URBAN WATER REUSE
HANDBOOK

URBAN WATER REUSE
HANDBOOK

Edited by
Saeid Eslamian

CRC Press
Taylor & Francis Group
Boca Raton London New York

CRC Press is an imprint of the
Taylor & Francis Group, an **informa** business

CRC Press
Taylor & Francis Group
6000 Broken Sound Parkway NW, Suite 300
Boca Raton, FL 33487-2742

© 2016 by Taylor & Francis Group, LLC
CRC Press is an imprint of Taylor & Francis Group, an Informa business

No claim to original U.S. Government works

Printed on acid-free paper
Version Date: 20151102

International Standard Book Number-13: 978-1-4822-2914-1 (Hardback)

Visit the Taylor & Francis Web site at
http://www.taylorandfrancis.com

and the CRC Press Web site at
http://www.crcpress.com

Contents

PART VIII Water Reuse Systems and Technologies

PART IX Water Reuse Risk and Impact Analysis

PART X Water Reuse Management and Monitoring

PART XI Natural Wetlands and Wetland Plants

PART XII Water Reuse and Artificial Wetlands

PART XIII Rainwater Harvesting in Cities

PART XIV Water Reuse Specific Applications

PART XV Water Reuse in Climatically and Physically Different Regions

PART XVI Water Reuse Case Studies

PART XVII Impacts of Climate Change, Mitigation, and Adaptation on Water Reuse

Preface

Water scarcity and pollution have long been a reality throughout much of the world's arid zones where the availability of water of suitable quality has a direct impact on growth and prosperity. Throughout much of the twenty-first century, dams, reservoirs, canals, and other measures provided the water needed to accommodate the region's growing population and economic needs. However, in communities all around the world, water supplies are coming under increasing pressure as rapid population growth coupled with drought, water-intensive energy development, climate change conditions, pollution, and changes in land use affect water quantity and quality, and a number of other factors are now placing additional stress on world water supplies. The increasing scarcity of water in the world along with rapid population increase in urban areas gives rise to concern about appropriate water management practices, while in urban areas, it is becoming difficult for authorities to manage water supply and wastewater. Not surprisingly, there is an increasing need and interest in many areas of the world to identify and develop alternative and sustainable water supplies. To address existing and anticipated water shortages, many communities are working to increase water conservation and are seeking alternative sources of water. In many countries throughout the world, water reuse has proved to be an effective and safe means to help satisfy growing water demands.

Water reuse is defined as the recycling of treated wastewater derived from domestic, agriculture, and industrial sources for beneficial purposes. Although water reuse activities are limited to nondrinking water purposes, a wide range of activities can occur, including irrigation of crops and pastureland, irrigation of urban landscapes, industrial cooling, dust control, street sweeping, and artificial groundwater recharge. The production of clean water from desalting is very energy-demanding, costly, and of concern; therefore, the reuse of water does for many purposes serve important roles. For reuse, the needed cleaning of the water to serve different purposes is of major importance and the need for advice is great.

This handbook provides the latest information on water reuse and attempts to increase attention on the opportunities for a majority of urban regions of the world. It includes authors from Australia, Bangladesh, Canada, Denmark, Germany, Ghana, Greece, France, India, Iran, Ireland, Italy, Japan, Jordan, Malaysia, Mexico, Nepal, the Netherlands, Nigeria, Pakistan, Palestine, Portugal, South Africa, Spain, Sri Lanka, Thailand, Turkey, UAE, United Kingdom, United States, and Zimbabwe.

The possibilities of water reuse in agriculture, industry, urban use, and environmental and ecological water enhancement are discussed with the assistance of practical examples.

The handbook focuses on the following titles, which present significant merits over the majority of previously published water reuse books:

- *Climate change impact and adaptation on water reuse*
- *Economic aspects of water reuse*
- *Environmental and ecological impacts/benefits*
- *Natural wetlands and wetland plants*
- *Rainwater harvesting in cities*
- *Social and cultural aspects of water reuse*
- *Sustainable water reuse and outlooks*
- *Urban water reuse and sanitation*
- *Urban water reuse options*
- *Water reuse and artificial wetlands*
- *Water reuse case studies*
- *Water reuse in climatically and physically different regions*
- *Water reuse management and monitoring*
- *Water reuse standards and regulations*
- *Water reuse risk and impact analysis*
- *Water reuse specific applications*
- *Water reuse systems and technologies*

This handbook is useful for academics, administrators, researchers, students, and practitioners in civil engineering, geosciences, geography, agricultural engineering, natural resources, environmental health, industrial engineering, chemical engineering, urban planning, and other related disciplines. It could be used in conjunction with institutional courses such as water quality, environmental engineering, solid wastes, water and wastewater treatment system, hazardous water management, an introduction to environment, water resources management, and groundwater engineering.

This handbook consists of 84 chapters by approximately 150 authors from more than 30 different climate countries across the world that have encountered water deficit and have performed water reuse in practice. Authors come from diverse sectors such as universities (civil engineering, environmental engineering, water engineering, mechanical engineering, bioresource engineering, sanitary engineering, agriculture engineering, chemical engineering, biochemical engineering, electrical engineering, applied sciences, social sciences, nanoscale sciences, physical sciences, biological sciences, marine sciences, geosciences, irrigation management, disaster management, biotechnology, information technology, food technology, urban planning, natural resources, ecosystem informatics, sustainable development, public policy, public health, hydrology, geography, agronomy, energy, pharmacy, microbiology, mines, arts, laws, architecture, economics, mathematics), research institutes, technology centers, consulting companies, and private foundations.

Well-known universities and organizations have contributed to this handbook. Some famous examples include CSIRO (Australia), MWH Global (Australia), SMEC (Australia), IWMI (Sri Lanka), ICARDA (Jordan), NanoH2O (USA), ITKI (Italy), OYO (Japan), MIWT (Mexico), Goadex (South Africa), University of Maryland (USA), Colorado School of Mines (USA), Worcester Polytechnic Institute (USA), San Diego State University (USA), Lamar University (USA), Georgia Institute of Technology (USA), Science and Public Policy Institute (USA), McGill University (Canada), Concordia University (Canada), University of Waterloo (Canada), University of Calgary (Alberta), United Nation University (Canada), University of Technology Sydney (Australia), University of Western Sydney (Australia), University of South Australia (Australia), Curtin University of Technology (Australia), Flinders University (Australia), Edith Cowan University (Australia), University of Brighton (UK), Delft University of Technology (Netherlands), University of Stuttgart (Germany), Institute for Sanitary Engineering (Germany), Brandenburg University of Technology (Germany), Umwelt und Informationstechnologie Zentrum (Germany), Eberswalde University for Sustainable Development (Germany), Technical University of Lulea (Sweden), University of Copenhagen (Denmark), University of Lisbon (Portugal), University of Beira Interio (Portugal), Limerick Institute of Technology (Ireland), Second University of Naples (Italy), University of Trento (Italy), CIHEAM/Mediterranean Agronomic Institute (Italy), Hellenic Open University (Greece), Aristotle University of Thessaloniki (Greece), University of Patras (Greece), Rhodes University (South Africa), University of Free State (South Africa), Tottori University (Japan), Nanyang Technological University (Singapore), Asian Institute of Technology (Thailand), TERI University (India), Jawaharlal Nehru University (India), Indian Institute of Technology (India), American University of Sharjah (UAE), University of Dhaka (Bangladesh), and Isfahan University of Technology (Iran).

- **Vijay P. Singh**, a distinguished professor at Texas A & M University says:

"*Urban Water Reuse Handbook* provides a comprehensive treatment of water reuse. The authors of chapters represent virtually all major countries of the world. The handbook has a global perspective and will be useful for practitioners and academics as well as researchers."

- **Soroosh Sorooshian**, a distinguished professor at the University of California, Irvine says:

"The handbook is a timely contribution to the field of water resources engineering and planning. With the growing world population, especially the growth of urban population in mega cities of Asia and Africa, the issues of safe urban water supply and adequate sanitation system will challenge most countries. The body of the contributions in this handbook should be of great value to most experts and planners."

- **Jose Salas**, a professor at Colorado State University, says:

"*Urban Water Reuse Handbook* includes a variety of topics that are of current interest to professionals involved in planning and management of water resources in urban environments. Technical, social, economic, legal, cultural, ethical, and climatic factors are discussed in various dimensions and perspectives."

- **Tsuyoshi Miyazaki**, a professor from the University of Tokyo, Japan, says:

"The newly published UWRH aiming to explain the current technique on water reuses in urban areas is an up-to-date and minute one, suitable to be read by researchers, graduate students, engineers, official administrators, and even by social activists."

- **Ove T. Gudmestad**, an adviser in Statoil on Marine Technology, Norway, says:

"This *Urban Water Reuse Handbook* is issued at an important point in time and will, hopefully, serve as guidance and useful information to select the best strategy to provide the growing population of the world with a much-needed water supply."

- **Olga Eugenia Scarpati**, IGU Water Sustainability Commission Secretary, says:

"*Urban Water Reuse Handbook* presents the different possibilities of water reuse which are discussed with the collaboration of experts in the field. It contains a broad overview of the matter and provides a basis understanding of the field. It can be used as a textbook for use in introductory graduate courses in water research and environmental science; it also can be used as a general reference for scholars or practicing engineers. It also covers the different ways that water reuse can help, from contaminants management to hydroelectric power generation. It will be useful for postgraduate courses or for professionals who wish to update their knowledge in this area of science."

I am deeply grateful to my family, Jacklin, Faezeh, and Alireza, for their continuous encouragement during the development of this handbook, as well as to others who also have directly and indirectly contributed.

I also acknowledge the assistance of the staff of Taylor & Francis, including Joe Clements, Jennifer Ahringer, and others.

Saeid Eslamian, PhD (University of New South Wales)
Full Professor in Environmental and Statistical Hydrology
(Isfahan University of Technology)

Editor

Saeid Eslamian is a full professor of hydrology and water resources engineering in the Department of Water Engineering at Isfahan University of Technology, where he has been since 1995. His research focuses mainly on statistical and environmental hydrology in a changing climate. In recent years, he has worked on modeling natural hazards, including floods, severe storms, wind, drought, pollution, water reuses, sustainable development and resiliency, etc. Formerly, he was a visiting professor at Princeton University, New Jersey, and the University of ETH Zurich, Switzerland. On the research side, he started a research partnership in 2014 with McGill University, Canada. He has contributed to more than 400 publications in journals, books, and technical reports. He is the founder and chief editor of both the *International Journal of Hydrology Science and Technology (IJHST)* and the *Journal of Flood Engineering (JFE)*. He is the author of approximately 90 book chapters and books.

Dr. Eslamian's professional experience includes being on editorial boards and a reviewer of approximately 40 Web of Science (ISI) journals, including the *ASCE Journal of Hydrologic Engineering, ASCE Journal of Water Resources Planning and Management, ASCE Journal of Irrigation and Drainage Engineering, Advances in Water Resources, Groundwater, Hydrological Processes, Hydrological Sciences Journal, Global Planetary Changes, Water Resources Management, Water Science and Technology, Eco-Hydrology, Journal of Hydrology, Journal of American Water Resources Association, American Water Works Association Journal*, etc. UNESCO also nominated him for a special issue of the *Eco-Hydrology and Hydrobiology Journal* in 2015.

Professor Eslamian was selected as an Outstanding Reviewer for the *Journal of Hydrologic Engineering* in 2009 and received the EWRI/ASCE Visiting International Fellowship in Rhode Island (2010). He was also awarded outstanding prizes from the Iranian Hydraulics Association in 2005 and Iranian Petroleum and Oil Industry in 2011. Professor Eslamian has been chosen as a distinguished researcher of Isfahan University of Technology (IUT) and Isfahan Province in 2012 and 2014, respectively.

He has also been the referee of many international organizations and universities. Some examples include the United States Civilian Research and Development Foundation (USCRDF), Majesty Research Trust Fund of Sultan Qaboos University of Oman, Royal Jordanian Geography Center College, and Research Department of Swinburne University of Technology of Australia. He is also a member of the following associations: American Society of Civil Engineers (ASCE), International Association of Hydrologic Science (IAHS), World Conservation Union (IUCN), GC Network for Drylands Research and Development (NDRD), International Association for Urban Climate (IAUC), International Society for Agricultural Meteorology (ISAM), Association of Water and Environment Modeling (AWEM), International Hydrological Association (STAHS), and UK Drought National Center (UKDNC).

Professor Eslamian finished Hakimsanaei high school in Isfahan in 1979. After the Islamic Revolution, he was admitted to IUT for a BS degree in water engineering, and graduated in 1986. After his graduation, he was offered a scholarship for a master's degree program at Tarbiat Modares University, Tehran. He finished his study in hydrology and water resources engineering in 1989. In 1991, he was awarded a scholarship for a PhD in civil engineering at the University of New South Wales, Australia. His supervisor was Professor David H. Pilgrim, who encouraged him to work on "Regional Flood Frequency Analysis Using a New Region of Influence Approach." After he earned his PhD in 1995, he returned to his home country and was employed by IUT upon arrival. In 2001, he was promoted to associate professor, and in 2014 to full professor. For the past 20 years, he has been nominated for different positions at IUT, including university president consultant, faculty deputy of education, and head of department.

Professor Eslamian had made two scientific visits to the United States and Switzerland in 2006 and 2008, respectively. In the former, he was offered the position of vice professor by Princeton

University and worked jointly with Professor Eric F. Wood from the School of Engineering and Applied Sciences for one year. The outcome was a contribution in hydrological and agricultural drought interaction knowledge by developing multivariate L-moments between soil moisture and low flows for northeastern United States streams.

Recently, Professor Eslamian has begun the editorship of several handbooks published by Taylor & Francis (CRC Press). The three-volume *Handbook of Engineering Hydrology* in 2014, *Urban Water Reuse Handbook* in 2015, and the three-volume *Handbook of Drought and Water Scarcity* (2016) are published/contracted handbooks.

Contributors

Lina Abu-Ghunmi earned a PhD in environmental engineering from Wagenining University in the Netherlands in 2009 and a MSc and BSc in civil engineering and chemical engineering, respectively, from the University of Jordan. Research in wastewaters attracted Dr. Abu-Ghunmi's attention during her MSc study, which encouraged her to carry out her MSc thesis on the treatment of textile wastewater using sequencing batch reactor technology. She has been very keen and motivated to complete further research in this area. She received the Dutch government scholarship "NUFFIC" to complete her PhD in the area of greywater. Her journey with the research in wastewaters and greywater, her favorite topics, did not end when she obtained her PhD. In fact, it began in earnest. Dr. Abu-Ghunmi is currently an assistant researcher at the University of Jordan. She succeeded in receiving a number of funded research projects from the United States Agency for International Development (USAID), from the Third World Academy of Science (TWAS), from the Jordanian Scientific Research Support Fund (SRSF), and from the University of Jordan. Furthermore, Dr. Abu-Ghunmi contributed to the development of an online course on greywater for the UNESCO-IHE Institute in 2009–2011, and she has several publications in the area of wastewaters in highly accredited journals.

Iván Rivas Acosta is a civil engineer specializing in water resources. Dr. Acosta attended the University of San Luis Potosi in Mexico, where he graduated in 1995 with a degree in civil engineering (with honors). He holds a master of science in water resources from Colorado State University (CSU) and in 2009, he earned a PhD also from CSU with the dissertation "Design and Implementation of Hydrologic Unit Watersheds for Stormwater Modeling in Urban Areas." After completing his graduate studies, he developed several drainage studies for land development for Bulloch Brothers Engineering, Inc. in Nevada, and later, he worked for PBSJ in Maryland, performing GIS floodplain mapping for the Federal Emergency Management Agency (FEMA) within the National Flood Insurance Program (NFIP). Currently, Dr. Acosta works as a senior hydrologist in The Mexican Institute of Water Technology, where his research is focused on assessing the climate change impacts on surface hydrology, developing long-term water resource planning strategies, and preparing environmental impact assessments.

Jan Adamowski is an associate professor of hydrology and water resources management in the Department of Bioresource Engineering at McGill University, Canada. At McGill, he is also the Liliane and David M. Stewart scholar in water resources, the director of the integrated water resources management program (which comprises a master of science program and an online certification program), and the associate director of the Brace Centre for Water Resources Management. Dr. Adamowski's teaching and research activities center on statistical hydrology and integrated and adaptive water resources management.

Catherine Adley has more than 25 years of experience in microbiology research. Her interests are in diverse areas of microbiology, including mobile elements in antibiotic resistance, microbial water quality, and rapid microbial detection methods for process analytical technology (PAT) applications. She is presently a senior lecturer in the Department of Chemical and Environmental Sciences at the University of Limerick, Ireland. She has worked in the UK at the London School of Hygiene and Tropical Medicine and at the University of Surrey. In the United States, she worked at Cold Spring Harbor Laboratory, New York, and at Boston University. She spent a sabbatical in the Pasteur Institute, Paris, France. She is an elected member of the Royal Dublin Society (RDS) Science and Technology Committee and a member of its board of management. She was appointed to the Scientific Committee of the Food Safety Authority of Ireland (2005–2010) and was an invited advisor to the EU FP6 Science and Society Directorate from 2002 to 2007.

Christos S. Akratos earned his diploma and doctorate from the Department of Environmental Engineering, Democritus University of Thrace, Greece, and he is now an assistant professor at the Department of Environmental and Natural Resources Management, University of Patras. He has extensive experience on constructed wetlands that commenced from his doctorate dissertation, and the majority of his publications deal with wastewater treatment with constructed wetlands. His current research interests focus on decentralized wastewater treatment systems aiming to develop innovative low-cost treatment systems for wastewaters of small industries, agro-industries, and remote villages and towns. He has published 25 refereed journal papers, and has participated as a research team member in several research programs, the majority of them focusing on the area of wastewater treatment in constructed wetlands.

António Albuquerque is currently an assistant professor in the Department of Civil Engineering and Architecture of the University of Beira Interior, Covilhã, Portugal, and senior researcher at the Research Unit FibEnTech, Covilhã, Portugal. He earned his B.Eng. in environmental engineering from the New University of Lisbon, Portugal (1991), MSc in environmental engineering from the Imperial College, London, UK (1993), and PhD in civil engineering from the University of Beira Interior, Covilhã, Portugal (2004). He has been involved in research related to nutrient removal in wastewater treatment processes, constructed wetland technology, water reuse in rural areas, water quality management, waste valorization, and sensors for water systems. He is the author or coauthor of approximately 180 publications, including books, book chapters, journal papers, conference proceedings papers, technical reports, and academic texts.

Hussein A. Amery is an associate professor in the Division of Liberal Arts and International Studies at the Colorado School of Mines. He joined the school in 1997 and has served as associate provost for academic affairs, division director, associate division director, and as the director of the division's graduate program. His current research interests are in the areas of water and food security in the countries of the Gulf Cooperation Council (GCC), including security of critical infrastructure and the security implications of labor migration to that region. He also has expertise in the geopolitical and security implications of transboundary water issues, and in Islamic perspectives on water.

Sayed Mohamad Amininejad is an alumnus of Isfahan University of Technology in Iran and earned his bachelor's degree in water engineering. Currently, he is a master's degree student at the University of Stuttgart. He has researched the photocatalytic and antimicrobial activity of nanostructures and has published in the field of environmental chemistry and pollution.

Seyedeh Matin Amininezhad earned her master of science in inorganic chemistry from Islamic Azad University of Shahreza in Iran. She has conducted several research projects on photocatalytic, antibacterial, and antifungal activity of nanostructures and Schiff-base complexes. Currently, she is working as a researcher on a series of projects. In addition, she has published several articles mainly on environmental chemistry.

Mohammad Javad Amiri earned his PhD in water engineering from Isfahan University of Technology in 2013. His research has been mostly about environmental issues. He has published 15 international journal papers and 25 papers in scientific conferences, mainly in environmental, pollution of surface and groundwater, nanotechnology, and artificial intelligence. He is currently an assistant professor of water engineering in Fasa University.

Zouboulis Anastasios is a professor of chemical and environmental technology, Department of Chemistry, Aristotle University of Thessaloniki, Greece. His research is in the field of water and wastewater treatment technologies, wastewater management, reclamation, and reuse. He is author/coauthor of more than 150 papers published in scientific journals and of more than 100 papers published in the proceedings of national and international conferences. He has supervised 8 PhD and 12 MSc theses and more than 50 diploma theses. He has participated in more than 50 national and international research and demonstration projects funded by the EU and from local sources (ministries, industries, companies, etc.), and was scientifically responsible for more than 30 of them, related mainly to the fields of wastewater treatment processes, environmental biotechnology, as well as the treatment of industrial solid toxic wastes by the application of appropriate stabilization methods. His international reputation was recognized by his election as a (foreign) member in the Russian Academy of Sciences (2003). He has received several prizes and scholarships from national and international organizations. He is also an active consultant for several local industries in the field of environmental protection, a national expert for the horizontal activities of FP6 and FP7 (EU), and he acts often as a reviewer/evaluator for scientific publications, as well as for several EU or national research projects. He is also a member of the editorial board of scientific publications (*Separation Science & Technology, Water Research, Journal of Hazardous Materials*, etc.), as well as a member of the organizing and scientific committees of several national and international conferences (more than 20).

Mohammad Arshadi is currently a postdoctoral researcher at Shiraz University. He earned his BS (2003) from Arak University and MS (2005) from Isfahan University of Technology (IUT). His research activities concern heterogeneous catalysts, immobilized organometallic compounds, immobilized nanoparticles as nanoadsorbent of dye, surface chemistry, and electrochemical sensors.

Elif Burcu Bahadır earned a BSc in biochemistry from Ege University (EU), an MSc in environmental engineering from Namık Kemal University (NKU), Institute of Sciences, Environmental Engineering Program, Turkey. She worked at a soil testing laboratory in Hayrabolu, Tekirdağ, between 2010 and 2012. She enrolled as a biochemistry specialist at the central laboratory of NKU in November 2012. The subject of her MSc thesis was "An İnvestigation on Color, COD and Priority Pollutants Removal with Ozone Technology in Treated Effluents from Two Textile İndustries." She took part in the NKU-funded "Characterization and Cost Effective Treatment Methods to Reclaim Rainwater and Stormwater" project. She has developed expertise on ICP-MS, ICP-OES, and HPLC to measure various groups of chemicals.

Bosun Banjoko is a senior lecturer in the Department of Chemical Pathology and the Institute of Public Health of Obafemi Awolowo University, Ile-Ife, Nigeria. He earned his MSc, PhD (biochemistry), and MPH (environmental health) degrees from University of Ibadan, Ibadan, Nigeria; MPA (public policy) degree from Obafemi Awolowo University, Ile-Ife,

Nigeria; and diploma in international environmental law from UNITAR, Geneva, Switzerland. He is currently registered for the LLM Environmental Law and Human Rights degree at Aberystwyth University, Aberystwyth, UK. He specializes in toxicology immunology, environmental technology, environmental law, and policy. He is an author of many articles and book chapters.

M. Dolores Hidalgo Barrio is currently the R&D projects scientific manager at the Environment and Biofuels Area in CARTIF Technology Centre and associate researcher of the ITAP Institute. Dr. Barrio studied chemical engineering at Valladolid University where she earned her doctorate in 1999 (in anaerobic digestion). She has a degree in environmental technology earned in 1993 and a degree in finances earned in 2004. After having held for one year the position of environmental consultant-in-expert, Bogotá (engineering consulting), she worked at the Unión Fenosa Company (thermal energy plant) from 1996 to 1998. Since then, she has worked for CARTIF. As coordinator of numerous research projects, Dr. Barrio has been involved in research related to waste and wastewater treatment, and water management in agriculture and atmospheric emissions control. She is the author of more than 40 articles and more than 65 papers at international conferences, organizer of a number of workshops, and inventor of three patents.

Olga Barron joined CSIRO as a hydrogeologist in 2003. She is a principal research scientist leading the Water in Resources Sector of CSIRO Land and Water. She predominately leads multidisciplinary projects, which incorporate biophysical science, social science, as well as application of advanced technologies in water management (e.g., desalination). In this research domain (in addition to research in urban water), she led a multiagency multimillion dollar project "Rural Towns—Liquid Asset," and a project investigating the opportunities of desalination in Australian agriculture, funded by the National Centre of Excellence in Desalination. Over the last 10 years, Dr. Barron has led research related to the climate change impact on water resources in Australia and its implication to future environmental water demands as well as the assessment of land use change on catchment water balance and water quality and their environmental significance. Recently, she extended her area of expertise to adaptation of remote sensing techniques to mapping and monitoring of groundwater-dependent ecosystems. She earned a doctorate degree in engineering geology and environmental geology from Moscow State University.

Mohammad Hadi Bazrkar earned his MSc in water resources engineering from Tarbiat Modares University, Tehran, Iran. He is interested in surface and groundwater resources modeling using various models. His thesis title was "SWAT Model Application in Simulation of Nutrients to Identify Pollutant Sources Contribution in Chamgordalan Reservoir Watershed in Wet Seasons." Furthermore, his publications in international conferences and journals mainly relate to water pollution, drought, flood, SWAT, and system dynamics.

Atul Bhargava is an associate professor at Amity University Lucknow, India, and has contributed significantly to the field of science. Dr. Bhargava is a botanist and has interest in environmental biotechnological research. He has published several research papers and some of which are extremely important and filled gaps in the field of botany.

Lee Blaney is an assistant professor in the Department of Chemical, Biochemical, and Environmental Engineering at the University of Maryland Baltimore County (UMBC). He earned his BS and MS degrees in environmental engineering from Lehigh University (2005, 2007). In 2011, he earned his PhD in civil engineering from the University of Texas at Austin. Dr. Blaney's current research focuses on measuring and treating trace concentrations of pharmaceuticals and personal care products (PPCPs) in water and wastewater. His lab uses advanced liquid chromatography, tandem mass spectrometry, and fluorescence detection methods to detect PPCPs in water. In general, the Blaney lab is focused on understanding the fate of PPCPs in physicochemical treatment processes. Dr. Blaney has been at UMBC for three years, and is a faculty member of the Marine Estuarine Environmental Sciences (MEES) program.

Floris Boogaard (professor of spatial transformations, PhD in stormwater handling, senior consultant at Tauw, and director of INDYMO) joined Delft University of Technology in 1992 and graduated in 1998 in water management with an additional graduation in working in third-world countries. After graduating as a civil engineer, he joined the consulting agency Tauw BV to do work on urban drainage and water management for municipalities, water authorities, project developers, special planners, universities, and other consulting agencies. In addition to his research, he has been a group manager since 2007 for the water management group at Tauw BV in Amsterdam dealing with urban drainage, hydrology, ecology, waste, and surface water.

His research and advice fields include stormwater drainage and infiltration, complex monitoring and optimizing sewer systems, design of drainage facilities, water supply and groundwater pollution, urban water quality management, and urban water management planning. He aims at making cities sustainable in terms of water management. In 2008, he joined the University of Delft to conduct PhD research on the quality of stormwater and optimization SUDSs in the international research group Skills Integration and New Technologies (SKINT) in order to integrate the worlds of spatial planning and water management and encourage the implementation of innovative technical and sustainable solutions.

Dr. Boogaard accepted a professorship in spatial transformations at the School of Architecture, Built Environment and Civil Engineering, Noorder Ruimte Centre of Applied Research and Innovation on Area Development at Hanze University of Applied Sciences in Groningen in 2013.

Babak Bozorgy is a chartered professional principal civil/water engineer (CPEng, NPER, RPEQ) of MWH Global with a PhD in water resources engineering and 20 years of professional experience in the water industry in different countries such as Australia, Qatar, UAE, Saudi Arabia, and Iran. His previous works vary from feasibility studies, conceptual, preliminary, and detailed design of dams, hydraulic structures, hydropower schemes, irrigation schemes, water supply and distribution network, and stormwater drainage and wastewater collection schemes, and studies on water resources planning and flood risk management. Dr. Bozorgy has published several technical and research papers in major scientific journals and international conferences. He has also been a university lecturer of undergraduate civil engineering courses, including fluid mechanics, hydrology, hydraulics, design of hydraulic structures, water and wastewater engineering, and environmental engineering.

Stewart Burn is a CSIRO principal research scientist with more than 30 years of experience primarily in water delivery and treatment technologies. Specifically, he has developed methodologies to allow the transition of existing systems to more sustainable states through the adoption of decentralized technologies and the development of water treatment technologies to recover resources from wastewater. He is currently the program leader of the Urban Water Systems Engineering group.

Elena Campos-Pozuelo is head of research of the R&D department of Valoriza Agua (SACYR). She earned a doctor of engineering from the University of Lleida and Polytechnic University of Madrid, Spain. Dr. Campos-Pozuelo has more than 20 years of experience in the treatment of drinking water, industrial water, sewage, and organic wastes. Her expertise focuses primarily on biological treatment, activated sludge, nutrient removal, anaerobic processes, and membrane bioreactors. She has also worked extensively in design and research of membrane desalination processes and tertiary treatment by ultrafiltration. She has worked in several private companies (Valoriza Agua Sadyt, ITT, and Tragsa). She has also worked as a researcher for public research centers (IRTA and University of Lleida). Dr. Campos-Pozuelo has participated in and coordinated several research projects and has published more than 40 papers in scientific journals, technical magazines, international and national conferences, and book chapters. She is also a lecturer in different master courses in several Spanish universities.

Zhuo Chen is a postdoctoral research associate at the graduate school of Shenzhen Tsinghua University, Shenzhen, China. She earned her PhD from the School of Civil and Environmental Engineering at the University of Technology, Sydney, Australia. Her research interest is mainly in environmental engineering and management with specialization in areas of wastewater treatment, water reuse, assessment, and management.

S. Chidambaram is a gold medalist in both BSc and MSc geology. He earned his PhD in applied geology from Annamalai University in the field of hydrogeochemistry. He is currently working as a reader in the Department of Earth Sciences, Annamalai University. He is an expert in the field of hydrogeochemistry, hydrogeochemical modeling, and the application of stable isotope techniques in groundwater. He developed a computer program, WATCLAST, in C++ for the classification and thermodynamics studies of water in 2003. He has completed a few projects, running several projects funded by various government organizations in India such as UGC, DST, BRNS, MoEn&F, and AERB. He developed collaboration with CWRDM, Kozhikode; JNU, New Delhi; BARC, Mumbai; IGCAR, Kalpakkam; and an international collaboration with the Russian Academy of Sciences. He has published research papers in several national and international journals and edited three books in the field of groundwater chemistry.

Didem Civancık is a teaching assistant in the Department of Environmental Engineering at Uludag University. She earned her undergraduate degree in environmental engineering at Middle East Technical University (METU) in 2012. She is working toward her master's degree in environmental engineering at METU.

Theodore C. Crusberg earned his BA from the University of Connecticut in 1963, his MS from Yale University in 1964, and his PhD from Clark University (Worcester, Massachusetts) in 1968, all in chemistry. After spending a year at Tufts University School of Medicine (Boston, Massachusetts) in the Department of Biochemistry, he joined the chemistry faculty of Worcester Polytechnic Institute and was instrumental in the formation of the new Life Sciences Department, which in time became the Department of Biology and Biotechnology. As professor emeritus, he occasionally teaches courses in environmental science and microbial physiology and carries out research dealing with water quality and other environmental issues.

Banti C. Dimitra is currently a PhD candidate at the Laboratory of Analytical Chemistry in the Department of Chemical Engineering at Aristotle University of Thessaloniki (AUTH). She works in a research program co-funded by the European Union; and national sources, at the Laboratory of Water Technology in the Department of Food Technology at the Technological Educational Institute of Thessaloniki. She earned her bachelor–master degree in environmental engineering and a second master's degree in hydraulic engineering at Democritus University of Thrace, Greece. She earned, with honors, her third master of science in environmental protection and sustainable development at AUTH. The research she has conducted thus far is related to environmental engineering, including optimization of a solid waste composting system and *in situ* remediation of groundwater aquifers, focused on graphic design, simulation, and optimization of permeable reactive barriers. At this moment, her PhD research is in progress and it concerns the enhancement of wastewater treatment systems (membrane bioreactors and conventional biological wastewater treatment systems) by studying the mechanisms of the formation of extracellular polymeric substances.

Armando Di Nardo earned a PhD in civil networks and environmental systems from the University of Napoli "Federico II."

He is an assistant professor of hydraulics in the Department of Civil Engineering, Design Building and Environment (DICDEA) of the Second University of Napoli (IT). Dr. Di Nardo is responsible for some degree courses, is a delegate of the Faculty of Engineering of the Second University of Naples for ERASMUS and internationalization projects; a research fellow in many Italian and international research projects; scientific adviser in water resources management for some Italian and international universities, research centers, and companies; and the author of many international papers. His main research topics include management optimization and protection of water resources; optimal water supply network partitioning; optimal management of artificial reservoirs; and permeable reactive barriers. The research is based on innovative techniques: fuzzy logic, heuristic optimization procedure (genetic algorithms), identification linear (AR, ARX, ARMAX, etc.) and nonlinear (neural networks) models, and graph theory.

Michele Di Natale is a full professor of hydraulic construction in the Department of Civil Engineering, Design Building and Environment (DICDEA) of the Second University of Naples (IT). Scientifically, he studies different arguments related to environmental and technical problems of the water in nature (hydraulic, hydrology, maritime and hydraulic structures, management of water system, etc.). His main research themes are design and management of the integrated urban water systems; remediation techniques of contaminated groundwater; phenomena of dispersion and diffusion of pollutants in marine and fluvial field; and modeling of incoherent seabed subjected to waves and currents actions. He is a member of various national and international scientific organizations (fellow of ASCE, member of IAHR, member of AII). He is the author of approximately 180 scientific publications on national and international journals and proceedings of congresses and also is scientific coordinator of many research projects. Furthermore, he has developed remarkable professional experiences in national and international consulting firms, taking part in important hydraulic engineering projects.

Mike Dixon is a senior applications engineer for NanoH$_2$O, a membrane manufacturing company based in Los Angeles, California. Before this, he formulated and managed the R&D plan requirements for the Adelaide Desalination Project alongside developing a membrane and desalination R&D strategy for South Australia Water Corporation. During this time, he was a co-investigator on eight National Centre of Excellence in Desalination Australia research projects. In addition to working in membrane and desalination research, he has experience in ion-exchange chemistry and activated-carbon adsorption, including the combination of these technologies with membrane processes. Dr. Dixon earned his PhD in chemical engineering at the University of Adelaide. With more than 45 publications in international journals and conference proceedings, he is also a reviewer for the *Journal of Membrane Science*,

Desalination, Water Research, and several other international journals. He is a senior adjunct lecturer at the University of South Australia and Flinders University. He is the Asia-Pacific special/technical program coordinator for IDA's Young Leaders Program and was national president of the Australian Water Association (AWA) Young Water Professionals in 2012. Dr. Dixon is the recipient of the several Australian Water Association awards and the International Desalination Association's Fellowship Award.

Aideen Dowling is a PhD candidate in environmental microbiology under the supervision of Dr. Catherine Adley in the Department of Chemical and Environmental Sciences at the University of Limerick. In 2010, Dowling earned her undergraduate honours degree in microbiology from Trinity College, Dublin. After graduation, she worked as a quality control research assistant in Alltect and in The State Laboratory, Backweston. Her current research focuses on the epidemiological spread of antibiotic-resistant bacteria, the risk to human health, and predictive trend analysis. She is a member of the Society of General Microbiology (SGM).

Pay Drechsel is a trained environmental scientist with 20 years of experience in integrated natural resources management and sustainable agricultural production in low-income countries. After several years with the International Board for Soil Research and Management, Drechsel joined in 2001 the International Water Management Institute (IWMI-CGIAR) where he leads the global theme on Resource Recovery, Water Quality, and Health. He coordinates a number of research projects on food safety, especially where untreated wastewater is used in urban and peri-urban agriculture in Africa and Asia. Drechsel served in scientific and technical advisory committees of different FAO, WHO, and EU projects and is author or coauthor of more than 250 publications, half of them in peer-reviewed journals and books.

Ali Elkamel is professor of chemical engineering at the University of Waterloo. He holds a BSc in chemical engineering and a BSc in mathematics from Colorado School of Mines, an MS in chemical engineering from the University of Colorado Boulder, and a PhD in chemical engineering from Purdue University, West Lafayette, Indiana. His specific research interests are in computer-aided modeling, optimization, and simulation with applications to energy production planning, sustainable operations and product design. He has supervised more than 70 graduate students (of which 30 are PhDs) in these fields and all his graduate students obtained good jobs in the chemical process industry and in academia. He has been funded for numerous research projects from government and industry. His research output includes more than 190 journal articles, 90 proceedings, more than 240 conference presentations, and 30 book chapters. He is also a coauthor of four books; two recent books were published by Wiley and titled *Planning of Refinery and Petrochemical Operations* and *Environmentally Conscious Fossil Energy Production*.

Amanda V. Ellis is currently a professor in chemistry/nanotechnology at Flinders University of South Australia. She graduated with a PhD (applied chemistry) from the University of Sydney in 2003. She then undertook two postdocs in the United States, including Rensselaer Polytechnic Institute and New Mexico State University. After these, she returned to New Zealand as a prestigious Foundation of Research Science and Technology Postdoctoral Research Fellow at Industrial Research Ltd. (now Callaghan Innovations). In 2006, Professor Ellis commenced at Flinders University as a teaching/research academic. Since then she has secured over $10M in funding on projects involving novel polymer coatings, functionalized carbon nanotubes, microfluidics, and desalination. She has published more than 100 journal publications in these areas. Currently, she is the chair of the Royal Australia Chemical Institute (RACI), National Polymer Division, a research leader in the Flinders Centre for nanoscale Science and Technology, and an Australian Research Council (ARC) Future Fellow.

Adebayo Oluwole Eludoyin is lecturer with the Department of Geography in Obafemi Awolowo University, Ile-Ife, Nigeria. He is an associate researcher with the Institute of Ecology and Environmental Studies, and the Centre for Science and Technology Education in the same university. He earned his doctorate in geography (hydrological science) at the University of Exeter, United Kingdom, after previous postgraduate and undergraduate programs at the Obafemi Awolowo University, Federal School of Surveying and Adeyemi College of Education in Nigeria. His research focuses on physical geography and geographic information systems (GIS). His expertise and practical experience cover the areas of hydrology, soil science, climate science, disasters, and GIS applications.

Saleh Tarkesh Esfahany obtained his BSc from Isfahan University of Technology (IUT), Isfahan, Iran, in 2012 and is currently a graduate student of environmental engineering in Science and Research Branch, Islamic Azad University of Tehran. He has contributed to three research projects and was introduced as the distinguished researcher of the Iranian Gas Industry in 2011. He has designed more than 50 projects about canal systems, modern irrigation systems, water and wastewater collection systems, and treatment. He has jointly written a book on water reuse in Iran as well as more than 30 articles.

Alireza Eslamian earned his bachelor's degree in civil engineering from Isfahan University of Technology, Isfahan, Iran, and is currently pursuing his MSc in Concordia University, Montreal, Canada. He is a research assistant and his area of research is water resources and hydraulics. He is working on modeling urban water use in Montreal for his master's thesis.

Faezeh Eslamian is currently a PhD student and research assistant of bioresource engineering in the Department of Environmental Sciences at McGill University, Montreal, Canada. She earned her bachelor's and master's degrees in civil and environmental engineering from Isfahan University of Technology, Iran. Eslamian has several publications and projects in the areas of environmental remediation, wastewater treatment, water reuse, hydrology, and drought. She is presently working on modeling phosphorus pollution in subsurface drained farmlands.

Kirsten Exall earned a PhD in chemistry at Queen's University in Kingston, Canada. As a former research scientist with Environment Canada, her research focused on urban water quality and treatment, including processes in drinking water, stormwater and wastewater treatment, water reuse and recycling, and water quality impacts of urban runoff and snow disposal sites. She has contributed over 45 publications and served as a member of the first international editorial board for the *Journal of Water Reuse & Desalination* (IWA Publishing). She is currently a research advisor at the University of Calgary in Alberta, Canada.

Jannatul Ferdaush is an assistant professor in the Department of Urban and Regional Planning (DURP) at Jahangirnagar University, Savar, Dhaka, Bangladesh. She has taught at the university since 2011. She has also worked in the National Development Organization—Unnayan Onneshan in Bangladesh. She is working on urban ecotourism, gender issues, urban development theories, applicability of pragmatic governmental tourism development policy, recurring municipal finance, middle-income housing affordability, riverbank erosion and population displacement, coastal urban drinking water, water reuse, and urban ethics. She has experience in applied research and leadership capacity in urban development project implementation. The main focus of her research interest is interdisciplinary research approach, as well as sustainable urban development theories. She has 12 publications in journals and books; in addition, she has experience in academic projects operation. She has recently completed two university faculty projects funded by University Grants Commission (UGC) of Bangladesh. She is currently pursuing higher studies in the United States and doing some projects on urban planning related issues. Beside these, since August 2014 she has worked for Savannah State University, Georgia, as a teaching assistant.

Cristina Fernandez is a technologist at the Waste, Mobility, and Works Department of Valoriza Servicios Medioambientales (SACYR Group). She is a chemical engineer from the University of Alicante, Spain, specialized in water treatment and waste management areas. She worked for two years in the R&D Department of Valoriza Agua. Her expertise primarily focuses on wastewater treatment by biofiltration, flocculator design for wastewater treatment, high load water treatment by forced evaporation and reverse osmosis for its reuse, and sewage sludge management and its application in the agriculture.

Paola Foladori is currently an associate professor of environmental and sanitary engineering at the Department of Civil, Environmental and Mechanical Engineering, University of Trento, Italy. She earned her PhD in 2002 from the Politecnico of Milano. Since then, she has been involved in research and teaching in the water and wastewater process engineering sector. In various research projects, several treatment processes of urban and industrial wastewater have been investigated: activated sludge plants, biofilm reactors, constructed wetlands, and microbial fuel cells. Innovative tools for advanced monitoring (respirometry and flow cytometry) have been developed for in-depth investigations in the field of water and wastewater treatment. She is the author or coauthor of 6 books, including *Sludge Reduction Technologies in Wastewater Treatment Plants*, 29 chapters in books, 35 papers in peer-reviewed journals, and some 100 other scientific publications and conference proceedings papers.

Richard M. Gersberg is currently a professor (and head of the division) of environmental health in the Graduate School of Public Health at San Diego State University (SDSU). Dr. Gersberg specializes in water quality research and limnology, and has broad experience working with both chemical and microbiological pollutants and risk assessments. He has more than 70 scientific publications in these areas. Dr. Gersberg has conducted a number of studies on the fate of bacterial indicators and toxics in both freshwater and estuarine ecosystems.

Ehsan Goodarzi is currently a research scientist in the School of Civil and Environmental Engineering at the Georgia Institute of Technology. He earned his PhD from Universiti Putra Malaysia (UPM) in water resource engineering with a focus on risk and uncertainty analysis in hydrosystem engineering. His research interests cover various fields of water resources engineering, including hydraulic structures, surface water hydrology, watershed management, optimization analysis, risk and uncertainty analyses, as well as climate change impacts on the water quantity and quality.

Nigel Goodman is a CSIRO scientist who holds a master's degree in environmental engineering. He has more than 15 years of experience working on a number of projects related to environmental chemistry, wastewater treatment, nutrient recovery, and water supply. His work is currently focused on municipal wastewater desalination using electrodialysis reversal to provide recycled water for horticultural irrigation.

Ritusmita Goswami earned her PhD in environmental science from Tezpur University, India. She is now working as a young scientist at Tezpur University under the DST-SERB Fast Track Young Scientist Scheme. Her research focuses on groundwater contamination, exposure impacts, and remediation technologies. She has published her research work in reputed journals such as the *Journal of Hydrology, Separation Science & Technology, International Journal of Nanoscience*, and others.

Manish Kumar Goyal works as faculty in the Department of Civil Engineering, Indian Institute of Technology, Guwahati, India. Earlier, he worked at McGill University, Montreal, Canada, and Nanyang Technological University (NTU), Singapore. He has published more than 35 international papers in peer-reviewed journals and presented his work in several national and international conferences/workshops. He was awarded a Japan Society for the Promotion of Science (JSPS) fellowship and the Commonwealth Scholarship Program.

Wenshan Guo is a senior lecturer in the School of Civil and Environmental Engineering at the University of Technology, Sydney. Her research focuses on innovative water and wastewater treatment and reuse technologies. Her expertise and practical experience cover the areas of water and wastewater engineering such as membrane technologies (e.g., membrane bioreactor, microfiltration, membrane hybrid system, and PAC-submerged membrane bioreactor), advanced biological wastewater treatment technologies (e.g., suspended growth reactors and attached growth reactors), and physicochemical separation technologies (e.g., adsorption, column, and flocculation). She also has the capability to work in solid waste management, life-cycle assessment, and desalination.

Deepa Gurung is a research associate at the Environmental Research Department at Thakur International, Kathmandu, Nepal. She earned her master's degree in environmental sciences from Jawaharlal Nehru University, New Delhi (2013).

Marwan Haddad is a full professor of environmental engineering and directs the Water and Environmental Studies Institute (WESI) at An-Najah National University (ANU) in Nablus, Palestine. Haddad's main research area is in water quality and resource management. He has published more than 190 papers in his field, and edited more than 10 international conference proceedings and refereed books. Haddad has directed and acted as a team leader of numerous major projects. He has served as an editorial board member and a reviewer for several local and international journals in his field.

Atef Hamdy is an emeritus professor in water resources management, CIHEAM/Mediterranean Agronomic Institute, Bari, Italy. He earned a PhD in soils and water science (1972) from Ain Shams University, a diploma in the use of isotopes and radiation in agriculture ITAL from Wageningen, the Netherlands (1970), and a diploma in irrigation sciences from CIHEAM/IAM, Bari, Italy (academic year 1978–1979). He has 40 years of professional experience and has worked with the Atomic Energy Authority, the Land Resources Department, the Soil and Water Research Department, and the Nuclear Research Center in Cairo, Egypt; he has also been the director of research at CIHEAM/Mediterranean Agronomic Institute in Bari, Italy. His fields of specialization are soil and water science (irrigation), water resources management, and, in particular, nonconventional water resources. Professor Hamdy is the editor and author of many books and special publications.

Armando Rivas Hernández is a biologist at Michoacan University, Mexico (1979). He holds a specialty on biological treatment of polluted waters from Shimane University, Japan (1989); a master's of environmental engineering from the National Autonomous University of Mexico (1997); and a PhD in science and environmental engineering from Metropolitan Azcapotzalco Autonomous University, Mexico. He has 25 years of experience on wastewater treatment plants at the Mexican Institute of Water Technology. His focus is on design, construction, evaluation, training, and technological research of natural systems, stabilization ponds, and treatment wetlands. He has written 43 technical papers, 7 reviewed papers, 7 book chapters, 14 technical notes; organized and participated in 26 national and international workshops on stabilization ponds, 78 national and international conferences, 14 international workshops on wastewater treatment by natural systems, training collaboration between the Japanese International Cooperation Agency and the Mexican Institute of Water Technology (1995–2014), and 26 national and international congresses on wastewater treatment systems.

Zareena Begum Irfan is currently serving as an associate professor of environmental economics at the Madras School of Economics, Chennai, Tamil Nadu, India. She earned her master's degree in integrated science program from the Pondicherry Central University (2003) and PhD (2007) from the Indian Institute of Technology Roorkee, India. Dr. Irfan has been involved in research related to water pollution, urban hydrology, the arsenic-related environment–health–development nexus, environmental valuation, and ecotourism. She has been a faculty member at the Madras School of Economics since 2007. She is the author or coauthor of more than 30 international journal publications and conference proceeding papers. She has served as a project investigator of Deutsche Gesellschaft für Internationale Zusammenarbeit (GIZ) GmbH, Indo-German Biodiversity Programme, Ministry of Environment and Forest, Government of India–sponsored projects on anthropogenic activities impacting environmental resource via valuation tool, cost–benefit analysis, and impact assessment. Her latest publications deal with the conservation of biodiversity and resource conservation. Her publications mainly deal with the interdisciplinary aspect of any environmental resource.

Kamrul Islam earned his undergraduate and postgraduate degrees in geology from the University of Dhaka before spending four years as a hydrogeologist and consultant at the Irrigation Management Division, Institute of Water Modeling (IWM), Dhaka, Bangladesh. His thesis research for his MS degree focused on wastewater reuse in agriculture. Presently, he is working on irrigation management for drought-prone areas of Northern Bangladesh.

Shafi Noor Islam is a lecturer in Euro hydro-informatics and water management, at the Brandenburg University of Technology Cottbus-Senftenberg, Germany. He holds a PhD in environmental and resource management, and worked as a lecturer in the master's program of Urban Management Studies at the Technical University of Berlin, Germany. He also worked as a visiting lecturer for the master's program of environmental and applied science at Halmstad University, Sweden. He has

postdoctoral research experience on food security and is presently a candidate for the habilitation (Dr.rer.nat.habil.) research degree on the topic "Ecosystem Services and Food Security under Threat in the Mega Deltas of Asian Coastal Regions: An Analysis on the Ganges–Brahmaputra–Meghna Delta." He is currently the chair of ecosystems and environmental informatics at the Brandenburg University of Technology Cottbus-Senftenberg, Germany. His focus research fields include interdisciplinary water resource research on water salinity and mangrove ecosystems analysis and wetlands and coastal resource management as well as climate change impacts on food security and ecosystem services. Transboundary water and char-land management strategies, encompassing groundwater quality and urban water supply, ecology management are some subjective interests. He has published 50 articles in international journals and books, and has research experiences in China, Malaysia, Indonesia, Germany, Poland, Denmark, Italy, Spain, France, the Netherlands, Sweden, and Mexico.

S.M. Tariqul Islam is a lecturer in the Institute of Disaster Management (IDM) at Khulna University of Engineering and Technology (KUET), Bangladesh. He also works as an editor and conference organizing secretary for WasteSafe International Conference (www.wastesafe.info). He has more than 10 years of experience working in the area of waste and wastewater treatment at different urban areas, especially in the coastal climatic refuges of Bangladesh. He has the life-long dream of low-cost potable drinking water supplies in urban poor areas.

Nidhi Jha is a doctoral scholar at the Department of Natural Resources at TERI University, Delhi, India. Her research interests are mainly in groundwater management and she has practical experience of more than two years working on various groundwater issues. Her doctoral research focuses on integrated water resource management by using cutting-edge techniques, remote sensing, and geographical information systems.

Amos T. Kabo-bah is an assistant professor in the Department of Energy and Environmental Engineering, University of Energy and Natural Resources, Sunyani, Ghana. In 2010, he earned his second MSc in geo-information science and earth observation for water resources and environmental management from the Faculty of International Institute for Geo-Information Science and Earth Observation (ITC), Universiteit Twente, Enschede, the Netherlands. Upon completion of these degrees, he pursued his doctorate at Hohai University, Jiangsu, China. Currently, he is responsible for water reuse research at University of Energy and Natural Resources, Sunyani, Ghana.

Ioannis K. Kalavrouziotis earned his PhD in environmental geochemistry from the Department of Geology, University of Patras, Greece, in 1999. He was as an associate professor in the Department of Environmental and Natural Resources Management in Agrinio, University of Western Greece, from 2000 to 2013. He is currently an associate professor in wastewater management in the School of Science and Technology, Hellenic

Open University, since July 2013. He is also the director of education in the wastewater management master's program.

Dr. Kalavrouziotis successfully completed his administrative responsibilities as an agronomist in the Ministry of Agriculture from 1988 to 2000, director of the Western Greece Region Administration from January 1993 to November 1993, member of the administrative board of the National Agricultural Research Foundation (NAGREF) from 2006 to 2009, and president of the Sector for the Management of Messologion Lagoon (Ministry of Environment, Physical Planning and Public Works) from June 2006 to January 2009.

He is an editorial board member of several reputed journals such as the *Journal of Water Reuse and Desalination* (IWA), *Environment and Pollution* (Canadian Center of Science and Education), *Journal of Environmental & Analytical Toxicology*, and *Frontiers in Green and Environmental Chemistry*.

Dr. Kalavrouziotis is a member of the IWA. He was the president of the Organizing and Scientific Committee of the IWA symposium on Water, Water and Environment: Traditions and Culture, March 22–24, 2014, Patras, Greece. He is a member of the interim management committee at the IWA specialist group on water and wastewater in ancient civilizations, WATERWiki group space, and WG on environment.

He has published 5 books and book chapters, 70 peer-reviewed full research papers in international journals, 3 papers in Greek national journals, 40 papers in international conferences, 38 papers in national conferences, and 90 articles in journals and newspapers.

He was a member of the administrative board of the Geotechnical Chamber of Greece from 2002 to 2003; a member of the administrative board of the State Agronomists Pan-Hellenic Union (PUSA) during the periods 1991–1993, 1993–1995, and 1995–1997; secretary general of the State Agronomists Pan-Hellenic Union in 1997; a member of the administrative board of the Organization for the Authentication (certification) and Supervision of Agricultural Products (OPEKEPE), representing the Geotechnical Chamber of Greece from 2004 to 2005; and a member of the administrative board of the National Agricultural Research Foundation (NAGREF) from 2005 to 2009.

Shams Kandhro is currently a graduate student in chemical engineering at the University of Waterloo. He has several years of industrial experience in process engineering, laboratory analysis, chemical treatment, waste treatment, and petrochemical processes.

Sweety Karmacharya is a research associate at UIZ Umwelt und Informationstechnologie Zentrum, Neue Grünstraße, Berlin, Germany, jointly with Eberswalde University for Sustainable Development, Eberswalde, Brandenburg, Germany. She is involved in a water resource management research project on the global climate change environment.

Bernard Keraita is an irrigation and water engineer with more than 12 years of research experience on water, health, sanitation, and environment in low-income countries. He worked as a researcher with the International Water Management Institute (IWMI-CGIAR) from 2002 to 2011 in its Africa Office in Ghana. His research focused on safe wastewater use in agriculture. He also worked for IWMI on agricultural water management research in Tanzania and Ghana. Since 2011, still based in Africa, he works as an international researcher with the Copenhagen School of Global Health, University of Copenhagen. His current core tasks are research, research capacity strengthening, and curriculum development in Ghana and Tanzania/Zanzibar. Keraita is a regular reviewer in leading environmental health journals, frequent external examiner for graduate schools, and has authored more than 70 scientific publications.

Ganesh Keremane is a research fellow at the School of Law, University of South Australia. He earned his PhD at the University of South Australia for which he received the 2008 CRC Irrigation Futures Director's Award. He also has a master's degree in agricultural economics from the University of Agricultural Sciences, India. After completing his PhD, Dr. Keremane continued at the CCWPL and is currently working with his colleagues on different research projects at the Center, including institutional analysis of implementing desalination projects and an integrated urban water management strategy. Dr. Keremane's research interests lie in the field of natural resources management with an emphasis on institutional and policy analysis of surface and groundwater management. He is also interested in assessing community attitudes and perceptions toward alternative water sources and has published several peer-reviewed international journal papers and book chapters related to these topics.

Behnaz Khosravi is a master's student in the Department of Water Engineering at Isfahan University of Technology (IUT), Iran. She earned her BS in water engineering from IUT (2009–2013), Iran. She researched using wastewater in agriculture and the effect of it on soil. Khosravi's current research focuses are on evaluating the electro-Fenton method to reduce coliform and organic loads of treated municipal wastewater.

Maryam Khozaei earned her master of science at Shiraz University. She has published more than 3 peer-reviewed manuscripts and 5 conference papers from her research, mainly in surface and groundwater, soil, water, and plant relationships.

Mohammad Mahdi Kohansal earned his MSc degree in irrigation and drainage in 2014 from Isfahan University of Technology (IUT), Isfahan, Iran. He is interested in climate change and variability, reuse water and reclamation, environmental nanotechnology, evapotranspiration and sensitivity analysis, sustainable development, interbasin water transfer, vermin compost, GIS, and wastewater treatment. His publications in national and international conferences are mainly in biodrainage, interbasin transfer, climate change, Urmia Lake, wastewater treatment, vermin compost, and GIS.

Manish Kumar is a faculty member at Tezpur Central University, Assam, India. He earned his PhD in environmental engineering from the University of Tokyo, Japan, and has received many prestigious fellowships, such as the Japan Society for the Promotion of Science (JSPS) foreign research fellowship, Brain Korea (BK)-21 postdoctoral fellowship, Monbukagakusho scholarship, Linnaeus-Palme stipend from SIDA, Sweden, research fellowship from CSIR, India, and others. Dr. Kumar is active in the fields of hydrogeochemistry, diffuse pollution, urban and agricultural water management, and has more than three dozen publications in reputed journals, books, and referred conferences. He has been working as a referee for many reputed journals, such as *Chemosphere, Journal of Hydrology, Environmental Research Letter, Contaminant Hydrology*, and others. Currently, he is a member of the editorial board for two international journals, and his name appeared in the 2010 Edition Marquis' *Who's Who in the World*. Dr. Kumar has also coedited a book titled *Groundwater Monitoring and Management through Hydrogeochemical Modeling Approach*.

Pietro Laureano, an architect and urban planner, is the UNESCO expert for arid regions, water management, Islamic civilization, and endangered ecosystems. He lived for 8 years in the Sahara Desert (Algeria) and for 10 years in the caves of Sassi of Matera (Italy) where he has led the recovery of this troglodyte city and founded IPOGEA (www.ipogea.org), which coordinates EU projects in 10 countries all over the Mediterranean and research and landscape restoration works carried out by means of traditional techniques and their innovative use. He has worked on the UNESCO plan for Petra (Jordan), has restored monument and hydraulic systems in Lalibela, Ethiopia, and realized projects in Algeria, Yemen, and Morocco. He is a member of the scientific committee of several international bodies (FAO, UNCCD) and of the working group of the new UNESCO Landscape Convention. He has created the Traditional Knowledge World Bank (www.tkwb.org) and is the president of the International Traditional Knowledge Institute (ITKI) promoted by UNESCO in Florence to pursue this initiative. He has published several books, including *The Water Atlas—Traditional Knowledge to Combat Desertification (UNESCO)*, which has been translated into many languages.

Sophie C. Leterme is a senior lecturer at Flinders University, Adelaide, South Australia. She earned a master's in biological oceanography in 2003 at l'Universite Pierre and Marie Curie—Paris VI, France, and a PhD in marine biology in 2006 from the University of Plymouth, UK. She then undertook 2 years of postdoctoral work at Flinders University working on the ecology of the Coorong wetlands and focusing on the adaptation of diatoms to salinity fluctuations. In 2008 she took a position as a lecturer in biological oceanography at Flinders University. Since then, she has secured over $5M in funding on projects involving plankton biology and ecology, environmental impacts of desalination plants, and biofouling issues within desalination systems. She has published more than 20 journal publications in these areas. Since 2009, Dr. Leterme has been the deputy node leader of the South Australian Node of Integrated Marine Observing System (IMOS).

Catherine Diane Luyt completed her secondary school education from Stirling High School in East London, South Africa, in 2003. She then continued to complete her bachelor of pharmacy summa cum laude from Rhodes University in Grahamstown in 2009 and her PhD in 2013 from the same university. Her research interests include community service, pharmacy, microbiology, and water. She has coauthored 7 peer-reviewed publications.

Jiri Marsalek is professor of urban water at the Technical University of Lulea, Sweden. His research interests, including sustainable stormwater management, drainage adaptation in a changing climate, and control and treatment of combined sewer overflows, are documented in almost 400 publications, including 140 journal papers. He has served as secretary of the International Association for Hydro-Environment Engineering and Research (IAHR) and International Water Association (IWA) Joint Committee on Urban Drainage (a specialist group with >1200 members worldwide) and has worked extensively with UNESCO and NATO on urban water management. His awards from the last decade include Environment Canada's Citation for Excellence (2005), two honorary doctorates from Sweden (2006) and Denmark (2008), sharing the 2009 Canadian Society for Civil Engineering Award for the best paper in environmental engineering (2010), and the International Water Association (>10,000 members) Honorary Membership Award (2010) for outstanding contribution to the association and to the water sector.

Jesús M. Martin Marroquin is currently head of the Occupational Risk Prevention Department and Researcher at the Agrifood and Sustainable Process Division in the CARTIF Technology Center. He is also a PhD candidate in environmental engineering at the University of Valladolid, ITAP Institute. He earned a bachelor's degree in chemistry in the field of chemical engineering at Valladolid University (Spain) in 1994. He has a master's degree in occupational risk prevention, earned in 1998, in the specialties of safety at work, industrial hygiene, and psychosociology and applied ergonomics. From 1997 to 1999, he joined R&D Department of Ebro Agrícola (sugar company) where he studied the sugar decoloration process. Since 2000, he has developed his professional career in CARTIF where he has been involved in research related to chemical recycling of plastic waste, advanced oxidation processes, and water reuse. He is the author of more than 10 articles, 10 papers at international conferences, and is an organizer of a number of workshops.

José Saldanha Matos is currently a full professor and senior researcher at the Department of Civil Engineering, Architecture and GeoResources of the Technical Superior Institute of the University of Lisbon (Portugal) and managing partner of Hidra, Hydraulics and Environment Ltd. and Vlow.Ges, Advanced Management of Wastewater and Environmental Systems, Ltd. He earned his B.Eng. in civil engineering from the Technical University of Lisbon (Portugal, 1978), and MSc and PhD from the same university in 1987 and 1992, respectively. He is a member of the management committee of the European Water Association (EWA) and fellow of the International Water Association (IWA). He has been involved in research related to urban drainage, sewer systems and processes, wastewater treatment, and asset management. He is the author or coauthor of approximately 250 publications, including books, book chapters, journal papers, and conference proceedings papers, and is responsible for more than 200 engineering studies and projects involving water supply and sanitation, in Europe, Africa, South America, and Central Asia.

Süreyya Meriç holds BSc, MSc, and PhD degrees in environmental engineering from Istanbul Technical University (ITU), Turkey. She was appointed as a full professor in November 2010 at Namık Kemal University, Tekirdağ, Turkey, as the head of Environmental Engineering Department (EED) and responsible for the EED's laboratories. She collaborated with the Sanitary and Environmental Engineering Division (SEED) of the University of Salerno (Italy) from 2004 to 2010. She was a visiting scientist in Italy from 1998 to 2000 at the National Cancer Research Institute, University of Naples Federico, Naples, and from September to December 2003 at SEED.

Previously, Dr. Meriç worked as research assistant and assistant professor at ITU, EED from 1988 to 2004. She has coordinated or collaborated with many international and national projects, supervised a number of graduate theses in Turkish, Italian, and English, and edited special issues for SCI journals and book chapters. She was chosen as an expert to evaluate national and international projects. She was engaged in organizing scientific committees of national and international symposiums and congresses. She has developed a long-lasting expertise of 25 years mainly in advanced (waste)water treatment technologies, recycling and reuse, alternative water sources, and impact assessment. She has published more than 70 SCI papers with a factor of 21. Her current research is focused on impact assessment and removal of xenobiotics, mainly pharmaceuticals, personal care products, and nanoparticles in wastewater and stormwater. She is currently in the MC of "COST ENTER Action ES1205" (the transfer of engineered nanomaterials from wastewater treatment and stormwater to rivers). She is also involved in the COST Action Water 2020.

Christopher Monckton of Brenchley has a master of arts in classical architecture in the University of Cambridge. He is a former Fleet Street editor and policy adviser to Margaret Thatcher during her years as prime minister of the United Kingdom. He has lectured at the faculty level on climate change science and economics all over the world, and has advised governments and corporations on the subject. His publication journals include climate sensitivity reconsidered (*Physics & Society*, 2008); clouds and climate sensitivity (*Annual Proceedings, Seminars on Planetary Emergencies, World Federation of Scientists*, 2010); is CO_2 mitigation cost-effective? (*Annual Proceedings, Seminars on Planetary Emergencies, World Federation of Scientists*, 2012); agnotology and climate consensus (*Science and Education*, 2013); political science: the dangers of apriorism in intergovernmental

science (*Energy & Environment*, 2014); and why models run hot: results from an irreducibly simple climate model (*Science Bulletin*, 2015, Vol. 60, no. 1, Jan. 2015).

Never Mujere earned a master of philosophy in geography (MPhil) from the University of Zimbabwe (UZ). Currently, he is reading for a doctor of philosophy (DPhil) in hydrology and water resources management at the UZ. He is also a lecturer in the UZ's Department of Geography and Environmental Science. He is a reviewer of the *International Journal of Hydrological and Technology* (IJHST), has authored 2 books, contributed to 4 book chapters, published 7 papers in refereed journals, and presented papers at international workshops. His areas of research interests are water resources and environmental issues.

Blair E. Nancarrow is a director of a social science consulting partnership, Syme and Nancarrow Water. Now semiretired, she spent 21 years as a social scientist in CSIRO Land and Water Australia and 3 years as a visiting fellow at the Fenner School of Environment and Society at the Australian National University in Canberra. She led a team who researched the social aspects of water recycling nationally for more than 7 years from 2002, which challenged the then-international view of the drivers of behavioral intentions in relation to contact with recycled water. Her particular area of expertise has been in the design and implementation of large, community-based research programs and experiments in the people/environment interface. Nancarrow has a special interest in the development of methods and processes to ensure the meaningful incorporation of social science in triple-bottom-line analysis of environmental policy and developments. She was awarded a Biennial Medal for services to the Modelling and Simulation Society in Australia and New Zealand and also was made a fellow of the society.

Syed Amir Abbas Shah Naqvi earned his MSc chemistry degree from the University of Education, Township Lahore. Naqvi is currently serving as a lecturer in the Department of Chemistry at Pakistan Public School and College, Abbottabad, Pakistan, and is also perusing his MS in chemistry from COMSATS Institute of Technology, Abottabad, Pakistan. He has extensive experience and expertise in the field of analytical chemistry. He had the opportunity to take part in a joint study with the National Tea Research Institute Shinkiari-Mansehra, Pakistan, that involved optimization of soil pH for tea cultivation.

Sara Nazif is an assistant professor in the School of Civil Engineering, College of Engineering at the University of Tehran. She earned her BS and MS degrees in civil engineering and water engineering from the University of Tehran (2003, 2005). In 2011, she earned her PhD in civil engineering from the University of Tehran. She is an international expert and consultant in the application of systems engineering and computer modeling in urban water engineering and management. She has authored more than 100 research and scientific papers published in national and international journals and conferences, and has published two books: *Urban Water Engineering and*

Management and *Hydrology and Hydroclimatology: Principles and Applications* (CRC Press, 2010 and 2012). Her research and training interests include water distribution systems, disaster management, urban water demand management, stochastic hydrology, climate change impacts on urban water engineering and management, drought and flood management, water and wastewater treatment plant operation, and systems analysis.

Huu Hao Ngo is a professor of environmental engineering at the University of Technology, Sydney. He has more than 30 years of professional experience in Australia and in Asian countries. He is well known for his activities in the areas of water and wastewater treatment and reuse technologies, which include advanced biological wastewater treatment technologies, membrane technologies, and physical–chemical separation technologies as pretreatment or post-treatment. His expertise and practical experience also cover the areas of water and wastewater quality monitoring and management, water and wastewater treatment technology assessment, desalination, and solid waste management. Currently, his activities are also in the development of specific green technologies: water–waste–energy nexus and greenhouse gas emission control. He has published more than 300 technical papers, 2 books and a number of book chapters and patents. He is appointed as an editor of *Bioresource Technology*, Elsevier, and is also a founder and editor in chief of the *Journal of Water Sustainability* while being an editorial board member of numerous international journals such as *Science of the Total Environment*, Elsevier; *Environmental Nanotechnology, Monitoring and Management*, Elsevier; *Journal of Chemistry*, Hindawi; *Journal of Advances in Environmental Chemistry*, Hindawi; *Journal of Waste Resources*, OMICS; and *Journal of Energy and Environmental Sustainability* (the International Society for Energy, Environment, and Sustainability).

Vilas Nitivattananon earned his academic degrees in engineering, economics, and management from Thailand and the United States. He is currently an associate professor of urban environmental management in the School of Environment, Resources and Development at the Asian Institute of Technology, Thailand. His areas of specialization and interest include management of urban infrastructure and services, water and environmental management, disaster and climate change risk assessment, environmental assessment, and economic analysis. He has published more than 50 articles in international journals, book chapters, and conference proceedings, including in the *Journal of Water Resources Planning and Management, Resource Conservation and Recycling, Desalination, Environmental Impact Assessment Review,* and *Journal of Environmental Management*.

Jean O'Dwyer earned her PhD from the University of Limerick, Ireland, in 2015. In 2011, she earned her bachelor of science undergraduate degree in environmental science, graduating with honors. Her research areas include environmental microbiology and environmental management in relation to public health. Her PhD thesis, under the supervision of Dr. Catherine Adley, involved the modeling of groundwater-derived water

supplies in order to predict microbiological contamination as a function of hydrogeological and climatological parameters. The goal of the research was to drive policy-based remediation strategies pertaining to environmental health.

Saeid Okhravi earned his undergraduate degree in water engineering from Isfahan University of Technology (IUT), Iran, and currently is a distinguished graduate student in water structure engineering at IUT. He has contributed to more than 10 scientific publications, primarily on rainwater harvesting systems, groundwater–surface water interactions, and wastewater treatment. He also has experience in the construction of artificial wetlands and rainwater harvesting systems. He contributed to the *Handbook of Engineering Hydrology* (Volume 1, *Fundamentals and Applications, Groundwater–Surface Water Interactions*) by Taylor & Francis (CRC Press).

Can Burak Özkal is a PhD student in the environmental sciences and technologies program at Namık Kemal University. He is also employed as a research assistant in the Environmental Engineering Department of Namık Kemal University Faculty of Engineering (2010–2014). He is currently involved in COST ENTER Action ES1205 (the transfer of engineered nanomaterials from wastewater treatment and stormwater to rivers), and also in the COST Action Water 2020. He is also a PhD candidate in the European PhD School on advanced oxidation processes and plans to conduct his studies on thin film.

Declan Page earned his PhD in chemical technology from the University of South Australia, Adelaide, Australia. He earned his undergraduate BSc from the University of Adelaide, majoring in chemistry, and his B.App. Sci. (honors) from the University of South Australia, majoring in colloid and interface science. He also holds an MBA from Monash University in Victoria, Australia. After completing his PhD, Dr. Page worked as a consultant scientist for several water utilities across Australia and in international development in Indonesia and Thailand. He is currently a team leader in the Water Reuse and Environmental Process Engineering Program of CSIRO Land and Water. His major research interests focus on water reuse via aquifers, water treatment, quantitative human health risk assessment, and water reuse.

Johanny A. Perez Sierra graduated with an MSc in environmental engineering from the University of Stuttgart in Germany, as part of the WASTE program with the support of the International Postgraduate Studies in Water Technologies (IPSWAT) scholarship of the German Federal Ministry of Education and Research (BMBF). Perez Sierra is an alumna of EARTH University, where she finished her BSc in agriculture engineering in 2010, thanks to a scholarship from the Annenberg Foundation. Currently, Perez Sierra is a scientific staff and a PhD candidate in the Institute of Agricultural Policy and Markets at the University of Hohenheim, researching in the field of bioeconomy modeling, analyzing the potential of biomass to feed biogas plants in Germany up to 2050, incurred effects on land use change, and greenhouse gas emissions.

Samaras Petros, chemical engineer, is a professor at the Laboratory of Water Technology, at the Department of Food Technology, of the Technological Education Institute of Thessaloniki, with expertise in the application of physical, chemical, and biological processes for water and wastewater treatment. He has participated in more than 40 national and international research projects, funded by the EU and from national sources (ministries, industries, companies, etc.). He has more than 60 publications in journals; more than 100 presentations in international conferences; 6 chapters in books; and holds 1 EU patent. He is a member of the organizing committee of 14 international conferences; member of the scientific committee of 18 international conferences; reviewer of research project proposals in EU and Greece; reviewer of papers submitted for publication to international scientific journals; instructor for continuing education courses on wastewater treatment for various private and public organizations; and consultant in private enterprises related to the examination of water–wastewater quality and the preparation of environmental impact assessment reports.

Manzoor Qadir is an assistant director of the Water and Human Development Program at the United Nations University Institute for Water, Environment and Health (UNU-INWEH). Qadir is a water and soil management scientist with interests in water recycling and safe and productive use of wastewater, water quality and environmental health, and amelioration of salt-induced land degradation. He has implemented multidisciplinary projects and directed research teams in Central and West Asia and North Africa. He previously held professional positions as senior scientist jointly appointed by the International Center for Agricultural Research in the Dry Areas (ICARDA) and the International Water Management Institute (IWMI); visiting professor at the Justus-Liebig University, Giessen, Germany; and associate professor at the University of Agriculture, Faisalabad, Pakistan. He is a fellow of the Alexander-von-Humboldt Foundation and serves on the editorial boards of several international journals.

Qin Qian earned a PhD in civil engineering. She is an associate professor in the Department of Civil Engineering, Lamar University, Beaumont, Texas. Dr. Qian joined Lamar University in June 2008, immediately after completing her PhD study in civil engineering at the University of Minnesota. Her research interests are in hydrology, hydraulics, and water resources. Her research goal is to advance process-based knowledge to allow better-informed land use planning, ecological restoration design, and preservation of aquatic ecosystems. Dr. Qian is a member of the American Society of Civil Engineers (ASCE), American Geophysical Union (AGU), Overseas Chinese Environmental Engineers and Scientists Association (OCEESA), Chi Epsilon Honor Society of ASCE, and Chinese American Water Resources Association (CAWRA). She served as a National Science Foundation Panel reviewer for the Water Quality/Pollution Control Unsolicited Spring Panel (P091361); a committee member of the ASCE Groundwater Hydrology Committee; technical assistant of the Technical Advisory Panel

for the Research Management Committee 5 (RMC-5) of the Texas Department of Transportation, and board member for the Gulf Coast Recovery and Protection District.

Ataur Rahman earned his PhD from Monash University in Australia and a master's in hydrology from University College Galway, Ireland. He completed his sabbatical at Cornell University and the University of Newcastle in 2010. He is currently an associate professor in water and environmental engineering in the University of Western Sydney, Australia. He has more than 20 years of experience in water research. His research interest includes flood hydrology, urban hydrology, and environmental risk assessment. He received the G. N. Alexander Medal from the Institution of Engineers Australia in 2002. He has published more than 260 research papers, book chapters, and technical reports in water and environmental engineering fields. He has been heavily involved in the preparation of the forthcoming revised version of *Australian Rainfall and Runoff—A Guide to Flow Estimation.*

A.L. Ramanathan is a professor in the School of Environmental Sciences, Jawaharlal Nehru University, New Delhi, India. His research area is in the field of hydrogeochemistry from inland and coastal surface and groundwater and their resource management. He has 18 years of teaching experience (PG) and research experience in this subject. He has taught at numerous universities in India and abroad. He has guided quite a number of PhD research students on groundwater quality and modeling aspects. He has spoken all over the world on the advanced research work in groundwater. He is also the recipient of various international and national scholarships and collaborations with institutes and universities reputed in groundwater research in India and abroad. He published two dozen articles in reputed refereed journals and authored five books on these aspects. He has also completed and continues his research work on groundwater in India and with international agencies.

Sandra Reinstädtler is a lecturer and doctoral research associate in the Department of Environmental Planning at the Brandenburg University of Technology Cottbus-Senftenberg in Germany. Besides an engineering degree as a landscape architect from Technical University of Dresden (TU Dresden), Reinstädtler is specialized in landscape planning combined with her doctoral research fields and researcher's mission in sustainable landscape, land use, and water management and its instrumentations in dealing with climate change. Her diploma thesis focused on regional scaled strategies for rural land use and the sustainable development of the cultural landscapes in a commune in Switzerland. She taught the Cultural Landscapes and Historical Gardens module in the World Heritage Master's program for three years. She currently holds lectures in the field of environmental planning in the land use and water management study program as well as strategic environmental assessment and environmental impact assessment for master's degree students in the environmental engineer's study program within the Brandenburg University of Technology Cottbus-Senftenberg. She also worked as a visiting lecturer in the master's environmental and applied science program at Halmstad

University, Sweden. She has thus far 7 publications in different international journals, conference proceedings, and books. She gained international work, scientific research and field study experiences in Germany, Canada, the United States, Great Britain, Sweden, and Switzerland.

Mohammad Naser Reyhani is completing his MSc degree in water engineering at Isfahan University of Technology (IUT) under the supervision of Professor Saeid Eslamian. The title of his thesis is "Assessing the Trend of Regional Development Based on Patterns of Sustainable Development in Water Resources by System Dynamics Approach." He has published 5 scientific publications in national and international journals, and has contributed to 3 chapters in the *Handbook of Drought and Water Scarcity*, published by Taylor & Francis (CRC Press). His research and publications are focused on sustainable development, integrated water resource management, social–ecological resilience, systems thinking, climate change, and drought assessment.

Kumari Rina is currently an assistant professor in the Central University of Gujarat, Gandhinagar, India. She completed her BSc and MSc in botany, and earned her MPhil and PhD in environmental sciences from the School of Environmental Science, Jawaharlal Nehru University, New Delhi. During her doctoral research, she worked on various aspects of groundwater: groundwater potential, groundwater quality, impact of land use/land cover change, and sea water intrusion. She has been involved as a scientist in various projects of prestigious organizations in the government of India, such as the Department of Science and Technology and the National Remote Sensing Centre. In June 2013, she joined as an assistant professor in the Biodiversity and Natural Resource Department of the Central University of Orissa, India. Her work is currently focused on groundwater contaminants, groundwater modeling, isotope hydrology, and remote sensing and GIS.

Giorgio Rovero earned his master's degree in chemical engineering at Politecnico di Torino before being appointed an assistant professor. He was then enrolled by the University of British Columbia as a visiting professor and research associate for about three years on several sprouted bed projects. He was appointed as an associate professor in Politecnico di Torino in 1986 and dedicated his research activity to fluidization, textile wet processing, and wastewater treatment. Rovero proposed and managed an extensive project in collaboration with about 30 industries to devise optimal technologies for water treatment, ultimately aiming at freshwater economization, exhausted liquor reuse, and water recycling. He currently acts as the head of the textile engineering group of Politecnico at the Biella Campus.

Sara Saadati is an MSc graduate in watershed engineering from Isfahan University of Technology, Iran. She is interested in advanced hydrology, sustainable development, environmental assessment, water and wastewater reuse, environmental flows, and climate change. Her publications and presentations in national and international journals and conferences are

mainly in drought, wastewater treatment, environmental flow, and IWRM. She is currently working as a water resources management specialist in a national project titled "Water Recycling Science and Technologies for Municipal, Industrial and Agricultural Wastewater," sponsored by the Ministry of Energy.

Subrota Kumar Saha is a professor at the Department of Geology, University of Dhaka. Dr. Saha is involved in research and teaching in environmental and geological interactions. He has published a good number of research papers at home and abroad. Dr. Saha is the author of *Environmental Auditing—A Road Map to Registration*, a manual on reducing vulnerability on climate change (RVCC); *Environmental Impact Assessment for Changing World*, published by AH Development Publishing House, Dhaka, 2007; *Poribesh Bijnan* ("Environmental Science" in Bengali), published by Bangla Academy, Dhaka, 2007; and *Paribesh Samaj Bigyan* in Bengali by Millenium Pub, Dhaka. He is a member of the Asiatic Society of Bangladesh and the Geological Society of Bangladesh.

R.K. Saket has worked as an associate professor in the Department of Electrical Engineering, Indian Institute of Technology (Banaras Hindu University), Varanasi, Uttar Pradesh, India, since 2005. Previously, he was a faculty member at the Birla Institute of Technology and Science, Pilani, Rajasthan, India; Sam Higginbottom Institute of Agriculture, Technology and Sciences—Deemed University, Allahabad, Uttar Pradesh, India, and University Institute of Technology, Rajiv Gandhi University of Technology, Bhopal, Madhya Pradesh, India. He has contributed more than 60 scientific articles, including prestigious book chapters and research papers in reputed international journals and conference proceedings. His research interests include reliability engineering, electrical machines and drives, power system reliability, self-excited induction generator/doubly fed induction generator, micro-hydro power generaration using municipal wastewater, and renewable energy systems. Dr. Saket is a life member of the Indian Society of Technical Education, New Delhi, India. He is an editorial board member of the *International Journal of Research and Reviews in Applied Sciences*; *International Journal of Management, Modern Science and Technology (UAE)*; and *Engineering, Technology and Applied Science Research (Greece)*. He is a reviewer of the scientific article and research papers of the *IEEE Transactions on Industry Applications, IET Renewable Power Generation, International Journal of Modeling and Simulations, International Journal of Electric Power and Energy Systems* and *International Journal of Global Energy Issues*. The Nehru Encouragement Award was conferred upon him on January 26, 1989, and January 26, 1991, by Madhya Pradesh, state government of India, for academic and research excellences toward his higher technical education.

Md Salequzzaman is an expert on sustainability and environmental technology specialized to work with sustainable development for society, including climate change impact on water supply and sanitation; planning, environmental, and social development for the decision-making process; and participatory community movement. He has 25 years of experience in the field of sustainability applications in education and industrial and natural resources management, especially for water supply and sanitation. He also taught water supply and sanitation; sustainable technology and policy; global environmental issues; sustainability rules, legislation and protocol; ecologically sustainable development; environmental economics; natural resources management; environmental and social impact assessment; and environmental planning at the university level where he worked with various local and international organizations on different sustainable community issues.

His research interest focuses on the optimization of sustainability in the water supply and sanitation especially in industrial and natural resources management and decision-making processes using sustainability indicators. His tertiary education includes a postdoc in sustainability integration; a PhD in sustainability (2004); an MSc in environmental engineering (1996); an MSc in natural resources management (1988); and an MSc in industrial resources management (2007). In addition, very recently, he completed a further study on project management, laboratory technology, occupational health and safety and education from Western Australia. In Australia, he worked on sustainability in the Centre for Advanced Studies in Australia, Asia and the Pacific at Curtin University of Technology and Royal Perth Hospital. He has substantial research and working experience in various forms of waste and wastewater treatment and reuse. Currently, he is working in environmental science at Khulna University of Bangladesh, as well as in a few universities in Western Australia in the area of waste management.

Nitirach Sa-nguanduan earned a PhD in urban environmental management from the Asian Institute of Technology, Thailand, with research focusing on water reuse. She is currently a lecturer at Kasetsart University, Thailand. Her academic and research interests are in environmental engineering and management with specialization areas in water reuse, wastewater treatment, and water quality management.

Toshio Sato, based in Tokyo, works in the Global Environment Business Division of the Japan-based OYO Corporation. He holds a PhD in global arid land science from Tottori University, Japan. Dr. Sato has more than 5 years of experience in working at national and international levels with a focus on environmental implications stemming from the use of wastewater in agriculture in untreated or partly treated forms. His postgraduate study and research work undertaken in 2008–2010 was part of the International Joint Master's degree program on integrated land management in drylands, supported by the United Nations University (UNU); Institut des Régions Arides (IRA), Institut National Agronomique de Tunisie (INAT); International Center for Agricultural Research in the Dry Areas (ICARDA); Cold and Arid Regions Environmental and Engineering Research Institute (CAREERI); and Arid Land Research Centre, Tottori.

In addition to research, he has also worked and published on global and regional assessments of wastewater production, treatment, and use.

Majedeh Sayahi is a master's degree student in the Department of Water Engineering at Isfahan University of Technology (IUT), Iran. She earned her BS in water engineering from Shahrekord University, Iran (2009–2013). Sayahi's current research focuses on recharging groundwater using urban wastewater.

Michael Schmidt (Prof. Dr.-Eng. Dr. h.c. (NMU Dnepropetrovsk)) is a professor of environmental planning at Brandenburg University of Technology Cottbus-Senftenberg at Cottbus, Germany. He is one of the prominent experts on land use and environmental planning and its sustainable national sectored implementation of legal frameworks for the instruments of Environmental Impact Assessment (EIA) in Syria in cooperation with GIZ (Germany) and Yemen in cooperation with the EU-FSMU program. He is also a prominent expert in the development of guidelines for transboundary EIA and Strategic Environmental Assessment (SEA). He has written several popular books on land use and environmental planning, and has published several articles in journals and books. His research focuses on multivariate environmental scientific works in interdisciplinary and transboundary fields in world heritage context, regional as well as urban planning and the environment, mitigation and adaptation to climate change impacts, human geography, soil protection, landscape archaeology and soil geography, as well as avoided deforestation and REDD. He earned his doctoral degree from the University of Göttingen, Germany. After serving in the Federal Agency for the Environment, he became head of the soil protection unit at the Brandenburg State Agency of the Environment and was appointed governmental executive director in 1992. In 1994, he became a professor at the newly established chair of environmental planning at BTU Cottbus. In 2002, he received the Award for Excellence in International University Cooperation from the German State Federal Ministry of Education and Research and in 2005 the Dr. h.c. of the National Mining University, Dnepropetrovsk, Ukraine. He was vice president for international affairs from 2000 to 2006. Since 2009, he has been the head of the world heritage studies master's program, and since 2011 he has been the dean of the Faculty of Environmental Science and Process Engineering at BTU Cottbus-Senftenberg (second term). In 2012, he initiated the joint master program, heritage conservation and site management, in cooperation with Helwan University, Egypt, and in 2013, the double master program, world heritage studies, at BTU Cottbus-Senftenberg in cooperation with Deakin University, Australia. He has more than 25 years of experience in intercultural teamwork and scientific research in Germany, United States, Belgium, Jordan, Yemen, Syria, Lebanon, Ukraine, and Vietnam.

Peter W. Schouten is a postdoctoral scientist working as a visiting fellow with the Australian Commonwealth Scientific and Industrial Research Organization (CSIRO). His research is currently focused on the online measurement of non-CO_2 greenhouse gas emissions from modern decentralized wastewater treatment processes. By performing this research, Dr. Schouten will determine if decentralized wastewater treatment plants (WWTPs) have a measurable influence on positive atmospheric radiative forcing, and ultimately, climate change.

Zarook Shareefdeen is an associate professor at the American University of Sharjah, UAE. Previously, he worked at the University of Waterloo, Canada; BIOREM Inc., Ontario, Canada; University of California; and King Fahd University of Petroleum and Minerals, Saudi Arabia. He is a registered professional engineer in Ontario, Canada, and earned his PhD in chemical engineering from the New Jersey Institute of Technology. His research areas include biofiltration, bioreactor modeling, waste management, waste to energy, and pollution control. He holds two U.S. patents on bioreactor operations and bioreactor media, published several book chapters, edited a book, and has published in refereed journal publications.

Ashok Sharma is a CSIRO principal research scientist and project leader. He is a professional chartered engineer with more than 25 years of experience in wastewater and stormwater infrastructure planning and design, water pollution evaluation and mitigation, water supply management, and sustainability/environmental footprint assessment. Presently, Sharma leads a number of projects investigating the long-term environmental, economic, and social impacts of the uptake of decentralized and alternative water supply and reuse technologies in Australian urban areas.

Mostafa Shiddiquzzaman is a young educationist and researcher in the area of environmental safety principles. Currently he is completing his undergraduate study in the area of marine science. In the last few years, he has completed several courses in molecular biology and biomedical science along with urban and regional planning. As a young fellow, he has a great passion to conduct research for humanity.

Jonathan Sierke is currently a PhD candidate in chemistry at Flinders University, Adelaide. He earned his bachelor of science and a first-class bachelor of science (honors) the following year. His honors research focused on the synthesis and thermal properties of silicone copolymers with different monomer distributions. For his PhD research, he moved to studying fillers and polymeric coatings for UF and MF polymeric membranes, particularly those used for pretreatment in desalination. The goal of his research is to produce compaction and biofouling resistant membranes for use in the water industry to improve the efficiency of the water purification process.

Chander Kumar Singh is an assistant professor in the Department of Natural Resources at TERI University, New Delhi, India. He earned his MSc degree in analytical chemistry from Banaras Hindu University and MPhil and PhD in remote sensing, GIS, and geophysical applications in hydrology from Jawaharlal Nehru University, New Delhi. He has worked for many premier organizations in India. He has published several international scientific papers in peer-reviewed journals on topics ranging from water pollution and remote sensing to urban heat islands. His research interests are in hydrogeochemistry, geogenic pollutants, hydrology, urban sprawl, urban heat islands, and public health. He has been using a very integrated approach to remote sensing, GIS, for water resources management. He has been actively involved in research with several universities across the world. Currently, he is leading a National Science Foundation study on arsenic and fluoride contamination under a PEER Science Grant. He has also conducted a first-of-its-kind sociobehavioral study on an arsenic-affected region in India.

Neha Singh is a doctoral student in the School of Environmental Sciences, Jawaharlal Nehru University. She studied at the graduate and postgraduate levels at Banaras Hindu University, Varanasi, India. Her doctoral research mainly emphasizes groundwater arsenic contamination.

Prafull Singh is an assistant professor in the Amity Institute of Geoinformatics and Remote Sensing, Amity University, Noida, India. He earned his BS and MS degrees in environmental geology from Jiwaji University, Gwalior (2000, 2003). In 2009, he earned his PhD from Jiwaji University, Gwalior, in groundwater potential and recharge modeling using geophysical, remote sensing, and GIS techniques. Dr. Singh's current research focuses on groundwater recharge and potential modeling, groundwater pollution and prediction modeling, aquifer characterization using geophysical survey, watershed modeling using remote sensing, and GIS.

Vishal Singh is a research scholar in the Department of Civil Engineering, Indian Institute of Technology, Guwahati, India. His research interests include hydrological modeling, remote sensing, and climate change. He has published several national and international peer-reviewed research articles.

Seval Kutlu Akal Solmaz is a professor in the Department of Environmental Engineering at Uludag University. She earned her BS and MS degrees in environmental engineering from Istanbul Technical University (ITU) (1986, 1990). In 1997, she earned her PhD in environmental engineering from ITU. Dr. Solmaz currently focuses on treatment of micropollutants with advanced oxidation processes and wastewater treatment, recycle, and reuse.

Mynepalli Kameswara Chandra Sridhar is a professor of environmental health sciences at the College of Health Sciences, Niger Delta University, Wilberforce Island, Bayelsa State, Nigeria, and is currently a professor on contract in the Department of Environmental Health Sciences, Faculty of Public Health, College of Medicine, University of Ibadan, Ibadan, Nigeria, where he spent his academic career before retiring. He earned his BSc (biological sciences) in 1961 (first class) from Andhra University, Waltair, India; MSc (biochemistry) (first class) in

1964 from Maharaja Sayajirao University of Baroda, Baroda, India; and PhD (biochemistry) in 1971 from the Indian Institute of Science, Bangalore, India. A recipient of many awards and distinctions, he is a fellow of the Royal Society of Health, UK (FRSH) (1978), member of the Chartered Institution of Water and Environmental Management, UK (MCIWEM) (1984), chartered chemist and member of Royal Society of Chemistry, UK (C.Chem., MRSC, 1985), member of the International Society of Environmental Epidemiology, USA (1996), and Fellow of the Institute of Waste Management, Nigeria (2012). His areas of specialization include water quality, water and waste management (domestic, industrial, and hazardous), waste recycling, food safety and hygiene, vector control, indoor and outdoor air pollution, climate change, monitoring/surveillance, environmental impact assessment/risk assessment/environmental audit, environmental education, and community mobilization in environmental management. He is also the author of many book chapters and research papers, and holds many patents. Over the years, he has inspired research and mentored many graduate students in the area of environmental health sciences.

Jatin Kumar Srivastava is an environmentalist based in Lucknow, India, who works as an environmental coordinator at Global Environment Engineering Service, Lucknow. Dr. Srivastava has 14 years of experience and is currently involved in nonconventional methods of improvement of urban water quality. He has worked in several national laboratories and has published several research papers and reviews on major issues in environmental sciences having substantial impact in the field of knowledge of science.

Carlo Stagnaro is head of the Minister's Technical Secretariat at the Italian Ministry of Economic Development. He is also a senior fellow at Istituto Bruno Leoni (IBL), a Milan-based think tank. He was the IBL research and studies director until March 2014. Dr. Stagnaro earned a laurea degree in environmental engineering from the University of Genoa, Italy, and a PhD in economics, markets, and institutions from IMT Alti Studi—Lucca, Italy. He is a member of the editorial board of the quarterly journal *Energia*. Since 2007, he has edited the IBL's *Indice delle liberalizzazioni* (*Index of Liberalizations*), an annual report on the degree of market openness in 10 sectors of the economy in the EU15 member states. His research interests are in the fields of energy economics and public service economics. He is currently writing a book on the impact of renewable energy subsidies on competition in the EU's electricity markets.

Alexandros I. Stefanakis is an environmental engineer with a diploma and doctoral degrees from the Department of Environmental Engineering, Democritus University of Thrace, Greece. He also holds an MSc degree in the field of hydraulic engineering from the Department of Civil Engineering of the same university. He has approximately 10 years of experience in the fields of natural systems for wastewater treatment with emphasis on constructed wetlands and water resources engineering. His diploma thesis investigated the use of ultrafiltration membranes

and activated carbon filters as an advanced stage in the wastewater treatment chain. His MSc and PhD theses focused on ecological engineering, that is, ecological treatment of wastewater and sludge dewatering using constructed wetlands. He has published several papers in international high-ranking peer-reviewed journals, papers in conference proceedings, a book on constructed wetlands, technical reports, lecture notes, and book chapters on environmental and ecological engineering. His research work includes numerous experiments on different treatment methods. He has participated in national and EU research projects in Greece (Democritus University of Thrace, University of Patras), Portugal (University of Beira Interior), Germany (UFZ-Helmholtz Centre for Environmental Research), and the UK (University of Brighton). He has taught at the undergraduate level on the topic of environmental protection, delivered seminars in postgraduate studies programs, and has been involved in the organization of two international IWA conferences. He has been awarded twice during his career for his research work. Currently, Dr. Stefanakis is employed as a lecturer of water engineering at the University of Brighton in the United Kingdom. He is also a practicing environmental engineer in Greece, dealing mainly with the design of constructed wetlands systems.

Aeysha Sultan earned a master of science with specialization in organic chemistry from the University of Sargodha. She earned her PhD from the University of Sargodha and had the opportunity to work as an exchange student at the School of Chemistry, University of New South Wales (UNSW), Sydney, Australia. She is currently a lecturer at the University of Sargodha, Sargodha, Pakistan. Although she is an organic chemist with interest in synthesis, she has undertaken advanced courses and has an intense interest in analytical and environmental chemistry.

Geoffrey J. Syme is currently a professor of planning at Edith Cowan University in Perth, Western Australia. He has spent 35 years researching and consulting in the area of social and policy aspects of water resource management at CSIRO Land and Water. He has been the co-winner of the AWA Peter Hughes Water Award and the CSIRO Medal for Research Excellence in urban water conservation. As a social scientist, he is currently an editor in chief of the *Journal of Hydrology* and associate editor of the *Journal of Water Reuse and Desalination*. He has been a regular consultant to water utilities and regulators in all states and territories. He has published approximately 100 articles in international journals, 20 book chapters, and 125 consultancy reports in the water resource management fields.

Roman Tandlich earned his bachelor and master of science degrees in biochemistry and biotechnology from the Slovak University of Technology in Bratislava in 1998. In 2004, Dr. Tandlich earned his PhD in pharmaceutical sciences from North Dakota State University in Fargo. After two postdoctoral fellowships in Israel and South Africa, he became a (senior) lecturer in pharmaceutical chemistry and biochemistry at Rhodes University in 2008. His research interests include wastewater treatment, microbial water quality, and disaster management. Dr. Tandlich has published 52 peer-reviewed papers, 7 technical reports, and 3 book chapters.

Jay Krishna Thakur is a director at UIZ Umwelt und Informationstechnologie Zentrum, Berlin, Germany. He earned his BSc in environmental science and BEd in educational administration and supervision from Tribhuvan University, Kathmandu, Nepal. In 2008, he earned his first MSc in environmental sciences from the School of Environmental Sciences, Jawaharlal Nehru University, India. In 2010, he earned his second MSc in Geo-Information Science and Earth Observation for water resources and environmental management from the Faculty of the International Institute for Geo-Information Science and Earth Observation (ITC), Universiteit Twente, Enschede, the Netherlands. Upon completion of these degrees, he pursued his doctorate in natural sciences (Dr. nat. sc.) at the Department of Hydrogeology and Environmental Geology, Institute for Geosciences and Geography, Martin Luther University, Halle, Germany, with the HIGRADE Scholarship from the Department of Groundwater Remediation, Helmoltz Centre for Environmental Research, UFZ, Leipzig, Germany. Dr. Thakur's current research focuses on the holistic approach of water resources mapping and management.

Josephine Treacy earned her undergraduate degree in chemical instrumentation at the Limerick Institute of Technology, Ireland. After graduating, she worked with the Cork County Council environmental sector in the area of pollution control and management. She earned her MSc and PhD at University College Cork, Ireland, in the area of environmental analytical science while working with the Cork County Council.

University College Cork, Ireland, then awarded Dr. Treacy a postdoctorate scholarship researching remediation techniques using supercritical fluid technology. Presently, Dr. Treacy holds a lecturing position with the Limerick Institute of Technology, Ireland, lecturing in the following subject areas: environmental studies, pollution control, environmental forensics, green technology and innovation, advanced pharmaceutical technology, spectroscopic and complementary methods, instrumentation, inorganic and physical chemistry, and real-time monitoring.

Her research interests include environmental ecosystems, monitoring and management, drinking water treatment, wastewater treatment, and river catchment surveys. Other research interests include sensor deployment and validation for air and water applications, including biofouling elimination and prevention on sensors. Recently, Dr. Treacy earned an MEd in adult and further education at Mary Immaculate College University of Limerick.

Anthony Richard Turton is a professor at the Centre for Environmental Management, University of Free State, Bloemfontein, South Africa. He is a transboundary water resource specialist with a DPhil from the University of Pretoria. His current work focuses on the strategic issues arising from the management of mine water after more than a century of largely unregulated mining in the South African gold and coal fields. Central to this is the vexing challenge of acid mine drainage (AMD) and uranium contamination arising from the many mine tailings dams that characterize the landscape around the city of Johannesburg. He is the past vice president of the International Water Resources Association (IWRA) and served as a deputy governor at the World Water Council. He serves either as an editor or on the editorial board of a number of publications, including *Water Policy* and *Water International, Water Alternatives,* and the *International Journal of Water Governance.* He is a founding trustee of the Water Stewardship Council of Southern Africa and serves as an advisor to executive management in the mining sector. He is a regular contributor to various TV, radio, and print media and holds a number of awards for his work as an environmentalist.

Geetika Tyagi completed her graduate degree from Banasthali University, India, and postgraduation from Banaras Hindu University, Varanasi. Currently, she is pursuing her MPhil from the Central University of Gujarat. Her research interest is on waste to energy generation.

Gökhan Ekrem Üstün is an associate professor in the Department of Environmental Engineering at Uludag University. He earned his undergraduate degree in environmental engineering at Uludag University in 1998. He then earned his MS and PhD in environmental engineering from the Uludag University (2001, 2006). Dr. Üstün's current studies focus on color removal with advanced oxidation processes, treatment of micropollutants, and wastewater treatment, recycle, and reuse.

Amanda Van Epps earned her undergraduate degree in chemical engineering at Stanford University before spending four years as an environmental engineer and consultant on soil and groundwater remediation projects in Northern California. She then enrolled at the University of Texas at Austin to pursue master's degrees in both environmental and water resources engineering and public affairs. Her thesis research focused on the management of used oil in the state of Texas. Upon completion of these degrees, she pursued her doctorate in civil engineering at the University of Texas, focusing her doctoral research on household greywater treatment and reuse. She is currently a Robert Bosch Foundation Fellow, spending a year learning from the German experience with water reuse.

Joanne Vanderzalm earned her PhD in geochemistry from Flinders University, South Australia. Prior to this, Vanderzalm completed a bachelor of applied science and master of applied science, majoring in chemistry from Monash University, Victoria. Dr. Vanderzalm worked in industrial and environmental chemistry within the electricity sector for several years. She is currently a research scientist within the CSIRO Sustainable Water Systems Stream in the Water for Healthy Country Flagship. Her research area is geochemical processes associated with water reuse via managed aquifer recharge, in particular arsenic mobility, aquifer treatment processes, and using environmental tracers to assess reaction processes.

Zhifang Wu is currently a research fellow at the Centre for Comparative Water Policies and Laws (CCWPL), University of South Australia. She earned her undergraduate in economics at Changchun Taxation Institute in China. After completing her PhD at the University of South Australia, Dr. Wu was employed by the CCWPL and worked on research projects with colleagues in the Center. Her research interests lie in the field of exploring community attitudes and perceptions regarding using nonconventional water sources, such as stormwater, reclaimed water, and so on. Dr. Wu also has interest in water governance, and in particular institutional and policy analysis in relation to diversifying a sustainable water supply portfolio.

Yohannes Yihdego is a senior hydrogeologists/scientist at the Snowy Mountains Engineering Corporation (SMEC), Australia, which is one of the leading international engineering consulting services. Prior to his current position, Dr. Yihdego served as senior hydrogeologists/modeler at Cardno Lane Piper and GHD. His duties include working as an engineering geologist and consulting hydrogeologist. He has worked extensively with the mining, infrastructure, agriculture, and civil construction industries on hydrogeological, contamination, and environmental assessments in multidisciplinary teams as a team member and project manager. The majority of his work has involved hydrogeological investigations, resources development, evaluation and management, impacts of mining, land use change and groundwater extraction on water resources, beneficial users and dependent ecosystems, water supplies for agriculture and industry, civil construction dewatering, water and soil contamination assessments, and remediation designs. His expertise includes contaminated land hydrogeology and his clients have included local government, government departments, private developers, investors, financiers, and land owners. Dr. Yihdego is involved with drought monitoring, assessment, and prediction. In addition, he has extensive experience working with international drought, climate, and water-related issues via project collaboration and consultation with more than 15 countries. Dr. Yihdego earned his

bachelor's in geology from Addis Ababa University, Ethiopia, and his master's with distinction in hydrogeology from ITC, University of Twente, the Netherlands. He earned his PhD in the field of hydrogeology from La Trobe University, Australia. He has authored or coauthored more than 100 journal articles, conference papers, and technical reports. He is an associate editor of an international journal. Dr. Yihdego was named by the Australian government as one of the most influential African-Australians of the year 2013.

Negar Zamani earned her degree in civil engineering from Guilan University. She is interested in water resources and environmental research.

Dongqing Zhang is currently serving as a senior research fellow at the Nanyang Environment and Water Research Institute, School of Civil Environmental Engineering, after earning her PhD from the Dortmund University of Technology, Germany. Her main research field is the removal of emerging contaminants (e.g., pharmaceuticals and personal care products, nanoparticles) from wastewater.

Lotfollah Ziaei is a water resources engineer in Zayandab Consulting Engineers Co., Isfahan, with more than 30 years of experience in water resources evaluation, planning, and watershed management. He has worked on numerous projects in water management, water supply development, water resources optimization, watershed protection and stormwater master planning, integrated surface and groundwater modeling, water quality assessment and modeling, program and project management, and flood forecasting/flood mitigation.

Bongumusa M. Zuma undertook and earned a bachelor of science in biochemistry summa cum laude from the University of Zululand, and his bachelor of science with honors in biotechnology and master of science in water resource sciences from Rhodes University, South Africa, between 2006 and 2009. He also holds a PhD in biotechnology from Rhodes University, South Africa (2012). He specializes in water and wastewater management and treatment, water quality monitoring, and onsite sanitation systems including greywater treatment. He has (co-)authored 13 peer-reviewed papers and 2 book chapters.

I

Introduction to Water Reuse

1

Water Shortages

Zhuo Chen
University of Technology Sydney
Graduate School at Shenzhen
Tsinghua University

Huu Hao Ngo
University of Technology Sydney

Wenshan Guo
University of Technology Sydney

Saeid Eslamian
Isfahan University of Technology

PREFACE

Growing environmental problems, including diminishing natural water resources, greater water demand triggered by population growth and urbanization, deteriorated water quality, and highly changing climate, have highlighted the importance of considering the sustainability of water resources and supply systems in both urban and rural areas. Specifically, the world population continues to expand with a growth rate of 1.2% each year, resulting in increased pressure on water quality, safety, and health.

This chapter overviews the current situations in water availability over the world, which is based on information from literature review, recent research studies, and analyses of case studies. It focuses on the issues related to the deficiency of water quantity and deterioration of water quality. With identified problems and challenges, this chapter also discusses some potential responses, opportunities, and solutions to combat water scarcity.

1.1 Introduction

Water is vital to human health and well-being, and to the support of ecosystems, agricultural and industrial development, and the environment. It also has crucial cultural value and social significance [2]. Because of climate change (e.g., flooding, prolonged drought, and severe cold), increasing population, rapid urbanization, and deteriorating water quality, water scarcity is considered one of the most important threats to society and a constraint for sustainable development. Within the next decades, due to continuous economic and population growth, water may become the most strategic resource in many areas of the world, especially the arid and semi-arid regions [28,60]. Based on the International Water Management Institute definition, a very significant portion of the world is projected to suffer from both physical (1000 m³ per person/annum renewable water supply) and economic water scarcity by 2025.

In Australia, because of water scarcity issues, population losses have been observed in inland rural areas (e.g., northeastern, south-eastern, and western Australia), which led to a further decline in irrigation services, failure of local businesses, and growing unemployment [34]. In the Middle East, water-related issues have already played an important part in the wide disputes between the regional countries, which could also become the case in Central Asia and parts of Southeast Asia. It is predicted that growing conflicts over access to water may even lead to armed conflicts among nations over control of rivers and aquifers in the future [10,35,51]. Additionally, the United Nations stated that water shortages in terms of reduced water availability, desertification, recurrent flooding, and increased salinity in coastal zones could indirectly lead to internal instability in many areas. Such deterioration may force movement, migration, and/or displacement, triggering social and ethnic tensions in cities and local areas [40]. It is worth noting that

3

the number of "water refugees" around the world has increased from 25 million in 1995 to 50 million in 2010, and the figure will continue to rise steeply over the next decades [30,35]. In addition, an Australian report pointed out that drought is also a significant cause of suicide and mental illness for Australian farmers and rural communities [34].

In light of potential water shortages, some cities have increasingly recognized the significance of considerable water conservation and water demand management as a long-term water supply strategy. However, as water conservation is inadequate to close the water supply–demand gap in some cases, sustainable management solutions and technological options in driving green growth for augmenting existing water supplies should be considered, including the exploitation of alternative water resources such as desalinated and recycled water, and the development of aquifer storage and treatment of unusable water [53]. Moreover, other innovations should also be taken into account, which can promote the water market toward increased efficiency, productivity, and enhanced environmental outcomes to balance the environmental, economic, and social issues [2].

1.2 Current Situations in Water Availability

The availability of water in sufficient quantity and adequate quality is indispensable for human beings and the functioning of the biosphere [37].

1.2.1 Deficiency of Water Quantity

The surface water and groundwater constitute the major constant water resources for water supply in both urban and rural areas. As shown in Figure 1.1, in a natural water cycle, rainfall normally enters the surface water system via precipitation, namely rainwater, or from the upper catchment or rivers, namely stormwater, and then leaves the system through evaporation, transpiration, downstream outflow, or surface runoff. At the same time, surface water can contribute to groundwater replenishment through seepage and infiltration [56]. Although the

hydrological cycle links all waters, surface water and groundwater are usually studied separately and represent different development opportunities.

As can be seen from Figure 1.2, the total water resources in the globe are estimated to be 43,000 cubic kilometers per year (km^3/year), which are distributed throughout the world unevenly on the basis of climate and physiographic structures [23].

When it comes to the continental level, America (including North America, Central America and the Caribbean, and Southern America) has the world's most abundant freshwater resources (45%), followed by Asia (28%), Europe (15.5%), and Africa (9%). It is widely accepted that the per capita water resource can be used as a parameter for a reasonable evaluation of water supply conditions. Thus, from the perspective of per capita resources in each continent, the Thelon River basin in Canada's Northwest Territories has the highest per capita water resource at around 15,000,000 m^3/person/year, while the Yaqui River in Sonora, Mexico has the lowest per capita water resource at only 173 m^3/person/year. The Yellow River basin in China is also among the watersheds with very low per capita water resource at 361 m^3/person/year [65,66]. Africa and Asia only occupy 5000 and 3400 cubic meters per year (m^3/year) of water resources per inhabitant, respectively, compared with 24,000 m^3/year of that in America (including North America, Central America and the Caribbean, and Southern America) [23]. In highly populated countries such as China and India, the per capita water availability is as low as 1700–2000 m^3/year [27].

Once a country's available water resources drop below 1700 m^3 per person per year, the country can be expected to experience regular water stress, a situation in which disruptive water shortages can frequently occur. When the available water resource drops below 1000 m^3/person/year, the situations can be more severe and lead to problems with food production and economic development. Moreover, if the amount of water available per capita drops below 500 m^3/person/year, countries face conditions of absolute water scarcity. By 2025, it is projected that, assuming current consumption patterns continue, at least 3.5 billion people, or 48% of the world's population, will live in water-stressed river basins. Of these, 2.4 billion will live under high water stress conditions [65,66].

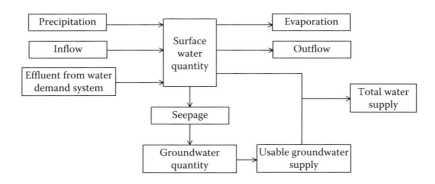

FIGURE 1.1 Schematic diagram of a generic water supply system. (Reprinted from Tan, Y.Y. and Wang, X., An early warning system of water shortage in basins based on SD model, *Procedia Environmental Sciences*, 2, 399–406. Copyright 2010, with permission from Elsevier.)

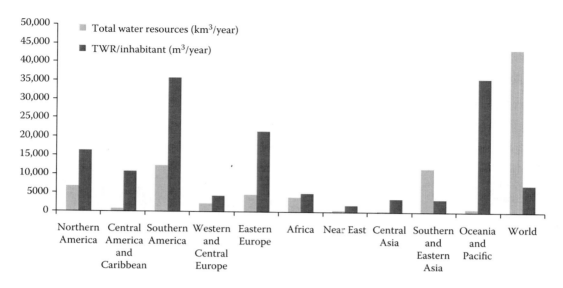

FIGURE 1.2 World water resources by region. (Data adapted from Food and Agricultural Organization. 2003. *Review of World Water Resources by Country.* Food and Agricultural Organization of the United Nations, Rome, Italy.)

It is recognized that water stress and sustainability are a direct function of the water resources available and of water withdrawal and consumption [24]. The increasing water needs/demands because of population growth, rapid economic development with land overuse, and climate change are the major drivers of increasing global water stress [24]. The world population is predicted to grow from 6.9 billion in 2010 to 8.3 billion in 2030 and 9.1 billion in 2050. By 2030, food demand is predicted to increase by 50% while energy demand from hydropower and other renewable energy resources will rise by 60%. These issues are interconnected and likely to lead to increased competition for water between the different water-using sectors. Currently, agricultural irrigation continues to play an important role in food production, which accounts for 70% of all water withdrawn

by the agricultural, municipal, and industrial sectors. To meet the future water and food demand in this sector, innovative technologies will be needed to improve crop yields and drought tolerance, produce smarter ways of using fertilizer and water, new pesticides and nonchemical approaches to crop protection, reduce postharvest losses, and ensure more sustainable livestock and marine production [61].

Furthermore, the rapid global urbanization has led to increasing growth of urban water use, especially domestic water consumption, giving rise to tension in urban water supply and demand (Figure 1.3). The world's cities are growing at an exceptional rate and urbanization is a continuum. Ninety-three percent of urbanization occurs in poor or developing countries [14]. The accelerating process of urbanization also translates

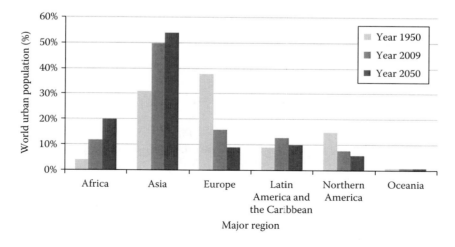

FIGURE 1.3 Distribution of the world urban population by major regions. (Data adapted from Department of Economic and Social Affairs. 2010. *World Urbanization Prospects the 2009 Revision.* United Nations, New York.)

into increased exposure to poorly designed or managed water systems and poor access to hygiene and sanitation facilities in public settings. For instance, in China, rapid urbanization is the dominant social and economic phenomena since the late 1970s. The urban population in China increased from 172.45 million in 1978 to 665.58 million in 2010, and its proportion in total population increased from 17.92% to 49.68%. Because of the growing urban population and industries, about two-thirds of China's 661 cities are short of water, of which more than 110 are suffering from severe water shortage. Hence, it is necessary to study the interaction of urbanization and water utilization so that those areas with water problems could achieve a sustainable urbanization [71].

Climate change is posing an additional challenge to urban water supplies. It may lead to increases in temperature, higher evaporation rates, and lower rainfall, exacerbating water-related disasters. For instance, this is likely to result in increased rainfall variability, increased risk of drought, and reduced availability of water for inland regions [29]. Consequently, surface water and groundwater resources are likely to be affected by these changes. Ceres [9] listed the observed changes in North American water resources during the past century. These include (i) 1–4 week earlier peak streamflow due to earlier warming-driven snowmelt; (ii) decreased proportion of precipitation falling as snow; (iii) reduced duration and extent of snow cover and ice cover; (iv) increased annual precipitation in most of North America but decreased annual precipitation in places like the Central Rockies and southwestern United States; (v) declined mountain snow water equivalent; (vi) increased frequency of heavy precipitation events; (vii) reduced runoff and streamflow; (viii) increased temperature of lakes; (ix) salinization of coastal surface waters; and (x) increased periods of drought. As can be seen, the climate change contributed to increases in intensity, duration, and spatial extent of droughts associated with higher temperatures, warmer sea surface temperatures, changes in precipitation patterns, and diminishing glaciers and snow pack [9].

Similarly, in Australia, climate change is likely to increase the difficulties that the country faces to secure adequate water supplies for cities and irrigation. Specifically, climate change projections showed increased potential evaporation, a tendency for decreases in winter–spring rainfall over the southern half of the continent, and a tendency for increased summer/autumn rainfall in northern Australia. In southern Australia, where winter and spring rainfalls are more important and competition for water between natural and human uses is already high, reductions in water supply appear to be much more severe [10].

With respect to surface water, water may be abstracted from artificial reservoirs, dams, rivers, lakes, and household water tanks. Notably, the rivers and surrounding watersheds are the major sources of water for agricultural irrigation, industrial and residential purposes, ecological/environmental flows, drinking, and other uses. Table 1.1 lists the major rivers in the world together with their drainage area and average annual runoff. Nevertheless, major river levels have dropped significantly at

TABLE 1.1 Major Rivers in the Globe with Their Average Annual Runoff

River	Length (km)	Drainage Area (km²)	Average Annual Runoff (km³)	References
Murray-Darling, Australia	3780	1,056,000	530.0	[42]
Yangtze, China	6300	1,808,500	951.3	[44]
Yellow, China	5464	752,433	66.1	[44]
Mekong, China	4350	810,000	505.0	[67]
Volga, Russia	3692	1,380,000	254.4	[67]
Yenisei, Russia	5539	2,580,000	618.1	[67]
Ob-Irtysh, Russia	5410	2,990,000	403.7	[67]
Lena, Russia	4400	2,490,000	539.2	[67]
Nile, Africa	6650	3,254,555	161.0	[67]
Congo-Chambeshi, Africa	4700	3,680,000	1318.0	[67]
Niger, Africa	4200	2,090,000	301.8	[67]
Mississippi, North America	6270	2,980,000	530.0	[67]
Mackenzie, North America	4241	1,790,000	324.8	[67]
Amazon, South America	6400	7,050,000	6911.6	[67]
Paraná, South America	4880	2,582,672	567.6	[67]

present and the existing water supply is often insufficient to meet the increasing water demands particularly in times of low rainfall, in both rural and urban areas. For instance, in Australia, the recent drought conditions across much of southeastern Australia during 2008–2009 have directly led to a 30% reduction in fresh water supply volumes compared to 2006–2007. It is estimated that the rivers and streams in the Murray-Darling Basin may have flows reduced by up to 30% by 2050, and 45% by 2070, which will further increase competition for already scarce water resources [29].

As a result, water restrictions across Australia have been invoked more often and at deeper levels, affecting many irrigation supply services. More seriously, the possibility of acute supply failure has become increasingly apparent and some smaller towns have actually run out of water [68]. Likewise, Northern China's largest natural freshwater lake, Lake Baiyangdian, is both disappearing and grossly contaminated [27], and water crises have occurred in more than 400 Chinese cities at the beginning of 2000 [74].

Moreover, groundwater is a precious and indispensable water resource that has become the principal resource in many cities around the world. Due to advances in technology that decline extraction costs and offer more reliable water delivery, as well as subsidies from the government and local authorities for power and pump installation, the privately owned and managed tube wells are also widely distributed. In Asia, more than 1 billion people rely on groundwater for drinking. Besides, in India and China, which account for 40% of the world's irrigation, the area irrigated by groundwater rose from about 25% in the 1960s to over 50% in the 1990s. This has led to the overexploitation of groundwater resources in the semi-arid regions where water tables have been falling at an alarming rate, often 1–3 m per year [6]. Particularly, in northern part of China, over-pumping and contamination of groundwater has forced cities and businesses to dig as deep as 120–200 m to find clean adequate supplies, compared with only 20–30 m deep a decade ago. Apart from additional costs caused by digging and maintenance, groundwater funnels have been reported and a remarkable one in the northern area of China, with an area of 50,000 km², is considered to be the largest in the world [27,74].

In Europe, groundwater takes up approximately 65% of total water supply. Particularly, in Berlin, Germany, approximately 70% of the water for domestic and industrial uses comes from groundwater. In the Middle East and Africa, many countries such as Saudi Arabia, United Arab Emirates, Libya, and Oman also depend heavily on groundwater [36,49]. Nonetheless, over-extraction has triggered groundwater depletion and related environmental problems. It was estimated that in some regions of the Mediterranean, 58% of coastal aquifers suffer from saline ingress so that agricultural industries were severely affected [18]. For instance, the Disi Aquifer between Jordan and Saudi Arabia is being developed by Jordan, with plans to extract unsustainable amounts of a nonrecharging water resource. The aquifer is likely to be depleted within 50 years [50]. Furthermore, ground subsidence has been observed in some big cities such as Shanghai and Mexico City, which can be a huge threat to construction and buildings [6,13].

To optimize the surface water and groundwater management and match the water resources with the water demands over the short and long term, the linkages between precipitation, surface water runoff, aquifer recharge, and pumping and the correlations between urban water demand, population variation, and climate change should be established [73].

1.2.2 Deterioration of Water Quality

Water quality is inextricably linked with water quantity as both are key determinants of supply. Poor water quality can impact water quantity in a number of ways. For example, polluted water that cannot be used for drinking, bathing, industry, or agriculture may largely reduce the amount of water available for use in a given area. When the water is heavily polluted, some additional costs are required to treat and purify the water for meeting the quality standard of discharge and reuse [61].

It is significant that water quality deterioration has been observed in rural and urban areas throughout Africa, Asia, and Latin America [58]. The degradation of water resources can have a far-reaching effect on environmental quality, human health, and even on global warming [48]. The deteriorating quality of water resources is an immediate result of wastewater discharge from cities and industry. Half of the world's rivers and lakes are polluted due to a lack of proper sanitation. For instance, industrial wastewater discharge can be chronic or accidental. Some chemicals that are used for health care such as endocrine disrupting compounds and pharmaceutical active compounds are increasingly discharged to municipal sewage. These substances are synthetic and have received much concern due to their uncertainty, toxicity, and persistence. They are difficult to trap using conventional treatment methods. Thus, the potential risks are sometimes underestimated [19,41,62].

For surface water such as rivers and lakes, overuse can lead to increased concentration of harmful substances present in the water due to pollution or mineral leaching [61]. For instance, India's diversion of water from the Ganges River over the past 30 years has reduced the water available to its other neighbor, Bangladesh, resulting in increased salinity, a decline in river fish, droughts, declining sedimentary deposits, decreased agricultural productivity, and loss of land to the sea [50]. Additionally, water resources have also been contaminated by pollution from agriculture. A major cause could be the indiscriminate use of chemical fertilizers. Particularly, some inorganic chemical pollutants (e.g., sodium, potassium, calcium, chloride, bromide, and trace heavy metals) are of high concern, as highly saline irrigation water can severely degrade the soil and the accumulation of heavy metals in soil can pose threats to the food chain [13,48]. The unsustainable irrigation activities might also result in groundwater contamination and environmental degradation. Moreover, owing to inappropriate management, water quality deterioration has also been detected in a wide variety of water supply types including piped supplies, boreholes, hand-dug wells, and even traditional water collection sources such as plastic and metal buckets [58].

Besides, water quality can be affected by excessive groundwater pumping through increased concentrations of naturally occurring compounds that become dangerously high as the amount of water dwindles. On the other hand, water quality can be affected by increasing salinity levels as a result of saltwater intrusion into coastal aquifers. Saltwater intrusion will not only reduce the amount of water available for human consumption, but also impact the other uses including agricultural and industrial purposes [61].

It is worth noting that in developing countries, unclean water causes severe health threats to millions of people, particularly causing challenges for children. More than 2.6 billion people (42% of the total population) lack access to basic sanitation facilities at present and are exposed to contaminated water sources, which contain excessive chemical compounds and high concentrations of pathogens (e.g., parasites, bacteria, and viruses). These water-borne substances can lead to a wide variety of diseases (e.g., diarrhea, typhus, guinea worm, malaria, and others). It is reported that 2.2 million deaths per year, 1.46 million of which are children, are caused by sanitation-related diseases, poor hygiene conditions, and unclean water [57,72].

Therefore, it is essential to plan coping strategies effectively together with interventions within minimized investments, especially for rural and regional areas [55]. Baguma et al. [5] conducted a social survey study in Uganda, Africa to examine the efficient water management in developing countries and found that the level of per capita income and involvement of women in water-related operation and maintenance partly influenced the safe water supply in households. They also indicated that the large size of a family could play an important role in the provision of safe water as a large number of active laborers in a family can help to remove unhygienic items from water storage systems or assist in removing waste that could facilitate the growth of pathogens. In addition, Ujang and Henze [59] addressed the use of practical measures for preservation of public health associated with water supply, sewage, solid waste management, and other public health services. By June 2010, there were 1519 centralized municipal wastewater treatment plants (WWTPs) in China and 18 plants were added each week.

Nevertheless, other researchers claim that the conventional management approaches using extensive sewer networks and big sewage treatment plants (STPs) require a high degree of legal enforcement, high capital, operating, maintenance cost, and in-house water and sewer connections and will not be sustainable in nature. There is a need for sustainable sanitation, which is technically manageable, sociopolitically appropriate, systematically reliable, and economically affordable which utilizes minimal amounts of energy and resources with the least negative impact and recovery of useable matters. The concepts of sustainable sanitation emphasize three major components: separation of pollutants at the source, decentralization of facilities, and reuse of by-products, especially treated wastewater and sludge. Since in many rural regions of China, no public wastewater treatment facilities are available and water supply and sanitation infrastructures are still in development stages, it is an ideal situation to start a new infrastructure with a new framework [59]. The components of sustainability in water supply and sanitation include policy and institutional, economic, environmental, technical, and sociopolitical aspects.

1.3 Responses and Opportunities Related to Water Shortage

To combat the water scarcity issues over the world, some incentives that improve water use efficiency and productivity such as technical innovations in developing alternative water resources should be considered. Additionally, the sustainable water management approaches have also been increasingly applied to explore and analyze existing and future water-related issues, as well as to support water managers and decision-makers to put forward solutions for potential situations [16]. The sustainable management generally incorporates the environmental, economic, social, and energy and resource sustainability [15]. To particularly address water shortage problems, some management strategies on demand–supply estimation, integrated water resources management and pricing, and policy and investment decisions should be increasingly discussed and established.

1.3.1 Identification and Evaluation of Possible Water Resources

Previously, the traditional solutions of increasing water supply and providing water security mainly focused on the importation of distant water resources or building additional water dams. For instance, in Asia, Singapore has been importing water from Malaysia under bilateral agreements since 1927. The imported water makes up around 30% of Singapore's total supply [17]. However, increasing supply from traditional sources can be very costly and tedious due to regulations, negotiations, competing users, and high marginal costs of extraction and pumping. Moreover, high infrastructure and energy costs can also increase the difficulty of importing water from distant sources [53,73].

In addition, building large dams for adding additional water supply has been considered unsustainable, which has not only presented numerous issues to the environment such as flooding, deforestation, and reduced pastureland, but also exerted negative impact on the livelihoods of local populations. It is estimated that 40–80 million people have been displaced by the construction of large dams worldwide [21]. Currently, owing to the technical possibility and economic affordability of previously prohibitively expensive technological alternatives, many countries, regions, and cities have increasingly recognized the significance of water conservation, water recycling and reuse, and desalination of brackish or ocean water as long-term water supply strategies [53].

1.3.1.1 Water Conservation

Water conservation is regarded as an environmentally friendly solution. The major advantages of local potable water use reduction, minimized cost of water supply network expansion, and cross-connection as well as leakage checking and repair have been widely acknowledged [47]. Municipalities should routinely promote actions to lower domestic water consumption and cities should require water-saving fixtures in all newly constructed homes, offices, and public buildings and should offer incentives to those who retrofit with these technologies [3].

In industrial sectors, there is a significant variation in the per-production-unit water consumption by different enterprises. The improvement of productivity and efficiency under water scarcity situations requires an acceleration of effort to reduce demand, including water withdrawals by industry and energy producers [3]. With respect to public sports and recreational facilities such as swimming pools, the implementation of water saving strategies such as dual flush toilet systems, water saving and flow regulation devices in shower heads, and filtration systems and pool covers to reduce evaporation are likely to minimize water consumption and the environmental footprint. These actions are being successfully implemented in many newly constructed aquatic centers [33].

Notably, as residential areas are also the main users of urban water, the development of effective strategies and policies to obtain detailed knowledge of water consumption regarding how and where residential water is consumed can provide a greater understanding of the key determinants of each and every water end use and, in return, will allow for the development of improved long-term forecasting models [38]. By implementing water use restrictions as well as regulations and reducing losses due to leaks at the household level, water consumption has been successfully reduced by approximately 22%–40% in many countries such as Australia, France, Canada, and Jordan [39].

Generally, showering is reported as the highest indoor consumption category, being 33% or almost 50 (L/p/d) Litres of water per person per day [39,69]. This water end use category has the potential to be substantially reduced in drought periods. Specifically, the replacement of low efficiency showerheads with higher ones will result in a considerable reduction in average daily shower consumption in the household. From an economic perspective, showerhead retrofit programs and conservation awareness campaigns are considered as least cost potable water saving measures that can be easily implemented by the water business or authority. In addition, laundry is regarded as the second largest indoor user of water [4]. The choice of washing machine type is the main factor affecting the annual water consumption in laundry, and front loaders typically consume less than half as much water per wash as top loaders. The higher occupancy (large families) can lead to lower total per capita water use compared with single person households [12,25,70,69].

1.3.1.2 Water Recycling and Reuse

With the rapid development of advanced wastewater treatment technologies such as membrane filtration in the last 10–15 years, many cities and regions of developed countries including North America, Australia, the Middle East, the Mediterranean, Asia, and Africa have considered water recycling and reuse as an alternative water resource to combat water scarcity issues. Some planned recycled water schemes have also been observed in many developing countries, especially in intensive agricultural areas [11,22]. These are due to the increased knowledge and understanding on the merits of recycled water (e.g., alleviation of the pressure on existing water supplies, reduction of effluent disposal to surface and coastal waters, and provision of more constant volume of water than rainfall-dependent sources). As more and more countries begin to notice the prospect of the fit-for-purpose recycled water reuses, the global water reuse capacity is projected to rise from 33.7 (GL/d) Giga Litres per day in 2010 to 54.5 GL/d in 2015 [44].

However, the real applications of recycled water vary in different countries which depend heavily on available treatment levels, current water supply status, environmental conditions, and public perceptions. Notably, agricultural irrigation currently represents the largest use of recycled water throughout the world. With respect to reuse categories, approximately 91% of the water was used for irrigating crops and pastures (e.g., cotton and vegetable growing, grain farming, fruit and tree nut growing), while 9% was used for other agricultural purposes such as stock drinking water and dairy and piggery cleaning [46].

Moreover, landscape irrigation is the second largest user of recycled water in the world currently, although the particular water demand for different countries and regions varies greatly by geographical location, season, plants, and soil properties. The specific applications include golf course uses, public parks, schools and playgrounds, and residential area landscape uses. Apart from irrigation purposes, recycled water has also been successfully applied to industrial sectors in Japan, the United States, Canada, and Germany since the Second World War for more than 70 years. Recently, industrial use is the third biggest contributor to recycled water consumption. The major categories associated with substantial water consumption include cooling water, boiler feed water, and industrial process water [11]. Furthermore, other applications on environmental flow regulations and residential purposes (e.g., fire protection, toilet flushing, car washing, and clothes washing) are being widely practiced. Residential applications are observed mostly in well-developed countries and regions, especially in highly urbanized areas where dual pipe systems have been established [1].

Notably, some sustainable management or control approaches can be carried out in existing or future recycling schemes to reduce water consumption and overall environmental footprint further. In addition to wastewater reuse, capturing and storing rainwater and stormwater can also reduce the demand on freshwater significantly. For instance, it is suggested to construct micro-dams along major waterways to store floodwater during winter seasons to reuse it again in the summer farming seasons as complementary irrigation water. This action will benefit the farmers and raise the national food sufficiency [31]. However, the rainfall-dependent water resources are highly variable and subject to climate conditions.

1.3.1.3 Desalination

The desalination of saline water has been increasingly recognized as one of the new water resource alternatives to provide freshwater for many communities and industrial sectors in a number of countries around the world. In desalination technologies, saline water is separated into two parts using different forms of energy. The potable water with a low concentration of dissolved salts is produced in one part. In the other part, the brine concentrate is collected, which has a much higher concentration of dissolved salts than the original feed water [7,52]. Remarkably, the desalination market is growing very rapidly with a growth rate of 55% per year. In some arid, semi-arid, and remote regions, desalination has now been able to successfully compete with conventional water resources and water transfers for potable water supply. The large centralized or dual purpose desalination plants are shown to be more economical and suitable for large density population areas [52,26].

Regarding the treatment approaches, the majority of water desalination processes can be divided into two types: phase change thermal processes and membrane processes, both encompassing a number of different processes. All are operated by either conventional or renewable energy sources to produce fresh water [52]. On the other hand, thermal desalination that is based on the principles of evaporation and condensation accounts for about 30%–50% of the entire desalination market [26,20].

TABLE 1.2 Market Historical and Forecast of Desalination Systems

Project Type	2006–2010 (billion USD)	2011–2015 (billion USD)
Seawater reverse osmosis (RO)	9.92	15.48
Seawater multieffect distillation (MED)	3.03	4.04
Seawater multistage flash distillation (MSF)	8.39	7.07
Small thermal	2.06	2.33
Brackish RO	1.43	2.18
Ultrapure RO	0.21	0.30
Total (billion USD)	25.04	31.40

Source: Reprinted from Ghaffour, N., Missimer, T.M., and Amy, G.L., *Desalination*, 309, 197–207, 2013. Copyright 2013, with permission from Elsevier.

In the near future, desalination has great development potential on a global scale. This is attributed to the fact that desalinated water is a new water source that has an essentially unlimited capacity. Table 1.2 shows the overall desalination historical development and market potential in the following years. As can be seen, about 50% of the investments are for reverse osmosis (RO) projects because of the lower capital and water costs and smaller footprint sizes [8]. Thermal processes will also continue to be utilized, especially where energy is available at low cost, and multiple-effect distillation (MED) is likely to replace multi-stage flash desalination (MSF) in future projects [26]. However, desalination processes normally consume a large amount of energy [20]. As many areas often experience a shortage of fossil fuels and inadequate and unreliable electricity supply, the integration of renewable energy resources such as geothermal, solar, or wind energies in desalination and water purification is becoming more viable and promising. Besides, the development of new low cost technologies such as adsorption desalination, improvements in membrane design, and system integration and utilization of low grade waste energy will likely decrease the desalinated brackish water cost. The cost reduction as well as minimization in energy consumption will further benefit the smooth expansion of the desalination market.

1.3.2 Water Demand and Supply Analysis

To calculate the water resources amounts and utilization needs, a series of analyses including the water consumption balance analysis, available water resources analysis, supply–demand water resources balance analysis, and over water resources security analysis can be performed. Based on Reference 56, the total water demand in a socioeconomic system can be generally figured out by the following equation:

$$D = D_I + D_A + D_D + D_E - WC - RW - DE \qquad (1.1)$$

where D_I is the monthly total industrial water demand, D_A is the monthly total agricultural water demand, D_D is the monthly total domestic water demand, D_E is the monthly minimum ecological flow, WC is the total water savings from conservation

strategies, WR is the total recycled water, and DE is the total desalinated water. Accordingly, the early warning level for water shortage can be classified according to the supply–demand ratio (R_{sd}) as follows:

$$R_{sd} = TS/D \qquad (1.2)$$

where R_{sd} is the supply–demand ratio, TS is the monthly total water supply, D is the total water demand and $R_{sd} \geq 1$ indicates the safety level [56].

Moreover, sensitivity analyses using extreme hypothetical socioeconomic and water management scenarios (e.g., population control, economic recession, industrial watershed, no surface water withdrawal, and no groundwater withdrawal) can provide insights to identify key drivers and effective strategies [28].

1.3.3 Integrated Water Resources Management

There is growing evidence that water scarcity issues can be created or intensified by unsustainable decisions to meet increasing water demands. Thus, the integrated water resources management (IWRM) has become an important framework in sustainable development and management of water resources. It involves the consideration of various possible water sources to satisfy the demands of different users, environment protection, and land and urban planning. This allows the capture of added aesthetic, ecological, economic, energy production and conservation, recreational, social, and other benefits in ways that have never been realized before [54]. The key principles in IWRM include

- Freshwater is a finite and vulnerable resource, essential to sustain life, development, and the environment.
- Water development and management should be based on a participatory approach involving users, planners, and policy makers at all levels.
- Women play a central part in the provision, management, and safeguarding of water.
- Water has an economic value in all its competing uses and should be recognized as an economic good.
- The important components in IWRM include economic efficiency, equity, and environmental sustainability [32].

Presently, based on the IWRM concept, some models have been developed such as the dynamic models, linear programming, simulation-based models, multilevel optimization techniques, and nonlinear programming. Particularly, the system dynamics models can facilitate understanding of the interactions among diverse and interconnected subsystems that drive dynamic behaviors. With identified problematic trends and their root drivers, these models are shown to be beneficial in water resources planning and management [28]. Besides, Wang and Huang [63,64] have utilized stochastic fuzzy programming approaches for water resources allocation problems. The developed methodology was applied to case studies and a set of solutions under different feasibility degrees was estimated based on economic efficiency, degree of satisfaction, and risk of constraint violation. Likewise, Pingale et al. [45] have applied the integrated urban water management model considering climate change (IUWMCC) for the optimum allocation of water from various water supply sources to satisfy the water requirements of different users along with considering various system and geometric constraints. Consequently, the application of IWRM concepts and subsequent models would be useful for decision makers or local authorities in optimum planning and utilization of water resources.

1.3.4 Water Pricing, Policy, and Investment

Water pricing reforms and incentives need to be performed to encourage economic efficiency in service delivery, investment, and water use. This involves regulations to ensure that

- Pricing sends a signal for efficient water use and investment.
- Prices are high enough to recover costs of operating and renewing water infrastructure and to ensure that autonomous water businesses are financially viable and able to invest.
- There are strong incentives and disciplines to ensure that costs are not excessively high [43].

At present, less than full cost recovery is the common feature of water utilities servicing residential areas. Many local utilities are not coupling the costs of supplying water to price increases but charging prices significantly lower than those in major urban areas. For long-term sustainable management, they should reduce nonrevenue water, improve water services by establishing autonomous and accountable providers, and increase price signals by improving metering and collection of service charges [3]. The movement to full cost recovery will allow many water businesses to fund major new investments from their customers substantially more than they would have otherwise been. Furthermore, wherever feasible, for IWRM, tariffs should be differentiated to reflect key cost drivers in meeting environmental objectives. Well-designed tariff structures with effective information, education, and communication programs can reduce water demand substantially. In addition, ensuring that pricing of access to water, wastewater, and desalination infrastructure by third parties is efficient will be important in promoting innovation in IWRM [13,43]. Without sufficient incentives and pricing reforms, water utilities, even the larger ones, will become unsustainable and water quality and security will suffer as a result. At the same time, government and local authorities should also invest in inclusive public awareness programs to highlight water issues [3].

1.4 Summary and Conclusions

The need for water continues to become more acute with the changing requirements of an expanding world population. To combat water scarcity issues, sustainable water management has received great attention in recent years. This requires water-related decision making to be performed on a holistic view of the problems due to the multitude of complex, interlinked, socioeconomic, and biophysical subsystems within watershed systems. In addition to rational planning and management of existing water supplies from traditional water resources, the significance and feasibility of potential alternative water resources and long-term water supply strategies should also be highlighted. The solutions may include water conservation, water recycling and reuse, and desalination. However, the smooth establishment and implementation of a sustainable water management framework, principles, and approaches are still challenging, especially in less developed areas. Consequently, before the implementation processes, additional assessment and reforms on water demand–supply analysis, integrated water resources management, and water pricing and corresponding governmental policies are highly recommended.

References

1. Asano, T., Burton, F.L., Leverenz, H.L., Tsuchihashi, R., and Tchobanoglous, G. 2007. *Water Reuse–Issues, Technologies and Applications*, McGraw-Hill, New York.
2. Australian Academy of Technological Sciences and Engineering. 2012. *Sustainable Water Management Securing Australia's Future*, Australian Academy of Technological Sciences and Engineering, Melbourne, Australia.
3. Asian Development Bank. 2013. *Asian Water Development Outlook 2013, Measuring Water Security in Asia and the Pacific*, Asian Development Bank, Manila, Philippines.
4. Babin, R. 2005. *Assessment of Factors Influencing Water Reuse Opportunities in Western Australia*, Dissertation. University of Southern Queensland, Queensland, Australia.
5. Baguma, D., Hashim, J.H., Aljunid, S.M., and Loiskandl, W. 2013. Safe-water shortages, gender perspectives, and related challenges in developing countries: The case of Uganda. *Science of the Total Environment*, 442, 96–102.
6. Barker, R., Scott, C.A., de Fraiture, C., and Amarasinghe, U. 2000. Global water shortages and the challenges facing Mexico. *International Journal of Water Resources Development*, 16(4), 525–542.
7. Buros, O.K. 2000. *The ABCs of Desalting*, 2nd ed. International Desalination Association, Massachusetts.
8. Campos, C. 2010. *The Economics of Desalination for Various Uses*. CETAQUA Water Technology Centre. http://www.rac.es/ficheros/doc/00731.pdf.
9. Ceres. 2009. *Water Scarcity and Climate Change: Growing Risks for Businesses and Investors*. Ceres, Boston, MA.
10. Chartres, C. and Williams, J. 2006. Can Australia overcome its water scarcity problems? *Journal of Developments in Sustainable Agriculture*, 1, 17–24.
11. Chen, Z., Ngo, H.H., and Guo, W.S. 2012. A critical review on sustainability of recycled water schemes. *Science of the Total Environment*, 426, 13–31.
12. Chen, Z., Ngo, H.H., Guo, W.S., Listowski, A., O'Halloran, K., Thompson, M. et al. 2012. Multi-criteria analysis towards the new end use of recycled water for household laundry: A case study in Sydney. *Science of the Total Environment*, 438, 59–65.

13. Chen, Z., Ngo, H.H., and Guo, W.S. 2013. A critical review on the end uses of recycled water. *Critical Reviews in Environmental Science and Technology*, 43, 1446–1516.

14. Department of Economic and Social Affairs. 2010. *World Urbanization Prospects the 2009 Revision*. United Nations, New York.

15. Dincer, I. and Rosen, M.A. 2005. Thermodynamic aspects of renewables and sustainable development. *Renewable and Sustainable Energy Reviews*, 9, 169–189.

16. Dong, C., Schoups, G., and van de Giesen, N. 2013. Scenario development for water resource planning and management: A review. *Technological Forecasting and Social Change*, 80, 749–761.

17. Duerr, R.I. 2013. Water scarcity in Singapore pushes 'toilet to tap' concept. http://www.dw.de/water-scarcity-in-singapore-pushes-toilet-to-tap-concept/a-16904636.

18. Durham, B., Rinck-Pfeiffer, S., and Guendert, D. 2002. Integrated water resource management—through reuse and aquifer recharge. *Desalination*, 152, 333–338.

19. Eriksson, E., Auffarth, K., Henze, M., and Ledin, A. 2002. Characteristics of grey wastewater. *Urban Water*, 4, 85–104.

20. Ettouney, H. 2009. Chapter 2—Conventional thermal processes. In: *Seawater Desalination, Green Energy and Technology*, Cipollina, A. (ed.), Springer-Verlag, Berlin.

21. Fatoorehchi, C. 2012. World water forums expose large dams as "unsustainable". http://www.counterbalance-eib.org/?p=1682.

22. Fatta-Kassinos, D., Kalavrouziotis, I.K., Koukoulakis, P.H., and Vasquez, M.I. 2011. The risks associated with wastewater reuse and xenobiotics in the agroecological environment. *Science of the Total Environment*, 409, 3555–3563.

23. Food and Agricultural Organization. 2003. *Review of World Water Resources by Country*. Food and Agricultural Organization of the United Nations, Rome, Italy.

24. Gallopín, G.C. 2012. *Five Stylized Scenarios*. United Nations Educational, Scientific and Cultural Organization, Paris, France.

25. Gato-Trinidad, S., Jayasuriya, N., and Roberts, P. 2011. Understanding urban residential end uses of water. *Water Science and Technology*, 64, 36–42.

26. Ghaffour, N., Missimer, T.M., and Amy, G.L. 2013. Technical review and evaluation of the economics of water desalination: Current and future challenges for better water supply sustainability. *Desalination*, 309, 197–207.

27. Gleick, P.H. 2009. China and water. In: *The World's Water 2008–2009: The Biennial Report on Freshwater Resources*. Island Press, Washington, DC.

28. Gohari, A., Eslamian, S., Mirchi, A., Abedi-Koupaei, J., Bavani, A.M., and Madani, K. 2013. Water transfer as a solution to water shortage: A fix that can backfire. *Journal of Hydrology*, 491, 23–39.

29. Government of South Australia. 2008. Water quantity fact sheet. http://www.epa.sa.gov.au/soe_resources/education/water_quantity.pdf.

30. Guler, C. 2009. The climate refugee change. http://www.isn.ethz.ch/Digital-Library/Articles/Detail/?lng=en&id=98861.

31. Hadadin, N., Qaqish, M., Akawwi, E., and Bdour, A. 2010. Water shortage in Jordan—Sustainable solutions. *Desalination*, 250, 197–202.

32. Hassing, J., Ipsen, N., Clausen, T.J., Larsen, H., and Lindgaard-Jørgensen, P. 2009. *Integrated Water Resources Management in Action*. United Nations Educational, Scientific and Cultural Organization, Paris, France.

33. Hazell, F., Nimmo, L., and Leaversuch, P. 2006. *Best Practice Profile for Public Swimming Pools—Maximising Reclamation and Reuse*. Royal Life Saving Society (WA Branch), Perth, Western Australia.

34. Hogan, A. and Young, M. 2012. Visioning a Future for Rural and Regional Australia, the National Institute for Rural and Regional Australia. The Australian National University, Canberra, Australia.

35. International Institute for Strategic Studies. 1999. Global water shortages, root of future conflicts? *Strategic Comments*, 5–6, 1–2.

36. Jimenez, B. and Asano, T. 2008. *Water Reuse: An International Survey of Current Practice, Issues and Needs*. IWA Publishing, London, UK.

37. Lehr, J.H. and Keeley, J. 2005. *Water Encyclopedia, Vol. 1, Domestic, Municipal, and Industrial Water Supply and Waste Disposal*. John Wiley & Sons, New York.

38. Makki, A.A., Stewart, R.A., Panuwatwanich, K., and Beal, C. 2013. Revealing the determinants of shower water end use consumption: Enabling better targeted urban water conservation strategies. *Journal of Cleaner Production*, 60, 129–146.

39. Marleni, N., Gray, S., Sharma, A., Burn, S., and Muttil, N. 2012. Impact of water source management practices in residential areas on sewer networks—A review. *Water Science and Technology*, 65(4), 624–642.

40. McAdam, J. 2011. Climate Change Displacement and International Law: Complementary Protection Standards. United Nations High Commissioner for Refugees, Bellagio, Italy.

41. Melin, T., Jefferson, B., Bixio, D., Thoeye, C., Wilde, W.D., Koning, J.D., Van der Graaf, J., and Wintgens, T. 2006. Membrane bioreactor technology for wastewater treatment and reuse. *Desalination*, 187(1–3), 271–282.

42. Murray-Darling Basin Authority. 2013. Surface water in the basin. http://www.mdba.gov.au/what-we-do/water-planning/surface-water-in-the-basin.

43. National Water Commission. 2011. *Review of Pricing Reform in the Australian Water Sector*. National Water Commission, Canberra, Australia.

44. Pearce, G.K. 2008. UF/MF pre-treatment to RO in seawater and wastewater reuse applications: A comparison of energy costs. *Desalination*, 222, 66–73.

45. Pingale, S.M., Jat, M.K., and Khare, D. 2014. Integrated urban water management modelling under climate change scenarios. *Resources, Conservation and Recycling*, 83, 176–189.

46. Primary Industries Standing Committee. 2011. Water use in agriculture RD&E strategy. http://www.npirdef. org/files/resourceLibrary/resource/19_Water_Use_in_ Agriculture_RDE_Strategy.pdf.

47. Radcliffe, J. 2010. Evolution of water recycling in Australian cities since 2003. *Water Science and Technology*, 62(4), 792–802.

48. Raghupathi, H.B. and Ganeshamurthy, A.N. 2013. Deterioration of irrigation water quality. *Current Science*, 105(6), 764–766.

49. Ramsar Convention on Wetlands. 2010. Groundwater replenishment. http://www.ramsar.org/pdf/info/services_ 02_e.pdf.

50. Saul, B., Sherwood, S., McAdam, J., Stephens, T., and Slezak, J. 2012. *Climate Change and Australia, Warming to the Global Challenge*. The Federation Press, Leichhardt, Australia.

51. Seckler, D., Amarasinghe, U., David, M., de Silva, R., and Barker, R. 1998. *World Water Demand and Supply, 1990–2025: Scenarios and Issues, Research Report 19*. International Water Management Institute, Colombo, Sri Lanka.

52. Shatat, M., Worall, M., and Riffat, S. 2013. Opportunities for solar water desalination worldwide: Review. *Sustainable Cities and Society*, 9, 67–80.

53. Smith, T. 2011. *Overcoming Challenges in Wastewater Reuse: A Case Study of San Antonio, Texas*. Bachelor dissertation, Havard University, Cambridge, MA.

54. Struck, S.D. 2012. *Visions of Green Technologies in 2050 for Municipal Resource Management*. Toward a Sustainable Water Future, ASCE, Reston, VA.

55. Surinkul, N. and Koottatep, T. 2010. Advanced sanitation planning tool with health risk assessment: Case study of a peri-urban community in Thailand. *Human and Ecological Risk Assessment: An International Journal*, 15(5), 1064–1077.

56. Tan, Y.Y. and Wang, X. 2010. An early warning system of water shortage in basins based on SD model. *Procedia Environmental Sciences*, 2, 399–406.

57. Törnqvist, R., Norström, A., Kärrman, E., and Malmqvist, P.A. 2008. A framework for planning of sustainable water and sanitation systems in peri-urban areas. *Water Science and Technology*, 58(3), 563–570.

58. Trevett, A.F., Carter, R.C., and Tyrrel, S.F. 2004. Water quality deterioration: A study of household drinking water quality in rural Honduras. *International Journal of Environmental Health Research*, 14(4), 273–283.

59. Ujang, Z. and Henze, M. 2006. *Municipal Wastewater Management in Developing Countries: Principles and Engineering*. IWA Publishing, Lyngby, Denmark.

60. UN Water. 2008. *Status report on integrated water resources management and water efficiency plans*. In: The 16th Session of the Commission on Sustainable Development. Economic and Social Council, New York, USA.

61. United Nations Educational Scientific and Cultural Organization. 2012. *Managing Water under Uncertainty and Risk, the United Nations World Water Development Report 4*, Volume 1. World Bank, Washington, DC.

62. Veolia Environment. 2012. Water quality is deteriorating. http://www.thecitiesoftomorrow.com/solutions/water/ background/water-quality.

63. Wang, S. and Huang, G.H. 2011. Interactive two-stage stochastic fuzzy programming for water resources management. *Journal of Environmental Management*, 92, 1986–1995.

64. Wang, S. and Huang, G.H. 2014. An integrated approach for water resources decision making under interactive and compound uncertainties. *Omega*, 44, 32–40.

65. Wang, X.C. and Jin, P.K. 2006. Water shortage and needs for wastewater re-use in the north China. *Water Science & Technology*, 53, 35–44.

66. Water Resources Institute. 2001. Major watersheds of the world. http://www.wri.org/publication/watersheds-world.

67. Wikipedia. 2013. List of rivers longer than 1000 km. http://en.wikipedia.org/wiki/List_of_rivers_by_length.

68. Whiteoak, K., Boyle, R., and Wiedemann, N. 2008. National Snapshot of Current and Planned Water Recycling and Reuse Rates, Final Report Prepared for the Department of the Environment, Water, Heritage and the Arts. Marsden Jacob Associate, Melbourne, Australia.

69. Willis, R.M., Stewart, R.A., Giurco, D.P., Talebpour, M.R., and Mousavinejad, A. 2013. End use water consumption in households: Impact of socio-demographic factors and efficient devices. *Journal of Cleaner Production*, 60(1), 107–115.

70. Willis, R.M., Stewart, R.A., Panuwatwanich, K., Williams, P.R., and Hollingsworth, A.L. 2011. Quantifying the influence of environmental and water conservation attitudes on household end use water consumption. *Journal of Environmental Management*, 92(8), 1996–2009.

71. Wu, P. and Tan, M. 2012. Challenges for sustainable urbanization: A case study of water shortage and water environment changes in Shandong, China. *Procedia Environmental Sciences*, 13, 919–927.

72. Wuyts, K. 2010. Water and sanitation—A remaining priority. *Water*, 37(5), 34.

73. Xu, P., Brissaud, F., and Salgot, M. 2003. Facing water shortage in a Mediterranean tourist area: Seawater desalination or water reuse? *Water Science and Technology: Water Supply*, 3(3), 63–70.

74. Zhu, Z., Zhou, H., Ouyang, T., Deng, Q., Kuang, Y., and Huang, N. 2001. Water shortages: A serious problem in sustainable development of China. *International Journal of Sustainable Development and World Ecology*, 8(3), 233–237.

2

Urbanization and Water Reuse

Vilas Nitivattananon
Asian Institute of Technology

Nitirach Sa-nguanduan
Kasetsart University

PREFACE

The world urban population is expected to highly increase by 72%, during 2011–2050 in both developed and developing countries. Urbanization promotes rapid social and economic development, but at the same time, leads to many problems, such as concentration of the population, traffic jams, housing shortages, resource shortages, biodiversity reductions, "heat island" effects, noise, and air and water pollution. In the water sector, urban growth generally increases water demand and, consequently, generates more wastewater and related adverse environmental impacts. Limited land is available in urban areas; therefore, the water supply is typically transported from sources in rural areas, and end-of-pipe technology is normally applied for wastewater management. This situation leads to possible unsustainability of water resource systems in urban areas, particularly in the case of rapid urbanization without proper planning.

Water reuse (WR) has been introduced as an eco-efficient way to reduce water stress, and is a growing practice in many cities aiming for improved sustainable water management in their urban areas. While there is a limited understanding of the relationship between urbanization and WR applications, this chapter presents the trend, concept, and challenge of urbanization, and explores roles and applications of WR in urban areas, including a case study of WR from urbanization in Thailand, representing a rapidly urbanizing country. It is based on information from literature review, recent research studies, and analysis and discussions of the case study. WR is expected to provide an alternative water source. Urbanization can provide opportunities to reuse treated wastewater. However, public acceptance, potential adverse impacts, and adequate financing can be key challenges and constraints of WR applications.

2.1 Introduction

Urbanization is the transition from a rural–agrarian to an urban–industrial society. Globally, the level of urbanization is expected to rise from 47% in 2000 to 69% in 2050 (Figure 2.1). Thus, the world urban population is expected to increase by 72%, from 3.6 billion in 2011 to 6.3 billion in 2050 [32] in both developed and developing countries. Urbanization promotes rapid social and economic development, but at the same time, leads to many problems, such as concentration of the population, traffic jams, housing shortages, resource shortages, biodiversity reductions, "heat island" effects, noise, and air and water pollution [13].

In the water sector, urban growth generally increases water demand, and consequently generates more wastewater and related adverse environmental impacts. With a limited supply of land in urban areas, the water supply is typically transported from sources in rural areas, and end-of-pipe technology is normally applied for wastewater management. This situation leads

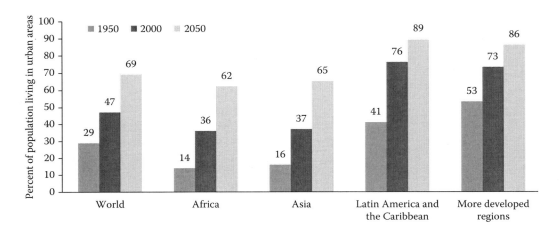

FIGURE 2.1 World and regional percent of population living in urban areas. (Adapted from Population Reference Bureau. 2010. *Trends in Urbanization, by Region.* From http://www.prb.org/Publications/GraphicsBank/PopulationTrends.aspx [retrieved 2010].)

to possible unsustainability of water resource system in urban areas, particularly in case of rapid urbanization without proper planning.

WR has been introduced as an eco-efficient way to reduce water stress, and is a growing practice in many cities aiming for improved sustainable water management in their urban areas [9,24,29]. In addition, WR has a significant impact on relieving the pressure caused by insufficient water resources and suffering from severe environmental problems due to rapidly increasing urbanization in both developed and developing countries. While there is limited understanding of the relationship between urbanization and WR applications, this chapter presents the trend, concept, and challenge of urbanization, and explores the roles and applications of WR in urban areas, including a case study of WR from urbanization in Thailand, representing a rapid urbanizing country. It is based on information

from literature review, recent research studies, and analysis and discussions of the case study.

2.2 Urbanization Trend and Challenges in Relation to Sustainable Water Management

2.2.1 Urbanization

2.2.1.1 Trend of Urbanization

Urbanization and city growth are caused by a number of different factors including rural–urban migration, natural population increase, and annexation [3]. Urbanization is also projected to highly increase in less developed countries in the next decades, including Asia (Figure 2.2). In the less urbanized regions of the world, namely, Africa and Asia, the proportion of the urban

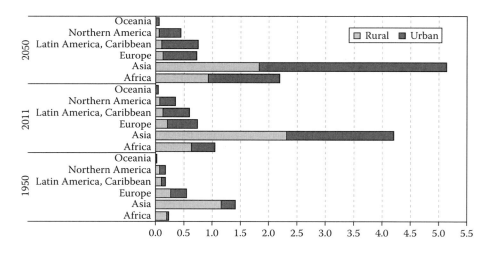

FIGURE 2.2 Rural and urban population (billions) by major regions in 1950, 2011, and 2050, world and regional percentage of population living in urban areas. (Adapted from United Nations. 2012. *World Urbanization Prospects: The 2011 Revision.* United Nations, Department of Economic and Social Affairs, Population Division.)

population is expected to increase to 62% and 65%, respectively, by 2050.

2.2.1.2 Sustainable Water Management

Per limitation of resources, sustainable development (SD) is a concern as the core concept in development. The most popular definition of SD is the one given in the Brundtland Report (1987, p. 43): "development that meets the needs of the present without compromising the ability of future generations to meet their own needs." SD emphasizes the need for understanding the interconnections between the environment, economy, and society. Urbanization has been recognized as one of the major challenges toward SD. Sustainable access to drinking water is one of the Millennium Development Goals (MDGs). While many countries have made significant progress in achieving the goal, several pressures, including rapid urbanization, require significant and more innovative efforts for sustainable water management.

Traditionally, the water sector mainly focuses on sufficiency of a water supply. However, sustainable water management should also be concerned with resource use, environmental effects, and access to water services, service quality, and other sustainable aspects. In urban areas with a focus on water and wastewater, the objectives of sustainable urban water systems management are to satisfy the water-related needs of the community at the lowest cost to the society while minimizing environmental and social impacts, both now and in the future [23]. Demand management and other proactive approaches through conservation, prevention, and closed-loop solutions for the water cycle have been considered as a framework for water management toward sustainability. For example, water-quality regulation and user-pay principal are expected to contribute to both water and wastewater services.

2.2.1.3 Challenges of Water Management under Rapid Urbanization

Typical urban water management is to seek the cheapest source of water, to treat the wastewater, and to dispose of the wastewater. Thus, challenges on water management under rapid urbanization toward sustainability can be identified in three dimensions:

- On water source: To deal with water limitations in urban areas, transportation of raw water from rural areas has been applied, including from nearby countries, such as import of water from Singapore to Malaysia. Thus, new or alternative water sources is a challenge for sustainable urban development.
- On water use: Urbanization causes an increase in water demand and water-quality stress in relation to a better standard of living. Consequently, potable water has been supplied for households, but a large amount has different uses, such as garden irrigation or toilet flushing.
- On wastewater: Wastewater pollution has increased not only from rapid urbanization, but also from end-of-pipe technology, which has been implemented by achieving

the minimum requirement of effluent standards without concern for possible contamination of water sources in the downstream and groundwater pollution. In the case of Asia, only a few cities have the capacity or resources to set up large-scale wastewater treatment facilities. This is a serious problem and improved sanitation and wastewater treatment are major issues in water management in this region [34].

2.2.2 WR in Urban Areas

2.2.2.1 Roles of WR in Sustainable Urban Water Management

WR generally refers to projects or actions that use technology to speed up the natural water cycle processes. A common example of unplanned water recycling occurs when cities draw their water supplies from rivers that receive wastewater discharges upstream from those cities. Water from these rivers has been reused, treated, and piped into the water supply a number of times before the last downstream user withdraws the water [35].

Although there is a long history of reusing wastewater throughout the world, increasing urbanization, particularly in developing countries, is posing a new role to WR as a new water resource for self-sufficient and sustainable urban development. WR has become a more significant component of the hydrological cycle by changing from end-of-pipe to loop systems. In addition, now, recycled water is regarded as a water resource and not as waste [28]. In addition, Asano [1] has also expressed that water reclamation, recycling, and reuse make the best use of existing water recourses by

- Conserving high-quality water supplies by substituting reclaimed water for applications that do not require that quality
- Augmenting potable water sources and providing an alternative source of supply to assist in meeting both present and future water needs
- Protecting aquatic ecosystems by decreasing the diversion of freshwater and reducing the need for water control structures such as dams and reservoirs
- Complying with environmental regulations and liability by better managing water consumption and wastewater discharges to meet regulatory limitations.

According to Miller [16], one of the most significant benefits of WR is the value created by the inclusion of WR in integrated water resources management and other aspects of water policy and the implementation of water projects resulting in the long-term sustainability of our water supplies. The true value of the WR system emerged over time when it permitted urban growth while reducing water supply demands [19]. The benefits (and disbenefits) of WR, as produced externalities, can be classified according to environmental, economic, and social points of view as shown in Table 2.1.

TABLE 2.1 Advantages and Disadvantages of WR

Aspects	Advantages	Disadvantages
Environment	Decrease the diversion of water from sensitive ecosystem Reduce water consumption Reduce and prevent pollution Create or enhance wetlands and riparian habitats Provide an appropriate and nutrient-rich source for agricultural and landscape irrigation More sustainable resource utilization	Probable contamination of groundwater Built-up chemical pollutants in the soil
Economic	Cost savings from reducing cost of water/wastewater treatment, disposal, and new water supply investment Increased economic development through additional water being available for agriculture and industrial expansion	More investment for WR technology
Social	Provide a secure long-term supply of water Assist in securing long-term supplies of drinking water to meet population growth in urban and regional centers	Unsure of human health risks

Source: Adapted from Asano, T. 2005. *Water Science & Technology*, 51(8), 83–89; EUREAU. 2004. *EUREAU Position Pater on Water Reuse.* Brussels, European Union of National Association of Water Supplies and Waste Water Services: 1–7; Marsden Jacob Associates. 2005. *National Guidelines on Water Recycling—Managing Health and Environmental Risks—Impact Assessment*; Miller, G.W. 2006. *Desalination*, 187, 65–75; UNEP and GEC. 2005. *Water and Wastewater Reuse: An Environmentally Sound Approach for Sustainable Urban Water Management.* UNEP.

2.2.2.2 WR Applications in Urban Areas

On a wider international concern, direct nonpotable WR is currently the dominant mode for supplementing public water supplies for landscape irrigation, industrial cooling water, river flow augmentation, and other applications. From urbanization, WR applications have been primarily focused as supplied sources for nonpotable water demand, except the urban areas in some arid countries such as Saudi Arabia, Kuwait, and Yemen, the WR would be enforced for potable water demand. The principle categories of WR applications in urban areas are listed in Table 2.2 in descending order of projected volume of use along with examples and issues for their applications. In addition, the WR of urban areas may be used for agricultural irrigation in peri-urban or nearby rural areas.

2.3 Drivers and Opportunities of WR in the Context of Urbanization

Urbanization is projected to highly increase, particularly in developing countries and potentially relates with WR development, due to pressure on water demand and limited sources. Based on a literature review, relationships between drivers and opportunity of WR and urbanization are discussed in this section.

2.3.1 Drivers to Reuse of Treated Urban Wastewater

Drivers of WR that have been addressed in past research are summarized in Table 2.3. Most of these drivers are accompanied by urbanization, particularly in environmental and economic factors. Trends toward SD also push the WR practicing in urban areas. Consequently, WR has been paid more attention in the context of rapid urbanization for reliability of water supply. This

is confirmed by Table 2.4, which shows that urbanized population growth rate has a positive relationship in proportion to WR. In addition, a case of high urbanization rate in China has also followed high growth in WR. In Thailand, WR has been paid more attention since experience with an unusually prolonged dry season in 2003–2004.

2.3.2 Opportunities of Urban WR

In addition to the challenges of urban water management, urbanization could also provide an opportunity to reuse treated wastewater with an aim at sustainable urban development. For example, WR would be directly beneficial to water management by reducing water demand, and to wastewater management by reducing effluent discharged to the natural environment.

The opportunities of WR from urbanization could be defined in terms of two areas, namely sources and areas.

2.3.2.1 Significant Sources of Treated Wastewater

As wastewater is typically about 80% of the water supply use, urban areas present a large amount of potential sources for WR from their wastewater. In addition, constructing wastewater treatment plants and regulating effluent quality standards have been typically carried out in urban areas due to tangible negative effects from wastewater problems. Consequently, treated wastewater from both central and on-site wastewater treatment plants would be significant potential sources for WR.

2.3.2.2 Compacted Areas for Distribution System

According to Jia et al. [10], while small WR systems (as individual household systems reusing effluent from on-site wastewater treatment) might be flexible, large WR systems reusing effluent from central wastewater treatment plants are safer, more reliable, and acceptable to the public. However, diseconomies of

TABLE 2.2   Categories, Examples, and Issues of WR in Urban Areas

Categories	Examples	Issues
Landscape irrigation: Parks, school yards, freeway medians, golf courses, cemeteries, greenbelts, residential areas	In United States, Australia, Spain, China	Effect of water quality on soils and crops Public health concerns related to pathogens (e.g., bacteria, viruses, and parasites) Using area control including buffer zone may result in high user costs
Industrial application: Cooling water, boiler make-up water, industrial process water—for example, in power plants and iron and steel production	POSTECH Steel Company in Pohang, Korea Sugar industry in Thailand	Constituents in reclaimed water related to scaling, corrosion, biological growth, and fouling Public health concerns, particularly aerosol transmission of pathogen in cooling water Cross-connection of potable and reclaimed water lines
Groundwater recharge: Groundwater replenishment, salt water intrusion control, control or prevention of ground subsidence	Groundwater recharge program in Los Angeles and in Northern Virginia	Possible contamination of groundwater aquifer used as source of potable water Organic chemicals in reclaimed water and their toxicological effects Total dissolved solids, nitrates, and pathogens in reclaimed water
Recreational/environmental use: Lakes and ponds, marsh enhancement, wetland enhancement and restoration, stream-flow augmentation, fisheries, snow making	Pond maintained with reclaimed wastewater in Los Angeles	Health concerns related to presence of bacteria and viruses (e.g., enteric infections and ear, eye, and nose infections) Eutrophication due to nitrogen and phosphorus in receiving water Toxicity to aquatic life
Nonpotable urban application: Fire protection, toilet flushing, air conditioning, commercial uses—for example, vehicle washing, laundry, construction water	On-site WR within buildings in Australia, Canada, and Japan NEWater in Singapore	Public health concerns about pathogens transmitted by aerosols Effects of water quality on scaling, corrosion, biological growth, and fouling Cross-connection of potable and reclaimed water lines
Potable reuse: Blending in water supply reservoir, pipe-to-pipe water supply	Discharge of highly treated municipal wastewater to the water supply source in Virginia	Constituents in reclaimed water, especially trace organic chemicals and their toxicological effects Aesthetics and public acceptance Health concerns about pathogen transmission, particularly enteric viruses

Source: Adapted from Levine, A.D. and Asano, T. 2004. *Environmental Science & Technology*, 201A–208A; Noh, S. et al. 2004. *Water Science & Technology*, 50(2), 309–314; Rowe, D.R. and Abdel-Magid, I.M. 1995. *Handbook of Wastewater Reclamation and Reuse*. CRC Press, Inc., Boca Raton, FL; Tchobanoglous, G. and Burton, F.L. 1991. *Wastewater Engineering: Treatment, Disposal, and Reuse*, 2nd edition. McGraw-Hill, Inc.; Tchobanoglous, G., and Burton, F.L. 2003. *Wastewater Engineering: Treatment and Reuse*, 4th edition. McGraw-Hill, Inc.; Trang, D.T. et al. 2006. *Selected Papers from the First International Symposium on Southeast Asian Water Environment (Biodiversity and Water Environment)*. IWA Publishing. 167–174; UNEP and GEC. 2005. *Water and Wastewater Reuse: An Environmentally Sound Approach for Sustainable Urban Water Management*. UNEP; U.S. EPA. 2004. *Guidelines for Water Reuse*. U.S. Environmental Protection Agency; Wei, C., Xingcan, Z., and Tao, L. 2007. *Water Science & Technology*, 55(1), 387–395.

TABLE 2.3   WR Drivers

Categories	Drivers
Environmental factors	Water scarcity and droughts Environmental protection and enhancement
Economic factors	Water charging Wastewater treatment cost Economic incentives Increased costs of upgrading wastewater treatment facilities
Social factors	Increasing water demands Public health protection Increasing awareness of the environmental impacts
Political factors	Public policy direction New regulations
Other factors	Growing number of successful WR projects in the world Growing recognition of benefit of using recycled water

Source: Summarized from Asano, T. 2005. *Water Science & Technology*, 51(8), 83–89; Hermanowicz, S.W. 2006. Is scarcity a real driver for water reuse? http://repositories.cdlib.org/wrca/wp/swr_706 2007. Water Resources Center Archives, Working Paper; Lundin, M. and Morrison, G.M. 2002. *Urban Water*, 4, 145–152; Marsden Jacob Associates. 2005. *National Guidelines on Water Recycling—Managing Health and Environmental Risks—Impact Assessment*; UNEP, WHO, HABITAT, and WSSCC. 2004. *Guidelines on Municipal Wastewater Management*, The Hague, The Netherlands, UNEP/GPA coordination office; U.S. EPA. 2004. *Guidelines for Water Reuse*. U.S. Environmental Protection Agency.

scale are apparent in all pipe networks and the extent of this dis-economy varies with urban density and is affected by the layout of the piping network [5].

Strong empirical evidence confirms that the concentration of people and productive activities in cities generates economies of scale and proximity that stimulate growth and reduce the costs of production, including the delivery of collective basic services such as piped water, sewers, and drains [33]. Consequently, urbanization through compacted households would provide nearby users with better economic feasibility on installation of WR pipe as a dual distribution system.

Regarding the national guidelines on water recycling of Australia [15], the economics of greywater recycling for domestic use are attractive, and past studies suggest that medium-sized recycling schemes servicing 1200–12,000 households can be the most economical. In addition, WR will continue to be most attractive in serving new (where the installation of dual distribution systems would be far more economical) than in already developed areas. Consequently, it would be a substantial matter for WR in urbanizing areas where there are many new or expanding areas. However, as mentioned earlier, this is subject to proper planning and implementation of WR schemes.

TABLE 2.4 Water Availability and WR Proportion in Selected Countries

	Australia	U.S.	Japan	China	Thailand
Urban population (% of total population)					
2004[a]	92	80	66	40	32
2010[b]	89	82	67	47	34
Urbanization growth rate (%), 2010–2015[b]	1.2	1.2	0.2	2.3	1.8
TARWR per capita (m³/year)					
2000[c]	25,708	10,837	3383	2259	6527
2005[c]	24,710	10,270	3360	2140	6460
2006[d]	24,158	10,135	3354	n/a	6330
Total use (% TARWR)[b]	5	16	21	n/a	21
WR (% total water use)	9.1 (2002)[e]	7.4 (2000)[e]	1.5 (2001)[f]	n/a	n/a
Growth in reuse flow rate treated per annum (%)[g]	42 in 42 years (1.00%/year)	2.5 in 11 years (0.23%/year)	25 in 30 years (0.83%/year)	67 in 4 years (16.75%/year)	n/a

Note: TARWR means total actual renewable water resource; n/a means data not available.

[a] World Bank. 2006. *2006 World Development Indicators*. The International Bank for Reconstruction and Development, The World Bank.

[b] Central Intelligence Agency. 2012. *The World Factbook*. Central Intelligence Agency. From https://www.cia.gov/library/publications/the-world-factbook/index.html.

[c] Lee, S., Cha, D., and Park, H. 2006. *Water Science & Technology*, 53 (6), 75–82.

[d] FAO-AQUASTAT. 2009. *Water Resources by Country/Territory and by Inhabitant, and MDG Water Indicator*. From http://www.fao.org/nr/water/aquastat/maps/ AQUASTAT_water_resources_and_MDG_water_indicator-March_2009.pdf [retrieved 2011].

[e] Miller, G.W. 2006. *Desalination*, 187, 65–75.

[f] Minamiyama, M. 2001. *Present State of Treated Wastewater Reuse in Japan*. National Institute for Land and Infrastructure Management, Japan. From http://www.nilim.go.jp/lab/bcg/siryou/tnn/tnn0264pdf/ks0264029.pdf [retrieved 2011].

[g] EUREAU. 2004. *EUREAU Position Pater on Water Reuse*. Brussels, European Union of National Association of Water Supplies and Waste Water Services: 1–7.

2.4 A Case of Urbanization and WR

The relationship between urbanization and WR, as well as its drivers and opportunities, has been initially discussed in previous sections. More specific and additional information is presented in this section with a case study from Thailand. Particular challenges and constraints of WR practices are also investigated.

Thailand's urbanization rate has been significant with an increase in its urban population from 31.1% in 2005 to 44.1% in 2010. Two urban areas of Thailand—Pattaya City (PC) and Chacheongsao Town Municipality (CTM)—were selected as study areas. The PC, a tourism area, was selected as representative of rapid urbanization, whereas the CTM represents a typical urban area, both of which have been urbanized due to population and economic growth. Review of official documents, key informants' interviews, and a household questionnaire survey were used to explore challenges and constraints for actual WR practices in both areas.

2.4.1 Background of the Study Areas

2.4.1.1 Urbanization Characteristics

The PC has comparatively had a higher urbanization level due to rapid economic growth, particularly from the tourism industry. Urbanization has also affected the environment. Consequently, PC was established as a Pollution Control Zone according to the National Environment Quality Act of 1992. This is different from CTM, which is a commercial business base with low population growth. The majority of people in CTM are from a local community, whereas PC is populated with visitors and employees.

Additionally, the PC is different from the CTM in governing structure and population density. Table 2.5 summarizes key information on these two study areas.

2.4.1.2 Water Management

Both PC and CTM are located in the Bang Pakong River Basin, which was predicted to have a scarce water supply during the dry season by 2021. This is confirmed by a water crisis record in 2003–2004 due to irregular rainfall. The major source of water in both study areas is tap water. With the high population density of PC, it has been difficult to find an area for large water supply storage. Therefore, PC has drawn raw water for tap water production from outside the city. Part of the raw water is provided by EastWater* at USD† 0.25 per cubic meter during the dry season. Conversely, the CTM is located in a downstream area of the basin that connects to the Gulf of Thailand. Thus, in this area, water shortage in the dry season (November to February) and salt-water intrusion are major problems. In addition, the amount of water use in PC is about 1.72 times of that in CTM, showing a higher water demand and wastewater generation in PC.

2.4.1.3 Wastewater Management

The main household wastewater system is a septic tank, which is a typical system in Thailand, before discharging to a sewer

---

* EastWater: Eastern Water Resources Development and Management Public Company Limited is a company responsible for supplying raw water in the eastern seaboard of Thailand.

† In 2013, 1 USD = 32 Thai Baht.

TABLE 2.5 Key Information of the Study Areas

Items	PC	CTM
Governing structure	Special governing	Normal governing
Population density, 2011 (persons/km²)	10,456 (incl. nonregistered inhabitants)	3101
Annual rate of change of the population (%)	2.61	0.09
GPP per capita in 2006 (Baht/person)	368,369	256,448
Nonagricultural GPP (%)	95.93	93.53
Water use (m³/connection/day)	0.19	0.11
Number of central wastewater treatment plant	2	1
Total capacity of central wastewater treatment plants (m³/d)	85,000	24,000
Wastewater tariff collection	Yes	No

system. Both areas have their central wastewater treatment plants utilizing an activated sludge process. The PC is the first city in Thailand to collect a wastewater tariff from households and commercial businesses. This differs from the CTM in that wastewater operation and maintenance costs are fully subsidized by local government without collection of wastewater fees.

2.4.2 Current WR Practices

WR in PC has been practiced in both public and private sectors. The private sector uses their effluent as individual systems typically for landscape irrigation and recreation (lake) especially in the hotel business. The effluent from the central wastewater treatment plant has been reused by using their municipal budget to develop a piping system for irrigation in public parks and roadway medians. Similarly, current practices of WR in the CTM are reusing effluent from the central wastewater treatment plant for irrigation in nearby public parks and roadway medians by a

watering truck. However, the purposes of WR practices from the central wastewater treatment plants of these two study areas are dissimilar in that PC implements WR for saving the water supply and reducing water pollution, while CTM implements WR following a big promotion and support of the central government. More specific characteristics of the current WR practices in the two areas are discussed in the following subsections.

2.4.2.1 Environmental and Social Concerns

Urbanization, especially unplanned urbanization, almost comes with environmental problems that also bring environmental concerns and awareness to a community. This enhances public acceptance of WR. This is confirmed by the results of a household questionnaire survey on environmental problems in the PC and the CTM and the level of public acceptance of WR. With different urbanization levels, it is indicated that there are more significant water and wastewater problems in PC than in CTM (Figure 2.3). In addition, households in PC (90%) have a much

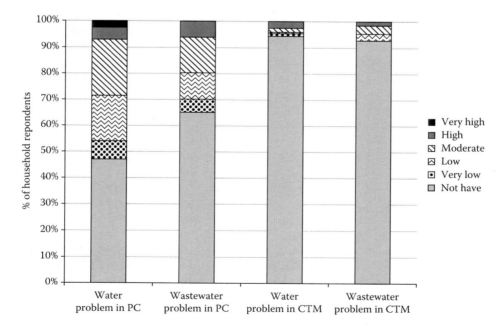

FIGURE 2.3 Levels of water shortage and wastewater problems in PC and CTM. (Adapted from Sa-nguanduan, N. 2011. Strategic decision-making system for urban water reuse application. PhD dissertation, Asian Institute of Technology, Bangkok.)

TABLE 2.6 Example of Internal Benefits of WR

WR Case	Benefits	Benefit (USD/m³)	O&M Cost (USD/m³)	B/C Ratio
Landscape irrigation in the hotel in PC	Internal benefits—savings of water cost for landscape irrigation	0.625	0.021	29.76
	External benefits—savings of wastewater fee	0.039		31.62
Toilet flushing and irrigating the food crops of residential schools (schools) in Madhya Pradesh, India [7]	Internal benefits—reduction in tankered water for toilet flushing and irrigating the food crops	0.176	0.034	5.18
	External benefits from avoidance of water infrastructure, environmental benefits, and health benefits	5.270		160.18
WR of the Ferrara wastewater treatment plant, Po Valley, Italy [20]	Internal and external benefits from continuous availability of water, quality of the Po di Volano canal, reduction in energy consumption, and creation of green areas and ponds	0.097	0.027	3.59

higher acceptance of WR practices than in CTM (71%). As a result, the urbanization level is an important driver for WR from environmental and social points of view.

2.4.2.2 Economical and Technical Concerns

From the review of current practices in the study areas, WR in PC is more innovative by implementing in both public and private sectors, whereas in CTM, reuse of effluent has only been from the central wastewater treatment plant. Based on the results of key informant interviews, an influencing factor on WR practice in PC is a water shortage problem related to cost savings from buying water from private water truck sellers at a price 2–3 times that of typical tap water. In addition, savings from wastewater management fees and improving the company's image to be more environmentally friendly are major factors to WR in the private sector in PC. The economic feasibility of a WR project in PC has been compared with cases in other countries as shown in Table 2.6.

In both areas, urban expansion not only provides an increased amount of sources of WR and potential areas for installing pipe of WR, but also allows improved capacity building of government authority in terms of economic and technical concerns for WR management. High economic and technical capacity would be provided for effective wastewater treatment and related systems. Consequently, these would ensure public health and environment protection with a high degree of reliability for the wastewater treatment process and operations.

2.4.3 Challenges and Constraints of WR Practices

Based on the interview results of key informants in both areas, public acceptance trends will increase according to public concern about limited water sources and environmental problems. Urbanization could potentially enhance economic opportunities of WR; however, failure to actually recognize the economic values of treated wastewater for reuse would be a challenge in the study areas. In addition, a challenge of large-scale WR systems is institutional cooperation, particularly between waterworks

authorities and municipalities responsible for wastewater management. For example, in the case of PC (Figure 2.4), if 100% of current effluent from the central wastewater treatment plant were reused, the drought period would be shortened. In addition, in an extreme case, if WR were implemented with the cooperation of waterworks authorities and municipalities, the waterworks authorities could reduce buying raw water from EastWater in a drought period, and could avoid buying raw water if 81% of the total capacity of the central wastewater treatment plant was reused. However, to expand WR application, implementing a government authority that addresses a controlling and monitoring system, particularly the WR system in the private sector, would be a major challenge.

According to the results of a household questionnaire survey (Table 2.7), key constraints of WR are typically negative impacts on public health and limited development of WR for other applications such as landscape irrigation. These are also related to efficiency and reliability of wastewater treatment systems, and are considered key constraints in both areas. However, financial issues regarding investment, operation, and maintenance costs are major constraints of WR implementation in CTM, which had a lower economic development level than PC. In addition, lack of laws and regulations to control WR applications is another constraint. Consequently, standards of WR quality and related monitoring and evaluation mechanisms are required to ensure no negative impacts from WR practices, particularly in relation to public health. It is a challenge to avoid (or reduce) adverse effects of WR and increased costs of wastewater treatment systems with the established standards and related control schemes.

2.5 Summary and Conclusions

Urbanization is a global development trend that can potentially face serious problems regarding limitation of water sources, increase in water demand, and wastewater pollution, and hence would be a challenge for urban and water management sustainability. The drive behind WR has been to provide alternative water sources. Opportunities from sources of treated wastewater

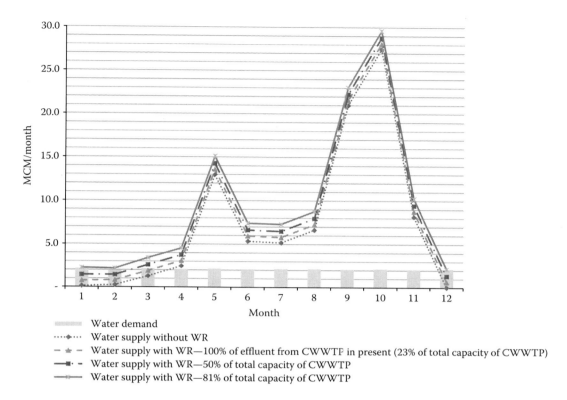

FIGURE 2.4   Scenario of water demand and supply in PC.

to reuse and users of WR, including economic feasibility, can be provided from urbanization. However, public acceptance, potential adverse impacts, and adequate financing can be key challenges and constraints of WR applications. Furthermore, the case from Thailand has shown that the level of urbanization would also influence WR applications, in economic, social, environmental, and technical issues. Higher levels of urbanization provide greater public acceptance, and better opportunity for the needs of and strengthened capacity building for WR.

Although urbanization provides good opportunities, WR requires long-term planning and has several constraints. Therefore, effective planning of WR under rapid urbanization

TABLE 2.7   Key Constraints of WR in PC and CTM

Constraints	Case Studies	Rank (with % of Respondents)						
		1	2	3	4	5	6	7
Adverse effects on human health	PC	34.7	16.1	18.6	11.6	4.0	5.5	9.5
	CTM	16.8	11.1	13.2	9.5	11.6	11.1	26.8
Efficiency of wastewater treatment technology	PC	41.2	41.2	13.6	2.0	2.0	0.0	0.0
	CTM	12.6	17.9	13.7	12.6	14.2	12.1	16.8
Reliability of wastewater treatment operators	PC	20.6	26.6	29.1	14.6	6.5	2.0	0.5
	CTM	19.7	12.8	14.9	14.4	14.9	12.2	11.2
Investment costs	PC	2.0	2.0	8.5	21.6	28.6	21.1	16.1
	CTM	17.6	12.8	18.1	17.0	11.7	16.0	6.9
Operation and maintenance costs	PC	2.0	5.0	7.5	25.1	30.7	22.1	7.5
	CTM	14.0	21.0	14.5	12.4	17.2	11.3	9.7
Laws and regulations	PC	0.0	2.0	7.5	7.0	16.1	29.6	37.7
	CTM	7.5	14.4	13.9	15.5	18.7	15.5	14.4
Public acceptance of WR	PC	3.5	9.0	17.6	15.6	12.1	17.1	25.1
	CTM	14.4	11.2	13.3	17.0	10.1	20.7	13.3

Source: Analyzed from Sa-nguanduan, N. 2011. Strategic decision-making system for urban water reuse application. PhD dissertation, Asian Institute of Technology, Bangkok.

would be required for developers and government authorities when integrating WR in an urban water management plan, in both developed and developing countries. Enhancing WR benefits from positive externalities would be helpful especially on strategic decision-making. More efforts should be made in developing and disseminating scientific information, such as the relationship between stages of urbanization and applications of WR, information support tools for decision-making system in terms of WR scales and applications, and good practices on control, monitoring, and evaluation of WR practices.

References

1. Asano, T. 2005. Urban water recycling. *Water Science & Technology*, 51(8), 83–89.
2. Central Intelligence Agency. 2012. (Retrieved 2012). *The World Factbook*. Central Intelligence Agency. From https://www.cia.gov/library/publications/the-world-factbook/index.html.
3. Cohen, B. 2006. Urbanization in developing countries: Current trends, future projections, and key challenges for sustainability. *Technology in Society*, 28(1–2), 63–80.
4. EUREAU. 2004. Why is water reuse so important to the EU?: Drivers, benefits and trends. *EUREAU Position Pater on Water Reuse*. Brussels, European Union of National Association of Water Supplies and Waste Water Services: 1–7.
5. Fane, S.A., Ashbolt, N.J., and White, S.B. 2002. Decentralised urban water reuse: The implication of system scale for cost and pathogen risk. *Water Science & Technology*, 46(6–7), 281–288.
6. FAO-AQUASTAT. 2009. (Retrieved 2011). *Water Resources by Country/Territory and by Inhabitant, and MDG Water Indicator*. From http://www.fao.org/nr/water/aquastat/maps/ AQUASTAT_water_resources_and_MDG_water_indicator-March_2009.pdf.
7. Godfrey, S., Labhasetwar, P., and Wate, S. 2009. Greywater reuse in residential schools in Madhya Pradesh, India—A case study of cost-benefit analysis. *Resources, Conservation and Recycling*, 53(5), 287–293.
8. Hermanowicz, S.W. 2006. Is scarcity a real driver for water reuse? http://repositories.cdlib.org/wrca/wp/swr_706 2007. Water Resources Center Archives, Working Paper.
9. Hochstrat, R., Wintgens, T., and Melin, T. 2008. Development of integrated water reuse strategies. *Desalination*, 218(1–3), 208–217.
10. Jia, H., Guo, R., Xim, K., and Wang, J. 2002. Research on wastewater reuse planning in Beijing Central region. *Water Science & Technology*, 51(10), 195–202.
11. Lee, S., Cha, D., and Park, H. 2006. International standards for services activities relating to drinking water supply systems and wastewater systems: Implications for developing countries. *Water Science & Technology*, 53(6), 75–82.
12. Levine, A.D. and Asano, T. 2004. Recovering sustainable water from wastewater. *Environmental Science & Technology*, 201A–208A.
13. Liu, F., Li, X., Wang, D., Hu, R., Yang, W., Li, D., and Zhao, D. 2009. Measurement indicators and an evaluation approach for assessing urban sustainable development: A case study for China's Jining City. *Landscape and Urban Planning*, 90(3–4), 134–142.
14. Lundin, M. and Morrison, G.M. 2002. A life cycle assessment based procedure for development of environmental sustainability indicators for urban water systems. *Urban Water*, 4, 145–152.
15. Marsden Jacob Associates. 2005. *National Guidelines on Water Recycling—Managing Health and Environmental Risks—Impact Assessment*.
16. Miller, G.W. 2006. Integrated concepts in water reuse: Managing global water needs. *Desalination*, 187, 65–75.
17. Minamiyama, M. 2001. (Retrieved 2011). *Present State of Treated Wastewater Reuse in Japan*. National Institute for Land and Infrastructure Management, Japan. From http://www.nilim.go.jp/lab/bcg/siryou/tnn/tnn0264pdf/ks0264029.pdf.
18. Noh, S., Kwon, I., Yang, H.M., Choi, H.L., and Kim, H. 2004. Current status of water reuse systems in Korea. *Water Science & Technology*, 50(2), 309–314.
19. Okun, D.A. 2002. Water reuse introduces the need to integrate both water supply and wastewater management at local and regulatory level. *Water Science & Technology*, 46(6–7), 273–280.
20. Population Reference Bureau. 2010. (Retrieved 2011). *Trends in Urbanization, by Region*. From http://www.prb.org/Publications/GraphicsBank/PopulationTrends.aspx.
21. Rowe, D.R. and Abdel-Magid, I.M. 1995. *Handbook of Wastewater Reclamation and Reuse*. CRC Press, Inc., Boca Raton, FL.
22. Sa-nguanduan, N. and Nitivattananon, N. 2006. Sustainable urban water management system: Review of status with focus on southeast Asian countries. *Proceeding of Regional Conference on Urban Water and Sanitation in Southeast Asian Cities*, 235–250.
23. Sa-nguanduan, N. 2011. Strategic decision-making system for urban water reuse application. PhD Dissertation, Asian Institute of Technology, Bangkok.
24. Sa-nguanduan, N. and Nititvattananon, N. 2011. Strategic decision making for urban water reuse application: A case from Thailand. *Desalination*, 268(1–3), 141–149.
25. Tchobanoglous, G. and Burton, F.L. 1991. *Wastewater Engineering: Treatment, Disposal, and Reuse*, 2nd edition. McGraw-Hill, Inc.
26. Tchobanoglous, G. and Burton, F.L. 2003. *Wastewater Engineering: Treatment and Reuse*, 4th edition. McGraw-Hill, Inc.
27. Trang, D.T., Molbak, K., Dalsgaard, A., Cam, P.D., and Hoa, N.V. 2006. Impacts of wastewater reuse in agriculture and aquaculture in Hanoi, Vietnam. *Selected Papers from the First International Symposium on Southeast Asian Water Environment (Biodiversity and Water Environment)*. IWA Publishing. 167–174.

28. Tsagarakis, K.P. 2005. Recycled water valuation as a corollary of the 2000/60/EC water framework directive. *Agricultural Water Management*, 72(1), 1–14.

29. UNEP and GEC. 2005. *Water and Wastewater Reuse: An Environmentally Sound Approach for Sustainable Urban Water Management*. UNEP.

30. UNEP, WHO, HABITAT, and WSSCC. 2004. *Guidelines on Municipal Wastewater Management*. The Hague, The Netherlands, UNEP/GPA coordination office.

31. United Nations. 2012. *World Urbanization Prospects: The 2011 Revision*. United Nations, Department of Economic and Social Affairs, Population Division.

32. United Nations. 2012. *World Urbanization Prospects: The 2011 Revision*.

33. UN-HABITAT. 2008. *State of the World's Cities 2010/2011: Bridging the Urban Divide*. UN-HABITAT.

34. UN-HABITAT. 2012. *Sustainable Urbanization in Asia: A Sourcebook for Local Governments*. UN-HABITAT.

35. U.S. EPA. (Retrieved 2007). *Water Recycling and Reuse: The Environmental Benefits*. From http://www.epa.gov/region9/water/recycling/index.html.

36. U.S. EPA. 2004. *Guidelines for Water Reuse*. U.S. Environmental Protection Agency.

37. Verlicchi, P., Aukidy, M. Al., Galletti, A., Zambello, E., Zanni, G., and Masotti, L. 2012. A project of reuse of reclaimed wastewater in the Po Valley, Italy: Polishing sequence and cost benefit analysis. *Journal of Hydrology*, 432–433, 127–136.

38. Wei, C., Xingcan, Z., and Tao, L. 2007. Consideration on the issue of water reuse in eastern China plain brooky regions. *Water Science & Technology*, 55(1), 387–395.

39. World Bank. 2006. *2006 World Development Indicators*. The International Bank for Reconstruction and Development, The World Bank.

3

Energy and Water Reuse

Carlo Stagnaro
*Italy's Ministry of
Economic Development*

PREFACE

Several activities within the water cycle, and particularly many water reuse-related ones, are energy intensive. At the same time, water and wastewater have an energy content under different forms. Energy consumption is one of the major sources of environmental impact from water systems. The problems underlying the energy footprint of water, the water footprint of energy, and the environmental impacts of both are becoming increasingly important, as global water consumption grows under demographic and GDP pressures. Indeed, world population is expected to grow from 7.2 billion in 2012 to 8.3–19.9 billion in 2050, with a medium value of 9.6 billion [24]. In the same period, world GDP is expected to grow at an average of about 3%, nearly tripling by 2050 [18].

This chapter focuses on energy use in water reuse in urban areas as well as on the potential for energy recovery from the water cycle—particularly from wastewater. The chapter surveys the most relevant, available technologies and emphasizes, on one hand, the site-specific aspects of technological choices, and on the other hand, the underlying institutional and financial dimensions. The major sources of environmental impacts and the subsequent trade-offs—both cost-, energy-, and environment-wise—are reviewed. The chapter also shows how design choices may influence the outcome in terms of energy, economic, and environmental performance of water systems.

3.1 Introduction

The interaction between energy and water in water reuse is bidirectional. Energy is used to make water available and usable; and water is required in most energy-generating technologies, regardless of whether they are water-based. Therefore, a double challenge is to be met: on one hand, energy use in water systems should be minimized; on the other hand, energy recovery from wastewater discharges should be maximized [26]. Either goal is dependent on the other—choices made on one end of the chain may have impacts on what happens on the other end. Therefore, the problem to be solved is not one of mere maximization, but one of optimization in technological, behavioral, and regulatory decisions. Moreover, social and environmental impacts should also be considered. Both energy use in water recycling

and the technologies that are employed (if any) for energy recovery from wastewater may have adverse environmental impacts on several dimensions, such as the production of pollutants, the emission of CO_2 as well as other greenhouse gases (GHGs), etc. It may—and does—happen that a setting that minimizes a given environmental impact also results in an increase in some other environmental impacts. For example—as it will be shown—it happens that, under specific circumstances, the reduction of CO_2 increases the release in the atmosphere of other GHGs. Even from this point of view, decision makers face important trade-offs, under significant information asymmetries, lack of information, and limited ability to forecast future trends.

To make things even more complex, there is no way of assessing what is the most convenient organizational solution for a water system in principle. The choice of technologies, organizational settings, financial arrangements, and supply- and demand-side measures strongly depends on site-specific circumstances. To make only a few examples, it does matter a lot whether freshwater is readily available; whether supply and demand are subject to huge fluctuations (e.g., due to seasonal trends); whether water demand is driven by uses that require high- or low-quality standards; how energy is provided; whether an urban area is densely populated or sprawled out; whether wastewater treatment plants are already available and, more generally, how the existing systems—both with regard to water conveyance to its final consumption places and to wastewater collection and treatment—are organized; whether energy is easily and cost-effectively available; whether the general quality of institutions is such that investments with a long recovery time are easily made; etc. In other words, there is no "one size fits all" kind of solution—several alternatives are available, including the null hypothesis of *not* reusing water if abundant freshwater is available or if other alternative sources of water are more competitive (e.g., water desalination). Extreme options are hardly convenient, though. In most cases, the decision maker will have to find the right balance between technologies, approaches, and use regulations that lead to the best outcome. In this respect, it can be said that, while a "theory" of energy–water interactions in water reuse is impossible to develop, a number of case studies may provide the decision maker with the most valuable information [11].

This chapter is organized as follows. Section 3.2 deals with energy use in water reuse. Both the provision of water and the processes needed to reclaim and treat it to make it usable again are energy intensive. The choice of the best technologies for wastewater treatment, as well as the correct scale for treatment plants, is crucial in order to ensure that energy is not wasted (and hence the environment is not pointlessly harmed) either from an operational perspective or from the point of view of life-cycle assessment. Depending on site-specific features—including, but not limited to, freshwater availability, the possibility of reclaiming greywater, blackwater, and perhaps yellow water independently from each other, the origins of discharges, the existence of incentive schemes for renewable energies, etc.—it may be possible both to reduce energy demand and to increase the share of carbon-free and pollution-free energy.

Section 3.3 looks at the opposite end of the problem: energy recovery from wastewater. Energy is stored in wastewater in a variety of forms, from the potential energy of it to its heat content and the chemical energy embodied in the discharge flow. Recovering energy is therefore technically possible, although not always cost-effective. Environmental impacts of energy recovery should also be considered: often pollution is not abated, it is merely shifted from one form to another or from a recipient body to another. Sometimes this may be—on balance—environmentally benign, sometimes not. Much depends on the available technologies. In principle, the more energy recovered from wastewater, the more energy demand from wastewater treatment and water provision is reduced, and the more energy-related environmental impacts are reduced (but not necessarily *all* environmental impacts).

Section 3.4 is concerned precisely with environmental impacts and the much-challenging trade-offs that the decision maker has to solve. Such trade-offs involve both the opportunity cost of the choices being made, and the choice between different pollutants that may be caused by the transformation of each other (e.g., non-CO_2 GHGs might be emitted as CO_2 is abated). Such decisions have both an economical dimension and an environmental one.

Section 3.5 puts all the above against the feasibility test: all the solutions that have been presented make energy and environmental sense, but are they financially sustainable, too? Occasionally, it may be the case that local conditions make an investment worthwhile because savings of future operational expenses are evident that anyone who is in charge would go for it. However, often reshaping water provisions and reclamation and treatment systems requires major upfront investments. Their bankability depends on the degree of certainty of the regulatory environment. Therefore, a discussion of the political, legal, and regulatory conditions behind water systems is unavoidable.

Summary and conclusions are provided in Section 3.6.

3.2 Energy Use in Water Reuse

The amount of energy used in water provision and reuse varies significantly depending on a number of variables, of which the most important ones are site-specific and include (but are not limited to) the availability of freshwater, the availability of indigenous, carbon-free sources of energy, and the general setting of the place (urban vs. rural or industrial). With regard to this latter point, this chapter mostly deals with urban contexts. It also makes a major difference whether a water system is being designed within a developing context or it is being adjusted within an existing context. Especially historical settings may pose significant challenges in finding the right balance between the optimization of the water system and the preservation of the existing architectures, buildings, and urban structures.

The use of energy in traditional water systems used to be relatively limited: energy was needed for moving water through pipes, treating water before use, and treating wastewater in order to release it in the environment without spreading pollution [23]. Figure 3.1 shows the breakdown of energy consumption of a typical water utility.

The energy footprint of water use, however, significantly increases as alternative water sources are employed. For example, water reuse—hence the need to treat wastewater up to drinking water standards—and desalination are typically energy-intensive activities. Subsequently, the environmental footprint of alternative water sources may also be nonnegligible. However, relying on alternative water sources may become ever more important as the world population grows, as does the average income, driving up water demand, and setting up existing or potential problems of water scarcity. In fact, as the population grows larger and wealthier, a process that is typically associated with urbanization, water demand follows. As Figure 3.2 shows, the relationship between average income levels and water consumption per capita is both straightforward and clear.

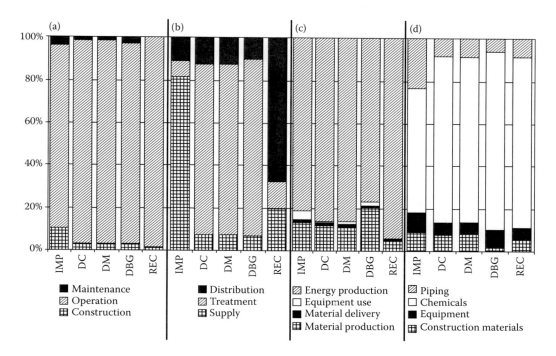

FIGURE 3.1 Breakdown of energy consumption of a water utility: (a) life-cycle phase, (b) water supply phase, (c) life-cycle activity, and (d) material production category. (Adapted from Stokes, J.R. and Horvath, A. 2009. *Environmental Science and Technology*, 43(8), 2680–2687.)

It can be inferred that both population growth—which is higher in the developing world than in the developed world [24]—and economic growth will drive per capita water demand in today's low-demand countries up toward the levels that are observed in higher-income countries. Water reuse becomes a crucial element to meet the growing demand while preventing water conflicts. Smart regulation—as it will be shown in Section 3.5—will also play a key role, as it crucially depends on the quality of institutions whether the needed investments will be performed.

In order to minimize the energy footprint of alternative water sources, three questions should be properly answered: (1) What alternative water source is best suited to meet the demand, both from an operational and from a life-cycle point of view? (2) What technological choices should be made, under (1)? (3) Given (1) and (2), how should the water system be designed and what kind of modifications are needed to the existing infrastructures?

Providing a correct answer to question (1) is particularly important in dry areas, where water scarcity poses major constraints to

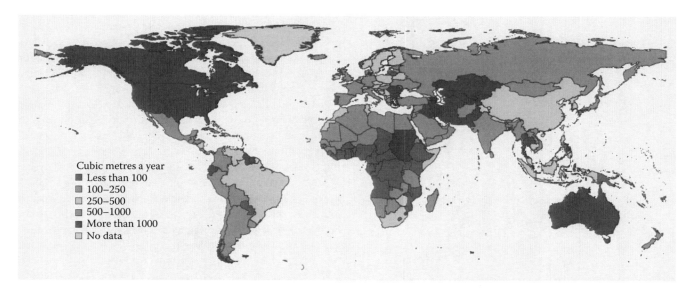

FIGURE 3.2 Annual water withdrawals per person by country (2000). (Adapted from UNESCO 2009. Water in a Changing World. The United Nations World Water Development Report 3, UNESCO Publishing, Paris, France.)

the ability of the population—particularly low-income groups—to have access to reliable and affordable water distribution and wastewater collection systems. Providing a well-functioning water cycle system is not just about enforcing the most basic human of all rights, but also about setting in place conditions for a number of epidemic diseases to be eradicated [9,12].

When the decision of relying on alternative water sources is made (or when there is no alternative to it), the decision maker is left with three options: (1) desalination (if seawater is at a reasonable distance), (2) water import (where possible), and (3) water reuse. Even though the balance may vary significantly depending on site-specific conditions, water reuse is often the least energy-intensive solution [17,22,28].

In the past, water reuse—especially for drinking purposes—had to overcome two major challenges: cost and public acceptance. Significant technological progresses in membrane filtration (with regard to both microfiltration and ultrafiltration) as well as in reverse osmosis (RO) technologies helped to tackle the former problem. Public acceptance is gradually being achieved, in part because populations around the world are becoming more aware of the size of the water challenge and the importance of not wasting water resources, especially in water-poor areas. Energy is an important input for water and wastewater treatment. Given the wide spectrum of available technologies—as well as the large differences in the quality, quantity, and regularity of the feedstock—the drivers of energy consumption may vary significantly. Figure 3.3 shows the breakdown of energy consumption for the typical treatment plant.

From Figure 3.3, it is evident that the most energy-intensive features of a conventional wastewater treatment plant are aeration of the activated sludge, which accounts for 45% of total energy consumption, and then—much less relevant—odor treatment, auxiliary equipment, and pumping, which account for 12%, 10%, and 8%, respectively. Three quarters of total energy demand is connected with the above-mentioned components. It follows that energy-saving efforts in wastewater treatment should be focused on these parts of the plant. Beyond the specific available technologies [2,3,9], attention should be paid to the general design of the system, of which the wastewater treatment plant is but the last step. In fact, the efficiency of the aeration system (and more broadly of the whole plant) may be increased if the wastewater inflow is more foreseeable and regular; on the other hand, energy consumption may be lower if the outflow of water may be directed toward nondrinking uses, whereby lower quality standards are acceptable. Clearly, under this latter hypothesis, fewer treatments are necessary and energy requirements are accordingly lower.

In other words, energy consumption in wastewater treatment may be reduced both by employing more efficient technology, and by improving the quality of the input (as well as reducing the required quality of the output). In either case, a different design is required from that which has historically characterized water distribution and wastewater collection systems [5]. The key concept is that multiple sets of pipes should be put in place, in order to better discriminate between alternative uses of water. This would also help to solve the social acceptance dilemma. If households could rely on freshwater for drinking, kitchen, and bathing purposes, and on recycled water for other uses (most notably toilet flushing and perhaps washing machines), they might have less to object because they would feel less unsafe (although there is no reason to believe recycled water is less safe than freshwater).

By the same token, what in most cities of today is one single flow of wastewater might be divided into two or even three separate flows: greywater (i.e., water used for bathing, laundry, etc.), blackwater (which includes feces and water from the kitchen and the like), and perhaps yellowater (urine). Greywater is the largest contribution to wastewater, and also the least contaminated one; blackwater has the highest content of biodegradable organic matter that needs to be depurated, but also has the highest potential for energy recovery; and yellowater contains the vast majority of such nutrients as nitrogen, phosphorus, and potassium. Separate collection of yellowater may be very costly or even impossible in historical settings, but it may be a useful improvement in new development areas.

Dual distribution systems and separate collection might be necessary to achieve two distinct, but interrelated, results: on one hand, minimizing energy consumption in water reuse, and on the other hand, maximizing energy recovery from wastewater. This is the subject of the next section.

3.3 Energy Recovery from Wastewater

Energy is used at every step of the water system. Energy is needed for water abstraction, purification, and distribution;

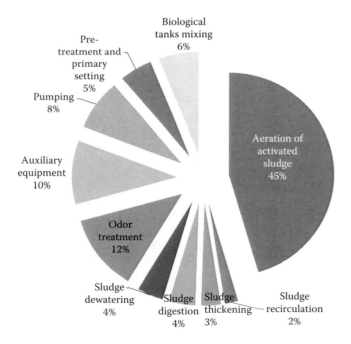

FIGURE 3.3 Typical energy consumption breakdown of a conventional wastewater treatment plant. (Elaboration on data from Lazarova, V., Choo, K.-H., and Cornel, P. (eds.) 2012. *Water–Energy Interactions in Water Reuse*, IWA Publishing, London and New York.)

energy is consumed to reclaim, treat, and release wastewater into the environment; and energy is needed to make water reuse possible when and if it seems convenient based on general or site-specific conditions, as well as on policies oriented at saving water and, when applicable, energy (in fact, it is not obvious that water reuse implies energy savings, and vice versa, while it is generally true that a well-designed water system allows for both water and energy savings).

Energy is not just spent on the water cycle: it is also transported by water and wastewater themselves, and on some occasions it is even injected into water (e.g., when cold water is heated—for kitchen or bathing uses—and then flushed, or when organic matter—such as feces—is added to water and then sent down to the sewage).

In particular, as Figure 3.4 shows, energy is present in—and can theoretically be reclaimed from—water and wastewater in four forms [13]:

- Kinetic energy
- Potential energy
- Thermal energy
- Chemically bound energy

The amount of energy attributable to each of these forms varies significantly; it depends on local conditions and it cannot always be recovered in a cost-effective way, leaving aside the technical aspects of energy recovery from wastewater.

Kinetic energy depends on the speed of water and wastewater flows. As these are generally very slow, the amount of recoverable energy is accordingly low, and generally it is not worth the cost. However, at particular points within the system, the speed of water flows may be higher, in which case—based upon site-specific features—energy recovery may be economically as well as energetically convenient.

Potential energy depends on the height. Even this potential source of energy does not appear very promising. A simple

calculation will explain why. Potential energy follows the following well-known formula:

$$U = mgh$$

where U is the potential energy; m is the mass of water or wastewater; g is standard gravity, that is, an acceleration of approximately 9.8 m/s² due to the Earth's gravitational attraction; and h is the height.

It follows that, under the assumption of a standard wastewater flow rate of 0.15 m³ per person per day, heights of as much as 10, 30, or 50 m imply a potential energy content (ignoring losses) of as much (or as little) as 14.7, 44.1, or 73.5 J per person per day, respectively, or 4.1, 12.2, 20.4 Wh per person per day, respectively.

Thermal energy depends on the heat that is embodied in wastewater flows. The heat, in turn, depends on wastewater's temperature, which derives from the amount of hot water that is mixed with cold water from, largely, kitchen and bathing uses. While thermal energy is the most important form of energy that is bound into wastewater, it may be difficult to recover nevertheless. In order to maximize the possibility of reclaiming thermal energy, one or more of the following conditions must be met: (1) as thermal energy is easily lost to the surrounding environment, it should be reclaimed as close as possible to the source; (2) in order to reduce losses, pipes should be insulated; and (3) in order to maximize wastewater's heat content greywater should be collected separately from other sources.

Finally, chemically bound energy—while less important than thermal energy for its size relative to the total energy content of wastewater—is the form that is most easily recovered and turned into usable energy. On one hand, wastewater can be moved wherever treatment plants are without compromising the amount of recoverable energy. On the other hand, several technologies are available to maximize energy recovery as well as to make it more cost-effective. Chemically bound energy depends on the organic content of wastewater, which is conventionally estimated through its chemical oxygen demand (COD). Every kilogram of COD embodies 3.49 kWh, which—with a COD load of 110–120 g per person per day—leads to a theoretical maximum potential of chemically bound energy of 400 Wh per person per day [13]. How much of it can be recovered, and at what cost, entirely depends on local conditions as well as on the technologies that are (or may be) employed. More energy can be recovered if the organic content is more concentrated; that is, systems that allow for a separate collection of blackwater are more responsive to this goal.

How much of this energy can be recovered, as well as at what cost, depends on (1) site-specific conditions; (2) the water system design; and (3) the available technologies. As to (1), some sites may be better suited to allow for energy recovery than others: higher systems can create a larger potential for exploiting potential energy; places where wastewater production follows more regular patterns may allow for a more efficient use of wastewater treatment plants because of a more stable load; etc. With regard to (2), places where blackwater, greywater, and perhaps

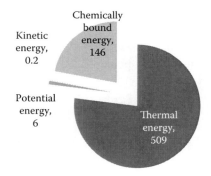

FIGURE 3.4 Theoretical energy potential of wastewater. Assumptions: water consumption = 122 L per person per day; flow rate = 5 m/s; greywater = 40 L per person per day; ΔT ~ 15 K; organic content = 115 g COD per person per day. (Elaboration on data from Meda, A. et al., 2012. In Lazarova, V., Choo, K.H., and Cornel, P. (eds.), *Water–Energy Interactions in Water Reuse*, IWA Publishing, London and New York.)

yellowater are collected separately are usually more manageable than those where these flows are mixed. As far as (3) is concerned, the choice of technologies—which is strongly influenced by the features *sub* (1) and (2)—is crucial in order to make energy recovery effective both energy- and cost-wise.

Whenever a technological, organizational, or policy choice is made with regard to energy recovery from wastewater, a number of variables should be considered. Such variables have to do with site-specific features, the final goals that are to be pursued—for example, cost-effectiveness and/or sustainability of the water cycle, which may not always be aligned—, and direct as well as indirect impact of the choices being made. Such impacts should be made not only with reference to the operational costs and impacts, but also with regard to life-cycle analysis. Life-cycle analysis reveals that the amount of energy that is bound into the creation of water facilities and water distribution is about 30% of its operational life-cycle cost [20].

A review of the literature on biosolids management [27] revealed that, while agricultural land application is still the most common usage of wastewater's organic content, a growing number of energy-related options are emerging and spreading out. Among them, the most promising are biosolids combustion and anaerobic digestion.

As far as biosolids combustion [21] is concerned, while promising for its ability to reduce volumes and the potential for energy recovery, it still poses some questions. Particularly, biosolids resulting from wastewater tend to have very low calorific power as well as high moisture and ash content. However, energy recovery from incineration can still be a viable option both to reduce energy consumption from the water system as a whole and as a means of pollution abatement.

Anaerobic digestion [19,20] refers to a wide number of different techniques: "combined carbon and ammonia oxidation, activated sludge, biological nutrient removal, aerobic digestion, anaerobic processes, lagoons, trickling filters, rotating biological contactors, fluidized beds, and biologically aerated filters to provide a comprehensive understanding of the field of biological wastewater treatment" [7].

3.4 Environmental Challenges

The choices that are being made on the water system design and water reuse technologies are not neutral with regard to environmental impacts. Some of them may result in reduced environmental impact—particularly if they lead to energy savings—but often they merely change the kind of environmental impact that is produced. For example, they may reduce carbon emissions while increasing the production of noncarbon GHGs, particularly methane, which may have a similar (or higher) effect on the planet's warming. In other cases, they may remove GHGs while creating other byproducts that cannot be released into the environment without proper treatment or abatement.

The most instructive approach to measure actual environmental impacts is—again—life-cycle analysis, with reference, respectively, to wastewater treatment [16] and reclamation treatment [22].

The former shows that the highest environmental impacts from wastewater treatments are linked to energy consumption. In particular, 95% of the biogas resulting from the treatment is burned in torch, while the sludge is used largely for agricultural use (98.6%), with the remaining amount being employed for compost. The most effective ways to abate pollution, therefore, rely on electricity generation from biogas, which in turn would reduce energy demand (and carbon footprint) of the water cycle itself, although it is very unlikely that it—or significant parts thereof—may become fully self-sufficient energy-wise.

With regard to reclamation treatments, it has been found that recycling water may or may not be energy-efficient, depending on local conditions. For example, a study [22] examined two case studies whereby energy demand from recycling was in one case higher and in the other case lower than energy demand from importing water. Desalination was the most energy-intensive water source. It should be emphasized that, as far as water recycling is concerned, distribution proved to be the most relevant source of energy demand (explaining 61%–74% of total energy demand). This suggests that (1) the more energy that can be recovered, the less environmental impact, and (2) any design measure that reduces the energy cost of collecting wastewater and distributing treated water will significantly reduce the life-cycle environmental impact of the water system.

Energy is by far the most important source of environmental impact from water recycling, so it makes sense to focus on it. Yet, two further elements should be considered. First, water quality may be a very relevant variable: the higher the expected quality of the output (and the lower the quality of the input), the more energy intensive the treatment. Therefore, where the discrimination between potable and nonpotable water is possible, it may be very meaningful to employ recycled water for nondrinking uses in order to reduce energy consumption. Second, other impacts—in terms of treatment byproducts—may derive from water recycling. Therefore, focusing on carbon, while providing a reliable proxy, is not enough as it comes to comparing not just economic costs, but also environmental costs and externalities [1].

Finally, depending on the technologies employed to achieve energy recovery from wastewater, the goal of reducing carbon emissions (through a lower production of CO_2) may come at the cost of increasing noncarbon GHGs, such as nitrous oxide (N_2O) or methane (CH_4). If not properly abated (or recycled themselves), they may result in an increased, not reduced, greenhouse effect. In fact, the greenhouse power of N_2O and CH_4 is as much as 310-fold and 21-fold of that of CO_2, respectively. This is a frequent problem with wastewater treatment plants.

In fact, the most straightforward way to reduce CO_2 emissions is reduced energy consumption, which—beyond investments in making the plants as well as the whole system more energy-efficient—may be achieved by relying on self-produced energy. However, the process of energy recovery produces byproducts at several stages. The amount of CH_4 that is produced depends mainly on the residence time and the temperature of wastewater in the sewage system [8]. The major sources of N_2O are the nitrification and denitrification stages in the plant [10]. Paradoxically, measures aimed at reducing energy demand may focus on the aeration strategy that, in turn, might increase N_2O emissions. Even here, there is no general rule except that all aspects should be properly considered, and all potential sources of environmental impact should be properly accounted for, as the water system is designed.

3.5 Role of Smart Regulation

In order to make the best use of existing technologies—and to reduce costs as well as environmental impact from water consumption—massive investments should be undertaken. Both from the point of view of the investment profile and from that of life-cycle environmental impacts, much depends on the conditions under which the system is designed and the degree of the investor's confidence in the system itself. This is especially true when private capitals are involved which, as we shall see, is also the most convenient case.

Underlying this issue are many questions regarding how the water system is organized and regulated, from both a financial and an institutional perspective. In fact, in this case as well as in many others where network infrastructures are involved, institutional variables—that affect both the attractiveness of the water system for private investors and the effectiveness of the public decision-making process with regard to resource allocation and design choices—are in a way the most relevant ones. Institutions allow financing in a cost-effective way the required investments, institutions drive the criteria under which the system is designed, and institutions may prevent rent-seeking activities and misallocation of resources.

An OECD report [14] enumerates seven dimensions under which institutional gaps may hinder water-energy policy coordination, resulting in suboptimal results with regard to either water- or energy-related goals, or both. Such dimensions are

- *Policy framework*: Different political agendas, visibility concerns, and power rivalries across ministries and agencies at a central level as well as problems from national ministries dictating vertical approaches to cross-sectoral policies that would benefit from codesign at the local level
- *Administrative roles*: Unclear and overlapping roles and responsibilities among government ministries as they relate to economic, social, and physical boundaries of water and energy flows
- *Capacity resources*: A lack and/or asymmetry of knowledge, enforcement capacity, and infrastructural resources within all levels of government
- *Funding resources*: Asymmetry of revenues and distribution of resources across ministries and levels of government
- *Informational challenge*: Data gaps and inconsistencies between and within the levels and ministries of government
- *Time frame and strategic planning*: Different schedules and deadlines occur between ministries

- *Evaluation*: Without evaluation, governance practices cannot be assessed, but very often feasibility is limited [14, p. 42]

Even under this respect, there is no clear evidence that a given setting is more effective than others are. However, institutional settings whereby liabilities are clearly allocated, regulatory choices are regarded as credible and stable over time, and externalities are properly internalized tend to outperform those that fail to meet such targets. Transparency is also important, especially in those countries where corruption is widespread.

Moreover, in several instances, water is—or may be treated as—a common-pool resource, that is, a resource that is virtually accessible to everyone (or, to be more precise, a resource such that excluding potential users is either impossible or too costly). Often, it is better to rely on bottom-up institutional arrangements to avoid the "tragedy" of growing scarcity, rather than on top-down policies. Bottom-up approaches may also make more evident the benefits (if any) of water reuse as opposed to other conventional or alternative water sources. If this is true, then Elinor Ostrom's lesson on how to manage the commons should not be ignored [15].

One specific issue regarding water/energy management and the underlying institutional setting has to do with the nature of the relevant stakeholders. The following are particularly important: (1) the understanding of who should pay for the investment (the government, water consumers, etc.); (2) who is in charge of making investments; and (3) how the compliance between investments actually being made and projected investments/costs/financing mechanisms is assured (and *a fortiori* how the monitoring of all the activities within the water system is ensured).

In this respect, it is particularly important that a solid institutional framework is settled when water reuse comes into consideration. Among the various, water-related business opportunities, water reuse is—on average—one with low expected returns. In fact, returns on water reuse investments are expected to be below 10% in a business-as-usual scenario, and slightly above 10% when a carbon-pricing scheme is set into operation [6]. This makes the reliability of the relevant set of rules absolutely crucial: it is only by virtue of the subsequent certainty that substantial resources may be confidently invested.

Strong institutions may be useful also to address the low (although growing) confidence of public opinion in using recycled water. This can be partly offset by using recycled water for nondrinking uses, but ultimately it is fundamental that people believe that those in charge of monitoring and controlling water quality are trustworthy. A major contribution can and should come from investments in information and communication to promote the positive sides of water reuse [4,5].

3.6 Summary and Conclusions

Energy is consumed in the water cycle—particularly when water is reused—and water is a relevant input in energy production. Moreover, energy is embodied in, and can be recovered from, wastewater. Therefore, much value can be created, and a lot can be done to increase sustainability, if water reuse technologies are employed and, more generally, if water systems are designed with attention to the two dimensions of cost-effectiveness and reduction of environmental impacts.

Many choices—particularly those regarding the preferred technologies and, more fundamentally, those related to relying on water recycling or other alternative water sources, such as desalination—depend on local, site-specific variables; for example, whether an area is urban or rural, whether abundant water is available and accessible, whether water demand and wastewater production patterns are more or less variable over time, etc. [25].

There are, however, three general features that, once the decision of recycling water has been made, have a fairly broad scope, to the point that they can be generalized.

First, the system should be designed to separate as much wastewater flows as possible. It is less costly, less energy intensive, and less environmentally harmful to treat blackwater, greywater, and yellowater separately rather than conveying them together.

Second, and by the same token, the separation of potable and nonpotable water can provide a major contribution to the effectiveness and efficiency of the whole water system: in that case, wastewater should be treated up to a quality standard that is lower than the drinking standard, and subsequently is less costly, simpler, and less energy intensive.

Third, in order to deploy a well-functioning water system, especially if it encompasses water reuse, massive investments are needed. Resources to be invested may be either public or private—depending on the cases and social as well as political variables—but, in either case, a credible institutional framework should be set in place. This is particularly important for the kind of investments that have relatively low expected returns (such as water recycling infrastructure), and whose ability to attract capital critically depends on the credibility of the expected cash flow that would be generated by the system operation over the next decades.

A consequence of the relevance of the upfront investment vis-à-vis operational costs is that a relevant part of the energy consumption is embodied in the construction of the water system itself. The appropriate tool to assess economic as well as environmental performance of the various technological alternatives and designs is, therefore, life-cycle analysis.

References

1. Asano, T., Burton, F.L., Leverenz, H.L., Tsuchihashi, R., and Tchobanoglous, G. 2007. *Water Reuse—Issues, Technologies, and Applications*, Metcalf & Eddy/WECOM, New York.
2. Barillon, B., Martin Ruel, S., and Lazarova, V. 2011. Full scale assessment of energy consumption in MBRs. *6th IWA Specialist Conference on Membrane Technology for Water & Wastewater Treatment*, Aachen, Germany, October 4–7, 2011.
3. Besnault, S., Martin Ruel, S., Carrand, G., and Dauthille, P. 2011. Towards the positive energy WWTP: Increase the efficiency of the overall aeration system. *Proceedings of Singapore International Water Week: Water Convention*, July 4–11, 2011.
4. Ching, L. 2010. Eliminating 'yuck': A simple exposition of media and social change in water reuse policies. *Water Resource Development*, 26(1), 113–126.
5. Daigger, G.T. 2009. Evolving urban water and residuals management paradigms: Water reclamation and reuse, decentralization, resource recovery. *Water Environment Research*, 81(8), 809–823.
6. Dobbs, R., Oppenheim, J., Thompson, F., Brinkman, M., and Zornes, M. 2011. Resource revolution: Meeting the world's energy, materials, food, and water needs. McKinsey Global Institute—McKinsey Sustainability and Resource Productivity Practice, November 2011.
7. Grady, C.P.L. Jr., Daigger, G.T., Love, N.G., and Filipe, C.D.M. (eds.) 2011. *Biological Wastewater Treatment*. IWA Publishing, London (UK) and New York (US).
8. Guisasola, A., Haas De, D., Keller, J., and Yuan, Z. 2008. Methane formation in sewer systems. *Water Research*, 42/6–7, 1421–1430.
9. Hunter, P.R., Macdonald, A.M., and Carter, R.C. 2010. Water supply and health. *PLoS Medicine*, 7(11), e1000361.
10. Kampschreur, M.J., Temmink, H., Kleerebezem, R., Jetten, M.S.M., and Van Loosdrecht, M.C.M. 2009. Nitrous oxide emission during wastewater treatment. *Water Research*, 43(17), 4093–4103.
11. Lazarova, V., Choo, K.-H., and Cornel, P. (eds.) 2012. *Water-Energy Interactions in Water Reuse*. IWA Publishing, London (UK) and New York (US).
12. Mara, D., Lane, J., Scott, B., and Trouba, D. 2010. Sanitation and health. *PLoS Medicine*, 7(11), e1000363.
13. Meda, A., Lensch, D., Schaum, C., and Cornel, P. 2012. Energy and water: Relations and recovery. In: Lazarova, V., Choo, K.H., and Cornel, P. (eds.), *Water-Energy Interactions in Water Reuse*, IWA Publishing, London (UK) and New York (US), 21–35.
14. OECD 2012. Meeting the water coherence challenge. In *OECD, Meeting the Water Reform Challenge*, OECD Studies on Water, OECD Publishing, Paris (France), 129–171.
15. Ostrom, E. 1990. *Governing the Commons: The Evolution of Institutions for Collective Action*. Cambridge University Press, New York.
16. Pasqualino, J., Meneses, M., Abella, M., and Castells, F. 2009. LCA as a decision support tool for the environmental improvements of the operation of a municipal wastewater treatment plant. *Environmental Science and Technology*, 43(9), 3300–3307.
17. Pearce, G.K. 2012. Desalination vs. water reuse: An energy analysis illustrated by case studies in Los Angeles and London. In: Lazarova, V., Choo, K.H., and Cornel, P. (eds.), *Water-Energy Interactions in Water Reuse*, IWA Publishing, London (UK) and New York (US) 257–267.
18. PWC. 2013. *The World in 2050*. PwC, London, UK.

19. Rajeshwari, K.V., Balakrishnan, M., Kansal, A., Lata, K., and Kishore, V.V.N. 2000. State-of-the-art of anaerobic digestion technology for industrial wastewater treatment. *Renewable and Sustainable Energy Reviews*, 4(2), 135–156.

20. Reiner, M., Pitterle, M., and Whitaker, M. 2007. Embodied energy considerations in existing LEED credits. *Symbiotic Engineering*, Boulder, CO.

21. Roy, M.M., Dutta, A., Corscadden, K., Havard, P., and Dickie, L. 2011. Review of biosolids management options and co-incineration of a biosolid-derived fuel. *Waste Management*, 31(11), 2228–2235.

22. Stokes, J.R. and Horvath, A. 2006. Life cycle energy assessment of alternative water supply systems. *The International Journal of Life Cycle Assessment*, 11(5), 335–343.

23. Stokes, J.R. and Horvath, A. 2009. Energy and air emissions effects of water supply. *Environmental Science and Technology*, 43(8), 2680–2687.

24. UN 2013. World Population Prospects: The 2012 Revision. Working Paper no.ESA/P/WP.228.

25. UNESCO 2009. *Water in a Changing World. The United Nations World Water Development Report 3*, UNESCO Publishing, Paris (France).

26. UNESCO 2014. *Water and Energy. The United Nations World Water Development Report 2014*. Volume 1, UNESCO Publishing, Paris (France).

27. Wang, H., Brown, S.L., Magesan, G.N., Slade, A.H., Quintern, M., Clinton, P.W., and Payn, T.W. 2008. Technological options for the management of biosolids. *Environmental Science and Pollution Research*, 45(4), 308–317.

28. Yüce, S., Kazner, C., Hochstrat, R., Wintgens, T., and Melin, T. 2012. Water reuse versus seawater desalination—evaluation of the economic and environmental viability. In Lazarova, V., Choo, K.H., and Cornel, P. (eds.), *Water–Energy Interactions in Water Reuse*, IWA Publishing, London (UK) and New York (US), 243–256.

4

Water Reuse and Sustainable Urban Drainage Systems

Floris Boogaard
Delft University of Technology

*Hanze University of
Applied Sciences*

*Tauw
INDYMO*

Saeid Eslamian
Isfahan University of Technology

PREFACE

In this chapter, cost-effective recycling is discussed. The processes followed are considered to be noncomplex and cost-effective on recycling water that is "moderately clean" for the processes that do not require water from a high-quality standard.

4.1 Introduction

The increasing scarcity of water will promote the recycling of water from hydrologic cycle. Since the water cycle consists of freshwater (rainwater, surface water, and groundwater) and saltwater (seawater) as well as used water, there are multiple options for our resources.

It is possible to deliver water of almost any quality to meet public health standards. Industrial water use includes water for purposes such as

- Drinking
- Processing
- Cleaning
- Cooling

Water recycling is normally an integral part of the process and treated and reclaimed water is recycled to conserve water and avoid stringent discharge requirements. The processes that produce the required water quality are as follows:

- Sedimentation
- Filtration
- Coagulation
- Aeration
- Micronano filtration

If water of the highest quality standard is needed, for example, for consumption, then mostly water is treated by specialized companies using the processes from sedimentation to nanofiltration. The use of groundwater, surface water, or stormwater is considered to be cost-effective for industrial water use as cooling and washing with processes such as sedimentation and filtration. Surface water quality is highly dependent on place and time (quality can depend on seasons and even day and night rhythm). Groundwater quality is relatively stable; the quality is less dependent on place and time. Stormwater quality can depend on place. Stormwater quality data for various locations of the world is shown in Table 4.1.

This chapter describes some sustainable urban drainage systems (SUDS) that are noncomplex and cost-effective for recycling water.

4.2 Bound and Dissolved Substances in Rainwater Runoff

Dissolved substances and nondissolved or solid substances are present in water (Figure 4.1). Dissolved substances are substances

TABLE 4.1 International Stormwater Quality Data from Residential Areas in United States, Australia, and Europe

		Dutch[a]	USA NSQD [7][b]	Europe/ Germany ATV Database [6][c]	Worldwide [3][d]
		Mean	Median	Mean	Mean
TSS	mg/L	17	48	141	150
BOD	mg/L	5.7	9	13	
COD	mg/L	32	55	81	
TKN	mg N/L	1.9	1.4	2.4	2.1
TP	mg P/L	0.4	0.3	0.42	0.35
PB	μg/L	18	12	118	140
Zn	μg/L	102	73	275	250
CU	μg/L	19	12	48	50
E. coli	kve/100 mL	1.9E + 04			

Source: Boogaard, F., Ven van de, F., Langeveld, J., and Giesen van de, N. 2014. *Challenges*, 5(1), 112–122.

[a] Dutch STOWA database (version 3.1, 2013) [1], based on data monitoring projects in the Netherlands, residential and commercial areas, with *n* (amount of monitored stormwater events) ranging from 26 (SS) to 684 (Zn).

[b] NSQD monitoring data collected over nearly a 10-year period from more than 200 municipalities throughout the United States. The total number of individual events included in the database is 3770, with most in the residential category (1069 events).

[c] ATV database, like [5] partly based on the US EPA nationwide runoff program (NURP), with n ranging from 17 (TKN) to 178 (SS).

[d] Typical pollutant concentrations based on review of worldwide [5] and Melbourne [9] data.

that degrade in water into individual molecules. The nondissolved substances can be classified as follows:

- Colloidal substances: These are substances with a diameter between 10^{-9} and 10^{-7} m and densities between 1000 and 2000 kg/m^3. These particles can cloud the water or cause an unpleasant color, for example, the brown color of water by humus-like substances. These substances form a transition between dissolved and suspended solids.
- Suspended solids are substances with a diameter greater than 10^{-6} m and a density greater than water. These particles reach the suspended state by the turbulence of running water. The relatively larger and heavier particles will slide, abrade, and settle, while the smaller particles keep their floating state through the turbulence. In stagnant water, suspended solids will sediment under the influence of gravity.
- Floating substances: Substances with a density less than water suspended matter are composed of an organic fraction and an inorganic fraction. Organic micropollutants bind themselves particularly to the organic fraction while the inorganic micropollutants bind to the inorganic fraction (e.g., lutite particles). Substances that are bounded to the floating fractions are less easily included in the ecosystem than dissolved substances.

From the considered measurements, as presented in Figure 4.1, it can be concluded that from the metals, lead (92%) and iron (98%) were bound most frequently and that nickel (55%) is the least bounded metal. On average, 72% of the metal has been

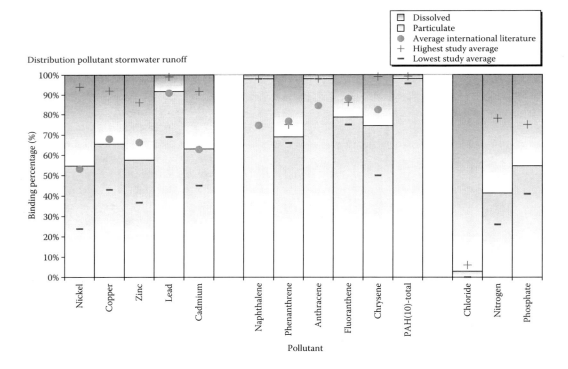

FIGURE 4.1 Percentage of bound and dissolved substances. (Adapted from Boogaard, F., Ven van de, F., Langeveld, J., and Giesen van de, N. 2014. *Challenges*, 5(1), 112–122.)

bounded and 28% of the metal has been dissolved. Polycyclic aromatic hydrocarbon (PAH) is bound for 98%. Of the undissolved oil, 93% is bound and nitrogen and phosphate are bounded, respectively, 40% and 50% to suspended matter. The most common form of nitrogen in water is nitrate.

The binding of the main contaminants in rainwater runoff (PAH and heavy metals) is relatively high, causing a high probability of capturing these bounded contaminants. Within the substances, the bonds can vary such as with the heavy metal cadmium (a relatively toxic substance for flora and fauna) that binds to a lesser degree than the heavy metal lead. For nutrients and other substances, the bond seems to be lower, which implies that these substances are less important in rainwater (relative to other sources).

The considered main "problem substances" in water are copper, lead, zinc, PAH, and mineral oils. Of these substances, lead, PAHs, and oils can be bound for more than 90%. Zinc and copper, on the other hand, bind less easily 58% and 65%, respectively (however, large ranges are found in international literature). It will be relatively easier to get a higher removal efficiency with SUDS for substances that are relatively more bound to suspended solids.

4.3 Purification: SUDS

Methods that can be used to get the desired quality for reused water can be described as (BMPs and SUDS . These systems will be discussed as follows.

4.3.1 Best Management Practices

The term SUDS or stormwater BMPs refers to a wide range of stormwater control systems, which enable the planning, design, and management of stormwater to be tackled equally from hydrological, environmental, and public amenity perspectives [4]. These systems can be used individually or in combination with each other (as a treatment train) and both as an alternative to or in combination with conventional piped stormwater drainage systems. The following sections provide a brief description of the main types of BMPs/SUDS followed by measured pollutant efficiency performances.

4.3.2 Physical Treatment Processes

The physical treatment processes are as follows:

- Litter and debris removal (using screens, floatation, and separation to remove gross solids)
- Oil and grease removal (using floatation and separation)
- Sedimentation (to remove fine solids and attached pollution)
- Filtration (to capture very fine particles)

Some of the widely used facilities are

- Constructed wetlands
- Subsurface flow constructed wetlands
- Detention ponds/basins (dry ponds)

- Extended detention basin (EDB)
- Retention ponds/basins (wet ponds)
- Sedimentation tank
- Bioswales

In the next sections, these SUDS are described with definitions from CIRIA, BMPs,[*] and SUDS Susdrain with recent information on the monitoring of SUDS and photos.

4.3.3 Constructed Wetlands

Constructed wetlands are artificial, designed complex vegetative water bodies that can provide treatment (and recycling) of both wastewater effluent and stormwater runoff. Surface flow systems (also known as free water systems) are wetlands in which water primarily flows above the ground surface and through the litter layer. They simulate natural marshes, employing shallow channels and basins planted with emergent, submergent, and/or floating vegetation through which water flows at shallow depths and low velocities.

A constructed subsurface flow system is a wetland in which wastewater flows through a lined basin or channel that is filled with a permeable substrate. This is planted with wetland plants and flow remains below the media surface.

4.3.4 Detention Ponds/Basins (Dry Ponds)

Depressed basins are normally dry but temporarily store and attenuate a portion of stormwater runoff following a storm event [10]. Water is controlled by means of a hydraulic control structure to restrict outlet discharge according to the required detention time. Such dry basins offer public open space for recreational uses but are of limited habitat value.

4.3.5 Extended Detention Basin

This typically consists of a two-stage design providing a dry upper level and a smaller lower stage containing permanent water and/or a shallow marsh. To serve as an effective BMP, EDBs need to hold stormwater in the lower stage basin for relatively long periods.

4.3.6 Retention Ponds/Basins (Wet Ponds)

These posses a permanent pool of water incorporated into the design and are also known as balancing ponds or flood storage basins. They principally function as sedimentation facilities with soluble pollutants being removed by biological processes, which can be enhanced by marginal planting. Such wet ponds/basins can have substantial aesthetic, amenity, and ecological

---

[*] Priority pollutant behavior in stormwater best management practices (BMPs), Deliverable No. D5.1, 18/09/07, Revised version: 15/02/08, Dissemination level: PU, Lian Scholes, Mike Revitt, Johnny Gasperi, Erica Donner, Middlesex University, Queensway, Enfield, UK.

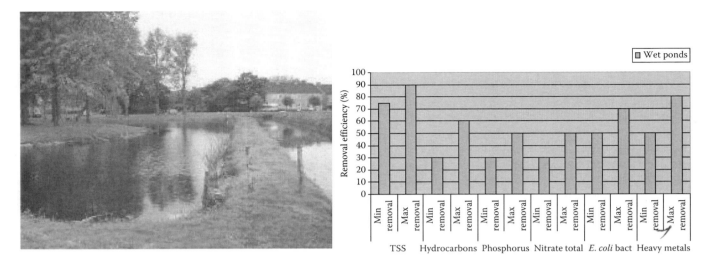

FIGURE 4.2 Example of purification for possible water reuse at stormwater sewer in Limmen and measured efficiency rates (photograph: Floris Boogaard); removal efficiencies of wet ponds. (Adapted from Wilson, S., Bray, R., and Cooper, P. 2004. *Sustainable Drainage Systems, Hydraulic, Structural and Water Quality Advice*, London, CIRIA C609.2004 RP663, ISBN 0-86017-609-6.)

benefits in addition to their flood and water quality control benefits (Figure 4.2).

4.3.7 Sedimentation Tank

These structures are intended to intercept and retain coarse sediment and litters carried in stormwater runoff by means of a bed load mechanism and are often located at the front end of a treatment train system. In Figure 4.3, an outlet of the stormwater system in Amsterdam with a catchment area of 12 hectares. The picture on the left shows the sediment load after 8 years (Figure 4.3).

The removal efficiency of sediment basins is highly dependent on the characteristics of the pollutants and dimensions of the

structure. Being a sedimentation facility, only the (larger) bound particles will be removed.

4.3.8 Bioswales

A properly designed bioswale system buffers rainwater and allows it to infiltrate; this improves the quality of rainwater, which can be reused for different purposes. A bioswale is a ditch with vegetation and a porous bottom. The top layer consists of enhanced soil with plants. Below that layer, different infrastructure can be constructed such as a layer of gravel, scoria, or baked clay pellets packed in geotextile. These materials have large empty spaces, allowing large quantities of the rainwater to be stored for reuse. An infiltration pipe/drainpipe is situated

FIGURE 4.3 Sedimentation basin in Amsterdam after 8 years of implementation.

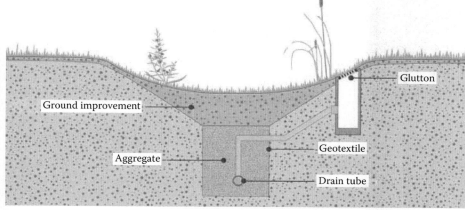

Section scheme of a bioswale when it is dry (© image by atelier GROEBLAUW, Marlies van der linden) (based on: Boogaard et al., 2006).

FIGURE 4.4 First large implementation of swales in Enschede, the Netherlands.

below the second layer. To prevent the bioswale from overflowing its banks during heavy rainfall, overflows are added that are connected directly to the infiltration pipe/drainpipe. With the filtration and adsorbtion of bound pollutants, high removal efficiencies are achieved with bioswales (Figure 4.4).

4.4 Summary and Conclusions

SUDSs are a collection of management practices, control structures, and strategies to efficiently and sustainably drain stormwater, while minimizing pollution and managing the impact on water quality of local water bodies. Sedimentation devices can be part of such a local treatment strategy to minimize the impact of local stormwater discharges on receiving surface water bodies. That is why these structures are referred to as sedimentation SUDS.

The settling efficiency of sedimentation devices depends highly on the characteristics of the pollution load, the dimensions of the facility, and implementation in the field. For detailed determination of the pollution removal efficiency of these sedimentation devices, information is needed on

- Quality of stormwater, as the concentration of the pollutants determines the need for stormwater treatment and acquired removal efficiencies [1]
- Suspended solids, pollutant adsorption behavior, particle size distribution, and settling velocities of particles, which determine the potential of treatment techniques based on settling [12]
- Hydraulic loading and geometry of the facility, as this determines local flow velocity, which affects the settling process of particles

Urban water managers can use this information to identify the most appropriate stormwater management strategy for a specific drainage area. Some SUDSs described in this chapter are as follows: constructed wetlands (horizontal, vertical), subsurface flow constructed wetlands, detention ponds/basins (dry ponds), EDB, retention ponds/basins (wet ponds), sedimentation tank, and bioswales.

References

1. Boogaard, F. and Lemmen, G. 2007. *STOWA Stormwater Database: The Facts about the Quality of Stormwater Runoff, STOWA 2007–21*, The Netherlands (in Dutch).
2. Boogaard, F., Ven van de, F., Langeveld, J., and Giesen van de, N. 2014. Selection of SUDS based on stormwater quality characteristics, *Challenges*, 5(1), 112–122.
3. Bratieres, K., Fletcher, T.D., Deletic, A., and Zinger, Y. 2008. Nutrient and sediment removal by stormwater biofilters: A large-scale design optimisation study. *Water Research*, 42, 3930–3940.
4. CIRIA. 2001. *Sustainable Urban Drainage Systems: Best Practice Manual for England, Scotland*, CIRIA Paperback, Wales and Northern Ireland.
5. Duncan, H.P. 1999. *Urban Stormwater Quality: A Statistical Overview; Cooperative Research Centre for Catchment Hydrology*, Melbourne, Australia.
6. Fuchs, S., Brombach, H., and Wei, B.G. New database on urban runoff pollution. In *Proceedings of NOVATECH 5th International Conference Sustainable Techniques and Strategies in Urban Water Management*, Lyon, France, June 6–10 2004; pp. 145–152.
7. Pitt, R. 2004. *National Stormwater Quality Database (NSQD)*, Department of Civil and Environmental Engineering, University of Alabama, Tuscaloosa.
8. Susdrain. http://www.susdrain.org/delivering-suds/using-suds/background/sustainable-drainage.html. Accessed December 5, 2014.

9. Taylor, G.D., Fletcher, T.D., Wong, T.H.F., Breen, P.F., and Duncan, H.P. 2005. Nitrogen composition in urban runoff—implications for stormwater management, *Water Research* 39, 1982–1989.

10. Wilson, S., Bray, R., and Cooper, P., 2004. *Sustainable Drainage Systems, Hydraulic, Structural and Water Quality Advice*, London, CIRIA C609.2004 RP663, ISBN 0-86017-609-6.

11. Xuheng, K. 2014. Stormwater modeling and management. In S. Eslamian, (ed.), *Handbook of Engineering Hydrology*, Vol. 3, Chapter 16. Environmental Hydrology and Water Management. Taylor & Francis, CRC Press, Boca Raton, FL, 328–346.

12. Boogaard, F.C. 2015. *Stormwater Characteristics and New Testing Methods for Certain Sustainable Urban Drainage Systems in The Netherlands*, Delft. ISBN: 978-94-6259-745-7.

II

Urban Water Reuse and Sanitary

5

Reused Urban Water Microbiology

Roman Tandlich
Rhodes University

Catherine Diane Luyt
Rhodes University

Bongumusa M. Zuma
Rhodes University

PREFACE

Guidelines for microbial quality of reclaimed water and wastewater have been published by organizations such as the World Health Organization (WHO) and they are generally based on concentrations of indicator microorganisms, protozoans, and helminths. The maximum allowable concentrations are determined by the source and intended use of the reclaimed water and the infectious dose of the given microorganism. To achieve the microbial quality of the reclaimed water for reuse, a minimum of secondary treatment must be performed, for example, the lagoon-type systems with hydraulic residence time of 25 days or more. Specific treatment such as disinfection with ultraviolet (UV) radiation or chlorine dioxide must be performed to remove protozoa from laundry greywater. In decentralized settings, greywater treatment can be performed using low-cost systems such as fly-ash lime filter tower with fecal coliforms (FC) and total coliforms (TC) removal shown to range from 35% to 71% at the hydraulic residence time of approximately 18–24 min. For rainwater recycling, passive treatment by rainwater contact with roofing material can be used, or the first-flush device can be installed onsite. Removal of helminth pathogens can be achieved using coarse filtration systems such the Hydrotech Discfilter process system.

5.1 Introduction

Reclaimed water has been used for various purposes [67], at the household and city levels. Microbial composition of reclaimed water and wastewater has been studied extensively and guidelines with respective microbiological limits have been published by international bodies [67] and national governments [17]. Characterization has been done using the indicator microorganism approach, as well as by enumeration of individual pathogens. The main four groups of microorganisms of public health concern in reclaimed water reuse are viruses, bacteria, protozoa, and helminths. Various types of reclaimed or recycled water are relevant in urban settings such as greywater; that is, domestic wastewater without any toilet input. Its treatment and reuse has been investigated as one of the ways to decrease potable water use in activities such as irrigation and toilet flushing [18,64,69]. Public health threats of reusing raw greywater are associated with the presence of indicator microorganisms and

(opportunistic) pathogens [9,18,46,50,65,66,68]: *Escherichia coli* (*E. coli*), FC, *Salmonella* spp., *Cryptosporidium* spp., *Shigella* spp., *Legionella* spp., and enteric viruses. Tandlich et al. [64] and Zuma et al. [69] reported indicator microorganism that indicate substantial risk to public health [59,60]; that is, microbial greywater quality must be known in detail before any reclamation can commence.

Onsite treatment and reclamation/recycling of greywater and other wastewaters gain significance during potable water supply interruptions, which is a common problem in developing countries, for example, South Africa. In 2010, up to 47.6% of all households in South Africa experienced water supply interruptions [62]. Considering local data, up to 63.4% of all interruptions resulted from insufficient maintenance and, in many cases, they lasted for 15 days or more [32,62]. Pipe breaks and inadequate maintenance lead to water pressure fluctuations, which in turn increases risk of microbial contamination of drinking water distribution systems and possibly also of waterborne disease outbreaks [8,43]. The dissemination of news about the occurrence of such problems can increase population dissatisfaction with potable water quality [61], which brings its safety into question [16]. Under such conditions, the population can seek alternative supplies of drinking water to serve during municipal supply interruptions. One such source is rainwater harvesting, taking place in rainwater tanks at the household level.

Rainwater tanks can be installed in such places as family homes, schools, and community centers [52]. After installation, microbial quality of the collected rainwater must be investigated. The significance of regular microbial rainwater quality examinations is further stressed by literature reports that raw and freshly harvested rainwater is commonly unsuitable for potable purposes in many developing countries [28,60]. Other domestic uses such as laundry and toilet flushing are, however, feasible [7]. If microbial contamination of rainwater is detected during routine monitoring [38], then rainwater should be subjected to minimum treatment before human consumption and such minimum treatment can include boiling, addition of bleach, or sand filtration [42]. The treatment of water will decrease the risk of a disease outbreak upon human consumption [40]. However, the extent and frequency of such minimum treatment can be hindered by limited financial resources in poor households [40]. A combination of these factors increases the chances of waterborne disease outbreaks if guidelines for safe use of rainwater are not followed [63]. The microbiology of recycled/reclaimed water, greywater, and rainwater is summarized in this chapter.

5.2 Microbial Quality of Recycled/ Reclaimed Water

Use of recycled/reclaimed water must be based on legally binding agreements between the recycled water producer or provider and the end-user. Such agreements have to contain specifications on the quality of the reclaimed water, the volume supplied, and exact applications in which the recycled or reclaimed water will be used (see page 9 in Reference 17). Water authorities are responsible for supply and microbial quality of water provided to the public (see page 10 in References 17 and 51). Therefore, they take charge of identifying risks and the routes of human exposure to recycled water (see page 10 in Reference 17) and set the relevant microbial standards that recycled water needs to meet (see Table 5.1 for details). Routes of human exposure to microbes contained in reclaimed water can be divided into two groups. The first one is "direct routes" such as consumption of fresh produce irrigated with reclaimed water or human contact with soils previously irrigated with the reclaimed water in question [17]. The second group of exposure routes is called "indirect routes" and relevant examples include exposure of agricultural workers to reclaimed water during milking of dairy cows [2].

Microbial considerations and routes of human exposure are of particular significance for several reasons [17]. First, there are links between the intended reclaimed water use and human consumption, for example, use of reclaimed water for irrigation of produce eaten raw or potable purposes. Second, aerosols form during application of reclaimed water and the resulting aerosolized particles can contain the causative agents of communicable diseases. Once irrigation has taken place, aerosols with infectious disease agents can be carried by wind as so-called "drift," resulting in human exposure via the atmosphere and inhalation downwind from the irrigation site. The third significance of the routes of human exposure to microbial components of reclaimed water is in aquaculture and fish farming, sports activities, and industrial operations. These will be important in the context of recreational activities, industrial production, and related activities. Direct and indirect routes of human exposure differ among the four groups of human pathogens (see Section 5.1). Relative importance of individual routes is a function of the structure of the basic infectious unit and the severity of the threat of the

TABLE 5.1 Classes of Reclaimed Water Based on the Concentrations of *E. coli*, Indicator Microorganisms, and Level of Treatment Required

	Required Levels of Wastewater Treatment Based on the Indicator of the Microorganisms	Possible Uses of the Reclaimed Water Class
Class of reclaimed water		
A	<10 *E. coli* cells or CFUs/100 mL <1 nematode egg/1 L <1 protozoan cell or oocysts/50 L <1 virion/50 L	Irrigation of areas with unlimited access for the general public Industrial applications with no limitations for employees Irrigation of agricultural crops that the target human population consumes raw
B	<100 *E. coli* cells or CFUs/100 mL Nematode egg concentrations should allow for livestock grazing	Application of the reclaimed water in areas involved in dairy livestock grazing Industrial applications such as rough floor washing
C	<1000 *E. coli* cells or CFUs/100 mL Nematode egg concentrations should allow for livestock grazing "schemes"	Nonpotable use for areas where public access can be controlled Irrigation of agricultural produce which will undergo cooking and heat treatment before human consumption; also can be applied to grow fodder crops for livestock Application in industrial operations where no human contact with the reclaimed water is guaranteed
D	<10,000 *E. coli* cells or CFUs/100 mL	Irrigation of agricultural crops that are not aimed for human consumption

Source: Adapted from Environmental Protection Agency. 2003. *Guidelines for Environmental Management: Use of Reclaimed Water.* Environmental Protection Agency Publication 464.2. Southbank, Victoria, Australia.

infectious disease outbreak from the given human pathogen in the reclaimed/recycled urban water.

Basic structural and infectious units can be summarized as follows: virion is the basic structural/infectious unit of viruses; the prokaryotic cell or spore is the infectious or structural unit of bacteria; the eukaryotic cell or spore is the basic structural/infective unit of fungi; oocysts or eukaryotic cells are the basic structural units and infectious particles for protozoa; and, finally, helminth infections are caused by eukaryotic cells or ova and cysts. From the viral causative agents, the highest threats in reclaimed water reuse originate from hepatitis E virus and rotavirus in adults [20,21,22,24,54]. Outbreaks with a bacterial causative agent can include *E. coli* and *Pseudomonas aeruginosa* [68]. Protozoans as human pathogens that have been detected in recycled water are *Cyclospora* spp. [26], *Giardia* spp., and *Cryptosporidium* spp. (see page 12 in Reference 17). The main helminth pathogens in reclaimed water include *Taenia* spp., *Trichuris* spp., and *Ascaris* spp. (see page 12 in Reference 17). Each of the four groups of infectious agents poses a risk to public health and this risk is quantified using the infectious dose concept.

Infectious dose is the number of infective particles that must enter the human body via direct or indirect routes of exposure for the affected individual to exhibit the clinical symptoms of a particular infectious disease [29]. This parameter depends on the type of pathogen, route of exposure, immunity of the affected individual, and environmental factors; thus, it is paramount in setting the guidelines for recycled water use. Examples of such guidelines are those set by WHO, which recommends that if recycled water is applied in "unrestricted" irrigation [53], then the maximum helminth concentration is 1 helminth egg/L [53,67]. If reclaimed water is used for irrigation and the produce is to be consumed by children 15 years of age or younger, then the maximum helminth concentration should not exceed 1 helminth egg in 10 L of recycled wastewater [53,67].

Testing for all pathogens is cost-prohibitive and guidelines for microbial quality of recycled water are often based on the indicator microorganism concentrations [34,68]. Combining the obtained concentrations with the infectivity of individual microbial pathogen classes gives rise to different classes of recycled water as shown in Table 5.1 [17,67]. In Table 5.1, the unit CFUs/100 mL stands for the colony-forming units per 100 mL of the recycled water sample. The most common indicator microorganisms for bacteria are FC and TC, the heterotrophic plate count (HPC), and *E. coli*. They can be used to evaluate the presence of fecal contamination or treatment problems in the distribution system of recycled/reclaimed water [34,67]. Enumeration of FC and *E. coli* can be achieved using selective media m-FC agar and m-TEC agar; and inoculation by spread-plating or by using membrane filtration [34]. The most-probable number technique can also be used, for example, using the Colilert system [42]. The same can be used for the TC enumeration on the MacConkey agar or the m-Endo agar [34]. HPC is enumerated on the R2A agar [34].

After inoculation, incubations are conducted at $35 \pm 2°C$ and $44.5 \pm 0.2°C$; and the concentrations of *E. coli*, FC, and TC are counted after 24 h. Low-cost indication of fecal contamination can be performed using the hydrogen-sulfide test [31,41]. This test can be used in remote areas or when laboratory equipment and qualified staff are not available. It detects bacteria of fecal origin, for example, *Citrobacter* spp. and *Salmonella* spp., in the water sample in question through the reduction of sodium thiosulfate into hydrogen sulfide and the following *in situ* precipitation of this compound with Fe^{3+} [31,41]. Examples of concentration ranges can be seen in Tables 5.2 (greywater) and 5.3 (rainwater).

The concentration of *E. coli* should be below 0 CFUs/100 mL of recycled water, if it is to be used for human consumption or in irrigation of produce that is consumed raw by humans, that is, to prevent infection through direct routes of human exposure [58]. The *E. coli* concentrations in greywater contaminated with feces has been shown to range from 1.58×10^3 to 1.59×10^7 CFUs/100 mL [68]. *E. coli* is part of the FC group and some guidelines are based on the FC concentrations instead of the *E. coli* levels. The FC concentration in recycled water should be below 0 CFUs/100 mL to avoid any risk of waterborne infection through direct routes of exposure [59,60]. If the relevant FC concentrations range from 0 to 10 CFUs/100 mL, then a slight risk of microbial infection exists through a continuous and direct human exposure [59,60]. However, short-term exposure is not expected to result in any infections [59,60].

If the concentrations of FC in recycled water range from 10 to 20 CFUs/100 mL, then there is a definite risk of human infections upon direct exposure to the recycled water on a continuous basis [59,60]. A slight risk of human infections will result from short-term exposure to recycled water through direct routes [59,60]. FC concentrations above 20 CFUs/100 mL indicate a significant risk of microbial infection among the recycled water end-users through direct and continuous exposure [59,60].

The older guidelines might still be based on TC concentrations. If this parameter lies inside the interval from 0 to

TABLE 5.2 Overview of the Indicator Microorganism Concentrations in Greywater Contaminated by Feces

Wastewater	TC Concentrations ($\log_{10}$ CFU/100 mL)	FC Concentrations ($\log_{10}$ CFU/100 mL)
Laundry	3.4–5.5	2.0–3.0
Shower, hand basin	2.7–7.4	2.2–3.5
Greywater	7.9	5.8
Shower, basin	1.8–3.9	0–3.7
Laundry wash	1.9–5.9	1.0–4.2
Laundry rinse	2.3–5.2	0–5.4
Greywater, 79% shower	7.4	4.3–6.9

Source: Collated from Ottosson, J. 2003. Hygiene aspects of greywater and greywater reuse. Licentiate thesis, Department of Water and Environmental Microbiology of the Swedish Royal Institute of Technology, Stockholm, Sweden; Ottosson, J. and Stenström, T.A. 2003. *Water Research*, 37(3), 645–655; Zuma, B.M. 2012. The efficacy and applicability of low-cost adsorbent materials in wastewater treatment. PhD thesis, Rhodes University, Grahamstown, South Africa.

TABLE 5.3 Microbiological Rainwater Quality in the Makana Municipality, South Africa

Location/Tank Site	GPS Coordinates	Sampling Date in 2011	HPC (CFU/mL)	FC (CFU/100 mL)	TC (CFU/100 mL)	H₂S Strip Test
Grounds and Gardens (RU)	S33°18′49.6″ E26°30′56.2″	March 25	4880	8	7900	Positive at 48 h
Continuing Education Centre (RU)	S33°19′01.9″ E26°30′46.7″	March 25	1000	16	500	Positive at 48 h
Victoria School for Girls	S33°17′42.8″ E26°33′17.6″	April 19	4660	224	7400	Positive at 48 h
Continuing Education Centre	S33°19′01.9″ E26°30′46.7″	April 19	150	138	600	Positive at 48 h
Eluxolweni Children's Shelter	S33°18′22.5″ E26°32′006 ″	April 19	800	1648	700	Positive at 48 h
Samuel Ntsiko Primary School	S33°18′10.3″ E26°32′48.1″	May 6	70	0	200	Positive at 48 h
GADRA School for the Blind	S33°18′29.2″ E26°32′01.4″	May 6	5620	66	700	Positive at 48 h
Andrew Moyake Tank 1	S33°18′37.6″ E26°32′58.1″	May 6	7870	0	4100	Negative
Andrew Moyake Tank 2	S33°18′37.6″ E26°32′58.1″	May 6	1010	4	<5	Positive at 48 h
Eluxolweni Children's Shelter	S33°18′22.5″ E26°32′006 ″	May 13	3520	254	4000	Positive at 24 h
Grounds and Gardens (RU)	S33°18′49.6″ E26°30′56.2″	May 13	2630	554	<5	Positive at 24 h
Continuing Education Centre (RU)	S33°19′01.9″ E26°30′46.7″	May 13	870	40	500	Positive at 24 h
Victoria School for Girls	S33°17′42.8″ E26°33′17.6″	May 13	2360	408	400	Positive at 24 h
Andrew Moyake Tank 1	S33°18′37.6″ E26°32′58.1″	May 13	4050	180	2000	Positive at 48 h
Andrew Moyake Tank 2	S33°18′37.6″ E26°32′58.1″	May 13	4320	208	2100	Positive at 48 h
Samuel Ntsiko Primary School	S33°18′10.3″ E26°32′48.1″	May 13	3930	644	30,700	Positive at 48 h
GADRA School for the Blind	S33°18′29.2″ E26°32′01.4″	May 13	6580	>500	13,700	Positive at 48 h
Archie Mbolekwa 1	S33°17′44.0″ E26°33′16.9″	May 27	2810	>500	16,200	Positive at 48 h
Archie Mbolekwa 2	S33°17′44.0″ E26°33′16.9″	May 27	650	>500	4100	Positive at 48 h
Victoria School for Girls	S33°17′42.8″ E26°33′17.6″	May 27	1270	>500	13,500	Positive at 48 h
TEM Mrwetyana Senior Secondary School	S33°17′35.2″ E26°33′26.5″	May 27	890	>500	4900	Positive at 48 h
GADRA School for the Blind	S33°18′29.2″ E26°32′01.4″	May 27	290	<4	1100	Negative
TEM Mrwetyana Senior Secondary School	S33°17′35.2″ E26°33′26.5″	August 17	1770	<4	500	Negative
Andrew Moyake Tank 1	S33°18′37.6″ E26°32′58.1″	August 17	80	<4	0	Negative
Andrew Moyake Tank 2	S33°18′37.6″ E26°32′58.1″	August 17	270	<4	600	Negative
Eluxolweni Children's Shelter	S33°18′22.5″ E26°32′006 ″	August 17	1190	<4	6200	Negative
Rhodes University Grounds and Gardens (RU)	Opposite Rhodes University Smuts Hall	August 17	10	<4	3900	Negative
Archie Mbolekwa 2	S33°17′44.0″ E26°33′16.9″	August 17	1250	<4	1300	Negative

Source: Luyt, C.D. 2013. Faecal source tracking and water quality in the Eastern Cape, South Africa. PhD thesis, Grahamstown, Rhodes University, South Africa.

5 CFUs/100 mL, then there is negligible risk of a waterborne disease outbreak from direct human exposure to recycled water [59,60]. TC concentrations from 5 to 100 CFUs/100 mL indicate that the reclaimed water has been inadequately treated, or that microbial contamination has occurred after the treatment and microbial regrowth has taken place in the distribution system or during storage [59,60]. As a result, there is a slight risk of an infectious disease outbreak during occasional domestic use of such reclaimed water, while a high risk is associated with continuous and direct human exposure [59,60]. Finally, if the TC concentrations in reclaimed water increase above 100 CFUs/100 mL, then analogical problems occur as with the previous concentration interval, except there is a significant risk of a disease outbreak associated with any direct human exposure [59,60]. The reader can consult relevant literature reviews, for example, [33,34,47] for further details on the use of indicator microorganisms.

The need to perform chlorination and general maintenance can be estimated using HPC levels [59,60]. There is no threat to human health if the HPC concentration in recycled water is inside the interval of 5–100 CFUs/mL [6,59,60]. Intermediate risk for direct routes of human exposure is encountered if HPC ranges from 100 to 1500 CFUs/mL in the reclaimed water [59,60]. Finally, high risk is reported if HPC increases above 1500 CFUs/mL [59,60]. Routine microbiological examination of recycled water should contain the enumeration of helminth eggs and the infectivity assessment [51], using standard methods [6]. Flotation isolation and fluorescence microscopy counting based on antibodies should be done for *Cryptosporidium parvum* and other protozoa [15,44]. The transmission routes for humans need to be understood in detail for the wide variety of recycled water types [45,51].

Helminth guidelines in Table 5.1 are supported by literature data on helminth infections among farm workers who ate fresh produce previously irrigated with helminth-contaminated wastewater [23]. Human helminth infections from exposure to recycled/reclaimed water are rare, but the patient's quality of life can be severely compromised if they occur [35]. Thus, infectious particles of helminths must be eliminated if reclaimed water is to be used for livestock watering (see page 27 and Section 4.1 in Reference 17). If irrigation is performed with reclaimed water that has not been subjected to helminth removal, then the pastureland cannot be used for cattle grazing until the helminth concentrations have decreased due to natural attenuation, that is, 2 years [17]. Reclaimed human sewage cannot be used as a source of drinking water for pigs due to potential transmission of *Taenia solis* [17]. If the class C reclaimed water is used in animal husbandry with sheep, goats, and horses, then helminth concentrations are of no concern (see page 28 in Reference 17). Compliance sampling to ascertain the reclaimed microbial water quality follows the hazard analysis and critical control points (HACCP) principles with frequency ranging from weekly for classes A and B to monthly for classes C and D (see Tables 5 and 6 in Reference 17).

Abbott et al. [1] found that up to 70% of the rainwater samples collected around the North Island of New Zealand contained more than 60 CFUs/100 mL of TC or *E. coli*. Thirty-two percent for TC and 36% for *E. coli* of the samples taken had concentrations of the two indicator microorganisms between 0 and 10 CFUs/100 mL, that is, indicating low to intermediate risk waterborne disease outbreak upon water consumption. Questionnaires were given to household residents where rainwater sampling took place, and 3%–18% did not know the frequency of gutters or water tank cleaning [1]. Thirty-three percent of gutters were only cleaned yearly and 30% of the tanks were said to be never cleaned [1]. Up to 52% of the rainwater tanks did not have any treatment facilities in place, for example, first-flush filters, while no maintenance could be performed in 32% of the cases as the physical barrier of debris screens prevented this [1]. Rainwater was only filtered in 10% of the sampled households, with 71% of these samples contaminated with indicator microorganisms [1]. Results of Reference 1 are in line with the findings of the previous studies in New Zealand [19,57].

The microbial quality of rainwater will depend on the harvesting method [39]. Many of the in-field harvesting methods will be applied in conjunction with a fertilizer such as manure and thus the microbial water quality of harvested rainwater in these cases is likely to be governed by similar factors as the flood waters as summarized by Luyt et al. [33]. Generally speaking, simple treatment methods can help improve the microbial rainwater quality and they include gutter and rainwater tank cleaning [31], filtration using devices such as first-flush devices, and rainwater boiling [11]. Microbiological quality is directly proportional to the rainfall intensity during the sampling period [14]. High levels of indicator microorganisms were suggested in the initial 50 mm of rainwater collected after the onset of a rain event [10]. Regrowth of indicator bacteria in rainwater tanks has been reported if organic matter is present, allowing for growth and cell division [5] and rotting vegetation that settles in rainwater tanks [1].

Anecdotal evidence on the link between the presence of fecal contamination in rainwater and the outbreak of diarrheal diseases exists [63]. Rainwater sampling in Australia has shown the presence of the enteropathogenic and extraintestinal virulence genes can be present in *E. coli* isolates from rainwater [3]. Examples include *eaeA, ST1, cdtB, cvaC, ibeA, kpsMT allele III, PAI, papAH,* and *traT* [3]. In other studies, opportunistic pathogens belonging to the species *Enterococcus* spp. have been detected in harvested rainwater [2,12,13]. Some studies have also contained references to the detection of zoonotic pathogens in rainwater [4]. These results further strengthen the requirements for regular microbial monitoring of reclaimed water quality [34].

Page et al. [48] assessed data from four case studies from Australia, South Africa, Belgium, and Mexico on the possibility of using aquifers as barriers to human infection by pathogenic microorganisms. The recycling was done for stormwater and reclaimed wastewater [48]. The final use of the water after aquifer storage was drinking. Data from South Africa and Australia indicate that the residence time of the water in the aquifer played an important role in the minimization of the human health risk from the consumption of the treated reclaimed water and stormwater [48]. Pre and posttreatment were required in

Belgium, while high risk to human health could be inferred from the microbial risk assessment for the Mexican data [48]. This was the result of the low residence times and virtual lack of any pretreatment of the reclaimed water and stormwater in the aquifer [48]. The pretreatment is especially critical for rotavirus and *Cryptosporidium* spp. as these pathogens have the ability to survive outside of the human host for extended periods of time [48]. The aquifer residence time should be maximized if the final effluent is to be used for potable purposes [48].

5.3 Removal of Microorganisms from Recycled Water

To achieve the microbial guidelines for individual reclaimed water classes in Table 5.1, a minimum of a secondary treatment needs to be performed [17]. An example of secondary treatment methods includes lagoon-type systems [37]. Reclaimed water monitoring must focus on the indicator microorganisms mentioned in the previous section, as well as the compliance of the 5-day biochemical oxygen demand (BOD5) below 20 mg/L and the total suspended concentrations (TSS) below 30 mg/L [17]. If the reclaimed water is going to be used for nonpotable uses with high probability of human contact, then tertiary treatment and chlorination should also be included [36]. Residence time of the reclaimed water in the lagoons should be at least 25 days to achieve helminth egg removal to comply with all the microbial limits in Table 5.1 (see above and in Reference 17). Additional secondary treatment processes for achievement of the microbial targets in Table 5.1 include sand filtration or membrane filtration and can be considered on a case-by-case basis [68]. Quality of the reclaimed water must be the same at the point of production and the point of use [17].

In order to manage and safely apply recycled water, sources of microorganisms in it must be known. O'Toole et al. [45] measured the transfer of microorganisms during a washing cycle in a top-loading washing machine. They spiked several materials with 340 cells/cm² of *E. coli*, 199,500 oocysts/cm² of *Cryptosporidium* spp., 563,000 PFUs/cm² for PDR-1 virus, and 871,000 PFUs/cm² for MS-2 virus [45]. The percentage of transfer onto the clothing materials ranged from below 0.01% for the two viral indicators to 0.09% for the *E. coli* cells [45]. Negligible transfer of other indicator microorganisms from the fabric to the user's hand was observed and limited contamination of the hand from the recycled water was also reported [45].

The total number of oocysts of *Cryptosporidium parvum* transferred to the clothing during washing ranged from 5130 to 12,020 [45]. The transferred amounts are significant only with the microorganism with the lowest infectious dose, that is, *Cryptosporidium parvum*, which ranges from 10 to 30 oocysts [15,44]. Therefore, the protozoans can be transferred into the laundry greywater and the specific disinfection precautions must be performed before any reuse. Examples include the application of UV radiation and disinfection with chlorine dioxide. Various authors have studied these methods. Keegan et al. [27] found that UV radiation is an effective disinfection tool for

the removal of *Cryptosporidium parvum* from wastewater with radiation doses up to 1000 mJ/cm².

The easiest pathogens to remove are helminths using coarse filtration systems such as the commercially available and patented Hydrotech Discfilter process, where rotating stainless steel filters with pore size of 10 μm strain helminth eggs from treated recycled water [49,53]. Its removal efficiency for *Trichuris suis* was tested by Sanz et al. [53] in a closed-circuit wastewater treatment system. After flocculation and primary settling, the influent was seeded at initial concentrations from 55 to 15,990 *Trichuris suis* eggs/100 L [53]. Complete removal of the helminth eggs was observed in all but one sample with the maximum reduction between influent and effluent concentrations of 4.2 log units [53]. The stainless steel filter material could be used for 5 years without any treatment problems, provided efficient backwashing and leakage protocols are in place and online turbidity monitoring is performed [53].

A major challenge in the current water scare countries is the treatment and recycling of wastewater under conditions of the decentralized potable water supply and lack of treatment facilities. Here, low-cost treatment systems are required. In recent years, a significant amount of research focused on the development of low-cost treatment systems has gone into their development. Tandlich et al. [64] used a controlled-release type chlorination device to eliminate 109 CFUs/100 mL of fecal and TC from the mixed-origin greywater within 65 h. The effluent concentrations of both indicator microorganism groups were below 800 CFU/100 mL. The system was unsustainable due to maintenance and operation challenges. To address these issues, Zuma [68] developed the fly-ash lime filter tower with the schematic representation of the system as shown in Figure 5.1.

Pulp and paper mill fly ash was obtained from SAPPI (Mandeni, South Africa) with an optional addition of mulch (Makana Meadery, Grahamstown, South Africa). The first reactor setup was referred to as FLFT A and it consisted of the biological and physicochemical treatment components as follows: at the top of the reactor, 10 cm free/empty space was allowed to contain 1.7 L of greywater influent. It provided a secondary treatment as required to meet Table 5.1 targets (see above). The mulch layer was then added to provide the substrata surface for promotion of the biofilm development to facilitate organic matter biodegradation, while permitting airflow in the column. In a reactor, the biofilm acts as a biological barrier because of the extracellular polymeric substance secreted by the microbial community [55].

Various biodegraders develop because of substrate availability in the biobarrier, leading to increased biodegradation of easily degradable components and, in some cases, even the recalcitrant compounds [25]. The fly ash/lime mixture layer was added to enable removal of the inorganic and organic components and for sterilizing bacteria [30]. The layer of fine sand mixed with silica sand was prepared to increase porosity while providing additional effluent treatment. The stone layer was added to increase porosity at the bottom of the reactor. The system in Figure 5.1 was tested for the treatment of greywater from

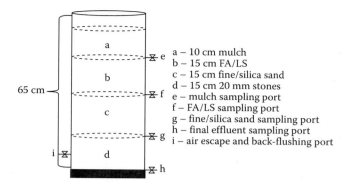

FIGURE 5.1 The schematic representation of the fly-ash/lime filter tower greywater treatment. (Adapted from Zuma, B.M., Tandlich, R., Burgess, J.E., and Whittington-Jones, K. 2009. *Desalination*, 242(1–3), 38–56.)

the kitchens and bathrooms from three donor households in Grahamstown, South Africa [68]. The average influent FC concentration ranged from 8616 to 28,356 CFU/100 mL, with the overall range spanning from 10 to 176,000 CFU/100 mL [68]. An average removal ranged from 58% to 71%. The mean influent TC averaged at 119,235 CFU/100 mL, with the overall range varying from 1000 to 484,000 CFU/100 mL. The TC removal percentage ranging from 35% to 64% and the effluent met the general requirements for recycled water quality for use in irrigation as outlined in Table 5.1 and listed in other regulatory documents [59,60].

Compared to greywater, rainwater treatment does not necessarily require installation of a designated onsite treatment system, as it has been shown that contact of rainwater with roofing materials during collection can provide a simple form of water treatment [56]. The main sources of microbial contamination of rainwater include leaves, branches, and bird feces near the rainwater-harvesting tank from the storage tank [56]. Thus, cleaning the roofs regularly and cleaning the gutters can significantly improve rainwater quality [63]. If the roof material is galvanized iron sheeting, then rainwater harvested from such roofs is reported to have a lower concentration of indicator bacteria [63]. This is the result of the roof's relatively smooth surface, which prevents dust build-up and thus eliminates the habitats for possible microbial regrowth [69]. At the same time, exposure of the galvanized iron to sunlight leads to high temperatures, which sterilize the harvested rainwater [56,63].

If the roof material is cement, then similar levels of sterilization can be observed as with galvanized iron sheeting, but the roof material also increases chances of the accumulation of leaves and animal droppings in the tile joints, which in turn compromises microbial water quality [56]. Tile roofs could be a potential source of contamination as leaves and bird feces can be caught and remain in the tile joints [56,63]. In such cases, regular (monthly) inspections of the roof and gutter cleanliness are required, along with the need for additional (secondary) treatment before rainwater meets the requirements in Table 5.1. Such treatment methods include any secondary treatment or

installation of the first-flush devices on the rainwater-harvesting tank [56].

5.4 Summary and Conclusions

To achieve the microbial quality of reclaimed water for reuse, a minimum of secondary treatment must be performed such as the lagoons with hydraulic residence time of 25 days or more. Specific treatment such as disinfection with UV radiation or chlorine dioxide must be performed for the removal of protozoa from laundry greywater. In decentralized settings, greywater treatment can be performed using low-cost systems such as fly-ash lime filter tower with the FC and TC removal shown to range from 35% to 71% at the hydraulic residence time of approximately 18–24 min. For rainwater recycling, passive treatment by rainwater contact with roofing material can be used, or the first-flush device can be installed onsite. Removal of helminth pathogens can be achieved using coarse filtration systems such the Hydrotech Discfilter process system.

References

1. Abbott, S., Caughley, B., and Douwes, J. 2007. *The microbiological quality of roof-collected rainwater of private dwellings in New Zealand*. Available at http://www.ctahr.hawaii.edu/hawaiirain/Library/papers/Abbots,%20Leder,%20Heyworth%20papers/Abbott1.pdf. Accessed on February 13, 2014.

2. Ahmed, W., Brandesa, H., Gyawalia, P., Sidhua, J.P.S., and Toze, S. 2014. Opportunistic pathogens in roof-captured rainwater samples, determined using quantitative PCR. *Water Research*, 53(1), 361–369.

3. Ahmed, W., Hodgers, L., Masters, N., Sidhu, J.P.S., Katouli, M., and Toze, S. 2011. Occurrence of intestinal and extraintestinal virulence genes in *Escherichia coli* isolates from rainwater tanks in Southeast Queensland, Australia. *Applied and Environmental Microbiology*, 77(20), 7394–7400.

4. Ahmed, W., Hodgers, L., Sidhu, J.P.S., and Toze, S. 2012. Fecal indicators and zoonotic pathogens in household drinking water taps fed from rainwater tanks in Southeast Queensland, Australia. *Applied and Environmental Microbiology*, 78(1), 219–226.

5. Ahmed, S.A., Hoque, B.A., and Mahmud, A. 1998. Water management practices in rural and urban homes: A case study from Bangladesh on ingestion of polluted water. *Public Health*, 112(5), 317–312.

6. American Public Health Association, American Water Works Association and Water Environment Federation (APHA). 2012. *Standard Methods for the Examination of Water and Wastewater*, 22nd edition. American Public Health Association, Washington, DC.

7. Amin, T.M. and Han, M. 2011. Microbial quality variation within a rainwater storage tank and the effects of first flush in rainwater harvesting (RWH) system. *Australian Journal of Basic and Applied Sciences*, 5(9), 1804–1813.

8. Besner, M.C., Prevost, M., and Regli, S. 2011. Assessing the public health risk of microbial intrusion events in distribution systems: Conceptual model, available data, and challenges. *Water Research*, 45(3), 961–979.

9. Carden, K., Armitage, N., Winter, K., Sichone, O., and Rivett, U. 2007. Understanding the use and disposal of greywater in the South Africa. Water Research Commission of South Africa, Report No. 1524/1/07, Pretoria, South Africa.

10. Coombes, P., Spinks, A., Evans, C., and H. Dunstan. 2004. Performance of rainwater tanks at an inner city house in Carrington NSW during drought. *WSUD*, 504–515. http://waterplex.com.au/download.php?file=media/Research/Rainwater_Tank_Research__Rainwater_Tank_in_Carrington_House.pdf (accessed on July 7, 2015).

11. Cunliffe, D.A. 2004. *Guidance on the Use of Rainwater Tanks*. EN Health Council, Department of Health and Ageing, Australian Government, Canberra, Australia.

12. de Kwaadsteniet, M., Dobrowsky, P.H., van Deventer, A., Khan, W., and Cloete, T.E. 2013. Domestic rainwater harvesting: Microbial and chemical water quality and point-of-use treatment systems. *Water, Air and Soil Pollution*, 224(7), article number 1829.

13. Dobrowsky, P.H., de Kwaadsteniet, M., Cloete, T.E., and Khan, W. 2014. Distribution of indigenous bacterial and potential pathogens associated with roof-harvested rainwater. *Applied and Environmental Microbiology*, 80(7), 2307–2316, doi: 10.1128/AEM.04130-13.

14. Dobrowsky, P.H., van Deventer, A., de Kwaadsteniet, M., Ndlovu, T., Khan, S., Cloete, T.E., and Khan, W., 2013. Prevalence of virulence genes associated with pathogenic *Escherichia coli* strains isolated from domestically harvested rainwater during low- and high-rainfall periods. *Applied and Environmental Microbiology*, 80(5), 1633–1638.

15. DuPont, H.L., Chappell, C.L., Sterling, C.R., Okhuysen, P.C., Rose, J.B., and Jakubowski, W. 1995. The infectivity of *Cryptosporidium parvum* in healthy volunteers. *New England Journal of Medicine*, 332(13), 855–859.

16. Duse, A., Da Silva, M., and Zietsman, I. 2003. Coping with hygiene in South Africa, a water scarce country. *International Journal of Environmental Health Research*, 13(Suppl 1), S95–S105.

17. Environmental Protection Agency. 2003. *Guidelines for Environmental Management: Use of Reclaimed Water*. Environmental Protection Agency Publication 464.2. Southbank, Victoria, Australia.

18. Eriksson, E., Auffarth, K., Henze, M., and Ledin, A. 2002. Characteristics of grey wastewater. *Urban Water*, 4(1), 85–104.

19. Fleming, J. 2000. Risk perception and domestic roof-collected rainwater supplies. Master of Public Health thesis, Auckland University, Auckland, New Zealand.

20. Gerba, C.P. and Rose, J.B. 2003. International guidelines for water recycling: Microbiological considerations. *Water Science and Technology: Water Supply*, 3(4), 311–316.

21. Gerba, C.P., Rose, J.B., and Haas, C.N. 1996. Sensitive populations: Who is at the greatest risk? *International Journal of Food Microbiology*, 30(1), 113–123.

22. Gerba, C.P., Rose, J.B., Haas, C.N., and Crabtree, K.D. 1997. Waterborne rotavirus: A risk assessment. *Water Research*, 30(12), 2929–2940

23. Havelaar, A., Blumenthal, J., Strauss, M., Kay, D., and Bartram, J. 2001. Guidelines: The current position. In: Fewtell, L. and Bartram, J. (eds), *Water Quality Standards and Health*, IWA Publishing, London, UK, pp. 16–42.

24. Hung, T., Chen, G.M., Wang, C.G., Yao, H.L., Fang, Z.Y., Chao, T.X., Chou, Z.Y., Ye, W., Chang, X.J., and Den, S.S. 1984. Waterborne outbreak of rotavirus diarrhoea in adults in China caused by a novel rotavirus. *Lancet*, 1(8387), 1139–1142.

25. Kästner, M. and Mahro, B. 1996. Microbial degradation of polycyclic aromatic hydrocarbons in soils affected by the organic matrix of compost. *Applied Microbiology and Biotechnology*, 44(5), 668–675.

26. Katz, D., Kumar, S., Malecki, J., Lowdermilk, M., Koumans, E.H., and Hopkins, R. 1999. Cyclosporiasis associated with imported raspberries, Florida, 1996. *Public Health Reports*, 114(5), 427–428.

27. Keegan, A.R., Fanok, S., Monis, P.T., and Saint, C.P. 2003. Cell culture-Taqman PCR assay for evaluation of *Cryptosporidium parvum* disinfection. *Applied and Environmental Microbiology*, 69(5), 2505–2511.

28. Lee, E.J. and Schwab, K.J. 2005. Deficiencies in drinking water distribution systems in developing countries. *Journal of Water and Health*, 3(2), 109–127.

29. Leggett, H.C., Cornwallis, C.K., and West, S.A. 2012. Mechanisms of pathogenensis, infective dose and virulence in human parasites. *PLOS Pathogens*, 8(2), 1–5.

30. Linstedt, K.D., Houck, C.P., and O'Connor, J.T. 1971. Trace element removals in advanced wastewater treatment processes. *Journal of Water Pollution Control Federation*, 43(7), 1507–1513.

31. Luyt, C.D. 2013. Faecal source tracking and water quality in the Eastern Cape, South Africa. PhD thesis, Grahamstown, Rhodes University, South Africa.

32. Luyt, C.D., Muller, W.J., and Tandlich, R. 2011. Low-cost tools for microbial quality assessment of drinking water in South Africa. *HealthMed*, 5(6 Supplement 1), 1868–1877.

33. Luyt C.D., Muller, W.J., Wilhelmi, B.S., and Tandlich, R. 2011. Health implications of flood disaster management in South Africa. *Proceedings of the 18th Annual Conference of the International Emergency Management of Society* held in Bucharest, Romania from June 7–10, 2011, The International Emergency Management Society, Brussels, Belgium, pp. 376–385.

34. Luyt, C.D., Tandlich, R., Muller, W.J., and Wilhelmi, B.S. 2012. Microbial monitoring of surface water in South Africa. *International Journal of Environmental Research and Public Health*, 9(8), 2669–2693.

35. Mahmoud, A.A. 2000. Diseases due to helminths. In: Mandell, G.L., Bennett, J.E., and Dolin, R. (eds), *Principles and Practice of Infectious Diseases*, 5th edition. Churchill Livingstone, Philadelphia, PA, pp. 2937–2986.

36. Martinez, C.J. and Clark, M.W. 2013. *Reclaimed Water and Florida's Water Reuse Programme*. University of Florida IFAS Extension, Gainesville, FL.

37. Maynard, H.E., Ouki, S.K., and Williams, S.C. 1999. Tertiary lagoons: A review of removal mechanisms and performance. *Water Research*, 33(1), 1–13.

38. Meays, C.L., Broersma, K., Nordin, R., and Mazumder, A. 2004. Source tracking fecal bacteria in water: A critical review of current methods. *Journal of Environmental Management*, 73(1), 71–79.

39. Monde, N., Botha, J.J., Joseph, L.F., Anderson, J.J., Dube, S., and Lesoli, M.S. 2012. Sustainable techniques and practices for water harvesting and conservation and their effective application in resource-poor agricultural production. Water Research Commission of South Africa, Report No. 1478/1/12, Pretoria, South Africa.

40. Monyai, P. 2004. Health-related water quality and surveillance model for the Peddie district in the Eastern Cape, Pretoria, South Africa. Water Research Commission of South Africa, Report No. 727/1/04.

41. Mosley, L.M. and Sharp, D.S. 2005. The hydrogen sulphide paper strip test. South Pacific Applied Geoscience Commission, *South Pacific Applied Geoscience Commission and WHO Technical Report* 373, Suva, Fiji.

42. Murray, K., du Preez, M., Kuhn, A.L., and van Niekerk, H. 2004. A pilot study to demonstrate implementation of the national microbial monitoring programme. Water Research Commission of South Africa, Report No. 1118/1/04, Pretoria, South Africa.

43. Nygård, K., Wahl, E., Krogh, T., Tveit, O.A., Bøhleng, E., Tverdal, A., and Aavitsland, P. 2007. Breaks and maintenance work in the water distribution systems and gastrointestinal illness: A cohort study. *International Journal of Epidemiology*, 36(4), 873–880.

44. Okhuysen, P.C., Chappell, C.L., Crabb, J.H., Sterling, C.R., and DuPont, H.L. 1999. Virulence of three distinct *Cryptosporidium parvum* isolates for healthy adults. *Journal of Infectious Diseases*, 180(4), 1275–1281.

45. O'Toole, J., Sinclair, M., and Leder, K. 2009. Transfer rates of enteric microorganisms in recycled water during machine clothes washing. *Applied and Environmental Microbiology*, 75(5), 1256–1263.

46. Ottosson, J. 2003. Hygiene aspects of greywater and greywater reuse. Licentiate thesis, Department of Water and Environmental Microbiology of the Swedish Royal Institute of Technology, Stockholm, Sweden.

47. Ottosson, J. and Stenström, T.A. 2003. Faecal contamination of greywater and associated microbial risks. *Water Research*, 37(3), 645–655.

48. Page, D., Dillon, P., Toze, S., Bixio, D., Genthe, B., Cisneros, B.E.J., and Wintgens, T. 2010. Valuing the subsurface pathogen treatment barrier in water recycling via aquifers for drinking supplies. *Water Research*, 44(6), 1841–1852.

49. Persson, E., Ljunggren, M., la Cour Jansen, J., Strube, R., and Jönsson, L. 2006. Disc filtration for separation of flocs from a moving bed bio-film reactor. *Water Science and Technology*, 53(12), 139–147.

50. Rose, J.B., Gwo-Shing, S., Gerba, C.P., and Sinclair, N.A. 1991. Microbial quality and persistence of enteric pathogens in graywater from various household sources. *Water Resources*, 25(1), 37–42.

51. Salgot, M., Vergés, C., and Angelakis, A.N. 2003. Risk assessment in wastewater recycling and reuse. *Water Science and Technology: Water Supply*, 3(4), 301–309.

52. Salukazana, L., Jackson, S., Rodda, N., Smith, M., Gounden, T., McLeod, N., and Buckley, C. 2005. Plant growth and microbiological safety of plants irrigated with greywater. *Proceedings from the 3rd International Conference on Ecological Sanitation*, May 23–26, 2005. University of KwaZulu-Natal, Durban, South Africa.

53. Sanz, J., Strube, R., Quinzaños, S., Dahl, C.P., Montoliu, I., Gracenea, M., and Mujeriego, R. 2009. Helminth eggs removal in water reclamation: Disc filtration as an effective barrier. *Proceedings from IDA World Congress, Atlantis, The Palm, Dubai, UAE*, November 7–12, 2009, pp. 1–14.

54. Sattar, S.A., Springthorphe, V.S., and Tetro, J.A. 2001. In: Hui, Y.H., Sattar, S.A., Murrell, K.D., Nip, W.K., and Stanfield, W.K. (eds), *Rotavirus. Foodborne Disease Handbook. Volume 2: Viruses, Parasites, Pathogens, and HACCP*, 2nd edition. Marcel Dekker, New York, pp. 99–125.

55. Seo, Y. and Bishop P.L. 2008. The monitoring of biofilm formation in a mulch biowall barrier and its effect on performance. *Chemosphere*, 70(3), 480–488.

56. Sharma, S.K. 2007. *Rainwater Harvesting*. UNESCO-IHE Institute for Water Education, Delft, The Netherlands.

57. Simmons, G., Gould, J., Gao, W., Whitmore, J., Hope, V., and G. Lewis. 2000. The design, operation and security of domestic roof-collected rainwater in rural Auckland. *Proceedings of the Water 2000 Conference Guarding the Global Resource, New Zealand Water and Wastes Association*, Auckland, New Zealand, pp. 1–16.

58. South African Bureau of Standards (SABS). 2006. *South African Standard Specifications for Water for Domestic Supplies* (*SANS 2009*). South African Bureau of Standards, Pretoria, South Africa.

59. South African National Department of Water Affairs and Forestry (DWAF). 1996a. *South African Water Quality Guidelines: Domestic Use* (Vol. 1). Department of Water Affairs and Forestry, Pretoria, South Africa.

60. South African National Department of Water Affairs and Forestry (DWAF). 1996b. *South African Water Quality Guidelines: Irrigation*, Vol. 4. Department of Water Affairs and Forestry, Pretoria, South Africa.

61. Statistics South Africa. 2010. *General Household Survey 2010 (Revised Version)*. Statistics South Africa, Pretoria, South Africa.

62. Statistics South Africa. 2011. *GHS Series Volume III Water and Sanitation 2002–2010: In-Depth Analysis of the General Household Survey Data.* Statistics South Africa, Pretoria, South Africa.

63. Tandlich, R., Luyt, C.D., Gordon, A.K., and Srinivas, C.S. 2012. Concentrations of indicator organisms in the stored rainwater in the Makana Municipality, South Africa. *Proceedings from the Air and Water Components of the Environment Conference* held in Cluj, Romania, March 23–24, 2012, pp. 89–96, University of Cluj Press, Cluj, Romania.

64. Tandlich, R., Zuma, B.M., Burgess, J.E., and Whittington-Jones, K. 2009. Mulch tower treatment system for greywater re-use. Part II: destructive testing and effluent treatment. *Desalination*, 242(1–3), 57–69.

65. Whittington-Jones, K., Tandlich, R., Zuma B.M., Hoossein, S., and Villet, M.H. 2011. Performance of the pilot-scale mulch tower system in treatment of greywater from a low-cost housing development in the Buffalo City, South Africa. *International Water Technology Journal*, 1(2), Paper 7, http://iwtj.info/wp-content/uploads/2011/12/v1-n2-p7.pdf.

66. Winward, G.P., Avery, L.M., Frazer-William, R., Pidou, M., Jeffrey, P., Stephenson, T., and Jefferson, B. 2008. A study of the microbial quality of greywater and an evaluation of treatment technologies for reuse. *Ecological Engineering*, 32(2), 187–197.

67. World Health Organization. 2006. *Guidelines for Safe Use of Wastewater, Excreta and Greywater.* World Health Organization Press, Geneva, Switzerland.

68. Zuma, B.M. 2012. The efficacy and applicability of low-cost adsorbent materials in wastewater treatment. PhD thesis, Rhodes University, Grahamstown, South Africa.

69. Zuma, B.M., Tandlich, R., Burgess, J.E., and Whittington-Jones, K. 2009. Mulch tower treatment system for greywater re-use. Part I: Overall performance in greywater treatment. *Desalination*, 242(1–3), 38–56.

6

Pharmaceuticals and Personal Care Products in Wastewater: Implications for Urban Water Reuse

Amanda Van Epps
Berlin, Germany

Lee Blaney
University of Maryland Baltimore County

PREFACE

The presence of pharmaceuticals and personal care products (PPCPs) in water and wastewater sources has been well documented over the last 15 years. These compounds represent an ecological and public health concern in the context of water reuse, due to the potential accumulation of PPCPs in municipal water sources. Furthermore, planned indirect water reuse may introduce PPCPs to water supplies that were previously uncontaminated by xenobiotics. In this chapter, a short introduction to the importance of PPCPs in the context of water reuse is provided, and the classes of PPCP compounds that comprise the most significant concerns are discussed. The bulk of this chapter focuses on the state-of-the-knowledge regarding PPCP removal in conventional and advanced wastewater treatment and water reclamation processes. Finally, the ecological and public health concerns associated with these compounds are highlighted to demonstrate the importance of achieving higher removal efficiencies of PPCPs within treatment plants.

6.1 Introduction

Over the past 30 years, the presence of pharmaceuticals and personal care products (PPCPs) has been documented in wastewater supplies and the environment [74,85,135,171,190]. Unlike many traditional contaminants for which the toxicological activity is separate from the intended use, pharmaceuticals have been specially designed to elicit a biochemical response in organisms. For this reason, pharmaceuticals represent a unique threat to ecological and public health, even at trace concentrations. Given this threat, increased understanding of the removal of PPCPs in wastewater treatment processes is required to ensure ecological well-being and sustained protection of public water supplies.

Unplanned potable water reuse has long been a reality in many locations, as wastewater effluent from upstream cities comprises some fraction of the drinking water supply for downstream cities. However, as per capita freshwater supplies continue to decrease due to increasing global population and climate change, planned water reuse is becoming a deliberate strategy to cope with pressures on water supplies in many parts of the world. Windhoek, Namibia has been practicing direct potable water reuse since the 1980s [194]. Many other municipalities already use wastewater effluent as a source of irrigation water for highway medians and golf courses or makeup water in cooling towers, but more are now actively pursuing indirect potable water reuse by injecting wastewater effluent into groundwater aquifers or pumping wastewater effluent into upstream reservoirs [106]. These utilities include Los Angeles (CA), San Diego (CA), Franklin (TN), Mexico City (Mexico), Gwinnett County (GA), Fairfax County (VA), Scottsdale (AZ), Miami-Dade (FL), Sulaibiya (Kuwait), Willunga (Australia), El Paso (TX), and Adelaide (Australia), among others [46]. The U.S. Environmental Protection Agency (EPA) has developed a comprehensive document entitled *Guidelines for Water Reuse*, which provides an overview of issues, examples, and treatment guidelines for water reuse operations [46]. The presence of trace concentrations of PPCPs in reclaimed water prompts not only ecological concerns for receiving waters, but also potential public health consequences in direct and indirect potable water reuse scenarios. Some of these concerns include reproductive abnormalities [77,114], antibiotic resistance [117,136], and developmental impacts [49,51].

Water reuse scenarios involve engineering and science facets, many of which have been described in detail in earlier chapters of this book. In this chapter, the fate of PPCPs is explored in wastewater treatment, and a context for the ecological and public health concerns is provided. The objective of this chapter is to provide a snapshot of the removal mechanisms for PPCPs in wastewater treatment processes to build greater understanding of this emerging area of interest. As there are over 7700 pharmaceuticals approved for human use in the world [63], PPCPs demonstrate a wide range of physicochemical properties and biodegradability. For that reason, this chapter provides an insight into the behavior of select PPCPs through conventional and advanced wastewater treatment processes. With mounting public pressure and ecological evidence, regulations for select PPCPs in wastewater effluent are expected in the future. Ultimately, this chapter aims to provide a context for considering treatment of such waters.

6.2 Snapshot of PPCPs in Wastewater

Hundreds of PPCPs have been detected in raw wastewater at concentrations ranging from nanograms per liter (ng/L) to hundreds of micrograms per liter (µg/L) [109,110,184]. Wastewater effluent may contain PPCP concentrations as high as tens of µg/L [110,170]; furthermore, these measurements likely underestimate the presence of PPCPs in wastewater due to limited access to advanced analytical instruments/methods and high limits of quantitation. PPCPs are present in wastewater largely because of human consumption. Some fraction of the pharmaceutical mass consumed by humans is excreted unchanged; furthermore, improper disposal of pharmaceuticals down the drain

also contributes to detectable concentrations in raw wastewater. Personal care product ingredients typically enter wastewater supplies through rinse water from laundry, hand washing, or showering [173]. Chemicals in both categories can also originate from manufacturing effluent, stormwater runoff, or landfill leachate [88,116,173].

The presence of PPCPs in reclaimed water is influenced by trends in local water use and PPCP consumption. For example, Clara et al. [27] analyzed raw wastewater and wastewater effluent in Austria for diazepam, but the compound was not detected. The authors attributed the absence of diazepam in wastewater to the relatively low consumption of that pharmaceutical in Austria. Similarly, iopromide, an x-ray contrast agent, was not detected in raw wastewater, presumably due to the lack of hospitals connected to the corresponding wastewater collection system. Note that these compounds have both been detected in other studies [79,169,203]; therefore, select PPCPs may be relevant to some systems, but not others.

A fundamental challenge associated with measuring PPCPs in wastewater, and designing treatment schemes to remove or degrade these emerging contaminants, is the breadth of compounds that fall into these categories. No central strategy for prioritization of PPCPs has been adopted; however, the corresponding approach should include a number of factors such as mass consumption rates, ecological or human health risk factors, physicochemical properties, biodegradability, pharmacological class, and a sustainability index [21,170]. The following classes of compounds have been found to be among the most frequently detected in environmental matrices, including wastewater and surface waters [110,116,160]: antibiotics, anti-inflammatory/analgesic drugs, lipid regulators and beta blockers, hormones, antiepileptics, various personal care product ingredients, and endocrine-disrupting chemicals (EDCs). Given the frequency of detection in environmental media, these compounds are expected to play a significant role in the urban water cycle. Properties of representative compounds for each category are provided in Table 6.1.

Antibiotics are prescribed to treat bacterial infections in humans and animals. In addition, antibiotics are extensively used as growth promoters in concentrated animal feeding operations, indicating that these compounds are important components of both urban and rural water cycles. More than 400 such compounds are registered for medical use, and 100,000–200,000 metric tons of antibiotics are manufactured every year [88,195]. Urban wastewater contains high antibiotic loads because a large fraction of the consumed mass (e.g., up to 98% in the case of amikacin [18]) is excreted unchanged [88]. More than 50 different antibiotics have been detected in raw wastewater, and typical concentrations are tens of ng/L to tens of µg/L [88,100,109,110,184]; the corresponding concentrations in wastewater effluent are as high as µg/L [67,88,100,109,110]. The antibiotic potency of metabolites has not been widely considered from an environmental context; however, these species may also be environmentally relevant because many antibiotics are metabolized into their active form.

TABLE 6.1 Summary of Chemical Properties of a Representative Compound from Each Prominent PPCP Category

Category	Compound	Chemical Formula	Chemical Structure	Molecular Weight (Da)	K_H (atm m³/mol)[a]	Log k_{ow}^a	pk_a^a
Antibiotics	Ciprofloxacin	$C_{17}H_{18}FN_3O_3$	*(chemical structure)*	331.35	5.09×10^{-19}	0.28	6.09 8.82
Antiepileptic agents	Carbamazepine	$C_{15}H_{12}N_2O$	*(chemical structure)*	236.28	1.08×10^{-10}	2.45	–
Anti-inflammatory drugs/analgesics	Ibuprofen	$C_{13}H_{18}O_2$	*(chemical structure)*	206.29	1.50×10^{-7}	3.97	4.91
Beta blockers	Propranolol	$C_{16}H_{19}NO_2$	*(chemical structure)*	257.34	7.98×10^{-13}	3.48	9.42
Hormones	17β-Estradiol	$C_{18}H_{24}O_2$	*(chemical structure)*	272.39	3.64×10^{-11}	4.01	–
Lipid regulators	Gemfibrozil	$C_{15}H_{22}O_3$	*(chemical structure)*	250.34	1.19×10^{-8}	4.77	4.50
Musk fragrances	Galaxolide	$C_{18}H_{26}O$	*(chemical structure)*	258.41	1.32×10^{-4}	5.90	–

[a] Reference 155.

Anti-inflammatory drugs and analgesics are widely used to treat inflammation and pain. These pharmaceuticals are available as prescription drugs and over-the-counter remedies, such as ibuprofen. At least 20 analgesic drugs have been detected in raw wastewater with typical concentrations ranging from tens of ng/L to hundreds of µg/L; the concentrations in wastewater effluent are approximately one order of magnitude lower [110,184].

Lipid regulators and beta-blockers are prescribed to treat cardiovascular disease. At least six lipid regulators and seven beta-blockers have been detected in raw wastewater at concentrations of hundreds of ng/L to µg/L [110,184]. Interestingly, more compounds within these classes have been detected in wastewater effluent. In particular, at least 9 lipid regulators and 12 beta-blockers have been detected in wastewater treatment plant (WWTP) effluent at concentrations similar to those detected in raw wastewater [110,184].

The most commonly studied antiepileptic drug is carbamazepine. In one review, carbamazepine was detected in 100% of the investigated raw wastewater samples and 98% of wastewater effluent samples [110]. Concentrations of carbamazepine in raw wastewater typically range from tens of ng/L to µg/L [110,184]; effluent concentrations remain in the range of tens of ng/L to µg/L [110,116,184]. As indicated in the following, this compound may be a good indicator of PPCP presence in wastewater due to the high frequency of detection and low removal during biological treatment processes.

Personal care product ingredients include a wide range of compounds including musk fragrances, ultraviolet (UV) filters, surfactants, preservatives, and plasticizers. In the Miège et al. [110] review, galaxolide, a musk fragrance, was detected at µg/L concentrations in every raw wastewater sample for which it was analyzed. Galaxolide was also detected in 100% of wastewater effluent samples at concentrations of hundreds of ng/L to µg/L

[110]; therefore, this compound may also serve as a robust indicator compound for personal care products in wastewater.

Hormones detected in wastewater can include human hormones that are naturally excreted and natural or synthetic hormones prescribed for medical purposes. To date, at least five different hormones have been detected in raw wastewater at concentrations of ng/L to hundreds of ng/L [100,110,184]. These compounds are potent at concentrations as low as 1–10 ng/L, as demonstrated by Kidd et al. [77]; therefore, measured wastewater effluent concentrations of hundreds of picograms per liter (pg/L) to tens of ng/L are environmentally relevant.

Endocrine-disrupting chemicals are often included in studies of PPCPs but are not limited to a particular treatment class or category. Chemicals as broadly distributed as birth control compounds (e.g., ethinyl-estradiol), reproductive hormones (e.g., estrone), replacement hormones (e.g., equilenin), plant steroids (e.g., stigmastanol), alkylphenols (e.g., nonylphenol), and plasticizers (e.g., bisphenol A) exhibit estrogenic activity. Thus, potential endocrine disruption effects in urban water reuse must be considered for a broad range of PPCPs that may demonstrate temporal and geographical consumption patterns.

Several review articles have provided exhaustive data on detections of PPCPs in raw and treated wastewater [88,100,109,110,184]. The breadth of compounds detected in urban wastewater, and the associated variability in physicochemical properties, is captured in Table 6.1. For the PPCP classes of interest, a significant fraction of the influent mass load is discharged in wastewater effluent, a phenomenon that is not particularly surprising given that awareness of these compounds and their potentially deleterious effects has only recently developed. Furthermore, existing WWTPs were not designed to consider removal of these emerging contaminants. In Section 6.3, the knowledge base that has developed over the last decade is considered to better understand removal of PPCPs in conventional and advanced wastewater treatment processes.

6.3 Removal of PPCPs in WWTPs and Water Reclamation Plants

As discussed in Section 6.2, a large fraction of consumed PPCPs enter the wastewater system through excretion, rinsing, or improper disposal; furthermore, conventional WWTPs are not designed, or optimized, for removal of these emerging contaminants. For this reason, WWTPs comprise the largest point sources for PPCP introduction to the environment and represent an important consideration in the urban water cycle. To address the occurrence of PPCPs in the urban water cycle, baseline removal of PPCPs in conventional wastewater treatment processes must be understood to optimize operation and improve removal efficiencies for emerging contaminants. This information will also prove useful in the design of WWTPs and water reclamation plants, especially as future regulations associated with particular PPCPs are expected.

6.3.1 Adsorption of PPCPs to Solids in Primary Treatment

For most municipal WWTPs, the initial treatment process involves settling of large particles present in raw wastewater in primary clarifiers. In this stage, PPCP removal is not expected to be high, especially as PPCPs discharged into the wastewater system will have already been in contact with these solids in the wastewater collection system. Regardless, low fractional removals can be expected as PPCPs may sorb to primary solids. Prediction of PPCP sorption to primary solids based on physicochemical properties is challenging because two underlying mechanisms are involved with this process [158,164,170,172]:

- Partitioning of PPCPs into primary solids through interactions between hydrophobic moieties of PPCP molecules with lipid and lipophilic fractions of sludge particles
- Sorption of PPCP molecules due to electrostatic interactions between cationic PPCP species (usually due to protonated amine functionalities) and negatively charged particles

Therefore, while the octanol–water partition coefficient (K_{ow}) might be expected to describe hydrophobic interactions, the corresponding acid dissociation constants (K_a) are needed to predict electrostatic interactions. One approach is to consider both sorption processes using the pH-dependent octanol–water partition coefficient D_{ow}. Note that D_{ow} is equal to K_{ow} for PPCPs without an acidic or basic moiety. For compounds that demonstrate acid/base speciation, D_{ow} is defined as shown in Reference 145:

$$D_{ow} = \frac{K_{ow}}{1 + 10^{pH - pK_a}} \quad (6.1)$$

In spite of its relevance to these scenarios, D_{ow} has not been reported for a wide range of PPCPs to date. More commonly, experimentally derived sorption coefficients (K_D) are used to describe sorption of each PPCP to primary sludge. This approach remains challenging because the empirical value of K_D is dependent on the conditions for which it was determined, including water quality, the characteristics of sludge, pH, and the initial PPCP concentration; therefore, this parameter may not be universally applicable. Nevertheless, considerable efforts have focused on experimental determination of sorption coefficients for PPCPs, as summarized in Table 6.2.

Dargnat et al. [34] demonstrated the relevance of hydrophobicity by monitoring the fate of five phthalates through a wastewater treatment train in France. The overall removal of phthalates ranged from 78% to 99% for the entire WWTP, but the majority of that removal occurred in the primary clarifier for four of the five compounds. Only 5% of the influent dimethyl phthalate was removed in the primary clarifier; note that the K_{ow} value of dimethyl phthalate is at least an order of magnitude lower than that of the other studied compounds. On the other hand, Golet

TABLE 6.2 Summary of Experimentally Determined Sorption Coefficients for PPCPs onto Primary and Secondary Sludge

Compound	Category	Log k_{ow}^a	K_D (L/kg)			
			Primary Sludge	Reference	Secondary Sludge	Reference
17α-Ethinylestradiol	Hormone	3.67	–		316	[28]
					584	[7]
17β-Estradiol	Hormone	4.01	10,400	[22]	631	[28]
					35,000	[22]
4-Methylbenzylidene camphor	UV filter	–	1220	[90]	5700	[90]
Acetaminophen	Analgesic	0.46	6.7	[132]	1160	[132]
Atenolol	β-blockers	0.16	95	[132]	64	[132]
Azithromycin	Antibiotic	4.02	–		376	[55]
Benzophenone-1	UV filter	2.96	–		260	[193]
Benzophenone-2	UV filter	2.78	–		1300	[193]
Benzophenone-3	UV filter	3.79	–		1300	[193]
Benzophenone-4	UV filter	0.37	–		8	[193]
Bisphenol-A	Plasticizer	3.32	–		1000	[28]
Carbamazepine	Antiepileptic	2.45	314	[132]	1.2	[172]
					135	[132]
Celestolide	Fragrance	–	5300	[90]	8800	[90]
Chlorophene	Preservative	3.60	–		2000	[193]
Ciprofloxacin	Antibiotic	0.28	2512	[56]	417	
					19,953	[158]
Clarithromycin	Antibiotic	3.16	–		262	[55]
Clofibric acid	Metabolite of lipid regulator	2.57	–		4.8	[172]
Cyclophosphamide	Chemotherapeutic agent	0.63	55	[172]	2.4	[172]
Diazepam	Anxiolytic	2.82	44	[172]	21	[172]
Diclofenac	Anti-inflammatory agent	4.51	194	[132]	16	[172]
			459	[172]	118	[132]
Erythromycin	Antibiotic	3.06	309	[132]	74	[132]
Estrone	Hormone	3.13	–		402	[7]
Galaxolide	Fragrance	5.90	4920	[172]	1810	[172]
			5050	[90]	9600	[90]
			20,600	[22]	14,300	[22]
Gemfibrozil	Lipid regulator	4.77	23	[132]	19.3	[132]
Glibenclamide	Antidiabetic	4.79	282	[132]	239	[132]
Hydrochlorothiazide	Diuretic	−0.07	25.8	[132]	20.2	[132]
Ibuprofen	Anti-inflammatory agent	3.97	9.5	[132]	0.0	[132]
					7.1	[172]
					251	[158]
					356	[22]
Ifosfamide	Chemotherapeutic agent	0.86	22	[172]	1.4	[172]
Iopromide	Contrast medium	−2.05	5	[172]	11	[172]
Ketoprofen	Anti-inflammatory agent	–	226	[132]	16	[132]
Loratadine	Antihistamine	5.20	2336	[132]	3321	[132]
Mefenamic acid	Anti-inflammatory agent	5.12	294	[132]	434	[132]
Norfloxacin	Antibiotic	−1.03	2512	[56]	15,848	[56]
Octocrylene	UV filter	6.88	14,000	[90]	–	
Octyl-methoxy-cinnamate	UV filter	5.80	1000	[90]	11,000	[90]
Oxytetracycline	Antibiotic	−0.90	–		3020	[158]
Phantolide	Fragrance	–	6500	[90]	7500	[90]
Propranolol	β-blockers	3.48	641	[132]	366	[132]

(Continued)

TABLE 6.2 (*Continued*)   Summary of Experimentally Determined Sorption Coefficients for PPCPs onto Primary and Secondary Sludge

Compound	Category	Log k_{ow}^{a}	K_D (L/kg)			
			Primary Sludge	Reference	Secondary Sludge	Reference
Sulfamethoxazole	Antibiotic	0.54	3.2	[132]	77	[132]
					256	[55]
Tetracycline	Antibiotic	−1.30	8400	[78]	–	
Tonalide	Fragrance	5.70	5000	[90]	2400	[172]
			5240	[22]	11,500	[90]
			5300	[172]	18,000	[22]
Traseolide	Fragrance	–	5600	[90]	17,000	[90]
Triclocarban	Antimicrobial	4.90	–		40,000	[193]
Triclosan	Antimicrobial	4.76	–		17,000	[193]
Trimethoprim	Antibiotic	0.91	427	[132]	208	[55]
					253	[132]

   [a] Reference 155.

et al. [56] demonstrated the importance of both hydrophobic and electrostatic interactions in an evaluation of the removal of two fluoroquinolone antibiotics in a WWTP with tertiary treatment in Switzerland. Both compounds (i.e., ciprofloxacin and norfloxacin) have negative log K_{ow} values, indicating the relatively hydrophilic nature of these antibiotics; additionally, these compounds are zwitterionic at near-neutral pH due to the presence of a protonated amine and a deprotonated carboxylate. For these reasons, the sorption of fluoroquinolone antibiotics onto primary sludge is expected to occur via electrostatic interactions and will, therefore, vary with wastewater pH. Consistent with these expectations, Golet et al. [56] found sorption coefficients at pH 7.5–8.4 to be approximately 2500 L/kg, which are larger than expected given the hydrophilicity of these molecules.

Table 6.2 contains a summary of experimentally determined sorption coefficients for various PPCPs. Generally, sorption is negligible if K_D is lower than 100 L/kg; on the other hand, substances with K_D values greater than 10,000 L/kg demonstrate significant removal in primary, as well as secondary, treatment [28]. As shown in Table 6.2, only a small fraction of PPCPs is expected to demonstrate substantial removal via sorption to primary sludge; these PPCPs include hydrophobic compounds such as 17β-estradiol (estrogenic hormone) and galaxolide (musk fragrance), as well as zwitterionic species like tetracycline and norfloxacin.

6.3.2 Removal of PPCPs in Biological Processes

The heart of wastewater treatment trains is the biological process, which effectively converts organic matter to carbon dioxide and biomass. Removal of PPCPs has been investigated in a variety of biological reactor configurations and operating conditions. Much like sorption of PPCPs to primary solids, prediction of biodegradation rates by chemical class or structure is challenging [72,120,170]. For this reason, biodegradation rate constants must be individually determined for each compound. Pseudo-first-order reaction kinetics have been used to describe degradation of micropollutants in biological processes. As shown in Equation 6.2, the biodegradation rate is directly proportional to both the aqueous PPCP concentration and the mixed liquor suspended solids (MLSS), but the MLSS concentration is assumed constant over the course of the experiment [73]:

$$\frac{dC}{dt} = -k_{biol}SX_{ss} \tag{6.2}$$

In Equation 6.2, C is the total PPCP concentration (μg/L), k_{biol} is the biodegradation rate constant (L/g_{ss}-d), S is the soluble PPCP concentration (μg/L), and X_{ss} is the MLSSs concentration (g_{ss}/L). A selection of published biodegradation rate constants is shown in Table 6.3.

Ternes et al. [170] and Joss et al. [73] asserted that PPCPs could be classified with respect to biodegradation potential by the magnitude of the corresponding rate constant. Compounds with rate constants less than 0.1 L/g_{ss}-d are not significantly degraded in conventional biological processes, while compounds with rate constants greater than 10 L/g_{ss}-d generally demonstrate more than 95% removal during wastewater treatment. Using this classification, Joss et al. [73] concluded that only four compounds (i.e., ibuprofen, paracetamol, 17β-estradiol, and estrone) from a study of 35 PPCPs would demonstrate more than 90% removal, while negligible removal was expected for 17 others. These results are generally consistent with the conclusions of Oulton et al. [120], who reviewed published data for removal of 140 PPCPs at pilot- and full-scale wastewater treatment facilities and found that a combination of primary treatment and conventional activated sludge (CAS) tended to result in no more than 1-log (or 90%) removal of PPCPs, regardless of influent concentrations.

6.3.2.1 Effect of Reactor Configuration on PPCP Removal

For compounds that demonstrate biodegradation rate constants in the 0.1–10 L/g_{ss}-d range, the biological reactor configuration has important consequences for removal efficiencies [73,170].

TABLE 6.3 Summary of Experimentally Determined Rate Constants Describing Suspended Growth Biodegradation of PPCPs

Compound	Category	Biodegradation Rate Constant L/g_{ss}-d	Reference
(Anhydro-)erythromycin	Antibiotic	<0.12	[73]
5-Amino-2,4,6-triiodo-2,3-dihydroxypropyl-amid-phthalic acid	Contrast agent	1.3–1.9	[73]
Azithromycin	Antibiotic	<0.13	[73]
Bezafibrate	Lipid regulator	−0.3–8.0 2.1–3.0	[26]
Bisphenol-A	Plasticizer	0.4–58	[26]
Carbamazepine	Antiepileptic agent	−0.1–0.21 <0.06	[26] [161]
Cefalexin	Antibiotic	2.82	[97]
Celestolide	Fragrance	75	[161]
Citalopram	Antidepressant	3	[161]
Clarithromycin	Antibiotic	<0.5	[73]
Clofibric acid	Metabolite of lipid regulator	0.3–0.8	[73]
Desmethoxyacetyliopromide	Contrast agent	1.9–4.9	[73]
Diatrizoate	Contrast agent	<0.1	[73]
Diazepam	Tranquilizer	<0.4 −0.7–0.6	[161] [26]
Diclofenac	Anti-inflammatory agent	<0.1 1.2	[73] [161]
Erythromycin	Antibiotic	6 0.0–18.8	[161]
Ethinylestradiol	Hormone	20	[26] [161]
Fenofibric acid	Lipid regulator	7.2–10.8	[73]
Fenoprofen	Anti-inflammatory agent	10–14	[73]
Fluoxetine	Antidepressant	9	[161]
Galaxolide	Fragrance	170	[161]
Gemfibrozil	Lipid regulator	6.4–9.6 − 0.1–33.6	[73]
Ibuprofen	Anti-inflammatory agent	21–35 20	[73] [161]
Indomethacin	Anti-inflammatory agent	<0.3	[73]
Iohexol	Contrast agent	1.8–2.4	[73]
Iomeprol	Contrast agent	1.2–1.6	[73]
Iopamidol	Contrast agent	<0.36	[73]
Iopromide	Contrast agent	1.6–2.5	[73]
Iothalamic acid	Contrast agent	<0.24	[73]
Ioxithalamic acid	Contrast agent	0.2–0.7	[73]
N4-Acetyl-sulfamethoxazole	Antibiotic	5.9–7.6 1.0–1.9	[73] [73]
Naproxen	Anti-inflammatory agent	9	[161]
Paracetamol	Anti-inflammatory agent	58–80 <0.2	[73]
Roxithromycin	Antibiotic	9	[73] [161]
Sulfadiazine	Antibiotic	0.174 0.3	[97] [161]
Sulfamethoxazole	Antibiotic	0.12	[97] [161]
Tonalide	Fragrance	115	[161]
Trimethoprim	Antibiotic	0.15	[161]

Extensive studies have focused on CAS systems and membrane bioreactors (MBRs). In MBRs, biological treatment by activated sludge and separation of solids from secondary effluent are combined in one treatment step using a membrane to retain sludge in the reactor. Conflicting evidence exists as to whether biodegradation of PPCPs in MBRs is superior to CAS. The lower sludge loadings in MBRs may induce a broader range of metabolic pathways, and smaller flocs will reduce mass transfer limitations [73]. Oulton et al. [120] found that PPCPs exceeded the 1-log removal threshold in MBRs for 49% of studies, while the same threshold was only met in 25% of CAS studies. Sui et al. [163] and Lesjean et al. [95] reported higher PPCP removal efficiencies in an MBR compared to CAS and attributed this difference to decreased sensitivity to temperature and variable influent conditions. On the other hand, Joss et al. [73] found that PPCPs with biodegradation rate constants less than 50 L/g_{ss}-d demonstrated slightly lower degradation rate constants in MBRs than in CAS; the authors hypothesized that this difference was attributable to a higher fraction of inert solids in MBR sludge. Other authors have found comparable rates of PPCP removal in CAS and MBRs when similar solids retention times (SRTs) are employed [26,72].

Fewer data exist with respect to the performance of other reactor types in removing PPCPs from wastewater. Kasprzyk-Hordern et al. [75] compared removal of 37 pharmaceuticals and 15 personal care product ingredients in two full-scale WWTPs in the United Kingdom. One plant used an extended aeration oxidation ditch, while the other plant utilized primary treatment followed by a trickling filter and secondary clarifier. On average, removal efficiencies in the plant with the extended aeration oxidation ditch were greater than 85%, while the corresponding removal efficiency in the trickling filter plant was less than 70%. However, Kasprzyk-Hordern et al. [75] did find enhanced removal of a subset of the studied compounds in trickling filters compared to CAS. The authors hypothesized that the attached growth in a trickling filter allows for colonization of heterogeneous microbial populations capable of effectively degrading micropollutants; such populations may not be able to thrive in suspended growth reactors. Others have reported no significant differences in PPCP removal for trickling filter systems and CAS [142].

6.3.2.2 Impact of Operating Parameters on PPCP Removal

Within conventional wastewater treatment and water reclamation plants, operating conditions are routinely varied to ensure that treatment goals for parameters such as biochemical oxygen demand or nutrient concentrations are met. These parameters are expected to impact PPCP removal as well. In the following subsections, the impacts of four important parameters are discussed.

6.3.2.2.1 Solids Residence Time

Many reports have concluded that SRT is a major determinant of the extent of PPCP removal in biological processes [119,170].

Longer SRTs facilitate increased microbial diversity in biological reactors by allowing slower-growing microorganisms to thrive and contribute to metabolism of PPCPs [27,87,170]. However, at longer SRTs, suspended solids contain a greater fraction of inert matter, indicating that biodegradation rates may be lower [170]. Oulton et al. [120] examined treatment data for plants with SRTs ranging from 5 h to 100 days, but did not determine a clear relationship between SRT and PPCP removal. Vieno et al. [186] found no apparent correlation between PPCP removal and SRT in an evaluation of 12 full-scale WWTPs with SRTs ranging from 2 to 20 days. Similarly, Sui et al. [164] compared removal of PPCPs at four different WWTPs with SRTs varying from 12 to 20 days and found no relationship; however, the authors concluded that these results might have been masked by the fact that the SRTs were already relatively long.

In general, Kreuzinger et al. [87] found that no PPCP removal was achieved in lab- and full-scale plants with SRTs less than 1 day. In examining removal of 20 different PPCPs during secondary treatment at eight different WWTPs, Oppenheimer et al. [119] found that each PPCP had a different critical SRT, above which 80% removal was consistently achieved. The critical SRT was between 5 and 15 days for the majority of PPCPs; however, a small number of compounds were determined to have a critical SRT of longer than 15 days. Similarly, Clara et al. [26] determined critical SRTs ranging from 5 to 10 days for four PPCPs. The authors indicated that an SRT of 10 days corresponds to typical design criteria for achieving nitrification. This assertion is consistent with studies that found nitrifying activated sludge enhanced removal of certain compounds compared to CAS [11,122]. The authors attributed this difference to changes in the microbial community related to the survival of slower-growing nitrifying bacteria. Batt et al. [11] reported inhibition of nitrifiers in activated sludge reduced removal of select PPCPs, suggesting that the microbial populations responsible for nitrification and PPCP biodegradation may be related. Other work reiterates that the impact of SRT variation on PPCP removal will differ for individual compounds [27,87,95,157].

6.3.2.2.2 Hydraulic Residence Time

The literature has demonstrated that the hydraulic residence time (HRT) of biological reactors does not significantly impact biodegradation of PPCPs [72], although HRT may be important for compounds that are removed by adsorption to secondary sludge by precluding attainment of equilibrium conditions [87]. Oulton et al. [120] examined treatment data for plants with HRTs ranging from 1 h to 10 days and did not ascertain correlations between HRT and PPCP removal.

6.3.2.2.3 Temperature

Clear relationships between wastewater temperature and biodegradation of PPCPs have also not been established. At higher temperatures, higher microbiological activity is expected and has been observed [95,163]. Similarly, Vieno et al. [187] found the highest effluent PPCP concentrations in a WWTP encompassing CAS with phosphorus precipitation during the winter, which the authors attributed to lower rates of biodegradation in colder temperatures. In addition, temporal variations in PPCP and water consumption may also contribute to differences in wastewater composition between winter and summer months. Reyes-Contreras et al. [134] found high removal efficiencies of PPCPs in constructed wetlands during summer months; however, the authors also reported that removal efficiencies for select PPCPs in an upflow anaerobic sludge blanket reactor improved in winter. Regardless, other authors found no correlation between operating temperature and PPCP biodegradation rates [54,99,202].

6.3.2.2.4 Redox Conditions

Several authors have found that biodegradation rates under aerobic conditions exceed those corresponding to anoxic and anaerobic conditions [30,71,163]; nevertheless, limited data exists on the effect of redox conditions or dissolved oxygen concentrations on PPCP removal in the aqueous phase. However, Zwiener and Frimmel [209] did find improved degradation for diclofenac and clofibric acid under anaerobic conditions, suggesting that certain compounds may exhibit more favorable anaerobic degradation pathways. These findings, as well as those discussed for other operating conditions, indicate that the physicochemical diversity of the thousands of PPCPs expected to be present in wastewater hampers efforts to find singular treatment solutions.

6.3.2.3 Partitioning of PPCPs into Sludge

Sorption of PPCPs to sludge in wastewater treatment varies between the primary and secondary clarifiers because of the difference in sludge composition. Primary sludge has a greater lipid fraction such that hydrophobic compounds are more likely to partition into sludge and be removed in the primary clarifier. PPCPs retaining a positively charged functionality at wastewater pH are expected to sorb to secondary sludge [174]. In an evaluation of two fluoroquinolone antibiotics, Golet et al. [56] found sorption coefficients of approximately 2500 L/kg in primary sludge, but 15,000–20,000 L/kg in secondary sludge. This difference stems from the speciation behavior of fluoroquinolone antibiotics, which are predominantly zwitterionic at pH 6.1–8.8. Sorption of fluoroquinolones to secondary solids occurs via electrostatic interactions, while hydrophobic interactions drive sorption of fluoroquinolones to primary sludge. Joss et al. [72] concluded that PPCP sorption in secondary treatment processes is negligible when the sorption coefficient is less than 300 L/kg.

As with primary sludge, solution pH plays a major role in determining the extent of PPCP sorption to secondary sludge. Urase and co-workers [182,183] evaluated the removal of 15 PPCPs and found a low degree of partitioning into sludge at near-neutral pH. However, when the pH was lowered to 5.6, the sorption coefficient for acidic pharmaceuticals increased. This difference was attributed to acidic pharmaceuticals existing as neutral molecules at acidic conditions, thereby increasing

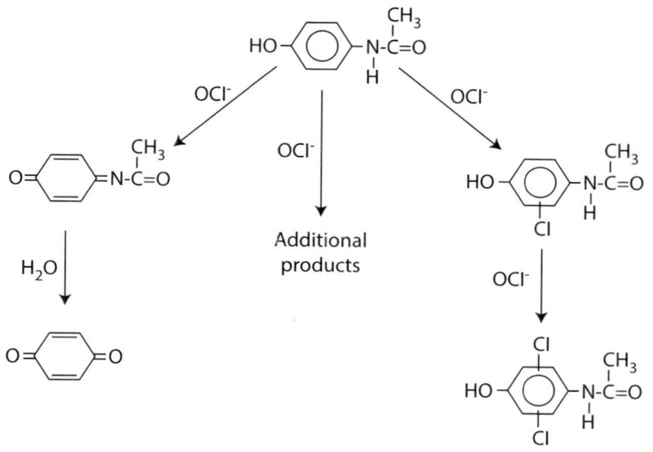

FIGURE 6.1 Reaction pathway for acetaminophen with hypochlorite. (Reprinted with permission from Bedner, M. and MacCrehan, W.A. 2005. *Environmental Science and Technology*, 40(2), 516–522. Copyright 2005, American Chemical Society.)

sorption to sludge particles. Conversely, Ternes et al. [172] found a sorption coefficient for diclofenac in secondary sludge that was an order of magnitude lower than the corresponding value in primary sludge (pH 6.6). The authors attributed this finding to the pH of the secondary reactor (pH 7.5), which caused the acidic pharmaceutical to be more deprotonated in this reactor. Urase et al. [182] also reported that sorbed PPCPs underwent a higher degree of metabolism. Ultimately, sorption of PPCPs to primary or secondary sludge reduces concentrations in the aqueous phase but may have ramifications for reuse of digested sludge.

6.3.3 Transformation during Disinfection Processes

The final stage of wastewater treatment involves disinfection to prevent pathogens from entering receiving bodies of water. Typically, wastewater disinfection employs chlorine or UV light. Chlorine is bactericidal and disinfects organisms through oxidation of the cell wall. UV light is bacteriostatic and operates by forming thymine and cytosine dimers in DNA, thus preventing replication. In both cases, disinfectants have a potential to cause chemical transformations in PPCP molecules, although the kinetics of these reactions are usually slower than disinfection. An understanding of the chemistry and reaction kinetics of these processes provides insight into baseline transformation of PPCPs in wastewater treatment. In this section, a focus will be placed on chlorine- and UV-based disinfection processes as these are widely employed in WWTPs.

6.3.3.1 Transformation of PPCPs by Chlorine

The mechanisms of action for chlorine-based disinfection include oxidation, chemical reactions between microorganisms

and chlorine, precipitation of proteins, increased cell wall permeability, and hydrolysis [168]. Many of these mechanisms are also relevant to understanding reactions between chlorine and PPCPs. For example, consider the reaction between chlorine and acetaminophen illustrated in Figure 6.1. In this case, acetaminophen reacts with hypochlorite to form *N*-acetyl-*p*-benzoquinone imine and 1,4-benzoquinone [13], both of which are known toxicants. The formation of toxic or biochemically active products during chlorination of PPCPs is an ongoing issue of interest and may have important consequences for evaluating the treatment efficiency of select pharmaceuticals.

Typically, chlorine is introduced as hypochlorous acid (HOCl) or hypochlorite (OCl⁻); the pK_a of the hypochlorite system is 7.60 (see Figure 6.2). This speciation is important because HOCl is a stronger disinfectant compared to OCl⁻. Together, HOCl and OCl⁻ are termed free chlorine, as opposed to combined chlorine (i.e., chloramines). PPCPs are expected to react with both HOCl and OCl⁻, but each of these reactions will exhibit distinct reaction kinetics (Reactions 6.3 and 6.4).

$$PPCP + HOCl \xrightarrow{k''_{HOCl,PPCP}} products \tag{6.3}$$

$$PPCP + OCl^- \xrightarrow{k''_{OCl^-,PPCP}} products \tag{6.4}$$

In Reactions 6.3 and 6.4, $k''_{HOCl,PPCP}$ and $k''_{OCl^-,PPCP}$ are the second-order rate constants for transformation of an individual PPCP with HOCl and OCl⁻, respectively. Many authors have reported the reaction kinetics of various PPCPs with chlorine [2,3,20,44,93,153,154,191]. For example, Soufan et al. [153] measured the reaction kinetics of carbamazepine, an antiepileptic that is recalcitrant in biological processes, with aqueous chlorine. The second-order rate constants for transformation of neutral carbamazepine by hypochlorous acid and hypochlorite were measured as $<1.0 \times 10^{-4}$ and $2.7(\pm0.4) \times 10^{-2}$ M⁻¹ s⁻¹, respectively [153]; these rate constants suggest that chlorine reactions with the neutral form of carbamazepine are negligible. The rate constant for the anionic form of carbamazepine with HOCl was eight orders of magnitude higher (i.e., $1.8(\pm0.3) \times 10^6$ M⁻¹ s⁻¹ [153]), indicating that this reaction will drive the majority of carbamazepine transformation in chlorination processes. Overall, hypochlorite is a considerably weaker electrophile than hypochlorous acid [53] and, therefore, does not contribute much to oxidation reactions for PPCPs. For that reason, most authors have not reported rate constants for transformation of PPCPs with hypochlorite. Second-order rate constants for transformation of select PPCPs by hypochlorous acid are provided in Table 6.4.

Clearly, these rate constants (Table 6.4) can vary considerably depending on the chemical structure and valence of PPCP molecules. Consider that the second-order rate constant corresponding to neutral molecules varies from $<1.0 \times 10^{-4}$ M⁻¹ s⁻¹ for carbamazepine to 3.23×10^3 M⁻¹ s⁻¹ for antipyrine [20,153]. The rate constant for anionic species can also vary over orders of magnitude (e.g., <0.1 M⁻¹ s⁻¹ for ibuprofen and 2.83×10^6 for

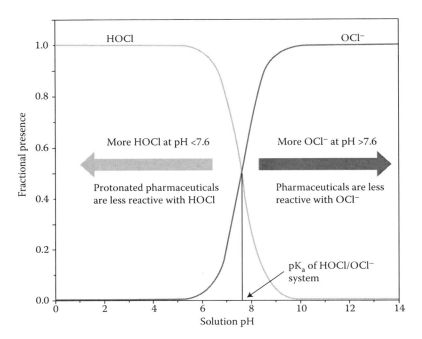

FIGURE 6.2 Speciation of HOCl and OCl⁻ in solution. This speciation drives the apparent reaction kinetics of pharmaceuticals with chlorine in solution.

tetracycline) [93,191]. The rate constants presented in Table 6.4 demonstrate that anionic species are generally more reactive with hypochlorous acid than cationic or neutral molecules. Some authors have proposed quantitative–structure–activity relationships (QSARs) for transformation of PPCPs during chlorination [94]; however, these models have not been widely successful for chlorine due to the abundant reaction mechanisms available in the chlorine system. Ultimately, the inherent differences in PPCP reactivity with HOCl indicate that a wide range of transformation efficiencies will be observed in disinfection processes.

As chlorine contact basins are designed to be plug-flow reactors, the relevant steady-state mass balance can be determined using

$$Q\frac{\delta[PPCP]}{\delta V} = -k''_{HOCl,PPCP}[PPCP][HOCl] \qquad (6.5)$$

In Equation 6.5, Q is the flow rate and V is the reactor volume. With known second-order rate constants, PPCP concentrations, and reactor design, the transformation of PPCPs in chlorine-based disinfection processes can be predicted.

Typical wastewater treatment designs for disinfection processes are well developed. For 3-log inactivation of bacteria at pH 7°C and 20°C, the maximum design chlorine exposure is approximately 3 mg-min/L [168]. Given HOCl/OCl⁻ speciation, the corresponding HOCl exposure is 2.40 mg-min/L or 2.74×10^{-3} M−s. For the PPCP molecules identified above, the corresponding expected transformation efficiencies can be calculated: 2.7×10^{-5}% (neutral carbamazepine), 99.99% (neutral antipyrine), 0.027% (ibuprofen anion), and ~100% (tetracycline anion). Calculating these efficiencies for disinfection-level

chlorine doses is extremely useful to build an understanding of PPCP fate in conventional wastewater treatment processes. As chlorine-contact basins are designed as plug-flow reactors and chlorine exposure is rigorously measured (as the product of concentration (C) and duration in time (T) of exposure, also known as CT values), a kinetic understanding provides a good deal of insight into the expected transformation efficiency and what CT values would be required for better transformation of PPCPs.

6.3.3.2 UV-Based Transformation of PPCPs

Water and wastewater utilities are conservative, and implementation of new or emerging technologies is often a long process. Regardless, UV-based disinfection has been extensively implemented by water and WWTPs over the past decade. Currently, about 25%–30% of U.S. WWTPs are using UV technologies [178]. The major advantage of UV systems over chlorine disinfection is that there is no need for subsequent treatment. That is, in chlorine-based disinfection, the chlorine residual must be quenched before discharge.

The majority of UV disinfection processes employ low- or medium-pressure lamps. Low-pressure lamps emit UV light at 253.7 nm, while medium-pressure lamps emit polychromatic light across the UV spectrum. Examples of the relevant emission spectra are shown in Figure 6.3; in this case, the molar extinction coefficients of H_2O_2 and N-nitrosodimethylamine are overlaid on the spectra [148]. PPCPs can absorb that light, become excited, and undergo chemical transformations; additionally, UV light interacts with background organic matter to generate reactive oxygen species, including hydroxyl radicals, singlet oxygen, and triplet state excited dissolved organic matter (DOM).

TABLE 6.4 Second-Order Rate Constants for Transformation of Select Pharmaceuticals by HOCl

Pharmaceutical	pK_a[a]	Structure	$k''_{HOCl,PPCP}$ $(M^{-1}s^{-1})$	Reference
17α-Ethinylestradiol	$pK_{a1} = 10.33$		$k_{neutral} = 4.33\ (\pm0.53)$ $k_{anion} = 3.52\ (\pm0.10) \times 10^5$	[36]
Antipyrine	$pK_{a1} = 0.37$		$k_{cation} =$ not presented $k_{neutral} = 3.23 \times 10^3$	[20]
Carbamazepine	$pK_{a1} = 15.96$		$k_{neutral} = <1.0 \times 10^{-4}$ $k_{anion} = 1.8\ (\pm0.3) \times 10^6$	[153]
Diclofenac	$pK_{a1} = 4.00$		$k_{neutral} = 0.37$ $k_{anion} = 2.41$	[154]
Ibuprofen	$pK_{a1} = 4.85$		$k_{neutral} = <0.1$ $k_{anion} = <0.1$	[93]
Naproxen	$pK_{a1} = 4.19$		$k_{neutral} = 300\ (\pm60)$ $k_{anion} = 9\ (\pm3)$	[2]
Sulfamethoxazole	$pK_{a1} = 1.97$ $pK_{a2} = 7.66$		$k_{cation} =$ not presented $k_{neutral} = 1.1 \times 10^3$ $k_{anion} = 2.4 \times 10^3$	[38]
Tetracycline	$pK_{a1} = 3.32$[b] $pK_{a2} = 7.78$ $pK_{a3} = 9.58$		$k_{cation} =$ not presented $k_{zwitterion} =$ not presented $k_{anion} = 2.83\ (\pm0.24) \times 10^6$ $k_{dianion} = 1.01\ (\pm4.96) \times 10^6$	[191]

[a] pK_a values were collected from www.chemicalize.org [23].
[b] pK_a values from Qiang and Adams [130].

These species may also contribute to transformation of PPCPs in UV-based disinfection depending on the light source and water quality. These two mechanisms are shown in Reactions 6.6 and 6.7, where $h\nu$ represents a photon and $k''_{HO\bullet,PPCP}$ is the second-order rate constant for PPCP transformation by hydroxyl radicals; note that reactions involving singlet oxygen and triplet state excited DOM were not included here for the sake of brevity:

$$PPCP \xrightarrow{h\nu} products \qquad (6.6)$$

$$PPCP + HO\bullet \xrightarrow{k''_{HO\bullet,PPCP}} products \qquad (6.7)$$

As above, these reactions can be substituted into the general mass balance for plug-flow reactors to better understand how PPCPs are transformed in UV processes:

$$Q\frac{\delta[PPCP]}{\delta V} = -k'_{p,PPCP}[PPCP]H' - k''_{HO\bullet,PPCP}[PPCP][HO\bullet]$$

$$(6.8)$$

In Equation 6.8, $k'_{p,PPCP}$ is the fluence-based pseudo-first-order rate constant for transformation of PPCPs by photons and H' is the fluence or UV dose. For disinfection processes, the fluence is generally 30–50 mJ/cm². Given the uniformity of UV doses in wastewater treatment, the fluence-based pseudo-first-order rate

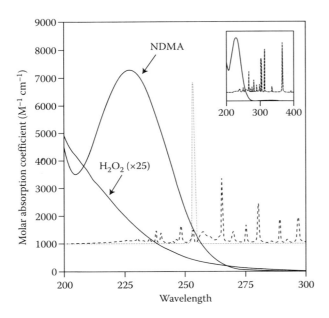

FIGURE 6.3 Comparison of low- and medium-pressure UV lamp spectra. (Reprinted with permission from Sharpless, C.M. and Linden, K.G. 2003. *Environmental Science and Technology*, 37(9), 1933–1940. Copyright 2003, American Chemical Society.)

constant determines whether significant transformation occurs for particular PPCPs. Table 6.5 provides a snapshot of fluence-based rate constants for several PPCPs with UV_{254} [10,128,143]. Unlike chlorine-based kinetics, few authors have deconvoluted the UV transformation kinetics to calculate specific fluence-based pseudo-first-order rate constants for UV-based transformation of PPCPs. Regardless, this approach can be advantageous for understanding the baseline UV transformation kinetics as a function of pH as shown in Figure 6.4, which describes the apparent fluence-based pseudo-first-order rate constant for tetracycline with UV_{254}. To contextualize these values, the corresponding transformation efficiency for PPCPs at 50 mJ/cm² is also included in Table 6.5. From the magnitude of transformation efficiencies, it is clear that PPCPs demonstrate a range of reactivity with UV_{254}; however, this reactivity varies over fewer orders of magnitude than reactions with hypochlorous acid. Regardless, transformation efficiencies of PPCPs are lower than the inactivation efficiencies for pathogens and other microbial contaminants at 50 mJ/cm². For that reason, coupling UV disinfection processes with chemical or catalytic processes to achieve advanced oxidation may be an effective technique to improve transformation of PPCPs, while continuing to meet disinfection regulations.

6.3.3.2.1 Transformation of PPCPs by Hydroxyl Radicals

Second-order rate constants for transformation of PPCPs with hydroxyl radicals are generally in the 10^8–10^{10} M^{-1} s^{-1} range. The UV system and water quality (i.e., carbonate system, DOM, etc.) sets the corresponding hydroxyl radical exposure term, providing a measure of the relative importance of direct (UV) and indirect (HO•) transformation mechanisms. In fact, the $R_{OH,UV}$ term proposed by Rosenfeldt and Linden [138] provides unique insight into the relative importance of these mechanisms for select water chemistries; this term derives from the original R_{ct} term coined by Elovitz and von Gunten [45], which was used to describe the ratio of hydroxyl radical- and ozone-exposure in ozone-based oxidation processes. Rate constants for reactions between select pharmaceuticals and hydroxyl radicals are included in Table 6.6; additionally, second-order rate constants for hydroxyl radical-based transformation of most PPCPs of concern can be found in the literature [14,37,58,64,70,121]. QSARs have been developed to predict second-order rate constants for transformation of PPCPs by hydroxyl radicals [91,118,162,192]. These tools have been refined in the past decade and represent an important method for assessing the reactivity of the full set of PPCPs in hydroxyl radical-driven processes. As the incorporation of advanced oxidation processes (AOPs) in drinking water and WWTPs becomes more common, an understanding of the baseline chemistry is important.

6.3.4 Removal in Advanced Treatment Processes

As WWTPs continue to evolve and the practice of water reuse becomes more widespread and more planned, advanced treatment processes will be more directly incorporated into conventional utilities. This process has already started with the advent of indirect planned water reuse operations around the world [46]. These facilities are gaining momentum for two reasons: (1) decreasing per capita freshwater supplies due to increasing population density and accelerated climate change and (2) increasingly stringent wastewater permits for discharge into sensitive ecosystems, such as the Florida Everglades and the Chesapeake Bay. To achieve increased treatment efficiencies, many plants have begun installing membrane systems and AOPs. The following sections discuss recent work in these areas with respect to the ability of such processes to separate PPCPs from effluent streams or chemically transform PPCP molecules.

6.3.4.1 Rejection of PPCPs in Membrane Processes

Much of the focus on membrane processes for wastewater treatment of PPCPs has focused on MBRs [1,40,41,86,101,103,131,141,150]. MBRs effectively combine biological processes, such as activated sludge or the A²O (anaerobic–anoxic–oxic) process, with micro- or ultrafiltration membranes to retain suspended solids. In MBRs, the HRT and the SRT can be effectively separated to achieve higher solids concentrations, more diverse microbial populations, and better metabolism of some PPCPs. As Section 6.3.2 has already provided discussion of biological degradation mechanisms, this section focuses directly on the interactions of PPCPs with membranes in wastewater treatment.

Due to the size of PPCP molecules, significant removal by micro- or ultrafiltration membranes is not expected. For example, consider that the maximum length of sulfamethoxazole and carbamazepine molecules are on the order of 12 Å [23]; in

TABLE 6.5 Fluence-Based Pseudo-First-Order Rate Constants for Transformation of Select Pharmaceuticals by UV Light at 254 nm

Pharmaceutical	pK_a[a]	Structure	$k''_{p,app}$ (cm²/mJ)	Reference	Transformation Efficiency at 50 mJ/cm² (%)
Atenolol	$pK_{a1} = 1.97$ $pK_{a2} = 7.66$		5.73×10^{-5} (pH 6.2)	[143]	0.29
Diclofenac	$pK_{a1} = 4.00$		$5.33\ (\pm0.13) \times 10^{-3}$ (pH 7.85) 3.63×10^{-4} (pH 6.2)	[10] [143]	23.4 ($\pm$0.65) 1.80
Ketoprofen	$pK_{a1} = 15.96$		4.39×10^{-3} (pH 6.2)	[143]	19.7
Metronidazole	$pK_{a1} = 3.09$		$3.56\ (\pm0.19) \times 10^{-5}$ (pH 5–7)	[128]	0.18 ($\pm$0.01)
Sulfamethoxazole	$pK_{a1} = 1.97$ $pK_{a2} = 7.66$		9.76×10^{-3} (pH 3.60) 2.40×10^{-3} (pH 7.85)	[10]	38.6 11.3
Trimethoprim	$pK_{a1} = 7.16$		$1.4\ (\pm0.7) \times 10^{-5}$ (pH 3.60) $1.7\ (\pm0.2) \times 10^{-5}$ (pH 7.85) 1.9×10^{-5} (pH 9.7)	[10]	0.07 ($\pm$0.03) 0.08 ($\pm$0.01) 0.09

[a] pK_a values were collected from www.chemicalize.org [23].

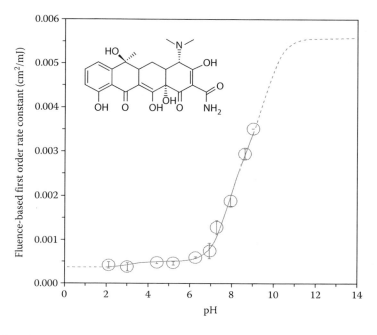

FIGURE 6.4 Apparent fluence-based pseudo-first-order rate constant for tetracycline with UV₂₅₄ in lab-grade water.

Table 6.6   Second-Order Rate Constants for Transformation of Select Pharmaceuticals with Hydroxyl Radicals

Pharmaceutical	Structure	$k''_{HO\bullet,PPCP}(M^{-1}s^{-1})$	Reference	Required Contact Time (s) for 2-Log Transformation If $[HO\bullet]_{ss} = 1.5 \times 10^{-11}$ M
17α-Ethinylestradiol		$9.8\ (\pm1.2) \times 10^9$	[64]	32
Atenolol		$8.0\ (\pm0.5) \times 10^9$	[14]	40
Carbamazepine		$8.8\ (\pm1.2) \times 10^9$	[64]	36
Diclofenac		$7.5\ (\pm1.5) \times 10^9$	[64]	42
Ibuprofen		$7.4\ (\pm1.2) \times 10^9$	[64]	43
Metronidazole		1.98×10^9	[70]	161
Naproxen		$9.6\ (\pm0.5) \times 10^9$	[121]	33
Sulfamethoxazole		$5.5\ (\pm0.7) \times 10^9$	[64]	58
Tetracycline		$7.7\ (\pm1.2) \times 10^9$	[37]	41
Trimethoprim		$6.9\ (\pm0.2) \times 10^9$	[37]	46

comparison, the minimum pore size of micro- and ultrafiltration membranes are approximately 0.1 and 0.01 μm, respectively. Consequently, more than 80 sulfamethoxazole molecules and 8 carbamazepine molecules can concurrently pass through these pores, negating the importance of rejection mechanisms such as steric exclusion, hydrophobic adsorption and partitioning, and electrostatic repulsion. Regardless, ultrafiltration (UF) studies have been conducted to assess rejection of PPCPs [79,140,151,202]. Snyder et al. [151] investigated removal of more than 30 PPCPs in a pilot-scale UF plant. The removal efficiencies of compounds varied widely: ethinylestradiol (78 ng/L → <1 ng/L), naproxen (24 ng/L → 21 ng/L), and triclosan (32 ng/L → 4 ng/L) [151]. However, UF of drinking water in Seoul, South Korea demonstrated slightly elevated permeate concentrations of PPCPs [79].

Much of the research concerning membrane-based treatment of PPCPs has focused on nanofiltration (NF; minimum pore size = 0.001 μm) or reverse osmosis (RO). In these cases, PPCP molecules must diffuse across the membrane. As the mass transfer kinetics are limited by diffusion of PPCPs in the polymeric membranes, relatively high removal efficiencies can be attained. Previous authors have suggested other operationally relevant mechanisms for pharmaceutical transport through membranes [151]. For NF and RO applications, rejection of pharmaceuticals is controlled by a number of membrane and solute properties, including physicochemical properties of the solute, membrane characteristics, and operating conditions. Specific properties of interest are highlighted:

- *Solute*: Molecular size, molecular shape, molecular charge, and reactivity with dissolved/particulate species
- *Membrane*: Salt rejection, molecular weight cut off (MWCO), hydrophobicity, extent of biofouling, and pure water permeability
- *Operating conditions*: Pressure, cross-flow velocity, recovery, and background water quality

As expected, NF and RO membranes provide better removal efficiencies compared to microfiltration and UF systems. The Snyder et al. [151] study cited previously about UF processes also investigated the performance of RO membranes for the same wastewater matrix. The corresponding removal of the same suite of compounds was high: ethinylestradiol (125 ng/L → <25 ng/L), naproxen (118 ng/L → <25 ng/L), and triclosan (246 ng/L → <25 ng/L) [151]. Ahel et al. [4] investigated treatment of pharmaceutical metabolites (i.e., diacetone xylose, diacetone sorbose, and diacetone α-ketogulonic acid) present in landfill leachate using NF membranes with MWCOs of 200, 300, and 400 Da. Rejection of the pharmaceutical metabolites was greater than 90% [4]. The differences between the abilities of NF and RO membranes were highlighted by Xu et al. [201], who found high rejection efficiencies (e.g., >90%) for naproxen, diclofenac, and ibuprofen, among others, using RO membranes (MWCO = 100 Da) and low rejection (e.g., 10%–65%) with NF membranes (MWCO = 400 Da). With the exception of tris (2-chloroethyl) phosphate, over 15 PPCPs were removed to below the analytical limits of detection in South Korean wastewater using NF and RO membranes [79]. Other work with NF and RO membranes reported similar results [80,104,140,204,205], although some studies did show significant removal capacities for select PPCPs with NF membranes [80]. In general, those membranes had lower MWCOs (i.e., 150–200 Da).

Water quality and membrane conditioning also affects the rejection efficiency of PPCPs in membrane processes [80,151]. This phenomenon stems from interactions of PPCPs with DOM, including the negative surface charge associated with DOM accumulated at the membrane surface due to concentration polarization [140]. Regardless, many PPCPs demonstrate a positive molecular charge at pH 7, indicating potential for facilitated transport through the membrane via complexation with DOM or other dissolved species. The impact of molecular charge was demonstrated by Botton et al. [17] through documentation of the rejection in virgin and biofouled membranes; however, rejection efficiencies were still higher than 80% for both scenarios. Similar effects have been documented for forward osmosis processes [197]. While fouling may be an issue, advancements in MBR technologies coupled with the enhanced rejection efficiencies observed with biofouled membranes indicate a delicate balance between maintaining flux and achieving high rejection of PPCPs.

6.3.4.2 Transformation of PPCPs in AOPs

In many ways, AOPs are similar to disinfection processes; however, the explicit goal in advanced oxidation is to generate hydroxyl radicals. Hydroxyl radicals are an effective disinfectant [25,66,208] due to oxidative attack of cell walls. Within the context of wastewater treatment and water reclamation, several AOPs have been investigated. Some of these processes include UV-H_2O_2, O_3-H_2O_2, UV-TiO_2, and Fe^{2+}-H_2O_2, among others [9,96,127,159]. The mechanism for producing hydroxyl radicals varies for each process. For example, the UV-H_2O_2 process operates through photolytic decomposition of hydrogen peroxide. Alternatively, hydrogen peroxide reacts with ozone to instigate rapid ozone decomposition. The energy/chemical costs associated with AOPs, as well as the achievable steady-state hydroxyl radical concentrations, drive usability. Ozone- and UV-based processes have seen the largest potential for introduction at full-scale water and WWTPs, presumably due to the current use of such processes for disinfection purposes.

The second-order rate constants for transformation of PPCPs by hydroxyl radicals were discussed in Section 6.3.3. In the case of AOPs, some reactivity with the base mechanism (e.g., UV or O_3) may occur, but the process is designed around a specific hydroxyl radical exposure (i.e., $\int_0^t [HO\bullet] dt$), and so the base mechanism does not usually contribute much to the transformation efficiency of PPCPs. The peroxone process is a good example of this phenomenon. Consider that the second-order rate constants for transformation of sulfamethoxazole with ozone and hydroxyl radicals at pH 7 are 5.5×10^5 and 5.5×10^9 M^{-1} s^{-1}, respectively [37,64]. The second-order rate constants for transformation of peroxide with ozone and hydroxyl radicals are 2.2×10^6 M^{-1} s^{-1} [175] and 2.7×10^7 M^{-1} s^{-1} [19], respectively. For this example using the PPCP sulfamethoxazole (SMZ), let us assume that [H$_2$O$_2$] = 1 mg/L (or 2.94×10^{-5} M) and [SMZ] = 500 ng/L (or 1.97×10^{-9} M). The ozone scavenging potential of peroxide—the reaction that leads to formation of hydroxyl radicals—can be determined from

$$\text{O}_3 \text{ scavenging efficiency} = \frac{k''_{\text{O}_3,\text{app},\text{H}_2\text{O}_2}[\text{H}_2\text{O}_2]}{k''_{\text{O}_3,\text{app},\text{H}_2\text{O}_2}[\text{H}_2\text{O}_2] + k''_{\text{O}_3,\text{app},\text{SMZ}}[\text{SMZ}]}$$

(6.9)

For the chemical conditions described above, hydrogen peroxide will react with more than 99.99% of ozone molecules. Assuming that the molar ratio of ozone to hydrogen peroxide is set to the stoichiometric value (2 mol O$_3$/mol H$_2$O$_2$) and the hydroxyl radical yield is approximately 50% of the ozone consumption [50], then the reaction between ozone and hydrogen peroxide should result in approximately 2.94×10^{-5} mol/L HO•. Using the equivalent of Equation 6.9, but for hydroxyl radicals, the hydroxyl radical reaction efficiency for sulfamethoxazole is 1.35%. That efficiency indicates that the 1.97×10^{-9} M of sulfamethoxazole (or 500 ng/L) will react with approximately 3.97×10^{-7} M of hydroxyl radicals, leading to relatively high transformation efficiencies. Thus, even though the majority of hydroxyl radicals are scavenged by hydrogen peroxide, excellent transformation efficiencies of PPCPs can still be attained.

From Table 6.6, the second-order rate constants for PPCP transformation by hydroxyl radicals generally ranges from 10^9 to 10^{10} M^{-1} s^{-1} [64,70]. Therefore, for a design hydroxyl radical exposure (or steady-state hydroxyl radical concentration), transformation of PPCPs will be relatively similar regardless of structure or class. In comparison, the corresponding rate constants for transformation by chlorine, ozone, and UV can vary over orders of magnitude [93,149,188], resulting in wide-ranging transformation efficiencies for different PPCPs. For this reason, AOPs effectively level the playing field among PPCPs. Regardless, the source of hydroxyl radicals (e.g., as in the H$_2$O$_2$ case demonstrated previously) and high chemical oxygen demand (COD) of wastewater can be a significant sink for hydroxyl radicals. Consider that COD levels for wastewater effluent are generally on the order of 10–100 mg/L. As pharmaceuticals are present on the ng/L to μg/L level [85,185], the elevated COD concentrations scavenge a large fraction of the hydroxyl radicals that are produced. The hydroxyl radical efficiency term for pharmaceuticals can be determined using

$$\text{HO• efficiency} = \frac{\sum k''_{\text{HO•},\text{PPCP}}[\text{PPCP}]}{\begin{bmatrix} k''_{\text{HO•},\text{H}_2\text{O}_2}[\text{H}_2\text{O}_2] + \sum k''_{\text{HO•},\text{PPCP}}[\text{PPCP}] \\ + \sum k''_{\text{HO•},\text{CO}_3}[\text{CO}_3] + k''_{\text{HO•},\text{COD}}[\text{COD}] \end{bmatrix}}$$

(6.10)

In Equation 6.10, the summation symbols indicate that multiple PPCPs may be present; furthermore, the carbonate system (i.e., HCO$_3^-$ and CO$_3^{2-}$) needs to be considered for hydroxyl

radical scavenging. Any other inorganic or organic compounds that demonstrate considerable reactivity with hydroxyl radicals should also be included in Equation 6.10. As written, the equation demonstrates that species at much higher concentrations will drive the steady-state concentration of hydroxyl radicals, limiting the effective transformation of PPCPs. Note that at high concentrations, hydrogen peroxide can also scavenge hydroxyl radicals; therefore, AOPs employing H$_2$O$_2$ have an optimal operating range.

Ultimately, the transformation efficiency of PPCPs in AOPs is a function of the hydroxyl radical exposure (often written as a steady-state hydroxyl radical concentration) and the second-order rate constants for transformation of PPCPs by hydroxyl radicals. With these parameters, advanced oxidation of PPCPs in wastewater can be successfully modeled. Several authors have investigated the efficacy of AOPs to treat PPCPs in wastewater [10,16,47,82,105,108,113,133,137,144]. For example, Rosario-Ortiz et al. [137] investigated the transformation of six PPCPs in three tertiary wastewater samples using the UV-H$_2$O$_2$ process. The concentrations of PPCPs in this study ranged from 38 to 2600 ng/L and the total organic carbon (TOC) of the tertiary wastewater varied from 6.6 to 10.3 mg/L. The overall transformation efficiency was highest in the low TOC wastewater; furthermore, transformation efficiency generally increased with the applied H$_2$O$_2$ dose. The authors were able to correlate transformation efficiencies to the reduction in UV$_{254}$ absorbance, indicating that the background organic matter served as an important scavenger of hydroxyl radicals. Like biological processes and disinfection mechanisms, AOPs cause chemical transformations in PPCPs. The identification and characterization of the resultant transformation products remains a concern from an effluent quality perspective.

6.3.5 Removal during Infiltration and Riverbank Filtration

Although not explicitly part of the treatment train, posttreatment filtration during recharge of drinking water aquifers provides a final opportunity to prevent PPCP introduction into water supplies. Minimal PPCP degradation is expected in the aqueous subsurface as microbial populations are more limited and less active [123]. In addition, the long residence time of water in the subsurface (e.g., indirect water reuse guidelines stipulate at least 2 months) may result in PPCP accumulation in groundwater [123]. For these reasons, recent work has focused on infiltration or riverbank filtration, which provides the potential for removal through adsorption, dispersion, and biodegradation mechanisms, before reaching groundwater [123]. The overall efficacy of such treatment for PPCPs depends on local characteristics including groundwater depth at the point of application and soil type, porosity, and permeability [123]. The corresponding redox conditions, bulk organic carbon concentrations, and nutrient availability also affect the microbial community structure and function, thereby influencing the extent to which PPCPs will be biodegraded [62,123]. Hoppe-Jones et al. [62] also found that microbial adaptation over the course of several months of infiltration improved PPCP removal rates. As with other treatment technologies described in this section, the removal efficiency can vary widely for individual PPCPs. For example, carbamazepine (an antiepileptic) demonstrated almost no removal, while the antibiotic clindamycin was almost completely removed [42,123].

6.4 Concerns about PPCP Presence in Reclaimed Water

As PPCPs are not completely removed or transformed during wastewater treatment, trace concentrations of PPCPs are continuously discharged to the environment. The toxicological significance of long-term exposure to trace concentrations of these molecules is not yet well understood, but preliminary research has indicated that these compounds can impact aquatic ecological systems. Furthermore, in the context of direct/indirect water reuse, a potential for impacts to human health exists. These topics are discussed in more detail.

6.4.1 Ecological and Human Health Concerns

Sections 6.2 and 6.3 demonstrated that PPCPs are widely present in raw wastewater and wastewater effluent. This presence is a concern for downstream ecological systems as PPCPs have been specially designed to instigate a specific biochemical response in organisms. Furthermore, in the context of indirect planned water reuse, these issues apply directly to human health. The most well-understood threat stems from the estrogenic/androgenic activity associated with wastewater effluent. A well-controlled study in the Canadian experimental lake system demonstrated that wastewater effluent containing 5 ng/L of 17α-ethinylestradiol can overwhelmingly alter the gender distribution of fathead minnows [77] leading to population collapse. In 1989, Aherne and Briggs [5] raised similar questions about human health after detecting synthetic steroids in reservoirs. Concerns associated with other PPCP classes are detailed in the following subsections.

6.4.1.1 Endocrine-Disrupting Chemicals

In 1994, Purdom et al. [129] demonstrated that wastewater effluent from the United Kingdom was estrogenic by reporting vitellogenin concentrations in fish. EDCs, and their corresponding estrogenic activity, have been widely detected in wastewater effluent, reclaimed water, and surface waters [12,98,107]. Feminization of fish has been tied to the presence of dozens of chemicals that are commonly present in wastewater effluent at trace concentrations. Indeed, endocrine-disrupting effects have been identified throughout the entirety of the aquatic food web from mollusks to polar bears [181,189,206]. As exposure to EDCs can alter the reproductive system of organisms, widespread ecological impacts are expected. With feminization and intersex development, the major concern is that vulnerable aquatic wildlife populations will be decimated; however, the long-term impacts of subtle changes in reproductive systems may have other consequences [15,167,198].

Current epidemiological information on the impacts of environmentally relevant concentrations of EDCs typically associated with PPCPs on human health is lacking. Regardless, the human health effects associated with other EDCs, including dioxin, polychlorinated biphenyls, and select pesticides have been well documented in humans [29,83,152]. Prescription of select estrogenic compounds to pregnant women has also resulted in apparent health issues (e.g., cancer, reproductive abnormalities, and fertility problems, among others) for offspring [31,115,176]. In addition, estrogenic compounds have been widely linked to changes in the global male/female sex ratio [35,112]. These controversial issues maintain a high level of public awareness, as evidenced by recent media interest in bisphenol A. Given these concerns and the relatively unknown human health effects of long-term and multigeneration exposure to trace concentrations of EDCs, the monitoring of EDC concentrations, and the associated toxicological activity, in reclaimed water is a highly important issue.

6.4.1.2 Antibiotics

Discharge of antibiotics in wastewater effluent represents a unique concern to ecological as well as public health. Direct toxicological impacts of antibiotics in receiving water bodies are not expected in most scenarios because antibiotics generally show antimicrobial activity at the mg/L range, which is outside the range of environmentally relevant concentrations. However, some classes of antibiotics, such as fluoroquinolones, can inhibit select microbial populations at concentrations as low as 1–10 μg/L. In this case, a potential for direct toxicity exists, especially in locations that have demonstrated high concentrations of fluoroquinolones in municipal or industrial wastewater effluent [92].

Recent review articles have highlighted discharge of antibiotics, antibiotic resistant bacteria, and antibiotic resistance genes from urban WWTPs [109,136]. Hundreds of antimicrobial resistance genes have been detected in WWTPs and water reclamation plants, as well as reclaimed water [33,165]. Discharge of antimicrobial resistance into reservoirs and aquifers may serve as a mechanism for increased public exposure. Regardless, the impact of trace concentrations of antibiotics present in surface water on development of multidrug antibiotic resistance is not yet known [65]; however, as antibiotic resistance represents a global health challenge [43,57], this issue is becoming more important and may have considerable human health and economic impacts. Furthermore, as generic antibiotics are widely available in many developing countries, gathering environmental data on antibiotic concentrations in those settings, as well as corresponding public health records documenting antimicrobial resistance will be vital to better understanding this important topic.

6.4.1.3 Chemotherapy Agents

The mechanism of action for chemotherapy agents, such as 5-fluorouracil, cyclophosphamide, and cisplatin, involves DNA alkylation. DNA strands are covalently bonded by the active metabolite of these drugs, inhibiting replication and preventing growth of new cells. The efflux of chemotherapy agents in wastewater has triggered concerns that these highly toxic compounds could affect organisms in aquatic ecosystems [68,69,207] and humans in water reuse scenarios [6,177]. Many of these compounds are carcinogenic, teratogenic, and mutagenic. For these reasons, concerns over the handling of these species in occupational scenarios have drawn recent attention [179,180], and subsequent attention to the presence of chemotherapy agents in wastewater and reclaimed water is likely. Indeed, Rowney et al. [139] recently indicated that although the risk associated with chemotherapy agents in water was low for adults, concerns associated with vulnerable populations, including pregnant women, fetuses, and breastfeeding babies, exist.

6.4.1.4 UV Filters

While UV filters are usually associated with sunscreens, these compounds are extensively employed in various personal care products ranging from lotions to cosmetics. Unlike pharmaceuticals, which are designed to achieve a specific biochemical response, UV filters have been shown to demonstrate endocrine activity, developmental toxicity, and up-/down-regulation of gene expression associated with the reproductive system [89,102,146,147]. Furthermore, recent studies have indicated that UV filters can affect the growth, cell viability, and oxidative stress in protozoa [52]. Human health concerns associated with these compounds have not yet been identified, but the endocrine activity and developmental toxicity of UV filters in animal models suggest that screening for human toxicity is warranted. In water reuse scenarios, UV filters may be present at elevated concentrations due to heavy household use of personal care products. For that reason, building and residential systems for water reclamation may exhibit characteristically higher concentrations of UV filters than municipal systems.

6.4.1.5 Other PPCPs of Concern

Hundreds of PPCPs have been identified in wastewater and reclaimed water. These compounds all have specific biochemical properties that can be tied to specific ecological and public health concerns. Some of the classes of compounds that have received significant attention, due to high consumption, include analgesics, antidepressants, nonsteroidal anti-inflammatory drugs, antiepileptics, antimicrobials, surfactants, bronchodilators, perfluorinated species, musk fragrances, antacids, antipsychotics, antidiabetics, anticoagulants, and antihypertensives, among many others. These compounds are being increasingly detected in wastewater effluent, and toxicological studies are being pursued at environmentally relevant concentrations. Pomati et al. [124–126] reported inhibitory impacts from trace concentrations of 11 pharmaceuticals on human, *E. coli*, and zebrafish cells. Regardless, the ultimate toxicity associated with mixtures of PPCPs at environmentally relevant concentrations in wastewater and surface water matrices is not yet well understood. The constantly changing composition of PPCPs in reclaimed water, especially due to seasonal and temporal variations, may shift toxicity concerns during the course of the year. In this regard, more toxicological data is required; however, the difficulties in assessing these effects in humans makes this a long-term challenge, especially as animal models may not directly apply to human health concerns. A systematic approach to addressing these concerns in the context of planned indirect water reuse is required to continue protecting human health and ensuring that PPCPs do not accumulate in the aquatic environment.

6.4.2 Application-Specific Concerns

Wastewater supplies have been used for agricultural irrigation for millennia. Today, that use has been expanded to include irrigation of golf courses, highway medians, and other public landscapes. In recent years, a number of studies have examined the general impact of long-term irrigation with both treated and untreated municipal wastewater sources. Soils irrigated with treated municipal wastewater demonstrated lower pH; higher sodium adsorption ratios (SARs), boron concentrations, microbial densities, electrical conductivity, metal concentrations, organic matter, total carbon, and total nitrogen; and changes in the indigenous bacterial and fungal communities [61,156,200]. Multiple studies have also focused on the particular impacts of PPCPs in treated wastewater on soil, plant growth, and microbial ecology. Chen et al. [24], Tamtam et al. [166], and Kinney et al. [81] investigated accumulation of PPCPs in soil irrigated with municipal wastewater effluent containing 28 different PPCPs. Concentrations as high as 212 µg/kg (of oxytetracycline) were detected in irrigated soils, and PPCPs were detected as deep as 75 cm below the soil surface, indicating a potential for groundwater contamination.

Fewer data exist regarding the effects of irrigation with wastewater containing PPCPs on soil microbial ecology. Xu et al. [199] investigated the sorption of five PPCPs to agricultural soils irrigated with reclaimed wastewater, but also found that degradation rates diminished with increasing chemical concentrations. These results indicate that higher PPCP concentrations inhibit microbial activity. Concerns also exist about the effects of antimicrobial compounds on microbial diversity and the spread of antimicrobial resistance in soil. Dalkmann et al. [33] detected eight pharmaceuticals in soils that had been irrigated with untreated wastewater for 100 years at concentrations up to 8.4 µg/kg; three of the pharmaceuticals in that study exhibited increasing soil concentrations over time, suggesting active accumulation of PPCPs. Soils irrigated with wastewater also demonstrated number concentrations of the sulfonamide-resistance genes *sul*1 and *sul*2 (normalized by 16S rDNA) 20 times higher than soils irrigated with precipitation. After 10 weeks of irrigation with synthetic greywater containing 2 mg/L of triclosan, Harrow et al. [59] found significantly lower numbers of culturable microorganisms and a less diverse microbial community than soil irrigated with synthetic greywater without triclosan. Furthermore, the authors reported that 40%–55% of viable heterotrophs isolated from soil treated with triclosan-containing greywater demonstrated resistance to four antibiotics, while only 15%–25% of viable heterotrophs isolated from soil irrigated with greywater without triclosan were resistant.

Plants grown in soils irrigated with reclaimed wastewater have also been shown to accumulate PPCPs. Wu et al. [196] studied growth of soybean plants irrigated with water containing five PPCPs at concentrations of 10 μg/L. In less than 4 months, PPCPs were detected in the root, stem, leaf, and bean at concentrations as high as 80 μg/kg. Herklotz et al. [60] and Migliore et al. [111] also confirmed accumulation of pharmaceuticals in six plant species at concentrations as high as 457 μg/kg when exposed to reclaimed water containing PPCPs at concentrations of 50–233 μg/L. PPCP presence in irrigation water also demonstrated inhibitory effects on plant growth. D'Abrosca et al. [32] investigated the phytotoxicity of eight pharmaceuticals on eight different plant species and found inhibitory effects on both seed germination and root lengths at concentrations lower than 1 μg/L in cultivated species such as lettuce and carrot plants. Migliore et al. [111] and Aristilde et al. [8] demonstrated that plant exposure to antibiotics at concentrations greater than 100 μg/L reduced root length and leaf number. In similar studies, Kong et al. [84] reported that oxytetracycline concentrations of 920 μg/L resulted in reduced shoot mass and root growth in alfalfa plants. Even more dramatically, Herklotz et al. [60] found that two types of plants would not grow in solutions containing 440 μg/L of triclosan. Because of these findings, implementation of water reuse for irrigation purposes should consider the presence and concentration of PPCPs due to the potential for inhibitory or toxic effects resulting from plant exposure to water containing these chemicals.

6.5 Summary and Conclusions

Existing WWTPs are primarily designed for bulk organic and nutrient removal; therefore, the data presented in Section 6.3 demonstrating that traditional WWTP processes are often ineffective at removing PPCPs are not surprising. If effluent water is to be reused in the urban water cycle, the breadth of compounds, with respect to chemical class, molecular structure, expected concentrations, and recalcitrance, indicates the need for redundant treatment processes. In this manner, removal of the broad range of PPCPs that represent threats to ecological and public health can be effectively treated.

Few studies demonstrate complete mineralization of PPCPs during wastewater treatment or water reclamation, and fewer studies focus on metabolic or transformation products within the treatment plant; therefore, while a parent compound may be removed, the corresponding metabolic/transformation compounds may still exhibit adverse effects. For this reason, monitoring of PPCP removal in urban water cycles must involve a comprehensive approach that focuses on reducing concentrations of pharmaceuticals as well as byproducts formed through human metabolism, biodegradation, or oxidation processes. The sheer number of compounds present in raw wastewater

represents a grand challenge to environmental engineers, especially as monitoring of all of these compounds is not economically feasible on a large scale. In addition, such an approach will not capture potential synergistic or antagonistic effects elicited by mixtures of PPCPs and other water quality components. Instead, a more reasonable approach may be to monitor specific anticipated effects, such as antimicrobial activity or endocrine-disrupting effects [39,107].

As the urban water cycle becomes shorter, awareness of the potential issues associated with PPCPs in urban water reuse is growing. While water treatment is an important part of the solution, greater consideration must be paid to reducing influent loads to WWTPs by use of appropriate source control. Recent efforts by the U.S. Food and Drug Administration (FDA) to work with industry to reduce antibiotic usage in farm animal feed is one example of such a move that would lead to reduced antibiotic loads to WWTPs receiving agricultural runoff [48]. Efforts to design green pharmaceuticals are also an innovative approach to this issue but would require cross-disciplinary research and collaboration potential between the medical and environmental engineering fields [76].

The need for increased sustainability in our urban infrastructure, along with practical concerns about availability of adequate water sources in both quantity and quality, demands greater levels of urban water reuse. Now, the potential for PPCP presence in urban water reuse and the corresponding impacts on human and ecological health require an intensive analysis of the trade-off between sustainability issues and risk factors associated with micropollutant presence in reclaimed water. Continued research and innovation in both the treatment technologies and monitoring techniques are essential to ensure sustainable water practices and sustained public health.

References

1. Abegglen, C., Joss, A., McArdell, C.S., Fink, G., Schlüsener, M.P., Ternes, T.A., and Siegrist, H. 2009. The fate of selected micropollutants in a single-house MBR. *Water Research*, 43(7), 2036–2046.

2. Acero, J.L., Benitez, F.J., Real, F.J., and Roldan, G. 2010. Kinetics of aqueous chlorination of some pharmaceuticals and their elimination from water matrices. *Water Research*, 44(14), 4158–4170.

3. Acero, J.L., Benitez, F.J., Real, F.J., Roldan, G., and Rodriguez, E. 2013. Chlorination and bromination kinetics of emerging contaminants in aqueous systems. *Chemical Engineering Journal*, 219, 43–50.

4. Ahel, T., Mijatovic, I., Matosic, M., and Ahel, M. 2004. Nanofiltration of a landfill leachate containing pharmaceutical intermediates from vitamin C production. *Food Technology and Biotechnology*, 42(2), 99–104.

5. Aherne, G.W. and Briggs, R. 1989. The relevance of the presence of certain synthetic steroids in the aquatic environment. *Journal of Pharmacy and Pharmacology*, 41(10), 735–736.

6. Aherne, G.W., Hardcastle, A., and Nield, A.H. 1990. Cytotoxic drugs and the aquatic environment: Estimation of bleomycin in river and water samples. *Journal of Pharmacy and Pharmacology,* 42(10), 741–742.

7. Andersen, H.R., Hansen, M., Kjølholt, J., Stuer-Lauridsen, F., Ternes, T., and Halling-Sørensen, B. 2005. Assessment of the importance of sorption for steroid estrogens removal during activated sludge treatment. *Chemosphere,* 61(1), 139–146.

8. Aristilde, L., Melis, A., and Sposito, G. 2010. Inhibition of photosynthesis by a fluoroquinolone antibiotic. *Environmental Science and Technology,* 44(4), 1444–1450.

9. Autin, O., Hart, J., Jarvis, P., MacAdam, J., Parsons, S.A., and Jefferson, B. 2012. Comparison of UV/H2O2 and UV/TiO2 for the degradation of metaldehyde: Kinetics and the impact of background organics. *Water Research,* 46(17), 5655–5662.

10. Baeza, C. and Knappe, D.R.U. 2011. Transformation kinetics of biochemically active compounds in low-pressure UV Photolysis and UV/H₂O₂ advanced oxidation processes. *Water Research,* 45(15), 4531–4543.

11. Batt, A.L., Kim, S., and Aga, D.S. 2006. Enhanced biodegradation of iopromide and trimethoprim in nitrifying activated sludge. *Environmental Science and Technology,* 40(23), 7367–7373.

12. Bayen, S., Zhang, H., Desai, M.M., Ooi, S.K., and Kelly, B.C. 2013. Occurrence and distribution of pharmaceutically active and endocrine disrupting compounds in Singapore's marine environment: Influence of hydrodynamics and physical–chemical properties. *Environmental Pollution,* 182, 1–8.

13. Bedner, M. and MacCrehan, W.A. 2005. Transformation of acetaminophen by chlorination produces the toxicants 1,4-benzoquinone and *N*-acetyl-*p*-benzoquinone imine. *Environmental Science and Technology,* 40(2), 516–522.

14. Benner, J., Salhi, E., Ternes, T., and von Gunten, U. 2008. Ozonation of reverse osmosis concentrate: Kinetics and efficiency of beta blocker oxidation. *Water Research,* 42(12), 3003–3012.

15. Bertanza, G., Papa, M., Pedrazzani, R., Repice, C., Mazzoleni, G., Steimberg, N., Feretti, D., Ceretti, E., and Zerbini, I. 2013. EDCs, estrogenicity and genotoxicity reduction in a mixed (domestic + textile) secondary effluent by means of ozonation: A full-scale experience. *Science of the Total Environment,* 458–460, 160–168.

16. Biń, A.K. and Sobera-Madej, S. 2012. Comparison of the advanced oxidation processes (UV, UV/H₂O₂ and O₃) for the removal of antibiotic substances during wastewater treatment. *Ozone: Science & Engineering,* 34(2), 136–139.

17. Botton, S., Verliefde, A.R.D., Quach, N.T., and Cornelissen, E.R. 2012. Influence of biofouling on pharmaceuticals rejection in NF membrane filtration. *Water Research,* 46(18), 5848–5860.

18. Brunton, L., Lazo, J., and Parker, K. (eds.), 2006. *Goodman & Gilman's The Pharmacological Basis of Therapeutics,* 11th edition, McGraw-Hill, New York.

19. Buxton, G.V., Greenstock, C.L., Helman, W.P., and Ross, A.B. 1988. Critical review of rate constants for reactions of hydrated electrons, hydrogen atoms and hydroxyl radicals (OH/O⁻) in aqueous solution. *Journal of Physical and Chemical Reference Data,* 17(2), 513–886.

20. Cai, M.-Q., Feng, L., Jiang, J., Qi, F., and Zhang, L.-Q. 2013. Reaction kinetics and transformation of antipyrine chlorination with free chlorine. *Water Research,* 47(8), 2830–2842.

21. Carballa, M., Omil, F., and Lema, J.M. 2005. Removal of cosmetic ingredients and pharmaceuticals in sewage primary treatment. *Water Research,* 39(19), 4790–4796.

22. Carballa, M., Omil, F., and Lema, J.M. 2007. Calculation methods to perform mass balances of micropollutants in sewage treatment plants. Application to pharmaceutical and personal care products (PPCPs). *Environmental Science and Technology,* 41(3), 884–890.

23. ChemAxon. 2013. www.chemicalize.org. Accessed on June 26, 2013.

24. Chen, F., Ying, G.-G., Kong, L.-X., Wang, L., Zhao, J.-L., Zhou, L.-J., and Zhang, L.-J. 2011. Distribution and accumulation of endocrine-disrupting chemicals and pharmaceuticals in wastewater irrigated soils in Hebei, China. *Environmental Pollution,* 159(6), 1490–1498.

25. Cheng, Y.W., Chan, R.C.Y., and Wong, P.K. 2007. Disinfection of *Legionella pneumophila* by photocatalytic oxidation. *Water Research,* 41(4), 842–852.

26. Clara, M., Kreuzinger, N., Strenn, B., Gans, O., and Kroiss, H. 2005. The solids retention time—A suitable design parameter to evaluate the capacity of wastewater treatment plants to remove micropollutants. *Water Research,* 39(1), 97–106.

27. Clara, M., Strenn, B., Gans, O., Martinez, E., Kreuzinger, N., and Kroiss, H. 2005. Removal of selected pharmaceuticals, fragrances and endocrine disrupting compounds in a membrane bioreactor and conventional wastewater treatment plants. *Water Research,* 39(19), 4797–4807.

28. Clara, M., Strenn, B., Saracevic, E., and Kreuzinger, N. 2004. Adsorption of bisphenol-A, 17β-estradiole and 17α-ethinylestradiole to sewage sludge. *Chemosphere,* 56(9), 843–851.

29. Colborn, T., vom Saal, F.S., and Soto, A.M. 1994. Developmental effects of endocrine-disrupting chemicals in wildlife and humans. *Environmental Impact Assessment Review,* 14(5–6), 469–489.

30. Conkle, J.L., Gan, J., and Anderson, M.A. 2012. Degradation and sorption of commonly detected PPCPs in wetland sediments under aerobic and anaerobic conditions. *Journal of Soils and Sediments,* 12(7), 1164–1173.

31. Crews, D., and McLachlan, J.A. 2006. Epigenetics, evolution, endocrine disruption, health, and disease. *Endocrinology,* 147(6), s4–s10.

32. D'Abrosca, B., Fiorentino, A., Izzo, A., Cefarelli, G., Pascarella, M.T., Uzzo, P., and Monaco, P. 2008. Phytotoxicity evaluation of five pharmaceutical pollutants

detected in surface water on germination and growth of cultivated and spontaneous plants. *Journal of Environmental Science and Health, Part A,* 43(3), 285–294.

33. Dalkmann, P., Broszat, M., Siebe, C., Willaschek, E., Sakinc, T., Huebner, J., Amelung, W., Grohmann, E., and Siemens, J. 2012. Accumulation of pharmaceuticals, *Enterococcus,* and resistance genes in soils irrigated with wastewater for zero to 100 years in Central Mexico. *PLoS One,* 7(9), e45397.

34. Dargnat, C., Teil, M.-J., Chevreuil, M., and Blanchard, M. 2009. Phthalate removal throughout wastewater treatment plant: Case study of Marne Aval station (France). *Science of the Total Environment,* 407(4), 1235–1244.

35. Davis, D., Gottlieb, M.B., and Stampnitzky, J.R. 1998. Reduced ratio of male to female births in several industrial countries: A sentinel health indicator? *Journal of the American Medical Association,* 279(13), 1018–1023.

36. Deborde, M., Rabouan, S., Gallard, H., and Legube, B. 2004. Aqueous chlorination kinetics of some endocrine disruptors. *Environmental Science and Technology,* 38(21), 5577–5583.

37. Dodd, M.C., Buffle, M.O., and von Gunten, U. 2006. Oxidation of antibacterial molecules by aqueous ozone: Moiety-specific reaction kinetics and application to ozone-based wastewater treatment. *Environmental Science and Technology,* 40(6), 1969–1977.

38. Dodd, M.C. and Huang, C.-H. 2004. Transformation of the antibacterial agent sulfamethoxazole in reactions with chlorine: Kinetics, mechanisms, and pathways. *Environmental Science and Technology,* 38(21), 5607–5615.

39. Dodd, M.C., Kohler, H.P.E., and von Gunten, U. 2009. Oxidation of antibacterial compounds by ozone and hydroxyl radical: Elimination of biological activity during aqueous ozonation processes. *Environmental Science and Technology,* 43(7), 2498–2504.

40. Dolar, D., Gros, M., Rodriguez-Mozaz, S., Moreno, J., Comas, J., Rodriguez-Roda, I., and Barceló, D. 2012. Removal of emerging contaminants from municipal wastewater with an integrated membrane system, MBR-RO. *Journal of Hazardous Materials,* 239–240, 64–69.

41. Dorival-García, N., Zafra-Gómez, A., Navalón, A., González, J., and Vílchez, J.L. 2013. Removal of quinolone antibiotics from wastewaters by sorption and biological degradation in laboratory-scale membrane bioreactors. *Science of the Total Environment,* 442, 317–328.

42. Drewes, J.E., Heberer, T., Rauch, T., and Reddersen, K. 2003. Fate of pharmaceuticals during ground water recharge. *Ground Water Monitoring & Remediation,* 23(3), 64–72.

43. Eggleston, K., Zhang, R., and Zeckhauser, R.J. 2010. The global challenge of antimicrobial resistance: Insights from economic analysis. *International Journal of Environmental Research and Public Health,* 7(8), 3141–3149.

44. El Najjar, N.H., Deborde, M., Journel, R., and Vel Leitner, N.K. 2013. Aqueous chlorination of levofloxacin: Kinetic and mechanistic study, transformation product identification and toxicity. *Water Research,* 47(1), 121–129.

45. Elovitz, M.S. and von Gunten, U. 1999. Hydroxyl radical/ozone ratios during ozonation processes. I. The Rct concept. *Ozone: Science & Engineering,* 21(3), 239–260.

46. EPA. 2012. *Guidelines for Water Reuse.* EPA document #EPA/600/R-12/618.

47. Fatta-Kassinos, D., Vasquez, M.I., and Kümmerer, K. 2011. Transformation products of pharmaceuticals in surface waters and wastewater formed during photolysis and advanced oxidation processes—Degradation, elucidation of byproducts and assessment of their biological potency. *Chemosphere,* 85(5), 693–709.

48. FDA 2013. Phasing out certain antibiotic use in farm animals. http://www.fda.gov/ForConsumers/Consumer Updates/ucm378100.htm.

49. Fent, K., Weston, A.A., and Caminada, D. 2006. Ecotoxicology of human pharmaceuticals. *Aquatic Toxicology,* 76(2), 122–159.

50. Fischbacher, A., von Sonntag, J., von Sonntag, C., and Schmidt, T.C. 2013. The •OH radical yield in the $H_2O_2 + O_3$ (peroxone) reaction. *Environmental Science and Technology,* 47(17), 9959–9964.

51. Flaherty, C.M. and Dodson, S.I. 2005. Effects of pharmaceuticals on Daphnia survival, growth, and reproduction. *Chemosphere,* 61(2), 200–207.

52. Gao, L., Yuan, T., Zhou, C., Cheng, P., Bai, Q., Ao, J., Wang, W., and Zhang, H. 2013. Effects of four commonly used UV filters on the growth, cell viability and oxidative stress responses of the *Tetrahymena thermophila. Chemosphere,* 93(10), 2507–2513.

53. Gerritsen, C.M. and Margerum, D.W. 1990. Non-metal redox kinetics: Hypochlorite and hypochlorous acid reactions with cyanide. *Inorganic Chemistry,* 29(15), 2757–2762.

54. Göbel, A., McArdell, C.S., Joss, A., Siegrist, H., and Giger, W. 2007. Fate of sulfonamides, macrolides, and trimethoprim in different wastewater treatment technologies. *Science of the Total Environment,* 372(2–3), 361–371.

55. Göbel, A., Thomsen, A., McArdell, C.S., Joss, A., and Giger, W. 2005. Occurrence and sorption behavior of sulfonamides, macrolides, and trimethoprim in activated sludge treatment. *Environmental Science and Technology,* 39(11), 3981–3989.

56. Golet, E.M., Xifra, I., Siegrist, H., Alder, A.C., and Giger, W. 2003. Environmental exposure assessment of fluoroquinolone antibacterial agents from sewage to soil. *Environmental Science and Technology,* 37(15), 3243–3249.

57. Gootz, T.D. 2010. The global problem of antibiotic resistance. *Current Reviews in Immunology,* 30(1), 79–93.

58. Haag, W.R. and Yao, C.C.D. 1992. Rate constants for reaction of hydroxyl radicals with several drinking water contaminants. *Environmental Science and Technology,* 26(5), 1005–1013.

59. Harrow, D.I., Felker, J.M., and Baker, K.H. 2011. Impacts of triclosan in greywater on soil microorganisms. *Applied and Environmental Soil Science,* 2011, 1–8.

60. Herklotz, P.A., Gurung, P., Vanden Heuvel, B., and Kinney, C.A. 2010. Uptake of human pharmaceuticals by plants grown under hydroponic conditions. *Chemosphere*, 78(11), 1416–1421.

61. Hidri, Y., Bouziri, L., Maron, P.-A., Anane, M., Jedidi, N., Hassan, A., and Ranjard, L. 2010. Soil DNA evidence for altered microbial diversity after long-term application of municipal wastewater. *Agronomy for Sustainable Development*, 30(2), 423–431.

62. Hoppe-Jones, C., Dickenson, E.R.V., and Drewes, J.E. 2012. The role of microbial adaptation and biodegradable dissolved organic carbon on the attenuation of trace organic chemicals during groundwater recharge. *Science of the Total Environment*, 437, 137–144.

63. Huang, R., Southall, N., Wang, Y., Yasgar, A., Shinn, P., Jadhav, A., Nguyen, D.T., and Austin, C.P. 2011. The NCGC pharmaceutical collection: A comprehensive resource of clinically approved drugs enabling repurposing and chemical genomics. *Science Translational Medicine*, 3(80), 38.

64. Huber, M.M., Canonica, S., Park, G.-Y., and von Gunten, U. 2003. Oxidation of pharmaceuticals during ozonation and advanced oxidation processes. *Environmental Science and Technology*, 37(5), 1016–1024.

65. Huerta, B., Marti, E., Gros, M., López, P., Pompêo, M., Armengol, J., Barceló, D., Balcázar, J.L., Rodríguez-Mozaz, S., and Marcé, R. 2013. Exploring the links between antibiotic occurrence, antibiotic resistance, and bacterial communities in water supply reservoirs. *Science of the Total Environment*, 456–457, 161–170.

66. Ikai, H., Nakamura, K., Shirato, M., Kanno, T., Iwasawa, A., Sasaki, K., Niwano, Y., and Kohno, M. 2010. Photolysis of hydrogen peroxide, an effective disinfection system via hydroxyl radical formation. *Antimicrobial Agents and Chemotherapy*, 54(12), 5086–5091.

67. Jelić, A., Gros, M., Petrović, M., Ginebreda, A., and Barceló, D. 2012. Occurrence and elimination of pharmaceuticals during conventional wastewater treatment. In: Guasch, H., Ginebreda, A., and Geiszinger, A. (eds.), *Emerging and Priority Pollutants in Rivers*. Springer Berlin, Heidelberg, pp. 1–23.

68. Johnson, A.C., Juergens, M.D., Williams, R.J., Kuemmerer, K., Kortenkamp, A., and Sumpter, J.P. 2007. Do cytotoxic chemotherapy drugs discharged into rivers pose a risk to the environment and human health? An overview and UK case study. *Journal of Hydrology*, 348(1–2), 167–175.

69. Johnson, A.C., Oldenkamp, R., Dumont, E., and Sumpter, J.P. 2013. Predicting concentrations of the cytostatic drugs cyclophosphamide, carboplatin, 5-fluorouracil, and capecitabine throughout the sewage effluents and surface waters of europe. *Environmental Toxicology and Chemistry*, 32(9), 1954–1961.

70. Johnson, M.B. and Mehrvar, M. 2008. Aqueous metronidazole degradation by UV/H2O2 process in single-and multi-lamp tubular photoreactors: Kinetics and reactor design. *Industrial & Engineering Chemistry Research*, 47(17), 6525–6537.

71. Joss, A., Andersen, H., Ternes, T., Richle, P.R., and Siegrist, H. 2004. Removal of estrogens in municipal wastewater treatment under aerobic and anaerobic conditions: Consequences for plant optimization. *Environmental Science and Technology*, 38(11), 3047–3055.

72. Joss, A., Keller, E., Alder, A.C., Göbel, A., McArdell, C.S., Ternes, T., and Siegrist, H. 2005. Removal of pharmaceuticals and fragrances in biological wastewater treatment. *Water Research*, 39(14), 3139–3152.

73. Joss, A., Zabczynski, S., Göbel, A., Hoffmann, B., Löffler, D., McArdell, C.S., Ternes, T.A., Thomsen, A., and Siegrist, H. 2006. Biological degradation of pharmaceuticals in municipal wastewater treatment: Proposing a classification scheme. *Water Research*, 40(8), 1686–1696.

74. Karthikeyan, K.G. and Meyer, M.T. 2006. Occurrence of antibiotics in wastewater treatment facilities in Wisconsin, USA. *Science of the Total Environment*, 361(1–3), 196–207.

75. Kasprzyk-Hordern, B., Dinsdale, R.M., and Guwy, A.J. 2009. The removal of pharmaceuticals, personal care products, endocrine disruptors and illicit drugs during wastewater treatment and its impact on the quality of receiving waters. *Water Research*, 43(2), 363–380.

76. Khetan, S.K. and Collins, T.J. 2007. Human pharmaceuticals in the aquatic environment: A challenge to green chemistry. *Chemical Reviews*, 107(6), 2319–2364.

77. Kidd, K.A., Blanchfield, P.J., Mills, K.H., Palace, V.P., Evans, R.E., Lazorchak, J.M., and Flick, R.W. 2007. Collapse of a fish population after exposure to a synthetic estrogen. *Proceedings of the National Academy of Sciences*, 104(21), 8897–8901.

78. Kim, S., Eichhorn, P., Jensen, J.N., Weber, A.S., and Aga, D.S. 2005. Removal of antibiotics in wastewater: Effect of hydraulic and solid retention times on the fate of tetracycline in the activated sludge process. *Environmental Science and Technology*, 39(15), 5816–5823.

79. Kim, S.D., Cho, J., Kim, I.S., Vanderford, B.J., and Snyder, S.A. 2007. Occurrence and removal of pharmaceuticals and endocrine disruptors in South Korean surface, drinking, and waste waters. *Water Research*, 41(5), 1013–1021.

80. Kimura, K., Iwase, T., Kita, S., and Watanabe, Y. 2009. Influence of residual organic macromolecules produced in biological wastewater treatment processes on removal of pharmaceuticals by NF/RO membranes. *Water Research*, 43(15), 3751–3758.

81. Kinney, C.A., Furlong, E.T., Werner, S.L., and Cahill, J.D. 2006. Presence and distribution of wastewater-derived pharmaceuticals in soil irrigated with reclaimed water. *Environmental Toxicology and Chemistry*, 25(2), 317–326.

82. Klavarioti, M., Mantzavinos, D., and Kassinos, D. 2009. Removal of residual pharmaceuticals from aqueous systems by advanced oxidation processes. *Environment International*, 35(2), 402–417.

83. Kogevinas, M. 2001. Human health effects of dioxins: Cancer, reproductive and endocrine system effects. *Human Reproduction Update*, 109(S103), S223–S232.

84. Kong, W.D., Zhu, Y.G., Liang, Y.C., Zhang, J., Smith, F.A., and Yang, M. 2007. Uptake of oxytetracycline and its phytotoxicity to alfalfa (*Medicago sativa* L.). *Environmental Pollution*, 147(1), 187–193.

85. Kostich, M.S., Batt, A.L., and Lazorchak, J.M. 2014. Concentrations of prioritized pharmaceuticals in effluents from 50 large wastewater treatment plants in the US and implications for risk estimation. *Environmental Pollution*, 184, 354–359.

86. Kovalova, L., Siegrist, H., Singer, H., Wittmer, A., and McArdell, C.S. 2012. Hospital wastewater treatment by membrane bioreactor: Performance and efficiency for organic micropollutant elimination. *Environmental Science and Technology*, 46(3), 1536–1545.

87. Kreuzinger, N., Clara, M., Strenn, B., and Kroiss, H. 2004. Relevance of the sludge retention time (SRT) as design criteria for wastewater treatment plants for the removal of endocrine disruptors and pharmaceuticals from wastewater. *Water Science & Technology*, 50(5), 149–156.

88. Kümmerer, K. 2009. Antibiotics in the aquatic environment—A review—Part I. *Chemosphere*, 75(4), 417–434.

89. Kunz, P.Y., Galicia, H.F., and Fent, K. 2006. Comparison of *in vitro* and *in vivo* estrogenic activity of UV filters in fish. *Toxicological Sciences*, 90(2), 349–361.

90. Kupper, T., Plagellat, C., Brändli, R.C., de Alencastro, L.F., Grandjean, D., and Tarradellas, J. 2006. Fate and removal of polycyclic musks, UV filters and biocides during wastewater treatment. *Water Research*, 40(14), 2603–2612.

91. Kušić, H., Rasulev, B., Leszczynska, D., Leszczynski, J., and Koprivanac, N. 2009. Prediction of rate constants for radical degradation of aromatic pollutants in water matrix: A QSAR study. *Chemosphere*, 75(8), 1128–1134.

92. Larsson, D.G.J., de Pedro, C., and Paxeus, N. 2007. Effluent from drug manufactures contains extremely high levels of pharmaceuticals. *Journal of Hazardous Materials*, 148(3), 751–755.

93. Lee, Y. and von Gunten, U. 2010. Oxidative transformation of micropollutants during municipal wastewater treatment: Comparison of kinetic aspects of selective (chlorine, chlorine dioxide, ferrateVI, and ozone) and non-selective oxidants (hydroxyl radical). *Water Research*, 44(2), 555–566.

94. Lei, H. and Snyder, S.A. 2007. 3D QSPR models for the removal of trace organic contaminants by ozone and free chlorine. *Water Research*, 41(18), 4051–4060.

95. Lesjean, B., Gnirss, R., Buisson, H., Keller, S., Tazi-Pain, A., and Luck, F. 2005. Outcomes of a 2-year investigation on enhanced biological nutrients removal and trace organics elimination in membrane bioreactor (MBR). *Water Science & Technology*, 52(10–11), 453–460.

96. Lester, Y., Avisar, D., and Mamane, H. 2010. Photodegradation of the antibiotic sulphamethoxazole in water with UV/H2O2 advanced oxidation process. *Environmental Technology*, 31(2), 175–183.

97. Li, B. and Zhang, T. 2010. Biodegradation and adsorption of antibiotics in the activated sludge process. *Environmental Science and Technology*, 44(9), 3468–3473.

98. Li, J., Fu, J., Zhang, H., Li, Z., Ma, Y., Wu, M., and Liu, X. 2013. Spatial and seasonal variations of occurrences and concentrations of endocrine disrupting chemicals in unconfined and confined aquifers recharged by reclaimed water: A field study along the Chaobai River, Beijing. *Science of the Total Environment*, 450–451, 162–168.

99. Lishman, L., Smyth, S.A., Sarafin, K., Kleywegt, S., Toito, J., Peart, T., Lee, B., Servos, M., Beland, M., and Seto, P. 2006. Occurrence and reductions of pharmaceuticals and personal care products and estrogens by municipal wastewater treatment plants in Ontario, Canada. *Science of The Total Environment*, 367(2–3), 544–558.

100. Liu, J.-L. and Wong, M.-H. 2013. Pharmaceuticals and personal care products (PPCPs): A review on environmental contamination in China. *Environment International*, 59, 208–224.

101. Maeng, S.K., Choi, B.G., Lee, K.T., and Song, K.G. 2013. Influences of solid retention time, nitrification and microbial activity on the attenuation of pharmaceuticals and estrogens in membrane bioreactors. *Water Research*, 47(9), 3151–3162.

102. Maerkel, K., Durrer, S., Henseler, M., Schlumpf, M., and Lichtensteiger, W. 2007. Sexually dimorphic gene regulation in brain as a target for endocrine disrupters: Developmental exposure of rats to 4-methylbenzylidene camphor. *Toxicology and Applied Pharmacology*, 218(2), 152–165.

103. Maletz, S., Floehr, T., Beier, S., Klümper, C., Brouwer, A., Behnisch, P., Higley, E. et al. 2013. *In vitro* characterization of the effectiveness of enhanced sewage treatment processes to eliminate endocrine activity of hospital effluents. *Water Research*, 47(4), 1545–1557.

104. Martínez, F., López-Muñoz, M.J., Aguado, J., Melero, J.A., Arsuaga, J., Sotto, A., Molina, R. et al. 2013. Coupling membrane separation and photocatalytic oxidation processes for the degradation of pharmaceutical pollutants. *Water Research*, 47(15), 5647–5658.

105. Medellin-Castillo, N.A., Ocampo-Pérez, R., Leyva-Ramos, R., Sanchez-Polo, M., Rivera-Utrilla, J., and Méndez-Díaz, J.D. 2013. Removal of diethyl phthalate from water solution by adsorption, photo-oxidation, ozonation and advanced oxidation process (UV/H2O2, O3/H2O2 and O3/activated carbon). *Science of the Total Environment*, 442, 26–35.

106. Metcalf and Eddy, Asano, T., Burton, F., Leverenz, H., Tsuchihashi, R., and Tchobanoglous, G. 2007. *Water Reuse: Issues, Technologies, and Applications*. McGraw Hill Professional, New York.

107. Metcalfe, C.D., Kleywegt, S., Letcher, R.J., Topp, E., Wagh, P., Trudeau, V.L., and Moon, T.W. 2013. A multi-assay screening approach for assessment of endocrine-active contaminants in wastewater effluent samples. *Science of the Total Environment*, 454–455, 132–140.

108. Michael, I., Hapeshi, E., Aceña, J., Perez, S., Petrović, M., Zapata, A., Barceló, D., Malato, S., and Fatta-Kassinos, D. 2013. Light-induced catalytic transformation of ofloxacin by solar Fenton in various water matrices at a pilot plant: Mineralization and characterization of major intermediate products. *Science of the Total Environment,* 461–462, 39–48.

109. Michael, I., Rizzo, L., McArdell, C.S., Manaia, C.M., Merlin, C., Schwartz, T., Dagot, C., and Fatta-Kassinos, D. 2013. Urban wastewater treatment plants as hotspots for the release of antibiotics in the environment: A review. *Water Research,* 47(3), 957–995.

110. Miège, C., Choubert, J.M., Ribeiro, L., Eusèbe, M., and Coquery, M. 2009. Fate of pharmaceuticals and personal care products in wastewater treatment plants—Conception of a database and first results. *Environmental Pollution,* 157(5), 1721–1726.

111. Migliore, L., Cozzolino, S., and Fiori, M. 2003. Phytotoxicity to and uptake of enrofloxacin in crop plants. *Chemosphere,* 52(7), 1233–1244.

112. Mocarelli, P., Gerthoux, P.M., Ferrari, E., Patterson Jr, D.G., Kieszak, S.M., Brambilla, P., Vincoli, N. et al. 2000. Paternal concentrations of dioxin and sex ratio of offspring. *The Lancet,* 355(9218), 1858–1863.

113. Mohapatra, D.P., Brar, S.K., Tyagi, R.D., Picard, P., and Surampalli, R.Y. 2013. A comparative study of ultrasonication, Fenton's oxidation and ferro-sonication treatment for degradation of carbamazepine from wastewater and toxicity test by Yeast Estrogen Screen (YES) assay. *Science of the Total Environment,* 447, 280–285.

114. Nash, J.P., Kime, D.E., Van der Ven, L.T.M., Wester, P.W., Brion, F., Maack, G., Stahlschmidt-Allner, P., and Tyler, C.R. 2004. Long-term exposure to environmental concentrations of the pharmaceutical ethynylestradiol causes reproductive failure in fish. *Environmental Health Perspectives,* 112(17), 1725–1733.

115. Newbold, R.R., Padilla-Banks, E., and Jefferson, W.N. 2006. Adverse effects of the model environmental estrogen diethylstilbestrol are transmitted to subsequent generations. *Endocrinology,* 147(6), s11–s17.

116. Nikolaou, A., Meric, S., and Fatta, D. 2007. Occurrence patterns of pharmaceuticals in water and wastewater environments. *Analytical and Bioanalytical Chemistry,* 387(4), 1225–1234.

117. Novo, A., André, S., Viana, P., Nunes, O.C., and Manaia, C.M. 2013. Antibiotic resistance, antimicrobial residues and bacterial community composition in urban wastewater. *Water Research,* 47(5), 1875–1887.

118. Öberg, T. 2005. A QSAR for the hydroxyl radical reaction rate constant: Validation, domain of application, and prediction. *Atmospheric Environment,* 39(12), 2189–2200.

119. Oppenheimer, J., Stephenson, R., Burbano, A., and Liu, L. 2007. Characterizing the passage of personal care products through wastewater treatment processes. *Water Environment Research,* 79(13), 2564–2577.

120. Oulton, R.L., Kohn, T., and Cwiertny, D.M. 2010. Pharmaceuticals and personal care products in effluent matrices: A survey of transformation and removal during wastewater treatment and implications for wastewater management. *Journal of Environmental Monitoring,* 12(11), 1956–1978.

121. Packer, J., Werner, J., Latch, D., McNeill, K., and Arnold, W. 2003. Photochemical fate of pharmaceuticals in the environment: Naproxen, diclofenac, clofibric acid, and ibuprofen. *Aquatic Sciences,* 65(4), 342–351.

122. Pérez, S., Eichhorn, P., and Aga, D.S. 2005. Evaluating the biodegradability of sulfamethazine, sulfamethoxazole, sulfathiazole, and trimethoprim at different stages of sewage treatment. *Environmental Toxicology and Chemistry,* 24(6), 1361–1367.

123. Petrovic, M., de Alda, M.J.L., Diaz-Cruz, S., Postigo, C., Radjenovic, J., Gros, M., and Barcelo, D. 2009. Fate and removal of pharmaceuticals and illicit drugs in conventional and membrane bioreactor wastewater treatment plants and by riverbank filtration. *Philosophical Transactions of the Royal Society A: Mathematical, Physical and Engineering Sciences,* 367(1904), 3979–4003.

124. Pomati, F., Castiglioni, S., Zuccato, E., Fanelli, R., Vigetti, D., Rossetti, C., and Calamari, D. 2006. Effects of a complex mixture of therapeutic drugs at environmental levels on human embryonic cells. *Environmental Science and Technology,* 40(7), 2442–2447.

125. Pomati, F., Cotsapas, C.J., Castiglioni, S., Zuccato, E., and Calamari, D. 2007. Gene expression profiles in zebrafish (*Danio rerio*) liver cells exposed to a mixture of pharmaceuticals at environmentally relevant concentrations. *Chemosphere,* 70(1), 65–73.

126. Pomati, F., Orlandi, C., Clerici, M., Luciani, F., and Zuccato, E. 2008. Effects and interactions in an environmentally relevant mixture of pharmaceuticals. *Toxicological Sciences,* 102(1), 129–137.

127. Poyatos, J.M., Muñio, M.M., Almecija, M.C., Torres, J.C., Hontoria, E., and Osorio, F. 2010. Advanced oxidation processes for wastewater treatment: State of the art. *Water, Air, and Soil Pollution,* 205(1–4), 187–204.

128. Prados-Joya, G., Sánchez-Polo, M., Rivera-Utrilla, J., and Ferro-garcía, M. 2011. Photodegradation of the antibiotics nitroimidazoles in aqueous solution by ultraviolet radiation. *Water Research,* 45(1), 393–403.

129. Purdom, C.E., Hardiman, P.A., Bye, V.V.J., Eno, N.C., Tyler, C.R., and Sumpter, J.P. 1994. Estrogenic effects of effluents from sewage treatment works. *Chemistry and Ecology,* 8(4), 275–285.

130. Qiang, Z. and Adams, C. 2004. Potentiometric determination of acid dissociation constants (pK_a) for human and veterinary antibiotics. *Water Research,* 38(12), 2874–2890.

131. Radjenovic, J., Petrovic, M., and Barceló, D. 2007. Analysis of pharmaceuticals in wastewater and removal using a membrane bioreactor. *Analytical and Bioanalytical Chemistry,* 387(4), 1365–1377.

132. Radjenović, J., Petrović, M., and Barceló, D. 2009. Fate and distribution of pharmaceuticals in wastewater and sewage sludge of the conventional activated sludge (CAS) and advanced membrane bioreactor (MBR) treatment. *Water Research,* 43(3), 831–841.

133. Razavi, B., Song, W., Santoke, H., and Cooper, W.J. 2011. Treatment of statin compounds by advanced oxidation processes: Kinetic considerations and destruction mechanisms. *Radiation Physics and Chemistry,* 80(3), 453–461.

134. Reyes-Contreras, C., Matamoros, V., Ruiz, I., Soto, M., and Bayona, J.M. 2011. Evaluation of PPCPs removal in a combined anaerobic digester-constructed wetland pilot plant treating urban wastewater. *Chemosphere,* 84(9), 1200–1207.

135. Richardson, M.L. and Bowron, J.M. 1985. The fate of pharmaceutical chemicals in the aquatic environment. *Journal of Pharmacy and Pharmacology,* 37(1), 1–12.

136. Rizzo, L., Manaia, C., Merlin, C., Schwartz, T., Dagot, C., Ploy, M.C., Michael, I., and Fatta-Kassinos, D. 2013. Urban wastewater treatment plants as hotspots for antibiotic resistant bacteria and genes spread into the environment: A review. *Science of the Total Environment,* 447, 345–360.

137. Rosario-Ortiz, F.L., Wert, E.C., and Snyder, S.A. 2010. Evaluation of UV/H2O2 treatment for the oxidation of pharmaceuticals in wastewater. *Water Research,* 44(5), 1440–1448.

138. Rosenfeldt, E.J. and Linden, K.G. 2007. The ROH, UV concept to characterize and the model UV/H2O2 process in natural waters. *Environmental Science and Technology,* 41(7), 2548–2553.

139. Rowney, N.C., Johnson, A.C., and Williams, R.J. 2009. Cytotoxic drugs in drinking water: A prediction and risk assessment exercise for the thames catchment in the United Kingdom. *Environmental Toxicology and Chemistry,* 28(12), 2733–2743.

140. Sahar, E., David, I., Gelman, Y., Chikurel, H., Aharoni, A., Messalem, R., and Brenner, A. 2011. The use of RO to remove emerging micropollutants following CAS/UF or MBR treatment of municipal wastewater. *Desalination,* 273(1), 142–147.

141. Sahar, E., Messalem, R., Cikurel, H., Aharoni, A., Brenner, A., Godehardt, M., Jekel, M., and Ernst, M. 2011. Fate of antibiotics in activated sludge followed by ultrafiltration (CAS-UF) and in a membrane bioreactor (MBR). *Water Research,* 45(16), 4827–4836.

142. Salgado, R., Noronha, J.P., Oehmen, A., Carvalho, G., and Reis, M.A.M. 2010. Analysis of 65 pharmaceuticals and personal care products in 5 wastewater treatment plants in Portugal using a simplified analytical methodology. *Water Science & Technology,* 62(12), 2862–2871.

143. Salgado, R., Pereira, V.J., Carvalho, G., Soeiro, R., Gaffney, V., Almeida, C., Cardoso, V.V. et al. 2013. Photodegradation kinetics and transformation products of ketoprofen, diclofenac and atenolol in pure water and treated wastewater. *Journal of Hazardous Materials,* 244–245, 516–527.

144. Sánchez-Polo, M., Abdel daiem, M.M., Ocampo-Pérez, R., Rivera-Utrilla, J., and Mota, A.J. 2013. Comparative study of the photodegradation of bisphenol A by HO, SO_4^- and CO_3^-/HCO_3 radicals in aqueous phase. *Science of the Total Environment,* 463–464, 423–431.

145. Scheytt, T., Mersmann, P., Lindstädt, R., and Heberer, T. 2005. 1-Octanol/water partition coefficients of 5 pharmaceuticals from human medical care: Carbamazepine, clofibric acid, diclofenac, ibuprofen, and propyphenazone. *Water, Air, and Soil Pollution,* 165(1–4), 3–11.

146. Schlumpf, M., Cotton, B., Conscience, M., Haller, V., Steinmann, B., and Lichtensteiger, W. 2001. *In vitro* and *in vivo* estrogenicity of UV screens. *Environmental Health Perspectives,* 109(3), 239–244.

147. Schlumpf, M., Durrer, S., Faass, O., Ehnes, C., Fuetsch, M., Gaille, C., Henseler, M. et al. 2008. Developmental toxicity of UV filters and environmental exposure: A review. *International Journal of Andrology,* 31(2), 144–151.

148. Sharpless, C.M. and Linden, K.G. 2003. Experimental and model comparisons of low- and medium-pressure Hg lamps for the direct and H_2O_2 assisted UV photodegradation of *N*-nitrosodimethylamine in simulated drinking water. *Environmental Science and Technology,* 37(9), 1933–1940.

149. Shu, Z., Bolton, J.R., Belosevic, M., and Gamal El Din, M. 2013. Photodegradation of emerging micropollutants using the medium-pressure UV/H_2O_2 advanced oxidation process. *Water Research,* 47(8), 2881–2889.

150. Sipma, J., Osuna, B., Collado, N., Monclús, H., Ferrero, G., Comas, J., and Rodriguez-Roda, I. 2010. Comparison of removal of pharmaceuticals in MBR and activated sludge systems. *Desalination,* 250(2), 653–659.

151. Snyder, S.A., Adham, S., Redding, A.M., Cannon, F.S., DeCarolis, J., Oppenheimer, J., Wert, E.C., and Yoon, Y. 2007. Role of membranes and activated carbon in the removal of endocrine disruptors and pharmaceuticals. *Desalination,* 202(1–3), 156–181.

152. Solomon, G.M. and Schettler, T. 2000. Environment and health: 6. Endocrine disruption and potential human health implications. *Canadian Medical Association Journal,* 163(11), 1471–1476.

153. Soufan, M., Deborde, M., Delmont, A., and Legube, B. 2013. Aqueous chlorination of carbamazepine: Kinetic study and transformation product identification. *Water Research,* 47(14), 5076–5087.

154. Soufan, M., Deborde, M., and Legube, B. 2012. Aqueous chlorination of diclofenac: Kinetic study and transformation products identification. *Water Research,* 46(10), 3377–3386.

155. SRC. 2013. FatePointers Search Module, http://esc.syrres.com/fatepointer/search.asp.

156. Stevens, D.P., McLaughlin, M.J., and Smart, M.K. 2003. Effects of long-term irrigation with reclaimed water on soils of the Northern Adelaide Plains, South Australia. *Soil Research,* 41(5), 933–948.

157. Strenn, B., Clara, M., Gans, O., and Kreuzinger, N. 2004. Carbamazepine, diclofenac, ibuprofen and bezafibrate-investigations on the behaviour of selected pharmaceuticals during wastewater treatment. *Water Science & Technology*, 50(5), 269–276.

158. Stuer-Lauridsen, F., Birkved, M., Hansen, L.P., Holten Lützhøft, H.C., and Halling-Sørensen, B. 2000. Environmental risk assessment of human pharmaceuticals in Denmark after normal therapeutic use. *Chemosphere*, 40(7), 783–793.

159. Su, C.-C., Chang, A.-T., Bellotindos, L.M., and Lu, M.-C. 2012. Degradation of acetaminophen by Fenton and electro-Fenton processes in aerator reactor. *Separation and Purification Technology*, 99, 8–13.

160. Suárez, S., Carballa, M., Omil, F., and Lema, J. 2008. How are pharmaceutical and personal care products (PPCPs) removed from urban wastewaters? *Reviews in Environmental Science and Bio/Technology*, 7(2), 125–138.

161. Suarez, S., Lema, J.M., and Omil, F. 2010. Removal of pharmaceutical and personal care products (PPCPs) under nitrifying and denitrifying conditions. *Water Research*, 44(10), 3214–3224.

162. Sudhakaran, S. and Amy, G.L. 2013. QSAR models for oxidation of organic micropollutants in water based on ozone and hydroxyl radical rate constants and their chemical classification. *Water Research*, 47(3), 1111–1122.

163. Sui, Q., Huang, J., Deng, S., Chen, W., and Yu, G. 2011. Seasonal variation in the occurrence and removal of pharmaceuticals and personal care products in different biological wastewater treatment processes. *Environmental Science and Technology*, 45(8), 3341–3348.

164. Sui, Q., Huang, J., Deng, S., Yu, G., and Fan, Q. 2010. Occurrence and removal of pharmaceuticals, caffeine and DEET in wastewater treatment plants of Beijing, China. *Water Research*, 44(2), 417–426.

165. Szczepanowski, R., Linke, B., Krahn, I., Gartemann, K.H., Gotzkow, T., Eichler, W., Puhler, A., and Schluter, A. 2009. Detection of 140 clinically relevant antibiotic resistance genes in the plasmid metagenome of wastewater treatment plant bacteria showing reduced susceptibility to selected antibiotics. *Microbiology*, 155(7), 2306–2319.

166. Tamtam, F., van Oort, F., Le Bot, B., Dinh, T., Mompelat, S., Chevreuil, M., Lamy, I., and Thiry, M. 2011. Assessing the fate of antibiotic contaminants in metal contaminated soils four years after cessation of long-term waste water irrigation. *Science of the Total Environment*, 409(3), 540–547.

167. Tang, J.Y.M., McCarty, S., Glenn, E., Neale, P.A., Warne, M.S.J., and Escher, B.I. 2013. Mixture effects of organic micropollutants present in water: Towards the development of effect-based water quality trigger values for baseline toxicity. *Water Research*, 47(10), 3300–3314.

168. Tchobanoglous, G., Burton, F.L., Stensel, H.D., and Metcalf & Eddy, Inc. 2003. *Wastewater Engineering: Treatment and Reuse*. McGraw-Hill Education, New York.

169. Ternes, T., Bonerz, M., and Schmidt, T. 2001. Determination of neutral pharmaceuticals in wastewater and rivers by liquid chromatography–electrospray tandem mass spectrometry. *Journal of Chromatography A*, 938(1–2), 175–185.

170. Ternes, T., Joss, A., Kreuzinger, N., Miksch, K., Lema, J.M., Gunten, U.v., McArdell, C.S., and Siegrist, H. 2005. Removal of pharmaceuticals and personal care products: Results of the Poseidon Project. *Proceedings of the Water Environment Federation*, 2005(16), 227–243.

171. Ternes, T.A. 1998. Occurrence of drugs in German sewage treatment plants and rivers. *Water Research*, 32(11), 3245–3260.

172. Ternes, T.A., Herrmann, N., Bonerz, M., Knacker, T., Siegrist, H., and Joss, A. 2004. A rapid method to measure the solid–water distribution coefficient (K_d) for pharmaceuticals and musk fragrances in sewage sludge. *Water Research*, 38(19), 4075–4084.

173. Ternes, T.A., Joss, A., and Siegrist, H. 2004. Peer reviewed: Scrutinizing pharmaceuticals and personal care products in wastewater treatment. *Environmental Science and Technology*, 38(20), 392A–399A.

174. Thompson, A., Griffin, P., Stuetz, R., and Cartmell, E. 2005. The fate and removal of triclosan during wastewater treatment. *Water Environment Research*, 77(1), 63–67.

175. Tomiyasu, H., Fukutomi, H., and Gordon, G. 1985. Kinetics and mechanism of ozone decomposition in basic aqueous solution. *Inorganic Chemistry*, 24(19), 2962–2966.

176. Toppari, J., Larsen, J.C., Christiansen, P., Giwercman, A., Grandjean, P., Guillette Jr, L.J., Jégou, B., Jensen, T.K., Jouannet, P., and Keiding, N. 1996. Male reproductive health and environmental xenoestrogens. *Environmental Health Perspectives*, 104(Suppl 4), 741–803.

177. Touraud, E., Roig, B., Sumpter, J.P., and Coetsier, C. 2011. Drug residues and endocrine disruptors in drinking water: Risk for humans? *International Journal of Hygiene and Environmental Health*, 214(6), 437–441.

178. Trojan. 2012. Ultraviolet disinfection for wastewater. The Operator Training Committee of Ohio, Inc. 49th Annual Wastewater Workshop and Exhibition, Columbus, Ohio, March 8, 2012.

179. Tuerk, J., Kiffmeyer, T.K., Hadtstein, C., Heinemann, A., Hahn, M., Stuetzer, H., Kuss, H.-M., and Eickmann, U. 2011. Development and validation of an LC–MS/MS procedure for environmental monitoring of eight cytostatic drugs in pharmacies. *International Journal of Environmental Analytical Chemistry*, 91(12), 1178–1190.

180. Turci, R., Sottani, C., Spagnoli, G., and Minoia, C. 2003. Biological and environmental monitoring of hospital personnel exposed to antineoplastic agents: A review of analytical methods. *Journal of Chromatography B*, 789(2), 169–209.

181. Tyler, C.R., Jobling, S., and Sumpter, J.P. 1998. Endocrine disruption in wildlife: A critical review of the evidence. *Critical Reviews in Toxicology*, 28(4), 319–361.

182. Urase, T., Kagawa, C., and Kikuta, T. 2005. Factors affecting removal of pharmaceutical substances and estrogens in membrane separation bioreactors. *Desalination,* 178 (1–3), 107–113.

183. Urase, T. and Kikuta, T. 2005. Separate estimation of adsorption and degradation of pharmaceutical substances and estrogens in the activated sludge process. *Water Research,* 39(7), 1289–1300.

184. Verlicchi, P., Al Aukidy, M., and Zambello, E. 2012. Occurrence of pharmaceutical compounds in urban wastewater: Removal, mass load and environmental risk after a secondary treatment—A review. *Science of the Total Environment,* 429, 123–155.

185. Verlicchi, P., Galletti, A., Petrovic, M., Barceló, D., Al Aukidy, M., and Zambello, E. 2013. Removal of selected pharmaceuticals from domestic wastewater in an activated sludge system followed by a horizontal subsurface flow bed—Analysis of their respective contributions. *Science of the Total Environment,* 454–455, 411–425.

186. Vieno, N., Tuhkanen, T., and Kronberg, L. 2007. Elimination of pharmaceuticals in sewage treatment plants in Finland. *Water Research,* 41(5), 1001–1012.

187. Vieno, N.M., Tuhkanen, T., and Kronberg, L. 2005. Seasonal variation in the occurrence of pharmaceuticals in effluents from a sewage treatment plant and in the recipient water. *Environmental Science and Technology,* 39(21), 8220–8226.

188. Von Sonntag, C. and von Gunten, U. 2012. *Chemistry of Ozone in Water and Wastewater Treatment: From Basic Principles to Applications.* IWA Publishing, London.

189. Vos, J.G., Dybing, E., Greim, H.A., Ladefoged, O., Lambre, C., Tarazona, J.V., Brandt, I., and Vethaak, A.D. 2000. Health effects of endocrine-disrupting chemicals on wildlife, with special reference to the European situation. *Critical Reviews in Toxicology,* 30(1), 71–133.

190. Waggott, A. 1981. *Trace Organic Substances in the River Lee* [*Great Britain*]. Ann Arbor Science, Ann Arbor, MI.

191. Wang, P., He, Y.-L., and Huang, C.-H. 2011. Reactions of tetracycline antibiotics with chlorine dioxide and free chlorine. *Water Research,* 45(4), 1838–1846.

192. Wang, Y.-N., Chen, J., Li, X., Wang, B., Cai, X., and Huang, L. 2009. Predicting rate constants of hydroxyl radical reactions with organic pollutants: Algorithm, validation, applicability domain, and mechanistic interpretation. *Atmospheric Environment,* 43(5), 1131–1135.

193. Wick, A., Marincas, O., Moldovan, Z., and Ternes, T.A. 2011. Sorption of biocides, triazine and phenylurea herbicides, and UV-filters onto secondary sludge. *Water Research,* 45(12), 3638–3652.

194. Windhoek. 2015. *The City of Windhoek: Water Reclamation.* http://www.windhoekcc.org.na/info_facts.php.

195. Wise, R. 2002. Antimicrobial resistance: Priorities for action. *Journal of Antimicrobial Chemotherapy,* 49(4), 585–586.

196. Wu, C., Spongberg, A.L., Witter, J.D., Fang, M., and Czajkowski, K.P. 2010. Uptake of pharmaceutical and personal care products by soybean plants from soils applied with biosolids and irrigated with contaminated water. *Environmental Science and Technology,* 44(16), 6157–6161.

197. Xie, M., Nghiem, L.D., Price, W.E., and Elimelech, M. 2013. Impact of humic acid fouling on membrane performance and transport of pharmaceutically active compounds in forward osmosis. *Water Research,* 47(13), 4567–4575.

198. Xu, H., Yang, M., Qiu, W., Pan, C., and Wu, M. 2013. The impact of endocrine-disrupting chemicals on oxidative stress and innate immune response in zebrafish embryos. *Environmental Toxicology and Chemistry,* 32(8), 1793–1799.

199. Xu, J., Wu, L., and Chang, A.C. 2009. Degradation and adsorption of selected pharmaceuticals and personal care products (PPCPs) in agricultural soils. *Chemosphere,* 77(10), 1299–1305.

200. Xu, J., Wu, L., Chang, A.C., and Zhang, Y. 2010. Impact of long-term reclaimed wastewater irrigation on agricultural soils: A preliminary assessment. *Journal of Hazardous Materials,* 183(1–3), 780–786.

201. Xu, P., Drewes, J.E., Bellona, C., Amy, G., Kim, T.U., Adam, M., and Heberer, T. 2005. Rejection of emerging organic micropollutants in nanofiltration-reverse osmosis membrane applications. *Water Environment Research,* 77(1), 40–48.

202. Yang, X., Flowers, R.C., Weinberg, H.S., and Singer, P.C. 2011. Occurrence and removal of pharmaceuticals and personal care products (PPCPs) in an advanced wastewater reclamation plant. *Water Research,* 45(16), 5218–5228.

203. Yoon, Y., Ryu, J., Oh, J., Choi, B.-G., and Snyder, S.A. 2010. Occurrence of endocrine disrupting compounds, pharmaceuticals, and personal care products in the Han River (Seoul, South Korea). *Science of the Total Environment,* 408(3), 636–643.

204. Yoon, Y., Westerhoff, P., Snyder, S.A., and Wert, E.C. 2006. Nanofiltration and ultrafiltration of endocrine disrupting compounds, pharmaceuticals and personal care products. *Journal of Membrane Science,* 270(1–2), 88–100.

205. Yoon, Y., Westerhoff, P., Snyder, S.A., Wert, E.C., and Yoon, J. 2007. Removal of endocrine disrupting compounds and pharmaceuticals by nanofiltration and ultrafiltration membranes. *Desalination,* 202(1–3), 16–23.

206. Young, R.B. and Borch, T. 2009. Sources, presence, analysis, and fate of steroid sex hormones in freshwater ecosystems—A review. In: Nairne, G.H. (ed.), *Aquatic Ecosystem Research Trends.* Nova Science Publishers, Hauppage, New York.

207. Zhang, J., Chang, V.W.C., Giannis, A., and Wang, J.Y. 2013. Removal of cytostatic drugs from aquatic environment: A review. *Science of the Total Environment,* 445–446, 281–298.

208. Zhang, L.-S., Wong, K.-H., Yip, H.-Y., Hu, C., Yu, J.C., Chan, C.-Y., and Wong, P.-K. 2010. Effective photocatalytic disinfection of *E. coli* K-12 using AgBr–Ag–Bi2WO6 nanojunction system irradiated by visible light: The role of diffusing hydroxyl radicals. *Environmental Science and Technology,* 44(4), 1392–1398.

209. Zwiener, C. and Frimmel, F.H. 2003. Short-term tests with a pilot sewage plant and biofilm reactors for the biological degradation of the pharmaceutical compounds clofibric acid, ibuprofen, and diclofenac. *Science of the Total Environment,* 309(1–3), 201–211.

Escherichia coli in Urban Water

Süreyya Meriç
Namık Kemal University

Can Burak Özkal
Namık Kemal University

PREFACE

Antibiotics are a factor in developing antibiotic resistance in the environment. Outbreaks due to pathogens and resistant bacteria are an emerging issue in this decade. Resistance of *Escherichia coli* to two groups of antibiotics has been revised recently by the World Health Organization (WHO). These data showed that bacteria have already developed resistance to third and fourth generation antibiotics. The WHO report on surveillance and antibiotics consumption evaluation showed that antibiotic consumption varies in EU countries. Outbreaks have increased in parallel to these data depending on country, season, sex, and age group.

This chapter revisits the routes of spreading and surveillance of *E. coli*. There is a particular focus on water sources including hospitals, urban wastewater treatment plants (UWTPs), diffuse sources, and water reuse. Extensively revised data are given on the control techniques by biological and advanced processes. The emerging issue of gene transfer control in parallel to the control of bacteria is expressed. A detailed literature survey of emerging technologies of photocatalysis and nanoparticles is given.

7.1 Introduction

Escherichia coli O157 is an uncommon but serious cause of gastroenteritis. This bacterium is noteworthy because a few, but significant, number of infected people develop hemolytic uremic syndrome (HUS), which is the most frequent cause of acute renal failure in children in the Americas and Europe. Many infections of *E. coli* O157 could be prevented by the more effective application of evidence-based methods, which is especially important because once an infection has been established, no therapeutic interventions are available to lessen the risk of the development of HUS [41].

Fecal to oral transmission occurs by many routes; therefore, many barriers are needed to prevent infection. Some of these barriers, such as milk pasteurization and water chlorination, protect the bulk of the population effectively. However, hand washing and the working practices that prevent cross-contamination rely heavily on human behavior. The organism can escape detection by the traditional visual inspection systems still in use in European and North American slaughterhouses. Therefore, risk reduction and mitigation strategies—available after transmission to human beings, such as outbreak control and rapid diagnosis with timely supportive treatment—are all that can be expected [41].

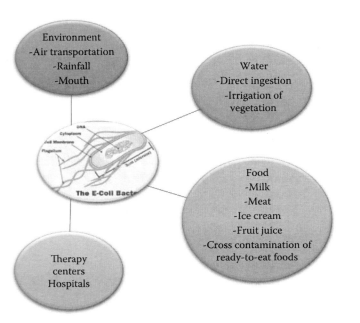

FIGURE 7.1   Contact routes of *E. coli* with humans.

The common contact routes of human beings with *E. coli* are demonstrated in Figure 7.1. This work aims to revise the recent literature on sources, outbreaks, and resistance development and control technologies of *E. coli* in urban water.

7.2 Sources of *E. coli* in Water Origin

The pathogens, as do any other pollutants, reach urban wastewater via different sources as indicated in Figure 7.2. These main sources are explained in the following sections.

7.2.1 Hospitals

Urinary tract infections (UTIs) are among the most common infectious diseases diagnosed in outpatients. *E. coli* is the main causative pathogen of UTIs in both out- and inpatients [1]. A variety of antibiotics is used to treat patients for these infectious diseases. However, hospital wastewaters become important sources of a large variety of pharmaceuticals including antibacterial agents, evidenced by the fact that they occur in these wastewaters at higher concentrations than in wastewater from household effluents, which is due to high usage and low dilution [14].

7.2.2 Urban Wastewater Treatment Plants

One of the most commonly suspected sources of fecal contamination of water resources is onsite wastewater treatment systems (OWTS), particularly septic tank–soil adsorption systems. Increased urbanization in the fringes of metropolitan areas has led to the reliance on OWTS for the treatment and dispersal of sewage effluent. Blaak et al. [5] and Carrolla et al.

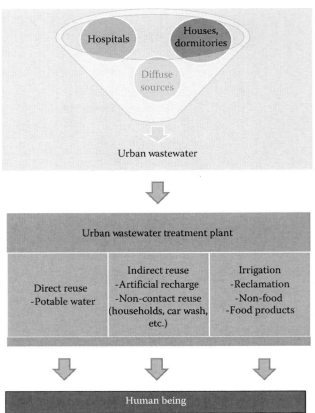

FIGURE 7.2   Pathways of *E. coli* contamination from contaminated water origin.

[7] investigated prevalence and characteristics of extended spectrum β-lactamases (ESBL)-producing *E. coli* in four Dutch recreational waters and the possible role of nearby wastewater treatment plants (WWTP) as contamination sources. Isolates from recreational waters were compared with isolates from WWTP effluents, from surface water upstream of the WWTPs, at WWTP discharge points, and in connecting water bodies not influenced by the studied WWTPs. ESBL-producing *E. coli* were detected in all four recreational waters, with an average concentration of 1.3 colony-forming units/100 mL, and in 62% of all samples.

7.2.3 Diffuse Pollution

Floods can bring pathogens into homes and cause lingering damp and microbial growth in buildings, with the level of growth and persistence dependent on the volume and chemical and biological content of the flood water, the properties of the contaminating microbes, and the surrounding environmental conditions, including the restoration time and methods, the heat and moisture transport properties of the mesh design, and the ability of the construction material to sustain the microbial growth [47]. Kramer et al. [28], reported that flood-borne *E. coli* bacteria may display a total survival time of 1.5 h–16 months on inanimate surfaces.

The increased emphasis on rapid growth and disease prevention in animal production has led to the increased use of subtherapeutic levels of antibiotics in veterinary practices. Suggestion that this type of antibiotic administration may promote the emergence of resistant bacteria has elicited a growing concern among consumers as well as medical experts [33]. Both surface soil and subsurface soil can be exposed, even with strongly sorbed veterinary antibiotics, which surely end up in the receiving water bodies after severe rainfall [27].

7.2.4 Reuse of Treated Wastewater and Greywater

Epidemiological evidence from recent outbreaks of *E. coli* O157:H7 and *Salmonella* spp. point to irrigation water as the vector in the production environment of spinach [9], lettuce [45], and pepper [8].

The reuse of domestic greywater has become common in Australia, especially during periods of extreme drought. Greywater is typically used in a raw, untreated form, primarily for landscape irrigation, but more than a quarter of greywater users irrigate vegetable gardens, despite government advice against this practice. Greywater can be contaminated with enteric pathogens and may therefore pose a health risk if irrigated produce is consumed raw. A quantitative microbial risk assessment (QMRA) model was constructed to estimate the norovirus disease burden associated with consumption of greywater-irrigated lettuce. It was shown that based on *E. coli* data evaluation, greywater use across the Melbourne population had a median annual disease burden of $<10^{-6}$ disability adjusted life-years (DALY; this is a measure of overall disease burden and is expressed as the number of years lost due to illness, disability, or premature death per person), while among those using greywater to irrigate home-grown vegetables, median annual disease burdens ranged from 10^{-10} to 10^{-4} depending on the source of greywater and vegetable washing behaviors [2].

7.3 Outbreaks and Incidences Related to Pathogenic Organisms in Urban Water

7.3.1 Occurrence and Outbreaks

Human infection with Shiga toxin/verocytotoxin-producing *E. coli* (STEC/VTEC) is characterized by an acute onset of diarrhea, which may be bloody, and is often accompanied by mild fever and/or vomiting. The infection may lead to potentially fatal HUS, affecting renal function and requiring hospital care. Infection is mainly acquired by consuming contaminated food, such as undercooked or contaminated beef or vegetables, or water, but person-to-person and direct

transmissions from animals to humans may also occur. The main reservoirs for STEC/VTEC bacteria are ruminants such as cattle, goats, and sheep. In 2011, 9534 confirmed cases of STEC/VTEC were reported by 27 EU/EEA countries [15]. This represents 2.5 times the number of confirmed cases reported in 2010 ($n = 3715$). The overall notification rate was also higher in 2011: 2.54 cases per 100,000 compared with 1.00 cases per 100,000 in 2010. This marked an increase in outbreak of VTEC O104:H4, which occurred in Germany in the early summer of 2011. Germany accounted for 58.6% ($n = 5558$) of all confirmed cases reported and also had the highest notification rate in 2011 (6.80 per 100,000 population). Overall, the number of confirmed cases reported increase in 18 Member States compared to 2010. Between 2009 and 2011, the Netherlands showed a steady rise in the number of reported confirmed cases, resulting in a 169% increase since 2009. Since 2007, the trend of confirmed STEC/VTEC cases had been stable in the EU until a sharp increase in 2011 due to the VTEC O104:H4 outbreak (Figure 7.3a).

Among the 27 EU/EEA countries with known data on sex, 24.4% more female than male cases were reported, with the female-to-male ratio 1.32:1. However, the highest rate of confirmed cases was reported in 0–4-year-old males (8.92 cases per 100,000 population) (Figure 7.3b). The number of reported cases of STEC/VTEC showed a sharp peak in May–June 2011 due to the VTEC O104:H4 outbreak in Germany (Figure 7.3c). There is a clear seasonality, indicating that STEC/VTEC infections are mainly acquired and reported in the summer months between June and September. Data on HUS were reported by 15 EU/EEA countries. A total of 1006 (11%) confirmed VTEC cases ($n = 9672$) developed HUS in 2011. Only 318 of these cases were reported to be due to STEC/VTEC O104, but of the 411 HUS cases with unknown serogroups reported from Germany, the majority are expected to have been caused by the outbreak. Twenty-eight percent of HUS cases ($n = 162$) were reported in 0–4-year-old children with O157 and O26 as the dominant serogroups, followed by 25–44-year-old adults with O104 as the dominant serogroup (91%). VTEC O104 was the predominant serogroup in HUS cases for all age groups above 15 years in 2011 (Figure 7.3d).

On May 25, 2012, the Netherlands issued an Early Warning and Response System (EWRS) message reporting a case of VTEC O104:H4 for one of the staff working at a laboratory. The person carried out cleaning tasks in the laboratory. The clinical signs were gastroenteritis with thrombocytopenia and renal function failure. The laboratory analysis identified the same genetic profile as VTEC O104:H4, which caused the STEC outbreak in Germany in 2011. The patient reported not having consumed raw vegetables (or fenugreek/sprouted seeds) but had eaten fresh mint leaves. Polymerase chain reaction (PCR) results on the leaves obtained from three shops where the patient bought this food item were positive for VTEC O104. Further laboratory diagnostic analysis was able to differentiate the VTEC O104 isolated strain from the German VTEC O104 strain [12].

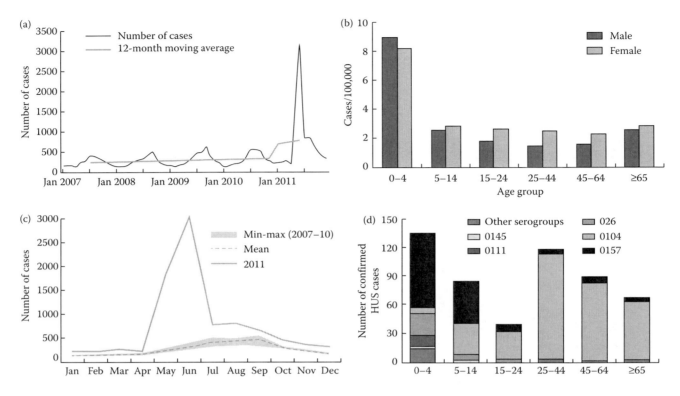

FIGURE 7.3 (a) Trend and number of confirmed cases of STEC/VTEC reported in the EU/EEA, 2007–2011. (b) Rates of confirmed STEC/VTEC cases reported in the EU/EEA, by age and sex, 2011. (c) Seasonal distribution: Number of confirmed cases of STEC/VTEC by month, EU/EEA, 2007–2011. (d) Number of confirmed STEC/VTEC HUS cases, by age and most common O-serogroups, 2011. (Adapted from ECDC. 2013. *Surveillance Report. Reporting on 2011 Surveillance Data and 2012 Epidemic Intelligence Data,* 2013, www.ecdc.europa.eu.)

7.3.2 Antibiotic Resistance

Antibacterial resistance has been an increasingly serious problem since the advent of antibiotics. With the introduction of newer β-lactam agents and other groups of antibiotics, natural selection was operating so that bacteria often managed to combat antibiotics. *E. coli*, a dominant facultative aerobe in the gut and an important human pathogen, may be a major reservoir of antibiotic resistance determinants on conjugative plasmids. Multiresistant *E. coli* have been isolated from a variety of water sources [25] and WWTP effluents [24,44].

The latter phenotypes are strongly influenced by multiple factors such as extensive and improper use of antimicrobial agents. Producing ESBLs is one of the major mechanisms of antibacterial resistance in general. These enzymes can hydrolyze oxyimino-cephalosporins (ceftriaxone, cefotaxime, ceftazidime, and cefepime) and monobactams (aztreonam) but not cephamycins or carbapenems. Most ESBLs are derived from classic Temoneira (TEM) and sulfhydryl reagent variable (SHV) groups of β-lactamases, thus belonging to Ambler Class A. Clonal and horizontal transfer of resistance genes aids the spread of antibiotic resistance. Thus, ESBL genes can be transmitted between different gram-negative bacteria. ESBL genes (blaESBL) are often associated with conjugative plasmids and

mobile genetic elements (e.g., integrons and transposons), which in turn can facilitate the capture and expression of such genes. There are more than 700 types of ESBLs, and they have been detected all over the world. A real medical challenge in this context is that ESBL-producing bacteria can show cross-resistance to other classes of antibiotics such as aminoglycosides, fluoroquinolones, cotrimoxazole, and tetracycline. It is reasoned that plasmids harboring ESBL genes also contain other antibiotic resistance genes, four of which may have varying clinical consequences depending on the local context [1].

Data from the European Antimicrobial Resistance Surveillance Network (EARS-Net) show large variations in the occurrence of antimicrobial resistance (AMR) in Europe depending on microorganism, antimicrobial agent, and geographical region. Over the last four years, there has been a significantly increasing trend of multidrug resistance (combined resistance to multiple antibiotics) in both *E. coli* and *Klebsiella pneumoniae* in more than one-third of the reporting EU/EEA countries. According to EARS-Net data on invasive bacterial isolates from more than 900 public health laboratories serving over 1400 hospitals in Europe, between 2008 and 2011, the percentages of *E. coli* isolates resistant to third-generation cephalosporins significantly increased in 18 of 28 reporting countries (Figure 7.4). No country showed a decreasing trend during this

FIGURE 7.4 *Escherichia coli*: Percentage (%) of invasive (blood and cerebrospinal fluid) isolates resistant to third-generation cephalosporins, EU/EEA, 2011. (Adapted from ECDC. 2013. *Surveillance Report. Reporting on 2011 Surveillance Data and 2012 Epidemic Intelligence Data,* 2013, www.ecdc.europa.eu.)

period. A majority of the isolates that were resistant to third-generation cephalosporins were ascertained as being ESBL-positive, ranging between 71% and 100%, depending on the reporting country.

WHO Surveillance report [52] showed that a global spread of antibiotic resistance has increased in recent years. The resistance of *E. coli* to two types of antibiotics is reported in Table 7.1.

7.4 Control Techniques

According to the source of occurrence of *E. coli,* mitigation measures would be indicative as shown in Table 7.2 as adopted from Wellington et al. [51]. Apart from other resources, the control of resistant *E. coli* should be handled accurately because transfer of genes play a very important role in spreading resistance [44]. The following sections approach the occurrence and control of *E. coli* in treatment units in a manner of securing environmental and human health.

7.4.1 Biological Processes

In the effluent of urban wastewater treatment plants (UWTP), 10^2 CFU/mL of *E. coli* may remain. According to the study by Duong et al. [14], fluoroquinolones (FQ) removal from the

wastewater stream was between 80% and 85%, probably due to sorption on sewage sludge. Simultaneously, the numbers of *E. coli* were measured and their resistance against ciprofloxacin and norfloxacin was evaluated by determining the minimum inhibitory concentration. Biological treatment lead to a 100-fold reduction in the number of *E. coli* but still more than 1000 *E. coli* colonies per 100 mL of wastewater effluent reached the receiving water. The highest resistance was found in *E. coli* strains in raw wastewater and the lowest in isolates of treated wastewater effluent. Thus, wastewater treatment is an efficient barrier to decrease the residual FQ levels and the number of resistant bacteria entering ambient waters [14].

It was demonstrated in previous studies that in an activated sludge WWTP, treated wastewater presented significantly higher percentages of *E. coli* and *Enterococcus* spp. resistant to the fluoroquinolone ciprofloxacin than the raw influent [11,12]. Further studies are summarized in Figure 7.5, showing the resistance of *E. coli* variation from one treatment plant to another one treating different antibiotics.

7.4.2 Advanced Treatment

Advanced treatments aim at improving the quality of the secondary effluent before disposal or reuse. Sand filtration,

TABLE 7.1 *E. coli* Surveillance in the World

Resistance to Third-Generation Cephalosporins[a] (Summary of Reported or Published Proportions of Resistance by WHO Region)			Resistance to Fluoroquinolones[a]		
Data Sources Based on at Least 30 Tested Isolates[b]	Overall Reported Range of Resistant Proportion (%)	Reported Range of Resistant Proportion (%) in Invasive Isolates[c] (Number of Reports)	Data Sources Based on at Least 30 Tested Isolates[b]	Overall Reported Range of Resistant Proportion (%)	Reported Range of Resistant Proportion (%) in Invasive Isolates[c] (Number of Reports)
African Region					
National data ($n = 13$ countries)	2–70	28–36 ($n = 4$)	National data ($n = 14$ countries)	14–71	34–53 ($n = 2$)
Publications ($n = 17$) from 7 additional countries	0–87	0–17 ($n = 5$)	Publications ($n = 23$) from 8 additional countries	0–98	0–10 ($n = 4$)
Region of the Americas					
National data or report to ReLAVRA ($n = 14$ countries)	0–48		National data or report to ReLAVRA ($n = 16$ countries)	8–58	
Publications ($n = 10$) from 5 additional countries	0–68		Publications ($n = 5$) from 4 additional countries	2–60	
Eastern Mediterranean Region					
National data ($n = 4$ countries)	22–63	41 ($n = 1$)	National data ($n = 4$ countries)	21–62	54 ($n = 1$)
Surveillance network in one country[d]	39 (caz)–50 (cro)		Surveillance networked, one additional country	35	
Publications ($n = 44$) from 11 additional countries	2–94	11–33 ($n = 6$)	Publications ($n = 32$) from 10 additional countries	0–91	15–53 ($n = 5$)
European Region					
National data or report to EARS-Net ($n = 35$ countries)	3–82	3–43 ($n = 32$)	National data or report to EARS-Net ($n = 35$ countries)	8–48	8–47 ($n = 33$)
Publications ($n = 5$) from 2 additional countries	0–8	0–8 ($n = 2$)	Publications ($n = 3$) from 2 additional countries	0–18	0–18 ($n = 2$)
South-East Asia Region					
National data ($n = 5$ countries)	16–68		National data ($n = 5$ countries)	32–64	
Publications ($n = 26$) from 2 additional countries	19–95	20–61 ($n = 2$)	Publications ($n = 19$) from 2 additional countries	4–89	
Western Pacific Region					
National data ($n = 13$ countries)	0–77		National data ($n = 16$ countries)	3–96	7 ($n = 1$)
Institute surveillance (data from 3 hospitals in one country)	4–14		Institute surveillance (data from 3 hospitals in one country)	0–14	
Publications ($n = 4$) from 2 additional countries	8–71		Publications ($n = 5$) from 3 additional counties	0.2–65	31 ($n = 1$)

EARS-Net = European Antimicrobial Resistance Surveillance Network; ReLAVRA = Latin American Antimicrobial Resistance Surveillance Network. (See Section 2. Tables A2.1–A2.6 on 2014 Antimicrobial Resistance Global Report on Surveillance.)

[a] Based on antibacterial susceptibility testing with caz, ceftazidime; cefotaxime or cro, ceftriaxone.

[b] Reported proportions may vary between compound used for testing and some countries report data for several compounds, or data from more than one surveillance system.

[c] Invasive isolates are deep infections, mostly bloodstream infections and meningitis.

[d] U.S. Naval Medical Research Unit No 3, Global Disease Detection Program, Egypt.

EARS-Net = European Antimicrobial Resistance Surveillance Network; ReLAVRA = Latin American Antimicrobial Resistance Surveillance Network. (For details, see Annex 2, Tables A2.7–A2.12).

[a] Based on antibacterial susceptibility testing with ciprofloxacin, gatifloxacin, levofloxacin, moxifloxacin, norfloxacin, ofloxacin, peflox-acin, refloxacin, or sparfloxacin. Where the fluoroquinolone was not specified, ciprofloxacin was used.

[b] Reported proportions may vary between compound used for testing and some countries report data for several compounds, or data from more than one surveillance system.

[c] Invasive isolates are deep infections, mostly bloodstream infections and meningitis.

[d] U.S. Naval Medical Research Unit No 3, Global Disease Detection Program, Egypt.

Source: Adapted from World Health Organization (WHO). 2014. *Antimicrobial Resistance Global Report on Surveillance.*

TABLE 7.2   Mitigation Strategies to Control Pathogens in the Environment

Control Point	Mitigation Strategies
Wastewater treatment	• Reduced microbial pollution, the present best practice in wastewater treatment (UV treatment). Only undertaken in so-called sensitive areas. • Dedicated hospital wastewater treatment. • Adoption of new technologies to remove pharmaceuticals from wastewater such as ozonation and membrane technology. • Further investment in wastewater system to reduce combined storm overflow discharges of raw sewage.
Farming	• Observe best practice in reducing livestock access to water courses. Implementation of buffer zones to reduce run-off. • Treatment of animal wastes to reduce microbial pollution. • Reduction in antibiotic use in agriculture—prophylactic use supporting unsustainable farming practice. Use alternative therapies where possible (probiotics).
Medical	• Ensure best practice in prescription of antibiotics, use alternative therapies such as bacteriophage or probiotics where possible. • Green drug choice: Use of degradable pharmaceuticals rather than environmentally persistent compounds. • Ethical procurement: Purchase from sources not polluting the environment with pharmaceuticals.

Source: Adapted from Wellington, E.M.H., Boxall, A.B.A., Cross, P., Feil, E.J., Gaze, W.H., Hawkey, P.M., Johnson-Rollings et al. 2013. *The Lancet Infectious Diseases*, 13(2), 155–165.

adsorption, membranes, and advanced oxidation processes (AOPs) are among the most applied and studied advanced treatment processes/technologies. Despite many available studies regarding the effect of advanced treatment technologies on bacteria inactivation [3,4,36,46,51], these processes have not been well documented for the control of gene transfer for antibiotic resistance control [16,17]. This section focuses mainly on innovative processes such as AOPs and nanoparticles/nanocomposites adsorption, which have been found effective for inactivation of both *E. coli* bacteria and genes provoking *E. coli* growth. Main outlines of those processes are given in Figure 7.6.

7.4.2.1 Inactivation of *E. coli* by AOPs

Cengiz et al. [10] compared Fenton and ozone oxidation processes in the removal of tetM gene and its host *E. coli* HB101 from synthetically contaminated cow manure. PCR-based monitoring assays showed that the band intensity of the tetM gene gradually decreased by increasing the Fenton reagent and the applied ozone dose.

Ozonation and TiO_2 heterogeneous photocatalysis were compared with conventional chlorination in terms of effects on DNA structure and integrity [38]. Öncü et al. reported that chlorine did not affect plasmid DNA structure at the studied doses. Ozone and photocatalytic treatment resulted in increased damage with increasing oxidant doses.

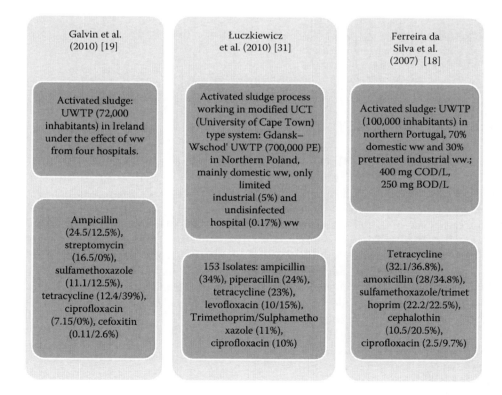

FIGURE 7.5   Examples of presence of *E. coli* resistant to antibiotics treated in biological treatment. (Adapted from Galvin, S., Boyle, F., Hickey, P., Vellinga, A., Morris, D. and Cormican, M. 2010. *Applied and Environmental Microbiology*, 76(14), 4772–4779; Łuczkiewicz, A., Jankowska, K., Fudala-Książek, S., and Olańczuk-Neyman, K. 2010. *Water Research*, 44(17), 5089–5097; Ferreira da Silva, M., Vaz-Moreira, I., Gonzalez-Pajuelo, M., Nunes, O.C., and Manaia, C.M. 2007. *FEMS Microbiology Ecology*, 60(1), 166–176.)

Fenton oxidation

- Fe (II) or Fe (III) and H_2O_2 are used. The process may be integrated with solar energy which has been effective one for inactivation.

UV

- UVC can be used alone. Its efficiency was increased by adding H_2O_2 or ozone or other catalysts.

Ozonation

- Alone or combined with UV, H_2O_2, H_2O_2
- At a moderate ozone dose of 2.5 mg/L (1 = 4 0.55 mg O_3 mg/L DOC), 6 logs inactivation of _E. coli_ are predicted, ozone has entirely reacted before 20 s.

Photocatalysis

- Slurry system or thin film reactors.
- TiO_2 and various catalysts have been investigated for increasing process efficiency.

Nanotechnology (Adapted from Pelgrift et al., 2014) [40]

- Nitric oxide-releasing nanoparticles (NO NPs).
  Hydrogel/glass composite NO NP platform by Friedman et al.
  Chitosan-containing nanoparticles (chitosan NPs).
  Metal-containing nanoparticles.
  Silver-containing nanoparticles (Ag NPs).
  Zinc oxide-containing nanoparticles (ZnO NPs).
  Copper-containing nanoparticles.
  Titanium dioxide-containing nanoparticles (TiO_2 NPs).
  Magnesium-containing nanoparticles.
  Gold-containing nanoparticles (Au NP).
  Bismuth-containing nanoparticles (Bi NPs).
  Aluminum oxide-containing nanoparticles (Al_2O_3 NPs).
  Resistance to metal-containing nanoparticles in certain bacteria.

FIGURE 7.6 Advanced oxidation processes used for inactivation of _E. coli_ and genes. (Adapted from Pelgrift, R.Y. and Friedman, A.J. 2013. _Advanced Drug Delivery Reviews_. Doi: 10.1016/j.addr.2013.07.011.)

A different approach was used by Paul et al. [39] to evaluate the effects of photolytic and TiO_2 photocatalytic treatment processes on the antibacterial activity of ciprofloxacin. In particular, quantitative microbiological assays with a reference _E. coli_ strain showed that for each mole of ciprofloxacin degraded, the antibacterial potency of irradiated solutions decreased by approximately 1 "mole" of activity relative to that of the untreated ciprofloxacin solution. The authors inferred that the ciprofloxacin photocatalytic transformation products retain negligible antibacterial activity compared to the parent compound. Moreover, according to their experimental system, the lower energy demand (20 J/cm²) to reduce antibacterial activity by one order of magnitude was achieved by UVA-TiO_2 photocatalysis. Table 7.3 compares slurry and thin film-based studies for inactivation of _E. coli_ and genes.

7.4.2.2 Nanoparticles

Among their different functions, nanoparticles can overcome drug resistance mechanisms of microbes, including decreased uptake and increased efflux of drug from the microbial cell [21,53], biofilm formation, and intracellular bacteria [6,21]. Nanoparticles have been used to target antimicrobial agents at the site of infection, so that higher doses of drug can be given at the infected site, thereby overcoming resistance with fewer adverse effects upon the patient [29]. Recent studies are indicative of solar photocatalysis potential for composites such as titanium dioxide graphene [13,20]. The main group of nanoparticles applied as antimicrobial agent has been specified by Pelgrift et al. [40], and shown schematically in Figure 7.6. The mode of actions and application fields of the most commonly used nanoparticles for bacteria control are given in Table 7.4.

7.4.3 Disinfection

Many factors must be considered in order to develop and implement treatment systems to improve the microbial quality of surface water and prevent the accidental introduction

TABLE 7.3 Removal of *E. coli* by TiO$_2$-Based Photocatalysis Process

Reference	Reactor	Strain	Time (h)	Light Emission	TiO$_2$
[30]	Suspended system		2	69 m Einstein/s · m^2	0.4 mg/L
[31]	Suspended system	HB 1001 (10^3 cells/mL)	1	67.9 µ Einstein/s · m^2)	1 mg/mL
[32]	Full-scale continuous flow solar parabolic and V-grove reactors	K-12 (1 · 10^6 CFU/mL)	20.5 min, 26.1 min, and 25.9 min for the CP, P, and V reactors (at 30 W/m^2)	22 and 44 W/m^2 and the average UV intensity during the trials was calculated to be 38.8 W/m^2	Concentration of TiO$_2$ coating per unit illuminated volume to be 94 mg/L
[33]	Suspended system	K-12 wild type, 4 × 10^7 CFU/mL	30–60 min	120 mJ/pulse, 1500 mW/cm^2	1–2.5 g/L
[34]	Suspended system	K-12, 10^7–10^8 and 10^5–10^6 CFU/mL	3 h	4–7 UV-A, 13.5–8.5 UV-B, 9–13 UV-C mW/cm^2	0.25–2.5 g/L
[35]	Suspended system	K-12, 10^7 CFU/mL	225 min	36 W	0.2–1.5 g/L
[36]	Suspended + TiO$_2$–coated composite catalyst	J109 *E. coli* 10^7 CFU/mL	60 s	44 W	0.2 g/L
[11]	Suspended system	K-12 *E. coli*, 7 × 10^7 cells/mL	105 min	72.8 mJ/m^2	30 mg/L
[40]	Suspended system	K-12 *E. coli*, 10^6 CFU/mL	120 min	2.8 × 10^{-6} Einstein/s, UV-A	Degussa, P25 0.1 g/L
[37]	Suspended system	K-12 108 CFU/mL	60 min adsorption equilibrium, total 150 min	800 mW/cm^2	Degussa, P25 0.5 g/L
[38]	Suspended, thin-film coated	*E. coli*, CIP 53126, 1.3 × 10^7 CFU/mL	120 min	1.1 × 10^{-5} Einstein/s, UV-A	1 g/L TiO$_2$, commercial, support, PC500 TiO$_2$ 18 g/m^2, 2.2 g/m^2 P25 Degussa
[39]	Coated on substrates, electrodes	K-12 *E. coli*	120 min	1.5–4 × 10^{-8} Einstein/cm^2 · s	P25 and Aldrich TiO$_2$, 0.8 mg/cm^2
[36]	Suspended + TiO$_2$–coated composite catalyst	J109 *E. coli* 10^7 CFU/mL	60 s	44 W	0.2 g/L
[41]	Suspended system	*E. coli* NCIM 2066, 10^8 CFU/mL	120 min	0.5 mW/cm^2, solar	Cu doped, 200 mg/L
[42]	Suspended system	K-12 *E. coli* 10^6 CFU/mL	5 h	Solar and UV-A filtered solar irradiation, 23.3 W/m^2	10–500 mg/L TiO$_2$ graphene composites versus P25 Degussa
[42]	Suspended system, varying TiO$_2$ catalysts	K-12 *E. coli*, 10^6 CFU/mL	90 min	10 kJ/L for Ruana and 7 kJ/L for others, based on Quv approach	Evonik, P25, PC500 and Ruana, 0.05 g/L and 0.5 g/L for others
[40]	Suspended, fixed-bed, fixed-wall	K-12 strains (CECT 4624, corresponding to ATCC 23631, 10^6 CFU/mL)	120 min	1.2 × 10-5 Einstein/s	Equivalent to 0.1 g/L, 1–3 times coating on fixed-bed and wall reactor
[43]	Suspended system	K-12 *E. coli* MG1655, 10^8 CFU/mL	6 h	4–8 W/m^2	0.25 g/L P25 Degussa, combustion synthesized TiO$_2$

of plant and human pathogens into vegetable crops. The most applied disinfection process in wastewater treatment is chlorination, but UV radiation also finds extended applications. Germicidal effects of chlorine (as chlorine gas or hypochlorites) include the following mechanisms: oxidizing the germ cells, altering cell permeability, altering cell protoplasm, inhibiting enzyme activity, and damaging the cell DNA and RNA. Chlorine appears to react strongly with the lipids of the membrane and the membranes that have high lipid concentrations appear to be more susceptible to destruction. The predominant disinfection mechanism depends on the microorganism (the resistance of a particular strain), the wastewater characteristics, and chlorine dose. Unfortunately, bacteria injured by disinfection processes can survive and regrow at low chlorine doses. UV radiation can also damage DNA, resulting in inhibition of cell replication and, in the case of lethal doses, in a loss of reproducibility. The effectiveness of a UV disinfection system depends on the characteristics of the wastewater, the intensity of UV radiation (optimum wavelength to effectively inactivate microorganisms is 250–270 nm), the time the

TABLE 7.4 Current and Potential Applications of Antimicrobial Nanomaterials

Nanomaterial	Antimicrobial Mechanism	Current Applications
Chitosan	Membrane damage, chelation of trace metals	Personal care products, microbicide in agriculture and biomedical products, food wraps, biomedical, flocculants in water and wastewater treatment
nAg	Release of Ag+ ions, disruption of cell membrane and electron transport, DNA damage	Portable water filters, clothing, medical devices, coatings for washing machines, refrigerators, and food containers
TiO$_2$	Production of ROS, cell membrane, and cell wall damage	Air purifiers, water treatment systems for organic contaminant Degradation
CNT	Physically compromise cell envelope	None
ZnO	Intracellular accumulation of nanoparticles, cell membrane damage, H$_2$O$_2$ production, release of Zn^{+2} ions	Antibacterial creams, lotions, and ointment, deodorant, self-cleaning glass, and ceramics

Source: Li, Q., Mahendra, S., Lyon, D.Y., Brunet, L., Liga, M.V., Li, D., and Alvarez, P.J.J. 2008. *Water Research*, 42, 4591–4602.

microorganisms are exposed to the radiation, and the reactor configuration [37,49,50].

Antibiotic resistant *E. coli* and other coliforms were investigated in an UWTP in Tokyo Metropolitan Prefecture [23]. *E. coli* strains, randomly isolated from wastewater samples, were tested for their sensitivity to seven antimicrobial agents in three different UWTP locations: the inflow, before chlorination, and after chlorination. Chlorination treatment did not significantly affect the percentage of resistance in *E. coli* to one or more antibiotics (from 14.7% to 14.0%) or specifically to ampicillin (constant at 7.3%) and tetracycline (from 8.0% to 6.7%).

Templeton et al. [48] investigated the effect of free chlorine and ultraviolet (UV intensity 0.247 mW/cm^2) disinfection on *E. coli* strains resistant to ampicillin and trimethoprim, in comparison to an antibiotic-susceptible strain of *E. coli* isolated from sewage sludge. Trimethoprim-resistant *E. coli* was found to be slightly more resistant to chlorine than the antibiotic-susceptible isolate and the ampicillin resistant *E. coli* under the studied conditions (95% confidence). Moreover, no statistically significant differences between the UV dose–response profiles of the antibiotic-resistant and antibiotic-susceptible *E. coli* strains over the UV dose range tested were observed.

When 2.0 mg/L chlorine dose was used to investigate the inactivation of multidrug resistant *E. coli* strains selected from

a UWTP effluent, the number of colonies decreased by 99.999% after 60 min of contact time [43]. However, the minimum inhibitory concentration to the antibiotics amoxicillin, ciprofloxacin, and sulfamethoxazole was not altered for the surviving cultures.

Ivey et al. [22] evaluated the efficacy of chlorine gas (Cl$_2$[g]) and chlorine dioxide (ClO$_2$) injection systems in combination with rapid sand filtration (RSF) in killing fecal indicator microorganisms in irrigation water in a vegetable-intensive production area. Sampling date and sampling point also had a significant effect on the abundance of generic *E. coli* in Cl$_2$(g)-treated water but only sampling point was significant in ClO$_2$-treated water. Accordingly, injection of ClO$_2$ and Cl$_2$(g) into surface water prior to RSF is inadequate in reducing fecal indicator microorganism populations and ClO$_2$ ineffectively kills infectious propagules of *Phytophthora capsici*.

7.5 Summary and Conclusions

Escherichia coli has globally spread antibiotic-resistant bacteria to many antibiotics. Recent surveillance reports showed that the data banks for occurrence of *E. coli* [1] are growing and this will help to map this information for more effective control of outbreaks and resistance. The origin and pathways of *E. coli* to human contact are clear and among these sources, urban water needs to be dealt with accurately to protect human and environmental health. This work outlines the following issues for urban water safety:

- Conventional UWTPs may positively affect antibiotic resistant bacteria (ARB) spread and selection as well as antibiotic resistance gene (ARG) transfer.
- All known types of antibiotic resistance mechanisms are represented in UWTP, suggesting the relevance of these facilities as reservoirs and environmental suppliers of genetic determinants of resistance.
- AOPs such as ozone, UV, and photocatalysis seem to be very effective for bacteria inactivation. However, reactivation is not to be ignored and a final disinfection seems unavoidable for long-term safety.
- Effect of conventional (e.g., chlorination and UV radiation) and new/alternative disinfection processes on the inactivation of specific ARB as well as the capacity to control resistance spread into the environment are strongly recommended because the few studies available show that, despite an effective decrease of the total number of bacteria, they may simultaneously promote the selection of ARB.

Acknowledgment

This work has been performed in the context of Research and Development Fund of NKU (NKUBAP.00.17.AR.13.13).

References

1. Al-Assil, B., Mahfoud, M., and Hamzeh, A.R. 2013. Resistance trends and risk factors of extended spectrum b-lactamases in *Escherichia coli* infections in Aleppo, Syria. *American Journal of Infection Control*, 41, 597–600.

2. Barker, S.F., O'Toole, J., Sinclair, M.I., Leder, K., Malawaraarachchi, M., and Hamilton, A.J. 2013. A probabilistic model of norovirus disease burden associated with greywater irrigation of home produced lettuce in Melbourne, Australia. *Water Research*, 47, 1421–1432.

3. Bekbölet, M. 1997. Photocatalytic bactericidal activity of TiO_2 in aqueous suspensions of *E. coli*. *Water Science Technology*, 35(11), 95–100.

4. Benabbou, A.K., Derriche, Z., Felix, C., Lejeune, P., and Guillard, C. 2007. Photocatalytic inactivation of *Escherichia coli*: Effect of concentration of TiO_2 and microorganism, nature, and intensity of UV irradiation. *Applied Catalysis B: Environmental*, 76(3–4), 257–263.

5. Blaak, H., de Kruijf, P., Hamidjaja, R.A., van Hoek, A.H.A.M., de Roda Husman, A.M., and Schets, F.M. 2014. Prevalence and characteristics of ESBL-producing *E. coli* in Dutch recreational waters influenced by wastewater treatment plants. *Veterinary Microbiology*, 171(3), 448–459.

6. Blecher, K., Nasir, A., and Friedman, A. 2011. The growing role of nanotechnology in combating infectious disease. *Virulence*, 2(5), 395–401.

7. Carrolla, S.P., Dawes, L., Hargreaves, M., and Goonetilleke, A. 2009. Faecal pollution source identification in an urbanising catchment using antibiotic resistance profiling, discriminant analysis and partial least squares regression. *Water Research*, 43, 1237–1246.

8. CDC. 2006. Update on multi-state outbreak of *E. coli* O157:H7 infections from spinach, October 6, 2006. Centers for Disease Control and Prevention, Atlanta, GA. http://www.cdc.gov/ecoli/2006/september/updates/100606.htm.

9. CDC. 2008. Investigation of outbreak of infections caused by *Salmonella* Saintpaul. Centers for Disease Control and Prevention, Atlanta, GA. http://www.cdc.gov/salmonella/saintpaul/jalapeno/index.html.

10. Cengiz, M., Uslu, M.O., and Balcioglu, I. 2010. Treatment of *E. coli* HB101 and the tetM gene by Fenton's reagent and ozone in cow manure. *Journal of Environmental Management*, 91, 2590–2593.

11. Da Silva, F., Tiago, M., Veríssimo, I., Boaventura, A., Nunes, A.R., and Manaia, C.M. 2006. Antibiotic resistance of enterococci and related bacteria in an urban wastewater treatment plant. *FEMS Microbiology, Ecology*, 55, 322–329.

12. Da Silva, F., Vaz-Moreira, M., Gonzalez-Pajuelo, I., Nunes, M.O.C., and Manaia, C.M. 2007. Antimicrobial resistance patterns in *Enterobacteriaceae* isolated from an urban wastewater treatment plant. *FEMS Microbiology, Ecology*, 60, 166–176.

13. Dunlop, P.S.M., Byrne, J.A., Manga, N., and Eggins, B.R. 2002. The photocatalytic removal of bacterial pollutants from drinking water. *Journal of Photochemistry and Photobiology A: Chemistry*, 148(1–3), 355–363.

14. Duong, H.A., Pham, N.H., Nguyen, H.T., Hoang, T.T., Pham, H.V., Pham, V.C., Berg, M., Giger, W., and Alder, A.C. 2008. Occurrence, fate and antibiotic resistance of fluoroquinolone antibacterials in hospital wastewaters in Hanoi, Vietnam. *Chemosphere*, 72, 968–973.

15. ECDC. 2013. *Surveillance Report. Reporting on 2011 Surveillance Data and 2012 Epidemic Intelligence Data*, www.ecdc.europa.eu.

16. Faure, M., Gerardin, F., André, J.C., Pons, M.N., and Zahraa, O. 2011. Study of photocatalytic damages induced on *E. coli* by different photocatalytic supports (various types and TiO_2 configurations). *Journal of Photochemistry and Photobiology A: Chemistry*, 222(2–3), 323–329.

17. Fernandez, P., Dunlop, P.S.M., D'Sa, R., Magee, E., O'Shea, K., Dionysiou, D.D., and Byrne, J.A. 2015. Solar photocatalytic disinfection of water using titanium dioxide graphene composites. *Chemical Engineering Journal*, 261, 36–44.

18. Ferreira da Silva, M., Vaz-Moreira, I., Gonzalez-Pajuelo, M., Nunes, O.C., and Manaia, C.M. 2007. Antimicrobial resistance patterns in *Enterobacteriaceae* isolated from an urban wastewater treatment plant. *FEMS Microbiology Ecology*, 60(1), 166–176.

19. Galvin, S., Boyle, F., Hickey, P., Vellinga, A., Morris, D., and Cormican, M. 2010. Enumeration and characterization of antimicrobial-resistant *Escherichia coli* bacteria in effluent from municipal, hospital, and secondary treatment facility sources. *Applied and Environmental Microbiology*, 76(14), 4772–4779.

20. Guillard, C., Helali, S., Polo-López, M.I., Fernández-Ibáñez, P., Ohtani, B., Amano, F., and Malato, S., 2014. Solar photocatalysis: A green technology for *E. coli* contaminated water disinfection. Effect of concentration and different types of suspended catalyst. *Journal of Photochemistry and Photobiology A: Chemistry*, 276, 31–40.

21. Huh, A. J. and Kwon, Y.J. 2011. Nanoantibiotics: A new paradigm for treating infectious diseases using nanomaterials in the antibiotics resistant era. *Journal of Controlled Release*, 156(2), 128–145.

22. Ivey, M.L. and Miller, S.A. 2013. Assessing the efficacy of pre-harvest, chlorine-based sanitizers against human pathogen indicator microorganisms and *Phytophthora capsici* in non-recycled surface irrigation water. *Water Research*, 47, 4639–4651.

23. Iwane, T., Urase, T., and Yamamoto, K. 2001. Possible impact of treated wastewater discharge on incidence of antibiotic resistant bacteria in river water. *Water Science and Technology*, 43(2), 91–99.

24. Jurry, K.L., Khan, S.J., Vancov, T., Stuetz, R.M., and Ashbolt, N.J. 2011. Are sewage treatment plants promoting antibiotic resistance? *Critical Reviews in Environmental Science and Technology*, 41, 243–270.

25. Kamruzzaman, M., Shoma, S., Naymul Bari, S.M., Ginn, A.N., Wiklendt, A.M., Partridge, S.R., Faruque, S.M., and Iredell, J.R. 2013. Genetic diversity and antibiotic resistance in *Escherichia coli* from environmental surface water in Dhaka City, Bangladesh. *Diagnostic Microbiology and Infectious Disease*, 76, 222–226.

26. Khalil, A., Gondal, M.A., and Dastageer, M.A. 2011. Augmented photocatalytic activity of palladium incorporated ZnO nanoparticles in the disinfection of *Escherichia coli* microorganism from water. *Applied Catalysis A: General*, 402(1–2), 162–167.

27. Kim, S.C., Davis, J.G., Truman, C.C., Ascough II, J.C., and Carlson, K. 2010. Simulated rainfall study for transport of veterinary antibiotics—Mass balance analysis. *Journal of Hazardous Materials*, 175, 836–843.

28. Kramer, A., Schwebke, I., and Kampf, G. 2006. How long do nosocomial pathogens persist on inanimate surfaces? A systematic review. *BMC Infectious Diseases*, 6, 130–138.

29. Leid, J.G., Ditto, A.J., Knapp, A., Shah, P.N., Wright, B.D., Blust, R., Christensen, L. et al. 2012. In vitro antimicrobial studies of silver carbene complexes: Activity of free and nanoparticle carbene formulations against clinical isolates of pathogenic bacteria. *Journal of Antimicrobial Chemotherapy*, 67(1), 138–148.

30. Li, Q., Mahendra, S., Lyon, D.Y., Brunet, L., Liga, M.V., Li, D., and Alvarez, P.J.J. 2008. Antimicrobial nanomaterials for water disinfection and microbial control: Potential applications and implications. *Water Research*, 42, 4591–4602.

31. Łuczkiewicz, A., Jankowska, K., Fudala-Książek, S., and Olańczuk-Neyman, K. 2010. Antimicrobial resistance of fecal indicators in municipal wastewater treatment plant. *Water Research*, 44(17), 5089–5097.

32. Madras, G., Sontakke, S., Mohan, C., and Modak, J. 2012. Visible light photocatalytic inactivation of *Escherichia coli* with combustion synthesized TiO$_2$. *Chemical Engineering Journal*, 189–190, 101–107.

33. Mathew, A.G., Arnett, D.B., Cullenc, P., and Ebner, P.D. 2003. Characterization of resistance patterns and detection of apramycin resistance genes in *Escherichia coli* isolated from swine exposed to various environmental conditions. *International Journal of Food Microbiology*, 89, 11–20.

34. Matsunaga, M. and Okochi, M. 1995. TiO$_2$ mediated photochemical disinfection of *Escherichia coli* optical fibres. *Environmental Science and Technology*, 29, 501–505.

35. McLoughlina, O.A., Kehoe, S.C., McGuigan, K.G., Duffy, E.F., Al Touati, F., Gernjak, W., Oller Alberola, I., Malato Rodriguez, S., and Gill, L.W. 2004. Solar disinfection of contaminated water: A comparison of three small-scale reactors. *Solar Energy*, 77, 657–664.

36. Meric, S. and Fatta-Kassinoss, D. 2009. Water treatment, municipal. In: M. Schaechter (ed.), *Encyclopedia of Microbiology*. Elsevier, Oxford, pp. 587–599.

37. Michael, I., Rizzo, L., McArdell, C.S., Manaia, C.M., Merlin, C., Schwartz, T., Dagot, C., and Fatta-Kassinos, D. 2013. Urban wastewater treatment plants as hotspots for the release of antibiotics in the environment: A review. *Water Research*, 47, 957–995.

38. Öncü, N.B., Menceloğlu, Y. Z., and Balcioğlu, I. A. 2011. Comparison of the effectiveness of chlorine, ozone, and photocatalytic disinfection in reducing the risk of antibiotic resistance pollution. *Journal of Advanced Oxidation Technologies*, 14(2), 196–203.

39. Paul, T., Dodd, M. C., and Strathmann, T. J. 2010. Photolytic and photocatalytic decomposition of aqueous ciprofloxacin: Transformation products and residual antibacterial activity. *Water Research*, 44(10), 3121–3132.

40. Pelgrift, R.Y. and Friedman, A.J. 2013. Nanotechnology as a therapeutic tool to combat microbial resistance. *Advanced Drug Delivery Reviews*, 65(13), 1803–1815. Doi: 10.1016/j.addr.2013.07.011

41. Pennington, H. 2010. *Escherichia coli* O157. *The Lancet*, 376(9750), 1428–1435.

42. Rincón, A.G. and Pulgarin, C. 2005. Use of coaxial photocatalytic reactor (CAPHORE) in the TiO$_2$ photo-assisted treatment of mixed *E. coli* and *Bacillus* sp. and bacterial community present in wastewater. *Catalysis Today*, 101(3–4), 331–344.

43. Rizzo, L., Fiorentino, A., and Anselmo, A. 2012. Effect of solar radiation on multidrug resistant *E. coli* strains and antibiotic mixture photodegradation in wastewater polluted stream. *Science of the Total Environment*, 427–428, 263–268.

44. Rizzo, L., Manaia, C., Merlin, C., Schwartz, T., Dagot, C., Ploy, M.C., Michael, I., and Fatta-Kassinos, D. 2013. Urban wastewater treatment plants as hotspots for antibiotic resistant bacteria and genes spread into the environment: A review. *Science of the Total Environment*, 447, 345–360.

45. Schwegmann, H., Ruppert, J., and Frimmel, F.H. 2013. Influence of the pH-value on the photocatalytic disinfection of bacteria with TiO$_2$—Explanation by DLVO and XDLVO theory. *Water Research*, 47(4), 1503–1511.

46. Sun, D.D., Tay, J.H., and Tan, K.M. 2003. Photocatalytic degradation of *E. coliform* in water. *Water Research*, 37(14), 3452–3462.

47. Taylor, J., Lai, K., Davies, M., Clifton, D., Ridley, I., and Biddulph, P. 2011. Flood management: Prediction of microbial contamination in large-scale floods in urban environments. *Environment International*, 37, 1019–1029.

48. Templeton, M.R., Oddy, F., Leung, W.K., and Rogers, M. 2009. Chlorine and UV disinfection of ampicillin-resistant and trimethoprim-resistant *Escherichia coli*. *Canadian Journal of Civil Engineering*, 36(5), 889–894.

49. van Grieken, R., Marugán, J., and Sordo, C.C. 2009. Comparison of the photocatalytic disinfection of *E. coli* suspensions in slurry, wall and fixed-bed reactors. *Catalysis Today*, 144(1–2), 48–54.

50. Wang, C., Li, H., and Zhuang, J. 2013. Photocatalytic degradation of methylene blue and inactivation of gram-negative bacteria by TiO_2 nanoparticles in aqueous suspension. *Food Control*, 34(2), 372–377.

51. Wellington, E.M.H., Boxall, A.B.A., Cross, P., Feil, E.J., Gaze, W.H., Hawkey, P.M., Johnson-Rollings, A.S. et al. 2013. The role of the natural environment in the emergence of antibiotic resistance in gram-negative bacteria. *The Lancet Infectious Diseases*, 13(2), 155–165.

52. World Health Organization (WHO). 2014. *Antimicrobial Resistance Global Report on Surveillance.*

53. Yadav, H.M., Otari, S.V., Bohara, R.A., Mali, S.S., Pawar, S.H., and Delekar, S.D. 2014. Synthesis and visible light photocatalytic antibacterial activity of nickel–doped TiO_2 nanoparticles against gram–positive and gram–negative bacteria. *Journal of Photochemistry and Photobiology A: Chemistry*, 294, 130–136.

III

Urban Water Reuse Options

8

Water Quality Issues in Urban Water

Syed Amir Abbas
Shah Naqvi
COMSATS

Aeysha Sultan
University of Sargodha

Saeid Eslamian
Isfahan University of Technology

PREFACE

Water pollution is the contamination of water bodies, which occurs when contaminants/pollutants are discharged into water bodies without prior treatment to remove hazardous wastes. The contaminated water causes damage to humans, animals, and plants, which in most cases is irreversible. The problem of water contamination is more pronounced in urban areas than in rural areas. The street runoff carrying oil and heavy metals from industries and automobiles contributes to the major factors of water pollution in these areas.

This chapter will cover the different sources of water contaminants in urban areas and the issues that originate because of these issues. Various potential solutions to the problems are also discussed briefly along with the role of different agencies in monitoring the quality control of urban water bodies.

8.1 Introduction

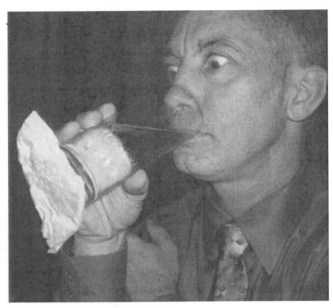

Great reasons to drink water, and how to form the water habit. (Adapted from www.dumblittleman.com, accessed on May 15, 2013.)

From the minute we turn the lights on and brush our teeth in the morning, we rely on energy and water for our quality of life. These two critical resources are intimately connected: it takes energy to treat and transport water and most power generation requires water. Climate change and other trends such as population growth are likely to affect both; the shift toward greener energy sources is leading toward diminished water resources.

Water is the foundation of life on Earth. Water makes up more than 60% of the human body, comprising 83% of our blood, 70% of our brain, and 90% of our lungs. The quality of water resources profoundly affects our quality of life. On Earth, 1.1 billion people lack access to clean water and 2.6 billion people lack basic sanitation. This leads to significant morbidity and mortality. According to UN statistics, after every 15 s a child dies of a preventable waterborne disease (more than 2 million children a year) [11].

Water pollution is a major cause of waterborne diseases, which is in turn responsible for poor health conditions as well as increased death rate in urban areas. Urban sources of water pollution have often been cited as the primary cause of poor water quality. The aim of this chapter is to provide an overview of the causes/sources of water pollution as well as the current technologies being used in this regard. The role of various agencies in controlling/monitoring water quality is also briefly discussed.

8.1.1 Why Study Urban Water?

Urban watersheds are complex environments in which people live, work, and recreate. At the same time, they are subject to natural and anthropogenic inputs that alter ecosystem dynamics and affect water quality. As the twenty-first century began, 50% of the world's population (6 billion) was living in urban environments, and this number is expected to increase significantly over the next 100 years. High population density in urban areas is straining existing water and sewer infrastructure, water resources, and natural systems that support human life.

8.1.2 What Is Urban Water?

Urban water is the water in our built environment that includes

- Water in pipes that travel to our homes for drinking and household use
- Waste water in pipes that travel from our homes to treatment plants
- Water that runs through the streets and into stormwater systems
- Wastewater from industrial activity

All this water represents engineering challenges and management issues related to the dense population associated with urban centers. Provision of sufficient, high-quality water and removal of wastewater are critical issues of urban populations, which have increased in the world [21].

8.1.3 Watershed Management

Urban areas are important contributors in generating large amounts of nonpoint pollution from runoff and storm-sewer discharge. The first step toward controlling the water quality in urban systems is to improve the scientific basis for management of inputs and inventories of contaminants in urban watersheds. This involves the following:

1. Primary research on environmental sources and fate of contaminants of concern in urban systems (e.g., pathogens, persistent toxicants)
2. Development of necessary monitoring and modeling capabilities for urban watersheds
3. Development of technologies and modeling tools for improved management of water quality in urban systems
4. Outreach and education to develop a citizenry ready for the challenges of managing urban water systems

8.2 Contaminants in Urban Water, Their Environmental Sources, and Fate

Water quality standards serve as the foundation for the water-quality based approach to pollution control. Untreated or poorly treated sewage can be low in dissolved oxygen (DO) and high in pollutants such as *fecal coliform* bacteria, nitrates, phosphorus, chemicals, and other bacteria. Treated sewage can still be high in nitrates. Groundwater and surface water can be contaminated from many sources such as garbage dumps, toxic waste and chemical storage and use areas, leaking fuel storage tanks, and intentional dumping of hazardous substances. Air pollution can lead to acid rain, nitrate deposition, and ammonium deposition, which can alter the water chemistry of lakes. The water quality in urban areas is quite poor as compared to that in rural areas because of factors such as industrialization, increased traffic, and poor waste disposal protocols.

8.2.1 Water Quality in Urban Areas and Sources of Contamination

The rapid urbanization and poor consideration in land management has resulted in serious ecological problems, water contamination being an important one. Increased urbanization has several serious environmental implications, which include an increase in flow, nutrients, heavy metals, sediments, and bacteria. The sources of water contamination can be broadly classified as natural and anthropogenic; however, sometimes it becomes more convenient to classify the sources as

1. Surface water contamination
2. Ground water contamination

Both of these categories are interrelated and at times it becomes quite difficult to separate the two. The surface water contamination or pollution refers to the pollution of water available on the Earth's crust whether in the form of lake, rivers, etc. or in the form of pipe or ditch. The surface water finds its way to the underground water by a seepage process. Therefore, contaminated surface water also pollutes underground water reserves.

8.3 Urban Watershed Monitoring and Modeling

The worldwide diminishing of pure water deposits has resulted in an increased pressure on quantity as well as quality of water resources. Therefore, management of the urban water cycle is needed in order to sustain the existing water resources and to attain/maintain high levels of public health, food production/protection, and improvement of the aquatic ecosystem.

8.3.1 Monitoring of Physicochemical Characteristics

A number of parameters may be used to indicate the physicochemical status of a water body. Some other parameters such as pH may be measured directly, while for others such as salinity indirect measures such as conductivity are used. Nutrients such as ammonium, nitrite, nitrate, phosphate, and more generally total nitrogen and phosphorus may be useful indicators in monitoring programs as they are involved in eutrophication processes, and some (e.g., nitrates) may contaminate groundwater after fertilizer applications. Other parameters may be used to characterize the degree of oxygenation of a water body, and include DO, the chemical oxygen demand (COD), the biochemical oxygen demand (BOD), redox conditions, or respirometry measurements. In turn, the presence and levels of organic matter strongly influence the COD of a water sample. Different types are found based on a range of specific electrodes, optical sensors, ultraviolet (UV), visible spectroscopy, colorimetry, chemiluminescence, titrimetric methods, or ion chromatography. In most legislation governing this area, there are lists of priority pollutants that are regarded as being harmful or potentially harmful. These are normally classified under three major categories: nonpolar organic compounds (e.g., some pesticides, and some industrial chemicals such as polycyclic aromatic hydrocarbons [PAHs]), polar organics (some pesticides, and pharmaceuticals) and heavy metals (e.g., mercury and cadmium). The methods for monitoring these pollutants vary a lot, but following steps are common to all classes of chemicals:

- Sampling
- Stabilization
- Transport
- Storage
- Preparation for analysis
- Analysis

Depending on the monitoring methods used, some of these steps may be omitted. For instance, on-site methods obviate the need for transport and storage. The best established and currently most tightly monitored in terms of quality assurance and control are chromatographic methods (gas chromatography or liquid chromatography) linked to a sensitive detector (e.g., flame ionization, electron capture, mass spectrometer, fluorescence spectrometer, UV spectrometer), and for metals methods such as inductively coupled plasma mass spectrometry or graphite oven atomic absorption spectrometry. However, a range of other methods based on chemical, optical, or electrochemical of biological techniques are available for carrying out the analytical step, depending on the chemical to be measured. All of the methods used for monitoring chemical parameters have one feature in common—a recognition system that is specific for the target molecules/parameter. Some methods are based on specific complexation reactions whose products (usually colored) can be quantified in a colorimeter or a spectrophotometer.

Electrochemical techniques are based on the measurement of changes in a potential (or a current) generated between two electrodes placed in a solution to be analyzed. These methods are for measuring either directly (e.g., pH, concentrations of some metals) or indirectly by using an intermediary substrate that recognizes the analyte of interest (as used for some organic molecules). Immunoassays comprise a further set of assays or methods for measuring levels of different chemicals in water samples. They rely on the interaction between a biological material (the recognition step) called an antibody, and this is specific to the analyte under study. These assays are available in a range of formats such as bound to magnetic particles in 96-well plates or coated at the bottom of test tubes. An amplification system is incorporated and is often provided by a linked-enzyme system that provides a colored product from a colorless substrate.

Other systems for quantifying the bound analyte include the use of fluorescent or radiolabeled tags. These methods are very sensitive, and are applicable to a wide range of compounds, for example, PAHs, pesticides, phenols, surfactant residues, heavy metals, mutagens and other polychlorinated biphenyls (PCBs). Many of the assays described here are commercially available in kit form. There is a proportional relationship between the concentration of an analyte in the water and its impact on a living organism. This bio monitoring can be based on changes measured at the level of the whole organism, tissues, cells, or isolated biochemical mechanisms. The data can be qualitative, semi-quantitative, or quantitative [15].

The biological monitoring techniques include bioassays, biomarkers, and biological early warning systems (BEWS). Bioassays are based on the use of whole organisms (including a range of animals [both vertebrates and invertebrates], yeast, algae, and bacteria) or of isolated parts of organisms (including isolated cells and tissues, and enzyme systems). Toxicity depends in part on the bioavailability of the toxicants, and this is affected by factors such as the presence of suspended matter and the concentration of dissolved organic carbon. It is also affected by physicochemical variables such as pH, temperature, water hardness, oxygen tension, and salinity.

Information on limits of detection for various chemicals and standard operating procedures (SOPs) are usually supplied by the producers. Most of these assays provide a measure of acute toxicity only, and do not provide information on chronic toxicity produced by long-term exposure to toxicants (possibly present in very low concentrations).

Table 8.1 gives a list of some of these technologies that are either commercially available or still at the prototype stage. Some of these systems may be able to detect microbiological pathogens as well as chemical toxicants [12].

8.3.2 Modeling of Urban Water Quality

Urban sources of water pollution have often been cited as the primary cause of poor water quality in receiving water bodies (RWB), and recently many studies have been conducted to

TABLE 8.1 Some of the Commercially Available Bioassays

System	Organism Used	Stimuli
AquaTox Control, behavioQuant®, Truitosem™	Freshwater fish	Swimming capacity behavior
Truitel™	Freshwater fish	Swimming capacity behavior
Fish taximeter	Zebra fish	Movement rapidity
Daphnia test®	Daphnia	Daphnia activity, swimming behavior
MosselMonitor®	Mussel	Valve movement
ToxAlarm, Toxiguard®	Algae and bacteria	Respirometry, oxygen production and consummation
Algae taximeter	Algae	Photosynthetic activity
Fluotox	Algae	Fluorescence natural emission
Microtox®, Toxscreen-II test, ToxAlert®, Vititox®, Biotox®, GreenScreen®	Bacteria	Natural bioluminescence inhibition

Source: Adapted from ISO, CD17994. 2001. *Water Quality—Criteria for the Establishment of Equivalence between Microbiological Methods*, Final Version, June 15, 2001.

investigate both continuous sources, such as wastewater treatment plant (WWTP) effluents, and intermittent sources, such as combined sewer overflows (CSOs).

The wetlands are conveniently categorized as surface flow wetlands and subsurface wetlands. Both of these are quite interrelated and most of the hydrologic processes occurring in these systems can be represented by five model components: precipitation, precipitation excess (runoff), surface flow through wetland, infiltration, and evapotranspiration (ET). Different methods being used commonly for modeling the wetland systems include black-box models (BBM), process-based models (PBM), and integrated urban water management (IUWM).

8.3.3 BBM Is Often Considered an Inadequate Wetland Process

The PBM utilize equations to model the processes of a wetland and, therefore, are often quite complicated and require intensive data collection. The PBM include the FITOVERT model, the constructed wetland two-dimensional (CW2D) model, the structural thinking experimental learning laboratory with animation (STELLA) software, the 2D mechanistic model, hydrological simulation program—Fortran (HSPF), and the constructed wetland model no. 1 (CWM1) [7]. The IUWM requires the individual components of the urban water cycle to be designed and managed together, rather than individually. An urban drainage system must, therefore, be considered jointly, that is, by means of an integrated approach. Integrated modeling of urban wastewater systems is of growing interest, mainly as a result of the recent adoption of the European Union (EU) Water Framework [3]. An integrated modeling approach is also required due to the concurrently growing awareness that optimal management of the individual components of urban wastewater systems (i.e., sewer systems, WWTPs and RWBs) does not lead to optimum performance of the entire system. However, although the benefits of an integrated approach have been widely demonstrated, several aspects have prevented its wide application. These factors, along with the high complexity level of the currently adopted approaches, introduce uncertainties in the modeling process that are not always identifiable [18].

8.4 Public Awareness and Educational Measures Regarding Urban Water Quality

Good environmental and natural resource management depends fundamentally on understanding human behavior. Participation, within this broad argument, also seeks to maintain commitment to social and democratic justice by being concerned and aware of social equity, access to information and rights to participate, and being heard and ultimately influencing decision-making processes.

One of the main differences between Europe and the United States is the development of volunteer networks for water monitoring, working in a complementary way with water authorities or state agencies. Volunteer actions are numerous and increasing, involve a wide range of citizens, including students and seniors, and are facilitated by a variety of programs some of which are supported by the United States Environmental Protection Agency (USEPA). The volunteers' work is essentially informing communities and citizens of water quality issues, getting data on water bodies that otherwise may go unmonitored, and helping monitoring programs for better water quality [10].

8.4.1 Monitoring and Assessment of Water

Clean Water Act (CWA) implementation needs the involvement of multilevel stakeholders for water monitoring (USEPA, www.epa.gov) in order to characterize waters and identify changes or trends in waters over time:

- Identify specific or existing or emerging quality problems
- Gather information to design specific remediation for identified pollutants
- Determine whatever program goals are being met
- Respond to emergencies, such as spills and floods

At the federal level, USEPA conducts limited monitoring surveys for specific issues. The regional offices have to comply with these and inspect the monitoring of industrial and municipal discharges. Since 1974, the U.S. Geological Survey (USGS) has been in charge of the National Stream Accounting Network (NASQAN) (http://water.usgs.gov/nasqan), which covers extensive chemical monitoring in large rivers. Since 1991, it has also piloted the National Water Quality Assessment Program (NAWQA), which focuses on status and trends in water, sediment, and biota at the regional level (http://water.usgs.gov/nawqa). Other federal agencies conduct water quality monitoring programs such as the U.S. Fish and Wildlife Service, the National Oceanic and Atmospheric Administration, the U.S. Army Corps of Engineers, and some water authorities such as the Tennessee Valley Authority. At the regional level, states and Indian tribes have the main responsibilities for conducting monitoring programs, and these activities are supported by grants from USEPA. They have to develop and conduct the state water monitoring and assessment program. Moreover, local governments (city and county environmental offices) can also implement monitoring programs.

8.5 Introduction and Highlighting Urban Water Quality Issues

At the local level, many private organizations (universities, watershed associations groups, permitted dischargers) are becoming increasingly important in water quality monitoring. They also have a great importance in the development and validation of new screening methods. Water quality indicators are particularly important in this depending on the monitoring design and sample site (Table 8.2).

As part of this monitoring, the chemical quality of waters, sediments, and the health of organisms can be analyzed. There are additional core indicators for particular media, such as lakes and wetlands. Finally, data quality is also a priority as well as their management by various local or national databases such as STORET (Data Storage and Retrieval System, www.epa.gov/storet/). These mechanisms allow the timely production of reports as required under the CWA. The design of prospective actions has an influence on the development of new monitoring tools and methodologies. In order to estimate these indicators, it is necessary to use various analytical methods, generally approved by USEPA or standardized by other national or international organizations such as American Public Health Association (APHA) ASTM, AOAC-International, or ISO [5]. In order to approve and validate new test procedures, confirm laboratory performance, and update approved methods (Table 8.3), USEPA publishes both approved test methods and procedures and those that have not been promulgated.

In order to ensure quality control, USEPA has listed regulations, rules, and guidance on approved analytical methods (test procedures) in the 40 CFR 136. This is especially designed for use by industries and municipalities for the measurement of chemical and biological components of wastewaters, drinking waters, sediments, and other environmental samples. One way for state agencies to face increasing needs for water monitoring is to collaborate with volunteers to encourage all citizens to improve their knowledge of water resources. In turn, this supports the implementation of monitoring activities, and the volunteer/state partnerships are provided with access to equipment and with a transfer of expertise from professionals to volunteers who are participating in programs.

TABLE 8.2   Examples of Recommended Core and Supplemental Water Quality Indicators for General Designated Use Categories

	Aquatic Life and Wildlife	Recreation	Drinking Water	Fish/Shellfish Consumption
Core indicators	Conditions of biological communities (EPA recommends use of at least two assemblages) DO temperature conductivity pH Habitant assessment	Pathogen indicators (*E. coli*, *Enterococci*) Nuisance plants growth Flow Nutrients, chlorophyll, landscape conditions (e.g., percentage cover of land uses)	Trace metals Pathogens Nitrates Salinity Sediments/TDS Flow Landscape conditions (e.g., percentage cover of land uses)	Pathogens Mercury Chlordane DDT PCBs Landscape conditions (e.g., percentage cover of land uses)
	Flow Nutrients Landscape conditions (e.g., percentage cover of land uses) Additional indicators for lakes: Eutrophic condition Additional indicators for wetlands: Wetland hydrogeomorphic settings and functions	Additional indicators for lakes: Secchi depth Additional indicators for wetlands: Wetland hydrogeomorphic settings and functions		
Supplemental indicators	Ambient toxicity Sediment toxicity Other chemicals of concern in water column or sediment Health of organisms	Other chemicals of concern in water column or sediment Hazardous chemicals Aesthetics	VOCs (in reservoirs) Hydrophilic pesticides Nutrients Other chemicals of concern in water column or sediment algae	Other chemicals of concern in water column or sediment

Source: USEPA. 2003.

TABLE 8.3   Different U.S. Methods and Procedures Relative to Analytical Test Methods

Approved methods	Compliance methods to measure pollutants in various media for many CWA applications (Section 304(h), 40 CFR 136)
Other approved methods	Compliance methods (40 CFR 136.3, Table IF and IG) specific to pharmaceutical manufacturing and pesticide chemical point source catagories (40 CFR 439 and 435)
Other methods	Methods published by USEPA but not promulgated
Alternative test procedure (ATP)	Approval and validation process for submitting new test methods or procedures for USEPA approval (40 CFR 136.4, 136.5 and 141.27)
Detection and quantification	Procedures to confirm laboratory performance, managed by the Federal Advisory Committee on Detection and Quantification (FACQD)
Method updates	Revision of the list of approved analysis and sampling procedures (new methods added, other methods withdrawn, new sample collection procedures, general analytical requirements for multi-analyte methods, method flexibility requirements)

Source: Adapted from ISO, CD17994. 2001. *Water Quality—Criteria for the Establishment of Equivalence between Microbiological Methods*, Final Version, June 15, 2001.

8.6 Technologies and Tools for Urban Water Improvement

States and regulatory agencies do not accept most data that have been obtained by analysis conducted in the field by technologies such as test kits or portable meters. The main reason is that field equipment has to be meticulously and accurately calibrated and maintained. Another reason is that most portable kits do not have the ability to detect low concentrations of chemicals in surface waters because they are mainly designed for monitoring pollutants in high concentrations such as are found in effluents from wastewater treatment plants or in highly polluted water. However, these sorts of equipment can be used depending on the needs and ability of the end user. Field portable test kits and handheld instruments can produce accurate and reliable data but they must be calibrated properly and be used in appropriate conditions. These kits or devices use prepackaged sets of chemicals. The USEPA has identified five major criteria for data from measurements of water quality data:

Precision: Measured by the difference between samples taken from the same place at the same time

Accuracy: Given by comparing analysis of a known standard or reference sample with its actual value

Representativeness: How closely samples represent the true environmental condition or population at the time a sample was collected

Completeness: Whether enough valid or usable data has been collected (comparing what was originally planned and what has actually been collected)

Comparability: How data compares between sample locations or periods of time within a project or between volunteers

SOPs are step-by-step directions for methods including calibration and maintenance procedures for field and laboratory analytical instrumentation [13].

8.7 Poor Quality in Urban Water Supply Chain and Contributing Factors

8.7.1 Deregulation Global Crisis

Global crisis concerning food, water, and fuel instrumented in a detached fashion through computer program trading on the New York and Chicago mercantile exchanges, where the global prices of rice, wheat, and corn are decided upon. Poor water supply is not solely the result of policy failures at a national level. The process of global impoverishment leads to the simultaneous impingement on three fundamental necessities of life: food, water, and fuel. A small number of financial institutions and global corporations have set standards of living for millions of people around the world through market manipulation.

We are at the crossroads of the most serious water crisis in modern history.

8.7.2 Where Do These Problems Come From?

The main intergovernmental bodies including the United Nations, the Bretton Woods Institutions, and the World Trade Organizations (WTO) have endorsed the New World Order on behalf of their corporate sponsors. Governments in both developed and developing countries have abandoned their historical role of regulating key economic variables as well as ensuring a minimum livelihood for their people.

8.7.3 Global Mall Policies Effect

Protest movements are going on in different regions of the world. The conditions are particularly critical in Haiti, Nicaragua, Guatemala, Bangladesh, and India. These masses have been coupled with severe water shortages. An equally serious situation prevails in Ethiopia.

Other countries affected by food, water, and fuel crisis include Liberia, Egypt, Sudan, Mozambique, Zimbabwe, Kenya, Eritrea, and a long list of impoverished countries, not to mention those under foreign military occupation including Iraq, Afghanistan, and Palestine [17].

8.7.4 Water, Food, and Fuel: Necessities of Life in Jeopardy

The provision of water, fuel, and food have no longer been the core object of intergovernmental regulations or governmental interventions nowadays. With a view to averting the outbreak of famines or alleviating poverty from the community, these agencies have become blunt in their edge-cutting activities.

Recently, many studies have been conducted to investigate urban sources of water pollution such as WWTP effluents, CSOs,

and continuous and intermittent sources. These have often been cited as the primary causes of poor water quality in RWB.

An urban drainage system must be considered by means of an integrated approach which, although, have benefits but several aspects have prevented its wide application, one such aspect is the scarcity of field data (both input and output variables) and parameters that govern intermediate stages of the system, which are useful for robust calibration. The study showed the ease in model simplification, potential of the identification analysis for selecting the most relevant parameters in the model, and in assessing the impact of data sources for model reliability. Further, the analysis showed some critical points in integrated urban drainage modeling that can prevent the identification of some of the related parameters such as the interaction between water quality processes on the catchment, in the sewer, etc. [4].

8.8 Major Sources of Pollution in Urban Water Supply Systems

Urban water is being polluted by a variety of anthropogenic factors. Here is one of the real-time examples. There are many others like it that can be attributed to urban water quality consensus. We have water distribution related issues in Pakistan and probably South Asia is suffering as well. The water distribution lines to consumers are controlled by a valve at the start of each street, which is located in a chamber 4–5 ft deep. Mostly the valves leak and the pit is full of dirty water. Supply is intermittent and when stopped this dirty water is sucked in and mixes with clean water. We need to have some easy and viable solutions for this problem, which is polluting the majority of our urban water supply water.

8.8.1 Management and Best Solutions for the Improvement of Water Quality in Urban Areas

The EPA has advised to lessen detrimental effects on urban water quality.

Everyone can help in the management and quality standards maintenance of urban water. Here are a few practical measures that can be done:

Cleaning of spilled brake fluid, grease, oil, and antifreeze must be strictly watched by governing agencies. Otherwise, all the waste eventually reaches local streams, rivers, other water bodies, and lakes.

Street gutters and storm drains must be free of litter, pet wastes, leaves, and debris because these outlets drain directly to wetlands, lake, rivers, and streams.

Application of garden and lawn chemicals must be done sparingly and according to standard directions given by agricultural extension officers.

Disposing properly of all used antifreeze, oil, paints, and other household chemicals into prescribed chambers, not in storm sewers or drains because of sheer carelessness or due to mere ignorance. It is a collective responsibility to abide by these rules to keep the community clean.

Soil erosion can be controlled by planting ground cover on your property and stabilizing erosion-prone areas to mitigate sealing issues occurring into water runoff channels.

Community appraisal is inevitable to encourage local government officials to develop ordinances for the construction of the sediment/erosion control; there should be implementation on legislation and recuing protocols in the cities should be a top priority.

Proper operation of a septic system is only possible by its timely inspection, maintenance, and pumping every 3–5 years at a minimum.

Phosphorus-deficient detergents and cleaners should be introduced to the public to reduce the amount of nutrients discharged into our lakes, streams, and coastal waters because higher contents of this promotes eutrophication and blockages in the water pathways.

All water agencies and other social corporate departments should be aware of the existing problems and their minimization should be among their top priorities [6].

8.9 Various Agencies Dealing with the Issues of Urban Water Quality

According to UN sources, approximately 1 billion people around the globe, that is, 15% of the total world population have no access to clean drinking water. Almost 6000 children die every day because of infections linked to unclean water [9] (Figure 8.1).

8.9.1 Privatization of Water: Nestlé Denies That Water Is a Fundamental Human Right

A handful of global agencies and corporations including Suez, Bechtel-United Utilities, Veolia, Thames Water, and Germany's

FIGURE 8.1 The privatization of water: Nestlé denies that water is a fundamental human right. (Adapted from Global Research [Activist post by Kevin Samson, June 27, 2013].)

RWE-AG are acquiring ownership and control of public water utilities and waste management mechanisms worldwide. Approximately 70% of the total water resources management and all of the quality control systems that Veolia and Suez control are being privatized worldwide. This centralized management has its own importance for the stakeholders, but at the same time, people who cannot afford these standards set by advanced and developed countries of the world are suffering.

Here is the sad and bitter story about the regulatory and owner agencies. Privatization of water networks under World Bank auspices feeds on the collapse of public distribution and supply chains of safe tap drinking water: "The World Bank is serving the interests of water companies both through its regular loan programs to governments, which often comes up with new set conditions that quite explicitly have need of the privatization of water provision and public access." In fact, it's the reason of complete exploit of all and sundry [16]. "The modus operandi [in India] is clear—neglect development of water resources [under World Bank budget austerity measures], claim a 'resource crunch' and allow existing systems to deteriorate" [1].

Meanwhile, the markets around the globe for bottled water have been appropriated and governed by a handful of corporations including Coca-Cola, PepsiCo, Danone, and Nestlé. Is there a special reason for this trend the world over? These companies are linked with many other related agencies as they interact equally with the agribusiness-biotech companies. Currently, these corporations not only work hand in glove with the water utility, but trying the real resources crunch as a whole, they are involved in the food industry as well. It has been reported that simple tap water is purchased by agencies like Coca-Cola from a municipal water facility, even though it is free, and then resold on a retail basis to the masses. There are no checks and balances. From all of the previously mentioned situations, we conclude that privatization is the root cause of many problems to the public of developing countries. Let us suppose you have no access to these products and you cannot drink clean water or you have no governmental rights regarding the acquisition of the fundamental necessities of life. The presence of these companies altogether is a ban on governmental moves toward solving these problems within the community. It is estimated and reported that in the United States, 40% of bottled water is tap water [14].

One more example in the advanced phase of water resources privatization is the contemplation of ownership of lakes and rivers by private corporations. Mesopotamia, which was the valley of the two rivers (Tigris and Euphrates), was not only invaded for its extensive oil resources.

8.9.2 Concluding Remarks on Water-Related Agencies

Today we are dealing with a centralized constellation and complicated system of economic powers that have a direct bearing on the lives of millions of people. The prices of food, water, and fuel are determined beyond the approach of national government policies and concerns, but at the global level. Hikes in the prices of food, water, and fuel are contributing to "eliminating the poor" through "starvation deaths." The act to kill is instrumented in a detached fashion electronically on the commodity exchanges. This trend is universal in the exploitation around the world. If we are to check the issues and challenges in urban water quality, it is not a matter to be overlooked to make real-time reforms in the urban water quality network.

8.9.3 Other Agencies Dealing with Water

The Water Framework Directive (WFD) is a legislative framework to protect and improve the quality of all water sources, including lakes, rivers, transitional and coastal waters, and groundwater in the EU.

The success of the implementation of the WFD will depend on the information available to those charged with managing water quality being fit for purpose in both quantity and quality. Three modes of monitoring are specified in the directive:

- Surveillance monitoring to assess long-term changes
- Operational monitoring to provide extra data on water bodies at risk or failing to meet the environmental objectives of the WFD
- Investigative monitoring to determine the causes of such failure where they are unknown

U.S. water regulation is mainly based on three laws, respectively, the CWA [8], the Safe Drinking Water Act, and the Marine Protection, Research, and Sanctuaries Act. USEPA is a regulatory agency that proposes complementary regulations and rules that explain the critical technical, operational, and legal details necessary to implement laws.

The dynamism of all these stakeholders allows both the improvement of knowledge exchange and experience and promotion of the use of alternative screening methods to improve U.S. water monitoring. The CWA implementation allows the participation not only of the USEPA and its 10 regional offices, other federal agencies, states, and Indian tribes, but also other stakeholder groups (including associations, universities, and volunteers networks) more particularly for efficient monitoring of water bodies (USEPA, website). Actually, screening methods are defined as qualitative or semiquantitative methods, but must also include rapid methods. They are mainly used for field measurements and are also called alternative methods [20].

8.9.4 Clean Water Act

The Federal Water Pollution Control Act (FWPCA), called CWA since 1977, aims at the protection of surface water quality in the United States. It was written in 1948, but largely reorganized and expanded in 1972, and covers both the chemical aspects for the "integrity" goal, and the point source pollution issue (municipal and industrial wastewaters mitigation). Moreover, voluntary

TABLE 8.4 Key CWA Elements to Improve Quality of Threatened and Impaired Waters

Name	Targets
Section 402	Point sources of pollution discharging into surface water body; also called national pollutant discharge elimination system (NPDES) permit program
Section 319	Nonpoint sources of pollution (farming, forestry operations, etc.)
Section 404	Placement of dredged or filled materials into wetlands and other waters
Section 401	Certification from state, territory, or Indian tribes of no violation of water quality standard (WQS) for a construction or other activity before federal issuance of permits or licenses
State resolving funds (SRF)	Money for municipal point sources, nonpoint sources, and other activities

Source: USEPA. 2008.

programs and a regulatory approach have been used for, respectively, "nonpoint" runoff and "wet weather point sources" (USEPA, website). The key CWA elements are summarized in Table 8.4 (Figures 8.2 and 8.3).

States must assess their waters in order to know if the quality standards for the designated use of water bodies are met. The information provided is

- The status of the water bodies in the state, tribal land, or other territory, according to their designated use (healthy, threatened, impaired)
- The presence of pollutants (chemicals, nutrients, sediments, metals, temperature, pH) or other stressors (altered flows, invasive species), which causes impairment to water bodies
- The sources of pollutants and stressors

8.10 Environmental, Anthropogenic, Industrial, and Other Sources

8.10.1 Urbanization and Water Quality

One can understand why the water quality of our urban water supplies is so important, as the U.S. population now lives in or near cities rather than in the countryside. Big cities have an impact on the local water supply, which means big development over large areas. In fact, things happen that can harm the quality of all local water resources. Here are some urban development related water-quality issues:

- Ever-multiplying rapid rates of population especially in the developing nations of the world
- Erosion and sedimentation channelized by a variety of sources
- Urban runoff and sewage
- Others like pesticides, nitrogen, phosphorus, sewage overflows, and waterborne pathogens [6]

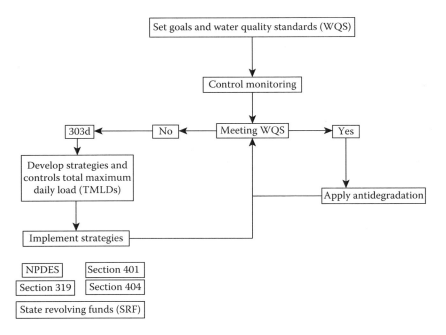

FIGURE 8.2   Key CWA elements. (From www.epa.gov/watertrain/cwa/cwa1.html.)

8.10.2 Contamination Emerging Concerns

Chemicals are being discovered, especially those referred to as "contaminants of emerging concern" (CECs), in water that previously had not been detected or reported as fata USEPA is conducting research to improve its understanding of a number of pharmaceuticals and personal care products (PPCPs) (Figure 8.3).

Interesting research being discussed is part of USEPA's ongoing endeavors to better understand its environmental occurrence, potential risks and threats, and is related to CECs. It summarizes technical information on pathogens and hormones that may affect water quality. It will identify information gaps that may help define research needs for USEPA and its federal agency, state, and local partners to better understand these issues [16].

8.10.3 Guide for the Identification of Helminth Eggs Present in Water Bodies

It has been extensively studied that wastewater frequently contains the eggs of parasites of animals, for example, domestic animals such as dogs, pigs, birds, and rats. Although it is not necessary to identify and classify these positively, it is still important to recognize that they are not of human origin and may cause risks of fatality. Plates I–XVII show a number of the egg models from human parasitic helminthes most frequently encountered in wastewater samples. Plate IV shows a clear view of the eggs [19] (Figure 8.4).

8.11 Quality Control Measures for Urban Water Quality

Here are some of the highlights from the "Announcement of 2013 Urban Water Quality Grant" that will help in understanding and establishing standards for the quality of urban water.

Dane County Urban Water Quality Grant proposed in Wisconsin states vividly the goals of the Urban Water Quality Grant (UWQG) are to improve the quality of urban stormwater runoff entering Dane County streams, lakes and rivers, and other water receiving bodies and channels. This will be helpful in increasing public awareness of urban water quality issues because humans are a key variable and focal point in the urban water quality "cause and effect relationship." This movement

FIGURE 8.3   Contamination emerging concerns.

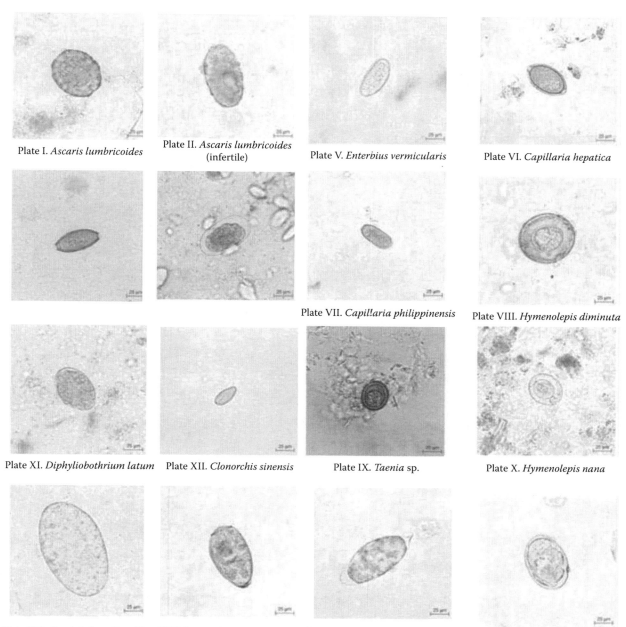

Plate I. *Ascaris lumbricoides*

Plate II. *Ascaris lumbricoides* (infertile)

Plate V. *Enterbius vermicularis*

Plate VI. *Capillaria hepatica*

Plate VII. *Capillaria philippinensis*

Plate VIII. *Hymenolepis diminuta*

Plate XI. *Diphyliobothrium latum*

Plate XII. *Clonorchis sinensis*

Plate IX. *Taenia* sp.

Plate X. *Hymenolepis nana*

Plate XIII. *Fasciola hepatica*   Plate XIV. *Paragonimus westermani*   Plate XV. *Schistosoma haematobium*   Plate XVI. *Schistosoma japonicum*

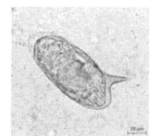

Plate XVII. *Schistosoma mansoni*

FIGURE 8.4   Guide for the identification of helminth eggs present in water bodies.

will provide public education for urban stormwater quality improvement practices and remedial measures adaptable by local agencies and other treatments concerning personnel. The project goals will be achieved through the construction of best management practices that will provide efficient, cost-effective treatment of urban runoff.

8.11.1 Financial Assistance Indicator

Funding should reimburse the total costs. Financial assistance is available to municipalities in the form of cost-sharing up to 50% of the total project cost, not to exceed $100,000.

8.11.2 Supervision of Ongoing Programs

All agencies carrying practices must meet the following eligibility requirements for urban water quality to be considered for funding.

All practical protocols must be designed to improve the quality of stormwater runoff from a developed drainage area with inadequate stormwater controls. It is a prerequisite demand of the existing systems on urban water quality maintenance executions. These should treat urban runoff draining to a lake, river, or stream in order to avoid all previously mentioned hazards to all species.

Practice performance should meet these factors as well. Annual sediment delivery processes must submit their activity log in a presentable layout at any international forum, sediment removal efficiency based on the impurities fate should be profunctional, contributing watershed area and the nature of hydrological cycles and the proximity to a beach are the main things under consideration. Demonstration value is mainly concerned with these key factors; location extent of the set quality maintenance station, public visibility is the fundamental desire of each sustainable sanitation area. Public accessibility at any time is a major requirement and it should impart some educational value to the public about the urban water quality. It should serve as the motivation point and ever-updated source of public awareness regarding everyday changes to the existing quality control agencies and novel technologies [2].

Figure 8.5 displays a schematic set up recommended by World Health Organization (WHO) to clean up wells. It is self-explanatory in nature (Figure 8.5).

8.11.3 Monitoring Groundwater Levels in Major Cities

It was studied that the rapid growth in well installation is causing adverse effects on the sustainability of groundwater potential and the water table in urban areas. Planning and management is inevitable; however, there should be development of a permanent water table monitoring network in major cities of the country. This monitoring is suggested to be done twice a month. A diagnostic survey of the main city of Pakistan is shown in Figure 8.6 [6].

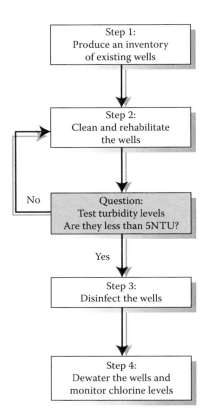

FIGURE 8.5 Steps for cleaning and disinfecting wells.

8.12 Summary and Conclusion

Water is a necessity that is required not only for driving life forces but also for various processes such as maintenance of a cleaner environment, generation of energy, and many others. With a growing urban population, water has suffered not only in quantity but also in quality. The reason for this chaos is poor management of resources and bad government policies.

Although scientific and technological developments in the field of water quality analyses has been very rapid, because these cannot be applied on a large scale they remain to be of lesser use. The portable kits and field equipments are often not sensitive enough or require calibration to determine trace levels of contaminants.

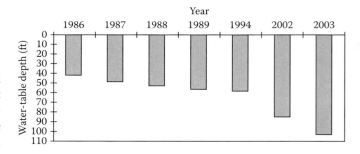

FIGURE 8.6 Groundwater level behavior in Islamabad, Pakistan.

A number of agencies and NGOs are working for maintenance/management of urban water resources at both national and international level; however, without the contribution of the public, the task of "pure water for all" cannot be accomplished. The governments, WHO, and responsible bodies must play their honest and sincere roles to eliminate the water crunch. Policies must be devised that are of worldwide acceptance and are beneficial to all people irrespective of their region, race, and color.

References

1. Ann, N., *Private Water, Public Misery, India Resource Center,* April 16, 2003.
2. Ayres, R.M. 1991. Comparison of techniques for the enumeration of human parasitic helminth eggs in treated wastewater. *Environmental Technology*, 12, 617–623.
3. Commission, 2000. Directive 2000/60/EC of the European Parliament and of the council establishing a framework for the community action in the field of water policy.
4. Chossudovsky, P.M. 2008. Global economy, poverty & social inequality. http://www.globalresearch.ca/the-global-crisis-food-water-and-fuel-three-fundamental-necessities-of-life-in-jeopardy/9191.
5. Csrees, K.A., Green, L., Herron, E., and Stepenuck, K. 2002. Why volunteer water quality monitoring makes sense. Factsheet II. http://www.usawaterquality.org/volunteer/.
6. Dane County Land and Water Resources Department (D.C.L.W.R.), Announcement of 2013 Urban Water Quality Grant, 2013, Madison.
7. Eslamian, S. (ed.). *Handbook of Engineering Hydrology and Water Management: Modeling of Wetland Systems*, CRC press, Taylor & Franscis, p. 241.
8. Final Revisions to the Clean Water Act Regulatory Definitions of "Fill Material" and "Discharge of Fill Material," USEPA, 2002.
9. Free water for all in Africa? BBC News, March 24, 2005.
10. Freni, G., Mannina, G., and Viviani, G. 2011. Assessment of the integrated urban water quality model complexity through identifiability analysis. *Water Research*, 45(1), 37–50.
11. Global Water Supply and Sanitation Assessment 2000 Report (UNICEF-WSSCC-WHO, 2000, 90 p.).
12. Gonzalez, C., Greenwood, R., and Quevauviller, P. 2009. Rapid chemical and biological techniques for water monitoring, 44–45.
13. ISO, CD17994. 2001. *Water Quality—Criteria for the Establishment of Equivalence between Microbiological Methods*, Final Version, June 15, 2001.
14. Jared, B. and Susan, L. The real cost of bottled water, *San Francisco Chronicle*, February 18, 2007.
15. Kumar, J.L.G. and Zhao, Y.Q. 2011. A review on numerous modeling approaches for effective, economical and ecological treatment wetlands. *Journal of Environmental Management*, 92(3), 400–406.
16. Maude, B., and Tony, C. *Water Privatization: The World Bank's Latest Market Fantasy,* Polaris Institute, Ottawa, 2004.
17. Picotte, A. and Boudette, L. 2005. *Vermont Volunteer Surface Water Monitoring Guide, Vermont Department of Environmental Conservation*, Water Quality Division.
18. Rauch, W., Bertrand-Krajewski, J.-L., Krebs, P., Mark, O., Schilling, W., Schuetze, M., and Vanrolleghem, P.A. 2002. Deterministic modelling of integrated urban drainage systems. *Water Science & Technology*, 45(3), 81–94.
19. Thomas, O. 2006. Alternative methods. In: Quevauviller, P., Thomas, O., and Van der Beken, A. (eds.), *Wastewater Quality Monitoring and Treatment*. John Wiley & Sons, Chichester, pp. 53–66.
20. U.S. Government Survey. 1996. *Water-Resources Investigations Report*, Everyone Lives Downstream (poster). Poster is available by calling 770-903-9100, p. 4302.
21. VanBriesen, J. 2013. Water Quest (Water Quality in Urban Environmental Systems). http://www.ices.cmu.edu/waterquest/home.asp.

9

Reuse, Potable Water, and Possibilities

Chander Kumar Singh
TERI University

Nidhi Jha
TERI University

Saeid Eslamian
Isfahan University of Technology

PREFACE

The growing population and exponentially expanding urbanization and development have pushed water demands beyond limits resulting in uncertainties for future generations. These uncertainties and scarcity of water have resulted in the development of many nonconventional water resources and development of various technologies so that used water can be reused and reclaimed.

This chapter reviews the present notion of water reuse, the various techniques and technologies of reclaiming water from different water resources, and the present extent of utilization of these reused water. Its focal point is the development of nonconventional water resources so that demand of the present generation is met without compromising the future generation. The chapter is based on the information obtained from various literature reviews of recent research studies, books, and case studies.

9.1 Introduction

9.1.1 Background

Earth is known as the "blue planet," with almost 70% of the Earth's surface covered with water, but in reality how much of that water is available to us? About 97% of the total water is saline water and only 3% of this is fresh water, of which only 33% is available to us in the form of lakes, groundwater, etc. [38]. This means water is a scarce, limited, but renewable resource. Unpredictable rainfall patterns due to climate change, depleting water resources, overdevelopment of groundwater, contamination of available surface/groundwater resources due to discharge of untreated sewage, and industry effluents add to water problems.

India is already experiencing water shortages in most blocks of the districts and this problem will become so acute in the near future that it will collapse the system unless we move toward conservation measures on a massive scale [34].

A notion about using recycled water and reclaimed water is that it is considered unsafe for potable or nonpotable uses. However, a water quality study done by Helgeson in 2009 compared the difference in quality of reclaimed/recycled water, groundwater, and surface water [16]. Results indicated that they are quite similar when one talks about its constituents. In the same research, the largest difference that appears between other water and reclaimed water is that the latter is disinfected and thus contains byproducts from disinfection [16].

9.1.2 General

Water is the most essential element for all living beings on Earth, so in order to preserve the biological functions and hydrological balance, it is important to develop adequate supplies of water. Effort should be made so that development of water sources will be within the capacity of nature to replenish. The latest technological intervention is therefore required for management of existing water sources so that sustainability is well ensured, safeguarding different water resources from getting polluted and to reuse, reclaim/treat the used water resources.

Water shortage is the most perpetual problem that is being faced by every continent. Almost one-fifth of the world's populations are living in areas suffering from physical scarcity of water, and by the year 2025, 1800 million people will be living in countries or areas with absolute water scarcity and two-thirds will be under stress conditions [18]. In addition to this, climate change produces variations in rainfall; these challenges could intensify in nearly every region of the world.

Resilience to water scarcity will require a large range of solutions, like using alternative water resources, depending less on groundwater resources, reusing water, treatment technologies, etc.

9.2 Definition of Nonconventional Water

Several different types of harvesting of water and its reclamation will have to be done to meet the huge demand of water and to decrease the gap between demand and supply. These new water resources are called nonconventional sources of water [19]. Nonconventional water includes all water that we get from reuse, treatment of wastewater, recycling of rainwater, and harvesting of different water resources.

There are three R's to save and conserve water. These three R's are Recycle, Reuse, and Reclaim. These practices will extend our limitation of finite water resources because the less we use now, the more we will have available for the future [11].

9.2.1 Recycle

The water recycling process is one in which gray- or wastewater is treated so that it can be used again either for potable use or for others (domestic, agricultural, and industrial). The recycling technique utilizes the very basic process of physical, biological, chemical, and disinfection principals to remove contaminants from wastewater. These processes can be modified into primary treatment (physical), secondary treatment (biological), and tertiary treatment (chemical). Tertiary treatment is expensive and not widely practiced, so different processes applied strictly depend on the demand for which water is required [26].

Reusing and recycling industrial water can ease the pressure on our water resources and avoid the need to discharge to the sewer or the environment. With appropriate management, which may include treatment, industrial water can be used for a wide range of purposes including industrial uses (e.g., cooling or material washing) or nonindustrial uses (e.g., irrigation or toilet flushing) [35].

The different processes of wastewater treatment are given here:

  i. *Primary treatment:* Half of the contaminants in wastewater are removed in the primary treatment process. This is the first step in the water recycling process in which raw sewage is made to pass through screens, coagulation, flocculation, floatation, etc., which separates large objects such as rags, plastic, oil fatty acids, and suspended solids.
  ii. *Screens:* This is the first process in the treatment to prevent large objects that are present in the wastewater. A screen can be designed depending on the contaminant to be removed, like course screening is done with spacing of more than 50 mm of screen bars to remove bigger contaminants. Medium screening is for medium-sized contaminants and the bar spacing is from 10 to 40 mm. Fine screening is for very small contaminants and the bar screening spacing is anywhere less than 10 mm. Even after screening, water still has physical compounds such as suspended solids and colloidal particles (>1 μm).
  a. *Grit chamber:* The semi-filtered water is then passed to the grit chamber where the influent flow is slowed down so that the suspended particles like sand and gravel fall down to the bottom, and fats, oil, or greases float on the top.
  b. *Coagulation:* It is the process by which destabilization of the colloidal particles are done by addition of coagulants. Addition of the coagulants to water results in

coming together of the different colloidal practices and suspended particles. Aluminum sulfate (alum) is the most common utilized coagulant.

c. *Flocculation:* Flocculation is the formation of agglomerates of destabilized colloidal particles (floc) through the addition of the reagents (flocculants). Most commonly used flocculants are activated silica and starch.

d. *Flotation:* It is the solid–liquid or liquid–liquid separation when the density of particles is less than the density of the liquid.

iii. *Secondary treatment:* This treatment process utilizes the microorganisms to convert most of the contaminants from the soluble and nonsettling solids into settleable solids, which are later captured in the final clarifiers, forming wastewater biosolids [16]. Microorganisms consume the organic matter as food, which greatly reduces the biochemical oxygen demand (BOD) in the water. First the air is mixed with the partially treated wastewater so that aerobic microbes can survive in the system and consume the organic matter. Since the process uses microbes, this process is always followed by a disinfection system, mainly chlorination. Then chlorine can be removed using sulfur dioxide (SO_2) to protect the aquatic life from highly toxic chlorine gas.

Most of the wastewater treatment is only up to this point; after this, the water is discharged in a stream or any water body so that dilution takes place. However, sometimes depending upon the usability, it is allowed to pass through the advanced stage of treatment process called tertiary treatment.

iv. *Tertiary treatment:* Tertiary treatment is used to separate salts and other fine mixed solids and helps in the disinfection of the water. These can be done using chemicals, membrane system (reverse osmosis [RO]), and using disinfection systems like that of ultraviolet disinfection, ozone disinfection, and later a cooling tower. This treatment process is costly and is only utilized if the water is required for potable use. Therefore, the factor to be considered for designing a water-recycling unit is the water quality requirements.

9.2.2 Reuse

Reuse of water is the repeated use of water so that water is not discarded after the first use. Domestic wastewater from the kitchen and bathroom can be reused for many purposes before discarding it [16]. Domestic wastewater is usually free of toxic materials and hence it can be reused in landscaping irrigation, plantation, etc.:

Urban reuse: The irrigation of parks, lawns, schoolyards, highways, and residential landscaping as well as for fire protection and toilet flushing for residential and commercial purposes. This does not require high-quality water and hence can be reused.

Agricultural reuse: Irrigation of nonfood crops like fiber and fodder, plantation, etc. can be done using domestic waste water. The irrigation of food crops requires high-quality reclaimed water.

Recreational impoundment: It is the body of reclaimed water made for aesthetic beauty. These include ponds, lakes, and landscape impoundments.

Wetlands: Creating artificial wetlands, enhancing natural wetlands, and sustaining stream flows can be done using domestic wastewater.

Industrial reuse: Some water requirements in industries do not require high-quality water like that of a cooling tower and hence can be reused.

9.2.3 Reclaim

Reclaimed water is the final product of wastewater treatment, which meets all of the requirements of water quality, that is, chemical, biological, and physical [35]. The major difference between recycled and reclaimed water is that recycled water only requires treatment up to the secondary treatment, but for reclamation, a high degree of tertiary treatment is required [16,17]. Hence, reclaimed water is a better version that is considered reliable and safe. The quality of this kind of water becomes more predictable than many groundwater resources and existing surface resources.

9.3 Sources of Nonconventional Water

9.3.1 Rainwater

Rainwater is the first form of water and is the primary source of water in the entire hydrological cycle. Rainwater is the source of all water forms like groundwater, reenter lakes, rivers, and oceans, which is being cycled via various processes of hydrogeological cycles. In India, monsoon season is from June to October, and the annual precipitation including snowfall is in the order of 4000 billion cubic meters [14] and much of this rainwater is actually left as runoff. To utilize the rainwater, low cost rainwater harvesting (RWH) can be installed to prevent runoff. Water harvesting is the collection of runoff for productive purposes. RWH can actually help in understanding the real value of rain and can help to overcome the water scarcity problem by storing rainwater for secondary uses and to recharge groundwater. The total mechanism of rainwater will be studied in detail later.

9.3.2 Groundwater

Groundwater is all the water that is present in the soil pores and rocks beneath the Earth's surface. These rocks bearing water are known as aquifers. The groundwater occurred in two zones—zones of aeration and zones of saturation. The zone of aeration comprises interstices partially filled by water and partially by air, whereas all the interstices of the zone of saturation are filled with water. The water table is the uppermost surface of the ground water, below which the entire pore spaces in soil or rocks are saturated with water.

Groundwater is the most precious form of water. Of all the fresh water on Earth, that is, about 3%, about 30.1% is present in groundwater. Groundwater occurrence in the country is highly uneven due to diversified geological formations with considerable lithological and chronological variations, complex tectonic framework, and climatological variations.

Groundwater resource comprises two parts—static groundwater resource, which remains perennially saturated, and the dynamic resources that reflect how seasonal recharge and aquifer discharge reflects water table fluctuations in major regions [24].

Thus, dynamic groundwater resource is basically the exploitable quantity of groundwater that is recharged annually and hence termed annually replenishable groundwater resource. In India, dynamic groundwater resources are estimated jointly by the Central Ground Water Board (CGWB) and state governments at periodical intervals.

9.3.3 Wastewater

Wastewater is water that is contaminated by different human activity such as bathing, dishwashing, laundry, fertilizing crops, and flushing the toilet. The contaminants include soaps and detergents, cooking oils, pesticides, paint, gasoline, seawater, pharmaceuticals, solid waste, and, of course, human waste (feces and urine along with used toilet paper and wipes). There are three types of wastewater depending on the source of their generation—domestic, sewage, and industrial wastewater. Domestic wastewater is the least contaminated of all. It can be reused for various purposes directly without any treatment, while the other types need a proper treatment before using. In a developing urban society, wastewater generation usually averages 30–70 m³ per person per year. In a city of 1 million people, the wastewater generated would be sufficient to irrigate approximately 1500–3500 ha. If we are able to treat wastewater and reuse it, then this wastewater will be a great resource [21].

9.3.4 Mine Water

Mine water treatment is a challenge that is being faced by many industries. Mine operators work to improve their water management system by reusing water, increasing the efficiency, and by using low quality and recycled water. However, acid mine drainage is the biggest problem that leads to contamination of surface and groundwater.

9.3.4.1 Acid Mine Drainage

Chemical reactions between water and rocks containing sulfur results in the formation of metal-rich water called acid mine drainage. It is through the mechanism that the heavy metals can be leached from rocks that come in contact with the acid; this process is accelerate by some of the microbes. This acid-rich water is so strong that it is able to dissolve heavy metals like copper, lead, and mercury into groundwater and surface water resulting in pollution [33]. Water from acid mine drainage can be reclaimed and remediated through biotic and abiotic treatments:

a. *Abiotic treatment*: It involves the passive and active system of neutralizing an acid mine drainage with an alkaline with high pH. This can be done by aeration and addition of lime and by forming passive limestone drains. Then it is passed through settling tanks to remove the sediment and particulate metals.

b. *Biotic treatment*: This involves the use of various bacteria to neutralize the acidic water. Biotic treatment is also categorized as acidic and passive processes. The most common biotic treatment involves formation of wetlands. Wetlands can reduce acid mine drainage through chemical, physical, and biological processes. Wetlands can remove metals through adsorption to organic substrates, and through bacteria, aerobic or anaerobic depending on the requirement. The residence time for which the water needs to be held in the wetland should be anywhere between 20 and 40 h for the acidic water with a pH of more than 5. If the mine water pH is below 5, the residence time should be 40 h [3].

9.3.5 Seawater and Brackish Groundwater

Seawater or brackish water was considered useful because of high total dissolved solids (TDS) content, but with advancing technology and increasing water demand, the possibility of utilization of brackish water and its management is possible. Studying of seawater and its movement is considered an important aspect as the overdevelopment of groundwater results in declining groundwater levels and these cause gradient so seawater enters the fresh groundwater zone, making the whole water saline. These cause serious implications for the environment.

9.4 Current Extent of Reuse of Water

The concept of reuse has been known for a long time in India; however, recycled water is most commonly used only for nonpotable (not for drinking) purposes, such as agriculture, landscape, public parks, and golf course irrigation. Other nonpotable applications include cooling water for power plants and oil refineries, industrial process water for such facilities as paper mills and carpet dyers, toilet flushing, dust control, construction activities, concrete mixing, and artificial lakes [8].

9.5 Rainwater

9.5.1 Rainfall Enhancement

Rainwater is the primary source of all of the sources of water present on Earth, but rain is uncertain and spread over time and space. In order to get continuous rain, many rainfall enhancement techniques are used. One of the most commonly used is cloud seeding.

Cloud seeding is the process of weather modification using application of scientific technology to enhance the cloud's ability to produce precipitation over an area. Cloud seeding is done by the process of spreading dry ice, silver iodide aerosols into the upper parts of the cloud to form new artificial ice particles. Cloud seeding can be done by static, dynamic, and hygroscopic cloud seeding methods. Static cloud seeding involves spreading the chemicals like silver iodide on the upper parts of the clouds and these chemicals form a crystal around which moisture can condense by dispensing the water. Dynamic cloud seeding involves boosting the currents in the air, which further encourages even more water to actually pass through the cloud, which gets converted into more air. This process is very complex and many meteorological conditions are required for using this methodology. Hygroscopic cloud seeding is done by dispersing salts through flares in the lower portions of clouds. It is also found that these salts actually grow in size when they come in contact with water [5].

Although in past this process has experienced some uncertain results because in which area this induced rain is going to happen is not certain, as clouds move it will be never be known whether a cloud that rains after seeding could have rained anyway. Therefore, the effectiveness of cloud seeding has to be increased in order to avoid uncertainties. Therefore, a key challenge for cloud seeding is to know how much precipitation would have happened had the clouds not been artificially seeded [32].

9.5.2 Rainwater Harvesting

Rainwater harvesting is the process of collecting rain where it falls or from the catchment area directly or by recharging water into the ground at a rate exceeding natural replenishment. It aims to prevent runoff by collecting it in reservoirs, which can be surface and subsurface [6]. Rainwater is the answer to various problems like the decline of the water level, seawater intrusion in coastal areas, improves the quality of groundwater and is a sustainable solution to water problems.

There are two main techniques of rainwater harvesting:

1. Storage of rainwater on the surface for future use
2. Recharge of aquifers/groundwater

These can be done through pits, trenches, dug wells, hand pumps, recharge wells, recharge shafts, rooftop rainwater harvesting, spreading techniques, lateral shafts, etc.

9.5.2.1 Quantification

The quantification of the rainwater that can be harvested in a region can be quantified very easily. The rainwater harvesting system mainly depends on the annual rainfall, the catchment area (the total area that catches rain), and the soil characteristics or collection efficiency (runoff coefficient). The total amount of rainfall received in an area is called rainwater endowment of that area, and when it is multiplied by the efficiency coefficient it gives the potential of rainwater harvesting [15].

Potential for rainwater harvesting (m³) = area (m²) × annual rainfall (m) × collection efficiency

Collection efficiency will mainly depend on

1. Runoff coefficient of the surface (depends on the type of surface as per Table 9.1)
2. Constant coefficient of evaporation, spillage, and wastage (always taken as 0.80) [25]

9.5.2.2 Factors

We observe that RWH potential will primarily depend on these factors and they should be taken care of when designing RWH:

1. Precipitation: Intensity of rainfall, duration of rainfall, and distribution of rainfall
2. Characteristics of the catchment: Size and shape, area, topography, and land use
3. Geological characteristics: Soil type
4. Meteorological characteristics: Temperature, humidity, and duration of sunshine

9.5.2.3 Designing

9.5.2.3.1 Subsurface storage

Designing the RWH structure for surface storage mainly depends on designing the storage/settlement tanks [25]. The storage requirements depend on local rainfall data, collection surface, runoff coefficient, user numbers, and consumption rates or water needs for productive use. The simplest method to calculate the required water volume is to use the following formula:

$$V = (t \times n \times q)$$

where

V: volume of the tank
t: number of days in dry period
n: number of people using the tank
q: total average consumption per capita per day (including all uses) [2]

TABLE 9.1 Values of Runoff Coefficient

Surface Type	Coefficient of Runoff
Bituminous streets	0.70–0.95
Concrete streets	0.80–0.95
Driveways, walks	0.75–0.85
Roofs	0.75–0.95
Lawns: sandy soil	
Flat, 2%	0.05–0.10
Average 2%–7%	0.10–0.15
Steep, 7%	0.15–0.20
Lawns: heavy soil	
Flat, 2%	0.13.0.17
Average 2%–7%	0.18–0.22
Steep, 7%	0.25–0.35

Source: Joint Committee of the American Society of Civil Engineers and the Water Pollution Control Federation (ASCE-WPCF). 1970. Design and construction of sanitary and storm sewers, ASCE Manual of Engineering Practice No. 37, WPCF Manual of Practice No. 9, 1st Revised Ed., New York.

As a safety factor, the tank should always be built 20% larger than the volume calculated from the equation.

9.5.2.3.2 Aquifer recharge structure

Harvested rainwater can be directed to the groundwater to augment depleted aquifers. The various types of recharge structures constructed are through abandoned dug wells, hand pumps, recharge pits, trenches, gravity head recharge tubewells, and recharge shafts. It is discussed in detail in groundwater recharge Section 9.6.2 [25].

9.6 Groundwater

9.6.1 Brackish Groundwater

Groundwater is classified as brackish water when the TDS ranges from 1000 to 10,000 mg/L [13]. The geographical distribution of brackish and saline groundwater is not fixed in time and subject to change because depletion of groundwater can lead to the concentration of salt and hence conversion of brackish water to saline takes place and from saline to brine. Another effect of groundwater depletion is seawater intrusion. When the fresh water from shallow aquifers near the sea is abstracted, it results in groundwater head decrease. This can cause up-coning of the saline sea water and even fresh water can be converted into saline water. Salinity of the groundwater can be reduced by recharging the aquifers and hence diluting the salt [37].

To deal with saline groundwater and to make it usable, several management and scientific techniques should be taken up:

a. *Conjunctive use of water:* When there is marginal quality groundwater, it makes sense to use this water in conjunction with better types of water. The brackish quality of groundwater could physically be blended with more freshwater to provide water with comparatively less TDS levels, which can be used for many applications. Alternatively, poor quality groundwater could be applied in an alternating fashion with better quality water.

b. *Desalination of groundwater:* There are various techniques to treat the saline groundwater in situ as well as ex situ. The desalination process separates dissolved solids from the water by phase change process (distillation, freezing) and also be membrane processes (RO, electrodialysis, ultrafiltration). However, this process can be costly because advanced technologies are used.

c. *Carbon dioxide sequestration:* Carbon dioxide sequestration in deep saline aquifers in order to reduce climate change has been researched. This way, saline aquifers can act as storage houses for the carbon dioxide and hence can be used.

Other methods include reducing the effluents from agriculture and industries, aquifer storage and recovery (ASR), and desalination using renewable energy

9.6.2 Direct Aquifer Recharge

There are two methods by which groundwater is being recharged. It is either by direct aquifer recharge methods or by indirect aquifer recharge methods [15]. We will study the indirect aquifer recharge technique in the next subsection.

Direct aquifer recharge is the simplest form of aquifer recharge. Here water from the land surface moves to the aquifers by means of percolation and filtration through soil. Most of the aquifer recharging is done from this method. The direct aquifer recharge method can be further subdivided into surface spreading method and subsurface techniques [30]. The schematic representation of the types of direct aquifer recharge methods is shown in Figure 9.1.

9.6.2.1 Surface Spreading Techniques

In the surface spreading technique, the water falls on the soil surface and directly is percolated to the aquifer through soil. The surface spreading technique does not require any engineering design, is very easy to implement, and is less costly [28]. The water that enters the system through this technique is always better in quality because it passes naturally through various unsaturated zones of soil. Quantity-wise, it does not give good results. The quantity of water recharged through the surface spreading technique is always less due to the infiltration capacity of soil. Another disadvantage of this technique is that the length of time water is in contact with the soil and area of recharge is always large in order to get a good recharge rate. Through this technique, only shallow aquifers can be recharged. Some of the examples of surface spreading techniques are flooding techniques, by making ditch and furrow to arrest water, by making artificial recharge basins, runoff conservation structure, stream modification, surface irrigation, etc.

9.6.2.2 Subsurface Recharge Techniques

In the subsurface spreading technique, the water is taken directly to the aquifers. For this method, the water is collected through various runoff conservation structures or stream modifications and stormwater is directed toward the recharge or injection well, which takes the water in the aquifers. Through this technique, deeper aquifers can be recharged and it takes comparatively less area. However, construction and maintenance of such a structure are costly and require proper planning and designing of the recharge unit. The general conceptual design of the subsurface recharge structure connected to stormwater is shown in Figure 9.2.

A subsurface recharging system will always have a PVC pipe that will take the stormwater to the recharge cell. In recharge cells, different permeable materials are organized in layers to do the work of filtration and oxidation before water reaches the injection well or recharge well. The first layer is sand, then gravel, followed by boulder at the bottom. This acts as a natural filter to the storm runoff water. All subsurface recharge techniques are susceptible to clogging and choking by suspended

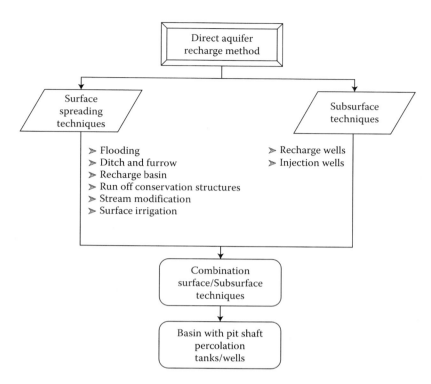

FIGURE 9.1 Flow diagram showing direct aquifer recharge methods.

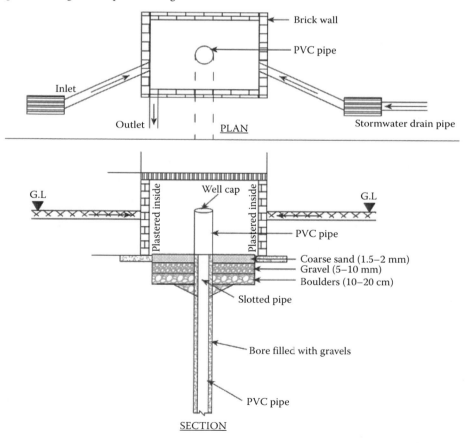

FIGURE 9.2 Conceptual design of artificial recharge structure. Note: Length and placement of slotted pipe shall be based on the aquifer zones encountered. Figure not to scale.

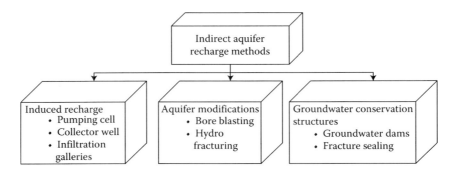

FIGURE 9.3 Flow chart showing indirect aquifer recharge methods.

solids, biological activities, and chemical impurities, so proper maintenance and cleaning of the materials are required.

9.6.2.3 Combination of Surface and Subsurface Techniques

A combination of surface and subsurface techniques gives the best result. The water is collected through the surface spreading techniques like through stream modification, by making furrow and ditch, and this water is collected with pipes to the subsurface recharge shafts or injection wells to recharge aquifers that are relatively deeper [20]. These can also recharge aquifers that are shallow or superficial aquifers formations that are impermeable or clayey.

9.6.3 Dams/Indirect Aquifer Recharge

Indirect aquifer recharge methods are all those methods that involve modification of aquifers or artificially inducing recharge and by groundwater conservation structures. The schematic representation of the types of indirect aquifer recharge is given in Figure 9.3. The effectiveness of this recharge type will depend on a number of factors like the nearness of water bodies to the surface or, perhaps, the hydraulic transmissivity of said aquifer. Some other factors include, but are not limited to, the permeability and area of the lake bottom or streambed [10]. The latter technique can be used to modify structures that are impeding the groundwater flow by the use of aquifers. Creating additional storage capacity is also another method to do this. There are a number of dams or groundwater barriers that have been constructed on these riverbeds in the recent past. This also includes India, which further helps in detaining and obstructing the groundwater to sustain the capacity of the storage of said aquifer. This also helps in meeting water demands during periods of need. Here, constructing aquifers on a small scale also seems a feasible solution.

a. *Induced recharge:* This is done by artificially installing a groundwater pumping facility near the hydraulically connected streams or lakes to lower the groundwater level and induce infiltration elsewhere in the drainage basin. It can be done using a pumping cell, collector well, and infiltration galleries. The advantage of this type of recharge is

that the quality of the surface water generally improves before it reaches the ground, due to its path through the aquifer material [9].

b. *Aquifer modification:* When there is an impermeable layer over an aquifer, sectional blasting of boreholes with suitable techniques is done on that layer to interconnect the fractures and to increase recharge. This can be done through bore blasting and hydrofracturing.

c. *Groundwater conservation structures:* To arrest the subsurface flow, certain structures are made to arrest the water in an area so that water is percolated down in the aquifers. The groundwater conservation structures are dams or subsurface dykes (locally termed Bandharas).

A groundwater dam barrier across any subsurface water stream stores the water to one side of the aquifer. The main purpose of a groundwater dam is to seize the flow of groundwater out of the subbasin and increase the storage within the aquifer. Underground dams (Figure 9.4) are advantageous because the loss through evaporation from a reservoir is not there because it is subsurface, no siltation in the reservoir takes place, and disaster like collapse of the dams can be avoided.

9.7 Wastewater

Wastewater is no longer considered waste. Various technologies are used to reclaim this wastewater and reuse it. Generally, wastewater is defined as any liquid and waterborne solids that have been used and fouled by different human activities and hence the quality has been degraded and usually is discharged to a sewage system. Wastewater is broadly classified into domestic wastewater, storm sewage, and industrial wastewater.

9.7.1 Domestic Wastewater

Domestic wastewater is also called greywater and covers all the non-toilet household wastewater from industries, commercial and industrial bathroom sinks, bathtubs, kitchen waste, clothes washing, etc. Greywater can be used for various purposes like landscaping irrigation, plantation, etc. because it is nontoxic. Domestic wastewaters are usually of a known quality and hence can be reused for many purposes including

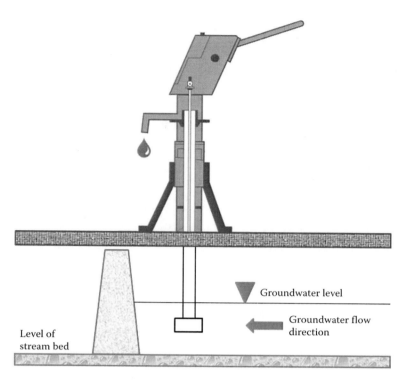

Level of
stream bed

Groundwater level

Groundwater flow
direction

FIGURE 9.4 Diagram showing a groundwater dam.

Urban reuse: The irrigation of parks, lawns, schoolyards, highways, and residential landscaping as well as for fire protection and toilet flushing for residential and commercial purposes. All these do not require high-quality water and hence can be reused.

Agricultural reuse: Irrigation of nonfood crops like fiber and fodder, plantation, etc. can be done using domestic wastewater, while the irrigation of food crops requires high-quality reclaimed water.

Recreational impoundment: It is the body of reclaimed water made for aesthetic beauty. These include ponds, lakes, and landscape impoundments.

Wetlands: Creating artificial wetlands, enhancing natural wetlands, and sustaining stream flows can be done using domestic wastewater.

Industrial reuse: Some water requirement in industries does not require high-quality water like that of cooling towers and hence can be reused.

9.7.2 Industrial Wastewater

Industrial wastewater includes all the wastewater that is not defined as domestic wastewater. These are the discharges from industrial plants and manufacturing units. The industries that give out a large amount of wastewater include paper and fiber industries, chemical and fertilizers units, refining and petrochemical operations, vegetable and fruit packing, processed food like meats, etc. Industrial wastewater varies widely in composition of chemicals, strength, pH, flow, and volume.

In general, industrial discharge mostly consists of strong organic wastewater with an increasingly high oxygen demand or may also contain undesirable chemicals. This can further damage many sewers and other structures. They generally contain compounds that resist biological degradation and are therefore harmful to the satisfactory operation of the wastewater treatment plant. As per government norms, industries should provide a pretreatment or partial treatment of the wastewater coming out of industries to certain permissible conditions prior to discharge to municipal sewers. Treatment of these units is a tedious exercise for all industrial units, so many industries work together to treat their wastewater. In location of industrial units, wastewater discharge from different industries is collected and mixed wastewater is treated together in a treatment unit; this is called a common effluent treatment plant (CETP).

In CETP, the industrial wastewater is first passed through a combination of physical, biological, and chemical processes that remove some or most of the pollutants. For meeting the standards of effluent wastewater, CETP has provided physicochemical and biological treatment. After biological treatment, tertiary treatment with an oxidation tank is also provided in the treatment scheme. Physical treatment involves a screen chamber, grit chamber, oil and grease trap, and equalization tank. Biological treatment consists of diffused aeration and a settling tank. Tertiary treatment provided is chemical oxidation and filtration. A general flow diagram of the CETP at Five Star MIDC, Kagal is given in Figure 9.5 [29].

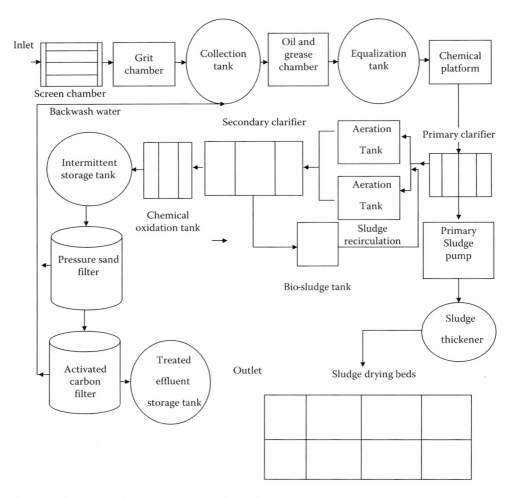

FIGURE 9.5   Flow diagram of common effluent treatment plant (CETP).

9.7.2.1 Preliminary Treatment

Primary treatment starts when the industrial wastewater is taken in the screen chamber where larger particles like debris, stones, leaves, etc. are separated here. After that, it is passed through the grit chamber where sedimentation and suspended large particles are removed [26]. Water from this is collected and sent to the oil and grease trap, which is a baffled wall channel where wastewater is subjected to up and down flow for the removal of floating particles at the top surface. Separated oil and grease is separated from the top layer by skimming and it is sent to the equalization tank. The equalization tank ensures the complete mixing of quality and quantity. Equalization is provided with submerged mixers in order to provide complete mixing of the water [26].

9.7.2.2 Chemical Treatment

Effluents from the primary treatment are sent to the physio-chemical treatment section of CETP. Here treatment of wastewater is done using various chemicals like coagulants and flocculants mainly to precipitate and coagulate the suspended solids and to remove the big coagulated mass using gravity settling in the primary clarifiers. From primary clarifiers, sludge is

removed and sent for processing to the sludge thickening unit and then to the sludge drying beds.

9.7.2.3 Secondary Treatment/Biological Treatment

Clear overflow from primary clarifier is conveyed to aeration tanks for biological treatment. Biological treatment is achieved by providing aeration to facilitate growth of aerobic microbes. Nutrient and food is supplied to microorganisms for enhancing their growth and after microbes fully decompose the chemicals, it is sent to the secondary clarification where bottom sludge is recirculated back to the aeration tank through a bio-sludge tank [26].

9.7.2.4 Tertiary Treatment

Tertiary treatment consists of chemical oxidation, pressure sand filters, and activated carbon filters. Effluents from the biological treatment are passed to the chemical oxidation where hydrogen peroxide dosing is done. After that, it is collected in the intermittent storage tank; from there, effluent is subjected to different filtering systems like pressure sand filter and activated carbon filter. Suspended solids are completely removed in pressure sand filters and activated carbon filters provide treatment of removal of color and chemical oxygen demand (COD) so that the final outlet meets the norms of the government.

Wastewater treatment processes differ in reducing the concentration of parameters of concern such as BOD or suspended solids, etc. and the standards of discharge determine whether a given combination of treatment processes provides an acceptable level of treatment. Thus, CETP measures the quality of the effluents at the inlet and at the outlet of the plant to meet the standard.

9.8 Social and Traditional Aspects

9.8.1 Religious

Water is regarded as precious in the Hindu religion and rivers are considered as goddesses and are worshiped in many parts of India. Hence, water conservation and preservation was sanctioned in Vedas and our religious texts. Various religious, cultural, and social rituals involved purification with water. Some of our ancient texts and Vedas give good examples of water storage and conservation [4]. There are examples where sage Narad during his visits to different kingdoms would regularly enquire about the state of ponds and other water bodies and if the water was enough to cater to the needs of the people. Stories of childhood and early life of Shri Krishna showed the importance of life-sustaining water of Jamuna (Yamuna) River where he grew up.

In fact, in Islam, water conservation is crucial because this religion originated in the Arabian Peninsula, which is a desert area, so scarcity of water has always prevailed in those areas and hence its water conservation techniques were crafted in their customs and behavior. It is evident from numerous verses in the Quran that water is a major theme in Islamic cosmogony and iconography as well as a recurrent topic in liturgy and daily life. Some hadiths include sayings of Mohamed prohibiting the wasteful use of resources, most importantly water, which is a very valuable resource in Arab lands [36].

9.8.2 Traditional Aspect

India has the proud traditional wisdom of many thousands of years in development of technologies for water conservation and groundwater recharge:

1. In order to protect water conservation structures like wells, etc. from any form of pollution, religious paintings or inscriptions were done on these structures so that only bare feet were allowed to keep them pure and clean.
2. The water conservation structures have been a continuous process for ages in India. In the eleventh and twelfth century AD, major revolution took place as one of the major existing lakes in Bhopal was constructed in an area of 647 km². The development of various structures for water management from third century BC to eighteenth century AD is briefly given in Table 9.2.

TABLE 9.2 Water Harvesting Practices in India over the Years

Time	Place/Structure
Third millennium BC	Dams built of stone rubble were found in Baluchistan and Kutch
3000–1500 BC	Indus–Sarasvati civilization had several reservoirs to collect rainwater runoff. Each house had an individual well
Third century BC	Kautilya's Arthasastra mentions irrigation using water harvesting systems
First century BC	Sringaverapura near Allahabad had a sophisticated water harvesting system using the floodwaters of the Ganges
Second century AD	Grand Anicut or Kallanai built by Karikala Chola across the River Cauvery to divert water for irrigation is still functional
Eleventh century AD	King Bhoja of Bhopal built the largest artificial lake (65,000 acres) in India fed by streams and springs
Twelfth century AD	Rajatarangini by Kalhana describes a well-maintained irrigation system in Kashmir

3. Design and construction of Baoli that is step wells was a revolutionary development in water conservation structures. These were mostly constructed in the arid and semi-arid areas of North and West India between the ninth and seventeenth centuries. In the Baolis, the steps were constructed beyond the occurrence of the water table and the paved areas of the Baolis acted as catchment to create enough runoff and diverting it to the step well. Some of the Baolis are still functional and in use. These are well-designed ethnic structures showing high competency in architectural design and technology [1].
4. Another important technology evolution was during the twelfth century when the tankas (a cylindrical underground structure with variable depths) were constructed in arid areas for water storage. The tankas were typical of the desert area where runoff is not created due to sand dunes; therefore, the artificial catchment area was created by constructing a large diameter cemented structure around the tankas.

9.9 Impacts

9.9.1 Impact on Groundwater Quality

Groundwater contains a wide variety of dissolved inorganic chemical constituents in various concentrations, resulting from chemical and biochemical interactions between water and the geological materials. Inorganic contaminants including salinity, chloride, fluoride, nitrate, iron, and arsenic are important in determining the suitability of groundwater for drinking purposes. Whenever any of these compounds are present in undesirable conditions, it then affects the health of the water and results in contamination.

Groundwater quality can be influenced by the atmosphere through rainwater and also from surface water bodies. Quality of groundwater is mainly influenced by anthropogenic factors like overexploitation of groundwater, excessive use of fertilizers, industrial effluents, etc. [7].

Groundwater recharge through an artificial recharge system can enhance the water quality of the groundwater by removing suspended solids by filtration through the soil or by dilution by mixing with naturally occurring groundwater. Reclaimed water, which is actually planned to be used in recharged aquifers or further for augmenting the surface water, actually receives a number of adequate treatments before really mixing it with naturally occurring water or subjecting it to natural restoration process [23].

9.9.2 Environment

Impact of reuse and recycle of nonconventional water resources is always positive. In order to study the environmental issues, a risk assessment study for possible health risks of recycled water and comparisons to conventional pharmaceuticals and personal care product (PPCP) exposures was conducted by the WateReuse Research Foundation [12]. For each of four scenarios in which people come into contact with recycled water used for irrigation—children on the playground, golfers, and landscape and agricultural workers—the findings from the study indicate that it could take anywhere from a few years to millions of years of exposure to nonpotable recycled water to reach the same exposure to PPCPs that we get in a single day through routine activities.

9.10 Summary and Conclusions

Expanding population, climate change, and exponential urbanization has led to tremendous pressure on water resources. In fact, water use has been growing at more than twice the rate of population increase in the last century and hence the biggest problem arising is scarcity of water [31]. As discussed, technological interventions, exploiting of nonconventional water resources, maximum utilization of rainfall, treatment technologies of wastewater, etc. will play a significant role in safeguarding the long-term water supply in order to maintain increasing water demand [27]. This type of sustainable management of water resources can bring economic, social, and environmental benefits. By wasting less water, polluting less, reusing more, managing effectively, and becoming more efficient in all uses of water, we can help in the reduction of water stress.

References

1. Agarwal, A. and Narain, S. 1997. *Dying Wisdom: Rise, Fall and Potential of India's Water Harvesting Systems*. Centre for Science and Environment, New Delhi.
2. Anjaneyulu, L., Arun Kumar, E., Sankannavar, R., and Kesava Rao, K. 2012. Defluoridation of drinking water and rainwater harvesting using a solar still. *Industrial & Engineering Chemistry Research*, 51(23), 8040–8048.
3. Bolis, J.L., Wildeman, T.R., and Cohen, R.R. 1991. The use of bench-scale parameters for preliminary analysis of metal removal from acid mine drainage by wetlands. In: *Wetlands Design for Mining Operations. Proceedings of the Conference of American Society for Surface Mining and Reclamation*, May 14–17, Durango, CO, pp. 123–126.
4. Chuvieco, E. 2012. Religious approaches to water management and environmental conservation. *Water Policy*, 14(2012), 9–20.
5. Cotton, W.R. 1998. *Weather Modifications by Cloud Seeding—A Status Report* 1989–1997, Colorado State University, Department of Atmospheric Science.
6. *Criteria and Guidelines for the Rainwater Harvesting*. 2010. Pilot Project Program, Colorado Water Conservation Board (CWCB).
7. Das, S. 2006. Groundwater overexploitation and importance of water management in India—Vision 2025. *Tenth IGC Foundation Lecture*, The Indian Geological Congress, Roorkee.
8. Dhiman, S.C. and Gupta, S. 2011. *Select Case Studies Rain Water Harvesting and Artificial Recharge*. Central Ground Water Board, Ministry of Water Resources, New Delhi.
9. Dillon, P.J. (ed.). 2002. *Management of Aquifer Recharge for Sustainability*. A.A. Balkema Publishers, Lisse, the Netherlands, 567pp.
10. Dillon, P., Fernandez Escalante, A.E., and Tuinhof, A. 2012. Management of aquifer recharge/discharge processes and aquifer equilibrium states. GEF-FAO Groundwater Governance, Thematic Paper.
11. European Commission. 2012. Science for environment policy. Thematic issue: Managing water demand, reuse and recycling, Issue 33.
12. Falkenmark, M., Lundqvist, J., and Widstrand, C. 1989. Macro-scale water scarcity requires micro-scale approaches. *Natural Resources Forum*, 13(4), 258–267.
13. Freeze, R.A. and Cherry, J.A. 1979. *Groundwater*. Prentice-Hall, Inc., Englewood Cliffs, NJ.
14. Freshwater Year. 2003. *Rain Water Harvesting Techniques to Augment Ground Water*. Central Ground Water Board, Ministry of Water Resources, New Delhi.
15. *Guide on Artificial Recharge to Ground Water*. 2000. Central Ground Water Board, Ministry of Water Resources.
16. Hartley, T.W. 2006. Public perception and participation in water reuse. *Desalination*, 187(1), 115–126.
17. Helgeson, T. 2009. *A Reconnaissance-Level Quantitative Comparison of Reclaimed Water, Surface Water, and Groundwater*. Water Reuse Research Foundation, Alexandria, VA. p. 141.
18. *Human Development Report*. 2006. United Nations Development Programme.
19. Jain, S.K. 2012. Sustainable water management in India considering likely climate and other changes. *Current Science*, 102(2), 177–188.

20. Jha, B.M. and Sinha, S.K. *Towards Better Management of Ground Water Resources in India.* Central Ground Water Board.

21. Jhansi, S.C. and Mishra, S.K. 2013. Wastewater treatment and reuse: Sustainability options. *Consilience: The Journal of Sustainable Development*, 10(1), 1–15.

22. Joint Committee of the American Society of Civil Engineers and the Water Pollution Control Federation (ASCE-WPCF). 1970. Design and construction of sanitary and storm sewers, ASCE Manual of Engineering Practice No. 37, WPCF Manual of Practice No. 9, 1st Revised Ed., New York.

23. Llamas, M.R., Mukherjee, A., and Shah, T. 2006. Guest editors' preface on the theme issue "Social and economic aspects of groundwater governance". *Hydrogeology Journal*, 14(3), 269–274.

24. *Manual on Artificial Recharge of Groundwater.* 2007. Central Ground Water Board, Ministry of Water Resources.

25. *Manual on Rainwater Harvesting and Conservation*, 2002. Central Public Works Department, New Delhi.

26. Metcalf, L. and Eddy, H.P. 1930. *Sewerage and Sewage Disposal: A Textbook*, 2nd edition. McGraw-Hill Book Company, Inc., New York.

27. *Ministry of Water Resources.* 2002. National Water Policy, New Delhi.

28. National Institute of Hydrology, Roorkee and Central Ground Water Board. 2006. An operational model for groundwater pumping at Palla Well Fields, Project Report, NCT Delhi.

29. Powar, M.M., Kore, V.S., and Kore, S.V. 2012. A case study on common effluent treatment plant at five star MIDC, Kagal. *World Journal of Applied Environmental Chemistry*, 1(1), 1–6.

30. Raju, K.C.B. 1990. Rainwater harvesting through subsurface dam. In: *Proceedings of Seminar on "Modern Techniques on Rainwater Harvesting,"* GSDA, Government of Maharashtra, Pune.

31. Romani, S. 2006. Groundwater management—Emerging challenges. Groundwater governance—Ownership of groundwater and its pricing. *Proceedings of the 12th National Symposium on Hydrology*, New Delhi, November 14–15.

32. Sibal, S.K. 2005. "Cloud Seeding" (http://dst.gov.in/admin_finance/un-sql207.htm). Department of Science and Technology, Government of India.

33. Skousen J. and Ziemkiewicz, P. 2005. Performance of 116 passive treatment systems for acid mine drainage. *National Meeting of the American Society of Mining and Reclamation*, Breckenridge, CO.

34. Speidel, D.H. and Agnew, A.F. 1988. The world water budget. In: Speidel, D.H., Ruedisili, L.C., and Agnew, A.F. (eds.), *Perspectives on Water: Uses and Abuses.* Oxford University Press, New York, pp. 27–36.

35. Wade Miller, G. 2006. Integrated concepts in water reuse: Managing global water needs. *Desalination*, 187(1), 65–75.

36. Water Harvesting—Our Age Old Tradition, Chapter II. As read in http://megphed.gov.in/knowledge/Rainwater Harvest/Chap2.pdf.

37. Weert, F.V., Gun, J.V., and Reckman, J. 2009. Global overview of saline groundwater occurrence and genesis, International Groundwater Resources Assessment Centre (IGRAC), Utrecht.

38. World Resources Institute. 2000. World Map, World Resources Institute, Washington, DC, available at http://earthtrends.wri.org/images/maps/2-4_m_Water Supply2025_lg.gif (verified April 16, 2008).

10

Urban Water Reuse in Agriculture

M. Dolores
Hidalgo Barrio
CARTIF Technology Center
ITAP—University of Valladolid

Jesús M. Martin
Marroquin
CARTIF Technology Center
ITAP—University of Valladolid

PREFACE

Water is an environmental, social, and economic asset and as such, it needs to be managed with the objective of conserving a common patrimony in the interests of the community at large.

Many countries have the problem of a severe water imbalance. This imbalance in water demand versus supply is due mainly to the relatively uneven distribution of precipitation, high temperatures, increased demands for irrigation water, and the impacts of tourism. To alleviate water shortages, serious consideration must be given to wastewater reclamation and reuse because it is necessary and important to guarantee water availability over time by means of sustainable forms of management, which will allow countries to cope with present demands without jeopardizing environmental balance and the needs of future generations.

This chapter refers to the collection and analysis of information regarding previously implemented methodologies, schemes, and good practices with respect to urban wastewater treatment and effluent reuse in agricultural production. The aim is to reveal the problems associated with the water resources available, the wastewater treatment capacity, and the treated wastewater reused for irrigation purposes. Furthermore, some good examples of the potential benefits of wastewater reuse in different countries are presented, giving special consideration to the costs associated with the processes. Naturally, they are most obvious for arid areas but the general increasing pressures on water resources all over the world should also make wastewater reuse attractive in other areas. The criterion for selecting these cases has been the contribution of the system into the overall increase of wastewater reuse in the country implemented. The effect of potential success factors and parameters has been looked into in order to realize why the implementation of these systems was proven highly successful.

10.1 Introduction

Water availability is central to the success of agricultural enterprises, domestically and globally, and cuts across multiple disciplines related to human health, food safety, economics, sociology, behavioral studies, and environmental sciences [21]. As such, almost 60% of all the world's freshwater withdrawals go toward irrigation uses. Farming could not provide food for the world's current populations without adequate irrigation [14].

Urban wastewater is used to irrigate agricultural land in many countries. The use of wastewater for irrigation is a way of disposing of urban sewage with several advantages. Wastewater contains many nutrients, which make the crop yields increase without using fertilizer. Furthermore, sewage is an alternative water source in arid and semi-arid areas where water is scarce. Besides these advantages, wastewater can contain heavy metals, organic compounds, and a wide spectrum of enteric pathogens that could have a negative impact on the environment and human health.

Focusing on arid countries, these are characterized by the low level and irregularity of water resources, both through time (summer drought, interannual droughts) and location (dry in the South). Furthermore, the rapid growth in urbanization, tourism, irrigation, and population can only increase tensions in these countries and regions where consumption has already reached the available amount.

Countries of the Southern Mediterranean and Middle East region, for example, are facing increasingly more severe water shortage problems [4,5]. Some of them have few naturally available freshwater resources and rely mainly on groundwater. Surface waters have already been, in most cases, utilized up to their maximum capacity. Groundwater aquifers are often overdrafted, and brackish and seawater intrusion in coastal areas has reached threshold limits in some locations.

Nonrenewable deep or fossil aquifers are being tapped to varying degrees. Exploitation of nonrenewable resources of Saharan aquifers is intensive in Libya, Egypt, Tunisia, and Algeria. Desalination of brackish and seawater has already been under implementation or planned in some countries despite its high cost. National exploitation ratios over 50%, or even approaching 100%, in several Mediterranean countries (Egypt, Gaza, Israel, Libya, Malta, Tunisia) show that the actual water consumption exceeds the renewable conventional water resources. Consequently, several problems appear all around the basin such as water and soil salinization, desertification, increasing water pollution, and unsustainable land and water utilization.

On the other hand, the volume of wastewater is also increasing in these arid regions, so irrigational reuse appears here as an adequate strategy to dispose of the effluents of conventional wastewater treatment plants. Large areas may be supplied with recycled water, which may also be used for other different purposes apart from irrigation, depending on the demand, the water characteristics, its suitability, and so on. Consequently, there is a potential for use of recycled water. However, it is essential that the developments of water reuse in agriculture and other sectors be based on scientific evidence of its effects on environment and public health.

10.2 Problems to Be Addressed

By 2050, rising population and incomes are expected to demand 70% more production, compared to 2010 levels. Increased production is projected to come primarily from intensification on existing cultivated land, with irrigation playing an important role [6].

However, many countries around the world are characterized by a severe water imbalance, mainly in the summer months, so to find available water sources for increasing cultivated areas is not always an easy task. This imbalance in water demand versus supply in these arid areas is mainly due to the relatively and uneven distribution of precipitation and high temperatures. Consequently, and in order to alleviate water shortages, serious consideration must be given to wastewater reclamation and reuse.

Concern for human health and the environment are the most important constraints in the reuse of wastewater. Sometimes the wastewater is not properly treated because the construction cost of efficient treatment systems is very high, especially for small and medium size communities. Of course, many alternative solutions have been developed with scientific and technological progress during the last years. However, the selection of the appropriate treatment technique that is tailored to the needs of each community usually means the involvement of qualified specialists, which are not always available. Moreover, in some cases, the outflow of the wastewater treatment systems could not have a standard quality either because standard operating procedures are not followed or because there is no qualified personnel able to overcome usual problems and to control/monitor the whole treatment procedure.

When wastewater is not properly treated, the problems that might be related to its repeated release in the environment for reuse applications are heavy metals, boron, chloride, and sodium accumulating in soil and plants and especially in their edible parts, xenobiotic compounds, including endocrine-disrupting compounds (EDCs), pharmaceuticals and personal care products, drugs' metabolites, illicit drugs, transformation products, and also genes resistant to antibiotics [5]. In addition, the creation of habitats for disease vectors or an excessive growth of algae and vegetation in canals carrying wastewater (eutrophication) can be negative aspects related to irrigation when using not properly treated wastewater [11]. On the other hand, the use of untreated or poorly treated wastewater for irrigation may affect crop growth and quality [27].

While the risks do need to be carefully considered, the need to establish monitoring and control systems to ensure reuse of suitably treated water is quite essential. Currently, the need for capacity building and for intra-regional transfer of knowledge and experience between countries is apparent. All actors involved in the water management cycle are not equipped with the necessary tools, methodologies, and know-how in order to be able to promote those urban wastewater treatment technologies (including innovative ones) that could be easily implemented and combined with recovery and reuse systems, the aim being the safe reuse of the treated effluents in the agricultural production. Sometimes the personnel of the governmental competent authorities do not have the knowledge and technical skills needed for the effective implementation of best practices, control, and monitoring of the treatment plants operation as well as experience concerning socioeconomic instruments, standards, and criteria that can facilitate the enforcement of sustainable schemes. Farmers and the public at large are not familiar with the hazards that are related to the uncontrolled direct usage of raw sewage in land and crops irrigation. The special characteristics of the countries, including gender, religious and cultural peculiarities are not elaborated and integrated in the overall management cycle to the necessary degree.

As a result, the main problem that can create significant obstacles in the safe reuse of treated wastewater in agriculture is the lack of information of all the involved actors, namely

- Governmental authorities: Lack of legislation and guidelines on the reuse of treated wastewater
- Local authorities and authorities responsible in wastewater treatment: (i) Lack of information on innovative cost-effective technologies for wastewater treatment, (ii) difficulties in the development of technical specifications for the construction and operation of appropriate wastewater treatment systems (in terms of technology, size, quality of the outflow), (iii) difficulties in the development of specifications for the proper use of the final outflow, and (iv) difficulties in finding the appropriate funds for the improvement of the wastewater treatment system
- Citizens, industries, and so on: Lack of information on the quality of wastewater that can be rejected in the wastewater treatment network
- Operators: Lack of knowledge for the efficient operation, control, and monitoring of the wastewater treatment system
- Farmers: Lack of information on the health risks related to the use of treated wastewater and the appropriate management procedures

Another factor to be taken into consideration is that water reuse for irrigation is frequently conducted irrespective of the perception of the farmer toward the resource. However, there are aspects of reuse that lead farmers to a more positive or negative view of reclaimed water, which may have implications for their acceptance of the resource. Farmer perceptions do not necessarily reflect the quality of water delivered to the farm. Additional factors influence farmer perceptions, such as the farmer's actual and perceived capacity to control water quality, and the farmer's ability to manage the negative aspects of the water that jeopardize productive agriculture [4].

Concluding, the main concerns that will be dealt with in the framework of sustainable urban wastewater reuse in the agricultural production, mainly in developing countries but also in some developed ones, are shown in Figure 10.1 and summarized here:

- The nonregulated use of treated water in agriculture
- The nonexisting reuse criteria related to hygiene, public health, and quality control
- The nonexisting reuse criteria related to irrigation techniques, degree of wastewater treatment, and choice of areas and types of crops to be irrigated

- The lack of efficient control and monitoring of urban wastewater treatment plants
- The lack of trained personnel both in the competent authorities and in the treatment plants
- The low level of awareness of the farmers and the public at large

10.3 Regulation of Urban Water Reuse in Agriculture

Different regions and governmental agencies have adopted a variety of standards for use of reclaimed water for irrigation of crops to provide effective measures to protect against risk to public health and the environment. These rules and regulations have been developed primarily to protect public health and water resources; specific crop water quality requirements must be developed with the end users [25].

The World Health Organization (WHO) guidelines [26] for irrigation with reclaimed water, widely adopted in Europe and other regions, are a science-based standard that has been successfully applied to irrigation reuse applications throughout the world.

In the United States, the California Water Recycling Criteria (Title 22 of the state Code of Regulations) require the most stringent water quality standards with respect to microbial inactivation (total coliform <2.2 cfu/100 mL). California Water Recycling Criteria require a specific treatment process train for production of recycled water for unrestricted food crop irrigation that includes, at a minimum, filtration and disinfection that meets the state process requirements.

China has established quality requirements for utilizing reclaimed water to regulate and ensure the safety of reclaimed water uses. The government decree, Urban Wastewater Reuse Category (GB/T 189198-2002), divided wastewater reuse into five use categories, and correspondingly five recommended national standards, one of them being "Urban Wastewater Reuse—Water Quality Standard for Farmland Irrigation Water" that stipulated water quality control programs, requirements, and analysis methods of reclaimed water for farmland irrigation [28].

However, irrigation of crops (both food and non-food) with untreated wastewater is widely practiced in many parts of the developing world with accompanying adverse public health outcomes. Nonetheless, this practice represents an economic necessity for many farming communities and for the rapidly expanding population at large, much of which is dependent on locally grown crops. Various international aid organizations have mobilized to improve upon these irrigation practices and provide barriers against transmission of disease-carrying agents.

For diverse reasons, many developing countries are still unable to implement comprehensive wastewater treatment programs. Therefore, in the near term, risk management and interim solutions are needed to prevent adverse impacts from wastewater irrigation. A combination of source control and farm-level and post-harvest measures can be used to protect farm workers and consumers. The WHO guidelines for

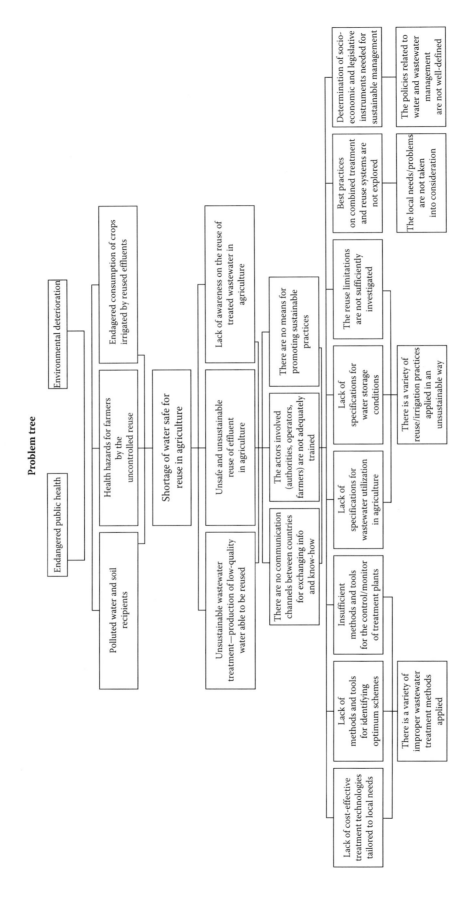

FIGURE 10.1 Main problems to be dealt with in the framework of sustainable urban wastewater reuse in agricultural production.

wastewater use suggest measures beyond the traditional recommendations of producing only industrial or nonedible crops, as in many situations it is impossible to enforce a change in the current cash crop pattern, or provide an alternative vegetable supply to urban markets.

According to Qadir et al. [23], in many Asian and African cities, population growth has outpaced improvements in sanitation and wastewater infrastructure, making management of urban wastewater a tremendous challenge. Some specific examples include India where only 24% of wastewater from households and industry is treated, or Pakistan where only 2% is treated. In West African cities, usually less than 10% of the generated wastewater is collected in piped sewage systems and receives primary or secondary treatment. In many developing countries, large centralized wastewater collection and treatment systems have proven difficult to sustain. Decentralized systems that are more flexible for long-term operation and financial sustainability and compatible with demands for local effluent use have been promoted in many areas, although not without challenges. In Ghana, for example, only 7 of 44 smaller treatment plants are functional and probably none meets the designed effluent standards.

Economics is a key factor in the choice of appropriate solutions [19]. The developed countries have tended to adopt an approach leading to conservative high technology/high cost/low risk guidelines or regulations among which California's water recycling regulations are the best-known example. Some countries have endeavored to follow this regulatory approach to guidelines, but have not always achieved low risk in practice because of insufficient money, experience, or regulatory controls. Limits of affordability have led some developing countries to follow the low technology/low cost/controlled risk path of the attributable risk approach that is embodied in the WHO guidelines.

In general, the adoption of standards for wastewater reclamation and reuse follows the problems encountered in each country. For example, the legal status of wastewater reuse is not uniform across Europe. Many European countries and most northern European countries (e.g., United Kingdom, Belgium, and the Netherlands) do not have any specific legislation on the matter. Regarding European Mediterranean countries, France has national recommendations, Italy a national law, and Spain various regional regulations. Portugal and Greece are considering developing national guidelines. Because of the absence of comprehensive international guidelines and of a scientific consensus, Kamizoulis et al. proposed in 2003 [13] common guidelines on water reuse in all Mediterranean countries. These guidelines were based on the consideration that

- An agricultural Mediterranean market was developing with large amounts of agricultural products (vegetables, fruits, etc.) imported and exported among European and other Mediterranean countries.
- Tourism is an essential part of the economic activity of the region; its development might be jeopardized in the long term by disease outbreaks linked to wastewater mismanagement.
- There is a growing concern among consumers about the food quality and health hazards.
- Unfair competition among farmers should be avoided (according to the European Common Market regulations).

Europe has focused on environmental improvement and indirect incentives through fiscal policy (effluent charges and water abstraction taxes). These policies aim to discourage the discharge of effluent, reduce over-abstraction of groundwater, and encourage sustainable innovative alternative solutions. The use of taxes is becoming more widespread in EU, but they have often been implemented on a case-by-case basis, and generally do not appear to be part of a broader strategy of encouraging alternative water resources. Furthermore, taxes have been frequently focused on revenue generation and not explicitly on providing incentives to change behavior.

10.4 Water Quality Requirements in Irrigation with Reclaimed Water

The water quality of treated wastewater depends largely on the quality of the municipal water supply, the nature of the wastes added during use, and the degree of treatment the wastewater has received [22].

The approach often used for evaluating irrigation water quality is given in Table 10.1 [3]. The "potential restrictions in use" are somewhat arbitrary because water quality changes occur gradually and there is no clear-cut break point. It is also necessary to consider that is not possible to cover all local situations when preparing water quality guidelines.

In the following subsections, the most common water quality problems are discussed with reference to the guidelines given in Table 10.1 [22].

10.4.1 Salinity

Historically, the quality of irrigation water has been determined by the quantity and kind of salt present in these water supplies. Although crops vary considerably in their ability to tolerate saline conditions, in general, as salinity increases in the treated wastewater used for irrigation, the probability for certain soil, water, and cropping problems increases.

The salts associated with salinity can affect the growth and health of plants in several ways. One is referred to as the "osmotic effect," in which salts attract water and compete with plants for it, meaning that as salinity rises, plants have to expend more energy to take in water (energy that would otherwise be used for growing, flowering, or fruiting). The result is negative: stunted growth, wilting, and other damage. Another major concern is that constituents making up salinity; for example, sodium and chloride ions as well as boron can be toxic to sensitive plants, causing severe damage such as leaf burn, leaf drop, and plant death.

TABLE 10.1 Guidelines for Interpretation of Water Quality for Irrigation

Potential Irrigation Problem	Units	Degree on Restriction on Use		
		None	Moderate	Severe
Salinity				
EC	dS/m	≤0.7	0.7–3.0	≥3.0
TDS	mg/L	≤450	450–2000	≥2000
Permeability (effects of infiltration rate of water into the soil; evaluate using EC and SAR together)				
	SAR = 0–3	EC ≥0.7	EC = 0.7–0.2	EC ≤0.2
	SAR = 3–6	EC ≥1.2	EC = 1.2–0.3	EC ≤0.3
	SAR = 6–12	EC ≥1.9	EC = 1.9–0.5	EC ≤0.5
	SAR = 12–20	EC ≥2.9	EC = 2.9–1.3	EC ≤1.3
	SAR = 20–40	EC ≥5.0	EC = 5.0–2.9	EC ≤2.9
Specific ion toxicity				
Sodium (Na)				
Surface irrigation	mg/L	≤3	= 3–9	≥9
Sprinkler irrigation	mg/L	≤70	>70	
Chloride (Cl)				
Surface irrigation	mg/L	≤140	= 140–350	≥350
Sprinkler irrigation	mg/L	≤100	>100	
Boron (B)				
Surface-sprinkler irrigation	mg/L	≤0.7	= 0.7–3	≥3
Miscellaneous effects				
Nitrogen (total N)	mg/L	≤5	= 5–30	≥30
Bicarbonate (overhead sprinkling only)	mg/L	≤90	= 90–500	≥500
Residual chlorine (overhead sprinkling only)	mg/L	≤1	= 1–5	≥5
pH		Normal range 6.5–8.4		

EC: electrical conductivity; TDS: total dissolved solids; SAR: sodium adsorption ratio.

10.4.2 Nutrients

The nutrients in treated municipal wastewater provide fertilizer value to crop or landscape production but in certain instances are in excess of plant needs and cause problems related to excessive vegetative growth, delayed or uneven maturity, or reduced quality. Nutrients occurring in important quantities include nitrogen and phosphorus and occasionally potassium, zinc, boron, and sulfur. The nutrients in reclaimed wastewater can contribute to crop growth, but periodic monitoring is needed to avoid imbalanced nutrient supply. In addition, care needs to be taken in the concentrations of nutrients in recycled water to avoid detrimental impacts on the porosity of soils.

10.4.3 Microbiological Content

The most common human microbial pathogens found in recycled water are enteric in origin. Enteric pathogens enter the environment in the feces of infected hosts and can enter water directly through defecation into water, contamination with sewage effluent, or from runoff from soil and other land surfaces. The types of enteric pathogens that can be found in water include viruses, bacteria, protozoa, and helminthes. The risk of water-borne infection from any of these pathogens is reliant on a range of factors including pathogen numbers and dispersion in water, the infective dose required and the susceptibility of an exposed population, the chance of fecal contamination of the water, and the amount of treatment undertaken before potential exposure to the water [24].

10.4.4 Other Aspects

Wastewater often contains a variety of pollutants: metals, met-alloids, residual drugs, organic compounds, EDCs, and active residues of personal care products [5]. Any of these components can harm human health and the environment.

Another possible problem of wastewater reuse is the excessive residual chlorine in treated effluent. Residual chlorine causes plant damage when sprinklers are used if the high chlorine residual exists at the time the effluent is sprinkled on plant foliage. Residual chlorine less than 1 mg/L should not affect plant foliage, but when chlorine residual is in excess of 5 mg/L, severe plant damage can occur.

Clogging problems with sprinkler and drip irrigation systems have been reported when treated municipal wastewater is used. The most frequent clogging problems occur with drip irrigation systems.

10.5 Good Practices in Agricultural Reuse of Urban Wastewater

The largest scheme in the world of wastewater use in agriculture is in the Mezquital Valley in Mexico. Starting around 1890, drainage canals were built to take wastewater from Mexico City to provide irrigation water for agricultural lands. The scheme now irrigates up to 90,000 ha of agricultural crops [16].

The first engineered systems of spreading urban wastewater for agricultural irrigation were implemented in Paris, France and Milan, Italy urban areas. The first regulation on water reuse by spreading was published in 1899 and by the end of the nineteenth century, almost all sewage from Paris was used to irrigate about 5000 ha of land.

In the twentieth century, the development of effective treatment systems opened the way to the safe use of reclaimed water for agricultural irrigation. In recent times, water shortages and increasing competition for available fresh water supplies has prompted the development of numerous large agricultural reuse schemes around the world.

In the United States, agricultural irrigation totals about 5.6 M L/s [14], which represents approximately 37% of all freshwater withdrawals. Confounding the agricultural water supply issue are the recent increases in Midwestern and Southeastern inter-annual climate variability that has led to more severe droughts, making issues of agricultural water reliability a greater national challenge. In many regions of the United States, expanding urban populations and rising demands for water from municipal and industrial sectors now compete for water supplies traditionally reserved for irrigated agriculture. In other areas, irrigation water supplies are being depleted by agricultural use. These shifts in the availability and quality of traditional water resources could have dramatic impacts on the long-term supply of food and fiber in the United States [25].

The use of reclaimed water for agriculture has been widely supported by regulatory and institutional policies. In 2009, for example, California adopted both the Recycled Water Policy and "Water Recycling Criteria." Both policies promote the use of recycled water. In response to an unprecedented water crisis brought about by the collapse of the Bay-Delta ecosystem, climate change, continuing population growth, and a severe drought on the Colorado River, the California State Water Resources Control Board (SWRCB) was prompted to "exercise the authority granted to them by the Legislature to the fullest extent possible to encourage the use of recycled water, consistent with state and federal water quality laws." As a result, future recycled water use in California is estimated to reach 2500 Mm³/yr by 2020, and 3700 Mm³/yr by 2030. Consequently, California presently recycles about 800 Mm³/yr, an amount that has doubled in the last 20 years, with agriculture as the top recycled water user.

In Florida, promotion of reclaimed water began in 1966; currently, 63 of 67 counties have utilities with reclaimed water systems. One of the largest and most visible reclaimed water projects is known as Water Conserv II in Orange County, Florida, where farmers have used reclaimed water for citrus irrigation since 1986. Another long-serving example of reclaimed water use in the United States is the city of Lubbock, Texas, where reclaimed water has been used to irrigate cotton, grain sorghum, and wheat since 1938. In addition, reclaimed water is a significant part of the agricultural water sustainability portfolio in Arizona, Colorado, and Nevada.

With population growth, accelerated industrialization and urbanization, and global climate change, China's water crisis is exacerbated. Water shortage has become a major obstacle restricting China's economic development [12].

The majority of available water is concentrated in the south, leaving northern and western China to experience perpetual droughts. Rivers, lakes, and underground aquifers are literally drying up due to overdrafts. Most of the remaining surface waters are so polluted that they are no longer suitable for human contact [28].

According to the Chinese Ministry of Construction, 1660 Mm³ reclaimed wastewater was used in 2008, amounting to 8% of the total treated municipal wastewater. The reclaimed wastewater was used mainly for agricultural irrigation (480 Mm³).

Spain shows by far the highest agricultural water reuse potential in Europe. The calculations suggest a value of over 1300 Mm³/yr. In addition, all Mediterranean countries exhibit estimated high reuse potentials of 550 Mm³/yr (Italy) and 120 Mm³/yr (France). Wastewater reuse appraisals for Germany amount to 150 Mm³/yr, whereas Portugal and Greece account for reuse potentials of less than 60 Mm³/yr [2]. The estimates suggest a treated wastewater reuse potential of 2455 Mm³/yr for EU. It accounts for 2% of irrigation water used in European countries, which is over 60% of total water use. However, considering only the seven Mediterranean countries (including Portugal), their reuse potential of 2150 Mm³/yr accounts for 3.5% of the irrigation water used in these countries. All of EU-Mediterranean states (with the exception of Malta) have recently established new regulations or they have revised existing ones.

Water reuse has recorded indisputable progress in recent years in the Mediterranean [7]. Some concrete recent noteworthy examples can be pointed out in some Mediterranean countries, both nationally and locally. These examples show the considerable benefits of wastewater reuse. These benefits are both financial (savings in very costly heavy infrastructures), economic and social (wastewater reclamation and reuse processes, if they are well managed, may, for example, allow agricultural incomes to be improved and give better access to water for the least well-off), and environmental (reducing pressure on, and even restoring, ecosystems and resources) [8]. An example on the above-mentioned information is the strategy of irrigation water saving in Tunisia. In Tunisia, a gradual approach has been adopted to expand reuse since the mid-1960s. Nowadays, out of 61 wastewater treatment plants that treat 140 Mm³/yr, 41 have a capacity less than 3500 m³/d and 10 above 10,000 m³/d [10].

In Israel, municipal sewers collect about 92% of the wastewater. Subsequently, 72% is used for irrigation (42%) or groundwater recharge (30%) [15]. Local, regional, and national authorities must approve the use of recycled wastewater. Effluent used for irrigation must meet water quality criteria set by the Ministry of Health. The trend is toward bringing all effluents to a quality suitable for unrestricted irrigation with wider crop rotation, which will require more storage and higher levels of treatment in the future. Cost-benefit analysis indicates that recycled wastewater is a very low cost source of water in Israel. As a result, treated wastewater within the overall water supply, particularly for irrigation, has risen to 24.4% of the allocations. The water crisis in Israel and the relatively low cost of treated wastewater, rather than pure environmental considerations, are the main driving forces behind the high percentage of reuse.

10.6 Cost of Water Reclamation and Agricultural Reuse

Economic considerations are of high importance when assessing the potential of water-reuse projects [17]. Furthermore, the Water Framework Directive (WFD) assigns significance to economic analysis to achieve suitable water resource management. However, economic feasibility of research on wastewater regeneration and reuse remains the least studied component. In part, this is because internal and external economic impacts of environmental-based projects should be identified and quantified when analyzing economic feasibility, such as for water reuse. While internal impacts may be easily translated into monetary units, external effects (or externalities) are not considered by the market, thus requiring economic valuation methods for their quantification. As a result, a series of statements about the advantages of wastewater regeneration and reuse are often presented without supporting economic quantification. Consequently, the true benefits and costs of many projects are not properly evaluated [20].

Reclaimed wastewater can be used for a number of options including agricultural irrigation, landscape irrigation, groundwater recharge, and industrial processes. Water quality requirements for reuse alternatives vary extensively depending on the extent of potential public exposure and, consequently, wastewater reclamation costs are usually not well documented, calculated, or even known. A problem associated with wastewater reclamation projects is that the actual cost is usually considerably higher than the estimated cost of the design stage. This is a result of insufficient planning before design and construction of water reclamation projects. Once the wastewater reclamation and reuse system has been built and operated, it is also necessary to take into consideration a code of good practices in order to avoid any possible problems related to the use of this water. First, it must be taken into account that reclaimed wastewater must only be reused for the purposes permitted. When reclaimed water quality does not meet the fixed standards, reuse must be ceased. In all cases, quality monitoring and process controls should be supported [10].

The development of a cost estimate must include projections of capital costs, annual operation and maintenance costs, and life-cycle cost. Life-cycle costs are useful in comparing the economic feasibility of alternative water resources projects over a specific time period.

Wastewater reclamation system costs can be presented as a function of facility capacity, end-use option, and treatment process configuration. Costs should be identified estimating facility construction costs, equipment purchases, and operation and maintenance fees. Reclamation system annual cost is comprised of treatment and distribution facility personnel salaries, operation (recurring power and chemical cost), and maintenance costs (equipment repairs and replacements). Personnel requirements are a function of the size and complexity of the plant. Operating cost depends on energy usage and chemical consumption. Maintenance costs (spare parts, replacements) are estimated generally as a percentage of equipment first cost (e.g., 5%). For pipelines and storage tanks, maintenance costs can be projected as 2% of capital costs. Each of these annual cost components should be analyzed for various reclamation options based on the guidelines presented in technical literature and actual operating experience [1,18,27].

Capital costs include cost incurred for wastewater treatment, storage, and distribution. Capital costs for wastewater treatment and water reclamation can be estimated by summing the individual unit process cost with required supporting facilities fees. Operation and maintenance costs for a total reclamation system can be analyzed by means of capital costs.

Referring to disinfection, agricultural food crops, parks, playgrounds and schoolyard irrigation, and nonrestricted recreational impoundments reuse options require a pathogen-free effluent. To upgrade the secondary wastewater treatment plants to meet the desired effluent quality, chemical addition, coagulation, filtration, and disinfection facilities must be provided.

Summarizing, the cost of water reuse involves

- The extra treatment needed to reach the reuse quality requirement above and beyond the mandatory baseline treatment required to protect the community safety, health, and environment
- The extra conveyance of the effluent to the reuse site

To illustrate how water reuse and exchange can be a cost-effective approach to managing water scarcity at the basin level, individual case studies must be examined. The cost-benefit analysis revealed that one project in the Llobregat Delta in Spain would be feasible with a total water-exchange cost of €5.2 million per year. The city would benefit by €14.4 million per year from an additional 13 Mm³ of freshwater a year released from agricultural use (saving the same volume of freshwater being extracted from river sources, which also benefits the environment). Farmers' incomes would collectively rise by €351,000 per year in the area because of reduced costs of pumping water and fertilizer use. Overall, the benefits were calculated to outweigh the costs by €9.5million per year [9].

As a general conclusion, it can be pointed out that when the economic feasibility of an agricultural water-reuse project is assessed, water management authorities and companies should not only consider the benefits of market value because the development of such projects may also be justified by other reasons, such as environmental benefits or the increase in the availability of a scarce resource. Hence, for the objective evaluation of agricultural water-reuse projects, economic feasibility studies should incorporate all parameters including economic, environmental, and resource availability [20].

The economics of water-reuse projects are a major barrier for their implementation. The objectives of pricing for water demand management, pricing for encouraging the use of regenerated water, and pricing for cost recovery are not simultaneously achievable. Molinos-Senante et al. [19] proposed a two-part tariff with a combination of a decreasing and an increasing rate structure for the variable charge as a partial solution to improve water-reuse pricing.

10.7 Summary and Conclusions

Increasing water productivity is often recommended as an easy way to address water scarcity problems in agriculture. The increasing scarcity of water in the world along with rapid population increases in urban areas are reasons for concern and the need for appropriate water management practices. Inadequate investment has been made in the past on sewage treatment facilities; water supply and treatment often have received more priority than wastewater collection and treatment. However, due to the trends in urban development, wastewater treatment needs greater emphasis. Currently, there is an increasing awareness of the impact of sewage contamination on rivers and lakes; wastewater treatment is now receiving greater attention from both the World Bank and the various international regulatory bodies.

According to the World Bank, "The greatest challenge in the water and sanitation sector over the next two decades will be the implementation of low cost sewage treatment that will at the same time permit selective reuse of treated effluents for agricultural and industrial purposes." It is crucial that sanitation systems have high levels of hygienic standards to prevent the spread of diseases.

There still exists a need for research in this area to improve or optimize the current methods of wastewater treatment. More emphasis on the topic will improve health, economy, and agricultural factors of developing countries. Successful development of a water reclamation and reuse project requires careful planning and depends on many factors, not all of which are under the control of the project owner or manager. Identification of large customers of reclaimed water and a realistic assessment of their water demand is crucial for project development. Water quality requirements of all customers must be evaluated. If these requirements cannot be met, especially for large potential customers, project feasibility may be doubtful.

The benefits of reusing and exchanging water between farmers and municipalities can be summarized as follows:

- For cities, potential benefits arise from the extra availability of freshwater that is not being used by farmers, which saves water extraction, desalination, and water treatment costs, in addition to transfer costs from remote water extraction sites.
- For the environment, potential benefits include a reduction of contaminants (such as salts and metals) and nutrients released into rivers and coastal waters; reduced freshwater extraction; renewed river flows; conservation of wetlands; and using reclaimed water as a barrier to prevent the intrusion of seawater into aquifers.

Set against these benefits are the costs associated with measures to minimize health risks from reusing wastewater and possible risks to the environment from any contaminants. Other costs include those for building new infrastructure to take the reclaimed water to agricultural areas.

Public awareness campaigns are highly needed to address the legal, social, economic, and institutional consideration for treated wastewater reuse. Participation of farmers in developing guidelines, standards, policies, and plans for agricultural reuse is very important for the sustainability of treated wastewater reuse.

Acknowledgments

The authors gratefully acknowledge support of this work by the LIFE+ Program under the responsibility of the Directorate General for the Environment of the European Commission through the agreement LIFE 12 ENV/ES/000727-REVAWASTE project. The authors also wish to thank all the MEDAWARE project partners for their contribution to the preparation of this document.

References

1. Alcon, F., Martin-Ortega, J., Pedrero, F., Alarcon, J.J., and de Miguel, M.D. 2013. Incorporating non-market benefits of reclaimed water into cost-benefit analysis: A case study of irrigated mandarin crops in southern Spain. *Water Resources Management*, 27(6), 1809–1820.

2. Angelakis, A.N. 2011. Water recycling and reuse in EU: Necessity and perspectives for establishing EU Legislation. In: *Book of Abstract of SmallWat11, 3rd International Congress on Wastewater in Small Communities*, Fundación Centro de las Nuevas Tecnologías del Agua (CENTA), Seville, Spain, p. 39.

3. Asano, T., Burton, F.L., Leverenz, H.L., Tsuchihashi, R., and Tchobanoglous, G. 2006. Agricultural irrigation with reclaimed water: On overview. In: Asano, T. (ed.), *Water Reuse. Issues, Technologies and Applications*. Metcalf & Eddy/AECOM, McGraw Hill, New York.

4. Carr, G., Potter, R.B., and Nortcliff, S. 2011. Water reuse for irrigation in Jordan: Perceptions of water quality among farmers. *Agricultural Water Management*, 98(5), 847–854.

5. Fatta-Kassinos, D., Kalavrouziotis, I.K., Koukoulakis, P.H., and Vasquez, M.I. 2011. The risks associated with wastewater reuse and xenobiotics in the agroecological environment. *Science of the Total Environment*, 409(19), 3555–3563.

6. Food and Agriculture Organization of the United Nations (FAO). 2011. *Executive Summary. Thirty-Seventh Session Rome 25 June–2 July 2011. The State of the World's Land and Water Resources for Food and Agriculture (SOLAW).* C 2011/32, pp. 1–12. http://www.fao.org/docrep/meeting/022/mb213e.pdf. Accessed July 2015.

7. Ghaffour, N., Missimer, T.M., and Amy, G.L. 2013. Combined desalination, water reuse, and aquifer storage and recovery to meet water supply demands in the GCC/MENA region. *Desalination and Water Treatment*, 51(1–3), 38–43.

8. Hafeez, M., Bundschuh, J., and Mushtaq, S. 2014. Exploring synergies and tradeoffs: Energy, water, and economic implications of water reuse in rice-based irrigation systems. *Applied Energy*, 114, 889–900.

9. Heinz, I., Salgot, M., and Mateo-Sagasta Dávila, J. 2011. Evaluating the costs and benefits of water reuse and exchange projects involving cities and farmers. *Water International*, 36(4), 455–466.

10. Hidalgo, D., Irusta, R., and Fatta, D. 2006. Sustainable and cost-effective municipal wastewater reclamation: Treated effluent reuse in agricultural production. *International Journal of Environment and Pollution*, 28(1), 2–15.

11. Jang, T., Lee, S.B., Sung, C.H., Lee, H.P., and Park, S.W. 2010. Safe application of reclaimed water reuse for agriculture in Korea. *Paddy and Water Environment*, 8(3), 227–233.

12. Jiang, Y. 2009. China's water scarcity. *Journal of Environmental Management*, 90(11), 3185–3196.

13. Kamizoulis, G., Bahri, A., Brissaud, F., and Angelakis, A.N. 2003. Wastewater recycling and reuse practices in Mediterranean region: Recommended Guidelines. http://www.a-angelakis.gr/files/pubrep/recycling_med.pdf. Accessed July 2015.

14. Kenny, J.F., Barber, N.L., Hutson, S.S., Linsey, K.S., Lovelace, J.K., and Maupin, M.A. 2009. Estimated Use of Water in the United States in 2005. United States Geological Survey (USGS). http://pubs.usgs.gov/circ/1344/pdf/c1344.pdf. Accessed March 2014.

15. Kumar, S. 2014. Study on waste water recycling and reuse. *International Journal of Applied Engineering Research*, 9(3), 329–334.

16. Lazarova, V., Asano, T., Bahri, A., and Anderson, J. (eds.). 2013. *Milestones in Water Reuse: The Best Success Stories.* IWA Publishing, London, UK.

17. Lazarova, V., Choo, K.H., and Cornel, P. (eds.). 2012. *Water-Energy Interactions in Water Reuse.* IWA Publishing, London, UK.

18. Mizyed, N.R. 2013. Challenges to treated wastewater reuse in arid and semi-arid areas. *Environmental Science and Policy*, 25, 186–195.

19. Molinos-Senante, M., Hernández-Sancho, F., and Sala-Garrido, R. 2011. Cost–benefit analysis of water-reuse projects for environmental purposes: A case study for Spanish wastewater treatment plants. *Journal of Environmental Management*, 92(12), 3091–3097.

20. Molinos-Senante, M., Hernandez-Sancho, F., and Sala-Garrido, R. 2013. Tariffs and cost recovery in water reuse. *Water Resources Management*, 27(6), 1797–1808.

21. O'Neill, M.P. and Dobrowolski, J.P. 2011. Water and agriculture in a changing climate. *Hort Science*, 46, 155.

22. Pedrero, F., Kalavrouziotis, I., Alarcón, J.J., Koukoulakis, P., and Asano, T. 2010. Use of treated municipal wastewater in irrigated agriculture—Review of some practices in Spain and Greece. *Agricultural Water Management*, 97(9), 1233–1241.

23. Qadir, M., Wichelns, D., Raschid-Sally, L., McCornick, P.G., Drechsel, P., Bahri, A., and Minhas, P.S. 2010. The challenges of wastewater irrigation in developing countries. *Agricultural Water Management*, 97(4), 561–568.

24. Toze, S. 2006. Reuse of effluent water-benefits and risks. *Agricultural Water Management*, 80(1), 147–159.

25. U.S. Environmental Protection Agency (EPA). 2012. *Guidelines for Water Reuse.* U.S. Environmental Protection Agency (USEPA), Washington D.C.

26. World Health Organization (WHO). 2006. *WHO Guidelines for the Safe Use of Wastewater, Excreta and Greywater.* United Nations Environment Program, Paris.

27. Xu, J., Wu, L., Chang, A.C., and Zhang, Y. 2010. Impact of long-term reclaimed wastewater irrigation on agricultural soils: A preliminary assessment. *Journal of Hazardous Materials*, 183(1), 780–786.

28. Yi, L., Jiao, W., Chen, X., and Chen, W. 2011. An overview of reclaimed water reuse in China. *Journal of Environmental Sciences*, 23(10), 1585–1593.

11

Urban Water Reuse in Industry

Mohammad Mahdi Kohansal
Isfahan University of Technology

Sara Saadati
Isfahan University of Technology

Saleh Tarkesh Esfahany
Isfahan University of Technology

Saeid Eslamian
Isfahan University of Technology

PREFACE

According to predictions, one of the conflicts in the twenty-first century is a water war. There are water-related issues such as water scarcity, water pollution, and damages caused by floods and droughts in many parts of the world, especially in developing countries. Some issues include population growth, agricultural development in arid and semiarid regions of the world, growing industries, and uneven spatiotemporal distribution of freshwater. Water quality problems have led the water supply to be a major global challenge of the twenty-first century. Water resource limitations and water pollution caused by municipal, industrial, and agricultural wastewater have made nations pay extra attention to wastewater management and use new water resources in the context of water resource management. The use of reclaimed wastewater for a variety of purposes such as industrial uses, agricultural and landscape irrigation, and other applications is becoming very common because of a global water shortage particularly in arid zones. The purpose of this chapter is to study urban water reuse for different industries and the associated methods including its advantages and disadvantages. In order to enrich the chapter, a case study is investigated at the end of the chapter.

11.1 Introduction

The economic, social, and environmental impacts of past water resources development and inevitable prospects of water shortage are leading to new patterns in water resources management. Now, new techniques have to consider the principles of sustainable development, environmental ethics, and public participation in project development.

In many communities and countries approaching the limits of their available water resources, water reclamation and reuse have become an important and attractive option to conserve and extend available water supply. Water reuse is very important in the current situation where available water supply is already overcommitted and expanding water demands in a growing society cannot be impeded [8].

Countries in the arid belt of the Earth are facing water shortage. Population growth and limitations of water resources have made the issue a major challenge in these communities. On a global scale, by 2025, these countries will have the lowest ratio of available freshwater to population size. Anticipated shortage of water resources in the coming years can be seen in the studies conducted in different countries [14].

Specifically, the arid part of the central Iranian plateau in recent years has lost water resources and faces drought. Therefore, using the methods of sustainable management of water resources in these countries is essential to deal with drought and maintain the principles of sustainable development.

The principle of sustainability, which was the cornerstone in the Brundtland Commission's report entitled "Our Common Future" is defined as follows [27]:

> Humanity has the ability to make development sustainable to ensure that it meets the needs of the present without compromising the ability of future generations to meet their own needs

The new and emerging paradigm of sustainable water resources management has been interpreted in different ways by different stakeholders. The American Society of Civil Engineers (ASCE) suggests the following definition for sustainable water resources systems [3]:

> Sustainable water resources systems are those designed and managed to fully contribute to the objectives of society, now and in the future, while maintaining their ecological, environmental, and hydrological integrity

Reuse of reclaimed water is one of the strategies for sustainable management. Water reuse can be defined as the use of treated wastewater for a direct beneficial purpose such as agricultural irrigation, municipal uses, and industrial cooling [27].

Reuse of municipal water for industry has been prominent and increased strongly since the early 1990s for many of the same reasons, specifically in arid regions, regarding new rules on water resources conservation and environmental compromises. Utility power plants are ideal facilities for reuse because of their high water requirements for cooling systems, ash sluicing, waste dilution, and flue gas scrubber. Oil refineries, chemical production, and metalworking facilities require reclaimed water not only for cooling but also for other processes among other industrial facilities. In many industries, the largest use of reclaimed water is cooling water because improvements in water treatment technologies have allowed industries to use successfully lesser quality waters. These advancements have caused better control of deposits, corrosion, and biological problems related to the use of reclaimed water in a concentrated cooling water system. The most common problems facing water quality in cooling water systems are corrosion, biological growth, and scaling. These problems are caused by contaminants in potable water as well as in reclaimed water, but contaminant concentrations in reclaimed water may be higher than in potable water [6].

The use of reclaimed water for industrial use has been proposed in many industrial sectors in the United States, Europe, and other developed countries.

11.2 Urban Water

Unlike municipal water, municipal wastewater might be seen as dirty and unpleasant, but it could be the best option for industry due to no heavy metals or low content of heavy metals. Heavy metals are among the major pollutants of the water industry. Meanwhile, there is very little or no chemical fertilizer in municipal water, which is considered an advantage in the reuse of municipal water. However, it would be a problem if urban water is reused for agriculture.

Perhaps outlet wastewater of cities would be the best definition for urban water. However, most cities and towns are not served by municipal sewerage systems and in some cities, which have sewerage systems, a large part of the municipal water is not connected. An example is the wastewater of landscape irrigation [23].

The increasing scarcity of freshwater in the world along with rapid growth of population in urban areas has raised concerns about problems including not enough water and heavy reliance on groundwater. Therefore, an important factor that makes urban wastewater valuable is that it is an alternative and a reliable source of water supply and it is available all year round.

Since water scarcity problems lead to imbalances in water supply and demand and urban population growth has not only increased the freshwater demand but also increased the volume of urban wastewater produced, it would be a new water resource that can play an important role in integrated water resources management addressing both water demand and supply.

Urban water pollution includes only a group of pollutants (such as fecal coliform bacteria, nitrates, phosphorus, chemicals, and other bacteria from many avenues such as runoff from streets carries oil, rubber, heavy metals, and other contaminants from automobiles) that would shorten and simplify the treatment process and reduce refining costs. In addition, the treatment of these contaminations is easier and faster. The volume of urban water in large cities in which manufacturing and industrial companies are located is more. It causes more attention to the importance of urban water and reusing it in the industrial sectors as alternative water sources for cooling towers and cooling water applications at power plants. Urban water reuse in developed countries is much more common than in most developing countries. Municipal sewerage systems in many developed countries are leading to water treatment plants and urban water is used for different purposes after treatment.

11.3 Water in Industry

Water is one of the most important parameters in many activities and industrial activities are no exception in this regard. There are different uses of water in industry; major industrial uses of water include cooling towers. Cooling towers are probably one of the most important applications of water in industries. Today's water system is an important part of the industrial park because it is used for domestic purposes, such as drinking and personal hygiene, and solid waste transfer and cooling [20].

Industries that produce metals, wood, paper, chemicals, oil and gas, food, and most other known products all use water in some part of their production process. The high percentage of water consumption of industrial and developing countries is allocated to the industrial sector while the largest consumer of water is the agriculture sector in underdeveloped or developing countries. Regarding the tendency of countries toward industrialization, planning would be necessary for water uses in industry [1].

Water can be used as a raw material in some industries. For example, in addition to using water for cooling purposes in the canning industry, it is used as raw material in making canned goods. Industrial water use includes water used for purposes such as processing, washing, dilution, cooling, transporting, incorporating water into a product, or for sanitation.

Surface water, groundwater, or unconventional water can be used for industry. As shown in Figure 11.1, the majority of water used for industrial purposes came from surface-water sources in 2005. In addition, according to the figure, the percentage of industrial water allocation is observed. Although this value is very low, it would be undeniable due to two reasons.

As countries become more developed, the industries' tendency is more, so, the total amount of industrial water use will be increased. However, the amount of water used in all sectors including agriculture and industry in underdeveloped or developing countries is much higher than the per capita water consumption compared to developed countries due to improper water use and low water use efficiency.

Today, researchers are looking for new methods for cooling in industry, which do not require water and could save a lot of water. For example, one of these methods is cooling by air. However, these methods require less water than others do but they still require water. In addition, all these methods are only at the introduction and testing stage and there is no alternative for water [16].

11.4 Quality of Water in Industry

Water is an important part of the manufacturing process in the industry. Water that is used in industry and agriculture sectors shall be at permitted values based on quality as well as drinking water. This range is different for various applications as well as in various industries. For example, water that is used in cooling towers has to be quite controlled about some factors such as heavy metals and microorganisms because heavy metals cause corrosion and scale deposits on pipes and heat exchangers in the towers and microorganisms lead to algae growth [22].

Other parameters such as hardness, pH, chemical oxygen demand (COD), and biochemical oxygen demand (BOD$_5$) are very important. Also, coliform indicators in the food industry should also be reviewed. If the water that is used for food production and processing is out of range of the coliform index, it will cause various diseases in humans [25].

According to the above, it has been concluded that it is not possible to consider a fixed index for water reuse in all industries.

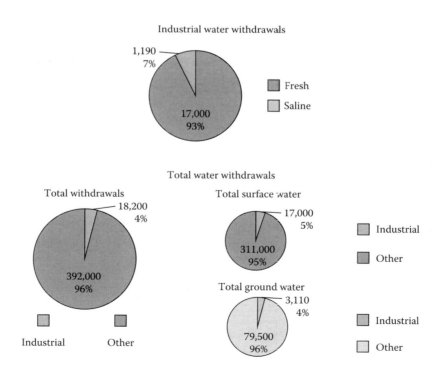

FIGURE 11.1   Water use in the United States.

TABLE 11.1   Water Quality Standards and Criteria for Industrial Uses Based on the Guidelines for Canadian Drinking Water Quality

Index	Permitted Values (mg/L)		
	Group (1)	Group (2)	Group (3)
Iron	<0/3	<1	<1
pH	6–9	6–9	6–9
COD	<20	<75	<75
Hardness	<250	<500	<500
Alkalinity	<150	<500	<500
Sulfate	<250	<500	<500
Silica	<20	<50	<50
Suspended solids	<50	<100	<100
TDS	<500	<1000	<1000
Chloride	<200	<500	<500

Source: Adapted from Government of Canada. 2008. Technical guidance document for Water Quality Index practitioners reporting under the Canadian Environmental Sustainability Indicators (CESI) initiative. Environment Canada and Statistics Canada.

Therefore, it is better to prepare the tables for reviewing water quality standards including maximum and minimum and the ranges allowed in important parameters in various industries.

These tables may be changed each year according to new findings and various conditions, and the values of the parameters might be low or high.

Water for industrial uses can be divided into three groups. The three groups are as follows [28]:

- *Group 1*: This group includes water in the industry used for processes that do not require high water quality and can be used without treatment or minimal treatment. In addition, it would be necessary to be treated for processes with high sensitivity. This group has a good quality for industrial applications.
- *Group 2*: This group includes water that is used without treatment or minimal treatment for processes with the least sensitivity. However, implementation of physical and chemical processes depending on use cases is necessary for relatively insensitive processes. This group has an average quality for industrial applications.
- *Group 3*: This group includes water that requires treatment for each application and has a poor quality. Therefore, it is recommended to use more for cooling applications, which do not require high treatment. Regarding the high levels of treatment for insensitive processes, the use of these kinds of waters is not recommended. Indicators and parameters for each of these groups for use in industry are shown in Table 11.1.

11.5 Wastewater Treatment

Factors such as successive droughts, population growth, climate change, and other factors have caused new problems facing humanity every day. One of these problems is water resources limitation, as far as it is predicted that a future world war (Third World War) occurs over water resources. These factors have led humans to seek unlimited water resources. Water resources are always known as limited resources; however, water is not an exception from the law of conservation. Today, using unconventional water sources is causing a discussion of unlimited water resources to be realized. Freshening of seawater and wastewater treatment resources can be identified as the ultimate water resources. The amount of wastewater increases during droughts and population growth; hence, it can resemble an unlimited resource [7].

Wastewater treatment generally includes physical and chemical processes to remove physical and chemical contaminants. During physical processes of wastewater treatment, solid particles can be separated from water using a filter with different diameters, but the chemical method is the process of removing contaminants by adding chemicals like chlorine to be deposited and removed. Treatment generally involves two stages—primary and secondary. During primary treatment, all objects floating on the top of the water and settable solids are separated by passing the wastewater stream through a mesh screen or settling basin. During secondary wastewater treatment, organic materials are consumed by bacteria (using trickling filters or activated sludge). Secondary treatment removes all suspended solids, settled materials, and reduces the amount of total suspended solids, BOD_5, COD, and all other pollution parameters up to 90%. During primary and secondary treatments, which are conventional treatments, synthetic organic and inorganic ions have not been removed; therefore, an advanced treatment is required. That is why sometimes advanced treatment and tertiary treatment are necessary. Tertiary treatment involves removing phosphorus concentrations such as phosphorus and nitrogen and a high percentage of suspended solids from wastewater. Advanced treatment of wastewater is known as clarification and results in high quality wastewater. In addition, other technologies are used in the treatment plant of industrial centers, power plants, and refineries for the removal of metal and persistent organic contamination that is very costly and requires a specialized and skilled workforce. Some effective processes in these technologies include the following [13,18]:

1. Flotation
2. Thermal hydrolysis
3. Wet oxidation
4. Adsorption and ion-exchange
5. The use of nano-adsorbents
6. The use of nano-filters
7. RO
8. The use of algae
9. The use of bacteria

11.6 Water Reuse

Water is another critical ingredient playing a key role in all industrial activities as a main resource. However, many problems are linked to the global water shortage. Owing to the complexity of the problem, many solutions have been proposed in order to provide water resources such as dams, use of groundwater, and desalination of seawater.

In particular, water reuse has been suggested as one of the promising solutions for providing an alternative resource in the future [2].

Wastewater can be considered as an alternative water source that is sustainable and unlimited. It contains a significant amount of resources that will allow sustainable development as a stable resource.

In order to reuse and recycle water in industry, the following should be considered [11,21]:

- Water should not cause precipitation on heat transfer surfaces.
- It should not cause corrosion in the structural materials of equipment.
- It does not contain nutrients for growing microorganisms.
- It does not create too much foam.

This unlimited source of water can be used for different purposes. For example, it can be used in agriculture for landscape irrigation, as cooling water in cooling towers for industrial purposes, for toilet flushing and other uses that are not a threat to humanity for municipal purposes.

In addition, reclaimed water can be used for fountains and other similar cases if it contains no harmful ingredients and does not contact the body. There are many methods of wastewater treatment and reuse, and the amount of reused water has been increased daily due to the advancement of science and technology [19].

It can also be achieved using methods such as RO and nanowater with high quality such as distilled water.

Although much attention has been paid to water reuse, this amount is still low and it is predicted to develop a tendency in water reuse with shortage of water resources and higher price of water in the not too distant future. This becomes more important considering overpopulation [17].

It should be noted that water reuse has been long promoted in different countries. In most developed countries, only treated wastewater is used regarding standards related to various applications. However, in most developing countries, raw wastewater has been used for different purposes such as agriculture and has caused health and environmental problems [24].

11.7 Urban Water Reuse in Industry

As noted earlier, in recent years, due to limited water resources, humans seek to achieve an unlimited water supply. This has led humans to focus on water reuse. Municipal water is safe to use in industry because the basic standards needed for industrial water from municipal wastewater are within the allowable limit and do not have any problems relating to reuse.

This led to a decrease in treatment processes and following them, the cost and time required to treat wastewater for water reuse decreased as well [26].

Most industrial water is mostly used as cooling water, boiler water, and transport of materials and utilization of the process.

The remarkable thing is that there are no regulations governing water reuse for industrial cooling in some countries such as South Korea, whereas water that is used in cooling towers should adhere to certain standards.

Today, water reuse is considered especially in industry because the water used in industry has no direct relation to humans, especially for cooling, and is much safer and more reliable in environmental problems.

Municipal water is more problematic for humans but creates fewer problems in industry and therefore, it is considered more. Two important factors that should be considered in order to reuse water in industrial cooling are sedimentation and corrosion.

In an inverse variation, the values of these two factors change in an opposite manner. As sediment is produced, corrosion does not occur and then sediment is not generated. These two indices are calculated by the saturation index (SI).

Even though recycled water in industrial processes is based on parameters such as microorganisms, nitrogen, and the amount of dissolved organic carbon, it should also be evaluated using the SI to confirm the role of sedimentation and corrosion.

Recycled water can be used for a wide range of industries and applications. Industrial cooling towers are the most important use of recycled water.

Sedimentation and corrosion are two main parameters for using water in cooling towers.

Considering the above-mentioned, it can be concluded that water reuse is essential and undeniable. It should be also noted that municipal wastewater is more suitable for industrial applications than other wastewaters.

11.8 Why Urban Water Reuse in Industry

Today, industry is an important part of nations and the needs for water and use of it in industry are undeniable. As mentioned previously, much attention has been paid to water reuse due to lack of sufficient available water resources, population growth, and other reasons. In the meantime, water reuse in industry has received more attention than any other application.

In addition to the large amounts of water consumed in industry (even in developing countries), it has no direct contact with humans and is not a leading cause of death and disease. In addition, the other benefit of urban water reuse in industry is that urban water meets the majority of water quality standards for industry uses and does not require different complex treatment processes and can be reused for industry only by passing through a simple process of treatment. If municipal water is considered to be reused as municipal water, it should go through different stages of treatment because it will be expected to reach standards before use. That means all of the changes that have been made in standards should return to initial status through treatment.

Containing small amounts of heavy metals is one of the main characteristics of municipal water that causes it to be more reliable

TABLE 11.2 Performance of Treatment Method for Quality Improvement of Municipal Wastewater

| Wastewater Treatment System | Activated Sludge | | | | | Stabilization Pond | |
Parameters	Conventional	Extended Aeration	Oxidation Ponds	Aerated Lagoon	Trickling Filter	Aerated	Other Methods
BOD removal	Moderate	Moderate	Good	Good	Moderate	Good	Good
Fecal coli form removal	Weak	Moderate	Moderate	Moderate	Weak	Good	Good
Removal of suspended solids	Good	Good	Good	Moderate	Good	Moderate	Moderate
Removal of parasitic egg	Moderate	Weak	Good	Moderate	Weak	Moderate	Good
Virus removal	Moderate	Weak	Good	Moderate	Weak	Good	Good

than other types of wastewater in terms of sedimentation and corrosion. In addition, municipal water due to no direct contact with humans can be a reliable source of water supply for industrial use.

Regarding the fact that municipal wastewater contains many harmful parameters relevant to human health has received increasing interest.

Another characteristic of municipal water, which made it suitable for industrial applications, is that municipal water has more values of water in comparison with industrial water (approximately double), which is leading to supplying industrial water use by municipal water. This is observed and understood in developing countries because the proportion of municipal water compared to the proportion of water used in industry in developing countries is more than in other developed countries. It is very important because if municipal water is less than the water requirement of industry, it will face a water shortage and if it is more than the water requirement of industry, it will be required to be reschedule for use or disposal.

11.9 Methods of Urban Water Reuse in Industry

Water treatment methods generally are divided into physical and chemical methods. Physical methods are physical filters that are used to separate particles from water. Chemical methods of water treatment use chemicals for coagulation, sedimentation, flocculation, and total separation of particles and contamination from water [4].

These two methods are generally used in wastewater treatment and municipal water reuse, but they are also performed in different ways. Some of the ways that are often used include

1. Activated sludge: This method is a process for treating wastewater divided into three treatment methods, including conventional activated sludge, extended aeration activated sludge, and oxidation ponds
2. Aerated lagoon
3. Trickling filter
4. Stabilization pond: A pond may be aerated or other methods should be used for sedimentation, flocculation, or other treatment operations

Assessment of the latest statistics of treatment plant shows that three methods in order of propagation include activated sludge

process, stabilization ponds, and aerated lagoons used for municipal wastewater [5]. The performance of different treatment methods for quality improvement of municipal wastewater by different sources is presented in Table 11.2.

Today, with advances in technology and science, so many methods have been introduced for wastewater treatment. Some of these methods have only been experiments, and some of them have been marketed.

The uses of nanotechnology in wastewater treatment has made a huge change in this area and led to improve wastewater quality to distilled water. Nano filters and nano particles are two forms of nanotechnologies that are used in wastewater treatment. Also, biological methods of wastewater treatment have recently been regarded. There are two types of biological treatment process including aerobic (in the presence of oxygen) or anaerobic (in the absence of oxygen).

The use of media and biofilms in both aerobic and anaerobic wastewater treatments has garnered attention.

Determination of the suitable method for wastewater treatment in order to reuse municipal wastewater in industry must be considered with respect to the region and the industry.

This is possible only by the different types of testing. After performing necessary testing, a method is selected as the most appropriate method for the output of the treatment plant, which has more permissive standards for use in industry.

11.10 Alternative Methods of Urban Water Reuse in Industry

Access to unlimited water sources is very pleasant and perhaps unlimited water sources may be an appropriate source for utilization.

However, a few questions remain. The answers to these questions are very important.

Some of the questions are as follows:

1. Is there another source of unlimited water?
2. Is municipal water reuse the best available source for industrial uses?
3. Is water reuse appropriate and affordable for industry by going through a treatment process?
4. Is there a less expensive way than municipal water reuse?
5. Are other waters more suitable for reuse (apart from municipal water)?

Asking such questions as above has led humans to seek other alternative methods and even comparing this method to other methods that have existed in the past. Some even believe municipal water reuse is not a suitable alternative for previous methods. Some also believe that municipal water will cause more environmental problems in the long term.

The question now is what methods are there for providing required water for industrial applications apart from municipal water reuse? Some of these methods are as follows:

Using groundwater or surface water (traditional water supplies): The quality of groundwater and some surface water is much better and more reliable. However, this method is not acceptable due to water resources limitation and it is not recommended.

Industrial water reuse: This method has a high risk because the level of contaminants in water increases during water recirculation in the system. This leads to an increase in the risk of environmental problems and more complete treatment of wastewater is needed to solve this problem. This causes a sharp increase in the cost that may make it unaffordable. However, it is recommended to reuse the output from one industry to another in some cases.

Seawater desalination: Seawater desalination is one of the new methods that in recent years has gained attention. Seawater is an unlimited source due to the extremely high volume of it. This issue has led to consideration of achieving affordable ways to harvest water from seawater by many scientists in water-related disciplines. Today, seawater desalination is done by methods such as reverse osmosis (RO), but there are two major problems. One of them is the very high cost of treatment and another is the large amount of materials that are separated from seawater as contamination that will face a huge disposal problem. Meanwhile, seawater desalination can be a suitable source to meet the water requirements of industries.

Reuse of agricultural wastewater: The agriculture sector is the largest consumer of water in all regions of the world. We need to pay particular attention to water used in agriculture. There are many studies on increasing irrigation efficiency and water use in agriculture, but a lot of water is lost in agriculture as runoff or infiltration. The losses in developing countries with lower irrigation efficiency have been very high and a lot of water is wasted in the agricultural sector. Very large amounts of water used in agriculture and its wastewater can be a suitable source for reuse. It is necessary to consider that agricultural wastewater contains large amounts of chemical fertilizers and some pesticides and herbicides that require strong management and a lot of money to control the treatment and disposal of wastewater. Is agricultural water reuse appropriate and affordable?

The mixing of high quality water with lower quality water: Although mixing of water is a good way to use poor quality water, the mixing of high quality water with lower quality water has not been paid too much attention. It is likely that in the not too distant future this approach will be highly regarded.

Using methods to increase water resources like cloud seeding: There are ways to increase water resources including cloud seeding to increase precipitation. Artificial recharge and so on are such suitable methods that much attention has been recently paid. These methods require strong management and are costly. Another problem with these methods is that implementation of the method needs to consider a way to harvest water because the amount of water in the area (including surface water and groundwater) is only added.

Applying interbasin water transfer: It should be mentioned that interbasin water transfer, as a way to alleviate water shortages, is facing too much criticism because the climate of two areas is affected by water transfer and may even cause loss of plant and animal species in the basins.

It should also be noted that interbasin water transfer is not an unlimited water resource; water is moved from one basin (or watershed) to another. Despite these explanations, interbasin water transfer can also be considered as an urgent method to provide water.

The use of alternative methods that do not need water: These methods can be used in some sectors of industry. For example, air can be used for cooling. Some of these systems have been only suggested and others have been tested for years and now are practically used. It may reduce the amount of water needed, but cannot reduce it to zero. It should also be kept in mind that alternative methods which do not require water are only used in some industrial sectors and alternative methods cannot be found in many sectors of industry. It is also worth noting that the use of alternative methods is not possible and probably affordable everywhere. In some cases, it is not recommended to use alternative methods due to too much cost and challenging environmental problems caused by applying the above-mentioned methods.

11.11 Problems of Urban Water Reuse in Industry

It should be considered that there is a problem in every opportunity. Therefore, there are some problems with municipal water reuse [10].

Of course, the challenges associated with municipal water reuse in industry are much lower than the challenges for agricultural and municipal uses. However, there are some problems with municipal water reuse that are very important to consider. Some of the problems of municipal water reuse are as follows:

1. Some parasites and pathogens exist in municipal wastewater that may be returned to circulation through reuse of municipal water and cause disease in a particular group of organisms (even humans) in a region.
2. After municipal wastewater treatment, some contaminated wastewater remains. It is difficult to manage the disposal in the environment, and accuracy and careful planning are needed.
3. In order to reuse municipal water, the process requires spending a lot of time and energy. It should also be kept in mind that in addition to very high initial costs, operation and maintenance expenses are also important. In

addition, it is important to mention that energy requirements for wastewater treatment are unavoidable.

4. Municipal water pollution is not always the same and may be even different in type and amounts of pollutions day to day. Therefore, this means that the exiting wastewater is not the same and the parameters are different during various days. This led treatment management to be more difficult and causes difficulty in selecting an appropriate method for wastewater treatment.

5. It must be noted that municipal water reuse requires the construction of wastewater treatment plants and it, therefore, requires a dedicated place for wastewater treatment plants. In areas where the land is scarce and expensive, allocating land for wastewater treatment plants is considered a major problem.

It is worth noting that the cases, as mentioned above, are only part of the problems that municipal wastewater reuse is faced with. Meanwhile, some of these cases are not recognized as a problem of municipal wastewater reuse for industrial purposes. For example, because the water used in industry does not have direct contact with organisms, the existence of pathogens in wastewater will not cause problems while using this unlimited source for industrial uses such as cooling tower makeup, boiler feed water, and other industrial applications [15].

11.12 Benefits of Urban Water Reuse in Industry

It should be emphasized that municipal water reuse is linked to many benefits. Having consecutive drought periods, population growth, and many problems of a required supply water, the most important advantage of municipal water reuse is saving water and increasing water resources. Water saving is not the only advantage of municipal water reuse. Other benefits of municipal water reuse are

- Reuse of municipal water causes much less environmental problems due to municipal wastewater discharged to the natural environment.
- Control and management of municipal wastewater becomes easier due to the volume of wastewater that must be discharged directly to the environment.
- Increasing land use according to reducing the volume of contaminated wastewater and available wastewater harvesting.
- Employment: Water treatment plants typically need a labor force. It should also be kept in mind that increasing water resources may lead to agriculture and industry sectors thriving.
- Reduction in social problems and challenges due to reduction in unemployment.

The benefits of municipal water reuse are not only limited to the above-mentioned cases; there are also additional benefits.

Municipal water reuse has an advantage over the rest of the parts in each sector. For example, regarding small amounts of heavy metals in municipal wastewater, application of it in industry can be a remarkable advantage. Other advantages in industry that can be noted include that the water used in industry has no direct contact with animals and, most importantly, humans and it requires less treatment.

According to explanations provided on benefits offered by the use of municipal wastewater for industrial applications, municipal wastewater may be considered the best option for industrial use and industry is the best option for municipal water reuse.

11.13 Case Study: Esfahan, Iran

Due to the high importance of urban wastewater reuse in industrial applications such as boilers and cooling, many studies have been conducted on these areas. The Eslamian et al. [9] study can be noted as an example. They did their study in Isfahan, which is located in the central Iranian plateau in more than 10.5 million acres of land bounded by the Zagros Mountains containing arid and semiarid climate which averages precipitation of about 187 mm and averages evapotranspiration of about 2000 mm/yr. This study was done in Borkhar, which is located 10 km north of Esfahan City. This city is shown in Figure 11.2.

Eslamian et al. [9] used treated wastewater of the Shahinshahr Wastewater Treatment Plant (WWTP1). This treatment plant consists of two similar phases that are planned to have a capacity of 36,000 m³/day. This area is divided into seven categories based on their plans.

These categories are WWTP 1, North Wastewater Treatment Plant (WWTP 2), Esfahan Oil Refining Company (Plant 1), Shahid Montazeri Power Plant (Plant 2), Esfahan Petrochemical Co. (Plant 3), Esfahan Chemical Industries Co. (Plant 4), and Mahmoudabad Industrial City (Zone 1), which are shown in Figures 11.3 and 11.4.

There are four major industrial units that need water in the north of Esfahan containing Esfahan Oil Refining Company (EOR), Shahid Montazeri Power Plant (SPP), National Petrochemical Company (NPC), and Esfahan Chemical Industries Company (ECI), which use water for several purposes such as cooling towers, boilers, firefighting, and so on.

Due to recent frequent droughts and reducing the amount of water in Zayandehrood River, reclaiming water and meeting its quality with Zayandehrood River was considered. Finally, it is concluded that reusing wastewater from a treatment plant (WWTP1 is considered in this project) is an appropriate method due to changing patterns of water consumption in the Borkhar plain and drought-related problems. This method can be considered a model for sustainable development strategies for other countries in the Middle East region and even many other countries [9].

Some parameters of water quality of Zayandehrood compared to effluent of wastewater treatment plant are shown in Table 11.3. As can be seen from the table, the effluent quality of the wastewater treatment plant is very close to Zayandehrood's water quality.

FIGURE 11.2 Location of Esfahan on the map of Iran.

FIGURE 11.3 Satellite picture of the study area, which is located in the north of Isfahan.

FIGURE 11.4   Mahmoudabad plain region, which is selected as the industrial area under study.

TABLE 11.3   Comparison between Water Quality Parameters of Zayandehrood and the Effluent of Shahinshahr Wastewater Treatment Plant

Parameters	Dimensions	Zayandehrood Water	Effluent of Shahinshahr Wastewater Treatment Plant
Turbidity	NTU	0.1	17
pH	–	7.81	7.67
EC	μmhos/cm	892	970
Total hardness (TH)	mg/L CaCO$_3$	259.77	183.02
Total suspended solids (TSS)	mg/L	–	4
Total dissolved solids (TDS)	mg/L	467	510
BOD$_5$	mg/L O$_2$	2	20
COD	mg/L O$_2$	5	45
Nitrite (NO$_2$)	mg/L NO$_2^-$	0.01	0.03
Nitrate (NO$_3$)	mg/L NO$_3^-$	6.64	27.00
Carbonate (CO$_3^{2-}$)	mg/L CaCO$_3$	0	0/00
Bicarbonate (HCO$_3^-$)	mg/L CaCO$_3$	140	220
Sulfate (SO$_4$)	mg/L SO$_4^{2-}$	78	70
Phosphate (PO$_4$)	mg/L P	0	5.80

11.14   Summary and Conclusions

According to increasing scarcity of freshwater in the world with rapid growth of population in urban areas, urban wastewater can be an alternative and a reliable source of water supply year round. Water reuse for different purposes requires an appropriate treatment regarding quality of the inlet wastewater and the expected quality of it for a variety of applications. Water reuse is one of the most important resources for industrial sectors such as cooling towers, boilers, and other sectors. Reusing water in industry has several advantages like no direct contact with humans, improving water use efficiency, reducing the use of scarce water resources, and so on. Industrial

operations require various water qualities for different applications. Therefore, the levels of treatment must be different for each purpose.

Based on this study, it can be concluded that municipal water as a sustainable and unlimited source of water can be considered the best option for industrial uses, and industry as the best option for municipal water reuse.

References

1. Abdel-Shafy, H.I., Mona, S., and Mansour, M. 2013. Overview on water reuse in Egypt: Present and future. *J. Sustainable Sanitation Practice*, 14, 17–25.

2. Asano, T., Burton, F.L., Leverenz, H.L., Tsuchihashi, R., and Tchobanoglous, G. 2007. *Water Reuse—Issues, Technologies and Applications.* McGraw-Hill, New York.

3. ASCE. 1998. *Sustainability Criteria for Water Resources Systems, Prepared by the Task Committee on Sustainability Criteria.* Water Resources Planning and Management Division, Washington, DC.

4. Babin, R. 2005. *Assessment of Factors Influencing Water Reuses Opportunities in Western Australia.* USQ Project, Faculty of Engineering and Surveying, University of Southern Queensland, Queensland, Australia.

5. Carr, G., Potter, R., and Nortcliff, S. 2010. Water reuse for irrigation in Jordan: Perceptions of water quality among farmers. *Agricultural Water Management,* 98(5), 847–854.

6. Environmental Protection Agency. 1981. *Process Design Manual: Land Treatment of Municipal Wastewater.* EPA 625/1-81-013. EPA Center for Environmental Research Information. Cincinnati, OH.

7. Eriksson, E., Auffarth, K., Henze, M., and Ledin, A. 2002. Characteristics of grey wastewater. *Urban Water,* 4, 85–104.

8. Eslamian, S.S. and Tarkesh Isfahany, S. 2011. Industrial reuse of urban wastewaters, a step towards sustainable development of water resources. *1st International Conference on Desalinization and Salinity, A Water Summit,* Abu Dhabi, UAE.

9. Eslamian, S.S., Tarkesh Esfahany, S., Nasri, M., and Safamehr, M. 2013. Evaluating the potential of urban reclaimed water in area of north Isfahan, Iran, for industrial reuses. *International Journal of Hydrology Science and Technology (IJHST),* 3(3), 257–269.

10. Fatta-Kassinos, D., Kalavrouziotis, I.K., Koukoulakis, P.H., and Vasquez, M.I. 2011. The risks associated with wastewater reuse and xenobiotics in the agroecological environment. *Science of the Total Environment,* 409, 3555–3563.

11. Gohari, A., Eslamian, S., Mirchi, A., Abedi-Koupaei, J., Bavani, A.M., and Madani, K. 2013. Water transfer as a solution to water shortage: A fix that can backfire. *Journal of Hydrology,* 491, 23–39.

12. Government of Canada. 2008. Technical guidance document for Water Quality Index practitioners reporting under the Canadian Environmental Sustainability Indicators (CESI) initiative. Environment Canada and Statistics Canada.

13. Grönlund, E. 2014. Sustainable wastewater treatment. In: Eslamian, S. (ed.), *Handbook of Engineering Hydrology: Environmental Hydrology and Water Management.* Taylor & Francis, Boca Raton, FL. 387–400.

14. IWMI. 2000. *World Water Supply and Demand: 1990 to 2025.* International Water Management Institute, Colombo, Sri Lanka.

15. Jimenez, B. and Asano, T. 2008. *Water Reuse: An International Survey of Current Practice.* Issues and Needs. IWA Publishing, London, UK.

16. Lazarova, V. 2001. Role of water reuse in enhancing integrated water resource management. *Final Report of the EU Project CatchWater,* EU Commission.

17. Lazarova, V. and Bahri, A. 2005. *Water Reuse for Irrigation Agriculture, Landscapes, and Turfgrass.* CRC Press, Boca Raton, FL.

18. Metcalf and Eddy. 2003. *Wastewater Engineering: Treatment and Reuse.* 4th edition. McGraw-Hill Inc., New York.

19. Metcalf and Eddy. 2007. *Water Reuse: Issues, Technologies, and Applications.* McGraw-Hill Professional, New York.

20. Mohamadi, O., Heidarpour, M., and Kohansal, M.M. 2014. *Evaluation of the BOD and Phosphorus Removal in Industrial Wastewater Using Anaerobic—Aerobic Treatment. The 4th International Conference on Environmental Challenges and Dendrochronologoy,* Sari, Iran.

21. Nazari, R., Eslamian, S.S., and Khanbilvardi, 2012. Water reuse and sustainability, Chapter 11. In: Voudouris, K. and Vousta, D. (eds.), *Ecological Water Quality-Water Treatment and Reuse.* pp. 241–254. InTech, Croatia.

22. Pearce, G.K. 2008. UF/MF pre-treatment to RO in seawater and wastewater reuse applications: A comparison of energy costs. *Desalination,* 222, 66–73.

23. Pingale, S.M., Jat, M.K., and Khare, D. 2014. Integrated urban water management modelling under climate change scenarios. *Resources, Conservation and Recycling,* 83, 176–189.

24. Smith, T. 2011. *Overcoming Challenges in Wastewater Reuse: A Case Study of San Antonio.* Texas. Bachelor Dissertation, Harvard University, Cambridge, MA.

25. USEPA. 2012. Guidelines for water reuse. 600/R-12/618, Washington, DC.

26. Wang, X.C. and Jin, P.K. 2006. Water shortage and needs for wastewater reuse in the north China. *Water Science Technology,* 53, 35–44.

27. WCED. 1987. *Our Common Future (The Brundtland Commission's Report).* World Commission on Environment and Development, Oxford University Press, Oxford, UK.

28. WHO. 2006. *Guidelines for the Safe Use of Wastewater. Excreta and Greywater,* 3rd edition, Volume II, Wastewater Use in Agriculture, World Health Organization, Geneva, Switzerland.

12

Water Recycling in the Textile Industry: A Holistic Approach from In-Field Experimentation to Impact on Production

Giorgio Rovero
Politecnico di Torino

PREFACE

This chapter summarizes the results of a comprehensive research project aimed at water recycling for wet operations in the textile industry. The ultimate goal of this project is the validation of a modular process scheme for wastewater treatment, based on results obtained from very extensive in-field experimentation with pilot plants. As the textile industry is characterized by small medium size enterprises (SMEs) involved in quite different productions, a thorough process analysis was carried out in order to collect data on the operating procedures, requirements, and medium-term expectations regarding technological improvements. An advanced treatment scheme (implying primary, secondary, and tertiary operations) related to wastewater generated from several dye-houses and finishing departments was based on a process design procedure. Consequently, several continuously operating small units were operated on field to produce a number of water batches, which were delivered to the laboratories of the industrial partners of this project. After tests concerning textile upgrading (i.e., dyeing or finishing), it was demonstrated that recycled water with defined quality specifications can be utilized in these operations to obtain a high standard according to consolidated production protocols.

12.1 Introduction

The water requirement of the textile sector to carry out wet operations is massive: the need ranges from a minimum of about 10 L/kg of finished product to a maximum close to 600 L/kg, the broad variability depending on production steps, technologies, and process strategies. The cost of industrial water throughout Europe varies largely and probably reaches the highest value in Germany (0.6 €/m³) with a minimum in Italy (0.004 €/m³). This irrational spread has been originated by political criteria, rather than by the actual value of freshwater. A very low cost may give origin to an excessive generation of wastewater instead of stimulating operators to save this natural resource. However, the general political trend is directed toward increasing the cost of good quality freshwater to guarantee a proper availability for human consumption as a social priority.

The Biella District in Northwest of Italy is an example of a high concentration of textile manufactures because about 80 dye-houses and 3 wool scouring processes operate in this area. The leading manufacturing is wool milling, with important polyester yarn production and several leading third-party operators. Additionally, new wet processes are being developed to attain knowledge oriented to technical fiber upgrading. This scenario has offered an ideal location to undertake a comprehensive project aimed at advanced wastewater treatment and recycling. As a symptomatic example of the issue in this area, competition between the two main users may occur in drought periods, as industries require about 30×10^6 m³/yr of freshwater, compared to about 20×10^6 m³/yr needed for domestic uses.

A study on a holistic approach to wastewater minimization in process industries was presented in the literature [22]: the reuse and recycling of water were defined therein. A complete hydraulic analysis and modeling on dyeing equipment was presented [5] to provide evidence that operating methods can actually reduce freshwater intake, without affecting product quality. Technological suggestions on how to achieve recycling and water conservation in the textile sector have been formulated, although the application of these concepts has had little success because of process bottlenecks related to strict quality specifications of the finished products. Empirical trials in wool dyeing and finishing have been mentioned in the technical literature, while a systematic recycling of purified wastewater was carried out in treatment consortia in Como and in Prato [4,19].

A comprehensive vision of wet textile productions integrated with wastewater treatment is illustrated in Figure 12.1, which implies the basic concept that every process must have a purging in order to avoid the accumulation of undesired substances; in this case, it was assumed to range between 20% and 30% of the total process water flow. In addition, the evaporation due to products drying was considered to close the water mass balance, according to actual process data.

Treatment costs and orographic constrains, as well as strict quality control on the water used to supply a textile process, have recommended local treatment at a single industry site [15,16], rather than a consortium operation. As far as the textile sector is concerned, wastewater treatment is often limited to a primary step (sedimentation and balancing) followed by a secondary operation (aerobic sludge digestion and consequent settling). If the wet departments of a textile industry are carefully managed, the resulting raw wastewater can be assimilated largely to domestic sewage in terms of biodegradability. The above sequence of treatment operations is recommended to meet the limits imposed by law for most parameters, with the exception of residual coloration and surfactant concentration. It is worthwhile to point out that the quality specifications required for wet textile processes for safe water recycling are much stricter compared to the legal bounds required for discharging into the environment. A specific result reached with this research project is to make these specifications explicit. This technological problem requires a very careful evaluation to implement wastewater treatment and recycling, and not to introduce additional problems to dyeing operations. Conversely, at the present time, quality specifications for the freshwater used in textile wet processes have been generally defined in a simplified way (hardness, clearness, and reduction potential).

The goals of this ambitious study, by operating with pilot-scale treatment units at many industrial sites, have been

1. To provide guidelines useful to optimize wastewater treatment in terms of biodegradation and overall effectiveness

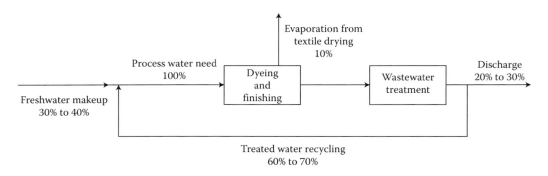

FIGURE 12.1 Conceptual block diagram of a wet textile process considering water recycling.

2. To compare the effectiveness of several tertiary operations with a special emphasis on the removal of residual coloration, surfactants, refractory chemical oxygen demand (COD), and ionic content

3. To run small-scale industrial tests and assess the quality of textile products dyed and finished with nondiluted recycled water

4. To establish water quality specifications in order to match the consolidated wet process reliability, where good quality freshwater is used as a standard

The project strategy implied sharing responsibilities between academy personnel (responsible for running the equipment and preparing batches of treated water) and textile industry technicians (responsible for the ultimate use of the purified water, either in dyeing or wash-off, as well as the final evaluation of textile wet upgrading).

Due to the most usual residual pollutants, the tertiary treatments, usually applied downstream of the secondary sedimentation, involve granular activated carbon adsorption or ozonization. The usual sequence advises to apply the oxidative process as a refining treatment to minimize O_3 generation. However, an example of a reverse order in the process scheme was also presented in the literature [2]. Decoloration kinetics and total organic carbon (TOC) removal efficiency were considered in a research project [6]. Single and multistage bubble contactors demonstrated the reduction profiles of TOC and COD [14]. The characterization of the hydraulic behavior of a pilot unit for ozone absorption, the determination of O_3 mass transfer efficiency, and the mixing of the continuous liquid phase were demonstrated in a fundamental study [9]. A similar approach was considered to define the concentration profile of residual O_3 in the gas phase along a bubble column [23]. A recent comprehensive evaluation has been carried out to identify the optimal hydrodynamics of O_3 contactors, where the gas to liquid mass transfer may gain benefit from ultrasounds or cavitation [1].

Membrane separation processes (ultrafiltration [UF], nanofiltration [NF], and reverse osmosis [RO]) have received considerable attention over the past two decades for advanced treatments of municipal and industrial wastewater to remove pollutants at low concentration and, specifically, to remove the ionic load. Colloid separation by UF, as an alternative to coagulation, was investigated as a preliminary treatment to textile productions [17] to achieve high quality dyeing. The above technologies were applied for wastewater reuse [4,19]. NF has a specific application to quantitatively reject bivalent ions and dyes having a molecular weight ranging between 400 and 800 Dalton. Instead, RO can concentrate monovalent and upper charged ions (NaCl and Na_2SO_4 are typical salts used for textile dyeing and finishing) to a degree such that downstream thermal processes become energetically sustainable to further concentrate the brine and then directly reuse it in the required production steps.

There is emerging scientific evidence that anaerobic processes (such as the upflow anaerobic-sludge blanket [UASB] technology [10]), would produce potentially harmful aromatic amines originated by dyestuff treatment [8,20]. However, coagulation/sedimentation physicochemical pretreatment could remove most dyestuff by adsorption, thus making the integrated process very interesting for the textile industry.

12.2 Process Design and Freshwater Need Mitigation

In order to overcome the variability of wet operations carried out in the textile industry, an ideal process was identified to provide a general vision of real cases. The basic approach of such process has implied: (a) considering the best available technologies (BAT) to conceive the so-called "Factory of the Future" according to a modular design. Thanks to this criterion, a process simplification is possible whenever a given operation is not required; (b) a hydraulic scheme representing both wet production steps and wastewater treatment to consider freshwater and recycled water utilization; this scheme was elaborated analyzing a relevant number of textile manufacturing sites in order to facilitate any technological transfer; and (c) quantifying the requirement of water and chemicals according to selected production processes, representative of real wet textile manufacturing.

12.2.1 Process Flow Diagram for Wastewater Treatment in a Wet Textile Industry

Figure 12.2 presents a complete process flow diagram (PFD) based on a water system existing in a standard textile industry, added of the operations needed to implement water recycling. This diagram includes all the unit process steps subsequently considered in the experimental program. The scheme also considers the possibility of reusing process water having different quality specifications required by dissimilar dye-house operations (purging, dyeing, wash-off, and textile finishing, in that order).

Dyeing and finishing are usually conclusive operations that upgrade textile products with well-defined properties, that is, coloration and color fastness, touch, etc. In particular, pale and pastel nuances may require extremely limpid water provided by a UF pretreatment [17]. Moreover, dyeing is very water demanding according to the liquor ratio of equipment and dyeing recipe used, as dyes are fixed at different liquor exhaustions, depending on the coloristic class and recipe formulation. A comprehensive overview of textile dyeing appears in the literature [21], even though, considering the variability of processes, a detailed evaluation can be given only through a case-by-case process analysis. It is quite a general opinion that a large liquor ratio helps in achieving optimal coloration evenness. On the contrary, it can be demonstrated that a well-designed low-liquor-ratio machinery, properly formulated dyeing recipes, and a sound product rinsing method both maximize dyeing quality and reduce process water needs, thus making wastewater treatment more effective and economical. Ultimately, recycling is more feasible in these instances. In this view, an example for increasing wastewater

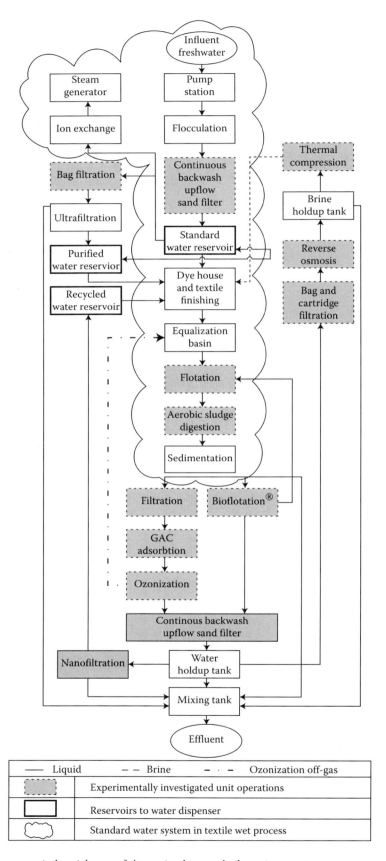

FIGURE 12.2 Process flow diagram at industrial state-of-the-art implemented pilot unit operations.

biodegradability by replacing standard dyeing auxiliaries with low molecular weight alcohols is given [7] both at the laboratory and industrial scale.

12.2.2 Dyeing Diagrams as Hydraulic Tool for Water Need Estimation

As most of dyeing is run batchwise, evaluating the minimum freshwater amount needed to carry out a given dyeing process implies knowing the type of machinery (i.e., its nominal liquor ratio), as well as the sequence of steps through which the process is accomplished. A clear graphical representation is given by the so-called dyeing diagram, which typically represents the temperature evolution as a function of time, together with the relative time-dependent feeding of chemicals. A useful completion of this diagram requires adding the complementary steps given by textile material pretreatment (or purge), wash-off operations (to remove soluble contaminants, usually represented by unfixed dyestuff), and textile finishing (application of softeners, stain repellents, or waterproof substances). An example of a complete diagram is given in Figure 12.3, which gives evidence that several draining steps take place during each batch-dyeing operation.

12.2.3 Textile Rinsing Methods and Water Economization

Considering batch operations, wash-off can be carried out according to two methods: overflowing the exhaust liquor by continuously feeding freshwater, or step operations by draining and refilling the system with freshwater. The hydrodynamics of a standard apparatus is favorable for both strategies (high liquor circulation, effective fiber-to-liquor mass transfer, and good hydraulic mixing). However, by applying the mass balances given next, it can be demonstrated that a step procedure is more effective in removing any soluble contaminant from the textile substrate, thus generating an overall rinsing water conservation of the order of 50% with respect to an overflowing technique. Industrial validation of this procedure for hank dyeing is given in the literature [5]. Moreover, it should be additionally considered that the step rinsing procedure allows also a partial refilling in several types of dyeing equipment (namely textile packages in vertical or horizontal axis kier) giving additional advantages in terms of the overall water requirement for rinsing.

The step rinsing procedure can be described by

$$C_n = C_0 \cdot \left(\frac{V_{hu}}{V} \right)^n \tag{12.1}$$

The overflow rinsing method is given by

$$C = C_0 \cdot e^{-t/\tau} \tag{12.2}$$

where

C_n = concentration of pollutant after n rinsing steps
C_0 = concentration of pollutant at the end of a dyeing step
V_{hu} = hold-up volume originated by equipment and textile
V = equipment volume
C = concentration at any time t
τ = equipment filling time

Again, industrial practice and model predictions match in terms of water need if the initial boundary condition is given by a dye concentration corresponding to the dye liquor exhaustion; then, the final boundary condition is assumed with a dye concentration equal to 0.1% of the dye-liquor recipe. The number of wash-off steps, both according to theory and practice, varies between one and five, depending on dyestuff exhaustion. Considering the mean step number, it follows that rinsing accounts for about 60% of the total water requirement in the process.

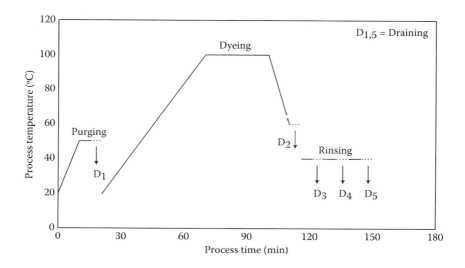

FIGURE 12.3 Complete dyeing diagram including wash-off draining steps (D_1 to D_5).

12.2.4 Specific Wastewater Treatment Goals for Recycling to Wet Textile Processes

The main objectives of the complete treatment scheme are: (1) decreasing the COD in the outgoing stream to such a low concentration at which the consequent reducing potential does not affect the coloristic result; (2) eliminating any residual coloration to avoid interference with dyeing recipes; (3) clearing the treated water from any suspended fine solids because many textile substrates act as a filter of impurities; and (4) regulating the total salinity in the treated water because the ionic content has a plurality of effects on dyeing equilibrium and has to be properly calibrated in terms of amount and timing throughout dyeing. As a conclusive remark, it can be stated that dyeing is a complex multiphase hydraulic and physicochemical operation, where water acts as a process fluid, which must have strict quality specifications.

The capacity of the described scheme was targeted for water treatment flow rates varying from several hundred to a few thousand m³/day. Considering exhaustion dyeing processes, the minimum specific water consumption target is estimated to be 30 L of freshwater per kilogram textile product. This goal can be reached by applying all the measures suitable to achieve a freshwater intake reduction, namely dyeing equipment selection and wash-off policy.

12.3 Process Modeling and Dynamics

A mathematical model has been conceived to predict the release dynamics of ionic substances, which cannot be removed by the standard technologies included in a wastewater treatment layout. The sole operation to consider for this purpose is membrane filtration. According to many consolidated evaluations of experienced dyers, the saline content seems to represent the most serious barrier to water recycling in wet textile processing.

12.3.1 Model Structure and Inputs

The inputs to the model were given by: (1) assuming a determined textile production, both in terms of mass and fiber types; (2) choosing referenced dyeing and wet finishing recipes; (3) defining the equipment type and its operating liquor ratio; and (4) engineering the whole process, according to Figure 12.3. As a mere example, referring to industrial data (30 tons of textile material per day, subdivided as 50% wool, 30% polyester, 15% acrylic, and 5% cotton, with a liquor ratio = 5 L/kg for most common dyeing recipes), the specific water consumption amounted to 30 L/kg$_{textile}$ and the overall freshwater requirement reached 900 m³/day.

Depending on recipes, the crystal NaCl salt feed was assumed to fluctuate between 400 kg/day (for 3 days a week) to 700 kg/day (for 4 days a week), in order to generate a periodic disturb to the ionic content and to analyze a potentially critical dynamics, its maximum allowable concentration being that consented by law. All the relevant equipment capacitances appearing in Figure 12.2 were included into the model scheme represented in Figure 12.4 to account for accumulation, dumping of concentration peaks, and hydraulic delay.

12.3.2 Flow Rate and Ionic Concentration Predictions from the Model

One model output is given in Figure 12.5: the concentration peak of salt at the discharge was limited by RO to 1200 mg/L of Cl⁻ (this value is typically set by law) when the system worked at its maximum load and the freshwater intake was reduced to about 360 m³/day, thanks to a desalinated flow rate of about 250 m³/day produced by the RO coupled with thermal compression to concentrate and recycle salt brine. A weekly membrane reactivation was also taken into account and appears in the plot.

The correlation of Figure 12.6 generated by the model demonstrates that treated water cannot be recycled over about 30% when NaCl crystal salt is continuously introduced into the process at a rate of 700 kg/day. If brine is reused or otherwise disposed, higher water recycling ratios are possible without exceeding the maximum allowable concentration of chloride ions.

This modeling procedure was proved to be effective to design the membrane filtration unit and predict the dynamic output of any substance that cannot be retained by standard treatment units. As a conclusive remark to simulate a chemical accident, also an abrupt 100 kg salt spilling was considered in the dyehouse to test the robustness of the model, as well as the damping capacity of the whole wastewater treatment process.

12.4 Industrial Experimentation

As the textile industry offers a wide spectrum of operating conditions and wastewater treatments, great emphasis and attention were given to the pilot equipment selection. Then, all dyeing and finishing steps in the industrial experimentation were carefully evaluated. An effective cooperation between the textile industries operating in the Biella District and Politecnico di Torino allowed planning and carrying out such a complex research project. A broad group of textile industries (18 enterprises) and technology providers (10 units) offered an adequate support to generate a sound overall result. The scientific group took charge of designing and operating the pilot units and preparing the water lots; the production industries carried out the textile operations according to best practice criteria. The results were jointly interpreted.

Several unit operations were specifically chosen and in-field interconnected to the existing industrial treatment plants to implement the process scheme given in Figure 12.2. The pilot equipment experimented on throughout the research project are listed in Table 12.1. All these units had a nominal treatment capacity of 100 L/h, which is a suitable size to generate water lots for industrial experimentation, while contemporarily maintaining all features of easy to operate equipment. Details of the single apparatuses, operation strategies, and effectiveness are given in the literature [18]. A summarization of the results is provided next.

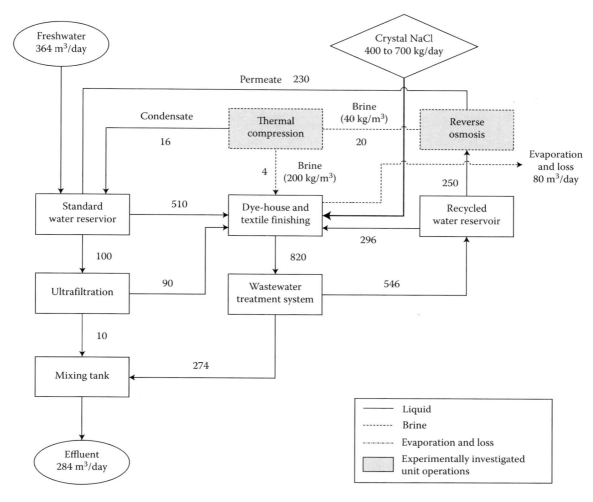

FIGURE 12.4 Model diagram for evaluating the ionic concentration dynamics (water overall utilization rate for dyeing: 900 m³/day; load range of crystal salt: 400–700 kg/day).

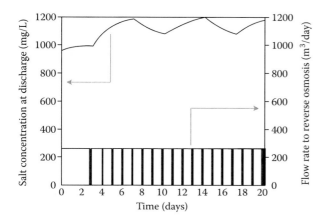

FIGURE 12.5 Dynamical output from the model, considering crystal salt utilization ranging from 400 to 700 kg/day.

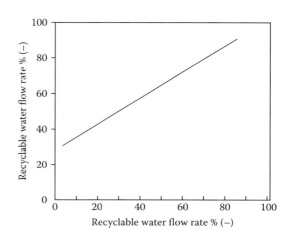

FIGURE 12.6 Percentage recycling of wastewater versus brine reuse at steady state.

TABLE 12.1 Pilot Equipment List

Technology	Number of Units
Automatic controlled downflow sand filter	1
Continuous backwash upflow sand filter	1
Ozonization	2 bubble columns + 1 ejector mixer
Granular activated carbon (GAC) adsorption	3
Mobile bed biofilm reactor (MBBR)	1
Fixed bed biofilm reactor (FBBR)	1
Electroflotation	1
Dissolved air flotation (DAF)	1
Settler	1
Thermal compression	1
Nanofiltration (NF)	1
Reverse osmosis (RO)	1
Comparative aerobic digestion set-up: Bioflotation® + flow-jet aeration + activated sludge + FBBR	1
Wastewater treatment platform at FILIDEA industrial site	1

12.4.1 Primary Treatment

Dissolved air flotation (DAF) and electroflotation were extensively tested to compare their performance. The specific goal of these physicochemical operations is to remove the fiber lubricant oils (mineral, silicon, or vegetable). It was proved that a high oil concentration seriously affects the biological action in the aerobic sludge digestion step. As an additional consideration, any residual oil, downstream of the secondary settling, seriously affects the operations in the tertiary treatments (lifetime of adsorbent media, ease of membrane cleaning, or gas-to-liquid mass transfer, which are the cases given by aeration and ozonization). An electroflotation equipment was specifically

conceived, built, and tested for continuous operations: this process is illustrated in the literature [12,13] and is schematically shown in Figure 12.7.

Two advantages given by electroflotation can be mentioned: the oil yield removal in the electroflotation unit easily reached values close to 80%, compared to an average value of 40% obtained by standard DAF. Electroflotation does not require any dosage of chemicals because the coagulant (Al(OH)$_3$ · nH$_2$O) is generated in situ by anodic corrosion. The COD removal also reached interesting yields close to 70%. The drawback of this technology is originated by the anodic corrosion of aluminum electrodes, which is rather costly in terms of material and energy.

12.4.2 Secondary Treatment

Most wet processes in textile industries operate with wastewater treatment schemes that include aerobic digestion having different features. The key characteristics of aerobic digestion technologies are linked to the biomass hydraulics (suspended or segregated), the wastewater residence time, and the aeration devices. The ultimate goals of this technology are given by the maximum removal of biodegradable matter, while minimizing the sludge production, and the highest oxygen transfer efficiency. Important concern is about the installation capital cost (due to size and technological complexity) and the operative cost (mainly pumping and air blowing). Several comparative experimentations [11] with long lasting campaigns were scheduled to compare standard activated sludge digestion to Bioflotation®, flow-jet aeration and an immobilized biomass digestion (see Figure 12.8).

These technologies are characterized by dissimilar oxygen transfer efficiencies and biomass hold-up values. All these units were fed in parallel to guarantee the same wastewater residence time. Isothermal conditions were obtained over time by interconnecting the biological reactors by means of a heat exchange network.

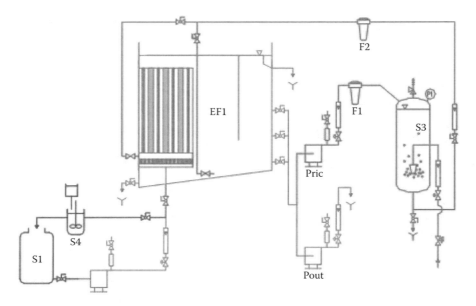

FIGURE 12.7 Scheme of the pilot unit for dissolved air flotation (DAF) and electroflotation.

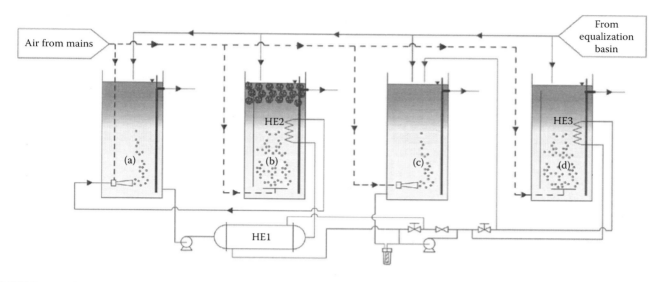

FIGURE 12.8 Comparative aerobic digestion set-up: (a) bioflotation, (b) FBBR, (c) flow jet aeration, (d) activated sludge as a reference. HE1, 2, 3: heat exchanger network.

Bioflotation and the fixed bed biofilm reactor (FBBR) offered the best performance in terms of TOC and COD removal and as far as the sludge production was concerned, neither technology is affected by sludge washout nor do they need sludge recirculation.

12.4.3 Tertiary Treatments

Since it is advisable to protect water-refining technologies from the suspended matter leaving the secondary settler, a continuous backwash upflow sand filter by OMC (Italy) was investigated to comprehend its effectiveness in retaining sludge particles. Removal yields of about 80% were steadily obtained by feeding wastewater having a sludge concentration in the range between 10 and 100 mg/L. Figure 12.9 reports the silhouette of this unit, which demonstrated excellent continuous performance both at pilot scale as well at full scale in two industrial sites.

12.4.3.1 Adsorption on Granular Activated Carbon

Granular activated carbon (GAC) was used in several units (see Figure 12.10) to determine the removal of contaminants (COD, color, and surfactants) versus operation time at different operating conditions. Nonpolar organic substances were retained, with yields decreasing with time from about 95% for the freshly charged adsorbent to values adequate to guarantee the water quality specifications (see the following section). The GAC saturation load was found to range between 0.4 and 0.6 kg_{COD}/kg_{GAC}. As a mean evaluation, the color removal correspondingly decreased from 99% to about 35%. As far as this technology is concerned, it was demonstrated that a critical operating condition is given by fiber lubricating oils because this organic pollutant contaminated the active carbon irreversibly with a much lower load, close to $1.5 \times 10^{-3}\ kg_{oil}/kg_{GAC}$ [3].

According to a combined application of GAC, it is worthwhile mentioning that the FBBR technology was subsequently chosen

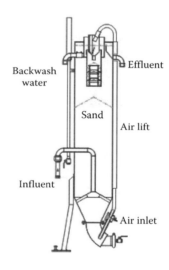

FIGURE 12.9 Scheme of the pilot unit for backwash continuous upflow sand filter.

to assemble a complete pilot unit treatment, as illustrated in Figure 12.11, as this digester requires minimum operating care and displays a high oxygen transfer efficiency. The main goal of this set-up was to demonstrate the effect of an efficient electroflotation on biological digestion and lifetime of a GAC unit. In these experiments, the specific treating capacity of the adsorbent medium was as high as 25 $m^3_{wastewater}/kg_{GAC}$ to reduce the COD from about 60–25 ppm (the latter being a water quality specification for recycling to dyeing or wet finishing).

12.4.3.2 Ozone Treatment

Decoloration treatments by ozone absorption were carried out with two kinds of contactors (see Figure 12.12), namely a bubble column (gas to liquid countercurrent characterized by plug

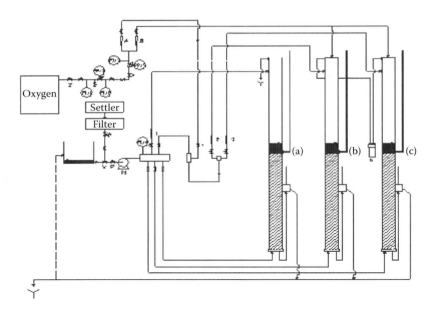

FIGURE 12.10 Scheme of the pilot unit with granular activated carbon and expanded clay columns to compare contaminant removal: (a) GAC with wastewater from secondary treatment, (b) GAC with O_2-saturated wastewater from secondary treatment, and (c) expanded clay with wastewater from secondary treatment.

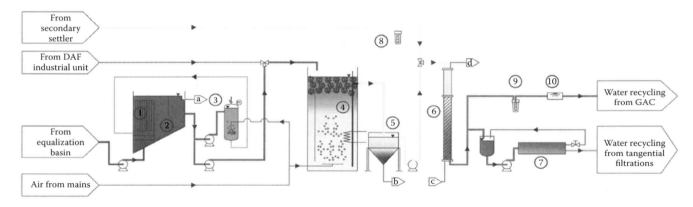

FIGURE 12.11 Complete pilot unit assembly for wastewater treatment in a polyester mill: (1) electroflotation, (2) DAF, (3) pressurized water saturation, (4) FBBR aerobic digester, (5) secondary settler, (6) GAC column, (7) tangential filtration module, (8) screen cartridge filter, (9) package cartridge filter, and (10) UV treatment.

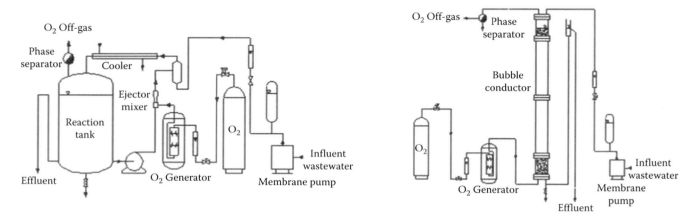

FIGURE 12.12 Scheme of the pilot units for ozone treatment: ejector well-mixed reactor on the left, plug-flow bubble column on the right.

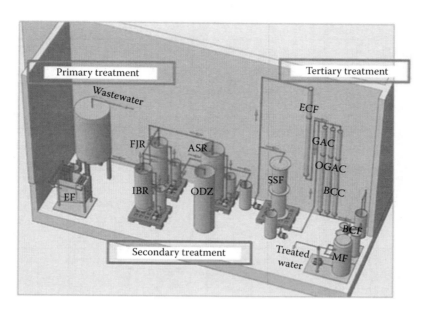

FIGURE 12.13 Industrial onsite wastewater treatment platform for the FILIDEA project.

flow-like hydraulics) and a recirculation contactor through an ejector mixer (well-mixed unit). The first technology, operating in a 3-m deep unit, with a 30-L capacity and 0.5-h residence time of the effluent, represents the industrial conditions very closely in terms of gas/liquid hydraulics. From these experimental conditions, the process scale-up is straightforward by maintaining identical superficial velocities. Parallel runs showed that the other more complex technology did not provide any advantage, both considering the oxidation kinetics and ozone dosage. The main advantages of ozone treatment consist of a complete decoloration, high operating flexibility, and prompt O_3 dosage to control the reaction.

12.4.3.3 Separation by Tangential Filtration

Membrane filtration (NF and RO) was selected to separate dissolved ions, nonbiodegradable matter, and refractory pollutants not otherwise retained by the chosen upstream steps. NF plays a specific role in textile wastewater softening for the separation of bivalent ions (mainly SO_4^{2-} and Ca^{2+}) introduced into dyeing liquors or during the primary physicochemical treatment. Low and medium molecular weight dyes are quantitatively separated by NF with a high rejection degree. RO was specifically considered to reduce the total saline load (monovalent and bivalent ions) and generate a recyclable brine. Additionally, it should be taken into account that anti-shrinking wool processing generates a high concentration of NH_4^+, which is quite noxious for dyeing results: in this view, RO is essential to conceive and design a wastewater recycling plant.

12.4.3.4 Design of a Complete Pilot-Scale Wastewater Platform

As a final example of industrial experimentation, a wastewater treating platform designed and built for the FILIDEA project (a settling entrepreneurship in the Biella area financed by the Piemonte regional government in 2009), which is a joint venture between Italy and Turkey, has to be mentioned. The graphic of this complex assembly is given in Figure 12.13.

One of the goals of this platform is to test the effect of novel dyeing recipes for technical fibers on all the steps of the effluent treatment and the consequent water recyclability. Considering this peculiar aspect, no final assessment can be provided by now.

12.5 Dyeing and Finishing Results

Over 550 industrial tests on dyeing and finishing were carried out by experts in the field with the water treated in the pilot units. The strategy to run each test was chosen by (a) defining the integration of the pilot units into an existing industrial wastewater treatment scheme; (b) defining the volume of water to treat according to downstream uses in the textile operations, namely purging, dyeing, wash-off, or finishing of a given substrate; and (c) running a parallel trial with standard freshwater to correlate the results in terms of coloration and other key textile properties (color fastness, touch, etc.)

A database to which the industrial partners in this project could refer to for future strategies was then implemented. Table 12.2 gives a brief account of the results obtained: it considers treatment technology, industrial partners ID, textile substrate used (type, fiber, and lot size) and general information about the dyeing recipe. Apart from a traditional visual assessment, reflectance spectrophotometry was used to evaluate each dyed material, thus providing an objective judgment. The ΔE parameter is generated by a linear combination of ΔC (difference of chroma), ΔH (difference of hue), and ΔL (difference of lightness) and was obtained by comparing specimens to corresponding reference samples. Any discrepancy could have been caused by solid impurities, dyestuff modification at the molecular scale, dye insolubility (or precipitation), or inadequate dyeing kinetics

TABLE 12.2   Dyeing Tests with 100% Recycled Water

Test Number	Technology	Industry ID	Substrate	Material	Lot Size (kg or m)	Color	ΔE	Result	COD (mg/L)
1	GAC	B	y.p.	W	2	Sky blue	0.89	+	<30
2	GAC + NF	B	y.p.	W	2	Sky blue	1.90	−	<10
3	GAC	S	y.p.	PET	2	Pink	0.96	+	<30
4	NF	M	y.p.	W/AN	24	Sky blue	1.18	+/−	12
5	O_3	L	y.p.	W	0.7	Sky blue	0.65	+	24
6	O_3	L	y.p.	W	0.7	White	0.12	+	18
7	FBBR + O_3	Z	Fabric	W	50	Pale gray	0.2	+	32
8	O_3 + RO[a]	TS	Hank	W	1	Sky blue	vis	+	<10
9	GAC + NF	T	Fabric	W/PET	17	Pink	vis	+	<10
10	GAC + O_3	TT	l.f.	C	3	Pale bue	vis	+	45
11	GAC + O_3 + NF	TT	l.f.	AN	3	Beige	vis	+	31
12	NF + RO + TC[b]	M	y.p.	W/AN	2	Black	1.9	−	—

Note: Representative results out of 550 industrial tests. y.p. = yarn package; l.f. = loose fiber (staple); W = wool; C = cotton; PET = polyester; AN = acrylic; vis = visual evaluation of dyeing result; ΔE = spectrophotometric color difference.

[a] Two RO stages.

[b] NaCl reuse 80 g/L brine.

caused by extraneous substances brought into the textile process by the recycled water. In the textile field, the discrimination criterion is conventionally assumed to be ΔE > 1.

Finishing quality was judged according to the experience of technologists.

An overall rate of success of the trials exceeding 90% was reached, which is fully acceptable by industrial practice. It is worthwhile to note that any dyeing that had a higher quality than the usual results (i.e., very brilliant or highly pure coloration) was considered negative because it was not in conformity with commercialization standards.

Some tests from Table 12.2 are specifically commented:

- Runs 2 and 4: The very pure water generated by the NF unit produced such a brilliant coloration, which was rejected according to the ΔE criterion and for commercial strategies of the industry
- Run 12: Besides water recycling, brine recycling was also carried out; the partially negative output (ΔE moderately exceeded unity) would suggest further and more detailed trials aimed at resolving this specific aspect, which deals with the recycling of process chemicals
- All the other positive results allowed to draw a general acceptability criterion (given by COD <25 ppm) taken as a clue specification for water recycling

12.6 Quality Specifications for Water Recycling to Textile Dyeing and Finishing

Dyeing and finishing tests were always carried out with 100% recycled water in order to run each trial at the most severe conditions. Besides the quality parameters defined hereafter, it is worthwhile mentioning that textile industries require the steadiest chemical conditions to run processes at the highest confidence

and then attain the highest quality of products. As dyeing and finishing are crucial operations to qualify the commercial value of a textile product, textile technologists really "dream" to minimize the number of process variables: In their opinion, wastewater recycling could be a critical issue to be faced on the basis of freshwater cost.

The experimentation on the pilot units generated the following fundamental data, definitely assessed thanks to the very large number of industrial runs.

1. The recycled water must have a minimal reduction potential. Additionally, inorganic ions characterized by reducing properties (such as NH_4^+) must be removed not to affect the function of the chromophore groups of a given dye.
2. A COD <25 mg/L was chosen as a safe overall parameter to qualify the chemical reducing activity.
3. Total suspended solids (TSS) <1 mg/L did not affect dyeing even when operating with the most compact yarn structures.
4. Nondetectable water coloration with a 200-mm optical path-length was a safe measure of noninterference on pale color dyeing.
5. Hardness (expressed as $CaCO_3$) <50 mg/L did not affect dyeing and touch of natural or synthetic fibers.
6. Salinity <50 mg/L was successfully tested and given as a process specification for process modeling and membrane filtration design.

12.7 Wastewater Treatment Cost and Additional Industrial Considerations

The problem of defining the total cost is of major significance when discriminating between various process alternatives. A modular design of the process flow diagram was selected to

TABLE 12.3 Unit Cost for Each Technology Considered in the On-Field Experimentation Program

Technology	Unit Cost (€/m³)
Primary and secondary treatment	0.60
Sand filtration	0.01
Bag and cartridge filtration	0.04
Electroflotation	0.12
Bioflotation	0.80
GAC	0.15
Ozonization	0.15
UF	0.16
NF	0.22
RO	0.25

confer flexibility to the process scheme. A process scheme with 60% recycle was conceived to evaluate the actual flow rates, the conceptual planning, and designing the units. This relevant recycling rate was assumed to provide economical significance to the more complex structure of a wastewater treatment plant, which has to be supported by the factory mains. The total unitary costs for the various technologies considered (given as €/m³ of recycled water) are listed in Table 12.3, following a joint estimation with the technology providers participating in the project. Each of these values considers the actualized capital and operating costs for a standard mean life span of the investment.

The combination of technologies included in the primary and secondary treatments by a survey in the Biella District has led to an appraisal in the range of 0.7–0.8 €/m³, which is the economical duty to environmental conservation. The additional cost for the refining treatments required by recycling goals is in the order of 0.6–0.7 €/m³, which has to be compared to the cost of saved freshwater.

Following the experimentation at pilot scale, it has been proven at full scale that an additional flotation, placed downstream of the secondary settler, generates great advantages to the following tertiary treatment steps. Namely, lifetime of GAC units was extended to such a degree that this technology has become very competitive.

It has emerged that a rational recycling rate (set at about two-thirds of wastewater entering the treatment facility) corresponds to the flow rate typically required by wash-off operations in a standard dye-house (about 60% of the total freshwater need). This consideration offers the additional possibility to address recycled water to rinsing instead of dyeing, as it is a much less critical production step.

12.8 Summary and Conclusions

A systematic research project was initiated by designing and modeling a wastewater system connected to wet textile processes. A preliminary analysis was based on the evaluation of the state-of-the-art in a large number of textile industries and the possible structural completion of their wastewater treatment facilities by aiming at water recycling. The conclusive results have provided a sound selection of tertiary technologies as well as water quality specifications of recyclable process water. Due to the complex case study, the experimental results were obtained by operating on field at the industrial sites and running a complete sequential arrangement of pilot plants to demonstrate the feasibility of water recycling. Several hundreds of dyeing and finishing industrial tests on textile products have proven a positive response for over 90% of the tests as a "go-no-go" quality evaluation of the textile product. A few lots of recycled water, produced by tangential filtration (NF or RO), generated a coloration of the textile specimen with a chroma even purer than the reference dyeing, run with freshwater; these unexpected results were most likely due to the very controlled process conditions in the wastewater treatment system.

The quality specifications for water utilized in textile processing have been reliably defined and proven on a very large statistical basis. The objective evaluation of the treatment costs carried out by a public institution (Politecnico di Torino) now provides a sound basis for future industrial developments. The implementation of this technological know-how, promptly transferable to full-scale facilities, depends only on the industrial freshwater cost charged to industries. Therefore, great benefits are foreseen for the environment both in the developed countries, where high-quality textiles are produced, and in the emerging countries, where commodity textiles are now manufactured with heavy environmental burdens. By implementing water recycling, freshwater can be reserved for domestic uses to a larger extent, as advised by transnational directives, and the release of non-biodegradable pollutants reduced.

Acknowledgments

The author thanks Professor Franco Ferrero for his fundamental advice and Professor Silvio Sicardi for his continuous support in this challenging project, addressed to water recycling, in the Biella Textile District.

References

1. Actis Grande, G. 2014. Treatment of wastewater from textile dyeing by ozonization. PhD thesis, Politecnico di Torino, Italy.
2. Boere, J.A. 1991. Combined use of ozone and granular activated carbon (GAC) in potable water treatment; effect on GAC quality after reactivation. *Ozone Science and Engineering*, 14, 123.
3. Ceresa Gianet, M., Rovero, G., Beltramo, C., and Sicardi, S. 2007. Wastewater recycling from polyester dyeing: An experimental study by pilot equipment units. *7th World Textile Conference AUTEX*, Tampere, Finland.
4. Ciardelli, G., Corsi, L., and Marcucci, M. 2000. Membrane separation for wastewater reuse in the textile industry. *Resources, Conservation and Recycling*, 31, 189.

5. Cignolo, S., Rovero, G., Banchero, M., and Ferrero, F. 2004. Industrial experimentation on hank-dyeing: Modelling of equipment and water economisation during rinsing. *AUTEX Research Journal*, 4(4), 192–203.

6. Ferrero, F., Iennaco, M., and Rovero, G. 1992. Degradazione ossidativa dei coloranti con ozono (oxidative degradation of dyes by ozone). *Inquinamento (Pollution)*, 4, 66.

7. Ferrero, F., Periolatto, M., Rovero, G., and Giansetti, M. 2011. Alcohol-assisted dyeing processes: A chemical substitution study. *Journal of Cleaner Production*, 19(12), 1377–1384.

8. Isik, M. and Sponza, D.T. 2008. Anaerobic/aerobic treatment of a simulated textile wastewater. *Separation and Purification Technology*, 60, 64–72.

9. Le Sauze, N., Laplanche, A., Orta De Velasquez, M.T., Martin, G., Langlais, B., and Martin, N. 1992. The residence time distribution of the liquid phase in a Bubble column and its effect on ozone transfer. *Ozone Science and Engineering*, 14, 245.

10. Metcalf and Eddy Inc., 2003. *Wastewater Engineering. Treatment, Disposal and Reuse*, 4th edition. McGraw-Hill Co., New York.

11. Papadia, S., Rovero, G., Fava, F., and DiGioia, D. 2011. Comparison of different pilot scale bioreactors for the treatment of a real wastewater from the textile industry. *International Biodeterioration and Biodegradation*, 65(3), 396–403.

12. Percivale, M. 2005. Rottura di emulsioni generate dall'industria tessile mediante elettroflottazione. PhD thesis, Politecnico di Torino, Italy.

13. Percivale, M., Rovero, G., and Ferrero, F., 2005. Electroflotation of wastewater from synthetic fibre dyeing. *5th World Textile Conference AUTEX*, Portoroz, Slovenia.

14. Rakoczi, F. 1990. Multistage ozone treatment of dye waste. *Ozone Science and Engineering*, 11, 11.

15. Rovero, G. 2001. Water recycling in textile industry: Part I. *Industria Laniera Tessile Abbigliamento CXV*, (2), 112.

16. Rovero, G. 2001. Water recycling in textile industry: Part II. *Industria Laniera Tessile Abbigliamento CXV*, (3), 194.

17. Rovero, G., Pagliai, A., Manna, L., Sicardi, S., Xotta, C., and Baltera, V. 1995. Separation of colloid particles from feed water for package-dyeing. *9th International Wool Textile Research Conference,* International Wool Secretariat, Biella.

18. Rovero, G., Percivale, M., and Beltramo, C. 2004. Water recycling for wet textile productions: An example of collective research between SMEs and University. *2nd CIWEM National Conference,* ovember 13-15, Leeds, UK.

19. Rozzi, A., Malpei, F., Bonomo, L., and Bianchi, R. 1999. Textile wastewater reuse in Northern Italy (Como). *Water Science and Technology*, 39(5), 121.

20. Sandhya, S., Sarayu, S., and Swaminathan, K. 2008. Determination of kinetic constant of hybrid textile wastewater treatment system. *Bioresource Technology*, 99, 5793–5797.

21. Textile dyeing. *Ullmann's Encyclopedia of Industrial Chemistry*. 1995. 26, 351–477.

22. Wang, Y.P. and Smith, R. 1994. Wastewater minimization. *Chemical Engineering Science*, 49(7), 981.

23. Zitella, P. 2003. Riciclo delle acque nell'industria tessile—Ottimizzazione dei trattamenti terziari a ozono e carbone attivo. PhD thesis, Politecnico di Torino, Italy.

13

Criterion, Indices, and Classification of Water Quality and Water Reuse Options

Manish Kumar
Tezpur University

S. Chidambaram
Annamalai University

A.L. Ramanathan
Jawaharlal Nehru University

Ritusmita Goswami
Tezpur University

Saeid Eslamian
Isfahan University of Technology

PREFACE

Water is an essential but scarce resource in India. Presently, the quality and the availability of freshwater resources is one of the most challenging issues facing the country. Overextraction of groundwater for different purposes has triggered groundwater depletion and related environmental problems in many parts of the world. In India, population increase coupled with rapid urbanization, industrialization, and agricultural development has resulted in high impact on the quality and quantity of water. The situation warrants immediate redressal through safe water resource and improved water quality management strategies. Water reuse allows communities to become less dependent on groundwater and surface water sources and can decrease the diversion of water from sensitive ecosystems. Additionally, water reuse satisfying most of the water demands depending on the level of treatment may reduce nutrient loads from wastewater discharges into waterways, thereby reducing and preventing pollution. This chapter highlights the basic concepts and issues involved in water reclamation and reuse.

13.1 Introduction

In view of an international perspective of <1700 m³/capita/year (0.45 mg/capita/year) as water stressed and <1000 m³/capita/year (0.26 mg/capita/year) as water scarce, today India is water stressed and is likely to be water scarce by 2050 [10]. It is now projected that by 2020, the number of people living in water-scarce countries will increase from about 131 million to more than 800 million [9]. It is reported that India supports more than 19% of the world's population with only 4% of the world's freshwater resources [19]. According to some experts, below 500 m³/capita/year (0.13 mg/capita/year), countries experience absolute water stress and the value of 100 m³/capita/year (0.026 mg/capita/year) is the minimum survival level for domestic and commercial use [7,11]. Projections predict that in 2025, two-thirds of the world's population will be under conditions of moderate to high water stress and about half of the population will face real constraints in their water supply.

Overextraction of groundwater for different purposes has triggered groundwater depletion and related environmental problems. Industrial consumption of water is low as compared to agricultural consumption, but it is raising with time due to a wave of industrialization and it may go up 193 km²/year by 2050 in India [23]. There are two types of problems with water

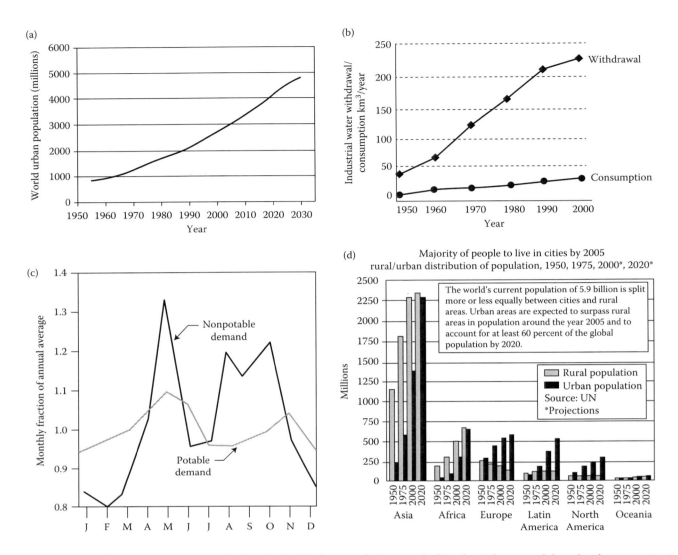

FIGURE 13.1 Describing current water scenario showing in (a) urban population growth, (b) industrial water withdrawal and consumption in Asia during 1950–2000 [22], (c) potable and nonpotable demand throughout the year, and (d) rural and urban population difference. (Adapted from UNESCO-WWAP. 2006. *The United Nations World Water Development Report 2.* UNESCO Publishing, Paris, and Earthscan, London.)

demand for industrial consumption; first, it is needed in huge quantities and, second, industries release of highly contaminated effluent. The situation is very alarming in India, which is already facing a moderate (39.9%) freshwater crisis [23]. The large gap between water withdrawal and actual consumption is a silent feature of our consumption pattern. Figure 13.1 shows the current water scenario in (a) urban population growth, (b) industrial water withdrawal and consumption in Asia during 1950–2000 [21], (c) potable and nonpotable demand throughout the year, and (d) rural and urban population difference.

Continued population growth, deterioration of water quality due to contamination of both surface water and groundwater, and uneven distributions of water resources have forced water agencies to search for new sources of water supply. In many parts of the world, water reuse is an important option in water resources planning and implementation. Water reuse

allows communities to become less dependent on groundwater and surface water sources. Additionally, water reuse may reduce the nutrient loads from wastewater discharges into waterways, thereby reducing and preventing pollution. While water reuse is a viable option, conservation of water resources, efficient and economic use of existing water supplies, and new water resources development and management are other alternatives that must be evaluated. Figure 13.2 illustrates water reuse by type in California and Florida, while Table 13.1 summarizes the top seven countries that are using reclaimed water.

The purpose of this chapter is to (1) discuss water quality in terms of drinking, irrigation, and industrial purposes, (2) introduce the basic concepts and issues involved in water reclamation and reuse, and (3) review briefly the need for and status of water reuse.

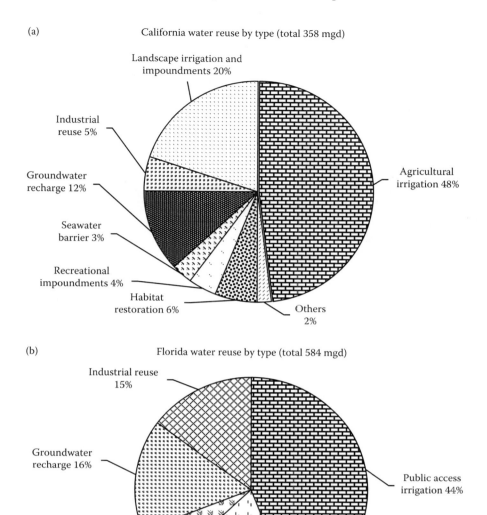

(a)   California water reuse by type (total 358 mgd)

Landscape irrigation and impoundments 20%

Industrial reuse 5%

Groundwater recharge 12%

Seawater barrier 3%

Recreational impoundments 4%

Habitat restoration 6%

Agricultural irrigation 48%

Others 2%

(b)   Florida water reuse by type (total 584 mgd)

Industrial reuse 15%

Groundwater recharge 16%

Wetlands and other 6%

Agricultural irrigation 19%

Public access irrigation 44%

FIGURE 13.2 Water reuse by type: (a) California. (Adapted from California Environmental Protection Agency.) (b) Florida. (Adapted from 2001 Florida Water Reuse Inventory.)

TABLE 13.1 Top Seven Countries Using Reclaimed Water

Country	Total Annual Water Withdrawal			Annual Reclaimed Water Usage			Reclaimed Water as Percentage of Total (%)
	Year	Mm	MG³	Year	Mm	MG³	
Kuwait	1994	538	142,140	1997	80	21,136	15
Cyprus	1993	211	55,746	1997	23	6077	11
Israel	1995	2000	528,400	1995	200	52,840	10
Qatar	1994	285	75,297	1994	25	6605	9
Turkmenistan	1995	2108	556,934	1999	185	48,877	9
Bahrain	1991	239	63,144	1991	15	3963	6
Jordan	1993	984	259,973	1997	58	15,324	6

13.2 Water Reuse Options

Reuse and recycling alternative water supplies is a key part of reducing the pressure on water resources and the environment. When considering alternative water supplies, one should choose the most appropriate water source, taking into account end use, risk, and resource and energy requirements. An understanding of the potential of wastewater for reuse to overcome shortage of freshwater existed in Minoan civilization in ancient Greece, where indications for utilization of wastewater for agricultural irrigation date back 5000 years [3]. Irrigation with sewage and other wastewaters has a long history also in China and India. In more recent history, the introduction of waterborne sewage collection systems during the nineteenth century, for discharge of wastewater into surface water bodies, led to indirect use of sewage and other wastewaters as unintentional potable water supplies [3].

Wastewater can be classified into two main types. The first type is domestic wastewater. This type of wastewater is commonly known as sanitary wastewater. "Domestic wastewater includes discharge from residence, commercial, institutional, and similar facilities" [20]. The second type of wastewater is industrial waste. Industrial wastewater contains different types of contaminants than domestic. Reuse of wastewater for domestic purposes has been occurring since historical times. However, planned reuse has

gained importance only two or three decades ago, as the demands for water dramatically increased due to technological advancement, population growth, and urbanization, which put great stress on the natural water cycle. Untreated sewage poses an acute water pollution problem that causes low water availability. Development of human societies is heavily dependent on availability of water with suitable quality and in adequate quantities, for a variety of uses ranging from domestic to industrial supplies. An estimate infers that every year, wastewater discharges from domestic, industrial, and agricultural practices pollute more than two-thirds of the total available runoff through rainfall, thereby, creating what can be called a "man-made water shortage." Thus, in spite of seeming abundance, water scarcity is endemic in most parts of the world. Nowadays, a number of methods have been adopted for recycling of wastewater. Figure 13.3 shows the water conservation hierarchy and inherent risk for alternative water sources.

Major reuse applications include

- Urban
- Industrial
- Agricultural
- Environmental and recreational
- Groundwater recharge

For each reuse application, quantity and quality requirements are generally considered.

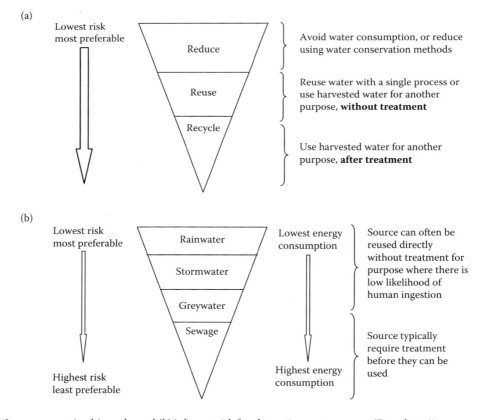

FIGURE 13.3   (a) Water conservation hierarchy and (b) inherent risk for alternative water sources. (From http://www.epa.vic.gov.au)

13.2.1 Urban Reuse

Urban reuse systems provide reclaimed water for various nonpotable purposes including

a. Irrigation of public parks and recreation centers, athletic fields, schoolyards and playing fields, highway medians and shoulders, and landscaped areas surrounding public buildings and facilities

b. Irrigation of landscaped areas surrounding residential areas, general wash down, and other maintenance activities

c. Irrigation of landscaped areas surrounding commercial, office, and industrial developments

d. Commercial uses such as vehicle washing facilities, laundry facilities, window washing, and mixing water for pesticides, herbicides, and liquid fertilizers

e. Ornamental landscape uses and decorative water features, such as fountains, reflecting pools, and waterfalls

f. Dust control and concrete production for construction projects

g. Fire protection through reclaimed water fire hydrants

h. Toilet and urinal flushing in commercial and industrial buildings

13.2.2 Industrial Reuse

Industrial reuse has increased substantially since the early 1990s for many of the same reasons urban reuse has gained popularity, including water shortages and increased populations, particularly in drought areas, and legislation regarding water conservation and environmental compliance. To meet this increased demand, many countries have increased the availability of reclaimed water to industries and have installed the necessary reclaimed water distribution lines.

13.2.3 Environmental and Recreational Reuse

Based on the inventory, current regulations and guidelines of water reuse may be divided into the following categories:

I. *Unrestricted urban reuse*—Irrigation of areas in which public access is not restricted, such as parks, playgrounds, schoolyards, and residences; toilet flushing, air conditioning, fire protection, construction, ornamental fountains, and aesthetic impoundments.

II. *Restricted urban reuse*—Irrigation of areas in which public access can be controlled, such as golf courses, cemeteries, and highway medians. Agricultural reuse on food crops—irrigation of food crops that are intended for direct human consumption.

III. *Agricultural reuse on non-food crops*—Irrigation of fodder, fiber, and seed crops, pastureland, commercial nurseries, and sod farms.

IV. *Unrestricted recreational reuse*—An impoundment of water in which no limitations are imposed on body contact water recreation activities.

V. *Restricted recreational reuse*—An impoundment of reclaimed water in which recreation is limited to fishing, boating, and other noncontact recreational activities.

VI. *Environmental reuse*—Reclaimed water used to create manmade wetlands, enhance natural wetlands, and sustain or augment stream flows.

VII. *Industrial reuse*—Reclaimed water used in industrial facilities primarily for cooling system make-up water, boiler-feed water, process water, and general.

Table 13.2 shows the suggested guidelines for water reuse in different categories.

A summary of the unit operations and processes commonly used in wastewater reclamation and the principal contaminants removed (Rein Munter) are presented in Table 13.3.

In wastewater reclamation process, the removal performance by microfiltration influent reduction and reverse osmosis (RO) reduction (based on literature) is shown in Table 13.4.

13.3 Treatment of Toxic Materials

Toxic materials including many organic materials, metals (such as zinc, silver, cadmium, thallium, etc.) acids, alkalis, and non-metallic elements (such as arsenic or selenium) are generally resistant to biological processes unless very dilute. Metals can often be precipitated out by changing the pH or by treatment with other chemicals. Many, however, are resistant to treatment or mitigation and may require concentration followed by land filling or recycling. Dissolved organics can be incinerated within the wastewater by the advanced oxidation process. Table 13.5 shows the inorganic and organic constituents of concern in water reclamation and reuse.

Industrial wastewater treatment has taken place in a series of development phases starting from direct discharge to recycling and reuse. This development has been slow considering the growing awareness of environmental degradation, public pressure, implementation of increasingly stringent standards, and industrial interest in waste recycling. The declining supply and higher cost of raw water is also forcing industry to implement recycling technologies. Many industries are now concentrating on methods to abate potable water intake and reduce discharge of pollutant and effluent. The move toward wastewater is reflected in different cleaner production approaches such as internal wastewater recycling, reuse of treated industrial or municipal wastewater, and reuse of treated wastewater for other activities.

13.4 Measures of Water Quality

The chemical characteristics of groundwater are determined by the chemical and biological reactions in the zones through which the water moves. Consequently, soil zone, aquifer gases, and the most water-soluble minerals and salts in an aquifer generally determine the chemical composition of groundwater in an aquifer. The characteristics of water that affect water quality

TABLE 13.2 Suggested Guidelines for Water Reuse[a]

Types of Reuse	Treatment	Reclaimed Water Quality[b]	Comments
Urban reuse All types of landscape irrigation (e.g., golf courses, parks, cemeteries); also vehicle washing, toilet flushing, use in fire protection systems and commercial air conditioners, and other uses with similar access or exposure to the water	Secondary[c] Filtration[d] Disinfection[e]	pH = 6–9 (weekly) ≤10 mg/L BOD[f] (weekly) <2 NTU[g] (continuous) No detectable fecal coli/100 mL[h,i] (daily) 1 mg/L Cl_2 residual (minimum)[j]	At controlled-access irrigation sites where design and operational measures significantly reduce the potential of public contact with reclaimed water, a lower level of treatment, for example, secondary treatment and disinfection to achieve <14 fecal coli/100 mL, may be appropriate. Chemical (coagulant and/or polymer) addition prior to filtration may be necessary to meet water quality recommendations. The reclaimed water should not contain measurable levels of viable pathogens.[k] Reclaimed water should be clear and odorless. A higher chlorine residual and/or a longer contact time may be necessary to assure that viruses and parasites are inactivated or destroyed. A chlorine residual of 0.5 mg/L or greater in the distribution system is recommended to reduce odors, slime, and bacterial regrowth.
Restricted access area irrigation Sod farms, silviculture sites, and other areas where public access is prohibited, restricted, or infrequent	Secondary[c] Disinfection[e]	pH = 6–9 ≤30 mg/L BOD[f] ≤30 mg/L TSS ≤200 fecal coli/100 mL[h,l,m] 1 mg/L Cl_2 residual (minimum)[j]	If spray irrigation, TSS less than 30 mg/L may be necessary to avoid clogging of sprinkler heads.
Agricultural reuse—food crops not commercially processed[n] Surface or spray irrigation of any food crop, including crops eaten raw	Secondary[c] Filtration[d] Disinfection[e]	pH = 6–9 ≤10 mg/L BOD[f] ≤2 NTU[g] No detectable fecal coli/100 mL[h,i] 1 mg/L Cl_2 residual (minimum)[j]	Chemical (coagulant and/or polymer) addition prior to filtration may be necessary to meet water quality recommendations. The reclaimed water should not contain measurable levels of viable pathogens.[k] A higher chlorine residual and/or a longer contact time may be necessary to assure that viruses and parasites are inactivated or destroyed. High nutrient levels may adversely affect some crops during certain growth stages.
Agricultural reuse—food crops commercially processed[n] Surface irrigation of orchards and vineyards	Secondary[c] Disinfection[e]	pH = 6–9 ≤30 mg/L BOD[f] ≤30 mg/L TSS ≤200 fecal coli/100 mL[h,l,m] 1 mg/L Cl_2 residual (minimum)[j]	If spray irrigation, TSS less than 30 mg/L may be necessary to avoid clogging of sprinkler heads. High nutrient levels may adversely affect some crops during certain growth stages. Milking animals should be prohibited from grazing for 15 days after irrigation ceases. A higher level of disinfection, for example, to achieve <14 fecal coli/100 mL, should be provided if this waiting period is not adhered to.
Recreational impoundments Incidental contact (e.g., fishing and boating) and full body contact with reclaimed water	Secondary[c] Filtration[d] Disinfection[e]	pH = 6–9 ≤10 mg/L BOD[f] ≤2 NTU[g] No detectable fecal coli/100 mL[h,i] 1 mg/L Cl_2 residual (minimum)[j]	Dechlorination may be necessary to protect aquatic species of flora and fauna. Reclaimed water should be nonirritating to skin and eyes. Reclaimed water should be clear and odorless. Nutrient removal may be necessary to avoid algae growth in impoundments. Chemical (coagulant and/or polymer) addition prior to filtration may be necessary to meet water quality recommendations. The reclaimed water should not contain measurable levels of viable pathogens.[k] A higher chlorine residual and/or a longer contact time may be necessary to assure that viruses and parasites are inactivated or destroyed. Fish caught in impoundments can be consumed.

(Continued)

TABLE 13.2 (*Continued*) Suggested Guidelines for Water Reuse[a]

Types of Reuse	Treatment	Reclaimed Water Quality[b]	Comments
Landscape impoundments Aesthetic impoundment where public contact with reclaimed water is not allowed	Secondary[c] Disinfection[e]	≤30 mg/L BOD[f] ≤30 mg/L TSS ≤200 fecal coli/100 mL[h,l,m] 1 mg/L Cl$_2$ residual (minimum)[j]	Nutrient removal may be necessary to avoid algae growth in impoundments. Dechlorination may be necessary to protect aquatic species of flora and fauna.
Construction use Soil compaction, dust control, washing aggregate, making concrete	Secondary[c] Disinfection[e]	≤30 mg/L BOD[f] ≤30 mg/L TSS ≤200 fecal coli/100 mL[h,l,m] 1 mg/L Cl$_2$ residual (minimum)[j]	Worker contact with reclaimed water should be minimized. A higher level of disinfection, for example, to achieve <14 fecal coli/100 mL, should be provided when frequent work contact with reclaimed water is likely.
Industrial reuse Once-through cooling	Secondary[c] Disinfection[e]	pH = 6–9 ≤30 mg/L BOD[f] ≤30 mg/L TSS ≤200 fecal coli/100 mL[h,l,m] 1 mg/L Cl$_2$ residual (minimum)[j]	Windblown spray should not reach areas accessible to workers or the public.
Recirculating cooling towers	Secondary[c] Disinfection[e] (chemical coagulation and filtration[d] may be needed)	Variable depends on recirculation ratio pH = 6–9 ≤30 mg/L BOD[f] ≤30 mg/L TSS ≤200 fecal coli/100 mL[h,l,m] 1 mg/L Cl$_2$ residual (minimum)[j]	Windblown spray should not reach areas accessible to workers or the public. Additional treatment by user is usually provided to prevent scaling, corrosion, biological growths, fouling, and foaming.

Other Industrial Uses

Types of Reuse	Treatment	Reclaimed Water Quality[b]	Comments
Environmental reuse Wetlands, marshes, wildlife habitat, stream augmentation	Variable Secondary[c] and disinfection[e] (minimum)	Variable, but not to exceed ≤30 mg/L BOD[f] ≤30 mg/L TSS ≤200 fecal coli/100 mL[h,l,m]	Dechlorination may be necessary to protect aquatic species of flora and fauna. Possible effects on groundwater should be evaluated. Receiving water quality requirements may necessitate additional treatment. The temperature of the reclaimed water should not adversely affect ecosystem. See Section 3.4.3 for recommended treatment reliability.
Groundwater recharge By spreading or injection into aquifers not used for public water supply	Site-specific and use-dependent Primary (minimum) for spreading Secondary[c] (minimum) for injection	Site-specific and use-dependent	Facility should be designed to ensure that no reclaimed water reaches potable water supply aquifers. For spreading projects, secondary treatment may be needed to prevent clogging. For injection projects, filtration and disinfection may be needed to prevent clogging.
Indirect potable reuse Groundwater recharge by spreading into potable aquifers	Secondary[c] Disinfection[e] May also need filtration[d] and/or advanced wastewater treatment[o]	Secondary[c] Disinfection[e] Meet drinking water standards after percolation through vadose zone	The depth to groundwater (i.e., thickness to the vadose zone) should be at least 6 ft (2 m) at the maximum groundwater mounding point. The reclaimed water should be retained underground for at least 6 months prior to withdrawal. Monitoring wells are necessary to detect the influence of the recharge operation on the groundwater. The reclaimed water should not contain measurable levels of viable pathogens after percolation through the vadose zone.[k]
Indirect potable reuse Groundwater recharge by injection into potable aquifers	Secondary[c] Filtration[d] Disinfection[e] Advanced wastewater treatment[o]	Includes, but is not limited to, the following: pH = 6.5–8.5 ≤2 NTU[g] No detectable total coli/100 mL[h,i] 1 mg/L Cl$_2$ residual (minimum)[j] ≤3 mg/L TOC ≤0.2 mg/L TOX Meet drinking water standards	The reclaimed water should be retained underground for at least 9 months prior to withdrawal. Monitoring wells are necessary to detect the influence of the recharge operation on the groundwater. Recommended quality limits should be met at the point of injection. The reclaimed water should not contain measurable levels of viable pathogens after percolation through the vadose zone.[k] Higher chlorine residual and/or a longer contact time may be necessary to assure virus and protozoa inactivation.

(Continued)

TABLE 13.2 (*Continued*) Suggested Guidelines for Water Reuse[a]

Types of Reuse	Treatment	Reclaimed Water Quality[b]	Comments
Indirect potable reuse Augmentation of surface supplies	Secondary[c] Filtration[d] Disinfection[e] Advanced wastewater treatment[o]	Includes, but is not limited to, the following: pH = 6.5–8.5 ≤2 NTU[g] No detectable total coli/100 mL[h,i] 1 mg/L Cl$_2$ residual (minimum)[j] ≤3 mg/L TOC Meet drinking water standards[q]	Recommended level of treatment is site-specific and depends on factors such as receiving water quality, time and distance to point of withdrawal, dilution, and subsequent treatment prior to distribution for potable uses.[p] The reclaimed water should not contain measurable levels of viable pathogens.[k] A higher chlorine residual and/or a longer contact time may be necessary to assure virus and protozoa inactivation.

[a] These guidelines are based on water reclamation and reuse practices in the United States, and they are especially directed at states that have not developed their own regulations or guidelines. While the guidelines should be useful in many areas outside the United States, local conditions may limit the applicability of the guidelines in some countries. It is explicitly stated that the direct application of these suggested guidelines will not be used by USAID as strict criteria for funding.

[b] Unless otherwise noted, recommended quality limits apply to the reclaimed water at the point of discharge from the treatment facility.

[c] Secondary treatment processes include activated sludge processes, trickling filters, rotating biological contactors, and may include stabilization pond systems. Secondary treatment should produce effluent in which both the BOD and TSS do not exceed 30 mg/L.

[d] Filtration means the passing of wastewater through natural undisturbed soils or filter media such as sand and/or anthracite, filter cloth, or the passing of wastewater through microfilters or other membrane processes.

[e] Disinfection means the destruction, inactivation, or removal of pathogenic microorganisms by chemical, physical, or biological means. Disinfection may be accomplished by chlorination, UV radiation, ozonation, other chemical disinfectants, membrane processes, or other processes. The use of chlorine as defining the level of disinfection does not preclude the use of other disinfection processes as an acceptable means of providing disinfection for reclaimed water.

[f] As determined from the 5-day BOD test.

[g] The recommended turbidity limit should be met prior to disinfection. The average turbidity should be based on a 24-h time period. The turbidity should not exceed 5 NTU at any time. If TSS is used in lieu of turbidity, the TSS should not exceed 5 mg/L.

[h] Unless otherwise noted, recommended coliform limits are median values determined from the bacteriological results of the last 7 days for which analyses have been completed. Either the membrane filter or fermentation-tube technique may be used. The number of fecal coliform organisms should not exceed 14/100 mL in any sample.

[i] The number of fecal coliform organisms should not exceed 14/100 mL in any sample.

[j] Total chlorine residual should be met after a minimum contact time of 30 min.

[k] It is advisable to fully characterize the microbiological quality of the reclaimed water prior to implementation of a reuse program.

[l] The number of fecal coliform organisms should not exceed 800/100 mL in any sample.

[m] Some stabilization pond systems may be able to meet this coliform limit without disinfection.

[n] Commercially processed food crops are those that, prior to sale to the public or others, have undergone chemical or physical processing sufficient to destroy pathogens.

[o] Advanced wastewater treatment processes include chemical clarification, carbon adsorption, reverse osmosis and other membrane processes, air stripping, ultrafiltration, and ion exchange.

[p] Setback distances are recommended to protect potable water supply sources from contamination and to protect humans from unreasonable health risks due to exposure to reclaimed water [6].

[q] Monitoring should include inorganic and organic compounds, or classes of compounds, that are known or suspected to be toxic, carcinogenic, teratogenic, or mutagenic and are not included in the drinking water standards [6].

TABLE 13.3 A Summary of the Unit Operations and Processes Commonly Used in Wastewater Reclamation and the Principal Contaminants Removed (Rein Munter)

	BOD	COD	TOC	Turbid	Color	Coli	NH –N
Primary treatment	x	x	x	x	o		o
Activated sludge	+	+	+	+	x	+	+
Nitrification	+	+	+	+	x	+	+
Denitrification	o	o	o	o			x
Trickling filter	+	+	x	x	o	o	
Coag.-floc.-sed.	+	+	+	+	+	+	o
Filtration A/S	x	x	x	x	x		x
GAC Adsorption	+	x	+	+	+	+	x
Ion exchange	x	x	o	o			+
Chlorination						+	+
Reverse osmosis	+	+	+	+	+		+
Ozone	o	+	x			+	

Symbols: o = 25% removal of influent concentration, x = 25%–50%, + = >50%. Blank denotes no data.

TABLE 13.4 Removal Performance Data

Constituent	Microfiltration Influent Reduction Reported in Literature (%)	Reverse Osmosis Reduction Reported in Literature (%)
TOC	45–65	85–95
BOD	75–90	30–60
COD	70–85	85–95
TSS	95–98	95–100
TDS	0–2	90–98
NH_3–N	5–15	90–98
NO_3–N	0–2	65–85
PO_4^-	0–2	95–99
SO_4^{2-}	0–1	95–99
Cl^-	0–1	90–98
Turbidity	–	40–80

depend both on substances dissolved in water and on certain properties. Natural inorganic constituents commonly dissolved in water that are most likely to affect water use include bicarbonate, sulfate, carbonate, calcium, magnesium, chloride, fluoride, manganese, sodium, and iron. The groundwater in natural systems generally contains less than 1000 mg/L dissolved solids. Natural groundwater generally acquires dissolved constituents by dissolution of aquifer gases, minerals, and salts. Properties of groundwater evaluated in a physical analysis include temperature, color, turbidity, taste, and odor. Biological analysis includes tests to detect the presence of coliform bacteria that indicted the quality of water in terms of sanitation for human consumption.

13.5 Groundwater Quality Criteria for Irrigation

The suitability of groundwater for irrigation is dependent on the effects of the mineral constituents of the water on both the plant and the soil [17,22]. For a suitable irrigation practice, water quality, soil types, and cropping practices play important roles. The main chemical constituents that affect the suitability of water for irrigation are the total concentration of dissolved salts, relative proportion of bicarbonate to calcium, magnesium, and relative proportion of sodium to calcium. Water quality problems in irrigation include salinity and alkalinity.

To ascertain the suitability of groundwater for any purposes, the total dissolved solids (TDS) should be below 500 mg L^{-1} [4,8]. Excessive sodium content relative to the calcium and magnesium reduces the soil permeability and thus inhibits the supply of water needed for crops [11]. The excess sodium or limited calcium and magnesium are evaluated by the sodium absorption ratio (SAR), which is expressed as

$$SAR = \frac{Na^+}{\sqrt{(Ca^{2+} + Mg^{2+})/2}} \qquad (13.1)$$

Sodium content expressed in terms of sodium percentage or soluble sodium percentage defined as

$$\%Na = \frac{(Na^+ + K^+) \times 100}{(Ca^{2+} + Mg^{2+} + Na^+ + K^+)} \qquad (13.2)$$

where all concentrations are expressed in meq L^{-1}

TABLE 13.5 Inorganic and Organic Constituents of Concern in Water Reclamation and Reuse

Measured Parameters	Constituent	Reasons for Concern
Suspended solids (SS), including volatile and fixed solids	SS	Organic contaminants, heavy metals, etc. are absorbed on particulates. Suspended matter can shield microorganisms from disinfectants. Excessive amounts of SS cause plugging in irrigation systems.
Biochemical oxygen demand	Biodegradable	Aesthetic and nuisance problems. Organic chemical oxygen demand, total organic carbon provide food for microorganisms, adversely affect disinfection processes, make water unsuitable for some industrial or other uses, consume oxygen, and may result in acute or chronic effects if reclaimed water is used.
Nitrogen, phosphorus, potassium	Nutrients	Nitrogen, phosphorus, and potassium are essential nutrients for plant growth and their presence normally enhances the value of the water for irrigation. When discharged to the aquatic environment, nitrogen and phosphorus can lead to the growth of undesired species.
Specific compounds (e.g., pesticides, chlorinated hydrocarbons)	Stable organics	Some of these organics tend to resist conventional methods of wastewater treatment. Some organic compounds are toxic in the environment, and their presence may limit the suitability of reclaimed water for irrigation or other uses.
pH	Hydrogen ion concentration	The pH of wastewater affects disinfection, coagulation, and metal solubility, as well as alkalinity of soils. Normal range in municipal wastewater is pH = 6.5–8.5, but industrial waste can alter pH significantly.
Specific elements (e.g., Cd, Zn, Ni, and Hg)	Heavy metal	Some heavy metals accumulate in the environment and are toxic to plants and animals. Their presence may limit the suitability of the reclaimed water for irrigation or other uses.
Total dissolved solids (TDS), electrical conductivity, specific elements (e.g., Na, Ca, Mg, Cl, and B)	Dissolved inorganics	Excessive salinity may damage some crops. Specific inorganics electrical conductivity ions such as chloride, sodium, and boron are toxic to specific elements (e.g., in some crops, sodium may pose soil permeability, Na, Ca, Mg, Cl, and B problems).
Free and combined chlorine	Residual chlorine	Excessive amounts of free available chlorine (>0.05 chlorine mg/L) may cause leaf-tip burn and damage some sensitive crops. However, most chlorine in reclaimed water is in a combined form, which does not cause crop damage.

Using both classifications may help to understand the evolution of different criteria to classify the quality of different types of irrigation water. In addition to the SAR and %Na, the excess sum of carbonate and bicarbonate in groundwater over the sum of calcium and magnesium also influences the suitability of groundwater for irrigation [19]. An excess quantity of sodium bicarbonate and carbonate is considered to be detrimental to the physical properties of soils as it causes dissolution of organic matter in the soil, which in turn leaves a black stain on the soil surface on drying. This excess is denoted by residual sodium carbonate (RSC) and is calculated as follows [16]:

TABLE 13.6 Suitability of Groundwater for Irrigation Based on Several Classifications

Based on EC (μS cm^{-1}) Values of Water	
<250	Excellent
250–750	Good
750–2000	Permissible
2000–3000	Doubtful
>3000	Unsuitable
Based on TDS (mg L^{-1}) Value of Water	
<300	Excellent
300–600	Good
600–900	Fair
900–1200	Poor
>1200	Unacceptable
Based on Total Hardness as CaCO$_3$ (mg L^{-1}) after Sawyer and Mc Cartly [19]	
<75	Soft
75–150	Moderately hard
150–300	Hard
Based on Alkalinity Hazard (SAR) after Richards [18]	
<10	Excellent
10–18	Good
18–26	Doubtful
>26	Unsuitable
Based on Percent Sodium after Wilcox [23]	
<20	Excellent
20–40	Good
40–60	Permissible
60–80	Doubtful
>80	Unsafe
% Na [5]	
<60	Safe
<20	Excellent
20–40	Good
40–60	Permissible
Based on Residual Sodium Carbonate (RSC) after Richards [18]	
<1.25	Good
1.25–2.50	Doubtful
>2.50	Unsuitable

$$RSC = (CO_3^{-2} + HCO_3^-) - (Ca^{2+} + Mg^{2+}) \quad (13.3)$$

where concentrations are expressed in meqL^{-1}

Table 13.6 shows the suitability of groundwater for irrigation based on several classifications.

Although all the criteria are very important in determining water quality, they are not feasible as there are different parameters like electrical conductivity (EC), total dissolved solids (TDS), etc., which have different values. Therefore, a holistic approach, as shown in Table 13.7, has been prepared based on guidelines given for irrigation water with its potential damage in Reference 13, which is adapted from References 2 and 14. This table shows the percentage of groundwater with its potential threat based on the irrigation method, that is, surface irrigation or sprinkler irrigation. It clearly shows that sprinkler irrigation method is much more suitable than that of field irrigation.

The entire previous discussion can be represented diagrammatically as well. Figure 13.4 describes the analytical data plotted on the U.S. salinity diagram [17], which illustrates the suitability of groundwater for irrigation. Groundwater samples that fall in

TABLE 13.7 Evaluation for the Suitability of Groundwater for Irrigation Based on the Guidelines

	None	Slight to Moderate	Severe
	(N)	(S-M)	(S)
Salinity (Affects Crop Water Availability)			
EC (μScm^{-1})	<700	700–3000	>3000
TDS (mg L^{-1})	<450	450–2000	>2000
Permeability (Affects Infiltration Rate of Water into the Soil)[a]			
SAR = 0–3	EC >700	700–200	<200
3–6	>1200	1200–300	<300
6–12	>1900	1900–500	<500
12–20	2900	2900–1300	<1300
Specific Ion Toxicity (Affects Sensitive Crop) (Sodium) Na$^+$			
Surface irrigation	SAR <3	3–9	>9
Sprinkler irrigation (mg L^{-1})	<70	>70	–
Cl$^-$ (mg L^{-1})			
Surface irrigation	<140	140–350	>350
Sprinkler irrigation	<100	>100	–
Miscellaneous Effects (Affects Susceptible Crops) Overhead Sprinkling Only			
HCO$_3^-$ (mg L^{-1})	<90	90–500	>500

Source: Adapted from Ayers, R.S. and Westcot, D.W. 1985. Water Quality for Agriculture, FAO Irrigation and Drainage Paper 29, Rev.1, Food and Agriculture Organization of the United Nations, Rome, Italy; Metcalf and Eddy. 2003. *Wastewater Engineering-Treatment and Reuse*, 4th edition, McGraw-Hill; Pettygrove, G.S. and Asano T. (eds) 1985. *Irrigation with Reclaimed Municipal Wastewater—A Guidance Manual*. Lewis Publishers, Inc., Chelsea, MI.

[a] Evaluated using EC and SAR of the groundwater.

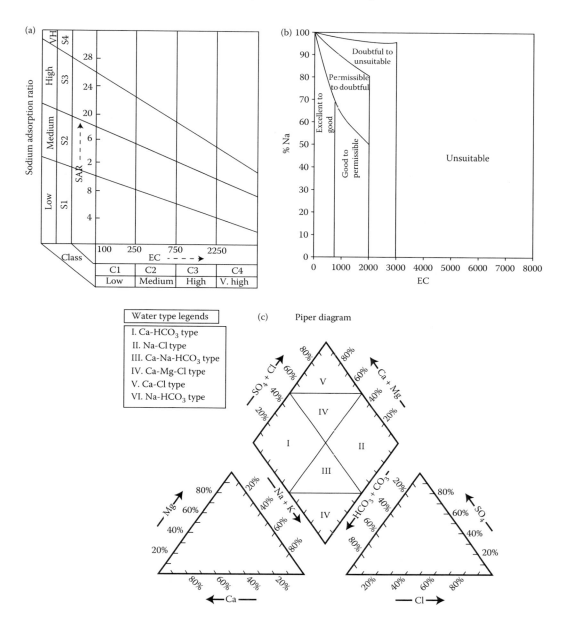

FIGURE 13.4 Showing diagrammatic representation of different water quality measuring criteria: (a) salinity diagram, (b) suitability of groundwater for irrigation in the Wilcox diagram, and (c) piper diagram showing type of water.

the low salinity hazard class (C1) can be used for irrigation of most crops and the majority of soils. However, some leaching is required, but this occurs under normal irrigation practices except in soils of extremely low permeability. Groundwater samples that fall in medium salinity hazard class (C2) can be used if a moderate amount of leaching occurs. High salinity water (C3, C4, and C5) cannot be used in soils with restricted drainage. Even with adequate drainage, special management for salinity control is required and crops with good salt tolerance should be selected. Such areas need special attention as far as irrigation is concerned. The geochemical evolution of groundwater can be understood by plotting the concentrations of major cations and anions in the tri-linear diagram [15].

13.6 Industrial Water Criteria

The quality requirements of water used in different industrial processes vary widely. Thus, makeup water for high-pressure boilers must meet extremely exacting criteria. Even within each industry, fixed criteria cannot be established; instead, only recommended limiting values or ranges can be stated [1,12]. Salinity, hardness, and silica are the three main parameters that are usually important for industrial water. Of almost equal importance for industrial purposes as quality of a water supply is the relative constancy of the various constituents. It is often possible to treat poor quality water or adapt to it so that it is suitable for a given process, but if the quality fluctuates widely, continued attenuation and expense may be involved. For industrial applications, fluctuations of water temperature can be equally troublesome. From this standpoint, groundwater supplies are preferred to surface water supplies, which commonly display seasonal variations in chemical and physical quality. An adequate groundwater supply of suitable quality often becomes a primary consideration in selecting new industrial plant locations.

13.7 Summary and Conclusions

The need for water continues to become more acute with the changing requirements of ever-increasing world population. In recent years, sustainable water management has received great attention to combat water scarcity issues. This implies conservation and channeling of water from areas where it is plentiful to deficient and scarce areas. In addition to rational planning and management of existing water supplies from traditional water resources, the significance and feasibility of potential alternative water resources and long-term water supply strategies should also be taken for consideration. The solutions may include water conservation, water recycling and reuse, and desalination. However, for implementation of sustainable water management framework, principles and approaches are still challenging, especially in less developed areas. In urban areas, it is becoming difficult for the authorities to manage water supply and wastewater. Strategies for wastewater reuse can successfully improve urban water management. Consequently, before the implementation processes, additional assessment and reforms on water demand-supply analysis, integrated water resources management, water pricing, and corresponding governmental policies are highly recommended. An efficient management of available water resources should be the main thrust for economic development and improvement of groundwater quality in India.

References

1. American Society for Testing Materials. 1966. Testing malts. In: *Manual on Industrial Water and Industrial Waste Water*, 2nd edition, Philadelphia.
2. Ayers, R.S. and Westcot, D.W. 1985. Water Quality for Agriculture, FAO Irrigation and Drainage Paper 29, Rev.1, Food and Agriculture Organization of the United Nations, Rome, Italy.
3. Beychok, M.R. 1971. Wastewater treatment. *Hydrocarbon Processing*, 50, 109–112.
4. Catroll, D. 1962. Rain water as a chemical agent of geological process—A view. *USGS Water Supply*, 1533, 18–20.
5. Eaton, F.M. 1950. Significance of carbonate in irrigation water. *Soil Sciences*, 69, 127–128.
6. Exall, K., Jimenez, B., Marsalek, J., and Schaefer, K. 2008. Water reuse in the United States and Canada. In: *Water Reuse—An International Survey of Current Practice, Issues and Needs*, Jimenez, B. and Asano, T. (eds.), IWA publishing, London, UK. pp. 88–89.
7. Falkenmark, M. and Widstrand, C. 1992. Population and water resources: A delicate balance. *Population Bulletin*, Population Reference Bureau.
8. Freeze, R.A. and Cherry, J.A. 1979. *Groundwater*, Prentice Hall, Englewood Cliffs, NJ.
9. Gardner-Outlaw, T. and Engelman, R. 1997 Sustaining Water, Easing Scarcity: A Second Update, Revised data for the Population Action International Report, Sustaining Water: Population and the Future of Renewable Water Supplies.
10. Gupta, S.K. and Deshpande, R.D. 2004. Water for India in 2050: First-order assessment of available options. *Current Science*, 86, 1216–1223.
11. Kumar, M., Kumari, K., Ramanathan, A.L., and Saxena, R. 2007. A comparative evaluation of groundwater suitability for irrigation and drinking purposes in two intensively cultivated districts of Punjab, India. *Environmental Geology*. 53, 553–574.
12. Lazarova, V., Levine, B., Sack, J., Cirelli, G., Jeffrey, P., Muntau, H., Salgot, M., and Brissaud, F. 2001. Role of water reuse for enhancing integrated water management in Europe and Mediterranean countries. *Water Science and Technology*, 43(10), 25–33.
13. McKee, J.E. and Wolf, H.W. 1963. Water quality criteria. Publication no. 3-A, California State Water Resources Control Board, Sacramento.
14. Metcalf and Eddy. 2003. *Wastewater Engineering—Treatment and Reuse*, 4th edition, McGraw-Hill, New York.
15. Pettygrove, G.S. and Asano T. (eds) 1985. *Irrigation with Reclaimed Municipal Wastewater—A Guidance Manual*. Lewis Publishers, Inc., Chelsea, MI.
16. Piper, A.M. 1944. A graphical procedure in the geochemical interpretation of water analysis. *American Geophysical Union Trans*, 25, 914–928.
17. Ragunath, H.M. 1987. *Groundwater*. Wiley Eastern Ltd., New Delhi, p. 563.
18. Richards, L.A. 1954. *Diagnosis and Improvement of Saline Alkali Soils*. U.S. Department of Agriculture, Hand Book 60.
19. Sawyer, G.N. and McCartly, D.L. 1967. *Chemistry of Sanitary Engineers*, 2nd edition. McGraw-Hill, New York, p. 518.
20. Singh, A.K. 2003. Water resources and their availability. In: *Souvenir, National Symposium on Emerging Trends in Agricultural Physics*, Indian Society of Agrophysics, New Delhi, April 22–24, 2003, pp. 18–29.
21. Tchobanoglous, G. and Burton, F. 1991. *Wastewater Engineering: Treatment, Disposal, and Reuse*, 3rd edition. Metcalf and Eddy, Inc., McGraw-Hill, Inc., New York.
22. UNESCO-WWAP. 2006. *The United Nations World Water Development Report 2*. UNESCO Publishing, Paris, Earthscan, London.
23. Wilcox, L.V. 1955. *Classification and Use of Irrigation Waters*. USDA, Circular 969, Washington, DC.
24. WWF. 2008. Living Planet Report.

IV

Water Reuse Regulations

14

Water Reuse Guidelines
for Agriculture

Faezeh Eslamian
McGill University

Saeid Eslamian
Isfahan University of Technology

Alireza Eslamian
Concordia University

PREFACE

Water deficiency has become a crucial issue in many countries. One of the approaches for confronting this problem is to reuse the water for agricultural purposes and irrigation. The effluent can be considered a source containing plant food nutrients and, therefore, the water reuse approach could lead to a decrease in the natural water resources consumption. However, because the reclaimed water contains undesirable materials that are threatening to human and environmental health, these effluents need to meet certain criteria prior to reuse. In this chapter, various guidelines of different countries and organizations regarding water reuse in agriculture are presented. Proper understanding and respecting of these guidelines are a major step in the practice of water reuse for agricultural purposes.

14.1 Introduction

In recent years, due to drought, population growth, and advances in technology, water demand along with wastewater production has consequently increased and, currently, many countries have faced critical water deficiency issues. One of the approaches for confronting this problem is to reuse the water for agricultural purposes and irrigation.

Currently, water reuse is used as an alternative for water resources. By reusing reclaimed water, not only are valuable drinking water sources preserved, but also the massive discharge of rich wastewaters containing nutrients into surface water resources will be avoided and reduced. The effluent (treated wastewater) can be considered a source containing plant food nutrients such as nitrogen, phosphorus, and potassium for plant growth and agriculture, which would otherwise be considered a pollutant for potable water. Therefore, the water reuse approach could lead to a decrease in natural water resources consumption.

However, since reclaimed water contains undesirable materials such as suspended solids, heavy metals, biodegradable organics, and pathogens, all threatening to human and environmental health, these effluents need to meet certain criteria prior to reuse. Various guidelines have been set for the reuse of water by different countries and organizations depending on the type of reuse, the plant receiving reclaimed water, land and meteorological conditions, and other effective factors.

This chapter mainly deals with the guidelines for water reuse in agriculture, and the various guidelines and allowable limits recommended and practiced by different countries or organizations will be comprehensively presented here.

The general measures that need to be taken in order to reuse water for agriculture purposes are as follows [6]:

- Wastewater treatment
- Restrictions for crops irrigated with reused water
- Water reuse application control
- Human contact control and respecting sanitary and health standards

In the past, the most attention was given to treatments measures; however, recently some limitations have been set for the type of crop cultivated. One of the most important issues associated with water reuse is health problems to which people in contact are exposed. According to Food and Agriculture Organization of the United Nations (FAO), four types of individuals are at risk [6]:

- Agricultural workers and their families
- Crop handlers
- Consumers of crops, milk, and meat
- Individuals living near the areas irrigated with wastewater

Therefore, crop handlers should wear protective clothing, maintain good hygiene, and need to be immunized against infectious diseases. Examples of these measures are given in the World Health Organization (WHO) Technical Report [11]. Furthermore, it is recommended that crops irrigated with reused water be cooked prior to consumption. In addition, households living near such agricultural fields should be fully informed so that they avoid these areas. Sprinklers should not be used within 100 m of houses or roads. All wastewater channels, pipes, and outlet have to be marked and distinctively colored so that workers do not mistake it for drinking water [6].

14.2 Guidelines for Water Reuse in Agriculture

The required quality for agriculture depends on the soil's physical and chemical properties, the crops resistance against salinity, and management methods. Therefore, water quality for agriculture is always relative and dependent on regional conditions.

14.2.1 World Health Organization

The recommended microbiological quality guidelines for water reuse in agriculture by WHO is presented in Table 14.1. In this table, the allowable coliform in wastewater and the required treatment is suggested with regard to three reuse conditions and taking into account the group of people exposed and the crop type. As it can be observed, category A, which involves crops likely to be eaten uncooked, has the most restricted limits with regard to its microbiological quality.

14.2.2 Food and Agriculture Organization

FAO [6] categorized water quality for reuse based on three degrees of restriction (Table 14.2). This guideline helps to identify the potential issues about crop production with doubtful waters and can be used for irrigation or agricultural purposes.

14.2.3 United States

The United States Environmental Protection Agency (USEPA) suggested limits for trace elements and heavy metals, given in Table 14.3. Furthermore, the USEPA recommended water quality criteria for irrigation as presented in Table 14.4, where agricultural reuse has been categorized based on crop type, irrigation method, and its distribution. This table also includes the required monitoring intervals as well as the minimum obligatory distance to potable water sources [8,9].

TABLE 14.1 Recommended Microbiological Quality Guidelines for Wastewater Use in Agriculture

Category	Reuse Condition	Exposed Group	Intestinal Nematodes[a] (Arithmetic Mean Number of Eggs per Liter)	Fecal Coliforms (Geometric Mean Number per 100 mL[b])	Wastewater Treatment Required
A	Irrigation of crops likely to be eaten uncooked, sports fields, public parks[c]	Workers, consumers, public	≤1	≤1000[c]	A series of stabilization ponds
B	Irrigation of cereal crops, industrial crops, fodder crops, pasture, and trees[d]	Workers	≤1	No standard recommended	Retention in stabilization ponds for 8–10 days or equivalent helminth and fecal coliform removal
C	Localized irrigation of crops in category B if exposure of workers and the public does not occur	None	Not applicable	Not applicable	Pretreatment as required by the irrigation technology, but not less than primary sedimentation

Source: Adapted from World Health Organization (WHO). 1989. *Health Guidelines for the Use of Wastewater in Agriculture*. Technical Report No. 778. WHO, Geneva.

[a] *Ascaris* and *Trichuris* species and hookworms.

[b] During the irrigation period.

[c] A more stringent guideline (<200 fecal coliforms per 100 mL) is appropriate for public lawns, such as hotel lawns, with which the public may come into direct contact.

[d] In the case of fruit trees, irrigation should cease 2 weeks before fruit is picked, and no fruit should be picked off the ground. Sprinkler irrigation should not be used.

TABLE 14.2 Guidelines for Interpretation of Water Quality for Irrigation

Potential Irrigation Problem		Units	Degree of Restriction on Use		
			None	Slight to Moderate	Severe
Salinity					
EC_w^a		dS/m	<0.7	0.7–3.0	>3.0
or					
TDS		mg/L	<450	450–2000	>2000
Infiltration					
SAR^b = 0–3	and EC_w =		>0.7	0.7–0.2	<0.2
3–6			>1.2	1.2–0.3	<0.3
6–12			>1.9	1.9–0.5	<0.5
12–20			>2.9	2.9–1.3	<1.3
20–40			>5.0	5.0–2.9	<2.9
Specific ion toxicity					
Sodium (Na)					
Surface irrigation		SAR	<3	3–9	>9
Sprinkler irrigation		me/L	<3	>3	
Chloride (Cl)					
Surface irrigation		me/L	<4	4–10	>10
Sprinkler irrigation		m³/L	<3	>3	
Boron (B)		mg/L	<0.7	0.7–3.0	>3.0
Trace elements (see Table 14.3)					
Miscellaneous effects					
Nitrogen (NO_3-N)^c		mg/L	<5	5–30	>30
Bicarbonate (HCO_3)		me/L	<1.5	1.5–8.5	>8.5
pH			Normal range 6.5–8		

Source: Adapted from Food and Agriculture Organization of the United Nations (FAO). 1985. In Ayers, R.S. and Westcot, D.W. (eds.), *Water Quality for Agriculture.* Irrigation and Drainage Paper 29 Rev. 1. FAO, Rome.

^{a} EC_w means electrical conductivity in deciSiemens per meter at 25°C.

^{b} SAR means sodium adsorption ratio.

^{c} NO_3-N means nitrate nitrogen reported in terms of elemental nitrogen.

14.2.4 Canada

According to the Canadian Council of Minister of the Environment (CCME), uptake of harmful amounts of toxic heavy metals by plants is not considered a potential risk in use of municipal wastewater, as most metals are removed from the wastewater during the primary treatment process. However, as a precautionary measure, all wastewater should meet the allowable concentration limits presented in Table 14.5 [3]. In addition, the wastewater should be analyzed both before and after irrigation to meet the standards specified in Table 14.6 [4].

TABLE 14.3 Recommended Water Quality Criteria for Irrigation

Constituent	Maximum Concentrations for Irrigation (mg/L)	Remarks
Aluminum	5.0	Can cause non-productiveness in acid soils, but soils at pH 5.5–8.0 will precipitate the ion and eliminate toxicity
Arsenic	0.10	Toxicity to plants varies widely, ranging from 12 mg/L for Sudan grass to less than 0.05 mg/L for rice
Beryllium	0.10	Toxicity to plants varies widely, ranging from 5 mg/L for kale to 0.5 mg/L for bush beans
Boron	0.75	Essential to plant growth; sufficient quantities in reclaimed water to correct soil deficiencies. Optimum yields obtained at a few-tenths mg/L; toxic to sensitive plants (e.g., citrus) at 1 mg/L. Most grasses are tolerant at 2.0–10 mg/L
Cadmium	0.01	Toxic to beans, beets, and turnips at concentrations as low as 0.1 mg/L; conservative limits are recommended
Chromium	0.1	Not generally recognized as an essential element; due to lack of toxicity data, conservative limits are recommended
Cobalt	0.05	Toxic to tomatoes at 0.1 mg/L; tends to be inactivated by neutral and alkaline soils
Copper	0.2	Toxic to a number of plants at 0.1–1.0 mg/L
Fluoride	1.0	Inactivated by neutral and alkaline soils
Iron	5.0	Not toxic in aerated soils, but can contribute to soil acidification and loss of phosphorus and molybdenum
Lead	5.0	Can inhibit plant cell growth at very high concentrations
Lithium	2.5	Tolerated by most crops up to 5 mg/L; mobile in soil. Toxic to citrus at low doses—recommended limit is 0.075 mg/L
Manganese	0.2	Toxic to a number of crops at a few tenths to a few mg/L in acidic soils
Molybdenum	0.01	Nontoxic to plants; can be toxic to livestock if forage is grown in soils with high molybdenum
Nickel	0.2	Toxic to a number of plants at 0.5–1.0 mg/L; reduced toxicity at neutral or alkaline pH
Selenium	0.02	Toxic to plants at low concentrations and to livestock if forage is grown in soils with low levels of selenium
Tin, tungsten, and titanium	–	Excluded by plants; specific tolerance levels unknown
Vanadium	0.1	Toxic to many plants at relatively low concentrations
Zinc	2.0	Toxic to many plants at widely varying concentrations; reduced toxicity at increased pH (6 or above) and in fine-textured or organic soils

Source: Adapted from United States Environmental Protection Agency (USEPA). 2012. *Guidelines for Water Reuse,* 600/R-12/618, Washington, DC.

TABLE 14.4 Suggested Guidelines for Water Reuse in Agriculture

Types of Reuse	Treatment	Reclaimed Water Quality	Reclaimed Water Monitoring	Setback Distances
Food crops not commercially processed Surface or spray irrigation of any food crop including crops eaten raw	• Secondary[a] • Filtration[c] • Disinfection[e]	• pH = 6–9 • ≤10 mg/L BOD[b] • ≤2 NTU[d] • No detectable fecal coli/100 mL[f,g] • 1 mgL Cl₂[h]	• pH: Weekly • BOD: Weekly • Turbidity: Continuous • Coliform: Daily • Cl₂ residual: Continuous	50 ft (15 m) to potable water supply wells
Food crops commercially processed Surface irrigation of orchards and vineyards	• Secondary[a] • Disinfection[e]	• pH = 6–9 • ≤30 mg/L BOD[b] • ≤30 mg/L TSS • 200 fecal coli/100 mL[f,i,j] • 1 mg/L Cl₂[h]	• pH: Weekly • BOD: Weekly • TSS: Daily • Coliform: Daily • Cl₂ residual: Continuous	300 ft (90 m) to potable water supply wells 100 ft (30 m) to areas accessible to the public (if spray irrigation)
Non-food crops Pasture for milking animals; fodder, fiber, and seed crops	• Secondary[a] • Disinfection[e]	• pH = 6–9 • ≤30 mg/L BOD[b] • ≤30 mg/L TSS • 200 fecal coli/100 mL[f,i,j] • 1 mg/L Cl₂[h]	• pH: Weekly • BOD: Weekly • TSS: Daily • Coliform: Daily • Cl₂ residual: Continuous	300 ft (90 m) to potable water supply wells 100 ft (30 m) to areas accessible to the public (if spray irrigation)

Source: Adapted from United States Environmental Protection Agency (USEPA). 2012. *Guidelines for Water Reuse,* 600/R-12/618, Washington, DC.

[a] Secondary treatment processes include activated sludge processes, trickling filters, rotating biological contractors, and may include stabilization pond systems. Secondary treatment should produce effluent in which both the BOD and TSS do not exceed 30 mg/L.

[b] As determined from the 5-day BOD test.

[c] Filtration means the passing of wastewater through natural undisturbed soils or filter media such as sand and/or anthracite, filter cloth, or the passing of wastewater through microfilters or other membrane processes.

[d] The recommended turbidity limit should be met prior to disinfection. The average turbidity should be based on a 24-h time period. The turbidity should not exceed 5 NTU at any time. If TSS is used in lieu of turbidity, the TSS should not exceed 5 mg/L.

[e] Disinfection means the destruction, inactivation, or removal of pathogenic microorganisms by chemical, physical, or biological means. Disinfection may be accomplished by chlorination, UV radiation, ozonation, other chemical disinfectants, membrane processes, or other processes. The use of chlorine as defining the level of disinfection does not preclude the use of other disinfection processes as an acceptable means of providing disinfection for reclaimed water.

[f] Unless otherwise noted, recommended coliform limits are median values determined from the bacteriological results of the last 7 days for which analyses have been completed. Either the membrane filter or the fermentation tube technique may be used.

[g] The number of fecal coliform organisms should not exceed 14/100 mL in any sample.

[h] Total chlorine residual should be met after a minimum contact time of 30 min.

[i] The number of fecal coliform organisms should not exceed 800/100 mL in any sample.

[j] Some stabilization pond systems may be able to meet this coliform limit without disinfection.

14.2.5 Australia

The Australian health guidelines for effluent reuse are presented in Table 14.7. This table suggests the required water quality and treatment level for different types of agricultural reuse. The maximum concentration of metals in irrigation waters is given in Table 14.8 [5].

The soil quality is also one of the important factors in considering irrigation with reused effluent. Table 14.8 presents the soil limitation guidelines for water reuse based on the soil property. As observed, soil limitations have been categorized into three groups of slight, moderate, and severe where soils with severe limitation are not recommended for effluent irrigation and soils with moderate limitations require careful design and monitoring [5] (Table 14.9).

14.2.6 France

Agence Française de Sécurité Sanitaire des Aliments (afssa), the French food safety agency, has guidelines regarding effluent

reuse for agriculture, which are presented in Tables 14.10 to 14.13. These guidelines are based on the public sanitation codes of France. Four water quality limits are defined in Table 14.10 and the reuse constraints for each level are presented in Table 14.11. For instance, in order to apply reused water for the irrigation of fruits and vegetables, which are not transformed during cooking, water quality level of A is required. Furthermore, the detailed specifications of the materials used in sprinkles for reused water irrigation such as type and jet angle can be observed in Table 14.12. Distance constraints according to the type of activity to be protected as well as the water quality level are summarized in Table 14.13 [2].

14.2.7 Other

Egypt has widely used effluent reuse for irrigation in comparison to other countries in the region. Unlike other countries, which have some guidelines and recommendations for reused water, Egypt has presented these standards in forms of national codes, which need to be respected.

TABLE 14.5 Canadian Water Quality Guidelines for the Protection of Agricultural Reuse

Parameters	Concentration (mg/L)
Aluminum	5
Arsenic	0.1
Boron	0.5–6
Cadmium	0.0051
Chromium	
• Trivalent Cr (iii)	0.0049
• Hexavalent Cr (vi)	0.008
Cobalt	0.05
Copper	0.2–1
Fluoride	1
Iron	5
Lead	0.2
Lithium	2.5
Manganese	0.2
Molybdenum	0.01–0.05
Nickel	0.2
Selenium	0.02–0.05
Uranium	0.01
Vanadium	0.1
Zinc	1–5

Source: Adapted from Canadian Council of Ministers of the Environment (CCME). 1999. *Canadian Guidelines for Municipal Wastewater Irrigation.* Alberta Environment.

TABLE 14.6 Treated Effluent Water Quality Standards for Wastewater Irrigation

Parameter	Standard	Type of Sample	Comments
Total coliform	<1000/100 mL	Grab	Geometric mean of weekly samples (if storage is provided as part of the treatment) or daily samples (if storage is not provided), in a calendar month
Fecal coliform	<200/100 mL	Grab	Geometric mean of weekly samples (if storage is provided as part of the treatment) or daily samples (if storage is not provided), in a calendar month
BOD	<100 mg/L	Grab/composite[a]	Samples collected twice annually, prior to and on completion of a major irrigation event
COD	<150 mg/L	Grab/composite[a]	Samples collected twice annually, prior to and on completion of a major irrigation event
TSS	<100 mg/L	Grab/composite[a]	Samples collected twice annually, prior to and on completion of a major irrigation event
EC	<1.0 dS/m for unrestricted use 1.0–2.5 dS/m for restricted use >2.5 dS/m unacceptable	Grab/composite[a]	Samples collected twice annually, prior to and on completion of a major irrigation event
SAR	<4 for unrestricted use 4–9 for restricted use when EC >1 dS/m >9 unacceptable	Grab/composite[a]	Samples collected twice annually, prior to and on completion of a major irrigation event
pH	6.5–8.5	Grab/composite[a]	Samples collected twice annually, prior to and on completion of a major irrigation event

Source: Adapted from Canadian Council of Ministers of the Environment (CCME). 1999. *Canadian Guidelines for Municipal Wastewater Irrigation.* Alberta Environment.

[a] Grab sample would suffice if storage is provided; composite sample is required if storage is not provided.

TABLE 14.7   Australian Health Guidelines for Effluent Reuse in Agriculture

Type of Reuse	Suggested Level of Treatment	Reclaimed Water Quality	Monitoring <3 mL	Monitoring >3 mL
Pasture and fodder for grazing animals (except pigs)	Secondary + pathogen reduction by disinfection or detention in ponds or lagoons	Thermo-tolerant coliforms: Median value of <1000 cfu/100 mL	Weekly	Weekly
		Disinfection systems	Weekly	Daily
		pH: 6.5–8.0 (90% compliance)	Weekly	Weekly
Silviculture, turf, and non-food crops	Secondary	Thermo-tolerant coliforms: Median value of <10,000 cfu/100 mL	Monthly	
		pH: 6.5–8.0 (90% compliance)	weekly	
Food crops in direct contact with water, for example, sprays	Secondary + filtration + pathogen reduction	Thermo-tolerant coliforms: Median value of <10 cfu/100 mL	Weekly	Weekly
		≥1 mg/L chlorine residual after 30 min or equivalent level of pathogen reduction	Daily	Daily
		pH 6.5–8.0 (90% compliance)	Weekly	Weekly
		≤2 NTU	As required	Continuous
Food crops not in direct contact with water (e.g., flood or furrow) or which will be sold to consumers cooked or processed	Secondary + pathogen reduction	Thermo-tolerant coliforms: Median value of <1000 cfu/100 mL	Weekly	Weekly
		BOD/SS	Monthly	Weekly
		pH 6.5–8.0 (90% compliance)	Weekly	Weekly

Source: Adapted from Environment Protection Policy. 1999. *Act Wastewater Reuse of Irrigation*, Canberra, Australia.

TABLE 14.8   Maximum Concentration of Metals in Irrigation Waters

Parameters	Concentration (mg/L)
Aluminum	5
Arsenic	0.1
Beryllium	0.1
Cadmium	0.01
Chromium	0.1
Cobalt	0.05
Copper	0.2
Iron	1
Lead	0.2
Lithium	2.5
Manganese	0.2
Molybdenum	0.01
Nickel	0.2
Selenium	0.02
Zinc	2

Source: Adapted from Environment Protection Policy. 1999. *Act Wastewater Reuse of Irrigation*, Canberra, Australia.

TABLE 14.9 Soil Limitations for Effluent Irrigation

Soil Property	Limitation			Restrictive Aspects
	Slight	Moderate	Severe	
Hydraulic connectivity (mm/h)				
• Topsoil	20–80	5–50	<5	Excess runoff
• Subsoil to 1 m	20–80	1–20	<1	Water logging
Phosphate sorption index $[(mg/kg)/log_{10}\mu g/L)]$	>100	10–100	<10	Acts as a poor filter, leaching of phosphorus to underground
pH (in $CaCl_2$)	5–8	4–8	<4, >8.5	Reduces optimum plant growth

Source: Adapted from Environment Protection Policy. 1999. *Act Wastewater Reuse of Irrigation*, Canberra, Australia.

TABLE 14.10 Quality Limit Values

Quality Level	A	B	C	D
Suspended solids	≤35	≤35	≤35	≤35
• In the case of natural lagoons	≤150	≤150	≤150	≤150
BOD (mg/L)	≤125	≤125	≤125	≤125
E. coli/L	≤10,000	≤10,000	≤100,000	–
Salmonella/L	Absent	–	–	–
Taenia coli/L	Absent	–	–	–

Source: Adapted from Agence Francaise de Sécurité Santaire des Aliments (afssa). 2008. Réutilisation des eaux usées traitées pour l'arrosage ou l'irrigation.

TABLE 14.11 Reuse Constraints

Application and/or Type of Activity	Required Quality Level	Restrictions or Intervention Methods
Non-arboriculture market crops, fruits, and vegetables Which are not transformed by means of cooking, pasteurization, or radiation *Arboricultural fruits* Pasture *Public green open spaces* (golf courts, sport fields, etc.)	A	With sprinkler Sprinkler outside the hours open to public
Non-arboriculture market crops, fruits, and vegetables Which are transformed by means of cooking, pasteurization, or radiation *Floral crops, nursery trees, and bushes* *Fodder and cereal crops*	B	With sprinkler
Floral crops, nursery trees, and bushes *Arboricultural fruits* *Fodder and cereal crops*	C	Restricted sprinkler use Not during fall and without sprinkler
Exploitation forests with open access to public	D	Restricted sprinkler use

Source: Adapted from Agence Francaise de Sécurité Santaire des Aliments (afssa). 2008. Réutilisation des eaux usées traitées pour l'arrosage ou l'irrigation.

TABLE 14.12 Required Adaption for Parts and Materials Used in Sprinkler Irrigation with Reused Water

Irrigation Component		Jet Attack Angle	Materials' Characteristics
Bubbler on nonmoving materials		–	Tools to be used on rotating and frontal ramps need to modified with descending canes
Nozzles (180°, 360°, rotating) on nonmoving materials (pulling, rotating, or frontal ramp)		For 360° nozzles, use a convex conical deflector	Should preferably be installed on descending canes, for the case of rotating and frontal ramps, the height needs to be compatible with crop in place
Sprinkler	Small sprinkler	Preferably use sprinklers with a low angle (12°) to avoid a very strong air inlet	To be used on a fixed position or on moving materials
	Irrigation cannon	Jet attack angle of 17° should preferably be used to avoid air inlet	For rotation, turbine is preferred to spoon and the slow returning cannon to the fast ones

Source: Adapted from Agence Francaise de Sécurité Santaire des Aliments (afssa). 2008. Réutilisation des eaux usées traitées pour l'arrosage ou l'irrigation.

TABLE 14.13   Distance Constraints

Activities to Be Protected	Quality Level		
	A	B with Sprinkler Using Aerosols	C and D
Habitation	50 m in case of sprinkler	100 m	–
Circulation lanes	50 m in case of sprinkler	50 m	–
Surface hydraulic centers	20 m	50 m	100 m
Shellfish farming and aquaculture	50 m	200 m	300 m
Bathing	50 m	100 m	200 m
Drinking water protected areas	Prohibition	Prohibition	Prohibition

Source: Adapted from Agence Francaise de Sécurité Santaire des Aliments (afssa). 2008. Réutilisation des eaux usées traitées pour l'arrosage ou l'irrigation.

According to Egypt's code, the application of reused effluent for the irrigation of vegetables consumed raw or cooked, exported products (such as cotton, rice, onion, potato, and aromatic plants), citrus trees, and schools' open spaces are prohibited [1].

Plants and crops irrigated by effluent are categorized into three groups based on the required treatment level. This code also has some restrictions for the crop type, irrigation methods, as well as safety measures for farm workers and the adjunct habitants. The three treatment levels are as follows [1]:

- Level A represents advanced treatment including sand filtration, disinfection, and other processes.
- Level B represents secondary treatment, which includes activated sludge, trickling filters, and the stabilization pond.
- Level C is the primary treatment level, which is limited to screening, deoiling, and settling pond.

Tables 14.14 and 14.15 present the types of plants irrigated with reused effluent and the allowable limits for each water treatment level.

Table 14.16 shows the guidelines presented by Syria for the reuse of water in agriculture. This table categorizes the types of irrigated crops into three groups of A, B, and C as follows [7]:

A. Fresh vegetables, urban grassland, and trees
B. Vegetables not consumed raw, suburb grassland, and trees
C. Forests

The Jordanian standards for effluent discharge and reuse are given in Tables 14.17 and 14.18, respectively. This standard

TABLE 14.14   Classification of Plants and Crops Irrigable with Treated Wastewater

Grade	Agricultural Group	Comments
A	G1-1: Plants and trees grown for greenery at touristic villages and hotels	Palm, Saint Augustine grass, cactaceous plants, ornamental palm trees, climbing plants, fencing bushes and trees, wood trees, and shade trees
	G1-2: Plants and trees grown for greenery inside residential areas at new cities	Palm, Saint Augustine grass, cactaceous plants, ornamental palm trees, climbing plants, fencing bushes and trees, wood trees, and shade trees
B	G2-1: Fodder/feed crops	Sorghum sp.
	G2-2: Trees producing fruits with epicarp	On condition that they are produced for processing purposes such as lemon, mango, date palm, and almonds
	G2-3: Trees used for green belts around cities and forestation of highways or roads	Casuarina, camphor, athel tamarix (salt tree), oleander, fruit-producing trees, date palm, and olive trees
	G2-4: Nursery plants	Nursery plants of wood trees, ornamental plants, and fruit trees
	G2-5: Roses and cut flowers	Local rose, eagle rose, onions (e.g., gladiolus)
	G2-6: Fiber crops	Flax, jute, hibiscus, sisal
	G2-7: Mulberry for the production of silk	Japanese mulberry
C	G3-1: Industrial oil crops	Jojoba and jatropha
	G3-2: Wood trees	Caya, camphor, and other wood trees

Source: Adapted from Abdel-Shafy H.I., Mansour, M.S.M. 2013. *J. Sustainable Sanitation Practice*, 14, 17–25.

divides the application type into the following three groups of products [10]:

- Cooked vegetables, parking areas, playgrounds, sides of roads, and inside cities
- Plenteous trees and green areas, sides of roads outside cities
- Field crops, industrial crops, and forestry

TABLE 14.15   Requirements for Treated Wastewater Reused in Agriculture

Treatment Grade Requirements (in mg/L)		A	B	C
Effluent limit values for BOD and suspended solids (SS)	BOD_5	<20	<60	<400
	SS	<20	<50	<250
Effluent limit values for fecal coliform and nematode cells of eggs (per liter)	Fecal coliform count (2) in 100 cm³	<1000	<5000	Unspecified

Source: Adapted from Abdel-Shafy H.I., Mansour, M.S.M. 2013. *J. Sustainable Sanitation Practice*, 14, 17–25..

TABLE 14.16 Effluent Characteristics for Irrigational Reuse

Parameter	Unit	Crop Type A	B	C
pH	–	6–9	6–9	6–9
BOD	mg/L	30	100	150
COD	mg/L	75	200	300
TSS	mg/L	50	150	150
Coliform	1/mL	<100,000	<100,000	1000

Source: Adapted from Syria National Report. The actual statues of astewater and reuse for agriculture in Syria. http://www.ais.unwater.org/ais/pluginfile.php/356/mod_page/content/111/Syria%20Reusing%20Sewage%20and%20treated%20Water%20for%20Irrigation%20and%20its%20Effects.pdf. Retrieved on September 23, 2013.

TABLE 14.17 Jordanian Standards for Maximum Acceptable Limits for Effluent Discharge Characteristics

Parameter	Unit	Discharge	Groundwater Recharge
BOD_5	mg/L	60	15
COD	mg/L	150	50
Dissolved oxygen	mg/L	1	2
Total suspended solids	mg/L	60	50
pH	Unit	6–9	6–9
Turbidity	NTU	ND	2
NO_3	mg/L	45	30
NH_4	mg/L	ND	5
T-N	mg/L	70	45
E. coli	MPN/100 mL	1000	2.2
Helminth eggs	Egg/L	1	1
Fat, oil, and grease	mg/L	8.0	ND
Group B			
Phenol	mg/L	0.002	0.002
Methelene blue active substance	mg/L	25	25
Total dissolved solids	mg/L	1500	1500
Total PO_4	mg/L	15	15
Chlorine	mg/L	350	350
SO_4	mg/L	300	300
HCO_3	mg/L	400	400
Sodium	mg/L	200	200
Magnesium	mg/L	60	60
Calcium	mg/L	200	200
SAR	mg/L	6.0	6.0
Aluminum	mg/L	2.0	2.0
Arsenic	mg/L	0.05	0.05
Beryllium	mg/L	0.1	0.1
Copper	mg/L	0.2	0.2
Fluorine	mg/L	1.5	1.5
Iron	mg/L	5	5
Lithium	mg/L	2.5	2.5
Manganese	mg/L	0.2	0.2
Molybdenum	mg/L	0.01	0.01
Nitrogen	mg/L	0.2	0.2
Lead	mg/L	0.2	0.2
Selenium	mg/L	0.05	0.05
Cadmium	mg/L	0.01	0.01
Zinc	mg/L	5.0	5.0
Chromium	mg/L	0.02	0.02
Mercury	mg/L	0.002	0.002
Vanadium	mg/L	0.1	0.1
Cobalt	mg/L	0.05	0.05
Boron	mg/L	1.0	1.0

Source: Adapted from World Health Organization (WHO). 2005. *A Regional Overview of Wastewater Management and Reuse in the Eastern Mediterranean Region.* WHO, Regional Office for the Eastern Mediterranean, Cairo, Egypt.

TABLE 14.18　Jordanian Standards Characteristics of Effluent Used for Irrigation of Crops

Parameter	Unit	Cooked Vegetables, Parking Areas, Playgrounds, Sides of Roads, and Inside Cities	Plenteous Trees and Green Areas, Sides of Roads Outside Cities	Field Crops, Industrial Crops, and Forestry
BOD$_5$	mg/L	30	200	300
COD	mg/L	100	500	500
DO	mg/L	2	Not detectable	Not detectable
TSS	mg/L	50	150	150
pH	Unit	6–9	6–9	6–9
Turbidity	NTU	10	ND	ND
NO$_3$	mg/L	30	45	45
T-N	mg/L	45	70	70
E. coli	MPN/100 mL	100	1000	ND
Helminth eggs	Egg/L	1	1	1

Source: Adapted from World Health Organization (WHO). 2005. *A Regional Overview of Wastewater Management and Reuse in the Eastern Mediterranean Region.* WHO, Regional Office for the Eastern Mediterranean, Cairo, Egypt.

14.3　Summary and Conclusions

Effluent reuse for irrigation and agricultural purposes is one of the common and widely used reuse techniques. This chapter attempted to gather some of the most important guidelines regarding effluent reuse in agriculture. Guidelines from various countries and organizations such as USEPA, WHO, FAO, Australia, Canada, France, Egypt, Syria, and Jordan were presented. Comparison shows that these guidelines have some similarity in values; however, they are often diversely presented. These differences originate from the type of application and climate conditions of each region where the guideline is used. Since countries have different demands and sources, the type of effluent reuse and to what extent it is applied often differs from one region to another.

As observed in the previous section, the guidelines divided the crops irrigated with effluent into several groups according to the level of treatment required. This is a desirable. In addition, it is proposed that the guidelines be adopted as codes to guarantee proper execution.

References

1. Abdel-Shafy, H.I., Mansour, M.S.M. 2013. Overview on water reuse in Egypt: present and future. *J. Sustainable Sanitation Practice,* 14, 17–25.

2. Agence Francaise de Sécurité Santaire des Aliments (afssa). 2008. Réutilisation des eaux usées traitées pour l'arrosage ou l'irrigation. (The French Food Safety Agency (afssa). Effluent reuse for irrigation.)

3. Canadian Council of Ministers of the Environment (CCME). 1999. *Canadian Guidelines for Municipal Wastewater Irrigation.* Alberta Environment.

4. Egyptian Environmental Association Affair (EEAA). 2000. *Law 48, No.61-63, Law of the Environmental Protection (1994)—Updating No. 44,* Cairo, Egypt.

5. Environment Protection Policy. 1999. *Act Wastewater Reuse of Irrigation,* Canberra, Australia.

6. Food and Agriculture Organization of the United Nations (FAO). 1985. In: Ayers, R.S. Westcot, D.W. (eds.), *Water Quality for Agriculture.* Irrigation and Drainage Paper 29 Rev. l. FAO, Rome.

7. Syria National Report. The actual statues of wastewater and reuse for agriculture in Syria. http://www.ais.unwater.org/ais/pluginfile.php/356/mod_page/content/111/Syria%20Reusing%20Sewage%20and%20treated%20Water%20for%20Irrigation%20and%20its%20Effects.pdf. Retrieved on September 23, 2013.

8. United States Environmental Protection Agency (USEPA). 2004. *Guidelines for Water Reuse.* Report 625/R-04/108, Washington, DC.

9. United States Environmental Protection Agency (USEPA). 2012. *Guidelines for Water Reuse.* 600/R-12/618, Washington, DC.

10. World Health Organization (WHO). 2005. *A Regional Overview of Wastewater Management and Reuse in the Eastern Mediterranean Region.* WHO, Regional Office for the Eastern Mediterranean, Cairo, Egypt.

11. World Health Organization (WHO). 1989. *Health Guidelines for the Use of Wastewater in Agriculture.* Technical Report No. 778. WHO, Geneva.

15

Water Reuse Guidelines for Industry

Alireza Eslamian
Concordia University

Faezeh Eslamian
McGill University

Saeid Eslamian
Isfahan University of Technology

PREFACE

Industrial water reuse is yet another way to recycle water and is gaining great importance nowadays. However, industrial equipment is sensitive and prone to destructive effects of low quality water such as corrosion, contamination, chemical reaction, etc. Therefore, the reclaimed water used in industries needs to follow minimum criteria to avoid such problems. The water quality guidelines for two major processes (cooling towers and boilers) in industries such as paper, textile, petroleum, and iron and steel are presented in this chapter.

15.1 Introduction

Over the past few years, industrial water reuse has gained great importance. Traditionally, reused water was utilized solely in paper and textile industries in cooling towers. However, with the growth in technology, water demand and the availability of proper guidelines, other procedures in a variety of industries are using this method [8].

In order to use reclaimed water in such industries, the effluent needs to meet a certain water quality criteria. Since most industrial equipment and machinery are either sensitive themselves or produce a highly sensitive material, failure to respect the allowable water quality limits could cause major damage to the equipment as well as consumers. The quality of water may affect the equipment or product by corrosion, contamination, scale, chemical reaction, or biological degradation [2].

The destructive effects of some water quality parameters on heating systems are summarized in Table 15.1. These effects outline the importance of existing guidelines and respecting them.

The cost effectiveness of water reuse needs to be analyzed as well. Since boilers and cooling towers require a lower water quality but higher volumes of water, the use of reused water for these systems is the most effective.

In this chapter, the water qualities required for different industrial sections and procedures will be presented.

15.2 Guidelines for Water Reuse in Industry

As stated before, cooling towers and boilers are the two main sections that are most desirable for the reuse of water. These systems need large volumes of medium to low quality water. The devices used in these two sections are less sensitive than that of others. However, the reclaimed water needs to meet the criteria set by the organization or country in which the industry operates.

15.2.1 Cooling Towers

Cooling towers are recirculating evaporative cooling systems that use reclaimed water to absorb process heat and then transfer

TABLE 15.1 Effects of Some Water Quality Parameters on Heating Equipment

Parameter	Effects on Heating Equipment
Color	The color-causing constituents may cause foaming in boilers. Hinders precipitation methods such as iron removal and softening.
Hardness	Chief source of scale in heat-exchange equipment, boilers, pipelines, etc.
Alkalinity	Foaming and carry-over of solids with steam. Embrittlement of boiler steel. Bicarbonate and carbonate produce CO_2 in steam, a source of corrosion in condensate lines.
Free mineral acids	Corrosion.
Carbon dioxide	Corrosion in water lines and particularly steam and condensate lines.
Sulfate	Adds to solids content of water but in itself is not usually significant. Combines with calcium to form calcium sulfate scale.
Chloride	Adds to solids content and increases corrosive character of water.
Nitrate	Adds to solids content but is not usually significant industrially. Useful for control of boiler-metal embrittlement.
Silica	Scale in boilers and cooling water systems. Insoluble turbine blade deposits because of silica vaporization.
Iron	Discolors water on precipitation. Source of deposits in water lines, boilers, etc.
Manganese	Discolors water on precipitation. Source of deposits in water lines, boilers, etc.
Oxygen	Corrosion of water lines, heat-exchange equipment, boilers, return lines, etc.
Hydrogen sulfide	Cause of "rotten egg" odor. Corrosion.
Ammonia	Corrosion of copper and zinc alloys by formation of complex soluble iron.
Dissolved solids	Process interference and causes foaming in boilers.
Suspended solids	Causes deposits in heat-exchange equipment, boilers, water lines, etc.

Source: Adapted from American Society of Mechanical Engineers. 1979. Consensus on Operating Practices for the Control of Feedwater and Boiler Water Quality in Modern Industrial Boilers. *Prepared by Feedwater Quality Task Group for Industrial Boiler Subcommittee. The ASME Research Committee on Water in Thermal Power Systems.* American Society of Mechanical Engineers, New York.

the heat by evaporation [8]. Reclaimed water is used to recompense for the loss of water due to evaporation in each cooling cycle. After several cycles, the water needs to be completely discharged and renewed to prevent a buildup of dissolved solids in the cooling system. Table 15.2 presents the recommended reused water quality in cooling towers. It can be interpreted from Table 15.2 that the reclaimed water recirculating a few times in the system requires a higher level of pretreatment and a better water quality compared to that of once-through because the water is in contact with the equipment for a longer time and could cause more serious problems.

15.2.2 Boilers

The use of reclaimed water in a boiling system requires extensive pretreatment procedures. The required water quality depends on the pressure at which the boiler functions. Higher pressures generally require higher water qualities [8].

The major issue in this case is corrosion. Hardness needs to be controlled or removed prior to use in boilers. Furthermore, calcium, magnesium, silica, aluminum, and alkalinity should be controlled [8]. The allowable limits for reused water in boilers are presented in Table 15.3.

The water quality for several generic processes and industries will be presented here. The water quality for generic systems such as cooling and heating depends on whether the process is once-through or recirculating. Table 15.4 shows the water quality guidelines for once-through cooling and makeup for recirculation limits; in addition, Table 15.5 presents the minimum and maximum numerical limits for water reuse in cooling towers.

Furthermore, in Table 15.6, the water quality limits for power generation stations are presented. The allowable concentration is given for three categories of cooling towers, boiling systems, and miscellaneous uses.

In Tables 15.7 through 15.10, the water quality guidelines for the reuse of water in industries such as iron and steel, pulp and paper, petroleum and textile are presented, respectively.

TABLE 15.2 Recommended Reused Water Quality in Cooling Towers

Application Type	Required Treatment	Reused Water Quality	Monitoring	Protected Distance
Once-through	Secondary	• pH: 6–9 • BOD: 30 mg/L • TSS: 30 mg/L • Fecal coliform: 200 in 100 mL • Cl_2: 1 mg/L	• pH: Weekly • BOD: Weekly • TSS: Weekly • Coliform: Daily • Cl_2: Continuous	90 m to public access
Recirculating	Secondary disinfection (chemical coagulation and filtration may be required as well)	• pH: 6–9 • BOD: 30 mg/L • TSS: 30 mg/L • Fecal coliform: 200 in 100 mL • Cl_2: 1 mg/L	• pH: Weekly • BOD: Weekly • TSS: Weekly • Coliform: Daily • Cl_2: Continuous	90 m to public access (in the case of disinfection, this distance can be reduced)

Source: Adapted from United States Environmental Protection Agency (USEPA). 2004. *Guidelines for Water Reuse, Report 625/R-04/108*, Washington, DC.

TABLE 15.3 Recommended Reused Water Quality in Boilers

Drum Operating Pressure (psig)	0–300	301–450	451–600	601–750	751–900	901–1000	1001–1500	1501–2000	OTSG
					Steam				
TDS max (ppm)	0.2–1.0	0.2–1.0	0.2–1.0	0.1–0.5	0.1–0.5	0.1–0.5	0.1	0.1	0.05
					Boiler Water				
TDS max (ppm)	700–3500	600–3000	500–2500	200–1000	150–750	125–625	100	50	0.05
Alkalinity max (ppm)	350	300	250	200	150	100	n/a	n/a	n/a
TSS max (ppm)	15	10	8	3	2	1	1	n/a	n/a
Conductivity max (µmho/cm)	1100–5400	900–4600	800–3800	300–1500	200–1200	200–1000	150	80	0.15–0.25
Silica max (ppm SiO$_2$)	150	90	40	30	20	8	2	1	0.02
				Feed Water (Condensate and Makeup, after De-aerator)					
Dissolved oxygen (ppm O$_2$)	0.007	0.007	0.007	0.007	0.007	0.007	0.007	0.007	n/a
Total iron (ppm Fe)	0.1	0.05	0.03	0.025	0.02	0.02	0.01	0.01	0.01
Total copper (ppm Cu)	0.05	0.025	0.02	0.02	0.015	0.01	0.01	0.01	0.002
Total hardness (ppm CaCO$_3$)	0.3	0.3	0.2	0.2	0.1	0.05	ND	ND	ND
pH at 25°C	8.3–10.0	8.3–10.0	8.3–10.0	8.3–10.0	8.3–10.0	8.8–9.6	8.8–9.6	8.8–9.6	n/a
Nonvolatile TOC (ppm C)	1	1	0.5	0.5	0.5	0.2	0.2	0.2	ND
Oily matter (ppm)	1	1	0.5	0.5	0.5	0.2	0.2	0.2	ND

Source: Adapted from United States Environmental Protection Agency (USEPA). 2012. *Guidelines for Water Reuse, 600/R-12/618*, Washington, DC.
n/a = not applicable.
ND = not defined.

TABLE 15.4 Water Quality Guidelines for Once-Through Cooling and Makeup Water Systems

	Once-Through[a]		Makeup for Recirculation[a]	
Parameter	Fresh	Blackish[b]	Fresh	Blackish[b]
Silica	<50	<25	<50	<25
Aluminum	NS[c]	NS	<0.1	<0.1
Iron	NS	NS	<0.5	<0.5
Manganese	NS	NS	<0.5	<0.02
Calcium	<200	<420	<50	<420
Bicarbonate	<600	<140	<24	<140
Sulfate	<680	<2700	<200	<2700
Chloride	<600	<19,000	<500	<19,000
Dissolved solids	<1000	<35,000	<500	<35,000
Hardness	<850	<6250	<130	<6250
Alkalinity	<500	<115	<20	<115
pH	5.0–8.3	6.0–8.3	NS	NS
Organic material:				
Methylene blue active substances	NS	NS	<1	<1
Carbon tetrachloride extract	NS[d]	NS[d]	<1	<2
Chemical oxygen demand	<75	<75	<75	<75
Suspended solids	<5000	<2500	<100	<100

Source: Adapted from Krisher, A.S. 1978. *Chemical Engineering*, 85, 78–98.
[a] Unless otherwise specified all units are mg · L^{-1}.
[b] Brackish water—dissolved solids concentration >1000 mg · L^{-1}.
[c] NS = not specified.
[d] No floating oil.

TABLE 15.5 Water Quality Guidelines for Cooling Towers

	Numerical Limits (mg · L⁻¹)	
Parameter	Minimum	Maximum
Langelier saturation index[a]	+0.5	+1.5
Ryzner stability index	+6.5	+7.5
pH units	≥ 6.0	≤ 8.0
Calcium (as $CaCO_3$)	>30	<300
		<400
Total iron	–	<0.5
Manganese	–	<0.5
Copper	–	<0.08
Aluminum	–	<1
Sulfide	–	<5
Silica	–	<150
	–	<200
$[Ca] \times [SO_4]$	–	<500,000
Total dissolved solids	–	<2500
Conductivity ($\mu S \cdot cm^{-1}$)	–	<4000
Suspended solids	–	<150

Source: Adapted from Krisher, A.S. 1978. *Chemical Engineering,* 85, 78–98.

TABLE 15.6 Water Quality Guidelines for Power Generation Stations

	Concentration (mg · L⁻¹)			
	Cooling Once-Through		Boiler Feedwater	Miscellaneous
Parameter	Fresh	Blackish[a]	(10.35–34.48 MPa)	Uses
Silica	<50	<25	<0.01	–
Aluminum	NS[b]	NS	<0.01	–
Iron	NS	NS	<0.01	<1.0
Manganese	NS	NS	<0.01	–
Calcium	<200	<420	<0.01	–
Magnesium	NS	NS	<0.01	–
Ammonia	NS	NS	<0.07	–
Bicarbonate	<600	<140	<0.5	–
Sulfate	<680	<2700	NS[c]	–
Chloride	<600	<19,000	NS[c]	–
Dissolved solids	<1000	<35,000	<0.5	<1000
Copper	NS	NS	<0.01	–
Hardness	<850	<6250	<0.07	–
Zinc	NS	NS	<0.01	–
Alkalinity (as $CaCO_3$)	<500	<115	<1	–
pH units	5.0–8.3	6.0–8.3	8.8–9.4	5.0–9.0
Organic material:				
Methylene blue active substances	NS	NS	<0.1	<10
Carbon tetrachloride extract	NS[d]	NS[d]	NS	<10
Chemical oxygen demand (COD)	<75	<75	<1.0	–
Dissolved oxygen	–	–	<0.007	–
Suspended solids	<5000	<2500	<0.05	<5

Source: Adapted from Krisher, A.S. 1978. *Chemical Engineering,* 85, 78–98.

[a] Brackish water—dissolved solids more than 1000 mg · L⁻¹.

[b] NS = not specified; the parameter has never been a problem at concentrations encountered.

[c] Controlled by treatment for other constituents.

[d] No floating oil.

TABLE 15.7 Water Quality Guidelines for the Iron and Steel Industry

Parameter	Concentration (mg · L^{-1})				
	Hot-Rolling, Quenching, Gas Cleaning	Cold-Rolling	Rinse Water		Steel Manufacturing
			Softened	Demineralized	
pH	5.0–9.0	5.0–9.0	6.0–9.0	–	6.8–7.0
Suspended solids	<25	<10	ND[a]	ND	–
Dissolved solids	<1000	<1000	ND	ND	–
Settleable solids	<100	<5.0	ND	ND	–
Dissolved oxygen	(Minimum for aerobic conditions)				
Temperature (°C)	<38	<38	<38	<38	<38
Hardness	NS[bc]	NS[b]	<100	<0.1	<50
Alkalinity	NS[c]	NS[c]	NS[c]	<0.5	–
Sulfate	<200	<200	<200	–	<175
Chloride	<150	<150	<150	ND	<150
Oil	NS	ND	ND	ND	ND
Floating material	NS	ND	ND	ND	ND

Source: Adapted from Canadian Council of Ministers of the Environment (CCME). 2008. *Canadian Water Quality Guidelines*, Canada.

[a] ND = not detected.

[b] Controlled by other treatments.

[c] NS = not specified; the parameter has never been a problem at concentrations encountered.

TABLE 15.8 Water Quality Guidelines for the Pulp and Paper Industry

Parameter	Concentration (mg · L^{-1})					
	Fine Paper	Ground-Wood	Kraft		Chem. Pulp and Paper	
			Bleached	Unbleached	Bleached	Unbleached
pH	–	6–8	–	–	6–8	6–8
Color (HU)	<40	<100	<25	<100	<50	<100
Turbidity (NTU)	<10	<20	<40	<100	<10	<20
Calcium	<20	<20	–	–	<20	<20
Magnesium	<12	<12	–	–	<12	<12
Iron	<0.1	<0.1	<0.2	<1.0	<0.1	<1.0
Manganese	<0.3	<0.1	<0.1	<0.5	<0.05	<0.5
Chloride	–	25–75	<200	<200	<200	<200
Silica	<20	<100	<50	<100	<50	<50
Hardness	<100	<100	<100	<100	<100	<100
Alkalinity	40–75	<150	<75	<150	–	–
Dissolved solids	<200	<250	<300	<500	<200	<250
Suspended solids	<10	–	–	–	<10	<10
Temperature (°C)	–	–	–	–	<36	–
CO$_2$	<10	<10	<10	<10	–	–
Corrosion tendency	Nil	Nil	Nil	Nil	Nil	Nil
Residual chloride	<2.0	–	–	–	–	–

Source: Adapted from Canadian Council of Ministers of the Environment (CCME). 2008. *Canadian Water Quality Guidelines*, Canada.

TABLE 15.9 Water Quality Guidelines for the Petroleum Industry

Parameter	Concentration (mg L^{-1})[a]
pH units	6.0–9.0
Color	NS[b]
Calcium	<75
Magnesium	<25
Iron	<1
Bicarbonate	NS
Sulfate	NS
Chloride	<200
Nitrate	NS
Fluoride	NS
Silica	NS
Hardness (as CaCO$_3$)	<350
Dissolved solids	<750
Suspended solids	<10

Source: Adapted from Ontario Ministry of the Environment. 1974. *Guidelines and Criteria for Water Quality Management in Ontario.* Water Resources Branch, Toronto, Ontario, Canada.

[a] Unless otherwise indicated.

[b] NS = not specified. The parameter has never been a problem at concentrations encountered.

15.3 Summary and Conclusions

Industrial water reuse is a growing category of water reuse. Since industrial equipment is sensitive and prone to destructive effects, the reclaimed water for use in industries needs to follow the guidelines set by organizations and countries where the industries operate to avoid environmental and human harm. Cooling towers and boilers are the two major operating systems in every industry, which require large quantities of water. The water quality guidelines for the use of reclaimed water for these two systems recommended by USEPA were presented here. Also, the required water qualities for four major industries were given as well. It can be concluded that the quality of the reclaimed water used in industries highly depends on the procedure and the type of industry. The more sensitive procedures require better water quality. A lower level of water quality is required for cooling towers, and they could be a major application for water reuse. Furthermore, a certain level of water quality parameters is required for each industry, depending on its specific characteristics and materials.

Close consideration should be given to these guidelines prior to reuse to avoid potential environmental, industrial, or human harm.

TABLE 15.10 Water Quality Guidelines for the Textile Industry

	Concentration (mg · L^{-1})					
	Cotton, Wool, Synthetics			Viscose, Rayon		
Parameter	Sizing Suspension	Scouring	Bleaching	Dyeing	Pulp Manufacture	Manufacture
Iron	<0.3	<0.1	<0.1	<0.1	<0.05(Fe+Mn)	ND[a]
Manganese	<0.05	<0.01	<0.01	<0.01	<0.03	ND
Copper	<0.05	<0.01	<0.01	<0.01	<5	–
Dissolved solids	<100	<100	<100	<100	<100	–
Suspended solids	<5	<5	<5	<5	–	–
Hardness (as CaCO$_3$)	<25	<25	<25	<25	<8	<55
pH: Cotton	6.5–10.0	9.0–10.5	2.5–10.5	7.5–10.0	–	–
Synthetics	6.5–10.0	3.0–10.5	NA[b]	6.5–7.5	–	–
Wool	6.5–10.0	3.0–5.0	2.5–5.0	3.5–6.0	–	–
Viscose, rayon	–	–	–	–	–	7.8–8.3
Color (relative units)	<5	<5	<5	<5	<5	—
Turbidity (NTU)	–	–	–	<15	<5	<0.3
Aluminum	–	–	–	–	<8	–
Silica	–	–	–	–	<25	–
Alkalinity (as CaCO$_3$)	–	–	–	–	50–75	50–75

Source: Adapted from Hart, B.T. 1974. *A Compilation of Australian Water Quality Criteria,* Australian Water Resources Council, Department of Environment and Conservation, Australian Government Public Service, Technical Paper No. 7, Canberra, Australia; McKee, J.E. and Wolf, H.W. 1963. *Water Quality Criteria,* 2nd edition, State Water Quality Control Board, Resources Agency of California, Publ. No. 3-A, Sacramento, California.

[a] ND = not detected.

[b] NA = not applicable.

References

1. American Society of Mechanical Engineers. 1979. Consensus on operating practices for the control of feedwater and boiler water quality in modern industrial boilers. *Prepared by Feedwater Quality Task Group for Industrial Boiler Subcommittee. The ASME Research Committee on Water in Thermal Power Systems.* American Society of Mechanical Engineers, New York.

2. Canadian Council of Ministers of the Environment (CCME). 2008. *Canadian Water Quality Guidelines*, Canada.

3. Hart, B.T. 1974. *A Compilation of Australian Water Quality Criteria.* Australian Water Resources Council, Department of Environment and Conservation, Australian Government Pubic. Service, Technical Paper No. 7, Canberra, Australia.

4. Krisher, A.S. 1978. Raw water treatment in the CPI. *Chemical Engineering*, 85, 78–98.

5. McKee, J.E. and Wolf, H.W. 1963. *Water Quality Criteria*, 2nd edition, State Water Quality Control Board, Resources Agency of California, Publ. No. 3-A, Sacramento, California.

6. Ontario Ministry of the Environment 1974. *Guidelines and Criteria for Water Quality Management in Ontario.* Water Resources Branch, Toronto, Ontario, Canada.

7. United States Environmental Protection Agency (USEPA). 2004. *Guidelines for Water Reuse, Report 625/R-04/108*, Washington, DC.

8. United States Environmental Protection Agency (USEPA). 2012. *Guidelines for Water Reuse, 600/R-12/618*, Washington, DC.

16

Water Reuse Guidelines for Recreation

Saeid Eslamian
Isfahan University of Technology

Faezeh Eslamian
McGill University

Alireza Eslamian
Concordia University

PREFACE

Worldwide water demand is increasing due to current population growth, and commercial and industrial development. Recreational water reuse is considered one of the major water reuse applications. Microbiological indicators are the most important parameters of concern on which most guidelines have emphasized. The water quality guidelines for recreational reuse also depend on whether the water body is in direct contact with humans. This chapter summarizes the various guidelines suggested in this area.

16.1 Introduction

Worldwide water demand is increasing due to current population growth, and commercial and industrial development. In order to maintain our current and limited water resources, efficient and supplementary sources of water are needed to recompense for the loss. Water reuse is widely used around the world in three major categories of agriculture, industry and urban, and recreation.

While in many Middle Eastern countries agricultural reuse is the major reuse category, in the United States and Canada, urban and recreational reuse is one of the highest volume uses [16].

Recreational water often refers to surface waters that are used primarily for sports in which the user comes into frequent direct contact with the water, either as part of the activity or incidental to the activity: for example, swimming and boardsailing. Other recreational uses include boating, canoeing, and fishing, which generally have less frequent body contact with water [1]. Therefore, water reused and discharged to these types of waters needs to meet certain criteria for environmental health. Furthermore, reuse applications such as recreational field and golf course irrigation, landscape irrigation, and other applications, including fire protection and toilet flushing, are important components of the recreational reclaimed water programs [16].

Urban and recreational reuse is often divided into applications that are either accessible to the public or have restricted access. In the case of restricted or controlled access, fences are used to avoid or reduce human contact [16].

In this chapter, the guidelines for water reuse in recreational purposes are presented.

16.2 Water Reuse Guidelines for Recreation

Water used for recreational purposes should be sufficiently free from microbiological, chemical, and physical hazards such as poor visibility, to ensure that there is negligible risk to the health and safety of the user [6]. Also, aesthetical aspects are of great concern. Table 16.1 shows water quality characteristics of importance to recreational water use presented by the Canadian Council of Ministers of the Environment (CCME) [1]. The most limiting factors for recreational water reuse are microbiological indicators, pathogens, and physical and chemical parameters such as taste, odor, color, and turbidity. The most important microbiological indicators are as follows: *Escherichia coli*, fecal coliforms, *Enterococci*, and coliphages. The Canadian guidelines for recreational water quality are summarized in Table 16.2.

TABLE 16.1   Water Quality Characteristics of Importance to Recreational Water Use

Characteristic	Swimming[a]	Non-Swimming
Microbiological		
Indicator organisms		
Escherichia coli	x	
Fecal coliforms	x	
Enterococci	x	
Coliphages	x	
Pathogens		
Pseudomonas aeruginosa	x	
Staphylococcus aureus	x	
Salmonella	x	
Shigella	x	
Aeromonas	x	
Campylobacter jejuni	x	
Legionella	x	
Viruses	x	
Giardia lamblia	x	
Cryptosporidium	x	
Nuisance organisms		
Vector and nuisance organisms	x	x
Aquatic vascular plants	x	x
Phytoplankton	x	x
Physical and chemical		
pH	x	
Temperature	x	
Aesthetics	x	x
Oil, debris	x	x
Clarity	x	
Turbidity	x	
Color	x	

Source: Adapted from Canadian Council of Ministers of the Environment (CCME). 2008. *Canadian Water Quality Guidelines*, Canada.

[a] Swimming also includes board-sailing, water skiing, white water sports, scuba diving, and dinghy sailing.

TABLE 16.2   Summary of the Guidelines for Recreational Water Quality

Parameter	Guideline
	Bacteriological
Escherichia coli and fecal coliforms	The geometric mean of at least five samples taken during a period not to exceed 30 d should not exceed 2000 *E. coli* per liter. Resampling should be performed when any sample exceeds 4000 *E. coli* per liter.
Enterococci	The geometric mean of at least five samples taken during a period not to exceed 30 d should not exceed 350 enterococci per liter. Resampling should be performed when any sample exceeds 700 enterococci per liter.
Aesthetics	All water should be free from • Materials that will settle to form objectionable deposits. • Floating debris, oil, scum, and other matter. • Substances producing objectionable color, odor, taste, or turbidity. • Substances and conditions or combinations thereof in concentrations that produce undesirable aquatic life.
Temperature	The thermal characteristics of water should not cause an appreciable increase or decrease in the deep body temperature of bathers and swimmers.
Clarity	The water should be sufficiently clear that a Secchi disc is visible at a minimum of 1.2 m.
pH	5.0–9.0, provided that when the pH is near the extremes of this range, the buffering capacity of the water is very low.
Turbidity	The turbidity of water should not be increased more than 5.0 NTU over natural turbidity when turbidity is low (<50 NTU).
Oil and grease	Oil or petrochemicals should not be present in concentrations that • Can be detected as a visible film, sheen, or discoloration on the surface. • Can be detected by odor. • Can form deposits on shorelines and bottom deposits that are detectable by sight and odor.
Aquatic plants	Rooted or floating plants that could entangle bathers should be absent; very dense growths could affect other activities such as boating and fishing.

Source: Adapted from Canadian Council of Ministers of the Environment (CCME). 2008. *Canadian Water Quality Guidelines*, Canada.

NTU = Nephelometric turbidity units.

Should the water be reclaimed for recreational purposes, it needs to meet the following criteria.

The recreational water quality guidelines suggested by the US Environmental Protection Agency (USEPA) are summarized in Table 16.3. This table is subcategorized into urban reuse and recreational reuse. Also, the water quality limits for each application are dependent on whether the application is restricted or unrestricted. This table also suggests the required monitoring frequencies for each water quality parameter.

The water quality requirements for bathing waters of the European Economic Community [6] are presented in Table 16.4.

Table 16.5 summarizes the different microbiological water quality guidelines for recreational reuse of approximately 20 countries and organizations. This table has been categorized based on shellfish harvesting, primary contact recreation, and protection of indigenous organisms. Table 16.5 shows that most countries have adopted the internationally accepted microbiological criterion for shellfish water quality of 70 total coliforms

TABLE 16.3 Suggested Guidelines for Recreational Water Reuse

Application Type	Required Treatment	Reused Water Quality	Monitoring	Protected Distance
Urban reuse				
Unrestricted: The use of reclaimed water in nonpotable applications in municipal settings where public access is not restricted.	• Secondary • Filtration • Disinfection	• pH: 6–9 • BOD <10 mg/L • Turbidity <2 NTU mg/L • Fecal coliform: Not detected in 100 mL • Cl$_2$ <1 mg/L	• pH: Weekly • BOD: Weekly • Turbidity: Continuous • Coliform: Daily • Cl$_2$: Continuous	50 ft (15 m) to potable water supply wells; increased to 100 ft (30 m) when located in porous media
Restricted: The use of reclaimed water in nonpotable applications in municipal settings where public access is controlled or restricted by physical or institutional barriers, such as fencing, advisory signage, or temporal access restriction	• Secondary • Disinfection	• pH: 6–9 • BOD <30 mg/L • TSS <30 mg/L • Fecal coliform: 200 in 100 mL • Cl$_2$ <1 mg/L	• pH: Weekly • BOD: Weekly • TSS: Daily • Coliform: Daily • Cl$_2$: Continuous	300 ft (90 m) to potable water supply wells 100 ft (30 m) to areas accessible to the public (if spray irrigation)
Recreational impoundment				
Unrestricted: The use of reclaimed water in an impoundment in which no limitations are imposed on body contact.	• Secondary • Filtration • Disinfection	• pH: 6–9 • BOD <10 mg/L • Turbidity <2 NTU mg/L • Fecal coliform: Not detected in 100 mL • Cl$_2$ <1 mg/L	• pH: Weekly • BOD: Weekly • Turbidity: Continuous • Coliform: Daily • Cl$_2$: Continuous	500 ft (150 m) to potable water supply wells (min.) if bottom not sealed
Restricted: The use of reclaimed water in an impoundment where body contact is restricted.	• Secondary • Disinfection	• BOD <30 mg/L • TSS <30 mg/L • Fecal coliform: 200 in 100 mL • Cl$_2$ <1 mg/L	• pH: Weekly • BOD: Weekly • TSS: Daily • Coliform: Daily • Cl$_2$: Continuous	500 ft (150 m) to potable water supply wells (min.) if bottom not sealed

Source: Adapted from USEPA. 2012. *Guidelines for Water Reuse, 600/R-12/618*, Washington, DC; USEPA. 2004. *Guidelines for Water Reuse, Report 625/R-04/108*, Washington, DC.

TABLE 16.4 Water Quality Requirements for Bathing Waters

Microbiological Parameter	Guide[a]	Mandatory[b]	Minimum Sampling Frequency
E. coli/100 mL	100	2000	Every two weeks
Fecal streptococci/100 mL	100	400	Every two weeks[c]
Enteroviruses PFU/10 L	–	0	Monthly[c]

Source: Adapted from Salas, H.J. 2000. *History and Application of Microbiological Water Quality Standards in the Marine Environment, World Health Organization (WHO).*

[a] 80% of samples less than.

[b] 95% of samples less than.

[c] Concentration to be checked by the competent authorities when an inspection in the bathing area shows that the substance may be present or that the quality of the water has deteriorated.

TABLE 16.5   Microbiological Water Quality Guidelines/Standards 100 mL

Country	Shellfish Harvesting Total Coliform	Fecal Coliform	Primary Contact Recreation Total Coliform	Fecal Coliform	Other	Protection of Indigenous Organisms Total Coliform	Fecal Coliform
USEPA, United States		14[a] 90% <43			*Enterococci* 35[a]		
California, United States	70[e]		80% <1000[i,j] 100% <10,000[k]	200[a,j] 90% <400[l]			
EEC,[b] Europe [5]			80% <500[c] 95% <10,000[d]	80% <100[c] 95% <2,000[d]	Fecal streptococci 100[c] *Salmonella* 0/L[d] Enteroviruses 0 PFU/L[d] *Enterococci* 90% <100		
UNEP [2,14]		80% <10 100% <100		50% <100[n] 90% <1000[n]			
Brazil [10]			80% <5000[m]	80% <1000[m]			
Colombia [7]			1000	200			
Cuba [8]			1000[a]	200[a] 90% <400			
Ecuador [11]			1000	200			
Mexico [12]	70[e] 90% <230		80% <1000[f] 100% <10,000[k]			10,000[e] 80% <10,000 100% <20,000	
Peru [9]	80% <1000	80% <200 200% <1000	80% <5000[f]	80% <1000[f]		80% <20000	80% <4000
Puerto Rico [4]	70[h] 80% <230			200[h] 80% <400			
Venezuela [17]	70[a] 90% <230	14[a] 90% <43	90% <1000 100% <5000	90% <200 100% <400			
France [18]			<2000	< 500	Fecal streptococci <100		
Israel			80% <1000[g]				
Japan [3]	70		1000			1000	
Poland					*E. coli* <1000		
Yugoslavia			2000				
People's Republic of China	Coli index <50[n]	14		<200[i]	Coli index <1000[i]		

Source: Adapted from Salas, H.J. 2000. *History and Application of Microbiological Water Quality Standards in the Marine Environment, World Health Organization (WHO).*

[a] Logarithmic average for a period of 30 days of at least 5 samples.
[b] Minimum sampling frequency—every two weeks.
[c] Guide.
[d] Mandatory.
[e] Monthly average.
[f] At least 5 samples per month.
[g] Minimum 10 samples per month.
[h] At least 5 samples taken sequentially from the waters in a given instance.
[i] Period of 30 days.
[j] Within a zone bounded by the shoreline and a distance of 1000 ft from the shoreline or the 30-ft depth contour, whichever is further from the shoreline.
[k] Not a sample taken during the verification period of 48 h should exceed 10,000/100 mL.
[l] Period of 60 days.
[m] "Satisfactory" waters, samples obtained in each of the preceding 5 weeks.
[n] Maximum permitted value in unit item determination.
[i] Sampling frequency no less than once a month. More than 95% of the samples in a year should accord with the standard.

per 100 mL, using a median most probable number (MPN), with no more than 10% of the values exceeding 230 total coliforms per 100 mL [13].

16.3 Summary and Conclusions

The guidelines for recreational water reuse recommended by three major organizations and several countries were presented in this chapter. As previously mentioned, the most important water quality parameters for recreational waters are the microbiological indicators. A wide variety of microbiological guidelines have been suggested and adopted by different countries and organizations for recreational waters. However, the number of epidemiological studies to justify these total and fecal coliform standards is very limited. Therefore, simple adaptation of a particular set of standards is inappropriate without a thorough review of their origin and taking into consideration the regional and social conditions.

In conclusion, as recreational water reuse is one of the major and growing water reuse applications, careful consideration should be given to existing guidelines and new and more detailed standards need to be adopted to assure environmental and human health, especially in cases with direct contact.

References

1. Canadian Council of Ministers of the Environment (CCME). 2008. *Canadian Water Quality Guidelines, Canada.*
2. Caribbean Environmental Programme (CEPPOL) and United Nations Environnent Programme (UNEP). 1991. *Report on the CEPPOL Seminar on Monitoring and Control of Sanitary Quality of Bathing and Shellfish-Growing Marine Waters in the Wider Caribbean.* Kingston, Jamaica, April 8–12, 1991. Technical Report No. 9.
3. Environmental Agency, Japan. 1981. *Environmental Laws and Regulations in Japan (III) Water.*
4. Environmental Quality Board, Puerto Rico. 1983. Regulation of Water Quality Standards, February 28, 1983.
5. European Economic Community (EEC). 1976. Council directive of December 8, 1975 concerning the quality of bathing water. *Official Journal of the European Communities.* 19, L 31.
6. Health and Welfare Canada. 1990. Guidelines for Canadian recreational water quality. *Federal-Provincial Advisory Committee on Environmental and Occupational Health,* Ottawa.
7. Ministry of Health. Colombia. 1979. Sanitary Water Provisions—Article 69 Law 05.
8. Ministry of Health, Cuba. 1986. Community hygiene, bathing places in coastal and inland bodies of water, sanitary hygienic requirements. 93–97. Havana, Cuba.
9. Ministry of Health, Peru. 1983. Amendments to Articles 81° and 82° Regulations Parts I, II and III of the General Water Law. Supreme Decree No. 007-83 -SA.
10. Ministry of Interior, Brazil. 1976. Bathing waters, Portoria No. 536.
11. Ministry of Public Health, Ecuador. 1987. Ecuadorian Institute of Sanitary Works, Draft regulations to implement the Act.
12. Ministry of Urban Development and Ecology (SEDESOL), Mexico. 1983. Ecological Legal breviary. Undersecretary of Ecology.
13. Salas, H.J. 2000. *History and Application of Microbiological Water Quality Standards in the Marine Environment,* World Health Organization (WHO).
14. United Nations Environment Programme (UNEP)/World Health Organization (WHO). 1983. *Assessment of the State of Microbial Pollution of the Mediterranean Sea and Proposed Measures,* Document UNEP/WG. 91/6, 1983.
15. USEPA. 2004. *Guidelines for Water Reuse, Report 625/R-04/108,* Washington, DC.
16. USEPA. 2012. *Guidelines for Water Reuse, 600/R-12/618,* Washington, DC.
17. Venezuela. 1978. Partial Regulations No. 4 of the organic law of the environment on water classification.
18. WHO. 1998. *Guidelines for Safe Recreational-Water Environments. Volume 1: Coastal and Freshwaters. Draft for Consultation.* WHO/EOS/98.14, Geneva.

17

State Legislation on Water Reuse

Jay Krishna Thakur
Environment and Information Technology Center

Amos T. Kabo-bah
University of Energy and Natural Resources

PREFACE

Water reclamation, recycling, and reuse have revolved over many years because of the growing pressure on water resources. Modern water reclamation strategies take into consideration efforts to minimize environmental and health risks especially with reuse applications. The future for sustainable water resources management depends on some feasibility on water reuse options for multiple water use objectives for domestic, industrial, and agricultural purposes. In order to ensure safe and efficient consideration of wastewater reclamation, recycling, and reuse, strong legislation on water reuse is necessary. This chapter researched the legislation on water reuse in selected countries and continents across the globe. The chapter envisaged a contribution to understanding the various emphases across the globe regarding wastewater reclamation, recycling, and reuse.

17.1 Introduction

The evolution of water reuse dates back as far as 1918 when there were fewer legislative issues on wastewater reuse. Perhaps, the call to attention to wastewater reclamation started when cholera outbreaks occurred across the globe [7,8,11,19,23]. This led to development of new systems and adoption of new policies to ensure that adequate and clean sanitation was acquired in all cases. Although policies and strategies were immediately put in place during this time, in some cases there were mere legal references on reclamation and reuse.

In the State of California, for example, policies on wastewater reclamation and reuse became a benchmark for water reuse. In the 1970s and 1980s, several revolutions occurred across different states in the United States and at the same time across the world. International organizations such as World Health Organization (WHO) and the World Bank began to play active roles in ensuring that countries and organizations considered the role of wastewater reclamation, recycling, and reuse as a measure toward safe and clean sanitation and a contribution toward sound public health. This prompted several sponsored studies to

compare existing laws with growing needs on wastewater reclamation and to recommend solutions [1,28,32]. Therefore, studies on progress of establishing guidelines and regulations in parts of Europe were implemented.

In the world today, challenges such as rapid population growth, industrial sprawl, and repercussions from climate change trigger crucial examination on sustainable management across both developing and developed economies to ensure every drop of water is reclaimed, recycled, and reused [5,7,8,11–19,23,24,37]. In that case, advances in the effective and reliable use of wastewater treatment systems and technologies that use reclaimed wastewater can serve as important tools for supplementing water sources as well as improving water quality levels and abating pollution problems. In developing economies, particularly in arid regions, reliable low-cost and affordable technologies for treatment and reuse are needed for protecting existing water schemes and water supplies to protect them from pollution [9]. This implies the implementation of wastewater reclamation; reuse and recycling techniques promote the preservation of scarce water resources while maintaining water conservation and sustainable watershed management. In implementing and planning water reclamation and reuse schemes, the intended water reuse program dictates the extent to which wastewater treatment plant (WWTP) quality and methods of distribution are applied [2].

In most developed economies such as the United States, Australia, and the European Union, water reclamation and reuse are well established to the extent that value of reclaimed water is fully recognized. This is because laws and citizenry have come to agree that water reuse is a paramount entity toward ensuring water quality and sustainable use among the society. For instance, several states conducted a survey to investigate the possibility of using reclaimed water for potable water purposes and freshwater [1]. Also, 18 states adopted regulations regarding the use of adopted guidelines and standards for reclaimed water [27,28]. This chapter broadly examine various legislation on water reuse across major economies across the world and makes recommendations on some of the future issues to consider in water reuse in the spate of large demand on water as a result of urban sprawl, industrialization, and scarcity of the resource.

17.2 Water Reuse and Historical Perspective

To provide a common basis for understanding what is water reuse and further to understand the term in the context of water reclamation and reuse, it is vital to expound some key terms. Water reclamation is meant to imply all the processes involved in the processing of wastewater to make it useable, and water reuse is the use of the treated wastewater for beneficial purposes including irrigation, household chores, and industrial cooling [3]. There are two ways of water reuse—direct and indirect. Indirect use is the discharge of an effluent to a receiving water body for withdrawal and assimilation downstream. Usually this type is not part of the usual direct water reuse. Direct water reuse is whereby water recycling involves one use, purpose, or user, with effluents from the user captured and redirected back into the use scheme.

The early development in the field of water reuse is parallel with the historical revolution in the practice of disposal of wastewater. With the advent of sewerage systems in the nineteenth century following severe public health crises (i.e., outbreak of cholera), domestic sewage farms were used for wastewater. The prime purpose of sewage farms was the disposal of wastewater, but later, other uses such as agricultural production were realized. Some examples of historical developments in water reuse across the globe are mentioned in Table 17.1.

The integration of planned water reclamation, recycling, and reuse in water resource management is part of the growing awareness of the scarcity of the water sources to meet ever-challenging societal demands, industrial advancement, effects of climate change, and improved knowledge of public health and risks [3,17]. It is imperative to note that the link between wastewater, reclaimed water, and reuse is becoming well understood and, hence, shorter recycle loops are possible [3]. The hydrologic cycle is the conceptual transport of water in the environment. The cycle consists of freshwater, saltwater, subsurface groundwater, water connected to land use functions, and atmospheric water vapor. In this cycle, many subcycles, including engineered transport of water, exist. Wastewater reclamation, recycling, and reuse constitute components of the hydrological cycle in industrial, urban, and agricultural areas. The major routes of water reuse include industrial use, surface water replenishment, irrigation, and groundwater recharge. Groundwater and surface water replenishment occur naturally in drainage systems and storm water runoff. The potential of reclaimed water for potable purposes is shown in Figure 17.1. The quantities of water transferred via each channel depend on the watershed properties, climatic and hydrogeological factors, and the various uses of direct and indirect water reuse. In this respect, water reuse regulations and legislations would promote safe hydrological cycles for protecting water systems against epidemics.

17.3 Overview of Legislation on Water Reuse

A brief recap on international standards is important to understand the state of legislation on water reuse in the world. In some countries such as the United States and Europe, there are some regional regulations and standards. Notwithstanding, within regional blocs such as the European Union, there are specified regulations and guidelines on wastewater reclamation, reuse, and recycling in some countries and the degree of implementation and strictness vary per country. The United States has its technical standards specified for individual limits for microorganisms and chemicals. This differentiation was well noted by the State of California, which as early as 1918 initiated efforts concerning the reuse of wastewater [1].

In most developing economies, the emphasis of legislation is on the use restriction. For instance, such regulations are a ban on the use of wastewater for irrigation for vegetables eaten raw and hence require a shorter time interval between irrigation and crop harvest. In such instances, functional agencies are required to properly ensure implementation of the legislation. There are, however, challenges involved in the regulation of use restrictions. This has led to the integration of use restrictions and easy-to-measure limits for biological and chemical parameters (i.e., biochemical oxygen demand [BOD] and chemical oxygen demand [COD]) and microorganisms. This practice has given rise to many discussions among scholars to the conclusive agreed use for wastewater.

WHO recognizes the potential and the risk involved in the use of untreated wastewater and as a result developed guidelines for policy makers to legislate permission for safe reuse of wastewater for agriculture [32,33]. The standards and guidelines from WHO were adopted and used by several countries especially in developed countries. However, in developing countries, several challenges still affect the adoption of these guidelines and regulations for using wastewater for irrigation [22].

TABLE 17.1 Example of Water Reuse between 1900 and 1990

Year	Location	Example of Water Reuse
1912–1985	Golden Gate Park, San Francisco, California	Watering lawns and supplying ornamental lakes.
1926	Grand Canyon National Park, Arizona	Toilet flushing, lawn sprinkling, cooling water, and boiler feed water.
1929	Pomona, California	Irrigation of lawns and gardens.
1942	Baltimore, Maryland	Metals cooling and steel processing at the Bethlehem Steel Company.
1960	Colorado Springs, Colorado	Landscape irrigation for golf courses, parks, cemeteries, and freeways.
1961	Irvine Ranch Water District, California	Irrigation, industrial and domestic uses, later including toilet flushing in high-rise buildings.
1962	County Sanitation Districts of Los Angeles County, California	Groundwater recharge using spreading basins at the Montebello Forebay.
1962	La Soukra, Tunisia	Irrigation with reclaimed water for citrus plants and to reduce saltwater intrusion into groundwater.
1968	Windhoek, Namibia	Advanced direct wastewater reclamation system to augment potable water supplies.
1969	Wagga Wagga, Australia	Landscape irrigation of sporting fields, lawns, and cemeteries.
1970	Sappi Pulp and Paper Group, Enstra, South Africa	Industrial use of reclaimed municipal wastewater for pulp and paper processes.
1976	Orange County Water District, California	Groundwater recharge by direct injection into the aquifers at Water Factory 21.
1977	Dan Region Project, Tel Aviv, Israel	Groundwater recharge via basins. Pumped groundwater is transferred via a 100-km long conveyance system to southern Israel for unrestricted crop irrigation.
1977	St. Petersburg, Florida	Irrigation of parks, golf courses, schoolyards, residential lawns, and cooling tower make-up water.
1984	Tokyo Metropolitan Government, Japan	Water recycling project in Shinjuku District of Tokyo providing reclaimed water for toilet flushing in 19 high-rise buildings in highly congested metropolitan area.
1985	El Paso, Texas	Groundwater recharge by direct injection into the Hueco Bolson aquifers, and power plant cooling water.
1987	Monterey Regional Water Pollution Control Agency, California	Monterey Wastewater Reclamation Study for Agriculture—agricultural irrigation of food crops eaten uncooked including artichoke, celery, broccoli, lettuce, and cauliflower.
1989	Shoalhaven Heads, Australia	Irrigation of gardens and toilet flushing in private residential dwellings.
1989	Consorci de la Costa Brava, Girona, Spain	Golf course irrigation.

Source: Adapted from Asano, T. 2002. *Water Science and Technology*. 45(8), 24.

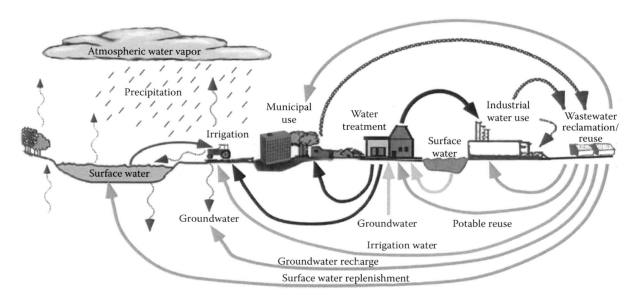

FIGURE 17.1 The role of engineered reclaimed water in the hydrologic cycle. (Adapted from Asano, T. and Levine, A.D. 1996. *Water Science and Technology*, 33(10–11), 1–16; Asano, T. 2002. *Water Science and Technology*. 45(8), 24.

In contrast to the WHO guidelines that focus mainly on the protection of human and public health, the Food and Agriculture Organization (FAO) has developed a field guide for evaluating the suitability of water for irrigation. Guideline values given identify potential problem water based on possible restrictions in use related to salinity, rate of water infiltration into the soil, specific ion toxicity, or to some other miscellaneous effects. The guide is intended to provide guidance to farm and project managers, consultants, and engineers in evaluating and identifying potential problems related to water quality. It discusses possible restrictions on the use of water and presents management options that may assist in farm or project management, planning, and operation. However, the FAO guidelines must be seen as orientation values that are in no way intended to replace case-by-case assessments.

17.3.1 Europe

Guidelines regarding reuse are made in article 12 of the European Wastewater Directive (91/271/EEC) stating: "Treated wastewater shall be reused whenever appropriate" [10]. This quest for water reuse is being driven by the need for alternative resources to protect receiving water bodies and the need to ensure stringent quality rules for discharged water. Water reuse has become a growing popular norm in populated Northern European countries such as England, Belgium, and Germany, and Western and Southern Europe [18].

The EU Water Framework Directive (WFD) noted as water scarcity increases worldwide, people would definitely look for options for water reuse [1]. This is because costs in pumping and wastewater treatment are so high that the issue of reuse becomes an alternative important way to natural management of water drainage. The WFD makes recommendations for Member States within each river basin to adopt as part of the program to promote water-efficient technologies in the industry and water-saving irrigation techniques [31].

Although the WFD does not include reuse in the body of the directive, it introduces a quantitative dimension to water management, on top of the usual qualitative dimension, which may stimulate the consideration of wastewater reuse. It also states, "Water resources should be of sufficient quality and quantity to meet other economic requirements." The Integrated Pollution Prevention Control (IPPC) legislation does encourage water reuse and is included in the legislation within the WFD [20].

17.3.2 Canada and United States

The well-known California criteria "California Code Regulations" stipulate conventional biological wastewater treatment by tertiary treatment, filtration, and chlorine disinfection to produce wastewater that is suitable for irrigation. In support of this approach, Asano and Levine [4] have reported two major epidemiological studies that were conducted in California during the 1970s and 1980s. These studies scientifically demonstrate that food crops that were irrigated with municipal wastewater reclaimed according to the California approach could be consumed uncooked without adverse health effects. However, the nutrients removed by the tertiary treatment are not available for fertilizing. Recently, California regulations have been revised to put measures that are more stringent on the domestic water supply, irrigation water quality requirements, and recycle water use specifications [25]. The State of California has provided a solid example for other states within the United States to emulate. However, the setting of actual standards, regulations, and guidelines is the responsibility of individual states and hence each state takes its own approach in specifying water quality standards for water reuse [6].

There is growing interest in water reuse applications in regions across Canada experiencing quantity and quality concerns. Some regions such as Alberta and British Columbia have produced guidelines relating to water reuse. There is, however, no clear national guidelines on wastewater reclamation and water reuse.

17.3.3 Australia

Australia is one of the driest inhabited continents on Earth. The national water consumption rates are now generally considered unsustainable due to either exceeding or nearing extraction limits. During the last decade, the country has begun to shift its focus of water management toward total water cycle management. This led to the development of new strategies to reduce the overall levels of wastewater discharged to rivers, seas, and oceans.

This has led to the development of strategies to reduce the overall amount of wastewater discharged to the ocean and rivers. Several citations demonstrate the introduction of water reuse in Australia.

The successful implementation of any reuse project hinges on public acceptance. As a concept for water management, water reuse is indeed widely accepted by the Australian community. In Western Australia, a recent focus group held by the Water Corporation of Western Australia [30] indicated that people rated the idea of using recycled water very positively. The same findings were obtained in studies on community perceptions of water reuse [26,29].

However, the widespread acceptance for water reuse in the Australian community does not automatically assume that any reuse projects will be readily accepted by the community. The general community often saw the logic in the move toward using recycled water, but felt that they themselves could not use the water. This view is again shared by the focus group participants in discussions held separately by Melbourne Water [29] and Water Corporation of Western Australia [30].

17.3.4 Africa

Over 85% of the urban dwellers in Africa, for example, Ghana, Ethiopia, Mali, and Tanzania, are served with onsite sanitation systems. Septic tank sludge regularly pollutes the environment in many cities while stormwater gutters also receive and channel greywater and other wastewater to larger drains and inner-city streams [16]. These appear in many cities as large wastewater drains absorbing in addition all kind of plastics and solid wastes.

Few cities within the Sub-Saharan Africa are sewered and about 1% of wastewater is treated [34]. The predominant challenge is low financial support, and lack of technical and managerial capacity. Existing wastewater treatment plants are often not functional, overloaded, and thus discharge into the environment. In this case, effluents are not safe for reuse. As population growth in Africa is on the rise, volumes of untreated wastewater would continue to increase and worsen the water bodies' ability to support irrigation agriculture. Industrial pollution is still limited especially with large industries. This is because of the large industries located at the seaports.

There are many challenges to wastewater reclamation, recycling, and reuse, which are varied, mixed, and differ from country to country. For instance, Namibia, Tunisia, and South Africa stand apart from the rest of the continent in treating sewage sludge through a range of conventional and nonconventional systems and having national guidelines and regulations [5].

17.3.5 China

There is increasing scarcity of freshwater resources as a result of the rising population growth and industrial activities in China. There have been efforts by the government to develop suitable reclamation processes related to water reuse to adapt to the cases of water scarcity [21]. Relevant national and local standards for reclaimed water quality are expected in China. The main users of treated wastewater include the urban recreation and agricultural irrigation sectors. Wastewater reclamation has great potential for all cities in China and other developing countries with similar conditions [35]. Due to the challenges China faced in terms of water scarcity and pollution during the last three decades, the government has launched nationwide efforts to optimize the use of reclaimed water through laws, policies, and legislations [36]. For instance, Beijing and Tianjin are some of the key cities in China using reclaimed water. It is envisaged that most cities in China would emulate the examples of these cities in wastewater reclamation, recycling, and reuse.

17.4 Summary and Conclusions

Water scarcity and water pollution are growing issues across the globe. These challenges have prompted the need to regulate and control the management of water through options such as wastewater reclamation, recycling, and reuse. This chapter examines some of the examples across the world on the legislation of water reuse. The research found that in most developed countries, efforts have been made in terms of policy and laws to regulate and promote water reuse. However, across developed economies, the emphases on the limits of effluents differ and the overall acceptability of users to water reuse also differs. In emerging developed economies such as China, wastewater reclamation, recycling, and reuse have become a core issue for water management especially in large cities such as Beijing. In Africa, wastewater

reclamation and issues of water reuse are limited often due to the lack of adequate technical and financial resources to support the management of such systems. Some countries within Africa—Namibia, South Africa—have advanced in the area of water reuse. Generally, it is strongly recommended that water reuse legislations be implemented in every country to promote sustainable water management.

References

1. Angelakis, A.N. and Durham B. 2008. Water recycling and reuse in EUREAU countries: Trends and challenges. *Desalination*, 218(1–3), 3–12.
2. Asano, T. 1998. *Wastewater Reclamation and Reuse: Water Quality Management Library*. CRC Press, Boca Raton, FL.
3. Asano, T. 2002. Water from (waste) water—The dependable water resource. *Water Science and Technology*. 45(8), 24.
4. Asano, T. and Levine, A.D. 1996. Wastewater reclamation, recycling and reuse: Past, present, and future. *Water Science and Technology*, 33(10–11), 1–16.
5. Bahri, A., Drechsel, P., and Brissaud, F. 2008. Water reuse in Africa: Challenges and opportunities: First African water week, *Accelerating Water Security for Socio-Economic Development of Africa*. Tunis, Tunisia.
6. Blumenthal, U.J., Mara, D.D., Peasey, A.A., Ruiz-Palacios, G., and Stott., R. 2000. Guidelines for the microbiological quality of treated wastewater used in agriculture: Recommendations for revising WHO guidelines. *Bulletin of the World Health Organization: The International Journal of Public Health*, 78(9), 1104–1116.
7. Colwell, R.R. 1996. Global climate and infectious disease: The cholera paradigm. *Science*, 274(5295), 2025–2031.
8. Cook, G.C. 1996. The Asiatic cholera: An historical determinant of human genomic and social structure. In: *Cholera and the Ecology of Vibrio Cholerae*, Drasar, B.S. and Forrest, B.D. (Eds.), Chapman and Hall, London, pp. 18–53.
9. Do Monte, M.H.F., Angelakis, A.N., and Asano, T. 1996. Necessity and basis for establishment of European guidelines for reclaimed wastewater in the Mediterranean region. *Water Science and Technology*, 33(10), 303–316.
10. EU. 1991. Council directive concerning urban wastewater treatment. *Official Journal of the European Communities*, No L 135/40, European Union.
11. Evans, R.J. 1988. Epidemics and revolutions: Cholera in nineteenth-century Europe. *Past and Present*, 120(1), 123–146.
12. Featherstone, D.J. 2013. The contested politics of climate change and the crisis of neo-liberalism. *ACME: An International E-Journal for Critical Geographies*, 12(1), 44–64.
13. Geels, F.W. 2013. The impact of the financial—Economic crisis on sustainability transitions: Financial investment, governance and public discourse. *Environmental Innovation and Societal Transitions*, 6, 67–95.

14. Hamidi, S. and Ewing, R. 2014. A longitudinal study of changes in urban sprawl between 2000 and 2010 in the United States. *Landscape and Urban Planning*, 128, 72–82.

15. Karanikolos, M., Mladovsky, P., Cylus, J., Thomson, S., Basu, S., Stuckler, D., Mackenbach J.P., and McKee, M. 2013. Financial crisis, austerity, and health in Europe. *The Lancet*, 381(9874), 1323–1331.

16. Keraita, B., Drechsel, P., and Amoah, P. 2003. Influence of urban wastewater on stream water quality and agriculture in and around Kumasi, Ghana. *Environment & Urbanization*, 15(2), 171–178.

17. Lamnisos, D., Anastasiou, C., Grafias, P., Panayi, A., Larkou, A., Georgiou, E., and Middleton, N. 2013. Awareness, attitudes towards wastewater reuse and perceptions of public health risks among the general public in Cyprus Demetris Lamnisos. *The European Journal of Public Health*, 23(suppl 1), ckt123.015.

18. Lazarova, V., Levine, B., Sack, J., Cirelli, G., Jeffrey, P., Muntau, H., Salgot, M. et al. 1997. Role of water reuse for enhancing integrated water management in Europe and Mediterranean countries. *Water Science and Technology*, 43(10), 25–33.

19. Lee, K. and Dodgson, R. 2000. Globalization and cholera: Implications for global governance. *Global Governance*, 6(2), 213–236.

20. O'Malley, V. 1999. The Integrated Pollution Prevention and Control (IPPC) Directive and its implications for the environment and industrial activities in Europe. *Sensors and Actuators B: Chemical*, 59(2), 78–82.

21. Pinjing, H., Phan, L., Guowei, G., and Hervouet, G. 2001. Reclaimed municipal wastewater—A potential water resource in China. *Water Science & Technology*, 43(10), 51–58.

22. Qadir, M., Wichelns, D., Raschid-Sally, L., McCornick, P.G., Drechsel, P., Bahri, A., and Minhas, P.S. 2010. The challenges of wastewater irrigation in developing countries. *Agricultural Water Management*, 97(4), 561–568.

23. Rosenberg, C.E. 2009. *The Cholera Years: The United States in 1832, 1849, and 1866.* University of Chicago Press, Illinois, USA.

24. Ruckelshaus, M., Doney, S.C., Galindo, H.M., Barry, J.P., Chan, F., Duffy, J.E., English C.A. et al. 2013. Securing ocean benefits for society in the face of climate change. *Marine Policy*, 40, 154–159.

25. Force, R.W.T. 2003. Recycled Water Task Force White Paper of the Public Information, Education, and Outreach Workgroup on Better Public Involvement in the Recycled Water Decision Process—DRAFT, Department of Water Resources, State Water Resources Control Board, Department of Health Services, Feb 4, 2003, p. 4.

26. Sydney Water. 1999. *Community Views on Re-cycled Water.* Research Report, Syndey Water (internal report), Syndey, Australia.

27. U.S. Environmental Protection Agency (EPA). 1992. *Guidelines for Water Reuse.* 625/R92004. Environmental Protection Agency, Washington, DC.

28. van Naersen, E., Mulder, J.W., Kramer, J.F., and Pluim, J. 2001. Micro- and hyperfiltration of WWTPeffluent; a new challenge in boiler feed water production. In: *Membrantechnik in der Wasseraufbereitung und Abwasserbehandlung.* 4. Aachener Tagung Siedlungswasserwirtschaft. Dohmann, M. and Melin, T. (Eds.), W9.

29. Water, M. 1998. Exploring community attitudes to water conservation and effluent reuse: A consultancy report prepared by Open Mind Group. St. Kilda, Victoria.

30. Water Corporation of Western Australia. 2003. Community attitudes and public perceptions. Paper presented at the Water Recycling Workshop, June 25–26, 2003, Perth, Australia.

31. WFD Directive 2000/60/EC. 2000. Establishing a framework for community action in the field of water policy. Annex V1, Part B, p. 64.

32. WHO. 1989. Health Guidelines for Use of Wastewater in Agriculture and Aquaculture. *Technical Report Series*, 778, World Health Organization, Geneva.

33. WHO. 2006. *Guidelines for the Safe Use of Wastewater, Excreta and Greywater—Volume 2: Wastewater Use in Agriculture.* WHO Press, Switzerland.

34. WHO & UNICEF. 2006. *Meeting the MDG Drinking Water and Sanitation Target: The Urban and Rural Challenge of the Decade.* WHO Press, Switzerland, 41 p.

35. Yang, H. and Abbaspour, K.C. 2007. Analysis of wastewater reuse potential in Beijing. *Desalination*, 212(1), 238–250.

36. Yi, L., Jiao, W., Chen, X., and Chen, W. 2011. An overview of reclaimed water reuse in China. *Journal of Environmental Sciences*, 23(10), 1585–1593.

37. Yue, W., Liu, Y., and Fan, P. 2013. Measuring urban sprawl and its drivers in large Chinese cities: The case of Hangzhou. *Land Use Policy*, 31, 358–370.

V

Environmental and Ecological Impacts/Benefits

18

Water Reuse Environmental Benefits

Zhuo Chen
University of Technology Sydney

Graduate School at Shenzhen Tsinghua University

Huu Hao Ngo
University of Technology Sydney

Wenshan Guo
University of Technology Sydney

PREFACE

With the socioeconomic development and population increase, freshwater consumption has increased beyond sustainable levels in many parts of the world. To combat water scarcity issues and environmental-related problems, recycled water has been increasingly considered an important alternative water resource. It can provide a viable opportunity to supplement existing freshwater supplies, alleviate contaminant loads exerted by effluent discharge, augment and/or improve the stream flow and wetlands, etc. This chapter examines the alternative water resources and identifies the importance as well as historical and current development of water reuse. It focuses on the environmental benefits associated with various recycled water applications. The illustrated case studies of recycled water in a number of countries, including Australia, Japan, the United States, Europe, and the Middle East, could be good examples for future projects. Notably, there is also a potential to exploit and develop new end uses of recycled water further in both urban and rural areas. This can contribute largely to freshwater savings, wastewater reduction, and water sustainability.

18.1 Introduction

Freshwater has become a scarce and overexploited natural resource in many parts of the world [34]. It has been recognized as a critical global issue and may become the most strategic resource in many areas, especially the arid and semi-arid regions. According to the International Water Management Institute's report, Australia, California, the Middle East, and the Mediterranean have been regarded as high water stress regions [26]. Likewise, the situation of water pollution and overextraction in Asia and Africa is far from optimistic. Within the next decades, due to continuous economic and population growth, climate change, rapid urbanization, and deteriorating water quality, water scarcity is likely to be a big constraint to the future food security and environmental sustainability [22,45]. The scientific definitions of water scarcity concepts include [38]: *Water scarcity* is the general collective term when water is scarce for whatever reason. *Water stress* is linked to difficulties in water use due to accessibility or mobilization problems (e.g., water infrastructures, flow control, costs). Normally, the use-to-availability ratio with values larger than 40% denotes high water stress. *Water shortage* refers to population-driven physical shortage of water when seen in relation to principal water requirements:

- Green water shortage, relating to deficiency in relation to crop water requirements

- Blue water shortage, when the number of people competing for a limited resource quantity, that is, water crowding is "high" (1000–1700 m³ per year supply per person shows water stress; less than 1000 m³ per year supply per person indicates extreme water scarcity) [38,51]

By 2025, it is projected that, assuming current consumption patterns continue, at least 3.5 billion people or 48% of the world's population, will live in water-stressed river basins. Of these, 2.4 billion will live under high water stress conditions [52,53]. Moreover, water quality deterioration has been widely observed in rural and urban areas through Africa, Asia, and Latin America [44]. More than 2.6 billion people (42% of the total population) lack access to basic sanitation facilities at present and are exposed to contaminated water sources, which contain excessive chemical compounds and high concentrations of pathogens (e.g., parasites, bacteria, and viruses) [43].

18.2 Coping with Freshwater Scarcity by Water Reuse

To meet the future water and food demand, some countries have increasingly recognized the significance of water conservation and water demand management as a long-term water supply strategy. Additionally, other sustainable management solutions and technological options in driving green growth for augmenting existing water supplies should be considered, including the exploitation of alternative water resources and the identification of the significance of water recycling and reuse.

18.2.1 Alternative Water Resources

With traditional water resources (e.g., surface water and groundwater) being depleted, it is essential to look for alternative water resources to meet the current water demand. The alternatives may include the capture and use of rainwater and stormwater as well as the exploitation of greywater, desalinated water, and recycled water.

- Desalinated water, that is, saline water, is separated into two parts using different forms of energy in desalination technologies; the potable water with a low concentration of dissolved salts is produced in one part; in the other part, the brine concentrate is collected which has a much higher concentration of dissolved salts than the original feed water [14,41].
- Greywater refers to urban wastewater that includes water from household kitchen sinks, dishwashers, showers, baths, hand basins, and laundry machines but excludes any input from toilets [29,37]; generally, it is less polluted and low in contaminating pathogens, nitrogen, suspended solids, and turbidity compared with municipal and industrial wastewaters [16]; greywater can be used immediately or treated and stored, the applications include irrigation, toilet flushing, and other purposes [30].

- Rainwater and stormwater, which can be harvested and treated locally; the harvested water can be used in place of drinking water in many functional areas for different purposes, including toilets, gardens, fire fighting, cooling towers, and manufacturing.
- Recycled water, that is, former wastewater that has been treated to remove solids and certain impurities; it is intended to be used for nonpotable uses (e.g., irrigation, industrial uses, recreational and environmental purposes, toilet flushing, car washing, dust control, and fire suppression); with more advanced treatment, it can be used for indirect potable reuse (i.e., discharge into a water body before being used in the potable water system) [4,24].

18.2.2 Importance of Water Reuse

The current severe water supply and demand situations have forced many water authorities and local councils to increasingly consider recycled water as a supplementary water supply [4]. This resource can help to alleviate the pressure on existing water supplies while protecting remaining water bodies from being polluted [24]. A distinct benefit of recycled water is the reliability of water supply all through the year for both household and local industries, which is available even in a drought. This is superior to rainfall-dependent water supplies (e.g., rainwater and stormwater) that are vulnerable to drought and infrastructure delivery problems. Ensuring water supply reliability supports the public health, quality of life, and the economic sustainability of the region. Moreover, Clean Ocean Foundation (COF) [19] has conducted a comparison between water recycling (Eastern Treatment Plant) and desalination (Wonthaggi) projects, which are both designed to deliver a similar volume of potable water to the city of Melbourne, Australia. With advanced water treatment facilities (e.g., biological filters, reverse osmosis [RO], and UV disinfection systems), both options can supply clean water of potable water quality. It was found that the capital cost of the water recycling approach is only \$1.86 billion, compared with \$3.1 billion of the desalination approach. Besides, the treatment energy cost is only 10% of desalination. With respect to environmental considerations, water recycling also produces lower CO_2 equivalent gas emission and less end of ocean outfall discharge. Likewise, Pasqualino [35] also showed that replacing potable and desalinated water with recycled water for nonpotable purposes (e.g., irrigation, industry, urban cleaning, and fire fighting) could result in lower environmental impacts in terms of acidification potential, global warming potential, and eutrophication potential.

Moreover, compared to other water resources, recycled water can contribute to a considerable wastewater reduction through reduced effluent discharges to the aquatic environment. It can also introduce some economic benefits to local government or private sectors. For instance, in Australia, irrigating vineyards at McLaren Vale with recycled water, which contains some amount of nutrients, has already brought \$120 million to the South Australia government [20]. To some extent, excessive costs on water infrastructure and energy consumption could be avoided as well [2,24]. Furthermore, when bringing recycled water and other water resources together in management, the ecological footprint of water, sewage, and drainage system could be potentially reduced by over 25% [1]. In a broader sense, water management can be further incorporated into climate change adaptation and environmental sustainable development [5,17].

18.2.3 Historical and Current Development of Water Reuse

The modern birth of recycled water application was in the mid-nineteenth century along with the prosperity of wastewater treatment technologies. Before the 1990s, 70% of reused wastewater was processed to a secondary treatment level by conventional activated sludge (CAS) methods and the effluent was only suitable for agricultural uses in less developed areas. With the rapid development of advanced wastewater treatment technologies such as membrane filtration in the last 10–15 years, the application of recycled water has been broadened from nonpotable uses (e.g., irrigation, industry, environmental flow, and residential uses) to indirect and direct potable reuses. Currently, thousands of water recycling schemes and pilot studies are being carried out worldwide with many more in the planning and construction stages [16,39]. The global water reuse capacity is projected to rise from 33.7 GL/d in 2010 to 54.5 GL/d in 2015 [36].

In developed countries, especially the cities and regions where freshwater resources are approaching the sustainable limit, recycled water would continue to be an important alternative water resource, especially for nonpotable purposes [16]. More stringent water treatment standards (e.g., tertiary treatment and additional nutrient removal) are expected to be required in most recycled water schemes. As highly advanced technologies are available for producing clean water from wastewater without adverse health effects, the focus of motivating water reuse should shift away from technological issues to environmental, social, and economic concerns [48]. While agricultural and industrial purposes are presently the dominant end uses of recycled water, urban and residential applications such as landscape irrigation, toilet flushing, and car washing are experiencing rapid development, the amount of which is likely to be as high as or much higher than that of agricultural irrigation schemes [13]. High value end uses with potential close human contact such as groundwater recharge and indirect potable reuse would be promising but still somewhat ambiguous due to strong public misgivings [17].

Comparatively, in less developed countries, owing to technical and economic constraints, a large proportion of water reuse activities still involve secondary wastewater treatment. There would be a tendency in recycled water market toward higher levels of treatment. With respect to end uses, apart from agricultural irrigation that will continue to be the major user of recycled water, other agricultural activities such as livestock consumption, using recycled water can be beneficial to alleviate freshwater stress and maintain economic development. According to these recent trends in both developed and developing areas [16], current end uses are mostly limited to a few nonpotable purposes. To meet aggressive water recycling targets, beyond the implementation of more recycled water schemes, the development of new end uses might be prospective and should be realized accordingly [17].

18.3 Environmental Benefits of Water Reuse

18.3.1 Freshwater Savings

Recycled water can satisfy many water demands, as long as it is adequately treated to ensure water quality is appropriate for use. In the United States, recycled water reuse accounts for 15% of the total water consumption, which is tantamount to save approximately 6.4 Gigaliters per day (GL/d) of fresh water [16]. Generally, the advanced treatment is required where there is a greater chance of human exposure to the water. When the water is not properly treated, health problems could arise from being exposed to recycled water, which might contain disease-causing organisms or other contaminants. Table 18.1 illustrates the possible recycled water end use categories associated with different treatment levels [47].

Notably, agricultural irrigation currently represents the largest use of recycled water throughout the world. In Australia, there are about 270 different agricultural irrigation schemes across the country, using 45–126 GL of recycled water per year. Nonetheless, considering the annual total water consumption in agriculture (6240 GL in 2011–2012), the contribution of recycled water was still small, which only accounted for 2% [9]. If higher amounts of recycled water can be properly reused, considerable freshwater savings would be achieved, especially in intensive farming systems. It is worth noting that the Shoalhaven Water's Reclaimed Water Management Scheme in New South Wales (4 GL/yr) has converted the region from dry land to dairy farm without introducing extra charges and environmental problems. Besides, the Wider Bay Water recycling scheme in rural Queensland, which used recycled water on 400 Ha of sugar cane in 2007, has resulted in the highest producing property in the district [8]. Moreover, landscape irrigation is the second largest user of recycled water in the world currently despite that the particular water demand varies greatly by geographical location, season, plants, and soil properties. The specific applications include golf courses, public parks, schools and playgrounds, and residential area landscape uses. A successful example is the Darwin Golf Course in Tasmania, Australia, where 450 Megaliters per year (ML/yr) of effluent provided by Darwin Golf Course Sewage Treatment Plant (STP) has well connected with the golf course irrigation. Part of the effluent sent to the golf course pond can be further utilized in sports fields such as Marrara Sports Complex, thereby great water savings can be achieved [8,16].

Apart from irrigation purposes, recycled water has also been successfully applied to industrial sectors, which became the third biggest contributor to recycled water consumption

TABLE 18.1 Types of Treatment Processes and Suggested Uses at Each Level of Treatment

	Types of Treatment Processes	Suggested End Uses	
Suggested water recycling treatment and reuse	Water collection system	–	Increasing level of treatment; Increasing level of human exposure (↓)
	Primary treatment sedimentation	• No uses recommended at this level	
	Secondary treatment: Biological oxidation, disinfection	• Surface irrigation of orchards and vineyards • Nonfood crop irrigation • Restricted landscape impoundments • Groundwater recharge of nonpotable aquifer • Wetlands, wildlife habitat, stream augmentation • Industrial cooling processes	
	Tertiary/advanced treatment: Chemical coagulation, filtration, disinfection	• Landscape and golf course irrigation • Toilet flushing • Vehicle washing • Food crop irrigation • Unrestricted recreational impoundment • Indirect potable reuse: Groundwater recharge of potable aquifer and surface water reservoir augmentation	

recently. The major categories associated with substantial water consumption include cooling water, boiler feed water, and industrial process water [16]. With respect to cooling water, the thermal power generation plants of MahaGenco Company at Koradi and Khaparkheda, India, reuse 110 ML/day of treated water for cooling purposes predominantly. This has become India's largest water reuse project and the company is going to use treated water constantly for the next 30 years, which will directly benefit 1 million people due to significant amount of freshwater savings [45]. In the pulp and paper mill industry, the Mondi Paper Company in Durban, South Africa, uses 47.5 ML/day of recycled water from the Durban Water Recycling Plant. As a result, great water savings in Mondi have been achieved and the water tariff has been reduced by 44% [50].

As for the metallurgical industry, the Port Kembla Steelworks in Australia, which belongs to BlueScope Steel Company, used 20 ML/day of recycled water from the Wollongong STP. The project has reduced the Port Kembla Steelworks' freshwater use by more than 50%. The new partnership with Sydney Water will further reduce the draw on freshwater from Avon Dam from 37 ML/day to 17 ML/day [12]. Similarly, the Port Kembla Coal Terminal also receives recycled water from the Wollongong STP and has been using it for dust suppression since 2009, reducing 70% of freshwater consumption [16]. Regarding the food processing industry, Matsumura and Mierzwa [31] reviewed water reuse for nonpotable applications in a poultry processing plant in Brazil. They found that prechiller effluent including continuous discharged effluent and batch-discharged effluent could be reused during chilling processes or for other nonpotable applications after ultrafiltration (UF). The water from a gizzard machine was able to be reused in inedible viscera flume as cascade water without pretreatment. Besides, wastewater from a thawing process and filer wash process might also be reused after filtration. By adopting water reuse programs, freshwater consumption was reportedly reduced by 21.9%.

Furthermore, other applications on residential and/or commercial purposes (e.g., fire protection, toilet flushing, car washing, gardening, and clothes washing) are being widely practiced. By constructing dual-reticulation pipe systems for water supply in several urban residential areas, substantial water savings has been achieved in many developed countries, including Australia, Japan, the United States, the United Kingdom, and Germany [6]. In Australia, one representative example is the Water Reclamation and Management Scheme (WRAMS) in Sydney Olympic Park. It has extended the urban water recycling concepts to integrated water management by incorporating both stormwater and recycled water in recycled water delivery systems. The novel stormwater reservoir design enabled stormwater from the Olympic Park and excess secondary effluent from STP to be stored and regulated so that the subsequent Water Reclamation Plant (WRP) can be operated at any rate to cope with large events. Up to 7 ML/day of recycled water under microfiltration (MF), UV, and super-chlorination was used for toilet flushing and open space area irrigation at sporting venues in Olympic Park, saving 850 ML/yr of Sydney's freshwater supply. The additional recycled water also served 2000 residential houses in Newington in terms of toilet flushing and garden watering. Recently, the end uses have been expanded to over 11 types, including swimming pool filter backwash and ornamental fountains [15].

In addition to centralized dual pipe systems, in the United States, the first large-scale onsite water recycling system was conducted at the Solaire Building (293 units) in New York City.

The wastewater treatment system, located in the basement, uses membrane bioreactor and UV disinfection to treat more than 95 ML/d of wastewater, of which 34 ML/d is for toilet flushing, 43.5 ML/d is used as makeup water for the building's cooling towers, and 22.7 ML/d is for landscape irrigation. The system has reduced the freshwater and energy consumption by 75% and 35%, respectively [3]. Remarkably, some sustainable management or control approaches (e.g., selection of advanced irrigation methods, adoption of additional treatment methods, and/or increase of capital and maintenance costs) in existing or future recycling schemes can be carried out to reduce the freshwater consumption and overall environmental footprint further.

18.3.2 Pollution Load Reduction

In some cases, the main impetus for water reuse comes not from a water supply need, but from a need to eliminate or decrease environmental loads exerted by effluent discharge to the ocean, an estuary, or a stream [47]. This strength of recycled water is distinct as many studies have already demonstrated massive adverse effects on aquatic sensitive ecosystems from wastewater effluent in terms of nutrients pollution, temperature disturbance, and salinity increase. For instance, the Rouse Hill Water Recycling Scheme, located in northwestern Sydney, is one of the most successful schemes in Australia. The primary objective of the scheme is to reduce the nutrient loads on the Hawkesbury-Nepean River system, which was caused by discharge of treated wastewater. It started in 2001 and uses up to 2.2 GL/yr of recycled water, serving over 19,000 homes. At the same time, it also helps to care for the environment and reduce impacts on waterways. The amount of treated wastewater discharged to the Hawkesbury-Nepean River has been largely reduced. In 2009, the upgrade of the plant enabled up to 4.7 GL/yr of wastewater to be recycled for residential use and can eventually serve 36,000 homes in the area [32]. Besides, the Illawarra wastewater recycling plant in Wollongong, Australia, is one of the most advanced coastal treatment plants in the world. The project was designed to improve water quality at Illawarra beaches, particularly those near STPs, and supply high-quality recycled water for industrial reuse. Presently, the plant adopts microfiltration and RO technologies to recycle 20 ML/day of sewage for reuse by BlueScope Steel at Port Kembla. The scheme facilitates protection of coastal waters, reduction of impact on sensitive marine ecosystems, and minimization of the effluents' negative impacts when released to the environment [49].

Shiratani [42] performed scenario analyses of recycling of sewage treated water into agricultural in order to reduce pollutant load discharged into the Aburagafuchi Lake in Japan. Since the lake was heavily polluted with organic matter, reducing pollutant load discharged into the lake has become the most important measure to improve the water environment. The results showed that irrigating paddy fields with the sewage-treated water could contribute to conserving water and reducing pollutant load, with reduction rate in BOD, nitrogen, and phosphorus ranging from 6%–36%, 16%–46%, and 18%–51%, respectively. Particularly, the results indicated that irrigating paddy fields with the treated water during noncropping periods and the accompanying reduction in withdrawn water from the river were more effective in reducing pollutant loads discharged into the lake.

Additionally, in the United States, the city of San Jose began implementing the South Bay Water Recycling (SBWR) Program in order to comply with the San Jose/Santa Clara Water Pollution Control Plant (SJ/SC WPCP)'s National Pollutant Discharge Elimination System Permit. Since most of the Bay Area's treated wastewater is discharged to the San Francisco Bay, which is a 303(d) listed impaired water body, the area's natural saltwater marsh might be threatened. Water reuse will result in direct water quality benefits for the Bay. The project can mitigate the effluent flows from the SJ/SC WPCP, thereby reducing the load of pollutants that enter the Bay. In 2008, the SBWR program had the capacity to provide 3363 million gallons per year of recycled water to over 500 customers for nonpotable purposes, such as agriculture, industrial cooling and processing, and irrigation of golf courses, parks, and schools. By avoiding the conversion of saltwater marsh to brackish marsh, the habitat for two endangered species can be protected [25,47].

Moreover, groundwater recharge with recycled water can reduce the decline of groundwater levels, dilute, filtrate, and store recycled water, partially prevent saltwater intrusion, and mitigate subsidence [7]. Presently, aquifer recharge with recycled water has been implemented in some areas in order to alleviate problems of falling groundwater tables. For example, in Mosman Peninsula, Western Australia, two golf courses are impacted by saltwater intrusion into their irrigation bores. Local government and residential groundwater users have also been affected. There has been consistent demand, particularly from the Town of Mosman Park, for a good and sustainable source of water. Consequently, a pilot study named Mosman Peninsula aquifer recharge scheme has been conducted, in which the superficial aquifer on the peninsula could be used to store reclaimed water over winter for maintaining the saltwater interface as well as summer irrigation use [11].

In the United States, one of the largest groundwater replenishment (GWR) systems in the world has been established in the Orange County Water District (OCWD), California, in 2007. The GWR system purifies highly treated wastewater through microfiltration, RO, UV disinfection, and hydrogen peroxide technologies. Half of the treated water is injected into OCWD's seawater intrusion barrier wells along the Pacific coastline; the other half is provided to groundwater spreading basins in Anaheim. The project has three growth stages with production rates of 265, 321, and 474 ML/day in 2008, 2010, and 2020, respectively. By 2020, GWR will be capable of supplying approximately 22% of total water demand in OCWD. Other environmental benefits of GWR system include protection from seawater intrusion, elimination of the need for additional ocean outfall, improvement of groundwater quality by decreased mineral levels, and more cost-effective and energy-efficient compared to water importation from northern California [33]. Similarly, in the Middle East, a

TABLE 18.2 Examples of Water Reuse for the Purpose of Stream Flow Augmentation

Location	Wastewater Treatment	Description	Motivation
San Antonio River (San Antonio, Texas), 2000–present	San Antonio Water System operates three WRCs. Dos Rios and Leon Creek WRCs are conventional activated sludge facilities. Medio Creek WRC employs an extended aeration process. Tertiary treatment includes filtration and disinfection (chlorination and dechlorination, except for Medio Creek WRC, which uses UV disinfection).	Recycled water replaces the use of groundwater for instream flow at the downtown River Walk attraction, which also flows through a city park and zoo, at three new discharge locations. Monitoring shows improved water quality and the return of sensitive, pollutant-intolerant species.	By the mid-1950s, headwater reaches near downtown were dry due to groundwater pumping. Reach downstream of WWTP discharge was considered a 40-mile (64-km) "dead zone" due to poor water quality. A city water recycling goal was to improve area streams by maintaining flows.
Salado Creek (San Antonio, Texas), 2001–present		Water quality monitoring and fish surveys were conducted before and after augmentation began, and improved water quality was observed. Creek was removed from the 303(d) List of impaired and threatened waterbodies for DO impairment.	See above. Impaired stream with low DO levels and occasional high fecal coliform. Community desired reliable base flow. Future discharge is under consideration at San Pedro Creek.
Bell Creek (Sequim, Washington), 2001–present	Sequim WRF. Includes tertiary treatment with UV disinfection and aeration via cascade structure.	Recycled water is discharged to maintain benthic species and improve salmon habitat (0.06 mgd or 250 m³/day).	City Council Water Reuse Task Force identified enhancement of Bell Creek as the number one alternative (followed by irrigation). Flow is committed to improve salmon habitat year-round.
Hillsborough River, Tampa, Florida (not implemented)	Howard F. Curren Advanced WWTP. Proposed tertiary treatment with aeration and UV disinfection.	Tampa Bay Downstream Augmentation Project was not implemented in part due to public concerns about discharge quality, including PPCPs. Future augmentation (Alafia River) is being considered.	To use recycled water from the City of Tampa to augment river flows and allow upstream potable withdrawal to increase by an equal amount.
San Luis Obispo Creek (San Luis Obispo, California), 1994–present	San Luis Obispo WRF. Primary, secondary with nitrification, and tertiary treatment with filtration and chlorination. Dechlorination and cooling tower prior to discharge.	WRF produces 3.6 mgd (14,000 m³/day) recycled water, of which a minimum of 1.6 mg/d (6000 m³/day) is released to creek at historical outfall. Creek habitat depends on recycled water discharge, which dominates summer flow.	Not originally intended as environmental enhancement. Observation of improved water quality following recycled water discharge to creek and presence of endangered species led to greater use for stream flow over landscaping and industrial uses.
Tossa de Mar Creek (Tossa de Mar, Spain), 1997–present	Tossa de Mar Water Reclamation Plant (coagulation, flocculation, sedimentation, filtration, UV + chlorine disinfection).	Tertiary-treated recycled water from an artificial pond percolates through soil to the creek, preventing the creek from becoming dry in the summer and providing ecological benefits.	To use recycled water from the WWTP to establish vegetation and create a park using marginal land located between the WWTP and Tossa de Mar Creek.
Nobidome Stream (Tokyo, Japan), 1984–present	Tamagawa-Johryu WWTP includes rapid sand filtration, partial P removal; chemical coagulation and ozonation added in 1989–1991.	Recycled water is viewed as an attractive water supply for stream augmentation in urban areas, as well as for creation of artificial streams.	Once an attractive riverine area of a Tokyo suburb, the stream dried when the headwaters were diverted in 1976.
Multiple rivers (Tokyo, Japan), 1995–present	Ochiai WWTP process includes tertiary treatment by rapid sand filtration.		Low flow or dry rivers (Shibuyahawa, Furukawa, Nomikawa, and Megurogawa Rivers) due to rapid urbanization.

Source: Adapted from Plumlee, M.H., Gurr, C.J., and Reinhard, M. 2012. *Science of the Total Environment*, 438, 541–548; Eckhardt, G. 2004. *Water Environment Federation*, 122–142; Latino, B.R. and Haggerty, A. 2007. *NC AWWA-WEA 87th Annual Conference. North Carolina American Water Works Association and North Carolina Water Environment Association*, Charlotte, NC; Asano, T., Burton, F.L., Leverenz, H.L., Tsuchihashi, R., and Tchobanoglous, G. 2007. *Water Reuse—Issues, Technologies and Applications*. McGraw-Hill, New York; Sala, L. and de Tejada, S.R. *Water Practice & Technology*, 3(2), doi:10.2166/wpt.2008.045; Yamada, K., Matsushima, O., and Sone, K. 2007. *6th IWA Specialist Conference on Wastewater Reclamation and Reuse for Sustainability*. International Water Association, Antwerp, Belgium.

DO = dissolved oxygen; mgd = million gallons per day; PPCPs = pharmaceuticals and personal care products; UV = ultraviolet; WRC = water recycling center; WRF = water reclamation facility; WWTP = wastewater treatment plant.

groundwater recharge project for seawater intrusion barrier as well as groundwater replenishment for agricultural irrigations is presently implemented in Salalah, Oman, where 20 ML/day of tertiary treated effluent is discharged to a series of recharge wells to form a barrier against seawater intrusion [16,27].

18.3.3 Augmentation of Stream Flow and Enhancement of Wetlands

Water reuse for stream flow augmentation (where "stream" refers to, in order of decreasing flow, a river, creek, or brook) or for constructed wetlands, ponds, or lakes has the potential to improve stream and wetland habitat and increase potable water supply. In contrast to traditional discharge of wastewater in which a site is selected for the purpose of disposal and any benefits are incidental, water reuse for stream flow restoration or augmentation is sited and designed to renew urban streams for environmental, ecological, societal, or other community benefit [37]. Using recycled water for environmental applications may allow utilities to obtain the full benefit of current and future recycled water supplies.

An increasing number of cities are considering or implementing these projects for a variety of reasons. In Australia, 17 inland STPs in Sydney, New South Wales, discharge recycled water into the Hawkesbury-Nepean River System where water supply dams and weirs have been built in the upper catchment. To release reliable environmental flows and protect the health of the downstream river, these STPs have been upgraded to advanced tertiary standards since 2004. Typically, the new St. Marys Water Recycling Plant in the west of the city is now in operation as the first of its kind in the world. Tertiary-treated wastewaters from the Penrith, St. Marys, and Quakers Hill STPs are transferred to this plant and undergo additional UF, RO, decarbonation, and disinfection processes. The recycled water is released to the Hawkesbury-Nepean River, providing 18 GL/yr of water for environmental flow regulation. This represents a very large savings of freshwater, as this flow would otherwise have been provided by freshwater from the Warragamba Dam. Until now, due to the high water quality requirement and limited exposure to the public, most of the environmental-related schemes have been successfully implemented and neither adverse environmental impacts nor human health problems have been identified [18].

In the United States, Halaburka [23] assessed the economic and ecological merits of stream flow augmentation using tertiary-treated recycled water in a California coastal stream. Compared to a direct ocean outfall discharge of secondary-treated effluent with no beneficial water reuse, numerous benefits were found by inland discharge of recycled water, including recreation, aesthetics, and habitat for native or endangered species. For example, the pedestrian path following the rehabilitated section of Calera Creek provides recreational benefit to people who use it for walking, biking, bird watching, and dog walking. The recreational value was estimated at $10.20 per visitor per recreation day for the base-case. The restoration also improved the aesthetics of the neighborhood, which can have a significant positive effect on housing prices. Further, the creek provides habitat for a number of native plant and animal species. The value of habitat protection can be substantial and in some cases is the most valuable ecosystem service of a stream restoration. Table 18.2 summarizes other projects in the United States and internationally. Generally, the main motivation of these projects has been to provide necessary stream flow for ecological benefit and/or to restore an unsightly, dry, or low-flow creek for aesthetic and recreational benefit in an urban or semi-urban community. Meanwhile, they can also contribute to decreased use of potable water supplies, reduced volume of wastewater discharge, and outfall contaminant loads [37]. Noticeably, three common issues, namely temperature, nutrients, and trace metals and organic contaminants, are consistently associated with urban stream degradation and are key potential hindrances to the reuse of water for stream flow augmentation [10].

With respect to wetlands, they have many noteworthy functions, such as flood attenuation, wildlife and waterfowl habitat, aquifer recharge, fisheries breeding grounds and water quality enhancement. For wetlands that have been impaired or dried from water diversion, water flow can be augmented and/or enhanced with recycled water to sustain and improve the aquatic and wildlife habitat. For instance, in the United States, recycled water from the Iron Bridge Plant was supplied to a wetland, breeding hundreds of aquatic animals and plants. After that, it was further discharged into St. Johns River in Orlando, Florida [16,47].

18.4 Summary and Conclusions

Water reuse has shown to be effective and successful in creating a new and reliable water supply while not compromising public health. As water demands and environmental needs grow, recycled water becomes an important water supply in many countries. An apparent environmental benefit of recycled water is to offset the use of potable water supplies (e.g., surface water or groundwater released from reservoirs or impoundments) being used for agricultural, industrial, residential, or commercial purposes at the site. Moreover, recycled water is beneficial to reduce the volume of wastewater discharge at the local rivers or ocean outfall locations, helping to minimize outfall contaminant loads and meet regulatory or permit requirements. Additionally, recycled water for stream flow augmentation and wetland enhancement applications has provided substantial recreational and aesthetic benefits to the community as well as environmental benefits to the local habitats. By working together to overcome obstacles, water reuse, along with water conservation and other sustainable water management strategies such as the development of green technologies, can facilitate conservation of our vital environment.

References

1. Anderson, J. 2003. Walking like dinosaurs: Water, reuse and urban jungle footprints. Water Recycling Australia. *AWA 2nd National Conference*, September 1–2, Brisbane, Australia.

2. Anderson, J., Adin, A., Crook, J., Davis, C., Hultquist, R., Jimenez-Cisneros, B., Kennedy, W. et al. 2001. Climbing the ladder: A step approach to international guidelines for water recycling. *Water Science and Technology*, 43, 1–8.

3. Applied Water Management Group. 2010. The Solaire wastewater treatment system. http://www.amwater.com/files/AMER0158_Project%20Sheets_Solaire-2.22.pdf.

4. Asano, T. 2001. *Water from Wastewater—The Dependable Water Resource*. Stockholm Water Prize Laureate Lecture, Stockholm, Sweden.

5. Asano, T. and Bahri, A. 2011. Global challenges to wastewater reclamation and reuse. *On the Water Front*, 2, 64–72.

6. Asano, T., Burton, F.L., Leverenz, H.L., Tsuchihashi, R., and Tchobanoglous, G. 2007. *Water Reuse—Issues, Technologies and Applications*. McGraw-Hill, New York.

7. Asano, T., and Cotruvo, J.A. 2004. Groundwater recharge with reclaimed municipal wastewater: Health and regulatory considerations. *Water Research*, 38, 1941–1951.

8. Australian Academy of Technological Sciences and Engineering. 2004. Water Recycling in Australia. http://www.atse.org.au/Documents/reports/water-recycling-in-australia.pdf

9. Australian Bureau of Statistics. 2013. Water Account, Australia 2011–12. http://www.abs.gov.au/AUSSTATS/abs@.nsf/Lookup/4610.0Main + Features202011-12.

10. Bischel, H.N., Lawrence, J.E., Halaburka, B.J., Plumlee, M.H., Bawazir, A.S., King, J.P., McCray, J.E., Resh, V.H., and Luthy, R.G. 2013. Renewing urban streams with recycled water for streamflow augmentation: Hydrological, water quality, and ecosystem services management. *Environmental Engineering Service*, 30(8), 455–479.

11. Blair, P.M. and Turner, N. 2004. Groundwater: A crucial element of water recycling in Perth, Western Australia. In: *WSUD 2004: Cities as Catchments, International Conference on Water Sensitive Urban Design, Proceedings of Barton, ACT*. Engineers Australia, 2004, 451–460.

12. Bluescope Steel. 2006. Community, Safety and Environment Report 2006. http://csereport2006.bluescopesteel.com/richmedia/BlueScope_Steel_CSE_Report_2006.pdf.

13. Brissaud, F. 2010. Technologies for water regeneration and integrated management of water resources. In: Sabater, S. and Barceló, D. (eds.), *Water Scarcity in the Mediterranean: Perspectives Under Global Change*. Springer-Verlag, Heidelberg, Germany.

14. Buros, O.K. 2000. *The ABCs of Desalting*, 2nd Edition. International Desalination Association, Massachusetts.

15. Chapman, H. 2006. WRAMS, sustainable water recycling. *Desalination*, 188, 105–111.

16. Chen, Z., Ngo, H.H., and Guo, W.S. 2013. A critical review on the end uses of recycled water. *Critical Reviews in Environmental Science and Technology*, 43, 1446–1516.

17. Chen, Z., Ngo, H.H., Guo, W.S., Lim, R., Wang, X.C., O'Halloran, K., Listowski, A. et al. 2014. A comprehensive framework for the assessment of new end uses in recycled water schemes. *Science of the Total Environment*, 470–471, 44–52.

18. Chen, Z., Ngo, H.H., Guo, W.S., and Wang, X.C. 2013. Analysis of Sydney's recycled water schemes. *Frontiers in Environmental Science and Engineering*, 7, 608–615.

19. COF (Clean Ocean Foundation). 2008. Desalination vs. Recycling. http://www.watershedvictoria.org.au/content/wp-content/uploads/2008/11/clean-ocean-foundation-desalination-vs-recycling.pdf.

20. Department of Environment and Natural Resources. 2010. Water reuse. Department of Water, Land & Biodiversity Conservation, Fact sheet 2, South Australia, Australia.

21. Eckhardt, G. 2004. The San Antonio River: Environmental restoration through streamflow augmentation. Proceedings of the Water Environment Federation, WEFTEC 2004: Session 71 through Session 80, pp. 122–142(21). Water Environment Federation.

22. Gohari, A., Eslamian, S., Mirchi, A., Abedi-Koupaei, J., Bavani, A.M., and Madani, K. 2013. Water transfer as a solution to water shortage: A fix that can backfire. *Journal of Hydrology*, 491, 23–39.

23. Halaburka, B.J., Lawrence, J.E., Bischel, H.N., Hsiao, J., Plumlee, M.H., Resh, V.H., and Luthy, R.G. 2013. Economic and ecological costs and benefits of streamflow augmentation using recycled water in a California coastal stream. *Environmental Science and Technology*, 47, 10735–10743.

24. Huertas, E., Salgot, M., Hollender, J., Weber, S., Dott, W., Khan, S., Schafer, A., Messalem, R., Bis, B., Aharoni, A., and Chikurel, H. 2008. Key objectives for water reuse concepts. *Desalination*, 218(1–3), 120–131.

25. ICF Jones & Stokes. 2009. Draft environmental assessment/initial study-mitigated negative declaration for the South Bay advanced recycled water treatment facility. http://www.usbr.gov/mp/nepa/documentShow.cfm?Doc_ID=4828.

26. International Water Management Institute. 2006. *Insights from the Comprehensive Assessment of Water Management in Agriculture*. Paper presented at Stockholm World Water Week, Colombo, Sri Lanka.

27. Jimenez, B. and Asano, T. 2008. *Water Reuse: An International Survey of Current Practice, Issues and Needs*. IWA publishing, London, UK.

28. Latino, B.R. and Haggerty, A. 2007. Utilizing reclaimed water for stream augmentation in North Carolina. In: *NC AWWA-WEA 87th Annual Conference. North Carolina American Water Works Association and North Carolina Water Environment Association*, Charlotte, NC.

29. Li, F.Y. 2009. Review of the technological approaches for grey water treatment and reuses. *Science of the Total Environment*, 407, 3439–3449.

30. Leflaive, X. 2007. Alternative ways of providing water, emerging options and their policy implications. http://www.oecd.org/env/resources/42349741.pdf.

31. Matsumura, E.M. and Mierzwa, J.C. 2008. Water conservation and reuse in poultry processing plant—A case study. *Resources Conservation and Recycling*, 52, 835–842.

32. NSW Office of Water. 2010. *2010 Metropolitan Water Plan*. The Department of Environment, Climate Change and Water, Sydney, Australia.

33. Orange County Water District. 2008. Groundwater replenishment system. http://www.gwrsystem.com/about-gwrs.html.

34. Page, G., Ridoutt, B., and Bellotti, B. 2011. Fresh tomato production for the Sydney market: An evaluation of options to reduce freshwater scarcity from agricultural water use. *Agricultural Water Management*, 100, 18–24.

35. Pasqualino, J.C., Meneses, M., and Castells, F. 2011. Life cycle assessment of urban wastewater reclamation and reuse alternatives. *Journal of Industrial Ecology*, 15, 49–63.

36. Pearce, G.K. 2008. UF/MF pre-treatment to RO in seawater and wastewater reuse applications: A comparison of energy costs. *Desalination*, 222, 66–73.

37. Plumlee, M.H., Gurr, C.J., and Reinhard, M. 2012. Recycled water for stream flow augmentation: Benefits, challenges, and the presence of wastewater-derived organic compounds. *Science of the Total Environment*, 438, 541–548.

38. Rockström, J., Falkenmark, M., Karlberg, L., Hoff, H., Rost, S., and Gerten, D. 2009. Future water availability for global food production: The potential of green water for increasing resilience to global change. *Water Resources Research*, 45, 1–16.

39. Rodriguez, C., Cook, A., Buynder, P.V., Devine, B., and Weinstein, P. 2007. Screening health risk assessment of micropullutants for indirect potable reuse schemes: A three-tiered approach. *Water Science and Technology*, 56, 35–42.

40. Sala, L. and de Tejada, S.R. Use of reclaimed water in the recreation and restoration of aquatic ecosystems: Practical experience in the Costa Brava region (Girona, Spain). *Water Practice & Technology*, 3(2), 1–8, doi:10.2166/wpt.2008.045.

41. Shatat, M., Worall, M., and Riffat, S. 2013. Opportunities for solar water desalination worldwide: Review. *Sustainable Cities and Society*, 9, 67–80.

42. Shiratani, E., Munakata, Y., Yoshinaga, I., Kubota, T., Hamada, K., and Hitomi, T. 2010. Scenario analysis for reduction of pollutant load discharged from a watershed by recycling of treated water for irrigation. *Journal of Environmental Sciences*, 22(6), 878–884.

43. Törnqvist, R., Norström, A., Kärrman, E., and Malmqvist, P.A. 2008. A framework for planning of sustainable water and sanitation systems in peri-urban areas. *Water Science and Technology*, 58(3), 563–570.

44. Trevett, A.F., Carter, R.C., and Tyrrel, S.F. 2004. Water quality deterioration: A study of household drinking water quality in rural Honduras. *International Journal of Environmental Health Research*, 14(4), 273–283.

45. Uday, G.K. and Kalyanaraman, B. 2012. City of Nagpur and MSPGCL Reuse Project 2012. *Guidelines for Water Reuse*, Appendix E International Case Studies. http://www.reclaimedwater.net/data/files/229.pdf.

46. UN Water. 2008. Status report on integrated water resources management and water efficiency plans. In: *The 16th Session of the Commission on Sustainable Development*. Economic and Social Council, New York.

47. US EPA (United States Environmental Protection Agency). 1998. Water recycling and reuse: The environmental benefits. http://www.epa.gov/region9/water/recycling/brochure.pdf.

48. Van der Bruggen, B. 2010. The global water recycling situation. In: *Sustainable Water for the Future, Volume 2-Water Recycling versus Desalination*. Elsevier, Amsterdam.

49. Veolia Water. 2014. Illawarra wastewater recycling scheme, Sydney Water–Wollongong, NSW. http://www.veoliawaterst.com.au/medias/case-studies/case_illawarra.htm.

50. Victoria University. 2008. *Guidance for the Use of Recycled Water by Industry*. Victoria University and CSIRO, Land and Water, Melbourne, Australia.

51. Vörösmarty, C.J., Lévêque, C., and Revenga, C. 2005. Fresh water. In: R. Hassan, R. Scholes, and N. Ash (eds.), *Ecosystems and Human Well-Being: Current State and Trends*, Volume 1. Island Press, Washington.

52. Wang, X.C. and Jin, P.K. 2006. Water shortage and needs for wastewater re-use in the north China. *Water Science & Technology*, 53, 35–44.

53. Water Resources Institute. 2001. Major watersheds of the world. http://www.wri.org/publication/watersheds-world.

54. Yamada, K., Matsushima, O., and Sone, K. 2007. Reclaimed wastewater supply business in Tokyo and introduction of new technology. In: *6th IWA Specialist Conference on Wastewater Reclamation and Reuse for Sustainability*. International Water Association, Antwerp, Belgium.

19

Ecological Impact of Water Reuse

Alexandros I. Stefanakis
University of Brighton

PREFACE

Water reuse is considered today one of the most desirable and attractive management practices in urban water management, applied worldwide. Although the economic and environmental advantages of this practice have been well documented, the overall impact on the environment is not understood in detail. Water reuse provides a series of ecological benefits like water resources conservation, freshwater ecosystems conservation, and aquatic life protection. However, if not properly and carefully applied, it may cause adverse ecological impacts. The major risk has to do with the potential redistribution of various pollutants still present in treated water and wastewater. Among them, nutrients like nitrogen and phosphorus and, especially, emerging micropollutants represent the greatest threat for the ecosystems. An ecological risk assessment appears as a vital prerequisite for the evaluation of the potential ecological impacts of water reuse and the implementation of integrated water resources management.

19.1 Introduction

Water reuse appears today as one of the most desirable and attractive management practices in urban water management. Treated water and wastewater from urban areas is usually discharged to surface bodies, like lakes, rivers, and streams. Direct recharge of underground aquifers is also an option. Reuse and recycling of treated water and wastewater through surface and groundwater bodies has been proved a valuable additional water source around the world.

When dealing with water reuse, the main concern is usually public health protection, hence very stringent limits have been adopted (depending on the use) [38]. Each water reuse application is usually accompanied by a dual characterization: as an environmental friendly practice which serves the sustainable water management approach and as an economically beneficial practice [4]. However, such applications possess a certain level of ecological risks and the implementation of water reuse projects could potentially result in a negative environmental impact, which should always be taken into account. Therefore, water reuse applications should ensure that the ecological risk taken remains within acceptable limits and is outweighed by the respective benefits. This chapter focuses on the global ecological impact, that is, the variety of ecological benefits and risks, associated with the practice of water reuse.

19.2 Environmental Drivers for Water Reuse and Applications

Probably the most important driver toward water reuse today is water scarcity. Although there is a general belief that mostly developing countries located in tropical regions face water problems, this is partially true. More than half of the European countries today face water stress issues [6,17]. Thus, finding an additional source of water appears as a necessary alternative worldwide [25], which could enhance the available water supplies [19]. Moreover, most populated parts around the world suffer from critical impacts on freshwater bodies. Overabstraction of freshwater ecosystems for agricultural irrigation resulted in significant deterioration of the status of surface water bodies (lakes and rivers) and aquifers. Pollution of receiving water bodies is also a major problem.

Generally, the practice of water reuse exists worldwide. The starting point that enables water reuse applications could be different from country to country, depending on the specific economic, social, and political parameters [20]. Although there can be some differences between the drivers among developed and developing countries, the major factors remain the lack of sufficient water supplies, high local demands, and the need for a reliable water source.

These facts, coupled with the gradual increase of environmental awareness and responsibility in modern societies, highlight the challenge and the need to find a more sustainable and reliable way to cover human needs for water, with less damage to the environment. New water policies and regulations are formed or old policies are upgraded to present new directions for an integrated resource management, which serves the aspect of sustainable development and environmental protection. The European Water Framework Directive [10] and the Millennium Development Goals [24] aim exactly at these goals, that is, to establish the perception of water reuse as an effective and acceptable solution to water-related issues and as an environmentally friendly approach. To this, despite the fact that water reuse practices are applied in many countries worldwide, the reused water volume still represents only a small fraction.

19.3 Water Reuse Applications

The overall environmental impact of water reuse practices is strongly connected to the respective applications. Reuse possibilities include a variety of potential applications [3,20]:

- Nonpotable use in urban areas, for example, street cleaning, toilet flushing, fire protection, car washing, sport fields, etc.
- Agricultural use (irrigation)
- Irrigation of nurseries, golf courses, urban parks, etc.
- Industrial use (process water)
- Artificial recharge of groundwater
- Recreational and environmental use
- Restoration of surface water bodies and wetlands
- Landscape irrigation
- Enhancement of drinking water reservoirs

Among these, agricultural reuse appears as the most important and most widespread application in all aspects (volume used, worldwide extension) because this application itself is one of the most water consuming worldwide. All of these applications

include a series of important economic and social values. The challenge, however, that appears today is to overcome possible public health issues and ecological impacts that may arise from this practice.

19.4 Environmental Benefits of Water Reuse

Water reuse as a practice generally possesses a positive meaning; it aims at a series of environmental benefits that make it acceptable and desirable. A fundamental principle in water reuse is that the recycled water, no matter the application, should have a quality at least equal or even higher to that of the effluents of water/wastewater treatment facilities. Considering that this principle is always (more or less) fulfilled, water reuse offers various environmental benefits, which are summarized in Table 19.1.

Treated water/wastewater represents a valuable alternative water source, especially in areas suffering from water scarcity. In the urban environment, where water needs are high, water reuse serves as a substitution to potable water for nonpotable uses. Limiting the volumes of public water supplies used for activities taking place within the urban environment, for example, reuse in the industrial processes, urban irrigation, etc., not only decreases the potable water consumption, but also adds value to the water as a resource. Water reuse for irrigation purposes probably represents the most important application, due to the large and increasing volumes of water consumed. Nutrient-rich water reuse in agriculture or landscape irrigation can also limit

TABLE 19.1 Environmental Benefits of Water Reuse Applications

1.	The volume of reused water is considered an additional, new component of the water balance
2.	Reduction of net water volume consumption
3.	Reduction of potable water use for nonpotable uses
4.	Water resources conservation
5.	Conservation of freshwater ecosystems
6.	Saving part of the freshwater stored in dams and reservoirs
7.	Limits the need for groundwater abstraction
8.	Maintenance of physical and chemical status of underground and surface water bodies
9.	Reduces the pollution of receiving bodies
10.	Protection of aquatic species and terrestrial animals
11.	Restoration of natural habitats, for example, wetlands
12.	Reduces the use of artificial/chemical fertilizers
13.	Prevents saltwater intrusion to freshwater habitats, for example, to aquifers, lakes, etc.
14.	Reduces the need for desalination facilities in areas with intense water scarcity problems
15.	Energy conservation—reduced greenhouse gas emissions
16.	Ecological responsibility and awareness—improved water quality in receiving waters/habitat

Source: Adapted from Anderson, L. 2003. *Water Science and Technology,* 3(4), 1–10; Miller, G.W. 2006. *Desalination,* 187, 65–75; Davis, C.K. 2008. *Water Reuse: An International Survey of Current Practice, Issues and Needs.* IWA Publishing, London.

the use of artificial fertilizers. The reuse of treated water/wastewater becomes even more attractive in countries with arid climate, in areas facing population growth, and in regions under water stress conditions. Figure 19.1 schematically depicts the various environmental beneficial uses of treated water/wastewater within the hydrological cycle.

On the other hand, overabstraction of surface and underground water supplies has led to the shrinking of aquifers, which, in some cases, (coastal line) results in saltwater intrusion. Groundwater is used worldwide not only to cover increasing irrigation needs, but also to provide more than 30% of the urban water supplies. Extensive and uncontrolled water abstraction from freshwater ecosystems resulted in significant degradation of surface water bodies like lakes, rivers, wetlands and also aquifers. Lower water quality in rivers and lakes is not only connected with the pollutant load that is discharged there, but also with the extracted freshwater volume [1]. Through water reuse, the freshwater volume that is removed from surface bodies is reduced, as well as the discharged pollutant load, which protects and improves the water quality. Therefore, the primary benefit of water reuse is water conservation via the substitution of natural water resources with treated wastewater. This practice is considered a sustainable way to reduce the anthropogenic impacts on freshwater ecosystems [1,16].

Ecological benefits also occur, when water is reused for ecological applications, like ecological restoration of degraded ecosystems, wetlands, and other natural habitats. Reclaimed water is used to cover the needs of freshwater flows in suffering ecosystems, as well as for aquifer recharge for the enhancement of urban water supplies. Such a strategy was followed, for example, for the ecological restoration of the Everglades ecosystem [27]. Water reuse could also contribute to the reduction of greenhouse gas emissions, as it is shown that it can be an energy-saving practice through the limitation of desalination facilities or energy-demanding drinking water facilities [31]. Water reuse also assists in better handling of nutrient discharge to the environment because it reduces the nutrient load that ends up in water bodies and, thus, reduces the treatment needs for nutrient load removal. And this, then, contributes to reduced greenhouse gas emissions, too.

An interesting example of the potential benefits of water reuse in water-stressed areas comes from the Mediterranean basin and, specifically, from Greece [13]. The regions of the island of Crete and of the Aegean Islands suffer from limited available water resources, while water demands often exceed water availability. The geographical fractionation of the Aegean islands to dozens of small and bigger islands makes it difficult to implement an integrated water management plan. In these regions, agricultural irrigation represents the major water consumer (up to 63%–80%). Increased numbers of tourist facilities also raise the need for domestic water. The fact that most cities in the region are served by wastewater treatment plants makes water reuse technically feasible and very attractive for agricultural and landscape irrigation. This would significantly reduce the need for artificial fertilizers by about 5–7 kg of nitrogen per year, which corresponds to 25–35 kg of common commercial chemical fertilizer. Moreover, these areas face the problem of seawater intrusion to aquifers due to overabstraction. Reuse of treated water for aquifer recharge could effectively deal with this problem, too. Another advantage in this case is that the maximum wastewater flow appears during summer months, when the demands for recycled water are also high. Thus, the need for large storage facilities can be avoided, which decreases the environmental footprint (e.g., greenhouse gas emissions) of the application. It is calculated that water reuse application only in the island of Crete could result in more than 5% water savings of irrigation water [37].

Generally, the practice of water reuse and recycling offers significant environmental benefits and can contribute to a higher level of sustainability in the field of water resources management,

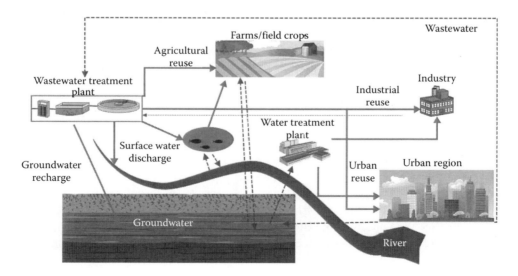

FIGURE 19.1 Water reuse applications and beneficial uses within the hydrological cycle.

water consumption, freshwater saving, ecosystems protection and restoration, and aesthetical upgrade. Furthermore, water reuse coupled with water conservation strategies represent an integrated urban water plan, which could exploit the respective environmental benefits.

19.5 Environmental Risks of Water Reuse

Despite the significant advantages of water reuse, this practice entails certain levels of risks. These risks are mainly associated with the insufficient treatment of the recycled water/wastewater and the pollutant load that results in surface and underground water bodies. Discharge of wastewater from urban settlements represents the main point-source pollution. Combined discharge of municipal treated effluent with untreated industrial wastewater is a major threat for aquatic ecosystems. Although in high-income countries of Europe and North America wastewater treatment takes place at a large extent, there are still fast growing areas and cities in both developed and developing countries with incomplete or even no wastewater infrastructure. Beside this, urban runoff is another nonpoint pollution source, which also adds pollutant load to the final water receiver.

The presence of organic and chemical pollutants in the environment could harm the ecosystems and the natural and wildlife habitat. General ecological risks include degradation of water and soil quality, negative effects on plant growth, disease transmission to aquatic life and animals, and death of fish species due to water pollution, among others [3]. Pollution generated by human activities has affected freshwater ecosystems like wetlands, lakes, rivers, and marshes, and has accelerated their degradation rate. Uncontrolled discharge of untreated or insufficiently treated wastewater usually creates problems to downstream areas and damages coastal ecosystems and habitats.

Although treated wastewater is considered a source rich in nutrients and necessary moisture for crop growth, if it is not properly treated and monitored, it could limit the yield and negatively affect the quality of the cultivated crops. Wastewater irrigation can give crops with higher yields and usually limits the use of chemical fertilizers; however, irrigation with treated wastewater represents a potential environmental risk due to possible improper treatment and contamination of surface and groundwater with nutrients, heavy metals, pathogenic microorganisms, organic chemicals, and salts.

19.5.1 Quality of Reused Water

Water quality is the most crucial parameter in water reuse applications, which defines the level of risk and acceptability of the use of reclaimed water. The pollutant load of treated water/wastewater discharge determines the impact on the environment, water bodies, and/or irrigated crops. Effluent discharge may contain residual concentrations of various pollutants, which represent a threat and may be toxic to the ecosystems. Usually, these constituents originate from domestic, municipal,

or industrial wastewater treatment plants. The impacts of the contaminant presence in aquifers and surface water bodies have an accumulative nature and the effects are usually long term. The various contaminant groups that could still be present in treated effluents and can be further redistributed with water reuse are microbial pathogens, nutrients, heavy metals, salinity, chlorine, and emerging contaminants.

19.5.1.1 Microbial Pathogens

Enteric pathogens are the most common pathogens in water and include bacteria, viruses, protozoa, and helminthes [36]. They enter the environment and infect living terrestrial and aquatic organisms by direct discharge and defecation into water or by runoff from land surfaces [15,28,36]. The presence of pathogenic microorganisms in treated effluents mainly depends on the level of previous treatment, the treatment technology applied, and the regulatory legislation [3]. These contaminants can cause waterborne diseases and infections. Reclaimed water and water reuse facilities should ensure that there is a minimal risk of pathogen exposure to the environment. Especially when the water reuse aims at indirect potable reuse, the wastewater should be treated using advanced treatment technologies to reach respective standards of drinking water quality, according to EPA and WHO. Common indicators of this group are bacteria like fecal (*Escherichia coli*) and total coliforms, *Salmonella typhi* and *Clostridium perfringens*, protozoa like *Cryptosporidium* and *Giardia lamblia*, enterococci, enteroviruses, and bacteriophages, among others.

19.5.1.2 Nutrients (Nitrogen and Phosphorus)

Nutrients are probably one of the oldest problems of water pollution, which still remains of interest [40]. Nitrogen (e.g., nitrate) and phosphorus (e.g., phosphate) represent the most common compounds used in fertilizers. The extensive use of fertilizers often results in wash-up of these compounds from the soil and their final discharge into lakes and rivers. Excessive availability of these pollutants in water bodies causes an increase in the microbial activity and primary productivity, which is known as eutrophication, a negative effect on the aquatic ecosystem [33]. This phenomenon affects the dissolved oxygen concentration and availability for the aquatic life and is usually accompanied by dead zones, death of fish species, algal blooms, and reduction of the submerged aquatic vegetation. Effects may also include a decline in the presence of water birds and reduction of biodiversity. Moreover, in irrigation applications, if the amount of nitrogen delivered to irrigated crops through water reuse exceeds the recommended dosage, it can cause a reduction in crop yield and negatively affect crop growth.

19.5.1.3 Heavy Metals

Heavy metals (lead, copper, cadmium, zinc, chromium, etc.) are present in wastewater, especially in that of industrial origin. Generally, common treatment technologies manage to remove the majority of trace metals, which usually end up in the sludge fraction of the wastewater. Treated effluents usually have low heavy metal concentrations [32], thus, they do not represent a major concern when treated effluents are reused. For industrial wastewater, though, or in cases of improper treatment, heavy metal concentration should be taken into account. When water is reused for irrigation applications, metals tend to accumulate in the soils and are assimilated in plant biomass [2]. Most of them can be toxic at higher amounts for plants and aquatic life and represent an environmental hazard [18], especially if they are in the dissolved or free form in water. Heavy metal presence in recycled water usually limits the suitability of this water for irrigation or other reuse purposes [29]. Hardness and pH values also affect heavy metal toxicity. Most heavy metals are usually regulated pollutants in wastewater effluents to protect aquatic life in cases of ecological applications of treated water [41].

19.5.1.4 Salinity

Water reuse could change the amount of dissolved solids and salts present in agricultural soils [43]. Generally, the total dissolved solids (TDS) content of treated wastewater is comparable to that of surface waters [3]. However, solid content of the final receiver—natural ecosystem—should always be measured and taken into account prior to the effluent discharge [7]. Possible accumulation of salts to plant roots may harm soil status (soil erosion due to increased sodium accumulation and respective decrease of the hydraulic conductivity) and crop yields (decreased productivity) due to osmotic stress or direct toxicity, while in free-draining soils downflow drainage could result in groundwater pollution [18]. High concentrations of sodium, mainly, and of calcium and magnesium, secondarily, results in sodic soils that are characterized by high sodicity values. High sodicity causes changes in the physical properties of the soil and dispersion of the soil aggregates, which could harm plants through reduced water and air permeability [11].

19.5.1.5 Chlorine

Irrigation with treated wastewater could result in the addition of chloride to groundwater [5]. Chlorine is a common disinfection chemical in wastewater treatment plants, used to eliminate pathogenic microorganisms. However, even at low concentrations, residual chlorine may react with organics in water and form a wide range of chlorinated by-products, for example, chlorinated hydrocarbons, which can cause ecotoxicity in the receiving water body [29]. Due to this, there is an increasing concern over the use of chlorine for recycled water disinfection. Chlorinated hydrocarbons are chemical compounds of chlorine, hydrogen, and carbon atoms, which can build blocks of other chemicals like pharmaceuticals, plastics, and solvents, which are discussed next. Some of them are ethylene dichloride, vinyl chloride monomer, and polyvinyl chloride. Chlorine is also produced from the chemical manufacturing industry (e.g., chlorine production via the chlor-alkali manufacturing process) or as a by-product of magnesium production and other processes. Moreover, besides chloride, specific ions like boron are also toxic to some crops and can cause serious damage [29].

19.5.1.6 Emerging Contaminants

Among the various constituents, chemicals of concern can cause the greatest environmental impact [36]. The term *emerging contaminants* is used for chemicals in water that have been recently identified. Most of them are still unregulated and originate from domestic and industrial wastewater, while most of them are used for various therapeutic purposes for humans and animals. This group includes contaminants like pharmaceuticals, antibiotics, personal care products (PCPs), endocrine disrupting chemicals (EDCs), and disinfection-byproducts (DBPs) [3,36]. Examples of EDCs are steroid estrogens (e.g., estrone, estriol, etc.) and xenoestrogens (e.g., alkylphenols and bisphenol A), while the group of pharmaceuticals includes antimicrobials (triclosan), carbamazepine, ibuprofen, amoxicillin, caffeine, and musk fragrances (galaxolide, tonalide), among others.

These substances can still be present in treated effluents and will eventually end up in lakes and rivers, thus creating a potential environmental threat. Aquatic organisms are as sensitive as or even more sensitive than humans to these chemical micropollutants. These chemicals can interfere with the normal function of the endocrine system of many aquatic and terrestrial organisms, which can result in reproductive dysfunction, reduced fertility, and population decline of the organisms [11]. Another concern related to pharmaceutical presence in water bodies is the potential development of antibiotic resistance in soil and water microorganisms due to continuous discharge of antibiotics to the environment [14]. However, the long-term impact of exposure to low concentrations of EDCs and PCPs and their metabolites is still not adequately known [22]. Test organisms like zebrafish have been used to investigate possible negative impacts [21]. In industrialized countries with extensive use of pharmaceuticals and PCPs, these micropollutants can be found in surface and groundwater bodies. Today, more than 3000 different pharmaceutical ingredients are known [9]. Continuous research and development of new measuring techniques reveal new micropollutants that represent a potential threat for the wellbeing of both ecosystems and humans.

The fate of these microcompounds in water and wastewater treatment is not well understood and detecting techniques with greater accuracy and lower detecting limits are currently under development. Conventional wastewater treatment plants (e.g., activated sludge) generally do not provide a sufficient treatment and removal of these compounds [3,26] because they were not designed for this type of constituents. This means that a residual concentration of these contaminants is discharged to the water bodies. Thus, these substances are increasingly found in surface waters, especially those receiving large volumes of treated effluents. They can be found in groundwater where aquifer recharge takes place using effluents from treatment facilities or where insufficiently treated water is used for irrigation. An efficient removal can be achieved only with the implementation of advanced tertiary treatment technologies (e.g., advanced oxidation and membrane technology), which—on the other hand—are high-energy-consuming methods, expensive to build and operate, and produce a final effluent of lower biological quality and not as biologically rich (as in the case of biologically based technologies) [22,34]. Although an additional level of treatment for the recycled water before the final reuse might be undesirable, current technological developments ensure a significant reduction of these compounds and a limited risk for the environment [36], with simultaneous minimum environmental impact. Overall, being a relatively new group of contaminants, the effects of these micropollutants are still under investigation. Current research activities focus on the fate of these pollutants, collecting transport data, development of predictive models, and determination of toxicological effects in water.

19.5.2 Further Treatment Demands

Another aspect of water reuse projects is that the promotion of this practice implies the construction of treatment and distribution facilities and the upgrade of existing facilities to meet treatment criteria for these compounds, which means that additional equipment, energy, and chemicals will be used [9]. Today there is a dispute over this assumption and the impact level it possesses. The use of sustainable treatment methods (e.g., natural treatment systems like constructed wetlands, pond systems, etc.) could reduce this environmental impact. These decentralized treatment facilities serve the concept of sustainable treatment with reduced construction and minimum operational environmental impacts [34]. For example, urban (stormwater) runoff is a common degradation factor of water quality in surface water bodies, especially for those located near densely populated areas. Flood control and pollutant removal from this water flow would provide an additional water source in urban areas. For this application, natural treatment systems like constructed wetlands appear as an alternative green technology, which helps with water quality improvement, pollutant removal, and control of large water volumes within a short time with an environmentally friendly nature of the treatment technology [30,34].

19.5.3 Greenhouse Gas Emissions

A relatively unknown and not thoroughly examined risk that may arise from water reuse is the potential increase in greenhouse gas emissions. Although water reuse is often characterized as an energy-saving practice [31], this is not a rule of thumb. The problem has to do with the fact that in water reuse projects there are usually large volumes of water to be transported. High needs for water pumping mean high rates of greenhouse gas emissions. Falconer et al. [11] report an example in Australia. This indicates that although water reuse for irrigation is considered a very desirable and beneficial application, in some cases potential side effects are ignored and not taken into account when estimating the overall environmental impact. This also counts not only for wastewater treatment and reuse but also for other alternatives like water desalination [8]. Therefore, a water reuse project should take into consideration all potential aspects that may affect the environmental footprint of the application.

Overall, the relatively limited information on the fate and effects of all these contaminants in the environment creates certain gaps in the risk taken when implementing water reuse applications. Potential impacts of water reuse projects include effects on the various environmental segments such as

- *Plants and soils*: Water reuse for irrigation may harm both plants and soils, if pollutants remain at relatively high concentrations in the recycled water. The presence of microorganisms could also be pathogenic for crops. Insufficient watering due to soil clogging (salt accumulation) or even extensive watering can have negative impacts. High rate of salts and dissolved solids accumulation (e.g., up to 2000 μS/cm) may be toxic to plants because salt ions can enter the plant structure and interfere with cellular processes and mechanisms.
- *Surface and groundwater*: Pollution of surface water bodies and groundwater reservoirs results in lower water quality and limits the potential beneficial use of these waters. Direct discharge to water bodies or uncontrolled runoff discharge can be sources of pollution. Excessive discharge of nutrients may result in the phenomenon of eutrophication, which negatively affects aquatic life, ecosystems, and drinking water supplies, and limits the beneficial use of water. Draining or irrigation water can reach groundwater and create pollution with nitrates and other contaminants (especially when treated effluents include industrial wastewaters). Agricultural runoff is also responsible for high concentrations of pesticides in surface waters.
- *Land*: It is possible that water reuse projects could stimulate urban growth and expansion of urban land. Increase of available water sources for nonpotable uses can be attractive for the development of residential, industrial, or agricultural activities which can, in the long term, have a negative effect on the initial environmentally friendly targets and ecological benefits.
- *Ecosystems*: Alteration in water quality and water flow path during water reuse projects could potentially cause adverse effects on ecosystem status and health. Implementation of such projects should be carried out carefully and be based on specific characteristics of the area. Treated wastewater should not represent a large fraction of the water flow volume of a river or stream. Ecological restoration could include increase of flow volume up to a minimum required level in order to protect aquatic life and downstream activities.

It is clear that the use of reclaimed water includes certain ecological risks for ecosystems and the aquatic and terrestrial life involved. These issues become even more important as water sources are gradually limited and population and agricultural needs increase. Under this scenario, water reuse (either directly or indirectly) gains increasing attraction and is expected to continue rising within the next years, taking into account that in many regions the available good quality water supply is already fully exploited. A critical issue of water reuse is that the redistribution of treated water/wastewater could also cause redistribution and further spreading of contaminants like heavy metals, salts, and emerging pollutants, with possible toxic and carcinogenic risks. Possible redistribution of these contaminants could take place not only via the treated water reuse but also via the wastewater sludge used as fertilizer, which is currently a very common practice. These contaminants tend to concentrate in the solid fraction of sludge and through extensive use of sludge as fertilizer they may be released back to the environment and end up in both crops and surface or groundwater [11].

19.6 Ecological Risk Assessment

Prior to any water reuse application, a respective assessment of ecological risk should be carried out. Although risk assessment and management is a relatively new field and is still evolving, it could be a significant tool for the estimation of the ecological impact of water reuse and for the better understanding of the various risks associated with this practice. Proper and actual consideration of all related risk aspects, quality restrictions and limits, and source-point measures (instead of simple end-of-the-pipe solutions) should be included. Main targets of risk assessment should be the minimization of risks for public health and the environment, the improvement of urban agriculture activities, and the integrated approach of treated wastewater as an element of water resources balance.

Especially for the evaluation of the ecological impact of water reuse applications, ecological risk assessment (ERA) is as a necessary tool. ERA is an evaluation process concerning the potential impacts of various factors-stressors, for example, chemicals-pollutants, land change, invasive species, climate change, etc., to the environment [39]. It is a flexible and helpful process for data organization and analysis, collection of relative information, and assumptions and uncertainties toward the evaluation of the possibility of adverse ecological impacts. It is based on the holistic approach of a water reuse application for environmental decisions, integrating not only available scientific information but also legal, economic, social, and political factors. It consists of three distinct phases: problem formulation, analysis of exposure and effects, and risk characterization [39].

In this process, key parameters are the end point, that is, habitat or endangered species, and the related sensitivity of the ecosystem under study, which could vary depending on the different reuse options. Based on these parameters, the evaluation of the exposure level and response to stressors is determined. Assessment of aquatic toxicity or water quality could be done through literature evaluation or additional measurements and tests. The final step is the calculation of the ratio of the predicted environmental concentration to the predicted no-effect concentration. Values of this ratio ≤1 or ≥1 indicate low or high environmental risk, respectively.

ERA meets growing interest worldwide as a tool for the estimation of the environmental impacts of water reuse applications. While the use of bacterial indicators has been widely used to demonstrate possible risks on human health, chemical risk on the environment is still developing. This evaluation process is particularly interesting when used for the estimation of the ecological risk of EDCs in recycled water. For example, Sun et al. [35] investigated the long-term ecological risk of eight typical estrogenic EDCs in effluents of wastewater treatment plants, which could affect the quality of recycled water, and found that among them steroidal estrogens had the highest ecological risk, followed by phenolic compounds and phthalate esters. This kind of study on ERA could be very useful toward the determination of priority pollutants that should be controlled in treatment plants. The number of micropollutants and organic chemicals is so big that prioritization is the key for the effective control and monitoring of these compounds, which allows for more reliable and efficient risk assessment. Current research activities investigate this topic with special focus on pollutants present in surface and groundwater and water reuse applications [12,23,42], indicating that the future challenges for effective, safe, and environmentally friendly water reuse applications is highly dependent on the effective control of emerging pollutants.

19.7 Summary and Conclusions

Water reuse today is recognized as an alternative and reliable source of water that could be a significant element toward the formation of sustainable water management and water policy. It is more often included in water resources plan development. Water reuse can be beneficial for all related segments of the ecosystems, offering a series of advantages like water resources conservation, reduction of potable water for nonpotable uses, ecological enhancement of ecosystems, decline of surface and groundwater abstraction, restoration of natural habitats, reduced use of artificial fertilizers, and protection of freshwater bodies from saltwater intrusion, among others. Especially for countries with serious water scarcity problems (e.g., Middle East region, Mediterranean countries), water reuse appears as an imperative practice.

However, if not properly implemented, that is, effective treatment and pollutant removal, safety measures, etc., the various environmental benefits can be switched and cause adverse impact on the natural habitat and the aquatic and terrestrial life. Improper management of treated water/wastewater reuse might result in enormous risks for the ecosystems, plants/crops, soils, and water bodies.

The presence of contaminants of different origin and composition represents the greatest threat to ecosystems. The level of risks, concerning water and soil contamination through water reuse, is strongly connected with the respective impacts of wastewater discharge to surface and underground water bodies and the level/effectiveness of treatment before the final discharge. Emerging organic micropollutants represent a relatively new source of concern when dealing with ecological enhancement applications via water reuse, due to the impacts they can cause to aquatic life. In contrast to other pollutants (e.g., nutrients, heavy metals) whose ecological impacts are more or less known and removal in wastewater treatment plants is satisfying, there is a gap in knowledge of emerging pollutants removal processes (e.g., pharmaceuticals, EDCs, PCPs, etc.) concerning their potential impact on ecosystems and natural habitats [34]. This group of pollutants is currently gaining increasing research interest and represents a scientific field with a great dynamic.

New scientific knowledge, technical (techniques, identification methods), and management tools (ecological risk assessment) are under continuous development and are proved to be very helpful and with increasing reliability for the proper implementation of water reuse projects. Integrated analysis of all related environmental risks and evaluation of potential multiple ecological impacts today is a prerequisite toward the development of best reuse practices for environmental protection. The adoption of new, upgraded international and national guidelines, which will include all potential risks and cover new sources of threat (i.e., emerging contaminants), is also almost a reality. These efforts will eventually result in a more holistic approach of water reuse practices.

References

1. Anderson, L. 2003. The environmental benefits of water recycling and reuse. *Water Science and Technology*, 3(4), 1–10.
2. Angelova, V., Ivanov, R., Delibaltova, V., and Ivanov, K. 2004. Bio-accumulation and distribution of heavy metals in fibre crops (flax, cotton and hemp). *Industrial Crops and Products*, 19, 197–205.
3. Asano, T., Burton, F.L., Leverenz, H.L., Tsuchihashi, R., and Tchobanoglous, G. 2007. *Water Reuse: Issues, Technologies, and Applications.* Metcalf & Eddy Inc., AECOM, McGraw-Hill, New York.
4. Atherton, J.G. 2011. Health and environmental aspects of recycled water. In: Doelle, H.W., Rokem, J.S., and Berovic, M. (eds.), *Biotechnology X, Encyclopedia of Life Support Systems (EOLSS),* Developed under the Auspices of the UNESCO, Eolss Publishers, Oxford, UK.
5. Babiker, I.S., Mohamed, M.A.A., Terao, H., Kato, K., and Ohta, K. 2004. Assessment of groundwater contamination by nitrate leaching from intensive vegetable cultivation using geographical information system. *Environment International*, 29(8), 1009.
6. Bixio, D., Thoeye, C., De Koning, J., Joksimovic, D., and Savic, D. 2006. Wastewater reuse in Europe. *Desalination*, 187, 89–101.
7. Brix, K.V., Gerdes, R., Curry, N., Kasper, A., and Grosell, M. 2010. The effects of total dissolved solids on egg fertilization and water hardening in two salmonids-Arctic Grayling (*Thymallus arcticus*) and Dolly Varden (*Salvelinus malma*). *Aquatic Toxicology*, 97(2), 109–115.
8. Cohen, I. 2006. Water-desalination versus recycling. Press release, Ian Cohen, The Greens, Member of the Legislative Council, New South Wales, Australia.
9. Davis, C.K. 2008. Ethical dilemmas in water recycling. In: B. Jiménez and T. Asano (eds.), *Water Reuse: An International Survey of Current Practice, Issues and Needs.* IWA Publishing, London.
10. EU. 2000. Establishing a framework for community action in the field of water policy. Directive 2000/60/EC of the European Parliament and of the Council of 23 October 2000. *Official Journal of the European Communities* L 327/1.
11. Falconer, I.R., Chapman, H.F., Moore, M.R., and Ranmuthugala, G. 2006. Endocrine- disrupting compounds: A review of their challenge to sustainable and safe water supply and water reuse. *Environmental Toxicology*, 21(2), 181–191.
12. Gavrilescu, M., Demnerová, K., Aamand, J., Agathos, S., and Fava, F. 2014. Emerging pollutants in the environment: Present and future challenges in biomonitoring, ecological risks and bioremediation. *New Biotechnology*, 32(1), 147–156. http://dx.doi.org/10.1016/j.nbt.2014.01.001.
13. Gikas, P. and Angelakis, A.N. 2009. Water resources management in Crete and in the Aegean Islands, with emphasis on the utilization of non-conventional water sources. *Desalination*, 248, 1049–1064.
14. Guardabassi, L., Petersen, A., Olsen, J.E., and Dalsgaard, A. 1998. Antibiotic resistance in *Acinetobacter* spp. isolated from sewers receiving waste effluent from a hospital and a pharmaceutical plant. *Applied and Environmental Microbiology*, 64, 2499–3502.
15. Guillaume, P. and Xanthoulis, D. 1996. Irrigation of vegetable crops as a means of recycling wastewater: Applied to Hesbaye Frost. *Water Science and Technology*, 33, 317–326.
16. Hamilton, A.J., Boland, A.M., Stevens, D., Kelly, J., Radcliffe, J., Ziehrl, A., Dillon, P.J., and Paulin, R. 2005. Position of the Australian horticultural industry with respect to the use of reclaimed water. *Agricultural Water Management*, 71, 181–209.
17. Hochstrat, R. and Wintgens, T. (eds.), 2003. AQUAREC, Report on Milestone M3.I, Draft of wastewater reuse potential estimation, Interim report.
18. Hussain, I., Raschid, L., Hanjra, M.A., Marikar, F., and Van der Hoek, W. 2002. Wastewater use in agriculture: Review of impacts and methodological issues in valuing impacts. Working Paper 37. International Water Management Institute, Colombo, Sri Lanka.
19. Janosova, B., Miklankova, J., Hlavinek, P., and Wintgens, T. 2006. Drivers for wastewater reuse: Regional analysis in the Czech Republic. *Desalination*, 187, 103–114.
20. Jiménez, B. and Asano, T. 2008. Water reclamation and reuse around the world. In: Jiménez, B. and Asano, T. (eds.), *Water Reuse: An International Survey of Current Practice, Issues and Needs.* IWA Publishing, London.
21. Knacker, T., Liebig, M., and Moltmann, J. 2006. Environmental risk assessment. In: *Human Pharmaceuticals, Hormones and Fragrances: The Challenge of*

Micropollutants in Urban Water Management. IWA Publishing, London, 121–135.

22. Li, Y., Zhu, G., Ng, W.J., and Tan, S.K. 2014. A review on removing pharmaceutical contaminants from wastewater by constructed wetlands: design, performance and mechanism. *Science of the Total Environment*, 468–469, 908–932.

23. López-Serna, R., Postigo, C., Blanco, J., Pérez, S., Ginebreda, A., de Alda, M.L., Munné, A., and Barceló, D. 2012. Assessing the effects of tertiary treated wastewater reuse on the presence emerging contaminants in a Mediterranean river (Llobregat, NE Spain). *Environmental Science and Pollution Research*, 19, 1000–1012.

24. MEU. 2013. *The Millennium Development Goals Report.* United Nations, New York.

25. Miller, G.W. 2006. Integrated concepts in water reuse: Managing global water needs. *Desalination*, 187, 65–75.

26. Murray, K.E., Thomas, S.M., and Bodour, A.A. 2010. Prioritizing research for trace pollutants and emerging contaminants in the freshwater environment. *Environmental Pollution*, 158, 3462–3471.

27. NRC (National Research Council). 2011. *Progress toward Restoring the Everglades: The Third Biennial Review – 2010.* The National Academies Press, Washington, DC.

28. Rose, J.B. 2007. Water reclamation, reuse and public health. *Water Science and Technology*, 55(1–2), 275–282.

29. Salgot, M., Vergés, C., and Angelakis, A.N. 2002. Risk assessment for wastewater recycling and reuse. *IWA Regional Symposium on Water Recycling in Mediterranean Region*, Iraklio, Greece, September 26–29.

30. Scholz, M. 2006. *Wetland Systems to Control Urban Runoff.* Elsevier Publishing, the Netherlands.

31. Serra, M. and Sala, L. 2003. Energy balance in several municipalities on the Costa Brava (Girona, Spain). In: *Proceedings of the Second International Conference on Efficient Use and Management of Water in Urban Areas*, Tenerife, Canary Islands (Spain), April 2–4.

32. Sheikh, B., Jaques, R.S., and Cort, R.P. 1987. Reuse of tertiary municipal wastewater effluent for irrigation of raw-eaten food crops: A five year study. *Desalination*, 67, 245–254.

33. Smith, V.H., Tilman, G.D., and Nekola, J.C. 1999. Eutrophication: Impacts of excess nutrient inputs on freshwater, marine, and terrestrial ecosystems. *Environmental Pollution*, 100, 179–196.

34. Stefanakis, A.I., Akratos, C.A., and Tsihrintzis, V.A. 2014. *Vertical Flow Constructed Wetlands: Eco-engineering Systems for Wastewater and Sludge Treatment.* Elsevier Publishing, the Netherlands.

35. Sun, Y., Huang, H., Sun, Y., Wang, C., Shi, X-L., Hu, H-Y., Kameya, T., and Fijie, K. 2013. Ecological risk of estrogenic endocrine disrupting chemicals in sewage plant effluent and reclaimed water. *Environmental Pollution*, 180, 339–344.

36. Toze, S. 2006. Water reuse and health risks—Real vs. perceived. *Desalination*, 187, 41–51.

37. Tsagarakis, K.P., Dialynas, G.E., and Angelakis, A.N. 2004. Water resources management in Crete (Greece) including water recycling and reuse and proposed quality criteria. *Agricultural Water Management*, 66, 35–47.

38. UKWIR. 2004. Framework for developing water reuse criteria with reference to drinking water supplies. UKWIR, WATEREUSE Foundation & AWWA Research Foundation. Draft Report Ref. No. 05/WR/29/1.

39. USEPA. 1998. Guidelines for Ecological Risk Assessment. United States Environmental Protection Agency, Office of Water, Washington, DC, EPA/630/R-95/002F, April.

40. USEPA. 2011. Memorandum: Working in partnership with states to address phosphorus and nitrogen pollution through use of a framework for state nutrient reductions. United States Environmental Protection Agency, Office of Water, Washington, DC.

41. USEPA. 2011. Water quality standards: Protecting human health and aquatic life. United States Environmental Protection Agency, EPA/820/F-11/001. Aquatic Life criteria table available at http://water.epa.gov/scitech/swguidance/standards/criteria/current/index.cfm.

42. von der Ohe, P.C., Dulio, V., Slobodnik, J., de Deckere, E., Kühne, R., Ebert, R-U., Ginebreda, A., de Cooman, W., Schüürmann, G., and Brack, W. 2011. A new risk assessment approach for the prioritization of 500 classical and emerging organic microcontaminants as potential river basin specific pollutants under the European Water Framework Directive. *Science of the Total Environment*, 409, 2064–2077.

43. Weber, B., Avnimelech, Y., and Juanico, M. 1996. Salt enrichment of municipal sewage: New prevention approaches in Israel, *Environmental Management*, 20(4), 487–495.

20

Environmental Impact Assessment: An Application to Urban Water Reuse

Bosun Banjoko
Obafemi Awolowo University

Saeid Eslamian
Isfahan University of Technology

PREFACE

Water is a precious and increasingly scarce resource, essential for both ecosystems and humans. Current global water needs rise due to several factors, which include population growth in areas with low freshwater resources and the resulting increase in water consumption. Pollution of surface water and groundwater and long-term changes in the hydrological cycle due to climate change underpin the requirement for water reuse. Although water reuse strategies are intended to address problems of water scarcity, this should not come at the price of increasing other environmental impacts because water reuse and reclamation technologies including wastewater treatment plants, drinking water treatment plants, desalination, and constructed wetlands are projects with environmental implications and the issue of water reuse may have transboundary implications. Environmental impact assessments (EIA) are inevitable and compulsory.

20.1 Introduction

Early practices in project assessment were limited and often based on technical feasibility studies and cost–benefit analysis. Until recently, environmental factors were not taken into consideration while examining the viability of a project. The growing number of cases globally of adverse environmental consequences as a result of interactions between socioeconomic and industrial development activities led to a serious rethinking that environmental consequences have to be explicitly considered in the decision-making process. It is therefore expedient that Earth's natural resources need to be conserved and utilized with great caution particularly in view of global population growth. The growing awareness and concern about the need for environmental protection has resulted in the introduction of related legislation and institutional arrangements globally. It is of importance to note the desire of nations to industrialize and pursue economic developmental goals for providing satisfactory quality of life; however, these desires should be linked with environmental protection and sustainability of existing natural resources for future generations [6].

Worthy of note is the fact that international trade, loan requests for development from such organizations like International Monetary Fund (IMF) and World Bank are tied to sound environmental management of proposed projects, which require environmental impact analysis and report to support such loan applications [22]. In this context, EIA has come to be recognized as an important process for incorporating the objectives of environmental concerns with the requirements of economic growth and social development. EIA helps in examining the options for choosing an environmentally acceptable course of action including site selection, choice of technology, resource conservation, etc. Furthermore, the international financing institutions have also introduced the procedure/development of EIA as a prerequisite for project funding. In addition, various laws have been enacted in different countries to make EIA essential for the approval of all the new development projects [1,2].

20.1.1 Environmental Impact Assessment

EIA is the process of identifying, predicting, evaluating, and mitigating the biophysical, social, and other relevant effects of development proposals prior to major decisions being taken and commitments made [27]. EIA is a formal process used to predict the environmental consequences (positive or negative) of a plan, policy, program, or project prior to the implementation of the decision. It proposes measures to adjust impacts to acceptable levels or to investigate new technological solutions. Although an assessment may lead to difficult economic decisions and political and social concerns, EIAs protect the environment by providing a sound basis for effective and sustainable development. The purpose of the assessment is to ensure that decision makers consider environmental impacts when deciding whether to proceed with a project. EIAs are unique in that they do not require adherence to a predetermined environmental outcome, but rather they require decision makers to account for environmental values in their decisions and to justify these decisions in light of detailed environmental studies and public comments on the potential environmental impacts [15]. The product of EIA is the preparation of an environmental impact statement (EIS). The adequacy of an EIS can be challenged in a federal court [8].

20.1.2 Environmental Assessment

The environmental assessment (EA) is a quasi-EIA public document made popular in the United States prepared pursuant to the National Environmental Policy Act to determine whether a federal action would significantly affect the environment and thus require a more detailed EIS. The certified release of an EA results in either a finding of no significant impact (FONSI) or requirement for an EIS. The EA is less complicated to prepare than an EIS and is thus a simple document that includes a brief description of the purpose and need of the proposal and of its alternatives as required by National Environmental Protection Agencies (NEPAs) and of the human environmental impacts resulting from and occurring to the proposed actions and alternatives considered practicable, plus a listing of studies conducted and agencies and stakeholders consulted to reach these conclusions. The action agency must approve an EA before it is made available to the public. The structure of a generic EA is as follows:

1. Introduction
   - Background
   - Purpose and need of action
   - Proposed action
   - Decision framework
   - Public involvement
   - Issues
2. Alternatives, including the proposed action
   - Alternatives
   - Mitigation common to all alternatives
   - Comparison of alternatives
3. Environmental consequences
4. Consultation and coordination

The EA becomes a draft public document when notice of it is published, usually in newspapers of general circulation in the area affected by the proposal. There is a 15-day review period required for an EA and 30 days in exceptional circumstances while the document is made available for public commentary, and a similar time for any or inadequate waste management operations. There are several different types and purposes of environmental audits. Examples include waste minimization audit and the transactional audit, which is performed prior to the sale or refinancing of manufacturing or commercial activities and is typically required by lending institutions, insurance companies, buyers, and state regulatory agencies before the transaction is finalized. Another type of environmental audit is the EIA audit. This is a follow-up process that evaluates the accuracy of the EIA by comparing actual to predicted impacts. The objective in this scenario is to make future EIAs more valid and effective. Two primary considerations underpin such audits: (1) Scientific—to examine the accuracy of predictions and to explain errors. (2) Management—to assess the success of mitigation in reducing impacts. Audits can be performed as a rigorous assessment of the mill objection to improper process. Commenting on the draft EA is typically done in writing or e-mail, submitted to the lead action agency as published in the notice of availability. An EA does not require a public hearing for verbal comments. Following the mandated public comment period, the lead action agency responds to any comments, and certifies either a FONSI or a notice of intent (NOI) to prepare an EIS in its public environmental review record. The preparation of an EIS then generates a similar but more lengthy, involved, and expensive process.

20.1.3 Historical Perspectives and Legal Framework

EIA commenced in the 1960s as part of increasing environmental awareness. EIAs therefore form the basis of a technical evaluation to contribute to more objective decision making with respect to projects and the environment. In the United States, EIA obtained formal status in 1969 with the enactment of the National Environmental Policy Act. The EIA is an activity that is implemented to find out the impact that is done before development will occur. EIA later become increasingly popular with many other countries incorporating EIA as requirement for project developments. These include Australia (1974), Canada (2012), the Netherlands (1987), Malaysia (1974), Nepal (1997), New Zealand (1974), Russian Federation (2004), and China (2004) [5,10].

The international laws supporting EIA include the following:

1. The 1991 Espoo Convention on Environmental Impact Assessment in a Transboundary Context "states that the state under whose jurisdiction a proposed activity is envisaged to take place shall provide an opportunity to the public in the areas likely to be affected to participate in relevant environmental impact assessment procedures regarding proposed activities and shall ensure that the opportunity provided to the public of the affected state is equivalent to that provided to its own public (Article 2(6))" [40]. The Convention on Environmental Impact Assessment in a Transboundary Context Espoo 1991, Convened by United Nations Economic Commission for Europe (UNECE) on the premise that environmental threats do not respect national borders. Governments have realized that to avert this danger, they must notify and consult each other on all major projects under consideration that might have adverse environmental impact across the borders. The Espoo Convention is a key step to bringing together all stakeholders to prevent environmental damage before it occurs. The convention entered into force in 1997.
2. Protocol on Strategic Environmental Assessment (SEA) Kyiv 2003. The SEA Protocol augments the Espoo Convention by ensuring that individual parties integrate environmental assessment into their plans and programs at the earliest stages, helping to lay the groundwork for sustainable development. The protocol also provides for extensive public participation in the governmental decision-making process. The protocol entered into force on July 11, 2010 [46,47].
3. Aarhus Convention on Access to Information, Public Participation, and Access to Justice in Environmental Matters June 25, 1988 [40].
4. Rio Declaration Principles 15. The precautionary principle: In order to protect the environment, the precautionary principle shall be widely applied by states according to their capabilities. When there are threats of serious or irreversible damage, lack of full scientific certainty shall not be used as a reason for postponing cost-effective measures to prevent environmental degradation.

Furthermore, EIA has legal backing in many countries and also in international environmental law. Examples of international laws supporting EIA will thus be enumerated.

1. Nigeria: The EIA Act 86, 1992
2. United States: National Environmental Policy Act of 1969
3. Australia: Environmental Protection Impact of Proposals Act 1974 superseded by the Environment Protection and Biodiversity Conservation Act 1999 [42]
4. Canada: Canadian Environmental Assessment Act 1995 superseded by the Canadian Environmental Assessment Act of 2012
5. China: Environmental Assessment Law 2004
6. Egypt: Environmental Law of the Environment 4/1994
7. EU: EIA directive 1985, superseded by Strategic Environmental Assessment (SEA) Directive (2001/42/EC)
8. Hong Kong: EIA Assessment Ordinance 1997
9. India: EIA Act originally enacted in 1994 under the Environmental Protection Act of 1986, amended in 2004, and improved to notification in 2006 by the Ministry of Environment and Forestry
10. Malaysia: Environmental Quality Act 1974
11. Nepal: Environmental Protection Act 1997
12. New Zealand: Resource Management Act 1991
13. Russian Federation: Federal Law on Ecological Expertise 1985 Regulations on Assessment of Impact from Intended Business and Other Activity on Environment 2000

20.1.4 Transboundary Context

The environmental impact of any activity can transcend boundaries of nations particularly with respect to water and air. International pollution can have detrimental effects on the atmosphere, oceans, rivers, aquifers, farmland, the weather, and biodiversity. Global climate change is transnational. Specific pollution threats include acid rain, radioactive contamination, debris in outer space, stratospheric ozone depletion, and toxic oil spills. Environmental protection is inherently a cross-border issue and has led to the creation of transnational regulations via multilateral and bilateral treaties. The United Nations Conference on the Human Environment (UNCHE), the Stockholm Conference of 1972, the United Nations Conference on the Environment and Development (UNCED), the Rio Summit Conference, and the Earth Summit held in Rio de Janeiro in 1992 were key to the creation of about 1000 international instruments that include at least some provisions related to the environment and its protection [44]. Furthermore, the UNECE's convention on EIA in a transboundary context held in Espoo, Finland in 1991 was negotiated to provide an international framework for transboundary EIA [43].

20.1.5 Limitations

EIA is used as a decision-aiding tool rather than decision-making tool [19]. There is a growing dissent about them as their influence on decisions is limited. Other areas of contention are determination of boundaries and scope of EIA. The boundary generally refers to the spatial and temporal boundary of the proposal effects. This boundary is determined by the application and the lead assessor, but in practice, almost all EIAs address only direct and immediate on-site effects [21]. The need for improved training for practitioners, guidance on best practice, and continuing research has been proposed [19]. However, there is now software that is ISO certification-based, which is apt to standardize and improve EIAs. An example is the Si SOSTA QUA, an environmental management tool adapted from the life-cycle analysis (LCA) manager tool (SIMPLESL, www.simple-sw.com) [35].

Even though EIA is potentially one of the most valuable, intra-disciplinary, objective, decision taking exercises, it is unable to quantify the values of various parameters to indicate their effect on the quality of the environment. In its present form, EIA suffers from several limitations at conceptual, methodological, and procedural levels. Presently, EIA is conceived merely as a project level tool and does not address developmental programs at the policy and planning levels. Many times, project level decisions are constrained by existent. In addition, there are no objective screening criteria to decide the type or scale of projects that should be subjected to an EIA. EIA should ideally be undertaken at the policy and planning levels as the environmental consequences of the project often arise due to higher-level decisions. However, the EIA policy is extremely complex mainly because the potential range of alternatives to achieve a desired goal would be unlimited. This problem can be solved to some extent by adopting a hierarchical approach. Only the number of alternatives are reduced by defining the problem in terms of a series of choices [25,26].

The most appropriate stage for implementing EIA is at the level of district planning because at this stage a reasonable number of alternatives are available to the developer. Another conceptual limitation of EIA is that it does not incorporate the strategies of preventive environmental intervention. The issues of resource conservation, waste minimization, project recovery, improvement in efficiency of equipment, etc. should be pursued as explicit goals in EIA. Experience over the years has shown that EIAs are always conducted with severe limitations of time, labor, financial resources, and data. India has unique problems resulting in widely divergent lifestyle, varying nature of terrain, flora, and fauna. No reliable environmental information base exists to cater to the needs of a comprehensive EIA study. Therefore, data for EIA could be merely reduced by focusing the study on a limited number of relevant issues rather than an elaborate listing of values for all environmental parameters. A major limitation of the present environmental process is the lack of objective criteria to decide whether a project requires a comprehensive EIA. Screening is a method of selection that allows elimination from the review process all those projects that do not require detailed EIA, thus avoiding unnecessary expense and delays in project clearance. Incorporation of screening in the environmental review process requires formulation of project- and site-related criteria for various types of developmental activities [18,20].

Most ecological problems are the cumulative result of environmental and social impacts of human activity in the region. Planning for sustainable development in the context of an ecosystem, carrying capacity thus requires systematic identification, quantification, and management of cumulative trends in significant environmental variables on a regional basis. Functional planning regions need to be identified based on ecological criteria such as climatic and vegetation patterns, soil classification, and watershed boundaries rather than political justifications. Within the context of sustainable development, regional EIA could provide the means for estimation of developmental limits imposed by regional carrying capacity.

20.1.6 International Association for Impact Assessment

Success of global environmental issues could not be said to be complete without the activities of international associations and nongovernmental organizations that are influential in negotiations and in the establishment of laws and policies. The International Association for Impact Assessment (IAIA) is an organization of researchers, practitioners, and users of various types of impact assessment from all parts of the world whose purpose is to advance the state of the air and science of impact assessment in applications ranging from local to global. IAIA members include corporate planners and managers, public interest advocates, government planners and administrations, private consultants and policy analysts, and university and college teachers and students.

20.1.7 List of Examples of Activities Requiring EIA Studies

1. Agriculture
   a. Land development schemes covering an area of 500 ha or more to bring forest into agricultural production
   b. Agricultural programs necessitating the resettlement of 100 families or more
   c. Development of agricultural estates covering an area of 500 ha or more involving changes in type of agricultural use
2. Airport
   a. Construction of airports (having airstrips of 2500 m or more
   b. Airstrip development in state and national parks
3. Drainage and irrigation
   a. Construction of dams and manufactured lakes and artificial enlargement of lakes with surface areas of 20 ha or more

b. Drainage of wetland, wildlife habitat, or virgin forest covering an area of 100 ha or more

c. Irrigation schemes covering an area of 5000 ha or more

4. Land reclamation
   a. Coastal reclamation involving an area of 5 ha or more

5. Fisheries
   a. Construction of fishing harbors
   b. Harbor expansion involving an increase of 50% or more in fish landing capacity per annum
   c. Land-based aquaculture projects accompanied by clearing of mangrove swamp forests covering an area of 50 ha or more

6. Forestry
   a. Conversion of hill forestland to other land use covering an area of 50 ha or more
   b. Logging or conversion of forestland to other land use within the catchment area of reservoirs used for municipal water supply, irrigation, or hydropower generation or in areas adjacent to state and national parks and national marine parks
   c. Logging covering an area of 500 ha or more
   d. Conversion of mangrove swamps for industrial, housing, or agricultural use covering an area of 50 ha or more
   e. Clearing of mangrove swamps on islands adjacent to national marine parks

7. Housing

8. Industry
   a. Chemical
      Where production capacity of each product or of combined products is greater than 100 tonnes/day
   b. Petrochemicals all sizes
   c. Nonferrous primary smelting
      Aluminum—all sizes
      Copper—all sizes
      Others—producing 50 tonnes/day and above of product
   d. Nonmetallic
      • Cement—For clinker throughput of 30 tonnes/h and above
      • Lime—100 tonnes/day and above burnt lime rotary kiln or 50 tonnes/day and above vertical kiln
   e. Iron and steel
      • Require iron ore as raw materials for production greater than 100 tonnes/day
      • Using scrap iron as raw materials for production greater than 200 tonnes per day
   f. Shipyards
      • Dead weight tonnage greater than 5000 tonnes
   g. Pulp and paper industry
      • Production capacity greater than 50 tonnes/day

9. Infrastructure
   a. Construction of hospitals with outfall into beachfronts used for recreational purposes

b. Industrial estate development for medium and heavy industry covering an area of 50 ha or more

c. Construction of expressways

d. Construction of national highways

e. Construction of new townships

10. Ports
    a. Construction of ports
    b. Port expansion involving an increase of 50% or more in handling capacity per annum

11. Mining
    a. Mining of materials in new areas where the mining lease covers a total area in excess of 250 ha.
    b. Ore processing, including concentrating for aluminum, copper, gold, or tantalum
    c. Sand dredging involving an area of 50 ha or more

12. Petroleum
    a. Oil and gas field development
    b. Construction of offshore pipelines in excess of 50 km in length
    c. Construction of oil and gas separation, processing, handling, and storage facilities
    d. Construction of oil refineries
    e. Construction of product depots for the storage of petrol, gas, or diesel (excluding service stations), which are located within 3 km of any commercial, industrial, or residential areas and which have a combined storage capacity of 60,000 barrels or more

13. Power generation and transmission
    a. Construction of steam-generated power stations burning fossil fuels and having a capacity of more than 10 MW.
    b. Dams and hydroelectric power schemes with either or both of the following:
       i. Dams over 15 m high and ancillary structures covering a total area in excess of 40 ha
       ii. Reservoirs with a surface area in excess of 400 ha
    c. Construction of combined cycle power stations
    d. Construction of nuclear-fueled power stations

14. Quarries
    Proposed quarrying of aggregate, limestone, silica, quartzite, sandstone marble, and decorative building stone within 3 km of any existing residential, commercial, or industrial areas, or any area for which a license, permit, or approval has been granted for residential, commercial, or industrial development.

15. Railways
    a. Construction of new routes
    b. Construction of branch lines

16. Transportation

17. Resort and recreational development
    a. Construction of coastal resort—facilities or hotels with more than 80 rooms
    b. Hill station resort or hotel development covering an area of 50 ha or more

c. Development of tourist or recreational facilities in national parks

d. Development of tourist or recreational facilities on islands in surrounding waters that may be declared as national marine parks

18. Waste treatment and disposal
    a. Toxic and hazardous waste
       i. Construction of incineration plant
       ii. Construction of recovery plant (off-site)
       iii. Construction of wastewater treatment plant (off-site)
       iv. Construction of secure landfill facility
       v. Construction of storage facility (off-site)
    b. Municipal solid waste
       i. Construction of incineration plant
       ii. Construction of composing plant
       iii. Construction of recovery/recycling plant
       iv. Construction of municipal solid waste landfill facility
    c. Municipal sewage
       i. Construction of wastewater treatment plant
       ii. Construction of marine outfall

19. Water supply
    a. Construction of dams, impounding reservoir with a surface area of 200 ha or more
    b. Groundwater development for industrial, agricultural, or urban water supply of greater than 4500 m³ per day

20.2 Principles and Objectives of EIA

Certain principles guide the successful implementation of an EIA study; in the same vein, there are certain aims and objectives that should be accomplished to produce an EIS.

20.2.1 Principles of an EIA Study

EIA is a procedure designed to identify the potential impacts (positive/adverse) of a development project on the surrounding environment.

Environmental clearance of these activities is carried out by the central or the state governments, with the following objectives:

1. Optimal utilization of finite natural resources through use of better technologies and management packages.
2. Incorporating suitable remedial measures at the project formulation stage to ensure minimum harm to the surrounding environment.

The objective of an EIA is to ensure that environmental problems are foreseen and properly addressed. To achieve this aim, decision makers must fully understand the EIA's conclusions. Most decision makers overlook information unless it is presented in terms of its immediate use in a concise form. Therefore, the format of EIA may include

1. "Hard" factors and prediction about impacts, comments on the reliability of this information, and summarized consequences of each of the proposed options.
2. Terminology and vocabulary used by the decision makers should communicate the right message to the community affected by the project.
3. Essential findings should be in a concise document, supported by separate background materials wherever it is necessary.
4. The document should be made easy to use, providing visuals wherever possible.

Principle 1: Focus on the Main Issues

The scope of the EIA should be limited only to the most likely and most serious of the possible environmental impacts. Some EIAs have resulted in large and complex reports running to several thousand pages. Such extensive work is immediately useful to decision makers and project planners. While suggesting the mitigating measures, it is important to focus on the study only on workable, acceptable solutions to the problems rather than wasting time on considering measures that are impractical or totally unacceptable to the developer or to the government. It is desirable to provide a summary of information relevant to each decision making group.

Principle 2: Involve the Appropriate Persons and Groups

It is important to be selective while involving various groups/people in EIA approaches. The following persons may be associated with this process:

1. Those appointed to manage and undertake the EIA process
2. Those who can contribute facts, ideas, or concerns to the study, including scientists, economists, engineers, policy makers, and representatives of interested or affected groups/people
3. Those who have direct authority to permit, control, or alter the project; that is, the decision makers including the developers, aid agencies, or competent authorities, regulators, and politicians

Principle 3: Link Information to Decisions about the Project

An EIA should be organized so that it directly supports the decisions that need to be taken about the proposed project. It should start early enough to provide information to improve the designs, and should progress through the several stages of project planning by providing the relevant information necessary for arriving at a particular decision for the project.

Principle 4: Present Clear Options for the Mitigation of Impacts and for Sound Environment Management

To help decision makers, the EIA must be designed to put forward clear choices on the planning and implementation of the project. It should clearly mention the positive as well as adverse impacts of the project on the environment. To mitigate adverse impacts, the EIA could propose the following measures:

1. Pollution control technology or design features
2. The reduction, treatment, recycling, or disposal of wastes
3. Optional utilization of natural resources
4. Compensation or concessions to affected groups/people

To enhance environment compatibility, the EIA should suggest

1. Several alternative sites
2. Changes to the project design and operation
3. Limitations to its initial sizes or growth
4. Separate programs that contribute in a positive way to local resources or to the quality of the environment

In addition, to ensure that the implementation of an approved project is environmentally sound, the EIA may prescribe

1. Monitoring programs or periodic impact reviews
2. Contingency plans for regulatory action
3. The involvement of the local community before taking final decisions

Principle 5

Details about the consequences of considered alternatives along with reasons to select the particular alternative should be provided to the decision makers.

20.2.2 Objectives of EIA Study

When it is established that EIA is necessary for clearance of the project after preliminary assessment, the developer should go to a full-fledged study. The various steps involved in an EIA study include the following:

1. To commission the project to an independent project coordinator with an expert team to carry out the detailed study for collection of baseline data
2. Identification of the key decision makers who will plan, finance, permit, and control the proposed project
3. To understand existing laws and regulations that will affect the decision about the project
4. Making contact with each of the various decision makers
5. Determining how and when the EIA's findings will be communicated

To carry out the detailed study, the study team can pursue the following steps:

1. The EIA study team should ensure that all the issues of importance are taken into account.
2. The team should initiate discussion with the project developers, decision makers, regulatory agencies, and local community on all the possible issues and concerns raised by the various groups.
3. The team should collect required information along with supportive baseline data to formulate the project profile.

4. To prepare the EIA, study reports taking into account the identification, predictions, or various factors of project development, evaluating and suggesting mitigative measures to save environmental degradation for the successful implementation of the proposed project.

The last step in the EIA process is documentation of the process and the conclusions. Many technically first-rate EIA studies fail to exert their importance and usefulness because of poor documentation. The EIA can achieve its purpose only if its findings are well communicated to decision makers.

To summarize, an EIA report typically contains

1. An executive summary of the EIA findings
2. A description of the proposed development project
3. The major environmental and natural resources issues
4. The project's impacts on the environment (in comparison with a baseline environment as it would be without the project) and how these impacts were identified and predicted
5. A discussion of options for mitigating adverse impacts and for shaping the project to suit its proposed environment, and an analysis of the trade-offs involved in choosing between alternative actions
6. An overview of gaps or uncertainties in the information
7. A summary of the EIA for the general public

All of this should be contained in a very precise, easy-to-read document, with cross-references to background documentation, if any, which may be provided in an appendix.

20.3 EIA Process

The EIA process involves both preliminary and specific processes.

20.3.1 Screening

Screening is the first and simplest stage of project evaluation. Screening helps to clear the type of projects that in past experience are not likely to cause any serious environmental problems/degradation.

20.3.2 Preliminary Assessment

If screening does not automatically clear a project, the developer may be asked to undertake a preliminary assessment. This involves sufficient research expert advice to

1. Identify the project's key impacts on the local environment
2. Generally describe and predict the extent of the impacts
3. Evaluate their importance to decision makers

The preliminary assessment can be used to assist early project planning, for instance, to narrow down the discussion of possible sites, to serve as an early warning that the project may have serious environmental problems, etc. It is in the developer's interest to do a preliminary assessment, as this step can clear some of the projects that do not require a full EIA.

20.3.3 Methods and Procedures

Many methods and procedures have been developed in EIA over the years. They all share the basic goal of providing a comprehensive and systematic environmental evaluation of the project, with the greatest degree of objectivity. These methods include the following.

20.3.3.1 Methods

1. *Good manufacturing practice (GMP)*: RAM and INOVA are EIA systems applicable to genetically modified plants' impact on the environment [14].

2. *Fuzzy logic*: EIA methods need measurement data to estimate values of impact indicators. However, many of the environmental impacts cannot be qualified, for example, landscape quality, lifestyle quality, and social acceptance. Therefore, the need to cater to this factor is created. Thus, information from similar EIAs, expert judgment, and community sentiment are employed. In addition, approximate reasoning methods known as fuzzy logic can be employed [36].

3. *The checklist method*: In the checklist method, all potential environmental impacts for the various project alternatives are listed and the anticipated magnitude of each impact is described quantitatively. For example, negative impacts can be indicated with a minus (−) sign. A more or moderate impact can be shown by two minus signs (− −), while a more severe impact can be shown with four minus signs, that is, (− − − −). A beneficial or positive impact can be shown with plus (+) signs. If the environmental impact is not applicable for a particular project, a zero (0) sign is used. Such a list presents an easy visual overview of the assessment.

4. *The matrix method*: The matrix method attempts to quantify or grade the relation impacts of the project alternatives and provide a numerical basis for comparison and evaluation. The anticipated magnitude of each potential impact may be rated on a scale of 0 to 10. The higher numbers may represent severe adverse effects, whereas the lower numbers represent minor or negligible effects, and zero (0) would indicate no expected impact for a particular activity or environmental component.

5. *Environmental quality index (EQI)*: EQI can be calculated for each project alternative. The alternative with the least index is the one that will probably cause the least harmful environmental impact overall. EQI can be estimated within the matrix model of EIA where numerical weighting factors are used to indicate the relative importance of a particular impact. These weighting factors are agreed on by the assessment team and are site and project specific. For example, the impact on groundwater quality may be considered more important in a particular area than impacts on air quality, particularly if the groundwater is a sole source of potable water. Therefore, groundwater quality could be assigned a relative importance or weight of say 0.5 compared to 0.2 for air quality. Weighting factors can be multiplied by the respective impact magnitudes to put each impact in perspective [30].

6. *Life cycle analysis (LCA)*: LCA is usually applied in the case of industrial products and it is used for identifying and measuring the impact of industrial products on the environment. These EIAs consider activities related to extraction of raw materials, ancillary materials equipment, production, use, and disposal [7]. LCA can also be used in the water cycle sector to evaluate the environmental profile of existing wastewater treatment plants and to propose alternative and economically sound practices [1,31,32,34].

7. *Computation, simulation, and modeling*: EIA is currently carried out with the aid of computers with various software, algorithms that are based on ISO standards. An example of an environmental management tool is SiSOSTAQUA LCA management tool (SIMPLE) SL Ecomvent 2.1 database (SCLCI 2009). It is also possible to model and simulate environmental impact using MATLAB® and Simulink® programs developed by Matrix Laboratories [34].

8. *Environmental audit*: An environmental audit is the inventory assessment of activities in relation to compliance with regulatory requirements. An environmental audit is usually applicable to industrial activities. Environmental auditing is a management tool that enhances the overall environmental performance of manufacturing activities and is now generally a requirement for property transfers and reduction of legal liabilities due to improper hypothesis or with a simpler approach. A comparison of outcome of the audit exercise with predictions in the EIA document is also performed [45]. After an EIA, precautionary and polluter pays principle may be applied to decide whether to reject, modify, or require strict liability or insurance coverage to a project based on predicted harms [45].

9. *Environmental impact statements (EIS)*: EISs are documents prepared to summarize the impact analysis of the EIA study. It is expedient for EIS to have the following sections for clarifications of the objectives:

   - Mitigating environmental damage
   - Secondary or indirect impacts
   - Irreversible and reversible commitments of resources
   - Project alternatives
   - No action alternatives
   - Public input and participation

20.3.3.2 General Procedures

General procedures for an environmental impact assessment involve the following:

1. Notification
2. Preparation and distribution of EIA documents
3. Consultation
4. Final decision

5. Postproject analysis
6. Public address

1. *Notification*

   With respect to such projects as listed earlier that require an EIA, notification of intention is made in a popular newspaper in the affected area in which the proposed project is to be sited. Thereafter, the appropriate assessment committee includes suitable members of the public residing in the affected area, representatives of nongovernmental organizations, sometimes foreign technical partners, and experts in EIA as constituted by the designated agency. The committee is headed by a lead expert in the subject (Figure 20.1) [38].

2. *Preparation and Distribution of EIA Document*

   After the various studies and analysis have been done, an EIS is prepared and distributed to the concerned people including the local and state government or counties as the case may be. The purpose of the distribution of the document is for response by the public and various interested parties to respond. There is usually a time frame of about 30 days before a consultation is done and a final decision is made [41].

3. *Consultation*

   The response will precede consultation by the expert panel, which will review the public response to benefits and alternatives for a final decision.

PUBLIC NOTICE
FEDERAL MINISTRY OF ENVIRONMENT

PUBLIC DISPLAY EXERCISE ON THE ENVIRONMENTAL IMPACT ASSESSMENT (EIA) OF THE PROPOSED DEVELOPMENT OF 50MW SOLAR FARM AT KADO, IGABI LGA, KADUNA STATE BY SYNERGENT POWERSHARE NIGERIA LIMITED.

In accordance with the Environmental Impact Assessment (EIA) Act No. 86 of 1992. Which makes it mandatory for proponents of all new major development activities to carry out Environmental Impact Assessment of their proposed projects, the Federal Ministry of Environment hereby announces a twenty – one (21) working day Public Notice for information and comments on the draft EIA report submitted by Synergent PowerShare Nigeria Limited.

The Display Centers Are:

- Headquarters Igabi Local Government Area, Turunku, Kaduna State.
- Kaduna State Ministry of Environment, Kaduna, Kaduna State.
- Federal Ministry of Environment, Federal Secretariat, Kaduna, Kaduna State.
- Federal Ministry of Environment, Environment House, Independence Way, Central Area Abuja.
- Federal Ministry of Environment, Conservation (Green) Building, Plot 444 Aguiyi Ironsi Street, Maitama, F.C.T Abuja.
- www.ea-environment.org

Project Description:

The proposed project will entail the design, development and installation of a **50MW** solar farm station which will be built in five (5) phases consisting of **10MW** of peak capacity for each phase on about **60** hectares of land in **Kado-Mando** community, **Igabi** Local Government Area, Kaduna State.

Duration of Display:

Date: 24ᵗʰ June to 22ⁿᵈ July, 2013, Time: 8:00am -4:00pm Daily.

ALL COMMENTS RECEIVED SHOULD BE FORWARDED TO THE HONOURABLE MINISTER, FEDERAL MINISTRY OF ENVIRONMENT ON OR BEFORE **30ᵀᴴ JULY, 2013.**

SIGNED
PERMANENT SECRETARY
For: Honourable Minister

FIGURE 20.1 Sample of environmental impact assessment public notification in Nigeria.

4. *Final Decision*

   The final decision leads to preparation of an EIS document. The EIS report is meant to be used as a planning and decision-making tool. It is supposed to be objective and unbiased and is not meant to either promote or block the implementation of a proposed project.

5. *Post-Project Analysis*

   The post-project analysis is an important development in EIA. It enables a comparison between environmental impact projection and the actual impact the project may have on the environment. The post-project analysis is usually periodic in which short-term and long-term effects are characterized.

6. *Public Hearing*

   Recently, a new dimension has been added in the EIA process known as a public hearing. The designated agency like a State Environmental Pollution Agency (SEPA) shall give notice for an environmental public hearing mentioning the date, time, and venue, which shall be published in at least two newspapers widely circulated in the region around the project. Suggestions, views, comments, and objections of the public on the project shall be invited within 30 days from the date of publication of the notice. All persons including residents, environmental groups, and others located at the project site or displacement sites likely to be affected can participate in the public hearing. They can also make oval/written suggestions to the SEPA for consideration during environmental clearance of the project. A public hearing is usually required in the case of projects involving large displacement or having severe environmental impacts [13].

20.3.3.3 Specific Procedures

Specific procedures involving EIA include the following:

1. Description of the existing environment
2. Description of the proposed project
3. Assessment of environmental impacts
4. Unavoidable adverse environmental impacts
5. Methods for reducing adverse impacts
6. Secondary or indirect impacts
7. Alternatives to the proposed project
8. Irreversible commitments of energy and resources
9. Public input and review

1. *Description of the Existing Environment*: It is quite desirable to have an accurate picture of what the pre-development (existing) environmental conditions are at and near the proposed site. The interdisciplinary team constituted to carry out the EIA assesses this. The inventory of the existing natural resources and urban facilities in the vicinity of the project sites typically includes the following: geology, soil, topography, water resources, vegetation, wildlife, air quality and noise, transportation, public utilities, population, and historical or unique cultural features.

2. *Description of the Proposed Project*: The developer's consulting engineers, architects, and planners are expected to provide information on the total area of the project, number of building lots, and relative distribution of residential, commercial, and industrial facilities. A proposed stormwater drainage including underground pipelines and stormwater detention basins would be shown. Some other information that could be useful in an EIA study includes type of construction to be undertaken, landscaping, and the expected market value of the constructed facilities. Plans for proposed water supply and wastewater collection system would be submitted showing the location and capacities of pipelines and other utilities.

3. *Assessment of Environmental Impacts*: Assessment of environmental impact of a project is aimed to predict any adverse or beneficial effects of a proposed project on the natural and urban environment and to offer alternative options. This is done so that measures can be taken to minimize or eliminate the harmful impacts when the project is implemented. The team to conduct the impact is usually multidisciplinary and may include engineers, ecologists, geologists, urban planners, social scientists, architects, and archaeologists, depending on the nature and scope of the project. Environmental impacts can be evaluated directly or indirectly. Use of mathematical models, computation, simulation, and modeling are making EIA more standardized and more reliable air quality impact can be assessed using existing databases, simulation, and mathematical modeling. However, impacts on vegetation and wildlife are more difficult to evaluate objectively and it is important to distinguish between long-term and short-term impacts [39].

4. *Unavoidable Adverse Environmental Impacts*: Some environmental impacts are unavoidable and these include factory noise, discharge of industrial waste effluents, gas emissions, etc. These should be recognized and well documented.

5. *Secondary or Indirect Impacts*: Some impacts are secondary or indirect. These impacts are not immediately apparent and are not directly caused by the project itself.

6. *Methods for Reducing Adverse Impacts*: It is quite expedient for methods to mitigate adverse impacts to be enumerated. This quality makes an EIA objective.

7. *Alternatives to the Proposed Project*: The alternatives to the proposed project should be proposed and well considered in the EIA and itemized in preparation of the EIA statement [11].

8. *Irreversible Commitments of Energy and Resources*: Consumption of energy has an environmental impact; therefore, energy consumption and material and natural resources requirements should be fashioned with cost analysis.

9. *Public input and review*: Public participation in EIA in the form of actual impact assessment and public hearings is a very important aspect of EIA and should be well documented in EIS to conform to the global EIA requirements.

20.4 Applications in Urban Water Reuse

Having put into consideration the principles, process, and methods previously discussed, applications of EIA in urban water reuse will be enumerated with specific procedures. First, the components of urban water reuse need to be identified and these include

1. Stormwater system
2. Wastewater system
3. Hazardous wastewater system
4. Wetlands: Natural and constructed
5. Drinking water system
6. Industrial effluents
7. Desalination system

The environmental impact categories will thereafter be identified and enumerated, for example, acidification potential, global warning potential, ozone depletion potential, ecotoxicity potential, etc. (Table 20.1).

20.4.1 Stormwater Management

Surface runoff and in uncontrolled stormwater can cause significant environmental damage which include flooding and water pollution. Therefore, in the design and analysis of stormwater control facilities, an estimate of the rate and value of surface runoff to be controlled is important. Computers are frequently used to analyze complex mathematical models of stormwater control systems and the use of on-site storage or detention basis is a common method for controlling stormwater [21].

20.4.2 Wastewater Management System

All the processes involved in wastewater management are assessed for their environmental impact and these include primary, secondary, and tertiary treatment facilities. The grit-removed and pretreatment waste that are buried in the ground must not pollute groundwater. The chemicals used for disinfection, like ozone and chlorine, will be assessed for potential impact in air, land, and water. Impact of sludge disposal and transportation to landfills will be documented. Energy consumption, sewers, and sewage systems must be shown to be safe and of good quality. Noise from generators and plants must meet acceptable limits [16,27,29].

20.4.3 Hazardous Wastewater System

Tanks for storage of hazardous waste and pipeline systems must meet international standard requirements to prevent leaking. It is recommended that fiberglass or cathodically protected steel be used to prevent corrosion and in addition double-walled tanks and cutoff walls with impervious underlayment be used for temporary storage of hazardous waste. Automatic leak detection systems and alarms must be installed as well as devices to prevent overfill. EIA concerning hazardous wastewater includes

TABLE 20.1 Example of Environmental Impact Categories for Wastewater Treatment for Urban Reuse

Impact Type	Measurable Parameter	Sources of Impact
Acidification potential (AP)	kg SO_2 eq	Contribution from substances that produce sulfuric acid when they are in contact with water. When these substances are present in the environment, they produce acid rain, causing terrestrial and aquatic species to degrade.
Global warming potential (GWP)	kg CO_2 eq	Contribution of the various emissions to the increase in global warming. The most important substances are CO_2, CH_4, N_2O, and the halogenated hydrocarbons.
Eutrophication (EP)	kg PO_4 eq	Contribution of the various emissions to the accumulation of nutrients in the environment. When nutrients accumulate in aquatic ecosystems, plant growth increases and depletes oxygen levels.
Photochemical oxidation (PHO)	kg formed ozone	Contribution of the various emissions to the formation of photo-oxidant substances (particularly ozone and peroxy-acetyl nitrate) through the photochemical oxidation of volatile organic substances and carbon monoxide.
Depletion of abiotic resources (DAR)	kg antimony eq	Contribution of the various emissions to the extraction of resources, including their availability, energy content, concentration, and rate of use.
Ozone depletion potential (ODP)	kg CFC 11 eq	Contribution of substances that deplete the ozone stratospheric layer. The most important substances are chlorated and bromated halocarbons, particularly trichlorofluoromethane (CFC 11, also known as Freon-11).
Ecotoxicity potential (ETP)	kg 1,4 DCB eq	Combined result of freshwater aquatic and sediment ecotoxicity, marine aquatic and sediment ecotoxicity, human toxicity, and terrestrial ecotoxicity. These substances affect the health of humans, flora, and fauna in different environments. The most important substances are heavy metals, persistent organic pollutants, and volatile organic compounds.
Fresh water use (FWU)	m^3	Reflects the consumption of freshwater from different sources (surface water, groundwater, etc.) throughout the whole life cycle.
Cumulative energy demand (CED)	MJ	Expresses the accumulated renewable and nonrenewable energy consumption throughout the whole life cycle.

Source: Adapted from Pasqualino, J.C., Meneses, M., and Castells, F. 2010. *Industrial Ecology* 15(1), 49–63.

the study of the nature of hazardous wastes whether they are infective, corrosive, or volatile. The land treatment for hazardous liquid waste must be shown not to have adverse effect on vegetation and groundwater.

20.4.4 Drinking Water System

The treatment of drinking water involves the use of chemicals. The potential environmental chemicals would be enumerated. The distributive system involves pipelines that should not affect the soil where they are laid.

20.4.5 Wetlands

Natural and constructed wetlands are becoming increasingly popular in reclamation and reuse of wastewater, particularly industrial water, stormwater, and water contaminated with toxic metals. While not all plants are resistant to toxic metals and pesticides, plants that are well known to abstract pollutants like the poplar trees, sunflower, and water hyacinth can be introduced.

20.4.6 Industrial Effluents

Pretreatment of industrial effluents should be made mandatory and the nature of the industrial effluents should be documented. Recent observations have shown the presence of pharmaceutical residues in many wastewaters, which may impact the ecosystem. Since most wastewater treatment plants do not have facilities for tertiary treatment, which can remove pharmaceutical residues, pharmaceutical companies are being encouraged to include

tertiary treatment for their wastewater before the effluent is discharged into water bodies [23,24].

20.4.7 Post-Project Monitoring

In view of the fact that different components constitute the system for urban water reuse, it is expedient to continue to monitor the environmental impact for comparison of the proposed impact as documented in the EIS and the actual impact observed as the services commenced.

20.5 Discussion

EIA is a procedure designed to identity the potential impacts, that is, positive or negative impacts of a development project on the surrounding environment. An EIS for a proposed project is a written report that summarizes the finding of a detailed review process of an EIA. Studies on impact of developmental activities could be said to have commended since the 1960s and attained formal status in 1969 in the United States with the enactment of the National Environmental Policy Act. Other countries later legislated acts on EIAs and these include Australia with the Environment Protection Impact of Proposals Act of 1974, which was later superseded by the Environment Protection and Biodiversity Conservation Act (EPBC) of 1999, Environmental Impact Assessment Law of China of 2004, the Egyptian Law 4 of 1994, Environmental Impact Assessment Ordinance of 1997 of Hong Kong, Malaysia Environmental Quality Act of 1974, New Zealand Resource Management Act of 1991, the Russian Federal Law on Ecological Expertise of 1995, the Regulation on Assessment of Impact from Intended Business and Other Activity on Environment in the Russian Federation of 2000, the Nigerian EIA Act 86 of 1992, and Nepal's Environmental Impact Assessment Guideline of 1997.

The EIS is meant to be used as a planning and decision-making tool. It is supposed to be objective and unbiased and it is not meant to promote or block the implementation of a proposed project. The greatest benefit of the EIS process is that environmental concerns must be examined thoroughly. Unexpected damage due to a construction project is significantly minimized. Unfortunately, EIS reports are sometime manipulated by developers to promote a construction project or they are misused by special interest groups to shop a project completely. A criticism of the EIS is that it may be imposed on small projects that might not warrant so much concern. An independent inventory of the natural resources and ecosystem of a proposed project environment should be taken before an EIA study to ensure transparency of the report [9]. With reference to urban water reuse, the various components of urban water reuse systems would be studied for their environmental impacts. These include the stormwater system, the wastewater treatment system, hazardous wastewater treatment system, sludge management, natural and constructed wetlands, desalination system, and municipal water treatment system [37]. Hazardous wastewater treatment system is a major environmental concern. Therefore, design and construction of tanks used in storing hazardous waste must ensure protection of the environment. There have been many cases of environmental damage particularly groundwater contamination caused by leaking pipes or storage tanks. It has been advised that fiberglass or cathodically protected steel be used for corrosion control. Other recommendations include double takes and thick walls with impervious under laymen. An EIA team is multidisciplinary and consists of different professionals including engineers, environmental scientists, ecologists, nongovernmental organizations with environmental interests, the public, and EIA experts. The various impacts to be considered include impact on air, water, and land.

In the case of urban water reuse, some chemicals used in disinfection like chlorine, ozone, etc. will be listed and the potential impact on the environment will be documented. Perhaps of great importance is sludge management whereby different alternatives can be proposed including sludge disposal or polluting for agricultural uses [33]. There is also concern for energy consumption because most processes in water treatment are dependent on electrical energy [28,30]. Wetland use for urban water reuse poses a challenge for harvested plants, which might abstract pesticides, chemicals, and toxic metals. Usually, incineration of such plants is recommended. There should be measures in preventing toxic metals from entering the food chain. Wetlands should also be monitored for species of the ecosystem. The impact of construction of treatment plants on local vegetation and wildlife should not be overlooked. The destruction of woodlands and meadows to make room for plants and roads can lead to significant ecological problems particularly if there are any rare or endangered species in the area. Cutting down trees

and paving over meadows cause soil erosion and stream sedimentation and long term may cause displacement of wildlife [11].

In developing countries where the electricity supply is usually inadequate, major sources of air pollution include final combination for power generation. An EIA report should know this impact can be integrated. Public participation and dissemination of information is usually a requirement for an EIA and should be well noted. Computation, simulation, and modeling continue to aid in improvement of EIAs and in standardization of impact assessments and reports.

20.6 Summary and Conclusions

EIA has become a requirement globally in any developmental project including systems that contribute to urban water reuse. While urban water reuse has the advantage of conserving potable water, which might require being transferred over a distance, the environmental impact of various processes and alternatives need to be considered. As the need for water increases globally with rising population and urban growth, water reuse will become increased as well as EIAs. It is therefore important that EIA studies become more refined for best practices.

References

1. Arpke, A. and Hutzler, N. 2006. Domestic water use in the United States: A life-cycle approach. *Journal of Industrial Ecology*, 10(1–2), 169–118.
2. Beavis, P. and Lundie, S. 2003. Integrated environmental assessment of tertiary and residuals treatment: LCA in the wastewater industry. *Water Science and Technology*, 47(7–8), 109–116.
3. Bhatia, S.C. (ed.) 2007. *Environmental Chemistry*. CBS, New Delhi, pp. 505–527.
4. Carroll, B. and Turpin, T. 2009. *Environmental Impact Assessment Handbook*, 2nd Edition. Thomas Telford Ltd., Telford, Shrophire.
5. CEAA. 2012. *What Is the Canadian Environmental Assessment Act, 2012?* Basics of Environmental Assessment. Canadian Environmental Assessment Agency. www.canada/en/environmental assessment_agency/index.html.
6. CML Institute of Environmental Studies. 2001. *CML 2 Baseline Method 2000*. University of Leiden, Leiden, the Netherlands.
7. Dixon, A., Simon, M., and Burkitt, T. 2003. Assessing the environmental impact of two options for small scale wastewater treatment: Comparing a reedbed and an aerated biological filter using a life cycle approach. *Ecological Engineering*, 20(4), 297–308.
8. Eccleston, C. and Doub, J.P. 2012. *Preparing NEPA Environmental Assessments: A User's Guide to Best Professional Practices*. CRC Press, Boca Raton, FL.
9. Elliott, M. and Thomas, I. 2009. *Environment Impact Assessment in Australia: Theory and Practice*, 5th Edition. Federation Press, Sydney.

10. Emond, D.P. 1978. *Environmental Assessment Law in Canada*. Emond-Montgomery Ltd., Toronto, Ontario.

11. Fernandes, J.P. 2000. EIA procedure, landscape ecology and conservation management—Evaluation of alternatives in a highway EIA process. *Environmental Impact Assessment Review* 20(6), 665–680.

12. Glasson, J., Therivel, R., and Chadwick, A. 2005. *Introduction to Environmental Impact Assessment*. Routledge, London.

13. Hanna, K. 2009. *Environmental Impact Assessment: Practice and Participation*, 2nd Edition. Oxford University Press, Oxford.

14. Hitzschky, K. and Silviera, J. 2009. A proposed impact assessment method for genetically modified plants (As—GMP method). *Environmental Impact Assessment Review*, 29, 348–368.

15. Holder, J. 2004. *Environment Assessment: The Regulation of Decision Making*. Oxford University Press, New York.

16. Hospido, A., Moreira, M.T., Fernandez-Couto, M., and Feijoo, G. 2004. Environmental performance of a municipal wastewater treatment plant. *International Journal of Life Cycle Assessment*, 9(4), 261–271.

17. IAIA. 1999. *Principle of Environmental Impact Assessment Best Practice*. International Association for Impact Assessment. www.iaia.org.

18. ISO 14040. 2006. *Environmental Management—Life Cycle Assessment Principles and Framework*. International Organization for Standardisation (ISO), Geneva, Switzerland.

19. Jay, S., Jones, C., Slinn, P., and Wood, C. 2007. Environmental impact assessment: Retrospect and prospect. *Environmental Impact Assessment Review*, 27(4), 289–300.

20. Jeffery, M.I. 1989. *Environmental Approvals in Canada*. Butterworths, Toronto, Ontario.

21. Lenzen, M., Murray, S., Korte, B., and Dey, C. 2003. Environmental impact assessment including indirect effects a case study using input-output analysis. *Environmental Impact Assessment Review*, 23(3), 263–282.

22. Lim, S.R., Park, D., and Park, J.M. 2008. Environmental and economic feasibility study of a total wastewater treatment network system. *Journal of Environmental Management*, 88(3), 564–575.

23. Lundin, M., Bengtsson, M., and Molander, S. 2000. Life cycle assessment of wastewater systems: Influence of system boundaries and scale on calculated environmental loads. *Environmental Science and Technology*, 34(1), 180–186.

24. Meneses, M., Pasqualino, J., Cespedes, R., and Castells, F. 2010. Alternatives for reducing the environmental impact of the main residue from a desalination plant. *Journal of Industrial Ecology*, 14(3), 512–527.

25. Meneses, M., Pasqualino, J.C., and Castells, F. 2010. Environmental assessment of urban waste water reuse: Treatment alternatives and applications. *Chemosphere*, 81, 266–527.

26. Meneses, M., Pasqualino, J.C., and Castells, F. 2009. Poster MO237. Environmental assessment of water supply technologies. Paper presented at the 19th annual meeting of SETAC Europe, May 31 to June 4, Goteborg, Sweden.

27. Milai, C.L., Chenoweth, J., Chapagain, A., Orr, S., Anton, A., and Clift, R. 2009. Assessing freshwater use impacts in LCA: Part 1—Inventory modeling and characterization factors for the main impact pathways. *International Journal of life Cycle Assessment*, 14(1), 28–42.

28. Munoz, I. and Fernandez Aba, A.R. 2008. Reducing the environmental impacts of reverse osmosis desalination by using brackish groundwater resources. *Water Research*, 42(3), 801–811.

29. Munoz, I., Rodriguez, A., Rosal, R., and Fernandez Alba, A.R. 2009. Life cycle assessment of urban wastewater reuse with ozonation as tertiary treatment: A focus on toxicity related impacts. *Science of the Total Environment*, 407(4), 1245–1256.

30. Nathanson, J.A. (ed.) 2003. *Basic Environmental Technology*, 4th Edition. Prentice Hall, New Jersey, pp. 475–477.

31. Ortiz, M., Raluy, R.G., Serra, L., and Uche, J. 2007. Life cycle assessment of water treatment technologies: Wastewater and water-reuse in a small town. *Desalination*, 204 (1–3), 121–131.

32. Owens, J.W. 2001. Water resources in life-cycle impact assessment: Considerations in choosing category indicators. *Journal of Industrial Ecology*, 5(2), 37–54.

33. Palme, U., Lundin, M., Tillman, A.M., and Molander, S. 2005. Sustainable development indicators for wastewater systems: Researchers and indicator users in a co-operative case study. *Resources, Conservation and Recycling*, 43(3), 293–311.

34. Pasqualino, J., Meneses, M., Abella, M., and Castells, F. 2009. LCA as a decision support tool for the environmental improvement of the operation of a municipal wastewater treatment plant. *Environmental Science and Technology*, 43(9), 3300–3307.

35. Pasqualino, J.C., Meneses, M., and Castells, F. 2010. Life cycle assessment of urban waste water reclamation and reuse alternatives. *Industrial Ecology*, 15(1), 49–63.

36. Peche, R. and Rodriguez, E. 2009. Environmental impact assessment procedure: A new approach based on fuzzy logic. *Environmental Impact Assessment Review*, 29, 275–283.

37. Pfister, S., Koehler, A., and Hellweg, S. 2009. Assessing the environmental impacts of freshwater consumption in LCA. *Environmental Science and Technology*, 43(11), 4098–4104.

38. Ruddy, T.F. and Hilty, L.M. 2008. Impact assessment and policy learning in the European Commission. *Environmental Impact Assessment Review*, 28, 90–105.

39. Sala, L. and Mo Serra. 2004. Towards sustainability in water recycling. *Water Science and Technology*, 50(2), 1–7.

40. Sands, P. 1989. The environment, community and international law. *Harvard International Law Journal*, 30(2), 393–420.

41. Shepherd, A. and Ortolano, L. 1996. Strategic environmental assessment for sustainable urban development. *Environmental Impact Assessment Review, 7*, 285–292.

42. *The Environment Protection and Biodiversity Conservation Act.* 2010. The Department of the Environment, Water, Heritage and the Arts, Australia, retrieved September 9, 2010. www.environment.gov.au/

43. UNECE. 1991. Convention on Environmental Impact Assessment in a Transboundary Context. Espoo, Unece. org. Retrieved January 3, 2013.

44. Weiss, E. 1999. *Understanding Compliance with International Environmental Agreements: The Bakers Dozen Myths.* University of Richmond. www.environment. gov.au.

45. Wilson, L. 1998. A practical method for environmental impact assessment audits. *Environmental Impact Assessment Review,* 18, 59–71.

46. Wolfrum, R. and Matz, N. 2003. *Conflicts in International Environmental Law.* Max-Planck-Institut fur Auslandisches Offentliches Recht and Volkerrech, pp. 38–64.

47. Young, O. 1999. *The Effectiveness of International Environmental Regimes.* MIT Press, Boston.

21

Environmental Impact Assessment for Urban Water Reuse: Tool for Sustainable Development

Sandra Reinstädtler
Brandenburg University of Technology Cottbus-Senftenberg

Michael Schmidt
Brandenburg University of Technology Cottbus-Senftenberg

PREFACE

Water reuse is one of the most pending management and technological challenges and in observation highly interdisciplinary in correlation to smart blue-green city development, public human health [7], and environmental health as well as in general to drought, desertification, land degradation, resulting food insecurity, poverty, hunger, climate change insecurities, and its ((water) resource) management. The quantitative facts are concretizing the necessity for water reuse demands, control by landscape and environmental planning instrumentations such as the environmental impact assessment (EIA) and water management options: in general, about 96.5% of the world's water is captured as saltwater in the oceans and salt lakes, which is about 1338 million km^3. One percent or 13 million km^3 is tied up as saline groundwater and the remaining 2.5% or 35 million km^3 are freshwater resources [45].

Water reuse is affecting positively with its correlation to water scarcity and directly influencing the most vulnerable urban ecosystems and life as well as human and environmental health quality especially of the poorest affected populations in so-called slum areas worldwide. Therefore, environmental [25] and urban water security should be considered as being a most important aim and mission within the context of controlling by the planning instrument of the EIA, smart city development, and combined water (reuse) management as well as monitoring. Managing water scarcity, upgrading water quality, and therefore water resources for water and urban security requires observances in multifarious interactions of the different social, ecological, economical, cultural, land use, legal, and political systems as well as including its pressures and drivers to be acknowledged.

This chapter discusses the role of EIA in the context of urban water reuse, the responsibility for cultivating enduring and sustaining freshwater resources, and its chances and challenges within the instrumental proceeding. The environmental media of water is standing in the focus of the overall holistic approach in environmental assessment and in specific for urban purposes in efficient recycling processes of qualified water purity. Life and health quality are next to environmental and climate friendly usage of natural and water resources.

Already for many years, existing water crises have to be acknowledged within the fact of just happening and further on coming changes in precipitation rates, rising temperatures, and other climatic insecurities. By using the instrumentation of EIA for a controlled water reuse, it is exactly adapting as a tool for interacting in spatial and environmental planning, projecting, and planning permission hearings.

21.1 Introduction

"In an age when man has forgotten his origins and is blind even to his most essential needs for survival, water along with other resources has become the victim of his indifference" [6]. Water quality, amount, and place of available water as well as water pollution are some of the most important guiding factors to sustainable processes, which are secured in quality and quantity on a project level by the EIA procedures. Important agreements, charters, and legal inventions were developed, came into force, and are thus supporting EIA as well as strategic environmental assessments (SEA) on a political level for solid sustainable processes [31]: the UN Conference on the Human Environment (Stockholm, 1972) [39], the 1983 evolved Brundtlandt Commission Report (1983) [44], "Our Common Future" (1987), Earth Summit (1992) and its most important developed declarations—Rio Declaration on Environment and Development, Agenda 21 and the Convention on Biological Diversity—further on the ICPD Programme of Action (1994), the Earth Charter, Lisbon Principles (1997) [8,9], UN Millennium Declaration (2000), and the Millennium Ecosystem Assessment (2005). Thus, these agreements, charters, and legal inventions also strengthened the awareness process of water quality, availability, water management, water reuse, and water pollution [28].

EIA is a basis for decision makers in deciding whether a project should be implemented as proposed. The EIA process covers a description of the project as well as a description of the environmental situation in the project perimeter. In addition, an identification and analysis of potential impacts is done next to proposed mitigating measures for those impacts. Further, on a proposed environmental management plan during implementation and up to that, a proposal for public involvement, participation and *later post-project monitoring* is included.

This chapter discusses the role of EIA and its generally described processing in the context of urban water reuse and its chances and challenges within the instrumental proceeding. The environmental media of water is standing in the focus of the overall holistic approach in environmental assessment and specifically for urban purposes in efficient recycling processes of qualified water purity. Life and health quality are next to environmentally and climate-friendly usage of natural and water resources.

Water reuse is one of the most pending management and technological challenges and in observation highly interdisciplinary in correlation to smart city development as well as in general to drought, desertification, land degradation, resulting food insecurity, poverty, hunger, its management, and water resources availability as well as quality.

With acknowledging that the Earth's water resources within an amount of 13 million km³ and 1% saline groundwater, saltwater in oceans and salt lakes inhabit 1338 million km³ and 96.5%, only 35 million km³ or 2.5% freshwater amount [45] is remaining (Figure 21.1) as qualified for pure drinking water purposes. This amount is available only in a small partial way (Figure 21.1).

Water reuse is affecting positively with its correlation to water scarcity and directly influencing the most vulnerable urban ecosystems and life as well as public health quality of especially the poorest affected populations worldwide. Therefore, environmental and urban water health and security should be considered as being a most important aim and mission within the context of controlling by the planning instrument of the EIA, smart city development, and combined water (reuse) management as well as monitoring. Especially in forms of main freshwater exploitation by humankind, the increasing demand for sanitation, drinking, manufacturing, leisure purposes, and agricultural land use has to be assessed, monitored against water pollution, and more efficiently reused. Managing water scarcity, upgrading

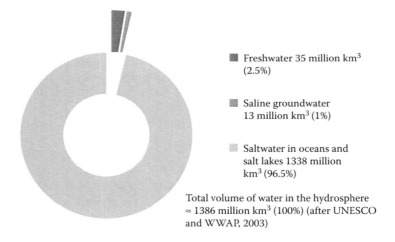

Freshwater 35 million km³ (2.5%)

Saline groundwater 13 million km³ (1%)

Saltwater in oceans and salt lakes 1338 million km³ (96.5%)

Total volume of water in the hydrosphere ≈ 1386 million km³ (100%) (after UNESCO and WWAP, 2003)

FIGURE 21.1 The world's water resources. (Adapted from Reinstädtler, S. 2014. *Conference Proceedings of the International World Heritage Studies (WHS)–Alumni—Conference "The Right to [World] Heritage,"* October 23–25, 2014, IAWHP [International Association for World Heritage Professionals], Cottbus, Germany; UNESCO and WWAP. 2003. *Water for People, Water for Life—The United Nations World Water Development Report (WWDR),* UNESCO Paris and Berghahn Books, New York.)

water quality, and therefore water resources for water and urban security requires observances in multifarious interactions of the different social, ecological, economical, cultural, land use, legal, and political systems as well as including its pressures and drivers to be acknowledged.

Water resources are most important for global management capacities combined to a sustainable, protective use for the most possible reduction of negative impacts to environmental and landscape resources in urban places. In fact, water resources are life protecting and life creating resources implementing standards of life and health quality, which have to be secured.

Changing negatively urban landscapes' and urban places' structures, values, identities, and functions such as ecological sustainability [24] and balance are therefore some of the results especially leading into less secured conditions. Water resources are one of the most affected balancing factors of ecological as well as urban sustainability [24]. The evolving challenges of more unsecured urban conditions in social, economical, cultural, ecological, and environmental life standards have to be directly prevented within proceeding EIA in urban water reuse.

Greatest engagement has to be fulfilled for the aims of water resource protection until 2050 and beyond for securing the worlds' urban and rural landscapes. Water resource security is the agar for an enduring sustainable future of human beings and life in honorable acknowledgment of approved and therefore balancing water availability as well as quality [28].

As stated by Smith [35] more than 200 years ago, "nothing is more useful than water; but it will purchase scarce anything; scarce anything can be had in exchange for it. A diamond, on the contrary, has scarce any value in use; but a very great quantity of other goods may frequently be had in exchange for it" [35].

21.1.1 Legal Situation for Protecting Water Resources

In many countries, water and sanitation are constitutionally guaranteed fundamental human rights. The fulfillment of instrumentations in different countries for securing the realistic proceeding of further guaranteeing this human right including water quality, amount, and bewaring the worldwide water cycle are dependent on international, national, and state-related water laws.

The existence of protective laws despite agreements (conventions), declarations, resolutions, agendas, programs, participation processes, functioning landscape and environmental planning instrumentations like EIA and SEA or informal as well as formal planning processes in general are tremendously important for developing frameworks for protecting natural and water resources and thus implementing water management strategies including sustainable water reuse.

But how is the legal situation of international agreements, declarations, resolutions, laws, or other important documents supporting forthcoming sustainable action for water resources and therefore water reuse?

International agreements such as the United Nations Convention on the Law of the Non-navigational Uses of International Watercourses (1997) [40] United Nations Convention to Combat Desertification (1994) [42] and Ramsar Convention on Wetlands of International Importance especially as Waterfowl Habitat (1971) [38] are strengthening the legal, administrative situation of water-related resource protection [28].

The International Declarations and Resolutions are helping to better the situation of water, specifically

- The United Nations Human Rights Council Resolution on Human Rights and Access to Safe Drinking Water and Sanitation, A/HRC/15/L.1 (2010) [23]
- The International Law Commission, Resolution on Confined Transboundary Groundwater (1994) [19]
- The Athens Resolution on the Pollution of Rivers and Lakes and International Law, done by the International Law Institute (1979) [20]
- The Heidelberg Resolution on International Regulation on River Navigation, done by the International Law Institute (1887) [15]

International Law Association Documents among others are the "Berlin Rules": International Law Association Berlin Conference on Water Resources Law (2004) and the Seoul Rules on International Groundwaters, International Law Association (1986) as well as the Helsinki Rules on the Uses of the Waters of International Rivers, done by the International Law Association (1966) [21,28].

Other international documents are the Resolution VIII. 40 [21], guidelines for rendering the use of groundwater compatible with the conservation of wetlands: Wetlands: water, life and culture within the 8th Meeting of the Conference of the Contracting Parties to the Convention on Wetlands (Ramsar, Iran, 1971) Valencia, Spain, November 18–26, 2002 [34,38] and the UN Conference on Environment and Development (Rio de Janeiro/Brazil, June 1992), Chapter 18: Protection of the Quality and Supply of Freshwater Resources: Application of Integrated Approaches to the Development, Management and Use of Water Resources.

The legal situation for optimized groundwater governance is showing that legal regulations have to be focused on

- Elaboration of water quality standards (not fulfilled in all countries yet)
- Implementing by law instrumentations like EIA
- Licensing of groundwater investigations [1] and water investigations in general
- Developing and improving national water laws and by-laws as well as international law with regard to groundwater use and protection [1]
- Establishment of transparent reporting systems [1] and framework systems such as already created throughout legal regulations by the EU Water Framework Directive (Figure 21.2) and its good functioning coordination system
- Enhancing international networking and information platforms

An example of integrated water management, the planning for sustainable use of water resources and transference of legal regulations is the renewal of the EU Water Framework Directive (2000) (Figure 21.2). It is a good functioning coordination system as well as an assessment throughout river basin districts (river shed units) *or also called water* catchment areas. The legal regulation to coordinate perimeters of river basin districts brought advantages especially for planning purposes including modeling, mapping, and realization of measurements. As unambiguous modeling and mapping are of great importance for verifying planning and management solutions on a spatial level [27,28] and specifically for integrated water management, regulating within river basin districts simplified these processing types. There are several planning instrumentations for analyzing, management, and decision making existing, such as mapping, modeling tools for communication, information exchange, and representing the relationships between a plan, its elements, measurements and development features. Apart from these, the example of the river basin districts with water catchment areas is already practiced for integrated water management in EU countries within the background of the EU Water Framework Directive (WFD) [13] (Figure 21.2). These observations within water catchment areas on regional to superregional scale present another working instrument, which can help despite the possibilities of EIA discussed in this chapter as an instrument on a project level for bettering the qualitative as well as quantitative situation of reused water resources.

Summarizing the legal situation, it can be stated that within the understanding and development for sustainable urban water management including urban water reuse as one tool and therefore protecting water resources, large numbers of international related organizations and governmental instances and policies (such as on the EU level) have to cooperate, communicate, [33] network, and create synergies as well as bring knowledge into commitments, consultations [33], and political processes. Further, international policy dialogues have to be supported by influencing global water policy and law with regard to sustainable urban water and especially groundwater management. In specific, the national and international water laws and policies are still inadequate to meet the challenges evolving from the global phenomenon of climate change as a main risk for freshwater resources. Adapting to the additional consequences of climate change induced impacts on water resources and water reuse purposes are inevitable, [12] but flexible planning instrumentations have to be developed or rearranged [27,30]. In addition, rare nations implemented yet view adequate measures fitting the impacts on climate change [12] and thus for the protection of the freshwater cycle.

21.1.2 Urban Water Demands

Presently, urbanization is the most important factor for human settlement within a modern territorial zone. The twenty-first century is the time for technological development in urban areas. All over the world, the urban areas are getting priority for human development and comfortable livelihoods. The growth of urban population rate is growing high very rapidly. Tropical and subtropical areas [8] are facing serious water scarcity for the last two decades and have witnessed growing water stress, both in terms of water scarcity and quality deterioration. By 2020, water use and reuse is

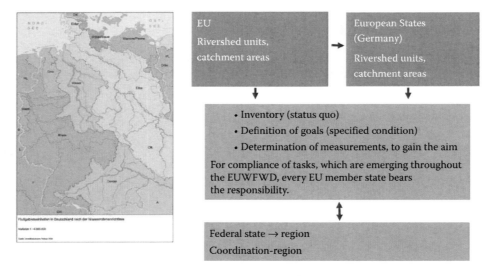

FIGURE 21.2 River basin districts (river shed units) as an example for functioning legal transference into coordination management in (global) European and national regions (WFWD = EU Water Framework Directive). (Adapted from EU Water Framework Directive (WFWD). 2000. Directive 2000/60/EC of the European Parliament and of the Council establishing a framework for the community action in the field of water policy. *Official Journal of the European Communities*, L327:1-L327:72. Appendix: Map River Basin Districts, 2012. http://eur-lex.europa.eu/legal-content/EN/TXT/?uri=CELEX:32000L0060. Accessed on October 6, 2011; Reinstädtler, S. 2011. *Conference Proceedings of the International World Heritage Studies (WHS)–Alumni—Conference "World Heritage and Sustainable Development,"* June 16–19, 2011, IAWHP (International Association for World Heritage Professionals), Cottbus, Germany. pp. 45–65.)

expected to increase by 40%, and 17% more water will be required for food production to meet the needs of the growing population. By 2025, 1.8 billion people will be living in regions with absolute water scarcity; about two out of three people in the world could be living under considerations of water stress [43].

Therefore, water use and reuse is a potential human demandable agenda to survive within a sustainable way. This is the most critical and potential subject of population distribution, settlement, urbanization, and agricultural development. The civilization, urban development, and agricultural cropping systems as well as industrial development are totally dependent on water availability. Soil fertility development also depends on water supply and the quality of surface and groundwater. Domestic and residential use, use for drinking water, industrial use, agricultural use, urban sanitation, recreational use, urban gardening, urban household, and hydroelectric power plants are even in concurrence with rare water resources and make it affordable to reuse water. The challenging fact is that due to huge amounts of water for drinking, irrigation, and industrial purposes by the last decades, water scarcity, and quality deteriorations, almost 70% of the population in cold, tropical, and subtropical climate regions are facing water stress challenges today [3].

The present population growth, socioeconomic activities, industrial development, and climate change impacts are the most critical threats for water resources uses and sustainable conservation. Most of the major cities in the world are facing critical potable drinking and irrigation water. Thus, urban water should be used and reused within scientific guidelines for better management and distribution of water resources in different climatic territories considering the value of ethical and cultural conditions. At a global level, water availability for 2006 was 8462 m³/capita-yr, but at a regional level, it went from as little as 1380 m³/capita-yr in the Middle East and North Africa to almost 53,300 m³/capita-yr in Oceania [22]. Water consumption in urban areas is roughly split, with 70% agricultural use, 22% industrial use, and 8% domestic use [45]. A fifth of the population lives in areas of water scarcity and one in eight lacks access to clean water especially in urban regions [46]. Most mega cities of the world are facing urban drinking water supply and management problems. It has been already projected by Habitat Unit of the United Nations that by 2050, 50% of the world population will live in urban regions. This development happened to get real much faster now in the years of 2000 with 47% and of approximated 50% in 2008 by the Habitat Agenda [41]. Therefore, water reuse in urban regions is very important for sustainable management of urban water resources and the protection of urban ecosystem services.

21.2 EIA as an Assessment Methodology for Sustainable Urban Water Reuse Infrastructure

This chapter discusses the role of the EIA as an assessment instrument in the context of sustainable urban infrastructure

with in-depth observations in the possibilities for qualitative water reuse procedures and protecting (drinking) water as well as groundwater quality for enabling an effective integrated water resource management. The combined view of the general purposes of the EIA process, such as protecting the environment throughout the prevention principle and sustainable management is to be set out. Some other common principles are observed within supporting sustainable urban infrastructure and water reuse purposes for an optimized water resource management as well as indirectly preserving desertification processes.

Land degradation by intensified land use especially in urban and suburbanized areas, less ground water accessibility or groundwater recharge, less precipitation as well as water scarcity in general are driving forces for enhancing the development of desertificated dry lands and an interrupted water cycle on a regional level. In general, sustainable, urban development and water-oriented planning are guiding concepts of so-called blue-green cities [5] or water-sensitive cities [4] for preventing degrading, unprofitable developments of water household in urban and suburban areas. During the first decade of the twenty-first century, instruments for water-centric urban planning are being developed in many countries [17]. Combined with green planning initiatives, the importance of EIA for screening and scoping before a project for water reuse purposes starts is highly important for the local and regional communities as well as for all environmental media and, specifically for water, its resource protection, and reuse quality.

Sustainable planning and management of urban infrastructure and especially of reusing water have to be supported throughout instrumental EIA processes. Some of the aspects, principles, and ways of procedure in EIA [48] (Figure 21.3) are for diminishing water scarcity while assessing disturbance grades and threshold lines within EIA report (Figure 21.3) in different kinds of projects [32]. Assessing water quality standards within projects of water purification plants, maintaining for receiving a sustaining water cycle, or drinking water quality, and therefore a functioning overall water resource management has to be monitored after finalizing the project (Figure 21.3) within a specific setup timeframe depending on the type of project with dependence and comparison on state-of-the-art and results of the established EIA report.

While there is overarching comparative research on EIA systems [15], the following principles are nevertheless seen in most comparisons of EIA systems around the world.

The principles of prevention and avoidance of adverse environmental consequences, proactive planning for involving options or alternatives and transferring as a decision tool, positive influence on decision making at the earliest possible opportunity and thinking proactively about options and alternatives are implemented within the steps of fulfillment of the project-cycle (Figure 21.3) and in accordance with EIA assessment and reporting. Up to that, flexibility in the project processing cycle for adapting to changes, robustness, and transparency within participation have to be implemented in principle by EIA procedures.

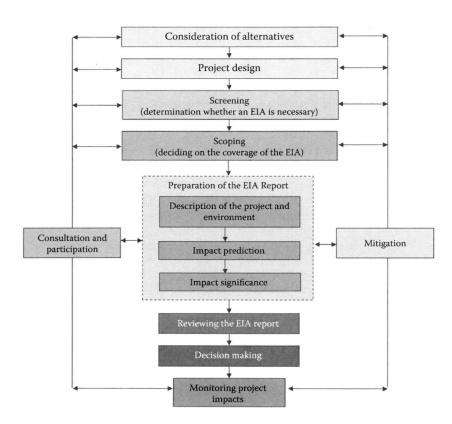

FIGURE 21.3 Generalized EIA procedure. (Adapted from Wood, C. 2003. *Environmental Impact Assessment: A Comparative Review.* Prentice Hall (Pearson Education), Harlow, United Kingdom.)

21.2.1 Integrating EIA in Urban Infrastructure Construction for Water Reuse

Sustainable urban infrastructure and especially water reuse, supported throughout instrumental EIA processes, are some of the aspects to diminish water scarcity and maintain for receiving a sustaining water cycle and therefore a functioning water resource management. Water reuse itself is one important part of an integrated water management strategy [11]. With outlining hydrological and urban related demands, water management inclusive of water reuse projects, flood control projects, and stormwater management have to be named as the most important pillars for managing urban water in general. Water supply, water conservation, irrigation (systems) and other appropriate technologies, hydrology, and urban planning as well as aquifers are some of the most important urban water demands and infrastructure constructions to be acknowledged in correlation with EIA processes.

The effective water reuse and the treatment of wastewater to meet water quality standards for water reuse applications in urban areas and for protecting public health is a critical element in water reuse systems to be assessed by EIA. Reuse also depends on the acceptance and willingness of the local population and farmers around urban areas. The treatment process would then have to be adopted to reuse when needed among others for irrigation.

The reuse or wastewater treatment consists of a combination of physical, chemical, and biological processes and operations to remove solids, organic matter, pathogens, metals, and nutrients from wastewater. In order to increase treatment levels, the different degrees of treatment are preliminary, primary, secondary, tertiary, and advanced treatment, which has been set up in a model and checked within EIA procedure. Household, industrial, and urban wastewater has to be treated for getting back quality standards for implementing again in a water cycle. On a microscale level, it means technically to treat in at least six different steps depending on the sort of wastewater including detergents or other nontoxic or toxic ingredients to be proved and assessed within EIA reporting. One of the first steps should include reducing or preventing waste to generate; Step 2, reducing waste toxicity; Step 3, including recycle waste; Step 4, reducing the process of waste; Step 5, treating waste before disposal; and Step 6, allocated to dispose in an environmental manner of wastewater treatment. After this procedure, wastewater may be treated as clean water for use in different purposes such as agricultural irrigation, industrial production, and gardening among others. Treated wastewater has to be approved more strongly for use as drinking water and so is frequently monitored for health quality.

A greater demand should also be set into the preventive program, plan, policy integrative instrument of the SEA, which will not be discussed in this chapter, but should be mentioned as being another very important environmental planning tool besides EIA for securing a more qualitative and quantitative reuse of water and therefore saving natural and water resources in urban as well as rural areas.

21.2.2 EIA in Context of Securing Urban Water Resources

One main aspect for using EIA in an urban water resource context is to get a better information basis about all environmental media involved in the specifics on the assessed project and the medias' management, protection status, and standards for supplying water quality, amount, availability, and surrounding of other environmental media interacting in a cumulative, synergistic way with the medium of water. Especially in urban background, public human health aspects can be better controlled within the procedure of EIA. One of the outcomes are better secured water resources by integrating water management, water resource management, water development, and sustainability as well as water supply in general.

Two-thirds of the groundwater reserve (Figure 21.4) lies below 800 m depth and is beyond human capacity to exploit [37]. The challenging fact is that due to huge amounts of water for drinking, irrigation, and industrial purposes in the last decades, water scarcity, and quality deteriorations, almost 70% of the population (in Europe) are facing water stress today [3].

Thus, objectives and long-term aims of water resource and combined wastewater-related urban projects could be strengthened throughout an EIA process. Specific objectives for the investor are, on the one hand, to establish a dependable treatment plant for wastewater, which is cost-effective and requires the minimum maintenance. It has to be combined to the recommendations of the responsible Ministries of Environment and so for managing the pathogenic risk inherent in wastewater to meet the effluent discharge standard sets. Up to that, the safe disposal of sludge has to be managed. On the other hand, other diversified interests, aims, and objectives of stakeholders and persons concerned have to be included, for which the processing throughout EIA is even more important.

For example, the project could aim at environmentally safe disposal of wastewater for upgrading the sanitary and health standards of city inhabitants. These aims and long-term objectives of such kinds of projects should especially include within the perimeter of developing urban areas:

- Preventing and controlling soil and groundwater pollution
- Controlling the contamination of water resources
- Preventing the spread of diseases, including the limitation of mosquito populations
- Preventing the prevalence of conditions offensive to sight and smell

Potential risks and impacts on water resources can be acknowledged in precautionary status throughout the holistic view and possibility of structured inclusion of different aims, objectives, risks, impacts, development recommendations, and monitoring as well as future perspectives by EIA.

With the example on geological, tectonically conspicuities, and a proposed and to be by EIA screened, scoped, and reviewed project of a collector line or wastewater treatment plant in a city, the location on a fault or fracture zone is easily found before investing a huge amount of money in a dangerous and nonlasting solution. The structural integrity of this implied system might be damaged by tectonic movements, for example, through earthquakes. Earthquake regions constitute, in general, a tectonically highly unstable zone. In case of existing numerous faults, some of them with eventually vertical tectonic displacements of several hundreds of meters could physically damage the wastewater facilities. These in stabilizing situations would lead into uncontrolled infiltration of untreated wastewater and thus contamination of the water resources. Therefore, EIA and its screening, scoping, assessing, reporting, public participation and administrative approval, decision making, and improvements as well

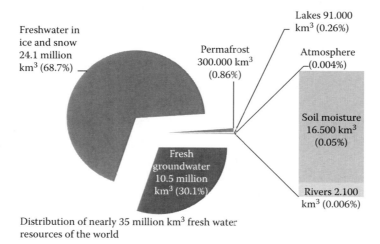

Distribution of nearly 35 million km³ fresh water resources of the world

FIGURE 21.4 The world's freshwater resources. (Adapted from UNESCO and WWAP. 2003. *Water for People, Water for Life—The United Nations World Water Development Report (WWDR),* UNESCO, Paris and Berghahn Books, New York.)

as post-project monitoring is giving precautionary structures with securing finances, approving engineering procedures, and protecting water resources. As economic growth, consumption of natural and water resources, and the generation of waste are in a greater imbalance than possible to be rejected by autarkic natural systems rehabilitation possibilities, these scenarios have to be changed by action [26]. Therefore, strong economic activities have to be coupled with sustainable use of natural resources, reducing waste, maintaining biodiversity, preserving ecosystems, and avoiding desertification [26].

Water resources are most important for global management capacities combined with a sustainable, protective use for a possible reduction of negative impacts to environmental and landscape resources in urban (and surrounding rural) places. In fact, water resources are life protecting and life creating resources implementing standards of life quality, which have to be secured. However, urban as well as rural landscapes are getting highly disturbed while getting threatened in the context of land degradation by intensified land use, less ground water accessibility, less precipitation beneath 250–80 mm/yr as well as water scarcity or low groundwater resource availability, in general, for

evolving to desertificated dry lands. Securing poverty reduction, sustainable land use and drought management, mitigation and adaptation for climate change, as well as specific groundwater management adapted to the quantity of existence (Figure 21.5) means to combat desertification effectively on 61-million km² dry land worldwide with a greater amount already including desertificated land masses.

The development of desertificated dry lands is already reflected by Figure 21.5: the map of groundwater resources is showing that already nearly half of the continental areas contain generally minor occurrences of groundwater. Those minor locations of groundwater occurrences are restricted to the near-surface unconsolidated rocks [1,45]. About 35% of relatively homogeneous aquifers are existing on the areas of continents (excluding the Antarctic) [1,45]. Eighteen percent of the areas are provided by groundwater, some of which are extensive especially in geologically complex regions [1,45].

It also implies securing landscapes, ecosystems, and urban (as well as rural) places ecological, economical, social, and cultural as well as environmental and human health [7] and wealth, which affects nearly 2 billion people living on 40%–41% of the

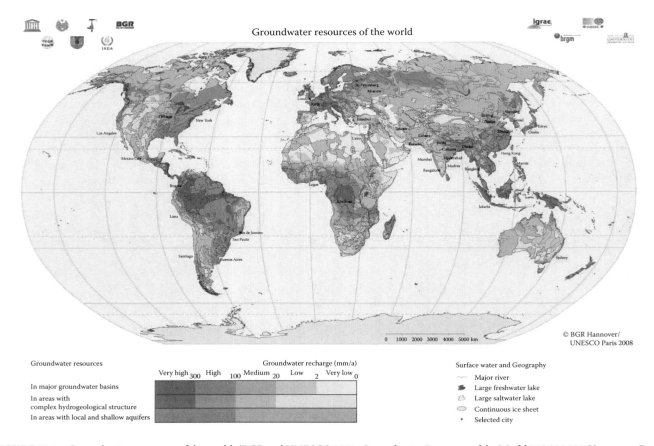

FIGURE 21.5 Groundwater resources of the world. (BGR and UNESCO 2008. *Groundwater Resources of the World 1:25 000 000*. Hannover, Paris; Richts, A., Struckmeier, W.F., and Zaepke M. 2011. WHYMAP and the Groundwater Resources of the World 1:25,000,000. In Jones J.A.A. (Ed.), *Sustaining Groundwater Resources*, pp. 159–173.)

total global land masses being dry lands already, which correlates also to the existing groundwater amount, quality, and level (Figure 21.5). It is estimated that 10%–20% of those dry lands are already destroyed while being turned to deserts. Especially regions in Africa south of Sahara and South Asia are facing these strong security problems in extreme poverty and hunger, desertification, and land degradation.

Changing landscapes and urban as well as rural places structures, values, identities, and functioning such as the ecological sustainability and balance [27] are therefore some of the results leading into less secure conditions. The evolving challenges of more unsecure conditions in social, economical, cultural, ecological, and environmental life standards have to be directly prevented—on a project level throughout EIA.

21.2.3 Importance of EIA for Urban Water Reuse

Urban water reuse is one part of waste management, which means to reuse, recycle, and dispose solid waste from manufacturing, commerce, and domestic sources. Assessing and processing EIA within urban water reuse and therefore within waste water treatment plants (WWTP) and its projects means to include project reviews and reporting of industrial, agricultural wastewater treatment, and sewage treatments with certain methods like biodegradation and other methods that are more effective than biodegradation with industrial wastewater including toxic chemicals [16]. In this context and within development it is increasingly preferred to be served by a single facility for treating water for all kinds of waste products [36], which has to be compared to EIA standards and limit values assessed in EIA.

However, what are the characteristics within a WWTP and its different sorts of wastewater to be treated and cleaned for reusing purposes and then screened and scoped by EIA?

Wastewater treatment works or a WWTP is an industrial object that is designed for removing chemical or biological waste products from water. The treated and cleaned water will be reused for other purposes again. Some of the functions of WWTP include industrial wastewater treatment, which is the treatment of wet wastes from manufacturing, industry, and commerce including heavy industries such as mining and quarrying. The agricultural wastewater treatment is one higher amount of treatment activities with the treatment and disposal of pesticide residues or animal waste and others from an agricultural background. The sewage treatment is treatment and disposal of human waste and other household waste such as liquid from kitchens, toilets, baths, showers, and sinks.

The EIA is one of the potential tools for assessing the environmental impacts in any developmental construction, large scale in fracture, and environmental projects in any type of climatic condition. The EIA procedure is introduced as an environmental assessing tool to the environmental scientists and other academic personalities. The general EIA procedure scheme has been introduced and shown in Figure 21.3. Figure 21.3 demonstrated the scientific view of the procedure where the scheme has shown the

detailed structural framework of EIA. The major functions of EIA procedures demonstrated in Figure 21.3 where the potential elements and factors of the scheme highlighted the important activities like project design, screening, scoping, reporting, mitigation, reviewing EIA project, decision support, and monitoring through consultation for alternative project planning, designing, and project implementation. The scheme can be used for water resources use and reuse processes and project verification in case of, for example, WWTP in urban territories in different climatic regions. For impact assessment and planning procedures, the general scheme and the proposed developed model (Figure 21.6) could be the appropriate model for urban water reuse and reuses in developed and developing countries of the world.

Water reuse is affecting positively with its correlation to water scarcity and directly influencing the most vulnerable urban ecosystems and life as well as human and environmental health quality of affected urban populations. Therefore, environmental [25] and urban water security should be considered as being a most important aim and mission within the context of controlling by the planning instrument of EIA, smart blue-green city development and combined water (reuse) management as well as monitoring system development. Managing water scarcity, upgrading water quality, and therefore water resources for urban water security requires observance in multifarious interactions of the different social, ecological, economical, cultural, land use, legal, and political systems as well as including its pressures and drivers to be acknowledged. The model shows the guideline framework of urban water reuse planning and implementation within [a] multifunctional divers and cyclic activities. Climate change impacts, urban land use system changes are further intensifying sealing, implementation of assessing multifarious drivers in cumulative and synergistic impacts on urban water resources. Figure 21.6 demonstrates the scenarios of methodological structure with interdisciplinary collaborative frameworks for implementing any kind of environmental assessment on urban environmental targets as well as urban water use and reuse practices.

Through this model of EIA in the context of urban water reuse, the responsibility for cultivating, enduring, and sustaining freshwater resources and its chances and challenges within the instrumental proceeding are becoming obvious. The environmental media of water is standing in the focus of the overall holistic approach in environmental assessment, specifically for urban purposes in efficient recycling processes of qualified water purity. Life and health quality are next to environmental and climate friendly usage of natural and water resources.

Already for many years existing water crises have to be acknowledged within the fact of just happening and further on coming changes in precipitation rates, rising temperatures, and other climatic insecurities. By using the instrumentation of EIA for a controlled projecting for water reuse, it is exactly adapting as a tool for interacting in spatial and environmental planning, projecting, and planning permission hearings. This is why EIA has so much potential in any kind of environmental project development, planning, and implementation within a

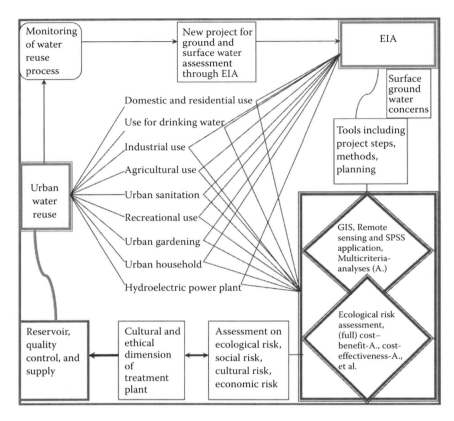

FIGURE 21.6   Model of urban water reuse through EIA practices. (Adapted from Reinstädtler, S. 2014. *Conference Proceedings of the International World Heritage Studies (WHS)–Alumni—Conference "The Right to [World] Heritage,"* October 23–25, 2014, IAWHP [International Association for World Heritage Professionals], Cottbus, Germany.)

critical urban environment. As it is covering the urban ecological risk assessment, cost-benefit approach through multicriteria analysis, (full) cost-benefit analysis, cost-effectiveness analysis, among others, EIA assessment itself forms for sustainability. Urban water supply, management, and planning is presently a critical target as urban population growth is getting higher; therefore, the urban drinking water crisis would be more harmful and a proper, integrative urban water management, planning, and implementation strategy is necessary. It is urgent for most of the medium, large, or even the mega cities of the world.

21.3 Managing Urban Security within Scoping Urban Water Reuse Cycles

The link between water and security on different levels and locations cannot be diminished. As water becomes scarcer, the importance of how it is managed grows vastly. Up to that, climate change will have major effects in directly impacting the Earth's freshwater systems [12]. A great amount of regions will suffer serious scarcity and other regions will face increased precipitation [12].

Therefore, water reuse especially of industrial wastewater treatment, agricultural wastewater treatment, and sewage treatment are some actions to be initialized within managing

water resources, water reuse, and therefore regional urban security. Following one of the three excellent convention results of the Rio Summit, the United Nations Convention to Combat Desertification [42], land degradation in correlation to scarce water availability and drought has to be effectively continued to be minimized in ecosystems, on landscape as well as urban and rural security level and for whole populations' social, natural, economic, and cultural wealth and health.

The vast area of tropical and subtropical as well as semi-arid and arid regions are recognized as warmer regions with higher and increasing temperatures. Even the continental (and partially cold climate) regions and in specific the so far moderate European regions face challenges because of higher and increasing temperatures. Water shortages attributed to climatic changes are other vast impacts. Shortages have resulted in record droughts, for example, in Australian regions, regions of south-eastern and western United States, southern Africa, and South America [12]. "Scientists are also blaming climatic changes for intensifying rain activities in northern Europe and North American by between 10% and 50% and for the related increase in flooding occurring in many of these regions" [12].

Except parts of semi-arid, arid, and some continental regions, many of those regions are thus far very fertile for human settlements and development activities as well as fertile for agriculture and biodiversity within specific productive landscapes. Population density, urbanization and water resources, natural productivities are the main resources for this huge population. However, the water resources are used within unlimited scales and scarcity of water next to adverse effects of changing weather patterns (floods, droughts) and their related economic costs. These main problems make it difficult to maintain natural resources in combination with the needs of agro- as well as industrial production for saving cultural, economic, and social needs. Maintaining landscapes, ecosystems, and ecosystem services in these regions is another challenge. Cold climate regions are more important for urban water reuse and feeding the overall global hydrological cycle [10] because cold climate regions are partially inhabiting permafrost-captured water and are industrially developed regions. High growth of urban population is distributed in the cold climate regions. The industries are using huge volumes of surface and groundwater, which is creating water scarcity in the regions. Therefore, for industrial, irrigation, and domestics uses, water reuse is more vital when climate change impacts are a threat to water scarcity.

Thus far, it can be stated within naming these different examples, that natural and water resources have to be managed more responsibly. Within the example of European Regions and its EU policy, an integrated product policy aimed at reducing resource use and the environmental impact of waste should be implemented in cooperation with business. In addition, biodiversity decline should be halted as set out in the 6th Environmental Action Programme [26].

Water reuses and cycle management systems in different climatic regions can be adopted within EIA process of project screening, scoping, and reporting. Figure 21.7 shows a general water cycle management, which is adaptable to different kind of regions. In specific climates and regions, detailed management options are possible. The water cycle management scheme has been recognized in scientific background of water and hydraulic research fields. Figure 21.7 demonstrates the general water cycle management model specifically to cold climate, tropical and subtropical climatic regions, and is adaptable to other regions of the world. The functions of this water cycle management model are also generalized and adaptable for urban and rural regions of different climatic regions. Wastewater reuse, when appropriately applied, is considered an example of environmental scientific technology (EST) applications.

The urban water cycle is representing the urban water management cycle, where human activities, urban agriculture, industries, forest, urban landscapes and wetlands, water shade, precipitation, and temperature are directly influencing the urban water cycle management. Briefly, it can be said that anthropogenic influences, rapid urban growth, and urban agglomeration are working as the driving force for urban water cycle disorder and crisis motive for urban inhabitants. Considering the urban water crisis scenarios, it could be stated that the urban security such as urban settlement, habitat, urban forests and water shades, urban gardens and urban agriculture, urban small entrepreneurs and industries, urban food security, urban landscapes,

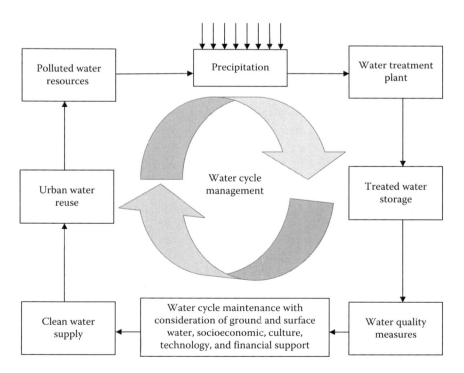

FIGURE 21.7 Water cycle management. (Adapted from Reinstädtler, S. 2014. *Conference Proceedings of the International World Heritage Studies (WHS)–Alumni—Conference "The Right to [World] Heritage,"* October 23–25, 2014, IAWHP [International Association for World Heritage Professionals], Cottbus, Germany.)

ecosystems and ecosystem services, as well as urban economic activities are dependent on the urban water quality and availability. Availability is dependent on the urban water cycle. Therefore, urban economic activities, agricultural practices and production, industrial quality products, cultural activities and environmental media are dependent on the water availability, which could ensure urban livelihoods, economic solvency, and urban security. All these factors are related to water supply availability and quality of urban drinking water and it is related to the health quality of urban inhabitants. As a result, urban water is considered the driving force of urban economy, urban food security, urban health, urban socioeconomic improvement, strength of urban economics, and urban security. To fulfill the demand of urban drinking, domestic and industrial water developmental planning, and decision support tools, the EIA methodology could be used to ensure urban security against risk of water scarcity and shifting water quality. Through the EIA procedure, the following potential sectors could be fulfilled for urban citizens such as

- Proper implementation of urban land use system changes or further intensifying sealing
- Implementation of assessing multifarious drivers in cumulative and synergistic impacts on urban water resources
- Implementation of climate change impacts in EIA and urban water reuse procedures

The modern water treatment plant should be implemented in all large and mega cities where urban water is treated as a critical commodity to the inhabitants. Before implementation of water treatment plants, the EIA practices could reduce the future conflicts and ecological negative impacts.

Subsequently a concept of sustainable urban water management and EIA for urban project assessment and monitoring is roughly needed. Further on, the concept of sustaining (urban) landscapes in the context of adapting to climate change [27] as well as avoiding political water dependencies are management recommendations that are strongly to be implemented.

21.4 Conclusions and Recommendations

"Anyone who can solve the problems of water will be worthy of two Nobel prizes—one for peace and one for science (John F. Kennedy, after Likhotal)" [23]. Looking back to historically managed water quality and quantity as well as wastewater treatment, it seems to be time is running faster in the present coupling multifarious pressures of population growth, food production, land use intensifications, and climate change insecurities as well as cultural, social, economical, and ecological impacts.

Historically, wastewater was used as an alternative water source for agricultural irrigation going back approximately 5000 years. By 2020, water use and reuse is expected to increase by 40%, and 17% more water will be required for food production to meet the needs of the growing population. By 2025, 1.8

billion people will be living in regions with absolute water scarcity; about two out of three people in the world could be living under considerations of water stress. Therefore, water use and reuse is a potential human demandable task for surviving in a sustainable way.

In general, the country's wetland natural resources as well as the ecosystems are degrading due to anthropogenic influences and natural calamities. The freshwater and saline water wetlands in Bangladesh are in an alarming situation. Similar challenges exist in diversified regions of the world, which have to communicate for proper management within networking. The methods for the generation of transformation knowledge to achieve and maintain natural resources, agro-farming system development, fishing development, and in general wise use of wetland natural resources could protect the biodiversity as well as self-cleaning-mechanisms in the freshwater wetlands and coastal saline mangrove wetland areas. The following recommendations focusing on EIA and water reuse could be followed in implementation and management stages, but should be seen as more generalized for the diverse multiple pressures coming up and in balancing the water resources.

Despite the fact that EIA is an important instrument for water reuse projecting and supporting the freshwater cycle and management, the overall surrounding of developing and enabling blue-green infrastructure and water sensitive urban design as well as urban planning will give even more opportunities for multiple benefits not only in water securing purposes [18].

Adopting blue-green infrastructure projects for sustainable and integrated water management in urban areas means also to couple the multiple benefits of blue-green infrastructure with qualitative EIA reporting as well as monitoring. Within these assessments and the approach of blue-green infrastructure, flood management, coastal (erosion) risk management, and water resource management could be assessed in a combined way throughout EIA and its in-depth analyses of medium of water as well as landscape in a multifunctional benefits seeking way. As flood risk reduction benefits better water quality in the viewpoint of the hydrological cycle, landscapes and ecosystems are provided by this integrative assessment. The local/regional and global/international scales should be touched not within these ongoing assessments of EIA reporting, but later should span both scales with overbridging throughout networking, communicating, and exchange of monitoring data. Further on the clarity of social and environmental consequences are enhanced in the decision-making process throughout the results of the EIA report.

One of the biggest concerns for our water-based resources in the future is the sustainability of the current and even future water resource allocation. As water becomes scarcer, the importance of how it is managed grows vastly. Finding a balance between what is needed by humans and what is needed in the environment is an important step in the sustainability of water resources. Attempts to create sustainable freshwater systems have been seen on a national level in countries such as Australia, and such commitments to the environment could set a model for the rest of the world [47]. Up to that, a platform could be created for better forwarding EIA processes, report, and project as well as monitoring examples to other projects and for a better, more cost-efficient as well as future-oriented sustainable planning or management manners. It is likely that ongoing climate change will lead to situations that have not been encountered. As a result, new EIA assessing and management strategies will have to be implemented in order to avoid setbacks in the allocation of water resources.

The present study has investigated the present scenarios of world urban water uses and management and sustainability issues in the context of EIA. Most of the regions of the world are continuously facing problems of water scarcity and climate change impacts have been added as new major threats to sustainable urban water use and distribution to the citizens. Based on the investigation of urban water availability and for new projects and developments, the EIA methods should be introduced and implemented properly including monitoring and combined networking platforms for enforcing that the world will have information and data for future project developments, EIA assessment, reporting, decision making, and monitoring as well as for the development of urban water use and reuse for a continuing sustainable hydrological cycle.

The methods for the generation of transformation knowledge to achieve and maintain urban water uses and in general wise use of water resources are combined with the EIA. The planning instrument of EIA and its practices could find out an alternative solution for urban water reuse methods that could ensure the sustainability of urban water reuse activities. The stated recommendations could be followed in implementation and management stages of urban water uses and reuses in different climatic regions of the world:

- People's participation for awareness of the importance of urban water use and reuse for the future generation and in specific also possible within participation processes of EIA processes. It could be a popular method for many peoples' participation in water conservation, harvesting, and sustainable uses within the ethics, culture, and religious values and practices.
- The requirement to ensure the provision of adequate institutional and project capacity for later implementing in decision making and policy development, delivery, and post-EIA monitoring; the importance of considering water sources as water management infrastructure rather than nature reserves and the need to consider the wise use and reuse of water within rural and beyond urban boundaries and understand the interconnectivity of watershed-scale issues.
- Traditionally, frameworks and spatial plans are area-based and structured in administrative boundaries. This approach fails to address to the required standard for the protection of natural and water resources, landscapes,

wetlands or ecosystem services. Because in many cases it is not possible to recognize that these boundaries are a functional advantage or commensurate with the environmental needs for balancing.

- Nations' cooperation instead of conflicts in correlation to an integrated water resource management.
- Tremendously destructuring processes of desertification, water scarcity (despite parallel contrary developments like storms, floods, and extreme weather events) are proceeding in diverse regions of the world. For that, developing as well as developed countries have to fulfill in greatest engagement the aims of UNCCD until 2015 and beyond for securing the worlds' landscapes, ecosystems, environmental, and rural as well as political security as being the agars for an enduring sustainable future of human beings and life on planet Earth—the one.

References

1. BGR. 2008. *Sustainable Groundwater Management for Tomorrow's Livelihoods—Strategies and Products*. Federal Institute for Geosciences and Natural Resources (BGR), Hannover, Germany.
2. BGR and UNESCO. 2008. *Groundwater Resources of the World 1:25 000 000*. Hannover, Paris.
3. Bixio, D., Thoeye, C., Koning, J.D., Joksimovic, D., Savic, D., Wintgens, T., and Melin, T. 2006. Water reuse in Europe. *Desalination,* 187: 89–101.
4. Brown, R., Keath, N., and Wong, T. 2008. Transitioning to water sensitive cities: Historical, current and future transition states. *Proceedings of the 11th International Conference on Urban Drainage*, Edinburgh, Scotland. http://web.sbe.hw.ac.uk/staffprofiles/bdgsa/11th_International_Conference_on_Urban_Drainage_CD/ICUD08/pdfs/618.pdf. Accessed on June 23, 2014.
5. Butterworth, J.A., McIntyre P., and da Silva Wells, C. (eds.). 2011. *SWITCH in the City: Putting Urban Water Management to the Test*. IRC International Water and Sanitation Centre, The Hague, Netherlands.
6. Carson, R. 1962. *Silent Spring*. Houghton Mifflin Company, New York.
7. Chorus, I. 2008. Assessing environmental impacts on human health—Drinking-water as an example. In: Schmidt, M., Glasson, J., Emmelin, L., and Helbron, H. (eds.). *Standards and Thresholds for Impact Assessment. Environmental Protection in the European Union,* Volume 3. Springer, Heidelberg, pp. 341–354.
8. Costanza, R. 2007. Lisbon principles of sustainable governance. In: Cleveland, C.J. (ed.), *Encyclopaedia of Earth*. Environmental Information Coalition, National Council for Science and the Environment, Washington DC Published in Encyclopaedia of the Earth, August 9, 2007; retrieved February 6, 2009. http://www.eoearth.org/view/article/154264. Accessed on June 23, 2014.

9. Costanza, R., Cumberland, J.H., Daly, H., Goodland, R., and Norgaard, R.B. 1997. *An Introduction to Ecological Economics*. St. Lucie Press and International Society for Ecological Economics, Boca Raton, Florida.

10. Crosby, R.L.J. 1997. A plan for community greenhouse, and waste treatment facility for cold climate regions. In: Etnie, C. and Guterstam, B. (eds.), *Ecological Engineering Wastewater Treatment*. Lewis Publishers, New York, pp. 145–152.

11. Daigger, G. (Author), Water Environment Federation (Publisher). 2009. Evolving urban water and residuals management paradigms: Water reclamation and reuse, decentralization, and resource recovery. *Water Environment Research*, 81(8), 809–823(15).

12. Eckstein, G. 2009. Water scarcity, conflict, and security in a climate change world: Challenges and opportunities for international law and policy. *Wisconsin International Law Journal*, 27(3), 2009, Texas Tech Law School Research Paper No. 2009–01. http://ssrn.com/abstract=1425796. Accessed July 4, 2014.

13. EU Water Framework Directive (WFWD). 2000. Directive 2000/60/EC of the European Parliament and of the Council establishing a framework for the community action in the field of water policy. *Official Journal of the European Communities*, L327:1-L327:72. Appendix: Map River Basin Districts, 2012. http://eur-lex.europa.eu/legal-content/EN/TXT/?uri=CELEX:32000L0060. Accessed on October 6, 2011.

14. Food and Agricultural Organization (FAO) Corporate Document Repository. 2014. *Sources of International Water Law Studies and Declarations Made by International Non-Governmental Organizations—Institute of International Law—International Regulation on River Navigation—Resolution of Heidelberg*, September 9, 1887. http://www.fao.org/docrep/005/w9549e/w9549e08.htm. Accessed on July 4, 2014.

15. Glasson, J., Therivel, R., and Chadwick, A. 2005. *Introduction to Environmental Impact Assessment*, 3rd Edition, Routledge, London.

16. Heldman, D.R. (ed.). 2003. *Encyclopedia of Agricultural, Food, and Biological Engineering*. Marcel Dekker, Inc., New York, p. 55.

17. Howe, C. and Mitchell, C. 2012. *Water Sensitive Cities*. IWA Publishing, London.

18. Hoyer, J., Dickhaut, W., Kronawitter, L., and Weber B. 2011. *Water Sensitive Urban Design—Principles and Inspiration for Sustainable Stormwater Management in the City of the Future*. Jovis, University of Hamburg, Berlin, Germany.

19. International Law Commission (ILC). 1994. Draft articles on the law of the non-navigational uses of international watercourses and commentaries thereto and resolution on transboundary confined groundwater. In: *Yearbook of the ILC* 1994. A/CN.4/SER.A/1994/Add. 1 (Part2), vol. II, United Nations Publications, Geneva, New York,

20. International Law Institute (ILI). 1979. Resolution on the Pollution of Rivers and Lakes and International Law. Athens, Greece, September 12, 1979. http://www.idi-iil.org/idiE/resolutionsE/1979_ath_02_en.PDF. Accessed on July 1, 2014.

21. International Water Law Project (IWLP). 2014. Addressing the future of water law and policy in the 21st century—International Documents. http://www.internationalwater-law.org/documents/intldocs/. Accessed on July 4, 2014.

22. Jiménez, B. and T. Asano (eds.). 2008. *Water Reuse: An International Survey of Current Practice, Issues and Needs*, IWA Publishing, London, Great Britain.

23. Likhotal, A. 2013. The future of water: Strategies to meet the challenge. *CADMUS Journal*, 2(1), 85–92. http://www.cadmusjournal.org/files/pdfreprints/vol2issue1/reprint-cj-v2-i1-the-future-of-water-alikhotal.pdf. Accessed on July 4, 2014.

24. Lundina, M. and Morrison, G.M. 2002. A life cycle assessment based procedure for development of environmental sustainability indicators for urban water systems. *Urban Water*, 4(2), 145–152.

25. Petrosillo, I., Müller, F., Jones, B. et al. 2008. *Use of Landscape Sciences for the Assessment of Environmental Security*. The NATO Science for Peace and Security Series–C: Environmental Security. Results of the NATO/CCMS Pilot Study on the Use of Landscape Sciences for Environmental Assessment 2001–2006. Springer, in cooperation with NATO Public Diplomacy Division, Dordrecht, Netherlands.

26. Presidency Conclusions—Göteborg, June 15 and 16, 2001 SN 200/1/01 REV 1 8EN. http://ec.europa.eu/smart-regulation/impact/background/docs/goteborg_concl_en.pdf.

27. Reinstädtler, S. 2011. Presentation of Sustaining landscapes—Landscape units for climate adaptive regional planning. In: H. Crescini and K. Sandberg (eds.), *Conference Proceeding of the International World Heritage Studies (WHS)–Alumni—Conference "World Heritage and Sustainable Development,"* June 16–19, 2011, IAWHP, Cottbus, Germany. pp. 45–65.

28. Reinstädtler, S. 2014. Sustaining UNESCO MAB Reserve Spree Forest—The Right for Preserving Landscape Values in the German Lusatia Region. *Conference Proceedings of the International World Heritage Studies (WHS)–Alumni—Conference "The Right to [World] Heritage,"* October 23–25, 2014, IAWHP (International Association for World Heritage Professionals), Cottbus, Germany.

29. Richts, A., Struckmeier, W.F., and Zaepke, M. 2011. WHYMAP and the groundwater resources of the World 1:25,000,000. In: Jones J.A.A. (ed.), *Sustaining Groundwater Resources*, London, Springer, pp. 159–173.

Switzerland, USA. http://legal.un.org/ilc/publications/yearbooks/Ybkvolumes%28e%29/ILC_1994_v2_p2_e.pdf. Accessed on July 1, 2014.

30. Roggema, R. 2009. *Adaptation to Climate Change: A Spatial Challenge.* Springer, Dordrecht.

31. Schmidt, M., João, E., and Albrecht, E. (eds.). 2005. *Implementing Strategic Environmental Assessment. Environmental Protection in the European Union,* Vol. 2. Springer, Heidelberg.

32. Schmidt, M., Glasson, J., Emmelin, L., and Helbron, H. (eds.). 2008. *Standards and Thresholds for Impact Assessment. Environmental Protection in the European Union,* Vol. 3. Springer. Heidelberg.

33. Schmidt, M., Albrecht, E., Helbron, H., and Palekhov, D. 2010. The proportionate impact assessment of the European Commission—Towards more formalism to backup "The Environment." In: Bizer, K., Lechner, S., and Führ, M. (eds.), *The European Impact Assessment and the Environment.* Springer, Berlin, Heidelberg, pp. 85–102.

34. Silk, N. and Ciruna, K. (eds.). 2005. *A Practitioner's Guide to Freshwater Biodiversity Conservation.* Island Press, Washington, DC.

35. Smith, A. 1776. *An Inquiry into the Nature and Causes of the Wealth of Nations,* Chapter 4, Book I, Wealth of Nations, Indianapolis, Indiana, USA.

36. Spellman, F.R. 2008. *Handbook of Water and Wastewater Treatment Plant Operations,* 2nd Edition, CRC Press, Boca Raton, USA, p. 8.

37. Thanh, N.C. and Tam, D.M. 1990. Water systems and the environment. In: Thanh, N.C. and Biswas, A.K. (eds.), *Environmentally Sound Water Management.* Oxford University Press, Delhi, pp. 1–30.

38. United Nations (UN). 1971. Multilateral Convention on wetlands of international importance especially as waterfowl habitat. Concluded at Ramsar, Iran, on February 2, 1971. https://treaties.un.org/doc/Publication/UNTS/Volume%20996/volume-996-I-14583-English.pdf. Accessed on July 1, 2014.

39. United Nations (UN). 1972. Report of the United Nations Conference on the Human Environment (A/CONF.48/14Ref.1), Stockholm, June 5-16, 1972, United Nations Publications, Geneva, New York, CH, USA. http://www.un-documents.net/aconf48-14r1.pdf. Accessed on June 23, 2014.

40. United Nations (UN). 1997. Convention on the Law of the Non-Navigable Uses of International Watercourses, adopted May 21, 1997, 36 ILM 700, entered into force August 17, 2014.

41. United Nations Centre for Human Settlement (Habitat). 2001. General Assembly. Special Session for an overall Review and an Appraisal of the Implementation of Habitat Agenda in New York, June 6-8, 2001, UNCHS, Nairobi. http://www.un.org/ga/Istanbul+5/bg10.htm. Accessed on June 1, 2014.

42. United Nations Convention to Combat Desertification (UNCCD). 1994. United Nations Convention to combat desertification in those countries experiencing drought and/or desertification, particularly in Africa from June 17, 1994, into force December 8, 1998.

43. United Nations Environment Programme (UNEP). 2007. Global Environment Outlook—Geo4 Environment for Development. United Nations Environment Programme, Nairobi, p. 116, http://www.unep.org/geo/geo4/report/04_water.pdf. Accessed on June 1, 2014.

44. United Nations Report of the Brundtland Commission. 1983. Process of Preparation of the Environmental Perspective to the year 2000 and beyond. General Assembly Resolution, 38/161, December 19, 1983. http://www.un.org/documents/ga/res/38/a38r161.htm. Accessed on June 23, 2014.

45. UNESCO and WWAP. 2003. *Water for People, Water for Life—The United Nations World Water Development Report (WWDR).* UNESCO Paris and Berghahn Books, New York.

46. Veil, J.A., Pudev, M.G., Elcock, D., and Redweik, R.J.J.R. 2004. A white paper describing produced water from production of crude oil, natural gas, and coal bed methane. Prepared for the US Department of Energy, Energy Technology Laboratory.

47. Walmsly, N. and Pearce, G. 2010. Towards sustainable water resources management: Bringing the strategic approach up-to-date. *Irrigation & Drainage Systems,* 24(3/4), 191–203.

48. Wood, C. 2003. *Environmental Impact Assessment: A Comparative Review.* Prentice Hall (Pearson Education), Harlow, United Kingdom.

VI

Social and Cultural Aspects

22

Water Recycling and Community

Mohammad Javad Amiri
Fasa University

Saeid Eslamian
Isfahan University of Technology

Mohammad Arshadi
Shiraz University

Maryam Khozaei
Shiraz University

PREFACE

Water resources are limited in both quantity and quality. As a result, industrial wastewater treatment, recycling, and reuse are important themes in today's context, not only to protect the environment from pollution, but also to conserve water resources so that water stress is reduced. The overview presented here should be useful in identifying key issues and facilitating selection of appropriate processes for treating wastewater. In this chapter, first the water reuse glossary is reviewed and then some benefits, applications, and risks of water recycling for the environment are described. Afterward, four different case studies regarding water recycling and environment that have been conducted in Australia as a developing country with semiarid zones are presented. At the end, the public perception of recycled water and its future are reported.

22.1 Introduction

Water scarcity is a fact of life in arid and semi-arid regions where agricultural, domestic, and industrial demands compete for limited water resources. Continuous population growth, rising standards of living, industrialization, urbanization, and climate changes have increased water demand significantly [20]. In the twenty-first century, recognition of water shortage and unpredictable impact of global warming on overall water scarcity postulates that the first and second decades should be referred to as "water crisis decades" [1]. Various countries in arid and semiarid regions are facing water shortage issues due to the low precipitation and high evaporation. This complicates the supply of water for domestic, industrial, and agricultural uses. Freshwater scarcity, especially in these regions, imposed humans to use poor quality water, such as saline or brackish water, for satisfying their demands [1]. Therefore, to overcome the shortage in water, the use of unconventional water resources can play an important role in satisfying the increasing water requirements [12]. Conventional water resources include surface and groundwater. Now, there is no uniform agreement for a definition of unconventional water resources. In general, it can be said resources other than good quality surface or groundwater are termed unconventional resources [15]. The main unconventional water resources are wastewater (high organic matter and microorganism content), seawater (high saline concentration), and water harvesting (rainwater collection and storage schemes on large or small scale). Wastewater can provide both an additional supply of water for irrigation and a low cost source of nutrients and organic material if a sufficient post-treatment in aerated ponds, soil gravel filters, or membrane ultrafiltration is provided [6]. Current water shortages and the cost associated with freshwater have caused us to conserve water and expand the use of recycled water. Recycled water comes from wastewater and stormwater that has been highly treated to make it suitable for use in a wide array of applications, which is currently applied for nonpotable purposes. Recycled water can be applied for municipal, industrial, and agricultural uses. However, recycled water cannot be used for human or animal drinking, household cleaning, cooking or other kitchen purposes, bathing and showering, swimming pools and spas, evaporative coolers, children's water toys, recreation involving water, and contact such as playing under sprinklers. Some parts of the world provide a great case study of the developments in maximizing water recycling opportunities from policy, regulatory, and technological perspectives [3]. Despite recycled water being a popular choice and being broadly embraced, the concept of indirect potable reuse schemes has lacked community and political support across the world to date.

22.2 Terminology

22.2.1 Water Recycling

Water recycling is the treating and managing of municipal, industrial, or agricultural wastewater to produce water that can be productively reused [5].

22.2.2 Recycled Water

Recycled water or reclaimed water, is former sewage that is treated to remove solids and impurities for sustainable landscaping irrigation, to recharge groundwater aquifers, and to meet commercial and industrial water needs. The purpose of these processes is sustainability and water conservation, rather than discharging the treated water to surface waters such as rivers and oceans [5].

22.2.3 Reused Water

Reused water is water used more than once or recycled [5].

22.2.4 Greywater

Greywater is defined as untreated household wastewater that has not come in contact with toilet waste and includes wastewater from bathtubs, showers, washbasins, clothes washing machines, and laundry tubs, but does not include wastewater from kitchen sinks, dishwashers, or laundry water from washing of materials soiled with human excreta [11,13].

22.2.5 Potable Water

Potable water is water that is fit for consumption by humans and other animals. It is also called drinking water, in a reference to its intended use. Water may be naturally potable, as is the case with pristine springs, or it may need to be treated in order to be safe. In either instance, the safety of water is assessed with tests that look for potentially harmful contaminants [5].

22.2.6 Nonpotable Water

Nonpotable water is water that is not of drinking water quality, but which may still be used for many other purposes, depending on its quality [5].

22.2.7 Potable Reuse

Potable reuse refers to reused water that you can drink [5].

22.2.8 Nonpotable Reuse

Nonpotable reuse refers to reused water that is not used for drinking, but is safe to use for irrigation or industrial purposes [5].

22.2.9 Indirect Potable Reuse (Planned)

Indirect potable reuse can be defined as the reclamation and treatment of water from wastewater (usually sewage effluent) and the eventual returning of it into the natural water cycle well upstream of the drinking water treatment plant. Planned reuse indicates that there is intent to reuse the water for potable use. The point of return could be into a major water supply reservoir, a stream feeding a reservoir, or into a water supply aquifer where natural processes of filtration and dilution of the water with natural flows aim to reduce any real or perceived risks associated with eventual potable reuse [22].

22.2.10 Indirect Potable Reuse (Unplanned)

Unplanned indirect potable reuse can be defined as wastewater entering the natural water system (creeks, rivers, lakes, and aquifers), which is eventually extracted from the natural system for drinking water with generally no awareness that the natural system contains treated wastewater [22].

22.2.11 Direct Potable Reuse

Direct potable reuse can be defined as either the injection of recycled water directly into the potable water supply distribution system downstream of the water treatment plant, or into the raw water supply immediately upstream of the water treatment plant. Injection could be either into a service reservoir or directly into a water pipeline. Therefore, the water used by consumers could be either undiluted or slightly diluted recycled water. In this definition, the key distinction with indirect potable reuse is that there is no temporal or spatial separation between the recycled water introduction and its distribution to consumers. Public perception of what extent of separation is required for reuse to become indirect may dictate the definition ultimately adopted [17,22].

22.3 Benefits of Recycled Water

An investment in recycled water could help solve many problems. It can be used to [8]

- Control water pollution
- Restore wetlands and marshes
- Forestall a water shortage by conserving freshwater
- Provide additional reliable local sources of water, nutrients, and organic matter for soil conditioning
- Provide drought protection
- Improve the economic efficiency of investments in pollution control and irrigation projects, particularly near urban areas
- Improve social benefits by creating more jobs and improving human and environmental health protection

In general, water recycling can help us find ways to develop new water sources, prevent water resource degradation, and improve efficiency of water consumption. Therefore, water recycling has a big potential to bring about environmental, economic, and financial benefits. The role of water recycling in improving health is shown in Figure 22.1.

22.4 Application of Recycled Water

Recycled water can satisfy most water demands, as long as it is adequately treated to ensure water quality appropriate for use. The most common treatment levels to produce recycled water is shown in Figure 22.2. As Figure 22.2 shows, improved and possible uses increase as the level of treatment increases. In uses where there is a greater chance of human exposure to the water, more treatment may be required. Reclaimed water can be treated at three different levels of increasing cleanliness/safety:

1. Primary: A physical process removes some of the suspended solids and organic matter from the wastewater. The remaining effluent from primary treatment will ordinarily contain considerable organic material and will have a relatively high biochemical oxygen demand (BOD).
2. Secondary: Biological processes involving microorganisms remove organic matter and suspended material. The effluent from secondary treatment usually has little BOD and few suspended solids (SS).

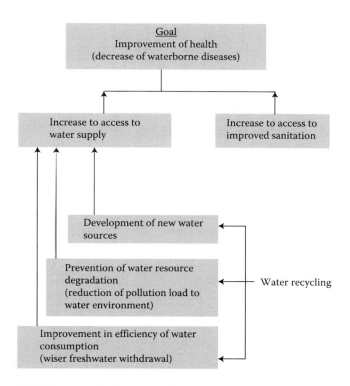

FIGURE 22.1 Role of water recycling.

3. Tertiary: This process further removes suspended and dissolved materials remaining after secondary treatment and often involves chemical disinfection and filtration of the wastewater [8].

Recycled water can be used in a variety of ways, including agricultural irrigation, landscape irrigation, groundwater recharge, industry, environment and recreation, and as an indirect potable water source. Direct and indirect recycled water use is illustrated in Figure 22.3. Direct and indirect recycled water use are classified as described in the following subsections.

22.4.1 Agriculture

Agriculture is the single greatest water consumer in the world. The agriculture sector controls the largest volumes of water and has a mandated responsibility to manage national and global water resources in a sustainable manner. The World Water Council, seeking to balance water supply with greater demand, suggested improved technologies to provide new water for agricultural use, such as the use of recycled water in agriculture [2]. The use of recycled water in agriculture is a growing practice that may help ensure safe and sustainable food crops. Recycled water is used in several states to irrigate both edible and nonedible crops. The overarching goals of water reuse in agriculture are to provide an adequate supply of high quality water for growers and to ensure food safety. In some parts of the world, recycled water is used to grow various kinds of vegetables and irrigate grapevines and citrus groves. Additionally, in many places,

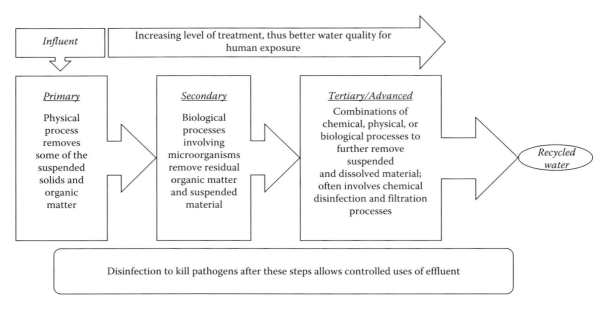

FIGURE 22.2 Treatment levels to produce recycled water. (From Department of Water Resources, State of California. 2004. *Water Facts*, No. 23, *Water Recycling*.)

recycled water might be the highest quality water available to farmers [7,16].

For example, in California, farmers use about 250,000 acre-feet of recycled water annually. A survey conducted in 1997 by the State Department of Water Resources and the Agriculture Committee of the Water Reuse Association indicated that 187,195 acre-feet of recycled water have been used on 61,553 acres of farm and ranch lands. It showed that recycled water in California is used to irrigate a wide variety of crops. In all, the survey found 52 different crops being grown with the help of recycled water.

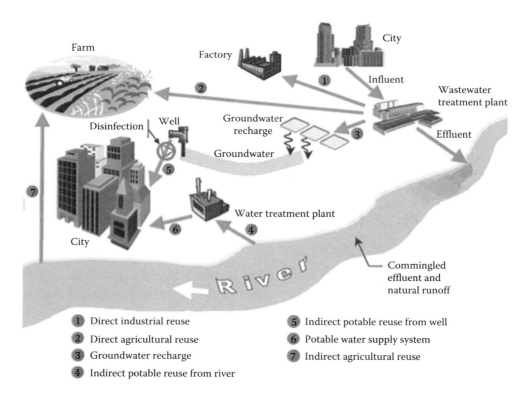

FIGURE 22.3 Direct and indirect recycled water use.

Its use for agriculture in California and worldwide has been shown to be economically and environmentally sound [8].

However, the type of irrigation is crucial for improving the quality and quantity of food production. Therefore, there is a need to identify the possible sources and problems involved with its use. The extent of contamination depends on the degree of treatment provided and the design of the irrigation system. Surface irrigation is the oldest and most common method (90% in the world and 60% in the United States) of applying water to croplands. This type of irrigation has a higher possibility of contaminating produce, especially when it is applied on creeping crops and where water drives through contaminated soil. In crops staked and raised from soil, such as those vegetables mostly grown for exportation, the risk of contamination is low. In addition, the sprinkler irrigation technique provides a rapid means to contaminate the product if the water is contaminated. Among the types of irrigation, drip irrigation is the most common type used on agriculture dedicated to exportation, reducing risk of contamination; however, this technique is used for approximately 15% of the irrigated land in the world [7,24].

22.4.2 Landscape Irrigation

Over time, recycled water could be the main source of irrigation for parks, playgrounds, golf courses, freeway landscaping, commercial and office building landscaping, and residential landscaping. Generally, recycled water will have a higher concentration of dissolved salts than drinking water [23]. Water with high levels of salts can have adverse effects on plant health and appearance; however, most recycled water produced does not have harmful levels of salts for most plants. Tertiary treated recycled water can be used for landscape; however, turf grasses, most annuals, and deciduous trees are more tolerant of saline water than evergreens [10].

22.4.3 Groundwater Recharge

Recycled water can be used to recharge groundwater aquifers. It can replenish, restore, or protect groundwater against saltwater intrusion. Water spreading (percolation from a basin) and injection are the common groundwater recharge practices. High quality water is necessary for injection. It may require advanced treatment such as reverse osmosis in addition to what is already required. There are several advantages in storing water underground via groundwater recharge including (1) the cost of artificial recharge may be less than the cost of equivalent surface water reservoirs; (2) the aquifer serves as an eventual natural distribution system and may reduce the need for transmission pipelines or canals for surface water; (3) water stored in surface reservoirs is subject to evaporation, taste, and odor problems due to algae and other aquatic productivity, and to pollution, which may be avoided by soil-aquifer treatment (SAT) and underground storage; (4) suitable sites for surface water reservoirs may not be available or may not be environmentally acceptable;

and (5) the inclusion of groundwater recharge in a wastewater reuse project may provide psychological and esthetic benefits as a result of the transition between reclaimed municipal wastewater and groundwater. However, at present, some uncertainties with respect to health risk considerations have limited expanding use of reclaimed municipal wastewater for groundwater recharge, especially when a large portion of the groundwater contains reclaimed wastewater that may affect the domestic water supply [4].

22.4.4 Industry

The potential use for recycled water in industry is high. Industry can use recycled water for heat dissipation, power generation, and processing. The advantages of recycling water in industry are as follows: (1) it allows recovering energy (as heat) from used effluents; (2) it reduces the production costs as it saves water [24]; (3) it allows recovery of prime matter; and (4) it reduces the cost of treating and disposing wastewater (discharging fees). However, the usage of recycled water in industry has different disadvantages that are classified as follows: (1) if the proper quality of water is not well defined when recycling, the background compounds contained in water are increased in each recycling cycle up to an extent in which the concentrations reached may impair the quality of the product. If no proper water quality criteria are set, the quality of the product might be decreased and the industrial processes may be affected or their efficiency reduced. (2) There are some health concerns because of the exposure to organic volatile compounds or microorganisms (such as *Legionella*) that might be disseminated through aerosols [21].

22.4.5 Environment and Recreation

Recycling not only can make extra water available, but also can protect sensitive water bodies from pollution. Recycled water can help protect and maintain the environment. It has helped in the development of recreational lakes, marsh enhancement, and stream flow augmentation. It also can be impounded for urban landscape development. In 1987, about 10,000 acre-feet of water were used for such purposes. By 2002, such volume had increased to more than 53,000 acre-feet used for recreational impoundment and wildlife habitat enhancement [8,24]. Some advantages of recycling water in environmental purposes are as follows: (1) protects water bodies from overexploitation, (2) surface water reuse can be integrated in urban areas as part of esthetic or recreational projects, (3) has the potential to be developed in multifunctional use projects considering treatment, storage, nature conservation, recreation, etc., and (4) can be combined with rainwater reclamation.

22.4.6 Greywater Reuse

In conventional sewer systems, greywater cannot be reused as it is discharged into the sewer system as well as blackwater. In onsite sanitation areas, greywater can be utilized, thereby reducing freshwater consumption, if it is separated from black-water (Figure 22.4) [14]. Greywater is generally reused without pretreatment for agricultural or landscape irrigation in household scale or in a larger scale. It is best to design a greywater system that prevents human contact and the potential for environmental contamination. Greywater reuse should be avoided for irrigation of root crops and edible parts of food crops. It is also needed to prevent groundwater contamination [11,13].

22.4.7 Nonpotable Urban Uses

Numerous nonpotable domestic and urban uses of recycled water can be identified. Examples of such urban uses include the use of recycled water for fire protection, air conditioning, toilet and urinal flushing, artificial snowmaking, concrete mixing, and dust control. Some of the advantages of recycling water for nonpotable urban uses are as follows: (1) it is a no consumptive activity providing water for a next reuse; (2) allows recovery of wastewater where it is produced; (3) reduces energy consumption by saving the net amount of first-use water supply; (4) provides water for a wide variety of uses, allowing flexible water reuse programs; (5) can be metered easily; and (6) in many cases, the cost of treating water for reuses demanding low quality is cheaper than bringing water from new sources [25].

22.4.8 Indirect Potable Uses

Indirect potable uses include the recharge of potable aquifers and the replenishment of surface reservoirs. Groundwater recharge with recycled water and indirect potable water reuse in general share many of the public health concerns encountered in drinking water withdrawn from polluted rivers and reservoirs. Four water quality factors are of special concern where recycled water is used in such applications: (1) enteric viruses and other pathogens; (2) organic and inorganic substances including industrial and pharmaceutical chemicals, residual home cleaning and personal care products, and other persistent pollutants; (3) salinity; and (4) heavy metals. The ramifications of many of these constituents in trace quantities are not well understood with respect to long-term health effects. For example, there are concerns about exposure to chemicals that may function as endocrine disrupters; also, the potential for development of antibiotic resistance is of concern. As a result, regulatory agencies are proceeding with extreme caution in permitting water reuse applications that affect potable water supplies [8,14].

22.5 Risks of Recycled Water

The risks associated with recycled water must be minimized to acceptable levels before recycled water can be used in any

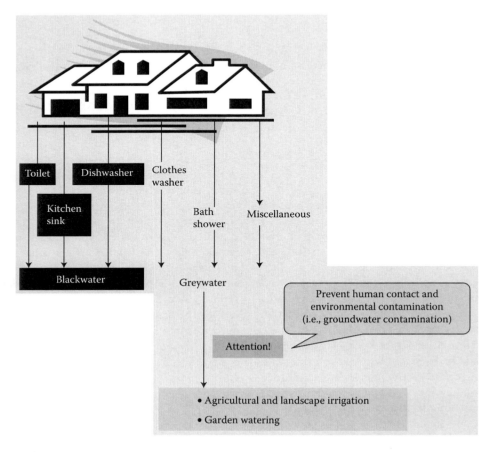

FIGURE 22.4 Outline of greywater reuse.

specific situation. In most cases, these environmental and health risks can be managed through the level of wastewater treatment or by the carefully managed use of recycled water. However, in some cases, these risks are too costly to manage and the reuse scheme may not be economically viable. Individual state environment and health-related authorities are generally responsible for ensuring the water recycled is fit for the intended use. Microbial pathogens in wastewater from sewage effluent are the major concern for human health [5]. The major groups of pathogens are [19]

- Bacteria (e.g., *Escherichia coli*, *Salmonella* spp.)
- Viruses (e.g., *Enterovirus*, *Rotavirus*, Hepatitis A)
- Protozoa (e.g., *Giardia lamblia*, *Cryptosporidium parvum*)
- Helminths (e.g., *Taenia* spp. [tapeworm], *Ancylostoma* spp. [hookworm])

However, not all infections make you sick. You must be exposed to a sufficient number of pathogens to become infected. If recycled water is fit for the intended purpose, exposure will be low and infection unlikely as it is related to the concentrations of pathogens in the recycled water and the amount of water ingested. Key potential environmental risks are shown in Table 22.1, together with their environmental endpoints and effects on the environment [5]. The review that identified the hazards listed in Table 22.1 did not cover environmental allocation of recycled (i.e., allocation directly to waterways or water bodies to benefit the environment) water detail because, in some states and territories, such allocations would be regulated under water allocation plans or under discharge rules for effluent from recycled water treatment plants. However, the review did include an initial screening-level risk assessment of chemical parameters in recycled water, with the aim of assessing whether these

TABLE 22.1 Key Environmental Hazards, Environmental Endpoints, and Common Effects on the Environment

Hazard	Environmental Endpoint	Effect or Impact on the Environment
Cadmium	A low risk with respect to cadmium concentrations in recycled water, but cadmium already in soils can be made more readily available to plants if chloride concentrations increase. Chloride can be measured indirectly, but reliably, as salinity (see the salinity section below).	
Chlorine disinfection residuals	Plants	Direct toxicity to plants
	Surface waters	Toxicity to aquatic biota
Boron	Accumulation in soil	Plant toxicity
Hydraulic loading	Soil	Waterlogging of plants
	Groundwater	Waterlogging of plants
	Groundwater	Soil salinity (secondary)
Nitrogen	Soils	Nutrient imbalance in plants
	Soils	Pest and disease in plants
	Soils	Eutrophication of soils and effects on terrestrial biota
	Surface water	Eutrophication
	Groundwater	Contamination
Phosphorus	Soils	Eutrophication of soils and toxic effects on phosphorus-sensitive terrestrial biota (native plants)
	Surface water	Eutrophication
Salinity	Infrastructure	Salinity may cause rising damp or corrosion of assets; this can also arise from excessive hydraulic load (secondary salinity)
	Soils	
	Soils	
	Surface water	Plants stressed from osmotic affects of soil salinity
	Groundwater	Contamination of soils by increasing plant availability of cadmium that is already in the soil
		Increasing the salinity of fresh groundwater
		Increasing the salinity of fresh surface waters
Chloride	Plants	Direct toxicity to plants when sprayed on leaves
	Soils	Plant toxicity via uptake through the root
	Surface water	Toxicity to aquatic biota
Sodium	Plants	Direct toxicity to plants when sprayed on leaves
	Soils	Plant toxicity via uptake through the root
	Soils	Soil structure decline due to sodicity

chemicals posed a hazard for environmental allocation of recycled water [19].

22.6 Risk Assessment

Assessment of risk is undertaken in relation to recycled water supplies for a number of reasons, including: (1) to predict the burden of waterborne disease in the community in both outbreak and nonoutbreak conditions; (2) to assist in setting target reference pathogen levels for recycled water supplies that will equate to tolerable levels of illness within populations exposed to that water; (3) to identify the most cost-effective method to reduce pathogen-related health risks to those exposed to recycled water; (4) to assist in determining the optimum balance in terms of pathogen kill versus the formation of disinfection by-products (DBPs); and (5) to provide a conceptual framework for consumers, organizations, regulators, and industry to understand the nature and risk to, and from, recycled water and how those risks can be managed. With respect to water recycling, health risk assessment components include consideration of chemical exposure, pathogen exposure, and system reliability/hazardous events analysis. The framework is depicted in Figure 22.5. This framework consists of a four-step process:

(1) hazard identification, (2) dose-response assessment, (3) exposure assessment, and (4) risk characterization. In Step (1), hazard identification includes identification of hazards that might be present and the associated effects on human health; this step also includes consideration of variability in hazard concentrations. In Step (2), dose response includes establishment of the relationship between the dose of the hazard and the incidence or likelihood of illness. In Step (3), exposure assessment includes determination of the size and nature of the population exposed to the hazard, and the route, amount, and duration of exposure. In Step (4), risk characterization includes integration of data on hazard presence, dose response, and exposure, obtained in the first three steps [5].

In Step (4), the magnitude of risk should be assessed on two levels: (a) maximum risk (risk in the absence of preventive measures) and (b) residual risk (risk that remains after consideration of existing preventive measures). Maximum risk is useful for identifying high-priority risks, identifying appropriate preventive measures, calculating performance targets, and preparing for emergencies should preventive measures fail. Residual risk provides an indication of the safety and sustainability of the recycled water scheme or the need for additional preventive measures.

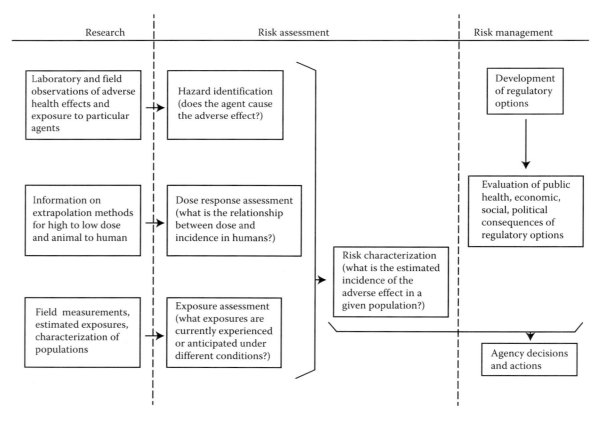

FIGURE 22.5 The risk assessment/risk management framework. (Modified from NRC. 1983. *Risk Assessment in the Federal Government: Managing the Process, Committee on the Institutional Means for Assessment of Risks to Public Health.* National Research Council: The National Academies Press, Washington, DC; Yokel, R.A. and MacPhail, R.C. 2011. *Journal Occupational Medical Toxicology,* 6(7), 1–27.)

22.7 Framework for Management of Recycled Water Quality and Use

The risk management framework is used to develop a management plan that describes the nature of a recycled water system and how it should be operated and managed. The plan is referred to as a risk management plan. The framework is a systematic approach to the management of recycled water quality from source through the treatment and end use to assure safety and reliability. The generic process in developing and implementing the preventive risk management systems for recycled water use can be applied to all combinations of water sources, treatment technologies, and end use. There is sufficient flexibility in the risk management framework to apply to all types of recycled water schemes, irrespective of size and complexity. The framework for management of recycled water quality incorporates 12 elements, organized within four general areas (Figure 22.6). These elements are interrelated, and each supports the effectiveness of the others. Because most problems associated with recycled water schemes are attributable to a combination of factors, the 12 elements need to be addressed together to assure a safe and sustainable recycled water supply [5]. The 12 elements are organized within four general areas that listed here:

1. Commitment to responsible use and management of recycled water. This requires the development of a commitment to responsible use of recycled water and to application of a preventive risk management approach to support this use. The commitment requires active participation of senior managers, and a supportive organizational philosophy within agencies responsible for operating and managing recycled water schemes.

2. System analysis and management. This requires an understanding of the entire recycled water system, the hazards and events that can compromise recycled water quality, and the preventive measures and operational control necessary for assuring safe and reliable use of recycled water.

3. Supporting requirements. These include basic elements of good practice, such as employee training, community involvement, research and development, validation of process efficacy, and systems for documentation and reporting.

4. Review. This includes evaluation and audit processes to ensure that the management system is functioning satisfactorily. It also provides a basis for review and continuous improvement.

The 12 elements within the framework are not discrete components, but interrelated activities to secure safe and sustainable recycled water supply. There must be an overarching commitment from all stakeholders to the responsible use and management of recycled water to ensure safe supply and use of recycled water for the life of the scheme. Having a clear understanding of the entire recycled water system, the hazards and events that can compromise recycled water quality, and the preventive measures is paramount for assuring safe and reliable use of recycled water. The supporting requirements of having training and education programs for the employee and the users of the recycled water, validation of process efficacy, and systems for documentation and reporting are essential for the continuous management of the scheme. Within the risk management plan, there also needs

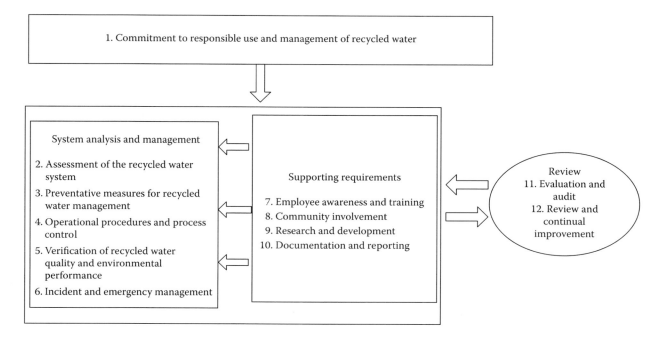

FIGURE 22.6 Framework for management of recycled water quality and use.

to be an evaluation and audit process to ensure that the management system is functioning satisfactorily and to provide a basis for review and continuous improvement.

22.8 Case Studies

In this section, four different case studies that have been conducted in Australia as a developing country with semi-arid zones that encountered water deficit problems are introduced. It is not surprising that Australia is a good case study for water recycling. Recognized as one of the driest continents on Earth with one of the most variable climates, Australia offers some valuable lessons to other countries looking at developing their wastewater management practices to maximize the benefits from this increasingly valuable source of water.

- Commercial crops irrigated with recycled water from a major metropolitan sewage treatment plant
- Use of greywater for toilet flushing and outdoor use
- Irrigation of municipal (landscape) areas with water recycled from a small community's sewage treatment plant
- Commercial laundry water recycling in Queensland, Australia

22.8.1 Commercial Crops Irrigated with Recycled Water from a Major Metropolitan Sewage Treatment Plant

In this case study, the source was treated sewage. Recycled water from a major metropolitan sewage treatment plant receiving domestic and industrial sewage was used for spray irrigation of commercial crops, including salad vegetables. The sewage treatment plant flow was 120 ML/day. The plant originally provided secondary treatment, followed by about 20 days of lagoon storage and polishing, before discharge of most of its treated sewage to the sea. The recycled water pipeline was commissioned in 1999 to supply treated sewage from the plant for irrigating commercial food crops. The aim was to supplement the existing use of groundwater, while substantially reducing discharge of nutrient-rich water to the sea. The users of the treated sewage were largely commercial market gardeners who required water that could be used to spray irrigate a range of crops, including salad vegetables such as lettuce. The human health risk assessment of the proposed recycled water scheme was performed in accordance with Element 2 of the framework for risk assessment and management. In addition, the environmental risk assessment of the recycled water scheme was performed in accordance with Element 2 of the framework for risk assessment and management [7].

22.8.2 Use of Greywater for Toilet Flushing and Outdoor Use

In this study, the source was treated greywater (laundry, bathroom, showers, and hand basins) from 100 units in an apartment complex. The greywater is used for toilet flushing, drip irrigation of

garden beds, and subsurface irrigation of grassed areas. Microbial hazards for human health include enteric bacteria, viruses, and protozoa. The total log reduction required for this scheme was determined and results from 6 months of greywater *Escherichia coli* monitoring at the development. The calculated targets were a 2-log reduction of protozoa, a 3.5-log reduction of viruses, and a 3-log reduction of bacteria. Exposure of chemicals in recycled water was too low to represent a health risk. Preventative measures included microfiltration, ultraviolet (UV), and chlorine disinfection. Onsite control measures were also applied. The membranes provided a 3-log reduction of bacteriophage and *E. coli*. This was discounted to 2 log for viruses, protozoa, and bacteria because membrane integrity monitoring (turbidity and particle counting) has limited sensitivity. A minimum UV dose of 25 mJ/cm^2 was applied. Based on published data, this provided a 1.5-log reduction of Cryptosporidium. Chlorine disinfection was expected to produce free chlorine residuals; however, the occasional presence of ammonia could result in production of chloramines, which are weaker disinfectants but persist for longer periods. A 2-log reduction of viruses was determined based on published data for free chlorine and chloramines. The combined impact of UV disinfection and chlorination would provide greater than a 4-log reduction of enteric bacteria. In total, the treatment processes provide a minimum 3.5-log reduction of protozoa, 4-log reduction of viruses, and 6-log reduction of viruses, which exceeded the calculated targets [5].

22.8.3 Irrigation of Municipal (Landscape) Areas with Water Recycled from a Small Community's Sewage Treatment Plant

In this case study, the source was treated sewage. Recycled water from a small community's sewage treatment plant was used for drip irrigation of trees and shrubs in municipal (landscape) areas of a local park. Microbial hazards for human health include enteric bacteria, viruses, and protozoa. The total log reduction required for this scheme is 3.5-log reduction of protozoa, 5-log reduction of viruses, and 4-log reduction of bacteria. Exposure of chemicals in recycled water is too low to represent a health risk. Limiting use to drip irrigation reduces the potential for public exposure to enteric protozoa, viruses, and bacteria by 4 logs. A further 1-log reduction of viruses is required. This can be achieved by secondary treatment and disinfection. Available data indicate that secondary treatment and disinfection provide the required log reduction. The preliminary risk assessment identified groundwater, landscape and garden plants, and specific soil types as potential environmental endpoints for hazards. Phase 1 of the risk assessment (preliminary screening) identified the hazards nitrogen and phosphorus as moderate to high risks. Phase 2 (maximal risk assessment) confirmed that phosphorus and nitrogen required preventive measures to lower the risk to acceptable levels. Phase 3 (residual risk assessment) identified a range of preventive measures available that should reduce the risks associated with key hazards identified above to acceptable levels [3,7].

22.8.4 Commercial Laundry Water Recycling in Queensland, Australia

The commercial laundry industry is one of the largest consumers of potable water, as well as being generators of huge quantities of polluted wastewater. A typical commercial laundry operation will consume on average 250 m³/day of potable water and discharge approximately 200 m³/day of polluted wastewater. Water analysis shows this discharged wastewater stream to be high in chemical oxygen demand (COD), BOD, grease/oil, lint, total suspended solids (TSS), total dissolved solids (TDS), detergents, and surfactants as well as being at temperatures well above ambient potable water. Australian Regulatory Authorities have stringent standards in regard to content, quantity, flow rates, and temperature of waste streams entering their infrastructure and most require some form of screening, dosing, flocculants, coagulates, detention, and cooling, as well as methods of measurement enabling them to assess the trade waste changes. Consequently, the recycling of commercial laundry wastewater has become a necessary requirement to aid in the reduction of operational costs and meet regulatory standards. Most recently, hollow fiber ceramic membrane technology has proven to be capable with the ability to treat wastewater at 80°C temperatures. The Gray Box is proving to be an effective and viable wastewater treatment solution, capable of removing grease, oil, and TSS, achieving government approved, A+ water quality, suitable for commercial laundry water reuse in continuous batch washers and washer extractors. The laundry wastewater is discharged to the raw water storage tank, and then pumped to a lint removal device, as direct steam injection maintains the desired hot water temperature of 60°C. The water is then feed via low-pressure transfer pumps (3bar) to the Gray Box process incorporating the Inocep, hollow fiber ceramic membranes. The water then flows through an activated carbon filter for polishing, then via UV treatment to the recycled water storage tank. Water quality in this tank is monitored and maintained using pH adjustment and ozone dosing. The process achieves a government approved A+ water quality standard. Due to the level of heat retention in the recycling process, and the high quality A+ recycled water produced by the Gray Box, it is extremely well suited for commercial laundry operations [5,22].

22.9 Water Recycling in Iran

Islamic Republic of Iran is in the midst of a serious water crisis. The looming crisis is being blamed on a number of factors including population growth and uneven distribution, natural phenomena such as droughts and changing climate patterns, and the mismanagement of existing water resources. The country's resources management is facing many subsequent challenges, including growing demand for water resources with proper quality, a considerable increase in the costs of supplying additional water, an urgent need to control water pollution, the uncontrolled exploitation of underground waters, and the necessity to conserve these valuable resources. If immediate mitigation measures are not taken, the situation could become even more disastrous in the years to come. Day by day, the water consumption by citizens is increasing and the volume of wastewater used for irrigation is also increasing [2]. One solution to this problem is the multiple use of wastewater by suitable treatment. By substituting the treated effluent for agricultural water from the reservoirs of the constructed dams and allocating reservoir water to the citizens, the problem could be alleviated. Therefore, the treatment of wastewater and its use for agriculture could provide an opportunity to allocate fresh river water for drinking purposes and for maintenance of health in the metropolitan cities of Iran. By recycling wastewater and preventing overexploitation of groundwater, the water table will come up and will be stabilized, and the risk of water, land, and soil pollution will be avoided. At present, wastewater reuse can be summarized as follows:

1. Indirect potable reuse. From 1000 cities in Iran, only 200 have wastewater collection networks operational or under design and construction. In many cities, wastewater from seepage pits infiltrates through layers of the ground ultimately reaching and recharging underground aquifers. The average distance between the water table and the bottom of these seepage pits should be above 20 m. Because of rapid industrialization and population increase, there are small communities and industries within city limits that draw their water from these underground strata.

2. Intentional groundwater recharge for nonpotable reuse. This is already practiced around the major city limits where underground aquifers are recharged with the seepage pits and wastewater treatment plant (WWTP) effluent to recharge a brackish groundwater aquifer through streambeds and channels and are used downstream through springs and qanats by farmers to irrigate their fields and for washing purposes.

3. Direct use. Direct use of untreated wastewater from sewage outlets, directly disposed of on land where it is used for crop production is not a common scene in Iran. Treated or partially treated wastewater used directly for irrigation without being mixed or diluted is more common. This is practiced in many plants. There is no exact estimate about the amount used by this method to irrigate fodders, cereals, fruit trees, and vegetables eaten cooked or uncooked.

4. Indirect use. Indirect use of untreated wastewater, where diluted wastewater is mixed with stormwater or small streams or tributaries of larger water bodies (polluted) are used for irrigation, is very common especially downstream of urban centers where treatment facilities are inadequate. In this way, a self-purification process takes place in full scale while effluents are flowing in streams, mixing with the stream water or diluted with qanat waters. This water is quite suitable for unrestricted irrigation.

5. Planned direct use. In this case, when the reclaimed water has been transported from the point of treatment to the point of use without an intervening discharge to waters, it is becoming more common especially in places where drinking water for cities is not adequate (i.e., city of

Mashhad during the recent drought). The planned direct use of reclaimed water is not administered by the authority responsible for managing wastewater treatment plants, that is, Water and Wastewater Consulting Engineers (WWCE), but is administered by special contract with the Wastewater Collection Division (WCD) and farmers formalizing their rights to use reclaimed water directly or by substitution of their rights.

22.10 Public Perception of Recycled Water versus Desalinated Water

As we know, water recycling is the treatment of municipal wastewater for the replenishment of available freshwater resources and consumption. Water recycling hence closes the water cycle on a more local level with the possibility of closing water cycles for individual households, buildings, factories, towns, or regions. The motivation for this activity is mostly the realization that human water consumption has increased beyond sustainable levels, resulting in extended periods of drought, depletion of environmental flows in natural water systems, and the decrease in healthy levels in drinking water reservoirs, including groundwater systems. However, the public often vehemently rejects water-recycling activities and as a result recycled water is available in countries with severe water restrictions, but clients for this recycled water often cannot be found. Several public consultation studies have been carried out to explore reasons for this resistance and how to gain community support [9]. In a different approach to circumvent the difficulties with recycled water acceptance, some countries are considering desalinating seawater to provide the shortfall in drinking water and avoid public acceptance problems. The results of investigations showed that acceptance levels of recycled water would be lower than acceptance levels of desalinated water. Interestingly, however, the results indicate that it is not possible to state that people generally perceived either desalinated water or recycled water as preferable. It appears that people discriminate by the nature of the water use where the likelihood of adoption for close to body uses is comparatively high for desalinated water as opposed to irrigation and cleaning the car and the house, for which recycled water is ranked higher in the adoption sequence. While these findings are derived from the aggregate of all respondents, future work should investigate whether personal characteristics, such as education level, prior experience with recycled or desalinated water, prior experience with drought, etc., impact the knowledge, perception, and likelihood of use. Further, other water resources such as stormwater, yellow water, and greywater need to be considered.

22.11 Summary and Conclusions

Water recycling has proven to be effective and successful in creating a new and reliable water supply without compromising

public health. Nonpotable reuse is a widely accepted practice that will continue to grow. However, in many parts of the world, the uses of recycled water are expanding in order to accommodate the needs of the environment and growing water supply demands. Advances in wastewater treatment technology and health studies of indirect potable reuse have led many to predict that planned indirect potable reuse will soon become more common. Recycling waste- and greywater requires far less energy than treating saltwater using a desalination system. While water recycling is a sustainable approach and can be cost-effective in the long term, the treatment of wastewater for reuse and the installation of distribution systems at centralized facilities initially can be expensive compared to such water supply alternatives as imported water, groundwater, or the use of greywater onsite from homes. Institutional barriers, as well as varying agency priorities and public misperception, can make it difficult to implement water-recycling projects. Finally, early in the planning process, agencies must reach out to the public to address any concerns and to keep the public informed and involved in the planning process. As water energy demands and environmental needs grow, water recycling will play a greater role in our overall water supply. By working together to overcome obstacles, water recycling, along with water conservation and efficiency, can help us to sustainably manage our vital water resources. Communities and businesses are working together to meet water resource needs locally in ways that expand resources, support the environment, and strengthen the economy. From the above case studies we will have excellent operational data to design new systems that protect the environment and reduce the carbon footprint of the water industry in a way that is acceptable to our communities. In the future, urban water-recycling schemes will be

- More energy efficient so that they do not contribute to GHG emissions
- Better integrated with urban planning
- More effective in substituting water imported to cities at either the regional or the household level

Industries will manage their water more efficiently within their premises and some may be co-located allowing for multiple uses of water before any release to the environment. There will be an increasing trend to recovering other high-value by-products from wastewater such as phosphorus, nitrogen, potassium, and other commodity chemicals. The term wastewater treatment plants may change to water management and nutrient and energy recovery plants.

References

1. Aghakhani, A., Mousavi, S.F., Mostafazadeh-Fard, B., Rostamian, R., and Seraji, M. 2011. Application of some combined adsorbents to remove salinity parameters from drainage water. *Desalination*, 275(1), 217–223.

2. Amiri, M.J. and Eslamian, S.S. 2010. Investigation of climate change in Iran. *Journal of Environmental Science Technology,* 3(4), 208–216.

3. Apostolidis, N., Hertle, C., and Young, R. 2011. Water recycling in Australia. *Water,* 3, 869–881.

4. Asanoa, T. and Cotruvo, J.A. 2004. Groundwater recharge with reclaimed municipal wastewater: Health and regulatory considerations. *Water Research,* 38(12), 1941–1951.

5. Australian Guidelines for Water Recycling: Managing Health and Environmental Risks (Phase 1). 2006. *National Guidelines for Water Recycling: Managing Health and Environmental Risks,* 1, 1–389.

6. Buchholz, M. 2008. Overcoming Drought, The Cycler Support Implementation Guide, A Scenario for the Future Development of the Agricultural and Water Sector in Arid and Hyper Arid Areas. Available at http://www.emwis.net/thematicdirs/news/2008/12/eu-project-cycler-suppport-implementation-guide.

7. Chaidez, C. and Soto, M. 2014. *Water: Waste, Recycling and Irrigation in Fresh Produce Processing.* Woodhead Publishing, Cambridge, UK.

8. Department of Water Resources, State of California. 2004. Water Facts, No. 23, *Water Recycling.* CA DWR Publishing, California.

9. Dolnicar, S. and Schafer, A.I. 2006. Public perception of desalinated versus recycled water in Australia. CD Proceedings of the AWWA Desalination Symposium. Available at http://ro.uow.edu.au/commpapers/138.

10. Ebrahimizadeh, M.A., Amiri, M.J., Eslamian, S.S., Abedi-Koupai, J., and Khozaei, M. 2009. The effects of different water qualities and irrigation methods on soil chemical properties. *Research Journal of Environmental Sciences,* 3(4), 497–503.

11. Eriksson E., Auffarth, K., Henze, M., and Ledin, A. 2002. Characteristics of grey wastewater. *Urban Water,* 4, 85–104.

12. Eslamian, S.S., Amiri, M.J., Abedi-koupai, J., and Shaeri Karimi, S. 2013. Reclamation of unconventional water using nano zero-valent iron particles: An application for groundwater. *International Journal Water,* 7(1/2), 1–13.

13. GESAP, Water and wastewater reuse. Available at http://nett21.gec.jp/GESAP/themes/themes2.html.

14. Godfrey, S., Labhasetwar, P., Swami, A., Parihar, G., and Dwivedi, H. 2007. Water safety plans for grey water in tribal schools. *Water Lines,* 5(3), 8–10.

15. Indelicato, S., Tamburino, V., and Zimbone, S.M. 1993. Unconventional water resource use and management. *CIHEAM, Options Mediterraneennes,* 1(1), 57–74.

16. Kirby, R.M., Bartram, J., and Carr, R. 2003. Water in food production and processing: Quantity and quality concerns. *Food Control,* 14, 283–299.

17. Leverenz, H.L., Tchobanoglous, G., and Asano, T. 2011. Direct potable reuse: A future imperative. *Journal of Water Reuse and Desalination,* 1(1), 2–10.

18. NRC. 1983. *Risk Assessment in the Federal Government: Managing the Process, Committee on the Institutional Means for Assessment of Risks to Public Health.* National Research Council: The National Academies Press, Washington, DC.

19. Pant, D. and Adholeya, A. 2007. Biological approaches for treatment of distillery wastewater: A review. *Bioresource Technology,* 98, 2321–2334.

20. Pearce, G.K. 2008. UF/MF pre-treatment to RO in seawater and wastewater reuse applications: A comparison of energy costs. *Desalination,* 222, 66–73.

21. Ranade, V.V. and Bhandari, V.M. 2014. *Industrial Wastewater Treatment, Recycling, and Reuse: An Overview.* Elsevier Ltd, Amsterdam, the Netherlands.

22. Recycled Water in Australia. Available at http://www.recycledwater.com.au/index.php?id=105.

23. Recycled Water Use in the Landscape. 2000. Available at http://www.leginfo.ca.gov.

24. Schroeder, E., Tchobanoglous, G., Leverenz, H.L., and Asano, T. 2012. *Direct Potable Reuse: Benefits for Public Water Supplies, Agriculture, the Environment, and Energy Conservation.* National Water Research Institute, Fountain Valley, CA.

25. USEPA. Water recycling and reuse: The environmental benefits. Available at http://www.epa.gov/region9/water/recycling/.

26. Yokel, R.A. and MacPhail, R.C. 2011. Engineered nanomaterials: Exposures, hazards, and risk prevention. *Journal Occupational Medical Toxicology,* 6(7), 1–27.

23

Ethical and Cultural Dimensions of Water Reuse: Islamic Perspectives

Hussein A. Amery
Colorado School of Mines

Marwan Haddad
An-Najah National University

PREFACE

Wastewater is often sewage water, which partly explains humans' historical negligence of this by-product of modern day living. However, in recent decades, developing countries have shown a growing interest in exploiting this underutilized resource by treating it and reusing it. The advantages of that include pollution mitigation and augmenting domestic water sources at a cost that is lower than desalination. This chapter explains wastewater reuse through a prism of Islamic water ethics. It outlines the critical value that fresh, pure water plays in the daily spiritual life of a Muslim, and invokes sustainability and intergenerational equity by addressing responsibilities of a Muslim toward protecting freshwater supplies from pollution. It also explains the role of science and technology in assessing purity of water and whether treated wastewater can be used for pre-prayer washing or ablution. Finally, the chapter argues that a culturally sensitive wastewater policy is likely to win popular acceptance in Muslim-majority countries.

23.1 Introduction

Ethics is a "normative science of the conduct of human beings," one that judges their "conduct to be right or wrong, to be good or bad or in some similar way" [32,36].

The word ethics is derived from the Greek and Latin to mean character and customs, or a "way of life." Islam instructs Muslims to "live by" God's commands in everything that they do; hence, Islam is intended to reach beyond the individual person and the mosque to guide and inform everything pious Muslims do whether they are political, economic, environmental, social, or other activities. Islamic water ethics can be defined as those direct and indirect interactions people have with actual or virtual water resources that are consistent with the teachings of Islam and humans' role on Earth as a khalifah or vicegerent, and masters on Earth, not of Earth [5,6]. Because ethics is part of culture and shapes society's values and principles, especially when it comes to human's relationship with water and nature at large, decision makers should incorporate local cultural values when crafting water policy, whether it pertains to pollution or conservation of freshwater or reuse of wastewater.

A Brazilian scientist offered a definition of the "environment as a whole," which includes physical and material factors, as well as economic and cultural aspects as well. He adds, "An accurate analysis of the environment must always consider the total impact of man and his culture on all the surrounding elements, and also the impact of ecological factors on every aspect of human life. Viewed in this perspective the environment includes biological, physiological, economic and cultural aspects, all linked in the same constantly changing ecological fabric" [24,30,44]. This understanding of the environment is broader than commonly considered because interactions between nature and humans are almost constant, as neither one exists separately from the other.

Water has life-giving characteristics and is an important part in people's cultural framework as it connects them with their past and influences their religious identity. The people are also entrusted with managing this fragile resource for current and future needs and are therefore required to protect it for the needs of future generations. For Nohad Seattle, a North American native, water that flows in nature is a lot more than water, "it represents the blood of our ancestors.... The murmur of water is the voice of the father of my father ... you have to teach to your children that rivers are our brothers, and yours too, and accordingly you have to treat them" [42]. Water is then a lot more than the hydrogen and oxygen that it is made of. In addition to its scientific dimension, water affects the economy, ecology, and cultural practices for religions around the world.

There is an Arabic proverb saying, "Rapid increase is fetid," and proved to be right in the case of present accelerating population growth and consequent water pollution and resource degradation. The rapid population increase of the last few decades along with the massive industrialization that often releases highly toxic pollutants into that natural environment have put unprecedented pressures on the water systems, and raised a host of new questions about the role religion and culture could play, and whether treated wastewater could be used for religious purposes. In different parts of the world, the term "water crisis" is often (over)used by people from all walks of life, yet many are ignorant of the fundamentals of the science or culture of water production, distribution, consumption, treating, and reusing wastewater. The latter is a particularly underutilized resource, partly because of the cultural barriers to using water that had become impure, befouled, and some communities may even consider it profaned. Therefore, treated wastewater can play an important role in augmenting freshwater supplies to help meet future water demand in countries as diverse as India, Indonesia, Saudi Arabia, and China. This chapter develops a framework of Islamic water ethics, and explains the central role that fresh, unpolluted water plays in the life of a Muslim. It also invokes sustainability and intergenerational equity by addressing Muslims' responsibilities toward protecting freshwater supplies from pollution and whether treated wastewater can be used for ablution. Finally, the chapter makes a few related contextual references to Hindu and Buddhist ethics toward wastewater.

All too often, wastewater flows untreated into seas, oceans, lakes, rivers, and to terrestrial destinations. Their environmental and health effects are local and sometimes transnational. Some 10% of the world's population consumes food that has been irrigated by raw or partially treated wastewater exposing them to serious health risks. Such irrigation practices affect 20 million

ha in around 50 countries [45]. In recent years, there has been growing interest in exploiting this resource by treating it and reusing it. In addition to having a supplementary water resource, studies showed that there is about 65% savings in actual fertilizer expenditure when irrigating crops with reclaimed wastewater compared to irrigating with fresh groundwater [7,10]. Wastewater reuse is often more expensive than existing freshwater from surface or subsurface sources. However, in order to encourage use, and to account for restrictions put on application of recycled water, its price would need to be lower than freshwater to encourage its use [35].

People generally accept the need for water withdrawal and view its treatment as a necessary step to make it fit to drink, but generally approach treated wastewater with far greater skepticism. For example, people in India's Tamil Nadu Water Supply and Drainage district have limited to low acceptability of using recycled sewage water [27]. People extend the "yuk" factor they associate with wastewater to "this foul liquid" is unhygienic, effectively discounting the role science and technology play in making it potable.

23.2 Approaches to Wastewater Reuse: Hinduism, Buddhism, and Islam

The psychological aversion toward reclaimed wastewater and crops irrigated with this water that some people had was based on "(i) the questionable origin of the reclaimed wastewater, (ii) health concerns, (iii) religious beliefs, and (iv) cultural values and traditions" [2]. A survey of the perceptions toward irrigation with treated wastewater in Jordan and Tunisia found that farmers with psychological aversion to using reclaimed wastewater were 16% and 50% in Jordan and Tunisia, respectively. It also found that farmers who did not have a firsthand experience irrigating with treated wastewater were "most likely to have negative prejudice." About one-fifth of the public cited religious reasons for their unwillingness to use reclaimed wastewater; this, despite the fact that Islam allows for such use if the impurities are removed from it [1,14,30]. Furthermore, survey data point to the fact that perceptions and attitudes toward irrigation with reclaimed wastewater can be conditioned and changed except for some fundamental assumptions and taboos [2].

Just as Islam allows for the use of science and technology to purify wastewater, Hindus have prohibitions against polluting sacred rivers. "In Hinduism, water, a symbol of purification, is used in all religious ceremonies" [41]. For Hindus, water is one of five intuitive elements of nature, which are Earth, water, fire, air, and ether (sky). Graphically, water is depicted as a circle, which signifies fullness (Nair). Hindus strive to achieve purity, which includes physical cleanliness and spiritual welfare. Water is imbued with physical and spiritually cleansing powers and therefore it has a "central place in the practices and beliefs of many a religious ritual. Physically and mentally clean person is enabled to focus on worship" (Nair). Temples are located near a water source; the faithful must use water for daily cleansing, and bathe before entering the temple [33].

For them, the Ganges, "which comes directly from a holy source in the Himalayas, is considered to be the most sacred of all the rivers." Hindus believe that a plunge into this river offers believers spiritual purification, frees them of sin, relieves their sickness and other ailments; morning cleansing with water is a daily obligation [11,41]. Rivers have been a vital component in the application of Hindu religious rites, and numerous hymns praise rivers, even ones that are now dry such as the Sarasvati River [33,41]. Pilgrims believe that the Ganges River has curative properties [11].

Rivers and water resources in general play a central role in religious rites of Hindus, Muslims, and Buddhists. Hindu doctrine states that the water must not be polluted, which implies that preserving its cleanliness is considered a good deed, and that polluters should be punished [16]. Similarly, Buddhist teachings denounce polluting of freshwater [16]. In a fatwa, Muslim scholar Abdel-Fattah Idrees [25] wrote that it is sinful to pollute water bodies [25]. Furthermore, Prophet Mohammad provided a pivotal legal rule through this pronunciation: "No harm shall be inflicted or reciprocated in Islam" (reported by Ahmad). Furthermore, the Qura'n itself makes it clear that people should avoid "mischief on earth" and those who abide by God's words and do good deeds will receive "the Mercy of Allah" (Qura'n, Al-'Araf: 56).

Water is endowed with more religious and cultural meaning and importance than any other natural resource. For Hindus, water is a "medium of purification and a source of energy" [30, pp. 294]. Muslims are required to wash with water before praying, an act that is referred to as ablution. Water is the only liquid that can make one pure. There are extensive guidelines on what makes water sufficiently pure for ablution [26]. Water quality is classified into *tahur* (purifier) and *najis* (impure). The former refers to water that is pure and purifying such as rain, snow, hail, springs, and water in seas, rivers, lakes, and wells. The Qura'n (25:48) states, "… And We sent down from the heaven tahur water."

Tahur water may become tahir or najis water. It will change into tahir water if it is used for ablution, or if something inherently clean is mixed with it, which changes any of its properties significantly. Tahur or tahir water may become najis (unclean) if mixed with pollutants such as feces, urine, or blood. If any volume of pure water is mixed with an impure substance (alcohol, pork, etc.) and one of its three qualities (color, taste, or odor) is changed, this water becomes najis and unfit for use in wudu, pre-prayer wash.

Water plays a strong role in the rituals of most religions. Just as Muslims need pure water to fulfill a central religious obligation, prayer, Buddhists also use water to bathe the Buddha daily. Over time, people developed cultural and scientific notions of what constitutes "clean water." These "socially constructed" understandings are based on long experiences and careful observations of the color, smell, and taste of water, and on the location of its source. In addition to this, cultural norms and traditions play an important role in shaping peoples' disposition, which is why, for example, the heavily polluted Ganges River remains holy to faithful Hindus. Furthermore, for Hindus and Buddhists, the

Bagmati River is holy. Their reverence coexists with the practice by some Hindus of bathing a corpse in the Bagmati, then cremating it and spreading the ashes on its banks. The river also receives large amounts of untreated wastewater and industrial sludge. For believers, the sacredness of the river is not shaken by its pollution load. In reality, however, many of the rivers were only holy in name, not in practice, and something had to be done.

"Humans transfer sin and pollution to the river and the river as a divine and holy body transforms the pollution into purity and thus the river remains pure. If the river was unable to transform impurity to purity, it was seen as proof that the river had lost its holiness. When engineers intervened and cleansed the deteriorated river, which it should have done by itself, numerous devotees perceived the river as being physically, but not ritually, pure" ("Water") [33].

Vinaya rules prevent monks from polluting green grass with impurities such a urine, saliva, and feces. Vinaya is a self-training code that was given by the Buddha for monks and nuns to observe. These common agents of pollution at that time would have aesthetic effects and would contaminate the grass that feeds animals. Therefore, adverse impacts of people's activities should be minimal [40].

23.3 Water in Islam: An Example

For Muslims, access to pure freshwater is critical for upholding two pillars of their faith: praying and pilgrimage (Qura'n 5:5, 72:16); hence, it is in their best interest to preserve and sustainably manage this vulnerable resource. In Islam, the water issue occupy, cover, and affect living things on Earth as they all contain water (Qura'n: 21:30, 24:45, 25:53, 25:54). Water use in Islam is part of the overall water management and interconnected with water resources creation development and consequences of human misbalance of use and development [18,19]. Water management in Islam addresses water issues at the personal, human level as well as the national level [15] and regulates state-personal relations. For example, the state is responsible for collecting, treating, and reusing wastewater; in return, the people are to pay their fair dues to the government. Individuals can expect abundance of resources, which would include treated wastewater, if they are faithful, fulfill their duties to God, and do righteous deeds (Qura'n 72:16, 77:41, 2:22, 6:99, 13:14, 13:17, 18:29) such as preventing (or minimizing) pollution levels affecting freshwater resources.

Water is used for different purposes and, in addition to natural conditions of availability and renewal, involves and is influenced by personal, social, legal, religious, cultural, economic, political, and other situational factors. Ethically and regardless of all previous issues and influences, it is important to note that water use is a vital matter and right to every human and non-human and should not be guided mainly by market forces and economic considerations; people should care for each other and the natural environment including ecology, water, land, air, and society as a whole regardless of the fluctuations of the economy.

It was stated by Jennings et al. [29] that (1) in a general sense, the significance of water for life and health is fundamental and can scarcely be overstated, and hence the pertinence of ethics to water utilization and management, (2) it is important for everyone involved in water resource management and in public health to have a well-reasoned understanding of the moral values and obligations that correspond to that significance, and (3) in the domain of ethics, questions of scientific knowledge come together with aspects of cultural meaning and perception; questions of conservation, sanitation, and health promotion come together with questions of justice, equity, and human rights; questions of sustainability and biodiversity come together with questions of democratic governance, law, and policy.

Ethics in Islam including that of water use is directly related to faith, behavior, and responsibility [23,47]. This means all Muslim's activities should be within and conforming to Islamic guidance and rules. Prophet Mohammad said in this regard:

1. I was sent to complete the best of ethics (Narrated by Abu Hurairah).
2. Faith is not by hoping, it is what settles in the heart and proved by actions/application (Narrated by Ibn Al-Najjar and Dailmy).
3. The most complete in faith believers are those having best ethics (Narrated by Tirmithi).

Making practical judgments about the types of uses that polluted water can be applied to have been taught and practiced for millenniums, and perhaps since the creation of mankind whereby people pass down knowledge that their elders had taught them. The water use issue relates to and deals with our behavior, conscience, and values during various acts and interventions. Water use ethics cannot be practiced without rules, principles, virtues, limits, and integrity [38] and this is why this chapter and discussion is important.

23.3.1 Islamic Approaches to Water Management

A distinct Islamic approach to environmental education and management including water was pinpointed as a result of (1) extracting by identification, listing, sorting, and grouping the verses in Qura'n (the holy book of Islam and Muslims) having commonalities related to environmental education and management, then (2) find the connection and relationships between groups according to mutual meanings and understandings [13,18-22]. This conceptualization is analogous to a balance tripod structure that relates the creation and development of water resources (control and source of water, creation, excellence, and directions for water use, and water as a testing matter) to its use for various living activities and needs (cleaning and wastes, the fulfillment of life needs, and worshiping) and to related impacts resulting from this use or misuse (degradation and perdition, protection and comfort). This approach and its guidelines will help in setting Muslims on a path of sustainable water use; the alternative route is one of depletion and degradation of water resources due to dumping or using of untreated wastewater with all the pathogens that are associated with that (see Reference 18 and Figure 23.1).

Tables 23.1, 23.2, and 23.3 include references found in the Qura'n regarding the three main approach sections, that is, source creation, use, and impacts, respectively. It is clearly stated in the Qura'n that worshiping Allah and contemplating the creation of environmental elements/resources—including mankind, Earth, water, and skies—to affirm His powers and promises—is necessary and required for directing Muslims to the right path in this life and in the hereafter. In Islam, water and other environmental resources were viewed by Allah as a blessing He bestowed on mankind. People will be put to the test later on how they used (or abused) that blessing.

Initially Allah sent down pure and purifying water (Qura'n 25:48, 8:11). One of the direct outcomes of human use of water for various purposes is the generation of wastewater, which, if mismanaged, would lead to environmental pollution adversely affecting the ecosystem as well as human beings. Any element in the natural environment has not been created in vain (Qura'n, 44:38–39). Accordingly, not all usable matters or objects are considered as waste in Islam, and throwing them away as waste is considered disobedience or mischief (Qura'n, 30:41, 7:56, 5:32, 2:205). This lead us to the definition of materialistic waste in Islam, which is any matter/state (gaseous, solid, or liquid) or object that is not usable, and/or spoiled in quality, and/or worthless in value or damaging to keep.

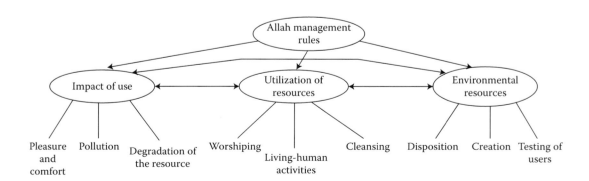

FIGURE 23.1 Schematic of Islamic approach to water resources management.

TABLE 23.1 Qura'nic References of Water Resource Creation and Development

Control and Source of Water	Creation, Excellence, and Directions for Water Use	Water as a Testing Matter
• Allah drives rain from the skies to earth for the benefit of mankind (Qura'n 78:14, 50:9, 31:10, 14:32). • Allah drives rain from the skies in specific quantities and distributes it over the earth (Qura'n 43:11). • Allah drives rain from the skies into the ground, where He can abate it (Qura'n 23:18), ensure it moves through the ground, emerging elsewhere as springs (Qura'n 39:21, 54:13), or cause it to flow in valleys (Qura'n 13:17), rivers and streams (Qura'n 14:32, 2:74). He could also cause it to fall as snow or hail (Qura'n 24:43).	• Allah created mankind and all living creatures from water (Qura'n 21:30, 24:45, 86:6, 32:8, 25:54). • Rain is created by and sent from the skies by Allah in abundance (Qura'n 42:28, 31:34, 31:48, 29:63). • Allah created water in specific quantities and caused it to be distributed among creatures, including mankind (Qura'n 23:18, 2:60). • Allah granted all creatures the right to drink (Qura'n 54:28).	• If man follows Allah's rule, Allah will send him plenty of rain (Qura'n 72:16). • Sending water and rain to give life to earth are signs from Allah to man to contemplate the creation (Qura'n 30:24) and declare the grace of Allah (Qura'n 35:3).

TABLE 23.2 Qura'nic References to Water Use for Various Purposes

Cleaning and Wastes	The Fulfillment of Life Needs	Worshiping
• Allah sends rain from the skies that channels into wadis and streams in specific quantities to assist in the process of waste disposal (Qura'n 13:17). • Allah sends rain from the skies for cleaning (Qura'n 8:11, 4:43) and for purity purposes (Qura'n 5:6).	• Allah provides water for creatures to drink (Qura'n 16:10, 15:22, 77:27). • Allah created agricultural species in males and females to grow agricultural products of various colors and produce food (Qura'n 39:21, 32:27, 22:5, 20:53, 31:10, 35:27). • Allah provides water so that man can grow gardens and produce beans (Qura'n 6:99, 27:60, 50:9). • Allah sends water with the purpose of greening the earth and giving it life (Qura'n 22:63, 43:11, 62:99). • Allah created the seas and provided ships for transport purposes (Qura'n 14:32, 8:11). • Allah created fish as food for mankind (Qura'n 16:14). • Allah described water as giving strength and strengthening hearts (Qura'n 8:11).	• Allah demanded that mankind ask Him for forgiveness of their sins and, in return, He would fill the skies with abundant rain for them (Qura'n 11:52). • Allah demanded that man praise Him and declare His glory for creating the skies and earth and for sending rain (Qura'n 27:60, 29:63, 45:12). • The creation of rain and driving it from skies to earth, along with all related matters, are a sign and illustration from Allah to mankind for thinking and consciousness (Qura'n 2:164, 56:68).

TABLE 23.3 Qura'n's References of Impacts of Water Resource Use or Misuse

Degradation and Perdition	Degradation and Depletion	Joy and Comfort
• Allah will cause the sky to withhold rain and the earth to swallow it in areas where man does that which is wrong and evil (Qura'n 11:44). • Allah will rain on those who conduct mischief showers of brimstones (Qura'n 24:58, 26:173). • Allah will send heavy rainstorms to land belonging to those who behave inappropriately (Qura'n 2:264, 2:265). • Allah, acknowledging that man needs to eat and drink, demands that people not waste excessively (Qura'n 7:31). • Allah demands that people eat and drink of the sustenance provided by Him but that they are not perditions (Qura'n 2:60, 7:31).	• Allah will punish the unjust and the unfaithful by drought and shortage of food (famine) (Qura'n 7:130). • For those who forget Allah's rules and way, Allah will degrade their water supply (Qura'n 18:45, 10:24). • Who will supply you with clear flowing water when the available one is degraded, Allah (Qura'n 67:30). • For those whose conduct is improper, Allah will destroy their resources and raise in their wake a new generation to succeed them (Qura'n 6:6).	• Allah promises that for those who remain on the path of righteousness He would bestow upon them rain in abundance (Qura'n 72:16, 24:39). • Allah encourages man to eat and drink with joy, pleasure, and health following and as a consequence of their good deeds (Qura'n 52:19, 77:43, 69:24). • Allah promises heaven on earth to those who have faith in Him (Qura'n 17:91).

An important juristic rule in Islamic Law states, "The averting of harm takes precedence over the acquisition of benefits" [9]. Therefore, minimizing waste production or generation was given a good deal of consideration in Islam and Islamic teachings. The following is an example of Islamic directions advising Muslims to initially minimize waste and optimize the use of available natural resources including water:

• Use material, crops, food either in eating or in development in specific quantities or as needed only and do not go in excess (Qura'n, 25:5, 20:81, 7:31, 6:141, 5:87, 4:6).

• Do not collect (wealth, resources, bounties, and others) and hide it from use or leave/store it without use (Qura'n, 70:18, 104:2).

• An example of the Islamic guiding principles for materialistic waste management (Qura'n, 8:37, 5:31, and 3:179) (Figure 23.2).

• Classification and separation: Classify then separate the impure from the pure or the usable (recyclable-good) from un-usable (Qura'n, 24:43). What could be recycled should be recycled and the unusable (non-recyclables, non-recoverable, bad in quality) would go into Step (2): disposal by either incineration, freezing or inland burial.

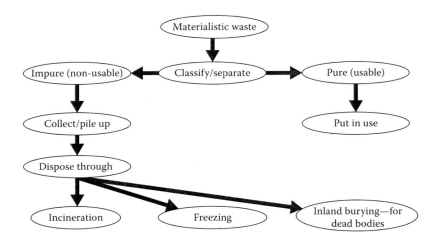

FIGURE 23.2 Schematic of Islamic waste management guiding principles.

- Compilation/heaping: Put the impure (being solid liquid or gaseous), one on another, heap them together (Qura'n, 24:43). Putting impure matter together here also means collecting it in one compilation.
- Waste disposal: It is better not to generate waste in the first place; however, in case waste already occurred, was collected, and needs to be disposed, disposal could take place by:
- Incineration or freezing: Cast impure matter or waste compiled from Step 1 and 2 into Hell (Qura'n, 24:43). Hell in Qura'n, contain two processes or extreme conditions: incineration by fire (Qura'n, 85:10, 72:23, 52:13) or exposure to excessive cold (Qura'n, 78:25, 38:57). Incineration will destroy waste and reduce it to ruins/ash—in reduced volume (Qura'n, 51:42). The residual ash from incineration is useful and stays or could be buried inland (Qura'n, 13:17). This is a unique solution to unusable-impure waste disposal. Dumping waste in frozen environment (soil, ice, artificial freezers, and others) in today's waste science and technology was not proposed or considered as a viable waste management option [37].
- Burying inland of dead bodies or tissues, which will decompose by time (Qura'n, 80:21, 81:8, 60:13). The degraded matter in land will be later (as nutrients) giving life to bare land through irrigating land and supporting crops growth (Qura'n, 77:25, 77:26, 27:67, 30:19, 10:31, 7:57).

After a group of Islamic scholars investigated and discussed with scientists the issue of wastewater treatment, the Council of Leading Islamic Scholars (CLIS)** in Saudi Arabia with which they are affiliated issued a fatwa that states:

Impure wastewater can be considered as pure water and similar to the original pure water, if its treatment using advanced technical procedures is capable of removing its impurities with

regard to taste, color, and smell, as witnessed by honest, specialized, and knowledgeable experts. Then it can be used to remove body impurities and for purifying, even for drinking. If there are negative impacts from its direct use on the human health, then it is better to avoid its use, not because it is impure but to avoid harming the human beings. The CLIS prefers to avoid using it for drinking (as possible) to protect health and not to contradict with human habits [1,4,14,34].

Similarly, the World Fatwa Management and Research Institute ruled, "If water treatment restores the taste, color, and smell of unclean water to its original state, then it becomes pure and hence there is nothing wrong to use it for irrigation and other useful purposes" [28]. Other scholars add to these perspectives that the removal of impurities needs to be witnessed by credible specialists [46]. These fatwas make it clear that there is no religious objection to Muslims using reclaimed wastewater for irrigation [3,23] and for other purposes.

Muslims as a people (umma) are balanced in their behavior and be in all aspects more or less in the middle (Qura'n, 2:143, 57:25, 55:8, 55:9, 42:17, 11:85). Muslims were instructed by Allah to be balanced in their activities and not to disrupt the natural scheme and rhythm (Qura'n, 28:77; Figure 23.3). In addition, Muslims were asked to do their best and avoid burdens that they cannot tolerate (Qura'n, 23:62, 7:42, 6:152, 2:286, 2:233). Therefore, the teachings are not about unrealistic ideals where nature is protected in its pristine state; in fact, they allow for some pollution and wastewater use as long as it does not harm humans nor irreversibly degrade nature whereby pollution levels from wastewater remain within the assimilative capacity of the natural system.

23.3.2 Women's Roles in Water Management

Water resource management is affected by people's attitudes, behavior, and activities. Women make up approximately half of any society, and play a pivotal role in water use and overall management especially in poorer, traditional societies. There,

---

* The fatwa was published in the *Daily Newspaper*, Al-Madina, Jeddah, on April 17, 1979.

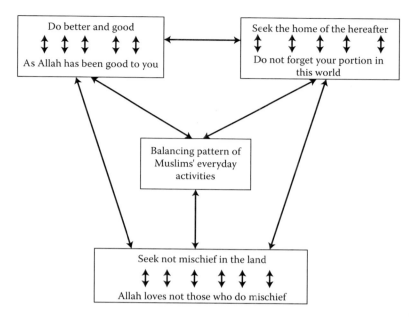

FIGURE 23.3 Schematic of Muslims' everyday activities balancing pattern.

women often make decisions regarding how much water to use and when, and on the quality of the water source from which to fill her container. In Islam, almost all commands in Qura'n that relate to water resources, its use, pollution, and/or depletion are directed to people in general or equally to both males and females (Qura'n 2:22, 5:6, 6:99, 7:57, 8:11, 10:24, 11:44, 13:4, 13:17, 14:32, 15:22, 16:10, 16:11, 16:65, 18:29, 18:45, 20:53, 22:63, 23:18, 27:60, 27:61, 29:63, 30:24, 31:10, 32:27, 35:27, 39:21, 43:11, 50:9, and others). Women's roles in water management are therefore acknowledged and are equally delegated and distributed among them and men. Consequently, women would need to be involved in water management decision-making processes, including when governments hold public participation hearings on how to manage wastewater in their communities [31].*

23.3.3 Considerations for Optimizing Water Management in Islam

According to the United Nations World Water Development Report [43], "Providing the water needed to feed a growing population and balancing this with all the other demands on water, is one of the great challenges of this century" [12]. This is the present and future main challenge: securing enough water in quantity, safe in quality, and affordable in cost at all times. As the Islamic approach listed previously explains, Muslims need to preserve and conserve water and use it as necessary, in balance, and in righteousness to society activities and projects, follow

Muslim's guidelines for allocation, minimize mischief and pollution, and think about developing and exploring means and tools for proper use of water.

Raising literacy levels, especially among females, is a challenge in the global South. Most Muslim countries are poor and the level of education, knowledge, and awareness about Islamic guidelines concerning water and wastewater use are limited or nonexistent, which hinders the implementation of a culturally sensitive water management strategy. Ironically, however, levels of religiosity among Muslims have been rising since the late 1960s. Wastewater treatment and reuse is not a leisure pursuit or an option for society; it is a necessity for sustainable development and a precondition for living in a clean, healthy environment. Once the people are aware of the multifaceted perspectives that Islam offers regarding humans' relations with the natural system, many will find it easier to embrace policies that are respectful of and sensitive to their cultural tradition. Muslim scholars as well as government officials and civil society groups need to make cultural education a priority and play a more proactive role in this endeavor.

The extent of proper wastewater collection, treatment, and reuse or disposal is still limited in Muslim countries; it needs a lot more attention, investment, awareness, legal frameworks, and appropriate technology, to extend coverage of sewerage to the entire country. This cannot be made by governments only; public-private partnerships in providing public services have been successful elsewhere.

The impacts of current wastewater management practices in Muslim countries on the environment (soil, water, plant, fish, ecosystem, public health, etc.) need to be monitored and evaluated. This monitoring and evaluation is needed to check Muslim's obedience to Allah's rules and guidance and upgrade their practices accordingly to get closer to the righteous path.

---

* The weaker status of some Muslim women and their inability to uphold their rights under (Shariah) law is based on their ignorance of their rights and about which procedures to follow to obtain them [8, p. 25], [17]. Interpretations of certain verses of the Qura'n have often been tailored to fit pre-existing traditions or to suit power relations between men and women [31].

23.4 Summary and Conclusions

Muslims take their cues from nature for their daily prayers and for their annual fasting in the month of Ramadan; hence, most are in tune with nature's rhythms and sensitive to environmental changes around them. Furthermore, the Qura'n invokes Muslims to contemplate the signs of God's creation as manifested in the natural environment, which affirm His powers; God also promises that if the people abide by His commandments, they will have abundance of resources including freshwater.

National policies that consider people's religious beliefs and cultural values are more likely to be embraced by citizens resulting in greater acceptance of reclaimed wastewater. However, field work done on this topic in a few Muslim countries reveals that most Muslims appear to have a narrow understanding of their faith (focused on routine acts of worship) and rarely do they connect its teachings with the natural environment or with wastewater treatment. The fact that the teachings of Islam support the preservation of water cleanliness and of reverting polluted water to its pure status is a useful finding. Invoking Islam's environmental and water ethics would pave the way for national policies that advocate capturing, treating, and reusing wastewater.

In summary, Islam views wastewater reuse as part of optimization of existing resources and as being supplementary; hence, misusing it is considered a disobedience to God. However, there are parameters to this reuse that need to be respected, and each type of waste needs to be considered by scientists and experts (Qura'n, 5:48). This chapter outlines an Islamic framework for water management, and develops a narrative on how wastewater should be managed, and what purified wastewater can be used for. It also shows that Muslim scholars acknowledge and respect the role that science and technology play in improving the lives of people through the mitigation of potential health threats that may arise from wastewater, the protection of freshwater sources from pollutants were wastewater to be released untreated, and through augmenting people's freshwater supplies by purifying wastewater and more.

The chapter calls attention to the challenges that face decision makers in implementing Islamic ethics with respect to wastewater primarily because many Muslims are uninformed about the environmental aspects of their faith. Therefore, a more substantial movement toward capturing, treating, and reusing wastewater would require a comprehensive approach, one that actively engages different stakeholders. Scientific data on the quality and health effects of treated wastewater need to be presented by qualified representatives. The data could convince the public that "impurities" have been removed from treated wastewater; this would remove a spiritual barrier from using that water. Note that psychological and other cultural barriers may remain. The input and engagement of men and women in the decision-making process is important to the success and expanded use of treated wastewater. Islamic law calls for the decision-making process to be inclusive of the people's wishes, a practice that the Qura'n calls *shoura*. The hurdles to Islamic water management are many but the intellectual roadmap and policy frameworks are becoming clearer by the growing body of academic literature on the subject.

References

1. Abderrahman, W.A. 2001. Water demand management in Saudi Arabia. In: Faruqui, N.I., Biswas, A.K., and Bino, M.J. (eds.), *Water Management in Islam*. UNU Press, New York.
2. Abu-Madi, M., Al-Sa'ed, R., Braadbaart, O., and Alaerts, G. 2008. Public perceptions towards wastewater reuse in Jordan and Tunisia. *Arab Water Council Journal*, 1(II), July.
3. Al Khateeb, N. 2001. Sociocultural acceptability of wastewater reuse in Palestine. In: Faruqui, N.I., Biswas, A.K., and Bino, M.J. (eds.), *Water Management in Islam*. United Nations University Press, New York.
4. Ali, I. 1987. Wastewater criteria for irrigation in arid regions. *Journal of Irrigation and Drainage Engineering*, 113(2), 173–183.
5. Amery, H.A. 2001. Islam and the environment. In: Faruqui, N.I., Biswas, A.K., and Bino, M.J. (eds.), *Water Management in Islam*. United Nations University Press, New York, pp. 39–48.
6. Amery, H.A. 2001. Islamic water management. *Water International*, 26(4), 481–489.
7. Asano, T. 2002. Water from (Waste) Water—The Dependable Water Resource. Delivered at the 11th Stockholm Water Symposium, August 12–18, 2001, Stockholm, Sweden. Available at http://www.iwaponline.com/wst/04508/wst045080023.htm.
8. Badri, B. 1989. Attitudes and behaviours of educated Sudanese women concerning legal rights in marriage and divorce. *The Afhad Journal: Women and Change*, 6(2), 21–26.
9. Bagader, A.A., El-Sabbagh, A.T., Al-Glayand, M.A., and Samarrai, M.Y.I. 1993. *Environmental Protection in Islam*, 2nd Edition. Mepa, Saudi Arabia. Available online at http://www.islamset.com/env/. Accessed December 17, 2006.
10. Bahri, A., Basset, C., Oueslati, F., and Brissaud, F. 2001. Reuse of reclaimed wastewater for golf course irrigation in Tunisia. *Water Science Technology*, 43(10), 117–124.
11. Dolnick, S. 2011. Hindus find a Ganges in Queens, to park rangers' dismay. *The New York Times*, April 21.
12. Eid, M. 2010. Social impact assessment of using treated wastewater in irrigation. In: Integrated Water Resource Management II: Feasibility of Wastewater Reuse, Report No. 14, Annex 13, pp. 1–9. International Resources Group. http://www.mwri.gov.eg/project/report/IWRMII/Report14Feasibility_of_WW_Reuse.pdf. Accessed July 2014.
13. Emrullah, A. and Hâdimî, M. 2001. *Ethics of Islam*. Waqf Ikhlâs Publications No. 17, 3rd Edition, İhlâs Gazetecilik A.Ş., Istanbul.
14. Farooq, S. and Ansari, Z. 1983. Wastewater reuse in Muslim countries: An Islamic perspective. *Environmental Management*, 7(2), 119–123.

15. Faruqui, N., Biswas, A., and Bino, M. (eds.) 2001. *Water Management in Islam*. United Nations University Press, New York.

16. Fisher-Ogden, D. and Saxer, S.R. 2006. World religions and clean water laws. *Duke Environmental Law and Policy Forum*, 17, 63–117.

17. Freedman, P. 1991. Women and the law in Asia and the Near East, Draft Paper, Development Law and Policy Program, Columbia University School of Public Health.

18. Haddad, M. 2000. An Islamic approach to the environment and sustainable groundwater management. In: M. Haddad and E. Feitelson (eds.). *Management of Shared Groundwater Resources: The Israeli-Palestinian Case with International Perspective*. Ottawa: IDRC and Kluwer Academic Publishers, pp. 25–42.

19. Haddad, M. 2006. Islam and Sustainable Development of Natural Resources. Lecture presented at the Center for Middle Eastern Studies, Department of Middle Eastern Studies, The University of Texas at Austin, May 25.

20. Haddad, M. 2006. Islamic approach towards environmental education. *Canadian Journal of Environmental Education*, 11(1), 57–73.

21. Haddad, M. 2011. The application of the Islamic approach as a way for the advancement of scientific research in the Arab World (in Arabic). Paper presented at the International Graduate Conference on Science, Humanities, and Engineering, May 4–5, 2011, An-Najah National University, Nablus, Palestine.

22. Haddad, M. 2012. An Islamic perspective of food security management. *Water Policy Journal*, 14, 121–135.

23. Husain, T. and Ahmed, A.H. 1997. Environmental and economic aspects of wastewater reuse in Saudi Arabia. *Water International*, 22(2), 108–112.

24. Hussain, I., Raschid, L., Hanjra, M., Marikar, F., and van der Hoek, W. 2002. Wastewater Use in Agriculture: Review of Impacts and Methodological Issues in Valuing Impacts. International Water Management Institute (IWMI), Working Paper 37. Colombo, Sri Lanka.

25. Idrees, A. No Date. Fatawa: Pollution in Islam. http://www.onislam.net/english/ask-the-scholar/health-and-science/environment/174913-pollution-islamic-view.html.

26. Imam Malik bin Anas. No date. Book of purification. In Arabic. http://ar.islamway.net.

27. Imranullah, M.A. 2013. Save water through recycling. *The Hindu*. March 26.

28. INFAD. 2012. Wastewater Treatment. World Fatwa Management and Research Institute, Islamic Science University of Malaysia. Retrieved on March 23, 2012 from http://www.onislam.net/english/ask-the-scholar/health-and-science/172614.

29. Jennings, B., Heltne, P., and Kintzele, K. 2009. Principles of water ethics. *Minding Nature*, 2(2), 25–29. http://www.humansandnature.org/principles-of-water-ethics-article-35.php. Accessed July 2014.

30. Jimenez, B. and Takashi A. 2008. *Water Reuse: An International Survey of Current Practice, Issues and Needs*. IWA Publishing, London.

31. Jütting, J. and Morrisson, C. 2005. Improve the Status of Women in Developing Countries. OECD Development Center, Policy Brief No. 27.

32. Lillie, W. 1971. *An Introduction to Ethics*, 3rd Edition, Methuen & Co. Ltd., p. 1.

33. Liu, D. et al. 2011. *Water Ethics and Water Resource Management*. UNESCO, Bangkok.

34. Mara, D.D. 2000. The production of microbiologically safe effluents for wastewater reuse in the Middle East and North Africa. *Water, Air, and Soil Pollution*, 123(1–4), 595–603.

35. Mekala, G.D., Davidson, B., Samad, M., and Boland, A. 2008. *A Framework for Efficient Wastewater Treatment and Recycling Systems*. International Water Management Institute, Colombo, Sri Lanka. (IWMI Working Paper 129).

36. Morgan, R.A. and Smith, J.L. 2012. Rethinking clean: Historicising religion, science and the purity of water in the twenty-first century, Conference Proceedings, Tapping the Turn Conference, November.

37. Nathanson, J.A. 2003. *Basic Environmental Technology: Water Supply, Waste Management, and Pollution Control*. Prentice Hall, Upper Saddle River, NJ.

38. Palmer, C. (ed.). 2006. *Teaching Environmental Ethics*. Koninklijke Brill NV, Leiden, the Netherlands.

39. Quran. 1998. Tahrike Tarsile Qur'an; Reissue edition. Translated by Abdullah Yusuf Ali.

40. Silva, P.D. 1984. Buddhism and behaviour modification. *Behaviour Research and Therapy*, 22(6), 661–678.

41. Singh, V.P. 2008. Water, environment, engineering, religion, and society 1. *Journal of Hydrologic Engineering*, 13(3), 118–123.

42. Torrecilla, N.T. and Martínez-Gil, J. 2005. *The New Culture of Water in Spain: A Philosophy Towards a Sustainable Development*. European Water Association, Brussels, Belgium.

43. UNESCO-WWAP. 2006. United Nations World Water Development Report No. 2. Water: A Shared Responsibility. Published jointly by the United Nations Educational, Scientific and Cultural Organization (UNESCO) and Berghahn Books, Paris.

44. Vidart, D. 1978. Environmental education—Theory and practice. *Prospects*, 8(4), 466–479.

45. WHO. 2012. WHO Guidelines for the Safe Use of Wastewater, Greywater, and Excreta in Agriculture and Aquaculture. World Health Organization.

46. WHO. 2006. Guidelines for the Safe Use of Wastewater, Excreta, and Greywater. Volume II—Wastewater Use in Agriculture. World Health Organization.

47. Wilkinson, J. 1978. Islamic water law with special reference to oasis settlements. *Arid Environments*, 1, 87–96.

24

Ethical and Cultural Dimensions of Water Reuse in Global Aspects

Jannatul Ferdaush
Savanna State University

Shafi Noor Islam
University of Brunei Darussalam (UBD)

Sandra Reinstädtler
Brandenburg University of Technology Cottbus-Senftenberg

Saeid Eslamian
Isfahan University of Technology

PREFACE

Water is essential for life and the total development of human societies in different climatic regions of the world. Geographically, the surface land of the world is divided due to geological tectonic aspects and climatic conditions. This is the most critical and potential aspect of population distribution, settlement, urbanization, and agricultural development. The civilization, urban development, agricultural cropping systems, and industrial development are totally dependent on naval communication and water availability. Soil fertility development depended on water supply and quality of surface and groundwater. Tropical, subtropical, and cold climate (CC) regions are facing serious water scarcity for the last two decades and have witnessed growing water stress, both in terms of water scarcity and quality deterioration. By 2020, water use and reuse is expected to increase by 40%, and 17% more water will be required for food production to meet the needs of the growing population. By 2025, 1.8 billion people will be living in regions with absolute water scarcity; approximately two out of three people in the world could be living under considerations of water stress. Therefore, water use and reuse is a potential human demandable issue to survive within a sustainable way.

Therefore, ethical and cultural subjects should be taken into consideration in water reuse and proper management strategies are very much potential factors for water management in urban and rural regions. The primary evaluation indicates that for an increased utilization of received wastewater, there needs to be cleaner arrangements. The ethical, cultural, and engineering aspects of water reuse guidelines should be considered in guideline development processes. The innovation of a water reuse approach would be the best practice in different climate regions toward protecting water resources management sustainability.

24.1 Introduction

Water is a vital natural resource of human beings and other species [63]. It is a critical resource for life and is essential for economic success and improvement of national and regional socioeconomy [21]. In general, about 73.3% of the Earth's water is in the oceans and bodies of surface water, of the remaining 2.7%, which is freshwater, 2.1% is tied up in the polar ice caps and glaciers, leaving only 0.6% to circulate [55]. At a global level, water availability for 2006 was 8462 m³/capita-yr, but at a regional level it went from as little as 1380 m³/capita-yr in the Middle East and North Africa to almost 53,300 m³/capita-yr in Oceania [59]. Water consumption worldwide is roughly split, with 70% agricultural use, 22% industrial use, and 8% domestic use [58]. One-fifth of the population lives in areas of water scarcity and one in eight lacks access to clean water [62]. Currently, properly treated produced water can be recycled and used for water flooding and other applications, such as crop irrigation, wildlife and livestock consumption, aquaculture and hydroponic vegetable culture, industrial processes, dust control, vehicle and equipment washing, power generation, and fire control [62]. The challenging fact is that due to the huge amount of water used for drinking, irrigation, and industrial purposes in the last decades, water scarcity, and quality deteriorations, almost 70% of the population in cold, tropical, and subtropical climate regions is facing water stress issues today [5].

Water supply, sanitation, and management variables to measures of "ecosystems services" and "ecosystems health" are in turn related to economic and human health benefits [12,36]. The question for sustainable development is what is the strongest driving force of the water sector [25]? Climatic condition is one of the potential factors of global and regional water use, reuse, and management. Cold, tropical, and subtropical climate regions are equally facing the problems of sustainable uses, reuses, and management. The tropical and subtropical climatic regions are the most hostile regions where people are facing severe environmental problems to maintain surface and groundwater uses and management. The cold climate (CC) region, in general, is the land of water; on the other hand, tropical and subtropical regions are considered water scarce regions. The scarcity of surface freshwater has created serious environmental problems in the tropical and subtropical climate regions. The location of the landward boundary of the coastal zone is a function of three basic geophysical processes: tidal fluctuations, salinity, and risk for cold storm surges. The coastal zone of CC, tropical climate, and subtropical climatic regions are affected by these processes and they cover an area almost 30% of the world (Figure 24.1). The coastal zones of different climatic zones are on both sides of the actual land–sea interface, where the influences of land and water on each side are still determining factors climatically, physiographically, and ecologically [19]. The climatic regions' water resources play an important role in socioeconomic development of all coastal regions in the CC regions as well as in Sweden and surroundings areas, the Bengal coastal region, and dry regions in Bangladesh [30]. The coastal region is composed of the land and the sea including estuaries and islands adjacent to the land–water

interfaces and the coast can multifariously divide into different distinctive features in terms of physiography, seasonality, and use patterns [46,47]. It is at the coast where the heavily sediment-laden river water with very little salinity meets the saline seawater. This mixing creates a unique ecosystem, which has given rise to the species of flora and fauna that exhibit their presence here. Among these, the most important is the northern Nordic region with its unique cold climatic landscape, vegetation, and wildlife; the fisheries are dominated by economically important species [47]. Water availability in cold and warm climate regions, supply to the coastal regions and surrounding areas plays a potential role in balancing climatic ecosystems [48].

Therefore, sustainable climate regional water resources management is essential for the ecosystem protection and crop production as well as cold and warm regions community development and livelihood sustainability in the geopolitical and climatic regions of the world. A comprehensive analysis of the various factors leading to cold, tropical, and subtropical surface, groundwater and water reuse management sustainability is made in this study. Produced water can potentially be treated to drinking water quality. Little research has been done on the feasibility and cost-effectiveness of direct or indirect potable reuse of product water from oil and gas production [15,54,65].

24.2 Objectives, Data, and Methodology

The present study was carried out based on secondary and primary data sources. The primary data was collected from field investigation in 2010 and 2013. Some group discussions on water issues were arranged with the scholar students who are working on water-related issues in the CC, tropical, and subtropical climatic regions. The WWW-YES 2010 was organized at the University of Paris, Ost, France. The information was collected from tropical and subtropical regions. In addition, some participatory rural appraisal (PRA) practices were arranged with the local students of environmental and applied science at Halmstad University in Sweden in the coastal region in 2013. The information was collected from the northern climatic region of Sweden and Tromsö City region of Norway. Besides these, many various reports and published articles in journals and conference proceedings have been used for this study. Published materials, reports, and journals were collected from different government and nongovernmental organizations, and university libraries in Germany, Sweden, Denmark, Poland, France, and Bangladesh. For secondary data collection, reports on coastal water resource reuse status and management have been used very openly. The research reports from Halmstad University Sweden, and University of Trömso, Norway, Brandenburg University of Technology Cottbus, Germany and Central Library of Jahangirnagar University, Dhaka, Bangladesh, and international organizational reports were also used for this study. In addition, some interviews were arranged with water engineers, environmentalists, geographers, geologists, sociologists, and experts on river systems and their ecology. The collected data were reconstructed, analyzed and visualized through MS EXCEL, VISIO 32, and ArcGIS 10.2 tools.

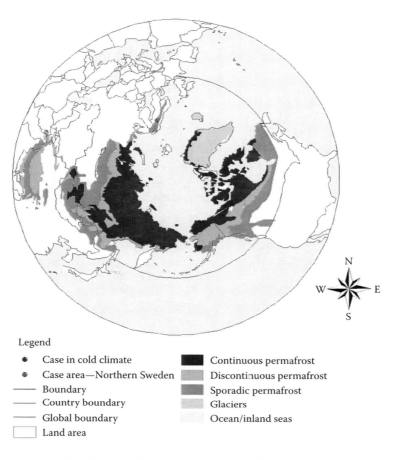

Legend

● Case in cold climate
● Case area—Northern Sweden
— Boundary
— Country boundary
— Global boundary
▢ Land area

■ Continuous permafrost
▨ Discontinuous permafrost
▨ Sporadic permafrost
▨ Glaciers
▢ Ocean/inland seas

FIGURE 24.1 Global water use and reuse locations in cold, tropical, and subtropical regions.

The specific objectives of this study are as follows:

- The objective of this chapter is to understand the critical characteristics of water reuse in different climatic regions of the world and their potential for socioeconomic and regional development.
- The study looks for the inadequate water reuse considering the ethical and cultural phenomena and barriers of sustainable adaptive management strategies.
- Introduction of geographical information system (GIS) and remote sensing (RS) and applied application for conservation measures and preparation of management plans for future development of agricultural production and services in different climatic regions for a community's livelihood and habitat sustainability.
- Preparation of recommendations for future sustainability of water reuse and management policies for CC, tropical, and subtropical regions for protection and management.

24.3 Geographical Characteristics of Water Reuse Regions

The world surface regions have been recognized based on their geographical, geological, hydrological, and meteorological characteristic conditions of the regions. In general, the temperature is very cold in the south and northern hemisphere and the middle of the surface region is tropical and subtropical where temperatures are comparatively 15–42°C in the summer time and humans have to face hardships in both environmental conditions. Most parts of the subtropical region are characterized by monsoon rainfall, which is the main source of water for human life and agricultural production. On the other hand, Smith [52] stated that the CC regions are recognized, where the mean temperatures for one month of the year are below +1°C and snow may stay on the ground for an extended period. The general temperatures range from as low as −50°C to as high as +40°C, and a total range of temperature is 90°C (−50°C + 40°C). The climate is harsh such as cold snow, wind, and rain, resulting in difficulties in water management and runoff conditions. The potential political land territories of the CC regions are Canada, northern United States, Greenland, Iceland, and north of Norway, Sweden, Finland, and the most northern region of Russia in the North Pole region. On the other hand, southern Argentina, the south part of Chile, the southeast corner of Australia, and New Zealand are the territorial part of the CC regions (Figure 24.1).

The above-mentioned political territorial boundaries are the permanent CC regions. Besides, the seasonal cold climate regions are extending gradually from the Northern pole to South pole.

It has been estimated that within the territorial boundaries of the surface areas of the world (cold, tropical, and subtropical regions) within different climatic regions, more than 7 billion people are settled and surviving within the cold and warm climate conditions [52]. Figure 24.1 shows global climatic geopolitical territory, which is categorized as CC regions such as northern and southern hemisphere counties, tropical climate region like the Middle East regions, and subtropical climate regions, which are mostly Asian countries, Pacific and Caribbean regions (Figure 24.1). In general, the industrialized regions and western regions are using the wastewater after treatment as freshwater for irrigation, industries, and drinking water. Most of the high-density populated regions, urban areas, and industrial regions are facing deficits of surface and groundwater quantity. Therefore, household and industrial wastewater are being considered as raw water for treatment for future use and reuse for human well-being and economic sustainability in the three different climatic regions.

The vast area of tropical and subtropical regions is recognized as warmer regions with higher and increasing temperatures. These regions are very fertile for human settlements and development activities as well as for agriculture and biodiversity with special productive landscapes. Population density, urbanization, and water resources are the main resources for this huge population. However, the water resources are used within unlimited scales and scarcity of water is the main problem to maintain natural resources, agricultural and industrial production as well as to maintain the ecosystems and ecosystem services in these tropical and subtropical regions. Figure 24.1 shows the climatic and geopolitical regional territories.

24.4 Historical Development of World Water Use and Reuse Technology

Water reuse and recycle management systems in the CC regions are more or less similar to the dry or semidry regions of the world.

The following cycle has been recognized in urban areas as one of the scholar models of water recycling (Figure 24.2). Historically, wastewater has been used as an alternative water source for agricultural irrigation, going back approximately 5000 years [1]. During the nineteenth century, the introduction of large-scale wastewater carriage systems for discharge into surface waters led to the inadvertent use of sewage and other effluents as components of potable water supplies. According to the historical water cycle management system, the water cycle flow (Figure 24.2) is applicable in the CC regions. Traditionally, this water cycle system (Figure 24.2) was implemented in three different regions. In addition, the modern water reuse systems are developing according to the demand and necessity in the communities.

Figure 24.2 demonstrates the general water cycle management model in CC, tropical, and subtropical climatic regions of the world. The functions of this water cycle management model are more or less similar for the urban and rural regions of different climatic regions. Wastewater reuse, when appropriately applied, is considered an example of environmentally sound technologies (EST) applications. According to the demand and criteria fulfillment of Agenda 21 as technologies that are indicated to achieve the following objectives: protect the environment, less polluting, use all resources in a more sustainable manner, recycle more of their wastes and products, and handle residual wastes in a more acceptable manner than the technologies for which they are substitutes. Water use and reuse is considered based on the necessity and demand of the community and regions. In general, it can be seen that in different climatic regions and religions, using and implementing strategies are different as the water reuse means it is already considered wastewater. Therefore, ethical values and cultural practices are considered in wastewater reuse in CC, tropical, and subtropical regions in the world. In Figure 24.2, the ethical and cultural dimension has been included as a major factor of these water use and reuse practices. In addition, water use and reuse practices are strongly followed in a religion's values and concepts.

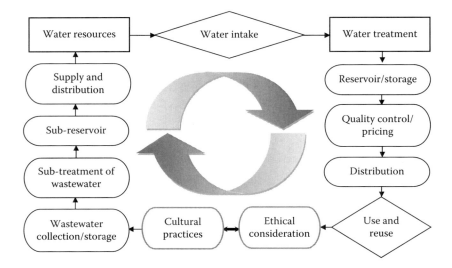

FIGURE 24.2 Urban and rural water cycle management in cold, tropical, and subtropical regions. (After Van Dijk, M.P. 2007. *Constitution to an International Seminar on Sustainable Urbanization in Tripoli*, Hotel Bab Africa, Libya, June 30–July 1.)

24.4.1 Global Water Reuse in Urban and Rural Areas

Surface water has advantages as a source of potable water. Surface water sources are usually easy to locate; unlike groundwater, finding surface water does not require a geologist or hydrologist. Normally surface water is not tainted with minerals precipitated from the Earth's strata. Ease of discovery aside, surface water also presents some disadvantages. Surface water sources are easily contaminated with microorganisms that can cause waterborne diseases and are polluted by chemicals that enter from surrounding runoff and upstream discharges. In general, surface water is highly variable for two main reasons: (1) human interference and (2) natural conditions. In some cases, surface water quickly runs off land surfaces [53]. Table 24.1 shows the different types of products content and water used in consideration of social ethics and cultural dimensions. The geocultural and morphological and climatic aspects are more important factors for water use and development.

Considering the international trade flows between all major countries of the world and looking at the major agricultural products being treated (285 crop products and 123 livestock products), the actual calculated water use for producing export products amounts to 1250 billion m³/yr. If the imported countries would like to produce the imported products domestically, they would require a total of 1600 billion m³/yr. This means that the global water saving by trade in agricultural products is 350 billion m³/yr [31]. Table 24.2 shows the per capita water availability and water intensity use (%) in geopolitical territories of the different climatic conditions of the world.

Now it is estimated that approximately 700 million people (11% of the total world population) in 43 countries live with less

TABLE 24.1 Global Average Virtual Water Content of Selected Products, per Unit of Product

Product	Virtual Water Content (L)
1 sheet of A4 paper (80 g/m²)	10
1 tomato (70 g)	13
1 potato (100 g)	25
1 slice of bread (30 g)	40
1 orange (100 g)	50
1 apple (100 g)	70
1 glass of beer (250 mL)	75
1 slice of bread (30 g) with cheese (10 g)	90
1 glass of wine (125 mL)	120
1 egg (40 g)	135
1 glass of orange juice (200 mL)	170
1 bag of potato crisps (200 g)	185
1 glass of apple juice (200 mL)	190
1 glass of milk (200 mL)	200
1 hamburger (150 g)	2400
1 pair of shoes (bovine leather)	8000

Source: Adapted from Hoekstra, A.Y. and Chapagain, A.K. 2008. *Globalization of Water-Sharing the Planet's Freshwater Resources.* Blackwell Publishing, Malden, MA.

TABLE 24.2 Water Availability and Water Intensity Use Indices

Geopolitical Regions	Water Availability Index in 2006 m³/Capita-yr	Water Intensity Use Index in 2000 (%)
Middle East and North Africa	1383	62.8
Asia (excluding Middle East)	3990	19.3
Mexico, Central America, and the Caribbean	6740	8.5
United States and Canada	19,649	9.3
Europe	10,660	6.4
Sub-Saharan Africa	7,200	3.1
Oceania	53,290	1.6
South America	45,400	1.3
Developed countries	11,392	9.0
Developing countries	7693	8.9
High-income countries	10,554	10.1
Middle-income countries	10,171	6.9
Low-income countries	5804	12.1
World	8462	8.9

Source: Adapted from Earth Trends, 2007. Vulnerability and adaptation to climate change, July 27, 2007.

than 1000 m³/capita-yr. By the year 2025, 38% of people (more than 3 billion) will live in such conditions, increasing the number to nearly half of the population and 149 countries by the year 2050 [59]. Water availability will certainly be reduced due to high population growth, agricultural extension, and irrigation and climate change impacts. Consequently, water use sustainability and wastewater reuse practices would be more effective for water use sustainability in different geopolitical and climatic regions.

24.5 Water Reuse from an Ethical Perspective

From our career perspective, we often see water as a common goods product that might be in good quality and will have efficient support for economic industrial development and maintaining public health. In addition to this we, who are working for water recycling, often go back against our ethics when our personal or self-interest took place in front of us. But in a world rife with competitive values, only a zealot or enthusiasm over water recycling can solve this ethical problem because no one wants to do anything wrong willingly; after all, we are all human beings. The technical, logistic, economical, and regulatory values are in such a tenet in our mind that we often try to neglect or oversimplify the ethical value of water recycling. We are biased to pretend that there is a scientific, economic, or legal basement for water recycling. We always feel comfortable talking about these issues, which involve various terms so that it creates a scholarly aptitude to our professional career development. It is very unlikely when we talk about our ethics, that our value to reserve or recycle water is for our own purpose, but rather for our own environment. Environment and humans, we have a inter dependency relationship, not a temporal relationship. Day by day due to population growth, we have to grow our ethical values regarding water recycling. Ethical value is a term that cannot be developed at once, it cannot be discussed in isolation of other values. Rather, we have to develop it from our early childhood. At that time, the growing population is increasing demands and many supplies are being degraded. According to the projection citation of UN, Water FAO (2007) [60], by 2025, two-thirds of the world population will face high-level water stress and about half of the population will face real constraints in their water supply.

24.6 Water Reuse from a Cultural Practices and Religious Perspective

As main references for the cultural theory, we use Thompson [56], Thompson et al. [57], and Schwarz and Thompson [50]. However, one of the roots of the cultural theory lies in the anthropological research of Mary Douglas. In her book *Natural Symbols* [16], Douglas introduced a group grid typology of cultures, based on a comparison of sociostructures and corresponding ideas about ritual, sin, and self. She argues that the character of social relations can be described along two axes: group and grid. The group axis represents the degree to which an individual is incorporated into confined units. A positive score on the group axis means that an individual strongly feels that he or she belongs to a group. The grid axis denotes the extent to which an individual's life is circumscribed by externally imposed prescriptions or, in other words, the extent to which external rules determine someone's behavior. A positive score on the grid axis indicates a high role definition, strong regulation of interactions between people, and little room for individual choice [57]. Based on the two dimensions, Douglas proposes to distinguish four types of social relations (Figure 24.3).

The combination of high group and high grid refers to social groups where individuals are involved with other people, but separated from them by numerous limits and boundaries. In the case of high group but low grid, all status is insignificant apart from one kind, the status involved in belonging to a defined group. Low group plus low grid means that individuals are free

from social constraints; group organization barely exists and fixed rules for behavior are lacking. In the last combination, low group and high grid, individuals do not belong to a circumscribed group, but they are nevertheless constrained in their relations with other people. The group grid typology reappears and has been further developed in the cultural theory, where the four types of social structures are called "ways of life." The four ways of life are described as the hierarchist, egalitarian, individualist, and fatalist (Figure 24.3).

To these, a fifth way of life has been added: the hermit, autonomous, or ineffectual way of life, where the individual withdraws from coercive or manipulative social involvement altogether. The hermit escapes social control by refusing to be controlled or to control others. Acknowledging that hermits are not involved in social transactions within the fourfold system of hierarchizes, egalitarians, individualists and fatalists (Figure 24.3), after Schwarz and Thompson [50] and Thompson et al. [57] only the first four ways of life will be considered and the way of hermits lifestyle as such can be ignored [3]. No substance in the world is endowed with more cultural and religious significance than water. Partow [45] point out that Hindu tradition considers water a medium of purification and a source of energy. "The water is the sky, the water of the rivers and water in the well whose source is the ocean, may all these sacred waters protect me" [39, p. 47]. Religions provide a reedy starting point for approaching cultural diversity in tackling environmental issues. Religions can be aligned with specific codified beliefs and values, and there are formal institutions and identifiable spokespeople, which can be taken as representative of large collectives of individuals [7].

In Islamic tradition, the Shariah is literally translated as "the source of water, contains legal rules and principles." The Quran has many references to water. It is the sole basis of creation by the will of God. Its scarcity or bounty often represent the status of relationship between humans and God. One of the principles is that water is proof of God's existence, proof of God's care, and proof of resurrection, as water restores life every day.

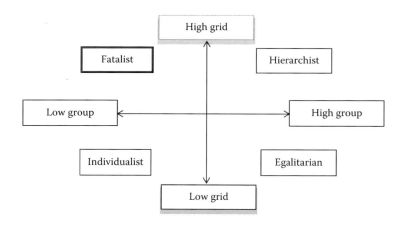

FIGURE 24.3 The framework model of group grid typology culture. (Adapted from Thompson, M., Ellis, R., and Wildavsky, A. 1990. *Cultural Theory*. Westview Press, Boulder, CO.)

The Goddess of freshwater is Oxum, who reigns over the rivers. African slaves from the Yoruba region brought their beliefs to Brazil, Cuba, Dominican Republic, and Puerto Rico, where the two water goddesses are still worshipped through song, dance, and gifts. Both the Jewish and Christian traditions use water as a symbol of cleaning. Ritually blessed water is representative of purification most obviously in baptism rituals. In Catholicism, it is more than symbolic that the water is actually purifying the baptized of original sin. In Buddhism, water is symbolic of human welfare and peace and prosperity [7].

24.7 Results and Discussion

The human-natural world intersection is vital but for many it has become a given. So, too, is the convention that the natural environment and human communities are facing potentially fatal disturbances [13,41]. On the other hand, environmental ethics or environmental psychology can be described as an exploration of the cosmos and humanity's relationship to it. Where environmental ethics could be said to differ from other explorations is in its explicit focus on the natural world. Environmental philosophy is the marriage of ecology and philosophy [41]. Therefore, water and natural resources are most important resources in consideration of environmental ethics, ethical values, cultural values, and religious values. Water is the most important good for life in general. Water use and reuse is now a subject to protect the resource and ensure sustainability. The present population growth, socioeconomic activities, industrial development, and climate change impacts are the most critical threats for water resource uses and sustainable conservation [41]. Most of the critical regions are now facing potable drinking and irrigation water loss. Thus, the natural sources of water should be used and reused within an ethical, cultural, and religious perspective. The history of water reuse is also very ancient, but the uses and practices are not very popular in different climatic regions [41]. This is why different climatic regions are facing challenges to use water in everyday life or for industrial use. The following subsections demonstrate the challenges and barriers of water use and reuse in different parts of the world.

24.7.1 Water Reuse and Challenges

Water use is very traditional, and common systems exist in every country and region as well as developed and developing countries. Due to high population and continuous increased demand, water reuse has come into focus. Anthropogenic influences, economic activities, technology development, and climatic conditions are more focal points. Water that is used a second time means it has been treated from a large plant or a microscale level plant for water purification and that water reuse is for many purposes. During normal production operations produced water, which is the aqueous liquid phase, is co-produced from a producing well along with the oil and/or gas phases. The following main contaminants of concern are ingredients in produced water, such as suspended solids, high level of total dissolved

solids (TDS), heavy metals, dissolved gases and bacteria, dissolved and volatile organic compounds, oil and grease, dispersed oil, radionuclides or chemicals (additives) used in production such as biocides, scale and corrosion inhibitors, and emulsion and reverse-emulsion breakers [27]. For purification, the contaminated water needs to be purified in several stages such as pretreatment, main treatment, polishing treatment, and tertiary treatment. Water reuse is a challenge for the construction and proper implementation of technical treatment plants. Finally, the ethical values and cultural practices and beliefs, and lack of technical knowledge are the more common challenges in water use and reuses in different geopolitical regions of the world.

24.7.2 Climate Change Is a Driving Force to Water Reuse for Agriculture

As global warming and population growth, urbanization, and industrial growth in different regions of the world as well as in the CC, tropical, and subtropical climatic regions continues, these areas are facing the challenge of clean water supply for drinking, agriculture, and industrial uses [8,49]. Due to the necessity of human well-being in the CC regions, England progressively introduced water filtration technology during the 1850s and 1860s. Climate change and sea level rise, induced by global warming, also compromise the ecological stability of the coastal zone; in sum, due to various natural and anthropogenic factors, the natural resource base of the zone is declining in Bangladesh and other low-lying countries in Asia and Pacific and Caribbean regions [43]. Failing ecosystem productivity further degrades the quality of life of the local population [14]. The relative sea-level rise (SLR) movement has an immediate and direct effect on the coastal inter-tidal ecosystems, particularly on vegetation. A rise of relative SLR decreases the influences of terrestrial processes and increases the influence of coastal marine processes [35,64]. The world's great deltas are among the most densely populated and most vulnerable of coastal areas; they are threatened by SLR [6]. Global warming, SLR, and vulnerability of coastal wetland ecosystems are factors that have to be considered for the long-term management strategy for dealing with the coastal mangrove wetland issues. The impacts of climate change in any given region depend on the specific climatic changes that occur in that region. Local changes can differ substantially from the globally averaged climate change [28]. Global warming and climate change cause a predicted SLR, which will cause a further "squeezing" of the natural tidal land. In Bangladesh, it has been projected by Integrated Pollution Prevention Control (IPCC) [34] and Ministry of Environment and Forest (MoEF) of Government of Bangladesh [23] that 3 mm/year SLR will occur before 2030 and 2500 km^2 land (2%) will be inundated. About 20% of the net cultivable land area is located in the coastal region in Bangladesh [29]. Present high saline water intrusion in the coastal regions is a threat for agricultural crops production. Therefore, water production, desalination, and water reuse would be an appropriate solution for agricultural sustainability in the coastal regions [23,34].

24.7.3 Water Reuse for Industrial Production and Peri-Urban Food Security

A very recent study on the coastal area in Bangladesh by Sing et al. [51] shows that the mean tidal level at Hiron Point is showing an increase of 4.0 mm/yr, which is higher than the global rate. Soils in this area are affected by different degrees of salinity [47]. It is estimated that about 203,000 ha very slightly, 492,000 ha slightly, 461,000 ha moderately, and 490,200 ha strongly salt affected soils are assessed in the southwestern part of the coastal area. The climate change impact issue is a new threat for the coastal agriculture and industrial productions [44]. For example, the Khulna paper mills and Hardwar mills are contributing raw materials to the Sundarbans mangrove forest wetlands. However, this forest is recently affected by scarcity of upstream freshwater supply. Huge amounts of freshwater is needed to keep the product quality of the coastal industries. The high saline water intrusion and shortage of upstream freshwater are damaging the quality of industrial products [24,37,38].

In addition, other environmental problems will arise in the coastal belt such as water pollution and scarcity, soil degradation, deforestation, solid and hazardous wastes, loss of biodiversity, estuary landscape damage, and riverbank erosion, which will create many new challenging problems for human livelihood in the coastal region. In such situations, it will further create unstable agricultural crop production, damaging fisheries and livestock and food security in the coastal riverine islands in Bangladesh. Some thousands of coastal towns and cities are located in the coastal regions as well as in landlocked areas. About 360 coastal towns are located in the coastal region of Bangladesh [24,37,38].

Most of the towns are facing drinking water supply and management problems. Almost all urban and peri-urban food security are dependent on freshwater supply. As most parts of the critical regions are facing the problem of freshwater supply, water production and reuse are essential to ensure peri-urban food security and ecosystems. Almost all sub-Saharan cities are in a big water supply crisis for drinking, irrigation, and industrial production. The peri-urban food security is also dependant on water supply and availability. Hence, new strategies and technical skills should be developed for water reuse in the case area in Bangladesh as well as other low-lying countries like Bangladesh in the Asia Pacific region, Africa, and Caribbean and Latin American countries [24,37,38].

24.7.4 Water Reuse for Biodiversity Protection

Biodiversity loss is a common scenario in the coastal and northern drought-prone areas in Bangladesh. As it has become a common environmental problem, water scarcity or less precipitation is the main reason of biodiversity damage. Similar problems have been seen in India, Pakistan, Sri Lanka, Nepal, China, Thailand, Vietnam, Philippines, Myanmar, Afghanistan, Azerbaijan, Mongolia, Iran, Iraq, Yemen, United Arab Emirates, Saudi Arabia, and other Asian countries [40,42]. The most solid

scenarios are available in the sub-Saharan countries where drought and water scarcity are common environmental problems. Deforestation of mangrove due to shrimp cultivation, salt farming, agricultural land extension, and settlement development adversely affects coastal fish production and leads to a loss of biodiversity and livelihood to over 3.5 million people who are dependent on mangrove resources in the coastal region of Bangladesh [2,17]. In Bangladesh, the mangrove forest areas were estimated (1963–1978) to be 685,000 ha and the present area is 587,000 ha, which is 86% covered. This means that 14% of mangrove has been lost or destroyed within 35 years. The damaged forests take a very long time to recover. Water scarcity, damage of soil fertility, and crops mortification are the main causes of biodiversity loss in Bangladesh and other Asian and African countries. The climate change impacts and droughts are new additional phenomena of biodiversity loss [18]. The annual natural calamity—global warming and its impacts—are new challenging threats for coastal food security and agro biodiversity [32]. The land management, landscapes, and protection of ecosystem services is very important for regional food security and biodiversity [22,26]. In such a critical situation, quality water plays the central role toward protection of biodiversity. Thus, water reuse could be the core alternative solution for biodiversity conservation and sustainable management in CC, tropical, and subtropical regions of the world.

24.7.5 Smart Plant for Water Reuse within Ethics and Cultural Practices

The climatic and geosphere condition, the uncontrolled disposal to the environment of municipal, industrial, and agricultural liquid, solid, and gaseous wastes constitutes one of the most serious threats to the sustainability of human civilization by contaminating the water, land, and air through global warming [4]. With continuous increasing population and economic growth and industrial development, treatment and safe disposal of wastewater is essential to preserve public health and reduce intolerable levels of environmental degradation in the CC regions. In addition, adequate wastewater management is also required for preventing contamination of water bodies to preserve the sources of clean water. Effective wastewater management planning is well established in developed cold regions, but is still limited and there are some limitations to water reuses. Now that the requirements for a sustainable wastewater treatment system have been presented, there are several options one can choose from in order to find the most appropriate technology for a particular region. This chapter will discuss a model (Figure 24.4) of microscale level water reuse treatment plant at the community level for the CC regions [9,11].

A plan for a community-scale integrated wastewater treatment facility is proposed for the CC, tropical, and subtropical regions for water reuse sustainability. Wastewater could be used in the waste treatment plant where the facilities would use biogas produced from the anaerobic digestion of organic waste to generate electricity, heat, light, and carbon dioxide for enhanced

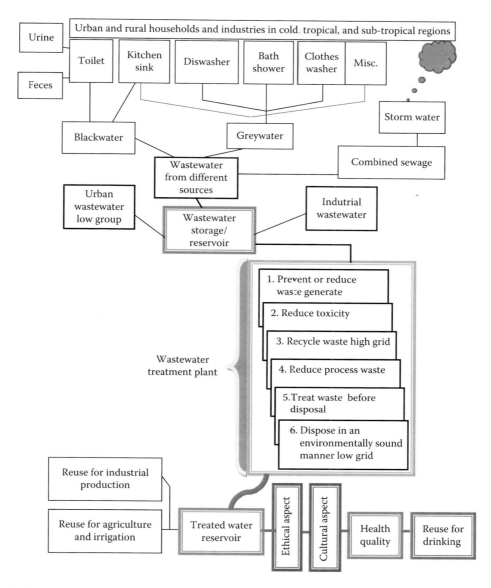

FIGURE 24.4 Model of wastewater treatment plant with consideration of ethical and cultural aspects (microscale level).

plant growth in a controlled environment [7,13]. In this study, the proposed model (Figure 24.4) could be a popular model that could be used in the community level because it has been designed based on the raw materials (wastewater) source at the household level. The importance of wastewater use and reuse in the CC, tropical, and subtropical regions is essential to demonstrate this potential message at the community level so that all levels of inhabitants could be involved in such an environmental program for the better management of their own regional water use and reuse activities. Communication about the treatment and quality of recycled water is particularly challenging. The technical language and tools used by the water industry to evaluate, quantify, and describe health risks is often neither comprehensible nor credible to the public. The early socialization of human beings to keep waste out of their mouths is often more persuasive than scientific analysis published by experts.

The present development of the microscale level model (Figure 24.4) implementation would be more popular and active at the community level if policy makers and environmental and climatic scientists have plan to involve the community and stakeholders in any kind of environmental program in the geopolitical climate regions. Figure 24.4 demonstrates a model flowchart of a wastewater reuse treatment plant at the household level in a community of the CC regions. The functionality, cost, and easy technology could be used at this model implementation. The effective treatment of wastewater to meet water quality objectives for water reuse applications and to protect public health is a critical element of water reuse systems. The wastewater treatment consists of a combination of physical, chemical, and biological processes and operations to remove solids, organic matter, pathogens, metals, and nutrients from wastewater [33].

In order to increase treatment levels, the different degrees of treatment are preliminary, primary, secondary, tertiary, and advanced treatment, which has been set up in the model. The household wastewater, industrial wastewater, and urban wastewater could be stored in the microscale level and that could be treated within this model in six different steps. Step 1 is to reduce or prevent waste generation. Step 2 is reduce waste toxicity. Step 3 is recycle the waste. Step 4 is reduce process of waste. Step 5 is treat the waste before disposal and Step 6 is to dispose in an environmentally sound manner low grid in the wastewater treatment (Figure 24.4). After this, wastewater could be treated as clean water for use in different purposes such as agricultural irrigation, industrial production, and gardening, and other portions of treated wastewater could be used for drinking as long as it has been monitored for quality of drinking water (Figure 24.4). This is the microscale level model of a wastewater treatment plant that could be implemented in the community household level and this type of environmental activity could ensure water reuse sustainability in the community level in CC, tropical, and subtropical climatic regions of the world.

24.8 Summary and Conclusions

Globally, water resources are potentially used for human life and socioeconomic and industrial development for human well-being. Water is a driving force for community, social, and economic development all over the world. The world geographical features do not consider the matter of water distribution, but climate issues are a matter of productivity and development. The CC, tropical, and subtropical climates are potential factors for quality water distribution, availability, uses, and future development issues. They have the ability to focus tremendous energy and to generate significant creative and economic betterment. In general, the country's wetland natural resources as well as the ecosystems are degrading due to anthropogenic influences and natural calamities. The freshwater and saline water wetlands in Bangladesh are in an alarming situation. The hydrological cycle of the wetland areas is losing the balance. The country needs an adequate interdisciplinary policy and strategies and political wills to implement it for sustainable management and protection of wetlands and ecologically sensitive wetland ecosystems in Bangladesh. The methods for the generation of transformation knowledge to achieve and maintain natural resources, agro-farming system development, fishing development, and, in general, wise use of wetland natural resources could protect the biodiversity in the freshwater wetlands and coastal saline mangrove wetland areas. The following recommendations could be followed in implementation and management stages:

- To build up people's awareness, there will be a yearly local cultural festival on international water day in different parts of the CC, tropical, and subtropical regions, where the local culture, heritage, and wetland ecosystem services and goods will be demonstrated and marketed to enforce the local capacity of income generation. It could

be a popular method for participation in water conservation, harvesting, and sustainable uses within the ethics, culture, and religious values and practices.

- The requirement to ensure the provision of adequate institutional capacity for policy development, delivery, and monitoring; the importance of considering water sources as water management infrastructure rather than nature reserves and the need to consider the wise use and reuse of water within rural and beyond urban boundaries and understand the interconnectivity of watershed-scale issues.

- Traditionally, frameworks and spatial plans are area based and structured on administrative boundaries. This approach fails to address or in many cases even recognize that these boundaries are not functional or commensurate with the environmental boundaries required for the protection of natural resources, wetlands, or ecosystem services.

For time series analysis, the satellite and remote sensing imageries could be used to compare the historical changes of water shades in CC, tropical, and subtropical climatic regions, agricultural cropping systems, and land use patterns, shrinking trends of wetlands and water bodies, and present scenario measures. As a whole, the developed water conservation planning and policy framework could be the fundamental output of applied research that could be implemented in the protected, environmental, and climate-sensitive regions in Bangladesh as well as others countries of the world.

References

1. Angelakis, A.N. and Spyridakis, S.N. 1996. The status of water resources in Minoan times a preliminary study. In: Angelakis, A.N. and Issar, A.S. (eds.), *Diachronic Climatic Impacts on Water Resources with Emphasis on Mediterranean Region*. Springer-Verlag, Heidelberg, Germany, pp. 161–191.

2. Anon. 1995. Integrated resource management plan of the Sundarbans Reserve Forest, Vol. 1, Draft final report of FAO/UNDP project BGD/84/056-Integrated Resource Management of the Sundarbans Reserve Forest, Khulna, Bangladesh.

3. Arjen, Y.H. 1999. *Appreciation of Water: Four Perspectives*. International Institute for Infrastructural, Hydraulic and Environmental Engineering (IHE), Delft, the Netherlands.

4. Beecher, J. 2001. *The Ethics of Water Privatization in Navigating Rough Water: Ethical Issues in the Water Industry*. American Water Works Association, Denver, CO, p. 247.

5. Bixio, D.T.C., Koning, J.D., Joksimovic, D.S., Savic, D., Wintgens, T., and Melin, T. 2006. Water reuse in Europe. *Desalination*. Elsevier, 187, 89–101.

6. Broadus, J.M. 1993. Possible impact of and adjustment to sea level rise: The cases of Bangladesh and Egypt. In: Warrick, R.A., Barrow, E.M., and Wigley, T.M.L. (eds.), *Climate and Sea-Level Change; Observations, Projections*

and Implication. Cambridge University Press, Cambridge, pp. 263–275.

7. Cheryl, K.D. 2008. Ethical dilemmas in water recycling. In: Jimenez, B. and Asano, T. (eds.), *Water Reuse—An International Survey of Current Practice, Issues and Needs.* Scientific and Technical Report No. 20, IWA Publishing.

8. Chowdhury, N.T. 2009. Water management in Bangladesh: An analytical review. *Journal of Water Policy,* IWA Publishing, London, 12(1), 32–52.

9. Cook, H.F. 1998. *The Protection and Conservation of Water Resources-A British Perspective.* John Wiley & Sons, Inc., New York.

10. Cooley, R.L. 1997. Confidence intervals for ground-water models using linearization, likelihood, and bootstrap methods, *Ground Water,* 35(5), 869–880.

11. Cooley, H. 2009. Water management in a changing climate. In: *The World's Water 2008–2009.* The biennial report on freshwater resources. Island Press, Washington, DC, pp. 39–56.

12. Costanza, R.L.J., Norton, B.G., and Halskell, B.D. (eds.) 1992. *Ecosystem Health: New Goals for Environmental Management.* Island Press, Washington, DC.

13. Crosby, M.P., Geenen, K., Laffoley, D., Mondor, C.I., and Sullivan, G.O. 1997. Proceedings of the Second International Symposium and Workshop on Marine and Coastal Protected Areas: Integrating Science and Management, NOAA, Silver Spring, MD, pp. 167.

14. Dasgupta, P. 2001. *Human Well-Being and the Natural Environment.* Oxford University Press, Oxford, UK.

15. Doran, G.F., Carini, F.H., Fruth, D.A., Drago, J.A., and Leong, L.Y.C. 1997. Evaluation of technologies to treat oil field produced water to drinking water or reuse quality. Paper SPE 38830 presented at the SPE Annual Technical conference and Exhibition, San Antonio, TX, October 5–8.

16. Douglas, M. 1970. *Natural Symbols: Explorations in Cosmology.* Barrie and Rockcli, London.

17. Earth Trends, 2007. Vulnerability and adaptation to climate change, July 27, 2007.

18. ECCD. 1991. Council Directive 91/676/EEC of December 12, 1991 concerning the protection of waters against pollution caused by nitrates from agricultural sources. *Official Journal of the European Union,* 375(31), 1–8.

19. Ferdra, K. and Feoli, E. 1998. GIS technology and coastal management, ICZM, Integrated coastal zone management. *EEZ Technology,* 3, 171–179.

20. FPCO (Flood Plan Coordination Organization). 1993. Southwest Area Water Resources Management Project. FAP4 Final Report. FPCO, Bangladesh Ministry of Irrigation, Water Development and Flood Control, Dhaka.

21. Gleick, P. 1998. *The World's Water.* Island Press, Washington, DC.

22. Gleick, P.H., Cooley, H., Cohen, M., Marikawa, M., Morrison, J., and Palaniappan, M. 2009. *The World's Water (2008–2009), The Biennial Report on Freshwater Reuses.* Inslet Press, Washington, DC, pp. 39–56.

23. GOB (Government of Bangladesh). 1999. *National Water Policy, Ministry of Water Resources.* Government of the People's Republic of Bangladesh, Dhaka.

24. Goodbred, S.L. and Kuehl, S.A. 2000. Late quaternary evolution of the Ganges–Brahmaputra River delta: Significance of high sediment discharge and tectonic process on margin sequence development. *Sedimentary Geology,* 133, 227–248.

25. Grigg, N.S. 1996. *Water Resources Management Principles, Regulation and Cases.* McGraw-Hill, Washington, DC, pp. 8–14.

26. Grigg, N.S. 2008. Total water management practices for a sustainable future. In: *American Water Works Association.* AWWA Publications, Denver, CO.

27. Guerra, K. Dahm, K., and Dundorf, S. 2011. Oil and Gas Produced Water Management and Beneficial Use in the Western United States. Report. U.S. Department of the Interior, Bureau of Reclamation, Denver Federal Center, Denver. https://www.usbr.gov/research/AWT/reportpdfs/report157.pdf. Accessed July 03, 2014.

28. Harvey, L.D.D. 2000. *Climate and Global Environmental Change.* Pearson Education Ltd., UK, pp. 187–207.

29. Hasan, S. and Mulamoottil, G. 1994. Natural resource management in Bangladesh. *AMBIO,* 23(2), 141–145.

30. Hidayati, D. 2000. *Coastal Management in ASEAN Countries. The Struggle to Achieve Sustainable Coastal Development.* UN University Press, Tokyo, pp. 1–74.

31. Hoekstra, A.Y. and Chapagain, A.K. 2008. *Globalization of Water-Sharing the Planet's Freshwater Resources.* Blackwell Publishing, Malden, MA.

32. Hussain, Z.K. 1995. The Farakka catastrophe reflections, In: H.J. Moudud (ed.), *Women for Sharing Water.* Academic Publishers, Dhaka, pp. 71–72.

33. ICWE (International Conference on Water and Environment). 1992. The Dublin statement on water and sustainable development. *International Conference on Water and Environment,* January 26–31, 1992, Dublin, Ireland.

34. IPCC (Intergovernmental Panel on Climate Change). 2007. Summary for policymakers, In: Parry, M.L., Canziani, O.F., Palutikot, P.J. et al. (eds.), *Climate Change 2007: Impact Adaptation and Vulnerability.* Contribution of working group II to the Fourth Assessment Report of the IPCC, Cambridge University Press, Cambridge, UK.

35. Islam, M.S. 2001. *Sea-Level Changes in Bangladesh: The Last Ten Thousand Years.* Asiatic Society of Bangladesh, Asiatic Military Press, Dhaka.

36. Islam, S.N. 2007. Salinity intrusion due to fresh water scarcity in the Ganges catchment: A challenge for urban drinking water and mangrove wetland ecosystem in the Sundarbans region, Bangladesh. At the 6th edition of the World Wide Workshop for Young Environmental Scientists (WWW YES 2007), April 24-27, 2007, Paris, France, pp. 20–30.

37. Islam, S.N. 2012. Degradation of coastal wetlands eco-systems and mangrove biodiversity in the Sundarbans

in Bangladesh. *Journal of the Bangladesh National Geographical Association*, 38(1–2), 1–13.

38. Islam, S.N. and Gnauck, A. 2009. Threats to the Sundarbans mangrove wetland ecosystems from transboundary water allocation in the Ganges basin: A preliminary problem analysis. *International Journal of Ecological Economics & Statistics (IJEES)*, 13(09), 64–78.

39. Jamieson, D. (ed.). 2003. *A Companion to Environmental Philosophy*. Wiley-Blackwell, Malden, USA.

40. Kundzewicz, Z.W., Mata, I.J., Arnell, N.W., Döll, P., Katat, P., Jimenez, B., Miller, K.A., Oki, T. Sen, Z., and Shiklomanov, I.A. 2007. Freshwater resources and their management, climate change 2007: Impacts, adaptation and vulnerability. Contribution of working group II to the fourth Assessment Report of the Intergovernmental Panel on Climate Change. Cambridge University Press, Parry, pp. 173–210.

41. Light, A. and Smith, J.M. (eds.). 1997. Philosophy and geography I: Space, place and environmental ethics. *A Peer Reviewed Annual*.

42. Liu, S.X. 2007. Innovative technologies for value-added substance/Energy recovery from wastewaters. In: *Food and Agricultural Wastewater Utilization and Treatment*. Blackwell Publishing, London, pp. 247–260.

43. Moeller, D.W. 2011. *Environmental Health*, 4th Edition. Harvard University Press, UK.

44. Nordell, E. 1961. *Water Treatment for Industrial and Other Uses*, 2nd Edition. Reinhold Publishing Corporation, New York.

45. Partow, H. 2001. The Mesopotamian marshlands: Demise of an ecosystem. Early Warning and Assessment Technical Report, UNEP/DEWA/TR.01-3, United Nations Environment Programme (UNEP). Nairobi, Kenya.

46. Pramanik, M.A.H. 1983. Remote Sensing Applications to Coastal Morphological Investigations in Bangladesh, Ph.D. Thesis, Jahangimagar University, Savar, Dhaka, Bangladesh.

47. Rahman, A.A. 1988. Bangladesh coastal environment and management. In: Hasna, M.J., Rashid, H.E., and Rahman, A.A. (eds.), *National Workshop on Coastal Area Resource Development and Management CARDMA*. Academic Publishers, Dhaka, pp. 1–22.

48. Rahman, M. and Ahsan, A. 2001. Salinity constraints and agricultural productivity in coastal saline area of Bangladesh. In: Rahman, M. (ed.), *Soil Resources in Bangladesh: Assessment and Utilization*, SRDI. ProkashMudrayan, Dhaka, pp. 2–14.

49. Rowe, D.R. and Abden-Magid, I.M. 1995. *Handbook of Wastewater Reclamation and Reuse*. CRC Taylor & Francis Group, New York.

50. Schwarz, M. and Thompson, M. 1990. *Divided Wee Stand: Re-defining Politics, Technology and Social Choice*. University of Pennsylvania Press, Philadelphia.

51. Sing, Y.F., Hui, S.S., and Rong, D.Z. 2000. Piezoelectric crystal for sensing bacteria by immobilizing antibodies on divinylsulphone activated poly-m-aminophenol film. *Talanta*, 51(1), 151–158.

52. Smith, K. 1996. Natural disasters: Definitions, data base and dilemmas. *Geography Review*, 10, 9–12.

53. Spellman, F.R. and Drinan, J.E. 2012. *The Drinking Water Handbook*, 2nd Edition. CRC Taylor & Francis Group, New York.

54. Tao, F.T., Curtice, S., Hobbs, R.D., Slides, J.L., Wiesser, J.D., Dyke, C.A., Tuohey, D., and Pilger, P.F. 1993. Reserve osmosis process successfully converts oil field brine into freshwater. *Oil and Gas Journal*, 91, 88–91.

55. Thanh, N.C. and Tam, D.M. 1990. Water systems and the environment. In: Thanh, N.C. and Biswas, A.K. (eds.), *Environmentally Sound Water Management*. Oxford University Press, Delhi.

56. Thompson, M. 1988. Socially viable ideas of nature: A cultural hypothesis. In Baark, E. and Svedin, U. (eds.), *Man, Nature and Technology: Essays on the Role of Ideological Perceptions*. Macmillan Press, London, pp. 80–104.

57. Thompson, M., Ellis, R., and Wildawsky, A. 1990. *Cultural Theory*. Westview Press, Boulder, CO.

58. UNESCO. 2003. *Water for People, Water for Life*. United Nations World Water Development Report.

59. United Nations. 2006. *Millennium Development Goals Report 2006*. UN, New York.

60. United Nations (UN), Water and Food and Agriculture Organization of the United Nations (FAO). 2007. *Coping with Water Scarcity: Challenge of the Twenty-First Century*. UN Water and FAO Report, New York, USA.

61. Van Dijk, M.P. 2007. Urban management and institutional change: An integrated approach to achieving ecological cities. *Constitution to an International Seminar on Sustainable Urbanization in Tripoli*, Hotel Bab Africa, Libya, June 30 –July 1.

62. Veil, J.A., Pudev, M.G., Elcock, D., and Redweik, R.J.J.R. 2004. A white paper describing produced water from production of crude oil, natural gas, and coal bed methane. Prepared for the U.S. Department of Energy, Energy Technology Laboratory.

63. Wei, S., Lei, A. and Islam, S.N. 2010. Modeling and simulation of industrial water demand of Beijing municipality in China. *Frontiers of Environmental Science and Engineering in China*, 4(1), 91–101.

64. White, I. 2010. *Water and the City: Risk, Resilience and Planning for a Sustainable Future*. Routledge, New York.

65. Xu, P., Drewes, J.E., and Heil, D. 2008. Beneficial use of co-produced water through membrane treatment: Technical-economic assessment. *Desalination*, 225(1–3), 139–155.

25

Community Attitudes and Behaviors and the Sociopolitics of Decision Making for Urban Recycled Water Schemes

Blair E. Nancarrow
Syme and Nancarrow Water, Perth

Geoffrey J. Syme
Edith Cowan University

PREFACE

It has been apparent for decades that communities support the concept of water reuse as a means of responsible water resources management. Reuse of stormwater and wastewater in urban developments for the irrigation of public open space and even private gardens receives little objection from communities. However, reactions from people when the use of recycled water involves close personal contact or ingestion are frequently quite different. Promoters of water recycling schemes have historically lamented the apparent emotive stand taken by communities in deciding if they will drink recycled water. The "yuck factor" has commonly been blamed for community objection to direct or indirect potable schemes. However, until recently, little had been known of how people actually made their decisions to accept or reject schemes or how to manage the "yuck factor."

This chapter discusses the Australian experience in planning and implementing urban reuse schemes with reference to international situations. It describes the psychology of community decision making and suggests that what has been commonly promoted as the key impediment to reuse schemes—community emotion—may not be that simple. The actuality is more complex and involves the relationship among the communities, proponents and regulators of the schemes, and politicians.

25.1 Introduction

Increasing urban water demand exacerbated by climate change has led to greater pressure among water managers to make efficient use of available resources. In the past, the solution to meeting demand was to build more dams or access more groundwater from aquifers. In the last 30 years, however, especially in developed countries, the easy sites for further dam construction have evaporated and those wishing to preserve key environmental assets often oppose new dams. Increased accession to groundwater has also been of concern in many areas due to the conservation of groundwater-fed environments and concerns about the potential collapse of aquifers.

Urban water managers have responded with demand management programs. However, increased urbanization will continue to create greater need for potable water and water for irrigation of green spaces and water features to support quality of life. Recycling water is being increasingly considered to augment supply. In Australia, the prolonged drought that commenced in 2002 and motivated an intergovernmental water reform program [45] resulted in the accelerated adoption of recycling programs for agriculture and industry, but only recently has potable recycled water been considered for urban areas.

In the past, the three water components of the urban water cycle have been managed separately, mostly in centralized systems. Wastewater and stormwater have been classified as "dirty" water and transported elsewhere for disposal. As a result, there have been calls for an integrated water cycle management that views recycling these waters as worthy sources of supply (e.g., References 35 and 55), which could make a significant contribution to sustainable urban water management. While seemingly a natural evolution of urban water management, the introduction of recycling systems has nevertheless created many institutional, engineering, community, economic, and political issues that need to be addressed if they are to become an integral part of urban water cycle management. In this chapter, we examine the social issues surrounding implementation of recycling and their relationship to decision making. Potable and nonpotable uses are considered separately, though the issues surrounding nonpotable water uses are less salient from the social science perspective.

The chapter begins with a review of the social and behavioral research associated with the introduction of recycling schemes. This review is then illustrated by case studies exemplifying the issues surrounding the rhetoric associated with the introduction of new recycling schemes and concludes with a discussion of the role of social science in decision-making and implementation.

25.2 Review of Literature

This review builds upon that of Po et al. [39], which demonstrates that prior to this time there had been relatively little social research in relation to recycling.

25.2.1 Nonpotable Reuse

Beginning with Bruvold and Ward's [9] U.S. survey, there has been a consistent finding that people tend to object to recycled water use the closer it comes to bodily contact and ingestion. For example, Syme and Nancarrow [52] in Perth, Western Australia, have shown that there is a relationship between initial acceptability of recycled water and the distance from personal contact or ingestion. In summary, there is strong support for recycled water to be used for outdoor uses and toilet flushing, but not for drinking (see Table 25.1).

Similar results have been found in other Australian studies [21,29,32]. Recent work in the United States has also shown the same trend [8] as did early work in the United Kingdom [22].

Half of water used by cities does not need to be of potable standards [44]. Given that a substantial proportion of a city's water demand can be met from nonpotable sources, it would seem then that major gains in supply can be made by recycling

TABLE 25.1 Community Acceptability of Uses of Recycled Water (in Descending Order of Acceptability)

Use of Recycled Water	% Acceptable
Reuse stormwater that has been treated to approved health standards on golf courses and ovals	97.5
Reuse stormwater that has been treated to approved health standards on your home garden	96.4
Reuse wastewater that has been treated to approved health standards on golf courses and ovals	94.7
Reuse stormwater that has been treated to approved health standards for toilet flushing	94.6
Reuse wastewater that has been treated to approved health standards for toilet flushing	91.9
Install a small, enclosed, and quiet wastewater treatment plant in your neighborhood to allow for reuse of wastewater in local parks and gardens	88.9
Reuse wastewater that has been treated to approved health standards on your home garden	88.4
Store stormwater that has been treated to approved health standards in wetlands for reuse at a later time	80.6
Store wastewater that has been treated to approved health standards in wetlands for reuse at a later time	70.2
Reuse stormwater that has been treated to approved health standards in the laundry for washing clothes, etc.	68.1
Reuse wastewater that has been treated to approved health standards in the laundry for washing clothes, etc.	51.0
Reuse stormwater that has been treated to approved health standards in the bathroom for personal washing, etc.	49.7
Reuse wastewater that has been treated to approved health standards in the bathroom for personal washing, etc.	30.8
Reuse stormwater that has been treated to approved health standards for drinking	28.9
Reuse wastewater that has been treated to approved health standards for drinking	15.8
Buy bottled drinking water, which would allow lesser quality water to be provided through the water supply system	10.7

Source: Adapted from Syme, G.J. 1995. *Acceptable Risks for Major Infrastructure.* A.A. Balkema, Rotterdam, the Netherlands, pp. 31–40.

fit-for-purpose quality water. Even at the household level, such substitutions can be made. For example, household toilet flushing can be supplied by connecting hand basins to toilet cisterns [12]. This would seem to be acceptable to the community [22].

At the large scale, numerous recycling projects serve industry and agriculture including retreated stormwater and wastewater for cooling in industry and irrigation for horticulture [45]. The products from these agricultural recycling schemes seem to be well accepted by communities (e.g., see Reference 56).

The provision of recycled water to households is more problematic. There have been a number of "third pipe" schemes trialed in Australia in which nonpotable water is delivered by a purple-colored pipe or tap to the household for external water use purposes. Mostly, these are operated at a development-by-development level in new suburbs. These smaller projects face economic, social, and behavioral challenges. There is the added expense of a third pipe in the developments and community issues of perceived inequity in paying the same or a higher cost for "poorer quality water" (see Reference 40). There is also a tendency among some householders to use the recycled water for unintended purposes.

For these reasons, it is seen by many urban water managers as more desirable to incorporate recycled water into the centralized potable system. This would seem to require fewer transaction costs, a smoother institutional pathway, less engineering innovation, and a noncontentious pricing policy. All it needs is community support. In many cases, that has been the stumbling block.

25.2.2 Potable Reuse

Most of the social science research on public perceptions, attitudes, and involvement in decision making has concentrated on drinking recycled sewage wastewater. The early work had a ready foundation in Rozin and colleagues' vivid research on the psychology of disgust [48], which later went on to be described in the popular press as well as the academic literature as the "yuck factor." There were, however, other factors influencing acceptance. Po et al.'s [39] review identified the following factors from the early literature that might influence acceptance of potable reuse:

- Disgust or "yuck factor"
- Perceptions of risk associated with using recycled water
- The specific uses of recycled water
- The sources of water to be recycled
- The issue of choice
- Trust and knowledge
- Attitudes toward the environment
- Environmental justice issues
- The cost of recycled water
- Sociodemographic factors

In general, this early review concluded that while there was general acceptance of the benefits accrued if recycled water was used for potable supply, there were personal concerns about the consequences. This finding has been replicated by Callaghan et al. [10], who termed the recognition of benefits a "normative" response to drinking and the concern about personal contamination a "functional" response.

Several of these factors have been the subject of later research efforts, although there have been a variety of research paradigms and decision-making contexts investigated. There has been more recent interest in the requirements for adequate decision-making processes and the rhetoric employed in community and political debate. Most of the research has been conducted in Australia and the United States.

25.2.2.1 Knowledge, Persuasion, and Acceptance

The implicit values of most social scientists seem to be that drinking recycled water is a positive thing and that if only people knew more about the topic, they would be accepting of the science involved and would support potable reuse [6]. The fact that Dolnicar and Hurlimann [13] found that the public did not know much about recycled water (at least in Australia) provided the opportunity for a direct knowledge intervention study. Dolnicar et al. [16] conducted a longitudinal study of the effects of information about the recycling treatment process. Acceptance was measured initially without information, and later with the same people with information when all potential uses showed an increase in acceptance. Unfortunately, there was no control group on the second measure to ensure that repeated measurement in itself did not have some influence on attitude. Dolnicar and Hurlimann[14] also examined the sources of information that were likely to be most significant in influencing people's acceptance of recycled water. The most influential were personal experience, values, and research findings. The preferred channels of information varied between groups.

The issue of persuasion was further researched by Kemp et al. [23] in a test of inoculation theory [31] to provide a theoretical basis for persuasion in this area. The theory was not supported but indicated that recency of communication was a key factor. The authors concluded that this indicated, "Continuous public communications are key to ensuring that community scare campaigns did not prevent implementation of water recycling projects."

This approach to persuasion is of necessity "top down" in that it is communication from the proponent to the community or stakeholders. During public or stakeholder discussions, however, the recipients actively process information as part of a dialogue when coming to a decision. This is illustrated by Price et al. [42] in their analysis of information use by supporters and opponents in a potable recycling proposal referendum in Australia. They showed clear "biases" in the way in which electors treated information with selective attention or preference being given to information that supported their position. This phenomenon has been well known for some time but is said to be exacerbated in times of threat. Evenhanded information seeking occurs more often when there is time to reflect and more than yes/no alternatives are available [24]. The finite date of the referendum and the discrete options offered were likely to have heightened selective attention.

25.2.2.2 Social Psychological Research

The phenomenon of disgust was a vivid starting concept for psychological research as was the hope that knowledge would be a basis on which acceptance could be achieved. The number of factors of interest soon moved beyond these two variables, as well as may have been expected from the Po et al. [39] review.

In two attitudinal modeling studies, Nancarrow and colleagues [36,37] developed a model of community decisions in the acceptance of recycled water for potable use. The model was developed in Australia in case studies in Perth and Melbourne, and confirmed in South East Queensland.

Ajzen's [1] Theory of Planned Behavior proposes that a person's behavior can be predicted from his or her behavioral intention. This intention is in turn determined by attitudes (toward the particular behavior), subjective norm (the influence of significant others to support the behavior), and perceived behavioral control (ease or difficulty of performing the behavior). Using Ajzen's behavioral model as a basis, a hypothesized model was developed which also incorporated additional factors from comparable literatures as follows: emotion, attitudes, subjective norms, risk perceptions, perceived control, knowledge, trust, responsibility, environmental obligation, and intended behavior. See Nancarrow et al. [36] for an explanation of the variables.

A range of measures were developed for each of the variables in the hypothesized model and tested for validity in a social experiment with a random sample of community members whereby they were asked to drink what they believed to be recycled waters. The measures were produced in a questionnaire format and the hypothesized model tested and then retested on two further occasions in circumstances where indirect potable reuse schemes were being proposed.

The final structural equation model, which accounted for 86% of the variation [37], is shown in Figure 25.1. The points of major note are that *Knowledge* was not a predictor of acceptance and that *Emotion* was not an overwhelming determinant. *Fairness* was roughly as influential as *Emotion*, indicating the

social context within which decisions are made. While *Health Risk* influenced acceptance behavior, it was determined by *Trust* in the authorities and *System Risk*. *System Risk* related not only to that associated with the technology, but also to the human capacity to manage it without error. There were obvious correlations between all of the predictor variables indicating the capacity for multiple influences on community.

The continued failure of *Knowledge* to emerge as a predictor of *Intended Behavior* in all test cases provides an explanation as to why persuasion has been ineffectual in the past to influence communities' support of reuse schemes (see also Reference 27).

In a later study, which examined determinants of acceptance of recycled water in a national sample, Dolnicar et al. [16] found that positive perceptions of recycled water was the dominant variable and that knowledge was the second greatest determinant. A variety of other factors was also found to be related to acceptance (such as influence of others [social norm] and experience with water restrictions).

Knowledge is of interest in that it was not an influential variable in the Nancarrow et al. [37] model. Nancarrow et al. [37] scaled perceived knowledge of water recycling generally, and

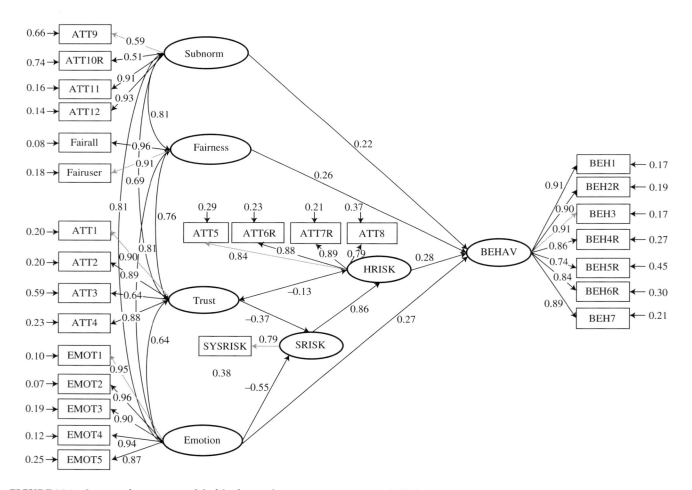

FIGURE 25.1 Structural equation model of the factors driving community intended behavior to drink recycled water. (Adapted from Nancarrow, B.E., Leviston, Z., and Tucker, D.I. 2009. *Water Science and Technology*, 60(12), 3199–3209.)

actual knowledge of the specifics of the scheme as stated by the respondents themselves. This was different from the more generalized issues and yes/no answers in the Dolnicar et al. [15] study. This may explain the difference. Nevertheless, it confirms the need to further consider knowledge and the nature of knowledge in conducting public discussions on recycled water acceptance. To quote Nancarrow et al. [37], "The absence of *Knowledge* as a predictor variable helps to explain why education campaigns have had limited influence in gaining public acceptance in the past. However, if the developers and planners engage in open and consistent conversations with the public on issues of concern to them, this should result in the earning of *Trust* and the model shows that this will assist in decreasing perceptions of both *System Risk* and *Health Risk*. This may also go some way to reducing the influence of *Emotion* ("yuck") on behavioral decisions, and also to reducing the influence of those who might try to use *emotion* to sway behavioral decisions."

Finally, from the Dolnicar et al. [15] study, it was concluded that acceptance would be maximized when people know there is no choice. In terms of information provision, it was concluded that nonthreatening ways for people to experience recycled water may be a useful method to encourage acceptance.

25.2.2.3 Cultural Studies and Discourse Analysis

Russell and Lux [49] voiced dissatisfaction with the psychologically based literature in terms of both its interest in disgust and the cross-sectional and noninteractive nature of attitudinal modeling work. In their view, the "yuck factor" had become too prominently considered by this literature and they felt that a better paradigm would be to take a sociological and cultural approach to the problem. The findings from this approach could be incorporated into well-designed public processes with an emphasis on knowledge gain through mechanisms such as deliberative decision-making.

Prior to Russell and Lux [49], probably the most comprehensive study of society and culture as it relates to recycled water was conducted by Marks and Zadoroznyi [30]. In this study, an adaptation of Szlompka's [53] model of "the social becoming of trust" was applied to four case studies in the United States and Australia. Using this theoretical approach, two kinds of components were identified: the first component consisted of five structural opportunities to build trust and the second was related to personal and collective characteristics. The five trust-building factors were

1. Coherence and consistency in regulation
2. Stability of social order or clarity in the functioning of institutions
3. Transparency of decision making
4. Familiarity through increasing knowledge and hence confidence in the new technology
5. Accountability through demonstration of properly constituted checks and balances

The second component was related to the personal and social characteristics of the main actors. That can be considered as

social capital but also as a form of interactive justice in which individuals and groups participating in decision-making are treated with dignity and respect [26].

Marks and Zadorozny's [30] multi-method study found that these components were integral for providing the basis for decision-making processes involving the community. In the large part, the findings suggested "best practice in residential water reuse should be coupled with institutional and structural arrangements that inform and involve the public and provide transparent governance."

There has not appeared to be a large response to Russell and Lux's [49] call for cultural and wider sociological studies since its publication. Hampton [18], however, in agreeing with the need for more interactive work with the community, conducted a discursive analysis of dialogue between differing groups as part of a larger nonpotable project for developing a deliberative public involvement strategy for water reuse. A key tenet of this work was that attitudes are context-based and that attitudes toward reuse need to be related clearly to the specific aspects of the proposal [41]. A two-stage focus group approach was used by Hampton [18]. In support of the discursive approach, it was found that item evaluation (i.e., evaluation of specific characteristics of the system under discussion) was preferred to category evaluation (general feelings about water reuse). The findings demonstrated the utility of a methodology and provided insights for future deliberative public programs. As to whether the same findings would have occurred for potable reuse is a moot point.

The discourse surrounding potential potable reuse was examined in Canberra, Australia, by Mikhailovich [34]. In her study, the "ways of talking" were examined through a variety of information sources. From the point of view of the relationship between different parties, Mikhailovich [34] notes, "In such consultations 'the community' is often described as though it is a distinct entity in contrast to 'policy makers' or 'scientific experts.' This differentiation is problematic and artificial when communities are not homogenous and contain within them sub-populations of experts and lay community members presenting multiple perspectives or worldviews on the issue of water recycling." Mikhailovich noted that the different discourses generated competing ways of thinking. Two distinct processes were observed: polarization of lay people and the professional community, and clashes in relation to the meanings attributed to water.

In summary, the more in-depth research and discourse analysis tends to reinforce the difference in requirements for public involvement for potable and nonpotable recycling.

25.2.2.4 Risk, Science, and Institutions in Decision-Making Approaches

The question of drinking recycled water can be seen as one about a low probability but high consequence risk. There has been much discussion about how public processes should be designed for such decisions. The conflict they have engendered for recycled water has led Dolnicar and Hurlimann [13] to ask the question as to whether it would be best to avoid public participation in decision making in such contexts. There is much literature to suggest this is not desirable, despite the difficulties involved (e.g., Reference 51).

There have been a number of studies describing best practice for public involvement in topics where risk is a prominent feature (for an early example, see Reference 5) and for recycled water in particular. Perhaps the first thing to note though is that the relationship between scientists and the community is key for success in a context such as this.

Hampton [19] noted that experts were often uncomfortable with questions from the lay community when challenged on technical questions and this was not conducive to social learning. Roth et al. [47] in a Canadian study of the interaction between scientific expertise and local knowledge found that scientists and residents differed in their assessment of water quality and quantity, "but also that there was a penchant for undercutting residents in their attempts to make themselves heard in the political process." In a Q-Method study, Leviston et al. [25] compared the structure of risk perceptions of experts and lay people in relation to a proposal for potable reuse through groundwater replenishment. The investigators found that there were five components to the lay perceptions of risk and four factors for the "expert" participants. While there were some differences between the two responses, there were also some commonalities. The most important aspect of the study was that the "experts" were then asked to respond in the way they thought the lay community would. In this case, only two factors emerged that "inaccurately caricatured (the community) as relatively non-accepting, emotion focused and driven, focused on health concerns and with a lack of trust and confidence in scientific, policy and management processes." It would seem that there is a need to research the structure of risk perceptions and how best to discuss the issues before any public programs begin.

Bagget et al. [4] in a UK study took this approach further in a social learning framework when they identified that effective collaboration between parties was dependent on the differing participating groups gaining a common language and shared understanding. For this reason, they surveyed people from all interests (e.g., management, research, domestic consumer) to establish their conception of the issue before the participatory process began. The questions encompassed expected levels of agreement about the nature of the risk, levels of trust in other stakeholder groups, and so on. Such information is almost never collected and would help in the design of engagement programs, whether or not it has a social learning framework.

Attwater and Derry [2] elaborate on this theme in their research on understanding how people with different associations (e.g., child care, education, industry) with a recycling scheme can be melded into a "community of practice" and share their experiences and thoughts as to how they can accommodate to the new system and move toward a continuous learning framework. The insights of "the community" could then be fed in a coherent way to the decision-making process as a whole rather than as fragments of thoughts and opinions, which are difficult for decision makers to accommodate. All this literature

emphasizes the need for preparatory social science research before public involvement processes are derived.

Marks [28] emphasized the need for appropriately structured institutions. This view is echoed by Stenekes et al. [50], who argue that the current decision-making systems in Australia tend to stigmatize the community as being risk averse and generally overwhelmed by the "yuck factor." According to Stenekes et al. [50], the difficulties for recycling schemes "lie in the complexity of achieving sustainability outcomes from the presence of old institutional structures." The problem definition for water recycling does not fit existing institutions easily, especially when integration of single-issue departments (e.g., health, water, land use planning) is required.

While there is clearly a need for institutional reform, it will have to be carefully planned, not only in the way it is structured but also in the way it behaves. In Hughes's [20] comparison of United States and Australian institutions of relevance to water reuse, she notes that while the United States has a less centralized system than Australia, both have encountered local resistance to the introduction of potable reuse. The political context, of course, will interact with institutional structures to govern success as will become evident in the next section.

25.3 The Sociopolitics of Implementing Urban Water Recycling: Some Case Studies

25.3.1 Nonpotable Residential Recycling Developments

There are numerous examples of the use of recycled water in urban areas in both dual pipe systems and for specific local circumstances, such as the irrigation of public open space. These dual pipe systems frequently occur in new developments mainly due to the cost of retrofitting in previously established areas. They reuse a range of wastewaters from single household greywater reuse systems, through stormwater to recycled sewage. Given that little or no personal contact is involved, there is little or no community opposition.

In Australia, the dual pipe systems occur mostly in capital cities and the waters are treated to high standards, though not to drinking water standards. Passing on the true cost of the high treatment of the recycled water to customers therefore becomes an issue as it costs more than the water in their drinking water pipes. There would be little incentive, other than environmental responsibility, for people to live in these areas. However, the incorporation of externalities in economic analyses, such as environmental protection of waterways, or ensuring long-term supplies of drinking water, obviously makes dual pipe developments attractive to water utilities. For example, Australia's largest dual pipe system in Rouse Hill, Sydney, has been operating since 2001 and now supplies 2.2 billion liters each year to 19,000 homes for garden irrigation and toilet flushing and has the potential to almost double in size [38, p. 28]. The cost of the recycled water is highly discounted to the customers. However, this is not always the case as is apparent with the City of Gold Coast's decision to decommission the dual pipe system in Pimpama Coomera, a system that only came online in December 2009. It was decided, "The high cost of the scheme for the Gold Coast community and the City outweighs the value to the city" [11]. Nevertheless, the development of residential dual pipe systems for outdoor purposes and toilet flushing is increasing in Australian capital cities.

25.3.2 Planned or Unplanned, Direct or Indirect, Potable Recycling

Unplanned water recycling occurs in river systems across the world. Wastewater, treated to various levels, is disposed from urban areas back into the rivers from which their drinking water was taken. That wastewater, then diluted by the river water, becomes part of the drinking water of the next urban area downstream. Numerous stormwater drains also flow to these rivers disposing of untreated waters. This unplanned water recycling is rarely challenged by communities, either through ignorance of the situation, or because "nature" is seen to be intervening and diluting. Treated wastewater does not make up a known percentage of the drinking water, as is the case in planned schemes.

Planned schemes that include the intervention of "nature" are more acceptable to communities than those schemes that do not have a natural buffer and hence there is currently little contemplation of direct potable schemes by water recycling proponents.

The best-known direct scheme is in Namibia where it has existed since 1969 as part of the drinking water supply of Windhoek. Surrounded by desert, with an arid climate and increasing populations, there was little choice. While the community has received education and awareness campaigns over the years, and is said to now be accepting of the scheme, they did not have a say in its introduction.

Similarly, Singapore introduced NEWater in 2003, an indirect potable scheme that originally supplied industry. By 2010, the NEWater plants could supply up to 30% of demand and plans to meet up to 55% of demand by 2060. With little catchment and storage space for freshwater and supply, and bilateral agreements with Malaysia ending in 2011 and 2061, Singapore had little choice but to introduce the scheme with no say being offered to its citizens. However, they did establish an education center that set a high standard for water utilities, and their subsequent community engagement conducted by the national water agency PUB aims "to get public buy-in, to have greater ownership of and to value our water resources" [43].

Orange County Groundwater Replenishment Scheme, California, opened in 2008 after many years of discussion with the community and awareness raising and education. While it originally provided a saltwater intrusion barrier, it now replenishes the groundwater aquifer that provides drinking water to the county.

A similar indirect potable recycling (IPR) scheme is planned for Perth, Western Australia through groundwater replenishment after many years of engagement with the community. An IPR scheme was planned to be introduced in 2008 in South East Queensland, Australia, by the introduction of purified recycled water (PRW) to Wivenhoe drinking water dam. In both these cases, social research established that while community support for the indirect potable recycling schemes was around 70%, this substantially reduced to around one-third when asked about support for a direct potable scheme. The additional treatment by "nature" and the dilution effect were provided as the major reasons for support of the IPR schemes over the direct schemes.

Both the Californian and Western Australian schemes avoided the DAD approach (Decide-Announce-Defend) that had consistently led to failure in the past (e.g., the San Diego water purification project, San Gabriel Valley groundwater recharge project). While the political circumstances and the water scarcity circumstances of Namibia and Singapore allowed the introduction of the schemes with no community comment, communities with a tradition of participation demand involvement in the decisions. Successful schemes are those that have allowed many years for this process.

25.3.3 Three Australian Case Studies

The following does not provide a comprehensive description of the three case studies. The purpose is to provide a brief outline that describes the different forces at play in attempting to implement indirect potable recycling schemes. As noted in the literature review, it is fashionable to say that the community is the major impediment to the implementation of these schemes: that they are emotive and their judgments are highly influenced by the "yuck factor." Replicated psychological modeling (including in two of the following case studies) has shown a more complex decision-making process as shown in Figure 25.1. The case study descriptions refer to these decision variables while also identifying other influential factors in the outcomes.

25.3.3.1 Toowoomba

Toowoomba is a regional city in Queensland. Due to drought and a critical water shortage, water use restrictions began in Toowoomba in 2003 (Stage 1) and became progressively more severe, being Stage 5 by 2006. In 2005, the Toowoomba City Council announced its "Water Futures Initiative," which was to involve comprehensive research into a range of water supply solutions and would include a three-year community engagement program. One of the solutions under investigation was to be an IPR scheme where treated wastewater would be piped into their drinking water supply dams. The Council had applied to the National Water Commission for funding for this initiative. It had considerable political support and so the launch, attended by key local, state, and federal politicians, received a lot of attention. Hence, a protest group, CADS (Citizens Against Drinking Sewage), formed shortly after as there was the impression of a DAD approach to potable recycling. As community unrest started to occur, a number of political figures withdrew their support. The mayor at that time noted that the focus had shifted from water to politics and vested interests [54].

In response to the growing unease, the federal government announced that a referendum would be held to ascertain if the community was supportive of an IPR scheme. If the result were positive, the federal government would provide considerable funding toward the project. Four months later the poll was held and the result was negative.

The process was highly destructive for the community, dividing friends and families [3]. People who voted YES considered the NO campaign circulated far more information and started much earlier than did the YES campaign. People who voted NO considered the YES campaign to be better funded and able to provide far more information supporting their campaign than could the NO campaign. Both the YES and the NO voters reported their decisions to be based principally on fear: fear of running out of water and fear of drinking recycled water.

A quick summary follows of what went wrong in Toowoomba in relation to what is known about the formation of acceptance decisions.

- The campaign was a short four months and based on persuasion and information. It forced people to make a quick decision without adequate time to reflect and ask their questions.
- Toowoomba people did not identify with people from Orange Country or other international examples with which they were provided. This then did not satisfy *Subjective Norm* as they considered their proposed scheme to be quite different.
- The process did not build *Trust*. There were statements from officials that the community *had* to trust them. This would heighten perceptions of *Risk*.
- It was seen to be *Unfair* in that many in the community saw it as being an Australian experiment. Impressions of Poowoomba were being promoted. Further, the YES campaign was publicly funded while the NO campaign had to find the funds themselves.
- All of the above allowed *Emotion* to play a significant role.

The referendum was a recipe for disaster from the beginning, which was brought on by a range of political influences. The community was caught in the middle. What might the result have been if the original Council proposal of a lengthy public engagement process had occurred?

25.3.3.2 South East Queensland

In January 2007, the then Premier of Queensland announced that an IPR scheme would be built to supply PRW to the South East Queensland (SEQ) region via the main water storage Wivenhoe Dam. This was in response to the record drought and dwindling dam levels to below 18% of their capacity. The IPR scheme was one aspect of a regional approach, which centralized water resources management away from local governments and connected all water sources in the region, including a desalination plant, on one major grid. The SEQ community was aware of the critical water shortage and had been on high level water use restrictions for some time.

While this was a DAD approach, it was occurring in conjunction with a major scientific research initiative with the collaboration of scientists from several universities, CSIRO, and the Queensland government. It was to integrate technical and social science and the original intention was to also provide the community with some input to the implementation process of the IPR scheme, which was to be completed by the end of 2008.

A major community survey, which included the decisional modeling mentioned previously, found that almost three-quarters of the community was accepting of the IPR scheme. The main reason was that there was little choice, given the water scarcity situation. Two-thirds of people advised that they would still support the scheme even if it rained, as they did not want to be in this situation again.

A Q-Sort research method followed to examine the structure of decisions about risks associated with PRW by both community members and technical personnel and showed a similar complexity in the ways both groups judged risk. However, the technical personnel perceived that the community made emotionally driven, risk averse judgments. Both groups emphasized the need for strict legal and legislative requirements surrounding the implementation of the scheme [7]. Recommendations from this study to follow up on the best ways to discuss and communicate risk between the technical personnel and the community were subsequently rejected by the government, and it decided to conduct a traditional education and persuasion campaign instead.

Political arguments started to occur in the media with potential politicians from an opposing political party to the government starting to question the safety of PRW as an election issue. In addition, a prominent health scientist from a national university was quoted in the media as saying that the safety of IPR schemes could not be guaranteed. These events were not advantageous in building community trust in the management of risk, or in maintaining community support. In addition, the health scientists who could have provided alternatives to the negative views did not do so.

Finally, even though a high degree of community support was maintained through to the end of 2008, the then Premier of Queensland announced that PRW would only be added to Wivenhoe Dam if the level dropped below 40%. With rains received in 2008, the level in the dam was above 40% and with record floods in the years since, the IPR scheme has slipped from public discussion. The management of water resources has also reverted to local governments. In 2012, the Liberal National Party defeated the governing Australian Labor Party and by April 2014, the media reported government statements that criticized water projects that they said were symbolic of Labor waste.

25.3.3.3 Perth

By 2001, it was evident that climate change was dramatically affecting Western Australia (WA). The inflows to Perth's surface water dams had reduced from an average of 338 GL/yr in the period 1935 to 1974, to 92.4 GL/yr for the period 2001 to 2005. The average to 2010 has since dropped to 73.3 GL/yr. The decreased recharge, together with the increased use of groundwater as the main source of Perth's drinking water, has resulted in stressed aquifers. Perth's population continues to grow from 1.3 m in 2001 to 1.7 m in 2010. WA's resources boom has produced continued growth. It was evident at the beginning of the century that ensuring Perth's drinking water into the future required rain-independent sources.

In 2003, reclaimed water for indirect use was proposed in the State Water Strategy for Western Australia [17]. The development of the strategy had included an extensive community engagement program in 2002. It was suggested that IPR through groundwater replenishment, as occurred in Orange County, could be feasible in WA.

In 2005, the EPA Strategic Advice Bulletin 1199—October 2005, suggested that managed aquifer recharge (MAR) had the potential to augment drinking water supplies but advocated a precautionary approach. It advised that a trial should be conducted outside the public drinking water supply areas before any large-scale development proposal.

The Premier of WA established the Premier's Water Foundation to fund multidisciplinary, integrated research. In 2005, a three-year, AU$3 million project, determining requirements for managed aquifer recharge in Western Australia began. It involved the collaboration of scientists from five organizations: the water utility, government, CSIRO, and two universities, also required the integration of social and technical science.

The social analysis informed the ongoing community engagement and the technical analysis provided the basis for a trial, funded by the federal government, which was conducted in 2011–2012. The IPR system has been approved and is due to come online by mid-2016. It will initially inject 7 GL to the aquifer and will most likely expand to 28 GL. The Water Corporation's modeling suggests that it could supply up to 20% of Perth's drinking water by 2060 [33].

The following is a quick summary of what went right in Perth:

- The three key areas of regulation, technical work, and community engagement received detailed attention.
- Regulation: The water utility worked with the Departments of Health, Conservation and Water, individually and together, for years until all concerns were addressed, a regulatory framework established, and approvals achieved.
- Technical: The technical work was rigorous by not only addressing the local issues, but also by addressing issues that had caused major setbacks in other projects internationally.
- Community engagement has been and continues to be comprehensive, including listening to people for more than 10 years, understanding and addressing their concerns, and providing an education center at the wastewater treatment plant in association with the trial.
- Politicians have been constantly briefed and bi-partisan support has been achieved and is being maintained.
- Potential detractors were identified, contacted, and briefed to their satisfaction.

- A respected "internal champion" addressed water utility concerns and ensured organizational support.
- Regular meetings were held to identify any potential problems or opposition and strategies devised to address these.
- Nothing was left to chance. Unlike the previous two case studies, in this case it was recognized that the community was not the only potential opponent to IPR. It is also recognized that there are still 2 1/2 years before the scheme comes online which also provided the opportunity for strategic party political negotiations.

25.4 Summary and Conclusions

The early social research related to recycled water started with the attention-grabbing theory of disgust and then moved to simple studies of acceptability of various uses. It was generally viewed that recycled water was a good thing and the role of social science would be to enhance adoption through the management of the public's misplaced "yuck factor." Survey work disagreed on the role of knowledge in acceptance decisions, but mostly in terms of its relationship to other variables. The social psychological work turned to formal modeling. This started to unpack the concept of risk perception, the roles of trust, and that of emotion. Emotion was only one factor in the public's decision-making. Despite the questionnaire design involving a lot of preliminary qualitative work, the benefit of this kind of work is that it can gain a broad cross-sectional insight as to what is important to the population as a whole. This can be significant when large-scale schemes with regional impact are proposed and when the public involvement programs associated with them need to be structured and evaluated.

Later research has called for more detailed culturally and sociologically orientated work that could assist in understanding the argumentation or rhetoric from differing points of view in order to frame social learning or deliberative public involvement. Inevitably, the research involved has gained a great depth of meaning for particular projects, but with limited breadth in terms of wider public or political perspectives. Unfortunately, most agencies do not have the resources, and sometimes the skills, to put into practice what has been learned. Nevertheless, this research (some with a social psychological paradigm) has clearly indicated the role of understanding the values, expectations, and meanings that will be brought to the table before public involvement programs begin. Techniques have been developed to plan public programs better and with clearer goals.

There has been a tendency to adopt an either/or judgment on the "cross-sectional" and "in-depth" approaches where in fact both are necessary for the introduction of technologies with low probability high consequence risk such as potable reuse. It is clear that many of the general concepts derived are similar. For instance, issues of trust, risk, and accountability occur in both research paradigms.

The cross-sectional work is required to see how relationships between key variables related to acceptability change over time and the reasons for that change. Qualitative work during questionnaire development can assist with quantitative interpretation, and the quantitative data can structure and evaluate following engagement programs. Quantitative analysis is useful for decision makers who like "hard numbers" and politicians who need to operate within their electorates.

The cultural and sociological work is clearly necessary to establish long-term goals and gain mechanisms for adaptive learning. It is highly important in gaining the trust of significant stakeholders and in assisting responsive decision makers at the project or regional level.

It is apparent that institutional work is underdone for recycled water. It is clear that if this is not rectified, the adoption of social science will be limited, as the wider questions that will be posed by it cannot be accommodated in the decision-making systems. This can be exacerbated by the behavior and attitudes of water professionals and scientists themselves. Radcliffe [46] commented on the results of a survey of the professional Australian water sector in 2013, where only 48% of respondents supported recycled water for potable use: "Have we no faith in our own capabilities? No wonder our politicians do not feel confident to take potable recycling forward to the community if industry professionals aren't comfortable with the idea."

The political psychology of decision making in this area needs to be urgently addressed. This would be very difficult to research but the discussion of the previous case studies may provoke thought in this regard.

References

1. Ajzen, I. 1985. From intentions to actions: A theory of planned behaviour. In: Kuhl, J. and Beckmann, J. (eds.), *Action Control: From Cognition to Behaviour.* Springer, Berlin, pp. 11–39.
2. Attwater, R. and Derry, C. 2005. Engaging communities of practice for risk communication in the Hawkesbury water recycling scheme. *Action Research*, 3(2), 193–209.
3. Australian Research Centre for Water in Society (ARCWIS). 2008. Unpublished research notes of discussions with groups of YES and NO voters in Toowoomba. CSIRO, Perth, WA.
4. Baggett, S., Jeffrey, P., and Jefferson, B. 2006. Risk perception in participatory planning for water reuse. *Desalination*, 187, 149–158.
5. Bidwell, R., Evers, F., De Jongh, J., and Susskind, L. 1987. Public perceptions and scientific uncertainty: The management of risky decisions. *Environmental Impact Assessment Review*, 7(1), 5–22.
6. Brown, J.D. 2010. Prospects for the open treatment of uncertainty in environmental research. *Progress in Physical Geography*, 34(1), 75–100.
7. Browne, A.L., Leviston, Z., Green, M.J., and Nancarrow, B.E. 2008. *Technical and Community Perspectives of Risks Associated with Purified Recycled Water in South East Queensland: A Q-Study.* Urban Water Security Research Alliance Technical Report No. 4.
8. Browning-Aiken, A., Ormerod, K.J., and Scott, C.A. 2011. Testing the climate for nonpotable water reuse: Opportunities and challenges in water scarce urban growth corridors. *Journal of Environmental Policy and Planning*, 13(3), 253–275.
9. Bruvold, W.H. and Ward, P.C. 1972. Using reclaimed water-public opinion. *Journal of the Water Pollution Control Federation*, 44(9), 1690–1696.
10. Callaghan, P., Moloney, G., and Blair, D. 2012. Contagion in the representational field of water recycling: Informing new environmental practice through social representational theory. *Journal of Community and Applied Psychology*, 22, 20–37.
11. City of Gold Coast. 2014. Pimpama Coomera waterfuture—residents and businesses. http://www.goldcoast.qld.gov.au/environment/pimpama-coomera-waterfuture-residents-and-businesses-7904.html. Retrieved January 11, 2014.
12. Comprasano, A. and Modica, C. 2010. Experimental investigation on water saving by reuse of washbasin grey water for toilet flushing. *Urban Water Journal*, 7(10), 17–24.
13. Dolnicar, S. and Hurlimann, A. 2009. Drinking water from alternative water sources differences in beliefs, social norms and factors of perceived behavioural control across eight Australian locations. *Water Science and Technology*, 60(6), 1433–1444.
14. Dolnicar, S. and Hurlimann, A. 2010. Water alternatives—who and what influences public acceptance? *Journal of Public Affairs*, 11(1), 49–59.
15. Dolnicar, S., Hurlimann, A., and Grun, B. 2011. What effects public acceptance of recycled and desalinated water? *Water Research*, 45(2), 933–943.
16. Dolnicar, S., Hurlimann, A., and Nghiem, L.D. 2010 The effect of information on public acceptance—the case of water from alternative sources. *Journal of Environmental Management*, 91, 1288–1293.
17. Government of Western Australia. 2003. *Securing our Water Future: A State Water Strategy for Western Australia.* http://www.water.wa.gov.au/PublicationStore/first/41070.pdf.
18. Hampton, G. 2010. Discursive evaluation of water recycling. *Qualitative Research Journal*, 10(2), 65–81.
19. Hampton, G. 2012. Social learning: Critical reflection and the perception of facticity in deliberation on water reuse. *Review of European Studies*, 4(5), 181–190.
20. Hughes, S. 2013. Authority structures and services reform in multilevel urban governance: The case of wastewater recycling in California and Australia. *Urban Affairs Review*, 49(3), 381–407.
21. Hurlimann, A. and Dolnicar, S. 2010. Acceptance of water alternatives in Australia. *Water Science and Technology*, 61(8), 2137–2142.
22. Jeffrey, P. and Jefferson, B. 2003. Public receptivity regarding "in-house" water recycling: Results from a UK survey. *Water Science and Technology: Water Supply*, 3(3), 109–116.

23. Kemp, B., Randle, M., Hurlimann, A., and Dolnicar, S. 2012. Community acceptance of recycled water: Can we inoculate the public against scare campaigns? *Journal of Public Affairs*, 12(4), 337–346.

24. Lavine, H., Lodge, M., and Freitas, K. 2005. Threat, authoritarianism and selective exposure to information. *Political Psychology*, 26(2), 219–244.

25. Leviston, Z., Browne, A.L., and Greenhill, M. 2013. Domain-based perceptions of risk: A case study of lay and technical community attitudes toward managed aquifer recharge. *Journal of Applied Social Psychology*, 43, 1158–1176.

26. Lukasiewicz, A., Bowmer, K.H., Syme, G.J., and Davidson, P. 2013. Assessing government intentions for Australian water reform. *Society and Natural Resources*, 26(11), 1314–1309.

27. Marks, J. 2004. Advancing community acceptance of reclaimed water. *Water*, 31, 46–51.

28. Marks, J. 2006. Taking the public seriously: The case of potable and nonpotable reuse. *Desalination*, 187, 137–147.

29. Marks, J., Martin, B., and Zadoroznyi, M. 2008. How Australians order acceptance of recycled water: National baseline data. *Journal of Sociology*, 44, 83–99.

30. Marks, J. and Zadoroznyi, M. 2005. Managing sustainable urban water reuse: Structural context and cultures of trust. *Society and Natural Resources*, 18, 557–572.

31. McGuire, W. 1961. The effectiveness of supportive and refutational defenses on immunizing and restoring beliefs against persuasion. *Sociometry*, 24, 184–197.

32. McKay, J. and Hurlimann, A. 2003. Attitudes to reclaimed water for domestic use. Part 1: Age. *Water*, 30(5), 45–49.

33. Mercer, D. 2014. Perth faces wastewater reality. *The West Australian*. https://au.news.yahoo.com/thewest/a/24218183/perth-faces-wastewater-reality/. Retrieved June 16, 2014.

34. Mikhailovich, K. 2009. Wicked water: Engaging with communities in complex conversations about water recycling. *EcoHealth*, 6, 324–330.

35. Mitchell, V.G. 2006. Applying integrated urban water management concepts: A review of the Australian experience. *Environmental Management*, 37(5), 589–605.

36. Nancarrow, B.E., Leviston, Z., Po, M., Porter, N.B., and Tucker, D.I. 2008. What drives communities' decisions and behaviours in the reuse of wastewater. *Water Science and Technology*, 57(4), 485–491.

37. Nancarrow, B.E., Leviston, Z., and Tucker, D.I. 2009. Measuring the predictors of communities' behavioural decisions for potable reuse of wastewater. *Water Science and Technology*, 60(12), 3199–3209.

38. Office of Water. 2010. *2010 Metropolitan Water Plan*. NSW Government, Sydney.

39. Po, M., Kaercher, J., and Nancarrow, B.E. 2004. *Literature Review of Factors Influencing Public Perceptions of Water Reuse*. Australian Conservation and Reuse Research Program. AWA Sydney, CSIRO, Perth.

40. Porcher, S. 2014. Efficiency and equity in two part tariffs: The case of residential water rates. *Applied Economics*, 46(5), 539–555.

41. Potter, J. 1998. Discursive social psychology: From attitudes to evaluative practice. *European Review of Social Psychology*, 9, 233–266.

42. Price, J., Fielding, K., and Leviston, Z. 2012. Supporters and opponents of potable recycled water: Culture and cognition in the Toowoomba referendum. *Society and Natural Resources*, 25, 980–995.

43. PUB. 2013. PUB, Singapore's National Water Agency website. http://www.pub.gov.sg/water/Pages/singaporewaterstory.aspx. Retrieved January 11, 2014.

44. Radcliffe, J. 2004. Water recycling options for the thirsty country. *Australasian Science*, 25(6), 38–42.

45. Radcliffe, J. 2010. Evolution of water recycling in Australian cities since 2003. *Water Science and Technology*, 62(4), 792–802.

46. Radcliffe, J. 2014. Preparing for the coming of direct potable water recycling. *Water*, 40(1), 8–9.

47. Roth, W-M., Riecken, J., Pozzer-Ardenghi, L., Mc Millan, R., Storr, B., Tait, D., Bradshaw, G., and Penner, T.P. 2004. Those who get hurt aren't always heard: Scientist-resident interactions over community water. *Science, Technology and Human Values*, 29(2), 153–183.

48. Rozin, P. and Fallon, A.E. 1987. A perspective on disgust. *Psychological Review*, 94, 23–41.

49. Russell, S. and Lux, C. 2009. Getting over yuck: Moving from psychological to cultural and sociotechnical analyses of responses to water recycling. *Water Policy*, 11, 21–35.

50. Stenekes, N., Colebatch, H.K., Waite, T.D., and Ashbolt, N. 2006. Risk and governance in water recycling: Public acceptance revisited. *Science, Technology and Human Values*, 31(2), 107–134.

51. Syme, G.J. 1995. Community acceptance of risk: Trust, liability and consent. In: Heinrichs, P. and Fell, R. (eds.), *Acceptable Risks for Major Infrastructure*. A.A. Balkema, Rotterdam, the Netherlands, pp. 31–40.

52. Syme, G.J. and Nancarrow, B.E. 2006. Social psychological considerations in the acceptance of reclaimed water for horticultural irrigation. In: Stevens, D. (ed.), *Growing Crops with Reclaimed Wastewater*. CSIRO Publishing, Collingwood, Victoria, pp. 189–196.

53. Szlompka, P. 1999 *Trust: A Sociological Theory*. Cambridge University Press, Cambridge, UK.

54. Thorley, D. 2007. Toowoomba recycled water poll. *Reform*, 89(Summer), 49–51.

55. Troy, P. 2011. The urban water challenge in Australian cities. In: Grafton, R.Q. and Hussey, K. (eds.), *Water Resources Planning and Management*. Cambridge University Press, Cambridge, pp. 463–482.

56. Wong, A.K. and Gleick, P.H. 2000. Overview to water recycling in California: Success stories. *Environmental Management and Health*, 11(3), 216–238.

26

Urban Water Reuse in Tourism Areas

António Albuquerque
University of Beira Interior

José Saldanha Matos
University of Lisbon

26.1 Introduction

Over the past decades, treated wastewater (reclaimed water) has been used as an alternative to potable water for a range of uses such as irrigation (landscapes, golf courses, and agricultural fields), aquifer recharge, industry applications, stream flow feeding, and nonpotable urban applications. In that period, there have been significant advances in reuse technologies, and an increase in the implementation of either rules or guidelines for water reuse.

The expected increase of population and changes in land use, sea level rise, and local climatic dynamics will continue demanding water supply challenges in many areas of the world, which can negatively affect the regions that have important economic benefits from tourism. These circumstances have forced planners to consider nontraditional water sources while maintaining environmental stewardship [9]. The water generated in tourist areas (namely stormwater and domestic wastewater), after treated and converted to reclaimed water, is seen as a valuable and alternative water resource to conventional water supplies for a range of applications.

According to the European Commission [13], droughts have increased in number and impact and in 2011 and 2012, water scarcity has affected large parts of Southern, Western, and even Northern Europe, with impact on economic activities connected with tourism. The number of areas and people affected is around 20% and the total costs of droughts amounted to 100 billion €. Tourism is a priority matter for the sustainable development in many countries and one of the concerns is the increase of water shortages as a consequence of the expected climate change [15,43].

Tourism generates one of the biggest pressures on water needs that coincides with the necessity to manage decreasing water resources more efficiently. Water consumption presents very atypical patterns compared with urban areas, which are related to seasonal people concentration (normally coincides with the period in which water resources are scarce), spatial concentration on coast areas (both urbanized and remote sites, normally presenting scarcity of local water resources and often located on sensitive natural sites), and use of facilities that consume large volumes of water (e.g., aquatic parks, swimming pools, golf courses, and spas). Additionally, the water and wastewater systems are designed for dealing with large variation of flow rate, which results in frequent problems for water management authorities.

As tourism areas continue to grow, pressure on local water sources will continue to increase and significant environmental, economic, and social impacts can arise where local freshwater supplies are limited or are available only with large capital investment. Mature tourist resorts over the world are developing activities that increase permanent water demand for facilities and leisure structures (e.g., golf courses, spas, aquatic parks, swimming pools, and gardens), as reported, for example, in Australia [21], French Polynesia [23], Portugal [41], Spain [18], Israel, Jordan, Morocco, and Tunisia [12].

The water-energy nexus points out that water and energy are mutually dependent [33]. This concept is also applied to water reuse because energy efficiency and sustainability are important keys for its application, which makes this type of water resource so essential to sustainable water management. The energy-water connection is particularly strong in regions with important water scarcity. Water reuse can significantly reduce energy consumption by eliminating some potable water treatment requirements and associated water conveyance (because reclaimed water can offset potable water use for some applications), protect dry areas from drought, lower consumers' utility bills, and reduce global warming pollution [35]. According to the California Energy Commission [5], approximately 20% of the electricity consumed in the state of California is associated with water-related energy use, including potable water conveyance, storage, treatment, distribution and wastewater collection, treatment, and discharge, and the energy required for treatment and transport of potable water is much greater than the amount required for polishing wastewater for reuse. Water reuse fits well the concept of sustainability because it allows other sources of water to remain in the environment and be preserved for future uses while satisfying some water demand of the present. The water-food nexus is also obvious in the face of the potential of reclaimed water to be used for farming purposes, increasing productivity, and contributing to employment and local economic development.

The integration of water reuse in water management strategies will contribute to reducing discharges to receiving waters and reducing reliance on natural water sources to meet water demands. In tourism areas, there are several opportunities for reusing reclaimed water produced from domestic wastewater, stormwater, and greywater. Reuse applications in tourist areas have been greatly developed over the past decade for landscape irrigation, urban applications (e.g., garden watering, car washing, and toilet flushing), wetland nature reserves, hydroponics applications, agricultural irrigation (e.g., turf farms, vineyards, and silviculture), and golf course irrigation [1,21]. Ecological treatment systems (e.g., constructed wetlands, sand filters, or rock filters) have been shown to be suitable for producing reclaimed water from domestic wastewater [25,30,37,50], stormwater [2,36,44] and greywater [6,31]. Therefore, the objective of this chapter is to address the potentialities and opportunities for reusing treated domestic wastewater, stormwater, and greywater in tourist areas, including aspects of risk control, regulation, and economical evaluation.

26.2 Urban Wastewater Characteristics and Challenges of Reuse

26.2.1 Urban Wastewater Characteristics

Urban waters produced at tourism areas are essentially composed of domestic wastewater and stormwater. These waters can be collected separately (by separate sewer systems) or jointly (by combined sewer systems). Older sewer systems typically constructed before the 1950s are combined, while more recent systems were designed as separate or partially separate systems. In

many countries, for example, in Mediterranean countries, the tourism season occurs mainly during summer, with very low average precipitation. Impact of tourism can lead to triplicate the average flows and aggravate the hourly peak factor, affecting sewer performance (e.g., risks of overflows) and the efficiency of treatment plants. In any case, the wastewater presents a variety of substances (Table 26.1), populated by diversity of microorganisms, many of which are of fecal origin and some are pathogenic, which can be considerably removed through secondary and tertiary treatment. However, these treatment levels do not remove all the harmful compounds and pathogens, and a residual load is usually discharged into water resources. Therefore, surface water and groundwater abstraction for producing drinking water normally configures a case of indirect and not planned water reuse.

Urban waters can be treated through a variety of physical, chemical, and biological processes, as presented in References 46 and 1, in order to produce final reclaimed water for discharging into a water stream or for reuse. The typical quality of reclaimed water after secondary treatment is presented in Table 26.2. Comparing Table 26.1 and Table 26.2, it can be noted that

TABLE 26.1   Urban Water Characteristics

Parameters	Domestic Wastewater + Stormwater[a,b,c]	Domestic Wastewater[b,c,d]	Stormwater[e,f,g]
pH	6.9–8.8	6.7–8	5.2–8
Conductivity (μS/cm)	820–1 320	540–1100	12–1480
TSS (mg/L)	200–450	390–1230	35–90
NO_3-N (mg/L)	0–3.3	0–4	0.4–2.2
NH_4-N (mg/L)	12–40	20–80	0.4–2
TN (mg/L)	20–60	36–88	0.9–4.5
TP (mg/L)	4–13	4–15	0.2–0.9
BOD_5 (mg/L)	110–400	250–1100	8–30
COD (mg/L)	250–800	450–1900	40–175
Oil and grease (mg/L)	90–475	40–150	3–60
Al (mg/L)	0.1–1.1	0.3–1	—
Cd (mg/L)	0.01–0.1	0.001–0.004	0.05–0.3
Cr (mg/L)	0.01–0.1	0.01–0.04	0.05–0.14
Cu (mg/L)	0.02–0.2	0.03–0.08	0.05–0.8
Fe (mg/L)	0.6–8	0.2–4	—
Ni (mg/L)	0.02–0.1	0.01–0.04	0.3–0.6
Pb (mg/L)	0.01–0.2	0.02–0.08	0.05–0.9
Zn (mg/L)	0.1–0.2	0.1–0.3	0.3–1,2
Total coliforms (MPN/100 mL)	10^7–10^9	10^{11}–10^{13}	10^2–10^6
Fecal coliforms (MPN/100 mL)	10^5–10^8	10^9–10^{11}	10^1–10^4
Fecal streptococci (MPN/100 mL)	10^4–10^6	10^6–10^8	10^2–10^4

[a] Marecos do Monte and Albuquerque [26].
[b] Asano et al. [1].
[c] Tchobanoglous et al. [45].
[d] von Sperling [47].
[e] EPA [6].
[f] Pitt [38].
[g] Hvitved-Jacobsen and Yousef [19].

TABLE 26.2 Typical Characteristics of Reclaimed Water after Secondary Treatment of Urban Waters

Parameters	Reclaimed Water[a,b,c,d,e]
pH	7.2–8
Conductivity (μS/cm)	300–900
TSS (mg/L)	10–50
NO_3-N (mg/L)	0–3
NH_4-N (mg/L)	10–20
TN (mg/L)	5–15
TP (mg/L)	5–10
BOD_5 (mg/L)	20–40
COD (mg/L)	80–140
Al (mg/L)	0.04–1
Cd (mg/L)	0.003–0.14
Cr (mg/L)	0.02–0.1
Cu (mg/L)	0.01–0.25
Fe (mg/L)	0.5–3.5
Ni (mg/L)	0.01–0.15
Pb (mg/L)	0.01–0.1
Zn (mg/L)	0.1–0.2
Total coliforms (MPN/100 mL)	10^4–10^7
Fecal coliforms (MPN/100 mL)	10^3–10^5
Fecal streptococci (MPN/100 mL)	10^2–10^4

[a] Tchobanoglous et al. [45].
[b] EPA [7].
[c] WHO [49].
[d] Asano et al. [1].
[e] Marecos do Monte and Albuquerque [26].

TABLE 26.3 Greywater Characteristics

Parameters	Bathroom Water[a,b]	Laundry Water[a,b]	Greywater[c]
pH	6.4–8.1	9.3–10	5–9.9
Conductivity (μS/cm)	82–250	190–1400	330–580
Color (Pt/Co)	60–100	50–70	–
Turbidity (NTU)	60–240	50–210	20–140
TSS (mg/L)	48–120	88–250	20–1500
NO_3-N (mg/L)	0.05–0.20	0.10–0.31	0–4.9
NH_4-N (mg/L)	0.1–15	0.1–1.9	0.1–8.1
TKN (mg/L)	4.6–20	1.0–40	0.6–50
TP (mg/L)	0.11–1.8	0.062–42	0.3–35
BOD_5 (mg/L)	76–200	48–290	33–620
Oil and grease (mg/L)	37–78	8.0–35	–
Alkalinity (mg/L)	24–43	83–200	125–382
Ca (mg/L)	3.5–7.9	3.9–12	4–824
Mg (mg/L)	1.4–2.3	1.1–2.9	1–15
Na (mg/L)	7.4–18	49–480	32–1090
K (mg/L)	1.5–5.2	1.1–17	4.5–13
Al (mg/L)	<1.0	<1.0–2.1	0.02–0.67
As (mg/L)	0.001	0.001–0.007	–
B (mg/L)	<0.1	<0.1–0.5	–
Cd (mg/L)	<0.01	<0.01	<0.01
Cl (mg/L)	9.0–18	9.0–88	3.1–136
Cu (mg/L)	0.06–0.12	<0.05–0.27	0.15
Fe (mg/L)	0.34–1.1	0.29–1.0	0.79–28
Se (mg/L)	<0.001	<0.001	–
Zn (mg/L)	0.2–6.3	0.09–0.35	0.38
Total coliforms (MPN/100 mL)	500–2.4×10^7	2.3×10^3–3.3×10^7	–
Fecal coliforms (MPN/100 mL)	170–3.3×10^3	110–1.09×10^3	17–1.6×10^5
Fecal streptococci (MPN/100 mL)	79–2.4×10^3	23–2.4×10^3	19–1.51×10^3

[a] Boal et al. [4].
[b] Lechte et al. [24].
[c] Jeppesen [20].

heavy metal and pathogen removal is not significant and polishing treatment may be needed (e.g., membrane technology and disinfection) if higher quality of reclaimed water is required. In remote resort areas, onsite wastewater treatment systems and ecological treatment processes are, in general, a more sustainable solution, using land treatment options, including wetlands and pond systems. Stormwater runoff presents higher quality than domestic wastewater or urban water, but can still need some treatment for removing trash racks, sediment traps, pathogens, and inorganic and organic compounds.

The wastewater produced by bathtubs, bathroom washbasins, clothes washing machines, and laundry tubs, known as greywater, presents advantages for reuse because it does not typically include kitchen and toilet wastewater, which have higher pollutant content and pathogens. Although feces and urine are not present, the quality of greywater is highly variable and can contain chemicals and pathogens that can be harmful for users and the environment (Table 26.3). Therefore, suitable treatment is also needed before its reuse.

According to Kavanagh [21], in Australian cities, stormwater runoff is equivalent to the amount of drinking water that is supplied. More than half of the drinking water is used for lower water quality purposes including garden watering and toilet flushing. Stormwater presents a high potential to be reused for non-drinking purposes and to markedly reduce the demand for drinking water [32].

Therefore, domestic wastewater, stormwater, and greywater are urban waters with potential to be reused after treatment in the form of reclaimed water.

26.2.2 Challenges of Reuse

The main constraints of water reuse in densely populated tourist areas are the greater probability of unknowingly exposing the public to reclaimed water and the lack of adequate regulations for allowing application in uses such as urban applications and golf course and landscape irrigation [23]. Many resorts will have to cope with increasing water demand and tourist flows, changing in climatic variables, and more droughts [18].

The Mediterranean region is one of the world's top tourism destinations and tourist flows to this region are constantly increasing (4% of the world total in 1990 and 6% in 2005, according to the European Commission) [12]. Water resources are limited and unequally distributed in space and time, thus exacerbating the pressures. Water consumption and water reuse by the tourism sector is not well documented by statistics and, therefore, it is difficult to assess the impact it has on water resources. Nevertheless, several water reservoirs are already under pressure and water supply increasingly relies on reclaimed water reuse or desalinized water, being necessary new tools and water management strategies for the tourism sector [12,15]. The expected trend for the near future is an increase in water consumption to satisfy the growth of tourists (396 million persons are expected in 2025 compared to 166 million in 1995), which look at comfort requirements and water consuming facilities such as swimming pools, aquatic sports, spas, and golf courses.

The residential sector has been grown around the tourist centers, with most houses having been built as second homes, often with high water needs for garden watering and swimming pools, which creates an increase in water consumption in the season of low rainfall and high evapotranspiration [10,43]. Additionally, the changes in garden design, using imported garden plants, and some garden irrigation practices are leading to significant differences in water consumption of tourist landscapes differentiated by land use pattern [18].

Remote tourist resorts normally are not connected to public water supply networks or sewer networks and wastewater treatment solutions normally involve natural systems such as septic tanks and filtration devices (typically filtration trenches or soak ways) or constructed wetlands, sometimes coupled with a disinfection system (e.g., UV or maturation ponds). This association of technology is normally sufficient for getting reclaimed water with the minimal quality for garden watering, toilet flushing, and car and boat washing.

Wastewater treatment technologies are now advanced enough to mitigate pathogens and chemical contaminants in reclaimed water, offering a multitude of process combinations that can be tailored to meet specific water quality objectives. Therefore, the concept of "fit for purpose" is highlighted to emphasize the efficiencies realized by designing reuse for specific end applications [34].

Health safety and economic viability are essential for some water reuse practices and can be achieved through suitable combinations of wastewater treatment technologies and best practices in the application sites. Two good examples are agricultural irrigation, which demands food security, and golf course and landscape irrigation, which allows contact between reclaimed water and users.

The public education and acceptance is essential for the success of setting up wastewater reuse projects in tourist areas. Periodic public presentations and discussion of projects should be organized with municipalities, professional organizations, end-users associations, consultants, public health authorities, and public representatives in order to evaluate the expected economic, social, and environmental benefits.

26.3 Potentialities and Opportunities for Reusing Reclaimed Water

Domestic wastewater, stormwater, and greywater produced at tourist areas, after converted in reclaimed water, are suitable for the irrigation of public parks, lawns and landscape, urban nonpotable applications (e.g., toilet flushing, washing of cars, municipal equipment, roads and pavements, and garden watering), and filling water nature reserves as well as for some agricultural activities.

According to Kavanagh [21], sustainable water management at tourist resorts should incorporate water reuse measures associated with water saving measures such as

- Rainwater should be reused as a water source.
- Extracted water should be returned to the source with no loss in quality.
- Treated wastewater should be reused for all nonpotable purposes.
- If possible, greywater should be kept separate from blackwater.
- Water saving devices should be used where possible.
- Dry composting toilets should be used.
- Xeriscape (low water) gardens should be employed.
- Wastewater should be treated according to the standards for the defined uses, utilizing a minimum of chemicals and energy.

Irrigation may use direct reuse, normally involving a reclaimed water distribution network, or indirect reuse, normally involving the discharge of reclaimed water into a water stream and then retrieved to be used again, providing additional benefits such as wildlife and wildfowl habitat, water quality improvement, flood diminishment, and fisheries breeding grounds. Indirect reuse also provides a polishing treatment because it allows an additional removal of residual pollutant load and pathogens elimination.

Golf courses represent another opportunity for water reuse. In many ways, it can be seen as agricultural fields, having a soil/plant/atmosphere consortium that is linked by the water needed for the plants to grow and presenting high evapotranspiration rates. The water consumption can rise up to 2500 m^3/day in a dry, hot season, depending on its dimension, local climate, water retention properties of substrata, and water requirements of turf [41]. A standard 18-hole golf course has an irrigated surface of 54 ha, and the annual average water consumption is approximately 0.3 Hm^3. Presently, water authorities in several countries (e.g., France, Spain, Portugal, and the United States) are already imposing the use of reclaimed water for the irrigation of golf courses and several projects submitted to environmental impact assessment evaluation have been approved conditionally to the use of reclaimed water.

The combination of water with low salinity and a high concentration of exchangeable sodium (which is common in desalinated regenerated water) can have a negative impact on the structural stability of soil, which loses fertility over the medium term because of losing its capacity to drain away water. Estévez et al. [11] have analyzed data and practices of reclaimed water reuse in golf courses of Canary Islands (Spain) for over 25 years, having concluded that is better to adapt the local plant species and varieties watered, instead of choosing species that are more tolerant of salinity. This practice would reduce the polishing requirements for high-quality reclaimed water producing. Other measures for getting water savings would include the adjustment of the volume and frequency of watering according to plant needs and evapotranspiration.

Other projects have considered the production of high-quality reclaimed water for several applications. The island of Bora Bora has implemented a reuse project [23], which has involved the upgrading of the wastewater treatment plant (WWTP) of Povai with an advanced membrane treatment to produce high-quality reclaimed water for toilet flushing in hotels, public spaces cleaning and landscape irrigation, boat and sea plane washing, filling of fire reservoirs of fire protection boats, washing of construction engines and pressure tests of concrete, and fire protection in buildings. The project has generated a good cooperation between the stakeholders, which was essential for the success of the project.

A study set up in the Llobregat Delta in Spain [16] has shown how water reuse and exchange can be a cost-effective approach to managing water scarcity at the basin level. The cost-benefit analysis has pointed out a total water-exchange cost of 5.2 million € per year, with an annual benefit of 14.4 million € from an additional 13 million m³ of freshwater released from agricultural use. Farmers would save about 351,000 € per year due to the reduction of costs with water pumping systems and fertilizer use. The annual global savings would be approximately 9.5 million €.

26.4 Regulation and Risk Control

Water reuse practices require strict control, which can be set up through the quality control of the reclaimed water or the risk management of the reuse procedures. Before reusing reclaimed water, it is necessary to evaluate if that water resource is safe for reuse, which can be done by applying traditional methods of comparing it with standards or using hazard/risk assessment tools as mentioned by Salgot et al. [41].

Regulations have been created for helping the control of reclaimed water quality and the reuse procedures. However, their efficiency depends on economic and social circumstances, legal capacitation by implicated entities and administrative bodies, human health/hygiene level (endemic illnesses, parasitism), technological capacity, previously existing rules and/or criteria, analytical capacity, risk groups possibly affected, and technical and scientific opinions [40]. Therefore, three groups of factors can be considered: technical and technological issues (e.g., analytical, treatment methods, capacity, and knowledge), legislative and economic issues (e.g., criteria, socioeconomic, and legal

competence), and health-related issues (e.g., sanitary state, diseases, and risk groups).

Regulations have been based traditionally on microbiological quality considerations [41]. More recently, chemical and toxicological considerations have been integrated, but qualitative aspects still need to be adjusted, not necessarily by increasing the number of analyses or adopting more restricted values, but rather by appropriately implementing complementary tools, such as risk assessment or good reuse practices.

Although most of the world regulations on water reuse are concerning agricultural applications, several experts and international entities have published standards for urban applications, and golf course and landscape irrigation [1,3,8,9,26,45,49], which can be used in tourist areas.

Reclaimed water that complies with reuse standards for a given use should have enough quality to reduce risks to an acceptable level. Nevertheless, as standards change for different countries, it is not generally feasible to apply uniform methodologies and analytical control worldwide or even at a regional level, and risk-related tools are recommended to apply for risk control. According to Salgot et al. [41], management systems based on the hazards/risks associated with wastewater treatment practices, recycling, and application are being developed for some reuse applications such as irrigation. The in situ risk evaluation normally follows a specified pattern, which can include real site observations, interpretation of quality analysis, and health risk assessment, including the hazard analysis and critical control points (HACCP).

The HACCP system allows the evaluation of hazards associated with all the steps in the water reuse system, its control, and the identification of the critical points. It can help the water authorities to undertake water-related inspections, while promoting self-control activities, and requires a compromise and involvement of stakeholders. It involves the following steps [48,49]: conduction of hazard analysis, identification of the critical control points (CCP; Figure 26.1), establishment of target and tolerance levels (e.g., identification of hazards that must be controlled), setting up a monitoring system (e.g., controls, critical limits, and procedures for monitoring), definition of corrective actions and verification procedures (validation of the HACCP plan), and production of documentation.

26.5 Economical Evaluation and Justification of Reuse Options

The economics of water reuse normally involve calculations based on different approaches, but consider the key variables, such as the cost of wastewater treatment, storage, and transport (investment and operational and maintenance costs). However, most focus on cost-benefit tools as presented in the works of Heinz et al. [16] and Hernández-Sancho et al. [17]. Costs for reclaimed water production from domestic wastewater are normally higher than the costs associated with stormwater recycling, rainwater collection, and treated greywater [42]. Costs

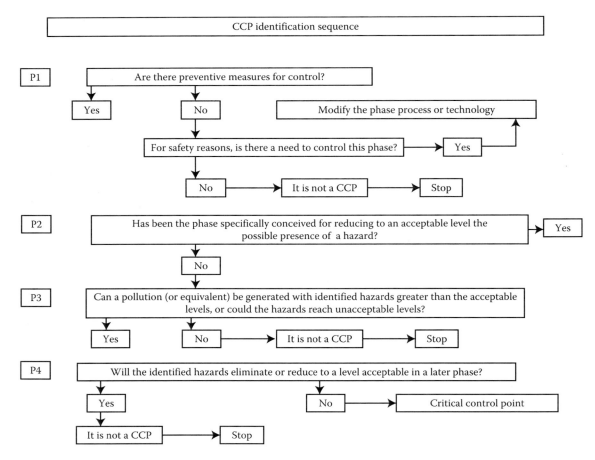

FIGURE 26.1 Procedure for the identification of Critical Control Points. (From Salgot, M., Priestley, G. and Folch, M. 2012. Golf course irrigation with reclaimed water in the Mediterranean: A risk management matter. *Water*, 4, 389–429.)

differ considerably, according to the size and sophistication of the reuse options and the seasonality of its use. For example, the costs of water reuse in golf courses can range from several cents to over 1 € per cubic meter, including analytical costs and mortgage of the treatment facilities [39].

Economic instruments, policy on water management, clearer regulations, and financial incentives are essential for enrolling stakeholders and the viability and cost competitiveness of water reuse projects. Water reuse is still suffering from the under-evaluated and subsidized conventional water resources and constraints for the application of "full-cost recovery" and "polluter-pays" principles.

Operation and maintenance costs of recycling facilities normally include costs with treatment technology (e.g., replacement of membranes and/or UV lamps) and of equipment of reservoirs, pumping systems, and distribution networks), being labor, chemicals, and energy. A project developed in Bora Bora [23] has shown that the main components of operation costs were labor (46%), repair and maintenance (21%, including membrane replacement), energy consumption (14%), chemicals (12%), and water quality monitoring (4%). The average energy consumption was 0.3 kWh/m³ at nominal flow and 0.95 kWh/m³ at 30%

of the hydraulic capacity of membrane treatment. Labor costs were close to the typical values in conventional activated sludge systems (45 ± 5%) and energy costs were lower compared to typical values of secondary treatment 25 ± 5% and tertiary treatment with microfiltration (MF) and reverse osmosis (RO) treatment 26%–32% [22]. This project has brought economic benefits for maintaining a number of economic activities such as construction, boat cleaning, and pleasant landscape in conditions of severe drought with strong restrictions on potable water consumption. For luxury hotels, the unit cost of high-quality reclaimed water was 2.5–3 times less expensive than the drinking potable water.

26.6 Case Studies in a Mediterranean Country

26.6.1 Overview

In this section, adapted from Viegas et al. [46], some reflections are presented about water reuse in Portugal, a Mediterranean country with a significant tourism industry, focusing on two representative case studies in water stressed regions: the Algarve, a very popular tourist destination principally during summertime, densely populated with urban and rural parishes, and Lisbon, the capital of the Portugal with tourism spread throughout the year.

Portugal has abundant potential freshwater resources. However, there is a shortage of its availability given the irregularity of their occurrence, both along the territory and over the year. The climate variability along the Portuguese territory is the first reason for the pronounced spatial differences that are observed in the freshwater availability. In a great part of the country, the water balance is deficient with the evapotranspiration exceeding the sum of infiltration and runoff. Moreover, the consequences of the climate diversity on freshwater availability are greatly amplified by other factors including the demographic characteristics and the pressure of some important economic activities, namely tourism.

Considering the population distribution along Portugal, a high density in major urban coastal areas is observed. The high potential for water reuse in large urban systems, namely for urban nonpotable uses, is then clear.

Tourism has been increasing heavily its weight in the economy of Portugal. With a high concentration during the summer, it significantly contributes to the seasonal stress of water scarcity, in particular in the Great Lisbon coastal area and in the Algarve. Also, the development of the golf industry, demonstrated by the growing number of golf courses being created, in addition to the high water needs for their maintenance and the current environmental policy, make golf course irrigation with treated wastewater a strategic goal relating resource sustainability. Actually, the Environmental Impact Declarations of new golf courses often impose its irrigation with treated wastewater. The case studies of these two Portuguese regions are given as examples.

26.6.2 Algarve Region Case Study

The Algarve region is a paradigmatic example in Portugal because tourism and golf are the structural basis of its economy while it suffers severe or extreme droughts from time to time. The wastewater utility for the region (Águas do Algarve, S.A.) explores 58 WWTPs, accounting for the treatment of 26 hm³ in 2007, 24 hm³ with secondary treatment, and 2 hm³ with tertiary treatment [14,27]. With a total of 31 golf courses in operation and 52 foreseen in the near future, the maintenance of the Algarve as a golf destination struggles with the high water consumption of this activity. However, the current environmental policy of treated wastewater use and the vicinity of 15 WWTPs to the golf courses open the opportunity of supplying the irrigation demand with treated wastewater, which proves to be sufficient except for the months of April, May, and October [28]. Studies performed [14,27–29] show that the available technology (traditional biological treatment) is able to provide the quality needed, except for a couple of wastewater systems with salt intrusion, which can increase the water salinity to 3 dS/m, cases for which desalination post-treatment is needed. The bulk supply of treated wastewater for golf course and landscape irrigation

in the Algarve region is still not fully implemented mainly due to economic reasons [14].

26.6.3 Lisbon Region Case Study

Lisbon is the largest and most populated city in Portugal, with a territory of 84 km², about 600,000 inhabitants and a fluctuating population of about 1.2 million people. SimTejo, S.A. is the wastewater utility responsible for the collection, treatment, and disposal of urban wastewater from six municipalities of the Lisbon region: Lisbon, Loures, Vila Franca de Xira, Amadora, Mafra, and Odivelas, covering approximately 1000 km². Today, SimTejo serves a population of about 1.5 million inhabitants, and operates 29 WWTPs, 74 pumping stations, and 245 km of sewerage systems, treating an average flow rate of 110 hm³/yr.

Given its urban and peri-urban profile, there are many potential applications for water reuse in the region, namely: (a) agricultural irrigation: nurseries and crop irrigation; (b) landscape irrigation: golf courses, industrial parks, public parks, cemeteries, and roadside plantings; (c) industrial: cooling water, fire protection, heavy construction, and process water; (d) nonpotable urban uses: air conditioning, car wash and laundries, decorative fountains, municipal urban services, toilet flushing, and sewer flushing; and (e) recreational/environmental uses: artificial lakes, fisheries, and wetlands. The company started using reclaimed water in 2001, first for internal use (e.g., polymer dilution, facilities cleaning, small gardens, scum and foam control in most WWTPs) and later for street cleaning. Today, SimTejo provides water reuse to three municipalities and the goal is to extend this practice to all the municipalities served.

In Lisbon (Chelas WWTP), besides internal use, treated wastewater is delivered to the municipality for landscape irrigation and street cleaning. Mafra municipality uses treated wastewater for landscape irrigation. In Loures, the vicinity of Frielas WWTP to a big commercial area (IKEA Loures) has created the opportunity to supply treated wastewater for cooling purposes in the air conditioning system (3200 m³/day in the summer and 1280 m³/day in the winter), representing the most impacting case of water reuse in the Lisbon area thus far.

The complete rehabilitation and upgrading of Alcântara WWTP, which now includes filtration and UV disinfection, together with the rehabilitation and expansion of the trunk sewer network allowed, through a cost synergy, the installation of a water reuse supply system to Lisbon downtown (Figure 26.2). This system offers new possibilities for treated wastewater use, namely for landscape irrigation, street cleaning, combined sewer flushing, and service water in wastewater pumping stations.

Next to Beirolas WWTP is the renovated east part of Lisbon—the Parque Expo area. When the renovation works took place, a dual plumbing system was installed, already prepared for landscape irrigation using treated wastewater from Beirolas WWTP, planned for the near future.

There is a growing trend toward the use of treated wastewaters in different countries of the world, with the major drivers being water scarcity and seasonal stress, in part due to tourism,

FIGURE 26.2   Alcântara WWTP and the treated wastewater supply system to Lisbon downtown.

and one of the major barriers is the economical feasibility of the projects.

26.7 Summary and Conclusions

Weather changes and population increases in touristic regions clearly dictate higher demands for water, which in arid and semi-arid regions may be considered a limited resource, usually with groundwater resources under heavy pressure. In most of those zones, future demands are not expected to be met by traditional water resources like surface and groundwater. In order to handle increased water demand, the treated wastewater originating from municipal WWTPs should be studied and developed, to be available to farmers and other end-users, for agricultural irrigation and for other compatible uses.

A significant number of tourist regions regularly experience severe water supply and demand imbalances, particularly in the summer months. Tourism is a very important economic activity and is pushing water demand particularly in regions suffering occasional water deficit, like the southern half of Portugal's mainland, namely the Algarve and Lisbon regions, which are presented in this chapter as case studies. Water reuse should be assumed an important management strategy in situations of water scarcity, being that there is a growing trend toward the use of treated wastewaters in different parts of the world, with the major drivers being water scarcity and seasonal stress, in part due to the tourism industry, and probably one of the major barriers is the economical feasibility and sustainability of the projects.

References

1. Asano, T., Burton, F.J., Leverenz, H.L., Tsuchihashi, R., and Tchobanoglous, G. 2007. *Water Reuse: Issues, Technology, and Applications.* McGraw-Hill Professional Publishing, New York.
2. Ávila, C., Salas, J., Martín, I., Aragón, C., and García, J. 2013. Integrated treatment of combined sewer wastewater and stormwater in a hybrid constructed wetland system in southern Spain and its further reuse. *Ecological Engineering*, 50, 13–20.
3. Bixio, D. and Wintgens, T. 2006. *Water Reuse System Management—Manual AQUAREC*, Directorate-General for Research, EC, Brussels, Belgium.
4. Boal, D., Eden, R., and McFarlane, S. 1996. An investigation into grey water reuse for urban residential properties. *Desalination*, 106(1–3), 391–397.
5. California Energy Commission. 2005. *California's Water-Energy Relationship*, Final Staff Report, CEC-700-2005-011-SF. August 2012, Sacramento, CA.
6. Devotta, S., Wate, S., Godfrey, S., Labhasetwar, P., Swami, A., Saxena, A., Dwivedi, H., and Parihar, G. 2007. *Greywater Reuse in Rural Schools.* Guidance Manual. National Environmental Engineering Research Institute, Nagpur, India.
7. EPA. 1993. *Natural Wetlands and Urban Stormwater: Potential Impacts and Management.* Technical Report, Washington, DC.
8. EPA, 2004. *Guidelines for Water Reuse.* Report EPA/625/R-04/108, Environmental Protection Agency, Washington DC.
9. EPA, 2012. *Guidelines for Water Reuse.* Technical Report, EPA/600/R-12/618, September 2012, Washington, DC.
10. Essex, S., Kent, M., and Newnham, R. 2004. Tourism development in Mallorca: Is water supply a constraint? *Journal of Sustainable Tourism*, 12, 4–28.
11. Estévez, E., Cabrera, M., Fernández-Vera, J., Hernández-Moreno, J., Mendoza-Grimón, V., and Palacios-Díaz, M. 2010. Twenty-five years using reclaimed water to irrigate a golf course in Gran Canaria. *Spanish Journal of Agricultural Research*, 8(2), 95–101.
12. European Commission. 2009. *MEDSTAT II: Water and Tourism Pilot Study.* Eurostat, Luxembourg.
13. European Commission. 2012. *Review of the European Water Scarcity and Droughts Policy.* Technical report, Brussels, Belgium.
14. Freire, J. 2010. Plans for bulk supply of TWW for golf course and landscape irrigation in the Algarve region. Workshop on Treated Wastewater Use in Portugal, October 2013, Lisbon, Portugal.
15. Hein, L., Metzger, M., and Moreno, A. 2009. Potential impacts of climate change on tourism; a case study for Spain. *Current Opinion in Environmental Sustainability*, 1, 170–178.

16. Heinz, I., Salgot, M., and Mateo-Sagasta Dávila, J. 2011. Evaluating the costs and benefits of water reuse and exchange projects involving cities and farmers. *Water International*, 36(4), 455–466.

17. Hernández-Sancho, F., Molinos-Senante, M., and Sala Garrido, R. 2011. Eficiencia técnica y económica en la depuración de aguas residuales: Aplicación de herramientas de benchmarking para su análisis dinámico. *Tecnología del Agua*, 31, 36–41.

18. Hof, A. and Schmitt, T. 2011. Urban and tourist land use patterns and water consumption: Evidence from Mallorca, Balearic Islands. *Land Use Policy*, 28(4), 792–804.

19. Hvitved-Jacobsen, T. and Yousef, Y.A. 1991. Highway runoff quailty, environmental impacts and control. In: Hamilton, R.S. and Harrison, R.M. (eds.), *Highway Pollution*, Vol. 44. Studies in Environmental Science, Pergamon Press, Amsterdam, the Netherlands pp. 165–208.

20. Jeppesen, B. 1994. Domestic grey water reuse: Australia's challenge for the future. Localised Treatment and Recycling of Domestic Wastewater Conference, Murdoch University, Australia, pp. 52–59.

21. Kavanagh, J. 2002. *Water Management and Sustainability at Queensland Tourist Resorts*. CRC Tourism's research report, Queensland Government, Brisbane, Australia.

22. Lazarova, V., Rougé, P., and Sturny, V. 2006. Evaluation of economic viability and benefits of urban water reuse and its contribution to sustainable development. *Proceedings of IWA Water Congress*, Beijing, China.

23. Lazarova, V., Carle, H., and Sturny, V. 2008. Economic and environmental benefits of urban water reuse in tourist areas. *Water Practice & Technology*, 3(2), 8.

24. Lechte, P., Shipton, R., and Boal, D. 1995. Installation and evaluation of domestic grey water reuse systems in Melbourne. 16th Australian Water and Wastewater Association Federal Convention, Australia, pp. 91–98.

25. Marecos do Monte, H. and Albuquerque, A. 2010. Of constructed wetland performance for irrigation reuse. *Water Science and Technology*, 61(7), 1699–1705.

26. Marecos do Monte, H. and Albuquerque, A. 2010. *Wastewater Reuse*. Technical Guide No. 14, ERSAR, Lisbon, Portugal (in Portuguese).

27. Martins, A. and Freire, J. 2007. Water reuse for irrigation of golf courses and landscapes in Algarve. *Proceedings of the 6th Conference on Wastewater Reclamation and Reuse for Sustainability*. October 9–12, 2007, Antwerp, Belgium.

28. Martins, A., Freire, J., Sousa, J., and Ribeiro, A. 2006. Potential treated wastewater use for golf courses and landscape irrigation in the Algarve region. *Proceedings of the 12th ENaSB–Encontro Nacional de Saneamento Básico*, October 24–27, 2006, Cascais, Portugal (in Portuguese).

29. Martins, A., Soares, M., Freire, J., Coelho, R., and Baptista, R. 2010. Proposed framework for treated wastewater quality for reuse in irrigation. *Proceedings of the 10th Congresso da Água*, March 21–24, 2010, Faro, Portugal (in Portuguese).

30. Masi, F. and Martinuzzi, N. 2007. Constructed wetlands for the Mediterranean countries: Hybrid systems for water reuse and sustainable sanitation. *Desalination*, 215, 44–55.

31. Morel, A. and Diener, S. 2006. *Greywater Management in Low and Middle-Income Countries, Review of Different Treatment Systems for Households or Neighborhoods*. Swiss Federal Institute of Aquatic Science and Technology (Eawag), Dübendorf, Switzerland.

32. National Capital Planning Authority. 1993. Designing subdivisions to save and manage water. Better Cities Program. Canberra, Australia.

33. NCSL. 2009. Overview of the water-energy nexus in the United States. *National Conference of State Legislatures*, Denver, CO. http://www.ncsl.org/issues-research/env-res/overviewofthewaterenergynexusintheus.aspx. Retrieved July 2014.

34. NRC. 2012. *Water Reuse: Potential for Expanding the Nation's Water Supply through Reuse of Municipal Wastewater*. National Research Council, The National Academies Press, Washington, DC.

35. NRDC. 2009. *Water Efficiency Saves Energy: Reducing Global Warming Pollution through Water Use Strategies*. Water Facts, Natural Resources Defense Council, New York.

36. NSW. 2006. *Managing Urban Stormwater: Harvesting and Reuse*. New South Wales Government, Department of Environment and Conservation, Sydney, Australia.

37. Pedrero, F., Albuquerque, A., Amado, L., Marecos do Monte, H., and Alarcón J. 2011. Analysis of the reclamation treatment capability of a constructed wetland for reuse. *Water Practice & Technology*, 6(3), 9.

38. Pitt, A. 2005. *The National Stormwater Quality Database*, Version 1.1. EPA, Washington, DC.

39. Salgot, M. and Folch, M. 2010. La reutilización del agua en la Región Mediterránea: Realidad y perspectivas. In: Navarro, T.M. (ed.), *Reutilización de Aguas Regeneradas. Aspectos Tecnológicos y Jurídicos*. Fundación IEA, Murcia, Spain.

40. Salgot, M. and Angelakis, A. 2001. Guidelines and regulations on wastewater Reuse. In: Lens, P., Zeeman, G., and Lettinga, G. (eds.), *Decentralised Sanitation and Reuse: Concepts, Systems and Implementation*. IWA Publishing, London, UK, Chapter 23.

41. Salgot, M., Priestley, G., and Folch, M. 2012. Golf course irrigation with reclaimed water in the Mediterranean: A risk management matter. *Water*, 4, 389–429.

42. Schwecke, M., Simmons, B., and Maheshwari, R. 2007. Sustainable use of stormwater for irrigation case study: Manly Golf Course. *Environmentalist*, 27, 51–61.

43. Scott, D. and Becken, S. 2010. Adapting to climate change and climate policy: Progress, problems and potentials. *Journal of Sustainable Tourism*, 18, 283–295.

44. Shutes, R., Brian, E., Revitt, D.M., and Scholes, L.N.L. 2010. Constructed wetlands for flood prevention and water reuse.

12th International Conference on Wetland Systems for Water Pollution Control, October 4–8, 2010, Venice, Italy.

45. Tchobanoglous, G., Burton, F., and Stensel, H. 2003. *Wastewater Engineering, Treatment, Disposal, and Reuse.* Metcalf & Eddy, McGraw Hill, New York.

46. Viegas, R., Sousa, A., Póvoa, P., Martins, J., and Rosa, M. 2011. Treated wastewater use in Portugal: Challenges and opportunities. *Proceedings of the 8th International Conference on Water Reclamation & Reuse*, September 26–29, 2011, Barcelona, Spain.

47. von Sperling, M. 2007. Wastewater characteristics, treatment and disposal. In: *Biological Wastewater Treatment Series*, Cap. 1, IWA, London, UK.

48. WHO. 2004. *Guidelines for Drinking Water Quality*, 3rd Edition. World Health Organization, Geneva, Switzerland.

49. WHO. 2006. *Guidelines for the Safe Use of Wastewater, Excreta and Greywater. Vol. 2: Wastewater Use in Agriculture.* World Health Organization, Geneva, Switzerland.

50. Zhai, J., Xiao, H., Kujawa-Roeleveld, K., He, Q., and Kerstens, S. 2011. Experimental study of a novel hybrid constructed wetland for water reuse and its application in Southern China. *Water Science and Technology*, 64(11), 2177–2184.

VII

Economical Aspects
of Water Reuse

27

Household Water, Wastewater Minimization, and Cost Analysis: A Case Study

Gökhan Ekrem Üstün
Uludag University

Didem Civancık
Uludag University

Seval Kutlu Akal Solmaz
Uludag University

PREFACE

This work studies ways of minimizing household water consumption. Calculations are performed to understand household water consumption in different water use areas for a representative family of four living in a house of 90 m². In this way, the family's water inventory is determined. Some advice and strategies for household water use are evaluated to minimize water consumption. Various types of water-saving equipment are proposed for household use to encourage water conservation, and this equipment's payback period is calculated by determining its cost. Because household water consumption can play an important role in water management strategies, this study focuses on the minimization of water consumption.

27.1 Introduction

Water, the main source of life, is required for the continuation of both ecosystems and humankind. Today, water is used in many different areas. Water is used mainly for irrigation in agricultural activities, for processes such as heating and cooling systems in industry, and for cleaning, washing, and drinking in households. However, factors such as increasing population, water pollution, urban development, agricultural irrigation, climate change, and drought directly affect the availability of water resources for increasing demand. Therefore, ensuring water security against this increasing demand has become a very important issue [18].

Between the years of 1990 and 2008, the amount of the population that had access to drinking water increased from 77%–87%. However, it was predicted that in 2008 there were 884 million people who did not have access to drinking water resources. According to current increasing demands for water resources, it is estimated that in 2015, 672 million people will not be able to reach drinking water resources [4].

Even Turkey, which is surrounded by the sea and rich in terms of rivers, streams, and water resources, is considered among the water-poor countries. The Wildlife Conservation Foundation predicts that Turkey will face water shortages in the coming years. The amount of water obtained from Turkish water resources is 110 billion m³. Of this amount, 16% is used as drinking water, 72% as agricultural irrigation, and 12% in industry. Rapid population growth, industrialization, water pollution, and low average annual rainfall threaten Turkey's water resources [14]. Thus, water must be used efficiently so that these water resources can be passed on to future generations with continuity.

Water conservation is an important factor in water management. According to Willis et al. [28], there is major potential for water conservation in the area of residential water consumption. Various institutions and organizations define practices and strategies to be used indoors to achieve water savings [21]. Studies show that water-saving applications have a great influence on water conservation [15,19,22–25].

According to Jergensen et al. [18], the key to understanding water demand management strategies is to understand how people think about water and its use. Studies show that water conservation is more successful when people recognize the problem of water scarcity and realize that other people also care about water conservation (Corral-Verdugo et al., 2002 as cited in Jorgensen et al. [18]). However, the effects of awareness and water pricing are not clear. Again, the factors affecting household use must be understood to reduce water demand and create a better water management strategy [18].

There are different approaches to household water conservation, and the greatest savings can be achieved when using special equipment and consuming water efficiently. The main goal of this study is to quantify the amount of water consumed in households and calculate the cost and payback period of equipment used for household water conservation.

27.2 Distribution of Household Water Use

The water use of a typical family living in a house without water-saving measures is portrayed in Figure 27.1. Comparable water use with water-saving measures is given in Figure 27.2, and a comparison of the two scenarios is shown in Figure 27.3 [16].

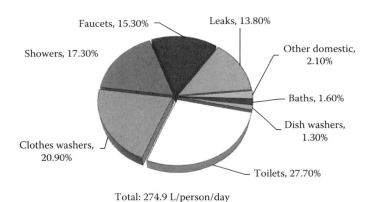

Total: 274.9 L/person/day

FIGURE 27.1 Water use of a typical family in a house without water conservation. (Adapted from Harris, J.L. and Ed M., CRS. 1999. *Texas AgriLife Extension*, p. 29.)

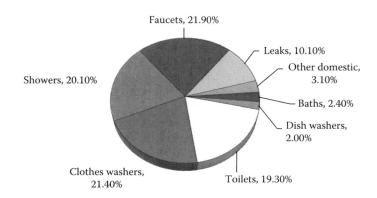

FIGURE 27.2 Water use of a typical family in a house with water conservation. (Adapted from Harris, J.L. and Ed, M., CRS. 1999. *Texas AgriLife Extension*, p. 29.)

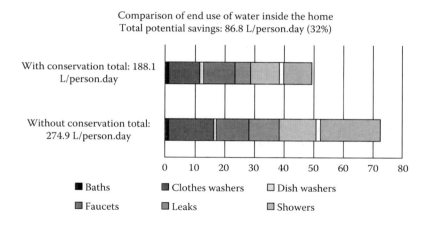

Comparison of end use of water inside the home
Total potential savings: 86.8 L/person.day (32%)

FIGURE 27.3 Comparison of water use in houses with and without water conservation. (Adapted from Harris, J.L. and Ed M., CRS. 1999. *Texas AgriLife Extension*, p. 29.)

27.3 Calculation of Household Water Use

A practical method exists to calculate water use for residential housing over an average of 320 days per year. This method assumes 150 L of water consumption per person per day in a residential building, excluding water consumption for irrigation.

A family of three in a four-bedroom house uses $3 \times 150 = 450$ L water per day. Assuming 320 days of use per year, this amount becomes $320 \times 450 = 144$ m^3 per year.

In the case of a house with a garden, the water used for irrigation must be included in calculations. It can be assumed that a person living in a house with a garden consumes 200–250 L water per day. In this case, the number of days the house is used throughout the year and the period of time over which the garden is irrigated become important issues.

It is assumed that a house with a garden and a family of four living in it irrigates approximately 300 m^2 for 80 days a year. The city of Istanbul is neither rainy nor arid, so 5 L of water consumption is assumed per m^2 of garden. Therefore, the consumption for 80 days is

$$300 \times 5 \times 80 = 120,000 \text{ L} = 120 \text{ m}^3$$

Assuming the water consumption for each family member as 150 L per day for a year:

$$150 \times 4 \times 320 = 192,000 \text{ L} = 192 \text{ m}^3$$

The total water consumption of a house with garden is then

$$120 + 192 = 312 \text{ m}^3$$

If we assume 200 L water consumption per person per day, the annual water consumption of the house, including irrigation water, is

$$200 \times 4 \times 320 = 256,000 \text{ L} = 256 \text{ m}^3$$

If 250 L of water consumption per person per day is assumed, the annual water consumption is

$$250 \times 4 \times 320 = 320,000 \text{ L} = 320 \text{ m}^3$$

When 250 L of water consumption is assumed per person per day, the results are closer to the previously calculated numbers. Therefore, it is reasonable to assume that the water consumption in houses with gardens per person per day is approximately 250 L [20].

27.4 Advice for Household Water Conservation

27.4.1 Kitchen

Independent of the type of house one lives in, the kitchen can be an effective place to reduce a house's water consumption. There are some easy ways to trim water consumption in the kitchen. In this way, both the reduction of water bills and water conservation can be achieved.

Dishwashers and washing machines should be operated when they are full and their most efficient water and energy settings should be chosen for operation [7]. Instead of leaving the tap running, a sink of water can be used to wash. For example, fruits and vegetables should be washed in a bowl instead of under running water [9]. The water used for washing can be reused to water flowers. In general, water should be reused as much as possible. Frozen food should be allowed to defrost in a refrigerator the night before instead of being placed under running water. When hot water is needed, only the amount needed should be heated in a coffee pot, teapot, or kettle [26]. The thermostat settings on hot water apparatus should not be excessively high because further water will be wasted in cooling very hot water. Moreover, water consumption can be reduced by 80% when using a special apparatus similar to that providing spray flow and increased water pressure [9]. Vegetables can be cooked with a pressure cooker instead of boiling them over an open fire. In this way, the nutritional value of vegetables is preserved and water is also saved [26]. Water should be kept in a refrigerator or ice can be added to make a drink cold instead of letting excess water run down the drain.

27.4.2 Dishwashing

Large amounts of water can be wasted during dishwashing if care is not taken. By applying some techniques and using efficient dishwashers, however, water conservation can be achieved during dishwashing.

Dishes should be washed as soon as possible to prevent food particles from hardening because removing hardened particles requires more water. Another technique is to consider order while hand washing dishes (low-fat substances first, followed by fats). After foaming the dishes, they should be placed under the tap over and over when rinsing. Alternatively, a pan of soapy water for washing and hot water for rinsing can be used. These methods help to save water compared to soaping and rinsing under running water. A minimal amount of detergent should be used when washing dishes because it is easier to rinse a smaller amount of detergent.

A dishwasher is preferred to hand washing if possible. Today, dishwashers use 12–18 L of water to wash 12 place settings, but hand washing consumes 40 L of hot water. Dishwashers and washing machines consume 16% of the total volume of water consumed in a typical household [1]. With water consumption at an all-time high, it is very important to choose an efficient dishwasher to preserve our most precious resource. Dishwashers providing 51% water savings are now available [8], and this type of machine should be preferred. The manufacturer's directions should also be followed to save water and energy. For example, the machine should be operated when it is full [8]. Dishes should not be rinsed in dishwater before being placed in the dishwasher. Instead, they should be scraped and the food remains composted [26].

27.4.3 Laundry

Doing laundry is a necessity, but it uses 22% of the water consumed in a household. Thus, individuals aware of this fact should ask themselves "Does this need washing?" before doing laundry [10].

Today's washing machines consume half of the water that a 10-year-old machine uses. New washing machines typically consume 50 L of water per 6 kg wash, and some even consume 40 L water per 6 kg wash. Washing machines have energy and water consumption labels for comparison. Some machines have 10 kg capacities and consume even less water for larger amounts of clothing [1]. These types of efficient washing machines should be preferred. Machines should be operated only when full and the water volume used should be reduced by choosing the most efficient cycle. Because 90% of a machine's life cycle impact occurs during its operation, the replacement of an old machine with a newer one can be more economical and environmentally friendly [1]. Before machine-washing, heavily polluted clothes should be soaked in detergent in another container. The used water from a washing machine can also be used for balcony and toilet cleaning.

27.4.4 Bathing

Approximately 40% of daily life is spent in the bathroom. New houses have showers and baths with water savings of 45%. Modern apparatus and lifestyle changes make it possible to conserve water in the bathroom [1].

Showers should be preferred over baths. During a shower, a person consumes one-third of the water consumed during a bath, and people usually to tend to take showers more often than baths [1]. When a tub is filled with water, 70–150 L of water is consumed [26]. On the other hand, a 5-min shower uses 35–60 L of water. If each member of a family of four decreases his or her shower time by 1 min, approximately 18–20 m^3 water can be saved [13]. More efficient and water-saving showerheads should be used to conserve even more water during showers. Before buying a new head, the flow of the existing head should be measured by timing how long it takes to fill a bucket [1]. Adjustable water taps consume a maximum of 9–12 L of water, while some can consume 25–30 L. Therefore, a 5–6 min shower might use 50–100 L of hot water, and energy savings could be achieved. If it is assumed that a family of four takes 3 showers a day, 55 m^3 of water savings could occur over the course of a year. The water should be turned off during soaping because there is no need for water during that time [5]. The amount of water consumed during a shower also depends on the person taking the shower because some people use a high flow of water during showering, while others use less. People should be aware of this difference and use water efficiently [1]. Water can also be collected in a bucket while waiting for the hot water to appear, and then used for other purposes.

27.4.4.1 Calculation of Water Consumed while Brushing Teeth

Brushing one's teeth usually takes 3 min. If the water tap is left open during brushing, a person consumes 2 L of water per minute. However, if it is opened only to rinse the mouth and clean the brush, only 1 L of water is consumed for 3 min of brushing.

The water consumption for a family of four is as follows when the tap is left open.

Because a person consumes $3 \times 2 = 6$ L of water during brushing,
Water consumed in a day assuming two brushes is $6 \times 2 = 12$ L
Water consumed in a day assuming two brushes for a family of four is $4 \times 12 = 48$ L
Water consumed in a year assuming two brushes for a family of four is $365 \times 48 = 17,520$ L [23]

The water consumption for a family of four is as follows when the tap is opened only to rinse the mouth and clean the brush.

Because a person consumes 1 L of water during brushing,
Water consumed in a day assuming two brushes is $1 \times 2 = 2$ L

Water consumed in a day assuming two brushes for a family of four is $4 \times 2 = 8$ L

Water consumed in a year assuming two brushes for a family of four is $365 \times 8 = 2920$ L [23]

Water saved: 17,520–2920 = 14,600 L [23]

27.4.4.2 Calculation of Water Consumed during Shaving

Shaving usually takes 6 min. If the water tap is left open, a person consumes 2 L of water per minute. However, if it is opened only to rinse the face and clean the razor blade, only 1 L of water is consumed during shaving.

Assuming half of the family is male and those members shave every morning, the water consumption for a family of four is as follows when the tap is left open.

Because a person consumes 2 L of water per minute during shaving,
Water consumed by a person is $6 \times 2 = 12$ L
Water consumed in a day by two people is $12 \times 2 = 24$ L
Water consumed in a year is $365 \times 24 = 8760$ L [23]

Assuming half of the family is male and those members shave every morning, the water consumption for a family of four is as follows when the tap is opened only to rinse the face and clean the razor blade.

Because a person consumes 1 L of water per minute during shaving,
Water consumed in a day by two people is $1 \times 2 = 2$ L
Water consumed in a year is $365 \times 2 = 730$ L [23]

Water saved: 8760–730 = 8030 L [23]

27.4.5 Toilet

Toilets are the main consumers of water in houses and constitute up to 30% of the total water consumed by a household. Old, inefficient toilets in terms of water consumption are the main reason water is wasted in most houses [27].

Assuming 6–16 L of water consumption per flush [26], a family of four consumes 7 m³ of water in a month if a toilet has a 16-L reservoir. To prevent this waste, water consumption can be reduced to 2.5–3 m³ by using a reservoir holding only 7 L [5]. Moreover, the overuse of siphons and flushes increases water consumption and wastewater production [26]. Thus, these mechanisms should not be used unnecessarily. If a member of a family of four uses a siphon unnecessarily once a day, 16 m³ of water is consumed in a year [6]. Water-saving closets should be preferred rather than standard models [26], and 60% water-saving closets are now available thanks to technology advances. Moreover, reservoir volume can be reduced to 6–7 L instead of 12–20 L, and stepped reservoirs can be used to reduce the amount of water used [12].

There are some practical ways to conserve water in toilets. Water savings can be achieved very easily, without the need for any special equipment, by placing a 1.5-L plastic bottle into the existing reservoir [6]. To reduce the amount of water inside the reservoir, a balloon can be placed to reduce its volume. In this way, the amount of running water can be reduced when flushing the toilet. For example, a method called the "hippo" exists in the United Kingdom and consists of putting a bag full of sand inside the reservoir to reduce its volume, achieving 45 L of water savings per day [2]. Old-fashioned siphons have the advantage of not leaking, but newer drop valve flushes can leak. Because the newer flushes are now mostly preferred, this leaking issue is becoming more of a problem. Leaking should be prevented because leaking reservoirs consume more water than that used during flushing [1].

27.4.6 Taps

Approximately 20% of household water flows through taps and sinks [1]. Excess water use can result from the use of old-fashioned, inefficient taps and leaks.

Water-saving fixtures should be chosen for households. Spray taps can save 80% of water by restricting its flow [1]. These systems can reduce the amount of running water while increasing its pressure. Dripping taps must be repaired because they waste 4–20 L of water per day [26]. Collecting dripping water from taps by putting pots under them is a good solution. Taps should also not be left open unnecessarily. The amount of water wasted in this way is approximately 15 m³ per person per year, or approximately 50 m³ for a family of four [6].

Taps should not be left open when brushing teeth and shaving [26]. Leaving the tap on during 3 min of brushing uses 5 L of water, while only 1 L of water is consumed when the tap is closed [3]. If teeth are brushed twice a day, 11–12 m³ of water is used [13]. Razor blades should be cleaned in a bowl of water instead of under running water.

27.4.6.1 Calculation of Water Wasted by Dripping Taps

A dripping tap wastes 1 L of water in an hour.
Therefore,

$1 \times 24 = 24$ L of water wasted in a day
$24 \times 365 = 8\,760$ L of water wasted in a year [23]

27.4.7 Garden

The amount of water used for irrigation during the growing season can constitute a high percentage of the total water used in a house. Water is required to provide a good landscape. If there is a sprinkler system, it should be set on a regular basis and leaks should be controlled. When watering a garden, a hose with a filter is preferred to water only the required area. Leaves that accumulate in a garden should be collected with a harrow instead of rinsing them away with water.

Native plants that can grow in a local climate should be chosen because they can grow with less water and care [6]. Conversely, a garden covered with grass in dry weather requires high amounts of water for irrigation. Only the amount of water required by the plants should be used for irrigation. Mornings and evenings should be chosen for irrigation because there is less evaporation compared to at other times of the day. During the middle of the day, more water is consumed and plants can be damaged [6]. Leaking hoses and drip irrigation can be used for the irrigation of trees and bushes. Plants that do not require water should not be watered. Trees and bushes should be watered longer and less often compared to shallow-rooted plants. Grass should not be cut too short, and grass cuttings, leaves, bushes, and mulch can be used to keep moisture in the ground. Organic matter such as decayed leaves and fertilizer can be added to soil so that it remains humid, reducing evaporation by approximately 70% [13]. Drinking water should not be used for garden irrigation. Rather, well water or collected rainwater can be chosen for this purpose, and recycled water should also be considered. For example, if there is an aquarium in the house, its water can be used for plant irrigation because it is rich in nitrogen and phosphorus, both of which are very beneficial to plant growth. The water used to boil eggs can be used to water plants instead of being discarded [26].

When washing a car, it should be parked so that the used water can go directly into a grassy area. In this way, the grass can be watered with increased efficiency.

27.4.8 Water Losses from Leaks

Regular maintenance of installments, siphons, and taps is crucial to saving water. Fixtures must be checked for leaks and repaired immediately. Unnoticed leaks can result in both water waste and increased cost. If water is assumed to leak from taps one drop at a time, then 1–10 m^3 of water is wasted in a year. Not only could this water be saved, but also its leakage creates wastewater and affects water bills.

Leaks can affect water consumption in a house. Water leakage from a reservoir can be identified by adding food coloring to it. If there is a leak, then colored water will be observed after a period of time (the siphon must be used after the test is done). Because reservoirs are cheap and can be easily installed, a new reservoir will pay for itself in a very short time by preventing water leaks.

It is also possible to determine leaks by checking a water meter when water is not being used [26]. Water leaks can also occur in sprinkler systems, and these leaks should be identified and repaired as soon as possible.

27.4.9 Water Conservation by Changing Habits

Changing daily habits can assist with water conservation. Some basic steps do not require extra money or energy to save water. A sweeper can be used to clean the front door, balcony, and terrace instead of a hose because washing these areas with a hose uses 200–250 L of water [6]. A bucket and sponge should be used to wash cars because only 20 L of water is used to wash

with a bucket, whereas a hose uses 150–550 L [12]. It should be noted that the cleaning detergents used in houses are ultimately discharged to rivers and seas. Water-soluble detergents that do not contain phosphorus should be used to protect the ecological balance. Because bleach is harmful to the bacteria responsible for the destruction of waste, it should be used as much as possible.

27.4.10 Water Conservation through Increased Awareness

Water conservation awareness is defined by the United Nations Environment Program as an understanding of the need to use water efficiently in every aspect of life and change use habits. Politicians, water managers, social marketers, and educators play an important role in spreading this idea to the public [11]. Reaching people of different backgrounds is critical to creating public awareness, and educating people about the importance of water conservation will make their children more aware in the future. This awareness can be achieved by warning other people about water resource scarcity and the importance of conserving water. For example, individuals should be encouraged to buy naturally grown vegetables and fruits because less water is used for this produce compared to processed agricultural products.

27.5 Household Water Use Calculations

This study calculates an inventory of the water consumed by a family of four living in a 90-m^2 house over a month. The water savings in this area are investigated, the equipment available to achieve water savings is defined, and this equipment's cost and payback period are calculated.

27.5.1 Bath

When showering, 5.3 L of water is used per minute. It is assumed that a person belonging to a family of four takes a 10-min shower three times in a week.

$$4 \text{ people} \times 3 \text{ showers/person.week} \times 4 \text{ weeks/month}$$
$$\times 10 \text{ min/shower} \times 5.3 \text{ L/min}$$
$$= 2544 \text{ L/month}$$

27.5.2 Toilet

If a person uses a toilet five times a day and 6 L of water is consumed for each use, the water use calculation for a family of four is as follows.

$$4 \text{ people} \times 6\,\text{L/use} \times 5 \text{ uses/person.day} \times 7 \text{ days/week}$$
$$\times 4 \text{ weeks/month}$$
$$= 3360\,\text{L/month}$$

27.5.3 Taps

If a person uses a washbasin eight times a day for different purposes such as washing hands and face, brushing teeth, and cleaning, and assuming 2 L of water is consumed each time, the calculation for a family of four is as follows:

$$8 \text{ uses/person.day} \times 2 \text{ L/use} \times 4 \text{ persons} \times 7 \text{ days/week}$$
$$\times 4 \text{ weeks/month}$$
$$= 1792 \text{ L/month}$$

27.5.4 Washing Machine

A washing machine uses 82 L of water for a wash. If a family of four uses a washing machine three times a week, then

$$82 \text{ L/use} \times 3 \text{ uses/week} \times 4 \text{ weeks/month}$$
$$= 984 \text{ L/month}$$

27.5.5 Dishwasher

A dishwasher uses 15 L of water for a wash. If a family of four is assumed to use a dishwasher four times a week, then

$$15 \text{ L/use} \times 4 \text{ uses/week} \times 4 \text{ weeks/month}$$
$$= 240 \text{ L/month}$$

27.5.6 Kitchen

Because water consumption in a kitchen is hard to measure, the amount of water used is calculated by estimating the water used outside of the kitchen, which is 8920 L/month, from the total household water use, which is 10,000 L a month.

$$10,000 \text{ L/month} - 8920 \text{ L/month} = 1080 \text{ L/month}$$

27.6 Equipment Providing Household Water Savings

The water saved by using efficient home appliances is determined and this equipment's payback periods are calculated. The appliances and their efficiencies and amortization periods are given in Table 27.1.

27.7 Summary and Conclusions

Because it is impossible to increase the Earth's freshwater resources, decreasing water consumption by using various techniques and conservation methods and water reuse has become a necessity. This study inventories the water consumption of a family of four living in a 90-m^2 house and investigates water conservation possibilities in those areas. Ideas for the use of water saving equipment are identified and their payback periods are calculated.

It is concluded that a family of four consumes 2544 L of water in the bath in a month, assuming that 5.3 L of water is consumed per minute and each person in the family takes a 10-min shower three times a week. The monthly water consumption by the toilet is calculated as 3360 L, assuming each person uses the toilet five times a day and consumes 6 L of water for each use. The water consumption with the use of washbasins is calculated as 1792 L per month. For this calculation, it is assumed that a person uses washbasins eight times in a day and consumes 2 L of water each time. Moreover, if 82 L of water is consumed by the family's washing machine, assuming the machine is operated three times a week, then 984 L of water is consumed for laundry per month. If a dishwasher consumes 15 L of water for each use, assuming four uses per week, the water consumed for washing dishes is calculated as 240 L per month. To calculate the amount of water used in the kitchen, the water used outside of the kitchen (8920 L/month) is subtracted from the total water used (10,000 L/month) by the household because it is difficult to estimate the water consumed in a kitchen. This subtracted number is 1080 L per month.

Household residents can achieve significant water savings [28], as cited in Lee and Tansel [21]. Special appliances can be used to save water. This study investigates these appliances, their savings ratios, cost and payback periods, and it is concluded that 1780.8 L of water can be saved in a month when using a 70% water-saving showerhead with a payback period of 3.2 months. With a 50% water-saving showerhead, 1272 L of water is saved in a month and the head pays for itself in 4.6 months. With a 50% water-saving doubled reservoir, 1680 L of water is saved and its payback period is 4.7 months. Ninety percent water-saving taps help to save 1612.8 L of water in a month and their payback period varies from 4.2–11.6 months. Seventy percent water-saving taps save 1254.4 L of water in a month and pay for themselves in 18.7 months. Using a 75%, 90%, or 60% water-saving faucet saves 1344 L, 1612.8 L, or 1075.2 L of water, and they can pay for themselves in 4.4, 4.4, or 7 months, respectively. In addition, 60% water-saving washing machines can save up to 590.4 L of water in a month and a 43% water-saving dishwasher can achieve savings of 103.2 L in a month.

When considering the world's water scarcity and insecurity, efficient use of water resources and their protection becomes increasingly important. Most of the water consumed in houses can be conserved by paying attention to daily habits and taking the precautions explained in this study. In this way, water resources can be conserved for future resources.

TABLE 27.1 Water-Saving Equipment, Amount Saved, and Amortization Period

Equipment	Cost (TL)[a]	Water Use without Conservation	Water Saved (%)	Water Use with Conservation (%)	Water Saved	Cost of Monthly Savings	Amortization Period
Showerhead							
	14.40	2544 L/m	70	763.2 L/m	1780.8 L/m	4.47 TL/m	3.2 months

Source: https://hergunozel.com/51/70-su-tasarrufu-saglayan-dus-basligi.

	14.90	2544 L/m	50	1272 L/m	1272 L/m	3.2 TL/m	4.66 months

Source: Metro Market Brochure. http://urun.gittigidiyor.com/ev-bahce/ac-kapa-tasarruflu-dus-basligi-universal-169610376.

Reservoir							
	19.99	3360 L/m	50	1680 L/m	1680 L/m	4.22 TL/m	4.74 months

Source: Metro Market Brochure. http://www.n11.com/basmali-rezervuar-ic-takimi-P55909807.

Faucet							
	16.90	1792 L/m	90	179.2 L/m	1612.8 L/m	4.05 TL/m	4.17 months

Source: http://www.hepsiburada.com/universal-su-tasarruf-kartusu-2-lt-p-HRPROFISU24X12.

| | 46.90 | 1792 L/m | 90 | 179.2 L/m | 1612.8 L/m | 4.05 TL/m | 11.6 months |

Source: http://urun.gittigidiyor.com/ev-dekorasyon-bahce/zaman-ayarli-tektus-otomatik-musluk-90-tasarruf-27559924.

Sensor Faucet							
	59.00	1792 L/m	70	537.6 L/m	1254.4 L/m	3.15 TL/m	18.7 months

Source: http://urun.gittigidiyor.com/ev-dekorasyon-bahce/aqua-sensor-ile-otomatik—su-tasarrufu-70-55512796.

Perlator							
	14.90	1792 L/m	75	448 L/m	1344 L/m	3.37 TL/m	4.42 months

Source: http://urun.gittigidiyor.com/ev-dekorasyon-bahce/volum-ayarli-su-tasarruf-perlatoru-54885784.

| | 18.00 | 1792 L/m | 90 | 179.2 L/m | 1612.8 L/m | 4.05 TL/m | 4.44 months |

Source: http://urun.gittigidiyor.com/ev-dekorasyon-bahce/cesme-tip-akis-kontrollu-su-tasarrufu-2-L-dk-51359470.

(Continued)

TABLE 27.1 (*Continued*) Water-Saving Equipment, Amount Saved, and Amortization Period

Equipment	Cost (TL)[a]	Water Use without Conservation	Water Saved (%)	Water Use with Conservation (%)	Water Saved	Cost of Monthly Savings	Amortization Period
Debi Sabitleyici Su Tasarruf Perlatörü	18.90	1792 L/m	60	716.8 L/m	1075.2 L/m	2.7 TL/m	7 months

Source: http://urun.gittigidiyor.com/ev-dekorasyon-bahce/eko-dus-debi-sabitleyici-su-tasarruf-perlatoru-57094909.

Washing Machine

9 kg.	999.90	984 L/m	60 (48L/82L)	393.6 L/m	590.4 L/m	1.48 TL/m	56.3 years

Source: http://www.hepsiburada.com/liste/vestel-cmh-xxl-8412te—8-kg-1200-devir-camasir-makinesi/productDetails.aspx?productId=mtvescmzxxl8412te&categoryId=155121.

	589.00	984 L/m	60 (49L/82L)	393.6 L/m	590.4 L/m	1.48 TL/m	33 years

Source: http://www.hepsiburada.com/Liste/arcelik-5063-5-kg-600-devir-camasir-makinesi-a-enerji-sinii/productDetails.aspx?productId=mtar5063&categoryId=155121.

Dishwasher

	1173.70	240 L/m	43 (6.5L/15L)	136.8 L/m	103.2 L/m	0.26 TL/m	188 years

Source: http://www.istanbulbilisim.com.tr/bosch-sms43d08tr-bulasik-makinesi-fiyati,65219.html.
[a] TL: Turkish Liras (1TL = 0.476 USD)

References

1. Anonymous. 2007. Conserving Water in Buildings: A Practical Guide, UK Environment Protection Agency, Bristol.
2. Anonymous. 2009. Water Save Methods. www.atikyonetim.com/evsel/ Verimli Su Kullanımı (in Turkish). Accessed on July 3, 2015.
3. Anonymous. 2009. http://www.talktalk.co.uk/money/features/water-saving.html. Accessed on July 3, 2015.
4. Anonymous. 2011. Drinking Water Equity, Safety and Sustainability, UNICEF and World Health Organization, JMP Thematic Report on Drinking Water 2011, pp. 11–14.
5. Anonymous. 2012. Water Use Awareness, İstanbul Water and Sewage Authority. http://www.iski.gov.tr/web/UserFiles/e-kutuphane/sukullanmabilinci.pdf (in Turkish). Accessed on May 8, 2012.
6. Anonymous. 2012. Water is life; do not waste it. Republic of Turkey Ministry of Forestry and Water Works, General Directorate of Environmental Management. http://www.semplastik.com.tr/pdf/su.pdf (in Turkish). Accessed on May 14, 2012.
7. Anonymous. 2013. Energy Saving Trust, Saving Water in Kitchen. http://www.energysavingtrust.org.uk/Take-action/Energy-saving-top-tips/Changing-your-habits-room-by-room/Saving-water-in-the-kitchen. Accessed on May 1, 2014.
8. Anonymous. 2013. Energy efficient appliances, Bosch. http://www.bosch-home.com/us/about-bosch/green-technology-inside.html. Accessed on April 28, 2014.
9. Anonymous. 2013. Water saving tips. http://www.highland.gov.uk/yourenvironment/sustainabledevelopment/greencouncil/WaterSavingTips.htm. Accessed on March 3, 2014.
10. Anonymous. 2013. http://environment.nationalgeographic.com/environment/freshwater/water-conservation-tips/. Accessed on July 3, 2015.
11. Anonymous. 2013. https://practicegreenhealth.org/topics/energy-water-and-climate/water/best-practices-water-conservation. Accessed on July 3, 2015.
12. Çetinavcı, İ.H. 2008. The responsibilities of the infrastructural organizations and people for water consumption, Water Consumption Treatment Reuse Symposium, September 3–5, Bursa, p. 9 (in Turkish).

13. Delen, G. 2010. *Practical Methods for Water Conservation.* İsmmmo Life Publication, 28, 44–45 (in Turkish).

14. Ergin, Ö. 2008. A project about water awareness, TMMOB 2. Water Policies Congressium, March 20–22, Ankara, pp. 531–538 (in Turkish).

15. Gilg, A. and Barr, S. 2006. Behavioral attitudes towards water saving? Evidence from a study of environmental actions. *Ecological Economics,* 57, 400–414.

16. Harris, J.L. and Ed M. CRS. 1999. Don't let the faucet run dry, water conservation in the home, family and consumer sciences. *Texas AgriLife Extension,* p. 29.

17. Heaton, L., Ilvento, T., and Taraba, J. 2013. Conserving water at home. http://www2.ca.uky.edu/agc/pubs/ip/ip2/ip2.htm. Accessed on May 2, 2014.

18. Jorgensen, B., Graymore, M., and O'Toole, K. 2009. Household water use behavior: An integrated model. *Journal of Environmental Management,* 91(1), 227–236.

19. Kolokytha, E.G., Mylopoulos, Y.A., and Mentes, A.K. 2002. Evaluating demand management aspects of urban water policy: A field survey in the city of Thessaloniki, Greece. *Urban Water Journal,* 4, 391–400.

20. Köktürk, U. 1995. Water consumption calculation in plumbing. *Installment Engineering Journal,* 21, 1–5 (in Turkish).

21. Lee, M. and Tansel, B. 2013. Water conservation quantities vs. customer opinion and satisfaction with water efficient appliances in Miami, Florida. *Journal of Environmental Management,* 128, 683–689.

22. Millock, K. and Nauges, C. 2010. Household adoption of water-efficient equipment: The role of socio-economic factors, environmental attitudes and policy. *Environmental Resources Economics,* 46, 539–565.

23. Özer, Z. (ed.) 2007. *Green Box Teacher's Handbook, Elements of Environment, Water,* TypoNova Kft., Macaristan, pp. 23–36 (in Turkish).

24. Randolph, B. and Troy, P. 2008. Attitudes to conservation and water consumption. *Environmental Science Policy,* 11, 441–455.

25. Russell, S. and Fielding, K. 2010. Water demand management research: A psychological perspective. *Water Resources Research,* 46, 1–12.

26. Şilliler T.D. 2011. Education Booklet for Wise Use of Water, Empowerment Project of Management of Forest Protected Areas, Küre Mountain National Park, pp. 1–52 (in Turkish).

27. U.S. Environmental Protection Agency. 2013. Water sense. Products. Toilets. http://www.epa.gov/WaterSense/products/toilets.html. Accessed on May 2, 2014.

28. Willis, R.M., Stewart, R.A., Panuwatwanich, K., Jones, S., and Kyriakides, A. 2010. Alarming monitors affecting shower end use water and energy conservation in Australian residential households. *Resources Conservation and Recycling,* 54, 1117–1127.

28

Evaluation of Socioeconomic Impacts of Urban Water Reuse Using System Dynamics Approach

Mohammad Hadi Bazrkar
Tarbiat Modares University

Negar Zamani
Guilan University

Saeid Eslamian
Isfahan University of Technology

PREFACE

The increasing demand for water and lack of sufficient and proper water in both quality and quantity have made the public and decision makers to investigate for new resources. One of the most available water resources is the reuse of water. Urban water reuse has been affected by social and economic drivers and vice versa. In this chapter, socioeconomic impacts of urban water reuse and the factors that influence application of this kind of water have been studied using a system dynamics approach.

28.1 Introduction

Population growth, development in urban areas and technology, as well as industry has brought an increase in water demand. In addition, climate change and land use have been steadily deteriorating Earth's freshwater resources in both quality and quantity [62]. The lack of insufficient freshwater in many countries has made the necessity of alternative water resources to meet high demand. Treating wastewater positively affects water resources in two ways: reducing wastewater discharge to the environment and wastewater reclamation and reuse in compensation of water deficit in arid and semi-arid areas. Water reuse has led to materials and energy recovery, and the life-cycle carbon emissions in the water supply process [22,33,34,35,42,53]. Treated wastewater can be considered for various applications such as irrigation, urban and recreational uses, groundwater recharge, aquaculture, and industrial consumptions. The various uses of wastewater and its importance indicate the necessity

of more studies on waste water reuse. Based on the complexity of the system, an integrated approach is needed. The system dynamics approach is a comprehensive method for studying wastewater and water reuse.

Climate change has threatened the global water sustainability by disturbing patterns of precipitation and increasing temperature and evapotranspiration [55]. Adapting to climate change will have a close resonance with adapting to water scarcity and is likely to require implementation of water demand management strategies, which may require capacity building and awareness raising across institutions and society. Adaptation measures on the supply-side include the ways to improve rain-harvesting techniques, increasing extraction of groundwater, water recycling, desalination, and improving water transportation. Climate change has many effects on the hydrological cycle and, therefore, on water resources systems. Global warming could result in changes in water availability and demand, as well as in the redistribution of water resources and in the structure and nature of water consumption, and exasperate the conflicts among water users.

Wastewater reuse could provide a mitigation solution to climate change through a reduction in greenhouse gases by using less energy for wastewater management compared to that for importing water, pumping deep groundwater, seawater desalination, or exporting wastewater. Reuse increases the total available water supply and reduces the need to develop new water resources and therefore provides an adaptation solution to climate change or population density induced water scarcity by increasing water availability. It may also contribute to desertification control and desert recycling.

As water affects human lives, humankind also has an effect on the hydrological cycle of the planet, in all dimensions from the very local to the global scale. Climate change is now a scientifically established fact. There is scientific consensus that the global climate is changing mainly due to manufactured emissions. The current water scarcity will be intensified by a further decrease in water availability due to reduced rainfall, which is projected to decrease by 20% over the next 50 years. Meanwhile, water demand will increase as a result of rising temperatures that lead to increase in evapotranspiration from irrigated agricultural zones and natural ecosystems. A decrease in rainfall and an increase in temperatures are projected to contribute to increased evaporation and decreased groundwater recharge.

It is expected that such dry conditions and rainfall decreases will put more pressure on available water resources, especially in the major river basins of some regions, which will also be influenced by the increase in water demands in the upstream areas of these rivers. This phenomenon will trigger more competition over water resources. Finally, increasing temperatures and the associated sea level rise will result in seawater intrusion in these rivers, deltas, and coastal groundwater aquifers.

28.2 System Dynamics Approach

Focusing on controls, feedbacks, and delays, system dynamics (SD) is a modeling approach used for studying and simulating complex systems while emphasizing policy analysis and design [26,29]. The relationships between system components as well as feedback affecting system regulations are depicted in causal loop diagrams (CLDs). Based on the positive and negative impacts of variables on each other forming the loop, feedback loops can be reinforcing or balancing [6,7].

Although water reuse is involved with a complex series of aspects such as social, economical, technological, environmental, and so on, SD rarely has been used to analyze such a complex system [15,35]. The aim of this research is to investigate the impacts of various drivers on water reuse systems in causal loop systems using SD.

In this research, SD has been applied in order to show the relationships of variables involved in water reuse systems. There are a number of dynamic hypotheses in water reuse systems such as social, economic, and cultural drivers [4,5] and the following dynamic hypotheses will be discussed.

28.3 Dynamic Hypotheses

In order to study the water reuse complex system, after describing the problem and necessity of an integrated approach, a holistic view is applied to find drivers that affect urban water reuse.

28.3.1 Social Impacts

Water scarcity in many regions is one of the most serious issues. A number of reasons are behind this situation, which include, but are not restricted to, the relatively uneven distribution of precipitation, high temperatures, increased demands for irrigation water, and impacts of tourism. Climate change is expected to aggravate the situation even more. The use of wastewater is one of the most sustainable alternatives to coping with water shortage. It has a number of advantages that include closing the gap between supply and demand, stopping the pollution of freshwater resources, and providing sound solutions to water scarcity and climate change.

Water is one of the most valuable resources on Earth. Water and sanitation have a great effect on human health, food security, and quality of life. Demands on water resources for household, commercial, industrial, and agricultural purposes are increasing greatly. Yet, water is becoming scarcer globally, with many indications that it will become even more scarce in the future. More than one-third of the world's population—roughly 2.4 billion people—live in water-stressed countries and by 2025 the number is expected to rise to two-thirds. Growing demand for water due to the growing world population is creating significant challenges to both developed and developing countries.

The world populations have grown 1.5 times over the second half of the twentieth century, but the worldwide water usage has been growing at more than three times the population growth. In most countries, human population is growing while water availability is not. World population growth is projected to reach over 8 billion in 2030 and to level off at 9 billion by 2050. The United Nations has challenged the international community to work together to improve the situation, and one of the main objectives of the Millennium Development Goal (MDG) is to halve the number of people without access to safe drinking water and adequate sanitation by 2015.

Human and natural systems sensitive to water availability and water quality are increasingly stressed, or coming under threat. Those countries will have to face water quality degradation and meet the increasing needs of environmental protection and restoration. Use of organic waste nutrient cycles, from point-of-generation to point-of-production, closes the resource loop and provides an approach for the management of valuable wastewater resources. Failing to recover organic wastewater from urban areas means a huge loss of life-supporting resources that, instead of being used in agriculture for food production, fill rivers with polluted water.

There is a concern about its social acceptance. However, using treated wastewater in agriculture will provide these benefits: (1) increasing crop production, (2) saving scarce freshwater resources for other high value uses, and (3) minimizing fertilizers due to the high nutrient contents of treated wastewater. Due to sanitary restrictions and social acceptance, the number of crops grown under irrigation using treated wastewater could be limited.

Increasing temperatures are speeding up the hydrological cycle, as evidenced by more intense rainfall patterns in some regions and more rapid return flows of water to the atmosphere. With higher temperatures, agricultural seasons are affected.

Similarly, an increased frequency of extreme events increases the risks for the users and managers of systems for water regulation and supply, for example, dam operators. Higher air temperatures increase vapor content, which adds to the severity of extreme events. Opportunities for reuse of water in agriculture are likely to be further curtailed, that is, within the time span of a season. Uncertainty is bound to increase. Societies are responding by implementing mitigation and adaptation measures. Some of these are slow, others are fast; some are going in one direction whereas others result in interactions.

Since the late 1970s and early 1980s, source controls and industrial wastewater pretreatment programs have been initiated to limit the discharge of industrial pollutants into municipal sewers, and these programs have resulted in a dramatic reduction of toxic pollutants in wastewater and sludge [43,52]. Several decades of research and operational experience have shown that through proper treatment and management, federal and state regulations can be met for reuse of wastewater and biosolids [17,30,45].

Furthermore, negative impacts such as contamination of groundwater and an increase in soil metal concentrations have been noted in some applications [5]. Over the past several decades, a number of surveys have been conducted to evaluate public attitudes toward the beneficial reuse of reclaimed wastewater and/or biosolids. Bruvold [10,11,12] summarized the results of several surveys that assessed public attitudes toward 27 uses of reclaimed water. His findings indicated that attitudes toward wastewater reuse were positive overall; however, specific options, such as reuse for drinking, were negatively perceived. Broad [9] noted a similar tendency in perception regarding wastewater reuse and further determined that females were less positive to personal use options and worried more about the health and safety aspects of reclaimed wastewater reuse. A survey by Lohman and Milliken [39] indicated an overall positive attitude with no significant change in general public attitudes between 1982 and 1985. Other surveys have also noted positive attitudes toward overall wastewater reuse [32].

28.3.1.1 People's Conception of Urban Water Reuse

Many factors that influence public attitudes of wastewater reuse and biosolids recycling have been identified such as demographics, knowledge, trust or belief, cost, health and environmental concerns, and information sources utilized. Bruvold [10,11,12] found that younger, well-educated people with higher incomes were associated with more positive attitudes than the reverse. Degree of contact was found to be a more reliable predictor than one comprised of indicators including environment, treatment,

distribution, and conservation ratings. Lohman and Milliken [39] also found an inverse relationship between age and acceptance. Males tended to be more supportive than females, and people from households with children were slightly more willing than average to drink reclaimed water; however, results also suggested that factors other than intimacy of contact appeared to be of importance in respondents' levels of support. For example, environmental impacts, conservation factors, and economic costs of various kinds all seem to be better predictors of public acceptance than a standard based on intimacy of contact. Broad [9] indicated that health and safety rather than cost were considered important in convincing people to drink and cook with reclaimed water. Four variables—trust, knowledge, antitechnology, and alienation were main factors in explaining attitudes. Additionally, trust in industry, government regulatory agencies, and science figured prominently in the decision to accept or reject reuse projects. Education and knowledge had direct effects, while the effect of age was reversed. Technical information about biosolids had little impact on acceptability. Lindsay [37] found that most socioeconomic characteristics of the respondents did not influence attitudes toward their acceptance or rejection of boisolids applications. Survey results suggested that as the volume of media information increases, support decreases, and that perception of the potential economic benefits and negative impacts from biosolids land application can be very influential in achieving public acceptance.

Factors influencing attitudes toward general environmental issues or specific issues (such as siting of landfills or nuclear storage/transport) may or may not be similar to those influencing attitudes toward wastewater reuse or biosolid recycling. Demographics (gender, age, education, income, household children, duration of resident, political status, or religion), and other factors such as knowledge, distance, and belief or trust have been identified in influencing attitudes concerning environmental issues [1,40,56]. However, the influence of these factors on attitudes toward wastewater reuse or biosolid recycling is inconclusive.

28.3.1.2 Regulation

28.3.1.2.1 Administrative Orders

Besides the policy solution of pricing instrument, administrative orders can result in similar effects on wastewater reclamation and reuse by means of orders, rules, and regulations that restrict the use of freshwater in particular areas or departments of society. Administrative orders may take the other form of encouraging or requiring the use of reclaimed wastewater in particular areas or departments. Individuals facing prolonged restrictions to the use of water may be willing to pay a higher price for recycled water than policy makers may anticipate [27].

28.3.1.2.2 Penalty for Violation of Policies

Even if designed well, policies may receive fewer positive responses than anticipated, whether from reclaimed wastewater suppliers or potential clients, for the ineffective encouragement or even violations of policies. Penalty provisions are the inherent constituent of policies to ensure the achievement of policy objectives. The water reuse in China has been stagnant despite widespread water shortages [60].

28.3.1.3 Population

It is foreseen that, within a couple of decades, water scarcity may affect about two-thirds of the world's population. In many countries, there is still a tendency to deal with water scarcity problems, mainly by augmenting the water supply, for example, by increasing surface and groundwater storage and allocation through the creation of new infrastructure, desalination of saltwater or brackish water, reuse of wastewater, or recharging aquifers. This tendency has prevailed over reducing water demand, for example, by stemming the losses in transport and distribution systems, implementing adequate tariff systems, which seek to encourage lower water demand levels, changing water use technologies, and, generally, increasing the efficiency of water use in domestic, industrial, and irrigation systems; in other words, seeking to increase overall water productivity [8].

28.3.1.4 Society Health Impacts

According to the World Health Organization (WHO), 1.1 billion people around the world lack access to "improved water supply" [59] and more than 2.4 billion lack access to "improved sanitation" [48].

Wastewater, if well treated, can be an important source of water and nutrients for irrigation in developing countries, particularly but not restricted to those located in arid and semi-arid areas. The use of wastewater is widespread and represents around 10% of the total irrigated surface worldwide, although varying widely at local levels [48]. While the use of wastewater has positive effects for farmers, it may also have negative effects on human health and the environment. The negative effects impact not only farmers but also a wide range of people.

There is no complete global data on the extent to which wastewater is used to irrigate land, mostly due to a lack of heterogeneous data and the fear that countries have about disclosing information; economic penalties can be imposed if produce is found to have been irrigated with low-quality water.

As wastewater reuse is becoming a necessity due to shortage in freshwater supply, it is important for governments to put in place wise but feasible management practices. In order to implement sustainable reuse of wastewater and to contribute to food security, reuse projects need to be planned and constructed for the long term, based on local needs.

28.3.1.5 Culture

Cultural factors have recently been identified as a key moderator in wastewater reuse for fish farming in Egypt [41]. Aside from the proposed Red Sea–Dead Sea Canal, desalination, dams, artificial recharge of the aquifers, microscale water management, improvements in irrigation technologies, and the reuse of treated wastewater were mentioned as areas for further development. However, it was also noted that it is imperative to invest in education: not only to educate the public, and future generations, about water shortages but also to raise awareness about implications of not conserving this resource and to provide effective methods for conservation [50].

The case of treated wastewater serves to highlight the disparity in knowledge and awareness between the public and the scientific community: the scientists and environmentalists are all in agreement that the reuse of treated wastewater should become standard procedure in agriculture whereas the local farmers were less sure of the concept, especially in Jordan. The perceived stigma of irrigating crops with treated wastewater is a concern among farmers who consider the practice to adversely affect the acceptance and price of their crops in overseas markets. Social understanding of the farmer's point of view needs to go hand in hand with investment to spread awareness about potential new water sources and to allay any concerns or questions the farmers and public may have.

A telephone survey instrument was developed and administrated to evaluate public attitudes and knowledge concerning wastewater reuse and biosolids recycling with respect to population demographics (gender, age, education level, and household income). Information from 300 Knox County area residents was collected and analyzed. The study indicated that 58% of respondents supported wastewater reuse and 75% of respondents supported biosolids recycling. Acceptance decreased with increasing possibility of contact for both wastewater reuse and biosolids recycling. For wastewater reuse, respondents were positive to applications not involving close personal contact (such as firefighting, car washing, lawn irrigation, and agricultural uses), while uses of wastewater for possible consumption (released into potable surface or groundwater supplies) or applications involving close personal contact (laundry) were unfavorable. No significant difference was found between females and males concerning wastewater reuse; however, females tended to be more resistant to options that involved close personal contact. Age was significantly associated with attitudes. In particular, the 65 or over age group was significantly less supportive of wastewater reuse. For biosolids recycling, participants responded favorably for all eight application uses (on farmland, grazing land, public parks, highway medians, home gardens, lawns, forestland, and as mulch after composting), except those 65 years of age and older. This cohort felt significantly less favorable toward five of the options than did other age groups. No significant difference was found between females and males concerning biosolids recycling. Education was not statistically associated with attitudes; however, the high school or less group tended to be less positive than other groups. This study also found that the test population was not very knowledgeable of wastewater reuse and biosolids recycling issues. The overall level of correct responses to six topical questions was less than 50%. Knowledge was found to be lower with increased age (≥65) and lower educational attainment level (≤ high school degree). Knowledge was significantly and positively associated with attitudes, so increasing knowledge was likely to increase public support. Additionally, this study indicated that television and newspaper were dominant information sources for environmental news and sources such as governmental agencies, local utilities, and environmental groups were likely to be negligible information sources [14]. Urban water reuse benefits as well as increasing people's awareness has led to a growing demand for reclaimed water in order to compensate for the effects of the main challenges of water in quantity and quality [2].

28.3.2 Economic Impacts

Desalination, water conservation, and reuse are the most common approaches in order to minimize the effects of water scarcity and increase water supply capacities [3,24]. Many countries invest in water reuse due to the adverse effects of desalination and its energy-intensive and expensive technology [21,24].

An integrated approach is associated with water quantity, quality, and pricing aspects and the benefits of which range from water conservation to energy and material efficiency.

If the reuse water price is set according to associated costs, the price may be higher than the potable water. Hence, considering the indirect savings and benefits of water reuse, the price is set much lower. There is a need for subsidies from government to compensate for the treatment-supply costs [3]. Since water reuse capacities increase, water withdrawal reduces in quantity and potable water treatment cost will decrease.

Where utilization is now approaching hydrological limits, and the combined effects of demographic growth, increased economic activity, and improved standards of living have increased competition for remaining resources. Water resources are already overexploited or are becoming so with likely future aggravation where demographic growth is strong.

Wastewater production is the only potential water source that will increase as the population grows and the demand on freshwater increases. Therefore, wastewater should be viewed as a resource that must be recovered and added to the water budget. If wastewater is recognized as part of the total water cycle and managed within the integrated water resources management (IWRM) process, this will help meet the requirement.

Reuse increases the total available water supply and reduces the need to develop new water resources and, therefore, provides an adaptation solution to climate change or population density induced water scarcity by increasing water availability. Water recycling and reuse is meant to help close the gap in the water cycle and therefore enable sustainable reuse of available water resources [8,46].

Decisions affecting future energy supplies need to consider water availability and costs just as decisions affecting future water supplies need to consider energy availability and costs. Both will impact our use of these resources as economic growth and protect our environments. For instance, the Mediterranean basin is nowadays depending on agriculture for its economic and social development, and secondarily on industry and other economic activities. Irrigated agriculture in competition with other sectors will face increasing problems of water quantity and quality, considering increasingly limited conventional water

resources, growing future requirements, and a decrease in the volume of freshwater available for agriculture. Agriculture will remain an important sector of economy in all Mediterranean countries. This is particularly true for the developing countries on the Mediterranean, which use export opportunities to neighboring countries and the European Union, but in order to satisfy the demand of these populations, agricultural production still has to be increased. This is not possible without available water resources for irrigation. Therefore, alternatives like the reuse of waste water in agriculture have to be seriously considered.

Wastewater reuse in the region can partially contribute to solving the problem of quality and quantity. Reusing of wastewater is considered one of the effective adaptation strategies in the water sector, which helps close the demand-supply gap in water resources, through sustainable reclamation of wastewater, especially in the agricultural sector. This would also have a major role in agricultural economy, both on a qualitative and a quantitative basis, and in the well-being and the health of society. All indicators point to an increase in environmental and water scarcity problems with negative implications toward current and future sustainability. As the demand for water continues to rise, it is imperative that this limited resource be used efficiently and in sustainable ways for agriculture and other purposes.

28.3.2.1 Pricing Instrument

Water-pricing policies have to provide adequate incentives for users to use water resources efficiently and thereby contribute to the environmental objectives [16]. Pricing instrument particularly means setting the price or pricing principles of reclaimed wastewater, usually compared with freshwater prices, for users. Generally, reclaimed wastewater price is discounted on the basis of freshwater price for the same water users. The willingness to reuse reclaimed wastewater may be promoted by pricing instrument. Valid pricing instruments are especially necessary when the urgent demand to develop an efficient reclaimed wastewater market is considered. Water pricing instruments can be further enhanced, if it is charged in a differentiated way [49].

28.3.3 Water and Wastewater Consumption

Uses for reclaimed municipal wastewater can be classified into the following categories: urban, industrial, agricultural, groundwater recharge and augmentation of potable supplies, recreational, and habitat restoration/enhancement. Urban reuse systems provide reclaimed wastewater for various nonpotable purposes including irrigation of public parks, recreation centers, athletic fields, schoolyards, playing fields, highway medians and shoulders, landscaped areas, and golf courses. Nonpotable wastewater reuse only requires the conventional water and wastewater treatment technologies that are readily available. Furthermore, because properly implemented reuse does not promote significant health risks, nonpotable applications have generally been accepted and endorsed by the public where introduced. Alternative nonpotable uses include commercial applications such as vehicle washing, window washing, mixing water for pesticides, herbicides, and liquid fertilizers, use in ornamental landscape and decorative water features (such as fountains, reflecting pools, and waterfalls), dust control and concrete production on construction projects, fire protection, and toilet/urinal flushing in commercial and industrial buildings [54].

Industrial wastewater reuse represents a significant potential market for reclaimed water in the United States and other developed countries. Prospective uses include evaporative cooling water, boiler-feed water, process water, and irrigation/maintenance of plant grounds. The Boeing Company's new park office in Renton, WA was the first facility in the Pacific Northwest to use a municipal wastewater treatment effluent-based cooling system [13]. Reclaimed municipal wastewater has also been used for cooling tower makeup water in Maryland and in cooling systems for office buildings in Japan [2].

Agricultural irrigation uses currently represent about 40% of the total water demand nationwide [51]. The use of reclaimed municipal wastewater in agriculture has been well accepted and is still the major application for water reuse. California uses over 40% of its 4.32×10^5 m³/yr reclaimed water for agriculture or landscape irrigation [2]. In Central Florida, use of reclaimed water for citrus irrigation [61] has occurred for years.

Ornamental and recreational reuse have also been carried out, including the discharge of treated municipal wastewater into city streams during dry seasons [44] and river flow augmentation [28]. Additional applications of municipal wastewater reuse include creation or enhancement of wetland habitat and direct potable reuse [4,23,25,31,47].

Industry in most countries is still emerging. Most of the waste disposed into the countries is composed of domestic and municipal sewage except in the case of some countries, where agricultural land drainage, mixed with both industrial and municipal wastewater, is also disposed to the sea.

Thanks to superior productivity, urban-based enterprises contribute large shares of gross domestic product (GDP). While in the past, industrial development has used as much water as might be available, today it is increasingly recognized through market signals that water has a value and that there is an opportunity cost associated with most uses of finite resources. As a result, the trend is clear. Less is withdrawn and more wastewater is treated, so that both the water and some of the elements that accumulate in the production process can be reused. Industries in many countries are now consuming less water per unit output and reducing pollution loads in their waste.

Water saving devices in households and elsewhere offer similar opportunities with lower net cost. For instance, domestic wastewater can be treated to separate valuable nutrients and be reused together with the treated water for urban agriculture [57].

28.3.4 The Role of Technology

28.3.4.1 Recycling, Reclamation, and Reuse

Water reclamation is an important source of water in water-scarce areas and its reuse can contribute significantly to meeting urban water needs [18]. Used water can be treated to meet any quality requirement, including for potable use or for industrial use, both requiring ultrapure water. Advanced treatment technologies are increasingly becoming realistic choices for water, wastewater, and stormwater treatment. They help cope with stringent standards, enhance capacities, and address contaminants that cannot be managed with conventional technologies [4]. An evolving approach is source separation, which recognizes that different water qualities are appropriate for different uses. This general concept has been used for several decades with dual distribution systems, where reclaimed water is distributed for urban agriculture with high value crops and is now being extended, recognizing the small proportion of water that actually is needed to meet potable water standards.

Given their better capabilities and performances, membrane-based technologies and membrane bioreactors are penetrating the markets in many water-scarce regions because they enable the recycling of wastes and the use of alternative sources (such as brackish water and seawater). Recent studies [19] suggest that a steady decrease in membrane costs has occurred during the last decade. Robust and durable membrane materials, as well as low-energy membrane systems (in some cases gravity driven) are being developed. Other technologies, such as photovoltaic systems with a renewable power source (solar driven) and oxidation processes, which can be enhanced with catalytic processes in combination with membrane systems, are coming onto the market. This trend will enable utilities to upgrade their systems [4].

Nanotechnology concepts are being investigated for higher performing membranes with less fouling properties, improved hydraulic conductivity, and more selective rejection/transport characteristics. Microbial fuel cells are emerging as a potential breakthrough technology that will be able to capture electrical energy directly from the organic matter present in the waste stream. Although these technologies are still in the early stages of development, and significant advances in process efficiency, demonstration, and production to commercial scale are necessary, they have the potential to enhance treatment-process performances and improve the efficiency of resources use [18].

28.3.4.2 Natural Treatment Systems

Natural treatment systems (NTSs) use natural processes to improve water quality, maintain the natural environment, and recharge depleted groundwater sources. These systems are, in general, cost-effective. Wetlands and oxidation ponds are examples of NTSs that are increasingly being used to treat and retain stormwater, wastewater, and drinking water flows. NTSs have the advantage of being able to remove a wide variety of contaminants at the same time, which makes them a total treatment system on their own, and they are increasingly being used for water reclamation.

28.3.4.3 Source Separation of Waste Streams

Key to the application of most of the new treatment technologies is the separation of the different flows of wastewater according to their pollution load. Most of the contaminants of concern in wastewater are contained in black water. For example, most of the organic and microbial contaminants are generated from fecal matter (which accounts for only 25% of domestic waste), while most of the nitrogen and the emerging contaminants, such as pharmaceutically active compounds and endocrine disrupting compounds, are present mainly in urine.

New technologies, such as vacuum sewage systems and urine separation toilets, which reduce most of the nitrogen and trace organic contaminants, have made it possible to handle a small and concentrated amount of waste.

28.3.4.4 Desalination

Desalination of brackish water and seawater is becoming increasingly economical, thanks to advanced membrane technologies and improved energy efficiency. In countries that have exhausted most of their renewable water resources, desalinated water meets both potable and industrial demands. However, its use in agriculture remains limited. Desalinated water is already being used for the cultivation of high value crops in greenhouses.

These technologies create opportunities for the reuse of greywater at the source and the recovery and reuse of nutrients. They also reduce the cost of extensive sewer systems and minimize, or may even avoid, the use of clean water to carry waste.

28.4 Integrated Approach

The review of previous studies and all aspects of water reuse show that water reclamation decision making in an integrated water resources management perspective is a complex process in comparison of societal, technological, economical, and environmental criteria [20,36,38].

SD approach is a useful tool in understanding nonlinear behavior of socioeconomic–environmental systems and policy implication assessment with feedbacks, controls, and delays

[30,58]. In this study, the impacts of involving drivers in water reuse such as system cost, water price, and, in particular, population have been captured. Furthermore, the feedbacks of water reclamation on the abovementioned drivers are investigated.

The CLDs in Figure 28.1 show the water reuse system. A "+" sign represents that an increase in the variable at arrow tail causes an increase in the arrow head variable and indicates a reinforcing relationship between two variables. A "−" sign indicates a balancing relationship between two variables at the ends of an arrow and represents one variable adversely affects the other.

Water withdrawal is related to water reuse as well as water demand. Therefore, by introducing the new resource of water, water reuse, water withdrawal will be reduced. In other words, the total water demand can be partially satisfied using water reuse for nonpotable water.

As it is shown in Figure 28.1, there are three loops in the water reuse system, all of which are Reinforcing (R). It is concluded that the governing mechanisms continue until an inhibitor driver is applied. In other words, population growth, water demand, water price, and urban water reuse cost will have an increasing trend. Inhibitor drivers that can be used to prevent this destructive mechanism are subsidies that belong to treated water, and improvement of technology in treated water in order to reduce the cost. Social conception of people is very important in enhancing demand for treated water and, consequently, decreases demand for freshwater. In this, freshwater resources will be preserved.

28.5 Case Study

Iran as an arid and semi-arid area is challenging with water scarcity and can be considered as a case study for this review study.

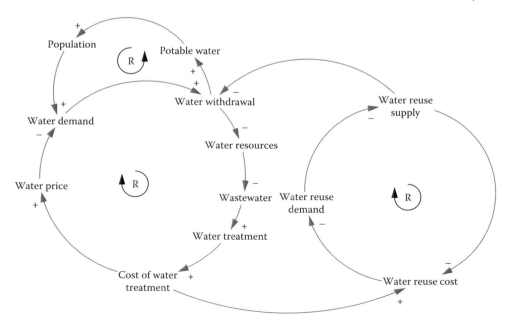

FIGURE 28.1   Interaction between drivers that influence urban water reuse.

With the support of the government, the population growth rate exceeded 4% during the Iran-Iraq war in the early 1980s. The result of this baby boom is a generation that accounts for 70% of unemployment and is ironically blamed by the government for lowering the current fertility level, which has brought down the population growth rate to 1.3%. Iranians are currently using more than 70% of their renewable freshwater resources, while using more than 40% of renewable freshwater resources means entering a water stress mode. To have sustainable water resources, Iran needs to apply urban water reuse using treatment plants. Recently, after drought periods start, decision makers were concerned about new water resources. Besides construction of huge dams, which impose noncompensable damages to the environment, construction of water treatment plants has been started. Yet, there are some concerns about using such kind of water resources for the following reasons.

28.5.1 People's Perception of Treated Water

Most populations of Iran are Muslim. Based on Islam principles, cleanness is frequently advised. Therefore, Iranians are very concerned about wastewater. In their mind, wastewater is not clean and any contact of clothes or body with wastewater will make them dirty and they cannot pray. Thus, they should remove impurity and wash it with clean water. This belief shows that there is a hindrance toward government in Iran for making water reuse prevalent.

By an increase in education level and knowledge of people about treated water, urban water reuse has been accepted by people for agricultural consumption and irrigation of landscapes. Yet, there are doubts for using water in household consumption.

28.5.2 Treated Water Cost

Water reuse cost in Iran is high due to import of treatment technology. Therefore, the government has to consider subsidies in order to reduce treated water. Since Iran is located in an arid area, the government has no way to supply water at a reduced price.

Rather than proactive management to prevent water problems, Iran's reactive management has focused on curing the symptoms, while the causes of the problems are becoming worse over time. If development is to be sustainable, Iranians need to replace symptom management with problem prevention. Hence, it is necessary to find a solution for application of treated water into recycled water.

28.6 Summary and Conclusions

It is concluded that considering regional and climatic parameters cannot lead to a true understanding of water reuse practices.

The world's freshwater resources are under strain. Reuse of wastewater, in concert with other water conservation strategies, can help lessen manufactured stresses arising from pollution of receiving waters. Reuse will provide relief to the Mediterranean

from the hazards of directing this water to its basin. Irrigation with treated municipal wastewater is considered an environmentally sound wastewater disposal practice compared to its direct disposal to the surface or groundwater bodies. There are concomitant environmental risks with wastewater reuse, such as transport of harmful contaminants in soils, pollution of groundwater and surface water, degradation of soil quality, for example, salinization, impacts on plant growth, and the transmission of disease via the consumption of wastewater-irrigated vegetables. The challenge facing wastewater reuse is to minimize such risks to maximize the net environmental gain.

One important change in the future is a likely shift toward a much higher water demand for household and industrial uses in urban areas. Much of the water use in these sectors is considerably less consumptive as compared to the open landscapes of agriculture, where the resource is depleted through evaporation and transpiration. However, storage capacities and conveyance systems are required. There are already many examples of huge transfers of water to urban and industrial uses. Even if a large fraction of the water used in urban centers returns to aquifers, lakes, or rivers, this will be in downstream areas affecting the quality. Further, many of the fast growing urban centers are located in coastal zones, which reduces options for reuse in other sectors at other locations.

Projects for sustainability require a holistic and integrated approach that takes into account the overall cultural context in which and by which water is used. Community measures (income, health, education) and resource measures (quality, quantity, consumption) coupled with the participation and empowerment of local communities should be the preferred methodological approach in water management.

References

1. Arcury, T.A. 1990. Environmental attitude and environmental knowledge. *Human Organization*, 49(4), 300–304.
2. Asano, T., Burton. F., Leverenz, H., Tsuchihashi, R., and Tchobanoglous, G. 2007. *Water Reuse: Issues, Technologies, and Applications*. McGraw-Hill Professional, New York.
3. AWWA (American Water Works Association). 2008. Water Reuse Rates and Charges: Survey Results. American Water Works Association, Denver, CO.
4. Bahri, A. 2012. Integrated Urban Water Management: Elanders, Global Water Partnership Technical Committee Elanders, Stockholm, Sweden.
5. Barry, G.A., Bloesch, P., Garden, E.A., Rayment, G.E., and Hogarth, D.M. 1998. Re-use of sewage biosolids on canelands. *Proceeding of the 20th Conference of the Australian Society of Surgar Cane Technologists Held at Ballina*, NSW, Australia, 69–75.
6. Bazrkar, M.H., Fathian, F., and Eslamian, S.S. 2013. Runoff modeling in order to investigate the most effective factors in flood events using system dynamic approach (case study: Tehran watershed, Iran). *Journal of Flood Engineering*, 4(1–2), 39–59.

7. Bazrkar, M.H., Tavakoli-Nabavi, E., Zamani, N., and Eslamian, S. 2013. System dynamic approach to hydropolitics in Hirmand transboundary river basin from sustainability perspective. *International Journal of Hydrology Science and Technology*, 3(4), 378–398.

8. Braga, B., Chartres, C., Cosgrove, W.J., Veiga da Cunha, L., Gleick, P.H., Kabat, P., Ait Kadi, M. et al. 2014. *Gulbenkian Think Tank on Water and the Future of Humanity, Water and the Future of Humanity: Revisiting Water Security*, Calouste Gulbenkian Foundation, UK.

9. Broad, P.A. 1996. *Community Views on Water Reuse*. Sydney Water, Australia.

10. Bruvold, W.H. 1981. Community evaluation of adapted use of reclaimed water. *Water Resources Research*, 17(3), 487–490.

11. Bruvold, W.H. 1985. Obtaining public support for reuse water. *Journal of American Water Works Association (AWWA)*, 77(7), 72–77.

12. Bruvold, W.H. 1988. Public opinion on water reuse options. *Journal of Water Pollution Control Federation*, 60(1):45–49.

13. Ceres and Pacific Institute. 2009. Water scarcity & climate change: Growing risks for businesses & investors. A Ceres Report. http://www.pacinst.org/reports/business_water_climate/full_report.pdf.

14. Chun, M. and Robinson, K.G. 2003. Assessment of Public Attitudes and Knowledge Concerning Wastewater Reuse and Biosolids Recycling, MSc thesis, University of Tennessee, Knoxville, TN.

15. Chung, G., Lansey, K., Blowers, P., Brooks, P., Ela, W., Stewart, S., and Wilson, P. 2008. A general water supply planning model: Evaluation of decentralized treatment. *Environmental Model Software*, 23, 893–905.

16. Correljé, A., François, D., and Verbeke, T. 2007. Integrating water management and principles of policy: Towards an EU framework? *Journal of Cleaner Production*, 15, 1499–1506.

17. Crook, J. 1999. Indirect potable reuse. *Water Environmental & Technology*, 11(5), 71–75.

18. Daigger, G.T. 2007. Creation of sustainable water resources by water reclamation and reuse. *Proceedings of the 3rd International Conference on Sustainable Water Environment: Integrated Water Resources Management—New Steps, 79–88*. Sapporo, Japan.

19. DeCarolis, J., Adham, S., Pearce, W.R., Hirani, Z., Lacy, S., and Stephenson, R. 2007. Cost trends of MBR systems for municipal wastewater treatment. In: Water Environment Federation (ed.), *WEFTEC 07 (3407–3418)*. Water Environment Federation, San Diego, CA.

20. Deviney, F.A. Jr., Brown, D.E., and Rice, K.C. 2012. Evaluation of Bayesian estimation of a hidden continuous-time Markov chain model with application to threshold violation in water-quality indicators. *Journal of Environmental Information*, 19(2), 70–78.

21. Exall, K., Marsalek, J., and Schaefer, K. 2004. A review of water reuse and recycling, with reference to Canadian practice and potential: Incentives and implementation. *Water Quality Resources Journal of Canada*, 39(1), 1–12.

22. Guest, J.S., Skerlos, S.J., Barnard, J., Brucebeck, M., Daigger, G., Hilger, H., Jackson, S.J. et al. 2009. A new planning and design paradigm to achieve sustainable resource recovery from wastewater. *Environmental Science & Technology*, 43, 6126–6130.

23. Gura, A. and Alsalem, S.S. 1992. Potential and present wastewater reuse in Jordan. *Water and Technology*, 26(7–8), 1573–1581.

24. GWI. 2009. *Municipal Water Reuse Markets 2010*, Global Water Intelligence, Oxford.

25. Hermanowicz, S.W. and Asano, T. 1999. Abel Wolman's the metabolism of cities revisited: A case for water recycling and reuse. *Water Science and Technology*, 40(4),29–36.

26. Huang, G.H. and Cao, M.F. 2011. Analysis of solution methods for interval linear programming. *Journal of Environmental Information*, 17(2), 54–64.

27. Hurlimann, A.C. 2009. Water supply in regional Victoria Australia: A review of the water cartage industry and willingness to pay for recycled water. *Resources, Conservation and Recycling*, 53, 262–268.

28. Juanico, M. and Friedler, E. 1999. Wastewater reuse for river recovery in semi-arid Israel. *Water Science and Technology*, 40(4–5), 43–50.

29. Khan, S., Yufeng, L., and Ahmad, A. 2009. Analyzing complex behavior of hydrological systems through a system dynamics approach. *Environmental Model Software*, 24, 1363–1372.

30. Lauer, W.C. 1993. *Denver's Direct Potable Reuse Demonstration Project Final Report Executive Summary*. Denver Water Department, Denver, CO.

31. Lauer, W.C. and Rogers, S.E. 1998. The demonstration of direct potable water reuse: Denver's landmark project. *Water Reclamation and Reuse*, 1269–1333.

32. LCRA (Lower Colorado River Authority). 1995. Lakeway area reuse public opinion survey results. http://twri.tamu.edu/ research/other/lakeway.html.

33. Leewongtanawit, B. and Kim, J. K. 2009. Improving energy recovery for water minimization. *Energy*, 34, 880–893.

34. Li, Y.P. and Huang, G.H. 2011. Planning agricultural water resources system associated with fuzzy and random features. *Journal of American Water Resources Association*, 47(4), 841–860.

35. Li, Y.P. and Huang, G.H. 2012. A recourse-based nonlinear programming model for stream water quality management. *Stochastic Environment Resources Risk Assessment*, 26, 207–223.

36. Li, Y.P., Huang, G.H., and Chen, X. 2009. Multistage scenario-based interval stochastic programming for planning water resources allocation. *Stochastic Environmental Resources Risk Assessment*, 23, 781–792.

37. Lindsay, B.E., Zhou, H., and Halstead, J. 2000. Factors influencing resident attitudes regarding the land application of biosolids. *American Journal of Alternative Agriculture*, 15(2), 88–95.

38. Liu, Z. and Tong, T.Y. 2011. Using HSPF to model the hydrologic and water quality impacts of riparian land-use

change in a small watershed. *Journal of Environmental Information*, 17(1), 1–14.

39. Lohman, L.C. and Milliken, G. 1984. Public attitudes toward potable wastewater reuse: A longitudinal case study. *Proceeding of the Water Reuse Symposium*, San Diego, CA.

40. MacGregor, D. and Slovic, P. 1994. Perceived risks of radioactive waste transport through Oregon: Results of a statewide survey. *Risk Analysis*, 14(1):5–14.

41. Mancy, K.H., Fattal, B., and Kelada, S. 2000. Cultural implications of wastewater reuse in fish farming in the Middle East. *Water Science and Technology*, 42(1–2), 235–239.

42. Mo, W., Nasiri, F., Eckelman, M.J., Zhang, Q., and Zimmerman, J.B. 2010. Measuring the embodied energy in drinking water supply systems: A case study in Great Lakes Region. *Journal of Environmental Science and Technology*, 44, 951–952.

43. NRC (National Research Council). 1996. *Use of Reclaimed Water and Sludge in Food Crop Production*. National Academy Press, Washington, DC.

44. Okun, D.A. 1997. Reclaimed water through dual systems. *AWWA*, 89(11), 52.

45. Oliver, A.W., Elsenberg, D.M. et al. 1998. City of San Diego health effects study on potable water reuse. *Wastewater Reclamation and Reuse*, 521–579.

46. Perrone, D., Murphy, J., and Hornberger, G.M. 2011. Gaining perspective on the water–energy nexus at the community scale. *Environmental Science and Technology*, 45, 4228–4234.

47. Reed, S., Parten, S., Matzen, G., and Pohern R. 1996 Water reuse for sludge management and wetland habitat. *Water Science and Technology*, 33(10/11), 213–219.

48. Rose, J.B. 2007. Water reclamation, reuse and public health. *Water Science and Technology*, 55(1/2), 275–282.

49. Shiferaw, B., Ratna Reddy V., and Wani, S.P. 2008. Watershed externalities, shifting cropping patterns and groundwater depletion in Indian semi-arid villages: The effect of alternative water pricing policies. *Ecological Economics*, 67, 327–340.

50. Shuval, H. and Dweik, H. 2007. *Water Resources in the Middle East: Israel-Palestinian Water Issues–From Conflict to Cooperation*. Springer, Berlin, 167–169.

51. Solley, W.B., Pierce, R.R., and Perlaman, H.A. 1993. *Estimate Use of Water in the United State in 1990. Circular 1081*. U.S. Geological Survey, Washington, DC.

52. Stehouwer, R.C., Wolf, A.M., and Doty, W.T. 2000. Chemical monitoring of sewage sludge in Pennsylvania: Variability and application uncertainty. *Journal of Environmental Quality*, 29(5), 1686–1695.

53. Strutt, J., Wilson, S., Shorney-Darby, H., Shaw, A., and Byers, A. 2008. Assessing the carbon footprint of water production. *Journal of American Water Works Association*, 100, 80–91.

54. U.S. Environmental Protection Agency. 2004. U.S. Agency for International Development. Guidelines for Water Reuse. Washington, DC. http://www.epa.gov/nrmrl/pubs/625r04108/625r04108.pdf.

55. UNESCO. 2009. *World Water Assessment Program. The United Nations World Water Development Report 3: Water in a Changing World,* Paris: UNESCO and London: Earthscan.

56. Van Liere, K.D. and Dunlap, R.E. 1980. The social bases of environmental concern: A review of hypotheses, explanation, and empirical evidence. *Public Opinion Quarterly*, 44, 181–197.

57. Vigneswaran, S. and Sundaravadivel, M. 2004. Recycle and reuse of domestic wastewater. In: *Vigneswaran Wastewater Recycle, Reuse, and Reclamation*. Encyclopedia of Life Support. Systems (EOLSS). Developed under the Auspices of the UNESCO, Eolss Publishers, Oxford, UK. http://www.eolss.net.

58. Winz, I., Brierley, G., and Trowsdale, S. 2009. The use of system dynamics simulation in water resources management, *Water Resources Management*, 23, 1301–1323.

59. World Health Organization, 1996. Water and Sanitation. WHO Information Fact Sheet No. 112. Geneva. http://www.who.int/inf-fs/en/fact112.html.

60. Yi, L., Jiao, W., Chen, X., and Chen, W. 2011. An overview of reclaimed water reuse in China. *Journal of Environmental Sciences*, 23(10), 1585–1593.

61. Zekrim, K. 1994. Treated municipal wastewater for citrus irrigation. *Journal of Plant Nutrition*, 17(5), 693–708.

62. Zimmerman, J.B., Mihelcic, J.R., and Smith, J.A. 2008. Global stressors on water quality and quantity. *Environmental Science and Technology Journal*, 42, 4247–4254.

29

Desalination Technologies: Are They Economical for Urban Areas?

Olga Barron
CSIRO

Elena Campos-Pozuelo
Valoriza Agua S.L.

Cristina Fernandez
Valoriza Agua S.L.

29.1 Introduction

The impacts of population growth and climate change on freshwater resources are forcing international scientific communities to look for alternative approaches to our current resource management. During the second half of the twentieth century, water withdrawals tripled worldwide (Figure 29.1) [46], but still 40% of the world's population is affected by a shortage of water. Currently, about 1.1 million people have limited access to safe drinking water and 2.4 million lack sanitation systems. Opportunities for generating cost-effective and potentially climate-independent water resources with controlled quality could be linked with desalination technologies [51]. Desalination allows a widening utilization of available resources by producing freshwater from saline or brackish natural water sources. However, currently, total desalinated water produced worldwide (77.4 million m^3/day; [25]) still comprises less than 1% of total worldwide water use.

Urban water demand is second only to agricultural water needs and accounts for more than 20% of total worldwide water use (Figure 29.1).

Until recently, desalination plants were limited to the water-scarce but oil-rich countries of the Middle East and North Africa, and to some tropical and subtropical islands. However, today, water sourced from desalinated seawater and brackish groundwater has become a commodity for many countries in order to satisfy their growing demand for water (Figure 29.2) [35]. Global desalination production includes 48% of water production in the Middle East (mainly in the Gulf States), 19% in Northern America, 14% in the Asia-Pacific region, 14% in Europe, and 6% in Africa [30].

The salinity of feedwater for desalination facilities ranges from approximately 1000 mg/L total dissolved salts (TDS) to 60,000 mg/L TDS, although feedwaters are typically labeled as one of two types: seawater or brackish water. Most seawater sources contain 30,000–4500 mg/L TDS, and seawater reverse osmosis (RO) membranes are used to treat waters within the TDS range 10,000–60,000 mg/L. Brackish water RO membranes are used to treat water (often groundwater) sources within a range of 1000–10,000 mg/L TDS. Feedwater type can dictate several design choices for a treatment plant, including desalination method, pretreatment steps, waste disposal method, and product recovery (the fraction of influent water that becomes product) [21]. Of the total global installed desalination capacity, 58.85% uses seawater as a feedwater source, followed by brackish water with 21.24% (Figure 29.3) [25].

By the 1990s, the use of desalination technologies for municipal water supplies had become commonplace with 70% of global desalinated water used by municipalities and 21% by industries (Figure 29.4) [30].

Over the years, as conventional water production costs have been rising in many parts of the world and costs of desalination declining, desalination has become more economically attractive and competitive. Lattemann et al. [30] estimated that by 2015 the costs of freshwater treatment, wastewater reuse, and desalination were likely to be similar, at least in the United States (Figure 29.5).

29.2 Desalination Processes and Desalination Cost

In general, desalination technologies may be grouped into thermal and nonthermal processes (Figure 29.6). Thermal methods are based on the separation of freshwater from saline water during water phase changes (i.e., distillation, freezing). Nonthermal methods include membrane processes based on forced filtration though semipermeable membranes (i.e., RO, nanofiltration) and ion removal from saline feedwater under the influence of an electrical potential difference (i.e., electrodialysis), which also uses a

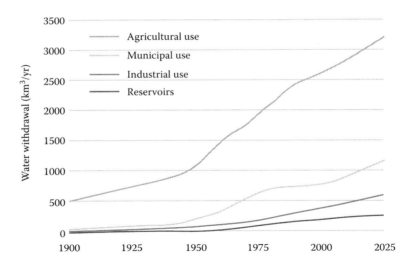

FIGURE 29.1 Evolution of freshwater demands worldwide. (Adapted from Desaldata. 2012. Global Desalination Market GWI. www.desaldata.com. Retrieved October 3, 2012.)

membrane technology for permeate and brine separation. There are also various combinations of these technologies available.

The installed world desalination capacity mainly uses the processes of RO and multistage flash distillation (MSF) (Figure 29.7) [12]. Other processes shown in Figure 29.7 include electrodeionization (EDI), electrodialysis (ED), and multiple-effect distillation (MED). Installed desalination capacity is roughly equally shared between thermal and membrane processes. However, older plants facing retirement are distillation units, while brackish water reverse osmosis (BWRO) or seawater reverse osmosis (SWRO) techniques are the fastest growing desalination techniques with the greatest number of

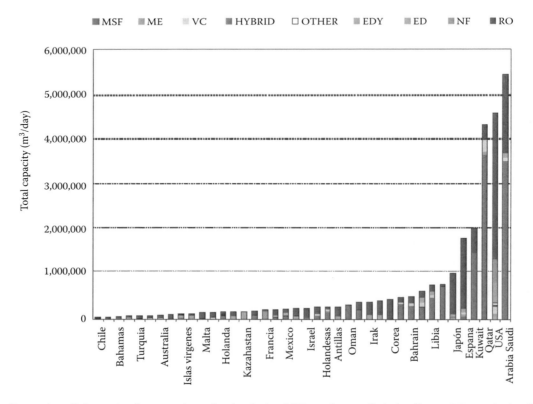

FIGURE 29.2 Current installed capacity (by countries and technologies: MSF—multistage flash distillation, ME—multiple-effect distillation, VC—vapor compression, EDI—electrodeionization, ED—electrodialysis, NF—nanofiltration, RO—reverse osmosis).

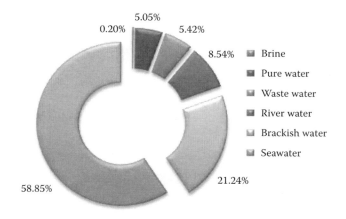

FIGURE 29.3 Total capacity installed in the world.

installations worldwide. It is likely that in the future, the total operating capacity of membrane units will increasingly exceed that of thermal units [9].

With upgraded designs and improved energy efficiencies, desalination units can currently deliver freshwater from the sea at costs that range from US$0.46 to US$0.80/m³, while freshwater from brackish water can now be produced at costs in the range of $0.10–0.20/m³, depending on feedwater salt content (this cost does not include the cost of water delivery to the consumers). Cost levels vary depending on local conditions, but there has been a significant cost reduction from the costs of the past (US$3.00 to US$5.00/m³ for seawater desalination).

By comparison, by the end of the twenty-first century the costs to withdraw freshwater from ground and/or surface water in western countries will range from US$0.40 to US$0.75/m³ before distribution [18]. For instance, domestic water price in Europe is on average US$0.70 with the highest prices in Germany (US$1.41), France (US$1.04), and the UK (US$0.93), and a lower price in Italy ($0.35) [6].

Desalination costs are dependent on the initial investment in the plant, the amortization period, and operating costs (which among other factors includes energy, labor employed, and maintenance). Distillation costs are commonly higher than RO costs [33]. The financial characteristics of a desalination plant are usually expressed in two ways: the capital costs and total annual operating water costs per unit of installed or process capacity, and are usually measured in units such as dollars per m³ (or kL) per day. Table 29.1 and Figure 29.8 show representative construction and operating costs on this basis for RO, MED, and electrodialysis reversal (EDR) desalination processes [2]. In addition to the desalination process itself, these costs include the cost for feedwater uptake and delivery to the desalination plant, feedwater pretreatment, product water posttreatment and storage, water delivery to the consumers, and brine (desalination byproduct) disposal.

29.2.1 Feedwater Quality

Feedwater quality largely influences desalination cost. The choice of desalination technologies, feedwater pretreatment needs, recovery rates (and hence volume of brine production), and energy requirements are largely controlled by feedwater salinity and feedwater chemical and biochemical composition.

Higher feedwater salt content leads to an increase in operating costs, as it requires more rigorous technological options, such as a larger membrane area and higher pressure for RO or a greater number of stages of distillation for MSF [9,28,43]. Typically, the cost of desalting seawater is three to five times the cost of desalting brackish water from the same size plant for both membrane and thermal methods [9,50]. Membrane processes achieve brackish water desalting most economically, with RO presently the cheapest process [19,29], which may provide a recovery rate (percentage of freshwater production from feedwater) of about 50% for seawater desalination and up to 90% for brackish water desalination [52].

Feedwater chemical composition may influence the scaling process in plants using thermal desalination techniques. Sulfate and bicarbonate ions in seawater can cause deposits of insoluble calcium and magnesium compounds at elevated temperatures and can lead to fouling of heat-transfer surfaces in the brine heater and heat-recovery stages. Biological composition of feedwater may lead to biological growth in pipelines and heat exchangers. High-temperature polymer additives, derived from polymaleic acid, are mostly used in conjunction with on-load sponge-ball cleaning to remove soft-scale depositions. Water can also be chlorinated to prevent biological growth.

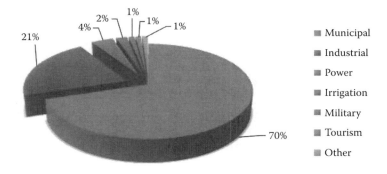

FIGURE 29.4 Global desalination capacities by user type.

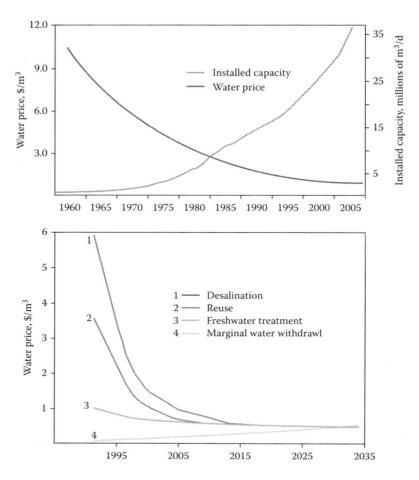

FIGURE 29.5 Water resource cost trends in the desalination market in US$ per cubic meter. Top: Total installed capacity and water price development. Bottom: Differentiated between water source types.

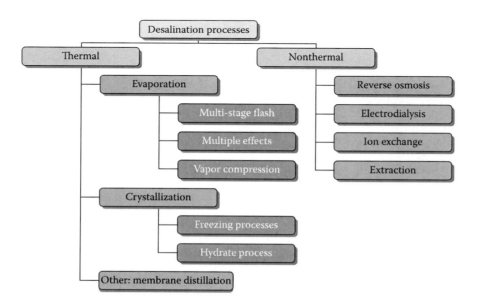

FIGURE 29.6 Commercially used technologies for water desalination.

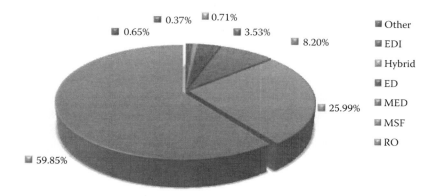

FIGURE 29.7 Total desalination capacity installed by technology.

TABLE 29.1 Construction and Operating Costs for RO, MED, and EDR

Parameter	Seawater RO	Brackish RO	MED	EDR
Capital cost (AU$/kL/day of product water)	1600–2500	600–1800	2500–3900	570–3250
Operating cost (AU$/kL/day of product water)	1.89–2.20	0.65–1.50	With waste heat: 0.55–0.95 Without waste heat: 1.8–2.80	1.00–2.80

Note: For construction costs, only direct capital costs associated with process works, including pretreatment and process treatment equipment, pumps, pipes, and control systems have been incorporated, and not the costs for delivery of the water to and from the plant, or associated posttreatment costs.

RO is generally more sensitive to feedwater quality, and becomes less productive at high water temperatures (up to 40°C); other factors that need to be considered include salinity level, bacterial activity, and water pollution level [8]. For instance, an increase in feedwater temperature results in an increased rate of salt and water diffusion across the membrane barrier at a rate of about 3%–5% per degree Celsius [52]. Fluctuation in feedwater quality may be particularly damaging for RO desalination units, which may cause unpredictable membrane fouling and affect RO membrane flux [20]. Hence, feedwater quality can influence the choice of membrane material (e.g., cellulose acetate or polyamide) and configuration (e.g., hollow fiber or spiral wound). Cellulose acetate membranes can be degraded by the formation of biological slimes on the membrane surface, while polyamide membranes cannot tolerate chlorine.

The cost of SWRO membranes contributes to the higher cost of seawater desalination; however, membrane cost has dropped

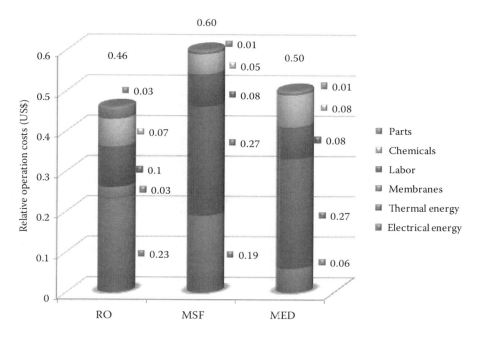

FIGURE 29.8 Relative operation costs in US$ of the main desalination processes.

by over 65% from 1985 to 2002 in real terms (more if adjusted for inflation), and membrane replacement now represents about 6% or less of the cost of produced desalinated water. The cost of chemicals for feedwater pretreatment and cartridge filter replacement has been reduced to 4% or less of the total cost of produced desalinated water [42]. Enhancement of membrane resistance to fouling (including biofouling) increases membrane flux; and membrane treatment and restoration result in a reduction in the energy required for desalination from a maximum of 25 kWh/m³ to as low as 3 kWh/m³.

Since desalination technologies, particularly RO, are sensitive to feedwater quality, the implementation of appropriate pretreatment processes facilitates a reduction in the overall cost of permeate production. The latest water pretreatment developments for RO desalination are based on integrated membrane systems (IMSs), as continuous microfiltration (CMF) or UF (ultrafiltration) in combination with RO and nanofiltration (NF). The IMSs are among the more expensive options for feedwater treatment, but they allow production of a better quality feedwater to be used for RO [48].

Use of NF as a water pretreatment option allows water softening and also removal of natural organic matter (NOM), viruses, pesticides and other micro pollutants, and heavy metal (arsenic) [47]. Among the new subjects that have been recently studied is the reduction of nitrate concentrations by NF.

NF technologies have been used in integrated desalination systems such as NF-SWRO and NF-MSF. The NF plant receives noncoagulated filtered seawater and reduces turbidity, microorganisms, and hardness. The concentration of monovalent salts can be reduced by 40%, and the overall concentration of TDS reduced by 57.7%, producing permeate far superior to seawater as a feedwater to SWRO or MSF. This makes it possible to operate a SWRO and MSF pilot plant at a high recovery rate (70% and 80%, respectively). The high water output in both integrated desalination systems, combined with a reduction of chemicals and energy (by about 25%–30%) allows producing freshwater from seawater at a 30% lower cost compared to conventional SWRO [4]. Furthermore, NF has also become a viable option for brackish water desalination.

29.2.2 Energy Requirements

As the energy component comprises a major cost, much effort in desalination research is related to minimizing the energy required for freshwater production. The requirement for thermal or electric energy input can represent 50%–75% of total operating costs [34]. Table 29.2 shows the energy use of various desalination processes.

The theoretical energy needed to remove salt from a 4% salt solution is 3.6 MJ/m³ or the equivalent of 1 kWh/m³, which is not likely to be achieved [45]. However, since the first installations, energy consumption by desalination plants has reduced by 60% to current consumption varying between 3 and 4.5 kWh/m³ [10].

RO seawater desalination currently consumes about 3.5–4.2 kWh/m³ of product in desalting seawater (3.5% salinity) at

TABLE 29.2 Energy Use by Various Desalination Processes

Process	Gain Output Ratio (GOR)[a]	Electrical Energy Consumption kWh/m³	Thermal Energy Consumption kWh/m³	Total Energy Consumption kWh/m³
MSF	8–12	3.25–3.75	9.75–6.75	13–10.5
MED	8–12	2.5–2.9	6.5–4.5	9–7.4
BWRO	N/A	1.0–2.5	N/A	1.0–2.5
SWRO	N/A	4.5–8.5	N/A	4.5–8.5

Source: Adapted from Water Corporation. 2000. *Desalination-Creating New Water Sources.* Water Corporation, Leederville, Australia.

[a] GOR: Gain Output Ratio; the ratio of freshwater output (distillate) to steam.

a recovery ratio of about 50%. Desalination of brackish water or slightly contaminated water requires less energy (1–2 kWh/m³) as the osmotic pressure for brackish water is significantly lower than for seawater (the osmotic pressure is 140 kPa for water of 1600 mg/L compared to 2800 kPa for seawater).

Desalination processes such as RO, EDR, and vapor compression (VC) require electrical power as the primary source of energy for both the desalination process itself and to drive ancillary equipment such as transfer pumps. Thermal processes such as MED and MSF use electrical energy as the secondary source of energy to drive recirculation and transfer pumps only.

Efforts toward reduction in desalination energy costs are mainly related to energy recovery systems, higher process efficiencies, new or improved construction materials, decreases in membrane prices, high-tech ultra- or microfiltration for pretreatment, the use of waste energy from other processes, and the use of low-grade energy from electricity generating plants, all of which contribute to substantially decreasing external energy use and subsequent product water cost.

Greenhouse gas emissions associated with fossil fuel use for energy also lead to a greater interest in the adaptation of renewable energy to desalination [17].

29.2.2.1 Renewable Energy Sources and Desalination

The cost, availability, and environmental implications of fossil energy use make renewable energy resources (RES) more attractive as an alternative for desalination energy. This is particularly relevant to freshwater production in remote areas, where both fuel and water supply are often expensive but where RES are in abundance. Desalination with renewable energy is more often adopted for smaller desalination plants and not commonly used for large capacity desalination. Thermal energy, electricity or mechanical (shaft) power generated from solar energy (as photovoltaic [PV], or solar thermal energy), wind energy, geothermal energy, and, to lesser extent, tidal and wave energy have been harnessed for desalination [41].

29.2.2.1.1 Solar Energy Desalination

Solar energy desalination is generally the collecting of solar thermal energy that is used for desalination directly in a solar still or that is converted to electricity first and then used in either thermal or membrane processes for desalination.

A solar still imitates part of the natural hydraulic cycle—humidification. Solar stills could provide the most economic supply of desalted seawater for a community in which there is abundant sunshine and a requirement for potable water in the range of 100 m³/day [31]. According to average estimates in Canada, about 1 m² of solar still is needed to produce 4.5 L of water daily [14]. Solar stills were developed as desalination units inherently have a major problem with energy loss in the form of the latent heat of the water condensate [3]. However, while solar desalination systems have low operation and maintenance costs, they require large installation areas and high initial investment.

Distillation solar collectors are used to concentrate solar energy to heat feedwater so that it can be used in the high temperature end of a standard thermal distillation process. Solar thermal energy can be used directly to operate an MED plant. MED systems are more flexible and inexpensive, less sensitive to scaling, and exhibit better performance than MSF systems. Another interesting solar distillation technology is the MED process powered by salinity gradient solar ponds. Several systems have been built based on either MED or MSF processes. Although this is not an efficient technology, the cost of the energy delivered by solar ponds is low. Moreover, the distillation system can be operated continuously throughout the year, thus resulting in a cost-effective technology [41].

Photovoltaic systems have been used as a source of electrical energy for operating RO or ED [14,36] and have been found to be most cost-efficient in Saudi Arabia [1].

As with other methods, larger distillation units are more economical [26]. The cost of water produced by a solar desalination plant in amounts of 20–400 m³/day is $1.58/m³, while much smaller installations (0.25–10 m³/day) generate permeate that is twice as expensive.

29.2.2.1.2 Wind

The maximum amount of wind power that can be extracted is 59.3% of the total wind energy. This theoretical maximum is not achieved by practical wind turbines, which typically are able to extract up to 50% of the wind's energy.

Wind energy is best suited to those desalination technologies that require electrical power rather than thermal energy input; as a result, wind turbines have been coupled with RO desalination units. Wind power cost ranges from $0.0234 to $0.1210/kWh between different locations and types of wind turbines [44], costs that do not account for energy storage.

As a standalone power system, wind was found to be more cost-effective than PV and hybrid wind/PV systems [27]. The cost of desalinated water production is less for wind-powered RO and VC (1.68–1.87 €/m³ and 2.42–2.74 €/m³, respectively) than for PV–RO (3.78–3.76 €/m³) [49] (see also Table 29.3).

29.2.2.1.3 Geothermal Energy

The amount of geothermal energy available on Earth is about 4×10^{17} KJ (Table 29.4), or 35 billion times the world's present

TABLE 29.3 Range of Product Water Cost Produced by Plants Operating on RES (A Compilation from Various Sources)

Desalination Method and Energy Source	Cost of Water Product ($/m³)
Solar thermal	0.87–5.48
PV–RO	0.56–3.14
W–VC	2.13–2.44
W–RO	1.5–1.77

total annual energy consumption. However, only a small fraction of this energy can be extracted. For the purpose of water desalination, thermal fluid of sufficiently high temperature can provide heat for thermal desalination technologies or, in the case of high fluid temperature, electricity can be generated to power membrane desalination technologies or to supply the electrical power required for the pumps in a thermal desalination process.

Although not as common as PV, solar, or wind, geothermal energy has been used for many years and can be considered a reasonably mature technology.

29.2.2.1.4 Ocean Energy

There are two main methods of energy generation using ocean tides: tidal turbines, similar to wind turbines but located beneath the surface of the sea; and tidal barrages along with low-head hydropower turbines, which generate electricity.

An ocean wave-powered RO desalination system consists of a wave pump and an RO module. There are small-scale systems in operation in Caribbean locations that are feasible where ocean water is shallow and under constant wave action [23]. A simple device to connect the hydro-ram to an existing desalination plant has been designed and patented at Salford University, which can be incorporated into existing conventional desalination plants producing cost savings [32].

29.2.2.1.5 Combinations and Limitations for RES/Desalination

According to publications of the Middle East Desalination Research Centre (MEDRC) [39], the choice for combination of RES/desalination technology is defined by feedwater quality (salinity) and the requirements for product water (Table 29.5). Overall, a wind-powered RO plant is considered the most cost-efficient.

TABLE 29.4 Capacity of Natural Energy Resources in the World

Classification	Capacity	Kind of Energy
Solar energy (kW)	0.4×10^{14}	Photon and heat
Wind power (kW)	9.7×10^{9}	Kinetic
Hydropower (kW)	3.0×10^{9}	Kinetic
Geothermal (kJ)	4×10^{17}	Thermal
Tide energy (kW)	6.7×10^{7}	Kinetic
OTEC (MW)	30×10^{9}	Thermal

Source: Adapted from Belessiotis, V. and Delyannis, E. 2001. *Desalination*, 139, 133–138.

TABLE 29.5   Recommended Renewable Energy—Desalination Technologies Combinations

Feedwater Available	Product Water	RE Resource Available	Small (1 − 50 m³ d⁻¹)	System Size Medium (50 − 250 m³ d⁻¹)	Large >250 m³d⁻¹	Suitable RE–Desalination Combination
Brackish water	Distillate	Solar				Solar distillation
	Potable	Solar	*			PV–RO
	Potable	Solar	*			PV–ED
	Potable	Wind	*	*		Wind–RO
	Potable	Wind	*	*		Wind–ED
Seawater	Distillate	Solar	*			Solar distillation
	Distillate	Solar		*	*	Solar thermal–MED
	Distillate	Solar			*	Solar thermal–MSF
	Potable	Solar	*			PV–RO
	Potable	Solar	*			PV–ED
	Potable	Wind	*	*		Wind–RO
	Potable	Wind	*	*		Wind–ED
	Potable	Wind		*	*	Wind–VC
	Potable	Geothermal		*	*	Geothermal–MED
	Potable	Geothermal			*	Geothermal–MSF

Source: Adapted from Oldach, R. 2001. Matching renewable energy with desalination plants. Middle East Desalination Research Centre, Report 97-AS-006a.

Despite the considerable capacity of natural energy resources (Table 29.4), there are certain limitations to RES use for desalination processes:

1. With the exception of geothermal energy, RES tend to have a variable rather than a constant power output. In order to provide constant power to the desalination plant, some form of energy storage is usually required.
2. RES and desalination plants are both relatively high-cost technologies. It is therefore imperative that all system aspects must be optimized, for example, the RES and the desalination system as well as the overall RES/desalination system integration.

3. The amount of product water required per day is an important factor, which will influence the overall system design.

Overall, RES are still more expensive than traditional resources. The most often used RES has been solar energy (70% of market), and RO has the majority (62%) of the renewable energy desalination market [21].

29.2.3 Economies of Scale

Economies of scale arise when increases in desalination plant size and hence water production lead to decreases in the unit freshwater cost (Figure 29.9) [16,22]. Economies of scale are

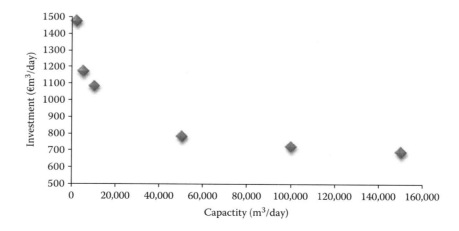

FIGURE 29.9   Investment costs of SWRO plant by size of plant (this figure shows investment costs, not water).

evident in all desalination processes, but to different extents. For example, RO has lesser scope for economies of scale when compared to distillation processes. However, operating and maintenance costs are not subject to economies of scale, but are directly affected by the water quality to be treated [37].

29.2.4 Brine Management

The desalination process results in a by-product stream of high salinity brine, which requires disposal and could have a significant environmental cost. The chemical composition of the brine is commonly similar to the feedwater: waste discharges from desalination plants consist primarily of 98.5% water with high salt and 1.5% wash water filters and cleaning products [10].

In the case of seawater desalination, the concentrated brine is commonly discharged back to the sea [13], where the impact of such a discharge on the marine environment is required to be assessed. Overall impacts depend on the physical and chemical properties of the reject streams and the hydrographical and biological features of the receiving environment [30]. These effects can be mitigated by adequate dispersal and mixing of concentrated brine wastes.

Brine disposal for inland plants is more challenging, ranging from simple evaporation to aquaculture or industrial reuse. The factors that influence the choice of brine-disposal method are illustrated in Figure 29.10 [38]. Disposal methods could include those listed in Table 29.6.

Given the high cost of building and operating desalination plants, brine water is often viewed as an asset, which when exploited is able to reduce the net cost of providing freshwater. Brine water value-adding enterprises are now accepted as an effective means to reduce the overall cost of providing freshwater while meeting environmental performance standards (Table 29.7) [2].

TABLE 29.6 Liquid Brine Disposal Options

Brine Concentrate Disposal Method	Key Issues
Blending with RO permeate feed	• Low capital cost • Technically easy to carry out • As long as the permeate conductivity does not increase over 1000 mS/cm for drinking water projects • If a two-pass process, then the first-pass reject is sent to the feed
Ocean/sewer	• Availability of discharge facilities with capacity • Environment habitats issues • Regulatory requirements/permits • Type of treatment available
Deep-well injection	• Land availability/seismic activity • Potential to use existing unused bores • Geologic/geohydrologic conditions • Potable water aquifers may get contaminated through leakage of wellhead due to corrosion • Need regulatory requirements/permits • Public perception/acceptance • Membrane brine concentrate compatibility • Need to condition the brine before discharge and associated costs • Typically used for plants with capacity >4 mL/day • Uncertainty of wellhead life, which can only be estimated
Land application (spray irrigation)	• Land availability and cost of land • Can be applied on lawns, parks, and golf courses • Percolation rates • Climate applicability • Salt drift • Water-quality tolerance of target vegetation to salinity • Aesthetic issues • Habitat issues if there is a large bird population • Long-term impacts on shallow groundwater

Source: Adapted from NSW Public Works. 2011. Brackish groundwater: A viable community water supply option? Waterlines Report Series No. 66.

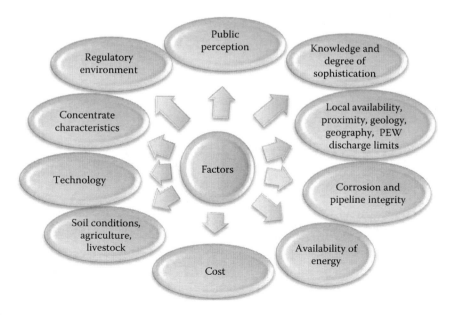

FIGURE 29.10 Factors affecting brine disposal options.

TABLE 29.7 Other Brine Disposal Options

Brine Concentrate Disposal Method	Key Issues
Salt production—SALPROC process	• Produces salts such as magnesium hydroxide, gypsum, and calcium chloride • High capital costs • Not economically viable for most because operating and selling salts are not core businesses for water desalination plant operation
Aquaculture—shrimp farming	• Can offset the cost of the desalination plant
Energy production—solar ponds	• Still experimental and requires government subsidies • A 100-ha solar-pond production cost of electricity is $0·30/kW h, which is around double the normal rate

TABLE 29.9 Concentrate Disposal Costs

Disposal Method	Construction Costs (US$ MM/MGD)
New outfall with diffusers	2.0–5.5
Power plant outfall	0.2–0.6
Sanitary sewer	0.1–0.4
Wastewater Treatment Plant (WWTP) outfall	0.3–2.0
Deep well injection	2.5–6.0
Evaporation ponds	3.0–9.5
Zero-liquid discharge	5.5–15.0

However, there are certain issues related to the brine concentrate disposal method (Table 29.8), one of which is related to its cost (Table 29.9) [37].

29.3 Urban Water Supply and Desalination in Australia

Australia is the world's driest inhabited continent. Over the last decade, the country has had to deal with a reduced

TABLE 29.8 Crystallization Brine Disposal Options

Brine Concentrate Disposal Method	Key Issues
Evaporation ponds/ mist sprays	• Land availability and cost of land • Climate applicability—limited to areas of dry, arid locations with high evaporation rates • Construction and maintenance requirements • Salt drift • Potential groundwater contamination from pond leakage; therefore, requires double lining with leakage-sensing probes • Aesthetic issues • Habitat issues if there is a large bird population • Disposal of crystallized salt • Excavating of salt without damaging liners
Wind-aided intensified evaporation	• Wind energy used to enhance evaporation • Small footprint compared to evaporation ponds by spraying the concentrate over vertical surfaces, thus increasing the evaporative area by a factor of 10 in comparison to evaporation ponds • Low energy costs • More suitable for climates with high natural evaporation
Forced circulation crystallizer	• Complexity of equipment • Mechanically intensive • Energy/power requirements • More suitable for power plants and other sites with a high degree of technical competence

availability of water due to drought, and debates about water rights are highly sensitive and nationally important. The production of desalinated water has significantly increased in Australia over the last decade, supplementing or even largely replacing more traditional water supply sources in many capital cities.

Water supply for the large capital cities in Australia (national population of 23 million) is mainly sourced from surface water reservoirs. Perth, the capital city of Western Australia, is an exception to this where, in addition to surface water, supply is also from local groundwater resources.

Municipal and industrial water use in Australia comprises over 30% of total water use, which is second only to agricultural water use (Figure 29.11) [12]. Total potable water consumption was greater than 17 GL/day in 2004–2005 [5], which makes Australians among the highest water users per capita worldwide.

The current urban water prices in the various states of Australia reflect the level of water use (sliding tariffs) and the type of water users (industry higher than domestic users). The water-related costs to households are exclusive of water service charge, water use, and wastewater collection and reticulation. Some example of water prices to domestic users are listed here:

- In Western Australia, the water price is calculated on a 2-month basis as $1.381/kL for water use less than 150 kL per two months, $1.841/kL for 51–500 kL, and $2.607/kL for water use more than 500 kL (www.watercorporation. com.au).

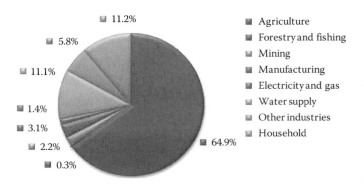

FIGURE 29.11 Sectoral water use in 2004–2005.

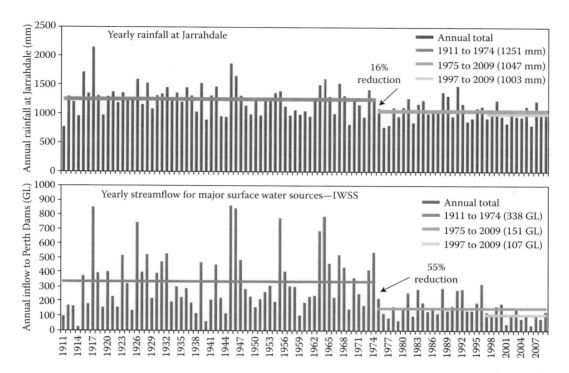

FIGURE 29.12 Rainfall and runoff to water supply dams in Perth. Note: A year is taken as May to April. (Data courtesy of Western Australian Water Corporation.)

- In South Australia, the water price is calculated quarterly as $2.26/kL for water use less than 30 kL, $3.23/kL for 30–130 kL, and $3.49/kL for more than 130 kL (www.sawater.com.au).
- In Victoria, the water price is calculated on daily basis as $2.3424/kL for the first 0.44 kL, $2.7486/kL for 0.44–0.88 kL, and $4.0609/kL for more than 0.88 kL of daily water consumption (www.citywestwater.com.au).
- In New South West, the water price is $2.168 (www.sydneywater.com.au).

It is common between the states to keep the same statewide price for water, regardless of the cost of getting that water to remote communities. This system is considered the fairest way to spread the cost of providing and maintaining basic water facilities across the community and is known as Community Obligation Service (COS).

There are also restrictions on water use, which can also be enforced during periods of water shortage.

Over the last decades, Australia has experienced a significant reduction in rainfall and hence a reduction in harvestable water resources. For example, Figure 29.12 shows a substantial reduction in rainfall (16%) in southwest Western Australia over the years 1975 to 2009, leading to an even greater reduction in runoff to water supply dams (55%).

Under such conditions, and considering the fast-growing population, maintaining a sufficient and reliable water supply is a key strategic element in ensuring adequate urban water supply for Australian cities and rural towns; this need has led to the establishment of a number of desalination plants. The amount of desalinated water production in 2013 was 4.3% of total water use with the total design desalination capacity of 1734 mL/day for potable and 461 mL/day for industrial supplies.

Currently there are six major seawater RO desalination plants in Australia, designed to supply potable water to major cities (Table 29.10) [11,15]. The total water production capacity of the six plants is 1.223 GL/day (ranging from 133,000 to 400,000 m³/day per plant). It appears that despite the relatively high costs of the unit water production, this cost is comparable to water prices adopted for urban water supply in Australia.

A series of small plants are also used by mining companies and in remote communities, where the cost of desalinated water production is much higher due to their locations. There are also examples of treated wastewater desalination and reuse. Overall, the proportion of water sources utilized as feedwater is 86% seawater, 1.2% brackish water, and 12% industrial effluent.

The particularly high cost of desalination in Australia (with a unit cost ranging from AU$1.00 to AU$2.27 per m³) is mainly attributed to the requirement of meeting strict environmental regulations and to high energy and labor costs.

Desperation in the face of drought has forced authorities to pay crisis prices ($12.9 billion) [40], passing the costs on to consumers through higher water bills. Uniquely, all of Australia's major urban seawater desalination plants buy wind and solar energy to offset their entire energy use, which makes the unit cost of water much higher as the cost of renewable energy is

TABLE 29.10 Major Desalination Plants in Australia and Their Cost

Plant	Perth Seawater Desalination Plant (Perth I)	Gold Coast Desalination Plant	Sydney Desalination Plant	Adelaide Desalination Plants I and II	Southern Seawater Desalination Plant (Perth II)	Victorian Desalination Project
Capacity (m³/day)	140,000 expandable to 250,000	133,000	250,000 expandable to 500,000	150,000 each	150,000 expandable to 300,000	400,000
Year	2007	2009	2010	2012	2011	2011
Capital cost	387 M AU$	943 M AU$	2 B AU$	1.83 B AU$	1450 M AU$	3.1 B AU$
ADS[a]	38.26 M AU$	86.66 M AU$	0.179 B AU$	0.153 B AU$	121.34 M AU$	0.259 B AU$
Unit cost (AU$/m³)	1.00	2.03	1.74	2.07[1]	1.81[1]	2.27[1]

[a] Calculated with an interest rate of 5.5% and an amortization period of 20 years; (1) estimated levelized cost.

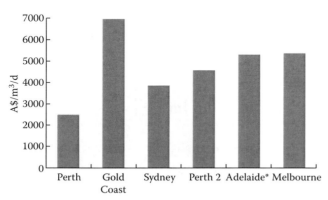

*Based on 140,000 m³/d of capacity.

FIGURE 29.13 Capital costs in Australian desalination plants.

still greater than the cost of conventional electricity (AU$0.12–0.17 kW/h) [24].

Design of Australia's desalination plants also aims to minimize their impact on the surrounding landscape. This leads to substantial increases in the cost of feedwater intakes and brine outfalls, which account for a substantial portion of a plant's capital costs; particularly at Queensland's Gold Coast plant (Figure 29.13). Meanwhile, the intakes and outfalls at the Perth II and Adelaide plants anticipate a future doubling of capacity at each site, and thus artificially inflating the initial costs.

Higher desalination cost is also related to Australia's higher costs for membranes and cartridge replacement, and chemicals consumption, which are much lower in other countries.

It is also not surprising that the desalination cost in remote country towns is even greater, reaching more than AU$5/m³ due to particularly remote locations and small plant sizes [38].

29.4 Summary and Conclusions

Continuous advancement in relevant technologies makes desalination more efficient and more cost-effective over the last decades. While more conventional water supply sources become increasingly stressed by overuse or by climate variability and changes, often leading to water quality deterioration and environmental degradation, desalination has grown to be a useful alternative for sourcing freshwater, particularly for urban water

supply. The following features of desalination illustrate the benefits of its adaptation:

- It can easily use renewable energy sources to ensure no or little greenhouse gass emissions.
- In a case of seawater desalination, the technologies provide drought-free water supply and also produce a totally new source, contrary to recycling.
- Produced water quality is not affected by fires, runoff, or recharge water quality or activities in catchments.
- Desalination is largely considered environmentally sustainable.
- It does not disturb natural water systems (rivers or groundwater) and associated habitat, particularly when seawater desalination is concerned. In this case, brine disposal can be environmentally managed, which was proven at numerous installations.
- Desalination plants have small environmental and terrestrial footprints. For instance, the first Perth SWRO plant acquired 6.5 ha of land and 2.5 ha of seabed.
- There is minimal greenhouse gas production during the manufacture of components and during the construction of the plant.
- The desalination plant and hence the potable water sources can be located near to where it is needed.

However, there are still some downsides or challenges of desalination technologies adaptation from urban water supply. Similar to freshwater resources, the water demand has to be close to the sources of brackish groundwater or seashore. Otherwise, the water delivery cost becomes a prohibiting factor for an effective water supply scheme.

For seawater desalination, the concerns are commonly related to low recovery rates (up to 50%), energy availability and cost, adequate site selection and design of seawater intake point, and the cost of produced water integration with the existing water supply reticulation systems.

When inland desalination is considered, a number of factors could lead to a high desalination cost.

- Feedwater quality may become a limiting issue, which may require additional feedwater treatment. Groundwater oxidation and mixing of groundwater from a number of production bores may result in certain compounds' precipitation (e.g., Fe and Mn). Silica concentration in groundwater is the main limiting factor in recovery rate for brackish water. Temporal variation in feedwater quality is also common, when groundwater is used.

- By-product (brine) disposal is commonly a challenge as it requires high capital and potentially environment costs.
- Economy of scale is not commonly applicable for relatively small inland plants.
- Brackish/saline water resources could be limited and could be climate dependent.

References

1. Abdul-Fattah, A.F. 1986. Supply of desalted water to remote arid zones. *Desalination*, 60, 161–189.
2. AFFA. 2002. *Introduction to Desalination Technologies in Australia*. Agriculture, Fisheries and Forestry Australia. www.affa.gov.au/content/publications.cfm.
3. Al-Salaymeh, A. and Zurigat, Y. 2006. A comprehensive study of greenhouse agriculture and water desalination for Jordan-Valley and Read-Dead Sea Region. *The 2nd International Conference on Water Resources & Arid Environment*, Riyadh, Kingdom of Saudi Arabia, November 26–29, 1–18.
4. Al-Sofi, M. 2001. Seawater desalination—SWCC experience and vision. *Desalination*, 135, 121–139.
5. Australian Bureau of Statistics. 2007. www.abs.gov.au/.
6. Barba, D., Caputi, P., and Cifoni, D. 1997. Drinking water in Italy. *Desalination*, 113, 111–117.
7. Belessiotis, V. and Delyannis, E. 2001. Water shortage and renewable energy (RE) desalination—Possible technological application. *Desalination*, 139, 133–138.
8. Birda, S.P. and Abosh, W. 2001. Recent developments in water desalination. *Desalination*, 136, 49–56.
9. Buros, O.K. 1999. *The ABC's of Desalting*. International Desalination Association, Massachusetts.
10. CONAMA. 2008. 9th Congreso nacional de medio ambiente (9th National Environment Congress). Nuevas fuentes de agua (New Water Sources).
11. Crisp, G.J. 2011. Desalination in Australia. Sustainably Drought Proofing Australia. Report.
12. Desaldata. 2012. Global Desalination Market GWI. www.desaldata.com. Retrieved October 3, 2012.
13. Dickie, P. 2007. Making Water—Desalination: Option or Distraction for a Thirsty World? assets.panda.org/downloads/desalinationreportjune2007.pdf.
14. El-Kady, M. and El-Shibini, F. 2001. Desalination in Egypt and the future application in supplementary irrigation. *Desalination*, 136, 63–72.
15. El Saliby, I., Okour, Y., Shon, H., Kandasamy, J.K., and Kim, I.S. 2009. Desalination plants in Australia, review and facts. *Desalination*, 247(1–3), 1–14.
16. Estevan, A. 2007. Desalación, energía y medio ambiente. (Desalination, Energy and Environment). University of Sevilla, Ministry of Environment.
17. Ettouney, H. and Rizzuti, L. 2007 Solar desalination: A challenge for sustainable fresh water in the 21st century. *Solar Desalination for the 21st Century*, 1–18. DOI: 10.1007/978-1-4020-5508-9_1.
18. Financial Times Global Water Report. January 26, 2001. Issue 114. 8 p.
19. Glueckstern, P. and Kantor, Y. 1983. Seawater versus brackish water desalting technology, operating problems and overall economics. *Desalination*, 44, 51–60.
20. Glueckstern, P., Nadav, N., and Priel, M. 2001. Desalination of marginal water: environmental and cost impact: Part 1: The effect on long-range regional development Part 2: Case studies of desalinated water vs. local desalination of marginal brackish water. *Desalination*, 138, 157–163
21. Greenlee, L.F., Lawler, D.F., Freeman B.D., Marrot, B., and Moulin, P. 2009. Reverse osmosis desalination: Water sources, technology, and today's challenges. *Water Research*, 43(9), 2317–2348.
22. Hammond, R.P. 1996. Modernizing the desalination industry. *Desalination*, 107, 101–109.
23. Hicks, D., Pleass, C., Mitcheson, G.R., and Salevan, J. 1989. DELBUOY: Ocean wave-powered seawater reverse osmosis desalination system. *Desalination*, 73, 81–94.
24. Hoang, M., Bolto, B., Haskard, C., Barron, O., Gray, S., and Leslie, G. 2009. Desalination plants: An Australia survey. *Water*, 36(2), 67–73.
25. IDA. 2012. *Desalination Yearbook*. International Desalination Association. www.idadesal.org. Accessed on May 8, 2013.
26. Kalogirou, S. 1997. Economical analysis of a solar assisted desalination system. *Renewable Energy*, 12, 351–367.
27. Kellogg, W.D., Nehrir, M.H., Venkataramanan, G., and Greez, V. 1998. Generation unit sizing and cost analysis for stand-alone wind Photovoltaic and hybrid wing/PV systems. *IEEE Transaction on Energy Conversion*, 13(1), 70–75.
28. Khan, A.H. 1986. *Desalination Processes and Multistage Flash Distillation Practice*. Elsevier, Amsterdam.
29. Larson, T.J. and Leitner, G. 1979. Desalting seawater and brackish water: A cost update. *Desalination*, 30, 525–539.
30. Lattemann, S., Kennedy, M.D., Schippers, J.C., and Amy, G. 2010. Sustainable water for the future: Water recycling versus desalination. In: I.C. Escobar and A.I. Schäfer (eds.), *Sustainability Science and Engineering*, Vol. 2. Elsevier, Amsterdam, the Netherlands, 7–39.
31. Malik, A., Hawlader, M.N.A., and Ho, J.C. 1996. Design and economics of RO seawater desalination. *Desalination*, 105, 245–261.
32. Maratos, D. 2002 A new device to allow wave power to be used to reduce the energy consumption on existing desalination plants. *EDS Newsletter*, 8.
33. Medina, J.A. 2004. Feasibility of water desalination for agriculture. Paper presented at the *Proceedings of the FAO Expert Consultation on Water Desalination for Agricultural Applications*, Rome. April 26–27, 2004.
34. Mesa, A.A., Gomez, C.M., and Azpitarte, R.U. 1996. Energy saving and desalination of water. *Desalination*, 108, 43–50.

35. Ministry of Environment, Rural and Marine. 2009. Desalination in Spain. Sectoral Notebooks 2009. Water Cycle 02, Report.

36. Mohsen M.S., and Jaber, J.O. 2001. A photovoltaic-powered system for water desalination. *Desalination*, 138, 129–136.

37. Morin, P.E. 1999. Desalting Plant Cost Update: 2000, International Desalination Association.

38. NSW Public Works. 2011. Brackish groundwater: A viable community water supply option? Waterlines Report Series No. 66.

39. Oldach, R. 2001. Matching renewable energy with desalination plants. Middle East Desalination Research Centre, Report 97-AS-006a.

40. Palmer, N. 2012. Current desalination in Australia. *Info Enviro Water*, 79.

41. Peñate, B. and García-Rodríguez, L. 2011 Current trends and future prospects in the design of seawater reverse osmosis desalination technology. *Desalination*, 284, 1–8.

42. Pique, G.G. 2002. Breakthroughs allow seawater desalination for less than $0.50/m³. *EDS Newsletter*, 16, 9–11.

43. Popkin, R. 1968 *Desalination: Water for the Worlds Future*. Frederick A Praeger Publishers, New York.

44. Rehman, S., Halawani, T.O., and Mohandes, M. 2003. Wind power cost assessment at twenty locations in the kingdom of Saudi Arabia. *Renewable Energy*, 28, 573–583.

45. Semiat, R., Sapoznik, J., and Hasson, D. 2010. Energy aspects in osmotic processes. *Desalination and Water Treatment*, 15(1–3), 228–235. doi: 10.5004/dwt.2010.1758.

46. UNESCO. 2010. Water in a Changing World: The 3rd report on the Development of Water Resources in the World.

47. Van der Bruggen B. and Vandecasteele, C. 2003. Removal of pollutants from surface water and groundwater by nanofiltration: Overview of possible applications in the drinking water industry. *Environmental Pollution*, 122, 435–445.

48. Visvanathafl, C., Boonthanon, N., Sathasivan, A., and Jegatheesanb, V. 2002. Pretreatment of seawater for biodegradable organic content removal using membrane bioreactor. *Desalination*, 153, 133–140.

49. Voivontas, D., Misirlis, K., Manoli, E., Arampatzis, G., Assimacopoulos, D., and Zervos, A. 2001. A tool for the design of desalination plants powered by renewable energies. *Desalination*, 133, 175–198.

50. Water Corporation. 2000. *Desalination—Creating New Water Sources*. Water Corporation, Leederville, Australia.

51. Water Desalination. 2004. *Proceedings of the FAO Expert Consultation on Water Desalination for Agricultural Applications*, Rome, April 26–27, 2004.

52. Wilf, M. and Klinko, K. 2001. Optimisation of seawater RO system design. *Desalination*, 138, 299–306.

VIII

Water Reuse Systems and Technologies

30

Water Reuse Traditional Systems

Pietro Laureano
President of International Traditional Knowledge Institute (ITKI)

PREFACE

Climate changes, ecosystems collapse, cataclysms, and the end of civilizations are conditions that humanity has had to face numerous times. People have had to deal with the unpredictability of the environment and the variability of the climate. The conditions of climate variability, ecosystem degradation, and natural disasters imposed a deep insight in managing resources and the invention of sound technologies and processes that were not invasive. In several climates and environments, incredibly tenacious cultures were able to use locally available materials and renewable resources. They applied the principles of nature—sun, wind, gravity, and humidity—to start positive interactive processes and amplify useful dynamics. These very conditions forged locally adapted knowledge capable of responding to adversity with appropriate techniques for water harvesting and distribution, soil protection, recycling, and optimal energy use. Such techniques constitute a huge reservoir of biological and cultural diversity and sustainable knowledge. This chapter aims to provide a general overview on the water management traditional techniques and local knowledge considered as part of cultural history. The evolution of these practices since the Paleolithic era, from nomad hunter-gatherers to agrarian societies in different parts of the world, is examined. Special attention is given to water management practices developed in the harsh and arid areas and notably to the small-scale societies such as the oases. The study of traditional knowledge is described as a contribution to a new water management paradigm and blue engineering in line with the sustainable development approach in order to promote the integration of technical, ethical, and aesthetical aspects. Several examples of innovative water use in traditional practices for agricultural, architectural, and urban development purposes are described.

30.1 Introduction

The ancient water traditional knowledge shows how to intervene in perfect agreement with the environment, highlighting its potential without exhausting it. Reusing water traditional knowledge does not mean directly reapplying the techniques from the past but understanding the logical reasoning underlying the knowledge system and reapplying it in a creative way: today's appropriate innovations constitute tomorrow's traditional knowledge.

30.2 Hydraulic Labyrinths and Stone Tumulus for Water Harvesting

From prehistory, humanity has elaborated on methods and techniques to provide drinkable water. The Paleolithic nomadic hunter-gatherers could move from place to place thanks to knowledge of the territory, particularly of the methods of water finding and supplies. The Paleolithic population used to harvest drinkable water in caves thanks to water dripping and percolation; moreover, they realized stone paving to harvest rainfalls and divert them to pits. They used dams, ditches, and stone arrangements to facilitate plant growing and fishing [11]. In steppes, in savannahs, and in deserts along the Karst Plateau or along interfluvial plains, human groups exploited favorable areas subject to becoming swampy and prone to drought, thanks to flow adjustment techniques. These techniques became imposing trapping systems for fishing as those found in Mount William in Australia [21] and in New Guinea where a complex system of draining canals developed from 9000 to 6000 years ago [9]. Such knowledge, which results from experiences gained in the end, is consolidated through the success of holders; it is recorded through the symbolic thought and art and is handed over to generations through oral tradition. Starting from the most archaic African sites, the knowledge is spread worldwide at the same time as the development of human groups. Labyrinthine spirals are reproduced in symbolic rock graffiti and we found the same shapes in the structures of the traps and the fences used for the first experiences of domestication. Water harvesting is associated with the origin of spirituality and art as is testified by drawings in caves. An artificial stone tumulus dating back 150,000 years ago in El Guettar in Tunisia, holding flints and Paleolithic handmaids, clearly shows functions linked to hydraulic practices [15]. Sedentariness in places without agricultural practices is a phenomenon that has existed from the late Paleolithic period. Imposing cult structures built up by nonagricultural groups 12,000 years ago have been recovered in the recent diggings of Göbekli Tepe (Anatolia) [24]. The quantity of people required to build the structures necessitates great availability of drinking water. Meaningfully, here we are in the same area where the first domestication and production of wheat started from the spontaneous mutation of a local wild species. In Jericho (Palestine) from 12,000–10,000 years ago, structures for support of the soil and water catchment were realized by building holding walls and clay platforms with coated surfaces to hold drinkable water

[5]. The recent anthropological acquisitions no longer view the Paleolithic as a less-developed stage of knowledge surpassed by evolution; instead, it is seen as an advanced and refined-level precursor of important acquisitions in the following areas: arts, sedentary lifestyles, environmental knowledge, symbolism, community organization, and the management of flora and fauna. The ancient hunter—gatherers' knowledge is a way of thinking troglodytic, with a lower waste of resources, labyrinthine, nomadic, passive, and slow, which can still provide indications for the careful and sustainable management of the environment. It persists as a sublayer shared by all populations that change, developed, or lost its importance according to socio-environment conditions. It reemerged in the shared similarities found in techniques, in architectural forms, and in myths among peoples in far-reaching places around the world.

30.3 Water Management in the First Societies

Since the eighth millennium in Africa, the Middle East, Anatolia, and the Pakistan-Indian area, original ways of cultivation developed in places where the warmth and the sun gave yields that justified the necessary engagement [6]. Nonetheless, due to the lack of rainfall, it was necessary to develop methods of water management. Before introducing irrigation techniques, populations used water directly and naturally available, which in arid conditions can be found as atmospheric moisture and in soil deposits. The moisture that condenses on the soil played a key role in the introduction of cultivations. Small orchards are implanted in the sites where the phenomenon was more important: areas situated near the basins, watercourses, or geological situations and stone mounds that favored the water steam and the dew condensation. By observing the best vegetative cycle of naturally grown plants, the most suited areas could be identified. Likewise, it was possible to establish where to exploit the water held in soil deposits. Alluvial soils, the loess, and wadi's dry courses were the most suitable places to preserve water resources inside the upper soil layers. This allowed the development of the first Neolithic societies that started from "nomadic cultivation," which relied on spreading seeds in favorable areas where the populations came back only for harvesting, switched to practices of agriculture.

30.3.1 The Neolithic Reuse of Wastewater with Ditches and Channels

Techniques of filters for the water and canalizations in order to fertilize the fields are realized in Anatolia, Syria, and Palestine from the first Neolithic societies and are diffuse to the east and west of this area. Successively, these methods become more and more elaborated. In the sixth and fifth millennium, before the development of the technologies necessary to control the great river courses and in the areas where these were not present, the villages are encircled by ditches that drain the water and it is reused for irrigation after it is used for the inhabitants and the cattle. In the trenched village of Murgia Timone near Matera in the south of Italy, there are reservoirs connecting each other in order to constitute a system for rainwater filtration and sanitation. Elliptical ditches encircling the village are also found in Germany in the site of Kol Lindenthal along the Rhine Valley and in Banpo, China, along the Yellow River. This site, near Xi'an, dated to 4500 BC, has ditches interrupted by stonewalls for storage and with leaking holes for water depuration.

Caves, characterized by natural water dripping, were excavated to follow the flows and better intercept them. They were artificially widened and deepened by creating openings on slope walls as well as draining and harvesting cisterns. These practices would then become the underground techniques of catchment tunnels and passive architecture, which in suitable geomorphologic situations would allow the troglodytic settlements and the development of stone towns [16]. The extraction activity in flint mines allowed the building of the first pit courtyards provided with radial tunnels. The pattern that can be found in Grimes Caves in England and in the mines of Gargano in the South of Italy [10] is reproduced in water harvesting devices and in underground dwellings with a central courtyard.

These techniques developed in dry karstic areas and on calcareous plateaus like the Murge in Italy, in Cappadocia (Turkish), in Iran, Israel, and Libya but also in the semi-arid clay plains of Tunisia and in the plains of loess in China. Elliptic or semicircular settlements, characterized by several perimeters of ditches, were established in rainy areas, or in karst areas, at the edges of high plateaus without superficial watercourses as well as in poorly drained plains like in Daunia, Italy. Under climate alternation conditions, they meet many functions linked to water balance: they drain water during rainfalls and preserve it for dry seasons; they are used as drinking troughs and ditches to collect useful sewage and waste to fertilize soils; they symbolically mark the places and strengthen social cohesion, the group identity, and the propensity to sedentariness. Ditches are multipurpose structures, resulting from the evolution of simple pits used to collect water and waste. This practice was useful to select domestic cultivated species and to identify seeding periods suited to any kind of plant. In fact, it was possible to observe that seeds carried from sewage into the ditches sprouted spontaneously in the most suited season. By sprinkling the fields with the water harvested in the ditches, it was possible to understand the fertilizing capacities of manure.

30.3.2 Large-Scale Water Management Techniques in Hydraulic Societies

In the areas surrounding great rivers, the realization of elaborated systems of canalization have been linked to the creation of the first cities. In the sixth millennium, the first techniques of flow diversion using water intakes and weirs that implemented irrigation by flooding were developed in the plains and interfluvial basins of Iraq and Mesopotamia. In the same period, in Baluchistan (Iran and Pakistan), the civilization preceding Harappa used heartened dams called *gabarband* to keep water

flows in the soil as well as dams accumulating sand that favored the sedimentation of slime and the control of alluvial sediments and floods. The settlement structures with an orthogonal texture developed on the matrix of the canals. The first noncircular constructions were realized with water and mud, the same slime of cultivations, like in the settlements of Çatal Hüyük and of Jarmo, in the plain of Konia in Anatolia and at the foot of Zagros Mountain in Iraq, which preceded urban organizations. The squared shape allowed diversified and complex solutions with greater evolution potentials. The dwellings could stretch and cluster gradually without leaving any free areas. In interfluvial plains, large clusters grew at the same time as the development of techniques supporting the soil with continuous wall curtains, and construction of platforms, banks, and canals. These practices allowed large-scale water management techniques to be carried out in alluvial deposits of slime, loess, and sand along Afro-asiatic fluvial basins, and also in karst areas of Mesoamerican rainforests, developed by important state-owned organizations rightly called hydraulic societies [26].

In the third millennium, these societies developed the first urban systems supplied by aqueducts, sewers, and water facilities along the course of the great rivers: the Nile in Egypt, the Tigris and Euphrates in Mesopotamia, the Karun in Iran, the Oxus (Amu Darya) and the Axartes (Syr Darya) in Central Asia, the Indus in the Indian subcontinent, and the Yangtze in China. The cities of Harappa and Mohenjo–Daro belonging to the Indus civilization had housing provided with running water and sanitary systems. Water supply systems were used in homes and techniques for disposal of wastewater and irrigation practices were comparable to those that would be used in Carthage and Rome more than 2000 years later [22]. In Harappa, we owe the first introduction of the wells provided with a stone structure that allows reaching groundwater at greater depths. In Iran, near the Ziggurat of Tchoga Zanbil dating back to 1.275–1.240 BC, the Elamite civilization, whose works to regulate the course of the Karun river are still visible in the town of Shushtar, realized a unique water filter system built of baked bricks [1]. The hydraulic Egyptian and Sumerian civilizations developed monumental architecture from the experience of constructing the banks and canals necessary for irrigating and fertilizing their lands. In Egypt, the first pyramids were made up of mud, which was the direct evolution of the Neolithic building techniques of riverbanks as well as of mud platforms in Africa, Mesopotamia, and Mesoamerica. The soil resulting from the excavation formed the first holy mounds. In hydraulic civilizations, the sovereigns gave a monumental interpretation of the techniques. In seasons when hydraulic work was not necessary, the labor shifted to the construction of great monuments. This pattern promoted urbanization and social identity. The powerful state control of hydraulic societies needs these big works that glorified and justified the despotism and the great administrative bureaucracy. The geographic size and the big hydraulic structures led to political despotism, bureaucratic hypertrophy, as well as state militarization. In Mesopotamia, the lack of the benefits of periodical floods, as occurred in Egypt, caused the loss of soil fertility and, together with the use of baked bricks and the consequent necessity of wood, brought the collapse of these societies.

Indeed, in more fragmented landscapes lacking in water coming from big rivers, the traditional techniques of water catchment and management developed thanks to the activities of small-scale communities having a more harmonious relationship with the environment [18].

30.4 Hydro Genesis in the Absence of Rivers

During the Metal Age, the populations organized in familiar clans could easily move thanks to carts and horses. By structuring paths in the mountain ridges and communications through the sea and the deserts, agro-pastoral and transhumant groups, great caravan nomads, trader, metal seekers, and cultivators exploited those areas that had not been previously colonized for the lack of perennial rivers and widespread techniques that would allow for the use of new tillable areas. Terracing was realized in inaccessible slopes and hanging gardens were introduced. The settlements were furnished with powerful systems for the supply of water and great reservoirs. Small-scale solutions for the production of water were managed in all the territory. Since the fourth millennium, the technique of excavating cisterns and water harvesting systems had developed in order to irrigate the fields from the settlements situated on high hills, thus exploiting the force of gravity. In the Bronze Age, the Edomite, then Nabataean, civilization emanating from the capital Petra in Jordan, in the Negev and disseminated in the Arabian Desert, used carries water sterilization methods consisting of layers of stone, sand, and charcoal. A systematic evolution of water supply management in ancient Greece began in Crete during the early Bronze Age, that is, the Early Minoan period (ca. 3500–2150 BC). In the Minoan civilization, extensive systems and elaborate structures for water supply, irrigation, and drainage were planned, designed, and built to supply the population and the agriculture [3].

Settlements placed on fortified hills developed in rough areas of the Middle East, Mediterranean Isles and Peninsulas, and on coast promontories of arid areas. Towns, citadels, and acropolises had to withstand the sieges and ensure a supply of drinkable water. In the Bronze Age, in the sites of Arad [2], Jawa, and Megiddo, in the north of the Arab desert [4], and of Qana, in the South of Yemen on the coast of the Indian Ocean [17], the area within the walls was used to harvest water and feed either open-air or excavated cisterns that could be reached through tunnels and staircases. Pipes fed tilled fields or the eventual urban development at the foot of the hill. This system is still functioning in Thula, Yemen. In the case of coastal settlements, they too supplied the harbor structures to provide ships with water, like in Aden, Yemen. During sieges, canals were cut down and defenders, who barricaded themselves on the summit, continued to produce the water that had been denied to attackers. Each rock or wall mass could produce water and protect the soils. The different thermal inertia combined with the atmosphere created colder surfaces

bringing about condensation. The walls intercepted winds and moisture. The interstices between blocks and the stone porosity withheld the water. The shadow protected it from evaporation. The rocks prevented the soils from dismantling, thus facilitating humus formation. The stone walls preserved the hydromorphic qualities of the soil, thus acting as thermoregulators and moisture balancers both in arid areas and under very cold conditions hindering the formation of ice. This is the reason why there are walls, stone circles, and stone arrangements that are generally used to catch moisture also found in those areas subject to heavy rains such as Ireland and the Orkney Islands. Here, thermoregulation counterbalances soil glaciations and permafrost formation.

In arid areas, a series of techniques developed ranging from simple stone tumulus, to half-moon mounds, to great stone-walls, thus evolving into complex double-wall curtain devices equipped with harvesting cisterns. In the Negev desert, archaeological research [13] has shown that century-old olive orchards and vineyards were irrigated thanks to a system of small stone-walls that collected dew. These walls are called *teleylat al 'anab* in Arabian, meaning mounds for the vineyard. The plants grew inside small fences whose stones, arranged with large interstices, caught the moisture from the wind. The vineyard and the olive tree could thus grow even without any source or water table. The walls, the stone mounds, the tumulus, the *trulli*, and the calcareous rock called *specchie*, in Southern Italy, the *talayotes*, in Baleares, Spain, the *nuraghi*, in Sardinia, Italy, the *telayet el anab*, in Arabia, all utilize water condensation. The stone mounds carry out their function during the day and at night. In hot seasons, the wind carries traces of moisture that seep into the interstices of the stone mounds. They have a lower temperature inside because it is not exposed to sunrays or it is cooled by the underground chamber. The decrease in temperature causes the condensation of drops that are absorbed by the soil or fall into the cavity. The collected water supplies further moisture and coolness, thus amplifying the efficiency of the condensation chamber. Overnight, the process is reversed and the condensation takes place externally, thus producing similar results. The cold surface of the stones condenses the moisture and dew slides into the interstices, thus wetting the soil, or it is harvested in the chamber of the cistern. Similar techniques, used in the Neolithic Saharan area, in the Arab desert and in Yemen, characterize a civilization of hidden waters developed by small-scale societies based on hydro genesis from air [23]. Their most imposing urban application is found in Petra, Jordan. The contacts with Southern Arabia through the Incense Route explain the similarities with the Sabean hydraulic techniques. The so-called Marib dam, which is actually a system of dividers and water intakes [8], boasts the most imposing example of soil formation practices by flooding control through complex hydraulic structures. In the highest sites characterized by rare rainfalls, these devices are associated with rain harvesting surfaces, which develop in a terraced or courtyard architecture organized for this purpose. Temples and worshipping monuments, successively like the mosques and the cloisters, were used to catch water. Likewise the tombs, the kurgan, the tholos, and the holy constructions

FIGURE 30.1 Graphic reconstruction of the water harvesting systems and reuse. Open air cisterns, underground cavities, and tunnels supplying water to the terraced gardens and the ablution rooms of the mosque, Thula, Yemen.

FIGURE 30.2 Reservoirs for water conservation on the acropolis of Thula, Yemen.

used the forms of the hydraulic structures for many reasons: (1) Water was actually used in religious and funerary ceremonies; (2) hydraulic knowledge was often handed over by holy or heroic personalities; and (3) the funerary mausoleum reproduced the architecture of the structures producing water because it is the source of life (Figures 30.1 and 30.2).

30.5 Water Technology's Evolution in Small-Scale Communities and in Hydraulic Societies

During the first millennium, hydraulic societies developed large-scale irrigation to connect interfluvial basins and to contrast the decreasing fertility of the soils. At the same time, in the less favored areas, from the margins of the great empires to small communities, perpetuated the knowledge tied to hydro genesis and water reuse developing a complex set of technologies to make areas, with no apparent water, livable. The separation between hydraulic societies and the small-scale communities is highlighted, from both a material and a symbolic standpoint, by the construction of the Imperial Canal started in the middle of the first millennium BC. The history of China is characterized by the course of the two great rivers that structure its geography. The ancient Chinese civilization flourished in the harsh conditions of the basin of the Yellow River (Huang He), which, situated at the end of the silk route, has promoted the spread of the water related traditional knowledge for hundreds of years. The second great river, the Blue River (Yang Tse), flows in large and favorable geographical conditions that allowed for great manufacturing and industry. The Imperial Canal was built to link the course of the Yellow River of the desert to that of the Blue River of the modernity and characterizes the unification of the Chinese Empire. Developed along a northern-southern direction for about 1700 km, it is the most extended canal in the world and, due to its geographic importance and to the impact on society, it can be considered an artificial Nile. Like the Great Wall, its realization promotes the territorial, social, and administrative creation of China. In an alternative to these great works, which unify wide territorial systems in a great empire, small communities in more difficult geographical situations perpetuate local solutions that create autonomy and independence. This is the way that the oases technology, humankind's most important experiment in survival in arid areas of the planet, was elaborated [19]. It is thanks to the oases, extended through the deserts, from Marrakesh to Xi'an, that the intercontinental communication network between Africa, Europe, and Asia was created [24].

30.6 The Hidden Waters Knowledge: Oases and Catchment Tunnels

In ancient Egypt, the word *oasis* appears in the late Old Kingdom on inscriptions of the Sixth Dynasty (2350–2200 bc) [14]. These show a Region of Oases joined to the Egyptian Kingdom by a route that, passing from one oasis to the next, permitted travel to the west and south. The ancient Egyptians clearly distinguished the areas of the western desert, which include the oases of Baharya, Dakhla, Farafra, Kharga, and Siwa, indicated by the term *oasis*, from the Nile Valley. The latter, in fact, although it depended on systems of irrigation and crop management, benefited from the plentiful water supply of a large river. The Egyptians considered as oases only those locations characterized by apparently nonexistent water resources. The oasis is an autocatalytic system whose initial supply of condensation and moisture is extended by planting date palm (*Phoenix dactylifera*) in the hot desert and the poplar (*Populos nigra*) in the cold desert. The trees produce shadows and attract organisms, thus forming the humus. The palm grove creates a wet microclimate that is fed by hydraulic catchment techniques and water management and reuse. An oasis is never a natural or casual creation. It is formed by small-scale local communities possessing environmental understanding specific to sites made habitable by applying techniques whose invention and preservation require considerable effort. The oases associate different skills and elements that already exist by using them in a new way. It is the fruit of the union of the environmental know-how of nomadic hunter-gatherers and herdsmen with the water techniques of farmers. The creation of the oases depends on the possession of qualified hydraulic expertise and the combined use of animals and plants suitable for the purpose, conditions that were first met in the early age of metals, around the third millennium bc. In this period, nomadic populations that had remained on the margins of the age's great city-building processes chose an agro-pastoral lifestyle and, driven by motives and pressures related to that choice, interacted, allied, established symbiosis with, or assimilated other groups, opening the whole package of specific concepts that will lead to a leap in complexity and establish the oasis as a complete system for the support of lives and livelihoods [7]. Through oases, these groups ensured physical and economic survival in hostile but mineral-rich areas that had become strategic in the Bronze and Iron Ages.

It is in this context that the technology of catchment tunnels was introduced. That these techniques are not the result of an imposition by a central power, but expressions of the knowledge of local populations, is demonstrated by their extreme variety and environmental adaptability, and by the diverse denomination used in different regions: *qanat, kareez, falaj, foggara, khettara*, etc. [24]. The catchment tunnels are constituted from an underground gallery dug semiparallel to the ground with vertical shafts. The underground gallery has a slight slope used for the water to flow down and arrive in the open air by gravity. The tunnel does not dip into the aquifer; therefore, it drains the upper part, often through its filtering walls, as it crosses that part of the soil where the exchanges between deep waters and surface-saturation waters are greatest. Continuous exchanges take place between the air above and below ground, and one consequence of this circulation is the condensation of water in the soil when the ground temperature is low enough. It is precisely in these exchanges and interactions that catchment tunnels intervene. The vertical shafts and the filtering tunnel walls work to absorb the humidity and to produce water. Between the extreme conditions—collecting water from a spring or from ground, or producing water by exploiting contributions from the atmosphere (humidity, occult precipitation, aerial sources) lies the full range of catchment tunnels. The operating mode of a catchment tunnel highly depends on environmental and topographical circumstances as well as the seasons, the alterations in climatic conditions, and the long-term climatic cycles. The catchment tunnels, point of arrival of diverse experiences developed in different areas and adapted to local geographical situations, operate in a diversified way in the different ecosystems to tap hidden water with an end product of free water [20] (Figures 30.3 through 30.5).

30.7 The Environmental Hydrology of Shushtar, Iran

The town of Shushtar in Iran, inscribed in the UNESCO World Heritage List and placed along the course of the Karun River, an affluent of the Euphrates, is an extraordinary example of a city designed completely for the management of the waters. The center of the city is constituted by a great depression completely artificially dug in a desert area. Here the waters of the river are conveyed through a system of weirs, tunnels, and channels to create an extraordinary landscape formed by little lakes and basins encircled by water cascades. The depression is dug exclusively in order to create a quota jump and to the cascades that supply energy. The water works of Shushtar are mentioned

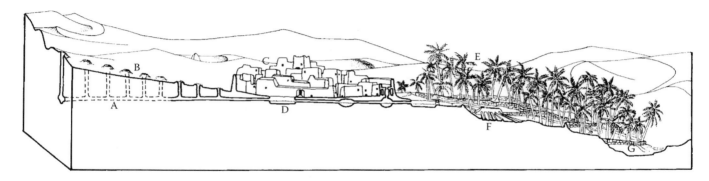

FIGURE 30.3 Structure of the oasis. Water produces in the catchment tunnel (a), which is visible thanks to the excavation shaft on the surface (b), runs beneath the adobe habitat (c), and gathers further along in decantation tanks (d), useful for drinking water, ablutions, and for cooling the dwelling. Once conveyed in open-air channels by means of the repartitions (kesria) (e), which serve to measure and distribute the water flow, water irrigates the palm groove (f) subdivided into tilled parcels by low mud walls (g).

in Zoroastrian texts, by Roman historians, and described with ample admiration by Arab travelers. They form a whole system that organizes and shapes, according to gravity and the required water flow, the entire conurbation, and the landscape as a whole, on a large territorial scale. From the archaic times of the Elamite water deity Anahita, to Zoroastrian cult practices of the Achemenian period, to Sabaean liturgies and to Islam, all the religions of the area have regarded water as the fundamental element in rituals. The historical hydraulic system of Shushtar consists of several devices that were built for the purpose of diverting the river course and thus utilizing its water resource in all possible ways. They have characterized this city from the most ancient times. Thanks to these works, the floods of the Karun River have been regulated and conveyed to cultivated fields in the desert, by means of banks and canals, thus supporting fertile oases, impoundments, and reservoirs, for the benefit

of agriculture and fisheries. Using the gravity, the water, passing through sluices, floodgates, superficial channels, tunnels, and feeding underground rooms of the houses, great cavities and industrial fabrics, is constantly used and reused. Water is diverted by means of underground canals, in such a way as to create waterfalls, which power a series of mills, essential within the local agricultural economy, and providing energy for the city. Water reaches town houses through underground conduits providing for not only domestic needs, but generating, at the same time, favorable climatic conditions acting as a thermal regulator. Waterworks consist of functional structures such as water intakes, bridges, dams and regulation weirs, water measuring and controlling towers, canals, underground conduits, penstocks, air vents, great basins, and also symbolic structures such as underground cult places, purification basins, baptism fonts, and ablutions halls.

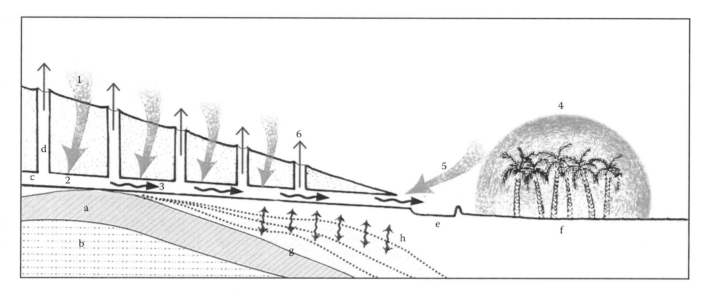

FIGURE 30.4 Water catchment tunnels hydro atmospheric cycle and water condensation in the soil and in the sub-soil: (1) atmospheric humidity; (2) condensation in the tunnels; (3) water runoff; (4) humidity in the cultivations; (5) absorption of the humidity; (6) output of dry air; (a) aquifer; (b) impermeable layer; (c) horizontal tunnel; (d) vertical shafts; (e) surface channels; (f) settlement and/or irrigated area; (g) aquifer level; (h) fluctuations in aquifer level.

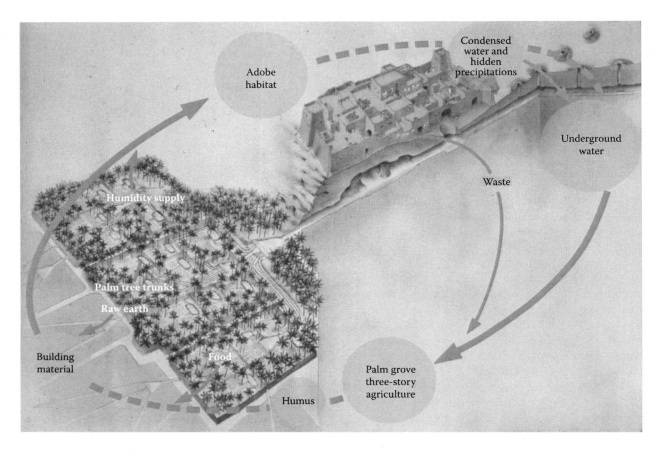

FIGURE 30.5 The oasis ecosystem. The oasis is a self-catalytic system in which the first supply of water condensation and moisture is increased by the installation of palm trees, which produce shade, attract organisms, and form humus. The palm grove determines a humid microclimate fed by hidden precipitations, water condensation, and underground drainage through the underground passageways of the catchment tunnel. The adobe habitat does not waste wood for firing bricks; it is kept cool by the underground water passageway and provides waste to fertilize the fields. The system runs the water resource in a cycle of use, which is not only compatible with the renewable quantities available but also increases them.

The age of such structures is attested by archaeological chronologies, but having been continually in use through time, they display succession of intervention works that overlap one another over extremely long periods. The oldest large-scale structural works date back to the Elamite civilization, which developed in the south and southwest of Iran from 3000 to 500 BC. The Elamite capital was Susa, today an important archaeological site, 60 km due northwest of Shushtar. Many authors believe, however, that the present site of Shushtar should be identified with the main Elamite cultural center in Khuzestan, linked with the great cult center of the Chogha Zambil ziggurat, situated 30 km southwest of Shushtar, and dating back to 1275–1240 BC. The earliest major earthworks channeling the waters of the Karun River date back to this period, as probably does the original network of the Shushtar canals. During the period of the Great Iranian Empire, which follows the Elamite civilization, dating from 550 BC to 330 AD, great waterworks were carried out. It is possible that the experiences in Shushtar made by the intense underground hydraulic engineers gave impulse to the development of the qanat technology. The imposing underground tunnel that conveys the waters of the Karun River to the city and to its fields dates back to the

Achaemenian King of Kings, Darius the Great (521–485 BC) and is still known to this day as Nahr e Dariun, or Dariun's Canal. Among these works, the most remarkable were the Polband-e Shâdorvân Bridge Dam and the Band e Mizan. They consist of a great bridge, for whose erection Roman captives were employed, among them was probably Emperor Valerian himself, captured in 259 AD, and of the weirs and dams across the Gargar canal, controlling and partitioning the waters, thus providing hydraulic power to operate watermills (Figure 30.6).

30.8 The Town Designed to Reuse Wastewater and Excrements: Shibam, Yemen

Shibam, Yemen, is constructed entirely of adobe. It is made up of very high towers made of unfired bricks, which have been left out in the sun to dry. Using adobe means enormous energy savings; adobe is also a better heat insulator in buildings. The city is right in the middle of a large wadi valley, the dry bed of a river, which only sporadically fills with water. Unusual swelling of the river

FIGURE 30.6 The hydraulic system of Shustar, Iran.

could completely overwhelm and destroy the city. The city protected itself with deflecting weirs and ditches that broadened the floodable area and dissipated the force of the water out over an enormous surface area that was thus made suitable for agriculture. Great depressions in the ground were dug around the city to collect and absorb the water. These artificial sand craters, protected along the rims by earth and shaded by the palms, could be tilled. The organic waste of the city was dumped into the depressions that, together with the water, changed the sterile sand of the loess into fertile soil.

It is indeed the very existence of the city of Shibam, with its supply of biological matter, which made the palm trees and farming possible creating a continuous positive feedback cycle. Not only foodstuffs feed the population and return to the soil as fertilizer but the entire city, with its forms and architecture, is founded on the eternal principle of complete reuse. The adobe bricks come from the soil of the gardens. Humus is continuously created and dug up in the craters and gives the soil its colloidal quality and cohesion, which in turn permits the bold architecture and solidly constructed buildings. The buildings followed a town plan and an architectural structure that were in harmony with the need to collect precious organic waste. All the tower houses had a facade along a dead-end street. Toilet drains situated on each floor of the building deposit the waste matter in the dead end street. The excrement was retrieved in woven straw

baskets that were kept at the bottom of the facades where the excrement dropped through trap doors. All of this was possible because liquid waste, which can damage adobe buildings, was kept separate from solid waste. Therefore, the solid waste dried quickly and was easily transported to the farm craters. The separation of liquid and solid waste was carried out thanks to a clever invention, a toilet, which had been used for centuries before the water closet cabinet came into use in modern time. The toilets had two outlets: a front outlet for urine and a back outlet for feces. Both are carried down by gravity, the liquid from a pipe and the solid from a shaft, to the collection baskets on the dead-end street. So its toilets, based on water reuse, can explain the entire town design and architecture of Shibam (Figure 30.7).

30.9 Unconventional Solutions to Improve Human Waste Management in Petra, Jordan

The archeological city of Petra, Jordan, the ancient capital of the Nabatean [28], was carved out of the desert canyon by nomad tribes thousands of years ago. It is now an endangered site. Petra is an environment undergoing a dynamic transformation. The erosion of the sandstone walls is part of a geological process. However, since Petra was abandoned, the crumbling of its surface

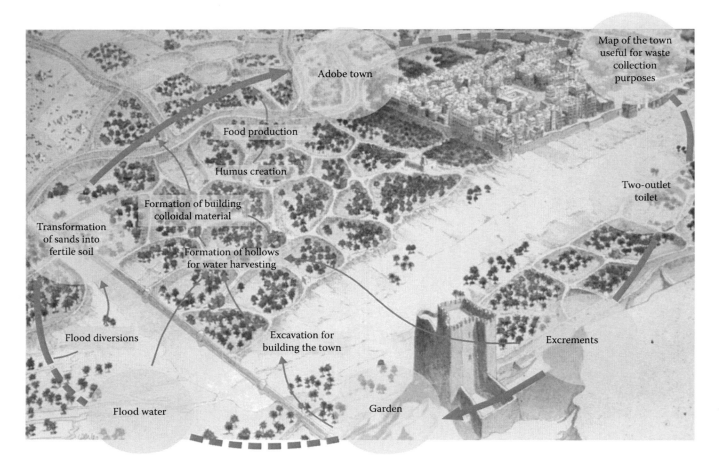

Map of the town useful for waste collection purposes

Adobe town

Food production

Humus creation

Formation of building colloidal material

Transformation of sands into fertile soil

Formation of hollows for water harvesting

Flood diversions

Excavation for building the town

Flood water

Garden

Two-outlet toilet

Excrements

FIGURE 30.7 The ecosystem of Shibam, Yemen. In Shibam, the habitat is important for the fertilization of the fields with which it interacts in an indissoluble cycle of careful use of the resources. The town is able to meet the need of collecting human excrements, thanks to the kind of closet, the fabric of the houses, and the whole urban plan. Excrement, essential in order to cultivate the desert, is dried in the sun. Thanks to the supply of floodwaters impounded by deviation dams, the excrement turns into humus and colloidal material, which is dug out and used for building and periodically renovating the tall adobe houses of the town. Depressions are made, surrounded by embankments and channels, and shaded by the palm-grove. Their function is that of providing agricultural foodstuffs and protecting the habitat from the floods by absorbing and storing quantities of water.

has sped up and is now proceeding at a catastrophic rate. The presence of human beings kept the stones of Petra from crumbling. The Nabatean peoples of Petra were able to make use of the rare rainfalls in order to create gardens and tilled land. The so-called Nabatean farms are exemplary in terms of water production by hydro genesis. As described in ancient writings, Petra, created with the technologies of the Nabatean agriculture [12], was a city of canals, basins, fountains, and gardens. An urban microclimate was created by means of what might be considered aesthetic measures such as waterfalls and gardens but which actually were the best protections for the architecture carved in the sandstone. Without recovering these ancient systems, it is not possible to save Petra. Today the number of visitors, the vastness of the park, and the climatic conditions require adequate sanitation facilities in the area of Petra. The park needs facilities for 3000–4000 people a day. The problem is important not only because toilets are necessary to provide a service to tourists and workers in the area, but also because it causes enormous damage

if left unsolved, as the absence of toilet facilities necessarily results in a release into the environment of human waste. Appropriate solutions must be found because one cannot fill Petra with toilets, especially the kind of plastic chemical toilets now located everywhere on the site. Nor is it sustainable to realize waterworks and sewer systems necessary to the installation of conventional flush toilets. An unconventional approach to the problem started from the acquisition that the sanitation systems conventionally used internationally are hardly the best example of sustainability. The enormous waste of water, in most cases drinking water, for toilets and the need for supply and disposal networks lead to an irresponsible attitude toward natural resources and to indifference to special geographical and environmental conditions. This situation is ethically and materially indefensible, particularly in arid areas. Toilets that do not require water are a necessity today not only for Petra and other places of human heritage but also to rethinking modern housing and urban design in a more sustainable way (Figures 30.8 and 30.9).

FIGURE 30.8 The actual situation of the so-called Palace Tomb, a monumental complex situated at the end of the long aqueduct of Wadi Al Mataha, Petra, Jordan.

FIGURE 30.9 Graphic reconstruction of the water systems, cascades, and tilled terraces of the Palace Tomb, Petra, Jordan.

30.10 Examples of Innovative Use of Traditional Water Systems

30.10.1 No-Flush Toilets

In so-called composting toilets, waste matter is processed onsite to produce compost for use as fertilizer. The result is aseptic and odorless. A wide range of solutions exists, ranging from simple devices equipped with aeration tubes to others with forced ventilation systems powered by solar panels, to technologically complex apparatus equipped with ultraviolet-lamp sterilizers. Composting toilets should be easy to build and maintain avoiding the use of water in order to limit the materials requiring disposal and, hence, facilitate their removal. One of the biggest problems of conventional toilets is the mixing of feces and urine, which makes the disposal of waste matter difficult in the absence of a sewage system. To remove and treat solid and liquid matter separately is in fact much easier, as the solid matter remains dry and more easily manageable. Traditionally this problem has been solved in the toilets used since antiquity in Shibam. It is a urine-diverting toilet that allows liquids to be stored and disposed of separately from feces. This solution is ideal for easy waste treatment. It has been realized in innovative ways with two-way ceramic or fiberglass bowls. Both solids and liquids are treated *in situ* or will be taken to remote areas where the composting will be done. Composting units can also be set up on local farms, which will use the compost as fertilizer.

30.10.2 Water Harvesting and Reuse

Prehistoric traditional techniques, which were used to build the agricultural landscape, today are reposed in agriculture as the best practices to replenish soils, save water, and combat hydro geological instability and desertification. The technique of drainage ditches spread in the Apulia district of Daunia 6000 years ago when Neolithic communities built more than 3000 villages surrounded by trenches. The ditches met environmental needs by draining water and drying some areas to be tilled during the humid season and by working as drinking troughs for cattle, humus collection, and water reserves during the dry season. After this practice has been replaced by mechanized agriculture, today these places are suffering terrible inundations in winter and extreme drought in summer. On the Ethiopian highlands and on the slopes of the Rift Valley ridges, there are many villages where multipurpose ditch systems are still used to store and manage water resources, gather sewage, and produce fertilizers. The techniques to harvest atmospheric water such as caves to condense humidity, mounds of stones, and limestone walls are used by all the ancient societies in arid areas. Today, authentic aerial wells, atmospheric condensers, water turbines, and textile surfaces of catchment producing water from vapor and fog are used. The practice of setting cistern jars full of water or calcareous masses close to plants to provide irrigation is reposed today with innovative techniques. Within the framework of this family of techniques, it is elaborated a product called *dry water*, which, set into the soil close to the roots, progressively furnishes the necessary water supply. The catchment tunnels, which produce only the right quantity of water that the environment itself can resupply, are proposed again as alternatives to the excavation of wells, which lower the groundwater and deeply perforate the soil, thus causing pollution and salinization on the surface.

30.10.3 Wastewater Reuse in Urban Settlements

Several innovative techniques coming from tradition are being experimented in urban areas. The building of most of the ancient towns followed the layout of the terracing and the water systems network. The rainwater harvesting techniques, the areas with the walled gardens, the use of organic remains for the production of humus, the passive architecture methods and climate control for food conservation and for energy saving, and the practices of recycling production and food residues used in the historical towns have been reproduced and innovated. This category includes all innovative techniques in the photovoltaic, sun warming, water catchment, composting, and waste recycling fields. In some advanced contexts, for example, Tokyo, the roofed-garden technique is being proposed by law in new houses with the vegetable covering on the terrace of modern buildings like the traditional hanging gardens. This keeps optimal climatic conditions inside the houses, harvests water, and becomes an area for entertainment and contemplation. The micro-solutions for city quarters and houses represent a large innovative sector in the waste-recycling field. Several mini-compost machines placed inside the gardens or in common areas of the quarters have been realized to directly absorb organic waste and supply the gardens with humus. Biomass mini-reactors, which transform waste into kitchen gases, as well as greater plants for heating the whole house, have been realized. Also small- and large-scale solutions for sewage water have been found. In Germany, modern houses have been equipped with a vertical marsh, a device that reproduces the processes of water decantation and filtration still existing naturally in marshlands. The process is reproduced along the wall of the building in glass interspaces where sewage waters seep into, infiltrate, and constantly recycle themselves by gravity. In Calcutta, an innovative traditional technique used on a very large scale solved the serious problem of used waters. Sewage waters, traditionally reused in rice fields, are today turned into a resource for irrigating and fertilizing rice fields by using proper innovative systems of sewage water filtration and sterilization.

30.10.4 Roof Water Harvesting and Two-Outlet Water Supply System

Today, the public water distribution networks are fed by excellent water, but only a very small quantity of it is used for drinking. Systems of meteoric water harvested off the roofs may supply the quantity of water necessary for nondrinking household use. To use rainwater, it is necessary to settle in the houses a two-outlet water supply system in order to separate the potable water supply network from water for sanitary use, shower, water closet, etc. A two-outlet network is a water supply system that allows distributing different quality waters according to their use. In fact, it is clear that the standards of water for human consumption are strongly secure and not necessary to other household uses (water closet, clothes washing, etc.). The possibility of distributing different water qualities would allow a substantial control of the costs of production and the distribution of waters with the best qualities for drinking use.

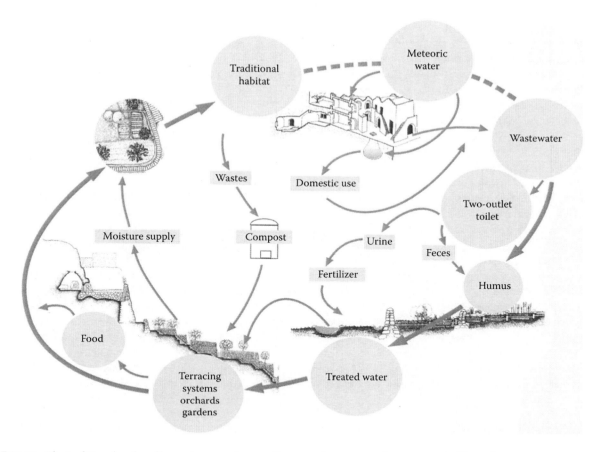

FIGURE 30.10 The traditional cycles of harvest, use, and reuse of water and wastewater for new sustainable settlements.

30.10.5 From Traditional Knowledge the Technologies of the Future

In Valais, the water catchment systems from the sources of springs and from glaciers that, through little surface canals called *bisse*, allow mountain slopes to be irrigated by gravity on a higher level than the stream's natural course. A similar technique used today is reproposed in Tibet with innovative methods to protect glaciers that are in danger because of climate change.

In Burkina Faso, *zai* is a particular traditional technique able to regenerate highly degraded soils. The soil is dug with holes that fill up with water in the humid season and are used as dumpsites for rubbish and manure in the dry season. This practice attracts termites that digest rubbish, thus its absorption by the plants' roots. Furthermore, the tunnels dug by the termites increase the soil's porosity. Seeds are then sown in the holes, giving very high crop yields. Innovative practices that promote original forms of symbiosis between humankind and animals or microorganisms are reproposed today to rehabilitate degraded soils or soils made suitable for human living in extreme areas. In the Balearic Islands, *feixes* are a traditional system of agricultural organizations according to which the plant roots are irrigated directly from underground without wasting water. Superficial drainage channels, into which water flows, separate the tilled fields. From

these a network of underground pipes made of porous materials and covered with a layer of branches and algae carries the water under the cultivation. Thus, channels release the quantity of water to the tilled soil according to seasonal and climatic needs. The technique is reproposed in hydroponic cultivations and for space stations (Figures 30.10).

30.11 Summary and Conclusions

Ecomimicry: The art of learning from the nature and the landscapes.

It is wrong to consider water traditional knowledge as marginal compared to the great economic and technological processes under way. Even from a quantitative point of view, their use still supports most of humankind that is distributed throughout the less industrialized countries. Paradoxically, in these places where traditional techniques are still used in a massive way, the modernist considers these as a phenomenon of backwardness, whereas in advanced countries, they create an image of desirability and provide added value. In the regions of Valais in Switzerland, in the Loire Valley in France, and in Tuscany in Italy, the maintenance of traditional techniques in agriculture has ensured the high value of the landscapes and the quality of the products that can be obtained with these techniques. Traditional water knowledge and oasis systems, such as Shushtar, Shibam, and Petra, show how archaic societies developed a sound economy, and survived thanks to the ingenuous management of natural resources. These experiences are now fundamental to completely rethinking the conventional system of water management, not only because of the growing needs of drinking water but also for the waste and destruction of resources that current methods involve. Among these is sufficient to consider that the discharge of wastewater into the sea causes the loss of nutrients on Earth and the hypertrophy and destruction of the marine ecosystem. The used waters are not a waste or a useless discharge, but a resource that, subjected to a continuous reuse and recycling, can provide freshwater, fertilizer, and energy. A recent discipline, *biomimicry*, creates innovative materials by copying desert plants and marine organisms. New aseptic textiles are invented studying the property of the crab carapaces. Innovative generations of nanomaterials like the graphene, that through microfiltration can return water to a drinkable state, are created by observing the membranes of natural organisms. Similarly, observing and studying the water traditional knowledge it is possible to elaborate the *ecomimicry*, a naturalistic and blue engineering: the art of learning from the great heritage of experience accumulated by humankind in building landscapes.

References

1. Abdollahi, K. 2006. Ancient hydro-structures for water management in Chogha Zanbil, Shushstar and Dezful, Iran. In: *Proceeding IWA Specialized Conference on Water & Wastewater Technologies in Ancient Civilizations*. Istanbul, pp. 239–245.
2. Amiran, R. 1970. The beginnings of urbanization in Canaan. In: *AA.VV. Near Eastern Archaeology in XXth Century*. Doubleday, New York.
3. Angelakis, A.N. and Spyridakis, S.V. 1996. Wastewater management in minoan times. In: *Proceedings of the Meeting on Protection and Restoration of Environment*. Chania, Greece, August 28–30, pp. 549–558.
4. Barrois, A.G. 1937. *Les Installations Hydrauliques de Megiddo*, extrait de la *Revue Syria*, n.3, Librairie Orientaliste Paul Geuthner, Paris.
5. Çauvin, J. 1994. *Naissance des divinités Naissance de l'agricolture*. CNRS, Paris.
6. Childe, V.G. 1954. Prime forme di società. In: *A History of Technology*. Clarendon Press, Oxford, V.I.
7. Cleuziou, S. and Laureano, P. 1998. Oases and other forms of agricultural intensification. In: M. Pearce and M. Tosi (eds.), *Papers from the EAA Third Annual Meeting at Ravenna 1997*. Volume I: Pre- and Protohistory. BAR International Series 717, Archaeopress, United Kingdom.
8. Dentzer, M. et al. 1989. *Contribution Francaise à l'archéologie Jordannien*. Ifapo, Amman.
9. Diamond, J. 1998. *Guns, Germs, and Steel, The Fates of Human Societies*. W.W. Norton & Company, New York.
10. Di Lernia, S. et al. 1990. *Gargano Prehistoric Mines Project: The State of Research in the Neolithic Mine of Defensola-Vieste (Italy)*, in Origini, XV. Gangemi, Italy, pp. 175–199.
11. Drower, M.S. 1954. Fornitura di acqua, irrigazione e agricoltura, In: *A History of Technology*. Clarendon Press, Oxford, V.I.
12. El Faïz, M. 2005. *Les maîtres de l'eau. Histoire de l'hydraulique arabe*. Actes Sud, Arles.
13. Evenari, M., Shanan L., and Tadmor, N. 1971. *The Negev the Challenge of a Desert*. Harvard University Press, Cambridge, MA.
14. Giddy, L. 1987. *Egyptian Oases*. Ares & Phillips, London.
15. Gruet, M. 1952. Amocellement pyramidal de spheres calcaires dans une source fossile moustérienne à El Guettar (Sud tunisien) in *Actes du II congrès panafricain de Préhistoire et de Protohistoire*. Crape, Alger, pp. 449–456.
16. Laureano, P. 1993. *Giardini di Pietra, i Sassi di Matera e la civiltà mediterranea*. Bollati Boringhieri, Torino.
17. Laureano, P. 1995. *La Piramide Rovesciata, il modello dell'oasi per il pianeta Terra*. Bollati Boringhieri, Torino.
18. Laureano, P. 2005. *The Water Atlas. Traditional Knowledge to Combat Desertification*. UNESCO, Laia Libros, Barcellona; *Water Conservation Techniques in Traditional Human Settlements*. Copal Publishing Group, India, 2013.
19. Laureano, P. 2012. Traditional water technology in dry land. In: *Afro-Eurasian Inner Dry Land Civilizations*, Association for Afro-Eurasian Inner Dry Land Civilisation Studies Graduate School of Letters, Nagoya University, Nagoya-shi, Vol. 1, pp. 11–26.
20. Laureano, P. 2012. Water catchment tunnels: Qanat, Foggara, Falaj. An ecosystem vision. In: *Proceeding IWA Specialized Conference on Water & Wastewater Technologies in Ancient Civilizations*, IWA, Istanbul.
21. Lourandos, H. 1980. Change or stability? Hydraulics, hunter-gatherers and population in temperate Australia. In: *World Archeology*, Vol. II N. 3, February. Routledge & Kegan Paul, London.
22. Mays, L.W. 2010. *Ancient Water Technologies*. Springer, London.
23. Pirenne, J. 1977. *La Maitrise de l'Eau En Arabie du sud Antique*. Geuthner, Paris.
24. Schimt, K. 2007. *Sie bauten die ersten Temple*. C. H. Beck Verlag, München.
25. Shimada, Y. 2012. *Afro-Eurasian Inner Dry Land Civilizations*, Vol. 1, 3–8.
26. Wittfogel, K.A. 1957. *Oriental Dispotism*. Yale University Press, New Haven.
27. Yazdi, S.A. and Khaneiki, M. 2010. *Veins of Desert—A Review on the Technique of Qanat/Falaj/Karez*. UNESCO, Yazd, Iran.
28. Zaydine, F. 1991. *La royaume des Nabatèens* in *Le Dossiers d'Archèologie*, n. 163, Faton, Quétigny, France.

31

Modern Water Reuse Technologists: Tertiary Membrane and Activated Carbon Filtration

Alexandros I. Stefanakis
University of Brighton

PREFACE

Water reuse is a widely desired practice around the world, mainly applied for water resources conservation, reduction of potable water supplies consumption, and ecosystems protection. Today, there is a variety of available effective technologies for water reuse, mainly as advanced treatment stage in wastewater treatment facilities. These technologies have different characteristics and are based on biological, chemical, mechanical, and natural processes. However, each technology has certain advantages and disadvantages, which should be taken into account when selecting the appropriate technological train for each water reuse application.

31.1 Introduction

Treatment technologies for water reclamation and reuse probably represent the most essential segment in integrated water resources management. A variety of technologies with different nature, processes, and means used exists for the treatment. A treatment technology can be employed singly or in combination with other technologies and processes for the optimum treatment result, which is also the usual practice. Due to the numerous available technologies and processes (Figure 31.1), there is a respectively high number of possible flow diagrams for the treatment train that can be adopted, depending on the specific characteristics of each reuse application. The basic principle of wastewater treatment plants is the optimum removal of the various pollutants present in wastewater. The necessary level of wastewater treatment is defined by the effluent limit concentrations, which needs to be fulfilled before the final discharge of the effluent, and by the option of water reuse of this treated effluent.

31.2 Treatment Train

A conventional treatment train usually includes up to two or three treatment stages: primary (preliminary), secondary, and tertiary treatment, as shown in Figure 31.1. Preliminary stage includes inflow-measuring devices, and screens for the removal of large solid materials, sand, grit, and oils. The primary stage includes a first sedimentation tank for the removal of settleable solids. The primary stage is followed by the second treatment stage, where removal of suspended solids, biodegradable organic matter, and—at a certain level—nutrient removal takes place. The most common method is that of aerated activated sludge. A treatment train of the conventional activated sludge system is shown in Figure 31.2 and includes aerated basins and sedimentation tanks in the secondary stage. Other methods for secondary treatment are trickling filters or rotating bio-contactors [14,83], which are based on the use of fixed-media. The secondary stage usually ends with a secondary settling tank or a combination of membrane filtration and disinfection [82].

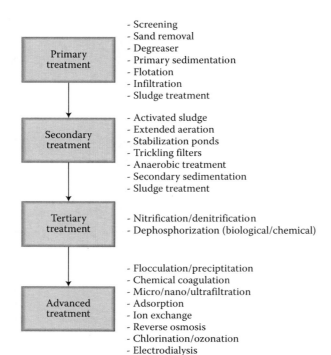

- Screening
- Sand removal
- Degreaser
- Primary sedimentation
- Flotation
- Infiltration
- Sludge treatment

- Activated sludge
- Extended aeration
- Stabilization ponds
- Trickling filters
- Anaerobic treatment
- Secondary sedimentation
- Sludge treatment

- Nitrification/denitrification
- Dephosphorization (biological/chemical)

- Flocculation/preciptitation
- Chemical coagulation
- Micro/nano/ultrafiltration
- Adsorption
- Ion exchange
- Reverse osmosis
- Chlorination/ozonation
- Electrodialysis

FIGURE 31.1 Treatment processes and technologies in wastewater treatment.

Until recently, a treatment train of two stages was common for conventional wastewater treatment. However, it was gradually realized that the effectiveness of this setup was no longer acceptable due to new challenges that came to the forefront in the wastewater treatment field: (1) adaptation of stricter regulated effluent limits for the final discharge for public health and environmental protection, (2) decline of available water resources, (3) rapid increase of population, (4) industrial development, (5) realization of treated water as a potential beneficial water source with multiple reuse applications, and (6) new groups of emerging pollutants with high health and environmental risks [71,98]. These challenges are dictated by the concept of integrated and sustainable management of water resources, which has been introduced to international and national legislation, for example, the Water Framework Directive 2000/60/EC in Europe [15].

31.3 Advanced Treatment Technologies

All these new challenges resulted in one common conclusion: the treated water from treatment facilities should be of higher quality

in order to cover all these new needs and to be appropriate for various reuse applications with minimum impact on human health and the environment. Conventional biological wastewater treatment cannot provide the required effluent quality, especially for discharge in ecologically sensitive areas and for the implementation of water reuse projects. Therefore, additional advanced steps should be added in the treatment train and advanced treatment technologies should be used to deliver a final effluent of higher quality. The main goal is to further treat and remove from the wastewater pollutants like organics, solids, nutrients, pathogens, and potentially toxic compounds like heavy metals or emerging pollutants (pharmaceuticals, endocrine-disrupting chemicals [EDCs], personal care products [PCPs], etc.) [2]. Thus, advanced treatment technologies could be classified according to the targeted pollutants as shown in Table 31.1.

Table 31.2 presents a general overview of various advanced treatment technologies' performance in the removal of pollutants.

31.3.1 Membrane Bioreactor

Membrane bioreactors (MBR) represent a system that combines biological and physical removal processes [11,88]. These systems are a development of the conventional activated sludge method, where ultrafiltration (UF) or microfiltration (MF) membranes are used for the separation and retention of the suspended sludge without the need of large secondary settling tanks [77]. A typical layout of a conventional MBR facility includes a pretreatment stage, for example, screening, and the aeration tank followed by the MBR unit, as shown in Figure 31.3a [5,9]. The aeration tank removes the biodegradable part of the influent compounds and the membrane unit separates the treated water from the mixed liquor. However, the aeration tank is often avoided and the pretreated wastewater is directed to the MBR unit; in this case, the membrane is submerged into the aeration tank. The configuration of the MBR unit can include a submerged membrane (Figure 31.4a) or a side-stream membrane (Figure 31.4b) [47]. In the first case, the membrane is submerged into the mixed liquor and usually operates under pressure. Periodical aeration or backflushing is applied for regular cleaning, while use of chemicals is also occasionally implemented [9]. In the second type, the mixed liquor is recirculated through an external membrane unit, which operates under the high cross-flow velocity along its surface.

Generally, their performance is good for the removal of solids, organics, nutrients, and turbidity, while a significant level of disinfection takes place. Especially for emerging pollutants like pharmaceuticals, PCPs and EDCs, membrane processes are considered an effective treatment technology for water and wastewater

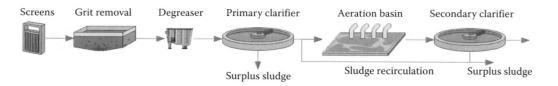

Screens   Grit removal   Degreaser   Primary clarifier   Aeration basin   Secondary clarifier

Surplus sludge   Sludge recirculation   Surplus sludge

FIGURE 31.2 Conventional wastewater treatment train (activated sludge system).

TABLE 31.1 Classification of Advanced Treatment Technologies Based on the Various Target Pollutants

Target Pollutant	Associated Risk	Technologies
Pathogenic microorganisms	Disease transmission	• Chlorination • Ozonation • UV radiation
Nutrients	Eutrophication in surface waters	• Nitrogen and phosphorus • Biological removal • NH_4^+ ion-exchange • Chemical precipitation • Adsorption • Constructed wetlands
Hardly biodegradable organics	Toxic for humans and the environment	• Activated carbon • Chemical oxidation with O_3, H_2O_2 or O_2 • Constructed wetlands
Heavy metals	Toxic for humans and the environment bioaccumulation	• Reverse osmosis • Nanofiltration • Chemical precipitation • Adsorption • Ion exchange • Constructed wetlands
Remained solids	Adverse effects on final effluent	• Tertiary treatment with filters (after activated sludge) • Ultrafiltration membrane • Membrane bioreactors • Constructed wetlands
Organic micro-pollutants and emerging contaminants		• Membrane filtration • Constructed wetlands

[36,57,70]. The hydraulic retention time in the MRB tanks is usually higher than that in conventional treatment systems, which makes the system more stable and, therefore, appropriate for relatively small facilities (<2000 m³/d) with fluctuations of the daily inflow rate [11]. Sludge production is also relatively low [88]. MBRs are generally smaller systems (compact) than conventional processes, which reduces the area demands [9]. The filtrate quality of MBR units can be appropriate for direct feed of reverse osmosis (RO) process. MBR systems are used in areas with limited water sources and limited available land for water reuse applications like small communities, commercial centers, resorts, hotels, etc. [18].

On the other hand, the basic disadvantage of this technology is the problem of membrane fouling due to gradual solid and colloidal matter accumulation on the external surface of the membrane [46,77,87]. Fouling affects system operation and can be a major failure parameter because it demands a frequent operation stop for cleaning and respective use of clean water and, possibly, chemicals. Also, total costs of the system, including both capital investment and high energy consumption for the system operation, is an obstacle toward the wider use of this technology [9,46,47,87]. Research on MBRs is still ongoing and focuses on limiting the impact of fouling on system performance and reducing operational costs.

31.3.2 Tertiary Membrane Filtration

Membrane units are also used as a tertiary treatment stage in existing conventional treatment facilities, for example, activated sludge systems (Figure 31.3b). Installation of a membrane unit is considered an upgrade of the facility to enhance its

TABLE 31.2 General Overview of Treatment Technologies' Performance in the Removal of Various Pollutants in Wastewater Treatment for Reuse Applications

Pollutant	BOD, COD	TSS	N	P	Alkalinity	Total Coliform	Viruses	TDS	TOC	Turbidity	Color	Emerging Micro-Pollutants (e.g., Pharmaceuticals, Pesticides, EDCs, etc.)
Conventional primary settling system	L	M	L	L					L	L	L	L
Conventional activated sludge system	H	H	L	L	M	L			H	H	L	L
Biological nutrient removal	H	H	H	H		L			H	H	L	L
MBR	H	H	H	H		M			H	H	H	M
Activated carbon	H	H	M	H		L	L	H	H	H	H	H
Tertiary filtration												
MF		H	L		M	M	L	L	M	H	M	L
UF		H	L		M	H	H	L	H	H	H	L
NF				H		H	H	H			H	H
Reverse osmosis			H		H	H	H	H	H		H	H
Chlorination						H		L				L
Ozonation						H	H					H
UV						H	M					
Constructed wetlands	H	M	H	M	H	M	M	L	M	M	M	M

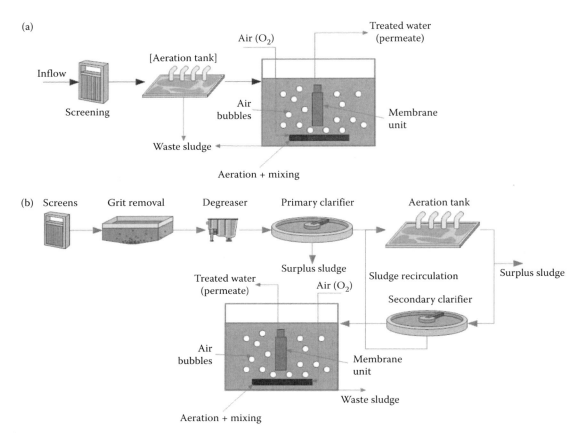

FIGURE 31.3 Typical treatment train of conventional membrane bioreactor facility.

performance. As for the MBR, the basic principle is the selective passage through the membrane unit, which results in the separation of specific compounds [98]. Membranes can be classified based on the size of the passing compounds, the chemical structure and composition, and the geometry of the unit. The most common membrane types are cross-flow pressure-driven processes such as MF, UF, nanofiltration (NF), RO, and electrodialysis (ED). Table 31.3 presents the basic characteristics of membrane processes.

31.3.2.1 Microfiltration

MF membranes remove fine particles and other suspended material from a liquid, such as suspended solids (sand, dirt), colloids, turbidity, and bacteria. A typical pore size is around 0.2 μm. This means that it is comparable with other filtration media such as sand, clay, etc. The MF membrane acts as a physical barrier for microorganisms. MF membranes are generally capable of achieving 90%–99.99% (1–5 logs) removal of bacteria and protozoa and 0%–99% (0–2 logs) of viruses [86]. MF can be

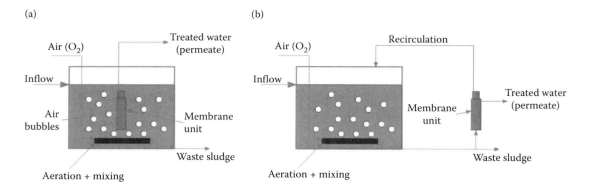

FIGURE 31.4 Membrane bioreactor configurations.

TABLE 31.3 Typical Characteristics of Membrane Filtration Processes

Membrane	Pore Size (nm)	Typical Transmembrane Pressure, ΔP (bar)	Removal
Microfiltration (MF)	0.05–2.0	0.2–3	Bacteria
			Colloids
Ultrafiltration (UF)	0.005–0.1	2.0–10.0	Large organic molecules
			Viruses
Nanofiltration (NF)	0.001–0.05	10.0–27.0	Small organic molecules
			Divalent ions
			Color
			Micro-pollutants
			Pesticides
			Herbicides
Reverse osmosis (RO)	0.0005–0.002	15.0–83.0	All dissolved species
			Aqueous salts

used as pretreatment to UF, NF, or RO membranes and is used in simple dead-end water filtration processes, sterile bottling of fruit juices and wine, pulp and paper wastewater, and aseptic use in the pharmaceutical industry [9].

31.3.2.2 Ultrafiltration (UF)

With UF membranes, salts, proteins, enzymes, gelatins, colloidal silica, manganese, iron, total organic carbon, bacteria, viruses, chlorine-resistant pathogens (e.g., *giardia*, *cryptosporidium*), emulsions, and, in general, large molecules are retained. A typical pore size is around 0.005–0.1 μm. Bacteria, viruses, and other fine particulate matter are also separated from the liquid. Common applications of UF membranes include water/oil separation, wood scouring, pulp mill effluents, tannery wastewater, fruit juice clarification, whey and milk processing, landfill leachate, cosmetic industry, pharmaceuticals purification, potable water production, and wastewater reuse [9].

31.3.2.3 Nanofiltration

NF membranes retain smaller organic molecules such as viruses, pesticides, herbicides, sugars, synthetic dyes, multivalent ions (e.g., Mg, Ca), chlorine resistant pathogens (e.g., *giardia*, *cryptosporidium*), dissolved organics, nitrates, antibiotics, and, generally, a wide range of micropollutants [94]. Almost all organic matter, a range of salts, and divalent ions are removed. A typical pore size is around 0.0001–0.005 μm. NF is often used for water softening (de-salting), that is, hardness removal, and color removal in various applications, such as lactose production from cheese whey [9]. Since they operate at a lower pressure, the drive-force through the membrane is smaller, which means that energy demand is respectively lower.

31.3.2.4 Reverse Osmosis

RO is the finest membrane separation process. It separates very fine particles and suspended material from a liquid, with a typical particle size up to 0.0001 μm. RO removes metal ions, monovalent ions (e.g., K, Na), acids, synthetic organic chemicals, emerging pollutants, and aqueous salts [65,71]. The basic principle of RO membranes is the natural tendency to be at equilibrium (osmosis). With RO, the osmotic process of a solution is reversed so that the water is driven away from dissolved molecules. The most common application of RO is the desalination of seawater and brackish water to produce freshwater suitable for human consumption. Other applications include the production of boiler feed water and ultra-pure water for municipal and industrial uses, like cheese whey concentration, fruit juice concentration, and car-wash water reclamation [9]. With RO, the goal is to produce a pure filtrate (water), reduce the wastewater volume to be discharged, and retain various constituents of the inflow stream as the product. As for NF, integrated membrane systems that will include both MF/UF followed by NF/RO are usually required in water reuse applications for the simultaneous removal of trace organic chemicals and dissolved solids. Generally, RO is more widely used than NF for water reuse.

31.3.2.5 Electrodialysis

ED is another, fundamentally different process than RO, used for desalination of brackish water and ion removal [55,71]. ED systems generally present some advantages over RO membranes like the absence of pretreatment necessity, enhanced membrane stability, higher treatment capacity, higher water recovery range, extended life cycle up to 10 years, and reduced costs [55]. Usually, in ED systems water is pumped at relatively low pressure (typical pressure 1–3 bar) between several flat, parallel, ion-permeable membrane sheets that are placed in series. In ED, transportation of ions takes place under the influence of an electric potential; membranes that allow cations to pass through are alternately placed with membranes that allow anions to pass through. Two electrodes placed at both ends of the membrane assembly create an electrical current and concentrate ions through the membranes. Alternate placement of membranes permeable to ions and cations results in gradual accumulation of ions in one membrane and reduction of ions in the following membrane. Membrane fouling effects are usually minimized by reversing the electrical current direction, that is, reversing the polarity of the electrodes, at 15–30 min intervals (ED reversal). Thus, ED produces two distinct flows: one with high salt content (brine waste) and the other with low salt content. ED has a wide use of applications in water treatment [7], refinery effluent treatment [43], and tannery wastewater treatment [78], among others.

31.3.3 Disinfection

Disinfection processes are used to eliminate the presence of pathogenic microorganisms in the effluent from water and wastewater treatment facilities. Increasing demands for sanitation and adoption of more stringent limit values for effluent discharge make necessary the implementation of advanced technologies to meet the requirements for effluent discharge to surface and groundwater, human health protection, environmental protection, and for water reuse [39]. Usually, targeted pathogenic pollutants are bacteria (e.g., *E. coli*, *Salmonella*, and *Shigella*), viruses (e.g., adenovirus and norovirus), and protozoa (e.g., *Giardia* and *Cryptosporidium*). Disinfection represents probably the most important stage in the treatment train because it prevents downstream contamination of surface and groundwater and allows for the safe reuse of treated water/wastewater.

31.3.3.1 Chlorination

The most widely applied method for disinfection is the use of chlorine, an agent known for its destructive activity over bacteria, viruses, and protozoa. Wastewater chlorination is common practice for the reduction of microbial contamination around the world. Chlorine can be used in many forms such as chlorine gas, hypochlorite solutions, and other chlorine compounds in liquid or solid state [85]. Generally, chlorination is a reliable and effective technique. Typical doses for municipal wastewater are 5–20 mg/L for a contact time of 30–60 min [39].

However, residual chlorine in the treated effluent might be present, which—even at low concentrations—is toxic to aquatic life and possesses a potential negative long-term impact on the ecosystem. Thus, often a dechlorination step is required prior to final discharge, which increases the total cost by about 20%–30% [39]. In addition, chlorine application results through the oxidation of organic compounds in the formation of various hazardous by-products like trihalomethanes (THMs) and haloacetic acids (HAAs) with adverse effects on the ecosystem and human health. It has to be noticed that chlorine probably transforms trace organic chemicals to new compounds rather than mineralizing them. If ammonia is also present, chloramines can also be formed like the carcinogen *N*-nitrosodimethylamine (NDMA) with the chlorination of amines [48]. Operating conditions, for example, daily dose, contact time, temperature, pH, etc., affect the formation of these compounds [79,96]; thus, a proper and careful management of the chlorination strategy is always necessary.

31.3.4 Advanced Oxidation Processes

Advanced oxidation processes (AOPs) are processes that are based on the production of hydroxyl radicals ($\bullet$OH) as the main oxidant agent for the majority of the chemicals present in water/ wastewater [1,10,24,98]. The aim of these processes is to deal with pollution in water and include a variety of particular processes and mechanisms. They are included in the treatment either as pretreatment, that is, to break down large organic molecules and improve their biodegradability in the following stages of biological treatment, or as a final polishing/disinfection stage [53,80].

Ozonation is a common disinfection technique that uses ozone (O_3), an agent with strong oxidizing capacity, for the elimination of bacteria, viruses, and some resistant protozoan parasites like *Giardia* spp. and *Cryptosporidium* spp. [39]. It is widely known for the inactivation of *E. coli* in drinking water. Its high effectiveness makes this process appropriate for effluent discharge and water reuse. Ozone is usually produced onsite from dry air or pure oxygen. It is very reactive with a series of ions present in water and possesses a high reduction potential. A critical parameter for its effective oxidation is the optimum solution to water because ozone has low water solubility [98]. Ozone diffusion within a gas film rather than in a liquid film is commonly used for the more effective mass transfer. Higher pH and pressure values, as well as increased contact time, enhance the oxidation effects, while high temperature limits the degradation efficiency [24]. Ozone has been applied not only for disinfection [95], but also for the removal of organic compounds, color, taste, odor, algae, Fe, and Mn [24].

The main disadvantage of ozonation is the relatively high energy demand and short life period of ozone, which can make this process economically unattractive especially for large wastewater treatment facilities. Also, the potential oxidation of bromate ions and the formation of harmful (carcinogenic) bromated organic compounds, coupled with the potential increase of downstream biological growth (due to breaking down of large organic molecules to lower ones), has led to the combination of ozone with activated carbon filtration for the decrease of these biodegradable compounds. Additionally, ozone is usually combined with other techniques like UV irradiation and hydrogen peroxide in hybrid methods.

- O_3/H_2O_2: It is also called peroxone process. The addition of hydrogen peroxide (H_2O_2) to the liquid solution enhances the ozone decomposition and forms $\bullet$OH [1]. It has been used for the oxidation of micro-pollutants and the removal of pesticides, odor, and taste [98]. The optimum rate of O_3/H_2O_2 varies between 0.3 and 0.6. This process has been applied in industrial effluents for the removal of phenolic compounds, cyanides, nitrite, and sulfide compounds [25].
- O_3/UV: This process is applied for the elimination of toxic and refractory organics [1]. During this process, water saturated with ozone is irradiated with UV light at 200–280 nm. It is considered a more complex method but with higher efficiency because the ozone activation by UV protons results in the formation of $\bullet$OH. However, this process is generally more expensive than O_3/H_2O_2. It has been used for the oxidation of aromatic and aliphatic chlorinated organic compounds, pesticides, and cyanide [98].
- H_2O_2/UV: In this process, UV irradiation (<280 nm) is applied for the photolysis of H_2O_2 and the respective formation of two $\bullet$OH, which react with organic compounds. The process is enhanced under alkalic pH values [1].
- Photocatalysis: Photocatalytic processes use some metal oxides as catalysts that produce $\bullet$OH when UV is absorbed. Oxygen is the oxidant agent [1]. The anatase form of TiO_2 appears to be the optimum catalyst that offers high stability and good performance at reasonable cost. The various

processes include TiO_2/UV and $TiO_2/H_2O_2/UV$. These processes are relatively new and have also been used for the oxidation of organic compounds [98].

- Fenton process: This is one of the oldest oxidation processes. This process consists of the combination of H_2O_2 with Fe(II) or Fe(III) oxalate ion for the production of $\cdot OH$. The photo-Fenton process includes UV irradiation, too. These processes are generally favored by the fact that iron is a nontoxic abundant element. Fenton processes, that is, activation of H_2O_2 with iron salts or ozone and UV, are capable of destroying various toxic organic compounds in wastewater, for example, phenols and herbicides [1,51].

The application of AOPs is mainly focused on disinfection potential and, lately, on the removal of emerging contaminants from water. Most of these processes have been found to be effective in this topic [17,29,37,54] and, along with activated carbon filtration, are considered best available techniques [10,80]. Despite their good effectiveness, AOPs have a high operational cost, which is still considered their main disadvantage [53].

31.3.5 Activated Carbon Filtration

Activated carbon (AC) has been widely used as filter media in filter units (columns) usually placed at the downstream end of a wastewater treatment facility. It is carbon activated with physical and/or chemical agents at high temperatures. Various raw materials have been used for the production of AC, like coal, peat, petroleum coke, wood, charcoal, nutshell, etc. [4]. AC filters can be effective in the reduction of organics (COD), total organic carbon (TOC), turbidity, and color when used at the tertiary treatment stage [4,38].

AC is also used in cases where removal of emerging contaminants is of concern. AC is known for its capacity to remove certain hydrophobic organic compounds and chlorine present in water, such as organic solvents and chlorinated organic solvents with low water solubility [12,49]. It has been characterized as one of the best available techniques for pharmaceutical and other organic chemicals removal [16]. These contaminants include solvents, pesticides (e.g., atrazine), benzene, chorolobenzenes, trichloroethylene, methylene chloride, vinyl chloride, phenolic compounds, etc. which can be found in contaminated surface and groundwater. In addition, disinfection with chlorine or chloramines can result in the formation of hazardous by-products like trihalomethanes, which can be removed in AC filters. AC has been proved effective in the removal of EDCs such as estrogens (e.g., estradiol, estrone, estriol, estradiol-17-glucuronide, etc.), pesticides (e.g., atrazine, simazine, desethylatrazine), disphenol A, phenols, and PCPs and pharmaceuticals like naproxen, ibuprofen, triclosan, caffeine, and diclofenac, among other [13,36,62,70,84,93]. A key parameter in AC filtration is the possibility to regenerate and reuse the material after its saturation with the adsorbed molecules, which makes its use economically attractive because AC regeneration restores the initial properties of the material [38].

31.3.5.1 Hybrid Filtration Systems

Combination of membrane filtration with activated carbon filtration is also a widely used alternative in tertiary treatment processes. This system usually includes MF or UF units in series with activated carbon filters. The combined process of both units further improves the performance in terms of soluble and dissolved organics, TOC, ammonia, bacteria, turbidity, color, and odor [21,34,35,42,64,67,68,81] when used in advanced wastewater treatment. The system performance is particularly enhanced in terms of emerging pollutants removal. The combination of an MRB unit with powdered or granular activated carbon has been widely and successfully tested for the removal of organic micro-pollutants such as pharmaceuticals, EDCs, PCPs, etc. [40,41,52,58,69]. Since sludge is retained in the MBR unit, fouling of the membrane is reduced and the retention time in the system is longer [59].

31.3.5.2 Activated Carbon/Ozonation

Combination of activated carbon filtration with ozonation processes is one alternative that rapidly attracts increasing interest. The need to produce effluents of high quality for water reuse applications means that disinfection for the elimination of pathogenic microorganisms as well as removal of chemical constituents should take place. Treated effluents still contain residual concentrations of several organic micro-pollutants, like pharmaceuticals, PCPs, pesticides, EDCs, etc. The combined effect of oxidation with ozone and activated carbon adsorption results in significant decrease of the micro-pollutant concentration. Activated carbon not only enhances the adsorption process but also acts as a catalyst in the ozonation reactions. It has been found that the ozonation process causes the increase of the pore size and the surface area of activated carbon, which respectively enhances the removal of phenolic compounds and COD [56]. It is also reported that the same combination can simultaneously remove color and organic matter from dyes solutions [19]. Trace organic chemicals (e.g., pharmaceuticals), dissolved organic carbon, toxicity, and bacteria removals are also improved in the combined system of ozonation followed by activated carbon filtration [23,61,60]. In all cases, the combined action of both processes is improved compared to the single application of each process.

31.3.6 Constructed Wetlands

Constructed wetlands (CWs) belong to the natural treatment systems category, which are based on the exploitation of naturally occurring processes and mechanisms under controlled conditions [74]. They are classified in free water surface CWs and subsurface flow CWs, vertical or horizontal [32,76,72–74]. The main characteristic of these systems is the use of plants (usually common reeds), which grow on natural filter media like sand and gravel. Plants develop a deep, dense, and complex root system. The surface of the filter media grains and the root surface act as available area for microbe attachment and the respective

TABLE 31.4 Pollutant Removal Processes in Constructed Wetlands

Pollutant	Removal and Transformation Processes		
	Physical	Chemical	Biological
Suspended solids (i.e., TSS)	Filtration Sedimentation		Bacterial decomposition
Organic matter (i.e., BOD$_5$ and COD)	Filtration and settling (particulate)	Oxidation	Bacterial degradation (soluble) Microbial consumption
Nitrogen	Volatilization	Ion exchange	Nitrification/denitrification Microbial uptake Plant uptake
Phosphorus	Filtration	Adsorption Precipitation	Plant uptake Microbial uptake
Pathogenic microorganisms	Filtration	Solar radiation Adsorption	Predation/biolytic processes Natural die-off
Heavy metals	Settling	Adsorption Precipitation	Biodegradation Phytodegradation Phytovolatilization Plant uptake
Emerging contaminants	Sedimentation Filtration	Adsorption Precipitation Solar radiation	Biodegradation Plant uptake

creation of the biofilm. Plants transfer oxygen through their roots to the deeper parts of the bed, which is then utilized by the various microorganisms for the biodegradation of the pollutants. CWs have a generally simple design and easy construction. CWs are an appropriate technology for decentralized wastewater treatment and water reclamation because they combine several environmental and economic benefits [74].

A typical CW facility includes a preliminary treatment (e.g., screening) followed by a primary sedimentation, which, however, can be avoided depending on the chosen design. The effluent criteria and the desire for reuse of the treated effluent determine the level of subsequent treatment. Usually, in the second and third stage, vertical and horizontal subsurface flow CWs are combined [32,74,89]. Besides secondary wastewater treatment, CWs application also includes tertiary treatment of municipal and industrial wastewater for water reuse [3,20,50] and urban runoff control and treatment for reuse [31,44,66].

Proper design and operation of a full-scale CW facility can produce an effluent of high quality that meets the standards for reuse [63]. The various pollutants present in wastewater such as organic matter, nitrogen, and phosphorus can be sufficiently removed by CWs through a variety of processes and mechanisms as shown in Table 31.4 [74,91,92], while elimination of pathogenic microorganisms also takes place [28,33,74,90]. The fate of emerging contaminants in CWs has also attracted increasing interest [22]. Currently, research focuses on the effectiveness of CWs as a remediation technology for various organic micropollutants like phenolic compounds [27,30,75], and pharmaceuticals, EDCs, and PCPs [6,8,26,45,97] with very promising results. Emerging pollutants are removed via a variety of complex biological, chemical, and physical processes and mechanisms, which could be optimized through appropriate design and operation.

The technology of CWs appears as an attractive alternative to conventional treatment technologies in terms of treatment efficiency and environmentally friendly character. Adoption of this technology increases rapidly around the world, especially in countries facing water scarcity problems. CWs have been successfully applied for water reclamation, ecosystem restoration, and recreational and environmental applications. They offer a decentralized solution not only to the problem of water/wastewater treatment but also to the option for safe water reuse.

31.4 Summary and Conclusions

Water reuse is a crucial issue around the world. Increasing realization of treated water as a valuable resource enhances the demand for water reuse, especially in the urban environment. Water reuse for nonpotable applications in urban areas significantly contributes to potable water supplies conservation. Today, water reuse applications take place mainly in large centralized treatment facilities. If water reclamation is the target, then advanced treatment technologies should be included in the treatment train. A variety of different technologies is now available to achieve an optimum effluent quality suitable for reuse applications. Microbial and chemical contamination and pollution can be effectively mitigated in reclaimed water to meet effluent regulated limits. However, when selecting the appropriate technology for water reuse, certain parameters should be taken into account including the operational scale, the desired end use of the treated water, the economic feasibility, the environmental impact, the social perspective, and local/regional customs and practices. Sustainability is a crucial factor in water reuse, meaning that water resources conservation dictates the use of technologies that are economically viable, technically and institutionally appropriate, socially acceptable, and can effectively protect the environment and the natural water resources.

Membrane technology today is one of the most attractive technologies available for potable and nonpotable water reuse applications, with many advantages like lower surface area demands, more or less process automation, effective pathogen removal, and good overall performance. However, membrane technology still faces limitations concerning higher investment and operational costs, high energy consumption, still unavoidable use of chemicals, complex and not quite easy to handle usage, limited useful lifetime of membranes, operational problems due to fouling, and the need for specialized personnel.

Furthermore, current practice and experience implies that for more effective and safe water reuse, various technologies should be combined like membrane filtration, activated carbon, ozonation, etc., which respectively enhances the above-mentioned limitations especially in terms of costing and reliable long-term operation. Ecological engineering developments like CW technology provide a sustainable solution as an economical, effective, environmentally friendly, socio-culturally acceptable, and simple to build and operate alternative. However, higher area demands are the main limitation for natural systems. Thus, it is clear that each water reuse application should be assessed separately and all related individual characteristics and parameters should be evaluated before the selection of the appropriate reuse technology.

References

1. Andreozzi, R., Caprio, V., Insola, A., and Marotta, R. 1999. Advanced oxidation processes (AOP) for water purification and recovery. *Catalysis Today*, 53, 51–59.

2. Asano, T., Burton, F.L., Leverenz, H.L., Tsuchihashi, R., and Tchobanoglous, G. 2007. *Water Reuse: Issues, Technologies, and Applications*. Metcalf & Eddy Inc., AECOM, McGraw-Hill, New York.

3. Ayaz, S.C. 2008. Post-treatment and reuse of tertiary treated wastewater by constructed wetlands. *Desalination*, 226(1–3), 249–255.

4. Bansode, R.R., Losso, J.N., Marshall, W.E., Rao, R.M., and Portier, R.J. 2004. Pecan shell-based granular activated carbon for treatment of chemical oxygen demand (COD) in municipal wastewater. *Bioresource Technology*, 94, 129–135.

5. Ben Aim, R.M. and Semmens, M.J. 2002. Membrane bioreactors for wastewater treatment and reuse: A success story. *Water Science and Technology*, 47(1), 1–5.

6. Berglund, B., Khan, G.A., Weisner, S.E.B., Ehde, P.M., Fick, J., and Lindgren, P.E. 2014. Efficient removal of antibiotics in surface-flow constructed wetlands, with no observed impact on antibiotic resistance genes. *Science of the Total Environment*, 476–477, 29–37.

7. Bernardes, A.M. and Rodrigues, M.A.S. 2014. Electrodialysis in water treatment. In: A.M. Bernardes, M.A.S. Rodrigues, and J.Z. Ferreira (eds.), *Electrodialysis and Water Reuse—Novel Approaches. Topics in Mining, Metallurgy and Materials Engineering*. Springer-Verlag, Berlin, Heidelberg.

8. Chapman, H. 2003. Removal of endocrine disruptors by tertiary treatments and constructed wetlands in subtropical Australia. *Water Science and Technology*, 47(9), 151–156.

9. Cicek, N. 2003. A review of membrane bioreactors and their potential application in the treatment of agricultural wastewater. *Canadian Biosystems Engineering*, 45, 6.37–6.49.

10. Comninellis, C., Kapalka, A., Malato, S., Parsons, S.A., Poulios, I., and Mantzavinos, D. 2008. Advanced oxidation processes for water treatment: Advances and trends for R&D. *Journal of Chemical Technology and Biotechnology*, 83, 769–776.

11. Cote, P., Masini, M., and Mourato, D. 2004. Comparison of membrane options for water reuse and reclamation. *Desalination*, 167, 1–11.

12. Crittenden, J.C., Sanongraj, S., Bulloch, J.L., Hand, D.W., Rogers, T.N., Speth, T.F., and Ulmer, M. 1999. Correlation of aqueous-phase adsorption isotherms. *Environmental Science and Technology*, 33, 2926–2933.

13. Dąbrowski, A., Podkościelny, P., Hubicki, Z., and Barczak, N. 2005. Adsorption of phenolic compounds by activated carbon—A critical review. *Chemosphere*, 58, 1049–1070.

14. Daigger, G.T. and Boltz, J.P. 2011. Trickling filter and trickling filter—suspended growth process design and operation: A state-of-the-art review. *Water Environment Research*, 83(5), 388–404.

15. EC. 2000. Establishing a framework for community action in the field of water policy. Directive 2000/60/EC of the European Parliament and of the Council, Official Journal of the European Communities, L 327/1, 23 October.

16. EPA. 2008. BAT guidance note for the pharmaceutical & other speciality organic chemicals. Environmental Protection Agency, Wexford, Ireland.

17. Esplugas, S., Giménez, J., Contreras, S., Pascual, E., and Rodriguez, M. 2002. Comparison of different advanced oxidation processes for phenol degradation. *Water Research*, 36, 1034–1042.

18. Fane, A.G. and Fane, S.A. 2005. The role of membrane technology in sustainable decentralized wastewater systems. *Water Science and Technology*, 51(10), 317–325.

19. Faria, P.C.C., Orfão, J.J.M., and Pereira, M.F.R. 2005. Mineralisation of coloured aqueous solutions by ozonation in the presence of activated carbon. *Water Research*, 39, 1461–1470.

20. Fibbi, D., Doumett, S., Lepri, L., Checchini, L., Gonnelli, C., Coppini, E., and del Bubba, M. 2012. Distribution and mass balance of hexavalent and trivalent chromium in a subsurface, horizontal flow (SF-h) constructed wetland operating as post-treatment of textile wastewater for water reuse. *Journal of Hazardous Materials*, 199–200, 209–216.

21. Gai, X-J. and Kim, H-S. 2007. The role of powdered activated carbon in enhancing the performance of membrane systems for water treatment. *Desalination*, 225, 288–300.

22. Garcia-Rodríguez, A., Matamoros, V., Fontás, C., and Salvadó, V. 2014. The ability of biologically based wastewater treatment systems to remove emerging organic contaminants—A review. *Environmental Science and Pollution Research*, 21 (20), 11708–11728.

23. Geritty, D., Gamage, S., Holady, J.C., Mawhinney, D.B., Quiñones, O., Trenholm, R.A., and Snyder, S.A. 2011. Pilot-scale evaluation of ozone and biological activated carbon for trace organic contaminant mitigation and disinfection. *Water Research*, 45, 2155–2165.

24. Gogate, P.R. and Pandit, A.B. 2004. A review of imperative technologies for wastewater treatment I: Oxidation technologies at ambient conditions. *Advances in Environmental Research*, 8, 501–551.

25. Gogate, P.R. and Pandit, A.B. 2004. A review of imperative technologies for wastewater treatment II: Hybrid methods. *Advances in Environmental Research*, 8, 553–597.

26. Haarstad, K., Bavor, H.J., and Maehlum, T. 2014. Organic and metallic pollutants in water treatment and natural wetlands: A review. *Water Science and Technology*, 65(1), 76–99.

27. Haberl, R., Grego, S., Langergraber, G., Kadlec, R.H., Cicalini, A-R., Dias, S.M., Novais, J.M., Aubert, S., Gerth, A., Thomas, H., and Hebner, A. 2003. Constructed wetlands for the treatment of organic pollutants. *Journal of Soils and Sediments*, 3(2), 109–124.

28. Hench, K.R., Bissonnette, G.K., Sexstone, A.J., Coleman, J.G., Garbutt, K., and Skousen, J.G. 2003. Fate of physical, chemical, and microbial contaminants in domestic wastewater following treatment by small constructed wetlands. *Water Research*, 37(4), 921–927.

29. Huber, M.M., Canonica, S., Park, G-Y., and von Gunten, U. 2003. Oxidation of pharmaceuticals during ozonation and advanced oxidation processes. *Environmental Science and Technology*, 37, 1016–1024.

30. Imfeld, G., Braeckevelt, M., Kuschk, P., and Richnow, H.H. 2009. Monitoring and assessing processes of organic chemicals removal in constructed wetlands. *Chemosphere*, 74, 349–362.

31. Jenkins, G.A., Greenway, M., and Polson, C. 2012. The impact of water reuse on the hydrology and ecology of a constructed stormwater wetland and its catchment. *Ecological Engineering*, 47, 308–315.

32. Kadlec, R.H. and Wallace, S.D. 2009. *Treatment Wetlands*, 2nd Edition. CRC Press, Boca Raton, FL.

33. Keffala, C. and Ghrabi, A. 2005. Nitrogen and bacterial removal in constructed wetlands treating domestic wastewater. *Desalination*, 185, 383–389.

34. Kim, H-S., Katayama, H., Takizawa, S., and Ohgaki, S. 2005. Development of a microfilter separation system coupled with a high dose of powdered activated carbon for advanced water treatment. *Desalination*, 186, 215–226.

35. Kim, H-S., Takizawa, S., and Ohgaki, S. 2007. Application of microfiltration systems coupled with powdered activated carbon to river water treatment. *Desalination*, 202, 271–277.

36. Kim, S.D., Cho, J., Kim, I.S., Vanderfold, B.J., and Snyder, S.A. 2007. Occurrence and removal of pharmaceuticals and endocrine disruptors in South Korean surface, drinking, and waste waters. *Water Research*, 41, 1013–1021.

37. Klavarioti, M., Mantzavinos, D., and Kassinos, D. 2009. Removal of residual pharmaceuticals from aqueous systems by advanced oxidation processes. *Environment International*, 35, 402–417.

38. Kuo, J-F., Stahl, J.F., Chen, C-l., and Bohlier, P.V. 1998. Dual role of activated carbon process for water reuse. *Water Environment Research*, 70(2), 161–170.

39. Lazarova, V., Savoye, P., Janex, M.L., Blatchley, E.R., and Pommepuy, M. 1999. Advanced wastewater disinfection technologies: State of the art and perspectives. *Water Science and Technology*, 40(4–5), 203–213.

40. Li, X., Hai, F.I., and Nghiem, L.D. 2011. Simultaneous activated carbon adsorption within a membrane bioreactor for an enhanced micropollutant removal. *Bioresource Technology*, 102(9), 5319–5324.

41. Lipp, P., Gross, H-J., and Tiehm, A. 2012. Improved elimination of organic micropollutants by a process combination of membrane bioreactor (MBR) and powdered activated carbon (PAC). *Desalination and Water Treatment*, 42(1–3), 65–74.

42. Ma, C., Yu, S., Shi, W., Tian, W., Heijman, S.G.J., and Rietveld, L.C. 2012. High concentration powdered activated carbon-membrane bioreactor (PAC-MBR) for slightly polluted surface water treatment at low temperature. *Bioresource Technology*, 113, 136–142.

43. Machado, M.d.B. and Santiago, V.M.J. 2014. Electrodialysis treatment of refinery wastewater. In: A.M. Bernardes, M.A.S. Rodrigues, and J.Z. Ferreira (eds.), *Electrodialysis and Water Reuse—Novel Approaches*. Topics in Mining, Metallurgy and Materials Engineering. Springer-Verlag, Berlin, Heidelberg.

44. Malaviya, P. and Singh, A. 2012. Constructed wetlands for management of urban stormwater runoff. *Critical Reviews in Environmental Science and Technology*, 42(20), 2153–2214.

45. Matamoros, V., Garcia, J., and Bayona, J.M. 2008. Organic micropollutant removal in a full-scale surface flow constructed wetland fed with secondary effluent. *Water Research*, 42, 653–660.

46. McAdam, E.L. and Judd, S.J. 2006. A review of membrane bioreactor potential for nitrate removal from drinking water. *Desalination*, 196, 135–148.

47. Melin, T., Jefferson, B., Bixio, D., Thoeye, C., de Wilde, W., de Koning, J., van der Graaf, J., and Wintgens, T. 2006. Membrane bioreactor technology for wastewater treatment and reuse. *Desalination*, 187, 271–282.

48. Mitch, W.A. and Sedlak, D.L. 2002. Formation of *N*-Nitrosodimethylamine (NDMA) from dimethylamine during chlorination. *Environmental Science and Technology*, 36, 588–595.

49. Mujeriego, R. and Asano, T. 1996. The role of advanced treatment in wastewater reclamation and reuse. *Water Science and Technology*, 40(4–5), 1–9.

50. Murphy, C. and Cooper, D. 2011. The evolution of horizontal subsurface flow reed bed design for tertiary treatment of

sewage effluents in the UK. In: J. Vymazal (ed.), *Water and Nutrient Management in Natural and Constructed Wetlands.* Springer Science and Business Media, Dordrecht, 103–119.

51. Neyens, E. and Baeyens, J. 2003. A review of classic Fenton's peroxidation as an advanced oxidation technique. *Journal of Hazardous Materials,* B98, 33–50.

52. Nguyen, L.N., Hai, F.I., Kang, J., Price, W.E., and Nghiem, L.D. 2012. Removal of trace organic contaminants by a membrane bioreactor–granular activated carbon (MBR–GAC) system. *Bioresource Technology,* 113, 169–173.

53. Oller, I., Malato, S., and Sánchez-Pérez, J.A. 2011. Combination of advanced oxidation processes and biological treatment for wastewater decontamination—A review. *Science of the Total Environment,* 409, 4141–4149.

54. Pera-Titus, M., Garcia-Molina, V., Baños, M.A., Giménez, J., and Esplugas, S. 2004. Degradation of chlorophenols by means of advanced oxidation processes: A general review. *Applied Catalysis B: Environmental,* 47, 219–256.

55. Pilat, B. 2001. Practice of water desalination by electrodialysis. *Desalination,* 139, 385–392.

56. Qu, X., Zheng, J., and Zhang, Y. 2007. Catalytic ozonation of phenolic wastewater with activated carbon fiber in a fluid bed reactor. *Journal of Colloid and Interface Science,* 309, 429–434.

57. Radjenovic, J., Petrovic, M., and Barceló, D. 2007. Analysis of pharmaceuticals in wastewater and removal using a membrane bioreactor. *Analytical and Bioanalytical Chemistry,* 387, 1365–1377.

58. Remy, M., van der Marel, P., Zwijnenburg, A., Rulkens, W., and Temmink, H. 2009. Low dose powdered activated carbon addition at high sludge retention times to reduce fouling in membrane bioreactor. *Water Research,* 43, 345–350.

59. Remy, M., Potier, V., Temmink, H., and Rulkens, W. 2010. Why low powdered activated carbon addition reduces membrane fouling in MBRs. *Water Research,* 44, 861–867.

60. Reungoat, J., Escher, B.I., Macova, M., Argaud, F.X., Gernjak, W., and Keller, J. 2012. Ozonation and biological activated carbon filtration of wastewater treatment plant effluents. *Water Research,* 46(3), 863–872.

61. Reungoat, J., Macova, M., Escher, B.I., Carswell, S., Mueller, J.F., and Keller, J. 2010. Removal of micropollutants and reduction of biological activity in a full scale reclamation plant using ozonation and activated carbon filtration. *Water Research,* 44(2), 625–637.

62. Rodriguez-Mozaz, S., de Alda, K.J.L., and Barceló, D. 2004. Monitoring of estrogens, pesticides and bisphenol A in natural waters and drinking water treatment plants by solid-phase extraction–liquid chromatography–mass spectrometry. *Journal of Chromatography A,* 1045, 85–92.

63. Rousseau, D.P.L., Lesage, E., Story, A., Vanrolleghem, P.A., and de Pauw, N. 2008. Constructed wetlands for water reclamation. *Desalination,* 218, 181–189.

64. Sagbo, O., Sun, Y., Hao, A., and Gu, P. 2008. Effect of PAC addition on MBR process for drinking water treatment. *Separation and Purification Technology,* 58(3), 320–327.

65. Sahar, E., David, I., Gelman, Y., Chikurel, H., Aharoni, A., Messalem, R., and Brenner, A. 2011. The use of RO to remove emerging micropollutants following CAS/UF or MBR treatment of municipal wastewater. *Desalination,* 273, 142–147.

66. Scholz, M. 2006. *Wetland Systems to Control Urban Runoff.* Elsevier Publishing, the Netherlands.

67. Seo, G.T., Suzuki, Y., and Ohgaki, S. 1996. Biological powdered activated carbon (BPAC) microfiltration for wastewater reclamation and reuse. *Desalination,* 106, 39–45.

68. Seo, G.T., Takizawa, S., and Ohgaki, S. 2002. Ammonia oxidation at low temperature in high concentration powdered activated carbon membrane bioreactor. *Water Science and Technology Water Supply,* 2(2), 169–176.

69. Serrano, D., Suárez, S., Lema, J.M., and Omil, F. 2011. Removal of persistent pharmaceutical micropollutants from sewage by addition of PAC in a sequential membrane bioreactor. *Water Research,* 45, 5323–5333.

70. Snyder, S.A., Adham, S., Redding, A.M., Cannon, F.S., de Carolis, J., Oppenheimer, J., Wert, E.C., and Yoon, Y. 2007. Role of membranes and activated carbon in the removal of endocrine disruptors and pharmaceuticals. *Desalination,* 202, 156–181.

71. Sonune, A. and Ghate, R. 2004. Developments in wastewater treatment methods. *Desalination,* 167, 55–63.

72. Stefanakis, A.I., Akratos, C.A., Gikas, G.D., and Tsihrintzis, V.A. 2009. Effluent quality improvement of two pilot-scale, horizontal subsurface flow constructed wetlands using natural zeolite (clinoptilolite). *Microporous and Mesoporous Materials,* 124(1–3), 131–143.

73. Stefanakis, A.I., Akratos, C.A., and Tsihrintzis, V.A. 2011. Effect of wastewater step-feeding on removal efficiency of pilot-scale horizontal subsurface flow constructed wetlands. *Ecological Engineering,* 37(3), 431–443.

74. Stefanakis, A.I., Akratos, C.A., and Tsihrintzis, V.A. 2014. *Vertical Flow Constructed Wetlands: Eco-Engineering Systems for Wastewater and Sludge Treatment.* Elsevier Publishing, the Netherlands.

75. Stefanakis, A.I., Seeger, E., Hübschmann, T., Müller, S., Sinke, A., and Thullner, M. 2013. Investigation of phenol and m-cresol biodegradation in horizontal subsurface flow constructed wetlands. *Proceedings of the 5th International Symposium on Wetland Pollutant Dynamics and Control,* October 13–17, Nantes, France.

76. Stefanakis, A.I. and Tsihrintzis, V.A. 2012. Effects of loading, resting period, temperature, porous media, vegetation and aeration on performance of pilot-scale vertical flow constructed wetlands. *Chemical Engineering Journal,* 181–182, 416–430.

77. Stephenson, T., Judd, S., Jefferson, B., and Brindle, K. 2000. *Membrane Bioreactors for Wastewater Treatment.* IWA Publishing, London, UK.

78. Streit, K.F., Rodrigues, M.A.S., and Ferreira, J.Z. 2014. Electrodialysis treatment of tannery wastewater. In: A.M. Bernardes, M.A.S. Rodrigues, and J.Z. Ferreira (eds.),

Electrodialysis and Water Reuse—Novel Approaches. Topics in Mining, Metallurgy and Materials Engineering. Springer-Verlag, Berlin, Heidelberg.

79. Sun, Y.-X., Wu, Q.-Y., Hu, H.-Y., and Tian, J. 2009. Effects of operating conditions on THMs and HAAs formation during wastewater chlorination. *Journal of Hazardous Materials,* 168(2–3), 1290–1295.

80. Suty, H., de Traversay, C., and Cost, M. 2004. Applications of advanced oxidation processes: Present and future. *Water Science and Technology,* 49(4), 227–233.

81. Suzuki, T., Watanabe, Y., Ozawa, G., and Ikeda, S. 1998. Removal of soluble organics and manganese by a hybrid MF hollow fiber membrane system. *Desalination,* 117, 119–130.

82. Tchobanoglous, G., Burton, F.L., and Stensel, H.D. 2002. *Wastewater Engineering: Treatment and Reuse,* 4th Edition. Metcalf & Eddy, McGraw-Hill, New York.

83. Tekerlekopoulou, A.G., Pavlou, S., and Vayenas, D.V. 2013. Removal of ammonium, iron and manganese from potable water in biofiltration units: A review. *Journal of Chemical Technology and Biotechnology,* 88(5), 751–773.

84. Ternes, T.A., Meisenheimer, M., McDowell, D., Sacher, F., Brauch, H-J., Haist-Gulde, B., Preuss, G., Wilme, U., and Zulei-Seibert, N. 2002. Removal of pharmaceuticals during drinking water treatment. *Environmental Science and Technology,* 36, 3855–3863.

85. USEPA. 1999. *Wastewater Technology Fact Sheet: Chlorine Disinfection.* EPA 832-F-99-062, United States Environmental Protection Agency, Office of Water, Washington, DC.

86. USEPA. 2001. *Low-Pressure Membrane Filtration for Pathogen Removal: Application, Implementation, and Regulatory Issues.* EPA 815-C-01-001, United States Environmental Protection Agency, Office of Water (4601), Washington, DC.

87. Van Nieuwenhuijzen, A.F., Evenblij, H., Uijterlinde, C.A., and Schulting, F.L. 2008. Review on the state of science on membrane bioreactors for municipal wastewater treatment. *Water Science and Technology,* 57(7), 979–986.

88. Visvanathan, C., Aim, R.B., and Parameshwaran, K. 2000. Membrane separation bioreactors for wastewater treatment. *Critical Reviews in Environmental Science and Technology,* 30(1), 1–48.

89. Vymazal, J. 2005. Horizontal sub-surface flow and hybrid constructed wetlands systems for wastewater treatment. *Ecological Engineering,* 25(5), 478–490.

90. Vymazal, J. 2005. Removal of enteric bacteria in Constructed Treatment Wetlands with emergent macrophytes: A review. *Journal of Environmental Science and Health, Part A: Toxic/Hazardous Substances and Environmental Engineering,* 40(6–7), 1355–1367.

91. Vymazal, J. 2007. Removal of nutrients in various types of constructed wetlands. *Science of the Total Environment,* 380(1–3), 48–65.

92. Vymazal, J. and Kröpfelová, L. 2009. Removal of organics in constructed wetlands with horizontal sub-surface flow: A review of the field experience. *Science of the Total Environment,* 407(13), 3911–3922.

93. Westerhoff, P., Yoon, Y., Snyder, S., and Wert, E. 2005. Fate of endocrine-disruptor, pharmaceutical and personal care product chemicals during simulated drinking water treatment processes. *Environmental Science and Technology,* 39, 6649–6663.

94. Wintgens, T., Gallenkemper, M., and Melin, T. 2002. Endocrine disrupter removal from wastewater using membrane bioreactor and nanofiltration technology. *Desalination,* 146, 387–391.

95. Xu, P., Janex, M-L., Savoye, P., Cockx, A., and Lazarova, V. 2002. Wastewater disinfection by ozone: Main parameters for process design. *Water Research,* 36, 1043–1055.

96. Yang, X., Shang, C., and Huang, J-C. 2005. DBP formation in breakpoint chlorination of wastewater. *Water Research,* 39(19), 4755–4767.

97. Zhang, D., Gersberg, R.M., Ng, W.J., and Tan, S.K. 2014. Removal of pharmaceuticals and personal care products in aquatic plant-based systems: A review. *Environmental Pollution,* 184, 620–639.

98. Zhou, H. and Smith, D.W. 2002. Advanced technologies in water and wastewater treatment. *Journal of Environmental Engineering and Science,* 1, 247–264.

32

Modern Water Reuse Technologies: Membrane Bioreactors

Zarook Shareefdeen
American University of Sharjah

Ali Elkamel
University of Waterloo

Shams Kandhro
University of Waterloo

PREFACE

Wastewater reclamation and reuse is of growing interest to the world due to increasing demand of water resources and environmental pollution. Since the last two decades, scarcity and deteriorated quality of water put pressure on many regions around the globe to search for an efficient use of water resources. In some parts of the world, water reuse is indispensable. The greater need of wastewater reclamation and reuse is not only to fulfill the increasing demand of water, but also it is important in terms of environmental protection and public health. Varieties of technologies currently available in the market are efficient to recapture the valuable resources and these technologies are environmentally friendly. Whereas use of reclaimed water sometimes can be risky due to the presence of microorganisms, it is important the quality of reclaimed water must match the current guidelines for reuse application. The fact is, "new water cannot be created easily" [1]; however, new approaches can guide us to a sustainable solution and make water and resources recovery possible. The main objective of this study is to understand new water reuse technologies and their benefits in terms of water reclamation and reuse application.

32.1 Introduction

Historically, water has been considered an abundant public commodity; whereas since the last decade, this attitude has changed due to the reduction of freshwater sources and growing water demand by an ever-increasing population. Big cities around the world are already densely populated and population will more than double

in coming decades. The greatest challenges for municipal authorities are to satisfy water and sanitation needs for their citizens and to minimize negative impacts on the environment. The United Nations Environmental Program (UNEP) [19] reported that about one-third of the world's population currently lives in the countries suffering high water stress, where reclaimed water consumption is more than 10% [19]. Lack of access to a safe water supply affects the health of millions of the people in African and Asian countries as reported by World Health Organization (WHO). Water quality has been deteriorated by industrial and municipal pollutants and overpumping of the ground water also has degraded the quality of water. Wastewater treatment, commonly known as water reclamation, has been the field of interest for some time. In order to improve the quality of the reuse—water as a sustainable alternative—several approaches have been used. However; there is a greater need to improve in the areas such as water consumption, water management, and most importantly, wastewater reuses [18]. Traditionally reclaimed wastewater has been used for agriculture irrigation, but with recent advancement in water treatment technologies, water quality has been improved and recycled water is used for many applications such as for cooling purposes in industries and toilet flushing. The health hazards have always been the greater concern with wastewater reuse and the risks mainly involved are classified as chemical and biological hazards. The local governments and international authorities for wastewater reuse propose several standards. Moreover, the greater need for wastewater reuse has forced the industry to put more efforts to innovate sophisticated tools, technologies, and methods to control the health hazards. As a result, the water treatment industry has progressed in numerous areas such as membrane separation technologies, membrane bioreactors, and disinfection processes [5]. Furthermore, efforts have been made to improve the efficiency of existing wastewater treatment plants to reduce the environmental impacts and provide sustainable solutions for water or wastewater treatment. Thus, the combination of process and modern tools and technologies has established new standards where wastewater is considered a valuable resource rather than a waste.

32.2 Innovations in Water Reuse and Resources Recovery

The vital importance of using modern technologies is to identify the maximum water sources to meet increased demand and to achieve the maximum benefits from a variety of wastewater resources. Conventional treatment facilities remove toxic wastes from wastewater before releasing the effluents into the surface water streams. The discarded effluent contains large amounts of nutrients such as nitrogen, phosphorus, and potassium. Before, very little thought was given to the valuable waste substances, which ended up in the surface water bodies. Industry slowly began to develop efficient technology to capture these nutrients from wastewater and, as a result, today millions of dollars of resources are converted into clean energy resources. Nutrient-rich fertilizer is one of the great sources that can potentially be used for agricultural irrigation. By utilizing modern technologies, most of the reclaimed water is used for a wide range of applications such as recreational, industrial, toilet flushing, and cooling water in industries [19]. Consequently, the freshwater sources are distributed only to areas where higher quality water is needed such as for drinking water. In fact, for some applications, treated wastewater can be economically viable and beneficial as compared to the freshwater resources. Thus, the

advancement in both technology and process offers an innovative approach to overall water treatment and its reuse applications. Another great advantage of emerging technology is that almost all the contaminants are removed from wastewater that would otherwise end up in surface water streams and affect the quality of surface water, which may increase the treatment cost for potable water filtration. In general, technologies are able to protect the environment, achieve better quality of recycle products, and use resources in a sustainable manner. Following are three main resources associated with wastewater.

32.2.1 Water Resources

Many parts of the world are facing serious challenges from an increased water demand. The ultimate purpose of the technology is to reclaim the reusable water from wastewater resources. Innovation in water treatment technology and advancement in existing water treatment plants are making a great deal of progress to meet the growing demand of water by different industries and households through reclaimed wastewater. Many African and Asian countries have extensively low water availability because of droughts and population increases. In many places, use of unsafe water is the main cause of death in humans and animals; more than 4% of diseases accounted for worldwide are waterborne diseases as reported by WHO [15]. This burden can be prevented by using innovative water treatment technologies to improve water quality and recover valuable substances from wastewater. With reclaimed water, some of the water demand can be fulfilled and consequently the lives of people can be improved in those regions as well as the environmental impact can be reduced.

32.2.2 Nutrient Resources

The agriculture sector is the largest freshwater user around the globe. Conventional wastewater treatment systems dispose tons of fertilizer into the environment, which leads to degradation of the soil and waste of valuable nutrients. On the other hand, processed fertilizer resources are shrinking; soon there will not be enough phosphorus supplies from mining resources to meet the agricultural demand. Food demand is rapidly growing especially in African and Asian countries where the population mostly depends on the agriculture industry and agriculture is dependent on the freshwater sources. For example, in the province Sindh (Pakistan), people have been greatly affected by drought, which causes a severe shortage of food, diseases, and death of thousands of inhabitants. Large bodies of freshwater resources are disappearing due to low rainfall and mismanagement of the water resources. Surface water has been polluted and this causes soil erosion to a historically agriculture land. Situations like this can easily be managed by the use of available water treatment technology in the market. Nowadays, modern technologies offer multiple opportunities including recovery of nutrients resources such as nitrogen, potassium, and phosphorous. Most of the organic matter plays an important role in the improvement of the soil fertility and even use of the direct nutrient-rich water may decrease or eliminate artificial fertilizer requirement. Advanced technologies provide innovative solutions to all sectors to recover productive resources and safe use of them as well. Above all, the nutrients recovered from wastewater have many benefits; a few of them are listed here [6].

1. By using modern technologies, nutrient-rich wastewater is used for irrigation and a number of other applications; thus, the freshwater sources can be allocated for drinking water.
2. Contribute to better crop production and agroforestry in order to meet the food demand particularly in areas under water stress.
3. Improve soil fertility by building up humus and prevention of land erosion.
4. Reduce or eliminate the requirements for processed fertilizers.
5. Reduce the surface water pollution.

Thus, recovering these resources and efficiently using them for agriculture is essential for sustainable environment and wastewater management.

32.3 Energy Resources

Energy consumption is constantly increasing all over the world. According to International Energy Agency (IEA), from 1990 to 2008 the average energy use per person increased 10% while world population increased 27%. It is expected that by the year 2035, energy consumption will be doubled [18]. On the other hand, potable water and wastewater treatment plants are also the largest energy consumers. According to the US Environmental Protection Agency (EPA), wastewater and drinking water treatment account for 4% of the energy use and emit over 45 million tons of greenhouse gases annually in the United States alone. Aerobic wastewater treatment is an energy-intensive process and emits nitrogen into the atmosphere. Nitrogen is generally used in the production of fertilizer and it takes enormous amounts of energy to recapture the nitrogen from the atmosphere [18]. Moreover, considering the increased fuel cost, global warming issues, and environmental pollutions, biosolids and biogas are great energy sources that can be derived from wastewater resources. Innovative technology is progressing in the direction of energy recovery, bio-solid management, odor reduction, and pathogen reduction in bio-solids. Anaerobic digestion for biogas generation is already used for industrial or domestic purposes such as producing fuel and waste management. Biogas produced from animal and human waste is a reliable source of energy and its use is growing in China, India, Nepal, and some western countries [2,19]. Similarly, bio-solids are rich in organic matter and nutrients, which are predominantly used as fertilizer. Dried sludge also can be used as a direct source of fuel in the industries. Furthermore, one of the important properties of the water is that it retains the temperature; water can transfer hot or cold temperature efficiently when it is in contact with other objects. By taking advantage of this characteristic of water, Japan

is saving 20%–30% of energy in the municipal government of Osaka. The temperature of the sewage treatment facility is relatively warmer in the winter and colder in the summer than the ambient temperature. Japan has developed a heating and cooling system with wastewater as an alternative source of energy [19].

Japan is also utilizing thermal energy from wastewater for snow melting. In the city of Sapporo, the snowfall average is 4.7 m/yr and the temperature of the effluent from sewage treatment facility reaches 13°C. With this unique approach, they are able to melt the snow between 600,000 m³ and 700,000 m³ per year [19]. While wastewater reuse has many advantages, it is essential to use appropriate technologies and processes in order to reduce the negative effects on human health and the environment. Water treatment processes largely depend on the level of protection; for pure quality products, an efficient treatment process is needed. In addition, modern technologies have the potential to protect the environment, use resources in a sustainable manner, and get maximum useable products from the wastes.

32.4 Modern Technologies in Water Reuse

Advancement in technologies has made wastewater reuse more feasible in terms of wastewater management, treatment, and prevention of the contaminants. Modern technology uses multiple barrier concepts in drinking water and wastewater treatment. Historically, conventional wastewater treatment methods consist of physical, biological, and chemical processes. In addition to the conventional treatment methods, modern technologies can be used as add-on technology or it can replace the traditional treatment system altogether. Generally, conventional treatment plants use three stages of treatment methods. The primary treatment method involves physically removing portions of suspended solids and organic matter. In the secondary treatment stage, biodegradable organic matter and suspended solids are removed using microorganisms. However, pathogens including bacteria, virus, and other contaminants present in the effluent can only be removed by tertiary treatment methods. In tertiary treatment methods, advanced techniques are used to remove the residual suspended solids and pathogens. Historically, after secondary treatment, effluents were used for irrigation purposes or released into the surface water bodies.

Due to water scarcity in arid climates and environmental issues, the reuse of treated wastewater has been getting more attention. As the numbers of reuse applications are increasing, so are the regulations becoming stricter. Research has shown that some virus and bacteria such as protozoa and *Giardia lamblia* are resistant to chlorine-based treatments offered by traditional conventional methods [16]; as a result, various schemes of tertiary and advanced methods have been proposed. Although there is not much change in primary and secondary methods, tertiary and advance treatment methods have changed significantly, which lowered the cost, health risk, energy use, and large infrastructure. The key benefits of modern water reuse technology are the increased level of environmental protection, improvement in water quality, and restoration of surface water sources, which revive the aquatic life. Mainly three water reclamation technologies are very popular because of their efficiency and increased reliability.

32.5 Membrane Filtration Technologies

Membrane technologies offer a great number of advantages over conventional water and wastewater treatment methods such as higher quality of water products, low environmental impact, and less infrastructure. The potential application involves removal of suspended solids and pathogens, wastewater disinfection, and desalination [5]. Membrane separation works as an add-on technology with conventional water treatment processes or it can replace the conventional treatment system. Membrane technology uses a selectivity permeable barrier mechanism that allows the water molecules to pass through a permeable barrier and retains a wide range of suspended solids, virus, and bacteria. Both system design and membrane technology offer a wide range of water recycle methods in drinking water filtration, wastewater treatment, municipal sewage treatment, agriculture waste treatment, and product recovery from effluents. Membrane filtration is classified by its pore size, which determines the size of particle that can be removed as shown in Table 32.1 [5].

It is also important to know that membranes are formed from different materials such as polymer membranes, metallic membranes, and ceramic membranes. Performance of the membrane is determined by its separation mechanism. This is due to the variability of membrane structure and the properties such as hollow fibers or flat sheets, which are mounted on different design housing modules. Following are the typical membrane configurations [20]:

- Hollow fiber ultrafiltration (UF) membrane is used for a wide variety of applications including municipal water and wastewater treatment, industrial biotechnology, and in the food, beverage, dairy, and wine industries.

TABLE 32.1 Membrane Filtration Matching Application

Filtration Type	Particle Size Removed	Contaminants Removed	Pressure Ranges
Microfiltration (MF)	0.1–10 μm	Suspended solids, bacteria, protozoa	1–50 psi
Ultrafiltration (UF)	0.01–0.1 μm	Colloids, proteins, polysaccharides, most bacteria, virus (partially)	30–150 psi
Nanofiltration (NF)	0.001–0.01 μm	Viruses, organic matter, multivalent ions, (hardness in water)	90–150 psi
Reverse osmosis (RO)	0.0001–0.001 μm	Almost all impurities, including monovalent ions	140–1000 psi

Source: Adapted from European Union (EU). 2010. Membrane technologies for water applications, highlights from a selection of European research projects, EUR 24552. http://ec.europa.eu/research/environment/pdf/membrane-technologies.pdf. Accessed April 10, 2014; Koch Membrane Systems. www.kochmembrane.com.

- A spiral-wound membrane is commonly used in reverse osmosis (RO) or nanofiltration (NF) technology and it effectively works at seawater desalination, brackish water treatment, water softening, dairy, food, and pharmaceutical applications, and organics removal.
- Tubular membrane works very well with UF and microfiltration (MF) technologies and easily removes suspended solids, colloidal haze particles, microorganisms, and undesirable proteins.

32.5.1 Microfiltration

MF use pore size from 0.01 to 10 μm for separation process, its molecular weight cut off (MWCO) is more than 100,000 Daltons and operating pressure ranges from 1 to 50 psi. MF is mostly used for water filtration, and the contaminants removed include suspended solids, clay, algae, *Giardia lamblia*, and some bacteria. Primarily MF is used for clarifying drinking water; it removes microorganisms and contaminants from the feed stream by retaining the contaminant particles on the membrane surface. Because of its larger pore size, it is called the "loosest" of all membrane processes and it can be operated under low pressure [17,20]. Conventional water treatment processes use chemicals and coagulants to remove the impurities from the wastewater, whereas MF physically removes the pathogens and significantly reduces the use of chemicals such as chlorine.

32.5.2 Ultrafiltration (UF)

UF pore size is 0.01 to 0.1 μm, its MWCO ranges from 10,000 to 100,000 Daltons, and typical driving force is from 30 to 150 psi. UF membranes are pressure-driven systems that typically remove colloids, emulsified oil, and suspended solids, and partially remove bacteria, humus materials, and some viruses. However, disinfection is recommended as a second barrier for complete removal of contaminants. UF membrane is generally made from polyethersulfone (PES), which is a hydrophilic material. By making the membrane surface hydrophilic and inside-out flow, the membrane performance is improved and flux is stabilized. UF membranes with small pore size are capable of removing bacteria, protozoa, and most viruses from the water; thus, UF membrane offers purer product with little use of a disinfection process such as ozone treatment or UV radiation. Some of the advantages of UF membrane processes compared with conventional treatment methods are listed here [17].

- UF membrane effectively removes microbial, turbidity, and silt
- Eliminates the use of chemicals (coagulants, flocculants)
- Offers pure quality of treated water
- Simple automation and periodic cleaning procedure to remove fouling layer
- Plant compactness and easy to operate

Some other uses of UF include fruit juice clarification, milk and whey processing, purification of pharmaceuticals, alcohol recovery, and oil and water separation [20].

32.5.3 Nanofiltration (NF)

Nanofiltration membrane pore size is from 0.001 to 0.01 μm, its MWCO ranges from 200 to 15000 Daltons. NF membrane usually requires higher operating pressure than MF or UF, and the pressure ranges from 90 to 150 psi. NF membrane is capable of removing all bacteria, viruses, cyst, humic acid materials, and divalent ions [20]. NF is used in between UF and RO membranes; it has various advantages such as increased flux rate, higher retention of organic matter including salts, and low operating and maintenance cost. Because of these advantages, NF can be an excellent choice for drinking water treatment or wastewater treatment and it can be used as a pretreatment cycle for a desalination process [5]. The primary advantages of NF filtrations for common applications are described next.

Surface water treatment: The main source of drinking water is surface water, whereas surface water contains numerous contaminants. Combination of both UF and NF can be a reliable method for surface water treatment. Most of the natural organic matter (NOM), turbidity, iron, manganese, and bacteria can be removed by UF, whereas NF removes viruses, bacteria, arsenic, and up to 96% of organic matter [7,17].

Groundwater treatment: NF membrane is used for softening the groundwater. It can remove up to 90% of multivalent ions, and 60%–70% of monovalent ions. There are several health benefits of using NF for drinking water treatment such as achieving pure quality of drinking water by removing disinfection by-products (DBP) precursors and micro-pollutants [7,17].

Wastewater treatment: A recent study demonstrates that NF membrane is a promising technology to treat wastewater generated from all kind of sources including industries. By integrating NF with RO membrane systems, a high rate of recovery and high quality of product can be achieved. In addition to removing most of the impurities from wastewater, more than 90% lead can be removed. NF offers almost 98% retention of dye, partially removes salts, and reduces chemical oxidation demand (COD) [2,7]. Thus, NF makes water quality reusable for many applications.

NF for desalination: Another great application of NF membrane is that it can be used as a pretreatment for the desalination process. In the integrated process, NF lowers the hardness by partially removing the salt contents and prevents scaling. It consequently minimizes the operating pressure for seawater reverse osmosis (SWRO) processes [7,17].

32.5.4 Reverse Osmosis (RO) Membrane Separation Technology

RO membrane separation technology is a fast-growing process; it has been used to treat different water sources including wastewater reclamation. The membrane used for RO contains the smallest pore size of all membrane types and ranges from 0.0001 to 0.001 μm. An RO system is extensively used to convert seawater or brackish water into drinking water and it is also used to recover dissolved salts in industrial processes. Currently, it has been used in homes to remove contaminants and other chemicals to improve drinking water quality and taste. Although the RO method can remove most of the contaminants, it is more effective for total dissolved solid removal compounds as shown in Table 32.2 [17,20].

RO, as it is reflected from its name, is involved with reversal of osmotic pressure, where water permeates through semipermeable membrane from the solution. In order to understand the RO mechanism, one needs to know the definitions of osmosis, osmotic pressure, and RO.

Osmosis: Osmosis is the diffusion of the fluid through a semipermeable membrane from a low solute concentration to higher concentration solution. Meanwhile, the semipermeable membrane prevents the dissolved matter from being transferred to low solute concentration as illustrated in Figure 32.1 [8].

Osmotic pressure: Similarly, the osmotic flow continues and low concentrated fluid defuses into the higher concentration until there is equal concentration of fluid on both sides of the membrane. Once the equilibrium state is achieved, the pressure difference in between fluid level on both sides is known as the osmotic pressure as shown in Figure 32.2 [8].

RO: When a pressure higher than osmotic pressure is applied to the concentrated solution, pure water flows through a semipermeable membrane from a concentrated solution to the pure water side. The semipermeable membrane retains dissolved

TABLE 32.2 Compounds Removed by RO Processes

Metals and Ions	Organic Chemicals	Particles
Fluoride (F), iron (Fe)	Benzene	Protozoan
Potassium (K), radium (Ra)	Trichloroethylene	Asbestos
Arsenic (As), aluminum (Al)	Trihalomethanes	
Calcium (Ca), carbonate (CO_3)	Toluene	
Barium (Ba), bicarbonate (HCO_3)	MtBE	Pesticides
Chloride (Cl), cadmium (Cd)	Carbon tetrachloride	Atrazine
Chromium (Cr), sodium (Na)	Dichlorobenzene	Pentachlorophenol
Sulfate (SO4), zinc (Zn)		1,2,4-Trichlorobenzene
Selenium (Se), silver, lead (Pb)		Endrin
Mercury (Hg), nitrate/nitrite		2,4-D
Magnesium (Mg), manganese (Mn)		Heptachlor
Copper (Cu)		Lindane

Source: Adapted from European Union (EU). 2010. Membrane technologies for water applications, highlights from a selection of European research projects, EUR 24552. http://ec.europa.eu/research/environment/pdf/membrane-technologies.pdf. Accessed April 10, 2014; Israel Desalination Society (IDS). 2009. Osmotic processes: Past, present and future (in memory of Prof. Sidney Loeb), http://w3.bgu.ac.il/ziwr/documents/11thIDSConference-Program.pdf. Accessed March 9, 2014.

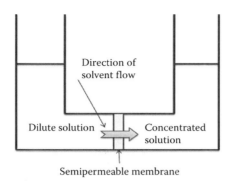

FIGURE 32.1 Osmosis through semipermeable membrane.

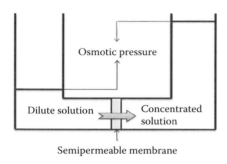

FIGURE 32.2 Osmotic pressure.

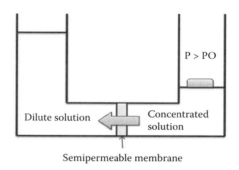

FIGURE 32.3 Reverse osmosis.

matter and salt contents, and produces a high quality of water as shown in Figure 32.3 [8].

32.5.4.1 Performance of RO Membrane Separation

Due to the advancement in technologies and the great contribution of Loeb and Sourirajan [8] in the development of the membrane separation technologies, RO is one of the well-established desalination processes. Innovation in membrane science and materials has improved salt rejection and pure water production. Recently, low operating pressures made RO less energy intensive and affordable, and a sound method for seawater desalination. Although RO is now a very popular technology, there are still some challenging areas such as cost, energy, equipment, membrane maintenance, and managing brine solution. However, research and development efforts are being made to improve the

efficiency of the membrane and associated hardware to reduce the feed pressure, increase flux rate, and make it a minimum energy-consuming operation [8,12].

32.5.4.2 Future Development of Osmotic Power

Growing demand for energy has created the need for renewable energy. During the earlier days of development of RO membrane separation technology, Sidney Loeb [8] pointed out the concept of osmotic power energy, which is generated by mixing freshwater and salt or saline water. Osmotic power is generated by taking advantage of pressure differences created by establishment of equilibrium between different concentrations of freshwater and seawater. That concept lead a power-generating company—Statkraft from Norway—to put its efforts in the new research area of osmotic power and convert it into electricity. As a result, today osmotic power is an important source of renewable energy and this source of energy is contributing more than 1600 TWh/yr worldwide. Obviously, the great benefit of osmotic power is to utilize renewable energy to convert sea or brine water into drinking water and to enable other emerging membrane separation technologies such as forward osmosis (FO) and membrane distillation (MD) to enhance the recovery of good quality drinking water and lower the cost [8]. Osmotic driven membrane technology has great potential in achieving a greater quality of drinking water, and enhances water recovery with low operating pressure. Osmotic power can become a great renewable energy source to meet the energy demand for the future RO separation processes.

32.6 Treatment Processes Selection

Different types of membranes are used for a wide variety of applications such as water, wastewater filtration, organic matter removal, disinfection, and desalination. Generally, one membrane filtration process is followed by another in order to increase the quality of the effluent water. For example, one type of membrane removes the contaminants that may otherwise affect the other equipment at a downstream level. In this multilayer mechanism, membrane technology offers complete recovery of resources [5]. General schematic process diagrams for water treatment are shown in Figure 32.4. As shown in Figure 32.4a, MF or UF membrane is used to remove microbial and small particles. Although treated water by this scheme can be used for many applications, this system is not capable of removing total NOM. In order to get pure quality drinking water, the system can work much better by adding absorbent or coagulant as illustrated in Figure 32.4b [9].

Figure 32.4c shows that NF/RO membrane is used with pretreatment, which works better for groundwater filtration where natural organic matter or salt needs to be removed. However, in the case of wastewater and surface water treatment, dual or multibarrier membrane mechanisms are required. Due to high pressure involved in downstream processes, a combination of the system and technology can be more effective as illustrated in Figure 32.4d [9].

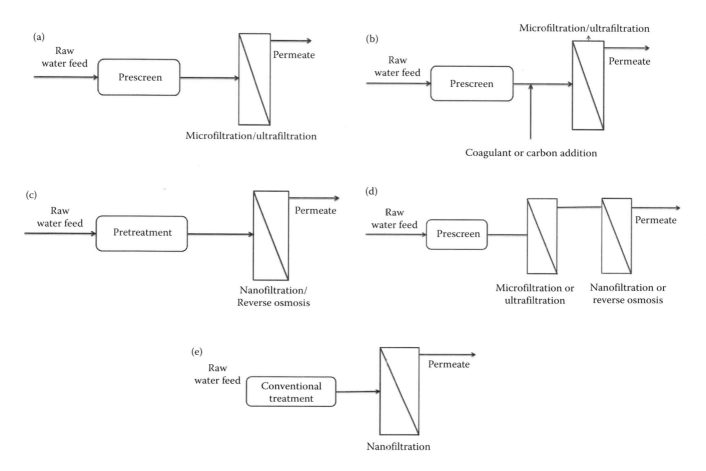

FIGURE 32.4 (a) MF/UF for removing microbial and small particles. (b) Addition of absorbent or coagulant. (c) NF/RO filtration for ground-water filtration. (d) Dual or multibarrier membrane for wastewater and surface water treatment. (e) Nanofiltration membrane used as add-on technology with conventional treatment plants.

Furthermore, the NF membrane as shown in Figure 32.4e can be used as a final treatment cycle with a relatively well-operating "conventional" treatment plant to remove residual NOM and micro-pollutants [9].

32.7 Membrane Bioreactors (MBR)

The membrane bioreactor (MBR) process is an advanced technology and plays a vital role in municipal wastewater and industrial wastewater treatment. MBR combines biological activated sludge processes with membrane separation technology. The membrane mechanism provides a physical barrier to suspended solids and other organic or inorganic matter, whereas in the biological treatment stage, the nutrients are removed by microbes. Compared to conventional wastewater treatment methods, an MBR system provides a simple, compact, and cost-effective alternative. Because of the innovation in membrane separation technologies where less chemical cleaning and low energy are required, the MBR is becoming an economically viable and environmentally sustainable treatment method [5,13].

32.7.1 MBR System Configuration

MBR process design is classified in two categories. The first category is referred to as pressure-driven membrane, where the membrane is externally attached to bioreactors as shown in Figure 32.5a.

In this method, wastewater is pumped through an external membrane module and the concentrated residual is returned back into a bioreactor compartment. Although this configuration is not used commonly, some of its advantages such as its higher flux rate and easy off-line cleaning or maintenance are very useful. Similarly, in the second configuration, membrane is immersed into the bioreactor compartment where MBR process is vacuum driven. In this method, the driving force across the membrane is achieved by creating a vacuum on the permeate side of the membrane as illustrated in Figure 32.5b.

The benefits of this kind of MBR configuration are that its energy consumption is significantly low, membrane packing density is higher, and less capital cost makes the design more attractive for common use [13]. Moreover, the design of the MBR system depends on the characteristics of individual project requirements.

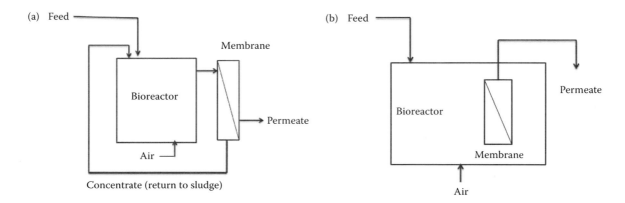

FIGURE 32.5 (a) Pressure-driven membrane externally attached to bioreactors. (b) Vacuum-driven membrane immersed into bioreactor compartment.

32.7.2 Pretreatment of Wastewater for MBR Processes

Usually, pretreatment is required for MBR processes; in this cycle of treatment, wastewater goes through primary screening or a clarification process where coarse particulate matter and large particles are removed by 0.5–3-mm screens. An efficient primary treatment method is essential for the membrane protection from the damage and for reducing nonbiodegradable substances [13].

32.7.3 Membrane Bioreactor Process

The MBR process is one of the most advanced treatment methods. MF or UF membranes are used for filtration, which produces high quality of permeate and eliminates the need for conventional retention ponds. The advantage of the membrane separation technology is the complete retention of the biomass and suspended solids, whereas conventional activation sludge treatment solely depends on the characteristics of the settling biomass. Major drawbacks of the MBR system are membrane fouling, which occurs from colloidal deposition, biofilm growth on the membrane surface, and inorganic matter precipitation. Because of the fouling, flux rate can be decreased significantly. However, the membrane cleaning is normally achieved by periodic back flushing or chemical cleaning. Besides a variety of membrane cleaning options, there are other alternatives recommended such as an equalization basin that may be helpful in case of low flux rate. Meanwhile, taking into consideration all the benefits offered by MBR technology, one has to make an extra effort to investigate or test other supplementary techniques such as using coagulants, flocculants, or adsorbent agents to minimize the fouling on the membrane surface, which can, in return, enhance the membrane performance [13].

32.7.4 Microbiological Treatment

In the microbial stage, the biodegradation of organic matter takes place in a biofilm reactor coupled with MF or UF membrane. Mostly, microbes remove the organic matter and nutrients from wastewater; these microbes grow and form micro-colonies, which

are an essential part of the MBR wastewater treatment process. An efficient treatment of activated sludge depends on the microorganisms' ability to form a flock structure. This biofilm flock structure is vital in the process of adsorption of colloids and macromolecule materials in wastewater. The microbial community in activated sludge consists of virus, bacteria, and eukaryotes, and plays an important role in the degradation processes. Recently, improvements have been made to existing MBR processes to optimize the performance in terms of reducing membrane fouling and sludge removing. Consequently, a biofilm reactor with membrane separation (BF-MBR) has been designed. The basic process concept is more or less the same except that the biomass reactor is divided into two separate entities as illustrated in Figure 32.6. Therefore, each operating unit works at its optimum level [11].

32.7.5 MBR Benefits and Future Developments

The largest application of MBR is the municipal wastewater treatment, where the nutrients, phosphorus, and nitrogen are separated from wastewater. Since MBR technology has been improved, it has been used for many industrial applications such as separation of oil from wastewater and treatment of wastewater from tanneries and textile industries. The key benefits of MBR technology are that these plants are relatively smaller in size but very efficient for tertiary wastewater treatment and potential for high quality of effluent discharge in the environment, which improves the quality of river and lake water and revives aquatic life. More importantly, the quality of the effluent is highly consistent due to the membrane separation barrier and meets most stringent water quality standards as shown in Table 32.3 [20].

Like other membrane separation process, use of MBR technology is increasing for wastewater treatment. The cleaning processes and membrane systems are now fully automated. Furthermore, MBR technology provides local decentralized and simpler solutions with minimum process components and maintenance requirement. The current market trend shows that the use of MBR technology will be increasing for wastewater treatment due to demanding applications such as agriculture irrigation, water parks, and for cooling purposes in industries [1].

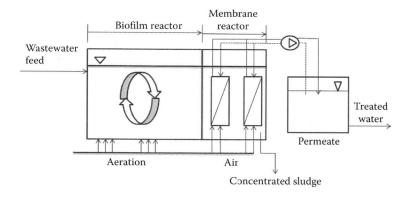

FIGURE 32.6 Moving-bed-biofilm reactor (MBBR) followed by submerged membrane modules. The outlet of the bioreactor is directed into the membrane reactor tank from where permeate is extracted through the membrane under vacuum.

TABLE 32.3 The Quality Standards of Treated Wastewater for Discharge into Class II Water and the Quality of Effluent Water from an MBR Plant

Parameters	Concentration Permitted for Discharge into Water Streams	MBR Values	MBR Plant Performance (%)
COD mg O$_2$/L	<125	<20	98
BOD mg O$_2$/L	<25	<2	96
Suspended solids mg/L	<35	<2	98
Total nitrogen	<21	<21	87
Total phosphorus	<1	<1	92
Bacteria removal	–	–	99
Turbidity NTU	<1	<1	99.9

Source: Adapted from Vlasic, A. and Cupic, D. 2009. Applying the MBR technology in wastewater treatment with reuse of wastewater on coastal areas in Croatia. Thirteenth International Water Technology Conference (IWTC), Hurghada, Egypt.

32.8 UV Disinfection

A combination of processes is being applied for water and wastewater treatment including clarification, nutrient removal, filtration, carbon adsorption, and disinfection. Disinfection is a process intended to stop the activity of viruses, bacteria, and protozoa. Disinfection procedures mostly depend on the delivered quality of water. There are a variety of disinfectant methods such as UV, ozonation, and chemical disinfection. In the chemical disinfection method, the commonly used chemicals are chlorine, chlorine dioxide, monochloramine, and ozone. Chlorine is considered a powerful disinfectant that is used for primary and secondary treatment of drinking water. Usually primary disinfection treatment is necessary to inactivate the pathogens and secondary disinfection treatment is used to maintain the disinfectant residual in the distribution system to protect the water from recontamination and decrease the bacterial regrowth. The effective applications of chlorine may be achieved by combining chlorine and ammonia, sodium hypochlorite, calcium hypochlorite, and chlorine dioxide. However, in the case of wastewater treatment, the danger with

chlorine-based treatment is that chlorine is highly toxic and it can react with organics present in wastewater and form DBP. UV irradiation is one of the greatest alternatives compared to chemical disinfection. The main benefits of UV include no use of toxic chemicals and therefore no chance of DBP. In addition, it is highly effective against bacteria, protozoa, and viruses. The application of (UV) light is usually used in combination with MBR technology to attain a water quality adequate for reuse. UV disinfection is rapid, efficient, safer, and comparatively an economical alternative than ozonation and chemical disinfection methods. Essentially, UV lamps are used for disinfection where UV light interacts with cellular components of microorganisms and destroys their DNA, consequently stopping the microorganism's ability to replicate itself. Thus, the use of the UV disinfection process controls the pathogen and eliminates the risk of DBP formation [3,4,14].

32.9 Summary and Conclusion

All the modern water treatment technologies have potential to reduce the pollution in the environment and provide sustainable solutions for water reuse. Traditionally, toxic waste from wastewater was removed by conventional treatment methods; thus, the water could be discharged safely into the surface water streams. However, these days increasing applications of wastewater reuse and production of profitable products at low cost from the waste have opened up a new era. The water treatment industry spends significant efforts into development of innovative technology where the substances in wastewater are being looked at as valuable resources. Reclaiming those profit-generating resources and converting them into useable forms of energy sources can actually save millions of dollars. Since those valuable products help to balance the equipment cost, the industry has developed a variety of technologies that can be utilized for the recovery of organic matter, fertilizers, and energy producing gases, and treat most of the wastewater quite easily. Therefore, water treatment technology is playing a crucial role in urban wastewater treatment. In general, membrane separation technology uses multiple barrier concepts in water and wastewater treatment. Microfiltration or ultrafiltration is used to remove particulate matter, while nanofiltration and reverse osmosis technologies are extensively used for removal of suspended solids and desalination purposes. The MBR system is a combination of both membranes coupled with a biological treatment system. With respect to the water treatment facilities, two options can be considered—central treatment plants and decentralized treatment plants. Decentralized wastewater treatment plants are practiced at a smaller scale such as industries, smaller residential areas, or an individual home. At a centralized plant location, an integrated system design is most effective because it utilizes all types of secondary, tertiary, and advance treatment processes. In contrast, nutrients recovery and energy management are more effective in a centralized system. In this chapter, new water reuse technologies and their benefits in terms of water reclamation and reuse application are discussed. The fact is, "new water cannot be created easily" [1]; however, new approaches can guide us to a sustainable solution and make water and resources recovery possible.

References

1. American Membrane Technology Association (AMTA). 2007. Membrane bioreactors (MBR). www.amtaorg.com. Accessed April 20, 2014.

2. Boussu, K., Van der Bruggena, B, Volodinb, A., Van Haesendonckb, C., Delcourc, J.A., Van der Meerend, P., and Vandecasteelea, C. 2006. Characterization of commercial nanofiltration membranes and comparison with self-made polyethersulfone membranes. *Desalination*, 191. www.elsevier.com/locate/desal. Accessed March 25, 2014. DOI: 10.1016/j.desal.2005.07.025.

3. Brahmi, M., Belhadi, N.H., Hamdi, H., and Hassen, A. 2010. Modeling of secondary treated wastewater disinfection, UV irradiation. *Journal of Environmental Sciences*, 22(8), 1218–1224. www.Sciencedirect.com. Accessed April 21, 2014.

4. Canadian Council of Ministers of the Environment (CCME). 2002. A workshop sponsored by the Canadian Council of Ministers of the Environment (CCME), linking water science to policy workshop series water reuse and recycling. http://www.ccme.ca. Accessed April 22, 2014.

5. European Union (EU). 2010. Membrane technologies for water applications, highlights from a selection of European research projects, EUR 24552. http://ec.europa.eu/research/environment/pdf/membrane-technologies.pdf. Accessed April 10, 2014.

6. Haering, K.C., Evanylo, G.K., Benham, B., and Goatley, M. 2009. Virginia cooperative extension; water reuse: Using reclaimed water for irrigation. https://pubs.ext.vt.edu/452/452-014/452-014_pdf.pdf. Accessed April 6, 2014.

7. Hilal, N., Al-Zoub, H., Darwish, N.A., Mohammad, A.W., and Abu Arabi, M. 2004. A comprehensive review of nanofiltration membranes: Treatment, pretreatment, modelling, and atomic force microscopy. *Desalination*, 170, 281–308. http://www.journals.elsevier.com/desalination. Accessed March 20, 2014.

8. Israel Desalination Society (IDS). 2009. Osmotic processes: Past, present and future (in memory of Prof. Sidney Loeb). 11th Annual IDS Conference. http://w3.bgu.ac.il/ziwr/documents/11thIDSConference-Program.pdf. Accessed March 9, 2014.

9. Jacangelo, J.G., Trussell, R.R., and Watson, M. 1997. Role of membrane technology in drinking water treatment in the United States. *Desalination*, 113, 119–127. www.sciencedirect.com/science/article/pii/S0011916497001203. Accessed March 19, 2014. DOI: 10.1016/S0011-9164(97)00120-3.

10. Koch Membrane Systems. www.kochmembrane.com.

11. Leiknes, T. and Degaard, H. 2007. The development of a biofilm membrane bioreactor, Norwegian University of Science and Technology (NTNU), Department of Hydraulic and Environmental. *Desalination*, 202, 135–143. www.elsevier.com/locate/desa. Accessed April 18, 2014.

12. New Ed, Nano for water cluster. http://www.new-ed.eu/. Accessed April 4, 2014.

13. Radjenovic, J., Matosic, M., Mijatovic, I., Petrovic, M., Barceló, D., and Heidelberg, B., 2007. Membrane bioreactor (MBR) as an advanced wastewater treatment technology, http://link.springer.com/chapter/10.1007/978-3-540-79210-9_2. Accessed April 16, 2014.

14. Safe Drinking Water Act. 2008. Procedure for disinfection of drinking water in Ontario, Ministry of Health and Long Term Care, http://www.health.gov.on.ca/english/public/program/pubhealth/safewater/docs/sdws_disinfection_manual_20081128.pdf. Accessed April 22, 2014.

15. Salgota, M., Huertasa, E., Weberb, S., Dottb, W., and Hollender, J. 2006. Wastewater reuse and risk: Definition of key objectives. *Desalination*, 187, 29–40.

16. Smith, T. 2011. Overcoming challenges in wastewater reuse: A case study of San Antonio, Texas. Thesis, Harvard College, Cambridge, MA. http://watersecurityinitiative.seas.harvard.edu/sites/default/files/Smith_Tiziana_Thesis_Final.pdf. Accessed April 8, 2014.

17. Tech Brief: Membrane Filtration, a national drinking water clearinghouse fact sheet. http://www.nesc.wvu.edu/pdf/dw/publications/ontap/2009_tb/membrane_DWFSOM43.pdf. Accessed April 12, 2014.

18. The International Water Association (IWA). 2013. For the UN-Water Task Force consultation on the sustainable development goal for water, Technological developments and innovations in water reuse. http://www.unece.org/fileadmin/DAM/env/water/meetings/Post_2015_Water_Thematic_Consultation/Technological_Developments_and_Innovations_in_Water_Reuse_Framing_Paper.pdf. Accessed July 5, 2015.

19. UNEP/DTIE/IETC. 2005. A division of technology, industry, and economics—International environmental technology center Japan. Water and Wastewater Reuse an Environmentally Sound Approach for Sustainable Urban Water Management, http://www.unep.org. Accessed April 4, 2014.

20. Vlasic, A. and Cupic, D. 2009. Applying the MBR technology in wastewater treatment with reuse of wastewater on coastal areas in Croatia. *Thirteenth International Water Technology Conference (IWTC)*, Hurghada, Egypt.

33

Blackwater System

Never Mujere
University of Zimbabwe

Saeid Eslamian
Isfahan University of Technology

PREFACE

In developed countries, around 97% of the population is connected to wastewater treatment works by sewers. However, in developing countries, the figure is less than 5% and most of the urban sewage is discharged into surface waters without any form of treatment. Domestic wastewater comprises concentrated blackwater from toilets and less concentrated greywater, which originates from showers, washing basins, and laundry among other sources. In most parts of the world, water is too valuable a resource to be thrown away; hence, wastewater needs to be recycled. The general principle in wastewater treatment is to remove pollutants from the water. Starting from the toilet, blackwater flows from the property into the plant to begin the treatment process. Treatment converts sewage into a liquid effluent, mainly water, leaving behind sludge. In treatment plants, four main stages are involved: preliminary, primary, secondary, and tertiary treatment.

Sewage treatment processes produce sludge and wastewater whose pathogen populations may be so low that they pose little or no harm to aquatic life. After treatment, this water can either be used for irrigation purposes or be directed into natural watercourses such as streams and rivers. On the other hand, the sludge can be incinerated or used as manure in surrounding farms.

This chapter attempts to answer the question: How do blackwater systems work? Arguments for and against blackwater recycling are presented. It focuses on blackwater management systems as described by four processes: generation, transport, treatment, and disposal. Treatment technologies include ponds, membranes, biological nutrient removal (BNR), and sequential batch removal (SBR).

33.1 Introduction

Domestic, industrial, commercial, institutional, and storm wastewater comprise concentrated blackwater or sewage and less concentrated greywater or sullage. Blackwater is wastewater containing human waste from toilet-flushing. Hence, it contains high concentrations of nitrogen and phosphorus, which are valuable nutrients. It differs from greywater, which is generated from showers, washing basins, washing machines, and laundry, among others [3]. The processes of wastewater management can be classified into three main aspects: collection, treatment, and disposal.

Many cities in developing countries lack sewer systems, let alone treatment plants. Sewage is often drained into rivers or lakes that may also be used as water sources with obvious problems for health. Contaminated water containing bacteria, parasites, and viruses derived from sewage is a major cause of death in most developing countries. Domestic and industrial sewage is discharged untreated into rivers and the sea. This is an acceptable method of disposal if the outfall is far enough offshore, if currents do not bring the sewage back to land, and if the discharge is not too great. In these circumstances, bacteria in the sea can break down the organic matter. Sea discharges are not always satisfactory because they compromise on bathing water quality. Discharge of raw sewage into watercourses leads to serious levels of water pollution [3,5,10]. Thus, it is necessary to treat sewage before it is allowed to flow into natural watercourses such as streams and rivers.

Suspended solids, total dissolved solids, electrical conductivity, alkalinity, heavy metals, and biochemical oxygen demand are regularly monitored by wastewater regulatory authorities to determine whether treatment plants comply with conditions of effluent disposal. Sewage treatment seeks to achieve three objectives, which are: to reduce the pathogen content of waste, to decrease its biochemical oxygen demand, and to reduce its solids content [1,3]. Efficient treatment of black or toilet water offers the possibility to recover energy and nutrients.

This chapter focuses on blackwater management systems as described by four processes: generation, transport, treatment, and disposal. It highlights the sources of blackwater before exploring arguments for and against blackwater recycling. Blackwater treatment technologies are also examined. Finally, common sludge disposal practices are highlighted.

33.2 Blackwater Generation, Collection, and Recycling

Toilet or human waste comprises feces and urine, which can be collected in containers such as buckets (night soil collection), vacuum tanks, and sewer systems, and transported to treatment plants. Sewer systems can either be combined (stormwater and wastewater are collected in one pipe) or separated (stormwater and wastewater are collected in separate pipes) systems.

A blackwater recycling system is a step beyond greywater recycling. Blackwater is generally not recycled because it contains sewage that is difficult to adequately clean for use. Although some systems make blackwater clean enough to be potable or drinkable, such systems are expensive, difficult to maintain, and people who drink recycled blackwater have to get over the stigma of drinking toilet water [9,10].

Blackwater recycling systems make up another category of blackwater treatment systems. In general, various types of physical, biological, and chemical methods for blackwater treatment are used and the treated water is then recycled back into the blackwater system loop rather than being discharged. The filtered blackwater is good for the environment and other benefits which include [4,6]:

Reduces stress on septic systems: A blackwater recycling system can take some of the stress off septic systems, which may be close to failing.

Conserves energy: Removing the harmful bacteria from blackwater costs a lot of money and uses energy from processing plants. Thus, recycling conserves energy.

Water conservation: Using recycled blackwater to water lawns, plants, and nonfood crops/gardens conserves freshwater.

Habitat protection: By recycling blackwater, there is less chance of wastewater seeping into natural habitats.

Plant growth: There are many nutrients left in the water after being treated that plants use. Thus, plants that are grown with recycled blackwater rarely need any fertilizer.

Although, there are advantages of blackwater recycling, there are also disadvantages that need to be considered when deciding to put in a blackwater recycling system. Some of these drawbacks include [4,6]:

Cost: Blackwater recycling systems are expensive to install, maintain, and fix if they break down.

Smell: A discernible smell from sewage and dying bacteria is produced from recycling plants.

Maintenance: The system usually needs frequent maintenance.

33.3 Blackwater Treatment Process

Blackwater treatment methods make use of physical, biological, and chemical methods to treat the solid and liquid organic and inorganic waste, commonly known as sewage, produced by humans via bodily excretion and domestic activities, as well as agricultural, industrial, and commercial activities. The goals are to remove solids, break down organic compounds, eliminate microorganisms that cause disease, remove harmful chemical substances, and prevent or eliminate offensive and harmful odors and soil discoloration. The different types of blackwater treatment include septic tanks, cesspools, soil drain fields, chemicals, bio-digesters, composting toilets, and blackwater recycling systems [7].

Cesspools, or cesspits, have long been used as an often-cheaper alternative because they are sealed tanks designed simply to hold sewage until it is pumped out and transported elsewhere for treatment. Septic tanks and soil drain fields have been the most common type of onsite blackwater treatment system for homes and small groups that are not connected to a centralized municipal sewage treatment system. They typically use physical and biological means for blackwater treatment. The former include percolation and filtration through gravel, sand, and activated charcoal. The latter take advantage of anaerobic and aerobic digestion by naturally occurring bacteria for blackwater treatment [11].

Alternatives that strive to mimic nature's way of treating blackwater, such as biodigesters, manufactured wetland and reed beds, composting toilets, and blackwater recycling systems are also used because they are considered less ecologically harmful.

Bio-digesters typically seek to make more efficient, effective use of anaerobic and aerobic digestion to treat blackwater. Plants that absorb nitrogen, phosphorous, and even metals might be planted in a manufactured wetland or reed bed.

Blackwater treatment also can involve the addition of chemicals, such as chlorine, to disinfect it. Chemicals can be added that cause reactions that result in the removal of harmful organic and inorganic substances.

Blackwater is treated to control the spread of disease by isolating the sewage so that viruses and other pathogens die, reduce the amount of oxygen consuming organic material before it is discharged into lakes or rivers, reduce biodegradable material and material in suspension, remove toxic materials and eliminate pathogenic bacteria, and break down the sewage into relatively harmless substances to protect the environment into which it is discharged [3].

The wastewater system relies on the force of gravity to move sewage from the sources to the treatment plants, which are often located on low ground, near rivers into which treated water can be released. If the plant is built above the ground level, the wastewater has to be pumped up to the aeration tanks. From the tanks, gravity moves the wastewater through the treatment process. There are four main stages of treatment in a typical sewage treatment plant; namely, preliminary, primary, secondary, and tertiary (Figure 33.1). However, the design or layout can vary from site to site.

33.3.1 Preliminary Treatment

33.3.1.1 Screening

The first stage or pretreatment stage is important to prevent damage to pumps and clogging of pipes. Sewage passes through mechanically raked bar screens to remove large debris such as pieces of wood, bottles, rags, plastics, wire, sticks, cans, stones, toilet paper, silt, pebbles, and other large objects. Insoluble material is efficiently reduced to a residue that is either discharged to sewer or de-watered and compacted for disposal as solid waste. The extracted material is burnt or buried in landfills. Unless they are removed, they can cause problems in the treatment process.

33.3.1.2 Flow Equalization

Wastewater seldom flows into wastewater treatment plants at the same rate throughout each day. Generally, the greatest flows reaching the wastewater treatment plants arrive mid-morning. Such uneven flow volumes reduce wastewater treatment plants' efficiency. To even out these periods of high and low flow, large basins are constructed at some wastewater treatment plants to store the wastewater flow from peak periods and release it for treatment latter. These basins require aeration and mixing to prevent odors and deposition of solids [5].

33.3.2 Primary Treatment

Once the large objects have been screened and removed during pretreatment, the sewage containing light organic suspended solids and small inorganic material goes for primary treatment. The primary treatment tank is where blackwater goes when it is removed from the home via gravity and pipes. Here, the sewage settles for some time (about 24 h) while an established colony of bacteria works to break down large particles. Some solids are crushed and settle down to form sludge or solid waste. The sludge is removed from the primary treatment tank by mechanical scrapers and pumps. Grease, oil, and other floating substances rise to the top, where they are removed by surface skimming equipment. Sewage containing small inorganic material flows slowly through grit tanks where particles of sand or gravel are removed by a grit removal system. Sand and grit settle

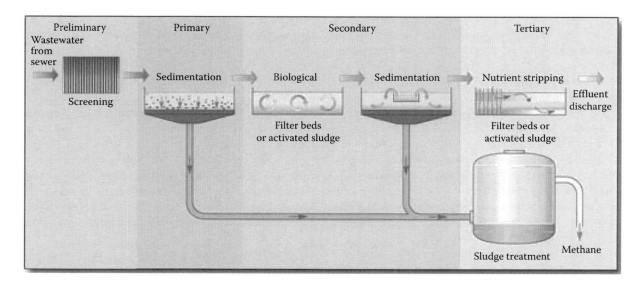

FIGURE 33.1 Flow diagram describing what happens at each stage of the blackwater treatment process.

out but light organic or fine particles remain suspended. Thus, sewage is passed to large primary sedimentation tanks or primary clarifiers where most of the remaining particles settle out by gravity to form sludge [6]. Primary treatment removes about 60%–70% of suspended solids. These settled solids, called primary sludge, are removed along with floating scum and grease, and are pumped to anaerobic digesters for further treatment. After 24 h, the settled blackwater is pumped to the secondary treatment tank. Liquid leaving the primary sedimentation tanks contains very fine solids and dissolved matter, so secondary treatment is required.

33.3.3 Secondary Treatment

The primary effluent from primary treatment is transferred to the biological or secondary stage. Secondary treatment is a biological process involving the oxidation of dissolved organic material by microorganisms. The process is aimed at restoring wastewater to a condition where it can sustain aquatic life such as fish, frogs, and turtles. Here, the wastewater is mixed with a controlled population of bacteria and an ample supply of oxygen [10]. The microorganisms digest the fine suspended and soluble organic materials, thereby removing them from the wastewater. In secondary clarifiers, biological solids or sludge is settled by gravity and then pumped to anaerobic digesters. The clear secondary effluent may flow directly to the receiving environment or to a disinfection facility prior to release. The secondary tank is separated into three chambers: blackwater aeration, sludge settling, and irrigation chamber.

Aeration chamber: This first chamber in the secondary treatment tank begins the aeration stage. Water and air are injected into the tank at timed intervals so that the tank contents are churned. Bacteria in the tank settle, multiply, and feed on sludge particles in the tank. After complete digestion of nutrients and oxygen from the sludge by bacteria, the water is piped or moved to the sludge-settling chamber.

Sludge-settling chamber: A bacteria biomass mechanism forces sludge downward and the partially treated water upward to be collected and sent on to the irrigation chamber stage. A colony of bacteria consumes most of the oxygen in the mix and breaks down any remaining solid particles.

Irrigation chamber: In the irrigation chamber, wastewater is clarified, chlorinated, and piped into ground irrigation systems for use in gardens.

There are several variations of secondary treatment, including trickling filters, activated sludge tanks, rotating biological contactors, lagoons, and oxidation (stabilization) ponds. The trickling filters and activated sludge tanks can be used anywhere while oxidation ponds are only applicable in warm climates such as the tropics [3,6]. The two most common secondary treatment methods are trickling filters and activated sludge.

Secondary treatment removes suspended solids and soluble materials that require oxygen for decay as well as further removal of about 70%–90% of the biochemical oxygen demand (BOD) in the sewage. Therefore, the effluent is usually sufficiently

purified to be discharged to a river, lake, or the sea. The process is speeded up by increasing the amount of oxygen available and is highly effective in reducing the BOD in wastewater. However, the secondary clarifiers used to settle out microorganisms in the secondary treatment process are not very effective. Some of these microorganisms remain in the wastewater after it leaves the secondary clarifier, and they add BOD because the decay of these microorganisms will exert their own oxygen demand. Sand filters are sometimes used for additional removal of microorganisms and other solids. The filters often use large, lightweight aggregates (such as coal) at the top to improve efficiency and facilitate cleaning. Partially treated wastewater usually contains higher concentrations of solids, so these filters must be designed for greater efficiency and for more frequent cleaning [1,2].

33.3.4 Tertiary Treatment

Tertiary treatment or advanced wastewater treatment (AWT) removes suspended and dissolved substances remaining after conventional secondary treatment. Some pollutants such as nitrogen, phosphorus, nonbiological *chemical oxygen demand* (COD), or heavy metals may be present and require removal so that the treated wastewater can be reused. AWT reduces plant nutrients or toxic compounds to acceptable levels by removing such things as color, metals, organic chemicals, and nutrients such as phosphorus and nitrogen. This may be accomplished using a variety of physical, chemical, or biological treatment processes to remove the targeted pollutants. However, this is very expensive and not commonly used. Some processes used in tertiary treatment include filtration, carbon adsorption, nutrient stripping, disinfection by ultraviolet (UV) light or filter membranes, phosphorus removal, and nitrogen removal [5].

Disinfection: Before the final effluent is released into the receiving waters, it may be disinfected to reduce the disease-causing microorganisms that remain in it. The most common method of disinfection is chlorination. Chlorine gas or a chlorine-based disinfectant such as sodium hypochlorite is injected into the wastewater and the wastewater is held in a basin for about 15 min to allow the chlorine to react with any remaining pathogens. Since excess chlorine is toxic to fish and the environment, the effluent is dechlorinated prior to discharge. Other disinfection options include UV light and ozone [4].

Filtration: The blackwater filtering system removes usable water from blackwater. Water is piped to it, and then it goes through a process before being used to water the lawn or non-food gardens via underground pipe systems. Water recycled from blackwater is rarely used for drinking or to water food crops because it may still contain harmful bacteria. Blackwater filtration can be achieved using the following processes: carbon adsorption, nitrification, ammonia stripping, phosphorus, and nitrogen removal.

Carbon adsorption: After secondary treatment and filtration, soluble organics (called refractory organics) may still be present in the wastewater. The most practical way to remove refractory organics is by adsorbing them on activated carbon. Adsorption is the accumulation of materials on an interface (in this case, the liquid/solid boundary layer). Carbon is activated by heating it in the absence of oxygen. The activation process creates many small pores in each carbon particle and increases the surface area of each carbon particle, thus making it more effective as an adsorption agent.

Phosphorus removal: Phosphorus, in wastewater, is considered a pollutant because it encourages the growth of algae. Phosphorus removal usually involves the addition of ferric chloride, alum, or lime to the wastewater, mixing it in a reaction basin, and then sending the mixture to a clarifier to allow the phosphorus-containing precipitate to settle out.

Nitrogen control: Nitrogen in any soluble form is a plant nutrient and may need to be removed from the wastewater to control the growth of algae. In addition, nitrogen in the form of ammonia exerts an oxygen demand and can be toxic to fish. Nitrogen can be removed from wastewater by both biological and chemical means. The biological process is called nitrification/denitrification and the chemical process is called ammonia stripping.

Nitrification/denitrification: The natural nitrification process can be forced to occur in the activated sludge process by maintaining a cell detention time of at least 15 days. Bacteria convert nitrates into water, nitrogen, and carbon dioxide. Small amounts of organic materials such as methanol and raw or settled sewage are added to provide a food source for the bacteria for the denitrification process.

Ammonia stripping: Nitrogen in the form of ammonia can be removed chemically by raising the pH (often, by adding lime) to convert the ammonium ion into ammonia, which can be stripped from the water by blowing large quantities of air through the water.

Primary and secondary treatments remove only 20%–40% of the phosphorus and nitrogen, and about half of the toxic compounds. Secondary treatment can remove over 85% of the BOD, suspended solids, and nearly all pathogens (Table 33.1).

Treated sewage from the secondary treatment is then passed for final clarification or filtration before discharge to the river or sea. The clarifier is a settling tank, similar to that used for primary treatment, and may be followed by a polishing filter.

33.4 Blackwater Treatment Systems

Treatment plants are categorized into one of three types, based on the method of secondary treatment, that is, oxidation (stabilization) ponds, trickling (biological) filter, activated sludge, BNR, sequencing batch reactors, membrane treatment, and Pasveer

TABLE 33.1 Percentage Removal (Cumulative, from Initial State) of Sewage Constituents or Characteristics of Sewage after Successive Stages of Treatment

Constituent	Primary	Secondary	Tertiary
Suspended solids	60–70	80–95	90–95
BOD	20–40	70–90	>95
Phosphorus	10–30	20–40	85–97
Nitrogen	10–20	20–40	20–40
E. coli bacteria	60–90	90–99	>99
Viruses	30–70	90–99	>99
Cadmium and zinc	5–20	20–40	40–60
Copper, lead, and chromium	40–60	70–90	80–89

ditch. The trickling filters and activated sludge tanks can be used anywhere while oxidation ponds are only applicable in warm climates such as the tropics. Biological filter and the Pasveer ditch systems, tend to be used on smaller, rural treatment plants.

33.4.1 Wastewater Stabilization Ponds

Stabilization or oxidation ponds are shallow manufactured impoundments for treating wastewater by entirely natural processes utilizing natural physical, biological, and chemical processes. They are suitable in tropical and subtropical climates, in developing countries where land is often available at reasonable cost, and skilled labor is in short supply. These sewage ponds need to be emptied every 5 years and their effluent is suitable for agriculture. They are simple, cost-effective, and highly efficient [3,4]. The systems comprise a series of ponds (anaerobic, facultative, and maturation) into which wastewater flows and from which, after a retention time of many days (compared to hours in conventional treatment processes), a well-treated effluent is discharged into water courses. Figure 33.2 shows the three main types of ponds joined in series.

Inlet works: Remove screenings, that is, rags, logs, stones, spoons, and grit—inorganic materials less than 10 mm in diameter.

Primary or anaerobic ponds: These are open septic tanks used to remove BOD for inoculation, break the scum, and treat large volumes of strong wastes. They remove 40% of BOD by sedimentation of suspended soils, which are rapidly digested by anaerobic bacteria. Retention time varies from 10 to 15 days. In most cases, there are two ponds in series and two ponds in parallel depending on quantity and strength of sewage. Ponds are 2–5 m deep and receive a high organic loading such that they contain no dissolved oxygen (DO) and algae.

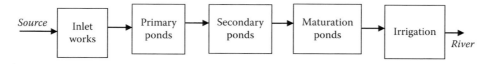

FIGURE 33.2 Layout of sewage ponds.

Secondary or facultative ponds: Comprise three or four ponds in series, each measures about 1.2–1.5 m deep to prevent vegetable growth and mosquito breeding. Facultative means that both anaerobic and aerobic conditions exist, resulting in a two-phase biological sewage treatment. In the aerobic upper layers, bacteria oxidize BOD using oxygen produced in the pond by photosynthesis of algae during the day. In the deep anaerobic layers, organic matter is digested by anaerobic bacteria.

Biological kinematic reactions remove 55%–57% BOD. The reactions depend on temperature. Retention time vary from 10 days in the first pond to 5 days in other ponds.

Maturation ponds: The ponds are wholly aerobic during the day, and are used primarily for pathogen (fecal bacteria and viruses) reduction, to ensure further BOD reduction, and nutrient removal (nitrates and phosphates). High pathogen removals are achieved because of high DO, high pH, and good light penetration. Pathogens are removed by UV radiation in sunlight. It is normal to provide three maturation ponds with retention time of 5 days and a depth of 1–1.5 m to allow sunlight to penetrate. They remove remaining BOD and pathogens are killed by UV radiation. Due to the algal and nutrient content of effluent, in Zimbabwe the effluent cannot be discharged into rivers but will have to be irrigated. Maturation ponds provide excellent conditions for fish breeding.

Irrigation pond: Phosphates and nitrates are removed by high nutrient uptake plants such as citrus plants, star grass, or Kikuyu grass.

33.4.2 Biological Filters System

This consists of a large circular tank or series of tanks containing stone or plastic pieces, which is sprayed with the screened sewage by a mechanical rotating arm moving over the surface of the bed. The medium is sufficiently loose to allow the liquid to permeate it and provide free access of air. A thin film of microorganisms is then supported on the bed, which provides biological decomposition of the sewage. The nature of the medium is critical to ensure an adequate retention time for maximum efficiency. The liquid leaving the bed passes to further settling or humus tanks, in which residual solids collect, and are withdrawn periodically for sludge disposal [9,10]. The disadvantage of the biological filter is mainly due to difficulties in control with variation in flow and concentration of the sewage, and its unsuitability for large-scale treatment (Figure 33.3).

The primary settling tank (PST) removes 40% BOD while filters remove 97% BOD. Humus and dead bacteria are removed in the humus tank. Final polishing is done in the holding ponds. Nutrients are removed during irrigation.

33.4.3 Activated Sludge Plants

The activated sludge method involves developing a culture of bacteria and other organisms in a large tank and on lanes containing the settled sewage. It is a process for treating sewage using air and a biological floc composed of bacteria and protozoa. Activated sludge comes from the biological mass formed when oxygen (in the form of air) is continuously injected into the wastewater. In this process, microorganisms are thoroughly mixed with organics under conditions that stimulate their growth. As the microorganisms grow and are mixed by the agitation of the air, the individual microorganisms clump (or flocculate) together to form a mass of microbes called activated sludge.

In the activated sludge process, wastewater flows continuously into an aeration tank where oxygen is injected into the wastewater to mix the wastewater with the activated sludge, and to provide the oxygen needed for the microorganisms to break down the organic pollutants. The mixed liquor (mixture of wastewater and activated sludge) flows to a settling tank (final clarifier or secondary tank) to allow the biological flocs (the sludge blanket) to settle, thus separating the biological sludge from the clear treated water. The activated sludge settles out. About 20%–30% of the settled sludge is returned to the aeration tank (and hence is called recycled or return sludge) to maintain a high population of microbes for breaking down the organics. Figure 33.4 shows the general arrangement of an activated sludge process.

Oxygen is forced into the sewage by mechanical means to sustain the oxidation process. It is provided by one of two methods:

Mechanical aeration: Agitators are used on the surface of the tanks; the rate of aeration can be controlled by varying the speed or depth of immersion of the agitator. In practice, many parallel lanes are often used with one or more agitator in each lane.

Air/oxygen diffusion: Oxidation is achieved using perforated pipes or domes called diffusers, positioned in the base of the tanks. Air or pure oxygen from compressors is pumped through these diffusers, producing small bubbles

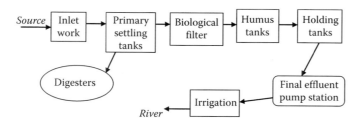

FIGURE 33.3 Biological filtration process.

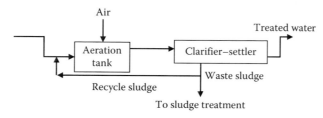

FIGURE 33.4 Activated sludge system.

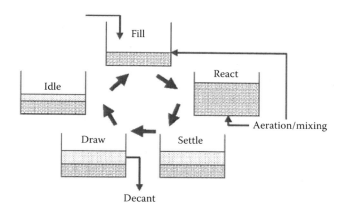

FIGURE 33.5 The Pasveer ditch system. (Adapted from Environmental Protection Agency (EPA). 2010. *Membrane Bioreactors: Wastewater Management Fact Sheet.* EPA, Washington, DC.)

in the sewage, providing very efficient oxidation. The rate of oxidation is then controlled by varying the speed of the compressors.

33.4.4 Sequencing Batch Reactors

SBR are more than an activated sludge system that operates in time rather than space [1]. In one unit, an SBR can undertake equalization, biological treatment, and secondary clarification with a timed, controlled sequence. Each of these processes occurs in separate units of a conventional activated sludge system (Figure 33.5).

A single-unit SBR can also include primary clarification and in a single chamber, there are multiple biological environments that can be created and maintained to nurture each specific type of bacteria for optimum performance at the right time and right stage of the process. For example, one type of bacteria absorbs excess phosphorous in the aerobic condition when aeration occurs, but the bacteria do nothing at all in the absence of oxygen or in the anaerobic stage. Bacteria quality and quantity is the key. With conventional systems, the bacteria are settled in the clarifier and pumped back to the head of the plant. The bacteria are recycled to maintain sufficient quantities. In SBRs, the bacteria continue to live and work in the unit as needed, when needed. These bacteria stay in the system longer and, therefore,

are stronger and more effective. The SBR system includes five steps: the idle, fill, react, settle, and draw steps.

During the fill step, static fill, mixed fill, or aerated fill can be used, depending on the facility and its goals. Static fill is when influent is mixed with existing biomass and no mixing or aeration has occurred when the mixing begins. Static fill can be compared to using "selector" compartments in a conventional activated sludge system [5]. In a mixed fill, there is mixing influent organics and biomass. An aerated fill consist of aerating the SBR contents to start an aerobic reaction.

During the react stage, biological reactions are finished so that nitrification or denitrification and phosphorous removal can be accomplished. In the settling stage, no influent or effluent can get in to disturb settling. Gentle mixing and settling time can produce a clearer effluent than a conventional system. In the draw step, a decanter is used to remove treated effluent out of the top of the tank while the bacteria settle at the bottom.

33.4.5 Pasveer Ditch System

This extended aeration system was developed in 1953 and is employed worldwide. It consists of an oval-shaped channel, approximately 2–3 m deep, into which sewage is passed after primary treatment. Based on the same biological processes as in activated sludge, the sewage is aerated and circulated around the ditch by means of one or more rotors mounted at different points around the ditch. The depth of immersion and speed of rotation are adjusted to suit the oxygen demand of the sewage [6]. The effluent is recirculated until adequate aeration is achieved before it is passed to the settling tank for tertiary treatment (Figure 33.6).

One variation of the oval Pasveer ditch design is an arrangement where the ditch is constructed as one long channel, often in a zigzag layout to save space. Rotors are used in the same way to aerate and propel the sewage along the ditch as the biological process takes place.

33.4.5.1 Aeration/Oxidation Control

In the case of the activated sludge and Pasveer ditch, the volumes of air or oxygen required to treat sewage is large. By measuring the dissolved oxygen in the aeration/oxidation lanes, careful control is kept of the volumes produced, limiting the value to

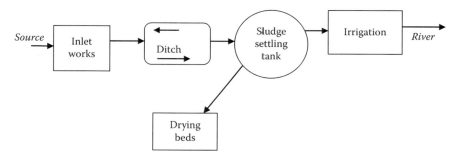

FIGURE 33.6 The BNR system.

around 2 mg/L in the sewage sludge. This is just sufficient for the degree of treatment required and ensures that complete nitrification of the ammonia is achieved [6].

33.4.6 Biological Nutrient Removal System

The BNR system is highly sophisticated concerning design criteria, process, operation, maintenance, and computer controls (Figure 33.7). It requires reliable power supply, oil, and spares for its operation [2,6,8]. BNR technology can treat phosphorous down to 1 mg/L. There is a significant chemical cost reduction using the BNR method. In plants with low alkalinity in the wastewater, BNR is advantageous in the denitrification step because it reduces oxygen required for treatment and restores alkalinity previously destroyed in the nitrification stage. BNR is an environmentally friendly process with no chemical additives. Plants discharging waste into receiving streams that merge with a major drinking water source need to discharge high quality effluent. The BNR can provide such conditions. The BNR system comprises seven or eight interrelated units; the screen, grit removal channel, skimming or sedimentation tank, digesters, return activated sludge (RAS) substation, clarifiers, thickeners, and the cascade [2,7,8].

The screen carries out preliminary treatment by removing large solid objects and heavy mineral particles. Grit removal channels also perform a similar function as they eliminate inorganic substances such as sand, glass, metals, and ashes from the wastewater. The screen and grit removal channel perform the same functions as in the trickling plant. The sedimentation or skimming tank separates solid from liquid waste at the bottom so that it can be disposed and eventually dried to produce manure for the farming community. The skimming tank is a small unit that traps some waste components of sewage, which may have by-passed the head works and tend to float over the surface of the tank. They include plastic wastes, coagulated oils, and greases, which interfere with subsequent sewage treatment processes. The bioreactor removes pollutants with biochemical reactions [2]. It is a sophisticated device comprising an anaerobic basin, aeration zone, and anoxic basin.

Primary or anaerobic digesters are used for the oxidation of raw sludge using microorganisms, such as bacteria. The trickling filter unit is an aerobic secondary treatment device, which is used to oxidize settled sewage with microorganisms. Humus tanks separate the final effluent from secondary sludge, which is then conveyed to disposal areas such as irrigation farms or pastures. Sludge drying beds perform two main functions; namely, sludge treatment and drying [2]. Once the sludge has dried, it can be sold to consumers in the farming community as sludge cakes or manure. In order for sewage treatment to succeed, all of the above components should operate efficiently. However, frequent electric power cuts and the breakdown of some parts in recent years have impacted negatively on the operation of the plant [8].

The RAS substation is a key component of the plant and no life can be sustained without it. It delivers sludge back to the bioreactor, supplies active microbes, and polishes/creates the desirable anaerobic conditions for fermentation.

Clarifiers are secondary sedimentation tanks that are designed to separate solid from liquid wastes. They thicken the accumulated solids content at the bottom of the tank and help to maintain suspension. Scum removal is achieved, separating it from the final effluent that is made to flow back to rivers and streams. Effluent from the clarifiers passes through it on its way to disposal channels such as rivers and streams.

Thickeners also play a significant role as they handle solids generated from the bioreactor and scum from the clarifiers. Once thickened, solids are conveyed to a nearby woodlot gum plantation for irrigation purposes [7]. The cascade is the last stage of the BNR plant.

33.4.6.1 The Anoxic/Oxic System Design

To address the nitrogen and phosphorus limits, innovative BNR with anoxic/oxic systems is added [2]. Anoxic/oxic systems are BNR processes that lower power costs because bacteria oxidized in the nitrate molecule is used. During the anoxic/oxic process, the facility uses oxygen that is available as a by-product of the treatment process itself. Without anoxic facilities, the oxygen produced is discharged with the effluent.

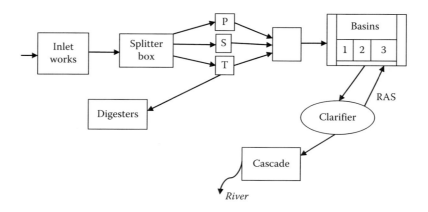

FIGURE 33.7   The SBR system (PST—primary settling tanks; basin 1—fermentation, 2—anoxic, and 3—aeration).

33.4.7 Membrane Treatment Technology

The technologies most commonly used for performing secondary treatment of wastewater rely on microorganisms suspended in the wastewater to treat it. Membranes provide a separation process using a wide variety of molecular sizes or pores to fit specific filtering or demineralization needs [12]. Membranes not only filter out most suspended solids but can also remove potentially dangerous bacteria. In a membrane plant, UV light can be used to effectively kill bacteria for disinfection. Membrane treatment technology completely eliminates the need for secondary clarifiers. Membrane technology is geared at specifics and not standardized; equipment can be specifically designed to meet a facility's specific effluent quality needs. The technology can be a more costly design feature than other filtration processes. The membrane's design life is usually between 7 and 10 years [1].

The advantages of membrane systems over conventional biological systems include better effluent quality, smaller space requirements, and ease of automation. Specifically, MBRs operate at higher volumetric loading rates, which result in lower hydraulic retention times. The low retention times mean that less space is required compared to a conventional system. Effluent from MBRs contains low concentrations of bacteria, total suspended solids (TSS), BOD, and phosphorus. This facilitates high-level disinfection. Effluents are readily discharged to surface streams or can be sold for reuse, such as irrigation.

The primary disadvantage of MBR systems is the higher capital and operating costs than conventional systems for the same throughput. Operation and maintenance (O&M) costs include membrane cleaning and fouling control, and eventual membrane replacement. Energy costs are also higher because of the need for air scouring to control bacterial growth on the membranes. In addition, the waste sludge from such a system might have a low settling rate, resulting in the need for chemicals to produce bio-solids acceptable for disposal (Figure 33.8).

Membrane treatment comprises [12]

Screening: The first step efficiently reduces insoluble material to a negligible residue. This residue is either discharged to sewer or it is dewatered and compacted for disposal as solid waste.

Biological treatment: Air is diffused into the water to create optimum conditions for bacteria to consume impurities. A sustainable biomass concentration is maintained, which metabolizes all incoming waste, resulting in negligible sludge.

Ultrafiltration: Ultrafiltration occurs through a special membrane of microscopic pores that prevents particles, bacteria, and viruses from passing through. The membranes are cleaned by air scouring, ensuring no wastewater is produced.

Ultraviolet disinfection: Ultraviolet lamps provide additional protection against pathogens.

TDS and nutrient removal: Proprietary technologies are employed for applications such as cooling tower reuse and discharge to sensitive environments.

Chlorination: Finally, chlorine residual is added to protect the water while in storage and the reticulation system. This is the only time any chemicals are used throughout the treatment process.

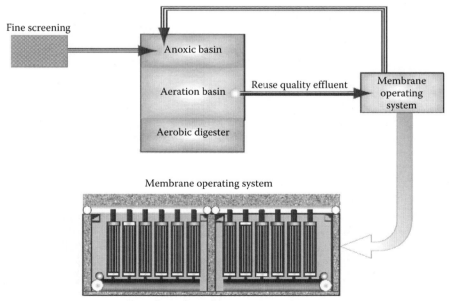

FIGURE 33.8 The MBR system. (Adapted from Environmental Protection Agency (EPA). 2010. *Membrane Bioreactors: Wastewater Management Fact Sheet.* EPA, Washington, DC.)

TABLE 33.2 Parameters Monitored at Each Stage during Blackwater Treatment

Stage in the Plant	Measurement	Purpose
Primary	pH and conductivity	To protect the process from unacceptable industrial discharges
	Ammonia and suspended solids	To monitor plant loading
	Suspended solids	To monitor discharge from settling tanks
Secondary	Dissolved oxygen and ammonia	To control aeration/oxidation
	Phosphate and nitrate	To monitor removal process, if these stages are incorporated in the design of the plant
Tertiary	Ammonia, nitrate, phosphate and suspended solids, BOD	To monitor the performance of the plant and the final treated sewage quality
	Suspended solids	To monitor discharge from sedimentation tanks and/or filtration

33.5 Blackwater Monitoring at Recycling Plants

Instrumentation on sewage plants is dependent on the size of the plant and the type of control required. Automatic monitoring and control instrumentation must be used to ensure that the optimum process conditions are maintained, and its use becomes essential with the introduction of more stringent controls on discharge into water courses as well as improving the control and the final sewage effluent quality from the plant. Operating costs can also be significantly reduced.

Dissolved oxygen monitors are essential to provide optimum control of the aeration/oxidation stage of the treatment process. Ammonia monitors are often installed on the final effluent stream to ensure the discharge limits have been met and as an indication of the process efficiency. Suspended solids monitors are invaluable throughout the plant to indicate load, efficiency, and correct operation of each stage of the process. pH is often used to monitor the raw sewage to guard against acidic or alkaline effluent entering the plant, which could upset, or in severe circumstances, kill the microorganisms used to carry out biological reaction [4,5]. Conductivity and pH systems could also be employed to monitor particular treatment processes. BOD, which gives an indication of the amount of organic matter entering the plant, is normally measured in the laboratory, but online analyzers are becoming more common (see Table 33.2).

33.6 Sludge Disposal and Utilization

Both the primary and the secondary treatment processes produce large amounts of sludge. Sludge is a nasty-smelling, thick liquid comprising about 96% water. Before final disposal from sewage treatment plants, sludge is sometimes held in closed tanks where, in the absence of oxygen, anaerobic bacteria further decompose the organic material, producing a relatively inoffensive digested sludge. Carbon dioxide and methane gases are produced. Methane gas, a by-product of anaerobic digestion, can be used as a fuel to heat the tanks or to generate electricity for the treatment works. The process takes 20–30 days at a temperature of around 30°C. Most of the sewage sludge is disposed in water bodies, farmland, landfills, incineration, land reclamation, composting, energy generation, and processed products [2,11].

Dumping in water bodies: This needs to be carefully controlled as the sea, like rivers and lakes, has only a limited capacity to absorb pollutants. It is particularly important to control pollution in enclosed seas such as the North Sea or the Mediterranean, where there is little water interchange with the larger oceans. The EU ended dumping at sea in 1998 [1].

Farmland: Digested sludge is a valuable fertilizer containing nitrogen, phosphates, potash, and organic matter. It converts into humus in the soil. However, it is important to investigate sludge for heavy metals before using it for agriculture. Sludge disposal on agricultural land is useful as a fertilizer and soil conditioner and it is a more convenient method of sludge disposal for sewage works that are not on the coast. Tankers are used to transport the large volumes of sludge to farms.

Landfill: Sludge containing toxic waste cannot be spread on fields. Instead, it is dumped in natural or artificial depressions in the ground or in trenches, where it dries, decomposes slowly (and may cause unpleasant smells), and covered with a layer of soil.

Incineration: Sludge may be dried and incinerated, leaving an inert ash. This is often the most expensive option.

33.7 Summary and Conclusions

In this chapter, various processes of blackwater treatment are described. Blackwater or sewage is wastewater containing human waste from toilet flushing. It is dangerous to people as it contains potentially harmful pathogenic organisms. Diseases such as cholera, typhoid, and dysentery as well as parasitic infestations are transmitted through waterborne contamination. If sewage is not adequately treated and disinfected, it becomes a source of community re-infection and therefore it is important that purified effluents be made biologically safe. Sewage water is treated to limit pollution and health risks, before being returned to the environment [5]. Some pollutants are easily removable while others must be converted to a settleable form before they can be removed.

The general principle in blackwater treatment is to remove pollutants by getting them either to settle or to float. Treatment facilities are designed in stages that include: preliminary/influent, primary treatment, secondary treatment, tertiary treatment, disinfection, and effluent discharge [6]. Each stage removes particles from the wastewater or changes dissolved and suspended material to a form that can be removed.

The goal of the primary first stage is to remove large objects from the wastewater, which include wood, bottles, stones, and toilet paper. Once the larger objects have been removed, the sewage goes further for primary treatment, which involves the crushing of some solids, which settle down to form sludge (solid waste). Secondary treatment is a biological process that is aimed at restoring wastewater to a condition where it can sustain aquatic life such as fish, frogs, and turtles. Treated sewage from the secondary treatment is passed for final clarification or filtration before discharge to the river or sea. The clarifier is a settling tank, similar to that used for primary treatment, and may be followed by a polishing filter.

Changing wastewater technology is providing improved efficiency in wastewater treatment. New technologies that make the systems work more effectively and efficiently include the membrane treatment technology and biological system enhancements such as BNR and SBR.

Sewage treatment processes produce sludge and wastewater whose pathogen populations are so low that they pose little or no harm to aquatic life. After treatment, this water can either be used for irrigation purposes or be directed into natural watercourses such as streams and rivers. On the other hand, the sludge can be incinerated or used as manure in surrounding farms.

References

1. Environmental Protection Agency (EPA). 2010. *Membrane Bioreactors: Wastewater Management Fact Sheet.* EPA, Washington, DC.
2. EPA. 2007. *Biological Nutrient Removal Processes and Costs.* EPA, Washington, DC.
3. Hoffmann, H., Rüd, S., and Schöpe, A. 2009. *Blackwater and Greywater Reuse System.* Sustainable Sanitation Alliance (SuSanA), Lima, Peru.
4. Hutton G., Haller, L., and Bartram, J. 2007. *Economic and Health Effects of Increasing Coverage of Low Cost Household Drinking-Water Supply and Sanitation Interventions to Countries Off-Track to Meet MDG Target 10.* World Health Organization, Geneva, Switzerland.
5. Katyal, T. and Satake, M. 2001. *Environmental Pollution.* Anmol Publications Pvt Ltd., New Dehli.
6. Lennartsson, M. and Ridderstolpe, P. 2011. *Guidelines for Using Urine and Blackwater Diversion Systems in Single-Family Homes.* Coalition Clean Baltic, Uppsala, Sweden.
7. Makoni, F.S., Ndamba, J., Mbati, P.A., and Manase, G. 2004. Impact of waste disposal on health of a poor urban community in Zimbabwe. *East African Medical Journal,* 81, 422–426.
8. Mapira, J. 2011. Sewage treatment, disposal and management problems and the quest for a cleaner environment in Masvingo City (Zimbabwe). *Journal of Sustainable Development in Africa,* 13(4), 353–363.
9. Nhapi, I., Siebel, M.A., and Gijzen, H.J. 2006. A proposal for managing wastewater in Harare, Zimbabwe. *Water and Environment Journal,* 20, 101–108.
10. Nyakutsikwa, H. 2010. *Assessment of Mhangura Water Supply and Sewage Systems.* UNICEF, Harare.
11. UNEP. 2002. *International Source Book on Environmentally Sound Technologies for Wastewater and Stormwater Management.* UNEP, UK.
12. Van Voorthuizen, E., Zwijnenburg, A., van der Meer, W., and Temmink, H. 2008. Biological black water treatment combined with membrane separation. *Water Research,* 42, 4334–4340.

34

Greywater

Lina Abu-Ghunmi
The University of Jordan

Saeid Eslamian
Isfahan University of Technology

PREFACE

The increasing global demand on water, along with the mounting amounts of polluted water, have revolutionized the conventional concept of domestic wastewater management. Consequently, a new direction in thinking has emerged that initiated the concepts of Ecological Sanitation (EcoSan) and Decentralized Sanitation and Reuse (DeSaR). The latter concept is the one behind the idea of separating domestic wastewaters into blackwater and greywater, which is in line with the sustainable management of wastewater. In repose to this, a plethora of publications and reports has been issued to cover different aspects of greywater management. As it is the case of every science, moving from theory to practice requires access to relevant information, which has been the motivation to develop a chapter on greywater as a part of this handbook. This chapter on greywater is an initiative to be used as a useful tool for academics, students, and practitioners and as a guide for decision-making processess. It introduces the concept of greywater and its importance, and further discusses characteristics of greywater and the many other issues that are important to running a sustainable greywater industry. Furthermore, the chapter reports on some of the technologies used for treating greywater, guidelines for designing such technologies, and results of their performances. Then it demonstrates the potential uses of (reclaimed) greywater guided by international standards and guidelines, which were developed for such purpose.

34.1 Introduction

Greywater is the water generated by domestic washing and cleaning activities, which include all household water-related activities except the water that is used in the toilet [13,18,45]. Greywater consists of water streams that are discharged in the appliances of shower, bath, washbasin, dishwasher, and washing machines, in addition to the water that is discharged in the sewer system of household cleaning activities.

Sources of greywater include any location that generates these kinds of waters, and they are residential places such as houses and apartments; commercial and recreational facilities that include offices, resorts, cafes, hotels, restaurants, airports, gyms, malls, and theaters; and institutional facilities that include

hospitals, mosques, prisons, and schools; in addition to industrial facilities that generate greywater as well [4,6,15,26,27,30,44].

Greywater is composed of water blended with other ingredients that affect water's physical properties and its chemical and biological characteristics. The ingredients are organic, inorganic, and biological constituents that are measured using the conventional domestic wastewater parameters, which are chemical oxygen demand (COD), biological oxygen demand (BOD), total suspended solids (TSS), phosphorous (P), nitrogen (N), total coliform (TC), and *Escherichia coli* (E.C), among others. The water component of greywater forms 50%–85% of the water of domestic wastewater [6,18,32] while the other ingredients of greywater form 40% of the COD, 10% of the nitrogen, and 10%–50% of the phosphorus that are contained in domestic wastewater [2,7,18]. The rest of the domestic wastewater components make blackwater [18,57]. In literature, greywater is written as grey water, greywater, gray water, or graywater where all are used interchangeably.

34.2 Characteristics of Greywater

Greywater is characterized in terms of quantity [6,10,21,35] and quality [6,13,18,21]. According to Metcalf and Eddy's [40,41] classification of domestic wastewaters, greywater quantity is defined in terms of its flow and flow pattern, while its quality is defined in terms of its physical properties and its chemical and biological characteristics.

34.2.1 Quantity Characteristics of Greywater

For a better understanding of the characteristics of greywater quantity, two measures (subdaily flow and daily flow) should be comprehended. Greywater subdaily flow is the average daily-flow per activity, which is the average daily volume of greywater that is generated per capita for a particular domestic activity such as shower, laundry, hand-washing, etc. (Table 34.1). Daily-flow is the average daily flow per capita from all domestic activities, which is the summation of the average daily volumes of greywaters that are generated per capita from all domestic activities; that is, summation of subdaily flows in Table 34.1. Subdaily flows (Table 34.1) could also be estimated using Equation 34.1 and data in Table 34.2, while daily flows could be estimated by Equation 34.2 [6,37].

Subdaily flow [L cap^{-1} d^{-1}] = Activity Daily flow (ADV)

= Activity Frequency per day per cap (F)

* Volume per frequency (V)   (34.1)

where subdaily flow or activity daily flow (L cap^{-1} d^{-1}) could be greywater flow from shower, bath, washbasin, or kitchen sink; F: mean daily frequency; V: mean water volume per use (m^3) (Table 34.2).

TABLE 34.2 Frequency and Volume of Greywater Subdaily Flow Generated per Household Activity per Person

Activities	Frequency of Activity (No. of Activity/cap/d)	Volume per Activity (L/Activity)	Reference
Toilet	2.29–4	15–19	[37,55]
Laundry	0.25–0.31	127–151	
Bath or shower	0.47–0.5	81–95	
Dishwashing	0.39–0.5	27–47	

$$\text{Daily flow (DV)[L cap}^{-1}\text{d}^{-1}] = \sum_{I=1}^{J} \text{ADV}_i \qquad (34.2)$$

where, I = 1, 2, 3 … J: 1 = shower, 2 = bath, 3 = washbasin, and 4 = kitchen sink.

In order to design greywater collection, storage, and treatment units for a particular location, such as a single house, a multi-story building, etc., greywater subdaily and daily flows' averages are multiplied by the number of residences in that location to arrive at the total volume of these flows.

Greywater subdaily flow and daily flow patterns are intermittent in nature [6,10]. These flows do vary among activities for the same location, [6], and among different locations [33,43]. The variations occur on an hourly, daily, weekly, and monthly basis, which is due to the impact of many factors including, among others, climate, geographical location, culture, habits, and economical and social situations [7,8]. Furthermore, hourly, daily, weekly, or monthly peak factors, which is the largest flow volume relative to the average volume, depends on the occurrence of a specific activity. For instance, hourly peak factor may occur due to showering activity; daily peak factor may occur due to laundry activity; and monthly peak factor could be due to it being summertime. For weekly peak factor, more than one activity is required. Peak factor could take the value of one to eight of the average daily flow. Therefore, flow and the peak factors are very important in designing of collection, storage [6,35], and treatment facilities [6].

TABLE 34.1 Greywater Subdaily Flow Generated per Household for a Number of Washing Activities in Different Countries

Location	Units	Kitchen Sink + Dishwasher	Washbasin	Bathroom Shower	Laundry	Reference
US	l/cap · day	27.8		41.7	30.5	[51,52]
Australia/Sydney	l/cap · day	29		41	30	[39]
Israel	l/cap · day	30	15	40	13	[22]
Jordan/Amman	l/cap · day	NA	28 (5)	21 (16)	10 (4)	[32]
India/cities[a]	l/cap · day	4		15	10	[1] cited by [23]
Oman	l/cap · day	34	7	51	8	[31]
Jordan/dormitory	l/cap · day		21 (5)	45 (6)		[6]
Yemen/Sana'a		37	32	18	13	[9]

[a] The study was conducted in Delhi, Mumbai, Ahmadabad, Madurai, Hyderabad, and Kolkata cited by Golda, A., Gopalsamy, P., and Muthu, N. 2014. *Applied Water Science*, (4), 39–49.

34.2.2 Quality Characteristics of Greywater

Greywaters are generated from shower, bath, washbasin, washing machine, kitchen sink, and dishwasher. Therefore, greywater components, other than water, are the residuals of vegetables, fruits, meat, food, detergents, disinfectants, clothes, fibers, glasses, plastics, sand, human cells, and saliva. It may also contain urine, feces, total coliform, *Escherichia coli*, and others [18,20,37]. These components, as domestic wastewater components [40,41], can be categorized into three major groups based on chemical characteristics, where this group consists of organics, inorganics, and gasses; physical properties, where this group consists of solids, which in turn are classified into settable, total, suspended, and dissolved. Solids are also classified into fixed and volatile. The third category is based on biological constituents, which consists of microorganisms such as bacteria, worms, and viruses. These three groups are very interrelated and their constituents are measured as a lump sum in terms of COD for organic and oxidized inorganic matter and BOD for biodegradable organic matter and solids fractions tests. Other measuring parameters include nitrogen, phosphorous, pH, alkalinity, heavy metals, dissolved oxygen, hydrogen sulfide, methane, turbidity, color, odor, and temperature.

In line with quantity characteristics, greywater quality parameters do vary for the same location, and among different locations (Tables 34.2). Therefore, based on Metcalf and Eddy's [40] classification of domestic wastewaters, greywaters could be classified into strong, medium, and weak. The variations in greywater quality are due to the same causes of the variation in its quantity characteristics.

Greywater loads that are reported per activity in Table 34.3 can be used to estimate greywater loads that result from combining two or more of the domestic activities using Equation 34.3.

$$\text{Ingredients load [IL] (mg L}^{-1}) = \frac{\sum_{I=1}^{J} V_I C_I}{\sum_{I=1}^{J} V_I}$$

where, IL is ingredients load of greywater from all streams; loads can be BOD, nitrogen, phosphorous, etc.; I denotes greywater stream, where 1 = shower, 2 = bath, 3 = washbasin, and 4 = kitchen sink; V_I is average daily volume of activity per capita; C_I is average daily concentration per activity.

34.2.3 Biodegradability of Greywater

A major fraction of greywater ingredients is in dissolved form, which is measured using physical parameters such as dissolved solids (TDS or DS) and using chemical parameters such as dissolved and colloidal CODs; namely COD_{col} and COD_{dis} (Table 34.4). This dissolved fraction is aerobically and anaerobically biodegradable. In fact, its biodegradability is a major contributor to the total biodegradability of greywater ingredients. Biodegradability is defined in Table 34.4 as the fraction of COD that is removable, in addition to BOD_5 and BOD_u. It is found that the fraction of COD that is anaerobically biodegradable is less than the fraction that is aerobically biodegradable (Table 34.4). Moreover, the biodegradation rate under aerobic conditions is faster than that under anaerobic conditions [4].

34.3 Management of Greywater

Greywater management could be defined as the process that brings together the technical, environmental, social, institutional, and other aspects that are interrelated and required to improve the performance of the greywater industry. Technical aspects cover issues such as separation, collection, transportation,

TABLE 34.3 Quality Characteristics of Greywater Generated from Different Locations in Different Countries

Parameters	Unit	Australia	US	Kuwait (Kuwait)	Sana'a (Yemen)	Sweden	Qebia (Palestine)	Karak[a] (Jordan)	Dormitory/ Jordan[b]	Amman[a]/ Jordan	Amman & Salt/Jordan Kitchen	Amman & Salt/Jordan Shower	Amman & Salt/Jordan Laundry
pH		7.3	6.8	7.5	6	–	6.60–6.86	5.93–7.82	7.95 (0.05)	7.81	6.83 (0.65)	7.15 (0.87)	9.6 (0.72)
TSS	mg/L	–	–	204	510.8	–	36–396	23–358	122 (78)	168	4101 (419)	750 (94)	3940 (2885)
Turbidity	NTU	112	–	120	618.6	–	–	–	122 (78)	49	NA	NA	NA
NO_3^-	mg/L	–	–	5	98.12	–	0–1.3	–					
$NH_4.N$	mg/L	–	–	4.8	11.28	–	25–45	–	8(6)		NA	NA	NA
TN	mg/L								10 (14)	9.2	25.9 (33.8)	2.4 (1.45)	2.8 (0.45)
PO_4^-	mg/L		7.8		16.1	–	–	–					
TP	mg/L								7 (7)	9	4.3 (39)	1.2 (1.1)	9 (0.7)
BOD_5	mg/L	159	164	40	518	196	941–997	110–1240	149 (46)	314	1850 (890)	120 (40)	1266 (232)
COD	mg/L	–	366	–	2000	–	1391–2405	92–3330	551 (202)	870	8071 (3535)	537 (165)	2500 (1865)
F.C	MPN/100 mL	–	8.8×10^5– 13×10^6		19×10^6	–	10^4–37×10^4	–					
References				Adapted from [9]					[6]	[32]		[6]	

[a] Karak and Amman are, respectively, the rural and the capital cities of Jordan.

[b] Standard deviation.

TABLE 34.4 Greywater COD Fractions and Their Aerobic and Anaerobic Biodegradations in Different Countries

		Dormitory, University of Jordan		Sneek, Groningen, Netherlands		Flintenbreite' Settlement, Luebeck, Germany	
COD_{tot}	mg/L	362 (100)		827 (204)		649 (124)	
COD_{ss}	%	28(9)		49		50	
COD_{col}	%	32 (8)		24		30	
COD_{dis}	%	40 (7)		27		20	
Biodegradability		Aerobic	Anaerobic	Aerobic	Anaerobic	Aerobic	Anaerobic
$COD_{28d\text{-}20°C}$	%	62 (6)	44(13)				74
$COD_{60d\text{-}20°C}$	%	86 (4)	70 (6)				70
COD_{ss}	%	78 (1)	60 (6)				84
COD_{col}	%	90 (3)	77 (6)				70
COD_{dis}	%	87 (2)	71 (1)				
$BOD_{5\text{-}20°C}/COD$	%	43 (4)					
$BOD_{U\text{-}20°C}/COD_{tot}$	%	75 (4)					
$BOD_{98D\text{-}35°C}/COD_{tot}$	%		57 (5)				
Sources		[3]		[26]		[17]	

storage, and treatment of greywater. Greywater uses are linked to the environmental and social aspects, which are partially reflected in standards and guidelines that administer the greywater industry. Implementation of these standards and guidelines are enforced through regulations and laws enacted and overseen by the related institutions. Greywater management could be applied to a single premise, cluster of premises, a small community, or an entire country. Premises include houses, apartments, commercial facilities, government buildings, universities, hospitals, schools, mosques, or any other premise that generates greywater. One of the major issues of concern in greywater management is whether there is a sewer system.

34.3.1 Separation, Collection, and Transportation of Greywater

Greywater is separated onsite to be collected manually or via household piping system. The manual collection means emptying the appliances such as kitchen sink, washbasin, bath, dishwasher, and washing machine into buckets. The household piping system is composed of two main parts; each is constructed based on a building code. One part is for greywater and the other is for blackwater, which is mainly toilet water. The collected greywater is then transported manually or via sewer system to be handled onsite or offsite.

Onsite handling of greywater is an option that can be used for a single premise or a cluster of premises, especially in unsewered areas. Offsite handling of greywater is an alternative to a cluster of premises, a small community that is not more than 1000 persons, or a large community. Greywater could be transported offsite using low-cost specialized sewer systems, such as that reported by Casey et al. [11] and Metcalf and Eddy [41]

for a small community. However, pretreatment of greywater is a prerequisite to prevent the clogging and damaging of these specialized sewer system's equipment [11,40,41]. A conventional pretreatment to remove settable and floatable solids from water that is transported by specialized sewer systems is a septic tank, which could be accompanied with a grinder [11,41].

The specialized sewer systems choices are (1) a small-diameter variable-slope gravity-flow sewer (SDGS), (2) a pressure sewer, and (3) a vacuum sewer. These three systems are promising options considering the intermittent nature and low volume of greywater flow in comparison with domestic wastewater flow. Taking into account the proper selection and design, the specialized sewer systems, especially the SDGS type, in comparison with a conventional sewer system is less expensive. It is important to note that geographical location, topography, and cost play crucial roles in the selection of the sewer system. For the design of pressure and vacuum sewer systems, consult Casey et al. [11] and Metcalf and Eddy [40,41] for the details.

The SDGS (Figure 34.1) is commonly used in Australia to transport greywater that is collected in septic tanks, where greywater is transported from its location, for example, a single house, through 2.5 to 3.75-cm lateral pipes, to be collected in septic tanks (Figure 34.1). Then the effluent of the septic tanks is transported through a 7.5 to 10-cm diameter plastic pipe that is buried at the same shallow depth below the frost line, regardless of variable grade [11,41]. There are two major design requirements for the SDGS, which are (1) the net slope from the inlet to outlet should be positive (Figure 34.1) and (2) the outlet point should be less than any of the houses' connection to the system are [41]. The lateral pipes are cleaned through cleanout or manholes [11] and there should be an annual inspection for both septic tanks and the SDGS in order to remove accumulated solids [11].

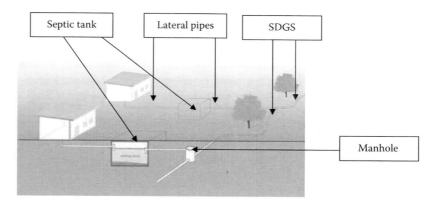

FIGURE 34.1 Photo for small-diameter variable-slope gravity-flow sewer system. (Adapted from http://www.grassrootswiki.org/images/thumb/4/47/Solids-Free_Sewer.png/400px-Solids-Free_Sewer.png.)

34.3.2 Storage of Greywater

The intermittent nature of raw greywater requires storage [35,50] to meet the demand for treatment and use in terms of timing and volume [6]. The storage options could be (1) a collection tank that collects, stores, and pretreats raw greywater and, therefore, regulates its flow to the treatment system; (2) a treatment unit that treats and stores raw greywater in one unit; or (3) a storage tank that collects and stores reclaimed greywater to meet the demand for use. The volume of greywater storage unit is determined based on whether greywater is raw or reclaimed, daily-flow volume and flow pattern of greywater, type of treatment, and its mode of operation; for example, batch, semi-batch, or continuous, type of uses, and demand for use; that is, time and volume.

Guidelines for designing a storage tank for raw or reclaimed greywater:

Step 1. Calculate the storage volume as the maximum accumulated daily inflow volume minus the amount that is released; that is, for use or for treatment.

Step 2. Calculate the minimum storage volume as the total volume that should be released at one time, such as the volume required for toilet cistern, crop irrigation, etc. [6], or the volume required as inflow to the treatment unit.

Step 3. Compare the two volumes from the above two steps and select the largest to be the volume of the storage tank.

Step 4. Make sure that the greywater residence time in the storage tank does not exceed 24 h for both raw [16] and reclaimed greywater [36]. The limited residence time is recommended in order to avoid quality deterioration of stored water; such as release of odors and mosquito breeding.

Step 5. Ensure that greywater flows to the storage tank by gravity.

34.3.3 Treatment of Greywater

Selection of greywater treatment system depends on the location; whether onsite or offsite, quality characteristics of raw greywater, cost, space, site geography, climate, and the intended use of reclaimed greywater, which is controlled by standards, regulations, and guidelines. The levels of greywater treatment are primary, secondary, and tertiary.

34.3.3.1 Primary Treatment

The primary treatment aims at removing settable and floatable materials to prevent clogging of pipes of the transporting and distributing system. Primary treatment applies physical operations that separate undesirable materials based on particle size and densities. These physical unit operations are low-cost and affordable and can be locally constructed, operated, and maintained. Primary units include mainly septic tanks, modified septic tanks, and filters.

34.3.3.1.1 Septic Tanks

A septic tank is a compact unit applicable to on-site and offsite treatment of greywater (see, e.g., Abu-Ghunmi et al. [4], who reviewed the works of Imura et al. [29], Shrestha et al. [47], and Li et al. [38] on using settling tanks in treatment of greywaters). Septic tanks serve a single and a cluster of premises in addition to a small community. Furthermore, septic tanks serve multiple functions: as a flotation tank, settler, anaerobic digester, and storage tank, all in one unit. The basic mechanism of separation is gravity settling and low-density flotation, where the latter is supported by the produced gases of anaerobic digestion of the settled solids. The function of the upper-part of the septic tank is to collect floatable materials, such as grease and oil, while the lower-part functions as a settler and an anaerobic digester that receives and digests settable solids in addition to fractions of suspended and colloidal particles. Septic tanks, depending on the residence time and the proper design, function as storage tanks for both greywater and settled solids. The removed solids can be food residual, rags, sands, etc.

Guidelines for selecting, constructing, maintaining, and designing a septic tank:

Step 1. Select the location of the septic tank that allows gravity flow of greywater from premises to septic tank.

Step 2. Select a material for constructing a watertight septic tank, such as concrete, fiberglass, polyethylene, or any other relevant local material.

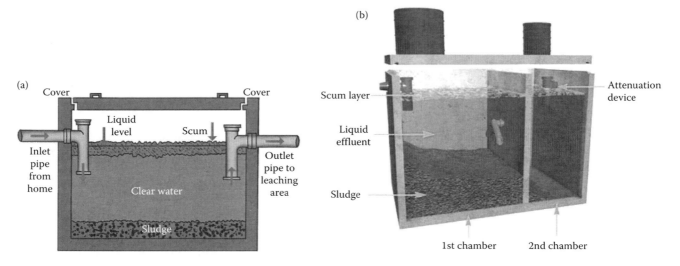

FIGURE 34.2 Two septic tanks: (a) A one-compartment septic tank (adapted from http://allcentexsepticaustin.com/wp-content/uploads/Septic-Tank-Pumping.jpg); (b) a two-compartment septic tank (adapted from http://genesbackhoeservice.com/web_images/tank-diagram.jpg).

Step 3. Select the number of compartments in the septic tank—one, two (Figure 34.2), or three compartments. The role of quantity and quality characteristics of greywater, such as the fraction of settable and suspended solids, flow volume, and flow pattern, are crucial in the selection of the number of septic tank compartments.

Step 4. Select the auxiliaries, if any. For example, in the case of one compartment, a filter vault, baffles, or inclined tubes are required to control the consistency of the quality of tank effluent.

Step 5. Maintain the septic tank by periodic discharge of the accumulated sludge to maintain the full capacity of the tank and its performance. The frequency of sludge discharge depends on the quality of the raw greywater and the residence time in the septic tank, considering that a properly designed septic tank produces anaerobically stable sludge.

Step 6. Apply Metcalf and Eddy's [41] criteria for designing domestic wastewater septic tanks that are included in Table 34.5; other sources can be consulted as well. There are two reasons for recommending using domestic wastewater criteria here; first, greywater is domestic wastewater excluding toilet stream. Second, septic tank functions have been tested for treatment of greywater under different names [5] (such as settling tank [43] [7], and a 1-mm fine screen followed by an equalization tank with 10 h residence time of Friedler et al. [19]). Although these studies reported positive performance for these tested systems, information about the design criteria and operation conditions is not available.

34.3.3.1.2 Modified Septic Tank

Abu-Ghunmi et al. [5] tested a cost-effective locally constructed anaerobic pretreatment unit. The applied unit is a modified septic tank that aims to improve both solids settling and the

TABLE 34.5 Designing Criteria for Septic Tanks

Design Parameter	Unit	Value Range	Typical
Liquid volume			
Minimum	gal	750–1.000	750
1–2 bedroom	gal	750–1.000	750
3 bedroom	gal	1.000–1500	1.200
4 bedroom	gal	1.000–2.000	1.500
5 bedroom	gal	1.200–2.000	1.500
Additional bedroom	gal	150–250	250
Compartments			
Number	No.	1–3	2
Volume distribution in multi-compartment tanks			
Two-compartment Tank	% 1st, 2nd	67, 33	67, 33
Three-compartment Tank	% 1st, 2nd, 3rd	33, 33, 33	33, 33, 33
Length to width	ratio	2:1–4:1	3:1
Depth	ft	1–6	4
Clear space above liquid	in	10–12	10
Depth of water surface below inlet	in	3–4	3
Inspection ports	No.	2–3	2

Source: Adapted from Metcalf and Eddy. 1999. *Wastewater Engineering; Treatment Disposal Reuse*, 3rd Edition, McGraw-Hill, New York.

anaerobic digestion of greywater. The system, as shown in Figure 34.3, consists of a circular tank that is internally constructed to have polyvinyl chloride (PVC) rods stretched from the bottom to the top. The PVC rods hold polyethylene corrugated discs that are spaced by PVC rings. The tank receives gravity flow of greywater on the up-flow mode. Details about the design criteria and operation conditions are reported in Table 34.6 and the treatment performance is reported in Table 34.7. The sludge discharged from this tank is characterized as stable [5].

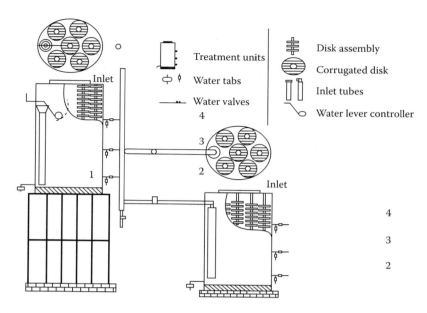

FIGURE 34.3 Sketch of the main items of the modified septic tank. (Adapted from Abu-Ghunmi, L. et al. 2010. *Bioresource Technology*, 101(1), 41–50.)

TABLE 34.6 Designing Criteria for the Modified Septic Tanks

Items	Unit	Values
Tank volume	m³	2.0
Depth	m	1.38
Depth/diameter	m	1.33
PVC rods		
Length	m	1.38
Diameter	cm	2.54
Thickness	cm	0.03
Polyethylene corrugated discs		
Diameter	cm	30
Diameter of center hole	cm	2.57
Thickness		3
PVC rings		
Length	cm	5
Diameter	cm	5
Thickness	cm	0.003
Port depths from bottom; for inspection and sludge discharge	m m m	
Operation		
Operating-mode		Batch in-flow/continues out-flow/variable volume
Flow-mode		Up-flow
Effective volume[a]	m³	1.18
HRT	hours	24
SRT_{min}	days	30
SRT_{max}	days	Infinite

Source: Adapted from Abu-Ghunmi, L., Zeeman, G., Fayyad, M., and van Lier, J.B. 2010. *Bioresource Technology*, 101(1), 41–50.

[a] Water volume.

34.3.3.1.3 *Filters*

Filters are compact units applicable to onsite and offsite treatment of greywater. Treatment of greywater in filters was tested by, for example, Shin et al. [49], Jefferson et al. [35], and Ramon et al. [46], among others, which were reviewed by Abu-Ghunmi et al. [4]. Filters serve a single premise, a cluster of premises, and a small community. The filter separates suspended, colloidal, and dissolved particles from water. The mechanism of separation includes settling, straining, interception, and coagulation–flocculation [42]. Filters can be classified into gravity filters [7,49], modified filters [15,56], and pressurized filters [28,46]. A pressurized filter is presented in Section 34.4 as an advanced treatment. Filters can also be classified, based on the operation mode, into batch, semi-batch, and continuous.

A gravity filter is a compact low-cost treatment technology in which greywater passes through filter media that varies in porosity from fine to course size. For the selection, construction, maintenance and design of a gravity filter, the following guidelines can be followed:

Step 1. Select the filter location that allows greywater to flow by gravity from the premises to the filter.

Step 2. Select filter media that can be any local material such as gravel, sand, plastics, glasses, quartz, or any other relevant material.

Step 3. Select the particle size of the filter media that ranges from a very fine to a course particle. Also, choose the number, order, and depth of the layers in the filter. The qualities of raw and reclaimed greywater are crucial in the selection of the filter media; for example, the more the dissolved particles in greywater, the finer should be the media. The operation mode of the filter is a determinant

TABLE 34.7 Treatment Performance of the Modified Septic Tank

Parameters		Units	tot	ss	col	dis	NH-N	P$_3$O$_4$-P
COD	Influent	mg L^{-1}	366 (165)	93 (66)	133 (58)	139 (64)		
	Removal	%	45 (5)	71 (4)	44 (7)	25 (5)		
BOD	Influent	Mg L^{-1}	150 (31)					
	Removal	%	37 (9)					
BOD/COD	Influent	%	55 (7)					
	Effluent	%	76 (8)					
Solids	Influent	g L^{-1}	1.081 (0.062)	0.169 (0.060)	0.094 (0.023)	0.797 (0.083)		
	Removal	%	24 (7)	81 (9)	42 (24)	7 (6)		
Volatile solids	Influent	g L^{-1}	0.237 (0.133)	0.077 (0.017)	0.044 (0.019)	0.11 (0.031)		
	Removal	%	33 (5)	57 (1)	15 (0.2)	4 (4)	8 (2)	
Nitrogen	Influent	mg L^{-1}	12 (0.6)	1.2 (0.1)	0.6 (0)	9 (0.5)	7 (3)	
	Effluent	mg L^{-1}	10 (5)	1.5 (1)	0.3 (0.2)	8 (4)	8 (2)	
Phosphorous	Influent	mg L^{-1}	11 (8)	1 (1)	3 (1)	7 (1)		6 (2)
	Effluent	mg L^{-1}	10 (4)	1 (2)	2 (2)	6 (3)		6 (1)
Total coliform	Influent	MPN/100 mL	7.1E + 7 (1.27E + 8)					
	Effluent	MPN/100 mL	1.9E + 5 (1.95E + 5)					
Escherichia coli	Influent	MPN/100 mL	1.4E + 6 (1.9E + 6)					
	Effluent	MPN/100 mL	1.0E + 5 (7.0E + 4)					

Source: Adapted from Abu-Ghunmi, L. et al. 2010. *Bioresource Technology*, 101(1), 41–50.

factor in choosing the number, depth, and order of the filter's layers.

Step 4. Choose the operation mode of the filter, which could be continuous, batch, or intermittent.

Step 5. Design a backwashed gravity sand filter for greywater using the design criteria in Table 34.8, which is adapted from Friedler et al. [19].

Step 6. Modify the filter design, based on trial and error, if the quality of raw greywater is different form that reported in Table 34.9. This is in order to keep or to improve the quality of treated water to the required level.

Step 7: Backwash the filter when observing deterioration in the reclaimed greywater quality and/or slowness in the filtration

TABLE 34.8 Designing Criteria of Backwashed Gravity Sand Filter

Item	Identification
Filter shape	Cylindrical
Operating media	Sand and quarts
Supporting media	Gravel
Ratio of operating-media depth to diameter	7
Ratio of operating-media to supporting-media	14
Porosity	36%
Operation	
Filtration rate	8.33 m/h
Operation mode	Intermittent: 11 times a day for 1 min each
Backwash frequency	Every 1.26 m^3 of filtration

Source: Adapted from Friedler, E. and Hadari, M. 2006. *Desalination*, 190(1–3), 221–234.

TABLE 34.9 Quality of Raw and Reclaimed Greywater from the Backwashed Sand Filter

Parameters	Raw Greywater (mg L^{-1})	Effluent (mg L^{-1})
SS	92 (115)	32 (13)
Turbidity	65 (68)	35 (25)
COD$_{tot}$	211 (141)	130 (37)
COD$_{dis}$	108 (47)	87 (28)
BOD	69 (33)	62 (21)
Fecal coliform	3.4E5 (4.2E5)	1.3E5 (1.4E5)

Source: Adapted from Friedler, E. and Hadari, M. 2006. *Desalination*, 190(1–3), 221–234.

rate. The frequency of backwashing depends on the characteristics of raw greywater and the size of the filter. The produced sludge is unstable and, therefore, a proper disposal is required in order to avoid health and aesthetic problems.

A modified filter includes, among other types, constructed wetlands that are planted gravity filters. The selection, design, and operation of the constructed wetlands depend on the location, volume of greywater flow, size of the filter, and the weather conditions. For further information on constructed wetlands and other modified filters that treat greywater, consult, for example, Shrestha et al., Dallas et al., Gross et al. [15,24,47,48], and others in Abu-Ghunmi et al. [4].

34.3.3.2 Secondary Treatment

Secondary treatment of greywater aims at separating the dissolved constituents, especially the biodegradable part, from water by biological processes. These processes are basically targeting the dissolved particles, in addition to other particles, with the purpose of converting solids into gases and sludge, where the latter is settled by gravity. The produced gases include carbon dioxide and, under anaerobic conditions, methane. The biological processes are commonly combined with a secondary sedimentation tank to separate sludge from treated greywater. Biological processes can be anaerobic or aerobic, where both are designed and operated as attached or suspended growth systems.

34.3.3.2.1 Anaerobic Processes

Greywater is treated anaerobically in an upflow anaerobic sludge blanket (UASB) reactor [17,26,27]. The UASB reactor (Figure 34.4) is a compact unit applicable to onsite and offsite treatment of greywater. The UASB serves a single premise and a cluster of premises, in addition to small and large communities. The UASB reactor is a suspended growth system that feeds in an upflow mode and relies on a sludge blanket to anaerobically treat greywater. The designing criteria for the UASB reactors for treating greywater are reported in Table 34.10 and the qualities of raw and reclaimed greywater by UASB are reported in Table 34.11. If the quality of raw greywater or the required quality for reclaimed water differs from that in Table 34.11, modify the operation conditions, by trial and error, to reach the target required effluent quality. Hernandez et al. [26,27] posttreated the the UASB effluent, aerobically, in a sequencing batch reactor technology to reach the standards of uses.

34.3.3.2.2 Aerobic Processes

Greywater treatment in an aerobic compact unit includes sequencing batch reactor (SBR) [49], rotating biological contractor (RBC) [19,22,43], fluidized bed [43,28], and modified activated sludge unit that combines both suspended and attached growth systems in one unit [5]. The aerobic units are

TABLE 34.10 Design Criteria of the UASB Reactor

Item	Identification
UASB	Cylindrical
Ratio of height to depth	28.6
Up-flow velocity	0.33 m h^{-1}
HRT	16 h
SRT	93–491

Source: Adapted from Elmitwalli, T. and Otterpohl, R. 2007. *Water Research*, 41(6), 1379–1387; Hernandez, L. et al. 2008. Comparison of three systems for biological greywater treatment. In *Proceedings of IWA Conference on Sanitation Challenges*. May 19–22, Wageningen, Netherlands.

TABLE 34.11 Influent Quality of Greywater and Performance of the UASB Reactor

Item	Influent Quality (mg L^{-1})	Removal (%)
COD$_{tot}$	618 (130)	64 (5)
COD$_{ss}$	308 (162)	83.5 (5.4)
COD$_{col}$	177 (114)	51.7 (19)
COD$_{dis}$	133 (63)	50.9 (8.9)
TN	27.1 (3.5)	29.8 (4.8)
Total phosphorus	9.9 (0.3)	15.2 (3.6)

Source: Adapted from Elmitwalli, T. and Otterpohl, R. 2007. *Water Research*, 41(6), 1379–1387; Hernandez, L. et al. 2007. *Water Science and Technology*, 56(5), 193–200.

applicable to onsite and offsite greywater treatment for a single premise and a cluster of premises as well as for small and large communities.

34.3.3.2.2.1 Sequencing Batch Reactor
A sequencing batch reactor (SBR) is a suspended growth system that combines all stages of biological treatment in one single unit (Figure 34.5). The SBR operates on timing-cycle phases, where treatment stages occur sequentially (Figure 34.5). The SBR operation cycle phases are filling and mixing, reacting, settling, decanting, and idle.

For the selection, construction, maintenance, and design of SBR, the following guidelines can be followed:

Step 1. Equalize greywater inflow to SBR in order to keep the performance of the system [49]. Therefore, SBR is recommended as a posttreatment of the effluents of physical unit operations and anaerobic process.

Step 2. Consult Shin et al. [49] for the designing criteria and operation conditions presented in Table 34.12 in order to design and operate the SBR for treating greywater.

Step 3. Posttreat SBR effluent for suspended solids (SS). Therefore, settling tank or filters are recommended as a posttreatment of the SBR effluent.

Step 4. Modify the operation conditions, by trial and error, to reach the required effluent quality, in case the quality of raw or reclaimed greywater differs from that in Table 34.13.

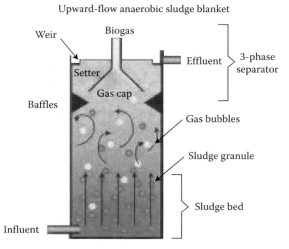

Upward-flow anaerobic sludge blanket

Weir
Biogas
Effluent
3-phase separator
Setter
Gas cap
Baffles
Gas bubbles
Sludge granule
Sludge bed
Influent

FIGURE 34.4 Sketch of a UASB reactor. (Adapted from https://microbewiki.kenyon.edu/images/thumb/5/59/Conceptual_diagram_of_UASB.png/400px-Conceptual_diagram_of_UASB.png.)

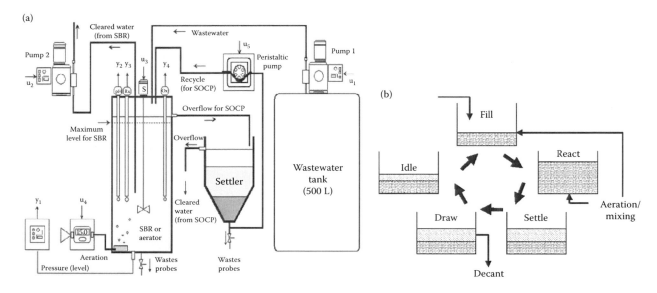

FIGURE 34.5 (a) Sketch of SBR technology components. (Adapted from http://www.ua.ia.polsl.pl/plants/pics/sbr1.jpg.) (b) Sketch of the time-cycle of SBR phases. (Adapted from http://api.ning.com/files/J51L4LBDbSQWHVlt4122jQaMixqlTjGfW5ULDI4POtQpppwOb3S7SfP-V1o4Lk7ncmovDxxh6ro*8dUDLl-EoBKq7UiyJvac/sbatchreactor.png.)

TABLE 34.12   Design Criteria of SBR Reactor

Item	Identification
SBR	
Sludge to liquid volume	2/3
MLSS	3579 mg L^{-1}
SVI	160 mL g^{-1}
Timing-cycle phases	12 h
Filling with mixing	1 h
Reaction	
Aeration	4 h
Anoxic	5 h
Settling	1 h
Decanting	0.5 h
Idle	0.5 h
Equalization basin (pretreatment)	
Volume	2.5 m^3

Source: Adapted from Shin, H. et al. 1998. *Water Science and Technology*, 38(6), 79–88.

TABLE 34.13   Greywater Quality of SBR's Influent and Effluent

Item	Units	Influent Quality	Effluent Quality
COD$_{dis}$	mg L^{-1}	30–194	20
SS	mg L^{-1}	185	20
NO$_3$-N	mg L^{-1}	6–23	<1

Source: Adapted from Shin, H. et al. 1998. *Water Science and Technology*, 38(6), 79–88.

34.3.3.2.2.2 Rotating Biological Contactors   RBC is a compact attached biological growth system applicable to onsite and offsite treatment of greywater. RBC treats greywater for a single premise and a cluster of premises, also for small and large communities. The RBC unit consists of a container that carries a horizontal, slowly rotating shaft, which holds up contactors in the shape of corrugated discs.

For the selection, construction, maintenance, and design of RBC, the following guidelines can be followed:

Step 1. Equalize greywater inflow to RBC in order to keep the performance of the system. Therefore, RBC is recommended as a posttreatment for physical unit operations' effluent. For example, Friedler et al. [19,22] recommend pretreating greywater inflow to RBC in an equalization basin.

Step 2. Consult Friedler et al. [19,22] for designing criteria and operation conditions presented in Table 34.14 in order to design and operate the RBC for treating greywater.

Step 3. Posttreat RBC effluent for SS. For example, Friedler et al. [19,22] recommend posttreating RBC effluent in secondary settling tanks and filters.

Step 4. Modify the operation conditions, by trial and error, to reach the target required effluent quality, in case the quality of raw or reclaimed greywater differs from that in Table 34.15.

34.3.3.3 Tertiary Treatment of Greywater

Tertiary treatment is an advanced treatment that aims at treating greywater to a high quality level for inside uses such as toilet flushing and laundry. Advanced treatment includes chemical and physical operations with particular attention to disinfection of greywater and separation of dissolved particles including removing mirco- and ultra-size particles.

34.3.3.3.1 Chemical Units

Chemical units make use of chemical properties and chemical reactions for treating greywater. Thus, chemical disinfectants

TABLE 34.14 Design Criteria of RBC Reactor

Item	Identification
RBC	
Stages	2
Volume/stage	15 L
Discs surface area/stage	0.5 m²
Discs submerged surface area	40%
Shaft rotation speed	1–1.5 rpm
HRT/stage	2 h
Settling tank (posttreatment)	
Volume	7.5 L
HRT	1 h
Equalization basin (pretreatment)	
Volume	330 L
HRT	1–10 h

Source: Adapted from Friedler, E. and Hadari, M. 2006. *Desalination*, 190(1–3), 221–234; Friedler, E., Kovalio, R., and Galil, N. 2005. *Water Science and Technology*, 51(10), 187–194.

TABLE 34.15 Greywater Quality of RBC's Influent and Effluent

Item	Units	Influent Quality	Effluent Quality
BOD_5	mg L⁻¹	59 (29.6)	6.6 (9.45)
COD_{tot}	mg L⁻¹	158 (60)	46 (19.4)
COD_{dis}	mg L⁻¹	110 (54)	47 (27)
SS	mg L⁻¹	43 (25.1)	16 (14.5)
Turbidity	NTU	33 (23.3)	1.9 (2.3)

Source: Adapted from Friedler, E. and Hadari, M. 2006. *Desalination*, 190(1–3), 221–234; Friedler, E., Kovalio, R., and Galil, N. 2005. *Water Science and Technology*, 51(10), 187–194.

that include chlorine and bromide in addition to titanium, which is combined with UV rays, are used in the disinfection process of greywater (Table 34.16). The selection of relevant disinfectant depends on cost, quality of reclaimed greywater, construction material of the storage tank, and the piping system. Furthermore, disinfectants' dosages and timing determine the efficiency of the disinfection process.

34.3.3.3.2 *Advanced Physical Operations*

Pressurized filter is an advanced physical unit operation that applies pressure to filter greywater through micro- or ultra-membranes made in the form of fibers or sheets. The high operational cost is the main obstacle of these technologies, especially for use for a single premise. To treat greywater in a pressurized filter, consult Shin et al. [49], Hills et al. [28], and Ramon et al. [46]; that is, for membrane type, pore sizes, and the quality of reclaimed greywater, which are reported in Table 34.17.

34.3.4 Greywater Use Options

Greywater onsite and offsite uses are many [33]. Selection of greywater use option depends on the needs, standards and regulations of use, characteristics of raw greywater, and applied treatment technology. The onsite uses are indoor and outdoor. For example, indoor uses include toilet flushing, laundry washing, and boiler feeding, while outdoor uses include irrigating gardens and olive and orchard trees, and car washing [28,34,43]. The offsite uses include groundwater recharge, irrigation of golf courses, cemeteries, and ornamental trees, in car washing stations, and cement mixing for building construction [7,12,25,34].

The standards for using greywater are either adopted from the standards of reclaimed domestic wastewater uses or developed especially for greywater uses. The standards still need to be revised and developed further in order to cover the many potential use options considering all factors of concern related to health, aesthetic, and environment. In 2006, the WHO [53,54] issued standards of greywater use that are accompanied with guidelines for safe use (Table 34.18). In 2008, the Jordanian Institution for Standards and Metrology issued standards and guidelines number (1776) [36] for using reclaimed greywater in rural areas. These standards and guideline were modified in 2013 (Table 34.18).

The standards are composed of quality parameters related to health and environment issues arising from using greywater.

TABLE 34.16 Summary of Disinfectants Applied to Greywater and the Corresponding Effluent Quality

			Qualities[a]				
			TC		FC		
Disinfection	Does	Comments	In	Out	In	Out	Reference
Chlorine	Solid or liquid	Slow released solid block or liquid solution					[35]
	Solid						[29,13]
Hypochlorite	0.2%–0.25%	For 1 mg/L residual after 1.0 min			5.1E4 ± 6.6E2	1.1–3.1E1	[21,19]
Bromide	Solid and liquid	Slow released solid block or liquid solution					[35]
UV	250 mj/m²						[14]
	150–400 j/m²		2.0E2–1.0E6	(2.0–4.0)E2	2.0E2–1.0E6	1.0E0–2.0E2	[43]

Source: Adapted from Abu-Ghunmi, L. et al. 2011. *Critical Reviews in Environmental Science and Technology*, 41(7), 657–698.
[a] TC: Total coliform; FC: Fecal coliform.

TABLE 34.17 Summary of Membrane Filters Applied to Greywater Treatment and the Corresponding Effluent Quality

Filter Type	Type	Pore size (μm)	Volume (m³)	Applied Pressure	Comment	SS	BOD	Turbidity (NTU)	Reference
						(mg/L)			
Micro-membrane filter	Hollow-fiber polypropylene	0.2	20		Backwashing with adding chemicals	<1.0			[49]
Ultra-membrane filter	Polyvinylslidenofloride	Reject down to 200 kDa			Tube membrane with area 0.22 m²		8.3		[28]
	Modified polyethersulphone	Reject down to 6 kDa					4.7		
	Tight polyethersulphone	Reject down to 4 kDa					5.8		
	Polyamine film	75% Cl₂ rejection					3.1		
	Polyamine film	80% Cl₂ rejection					2.4		
	Cellulose acetate	90% Cl₂ rejection					4.7		
	Polyacrilonitriel	Reject down to 400 kDa		1–2 bar	Member sheet			1.4 (0.4)	[46]
		Reject down to 200 kDa						1.0 (0.5)	
		Reject down to 30 kDa						0.8 (0.2)	
		Reject down to 200 Da		6–10 bar	Tubular nono-fitler, cross flow filtration 150 L/hr			29.5 (0.6)	

Source: Adapted from Abu-Ghunmi, L. et al. 2011. *Critical Reviews in Environmental Science and Technology*, 41(7), 657–698.

TABLE 34.18 Standards from Different Sources for Greywater Uses

No	Parameters	Units	Toilet Flushing	Irrigation of Edible Crops	Irrigation for Nonedible Crops, Landscaping, and Other Crops	Guidelines for Treated Greywater Quality for Cooling Tower Make Up	Restricted Irrigation	Unrestricted Irrigation
			Acceptable Limit—Jordanian Standards [36]				WHO for Use of Greywater [53]	
	DO	mg/L						
1	Odor	in hazen units				Nonoffensive		
2	Color	in hazen units				<15		
3	pH	Unit	6–9	6–9	6–9	6.0–9.0		
4	Turbidity	NTU	5≥	NA	NA	<2		
5	BOD₅	mg/L	10≥	60	60	<5		
6	COD	mg/L	20≥	120	120			
7	TSS	mg/L	10≥	100	100			
8	TDS	mg/L						
9	T-N	mg/L	50	50	50			
10	NO₃	mg/L	70	70	70			
11	T-P	mg/L						
12	Total coliform	CFU/100 mL				<10	≤1E5	≤1E3
13	*Escherichia coli*	MPN/100 mL	10≥	10³	10⁴	Nondetectable/100 mL		
14	Total *Legionella* count	CFU/mL				Nondetectable or below 1000 CFU/liter		
15	Standard plate count/ heterotrophic plate count	CFU/mL				Maximum 500		
16	Total residual chlorine	mg/L				0.5–2.0		
17	Intestinal helminths eggs	Egg/L	1≥	1≥	1≥			
18	Phenol	mg/L						
19	MBSA	mg/L						

The guidelines are concerned with the safety that is related to the technical, environmental, and social aspects, such as safe separation of greywater, collection, treatment, and use. For example, the guidelines recommend that the piping system used to transport and distribute greywater should be marked to distinguish it from the piping system of portable water. Furthermore using greywater onsite should be restricted for irrigation, toilet flushing, and tower water make up, but not for potable use [36,53]. In addition, the guidelines recommend that awareness should be increased about the purpose of separation of greywater as awareness contributes to the proper management of greywater. Such awareness would result in, for example, the prevention of the use of harmful materials that adversely affect health and environment, and the implementation of safety rules when using reclaimed water.

Furthermore, the guidelines recommend that reclaimed greywater for irrigation is to be used in drip irrigation systems and to cover the soil and the irrigation system with mulch. Stop irrigation to dry the soil before picking up the fruits and the fallen fruits should be disposed of. Furthermore, if there is not enough rainfall, it is recommended to frequently irrigate soil with freshwater for the purpose of soil washing.

34.4 Summary and Conclusions

Greywater is domestic wastewater excluding toilet wastewater. This greywater is characterized by variation in its quantity and quality characteristics, where its ingredients are biodegradable to different extents under aerobic and anaerobic conditions.

Greywater quantity and quality characteristics are important in the selection of the appropriate type of treatment, which can be classified into primary, secondary, and tertiary. Primary treatment is basically targeting the settable and suspended solids, while secondary treatment is mainly biological processes to remove the biodegradable ingredients. Finally, tertiary treatment is an advanced type of treatment that utilizes chemicals or physical operations to remove the ingredients that could not be tackled by primary and secondary treatments.

For each type of treatment, a number of technologies are presented and discussed. Septic tank, modified septic tank, and filters are discussed under primary treatment. UASB, SBR, and RBC are reported as secondary treatment options, while chemical disinfections and ultra- and microfiltration technologies are considered as tertiary treatment options. This is followed by looking at the use options of reclaimed greywater, which is governed by standards and guidelines developed by the relevant authorities.

References

1. Abdul, S. and Sharma, R.N. 2007. Water consumption pattern in domestic households in major Indian cities. *Economic and Political Weekly*, XLII (23), 2190–2197.
2. Abu-Ghunmi, L. 2009. Characterization and treatment of grey water; options and (re)use. PhD thesis, Wageningen University, the Netherlands.
3. Abu-Ghunmi, L., Zeeman, G., Fayyad, M., and van Lier, J.B. 2011. Grey water biodegradability. *Biodegradation*, 22(1), 163–174.
4. Abu-Ghunmi, L., Zeeman, G., Fayyad, M., and van Lier, J.B. 2011. Grey water treatment systems: A review. *Critical Reviews in Environmental Science and Technology*, 41(7), 657–698.
5. Abu-Ghunmi, L., Zeeman, G., Fayyad, M., and van Lier, J.B. 2010. Grey water treatment in a series anaerobic–Aerobic system for irrigation. *Bioresource Technology*, 101(1), 41–50.
6. Abu-Ghunmi, L., Zeeman, G., van Lier, J., and Fayyad, M. 2008. Quantive and qualitative characteristics of gray water for reuse requirements and treatment alternatives: The case of Jordan. *Water Science and Technology*, 58(7), 1385–1396.
7. Al-Jayyousi, O. 2003. Grey water reuse: Towards sustainable water management. *Desalination*, 156(1–3), 181–192.
8. Al-Jayyousi, O. 2002. Focused environmental assessment of grey water reuse in Jordan. *Environmental Engineering Policy*, 3(1–2), 67–73.
9. Al-Mughalles, M., Rahman, R., Suja, F., Mahmud, M., and Jalil, N. 2012. Household greywater quantity and quality in Sana'a, Yemen. *Electronic Journal of Geotechnical Engineering*, (17), 1025–1034.
10. Butler, D., Friedler, E., and Gatt, K. 1995. Characterizing the quantity and quality of demostic wastewater inflow. *Water Science and Technology*, 31(7), 13–24.
11. Casey, P., Ross, J., Winant, E., and Falvey, C. 1996. Alternative sewers: A good option for many communities. *Pipeline; National Small Flows*, 7(4): 1–7.
12. Centre for the Study of the Build Environment (CSBE). 2003. Grey water reuse in other countries and its applicability to Jordan. Project funded by the ministry of planning enhanced productivity program, Amman, Jordan.
13. Christova-Boal, D., Eden, R., and McFarlane, S. 1995. An investigation into grey water reuse for urban residential properties. *Desalination*, 106(1–3), 391–397.
14. Cui, F. and Ren, G. 2005. Pilot study of process of bathing wastewater treatment for reuse. *IWA Conference 2005*, May 18–20, 2005, Xi'an.
15. Dallas, S. and Ho, G. 2005. Subsurface flow reedbeds using lternative media for the treatment of demostic grey water in Monteverde, Costa Rica, Central America. *Water Science and Technology*, 51(10), 119–128.
16. Dixon A., Butler, D., Fewkes, A., and Robinson, M. 1999. Measurment and modelling of quality changes in stored untreated grey water. *Urban Water*, 1(4), 293–306.
17. Elmitwalli, T. and Otterpohl, R. 2007. Anaerobic biodegradability and treatment of greywater in upflow anaerobic sludge blanket (UASB) reactor. *Water Research*, 41(6), 1379–1387.
18. Eriksson, E., Auffarth, K., Henze, M., and Ledin A. 2002. Characteristics of grey water. *Urban Water*, 4(1), 85–104.
19. Friedler, E. and Hadari, M. 2006. Economic feasibility of on-site grey water reuses multi-story buildings. *Desalination*, 190(1–3), 221–234.

20. Friedler, E., Kovalio, R., and Ben-zvi, A. 2006. Comparative study of the microbial quality of grey water treated by three on-site treatment systems. *Environmental Technology*, 27(6), 653–663.

21. Friedler, E. 2004. Quality of individual domestic grey water streams and its implication for on-site treatment and reuse options. *Environmental Technology*, 25, 997–1008.

22. Friedler, E., Kovalio, R., and Galil, N. 2005. On-site grey water treatment and reuse in multi-story buildings. *Water Science and Technology*, 51(10), 187–194.

23. Golda, A., Gopalsamy, P., and Muthu, N. 2014. Characterization of domestic grey water from point source to determine the potential for urban residential reuse: Short review. *Applied Water Science*, (4), 39–49.

24. Gross, A., Shmueli, O., Ronen, Z., and Raveh, E. 2007. Recycled vertical flow constructed wetland (RVFCW)-a novel method of recycling grey water for irrigation in small communities and households. *Chemosphere*, 66(5), 916–923.

25. Halalsheh, M., Dalahmeh, S., Sayed, M., Suleima, W., Shareef, M., Mansour, M., and Safi, M. 2008. Grey water characteristics and treatment options for rural areas in Jordan. *Bioresource Technology*, 99(14), 6635–6664.

26. Hernandez, L., Zeeman, G., Temmink, H., and Buisman, C. 2007. Characterization and biological treatment of grey water. *Water Science and Technology*, 56(5), 193–200.

27. Hernandez, L., Zeeman, G., Temmink, H., Marques, A., and Buisman, C. 2008. Comparison of three systems for biological grey water treatment. In: *Proceedings of IWA Conference on Sanitation Challenges*. May 19–22, Wageningen, Netherlands.

28. Hills, S., Smith, P., Hardy P., and Briks, R. 2001. Water recycling at the millennium dome. *Water Science and Technology*, 43(10), 287–294.

29. Imura, M., Sato, Y., Inamori, Y., and Sudo, R. 1995. Development of a high efficiency household biofilm reactor. *Water Science and Technology*, 31(9), 163–171.

30. Itayama, T., Kiji, M., Suetsugu, A., Tanaka, N., and Saito, T. 2006. On site experiments of the slanted soil treatment systems for domestic grey water. *Water Science and Technology*, 53(9), 193–201.

31. Jamrah, A., Al-Futaisi, A., Prathapar, S., and Harrasi, A. 2008. Evaluating grey water reuse potential for sustainable water resources management in Oman. *Environmental Monitoring and Assessment*, 137(1–3), 315–27.

32. Jamrah, A., Al-Omari, A., Al-Qasem, L., and Niveen, A. 2006. Assessment of availability and characteristics of grey water in Amman. *Water International*, 31(2), 210–220.

33. Jefferson, B., Palmer, A., Jeffrey, P., Stuetz, R., and Judd, S. 2004. Grey water characterization and its impact on the selection and operation of technologies for urban reuse. *Water Science and Technology*, 50(2), 157–164.

34. Jefferson, B., Laine, A., Parsons, S., Stephenson, T., and Judd, S. 2001. Advanced biological unit processes for domestic water recycling. *Water Science and Technology*, 43(10), 211–218.

35. Jefferson, B., Laine, A., Parsons, S., Stephenson, T., and Judd, S. 1999. Technologies for domestic wastewater recycling. *Urban Water*, 1(4), 285–292.

36. JS. 1776. 2013. *Jordanian Standards for Water-Reclaimed Grey Water*. Jordan Standards and Metrology Organization, Amman, Jordan.

37. Legman, K., Hutzler, N., and Boyle, W.C. 1974. Household waste characterization. *Journal of the Environmental Engineering Dvision*, 100(1), 201–213.

38. Li, Z., Gulyas, H., Jahn, M., Gajyrel, D., and Otterphohl, R. 2003. Grey water treatment by constructed wetlands in combination with TiO2-based photocatalytic oxidation for suburban in rural areas without sewer system. *Water Science and Technology*, 48(11–12), 101–106.

39. Loh, M. and Coghlan, P. 2003. Domestic water use study. In: *Perth, Western Australia 1998–2001*. Water Corporation, Perth, Western Australia.

40. Metcalf and Eddy. 2003. *Wastewater Engineering; Treatment Disposal Reuse,* 4th Edition, McGraw-Hill, New York.

41. Metcalf and Eddy. 1999. *Wastewater Engineering; Treatment Disposal Reuse,* 3rd Edition, McGraw-Hill, New York.

42. McGhee, T. 1991. *Water Supply and Sewerage,* 6th Edition, McGraw-Hill, New York.

43. Nolde, E. 1999. Grey water reuses systems for toilet flushing in multi-story buildings-over ten years' experiences in Berlin. *Urban Water*, 1(4), 275–284.

44. Nolde, E. and Dott, W. 1992. Experimenteller wohnungs-und stadtebau–Forschungskonzept Block 103-Grauwasserprojekte in Berlin-Kreuzberg. Erstellet im Auftrag der Senatseverwaltung fur Bauund Wohnungswesen, Berlin (unpublished data) (cited by Nolde 1999).

45. Otterpohl, R., Albold, A., and Oldenburg, M. 1999. Source control in urban sanitation and waste management: Ten systems with reuse of resources. *Water Science and Technology*, 39, 153–160.

46. Ramon, G., Green, M., Semiat, R., and Dosortez, C. 2004. Low strength grey water characterization and treatment by direct membrane filtration. *Desalination*, 170(3), 241–250.

47. Shrestha, R., Haberl, R., and Laber, J. 2001. Constructed wetlands technology transfer to Nepal. *Water Science and Technology*, 43(11), 345–350.

48. Shrestha, R., Haberl, R., Laber, J., Manandhar, R., and Madar, J. 2001. Application of constructed wetlands for wastewater treatment in Nepal. *Water Science and Technology*, 44(11–12), 381–386.

49. Shin, H., Lee, S., Seo, S., Kim, G., Lim, K., and Song, J. 1998. Pilot-Scale SBR MF operation for the removal of organic compounds from grey water. *Water Science and Technology*, 38(6), 79–88.

50. Surendran, S. and Wheatly, A. 1998. Grey and roof water reclamation at large institutions–Loughborough experiences. *Presented at Water Recycling and Effluent Re Use.* April 26–27, 1999, London, UK.

51. USEPA. 1992. Guidelines for Water Reuse, EPA/625/R-92/2004, http://www.epa.gov/ORD/NRMRL/Pubs/625r04108/625r04108.pdf. Accessed on January 19, 2009.

52. USEPA. 2002. Water Supply and Demand in the United States. www.epa.gov/seahome/groundwater/src/supply.htm.

53. WHO. 2006. *Guidelines for the Safe Use of Wastewater, Excreta and Grey Water. Excreta and Grey Water Use in Agriculture,* Vol. 4. WHO Press, World Health Organization, Geneva, Switzerland.

54. WHO. 2001. Water Quality: Guidelines, Standards and Health, http://www.who.int/water_sanitationhealth/dwq/iwachap2.pdf. Accessed on January 19, 2009.

55. Witt, M., Siegrist, R., and Boyle, W.C. 1975. Small scale waste managment progect. University of Wisconsin-Madison. Puplication 1.5. Rural Household Wastewater Characterization.

56. Winward, G., Avery, L., Farzer-Williams, R., Pidou, M., Pidou, M., Jeffrey, P., Stephenson, T., and Jefferson, B. 2008. A study of the microbial quality of grey water and an evalution of treatment technologies for reuse. *Ecological Engineering,* 32(2), 187–197.

57. Zeeman, G., Kujawa, K., Mes de, T., Hernandez, L., Graff de, M., Abu Ghunmi, L., Mels, A. et al. 2008. Anaerobic treatment as a core technology for energy, nutrients and water recovery from source-separated domestic waste(water). *Water Science and Technology,* 57(8), 1207–1212.

35

Wastewater Operation Requirements and Distribution Systems

Josephine Treacy
Limerick Institute of Technology

PREFACE

Clean water is only as good as our wastewater operations and distribution systems. The challenge of today's society is to maintain the quality of our clean water. In line with this challenge, one must enact the three pillars of sustainability within our wastewater strategy; namely, environmental, social, and economical.

It is important to note water is not a commercial product like other commercial products but a heritage that has been given to us from previous generations and must be protected by us for future generations.

The modern paradigm of waste treatment is to move in the direction of resource recovery from waste rather than just treating the waste. This chapter outlines the importance of understanding the characteristics of wastewater and discusses emerging components present in wastewater. By having better control of the inputs to a wastewater treatment plant (WWTP), this can enable better control of both the outputs and the processes within the plant itself. This chapter discusses the legislative requirements governing WWTP operations. The WWTP operations, namely, preliminary, primary, secondary, tertiary, and sludge management and treatment will be discussed bearing in mind the operators' role.

The monitoring requirements and the emergence of real-time monitoring will also be addressed as part of this chapter.

Modern concepts such as moving in the direction of zero waste will be discussed. Other initiatives such as the multibarrier approach, BATNEEC, and life-cycle assessment as aids in empowering the efficiency of WWTPs are highlighted. The future concerns with regard to wastewater treatment such as climate change will also be addressed.

The summary and recommendations of this chapter outline the variables to consider in order to maintain the efficient operations of WWTPs. Our clean water is only as good as our wastewater treatment. Understanding and public awareness regarding the input to the plant can have an overall impact on the efficiency of the plant. The modern paradigm of resource recovery from waste will help ensure WWTPs become more sustainable, as financial constraints can have a large impact on the operations of the plant. Basic controls such as the correct time allowed for each process can enhance the performance of the plant. Sharing knowledge and developing key performance indicators are important for the future. WWTP operations are not static but dynamic and will continuously evolve to maintain the quality of our receiving waters.

35.1 Introduction

Wastewater operations and distribution systems are a fundamental concern in today's society. The production of waste from human activities has to be treated and disposed of efficiently in order to sustain the quality of our ecosystems. This chapter will discuss interesting variables around wastewater operations and distribution systems. Throughout this chapter, focus on the operations of the plant as an aid for the operator will be emphasized.

35.2 The Importance of Wastewater Treatment

Clean water is only as good as our wastewater treatment. In order to maintain and improve the quality of our clean water, society must invest in the technologies and infrastructure for the treatment and disposal of our wastewaters. An interesting paper by Somlyody summarizes the trends related to water quality management which include wastewater treatment, pollution control, monitoring, modeling, planning, environmental impact assessment, legislation, and sustainability development [53].

35.2.1 Wastewater Characteristics

Wastewater characteristics are influenced by the behavior and lifestyle of the producers of the waste. The characteristics of wastewater are influenced by water usage in the community, including industrial and commercial contributors. Seasonal factors also influence the characteristics of the wastewater. During wet weather, a significant quality of infiltration and inflow may also enter the collection systems. The quantity of inflow and infiltration depends on the condition of the sewage system in terms of cracks in piping systems and defectives in manholes, for example [63]. Sewer systems including sanitary sewers, stormwater sewers, and mixed sewers can also influence the quality of the wastewater. The constituents of wastewater are summarized in Table 35.1.

Indicator parameters used to study wastewater quality include biological oxygen demand (BOD_5), chemical oxygen demand (COD), total suspended solids (TSS), pH, total phosphorus, and

TABLE 35.1 Constituents Present in Wastewater

Characteristics	Composition
Physical	Color, odor temperature
	Suspended solids
Chemical	Nutrients
	Metals
	Organic materials
	Volatile and nonvolatile
	Other inorganic material
Biological	Pathogenic bacteria, virus, and worm eggs

Source: Adapted from EPA. 1997. *Waste Water Treatment Manuals Primary, Secondary and Tertiary Treatment.* Environmental Protection Agency, Ireland.

TABLE 35.2 Typical Indicator Parameter Levels in Wastewater

Parameter	Concentration (mg/L)
BOD	100–300
COD	250–800
Suspended solids	100–350
Total nitrogen (as N)	20–85
Ammonia (NH_3 as N)	10–30
Organic phosphorus (as P)	1–2
Inorganic phosphorus (as P)	3–10
Oils fats and grease	50–100
Total inorganic constituents (Na, Cl, Mg, S, Ca, K, Si, Fe)	100
Heavy metals (Cd, Cr, Cu, Pb, Hg, Ni, Ag, Zn)	<1 mg/L

Source: Adapted from EPA. 1995. *Waste Water Treatment Manuals, Preliminary Treatment.* Environmental Protection Agency, Ireland.

total nitrogen. Typical wastewater indictor level parameters are shown in Table 35.2.

35.2.2 Emerging Pollutants in Wastewater

In today's society, there are many challenges related to emerging micropollutants that can be potentially found in wastewater.

Pharmaceutical residuals, personal care products, and endocrine disrupting chemicals (EDCs) are classified as emerging micropollutants. Micropollutants can have a diverse effect on water ecosystems such as bioaccumulation and transboundary pollution. The levels of these micropollutants found in water can range from μg/L to ng/L. The challenges concerning micropollutants present in wastewater include the following: How can one analyze them efficiently? What evolving technologies are needed to treat these micropollutants? An interesting review shows the occurrence, fate, and transportation of micropollutants in the environment of 14 countries, which portrays the research interests in this topic globally [30]. Classes of emerging compounds and micropollutants in wastewater systems can be seen in Table 35.3.

Due to the debates surrounding emerging pollutants in wastewater treatment, one needs to ask the question, Is there a gap in knowledge in terms of what is in wastewater? The challenge for WWTP operators is to understand the inputs in order to control the outputs. Typical sources of pollutants entering our WWTPs can be seen in Figure 35.1.

Emerging pollutants can also bioaccumulate in the sludge formed during primary, secondary, and tertiary treatment. Common pollutants such as oil and grease can have major impacts on the secondary treatment process [7,37]. Pollutants can eventually enter drinking water if not treated efficiently due to the practically closed-loop system. As stated in the introduction, our clean water is only as clean as guided by the efficiencies of our wastewater treatment technologies. Control of our inputs to the treatment process and the output will enhance the quality of our clean waters.

Water ecosystems can be considered a partially closed loop as no water source is isolated from one another. The discharges from WWTPs are readily linked with groundwater sources and surface water sources. Due to the partially closed-loop system, wastewater quality has a major impact on our drinking water quality. The partially closed water loop is demonstrated in Figure 35.2.

35.2.3 Sustainability of Our Ecosystems

In this section, the three pillars of sustainability will be addressed in relation to the creation of awareness of the importance of monitoring the characteristics of wastewater (Figure 35.3).

Public awareness of wastewater operations is very important. A fundamental study showed the attitude of the general public (N = 90) in relation to the question, "What goes down your kitchen sink?" (n = 90) [11]. A graphical presentation of the findings can be seen in Figure 35.4.

Figure 35.4 shows there is still educational strategies needed to advise the general public on what does go down the kitchen sink and the disposal of waste.

An interesting guide showing the variables to consider for the sustainability of wastewater treatment can be seen in Table 35.4 [38].

The sustainability of our water systems remains a challenge in today's society [27].

35.3 Wastewater Operations

The operations of a WWTP should be considered interdisciplinary combining engineering, biology, mathematics, hydrology, chemistry, physics, analytical science, environmental science, social science, and areas such as validation and quality management.

35.3.1 Legislative and Policy Requirements

The regulation of water resources varies from country to country. With regard to legislation and policy requirements, specific

TABLE 35.3 Emerging Micropollutants Found in Wastewater Systems

Compound Class	Example
Pharmaceutical	
Veterinary and human antibiotics	Lincomycin, erythromycin
Analgesics and anti-inflammatory drugs	Diclofenac
Lipid regulators	Bezafilbrate, clofibric
Beta blockers	Metoprolol, propranolol
X-ray contrast media	Iopamidol, diatrizoate
Steroids and hormones	Estradiol, estrone
Personal care products	Macrocyclic musks
Fragrances	Benzophenone ethoxylates
Sunscreen agents	N,N-Diethyltoluamide
Insect repellents	
Antiseptics	Triclosan
Surfactants and surfactant metabolites	Alkylphenol
Flame retardants	Polybrominated diphenyl ethers (PBDEs)
Industrial additives and agents	Chelating agents (EDTA), aromatic sulfonates
Gasoline additives	Dialkyl ethers
Disinfection by-products	Iodo-THMs, bromoacids

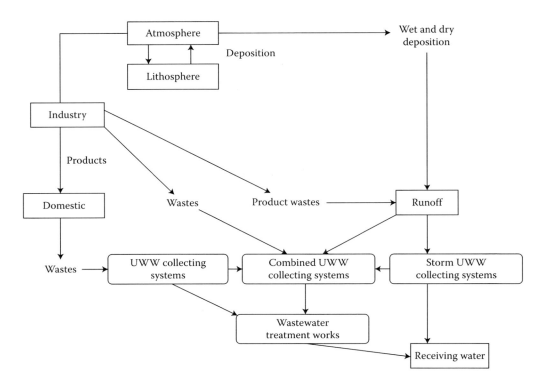

FIGURE 35.1 Sources of wastewater entering WWTPs. (Adapted from LC Consultants Ltd. London. 2001. *Pollutants in Urban Waste Water and Sewage Sludge*. Office of Official Publications of the European Communities, Luxemburg.)

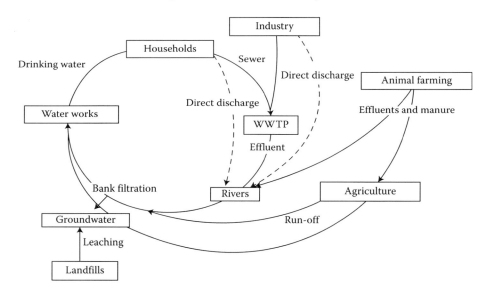

FIGURE 35.2 The partially closed water loop system. (Adapted from Petrovic, M., Gonzalez, S., and Barcelo, D. 2003. *Trends in Analytical Chemistry*, 22(10), 685–696.)

requirements apply to the collection systems, the treatment plant itself, and the monitoring of the inputs and outputs of the discharges. In terms of a European perspective, the Council Directive 91/271/EEC concerning urban wastewater treatment was adopted on May 21, 1991 [17]. Its objective is to protect the environment from the adverse effects of urban wastewater discharges and discharges from certain industrial sectors

(see Annex III of the Directive). The directive regulates the collection, treatment, and discharge of

- Domestic wastewater
- Mixture of wastewater
- Wastewater from certain industrial sectors (see Annex III of the Directive)

Specific articles within the directive for certain waste streams can be seen in Figure 35.5.

Agglomeration refers to the area where the population or economic activities are sufficiently concentrated for urban wastewater to be collected and transported to an urban WWTP. With regard to the sizes of WWTPs, measurement is in terms of the population served or people equivalent (PE). The fundamental principles of the directive (91/271/EEC) include planning, regulation, monitoring, and reporting. Other important variables pending on the directive include

- The collection and treatment of wastewater in all agglomerations >2000 PE
- Secondary treatment of all discharges from agglomerations >2000 PE and more advanced treatment for agglomerations >10,000 PEs
- A requirement for preauthorization of all discharges of urban wastewater, of discharges from the food-processing industry, and of industrial discharges into urban wastewater collection systems

- Monitoring of the performance of treatment plants and receiving waters
- Control of sewage sludge disposal and reuse, and treated wastewater reuse whenever it is appropriate [17]

The water frame work directive is the parent directive in which the urban WWTPs directive emerged. The basic requirement of the any water policy is to expand water protection to all waters including inland and coastal, surface water, and groundwater [14].

TABLE 35.4 Indicators Developed to Assess Sustainability of Wastewater Treatment Technologies

Economic
Capital costs
Operation and management
Environmental energy use
Biochemical oxygen demand (BOD) % removal
Total suspended solids (TSS) % removal
Phosphorus % removal
Nitrogen % removal
Pathogens % removal
Societal public participation
Community size served population
Aesthetics
Measured level of nuisance from odor
Staffing required to operate plant
Level of education
Operational requirements
(operator license)
Open space availability

Source: Adapted from Muga, H.E. and Mihelcic, J.R. 2008. *Journal of Environmental Management*, 88, 437–447.

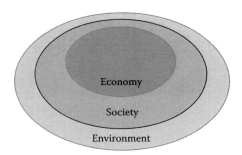

FIGURE 35.3 Three pillars of sustainability linking with the importance of understanding the characteristics of wastewater.

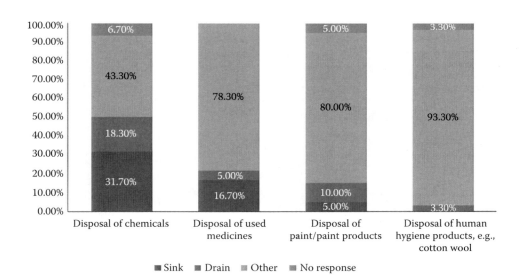

FIGURE 35.4 Percentage of responses from the general public on what goes down your kitchen sink (n = 90). (Adapted from Devane, N., Treacy, J., and McDonnell, K. 2008. *ESAI Environews*, Issue No. 16, p. 15, Winter 2008.)

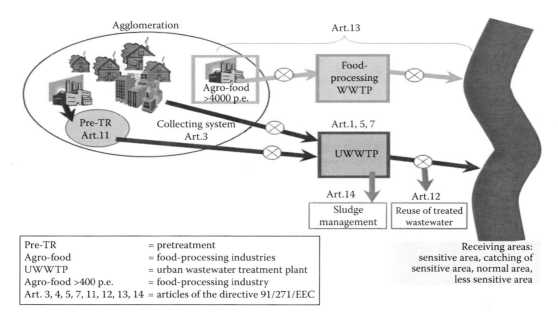

FIGURE 35.5 Systematic diagram of urban WWTP (UWWTP) and receiving waters and article implementations. (Adapted from EC Europe. 2014. Directive 2000/60/EC of the European Parliament. http://www.doeni.gov.uk/niea//pubs/publications/waterframeworkdirective.pdf. Accessed on July 2014.)

35.3.2 Loading Capacity

A critical consideration in the design of a WWTP is to understand the peak hydraulic loading and the maximum flow through the plant. The inflow and composition of wastewater are very important in order to optimize the design for the collection and the treatment of wastewater. For large treatment plants, it is generally recommended to design a plant for a maximum 3 times dry weather flow (DWF). For smaller treatment plants, the design is set for 4–6 times DWF. A typical 3DWF can be seen in Figure 35.6 [20].

The flow entering the plant can come from different sources; namely, domestic wastewater, industrial wastewater discharges from urban areas, infiltration, surface water and stormwater, domestic sewers, and combined sewers [6]. Seasonal variance can impact the flow. WWTPs can also receive inputs from private contractors. In Ireland, for example, the septic tank waste and landfill leachate are treated in WWTPs. Septic tank waste and landfill leakage enter the plant as separate loads. The loading

capacity is determined by populations served by a particular treatment plant.

The wastewater from inhabitants is expressed in unit population equivalent; person equivalent (PE). PE can be expressed in water volume or BOD.

The definitions used worldwide are as follows [21]:

$$1\,PE = 0.2\,m^3/d \tag{35.1}$$

$$1\,PE = 60\,g\,BOD/d \tag{35.2}$$

The daily load per person can be expressed as

$$\begin{aligned} \text{Daily load} &= \text{kg BOD/day} \\ &= (\text{BOD}(mg/L) \times \text{daily volume}\,m^3/\text{day})/1000 \end{aligned} \tag{35.3}$$

When designing a WWTP engineers should design to facilitate for extra space in the vicinity of the plant to allow for potential expansion of the plant. Each of the fundamental steps in the wastewater treatment operations will now be discussed in turn.

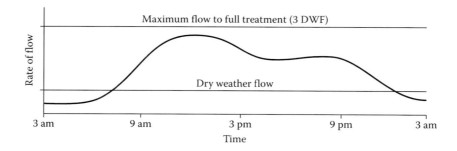

FIGURE 35.6 Typical WWTP input flow. (Adapted from EPA. 1995. *Wastewater Treatment Manuals, Preliminary Treatment.* Environmental Protection Agency, Ireland.)

35.3.3 Systematic Design of a Wastewater Treatment Plant

A systemic diagram of WWTP operations can be seen in Figure 35.7.

The range of technologies for preliminary, primary, secondary, and advanced tertiary treatment and sludge treatment are shown in Figure 35.7. The fundamental technologies will now be discussed in turn.

35.3.4 Technologies Related to Preliminary and Primary Treatment

Wastewater preliminary treatment processes essentially comprise physical processes to ensure the water flows easily through the other processes [20]. Preliminary treatment involves the coarse filtering of the water to remove sand and silt. Other components that can be removed include paper and plastic. Oils and grease can also be removed at this stage. The fundamental role of

the operator at this stage is to ensure the screens are not clogged. Routine cleaning and back flushing of the screens is essential to ensure maximum efficiency of the preliminary treatment process.

The purpose of primary treatment, which is known as primary clarification or sedimentation, is to remove the settable solids from the water and to remove oil and grease. If the primary clarification is working efficiently, it can remove 50%–70% total suspended solids and reduce the BOD by 50%. The efficiency of the primary settlement depends on the length of time the water is left to settle. If, for example, the time is too long, gas bubbles H_2S can form due to anaerobic conditions developing, which can affect the settling process [21]. The amount of solids in the wastewater can also influence the process. The settled sludge should also be regularly removed from the tanks. Oils and grease can be removed at this stage, for example, by dispersing air through the settled water. This causes the formation of microbubbles, which can attach themselves to the particles in the water and make them float. Other technologies related to this can be seen

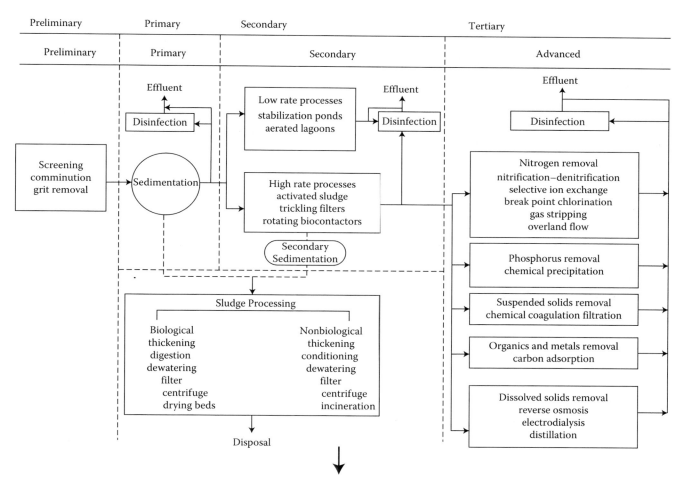

Sludge processing biological/nonbiological
Other advanced technologies for tertiary treatment include ultrafiltration and microfiltration
Figure information above (Google images)

FIGURE 35.7 Systematic diagram of WWTP technologies.

in the literature, such as vertical surface aerators [48]. The timing allowed for the primary treatment stage can affect the overall efficiency of the process.

35.3.5 Technologies Related to Secondary Treatment

The function of secondary treatment is to remove the soluble and colloidal waste that cannot be removed by primary treatment. The basic objective of the biological treatment is to feed the soluble and colloidal waste to a microbial culture. A large variety of microorganisms are being investigated for this role; namely, protozoa, rotifers, nematodes, fungi, and algae. Secondary treatment can be considered as supplying waste as a food source for the microbial culture. With secondary treatment, there are two fundamental technologies; namely, the activated sludge systems and the biofilm systems. Both technologies use a microbial culture [56].

35.3.5.1 Activated Sludge System

The activated sludge system in the wastewater is in an aeration tank, which contains microorganisms; namely, bacteria, fungi, yeast, protozoa, and worms. The microorganisms attach particles from the wastewater forming a floc. During the activated sludge process, the floc (known as sludge) is required to remain suspended in the water. This floc attaches both inorganic and organic components from the water by van der Waals forces, adsorption, absorption, and entrapment (coagulation followed by flocculation). Pathogens in the water also move toward the sludge. As more waste moves toward the sludge, the sludge increases in size. The challenge for the operator is to maintain the efficiency of the activated sludge system. In order to maintain its efficiency, the activated sludge system must be aerated. Aeration is achieved by using submerged or surface mechanical aerators. Much research in the design of these aeration devices can be seen in the literature such as microbubble aerators (Dionergy EL 400 Microbubble Technology) [12]. The aeration level should remain between 1 and 2 ppm dissolved oxygen (DO) for the efficient activity of the microbes. The aeration of the tank also enables the sludge to remain in suspension in the water. The aeration efficiency is measured in terms of the amount of oxygen transferred per unit of air introduced to the wastewater under conditions such as depth of immersion, temperature, and the chemical matrix of the wastewater [1]. The unit related to aeration efficiency is KgO_2/KWh.

The total oxygen demand on the activated sludge system is based on the BOD inflow and the BOD outflow.

Other factors to consider include

Ammonia inflow ammoniacal nitrogen (mg/L)
Ammonia outflow ammoniacal nitrogen (mg/L)
Inflow and outflow nitrate nitrogen (mg/L)

Other important variables to consider are the pH and the food:microbes ratio [44]. The food is the waste in the water that moves toward the microbes forming the sludge.

Similar to the primary treatment, the contact time of the wastewater in the activated sludge system is an important variable in the operations of the plant.

The next stage of the process involves the output from the activated sludge tank being feed into the secondary clarifier. The secondary clarifier works very similar to the primary clarifier. The settled sludge can be taken away for further treatment such as resource recovery, which will be discussed later.

In order to have efficient operations of the activated sludge system, a certain portion of the settled sludge is returned to the aeration tank. This is known as the "returned activated sludge" (RAS). In the secondary treatment, two terms are important for the operator to understand: mixed liquid suspended solids (MLSS) and the mixed liquor, which is the liquid suspension. The MLSS can be measured by gravimetric analysis. The organic and inorganic component can be determined by burning off the organic component at 500°C and reweighing. The purpose of the secondary clarifier is to enhance the clarity of the water and return some of the settled sludge to the aeration tank.

The sludge stability is measured in terms of sludge volume index (SVI) as can be seen in Equation 35.4 [21]

$$SVI = (\text{volume of settled sludge (mg/L)} \times 1000)/MLSS \text{ (mg/L)}$$

(35.4)

The sludge stability can impact the efficiency of the secondary clarifier.

The efficiency of the secondary clarifier can be influenced by the sludge blanket, concentration of suspended solids, the speed in which the sludge is returned to the activated sludge tank, the turbidity of the supernatant, and the concentration of DO.

Observations of the sludge by the operator can determine if the plant is performing correctly [46]. An experienced operator can physically observe for white foam, brown foam, or black foam on the surface of the activated sludge tank. The color can be an indicator of the quality of the sludge [47]. Regular examination of the mixed liquor should be carried out examining floc/sludge size and shape, the presence of filamentous bacteria, the presence of protozoans (flagellates, ciliates) and the presence of rotifers under a microscope [31]. The activity of the sludge can affect the overall performance of the secondary treatment process. The activity of the sludge can be influenced by floc size and shape, foam color, and microbiological characteristics.

35.3.5.2 Biofilm Systems

The second type of secondary treatment is known as the biofilm design. The biofilm approach can involve the following types of processes: percolated or trickling filters, rotating biological contractors (RBC), and submerged (aerated filters). The percolated filter is a bed of packed medium, which is highly permeable. The RBC are large diameter structure discs. In this design, the culture is on a substrate such as plastic and rock. In the submerged aerated filters, biomass is supported within a tank in which the wastewater flows. As the water passes through the biofilm, the waste in the water is absorbed by the biofilm layer forming a humus type sludge. Biofilm processes are aerobic. Oxygen must be supplied. In certain cases, a natural draught is created by the water passing through the biofilm. Activated sludge like blowers and fine bubble diffusers on some submerged aerated filters can be utilized to supply the oxygen. The biofilm can lose its thickness due to the passage of water through the biofilm. This is known as sloughing of the biofilm. Anaerobic conditions can also break down the biofilm. Heterotopic bacteria and fungi are the predominate bacteria used in the biofilm design. Recirculation of the outflow from the biofilm process dilutes the inflow and increases the hydraulic loading rate of the plant. Erosion of the biofilm due to the wall thickness being reduced gives rise to a humus material; this is due to the humus-like nature of the sludge produced by the biofilm systems. After the water passes through the biofilm, the next process is again a secondary clarifier where the humus-type sludge is settled out for further treatment. Modern technologies and research in the area of biofilm design can be seen in the literature [26].

35.3.5.3 Nutrient Removal and Wastewater

Nutrient removal is a very important concern with regard to wastewater treatment as excessive nutrients can give rise to eutrophication problems in the receiving waters. An interesting American approach to this is the nutrient removal challenge [63]. The American group is researching the best sustainable cost-effective processes to meet nutrient limits in wastewater treatment facilities. The research involves developing new and improving existing technologies [59,65]. An interesting one to mention is the mechanism of nitrification and denitrification for nutrient removal.

The nitrification process is as follows:

ammonia NH_3 to nitrite NO_2^- to nitrate NO_3^-

The denitrification process is as follows:

nitrate NO_3^- to nitrite NO_2^- to nitrogen monoxide NO to dinitrogen oxide N_2O to nitrogen N_2

The nitrite and nitrate produced during nitrification becomes the oxygen source for heterotopic bacteria, which utilized the available BOD in the wastewater as a carbon source. The oxygen created by the nitrification process utilized by the heterotopic bacteria is known as a denitrification process [21]. The combined nitrification and denitrification processes can reduce the overall BOD of the water. With regard to biological phosphate removal, this can be achieved using two approaches: stoichiometric coupling to microbial growth and enhanced storage in the biomass as polyphosphate (Poly-P) [36].

35.3.6 Technologies Related to Tertiary Treatment

More advanced nutrient removal is considered tertiary treatment. Removal of excessive phosphorous and nitrate and the disinfectant of the final discharge waters are examples of tertiary treatment. Precipitation of nutrients with metal salts can be used, but this can give rise to more salt residuals in the discharge waters.

Disinfection is mainly concerned with auxiliary directives such as the shellfish directive and the bathing water directive, where the discharge waters are entering a shellfish or bathing water area. UV disinfection using 254 nm is readily used [40,57]. Membrane technologies such as UF membranes can also be used [39]. The ozonation of the water is another method used for disinfection [43]. Chlorination prior to the discharge of the treated water can also be a consideration within the operations processes [3]. Chlorination is not used regularly in Ireland within WWTPs. There are many concerns related to disinfectant byproducts (DBP), which could potentially create more risk regarding the discharge waters [49].

There are many cost constraints concerning the implementation of tertiary treatment. As mentioned previously, water is a partially closed-loop system. Tertiary treatment can again minimize contamination of the receiving waters. In Ireland, for example, most treatment plants would be obliged to have preliminary, primary, and secondary treatment. Tertiary treatment is an extra bonus. Unfortunately, there are a number of discharge points that still have no treatment in European countries such as Ireland [50]. In an ideal situation, it would be great to have every treatment plant at the level of tertiary treatment in order to minimize the risk associated with waste streams on the receiving waters.

35.3.7 Sludge Management and Treatment

Sewage sludge is a byproduct of the wastewater treatment process and is often known as biosolids because environmentalists would like to consider biosolids as a resource rather than a waste [58]. Sludge treatment can involve treatment for disposal or further use. The source of sludge in wastewater treatment is formed from the primary clarification and secondary clarification processes. Precipitates from tertiary treatment can also give rise to sludge. Sludge from private contractors and small treatment plants that have only dewatering facilities can be transported to larger treatment plants for further treatment. The most common initial treatment of sludge is dewatering. The water collected from the dewatering process is usually returned to the treatment plant and this can impact the loading of the plant. Developments concerning resource recovery from the sludge are a very strong research area in today's society. If one can make plants more economically viable and self-economical, our WWTPs will become more efficient due to the self-financing of their operations.

The most common sludge stabilization is known as anaerobic digestion. The process transforms sludge to a biogas, which is a mixture of methane, carbon dioxide, and traces of other gases. Certain treatment plants can use the anaerobic digestion process and the methane liberated as an energy source for the plant [13]. Other technologies used for resource recovery from sludge can

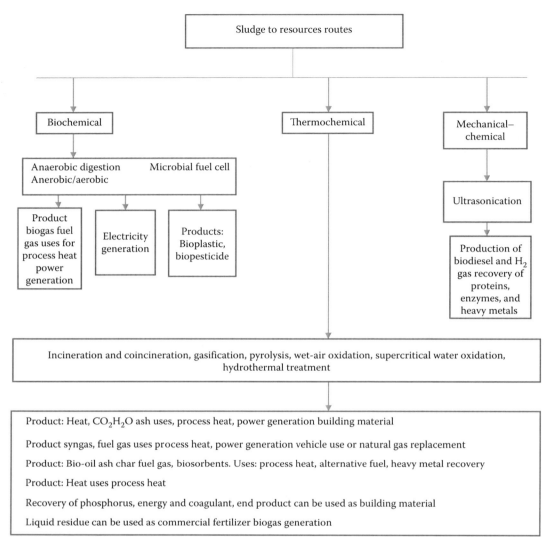

FIGURE 35.8   Technologies for resource recovery from sewage sludge. (Adapted from Tyagi, V.K. and Lo, S.L. 2013. *Renewable and Sustainable Energy Reviews*, 25, 708–728.)

be seen in Figure 35.8, which can also be considered treatment methods for sludge stabilization [60].

35.3.8 Discharge of Treated Water

The discharge points from WWTPs can be a river, bathing water, or estuary, for example. The management of the discharge points is also important to consider when designing WWTPs. One must consider the discharge point and the risk associated with the surrounding environment. Environmental impact assessments and licenses can minimize the risks [10].

35.4 Wastewater Treatment Plant Monitoring

The monitoring requirement of a WWTP will now be discussed.

35.4.1 Role of the Plant Operator and Monitoring Requirements

The role of the plant operator entails the following:

- Meeting the emission limits of the treatment plant (for air and water)
- Meeting the standards for sensitive waters (receiving waters)
- Preventing odors in the vicinity of the plant (occurring from lack of maintenance and rushing the water through the plant processes)
- Efficient operations and management of the plant
- Maintaining an effective maintenance program for the infrastructure of the plant, namely preliminary, primary, secondary, and tertiary treatment

In order for the operators to be involved in efficient management of the plants, it is important to regulate audits of the plants. Included in this are preventative maintenance, routine servicing, emergency response, equipment replacement, and quantity of inputs and outputs. The operators of WWTPs should maintain a schedule monitoring program including a frequency of analysis template. The object of an audit, either internal or external, is to evaluate the plant's performance.

35.4.2 Monitoring Requirements

The monitoring of WWTPs is required in order to ensure efficient performance of the plant; performance is measured in terms of efficiency or percentage (%) removal of waste. Water and air quality investigations can be divided into sampling, measurement, recording, and analysis. The monitoring and sampling requirement can be represented as routine, event, and spatially intensive [27]. Routine monitoring is the backbone of any treatment plant and will determine the overall performance of the plant. In cases of an unusual event such as an odor near the plant, event sampling and monitoring can be undertaken. The spatially intensive approach would involve taking samples in different locations around the plant under stable conditions and to potentially use the data obtained as a baseline in cases of unusual occurrence around the plant. The automation of WWTP operations is available for certain large treatment plants. SCADA systems are examples of system controllers for real-time data. The SCADA system can also be linked with pump control hydraulic flows entering, leaving, and around the plant to evaluate the performance of the plant. Real-time sensors around each of the treatment processes can enhance the operations of the plant. Sensors such as pH, DO, turbidity, and nutrients are beginning to be used in the monitoring of WWTPs [2,35]. Monitoring requirements with regard to the maintenance of the infrastructure of the plant are also very important.

35.4.3 The Benefits of Accreditation

Accreditation can enhance the operations of the plant. It is important to state factors such as poor design, inefficient monitoring and inspection of the plant operations, and the quality and quantity of the wastewater entering the plant. These can affect the overall performance of the plant. Countries all over the world have seen a need for environmental management, sustainable growth, and development. ISO is one such standard that is utilized all over the world. The ISO 14001 not only addresses the environmental aspects of organizations, but also its products and services. ISO 14001 is an international standard that details the requirements for an environmental management system. Other ISO standards applicable to WWTP operations include OHSAS 18001: Health & Safety Management System; ISO 17025:2005 Laboratory Competence; and ISO/IEC 27001:2005 Information and Data Security. Another interesting standard (ISO 50001) is based on the management system model of continual improvement similar to ISO 14001. The focus of ISO

50001 is to make it easier for organizations to integrate energy management into their overall efforts to improve quality and environmental management [28]. It is important even if a plant does not have ISO accreditation to maintain good manufacturing practice throughout the plant.

35.4.3.1 Documentation

Basic documentation that can be enhanced by accreditation include [22]

- Monthly influent monitoring
- Discharges from the agglomeration
- Ambient monitoring report
- Data collection and reporting requirements
- Pollution release and transfer register (PRTR) specific to Europe [23]
- Operational reports, complaints, reports on incident
- Infrastructural capacity, stormwater overflow identification and inspection, report on improvement program

A manual maintenance log for each tank throughout the process is important if no automatic system is available such as a SCADA system. SCADA systems can be linked to flows around the plant, alarms linked with pressure of flows, and maintenance needs. SCADA systems can also be linked with real-time sensors [64].

35.4.4 Real-Time Monitoring

Ireland and other countries need efficient, sustainable, real-time, high-quality environmental monitoring systems, which can be harmonized throughout the country. Data quality control is a key element in sustainable environmental management [34].

The importance of coordination of data monitoring in order to enhance the sustainability of WWTPs is seen in Figure 35.9. Bench marking and coordination are important in order to ensure good efficiency of WWTPs.

35.4.5 Conventional Sampling and Analysis

Conventional sampling and analysis can complement real-time monitoring. Grab and composite sampling compliment real-time monitoring. From an operator's point of view, conventional sampling can enable the operator to physically observe the performance of the plant such as mentioned earlier; the physical color of the sludge can tell the condition of the sludge, and how effective the sludge is at breaking down the waste. Visual observations of the coarse filters at preliminary treatment can inform an experienced operator of the need for maintenance.

35.4.6 Maintenance Monitoring Requirements

Odors can be an indicator of maintenance needs due to issues around plants such as infrequent removal of settled solids, inadequate sludge removal, and release of odorous gases dissolved in primary overflow. Other maintenance problems

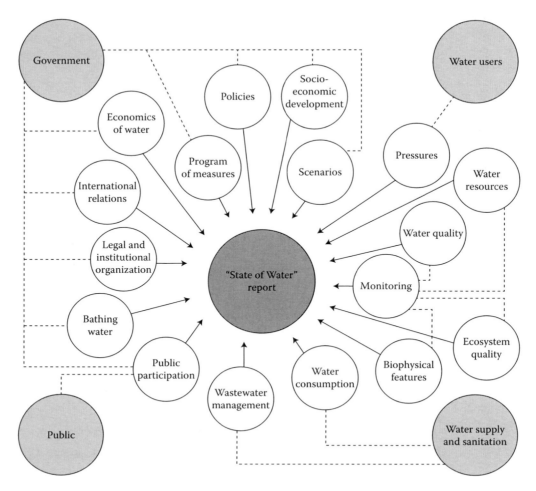

FIGURE 35.9 Coordination of data to improve water quality. (Adapted from EEA. 2011. *Europe's Environment—An Assessment of Assessments.* EEA, Copenhagen.)

include insufficient oxygen in the activated sludge system, and septic conditions as a result of hydraulic overloading or clogging. Other odor problems include capacity organic overload of the plant, inadequacy wetting of media, and inadequate ventilation [15].

In recent years, computer-based techniques such as SCADA systems have been developed for studying water quality management systems and providing related decision support [64].

35.5 Distribution Systems

The distribution system is often not considered part of the overall wastewater treatment process. This is an area that requires much financial support to ensure good maintenance and efficiency.

35.5.1 Infrastructure Maintenance and Efficiency

It is important to outline the most frequent types of defects that are present in wastewater pipe systems. These vary by the type of

pipe system and by the type of pipe material used. The common deflects can be summarized as follows [61]:

- Hydraulics of plant centrifugal flow pump and head loss
- Cracks
- Internal corrosion
- Grease build-up (plus grit and debris)
- Root intrusion
- Joint misalignment/separation/leakage
- Excessive pipe deflection
- Lateral connection/leakage grade and alignment

The financial constraints around infrastructure of WWTP are a major concern. If WWTPs can become more self-financing, this would support the intrastructural upgrading and maintenance issues of the plant.

35.6 New Directions and Concerns

Some interesting initiatives and concerns related to WWTP operations and distribution systems will now be discussed.

35.6.1 Global Warming Concerns

Global warming can have severe effects on society. One such effect is more extensive rainwater, giving rise to flash flooding and more stormwater problems. Stormwater is retained in stormwater holding tanks in WWTPs and treated sequentially to take the pressure off the loading capacity of the plant. In cities such as Limerick and Kilkenny in Ireland, the extensive constructive of roads, parking lots, and building reduces the area for land available to absorb runoff. The excessive runoff instead is diverted into a stormwater system. Stormwater can accumulate chemicals such as the residuals of oil, road salt, garden chemicals, and car exhaust chemicals. Sewers can consist of stormwater sewers, combined sewers, and sewage sewers. The sewage sewers are diverted directly to the plant's treatment process. Stormwater sewers and combined sewers can be diverted to holding tanks within the plant. In certain cities and towns, the stormwater holding tanks are overflowing and the water has to be directed into the received water without treatment. By creating greater awareness on how society can control the quality of stormwater and how remediation technologies can improve stormwater quality, certain pressures related to stormwater concerns can be eliminated. An interesting review examining climate change and water can be seen in the literature [52].

35.6.2 The Multibarrier Approach

The multibarrier approach is beginning to be placed in high regard in relation to water management. The multibarrier approach examines wastewater treatment not just in terms of plant operations but also in terms of controlling the inputs and the distribution of the outputs. The words *point* and *diffuse source of pollution* have a big role to play in the multibarrier approach. The multibarrier approach will enable WWTP operators to use a holistic approach to WWTP operations [54].

35.6.3 The Precautionary Principle (PP)

The implementation of the precautionary principle (PP) to domestic wastewater treatment systems, which has been strongly applied in world trade decisions and the REACH directive, could potentially enhance the operations of the plant. The PP applies to situations where scientific evidence is insufficient, inconclusive, or uncertain. Preliminary scientific evaluation indicates that there are reasonable grounds for concern that the potentially dangerous effects on the environment, human health, and animal or plant health may be inconsistent, with the high level of protection recommended by the EU. The PP is aimed at addressing problems of decisions under significant uncertainty where a course of action poses a risk that cannot be fully quantified scientifically. The variables of risk related to the PP can be categorized as follows: assimilative capacity, risk based on quantifying and analyzing problems rather than solving them, and acceptable risk. The PP is basically related to controlling and minimizing the risk and uncertainty surrounding the WWTP operations [9].

35.6.4 Life-Cycle Assessment

Life-cycle analysis involves studying a WWTP from the cradle to the grave approach [8,42]. This approach is being developed to improve the overall operations and performance of WWTPs.

35.6.5 BATNEEC Principle

The term BATNEEC means "the best available technology not entailing excessive costs." In terms of the operations of a WWTP, the "best" means the best approach to prevent pollution. "Available" means it is procurable by the industry concerned. The "technology" variable is based on the technologies available for use within the WWTP operations and the use of these techniques [24]. The BATNEEC approach can be used to prevent, eliminate, limit, abate, or reduce any emissions from an activity. The energy use in a WWTP can be categorized as electrical, manual, fuel assumption, and chemical use [51]. One of the big cost factors for WWTPs are its energy costs. Energy savings in a WWTP is an emerging research need in the vicinity of WWTPs due to climate change concerns and economical benefits [51]. In today's society, WWTPs need to reassess the operations of the plant in order to enhance the efficiency of the plants in terms of the BATNEEC approach. Areas to be addressed include understanding and controlling outputs and inputs, identification of any aspects of the WWTP that can influence the quality of the discharged wastewater such as maintenance shutdown, technologies used in the plant at different stages of treatment, and time for each stage of the wastewater treatment process [19,41].

35.6.6 Moving in the Direction of Zero Waste

The challenge is to move in the direction of zero waste. Wastewater recycling within the plant is one such initiative in terms of zero waste. Other initiatives can be seen in the literature [5,42,62]. Resource recovery from sludge has been previously mentioned in this chapter as a zero waste initiative.

35.6.7 Challenges in Wastewater Treatment Plant Operations

Failures in WWTPs should be eliminated by promoting the enhancement of the use of conventional sampling and analysis and real-time monitoring. Key equipment flow meters, telemetry, and alarms should be installed where possible. Training and revision courses for plant operators in order to improve the management of the operations of the WWTPs should be encouraged. Sharing knowledge such as demonstrated in Figure 35.9 is very important to enhance the operations of WWTPs [18]. Sharing knowledge and awareness involving a multidisciplinary team can empower WWTPs in Ireland, Europe, and globally.

35.7 Summary and Conclusions

The main issues constantly emerging with regard to wastewater operations and distribution systems include complying with regulation, coping with new regulation, and maintaining the infrastructure. In Ireland and other European countries, a number of WWTPs are contracted to private companies rather than directly run by local authorities [54]. Other concerns are bench marking the performance of the WWTPs and developing key performance indicators [4,25]. Upgrading and loading capacity upgrades is also a concern in WWTP operations [55]. Public awareness on the importance of understanding the operations of WWTPs and the consequences of waste disposal is very important for the future operations of our WWTPs in Ireland and globally.

Sharing knowledge on a global platform is important in the future to help sustain our water ecosystems [16,32]. Knowledge will help society move in the direction of the waste paradigm—resource recovery from waste rather than just treating our waste [29]. The paradigm will help society move in the direction of zero waste and sustainable water management.

References

1. American Society of Civil Engineering. 1998. *Aeration A Wastewater Treatment Process WPCF—Manual of Practice—No. FD-13 ASCE Manuals and Reports on Engineering Practice No. 68.* Water Pollution Control Federation, Virginia.
2. Anders, L.-J. 1999. Trends in monitoring of waste water systems. *Talanta*, 50(1999), 707–716.
3. Asano, T. and Levine, A.D. 1996. Wastewater reclamation recycling and reuse past, present and future. *Water Science and Technology*, 33(10–11), 1–14.
4. Balkema, A.J., Preisig, H.A., Otterpohl, R., and Lambert, F.J.D. 2002. Indicators for the sustainability assessment of wastewater treatment systems. *Urban Water*, 4(2), 153–161.
5. Buckhard, R., Deletic, A., and Craig, A. 2000. Techniques for water and waste water management: A review of techniques and their integration in planning. *Urban Water*, 3(2), 197–221.
6. Butler, D. and Davies, J.W. 2000. *Urban Drainage.* E. & F.N. Spon, London.
7. Butler, D., Friedler, E., and Gatt, K. 1995. Characterising the quantity and quality of domestic wastewater inflows. *Water Science and Technology,* 31(7), 13–24.
8. Corominas, L., Foley, J., Guest, J.S., Hospide, A., Larsen, H.F., Morera, S., and Shaw, A. 2013. Life cycle assessment applied to wastewater treatment; state of the art. *Water Research*, 47(15), 5480–5492.
9. Crawford-Brown, D. and Crawford-Brown, S. 2011. The precautionary principle in environmental regulation for drinking water. *Environmental Science and Policy*, 14, 379–387.
10. de los Ríos, A., Pérez, L., Ortiz-Zarragoitia, M., Serrano, T., Barbero, M.C., Echavarri-Erasun, B., Juanes, J.A., Orbea, A., and Cajaraville, M.P. 2013. Assessing the effects of treated and untreated urban discharges to estuarine and coastal waters applying selected biomarkers on caged mussels. *Marine Pollution Bulletin*, 77(1), 251–265.
11. Devane, N., Treacy, J., and Mc Donnell, K. 2008. Wastewater: Do you know where it goes? *ESAI Environews*, Winter (16), 15.
12. Dionergy. 2014. http://dionergy.com/contact-us. Accessed on October 23, 2014.
13. Dublin. 2014. http://www.dublincity.ie/main-menu-services-water-waste-and-environment-water-projects/dublin-bay-project. Accessed on October, 2014.
14. EC. 2000. http://ec.europa.eu/environment/water/water-urbanwaste/index_en.html. Accessed on July, 2014.
15. EC. 2005. The European Communities Wastewater Treatment (Prevention of Odours and Noise) Regulations 2005 SI No. 787 of 2005.
16. EC. 2006. Communication from the Commission to the European Parliament and the Council—Towards sustainable water management in the European Union—First stage in the implementation of the Water Framework Directive 2000/60/EC—[SEC(2007) 362] [SEC(2007) 363]/* COM/2007/0128 final */ [Regulation EC No 116/2006 of the European Parliament and the Council of 18th January 2006.
17. EC Europe. 2014. Directive 2000/60/EC of the European Parliament. http://www.doeni.gov.uk/niea//pubs/publications/waterframeworkdirective.pdf. Accessed on July, 2014.
18. EEA. 2011. *Europe's Environment—An Assessment of Assessments.* EEA, Copenhagen.
19. Ellis, J.B. 2000. Risk assessment approaches for ecosystem responses to transient events in urban receiving waters. *Chemosphere*, 41(2), 85–91.
20. EPA. 1995. *Waste water Treatment Manuals, Preliminary Treatment.* Environmental Protection Agency, Ireland.
21. EPA. 1997. *Waste Water Treatment Manuals Primary, Secondary and Tertiary Treatment.* Environmental Protection Agency, Ireland.
22. EPA. 2012. *Guideline on the Preparation and Submission of the Annual Environmental Report (AER) for Waste Water Discharge Licences for 2011.* Environmental Protection Agency, Ireland.
23. EPER. 2006. European Pollutant Release and Transfer Register Regulation EC No 166/2006 of the European parliament and of the council of 18th January 2006, concerning the establishment of a European pollution release and transfer register and amended council directives 91/689/EEC and 96/61/EEC[5] S.I. No. 722 of 2003. http://ec.europa.eu/environment/industry/stationary/eper/implementation.htm. Accessed on June 18, 2014.
24. Generowicz, A., Kulczycka, J., Kowalski, Z., and Banach, M. 2011. Assessment of waste management technology using BATNEEC options, technology quality method and multi-criteria analysis. *Journal of Environmental Management*, 92(2011), 1314–1340.

25. Gernaey, K.V., Jeppsson, U., Vanrolleghem, P.A., Copp, J.B., and Steyer, J.-P. (eds.). 2012. *Benchmarking of Control Strategies for Wastewater Treatment Plants*. IWA Scientific and Technical Report. IWA Publishing, London.

26. Grady, C.P.L., Daigger, G.T., Love, N.G., and Filipe, C.D.M. 2011. *Biological Waste Water Treatment*. IWA, UK.

27. Huang, G.H. and Xia, J. 2001. Barriers to sustainable water-quality management. *Journal of Environmental Management*, 61, 1–23.

28. ISO. 2014. Environmental management, the ISO 14000 family of international standards. http://www.iso.org/iso/home/about.htm. Accessed on May 13, 2014.

29. Jacobsen, Bo.N. 2014. *Performances of Water Utilities Beyond Compliance Sharing Knowledge Bases to Support Environmental and Resources Policies and Technical Improvements*. EEA Technical Report No 5. Publications Office of the European Union, Luxembourg.

30. Jiang, J.Q., Zhou, Z., and Sharna, V.K. 2013. Occurrence, transportation, monitoring and treatment of emerging micro-pollutants in waste water—A review from global views. *Microchemical Journal*, 110, 292–300.

31. Kale, V.V. and Bhusari, K.P. 2007. *Applied Microbiology*. Himalya Publishing, Himalya.

32. Larsen, T.A., Udert, K.M., and Lienert, J. (eds.). 2013. *Source Separation and Decentralization for Wastewater Management*. IWA, UK.

33. LC Consultants Ltd London. 2001. *Pollutants in Urban Waste Water and Sewage Sludge*. Office of Official Publications of the European Communities, Luxemburg.

34. Loucks, D.P. 1997. Quantifying trends in system sustainability. *Hydrological Sciences Journal*, 42, 513–530.

35. Lynggaard-Jensen, A. 1999. Trends in monitoring of waste water systems. *Talanta*, 50, 707–716.

36. Mino, T., van Loosdrecht, M.C.M., and Heijnen, J.J. 1996. Microbiology and biochemistry of the enhanced biological phosphate removal process. *Water Research*, 32(11), 3193–3207.

37. Moysa-Lobo, E. 2011. *Effects of Municipal Waste Water Containing Oil on Activated Sludge Under Aerobic Conditions*. Architecture Civil Engineering Environment, The Salesian University of Technology No. 4 147-150 (25-6-2011).

38. Muga, H.E. and Mihelcic, J.R. 2008. Sustainability of wastewater treatment technologies. *Journal of Environmental Management*, 88, 437–447.

39. Nicolaisen, B. 2003. Developments in membrane technology for water treatment. *Desalination*, 153(1–3), 355–360.

40. Oguma, K., Kita, R., Sakai, H., Murukami, M., and Takizawa, S. 2013. Application of UV light emitting diodes to batch and flow-through water disinfection systems. *Desalination*, 328, 24–30.

41. O'Reilly, A.J. 2000. Waste water treatment process selection: An industrial approach. *TransIChemE*, (78), 454–464.

42. Ortiz, M., Raloy, R., and Serra, L. 2007. Life cycle assessment of water treatment technologies waste water and water reuse in a small town. *Desalination*, 204, 121–131.

43. Papa, M., Pedrazzani, R., and Bertanza, G. 2013. How green are environmental technologies? A new approach for the global evaluation: The case of WWTP effluents ozonation. *Water Research*, 47(11), 3679–3687.

44. Petrasek, A.C., Kugelman, I.J., Austern, B.M., Pressley, A.T., Winslow, L.A., and Wise, R.H. 1983. Fate of toxic organic compounds in wastewater treatment plants. *Journal (Water Pollution Control Federation)*, 55(10), 1286–1296.

45. Petrovic, M., Gonzalez, S., and Barcelo, D. 2003. Emerging contaminants in wastewater and drinking water. *Trends in Analytical Chemistry*, 22(10), 685–696.

46. Petrovski, S., Dyson, Z.A., Eben, S., McIiroy, S.J., Tillett, D., and Sevior, R.J. 2011. An examination of the mechanisms for stable foam formation in activated sludge systems. *Water Research*, 45(5), 2146–2154.

47. Pujol, R., Duchene, P., Schetrite, S., and Canler, J.P. 1991. Biological foams in activated sludge plants—Characterization and situations. *Water Research*, 25(6), 1879–1448.

48. Response Group. 2014. http://www.response-group.ie. Accessed on October 16, 2014.

49. Richardson, S.D. 2005. New disinfectant by-products issues emerging DBPs and alternative routes of exposure. *Global Nest Journal*, 7(1), 43–60.

50. Shannon, D., Wall, B., and Flynn, D. 2014. *Focus on Urban Waste Water Treatment in 2012*. Environmental Protection Agency, Ireland.

51. Singh, P., Carliell-Marquet, C., and Kansal, A. 2012. Energy pattern analysis of a waste water treatment plant. *Applied Water Science*, 2(40), 221–226.

52. Smith, J. 2009. *Climate Change and Water: International Perspectives on Mitigation and Adaptation*. American Water Works Association (AWWA), London.

53. Somlyody, L. 1995. Water-quality management: Can we improve integration to face future problems? *Water Science and Technology*, 31, 249–259.

54. Spellman, F.R. 2009. *Handbook of Water and Wastewater Treatment Plant Operations*, 3rd Edition. CRC Press, Boca Raton, FL.

55. Stephens, A. and Fuller, M. 2009. *Waste and Waste Management: Sewage Treatment: Uses, Processes and Impact*. Nova Science, New York.

56. Surampalli, R.Y. and Tyagi, R.D. 2004. *Advances in Water and Wastewater Treatment*. Blackwell, North America.

57. Thompson, S.S., Jackson, J.L., Suvacastillo, M., Yanko, W.A., El Ziad, J., Kuo, J., Ching Lin Chen, William, F.P., and Schurr, D.P. 2003. Detection of infectious human adenoviruses in tertiary-treated and ultraviolet-disinfected waste water. *Water Environment Research*, 75(2), 163–170.

58. Treacy, J. 2009. Present and future technology for the treatment of sewage sludge in Ireland. 2005-WRM-EPA-Ds-e http://erc.epa.ie/safer/iso19115/displayISO19115.jsp?isoID=129. Accessed on July, 2015.

59. Trussell, R. and Hand, R. 2012. *MWH's Water Treatment Principles and Design*. Wiley, Hoboken, NJ.

60. Tyagi, V.K. and Lo, S.L. 2013. Sludge: A waste or renewable source for energy and resources recovery? *Renewable and Sustainable Energy Reviews*, 25, 708–728.

61. USEPA. 2009. *Rehabilitation of Wastewater Collection and Water Distribution Systems White Paper 600R09048*. National Service Center for Environmental Publications (NSCEP), Cincinnati, Ohio.

62. Verstraete, W. and Siegfried, E.V. 2011. Zero waste water short-cycling of waste water resources for sustainable of cities of the future. *International Journal of Sustainable Development and World Ecology*, 18(3), 253–264.

63. Water Environment Research Foundation (WERF). 2004. An examination of innovative methods used in the inspection of wastewater systems. http://www.werf.org/a/ka/Search/ResearchProfile.aspx?ReportId=01-CTS-7. Accessed on October, 2014.

64. WaterWorld. 2014. SCADA system monitors water transport, distribution, treatment. http://www.waterworld.com/articles/print/volume-27/issue-8/departments/automation-technology/scada-system-monitors-water-transport-distribution-treatment.html. Accessed on October 2014.

65. WEF/WERF. 2008. *Workshop W101 Demonstrated Processes for Limit of Technology Nutrient Removal Achievable Limits and Statistical Reliability*. WEFTEC 08 D. Parker (chair), Chicago, IL, October 18, 2008.

36

Greenhouse Gas Emissions from Decentralized Systems

Peter W. Schouten
CSIRO Land and Water

Ashok Sharma
Victoria University

Stewart Burn
CSIRO Land and Water

Nigel Goodman
CSIRO Land and Water

PREFACE

The emission of greenhouse gases (GHG) has gained a great deal of interest from scientists, policy makers, and the general public in recent history. Two GHGs that have come under increasing examination are nitrous oxide (N_2O) and methane (CH_4). N_2O and CH_4 are both known to be significant contributors to the greenhouse effect, at a level higher than that of carbon dioxide (CO_2). Not only can N_2O and CH_4 emissions be generated in industrial and agricultural systems, but also fugitive emissions of N_2O and CH_4 can occur in wastewater treatment processes. The magnitude and temporal variation of these fugitive emissions have been measured in investigations performed at numerous wastewater treatment systems using a variety of analysis techniques. This chapter provides a literature review summarizing how these investigations were performed and their most significant outcomes. In addition, recommendations are made on how this research can be developed to better understand wastewater-based GHG production, how to capture and reuse GHG emissions onsite, and how to improve the accuracy of N_2O and CH_4 measurements made at both large-scale (centralized) and small-scale (decentralized) wastewater treatment plants in order to deliver better GHG footprint reporting data for water management authorities.

36.1 Introduction

Ongoing emissions of greenhouse gases (GHG) have been under extensive examination by the scientific community and policy makers in recent history. The three GHGs that have been subject to the most scrutiny are carbon dioxide (CO_2), nitrous oxide (N_2O), and methane (CH_4). Once present in the atmosphere, CO_2, N_2O, and CH_4 can contribute to the greenhouse effect, with N_2O and CH_4 both found to be more efficient at trapping infrared radiation (heat) in comparison to CO_2. Specifically, CH_4 has 25 times and N_2O has 298 times greater atmospheric heating potential over 100 years in comparison to CO_2 [26]. From just before the Industrial Revolution (midway through the eighteenth century), up to the late twentieth century, atmospheric CH_4 increased by a factor of 2.5 up to 1745 parts per billion by volume (ppbv) and atmospheric N_2O increased by 45 ppbv up to 314 ppbv [66]. Along with CO_2, CH_4, and N_2O contribute close to 80% of the positive atmospheric radiative forcing caused by long-lived GHGs [66]. As a result, both CH_4 and N_2O are recognized and addressed by the Kyoto Protocol.

In the vast majority of the first world, sewage is removed via an extensive infrastructure of gravity and pressure-fed pipelines running from various industrial and domestic sources, and is collected at both centralized (large-scale sites servicing extensive populations and catchment areas) and decentralized (small-scale sites servicing smaller populations and catchment areas) wastewater treatment plants (WWTPs). At these WWTPs, a combination of physical, chemical, and biochemical mechanisms actively filter out or break down solid objects, chemicals, solid organic matter, and pathogens present in the influent. Organic solids in the wastewater are removed via a dynamic biochemical process in which freely roaming microorganisms utilize the solids as a food source. From this process, a biologically activated sludge is generated. This sludge can be recycled back into the treatment process in order to reseed more biochemical activity or it can be removed, after which it may be burnt off, buried, or reused in another application offsite (such as fertilizer). Various stages of this treatment cycle can take place under both aerobic (oxygen present) or anoxic (oxygen absent) conditions as defined by the specific requirements of site operators. Specifically, CH_4 gas is known to be produced and released into the atmosphere primarily during anoxic wastewater treatment processes and during sludge handling and storage [21]. In addition, N_2O gas is produced and emitted as a by-product of nitrification and denitrification taking place in treatment reactors. Along with CH_4 and N_2O, CO_2 emissions can also be readily released from a variety of processes within the wastewater treatment cycle. However, this CO_2 is regarded as being a short half-life biogenic gas produced by the breakdown of organic matter. As it is not generated by fossil fuel burning and does not contribute to the greenhouse effect, it is not included in any GHG footprint accounting.

A large number of water management authorities use desktop modeling (generally by employing a software spreadsheet package) to estimate their total GHG production from both energy use and from fugitive emissions of N_2O and CH_4. For this, most water management authorities use a number of semi-empirical models to make estimates of the overall annual GHG footprint for each of their WWTPs. The data used in these semi-empirical models is often obtained from real-world measurements made using water quality measurement instrumentation deployed at their WWTPs, from grab samples analyzed in a laboratory, or from inferences or extrapolations from previously published data. This data is generally averaged out over an extended time interval (usually over one year), and as such peaks and troughs in the daily data stream are smoothed out or removed entirely. Empirical emission and conversion factors are also inserted into models, which may or may not have been derived from

measurements taken from WWTPs using entirely different treatment and management practices to the WWTP under analysis. In addition, numerous basic assumptions are usually made during the modeling process, which may lead to further miss-estimations in modeled GHG data. As a result, the final emissions reports generated by water management authorities may be inaccurate. This could mean that water management authorities may end up paying either a smaller or a larger amount of carbon taxation than required (if they are required to do so). Specifically, due to the greater relative potential of N_2O and CH_4 to influence global climate change over time in comparison to CO_2, even relatively small inaccuracies in the reporting may lead to water management authorities being greatly under- or overcharged. As a result, an increasing number of water management authorities are measuring the fugitive GHG emissions from their WWTPs using appropriate online instrumentation, in order to improve their modeling systems and to make more accurate calculations of their yearly overall GHG footprint. By determining the extent and the location of fugitive GHG emissions within their WWTPs using online monitoring, water management authorities are also installing equipment to capture and store N_2O or CH_4 so it can be sold to other companies, or are reusing the emitted N_2O or CH_4 to power mechanical processes taking place within the WWTP.

The objective of this chapter is to provide a literature review on where and how fugitive N_2O and CH_4 emissions are generated and emitted during modern wastewater treatment practices. In addition, a treatise is provided on how these emissions can be accurately measured both in the field and in the laboratory using in-situ and ex-situ analysis techniques and how they can be mitigated and captured using new noninvasive technologies. A literature review detailing the outcomes from the most prevalent peer-reviewed GHG measurement studies performed at numerous types of wastewater treatment systems located across the world has also been detailed. Finally, several recommendations are made on how to further advance this research area in order to improve the accuracy and validity of N_2O and CH_4 measurements made at both centralized and decentralized wastewater treatment facilities in order to provide better fugitive GHG footprint reporting data. In addition, several gaps in the current worldwide wastewater treatment GHG inventory and knowledge base are identified and a series of suggestions are made on how to effectively remedy them.

36.2 Emission and Measurement of Fugitive Nitrous Oxide and Methane from Wastewater

36.2.1 Generation and Distribution of Fugitive Nitrous Oxide in Wastewater

In general, N_2O is produced in and released from wastewater that contains a concentration of nitrogen-based organic material, typically originating from waste matter produced by both humans and animals. At the cellular level, N_2O is ejected from microbial cells during the nitrification and denitrification process during wastewater treatment. Specifically, N_2O is produced via two pathways: the oxidation of ammonia to nitrate under aerobic conditions (nitrification) and the reduction of nitrate to dinitrogen gas under anaerobic conditions (denitrification). Predominantly nitrate (NO_3) and nitrogen gas (N_2) are provided by nitrification and denitrification, respectively. However, a small amount of available N can be released in the form of N_2O [45]. Specifically, nitrification occurs in aerobic conditions when ammonium-oxidizing bacteria (AOB) and ammonium-oxidizing archaea (AOA) convert ammonia into nitrite, and nitrite-oxidizing bacteria (NOB) converts nitrite into nitrate [29]. In wastewater treatment, nitrification is facilitated by autotrophic AOB and NOB using either ammonia or nitrite for energy and CO_2 for carbon [29]. AOB are known to produce N_2O, as they have the enzymes necessary to break down NO_2—N and NO leaving N_2O as a remainder [21]. Denitrification occurs in anaerobic conditions and is made possible by a variety of archaea, bacteria, and microorganisms that start the oxidation of organic or inorganic substrates to the reduction of nitrate, nitrite, NO, and N_2O [21]. As a result, as N_2O is released as an intermediate product during denitrification, incomplete denitrification can lead to the emission of N_2O gas into the local atmosphere.

The time required to eliminate N_2O from the atmosphere is much longer in comparison to CH_4. Specifically, less than 1% of atmospheric N_2O is eliminated each year via stratospheric oxidative reactions and photolysis [45]. Due to the fact that N_2O is produced during nitrification and denitrification, and aerated treatment processes are generally more like to emit N_2O gas than nonaerated treatment processes, the largest N_2O emissions within a WWTP generally take place in aeration tanks employing the activated sludge process [23,51]. N_2O emissions can also be measured at lower concentrations around anoxic/anaerobic tanks and in effluent holding lagoons [21]. In addition to this, measurements performed by Czepiel et al. [9] have shown that relatively low N_2O emissions can emanate from grit tanks and sludge storage tanks. Organic reactive nitrogen can also be stored in final effluent and in waste sludge. Both final effluent and waste sludge may be returned to the environment (via natural waterways for the effluent and via landfills, land application, or stockpiles for the sludge) and as such will be available for nitrification and denitrification resulting in further possible N_2O emissions from the WWTP [1]. In general, N_2O emissions have been shown to vary significantly between WWTPs, with the variations stemming from different plant sewage flow and treatment configurations and operational protocols along with dissimilar influent biochemical properties [36].

Bhunia et al. [4] has previously stated that there are numerous findings in the literature stating what biochemical conditions are the most likely to generate N_2O gas, with some studies [2,3,31,46,47,48]; showing that varying levels of dissolved oxygen (DO) within the aeration treatment cycle were the primary causative factor, while other investigations have detailed how denitrification is the main pathway from which N_2O is emitted

from raw sewage in a low oxygen environment [13,14,57]; or from wastewater with a low BOD to N ratio [33]. Gaseous N_2O emissions can be monitored over time using traditional gas sampling and measurement techniques, with a specific measurement protocol developed recently [7]. However, these N_2O emissions cannot be used as a final measure of total N_2O production because N_2O has high water solubility. As a result, the direct measurement of N_2O emissions from reactor surfaces most probably will not provide an accurate representation of the N_2O production rate in the sewage. This is especially the case in reactors with no stirring or mixing process, whereby the absence of agitation results in less gas stripping. However, dissolved measurements of N_2O can be made with handheld online sensors, such as the Unisense N_2O microsensor, which is capable of detecting minute N_2O concentrations in wastewater without being effected by oxygen.

Between 0.5% and 1% of all N arriving into wastewater treatment systems is believed to be ejected as N_2O gas. However, recent field sampling is showing that the real N to N_2O conversion percentage can fluctuate highly depending on a wide variety of environmental factors and treatment conditions [20,24,25,29,60], and may also be dependent on the organic load contained in the influent, which can be highly variable [13]. Recent continuous diurnal monitoring studies have shown that a peak in N_2O production and ejection can be expected during times of maximum N load occurring within a treatment plant [40]. Numerous strategies have been postulated and developed in order to minimize N_2O emissions from wastewater treatment processes. Unfortunately, none of these have been tested and proven at real-world WWTPs [36].

36.2.2 Generation and Distribution of Fugitive Methane Emissions in Wastewater

CH_4 is currently the second most prevalent GHG in the atmosphere behind CO_2. CH_4 is primarily produced in anaerobic wastewater treatment processes as a by-product of anaerobic digestion activated by methanogenic bacteria. Once produced, CH_4 is released to the atmosphere by either mechanical aeration or surface diffusion. Specifically, the anaerobic digestion process takes place in four separate stages, with each stage requiring facilitation by a specific group of microorganisms. These four stages are as follows:

Hydrolysis: The breakdown of nonsoluble biopolymers to soluble organic compounds
Acidogenesis: The breakdown of soluble organic compounds to volatile fatty acids and CO_2
Acetogenesis: The breakdown of volatile fatty acids to acetate and H_2
Methanogenesis: The breakdown of acetate, CO_2, and H_2 to CH_4 [44]

Anaerobic digestion and CH_4 production are both directly linked to water toxicity, temperature, pH, and chemical oxygen demand (COD). Generally, COD can be employed as a predictor for the potential of CH_4 production and emission to occur from a wastewater stream [44]. Sludge generated from various wastewater treatment processes can also be subject to anaerobic breakdown when in storage, and as a result CH_4 emissions can also be produced [65]. Investigations have shown that CH_4 is most readily emitted from WWTPs in processes where high biological oxygen demand (BOD) and low oxygen concentration conditions are present. These locations can include primary sedimentation tanks, secondary clarification tanks, and sludge storage [10]. Additionally, measurable CH_4 fluxes are known to be present in Imhoff tanks, anaerobic digesters, and aerobic tanks [21,34]. It is believed that the majority of the CH_4 emitted from aeration tanks is generated earlier in the treatment process or arrives from the rejection water returned from sludge handing [21]. In WWTPs using an anaerobic digester for sludge processing, the biogas produced by the digester is generally flared off in a burner that may allow for further CH_4 emissions to escape in the off-gas [11,38,68].

Recent investigations suggest that CH_4 emissions from WWTPs can be modulated by varying DO concentrations in aerobic grit chambers and anoxic tanks, and by temperature control in sedimentation tanks [65]. In general, environmental parameters that can be used as indicators of possible CH_4 production can include: pH, temperature, competition between methanogenic bacteria, sulfate reducing bacteria and toxicants, and wastewater treatment degree and retention time [17]. Specifically, the biophysical conditions most conductive for the emission of CH_4 within the wastewater treatment process are minimal to no aeration, high amounts of solid settling and buildup, accumulation of a sludge blanket within which anaerobic conditions can prevail, and a solids retention time (SRT) of less than 30 days (at ambient temperature) [58]. In addition, decomposition processes in wastewater treatment can demand greater amounts of oxygen than can be delivered by surface diffusion. In this scenario, when no supplementary artificial aeration is supplied, methanogenic bacteria will initiate anaerobic decomposition from which CH_4 gas can be produced and eventually released into the atmosphere [10]. Along with emissions produced directly from a WWTP, large amounts of CH_4 can be produced within sewer systems [22]. This CH_4 can be emitted into the atmosphere and, as a result, may have a potential GHG effect that is similar to the footprint that is associated with the energy generation necessary to power a WWTP [22].

36.2.3 Fugitive Nitrous Oxide and Methane Emissions Analysis Techniques

36.2.3.1 Sample Gas Collection and Infrared Analysis

Numerous invasive and noninvasive techniques have been employed to measure N_2O and CH_4 emissions from wastewater treatment processes at real-world WWTPs. The most common technique used has been the combination of staggered gas collection with a floating flux chamber/hood device with ex-situ laboratory gas analysis performed with a gas chromatography/mass spectrometry unit or similar device calibrated to a primary standard. In this technique, the flux chamber/hood is lowered onto the water surface, after which an airtight seal is created at the interface between the perimeter of the flux chamber/hood and the water. Gases are usually pumped out of the hood (at a constant flow rate) and captured in a Tedlar bag, syringe, or solid container situated inside or connected to the outside of the flux chamber/hood device. Desiccants and other similar filter materials are often situated in the off-gas pipeline and are used to remove any water vapor or other contaminants from the incoming gas before final collection. A fan is often placed inside the flux chamber/hood in order to improve air circulation above the water surface. In addition to this, various pulley and anchoring systems can be connected to the flux chamber/hood in order to better stabilize the position of the collection equipment, particularly on oscillating aerated water surfaces. Currently, gas chromatography/mass spectrometry is the most accurate equipment to determine gas concentrations within a grab sample. However, the collection of gases onsite with the flux chamber/hood has a number of inherent limitations [35]:

- The gas flux cannot be immediately determined on site, and as such it is not known if useable measurements have been obtained until returning to the laboratory. This may lead to multiple return trips to field locations, resulting in a loss of both time and money.
- There is minimal mixing of the air within the flux chamber/hood during the sampling process. This can result in the capture of inaccurate gas concentrations, particularly over extended sampling intervals.
- Loss of stability in the gas sample can occur over time when there is a substantial delay between gas capture in the field and concentration measurement in the laboratory.
- Depending on the efficiency of the gas extraction mechanism/pump, gas collection required for a single flux measurement may take a substantial amount of time. As a result, only a small number of treatment processes may be adequately evaluated over a single day.

Flux hoods used in conjunction with infrared gas analyzers have been used previously to measure GHG emissions from water surfaces in real-time. Tremblay et al. [61] is one such example employing flux chambers similar to one used originally by Carignan [6] in combination with both a nondispersive infrared (NDIR) gas analyzer and a Fourier transform infrared (FTIR) gas analyzer to measure GHG emissions from lakes and reservoirs around Canada. This measurement technique has recently been adopted and modified to measure GHGs released from large-scale and small-scale WWTPs [23]. Generally, the sampling technique works by trapping gases escaping from the water surface with a flux chamber/hood. These gases are pumped (and filtered to remove water and any contaminants that may skew the gas concentration readings) through to the gas analyzer. Once the constituent gas concentration is determined, the sampled gas is sent back into the flux chamber/hood so that the trapped air is continuously mixed, allowing for a much more accurate measure of gas concentration to be obtained [35,62]. The online use of either NDIR or FTIR gas analyzers has numerous advantages over the manual collection of gases onsite with ex-situ laboratory gas analysis. The most significant of these are [35]

- Flux measurements can be made rapidly. As such, many fluxes from multiple treatment processes can be calculated each day.
- Both NDIR and FTIR gas analyzers can be configured to allow for the simultaneous measurement of both N_2O and CH_4 concentrations along with any other gases of interest such as CO_2 and hydrogen sulfide (H_2S).
- Any inaccuracies with the gas concentrations and/or flux measurements caused by faulty equipment can be detected in real-time and fixed onsite.
- NDIR and FTIR gas analyzers are generally less expensive in comparison to gas chromatography/mass spectrometry instrumentation and require less maintenance and ongoing calibrations.

Unfortunately, the application of flux chambers to measure GHG emissions does have several inherent limitations that must be acknowledged. Numerous soil emissions studies have shown how flux chambers are susceptible to large increases in the chamber gas concentration over time, which can lead to a measurable forced decrease in the release of gas from the emitting surface. This often leads to the recording of a nonlinear gas concentration versus time relationship, resulting in an underestimation in the final flux calculation [39,49,64]. This is commonly referred to as the "chamber effect," and flux underestimations as large as 40% can occur from it [39,64]. This underestimation can be reduced by increasing the height of the flux chamber and reducing the total time taken for a single flux measurement [63,64]. One downside to using this error mitigation strategy is that the ratio between the gas concentration inside the flux chamber and the external ambient gas concentration will be reduced. As such, the gas analysis instrumentation employed must have a measurement range capable of accurately measuring the smaller gas concentration [64].

36.2.3.2 Eddy Covariance

CH_4 gas flux (along with other fluxes, such as sensible and latent heat) can be measured from water surfaces by utilizing gas exchange and atmospheric data taken from micrometeorological towers in combination with the eddy covariance technique. This particular method is very useful in locations where continuous flux measurements are required over a surface area spanning over several hundred square meters [52]. Such locations may include sludge drying pans, sewage treatment lagoons, and effluent basins. Jones et al. [27], Meijide et al. [41], and Schubert et al. [54] have shown that eddy covariance flux measurements do agree reasonably well with flux measurements made with a flux chamber/hood, as long as the chamber is within the fetch range of the eddy covariance system. The micrometeorological instrumentation required for eddy covariance is expensive, can be difficult to set up and requires a great deal of ongoing maintenance [16]. Also, eddy covariance and other online

continuous measurement systems may not be entirely useful for determining gaseous emissions released from relatively small and tightly confined surface areas, such as aeration and sedimentation tanks (especially those housed within decentralized systems and septic treatment sites), not only due to their size, but also due to the fact that emissions from treatment tanks do not have any influence upon the ambient GHG concentration in air more than several meters away from the tank [60]. Additionally, it must be ensured that the surface area under analysis must remain within the measurement footprint area. This can be difficult to maintain, as the shape of the measurement footprint can change rapidly, primarily due to variations in wind direction.

36.2.3.3 Other Methods

Several other GHG emissions measurement systems have been employed to a lesser extent than the aforementioned flux chamber/hood techniques. CH_4 and N_2O emissions can also be estimated from water samples removed from treatment tanks via glass bottles. Gases are extracted from the water in the laboratory using headspace or purge-and-trap techniques and are then fed through to a gas chromatography unit. These gas extraction methods yield high quality estimates, but each process can take several hours to convert all the gas from the dissolved phase [5]. Boundary layer flux equation (BLE) techniques combined with automated measurement systems have been employed to rapidly estimate GHG emissions at the air–water interface. Despite their simplicity, BLE techniques have been shown to be highly susceptible to wind fluctuations (especially at higher wind speeds), and may underestimate gas fluxes [16,54].

36.3 Previous Real-World Investigations

36.3.1 Fugitive Nitrous Oxide Measurements

To date, there have been a far greater number of studies investigating wastewater N_2O emissions in comparison to CH_4 emissions, with the majority of the emissions data obtained for N_2O varying greatly between each particular measurement study. The large variations in the published N_2O emissions data can be related to differences between the physical configurations of the WWTPs from site to site and the particular sewage environment created by the treatment processes under analysis. These differences include

- Different biochemical conditions in the primary and secondary treatment tanks (caused by varying aeration and anaerobic cycles, distinct sludge retention times along with dissimilar water quality parameter values)
- Different amounts and types of influent arriving from domestic and/or industrial sources, resulting in different levels of nitrogen, BOD, and COD being present in the wastewater prior to entry into the treatment process
- Varying atmospheric conditions (such as air temperature and pressure) occurring during the measurements

- Inherent differences between measurement systems (including accuracy and precision levels, primary and secondary calibration standards along with instrumentation susceptibility to external environmental factors such as air temperature and pressure)
- Data analysis protocols used by research groups may also play a part in the relatively large discrepancies seen in the current N_2O emissions inventory

The amount of time taken to measure the temporal variation of N_2O emissions also varies greatly between different studies (ranging from several hours over a single day up to several months), along with the frequency of the gas sampling (generally from a few minutes to several hours between collection/measurement). Naturally, estimates of annual per capita emissions, emission factors for N to N_2O conversion rates made using short-term flux calculations, are more likely to be less reliable than estimates made with flux calculations taken over extensive time periods.

The first published study measuring direct N_2O gas emissions from wastewater handling infrastructure was carried out by DeBruyn et al. [13,14]. This study involved off-gas sampling from three Belgian WWTPs in order to determine the primary biochemical reactions governing N_2O generation and to calculate general N_2O emissions factors for various types of wastewater. Since then, several other notable studies have been performed. The metadata and major findings from each of these studies are summarized in Table 36.1.

Aside from the research summarized in Table 36.1, the most wide-scale and detailed investigation evaluating N_2O emissions from centralized WWTPs was performed by a multinational research team for the Global Water Research Commission [21]. In this investigation, a widespread study was completed measuring fugitive N_2O emissions emitted from a wide variety of activated sludge WWTPs (using both BNR and non-BNR practices) at locations in Australia, the Netherlands, France, and the United States. N_2O emissions were found to be extremely variable from site to site. Several important findings emerged from the study: N_2O emission is produced most commonly via the nitrification pathway; high concentrations of ammonium nitrogen ($NH_4 + -N$) can facilitate the emission of N_2O if nitrification takes place; in aerobic treatment tanks, nitrite accumulation can lead to the formation of N_2O; treatment plants not using BNR that have nitrification occurring are more likely to produce N_2O emissions; high variations in organic loads can increase the probability of N_2O emissions taking place; and a high quality final effluent with a measured TN value of <5 mg L^{-1} can be regarded as an indicator of lower or minimal N_2O emissions occurring upstream at a WWTP. Further details pertaining to the N_2O measurements made by the Australian research team have been presented in References 18, 20.

36.3.2 Fugitive Methane Measurements

Numerous field investigations have been carried out reporting on CH_4 emissions from numerous natural and artificial sources.

However, there is only a small amount of wastewater-related CH_4 emission field studies presented in the literature. As noted in the previous section, far fewer investigations have been carried out on the measurement of CH_4 in comparison to N_2O. The very first field investigation measuring CH_4 emissions from wastewater treatment processes was performed by Kinnicutt et al. [32]. The next field study was carried out just over 80 years later by Czepiel et al. [10]. This study was undertaken at the same WWTP that was evaluated in the Czepiel et al. [9] N_2O emissions investigation. Following this initial research, only three more field CH_4 measurement campaigns have been completed and published. The metadata and the significant outcomes from these studies are detailed in Table 36.2.

In addition to the fieldwork described in Table 36.2, the Global Water Research Commission [21] has completed a large-scale CH_4 emission measurement program at several WWTPs located in Australia, the Netherlands, France, and the United States. The WWTPs analyzed used a variety of BNR and non-BNR treatment regimes. From this work, several new insights into how and where CH_4 is generated and emitted within the wastewater treatment process were presented, including the following: CH_4 is mainly generated in sewers, in influent works, and in the sludge handling process (after which it can be released in other locations within a WWTP such as aeration tanks via agitation); sulfuric odors are a good indicator of where CH_4 emissions may be occurring; and the combination of both CH_4 and N_2O fugitive emissions commonly constitute a large part of the total GHG footprint of a WWTP. Further information relating to the CH_4 emissions evaluated by the Australian research team can be found in References 18, 19.

36.4 Discussion

36.4.1 Appropriate Gas Measurement Scheme for Water Treatment Operators

It is known that GHG emissions can vary greatly between different WWTPs and within WWTPs over the space of a single day and throughout a season, depending on variations in influent organic load, aeration cycles, temperature, and other biochemical and environmental factors [21]. The use of uncalibrated models with predetermined or generic emission and conversion factors to calculate yearly emissions may not be entirely appropriate and may lead to sizeable under- or overestimations. Therefore, in order to more accurately determine the annual GHG footprint at WWTPs, water management authorities should consider permanently positioning appropriate online infrared gas monitoring equipment within each of their WWTPs, at locations where off-gases are expected to be generated and released. If this is not possible, daily grab water sampling or gas collection (during times of peak inflow and nutrient concentration) in conjunction with laboratory analysis will suffice. The spatial distribution of both N_2O and CH_4 emissions across the surface to air interface of a treatment tank/reactor may be nonhomogenous and could vary to a great extent across even a small surface area

TABLE 36.1 Metadata and Major Findings Summary from Previous Field Wastewater Treatment N_2O Gas Emissions Measurement Studies

Reference and Location	Processes/Systems Evaluated	Measurement Technique	Major Findings
[13,14], Belgium	Influent flow and settling tanks	Gas capture via airtight vessels Gas analysis performed by ex-situ gas chromatography	Nitrification was the primary reaction causing N_2O emission to occur from the influent and the settling tanks N_2O emissions from WWTPs may account for 0.6% of Belgium's annual total N_2O footprint
[10], United States (New Hampshire)	Grit tanks, primary settling, aeration tanks, secondary clarification, and sludge holding	Closed chamber gas capture and collection Gas analysis performed by ex-situ gas chromatography	Highest emissions were measured in the aeration tanks, where N_2O was produced via denitrification during primary settling and stripped by means of mechanical aeration A total yearly per capita emission of 3.2 g N_2O person^{-1} year^{-1} for the local population was calculated
[59], Sweden (Rya)	Sampled anoxic activated sludge	Gas analysis performed by gas chromatography	Temperature variations, carbon source, and nitrate or nitrite concentration did not influence generation of N_2O Changes in pH were linked to N_2O production No N_2O was detected at pH values greater than 6.8 The highest levels of N_2O production occurred when pH levels were between 5 and 6
[8], Germany (Bayreuth)	Sewer system	Closed chamber gas capture and collection Gas analysis performed by ex-situ gas chromatography	Postulated that N_2O was generated in the sewer biofilm by either nitrification or denitrification A per capita emission of 3.5 g N_2O person^{-1} year^{-1} was calculated
[31], Japan (Chiba Prefecture)	Intermittent aeration tank	Closed chamber gas capture and collection Gas analysis performed by ex-situ gas chromatography	Most N_2O was released during the aerobic period, with the per capita emission estimated to range between 0.43 g N_2O person^{-1} year^{-1} and 1.89 g N_2O person^{-1} year^{-1} Nitrification and denitrification took place during the intermittent regimes, allowing for a larger removal of N in comparison to a continuously aerated control regime
[50], France	Activated sludge aeration tanks at one high capacity WWTP and one low capacity WWTP	Closed chamber gas capture and collection Gas analysis performed by in-situ gas chromatography	221 kg and 42 kg N_2O year^{-1} were emitted from the low capacity and high capacity WWTPs, respectively, representing a per capita gas flux equivalent of 17.2 and 0.36 g N_2O person^{-1} year^{-1}
[28], Netherlands (Rotterdam)	Full-scale nitration and anammox reactors	Gases were sampled from the nitritation reactor headspace and from the anammox reactor gas recycle loop with a chemi-luminiscence gas analyzer	Low dissolved oxygen and high nitrite levels were postulated to be the main causation of N_2O emission by ammonia oxidizing bacteria The total yearly N_2O footprint for the system was calculated to be 8600 kg N_2O year^{-1}
[30], Netherlands	Single-stage nitritation–anammox reactor	Gases were sampled from the reactor headspace with a chemiluminiscence gas analyzer	1.2% of the total nitrogen load entering the reactor was converted to N_2O gas Limiting the oxygen concentration between 0 and 2 mg O_2 L^{-1} resulted in a decrease in N_2O emissions
[23], United States	Numerous systems from 12 BNR and non-BNR WWTPs	Closed chamber gas capture and collection Gas analysis performed by an online infrared gas analyzer	At BNR sites, aerobic zones emitted more N2O than anoxic zones Ammonium, nitrite, and DO levels were positively correlated with N2O emissions released in aerobic regions of activated sludge reactors Nitrite and DO levels were positively correlated with anoxic zone N2O emissions
[60], United States (Southern California)	A total of five WWTPs and water reclamation plants	Gas phase and liquid phase collection with stainless steel canisters and airtight bottles Gas analysis performed by ex-situ gas chromatography Stable isotope analysis was also performed	N_2O generation represented 1.2% of the total N removed from the inflow to the water reclamation plants In comparison, for the sites using traditional treatment processes for carbon oxidation only, N_2O generation represented 0.4% of the total N removed from the influent
[67], United Kingdom	WWTP ventilation system	Continuous off-gas monitoring in ventilation system	N_2O emissions from the activated sludge process (nitrification and denitrification) were found to provide 17.5% of the yearly total GHG emissions from the WWTP Decreasing DO led to higher N_2O emissions Recommended that the DO in aeration tanks be held above 1 mg^{-1} to minimize N_2O emissions during the nitrification process
[53], Australia	Aeration tank in a sewer mining system	Closed chamber gas capture and collection Gas analysis performed by an online infrared gas analyzer	Substantial N_2O emissions (up to 11.6 g N_2O m^{-2} day^{-1}) were measured with substantial fluctuations in flux occurring over short time periods Large variations in flux were also found to occur over short distances

TABLE 36.2   Metadata and Major Findings Summary from Previous Field Wastewater Treatment CH_4 Gas Emissions Measurement Studies

Reference and Location	Processes/Systems Evaluated	Measurement Technique	Major Findings
[9], United States (New Hampshire)	Grit tanks, primary settling, aeration tanks, secondary clarification, and sludge holding	Closed chamber gas capture and collection Gas analysis performed by ex-situ gas chromatography	Strong positive correlation between CH_4 flux and liquid temperature was found in aerated and nonaerated zones within grit tanks A per capita emission rate of 39 g CH_4 person^{-1} year^{-1} was calculated
[15,37], United States (California)	Eight different septic tank systems	Gas capture and collection done with a modified PVC closed chamber device Gas analysis performed by ex-situ gas chromatography	Total GHG emissions (CH_4, CO_2, and N_2O combined) measured from the septic tanks and from the septic tank vents were 276.5 and 327.1 g $CO_{2equivalent}$ person^{-1} day^{-1}, respectively
[65], China (Jinan)	Influent pump, aerated grit chambers, anaerobic/anoxic tanks, aeration tanks, final clarifier tanks, high density settler tanks, high efficiency fiber filter beds, and sludge processing units	Closed chamber gas capture and collection Gas analysis performed by ex-situ gas chromatography	A total annual per capita emission of 11.3 g CH_4 person^{-1} yr^{-1} was calculated The most prevalent source of CH_4 was the anaerobic tanks, with aerated grit chambers, aeration tanks, and sludge concentration tanks also releasing CH_4
[11], Netherlands (Kralingseveer)	Pipeline receiving pumped off-gases from primary and secondary treatment systems	Gas analysis performed by an online infrared gas analysis, mass balance, and the salting-out of dissolved CH_4	Average daily CH_4 measured across the entire WWTP was 302 kg CH_4 day^{-1} Close to 75% of the WWTPs CH_4 was generated by the anaerobic digestion of primary and secondary sludge Both the digested sludge buffer tank and the dewatered sludge storage tank produced the highest overall levels of CH_4 gas
[53], Australia	Aeration tank in a sewer mining system	Closed chamber gas capture and collection Gas analysis performed by an online infrared gas analyzer	Substantial CH_4 emissions (up to 1.1 g CH_4 m^{-2} day^{-1}) were measured Gas output changed significantly over both time and position across the tank

(especially in tanks with variable aeration and mixing profiles), and as a result single point measurements made over a limited time-frame are not sufficient in order to gain a detailed understanding of the total GHG footprint of a WWTP. As such, gas sampling should be performed over multiple evenly spaced sections over the entirety of a treatment tank/reactor in order to define the overall surface emissions distribution. Following this, an integration of the measured GHG emissions across the tank/reactor surface over time can be made to deduce the total GHG flux output. Supporting this, recent field research performed by Daelman et al. [12] has shown in detail that short-term point-scale measurements of N_2O cannot reliably evaluate the total N_2O emissions (and associated environmental footprint) released by a WWTP over an extensive time period, and that continuous online gas analysis is required to develop an understanding of the diurnal variations in fugitive gas emissions.

36.4.2 Fugitive Emissions from Decentralized Systems

The amount of real-world CH_4 and N_2O emission studies carried out at small-scale decentralized WWTPs treating sewage for small populations and catchment areas (generally in Greenfield and infill developments as well as self-sufficient eco-communities) is especially limited given that the uptake of decentralized WWTPs is rapidly increasing across the world [55].

Consequently, this represents a serious deficiency in the knowledge base. The effluent from these sites is commonly being employed to irrigate golf courses, public parks, and sports fields. The lack of information on the GHG footprint associated with these sites could be partly due to the reluctance of researchers to do field work in and around hazardous (and potentially infectious) raw and semi-processed wastewater or due to the inability to gain access to an appropriate cross-section of treatment facilities. However, it is important to gain a deeper understanding on the fugitive CH_4 and N_2O emissions released from these decentralized WWTPs, as some water management authorities are now operating as many as 10 sites (or even more in some cases) within an urban or suburban municipality.

Decentralized WWTPs running activated sludge treatment generally take in an annual total N load of between 10 and 300 tonnes and an annual total COD load of between 25 and 3000 tonnes. If the conversion percentage ratio of N to N_2O emission remains similar to an expected value of 1% and if the conversion percentage ratio of COD to CH_4 emission is consistent to an approximate maximum of 1% as reported in previous field studies conducted at centralized WWTPs [21], 0.1–3 tonnes of N_2O and 0.25–30 tonnes of CH_4 may be released from each decentralized WWTP per year. Consequently, over 10 WWTPs across an entire municipality, the total annual N_2O and CH_4 output may reach 30 tonnes and 300 tonnes, respectively. However, this estimate does not include emissions from sludge processing and the extra N_2O emissions that may occur beyond the final treatment stage, as the water solubility of N_2O is high, and as a result N_2O generated upstream in other treatment processes (such as in the aeration tanks) may remain in the effluent for an extended time interval. Additionally, reactive nitrogen that is held within the final effluent produced by decentralized wastewater treatment and recycling facilities may be applied across parks and playing fields, and as such will be available for nitrification and denitrification resulting in more long-term N_2O emissions, further exacerbating the site GHG footprint. This footprint will be increased if the treatment plant operators are required to build and run more decentralized WWTPs to service the constant population, housing, and infrastructure growth taking place in modern outer city and suburban regions.

36.4.3 Methods for Reducing Fugitive Emissions from WWTPs

The most direct way to reduce GHG emissions at both centralized and decentralized WWTPs could be to change and modify the day-to-day operating regime. Essentially, it may be possible to tune and optimize the wastewater treatment process to remove, limit, or increase GHG emissions, depending on the requirements and strategy of the plant operators. This is important, as some operators may wish to generate more GHG emissions in order to provide a consistent self-contained energy source for other processes taking place at the plant. There is a particularly high potential for this to be trialed at decentralized/smaller-scale WWTPs as they have a much smaller spatial footprint (with less treatment reactors and systems) and a smaller operational volume, which makes them far more controllable and manageable in comparison to their centralized/large-scale counterparts. Any measureable fluctuations in GHG emissions caused by deliberate changes made to treatment parameters should appear quicker at smaller WWTPs. Also, another advantage of using small-scale WWTPs is that any errors made by site operators during the GHG emission tuning process (which may inadvertently release excess amounts of GHG emissions into the atmosphere or deliver untreated nutrients into local waterways and storage ponds) are more likely to have a relatively low short-term environmental impact, relative to larger treatment plants servicing extensive catchment areas, due to their generally lower average daily inflow and organic loading. Operators may be able to control N_2O emissions via directly manipulating aeration levels along with altering the concentration of other materials in the sewage, such as nitrite and ammonium (if possible), and by ensuring the rapid changes in treatment system parameters are kept to an absolute minimum. However, depending upon operator requirements, BNR systems may be optimized to either promote or reduce complete nitrification and denitrification in order to decrease or increase N_2O emissions. CH_4 emissions may be varied as required by controlling temperatures and aeration levels in treatment tanks and by adequately optimizing sludge retention and digestion times along with the amount of burn-off flaring performed on off-gases emitted during anaerobic digestion and sludge handling.

36.4.4 Capture and Reuse of Fugitive Emissions

In order to minimize the cumulative effect that GHG emissions have on the natural environment, treatment systems at both small-scale and large-scale WWTPs could be enclosed (or have a permanent restrictive interface installed) in order to restrict the continuous venting of the fugitive N_2O and CH_4 into the surrounding atmosphere. In addition to this, a gas extraction system may be installed onsite in order to adequately remove and store the N_2O and CH_4 gases so they can be either destroyed or employed for another application either onsite or sold off for use elsewhere. For example, the Western Treatment Plant (WTP) in Melbourne, Australia, uses large-scale blankets spread over the surface of their treatment lagoons as a means to minimize odor and to reduce CH_4 emissions released into the atmosphere. These covers have helped to limit the total odor emissions from the lagoons by 90% since they were first installed in 1992, with the elimination of as much as an estimated 83,300 tonnes $C_2O_{equivalent}$ biogas emissions taking place each year [42,43]. After being blocked by the blankets, the CH_4 is captured and is used as a fuel source for an onsite power station, allowing the WTP to be self-sufficient for up to 95% of its total annual electricity requirement [42]. Collected N_2O gas can also be effectively used as a source of energy in a wide variety of applications. For instance, N_2O gases emitted from wastewater treatment can be burnt off and broken down in an engine, producing environmentally neutral oxygen and nitrogen along with heat. This heat can then be converted into energy for a multitude of uses, such as powering other functions across a WWTP [56].

36.5 Summary and Conclusions

The continuation of the monitoring and analysis of fugitive CH_4 and N_2O emissions made at various types of centralized and decentralized WWTPs is important, as there is still relatively very little data currently available on the total environmental footprint related to these sites. Further measurements are needed to calibrate and optimize existing modeling techniques in order to improve the accuracy of environmental footprint reporting performed by water management authorities and operators, which has been shown to be too general in application and at times highly inaccurate [21]. In addition, the detailed measurement of fugitive emissions at centralized and decentralized WWTPs not only help to better quantify the magnitude and extent of N_2O and CH_4 production over time, but can also assist in determining how N_2O and CH_4 gases are generated and released from wastewater. As a result, they can aid in the development of innovative strategies on how N_2O and CH_4 fugitive emissions can be reduced, captured, and recycled. Emissions data is also required to evaluate if centralized and decentralized WWTPs are capable of being continuously self-powered by the efficient conversion of their emissions into a sustainable energy supply. This is important to quantify, as it is necessary to determine if the increasing installation of WWTPs (most particularly decentralized WWTPs) will be ecologically and economically sustainable going into the future.

References

1. Barton, P. and Atwater, J. 2002. Nitrous oxide emissions and the anthropogenic nitrogen in wastewater and solid waste. *Journal of Environmental Engineering (ASCE),* 128(2), 137–150.

2. Beline, F. and Martinez, J. 2002. Nitrogen transformations during biological aerobic treatment of pig slurry: Effect of intermittent aeration on nitrous oxide emissions. *Bioresource Technology,* 83, 225–228.

3. Beline, F., Martinez, J., Chadwick, D., Guiziou, F., and Coste, C.-M. 1999. Factors affecting nitrogen transformations and related nitrous oxide emissions from aerobically treated piggery slurry. *Journal of Agricultural Engineering Research,* 73, 235–243.

4. Bhunia, P., Yan, S., LeBlanc, R.J., Tyagi, R.D., Surampalli, R.Y., and Zhang, T.C. 2010. Insight into nitrous oxide emissions from biological wastewater treatment and biosolids disposal. *Practice Periodical of Hazardous, Toxic, and Radioactive Waste Management,* 14(3), 158–169.

5. Boulart, C., Connelly, D.P., and Mowlem, M.C. 2010. Sensors and technologies for *in situ* dissolved methane measurements and their evaluation using technology readiness levels. *Trends in Analytical Chemistry,* 29(2), 186–195.

6. Carignan, R. 1998. Automated determination of carbon dioxide, oxygen, and nitrogen partial pressures in surface waters. *Limnology and Oceanography,* 43, 969–975.

7. Chandran, K. 2011. Protocol for the measurement of nitrous oxide fluxes from biological wastewater treatment plants. *Methods in Enzymology,* 486, 369–385. DOI: 10.1016/S0076-6879(11)86016-7.

8. Clemens, J. and Haas, B. 1997. Nitrous oxide emissions in sewer systems. *Acta Hydrochimica et Hydrobiologica,* 25, 96–99.

9. Czepiel, P., Crill, P., and Harriss, R. 1995. Nitrous oxide emissions from municipal wastewater treatment. *Environmental Science and Technology,* 29, 2352–2356.

10. Czepiel, P.M., Crill, P.M., and Harriss, R.C. 1993. Methane emissions for municipal wastewater treatment processes. *Environmental Science and Technology,* 27, 2472–2477.

11. Daelman, M.R.J., van Voorthuizen, E.M., van Dongen, U.G.J.M., Volcke, E.I.P., and van Loosdrecht, M.C.M. 2012. Methane emission during municipal wastewater treatment. *Water Research,* 46(11), 3657–3670.

12. Daelman, M.R.J., de Baets, B., van Loosdrecht, M.C.M., and Volcke, E.I.P. 2013. Influence of sampling strategies on the estimated nitrous oxide emission from wastewater treatment plants. *Water Research,* 47(9), 3120–3130.

13. DeBruyn, W., Lissens, G., Van Rensbergen, J., and Wevers, M. 1994. Nitrous oxide emissions from wastewater. *Environental Monitoring and Assessment,* 31, 159–165.

14. DeBruyn, W., Wevers, M., and Van Rensburgen, J. 1994. The measurement of nitrous oxide from sewer systems in Belgium. *Fertiliser Research,* 37, 201–205.

15. Diaz-Valbuena, L.R., Leverenz, H.L., Cappa, C.D., Tchobanoglous, G., Horwath, W.R., and Darby, J.L. 2011. Methane, carbon dioxide, and nitrous oxide emissions from septic tank systems. *Environmental Science and Technology,* 45, 2741–2747.

16. Duchemin, E., Lucotte, M., and Canuel, R. 1999. Comparison of static chamber and thin boundary layer equation methods for measuring greenhouse gas emissions from large water bodies. *Environmental Science and Technology,* 33, 350–357.

17. El-Fadel, M. and Massoud, M. 2001. Methane emissions from wastewater management. *Environmental Pollution,* 114, 177–185.

18. Foley, J. and Lant, P. 2009. *Direct Methane and Nitrous Oxide Emissions from Full-Scale Wastewater Treatment Systems.* Report No. 24. Water Services Association of Australia, Australia.

19. Foley, J., Yuan, Z., and Lant, P. 2009. Dissolved methane in rising main sewer systems: Field measurements and simple model development for estimating greenhouse gas emissions. *Water Science and Technology,* 60(11), 2963–2971.

20. Foley, J. de Haas, D., Yuan, Z., and Lant, P. 2010. Nitrous oxide generation in full-scale biological nutrient removal wastewater treatment plants. *Water Research,* 44, 831–844.

21. Global Water Research Commission. 2011. *N_2O and CH_4 Emission from Wastewater Collection and Treatment Systems.* Technical Report. Report of the GWRC research strategy workshop. Global Water Research Coalition, London.

22. Guisasola, A., de Haas, D., Keller, J., and Yuan, Z. 2008. Methane formation in sewer systems. *Water Research,* 42, 1421–1430.

23. Ho Ahn, J., Kim, S., Park, H., Rahm, B., Pagilla, K., and Chandran, K. 2010. N_2O emissions from activated sludge processes, 2008–2009: Results of a national monitoring survey in the United States. *Environmental Science and Technology,* 44, 4505–4511.

24. Intergovernmental Panel on Climate Change (IPCC). 1996. IPCC guidelines for national greenhouse gas inventories. http://www.ipcc-nggip.iges.or.jp/public/gl/invs1.html. Accessed August 17, 2012.

25. Intergovernmental Panel on Climate Change (IPCC). 2006. 2006 IPCC guidelines for national greenhouse gas inventories. In: H.S. Eggleston, L. Buendia, K. Miwa, T. Ngara, and K. Tanabe (eds.), *The National Greenhouse Gas Inventories Programme.* IGES, Hayama, Japan.

26. Intergovernmental Panel on Climate Change (IPCC). 2007. *Climate Change 2007: The Physical Science Basis, Contribution of Working Group I to the Fourth Assessment Report of the Intergovernmental Panel on Climate Change.* Cambridge University Press, Cambridge, UK.

27. Jones, S.K., Famulari, D., Di Marco, C.F., Nemitz, E., Skiba, U.M., Rees, R.M., and Sutton, M.A. 2011. Nitrous oxide emissions from managed grassland: A comparison of eddy covariance and static chamber measurements.

Atmospheric Measurement Techniques Discussion, 4, 1079–1112.

28. Kampschreur, M.J., Poldermans, R., Kleerebezem, R., van der Star, W.R.L., Haarhuis, R., Abma, W.R., Jetten, M.S.M., and van Loosdrecht, M.C.M. 2009. Emission of nitrous oxide and nitric oxide from a full-scale single-stage nitritation-anammox reactor. *Water Science and Technology*, 60(12), 3211–3217.

29. Kampschreur, M.J., Temmink, H., Kleerebezem, R., Jetten, M.S., and van Loosdrecht, M.C. 2009. Nitrous oxide emission during wastewater treatment. *Water Research*, 43(17), 4093–4103.

30. Kampschreur, M.J., van der Star, W.R.L., Wielders, H.A., Mulder, J.W., Jettena, M.S.M., and van Loosdrecht, M.C.M. 2008. Dynamics of nitric oxide and nitrous oxide emission during full-scale reject water treatment. *Water Research*, 42, 812–826.

31. Kimoshi, Y., Inamori, Y., Mizuochi, M., Xu K-Q., and Matsumura M. 1998. Nitrogen removal and N_2O emission in a full-scale domestic wastewater treatment plant with intermittent aeration. *Journal of Fermentation and Bioengineering*, 86(2), 202–206.

32. Kinnicutt, L.P., Winslow, C.E.A., and Pratt, R.W. 1910. *Sewage Disposal*. John Wiley & Sons, New York.

33. Kishida, N., Kim, H.J., Kimoshi, Y., Nishimura, O., Sasaki, H., and Sudo, R. 2004. Effect of C/N ratio on nitrous oxide emission from swine wastewater treatment process. *Water Science and Technology*, 49, 359–371.

34. Kozak, J.A., O'Connor, C., Granato, T., Kollias, L., Belluci, F., and Sturchio, N. 2009. Methane and nitrous oxide emissions from wastewater treatment plant processes. In: *Proceedings of the Water Environment Federation WEFTEC 2009*, 5347–5361.

35. Lambert, M. and Frechette, J-.L. 2005. Analytical techniques for measuring fluxes of CO_2 and CH_4 from hydroelectric reservoirs and natural water bodies. In: A. Tremblay, L. Varfalvy, C. Roehm, and M. Garneau (eds.), *Greenhouse Gas Emissions—Fluxes and Processes: Hydroelectric Reservoirs and Natural Environments*. Springer Publishing, Germany.

36. Law, Y., Ye, L., Pan, Y., and Yuan, Z. 2012. Nitrous oxide emissions from wastewater treatment processes. *Philosophical Transactions of the Royal Society B: Biological Sciences*, 367, 1265–1277.

37. Leverenz, H.L., Tchobanoglous, G., and Darby, J.L. 2010. *Evaluation of GHG Emissions from Septic Systems*. Report DEC1R09, Water Environment Research Foundation (WERF), Alexandria, VA.

38. Liebetrau, J., Clemens, J., Cuhls, C., Hafermann, C., Friehe, J., Weiland, P., and Daniel-Gromke, J. 2010. Methane emissions from biogas-producing facilities within the agricultural sector. *Engineering in Life Sciences*, 10(6), 595–599.

39. Livingston, G.P. and Hutchinson, G.L. 1995. Enclosure-based measurement of trace gas exchange: Applications and sources of error. In: P.A. Matson and R.C. Harriss (eds.), *Biogenic Trace Gases: Measuring Emissions from Soil and Water. Methods in Ecology*. Cambridge University Press, United Kingdom.

40. [My paper]Lotito, A.M., Wunderlin, P., Joss, A., Kipf, A., and Siegrist, H. 2012. Nitrous oxide emissions from the oxidation tank of a pilot activated sludge plant. *Water Research*, 46(11), 3563–3673.

41. Meijide, A., Manca, G., Goded, I., Magliulo, V., di Tommasi, P., Seufert, G., and Cescatti, A. 2011. Seasonal trends and environmental controls of methane emissions in a rice paddy field in Northern Italy. *Biogeosciences Discussions*, 8, 8999–9032.

42. Melbourne Water. 2012. *Capturing Biogas at Western Treatment Plant and Reusing It as a Renewable Energy Source Case Study: Electricity from Treatment by-Products Reducing Operating Expenses and Greenhouse Gas Emissions*. Report. Melbourne Water, Australia.

43. Melbourne Water (n.d.). Melbourne Water: Sewerage: Western treatment plant: Community and environmental benefits. http://www.melbournewater.com.au/whatwedo/treatsewage/wtp/Pages/Community-and-environmental-benefit.aspx. Accessed October 2, 2012.

44. Mes, T.Z.D., de Stams, A.J.M., and Zeeman, G. 2004 Methane production by anaerobic digestion of wastewater and solid wastes. In: J.H. Reith, R.H. Wijffels, and H. Barten (eds.), *Biomethane and Biohydrogen. Status and Perspectives of Biological Methane and Hydrogen Production*. Dutch Biological Hydrogen Foundation, the Netherlands.

45. Montzga, S.A., Dlugokencky, E.J., and Butler, J.H. 2011. Non-CO_2 greenhouse gases and climate change. *Nature*, 476, 43–50.

46. Okayasu, Y., Abe, I., and Matsuo, Y. 1997. Emission of nitrous oxide from high-rate nitrification and denitrification by mixed liquor circulating process and sequencing batch reactor process. *Water Science and Technology*, 36, 39–45.

47. Park, K.Y., Inamori, Y., Mizuochi, M., and Ahn, H.K. 2000. Emission and control of nitrous oxide from a biological wastewater treatment system with intermittent aeration. *Journal of Bioscience and Bioengineering*, 90, 247–252.

48. Park, K.Y., Lee, W.J., Inamori, Y., Mizuochi, M., and Ahn, H.K. 2001. Effects of fill modes on N_2O emission from the SBR treating domestic wastewater. *Water Science and Technology*, 43, 147–150.

49. Parkin, T.B., Venterea, R.T., and Hargreaves, S.K. 2012. Calculating the detection limits of chamber-based soil greenhouse gas flux measurements. *Journal of Environmental Quality*, 41, 705–715. DOI:10.2135/jeq2011.0394.

50. Peu, P., Beline, F., Picard, S., and Heduit, A. 2006. Measurement and quantification of nitrous oxide emissions from municipal activated sludge plants in France. In: *Proceedings of IWA World Water Congress*. International Water Association, Beijing.

51. Rassamee, V., Sattayatewa, C., Pagilla, K., and Chandran, K. 2011. Effect of oxic and anoxic conditions on nitrous oxide emissions from nitrification and denitrification processes. *Biotechnology and Bioengineering,* 108(9), 2036–2045.

52. Rinne, J., Pihlatie, M., Lohila, A., Thum, T., Aurela, M., Tuovinen, J.P., Laurila, T., and Vesala, T. 2005. Nitrous oxide emissions from a municipal landfill. *Environmental Science and Technology,* 39(20), 7790–7793.

53. Schouten, P.W., Sharma, A. Burn, S., and Goodman, N. 2014. Spatial and short term distribution of fugitive methane and nitrous oxide emission from a decentralised sewage mining plant: A pilot study. *Journal of Water and Climate Change,* 5(1), 1–12.

54. Schubert, C.J., Diem, T., and Eugster, W. 2012. Methane emissions from a small wind shielded lake determined by eddy covariance, flux chambers, anchored funnels, and boundary model calculations: A comparison. *Environmental Science and Technology,* 46, 4515–4522.

55. Sharma, A., Chong, M.N., Schouten, P.W., Cook, S., Ho, A., Gardner, T., Umapathi, S., Sullivan, T., Palmer, A., and Carlin, G. 2013. *Decentralised Wastewater Treatment Systems: System Monitoring and Validation.* Report No. 70, Urban Water Security Research Alliance, Australia.

56. Strain, D. and Schwartz, M. 2010 Stanford engineers use rocket science to make wastewater treatment sustainable. *Stanford University News.* http://news.stanford.edu/news/2010/july/waste-072610.html. Accessed October 3, 2012.

57. Tallec, G., Garnier, J., Billen, G., and Gousailles, M. 2008. Nitrous oxide emissions from denitrifying activated sludge of urban wastewater treatment plants, under anoxia and low oxygenation. *Bioresource Technology,* 99, 2200–2209.

58. Tchobanoglous, G., Burton, F.L., and Stensel, H.D. 2003. *Wastewater Engineering: Treatment and Reuse.* McGraw-Hill, New York.

59. Thorn, M. and Sorensson, F. 1996. Variation of nitrous oxide formation in the denitrification basin in a wastewater treatment plant with nitrogen removal. *Water Research,* 30(6), 1543–1547.

60. Townsend-Small, A., Pataki, D.E., Tseng, L.Y., Tsai, C-Y., and Rosso, D. 2011. Nitrous oxide emissions from wastewater treatment and water reclamation plants in southern California. *Journal of Environmental Quality,* 40, 1542–1550.

61. Tremblay, A., Lambert, M., and Gagnon, L. 2004. Do hydroelectric reservoirs emit greenhouse gases? *Environmental Management,* 33(1), S509–S517.

62. Tremblay, A., Bastien, J., Demarty, M., and Demers, C. 2009. Eastmain-1 net GHG emissions project—The use of automated systems to measure greenhouse gas emissions from reservoirs. In: *Proceedings of Waterpower XVI,* July 27–30, Spokane, WA.

63. Venterea, R.T. and Baker, J.M. 2008. Effects of soil physical nonuniformity on chamber-based gas flux estimates. *Soil Science Society of America Journal,* 72, 1410–1417.

64. Venterea, R.T., Spokas, K.A., and Baker, J.M. 2009. Accuracy and precision analysis of chamber-based nitrous oxide gas flux estimates. *Soil Science Society of America Journal,* 73, 1087–1093.

65. Wang, J., Zhang, J., Xie, H., Qi, P., Ren, Y., and Hu, Z. 2011. Methane emissions from a full-scale A/A/O wastewater treatment plant. *Bioresource Technology,* 102, 5479–5485.

66. Warneke, T., de Beek, R., Buchwitz, M., Notholt, J., Schulz, A., Velazco, V., and Schrems, O. 2005. Shipborne solar absorption measurements of CO_2, CH_4, N_2O and CO and comparison with SCIAMACHY WFM-DOAS retrievals. *Atmospheric Chemistry and Physics Discussions,* 5, 847–862.

67. Winter, P., Pearce, P., and Colquhoun, K. 2012. Contribution of nitrous oxide emissions from wastewater treatment to carbon accounting. *Journal of Water and Climate Change,* 3(2), 95–109.

68. Woess-Gallasch, S., Bird, N., Enzinger, P., Jungmeier, G., Padinger, R., Pena, N., and Zanchi, G. 2010. *Greenhouse Gas Benefits of a Biogas Plant in Austria.* Report. Resources Institute of Water, Energy and Sustainability, Graz, Austria.

37

MF, UF, and SWRO Membranes in the Desalination Process

Mike Dixon
Flinders University

Sophie C. Leterme
Flinders University

Jonathan Sierke
Flinders University

Amanda V. Ellis
Flinders University

PREFACE

In today's world, the accessibility to clean and inexpensive water is quickly becoming more difficult. It is inevitable that water shortages will only become worse due to climate change resulting in disruptions to society such as environmental epidemics, environmental disasters, and growth shortfalls. This has defined the reasons behind the building of high-powered, energy-efficient desalination plants. Desalination refers to any of several processes that remove some amount of salt and other minerals from saline water. Seawater reverse osmosis (SWRO) desalination is considered the simplest and most cost-effective method of freshwater production in comparison to other separation methods such as distillation, extraction, ion exchange, and adsorption. To overcome urban water shortages, polymer-based membrane filtration is the current leading technology for energy efficient desalination of recycled water and seawater. The primary benefit of membrane filtration is reliable pretreatment and the removal of high salt content. The greatest limitations of current membrane technologies are biofouling (where microbes and other particles build up on the surface of the membrane causing clogging), chlorine resistance, and compaction. All result in higher energy requirements, costly membrane clean up and replacement, as well as down-time. This chapter will discuss some of the conventional membrane systems used in SWRO membrane desalination, namely, microfiltration and ultrafiltration membranes for pretreatment and SWRO membranes for salt removal. Limitations of these and potential solutions are then discussed. In certain specialized cases, nanofiltration (NF) is used and this chapter will introduce the use of carbon nanotubes related to this emerging area.

37.1 Introduction

Water is the backbone of the global economy, being vital for agriculture, industry, recreation, energy production, and domestic consumption [29]. Since water demands and water shortage concerns have increased with population growth and climate change, seawater desalination has become an important alternative source to produce high-quality potable water.

There are three categories of water treatment technologies that have been used for seawater desalination, namely: membrane technologies, distillation processes (thermal technologies), and chemical approaches. Here, we focus on membrane technologies used in seawater reverse osmosis (SWRO) desalination plants.

The first major SWRO desalination plant, with a capacity of up to 450 m³ d⁻¹, was built in Kuwait in 1950 in response to a

15-year-long drinking water crisis [21]. While desalination was first viewed as an economically unviable method of freshwater production, rapid technological advancements continuously lowered the operational costs and increased productivity [13]. Around the world, the use of seawater desalination has been expanding in response to climate change and associated increasing temperatures, desertification, and drought [24,46,50]. In 2007, approximately 40% of the world's population was suffering from water supply insufficiencies, with future projections estimating an increase to 60% by the year 2025 [2].

SWRO desalination plants are now widely recognized as effective treatments of seawater or brackish water for the production of potable water, especially with the improvements made to membrane materials and elements [31,50]. Seawater desalination continues to represent the largest percentage of online global capacity at 59%, followed by brackish water at 22%, river water at 9%, and wastewater and pure water at 5% each. There are currently over 17,200 SWRO desalination plants around the world, with a total capacity of 80 million m³ d⁻¹. The biggest plants are generally built in the United Arab Emirates, Saudi Arabia, Spain, Kuwait, Algeria, and Australia. There is an expected market growth in desalination plants in India, Mexico, Jordan, South Africa, Chile, Libya, and China over the next five years. All of which are expected to be more than double their desalination capacity.

Figure 37.1 shows a simplified schematic of a typical desalination plant using membrane separation. In particular, the three predominant membranes used in industry are microfiltration (MF) and ultrafiltration (UF) for pretreatment and SWRO for the removal of salts.

37.2 Current Filtration Membranes

In the last 15 years, MF and UF pretreatments have gained widespread attention. Systems composed of UF or MF pretreatment for SWRO desalination are often termed "integrated membrane systems" or "dual membrane systems." This section will

briefly discuss current, MF, and UF pretreatment membranes, and SWRO membrane technologies and how they are used in the context of seawater desalination. Figure 37.2 shows the typical operating pressures, pore sizes, and rejection capabilities of these membranes. The following section describes each in more detail.

37.2.1 Microfiltration Membranes

MF membranes have a pore size range of 0.1–5 μm, which is the largest pore size of the three main membrane types (Figure 37.2). MF, like UF, is highly effective in decreasing turbidity, and removing bacteria [8,51] and protozoa [43]. In particular, the use of MF prior to RO allows for a good quality filtrate suitable to feed an RO system [26]. A variety of polymers, such as polyolefins [68], poly(vinylidene fluoride) (PVDF), and poly(ethersulfone) (PES) are used to synthesize these membranes because of their low cost and versatility. However, fouling of MF membranes cannot be avoided. A few studies have investigated the effect of ozone pretreatment on the fouling of MF membranes in the SWRO system [51]. The efficiency of the ozone treatment depends on the charge of the membrane and is more effective on hydrophobic rather than hydrophilic MF membranes. It has been found that MF membranes fabricated using thermally induced phase separation (TIPS) have been shown to provide better fouling control characteristics compared to extruded polymeric microfiltration membranes [73], and are the most commonly used.

37.2.2 Ultrafiltration Membranes

After microfiltration, the raw feedwater to RO membranes must be further adequately prepared or "pretreated." Any silt or turbidity must be removed to avoid rapid plugging of spiral wound SWRO membrane feed channels and organic fouling of the surface itself. To date, the majority of major RO membrane users employ sand or dual media filtration to pretreat water prior to

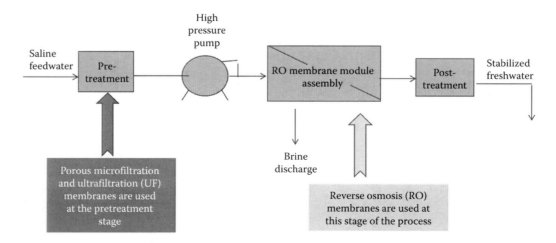

FIGURE 37.1 Simplified schematic representing a typical desalination process using membrane separation.

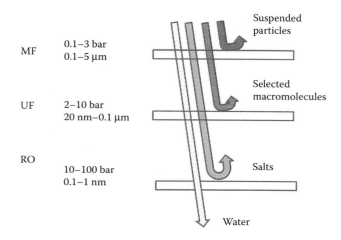

MF   0.1–3 bar   0.1–5 μm

UF   2–10 bar   20 nm–0.1 μm

RO   10–100 bar   0.1–1 nm

Suspended particles

Selected macromolecules

Salts

Water

FIGURE 37.2 Typical operating pressures, pore sizes, and rejection capabilities for microfiltration (MF), ultrafiltration (UF), nanofiltration (NF), and reverse osmosis (RO) membranes [47]. (Adapted from Kawaguchi, T. and Tamura, H. 1984. *Journal of Applied Polymer Science*, 29, 3359–3367.)

the RO membranes. While this pretreatment is adequate for turbidity removal, bacteria are still present after filtration. For this reason, modern RO plants use ultrafiltration membranes as a pretreatment. Ultrafiltration is a cross-flow separation process using membranes with an integral barrier of pore sizes of between 0.1 μm and 20 nm (Figure 37.2). Pores sizes are large enough to allow both mono- and divalent salts to pass through, but small enough to reject particles such as silica [12], metal ions [49], colloids [32], macromolecules, endotoxins, proteins, viruses [55], bacteria, algae [81], plastics, and fumes.

In recent years, the use of UF membranes as pretreatment to raw seawater has gradually become more common [22]. UF pretreatment is becoming increasingly used due to many advantageous properties such as more consistent product quality and smaller footprint, ease of design and operation, and lower environmental impact when compared to conventional processes [14]. However, the most challenging problem interfering with the UF performance is membrane fouling. In particular, one of the biggest issues is the fouling caused by organic materials, which are amplified during seawater algal blooms. Algae constantly secrete extracellular polymer substances (EPS) into the water, which are mainly composed of polysaccharides and proteins [25]. The membrane flux is considered the most important parameter in UF design, and it is closely related to membrane filtration characteristics and membrane fouling. Xu et al. [81] showed that various factors (i.e., dosage of humic acids, calcium concentration, and hydrodynamic conditions) would impact the UF performance in terms of the critical flux, rejection, and filtration resistance during the pretreatment of seawater.

37.2.3 Seawater Reverse Osmosis Membranes

SWRO membranes have a pore size range of 0.1–1 nm (Figure 37.2) and are semipermeable in that they allow for the removal

of salts, molecules, and ions but retain water molecules. They are operated in a cross-flow mode that achieves salt rejections greater than 99% [46]. SWRO membranes offer simultaneously high solute rejection, water permeability, chemical stability, and good chlorine resistance, particularly for aqueous separations. However, their small pore size means that a significant amount of osmotic pressure is required to force filtration. As the application of high pressure generates high operational costs, low pressure reverse osmosis membranes (LPROM) have been proposed as an alternative for water and wastewater industries [60] but are not often used. Cellulose acetate and aromatic polyamide type membranes are considered the two best polymer materials for RO applications. In particular, polyamide thin-film composite (TFC) membranes are currently the most widely used desalination membranes. TFC-RO membranes are made of an ultrathin polyamide film formed in situ by the polycondensation reaction of polyfunctional amine and acid chloride monomers over a porous polysulfone support membrane [63]. However, high feed pressures damage the membranes due to physical compaction of the porous support of the membrane and create an internal irreversible fouling that remains a serious concern for RO membranes. One of the challenges for successful operation in SWRO desalination is membrane fouling. For a successful RO desalination operation, consistent high-quality feedwater is important [57].

37.3 Current Issues with Membranes

37.3.1 Biofouling

Biofouling is an unwanted excessive accumulation of biological matter and is one of the most serious problems affecting membrane filtration systems as it impinges upon the quantity and quality of water produced and increases water production costs. Biofouling causes unacceptable increases in operating pressure or reductions in flow [16] increasing energy usage and decreasing salt rejection characteristics of the membrane [28]. This type of fouling is caused by the attachment of microorganisms such as bacteria to the membrane surface with subsequent growth of colonies to form a biofilm. The biofilm is then stabilized by weak interactions (physiochemical absorption), including hydrophobic and/or electrostatic interactions, hydrogen bonding and van der Waal's interactions. Formation of a biofilm involves four phases: (1) conditioning of the membrane surface with extracellular polymeric substances excreted by naturally occurring bacteria, (2) adhesion and attachment of bacteria, (3) growth of the bacteria, and (4) partial biomass detachment through erosion and sloughing [3,80]. The presence of biofouling may be observed after only a few minutes of operation time [67].

Biofilms have a number of effects on membranes. In particular, in spiral wound membranes (refers to a membrane configuration that is comprised of "flat sheet membrane—permeate channel spacer—flat sheet membrane—feed channel spacer" combinations rolled up around a product collection tube) there are two effects: feed channel pressure drop and transmembrane pressure increase (TMP). The feed channel pressure drop is the accelerated pressure differences between the feed and the concentrate while the TMP is the difference in pressure between the feed and the permeate. At constant flux, biofouling causes the TMP to be increased. It has also been demonstrated that, irrespective of flux (with and without permeation) as the biofilm concentration increases, feed channel pressure further increases [77].

In addition to this hydraulic resistance to permeation caused by surface biofilms, cake-enhanced osmotic pressure (CEOP) is induced [17,37], also known as biofilm enhanced osmotic pressure (BEOP) [18,34]. The CEOP/BEOP phenomenon occurs because the biofilm layer traps solutes thus hindering back-diffusion. This greatly increases osmotic pressure at the membrane surface and causes a reduction in net driving pressure for permeate production.

37.3.1.1 Mitigation Using Chlorination

It is common for shellfish and bacteria to grow in seawater desalination plant intakes as well as in sand filters and UF pretreatment systems [23]. For this reason, chlorine is a cost-effective method for control. Biofouling can easily be controlled in MF and UF applications due to the chemical resistance to chlorine of MF and UF membranes. In particular, surfaces such as PVDF and PES are not readily oxidized by chlorine. Two methods are used to control biofouling in microfiltration and ultrafiltration: (1) chlorine enhanced backwashes, where the normal backwash cycle is injected with chlorine to a low concentration, for example, 5 ppm and (2) clean in-place or shock chlorination, where membranes are exposed to greater concentrations while offline, for example, 50–300 ppm [23]. However, the majority of modern NF and RO membranes are readily oxidized by chlorine.

While the original RO membranes, developed by Sidney Loeb at the University of California, Los Angeles, had cellulose acetate surfaces that were chlorine resistant, these membranes did not have sufficient flux and rejection characteristics [54]. For this reason, TFC polyamide membranes were developed, which are more favorable for industry in the reduction of capital expenditure at large plants. However, they are not compatible with chlorine and this therefore necessitates the control of biofouling on RO membranes using methods other than chlorination, for example, the dosing of biocides [16], phosphorus limitation [76], and membrane modification [65].

37.3.1.2 Mitigation Using Biocides and Phosphorus Limitation

Due to its oxidizing nature, chlorine cannot be dosed into RO systems and so the use of nonoxidizing biocides needs to be employed. Common nonoxidizing biocides include glutaraldehyde, formaldehyde, sodium bisulfite, isothiazolin, and 2,2-dibromo-3-nitrilopropionamide acid. Periodic dosing of these chemicals can delay the onset of biofouling, but the removal of biofouling is more difficult, requiring extended downtime for chemical cleaning, which is not guaranteed to remove well-developed biofilms [76].

Another method of biofouling control is phosphorus limitation. Vrouwenvelder et al. [76] showed that bacteria, which commonly caused biofouling, could thrive when the concentration ratio of carbon:nitrogen:phosphorus is 100:20:10 and when this ratio is interrupted, bacterial growth is limited. One simple way of disrupting the ratio is to limit the concentration of phosphorus in the RO feed. It is necessary in many plants to dose antiscalant into the feedwater due to the presence of calcium. The process concentrates calcium salts toward the end of an RO pressure vessel causing salt deposition and an increase in feed channel pressure drop. Many common antiscalants are phosphonate-based; therefore, the use of alternate antiscalants such as dendripolymers is advised.

Studies focusing on the alteration of the membrane itself to make it resistant to biofouling are prevalent in the literature. Numerous growth-inhibiting surface modifications have been proposed, including the attachment of polycationic chains that disrupt the cellular membrane, silver and copper particle coatings, and TiO_2 nanoparticles that catalyze the production of cell-killing reactive oxygen species when excited by UV light [65].

37.3.1.3 Mitigation Using Antibiofouling Coatings: Surface Modification

Modification of the polyamide layer of RO membranes is by far the most studied. Ideal polymeric coating materials with reduced bacterial affinity should (1) be hydrophilic; (2) have a low surface roughness (microscale); (3) be charge neutral; (4) have high surface energy; and (5) have stable surface hydration. Truly biofouling-resistant membranes are yet to be realized. Poly(ethylene glycol) (PEG) or oligo(ethylene glycol) (OEG) coatings have been shown to improve resistance to nonspecific protein absorption [33,59]. Steric exclusion effects and hydration are considered as critical factors for PEG polymers to resist protein adsorption. Unfortunately, PEG polymers are not stable and are easily auto-oxidized in the presence of oxygen or transition metal ions. Similarly, phosphorylcholine (PC)-based polymers have been shown to decrease biopolymer adsorption, but these polymers are fragile; the phosphoester groups are readily hydrolysable [52,42]. Zwitterionic polymers, such as polysulfobetaines, are more stable under such conditions. In addition, they are biocompatible [85] and are promising materials for use as biofouling coating materials [44].

37.3.2 Chlorine Resistance

Chemical attack of TFC polyamide membranes by chlorine ultimately results in membrane failure as measured by enhanced passage of both salt and water [75]. Membrane failure is due to certain structural changes within the polymer in response to chlorine exposure. These changes result from chlorine attack on amide nitrogen and aromatic rings in the polyamide-type membranes (Figure 37.3). The resulting substitution products may cause deformation in the polymer chain or cleavage at amide linkages [30].

A strategy to make a more robust and durable material could be to select polymers that feature chemically resistant backbones; unfortunately, these materials are often very hydrophobic, thus reducing flux. Water must sorb into the polymer membrane to an adequate level to achieve high water flux; however, high water sorption typically reduces salt rejection. Manufacturing membranes that continue to reduce desalination plant sizes are critical to industry. Hence, any surface must maximize salt rejection and flux simultaneously. Specific strategies include (1) altering aromatic diamine groups with aliphatic or cycloaliphatic diamine moieties with secondary amino groups, (2) introducing polyamide derived from secondary diamines, (3) modifying functional groups at the ortho position of an aromatic diamine moiety to hinder rearrangement [56], and (4) sulfonation to introduce a hydrophilic character into an otherwise hydrophobic polymer [28]. The use of sulfonated pentablock copolymers has shown some promise in delivering a more chlorine-resistant membrane [28].

37.3.3 Compaction

Compaction, like fouling, is another factor that causes flux decline during pressure-driven filtration processes. Compaction refers to the physical compression of the membrane due to the applied pressure, which decreases the efficiency of filtration by constricting the pores in the membrane. Unlike fouling, compaction is an intrinsic part of the filtration process as the use of pressure is unavoidable.

Compaction has been proposed as a cause of flux decline for many years, with early literature focusing on cellulose acetate RO membranes [5,7,10,11,48]. Filtrations (typically involving differing concentrations of NaCl) both in short- and long-term were found to undergo flux decline over time [48]. Evidence for compaction being the cause came from reduction of membrane thickness after the filtration and was even found to alter the crystal structure of cellulose acetate [5]. Studies have long since moved to other types of membranes, such as UF and MF membranes, which operate under lower pressures but still showed similar behavior [71].

Direct evidence for compaction being the cause of flux decline did not come about until the introduction of ultrasonic time-domain reflectometry (UTDR) [64], which allowed simultaneous measurement of water flux and membrane thickness (to the micrometer level). Using UTDR, it was observed that flux decline correlated well with reduction of membrane thickness, for the first time providing direct evidence that compaction (compression strain) of the membrane caused flux decline (decrease in permeation rate) [64] (Figure 37.4).

37.3.3.1 Aspects of Compaction

Compaction, like most mechanical properties, has both a reversible and irreversible (unrecoverable) component [15,70]. Although the overall compaction decides the total amount of flux decline due to compression, the fact that some of the compaction is irreversible is significant for systems where the TMP is changed during use. This is because higher pressures not only induce higher overall compaction [36], but also higher levels of irreversible compaction [15,70,45]. This means that a membrane that was used at higher pressures will exhibit a lower flux than if the same membrane were only used at the lower pressure [15,70]. This has practical implications for desalination/water purification plants, which will increase TMP to maintain water output due to fouling, but then revert to lower pressures again once the membrane has been cleaned.

Another effect of compaction is potential alterations to the rejection properties of membranes, although this is still controversial. Studies have shown that membranes compacted at higher pressures can exhibit increased rejection, reduced rejection [4],

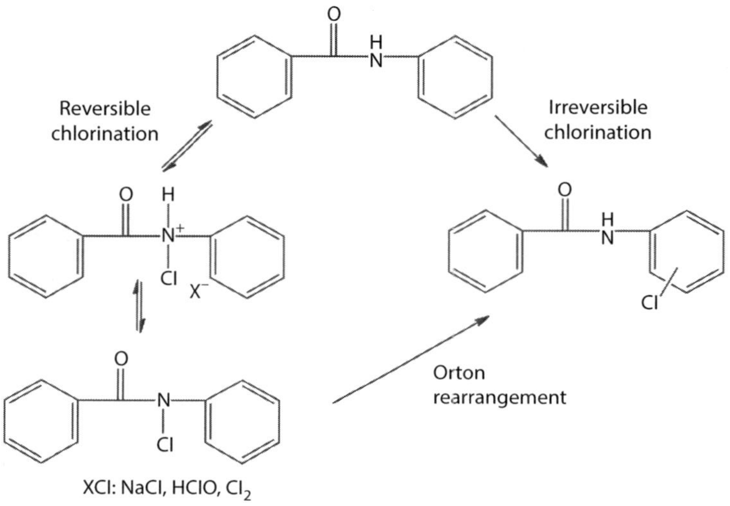

FIGURE 37.3 Polyamide rearrangement by chlorination [64]. (From Peterson, R.A. et al. 1998. *Desalination*, 116(2–3), 115–122. With permission.)

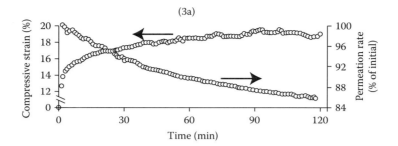

FIGURE 37.4 Variation in compressive strain and relative permeation rate with time of a commercial DOW BW-30 RO membrane at a constant pressure of 4.14 MPa. As it can be observed, there is a clear correlation between increasing compressive strain (or membrane compaction) and reduction in pure water permeance [56]. (From Misdan, N., Lau, W.J., and Ismail, A.F. 2012. *Desalination*, 287, 228–237. With permission.)

or show no change [71]. Thus far, studies have indicated that the major variables effecting compaction are the TMP, morphology of the membrane, and membrane material. Although the effects of the TMP and membrane morphology on compaction are well understood, currently there is little research on comparisons of different materials on compaction. The following sections will address methods of potentially controlling compaction.

37.3.3.2 Transmembrane Pressure

The TMP is the only factor that can easily be changed during a filtration process and is therefore the most studied. It is expected, following Hagen–Poiseuille's Law, that increasing the TMP should increase the flux proportionally. However, this is not the case and instead the hydraulic resistance of the membrane increases with increasing TMP [70,71] until increasing the TMP no longer produces significant gains in flux. Furthermore, exposure to higher TMP causes a greater irrecoverable flux decline due to irreversible compaction. Therefore, careful consideration should be undertaken before increasing the TMP as a way of increasing flux, as it may not be an economical choice.

37.3.3.3 Membrane Morphology

To understand the effect of membrane morphology on compaction, one must have a general understanding of the structure of a typical membrane. Membranes used in water filtration typically have an asymmetric structure—a dense skin layer sitting atop a much more porous underlayer. Membranes may contain multiple different layers of differing porosity and pore sizes, and each layer can be made of different materials. The pore morphology typically found in flat sheet water filtration membranes is displayed in Figure 37.5; common characteristics include a very thin selective skin layer sitting on a macroporous underlayer, with some membranes having an additional supportive backing to improve mechanical integrity. The differences in macroporous structure arise from differing conditions used to cast the membranes and the use of different polymers.

Although compaction is well known to occur in both the macroporous underlayers as well as the dense skin layer, the structure of the macroporous layer has been shown to have a profound effect on compaction. In particular, increasing size of the macrovoids in the macroporous layer has been correlated with

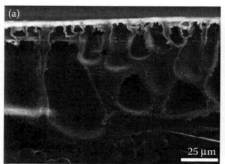 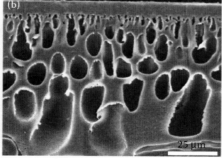 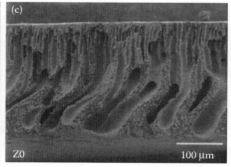

FIGURE 37.5 Scanning electron images. (a) Shows a UF membrane, regenerated cellulose with polyethylene terephthalate (PET) backing (commercial membrane from Microdyn-Nadir) [1]. (Reprinted from Aerts, P. et al. 2001. *Separation and Purification Technology*, 22–23, 663–669.) (b) Shows a UF membrane, permanently hydrophilic polyethersulphone with polyethylene/polypropylene backing (commercial membrane from Microdyn-Nadir) [62]. (Reprinted from Pendergast, M.T.M. et al. 2010. *Desalination*, 261(3), 255–263, Copyright 2010, with permission from Elsevier.) (c) Shows a hand-cast polysulfone membrane (no backing) (type of membrane not specified) [63]. (Reprinted from Petersen, R.J. 1993. *Journal of Membrane Science*, 83, 81–150, Copyright 1993, with permission from Elsevier.)

increasing susceptibility to compaction [58]. By having a less porous layer sitting in between the skin layer and the macrovoid layer, compaction of the macrovoids is suppressed and overall compaction of the membrane is reduced [70]. It should be noted that flux decline still occurs despite no continued decreases in membrane thickness, suggesting that the skin layers (and less porous layers) also undergo compaction. In that case, there is a limit to what effect morphology alone has on the overall compaction of the membrane and it may be more affected by the membrane material.

37.4 Emerging Membrane Technologies

37.4.1 Mixed Matrix Materials

Mixed matrix membrane refers to a membrane that consists of at least two different materials, typically a polymeric (organic) component and an inorganic component. Often, this is accomplished by blending or dispersing particles within a polymer matrix or coating inorganic particles on membrane surfaces. By doing this, several advantages are achieved compared to plain polymeric membranes. These include improved flux, improved fouling resistance and flux recoverability and, in the case of blended membranes, increased compaction resistance. Most interestingly, some additives have antibacterial properties which help reduce biofouling. A large amount of research has focused on using various filler materials (raw and/or modified) of various sizes. Typical fillers are comprised of metal or metal oxides, such as gold [74], silver [9], titanium dioxide (TiO_2) [66], silicon dioxide (SiO_2) [84], aluminium oxide (Al_2O_3) [82], zirconium oxide (ZrO_2) [6], and zeolites [27], but also includes carbon allotropes [72] including carbon nanotubes (CNT). CNTs are a special case and will be addressed in a separate section later in this chapter.

Typically, the blending of these fillers follows a similar trend; the addition of the filler improves hydrophilicity (decreases water contact angle), increases flux, increases fouling resistance, and increases mechanical properties until an optimal weight percentage (wt%) is reached. However, beyond the optimal weight percentage there can be a detrimental effect on properties, often due to aggregation of the particles. Most studies focus on the use of weight percentage ranges of filler up to 20 wt%, typically less than 10 wt%. However, it has been demonstrated that loadings up to 80–90 wt% can produce usable membranes with particles as large as 4.5 mm. Largely, this work has focused on the use of Fe_3O_4 [40] and ZrO_2 [6] particles. These exceptionally high weight percentages can cause a significant increase in flux (but sacrifice rejection in the process), but also show greater compaction resistance compared to unfilled membranes. In fact, flux profiles have been shown to be closer to that of ceramic membranes showing almost a linear trend between TMP and flux. Side effects of high loadings can result in surface defects and rough surfaces, which increases flux but rapidly decreases rejection [86] and increases the susceptibility to biofouling [39], respectively.

The application of TiO_2 particles, whether coated or blended, is very popular due to the photocatalytic properties of TiO_2, which can provide an additional antibacterial and antifouling mechanism. TiO_2 particles (nanoparticles [<25 nm], anatase crystals) produce radicals under UV light that can be used to degrade organic molecules or attack bacterial cells [20]. A 5 wt% inclusion of TiO_2 nanowires [79] has been shown to significantly increased hydrophilicity, improve tensile properties (tensile strength increased from 1.64 to 2.23 MPa; elongation at break increased from 36.91% to 59.01%), increase flux (both for pure water and bovine serum albumin [BSA] solution), and lower flux reduction due to BSA fouling (about 20% less flux reduction). This is particularly interesting as the membrane became less brittle with incorporation of the nanowires, whereas the opposite usually occurs with inorganic particle addition.

The incorporation of porous zeolites not only operates as other metal oxide nanoparticles, but can also provide additional benefits; one benefit is they can provide additional pathways for water flow through RO membranes, which can provide significant increases in flux (from 75% [72] up to 550% [62]) while still maintaining rejection of NaCl. Zeolites can also be used to give membranes antibacterial properties; through ion exchange, the zeolite can be loaded with Ag^+ ions, which can be reduced back to Ag(0). Incorporating silver-loaded zeolites into PVDF UF membranes was shown to inhibit the growth of *Escherichia coli* [53]. The incorporation of Ag(0) nanoparticles within polyethersulfone/polysulfone membranes has also been shown to give a similar result [9]. These antibacterial membranes show promise not only in reducing fouling but also as a method to resist biofilm growth.

The drawbacks to using mixed-matrix membranes include higher cost, cumbersome methodology, and leaching. The higher cost associated with using these fillers limits their commercial viability [61] as the improvements in membrane performance are not high enough. With the increasing awareness and production of nanomaterials within industry, it will become more affordable but until then most of these mixed-matrix membranes will remain an academic curiosity. In many cases, long and/or multi-step procedures are required, particularly if chemical modification is needed. Furthermore, the longevity, particularly of antibacterial membranes, needs to be accounted.

Leaching of the filler [9] is known to occur in these systems, which may both reduce the properties of these membranes overtime and represent a health hazard. Overall, however, the use of mixed-matrix membranes is still a promising avenue to improving upon current membrane technologies and continued research in the field will make this a more viable option.

37.4.2 Carbon Nanotube Membranes

One of the more unique nanomaterials that have recently been studied as an NF system in the desalination processes are CNTs. CNTs are allotropes of carbon in which the carbon atoms form a hexagonal sheet of sp^2 hybridized carbon bonds. These sheets of carbon, called graphene, are rolled into cylinders forming tube-shaped macromolecules called CNTs. Depending on the type of CNTs, they can have a radius as small as a few nanometers up to tens of nanometers and can be many microns in length [69]. There are many different types of carbon nanotubes; for example; single-walled carbon nanotubes (SWCNTs), double-walled carbon nanotubes (DWCNTs), and multi-walled carbon nanotubes (MWCNTs).

In 2001, fast water transport in SWCNTs was reported by Hummer [41] and Cicero et al. [19], opening new prospects for the world of water transport. CNTs are believed to be able to transport water through their narrow pores due to their very hydrophobic sidewalls [38]. Hummer et al. [41] found that for small diameter tubes (0.81 nm), diffusion occurs through a burst-like mechanism, producing single-files of water chains capable of moving with very little resistance. It was observed that a small perturbation of the interactions between water and the nanotube wall had a considerably large effect on the filling behavior of the CNT [41]. Water and CNTs have shown remarkable interactions where the Lennard-Jones interactions of the nanotube well were seen to be reduced from 0.11 to 0.065 kcal/mol, this resulted in a nanotube that fluctuated between empty and filled states [78]. These interactions have since resulted in an ordered chain of hydrogen-bonded water molecules spanning through the nanotube channel, indicating remarkable prospects for transport within such a small volume channel [78]. In 2004, Holt et al. [38] reported experimentally on the fabrication of CNT embedded membranes embedded in silicon nitride to investigate nanometer-scale mass transport [38]. The research showed results that supporting membranes are a robust platform for studies on molecule transport, where they calculated a water transport rate of 1.91 mol/m^2 s. Later in 2006, Hoek et al. [35] proposed the use of RO desalination using extremely high pressure to force saline or polluted water through pores of a semipermeable membrane. Pressure is required for the water molecules to actually pass through these pores; however, other impurities such as salt cannot pass through these, resulting in highly purified water [35].

More recently, Yang et al. [83] have reported that plasma-modified ultra-long carbon nanotubes exhibited ultrahigh specific adsorption capacity for salt (exceeding 400% by weight) that is two orders of magnitude higher than that found in the current state-of-the-art activated carbon-based water treatment systems. While these are not based on conventional RO membrane systems, they do offer some future opportunities for alternate desalination methods. These ultra-long carbon nanotube-based membranes may lead to next-generation rechargeable, point-of-use potable water purification appliances with superior desalination, disinfection, and filtration properties for both rural and urban use.

37.5 Summary and Conclusions

The desalination of water is highly reliant on filtration membranes of various types. Most are polymeric in nature and are involved in MF, UF (as a pretreatment stage), and RO membranes (as the salt exclusion stage). These membranes are prone to biofouling, chorine attack, and compaction, so strategies need to be put in place to mitigate these. Such strategies include chlorine enhanced backwashes and clean in-place or shock chlorination for MF and UF membranes. While for the more chlorine-sensitive RO membranes, biocides and phosphorus limitation can be used. Furthermore, increasingly research is focusing on surface modification of RO membranes with antifouling coatings and bulk modification of the polyamide layer during processing.

The future of desalination may be in emerging membrane technologies such as mixed membrane systems and forward osmosis based on carbon nanotubes for NF. These are still at their infancy and require changes in existing infrastructure to be implemented. Nonetheless, the benefits of such systems could provide significant cost and energy savings in the future. Other technologies on the horizon include graphene and zeolite membranes as well as hybrid systems with aquaporins and polyhedral oligomeric silsesquioxanes.

References

1. Aerts, P., Greenberg, A.R., Leysen, R., Krantz, W.B., Reinsch, V.E., and Jacobs, P.A. 2001. The influence of filler concentration on the compaction and filtration properties of Zirfon®-composite ultrafiltration membranes. *Separation and Purification Technology*, 22–23, 663–669.

2. Alameddine, I. and El-Fadel, M. 2007. Brine discharge from desalination plants: A modelling approach to an optimized outfall design. *Desalination*, 214, 241–260.

3. Allison, D.G. and Sutherland. I.W. 1984. A staining technique for attached bacteria and its correlation with extracellular carbohydrate production. *Journal of Microbiological Methods*, 2, 93–99.

4. Arkhangelsky, E. and Gitis, V. 2008. Effect of transmembrane pressure on rejection of viruses by ultrafiltration membranes. *Separation and Purification Technology*, 62(3), 619–628.

5. Arneri, G. 1981. The effect of pressure on the bulk polymer microstructure in cellulose acetate reverse osmosis membranes. *Desalination*, 36(1), 99–104.

6. Arthanareeswaran, G. and Thanikaivelan, P. 2010. Fabrication of cellulose acetate-zirconia hybrid membranes for ultrafiltration applications: Performance, structure and fouling analysis. *Separation and Purification Technology*, 74(2), 230–235.

7. Baayens, L. and Rosen, S.L. 1972. Hydrodynamic resistance and flux decline in asymmetric cellulose acetate reverse osmosis membranes. *Journal of Applied Polymer Science*, 16(3), 663–670.

8. Bae, H., Kim, H., Jeong, S., and Lee, S. 2011. Changes in the relative abundance of biofilm-forming bacteria by conventional sand-filtration and microfiltration as pretreatments for seawater reverse osmosis desalination. *Desalination*, 273, 258–266.

9. Basri, H., Ismail, A.F., and Aziz, M. 2011. Polyethersulfone (PES)-silver composite UF membrane: Effect of silver loading and PVP molecular weight on membrane morphology and antibacterial activity. *Desalination*, 273(1), 72–80.

10. Baum, B., Margosiak, S.A., and Holley, W.H. 1972. Reverse osmosis membranes with improved compaction resistance. *Product R&D*, 11(2), 195–199.

11. Bert, J.L. 1969. Membrane compaction: A theoretical and experimental explanation. *Journal of Polymer Science Part B: Polymer Letters*, 7(9), 685–691.

12. Bharwada, U. 2012. Silica removal projects in India use hollow-fibre membranes. *Membrane Technology*, 7, 6.

13. Bin Mohammed, T. and Al Sajwani, A. 1998. The desalination plants of Oman: Past, present and future. *Desalination*, 120, 53–59.

14. Brehant, A., Bonnelye, V., and Perez, M. 2002. Comparison of MF/UF pretreatment with conventional filtration prior to RO membranes for surface seawater desalination. *Desalination*, 144, 353–360.

15. Brinkert, L., Abidine, N., and Aptel, P. 1993. On the relation between compaction and mechanical properties for ultrafiltration hollow fibers. *Journal of Membrane Science*, 77(1), 123–131.

16. Characklis, W.G. and Marshall, K.C. 1990. *Biofilms*. John Wiley & Sons, New York.

17. Chong, F.S., Wong, F.S., and Fane, A.G. 2008. Implications of critical flux and cake enhanced osmotic pressure (CEOP) on colloidal fouling in reverse osmosis: Experimental observations. *Journal of Membrane Science*, 314, 101–111.

18. Chong, T.H., Wong, F.S., and Fane, A.G. 2000. The effect of imposed flux on biofouling in reverse osmosis: Role of concentration polarisation and biofilm enhanced osmotic pressure phenomena. *Journal of Membrane Science*, 325(2), 840–850.

19. Cicero, G., Grossman, J.C., Schwegler, E., Gygi, F., and Galli, G. 2008. Water confined in nanotubes and between graphene sheets: A first principle study. *Journal of the American Chemical Society*, 130(6), 1871–1878.

20. Damodar, R.A., You, S.J., and Chou, H.H. 2009. Study the self cleaning, antibacterial and photocatalytic properties of TiO_2 entrapped PVDF membranes. *Journal of Hazardous Materials*, 172(2–3), 1321–1328.

21. Delyannis, E. and Belessiotis, V. 2010. Desalination: The recent development path. *Desalination*, 264, 206–213.

22. Di Profio, G., Ji, X., Curcio, E., and Drioli, E. 2011. Submerged hollow fiber ultrafiltration as seawater pretreatment in the logic of integrated membrane desalination systems. *Desalination*, 269, 128–135.

23. Dixon, M.B., Ayala, V., Hijos, G., and Pelekani, C. 2012. Experiences from the Adelaide desalination project: Ultrafiltration cleaning optimisation—From pilot to fulls-cale operation. *Desalination and Water Treatment*, 1–10.

24. Dixon, M.B., Lasslett, S., and Pelekani, C. 2012. Destructive and non-destructive methods for biofouling analysis investigated at the Adelaide desalination pilot plant. *Desalination*, 296, 61–68.

25. Donlan, R. 2002. Biofilms: Microbial life on surfaces. *Emerging Infectious Diseases*, 8, 881–890.

26. Ebrahim, S., Abdel-Jawad, M., Bou-Hamad, S., and Safar, M. 2001. Fifteen years of R&D program in seawater desalination at KISR. Part I. Pretreatment technologies for RO systems. *Desalination*, 135, 141–153.

27. Fathizadeh, M., Aroujalian, A., and Raisi, A. 2011. Effect of added NaX nano-zeolite into polyamide as a top thin layer of membrane on water flux and salt rejection in a reverse osmosis process. *Journal of Membrane Science*, 375(1–2), 88–95.

28. Geise, G.M., Freeman, B.D., and Paul, D.R. 2010. Characterization of a sulfonated pentablock copolymer for desalination applications. *Polymer*, 51(24), 5815–5822.

29. Gewin, V. 2005. Industry lured by the gains of going green. *Nature*, 436(7048), 173.

30. Glater, J., Hong, S.K., and Elimelech, M. 1994. The search for a chlorine-resistant reverse osmosis membrane. *Desalination*, 95(3), 325–345.

31. Harif, T., Elifantz, H., Margalit, E., Herzberg, M., Lichi, T., and Minz, D. 2011. The effect of UV pre-treatment on biofouling of BWRO membranes: A field study. *Desalination and Water Treatment*, 31, 151–163.

32. Hébrant, M., Rose-Hélène, M., and Walcariusa, A. 2013. Metal ion removal by ultrafiltration of colloidal suspensions of organically modified silica. *Colloids and Surfaces A: Physicochemical and Engineering Aspects*, 417, 65–72.

33. Herrwerth, S., Eck, W., Reinhardt, S., and Grunze, M. 2003. Factors that determine the protein resistance of oligoether self-assembled monolayers—Internal hydrophilicity, terminal hydrophilicity, and lateral packing density. *Journal of the American Chemical Society*, 125, 9359–9366.

34. Herzberg, M. and Elimelech, M. 2007. Biofouling of reverse osmosis membranes: Role of biofilm-enhanced osmotic pressure. *Journal of Membrane Science*, 295(1–2), 11–20.

35. Hoek, E. 2006. *Nanotech Water Desalination Membrane, in Today's Seawater is Tomorrow's Drinking Water: UCLA Engineers Develop Nanotech Water Desal Membrane*. M. Abraham, UCLA, Los Angeles, CA.

36. Hoek, E.M.V., Allred, J., Knoell, T., and Jeong, B.H. 2008. Modeling the effects of fouling on full-scale reverse osmosis processes. *Journal of Membrane Science*, 314(1–2), 33–49.

37. Hoek, E.M.V. and Elimelech, M. 2003. Cake-enhanced concentration polarization: A new fouling mechanism for salt-rejecting membranes. *Environmental Science and Technology*, 37, 5581–5588.

38. Holt, J.K., Noy, A., Huser, T., Eaglesham, D., and Bakajin, O. 2004. Fabrication of a carbon nanotube-embedded silicon nitride membrane for studies of nanometer-scale mass transport. *Nano Letters*, 4(11), 2245–2250.

39. Huang, Z.Q., Chen, L., Chen, K., Zhang, Z., and Xu, H.T. 2010. A novel method for controlling the sublayer microstructure of an ultrafiltration membrane: The preparation

of the PSF–Fe$_3$O$_4$ ultrafiltration membrane in a parallel magnetic field. *Journal of Applied Polymer Science*, 117(4), 1960–1968.

40. Huang, Z.Q., Chen, K., Li, S.N., Yin, X.T., Zhang, Z., and Xu, H.T. 2008. Effect of ferrosoferric oxide content on the performances of polysulfone-ferrosoferric oxide ultrafiltration membranes. *Journal of Membrane Science*, 315(1–2), 164–171.

41. Hummer, G., Rasaiah, J.C., and Noworyta, J.P. 2001. Water conduction through the hydrophobic channel of a carbon nanotube. *Nature*, 414(6860), 188–190.

42. Iwasaki, Y. and Ishihara, K. 2005. Phosphorylcholine-containing polymers for biomedical applications. *Analytical and Bioanalytical Chemistry*, 381, 534–546.

43. Jacangelo, J.G., Adham, S.S., and Lainee, J.M. 1995. Mechanism of Cryptosporidium, Giardia, and MS$_2$ virus removal by MF and UF. *Journal of American Water Works Association*, 87, 107.

44. Jiang, S., Chen, S., and Zhang, Z. 2007. Super-low fouling sulfobetaine and carboxybetaine materials and related methods. WO/2007/02439, 01.03.2007.

45. Kallioinen, M., Pekkarinen, M., Mänttäri, M., Nuortila-Jokinen, J., and Nyström, M. 2007. Comparison of the performance of two different regenerated cellulose ultra-filtration membranes at high filtration pressure. *Journal of Membrane Science*, 294(1–2), 93–102.

46. Katebian, L. and Jiang, S.C. 2013. Marine bacterial biofilm formation and its responses to periodic hyperosmotic stress on a flat sheet membrane for seawater desalination pretreatment. *Journal of Membrane Science*, 425, 182–189.

47. Kawaguchi, T. and Tamura, H. 1984. Chlorine-resistant membrane for reverse osmosis. I. Correlation between chemical structures and chlorine resistance of polyamides. *Journal of Applied Polymer Science*, 29, 3359–3367.

48. Kimura, S. and Sourirajan, S. 1968. Performance of porous cellulose acetate membranes during extended continuous operation under pressure in reverse osmosis process using aqueous solutions. *Industrial & Engineering Chemistry Process Design and Development*, 7(2), 197–206.

49. Landaburu-Aguirrea, J., Pongráczb, E., Perämäkic, P., and Keiskia, R.L. 2010. Micellar-enhanced ultrafiltration for the removal of cadmium and zinc: Use of response surface methodology to improve understanding of process performance and optimisation. *Journal of Hazardous Materials*, 180, 524–534.

50. Lee, J. and Kim, I.S. 2011. Microbial community in seawater reverse osmosis and rapid diagnosis of membrane biofouling. *Desalination*, 273, 118–126.

51. Lee, J., Oh, B.S., Kim, S., Kim, S.J., Hong, S.K., and Kim, I.S. 2010. Fate of *Bacillus* sp. and *Pseudomonas* sp. isolated from seawater during chlorination and microfiltration as pretreatments of a desalination plant. *Journal of Membrane Science*, 349, 208–216.

52. Lewis, A.L., Hughes, P.D., Kirkwood, L.C., Leppard, S.W., Redman, R.P., Tolhurst, L.A., and Stratord, P.W. 2000.

Synthesis and characterisation of phosphorylcholine-based polymers useful for coating blood filtration devices. *Biomaterials*, 21, 1847–1859.

53. Liao, C., Yu, P., Zhao, J., Wang, L., and Luo, Y. 2011. Preparation and characterization of NaY/PVDF hybrid ultrafiltration membranes containing silver ions as antibacterial materials. *Desalination*, 272(1–3), 59–65.

54. Loeb, S. 1984. Circumstances leading to the first municipal reverse osmosis desalination plant. *Desalination*, 50, 53–58.

55. Madaeni, S.S., Fane, A.G., and Grohmann, G.S. 1995. Virus removal from water and wastewater using membranes. *Journal of Membrane Science*, 102, 65–75.

56. Misdan, N., Lau, W.J., and Ismail, A.F. 2012. Seawater reverse osmosis (SWRO) desalination by thin-film composite membrane—Current development, challenges and future prospects. *Desalination*, 287, 228–237.

57. Naidu, G., Jeong, S., Vigneswaran S., and Rice, S.A. 2013. Microbial activity in biofilter used as a pretreatment for seawater desalination. *Desalination*, 309, 254–260.

58. Ochoa, N.A., Masuelli, M., and Marchese, J. 2003. Effect of hydrophilicity on fouling of an emulsified oil wastewater with PVDF/PMMA membranes. *Journal of Membrane Science*, 226(1–2), 203–211.

59. Ostuni, E., Chapman, R.G., Holmlin, R.E., Takayama, S., and Whitesides, G.M. 2001. A survey of structure-property relationships of surfaces that resist the adsorption of protein. *Langmuir*, 17, 5605–5620.

60. Ozaki, H., Sharma, K., and Saktaywin, W. 2001. Performance of an ultra low pressure reverse osmosis membrane (ULPROM) for separating heavy metals: Effects of interference parameters. *Desalination*, 144, 287–294.

61. Pendergast, M.M. and Hoek, E.M.V. 2011. A review of water treatment membrane nanotechnologies. *Energy & Environmental Science*, 4(6), 1946–1971.

62. Pendergast, M.T.M., Nygaard, J.M., Ghosh, A.K., and Hoek, E.M.V. 2010. Using nanocomposite materials technology to understand and control reverse osmosis membrane compaction. *Desalination*, 261(3), 255–263.

63. Petersen, R.J. 1993. Composite reverse osmosis and nanofiltration membranes. *Journal of Membrane Science*, 83, 81–150.

64. Peterson, R.A., Greenburg, A.R., Bond, L.J., and Krantz, W.B. 1998. Use of ultrasonic TDR for real-time noninvasive measurement of compressive strain during membrane compaction. *Desalination*, 116(2–3), 115–122.

65. Rana, D. and Matsuura, T. 2010. Surface modifications for antifouling membranes. *Chemical Reviews*, 110(4), 2448–2471.

66. Razmjou, A., Mansouri, J., and Chen, V. 2011. The effects of mechanical and chemical modification of TiO$_2$ nanoparticles on the surface chemistry, structure and fouling performance of PES ultrafiltration membranes. *Journal of Membrane Science*, 378(1–2), 73–84.

67. Ridgway, H.F. and Flemming, H.C. 1996. Membrane biofouling. In: J. Mallevialle, P.E. Odenaal, and M.R. Wiesner (eds.), *Water Treatment Membrane Processes*. McGraw-Hill, New York.

68. Rincón, S.S. 2002. Nuevos desarrollos en PP: Copolimeros heterofasicos de alto modulo. *Revista Plasticos Modernos*, 83, 307.

69. Singh, P., Campidelli, S., Giordani, S., Bonifazi, D., Bianco, A., and Prato, M. 2009. Organic functionalization and characterization of single-walled carbon nanotubes. *Chemical Society Reviews*, 38, 2214–2230.

70. Stade, S., Kallioinen, M., Mikkola, A., Tuuva, T., and Mänttäri, M. 2013. Reversible and irreversible compaction of ultrafiltration membranes. *Separation and Purification Technology*, 118, 127–134.

71. Tarnawski, V.R. and Jelen, P. 1986. Estimation of compaction and fouling effects during membrane processing of cottage cheese whey. *Journal of Food Engineering*, 5(1), 75–90.

72. Taurozzi, J.S., Crock, C.A., and Tarabara, V.V. 2011. C60-polysulfone nanocomposite membranes: Entropic and enthalpic determinants of C60 aggregation and its effects on membrane properties. *Desalination*, 269(1–3), 111–119.

73. Vanegas, M.E., Quijada, R., and Serafini, D. 2009. Microporous membranes prepared via thermally induced phase separation from metallocenic syndiotactic polypropylenes. *Polymer*, 50, 2081–2086.

74. Vanherck, K., Hermans, S., Verbiest, T., and Vankelecom, I. 2011. Using the photothermal effect to improve membrane separations via localized heating. *Journal of Materials Chemistry*, 21(16), 6079–6087.

75. Voutchkov, N. 2013. *Desalination Engineering, Planning and Design*. McGraw Hill, New York.

76. Vrouwenvelder, J.S., Beyer, F., Dahmani, K., Hasan, N., Galjaard, G., Kruithof, J.C., and Van Loosdrecht, M.C.M. 2010. Phosphate limitation to control biofouling. *Water Research*, 44(11), 3454–3466.

77. Vrouwenvelder, J.S. and Graf von der Schulenburg, D.A. 2009. Biofouling of spiral-wound nanofiltration and reverse osmosis membranes: A feed spacer problem. *Water Research*, 43(3), 583–594.

78. Waghe, A., Rasaiah, J.C., and Hummer, G. 2002. Filling and emptying kinetics of carbon nanotubes in water. *Journal of Chemical Physics*, 117(23), 10789–10795.

79. Wei, Y., Chu, H.Q., Dong, B.Z., Li, X., Xia, S.J., and Qiang, Z.M. 2011. Effect of TiO$_2$ nanowire addition on PVDF ultrafiltration membrane performance. *Desalination*, 272(1–3), 90–97.

80. Wingender, J., Neu, T.R., and Flemming, H.C. 1999. *Microbial Extracellular Polymeric Substances: Characterization, Structure and Function*. Springer-Verlag. Berlin, Heidelberg.

81. Xu, J., Ruan, L.G., Wang, X., Jiang, Y.Y., Gao, L.X., and Gao, J.C. 2012. Ultrafiltration as pretreatment of seawater desalination: Critical flux, rejection and resistance analysis. *Separation and Purification Technology*, 85, 45–53.

82. Yan, L., Hong, S., Li, M.L., and Li, Y.S. 2009. Application of the Al$_2$O$_3$-PVDF nanocomposite tubular ultrafiltration (UF) membrane for oily wastewater treatment and its antifouling research. *Separation and Purification Technology*, 66(2), 347–352.

83. Yang, H.Y., Han, Z.J., Yu, S.F., Pey, K.L., Ostrikov, K., and Karnik, R. 2013. Carbon nanotube membranes with ultrahigh specific adsorption capacity for water desalination and purification. *Nature Communications*, 4, 2240.

84. Yu, L.Y., Xu, Z.L., Shen, H.M., and Yang, H. 2009. Preparation and characterization of PVDF-SiO$_2$ composite hollow fiber UF membrane by sol-gel method. *Journal of Membrane Science*, 337(1–2), 257–265.

85. Yuan, J., Chen, L., Jiang, X., Shen, J., and Lin, S. 2004. Chemical graft polymerization of sulfobetaine monomer on polyurethane surface for reduction in platelet adhesion. *Colloids and Surfaces B: Biointerfaces*, 39, 87–94.

86. Zhang, Y., Li, H., Lin, J., Li, R., and Liang, X. 2006. Preparation and characterization of zirconium oxide particles filled acrylonitrile-methyl acrylate-sodium sulfonate acrylate copolymer hybrid membranes. *Desalination*, 192(1–3), 198–206.

38

Upgrading Wastewater Treatment Systems for Urban Water Reuse

Bosun Banjoko
Obafemi Awolowo University

Mynepalli Kameswara
Chandra Sridhar
University of Ibadan

PREFACE

Steady increase in global population and increased movement of people from rural to urban centers worldwide in a quest for better economic opportunities has placed pressure on demand for water for domestic and industrial purposes. The realization that water is a finite entity has required the international community to rethink water economics and management. Therefore, water reuse and recycling are becoming much more common as demands for water exceed supply. Unplanned reuse occurs as the result of waste effluents entering receiving waters or groundwater and subsequently being taken into a water distribution system. Planned reuse utilizes wastewater treatment systems deliberately designed to bring water up to standards required for subsequent applications. Lack of adequate water supply and widespread deployment of modern water treatment processes, which significantly enhance the quality of water available for reuse, have continued to promote an urban water reuse culture. It is therefore inevitable that water recycling and reuse will continue to grow and this trend will increase the demand for water treatment, both for quantity and quality. However, for developing economies, elaborate waste treatment facilities that incorporate different advanced treatment technologies may not be economically prudent. The need arises for research into appropriate, cost-effective and sustainable technologies to meet the needs of developing countries. Therefore, technologies for upgrading wastewater treatment for urban reuse are apt to vary from place to place. Furthermore, there is a need for global cooperation in terms of transfer of technology, research, and training on wastewater reuse.

38.1 Introduction

After water is used in households, businesses, and industries, it is collected in a sanitary sewer system and sent to the local wastewater treatment facility. Industrial wastewater may be treated and released directly into the receiving body of water or into the municipal sanitary sewer system. In the latter case, the industrial effluent most oftentimes receives some pretreatment before it can be disposed of in the sanitary sewer system in line with guidelines stipulated by various countries. In the United States, regulation is under the Clean Water Act and a discharger is required to meet certain technology based effluent limits and perform effluent monitoring [24]. In Nigeria, industrial effluent discharge is regulated by the National Environmental Standard Regulatory Agency (NESREA). Furthermore, stormwater sewer systems that collect runoff from urban streets join the sanitary system and the combination of wastewaters flow to the municipal wastewater treatment plant (WWTP). These combined systems have their disadvantages. During rains, they often carry more wastewater than the local treatment plant can handle. When that happens, a portion of the flow, which includes raw sewage, must be diverted around the treatment plant and released directly into the receiving water at the cost of public health. Separating these combined systems is immensely expensive, a probable solution is to create massive reservoirs usually underground that store the combined flow until the storm passes after which time the reservoir is slowly drained back into the sanitary sewer system [24].

The purpose of water treatment systems is to bring raw water up to drinking water quality Municipal wastewater is typically over 99.9% water and only about 0.1% impurities, which are mostly organic material, pathogenic microorganisms, and many different substances including both suspended and dissolved solids. The total amount of organic matter represents the strength of the sewage. This is measured by the biochemical oxygen demand (BOD). Another important parameter for the strength of the sewage is the total suspended solids (TSS). On average, untreated domestic sanitary sewage has a BOD value of 200 mg/L and a TSS of about 240 mg/L. Industrial wastewater may have BOD and TSS values much higher than those for sanitary sewage. The composition is therefore source dependent. Also important in wastewater are the plant nutrients, namely, nitrogen (N) and phosphorus (P). On average, raw sewage contains about 35 mg/L of N and 10 mg/L of P. Equally of note in sewage is the amount of pathogens, which are evaluated in terms of fecal coliforms. The coliform concentration in raw sewage is roughly 1 billion per liter. Therefore, BOD, TSS, coliform concentration, N, and P are preferred parameters used in the assessment of water quality [9].

In planning and designing a wastewater disposal system, the texture of the soil into which the effluent is discharged is very important. A thorough investigation of site and soil conditions is quite necessary. Specifically, the permeability or hydraulic conductivity (k) of the soil must be within an acceptable range. For example, for sandy gravel, the range is 5.00E−04 and 5.00E−02 min(m/s). If it is too low, the wastewater will not be able to percolate fast enough for effective disposal, and if it is too high, there may not be sufficient time for purification of the effluent.

38.2 Wastewater Management Technologies

38.2.1 Onsite Wastewater Disposal Systems

The onsite waste disposal system is a subsurface waste disposal application used mostly in lightly populated suburban or rural areas where it is often uneconomical to build a public collection system and centralized treatment plant. It is also a very common system of wastewater disposal in developing countries. However, subsurface systems are deemed as a temporary kind of wastewater disposal because of frequent failure of such systems. They are often considered unreliable and undesirable in developed countries. However, if properly located, designed, and constructed, it can serve effectively for wastewater disposal on a long-term basis before it reaches the water table. In addition to hydraulic conductivity, the depth to groundwater and the depth to bedrock are important considerations in the siting and design of a subsurface system. Generally, bedrock and seasonally high groundwater table must be at least 3 m below the disposal system. These values will vary depending on local health department and environmental agency regulations. A simple practical assessment includes excavation of soil using a backhoe to a depth of 4 m to serve as a test pit. Then visual observations of soil texture, depth to groundwater, and depth to rock are recorded. Soil survey maps can also be obtained from local geological survey agencies or appropriate Ministries at the state level. Recommended soil types for subsurface disposal of wastewater are relatively coarse granular soils which contain a large percentage of sand and have high hydraulic conductivities. Fine-grained soils such as very fine sand and silts offer more resistance to the flow of water, while soils containing large fractions of clay have extremely low permeability and therefore are quite unsuitable for subsurface wastewater disposal [29].

38.2.2 Offsite Wastewater Disposal Systems

An offsite wastewater disposal system involves the transport of wastewater from sources of generation such as houses, business premises, and industries, conveyed by municipal sewers into publicly owned treatment works (POTWs). The POTWs are the centralized treatment plants used in urban cities. The offsite WWTPs have facilities for primary and secondary treatment technologies, which can be upgraded to include advanced treatment or tertiary treatment technologies for urban water reuse. The treatment of wastewater involves two or three levels. These include primary, secondary, and tertiary treatments. Primary treatment involves simple screening by removing large floating objects such as leaves and sticks that may damage the pumps or clog small pipes. After screening, the wastewater passes into a grit chamber for a few minutes (detention time) to allow sand, grit, and other materials to settle out. The wastewater from the grit chamber thereafter passes to the primary settling tank where the flow speed is reduced enough to allow most of the suspended solids to settle out by gravity. The settled solids from this process, called primary sludge, are removed and chlorinated for disinfection and further processed to the thickener, to anaerobic digester, and later to sludge dewatering chamber for removal of water, and finally, the decontaminated sludge is disposed on land. The secondary treatment, also referred to as biological treatment, further removes more suspended and dissolved solids beyond what is achievable by sedimentation. This is accomplished by three main methods utilizing the ability of microorganisms to covert organic waste into stabilized low-energy compounds. These methods are (1) trickling filters, (2) activated sludge, and (3) oxidation ponds.

38.2.3 Sanitary Sewerage System

38.2.3.1 The Septic Systems

The most common type of subsurface disposal system is the septic system, which includes (1) a septic tank and (2) an absorption or leaching field. The tank serves as a storage compartment for settled and floating solids and the leaching field serves as a distribution outlet for the effluent so that it can percolate through the soil. In this manner, decomposition or organic substances takes place under anaerobic conditions. A typical small rectangular concrete tank for an individual household is generally about 3800 L of liquid capacity. Septic tanks must be watertight and have adequate access for inspection and cleaning. The top of the tank is usually about 300 mm below the ground surface. The effluent from the septic tank flows into an absorption or leaching field, which distributes the liquid uniformly over a sizable area. From the leaching field, the effluent percolates downward to the water table. As it flows through the soil voids, microorganisms and other pollutants are removed from the effluent. Filtration, adsorption, and biological decomposition each play a role in the purification of the wastewater effluent before it is diluted in the groundwater. A very common type of leaching field consists of two or more separate trenches with pipes that serve to spread wastewater. Before reaching the trenches, the effluent flows into a distribution box, which serves to evenly divide the flow into each trench. The disposal leaches if the leaching fields are shallow excavations that are a minimum of 0.3 m wide and at least 0.6 m deep. About 150 m of gravel is placed on the bottom of the trench to support the line of perforated effluent distribution pipe. The pipe is covered with more gravel and then with straw or paper to prevent the final layer of backfill soil from penetrating the gravel voids.

38.2.3.2 Evapotranspiration Systems

An evapotranspiration (ET) system is an alternative onsite system of wastewater disposal when subsurface disposal is not feasible. This system can be used in conjunction with a septic tank. It consists of a sand bed, a network of perforated distribution pipes, and an impermeable liner that prevents the treatment effluent from reaching the water table. In some cases, the liner may be omitted in order to allow some of the effluent to seep into the soil. The basic objective of this type of system is to dispose of the wastewater into the environment and to avoid the need for discharge into either surface or groundwater. Effluent from a septic tank is distributed throughout the bed in the perforated pipe network. The effluent rises through the sand by capillary action and then evaporates into the air. Grass or other vegetation growing on the top of the bed serves to absorb some of the wastewater in the root zone and transpire it into the air through the leaves. Hence, the name evapotranspiration system. One of the most critical factors controlling the use and design of an ET system is the local climate, which affects the rate of evaporation. An important parameter to be considered in the design of an ET system is the hydraulic loading rate, which must be low enough to prevent the bed from filling completely with effluent. The hydraulic loading rate is determined from the difference between the rate of evaporation and the rate of precipitation. A typical ET system for a single-family residence would require about 464 m^3 of land area; however, this is not useful for water reuse.

38.2.3.3 Intermittent Sand Filters

This is one of the oldest methods of wastewater treatment. The sand bed is usually about 1 m deep and is under drained with gravel and collection pipes. The collected effluent may be disinfected with chlorine before discharge to land or surface waters. The filters may be built as open units at ground level or they may be buried in the ground. They provide efficient treatment while requiring a minimum of maintenance. The upper perforated distribution lines are level and are spaced about 2 m apart. The under drainpipes are perforated or open joint lines on a slope of about 0.5%. The hydraulic load on a buried filter is about 0.04 m^3/m^2. The finished grade of the top soil over the filter is mounded to direct runoff away from the bed. Small tanks called dosing chambers store the wastewater until it is applied to the filter by a pump or siphon device. Siphon chambers require very minimal routine maintenance.

38.2.3.4 Sludge Disposal

Sludge or bio-solids are suspended solids removed from wastewater during sedimentation and further treatment and disposal. Before it can be disposed of, sludge requires some form of treatment to reduce its volume and to stabilize it. Sludge forms initially as a 3%–7% suspension of solids and each person typically generates about 15 L of sludge per week. Because of the volume and nature of the material, sludge management is a major factor in the design and operation of all water pollution control plants [29]. A reduced sludge volume minimizes pumping and storage requirements and lowers overall handling costs. Several processes are available for accomplishing the two basic objectives of volume reduction and stabilization, and these include (1) sludge thickening, (2) sludge digestion, (3) sludge dewatering, and (4) co-composting.

38.2.4 Stormwater Management

Stormwater and surface runoff present a peculiar problem in wastewater management because storm runoff is a major non-point source of water pollution including fertilizers/pesticides, oil, organics, and other substances. This can be compounded by flooding. Best management practices of stormwater are reproduction of predevelopment flow conditions and pollutant control. Construction of stormwater storage basins after it has been collected from streets, parking lots, and other surfaces is being used to attenuate stormwater flow. It is also necessary to provide storm drains to remove excess water from streets and parking lots to prevent inconvenience of flood in localized areas. In newer cities, it is pertinent to completely separate municipal wastewater from stormwater sewer systems. Water pollution control is an important consideration in the design of stormwater management facilities, and computers and modeling are being used extensively in advanced countries in stormwater management [1].

38.2.5 Industrial Wastewater Treatment

To be treated, industrial wastewater must be fully characterized with a thorough chemical analysis of constituents and metabolic products and their biodegradability should be determined. It is mandatory that industrial wastewater be pretreated before the effluent is discharged into the water bodies. One of the major ways of removing organic wastes is biological treatment by an activated sludge or related processes. Consideration needs to be given to possible hazards of bio-treatment of sludge such as those containing excessive levels of heavy metal ions. The other major process for removing organics from wastewater is sorption by activated carbon, usually in columns of granular inactivated carbon. Activated carbon and biological treatment can be combined with the use of activated carbon in the activated sludge process. In some cases, simple density separation and sedimentation can be used to remove water–immiscible liquids and solids. Filtration is frequently required and floatation by gas bubbles generated on particle surfaces may be useful. Solutes in wastewater can be concentrated by evaporation distillation and membrane processes. Including reverse osmosis, hyperfiltration and ultrafiltration, organic constituents can be removed by solvent extraction, air stripping, or stream stripping. Furthermore, synthetic resin can be used in removing pollutant solutes from wastewater. For example, organophilic resins have proved useful for the removal of alcohols, aldehydes, ketones, hydrocarbons, chlorinated alkanes, alkenes, aryl compounds, and esters including phltalate esters and pesticides. In addition, carbon exchange resins are effective for the removal of heavy metals [5].

A variety of chemical processes has been found useful for treatment of industrial wastewater. These include acid/base neutralization, precipitation, and oxidation/reduction reactions. In some cases, those treatment steps precede biological treatment. For example, wastewater exhibiting extremes of pH must be neutralized in order for microorganism to thrive in it.

Cyanide in the wastewater may be oxidized with chlorine and organics with ozone while hydrogen peroxide is promoted with ultraviolet radiation or dissolved oxygen at high temperatures and pressures. Heavy metals may be precipitated with base, carbonate, or sulfide. It is therefore obvious that the treatment of industrial wastewater requires a different approach and tertiary treatment is mostly required. Calcium and magnesium salts, which are generally present in water as bicarbonates or sulfates, cause water hardness, which causes insoluble turbid solution soap to react with the calcium and magnesium ions in the hard water. Hence, calcium and magnesium must be complexed and removed from water for detergent to function properly. Another problem caused by hard water is the formation of mineral deposits containing insoluble calcium carbonate when the water is heated. One of the popular ways of removing calcium and magnesium ions is addition of lime soda.

Soluble iron and manganese can be removed from industrial wastewater by addition of chlorine and potassium permanganate as oxidizing agents. Furthermore, lime treatment can be used to remove heavy metals as insoluble hydroxides, basic salts, or coprecipitated with calcium of ferric hydroxide. Although this process does not completely remove mercury, cadmium, or lead, their removal can be accomplished by addition of sulfide because most heavy metals have affinity for sulfides. In addition, activated carbon adsorption effectively removes some metals from water. Sometimes, a chelating agent is sorbed to the charcoal to increase metal removal. In some cases, electrodeposition, that is, reduction of metal ions to metal by electrons at an electrode, reverse osmosis, and ion exchange are frequently employed for metal removal. Other methods of metal removal include solvent extraction using organic soluble chelating substances and cementation in which metal deposits are formed due to the reactions of their ions with a more readily oxidized metal [22].

Phosphorus as phosphate is commonly removed by precipitation and usually lime $(CaOH)_2$ is the choice chemical employed for phosphorus removal. Nitrogen removal can be achieved by a nitrification–gentrification process. The first step is an essentially complete to conversion of ammonia and organic nitrogen to nitrate under strongly aerobic conditions achieved by more extensive than normal aeration of the sewage. The second step is the reduction of nitrate to nitrogen gas. This reaction is also catalyzed by bacteria and requires a carbon source and a reducing agent such as methanol. The denitrification process may be carried out either in a tank or on a carbon column; it is possible to convert 95% of ammonia to nitrate and 86% of the nitrate to nitrogen gas by this method [1].

38.2.5.1 Primary Treatment

The bar screens, or racks as they are called, used in primary treatment of wastewater are made up of long narrow metal bars spaced between 1.5 and 6 mm apart. They retain floating debris such as wood, rags, and other bulky objects that could clog pipes or damage mechanical equipment in the rest of the plant. The collected debris or screenings are promptly disposed of usually by burial in the plant grounds.

38.2.5.2 Primary Sedimentation or Settling

The preliminarily processed water still contains suspended organic solids, which can be removed by plain sedimentation. The basic sedimentation or plain gravity settling is achieved by settling tanks called *primary clarifiers*. They provide between 1 and 2 h detention time and side water depth (SWD) is generally between 2.5 and 5 m. The tank may be circular or rectangular. In addition to mechanical sludge collections that continually scrape the settled solids along the bottom to a sludge hopper for removal, a surface skimming device is used to remove grease and other floating materials from the liquid surface. A standard treatment plant usually requires two or more clarifiers. The combination of preliminary treatment and gravity settling constitute what is called primary treatment. Primary treatment can remove up to 60% of the suspended solids and 35% of the BOD from wastewater.

38.2.5.3 Grit Removal

A portion of the suspended solids in raw sewage consists of gritty materials such as sand, coffee, eggshells, and other relatively inert materials. Suspended grit can cause excessive wear and tear on the pumps and other equipment in the plant. The grit is transported and collected in a long narrow tank called the *grit chamber*. The preliminary treatment of screening comminuting and grit removal prepares wastewater for secondary treatment.

38.2.6 Secondary or Biological Treatment

The limitation of primary treatment processes is that only those pollutants that will float or settle by gravity are removed. Therefore, the effluent contains suspended solids that did not settle out in the primary tanks and the dissolved BOD that is unaffected by physical treatment. Also known as biological treatment, secondary treatment ensures removal of 85% BOD and total suspended solids (TSS) [13]. Secondary treatment of sewage is termed biological treatment because it involves the activity of microorganisms including bacteria and protozoa that metabolize the biodegradable organic converting them into carbon dioxide, water, and energy for their growth and multiplication. This natural aerobic process requires oxygen; therefore, the design and operation of secondary treatment plants must take into account adequate provision of oxygen, adequate contact of microorganisms with the organic materials in the sewage, and suitable temperatures, which are all encompassed into a controlled artificial environment of steel or concrete. Two of the most common biological treatment systems are the trickling filter and the activated sludge process. The trickling filter or percolating filter is a type of fixed growth system whereby the microbes remain fixed or attached to a surface while the wastewater flows over the surface to provide contact with the organics. On the other hand, the activated sludge is a suspended-growth system whereby the microbes are thoroughly mixed and suspended in the wastewater rather than attached to a particular surface.

38.2.6.1 Land Treatment of Wastewater

Secondary effluent can be allowed to flow over the vegetated ground surface to percolate through the soil in what is known as land treatment of wastewater. This method provide moisture and nutrients for vegetative growth and also nutrients for ground water aquifers. The application of secondary effluent onto the land surface can provide an effective alternative to the expensive and complicated advanced treatment methods. Furthermore, a high quality polished effluent can be obtained by natural processes that occur as the effluent flows over the vegetative ground surface and percolates through the soil. In addition, land treatment of wastewater allows a direct recycling of water and nutrients for beneficial use whereby the sewage becomes a valuable natural resource that is not simply disposed of. The disadvantage of this mode of wastewater treatment is that large land areas are required and the success depends on soil type and climate. These factors also determine the mode of land treatment to be adopted. There are three basic types of land treatment: (1) slow rate, (2) rapid infiltration, and (3) overland flow.

38.2.6.1.1 Slow Rate

In the slow rate, also called irrigation, vegetation is very critical. In this situation, water and nutrient components are for vegetation growth. Wastewater is conserved when secondary effluent are used to irrigate lawns, landscaped areas, and golf courses. The slow rate method also provides good results in respect of tertiary treatment levels of pollutant removal. TSSs and bBOD are significantly reduced by filtration of the organics in the top few inches of soil. Nitrogen is removed by adsorption within the soil and pake by the plants.

38.2.6.1.2 Rapid Infiltration

The rapid infiltration mode of land treatment of wastewater is also referred to as percolation mode. It is primarily designed to recharge groundwater aquifers and in the process provide advanced treatment. The filtering and adsorption action of the soil remove BOD, TSS, and phosphorus and a little nitrogen. For this mode to be feasible, soil must be highly permeable to the rapid infiltration method. Usually the wastewater is applied in large ponds called recharge basins

38.2.6.1.3 Overland Flow System

In this mode, wastewater is sprayed on a sloped terrace and allowed to flow across the vegetated surface to a runoff collection ditch. Physical, biological, and chemical processes take place in order for purification to be achieved. Overland flow favorably removes BOD and nitrogen but only a little phosphorus. The water collected in the ditch is usually discharged to a nearby body of surface water [11,12].

38.2.6.2 Trickling Filters

Trickling filters consist of a structure containing a rotating distribution arm that sprays liquid wastewater over a circular bed of very small-sized rocks or other materials. The spaces between

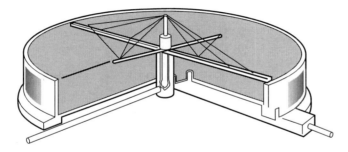

FIGURE 38.1 Cross section of trickling filter used in wastewater treatment.

the rocks allow air to circulate easily so that aerobic conditions can be maintained. Individual rocks in the bed are covered by a layer of biological slimes that absorb and consume the waste trickling through the bed. The slime consists of mainly bacteria, fungi, algae, protozoa, worms, and larvae of insect. The accumulating slime periodically slides off the rocks and is collected at the bottom of the filter along with the treated wastewater and passed on to the secondary settling tank where it is removed (Figure 38.1).

38.2.6.3 Activated Sludge

The aeration tank, which is the key biological unit in the activated sludge process, receives effluent from the primary classifier and recycled biological organisms from the secondary settling tank. Air or oxygen is led into the tank and the mixture is thoroughly agitated to maintain aerobic conditions. After about 6–8 h of agitation, the wastewater flows into the secondary settling tank, where the solids, mostly bacteria masses, are separated from the liquid by a portion of those solids that is relayed to the aeration tank to maintain proper bacterial population while the remainder are processed and disposed of. The resultant mixture of solids and water after the liquid effluent has been released into a nearby body of water is called sludge. The sludge is passed into two stages of anaerobic digester. The digested sludge is dewatered by pumping into large sludge drying beds. The digester and dewatered sludge can be used as a fertilizer or conveyed to a landfill.

38.2.6.4 Waste Stabilization Ponds

Waste stabilization ponds (also called oxidation ponds) consist of shallow manufactured basins comprising a single or several series of anaerobic, facultative, or maturation ponds. The primary treatment takes place in the *anaerobic pond*, which is mainly designed for removing suspended solids, and some of the soluble organic matter (BOD). During the secondary stage, in the *facultative pond*, most of the remaining BOD is removed through the coordinated activity of algae and heterotrophic bacteria. The main function of the tertiary treatment in the *maturation pond* is the removal of pathogens and nutrients (especially nitrogen).

Stabilization ponds are particularly well suited for tropical and subtropical countries because the intensity of the sunlight

and temperature are key factors for the efficiency of the removal processes. It is also recommended by the WHO for the treatment of wastewater for reuse in agriculture and aquaculture, especially because of its effectiveness in removing nematodes (worms) and helminth eggs. According to the IRC International Water and Sanitation Centre, stabilization pond technology is the most cost-effective wastewater treatment technology for the removal of pathogenic microorganisms. A World Bank study carried out in Sana'a, Yemen, in 1983 makes a detailed economic comparison of waste stabilization ponds, aerated lagoons, oxidation ditches, and trickling filters. According to this study, stabilization pond technology is a cheaper option up to a land cost of US$ 7.8/m². Above this cost, oxidation ditches become the cheapest option. However, often the main constraint against selecting this technology is not land cost but land availability. If land is available, stabilization ponds have the advantage of very low operating costs because they use no energy compared to other wastewater treatment technologies [39].

38.2.6.5 Oxidation Ditch

These are large, shallow ponds usually 2 m deep where raw or partially treated sewage is decomposed by microorganisms. They are cheap to construct and maintain. Although they can be used to treat raw sewage, they are usually used as an addition to secondary treatment in which case they are called polishing ponds. Oxidation ponds are usually referred to as facultative ponds because of the combined mechanism of aerobic digestion at the surface and anaerobic digestion near the bottom of the pond [4,17].

38.2.6.6 Secondary Effluent Disinfection

The last process in a secondary sewage treatment process is disinfection. The purpose of sewage disinfection is to destroy any pathogen in the effluent that may have survived the secondary treatment process, thereby protecting public health. While the removal of BOD and TSS serves primarily to protect the environment, the process of disinfection of secondary effluent is to ensure water quality for swimming or for further treatment for domestic uses.

Like drinking water, sewage is usually disinfected by chlorination. When chlorine is dissolved in pure water, it reacts with the H^+ ions and the OH^- ions in the water. Two of the products of this reaction are hypochlorous acid (HOCl) and the hypochlorite ions (OCl⁻), which are the actual disinfecting agents. If microorganisms are present in the water, HOCl and OCl⁻ penetrate the microbe cells and react with certain enzymes. This reaction disrupts the organisms' metabolism and kills them. Hypochlorous acid is a more effective disinfectant than the hypochlorite ion because it diffuses faster through the microbe cell wall. HOCl and OCl⁻ depend on the pH of the water; the more effective the chlorination, the better the disinfection process. When chlorine is first added to water containing some impurities, the chlorine immediately reacts with the dissolved inorganic or organic substances and is then unavailable for disinfection. If dissolved ammonia (NH_3) is present in the water, the chlorine

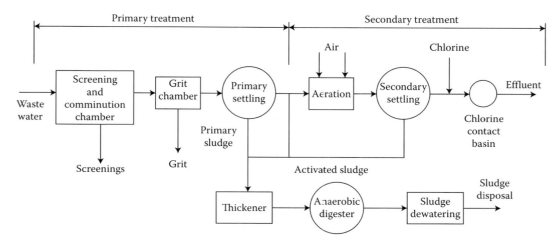

FIGURE 38.2 Schematic diagram of a primary wastewater treatment system.

reacts with it to form compounds called chloramines. The amount of chlorine used up in the initial reactions with impurities and dissolved ammonia is called chlorine demand of water. It is after the chlorine demand is satisfied and the reaction with all dissolved ammonia is complete that the residual chlorine is available in the form of $HOCl^-$ and OCl^-, called free available chlorine, which effects disinfection [29] (Figure 38.2).

38.2.6.7 Wetlands Ecosystems

Wetlands are areas of land that are covered with freshwater for at least part of the year. The two main types of freshwater wetlands are marshes and swamps. Marshes contain nonwoody plants such as cattails, while swamps are dominated by woody plants, such as trees and shrubs [3]. Wetlands perform many environmental functions and these include

1. Trapping and filtering sediments, nutrients, and pollutants, which keep these materials from entering lakes, reservoirs, and oceans
2. Reducing the likelihood of a flood, protecting agriculture, roads, buildings, and human health and safety
3. Buffering shorelines against erosion
4. Providing spawning grounds and habitat for commercially important fish and shellfish
5. Providing habitat for rare, threatened, and endangered species
6. Providing recreational areas for activities such as fishing, bird watching, hiking, canoeing, photography, and painting
7. Wetland vegetation traps carbon that would otherwise be released as carbon dioxide, which may be linked to rising atmospheric temperatures

In order to protect wetlands due to their global importance, the Ramsar Convention of 1972 was convened and a legal framework on protection of wetlands was established. Since natural wetlands act as a biofilter, removing sediments and pollutants such as heavy metals from water, constructed wetlands are being designed to emulate these features. A constructed wetland or wet park is an artificial wetland created as a new or restored habitat for native and migratory wildlife for anthropogenic discharge such as wastewater, stormwater runoff, or sewage treatment, for land reclamation after mining, refineries, or other ecological disturbances such as required mitigation for natural areas lost to a development. Vegetation in a wetland provides a substrate, namely, roots, stems, and leaves upon which microorganisms can grow as they break down organic materials. This community of microorganisms is known as periphyton. The periphyton and natural chemical processes are responsible for approximately 90% of pollutant removal and waste breakdown. The plans remove about 7%–10% of pollutants and act as a carbon source for the microbes when they decay. Different species of aquatic plants have different rates of heavy metal uptake and the knowledge of this is a consideration for plant selection in a constructed wetland designated for wastewater treatment. Constructed wetlands are of two basic types: subsurface flow and surface flow. The surface flow wetlands move effluent above the soil in a plant marsh or swamp and thus can be supported by a wider variety of soil types including mud and other salty clays [14,20]. Subsurface flow wetlands, which consist of horizontal flow and vertical flow types, are useful for treatment of household wastewater, mining runoff, and storm drains. Effluents are moved through a gravel or sand medium usually made of limestone or volcanic rock on which plants are rooted. An improvement in wetland construction is the tidal flow wetlands, which are found very useful in heavy load. In this system, organic carbon is primarily oxidized with nitrate, which is produced through a series of flood and drain cycles from one side of the wetland to the other [38]. Poplar plants used in wetland wastewater treatments include cattails (*Typha* spp.), water hyacinth (*Eichhornia crassipes*) and *Pontederia* spp., buck beans (*Menyanthes trifoliate*), and pendant grass (*Arctophila fulva*). Typha and phragmites are seen as invasive species. Physical, chemical, and biological processing combine in wetlands to remove contaminants from wastewater [19].

Theoretically, wastewater treatment within a constructed wetland occurs as it passes through the wetland medium and the plant rhizosphere. A thin film around each root hair is aerobic, caused by leakage of oxygen from rhizomes, roots, and rootlets. Along the line, aerobic and anaerobic microorganisms facilitate decomposition of organic matter. Microbial nitrification and subsequent denitrification release nitrogen as gas to the atmosphere. Phosphorus is coprecipitated with iron, alum, and calcium compounds located in the root bed medium. Suspended solids filter out as they settle in the water column in surface flow wetlands or are physically filtered out by the medium with subsurface flow wetlands. Harmful bacteria and viruses are reduced by filtration and adsorption by biofilms on the rock media in subsurface flow and vertical flow medium. The dominant forms of nitrogen in wetlands that are of importance to wastewater treatment include organic nitrogen, ammonia, ammonium, nitrate, nitrite, and nitrogen gases while the inorganic forms are essential to plant growth in aquatic systems and their scarcity can limit plant growth. Ammonia is toxic to fish; it is discharged into water courses and excessive nitrate can cause methemoglobinemia in children with subsequent decrease in blood oxygen supply capability [2,6,7].

38.2.6.8 Phytotechnologies

Phytoremediation is an emerging technology that involves the use of different types of plants to degrade, extract, contaminate, or immobilize contaminants from the soil and water [36]. Filtration in the rhizosphere or in rhizomes had been observed for water contaminated with radionuclides and metals. Volatilization or transpiration through plants into the atmosphere is another possible mechanism for removing contaminants from soil or water at a site. The fundamental basis of this remediation processes the accumulation of contaminants in roots by using hydroponic techniques. The hydroponic technique refers to growing plants in nutrient solutions. This is useful for separating inorganic contaminants from water.

38.2.7 Tertiary or Advanced Treatment

With current knowledge, the need arises for more advanced treatment of wastewater. This is because secondary treatment cannot achieve absolute removal of BOD and TSS in raw sanitary sewage. Generally, 85% and 95% of BOD and TSS, respectively, are removed by secondary treatment leaving about 30 mg/or less of BOD and TSS in the secondary effluent. Another limitation of secondary treatment is that it does not significantly remove the effluent concentrations of nitrogen and phosphorus, which are very important plant nutrients that promote algal bloom and accelerated lake aging or cultural eutrophication in lakes in which such effluents are discharged. In addition, the nitrogen in the sewage effluent may be present mostly in the form of ammonia compounds, which are toxic to fish in high concentrations. Furthermore, ammonia exerts a nitrogenous oxygen demand in the receiving water as it is converted to nitrates in the process of nitrification. Some of the current tertiary treatment

technologies include (1) effluent polishing, (2) nutrient removal, namely, phosphorus and nitrogen, and (3) membrane technologies, namely, membrane bioreactor process, ballasted floc reactor, and electrodialysis.

38.2.7.1 Effluent Polishing

Effluent polishing involves the additional removal of BOD and TSS from secondary effluents. This can be achieved by (1) granular media filtration and (2) microstraining or microscreening. The granular media filtration process is achieved by an automatic backwash tertiary filter which further removes BOD and TSS. Because of the organic and biodegradable nature of the suspended solids, in the secondary effluent, tertiary filters must be backwashed frequently to prevent decomposition, which could cause septic or anaerobic conditions to develop in the filter bed. In addition to the conventional backwash cycle, an auxiliary surface air wash is used to thoroughly scour and clean the filter bed. Filtration may be done by gravity in an open tank or by pressure in closed pressure vessels. On the other hand, the microstraining or microscreening method utilizes microstrainers or microscreeners, which are composed of specially woven steel wire cloth mounted around the perimeter of large revolving drums. The steel wire cloth acts as a fine screen with openings as small as 20 μm. The rotating drum is partially submerged in the secondary effluent, which flows into the drum and outward through the microscreen. As the drum rotates, captured solids are carried to the top, where a high velocity water spray flushes them into a hopper mounted on the hollow axle of the drum.

38.2.7.2 Nutrient Removal

Nutrient removal in wastewater treatment is regarded as an advanced treatment and the common nutrients removed are nitrogen and phosphorus.

38.2.7.2.1 Nitrogen Removal

During the decomposition of waste by bacteria, nitrogen bound up in complex organic molecules is released as ammonia nitrogen. Subsequent oxidation of ammonia requires oxygen, which, if it occurs in the receiving body of water, contributes to oxygen depletion problems. In addition, nitrogen is an important nutrient for algal growth and only about 30% is normally removed in a conventional secondary treatment facility. To avoid the oxygen demand and eutrophication problems, treatment plants need to be augmented to achieve higher rates of nitrogen removal. This can be affected utilizing aerobic bacteria such as *Nitrosomonas* spp. To convert ammonia (NH_4) to nitrate (NO_2) in a process called nitrification, it is followed by an anaerobic stage in which *Nitrobacter* spp. oxidizes nitrites to nitrates and nitrates to nitrogen gas in a process of denitrification. The overall process is referred to as nitrification/denitrification. Nitrification process of domestic wastewater becomes effective after 5–8 days; therefore, a detention time of 15 days is required by this method.

38.2.7.2.2 Phosphorus Removal

During conventional primary and biological treatments of wastewater, only about 30% of the phosphorus is removed. Phosphorus in wastewater exists in many forms but all of it ends up as orthophosphate (H_2PO_4, HPO_4, and PO_4^{3-}). Removing phosphorus is most often achieved by adding a coagulant usually alum $Al_2(SO_4)$ or lime $[Ca (OH)_2]$. Alum is usually added to the aeration tank when the activated sludge process is being utilized thus minimizing the need for additional equipment.

38.2.7.3 Membrane Technologies

Recent technological breakthroughs in wastewater treatment and reclamation for water reuse include membranes that have emerged as a significant innovation for treatment and reclamation as well as a leading process in the upgrade and expansion of WWTPs. Membranes may be an option when they enable the removal of contaminants that other technologies cannot. They are also more economical than other alternatives and require much less land area than competing technologies because they may replace several treatment processes with a single one. For wastewater treatment applications, membranes are currently used as a tertiary treatment for the removal of dissolved species, organic compounds, phosphorus, a nitrogen species, colloidal and suspended solids, and human pathogens including bacteria, protozoa, cysts, and viruses. Membrane technologies for wastewater treatment include membrane bioreactors, usually microfiltration (MF) or ultrafiltration (UF) membranes immersed in aeration tanks (vacuum systems) or implemented in external pressure-driven membrane units as a replacement for secondary clarifiers and tertiary polishing filters.

38.2.7.3.1 Low-Pressure Membranes

These are usually MF or UF membranes used as either a pressure system or an immersed system providing a higher degree of suspended solids removal following secondary clarifications. UF membranes are effective for the removal of viruses.

38.2.7.3.2 High-Pressure Membranes

This involves nanofiltration or reverse osmosis pressure treatment and production of high quality product water suitable for indirect potable reuse and as high purity industrial process water. Pharmaceuticals and personal care products can also be removed by high-pressure membranes. Despite the increasing popularity of membrane technologies, there are, however, some gray areas in the form of technical barriers including high potential for fouling caused by colloids, soluble organic compounds, and microorganisms, which are typically not well removed with conventional pretreatment methods. This caused increased feed pressure and requires frequent membrane cleaning, which can reduce the efficiency and life span of the membranes. Another technical barrier is the complexity and expense of disposal of the concentrate from high-pressure membrane. Nevertheless, the benefits of high-pressure membranes are numerous. It is more often a viable option for many plant upgrades and capacity expansions and it is beneficial for landlocked situations, urban, agricultural, and industrial reuse, groundwater recharge, salinity barriers, and augmentation of potable water supply meeting very low effluent nutrient limits.

38.2.7.3.3 Membrane Bioreactor Process

In this process, aeration, secondary clarification, and filtration occur within a single bioreactor rather than in three separate basins providing the required tertiary treatment within smaller land area compared to conventional treatment. This process utilizes cassettes in which hollow-fiber microfiltration membranes were bundled into modules and grouped together. Connected by a beader pipe to the effluent vacuum pumps, the cassettes are submerged in a bioreactor tank. The vacuum pumps pull the effluent through the membranes but leave the solids behind, eliminating the need for secondary clarification and return sludge pumping. Since activated sludge stays in the tank, the mixed liquor suspended solids levels are much higher than the conventional activated sludge systems, thus facilitating treatment within a smaller volume. Air is supplied through the coarse bubble diffusers below the membrane cassettes, providing oxygen for biological treatment and agitation to scour and clean the membranes. Fine bubble diffusers are also used to supply more air treatment. Automatic backwash cycles are used to clean and restore membrane permeability at regular intervals. An MBR system can be built new or retrofitted into an existing activated sludge tank (Figure 38.3).

38.2.7.3.4 Ballasted Floc Reactor

This is a physico-chemical process in which the settling rate of suspended solids is increased over that of a conventional primary sedimentation process. A coagulant is first mixed rapidly with the influent to promote flocculation. Sand and a polymer are than added to form heavier large floc particles, which settle rapidly. The polymer binds the sand to the organic floc particles. The clarified effluent is discharged over a weir and settled sludge is pumped to a hydroclone, where the sand is separated by centrifugal force. The sand is recycled back to the Ballasted Floc reactor (BFR) and the organic sludge is pumped to an appropriate treatment process. The MBR and BFR systems can be utilized in parallel to allow cost-effective designs (Figure 38.4).

38.2.7.3.5 Electrodialysis

Electrodialysis consists of applying a direct current across a body of water separated into vertical layers by membranes alternatively permeable to cations and anions. Cations migrate toward the cathode and anions toward the anode. Cations and anions both enter one layer of water and both leave the adjacent layer. Thus, layers of water enriched in salts alternate with those from which salts have been removed. The water in the brine-enriched layers is recalculated to a certain extent to prevent excessive accumulation of brine. Although the relatively small ions constituting the salts dissolved in wastewater readily pass through the membranes, large organic ions including proteins, for example,

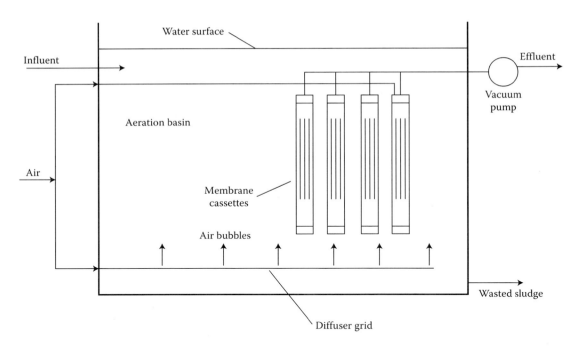

FIGURE 38.3 Schematic of a membrane bioreactor used in conjunction with activated sludge.

and charged colloids migrate to the membrane surfaces often causing fouling and plugging the membrane and therefore reducing the efficiency. In addition, growth of microorganisms on the membrane can also cause fouling and therefore requires regular cleaning. Electrodialysis possesses the potential to be a practical and economical method of removing up to 50% of the dissolved inorganics from secondary sewage effluent. Such a level of efficiency would permit repeated recycling of water without dissolved inorganic materials reaching unacceptably high levels [16].

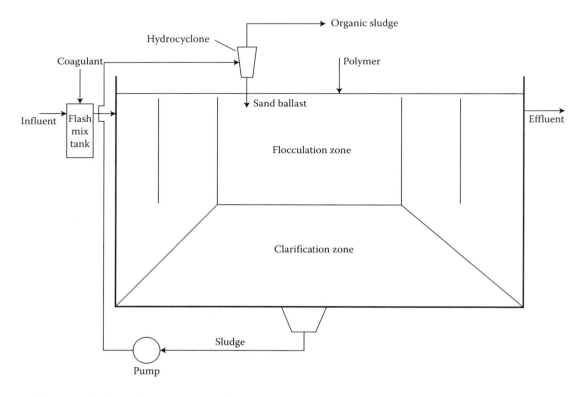

FIGURE 38.4 Schematic of ballasted flocculation reactor.

38.2.8 Desalination (Desalting)

Desalination is one of the processes of water reclamation for reuse from seawater into which treated wastewater effluents have been discharged containing approximately about 35,000 gm/L of salts and brackish water containing high levels of dissolved minerals or salts up to a concentration of about 1000 mg/L of salts. Brackish or seawater is unsuitable for domestic, agricultural, or industrial uses. Although expensive, desalination can be more economical than moving large quantities of freshwater over long distances. Desalinated or desalted water now provides the primary source of municipal water in many areas of the world particularly in the Caribbean, Middle East, North Africa, and other densely populated arid areas and coastal areas. There are two basic methods for desalination: thermal processes and membrane processes. The thermal processes involve the transfer of heat and a phase change of water into either vapor or ice, while membrane processes make use of thin sheets of special materials that allow freshwater to pass through but not salt.

However, both types of water desalination technology require high energy. The most common thermal desalination process in use is multistage flash distillation, which allows the production of relatively large quantities of desalted water. The mechanism of distillation involves the separation of freshwater by heating, evaporation, and condensation. The multistage flash distillation is carried in a series of closed vessels set at progressively lower pressures. The boiling temperature of water is lowered as the air pressure drops, reducing the amount of energy needed for vaporization. Heat is provided by steam from a boiler. When preheated saltwater enters a vessel that is at low pressure, some of it rapidly boils (flashes) into vapor, which condenses into freshwater on heat-exchange tubes and is then collected in trays under the tubes. The remaining saltwater flows into the next stage, which is set at even lower pressures, where some of the flashes continue until the process is completed.

The membrane technology utilizes two different types of processes for desalination: (1) reverse osmosis and (2) electrodialysis. However, these are usually limited to the treatment of brackish inland or wellwater supplies rather than seawater. However, reverse osmosis (also called UF) is next to multistage flash distillation, which is used primarily to treat seawater, in the global production of desalted water. Compared to multistage distillation, reverse osmosis requires only about half the energy to produce portable water. No heating is necessary and the major energy required is significantly reduced (Figure 38.5).

38.3 Hazardous Wastewater Treatment

Hazardous waste can be defined as any substance that because of its quality, concentration, physical, chemical, or infective characteristics may cause or significantly contribute to an increase in mortality, cause an increase in serious irreversible or incapacitating reversible illness, or pose a substantial present or potential hazard to human health and the environment when improperly treated, stored, transported, or disposed of or otherwise managed. A substance is described as hazardous if it possesses any of the following characteristics: reactivity, ignitability, corrosivity, toxicity, or infectivity [23].

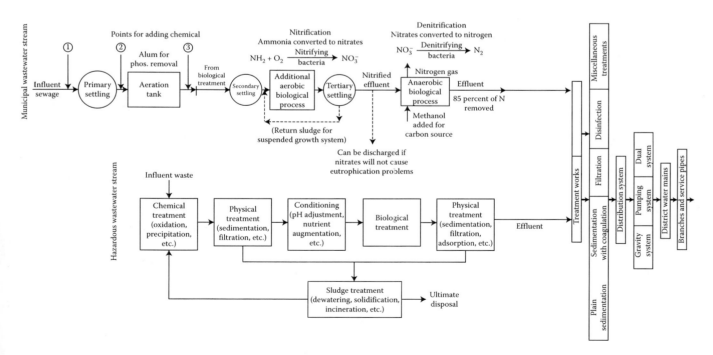

FIGURE 38.5 Schematic of an upgraded WWTP incorporating tertiary treatment facilities.

38.3.1 Legislation and Standards

Worldwide, there are legislations to guide safe provision of water for municipal and other uses. In the United States, one early effort at the federal level to guide the nation's clean water strategy was the Water Quality Act of 1965. This act was strengthened by the Federal Water Pollution Control Act Amendments of 1972. A basic goal of the amendments was to encourage individual states to clean up surface waters to the extent that they would once again be swimmable and fishable. The amendments were supported by programs to reduce water pollution from point sources, construction of sewage treatment plants, and Environmental Protection Agency (EPA) enforcement. The strategy of reducing pollution from point of sources was implemented by the National Pollution Discharge Elimination System (NPDES). Under the NPDES, all municipal or industrial treatment facilities that discharge wastewater effluents must obtain an NPDES discharge permit from EPA or a delegated state agency. The NPDES permits clearly state the allowable amounts of specific pollutants that a particular facility can discharge into the environment. The same act, amended in 1977, includes use of current technology that is economically viable. The Clean Water Act was also amended in 1986 with stricter control [35].

38.3.1.1 Primary Treatment

Primary treatment of hazardous wastewater includes dissolution, blending, slurring, and phase separation. The most straightforward means of physical treatment involves separation of components of a mixture that are already in two different phases. Separation and decanting can easily be accomplished with simple equipment. In many cases, the separation is facilitated by mechanical means, particularly filtration or centrifugation. The simplest physical process that can concentrate and reduce wastewater volume is evaporation, which may be facilitated by using mechanical sprayers. Other available processes utilized to separate hazardous waste from a liquid include sedimentation, floatation, and filtration [15].

38.3.1.2 Secondary Treatment

Secondary treatment of hazardous wastewater can be affected by chemical processes and biological treatment. Chemical processes include the following:

1. In exchange processes: In this case, the hazardous wastewater (usually industrial wastewater) is passed through a bed of resin that selectively adsorbs charged metal ions. This process is being utilized, for example, in the removal of waste, chronic acid from production rinse water in the metal finishing industry.
2. Neutralization: Neutralization is a process of pH adjustment to reduce the strength of acidity or alkalinity of waste. Limestone, for example, can be used to neutralize acids while compressed carbon dioxide can be used to neutralize strong bases. It is important to note that pH affects the effectiveness of chlorination and disinfection

in later purification processes of water. The lower the pH, the more effective the chlorination and disinfection process.
3. Precipitation: In a precipitation reaction, certain chemicals are made to settle out of a solution as a solid material. An example of its application is in the battery industry, where the addition of lime and sodium hydroxide to acidic battery waste causes lead and nickel to precipitate out of solution.
4. Oxidation and reductions: These are complementary chemical reactions involving the transfer of electrons among ions. These reactions have been used successfully in rendering toxic substances less hazardous; for example, oxidation of waste cyanide by chlorine.
5. Biological treatment: As stated earlier, biological treatment involves the action of living organisms whereby the microbes utilize the waste material as food, metabolize it, and convert it into simpler substances. Although biological treatment has common applications in stabilizing organic waste in municipal sewage, there is now emerging applications in petroleum industries whereby organic wastes are acted on by certain bacteria and genetically engineered species to do so.

38.3.2 Bioreactors

Certain organic hazardous wastes can be treated in a slurry form in an open lagoon or in a closed vessel called a bioreactor. A bioreactor encompasses a fine bubble diffuser whose function is to provide oxygen and a mixing device to keep the slurry solids in suspension. In principle, the bioreactor is a biological system.

38.3.3 Air Stripping

This involves the mass transfer of volatile contaminants from water to air. For groundwater remediation, this process is typically conducted in a packed tower or an aeration tank. The genetic packed tower air stripper includes a spray nozzle at the top of the tower to distribute contaminated water over the packing in the column, a fan to force air countercurrent to the water flow, and a sump at the bottom of the tower to collect decontaminated water. Auxiliary equipment that can be added to the basic air stripper includes automated control systems with sump level switches and safety features such as differential pressure monitors, high sump level switches, explosion-proof components, and discharge air treatment systems such as activated carbon units, and catalytic or thermal oxidizers. Packed tower air strippers can either be installed permanently on concrete pads on a skid or on a trailer (Figure 38.6).

38.3.4 Liquid Phase Carbon Adsorption

In this technology, hazardous wastewater, especially contaminated groundwater, is pumped through a series of vessels containing activated carbon onto which dissolved contaminants

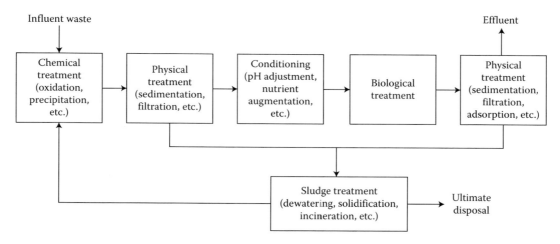

FIGURE 38.6 Schematic of a hazardous wastewater treatment system.

are adsorbed. When the concentration of contaminants in the effluents from the bed exceeds a certain level, the carbon can be regenerated in place, removed and regenerated at an offsite facility, or removed and disposed of. Adsorption by activated carbon has long played a big role in municipal, industrial, and hazardous waste management.

38.3.5 Oil/Water Separator and Recovery

This process is commonly used in sites where free product exists in the groundwater. Undissolved liquid-phase organics are removed from subsurface formations either by active methods such as pumping or by a passive collection system. In an active pumping system, the combined oil and water is pumped aboveground and passed through an oil/water separator, which separates out and skims the oil and stores it in an aboveground container. Following recovery, it can be disposed of, reused directly in an operation requiring high purity materials, or purified [32].

38.3.6 Underground Injection

The most popular way to dispose of liquid hazardous waste has been to force them underground through deep injection wells. This involves pumping the liquid down through a drilled well into a porous layer of rock. The liquid is then injected under high pressure into the pores and fissures of the rock. The layer or stratum of rock in which the waste is stored (usually limestone or sandstone) must lie between impervious layers of clay or rock. This injection zone can be about or above 700 m below the surface. The capacity of the geological strata to accept an injected waste depends on its porosity, permeability, and other factors. The injection well must be at least 0.4 km from an underground source of drinking water and the waste must be injected into a separate geological formation free from fractures. The chemical and petroleum refining industries are the largest users of

injection wells. The risk of hazardous waste injection wells to leak indicates they cannot be considered to be entirely safe; therefore, regulations on their operation are becoming more stringent and the reliance of technology is being discouraged prior to reuse. Some of these systems may be designed to recover only the product, mixed product and water, or separate streams of product and water.

38.4 Biomonitoring of Wastewater Effluents

38.4.1 Policies, Regulations, and Guidelines: Global Best Technologies

In the United States, the strategy for reducing pollution from point sources is implemented by the NPDES. Under the NPDES, all municipal or industrial treatment facilities that discharge wastewater effluents must obtain on NPDES discharge permit from the EPA or a delegated agency. In Nigeria, the Nigeria Environmental Standard Regulatory Agency (NESREA) is responsible for setting and enforcing minimal standards for wastewater effluents. Generally, the following values are internationally acceptable: BOD = 30 mg/L, TSS = 30 mg/L, pH 6.0–9.0, coliform count 200 per 100 mL. Efficiency of wastewater treatment can be monitored by assessing the effluents for the following: BOD TSS, pH, nitrogen, phosphorus, and *E. coli*.

38.4.1.1 Biochemical Oxygen Demand

When biodegradable organic matter is released into a body of water, microorganisms, especially bacteria, feed on the wastes breaking them down into simpler organic and inorganic substances. When this decomposition takes place in an aerobic environment, which is in the presence of oxygen, the end products are carbon dioxide (CO_2), sulfate (SO_4), orthophosphate (PO_4), and nitrate (NO_3). In a situation of insufficient oxygen, the resulting anaerobe composition is performed by completely different microorganisms where the end products include hydrogen sulfide (H_2S), ammonia (NH_3), and methane (CH_3). The methane produced is physically stable, biologically degradable, and is a potent greenhouse gas. It is also generated in the anaerobic environment of landfills, where it is sometimes collected and used as an energy source. The amount of oxygen required by microorganisms to oxidize organic wastes aerobically is called the biochemical oxygen demand (BOD) and is expressed in milligrams of oxygen per liter of wastewater (mg/L).

38.4.1.2 Total Solids

It is a water quality measurement listed as a conventional pollutant in the US Clean Water Act (USCWA). TS of a water sample or effluent sample is determined by pouring a carefully measured volume of water (typically 1 L but less if the particulate density is high or is as much as 2 or 3 L for very clean water) through a preweighed filter of a specified pore-size, and then weighing the filter again after drying to remove all water. Filters for TSS measurements are typically composed of glass fibers [25]. The gain in weight is a dry weight measure of the particles present in the water sample expressed in units derived or calculated from the volume of water filtered (typically milligrams per liter or mg/L). Although turbidity purports to measure approximately the same water quality property as TSS, the latter is more useful because it provides an actual weight of the particulate material present in the sample [10,28].

38.4.1.3 Total Coliform Count and *E. coli* Count

Coliform bacteria are hardy organisms that survive in water longer than most organisms and are well-known biological indicators of water quality and pollution. Often, nonpathogenic coliforms are always present in the intestinal tract of humans and millions are excreted with body wastes; consequently, water that has been contaminated with sewage will always contain coliforms. A particular species of coliforms found in domestic sewage is *Escherichia coli* (*E. coli*) and there are roughly 3 million *E. coli* bacteria in 100 mL volume of untreated sewage. Coliform estimation in effluents involves two methods; namely, membrane filtered and the multiple-tube fermentation. The membrane filter method takes less time and provides more of a direct coliform count then the multi-tube method and also requires less laboratory equipment. However, it is not applicable to turbid samples. The membrane filter method utilizes a special membrane that is flat-shaped in a paper-like disk with uniform microscopic pores small enough to retain the bacteria on its surface while allowing the water to pass through the filter. This is applied to a partial vacuum. After the sample is drawn through, the filter is placed in a Petri dish in which bacteria growth medium is embedded. The Petri dish holding the filter and nutrient is placed in an incubator at 35°C for 24 h. The bacteria grow on the medium as colonies and appear as characteristically green metallic dots. The filter has a grid printed on it to facilitate colony counting. A magnifying glass can be used to facilitate this. Small samples of polluted water or wastewater must be diluted with sterile water before filtering so that the filter is not overgrown with colonies,

making it impossible for an accurate count. Coliform concentrates are expressed in terms of the number of organisms per 100 mL of water. The following formula can be used to express the results of samples of various sizes:

$$\text{Coliforms per 100 mL count} = \frac{\text{number of colonies} \times 100}{\text{vol of sample}}$$

In the multitube fermentation method, the technique is based on the fact that coliforms possess the ability to utilize lactose as food and produce gas in the process (lactose fermentation). A measured volume of water sample is added to a tube that contains lactase broth nutrient medium. A small inverted vial in the lactose broth traps some of the gas that is produced as the coliform bacteria grow and reproduce. The gas bubble in the inverted vial along with a cloudy appearance of the broth provides visual evidence that coliforms may be present in the sample. However, if the gas is not produced within 48 h of incubation at 35°C, the test is said to be negative for coliform. Because some organism do produce gas in lactose, a confirmation test involves transferring the nutrient medium from a presumptive positive tube to another fermentation tube containing brilliant green bile, another nutrient medium. Coliforms are said to be present if gas is again formed within 48 h of incubation at 35°C. The fermentation tube method can be used to test for fecal coliforms as well as total coliforms, but the higher temperature of 44.5°C is required for incubation to grow fecal organisms.

pH signifies the degree of acidity or alkalinity of a medium. Industrial effluents usually have varying pH and therefore should normally be neutralized before being discharged. pH is determined using a pH meter. The pH of effluent is particularly important during disinfection with chlorine because the HOCl and OCl$^-$ ions, which are the active ions in chlorine, depend on the pH of the water to function. The lower the pH, the more effective the chlorination and disinfection of effluents and water. At lower pH, a smaller contact period is required for the same percentage of destroyed organisms [30].

38.4.1.4 Phosphorus Level

Effluent from advanced waste treatment is expected to have a low level of phosphorus below 0.5 mg/L. Advanced waste treatment normally requires removal of phosphorus to reduce algal growth. Algae may grow at PO_4^{3-} levels as low as 0.05 mg/L. The removal of phosphorus may occur in the sewage treatment process in the primary settler, in the aeration chamber of the activated sludge unit, and lastly after the secondary waste treatment. Therefore, effluent monitoring of phosphorus is important.

38.4.1.5 Nitrogen Level

Nitrogen level is also monitored in the effluent to assess quality. Nitrogen is the algal nutrient most commonly removed as part of the advanced waste treatment. In addition, ammonia is known to be toxic in fish while excessive nitrate can cause methemoglobin anemia in the infants.

38.4.2 Special Needs of Developing Countries

Success of wastewater treatment processes is assessed by biomonitoring of the effluent. Effluent quality is dependent on functional technologies. In the case of developing countries, there are challenges of funds and infrastructure, which may impact effluent quality. Despite the availability of advanced treatment technologies, cost benefit analysis should be considered before the adoption of advanced treatment technologies. Therefore, the primary focus of developing economies should be appropriate and sustainable technologies. For example, constructed wetlands may be quite appropriate in some instances than specialized treatment plants while drying and pelleting of dry sludge for use as fertilizer may reduce the cost of sludge management. In many instances of urban water reuse, public–private partnership has been discovered to be the way forward. Furthermore, international organizations such as World Bank, UNIDO, UNDP, UNICEF, and nongovernmental organizations can partner with developing countries to achieve successful wastewater treatment through funding, capacity development, infrastructure upgrade of electricity power, and development of appropriate and sustainable wastewater treatment. Some of the requirements of the developing countries include the following:

1. Institutional training

   Institutional training is an essential prerequisite to developing professional skills in the area of wastewater management. Accordingly, on a long-term basis, each nation should ensure that its colleges and universities are capable of imparting the type of training that is consistent with national needs. In most developing countries, however, appropriate universities or colleges where people could be properly trained in wastewater management do not exist. Not surprisingly, educational institutions share many of the same constraints of other public bodies. Such constraints include limited operating budgets, inadequate possibilities for career development of teaching staff, understaffing, inadequate library, computer, research, and other support facilities, inexperienced staff, and poor staff–students ratios. These factors generally contribute to theoretical and conventional orientations to teaching, which often are not appropriate for the solution of practical problems. Furthermore, in an area such as hazardous wastes management where the knowledge base is improving rapidly, the lack of training facilities available to teachers during their working lives is contributing to the obsolescence of the material being taught.

2. Short courses

   Well-designed short course (1–12 weeks) can be of immense benefit to all staff, improving their level of knowledge and complementing the formal education received during their earlier training. Such courses can be used to familiarize staff in new areas, such as environmental aspects of waste management, instrumentation, or use of computers and systems analysis. They can also serve periodically to update technical knowledge and skills and to broaden interdisciplinary understanding. These courses can be held either within a country or abroad.

3. Counterpart training

   In most multilateral or bilateral projects for hazardous wastes management, where expatriate professionals are used, counterpart staff are appointed with the objective of knowledge and experience transfer. Properly done, counterpart training can play an important role in developing countries. Generally, however, the training of counterpart personnel who are supposed to take over from expatriate staff is unsatisfactory. Expatriate staff are frequently used as substitutes for local work force and spend much of their time performing routine and mundane duties. The weakness of the national institutions often renders the foreign experts, even when they are suitably chosen, ineffective. Frequently there are not enough adequately qualified counterpart staff, and even when they exist, bureaucratic delays in their recruitment may not leave sufficient time for training.

4. Individually tailored programs

   Many multilaterally or bilaterally supported projects now contain specifically tailored programs for training of a limited number of select individuals. These training programs are invariably abroad. They may range from 2–8 weeks of study tours in various institutions in one or more countries in specific areas of specialization to more formal training at educational and research institutions leading to diplomas or degrees.

5. Conferences, symposia, and workshops

   Under certain conditions, participation at conferences, symposia, and workshops to present papers, discuss experiences, and share knowledge can have a training role. However, considering the fact that most conferences or symposia have durations of one week or less, they are not planned to play an explicitly educative role. If not suitably organized, their role in training may not be cost-effective. One could argue that the perusal of documentation from a conference in certain cases may be a more cost-effective training alternative.

6. Information and resource materials

   Pursuit of knowledge is seriously handicapped in all developing countries by lack of access to up-to-date books, manuals, journals, and databases. Inadequate funding means that many of the most recently published books and periodicals reporting new findings and methodologies are not available and the few publications that may have been purchased are not readily accessible because of poor documentation and library facilities. Similarly, the lack of microcomputers and audiovisual facilities means that staff cannot remain informed about the new developments in a rapidly changing technical area such as wastewater management. However, the

Internet, which is a good source for information, can be expensive for many developing countries to subscribe to on a long-term basis.

38.4.3 Ethics in Wastewater Treatment Upgrading

Wastewater operation is a professional occupation that requires education, training, experience, and attainment of state certificate to practice. In this profession, operators must be dedicated to the protection of public health and act skillfully and conscientiously. Operators have a direct impact on our island environment, the preservation and protection of which affects the quality of life for all residents and the experience of millions of annual visitors.

Rules of practice:

Operators, in fulfillment of their professional duties, shall:

1. Hold paramount the health and welfare of the public.
   a. Follow all procedures and guidelines designed to prevent pollution from occurring.
   b. Strive to increase public knowledge of the wastewater treatment field and its importance by leading tours of their treatment facilities.
2. Protect public property and the environment.
   a. Properly and conscientiously operate and maintain the treatment facilities with which they are publicly entrusted.
   b. Strive to maintain the aesthetics of the environment in and around their facilities.
3. Properly and accurately complete required records.
   a. Be objective and truthful in data collection and reporting.
   b. Acknowledge their errors and not distort or alter the facts.
4. Follow and comply with state and federal rules and regulations.
   a. Be familiar with all details of the permit requirements that apply to their facilities and understand the consequences of violations caused by inaction or negligence.
   b. Only undertake duties for which they are qualified.
5. Follow safe operating policies and ensure the best possible services.
   a. Always consider their personal safety, the safety of their fellow workers, and that of any other persons present at their facilities while performing their duties.
   b. Endeavor to increase their knowledge and skills through continuing education activities.
6. Avoid unprofessional practices and act honorably, responsibly, ethically, and lawfully to enhance the reputation of the profession.
   a. Accept personal responsibility for their professional actions.

b. Not misrepresent or exaggerate their responsibilities during prior work experience.
c. Not untruthfully criticize other operators to injure their professional reputation or employment.

38.4.4 Economics of Wastewater Treatment System Upgrade

There is need for cost analysis and cost–benefit ratio when considering the upgrade of a wastewater treatment system. All the components including EIA should be considered. It has been postulated that inclusion of advanced treatment in some cases can be twice the cost of initial facilities with secondary treatment capability. However, other experiences have revealed not only great improvement in effluent quality, but also reduction in power consumption, land use, emission, chemical use, better sludge management, and a better option than building a new treatment plant to accommodate urban growth [2,21].

38.4.5 Geographical Information System: Simulation and Modeling

Geographical information system (GIS) is an orderly assemblage of computer-based hardware, software, geographically referenced data, procedures and human ware (personnel) configured to handle all forms of spatial data to satisfy the geographic information needs of a user [34]. The technology, which is expanding, has been accommodating various programming languages and software for designs, simulation, and modeling with wide applications in environmental technology including wind turbines, water treatment plants, WWTPs, and upgrade. The MATLAB® and Simulink® systems are becoming increasing popular for these applications. MATLAB® (matrix laboratory) is a multiparadigm numerical computing environment and fourth generation programming language designed by Cleve Moler, Professor and Chairman of Computer Science Department of the University of New Mexico in the late 1970s to give his students access to LINPACK and EISPACK without having to learn Fortran and was later developed by MathWorks® and written in C language in 1984. MATLAB® allows matrix manipulations, plotting of functions, data implementation of algorithms, creation of user interfaces, and interfacing with programs within other languages including C, C++, Java, and Fortran. Although MATLAB® is primarily intended for numerical computing, an optional toolbox uses the MUPAD symbolic engine and allows access to symbolic computing capabilities. An additional package called Simulink® adds graphical multidomain simulation and model-based design for dynamic embedded systems. Simulink®, also developed by MathWorks®, is a data flow graphical programming language tool for modeling, simulating, and analyzing dynamic systems. Its primary interface is a graphical block diagramming tool and a customizable set of block libraries. It offers tight integration with the rest of the MATLAB® environment and can either drive MATLAB® or be scripted from it.

Simulink® is widely used in control theory and digital signal processing for multidomain stimulation and model-based design. A number of MathWorks® and third-party hardware products are available for use in Simulink® and include *State flow* extend simulation with a design environment for developing state machines and flow charts. Simulink® is typically used for quasi-continuous stimulation and event-controlled simulation. Typical systems that can be simulated include technical systems, biologic-technical systems, business processes, and many others. Simulink® modeling is becoming increasingly popular in WWTP monitoring and upgrade. Different water parameters can be used for simulation and modeling using MATLAB® and Simulink.® These include sand concentration and grain size, mass density, sludge value, turbidity, volumetric flow, and rate and concentrations of aluminum, dissolved organic carbon, chloride, dry matter, *E. coli*, and so on [27].

38.5 Discussion

Water reuse and recycling are becoming more common as demands for water exceed supply. Unplanned reuse occurs as the result of waste effluents entering receiving water or underground and subsequently being taken into a water distribution system. Planned reuse utilizes wastewater treatment systems deliberately designed to bring water up to standards required for subsequent application. Globally it has been accepted that water is a finite entity and therefore should be well managed. This thinking gave rise to the concept of integrated water management systems (IWMS) [8,18]. Reuse of water continues to grow due to a lack of adequate supply of water and availability and widespread deployment of modern water treatment processes, therefore significantly enhancing the quality of water available for use. These two factors underpin popularity of water reuse in semi-arid regions and in countries with advanced technology. Many countries, particularly in the Middle East, depend on irrigation for essentially all of their agriculture. For example, Israel reuses 66% of its sewage effluent for irrigation whereas in the United States, where water is more readily available, only about 2.4% of its sewage effluent is reused.

In order to meet new standards and regulations, it may be expedient to upgrade an existing WWTP instead of constructing a new plant. Other reasons for upgrading a WWTP for urban reuse include (1) population growth and increased requirements, (2) aging existing infrastructure, (3) new concept of integrated water management, (4) groundwater recharge, (5) cooling and process water in industrial applications, (6) irrigation of crop land and watering of golf courses and other plant-based requirements, (7) overall improvement in water quality, and (8) epidemics of infectious disease [37].

Standard WWTPs have facilities for primary and secondary treatments that can then discharge its effluents into the water courses. However, for urban purposes, it might be necessary to upgrade such facilities to include tertiary or advanced treatment with varying technologies. Tertiary treatment of sewage can remove more than 99% of the pollutant from raw sewage and can produce effluent of almost drinking water quality. However, the cost of tertiary treatment, construction, operation,

and maintenance could be very high. Sometimes doubling the benefit-to-cost ratio and usually is not always big enough to justify the addition expense. There are, however, situations when an upgrade is inevitable. There are different possibilities to upgrade a WWTP. There may be a requirement for activated sludge plant, precipitation and flocculation processes, influent balancing, increase of oxygenation capacity with pure oxygen, increase of capacity of final clarifiers and pretreatment of industrial effluents. Furthermore, an upgrade can include additional electrical and civil engineering processes, process control, new head works, new raw sludge pump, start, new, or additional primary clarifiers, new aeration air blowers, additional secondary clarifiers, and new chlorine contact tank (Figure 38.5).

In planning to upgrade a WWTP, the component to be upgraded should be clearly identified and compared with the cost benefits. There may be a need to include new processing lines of treatment or to separate existing combined systems. In this respect, the technology of GIS is becoming increasingly useful whereby a desired upgraded plant can be simulated and modeled using relevant wastewater parameters such as sand concentration, dissolved organic carbon, chloride, dry matter, coliform bacteria, volumetric flow rate, sludge volume, and so on as coordinates. In a particular model, the main results were expressed in terms of number and capacity and location of the WWTP and the length of the main sewers. The decision process concerning the location and capacity of a wastewater system has a number of parameters that can be optimized. Those parameters include the total sewer length, the number, capacity, and the location of the WWTP, the optimization of construction and operation cost of the integrated system. Therefore, GIS may represent an important role for the analysis of data and results especially in the preliminary stage of planning and design [21].

In another scenario, the main reason to upgrade was to curb an epidemic and improve public health safety. For example, an outbreak of cryptosporidosis in Milwaukee USA in 1993, required an upgrade of the wastewater treatment—and drinking water systems. The upgrade was designed to kill *Cryptosporidium* in wastewater and influent water entering the water treatment plant and to minimize the levels of disinfection by-product and reduce taste and odor problems caused by chlorine and other precipitation chemicals. Since application of ozone is the most cost-effective disinfectant against *Cryptosporidium*, ozone was generated at the two plants and mixed with the water using fire bubble diffusion technology. The upgrade was among the first to be done to inactivate *Cryptosporidium* with ozone. Since chlorine was no longer used, the generation of chlorine by-products was eliminated. The use of nonchlorine coagulants also reduced sludge production and sludge disposal cost. The filter run was lengthened and the volume of backwash water was reduced.

Another example of an upgrade was in the case of Olburgen, the Netherlands WWTP to improve effluent quality. In this scenario, separate and dedicated treatments were initially being utilized for industrial wastewater and reject water. The separate treatment of effluents proved to be more cost-effective and area and energy efficient than a combined treatment process. In the improved technology, the industrial wastewater was first treated in a UASB reactor; thereafter, the effluent was combined with the reject water and treated in a struvite reactor (phospaq process) followed by a one-stage granular sludge anammox process. For the first time, both reactors were operative at full capacity and have been upgrade stable over the years. The recovered struvite was also found to be a suitable substitute for commercial fertilizer. This public–private partnership project was found to be very successful because it requires a 17 times smaller reaction volume (20,000 m) (3 times less) and reduced sewer electricity power by approximately 1–5 Gigawatts hour (GWh) per year [1].

Another important component of the waste management system is the wetlands. With the potential of natural wetlands to remove pollutants, constructed wetlands are becoming increasingly popular. While subsurface flow wetlands and surface flow wetlands were the earlier types designed, the tidal flow wetlands are the latest evolution of wetland technology and are used to treat domestic, agricultural, and industrial wastewater, including heavy load. In this system, organic carbon is primarily oxidized with nitrate, which is produced through a series of flood and drain cycles from one side of the wetland to the other. This process holds a number of benefits over traditional subsurface and surface flow wetlands and these include reduced land requirements and increased denitrification capability because the treatment of heavy load constructed wetlands has been used to remove ammonia from nine drainage and extensively for the removal of dissolved metals and metalloids [26]. Use of plants in removal of pollutants called *phytoremediation* has found particularly great usefulness in constructed wetlands. Current knowledge of specific characteristics in abstracting pollutants from wastewater has made increased interest in this technology. Poplar trees and sunflowers are particular plants of note.

Although wastewater reuse is frequently mentioned as one of the main solutions for water scarcity, it is not widely applied in countries where water is a scarce response, for example, in the Mediterranean. All around the Mediterranean, however, several countries such as Italy, Spain, and Greece are experiencing droughts and, at the same time, difficulties in implementing water reuse systems. The major areas in which water is currently being reused are irrigation (agriculture), industry, urban cleaning, firefighting, recreation, surface water replenishment, and groundwater recharge. It should be taken into account that reclaimed water reduces the consumption of potable water (and thus the consumption of freshwater required to produce potable water) [18,31].

The immediate drivers behind water reuse may differ in each case; the overall goal is to close the hydrological cycle to a much smaller, local scale. In this way, used water (wastewater) becomes a valuable resource. Although reclaimed water can become an alternative source of water, social opposition to it is significant. This is because many end users and customers are unwilling to accept reclaimed water because of the risks that may be involved in using it. These fears are partly based on the potential effects of contaminants on human health and the environment (pollutants, e.g., pharmaceuticals and personal care products). To guarantee that wastewater is safely reused, effective tertiary treatment processes must be applied to the water lines of conventional WWTPs [37]. These issues imply that more knowledge is needed regarding the benefits of reclaimed water. One should also remember that although water reuse strategies are intended to address the problem of water scarcity, this should not come at the price of increasing other environmental impacts. Because of the need for balance, an environmental evaluation of water reuse alternatives is needed [33].

For urban reuse of treated wastewater, an upgrade may be inevitable. It is therefore being proposed that sewer system design must be integrated with WWTP design when moving toward a more sustainable urban water management (Figure 38.4). This integration allows an optimization of the design of both systems to achieve a better and more cost-efficient wastewater management. Hitherto, integrated process design has not been an option because the tools to predict sewer wastewater transformation have been inadequate. Simulation, modeling, and application-integrated treatment plant design are exemplified. However, it is expedient that industrial effluent should be pre treated onsite before transport to water treatment plants for reuse.

With respect to developing countries, for international cooperation, capacity development, funding and transfer of technology is needed for urban water reuse to be made feasible. The primary challenge of electricity sewer generation in developing countries should be put into consideration when planning the design of wastewater plants in such countries. Therefore, appropriate and sustainable technology should be the focus. International, government and nongovernmental organizations are particularly needed as partners in providing adequate water in land-locked and semi-arid zones. Such partners include the World Bank, IMF, UNICEF, UNIDO, and UNDP.

It is in evitable that water recycle and reuse will continue to grow and this trend will increase the demand for quantitative and qualitative wastewater treatment. In addition, it will require more careful consideration of the original uses of water to minimize water deterioration and enhance its suitability for reuse.

38.6 Summary and Conclusions

Global population growth, climate change, the finite nature of water, urban growth, and rapid industrialization underpin the necessity for better management of water resources. Increasing demand for water in urban centers in general and semi-arid zones in particular have made urban water reuse inevitable. In addition, current knowledge of water pollution, which includes contamination with active pharmaceuticals and household products and occasional epidemics with infectious diseases, has caused the establishment of new regulations and standards for water usage. Furthermore, improved technologies have resulted in safer treated wastewater effluent, which meets the Drinking Water Act Standards. However, cost and benefit analysis should be considered before upgrading WWTPs. The role of public–private partnership is a feasible upgrade of a WWTP particularly in developing economics and cannot be overemphasized. Simulation and modeling of an upgraded WWTP should be explored at the planning stage and this should incorporate EIA. Furthermore, in view of the fact that wetlands are very important components of waste management systems, constructed wetlands are increasingly becoming part of the upgrade and tidal flow wetlands are the latest evolution of wetland technology. In view of these reasons, WWTP upgrades are bound to be in demand in the near future.

References

1. Abma, W.R., Driessen, W., Haarhuis, R., and Van Loosdrecht, M.C. 2010. Upgrading of sewage treatment plant by sustainable and cost effective separate treatment of Industrial waste water. *Water Science and Technology*, 61(7), 1715–1722.
2. Arcata. 2012. California Constructed Wetland: A Cost-Effective Alternative for Wastewater Treatment. Ecotippingpoints.org. Retrieved May 23, 2012.
3. Arms, K. 2008. *Environmental Science*. Holt, Rinehart and Winston, Austin, TX, pp. 187–188.
4. Banjoko, B. 2014. Environmental engineering for water and sanitation systems. In: S. Eslamian (ed.), *Handbook of Engineering*. CRC Press, Boca Raton, FL, pp. 65–83.
5. Bhatia, S.C. (ed.). 2007. *Environmental Chemistry*. CBS, New Delhi.
6. Boubecar, B., Crosby, C., and Touchains, E. 2002. *Intensifying Rainfed Agriculture: South African Country Profile*. IWMI, Colombo, Sri Lanka.
7. Brix, H. 1993. Wastewater treatment in constructed wetlands: System design, removal processes, and treatment performance In: A.G. Moshiri (ed.), *Constructed Wetlands for Water Quality Improvement*. CRC Press, Boca Raton, FL.
8. CapNet. 2012. Tutorial on basic principles of integrated water resources management. www.cap-net.org. Accessed on May 15, 2012.
9. Chen, Y. and Regli, W.C. 2002. Disinfection practices and pathogen inactivation in ICR surface water plants. In: M.J. McGuire, J.L. McLain, and A. Obolensky (eds.) *Information Collection Rule Data Analysis*. AWWA, Denver, pp. 376–378.
10. Clescerl, L.S., Greenberg, A.E., and Eaton, A.D. (eds.). 1999. *Standard Methods for the Examination of Water and Wastewater*, 20th Edition. American Public Health Association, Washington, DC.
11. Critten Clan, J.C., Trusell, R.R., Hand, D.W., and Howe, K.J. 2005. *Water Treatment Principles and Design*, 2nd Edition. Wiley, New York.
12. Dutta, S. (ed.). 2002. *Environmental Treatment Technologies for Hazardous and Medical Wastes*, Tata McGraw-Hill, New Delhi.
13. Edzwald, J.K. 2011. *Water Quality and Treatment*, 6th Edition. McGraw-Hill, New York.
14. Fraser, L. and Keddy, P.A. (eds.). 2005. *The World's Largest Wetlands: Their Ecology and Conservation*. Cambridge University Press, Cambridge, p. 488.
15. Freeman, H.M. (ed.). 1989. *Standard Handbook of Hazardous Waste Treatment and Disposal*. McGraw Hill, New York.
16. Gese, T.P. 2001. *New Wastewater Treatment Technologies Public Works Manual, Ridgewood*. Public Works Journal Corporation, New Jersey.
17. Glaze, W.H. 1987. Drinking water treatment with ozone. *Environmental Science and Technology*, 21(3), 224–230.
18. Global Water Partnership. 2002. Toolbox. Integrated water resources management. www.gwp.org/Globa/04%20 Integrated%20water%20Resources%20. Accessed on June 12, 2014.
19. Hart, T.M. and Davis, S.E. 2011. Wetland development in a previously mined landscape of East Texas. *USA Wetlands Ecological Management*, 19, 317–329.
20. Keddy, P.A. 2010, *Wetland Ecology: Principles and Conservation*, 2nd Edition. Cambridge University Press, New York, p. 497.
21. Leitao, J.P., Mahos, J.S., Goncalves, A.B., and Matos, J.L. 2009. Contribution of geographic information systems and location models to planning of wastewater systems. *Water Science and Technology*, 52(33), 1–8.
22. Liikanen, A. 2009. Methane and nitrous oxide fluxes in two coastal wetlands in the northeastern Gulf of Bothnia, Baltic Sea. *Boreal Environment Research*, 14(3), 351–368.
23. Maltezom, S.P., Biswes, A.K., and Sulter, H. (eds.). 1989. *Hazardous Waste Management Selected Papers from an International Expert Workshop Convened by UNIDO, Vienna*, June 22–26, 1987, Tycooly, London.
24. Masters, G.M. (ed.). 1998. *Introduction to Environmental Engineering and Science*, 2nd Edition. Pearson Education, New Delhi.
25. Michaud, J.P. 1991. A citizens' guide to understanding and monitoring of lakes and streams No 94–149, Washington State Department of Ecology, Publications office, Olympia, WA.
26. Mitsch, W.J. and Gosselink, J.G. 2007. *Wetlands*, 4th Edition. John Wiley & Sons, New York.
27. Moler, C. 2004. The origins of MATLAB®. *MathWorks Newsletter*.
28. Moran, J.M., Morgan, M.D., and Wiersma, J.H. 1980. *Introduction to Environmental Science*, 2nd Edition. W.H. Freeman, New York.
29. Nathanson, J.A. (ed.). 2003. *Basic Environmental Technology*, 4th Edition. Pearson Education, New Delhi.
30. Park, K. (ed.). 2005. *Park's Textbook of Preventive and Social Medicine*, 18th Edition. Bhanot, Dapalpur.
31. Rees, J.A. 2002. *Risk and Integrated Water Management TAC Background Papers*. Global Water Partnership, Stockholm.
32. Richardson, J.L., Arndt, J.L., and Montgomery, J.A. 2001. Hydrology of wetland and related soils. In: J.L. Richardson and M.J. Vepraskas (eds.), *Wetland Soils*. Lewis Publishers, Boca Raton, FL.
33. Sharma, J.L. (ed.). 1986. *Public Health Engineering*. Satya Prakashan, New Delhi.

34. Uluocha, N.O. 2007. *Elements of Geographic Information Systems*. Sam Iroanusi, Lagos.

35. U.S. Clean Water Act, sec. 304(a)(4), 33 U.S.C. & 1314(a)(4).

36. USEPA. 2000. Introduction to Phytoremediation EPA/600/R-99/107, National Risk Management Research Laboratory; office of Research and Development, Cincinnati, OH.

37. Vollertsen, J., Hvitved–Jacobsen, T., Ujang, Z., and Talib, S.A. 2002. Integrated design of sewers and waste water treatment plants. *Water Science and Technology*, 46(19), 11–20.

38. Vymazal, J. and Kropfleova, L. 2008. Wastewater treatment in constructed wetlands with horizontal subsurface flow. *Science of the Total Environment*, 380, 48–65.

39. WHO. 2006 *Guidelines for the Safe Use of Wastewater, Excreta and Grey Water. Policy and Regulatory Aspects*, Vol. 1. World Health Organization, Geneva.

39

Contamination Warning System

Saeid Eslamian
Isfahan University of Technology

Seyedeh Matin Amininezhad
Islamic Azad University of Shahreza

Sayed Mohamad Amininejad
University of Stuttgart

PREFACE

Water distribution networks are a main area of vulnerability for contamination events (chemical, biological, and radioactive). Water utilities have traditionally concentrated on contamination from natural sources but in recent years, utilities have also considered issues associated with intentional contamination.

Therefore, it is important to consider the contamination events in source waters or treated water in time to allow an effective response that will considerably decrease or avoid harmful impacts on consumers or the environment. To achieve this purpose, water security research efforts have focused on the advancement of methods for mitigating contamination threats to drinking water systems. One of the favorable efforts for the mitigation of both accidental and intentional contamination is a contamination warning system (CWS), a system to deploy and operate online sensors, other surveillance systems, rapid communication technologies, and data analysis methods to provide an early indication of contamination.

This chapter reviews some technologies that are used for online monitoring of water and detecting contaminants.

39.1 Introduction

With the rapid development of the electronics industry, the philosophy of instrumentation and control systems has changed. Recently, the philosophy is based on system structure. In addition, the structures of systems are greatly dependent on the technologies that have been developed in recent years by the computer and telecommunications industries [13].

Computer-based control systems are used by many industries to monitor and control sensitive processes and physical functions. Control systems collect sensor measurements and operational data from the field, process and display this information, and relay control commands to local or remote equipment [17]. These systems are more properly considered management control systems, or decision control systems. The good news about the control systems is that the distinctions are becoming blurred between computer, smart components, controllers, and other components. It is no longer necessary to begin with the design of system hardware, but rather by outlining the functionalities desired. For any need, there are now multiple means of meeting it, and accomplished system integrators can design the hardware and software to meet requirements [33].

39.2 Instrument and Control System Terminology

For the use of instrument and control systems, it is vital to know the related terminology.

39.2.1 Telemetry

The definition of telemetry is primarily to measure the information at some remote location and then transmit that information to a central or host location. There, it can be monitored and used to control a process at the remote site.

A telemetry system is basically classified into three categories based on transmission medium:

1. Wire-link or wire telemetry system
2. Radio or wireless telemetry system, with two special types:
    a. Short-range radio telemetry system
    b. Satellite radio telemetry system
3. Optical fiber or fiber-optic telemetry system

The recent progress in electronics and telecommunications has made remote telemetry systems very reliable and cost-effective in water quality monitoring. The following advantages can be maintained by telemetry in a water quality monitoring project:

1. Environmental data can be continuously monitored at near real time.
2. Early detection and warning system (e.g., alerts) can be developed where and when a certain condition is favorable to occur.
3. A reduction of maintenance and project costs can be reached [7,28].

39.2.2 Data Acquisition Systems

Data acquisition systems measure, store, display, and analyze information collected from different devices. Most measurements should have a transducer or a sensor, a device that converts a measurable physical quantity into an electrical signal. Temperature, strain, acceleration, pressure, vibration, and sound are instances of this information. Data acquisition systems have evolved tremendously over time from electromechanical recorders containing typically from one to four channels to all electronic systems that can measure hundreds of variables simultaneously [15].

39.2.3 Supervisory Control and Data Acquisition

Supervisory Control and Data Acquisition (SCADA) is a technology that comprises supervision, control, and data acquisition. It enables the remote monitoring and control of the whole system, or parts of it, and processes information to generate alarms, reports, graphs, or other outputs essential to operation and maintenance. In water supply systems, SCADA can monitor and control various equipment and processes, from the water source to the customer's tap, including transmission pipes, treatment plants, tanks, and distribution networks. In the urban water cycle, SCADA systems can also be used to monitor and control wastewater and stormwater systems [6,27].

39.3 Alarm System

An alarm system is a type of electronic monitoring system that is used to detect and respond to specific types of events. In water/wastewater systems, alarms are also used to alert operators when process operating or monitoring conditions go out of pre-set parameters. These types of alarms are primarily integrated with process monitoring and reporting systems (e.g., SCADA systems) [24].

39.4 Access Control

A means of providing access control should be incorporated into all security systems. The practice of securing entry points into water and wastewater facilities is the foundation of prevention measures. Utilities need to limit access to their facilities only to those with an authorized business need. Access control systems include issuance of credentials to individuals necessary to enter specified secured facilities and computer networks with sensitive information.

Generally, access control systems are comprised of three parts:

1. The physical barrier. To physically restrict access to a building or location via such methods as
    a. Doors; secured by either a magnetic or strike lock or can be revolving or sliding
    b. Turnstiles and speed gates; designed to limit access to one person for one card presented
2. The identification device
3. There are a number of different technologies used to identify users of an access control system, such as
    a. A proximity card and reader using RFID—cards can either work at a short-read range or a long-read range
    b. A smart card and reader
    c. A swipe card and reader
    d. PIN pads
    e. Biometric (fingerprint, iris scanning, face recognition)
4. The door controller and software. The door controller and software are at the heart of the system and are used to decide who can gain access through which access point at

what time of the day. These can vary depending on the size of the system and how many readers or sites you are trying to control from one point. Some of the options include

    a. A standalone door controller linked to a single door with no software

    b. A number of door controllers all linked together to a single PC to control one site

  5. A number of sites all interlinked together over a wide network area [35].

39.5 Contamination Warning System

Recently, water security research efforts have focused on the advancement of methods for mitigating contamination threats to drinking water systems. A promising approach for the mitigation of both accidental and intentional contamination is a CWS, a system to deploy and operate online sensors, other surveillance systems, rapid communication technologies, and data analysis methods to provide an early indication of contamination. A CWS is a proactive approach that uses advanced monitoring technologies and enhanced surveillance activities to collect, integrate, analyze, and communicate information that provides a timely warning of potential contamination incidents [32].

39.6 Components of a Contamination Warning System

A contamination warning system consists of the following monitoring and surveillance components [19,29,30,31].

39.6.1 Online Water Quality Monitoring

The online monitoring component of a CWS is composed of multiple sensor stations that collect data continuously and transmit it to a central database in a control room, most commonly a SCADA database.

Online water quality monitoring systems consist of a platform of various online water quality parameter monitors (e.g., disinfectant residual and total organic carbon [TOC]) located at predetermined sites within the distribution system. Accompanying software analyzes data for abnormalities of the water quality to detect contamination events. Many contaminants affect various water quality parameters allowing the software to recognize abnormal water qualities.

39.6.2 Consumer Complaint Surveillance

Consumer complaints regarding unusual taste, odor, or appearance of the water are often reported to water utilities, which track the reports as well as steps taken by the utility to address these water quality problems. The WS Initiative is developing a process to automate the compilation and tracking of information provided by consumers. Unusual trends that might be indicative of a contamination incident can be rapidly identified using this approach.

39.6.3 Public Health Surveillance

Public health surveillance conducted by the public health sector, including information such as sales of over-the-counter medication, reports from emergency medical service logs, calls from 911 centers, and calls into poison control hotlines, could serve as a warning of a potential drinking water contamination incident. Information from these sources can be integrated into a CWS by developing a reliable and automated link between the public health sector and drinking water utilities.

39.6.4 Enhanced Security Monitoring

Security breaches, witness accounts, and notifications by perpetrators, news media, or law enforcement can be monitored and documented through enhanced security practices. This component has the potential to detect a tampering event in progress, potentially preventing the introduction of a harmful contaminant into the drinking water system. Operational security countermeasures such as adding a distribution system component to the facility's mission essential vulnerable area list, and physical security countermeasures such as an intrusion detection system (IDS) are examples of enhanced security.

39.6.5 Routine Sampling and Analysis

Water samples can be collected at a predetermined frequency and analyzed to establish a baseline of contaminants of concern. These samples will provide a baseline for comparison during the response to detection of a contamination incident. In addition, this component requires continual testing of the laboratory staff and procedures so that everyone is ready to respond to an actual incident.

39.7 Sensor for Online Monitoring of Chemical, Radiological, and Biological Contamination

The various sensors for online monitoring of chemical, radiological, and biological contamination are explained.

39.7.1 Chemical Sensor: Online Chlorine Measurement Systems

Residual chlorine is one of the most widely measured chemical parameters in the water distribution system monitoring. Chlorine monitoring verifies proper residual chlorine levels at all points in the system, enhances rate rechlorination when required, and quickly and accurately signals any unexpected increase in disinfectant demand. A significant decline of residual chlorine could be an indication of potential threats to the system. Online systems can be supplied with control, signal, and alarm systems that notify the operator of low chlorine concentrations, and some may be tied into feedback loops that automatically adjust chlorine concentrations in the system. Two factors related to residual chlorine monitoring for security purposes are

- Residual chlorine can be measured using continuous online monitors at fixed points in the system or by taking grab samples at any point in the system and using chlorine test kits or portable sensors to determine chlorine concentrations.
- Correct placement of residual chlorine monitoring points within a system is crucial to early detection of potential threats. Although dead ends and low-pressure zones are common trouble spots that can show low residual chlorine concentrations, these zones are generally not of great concern for water security purposes because system hydraulics will limit the circulation of any contaminants present in these areas of the system [23,25,].

39.7.2 Chemical Sensor: Total ORGANIC Carbon Analyzer

TOC analysis is a frequently used technique that measures the carbon content of dissolved and particulate organic matter present in water. Many drinking water utilities monitor TOC to evaluate raw water quality or to evaluate the effectiveness of treatment processes designed to remove organic carbon. Many analytical devices can be used to provide a gross measurement of the organic chemical content of water. These include TOC analyzers and ultraviolet-visible (UV-Vis) spectrometers [1,9].

39.7.2.1 TOC Analyzer

Online TOC analyzers can be placed at critical sites within a drinking water distribution system, at the intake of a drinking water treatment plant, or in a wastewater plant influent wet well, to detect potential organic chemical compounds.

39.7.2.2 UV-Vis Absorbance

A UV-Vis spectrometer will react with organic contaminants that absorb in the UV range. The alarm sensitivity for many organic contaminants is between 1 and 500 μg/L. Some instances of organic compounds detected include phenol, toluene, xylene, and many pesticides. The response time can be less than 1 min. In some types of compounds not detected, include short-chained aliphatics. UV–Vis spectrometry has some advantages over TOC measurement such as shorter response time, greater sensitivity, less maintenance, and lower instrument cost.

39.7.3 Chemical Sensor: Oil and Petroleum Detection

Many techniques can be used for online monitoring of oil in water. Some of these techniques are mentioned next.

39.7.3.1 Light Scattering

Light scattering was probably the most common technique for online measurement of oil in water.

The technique involves passing visible light through an oily water sample. Due to the presence of particles (oil droplets, solid particles, and gas bubbles), some light will be scattered, and a reduced amount will be transmitted. By measuring the amount of transmitted light together with the amount of light that is scattered at different angles, it is possible to distinguish the oil droplets from solid particles and gas bubbles, and determine the oil concentration [21].

39.7.3.2 UV Fluorescence

When aromatic hydrocarbons absorb UV light, they emit fluorescent light at a longer wavelength. By measuring the intensity of the UV fluorescent light, one can determine the amount of aromatic hydrocarbons, which is related to the total amount of hydrocarbons provided that the ratio of aromatics to total hydrocarbons remains relatively constant. UV fluorescent devices have been largely used by the oil and gas industry for the measurement of oil in produced water [18].

39.7.3.3 Focused Ultrasonic Acoustics

In this technique, a highly focused acoustic transducer is inserted directly into a produced water stream. The transducer focuses and a time window determines the measurement volume. Any particles, such as oil droplets, solid particles, and gas bubbles, that pass through the measurement volume will produce acoustic echoes. These signals are detected, classified, and used to work out particle size and size distribution. Oil concentration is then calculated from the size and size distribution obtained [3].

39.7.3.4 Photoacoustic Sensor

The principle of the photoacoustic sensor is simple. A pulsed laser light is focused onto a small sample of oily water. When oil (dissolved and dispersed) absorbs the optical energy, it causes sudden local heating and then produces thermal expansion, which generates high frequency pressure waves. These waves are detected and correlated with oil concentration [34].

39.7.4 Monitoring for Radiation to Detect Radionuclides

Drinking water may contain radioactive substances ("radionuclides") that can cause risks to human health. In terms of health risk assessment, the guidelines do not differentiate between radionuclides that occur naturally and those that arise from human activities. However, in terms of risk management, a differentiation is made because, in principle, human-made radionuclides are often controllable at the point at which they enter the water supply. Naturally occurring radionuclides, in contrast, can potentially enter the water supply at any point, or at several points, prior to consumption. For this reason, naturally occurring radionuclides in drinking water are often less amenable to control.

The common types of radiation are alpha, beta, and gamma. Alpha particles range in energy from 4 to 8 MeV, and have a relatively high atomic mass of 4. For these reasons, alpha particles travel in straight lines through matter and deposit their energy over short distances. They are unable to penetrate skin.

Beta particles are electrons emitted from an atom. In air, beta particles can travel a few hundred times farther than alpha particles (up to 6 ft [2 m] or more for the beta particles with higher energies).

Gamma particles are waves of energy that travel at the speed of light. These waves can have considerable range in air and have greater penetrating power (can travel farther) than either alpha or beta particles. Gamma waves can penetrate through a variety of objects including human skin and clothing. Measuring alpha and beta emissions in water is difficult because these short-range radiations are easily blocked by water before they reach the detector [24].

39.7.5 Monitoring for Biological Contaminant (Microbial Contaminant)

Detection of pathogenic bacteria in drinking water is an important issue for water utilities because they have critical impact on public health. The physical characteristics of microorganisms as a means to detect them are of high interest to many researchers [2,14,16,26]. This area of research is still trending, and the methodologies include turbidity measurement, vibrational spectroscopy, and multi-angle light scattering (MALS) technologies.

39.7.5.1 Turbidity Measurement

Turbidity is caused by suspended particles or impurities that interfere with the clarity of the water. Increased turbidity levels may indicate anomaly of water quality, and turbidity measuring is sometimes used to monitor microbial contamination; for instance, the Japanese Potable Water Quality Standard is a method that specifies turbidity level of 0.1 mg/L (kaolin turbidity standard) in finished drinking water as the standard turbidity value to prevent *Cryptosporidium* contamination. Turbidmeter technology is frequently used by many water utilities for monitoring water quality [4,8].

39.7.5.2 Vibrational Spectroscopy

This technology has to do with interpreting the spectra that are emitted from transitions between vibrational levels of a molecule following excitation by laser light. Molecules such as nucleic acids, cytoplasmic proteins, membrane lipids, or cell wall components are the building blocks of microorganisms, and their exact composition and distribution is unique for each organism. Vibrational spectroscopy is a noninvasive and reagent-free method. It has been successfully applied to identify, differentiate, and classify pathogenic microorganisms based on their unique spectroscopic signatures [5,10,11,20].

39.7.5.3 Multi-Angle Light Scattering

MALS is another form of turbidity measurements in which instead of using one light source, several light sources and angles of refraction are used. With proprietary algorithms, the shape, size, refraction index, and internal structure of a particle can be deduced from the light scattering patterns. With this technique, microorganisms can be accurately identified. MALS identification of microorganisms is less reliable than identification by vibrational spectroscopy. The sensor contains a laser beam that strikes individual cells or particles in water, resulting in unique light scattering patterns. Data obtained are compared to a computerized database of patterns from different kinds of microbes, which are then placed into four identifiable categories: rods, spores, protozoa, and unknown (applied for particles of an appropriate size for the prior categories but not classified specifically as being a member of the three categories) [12,20,21].

39.8 Summary and Conclusions

To ensure the safe supply of drinking water, the quality needs to be monitored online. The consequence of insufficient monitoring can result in significant health risks and economic damages. Therefore, these online sensors have to be placed at points throughout the distribution system.

A CWS is a proactive approach that uses advanced monitoring technologies and improved surveillance activities to collect, integrate, analyze, and communicate information that provides a timely warning of potential contamination incidents. These technologies can have clear and multiple benefits for water utilities.

In summary, CWA can allow water utilities to plan for and rapidly react to an accidental or intentional contamination of water distribution systems.

References

1. Allgeier, S.C., Hall, J.S., Rahman, M., and Coates, W. 2010. Selection of water quality sensors for a drinking water contamination warning system. *AWWA Water Quality Technology Conference*. AWWA, Savannah, GA.
2. Allmann, T.P. and Carlson, K.H. 2005. Modeling intentional distribution system contamination and detection. *Journal of American Water Work Association*, 97, 58–71.
3. Anaensen, G. and Volker, A. 2006. Produced water characterization using ultrasonic oil-in-water monitoring recent development and trial results. In: *A Paper Presented at NEL's 8th Oil-in-Water Monitoring Workshop*, September 21, 2006, Aberdeen, UK.
4. Chen, D., Huang, S., and Li, Y. 2006. Real-time detection of kinetic germination and heterogeneity of single Bacillus spores by laser tweezers Raman spectroscopy. *Analytical Chemistry*, 78, 6936–6941.
5. Connelly, J.T. and Baeumner, A.J. 2012. Biosensors for the detection of waterborne pathogens. *Analytical and Bioanalytical Chemistry*, 402, 117–127.
6. Daneels, A. and Salter, W. 1999. What is SCADA? *International Conference on Accelerator and Large Experimental Physics Control Systems*, Trieste, Italy.

7. Frank, C., Jedlicka, R.P., and Henry, R. 2002. *Telemetry Systems Engineering*. Artech House, Boston.

8. Greenwood, R., Mills, G.A., and Boig, B. 2007. Introduction to emerging tools and their use in water monitoring. *Trends in Analytical Chemistry*, 26, 263–267.

9. Hall, J., Zaffiro, A.D., Marx, R.B., Kefauver, P.C., Krishnan, E.R., Haught, R.C., and Herrmann, J.G. 2007. On-line water quality parameters as indicators of distribution system contamination. *American Water Work Association*, 99(1), 66.

10. Hasan, J., Goldbloom-Helzner, D., Ichida, A., Gibson, M., and States, S. 2005. Technologies and techniques for early warning systems to monitor and evaluate drinking water quality: State-of-the-art review. USEPA report.

11. Janzon, A., Sjoling, A., Lothigius, A., Ahmed, D., Qadri, F., and Svennerholm, A.M. 2009. Failure to detect *Helicobacter pylori* DNA in drinking and environmental water in Dhaka, Bangladesh, using highly sensitive real-time PCR assays. *Applied and Environmental Microbiology*, 75, 3039–3044.

12. King, K.L. and Kroll, D. 2005. Testing and verification of real-time water quality monitoring sensors in a distribution system against introduced contamination. *Proceedings of the AWWA, Water Quality Technology Conference*, Quebec City, Canada.

13. Krishna, K. 1997. *Handbook of Computer-Based Industrial Control*, Asoke K. Ghosh, PHI Learning Private Limited, Connaught Circus, New Delhi.

14. Li, F., Zhao, Q., Wang, C.A., Lu, X.F., and Li, X.F. 2010. Detection of *Escherichia coli 0157:H7* using gold nano particle labeling and inductively coupled plasma mass spectrometry. *Analytical Chemistry*, 82, 3399–3403.

15. Measurement Computing Corporation. 2004. *Data Acquisition Handbook*, United States of America.

16. Mikol, Y.B., Richardson, W.R., Vander Schalie, W.H., Shedd, T.R., and Widder, M.W. 2007. An online real-time biomonitor for contaminant surveillance in water supplies. *Journal of American Water Work Association*, 99, 107–115.

17. Radvanovsky, R. and Brodsky, J. (eds.). 2013. *Handbook of SCADA/Control System Security*. CRC Press, Boca Raton, FL.

18. Reeves, G. 2006. Alternative oil in water measurements using fluorescence and electrochemical nanotechnology. In: *A Paper Presented at TUV NEL's 8th Oil in Water Monitoring Workshop*, Aberdeen, UK, September 21, 2006.

19. Roberson, J.A., Morley, P.E., and Morley, K.M. 2005. Contamination warning system for water: An approach for providing actionable information for decision makers. American Water Works Association.

20. Samendra, P.S., Masaaki, K., Charles, P.G., and Ian, L.P. 2013. Evaluation of real-time water quality sensors for the detection of intentional bacterial spore contamination of potable water. *Biosensors & Bioelectronics*, 4(4), 1–5.

21. Samendra, P.S., Masaaki, K., Charles, P.G., and Ian, L.P. 2014. Rapid detection technologies for monitoring microorganisms in water. *Biosensors Journal*, 3(1), 1–8.

22. Schmidt, A.A. 2003. Oil-in-water monitoring using advanced light scattering technology—Theory and applications. In: *A Paper Presented at TUV NEL's 5th oil in Water Monitoring Workshop*, May 21–22, 2003, Aberdeen, UK.

23. Spellman, F.R. 2009. *Water and Wastewater Treatment Plant Operations*. CRC Press, Boca Raton, FL.

24. Spellman, F.R. 2009. *Handbook of Water and Wastewater Treatment Plant Operations*. CRC Press, Boca Raton, FL.

25. State, S. 2010. Security and emergency planning for water and wastewater utilities. American Water Work Association.

26. Straub, T.M. and Chandler, D.P. 2003. Towards a unified system for detecting waterborne pathogens. *Journal of Microbiology Methods*, 53, 185–197.

27. Temidoa, J., Sousab, J., and Malheiro, R. 2014. SCADA and smart metering systems in water companies. A perspective based on the value creation analysis. *Procedia Engineering*, 70, 1629–1638.

28. United States Environmental Protection Agency (EPA). 2002. Delivering timely water quality information to your community. The Chesapeake Bay and National Aquarium in Baltimore EMPACT Projects. Office of Research and Development. National Risk Management Research Laboratory Cincinnati. EPA625/R-02/018.

29. United States Environmental Protection Agency (EPA). 2005. *Water Sentinel System Architecture*. United States Environmental Protection Agency (EPA), National Homeland Security Research Center, Cincinnati, Ohio.

30. United States Environmental Agency (EPA). 2010. *Detection of Biological Suspension Using Online Detectors in a Drinking Water Distribution System Simulator*. United States Environmental Agency (EPA), National Homeland Security Research Center, Cincinnati, Ohio.

31. United States Environmental Protection Agency (EPA). 2007. *Water Security Initiative: Interim Guidance on Planning for Contamination Warning System*. United States Environmental Protection Agency (EPA), National Homeland Security Research Center, Cincinnati, Ohio.

32. United States Environmental Protection Agency (EPA). 2010. *Water Quality Event Detection System for Drinking Water Contamination Warning System*. United States Environmental Protection Agency (EPA), National Homeland Security Research Center, Cincinnati, Ohio.

33. Water Environment Federation. 2009. *Wastewater Solid Incineration Systems*. Water Environment Federation, Virginia.

34. Whitaker, T., Terzoudi, V., and Butler, E. 2001. Application of photoacoustic technology for monitoring the oil content in subsea separated produced water. In: *A Paper Presented at TUV NEL's 3rd Oil-in-Water Monitoring Workshop*, May 23, 2001, Aberdeen, UK.

35. A Guide to Access Control for the Utilities Sector, http://www.bsia.co.uk/Portals/4/Publications/119-ac-guide-utilities.pdf.

Water Reuse Risk and Impact Analysis

40

Urban Water Reuse Risk

Zareena Begum Irfan
Madras School of Economics

PREFACE

Water scarcity and water pollution pose a critical challenge in many developing countries. In urban areas, it is becoming difficult for the authorities to manage water supply and wastewater. Strategies for water and wastewater reuse can improve urban water management. This chapter provides an overview of global scenario of urban water reuse. The guidelines have been listed with respect to the application and treatment requirements of water reuse. The important aspects to minimize public health risks are identified. The possibilities of wastewater reuse in agriculture, industry, urban uses, and environmental water enhancement, including groundwater recharge, are discussed with the help of practical examples.

The economics of reuse system scale in urban water system has been analyzed with a response recovery framework development. The strategy to minimize the risk involved at different levels of urban water reuse management has been examined. The capacity-building policy-making, institutional strengthening, financial mechanisms, and awareness raising and stakeholder participation are vital to implement these strategies for wastewater reuse.

40.1 Introduction

During the last century, rapid urbanization and population growth have resulted in many environmental problems. Among those, the most serious are water shortage and pollution. People around the world are beginning to realize the interactions between human beings and the environment. Human activities are affecting the natural water ecological cycle in many ways, such as the reduction of forested areas, the shrinkage of grassland for grazing, and the spread of urban growth resulting in increased rainwater lost to runoff. Overexploitation of groundwater resources have decreased groundwater levels and caused problems of seawater intrusion; toxic industrial discharge and the extensive use of chemical fertilizers have polluted much of the water supply. Many regions in the world are facing the great challenge of water shortage and pollution, and the situation is getting worse. The United Nations Environment Program (UNEP) identified that water shortage and global warming are the two most worrying problems for the new millennium and the World Water Council believes that by 2020 the world will need 17% more water than is currently available [42].

As populations are generally concentrated in urban areas, urban demands on water resources have increased rapidly during the last few decades, and urbanization continues to worsen the situation. According to a United Nations report [31], over the last 50 years, the urban population has been tripled and now accounts for nearly half of the total population. The world water demand has also tripled since 1950 and is continuing to rise as 80 million more people are added each year. As urban populations place greater pressure on water supplies, expansion of water supplies is usually the only means employed to meet the growing water demands; at the same time, the economical and ecological limits of water supplies have generally been ignored. The fact is that in many regions of the world, water consumption is nearing or surpassing the limits of natural systems [21]. For example, Liaoning, the northeast province of China, is a traditional industrial base of China and has a large urban population (25.7 million of total population of 42.38 million) with heavy water consumption. In recent years, excessive extraction of water has resulted in the sinking of the ground in a 1500-km² area and intrusion of seawater in more than five cities [43]. The government has banned construction of new wells and other facilities for extracting groundwater in certain cities because it is realized that water resource management has to comply with the standards of sustainable development.

Like Liaoning province, many communities in the world are seeking a sustainable water management strategy. Sustainable development was defined by the World Commission on Environment and Development (WCED) as "development that meets the needs of the present without compromising the ability of future generations to meet their own needs" [38]. For several decades, urban water reuse has been recognized as a component of sustainable urban water management strategy and has been practiced by many communities. Water reclamation and reuse

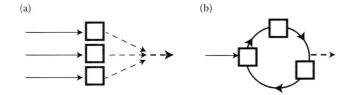

FIGURE 40.1 Once-through and looped systems.

are regarded as a "win-win" strategy for both reducing pollution and enhancing water supply resources [19]. In a traditional urban water system, after water use, wastewater is treated to certain legalized quality levels when discharged into receiving water bodies. Such a water use system is generally regarded as a once-through system [12], as illustrated in Figure 40.1a. In such a system, water is only used once, so the efficiency of water use is low. Figure 40.1b represents a looped system created when treated wastewater is reused for some applications that do not require high-quality drinking water, such as irrigation and sanitation. Wastewater reuse practices will help in satisfying more water demands while effluent discharge can be reduced. Although a looped system is relatively complex, it provides much higher water use efficiency.

40.1.1 Urban Water Reuse: World Overview

Urban water reuse has a long history dating back to ancient times. However, according to Asano and Levine [3], the development of programs for planned reuse of wastewater began in the 1910s and some of the earliest water reuse systems were developed during the 1920s. The first industrial water reuse was implemented in the 1940s and in the 1960s when Colorado and Florida developed urban water reuse systems. During the last 30 years, research works were extensively focusing on technical barriers and health risks associated with water reclamation [3,13].

During the last decade, wastewater reuse has gained much attention in many parts of the world as one means to alleviate the growing pressures for increasing water supplies for various applications. Technically speaking, modern wastewater treatment facilities are able to treat wastewater to the quality levels eligible for any purposes [3]. It has been recognized throughout the world that urban water reuse is an important factor in pursuing the optimal planning and efficient use of water resources.

Because of water scarcity, several countries are reusing wastewater. Countries worldwide use different types of reuse techniques. Concerning reuse options, agriculture is by far the most important in terms of volume, simply because it is the activity that demands the most water around the world. This reuse is expected to increase because the potential to reuse wastewater is still high (even agricultural reuse only represents <1% in volume of the total demand of water by the sector).

The tendency to reuse wastewater is most common among those activities demanding the most water. The following examples illustrate this:

1. Pakistan and Tunisia, with a water use for agricultural purposes of 96% and 86%, respectively, reuse a very large amount of their wastewater to irrigate (although in Pakistan it is nontreated wastewater while in Tunisia it is treated).
2. Namibia and Singapore, which use 29% and 45% of total water extracted for municipal purposes, respectively, are the two countries with the most important water for human consumption reclamation projects.
3. The United States, Singapore, and Germany, where 45%, 51%, and 69% of water is used for industrial purposes, respectively, have a large number of recycling and reuse projects across industries.

Water reuse is practiced in developed and developing countries in very different ways. In the former, it is mostly a planned activity while in the latter it frequently occurs unplanned. Nevertheless, both are wastewater reuse practices and should be acknowledged as such [14], and in all areas, the reuse of wastewater is considered economically interesting. In developed countries, reuse is considered viable for reasons including the application of stringent standards and the use of expensive technology and economic incentives. In contrast, in developing countries, the use of untreated wastewater to produce goods and beneficially recycle nutrients also makes reuse attractive. The reason why these different approaches converge is that wastewater is considered a resource. Nevertheless, this is not always officially recognized in national legislations or policies, making the implementation of reuse projects difficult in practice. Whether water reuse will reach its full potential depends on a number of variable factors in each country, such as economic consideration, potential uses for reclaimed water, the stringency of wastewater discharge requirements, and public policies for conservation and protection. Local strategies need to be flexible and promote wastewater reuse in the best, safest possible conditions.

40.1.2 Water Reuse Applications and Treatment Requirements

In an urban water system, reclaimed water can be used to replace potable standard water currently being used for many purposes. Figure 40.2 shows an example of such a looped system with water reuse activities, such as irrigation and groundwater recharge, etc. Entering the system from the surface or from groundwater sources to municipal water treatment plants, freshwater follows several steps, such as water distribution, water use, wastewater discharge, wastewater treatment, and then its return for certain forms of reuse. The water quality levels required by the specific water reuse applications and the water available for reuse determine the feasibility of the specific water reuse practice. Reuse water quality generally depends on the corresponding treatment processes.

40.2 Water Reuse Criteria: Environmental and Health Risk Based Standards and Guidelines

According to US EPA's *Guidelines for Water Reuse* [34], water reuse applications can be classified into certain categories. In the order of water quality requirements from the highest to the lowest, these categories are briefly introduced as follows:

1. Potable reuse. It is the water reuse application with the highest quality requirements. Due to the health risk concerns to the general public, potable reuse is not widely practiced.
2. Unrestricted urban and recreational uses, and agricultural irrigation of food crops. Very high levels of treatment are required for applications in this category. This represents the highest level of water reuse that is currently practiced in the world. Typical treatment processes

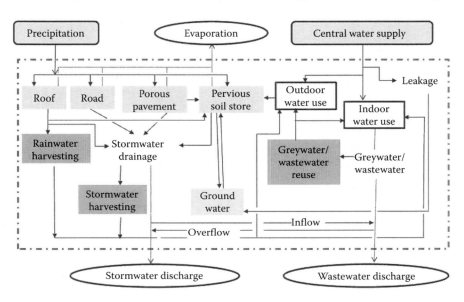

FIGURE 40.2 The concept of engineered urban water cycle with reuse.

include secondary treatment, filtration, and disinfection, with strict quality requirements on some parameters, such as effluent biochemical oxygen demand (BOD), turbidity, total and/or fecal coliforms, disinfectant residuals, pH levels, etc.

3. Restricted-access urban use, restricted recreational use, and agricultural irrigation of nonfood crops. This category defines the water reuse situation in which only limited populations have access to reclaimed water. In this category, reuse water for irrigation is beneficial because the nutrients in wastewater are good chemical fertilizers. As a result, using reclaimed water to irrigate golf courses and other landscapes is widely practiced.

4. Industrial reuse. Statistically, in primary resource and manufacturing industries, less than 20% of their water intake is consumed and, therefore, there are many opportunities for reuse between industries or other urban water use sectors. Treated municipal wastewater can be reused for industrial water supply. Furthermore, most industrial water reuse focuses on recycling and process modifications within the boundary of one plant [7]. In urban water reuse modeling studies, little attention has been given to study the opportunities of water reuse between industries. Typical reclamation treatment for municipal wastewater in this category includes secondary treatment and disinfection.

The most important consideration in developing a water reuse system is that the quality of the reclaimed water be appropriate for its intended use. Main reuse water quality parameters include pathogen and chemical constituents. Pathogen is the most common concern associated with nonpotable reuse of reclaimed municipal water due to the risk of transmission of infectious disease. Chemical constituents in municipal wastewater may affect the acceptability of such water for food crop irrigation, indirect potable reuse, and some industrial applications. Some typical water constituents and quality parameters include the following: suspended solid measured as total suspended solids (TSS); organics measured as BOD, chemical oxygen demand (COD), and total organic carbon (TOC); and dissolved inorganics measured as total dissolved solids (TDS) and specific elements (e.g., Ca and Mg). In water reuse applications, these constituents have certain effects on the specific applications. For example, suspended solids may absorb organic contaminants and heavy metals, and plug-up equipment and cause fouling, while organics may promote growth of slime-forming organisms, cause aesthetic and nuisance problems, and reduce heat transfer efficiency and water flow; dissolved inorganics may promote corrosion by increasing the electrical conductivity of the water and cause scaling [34]. Thus, these parameters are commonly used in water reuse studies and applications. Some studies, such as decision support system [23], contribute to the management and optimal selection of wastewater treatment trains that can produce the highest possible quality water for reuse at the lowest possible costs.

An important part of the process of achieving public acceptability of water reuse is demonstrating that the water is safe.

Even if advanced treatment is used, there is a need for criteria that can act as a benchmark against which to measure the safety of the final product within the context of the desired use. Such criteria also provide a technical benchmark for operators and for the development and deployment of treatment processes. These criteria may not be numbers relating to individual parameters but could be process standards. However, it is important that any criteria or standards are both scientifically sound and justifiable, that their derivation is transparent, all the options have been evaluated with the involvement of the stakeholders, and the selected solution has the support of the stakeholders. Guidelines and standards have been developed in many parts of the world; however, there is a need for a much wider agreement as to the criteria for both developing national or local standards and the approaches by which these can be achieved. UKWIR, Awwa Research Foundation (AwwaRF), and WateReuse Foundation commissioned this project to develop a consensus on criteria for water quality in reuse schemes and to further the potential for reuse of wastewater and provide a basis for discussion with regulators and potential users.

40.3 Water Reuse and Risk Management in the Urban Water Cycle

An urban water system includes natural water cycle, artificial water cycle, economic water cycle, and social water cycle, as shown in Figure 40.3. Artificial water cycle is comprised of water supply, water use, water drainage, and treatment. Water cycle operates by the consumption of part of water as well as treatment of the sewage generated. Part of urban water evaporates and becomes atmospheric water, while part of it becomes sewage. Part of the sewage will become new reuse water resources after environmental and ecological treatment, while part of the treated sewage will percolate through slowly and become groundwater resources or become urban water resources in the process of evaporation and precipitation with the rest draining into water bodies, which is how urban artificial water cycle repeats. Water environment recovery in artificial water cycle focuses on sewage treatment and water reuse. The urban water cycle refers to artificial water cycle in this study; that is, water circulates among the three processes of access, use and drainage, and relative water bodies. This is social reinforcement on water cycle in nature.

40.4 Reducing Risk from Wastewater Use in Urban Farming

Agricultural irrigation is crucial for improving the quality and quantity of production. Worldwide, agriculture is the largest user of water. Agriculture receives 67% of total water withdrawal and accounts for 86% of consumption in 2000 [32]. In Africa and Asia, an estimated 85%–90% of all the freshwater use is for agriculture. By 2025, agriculture is expected to increase its water requirements by 1.2 times [27]. Large-scale irrigation projects have accelerated the disappearance of water bodies, such as the

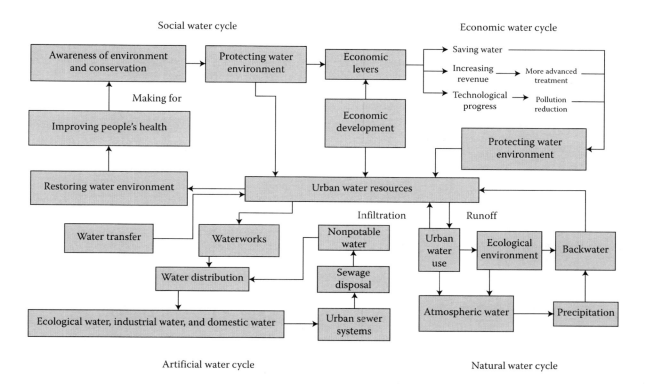

FIGURE 40.3 Water cycles in an urban system.

Aral Sea, the Iraqi Marshlands, and Lake Chad in West Africa. Thus, more efficient use of agricultural water through wastewater reuse is essential for sustainable water management.

40.4.1 Benefits

The ancient practice of applying wastewater containing human excreta to the land has maintained soil fertility in many countries of Eastern Asia and the Western Pacific for over 4000 years, and remains the only agricultural use option in areas without sewerage facilities [39]. Potential benefits of wastewater reuse for agriculture include the following:

- Conservation and more rational allocation of freshwater resources, particularly in areas under water stress.
- Avoidance of surface water pollution.
- Reduced requirements for artificial fertilizers and associated reduction in industrial discharge and energy expenditure.
- Soil conservation through humus build-up and prevention of land erosion.
- Contribution to better nutrition and food security for many households [39].

40.4.2 Potential Concerns

While wastewater reuse for agriculture has many benefits, it should be carried out using good management practices to reduce negative human health impacts. The guidelines include recommendations for crops to be cooked or used as feed, as well as for parks and localized irrigation. The guidelines are set to minimize exposure to workers, crop handlers, field workers, and consumers, and recommend treatment options to meet the guideline values [41]. The guidelines are focused on health-based targets and provide procedures to calculate the risks and related guideline values for wastewater reuse in agriculture.

In addition to ponds, the wastewater can be treated through reservoirs [16], wetlands, physicochemical process, soil-aquifer treatment (SAT), and other methods. However, the practice of stabilization ponds is more common in many developing countries. The guidelines state that "local epidemiological, sociocultural and environmental factors should be taken into account and the guidelines modified accordingly" [39]. The microbiological quality guidelines have been used as the basis for standard setting in several countries and regional administrations. In other situations, the quality guideline levels have been adopted with specifications of additional management practices and restrictions. Standard setting in other countries has been influenced by the WHO guidelines, but often with some modification of the microbiological guidelines before adoption as standards [40].

Wastewater intended for reuse should be treated adequately and monitored to ensure that it is suitable for the projected applications. If wastewater streams come from industrial sources and urban runoff, toxic chemicals, salts, or heavy metals in the wastewater may restrict agricultural reuse. Such materials may change soil properties, interfere with crop growth, and cause bioaccumulation of toxic materials in food crops. While separating household wastewater and runoff from industrial effluent is preferable, this may not be feasible. Thus, proper treatment and monitoring should be practiced. Wastewater reuse for agriculture needs to be planned with attention to target crops and existing water delivery methods. Nutrients in reclaimed water that are important to agriculture include nitrogen, potassium, zinc, boron, and sulfur [2]. However, excess nitrogen may cause overgrowth, delayed maturity, and poor quality of crops. While boron is an essential element for plant growth, excess boron becomes toxic [9]. Furthermore, proper care should be taken to control the saline problems caused by wastewater reuse [36].

40.4.2.1 Risk Assessment and Risk Reduction Measures

The key issue for assessing the potential risks of microorganisms and chemicals is whether there will be exposure of receptors such as plants and humans, and whether that exposure will be sufficient to be of concern in terms of both concentration, or numbers, and period of exposure. The questions that need to be addressed are as follows:

- Will the chemical or pathogen reach a receptor such as an animal, including humans, or plant, either directly or indirectly?
- Will a chemical or pathogen deposit on the leaves of plants? This may be of no direct concern for the plants, but if the plants are eaten by humans or livestock, it could be of concern.
- Will the pathogen be taken up by humans through contact during recreation or similar circumstances?
- Will the chemical accumulate in an animal or plant and, if so, will it accumulate in the parts of that animal or plant that will be used in food or food production?
- Is the chemical present in sufficient concentration to damage that animal or plant?
- Will the chemical leach or percolate into groundwater, or surface water, giving rise to subsequent problems? Will the chemical accumulate in the soil and damage crops at a later stage?
- Will the pathogen accumulate in shellfish or fish?
- Is there a potential for disease transmission or uptake of chemicals through aerosols associated with the use of wastewater?

With regard to drinking water, the position is relatively simple in terms of assessing human exposure, but with other uses of wastewater, for example, irrigation, aquaculture, stock watering, or industrial applications such as cooling water, assessing exposure may be more complex.

Exposure may also be specific to particular circumstances. This is of great importance in crop irrigation where the types of soil, the depth of groundwater, the types of crop grown, and the method of irrigation will all impact on the potential risks. Therefore, there will always be a requirement for consideration of local circumstances. Use on green spaces can lead to the potential presence of pathogens that could be taken up by hand-to-mouth transfer or through cuts. In other circumstances, the

use of reclaimed wastewater in fountains and in uses generating aerosols, such as toilet flushing, car washing, or cooling towers, have the potential to expose consumers through inhalation if pathogens are present in the reclaimed water.

40.4.2.2 Risk Reducing Measures

Improving safe and sustainable water reuse in areas of currently unplanned practice has been greatly influenced by the WHO guidelines [39,41]. In 2006, the WHO released a four-volume report titled *Guidelines for the Safe Use of Wastewater, Excreta and Greywater*. The first volume focuses on policy and regulatory aspects of wastewater, excreta, and greywater use; the second volume focuses on use of wastewater in agriculture; the third volume focuses on wastewater and greywater use in aquaculture; and the fourth volume focuses on excreta and greywater use in agriculture. The discussion in the WHO guidelines is limited to wastewater, excreta, and greywater from domestic sources that are applied in agriculture and aquaculture. Rather than relying on water quality thresholds as in past editions [39], the most current WHO guidelines [41] adopt a comprehensive risk assessment and management framework. This risk assessment framework identifies and distinguishes among vulnerable communities (agricultural workers, members of communities where wastewater-fed agriculture is practiced, and consumers) and considers trade-offs between potential risks and nutritional benefits in a wider development context. As such, the WHO approach recognizes that conventional wastewater treatment may not always be feasible, particularly in resource-constrained settings, and offers alternative measures that can reduce the disease burden of wastewater use. The specific approach utilized by the WHO [41] guidelines is to (1) define a tolerable maximum additional burden of disease, (2) derive tolerable risks of disease and infection, (3) determine the required pathogen reduction to ensure that the tolerable disease and infection risks are not exceeded, (4) determine how the required pathogen reductions can be achieved, and (5) put in place a system for verification monitoring.

The most effective health protection recommendation is the production of crops not eaten raw. However, this option requires appropriate monitoring capacity and viable crop alternatives for farmers. Other options include on-farm treatment and application techniques, as well as the support of natural die-off as described in two Africa case studies, Ghana-Agricultural and Senegal-Dakar, and natural attenuation in nonedible aquatic plants lining irrigation canals (Vietnam-Hanoi) [1]. There is reported success of blending of wastewater with higher-quality water to make it more suitable for production (Vietnam-Hanoi, Senegal-Dakar, India-Delhi, Jordan-Irrigation, and Israel/Palestinian Territories/Jordan-Olive Irrigation). In addition to the risks from pathogen contamination, wastewater may have chemical contaminants from industrial discharges or stormwater runoff. The WHO [41] guidelines provide maximum tolerable soil concentrations of various toxic chemicals based on human exposure through the food chain. For irrigation water quality, WHO refers to the FAO guidelines, which focus on plant growth requirements and limitations [4,20]. The guidelines do not specifically address how to reduce chemical contaminants from wastewater for use in irrigation. Resource-constrained countries may have historically been less prone to heavy metal contamination, which is usually localized and associated with industrial activities, but where industries are emerging, industrial source control measures are required to avoid potential contamination in food crops.

Likewise, where required, stormwater should be diverted and treated to remove pollutants. Alternative options for low-income countries to reduce the potential risk of chemical contamination, like through phytoextraction, crop selection, and soil treatment, are limited [28].

40.5 Economics of Reuse System Scale in Urban Water Systems

Economies of scale exist in both capital and operating costs of wastewater treatment. Diseconomies of scale are, however, apparent in all pipe networks [8]. This diseconomy is inherent and results because as the number of connections increase, the distance of pipe required per connection also increases, as does the need for larger pipes with greater volumes. The extent of this diseconomy varies with urban density and is affected by the layout of the piping network. Based on system replacement and operational costs for conventional wastewater infrastructure in Adelaide, Clark [8] reported little difference in system life-cycle cost (LCC) per household for wastewater systems between 500 and 1 million connections. The decrease in average LCC of wastewater treatment is balanced by an increasing average cost of sewering. Below 500 connections, treatment costs dominate and an economy of scale exists. A slight diseconomy of scale was evident above 10,000 connections. The actual optimum was dependent on both household density and the discount rate applied. In a separate Australian study estimating the economics of scale for greywater recycling systems, Booker [6] reported minimum cost per household within the range of 1200 to 12,000 connections depending on treatment scenarios. Little difference in LCC was seen to exist in a range from 120 to 120,000 connections with capital amortized over 20 years at 5%. The long-term economic viability of reuse projects also represents an important barrier to water reuse. Reclaimed water is often priced just below the consumer cost of drinking water to make it more attractive to potential users, but this may also affect the ability to recover costs [15]. Distortion in the market for drinking water supply complicates the pricing of reclaimed water, as does the lack of accounting for externalities, including water scarcity and social, financial, and environmental burdens of effluent disposal in the environment [25,37]. Although there is a movement toward increased or even full operations and maintenance cost recovery in the large market of agriculture water reuse (Morocco, Tunisia, Jordan), this is still the exception among many state-run service providers. However, there may be opportunities to set different tariff levels for different classes or types of users, thus subsidizing the resource for the poor while recovering costs from groups that are able to pay. Finally, financing of up-front costs remains an important barrier to introducing new reuse programs and often requires government intervention in the form of grants or subsidies combined with eventual revenues.

40.6 Drinking Water Potential Health Effects Caused by Infiltration of Pollutants from Solid Waste Landfills

In urban areas, the potential for introducing wastewater reuse is quite high, and reuse options may play a significant role in controlling water consumption and reducing its pollutant load on the environment. A large percentage of water used for urban activities does not need quality as high as that of drinking water. Dual distribution systems (one for drinking water and the other for reclaimed water) have been utilized widely in various countries, especially in highly concentrated cities of developed countries. This system makes treated wastewater usable for various urban activities as an alternative water source in the area, and contributes to the conservation of limited water resources. In most cases, secondarily treated domestic wastewater followed by sand filtration and disinfection is used for nonpotable purposes, such as toilet flushing in business or commercial premises, car washing, garden watering, parks or other open space planting, and firefighting.

40.6.1 Benefits

The benefits of wastewater reuse for urban applications include the following:

- High volume of wastewater generation, and a large number of potential applications and volume for water reuse, which may benefit from the economy of scale.
- Reduction in the wastewater volume to be treated by municipal wastewater treatment plants, which are over-extended and in need of expansion in many developing countries' mega-cities.
- Tokyo is one of the leading cities that are successfully implementing wastewater reuse, such as dual distribution systems and stream augmentation. In a water reuse project in the Shinjuku area of Tokyo, a dual distribution system has been adopted and sand-filtered water from the Ochiai Municipal Wastewater Treatment Plant is chlorinated and used as toilet-flushing water in 25 high-rise business premises and for stream augmentation. The system, which has been successfully operated since 1984, is supplying treated wastewater up to a maximum 8000 m^3/day [29].
- There is also a small-scale onsite system where the greywater is recycled as an in-building water resource, with a dual distribution system. Reclaimed water can be used for toilet flushing, car washing, stream augmentation, or landscape purposes.

40.6.2 Potential Concerns

One of the key concerns for wastewater reuse in urban applications is the protection of public health, as urban reuse has the potential to expose a large number of people to disease-causing microorganisms. Care should be taken to avoid contamination of drinking water by misconnection (cross-connection) between potable water pipes and reclaimed water pipes, and also to disinfect reclaimed wastewater properly.

In addition, the following problems have also been identified in wastewater reuse for toilet flushing:

- Corrosion of pipe.
- Blockage of pipe and strainer.
- Biofilm (slime) formation in reservoir tank due to reduction of residual chlorine in reclaimed water [11].

These problems, some of which are similar to concerns associated with industrial reuse, tend to occur because reclaimed water contains more salts and organics than drinking water. Pipe corrosion is seen, particularly at the joint part where galvanic corrosion occurs by an electrochemical process (Figure 40.4). In the end, it may result in blockage of a pipe or clogging of a strainer due to formation of iron oxide scale (e.g., Fe_2O_3) (Figure 40.5).

FIGURE 40.4   Corrosion of pipe joint.

FIGURE 40.5   Blockage of pipe joint.

Corrosion mitigation can be taken by several methods, such as application of protective coating, employing corrosion-resistant materials or adoption of screw-type fitting with inner sleeve, as shown in Figure 40.6.

Furthermore, an insufficient amount of residual chlorine in reclaimed water allows bacteria growth and biofilm formation in the reservoir tank. The reduction of residual chlorine occurs as the consumed chlorine reacts with salts and organics in reclaimed water. Therefore, the chlorine injection rate must be monitored carefully and should be kept at an appropriate level.

40.6.3 Groundwater Recharge

A groundwater aquifer is important for freshwater storage and water transmission. It provides water resources that can be withdrawn for various purposes. Three common methods for aquifer recharge are illustrated in Figure 40.7.

The use of a recharge basin requires a wide area with permeable soil, an unconfined aquifer with transmissivity, and an unsaturated (or vadose) zone without restricting layers. With this system, the vadose zone and aquifer work as natural filters and remove suspended solids, organic substances, bacteria, viruses, and other microorganisms. In addition, reduction of

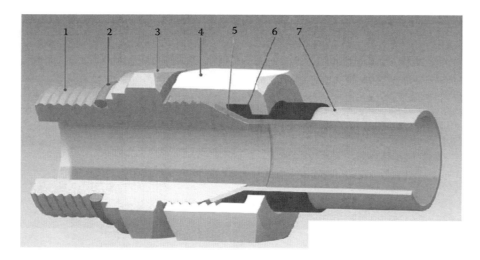

FIGURE 40.6   Screw-type fitting with inner-sleeve. Flare connection: (1) Screw thread, (2) o-ring, (3) body, (4) nut, (5) seal interface, (6) support ring (sleeve), (7) flared tubing.

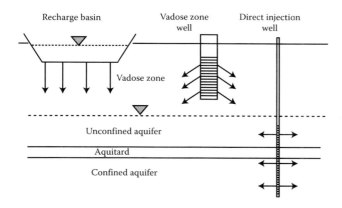

FIGURE 40.7 Method for aquifer recharge.

nitrogen, phosphorus, and heavy metals can also be achieved. This process is called soil aquifer treatment.

Direct injection of treated wastewater can access deeper aquifers through an injection well. Direct injection is utilized when aquifers are deep or separated from the surface by an impermeable layer. This method requires less land than the recharge basin methods, but it costs more to construct and maintain the injection well. A well wall is susceptible to clogging by suspended solids, biological activity, or chemical impurities. In this method, the soil aquifer treatment effect is not observed. The method requires advanced pretreatment of applied water, including sufficient disinfection. Without treatment, the injected wastewater may pollute the aquifer, causing health concerns.

Vadose zone injection is an emerging technology that provides some of the advantages of both recharge basins and direct injection wells. This method is used when a permeable layer is not available at a shallow depth, and a recharge well has a relatively large diameter [10].

40.6.4 Benefits

Groundwater recharge has been used to prevent the decline in groundwater level and to preserve the groundwater resource for future use. Compared to conventional surface water storage, aquifer recharge has many advantages, such as negligible evaporation, little secondary contamination by animals, and no algal blooming. It is also less costly because no pipeline construction is required. Furthermore, it protects groundwater from saltwater intrusion by barrier formation in coastal regions, and controls or prevents land subsidence.

Groundwater recharge site is one of the successful examples. For over 35 years, in the Montebello Forebay Ground Water Recharge Project, recycled water has been applied to the Rio Hondo spreading grounds to recharge a potable ground water aquifer in south-central Los Angeles County in California [35].

40.6.5 Potential Concerns

In any of the methods described previously, groundwater recharge with reclaimed water presents various health concerns

when water is extracted from a collection well and used for irrigation or other purposes. As the performance of soil aquifer treatment is uneven depending on hydraulic loading, each project should be carefully designed and adequate attention paid to reducing pathogens [10].

40.7 Exploding Sewers: The Industrial Use and Abuse of Municipal Sewers and Reducing the Risk

Industrial water use accounts for approximately 20% of global freshwater withdrawals. Power generation constitutes a large share of this water usage, with up to 70% of total industrial water used for hydropower, nuclear, and thermal power generation, and 30%–40% used for other, nonpower generation processes. Industrial water reuse has the potential for significant applications, as industrial water demand is expected to increase 1.5 times by 2025 [27].

40.7.1 Benefits

Industrial water reuse has the following specific benefits, in addition to the general environmental benefits discussed in earlier sections:

- Potential reduction in production costs from the recovery of raw materials in the wastewater and reduced water usage.
- Heat recovery.
- Potential reduction in costs associated with wastewater treatment and discharge.

Water reuse and recycling for industrial applications have many potential applications, ranging from simple housekeeping options to advanced technology implementation. Wastewater reuse for industry can be implemented through the reuse of municipal wastewater in industrial processes, internal recycling and cascading use of industrial process water, and nonindustrial reuse of industrial plant effluent, as summarized in Table 40.1.

In particular, cooling systems may consume 20%–50% of a facility's water usage, and also present a significant potential

TABLE 40.1 Types of Industrial Water Reuse with Examples

Types of Water Reuse	Examples
Reuse of municipal wastewater	Cooling tower make-up water
	Once-through cooling
	Process applications
Internal recycling and cascading use of process water	Cooling tower make-up water
	Once-through cooling and its reuse
	Laundry reuse (water, heat, and detergent recovery)
	Reuse of rinse water
	Cleaning of premises
Nonindustrial use of effluent	Heating water for pools and spas
	Agricultural applications

for reuse. Cooling systems remove heat from air-conditioning systems, power stations, oil refining, and other various industrial processes. Many facilities operate cooling towers, in which warm water is circulated and cooled continuously. Water, commonly referred to as make-up water, is added to replace evaporative loss and pollutant discharge. Some facilities also use once-through water to cool heat-generating equipment and discharge water after heat transfer. In both systems, adequately treated wastewater can be used as cooling water or make-up water, with or without mixing with tap water. Once-through cooling systems also present additional opportunities for water reuse, such as connection to a recirculating cooling system to reuse water, and cascading use of cooling water in other applications.

Water quality requirements for industry reuse differ according to application types. Obtaining the necessary quality may require secondary treatment, tertiary treatment, or specific methods to meet individual needs. For example, rinsing and cleaning for semiconductor wafer manufacturing requires ultra-pure water, which can be supplied from municipal wastewater that has undergone reverse osmosis and ultraviolet treatment [22]. Almost all well-managed cooling towers use a water treatment scheme, such as sulfuric acid treatment, side stream filtration, and ozonation to inhibit corrosion and scaling [17]. The cascading of process water of nonpotable quality without treatment may be sufficient for general office cleaning and rinse water. In the steel industry, the effluent from wet scrubbers of blast furnaces can be recycled after treatment to remove iron oxide, silica, carbon lime, and magnesium by coagulation or high-gradient magnetic separation. In the pulp and paper industry, water reuse is an important strategy for recovering fibers, chemicals, and heat from process effluent, as well as for reducing freshwater consumption and wastewater production [5]. Wastewater treatment, with respect to the industrial use, depends on the effluent, that is, if the effluent is coming from somewhere other than the industry that is reusing it, or if the industry is recycling its effluent for reusing within the same industry.

Cleaner production (CP) assessment is an analytical method of particular relevance for evaluating options for industrial water reuse. CP has been promoted by UNEP since the late 1980s, and is defined as the "continuous application of an integrated, preventive environmental strategy to processes, products, and services to increase overall efficiency, and reduce risks to humans and the environment." CP can be implemented to improve industrial processes, product performance, and various services provided in society. The CP assessment methodology is used for systematic identification and evaluation of CP opportunities, thereby facilitating their implementation.

The five areas of CP opportunities to be identified through CP assessment are the changes in raw materials, technological change, good operating practices, product changes, and onsite reuse and recycling [32]. Water reuse and recycling in the process can be achieved by implementing good operating practices, changing to technologies that require less process water and

onsite reuse and recycling. In addition, CP assessment can be applied to reformulate industrial products to use less water.

40.7.2 Potential Concerns

Potential concerns for industrial water reuse include scaling, corrosion, biological growth, and fouling, which may impact industrial process integrity and efficacy, as well as product quality. These concerns are often interrelated, and may be addressed by the options summarized in Table 40.2.

Salt concentrations can be affected by various factors, including process operating temperatures, sources of wastewater, and areas from which wastewater is collected (e.g., coastal areas may have higher concentrations).

Occupational health concerns include exposure to aerosols that contain toxic volatile organic compounds and bacteria, such as Legionella, which causes Legionnaire's disease.

Some of the important and practical aspects for industrial wastewater reuse are

- Usually, the industry itself decides the needs and extent of wastewater treatment for its reuse. The government does not decide how to reuse/recycle water in the industry; it only motivates industry through incentives such as the price of water or subsidies for the technology.
- Most industries select types of wastewater treatment processes that have a great level of reliability. This differs from wastewater reuse projects in municipalities, where the cost is a crucial factor to decide both the type of reuse and the type of treatment.

40.7.2.1 Wastewater Reuse for Environmental Water Enhancement

Another area where wastewater reuse is being applied is in environmental enhancement, such as the augmentation of natural/artificial streams, fountains, and ponds. In metropolitan areas, urbanization, and the resulting increase in surface area coverage by buildings and pavements, has resulted in decreased water retention capacity. In addition, stormwater is rapidly drained and discharged to a river and/or sea to prevent flooding, often leaving little water for environmental water usage.

TABLE 40.2 Concerns, Causes, and Treatment Options of Industrial Water Reuse

Concerns	Causes	Treatment Options
Scaling	Inorganic compounds, salts	Scaling inhibitor, carbon adsorption, filtration, ion exchange, blowdown rate control
Corrosion	Dissolved and suspended solids, pH imbalance	Corrosion inhibitor, reverse osmosis
Biological growth	Residual organics, ammonia, phosphorous	Biocides, dispersants, filtration
Fouling	Microbial growth, phosphates, dissolved and suspended solids	Control of scaling, corrosion, microbial growth, filtration chemical and physical dispersants

40.7.3 Benefits

The key benefit for environmental enhancement is the increased availability and quality of water sources, which provide public benefits such as aesthetic enjoyment and support ecosystem recovery. The restoration of streams or ponds with reclaimed water has been practiced in many cities, contributing to the revival of aquatic life, such as fish, insects, crawfish, and shellfish, and creating comfortable urban spaces and scenery. The recovery of water channels has great significance for creating "ecological corridors" in urban areas.

The landscape of Osaka Castle in Japan has been beautifully restored with a moat, which is filled with reclaimed water. In this case, 5000 m³/day of tertiary treated wastewater, by sand filtration and chlorine disinfection, is supplied from the sewage treatment works [18].

40.7.4 Potential Concerns

As with urban applications, public health concerns must be adequately addressed for environmental enhancement applications in order to avoid negative human health impacts. When treated wastewater is used for water augmentation in a water channel, proper water quality guidelines must be considered on the assumption that there will be human contact with the reused water, and sufficient disinfection must be carried out. Disinfection options may include chlorination or UV irradiation. In addition to the public health considerations, the removal of nutrients including nitrogen or phosphorus should be implemented because they may cause algal blooming, which spoils the appearance of streams, lakes, and reservoirs.

Care must also be taken to facilitate ecosystem recovery. In the case of the restoration of aquatic flora and fauna in a stream, ozone or UV disinfection is more preferable than chlorination because it generates fewer disinfection by-products with smaller residual effects to the flora and fauna.

40.8 Lessons Learned: A Response and Recovery Framework

Launching a wastewater reuse initiative requires careful consideration of the local conditions, and must be based on the sufficient and well-integrated analysis of technology options, financial implications, health risks mitigation, and other factors. In this section, key factors for establishing wastewater reuse initiatives are described.

40.8.1 Planning to Meet Specific Needs and Conditions

During a planning process, the appropriateness of water and wastewater reuse applications needs to be carefully evaluated against the volume of available wastewater, degree of water scarcity, availability of existing infrastructure, and receptivity of potential users. The purpose of the application, such as irrigation, industrial use, landscape, and household use, needs to be evaluated together with the water quality requirements and associated health risk. Such evaluation is useful in identifying necessary treatment and disposal technologies, as well as operational and maintenance requirements. Public perception and receptivity also need to be analyzed carefully. The public should be recognized as legitimate stakeholders, and their roles and responsibilities should be clearly defined in the planning process.

40.8.2 Selecting Options to Minimize Risk

One of the most important factors in water and wastewater reclamation projects is complying with water quality standards to minimize health risks or establishing them if they do not exist. While the WHO guidelines for agricultural applications of wastewater are available, there are no international guidelines or criteria for other types of wastewater reuse. Therefore, guidelines and standards need to be developed by each country, with health risks as well as technical and economic feasibility being taken account. Technological options should be selected to meet such guidelines and standards and ensure the protection of human health and the environment.

40.8.3 Utilizing Institutions and Organizations

Wastewater reuse involves many stakeholder institutions, such as utilities and private users that implement the initiative, local environmental authorities for permits and enforcement, financial institutions for provision of funding, the national environmental ministry for setting national standards and supporting local authorities, and so on. Their responsibilities and roles for facilitating reuse programs need to be identified and understood clearly. In many cases, institutions need to be supported or newly established.

40.8.4 Building Capacity

Capacity is needed to successfully plan and implement wastewater reuse initiatives. Such capacity encompasses human resources, policy and legal frameworks, institutional and organizational management, and financing, as well as public awareness and participation. The following sections explain in more detail the needs for various elements of capacity building, as well as concrete examples.

40.8.5 Meeting Standards and Guidelines

Standards and criteria for water reuse need to be complied with in order to protect human health and the environment. Applications need to be monitored to ensure that wastewater is being reused in a manner consistent with the intended applications and practice. WHO has established guidelines for agricultural reuse, which can serve as a basis for countries without their own standards or guidelines. Standards and guidelines for other applications have also been established in various localities and countries, for example by the US EPA and the State of California. They may also provide insight for other countries or localities.

40.9 Managing Urban Water Risks

40.9.1 Adaptation Case Study

40.9.1.1 Reviving of Flora and Fauna by Restoration of River Flow

The Meguro River, which flows through a residential area in Tokyo, had been abandoned by residents due to the decreasing flow of water and pollution with an unpleasant odor. To solve this issue, the Tokyo Metropolitan Government released water treated with UV radiation into the river. With the drastic improvement in water volume and quality, various living species have returned to the river. After the introduction of highly treated water, many insects and small animal populations have been reestablished, and fish such as Japanese trout, striped mullets, and gobies also returned to the river. Biodiversity and environmental amenities have thus been restored effectively with wastewater reuse.

40.9.1.2 Utilization of Thermal Energy of Wastewater (Japan)

One of the important characteristics of water is its ability to retain heat energy, and water can warm or cool other objects when it comes into contact with them. Japan implements a unique approach by utilizing the nature of water as a heat medium. The Japanese climate has a large variation in temperature, from below 0 in winter to nearly 40°C in summer, but the temperature of effluents from sewage treatment works stays relatively constant throughout the year, from 12°C to 30°C. The temperature of wastewater is therefore lower in summer and higher in winter than the ambient temperature. Based on this feature, a heating and cooling system through heat exchange with wastewater has been developed, which achieved energy savings of 20%–30% [18].

Another example of utilizing thermal energy from wastewater has been implemented in the city of Sapporo, where the average annual snowfall is 4.7 m. Snow removal from roads and road shoulders is very important to ensure the safety of drivers and pedestrians, particularly in downtown areas. Removed snow is transported and dumped by trucks on unused land. In Sapporo, snow melting by using effluent from sewage treatment works is implemented as an alternative approach. The temperature of effluent in winter reaches approximately 13°C [26], and 600,000 to 700,000 m³ of snow are melted per year by snow flowing conduits and snow melting tanks [24].

40.9.1.3 Artificial Snowmaking: Wastewater Reuse in Alpine Area (Australia)

Mount Buller Alpine Resort is located 200 km northeast of Melbourne, Australia. As one of Australia's most popular snow sports destinations, the resort is most active during the winter season. Approximately 70% of the entire annual effluent is produced in winter. A pilot study for making snow from treated water has been in operation since 2000. As the ultrafiltration system is used to remove pathogens in reclaimed water, the treated water quality has reached the US EPA's standards for unrestricted recreational use. The installation of a full-scale plant for artificial snowmaking will bring sustainable benefits to the resort in the future [30].

40.9.2 Economic Analysis

Economic and financial analyses are also needed to identify viable solutions and to access financial assistance when necessary. While wastewater reuse programs have many benefits and long-term cost effectiveness, they may have a high-initial cost associated with additional treatment and infrastructure needs, such as additional treatment, pumps, pipes, reservoirs, and so on. Alternatives to address this impediment, such as public assistance, incentives, and preferential private sector financing, must be explored. The decision makers and the users should be aware of the impact on water prices resulting from wastewater reuse projects. The quantifiable present worth costs and benefits were estimated for each option. The economic benefits included residents' willingness to pay for sewage service, reduction of sanitation costs at commercial establishments, tourists' willingness to pay for sewage service, value of water reuse, reduced beach erosion, avoidance of beach closures, enhanced tourism activities, and public health. The value of water reuse was determined as the cost of the water to be used if reclaimed water were not available. Thus, costs were determined for potable water, desalinated water for irrigation, groundwater for irrigation, and brackish water for cooling.

The costs and benefits of each option were discounted from their future values to an equivalent present value for comparison. A range of discount rates was used to test the sensitivity of the results to changes in the discount rate. The different discount rates affected the net project costs but did not change the ranking of the options in the quantifiable economic analysis.

40.9.3 Closing the Urban Water Cycle: An Integrated Approach toward Water Reuse

Water resources are the indispensable material base of life and the important safeguard for socioeconomic development. Strategies for water recycling and reuse promote circular economy and at the same time offer solutions to water pollution. Circular economy is centered on the relationship between pollution treatment and reuse of reclaimed water. The notions of waste reduction, even no waste, reuse, and turning waste into resources in a circular economy are consistent with those of economics. These methods not only reduce costs but also increase capital. Strategies for water recycling and reuse offer solutions to water scarcity and are a glaring contradiction between supply and demand in Tianjin. Such strategies are also positive approaches to the serious water pollution in Tianjin.

Health risk assessment of urban reclaimed water is aimed at assessing water quality as various pollutants may generate harmful effects in humans in the process of water reuse. The reclaimed water is primarily reused in agricultural irrigation, industry, groundwater recharge, landscapes, and recreation areas. As our reclaimed water management strategy depends primarily on the comparison of conventional water quality indexes and standard ones, the current assessment method, to some extent, weakened the potential impact of toxic and harmful factors stipulated by national water quality standards. If health risk assessment of reclaimed water becomes a part of routine environmental assessment, it will be easier to have a more scientific, objective, comprehensive, and immediate understanding of reclaimed water quality and safety to implement necessary pollutant priority control strategy and environmental protection strategy.

40.10 Summary and Conclusions

Recognizing that water-related problems are one of the most important and immediate challenges to the environment and public health, it is important to act now. Water scarcity and water pollution are some of the crucial issues that must be addressed within local and global perspectives. One of the ways to reduce the impact of water scarcity as well as minimize water pollution is to expand water and wastewater reuse. In this regard, ESTs are vital to implement wastewater reuse and recycling at the local level. The selection of the appropriate ESTs depends on the available quality and quantity of wastewater and the requirements for the wastewater reuse. The local conditions including regulations, institutions, financial mechanisms, availability of local technology, and stakeholder participation have a great influence over the decisions for wastewater reuse. Hence, the selection of appropriate ESTs should be paramount among those factors. In cities and regions of developed countries, for example, where wastewater collection and treatment have been established over the years, wastewater reuse is practiced with proper attention to sanitation, public health, and environmental protection. Unfortunately, in many places in developing countries, wastewater reuse is not being practiced in a way to protect the environment and public health. Three broad scenarios can be observed: in some places, untreated wastewater is being reused as some countries are obliged to resort to this despite the health risks; in other places, an intermediate quality is being reused, where some treatment has been given; and lastly, in few places, a level corresponding to developed-world practice is being implemented. This last level can be taken as a benchmark to meet the demand for water with minimum damage to the environment and protection of public health. The type of reuse will govern the roadmap for achieving the highest standards. Thus far, wastewater reuse in the agriculture sector is the highest, and stabilization ponds and artificial wetlands based on phytotechnologies are commonly used in developing countries to improve the water quality. Other methods including reservoirs and soil aquifer treatment are also in use in some places. The increasing awareness of food safety, and the influence of the countries, which import food, is influencing policy makers and agriculturists to improve the standards of wastewater reuse in agriculture. Wastewater reuse in industries is regulated mainly due to the water and wastewater charges and penalties on pollution levels. As the cost goes up, industries try to scale-up recycling and reuse. The role of government assistance, especially economic incentives and noncommercial credit for obtaining appropriate technology, has shown good success in some countries like Japan. Recently, the environmental awareness of consumers has been putting pressure on the producers (industries) to opt for environmentally sound technologies including those that conserve water and reduce the level of pollution. Urban applications are gaining ground due to water scarcity and the increasing cost of freshwater. Secondary uses including landscaping and gardening, which consume substantial quantities of freshwater, are now reusing greywater and reclaimed wastewater. The improvements in the technologies and their affordability at municipality as well as the household level will result in increasing wastewater reuse in the cities. To promote wastewater reuse on a sustainable basis and for wider applications, some key factors should be addressed. First, planning for wastewater reuse is important with reference to meeting specific needs and conditions. This could be facilitated by incorporating wastewater reuse into local plans for water management. Planning needs to take care of all the sensitive issues including public health, the role of stakeholders, and the viability of operation and maintenance. Second, economic and financial requirements are crucial, as less viable schemes for wastewater reuse will only create a social burden and will not last long. Cost-effectiveness should be given high priority. Partnerships with the private sector and the community may help to improve the level of investment and improve efficiency during operation and maintenance, thus reducing the overall cost and making it economically and financially viable. Third, local capacity, including human resources, policy and legal framework, and institutions, is very important in achieving sustainable targets of wastewater reuse plans. Capacity building should be an integrated part of the overall plan, and national and international agencies can actively assist in this.

In light of this, it may be observed that we have to move forward to implement strategies and plans for wastewater reuse. However, their success and sustainability will depend on political will, public awareness, and active support from national and international agencies to create an enabling environment for the promotion of environmentally sustainable technologies.

References

1. Amoah, P., Keraita, B., Akple, M., Drechsel, P., Abaidoo, R.C., and Konradsen, F. 2011. Low-Cost Options for Reducing Consumer Health Risks from Farm to Fork Where Crops are Irrigated with Polluted Water in West Africa. International Water Management Institute, Colombo, Sri Lanka.

2. Asano, T. (ed.). 1998. *Wastewater Reclamation and Reuse*, Vol. 10. CRC Press, Boca Raton, FL.

3. Asano, T. and Levine, A. 1996. Wastewater reclamation, recycling and reuse: Past, present and future. *Water Science and Technology*, 33(10–11), 1–14.

4. Ayers, R.S. and Westcot, D.W. 1985. Water quality for agriculture. FAO Irrigation and Drainage Paper 29. FAO, Rome.

5. Bedard, S., Sorin, M., and Leroy, C. 2000. *Application of Process Integration in Water Reuse Projects*, CANMET Energy Diversification Research Laboratory, Natural Resources Canada.

6. Booker N. 1999. Estimating the economic scale of greywater reuse systems. Program report FE-88, CSIRO Molecular Science.

7. Bowman, J.A. 1994. Saving water in Texas industries. *Texas Water Resources*, 20(1), 1–10.

8. Clark, R. 1997. Optimum scale for urban water systems. Report 5 in the water sustainability in urban areas series. Water Resources Group. Department of Environment and Natural Resources, Adelaide, July.

9. Food and Agricultural Organization of the United Nations. 1985. Water Quality for Agriculture. FAO Irrigation and Drainage Paper 29 Rev.1. FAO, Rome.

10. Fox, P. 1999. Advantages of aquifer recharge for a sustainable water supply, *Proceedings of the International Symposium on Efficient Water Use in Urban Areas*. IETC Report 9, pp. 163–172.

11. Fukuoka Municipal Government. 1999. *Report on Committee for Water Reuse in Housing Complex,* Japanese version.

12. Indigo Development. 2003. Center in the Sustainable Development Division of Sustainable Systems Inc. Oakland, California.

13. Jacques G. and Anastasia, P. 1996. Risk analysis of wastewater reclamation and reuse. *Water Science and Technology*, 33(10–11), 297–302.

14. Jimenez, B. and Asano, T. 2004. Acknowledge all approaches: The global outlook on reuse. *Water,* 21(December), 32–27.

15. Jimenez, B. and Asano, T. (eds.). 2008. *Water Reuse: An International Survey of Current Practice, Issues and Needs.* IWA Publishing, London, UK.

16. Juanico, M. and Shelef, G. 1991. The performance of stabilization reservoirs as a function of design and operation parameters. *Water Science Technology*, 23, 1509–1516.

17. North Carolina Department of Environment and Natural Resources (NCDENR). 1998. *Manual for Commercial, Industrial and Institutional Facilities*, North Carolina.

18. Osaka Municipal Government. 2003. *Sewage Works in Osaka.*

19. Parkinson, D.B. and J.V. Wodrich. 2000. Challenges in implementing the use of reclaimed water. *Water Management in the 90s: A Time for Innovation, Recycling.* Canadian Council of Ministers of the Environment, Winnipeg, Manitoba.

20. Pescod, M.B. 1992. Wastewater treatment and use in agriculture. FAO Irrigation and Drainage Paper, 47. FAO. Rome, Italy.

21. Postel, S. and Oasis, L. 1993. Facing Water Scarcity. Planning and Management Division, ASCE and Working Group UNESCO/IHP IV Project, Norton, New York, pp. 674–681.

22. Public Utilities Board of Singapore (PUB). 2003. *NEWater Sustainable Water Supply.* http://www.pub.gov.sg/NE Water_files/visitors/index.html.

23. Safaa A.A., Shadia, R., and Hala, A. 2002. Development and verification of a decision support system for the selection ofoptimum water reuse schemes. *Desalination,* 152, 339–352.

24. Sapporo Municipal Government Information on Sewage in the City of Sapporo: Application for Snow Melting (Japanese version).

25. Sheikh, B., Rosenblum, E., Kasower, S., and Hartling, E. 1998. Accounting for all the benefits of water recycling. *Proceedings, Water Reuse '98, a Joint Specialty Conference of WEF and AWWA.* Orlando, FL.

26. Shibuya, T. 1999. Sewage works for snow control. *Sewage Works in Japan*, 49–55.

27. Shiklomanov, I.A. 1999. World water resources and their use. http://webworld.unesco.org/water/ihp/db/shiklomanov/index.shtml.

28. Simmons, R., Qadir, M., and Drechsel, P. 2010. Farm-based measures for reducing human and environmental health risks from chemical constituents in wastewater. In: P. Drechsel, C.A. Scott, L. Raschid-Sally, M. Redwood, and A. Bahri (eds.), *Wastewater Irrigation and Health: Assessing and Mitigation Risks in Low-Income Countries.* Earthscan-IDRC-IWMI, United Kingdom.

29. Tokyo Metropolitan Government. 2001. *Sewage in Tokyo—Advanced Technology.* http://www.gesui.metro.tokyo.jp/english/technology.htm.

30. Tonkovic, Z. and Jeffcoat, S. 2002. Wastewater reclamation for use in snow-making within an alpine resort in Australia—Resource rather than waste. *Water Science and Technology*, 46(6–7), 297–302.

31. United Nations. 1989. World Population Prospects. Department of International Economic and Social Affairs. United Nations, New York. www.un.org.

32. United Nations Educational, Scientific and Cultural Organization (UNESCO). 2000. *Water Use in the World: Present Situation/Future Need.* http://www.unesco.org. United Nations World Commission on Environment and Development. Oxford University Press, New York.

33. United Nations Environment Program (UNEP). 2004. Saving Water through Sustainable Consumption and Production: A Strategy for Increasing Resource Use Efficiency, DTIE, Paris, France.

34. U.S. Environmental Protection Agency. 1992. Agency for International Development, Guidelines for Water Reuse. EPA/625/R-92/004; US EPA Technology Transfer, Cincinnati, OH.

35. US Environmental Protection Agency. 1998. *Water Recycling and Reuse: The Environmental Benefits*, EPA909-F-98–001, Washington, DC.

36. Weber, B. and Juanico, M. 2004. Salt reduction in municipal sewage allocated for reuse: The outcome of a new policy in Israel. *Water Science Technology*, 50(2), 17–22.

37. Wintgens, T. and Hochstrat, R. 2006. *Report On Integrated Water Reuse Concepts—Integrated Concepts for Reuse of Upgraded Wastewater*. AQUAREC. EVK1-CT-2002-00130 Deliverable D19. http://www.amk.rwth-aachen.de/fileadmin/files/Forschung/Aquarec/D19_final_2.pdf. Accessed on August 23, 2012.

38. World Commission on Environment and Development (WCED). 1987. Our Common Future: The Bruntland Report. UN Documents Gathering a body of global agreements.

39. World Health Organization. 1989. *Health Guidelines for the Use of Wastewater in Agriculture and Aquaculture*. World Health Organization, Geneva.

40. World Health Organization (WHO). 2001. Water Quality: Guidelines, Standards and Health Assessment of Risk and Risk Management for Water-related Infectious Diseases. http://www.who.int/water_sanitation_health/dwq/whoiwa/en/.

41. World Health Organization (WHO). 2006. Guidelines for the safe use of wastewater, excreta and greywater. Resource document. WHO. http://www.who.int/water_sanitation_health/wastewater/gsuww/en/index.html. Accessed on March 5, 2007.

42. World Water Council (WWC). 2000. Commission Report: A Water Secure World: Vision for Water, Life and the Environment. www.worldwatercouncil.org/publications.shtml.

43. Zhang X. 2000. Northeast China province to control use of groundwater. *Liao Ning Province*, China. www.acca21.org.cn 2000.

41

Agricultural Water Reuse in Low-Income Settings: Health Risks and Risk Management Strategies

Bernard Keraita
University of Copenhagen

Pay Drechsel
International Water Management Institute (IWMI)

PREFACE

Wastewater may be defined as the combination of liquid wastes discharged from domestic households, farms, institutions, and commercial and industrial establishments eventually mixed with groundwater, surface water, and stormwater. Wastewater is increasingly receiving global attention as it is seen as one of the alternative solutions to increasing global water scarcity. Indeed, wastewater is globally being reused in many applications including groundwater recharge, industrial reuse like for cooling, environmental and recreational uses, nonpotable urban uses, and indirect or direct potable reuse. However, agricultural irrigation and landscaping is by far the largest wastewater use sector. Indeed, millions of farmers worldwide are involved in wastewater irrigation activities.

However, wastewater contains a variety of pollutants and contaminants, which may pose health risks if not well managed. These pollutants include salts, metals, metalloids, pathogens, residual drugs, organic compounds, endocrine disruptor compounds, and active residues of personal care products. The kind and extent of health risks depend on many factors including the types and levels of contaminants as well as regional risk relevance. In low-income countries, risks from pathogens receive the most attention. This is because people in these countries are most affected by diseases caused by poor sanitation such as diarrheal diseases and helminth infections, so high loads of pathogenic microorganisms are often found in wastewater systems.

Focusing on low-income contexts, this chapter presents health risks posed by wastewater irrigation activities and some practical examples on how these risks could be managed.

41.1 Introduction

With increasing global water scarcity and pollution of water bodies, agricultural water reuse is becoming more common [17]. Globally, agricultural water reuse is by far the largest water reuse application portfolio [16]. Existing literature from global and country-level assessments that have been conducted show that wastewater irrigation systems are widespread, regardless of the development level and climate [21,35]. This literature shows that these systems are beneficial as they increase agricultural production, especially that of high-value crops like vegetables, leading to high-income generation for farmers and others in the supply chain. However, the literature also shows that wastewater can affect the quality of crops and pose public health and environmental risks. There are also productivity risks due to salinity, sodicity, and ion-specific toxicities [33]. This is especially so when untreated or only partially treated wastewater is used, which is a common practice in most low-income countries [24,37]. Of most concern are excreta-related pathogens associated with low sanitation coverage in these countries [7,43]. These risks jeopardize sustainability of agricultural water reuse systems, calling for a research and development focus on risk management in these systems.

41.2 Agricultural Water Reuse Systems in Low-Income Settings

Most wastewater irrigation systems in low-income countries use untreated or partially treated wastewater [24]. This is due to low levels of wastewater treatment, with median levels estimated at 35% in Asia, 14% in Latin America, and less than 1% in sub-Saharan Africa [46]. There are various use scenarios in the irrigation systems, but mainly farmers use wastewater directly for irrigation without being mixed or diluted by other water bodies (direct) or after dilution and mixing (indirect) [24].

41.2.1 Direct Use Irrigation Systems

Direct use of wastewater takes place often where alternative water sources are scarce, that is, usually in drier climates like in the Middle East and North Africa (MENA) region. Direct use of untreated wastewater in agriculture has been reported in Afghanistan, Algeria, Egypt, Iran, Iraq, Lebanon, Morocco, Palestine, Syria, and Yemen [5]. The situation is the same in drier parts of Latin American countries such as Mexico, Chile, and Brazil and in some Caribbean islands [21]. In other areas, some reasons such as salinity and cost implications restrict the use of alternative water sources. For example, in the arid climate of Haroonabad, Pakistan, untreated wastewater is directly used in farming because the groundwater, which is the only other alternative, is too saline for irrigation [14]. In the semiarid climate of the twin city of Hubli–Dharwad in Karnataka, India, farmers extract untreated wastewater from sewage nallas (open sewers) and underground sewer pipes [8]. Pumping wastewater from sewage nallas is cheaper than using groundwater from boreholes. This makes irrigation water more accessible to farmers with limited financial resources. In other cases, such as Cochabamba, Bolivia and Tamale, Ghana, farmers use wastewater from malfunctioning treatment, taking advantage of the already collected resource [2,20].

41.2.2 Indirect Use Systems

The indirect use of wastewater for irrigation is more common in wetter climates. Usually, untreated and/or partially treated wastewater from urban areas is discharged into drains, smaller streams, and other tributaries of larger water bodies, where it is mixed with storm- and freshwater (diluted wastewater) before it is used by farmers, most of them are traditional users of these water sources. This practice has been reported throughout sub-Saharan Africa, but also Nepal, India, and many other cities in Brazil, Argentina, and Colombia, which lack adequate sanitation facilities [24]. In West Africa, there is extensive cultivation of vegetables using this irrigation system as estimates show that up to 90% of vegetables consumed are grown within or near urban areas [11]. However, this system is not limited only to low-income countries that have no capacity for collecting and treating wastewater comprehensively, as similar cases have been reported in fast-growing economies such as China and the MENA region.

41.2.3 Other Systems

In other areas, irrigation infrastructure originally built for formal irrigation is now used for wastewater irrigation. Wastewater is pumped into the irrigation canals for supplementing irrigation water (surface/groundwater). For instance, in Vietnam, wastewater from Hanoi and other cities along the Red River delta is pumped into irrigation canals at certain times of the year for supplementing irrigation water [42]. However, in the dry season and at tail ends, wastewater ends up being the only water flowing in the canals such as in Haroonabad; Pakistan; and Hyderabad, India [13]. In some extreme cases, farmers rupture or plug sewage lines for wastewater. Such cases have been reported at Maili Saba, Nairobi, and Bhaktapur at the Katmandu Valley [19,36]. At Maili Saba, farmers have removed manhole covers and blocked the city's main sewer, causing raw sewage to rise up the manholes and flow out over the land [19]. In Northern Ghana, farmers use untreated human waste (fecal sludge) in rain-fed cropping systems [10].

41.3 Health Risks from Agricultural Water Reuse

Wastewater contains a variety of pathogens and pollutants. Extensive studies on human health risks posed by wastewater irrigation especially from pathogen contamination have been done [45]. Table 41.1 is a simplified presentation of wastewater related human health risks affected groups and exposure pathways. Other pollutants include salts, metals, metalloids, pathogens, residual drugs, organic compounds, endocrine disruptor compounds, and active residues of personal care products [41]. These components pose environmental and human health risks. Emphasis in discussions is often given to different types of pollutants depending on the regional risk relevance. For example, in low-income countries, risks from microbiological contaminants receive the most attention. This is because people in these countries are most affected by diseases caused by poor sanitation such as diarrheal diseases and helminth infections [34]. The situation changes significantly in transitional economies and is different in high-income countries, where microbiological risks are largely under control. In this context, chemical pollution (heavy metals, pesticides) and emerging pollutants (e.g., antibiotics) are a major public concern.

TABLE 41.1   Simplified Presentation of the Main Human Health Risks from Wastewater Irrigation

Kind of Risk	Health Risk	Who Is at Risk	Exposure Pathway
Occupational risks (contact)	• Parasitic worms such as *A. lumbricoides* and hookworm infections • Bacterial and viral infections • Skin irritations caused by infectious and noninfectious agents—itching and blister on the hands and feet • Nail problems such as koilonychias (spoon-formed nails)	Farmers/field workers Marketers of wastewater-grown produce	• Contact with irrigation water and contaminated soils • Contact with irrigation water and contaminated soils • Contact with contaminated soils during harvesting • Exposure through washing vegetables in wastewater
Consumption-related risks (eating)	• Mainly bacterial and viral infections, such as cholera, typhoid, enterotoxigenic *Escherichia coli* (ETEC), hepatitis A, and viral enteritis, which mainly cause diarrhea • Parasitic worms such as ascaris	Vegetable consumers	• Eating contaminated vegetables, especially those eaten raw
Environmental risks	• Similar risks such as those exposed to occupational and consumption risks, but decreasing with distance from the farm	• Children playing in wastewater-irrigated fields • People walking on or nearby fields	• Soil particle intake • Aerosols

Source: Modified from Abaidoo, R.C. et al., *Soil Biology and Agriculture in the Tropics*, 2009, pp. 498–535. From Springer Science+Business Media.

Humans are mainly exposed to wastewater-irrigation risks by (1) consuming irrigated produce (consumption-related risks), (2) coming into contact with wastewater when working in the farms (occupational risks), and (3) by getting exposed to wastewater and wastewater-irrigated soils when walking by or children playing on the fields (environmental risks). Constituents of most concern in wastewater are excreta-related pathogens and skin irritants [7,43]. For consumption-related health risks, the primary concern is uncooked vegetable dishes such as salad [18]. Several diarrheal outbreaks have been associated with wastewater-irrigated vegetables [39,45]. There is also strong epidemiological evidence for *Ascaris lumbricoides* infections for both adults and children consuming uncooked vegetables irrigated with wastewater [32]. Helminth infections, especially *A. lumbricoides* and hookworm, have higher importance in relation to occupation-related risks compared to bacterial, viral, and protozoan infections [7]. The most affected group is farm workers, owing to the long duration of their contact with wastewater and contaminated soils [45]. Recent studies from Vietnam, Cambodia, India, and Ghana have associated skin diseases such as dermatitis (eczema) to contact with untreated wastewater [24].

41.4 Management of Health Risks

41.4.1 Health Risk Management Approaches

Different approaches have been proposed for risk mitigation. For a long time, conventional wastewater treatment was regarded as the ultimate risk mitigation measure [4]. This approach put a strong emphasis on the use of water quality standards in wastewater irrigation systems and strict regulations as done in most high-income countries. However, the most recent World Health Organization (WHO) guidelines for wastewater irrigation recommend a shift from water quality standards to health-based targets, which can be achieved along a chain of multiple risk reduction measures [45]. Barriers are placed at critical control points along the food chain (from production to consumption), aiming at maximum risk reduction. For example, barriers can be placed at wastewater generation points, on farms, at markets, and even at the consumer level. A generic example of these barriers is shown in Figure 41.1. The guidelines set a health-based target of a tolerable additional disease burden of ≤10⁻⁶ DALYs* per person per year, translated to a performance target of 6–7 log units pathogen reduction at the point of exposure, however and wherever it can be achieved, between wastewater treatment and food intake. This new approach of health-based targets offers authorities more options and flexibility for reducing risks especially where conventional water treatment is not possible. However, as the multiple barrier approach is more complex than setting up water quality thresholds, thresholds remain globally the predominant thrust of wastewater reuse guidelines.

41.4.2 Examples of Risk Management Strategies

41.4.2.1 Wastewater Treatment

If all wastewater generated is adequately treated before it reaches farms, then the quality of irrigation water for direct and indirect agricultural reuse meets standards and health risks will only be limited to postharvest practices. Reviews from several studies show many options to improve irrigation water quality [31,45]. Table 41.2 shows various wastewater treatment practices and the ranges of pathogen removal.

41.4.2.2 Farm-Based Water Treatment

In cases where adequate wastewater treatment is not done at the municipal level and farmers still need to use this water for irrigation, as it is commonly observed in low-income contexts,

---

* The disability-adjusted life year (DALY) is a measure of overall disease burden, expressed as the number of years lost due to ill health, disability, or early death.

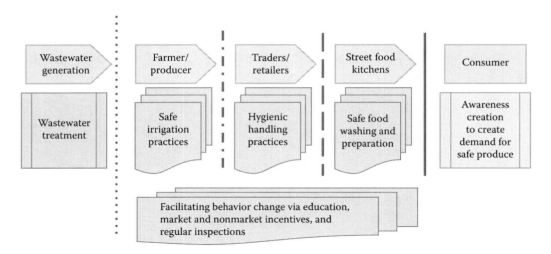

FIGURE 41.1   The multibarrier approach of good practices for consumption-related risks along the food chain as applied in wastewater irrigation.

some treatment options exist for improving water quality at the farm level.

Pond systems: They are widely used in irrigation systems for water storage and as water sources. Due to their larger sizes, helminth eggs and protozoa in ponds are mostly removed by sedimentation while bacteria and viruses are removed mostly from die-off in the ponds [29]. Sedimentation may also contribute to coliform bacteria and virus removal [22]. In water storage and treatment reservoirs, which are widely used in irrigation systems in Brazil, water quality can be improved by using three ponds in parallel and using batch-fed processes to ensure that all of the wastewater is retained for an adequate time, to reduce pathogen concentrations to the appropriate level prior to irrigation [30]. In West Africa, shallow dugout ponds (usually less than 1 m deep and 2 m wide) are widely used in irrigated urban vegetable farming sites. During the storage of water and gradual use in irrigation, sedimentation takes place. The extent of pathogen removal depends on the length of the undisturbed retention time. Studies conducted in Ghana showed that these ponds are very effective in removing helminths (reduced to less than 1 egg/L) when sedimentation is allowed for 2–3 days [25]. Reductions can be achieved with better pond designs (deeper, wedge-shaped beds) and training farmers on how to collect water [25]. In addition, measures that can enhance sedimentation, for example, using natural flocculants such as *Moringa oleifera* seed extracts in the ponds, seem to be promising in Ghana [38]. Furthermore, use of additional measures that influence pathogen die-off such as sunlight intensity, temperature, and crop type can help in lessening the pathogen load in irrigation water [27].

Filtration systems: There is a wide range of filtration systems but use of slow sand filters is probably the most appropriate to treat irrigation water at the farm level. Sand filters remove pathogenic microorganisms from polluted water, first by retaining them in the filtration media and then by promoting their elimination [40]. Retention is achieved mainly through straining whereby microorganisms are physically blocked as they move

through the filter media and through adsorption when they are attached to the filtration media. Generally, in filters, large-sized pathogenic microorganisms like parasites are retained mainly by straining while smaller-sized agents like bacteria and viruses are retained by adsorption. Elimination of pathogenic microorganisms is achieved mainly by their exposure to unfavorable environmental conditions such as high temperature, and through predation. Similarly, soils also can act as bio-filters. However, care should be taken, especially with the current thinking of using biochar as biochar in soil will only retain the pathogens and not remove or eliminate them, which may lead to concentration of pathogens in soils. The typical pathogen removal range reported by WHO based on a review of several studies for slow sand filters is 0–3 log units and 1–3 log units for bacteria and helminths, respectively [45]. Studies done in Ghana using column slow sand filters achieved more than 98% of bacteria removal, equivalent to an average of 2 $\log_{10}$ reduction units/100 mL, while 71%–96% of helminths were removed [26]. This removal was significant but not sufficient as irrigation water had very high initial levels of indicator organisms.

41.4.2.3 Irrigation Scheduling

Irrigation water scheduling, such as controlling the timing and frequency of irrigation, can minimize soil and crop contamination. One of the most widely documented water management measures used to reduce pathogen contamination is cessation of irrigation, where irrigation is stopped a few days before crops are harvested to allow for pathogen die-off due to exposure to unfavorable environmental factors such as sunlight [39]. As much as 99% elimination of detectable viruses has been reported after 2 days of exposure to sunlight, supporting regulations that a suitable time interval should be maintained between irrigation and crop handling or grazing time [15]. In another study, it was revealed that when trickling filter effluent with 6 log units of thermotolerant coliform per 100 mL was used to spray-irrigate lettuce, initial concentrations of indicator bacteria exceeded

TABLE 41.2 Log Unit Reduction or Inactivation of Excreted Pathogens Achieved by Selected Wastewater Treatment Processes

	Log Unit Pathogen Removal			
Treatment Process	Viruses	Bacteria	Protozoan (oo) Cysts	Helminth Eggs
Low-Rate Biological Processes				
Waste stabilization pounds	1–4	1–6	1–4	1–3
Wastewater storage and treatment reservoirs	1–4	1–6	1–4	1–3
Constructed wetlands	1–2	0.5–3	0.5–2	1–3
High-Rate Processes				
Primary treatment				
Primary sedimentation	0–1	0–1	0–1	0– <1
Chemically enhanced primary treatment	1–2	1–3	1–2	1–3
Anaerobic upflow sludge blanket reactors	0–1	0.5–1.5	0–1	0.5–1
Secondary treatment				
Activated sludge + secondary sedimentation	0–2	1–2	0–1	1– <2
Trickling filters + secondary sedimentation	0–2	1–2	0–1	1–2
Aerated lagoon + settling pond	1–2	1–2	0–1	1–3
Tertiary treatment				
Coagulation/flocculation	1–3	0–1	1–3	2
High-rate granular or slow-rate sand filtration	1–3	0–3	0–3	1–3
Dual-media filtration	1–3	0–1	1–3	2–3
Membranes	2.5– >6	3.5– >6	>6	>3
Disinfection				
Chlorination	1–3	2–6	0–1.5	0– <1
Ozonation	3–6	2–6	1–2	0–2
Ultraviolet radiation	1– >3	2– >4	>3	0

Source: Adapted from WHO. 2006. *Guidelines for the Safe Use of Wastewater, Excreta and Grey water: Wastewater Use in Agriculture* (Volume 2). Report of a WHO Scientific Group. WHO Technical Report Series 778. World Health Organization, Geneva, Switzerland.

5 log units of thermotolerant coliform per 100 g fresh weight. Once irrigation ceased, no salmonella could be detected after 5 days, and, after 7–12 days, the levels of thermotolerant coliform were comparable to those detected on lettuce irrigated with freshwater [44]. In Ghana, studies from field trials show an average daily reduction of 0.65 log units of thermotolerant coliforms on lettuce [27]. However, the studies from Ghana show that cessation of irrigation is accompanied by high-yield losses (1.4 tons/ha of fresh weight), which may dissuade farmers from adopting this practice [27].

41.4.2.4 Irrigation Methods

Under normal conditions, the type of irrigation method selected will depend on water supply, climate, soil, crops to be grown, cost of irrigation method, and the ability of the farmer to manage the system. However, when using wastewater as the source of irrigation, other factors, such as pathogen contamination of plants, harvested product, farm workers, and the environment need to be considered. Based on health impacts from wastewater, the WHO has classified irrigation methods in three distinct categories: flood and furrow, spray and sprinkler, and localized irrigation methods [45]. Flood and furrow irrigation methods apply water on the surface and pose the highest risks to field workers, especially when protective clothing is not used [7]. Spray and sprinkler are overhead irrigation methods and have the highest potential to transfer pathogens to crop surfaces, as water is applied on edible parts of most crops. They also promote wide movement of pathogens through aerosols. Localized techniques such as drip and trickle irrigation offer farm workers the best possible health protection and also ensure minimal pathogen transfer to crop surfaces because water is directly applied to the root [33].

However, localized techniques are comparatively the most expensive and are prone to clogging as polluted water has high-particulate levels. They can reduce contamination on crops by 2–4 log units [45]. Nevertheless, recently introduced low-cost drip irrigation techniques, like bucket drip kits from Chapin Watermatics, USA and International Development Enterprises (IDE-India), have more potential for use and adoption in low-income countries [23]. Studies done in Ghana using bucket drip kits show massive reduction in contamination (up to 6 log units), especially during the dry season [28]. These studies from Ghana also demonstrated that the traditional watering can system could be modified to diminish splashing of contaminated soils to the crops, which in turn would reduce crop contamination.

41.4.2.5 Crop Selection

Some crops are more prone than others are to contamination with pathogens. Thus, proper crop selection will contribute to a decrease in human health risks. For example, crops with their edible parts more exposed to contaminated soils and irrigation water, such as low-growing crops and root tubers, will be more prone to pathogen contamination. WHO, in its guidelines for safe use of wastewater in agriculture, advises for crop restrictions, especially targeting vegetables and other crops eaten raw [45]. In Mexico, farmers could shift from leafy vegetables to vegetables such as onions and garlic, which provide a less favorable environment for bacteria survival. However, a shift in crop choice is only feasible if the market value of the alternative crop is similar to that of the original crop. Crop restrictions can be hard to implement if necessary conditions such as law enforcement, market pressure, and demand of cleaner vegetables are not in place. So, while there have been successful crop restriction schemes in India, Mexico, Peru, and Chile [7,9], this has not been possible in sub-Saharan Africa and other countries where wastewater irrigation is informal.

41.4.2.6 Postharvest Measures

It is generally recognized that the above farm-based interventions can only reduce but not eliminate crop contamination at the farm level. Furthermore, several studies show that further contamination (mainly microbial) can occur during postharvest handling at markets and consumption points [3,12]. Therefore, postharvest interventions are equally important to ensure more comprehensive food safety. Improving stakeholders' knowledge and awareness of personal hygiene especially in relation to food safety has been advocated. It is conventional that vegetables be washed, but this only removes much of the adhering soil and some dirt. Studies show that the effect of washing in reducing the number of bacteria present is small with reductions of 0.1–1 log unit [6]. Other studies have reported that approximately 1 and 2 log reductions (depending on the nature of the surface) can be achieved when leaves are washed vigorously in tap water in the markets. A review done on the effectiveness of postharvest risk reduction options is presented in Table 41.3 [45].

In Ghana, perception studies show that food vendors are usually confident that their cleaning and treatment process is sufficient to eliminate contaminants. However, assessments done in West African cities show that only a few of the methods used in cleaning vegetables achieved some reductions [3]. For many of them, some adjustments were needed to achieve 2–3 log units' pathogen reductions as reported. Salt is considered the cheapest disinfectant and the most likely to be adopted, but it is known to cause deteriorating effects on lettuce at high concentrations of 23 and 35 ppm. A weak 7-ppm NaCl solution is recommended to achieve some pathogen reduction while preserving freshness of lettuce [3]. At an acetic acid concentration of 12,500 ppm (approximately one-part vinegar in five-parts water), a significant reduction in pathogen levels can be achieved with a contact time of at least 5 min. However, vinegar is considered expensive by most street food vendors and mainly used by restaurants and middle-/upper-class households. Locally promoted chlorine tablets are cheaper than vinegar and very efficient, allowing for 2–3 log unit reductions [3].

Various studies have shown that none of the available washing and sanitizing methods, including some of the newest sanitizing agents such as chlorine dioxide and ozone, can guarantee the microbiological quality of minimally processed vegetables without compromising their organoleptic quality [6]. This is probably due to insufficient knowledge of the proper application conditions of the sanitizing agent. Indeed, whereas the efficacy of a particular disinfection method is dependent on the type and physiology of the target microorganisms, characteristics of produce surfaces, exposure time, concentration of cleaners and sanitizers, pH, and temperature conditions also influence the effectiveness of washing. However, few studies have been conducted to optimize these parameters; hence, the varied and often-ineffective methods washing of vegetables.

41.5 Summary and Conclusions

Agricultural wastewater use poses human health risks, especially when adequate wastewater treatment is not done. Health risk levels vary widely depending on local sources of wastewater and farming practices in irrigation systems. Risk tends to be higher when untreated wastewater is used for irrigation as is common in many low-income contexts where peri-urban farmers have little or no choice apart from using surface water sources that are polluted with untreated wastewater. In such contexts, diarrheal diseases and helminth infections (especially hookworm and ascaris) have been found to pose the greatest human health risks. These diseases are transmitted mainly by consuming wastewater-irrigated produce or by coming into contact with contaminated soils and irrigation water. Although a number of risk management measures have been suggested, actual field testing and implementation is very limited.

Developing risk management measures in line with the multiple-barrier approach as supported by the latest WHO guidelines offers a variety of options to achieve a realistic chance for risk management from wastewater irrigation systems in low-income countries. This provides local health risk managers with flexibility to address wastewater irrigation risks with locally viable means, instead of not doing anything at all due to unattainable water quality threshold levels.

This chapter has highlighted some of the risk management strategies with the greatest potential in low-income contexts. The studies done thus far have shown that on-farm and postharvest risk mitigation measures provide more direct solutions to preventing contamination and decontamination of vegetables grown in wastewater irrigation systems. Although the effectiveness of individual measures may not be sufficient, they can be used in combination to complement each other in

TABLE 41.3 Postharvest Measures for Pathogen Reduction

Control Measure	Extent of Pathogen Reduction (Log Units)	Comments
Produce washing	1	Washing salad crops, vegetables, and fruit with clean water
Produce disinfectant	2–3	Washing salad crops, vegetables, and fruit with a disinfectant solution and rinsing with clean water
Produce peeling	2	Fruits, root crops
Produce cooking	6–7	Immersion in boiling or close-to-boiling water until the food is cooked for some period of time ensures pathogen destruction

Source: Adapted and modified from WHO. 2006. *Guidelines for the Safe Use of Wastewater, Excreta and Grey water: Wastewater Use in Agriculture* (Volume 2). Report of a WHO Scientific Group. WHO Technical Report Series 778. World Health Organization, Geneva, Switzerland.

order to achieve the acceptable risk levels. Combination can be done within and between operation levels, that is, farms, markets, and households. The measures presented, though not exhaustive, allow for flexibility to adaptation in different locations.

We recommend continued research focus on developing effective, appropriate, and easily adoptable risk management strategies for millions of farmers using wastewater especially in low-income countries. Development should involve actual field-testing and involve end users. Equally important is motivating, incentivizing, and encouraging innovations from farmers and institutionalization of appropriate guidelines by local authorities to manage the health risks.

References

1. Abaidoo, R.C., Keraita, B., Drechsel, P., Dissanayake, P., and Akple, M.S. 2009. Soil and crop contamination through wastewater irrigation and options for risk reduction in developing countries. In: Dion, P. (ed.), *Soil Biology and Agriculture in the Tropics*. Springer Verlag, Heidelberg, Germany, pp. 498–535.

2. Abdul-Ghaniyu, S., Kranjac-Berisavljevic, G., Yakubu, I.B., and Keraita, B. 2008. Sources and quality of water for urban vegetable production, Tamale, Ghana. *Urban Agriculture Magazine*, 8, 10.

3. Amoah, P., Drechsel, P., Abaidoo, R.C., and Klutse, A. 2007. Effectiveness of common and improved sanitary methods in West Africa for the reduction of coli bacteria on vegetables. *Tropical Medicine and International Health*, 12, 39–49.

4. Asano, T. and Levine, A. 1998. Wastewater reclamation, recycling, and reuse: An introduction. In: Asano, T. (ed.), *Wastewater Reclamation and Reuse*, Vol. 10, CRC Press, Boca Raton, FL, pp. 1–56.

5. Bahri, A. 2008. Middle East and North African countries. In: Jimenez, B. and Asano, T. (eds.), *Water Reuse: An International Survey of Current Practice, Issues and Needs*. IWA Publishing, London.

6. Beuchat, L.R., Nail, B.V., Adler, B.B., and Clavero, M.R.S. 1998. Efficacy of spray application of chlorine in killing pathogenic bacteria on raw apples, tomatoes, and lettuce. *Journal of Food Protection*, 61, 1305–1311.

7. Blumenthal, U.J., Peasey, A., Ruiz-Palacios, G., and Mara, D.D. 2000. Guidelines for wastewater reuse in agriculture and aquaculture: Recommended revisions based on new research evidence. *WELL Study*, Task No: 68 Part 1.

8. Bradford, A., Brook, R., and Hunshal, C.S. 2002. Crop selection and wastewater irrigation, Hubli–Dharwad, India. *Urban Agriculture Magazine*, 8, 31–32.

9. Buechler, S. and Devi, G. 2003. Household food security and wastewater dependent livelihood activities in Andhra Pradesh, India. Unpublished background document prepared for the WHO in preparation of Guidelines for Safe Use of Wastewater in Agriculture.

10. Cofie, O.O., Kranjac-Berisavljevic, G., and Drechsel, P. 2005. The use of human waste for peri-agriculture in northern Ghana. *Renewable Agriculture and Food Systems*, 20(2), 73–80.

11. Drechsel, P., Graefe, S., Sonou, M., and Cofie, O.O. 2006. *Informal Irrigation in Urban West Africa: An Overview*. IWMI Research Report 102. IWMI, Colombo.

12. Ensink, J.H.J., Mahamood, T., and Dalsagaard, A. 2007. Wastewater irrigated vegetables: Market handling versus irrigation water quality. *Tropical Medicine and International Health*, 12, 2–7.

13. Ensink, J.H.J. 2006. Water quality and the risk of hookworm infection in Pakistani and Indian sewage farmers. PhD thesis, London School of Hygiene and Tropical Medicine, University of London, UK.

14. Ensink, J.H.J., Van der Hoek, W., Matsuno, Y., Munir, S., and Aslam, M.R. 2002. *Use of Untreated Wastewater in Peri-Urban Agriculture in Pakistan: Risks and Opportunities*. Research Report 64. International Water Management Institute, Colombo.

15. Feigin, A., Ravina, I., and Shalhevet, J. 1991. *Irrigation with Treated Sewage Effluent: Management for Environmental Protection*. Springer, Berlin, Heidelberg, New York.

16. Global Water Intelligence (GWI). 2010. *Municipal Water Reuse Markets 2010*. Media Analytics Ltd., Oxford, UK.

17. Hamilton, A.J., Stagnitti, F., Xiong, X., Kreidl, S.L., Benke, K.K., and Maher, P. 2007. Wastewater irrigation the state of play. *Vedose Zone Journal*, 6(4), 823–840.

18. Harris, L.J., Farber, J.M., Beuchat, L.R., Parish, M.E., Suslow, T.V., Garrett, E.H., and Busta, F.F. 2003. Outbreaks associated with fresh produce: Incidence, growth, and survival of pathogens in fresh and fresh-cut produce. *Comprehensive Reviews in Food Science and Food Safety*, 2, 78–141.

19. Hide, J., Hide, C., and Kimani, J. 2001. Informal irrigation in the peri-urban zone of Nairobi, Kenya: An assessment of surface water quality for irrigation. Report OD/TN. HR Wallingford Ltd, Wallingford, UK.

20. Huibers, F.P., Moscoso, O., Duran, A., and van Lier, J.B. 2004. The use of wastewater in Cochobamba, Bolivia: A degrading environment. In: Scott, C.A., Faruqui, N.I., and Raschid-Sally, L. (eds.), *Wastewater Use in Irrigated Agriculture*. CABI Publishing, Wallingford, UK.

21. Jimenez, B. 2008. Water reuse overview for Latin America and Caribbean. In: Jimenez, B. and Asano, T. (eds.), *Water Reuse: An International Survey of Current Practice, Issues and Needs*. IWA Publishing, London.

22. Karim, M.R., Manshadi, F.D., Karpiscak, M.M., and Gerba, C.P. 2004. The persistence and removal of enteric pathogens in constructed wetlands. *Water Research*, 38, 1831–1837.

23. Kay, M. 2001. *Smallholder Irrigation Technology: Prospects for Sub Sahara Africa*. IPRTRID, FAO, Rome.

24. Keraita, B., Blanca, J., and Drechsel, P. 2008a. Extent and implications of agricultural reuse of untreated, partly treated, and diluted wastewater in developing countries. *CAB Reviews: Perspectives in Agriculture, Veterinary Science, Nutrition and Natural Resources*, 3(058) 1–15.

25. Keraita, B., Drechsel, P., and Konradsen, F. 2008b. Using on-farm sedimentation ponds to reduce health risks in wastewater irrigated urban vegetable farming in Ghana. *Water Science and Technology,* 57(4), 519–525.

26. Keraita, B., Drechsel, P., and Konradsen, F. 2008c. Potential of simple filters to improve microbial quality of irrigation water used in urban vegetable farming in Ghana. *Journal of Environmental Science and Health, Part A,* 43, 1–7.

27. Keraita, B., Konradsen, F., Drechsel, P., and Abaidoo, R.C. 2007a. Reducing microbial contamination on lettuce by cessation of irrigation before harvesting. *Tropical Medicine and International Health,* 12, 8–14.

28. Keraita, B., Konradsen, F., Drechsel, P., and Abaidoo, R.C. 2007b. Effect of low-cost irrigation methods on microbial contamination of lettuce. *Tropical Medicine and International Health,* 12, 15–22.

29. Mara, D.D. 2004. *Domestic Wastewater Treatment in Developing Countries.* Earthscan Publications, London.

30. Mara, D.D. and Pearson, H.W. 1992. Sequential batch-fed effluent storage reservoirs: A new concept of wastewater treatment prior to unrestricted crop irrigation. *Water Science Technology,* 26(7 – 8), 1459–1464.

31. Norton-Brandão, D., Scherrenberg, S.M., and van Lier, J.B. 2013. Reclamation of used urban waters for irrigation purposes: A review of treatment technologies. *Journal of Environmental Management,* 122, 85–98.

32. Peasey, A. 2000. Human exposure to Ascaris infection through wastewater reuse in irrigation and its public health significance. PhD thesis, University of London, England.

33. Pescod, M. 1992. Wastewater treatment and use in agriculture. Irrigation and drainage paper 47. FAO, Rome.

34. Prüss-Ustün, A. and Corvalan, C. 2006. Preventing disease through healthy environments, towards an estimate of the environmental burden of disease. WHO, Geneva.

35. Raschid-Sally, L. and Jayakody, P. 2008. Drivers and characteristics of wastewater agriculture in developing countries: Results from a global assessment. IWMI Research Report 127, International Water Management Institute, Colombo, Sri Lanka.

36. Rutkowski, T., Raschid-Sally, L., and Buechler, S. 2007. Wastewater irrigation in the developing world—Two case studies from Katmandu Valley in Nepal. *Agricultural Water Management,* 88, 83–91.

37. Scott, C., Drechsel, P., Raschid-Sally, L., Bahri, A., Mara, D., Redwood, M., and Jiménez, B. 2010. Wastewater irrigation and health: Challenges and outlook for mitigating risks in low-income countries. In: Drechsel, P., Scott, C.A., Raschid-Sally, L., Redwood, M., and Bahri, A. (eds.), *Wastewater Irrigation and Health: Assessing and Mitigation Risks in Low-Income Countries.* Earthscan-IDRC-IWMI, UK, pp. 381–394.

38. Sengupta, M.E., Keraita, B., Olsen, A., Boateng, O.K., Thamsborg, S.M., Palsdottir, G.R., and Dalsgaard, A. 2012. Use of *Moringa oleifera* seed extracts to reduce helminth egg numbers and turbidity in irrigation water. *Water Research,* 46(11), 3646–3656.

39. Shuval, H.I., Adin, A., Fattal, B., Rawitz, E., and Yekutiel, P. 1986. *Wastewater Irrigation in Developing Countries: Health Effects and Technical Solutions.* World Bank Technical Paper No. 51, Washington, DC.

40. Stevik, T.K., Aa, K., Ausland, G., and Hanssen, J.F. 2004. Retention and removal of pathogenic bacteria in wastewater percolating through porous media. *Water Research,* 38, 1355–1367.

41. Tchobanoglous, G., Burton, F.L., and Stensel, H.D. 1995. Wastewater treatment systems. In: *Wastewater Engineering: Treatment, Disposal and Reuse.* 3rd Edition, Metcalf and Eddy Inc., Tata McGraw-Hill, New Delhi, India, pp. 1017–1102.

42. Trang, D.T., van der Hoek, W., Tuan, N.D., Cam, P.C., Viet, V.H., Luu, D.D., Konradsen, F., and Dalsgaard, A. 2007. Skin disease among farmers using wastewater in rice cultivation in Nam Dinh, Vietnam. *Tropical Medicine and International Health,* 12, 51–58.

43. Van der Hoek, W., Tuan Anh, V., Dac Cam, P., Vicheth, C., and Dalsgaard, A. 2005. Skin diseases among people using urban wastewater in Phnom Penh. *Urban Agriculture Magazine,* 14, 30–31.

44. Vaz da Costa-Vargas, S., Bastos, R.K.X., and Mara, D.D. 1996. Bacteriological aspects of wastewater irrigation. In: *TPHE Research Monogram.* Department of Civil Engineering, University of Leeds, Leeds, UK.

45. WHO. 2006. *Guidelines for the Safe Use of Wastewater, Excreta and Grey Water: Wastewater Use in Agriculture* (Volume 2). Report of a WHO Scientific Group. WHO Technical Report Series 778. World Health Organization, Geneva, Switzerland.

46. WHO. 2000. *Global Water Supply and Sanitation Assessment 2000 Report.* World Health Organization (WHO), United Nations Children's Fund (UNICEF), Geneva and New York.

42

Choosing Indicators of Fecal Pollution for Wastewater Reuse Opportunities

Theodore C. Crusberg
Worcester Polytechnic Institute

Saeid Eslamian
Isfahan University of Technology

Preface

Under certain critical conditions, communities may have to commit to using domestic wastewater in order to supplement a potable water source to meet domestic needs. The safety of this water will be called into question by residents and local public officials, especially the water operations manager and public health director. Certifying or guarantying the safety of this alternate domestic water supply will require extensive testing of the treated water and vigilance within the community to ensure public confidence in their water. This chapter deals with the means used by local officials to ensure safe water when a community must employ wastewater reclamation/reuse to solve demand for a sufficient supply.

42.1 Introduction

Wastewater redirected through a treatment regimen making it safe and suitable for human consumption has taken the forefront in identifying novel methods of augmenting presently limited water supplies. This process known as wastewater reuse has many drawbacks, especially the general sense of the population that this is a "last resort" and only acceptable when all other methods of obtaining potable water have been exhausted. The public health officer in a community dealing with decisions regarding wastewater reclamation must be comfortable in assessing the science available to him or her to ensure that the public safety is given the highest level of consideration. The choice of what to measure to ensure this level of safety must be weighed in the sense of the capability of the laboratory and laboratory person-nel to ensure safety of those served by the water utility. The ben-efits of using certain indicators of human fecal pollution have been realized for decades but they may not in aggregate be suit-able for wastewater reuse.

42.1.1 Personal Observations

Imagine the surprise when this author learned that only about 100 m to the east ran a perennial river. After all, this was south-ern Arizona, one of the driest places in the United States. Where could water of this quantity actually come from? This was, in fact, the Santa Cruz River, surely ephemeral at its origins but now it ran impressively northward out of Sonora, Mexico, around 30 km to the south, to eventually disappear into the earth another 20 km to the north of Tumacacori, Arizona, where the author was spending a few weeks in the winter of 2014. A sign posted by the National Parks Service, which administers a trail system adjacent to the river, warned that the river water was not fit for human consumption. Indeed, this was true as the water is known to exhibit high bacteria counts and is actually derived in total from the outfall of an international wastewater treat-ment plant located just 15–20 km from the Mexican–U.S. bor-der, and serving the cities of Nogales, Sonora, Mexico (>300,000 residents), and Nogales, Arizona, U.S. (30,000 residents), releas-ing 55,000 m^3 daily. These cities derive their drinking water from subsurface wells, many in the Santa Cruz River aquifer. The water from the wastewater treatment plant meets discharge requirements most of the time, yet the fecal coliform levels at the Santa Gertudis Lane sampling site at the temporary residence of this author often exceeded the 200 colony-forming units (cfus) attaining a single sample maximum of 27,100 although exhibit-ing a geometric mean of 149 [39]. By the time the water reached Tumacacori, it had traveled many kilometers, yet at times the levels of fecal bacteria was augmented by runoff from agricul-tural lands, principally from cattle ranches, and perhaps other sources of human and animal excrement. The hydrology of the area (1400 km^2) was also described [43] and the subsurface is composed of relatively younger alluvium with transmissiv-ity values ranging between 400 and 288 m^3/day [43]. The Santa Cruz River rarely, if ever, reaches the Colorado River except per-haps during very infrequent flood periods. Yet in its 40-km trip northward from the wastewater treatment plant, it recharges the groundwater along its route and provides for an active riparian corridor for a variety of wildlife. Assuming that so much river water actually infiltrates the groundwater used eventually for domestic drinking water, one has to consider this recharge as an excellent example of what is now termed "indirect potable reuse" or IPR.

42.2 Indirect and Direct Potable Reuse

IPR refers to the process by which treated wastewater is allowed to enter an aquifer, injection wells, constructed wetlands, or other physical barrier prior to consumption [3,7,21,31,45] while

direct potable reuse (DPR) refers to a process of directly cou-pling the effluent of a wastewater treatment plant to a water treatment facility that will provide drinking water to a commu-nity. Unfortunately, there seems to be no well-documented effort to test the overall chemical or microbiological quality of water derived from private wells in the towns that benefit from the Santa Cruz River, namely Rio Rico, Tumacacori, and Tubac (all in Arizona). It is difficult to assess the real public health safety of the groundwater people in the area, including the author, are consuming. However, there appears to be no health issues reported in this area, so one may simply consider this an excel-lent example of how IPR should operate. Some concern has to be expressed, however, because a report from another area of Sonora, Mexico, does point to concerns over groundwater con-sumption as having a negative impact on the health of residents in the area [19].

42.3 Public Health Considerations

Public health officials express a philosophy of "better safe than sorry" and as a result tend to err on the side of safety over con-venience or profit. If there is going to be a general acceptance by the public of wastewater reuse, then there is going to have to be a strong scientific argument that the public health of the popula-tion being served will be protected "no matter what." For this reason, scientists must have at their disposal a way to ensure pub-lic health-related safety of the water supply provided. For most people living in the developed world, safe drinking water was sourced from well-protected upland supplies, from the ground, or from rivers and lakes that were somewhat compromised from a public health perspective, yet treated to a degree that satisfied the consumers. Even in the Third World, large segments of the population are served via a physical water distribution system as in Ghana [33]. Although even in developed countries there were assumptions that groundwater was usually safe, the formal documentation of waterborne disease outbreaks resulted in gov-ernmental regulations that ultimately identified many chemicals [28] both inorganic and organic, as well as a variety of biological agents that caused harm to the consumer [15,14]. Of the 733 out-breaks reported between 1971 and 2006, 110 were determined to be the result of groundwater contamination by municipal waste-water, caused by a group of seven enteric bacteria, three enteric viruses, and two protozoans. In addition, 10% of all outbreaks had a chemical, not a biological, origin.

42.4 Moves for Adoption of Wastewater Reuse

Wastewater reuse or what is also known as wastewater recla-mation is now a topic of great concern. Australia is taking one of the leads in this area [4]. In this report, the study commit-tee discussed and described IRP and DPR, providing a listing of the benefits of each process and the rationale for moving in this direction. The U.S. National Academy of Sciences also reflects on this issue in a positive note [32] but also suggests 14 goals needed to be explored before any significant action should be taken, including No. 13, "Identify better indicators and surrogates that can be used to monitor process performance and develop online real time or near real time analytical monitoring techniques for their measurement." The purpose of this chapter is to address the historical biological indicators that have been used to assess the public health safety of drinking water and to evaluate if they have served the public. In addition, using many of those same indicators but rather developing a suite consisting of both biological and chemical species may have a positive effect on attitudes of consumers who may be able to understand, appreciate, and even accept either IPR or DPR water entering their distribution system, their homes, and their lives. After all, it is likely that it is occurring already and in many cases is unknown to the recipients, as appears to be the case along the Santa Cruz River in southern Arizona.

42.5 Indicators of Fecal Pollution

Reclaimed wastewater, whether used for agricultural purposes, to prevent saltwater intrusion on coastlines, or for human consumption, must have some level of public health safety. Certainly, as a component of the drinking water supply the greatest level of safety must be recognized and upheld. Since wastewaters accumulate a wide and often unpredictable range of hazardous chemicals and biological materials often from many undefined sources until they receive treatment, it is mandatory that those materials of concern are removed or at least reduced to what is termed an "acceptable level" before being discharged back into the environment. Then again, should the treated wastewater need to be reclaimed, further treatment will be needed to ensure the safety of that water. Great concern is given to the biological entities that cause disease in effluents from wastewater treatment plants, those components called "pathogens." However, pathogens are often fleeting in their presence so testing for them is not a reliable means of determining drinking water safety. As far back as 1891, Franklands proposed the idea that organisms that are present in sewage could be employed to characterize water safety (reviewed in Ref. [33]). Rather than develop ways to identify specific pathogens, the use of indicators of pollution have been used and in fact have stood the test of time [6]. As has been already stated, the public health professional tries to balance safety with convenience yet chooses to err on the aide of the former. In 1885, Escherich [20] described the common colon bacillus, renaming it *Escherichia coli*, and bacteria resembling this species were identified as "coliform" and became an important indicator for the presence of human fecal pollution (reviewed in Ref. [33]). Today, many bacteria have been suggested as indicators and a few have been chosen. Although public health and safety are first and foremost of greatest concern, identifying fecal or even environmental pollution in both the environment and as it relates to drinking water means providing a level of vigilance over short time intervals and at a reasonable cost in terms of time and money. Therefore, a number of tests have been developed that meet these criteria and provide usually

ample information to allow a public health official to respond to an event and make corrections as needed, ensuring public safety. Sometimes, however, the system does break down resulting in public health emergencies.

42.6 Pathogens of Concern in Water

Pathogens of concern in water include a number of bacteria, viruses, protozoans, and some macroinvertebrates [8,23,37,46]. Although *E. coli* is usually not considered a pathogen, certain strains, notably *E. coli* O157:H7 is indeed a pathogen, but is usually found in prepared meats and not necessarily in water. Bacterial pathogens include *Salmonella* species, *Vibrio cholorae*, *Campylobacter jejuni* [25] to mention a few, while viral pathogens include the hepatitis A virus [17], noroviruses, and others, and protozoans of concern are *Giardia lamblia* and *Cryptosporidium parvum* [30]. Rather than rely on the presence of any single pathogen, public health officials rely on organisms that are considered always present indicators of fecal contamination. In essence, the role of any indicator is an "index" of the presence of fecal matter and thus the possible presence of a pathogen. An indicator of fecal pollution should have the attributes listed in Table 42.1 [37].

Studies have been reported originally with the intention of correlating the populations of indicator species with the populations of pathogens present in water usually to little or no success [37]. Given the criteria that would be assigned to an ideal indicator, it must be stated now that in nature, nothing is perfect and our choices of indicators will all have some flaws. It is important therefore to rely perhaps not on a single indicator but on several and perhaps it is best overall to choose a "suite" of technologies to assess water safety especially as reclaimed wastewater is chosen to supplement the traditional supplies of drinking water in the future. For this reason, a number of different indicators of fecal contaminations have been used to evaluate water safety. Table 42.2 lists these choices and their main purpose for the public health professional. Tests for these indicator organisms based upon the biochemistry inherent in the species or group have been developed and are commonplace in most public health laboratories.

As one explores the processes needed to provide safe drinking water from reclaimed wastewater, it is necessary to evaluate these

TABLE 42.1 Criteria for Selecting an Indicator of Fecal Contamination

1. The indicator should always be present in polluted water and found in significantly lower concentrations in unpolluted raw or untreated water.

2. The indicator should survive in the environment for periods of time equal to or greater than the longevity of pathogens also present. Again, this criteria has its faults.

3. The indicator should be present in numbers greater than any pathogen that may also be present.

4. The indicator should be responsive to treatment conditions in a manner like the pathogens that may also be present. As shall be seen, this condition is difficult to meet in all cases.

5. Simple and relatively inexpensive technologies should be available to identify the indicator and quantify or enumerate its presence.

6. The indicator should not be a pathogen in order to protect the health of those assaying for its presence.

TABLE 42.2 Current Indicators Used to Assess Water Contamination by Fecal Matter

1. *E. coli* is a species of fecal coliform bacteria that is specific to fecal material from humans and other warm-blooded animals. Several strains have been identified which are very pathogenic. The simple and inexpensive membrane filter technique and a USEPA approved and recommended protocol is used to evaluate the number of viable cells per 100 mL of sample.

2. Total coliform bacteria are widespread in nature and can occur in human feces, but some can be found in animal manure, soil, and on forest floor litter. For drinking water, total coliforms are still the standard test because their presence indicates contamination of a water supply by an outside source. This group is characterized by an inability to grow on specific bacteriological medium at temperatures above 37°C.

3. Fecal or thermotolerant coliforms are fecal-specific bacteria in origin but contain the genus *Klebsiella*, a species that is not necessarily of fecal origin. However, *Klebsiella* species is commonly found in paper mill wastes. Organisms belong to this group grow at higher incubation temperatures and are considered more likely to have come from the intestinal tract of a warm-blooded animal.

4. Fecal streptococci are Gram-positive bacteria inhabiting the digestive systems of humans and other warm-blooded animals. The Enterococci are a human-specific subgroup within the fecal streptococcus group distinguished by their ability to survive in saltwater.

5. *Clostridium perfringens* is a rod-shaped Gram-positive spore forming anaerobe found in both human and nonhuman intestinal tracts with toxigenic properties. It has been suggested as a surrogate for the infectious protozoan parasites as it is more resistant to water treatment, but it is indeed a pathogen and as such will serve in a limited capacity as an indicator species. Analytical tests for this organism require special laboratory conditions, which increases the cost of analysis.

6. Coliphages are viruses that infect *E. coli* specifically and are also strain specific. Rather than serve as an indicator of fecal pollution, these viruses have been occasionally used as a surrogate for the presence of other viral pathogens. Although coliphages should be regarded as true indicators, the methodology for their analysis is also much more difficult to carry out in the laboratory than the bacteriological analyses for the first four examples above.

and perhaps other possible biological and nonbiological entities as indicators of both system performance and the end product—drinking water that will be delivered to the public. Wastewater treatment most of the time involves a train of technologies linked in sequence, the first of which is primary settling to remove sediment and particulates, followed by aerobic or secondary treatment during which carbon-rich components are oxidized and the bacterial/fungal mass of the operation increases. Finally, tertiary treatment, if employed, involves nitrogen or phosphate removal and finally release of effluent into the environment, river, lake, or other body of water, injection into the subsurface occasionally, and finally disinfection usually with gaseous chlorine (Cl_2) to kill pathogens and other bacteria and viruses as well. If the wastewater is then to be directly diverted for reuse into a drinking water treatment operation, further treatment must be carried out. Technologies that are more sophisticated may be in order to satisfy the public health goals for the recycled water than would be the case for treating water from deep wells or from well-protected upland supplies [9]. This could involve membrane filtration technologies, ozone contact, flocculation and precipitation, irradiation with ultraviolet

light, or other technologies with chlorination following the last formal treatment step in the train. Chlorination of the water released into the distribution system is needed to both disinfect the raw or reclaimed water itself and to provide protection against infiltration of contamination caused by what are known as "cross contamination" events from leaky water pipes. Such a coupled wastewater treatment water reclamation process as already noted is termed direct water reuse or DPR. Release of treated wastewater into an aquifer, injected into the subsurface above the vadose zone, or other method involving percolation of the wastewater through an intermediate barrier is termed IPR. At each step in the treatment train, it is necessary to assess the removal of contaminants, thus justifying the actual value or merit of each of the steps. Such a value can be determined by quantifying the reduction of one or more indicator organisms listed in Table 42.2.

One may now ask if the microbiological indicators listed in Table 42.2 are sufficient to enable a public health official to determine the true potability of reclaimed wastewater. For example, treatment may have killed many of the microbial indicators in use, but one may ask, "Is the water now safe to drink?"

42.7 Limits of Bacteriological Pollution Indicators

Except for *E. coli*, the main key indicators represent not a single species of bacteria but a group. The primary test for these organisms is to obtain a 125 mL water sample in a sterile bottle from a faucet that has been first treated with a flame to sterilize the opening and then has been allowed to remain open for a couple of minutes. The water sample is kept at 10°C for no more than 30 h and then 100 mL of the water is transferred to a sterile vacuum filter funnel fitted into a holder with a 47-mm-diameter porous filter with pores of 0.2 μm diameter and the water allowed to pass through trapping any bacteria on the surface of the filter. Using sterile forceps, the filter was transferred to a plastic bacteriologic plate with special medium (m-FC for fecal coliform or m-endo for coliform) and incubated for 24–48 h at the appropriate temperature (37°C for m-endo and 44.5°C m-FC). Coliform organisms listed in Table 42.3 are seen as greenish colonies with a metallic sheen while fecal coliforms are blue in color.

As an example, water samples were obtained from a number of sampling sites throughout the drinking water distribution system in the city of Worcester, MA. In an attempt to identify organisms that appeared on these two types of agar, we tested 210 colonies and found that 81% tested Gram negative. Then 56 of these were selected for further analysis using the API-20E system (bioMerieux, Inc., Hazelwood, MO). A strip of 20 wells, each containing reagents to test for the presence of a specific enzyme, is then inoculated with bacteria from a single colony and after 24 h the colors in each of the wells are observed. Combining the various wells that yield a positive response, a seven-digit number is obtained, from which the probability of the identity of the organism is then proposed using a table provided by the company. Table 42.4 shows the results for those 56 colonies that tested Gram negative [18]. Background colonies (those that did

TABLE 42.3 Coliforms: Gram-Negative Bacteria, Which Produce the Enzyme β-Galactosidase

Genus	Source
Escherichia	Human digestive tract
Klebsiella	Human digestive tract and in the environment
Enterobacter	"
Citrobacter	"
Yersinia	"
Serratia	"
Hafnia	"
Pantoea	Found only in the environment
Kluyvera	"
Cedecea	"
Ewingella	"
Moellerella	"
Leclercia	"
Rahnella	"
Yokenella	"

Source: Adapted from Review of Coliforms as Microbial Indicators of Drinking Water Quality, National Health and Medical Research Council of Australia, 2009. Recommendations to change the use of coliforms as microbial indicators of drinking water quality: http://www.nhmrc.gov.au/_files_nhmrc/publications/attachments/eh32.pdf. Accessed May 25, 2014.

TABLE 42.4 Bacteria Identified in Worcester, MA Drinking Water Using the API20E System

Species	Colonies Identified and Origins of Organisms			
	Background on m-Endo Plates	Background on m-FC Plates	Plate Count Agar	Identified as Coliform on Either m-FC or m-Endo Plates
Enterobacter calcoaceticus	6	1	5	1
Enterobacter aerogenes	5			
Streptococcus liquefaciens	4		1	1
Enterobacter cloacae	4		1	1
Salmonella enteritidis	1			
Klebsiella pneumonia	4	1	3	2
Salmonella pullorum	11			
Escherichia coli	2	5	1	
Serratia marcescens	1			
Citrovorum freundii				2
Enterobacter agglomerans		1		

Source: Adapted from Crusberg, T.C. and Reynolds, J.T. 1974. Relationships between chlorine and microorganisms in a water distribution system. Presented at the *102 American Public Health Association Meeting*, New Orleans, LA, October 29.

Colonies were transferred using a sterile loop into wells on API20E strips, incubated as directed and identified from the seven-digit number derived from the results of inspecting the strips and referring to the serial number corresponding to the value as instructed by the manufacturer.

not test positive) were also tested and those organisms gave a negative response on the specified agar, and several colonies were also obtained from plate count agar, which is nonselective. Of those background colonies, 75% scored as being of possible enteric origin, and surprisingly seven of 35 colonies that did not score as coliforms or fecal coliforms gave a strong *E. coli* response. One can conclude that these two key tests used to identify enteric bacteria in drinking water yield a good many false negatives. The m-endo and m-FC agar both are selective for Gram-negative bacteria and each relies on a certain enzyme present in the Enterobaceriaceae to produce the color change indicative of the desired bacterial type. The take home message is that one must be careful using these two tests in interpreting a negative as the absence of coliforms or fecal coliforms. The USEPA has developed a test in which both coliforms and *E. coli* together are identified on the same bacteriological plate [44].

42.8 Proposed Alternative Pollution Indicators

Although Table 42.2 lists six indicators that are in use to evaluate water for the presence of potentially pathogenic fecal matter, a number of other candidate indicators have been proposed over the years mostly due to the fact that the principal protozoan pathogens, namely *Giardia lamblia* (Giardia) and *Cryptosporidium parvum* (Cryptosporidium) may persist in an environment where *E. coli* and other coliforms have been systematically removed by wastewater and water treatment processes. In fact, in chlorinated wastewater effluents, *E. coli* may be totally inactivated while viable protozoans persist. This is because lethal

concentrations and reaction (contact) times for gaseous chlorine are much longer to realize inactivation of Giardia and especially Cryptosporidium compared with the more fragile *E. coli*. Issues such as the fact that waterborne virus removal by disinfection [10] as well as the known fact that protozoans survive disinfection [1,13,30,40] attest to the need for alternative pollution indicators, some of which are listed in Table 42.5.

42.9 Disinfecting Potable Water

The ability of chlorine disinfectant to kill pathogens has its limits. The parameter used most in assessing the capability of this or any other disinfectant is the "Ct value" or the product of the concentration of the disinfectant in mg/L multiplied by the time the disinfectant and organism are in contact with one another in order to reach a level of 90% inactivation of the particular organism. For Cryptosporidium, this value is 5100 at pH 6.5 and 25°C [42], for Giardia the value is 15–20 at 20°C, for viruses the Ct is about 1.0, and for *E. coli* the Ct is 0.034–0.05 [16]. For chloramines, which are formed when chlorine and ammonia

TABLE 42.5 Alternate but Not Regulator-Approved Fecal Pollution Indicators

Candidate Indicator	How Manifested	Factor(s) Limiting Its Use or Demon-Strating Its Suitability	References
1. Epidemiology	Unusual incidence of enteric disease recognized by public health officials	Requires time to actually be able to identify a significant disease incidence in a population	[14,15,17,19]
2. Chlorine residual in water distribution system	*E. coli* and other fecal-derived coliforms are killed by low levels of chlorine, which is easily measured chemically or using simple instrumentation	Low levels of chlorine inactivate *E. coli* and other coliforms but not Giardia or Cryptosporidium	[2,35,38]
3. Anthropogenic chemicals ubiquitous in sewage that do not break down during sewage treatment	Acesulfame, sucralose, and carbamazepine	Requires special sophisticated instrumentation, usually gas chromatography and mass spectroscopy	[41,47]
4. *Bacteroides fragilis*	A normal anaerobic inhabitant of the human colon	Culturing anaerobes requires special facilities	
5. F-specific coliphages	Viruses that infect *E. coli*	Culturing bacterial viruses is very time consuming	[10,12]
6. Molecular markers	DNA probes bind with specific genomic sequences found only in certain indicator species and sequences of the ribosomal 16S small subunit RNA	Cannot distinguish between viable and nonviable organisms	[11]
7. Fluorescence excitation	The amino acid tryptophan found in proteins fluoresce at 340 nm when excited at 280 nm	Nonspecific but very easy to run analyses of the instrumentation is available	[5]

Note: Some of these may prove worthy for individual consideration and the local public health official must always be aware of his or her own community and the communicable diseases that are present at any moment in time as in no. 1 above. Chlorine residuals are relatively easy to measure and may be telling of certain conditions within the community water distribution system.

gas combine, the ct is raised upward by a factor of 400–500, but for ozone the Ct for Giardia is reduced by 100. Chloramines are often used to provide protection within a water distribution system but chloramine is a poor disinfectant overall. In contrast, the Ct value for Cryptosporidium oocyte inactivation at 20°C with ozone as disinfectant is about 2.0, meaning that to achieve a 3 log reduction of oocytes (99.9%) at this temperature, the contact time would be only 6 min if the ozone concentration was 1 mg/L. At 5°C, the Ct was 10 times greater. [13]. It is often desirable to assure that the pathogens of concern are subject to 2 or even 3 Ct values before the first consumer is served within the water distribution system.

42.10 Permissible Levels of Indicators

The total coliform rule mandated in the United States by the Environmental Protection Agency in 1990 states "Systems must not find coliforms in more than 5% of samples. When a system finds coliforms, the system must collect a set of repeat samples within 24 h. When a repeat sample tests positive for total coliforms, it must also be analyzed for faecal coliforms and *E. coli*. A positive result to this last test signifies an acute maximum contaminant level violation, which necessitates rapid state and public notification." Coliforms are a group of organisms that include *E. coli* and all produce the enzyme β-galactosidase, which hydrolyzes the sugar lactose forming galactose and glucose, which are used in metabolism. The enzyme is the key to a methodology for identifying this group of bacteria, the "coliforms" which include the genera listed in Table 42.3.

What is the public health official or advisor to do now that he or she must evaluate water destined for reuse by the public? What kinds of tests should be run before and after the

wastewater reclamation project is implemented? Clearly, there should be a consensus by all the parties concerned of how the project will be monitored in terms of the public health considerations assuming that the public itself has been properly educated and is willing to accept reclaimed wastewater to be admitted to the municipal water supply. Incorporating DPR into the drinking water will have one level of concern, but perhaps not as much as resorting to IPR, which provides what is called a public health consore that water is introduced into the municipal water distribution system. Every step in both the wastewater treatment train and the potable water treatment train must be quantified as to the degree of success in removal of both biological and chemical pollutants that will be acceptable for the water to meet its specific purpose. It must be recognized that chlorine disinfection alone will not provide sufficient mortality in Cryptosporidium oocytes and perhaps Giardia cysts to ensure public safety, but usually ozone treatment will. Ozone must be generated by electric discharge on-site and has little persistence in the water distribution system should fugitive pollution find its way into the piping through cross-connections or leaky pipes themselves. For that reason, a chlorine residual is desirable but the addition of ammonia to form chloramines may be less useful because the Ct values for these agents are very large and even *E. coli* is not all that sensitive to inactivation by this reagent. A public health program should be considered using a combination of different kinds of information that would alert public health officials of a possible problem and these include: (1) epidemiological surveillance, (2) analysis of coliform monitoring, (3) monitoring for *E. coli*, (4) monitoring for chlorine residual in the system, and (5) prescribed monitoring for specific pathogens using perhaps any of the other methodologies in the event any of the other approaches should be inadvertently failing at that time.

In any case, the public health official or advisor must always, if required by good conscience and common sense, err on the side of safety. Should an alternate pathway or decision be considered, then let that decision be made by others over strong objection!

42.11 General Public Acceptance of Bacteriological Indicators

Several recent publications suggest that *E. coli* alone are sufficient indicators of human fecal pollution [24,26,29,36] although there are others who have expressed great concern [27] based on the observation that chlorine disinfection of wastewater effluents can effectively kill all *E. coli* yet still allow protozoan pathogens to pass through unharmed. On the other hand, the public health issues must prevail and the capability of a community to monitor the safety of the drinking water provided to its consumers will dictate if that community can or cannot carry out proper water quality analysis. The trust of the public must be recognized and understood [22,34]. Most of the alternative candidate pollution indicators listed in Table 42.5 require very sophisticated instrumentation and highly skilled laboratory personnel. Compared with simpler technologies such as the use of microbiological plates (essentially determining *E. coli* and total coliforms), measurement of chlorine within the water distribution system and carrying out waterborne disease surveillance taken together as a composite process should be sufficient. Certainly, if the laboratory and personnel resources are present and the value of the reclaimed wastewater exceeds that of traditionally obtained water, there is no argument for exploiting such wonderful resources.

42.12 The Santa Cruz Wash: A Possible Model for IPR

If the Santa Cruz River is to serve as a model for IPR, then why is there very little data from local residents? The answer is simple enough. Requirements for sampling water supplies in the United States apply at least nationally for water supplies serving 25 people or more and a cursory look at the geography of the upper Santa Cruz River in Arizona clearly shows that most of those served by the river are individual domestic wells. It is up to the owners of properties to commission their own analyses and either few actually do or if they do, there is no public report of the analyses. Since this particular region is perfect for further study, it would be beneficial both locally and nationally for more work to be done evaluating the upper Santa Cruz River drainage for assessing the response of a groundwater supply to the intrusion of treated wastewater.

42.13 Summary and Conclusions

As populations increase and climate undergoes change, water supplies are subjected to greater demands, and often municipalities are stressed to a point that the only solution may be to consider reusing contaminated sewage or wastewater. Choosing to supplement a potable water supply applying technologies that are known to transform domestic wastewater into clean and safe water that can be consumed by humans requires constant vigilance to ensure that the resulting product is safe for all to use. Wastewaters are rich in pathogenic bacteria, viruses, protozoa, and macroinvertebrates, as well as a variety of chemicals that must be removed or inactivated before the effluent is released for human consumption. There is a wide variety of actions water system managers and public health officials can use to assess the quality and safety of recycled wastewaters. By applying a stringent set of standards or protocols, managers can ensure the safety of once-contaminated water that is released back into use by the public.

References

1. Agullóg-Barcelóarcelo, M., Oliva, M.F., and Lucena, F. 2013. Alternative indicators for monitoring Cryptosporidium oocysts in reclaimed water. *Environmental Science and Pollution Research*, 20, 4448–4454.
2. Al-Jassar, A.O. 2007. Chlorine decay in drinking water-transmission and distribution systems: Pipe service age effect. *Water Research*, 1, 387–396.
3. Asano, T. and Cotruvo, J.A. 2004. Groundwater recharge with reclaimed municipal wastewater: Health and regulatory considerations. *Water Research*, 38, 1941–1951.
4. ATSE, Australian Academy of Technological Sciences and Engineering. 2013. *Drinking Water through Recycling: The Benefits and Costs of Supplying Direct to the Distribution System*. Melbourne, Victoria, Australia.
5. Baker, A. 2001. Fluorescence excitation–emission matrix characterization of some sewage-impacted rivers. *Environmental Science and Technology*, 35, 948–953.
6. Berg, G. 1978. The indicator system. In: Berg, G. (ed.), *Indictors of Viruses in Water and Food*. Ann Arbor Science Publishing, Ann Arbor, MI, pp. 1–13, 948–953.
7. Brouwer, H. 2002. Artificial recharge of groundwater: Hydrogeology and engineering. *Journal of Hydrology*, 10, 121–142.
8. Cabral, J.P.S. 2010. Water microbiology. Bacterial pathogens and water. *International Journal of Environmental Research and Public Health*, 7, 3657–3703.
9. Campos, C. 2008. New perspectives on microbiological water control for wastewater reuse. *Desalination*, 218, 34–42.
10. Carducci, A., Battistini, R., Rovini, E., and Verani, M. 2009. Viral removal by wastewater treatment: Monitoring of indicators and pathogens. *Food and Environmental Virology*, 1, 85–91.
11. Cheri, R. and Mostaghim, M. 2013. Rapid, specific and sensitive coliforms/*Escherichia coli* detection in water samples through selectively pre enriching and an octaplex PCR. *Advances in Environmental Biology*, 7, 2681–2687.
12. Collins, K.E., Cronin, A.A., Rueedi, J., Pedley, S., Joyce, E., Humble, P.J., and Tellam, J.H. 2006. Fate and transport of bacteriophage in UK aquifers as surrogates for pathogenic viruses. *Engineering Geology*, 85, 33–38.

13. Corona-Vasquez, B., Samuelson, A., Rennecker, J.L., and Marinas, B.J. 2012. Inactivation of *Cryptosporidium parvum* oocysts with ozone and *Clostridium perfringens* spores by a mixed-oxidant disinfectant and by free chlorine. *Applied and Environmental Microbiology*, 63, 1598–1601.

14. Craun, G.F. 2012. The importance of waterborne disease outbreak surveillance in the United States. *Annali dellli del d superiore di sanita*, 48, 447–459.

15. Craun, G.F., Brunkard, J.M., Yoder, J.S., Roberts, V.A., Carpenter, J., Wade, T., Calderon, R.L., Roberts, J.M., Beach, M.J., and Roy, S.L. 2010. Causes of outbreaks associated with drinking water in the United States from 1971 to 2006. *Clinical Microbiology Reviews*, 23, 507–530.

16. Crusberg, T.C. 2014. Water supply and public health and safety. In: Eslamian, S. (ed.), *Handbook of Engineering Hydrology: Environmental Hydrology and Water Management*. CRC Press, Boca Raton, FL.

17. Crusberg, T.C., Burke, W., Reynolds, J.T., Morse, L.J., Reilly, J., and Hoffman, A.H. 1978. The reappearance of a classical epidemic of infectious hepatitis in Worcester, Massachusetts. *American Journal of Epidemiology*, 107, 545.

18. Crusberg, T.C. and Reynolds, J.T. 1974. Relationships between chlorine and microorganisms in a water distribution system. Presented at the *102 American Public Health Association Meeting*, New Orleans, LA, October 29.

19. Dominguez-Mariani, E., Carrillo-Chavez, A., Ortega, A., and Orozco-Esquivel, M.T. 2004. Wastewater reuse in Valsequillo agricultural area, Mexico: Environmental impact on groundwater. *Water Air and Soil Pollution*, 155, 251–267.

20. Escherich, T. 1885. Die Darmbakterien des Neugeborenen und Säuglings (Intestinal bacteria of the newborn and infant). *Fortschritte Journal of Medicine*, 3, 515–554.

21. Fetter, Jr. C.W. and Holzmacher, R.G. 2014. Groundwater recharge with treated wastewater. *Journal of the Water Pollution Control Federation*, 46, 260–270.

22. Foster, S.S.D. and Chilton, P.J. 2004. Downstream of downtown: Urban wastewater as groundwater recharge. *Hydrogeology Journal*, 12, 115–120.

23. Gibson, K. E. 2014. Viral pathogens in water: Occurrence, public health impact, and available control strategies. *Current Opinion in Virology*, 4, 50–57.

24. Harwood, V.J., Levine, A.D., Scott, T.M., Chivukula, V., Lukasik, J., Farrah, S.R., and Rose, J.B. 2005. Validity of the indicator organism paradigm for pathogen reduction in reclaimed water and public health protection. *Applied and Environmental Microbiology*, 71, 3163–3170.

25. Jones, K. 2001. Campylobacters in water, sewage and the environment. *Journal of Applied Microbiology*, 90, 68S–79S.

26. Kaya, D., Crowtherb, J., Stapletona, C.M., Wyera, M.D., Fewtrella, L., Edwardsa, A., Francisc, C.A., McDonaldd, A.T., Watkinsc, J., and Wilkinsone, J. 2008. Faecal indicator organism concentrations in sewage and treated effluents. *Water Research*, 42, 442–454.

27. Lalancette, C., Papoineau, I., Payment, P., Dorner, S., Servais, P., Barbeau, B., DiGiovanni, G.D., and Prevost, M. 2014. Change in *Escherichia coli* to *Cryptosporidium* ratios for various fecal pollution sources and drinking water intakes. *Water Research*, 55, 150–161.

28. Lapworth, D.J., Baran, N., and Stuart, M.E. 2012. Emerging organic contaminants in groundwater: A review of sources, fate and occurrence. *Environmental Pollution*, 163, 287–303.

29. McKay, L.D. 2011. Foreword: Pathogens and fecal indicators in groundwater. *Ground Water*, 49, 1–3.

30. Montemayor, M., Valero, F., Jofre, J., and Lucena, F. 2005. Occurrence of Cryptosporidium spp. oocysts in raw and treated sewage and river water in north-eastern Spain. *Journal of Applied Microbiology*, 99, 1455–1462.

31. MoratóoratóCodony, F., Sánchez, O., Martín-Pérez, L., García, J., and Mas, J. 2014. Key design factors affecting microbial community composition and pathogenic organism removal in horizontal subsurface flow constructed wetlands. *Science of the Total Environment*, 481, 81–89.

32. National Academy of Sciences U.S. 2013. *Water Reuse: Potential for Expanding the Nation's Water Supply through Reuse of Municipal Wastewater*. National Academies Press, Washington, DC.

33. Odonkor, S.T. and Ampofo, J.K. 2013. *Escherichia coli* as an indicator of bacteriological quality of water: An overview. *Microbiology Research*, 4(e2), 5–11.

34. Ormerod, K.J. and Scott, C.A. 2013. Drinking wastewater: Public trust inpotable reuse. *Science Technology Human Values*, 38, 351–373.

35. Panayiota, P., Andra, S.S., Charisiadis, P., Demetriou, G., Zambakides, N., and Makris, K.C. 2014. Variability of tap water residual chlorine and microbial counts at spatially resolved points of use. *Environmental Engineering Science*, 31, 193–201.

36. Paruch, A.M. and Maehlum, T. 2012. Specific features of *Escherichia coli* that distinguish it from coliform and thermotolerant coliform bacteria and define it as the most accurate indicator of faecal contamination in the environment. *Ecological Indicators*, 23, 140–142.

37. Payment, P. and Locas, A. 2011. Pathogens in water: Value and limits of correlation with microbial indicators. *Ground Water*, 49, 4–11.

38. Rebhun, M., Heller-Grossman, L., and Manka, J. 1997. Formation of disinfection byproducts during chlorination of secondary effluent and renovated water. *Water Environmental Research*, 69, 1154–1162.

39. Sanders, E.C., Yuan, Y., and Pitchford, A. 2013. Fecal coliform and *E. coli* concentrations in effluent-dominated streams of the upper Santa Cruz watershed. *Water*, 5, 243–261.

40. Savichtcheva, O. and Okabe, S. 2006. Alternative indicators for direct pathogen monitoring and future application perspectives. *Water Research*, 40, 2463–2476.

41. Scheurer, M., Storck, F.R., Graf, C., Brauch, H.-J., Ruck, W., Levc, O., and Lange, F.T. 2011. Correlation of six anthropogenic markers in wastewater, surface water, bank filtrate, and soil aquifer treatment. *Journal of Environmental Monitoring*, 13, 966–973.

42. Shields, J.M., Hill, V.R., Arrowood, M.J., and Beach, M.J. 2008. Inactivation of *Cryptosporidium parvum* under chlorinated recreational water conditions. *Journal of Water Health*, 6, 513–520.

43. Shamir, E., Mekong, D.M., Graham, N.E., and Georgakakos, K. 2007. Hydrologic model framework for water resources planning for the Santa Cruz River, southern Arizona. *Journal of the American Water Works Association*, 43, 1155–1170.

44. U.S. EPA. 2002. Method 1604: Total coliforms and *Escherichia coli* in water by membrane filtration using a simultaneous detection technique (MI medium). http://www.epa.gov/microbes/documents/1604sp02.pdf.

45. Voudouris, K. 2011. Artificial recharge via boreholes using treated wastewater: Possibilities and prospects. *Water*, 3, 964–975.

46. Winward, G.P., Avery, L.M., LeCorre, K.S., Fewtrell, L., and Jefferson, B. 2014. Pathogens in urban wastewaters suitable for reuse. *Urban Water Journal*, 6, 291–301.

47. Wolf, L., Zwiener, C., and Zemann, M. 2012. Tracking artificial sweeteners and pharmaceuticals introduced into urban groundwater by leaking sewer networks. *Science of the Total Environment*, 430, 8–19.

Water Reuse Management and Monitoring

43

Contaminant Management in Water Reuse Systems

Manzoor Qadir
*United Nation University
Institute for Water, Environment
and Health (UNU-INWEH)*

*International Water Management
Institute (IWMI)*

*International Center for
Agricultural Research in the
Dry Areas (ICARDA)*

Pay Drechsel
*International Water Management
Institute (IWMI)*

PREFACE

Although wastewater has been increasingly used to grow a range of crops for income generation and livelihood resilience in urban and peri-urban areas, irrigation with untreated or partially treated wastewater may result in negative impacts on irrigated crops, soils, and groundwater along with implications for human and environmental health through chemical and microbial risks. With the potential for environmental risks due to concentrations above the maximum allowable levels, the major chemical constituent groups that need to be addressed in wastewater-irrigated environments are metals and metalloids, essential nutrients, salts and specific ionic species, and persistent organic pollutants. To avoid potential negative impacts, conventional wastewater treatment options, which can control the release of these contaminants into the environment, remain the key to protecting water quality for beneficial uses in agriculture, aquaculture, and agroforestry systems. Effective legislation, monitoring, and enforcement are also essential and often neglected management strategies. At the farm level, some low-cost irrigation, soil, and crop management options, discussed in this chapter, are available to reduce the risk from contaminants added through wastewater irrigation.

43.1 Introduction

Wastewater generated by domestic, municipal, and industrial sectors has been increasingly used, particularly in dry areas, to provide water and nutrients to agriculture, aquaculture, and agroforestry systems. While such use helps in income generation and livelihood resilience amid water scarcity, use of wastewater in untreated or inadequately treated forms may result in negative impacts on human and environmental health through chemical and pathogenic risks [16,24,29].

Although farmers using untreated or partially treated wastewater provide a service by avoiding or decreasing effluent entering into polluted streams and applying it to soils, thus reducing the pollutant load in downstream locations, such irrigation also generates risk for farm communities and consumers of farm products. Polluted canals and ditches, and wastewater-irrigated

fields create hazards in which children and other residents are exposed to harmful pathogens and chemicals [10]. Consumers of farm produce also are at risk when they handle and ingest contaminated vegetables, particularly when the food is eaten raw or prepared with inadequate care toward reducing contamination risk.

Although wastewater treatment is the best choice in managing wastewater in agriculture, wastewater treatment in developing countries is limited, as investments in treatment facilities have not kept pace with persistent increases in population and the consequent increases in wastewater volume in many countries. A key challenge in almost all cases relates to the fact that the price of treated water does not justify its treatment cost. Thus, most of the treatment plants are either abandoned in the short-run or are highly subsidized to ensure their sustenance. Even where wastewater treatment plants are externally funded, they usually only treat a small fraction of the produced wastewater and can face, depending on their type, significant maintenance problems.

The average estimated rates of wastewater treatment are just 8% in low-income countries and 28% in lower-middle-income countries [32]. Thus, much of the wastewater generated in developing countries is not treated, and much of the untreated wastewater is used for irrigation by small-scale farmers. Where irrigation with untreated or inadequately treated wastewater cannot be avoided or is otherwise common, negative impacts on irrigated crops, soils, and groundwater are likely to affect human and environmental health [1,21,24,38]. However, some farm-based measures and low-cost treatment options can reduce the risk for environmental and human health [38]. This chapter elaborates the contaminant risks resulting from the use of untreated or partially treated wastewater and provides insight into the contaminant risk management strategies leading to safe and productive use of wastewater in agriculture.

43.2 Contaminants in Wastewater and Risk Management Options

Wastewater contains different types and levels of undesirable constituents, depending on the source from which it is generated and the level of its treatment. The nonpathogenic components of wastewater aside from organic chemicals, debris, and solutes, can comprise a range of elements that can be essential plant nutrients, undesirable salts, metals and metalloids, and pesticides, residual pharmaceuticals, endocrine disruptor compounds, and active residues of personal care products in toxic concentrations [33,35]. The pathogenic components could be viruses, bacteria, protozoa, and multicellular parasites. The

concentrations of these constituents above the permissible limits have bearing on human and environmental health [38].

With the potential for environmental risks due to concentrations above the maximum allowable levels, the chemical constituents that need to be addressed in wastewater-irrigated environments can be divided into: (1) metals and metalloids, such as cadmium, chromium, nickel, zinc, lead, arsenic, selenium, mercury, copper, and manganese, among others; (2) nutrients such as nitrogen, phosphorous, and magnesium, which in high concentrations might suppress other nutrients and/or affect plant growth otherwise negatively; (3) salts and specific ionic species such as sodium, boron, and chloride; and (4) persistent organic pollutants, such as pesticides as well as residual pharmaceuticals, endocrine disruptor compounds, and active residues of personal care products, among others. A generic framework for contaminant management while irrigating with wastewater is presented in Figure 43.1.

To avoid potential negative impacts, conventional wastewater treatment options, which can control the release of salts, metals and metalloids, nutrients, and emerging contaminants into the environment, remain the key to protect water quality for beneficial uses in agriculture, aquaculture, and agroforestry systems. In the case of metals, metalloids, nutrients, and emerging contaminants, pretreatment and/or segregation of industrial wastewater from the domestic and municipal wastewater stream is an important task [23]. Effective legislation, monitoring, and enforcement are also essential and often neglected management strategies. The sources of salts in wastewater can be reduced by using technologies in industrial sectors that reduce salt consumption vis-à-vis discharge into the sewage system. In addition, restrictions can be imposed on the use of certain products for domestic use that are major sources of salts in wastewater [17].

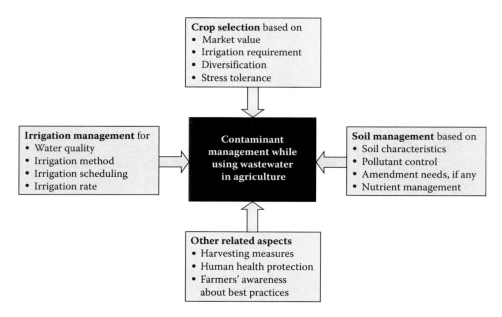

FIGURE 43.1 A framework for contaminant management while irrigating with wastewater, illustrated by four major categories, which are interrelated and implicate each other.

43.2.1 Metals and Metalloids

All of the potentially toxic metals are naturally present in the environment in trace amounts and are ingested with food, water, and air. Human bodies have the ability to deal with these background levels. The World Health Organization (WHO) has established guidelines on allowable consumption of various toxins and guidance values in irrigation water [38]. Several of these metals and metalloids are of particular concern due to their adverse effects on agricultural productivity as well as environmental and human health. In a review of using reclaimed water in the Australian horticultural production industry, Hamilton et al. [11] classified potentially phytotoxic metals in wastewater into four groups based on their retention in soil, translocation in plants, phytotoxicity, and potential risk to the food chain. They categorized cadmium, cobalt, selenium, and molybdenum as posing the greatest risk to human and animal health because they may appear in wastewater-irrigated crops at concentrations that are not generally phytotoxic, but posing risks to human and animal health. The WHO guidelines also consider cadmium with particular concern because of high levels of toxicity and bioaccumulation in crops [38].

Metals such as cadmium, mercury, and lead do not have any essential function but they are detrimental, even in small quantities, to plants, animals, and humans and accumulate because of their long biological half-life [9]. Other metals and metalloids, such as manganese, zinc, boron, or copper are, in small concentrations, essential micronutrients, but harmful to crops in higher concentrations. Some, like copper and zinc, become toxic to plants before they reach high enough concentrations to be toxic to humans; thus, plants function here as a barrier mitigating potential health risks [11,15].

In terms of risk management options with regard to metals and metalloids, the following are important for consideration: (1) soil-based treatment options; (2) crop-based management strategies; (3) crop restrictions; and (4) land retirement zones or other uses of land.

43.2.1.1 Soil-Based Management Options

Soil-based treatment options include the use of soil amendments such as gypsum, lime, phosphate materials, hydrous iron and manganese oxides, clay minerals, and organic matter. These soil amendments form insoluble complexes of metals and metalloids, thereby reducing their availability at low concentrations in the root zone and reducing their assimilation by plants. The amendments have been shown to immobilize metals and metalloids through the (1) formation of insoluble metal phosphate minerals; (2) sorption of contaminants on iron and manganese oxide surface exchange sites and coprecipitation; (3) sorption of contaminants on exchange sites of organic materials including manure, compost, and sewage sludge; and (4) sorption of contaminants on exchange sites at clay mineral surfaces or incorporation into the mineral structure of zeolites, natural aluminosilicate minerals, and aluminosilicate by-products.

Although soil-based management through the application of amendments offers an opportunity to immobilize or minimize bio-availability of metals and metalloids, there are practical limitations, which must be considered. These include the management of sites cocontaminated with several elements, cost and availability of amendments, cost of long-term monitoring programs, and suitability to particular soil and climatic conditions. In addition, there would be a need to follow up in the postmanagement phase, particularly in restricting the flush of acidic water, which may trigger the conversion of insoluble complexes of metals and metalloids into soluble and readily available forms at concentrations that may be phytotoxic and have implications for environmental, human, and livestock health.

43.2.1.2 Crop-Based Management Strategies

Soils contaminated with metals and metalloids can be improved through the use of certain plant species. This approach is broadly known as the phytoremediation of metal and metalloid contaminated soils [4,5,31]. As an important category of phytoremediation, phytoextraction involves the use of pollutant-scavenging plants to transport and concentrate metals and metalloids from the soil into above-ground biomass, which may be harvested to remove the elements from the field. The efficiency of phytoextraction for a given metal is the product of the dry matter yield of shoot and concentration of the metal in shoot.

The plants able to accumulate high concentrations of metals are known as hyperaccumulators. The concentrations of metals accumulated in such plants may be 100 times greater than those occurring in nonaccumulator plants growing on the same substrates [4]. More than 400 plant species have been categorized as hyper-accumulators of metals and metalloids. Following harvest of the metal-enriched plants, the weight and volume can be reduced by ashing. Metal-enriched plants can be disposed of as hazardous material or, if economically feasible, used for metal recovery.

Because the costs of growing a phytoremediation crop are minimal as compared to those of soil removal and replacement, the use of plants to remediate hazardous soils is seen as having great promise [4]. However, phytoremediation has certain limitations; the most crucial is the duration of its action, which may take years or even decades, thereby limiting its practical applicability under situations where more and more contaminants may be added to the soil than their removal through the phytoremediation process.

43.2.1.3 Crop Restrictions

Some crops are more prone than others to contamination with metals and metalloids and/or pose a greater risk to human health due to levels of dietary intake. For example, in the case of irrigation with untreated wastewater, leafy vegetables accumulate certain metals such as cadmium in greater amounts than nonleafy species (Figure 43.2). However, consideration must be given to the quantities of leafy vegetables actually consumed and hence their contribution to the dietary intake of heavy metals such as cadmium.

A shift in crop choice is only feasible and sustainable if there is a market and comparative market value for the alternative crop.

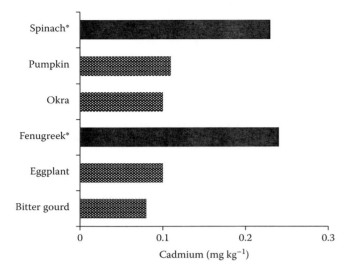

FIGURE 43.2 Cadmium concentration (mg kg⁻¹) in vegetables grown on urban agricultural soils irrigated with untreated wastewater. Spinach and fenugreek are leafy vegetables while others are nonleafy vegetables. (Based on data from Qadir, M., Ghafoor, A., and Murtaza, G. 2000. Cadmium concentration in vegetables grown on urban soils irrigated with untreated municipal sewage. *Environment, Development and Sustainability*, 2(1), 11–19.)

In addition, investing financial resources and time in acquiring the tools and skills to adopt alternative cropping patterns would be a challenge if farmers do not have secure land tenure to plant those plant species that take a long time to give financial returns, such as agroforestry systems. Therefore, crop restrictions can be hard to implement if necessary conditions such as market demand, land rights, and supporting extension services are not effectively in place. However, there are examples of successful or partly successful implementation of crop restriction in wastewater use schemes in several countries such India, Mexico, Peru, Chile, Jordan, and Syria [2,29].

43.2.1.4 Land Retirement Zones or Other Uses

Where there are no further options to maintain the farm, the affected areas might have to be mapped and taken out of production. Simmons et al. [34] developed a General Linear Regression Model to predict the spatial distribution of soil cadmium in a cadmium/zinc cocontaminated cascading irrigated rice-based system in Thailand. Preliminary validation indicated that the model can predict soil cadmium based on minimal soil sampling and the field's proximity to primary outlets from in-field irrigation channels and subsequent interfield irrigation flows. While cadmium is of high risk, soil sampling alone might not be a sufficient indicator of the actual health risk. However, zoning and taking contaminated areas out of food production for land retirement or other uses should be accompanied by adequate compensation for farmers or landowners or providing them income generating alternative livelihood opportunities, associated with training and assured markets.

43.2.2 Nutrient Elements

Although wastewater usually contains valuable nutrients essential for plant growth, the concentrations of the nutrients can vary significantly, and might reach levels that are in excess of crop needs or have antagonistic effect on certain nutrients when present at elevated levels. Therefore, maintaining appropriate levels of nutrients in wastewater is a challenging task and periodic monitoring is required to estimate the nutrient loads in wastewater and adjust fertilizer applications accordingly. Excessive nutrients can cause nutrient imbalances, undesirable vegetative growth, delayed or uneven maturity, and can also reduce crop quality while polluting groundwater and surface water supplies. The amount of nutrients applied via wastewater irrigation can vary considerably if it is raw, treated, or diluted with stream water.

To avoid excessive or unbalanced additions of particular nutrients to wastewater-irrigated soils and crops, farmers can select crops that are less sensitive to high nutrient levels or that can take advantage of high amounts of phosphorus and nitrogen. Land- and soil-based management options depend not only on the type of crop but also on the local soil and site characteristics. Medium- to fine-textured soils, for example, may hold more nutrients than sandy soils, thereby releasing fewer amounts of nutrients in the water percolating through the soil and adding to the groundwater. However, there is a need for groundwater quality monitoring when groundwater is shallow and used for drinking purposes.

Observations from larger urban settings in developed countries showed that effluent treatment by land application for cropping and forestry is often less economical than other treatment techniques. This might be due to the increasing economic land value near cities, but in particular the need in temperate climates to cater to the cold season when soils might be sealed by ice and plants not growing or in a dormant state [14]. In addition, where soils have only restricted internal drainage capacity, their degradation can occur through waterlogging and salinization [14,36]. Hence, most land disposal processes are dependent on freely draining soils and the existence of some diversion structure to store effluent during periods of low absorption capacity or crop water demand.

Riparian buffers can be used to remove sediments, nitrogen, and phosphorus, among other pollutants. Riparian buffers, both the grassed and forested portions, serve to slow water velocity, thus allowing sediment to settle out of the surface runoff water. The grassed portion of the buffer functions as a grass vegetated filter strip. In addition to acting as buffer zones, riparian buffers consisting of grasses, shrubs, trees, or mixed vegetation take up nutrients and thus reduce nutrient loads in water.

Based on data reported on the effects of different sizes of riparian buffers on the reduction of sediment and nutrient loads from field surface runoff [18], and economic valuation, based on shadow price of a pollutant, of the undesirable pollutant addition from wastewater to water bodies such as rivers [13], data reveal that forest-based riparian buffers are more efficient in reducing pollutant loads and increasing economic gains. The total benefits from the reduction in suspended solids, nitrogen, and phosphorus from field surface runoff into water bodies may range from €0.13 m^{-3} for grass-based systems to €0.49 m^{-3} for forest-based systems. These results suggest that options beyond the typical wastewater treatment plant could be used for water quality improvement for environmental and economic benefits.

43.2.3 Salts and Specific Ions

Wastewater contains more soluble salts than freshwater because salts are added to it from different sources. There are no economically viable means to remove the salts once they enter wastewater because of the prohibitively expensive techniques such as cation exchange resins or reverse osmosis membranes, which are only used to produce high-quality recycled water [25,37]. The amount and type of salts used in an industry and the relevant treatment affect the quality of wastewater it generates. For example, in the tannery industry, skins are usually salted with 50%–100% salt by weight and hides with 40%–50% salt [25]. These values suggest that each ton of salted skins contributes over a half ton of salt to the environment. In terms of salt concentration, wastewater from tanneries contains salt in the range of 10–50 g L^{-1}. For comparison, the domestic sector adds about 0.3–0.5 g of dissolved salt to each liter of its wastewater [25]. The situation is complicated when industrial or commercial brine waste streams are not discharged into separate waste sewers, but rather into main urban sewers that convey wastewater to the treatment plants or to disposal channels leading to farmers' fields.

Considering the presence of soluble salts in excessive concentrations, wastewater can be divided into: (1) saline wastewater containing excess levels of soluble salts; (2) sodic wastewater characterized by excess levels of sodium; and (3) saline-sodic wastewater having both salts and sodium in excess concentrations. For long-term irrigation with saline and/or sodic wastewater, there is a need for site-specific preventive measures and management strategies, which may include the following.

43.2.3.1 Crop Selection and Diversification

Considerable variation exists among crops in their ability to tolerate saline conditions [20]. Appropriate selection of crops or crop variety capable of producing profitable yield also with saline and/or sodic wastewater is vital at given levels of salt and sodium concentrations [25]. Such selection is generally based on the ability of the species to withstand elevated levels of salinity in irrigation water and soil [20] while also providing a saleable product or one that can be used on-farm [12]. The salt tolerance of a crop is not an exact value because it depends on several soil, crop, and climatic factors. This diversity can be exploited to identify local crops that are better adapted to saline and/or sodic soil conditions. Several field crops, forage grasses and shrubs, bio-fuel crops, and fruit-tree and agroforestry systems can suit a variety of salt-affected environments [28]. Such systems linked to secure markets should support farmers in finding the most suitable and sustainable crop diversifying systems to mitigate

any perceived production risks, while ideally also enhancing the productivity per unit of saline wastewater and protecting the environment.

43.2.3.2 Irrigation Management Strategies

There could be different options to irrigate crops with saline wastewater, such as surface or flood irrigation, manual irrigation with watering cans, furrow irrigation, sprinkler irrigation, and micro-irrigation such as drip or trickle irrigation. Flood irrigation is the lowest cost method with low water use efficiency. With a medium level of health protection, furrow irrigation needs land leveling. It is suitable when there is a greater leaching need to remove high levels of salts. Without the need of land leveling, irrigation with sprinklers involves medium to high cost and medium water use efficiency. Sprinkler irrigation systems have the advantage of reducing the amounts of water and salts applied to soil and crop [28]. As sprinkler irrigation may cause leaf burn from salts absorbed directly through wetted leaf surfaces where climatic conditions favor fast evaporation [1], irrigation at night can help avoid this. Alfalfa leaves, for example, are known for margin leaf-burn when sprinkled with saline water of 3–5 dS m^{-1}. Sprinkler irrigation of cotton when practiced during daytime with saline water of 4 dS m^{-1} may cause about 15% reduction in lint yield. Several other factors affect salt deposition on leaf surfaces when sprinkler irrigation is practiced, including leaf age, shape, angle, and position on plant; type and concentration of salt; ambient temperature; air velocity; irrigation frequency; and length of time the leaf remains wet [19]. Since the problem is also related more to the number than the duration of sprinkler irrigation, infrequent and heavy irrigations should be preferred over frequent and light sprinkler irrigations [27]. Drip irrigation systems are costly, but highly efficient in water use along with the highest levels of health protection. The clogging of drippers, on the other hand, may limit the use of drip irrigation systems for wastewater. Therefore, filtration is needed to prevent clogging of emitters.

43.2.3.3 Drainage Management for Root Zone Salinity Control

While using saline wastewater, the volume of irrigation water applied should be in excess of crop water requirement (evapotranspiration) or predictable rainfall should be taken into consideration to leach excess salts from the root zone. Salinity control by effective leaching of the root zone becomes an important option for farmers who do not have limited water allocations. In order to calculate leaching requirement, farmers will need assistance to analyze the electrical conductivity of their soils and irrigation water. The leaching required to maintain salt balance in the root zone may be achieved either by applying sufficient water at each irrigation to meet the leaching requirement or by applying, less frequently, a leaching irrigation sufficient to remove the salts accumulated from previous irrigations. The leaching frequency depends on the salinity status in water or soil, salt tolerance of the crop, and climatic conditions [27]. The amount of rainfall should be taken into consideration while estimating the leaching requirement and selecting leaching method. Although leaching is essential to prevent root zone salinity, leaching under saline wastewater irrigation may result in the movement of nitrates, metals and metalloids, and salts to the groundwater. Therefore, monitoring of groundwater levels and quality is an essential indicator of environmental performance [17]. In addition, adequate soil drainage is considered an essential prerequisite to achieve leaching requirement vis-à-vis salinity control in the root zone. Natural internal drainage alone may be adequate if there is sufficient storage capacity in the soil profile or a permeable subsurface layer occurs that drains to a suitable outlet [8].

43.2.3.4 Conjunctive Use with Freshwater

Saline wastewater can be used for irrigation in conjunction with freshwater, if available, through cyclic and blending approaches. These approaches allow a good degree of flexibility to fit into different situations. Guidelines pertaining to water quality for irrigation in terms of salinity and sodicity related parameters are available [1,38]. The cyclic strategy involves the use of saline wastewater and nonsaline irrigation water in crop rotations that include both moderately salt-sensitive and salt-tolerant crops. Typically, the nonsaline water is also used before planting and during initial growth stages of the salt-tolerant crop, while saline water is usually used after seedling establishment [22,30]. The cyclic strategy requires a crop rotation plan that can make the best use of the available good-quality water and saline wastewater, and take into account the different salt sensitivities among the crops grown in the region, including the changes in salt sensitivities of crops at different stages of growth. The advantages of a cyclic strategy include: (1) steady-state salinity conditions in the soil profile are never reached because the quality of irrigation water changes over time, (2) soil salinity is kept lower over time, especially in the topsoil during seedling establishment, (3) a broad range of crops, including those with high economic value and moderate salt sensitivity, can be grown in rotation with salt-tolerant crops, and (4) conventional irrigation systems can be used. Blending consists of mixing good- and poor-quality water supplies before or during irrigation. Saline wastewater can be pumped directly into the nearest irrigation canal or water channel. The amount of saline wastewater pumped into the canal can be regulated so that target salinities in the blended water can be achieved [22,30].

43.2.3.5 Crop Planting Techniques

Since most crops are salt sensitive at the germination stage, their establishment is important to avoid the use of saline wastewater during this critical growth stage. Under field conditions, it is possible to apply modifications to planting practices in order to minimize salt accumulation around the seed and to improve the stand of crops that are sensitive to salts during germination; for example, sowing near the bottom of the furrows on both sides of the ridges, raising seedlings with freshwater and their transplanting, using mulches to carry over soil moisture for longer periods, and increasing seed or seedling rate per unit area, thereby increasing plant density to compensate for possible decrease in germination and growth rates.

43.2.3.6 Soil and Water Treatment

Irrigation with sodic wastewater needs provision of a source of calcium to mitigate sodium effects on soils and crops. Gypsum is the most commonly used source of calcium; its requirement for sodic water depends on the sodium concentration and can be estimated through simple analytical tests. Gypsum can be added to the soil, applied with irrigation water by using gypsum beds, or placing gypsum stones in water channels. In the case of calcareous soils, containing precipitated or native calcite, the dissolution of calcite can be enhanced through plant root action to increase calcium levels in the root zone. Therefore, a lower rate of gypsum application may work well on calcareous soils. Plant residues and other organic matter left in or added to the field can also improve chemical and physical conditions of the soils irrigated with sodic wastewater. In addition, biological treatment of salt-prone wastewater by standard activated-sludge culture can be triggered by the inclusion of salt-tolerant organisms to improve treatment efficiency.

43.2.4 Persistent Organic Pollutants

In developing countries, the exposure of crops and farmers to organic contaminants is probably higher through pesticide application than organic contaminants in irrigation water. The challenge of any related risk mitigation starts with its assessment, which is costly if based on actual analysis. Thus, it is mostly based on observations [38]. An alternative between both extremes may be risk assessment based on easy to measure environmental factors and application practices [35]. The analysis of emerging contaminants, like residual pharmaceuticals or endocrine disruptor compounds would be more difficult and costly [38].

For residual pharmaceuticals, the approach for environmental risk assessments is, for example, regulated by the European Medicines Agency in their Guidelines on the Environmental Risk Assessment of Medicinal Products for Human Use (EMEA/CHMP/SWP/4447/00 corr 1*, 2006) and Environmental Impact Assessment for Veterinary Medicinal Products (CVMP/VICH/592/1998–CVMP/VICH/790/2003). These risk assessments start with an estimation of the exposure by calculating a predicted environmental concentration (PEC), based on dosage of pharmaceuticals or consumption data. These PECs are then compared to predict no effect concentrations (PNEC) in order to assess potential risks. Although pharmaceuticals and other emerging pollutants can accumulate in soil as a result of long-term irrigation with wastewater [6, 7] and may transfer from soils to crops, their amounts taken up by plants seem too small to cause acute toxic effects to humans [3]. However, little is known regarding health risks arising from the long-term uptake of small concentrations of mixtures of micropollutants with food and with drinking water use in poor communities in downstream areas.

Chemical stability and slow natural attenuation of certain persistent organic pollutants makes remediation of these pollutants a particularly intractable environmental challenge. The degree to which wastewater containing persistent organic pollutants needs to be treated depends on (1) pollutant loads, that is, concentration in wastewater × wastewater volume over time; (2) behavior of these compounds in the soil, which could be assessed through specific bioavailability tests to be performed before costly remediation strategies undertaken, for example, in the case of some compounds, a short time after their addition to the soil, diffusion and specific sorption processes lead to "aging" of these pollutants, a process that sequesters these compounds and inactivates their toxic effects; (3) soil properties as soils with large buffer capacities (adequate pH, high soil organic matter content, loamy clay texture, high cation exchange capacity, medium to deep profile) have the capacity to receive and filter larger pollutant loads. For the sites already contaminated with these compounds, the approach usually taken is to isolate the affected sites, and either remove the contaminated soil or rely on phytoremediation. In general, however, it remains crucial to ensure that industrial wastewater is treated at the source and/ or separated from other wastewater streams used for irrigation.

As pesticide contamination is more likely to reach significance through direct onsite application, farm-based measures like the use of alternative pesticides or integrated pest management remain the key for risk reduction. To avoid pesticides entry into streams used for irrigation or other purposes, buffer zones, run-off reduction, and use of wetlands for remediation could be considered [35]. Containment of contaminated water in dams or wetlands may provide time for pesticides to be removed by sedimentation or through degradation. Farming practices that reduce runoff such as the provision of vegetation cover or vegetation buffer strips can help reduce the environmental impacts. The key removal mechanisms for most organic substances are sorption and biodegradation [38]. Removal efficiencies are usually greater in soils rich in silt, clay, and organic matter.

43.3 Summary and Conclusions

To avoid or reduce the contaminant risks from irrigation with untreated or partially treated wastewater, the major chemical constituents that need to be addressed can be grouped into metals and metalloids, nutrients, salts and specific ionic species, and persistent organic pollutants. The strategies for contaminant risk management have been mainly focused on wastewater treatment options, which have been demonstrated to be beneficial when treated wastewater is used in agriculture. Some farm-based measures can also reduce contaminant risk for environmental and human health under situations where untreated or partially treated wastewater is used for irrigation. However, the number of strategies that have been economically assessed and proven to be cost-effective for contaminant risk management when irrigating with wastewater is rather limited, but all mention a positive impact.

The right combination of flexible policies, effective institutions, and wise financial planning can help in implementing contaminant risk management strategies as an interim measure to gradually reach a level when most wastewater in developing countries would be available in treated form for safe and productive reuse by all means.

References

1. Ayers, R.S. and Westcot, D.W. 1985. Water quality for agriculture, Food and Agriculture Organization of United Nations (FAO) Irrigation and Drainage paper 29 Rev 1., FAO, Rome, Italy.
2. Blumenthal, U.J., Peasey, A., Ruiz-Palacios, G., and Mara, D.D. 2000. Guidelines for wastewater reuse in agriculture and aquaculture: Recommended revisions based on new research evidence, WELL Study, Task No: 68 Part 1. http://www.lboro.ac.uk/well/resources/well-studies/full-reports-pdf/task0068i.pdf. Accessed on February 20, 2014.
3. Boxall, A.B., Johnson, P., Smith, E.J., Sinclair, C.J., Stutt, E., and Levy, L.S. 2006. Uptake of veterinary medicines from soils into plants. *Journal of Agricultural and Food Chemistry*, 54(6), 2288–2297.
4. Chaney, R.L., Angle, J.S., Broadhurst, C.L., Peters, C.A., Tappero, R.V., and Sparks, D.L. 2007. Improved understanding of hyperaccumulation yields commercial phytoextraction and phytomining technologies. *Journal of Environmental Quality*, 36(5), 1429–1443.
5. Cunningham, S.D., Berti, W.R., and Huang, J.W. 1995. Phytoremediation of contaminated soils. *Trends in Biotechnology*, 13(9), 393–397.
6. Dalkmann, P., Broszat, M., Siebe, C., Willaschek, E., Sakinc, T., Hübner, J., Amelung, W., Grohmann, E., and Siemens, J. 2012. Accumulation of pharmaceuticals, enterococcus, and resistance genes in soils irrigated with wastewater for zero to 100 years in central Mexico. *Plos One*, 7(12), e45397.
7. Durán-Alvárez, J.C., Prado-Pano, B., and Jiménez, B. 2012. Sorption and desorption of carbamazepine, naproxen and triclosan in soil irrigated with raw wastewater: Estimation of the sorption parameters by considering the initial mass of the compounds in the soil. *Chemosphere*, 88(1), 84–90.
8. Fisher, H. and Pearce, D. 2009. Salinity reduction in tannery effluents in India and Australia, ACIAR Project (AH/2001/005) Final Report, Australian Centre for International Agricultural Research, Canberra, Australia.
9. Göethberg, A., Greger, M., and Bengtsson, B.E. 2002. Accumulation of heavy metals in water spinach (*Ipomoea aquatica*) cultivated in the Bangkok region, Thailand. *Environmental Toxicology and Chemistry*, 21(9), 1934–1939.
10. Grangier, C., Qadir, M., and Singh, M. 2012. Health implications for children in wastewater-irrigated peri-urban Aleppo, Syria. *Water Quality, Exposure and Health*, 4(4), 187–195.
11. Hamilton, A.J., Boland, A.M., Stevens, D., Kelly, J., Radcliffe, J., Ziehrl, A., Dillon, P., and Paulin, B. 2005. Position of the Australian horticultural industry with respect to the use of reclaimed water. *Agricultural Water Management*, 71(3), 181–209.

12. Hamilton, A.J., Stagnitti, F., Xiong, X., Kreidl, S.L., Benke, K.K., and Maher, P. 2007. Wastewater irrigation: The state of play. *Vadose Zone Journal*, 6(4), 823–840.

13. Hernández-Sancho, F., Molinos-Senante, M., and Sala-Garrido, R. 2010. Economic valuation of environmental benefits from wastewater treatment processes: An empirical approach for Spain. *Science of the Total Environment*, 408(4), 953–957.

14. Jayawardane, N.S., Biswas, T.K., Blackwell, J., and Cook, F.J. 2001. Management of salinity and sodicity in a land FILTER system, for treating saline wastewater on a saline-sodic soil. *Australian Journal of Soil Research*, 39(6), 1247–1258.

15. Johnson, S. 2006. Are we at risk from metal contamination in rice? *Rice Today* (July-September), 12(3), Los Baños, Philippines.

16. Keraita, B., Drechsel, P., and Konradsen, F. 2010. Up and down the sanitation ladder: Harmonizing the treatment and multiple-barrier perspectives on risk reduction in wastewater irrigated agriculture. *Irrigation and Drainage Systems*, 24(1–2), 23–35.

17. Lazarova, V. and Bahri, A. 2005. *Water Reuse for Irrigation: Agriculture, Landscapes, and Turf Grass*. CRC Press, Boca Raton, FL.

18. Lowrance, R., Altier, L.S., Newbold, J.D., Schnabel, R.R., Groffman, P.M., Denver, J.M., Correll, D.L. et al. 1995. *Water Quality Functions of Riparian Forest Buffer Systems in the Chesapeake Bay Watershed*, U.S. Environmental Protection Agency, EPA 903-R-95-004/CBP/TRS 134/95, Washington, DC.

19. Maas, E.V. and Grattan, S.R. 1999. Crop yields as affected by salinity. In: R.W. Skaggs and J. van Schilfgaarde (eds.), *Agricultural Drainage*, ASA-CSSA-SSSA, Madison, WI, USA, pp. 55-108.

20. Maas, E.V. and Hoffman, G.J. 1977. Crop salt tolerance – current assessment. *Journal of the Irrigation and Drainage Division*, 103(2), 115–134.

21. Murtaza, G., Ghafoor, A., Qadir, M., Owens, G., Aziz, M.A., Zia, M.H., and Saifullah. 2010. Disposal and use of sewage on agricultural lands in Pakistan: A review. *Pedosphere*, 20(1), 23–34.

22. Oster, J.D. 1994. Irrigation with poor quality water. *Agricultural Water Management*, 25(3), 271–297.

23. Patwardhan, A.D. 2008. *Industrial Waste Water Treatment*. Prentice Hall of India, New Delhi, India.

24. Pescod, M.B. 1992. *Wastewater Treatment and Use in Agriculture*. FAO Irrigation and Drainage Paper No. 47, FAO, Rome, Italy.

25. Qadir, M. and Drechsel, P. 2010. Managing salts while irrigating with wastewater. *CAB Reviews: Perspectives in Agriculture, Veterinary Science, Nutrition and Natural Resources*, 5(016), 1–11.

26. Qadir, M., Ghafoor, A., and Murtaza, G. 2000. Cadmium concentration in vegetables grown on urban soils irrigated with untreated municipal sewage. *Environment, Development and Sustainability*, 2(1), 11–19.

27. Qadir, M. and Minhas, P.S. 2008. Wastewater use in agriculture: Saline and sodic waters. In: S.W. Trimble (ed.), *Encyclopedia of Water Science*, 2nd Edition. Taylor & Francis, Boca Raton, FL, pp. 1307–1310.

28. Qadir, M., Tubeileh, A., Akhtar, J., Larbi, A., Minhas, P.S., and Khan, M.A. 2008. Productivity enhancement of salt-affected environments through crop diversification. *Land Degradation and Development*, 19(4), 429–453.

29. Qadir, M., Wichelns, D., Raschid-Sally, L., Minhas, P.S., Drechsel, P., Bahri, A., and McCornick, P. 2007. Agricultural use of marginal-quality water—Opportunities and challenges. In: D. Molden (ed.), *Water for Food, Water for Life: A Comprehensive Assessment of Water Management in Agriculture*. Earthscan, London, UK, pp. 425–457.

30. Rhoades, J.D. 1989. Intercepting, isolating and reusing drainage waters for irrigation to conserve water and protect water quality. *Agricultural Water Management*, 16(1–2), 37–52.

31. Salt, D.E., Blaylock, M., Kumar, P.B.A.N., Dushenkov, S., Ensley, B.D., Chet, I., and Raskin, I. 1996. Phytoremediation: A novel strategy for the removal of toxic metals from the environment using plants. *Nature Biotechnology*, 13(5), 468–474.

32. Sato, T., Qadir, M., Yamamoto, S., Endo, T., and Zahoor, A. 2013. Global, regional, and country level need for data on wastewater generation, treatment, and use. *Agricultural Water Management*, 130, 1–13.

33. Siemens, J., Huschek, G., Siebe, C., and Kaupenjohann, M. 2008. Concentrations and mobility of human pharmaceuticals in the world's largest wastewater irrigation system, Mexico City-Mezquital Valley. *Water Research*, 42(8–9), 2124–2134.

34. Simmons, R.W., Noble, A.D., Pongsakul, P., Sukreeyapongse, O., and Chinabut, N. 2009. Cadmium-hazard mapping using a general linear regression model (Irr-Cad) for rapid risk assessment. *Environmental Geochemistry and Health*, 31(1), 71–79.

35. Simmons, R., Qadir, M., and Drechsel, P. 2010. Farm-based measures for reducing human and environmental health risks from chemical constituents in wastewater, In: P. Drechsel, C.A. Scott, L. Raschid-Sally, M. Redwood, and A. Bahri (eds.), *Wastewater Irrigation and Health: Assessing and Mitigating Risks in Low-income Countries*, Earthscan, London, International Development Research Centre (IDRC)-International Water Management Institute (IWMI), 209–238.

36. Su, N., Bethune, M., Mann, L., and Heuperman, A. 2005. Simulating water and salt movement in tile-drained fields irrigated with saline water under a serial biological concentration management scenario. *Agricultural Water Management*, 78(3), 165–180.

37. Toze, S. 2006. Reuse of effluent water—benefits and risks. *Agricultural Water Management*, 80(1–3), 147–159.

38. WHO (World Health Organization) 2006. *Guidelines for the Safe Use of Wastewater, Excreta and Grey Water: Volume 2. Wastewater Use in Agriculture*. WHO, Geneva, Switzerland.

44

Urban Meteoric Water Reuse and Management

Michele Di Natale
Second University of Naples

Armando Di Nardo
Second University of Naples

PREFACE

The increase in population along with the parallel per capita rise in demand for water has thrown many countries into a state of "water stress," as annual consumption exceeds natural capacities for resource renewal. A majority of countries have begun dedicating increasing attention to research and development of concrete actions aimed at reducing and optimizing drinking water consumption. The amount of water consumed in the home may be reduced both through targeting lifestyles and by modifying the equipment used. Another technique to reduce water consumption may be the collection, storage, and reuse of rainwater.

A system for the collection and reuse of rainwater consists of the following basic components: collection system, storage system, distribution system, and treatment system. With reference to the calculation of storage volume is proposed an original method that takes into account the net by the contribution of rainwater and the water requirement of service.

44.1 Introduction

If you could look at the Earth from above, you would see that most of it is blue in color. This is due to the fact that about 71% of the Earth's surface is covered by water, while only about 29% consists of land above sea level. Of the 510 million km² of the Earth's surface, a good 364 km² are occupied by water, which has a volume of 1.4 billion km³. Which is, of course, why Planet Earth is often referred to as "the blue planet."

It should be noted, however, that over 97% of all the water on Earth is saltwater (with an average salt content of 3500 mg/L), held in the oceans, seas, inland seas, brackish lagoons and in some of our groundwater. Freshwater, which has a maximum salt content of 500 mg/L, only accounts for 3% of all the water on Earth. Freshwater is distributed between groundwater, glaciers, the polar ice caps, rivers, and lakes. From all these sources, only a small amount—equivalent to 0.75% of all the water present on Earth—is to be found in the form of liquid freshwater in our lakes, rivers, and under the ground. An average of 70% of all our drinking water supplies is used in agriculture, while 20% goes to industry and just 10% of the remainder is for domestic use.

Numerous international studies were carried out on the evaluation of availability and use of water; a good summary can be found in References 20 and 22.

Water is distributed unevenly across the Earth's surface and as a result, although the world's existing water resources are sufficient to meet the needs of the whole of its population, certain regions, particularly Africa, the Middle East, eastern Asia, and some countries in Eastern Europe, are handicapped by a severe and chronic shortage of water.

The quantity of freshwater available for human consumption varies between 12,500 and 14,000 km³ per year. But as a result of the growth in the planet's population, water consumption has increased sixfold over recent years, while per capita availability has decreased from nearly 13,000 m³ per year available in 1970 to 9000 m³ in 1990, and down to less than 7000 in 2000. Among the causes of global water scarcity are the impacts of climate change such as global warming and desertification, but also the degradation of water quality due to pollution.

In tandem with this decrease in the amount of per capita renewable freshwater available, there has been an overall increase in consumption as a result of population dynamics and the resulting growth in demand for water for agricultural, industrial, and residential use. In 1900, per-capita global consumption was estimated at 350 L per person per day, but by the year 2000, this stood at 640 L per head. Over the course of one century, then, a halving of the availability of water has been accompanied by a doubling in demand [14].

This increase in population, along with the parallel per capita rise in demand for water has thrown many countries into a state of "water stress," as annual consumption exceeds natural capacities for resource renewal.

Faced with this generally disconcerting picture, a majority of countries have begun dedicating increasing attention to research and development of concrete actions aimed at reducing and optimizing drinking water consumption. If we focus specifically on water supplies used solely for domestic purposes, research and awareness-raising activities have branched out into three directions: (1) the dissemination of best practices; (2) the use of new technologies for water saving; and (3) rainwater harvesting and reuse [1,8,21].

After a brief glance at the first two types of initiatives, we will concentrate on analyzing systems used for the harvesting and reuse of rainwater.

44.2 Reduction of Domestic Water Consumption

Household drinking supplies represent about 75% of civil water use. Many "nonresidential" uses are, however, comparable to domestic water use—for example, water supplies to offices, schools, hotels, restaurants, and so on. These supplies perform the same functions as home water use, the only difference lying in the quantities involved. Just a tiny fraction of civil water consumption (of the order of between 1% and 2%) goes to distinctly different types of use such as traditional craft uses, the fire service, and similar.

The quantity of water consumed in the home can vary greatly with climate, social class, lifestyle, and the type of building construction involved.

Table 44.1 shows data on household water consumption in the world.

TABLE 44.1   Average Water Use per Person per Day in Liters

Country	L/(p × d)	Country	L/(p × d)	Country	L/(p × d)
US	575	Germany	193	Nigeria	36
Australia	493	Brazil	187	Burkina Faso	27
Italy	386	Peru	173	Niger	27
Japan	374	Philippines	164	Angola	15
Mexico	366	UK	149	Cambodia	15
Spain	320	India	135	Ethiopia	15
Norway	301	China	86	Haiti	15
France	287	Bangladesh	46	Rwanda	15
Austria	250	Kenya	46	Uganda	15
Denmark	210	Ghana	36	Mozambique	4

Source: From United Nations Development Program—Human Development Report 2006—www.data360.org. With permission.

Considerable variation can be seen between one city and another, while the lowest consumption rates occur in northern European cities, as well as in cities with hotter climates.

Table 44.2 shows an estimate of average domestic water consumption, based on an analysis of available data [6]. As can be seen, among those items labeled "domestic use," there appear some types of consumption that do not require drinking water. These include WC flushing, house cleaning, car washing, and watering the garden. For these uses, water that is purified and odorless, but not necessarily of potable quality, would suffice.

Uses that actually require drinking water may be limited to personal hygiene (32%), food preparation (12%), and dishwasher and washing machine (15%) use, even though, as we shall see, for both the dishwasher and the washing machine, using rainwater would bring the advantage of extending their mechanical durability, as rainwater contains less chalk than drinking water drawn from the mains supply. In short, it can be seen then, that "domestic uses" requiring drinking water amount to less than half of actual current domestic consumption [18].

The amount of water consumed in the home may therefore be reduced both through targeting lifestyles, that is, changing habits that have become ingrained over time, and by modifying the equipment used, that is, utilizing appliances designed to be more efficient, or no less wasteful of water and energy, than conventional models. The best results are attained when both

TABLE 44.2   Typical Household Use of Drinking Water in Europe

Type of Device	%
Bath and personal hygiene	32
WC	30
Washing machine	12
Kitchen	12
Irrigation and other uses	8
Cleaning the house	3
Dishwasher	3

Source: From Conte, G. 2008. *Nuvole e sciacquoni.* Ambiente, Italy. With permission.

these strategies are adopted in tandem; that is, when habits are changed and new equipment installed.

Numerous experiments conducted in Europe, as well as studies in the United States, Canada, and Australia, have shown that efforts to reduce drinking water consumption that are based on both factors have maintained the initially obtained saving effects into the long term [12,16]. The conceptual pairing of "technology and behavior change" therefore appears an essential one for the attainment of significant results and their maintenance over time.

44.2.1 Using "Best Practices"

Behaviors aimed at reducing water consumption affect various aspects of the civil use of this resource and have the purpose of improving and optimizing use without making substantial changes to installations.

Practices listed below provide interesting insights into some ways water consumption can be reduced by adopting water-sensitive behaviors [10].

Using a bathtub involves a consumption of 100–200 L of water; if one opts to have a shower rather than a bath, approximately 150–180 L of water is saved each time. Water consumption for a shower varies between 20 and 50 L. If, in addition to this, you use a shower with an air–water mixer, even less water (about 50% less) will be required.

Leaving a tap running causes considerable waste: the throughput of a tap is about 12 L of water per minute. In this way, a three-person household will save up to 7500 L of drinking water over the course of 1 year.

Good practice in washing food and washing the dishes is not to use running water from the tap, but to collect water for the purpose in a container. Furthermore, if the washing up is done soon after a meal, using the water from cooking to wash the dishes, savings will be obtained not just of water and detergents but of energy, too. Indeed, overuse of chemical products when washing dishes, pots, and pans as well as in cleaning around the house, pollutes our waterways. More water will also be consumed in rinsing soapy surfaces.

Washing machines and dishwashers consume a lot of water (80–120 L) regardless of the load of clothes or dishes being washed. A good way to save water and electricity, then, is to use these appliances at full load, thus decreasing the frequency of washing. Seven thousand to eleven thousand liters of water per household per year can be saved in this way.

The amount of water normally consumed when washing the car by using a garden hose, with the water left running during the wash, comes to 200–400 L per wash. Using a container will save more than 100 L of water.

It is good practice to position a hot water boiler as closely as possible to its point of use; in this way, one avoids the considerable wastage of energy and water that occurs during the waiting periods before the hot water starts to flow.

It is crucial that the toilet is never used as a discharge point for toxic substances (paints, varnishes, chemicals, cigarettes,

solvents) as doing so impairs the correct functioning of the sewage system.

Watering the garden in the evening hours, when temperatures are lower, means less evaporation and less waste. All of these measures will bring annual savings of approximately 6000 L of drinking water.

Taking these simple precautions, a household will save no fewer than 50,000 L of drinking water in 1 year. This is not only a significant step forward toward sustainable use of water resources, but also translates into cost-of-living savings for the family.

44.2.2 Using Technology to Save Water

Technologies are available today that offer substantial savings in water use; they feature very simple devices that halve the consumption of mains water [21]. Such technologies include

- Devices and components designed to reduce consumption in sanitary fittings (toilet bowls with reduced water consumption, devices to reduce flushing water requirements, low-flow flushes or flow-differentiated flushes, low-consumption taps, flow breakers, low-flow showers, pressure reducers, etc.), devices that reduce consumption in irrigation equipment for private or shared gardens (programmable watering timers, micro-irrigation systems, drip irrigation, water-efficient gardening techniques and practices, etc.).
- High-efficiency washing machines and dishwashers (Class A category) that reduce water and energy consumption.
- Regular maintenance of water and sanitary installations and equipment in private and shared households.
- Rainwater harvesting, reuse of greywater (water that has been used in food preparation or for rinsing clothes) and treated wastewater for appropriate uses.

For tap fittings, we show the main commercially available types of devices used to reduce water consumption that may be used in residential and public contexts:

- Flow limiters: Devices fitted with a flow-reducing valve, so that flows remain constant independently of mains pressure. This is achieved thanks to an internal variable equalizing valve. Savings in water consumption will depend on the degree of maximum flow adjustment.
- Flow-breaker/water saver: This device acts on the output from the tap by mixing air and water together, resulting in a powerful jet that uses much less water.
- Pressure limiters: Devices that reduce water pressure. Their installation is recommended in hotels, fitness centers, and on public premises where large quantities of water are consumed during certain periods of the day.
- Mechanical flow switches: Devices that can be opened and closed in a simple way by operating a lever. They function in practically the same way as a single-lever tap and are recommended for showers with two water inlets, as these

devices enable water flow to be interrupted when lathering and then allow the shower to be reactivated without any need to readjust water temperature settings.

- Single-lever taps: These devices offer significant advantages, as they allow you to adjust water flow and water temperature better and more quickly, thereby avoiding considerable wastage. Achievable savings will depend on the kind of device fitted to the tap.
- Tap timers: Timers are mechanisms that automatically shut off flow after a set period of time. Water savings can amount to 30%–40% for showers and 20%–30% for hand basins.
- Electronic taps: The flow stops automatically each time you withdraw your hands from the basin. These taps are costly, but can save 40%–50% of water consumption.
- Thermostatic taps: Thermostatic taps are fitted with a temperature selector that keeps the water at the selected temperature so that when you shut and then reopen the tap, the water flows at the same temperature. These taps, which are used primarily in showers, offer savings not only of water but also of energy.

Table 44.3 shows an overview of water savings achievable using an appropriate tap device.

When considering devices for reducing consumption via WC flush discharges, it should be noted that toilet flushing is the main single source of domestic potable water consumption. This amounts, on average, to more than 50 L per capita per day, or 30% of the total daily consumption of a typical water user who does not deploy water-saving devices. Conventional WC cisterns discharge between 9 and 16 L with each flush: considering that each person uses one on average four times a day, significant savings can be achieved. Considerable savings are also available in non-residential contexts: it is estimated that bathroom use in offices is approximately three times per day for women, while men will use the toilet once and the urinal twice a day on average [1].

Optimal functioning of the flush largely depends on three points:

- The flushing mechanism
- Speed (and therefore pressure) of water flow
- Water quantity

TABLE 44.3   Water Saving Using Devices Applicable to the Tap

Devices	Water Saving (%)
Flow limiters	30–40
Watersaver	30–70
Mechanical flow switches	50
Pressure limiters	10–40
Single-lever taps	30–40
Timed taps	30–40
Electronic taps	40–50
Thermostatic taps	50

Source: From Conte, G. 2008. *Nuvole e sciacquoni.* Ambiente, Italy. With permission.

The worse a system performs under the first two points, the greater will be the amount of water required to ensure efficient flushing.

Regarding the WC bowl, low-volume bowls are highly efficient in comparison to the huge amount of water (an average of 9–16 L) required for a single flush of a traditional toilet. Low-volume models are designed and manufactured with special shapes that exploit the jet of water and water flow to a maximum, thus necessitating on average just 6 L per flush.

The required flow-rate and the amount of water used will depend on the type of flushing system, which is usually a gravity-fed flushing cistern. If flushing cisterns are located immediately behind the bowl, they will need to hold at least 12–15 L of water, which will be completely emptied with each use. If, on the other hand, they are positioned higher up, they can be smaller (about 9 L), as the height will increase water pressure and speed of discharge. Without a doubt, the most effective way to achieve substantial savings in WC water consumption is through fitting a dual-action cistern with one flush lever of 3–4 L, while the other provides a 6–9 L flush.

These devices can bring about water savings of up to 60%, even though this does not generally exceed a savings of between 35% and 50% due to incorrect use. By way of an alternative, there are devices on the market that can interrupt the flow as required: while these are perfectly functional, they require an extra effort on the part of the user. Higher levels of water savings can be achieved using pressure discharging devices; these systems are more complex and expensive than the others and exploit an autoclave pump to flush the toilet more efficiently and with less water. The devices described above can provide high levels of water savings and economic efficiency when deployed in sanitary facilities for public access.

Looking, finally, at household appliances, among the many models available on the market we should identify those which, depending on intended use, ensure the lowest water consumption. For several years now it has been a requirement in the European Union that appliances carry an energy-efficiency rating label (Ecolabel). The label rates efficiency levels in six bands, from A (low consumption) to F (high consumption). In order to reduce consumption of water and energy, washing machines and dishwashers from category "A" or higher should always be chosen. Although these are more expensive, their savings on both resources is significant and appreciable over a short space of time. A typical washing machine uses around 100 L of water for a normal load of 5–6 kg of white wash, while the most efficient models offer consumptions of less than 12 L/kg, or about 60 L per wash. In this way, a smaller amount of detergent and fabric softener can be used, with further savings for the user and a reduction in pollution from detergents. In Australia, where very strict restrictions on the domestic use of water are in force, there are models on the market that consume about 35 L per wash. Similarly, outdated dishwashers consume 30–40 L per cycle, while the most efficient models get down to 10–14 L per cycle, thanks to nozzle systems and grid arrangements that reduce the amount of water needed to loosen and remove food particles, as well as the function of choosing the type of wash according to load and kind of food remains.

44.3 Systems of Rainwater Harvesting

The recovery and recycling of rainwater is an ancient practice. Archaeological evidence testifies how rainwater harvesting dates back at least 4000 years, while there are reports of findings that would date the presence of the first water tanks in China as far back as 6000 years ago [2].

In modern times, in Europe up until the mid-twentieth century, the practice of collecting rainwater was still widespread in many countries, as has been demonstrated by recent studies into traditional techniques for collecting rainwater conducted in the city of Matera, Italy. We now find that this great tradition has rapidly disappeared. Our ancient water-collection tanks fell into disuse toward the end of the 1960s and with disuse the old science of their management was lost. Of course, today, we could not do without our centralized supplies of mains water. Nonetheless, it is just as clear that we need to furnish ourselves with a broad-based water-collection capacity, creating new levels of stored volumes and recovering, above all, the lost science of constructing and maintaining rainwater storage systems [4,9,17].

The importance of collecting rainwater to meet a part of domestic consumption is now widely recognized throughout the world. In Australia, the United States, Europe, and South America, many public and private bodies have for some time been promoting and spreading rainwater harvesting awareness and technology, with many enterprises operating successfully in the sector [5,14,15,17].

The minimum purpose that a rainwater collection system should meet is to enable demand to be met for which drinking-quality water is not strictly necessary, but where purified and odorless water (e.g., harvested rainwater) would suffice. Specifically, this concerns uses related to WCs, washing machines, household cleaning, and irrigation of gardens, which amount to approximately 50% of daily per capita consumption.

It thus becomes clear how reuse and careful management of rainwater through an adequate system could halve levels of drinking water consumption.

A system for the collection and reuse of rainwater consists of the following basic components:

- Collection system
- Storage system
- Distribution system
- Treatment system

Figure 44.1 shows a classic example of a rainwater collection and reuse system, typically installed in single-household contexts.

44.3.1 The Collection System

This system consists of impervious water-shedding surfaces that are not subjected to vehicular transit (roofs, terraces, balconies, walkways) on which rain falls, and the fittings (gutters, gullies, downpipes, drainage wells, vertical pipes) necessary for the conveyance and storage of the intercepted water. In the collection phase, precautions should be taken to deal with "first-flush" waters, which are characterized by high concentrations

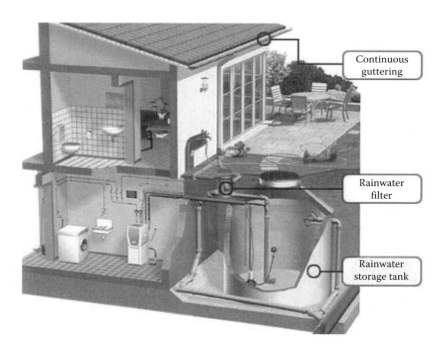

FIGURE 44.1 A model of an underground rainwater storage system. (From http://www.raincollectionsupplies.com/Rainwater_harvesting_books_s/29.htm. With permission.)

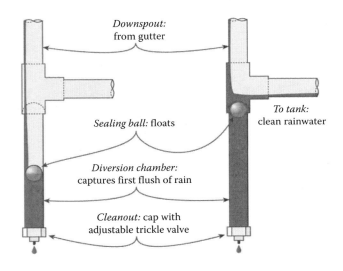

FIGURE 44.2 An example of first-flush diversion. (From Design for Water-New Society, 2007. With permission.)

of pollutants. The quality of this first-flush water is heavily influenced by specific features of the terrain under consideration, and in particular by the hydrological, climactic, and morphological characteristics of the area (frequency and intensity of precipitation, surface type, etc.)

As a rule, the first 2–5 mm of a given precipitation event are considered first-flush waters. The problem of first-flush pollution is solved inside the rainwater collector with the installation of a "first-flush diverter," whose function is basically to reduce the contamination peak by holding back the first part of the rainwater.

Various types of switches are available, such as the roof-washer; a downspout that is closed off at one end and runs alongside a pipe connecting the gutters to the water-collection tank. The pipes are configured in such a way that water coming from the roof is able to flow into the tank only once the roof-washer has been completely filled. Figure 44.2 shows an example of a roof washer with a ball valve. The system has been designed so that once the space for the first-flush water has been filled, the floating ball valve blocks the entry of any further water, which will therefore flow into the collecting tank.

Automatically or manually operated switches are also available. These are installed on the terminal portion of the downspout and block off a section of it; thereby deviating rainwater either toward an outflow or toward the tank.

Among the innovations aimed at solving the common problem of clogged-up gutters and downspouts (caused by an accumulation of leaves and the other debris that falls onto roofing profiles), we should mention special gutters equipped with fittings that close off the upper part of the gutter itself (see Figure 44.3).

Before storage, the collected water undergoes mechanical filtration. The filter can be installed at various points upstream of the collection tank (on the downpipes, aboveground, underground, integrated with the tank, etc.). Filters are equipped with automatic rinsing devices for removing the filtered-out material, which would otherwise deposit in layers and reduce efficiency.

Some examples of filters are shown in Figure 44.4.

44.3.2 The Storage System

The system consists of a tank whose volume has to be sized appropriately, as will be seen next. The range of tanks available varies according to their target location, capacity, shape, and material.

Among possible locations for the tank are aboveground, inside the building (basement, garage), and underground.

Aboveground tanks (Figure 44.5a) are generally preferred where water is collected for irrigation purposes (vegetable gardens, etc.) or for washing cars, etc. whereby the water distribution is affected by gravity feed from the tank without the use of pumps.

Indoor tanks (Figure 44.5b) are typically located in ground level rooms (garages, basements, etc.). This choice is typically driven by ease of installation, the nonavailability of outdoor sites, difficulties with excavating (rocky soil, high-water table, etc.) or by a need to avoid damage to areas around the building. Tanks are usually tall in profile in order to minimize the amount of floor space they occupy and sizes are usually modest in order to facilitate insertion into indoor spaces. Capacities can, however, be increased by having several tanks stacked in rows.

While involving more outlay, underground tanks (Figure 44.5c) offer a view uncluttered by features that may not match the functional and aesthetic requirements of the building. They

FIGURE 44.3 Gutters with prefiltering of rainwater. (From http://www.gutterguard.co.za/Products-and-Pricing. With permission.)

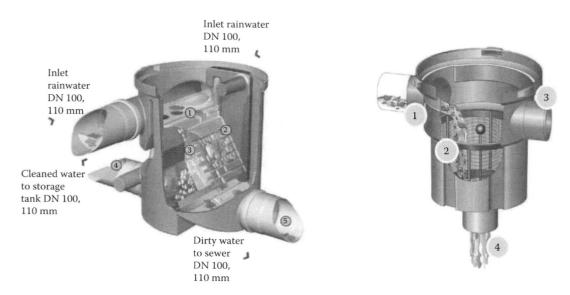

FIGURE 44.4 Examples of rainwater harvesting filter. (From http://www.freewateruk.co.uk/commercial-tank-filter.htm. With permission.)

FIGURE 44.5 (a) Storage systems—inner tank. (From http://www.capewatersolutions.co.za/photo-gallery. With permission.) (b) Storage systems—aboveground tanks. (From http://www.rainharvest.com/water-tanks-plastic/above-ground-tanks.asp. With permission.) (c) Storage systems—underground tank. (From http://www.stormsaver.com/write/Stormsaver_Commercial_Brochure.pdf. With permission.)

also permit structures of large dimensions to be installed. Each tank is fitted with a "manhole," a method for accessing the interior of the tank. This consists of an entry point in the upper part of the casing, which is finished with a watertight cover, removable when performing operations of maintenance and control. After it has been filtered, the water entering the tank has to pass via a stilling pipe. This is a vertical pipe as high as the tank itself and fitted with an upwardly curving end-fitting. Its function is to avoid the creation of turbulence within the tank that might stir up any layers of algae or other deposits present on the bottom, leaving them in suspension.

The tank also has to be fitted with an overflow device that allows water to enter the outflow system once the level of maximum tank capacity has been reached. The overflow tube has to be fitted with an airlock in order to avoid the emission of unpleasant odors from the waste disposal system.

Finally, the tank has to be fitted with a nonreturn valve. This is an essential fitting to avoid contamination of the water stored in the tank. This valve consists of a special device containing a self-closing gate (which may be operated manually in case of emergency or during maintenance). The gate prevents backflow of water originating from the disposal system. Normally this valve is fitted with a lattice filter that will prevent insects and other animals from accessing the tank from the discharge subsystems.

Tanks in production are generally cylindrical in shape with a horizontal or vertical axis.

Tanks have to be constructed out of materials that meet technical standards for the storage of water for human consumption. The material is usually high-density polyethylene (HDPE), fiberglass (GRP), concrete, or steel.

44.3.3 The Distribution System

The filtered and suitably stored water is now ready to be tapped off and reused through an appropriate distribution system. Where the water is intended simply for irrigation purposes, a pump of adequate capacity and delivery head can be either immersed in the tank or connected externally. Where the stored water is required for domestic use, additional precautionary measures need to be built into the delivery system. A lot of manufacturers recommend in these cases the use of central control units for automatic management. The central control unit is tasked with providing end users with a constant supply of water, even during prolonged periods of dry weather. This is achieved through the automatic management of the traditional and recovered water circuits, completely avoiding waste.

To this end, the first thing to determine is the minimum volume below which the stored volume of water should never fall. This will allow for simultaneous and prolonged draining of the systems by the end-users connected to them. The minimum volume is guaranteed by positioning a water-level sensor inside the tank; as the level of stored rainwater falls below the specified amount, this sensor opens a solenoid valve to the mains water supply and the tank is refilled via a submerged pump or via a pump external to the tank. It should be noted that the tank is

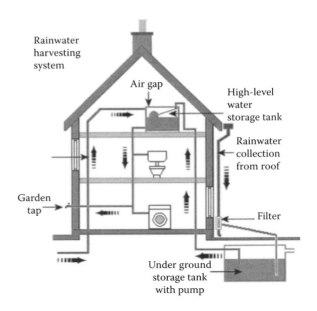

FIGURE 44.6 Example of the distribution network of a system of collection. (From http://www.lowenergyhouse.com/rainwater-harvesting.htm. With permission.)

replenished using the traditional mains water supply only for the purpose of maintaining the set minimum level (determined by the sensor). In this way, water from the next rainfall will not be wasted.

Figure 44.6 shows an example of a distribution network for harvested rainwater.

44.3.4 The Treatment System

Before it reaches the end user, the water collected in the tank must be treated in order to eliminate any risk to human health arising from its use. Treatment generally consists of filtration and disinfection processes. In an initial stage, the water is filtered to remove any suspended solids. This is done using cartridge filters of sand, activated carbon, or membrane filters. Following this, to eliminate any microorganisms present, the disinfection stage is usually performed using an ultraviolet germicidal sterilizer (see Figure 44.7), or by means of chemical treatments based on chlorine or ozone.

A typical treatment system comprises two in-line filters, one a 55-μm cartridge filter, followed by a 3-μm carbon filter, which are followed by an ultraviolet germicidal sterilizer.

44.4 Calculation Criteria of the Storage Volume

Assessment of the volume to be assigned to the system's storage tank must take into account the net flows of rainwater and the demand to be placed on the service supply. We will proceed below to provide a preliminarily indication of the criteria used

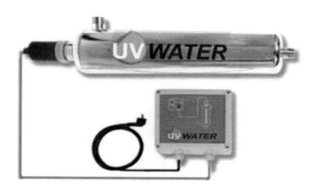

FIGURE 44.7 Sterilizer ultraviolet (UV-C). (From http://www.uv-water.com/modules/prodotti/uv-water-40h.php. With permission.)

TABLE 44.4 φ Values of Runoff Coefficient

Surface of Drainage	φ
Roof	0.90–0.95
Pavement	0.80–0.85
Porous asphalt/concrete and permeable pavers	0.60–0.70
Green roof with four or more inches of growing media	0.60–0.70
Synthetic turf athletic fields with subsurface gravel bed and underdrain system	0.60–0.65
Gravel parking lot	0.50–0.60
Undeveloped areas	0.30–0.40
Grassed and landscaped areas (including rain gardens and vegetated swales)	0.15–0.20

Source: From Strickland, C.H. and Bloomberg, R. 2012. *Guidelines for the Design and Construction of Stormwater Management Systems.* Developed by the New York City Department of Environment. With permission.

in calculating these quantities [3,24]. Following this, we will look at calculating the storage volume. To this end, a satisfactory design result is best achieved by treating instances of small- and medium-sized installations (single-family contexts) separately from large-scale installations [19].

44.4.1 Assessing Net Rainwater Capture

The net volume of theoretically collectable rainwater is given by the following formula:

$$V_o = V_p - V_{pp} - V_{pe} \qquad (44.1)$$

where

V_p = maximum amount of rainwater that can theoretically be collected

V_{pp} = amount of "first-flush" water

V_{pe} = amount of tank-stored water that is lost through evaporation

Calculating V_p: The maximum amount of rainwater that can theoretically be collected is calculated using the following:

$$V_p = 10^{-3} \varphi h S \qquad (44.2)$$

where

V_p = maximum volume of rainwater collectible [m³]

φ = runoff coefficient

S = summation of precipitation watershed surfaces [m²]

h = depth of rainfall [mm]

The runoff coefficient (φ) represents the ratio between the volume of net rainfall (actual precipitation) that one manages to divert toward the collection system and the gross total precipitation that falls on surfaces in the catchment area. It is a function of the type and quality of the areas exposed to precipitation and also the orientation of their exposed slopes. The values this coefficient can take on are given in Table 44.4.

Runoff collection surfaces (*S*) represent all the areas designated for the collection of rainwater, in horizontal projection. It is vital to make a careful choice of these catchment area surfaces for rainwater, as this choice will greatly influence the quality of the water stored. For example, areas where vehicular traffic is permitted yield levels of pollutants in stored rainwater that are significantly higher than is the case for surfaces consisting solely of the roofs of buildings.

With regard to meteoric precipitation (*h*), reference can be made to annual rainfall depth values (h_a), monthly (h_m), or daily (h_d) depending on whether the volume V_p calculated by means of Equation 44.2, is to produce an annual, monthly, or daily value.

Measurement of rainfall depth should be based on a series of historic rainfall data obtained from the monitoring stations of the National Hydrographic Service and distributed across the geographical area concerned.

As is known, the method of estimating hydrological magnitude attained using historic rainfall data, of amplitude N, may be based on two approaches:

- Deterministic approach
- Probabilistic approach

The deterministic approach consists of assuming that values corresponding to the historic series taken into consideration will be repeated in the future. In this case, the hydrological magnitudes for the project can be estimated by calculating the average values of a sample of length N:

$\overline{h_a}$ = annual mean rainfall depth h_a

$\overline{h_m}$ = monthly mean rainfall depth h_m

$\overline{h_d}$ = daily mean rainfall depth h_d

On the other hand, the probabilistic approach consists of predicting the values of rainfall depths that may occur in the future with a given probability. As will be known, this method consists of identifying an appropriate probability law that will be able to interpret a data sample of length N to determine with a given probability (return period) values for average or minimum precipitation depths.

Calculating V_{pp}: In Equation 44.1, apart from the maximum value of rainwater theoretically collectible (V_p), the corresponding volume of "first-flush" water (V_{pp}) is also given. So-called first-flush water is the volume of rainfall runoff during the first part of a precipitation event, which is characterized by high levels of pollutants. It is therefore necessary to quantify these "first-flush" volumes so that they can be discarded before calculating stored volumes. A first approximation value for V_{pp} may be obtained by assuming that it corresponds to a precipitation depth of 2–4 mm:

$$V_{pp} = 10^{-3}\varphi h_{pp}S \tag{44.3}$$

in which

V_{pp} = first-flush volume [m³]
h_{pp} = depth of first-flush precipitation [m]

Calculating V_{pe}: As shown in Equation 44.1, in order to determine the net volume of rainwater (V) that can be collected, it is also necessary to quantify the volume associated with evaporation losses (V_{pe}) that occur within the storage capacity. This loss is usually slight and is often passed over in simulation studies of manufactured tank functioning. At all events, if one wanted to take into account the value for the volume of water associated with evaporation losses, it should be calculated based on the thermodynamic properties of the air on the water surface area within the tank.

Before concluding, we note that, when calculating the collection capacity, the net volume (V) of water reaching the tank must also take into account the efficiency of the filter located upstream of the storage tank. We get, therefore:

$$V = \varepsilon V_o \tag{44.4}$$

coefficient ε is commonly assigned values of between 0.8 and 0.9.

44.4.2 Estimating Consumption

Water requirements for daily sanitary, irrigation, and cleaning uses (service uses) (F_d) can be determined by estimating the number of appliances present and the number of times they are used daily (WC, washing machine, dishwasher) along with the number of square meters to be supplied (irrigation, washing of rooms). The data obtained in this way is then assigned the consumption per unit values shown in Table 44.5.

Clearly, the annual consumption value (F_a) will then be given by the equation:

$$F_a = \sum_{i=1}^{365} F_{d,i} \tag{44.5}$$

44.4.3 Sizing a Plant of Small Dimensions

Small-sized systems for harvesting and reusing rainwater, which are commonly referred to as "domestic installations,"

TABLE 44.5 Estimated Consumption Compatible with the Quality of the Rainwater

Uses		Estimated Average Consumption
WC flushing	Traditional cistern	9–12 L
	Cistern with double button	6–9 L
	Cistern water saving	3–9 L
	Discharge with "pressed" system	2–4 L
Irrigation	English lawn	4–6 L/m² day
	Rustic lawn	1–3 L/m² day
	Hedges	6–8 L/m² day
Washing machine	Handwash	150 L
	Washing with pump	400 L
Washing areas	Washing floors	0.5 L/m²
Washing machine	Class A (low consumption)	60 L/load
	Class B (high consumption)	100 L/load
Dishwasher	Class A (low consumption)	30 L/load
	Class B (high consumption)	50 L/load

Source: Water Treatment Solutions, Lenntech. (www.lenntech.it)

are intended for meeting the requirements of a small number of users (Figure 44.8). For such installations, reference is usefully made to E DIN 1989-1: 2000-12.

The calculation for the storage volume is based on the net annual rainwater runoff (V_a) and the annual service-use water requirement (F_a). A value is found for this in the following way:

$$F_a = 365F_d \tag{44.6}$$

where F_d is the amount obtained for the daily water requirement.

Having calculated the two magnitudes, V_a and F_a, as shown above, we proceed to calculate the average monthly period of

FIGURE 44.8 Example of a small plant. (From http://www.seasons-matter.com/wp-content/uploads/2013/06/Keyport-Garden-Walk-pdf.pdf. With permission.)

dry weather (T_{MA}), which is the mean number of days per month during which there may be no precipitation. This value can be derived from the analysis of rainfall data or assessed by the following expression [9]:

$$T_{SM} = (365 - R)/12 \qquad (44.7)$$

where

T_{MA} = monthly average dry weather period [d]
R = number of rainy days in a year [d]

Volume W of the tank can now be determined analytically using the following equation:

$$W = T_{MA}F_a^*/365 \qquad (44.8)$$

The value of F_a^* for Equation 44.8 is to be derived as follows:

- if $V_a > F_a \rightarrow F_a^* = F_a$
- if $V_a < F_a \rightarrow F_a^* = (V_a + F_a)/2$

However, the method shown is not entirely satisfactory, as it fails to illustrate some important aspects. In particular, it does not provide a means of determining the efficiency of the system. Furthermore, it fails to take account of the actual distribution of daily consumption throughout the year. These deficiencies prevent us from being able correctly to estimate the level of convenience in constructing the system.

44.4.4 Sizing a Plant of Large Dimensions

The foregoing refers to single-family houses or buildings with few users. A different approach is necessary for large-scale rainwater harvesting systems, such as those for the reuse of rainwater for all compatible uses within public buildings such as hospitals, schools, and barracks (Figure 44.9).

We illustrate a method for sizing the storage volume of a large-scale system.

The method involves, as a first step, an estimate of the annual net theoretically collectible inflow of rainwater (V_a) and the annual service-use requirement of water (F_a).

V_a can be obtained from Equation 44.1. In detail, having estimated values for the surface area of the watershed and assigned a suitable value for the runoff coefficient (φ), we proceed to assessing the mean daily inflow V_{pd}, by means of Equation 44.2.

In this connection, we can turn to the deterministic approach referred to above. Given a historic series of N years of daily rainfall, we derive an average year of daily precipitation, estimating for each day of the mean year a depth of rainfall equal to the mean daily rainfall recorded over the N years taken into consideration.

With respect to the value of V_{pp}, we refer to the simplified criterion given above, which consists of reducing each mean rainfall depth value by 2–4 mm.

Evaporation losses (V_{ped}) have been ignored.

Having obtained the values of V_{pd}, V_{pp}, and V_{ped}, we now proceed to calculate V_{od}. Finally, from Equation 44.4, fixed a value for ε, we arrive at an estimate for daily mean values V_d.

The daily value for the volume of water requirement (F_d) is calculated, as shown previously, on the basis of unit consumption, number of appliances, and their daily utilization.

Having set out the above, it is now necessary to carry out a preliminary verification of the system on the basis of an annual reckoning of input volumes (V_a) and delivered volumes (F_a):

$$V_a = \sum_{i=1}^{365} V_{d,i} \qquad (44.9)$$

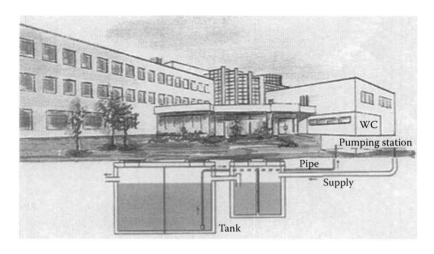

FIGURE 44.9 Example of a large plant. (From http://www.e-landscapellc.com/wp-content/uploads/2013/04/Rainwater-Harvest.jpg. With permission.)

$$F_a = \sum_{i=1}^{365} F_{d,i} \qquad (44.10)$$

Three possible scenarios can arise:

$$V_a = F_a; \quad V_a > F_a; \quad V_a < F_a$$

In the first scenario, the volume of rainwater collected in 1 year (V_a) equals the annual requirement (F_a). At the end of an average annual cycle, there will be no surplus or shortfall in water availability.

In the second scenario, the volume of rainwater that can potentially be collected in 1 year (V_a) turns out to be greater than the volume of maximum requirement (F_a) for the year in question. In this case, surplus water availability is removed from the storage tank by overflow.

In the third scenario, the volume of rainwater that can potentially be collected in 1 year (V_a) turns out to be less than the volume of maximum requirement (F_a) for the year in question. This

deficit in water availability will have to be made up for by drawing water from the mains water supply.

By way of example, Figure 44.10a–c shows cumulative values of the monthly inflow and the corresponding monthly requirement for three such scenarios as indicated.

Once the above preliminary evaluations for d and F_d have been attained, the volume W_o to be assigned to the storage tank can be calculated as follows.

We consider the 365 days of the reference mean year of rainfall and for each day the values of daily deviation, δ, are entered:

$$\delta = V_a - F_d \qquad (44.11)$$

The values shown for δ may, of course, turn out to be either positive (surplus of rainwater availability) or negative (a shortfall in availability).

The cumulative deviations over time are then calculated:

$$\Delta i = \sum_{j=1}^{i} \delta_j \qquad (44.12)$$

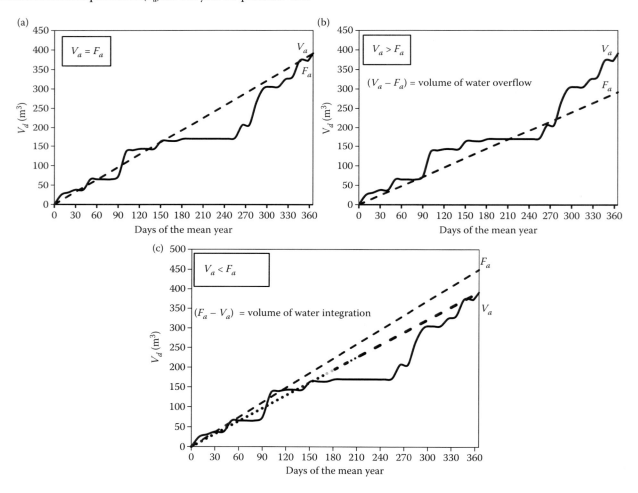

FIGURE 44.10 (a) Comparison between the values of V_d and F_d ($V_a = F_a$: compensation between availability and consumption of water). (b) Comparison between the values of V_d and F_d ($V_a > F_a$: surplus of water availability). (c) Comparison between the values of V_d and F_d ($V_a < F_a$: deficit in water availability).

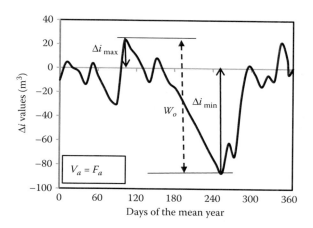

FIGURE 44.11 Typical trend of the values Δi in the days of mean year (case $V_a = F_a$).

where the ith cumulative deviation, Δi, is given by the sum of all deviations, δ, corresponding to the days of the year leading up to the ith, including the ith day itself.

The trend in the values of the cumulative deviation is characterized by an alternation between stretches of growth (accumulation phases) and stretches of shrinkage (emptying phase of the stored fillage).

From here, in order to calculate the volume W_o to be assigned to the tank, we have to proceed as follows:

Scenario 1: $V_a = F_a$

In this case, the histogram representing the trend in the volume of the tank is given by the equation:

$$W_o = \Delta i_{max} - \Delta i_{min} \qquad (44.13)$$

with Δi_{max} and Δi_{min} being, respectively, the maximum and minimum values of the cumulative deviations (Figure 44.11). The application of Equation 44.13 can be understood intuitively if you consider that any sequence of decreasing values of Δi represents in concrete terms a state of growing shortfalls in water availability that have to be made up for using stored water. For this reason, the sequence of shortfalls having the greatest amplitude gives us precisely a measurement for the maximum accumulation capacity that needs to be provided for.

It is helpful to note that during cyclic operation of the system, before reaching a critical point that requires having the full amount of stored water available, this volume will need to have been accumulated from the surpluses of the preceding days.

Scenario 2: $V_a > F_a$

In this scenario, which corresponds to a situation of a general surplus of water with an over-brimming tank, the useful compensating volume (W_o) to be assigned to the tank will again be calculated using Equation 44.13. With regard to the starting conditions, it can easily be shown how, following the first cycle, the system will go into equilibrium with the tank constantly returning to full

at the beginning of each cycle. It is then reasonable to begin simulations of daily water balances assuming the tank to be full.

Scenario 3: $V_a < F_a$

In this case, which corresponds to a situation of general water shortfall, the calculation of what useful volume W_o should be assigned to the tank proceeds by identifying an annual requirement F_a' ($<F_a$) equal to the net annual volume of input V_a in such a way as to simulate the condition $F_a' = V_a$.

Once this has been given, the volume W_o^* corresponding to this fictitious condition is once more calculated by means of Equation 44.13. This is the tank volume that will compensate for water requirement under the given conditions.

The water shortfall that will guarantee requirement F_d will now be obtained by drawing on the mains water supply, and for this function there is no need for a storage capacity.

Under real conditions, to volume W_o^*, a further capacity ΔW_o is added—(e.g., equal to the value for mean daily requirement)—which has a purely technical and noncompensatory function. The system would then have a tank whose total volume equals ($W_o = W_o^* + \Delta W_o$) and would be fitted with a ballcock/level detector, which activates external refilling as soon as the water level falls below capacity W_o^*.

It is worth mentioning that for each of the three scenarios analyzed, the planned system can be checked by running a simulation of its operational cycle.

To this end, it will suffice to add to the first day of the cycle a volume of water equivalent to Δi_{min} (for $V_a = F_a$), W_o (for $V_a > F_a$) and W_o^* (for $V_a < F_a$) and to check that the volume held in the tank at the end of the cycle is the same as the initial value.

With reference to scenario 1, Figure 44.12 shows how, in the tank W_o initially containing the volume of water Δi_{min}, the yearly operating cycle ends with the initial value of Δi_{min}. Furthermore, the tank during the operation has two specific moments: the first (point F in Figure 44.12), which is completely full; and the second (point E of Figure 44.12), which is completely empty instead.

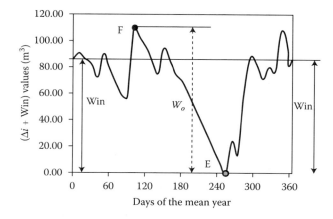

FIGURE 44.12 Simulation of the annual cycle of tank operation (case $V_a = F_a$; F = full tank; E = empty tank).

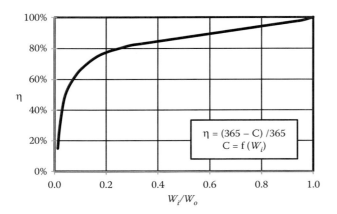

FIGURE 44.13 Typical trend of efficiency η versus W_i/W_o.

Finally, we can point out that in situation $(V_a > F_a)$, where one arrives at a situation of water overflow from the tank, the maximum volume of water overflow, W_{sf}, will be given by $(V_a - F_a)$.

44.4.5 Efficiency of Rainwater Reuse Systems

The efficiency η of the system is given by the ratio between the number of days during which the system is able to dispense stored rainwater and the total number of days of the chosen period of reference.

$$\eta = (365 - C)/365 \qquad (44.14)$$

where 365 represents the total number of days in the year and C indicates the number of days of the year in which the system cannot cope.

Figure 44.13 shows a typical efficiency trend η against variations in the volume W_o assigned to the system.

Once we have estimated a value for volume W_o to assign to the system, we can proceed to determining the efficiency (η) of the system by means of a series of simulated daily water balance calculations performed as shown above, changing each time capacity W_o.

More specifically, starting from the vale for W_o, we consider a series of values for $W_o1, W_o2 \ldots W_{oi}$ such that $W_o > W_o1 > W_o2 > \ldots W_{oi}$; thus, proceeding to carry out the simulation of system operation for each one of the values of W_{oi} assumed (i.e., to the initial day of the cycle one adds each time the value of W_{oi}). This simulation will show the negative values of cumulative deviations that correspond to days of the year in which it will be necessary to draw on external refurbishment; with the number of days during which the system cannot cope noted, you can determine, according to Equation 44.14, the efficiency η of the system.

44.5 Summary and Conclusions

The Earth's drinking water resources are generally sufficient, even if they are irregularly distributed in different countries. The remarkable growth of drinking water consumption leads us to predict water crisis scenarios for several countries. It is therefore necessary to intervene immediately with measures to minimize water losses, reduce consumption, and improve rainwater harvesting.

With reference to household water, about half of the entire daily drinking water consumption is actually wasted because we improperly use it for nondrinking purposes (toilet flushing, house cleaning, car washing, and garden irrigation). For these uses, it would be enough to have purified and odorless water but not necessarily of potable quality.

The approach to the problem of reducing consumption is carried out by means of educational programs aimed at the diffusion of best practices to avoid water wasting and through the application of appropriate technologies for domestic equipment (taps, toilet flushing, washing machines, dishwashers, etc.).

In this scenario, the adoption of appropriate plants aimed at household rainwater harvesting is of great interest. A plant for the reuse of rainwater is composed of several parts: collection system, storage system, distribution system, and treatment system.

With regard to the calculation of the volume to be assigned to the storage tank, we can distinguish the cases of small and large plants. In the first case, there are simple methods to determine the approximate volume of the tank. For the case of large systems, the problem is more complex.

In this chapter, we presented a criterion to determine the volume of the tank through a careful reconstruction of the heights of daily rainfall, which refer to an average meteorological year, and to the law of daily consumptions. The results obtained through the comparison between the two temporal laws allow examining the problem in detail under various practical scenarios. The technical and economical selection of the optimal value of the tank volume can be carried out by means of the calculation of the value of the efficiency of the plant.

References

1. AA. VV. 2002. *Acqua Sistemi e dispositivi per il risparmio e il riuso*. Edicom Edizioni (ed.), Gorizia, Italy, ISBN: 88-86729-31-6.
2. Abdulla, F.A. and Al-Shareef, A.W. 2008. *Roof Rainwater Harvesting Systems for Household Water Supply in Jordan*. Desalination, 243, 195–207.
3. Böse, K.H. 2012. *Recuperare l'acqua piovana per il giardino e la casa*. Terra Nuova Ed., Florence, Italy.
4. Cabell Brand Center. 2009. *Virginia Rainwater Harvesting Manual*, Version 2.0. Salem, VA. Ed Crawford, David Crawford, and Cabell Brand, www.cabellbrandcenter.org.
5. Campisano, A. and Modica, C. 2011. *Regional Evaluation of the Performance of Rooftop Rain Water Harvesting Systems for Domestic Use*, 2nd International Conference on Urban Drainage, Porto Alegre, Brazil, September 11–16, 2011.
6. Conte, G. 2008. *Nuvole e sciacquoni*. Ambiente, Italy.

7. Domenech, L. and Sauri, D. 2011. A comparative appraisal of the use of rainwater harvesting in single and multifamily buildings of Metropolitan Area of Barcelona (Spain). *Journal of Cleaner Production*, 19, 598–608.

8. European Environmental Agency. 2006. *The Problem of the Water Stress.* http://www.eea.europa.eu/publications/92-9167-025-1/page003.html.

9. Fanizzi, L. 2008. The systems for urban use of rainwater run-off. In: Ranieri, J. (ed.), *L'Ambiente*, vol. 1. Milano, Italy.

10. Fewkes, A. 1999. The use of rainwater for WC flushing: The field testing of a collection system. *Journal of Building and Environment*, 34, 765–772.

11. Foraste, J.A. and David, H. 2010. A methodology for using rainwater harvesting as a stormwater management BMP. *ASCE International Low Impact Development Conference, Redefining Water in the City.* San Francisco, CA, April 11–14, 2010.

12. Freni, G., Mannina, G., Torregrossa, M., and Viviani, G. 2007. *Sistemi localizzati di riuso delle acque reflue e meteoriche in ambiente urbano.* V Giornata di studio sul drenaggio, urbano sostenibile, riuso e risparmio acque reflue e meteoriche. Dicembre 13, 2007, Genova, Italy.

13. Ghisi, E., Lapolli Bressan, D. and Martini, M.. 2007. Rainwater tank capacity and potential for potable water saving by using rainwater in the residential sector of southeastern Brazil. *Building and Environment*, 42, 1654–1666.

14. Gleick, P.H. 2000. The changing water paradigm: A look at twenty first century water resources development. *Water International*, 25, 127–138.

15. Jones, M.P. and Hunt, W.F. 2010. Performance of rainwater harvesting systems in the southeastern United Sates. *Research Conservation and Recycling*, 54, 623–629.

16. Kinkade-Levario, H. 2007. *Design for Water: Rainwater Harvesting, Stormwater Catchment, and Alternate Water Reuse.* New Society Publisher Gabriola Island, BC, Canada, ISBN 978-0-86571-580-6.

17. Li, Z., Boyle, F. and Reynolds, A. 2010. Rainwater harvesting and greywater treatment systems for domestic application in Ireland. *Desalination*, 243, 195–207.

18. Maglionico, M. and Tondelli, S. 2003. *Gestione sostenibile delle risorse idriche e regolazione urbanistico edilizia.* DEI-Tipografia del Genio Civile, Roma, Italy.

19. Strickland, C.H. and Bloomberg, R. 2012. *Guidelines for the Design and Construction of Stormwater Management Systems.* Developed by the New York City Department of Environment.

20. UNESCO. 2006. Water a Shared Responsibility: The United Nations World Water Development Report 2 Parigi. http://www.unhabitat.org/programmes/water/documents/waterreport2.pdf.

21. Vickers, A. 2001. *Handbook of Water Conservation.* WaterPlow Press, Amherst, MA.

22. WHO/UNICEF. 2010. Joint Monitoring Program for Water Supply and Sanitation. Progress on Sanitation and Drinking-water. http://www.plaidoyercpu.org/documents/9789241505390_eng.pdf.

23. United Nations Development Program—Human Development Report 2006—www.data360.org.

24. Zini, L., Calligaris, C., Treu, F., Iervolino, D., and Lippi F. (a cura di) 2011. Risorse idriche sotterranee del Friuli Venezia Giulia: sostenibilità dell'attuale indirizzo. Edizioni EUT, 89 pp., Trieste, Italy, ISBN: 9-788883-033148.

45

Best Management Practices in Urban Water Cycle

Babak Bozorgy
MWH Global

PREFACE

This chapter provides an overview of best management practices in urban water cycle, with a focus on urban stormwater management, to achieve the objectives of sustainability, integrated urban water management, and water sensitive urban design.

First, the concepts and objectives of integrated water resources management, integrated urban water management, water sensitive urban design, and sustainable stormwater drainage, and their relationships are described.

Then, best practices of the application of these concepts in urban design and the management of stormwater drainage are described in brief and technical references and guidelines for design of the components of water sensitive urban design are introduced.

45.1 Introduction

According to the United Nations, "Sustainability calls for a decent standard of living for everyone today without compromising the needs of future generations." Since the early 1990s, when the concepts of "sustainability" and "sustainable development" became popular, scientists, engineers, and decision makers in different fields and sectors have been developing and adopting best management practices (BMPs) in an attempt to preserve the Earth's resources for future generations and building a better world.

In the context of "water resources," sustainability is defined as "coordinated development and management of water, land and related resources in order to maximize economic and social welfare in an equitable manner without compromising the sustainability of vital ecosystems" [3]. This is also called integrated water resources management (IWRM).

IWRM is a cross-sectoral policy approach, designed to replace the traditional, fragmented sectoral approach to water resources and management that has led to poor services and unsustainable resource use. It is based on the understanding that water resources are an integral component of the ecosystem, a natural resource, and a social and economic good [3].

In the urban context, IWRM is also called integrated urban water management (IUWM).

The world's urban population is estimated to grow by 2.6 billion over the next four decades, with most of that growth taking place in less developed regions. By 2025, cities of one million or more will be home to more than 47% of the urban population. Competition for water resources across all sectors will become fierce. At the same time, raw water sources risk becoming more contaminated through changes in land use patterns, poor solid waste management, inadequate wastewater treatment, and aging infrastructure. Consequently, the quantity and quality of water available to cities for agriculture, energy, industry, and human development needs is and will remain in constant flux [5].

With many sectors relying on the same water source, the competitive dynamics at play require an integrated approach to urban water management.

IUWM is a holistic mode of strategic planning. It takes a landscape view of water challenges by looking at competing water users in a given catchment or river basin. Through coordinated and flexible planning among water using sectors, IUWM allows for the optimal sequencing of traditional and new infrastructure with alternative management scenarios that leverage efficiencies and promote conservation. The emerging IUWM approach offers a more diverse and versatile set of options for dealing with larger and more complex urban water challenges [5].

IUWM provides a framework for planning, designing, and managing urban water systems. It is a flexible process that responds to change and enables stakeholders to predict the impacts of interventions. IUWM includes environmental, economic, social, technical, and political aspects of water management. It brings together fresh water, wastewater, stormwater, and solid waste, and enables better management of water quantity and quality.

IUWM calls for aligning urban development with basin management to ensure sustainable economic, social, and environmental relations along the urban–rural continuum [2].

Progressing from conventional urban water management toward IUWM requires a paradigm shift from resource use and waste management (trying to collect and dispose of stormwater as fast as possible) to resource management.

Development policies and strategies supported by financing strategies, technological developments, and tools for decision-making, in cooperation with both public and private sector partners, can facilitate putting IUWM into practice at all levels.

This chapter provides an overview of BMPs in urban water cycle, with a focus on urban stormwater management, to achieve the objectives of IUWM and urban water resource sustainability.

45.2 The Concept of Integrated Urban Water Management

IUWM is a concept designed to facilitate sustainable water use and management in urban areas. It calls for the alignment of urban development and basin management to achieve sustainable economic, social, and environmental goals. It brings together water supply, sanitation, and stormwater and wastewater management and integrates these with land use planning and economic development [2].

IUWM principles as defined by Global Water Partnership (GWP) are [2]

- Encompass alternative water sources.
- Match water quality with water use.
- Integrate water storage, distribution, treatment, recycling, and disposal.
- Protect, conserve, and exploit water resources at their source.

- Account for nonurban users.
- Recognize and seek to align formal and informal institutions and practices.
- Recognize relationships among water, land use, and energy.
- Pursue efficiency, equity, and sustainability.
- Encourage participation by all stakeholders.

The IUWM approach begins with clear national policies on IWRM, backed by effective legislation to guide local authorities. A successful IUWM plan requires a collaborative approach that involves all stakeholders in setting priorities, making decisions, and taking action and responsibility.

IUWM assesses both water quantity and quality, estimates future demands, anticipates the impacts of climate change, and recognizes the importance of efficiency, without which water operations cannot be sustainable. It also recognizes that different water sources can be used for different purposes; fresh water and desalinated water for domestic use; and treated wastewater for agriculture, industry, and the environment [2].

IUWM requires the development of planning and management for all components of urban water services.

Figure 45.1 illustrates the coordinating structure that will ensure communication between departments, levels of government, local communities, and stakeholders [2].

Urban planners can help governments overcome fragmented public policy and decision making by linking planning with other policy sectors like infrastructure, and adopting collaborative approaches that involve all stakeholders in determining priorities, actions, and responsibilities [2].

Figure 45.2 shows how different policy sectors are integrated under the concept of IUWM.

As shown in Figure 45.2, one of the main elements of IUWM is urban stormwater and drainage management. In recent years, the application of IUWM in urban stormwater management has been translated to the concept of sustainable stormwater and drainage management. Various practitioners, engineers, and scientists across the world have adopted the concept and developed it into practical principles and frameworks, all with the same overall concept and idea, but have used different terminologies, that is, sustainable drainage systems (SuDS) in the United Kingdom, low impact development (LID) and BMPs in the United States, and water sensitive urban design (WSUD) in Australia and New Zealand. However, the terminology of WSUD is now becoming more favorable and popular worldwide.

45.2.1 Water-Sensitive Urban Design

Conventional urban stormwater management has historically focused on providing efficient drainage infrastructure to rapidly collect and remove stormwater runoff. This has created problems such as huge investments on large infrastructure as well as increased environmental impacts on the downstream outlets of stormwater drainage systems, by transferring all the pollutants down the system [7].

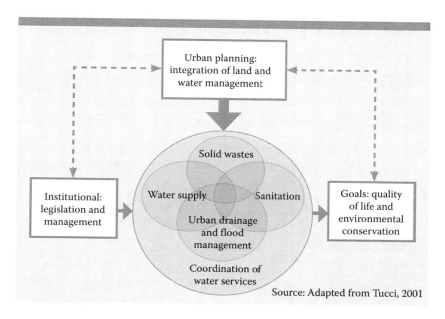

FIGURE 45.1 IWUM coordination structure. (From Global Water Partnership (GWP). 2013. *Policy Brief TEC 16, Integrated Urban Water Management (IUWM): Towards Diversification and Sustainability.* Global Water Partnership (GWP), Stockholm, Sweden.)

To tackle these issues, stormwater needs to be managed at the source, as much as practical, both in terms of quality and quantity. Management of stormwater runoff cannot be considered in isolation to the broader planning and design of the contributing urban area. Rather, stormwater management needs to be considered at all stages of the urban planning and design process.

The concept of WSUD is based on formulating development plans that incorporate multiple stormwater management objectives and involve a proactive process that recognizes the opportunities for urban design, landscape architecture, and stormwater management infrastructure to be intrinsically linked. At the building design level, innovative means for recycling stormwater for nonpotable usage can significantly reduce pressure placed on water resources development and drainage infrastructure by urban development [7].

The concept of WSUD provides the basis for a holistic approach to stormwater management at both regional and local levels. WSUD is the integration of urban planning and utilization of BMPs to achieve the objectives of sustainable drainage systems for urban areas.

At a particular site, sustainable drainage systems are designed both to manage the environmental risks resulting from urban runoff and to contribute, wherever possible, to environmental enhancement. Therefore, the objectives of such systems are to minimize the impacts from development on the quantity and quality of runoff, and maximize amenity and biodiversity opportunities.

The three-way concept, set out in Figure 45.3, shows the main objectives that the approach is attempting to achieve. The objectives should all have equal standing, and the ideal solution will achieve benefits in all three categories, although the extent to which this is possible will depend on site characteristics and constraints. The philosophy of sustainable drainage systems is to replicate, as closely as possible, the natural drainage from a site before development [8].

The framework of WSUD is also shown in Figure 45.4.

45.2.2 Sustainable Drainage Management Train

There are many pollutants present in stormwater runoff, such as oil, nutrients, and fine sediments, which need to be reduced

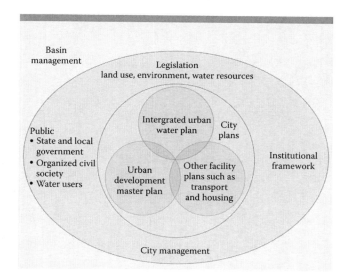

FIGURE 45.2 Integration of policy sectors under IUWM. (From Global Water Partnership (GWP). 2013. *Policy Brief TEC 16, Integrated Urban Water Management (IUWM): Towards Diversification and Sustainability.* Global Water Partnership (GWP), Stockholm, Sweden.)

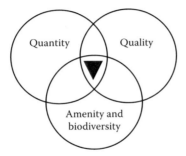

FIGURE 45.3 Sustainable drainage objectives. (Adapted from Woods-Ballard, B., Kellagher, R., Martin, P., Jefferies, C., Bray, R., and Shaffer, P. 2007. *CIRIA C697: The SUDS Manual*. CIRIA, London, United Kingdom.)

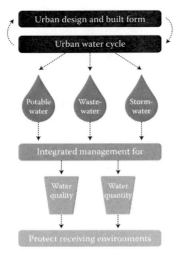

FIGURE 45.4 Framework of WSUD. (Adapted from Wellington City Council. 2014. *Water Sensitive Urban Design: A Guide for WSUD Stormwater Management in Wellington*. Wellington City Council, New Zealand.)

especially before the stormwater runoff is discharged into a downstream water body. This requires a number of measures used in sequence for treatment designed to meet the stormwater treatment needs of a particular environment. This sequence is called a "sustainable drainage management train (SDMT)."

To mimic natural catchment processes as closely as possible, a "management train" is required. This concept is fundamental to designing a successful WSUD scheme. It uses drainage techniques in series to incrementally reduce pollution, flow rates, and volumes. The hierarchy of techniques that should be considered in developing the management train is as follows [8]:

- *Prevention*: The use of good site design and site housekeeping measures to prevent runoff and pollution, for example, sweeping to remove surface dust and detritus from car parks, and rainwater reuse/harvesting. Prevention policies should generally be included within the site management plan.
- *Source control*: Control of runoff at or very near its source, for example, soakaways, other infiltration methods, green roofs, or pervious pavements.
- *Site control*: Management of water in a local area or site, for example, routing water from building roofs and car parks to a large soakaway, infiltration, or detention basin.
- *Regional control*: Management of runoff from a site or several sites, typically in a balancing pond or wetland.

Figure 45.5 shows the sustainable drainage management train. Wherever possible, stormwater should be managed in small, cost-effective landscape features located within small subcatchments rather than being conveyed to and managed in large systems at the bottom of drainage areas. The techniques that are higher in the hierarchy are preferred to those further down so that prevention and control of water at the source should always be considered before site or regional controls. However, where upstream control opportunities are restricted, a number of lower

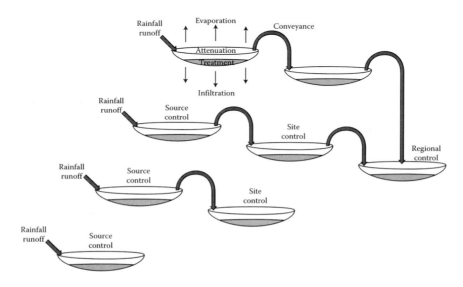

FIGURE 45.5 Sustainable Drainage Management Train. (Adapted from Woods-Ballard, B., Kellagher, R., Martin, P., Jefferies, C., Bray, R., and Shaffer, P. 2007. *CIRIA C697: The SUDS Manual*. CIRIA, London, United Kingdom.)

hierarchy options should be used in series. Water should be conveyed elsewhere only if it could not be dealt with on site [8].

Stormwater quality and quantity management strategies require a combination of measures. Some stormwater management measures have a dual function of quality and quantity control, while some other measures only have a single function, that is, they either control the flow and provide some attenuation or enhance the quality of the stormwater. Therefore the positioning and prioritization of the stormwater management measures within a sustainable drainage management train is of paramount importance and needs to take into account the site-specific characteristics, constraints, and opportunities.

45.3 Best Management Practices in Water Sensitive Urban Design

The previous section described the concepts and objectives of IUWM, WSUD, and SDMT. This section describes the best practices of the application of these concepts in urban design and the management of stormwater drainage.

45.3.1 WSUD System Components

Following the concept of SDMT and depending on the project requirements, environmental and urban planning legislation, and regulations and site constraints, different measures and system components can be included in the WSUD schemes.

A summary of the main WSUD measures and system components as per the BMPs developed and applied in different countries such as the United States, United Kingdom, Australia, and New Zealand, as the pioneers of the WSUD concept, is provided next.

For more details on the WSUD/SuDS measures and system components and their design criteria, refer to the following sources and the list of references at the end of this chapter.

- Stormwater Management Best Practices publication series by the US Environmental Protection Agency (EPA).
- SuDS publication series by the Construction Industry Research and Information Association (CIRIA) of the United Kingdom.
- Publications of the Australian Water Conservation and Reuse Research Program by the Australian Water Association (AWA) and the Commonwealth Scientific and Industrial Research Organization (CSIRO).
- WSUD guidelines of city councils in Australia and New Zealand.

45.3.1.1 Rainwater Tanks

The capturing or harvesting of rainwater for reuse can contribute to water conservation and water quality objectives. It complements other approaches to IUWM such as demand management and the reuse of effluent and greywater. Moreover, especially in wet regions, rainwater tanks can help in reducing water bills.

Rainwater tanks allow the collection and storage of rainwater from roofs and its usage for many purposes, including garden irrigation, washing cars, filling pools, and household tasks such as washing and toilet flushing.

Rainwater tanks are available in the form of under- or aboveground and the size is usually determined by vendor/manufacturer based on the available rainfall in the region and the surface area of the roof. An example of a rainwater tank is shown in Figure 45.6.

45.3.1.2 Gross Pollutant Traps

Gross pollutant traps (GPTs) are devices for the removal of solids conveyed by stormwater runoff that are typically greater than 5 mm. Examples of GPTs currently used in urban catchments include gully baskets, in-ground GPTs, trash racks, and pipe nets [4].

A schematic illustration of a GPT is shown in Figure 45.7.

GPTs should be considered at the scoping stage of any WSUD project. The primary purpose of GPTs is to remove gross pollutants and coarse sediments washed into the stormwater system before the stormwater enters the receiving waters. While most GPTs capture both categories of pollutants, there are some that

FIGURE 45.6 Example of a rainwater tank (City of Victor Harbor, SA Australia).

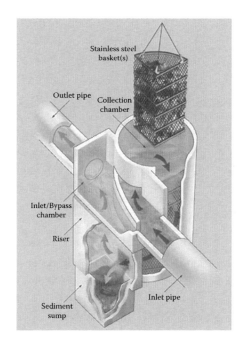

FIGURE 45.7 Schematic illustration of a GPT. (From Rocla.)

FIGURE 45.8 Pervious pavement (Manly, NSW Australia).

target litter and debris exclusively and others that are designed for sediment removal only. Generally, GPTs are used to provide primary treatment within a WSUD treatment train. GPTs do not contribute to flood control. Indeed, unmaintained inline GPTs can contribute to increased flooding by generating additional backwater effect [4].

GPTs require a considerable amount of maintenance to ensure that they continue to operate at the design level of performance. A maintenance and monitoring management plan is required to monitor the performance of and service the given GPT device; therefore, this should be developed during the design process.

45.3.1.3 Pervious Pavements

Pervious surfaces allow rainwater to infiltrate through the surface into an underlying storage layer, where water is stored before infiltration to the ground, reuse, or release to surface water.

Pervious pavement has two main advantages over impervious pavement, in terms of stormwater management [1]:

- Improvement in water quality, through filtering, interception, and biological treatment.
- Flow attenuation, through infiltration and storage.

However, it should be noted that pervious pavements might be expensive and prone to clogging.

Figure 45.8 shows an example of pervious pavement.

45.3.1.4 Vegetated Swales and Filter Strips

These are wide, gently sloping, and shallow channels covered by grass or other suitable vegetation, which provide some stormwater filtration and treatment prior to discharge to downstream drainage systems or receiving waters.

Figure 45.9 shows an example of a vegetated swale.

45.3.1.5 Infiltration and Bioretention Basins

Infiltration and bioretention basins are depressions in the surface that are designed to store runoff, infiltrate the water to the ground, and remove the pollutants to some extent. They may also be landscaped to provide aesthetic and amenity value.

Figure 45.10 shows an example of bioretention basin.

Infiltration and bioretention basins use vegetation and filter layers to remove pollutants from stormwater runoff. Infiltration basins

allow the stormwater runoff to infiltrate into the ground through the filter layer. Bioretention basins detain the stormwater runoff (primarily the first flush, which contains a higher concentration of pollutants) and slowly discharge it to the receiving stormwater drainage system, with removal of pollutants to some extent.

Maintenance of the vegetation in bioretention basins is critical as it contributes to biological uptake, maintaining porosity of the filter media, and enhancing sedimentation within the basin.

45.3.1.6 Ponds, Wetlands, and Sediment Basins

Ponds, wetlands, and sediment basins operate using similar mechanisms of flow attenuation, sedimentation, and in some cases, filtration to remove contaminants from urban stormwater [1].

Wet ponds are basins that have a permanent pool of water for water quality treatment. They provide temporary storage for additional storm runoff above the permanent water level. Wet ponds may provide amenity and wildlife benefits. Constructed wetlands are ponds with shallow areas and wetland vegetation to improve pollutant removal and enhance wildlife habitat. Extended detention basins are normally dry, though they may have small permanent pools at the inlet and outlet. They are designed to detain a certain volume of runoff as well as provide water quality treatment [8].

Figure 45.11 shows an example of a constructed wetland.

45.3.2 Modeling of WSUD Components

The performance of WSUD components needs to be assessed and estimated to ensure that the overall performance of the SDMT meets the water quality enhancement and water quantity control requirements of the particular project and/or regulating authority. Reference 8 of this chapter provides indicative estimates of the performance of WSUD components in pollutant removal.

To assess the performance of WSUD systems, two different sets of models, or a combination of both, can be used:

- Hydrologic/hydraulic models to assess the flow routing and attenuation and the hydraulic performance of the system.
- Water quality models to assess the pollutant removal performance of the system.

FIGURE 45.9 Vegetated swale.

FIGURE 45.10 Bioretention basin.

FIGURE 45.11 Constructed wetland.

There are a number of industry recognized software packages available for the assessment of the hydrologic, hydraulic, and water quality performance of WSUD systems. Examples of those are EPA-SWMM, InfoSWMM, InfoWorks ICM, MIKE URBAN, XP-SWMM, DRAINS, and MUSIC.

45.4 Summary and Conclusions

This chapter provides an overview of BMPs in urban water cycle, with a focus on urban stormwater management, to achieve the objectives of sustainability, IUWM, and WSUD.

It describes the concepts and objectives of IWRM, IUWM, WSUD, and sustainable stormwater drainage, and their relationships, briefly describes the best practices of the application of these concepts in urban design and the management of stormwater drainage, and introduces the technical references and guidelines for design of the components of WSUD.

WSUD shall be incorporated into and integrated with all other aspects of urban design in order to move toward sustainability in stormwater management.

One of the most important aspects of WSUD is the SDMT and its hierarchy and priority from prevention and source control to site control and regional control.

Incorporating WSUD components into the urban environment will help to reduce the concentration of pollutants in stormwater runoff and enhance the water quality of the receiving systems and water bodies and at the same time improve the urban amenities and landscapes.

References

1. Fletcher, T.D., Deletic, A.B., and Hatt, B.E. 2004. *A Review of Stormwater Sensitive Urban Design in Australia*. AWA & CSIRO, Monash University, Australia.
2. Global Water Partnership (GWP). 2013. *Policy Brief TEC 16, Integrated Urban Water Management (IUWM): Towards Diversification and Sustainability*. Global Water Partnership (GWP), Stockholm, Sweden.
3. Global Water Partnership (GWP). http://www.gwp.org/The-Challenge/What-is-IWRM/.
4. South Australia Department of Planning and Local Government. 2010. *Water Sensitive Urban Design Technical Manual for the Greater Adelaide Region*. Government of South Australia, Adelaide, Australia.
5. The World Bank. http://water.worldbank.org/iuwm.
6. Wellington City Council. 2014. *Water Sensitive Urban Design: A Guide for WSUD Stormwater Management in Wellington*. Wellington City Council, New Zealand.
7. Wong, T.H.F. 2000. Improving urban stormwater quality—From theory to implementation. *Water—Journal of the Australian Water Association*, 27(6), 28–31, Australia.
8. Woods-Ballard, B., Kellagher, R., Martin, P., Jefferies, C., Bray, R., and Shaffer, P. 2007. *CIRIA C697: The SUDS Manual*. CIRIA, London, United Kingdom.

46

Managing the Unintended Consequences of Mining: Acid Mine Drainage in Johannesburg

Anthony Richard Turton
University of Free State

PREFACE

The South African economy is based on mining, with the Witwatersrand Goldfields having provided 40% of all gold ever mined by humans in all of recorded history. Central to this is the city of Johannesburg, which owes its existence entirely to the gold mining industry. Unlike Kalgoorlie in Western Australia that is also a gold mining city, Johannesburg is a major urban center currently experiencing exponential population growth due to in-migration. The city is unique in many ways, most notably because of the intimate coexistence of a large human population, mostly in informal settlements, with mining waste. The South African gold mining sector is in rapid decline and is not expected to last as a viable industry for more than a decade. This raises the issue of managing the unintended consequences of mining in a city that is water-constrained and straddles a continental watershed divide.

This chapter deals with the issue of acid mine drainage (AMD) that became prominent in 2002 when the first of three major mining basins beneath the greater conurbation of Johannesburg flooded and began decanting into a small stream that had not flowed for more than a century because of active dewatering. The problem is complicated by the existence of uranium in vast quantities in the many mine tailings dams that litter the landscape. It is argued that it is in nobody's best interest for the mining industry to become insolvent, as this will leave the hundreds of tailings dams to collapse under the natural forces of wind and water erosion. The case is made for the transformation of conventional mining into what is called closure mining in which the treatment and reuse of AMD is a central element. The combined decant potential of the three mining basins is equivalent to 14% of the daily needs of the city, so the various options for the use of this water are analyzed.

46.1 Introduction

Johannesburg is a unique city from a water resource and water services perspective because it is not located on a river, lake, or waterfront. Instead, it straddles a continental watershed divide between two great river basins—the Orange that discharges into the Atlantic Ocean and the Limpopo that discharges into the Indian Ocean. This makes it an engineering masterpiece because water has had to be pumped uphill, from ever-distant basins, simply to sustain a growing city. As with all heroic battles against the persistence of gravity, however, the water simply flows away once used, this time as sewage or water contaminated by industry. Johannesburg is also the epicenter of the Witwatersrand Gold Mining Complex, which has produced a staggering 40% of all gold that has ever been mined in all of recorded history. The name for the city in the vernacular is "eGoli" (isiZulu) and "Gauteng" (Northern Sotho), both of which mean "place of gold." This chapter tells the story about the linkage between the engineering challenges to sustain a modern African city, in the face of the social challenge arising from the unintended consequences of externalities from more than a century of mining where state regulation was minimal.

46.2 The Problem and Its Setting

The Witwatersrand Gold Mining Complex is the result of the deposition of metals into a lagoon in what used to be an inland ocean. This drained a hinterland rich in metals, including gold and uranium, which flowed through high velocity braided rivers discharging into this lagoon, rich in filamentous algae that trapped these heavy particles [32]. This is the richest single gold deposit on Earth, yielding 40% of all gold mined in recorded history; this area is also rich in uranium [13]. In fact, depending on the reef band being mined, for every tonne of gold produced, between 10 and 100 tonnes of uranium was brought to the surface, mostly discarded as waste because it had limited commercial value at that time [33]. The legacy of gold mining is the existence of 600,000

tonnes of uranium in various species that currently lie discarded in these old dumps [9,30]. There is a swathe of land running just south of the center of the Johannesburg, approximately 98 km long and 2 km wide, that has been undermined and is thus geotechnically unstable [24,26]. With a complex surface striking reef band consisting of more than 12 clearly defined packages, there are openings to surface approximately every 100 m along the entire length of the strike, creating multiple ingress points for water and access points for illegal artisanal miners (see Figure 46.1).

The Witwatersrand mining basin beneath the city of Johannesburg and its immediate conurbation of satellite towns is divided into three distinct geohydrological units, each separated by clearly defined aquitards or impervious barriers associated with tectonic activities over geological timescales. These consist of the Eastern Basin (under the satellite conurbation towns of Brakpan, Springs, and Benoni), the Central Basin (under the city itself but stretching to Roodepoort), and the Western Basin (under the satellite towns of Krugersdorp and Randfontein).

The main gold reserves are largely depleted, with the life of the total resource based on existing economics and technologies expected to be at an end in the next decade [13], at least if current business models are used. The legislation that governs mine waste is complex [11], in essence removing it from the normal waste management regime by defining it as a resource that can be reused in the future. The unintended consequence of this is

FIGURE 46.1 Shallow undermining into gold-bearing reef. Thousands of these shallow workings exist in the Johannesburg area.

that all mine dumps in the Johannesburg region are not in their final resting place, so they cannot be rehabilitated without the significant investment of taxpayers' money into this process. The reason for this is that mineral resources have been closely linked to the national security of the country [12]. This meant that during the international isolation of the South African state arising from the imposition of comprehensive economic sanctions against the policy of Apartheid, the industry was protected as a matter of perceived national survival [29]. In effect, the state ceased to be a regulator from 1961 to 1994, becoming a partner in an elaborate collaboration that maximized profits to the mining houses by externalizing liabilities, but always assuring healthy revenue streams needed to keep the embattled pariah state afloat [27]. In all probability, had the gold sector not been as powerful as it was for such a long time, the transition to democracy would have occurred sooner than it actually did in 1994.

The strategic-level challenge that is now confronting the government is how to deliver potable water services and create jobs in a stagnating and water-constrained economy where the legacy of mining externalities have become constraints to future development.

46.3 Unintended Consequences

The economics of mining are driven on the supply side by three major costs—energy, labor, and ore grades. The price of energy is rising rapidly as a result of the failure by the state to recapitalize the national electricity utility called ESKOM, with year-on-year escalations of 16% over the last 3 years and 8% increases planned for the next 3 years. The price of labor is also rising rapidly, most notably as a result of demands by militant trade unions that are unrelated to increases in productivity, with year-on-year escalations in the order of 10% [1]. Ore grades are falling, with the majority of the resources considered to be economically viable having been depleted [13].

One of the biggest challenges facing the gold producing industry is the cost of dewatering deep mines. The gold bearing reef dips from the surface in a southerly direction, where it is overlain by one of the largest karstic aquifer systems in the world [3,16]. A number of mines coexist in a given mining basin, each linked in a geohydrological sense by virtue of their geographic location in a confined aquifer system. Because of the need to create multiple exit points from deep mines in the event of an emergency, all of these underground workings have been interconnected, further enhancing their hydrological connectivity. This means that under normal operational conditions, the volumes needed to be abstracted are large and variable, dependent on the ingress into the void during the rainy season. The volumes per basin needed to retain steady state operations once the void level has been drawn down to environmental critical level (ECL) are as follows: Eastern Basin—70–100 mega-liters per day (Ml/d); Central Basin—30–90 Ml/d; Western Basin—19–27 Ml/d [7]. When compared to the 1500 Ml/d supplied by the public utility Johannesburg Water, AMD decant from these three basins equals approximately 14% of the total daily needs of the city.

In order to ensure the economic viability of the industry, a number of independent mining companies share the cost of pumping. The unintended consequence of this approach is that when a given company ceases to be financially viable due to increasing costs of labor and energy in the face of falling ore grades, then it shuts off the pumps, placing an additional burden on the remaining companies in the given mining basin. This implies that the probability of insolvency for the remaining companies is inversely related to the number of mines still operating. This principle is known as "The Last Man Standing."

In the Western Basin, all underground mining ceased in the 1990s and the void started to fill with water. This became increasingly acidic as a result of exposure to pyrite, the majority of which is found in the surface tailings dams, which enters the void via multiple ingress points [28] (see Figure 46.2). This is exacerbated by the existence of many surface holings and cracks in the rock caused by shallow underground mining (see Figure 46.1). In 2002, the water level in the void reached the Black Reef Incline (BRI) Shaft, which is the lowest entry point into the Western Basin mine void, where decanting occurred into the Tweelopies Spruit, a small stream emerging from a karst structure that had been dry in living memory because of the dewatering. The decanting water (see Figure 46.3a and b) was highly acidic with a pH value of around 3, into which a large number of heavy metals (including uranium) had been dissolved along with sulfate loads of up to 4000 ppm [4,14,22,34].

Public interest was massive as the media picked up the story. The interest grew when scientists reporting on the subject were harassed and attempts were made to suppress their work [5,6,31]. Public interest was magnified when the cost to the taxpayer of derelict and ownerless mines was estimated to be ZAR 100 bn (US$ 10 bn) in 2007 [2], but this is, in all probability, a gross underestimation of the true value, given the known complexity of the problem [4]. This interest turned to anger as the media began presenting the story as an example of a greedy mining industry and an uncaring government [21,23,25].

46.4 Turning the Tide

In response to the growing tide of public anger, the government created the Inter-Ministerial Task Team on Acid Mine Drainage, which tabled a public report in December 2010 [4,19]. This report concluded with the following:

> Currently two plants are treating AMD to potable quality in South Africa at full scale. These are, however, not financially self-sustaining. This is similar to the experience internationally that has shown that AMD treatment is unlikely to be financially self-sustaining. The costs of this treatment are estimated at around R11 per cubic metre (approximately US$ 1.10/m³), with a capacity of treating 20 Ml/d (20,000 m³/d) at each plant, including amortisation of the capital costs of the plant (several hundred million Rand) over the projected 20-year design life of the plant. This is not economically self-sustaining and relies on a

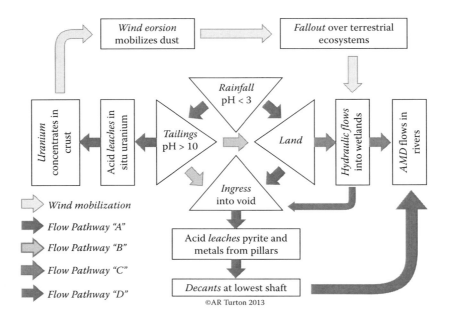

FIGURE 46.2   AMD flow pathways showing the existence of discreet subsystems.

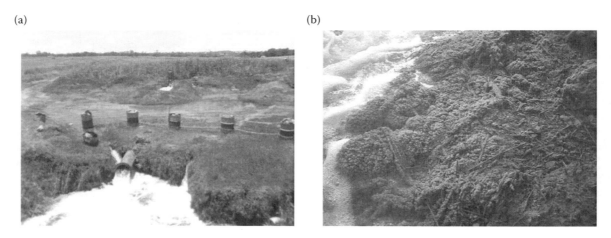

FIGURE 46.3   (a) Shows the decanting mine water under conditions of winter low flow. (b) Shows the precipitation of salts in the Tweelopiespruit, approximately a kilometer downstream of the decant point.

subsidy from the mining companies. Therefore, it is fore-seen that there will be a shortfall between the cost of clean water produced in a plant and the revenue recoverable from the sale of water [4].

What this report does not acknowledge is the growing tide of public anger, most notably as a consequence of perceptions that government is increasingly authoritarian and corrupt [8,15]. In Johannesburg, this anger has already resulted in the demise of an initiative by the mining sector known as Western Utilities Corporation (WUC), which sought to obtain mine closure cer-tificates by treating AMD to potable standards using the lowest cost technology, without adequate public consultation [27]. It is against this background that the potential for urban water reuse

as it pertains to the gold mining areas of Johannesburg needs to be understood.

46.5 Understating AMD as Discreet Hydrological Flow Pathways

There has been a lack of conceptual clarity needed to accurately understand what AMD is actually about, or more importantly, where interventions need to be made in the overall cycle. In the absence of this clarity, but in the face of growing public anger fuelled by environmental activists, all of the perceived evils of the mining industry have been placed at the door of AMD [23]. This has caused a shift in focus to one item only—the neutralization of acidic water at the decant point. We therefore need to better understand AMD as a complex hydrological flow in which a number of nested subsystems exist [28]. This is presented in Figure 46.2.

There are four discreet, but nested, flow pathways for AMD, each capable of being accurately modeled. The genesis of AMD is in Flow Pathway "A," which is driven by rain with a low pH (<4) falling on flat-topped tailings dams, where it attacks the hydroxide coating around quartzite particles that also contain pyrite. This oxidizes the pyrite, triggering the generation of additional acid, which in turns starts to leach the uranium in situ. The uranium concentrates in a crust that is eroded by wind when it desiccates. This flow pathway is thus limited in geographic scale to the mine dump itself, becoming the source of both acid and uraniferous dust that enters subsequent flow pathways.

Flow Pathway "B" is a gatekeeper, as it determines whether the water will either subsequently flow across the surface into aquatic ecosystems or into the underground void. The scale is limited to the footprint of the mine dump and the land immediately adjacent to it.

Flow Pathway "C" is driven by aquatic ecosystems with three main inputs, each of which can be numerically modeled. These are rainfall as an episodic series of events; fallout of uraniferous dust over large swathes of land that is mobilized hydraulically with each rainfall event; and direct runoff from tailings dams as a result of the surface flowing output from Flow Pathway "B." The scale of this subsystem ranges from quaternary catchment up to river basin level, with an increasingly large footprint as uraniferous dust deposition escalates from poorly maintained and collapsing tailings dams.

Flow Pathway "D" is entirely underground, with three unique but specific inputs: seepage of acidic water from beneath the footprint of the surface tailings dam (Flow Pathway "B"); infiltration of rainfall via multiple ingress points such as naturally

weathered rock and shallow underground workings (Figure 46.1); and direct ingress of surface flows from wetlands and rivers that intersect preferential pathways such as faults, fissures, and dykes. Once underground, this acidic water attacks the pyrite in the remaining pillars and stopes, aided by the bacterium *Thiobacillus feroxidans*, where it eventually decants from the void via the lowest shaft opening to the surface.

Armed with this enhanced conceptual clarity, it becomes evident that AMD is highly complex, so there is no single silver bullet solution. More importantly, it becomes increasingly evident that the predicted imminent demise of the gold mining industry, at least in the current format in which it is operating, will have catastrophic implications for water resource management in an area that will be defined by the fallout of uraniferous dust mobilized by rainfall events and deposited into aquatic ecosystems (wetlands and rivers). More importantly, the case for managing AMD as a desired outcome of closure mining (see Figure 46.4) is made more robust because it is only by means of such an integrated approach that management interventions can occur at appropriate points in the overall cycle. Ideally, these interventions should be aimed at removing surface tailings for placement back into the void while sequestering uranium as far as possible (Flow Pathway "A"); prevention of wind mobilization of uraniferous dust by means of adequate rehabilitation of remaining surface dumps; prevention of surface retention of rainfall and subsequent runoff from the dumps through reengineering the profile (Flow Pathway "B"); active management of wetlands as natural sinks for uranium and other elements (Flow Pathway "C"); active management of ingress through a directed approach designed to identify all intersections and develop appropriate solutions (Flow Pathway "D"); and finally the active pumping and treatment of the void water to prevent breach of the ECL.

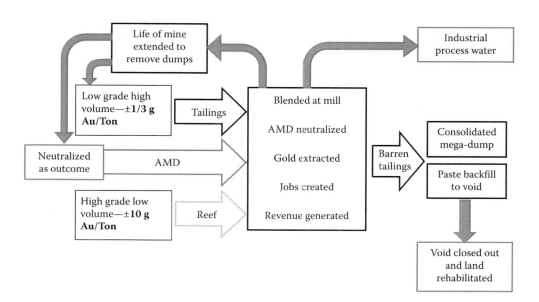

FIGURE 46.4 Closure mining as a conceptual model showing the core elements of the overall process.

46.6 Urban Water Reuse

The Inter-Ministerial Report noted that the following volumes would initially have to be dealt with in order to achieve the steady-state operation noted above: Eastern Basin—108 Ml/d; Central Basin—60 Ml/d; Western Basin—40 Ml/d [4]. As these volumes are quite significant and would require substantial infrastructure to treat and reticulate the water, focus is now shifting onto two distinct aspects:

- What is the most appropriate business model to sustain AMD treatment over time?
- What is the most appropriate process model that is socially acceptable?

With respect to the business models, there are two distinct versions—the Legacy Business Model and the Public Private Partnership (PPP) Business Model.

The Legacy Business Model is based on the assumption that AMD is the result of irresponsible mining. The few remaining mining companies are inevitably destined to become insolvent within the next decade, as a direct result of rising costs of labor, energy, and water treatment in the face of falling ore grades. Because of this, the mining industry is to be bypassed in solution seeking, with contracted third parties acting as service providers. The preferred technology is the high-density sludge (HDS) process, followed by desalination. The source of revenue for this solution would be taxation that will need to be provided in perpetuity. The disposal of HDS, rich in heavy metals and sulfate, and subsequently brine, is a major cost in this model. At a technical level, the core assumption is that AMD is generated in the mine void, so it is blind to the existence of the various flow pathways shown in Figure 46.2, focusing only on Flow Pathway "D," but without reference to that model. This means that surface tailings dams are not part of any remediation plan and would remain there in perpetuity. As a direct result of this, the 600,000 tonnes of uranium lying in those combined dumps will become a growing hazard as the dumps start to collapse through erosion by both rain and wind. The latter is inevitable as soon as revenues cease to flow because conventional mining operations include the constant maintenance of dumps needed to retain the structural integrity of these flat-topped stepped-sided structures. Of even greater significance, the mining void would remain open as this is not part of the overall management objective, which means that the long-term issue of illegal artisanal mining will, in all probability, grow in magnitude. Rehabilitation of mine impacted landscapes and ecosystems is not part of the management objective, which is focused only on the prevention of the breach of ECL and the neutralization of AMD before it causes environmental damage. The absence of a revenue stream means that not only does the taxpayer have to pay for the treatment cost, but also the water has to become potable in order to recover some of that cost. The current thinking has not factored in the possibility of public hostility, currently manifesting with some vigor in the form of opposition to

the recent introduction of electronic tolling to certain highways around Johannesburg without adequate consultation [17]. The known cost for the pretreatment phase only (neutralization of the acid by means of the HDS process and subsequent disposal of sludge) is ZAR 12.00/m^3 (US\$ 1.20/m^3).

The PPP Business Model has an inherently different logic underpinning it. As a point of departure, it acknowledges that the scale of complexity is so great that no single entity can resolve it unassisted. More importantly, it recognizes that in effect, it is the future of Johannesburg, the financial capital of the entire African continent, which is at stake if the transition to a postmining economy is inappropriately managed. Because of this logic, the ethos of partnership lies at the very heart of the solution. The mining sector is not regarded as the sole source of the problem, so remaining mining companies are encouraged to become rehabilitation-oriented as part of their adaptive response to the imminent closure of the resource. By adopting the concept of flow pathways (Figure 46.2), this model acknowledges that the majority of AMD is created on the surface, when acid rain attacks the hydroxide-coated tailings particles on the flat-topped tailings dams, exposing the pyrite to further oxidization (Flow Pathway "A"). Because of this bigger picture approach, it is accepted that AMD is a temporary manifestation only, with the closure of the void central to all management planning. This is to be achieved by embracing state-of-the-art mining practices that include the backfilling of high-density paste, sometimes with a binding agent like cement, back into the void as a deliberate part of rehabilitation mining [10]. Surface tailings are reprocessed to yield the remaining gold they contain. More importantly, this allows for the removal of uranium from the dumps, either as a sequestered Uranyl ion complex or as a chemically pure mineral recovered through an ion exchange process to be further beneficiated. By virtue of the fact that mining is still part of this solution, restructured mining companies become service providers to the state, locked into a PPP contract with all the necessary performance clauses needed to protect the public interest. With closure mining as the core element of the solution, the cost of rehabilitation and water treatment is subsidized by revenue generated from gold recovery. Given the known economics of uranium recovery, there is a case to be made for the possible subsidization by the fiscus, given that a public benefit would accrue from the safe removal of 600,000 tonnes of uranium from the environment. The interesting aspect of this model is that because rehabilitation drives the actual process, the end result is land that has been rendered safe for future economic and social use in a postmining economy. Calculations show that 5445 ha of land can be brought back into safe use this way, in a city that has seen exponential population growth since the collapse of Apartheid in 1994, where the population will be 20 million by 2020 with 4.7 million households that can only be built on mine residue areas [9,24].

One of the technologies in this model is called the tailings water treatment (TWT) process [18] (see Figure 46.5). This uses alkaline gold tailings (pH 10.5) to neutralize AMD (pH 3), with a range of benefits, which include the following:

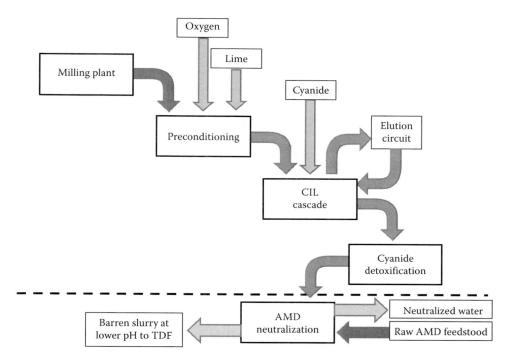

FIGURE 46.5 Schematic diagram of the tailings water treatment (TWT) technology that uses alkaline mine tailings to neutralize AMD.

- Uranium is sequestered in a chemically stable complex.
- There is no need for HDS disposal as the sludge naturally generated is codisposed with the barren tailing stream, so a major cost is removed from the model.
- The reprocessing of old tailings dams removes a source of long-term hazard, enabling the barren tailings to be placed into the void in order to close out future AMD generation underground, while also depositing to the surface as required in dumps that are engineered to twenty-first century design standards.
- All shallow surface striking reef is removed as part of the closure mining model (see Figure 46.4), which permanently closes out ingress points for water and access points for illegal artisanal miners (see Figure 46.1).
- Marginal mines can be rejuvenated as rehabilitation companies, creating a 20-year extension to the life of the mine, buying time to complete the necessary work while creating livelihoods in a sector currently shedding jobs.

The known cost of AMD neutralization using the TWT technology is ZAR 4.00/m³ (US$ .40 c/m³). This cost includes both CAPEX and OPEX items with water treated to the same standard as the HDS process used as an industry benchmark [18]. This water need not enter the potable water stream as it is ideally suited for industrial process water, most notably in the platinum mining industry, but it can also become feedstock for desalination processing if that is the final decision made by the regulatory authority in consultation with society. The TWT approach is thus suited to either the Centralized Process Model or Dual Stream Process Model, but it naturally favors the latter.

With respect to the operational models, there are also two distinct versions—the Centralized Process Model and the Dual Stream Process Model.

The Centralized Process Model is based on the core assumption that the state owns all of the water resources in the country, consistent with the National Water Act [20]. This means that the state has the sole right to treat and distribute water as it alone sees fit. The best analogy is the equivalent in the South African energy sector where the state-owned enterprise called ESKOM is the sole energy provider in the country. Within this model, all water is treated to one national standard and is sold at a national average cost, irrespective of the actual cost of reticulating it to the specific end user. This means that treated AMD is regarded as an integral component of the national water resource, so it will logically become potable water, irrespective of the wishes of the consumer. Public support for this model has been assumed, but not yet tested, so this manifests an as yet unquantified risk. Significantly, within this model, the reconciliation of localized demand and supply is managed at the national level through a centralized decision making entity. Given that this is essentially a monopoly, the overall process is somewhat price insensitive, with no real incentive to favor any least-cost maximum-benefit solution. Having said this, the Centralized Process Model is the product of current laws and policies, so there is no need for any reform, which makes the rollout relatively easy, assuming no public hostility.

The Dual Stream Process Model is based on a core assumption that when a national water resource approaches closure (where demand starts to exceed supply at a high level of assurance), then it becomes logical to treat water as a flux rather than a stock. Water moves in time and space so it is in effect a flux, which means that different quality water can be used for different purposes at different price structures. In essence then, a dual stream reticulation system is one that differentiates different water qualities at different prices for different users. In South Africa, this was pioneered in the late 1990s when the Southern Sewage Works in Durban entered into an agreement with Veolia as service provider, to treat sewage effluent to industrial standards for onward sale to bulk users in the paper-making and oil refining industries. In this model, the state retains the sole right to treat and distribute water, consistent with the National Water Act [20], but this is no longer done to one standard and one national average price. The analogy in the energy sector is the emergence of Independent Power Producers (IPPs) to augment supply by ESKOM. Because of the flexibility of this model, prices are based on local treatment costs and the needs of the end user. Of greater importance, the treated AMD need not necessarily become potable water, thereby negating one major risk silently embedded in the Centralized Process Model. However, if there is public acceptance of this practice, then it is technically feasible with the TWT process [18] (see Figure 46.5). Reconciliation of demand and supply is done at the localized level, freeing up potable quality water for use elsewhere in the national economy. An example of this is current practice in the platinum mining industry, based in the Crocodile West & Marico Water Management Area. Potable quality water is supplied into the platinum industry from the Orange River Basin by means of a high-pressure pipeline that literally flows within a few meters of the Western Basin decant point. This potable water is not suitable for the flotation process central to the metallurgy of platinum, so it is modified to the point where it meets the technical requirements. These requirements are identical to those inherent to the

treated AMD emerging from the TWT and HDS process, so by using this industrial water instead, potable water is spared for use elsewhere in the water-constrained national economy. With the emergence of independent water producers acting as agents on behalf of the state, this is nonmonopolistic, so it is inherently more price sensitive. It will require some policy and legal reform, so it is not that easy to roll out, but the precedent has already been created in the form of the Durban sewage works noted previously.

These two business models and process models can now be evaluated in the context of a matrix shown in Figure 46.6.

From Figure 46.6 it is obvious that the Legacy Business Model favors the Centralized Process Model, but this means that the mining sector is naturally excluded as a potential partner in solution-seeking. In effect then, this limits the range of solutions by naturally excluding the rehabilitation of mine-impacted landscapes and ecosystems. Given that it is centered on the HDS technology, not only are the surface dumps left unrehabilitated, but additional space is needed for sludge storage facilities. The economics of this combination means that the treated AMD has to become potable water, irrespective of what the end consumer feels about this inevitability. The strategic challenge in this combination is for the regulatory authority to make the case to an increasingly angry taxpaying public, that they will be expected to foot the bill in perpetuity, without the added benefits of a rehabilitated landscape, but with the added bitterness of being forced to drink the water. The Legacy Business Model when combined with the Dual Stream Process Model does not work well as the economics simply do not stack up.

From Figure 46.6 it is clear that the PPP Business Model tends to favor the Dual Stream Process Model. The advantage of this combination is that the last remaining mining companies are

	Centralized Process Model	**Dual Stream Process Model**
Legacy Business Model	• Mining is the *problem* • State to *contract third party service provider* (E-tolls) model) • Taxpayer to pay in *perpetuity* • *Landscape* not rehabilitated • **AMD** becomes *drinking water* • *HDS Storage* facilities needed • Cost = ≥ **R12.00 m3** (neutralized)	• *Economics* do not stack up for this combination • *Process* can support this combination • **HDS storage facilities not necessarily needed** *depending* on process selected
PPP Business Model	• *Economics* supports this combination • *Process* supports this combination • **HDS storage facilities not needed** *depending* on process selected	• **Closure mining** is the *solution* • *Mines* are service provider • *Mining revenues* cover majority of cost (partnership) • *Landscape* is rehabilitated • AMD *does not become drinking water* • HDS storage facilities *not needed* • Cost= ±**R4.00 m3** (neutralized)

FIGURE 46.6   Matrix showing the various combinations of the two business models and two process models.

given the chance to adapt by becoming rehabilitation companies instead of becoming insolvent, which is the inevitable outcome of retaining the business-as-usual approach. This means that closure mining becomes the solution (see Figure 46.4), which brings in mining companies as service providers capable of moving the vast quantities of tailings that lie in the surface dumps. The obvious advantage of this approach is that the surface dumps can be rehabilitated, either by being placed into consolidated megadumps engineered to twenty-first century standards, or by being backfilled into the void as high-density paste strengthened with a binding agent like cement [10]. This closes out the void as a potential future source of AMD, while increasing the geotechnical stability of the mine-impacted landscape and removing access points for illegal artisanal miners.

46.7 Lessons Learned

The major lesson learned from this whole AMD problem is the unintended consequence of unplanned mine closure. Given the significance of mining to the South African economy, the collapse of the gold sector has national implications, which seem not to have been grasped by the authorities in charge of the city of Johannesburg. The general rhetoric in the media, as a reflection of the discourse underpinning the issue, is that the mining industry is irresponsible, so what needs to be done is to neutralize the acidic water to prevent future decant. The simple fact of the matter is that Johannesburg is arguably the most uranium-contaminated city in the world, with the implications of this becoming apparent only after the mining industry ceases to function as a viable entity. Seen in this light, the AMD issue is in fact nothing more than the manifestation of a transition from an extractive to a postmining national economy, with the core challenge of needing to deal with environmental externalities now manifesting as constraints for future development and job creation. Unfortunately, very few people see it this way, with aggressive anti-mining activists shaping public opinion [23] and thus driving the response by government in a direction that is ultimately going to be unsustainable because it fails to recognize the need for an integrated response.

Another as yet dangerous lesson to be learned is the issue of human population migration into the city of Johannesburg. From 1900 to 1994, the population of Johannesburg grew to 6.2 million, but over the next decade (1994–2004) that population exploded with 25% growth to 7.7 million inhabitants living in 2.4 million households of which 0.6 million were in informal settlements [24]. In fact, 1.6 million people are now living in close proximity to uraniferous mine residues and that number is growing exponentially, so government has a hard choice to make—either move the people from the dumps or move the tailings from the people [9]. The population is set to grow at a rate of 6.7% per annum, which will yield a population of 20 million by the year 2020. More significantly, however, this will drive the demand for 4.7 million households, or 200% growth off the 1994 baseline. The only land left to settle is mining impacted, so the water resource management problem is intimately linked with

the mine rehabilitation issue, with uranium sequestration as a central feature to any viable strategy.

The biggest lesson is that closure mining is an appropriate adaptive response, so it is important to explain how this differs from conventional mining. Stated simplistically, conventional mining as practiced in South Africa maximized profits at the level of the shaft or pit, by externalizing costs in collusion with the Apartheid-era government, which essentially left the industry to self-regulate in order to survive as a pariah state [12,27,29]. This increasingly placed the outcome of mining—a highly impacted landscape in which uranium is a persistent hazard and water pollution a central feature—at odds with the broader interest of society. Closure mining is about realigning the outcome of the mining process—a rehabilitated landscape that is safe to use—with the broader interests of society, thereby reducing the conflict potential inherent to the conventional mining model. This is achieved by aligning the processes shown in Figure 46.4 in order to lower the breakeven point derived from falling grades of ore.

The essential idea behind closure mining is the recovery of gold from the existing tailings dams, which contain around 0.3 g of gold per tonne of tailings. This means that in order to recover 1 kg of gold, in excess of 3333 tonnes of tailings have to be reprocessed. This is a low-grade high-volume business with viability driven by the availability of low cost water and the constant throughput of tailings into the processing plant. However, for every tonne of tailings moved in the gold extraction process, between 50 and 200 g of uranium are also moved. This means that for every kilogram of gold recovered, up to 600 kg of uranium can also be processed or sequestered. The existence of significant bodies of ore, no longer in sufficient quantities to sustain conventional mining, but enough to change the economics of a marginal operation by increasing the grades through the milling circuit, remains a potential resource. Known reserves of surface striking reef (Figure 46.1) and pillars left in underground operations are of a grade that averages around 5 g/t, with sweet spots of higher value that can be selectively targeted. This is blended with the tailings stream at the mill in order to bring the run of mine grade up to 1 g/t, which is the breakeven point for the operation. In the process, the following key objectives are reached, thus representing a realignment of the outcome of closure mining with the broader interests of society:

- The remaining surface dumps are removed and consolidated using state-of-the-art processes that are designed to twenty-first century standards.
- The void is closed out as far as possible with high-density paste backfill in order to remove as much of the surface waste as possible.
- Geotechnical stability is ensured through the backfilling of high-density paste into the void, as well as the removal of surface striking reef in shallow undermined areas (Figure 46.1).
- Ingress points into the void are closed out, thereby reducing the future flow of surface water into subsurface flow pathways.

- Access points for illegal artisanal miners are permanently closed out, reducing the presence of criminal syndicates linked to the fencing of stolen goods and the laundering of money across international borders.
- Livelihoods are created in an industry currently shedding jobs, as the life of marginal mines is extended by up to 20 years, buying time to complete the rehabilitation process.
- The mining industry becomes part of the solution, so the burden on the state and taxpaying public is substantially reduced.
- An integrated rehabilitation process is enabled, with multiple intervention points at key parts in the complex hydrogeological and hydrochemical cycle underpinning AMD.

More importantly, in the context of this specific chapter, closure mining is centered on the reuse of highly contaminated industrial water because the very foundation of the process is hydraulic mining and aqueous metallurgical processes. In fact, inherent to the concept of closure mining is the neutralization of AMD by means of exposure to barren tailings at a naturally high pH, as shown in Figure 46.5.

Figure 46.5 shows the overall TWT technology in schematic format [18]. Blended paste enters the circuit from the mill with a grade value of ≥1 g/t. This is preconditioned by raising the pH through the addition of calcium hydroxide and oxygen in order to precipitate out the iron. This slurry is then pumped across to the carbon-in-leach (CIL) cascade where activated carbon and cyanide are both introduced. The cyanide captures the gold, which is adsorbed into the activated carbon matrix, to be stripped out and recovered during the elution phase. The barren slurry is then sent though a cyanide detoxification process, at which time the highly alkaline barren tailings, still at a pH value of ≥10.5, are exposed to raw AMD pumped in from the void. This exposure causes the low levels of residual cyanide to complex with other metals in the tailings stream, while also precipitating out metals in the form of HDS, which is seeded onto each individual tailings particle. The uranium in the AMD and tailings stream also becomes a chemically stable complex, thereby sequestering it onto the tailings particle. The neutralized AMD is used as process water for the hydraulic mining and related circuits, while the barren tailings are returned to their final resting place to be rehabilitated to twenty-first century engineering and environmental management standards.

46.8 Summary and Conclusions

South Africa was once the home of some of the most innovative mining engineering in the world. The result of over more than a century of mining has been a mixed blessing to the citizens of the country. On the one hand, the country has rapidly industrialized into a modern diversified economy capable of competing in a global market, with the transition from low-paid agricultural subsistence livelihoods into higher paid industrial jobs supported by a range of benefits. On the other hand, the

externalities of mining, mostly the result of the Apartheid-era survival tactics of an embattled pariah state, are now manifesting as significant constraints to future economic growth and job creation policies being rolled out by the first democratically elected government. Johannesburg is now one of the most uranium-contaminated cities in the world, but it remains the economic capital of the African continent. It remains to be seen how water resource management, most notably around the pressing need to mitigate the AMD risk, will become the catalyst for future economic wellbeing. The regulatory authority is now confronted by a difficult set of trade-offs, defined in this chapter as being the interaction between two different business models with two different operational process models. The decisions that are made in the near future will define the future of the city of Johannesburg. This chapter is offered in the sincere hope that it can inform that process of decision making.

References

1. Borrhalo, B. 2013. South African gold mining sector experiences volatility. *Mining Weekly*. August 2, 2013. http://www.miningweekly.com/article/volatility-in-the-south-african-gold-mining-sector-2013-08-02.
2. Brown, J. 2007. Derelict mines to cost state R100bn. *Business Report*, May 23, 2007.
3. Buchanan, M. (ed.), 2010. *The Karst System of the Cradle of Humankind World Heritage Site: A Collection of 13 Issue Papers by the South African Karst Working Group*. Water Research Commission, Pretoria.
4. Coetzee, H., Hobbs, P.J., Burgess, J.E., Thomas, A., and Keet, M. 2010. Mine water management in the Witwatersrand gold fields with special emphasis on acid mine drainage. Report to the Inter-ministerial Committee on Acid Mine Drainage, December 2010, pp. 1–128. http://www.dwaf.gov.za/Documents/ACIDReport.pdf.
5. Coetzee, H., Winde, F., and Wade, P.W. 2006. *An Assessment of Sources, Pathways, Mechanisms and Risks of Current and Potential Future Pollution of Water and Sediments in Gold-Mining Areas of the Wonderfonteinspruit Catchment*. WRC Report No. 1214/1/06. Water Research Commission, Pretoria.
6. CSIR. 2008. *High Confidence Study of Children Potentially Affected by Radionuclide and Heavy Metal Contamination Arising from the Legacy of Mine Water Management Practices on the Far West Rand of South Africa*. Project Concept Note. Council for Scientific and Industrial Research (CSIR), Pretoria.
7. DWA. 2013. Feasibility study for a long-term solution to address the acid mine drainage associated with the east, central and west rand underground mining basins, Gauteng Province. *Newsletter AMD FS LTS*, 2nd Edition. http://www.dwa.gov.za/Projects/AMDFSLTS.
8. Feinstein, A. 2007. *After the Party: A Personal and Political Journey Inside the ANC*. Jonathan Ball, Johannesburg.

9. GDARD. 2011. *Feasibility Study on Reclamation of Mine Residue Areas for Development Purposes: Phase II Strategy and Implementation Plan,* December 2011. Report No. 788/06/02/2011 (Final). Umvoto Africa (Chris Hartnady & Andiswa Mlisa) in association with TouchStone Resources (Anthony Turton). Gauteng Department of Agriculture and Rural Development (GDARD), Johannesburg.

10. Grice, A. 1998. Underground mining with backfill. *Paper presented to the 2nd Annual Summit of Mine Tailings Disposal Systems,* Brisbane, Australia.

11. Godfrey, L., Oelofse, S., Phiri, A., Nahman, A., and Hall, J. 2007. *Mineral Waste: The Required Governance Environment to Enable Reuse.* Report No. CSIR/NRE/PW/IR/2007/0080/C. Council for Scientific and Industrial Research (CSIR), Pretoria.

12. Gutteridge, W. 1984. Mineral resources and national security. In: *Conflict Studies,* No. 162, 1–25. Reprinted in W. Gutteridge (ed.). 1995. *South Africa: From Apartheid to National Unity, 1981–1994.* Dartmouth Publishing, Aldershot, Hants & Brookfield, VT, pp. 59–83.

13. Hartnady, C.J.H. 2009. South Africa's gold production and reserves. *South African Journal of Science,* 105, 328–329.

14. Holtzhauzen, L. 2004. Decanting minewater: Solving a 100-year-old problem. *Water, Sewage and Effluent,* 4(4), 18–20.

15. Johnston, S. and Bernstein, A. 2007. *Voices of Anger: Protest and Conflict in Two Municipalities. Report to the Conflict and Governance Facility (CAGE).* The Centre for Development and Enterprise, Johannesburg.

16. Jordaan, J.M., Enslin, J.F., Kriel, J.P., Havemann, A.R., Kent, L.E., and Cable, W.H. 1960. *Finale Verslag van die Tussendepartmentele Komitee insake Dolomitiese Mynwater: Verre Wes-Rand, Gerig aan sy Edele die Minister van Waterwese deur die Direkteur van Waterwese.* (In Afrikaans translated as, Final Report of the Interdepartmental Committee on Dolomitic Mine-water: Far West-Rand, Directed at His Excellency the Minister of Water Affairs by the Director of Water Affairs). Department of Water Affairs, Pretoria.

17. Magubane, K. 2012. Bigger turnout for second E-toll consultation. *Business Day,* November 15, 2012. http://www.bdlive.co.za/national/2012/11/15/bigger-turnout-for-second-e-toll-consultation.

18. Mintails. 2013. *Proposal for the Application of the Mintails Tailings Water Treatment Technology for the Mitigation of the Acid Mine Drainage Problem in the Western Basin of the Witwatersrand Gold Mining Complex.* Mintails (Ltd)., Krugersdorp. Submitted to the Department of Water Affairs in response to the public call for proposals.

19. Naidoo, B. 2009. Rising tides: Massive acid mine drainage project stimulus for local beneficiation. *Creamers Mining Weekly,* July 31–August 6, 8–9.

20. National Water Act. 1998. *The National Water Act.* Act 36 of 1998. Government Gazette, Pretoria.

21. Noseweek. 2009. Joburg's Poisoned Well. *Noseweek,* 118, August 2009.

22. Oelofse, S.H.H., Hobbs, P.J., Rascher, J., and Cobbing, J. 2007. The pollution and destruction threat of gold mining waste on the Witwatersrand: A west rand case study. *Paper presented at the 10th International Symposium on Environmental Issues and Waste management in Energy and Mineral Production (SWEMP, 2007),* Bangkok December 11–13, 2007.

23. Segar, S. 2013. Here comes the poison. *Noseweek,* 162, April 1, 2013. http://www.noseweek.co.za/article/2934/2013-Here-comes-the-poison.

24. Tang, D. and Watkins, A. 2011. *Ecologies of Gold: The Past and Future Mining Landscape of Johannesburg.* http://places.designobserver.com.

25. Tempelhoff, E. 2013. Miljoene om dié Water te Pomp: Jan Alleman Betaal vir Myne se Skade (Millions to pump the water: Joe Bloggs pays for mine's damage). *Beeld Newspaper.* November 21, 2013.

26. Toffa, T. 2012. Mines of gold, mounds of dust. Master's thesis for Urbanism and Strategic Planning, Catholic University of Leuven, Belgium.

27. Turton, A.R. 2010. The politics of water and mining in South Africa. In: K. Wegerich and J. Warner (eds.), *The Politics of Water: A Survey.* Routledge, London, pp. 142–160.

28. Turton, A.R. 2013. Debunking persistent myths about AMD in the quest for a sustainable solution. *SAWEF Paradigm Shifter* No.1. South African Water, Energy and Food Forum (SAWEF), Johannesburg.

29. Turton, A.R., Schultz, C., Buckle, H., Kgomongoe, M., Malungani, T., and Drackner, M. 2006. Gold, scorched earth and water: The hydropolitics of Johannesburg. *Water Resources Development,* 22(2), 313–335.

30. Van Deventer, P.W. 2009. *Conceptual Study on Reclamation of Mine Residue Areas for Development Purposes.* University of North West, Potchefstroom.

31. Weltz, A. 2009. The spy who came in from the gold: Has a leading water policy expert been suspended for being a former spook—or for treading on the mining industry's toes? *Noseweek,* 111, January 2009, 30–31.

32. Werdmüller, V.W. 1986. The central rand. In: E.S.A. Antrobus (ed.), *Witwatersrand Gold–100 Years.* The Geological Society of South Africa, Pretoria, 7–47.

33. Winde, F. 2013. Uranium Pollution of Water—A Global Perspective on the Situation in South Africa. Vaal Triangle Occasional Paper: Inaugural Lecture 10/2013. University of North West.

34. Winde, F. and Van Der Walt, I.J. 2004. The significance of groundwater-stream interactions and fluctuating stream chemistry on waterborne uranium contamination of streams—A case study from a gold mining site in South Africa. *Journal of Hydrology,* 287, 178–196.

Wastewater and Sludge Reuse Management in Agriculture

Ioannis K. Kalavrouziotis
Hellenic Open University

47.1 Introduction

As the technology of wastewater treatment progresses and possible contaminants are effectively removed, the treated effluent becomes less a threat to the environment and human health. Thus, the waste is in small or large toxic organic substances, micronutrients, pathogens, toxic organic chemicals, and dissolved minerals.

Application of sludge and treated municipal wastewater (TMWW) to soil, as well as sludge deposition to landfills and incineration, are some of the popular methods of wastewater and sludge disposal. Nearly 5.4 million dry tones are produced per year in the United States and comparable numbers can be found in Western European countries.

In a recent published research on the production, use, and disposal, United Kingdom reported that 45% of total wastewater of the state is applied to 0.3% of agricultural land, 30% is disposed into the ocean, 13% is deposited in dumps, 7% is incinerated, and 6% is used for various purposes [38].

If there were no toxic heavy metals and hazardous organic pollutants, the soil application would have been an ideal way for the disposal of wastewater because this use would be beneficial for both the residents and the farmers, as the soil provides a reliable outlet for assimilation of waste. Besides, the plant nutrients N, P, and organic matter added to the soil would create perfect conditions for plant growth.

Fifty years ago, wastewater included 160–400 mg Cu/kg, 80–320 mg Zn/kg, 930–1860 mg Pb/kg, and approximately 1400 mg Cr/kg of solid weight, respectively. These levels have been significantly increased due to the extensive use of heavy metals and other chemicals and their dispersion in the environment.

When the flow of municipal wastewater is mixed with industrial wastewater, then the content of the above toxic metals and chemicals is significantly increased. It is reported in a national bibliography concentrations of 41,000 mg Cr/kg, 12,000 mg Cu/kg, 26,000 mg Pb/kg, 62,000 mg Zn/kg, and 1500 mg Cd/kg [31].

Since the 1970s, the fate and transport in the food chain of micronutrients through the application of municipal wastewater in soil have been investigated [14].

The treated water used for irrigation must be of suitable quality; therefore, it should be subjected to the appropriate treatment, so the concentration of heavy metals does not exceed permissible limits.

The developing plants need a specific number of microelements, whose concentration cannot exceed the permissible limits. Thus, the wastewater disposed of in the soil is further purified by two ways:

1. By the plants that get the minerals for their development.
2. By the soil that favors the biodegradation of organic matter and retains them to a certain extent either by precipitation or adsorption in unavailable forms.

47.2 Quality Criteria of Irrigation Water

Quality of irrigation water is important in dry or semidry regions where high temperatures combined with low humidity create conditions for high evapotranspiration (ET). The water used for irrigation can vary, depending on the type and concentration of dissolved salts. Consequently, ET may lead to salt accumulation in the soil. The various chemical, physical, and mechanical soil properties such as the degree of dispersion of soil particles, stability of soil structure, and soil conductivity are parameters sensitive to ions that are contained in irrigation water (Table 47.1). This means

that if the treated wastewater is to be reused for crop irrigation, the design of the irrigation project must take into consideration the wastewater treatment plants (WWTP), crop yield, and soil properties.

The most important parameters of irrigation water is salinity, which is measured by electrical conductivity (EC), which is used for monitoring the concentration of total dissolved solids (TDS) in the wastewater. The following relation is used for the calculation of TDS, that is,

$$TDS(mg/L) = 640\,Ecw\,(mmhos/cm\,or\,ds/m) \qquad (47.1)$$

The presence of salts in irrigation water influences plant growth by the following two basic processes:

1. By osmotic effects causing the total concentration of salts in the soil.
2. By dispersion of soil particles caused by high concentration of sodium.

Furthermore, caution is needed in relation to the concentrations of Mn^{2+} and Cu^{2+}, when the wastewater is applied by sprinkling, as the accumulation of these ions on the leaf surface may damage the leaves due to their toxic effects on plants at the expense of their development. To avoid salt damage to plants in soil, enhance the vertical flow of water.

Securing conditions of a rational strategy will allow the continuous transfer of water and salts below the area of the root system. That is why it is necessary that under conditions of wastewater reuse irrigation of crops, the drainage of the soil must be very effective.

The use of wastewater for irrigation usually is controlled to avoid transporting diseases.

Forty-three of the 50 United States have regulations that control wastewater reuse for irrigation crops. [37].

The criteria of wastewater recycling that have been adopted by California can be used as a paradigm. This regulation does not

TABLE 47.1 Classification of the Quality of Irrigation Water

Irrigation Problem	Units	Without Problem	Increased Problem	Serious Problem
1. ECw	dS/m	<0.75	0.75–3.0	>3.0
2. RDS	mg/L	<450	450–2000	>2000
3. Filtration (calculated with ECw and SAR)				
SAR = 0–3	dS/m	>0.7	0.7–0.2	<0.2
SAR = 3–6	dS/m	>1.2	1.2–0.3	<0.3
SAR = 6–12	dS/m	>1.9	1.9–0.5	<0.5
SAR = 12–20	dS/m	>2.9	2.9–1.3	<1.3
SAR = 20–40	dS/m	>5.0	5.0–2.9	<2.9
Nutrient (total N)	mg/L	<5.0	5–30	>30
HCO_3	mg/L	<90	90–500	>500
Na^+	mg/L	<3.0	3–9	>9
Residual sodium for irrigation	mg/L	<1.0	1.0–5.0	>5.0
B^{3+}	mg/L	<0.7	7.0–3.0	>3.0

Source: Adapted from Ayers, R.S.S. and Westcot, D.W. 1985. Water quality of agriculture, FAO Irrigation and Drainage Paper 29:99–104 Rev. 1; Pettygrove, G.S. and Asano, T. 1984. Irrigation with reclaimed wastewater. Aquidance manual Califronia State Water Resources, Control Board, Report number 84–1, July.

set any numerical limit, apart from water bacteriological quality. It avoids setting a minimum level of treatment required for every water use, restricts people and grazing animals from using this water, and restricts where this water is used. Thereby, the risk of transferring diseases is minimized. If recycling wastes were applied in accordance to these requirements, the chemical constituents in the treated effluent would not be hazardous to human health and would not cause water contamination. This requirement may not apply if the industrial wastes are not confined.

Table 47.2 shows the maximum allowable concentrations of microelements in irrigation water, mentioning the toxicity that can be caused in plants due to excess use.

47.3 Application of Sludge and Wastewater to Soil

It has been generally accepted that municipal wastewater could be used in the soil according to the need of N and P of plants. If concentrations of the wastewater pollutant are high, the urban

contaminants in the soil would be evaluated, even when the wastewater is used according to agronomic rules.

The last two decades have formulated regulations to restrict the discharge of pollutants in the soil. The limits of the pollutant load are drastically different. Since 1973, the US Environmental Protection Agency (USEPA) has announced many regulations regarding the reuse of urban wastewater. The final regulation, which became law in November 1992, was significantly different from previous ones, reflecting the technical expertise and a change in the basic principles of the regulation [38]. Applying the maximum loading rate of pollutants for metal or other elements detection, the final regulation of USEPA for the use of wastewater in the soil is based on soil ability to absorb pollutants. Although more than 20 organic pollutants have been studied during the procedure of law formulation, none of these have been included in the directives due to

1. The concentrations were much higher than the permissible amounts of wastewater that will not be overcome, if the number of pollutants does not exceed the agronomic rate.

TABLE 47.2 Proposed Maximum Allowable Concentration of Metals in Irrigation Water

Elements	Proposed Maximum Concentration	Labeling
Al (aluminum)	5.0	May lower productivity in acidic soils (pH < 5.5) while in alkaline (pH > 7.0) causes precipitation of Fe, limiting the toxicity.
As (arsenic)	0.10	Plant toxicity varies widely, ranging from 12 mg/L, for grass (Sudan), to below 0.05 mg/L in rice (*Ozyza sativa*).
Be (verilio)	0.10	Toxicity in plants differs widely, ranging from 5 mg/L in sprouts to 0.5 mg/L in Nana beans.
Cd (cadmium)	0.01	Toxic in beans and sugar beet in concentration up to 0.10 mg/L in nutrient solutions.
Co (cobalt)	0.05	Toxic in tomatoes at 0.1 mg/L in nutrient solution. Tends to be inactivated at neutral and alkaline soils.
Cr (chrome)	0.10	It has not been identified as a basic in constituent for development. Precautionary limits proposed due to lack of knowledge of toxicity in plants.
Ca (copper)	0.20	Toxic for a great number of plants from 0.1 mg/L to 1.0 mg/L in nutrient solutions. Synergistic interaction with Zn.
F (fluorine)	1.0	Is inactivated in neutral and alkaline soils.
Fe (iron)	5.0	It is not toxic in plants in well-aerated soils. It decreases soil pH and compete the basic elements P and microelements Mn.
Li (lithium)	2.5	Tolerated by most crops up to 5 mg/L. Unstable to soil. Toxic in citrus in concentration (0.075 mg/L). Acting like boron.
Mn (magnanese)	0.20	Toxic in a great number of crops from 0.1 to 1.0 mg/L but in only dry soils.
Mo (molybdenum)	0.01	Nontoxic in plants where concentrations in irrigation water are normal it can be toxic in livestock if grazing is developed in soils with high concentrations of molybdemum.
Ni (nickel)	0.20	Toxic in a number of plants from 0.5 mg/L to 1.0 mg/L. Observed decreased toxicity in neutral and alkalic soils.
Pb (lead)	5.0	It can prevent the cellular plant growth in high concentrations.
Se (selenium)	0.02	Toxic in plants in low concentrations 0.025 mg/L and toxic in animals, if grazing is developed in soils with high levels of added selenium.
Sn (tin)		Effectively excluded from plants. Special tolerance unknown.
Ti (titanium)		It is not absorbed by plants yet special tolerance is unknown.
W (volthanio)		It is not absorbed by plants yet special tolerance is unknown.
V (vanadium)	0.10	Toxic in many plants in relatively high concentrations.
Zn (zinc)	2.0	Toxic in many plants, in concentrations widely differentiated. Reduced toxicity in pH > 6.0 and in well-treated and organic soils.

Sources: Adapted from Synergistic and antagonistic interaction have been referred in specific plants in relation with their resistence in heavy metals after irrigation with W.W.T.P. Described in Guidelines, WHO, 2006. *Excreta and Greywater,* 2, 178; Drakatos P.A., Kalavrouziotis, I.K., and Drakatu, S.P. 2000. *Journal of Land Contamination and Reclamation,* 8(3), 201–207 (EPP publications); Drakatos, P.A. et al. 2002. *International Journal of Environmental Studies,* 59(1), 125–132; Kalavrouziotis, I.K. and Drakatos, P.A. 2002. *International Journal of Water,* 2(4), 284–296.

TABLE 47.3 Allowable Heavy Metal Accumulation Levels in Soils of Various Countries Supplied with Sludge

Country	Year	Annual Limit (kg/ha/Year)					
		Cd	Cu	Ni	Pb	Zn	Hg
European Union	1986	0.15	12	3	15	30	0.1
Sweden	1986	0.015	3	0.5	0.3	10	0.008
Holland	1995	0.0025	0.15	0.076	0.45	0.6	0.0015
Germany	1986	0.15	6	1	6	15	0.125
Finland	1986	0.12	12	2	4.8	20	0.1
Norway	1986	0.03			0.6		0.014
Denmark	1986	0.015			0.06		
United States	1993	1.9	75	21	15	140	0.85

Source: Adapted from U.S. Environmental Protection Agency. 1993. *Federal Register*, 58(32), 9248–9415.

2. They are forbidden or no longer produced in the United States.

The maximum added limits of metal load, the detected elements, and the metal load pace are depicted in Table 47.3. Regulations of the wastewater reuse in the United States specify that wastewater cannot be used in rural land, if the concentration of any pollutants in wastewater is lower than "the levels of nonobserved adverse effect," which are based on an added assumed usage of 1000 metric tons (mT) dry sewage per hectare. These sewage categories can be used in soil with a small restriction on the total loads of metals, if the sewage load every year is below the limit (Table 47.4).

Regulation of wastewater use in the United States and Holland, Denmark, Norway, and Sweden represent two different approaches that nobody can use for the formulation of these regulations. Semantically both of those two approaches are applied. Due to a philosophical difference that was imposed, and the regulations goals, the numerical limits for the same pollutant sometimes can be significantly different and only large value differences can confuse us. In fact, the numerical values of European States represent the maximum load levels for no effect and those of United States represent the maximum permissible limit so as not to avoid any damages to humans and animals.

In Greece, there are guidelines regarding reuse, where besides all physicochemical parameters that have to be checked, the maximum permissible concentrations in priority and toxicity in reclaimed wastewater are determined.

47.4 Methods and Principles of Municipal Wastewater Disposal

The terrestrial disposal method that is offered for the reuse of wastewater of biological stations in agriculture is the method of slow disposal.

Besides the method of slow disposal, there are other two basic methods of terrestrial disposal, that is, rapid filtration and flooding.

These three methods as well as the principles and the terrestrial disposal of sewage are described in detail in the following sections.

47.4.1 Method of Slow Disposal

The system of slow disposal of the effluent flow on surfaces is covered by vegetation. A part of wastewater is filtered in deeper layers and another part of it is used by plants. The goals of the slow disposal system are

1. Disposal and treatment of wastewater
2. Economic exploitation of the sewage water and nutrients for cultivation and plant growth

TABLE 47.4 Critical Limits of Heavy Metals' Concentration in Soils of Various Countries

Element	United States	Canada	European Union	Denmark	France	Holland
As	41	15				0
Cd	39	4	1.25–6.25	0.2	3.75	1.25
Cu	1500	150	75–300		200	7.5
Cr	3000				250	7.5
Ni	420	36	12.5–125		62.5	3.8
Pb	300	100	0–625		125	22.5
Zn	2800	370	175–550		550	30
Hg	17	1	2.25–3.5		2.3	0.075

Source: Adapted from U.S. Environmental Protection Agency. 1993. *Federal Register*, 58(32), 9248–9415.

3. Saving water by using wastewater instead of natural irrigation or drinking water
4. Maintenance and enlargement of suburban green and open spaces

As recipients of such systems, the agricultural crops, forests, parks, groves, or fields can be used.

47.4.2 Method of Rapid Filtration

Rapid filtration systems are created when the greatest part of wastewater is filtered through soil in deeper layers and finally unites with the surface or groundwater. Of course, these systems of rapid filtration require a certain degree of purity of treated wastewater, which is often so refined, that it very quickly is mixed with soil water. The goals of the rapid filtration systems are

1. The increase of groundwater supplies
2. The recovery of clean wastewater from wells and water collection systems for reuse
3. The increase of surface water from groundwater restraint
4. Temporary wastewater storage in aquifers

47.4.3 Flooding Method

Flooding methods are created when the wastewater is left in a slope covered by plants and selects it in trenches downstream. The goals of these flood systems are

1. To accomplish the same clean water quality with that of the secondary treatment when the wastewater is derived from untreated wastewater or it has been subjected to only primary treatment.
2. To succeed high levels of nitrogen removal, BOD, and suspended particles.

47.5 Principles of Terrestrial Disposal

The principles of sewage disposal are all based on ecologically acceptable disposal and minimization of the ecological load, a fact that is the primary goal of geotechnical scientists.

The method of wastewater disposal is based on the operating principle of hydrological cycle in agricultural ecosystems.

Especially for forestry, together with ET and infiltration, the water retaining capacity of the tree foliage must be calculated for mature plant clusters.

Although the disposal of wastewater is accomplished according to the principles of hydrological cycle, the disposal of dry sewage is controlled by its nutrients, trace elements, and heavy metal content. The rate of application per hectare or acre is based on the concentration of nitrogen, phosphorus, and heavy metals. Consequently, the controlling factor is the recycling of the nutrients in the forestry ecosystem.

The principles of wastewater disposal are summarized as follows:

1. The disposal of wastewater must be consistent with the principle of minimum environmental load.
2. The disposal of wastewater must be regulated by hydrological cycle of the system.
3. The disposal of dry sewage must be regulated by the cycle of nutrients.
4. The disposal of wastewater should be done in accordance to the requirements of the WWTP, and the level of the function of its electrical installations.

47.6 Geotechnical Criteria for Site Selection of Terrestrial Sewage Disposal in Infertile and Problematic Soils

To select the best location for wastewater disposal cultivated with modern agricultural plantation, forest, or ornamental species, it should be made with scrupulous examination of the geotechnical characteristics of different available positions.

Criteria (characteristics) that should be examined are discussed in the following sections.

47.6.1 Climate–Meteorological Criteria

Climate and meteorological data that need to be assessed for slow disposal systems are

1. The frequency and quantity of precipitation of monthly and annual average, maximum and minimum in a time series of 10 years. This feature is used in the assessment of the water balance of aquatic, as mentioned previously.
2. Evaporation and transpiration, of monthly and annual average, in a time series of at least 10 years. Annual distribution of evaporation is used for ET assessment of water balance.
3. Temperature. The number of days where the temperature is below zero shows the number of days where we cannot use the system of terrestrial disposal. Therefore, it shows the need and the extent of wastewater storage.
4. Intensity, frequency, and duration of storms for assessment of the surface runoff of water.

All of the above with the exception of transpiration can be given by the local and meteorological stations.

47.6.2 Geological–Hydrological Criteria

Geological-topographical and hydrological characteristics of the regions that would be assessed for slow disposal systems are

1. Latitude and reliefs. Play the role in transport of wastewater.
2. Deviations. Strong slopes are prohibitive due to the hazard of territorial erosions and strong runoff. Slopes should be lower than 20%.

3. Petrographic data for a first assessment of water infiltration and territorial information.

4. Stratigraphic data for finding impermeable, permeable, and semipermeable zones in the subsoil.

5. Tracking and recording of the surface waters for later use in the detection of infiltration and draining of wastewater.

47.6.3 Tracking and Recording of Water Sources and Deep Wells for the Control of Water Quality: Soil Characteristics

Soil is the main zone where the processes take place between wastewater and environment. It is important to have as much information as we can about physical and chemical soil properties, which are going to be used for the establishment of plantations with modern disposal of wastewater and solid sewage (sludge).

Soil properties that will be used as criteria for site selection are

1. Mechanical composition of the soil.
2. Hydraulic conductivity (permeability) horizontally and vertically.
3. Soil structure, porosity, fluid distribution, and apparent wetness.
4. Cation exchange capacity (CEC). This shows the total charge of soil surface, and therefore the retention capacity of cations.
5. Organic matter.
6. Soil reaction (pH).
7. Inorganic, organic, and total nitrogen and phosphorus.

All of the above constitute the initial soil criteria for site selection. The soil will be monitored constantly after the site has been chosen to verify the accumulation of heavy metals and contaminants, including changes of the general soil properties.

47.6.4 Economic-Social Criteria

Last, another set of criteria that has to be taken into consideration for site selection is related to the economic and social conditions of the selected area.

1. Conditions of employment—workers availability for reforestation.
2. Land ownership—land registers.
3. Right to use and land and water chores.
4. Absorption—disposal of produced wood or biomass.

From the above, it can be seen that none of the criteria mentioned refer to the quality and quantity of sewage. This must be taken into consideration before one can begin to study the disposal site. If the quality of sewage is inconsistent with the requirements of the WWTP, we will have to study how to improve the quality of wastewater and the manner of disposal. Thus, the control of the electro-mechanics of installations of WWTP should always be preceded from the disposal study and be continued during the project [12].

If all proposed geotechnical criteria are carefully studied, then the best territorial disposal site will be selected and minimum environmental damage will be ensured.

47.7 Applicability of Disposal Model of Wastewater Treatment Plants and Biosolids in Reforestations

The laboratory of the Mechanical Engineering Department, University of Patras, has designed and implemented since 1991 a reforestation experiment. Seven acres of land near University of Patras were planted with 1213 forest trees of the species *Pinus brutia* and *Pinus maritime* in small groups. The *Pinus brutia* was of Greek origin and *Pinus maritima* was of Corsican origin.

The treatments applied were sludge plus TMWW, sludge, TMWW, and the control was well water.

Each tree was irrigated with 30 L of TMWW or well water, depending on the treatment. The total TMWW per year was 720 L. The experiment lasted from 1991 to 1994. The following results were obtained.

Based on the rate of mortality (Tables 47.5 and 47.6) *Pinus brutia* was most tolerant to the application of sludge and TMWW, and it is proposed to be used for reforestation and to be irrigated with TMWW or with the application of sludge. In contrast, the mortality of *Pinus maritima* was very high, almost up to complete failure (95%), especially under the effect of sludge due to the abnormal development of the root system.

Growth of *Pinus brutia* was satisfying, proving that the presence of sludge or wastewater did not slow down or prevent the normal development of the plants [7,8,10–13].

In Tables 47.7 and 47.8, the total increase in height (cm) of plants of *Pinus brutia* and *Pinus maritima* at the various treatments is shown.

Kalavrouziotis and Drakatos [16], studied capability of the Mediterranean plant species to absorb heavy metals that originate from wastewater treatment systems and the normal water irrigation enriched with Cu, Mn, and Zn. The plants that were studied were *Myoporum* sp., *Nerium oleander*, and *Geranim* spp.

Results showed that the *Myoporum* spp. had a maximum concentration of Cu, Mn, and Zn in its leaves without showing toxicity effects.

It was also found that metals were concentrated mainly in the root system and less in leaves. All the studied plant species except from *Nerium oleander*, which showed toxicity symptoms in leaves, showed great resistance to water irrigation enriched in heavy metals.

Drakatos et al. [9] studied the relevant absorption capability of Mn^{2+} and Fe^{2+}, under the influence of reuse of processed wastewater, as well as their resistance at high concentration levels of Mn^{2+}, that is contained in the water used for their irrigation. The experimental plants *Nerium oleander* and *Geranium* spp. were grown under greenhouse conditions, in special plastic bags, which were filled with soil from the region of the experimental site. The first group was irrigated with water that has

TABLE 47.5 Mortality among Manipulations of *Pinus brutia*

Experimental Period	Sludge and Wastewater	Sludge	Wastewater	Witness	All Treatments
March 1991	1 (1.2)	2 (2.4)	5 (8.1)	0 (0.0)	8 (2.1)
October 1991	2 (2.4)	2 (2.4)	3 (4.8)	6 (4.1)	13 (3.4)
March 1992	3 (3.5)	0 (0.0)	0 (0.0)	0 (0.0)	3 (3.4)
October 1992	1 (1.2)	0 (0.0)	0 (0.0)	10 (6.8)	11 (2.9)
October 1993	4 (4.7)	0 (0.0)	5 (8.0)	2 (1.4)	11 (2.9)
Total	11 (12.94)	4 (4.76)	13 (20.96)	18 (12.32)	46 (12.2)
Number (n) plants	85	84	62	146	377

TABLE 47.6 Mortality[a] in the Treatments of *Pinus maritime*

Experimental Period	Sludge and Wastewater	Sludge	Wastewater	Witness	All Treatments
March 1991	119 (64.7)	8 (3.1)	75 (36.6)	44 (22.9)	246 (29.4)
October 1991	52 (28.3)	25 (9.8)	74 (36.1)	16 (8.3)	167 (20.0)
March 1992	0 (0.0)	2 (0.8)	3 (1.5)	2 (1.0)	7 (0.8)
October 1992	3 (1.6)	21 (8.2)	30 (14.6)	43 (22.4)	97 (11.6)
October 1993	3 (1.6)	24 (9.4)	6 (3.0)	24 (12.5)	57 (6.8)
Total	177 (96.2)	80 (31.4)	188 (91.7)	129 (67.2)	574 (68.7)
Number (n) plants	184	255	205	192	836

[a] It is given the absolute number of detected dead plants of *Pinus brutia* and *Pinus maritime* and in parenthesis the calculated percentage by the number of plants at the given treatment.

TABLE 47.7 Mean Height (cm) and Increase of Plants: *Pinus brutia* at Various Treatments

Season Period	Sludge and Wastewater	Sludge	Sludge	Witness
March 1991	39.91	39.59	43.31	42.63
October 1991	77.22	101.29	106.38	102.65
Increase March–October 1991	37.31	61.92	62.51	60.09
Increase October 1991–March 1992	38.13	53.31	55.29	45.52
Increase March–October 1992	32.88	49.44	50.50	51.05
Increase October 1992–October 1993	82.12	114.95	113.14	101.98
October 1993–October 1994	65.65	94.68	85.86	87.57
Total increase				
March 1991–October 1994	256.1	374.3	367.3	346.21

TABLE 47.8 Mean Height (cm) and Increase of Plants of *Pinus maritima* at Various Treatments

Season Period	Sludge and Wastewater	Sludge	Wastewater	Witness
March 1991	19.07	19.98	23.45	22.37
October 1991	47.51	48.39	42.22	46.40
Increase March–October 1991	28.44	28.42	17.64	23.99
Increase				
October 1991–March 1992	30.65	32.99	20.80	27.03
Increase March—List October 1992	18.80	6.20	13.00	9.36
Increase				
October 1992–October 1993	62.71	47.17	49.58	43.72
October 1993–October 1994	41.14	32.54	38.36	21.96
Total increase				
March 1991–October 1994	181.7	149.20	122.00	137.13

undergone secondary treatment, while the second group was irrigated with natural water and the third one with water technically enriched with Mn^{2+}. Results showed low concentrations of metals in leaves and the roots of *Neriun oleander* and *Geranium* spp., where reuse of treated wastewater was used for irrigation.

Irrigation with technically enriched Mn^{2+} water decreased the corresponding concentration of Fe^{2+} in leaves. At 500 ppm, Mn^{2+} the proportion Mn/Fe was 1:2.43, 1:1.63, 1:0.86 for the above plants listed, respectively, while at 1000 ppm was 1:1.75, 1:2.45.1:0.92 for the three plants, respectively. The roots of three plants absorbed heavy metals. In the soil, the antagonistic effects between Fe and Mn elements were obvious [9]. In addition, the synergistic interaction of Cu and Zn of the three forest species were studied. Tzanakakis et al. [36] studied the influence of liquid urban wastewater in different forest species in Skalani Heraklion. The plants studied were *Eucalyptus camandulensis*, *Acacia cyanophylus*, and *Populus nigra*, which were transplanted in October 2000. Also, the plants *Arundo donax* were planted on February 2001. Irrigation with liquid treated wastewater started in June. It was found that *Populus nigra* trees had a higher increase in height, while *Acacia* plants had a higher diameter increase. The increase in production of biomass was the *Acacia* plant, followed by *Arundo donax*, while the lowest production was by *Eucalyptus* spp. and *Populus nigra* [36].

Mavrogiannopoulos and Kyritsis [32] conducted an experiment using 120 plants of *Buddleia variabilis* and 120 plants of *Medicago arborea*. The experiment involved three treatments. (1) No irrigation after planting, (2) irrigation with natural water, and (3) irrigation with TMWW. Results showed that up to September there were no differences in the biomass of *Eucalyptus* spp. between the studied treatments. The plants that were irrigated with TMWW had a significant increase in their height. In addition, significant differences in the growth of other plants were observed. The plants irrigated with municipal wastewater had the highest diameter increase. For *Buddleia variabilis*, significant differences regarding the diameter increase were found between those plants that were irrigated and those that were not irrigated, as well as those that were irrigated with natural water and with TMWW. For *Forsythia* sp., no significant differences in growth among plants of different treatments were found. For *Medicago arborea* and *Nerium oleander* plants, differences were found between plants irrigated with TMWW and nonirrigated ones and not with those plants irrigated with natural water. According to these results, generally, the plants that were irrigated with TMWW attained the highest growth, probably due to the TMWW content in essential nutrients [32].

Kalavrouziotis and Drakatos [17] studied the response of *Eucalyptus* spp., *Pinus maritime*, *Pinus Pinea*, and *Nerium oleander* to the TMWW on lands near the WWTP of Corfu.

The aim of this work was to explore possibilities for the installation of forest species in regions near the WWTP of Corfu, so that the lands could be environmentally restored and irrigated with the reuse of wastewater, ensuring the protection of the marine area of Corfu. Relevant experiments studying the wastewater reuse were conducted in the agricultural area of Agrinion

[27]. Also, in Patras [28], and especially in the area of vineyards of Metamorphosis, Attikis. Also, the planning of the wastewater reuse for the cultivated areas of western Greece was designed. The water requirements of crops of the referred areas were calculated and the possible effluent quantities were found that could be reused by taking advantage of the effluents produced by the WWTPs operating in western Greece.

Additionally, the irrigation problem in Argolida and Lakonia with municipal wastewater produced by the WWTPs of the two districts, and the application method in crops were studied [24–26]. Furthermore, research involving the study of the effect of the treated wastewater from WWTP of Agrinion and the accumulation of macro- and micronutrients of heavy metals in soils and plants *Alium cepa* (onion) and *Lactuca sativus* (lettuce) were investigated, and the following were found [19]:

1. Significant quantities of available P and Mn were accumulated in the soil of both crops.
2. Cu was accumulated in leaves while Zn accumulated in seed of the lettuce.
3. The treated wastewater of WWTP of Agrinion could be used for vegetable irrigation under certain conditions.
4. Continual testing of both liquid sewage and soil for potential presence of high concentrations of metals and other nutrients.
5. The appropriate measures under wastewater treatment for decrease of microbial load in permitted levels to ensure the safe use of treated sewage.
6. Testing of physical and chemical soil conditions (drainage, aeration, soil pH, and organic matter).

Furthermore, Kalavrouziotis et al. [24–26] studied the effect of the application of treated wastewater of WWTP of Mesolonghi on the following:

1. Behavior of nutrient elements P, K, Ca, Mg, Fe, Mn, Zn, and Cu in soil.
2. Heavy metals Pb, Ni, Co, Cd and the soil properties pH, $CaCO_3$, organic matter, and electrical conductivity (E.C.).
3. The accumulation of all of the above nutrients and metals in various parts of vegetables (roots, leaves and head: (a) *Brassica oleracea* var. *Italica* (Broccoli) and (b) *Brassica oleracea* var. *Gemmifera* (Brussels sprouts).
4. Interactions between the above nutrients, heavy metals, and soil properties.

There following were found in a series of experiments:

1. Zn, Cd, and P were accumulated in soil that was planted with broccoli, in levels that were within the critical permitted limits.
2. The statistically significant accumulation of Pb and Cd in heads and Fe in the roots, with combination of high-microbial load of studied wastewater, has been a prohibitive factor, for the present use of treated wastewater, for irrigation of studied vegetables, and all the other vegetables, generally.

3. Generally, in the plants' roots accumulation of nutrients Mn, Fe, Cu, and B, and heavy metals Ni and Pb were higher than in leaves and lower than in the heads.
4. The effect of treated wastewater in the interactions between the above nutrients elements and heavy metals in most cases were not statistically different from the corresponding of natural water irrigation and from this view, the studied TMWW can be used for vegetable irrigation.
5. Antagonistic interactions between some heavy metals and chemical soil properties, for instance, PbxpH, pHxCu, pHxMn, pHxNi, enable their utilization for treatment problems of soil pollution with the given heavy metals.

The reuse involves the application of a specific scientific methodology, which if it is applied in Greece visible benefits for agricultural production of the country will be possible.

47.8 National and Regional Planning for Reuse of Municipal Wastewater for Irrigation

The management of the reuse of municipal wastewater is an international practice. Eventual application of the TMWW reuse in Greece could indirectly contribute sufficient amounts of natural irrigation water, which could be used for the irrigation of arable and deserted areas.

Relevant efforts have been made in the direction of urban wastewater reuse. However, thus far, there has not been any progress in relation to implementation on a national and regional level for the reuse of the municipal wastewater in the irrigation of crops. For the time being, the matter of wastewater reuse in Greece is at the experimental level. The study of municipal wastewater reuse in soil, produced by all the operating WWTP in Greece, in terms of country planning, and environmental, geological, and geochemical terms, including those that are to be constructed in small and big municipalities in the near future, will allow the prevention of the disposal of liquid wastewater in streams, rivers, lakes, and seas, while contributing to the increase of irrigated areas and to the establishment of green urban sites.

Such application of municipal treated wastewaters in soils requires their safe reuse and will aim primarily at

1. Increasing the irrigated areas (today only 42% of planted areas are irrigated) [16].
2. Developing green urban sites.
3. Avoiding any future pollution in aquatic ecosystems mostly in tourist areas, where there is no safe rule in operation of WWTP [16].
4. Restoring problematic territories with resistant crops and, in turn, their irrigation with TMWW.

Safe reuse of municipal wastewater and solid byproducts that are produced by WWTP will maintain and increase the environmental quality in urban environments, while constantly controlling the water irrigation quality [4].

The reuse of municipal wastewater is based on the following premises [5]:

1. To provide reliable treatment of municipal wastewater to meet the requirements of irrigation water quality; this is needed for application in soils.
2. Public health protection.
3. Public acceptance.

Used as irrigation water are of particular geochemical important:

1. The total microbiological load that will probably bear municipal wastewater (*Salmonella* spp., *Escherichia coli*, *Shigella* spp., *Ascaris* spp., fecal coliforms)
2. The total content of wastewater in metals

The presence of toxic heavy metals such as Al, As, Ag, Ba, Be, B, Cd, Cr, Cu, Fe, Li, Mn, Mo, Se, V, and Zn may contribute to the accumulation of some of these metals in soils and plants under specific concentrations, which may be toxic for the aquatic world. Their presence in the municipal wastewater that is intended for irrigation can limit the natural irrigation water use. For this reason, it is necessary to continuously monitor the concentration of those elements in wastewater in all treatments stages, and also research their reuse in soils and crops in various plant species. Maximum concentrations of heavy metals in irrigation water have been reposted by Pettygrove and Asano [35].

The crucial question is for how long the reuse of wastewater could be applied without causing toxicity in soils and plant species. The answer to this question is the preparation of a special computer program that will allow us to foresee and forecast the possible creation of toxic conditions in soils to take the necessary measures and prevent the establishment of pollution in the soil.

47.9 Reuse of TMWW in Agriculture

The basic application of TMWW reuse should include the following:

1. Irrigation of agricultural farmlands
2. Irrigation of landscapes
3. Industrial recycling
4. For the enrichment of underground water
5. Fire prevention in urban centers
6. Protection and reformation of certain problematic areas
7. Application of wastewater in urban centers

The available water on a national level in Greece is 14.34×10^9 m³. Greece suffers from a significant water shortage every 40–45 years and presents a periodical water cycle every 5–7 years. From the total agricultural area of Greece, 40% is irrigated [16].

The total water quantity that is required for crops' irrigation annually amounts to 55×10^6 m³. On the other hand, 84%–86% of the total demands for water in the country go to irrigation of crops. This practically means that the largest water consumer in Greece is agricultural production.

The quantitative data that describe the conditions of water on a national level in Greece, are the following:

1. Atmospheric precipitations: 115,375 Mm³/annually.
2. Total water potential 69,000 Mm³/annually.
3. Total water usage: 8150 Mm³/annually.
4. Irrigation: (a) agricultural: 6900 Mm³/annually; (b) other uses: 1.25 Mm³/annually.
5. Irrigation water (%): 84.7.
6. Waste effluents: 700 Mm³/annually.
7. Reuse (2005): 6.00 Mm³/annually (0.07%–0.9%).
8. Predicted (2005): 60.00 Mm³/annually (0.75%).

From the above data, it is concluded that sooner or later it will be necessary to apply treated wastewater reuse in Greece especially in problematic and nonproblematic soils based on the research results, which have thus far been obtained in the country, as well as abroad [16].

47.10 Elemental Contribution to Soils and Plants of the Interactions between Heavy Metals and Micro Elements after Irrigation with TMWW

From the research conducted, regarding the interaction of heavy metal under the effect of TMWW in soil and vegetables, it was found that in leaves, roots, and seeds and generally in the heads of plants, a great number of interactions occur, amounting to hundreds, between heavy metals and nutrients. Application of the reuse of TMWW generally increases the number of statistically significant interactions.

In Brussels sprouts, the synergistic interactions were more than those of broccoli roots. In addition, the number of antagonistic interactions in roots, under the influence of TMWW, was found equal to that of leaves (Figure 47.1).

The total number of interactions that occurred in Brussels sprouts, regardless of the type of the interaction, for instance, antagonistic or synergistic, was higher in roots than in leaves. In Figure 47.2, the interactions between heavy metals, micronutrients, and macronutrients that occur in the roots of Brussels sprouts under the effect of TMWW are given.

Synergistic interactions contributed nutrients and heavy metals, while the antagonistic ones had a negative impact on the accumulation of heavy metals and nutrients. The total contribution of the interactions of nutrients were dependent on the number of different types of interactions (synergistic vs. antagonistic) and quantitative contribution of each interaction. Some antagonistic interactions, especially those that include heavy metals, had a positive impact on plant growth because they decreased toxic concentration of metals in plants. The antagonistic interactions were CaxCd, MnxDd, KxCd, and PxPb. These interactions can be used appropriately after their experimental validity in relation to controlling the pollution. However, other competitive interactions, which included heavy metals and nutrient elements, are able to have negative impacts on plants resulting in deficiency.

Additionally, other synergistic interactions contributed to the accumulation of heavy metals in plants. The synergistic

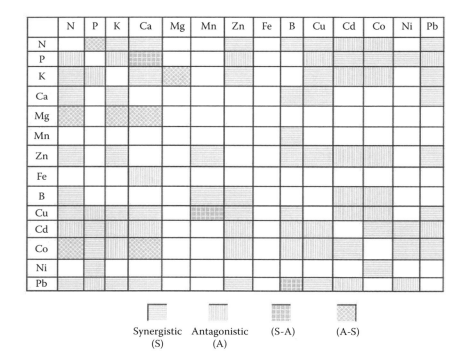

FIGURE 47.1 Interactions between heavy metals, macro-, and micronutrients that occur in brussels sprouts roots (*Brassica oleracea* var. *gemmifera*) under the influence of TMWW. (Adapted from Kalavrouziotis, I.K. and Koukoulakis, P.H. 2014. Unpublished data.)

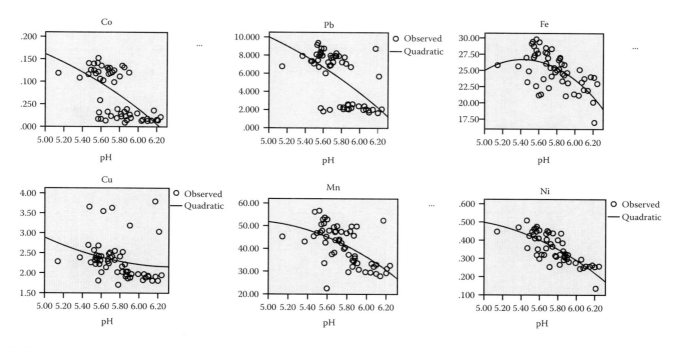

FIGURE 47.2 Antagonistic relationship of pH with metals Co, Pb, Fe, Cu, Mn, and Ni in mg/kg, respectively, under the influence of sludge. (Adapted from Ntzala, G. 2011. Investigation of Soil Pollution with Wastewater and Sludge and Their Impact on Soil and Lettuce Plants (*Lattuca stiva*, var. *congifolia*). MS thesis, Department of Environment and Natural Resources Management, University of Western Greece, pp. 1–165.)

interactions of PxCd, KxPb, ZnxPb, and especially ClxCd, which increased significantly the content of toxic Cd in vegetables leaves, as found lately in experimental vegetable irrigation with TMWW, indicated that it is very important because it may have health effects.

47.10.1 Quantification of Contributive Interactions in Heavy Metals and Nutrients

The regression analysis between analytical data of heavy metals and nutrient elements of both soil and plants enabled the quantification of the elemental contribution of the interactions in nutrient elements and heavy metals to soils and plants. It was found that the synergistic interaction of MnxP contributed to content increase of P in the heads of broccoli by 43%, while the antagonistic interaction of ZnxP decreased the content of P by 63% [22].

It was found that the final content of a given nutrient or heavy metal is dependent on the number of corresponding synergistic or antagonistic interactions, as well as the size of the contribution of each interaction in terms of element or metal.

As an example, in Table 47.9 the contribution of interactions in heavy metals and nutrients elements in various forest plant species is given.

The study of this table shows that some data have negative values. These values reflect the reduction of the content of the corresponding metal or nutrient because of the effect of antagonistic

interactions. It is highlighted here that these interactions are able to reduce to such an extent the concentration of the interacted metal, which in turn can result in plant deficiency with negative effects on plant growth and yields.

Statistically negative values were found between soil pH and heavy metals of Co, Pb, Fe, Cu, Mn, and Ni. This fact underlines the great environmental importance of these relationships (Figure 47.2). These relationships are practically applied to improve soils contaminated with metals, with increase of soil pH, by the application of liming.

In addition, organic matter of soil under the influence of sludge is related to negatively and statistically significant metals Mn, Cu, and Pb. In Figure 47.3, the negative relationships between the organic matter (OM%) and soil Mn, Cu, and Pb content are given.

47.11 Determination of Transfer Factors of Metals under the Influence of Irrigation with Treated Wastewater

The reuse of TMWW constitutes a necessary option for the arid conditions prevailing in many countries around the Mediterranean basin and generally in the world. Additionally, the reuse of TMWW is imposed by the necessity of producing irrigated crops, to address the nutritional needs of people. The implementation of reuse of TMWW can contribute to the

TABLE 47.9 Elemental Contribution of the Interactions between Heavy Metals and Micronutrients to Leaves, Stems, and Roots of Various Forest Species under the Influence of Sludge (Biosolids)

Forest Species	Plant Species	Cu	Fe	Mn	Zn	Ni	Cr	Co	Cd	Pb
Cupressus arizonica Green	Leaves	5.52	19.69	1.71	32.75	74.07	ns	7.62	ns	ns
	Plant stalk	−82.48	−84.99	ns	ns	−0.74	50.13	86.74	87.50	−17.91
	Roots	30.79	39.07	6.49	27.41	3.89	ns	ns	−74.45	22.67
Pinus pinea L	Leaves	41.91	30.41	−36.79	43.70	21.30	ns	34.49	−57.04	8.39
		−30.06	−78.16	ns	96.54	−37.52	−2.07	4.34	ns	69.69
	Roots	27.54	14.58	ns	34.56	39.72	ns	17.17	−45.94	41.80
Cotoneaster integerrimus Med	Leaves	−54.07	−61.67	ns	22.44	10.85	−17.16	−17.49	ns	−5.84
	Plant stalk	98.21	86.8	−2.09	−70.71	87.66	100.00	−27.57	53.69	ns
	Roots	ns	−96.86	ns	ns	79.88	ns	98.84	39.19	40.32

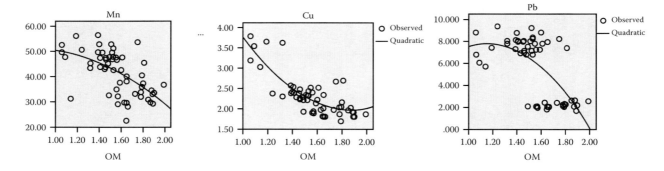

FIGURE 47.3 The competitive relationships between organic matter (OM%) and soil content in Mn (mg/kg), Cu (mg/kg), and Pb (mg/kg). The above negative relationships also can be practically applied for soil pollution with regard to heavy metals by simply increasing organic matter of soil.

accumulation of heavy metals in soil and plants with negative impact on human health and environment. The study of the accumulation of these metals in plants can be accomplished by taking into account the optimum "transfer factor" (TF) of metals, from soil to plants [20,21,23]. The TF value is given by the equation TF = Mp/Ms where Mp is metal content in the plant dry matter in mg/g and Ms is the metal content of soil in μg/g. One approach to formulate an optimum value of TF is to express it as a function of

1. The maximum yield of dry substance
2. The minimum content of dry matter in heavy metals, under the permissible limit
3. Minimum pollution load index (PLI) value

This approach can be accomplished by the application of multiple linear regression analysis using the experimental data of TF corresponding to the values of the above parameters.

For this purpose, an experiment was conducted applying different levels of wastewater to lettuce (*Brassica oleracea* var. *capitata*). The optimum TF values found for the various metals were as follows: ZnTF 3.44, CuTF 0.93, CdTF 4.32, CrTF 116.7, NiTF 2.86, and PbTF 0.56. TF values higher than the above values signify the transfer of high levels of the specific heavy metal from soil to plants. These values are tentatively considered as optimum or critical values of TF.

47.12 Relationship of the PLI with TF and Concentration Factor, under the Influence of Sludge

Sludge, a byproduct of the wastewater treatment, is used as an organic fertilizer and as a soil amendment. Besides the positive effects on soil and plants, sludge can also have toxic effects, given the fact that it includes heavy metals. Long-term use of sludge can contribute to the accumulation of heavy metals in soil [20], causing pollution and toxicity in plants, adversely affecting their development. The assessment of the level of soil contamination due to the long-term reuse of sludge is necessary for preventing health risk effects and environmental pollution.

PLI can be used as a relatively accurate determination of the level of soil contamination. For its calculation, the knowledge of the concentration factor (CF) of each metal, which is indicative for soil contamination, needs to be known. CF is equal to the ratio of the content of the soil heavy metal (Ms) to the reference value of a given metal (Mr): TF = Ms/Mr.

PLI is equal to the geometric mean of the concentration factor of each of the individual heavy metals as shown in

$$\text{PLI} = \sqrt[n]{CF_1 \times CF_2 \times CF_3 \dots \times CF_n} \qquad (47.2)$$

where CF_1, $CF_2 \ldots CF_n$ = concentration factor of the metals M_1, $M_2 \ldots M_v$, respectively.

The relationship between PLI with CF and TF was studied, likewise between CF with TF was studied, for the creation of mathematical models with the view to calculate one as a function of the other. The relationship of PLI with TF was found to be linear and normally negative (Figure 47.4), while the relationship of PLI with CF was found to be linear and positive (Figure 47.5),

and the relationship between CF and TF was curvilinear and positive (Figure 47.6).

To accurately assess the level of soil contamination, the knowledge of the optimum level of CF is necessary. This can be accomplished by the expression of CF as a function of the maximum yield of dry matter, optimum concentration of micronutrients in dry matter, and the minimum concentration value of heavy metals in dry matter [20,21,23].

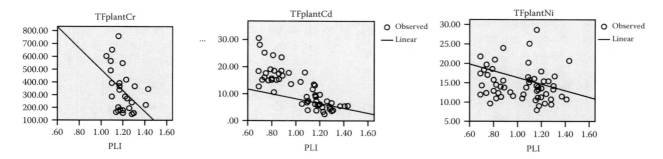

FIGURE 47.4 Relationship of PLI with transfer factor of Cr, Cd, and Ni of dry substance of lettuce. (Adapted from Ntzala, G. 2011. Investigation of Soil Pollution with Wastewater and Sludge and Their Impact on Soil and Lettuce Plants (*Lattuca stiva*, var. *congifolia*). MS thesis, Department of Environment and Natural Resources Management, University of Western Greece, pp. 1–165.)

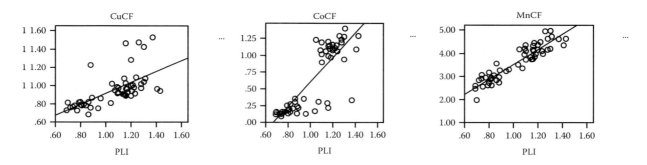

FIGURE 47.5 Relationship of PLI with concentration factor of Cu, Co, and M of soil, under the influence of sludge. (Adapted from Ntzala, G. 2011. Investigation of Soil Pollution with Wastewater and Sludge and Their Impact on Soil and Lettuce Plants (*Lattuca stiva*, var. *congifolia*). MS thesis, Department of Environment and Natural Resources Management, University of Western Greece, pp. 1–165.)

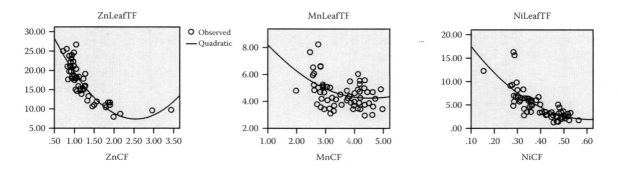

FIGURE 47.6 Relationship of concentration factor (CF) of Zn, Mn, and Ni with the relevant TF of the given metals of a dry substance of lettuce, under the influence of sludge. (Adapted from Ntzala, G. 2011. Investigation of Soil Pollution with Wastewater and Sludge and Their Impact on Soil and Lettuce Plants (*Lattuca stiva*, var. *congifolia*). MS thesis, Department of Environment and Natural Resources Management, University of Western Greece, pp. 1–165.)

1. From the application of linear regression analysis, the following optimum values of concentration factor of various heavy metals were found: ZnCF2, 93; CdCF, 0.39; CoCF, 1.47; and NiCF, 0.52.
2. Based on above values, the optimum PLI value was found to be equal to 0.97.

From the results obtained by our lab experiments, general conclusions are summarized in the next paragraphs.

The accumulation of heavy metals by the application of the wastewater reuse of the TMWW and sludge within the context of a national plan for the utilization of effluents (solid or liquid) of the WWTP will have to be done with specific experiments, such as that for the study of the interaction of ClxCd because of its direct relationship with the food chain and protection of human health and the environment.

The continuation of the study of the interactions between heavy metals and nutrients in soil and plants will have to be done in other crops besides vegetables (wheat, corn) to confirm their possible presence in all crops. The stability of the interactions must be assured during their mobility from plant roots to leaves, and from leaves to heads or from roots to leaves and to seeds. It is also necessary to study the fate of toxic contents in wastewater such as toxic organic substances, pesticides, and pharmaceutical compounds, and their accumulation in soil and absorption by plants, as well as the potential accumulation of heavy metals in the heads of the plants and finally in the food chain.

47.13 Development of Biosolids (Sludge) of Wastewater for Agriculture

The sludge processing of municipal wastewaters (biosolids) is derived from wastewater in WWTP. According to the instruction 86/278/EEC, the sludge derived from septic tanks is considered of equal importance (value) with the previous one; therefore, it is covered by the above directive.

The sludge is a renewable resource of enormous importance in economy and society generally. Undoubtedly, the use of sludge presents practical, technical, and financial difficulties. The choices for sludge utilization are composting, energy combustion, raw material in the brick-making industry, landfills in abandoned quarries, and for environmental restoration.

The management of municipal wastewater sludge for use in agriculture is one of the most important forms of environmental management, which positively contributes to restoring soil fertility with the enriched presence of organic matter and macronutrients [20].

It is worth mentioning that 1 tone of sludge contains 50 kg N, which corresponds to 250 kg ammonium sulfate (21-0-0). However, it should be noted that the positive utilization of sludge in agriculture, the overall management of sludge so that after appropriate treatment to be used as soil amendment and as a source of nutrients elements in agriculture, forestry, and landscape and soil restoration, poses risks from its potential content in harmful substances, leading to pollution and to the existence of biological parameters, which generate pollutions. To avoid these adverse impacts on crops, it is necessary to have exclusive control and to remove all harmful substances from wastewater. Any illegal connections to the drainage system of municipal wastewaters with industrial wastewaters, which are known to be loaded with heavy metals and other toxic substances, must be excluded. Particular attention should be given in areas of primary production that operate near WWTP, such as rural industrial units, such as olive oil factories, to avoid adverse impacts on the environment. This is strictly necessary given that 28% of the Greek territory consists of areas that are protected by RAMSAR and NATURA 2000 treaties. Any use of sludge in soils should be a constantly controlled process in order to avoid any risk of soil and aquifer pollution.

The use of sludge in soils is, in fact, a biochemical procedure that will have to be applied safely in order to be beneficial to soil restoration of low production, not only in Greece but also in all countries in the southern Mediterranean.

The application of sludge in soils will have to be done in a completely stable condition and under continual control [29,30]. In order to accomplish this goal, the following prerequisites are needed:

1. Informing properly the rural population for the usefulness of the sludge and for the risk involved.
2. The effective function of the WWTP practicing regular quality control.
3. The application of sludge to soils must be done only when there is agronomic interest, aimed at improving plant production and soil productivity.
4. The application of sludge of high quality in soils is judged in respect to the physicochemical and biological characteristics.
5. Special consideration should be given to the knowledge of concentration values of heavy metals, hazardous organic compounds, and pathogens of sludge and soils.
6. Avoidance of application of excessive quantities of sludge in soils, the addition of which should be supported by relevant soil analysis (physicochemical properties, macronutrients concentration, micronutrients, and trace elements). As generally recognized at the European level in relation to the use of sludge in soils, the application of 500 kg of sludge per acre and implementation period of 3 years is considered to be optimum for sludge reuse.
7. Avoidance of water pollution, when the sludge is to be applied in soils outside the basins of water drainage, and in these cases, the application of guidelines is deemed necessary (2000/60/E.K).
8. Each agricultural utilization of sludge is a process that should eventually be made under the sole responsibility of the producers after the relevant information is given by scientists.

47.14 Production of Sludge and Disposal in Europe and Greece

Today in Greece, about 450 WWTPs are functional, which covers 75% of the equivalent population (E.P) [2]. The capacity varies between those of Psitalia with a production quantity of sludge 35,000,000 and 700,000 E.P, respectively, but in periphery, significantly less sludge is produced (20,000 E.P up 190,000 E.P).

The production of sludge amounts to 19.8 g of dry sludge per head per day [1]. In Cyprus, the annual production of sludge amounts to 20,000 tones and in Europe annually 12 millions tones of sludge timber [3].

Depending on the stage of the wastewater treatment, the following types of sludge are recognized:

1. Primary sludge: Sludge produced during the primary treatment of wastewaters.
2. Biological sludge: Sludge produced during the secondary treatment of wastewaters.
3. Mixed sludge: A mix of primary and biological sludge.
4. Higher sludge: Produced during the higher or advanced treatment of wastewaters.

The sludge produced by WWTP is subject to additional processing, in order to reduce the water content for better economic management; attention is paid to the stability of the product and the removal of pathogens.

The quantities of sludge that are produced by WWTP in Greece, according to the official figures of YPEXODE amount to 103,865.80 t/year.

Agricultural use of sludge constitutes an important way of management of this byproduct, where in EU it amounts to 37% of the total produced sludge. In Greece, today, only small amounts have been used in agriculture (9357.90 t/year), either in pilot applications or in research and localized applications without central planning.

Since the end of 2008 in Greece, there is an approved manual by Ministry of Rural Development and Food issued by Agricultural Research Foundation for the appropriate utilization of municipal sludge, which aims to guide farmers to environmentally acceptable uses of sludge for agricultural purposes.

Based on statistical figures of EU for the period 2000–2006, 9.7 millions MG (1 ton = 1 MG) biosolids of sludge have been produced in EU annually, of which 8.7 million MG from EU 15 member states and an additional million MG from new member states. Approximately 80% of the total produced sludge was derived in Greece, Great Britain, Italy, Spain, and France. By 2020, the annual sludge production in EU is expected to increase by 30% and reach a height of 1 million MG (Figure 47.7).

In Figure 47.8, the annual mean production of sludge per capita in the EU member states is given, and on the basis of these data, the average planned annual production of sludge per capita in EU by the 27 member states is equal to 17 kgDS/a. According to these improvements, it is expected to be increased in 2020 at 25 kgDS/a.

For the period 2000–2006, the total produced quantity of sludge is recycled with great variability among the member states as to final method of disposal.

47.15 Other Methods of Sustainable Management of Biosolids (Sludge)

Management methods of sludge are listed as follows:

1. Thermal treatment. This method is effective in solving the problem because organic matter and phosphates are completely removed. Yet, problems of health risk from

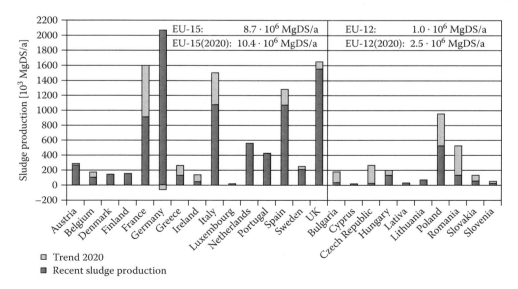

FIGURE 47.7 Recent sludge production in EU-15 and new member states EU-12 (database of 2002–2006) and 2012 prospects. (Adapted from Milieu Ltd, 2010. WRc, RPA: Environmental economic and social impacts of the use of sewage sludge on land. Final Report, Part III: Project Interim Reports, Report for the European Commission, D.G. environment, Study contact DG ENV. G4/ETU/2008/00762.)

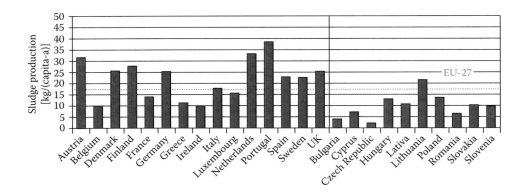

FIGURE 47.8 Per capita annual production of sludge in EU. (Adapted from Milieu Ltd, 2010. WRc, RPA: Environmental economic and social impacts of the use of sewage sludge on land. Final Report, Part III: Project Interim Reports, Report for the European Commission, D.G. environment, Study contact DG ENV. G4/ETU/2008/007662.)

hazardous substances, which are emitted such as Hg and Co may appear.

2. Improvement and restoration of degraded areas. The sludge that is derived from the stabilization procedure is enriched in organic matter and macronutrients. It can therefore be used for the improvement of areas that are poorly supplied with organic matter and nutrients for enhancing plant growth. This can be implemented in abandoned stadiums, camps, and quarries, all areas which undoubtedly exist in our country. It is simultaneously exploited as waste material for landfills and as a restorative material of landfills of sites of uncontrolled wastewater management. The sludge derived from composting is utilized in areas already contaminated.

3. Production of phosphoric fertilizers. The extraction of phosphorus-rich municipal sludge and wastewaters after crystallization procedure or thermal treatment.

4. Biogas production. Through anaerobic digestion process in WWTP, the production of biogas is possible, which could be used for the needs of WWTP or for motion of wheeled vehicles, and so on.

5. Production of building materials or others. Thermal treatment is a prerequisite for the production of building materials or asphalt products.

6. Production of combustible energy. Sludge can be used as a combustible material, with the reservation of potential problems regarding phenomena of pollution.

47.16 Summary and Conclusions

Large quantities of TMWW and sludge are produced the world over. In some countries with dry climate, a fraction of the TMWW is used for the irrigation of crops and the remaining is directed to the sea or oceans. In addition, a relatively small fraction of sludge is incorporated into soil, a larger one is added to dumps, and a smaller fraction is used for the production of biogas, building materials, and so on.

Generally, the agricultural reuse of both of these inputs is an attractive way of disposal. However, the quality of irrigation

wastewater in dry or semidry regions where high temperatures, combined with low humidity, create conditions for high ET is of extreme importance. In spite of the economic importance of wastewater and sludge, health and environmental risk problems are by necessity associated, which cannot be overlooked. This fact makes their management problematic. Thus, in the last two decades, regulations have been formulated by various committees such as USEPA for the controlled discharge of pollutants in the soil, and pollutant load limits have been enforced by law since 1973.

The terrestrial disposal is a method of "slow" disposal. However, there are two other basic methods such as rapid filtration and flooding.

The methods of sewage disposal are all based on ecologically acceptable practices aiming at the minimization of the ecological pollutant load of the environment.

The slow land disposal systems must be based on climate and meteorological data and the following factors must be taken into account:

The frequency and quantity of precipitation of monthly and annual rainfall, including mean monthly and annual ET, and annual distribution of ET for the assessment of water balance.

Temperature.

Intensity, frequency, and duration of rainfall for assessment of the surface runoff.

Geological–topographical and hydrological characteristics of the region.

Latitude and relief.

Topographic relief. Strong slopes are prohibitive due to the hazard of territorial erosions and strong runoff. Slopes should be lower than 20%.

Petrographic data for a first assessment of water infiltration.

Stratigraphic data for finding impermeable, permeable, and semipermeable zones in the subsoil.

Soil characteristics.

Mechanical composition of the soil.

Hydraulic conductivity (permeability) horizontally and vertically.

Soil structure, porosity, fluid distribution, apparent wetness.

Cation exchange capacity (CEC). This shows the total charge of soil surface, and therefore the retention capacity of cations.

Organic matter.

Soil reaction (pH).

Inorganic, organic, and total nitrogen and phosphorus.

The soil will be continuously monitored during the reuse for heavy metal accumulation, and presence of other organic contaminants including changes in soil properties.

In addition, the following, set of economic and social criteria must be taken into consideration for the site selection for land disposal, on the selected area.

- Conditions of employment—workers availability for reforestation.
- Land ownership.
- Right to use land and water chores.

The land disposal of TMWW should include the following practices:

- Irrigation of agricultural farmlands.
- Industrial recycling.
- For the enrichment of underground water.
- Fire prevention in urban centers.
- Protection and reformation of certain problematic areas.

The rational management of wastewater and sludge must be accompanied by continuous research, planning, and development related to the optimum reuse of these two inputs. The following aspects must therefore be investigated:

Methodology of reuse application, quantities of TMWW and sludge to be applied for optimum effects on plant growth and soil quality, effects on soil and plant quality, accumulation of heavy metals in soils and plant, and toxic organic substances in monitoring system for the control of soil toxicity problems due to long-term reuse, interactions between heavy metals, plant nutrients, and soil characteristics, quantification of elemental interaction's contribution to soils and plants, effective storing and distribution systems of TMWW and sludge, economic aspects of TMWW and sludge reuse, forecasting pollution problems and amelioration of polluted soils.

Establishment of a relevant computer program for the effective reuse of TMWW and sludge, effective management of sludge including thermal treatment, and production of biogas, phosphoric fertilizer, and energy from sludge combustion, exploitation of various kinds of water such as greywater, roof and street water by developing relevant collection systems, and so on.

Based on the aforementioned, the following conclusions could be drawn:

The present wastewater treatment technology has advanced to the extent of producing clean water for multiple use (urban, agricultural, industrial, etc.).

Among the methods of the TMWW disposal, the agricultural reuse seems to be most attractive.

The effective reuse of the TMWW could be accomplished by modern management practices, which must include the following: (i) quality criteria, (ii) economic and social factors, (iii) geotechnical, (iv) geological and hydrological criteria, and (v) national and regional planning.

All of the above must be accompanied by continuous research work on the various reuse aspects, such as methodology of application, impact on soil, and plants assessment of the soil pollution.

The pollution of heavy metals and other toxic compounds can be alleviated by the use of a model of reuse application to crops that will also help to decrease the fertilizer use forecast possible unfavorable effects on soil and on the environmental quality.

References

1. Andreadakis, A.D., Mamais, D.E., Gavalaki, E., and Kampylafka, S. 2002. Sludge utilization in agriculture: Possibilities and prospects in Greece. *Water Science and Technology*, 46(10), 231–238.
2. Angelakis, A. 1999. *Recording and Mapping of Wastewater Treatments Plants (WWTP) in Greece.* National Agricultural Foundation, Heraklion, Greece, p. 64.
3. Anonymous. 2002. EUREAU Position paper on Sewage Sludge. European Union of National Associations of Water Supplies and Wastewater Services, Brussels, p. 4.
4. Asano, T. 1998. *Wastewater Reclamation and Reuse*, Vol. 10. Technomic Publishing, Lancaster, PA.
5. Asano, T. 2002. Recycled water task force second meeting. Workgroup presentation and discussions, June 3, Manhattan Beach Marriott Hotel.
6. Ayers, R.S.S. and Westcot, D.W. 1985. Water quality of agriculture, FAO Irrigation and Drainage Paper 29:99–104 Rev. 1.
7. Drakatos, P.A., Fanariotou, I., and Kalavrouziotis, I.K. 1996. Impact of wastewater irrigation and sludge utilization on mechanical properties of Eucalyptus Wood. *Restoration and Protection of the Environment, Proceedings of the International Conference*, Chania, Greece, August 28–30, 1996, pp. 723–726.
8. Drakatos, P.A. and Kalavrouziotis, I.K. 1996. Impact of wastewater irrigation and sludge utilization on thermal and acoustic properties of Eucalyptus wood. *International Workshop, Sewage Treatment and Reuse for Small Unities: Mediterranean and European Experience*, November 23–26, 1996, Morocco, Africa.
9. Drakatos, P.A., Kalavrouziotis, I.K., Hortis, T.C., Varnavas, S.P., Drakatu, S.P., and Bladenopoulou, S. 2002. Antagonistic action of Fe and Mn in Mediterranean-type

plants irrigated with wastewater effluents. *International Journal of Environmental Studies*, 59(1), 125–132.

10. Drakatos, P.A., Kalavrouziotis, I.K., Skuras, D., and Drakatu, S.P. 1997. Impact of thermal properties of the trees cultivating by processed wastewater and sludge. *Proceedings of International Conference, Energy Week*, January 28–30, 1997, Houston, TX, pp. 401–403.

11. Drakatos, P.A., Kalavrouziotis, I.K., Skuras, D., and Drakatu, S.P. 1997. Thermal emissivity of leaves from trees cultivated using processed wastewater. *Proceedings of International Conference, Energy Week*, January 28–30, 1997, Houston,TX, pp. 404–407.

12. Drakatos, P.A., Lyon, R.H., Kallistratos, G., and Chrysolouris, G., 1992. Reliability and exploitation of wastewater treatment plants using diagnostic methods. *International Journal of Environmental Studies*, 40, 267–280.

13. Drakatos, P.A., Varnavas, S.P., and Kalavrouziotis, I.K. 1996. Irrigation of certain Mediterranean plants with wastewater. Tolerance in heavy metals. *Restoration and Protection of Environment, Proceedings of the International Conference, Chania*, Greece, August 28–30, 1996, pp. 412–418.

14. Hermite, L. 1991. *Treatment and Use of Sewage Sludge and Liquid Agricultural Wastes*. Elsevier Applied Science, London, pp. 559–601.

15. Kalavrouziotis, I.K. and Drakatos, P.A. 2001. The future of irrigation by processed wastewater in Greece. *Journal of Environmental and Waste Management*, 4(2), 107–110.

16. Kalavrouziotis, I.K. and Drakatos, P.A. 2002. Investigation of Corfu-Greece reclaimed Municipal wastewater suitability for irrigation. *International Journal of Water*, 2(4), 284–296.

17. Kalavrouziotis, I.K. and Drakatos, P.A. 2004. Investigation of Corfu-Greece reclaimed municipal wastewater suitability for irrigation. *International Journal of Water*, 2(4), 284–296.

18. Kalavrouziotis, I.K., Hortis, T., and Drakatos, P.A. 2004. The reuse of wastewater and sludge for cultivation of forestry trees in dessert areas in Greece. *International Journal of Environment and Pollution*, 21(6), 425–439.

19. Kalavrouziotis, I.K., Kanatas, P.I., Papadopoulos, A.H., Bladenopoulou, S., Koukoulakis, P.H., and Leotsinides, M.N. 2005. Effects of municipal reclaimed wastewater on the macro and microelement status of soil and plants. *Fresenius Environmental Bulletin*, 14(11), 1050–1057.

20. Kalavrouziotis, I.K. and Koukoulakis, P.H. 2012. Soil Pollution under the effect of treated municipal wastewater. *International Journal of Environmental Monitoring and Assessment*, 184, 6297–6305.

21. Kalavrouziotis, I.K., Koukoulakis, P.H., and Kostakioti, E. 2012. Assesment of metal transfer factor under irrigation with treated Municipal wastewater. *International Journal of Agricultural Water Management*, 103, 114–119.

22. Kalavrouziotis, I.K., Koukoulakis, P.H., and Mehra, A., 2010. Quantification of elemental interaction effects on Brussels sprouts under treated municipal wastewater. *Desalination Journal*, 254(1–3), 6–11.

23. Kalavrouziotis, I.K., Koukoulakis, P.H., Ntzala, G., and Papadopoulos, A. 2012. Proposed indices for assessing soil pollution under the application of sludge. *Water Air and Soil Pollution*, 223(8), 5189–5196. Doi: 10.1007/s11270-012-1270-x.

24. Kalavrouziotis, I.K., Koukoulakis, H., Robolas, P., Papadopoulos, A., and Pantazis, V. 2008. Interrelationship of heavy metals and properties of a soil cultivated with *Brassica oleracea* var. *Italica* (*Broccoli*), under the effect of Treated Municipal Wastewater. *International Journal of Air, Soil and Water Pollution*, 190(1–4), 309–321.

25. Kalavrouziotis, I.K., Koukoulakis, H., Robolas, P., Papadopoulos, A., and Pantazis, V. 2008. Macro and micronutrient interactions in soil under the effect of *Brassica oleracea var. Italica*, irrigated with treated municipal wastewater. *International Journal of Fresenius Environmental Bulletin*, 17(9a), 1270–1282.

26. Kalavrouziotis, I.K., Robolas, P., Koukoulakis, P., and Papadopoulos, A. 2007. Effects of municipal reclaimed wastewater on the macro and microelements status of soil and of *Brassica oleracea* var. *Italica*, and *Brassica oleracea* var. *gemmifera*. *International Journal of Agricultural Water Management*, 95(4), 419–426.

27. Kalavrouziotis, I.K., Sakellariou-Makrantonaki, M., Vagenas, I., Hortis, T., and Drakatos, P.A. 2005. The potential for the systematic reuse of the wastewater effluents by the biological treatment plant of Agrinio, Greece on soils and agriculture. *Freshenius Environmental Bulletin*, 14(3), 204–211.

28. Kalavrouziotis, I.K., Sakellariou-Makrantonaki, M., Vagenas, I.N., and Lemesios, I. 2006. Assessment of water requirements of crops for the reuse of municipal wastewaters from the WWTP. of Patras, Greece. *International Journal of Environment and Pollution*, 28(3–4), 485–495.

29. Kouloumbis, P. 2001. Possibilities of sludge exploitation of WWTP. *Proceedings Conference*, Xarokopio, University of Athens, pp. 179–189.

30. Kouloumbis, P. et al. 2000. Environmental Problems from the disposal of sewage sludge in Greece. *International Journal of Environmental Health Research*, 10(1), 77–83.

31. Mattheus, P.J. 1984. Control of metal application rates from sewage sludge utilization in agriculture. *CRC Critical Reviews in Environmental Control*, 14, 199–250.

32. Mavrogianopoulos, G. and Kyritsis, S. 1995. *Use of Municipal Wastewater for Biomass Production*. Agricultural University of Athens, Athens.

33. National Academy of Sciences and Engineering. 1973. *Water Quality Criteria EPA-R3-73-033*. U.S. Environmental Protection Agency, Washington, DC.

34. Ntzala, G. 2011. Investigation of Soil Pollution with Wastewater and Sludge and their Impact on Soil and Lettuce Plants (*Lattuca stiva*, var. *congifolia*). MS thesis, Department of Environment and Natural Resources Management, University of Western Greece, pp. 1–165.

35. Pettygrove, G.S. and Asano, T. 1984. Irrigation with reclaimed wastewater. Aquidance manual Califronia State Water Resources, Control Board, Report number 84–1, July.

36. Tzanakakis, V.E., Paranychianakis, N.V., Kyritsis, S., and Angelakis, A.N. 2002. Evaluation of wastewater treatment and biomass production by a slow rate system using different plant species. *IWA Regional Symposium on Water Recycling in Mediterranean Region*. Iraklio, Greece, September 26–29, pp. 353–360.

37. U.S. Environmental Protection Agency. 1992. *Guidelines for Water Reuse*. EPA/625R-92/004, Washington DC, p. 247.

38. U.S. Environmental Protection Agency. 1993. Standards for the use or disposal of sewage sludge. *Federal Register*, 58(32), 9248–9415.

39. Milieu Ltd. 2010. WRc, RPA: Environmental economic and social impacts of the use of sewage sludge on land. Final Report, Part III: project inerim Reports, Report for the European Commission, D.G. environment, Study contact DG ENV. G4/ETU/2008/007662.

40. WHO. 2006. Guidelines for the safe use of wastewater. *Excreta and Greywater*, 2, 178.

41. Drakatos P.A., Kalavrouziotis, I.K., and Drakatu, S.P. 2000. Synergism of Cu and Zn in the plants irrigated via processed liquid wastewater. *Journal of Land Contamination and Reclamation*, 8(3), 201–207 (EPP publications).

48

Wastewater Monitoring

Floris Boogaard
Delft University of Technology
Hanze University of
Applied Sciences
Tauw
INDYMO

Saeid Eslamian
Isfahan University of Technology

PREFACE

This chapter proposes a methodology for the selection of monitoring locations for wastewater quality monitoring, based on (pre-) screening, a quick scan monitoring campaign, and final selection of location and design of the monitoring setup.

48.1 Introduction

As described in Chapter 4, it is important to have continuous insight of the water quality that is being used in the (industrial) environment. For this reason, a monitoring campaign can be set up to ensure the quality of inflow and outflow of water. The challenge for the design of a monitoring campaign with a given budget is to balance detailed monitoring at a limited number of locations versus less detailed monitoring at a large number of locations.

The design of a wastewater monitoring campaign involves the selection of the monitoring setup, to be measured parameters, and the selection of the monitoring locations [6].

48.2 Method for Selection of Monitoring Locations

The method for the selection of monitoring locations comprises four steps: prescreening, screening, quick scan, and final selection [5].

48.2.1 Prescreening Monitoring Locations

The prescreening stage limits the total number of potential monitoring locations by applying the following criteria:

- General suitability for research
- Representativeness (e.g., in terms of catchment characteristics, such as construction period, number of inhabitants, average income, type of roads [high/low traffic intensity]), planned reconstruction, surcharged/non-surcharged storm sewers
- Geographic coverage of the total research area

In most cases, the prescreening can be based on information that is already available from field visits and general accessible information.

48.2.2 Screening Monitoring Locations

The screening stage further limits the total number of monitoring locations by assessing practical aspects such as

- Safety (traffic conditions, criminality, vandalism)

- Accessibility and available space for stormwater treatment and monitoring equipment
- Planned reconstruction of infrastructure

The screening can be based partly on available information and partly on information from a site visit. This data can be used later to select appropriate monitoring equipment.

48.2.3 Quick Scan

The quick scan aims at gathering information on

- The water quality to be expected at a potential monitoring location
- The system dynamics. This information is gathered using a very simple and relatively cheap approach
  - Installation of a continues loggers (such as, temperature, pressure, conductivity, and turbidity)
  - Use local radar weather forecast to timely detect significant storm events
  - Grab sampling of water for specific parameters
  - Laboratory analyses of samples

48.2.3.1 Parameters

The following water quality parameters, besides continuous monitoring (such as pH, EC, N, P, O_2, turbidity, see Table 48.1) [1], can be analyzed to determine the health issues in wastewater.

48.2.4 Final Selection Monitoring Campaign

The final selection of the monitoring campaign is the result of the previous steps and results and available budget. Most important for reusing water with a high quality ambition could be tracing possible pathogen microorganisms. In surface water and

TABLE 48.1 Water Quality Parameters Analyzed

Water Quality Problem Related	Parameter
Oxygen depletion	SS, BOD, COD
Eutrophication	NH3-NH4+, TKN, Total-P, Otho-P
Toxicity	Pb, Zn, Cu, PAH (16 compounds)
Hygienic quality	Total *Escherichia coli (E. coli), fecal streptococcus*, and thermal coli
Examples of sensors for online insight water quality	Temperature, oxygen, nitrate, phosphate, conductivity (EC), pH, suspended solids/turbidity

stormwater, these are well-documented [2]; however, (subsurface) groundwater contamination by pathogenic microorganisms should not be ruled out as a risk for health due to leakage of wastewater systems. When stormwater is being reused from stormwater drainage pipes, possible health problems can occur when foul connections are realized (wastewater from toilets to stormwater drainage pipes). An example is presented to cost-effectively trace these connections and get insight of the water quality of the setup from a monitoring campaign [5].

48.3 Monitoring Foul Connections: An Example

A temperature sensor is placed near the invert level of the conduit near the sewer manhole. The sensor also measures and records the water depth. Changes in temperature or water level indicate either a spill from a wrong connection or rainfall. Therefore, also the data from a local rain logger should be available. During dry periods, the water depth should be constant. Only after rainfall, the drainage system should discharge water but discharge

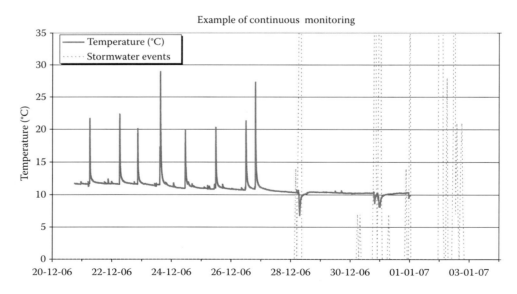

FIGURE 48.1 Example of information gathered with continuous monitoring. The temperature variations are an indication of a wrong connection in a stormwater drainage system and the effect of stormwater events on temperature. (Adapted from Kluck, J. and Boogaard, F. 2008. *11th International Conference on Urban Drainage*, Edinburgh, Scotland.)

Attached equipment:
1. *In situ* TROLL 9500 sensors:
   - Nitrate and ammonium ISE
   - Rugged dissolved oxygen
2. CTD diver:
   - Temperature
   - Pressure
   - Conductivity
3. Diving light
4. HD video camera (GoPro 3+)

FIGURE 48.2 Underwater drone with sensors attached for innovative dynamic monitoring (near the Floating Pavilion, Rotterdam).

off groundwater is also possible. A sudden temperature rise in a dry period indicates a foul spill and thus a wrong connection on the drainage system (Figure 48.1). During a warm day, a sudden temperature rise might be the result of relatively warm runoff water. The measurements indicate the upstream of the measurement location if the conduits are clean. The observed suspicious locations (measurement points) can be investigated in detail with visual or video inspections. Also at this point, the use of smoke and tracers can be used in the restricted area. The effects of illicit discharges on the surface water can be monitored continuously by using aquatic drones and sensors (Figure 48.2) [3].

These measurements give a quick and cost-effective insight in which sewers are likely to be clean and which part of the stormwater system needs to be restored to their primarily function: transporting stormwater only and reuse of water for different usage.

48.4 Summary and Conclusions

This chapter describes a short cost-effective methodology to select a number of suited monitoring locations for (waste) water quality assessment for reuse of water with a quick scan. As a result, the chance of failure of the subsequent research or monitoring project decreases significantly, especially as the quick scan reveals the onsite system dynamics. This not only enhances the selection of appropriate monitoring locations, but also the subsequent detailed design of the monitoring equipment and sampling strategy.

In addition, a quick scan will result in a first useful data set on water quality and a strong indication of time and place depending on factors such as illicit connections using (distributed) temperature sensing to detect illicit connections.

References

1. Balendonck, J., Boogaard F.C., Bruins, M.A., Ganzevles, P.P.G., and Gieling, Th. 2005. *Real Time Control (Storm) Water Quality (in Dutch)*. WUR, the Netherlands. https://www.wageningenur.nl/en/Publication-details.htm?publicationId=publication-way-333431383831.
2. Boogaard, F. and Lemmen, G. 2007. STOWA stormwater database: The facts about the quality of stormwater runoff. STOWA 2007-21, the Netherlands (in Dutch).
3. De Lima, R., Sazonov, V., Boogaard, F.C., de Graaf, R., and Dionisio, M. 2015. Monitoring the impacts of floating structures on the water quality and ecology using an underwater drone, *IAHR World Congress*, The Hague, the Netherlands. http://89.31.100.18/~iahrpapers/84564.pdf
4. Kluck, J. and Boogaard, F. 2008. Locating wrong connections in stormwater drainage systems by temperature logging. *11th International Conference on Urban Drainage*, Edinburgh, Scotland.
5. Langeveld, J., Boogaard, F., Liefting, H., Schilperoort, R.P.S., Hof, A., Nijhof, H., de Ridder, A.C., and Kuiper, M.W. 2013. *Selection of Monitoring Locations for Stormwater Quality Assessment*. NOVATECH.
6. Xuheng, K. 2014. Stormwater modeling and management. In: Eslamian, S. (ed.), *Handbook of Engineering Hydrology, Environmental Hydrology and Water Management*, Vol. 3. CRC Press, Boca Raton, FL, pp. 328–346.

49

Urban Stormwater Quality Monitoring

Paola Foladori
University of Trento

PREFACE

Urban stormwater runoff has a certain level of contamination affected by site-specific conditions such as land-use and associated human activities. During a rainfall event, large quantities of pollutant accumulated on the surfaces during dry periods are washed off and transported by stormwater runoff.

Due to the large variability in the quality of stormwater during a rainfall event and in the uses of land and thus its contamination levels, qualitative data acquisition from urban stormwater remains an open issue. An accurate site-specific monitoring of stormwater quality represents a key requirement to adequately explain the complex response to a rainfall event or site-to-site differences in stormwater quality. Decisions taken with an insufficient amount of information result in costs and time wastage and may cause water management problems in its treatment and utilization.

This chapter aims to provide a general overview of the main aspects affecting stormwater quality in urban areas and the approaches used for its monitoring. Methods and equipment currently used to monitor stormwater quality are presented focusing on the appropriate collection of water samples and the analysis of the most important pollutants. To this end, the use of monitoring stations equipped with automatic samplers, online sensors, and electronic instrumentation to record or transmit data can all help operators to obtain representative samples at reasonable costs. The collection of flow-weighted composite samplers is the correct approach for estimating the event mean concentrations for each pollutant and to calculate mass loads produced from stormwater. Both intra-storm and inter-storm samples should be included in a complete monitoring campaign to evaluate the short- and long-term effects on receiving water bodies or in water utilization.

49.1 Introduction

Urban stormwater runoff originates from the transformation of rainfall excess to a runoff hydrograph based on catchment characteristics such as impervious surface fraction [8] and the term usually defines wet weather discharges from separate sewer catchments.

Urban stormwater runoff is usually discharged into receiving water bodies, but it could play an important role in combating water shortages in urban areas, due to the increasing need for water in such places.

It is well known that urban stormwater runoff has a certain level of contamination affected by site-specific conditions such as land use and associated human activities. Large quantities of pollutant accumulated on surfaces during the antecedent dry days are washed off and transported by stormwater runoff, often causing the need for treatment before discharge or utilization.

A similarity exists between the characteristics of stormwater runoff from impervious or semi-impervious surfaces in urban areas (streets, parking lots, and roofs) and on highways: the pollutant contribution from traffic is central in both cases [4]. Indeed, in residential areas, the concentrations of pollutants such as organic matter or solids are higher in the stormwater from roads rather than from roofs [13]. However, the runoff from commercial and industrial areas generates more pollutants than residential areas [9,13].

Empirical approaches used to characterize stormwater quality require inventoried land use and estimated annual mass loads for a variety of pollutants and for each land use [9]. Although pollutant loads for the various land uses are referred to in the literature, a careful review of the pollutant loads for the same land use shows variability, which suggests the need for site-specific monitoring [9]. Poor water-quality characterizations are responsible for biased results in the estimation of stormwater pollutant loads, suggesting the need for improved monitoring to reduce the uncertainty of water quality [8,9].

Pollutant build-up and wash-off processes have been implemented in stormwater quality models to estimate stormwater pollution. However, these processes are more complex than what the process equations in the stormwater quality models take into account, and deficiencies in the current modeling approaches for urban stormwater quality assessment have been highlighted [8]. Although the hydraulic modeling is able to produce accurate results, the water quality estimation from models is not yet at a consolidated stage of knowledge and needs further advancement.

In this context, an accurate site-specific monitoring of stormwater quality remains a key requirement to adequately explain the complex response to a rainfall event or site-to-site differences in stormwater quality. Decisions taken with an insufficient amount of information represent costs and time wastage and may lead to water management problems [2].

This chapter aims to provide a general overview of the approaches currently used to monitor the quality of stormwater runoff produced in urban areas, by means of an appropriate collection of water samples and the measurement of the most important pollutants. Data acquisition in urban stormwater is an open issue [2] and there is not a unique procedure suitable for all monitoring studies. Thus, this work cannot be exhaustive of all aspects and details of the monitoring of urban stormwater quality because there is much to be said concerning this wide topic (*inter alia* Hvitved-Jacobsen [4,12]) and research is constantly bringing up new results and discussions [2]. This work is, rather, a synthesis of the most relevant and practical information that should be considered in the monitoring of stormwater quality and which can be easily understandable to practitioners at different levels of expertise. The better this practical information is understood, the more efficiently the monitoring data can be collected [12].

49.2 Factors Affecting Urban Stormwater Quality

Site specific, climatic, and other local variables play an important role in the qualitative characterization of stormwater runoff [2]. Among them, rainfall and catchment characteristics are frequently referred to as the principal factors affecting stormwater runoff quality.

49.2.1 Rainfall Characteristics

Precipitation is the phenomenon causing the formation of stormwater runoff. It is widely recognized that most of the pollutants found in stormwater runoff do not originate from natural rainwater but rather from the washing of catchment surfaces. As the volume or velocity of stormwater runoff increases, the capacity to wash off and transport soil particles and associated pollutants also increases. The quantitative characteristics of the rain event (amount, frequency, intensity, duration, etc.) thus significantly affect the quality of stormwater runoff. Rainfall measurements are important in a complete monitoring and the main collection parameters for describing a rain event are the following:

- Total rainfall depth, typically expressed in mm
- Rainfall duration, typically expressed in h
- Rainfall intensity, typically expressed in mm/min or mm/h

Concerning the influence of rainfall intensity on stormwater quality, a high rainfall intensity displays the highest pollutant concentrations in water samples, especially regarding solids and organic matter, because the relatively higher kinetic energy of a high-intensity rainfall event results in more pollutants being transported [8]. Conversely, it was observed [8] that low-intensity rainfall events, even of long duration, produce consistently lower pollutant concentrations [11].

The influence of rainfall duration on pollutant wash-off in low-intensity events is minimal. Compared to rainfall intensity, duration has, in general, a minor effect.

Another important parameter, which may help to understand runoff quality, is the antecedent dry weather period (ADWP) between two rain events, expressed in days. In general, the run-off pollution load is expected to be positively correlated to the ADWP parameter because the high ADWP allows more pollutants to accumulate and not be washed off [13], but this relationship is not always confirmed in the literature.

49.2.2 Rainfall–Runoff Relation

Stormwater runoff volume is highly related to the characteristics of the rainfall event, land use, and impervious surfaces. Runoff volume (V, expressed in m^3) can easily be estimated from a runoff coefficient, which accounts for the reduction in the runoff compared with the rainfall [4,9], according to the following equation:

$$V = \phi \cdot P \cdot A \qquad (49.1)$$

where ϕ is the runoff coefficient (adimensional), P is the rainfall depth (expressed in mm), and A is the contributing catchment area (expressed in m^2); a conversion factor of 1000 is used to convert mm to m.

In this equation, V is the runoff produced from the watershed during one storm event or the volume in a year (using the annual precipitation).

Equation 49.1 can be rewritten as

$$Q(t) = \phi \cdot i(t) \cdot A \qquad (49.2)$$

where $Q(t)$ is the runoff flow rate (expressed in m^3/s) and $i(t)$ is the rainfall intensity (expressed in m/s or m/min).

The relationship between $i(t)$ and the rainfall depth P is given by: $P = \int_0^T i(t) \cdot dt$.

The coefficient ϕ is a pragmatic but useful approach in place of the rather complicated rainfall–runoff process [4], which is held at a rather simple level because hydraulic aspects are not a major focus of this chapter. Standard ϕ values have been proposed for different types of urban surfaces (0.9–1.0 for roofs, 0.7–0.95 for asphalt and concrete pavement, pervious areas around 0.3).

The catchment, such as land use and type of urban surfaces, also has an important role in influencing runoff quality in addition to quantity. A different level of contamination is expected in the stormwater runoff originating from different functional areas (residential, commercial, and industrial) in an urban context. In general, the pollutant concentrations in these three functional areas are in the following order: industrial area>commercial area>residential area [13]. For instance, more pollutants are generally deposited in industrial areas, while the densities of vehicular traffic and human activities are higher in commercial rather than residential areas [13].

The quality of the runoff also depends on the type of surface material such as asphalt, cement, roofs, etc. In urban residential areas, the concentrations of organic matter or solids are higher in the runoff from roads rather than from roofs, due to the larger accumulation of pollutants on roads due to traffic [13].

Therefore, in order to include a high variability of catchment characteristics, direct monitoring of stormwater runoff is often advised. Furthermore, for a detailed monitoring of the quantity and quality of urban stormwater runoff, time-profiles are needed for both flow rate and for pollutant concentration, as described in the following section.

49.2.3 Hyetographs, Hydrographs, and Pollutographs

Rainfall events are characterized by a set of hydrographs and pollutographs for the various contaminants in stormwater runoff.

An example of the rainfall intensity over time (hyetograph) and the corresponding runoff flow rate (hydrograph) is shown in Figure 49.1a. There is a time delay between the initiation of the rainfall and the initiation of the corresponding runoff in the location of the monitoring station. This delay not only depends on some characteristics of the rainfall event but also on the catchment characteristics and transport time.

In urban areas, increasing urbanization causes a loss in vegetation and the replacement of pervious areas with impervious surfaces, thus resulting in changes to the characteristics of the surface runoff hydrograph, increasing stormwater runoff volumes, and peak flows [2].

The plot of the concentration of pollutants as a function of time during a runoff event and in correspondence to the monitoring station is defined as a pollutograph. A pollutograph can refer to either the concentration (expressed in mg/L) or the mass flow (expressed as g/s).

Figure 49.1b shows a typical profile of a pollutograph obtained from a discrete sampling during the runoff event, where the pollutant is expressed as concentration. In the example, the peak of concentration in the runoff appears in the first minutes, caused by the early runoff, which transports the accumulated pollutants from the urban surfaces. After the peak, the pollutant concentration decreases progressively and may remain approximately stable in the final part of the profile. In many cases, the pollutant concentration peak preceded the flow peak: this phenomenon is called first-flush and is explained in detail in Section 49.3.5.

The combined information that is given in terms of the hyetograph, the hydrograph, and the pollutograph for a rain event forms an essential basis for any engineering approach related to urban stormwater pollution [4].

49.3 Monitoring of Stormwater Quality

A monitoring program should be designed based on local characteristics and availability of budget and time and with a well-defined objective. The objective may be the need to build a treatment system, the evaluation of pollutant removal efficiency in a treatment, the assessment of impacts in receiving water bodies, or the potential of utilization.

A monitoring program focuses on the measurements of stormwater quantity and quality, taking into account the large variability during each event and during the different seasons. It

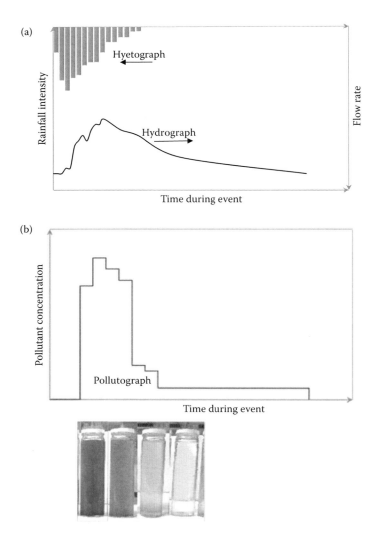

FIGURE 49.1 (a) Example of rainfall intensity versus time (hyetograph), compared to corresponding runoff flow rate (hydrograph); (b) example of profile of pollutant concentration in samples collected during the runoff event (pollutograph).

is well known that a few grab samples during one storm event during the year are not representative of the average conditions, and an appropriate design of intrastorm sampling (Section 49.3.1) and interstorm sampling (Section 49.3.2) is recommended. In the intrastorm sampling design, one sample representative of the entire storm event can be analyzed obtaining a flow-weighted concentration (Sections 49.3.1 and 49.3.3). Alternatively, the choice to analyze more samples along the event is more expensive, but permits to observe the occurrence of first-flush (Section 49.3.5) and variations in the pollutant concentration.

49.3.1 Sampling during a Storm Event

During a storm event, the level of contamination in the urban runoff depends on the time: the highest pollutant load is found in the first part of the runoff volume, while the following part has lower concentrations due to the progressive washing of urban surfaces. The different pollutant concentrations in both

the initial and the following runoff suggest that different measures have to be taken to deal with these fluxes. For example, the initial runoff should be collected and may need a separate treatment before discharge or utilization, while the following runoff can be discharged directly in order to reduce treatment costs during utilization or pollution control. The intra-storm sampling design should take into consideration this variability by collecting multiple samples throughout a storm event to incorporate changes in concentration and flow rate, with the aim to accurately represent the entire storm event. For this purpose, various types of storm sampling can be chosen: grab samples, time-weighted samples, or volume-weighted samples, described in more detail as follows.

Grab samples are taken manually by operators within a relatively short period of time (for instance, a few minutes) to form discrete grab samples (single samples stored in separated containers) or composite grab samples (samples stored in the same container). Grab samples can be collected any time during the storm event. However, the approach to form a composite sample using various grab samples is very time-consuming and thus generally impractical.

The timing of a grab sample is important because the first-flush effect (which can be observed in many stormwater events) creates a higher pollutant concentration at the beginning of run-off; thus, samples early in a storm event should be collected if the peak or maximum concentrations are desired [5]. Conversely, samples in the middle of the storm have to be collected to obtain samples that are more representative of the entire event [5]. Unfortunately, the duration of the initial phase and the period of time in the middle of the storm are hard to decide in practice because they may vary with the characteristics of the rainfall and catchment [13].

Grab samples collected manually suffer from some limitations due to: (1) the requirement of an operator to be onsite during actual stormwater sample collection; (2) the inability to accurately predict the starting time of a storm event with the risk of arriving too late and missing the early stage of the storm event or the first-flush, when pollutant concentrations are higher; and (3) the use of a single grab sample to estimate mass emission may lead to large errors when not collected at the appropriate time. However, grab samples are the only option for monitoring parameters that transform rapidly, such as bacteria and volatile organic compounds, or substances that adhere to containers such as oil and grease [12]. Furthermore, grab samples, which can be taken manually or with portable pumps, can be useful when the point of sampling is difficult to access or to reduce set-up costs when other solutions might be cost-prohibitive. In other cases, the use of automatic samplers is more appropriate for obtaining composite samples. In particular, the following sampling designs can be applied with automatic samplers:

- *Time-proportional sampling*: Samples of equal volume are collected at equal increments of time. For instance, a fixed increment of time in the range 5–20 min is suitable for the first hour of the storm, while it is around 0.5–1–2 h thereafter. Although the flow rate during the storm event varies over time, the time-proportional sample does not account for variations in flow. The samples can be collected as discrete samples or composite samples (Table 49.1).
- *Flow-proportional sampling*: Samples can be collected in several ways, but a continuous flow measurement device is required. The following methods can be used, depending on the types of programs installed in the automatic sampler [12]:
  1. *Constant volume – time proportional to flow volume increment* – Aliquots of equal volume are taken at equal increments of flow volume (for instance 1 L every 10 m³, but regardless of time)
  2. *Constant time – volume proportional to flow volume increment* – Aliquots are collected at equal increments of time and the volume of each aliquot is proportional to the volume of flow since the preceding aliquot was collected

3. *Constant time – volume proportional to flow-rate* – Aliquots are collected at equal increments of time and the volume of each aliquot is proportional to the flow rate at the time the aliquot was collected.

Methods 1 and 2 use the total volume of flow and thus are more accurate than method 3, which uses a single instantaneous measurement of flow rate over the entire sampling interval.

Flow-proportional samples can be collected as discrete samples or composite samples, depending on whether samples are stored separately or in a single container (Table 49.1).

- *Random sampling or user-defined sampling*: Aliquots are collected randomly during the storm event, and users may change the interval of sampling on the basis of real-time control of online instruments during the storm event.

In Table 49.1, samples are distinguished in discrete samples and composite samples. In the case of discrete samples, each sample is collected and analyzed separately and represents the average pollutant concentration for the volume of water to which it corresponds. Conversely, in the case of composite samples, all the samples collected are put in the same container; thus, only one aliquot is taken and analyzed from this sample, and it represents the average pollutant concentration of the entire composite sample.

However, in the case of "time-proportional sampling" and "composite samples" (Table 49.1), the average pollutant concentration measured in the single sample cannot be put in relation with the volume of water, which is variable over time. This means that pollutant concentrations in sample aliquots collected when the flow is lower are given the same "weight" as sample aliquots collected during higher flows [12]. Therefore, this sampling method is, in general, not recommended, while the various options of the "flow-proportional sampling" (Table 49.1) result as more suitable.

To represent the entire storm event, samples should be taken from the beginning of the runoff (which may not coincide with the initiation of the rain event) to the end of the event when the flow decreases below the dry weather water level. However, it may happen that the bottles in the sampler, which are normally up to 12 or 24 (which may correspond to up to 48–96 samples), are completely filled before the end of the storm event, resulting in the sampling of only a part of the storm event. Conversely, not every storm event may produce exactly 12 or 24 filled bottles. Because of the unpredictability of rainfall during a storm event, sampling will always involve some uncertainty.

With random or user-defined sampling, the user can vary the sampling time in real-time to optimize the sampling, even during the stormwater event, by using remote programming.

In the assessment of efficient sampling designs for urban stormwater monitoring, Leecaster et al. [7] demonstrated that single storms were most efficiently characterized (small bias and standard error) by taking 12 samples following a flow-proportional schedule. The observation that flow-weighted

TABLE 49.1 Synthesis of the Type of Samples Obtained Using Automatic Samplers

Type of Sampling	Collection of Aliquots	Storage in the Sampler	Type of Sample	Volume of Aliquots	Concentration in the Sample[a]
Time-proportional sampling	At constant time increments	Samples stored in separated containers	Discrete samples	Constant	Variable (C_i)[a]
	At constant time increments	Samples stored in a single container	Composite sample	Constant	Average
Flow-proportional sampling (constant volume and time proportional to flow)	At constant volume increments	Samples stored in separated containers	Discrete samples	Constant	Variable (C_i)[a]
	At constant volume increments	Samples stored in a single container	Composite sample	Constant	Average (EMC, see Section 49.3.3)
Flow-proportional sampling (constant time and volume proportional to flow)	At constant time increments	Samples stored in separated containers	Discrete samples	Variable (V_i)[a]	Variable (C_i)[a]
	At constant time increments	Samples stored in a single container	Composite sample	Variable (V_i)[a]	Average (EMC, see Section 49.3.3)

[a] i, aliquot number.

composite samples are the better choice for stormwater monitoring is widely confirmed in the literature (*inter alia* Lee et al. [5]).

The result from a flow-proportional composite sample is also called "event mean concentration" (EMC; see Section 49.3.3) and it is used to estimate pollutant mass load given by the product of EMC and runoff volume (Section 49.3.4).

49.3.2 Sampling of Different Storm Events

The decision as to how many and which storm events should be sampled (interstorm sampling design) is not easy.

Stormwater should be monitored for at least an entire year to take into account seasonal variations, especially in areas where precipitation is infrequent. Indeed, climatic conditions, such as long antecedent dry periods, may significantly increase the pollutant emissions from urban stormwater discharges, due to the long period of pollutant build-up. In this case, the first rainfall of the wet season may produce higher pollutant concentrations than later events, this phenomenon being called "seasonal first-flush" [5]. The presence of a seasonal first-flush is a dilemma in stormwater monitoring programs because this leads to identifying the highest mass emission but also to overestimating the central tendencies, such as annual mass loads [5].

Concerning the number of storms to be sampled, the decision is problematic because it depends on many factors. Stormwater data always has a very high intrinsic variability. The variance in the data of storm events will determine the assessment uncertainty, which needs to be controlled to obtain accurate and reliable results in a monitoring program. However, the challenge in developing better monitoring programs is to reduce the variability due to sampling errors and artifacts, while retaining the intrinsic variability of the rainfall phenomena.

A variety of sampling monitoring designs have been proposed and used in various countries, depending on objectives and availability of funds, personnel, and time. Many authors note that only one or two storms each year are unlikely to be representative [5]. Leecaster et al. [7] examined several monitoring programs in California and concluded that small confidence

intervals are obtained by monitoring at least seven storms per year, which are about 50% of the annual storm events. Hvitved-Jacobsen et al. [4] indicate that it is not possible to determine exactly the number of storm events to be monitored, and that a more pragmatic recommendation is to sample not less than 10–15 storm events in total. On the other hand, very large numbers of storm events or samples cause an increased burden of work and budget-costs without a reasonable benefit in terms of better statistical values.

Samples of storm events with different characteristics should be considered in a complete stormwater monitoring program during one year, including various rainfall intensities, storm durations, and seasons during the year. The selection of the storms for sampling is generally subjective, but in general, the preference is given to medium or high-intensity storms because signals of the installed equipment will certainly be above detection levels and concentrations will be higher. In terms of rain events that contribute to the runoff, rainfall events larger than a rainfall depth of 0.5–2 mm can be reasonably considered relevant in a stormwater monitoring program (*inter alia* Hvitved-Jacobsen [4]). Furthermore, storm events relevant for the assessment of pollution are those preceded by at least 72 h with less than 2 mm of precipitation [12].

When different storm events are sampled in a site during one year, the mean concentration of pollutants on the annual basis (called "site mean concentration") can be calculated as the mean or the median value of a number of EMCs at that site (see Section 49.3.3). Hvitved-Jacobsen et al. [4] indicate that the better calculations of the site's mean concentration are the weighted mean method and the median value method, in order to overcome the influence by outliers (e.g., events with high EMCs and small runoff volumes).

49.3.3 Event Mean Concentration of Pollutants

The concentration of pollutants varies greatly during a storm event and the flow rate varies to a large extent. Single stormwater events can be efficiently characterized (small bias and standard error) by using a flow-weighted estimation. The EMC (expressed in mg/L) of pollutants represents a flow-proportional average concentration during a storm event. It is defined as the total mass of a pollutant transported during an entire runoff event (M) divided by the total stormwater runoff volume (V), for an event of duration T, according to the following equation [6,9]:

$$EMC = \frac{M}{V} = \frac{\int_0^T c(t) \cdot q(t) \cdot dt}{\int_0^T q(t) \cdot dt} \qquad (49.3)$$

where $c(t)$ is the pollutant concentration variable over time (expressed in mg/L); $q(t)$ is the flow rate variable over time (expressed in L/min); the limits of integration $t = 0$ and $t = T$ refer to the times of initiation and cessation of runoff, respectively (expressed in min). Therefore, the EMC value refers to the entire runoff duration.

In practice, EMC is calculated based on a sampling procedure and thus refers to discrete time steps rather than the integrated form. Equation 49.3 can be presented as Equation 49.4: the product of the pollutant concentration and the runoff volume for each sample collected is divided by the sum of the total runoff volume.

$$EMC = \frac{\sum_{i=1}^N C_i \cdot V_i}{\sum_{i=1}^N V_i} \qquad (49.4)$$

where N is the total number of samples collected in the storm event; i (1,2,...,N) is the sampling sequence number in the event; C_i is the pollutant concentration in sample i; and V_i is the discharged runoff volume corresponding to sample i.

In a flow-proportional composite sample, the concentration corresponds to an average of all the samples taken, coinciding to EMC (Table 49.1). Therefore, in cases when time-profiles of pollutant concentrations are not needed, the flow-proportional sampling with a composite sample is a very straightforward method of sampling, as EMC can be immediately obtained and is presumed to be more accurate.

In the case of collecting grab samples, Lee et al. [5] observed that a single grab sample can be 10 times greater or smaller than EMC, leading to large errors in the estimation of mass loads from stormwater. Although averaging 12 grab samples is a much better estimate of the EMC, it can still lead to errors compared to EMC [5].

49.3.4 Total Load Calculation

The total load (M) for each pollutant in the stormwater runoff can be calculated as the product of the runoff volume and the concentration of contaminants represented by the EMC according to the following equation:

$$M = V \cdot EMC \qquad (49.5)$$

Using data of a discrete sampling, the above equation can be rewritten in the following form:

$$M = \sum_{i=1}^N C_i \cdot V_i \qquad (49.6)$$

where symbols are the same as indicated above.

The load can be calculated as event mass load when calculated for a single storm event or as annual mass load when the annual volume and the mean annual concentration are used.

49.3.5 First-Flush

The initial period of stormwater runoff during which a substantial concentration peak appears is called the "first-flush" phenomenon [4,6]. The occurrence of first-flush is evaluated from the relationship between the cumulative mass and the cumulative runoff volume [6].

First, the dimensionless cumulative mass of pollutant (f_m) is defined as the ratio between the cumulative mass up to interval j and the total cumulative mass in the entire event.

$$f_m = \frac{\sum_{i=1}^j M_i}{\sum_{i=1}^N M_i} = \frac{\sum_{i=1}^j C_i \cdot V_i}{\sum_{i=1}^N C_i \cdot V_i} \qquad (49.7)$$

where j (1,2,...,N) is the sampling sequence number in the event, i (1,2,...,j) is the sampling sequence number up to j, and M_i is the mass of pollutant in the runoff water corresponding to sample i; remaining symbols are defined previously.

Second, the dimensionless cumulative runoff volume (f_v) is defined as the ratio between the cumulative volume up to interval j and the total cumulative volume in the entire runoff event.

$$f_v = \frac{\sum_{i=1}^j V_i}{\sum_{i=1}^N V_i} \qquad (49.8)$$

An example of the relationship between f_m and f_v for a certain pollutant is shown in Figure 49.2 and the resulting curve is often referred to as an M(V) curve.

The first-flush phenomenon occurs for a pollutant when f_m exceeds f_v at all instants during the storm event [6]. In the example of Figure 49.2, the curve of f_m versus f_v indicates clearly that a first-flush occurs because data falls above the 45° line. Conversely, data on the 45° line indicates that pollutant concentration remains constant throughout the storm runoff, while a curve below the 45° line indicates a dilution. However, in the literature, other criteria have also been proposed for assessing the occurrence of first-flush and some are reviewed by Hvitved-Jacobsen et al. [4].

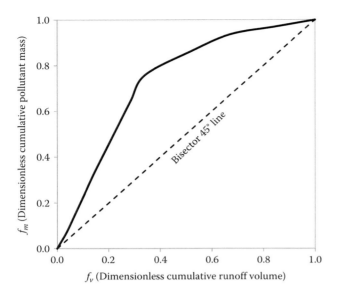

FIGURE 49.2 Relationship between dimensionless cumulative pollutant mass (f_m) and the dimensionless cumulative runoff volume (f_v) for the assessment of the first-flush phenomenon.

The occurrence of a first-flush and its strength depends on rainfall intensity and duration and on characteristics of the catchment such as land slopes, and the percentage of impervious area or watershed area. In particular, an inverse relationship was observed between the strength of first-flush and the watershed area [6].

When the first-flush effect occurs for a pollutant, it means that a large fraction of this pollutant is contained in the initial runoff. This has an immediate implication: the first-flush part of the flow should undergo treatment to reduce pollutant loads before utilization or discharge into receiving waters. Zhang et al. [13] report that the initial rainfall of 2 mm in residential areas, 5 mm in commercial areas, and 10 mm in industrial areas (considering pollutants such as COD and total suspended solids) should be collected and treated before direct discharge or utilization of stormwater.

The first-flush is not a universal and constant phenomenon and it may only take place for some pollutants [2]. Sansalone and Cristina[10] demonstrated that parameters such as suspended solids and particulate COD present a first-flush effect, while it was absent for soluble contaminants such as dissolved solids and soluble COD. This result can be explained with a difference in the response between the particulate fraction and the dissolved fraction of pollutants during stormwater runoff.

49.4 Methods and Equipment for Stormwater Monitoring

Monitoring of stormwater in a given location involves rainfall measurements, flow estimation, collection of a significant number of samples, and analyses of various contaminants. The

selection of the site, the appropriate equipment, and methods should be chosen taking into account local characteristics and availability of budget and time.

The selection of the site where locating a monitoring station is aimed at facilitating representative sampling and flow measurement during stormwater events. The following aspects can be considered: a separate storm drain system, open channels for measuring flow rate, sections with uniform flow conditions, a place that is safe to be reached by field personnel, and suitability for constructing additional channels or installing sampling equipment.

The equipment installed for the stormwater monitoring is based on various devices connected together, which include: rain gauges for *in situ* measurements of natural rainfall, online sensors for *in situ* monitoring of water quality, flow meters or water level probes for measuring runoff flow rate, automatic samplers for sample collection, and electronic instrumentation to record data or transmit data via cellular phone connection (Figure 49.3).

49.4.1 Rainfall Measurement

The monitoring of rainfall events is essential to help determine when to start sampling by automatic samplers. The importance of rainfall data, however, decreases as the accuracy of flow measurements is improved [12].

Rainfall gauges used for measuring local rainfall depth are relatively low-cost and they should be established as close as possible to the monitoring stations. Tipping bucket rain gauges or similar techniques are used. A rain gauge consists of a funnel that collects the rainwater and channels it into one of two compartments of a bucket (Figure 49.4). The water enters one compartment at a time and when it has filled to a specific volume (for instance, 0.1 mm of rainfall), the bucket tips and the water empties down a drainage hole. When the bucket tips, an electrical signal is recorded to give the cumulative rainfall (expressed in mm) or the rainfall intensity (expressed in mm/min). Therefore, the tipping bucket rain gauge is a useful method for online control. Rain gauges with a resolution of 0.1 mm (available on the market at accessible costs) are recommended.

Rainfall can also be monitored by radar, which permits detection of falling raindrops and the spatial variation of the intensity of the rain. Although calibration of radar against measurements from a rain gauge is possible, the radars do not ensure the same accuracy as a well-functioning rain gauge [4].

The measurement of rainfall depth can be used to start the stormwater sampling at the beginning of the storm event by connecting the rain gauge to an automatic sampler.

49.4.2 Flow Measurement

Accurate flow measurement is essential in a stormwater monitoring program for the determination of total mass loads. Continuous measurements of flow rate (typically in time steps of around 1 min) in an open channel can be performed with several

Input

Rain gauges for the *in situ*
measurements of natural rainfall

Output

Automatic sampler
for sample collection

Control module

Connection to electronic
instrumentation
to control sampler and record data
(datalogger)

Flow meters or water level probes
for measuring runoff flow rate

Composite of discrete
samples

Online sensors for *in situ*
monitoring of water quality

Record data
(datalogger) or transmit
data (phone connection)

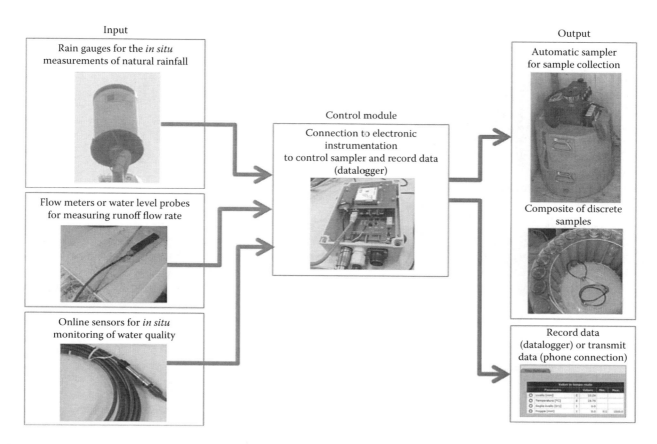

FIGURE 49.3   Equipment installed in a stormwater monitoring station.

approaches, but the most used methods are velocity-based methods and methods using weirs and flumes (Figure 49.5).

The velocity-based technique (such as using area–velocity sensors) is based on the simultaneous measurement of velocity and depth of flow. The method is useful but is considered in some cases to be difficult or expensive. With small flows, measurements may be intermittent and the probe may be susceptible to accumulation of sediment or litter.

It is possible to obtain a good monitoring of flow rate with indirect measurement of the flow through weirs or flumes.

With these devices, the need for expensive flow meters would be avoided, but construction of suitable obstructions or similar devices is needed to create an area of the channel where the hydraulics is controlled. The discharge through the device is directly related to the depth of the flow measured by a pressure transducer, an ultrasonic depth sensor, or similar systems. Flow rate is then calculated using curves based on the geometry of standard weirs.

49.4.3 Sample Collection Equipment

The installation of automatic samplers for stormwater sampling in the monitoring sites is preferable rather than taking grab samples when the measurement of EMC is needed.

An automated sampler is a programmable instrument capable of collecting a series of samples and of distributing them into separate bottles, which can be analyzed individually or composited.

An automatic sampler can be programmed to collect a sample at a specific time or on receipt of a signal from one of the following devices installed for continuous monitoring: rain gauges, flow meters, water level sensors, or online sensors such as conducibility or turbidity. In particular, the automatic sampler can be set to trigger the sampling of water (with aliquots typically from 0.25 to 1 L) based on some predetermined signals: (1) initiation of

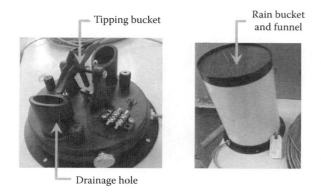

Tipping bucket

Rain bucket
and funnel

Drainage hole

FIGURE 49.4   Example of a tipping bucket rain gauge.

 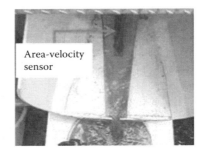

FIGURE 49.5 Examples of weirs and velocity-based sensors for flow rate measurement in stormwater.

rainfall, (2) raising of flow rate in open channels, (3) water level at a certain depth, or (4) achievement of a certain value of an online sensor such as conducibility or turbidity. For example, when the runoff flow in an open channel reaches a predetermined set point after the beginning of a rainfall event, the automatic sampler begins to sample.

Based on the observations discussed in Section 49.4.1, the more representative choice is to take composite samples collected with flow-weighted automatic samplers (or flow-proportional automatic samplers) to form a sample with a pollutant concentration equal to EMC.

As for the frequency of sample collection, samples are collected in general at 5-min, 10-min, or 15-min intervals. Sampling intervals can be extended to 30 min or an hour during tailing stormflows in extremely large storms when flow and runoff concentrations are changing slowly [7].

Unsampled volumes may occasionally occur as the result of signals below set points for triggering sampling, stormwater flow lower than the sampler pump intake, or accidental equipment malfunctions or breakages.

49.4.4 *In Situ* Sensors

Some pollutants can be measured *in situ* using online sensors installed in the stormwater channel, which permit continuous monitoring. The physico-chemical parameters, which can be analyzed in the field at reasonable costs, are the following:

temperature, turbidity (a surrogate parameter of suspended solids), conductivity, pH, dissolved oxygen, and some nutrients (phosphate, nitrate, and ammonium). Multiparameter probes are also available on the market.

The use of online sensors is advantageous because data can be collected frequently (every minute is common) and available remotely. On the other hand, probes require frequent maintenance in the field and in some cases exposure to air should be minimized [12].

49.4.5 Other Installations

Some instrumentation of a monitoring station (samplers, data loggers, etc.) can be placed in a dedicated structure suitable for equipment installation, routine servicing and operation, and to avoid any damage from the weather or vandalism (Figure 49.6).

The sampling station containing the automatic sampler, usually equipped with a peristaltic pump should not be more than 4–6 m above the water surface elevation to permit water pumping.

A data acquisition system for field use can be installed in the sampling station to monitor and store the signals acquired from a variety of sensors and in many cases, it represents the execution center of the monitoring station. Data stored may be downloaded to a portable PC or to a host PC via a modem. The data acquisition system can be used at various levels: (1) with user-selected intervals; (2) with an interface that permits onsite

FIGURE 49.6 Examples of dedicated structures for the equipment installation, routine servicing, and operation.

manipulation; and (3) with remote programming via a telephone modem.

The instrumentation may use a 12V DC battery power supply, solar battery, or AC power.

49.4.6 Management

A monitoring station equipped with automatic instrumentation, data acquisition system, and remote control does not need personnel to collect data or samples. Operators must visit the site periodically (every 1–2 weeks) for routine maintenance of the instrumentation. Onsite sampling with automatic samplers requires minimal efforts from operators, and is limited to the collection of automatically collected samples to carry them to a laboratory for physico-chemical analyses. Operators also have to manage, evaluate, and report the data acquired at the monitoring station during stormwater events.

49.4.7 Analytical Parameters

After collection, samples are transported and analyzed in a laboratory. Standard methods for the determination of various parameters are well defined [1].

Total suspended solids (TSS) is a common bulk parameter to determine the content of solids in stormwater, after filtration of samples on 0.45-μm membrane filters (for the separation of suspended and dissolved fractions). TSS measured in the laboratory is often correlated with the online measurement of turbidity.

Chemical oxygen demand (COD) is commonly analyzed to estimate the content of organic matter (in both particulate and soluble form), while dissolved organic carbon (DOC) refers only to the soluble fraction of organic matter. The biochemical oxygen demand (BOD_5), measured via respirometric methods, is the common analysis in estimating biodegradable organic matter. Total Kjeldahl nitrogen (TKN) and total phosphorus (P) concentrations are measured to quantify nutrients.

Heavy metals, such as copper, lead, zinc, cadmium, nickel, and chromium, are relevant in stormwater runoff because of their potentially toxic effects on biota. Other metals may be of interest in specific cases such as arsenic or mercury [4]. Iron is not generally considered a pollutant in stormwater. Because heavy metals are most toxic in their dissolved form, it is advised that the fraction of dissolved metals be determined [12].

Organic micropollutants include various organic substances (pesticides, polycyclic aromatic hydrocarbons, phenols, endocrine disrupting chemicals, and others) with potential toxic effects for the environment and humans, even if present in stormwater in trace amounts. In a literature survey [3], more than 600 organic micropollutants were found in stormwater runoff. However, it is generally neither feasible nor relevant to analyze such a number of micropollutants, it being preferable instead to analyze only some substances used as potential indicators for the group [4]. Specific chemical analyses should be included if there is any reason to believe they are present.

Concerning salinity, high concentrations of chloride and high electrical conductivity in stormwater runoff are due to the use of de-icing salts on roads during winter periods.

Microbiological analyses regard counts of cultivable bacteria or groups of bacteria, such as fecal indicators (*E. coli*, coliforms, *enterococci*, etc.). The microbial content in stormwater runoff is not as relevant as it is in wastewater, and relevance is limited to faulty connections in the sewer network and contributions from animals [4].

49.5 Characterization of Stormwater Pollutants

49.5.1 Type and Sources of Contamination

The parameters used for characterizing urban stormwater are in common with those used for municipal wastewater, but stormwater presents very different quality characteristics. For instance, urban stormwater is recognized as the most important source of heavy metals, whereas wastewater represents the main source of pollution from organic matter, bacteria, and nutrients [4, 2].

The pollutants in stormwater runoff can be divided into the following groups:

- Turbidity and suspended solids
- Organic matter and biodegradable fraction
- Nutrients
- Heavy metals
- Oils and greases
- Organic micropollutants
- Salinity
- Microorganisms

The main sources of these pollutants in stormwater are summarized in Table 49.2.

49.5.2 Levels of Pollutants in Stormwater Runoff

Urban stormwater pollutants impact in several ways on receiving water bodies or during utilization [2,4], depending on the type of contaminants present in water.

Table 49.3 shows a synthesis of typical EMCs estimated in urban stormwater for various pollutants and land uses as referred to by Park et al. [9]. A large variability in concentration is observed also for the same land use, which suggests that site-specific measurements are always preferable.

A strong relationship is, in general, observed between COD and TSS concentrations [13]. Most of COD in the runoff water is in particulate form and thus related to suspended solids; therefore, treatment methods such as sedimentation or filtration are, in general, very suitable in the removal of COD. Heavy metals, organic micropollutants, and PAHs are bonded to the smaller suspended particles. Organic matter in urban stormwater has a low biodegradability because it is dominated by plant material; therefore, it is also less problematic for the environment because

TABLE 49.2 Type and Sources of the Main Contaminants in Urban Stormwater Runoff

Group of Contaminants	Analytical Parameter	Main Sources
Suspended solids	TSS	Atmospheric fallout
		Pavement wear, construction sites, land erosion
Organic matter	COD	Vegetated areas (leaves and logs)
	DOC (Dissolved)	Contributions from animals
		Commercial activities
Biodegradable fraction of organic matter	BOD$_5$	Vegetated areas (leaves and logs)
		Contributions from animals
		Commercial activities
Nutrients	TKN	Atmospheric fallout
		Fertilizers, contributions from animals
	Total P	Atmospheric fallout
	Soluble P	Fertilizers, contributions from animals
Heavy metals	Copper	Traffic and vehicles: engine parts, tire wear, brake wear, leaded gasoline, diesel fuel, lubricating oil
	Lead	Road structures
	Zinc	Metal plating, copper roofs
	Cadmium	Industrial activities
	Nickel	
	Chromium	
Oils and greases		Traffic and vehicles: lubricating oil, fuel, vehicle spills
		Industrial activities
Organic micropollutants	Various	Atmospheric fallout
	(see Section 49.4.7)	Incomplete combustion processes
	Examples are PAHs, MTBE	Traffic and vehicles: abrasion of tires and asphalt pavement
		A large number of chemicals in household and industry
Salinity	Chloride	Deicing salts
	Electrical conductivity	
Microorganisms	Bacteria counts	Faulty connections in the sewer network
	Fecal indicators	Sewage overflow
		Inputs from diffuse urban sources
		Contributions from animals (bird and mammal feces)

Source: Adapted from Hvitved-Jacobsen, T., Vollertsen, J., and Nielsen, A.H. 2010. *Urban and Highway Stormwater Pollution: Concepts and Engineering.* CRC Press, Boca Raton, FL; Barbosa A.E., Fernandes J.N., and David, L.M. 2012. *Water Research,* 46, 6787–6798.

of the limited oxygen demand. The analysis of heavy metals is important especially in transportation land uses because these areas are a greater source of copper, lead, and zinc than other land uses [5]. The relevance of lead is decreasing in countries using unleaded gasoline.

It is important to relate the monitoring program, especially regarding interstorm sampling design, to the acute or cumulative impact produced by a given constituent [12]. Some parameters (TSS, COD, BOD, pH, metals, etc.) may cause an acute impact on the receiving water bodies in the short term. In these cases, a

TABLE 49.3 Values of Event Mean Concentrations (EMCs) Estimated in Urban Stormwater for Various Pollutants and Land Uses

Land Use ↓	EMC (mg/L)					EMC (µg/L)		
			Nutrients		Oil and greases	Heavy Metals		
	COD	TSS	TKN	Total P		Cu	Pb	Zn
Residential	46–226	42–414	0.2–4.5	0.2–0.7	4.0	1–33	2–144	46–141
Commercial	57–500	40–276	1.2–14	0.2–0.7	4.6	2–33	12–133	150–760
Industrial	50–118	50–221	1–7.2	0.2–0.4	4.8	1–46	17–133	149–326
Transportation	24–136	9–215	1–2	0.1–6.7	1–718	0–135	3–400	9–4274

Source: Adapted from Park, M-H., Swamikannu, X., and Stenstrom, M.K. 2009. *Water Research,* 43, 2773–2786.

short-term analysis on an intrastorm basis should be included in the monitoring, especially for extreme events. Under these circumstances, reduction of peak effluent concentrations may be more important than EMC reduction for impact reduction.

In the case of cumulative effects (accumulation in sediments and biota), when the negative impacts are expressed over a long time-scale, the intrastorm variability is not so important and mean or median of EMC over a certain period of time (for instance, one year) may be more appropriate. When contaminants may have both acute and cumulative effects, intra- and intersampling merit attention.

49.6 Summary and Conclusions

Due to the large variability of quantity and quality of stormwater during a rainfall event, the use of monitoring stations equipped with automatic samplers, online sensors, and electronic instrumentation to record or transmit data can help operators to obtain representative samples at reasonable costs. The collection of flow-weighted composite samplers is the correct approach for estimating the event mean concentrations for each pollutant, and is useful to calculate mass loads produced from stormwater. However, water quality measurement in urban stormwater is an open issue. A large variability in concentration of pollutants is observed in many studies referred to in the literature, also for the same land use, which suggest that site-specific measurements are always preferable.

References

1. APHA, AWWA, WERF. 2012. *Standard Methods for the Examination of Water and Wastewater*, 22nd Edition. American Public Health Association, Washington, DC.
2. Barbosa, A.E., Fernandes, J.N., and David, L.M. 2012. Key issues for sustainable urban stormwater management. *Water Research*, 46, 6787–6798.
3. Eriksson, E., Baun, A., Mikkelsen, P.S., and Ledin, A. 2005. Chemical hazard identification and assessment tool for evaluation of stormwater priority pollutants. *Water Science and Technology*, 51(2), 47–55.
4. Hvitved-Jacobsen, T., Vollertsen, J., and Nielsen, A.H. 2010. *Urban and Highway Stormwater Pollution: Concepts and Engineering*. CRC Press, Boca Raton, FL.
5. Lee, H., Swamikannu, X., Radulescu, D., Kim, S., and Stenstromd, M.K. 2007. Design of stormwater monitoring programs. *Water Research*, 41, 4186–4196.
6. Lee, J.H., Bang, K.W., Ketchum, L.H., Choe, J.S., and Yue M.J. 2002. First flush analysis of urban storm runoff. *Science of the Total Environment*, 293, 163–175.
7. Leecaster, M.K., Schiff, K., and Tiefenthaler, L.L. 2002. Assessment of efficient sampling designs for urban stormwater monitoring. *Water Research*, 36, 1556–1564.
8. Liu, A., Egodawatta, P., Guan, Y., and Goonetilleke, A. 2013. Influence of rainfall and catchment characteristics on urban stormwater quality. *Science of the Total Environment*, 444, 255–262.
9. Park, M.-H., Swamikannu, X., and Stenstrom, M.K. 2009. Accuracy and precision of the volume-concentration method for urban stormwater modeling. *Water Research*, 43, 2773–2786.
10. Sansalone, J.J. and Cristina, C.M. 2004. First flush concepts for suspended and dissolved solids in small impervious watersheds. *Journal of Environmental Engineering*, 130(11), 1301–1314.
11. Sansalone, J.J., Koran, J.M., Buchberger, S.G., and Smithson, J.A. 1997. Partitioning and first flush of metals and solids in urban highway runoff. *Journal of Environmental Engineering*, 123(2), 134–143.
12. USEPA. 2002. Urban storm water BMP performance monitoring, A guidance manual for meeting the national stormwater BMP database requirements. USEPA-821-B-02-001, April 2002.
13. Zhang, M., Chen, H., Wang, J., and Pan G. 2010. Rainwater utilization and storm pollution control based on urban runoff characterization. *Journal of Environmental Sciences*, 22(1), 40–46.

Natural Wetlands and Wetland Plants

50

Urban Wetland Hydrology and Water Purification

Never Mujere
University of Zimbabwe

Saeid Eslamian
Isfahan University of Technology

PREFACE

Wetlands are areas where the water table is at or near the surface of the land for at least a consecutive period of time or is covered by shallow water up to 2 m deep. The prefix "wet" in wetlands is the key. Various types of wetlands include flood plains, riparian wetlands, dambos, pans, swamps, and artificial impoundments. Wetlands perform an array of vital hydrological functions that require preserving rather than eliminating our wetland resources. These ecosystems provide numerous ecosystem values, goods, and services for human livelihood.

The benefits of wetlands are many: water purification, flood protection, shoreline stabilization, groundwater recharge, and streamflow maintenance. Wetlands form part of the catchment, thus holding water, which is recharged into main rivers and underground. Wetlands achieve water quality through trapping sediments and purifying water of biological pathogens and chemical pollutants. Wetlands filter suspended solids from water that comes into contact with wetland vegetation. Stems and leaves provide friction for the flow of the water, thus allowing settling of suspended solids and removal of related pollutants from the water column. Wetlands may retain sediment in the peat or as substrate permanently. Wetlands also provide habitat for fish and wildlife, including endangered species.

Since wetlands are located between uplands and water resources, many can intercept runoff from the land before it reaches open water. As runoff and surface water pass through, wetlands remove or transform pollutants through physical, chemical, and biological processes. Wetland processes play a role in the global cycles of carbon, nitrogen, and sulfur by transforming them and releasing them into the atmosphere. The values of wetland functions related to biogeochemical cycling and storage include water quality and erosion control.

Nevertheless, wetland ecosystems are under environmental siege due to unsustainable use of resources. With the rapid increase in human population, the pressures on natural resources have become intense. Urban agriculture and infrastructural development in wetlands are increasing. This chapter is an overview of urban wetland hydrology, along with a brief explanation of how wetlands function in terms of water quality and water quantity issues.

50.1 Introduction

Wetlands are natural and human-made infrastructures that receive, transport, clean, store, and deliver water to a wide range of users for domestic needs, agriculture, biodiversity, industry, and other economic production, as well as maintenance of social and cultural integrity. They provide important hydrological functions such as groundwater recharge, water quality improvement, and flood alleviation. However, these ecosystems are under environmental siege due to unsustainable use of resources [3]. With the rapid increase in human population, the pressures on natural resources have become intense, and hence, most wetlands are lost [7]. Recurrent food shortages, water scarcity, and general drought-induced desiccation have forced many societies, both urban and rural, to focus on wetland ecosystems as providers of food and water and as extremely valued natural resources [17]. Therefore, they may be considered to now form an integral part of urban livelihood. However, this may lead to degradation of these resources; hence, there is a need for exploitation systems that can be sustained over a long time period. There is need for wise use of these ecosystems for the benefit of humankind.

Wetlands have intrinsic values to communities living around them. The Ramsar Convention embodies commitments by member states to maintain the ecological character of wetlands and to plan for the wise use or sustainable use of all of the wetlands in their territories. Because of their importance, each year February 2 is World Wetlands Day. This day marks the date of the adoption of the Convention on Wetlands on February 2, 1971, in the Iranian city of Ramsar. The Ramsar Convention is an intergovernmental treaty to promote national action and international cooperation for the conservation and wise use of wetlands and their resources [23,24].

50.2 What Are Wetlands?

Wetlands are areas subject to permanent or temporary water logging, where water covers the soil, or is present either at or near the surface of the soil all year or for varying periods of time during the year, including during the growing season. Wetlands are areas of marsh, fen, peat-land or water, whether natural or artificial, permanent or temporary, with water that is static or flowing, fresh, brackish or salt, including riparian land adjacent to the wetland. Wetlands can be highly variable ecosystems that are characterized by fluctuating water levels and the prevalence of saturated soil conditions during the growing season Raisin et al. [23] define wetlands as areas of marsh, fen, peat land, or water, whether natural or artificial, permanent or temporary, with water that is static or flowing, fresh, brackish or salt, including areas of marine water the depth of which at low tide does not exceed 6 m. This definition, while emphasizing types and features of landscapes, is considered to have a wide applicability [9]. The definition of wetlands for ecological studies [6] is lands transitional between terrestrial and aquatic systems where the water table is usually at or near the surface or the land that is covered by shallow water. Wetlands must have one or more of these three attributes: (1) at least periodically, the land supports predominantly hydrophytes, (2) the substrate is predominantly undrained hydric soil, and (3) the substrate is non-soil and is saturated with water or covered by shallow water at some time during the growing season each year. Wetlands definition for management purposes, particularly regulation, refers to wetlands as areas that are inundated or saturated by surface or groundwater at a frequency and duration sufficient to support, and that under normal circumstances do support, a prevalence of vegetation typically adapted for life in saturated soil conditions [12].

50.3 Types of Wetlands

Wetlands are transitional zones between aquatic and terrestrial ecosystems all rich in plants and animals. Water is the primary factor that controls the environment, plants, soils, and animals. There is a wide range of wetland types with varying modes of formation and hydrology IUCN [12] recognizes

42 types of wetlands, including rivers and their tributaries and floodplains, lakes, estuaries, deltas, marshes, dambos, peat lands, oases, and coastal areas, together with mangroves and coral reefs. Five major wetland systems have been recognized, namely, marine, estuarine, lacustrine, riverine, and palustrine. Marine and estuarine systems are coastal wetlands such as tidal marshes and mangrove swamps. Lacustrine and riverine wetlands are associated with lakes and rivers, respectively. The palustrine system includes marshes, dambos, bogs, and swamps [5].

Dambos or vleis are most common wetlands in southern Africa [4]. Dambos are palustrine wetlands. A vlei or dambo is an important type of wetland in southern Africa. It is low-lying, marshy wet grassland, covered with water during the rainy season, and even though it may seem to be dry during the winter season and droughts, it is actually storing water under the ground, which it releases slowly into the streams and rivers.

A vlei ecosystem can be kilometers wide in extent with the drier fringes moving down to the wetter mid- and central areas. All areas have an essential role to play in the provision of water. Hence, dambos or vleis are natural regulators with shallow, seasonal, or permanently waterlogged depressions at or near the head of a drainage network, or alternatively occurring independent of a drainage system. Figure 50.1 shows the distribution of wetlands in Zimbabwe.

Zimbabwe's Environmental Management Act (CAP 20:27) and Statutory Instrument 7 of 2007 Environmental Management (EIA and Ecosystems Protection Regulations) govern wetland utilization in Zimbabwe. Section 113 of the Environmental Management Act (Chapter 20:27) gives the Minister of Environment powers to declare any wetland to be an ecologically sensitive area and may impose limitations on development in or around such an area. The law prohibits the reclamation or drainage, disturbance by drilling or tunneling

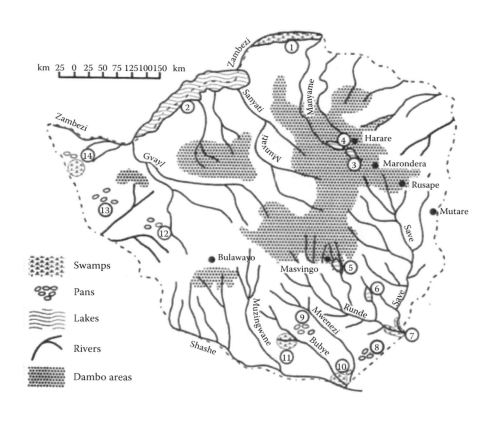

1 Mid Zambezi Valley-Mana Pools
2 Lake Kariba
3 Lake Chivero
4 Darwendale Dam/Manyame Lake
5 Lake Kyle/Mutirikwi
6 McDougal Dam
7 Save-Runde Confluence Floodplain
8 Gorhwana Pans
9 Bubye Pans
10 Majinji Pans
11 Kwaluzi Swamp
12 Mabomo-Joweni Pan Complex
13 Lememba-Shabashaba Complex
14 Tsamtsa

FIGURE 50.1 Wetlands distribution in Zimbabwe. (From Matiza, S., and Crafter, A. 1994. *Wetlands Ecology and Priorities for Conservation in Zimbabwe.* IUCN, Harare. With permission.)

in a manner that has or is likely to have an adverse impact on any wetland or adversely affect any animal or plant life therein. Failure to abide by the law is a crime that attracts a fine not exceeding US$500 or imprisonment not exceeding two years or both fine and imprisonment.

50.4 Wetland Hydrological Functions

The ecological goods and services of wetlands are, among other things,

- Provide important habitat for a wide variety of wildlife
- Ensure food security if sustainably utilized
- Trap moderate amounts of soil running off nearby uplands before they enter lakes and streams
- Maintain and improve water quality by filtering contaminants and excessive nutrients
- Renew groundwater supplies
- Help and control flooding and reduce flood damage
- Fire control
- Provide a source of economically valuable products such as wild rice and commercial fish
- Climate control
- Support recreational activities including fish, hunting, nature appreciation, bird watching, and so much more
- Provide opportunities to participate in outdoor educational activities and to enjoy the aesthetic qualities of wetlands

50.5 Degradation of Urban Wetlands

Wetlands are major water sources within and surrounding cities. However, they being affected by infrastructural development, polluted by waste, and degraded by informal urban agriculture. One model of wetlands alteration assumes three main factors that influence wetland ecosystems: water level, nutrient status, and natural disturbance. The exploitation of wetlands for agricultural purposes in urban areas is common as these areas were deliberately left out of residential development during town planning. The combined impacts of adverse weather (frequent dry spells and generally below-normal soil moisture conditions) and severe economic constraints have induced hardship and food insecurity among urban populations. This has resulted in urban communities shift to intensive wetlands utilization for food production as these areas have higher moisture

contents [13]. Therefore, the drivers of wetland degradation in urban areas include [11]:

- Agricultural activities can lead to wetlands loss and degradation. This gives rise to the need to sustainably utilize water provided by wetlands without tampering with their natural existence
- Commercial and residential development, road construction, resource extraction, industrial siting processes
- Waste and dredge disposal
- Cultivation of forest trees (silviculture)
- Commercial and housing construction projects are the greatest threat to wetlands

50.5.1 Change in Sizes of Harare Wetlands

A study conducted on three wetlands in Harare, namely Honeydew, Mabvuku, and Monavale, have shown that the wetlands significantly decreased in size from 1972 to 2008 [20]. However, the rate of change between each time period differed. A number of factors may have caused the changes in wetland size. Land use and climate changes were proposed as the drivers of the change. Other factors such as the topography may have been involved although their influence was not investigated in this study. Table 50.1 shows the changes in sizes of the three wetlands.

The largest decrease in wetland size was noted for Honeydew wetland (Table 50.1). Between the periods 1972–2008 there was an 88% decrease in the area of this wetland. Much of this change in size (78%) occurred between 1995 and 2008. This wetland was used by a commercial farm that had been in existence since the 1970s. Hence, agriculture was likely to have contributed to the shrinkage of the Honeydew wetland.

The decrease in total area of Mabvuku wetland was approximately 63% from the period 1972–2008. Initial changes of 64% from 1972 to 1984 may have been associated with agricultural activity and perhaps also the construction of sewage works. A change of 61% between 1995 and 2000 may have been a consequence of agricultural activity. The main crops grown in the Mabvuku wetland were maize and sweet potatoes. Wetlands in countries such as Malawi, Zambia, and Zimbabwe are mostly used for growing grain, tubers, vegetables, and fruit [18]. For these to grow well, the farmers create ridges to prevent water logging. A similar drainage system was observed in the Monavale and Honeydew wetlands. These ridges encourage runoff thereby

TABLE 50.1 Change in the Area (in m²) of Honeydew, Mabvuku, and Monavale Wetlands between 1972 and 2008

Wetland	1972	1984	1995	2008	% Change
Honeydew	1,560,496	1,072,370	822,724	180,368	88
Mabvuku	6,057,098	3,697,100	3,623,605	2,244,022	63
Monavale	3,383,220	2,981,482	2,360,621	2,162,178	36

Source: Adapted from Msipa, M. 2009. Land Use Changes between 1972 and 2008 and Current Water Quality in Wetlands of Harare, Zimbabwe. MSc thesis, University of Zimbabwe, Harare.

reducing seepage into the soil. Hence, over time, water levels in the wetland may go down causing wetland shrinkage. Estimates of the loss of wetlands in industrialized regions indicated that up to 60% of these have been destroyed in the last 100 years due to drainage, conversion, infrastructure development, and pollution [19].

The decrease in total area of Monavale wetland was approximately 36% from the period 1972–2008. The largest decrease in wetland size (26%) was between 1984 and 1995. The main cause of the decrease may have been increased in settlements. The settlement land use increased from 16% in 1972 to 44% in 2008. Monavale experienced the least reduction in wetland size. Using 1972 as a baseline, 64% of the wetland area remained in 2008. Subsistence agricultural activities flourished only recently between 1995 and 2008. Therefore, it may be a reason why the wetland has dried up at the slowest rate as compared to the other two wetlands (Honeydew and Mabvuku). Another crucial factor to which the smooth decrease in overall size of Monavale may be attributed was that about 6% of the 2008 total wetland area was conserved by some local residents. Within the conserved area, no agriculture or other forms of land conversion was allowed by the Monavale wetland committee. Results indicated that conservation of an area by limiting human activity may be critical to the conservation of the wetlands. Therefore, the use of the Monavale wetland for agriculture presents a threat to the existence of that wetland.

50.6 Urban Wetland Hydrology

50.6.1 Urban Hydrology

The physical environment of urban and industrial areas is completely different from that of forests, agricultural lands, and rangelands. Many observations and studies have shown greater precipitation in and around major urban areas than in the surrounding countryside due to the enormous number of condensation nuclei produced by human activities and atmospheric instability associated with the heat island generated by the city. The impervious surfaces created by urban development may cause an increase in flooding, soil erosion, stream sedimentation, and pollution of land and water bodies. The study and analysis of hydrology in urban areas and the hydrological problem associated with urbanization fall within the scope of urban hydrology [5].

50.6.2 Wetland Hydrology

Hydrology, which is the description of water level, flow, and frequency of water, is a very important aspect in wetlands studies because it is probably the single most important determinant of the establishment and maintenance of specific types of wetlands and wetlands processes [19]. Wetland hydrology studies the vegetation, flooding, and hydrologic characteristics of wetlands, streamflow, and sediments. It examines the impact of development projects on wetland ecosystems, groundwater fluctuations, water quality, nutrient removal and transformation, wetland construction and restoration, wetland management, soil characteristics, wetland delineation and classification, erosion control, etc. [19].

50.6.3 Importance of Wetlands Hydrology

Hydrologic conditions are extremely important for the wetland's structure and function because they affect many abiotic factors including soil anaerobics, nutrient availability, and salinity like in coastal wetlands. These in turn affect biota that develops in a wetland. In dry climate, for example, Southern Africa and most of Australia, wetlands are often restricted to river channels and depressions [26]. In many wetlands, organisms must survive or endure water-level fluctuations, ice formation, and even periodic absence of water. Finally, completing the cycle, biotic components are active in altering the wetland hydrology and other physiochemical features [19].

Wetlands are shallow; even small changes in water levels which would be inconsequential in large lakes and rivers can result in significant local environmental changes for sessile plants and animals [26]. Therefore, changes in water inputs and outputs in wetlands, in other words, their water budgets, and the water depths at any given place over time, and the local water regime, are important descriptors of the entire wetland and of the physical environment experienced by organisms found at different elevations within a wetland. On the other hand, past and current depth of flooding is the major determinant of the distribution of plants and animals within a wetland. Water levels in wetlands are also important in wetlands studies as they can be used to estimate evapotranspiration (ET). ET from wetlands has also been calculated from observing the diurnal cycles of groundwater or surface water in wetlands. The hydroperiod of a wetland has a significant effect on nutrient transformations on the availability of nutrients to vegetation, and on loss from wetland soil nutrients that have gaseous forms.

Furthermore, with records of water levels, the following hydrologic parameters can be determined: hydroperiod, frequency of flooding, duration of flooding, and water depth. Water levels in most wetlands are generally not stable but fluctuate seasonally (high order riparian wetlands), daily or semi-daily (types of tidal wetlands), or unpredictably (wetlands in low order streams and coastal wetlands with wind-driven tides). The amount of precipitation in most parts of the world varies seasonally and from year to year. This variation is reflected in the amount of water in a wetland. As such, a number of terms are commonly used to describe the general hydrology of wetlands, namely [6]: permanently flooded (standing surface water present throughout the year), intermittently exposed (standing water present throughout the year except in years of severe drought), semipermanently flooded (standing water present throughout the growing season in most years), seasonally flooded (standing water present for extended periods of time during the growing season), saturated (the soil or substrate is saturated with water [water logged] to the surface during the growing season, but standing water is rarely present), temporarily flooded (standing water is present only for brief periods during the growing season), and intermittently flooded (standing water is present periodically but without any seasonal pattern).

50.6.4 Urban Wetland Hydrology

Urban wetland hydrology provides the basic knowledge, principles, scientific evidence, and justification for managing water resources in urban watersheds. The hydrologic function affects many services cited in support of wetland protection both directly (floodwater retention, microclimate regulation) and indirectly (biogeochemical cycling, pollutant removal). It is vital to understand the links between condition and hydrologic function to test the hypothesis, embedded in regulatory assessment of wetland value, that condition predicts function. Condition is assessed using rapid and intensive approaches. Hydrologic function is assessed using hydrologic regime (mean, variance, and rates of change of water depth), and measurements of groundwater exchange and ET.

50.7 Modeling Urban Wetland Hydrology

Surface hydrology, in conjunction with groundwater hydrology and soil characteristics, controls the hydrology budget of a wetland. In highly urbanized locations, the input of stormwater runoff into local wetlands is a potentially critical component of the water budget. High amounts of impervious cover (roofs, road surfaces) in urban areas increase stormwater runoff velocities and volumes. These increased velocities produce water budgets that differ from those of wetlands in nonurban settings. Urban surface water inflows occur via both stream overbank flow and from storm drains that discharge directly into the wetland system. The basic hydrologic parameters of wetland water budgets include surface water influxes, precipitation, groundwater influxes, storage of water, percolation, and ET [21,25]. Precipitation is most likely the dominant factor in a hydrologic simulation of the wetlands.

Many water budgets have attempted to describe wetland hydrology, but models capable of describing urban wetland water flows are extremely few [8,21,22]. Koob et al. [15] identify three general approaches used to model wetland hydrology. These are (1) single event models, (2) stochastic models involving one parameter, and (3) complete or comprehensive water budget models.

50.7.1 Single Event Models

Single event models use rainfall of a given duration, frequency, and amount, coupled with site soils and land-use conditions to predict runoff. Such estimates of runoff are then used with general design criteria to construct the wetlands. For example, Koob et al. [15] propose that wetland volume should be based on hydraulic retention time and the runoff volume from the 90% storm.

50.7.2 Stochastic Models

The stochastic framework examines the probability of extended dry periods and daily seepage and ET losses in order to give some level of assurance to maintaining a specified wetland water volume. This method strives to provide the planner with some level of acceptable risk to associate with design [15]. Because urban hydrology may be subject to more highly fluctuating environmental conditions than a nonurbanized system, there is an advantage in applying a stochastic model to urban wetlands because this model type allows the incorporation of uncertainty into the model results. This approach contrasts with deterministic models, which produce identical results when provided with constant input parameters. Another alternative is to use a deterministic model with variable inputs to examine a range of conditions (e.g., dry conditions, wet conditions, average conditions).

50.7.3 Mass Balance Approach

A comprehensive water budget is necessary to characterize the hydrology of an urban wetland system. Thus, the mass balance approach provides a framework for developing a water budget, which seeks to incorporate the parameters that control a wetland's hydrology. This method involves use of models that attempt to simulate all or the majority of the components of the water budget. These models have been growing rapidly based on the decreasing costs of computer memory, increased computational speed, as well as the compilation and availability of regional databases from which to run and calibrate the models. Further generalizations or additional parameters, such as a proposed restoration design of the system's hydrology, may also be included in a model [21,29]. However, the approach is difficult to estimate the various components of urban hydrology or to create hydrologic simulations over extended time periods.

Water in the wetland is derived from (1) precipitation that falls directly on the wetland, (2) surface water runoff from the interstate, (3) surface water runoff and groundwater discharge that flows through the culverts that drain under the interstate, (4) sound water that discharges from seeps on the slope between the interstate and the wetland, and (5) groundwater that flows through the surficial aquifer and discharges to the wetland. The magnitude of these sources changes seasonally and annually. Water discharges from the wetland by (1) groundwater flow through the surficial aquifer, (2) surface water flow through the drainage ditch and surface depressions on the eastern side of the wetland, and (3) surface water flow across the extensive area of the wetland [1,8,27]. Figure 50.2 shows components of the wetland water balance.

50.7.4 Water Balance Calculations

The calculation of the water budget follows a mass balance approach provided by Mitsch and Gosselink [19] and Obropta et al. [21]. The general mass balance exists as (change in

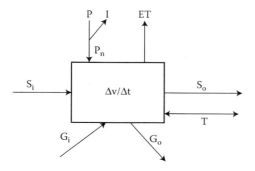

FIGURE 50.2 Components of the wetland water balance (where P = precipitation, P_n = net precipitation, ET = evapotranspiration, I = interception, S_i = surface water inflow, S_o = surface water outflow, G_i = groundwater inflow, G_o = groundwater outflow, T = tide, V = change in storage, and t = time). (From Mitsch, W.T., and Gosselink, J.G. 1993. *Wetlands*. Van Nostrand Reinhold, New York. With permission.)

storage = input − output). The mass balance applied to the wetland is derived from the water budget equation:

$$\Delta S = P + S_i + G_i - AET - I - S_o - G_o \pm T \qquad (50.1)$$

where

ΔS = change in storage volume
P = precipitation
S_i = surface water inflow
G_i = groundwater inflow
AET = actual evapotranspiration
I = infiltration
S_o = surface water outflow
G_o = groundwater outflow
T = tidal flow

Surface water inflows (S_i) and outflows (S_o) are in cubic meters or feet per second (Cumecs or CFS). The precipitation (P) input for the wetland itself is calculated by summing the hourly data (in mm or in.). ET is affected by land use, with higher rates in intensive (agriculture and urban) landscapes in response to higher leaf area. ET determines latent heat exchange, which regulates microclimate, a valuable service in urban heat islands. Higher ET also indicates higher productivity and thus carbon cycling. Groundwater exchange regularly reversed flow direction at all sites in response to rainfall. This buffering effect on regional aquifer levels, an underappreciated service of isolated wetlands, was provided regardless of condition. Intensive landscapes may benefit most from the hydrologic services that wetlands provide because that is where certain services (floodwater storage, microclimate regulation) are realized [19].

50.8 Hydrologic Flux and Storage

Wetlands play a critical role in regulating the movement of water within watersheds as well as in the global water cycle [18]. Wetlands, by definition, are characterized by water saturation in

the root zone, at or above the soil surface, for a certain amount of time during the year. This fluctuation of the water table (hydroperiod) above the soil surface is unique to each wetland type. Wetlands store precipitation and surface water and then slowly release the water into associated surface water resources, groundwater, and the atmosphere. Wetland types differ in this capacity based on a number of physical and biological characteristics, including landscape position, soil saturation, the fiber content/degree of decomposition of the organic soils, vegetation density, and type of vegetation. The change in storage of the system each month is calculated by subtracting the total outputs from the total inputs of the system. This represents the amount of water stored in or removed from the wetland system each month. To calculate the cumulative storage for the wetland system, the change in storage for each month is added to the previous month's cumulative storage, resulting in the cumulative storage plot [1].

50.9 Water Retention

Soil saturation and fiber content are important factors in determining the capacity of a wetland in retaining water. Like a sponge, as the pore spaces in wetland soil and peat become saturated by water, they are able to hold less additional water and are able to release the water more easily. Clay soils retain more water than loam or sand, and hold the water particles more tightly through capillary action because pore spaces are small and the water particles are attracted to the negatively charged clay. Pore spaces between sand particles are large and water drains more freely since less of the water in the pore is close enough to be attracted to the soil particle. Water drains more freely from the least decomposed (fibric) peat because pore spaces are large and the surface area for capillary action is small. Sapric peat (most decomposed, fibers unrecognizable) and hemic peat (intermediate) have very small pores. Water moves very slowly in such peats. Water in wetlands, as a result, flows over the surface or close to the surface in the fibric layer and root zone (acrotelm). Thus, wetlands with sapric peat and clay substrate will store water but will have no groundwater discharge (inflow) or outflow (recharge).

50.10 Water Supply

Wetlands act as reservoirs for the watershed. Wetlands release the water they retain (from precipitation, surface water, and groundwater) into associated surface water and groundwater. In Wisconsin, watersheds composed of 40% lakes and wetlands, spring stream outflows from the watersheds were 140% of those in watersheds without any wetlands or lakes [18]. Forested wetlands, kettle lakes, and prairie potholes have significant water storage and ground water recharge. Forested wetlands overlying permeable soil may release up to 100,000 gallons/acre/day into the ground water. A Minnesota bog released 55% of the entering water to stream and ground water [20].

The amount and source of water in a wetland are affected by landscape position. For example, wetlands that are near a topographical height, such as a mountain bog, will not receive as

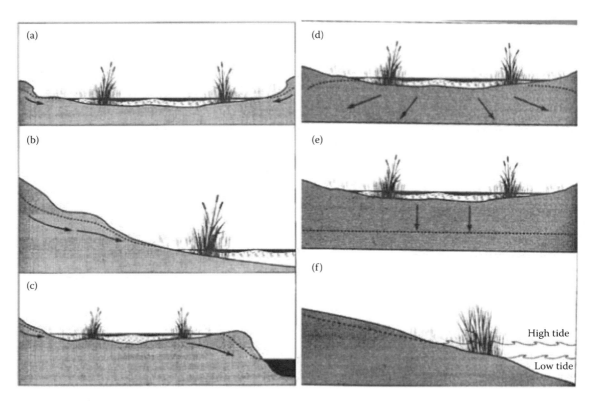

FIGURE 50.3   Discharge–recharge interchanges between wetlands and groundwater systems including: (a) marsh as a depression receiving groundwater flow, a "discharge" wetland. (b) Groundwater spring or seep wetland or groundwater slope wetland at the base of a steep slope. (c) Floodplain wetland fed by groundwater. (d) Marsh as a "recharge wetland" adding water to groundwater. (e) Perched wetland or surface water depressing wetland. (f) Groundwater flow through a tidal wetland. (From Mitsch, W.J., and Gosselink, J.G. 2000. *Wetlands*, 3rd edition. John Wiley & Sons, New York. With permission.)

much runoff as a marsh in a low area amid fields. Wetlands can be precipitation dominated, groundwater dominated, or surface flow dominated. Wetlands on local topographic heights are often precipitation dominated. Precipitation dominated wetlands may also be in flat or slightly elevated areas in the landscape, where they receive little or no surface runoff. Generally, such wetlands have a clay and peat layer that retains the precipitation and prevents discharge from ground water. Wetlands also form in landscape positions at which the water table actively discharges, particularly at the base of hills and in valleys. Such groundwater-dominated wetlands may also receive overland flow but they have a steady supply of water from and to groundwater. Most wetlands in low points on the landscape or within other water resources are dominated by overland flow. Such riverine, fringe (marsh), and tidal wetlands actively play a role in the landscape because they come in contact with, store, or release large quantities of water and act on sediments and nutrients. These wetlands may be recharged by groundwater as well, but surface water provides the major source of water (Figure 50.3).

50.11 Flow Velocity and Erosion Control

By virtue of their place in the landscape, riparian wetlands, salt marshes, and marshes located at the margin of lakes protect

shorelines and stream banks against erosion. Wetland plants hold the soil in place with their roots, absorb wave energy, and reduce the velocity of stream or river currents. Coastal wetlands buffer shorelines against the wave action produced by hurricanes and tropical storms [18]. The ability of wetlands to control erosion depends on vegetation density and type. Stems cause friction for the flow of the water, thus reducing water velocity. As density of vegetation increases, velocity decreases. A wetland with more vegetation will intercept more runoff and be more capable of reducing runoff velocity and removing pollutants from the water than a wetland with less vegetation. Wetland plants also reduce erosion as their roots hold the stream bank, shoreline, or coastline. Values associated with biological productivity of wetlands include water quality, flood control, erosion control, community structure and wildlife support, recreation, aesthetics, and commercial benefits.

50.12 Role of Wetlands in Shoreline Stabilization

Wetlands that occur along the shoreline of lakes or along the banks of rivers and streams help protect the shoreline soils from the erosive forces of waves and currents. The wetland plants act as a buffer zone by dissipating the water's energy and providing stability by binding the soils with their extensive root systems.

50.13 Hydrologic Flux and Life Support

Changes in frequency, duration, and timing of hydroperiod may impact spawning, migration, species composition, and food chain support of the wetland and associated downstream systems. Normal hydrologic flux allows exchange of nutrients, detritus, and passage of aquatic life between systems. Values of wetlands because of the functions of hydrologic flux and storage include water quality, water supply, flood control, erosion control, wildlife support, recreation, culture, and commercial benefits [29].

50.14 Flood Protection

Wetlands within and upstream of urban areas are particularly valuable for flood protection. The impervious surface in urban areas greatly increases the rate and volume of runoff, thereby increasing the risk of flood damage. Drainage of wetlands and the development in the floodplains cause damage to businesses, homes, crops, and property. Wetlands help protect adjacent and downstream properties from potential flood damage. The value of flood control by wetlands increases with: (1) wetland area, (2) proximity of the wetland to flood waters, (3) location of the wetland (along a river, lake, or stream), (4) amount of flooding that would occur without the presence of the wetlands, and (5) lack of other upstream storage areas such as ponds, lakes, and reservoirs [18].

Almost any wetland can provide some measure of flood protection by holding the excess runoff after a storm, and then releasing it slowly. The size, shape, location, and soil type of a wetland determine its capacity to reduce local and downstream flooding. While wetlands cannot prevent flooding, they do lower flood peaks by temporarily holding water and by slowing the water's velocity. Wetland soil acts as a sponge, holding much more water than other soil types. Even isolated wetlands can reduce local flooding. If the wetlands were not there to hold stormwater runoff, backyards and basements might end up under water.

50.15 Groundwater Recharge and Streamflow Maintenance

Aquifers and groundwater are recharged, that is, replenished with water by precipitation that seeps into the ground and by surface waters. Those wetlands connected to groundwater systems or aquifers are important areas for groundwater exchange. They retain water and so provide time for infiltration to occur. Groundwater, in turn, provides water for drinking, irrigation, and maintenance of streamflow and lake and reservoir levels. During periods of low streamflow (or low lake water levels), the slow discharge of groundwater often helps to maintain minimum water levels. In addition, wetlands located along streams, lakes, and reservoirs may release stored water directly into these

systems, thus also contributing to their maintenance. Wetlands' many intricate connections with groundwater, streamflow, and lake and reservoir water levels make them essential in the proper functioning of the hydrologic cycle [23].

Thus, wetlands help maintain the level of the water table and exert control on the hydraulic head. This provides force for groundwater recharge and discharge to other waters as well. The extent of groundwater recharge by a wetland is dependent on soil, vegetation, site, perimeter to volume ratio, and water table gradient. Groundwater recharge occurs through mineral soils found primarily around the edges of wetlands. The soil under most wetlands is relatively impermeable. A high perimeter to volume ratio, such as in small wetlands, means that the surface area through which water can infiltrate into the groundwater is high. Groundwater recharge is typical in small wetlands such as prairie potholes, which can contribute significantly to recharge of regional groundwater resources. Researchers have discovered groundwater recharge of up to 20% of wetland volume per season [20].

However, groundwater can be adversely affected by activities that alter wetland hydrology. Drainage of wetlands during urban development lowers the water table and reduces the hydraulic head providing the force for groundwater discharge. If a recharged wetland is drained, the water resources into which groundwater discharges will receive less inflow, potentially changing the hydrology of a watershed. It has been calculated that if 80% of a 5-acre Florida cypress swamp were drained, available groundwater would be reduced by an estimated 45% [16].

50.16 Evapotranspiration

Plants actively take up water and release it to the atmosphere through ET. This process reduces the amount of water in wetland soil and increases the capacity for absorption of additional precipitation or surface water flow. As a result, water levels and outflow from the wetland are less than when plants are dormant. Larger plants and plants with more surface area will transpire more. Most wetlands return over two-thirds of their annual water inputs to the atmosphere through RT. Wetlands may also act to moderate temperature extremes in adjacent uplands.

50.17 Urbanization and Water Quality in Wetlands

Every land use virtually has potential to change water quality. Land use changes are important drivers of water, soil, and air pollution. Urban agricultural practices, which include intensive inputs of nitrogen and phosphorus fertilizers substantially increase the pollution of surface water by runoff and erosion and the pollution of groundwater by leaching of excess nitrogen (as nitrate). Other agricultural chemicals, including herbicides and pesticides, are also released to ground and surface waters by agriculture and in some cases remain as contaminants [30].

Thus, wetlands are altered by pollutants upstream or local runoff and in turn change the quality of the water flowing out through them. The chemical nature and concentration of various substances dissolved in the water determine its pH, hardness, salinity, nutrient content, and other measures used to categorize water chemistry, and can have a significant impact on the flora and fauna of the wetland [28].

The major sources of water in an urban wetland are rainfall and groundwater discharge. The contribution of each source is determined by the geomorphic setting and local climatic conditions. In turn, the sources of the water in a wetland determine not only the amount present and when it is present, but also its chemistry. The water chemistry of wetlands whose primary source of water is precipitation will be very different from that of a wetland whose primary source of water is groundwater discharge. Seepage from sewer pipes also contributes to groundwater discharge into wetlands [14]. This can have a major effect on the species composition of the vegetation and its primary production.

Common and critical water quality impacts that result from different land uses are changes in suspended sediment load, organic matter, biological and chemical oxygen demand, bacteria and viruses, nutrient loads, temperature, heavy metals, toxins such as pesticides and herbicides, acidification (pH), salinization, and changes in the water flow itself. The physical parameters include conductivity, total dissolved solids, and pH. The chemical parameters selected for the study were total nitrogen and total phosphates. Microbiological analyses determine the presence of *Escherichia coli* and estimate the number of fecal coliform. Selection of water quality parameters is based on the effect they have on wetlands' water quality and productivity and the type of pollution likely to be received due to the land uses in wetlands. For example, presence of sewage ponds is indicated by fecal coliforms, while total nitrogen and total phosphates are indicators of agricultural activities. Total dissolved solids are selected to show the pollution levels at each point. The oxidation reduction potential (ORP) indicates quantity of decaying organic matter. Therefore, this parameter would vary with land uses. For instance, in agricultural land uses, ORP is expected to be lower than in areas adjacent to a sewage pond. ORP also has a bearing on the water-holding capacity of the soil, and hence water levels [28]. Finally, pH was considered a general parameter, which affects physical, chemical, and biological processes.

50.18 Role of Wetlands in Water Purification

Runoff waters often carry nutrients that can cause water quality problems. An example of such an occurrence is an algae bloom. Besides the aesthetic problems associated with algae blooms (a green, smelly slime), they result in low levels of oxygen in the water. This oxygen depletion can result in the death of fish and other aquatic life. Some algae release toxins that can kill pets and livestock when bloom conditions occur. Wetlands protect surface waters from the problems of nutrient overload by removing the excess nutrients, some of which are taken up and used by wetland plants, and some of which are converted to less harmful chemical forms in the soil [11].

Toxic chemicals reach surface waters in the same way as nutrients, and can cause disease, death, or other problems upon exposure to plants and animals (including humans). In a function similar to nutrient removal, wetlands trap and bury these chemicals or may even convert some of them to less harmful forms [26]. Disruptions of the wetland soils could release the toxins back into the aquatic environment.

Wetlands, and in particular marshes and riparian vegetation, contribute notably to the regular natural filtration of water and to the improvement of its quality when polluted. This service is especially important for human societies whose economic, social, and domestic activities inevitably lead to a substantial level of waste. Water pollution remains a key issue in urban areas. The main sources of water pollution are from agriculture (crops and livestock), sewage wastewater (industry and settlements), runoff from urban areas, and illegal dumping of solid and liquid waste [14].

Within certain limits ensuring their ecological functioning, wetlands in a good ecological state could help decision-makers reach the requirements of the new legal framework. It is obviously more vital than ever to reduce the release of contaminants in water and increase the use of technological equipment such as sewage plants and treatment centers. The natural ability of wetlands to filter and clean water is even reproduced to treat wastewater, in treatment plants using aquatic plants.

50.18.1 Nutrient Removal

50.18.1.1 *Nitrogen*

The biological and chemical process of nitrification/denitrification in the nitrogen cycle transforms the majority of nitrogen entering wetlands, causing between 70% and 90% to be removed [20]. In aerobic substrates, organic nitrogen may mineralize to ammonium, which plants and microbes can utilize, adsorb to negatively charged particles (e.g., clay), or diffuse to the surface. As ammonia diffuses to the surface, the bacteria *Nitrosomonas* can oxidize it to nitrite. During nitrification, the bacteria *Nitrobacter* oxidizes nitrite to nitrate. Plants or microorganisms can assimilate nitrate, or anaerobic bacteria may reduce nitrate (denitrification) to gaseous nitrogen (N_2) when nitrate diffuses into anoxic (oxygen depleted) water. The gaseous nitrogen volatilizes and the nitrogen is eliminated as a water pollutant. Thus, the alternating reduced and oxidized conditions of wetlands complete the needs of the nitrogen cycle and maximize denitrification rates [21].

50.18.1.2 Phosphorus

Significant quantities of phosphorus associated with sediments are deposited in wetlands. Phosphorus can enter wetlands with suspended solids or as dissolved phosphorus. Phosphorus removal from water in wetlands occurs through use of phosphorus by plants and soil microbes; adsorption by aluminum and iron oxides and hydroxides; precipitation of aluminum, iron, and calcium phosphates; and burial of phosphorus adsorbed to sediments or organic matter [16]. Riparian forest wetlands can reduce phosphate concentrations in runoff and floodwater by up to 50%. The estimated mean retention of phosphorus by wetlands is 45%. Wetlands with high soil concentrations of aluminum may remove up to 80% of total phosphorus [23,27]. Wetlands along rivers have a high capacity for phosphorus adsorption because as clay is deposited in the floodplain, aluminum (Al) and iron (Fe) in the clay accumulate as well. Thus, floodplains tend to be important sites for phosphorus removal from the water column, beyond that removed as sediments are deposited.

Dissolved phosphorus is processed by wetland soil microorganisms, plants, and geochemical mechanisms. Microbial removal of phosphorus from wetland soil or water is rapid and highly efficient; however, following cell death, the phosphorus is released again. Similarly, for plants, litter decomposition causes a release of phosphorus. Burial of litter in peat can, however, provide long-term removal of phosphorus. Harvesting of plant biomass is needed to maximize biotic phosphorus removal from the wetland system. However, wetland soils can reach a state of phosphorus saturation, after which phosphorus may be released from the system [16]. Phosphorus export from wetlands is seasonal, occurring in late summer, early fall, and winter, as organic matter decomposes and phosphorus is released into surface water.

50.18.2 Removal of Biological Oxygen Demand From Surface Water

Biological oxygen demand (BOD) is a measure of the oxygen required for the decomposition of organic matter and oxidation of inorganics such as sulfide. BOD is introduced into surface water through inputs of organic matter such as sewage effluent, surface runoff, and natural biotic processes. If BOD is high, low dissolved oxygen levels result. Low dissolved oxygen levels can lead to mortality of aquatic life. Wetlands remove BOD from surface water through decomposition of organic matter or oxidation of inorganics. BOD removal by wetlands may approach 100% [2].

50.18.3 Removal of Suspended Solids and Associated Pollutants From Surface Water

Suspended solids (such as sediment and organic matter) enter wetlands in runoff, as particulate litterfall, or with inflow from associated water bodies. Sediments, which are particles of soil, settle into the gravel of streambeds and disrupt or prevent fish from spawning, and can smother fish eggs. Sediment deposition in wetlands depends on water velocity, flooding regimes, vegetated areas of the wetland, and water retention time.

Sediment deposition in wetlands prevents a source of turbidity from entering downstream ecosystems. Typically, wetland vegetation traps 80%–90% of sediment from runoff. Less than 65% of the sediment eroded from uplands exits watersheds that contain wetlands [21]. Deposition of suspended solids, to which such substances are adsorbed, removes these pollutants from the water.

Sediments, nutrients, and toxic chemicals enter wetlands primarily by way of "runoff," a term used to describe the rain and stormwater that travels over land surfaces on its way to receiving waters. In urban areas, runoff washes over buildings and streets in industrial, commercial, and residential areas where it picks up pollutants and carries them to receiving waters. In rural areas, agricultural and forest practices can affect runoff. Where the runoff drains a freshly ploughed field or clear-cut area, it may carry too much sediment. Runoff may carry pesticides and fertilizers if these have been applied to the land.

Other pollutants, notably heavy metals, are often attached to sediments and present the potential for further water contamination [27]. Wetlands remove these pollutants by trapping the sediments and holding them. The slow velocity of water in wetlands allows the sediments to settle to the bottom where wetland plants hold the accumulated sediments in place.

Wetlands protect water quality by trapping sediments and retaining excess nutrients and other pollutants that impact water quality such as nutrients, organics, heavy metals, and radionuclides, which are often adsorbed onto suspended solids. These functions are especially important when a wetland is connected to groundwater or surface water sources (such as rivers and lakes) that are, in turn, used by humans for drinking, swimming, fishing, or other activities. These same functions are also critical for the fish and other wildlife that inhabit these waters.

50.18.4 Removal of Metals

Human activities have resulted in metal levels high enough to cause health or ecological risks in water resources. Metals may exist in wetland soils or enter wetlands through surface or groundwater flow. Wetlands can remove metals from surface and groundwater because of the presence of clays, humic materials (peats), aluminum, iron, or calcium. Metals entering wetlands bind to the negatively ionized surface of clay particles, precipitate as inorganic compounds (includes metal oxides, hydroxides, and carbonates controlled by system pH), complex with humic materials, and adsorb or occlude to precipitated hydrous oxides. Iron hydroxides are particularly important in retaining metals in salt marshes. Wetlands play an important role in removing metals from water resources, runoff, and groundwater [22].

Wetlands remove more metals from slow flowing water because there is more time for chemical processes to occur before the water moves out of the wetland. Burial in the wetland substrate keeps bound metals immobilized. Neutral pH favors metal immobilization in wetlands. With the exception of very low pH peat bogs, as oxidized wetland soils are flooded and reduced, pH converges toward neutrality (6.5–7.5) whether the wetland soils were originally acidic or alkaline. Forested wetlands remove 20%–100% of metals in the water downstream of urbanized areas. Obropta et al. [21] found that lead leaking from a Florida hazardous waste site was retained at high levels by a wetland; less than 20%–25% of the total lead in the soil and sediments was readily bioavailable. The majority of the lead was bound to soil and sediments through adsorption, chelation, and precipitation. Bioavailable lead was absorbed primarily by eelgrass, which had bioaccumulated the majority of the lead. In another case, researchers found that wetland vegetation and organic (muck) substrate retained 98% of lead entering the wetland [10].

50.18.5 Removal of Pathogens

Fecal coliform bacteria and protozoans, which are indicators of threats to human health, enter wetlands through municipal sewage, urban stormwater, leaking septic tanks, and agricultural runoff. Bacteria attach to suspended solids that are then trapped by wetland vegetation. These organisms die: after remaining outside their host organisms, through degradation by sunlight, from the low pH of wetlands, by protozoan consumption, and from toxins excreted from the roots of some wetland plants [10]. In this way, wetlands have an important role in removing pathogens from surface water.

50.18.6 Oxidation–Reduction

The fluctuating water levels (also known as hydrologic flux) that are characteristic of wetlands control the oxidation-reduction (redox) conditions that occur. These redox conditions governed by hydroperiod play a key role in nutrient cycling, availability, and export; pH; vegetation composition; sediment and organic matter accumulation; decomposition and export; and metal availability and export [26]. When wetland soil is dry, microbial and chemical processes occur using oxygen as the electron acceptor. When wetland soil is saturated with water, microbial respiration and biological and chemical reactions consume available oxygen. This shifts the soil from an aerobic to an anaerobic or reduced condition. As conditions become increasingly reduced, electron acceptors other than oxygen must be used for reactions. These acceptors are, in order of microbial preference, nitrate, ferric iron, manganese, sulfate, and organic compounds. Wetland plants are adapted to changing redox conditions. Wetland plants often contain arenchymous tissue (spongy tissue with large pores) in their stems and roots, which allows air to move quickly between the leaf surface and the roots. Oxygen released from wetland plant roots oxidizes the rhizosphere (root zone) and allows processes requiring oxygen, such as organic compound breakdown, decomposition, and denitrification, to occur [21].

50.19 The Limits of Wetlands

Besides their hydrological contributions, wetlands do have their limits. A partially filled or otherwise damaged wetland is one that only partially meets its potential for flood control, shoreline stabilization, or groundwater recharge. A badly degraded wetland can lose its capacity to remove excess sediments, nutrients, and other pollutants, and can lose its habitat value for fish and wildlife. Wetlands may have tremendous capacities to provide environmental benefits, but they are not indestructible. If we want wetlands to continue to perform their hydrological and ecological functions, then we have to do our part to protect them.

50.20 Summary and Conclusions

Wetlands are areas temporarily or permanently covered with water, which may or may not be flowing; it may be of natural or artificial origin. Their water tables occur at or near the land surface, and hence are saturated for long enough time to promote aquatic processes and various kinds of biological activities adapted to wet environments [17]. Values of wetlands include water quality, water supply, flood protection, habitat provision, microclimate regulation, floodwater storage, pollutant removal, erosion control, fish and wildlife habitat, recreation, aesthetics, culture, science, and commercial benefits. About water quality, wetlands help maintain and improve the water quality of streams, rivers, lakes, and estuaries. Due to rapid urbanization that occurred during the twentieth century, many wetlands in most urban areas have been cut off from their historic water sources. The hydrology of urban wetland systems, including the inflows from their surrounding watershed, are radically altered [10,13].

The wetlands function of maintaining water quality influences a much broader scale than that of the wetland ecosystem itself. Wetlands under favorable conditions remove organic and inorganic nutrients and toxic materials from water that flow through them because of the following attributes [11,12]:

- A reduction in water velocity as streams enter wetlands, causing sediments and chemicals sorbed to sediments to drop out of the water column.
- A variety of anaerobic and aerobic processes in close proximity promoting denitrification, chemical precipitation, and other chemical reactions that remove certain chemicals from the water.
- A high rate of productivity in many wetlands can lead to high rates of mineral uptake by vegetation and subsequent burial in sediments when plants die.
- A diversity of decomposers and decomposition processes in wetland sediments.
- A large contact surface of water with sediments because of the shallow water, leading to significant sediment–water exchange.
- An accumulation of organic peat in many wetlands that causes the permanent burial of chemicals.

References

1. Arnold, J.G., Allen, P.M., and Morgan, D.S. 2001. Hydrologic model for design and constructed wetlands. *Wetlands*, 21(2), 167–178.

2. Bhaduri, B., Minner, M., Tatalovich, S., and Harbor, J. 2001. Long-term hydrologic impact of urbanization: A tale of two models. *Journal of Water Resources Planning and Management*, 127(1), 13–19.

3. Breen, C.M., Quinn, N.W., and Mander, J.J. (eds.). 1997. *Wetland Conservation and Management in South Africa: Challenges and Opportunities*. IUCN, South Africa.

4. Bullock, A. 1988. Dambo and Discharge in Central Zimbabwe. PhD thesis, University of Southampton, United Kingdom.

5. Chang, M.C. 2013. *Forest Hydrology. An Introduction to Water and Forests*, 3rd Edition. CRC Press, Boca Raton, FL.

6. Cowardin, L.M., Carter, V.M., Golet, F.C. and La Roe, E.T. 1979. *Classification of Wetlands and Deep Water Habitats of the United States*. FWS/OBS-79/31, US Fish and Wildlife Service, Washington, DC.

7. Cunningham, W.P. and Cunningham, M.A. 2002. *Principles of Environmental Science: Inquiry and Applications*. McGraw-Hill, Boston.

8. Drexier, J.Z., Bedford, B.L., DeGaetano, A.T., and Siegel, D.I. 1999. Quantification of the water budget and nutrient loading in a small peatland. *Journal of the American Water Resources Association*, 35(4), 753–769.

9. Dugan, P.J. 1990. *Wetland Conservation: A Review of Current Issues and Required Action*. IUCN, Gland, Switzerland.

10. Ehrenfeld, J.G. 2000. Evaluating wetlands within an urban context. *Ecological Engineering*, 15, 253–265.

11. FAO. 2007. *FAO Global Information and Early Warning System on Food and Agriculture*. FAO, Rome.

12. IUCN. 1997. *Wetland Conservation and Management in South Africa: Challenges and Opportunities*. IUCN, Gland, Switzerland.

13. IUCN. 1999. *Wetlands and Climate Change*. IUCN, Gland, Switzerland.

14. Keddy, P.A. 1983. Freshwater wetland human induced changes: Indirect effects must also be considered. *Environmental Management*, 7, 299–302.

15. Koob, T., Barber, M.E., and Hathhorn, W.E. 1999. Hydrologic design considerations of constructed wetlands for urban stormwater runoff. *Journal of the American Water Resources Association*, 35(2), 323–331.

16. Mak, M. 2007. Development of an Urban Hydrological Model to Support Possible Urban Wetland Restoration. MS thesis, Rutgers University, New Brunswick, NJ.

17. Matiza, S. and Crafter, A. 1994. *Wetlands Ecology and Priorities for Conservation in Zimbabwe*. IUCN, Harare.

18. Mitsch, W.T. and Gosselink, J.G. 1993. *Wetlands*. Van Nostrand Reinhold, New York.

19. Mitsch, W.J. and Gosselink, J.G. 2000. *Wetlands*, 3rd Edition. John Wiley & Sons, New York.

20. Msipa, M. 2009. Land Use Changes between 1972 and 2008 and Current Water Quality in Wetlands of Harare, Zimbabwe. MSc Thesis, University of Zimbabwe, Harare.

21. Obropta, C., Kallin, P., Mak, M., and Ravit, B. 2001. *Modeling Urban Wetland Hydrology for the Restoration of a Forested Riparian Wetland Ecosystem*. Rutgers University, New Brunswick, NJ.

22. Owen, C.R. 1995. Water budget and flow patterns in an urban wetland. *Journal of Hydrology*, 169(1), 171–187.

23. Raisin, G., Bartley, J., and Croome, R. 1999. Groundwater influence on the water balance and nutrient budget of a small natural wetland in Northeastern Victoria, Australia. *Ecological Engineering*, 12(1), 133–147.

24. Ramsar Convention. 1971. *Convention on Wetlands*, Iran.

25. Ramsar Convention Bureau. 2000. *Directory of Wetlands of International Importance*. Ramsar, Gland, Switzerland.

26. Reinelt, L.E. and Horner, R.R. 1995. Pollutant removal from stormwater runoff by palustrine wetlands based on comprehensive budgets. *Ecological Engineering*, 4(2), 77–97.

27. Tsihrintzis, V.A. and Hamid, R. 1998. Runoff quality prediction from small urban catchments using SWMM. *Hydrological Processes*, 12(2), 311–329.

28. van der Valk, A.G. 2006. *The Biology of Freshwater Wetlands*. Oxford University Press, New York.

29. Yu, Z. and Schwartz, F.W. 1998. Application of an integrated basin-scale hydrologic model to simulate surface-water and groundwater interactions. *Journal of the American Water Resources Association*, 34(2), 1–7.

30. Zhang, L. and Mitsch, W.J. 2005. Modelling hydrological processes in created wetlands: An integrated system approach. *Environmental Modelling & Software*, 20(7), 935–946.

51

Urban Wetland Hydrology and Changes

Sara Nazif
University of Tehran

Saeid Eslamian
Isfahan University of Technology

PREFACE

Urbanization highly alters the natural hydrologic cycle. This will result in drastic changes in different components of the hydrologic cycle, such as increase in surface runoff and decrease in infiltration due to high percentage of impervious areas. Urbanization alters the quality of surface runoff as well. All of this will highly affect natural and manufactured wetlands hydrology located in urban areas. This chapter focuses on urban wetland hydrology and how urbanization affects wetlands as an important component of urban ecosystems.

    This chapter includes five main sections. Section 51.1 provides an introduction on wetlands and their importance and characteristics. Section 51.2 focuses on urban wetlands characteristics and value as well as effects of urbanization on wetlands. Why the urban wetland hydrology is important is discussed in Section 51.3. Section 51.4 provides a detailed description on urban wetlands hydrology. Hydrologic measures and classification of wetlands as well as description of wetlands' water budget components are included in this section. Finally, Section 51.5 discusses the factors that could alter the urban wetland hydrology.

51.1 Introduction

Wetlands are of high value in Earth life. They have different, irreplaceable hydrological and ecological functions, such as stabilization of water supplies, flood damping, improvement of water quality, soil erosion control, recharge of groundwater, and carbon sequestration [51]. These functions of wetlands could diminish the urbanization effects on aquatic resources and biota. Wetlands are well known because they function like a kidney even though these ecological services are rarely assessed quantitatively and usually not in urban contexts. The failure of wetlands in their function as a kidney, by exceeding the capacity of a wetland to assimilate additional hydrologic, nutrient, or sediments, is usually not discussed and there seems to be an implicit assumption in the common understanding of wetlands that their capacity to assimilate is unlimited [27].

Wetlands can also be degraded ecologically in the same manner as streams by stormwater, nutrient enrichment, sedimentation, and altered hydrologic cycles. However, wetlands have gained less attention historically and are considered waste areas. That is why numerous wetlands are drained and used for agricultural purposes [18]. As an example, about 70% of original wetland area within southern Ontario, Canada is drained for agricultural purposes, with the higher rate near major urban areas [12]. Due to the importance of wetlands and their valuable performance in urban areas, special attention should be paid to these systems. An important feature of these systems is their hydrology, which highly affects their function especially in constructed wetlands. It should be noted that the wetlands are commonly connected to other waterways and this should be considered in wetlands hydrology analysis. Therefore, this chapter deals with wetlands, their performance in natural and urban areas, as well as the hydrologic concepts of these systems.

51.1.1 Wetlands Characteristics

Generally, wetlands can be considered aquatic ecosystems composed of plants, animals, and soils. These components are well adapted to wet conditions, which can survive in permanent or periodic inundation. Wetlands are not continuously wet and could remain dry for years. Water in wetlands can be still or flowing, fresh or brackish. The main characteristics of wetlands are as follows:

1. Shallow water, either at the surface or within the root zone, continuously or for a period of time. The shallow and deep systems differ fundamentally in type of thermal stratification in the water column and degree of light attenuation and degree the light penetrates to bottom sediments [5]. Therefore, shallow hydrologic environment of wetlands creates unique biogeochemical conditions that distinguish it from aquatic and terrestrial environments.
2. Slow water movement or being static, the wetlands are called lentic environments.
3. Production of wetland soils that are reducing or at the least anaerobic because of water logging. This is unlike normal terrestrial soils that are oxic.
4. Hydrophytes (adapted vegetation to wet condition) availability in emergent or submerged forms. In wetlands, plants not tolerant to inundation are largely absent. This can be considered the main difference between shallow lakes and wetlands. In general, wetland is one that has at least some emergent vegetation but shallow lakes only have submerged vegetation and emergent vegetation are limited to the edges.

51.1.2 Wetlands Importance

Wetlands are a unique hydrologic feature of the landscape. An important characteristic of wetlands is that they share aspects of both aquatic and terrestrial environments. Wetlands can be considered a transition zone between aquatic and terrestrial ecosystems. Freshwater and marine aquatic environments are commonly distinguished by having permanent water but terrestrial environments are generally distinguished by having drier conditions, with an unsaturated zone present for most of the annual cycle. Since wetlands have both of these characteristics, they occupy the transition zone between predominantly wet and dry environments.

Wetlands provide habitat for plants and animals, water storage, and improve water quality as well as reduce pollution. They also protect against natural hazards through slowing floodwaters, their peak and volume, reducing the risk of fire, moderating strong winds, and reduction of river banks and coastline erosion. Wetlands and associated vegetation can cool surrounding areas in summer.

51.2 Urban Wetlands

The wetlands located within the boundaries of a city or town are called urban wetlands. The urban wetlands function differently from wetlands in nonurban lands because of numerous effects of urbanization on hydrology, geomorphology, and ecology. Furthermore, urban wetlands provide human-related values that they lack in nonurban areas, as they provide some contact with nature, and some opportunities for recreations that are otherwise rare in the urban landscape. There are varieties of urban wetlands that can be natural, constructed, or a mixture of both.

51.2.1 Characteristics of Urban Wetlands

The wetlands constructed and occurring naturally within an urban area suffer from different constraints that arise through anthropogenic changes within their catchment and surrounding environments [14]. A typical urban wetland is shown in Figure 51.1.

The urban wetlands are physically and biologically different from nonurban wetlands [37] as it is summarized in Table 51.1. As shown in this table, the regional changes in hydrology because of urbanization have indirect effects on wetland structure and function. In addition, urbanization directly affects wetland hydrology by filling, ditching, diking, draining, and damming. Physical changes to the shape of the land because of road and building construction also results in geomorphological changes both in wetlands and in the adjacent areas. Biological and ecological changes in urban wetlands happen along with the physical effects of urbanization. Both animal behavior and plant reproductive ecology may be strongly affected by the size, shape, and heterogeneity of habitat patches. Thus, the possible states for the flora and fauna in urban wetlands are qualitatively different from those of nonurban wetlands.

51.2.1.1 Natural and Constructed Wetlands

A general definition of natural wetlands can be natural areas that provide a transition zone between the terrestrial environment

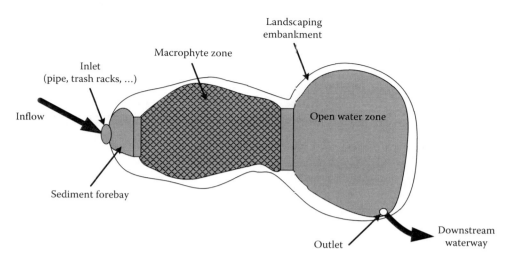

FIGURE 51.1 Schematic of an urban wetland.

TABLE 51.1 Urbanization Effects on Wetland Hydrology and Geomorphology

Type	Effect
Hydrology	• Increased surface water inflow because of increase in impervious area
	• Increased stormwater discharge relative to baseflow discharge, consequently increased erosive force within stream channels, therefore, increased sediment inputs to recipient coastal wetlands
	• Altered water quality (more turbidity, more nutrients, metals, organic pollutants, less O_2, etc.)
	• Replacing low-order streams with culverts, outfalls, etc. resulting in high variable baseflow and low-flow conditions
	• Less groundwater recharge results in less groundwater flow, resulting in decrease in baseflow and/or eliminating dry-season streamflow
	• Increased flood frequency and magnitude result in more scour of wetland surface, physical disturbance of vegetation
	• Increase in flow variability may deprive wetlands of water during dry weather
	• Greater regulation of flows decreases magnitude of spring flush
Geomorphology	• Decreased sinuosity of wetland/upland edge reduces amount of ecotone habitat
	• Decreased sinuosity of stream and river channels results in higher velocity of stream water discharge to receiving wetlands
	• Alterations in shape of slopes (e.g., convexity) affects water storage properties
	• Increased cross-sectional area of stream channels (due to erosional effects of increased flood peak flow) results in more erosion along banks
Vegetation	• Large numbers of exotic species present
	• Large and numerous sources for continuous reinvasion of exotics
	• Large amounts of land with recently disturbed soils suitable for weedy, invasive species
	• Restricted pool of pollinators and fruit dispersers
	• Chemical changes and physical impediments to growth associated with the presence of trash
	• Small remnant patches of habitat not connected to other natural vegetation
	• Human-enhanced dispersal of some species
	• Trampling along wetland edges and periodically unflooded areas
Fauna	• Large numbers of exotic species present; large and numerous sources for continuous reinvasion of exotics
	• Large amounts of land with recently disturbed soils suitable for weedy, invasive species
	• Restricted pool of pollinators and fruit dispersers
	• Chemical changes and physical impediments to growth associated with the presence of trash
	• Small remnant patches of habitat not connected to other natural vegetation
	• Human-enhanced dispersal of some species
	• Trampling along wetland edges and periodically unflooded areas

and deep open water and provide a habitat for aquatic flora and fauna. The nutrient rich inflows can degrade the wetlands water quality and make them eutrophic. Since wetlands are commonly placed in the downstream of catchments, their situation shows the watershed situation and health. Eutrophication highly affects the wetland health due to potential algal blooms and release of pollutants from sediments that declines in wetland biodiversity. Therefore, specific investigations should be done before wastewater release into natural wetlands or development plans around wetlands that affect the hydrology of the wetland

[20]. The following principles are provided by NSW Wetlands Management Policy [33]:

- The inflow of natural wetlands would have enough quality for keeping the wetlands quality.
- The construction of purpose-built wetlands instead of natural wetlands is not suggested.

The long-term plans for monitoring the quality and quantity of the wetland inflows should be prepared. The variables that are measured depend on the project constraints, sampling and test costs, and required instruments. However, usually the measurement of nutrients, pH, heavy metals, turbidity, conductivity, suspended solids, biochemical oxygen demand (BOD), fecal coliform contamination, and algae are considered.

Constructed wetlands are specifically constructed to reduce flows pollution before entrance into the receiving waters or underground aquifer. These wetlands are also fashionable for controlling the quantity of stormwater runoff, which has been generated by the increased impervious area common to urban development. For these constructed wetland treatment systems to be successful at retaining the range of pollutants in urban runoff, it is necessary to engineer a system that provides a suitable habitat for aquatic plants as well as enhancing the pollutant removal processes resident within natural wetlands.

High attention and efforts should be paid in determining the performance of a constructed wetland. The performance of these systems should be assessed against their design goals, even though they can provide some other advantages. These objectives should include an appreciation of the management constraints, the maintenance approach, the establishment of the wetland plants, and the changes in wetland by its aging.

When a wetland performance does not satisfy its objectives, this means that it is poorly designed or maintained, the considered objectives for its development are inappropriate, or the wetland has exceeded its designed life cycle. If it is possible, the construction of wetlands off-line from the main flow of water is preferred. Off-line constructed wetlands are more efficient because of better sedimentation, prevention from stripping epiphytes and biofilms from the plants by high flow velocities, and promotion of chemical reactions through water and sediments.

Commonly multiple objectives are considered in the design of constructed wetlands. It should be remembered that each component of a wetland has a specific function that is designed to achieve it, but an integration of these components provides a wetland that should meet wetland construction objectives [20].

51.2.1.2 Types of Constructed Wetlands

Constructed wetlands can be categorized into [20]

- Free water surface—In these wetlands a zone of free water is available as an open pond or between emergent aquatic vegetation
- Sub-surface—In these types of wetlands, the subsurface flows in different directions and these flows support aquatic plants

Some common wetlands design schemes based on Schueler [39] are as follows:

1. Shallow marsh system (Figure 51.2): These system developments need a large area, and permanent baseflow or groundwater supply for providing the required water level for emergent wetland plants support. Consequently,

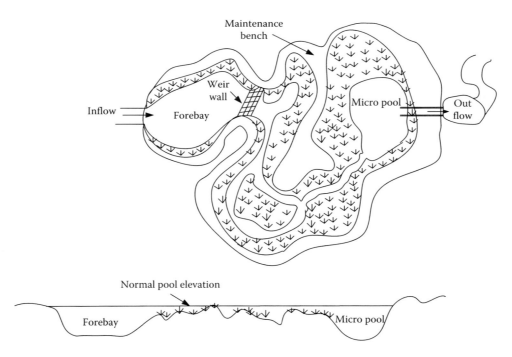

FIGURE 51.2 Schematic of a shallow marsh system.

the upstream watershed of these systems has a large area, commonly more than 10 ha. The water depth in the shallow marsh system is often less than 0.45 m, which provides appropriate conditions for emergent plant growth. Some deeper zones at the inlet and outlet are also provided for specific functions.

2. Pond/wetland system (Figure 51.3): This design is composed of a wet pond and a shallow marsh for stormwater treatment. The defined functions for the wet pond are sedimentation trap, reduction of runoff velocity, and pollutants removal. These systems need less space in comparison with the shallow marsh because the majority of treatment process is provided by the wet pool. An alternative of this system is providing a pretreatment system for sedimentation and nutrients (phosphorus) removal. This improves the pollutant removal efficiency of wet ponds.

3. Extended detention (ED) wetland (Figure 51.4): In these wetlands, extra runoff is temporarily detention above the shallow marsh. Therefore, the ED wetlands need less space because of temporary vertical storage. There is a growing zone along the side-slopes of ED wetlands, extending from the normal pool elevation to the maximum water level of the wetland. The water level can change up to 1 m after a storm event, and it may last about a day to return to normal level. About half of the treatment volume can be provided as ED storage, to protect downstream channels from erosion, and decrease the wetland's required area.

4. Pocket wetland (Figure 51.5): Pocket wetlands are commonly used in small size (often less than 0.05 ha) area and because of their small catchment, they do not have permanent baseflow and experience a high range of water level

fluctuations. The water levels in these wetlands are supported by groundwater. In drier areas, stormwater runoff supports the pocket wetland, and during extended periods of dry weather, the water level can go under the surface and they have only saturated soils. Because of small size and high water level variability of pocket wetlands, they have low aquatic life diversity and value.

51.2.2 Values of Urban Wetlands

Urban wetlands provide a variety of benefits and services to the community. In addition to providing habitat for plants and animals, wetlands provide water storage, improve surface runoff water quality, and reduce their pollution. Urban areas can also benefit from wetlands through protection against natural hazards such as urban flash flood and inundation, fire, strong winds, and soil erosion. Wetlands can also cool the surrounding area and moderate the heat island effect in urban areas.

Wetlands also contribute to the wellbeing of the public by acting as urban green spaces, which provide beautiful views, landscape diversity, and recreational opportunities. They can also contribute to cultural heritage and spiritual values. Furthermore, wetlands provide easily accessible educational opportunities to learn about the environment. Harvested stormwater in wetlands can also be used for irrigation of local sports grounds and green spaces instead of drinking water.

51.2.3 Urbanization Impacts on Wetlands

Urbanization is a major reason for wetlands damage. Urbanization directly results in wetland acreage loss as well as degradation. The degradation of wetlands is because of changes

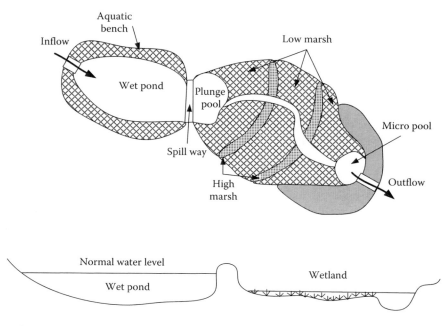

FIGURE 51.3 Schematic of a pond/wetland system.

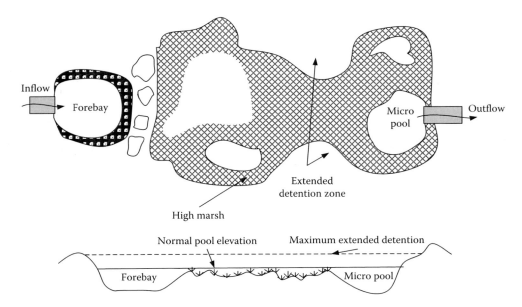

FIGURE 51.4 Schematics of an extended detention wetland.

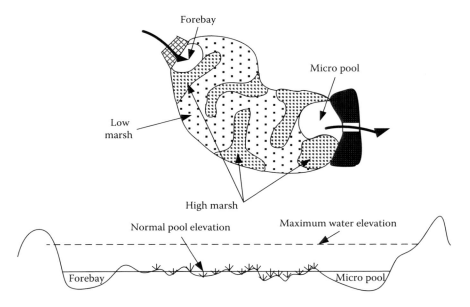

FIGURE 51.5 Schematic of a pocket wetland.

in water quality, quantity, and flow rates; entrance of huge amounts of different types of pollutants; and changes in species composition due to introduction of non-native species and disturbance. The major pollutants in wetlands resulting from urbanization are heavy metals, sediment, nutrients, road salts, hydrocarbons, bacteria, oxygen-demanding substances, and viruses. These pollutants enter wetlands from point sources or nonpoint sources. As an example, construction activities are a major source of suspended sediments in wetlands, which are carried through urban runoff.

The main potential risks that threaten urban wetlands are as follows:

- Direct habitat loss (because of development, land reclamation, roads, in-stream dredging, etc.)
- Altered water regime (because of dams/barriers construction, stream redirection, hard surfacing, water extraction, etc.)
- Pollution (with sources of garbage, sewage, oil and chemical spills, pesticides, airborne toxins, etc.)

- Biodiversity loss due to the introduction of exotic species (weeds, pests, and domestic pets)
- Other ecosystem modifications (e.g., altered fire regimes, dieback, and changes in salinity)

51.2.3.1 Changing Land Use

By development of roads, buildings, and parking lots, the pervious surfaces are changed into impervious surfaces. Therefore, the rainfall or snowmelt infiltration decreases and the runoff amount increases. The produced runoff carries sediments; organic matter; pesticides and fertilizers from lawns, gardens, and golf courses as well as heavy metals; hydrocarbons; road salts; and debris from urban impervious surfaces into urban streams and wetlands. Increased salinity, turbidity, and toxicity and decreased dissolved oxygen, resulting from pollutants entrance to wetlands affect aquatic life and, therefore, the food cycle. High loads of nutrients entrance can result in eutrophication or the release of pollutants from a wetland into connected water resources.

From another point of view, by runoff flow over warmed impervious surfaces (such as asphalt), the water temperature increases and results in less dissolved oxygen content of the runoff. Higher water temperature and resulting lower level of dissolved oxygen can cause stress or death of aquatic organisms. Furthermore, the increasing temperature can result in release of nutrients such as phosphorus from wetland sediment and intensify the eutrophication potential.

Due to connections between groundwater and wetlands, increase of impervious surfaces decreases ground water recharge over the watershed and reduces water flow into wetlands. Increases of runoff peak flow rates, and alternations in wetland hydrology as a result of land use changes, can cause erosion and channelization in wetlands, as well as alteration of species composition and decrease in pollutant removal efficiency. Changes in frequency, duration, and timing of the wetland hydroperiod (which will be discussed in the following sections) have adverse impacts on spawning, migration, species composition, and thus the food cycle in a wetland and associated ecosystems.

51.2.3.2 Wastewater and Stormwater Inflow to Wetland

Wastewater treatment plant effluent and urban stormwater are pollutant sources that can degrade wetlands. In case of filtering organic matter by wetlands, the "aging" of wetlands happens. "Aging" refers to the state in which the wetland is saturated by nutrients and heavy metals over time and the effectiveness of it is increased and the wetland is degraded [30]. Wastewater and stormwater can result in alternation of the ecology of a wetland ecosystem in case of extended eutrophication because of nutrients entrance or plant and aquatic organism toxicity because of metals [15]. Iron and magnesium, especially, can reach toxic concentrations, immobilize available phosphorus, and prevent nutrient uptake by coating roots with iron oxide.

Accumulation of heavy metals in estuarine wetlands may cause deformities, cancers, and death in aquatic animals and their terrestrial predators. Heavy metals are commonly used by benthic organisms in estuarine wetlands through the binding in sediments or the suspended solids that such organisms feed on or settling on the substrate where such organisms live.

The high level of nitrogen and phosphorus enters wetlands through urban and industrial stormwater, sludge, and wastewater treatment plant effluent, which can lead to algal blooms. Algal bloom results in depletion of dissolved oxygen and death of benthic organisms. Furthermore, some algae are toxic to aquatic life [23]. Excess algae also shades underwater grasses and prevents photosynthesis, resulting in underwater grass death [3]. Underwater grass meadows reduce turbidity by stabilizing sediments and provide critical food, refuge, and habitat for a variety of organisms; therefore, the death of these plants highly damages the estuarine ecosystem [3,13].

51.2.3.3 Development of Roads and Bridges among Wetlands

Because of low land value of wetlands, it is more economical to construct roads and bridges across them instead of around them [49]. Sediment loading to wetlands during road and bridge construction periods is increased [30]. Borrow pits during road construction besides the wetlands can degrade water quality through sedimentation and increase turbidity in the wetland [21]. Roads can impound a wetland, even if culverts are used. This impoundment and resulting hydrologic alteration affect the functions of the wetland [49]. Roads can also disrupt habitat continuity, driving out more sensitive, interior species, and providing habitat for hardier opportunistic edge and non-native species. Roads can slow down movement of special species or increase animals' mortality in crossing them.

The maintenance and use of roads contribute many pollutants into the surrounding wetlands. The main activities are

- De-icing roads by rock salt that damages or kills vegetation and aquatic life [52].
- Roads maintenance through use of herbicides, soil stabilizers, and dust palliatives, which damage wetland plants, accumulate in aquatic life, and can result in mortality.
- Bridge maintenance may result in direct entrance of lead, rust (iron), and other chemicals from paint, solvents, abrasives, and cleaners into wetlands.
- Bridge runoff can increase loadings of hydrocarbons, heavy metals, toxic substances, and other chemicals into wetlands.

51.2.3.4 Sanitary Landfill Development Near Wetlands

Sanitary landfills commonly receive hazardous wastes with household and industrial sources as well as sewage sludge. If landfills are not properly located, designed, or managed, these facilities can pose an ecological risk to wetlands by entering pollutants and changing the hydrology of nearby wetlands. Leachate from landfills often has high biological oxygen demand (BOD), and high concentration of ammonium, iron, and manganese that is toxic to plants and aquatic animals [25].

51.2.3.5 Non-native Plants and Animals

Wetlands can be attacked by aggressive, highly tolerant, non-native vegetation, such as purple loosestrife, water hyacinth, and salvinia, or can be dominated by a monoculture of cattails or common reed because of disturbance and habitat degradation [30]. This is more common in constructed wetlands, including restored wetlands, where non-native and tolerant native species out-compete other species resulting in less species diversity. In some cases, non-native species are introduced on purpose such as nutrients impoundment and wastewater purification. Carp and nutria are two introduced exotic animal species that degrade wetlands [30]. Carp, which are used for recreational fishing purposes, severely increase the turbidity of water resources. Nutria, which are used because of their pelts, voraciously eat and destroy wetland vegetation.

51.3 Urban Wetland Hydrology Importance

Hydrology is an important factor to be considered in studying wetlands. The wetland hydrology provides information about different wetland systems performance and is used for classification purposes. It is an important component of the cycle that determines the conditions and biota within a wetland (Figure 51.6). The wetland biotas are adapted to the hydrological characteristics and variability. Therefore, it is very important to study the wetland hydrology.

51.3.1 Driving Force of Wetland Hydrology

The spatial and temporal variability of energy controls the direction and rate of water movement into and out of wetlands. By changing the energy in two zones, a force is generated for water movement from zones of high energy to zones of lower energy. Water movements are mostly due to gravitational forces that force water to flow from higher to lower elevations. Viscous (friction) forces resist against the gravitational force and retard the fluid velocity. Momentum forces resist any change in flow velocity, causing water to move at a constant velocity, and in a straight line, unless additional energy is expended. Pressure change can also be a driving force and cause water movement from zones of high pressure to zones of low pressure. This is common in confined aquifers. Artesian flow from a confined aquifer to the surface occurs when an aquifer recharge zone is located at a higher elevation than the ground surface in the discharge zone.

Wetlands are commonly located in low-energy environments where water normally flows with a slow velocity. This can be because of the relatively flat topography of the wetland located areas [35]. Because of flat landscapes where wetlands are located, their surface expands by changing water stage changes and this allows for the storage of large volumes of water. Therefore, wetlands are moderators of hydrologic variability such as storing flood flows and velocities. Shallow depths and low slopes of land surface in wetlands play an important role in trapping nutrients and sediments.

51.3.2 Hydrology and Ecology in Wetlands

A simplified and idealized model of the wetland water regime is shown in Figure 51.7. This information can be used for real life situation identification. As an example, the area of growth of different types of plants is controlled by water level fluctuations and resulting wetting and drying regimes in different parts of a wetland (Figure 51.8).

Although hydrology plays an important role in the establishment and maintenance of wetlands and is determinant of plant types, wetlands are not completely passive players in this relationship. Vymazal [43] states that there are varieties of mechanisms that through them the biotic components control the wetland hydrology. These mechanisms are the generation of

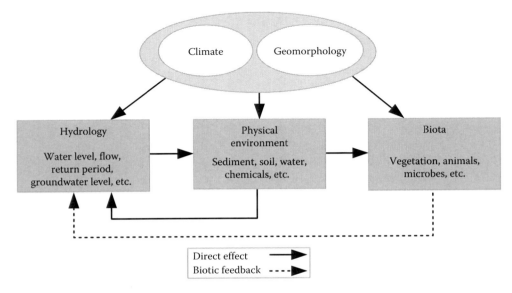

FIGURE 51.6 Dynamic of hydrology effect on wetland function and the biotic feedbacks that affect wetland hydrology.

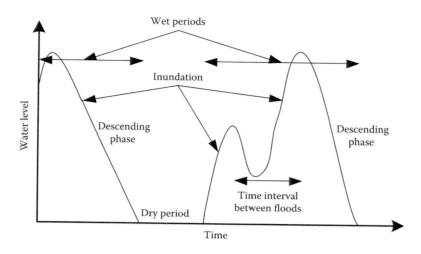

FIGURE 51.7 The main components of wetland hydrology.

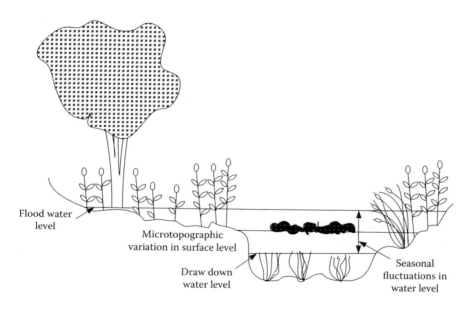

FIGURE 51.8 Effects of wetland water regime on the growth and presence of different zones of wetland plants. Completely different types of plants are available in different zones.

peat, trapping of sediment, vegetative shading, and altered rates of evapotranspiration because of the presence of large beds of emergent macrophytes.

Recent investigations show that emergent vascular plants in wetlands could modify the water regime as well as the same biogeochemical processes that operate in a wetland. This is done through creating micro-topographical relief in the profile of the bed of the wetland and in its sediments. These small-scale variations in topography help structure vegetation mosaics because certain plants grow better on the slightly elevated mounds [36,38]. The micro-topographic variations also affect biogeochemical processes such as sulfur metabolism and nitrogen cycling in wetlands [40,47,50].

The other aspects of the ecological structure and function of wetlands are also affected by hydrology. Water regime of wetlands highly influences the growth of submerged and emergent vegetation. It also affects the way plant material is entrained into wetlands from the upstream basin, and the relative importance of vascular plant tissue versus algal biomass to wetland food cycle. This process is of higher importance in arid-zone, Mediterranean, and wet-dry tropical systems, where temporal inundation results in pulsed inputs of vascular plant material from the floodplain and periods of autochthonous algal productivity in the pools retain water after water levels drop [9,46]. These periodic inundation and organic components entrance are reflected in pulsed rates of organic matter decay and methane emission [19,32].

The slow rise and fall of water levels in response to rainfall and evaporation moves the ponded, wet, damp, and dry sediments from the center of the wetland to its periphery. These different environments provide ecotones (zones of transition between adjacent ecological systems). By strong hydrological shift among different spatial or temporal scales, ecotonal environments are created. Seasonal rise and fall in groundwater tables has a vital role in mediating biogeochemical processes in wetlands, including rates of methane uptake and release, as well as in nitrogen and sulfate cycling [6,7].

51.3.3 Hydrology and Management of Urban Wetlands

The understanding of wetlands hydrologic and climatic characteristics plays an important role in urban wetlands management. The wet and dry periods that could result in severe droughts and floods should be considered a natural part of the environment. The wetlands are developed in the available climatic and hydrological situation and the biota is adapted to the resulted physico-chemical conditions. Therefore, the wetlands should be managed in a way that is consistent with their climatic and hydrological situation.

Based on Walters [45], the main issues in wetland management practices are protection, restoration, and operation of the system to provide the objective functions and values of a wetland with emphasis on both quality and acreage in a sustainable manner. Management of wetlands needs continuous monitoring as well as increased interaction among different stakeholders that affect or are affected by wetlands performance. The wetland management includes buffering wetlands from direct anthropogenic impacts that alter the wetlands normal functions as well as maintaining natural processes on which wetland functions are dependent.

Therefore, wetland management needs an integrated approach including planning, execution, and monitoring. An efficient management needs knowledge in different fields of ecology, economics, watershed management, and planners and decision makers. In this way, by better understanding wetlands, more comprehensive solutions are proposed for long-term conservation and management of wetlands. For all of these purposes, a deep understanding about the wetlands hydrology is necessary because, as mentioned before, the hydrologic condition highly affects the wetland functions.

51.4 Urban Wetland Hydrology

Wetland hydrology is the main feature describing the wetland function. Wetland hydrology is often described using a hydroperiod. Hydroperiod shows the water level fluctuation pattern calculated based on inflows and outflows balance, topography, and subsurface soil, geology, and groundwater conditions [30]. The application of wetlands for different functions that were described before, needs the wetland hydrology to be well understood and considered in their planning and management.

51.4.1 Urbanization and Wetland Hydrology

The main effect of urbanization on wetland's hydrology is alternation of the wetland hydroperiod. This alternation is usually observed as increase in the magnitude, frequency, and duration of wetland water levels. In other words, increased stormwater flows in urban area result in higher wetland water levels for longer periods of time, on more occasions during the wet season. Plant and animal communities living in wetlands were adapted to the preexisting hydrologic conditions. Therefore, changes in wetland hydroperiod will affect them. More detailed description on urbanization effect on wetlands' hydrology is given in the next sections.

51.4.2 Wetland Hydrologic Measures

Three hydrologic variables of the water level, the hydropattern, and the residence time are used for characterizing wetland hydrologic behavior. Elevation of wetland water levels relative to the soil surface is a general hydrologic descriptor. Open water is observed in deep areas with few emergent vegetation. The vegetation in this area is usually floating on the water surface and is not attached to the wetland bottom. An emergent zone is present in areas shallower than the open water zone. Considerable quantities of living or dead emergent vegetation are available in the emergent zone. However, some wetlands have large areas of exposed, saturated soil with green coverage. Therefore, water level is indicative of the vegetation types in different zones of a wetland.

The second measure of wetland hydrology is the temporal variability of water levels. The wetland hydropattern shows the timing, duration, and distribution of wetland water levels. The hydropattern fluctuation pattern is dependent on the system type. In some systems, such as tidal marshes, hydropattern highly fluctuates over short periods of time; but in seasonally flooded bottomland hardwood communities, the hydropattern fluctuations are more slowly over time. There are also some static wetland systems without any substantial short- or long-term variability. The wetland hydropattern is a function of the net difference between inflows and outflows from different sources of the atmosphere, groundwater, and surface water.

The third measure of wetland hydrology is the water residence time; that is, time of water movement through the wetland. Some wetland systems have a short resistance time and the wetland exchanges water quickly, while in some other systems water moves very slowly through the wetland. The residence time is determined as the ratio of the water storage volume within the wetland to the flow rate through the wetland. The residence time and hydropattern of a wetland are related to each other. Wetlands with large water level fluctuations such as in tidal marshes may have shorter residence times. However, some wetlands that have very long residence times due to slow loss rates may fluctuate rapidly because of large changes in inflow.

51.4.2.1 Water Level

An important feature of wetlands is the oxygen content in wetland soils. Anaerobic conditions are more probable and develop more quickly in saturated soils than in unsaturated soils because of

- Low oxygen solubility in water
- Slow rates of water advection
- Slow diffusion rates of oxygen through water

Anaerobic conditions in wetland soils result in adverse conditions for root survival and growth, and affect vegetation. Thus, the sustainable water level in wetlands highly affects soil oxygen concentrations as well as plant growth and survival. Furthermore, wetland is one of the most biologically productive ecosystems on Earth. Therefore, there are vegetative species with special physiological adaptations that enable them to survive and prosper in these harsh growing conditions. Many specific biochemical reactions also occur in low oxygen content area of soil. Therefore, wetland water level can be considered an indicator of the dissolved oxygen state of the soil-water system.

It should be noted that there is a difference between soil saturation and soil surface inundation. A system can be flooded for a period of time and have low pore water soil saturation, and vice versa. Furthermore, the saturated soil zone can be extended above the water table because of capillary rise in fine-grained soil types [4]. Since the capillary height of rise can be several meters above the water table in fine-grained soils, the soil can been saturated even when water levels are below the ground surface. Therefore, wetland water level is not a representative of soil saturation state.

51.4.2.1.1 Hydrograph

A hydrograph shows the water level (stage) variations over time. The water levels rapidly rise in response to precipitation and after that decline slowly over time. The time to peak water level is considered the time length between the peak precipitation and peak stage. In urban areas due to large impervious surfaces and modified channels with high stream velocities, times to peak are short. Forested areas because of few impervious surfaces and channels with many obstructions resulting in slow velocity of water have longer time to peak. Another important feature of wetland extracted from wetland hydrograph is the time of concentration. Time of concentration is defined as the time required for flow to travel from the most distant point on the watershed to the outlet and is related to the same factors that affect the time to peak.

51.4.2.1.2 Water Level Measurement

If the water level in wetland is above the ground surface, a staff gage can be used for wetland water level determination. The staff gage is an inexpensive tool including a vertical scale that determines the elevation of water with respect to a reference elevation [48]. The staff gage is placed in the wetland such that the base of the staff gage is always submerged in water. In case of high water level variations in a wetland, multiple staff gages are placed at different depths in a way that the nearest one is visible from the shoreline during the high water level periods. The other staff gages are used when the water level is below the nearer gage (Figure 51.9).

In some wetlands, a water line can be observed on periodically submerged vegetation and it can indicate a flood highwater level. In these cases, floating debris—such as leaves, trash, or branches—can be seen in the canopy of trees or bushes. In other cases, a horizontal line on the sides of trees is observed showing the normal water level in the wetland. It should be noted that this line is just visible during dry periods with water levels lower than normal.

In case of water level below the ground surface, piezometers are used for water level determination (Figure 51.10). A piezometer is a small-diameter perforated tube installed at a specified depth within the soil [4]. The piezometer should be narrow enough to minimize interference between layers, and located within a unique hydrogeologic unit.

Water levels can also be inexpensively determined by lowering a weighted, chalk-covered steel measuring tape into the piezometer until a part of it gets wet and determining the length of tape in the piezometer. A more expensive technique for water level determination is usage of a depth-to-water detector, which provides an audio or visual signal when facing the water surface. Another option is use of automated water level recorders such as float or pressure transducers. The automated techniques can be used in both conditions of water levels above and below the ground surface [4].

Time domain reflectometry (TDR) is used to measure the degree of soil saturation when water levels are below the ground surface. In TDR application, the electromagnetic properties of a wave pulse passing through a conducting set of rods placed in the soil are used for determining the water content and, therefore, there is no need for calibration. Additional measurements of the total porosity are required because the air porosity is not measured using TDRs. Then, the soil saturation is determined as the ratio of the water content to the total porosity. The basis of TDR

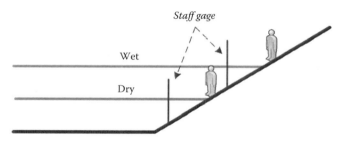

FIGURE 51.9 Multiple staff gages placement in a wetland.

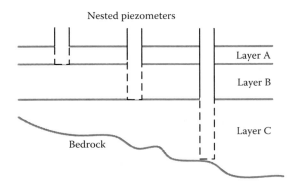

FIGURE 51.10   Nested piezometers location.

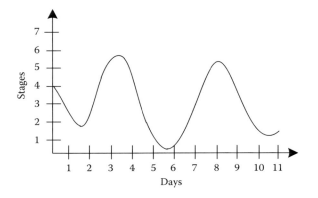

FIGURE 51.11   A sample wetland hydrograph.

work is that along a conductor, velocity of electromagnetic waves is adversely related to the dielectric coefficient of the surrounded media [41]. By increasing the dielectric constant, the wave velocity decreases resulting in longer travel times. Water has a dielectric constant of 80.2 at 20°C, while this value for quartz and air is 4.3 and 1.0, respectively. Therefore, by increasing the water content of a soil, the wave velocity is substantially decreased.

51.4.2.2 Hydropattern

The hydropattern shows the temporal and spatial variability of wetlands water level that is a result of dynamic changes in hydrologic inputs and outputs as well as changes of hydraulic controls within the wetland. The water level changes over the time are determinants for many aquatic flora and fauna. The reproductive success of these wetland species is highly dependent on correct synchronization of fluctuations with their developmental stages.

The hydropattern is a distinctive feature of the hydrologic variability [1,24]. Hydropattern is an extension of the traditional concept of hydroperiod (showing the frequency and duration of wetland saturation periods). In hydropattern, more information about the extension and duration of inundation are provided. The inundation extension is of high importance especially in large, complex wetlands including a variety of wetland features.

Different approaches are used for characterizing temporal changes in wetland stage (i.e., water levels). The simplest way is plotting wetland stage as a function of time, called the hydrograph. The hydrograph shows the water level variations for a period of time as well as the possible range of hydrologic variability (Figure 51.11). Using a hydrograph, the inter-annual, seasonal, event, and daily water level fluctuations are shown.

Based on the observed water levels, a plot of flooding duration versus wetland stage could be developed (Figure 51.12a). Using this graph, a descriptive summary about flood duration for each stage is provided. Higher water levels have shorter durations of flooding than lower water levels. Using this graph, a stage-duration relationship can be developed that quantifies the duration in time that a specified water level is exceeded. The seasonality of

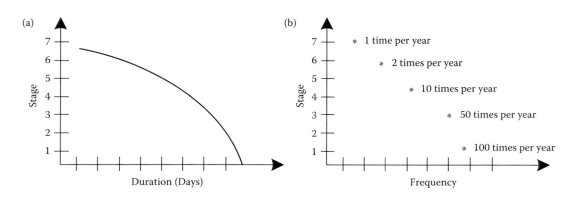

FIGURE 51.12   (a) The wetland hydroperiod, and (b) wetland stage–frequency plot.

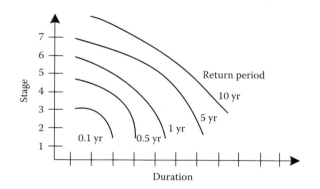

FIGURE 51.13 A sample wetland stage–frequency–duration (SDF) plot.

inundations should also be considered in this approach by dividing the data into specific time frames [30].

Even though the duration of time that the system is flooded can be recognized by the stage-duration approach, it does not provide any information about the frequency with which this occurs. Another approach is to quantify the frequency in time that the wetland exceeds a specified stage. This approach results in a cumulative frequency plot on which exceedance probabilities can be calculated (Figure 51.12b).

The main disadvantage of the exceedance probability approach is that the correlation between observations can or cannot be captured. The systems that vary slowly or quickly over time can provide the same frequency distribution of water levels. In other words, the magnitude of the fluctuations is shown without giving any information about their duration. Another shortcoming of the frequency diagram is that it can poorly convey daily and seasonal behavior of water level changes. Classifying data sets into seasonal or other periods can improve the representation of water level variations [30].

A stage-duration-frequency (SDF) curve can be used to overcome these shortcomings (Figure 51.13). The SDF curve is similar to intensity-duration-frequency (IDF) curve used in rainfall analysis [29]. SDF curves indicate the frequency that a depth-duration relationship is observed. The hydrodynamic behavior of large, complex wetlands is highly variable and characterizing a hydropattern for the whole wetland is considerably more difficult than a small, uniform wetland.

51.4.2.3 Residence Time

The hydrologic residence time shows the time required for a hydrologic input to pass through the wetland. The residence time, T, of a system with constant volume and flow rate is estimated as follows:

$$T = V/Q \tag{51.1}$$

where V is the water volume within the wetland, and Q is the flow rate. The estimated residence time in this way can be used only in the following conditions:

1. Plug flow
2. Steady state flow

3. Single locations of inflow and outflow
4. No atmospheric or groundwater exchanges

In other conditions, using Equation 51.1 only provides an estimation of the average residence time and the actual residence time would be spatially and temporally variable. In simple systems, functions are developed for describing the distribution of residence times. For example, for a fully mixed system with constant inputs over time, the residence time distribution is determined using an exponential function [26].

Different zones of a wetland could have different hydrologic residence times. Active, flow-through sections of a wetland may have shorter residence times than inactive, isolated parts of a wetland. Each section could have different hydropatterns but the flow is concentrated in one area and in other areas approximately the stagnant conditions exist. The same equation can be applied for residence time estimation in each section considering the volume of water present in the section, and the flow into the considered section.

It is more difficult to calculate the residence times for dynamic systems than steady flow systems. In dynamic wetland systems, the residence time is variable and increases when outflows exceed inflows, and decreases when inflows exceed outflows. Tracer tests can be used for confirmation of the hydraulic residence times calculations based on Equation 51.1. Conservative tracers are added at the wetland inflow point, and then tracer concentrations are monitored at the outflow point. A plot showing the tracer concentration at the outflow point of the wetland over time is generated. The time to median concentration (when the outflow concentration equals half of the input concentration) can be used as an estimation of the average residence time of the wetland.

51.4.3 Wetland Hydrologic Classification

Wetlands can be classified based on different features such as vegetation, geographical location, and hydrology. In hydrologic classification of wetlands, hydroperiod, illustrating the seasonal water level pattern of a wetland, and water level can be considered. Based on water level variation scheme, the wetlands are classified into tidal and nontidal wetlands. Each of these wetlands, based on their hydroperiod, are classified into different groups as given in Table 51.2. Hydroperiod is an important feature because it directly affects the physiochemical characteristics of a wetland due to changes in oxygen availability, soil and water chemistry, nutrients, toxicity, and sedimentation. These changes directly impact wetland biota through species composition, richness, and ecosystem productivity change [30].

Groundwater and wetlands relationship can also be considered for wetlands classification as it is shown in Table 51.3 [31]. Groundwater or surface water can be the determinant of wetland water level changes. Wetland and groundwater interactions occur in two ways of discharge and recharge. When the wetland water level is lower than groundwater table, the discharge occurs and vice versa. Different types of wetlands based on this classification are shown in Figure 51.14.

TABLE 51.2 Wetland Classification Based on Water Level and Hydric State

Class	Sub-class	Definition
Tidal wetlands	Subtidal	The land surface is always covered by tidal water.
	Irregularly exposed	The land surface is commonly covered by tidal water but is not exposed daily.
	Regularly flooded	Tidal water consecutively covers and daily exposes the land surface.
	Irregularly flooded	Tidal water covers the land surface less often than daily.
Nontidal wetlands	Permanently flooded	The land surface is covered by water all the time and vegetation is composed of obligate hydrophytes.
	Intermittently exposed	The land surface is covered by water throughout the year except in periods of extreme droughts.
	Semi-permanently flooded	The land surface is usually covered by water throughout the growing season. When the land surface is exposed, groundwater level is at or near the surface.
	Seasonally flooded	The land surface is covered by water for extended periods especially at the beginning of the growing season. The land surface is usually exposed by the end of the growing season and the groundwater level is at or near the surface at these times.
	Temporarily flooded	The land surface is covered by water for short periods during the growing season, and the groundwater level is usually below the surface. Plants that grow both in uplands and wetlands are present.
	Intermittently flooded	The land surface is covered by water in various periods of time throughout the year. The vegetation type changes regarding changes in soil moisture condition. Some areas may not exhibit hydric soils or support hydrophytes.
	Artificially flooded	The amount and duration of flooding is controlled by humans through application of pumps or siphons in combination with dikes or dams.

Source: Adapted from Topp, G.C. 1980. *Water Resources Research*, 16, 574–582.

TABLE 51.3 Wetland Classification Based on Water Level and Hydric State

Groundwater Flow Pattern Classification	Characteristics
Surface water depression wetland (Figure 51.14a)	• Controlled by surface runoff and precipitation • Limited groundwater outflow and recharge • Groundwater level usually below wetland
Surface water slope wetland (Figure 51.14c)	• Placed commonly beside a lake or stream • Fed mainly by overbank flooding from adjacent water body • Some inflow from runoff and precipitation • Groundwater recharge possible which usually discharges shortly back to water body • Groundwater level usually below wetland
Groundwater depression wetland (Figure 51.14b)	• Water level is low enough to intercept the groundwater level • Limited water-level fluctuations in comparison with surface-fed wetlands because of groundwater level stability
Groundwater slope wetland (Figure 51.14d)	• Located on slopes where groundwater discharges to surface through seeps or springs • Continuous or seasonal discharge of groundwater to wetland depending on geohydrology and evapotranspiration rates

51.4.4 Water Budgets of Urban Wetland

The hydrologic measures of wetland including water level, hydropattern, and residence times, are affected and controlled by hydrologic inputs and outputs of the system. To account for the inputs and outputs to the system, the wetland water budget is used. There can be exchanges with the atmosphere and ground or surface water, and can be affected by tidal actions. The wetland water level is affected by the sum of all exchanges. The water balance equation used for this purpose is formulated as follows:

$$\Delta Q = I - O = \Delta V / \Delta t \tag{51.2}$$

where ΔQ is the difference between inflows, I, and outflows, O, and the ΔV is change in the water storage volume over the time period of Δt. Based on this equation, the storage volume increases when the inflows exceed the outflows, and vice versa. Regarding the direct relationship between water level and storage volume, the changes in water are estimated as follows:

$$\Delta h / \Delta t = \Delta Q / A = (I - O)/A \tag{51.3}$$

where Δh is the water level change, and A is the wetland area. Equation 51.3 is obtained from Equation 51.2 considering that the change in wetland storage volume is equal to the product of the watershed area and the water level change. In cases when the water level falls below the ground surface, the water level changes relationship would be more complicated. In these cases, the amount of water release would be less because the mineral and organic soil materials retain water in their pores.

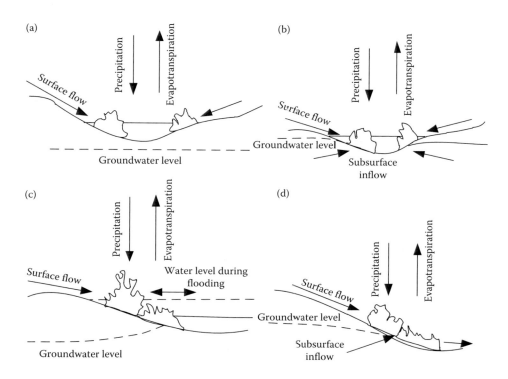

FIGURE 51.14 Wetlands classification based on the relative importance of surface water versus groundwater inputs: (a) surface water depression wetland, (b) groundwater depression wetland, (c) surface water slope wetland, and (d) groundwater slope wetland. (Adapted from Boon, P. 2013. *Work Book for Managing Urban Wetlands in Australia*. Sydney Olympic Park Authority, ebook available through www.sopa.nsw.gov.au\education\ WETebook\, ISBN 978-0-9874020-0-4.)

51.4.4.1 Inflows and Outflows Balance

The possible wetland inflow sources are as follows:

- Precipitation directly onto the wetland
- Direct overland flow
- Overbank flow
- Surface water inputs from related surface water bodies
- Inflow from groundwater sources including subsurface, lateral unsaturated and saturated flow from uplands to toe slope and flat landscapes

The possible outputs that balance the inputs are as follows:

- Evaporation
- Transpiration
- Surface water outflows
- Groundwater recharge

Water level falls over time when wetland outflows exceed inflows, and vice versa. The water level change based on Equations 51.2 and 51.3 is estimated as follows:

$$\Delta h = \frac{\Delta V}{A} = \frac{\Delta Q}{A} \Delta t \qquad (51.4)$$

where Δh is the water level change, ΔV is the net change in storage volume, ΔQ is the net change in inflows minus outflows, Δt is the calculation time step, and A is the wetland surface area. Hydrologic inflows and outflows are expressed in units of

volume per unit time. The unit of time of water level changes is not explicitly incorporated in water level determination. Water level changes in unit of time are commonly determined. Thus, observed water level is the result of accumulated water level changes over a period of time and is formulated as follows:

$$h(t_i) = \sum_{j=0}^{\infty} \Delta h(t_i - t_j) = \sum_{j=0}^{\infty} \frac{\Delta Q(t_i - t_j)}{A_e} \Delta t \qquad (51.5)$$

where $t_i - t_j$ is equal to Δt. This assimilation of water level changes means that wetland water level fluctuations are reduced by water storage during wet periods and release of the stored water during dry periods. Since wetland water levels are accumulators of spatial and temporal hydrologic changes, they would be sensitive to environmental condition changes. The mass balance of dissolved and suspended matter carried by the water can be considered in addition to the water balance equation.

$$\frac{\Delta M}{\Delta t} = \Delta L = L_1 - L_2 \qquad (51.6)$$

where ΔM is the change in mass of dissolved or suspended matter, Δt is the time interval, and ΔL is the change in load, all of them between two points. L_1 is the inflow load, and L_2 is the outflow load. The rate of mass change per unit time is related to the inflows and outflows balance. In water quality practices, the solute concentration is normally measured instead of solute load. The load, L, and the concentration, C, are related to each other as follows:

$$L = C \times Q \qquad (51.7)$$

where Q is the flow rate. This is because the concentration is

$$C = \frac{L}{Q} = \frac{M}{V} \qquad (51.8)$$

or mass per unit flow rate, which is just the mass, M, per unit volume, V.

51.4.4.2 Stage-Area-Volume Relationships

Changes in water depth must normally be converted to changes in water volume. For conversion of water depth to water volume, the effective storage area of the wetland, A, should be known

$$A = \Delta V / \Delta h \qquad (51.9)$$

where ΔV is the wetland water storage change and Δh is the water level change. Equation 51.9 is applicable when water level is entirely above the ground surface. The effective storage area is dependent on wetland stage, that is, the wetland area grows during high stage, wet periods. Therefore, a table or plot of wetland stage versus area should be developed.

51.4.4.2.1 Subsurface Storage

When wetland stage is below the ground surface, the situation would be more complicated and the specific yield of the organic and mineral sediments should be determined. The specific yield is defined as the released volume of water per unit area of wetland per unit decrease in water level. In some cases, a small volume of water is released as water level falls because the organic and mineral sediments remain almost saturated. Therefore, the effective storage area is determined by combining the storage above and below the ground surface as follows:

$$A = A_s + S_y \cdot A_e \qquad (51.10)$$

where A_s is the submerged area of wetland, A_e is the exposed area of wetland, and S_y is the specific yield of sediments, which is determined as the difference between the saturated water content and field capacity of the sediments. Specific yield of sediments is dependent on their particle size distribution and chemical composition. Clays and mineral soils have low specific yields while sands have large specific yields [29].

Based on the estimated effective areas, the stage-area and stage-volume relationships are developed for relating the water level changes to net changes in volume. There are two sharp changes in the stage-volume relationship. The first case is when water level falls below the ground surface and the second case is when water level overtops a natural bank or levee. Examples of developed stage-area and stage-volume curves are illustrated in Figure 51.15a and b, respectively.

51.4.4.2.2 Area and Volume Determination

Topographic maps with sufficient details, surface mapping, or aerial photographs can be used to determine the wetland area as a function of water elevation. A method for determining the volume of a wetland is to add a known mass of tracer to the wetland water storage and let the tracer completely mix with the stored water, and then measure the tracer concentration. Regarding Equation 51.8, the volume of water within the wetland is equal to the mass of tracer divided by the tracer concentration. In cases where there is no inflow and evapotranspiration is the only outflow, by decrease in the volume of water within the wetland, the tracer concentration would increase over time. If there are inflows to the wetland without any outflows, the tracer concentration will decrease over time and the wetland volume changes can be calculated. Water levels and tracer data conjunctively can be used to estimate the water balance and water budget components using Equations 51.2, 51.6, and 51.8.

51.4.5 Water Budget Components

Water budget is used as an important tool for characterizing the wetlands' behavior. The inflows and outflows from wetlands include four types.

- Atmospheric exchanges: rainfall, snow, evaporation, and transpiration
- Subsurface exchanges

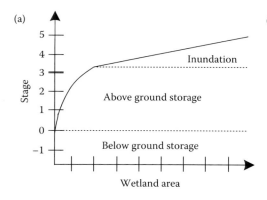
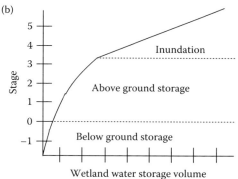

FIGURE 51.15   Examples of synthetic: (a) stage-area and (b) stage-volume curves.

- Surface exchanges: overland flow, interactions with rivers and streams
- Tidal exchanges that happen in marine systems

Brinson [8] characterized wetlands based on the type of hydrologic inflow; however, wetlands are affected by exchanges of water including both inflows and outflows. As an example, atmospheric exchanges include evapotranspiration components as an outflow as well as precipitation as an inflow. Surface water exchanges incorporate releases from wetlands as well as inflows to wetlands.

Brinson [8] states that wetlands follow different patterns of inflow and outflow. Some wetlands have simple exchanges with an adjacent water body. For example, a wetland could receive water from a river during a rising stage of a flood and return water to the river during the falling stage. The same relationship exists along the coast, where the water moves in and out of wetlands to its original source because of tidal effects. However, in some cases, exchanges have a hybrid character and are between different types of water bodies/sources of inflow and outflow. As an example, inflows can be in a type of subsurface exchange and outflows can be in a form of evapotranspiration. However, the key issue in understanding and managing wetlands is identifying the key wetland hydrologic inflow and outflow components.

51.4.5.1 Atmospheric Exchanges

Atmospheric exchanges including precipitation (including different forms of rain, snow, hail, sleet, freezing rain, fog drip, dew, and frost), evaporation, and transpiration are needed to be estimated for the water budget determination. The water level of wetlands is dependent on atmospheric exchange and, therefore, is highly affected by climatic signals. Precipitation occurs as discrete events described with their intensity, duration, frequency, and areal extent. Long-term variability of precipitation events is described using monthly and seasonal averages or summations. Even though the regional precipitation networks are used to estimate a watershed condition, because of high spatial variability of precipitation patterns, the onsite precipitation measurements are needed to estimate water budget components for water balance analysis.

The loss of wetland water through evaporation and transpiration (evapotranspiration) from the water surface and plant leaf and stem surfaces have considerable effects on water levels. This evaporation would be a dominant component when open water is present and vegetation is limited. The water loss through evaporation from saturated soils would be almost the same as open water in the absence of a litter or mulch layer. In wetland systems with large coverage of living vegetation and little open water, evapotranspiration rates are affected by a variety of factors including leaf and stem area, temperature of air, water, plant, atmospheric humidity, wind speed, and the water potential of exposed soils.

The evapotranspiration losses of close wetlands are generally similar because the source of vaporization energy (i.e., the sun) is uniform and the availability of water for vaporization is similar (if there is no water limitations in wetlands). Forested wetlands

due to higher leaf areas could have greater evapotranspiration rates. The wetlands with dead vegetation coverage also have less evapotranspiration rates because of lack of transpiration and less evaporation from shading and poorer wind exchange. Therefore, since the eutrophication could lead to increased plant leaf area, more evapotranspiration will be expected.

Evapotranspiration rates are determined based on evaporation rates measurements. An evaporation pan can be used for wetlands evaporation rate estimation. The potential evapotranspiration derived from pan estimates can be used for site condition estimation but it should be considered that the accuracy of the measurements is dependent on local effects of shading and wind shelter. Pan coefficients defined as the ratio of actual evapotranspiration to pan measurements are used to consider these effects.

Daily precipitation and evaporation data are plotted together. The difference between precipitation and evaporation can be compared to observed wetland water levels in systems where atmospheric exchanges dominate the wetland hydrology. In these systems, water levels rise during precipitation, and evaporation rates control the water level decrease.

51.4.5.2 Subsurface Exchanges

Groundwater discharge (subsurface inflows) to wetlands is provided by shallow, topographically induced drainage from nearby uplands, or through discharges of regional, confined aquifers [28]. On the other hand, groundwater recharge (subsurface outflows) from wetlands are provided by downward and lateral flow from the wetland to underlying unconfined aquifers, and to deeper, confined aquifers where the confining layer has been locally breached due to collapse or subsidence.

Shallow inflows may result from perched, or interflow, drainage on top of lower-permeability units within the unsaturated zone, such as clay beds, soil horizons, or even permafrost. Also, when the wetland water goes below the water table in the underlying unconfined aquifer, shallow subsurface inflows may arise. In this situation, the direction and magnitude of the hydraulic gradient is estimated by determining the aquifer water levels in the vicinity of the wetland through piezometers. The permeability of the aquifer and any organic and mineral benthic sediments also affect the hydraulic gradient.

There could be a concentrated or distributed (such as upward leakage) inflow within the wetland. If confined aquifers discharge into the wetland, subsurface inflows from deeper sources may be observed. Examples of these situations are observed in karst areas when the confining layer is breached due to subsidence or collapse. In this case, the piezometric surface in the confined aquifer controls the wetland water level. Confined aquifer discharges also happens when there is diffuse upward leakage from the confining layer into the overlying unconfined aquifer, and then to the wetland.

Discharges from deeper sources are more stable and are not affected by individual storm events, these discharges change in response to seasonal and longer-term changes. Shallow inflows commonly change more rapidly in response to individual storm events, as well as to seasonal and climatic changes. In other words, the interflow and water level in shallow aquifers are more sensitive to atmospheric flux changes (precipitation minus evapotranspiration) around the wetland.

If a layer of low permeability is placed beneath the wetland, it allows the water to perch and shallow subsurface outflows can occur. In these situations, the water can exit the wetland as overland flow, channel flow, or interflow in a low point on the perimeter of the wetland. In these systems, water level is perched above the regional water table, and there also can be an unsaturated (vadose) zone between the water table and the perched wetland. Recharge to the underlying, surficial aquifer also occurs through the low permeability layer.

Permanently frozen soils, that is, permafrost, prevent downward movement of water because the water is converted to solid form because of heat exchange in the underlying frozen unit. However, the ability of the frozen unit to freeze water can be overloaded over time, and in this situation, confining ability of the unit will decrease regarding the heat added to the unit.

When wetland water level is connected with water levels within the unconfined aquifer (Figure 51.9), flow through the surficial aquifer is affected by the wetland. The surficial aquifer water levels decrease in the direction of water flow, but the wetland water level is almost horizontal. Therefore, wetland water level would be below the water table in the up gradient direction, and above the water table in the down gradient direction. As a result, aquifer will be discharged at the wetland upstream, and will be recharged at the wetland downstream. This type of flow-through wetland is the reason for most of the flux of water through a wetland that has no readily apparent inflows or outflows.

Finally, the deeper, confined aquifers can be recharged when a subsidence or collapse has breached the confining layer that isolates the aquifer from the surface. In this situation, during wet weather periods that wetland water level increases, direct recharge to the deeper aquifer can be observed [30].

51.4.5.2.1 Groundwater Gradients

The flow of water in the subsurface is governed by the gradient of groundwater potentials that is a measure of the available energy for water movement. The direction of the gradient controls the movement direction. The gradient is calculated as follows:

$$G = (G_x, G_y, G_z) = \left(\frac{\Delta h}{\Delta x}, \frac{\Delta h}{\Delta y}, \frac{\Delta h}{\Delta z} \right) \qquad (51.11)$$

where G, the hydraulic gradient, is composed of three components in three directions named G_x, G_y, and G_z, which are determined using the head change, Δh, in each of the three directions, Δx, Δy, and Δz, respectively. For more simplicity in multilayered geologic deposits, the hydraulic gradient can be decomposed into horizontal and vertical components, with each layer having a unique horizontal flow pattern.

$$G_h = (G_x, G_y) = \left(\frac{\Delta h}{\Delta x}, \frac{\Delta h}{\Delta y} \right) \qquad (51.12)$$

and

$$G_v = G_z = \frac{\Delta h}{\Delta z} \qquad (51.13)$$

where G_h is the horizontal component and G_v is the vertical component of the hydraulic gradient. In case of horizontal layering, the magnitude of the horizontal gradient of each layer is determined based on the change in water levels with distance, while the vertical gradient between layers is determined based on the change in water level with each layer depth.

The groundwater flux, or rate of volume flow, is determined by a combination of the hydraulic gradient and the hydraulic conductivity. The groundwater flux is also separated into three components, two horizontal and one vertical as follows:

$$q = (q_x, q_y, q_z), \quad q_h = (q_x, q_y), \quad q_v = q_z \qquad (51.14)$$

The groundwater flux in different directions is determined using Darcy's law. The horizontal component of flow, q_h, is determined using the horizontal hydraulic conductivity, K_h, while the vertical component, q_v is determined based on the vertical conductivity, K_v

$$q_h = -K_h \cdot G_h \quad q_v = -K_v \cdot G_v \qquad (51.15)$$

The negative sign shows the flow direction from regions of higher hydraulic head to regions with lower head. The above equations estimate flow at a point. Flow across the area is determined by flows multiplication by the cross-sectional area of the considered unit. The total flow, Q, is calculated using

$$Q_h = A_h \cdot q_h \quad Q_v = A_v \cdot q_v \qquad (51.16)$$

where A_h is the cross-sectional area of flow, A_v is the map-view area of flow, and Q_h and Q_v are the horizontal and vertical components of total flow, respectively.

Groundwater flows are temporally and spatially variable. Therefore, taking multiple vertical and horizontal samples at a sufficiently frequent time interval is necessary in order to capture the system variability over time and space. Ground water flow monitoring is done by placing piezometers at different depths within and around the wetland. A nest network of piezometers, composed of multiple piezometers placed at different depths, is needed for complex layering systems.

Piezometers are also placed at different distances from the wetland to determine the water table pattern in the neighborhood of the wetland. This network of piezometers is used to determine the three-dimensional characteristics of water flows. Ideally, the locations of piezometers should form an equilateral triangle (Figure 51.16). Otherwise, the co-linearity of the wells interferes with the estimation of the gradient. As mentioned before, the total flow of water into or out of the wetland estimation requires the hydrologic conductivity to be known through aquifer tests or standard tables' application [16]. Commonly, core samples underestimate field hydraulic conductivity because of the difficulty in obtaining core samples from highest flow paths within the system.

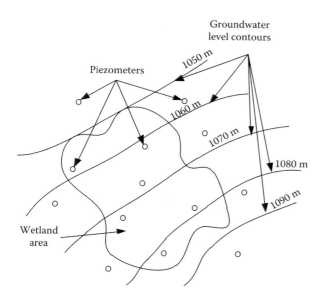

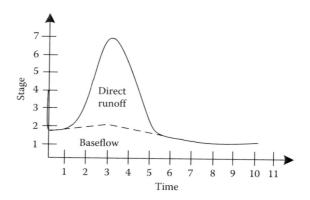

FIGURE 51.17 The water level changes during a storm.

FIGURE 51.16 The required network of piezometers for mapping water level in the vicinity of a wetland. (Adapted from Boon, P. 2013. *Work Book for Managing Urban Wetlands in Australia.* Sydney Olympic Park Authority, ebook available through www.sopa.nsw.gov.au\education\WETebook\, ISBN 978-0-9874020-0-4.)

After determining the hydraulic conductivity and gradients in different directions, the total flow within the system is calculated. There are often many uncertainties in this method because of high spatial variability of both gradients and conductivities. Different sources of water within wetlands can be determined through water quality sampling. For example, if groundwater is moving upward into the wetland from a deeper aquifer and shallow groundwater is also moving laterally into the wetland, then the geochemical characteristics of each source can be used to estimate the flow from each source.

51.4.5.3 Surface Water Exchanges

Surface water exchanges with wetlands have numerous sources and mechanisms such as

- Overland or sheet-flow
- Direct exchange with channel of a river or stream flows through the wetland
- Overbank flooding during wet weather when the channel is separated from the wetland by a levee or floodplain and along the edges of lakes, estuaries, and the ocean.

These surface water exchanges can be constant or happen occasionally between the surface water and the wetland. Streamflow is composed of two components of baseflow and stormflow (Figure 51.17). Baseflow refers to flow during low flow periods, while stormflow includes the response to precipitation events.

51.4.5.3.1 Flood Damping

Floods reduce flood velocities and flood wave velocity. Flood waves travel at different rates than water velocities. The flood wave velocity, c, is defined as the change in discharge per unit

change in stream cross-sectional area and is formulated as follows:

$$c = \frac{\Delta Q}{\Delta A} = \frac{\Delta(\bar{v}A)}{\Delta A} = \bar{v} + A\frac{\Delta\bar{v}}{\Delta A} \tag{51.17}$$

where $Q = \bar{v} \cdot A$ is the stream discharge, A is the stream cross-sectional area, and $\bar{v}$ is the mean stream velocity. Based on this equation, the flood wave velocity equals the water velocity plus a second term that is positive if the water velocity increases as the cross-water level (flow cross-section) increases, and is negative if the water velocity slows as the water level increases. In other words, the wave velocity is faster than the water velocity if the water velocity increases with stage, and vice versa. This concept can also be shown with kinematic ratio, k, which is defined as the ratio of the wave velocity to the water velocity:

$$k = \frac{c}{v} = 1 + \frac{A}{v} \cdot \frac{\Delta v}{\Delta A} = 1 + \frac{\Delta \ln v}{\Delta \ln A} \tag{51.18}$$

Based on this equation, the flood wave velocity is faster than the water velocity, $k > 1$, when the second term is positive, and vice versa. This term is a critical parameter that controls damaging flood waves. Wetlands impound flood waters resulting in decrease in average and incremental water velocities, flood waves travel times, and peak discharges. The flood wave velocity could actually be less than the water velocity if the velocity decreases with increasing depth.

51.4.5.3.2 Measurement of Surface Flows

Surface water flows are determined using flow measurement devices listed in the following:

- Weirs: Need a pool upstream and are not satisfactory in case of deep sedimentation
- Flumes: Flush sediments more effectively than weirs
- Culverts: Less accurate than other tools

The relationship between stage and streamflow discharge is called the rating curve. Using the rating curve, the stream discharge is estimated based on water level measurement.

In case of control structure's absence and no possibility of in-stream measurement, the Manning's equation can be used to indirectly measure water velocity:

$$\bar{v} = \frac{1}{n}R^{2/3}S^{1/2} \tag{51.19}$$

where n is the Manning's roughness coefficient, R is the hydraulic radius, and S is the gradient in total head. The Manning's roughness coefficient can be determined based on stream channel characteristics such as stream bed materials, amount of vegetation within the channel, variation in channel shape, and sinuosity using the standard tables available in different textbooks. Considering the hydraulic radius definition, as the ratio of the stream cross-sectional area to the wetted perimeter, in a shallow, wide channel, it would be approximately equal to the water depth. After determining the average water velocity, $\bar{v}$, the stream discharge, Q, can be estimated using

$$Q = \bar{v}A \tag{51.20}$$

where A is the cross-sectional area of the channel.

51.4.5.3.3 Estimating Overbank Flows

Wetlands located near the riverine systems are often affected by overbank flows during wet periods that the river spills out of its normal channel. The period of time that wetlands are affected by overbank flows is dependent on inundation duration and depth. The water level in wetlands adjacent to riverine systems is monitored using the water level measurement instruments mentioned before. The U.S. Geological Survey has also provided estimations of overbank flooding frequencies for ungagged sites [22].

51.4.5.4 Tidal Exchanges

Coastal wetlands are almost similar to freshwater wetlands, but they are transitional between marine and terrestrial environments. The unique attribute of coastal wetlands is the combination of flooding and soils near the water surface, that result in ecologic diversity and productivity in coastal wetlands. Different types of coastal wetlands exist. The tidal wetlands are dominated by the ebb and flow of the ocean due to tides. Because of the large magnitude of the daily tides and their regularity, a unique condition exists in these wetlands. There are also some marine wetlands that are less affected by tidal changes. Some others are affected by changes in water quality of freshwater tributaries.

51.4.5.4.1 Tidal Effects

Most of the coastal wetlands damp tidal and wave energies, therefore the fluctuation's magnitude generally diminishes with distance from the coast. The height of a harmonic wave with constant energy dissipation as a function of distance can be formulated as follows [10]:

$$h = h_0 e^{-\kappa x} \cos(\omega t - \kappa x) \tag{51.21}$$

where κ equals $\sqrt{\omega/2D}$ and is the wave or tide energy damping coefficient, h is the wave or tidal height, h_0 is the maximum magnitude of the fluctuation at the shoreline, x is distance from the shoreline, ω is the frequency of the fluctuation, t is time, and D is the hydraulic diffusion coefficient. The amplitude of the oscillation is obtained as follows:

$$|h| = h_0 e^{-\kappa x} \tag{51.22}$$

Based on this equation, the oscillation amplitude decreases with increasing frequency and distance from the shoreline. It also indicates that low-frequency waves, such as daily and twice-daily tidal fluctuations, in comparison with the high-frequency waves, propagate deeper into coastal regions, are attenuated less and lagged more. It should be noted that the harmonic wave height equation is not suitable for all coastal areas, but because of a combination of most of the important factors (e.g., time, distance, wave frequency, and hydraulic resistance) that affect water and energy movement, it is widely used.

In coastal wetlands, a good understanding of the local coastal morphology and vegetation and the energy inputs from marine sources and terrestrial surface water and groundwater inflows is important to know how coastal wetlands are affected by hydrologic tidal exchanges. The wetlands located many kilometers from any sources of saline water may be affected by tides. The hydrodynamic conditions imposed by a changing sea level decreases the flow velocity in rivers during high tides, and vice versa. This will result in changes in rivers upstream water level in tandem with the tides. The intensity of this effect decreases toward the river upstream depending on the local channel features.

51.4.5.4.2 Other Inputs

The coastal wetlands can be affected by adjacent freshwater inputs, especially in estuarine environments. As an example, temporal large stormwater inflows can result in rapid changes in the salinity, temperature, dissolved oxygen, and sediment concentration within the wetland. These inputs can be advantageous, such as historical sediment deposition in the Mississippi River delta region of southern Louisiana—in contrast to the current practice of diverting stormflows away from coastal wetlands, which has led to regional subsidence and salt water intrusion. However, these inputs can be destructive such as when including high amounts of urban wastewater and stormwater that alter the natural conditions.

Ground water inputs are also important in coastal wetlands and are in two forms of points such as springs and diffuse such as upward leakage from confined aquifers. Generally, characterizing the groundwater exchanges is more difficult than other sources. Radioisotopes can be used to differentiate between groundwater and other inputs [11].

51.4.5.4.3 Monitoring Coastal Wetlands

The geochemical information can be used to distinguish between tidal and freshwater and groundwater inputs. The tidal water inputs have high concentrations of sodium chloride but the freshwater inputs normally have lower specific conductivities and total dissolved solids. The geochemical features of groundwater inputs are dependent on their location but generally they are intermediate. For example, in carbonate aquifers, calcium signal can be observed. The water quality data can be used to determine the degree of tidal flushing and the residence times of water within the system. The fluxes estimations are used in the water balance equation.

51.4.6 Urban Wetland Hydrology Modeling

To manage and plan wetland systems within a highly urbanized watershed in a sustainable manner, a model of the site's existing hydrology is needed. Three approaches are commonly used for wetland hydrology modeling. These approaches include single event models, stochastic models, and comprehensive water budgets. The mass balance approach provides a framework for developing a water budget, which seeks to incorporate the parameters that control a wetland's hydrology.

Because urban wetland hydrology is subject to more highly fluctuating environmental conditions than a nonurbanized system, the application of a stochastic model to urban wetlands would be advantageous. In this way, the uncertainties can be incorporated in the modeling process. This approach contrasts with deterministic models that provide identical results using constant input parameters. Another alternative is to use a deterministic model with different inputs showing the range of variations to examine the range of possible conditions (e.g., dry conditions, wet conditions, average conditions).

In case of minimal or nonexistent groundwater interactions in urban wetland systems, nontraditional modeling approaches can be used. In these cases, the most effective modeling approach of wetland hydrology would be the application of a nonlinear reservoir method, such as the U.S. Environmental Protection Agency Storm Water Management Model (SWMM) [34].

51.5 Changes in Urban Wetland Hydrology

There are four types of alterations that can result in wetland degradation. These alternations include (1) geomorphic and hydrologic, (2) nutrient and contaminant, (3) harvest, extinction, and invasion, and (4) climate change [51]. These changes are further discussed in the following sections.

51.5.1 Natural Changes

Wetlands are the result of many geologic forces such as rivers flooding and flood plains inundation, glacier formation and movement, tectonic uplift and subsidence, and carbonate aquifers dissolution. Also, sediment can be moved out of natural channels by accelerated erosion, resulting in down-cutting and deepening of channels, which leads to a lowering of riparian water tables and the reduction of overland flows, both of which alter wetland saturation.

Wetlands can also modify their environment as they grow. Formed peats in wetlands can substantially alter the original landscape by filling in the depression in which they originally formed. There are other biological forces that result and help in wetlands formation. Beavers create impoundments in favorable habitats to their needs and form natural wetlands. The shallow wetlands can also be formed by large woody debris.

The wetlands can slowly age over time by filling in with external sources of materials (i.e., sediments from upland erosion) or detrital materials from wetland vegetation. The deposition rate of these materials can be slow (oligotrophic systems with small upstream catchment areas) or rapid (in nutrient-rich areas with large upstream areas with extensive erosion). Therefore, it can be said that natural wetlands are always being formed and lost regarding the natural forces balance.

Wetland hydrology is also changing during this time. The reduction of water storage volume within a wetland decreases the residence time. The depth and hydropattern are decreased by reduction in storage volume within the deep water areas that normally remain wet under drought conditions. The dynamic nature of wetlands hydrology means that wetlands cannot be studied apart from their regional environment and any hydrologic change upstream of the wetland affects wetland evolution.

51.5.2 Urbanization

The human development and urbanization have considerably increased hydrologic disturbances within watersheds [2] by increased sediment production and transport and increases in nutrient concentrations and loads. These changes naturally result in wetland stage reduction due to sediment trapping and nutrient uptake with subsequent deposition of organic sediments.

Urbanization has resulted in more surface water inflows to wetlands from urban, industrial, and agricultural areas and resulted in altered inflows because of effects of hydraulic structures, such as reservoirs, canals, levees, dikes, revetments, and jetties on natural hydrologic patterns.

Most of these effects are the result of efforts to drain wetlands. Outflows from wetlands are increased because of drainage ditches, channels and canals, construction or natural barriers, such as vegetation, removal, and by straightening streams. Some wetlands draining efforts use groundwater extraction techniques, such as underground tile drains and pumping wells that lower groundwater levels. By lowering the groundwater table, the subsurface drainage from the wetland is increased to the point of groundwater extraction. Ditches and tile drains increase the discharge of shallow groundwater, thus lowering water tables in the vicinity of the drain.

Water levels increase with distance away from the drains, reaching a maximum midway between the drains. Tile drainage systems increase the rate of shallow groundwater flow, thus favoring drier conditions within the wetland. Tile drainage systems are more effective for removing water resulting from low-intensity, long-duration storms, and are less effective for draining water resulting from short-duration, high-intensity storms [17].

Fields drainage for agricultural purposes decreases surface water inflows, lowers water tables, and reduces the soil saturation duration. Groundwater pumping near the wetland can lower shallow aquifer water levels, while irrigation increases groundwater levels, resulting in either decreases or increases in wetland water levels, respectively. The regional groundwater pumping can result in slow (even rapid in some cases) reductions in regional, confined aquifer groundwater levels. These changes may result in reductions in diffuse upward leakage or direct connections to wetlands and finally the reduction of wetland water level. The channels and canals alternation effect is two-fold. The new excavations increase the wetland outflow and, on the other hand, the accumulation of spoils (i.e., the materials removed from the excavated areas) near the excavations may concentrate or otherwise alter the natural wetland drainage system.

The wetlands clearly affect and are affected by biological factors such as vegetation, fish, and wildlife. Obstructions (such as beaver dams, roads, channels, and dams) to surface water exchanges can deprive the wetlands of natural flows. These changes may also result in requiring a higher stage to pass the same flow, which has both positive and negative effects on wetlands. Increased wetland water level alters the natural storage ability (a negative effect), but increases the residence time (a positive effect).

Some wetlands have an important function of flood flows mitigation by slowing the average water velocity as well as the flood peak velocity. This function is intensified by the additional friction to flow that wetlands provide and the increased water storage capacity associated with their area. This function of wetlands and damages of floods in urban areas have resulted in efforts to design and construct artificial wetlands [44]. For these purposes, the hydrodynamic models are used to achieve specific management goals that need reductions in nutrients, sediments, and peak flows.

Based on Manning's equation, several factors including the roughness of flooded ground, the hydraulic radius, and the water energy slope affect the water velocity. Wetlands with substantial macrophytic vegetation have higher hydraulic roughness and more decrease flow velocities. Shallow water bodies because of small hydraulic radius again decrease flow velocities. Increasing wetland loading rates results in less retention times. Therefore, the effectiveness of wetlands in storing water during flood periods is decreased.

If irrigation amount exceeds plant needs, it will result in water table increase. This situation happens in arid areas where surplus irrigation is required for removal of dissolved salts from the root zone. In these cases, water level may increase to the point where surface inundation results and saline wetlands form [31].

In coastal wetlands that are dominated by nearby freshwater inputs, occasional, large stormwater inflows can rapidly change the geochemical characteristics of the wetland such as the salinity, temperature, dissolved oxygen, and sediment concentration. Change in coastal morphology adversely affects natural wetlands

and increases the saltwater intrusion rates. Density-dependent stratification of estuarine waters prevents a saltwater presence in coastal areas. The salinity can be increased by construction of deep-water navigation channels.

Coastal ground water pumping reduces artesian pressures in underlying confined aquifers and results in less point and diffuse upward leakage. Furthermore, it results in considerable lowering of the ground surface relative to the sea level and intrusion of saline water into coastal wetlands. Pumping from shallow aquifers decreases the water level in the coastal zone and may result in local dewatering of coastal wetlands. Shallow disposal of septic wastes degrades the local groundwater quality and can affect local wetlands into which groundwater discharges.

51.5.3 Climate Changes

The climate change impacts should be considered in design and management of constructed wetlands. The main impacts would be on rainfall intensity, determinant of online wetlands peak flows, runoff volume, and sea level rise (for estuarine wetlands). The general investigations of climate change impacts on hydrology show that variability of rainfall intensities will increase, the average annual runoff volumes would remain almost constant, and due to temperature increase, the evaporation during the interevent rainfall periods would increase.

The recent investigations have demonstrated the increase in sea level due to climate change impacts. The sea level rise will result in

- More tidal inundation of land by seawater
- Recession of beach and dune systems
- Changes in tides behavior within estuaries
- Saltwater extending further upstream in estuaries
- Higher saline water tables in coastal areas
- Increased coastal flood levels due to a reduced ability to effectively drain low-lying coastal areas

The projections also show increasing intensity of storms, which their storm surge will result in further erosion. The integration of impacts of extreme weather conditions and the increased sea level is a complex problem and is dependent on local topographic and bathymetric conditions. The important point is that all of these changes would affect the wetlands performance and should be considered in their planning and management as well as deign of constructed wetlands [20].

51.6 Summary and Conclusions

Wetlands are aquatic ecosystems with different components of plants, animals, and soils adapted to survive in permanent inundation conditions even though they may remain dry for years. Wetlands water can be static or flowing, fresh or salty. Urban wetlands are those wetlands placed within a city's boundaries and could be natural, constructed, or a mixture of both. Previously, the high values of wetlands were not well understood and that is why numerous wetlands around the world drained.

In recent years, high attention is paid to different benefits and services that urban wetlands provide to a community.

Urban wetlands provide habitat for plants and animals, water storage, and improve water quality. They also protect against natural hazards such as flood and fire and erosion of river banks and coastlines. The stored water in wetlands during floods can be used for different purposes such as green space irrigation as well as groundwater recharge. They can also function as natural treatment systems especially for surface runoff if the required situations are provided.

Urbanization highly affects the wetlands through direct habitat loss, altered water regime, entrance of high loads of pollution including toxic compounds, introduction of exotic species, and different ways of ecosystem modifications such as altered fire regimes, dieback, and changes in salinity. The functions of wetlands are highly dependent on their hydrology; therefore, understanding of their hydrology is needed for efficient planning and management of these systems. For this purpose, intense and continuous monitoring of the hydro climatic condition of the wetlands is necessary. The wetlands management should be in a way that causes the least disturbance in natural functions and processes of wetlands along with satisfying the defined objectives for their operation in a sustainable fashion.

References

1. Acosta, C.A. and Perry, S.A. 2001. Impact of hydropattern disturbance on crayfish population dynamics in the seasonal wetlands of Everglades National Park, USA. *Aquatic Conservation-Marine and Freshwater Ecosystems*, 11(1), 45–57.

2. Azous, A.L. and Horner, R.R. (eds.) 2001. *Wetlands and Urbanization: Implications for the Future.* Lewis Publishers, Boca Raton, FL.

3. Batiuk, R.A., Orth, R.J., Moore, K.A., Dennison, W.C., Stevenson, J. C., Staver, L.W., Carter, V. et al. 1992. *Chesapeake Bay Submerged Aquatic Vegetation Habitat Requirements and Restoration Targets: A Technical Synthesis.* EPA, Annapolis, MD.

4. Black, P.E. 1996. *Watershed Hydrology*, 2nd Edition. Lewis Publishers, Washington, DC.

5. Boon, P. 2013. Hydrology of urban freshwater wetlands. In: Paul, S. (ed.), *Work Book for Managing Urban Wetlands in Australia.* Sydney Olympic Park Authority, ebook available through www.sopa.nsw.gov.au\education\WETebook\, ISBN 978-0-9874020-0-4.

6. Boon, P.I. 2006. Biogeochemistry, ecology and management of hydrologically dynamic wetlands. In: Batzer, D.P. and Sharitz, R.R. (eds.). *Ecology of Freshwater and Estuarine Wetlands.* University of California Press, Berkeley, pp. 115–176.

7. Bougon, N., Auterives, C., Aquilina, L., Marmonier, P., Derrider, J. et al. 2011. Nitrate and sulphate dynamics in peat subjected to different hydrological conditions: Batch experiments and field comparison. *Journal of Hydrology*, 411, 12–24.

8. Brinson, M.M. 1993. A Hydrogeomorphic Classification for Wetlands. Wetlands Research Program Technical Report WRP-DE-4, Army Corps of Engineers, Waterways Experiment Station, Vicksburg, MS.

9. Burford, M.A., Cook, A.J., Fellows, C.S., Balcombe, S.R., and Bunn, S.E. 2008. Sources of carbon fuelling production in an arid floodplain river. *Marine and Freshwater Research*, 59, 224–234.

10. Carslaw, H.S. and Jaeger, J.C. 1959. *Conduction of Heat in Solids*, 2nd Edition. Clarendon Press, Oxford University Press, New York.

11. Charette, M. A., Splivallo, R., Herbold, C., Bollinger, M.S., and Moore, W.S. 2003. Salt marsh submarine groundwater discharge as traced by radioisotopes. *Marine Chemistry*, 84, 113–121.

12. Credit Valley Conservation. 2010. Monitoring Wetland Integrity within the Credit River Watershed. Chapter 1: Wetland Hydrology and Water Quality 2006-2008. Credit Valley Conservation. vii + 79 p., Meadowvale, ON, Canada.

13. Dennison, W.C., Orth, R.J., Moore, K.A., Stevenson, J.C., Carter, V., Kollar, S., Bergstrom, P.W., and Batiuk, R.A. 1993. Assessing water quality with submersed aquatic vegetation. *BioScience*, 43, 86–94.

14. Duchatel, K. and Pratten, A. 2012. Maintenance regime around freshwater wetlands. In: Paul, S. (ed.). *Work Book for Managing Urban Wetlands in Australia.* Sydney Olympic Park Authority, ebook available through www.sopa.nsw.gov.au\education\WETebook\, ISBN 978-0-9874020-0-4.

15. Ewel, K.C. 1990. Multiple demands on wetlands. *BioScience*, 40, 660–666.

16. Fetter, C.W. 2001. *Applied Hydrogeology*, 4th Edition. Prentice Hall, New York.

17. Galatowitsch, S.M. and van der Valk, A.G. 1994. *Restoring Prairie Wetlands: An Ecological Approach.* Iowa State University Press, Ames, IO.

18. Grand River Conservation Authority. 2003. *Wetlands Policy.* Grand River Conservation Authority, Cambridge, Ontario.

19. Harms, T.K. and Grimm, N.B. 2012. Responses of trace gases to hydrologic pulses in desert floodplains. *Journal of Geophysical Research—Geosciences*, 117, G01035.

20. Hunter, G. 2013. Constructed wetlands: Design, construction and maintenance considerations. In: Paul, S. (ed.), *Work Book for Managing Urban Wetlands in Australia.* Sydney Olympic Park Authority.

21. Irwin, D.A. 1994. The regulatory framework used to evaluate the impacts of borrow pits on adjacent wetlands in Florida. *AWRA Symposium Effects of Human Induced Changes on Hydrological Systems*, Jackson Hole, WY. June 26–29, pp. 455–464.

22. Jennings, M.E., Thomas, W.O., and Riggs, H.C. 1994. Nationwide summary of US Geological Survey regional regression equations for estimating magnitude and frequency of floods for ungaged sites, 1993. U.S. Geological Survey, WRI 94-4002.

23. Kennish, M.J. 1992. *Ecology of Estuaries: Anthropogenic Effects.* CRC Press, Ann Arbor, MI.

24. King, R.S., Richardson, C.J., Urban, D.L., and Romanowicz, E.A. 2004. Spatial dependency of vegetation-environment linkages in an anthropogenically influenced wetland ecosystem. *Ecosystems*, 7(1), 75–97.

25. Lambou, V.W., Kuperberg, J.M., Moerlins, J.E., Herndon, R.C., and Gebhard, R.L. 1990. *Proximity of Sanitary Landfills to Wetlands: An Evaluation and Comparison of 1,153 in 11 States.* EPA/600/S4-90/012. EPA, Cincinnati, OH.

26. Law, A.M. and Kelton, W.D. 1991. *Simulation Modeling and Analysis*, 2nd Edition. McGraw-Hill, Inc., New York.

27. Mack, J.J. and Micacchion, M. 2007. An ecological and functional assessment of urban wetlands in central Ohio. Volume 1: condition of urban wetlands using rapid (level 2) and intensive (level 3) assessment methods. Ohio EPA Technical Report WET/2007-3A. Ohio Environmental Protection Agency, Wetland Ecology Group, Division of Surface Water, Columbus, OH.

28. Maley, M. and Peters, J. 1999. Hydrogeologic Evaluation of the Springtown Alkali Sink, California. In: Means, J.L. and Hinchee, R.E. (eds.), *Wetlands & Remediation: An International Conference, Proceedings of Conference Held in Salt Lake City, Utah,* November 16–17, 1999, Battelle Press, Columbus, OH, pp. 79–86.

29. McCuen, R.H. 1998. *Hydrologic Analysis and Design.* Prentice Hall, Upper Saddle River, NJ.

30. Mitsch, W.J. and Gosselink, J.G. 2000. *Wetlands*, 3rd Edition. John Wiley & Sons, Inc., New York.

31. Mitsch, W.J. and Gosselink, J.G. 2007. *Wetlands*, 4th Edition. John Wiley & Sons, Inc., New York.

32. Mitsch, W.J., Nahik, A., Wolski, P., Bernal, B., and Zhang, L. 2010. Tropical wetlands: Seasonal hydrological pulsing, carbon sequestration, and methane emissions. *Wetlands Ecology and Management*, 18, 573–586.

33. NSW Government. 1996. *The NSW Wetlands Management Policy.* Department of Land and Water Conservation, Sydney, Australia.

34. Obropta, C., Kallin, P., Mak, M., and Ravit, B. 2008. Modeling urban wetland hydrology for the restoration of a forested riparian wetland ecosystem. *Urban Habitats*, 5(1), 183–198.

35. Orme, A.R. 1990. Wetland morphology, hydrodynamics and sedimentation. In: Williams, M. (ed.), *Wetlands: A Threatened Landscape.* The Institute of British Geographers, The Alden Press Ltd., Osney Mead, Oxford, Great Britain, pp. 42–94.

36. Peach, M. and Zedler, J.B. 2006. How tussocks structure sedge meadow vegetation. *Wetlands*, 26, 322–335.

37. Platt, R.H., Rowntree, R.A., and Muick, P.C. (eds.). 1994. *The Ecological City.* University of Massachusetts Press, Amherst, MA.

38. Raulings, E., Morris, K., Roache, M., and Boon, P.I. 2010. The importance of water regimes operating at small spatial scales for the diversity and structure of wetland vegetation. *Freshwater Biology*, 55, 701–715.

39. Schueler, T.R. 1992. *Design of Stormwater Wetland Systems: Guidelines for Creating Diverse and Effective Stormwater Wetlands in the Mid-Atlantic Region.* Metropolitan Washington Council of Governments, Washington, DC.

40. Stribling, J.M., Cornwell, J.C., and Glahn, O.A. 2007. Microtopography in tidal marshes: Ecosystem engineering by vegetation? *Estuaries and Coasts*, 30, 1007–1015.

41. Topp, G.C. 1980. Electromagnetic determination of soil water content: Measurements in coaxial transmission lines. *Water Resources Research*, 16, 574–582.

42. U.S. EPA. 2008. *Methods for Evaluating Wetland Condition: Wetland Hydrology.* Office of Water, U.S. Environmental Protection Agency, Washington, DC. EPA-822-R-08-024.

43. Vymazal, J. 1995. *Algae and Element Cycling in Wetlands.* Lewis Publishers, Boca Raton, FL.

44. Walker, W.W. and Kadlec, R.H. 2002. Dynamic model for stormwater treatment areas. http://www.wwwalker.net/dmsta.

45. Walters, C. 1986. *Adaptive Management of Renewable Resources.* Macmillan, New York.

46. Warfe, D.M., Pettit, N.E., Davies, P.M., Pusey, B.J., and Hamilton, S.K. 2011. The 'wet-dry' in the wet-dry topics drives river ecosystem structure and processes in northern Australia. *Freshwater Biology*, 56, 2169–2195.

47. Wetzel, P.R., Sklar, F.H., Coronado, C.A., Troxler, T.G., Krupa, S.L. et al. 2011. Biogeochemical processes on tree islands in the Greater Everglades: Initiating a new paradigm. *Critical Reviews in Environmental Science and Technology*, 41, 670–701.

48. Williams, C., Firehock, K., and Vincentz, J. 1996. *Save Our Streams: Handbook for Wetlands Conservation and Sustainability.* Izaak Walton League of America, Gaithersburg, MD.

49. Winter, T.C. 1988. A conceptual framework for assessing cumulative impacts on the hydrology of nontidal wetlands. *Environmental Management*, 12, 605–620.

50. Wolf, K.L., Ahn, C., and Noe, G.B. 2011. Mictotopography enhances nitrogen cycling and removal in created mitigation wetlands. *Ecological Engineering*, 37, 1398–1406.

51. Zedler, J.B. and Kercher, S. 2005. Wetland resources: Status, trends, ecosystem services, and restorability. *Annual Review of Environment and Resources*, 30, 39–74.

52. Zentner, J. 1994. Enhancement, restoration and creation of freshwater wetlands. In: Kent, D.M. (ed.), *Applied Wetlands Science and Technology*, Chapter 7. CRC Press, Boca Raton, FL.

52

Wetland Systems and Removal Mechanisms

Banti C. Dimitra
Aristotle University
of Thessaloniki
Technological Education
Institute of Thessaloniki

Samaras Petros
Technological Education
Institute of Thessaloniki

Zouboulis Anastasios
Aristotle University
of Thessaloniki

PREFACE

Utilization of constructed wetlands for the last two decades indicates significant progress took place on the operation and efficiency of constructed wetland systems including free water surface constructed wetlands and sub-surface flow constructed wetlands; as a result, their application in wastewater treatment has been increased and expanded worldwide. Wetlands represent a cost-effective and technically reliable option for wastewater and stormwater treatment. In addition to acceptable water quality effluent, constructed wetlands are harmonized to the surrounding natural environment. However, constructed wetlands present a few drawbacks. Therefore, it is essential to understand their operation mechanisms, in order to optimize their efficiency for pollutants removal. This chapter constitutes a useful approach containing information about constructed wetland systems and the corresponding processes taking place by and due to the wetland plants. Emphasis is given on the reduction mechanisms of various parameters by the wetland plants, such as organic pollutants, nitrogen, phosphorus, pathogen microorganisms, and metals.

If one way be better than another, that you may be sure is nature's way.

Aristotle

52.1 Introduction

Significant work for development, implementation, and operation of constructed wetland systems (CWSs) has been performed in the last 20 years, resulting in their broad and increasing application: In 1996 more than 20,000 constructed wetland systems were operated worldwide, while recently (2011) their number is estimated to exceed 50,000 [62].

Constructed wetland systems are used for the treatment of wastewater from individual houses or groups of houses, while at a larger scale they are mainly used for the secondary treatment of domestic and municipal sewage [13]. Moreover, they can be used for the tertiary treatment of urban wastewater or for the treatment of stormwater runoff. Their use has been expanded for the treatment of other wastewater types, such as industrial, agricultural wastewaters, and landfill leachate. As many of these wastewaters are difficult to be treated in a single stage system, hybrid systems consisting of various types of constructed wetlands staged in series have been introduced [57]. Their rapid spread is because they constitute an environmentally friendly and cost-effective solution [13].

Research on wetlands is developing rapidly and the topics that are mainly examined by the researchers, toward their performance optimization, include the systems' operating parameters, such as flow characteristics (surface flow, subsurface vertical flow, subsurface horizontal flow), the loading and the substrate (soil or gravel) [32]. Other parameters include the role of various wetland plants in pollutants removal, the removal mechanisms, and the effect of climatic conditions [13,58,61].

52.2 Wetland Systems for Wastewater Treatment

52.2.1 Advantages of Wetlands for Wastewater Treatment

Wetlands constitute a cost-effective and technically reliable option for wastewater and stormwater treatment. Their great advantages include their simple, easy and inexpensive installation, and the low maintenance and operating costs, for both material supply and energy consumption [46,50]. Furthermore, they require intermittent and not continuous personnel employment to operate the facility. Nevertheless, wetlands can be efficiently operated under significant flow variations while they can be handled as hydrologic buffer or equalization tanks, for the simultaneous pathogen removal [50]. A good quality effluent is produced at the outlet of constructed wetlands, allowing the water reuse for various applications, for example, for irrigation. Other reasons favoring the spreading of wetlands are that they improve aesthetically the open spaces, fit harmoniously into the environment [25], and provide opportunities for environmental education, leisure, and exercise activities [34].

However, constructed wetlands display a number of disadvantages, such as low organic loadings are elaborated, requiring a greater land area compared to conventional activated sludge systems; and in winter or generally in cold regions their yield is often not satisfactory, as the low temperature reduces the rate of natural metabolic activities of microbes and macrophytes [36].

52.2.2 Natural and Constructed Wetlands

Natural wetlands are defined as transitional areas between the terrestrial and aquatic systems (Figure 52.1). These areas, due to their location, are humid for a long period during the year [51].

On the other hand, *constructed wetlands* have been manufactured by humans in order to serve specific purposes. The types of the constructed wetlands are the following [51]

- *Constructed habitat wetlands.* These wetlands are constructed to compensate for and help offset the rate of conversion of natural wetlands resulting from agriculture and urban development.
- *Constructed flood control wetlands.* These wetlands are constructed to act as a flood control facility.
- *Constructed aquaculture wetlands.* These wetlands are constructed to be used for production of food and fiber.
- *Constructed treatment wetlands.* These wetlands are constructed to act mainly as a wastewater treatment system and to improve water quality.

This chapter deals with constructed treatment wetlands, which hereafter will be called by the term constructed wetlands (CW).

CWs are large shallow tanks (with depths usually less than 1 m) filled by wastewater. The wastewater is treated by a combination of physical, chemical, and/or biological processes in order to produce a high-quality effluent that can be discharged safely to an aquifer or can be reused [11,22].

CW systems used for wastewater treatment consist of four major components [51]:

1. Wetland vegetation
2. Media or substrate supporting vegetation
3. Water column (in or above the media)
4. Living organisms

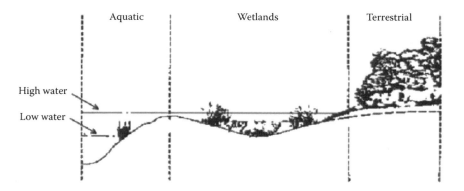

FIGURE 52.1 Formation of natural wetlands zone. (Adapted from Sundaravadivel, M. and Vigneswaran, S. 2004. *Wastewater Recycle, Reuse and Reclamation, Encyclopedia of Life Support Systems (EOLSS), Developed Under the Auspices of the UNESCO.* Eolss Publishers, Oxford, UK. http://www.eolss.net.)

52.2.3 Types of Constructed Wetland Systems

The basic types of constructed wetlands are

1. Free water surface constructed wetlands—FWS CWs
2. Sub-surface flow constructed wetlands—SSF CWs
   a. Sub-surface horizontal flow constructed wetlands—SHF CWs
   b. Sub-surface vertical flow constructed wetlands—SVF CWs

52.2.3.1 Free Water Surface Constructed Wetlands: FWS CWs

The free water surface wetland systems simulate the hydrological regime of natural wetlands, where wastewater flows above the surface of the wetland and evapotranspiration and infiltration is taking place in the "body" of the wetland. In these systems, the wastewater, which is going to be treated, is often mixed with surface water or purified effluent and, in general, flows through the system with a minimum residence time of 10 days [54]. In free water surface systems, various types of vegetation are used, such as free-floating, floating-leaved, submerged, and emergent ones [32,59].

CWs with free-floating macrophytes, for example, *Eichhornia crassipes* (Figure 52.2), consist of one or more shallow ponds, where plants float on the surface [59].

CWs with emergent macrophytes, for example, *Typha* spp., consist of one or more shallow sealed sequenced basins. They contain 20–30 cm of rooting soil and the depth of the water reaches 20–40 cm [59]. The vegetation is usually dense and covers more than 50% of the water surface. In addition to macrophytes, naturally occurring species may also be used [25]. The water flows in the horizontal direction on the surface of soil material. A free water surface wetland with emergent macrophytes is presented in Figure 52.3, while the configuration of a typical free water surface wetland system is shown in Figure 52.4.

52.2.3.2 Sub-Surface Flow Constructed Wetlands: SSF CWs

At the subsurface flow systems, the wastewater is discharged within the "body" of the wetland, which is always soil; influents are gradually processed by microorganisms lying around the rhizomes of plants. Plants have a natural mechanism that discharges the air from the atmosphere to the roots, creating aerobic conditions around the roots, and enhancing the establishment of communities of aerobic microorganisms, degrading the organic compounds [7].

There are two types of SSF constructed wetlands, depending upon the flow regime, the horizontal flow (HF) and the vertical flow systems (VF). SSFs are primarily built in HF configuration. The horizontal CWs show an efficient denitrification performance even at low C/N ratio [32]. However, the increased requirement for ammonia removal resulted in the development of VF wetlands, which are intermittently fed. They allow high oxygenation rate of the bed and consequently high nitrification capacity [32,59].

52.2.3.2.1 Sub-Surface Horizontal Flow Constructed Wetlands: SHF CWs

Systems of SHF wetlands are constructed in sealed basins with average depths of 40–60 cm. In these basins, inert materials are placed, with grain diameter sizes of 2.5–5 cm. Appropriate aquatic vegetation is planted and grown, usually consisting of reeds or rushes (Figure 52.5). The bottom slope is approximately 1%–2% and the wastewater enters into the system and flows in an underground horizontal direction (Figure 52.6). The required area for their orderly function ranges at about 6–7 m²/Person Equivalent-PE [14,30].

FIGURE 52.2 Constructed wetland with free-floating macrophytes (water hyacinth, *Eichhornia crassipes*) as a part of a treatment system in Yantian Industry Area in Baoan District, Shengzhen City, South China. (Adapted from Vymazal, J. 2008. *Proceedings of Taal 2007: The 12th World Lake Conference*, 28 October–2 November 2007, Jaipur, pp. 965–980.)

FIGURE 52.3 Free water surface constructed wetland with emergent vegetation for stormwater runoff treatment in Plumpton Park, NWS, Australia. (Adapted from Vymazal, J. 2008. *Proceedings of Taal 2007: The 12th World Lake Conference*, 28 October–2 November 2007, Jaipur, pp. 965–980.)

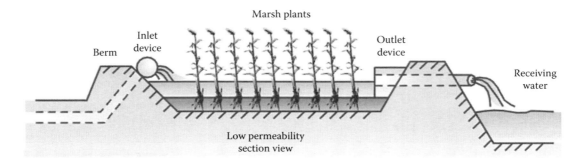

FIGURE 52.4 Configuration of a typical free water surface wetland system. (Adapted from Sim, C.H. 2003. *The Use of Constructed Wetlands for Wastewater Treatment*. Wetlands International, Malaysia Office, p. 24.)

FIGURE 52.5 Sub-surface horizontal flow CW for municipal sewage treatment at Gronfeld, Denmark. (Adapted from Vymazal, J. 2008. *Proceedings of Taal 2007: The 12th World Lake Conference*, 28 October–2 November 2007, Jaipur, pp. 965–980.)

52.2.3.2.2 Sub-Surface Vertical Flow Constructed Wetlands: SVF CWs

In VF CWs the basins have a depth of approximately 0.90–1.20 m and average bottom slope of about 1%. Flow takes place by means of gravity. The basins are sealed and filled with sand and gravel (Figure 52.7). The addition of wastewater is carried out by perforated pipes (usually under pressure), which are distributed, as a network, on the surface of the filling material. The wastewater is collected by perforated drainage pipe systems, distributed at the bottom of the reservoir (Figure 52.8), with approximately 1 m distance between the pipes. Vertical tubes are placed between the perforated pipes at 2 m distance, for oxygen supply. These systems are swamped periodically by a large sewage flow. The required area is approximately 1–2.5 m^2/PE [1,14,30].

52.3 Pollutants Removal Mechanisms

52.3.1 Removal of Organics

The quality control of the organic loading is performed by using standard measurement of organic compounds, such as

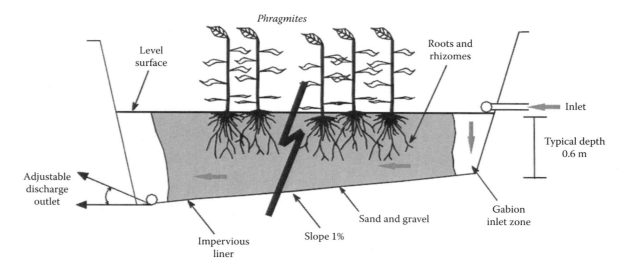

FIGURE 52.6 Configuration of a typical subsurface horizontal flow CW. (Adapted from Irish Environmental Protection Agency. 2000. *Wastewater Treatment Manuals-Treatment Systems for Single Houses*. EPA Publications, Wexford, Ireland.)

FIGURE 52.7 Subsurface vertical flow CW during its construction at Bojna, Slovakia. (Adapted from Vymazal, J. 2008. *Proceedings of Taal 2007: The 12th World Lake Conference*, 28 October–2 November 2007, Jaipur, pp. 965–980.)

biochemical oxygen demand (BOD₅), chemical oxygen demand (COD) and total organic carbon (TOC). In constructed wetlands, chemical and biological processes convert and/or disintegrate the organic contaminants and often lead to different types of substances. Therefore, the measurement of BOD₅ or COD at the effluent of the system does not represent the existence of the same organic components in the influent of the system [25].

Generally, BOD₅ removal efficiency is lower in FWS wetland systems than SSF wetland systems [30].

In SHF wetland systems, the organics are degraded by aerobic and/or anaerobic bacteria that are adherent to the underground bodies of the plants and on the surface of the filling material as well as by sedimentation and filtration of particulate organic matter [25,65].

Under high organic loading flows, continuous saturation of the filtration bed is taking place and anoxic/anaerobic processes are prevailing. Therefore, aerobic processes are restricted to short zones close to the plant roots and rhizomes and to a thin surface layer where oxygen diffusion from the atmosphere may occur [63]. In wetlands receiving low load, dissolved oxygen may be transferred by inflow wastewater.

Aerobic degradation of soluble organic matter is carried out by the aerobic heterotrophic bacteria according to the reaction:

$$C_6H_{12}O_6 + 6O_2 \rightarrow 6CO_2 + 6H_2O + 12e^- + \text{energy} \quad (52.1)$$

In most types of wastewaters, except some industrial wastewaters and runoff waters, the supply of dissolved organic matter is sufficient, and aerobic degradation is limited by dissolved oxygen concentration [56].

Organic matter is composed of a complex mixture of biopolymers [38]. Some of these compounds, such as proteins, carbohydrates, and lipids are easily degraded by microorganisms, while other compounds, such as lignin and hemicellulose, are resistant to decomposition [63]. Biopolymers are degraded by a multistep process (Figure 52.9). Firstin, a first stage, microorganisms degrade polymers to simplified monomers such as amino acids, fatty acids, and monosaccharides [38]. The primary endproducts of fermentation are fatty acids such as acetic (Equation 52.2),

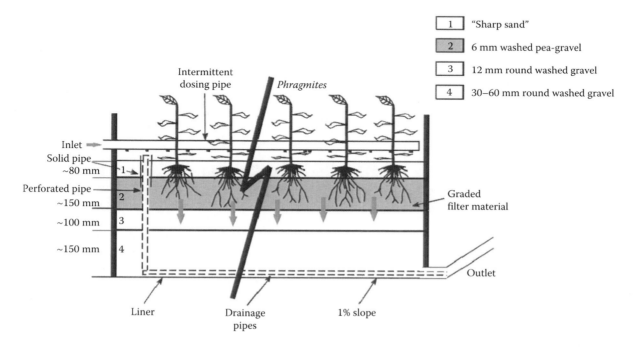

FIGURE 52.8 Configuration of a typical subsurface vertical flow CW. (Adapted from Irish Environmental Protection Agency. 2000. *Wastewater Treatment Manuals-Treatment Systems for Single Houses*. EPA Publications, Wexford, Ireland.)

butyric, and lactic acids, alcohols, and gases such as CO_2 and H_2 [38,41,45,55,63]:

$$C_6H_{12}O_6 \rightarrow 3CH_3COOH + H_2 \qquad (52.2)$$

Then, primary fermentation products are mineralized to CO_2 and CH_4, or they undergo secondary fermentation to smaller volatile fatty acids. Methane is formed primarily from the oxidation of H_2 coupled to CO_2 reduction or by the fermentation of acetate (Figure 52.9). Acetate is formed by primary fermentation, acetogenesis from H_2/CO_2 and by secondary fermentation of primary fermentation products [23,55,63].

It has been found that the highest removal BOD_5 and COD efficiencies are achieved in systems treating municipal wastewater ($BOD_5 = 80.7\%$ and COD $= 63.2\%$) while the lowest efficiency has been recorded for landfill leachate ($BOD_5 = 32.8\%$ and COD $= 24.9\%$) in SHF wetlands [63].

Macrophytes have been observed to efficiently affect the performance of SHF wetlands, between the various vegetation species examined [63]. With respect to the seasonal effect on BOD_5 and COD removal in SHF wetlands, conflicting studies have been reported, ranging from those indicating little or negligible seasonal effects, to those suggesting strong seasonal dependence [63].

In SVF wetland systems, the main mechanism for removal of organic pollutants is the following [44]: The *first step* is the

adsorption of organic pollutants on wetland substrate. The adsorption level depends mainly on the particular characteristics of the substrate and the organic loading. The substrate may be saturated; however, the biological degradation of adsorbed organic matter that follows usually regenerates the substrate's adsorptive capacity. The regeneration of substrate's adsorptive capacity depends on the good oxygenation conditions, and therefore, it is influenced by the recovery periods of the wetland. The identification of the best time intervals of charge and recovery regimes becomes an important parameter for the removal of organic pollutants by VF wetlands [21,42].

The *second step*, after the adsorption of the organics, is biological degradation, which takes place with the contribution of microorganisms. Microorganisms are developed in the form of thin biological films (biofilms) or they are excreted on the surfaces of the substrate and/or the underground parts of vegetation. Biological degradation is performed in two subsequent phases. Initially the hydrolysis of agglomerated organic matter takes place, followed by oxidation [21,25,42].

Toxic organics added to wetlands undergo similar pollutant removal processes as natural organic matter, such as aerobic and anaerobic microbial breakdown, vegetative uptake, volatilization, phytolysis, chemical hydrolysis, sorption, and burial in the soil. The extent of these processes depends on the type of compounds and the biological and chemical conditions in the soil water column, such as pH, temperature, light intensity, nutrient and electron acceptor availability, and organic matter content [44].

52.3.2 Removal of Nitrogen

Several processes are taking place for the removal or retention of nitrogen during treatment by CWs, including NH_3 volatilization, nitrification, denitrification, nitrogen fixation, plant and microbial uptake, mineralization (ammonification), nitrate reduction to ammonium (nitrate-ammonification), anaerobic ammonia oxidation (ANAMMOX), fragmentation, sorption, desorption, burial, and leaching (Table 52.1) [58]. The basic transformation mechanisms occurring in CWs are ammonification, nitrification, denitrification, uptake, and assimilation by plants [24].

However, only a few of the processes ultimately remove total nitrogen, while most processes convert nitrogen to its various forms. Mechanisms that ultimately remove nitrogen from wastewaters are ammonia volatilization, denitrification, plant uptake (with biomass harvesting), ammonia adsorption, ANAMOX, and organic nitrogen burial [58]. The rest of the processes convert nitrogen among other nitrogen forms, but do not actually remove nitrogen from wastewater. For instance, ammonification converts organic nitrogen to ammonia raising the amount of ammonia in the system. On the other hand, ammonia then becomes available to other processes, such as microbial uptake or assimilation of plants.

Removal of total nitrogen by several types of constructed wetlands varied between 40% and 55% corresponding to 250 and 630 g N/m² year depending on CW type and inflow loading [58].

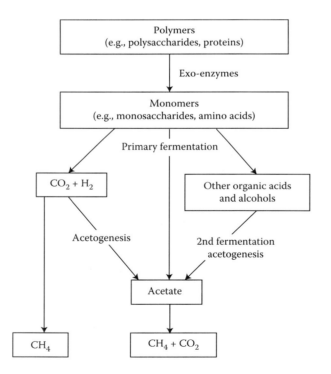

FIGURE 52.9 Metabolic scheme for the degradation of biopolymers, resulting in methanogenesis. (Adapted from Megonigal, J.P., Hines, M.E., and Visscher, P.T. 2004. *Biogeochemistry.* Elsevier-Pergamon, Oxford, UK, pp. 317–424.)

TABLE 52.1 Potential Magnitude of Nitrogen Transformations in Different Types of Constructed Wetlands

	Transformation	FFP	FWS	HSSF	VSSF
Volatilization	**Ammonia-N (aq) $\rightarrow$ ammonia-N (g)**	**Low**	**Medium**	**Zero**	**Zero**
Ammonification	Organic-N $\rightarrow$ ammonia-N	High	High	High	High
Nitrification	Ammonia-N (aq) $\rightarrow$ nitrite-N $\rightarrow$ nitrate-N	Low	Medium	Very low	Very high
Nitrate-ammonification	Nitrate-N $\rightarrow$ ammonia-N	?	?	?	?
Denitrification	**Nitrate-N $\rightarrow$ nitrite-N $\rightarrow$ gaseous N_2, N_2O**	**Medium**	**Medium**	**Very high**	**Very low**
N_2 fixation	Gaseous N_2 $\rightarrow$ ammonia-N (organic-N)	?	?	?	?
Microbial uptake (assimilation)	Ammonia-, nitrite-, nitrate-N $\rightarrow$ organic-N	Low	Low	Low	Low
Plant uptake (with harvest)		**Medium**	**Low**	**Low**	**Low**
Ammonia adsorption		Zero	Very low	Very low	Very low
Organic nitrogen burial		**Very low**	**Low**	**Low**	**Very low**
Fragmentation and leaching		?	?	?	?
ANNAMOX (anaerobic ammonia oxidation)	**Ammonia-N $\rightarrow$ gaseous N_2**	**?**	**?**	**?**	**?**

Source: Adapted from Vymazal, J. 2007. *Science of the Total Environment*, 380, 48–65.

Note: Words in bold indicate the processes that thoroughly remove total nitrogen from wastewater.

Abbreviations: FFP, free-floating plants CWs; FWS, free water surface CWs with emergent plants; HSSF, horizontal sub-surface flow CWs; VSSF, vertical sub-surface flow CWs; "?", Information does not exist.

The processes are heavily influenced by system's inflow loading and the alternation of feeding and relaxing periods in a CW [30]. Furthermore, they are affected in different extents among the variety of the systems, as shown in Table 52.1.

Single-stage constructed wetlands cannot achieve high removal rates of total nitrogen because they cannot provide simultaneously aerobic and anaerobic conditions. Specifically, free-floating plant CWs do not afford soil processes, free water surface CWs have very limited soil processes, and subsurface flow CWs present process deficiency in the free water zone. Subsurface horizontal flow systems provide the appropriate conditions for denitrification but they are disadvantageous in nitrifying ammonia [58]. On the other hand, subsurface vertical flow systems achieve high levels of nitrification due to their ability to transport oxygen [12]. They are also very efficient in organic nitrogen oxidation The presence of oxygen, in combination with wastewater flood feeding, contribute to the conversion of ammonia-N into nitrate via nitrifying bacteria; therefore, ammonia-N is successfully removed [52,58]. Due to the lack of anoxic zones with sufficient organic carbon for the activity of heterotrophic denitrification microorganisms, the biological denitrification is not favored, and nitrogen is released mainly as nitrate-N, resulting finally in limited overall removal of nitrogen [2]. Therefore, different types of CWs may be combined with each other, rather than exploit the advantages of the individual systems.

Removal of nitrogen via harvesting of aboveground biomass of emergent vegetation is low, but it could be substantial for lightly loaded systems, such as 100–200 g N/m^2 year [58]. Moreover, systems with free-floating plants may achieve higher removal of nitrogen via harvesting due to multiple harvesting schedules.

52.3.3 Removal of Phosphorous

Phosphorus is present in the water in the form of orthophosphates, organic and inorganic phosphorus [34,58]. It is found in wetlands as part of sediments. The phosphorus cycle in wetlands is shown in Figure 52.10. The soil phosphorus cycle is completely different from the N cycle, as there are no valency changes during biotic assimilation of inorganic P or during decomposition of organic P by microorganisms [58].

Transformation of phosphorus during wastewater treatment in CWs includes adsorption of dissolved phosphorus by porous media, desorption, precipitation with metals (e.g., Fe, Al, Mn), dissolution, plant and microbial uptake, fragmentation, leaching, mineralization, sedimentation (peat accretion), and burial [16,24,30,39,58].

The major phosphorus removal processes are sorption on antecedent substrates, precipitation, plant uptake, and the formation and accretion of new sediments and soils [30]. The potential magnitude of phosphorus removal in various types of CWs is given in Table 52.2. Among these processes, the first three present a maximum saturation capacity, that is, they have a finite capacity and therefore, long-term, sustainable P removal is not justified [17,30,40,58]. Nevertheless, soil accretion may be effective only in FWS CWs with emergent vegetation, where there is high production of biomass and water overlying the sediment [58].

The adsorption and precipitation of phosphorus take place mainly in subsurface flow CWs, as these mechanisms are effective in systems where wastewater is in contact with filtration substrate [16]. Between these systems, HF units have higher removal potential, as the substrate is constantly flooded and there is not much fluctuation in redox potential in the bed. On the other hand, VF systems, where wastewater is fed intermittently, present lower removal potential, as the oxygenation of the bed may cause the desorption and the subsequent release of phosphorus [58]. Nevertheless, materials that are often used for subsurface flow CWs, such as washed gravel or crushed rock, provide very low capacity for sorption and precipitation. For this reason, several filtration materials have been tested recently in constructed

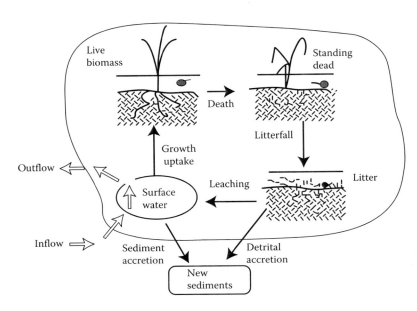

FIGURE 52.10 Phosphorus cycle in wetlands. (Adapted from Interstate Technology & Regulatory Council (ITRC), Wetlands Team. 2003. *Technical and Regulatory Guidance Document for Constructed Treatment Wetlands.* p. 199.)

wetlands, such as LECA (light weight clay aggregates) [58]. Thus, removal of phosphorus in various types of CWs is low, unless special substrates with high sorption capacity are used.

In general, removal of total phosphorus varies between 40% and 60% in all types of constructed wetlands, with load ranging between 45 and 75 g N/m² year depending on CW type and inflow loading [58].

Plant uptake is the primary phosphorus removal mechanism in free-floating macrophytes CWs. The development of an efficient harvesting frequency is important rather than keeping macrophytes at the optimum growth stage, for high phosphorus removal. Therefore, removal of phosphorus via harvesting of aboveground biomass of emergent vegetation may be substantial, for lightly loaded systems (cca 10–20 g P/m² year) [58].

The microbial uptake mechanism contributes only to temporary storage of phosphorus with very short turnover rate. Phosphorus that is taken up by microbiota is released back to the water after the decay of the organisms [58].

Nutrients removal depends on the degree of water logging, and the rate and the duration of nutrient loading [18]. The

removal of phosphorus increases if the substrate is nonreducing, which is in contrast to the condition required for efficient N removal. Furthermore, hydraulic loading and retention times are important in determining sedimentation rates, the contact time between nutrient load and wetland sediment and vegetation (Figure 52.11). In addition, N removal is negatively related to N loading, whereas P removal is positively related to P loading.

52.3.4 Pathogen Removal

The pathogen microorganisms existing in urban wastewater are mainly bacteria, viruses, protozoa, and helminthes. Natural

TABLE 52.2 Potential Magnitude of Phosphorus Removal Mechanisms in Different Types of Constructed Wetlands

	FFP	FWS	HSSF	VSSF
Soil accretion	Very low	High	Zero	Zero
Adsorption	Very low	Low	High[a]	High[a]
Precipitation[b]	Zero	Very low	Very low	Very low
Plant uptake [c]	Medium	Low	Low	Low
Microbial uptake	Low	Low	Low	Low

Source: Adapted from Vymazal, J. 2007. *Science of the Total Environment*, 380, 48–65.

[a] When special filtration materials are used.
[b] When washed gravel or crushed rock is used.
[c] With harvest.

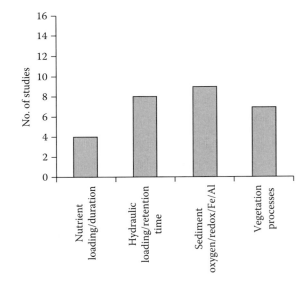

FIGURE 52.11 Most usual factors affecting nutrient reduction in CWs. (Adapted from Fisher, J., and Acreman, M.C. 2004. *Hydrology and Earth Systems Sciences*, 8(4), 673–685.)

removal mechanisms of bacteria take place, such as adsorption, precipitation, physicochemical (solar radiation, change of pH and temperature, presence of toxic substances produced by algae) and biological (physical elimination due to lack of nutrients, presence of microorganisms which consume pathogens, competition with other pathogens for food) [25,60].

Tertiary treatment wetlands are rather effective in removing pathogenic microorganisms [46]; treatment efficiencies vary between 0.5 and 3 log 10 units for bacteria, 3 log 10 units for helminths, 0.5–2 log 10 units for protozoa, and 1.5–2 log 10 units for viruses [31]. The rate of pathogen removal can be optimized by finer gravel size and longer HRT [19].

52.3.5 Metals Removal

CWs can efficiently remove heavy metals [20,35] and other low-concentration compounds such as pharmaceutical products [37]. The metals that enter the wetland system usually are Cu, Pd, and Zn in μg/L. More often, metals exist in the insoluble suspended solids form; according to pH and redox potential values, insoluble types can be redissolved. Metal removal mechanisms in CWs include mainly adsorption, microbiological mediation, chemical precipitation, ion exchange, filtration, and in a smaller extent uptake by plant roots [25].

It is noteworthy that long-term accumulation of heavy metals may render a CW into a black level. However, in the case of tertiary CWs for domestic wastewater, few adverse effects can be expected [46].

52.4 Design and Principles of Wetlands

52.4.1 Site Selection, Construction, and Start-Up of a Wetland

The preferred site for construction of a wetland should have the following characteristics: it should cover a relatively large available area, a source of year round water, an appreciable growing season, and a relatively large volume of water requiring treatment with low to moderate contaminant concentrations. Site selection is usually accomplished during a feasibility study [26].

During the construction of a wetland, the clearing and grubbing of the land takes place initially. The second step is the identification of general contours of the wetland, which is followed by the installation of water control structures. Subsequently, the wetland cell berms are constructed by compacting soil, and liners are set (e.g., clay layer, geomembrane). Finally, the wetland cell bottom is leveling to optimize the spreading of wastewaters in the wetland. The wetland cells are flooded to a "wet" mode and the planting begins. Wetland plants are transferred to the location and planted manually. After planting, water level is gradually increased to the design water level and the wetland is complete [25,49,53]. The stages of the construction of a wetland are presented in Figure 52.12.

The start-up periods are variable and they depend on the type of CW (FWS or SFS), the characteristics of the influent wastewater, and the season of the year. Start-up periods for FWS wetlands are significant because the establishment of flora and fauna should take place associated with the treatment processes, whereas the start-up period for SFS is less critical as their performance is less dependent on the planting [26,53].

52.4.2 Wetland Plants

The most frequently used vegetation in CWs is based on the type of attachment and includes [15,48]:

1. Emergent macrophytes
2. Submerged macrophytes
3. Floating leaved macrophytes (freely floating and rooted)

The role of macrophytes in constructed treatment wetlands is crucial. The aerial plant tissue of macrophytes attenuates sunlight and reduces subsequently the growth of phytoplankton; it provides insulation of the wetland during the winter and reduces wind velocity, reducing alongside the risk of resuspension; additionally, it stores nutrients. The plant tissue of macrophytes in water filters out the large debris. Furthermore, it reduces the current velocity, increasing the rate of sedimentation and reducing the risk of resuspension. It also provides surface area for attached biofilms and uptakes nutrients. Finally, the roots and rhizomes of the plants stabilize the sediment surface and thus decrease the erosion in the sediment, while they prevent clogging of the substrate in vertical flow CWs and they release antibiotics [8].

52.4.2.1 Emergent Macrophytes

Emergent macrophytes constitute the dominant macrophyte type in wetlands. They grow within a water table variation from 0.5 m below the soil surface to a water depth of 1.5 m or more (Figure 52.13). These plants transport oxygen to roots and rhizomes, as shown in Figure 52.14. Part of the oxygen may be diffused into the surrounding rhizosphere, creating oxidizing conditions. In this way, decomposition of organic matter and growth of nitrifying bacteria are enhanced. Nevertheless, CO produced by respiration of the root system and CH_4 produced in the sediment diffuse along the reverse gradient [9,63].

Most used types of emergent macrophytes include *Typha* spp. (cattails), *Scirpus* spp. (bulrushes), *Phragmites australis* (common reed), *Juncus* spp. (rushes), etc. [26,49]. However, native species should be preferred.

52.4.2.2 Submerged Macrophytes

Submerged macrophytes appear at all depths within the photic zone (Figure 52.15). The rooted submerged macrophytes receive phosphorus by direct uptake from the sediments or they can absorb it from water [3,4,5,43]. Likewise, they receive nitrogen from the sediments [10]; submerged macrophytes present bigger rates in ammonia assimilation by foliage than nitrate assimilation. The types of submerged macrophytes often used include *Elodea* spp. (waterweeds), *Potamogeton* spp. (pondweeds), etc.

FIGURE 52.12 Construction of a wetland. Clearing of the land, construction of berms and liners of wetland cells and planting. (Adapted from Sim, C.H. 2003. *The Use of Constructed Wetlands for Wastewater Treatment.* Wetlands International, Malaysia Office, p. 24.)

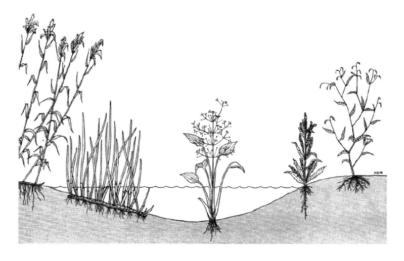

FIGURE 52.13 Emergent macrophytes. (Adapted from Sainty, G.R., and Jacobs, S.W.L. 1981. *Waterplants of New South Wales.* Water Resource Commission, N.S.W.; drawing by David Mackay.)

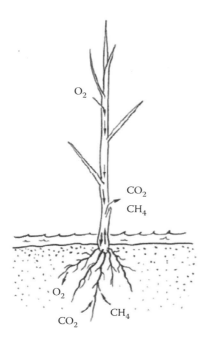

FIGURE 52.14 Passive diffusion of gases in the lacunar system of wetland plants. (Adapted from Brix, H. 1993. *Constructed Wetlands for Water Quality Improvement*. CRC Press, Boca Raton, FL, pp. 391–398.)

52.4.2.3 Floating Leaved Macrophytes (Freely Floating and Rooted)

The freely floating macrophytes (Figure 52.16) range from large plants with rosettes of aerial and/or floating leaves and well-developed submerged roots, for example, *Eichhornia crassipes* (water hyacinth), to minute surface-floating plants with few or no roots, for example, *Lemnaceae* (duckweeds) [9,63]. Freely

floating macrophytes are generally restricted to sheltered habitats and slow-flowing waters. Nutrient uptake is performed completely from the water [65].

The rooted floating-leaved macrophytes are primarily angiosperms appearing attached to submerged substrates at water depth from about 0.5–3.0 m (Figure 52.17). Types of rooted floating-leaved macrophytes include *Nuphar* or *Nymphaea*, *Brassenia*, and *Potamogeton natans* [65].

52.4.3 Process Design and Operation Parameters

A wetland is designed taking into account substrate types, pollutants loading rate, and retention time. The basic process design and operation parameters for FWS and SSF CWs are presented in Tables 52.3 and 52.4, respectively.

During the estimation of the required surface area of an SSF CW, it should be considered that the adoption of a maximum effective treatment area would reduce the short-circuiting problems. The following empirical formula is often used for the reduction of BOD_5 in sewage effluent:

$$A_h = KQ_d(\ln C_0 - \ln C_t) \qquad (52.3)$$

where

A_h: surface are of CW bed, m^2
K: rate constant, m/d
Q_d: average daily flow rate of wastewater, m^3/d
C_0: average daily BOD_5 of the influent, mg/L
C_t: required average daily BOD_5 of the influent, mg/L

In practice, most SSF systems operate on a basis of 3–5 m^2/PE [49].

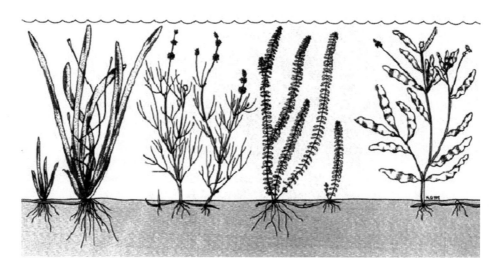

FIGURE 52.15 Submerged macrophytes. (Adapted from Sainty, G.R., and Jacobs, S.W.L. 1981. *Waterplants of New South Wales*. Water Resource Commission, N.S.W.; drawing by David Mackay.)

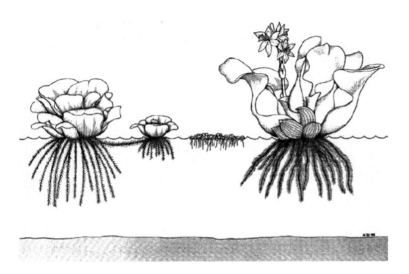

FIGURE 52.16   Free floating macrophytes. (Adapted from Sainty, G.R., and Jacobs, S.W.L. 1981. *Waterplants of New South Wales*. Water Resource Commission, N.S.W.; drawing by David Mackay.)

FIGURE 52.17   Floating-leaved rooted macrophytes. (Adapted from Sainty, G.R., and Jacobs, S.W.L. 1981. *Waterplants of New South Wales*. Water Resource Commission, N.S.W.; drawing by David Mackay.)

TABLE 52.3   Process and Operation Design Parameters for FWS CWs

Parameter	Recommended Design
Pretreatment	Oxidation ponds
Minimum number of cells	3 in each treatment train
Minimum number of trains	2 (unless it is very small)
Basin geometry (aspect ratio)	Optimum: 3:1 to 5:1
Water depth	0.6–0.9 m for fully vegetated zone
	1.2–1.5 m for open-water zone
Substrate porosity	0.65 for dense emergent species
	0.75 for less dense emergent species
	1.0 for open water areas
Minimum HRT (at Q_{max}) in 1	2 days (fully vegetated zone)
Maximum HRT (at Q_{ave}) in 2	2–3 days (open water)
Inlet structures	Uniform distribution across cell inlet
Outlet structures	Uniform distribution across cell outlet

Source: Adapted from U.S. Environmental Protection Agency. 2000. *Manual: Constructed Wetlands Treatment of Municipal Wastewaters*. EPA/625/R-99/010. U.S. EPA Office of Research and Development, Washington, DC.

TABLE 52.4   Process and Operation Design Parameters for SSF CWs

Parameter	Design Criteria
Pretreatment	Primary treatment is recommended. Oxidation ponds not recommended because of causing problems with clogging
Minimum number of SSFs	Recommended 2SSF wetlands in parallel
Substrate depth	0.5–0.6 m
Water depth	0.4–0.5 m
Bottom slope	0.5%–1%
TN Removal	Use another process in conjunction with SSF CW
TP Removal	SSF is not recommended
Inlet structures	Use adjustable inlet device to balance flow
Outlet structures	Use adjustable outlet device to balance flow

Source: Adapted from U.S. Environmental Protection Agency. 2000. *Manual: Constructed Wetlands Treatment of Municipal Wastewaters*. EPA/625/R-99/010. U.S. EPA Office of Research and Development, Washington, DC.

52.5 Summary and Conclusions

CWs use a combination of physical, chemical, and/or biological processes in order to produce high quality effluent. They consist of four major components: wetland vegetation, media or substrate supporting vegetation, water column, and living organisms. The basic types of CWs are free water surface CWs and subsurface flow CWs. The second type may be divided into subsurface horizontal flow CWs and subsurface vertical flow CWs. At free water surface wetland systems, wastewater flows above the surface of the wetland and evapotranspiration and infiltration is taking place in the "body" of the wetland. At subsurface flow systems, the wastewater is discharged within the "body" of the wetland, which is always soil; influents are gradually processed by microorganisms lying around the rhizomes of plants. These systems are primarily built with horizontal flow configuration because of its efficient denitrification performance.

Organics removal efficiency is higher in subsurface flow CWs than free water surface. In subsurface horizontal flow wetland systems, the organics are degraded by aerobic or anaerobic bacteria, which are adherent to the underground bodies of the plants and on the surface of the filling materials, as well as by sedimentation and filtration of particulate organic matter. In subsurface vertical flow wetland systems, the main mechanisms for organics removal are adsorption in wetland substrate and biological degradation.

The basic transformation mechanisms for nitrogen removal occurring in CWs are ammonification, nitrification, denitrification, uptake, and assimilation by plants. Single-stage constructed wetlands cannot achieve high removal rates of total nitrogen because they cannot provide aerobic and anaerobic conditions simultaneously. Therefore, different types of wetland systems are combined aiming to an efficient total nitrogen removal. Removal of dissolved phosphorus in CWs occurs by adsorption on porous media, desorption, precipitation with metals, dissolution, plant and microbial uptake, fragmentation, leaching, mineralization, sedimentation, and burial. Removal mechanisms of pathogen microorganisms may be physicochemical, such as solar radiation and change of pH and temperature, or biological, such as physical elimination due to lack of nutrients or presence of microorganisms, which consume pathogens. The rate of pathogen removal can be optimized by finer gravel size and longer HRT. Furthermore, CWs can efficiently remove heavy metals and other low-concentration compounds such as pharmaceutical products. According to pH and redox potential values, insoluble metal types can be redissolved. Metal removal could take place by adsorption, microbiological mediation, chemical precipitation, ion exchange, filtration, and in a smaller extent, uptake by plant roots.

Macrophytes role in wetland systems is crucial due to their multiple functions. Macrophytes insulate the wetland during the winter, reduce wind velocity, store nutrients, and provide surface area for attached biofilms and uptake nutrients. The most frequently used macrophytes include emergent macrophytes, submerged macrophytes, and floating leaved macrophytes, divided into freely floating and rooted species. Emergent macrophytes constitute the dominant macrophyte type in CWs, while freely floating ones constitute the less used, and they are restricted only to sheltered habitats and slow-flowing waters.

Construction of a treatment wetland commences by a feasibility study for site selection. The following steps during the construction of a wetland include land clearing, contours identification, installation of water control structures, construction of wetland cell berms, leveling of wetland cell bottom, planting, and finally increase of water level.

References

1. Akratos, C.S. 2006. Optimization of design parameters in subsurface flow constructed wetlands using pilot plants, PhD thesis, Department of Environmental Engineering, Faculty of Engineering, Democritus University of Thrace, p. 194. (In Greek.)
2. Arias, A.C., Brix, H., and Marti, E. 2005. Recycling of treated effluents enhances removal of total nitrogen in vertical flow constructed wetlands. *Journal of Environmental Science and Health*, 40, 1431–1443.
3. Barko, J.W. and Smart, R.M. 1980. Mobilization of sediment phosphorus by freshwater macrophytes. *Freshwater Biology*, 10, 229–238.
4. Barko, J.W. and Smart, R.M. 1981. Comparative influence of light and temperature on the growth and metabolism of selected submersed freshwater macrophytes. *Ecological Monographs*, 51, 219–235.
5. Barko, J.W. and Smart, R.M. 1986. Sediment related mechanisms of growth limitation in submerged macropyhtes. *Ecology*, 67, 1328–1340.
6. Brix, H. 1993. Macrophyte-mediated oxygen transfer in wetlands: Transport mechanisms and rates. In: A.G. Moshiri (ed.), *Constructed Wetlands for Water Quality Improvement*. CRC Press, Boca Raton, FL, pp. 391–398.
7. Brix, H. 1994. Functions of macrophytes in constructed wetlands. *Water Science Technology*, 29(4), 71–78.
8. Brix, H. 1997. Do macrophytes play a role in constructed treatment wetlands? *Water Science Technology*, 35(5), 11–17.
9. Brix, H. and Schierup, H.-H. 1989. The use of aquatic macrophytes in water pollution control. *Ambio*, 18, 100–107
10. Chambers, P.A. and Kalff, J. 1987. Light and nutrients in the control of aquatic plant community structure. I. *In situ* experiments. *Journal of Ecology*, 75, 611–619.
11. Chen, Y.T., Kao, M.C., Yeh, Y.T., Chien, Y.H., and Chao, C.A. 2006. Application of a constructed wetland for industrial wastewater treatment: A pilot-scale study. *Chemosphere*, 64(3), 497–502.
12. Cooper, P. 1999. A review of the design and performance of vertical-flow and hybrid reed bed treatment systems. *Water Science Technology*, 40(3), 1–9.
13. Cooper, P. 2009. What can we learn from old wetlands? Lessons that have been learned and some that may have been forgotten over the past 20 years. *Desalination*, 246, 11–26.

14. Crites, R., Middlebrooks, E., and Reed, S. 2006. *Natural Wastewater Treatment Systems*. CRC Press, Boca Raton, FL.

15. Daubenmire, R.F. 1947. *Plants and Environment. A Textbook of Plants Autecology*. John Wiley & Sons, New York.

16. Drizo, A., Frost, A.C., Grace, J., and Smith, A.K. 1999. Physico-chemical screening of phosphate-removing substrates for use in constructed wetland systems. *Water Research*, 33(17), 3595–3602.

17. Dunne, E.J. and Reddy, K.R. 2005. Phosphorus biogeochemistry of wetlands in agricultural watersheds. In: E.J. Dunne, R. Reddy, and O.T. Carton (eds.), *Nutrient Management in Agricultural Watersheds: A Wetland Solution*. Wageningen Academic Publishers, Wageningen, the Netherlands, pp. 105–119.

18. Fisher, J. and Acreman, M.C. 2004. Wetland nutrient removal: A review of the evidence. *Hydrology and Earth Systems Sciences*, 8(4), 673–685.

19. García, J., Vivar, J., Aromir, M., and Mujeriego, R. 2003. Role of hydraulic retention time and granular medium in microbial removal in tertiary treatment reed beds. *Water Research*, 37, 2645–2653.

20. Ghermandi, A., Bixio, D., and Thoeye, C. 2006. The role of free water surface constructed wetlands as polishing step in municipal wastewater reclamation and reuse. *Science of the Total Environment*, 380, 247–258.

21. Giraldo, E. and Zarate, E. 2001. Development of a conceptual model for vertical flow wetland metabolism. *Water Science Technology*, 44(11–12), 273–280.

22. Goulet, R.R., Pick, F.R., and Droste R.L. 2001. Test of the first-order removal model for metal retention in a young constructed wetland. *Ecological Engineering*, 17, 357–371.

23. Grant, W.D. and Long, P.E. 1985. Environmental microbiology. In: O. Hutzinger (ed.), *The Handbook of Environmental Chemistry. Vol. 1. Part D, The Natural Environment and Biogeochemical Cycles*. Springer-Verlag, Berlin, pp. 125–237.

24. Greenway, M. and Woolley, A. 1999. Constructed wetlands in Queensland: Perfomance efficiency and nutrient bioaccumulation. *Ecological Engineering*, 12, 39–55.

25. Brix, H., Kadlec, R.H., Knight, R.L., Vymazal, J., Cooper, P., and Haberl, R. 2000. *Constructed Wetlands for Pollution Control: Processes, Performance, Design and Operation. Scientific and technical report/IWA*, vol. 8, p. 156, IWA Publishing, London, UK, ISBNB number: 1900222051.

26. Interstate Technology & Regulatory Council (I.T.R.C.), Wetlands Team. 2003. *Technical and Regulatory Guidance Document for Constructed Treatment Wetlands*. p. 199. http://www.itrcweb.org/GuidanceDocuments/WTLND-1.pdf

27. Irish Environmental Protection Agency. 2000. *Wastewater Treatment Manuals-Treatment Systems for Single Houses*. EPA Publications, Wexford, Ireland.

28. Kadlec, R.H. 1994. Overview: Surface flow constructed wetlands. In: *Proceedings of the 4th International Conference on Water Science and Technology*, 32(3), 1–12, *Selected Proceedings of the 4th International Conference on Wetland Systems for Water Pollution Control*.

29. Kadlec, R.H. 2000. The inadequacy of first-order treatment wetland models. *Ecological Engineering*, 15(1–2), 105–119, doi:10.1016/S0925-8574(99)00039-7.

30. Kadlec, R.H. and Wallace, S.D. 2009. *Treatment Wetlands*. 2nd Edition. CRC Press, Boca Raton, FL.

31. Kamizoulis, G. 2005. The new draft WHO guidelines for water reuse in agriculture. *Proc. Technical Workshop on the Integration of Reclaimed Water in Water Resource Management*, Lloret de Mar, Spain, pp. 209–220.

32. Keffala, C. and Ghrabi, A. 2005. Nitrogen and bacterial removal in constructed wetlands treating domestic wastewater. *Desalination*, 185, 383–389.

33. Khanijo, I. 2007. *Nutrient Removal from Wastewater by Wetland Systems*. http://home.eng.iastate.edu/~tge/ce421–521/ishadeep.pdf.

34. Knight, R.L., Clarke, R.A., and Bastian, R.K. 2001. Surface flow (SF) treatment wetlands as a habitat for wildlife and humans, *Water Science Technology*, 44(11/12), 27–37.

35. Lesage, E., Rousseau, D.P.L., Tack, F.M.G., and De Pauw, N. 2006. Accumulation of metals in a horizontal subsurface flow constructed wetland treating domestic wastewater in Flanders. Belgium. *Science of the Total Environment*, 380, 102–115.

36. Liang, W., Wu, Z.B., Cheng, S.P., Zhou, Q.H., and Hu, H.Y. 2003. Roles of substrate microorganisms and urease activities in wastewater purification in a constructed wetland system. *Ecological Engineering*, 21, 191–195.

37. Matamoros, V., García, J., and Bayona, J.M. 2005. Behavior of selected pharmaceuticals in subsurface flow constructed wetlands: A pilot-scale study. *Environmental Science and Technology*, 39(14), 5449–5454.

38. Megonigal, J.P., Hines, M.E., and Visscher, P.T. 2004. Anaerobic metabolism: Linkage to trace gases and aerobic processes. In: W.H. Schlesinger (ed.), *Biogeochemistry*. Elsevier-Pergamon, Oxford, UK, pp. 317–424.

39. Merlin, G., Pajean, J.L., and Lissolo, T. 2002. Performances of constructed wetlands for municipal wastewater treatment in rural mountainous area. *Hydrobiologia*, 469, 87–98.

40. Meuleman, F.M.A., Van Logtestijn, R., Rijs, B.J.G., and Verhoeven, T.A.J. 2003. Water and mass budgets of a vertical-flow constructed wetland used for wastewater treatment. *Ecological Engineering*, 20, 31–44.

41. Mitsch, W.J. and Gosselink, J.G. 2000. *Wetlands*, 3rd Edition. John Wiley & Sons, New York, p. 920.

42. Prochaska, C.A. 2005. Study of wastewater processing using natural systems, PhD thesis, Department of Chemistry, Aristotle University of Thessaloniki, p. 325. (In Greek).

43. Rattray, M.R., Howard-Williams, C., and Brown, J.M.A. 1991. Sediment and water as sources of nitrogen and phosphorus for submerged rooted aquatic macrophytes. *Aquatic Botany*, 40, 225–237.

44. Reddy, K.R. and D'Angelo, E.M. 1997. Biogeochemical indicators to evaluate pollutant removal efficiency in constructed wetlands. *Water Science Technology*, 35(5), 1–10.

45. Reddy, K.R. and Graetz, D.A. 1988. Carbon and nitrogen dynamics in wetland soils. In: D.D. Hook et al. (eds.), *Ecology and Management of Wetlands*, vol. 1. Ecology of wetlands. Timber Press, Portland, OR, pp. 307–318.

46. Rousseau, D.P.L., Lesage, E., Story, A., Vanrolleghem, P.A., and De Pauw, N. 2008. Constructed wetlands for water reclamation. *Desalination*, 218, 181–189.

47. Sainty, G.R. and Jacobs, S.W.L. 1981. *Waterplants of New South Wales*. Water Resource Commission, N.S.W.

48. Sculthorpe, C.D. 1967. *The Biology of Aquatic Vascular Plants*. St. Martin's Press, New York.

49. Sim, C.H. 2003. *The Use of Constructed Wetlands for Wastewater Treatment*. Wetlands International, Malaysia Office, p. 24.

50. Siracusa, G. and La Rosa, A.D. 2006. Design of a constructed wetland for wastewater treatment in a Sicilian town and environmental evaluation using the energy analysis. *Ecological Modelling*, 197, 490–497.

51. Sundaravadivel, M. and Vigneswaran, S. 2004. Vol. I—Constructed wetlands for wastewater treatment. In: *Wastewater Recycle, Reuse and Reclamation, Encyclopedia of Life Support Systems (EOLSS), Developed Under the Auspices of the UNESCO*. Eolss Publishers, Oxford, UK. http://www.eolss.net

52. Tietz, A., Hornek, R., Langergraber, G., Kreuzinger, N., and Haberl, R. 2007. Diversity of ammonia oxidising bacteria in a vertical flow constructed wetland. *Water Science Technology*, 56(3), 241–247.

53. U.S. Environmental Protection Agency. 2000. *Manual: Constructed Wetlands Treatment of Municipal Wastewaters*. EPA/625/R-99/010. U.S. EPA Office of Research and Development, Washington, DC.

54. Verhoeven, J.T.A. and Meuleman, A.F.M. 1999. Wetlands for wastewater treatment: Opportunities and limitations. *Ecological Engineering*, 12, 5–12.

55. Vymazal, J. 1995. *Algae and Element Cycling in Wetlands*. Lewis Publishers, Chelsea, MI, p. 689.

56. Vymazal, J. 2001. Removal of organics in Czech constructed wetlands with horizontal sub-surface flow. In: J. Vymazal (ed.), *Transformations of Nutrients in Natural and Constructed Wetlands*. Backhuys Publishers, Leiden, the Netherlands, pp. 305–327.

57. Vymazal, J. 2005. Horizontal sub-surface flow and hybrid constructed wetlands systems for wastewater treatment. *Ecological Engineering*, 25, 478–490.

58. Vymazal, J. 2007. Removal of nutrients in various types of constructed wetlands. *Science of the Total Environment*, 380, 48–65.

59. Vymazal, J. 2008. Constructed wetlands for wastewater treatment: A review. In: M. Sengupta and R. Dalwani (eds.), *Proceedings of Taal 2007: The 12th World Lake Conference*, 28 October–2 November 2007, Jaipur, India, pp. 965–980.

60. Vymazal, J. 2008. Constructed wetlands, subsurface flow. In: S.E. Jørgensen and B. Fath (eds.), *Encyclopedia of Ecology*. Elsevier Science, Amsterdam, pp. 748–764.

61. Vymazal, J. 2009. The use of constructed wetlands with horizontal sub-surface flow for various types of wastewater. *Ecological Engineering*, 35, 1–17.

62. Vymazal, J. 2011. Constructed wetlands for wastewater treatment: Five decades of experience. *Environmental Science and Technology*, 45, 61–69.

63. Vymazal, J. and Kropfelova, L. 2008. Wastewater treatment in constructed wetlands with horizontal sub-surface flow. In: Alloway, B. and Trevors J. (eds.), *Environmental Pollution*, Vol. 14, Springer, Springer Science and Business Media B.V, Czech Republic, p. 566.

64. Vymazal, J. and Kröpfelová, L. 2009. Removal of organics in constructed wetlands with horizontal sub-surface flow: A review of the field experience. *Science of the Total Environment*, 407, 3911–3922.

65. Wetzel, R.G. 2001. *Limnology. Lake and River Ecosystems*, 3rd Edition. Academic Press, San Diego, CA.

53

Phytoremediation

Bosun Banjoko
Obafemi Awolowo
University Ile-Ife

Saeid Eslamian
Isfahan University of Technology

PREFACE

One of the greatest challenges of the twenty-first century is the environmental pollution of land air and water due to industrialization and population growth, which put a lot of pressure on natural resources. Conventional techniques for pollution control are typically very expensive, have high energy consumption, and generate large quantities of wastes that require disposal. Growing and, in some cases, harvesting plants on a contaminated site as a remediation method is an aesthetically pleasing, solar energy driven, passive technique that can be used to clean up sites with shallow, low to moderate levels of contamination. This technique, termed "phytoremediation," can be used along with or, in some cases in place of mechanical cleanup methods. Phytoremediation encompasses associated bioremediation in which vascular plants, bacteria, fungi, and algae are being utilized. Further genetic engineering is providing more information and transgenic plant products with more promising prospects for this technology. This chapter is therefore aimed at describing basic principles of phytoremediation and applications in general and water reuse in particular

53.1 Introduction

The quality of life on Earth is linked inextricably to the overall quality of the environment. In early times, it was believed that the Earth possessed an unlimited abundance of land and resources. Today however, recent events and audit of world resources have revealed misuse and careless management of the world's natural resources, which has caused a lot of concern. It is widely recognized that contaminated, land water and air are a potential threat to human health and their continual occurrences over recent years has led to international efforts to remedy many of these sites either as a response to the risk of adverse health or environmental effects caused by the contamination or to enable the site to be redeveloped for us [3,42]. The conventional techniques used for remediation have been to dig up contaminated soil and remove it to a landfill or to cap and contain the contaminated areas of a site. The methods have some drawbacks. The first method simply moves the contamination elsewhere and may create significant risks in the excavation, handling, and transport of hazardous materials. In addition, it is very difficult and increasingly expensive to find new land fill sites for the final disposal of the material. The cap and contain method is only an interim solution since the contamination remains on site, requiring monitoring and maintenance of the isolation barriers long into the future, will all the associated cost and potential inability. A better approach than these traditional method is to completely destroy the pollution if possible or at least to transform them to innocuous substances. Some technologies that have been used are high temperature incineration and various types of chemical decomposition, for example, base catalyzed dechlorination and ultraviolet ray oxidation. These methods can be very effective at reducing levels of contaminants but have several drawbacks which include high cost for small scale application and the lack of public acceptance, especially the incineration that may increase the exposure to contaminants for both workers at the site and nearby residents [137,202].

Worldwide, there is an increasing trend in areas of land, surface waters, and groundwater affected by contamination from industrial, military, and agricultural activities, either due to ignorance, lack of vision, or shared carelessness. The build-up of toxic pollutants including metals, radionuclides, and organic contaminants in soil, surface water, and groundwater, not only affects natural resources but also causes a major strain in the ecosystems [44]. The necessity to conserve natural resources including land and water underpins the need to develop and apply alternative, environmentally sound technologies (ESTS). It is against this backdrop that the process of phytoremediation emerged. Phytoremediation is the direct use of living green plants for *in situ*, or in place, removal, degradation, or containment of contaminants in soils, sludges, sediments, surface water, and groundwater. It is a low cost, solar energy driven cleanup technique has been found to be very useful for treating a wide variety of environmental contaminants particularly with shallow and low levels of contamination [30,37,38]. Phytoremediation has been found effective in conjunction with mechanical cleanup methods and in some cases is preferable over the latter [5].

Phytoremediation is a generic term derived from the Greek word phyto meaning "plant" and Latin word remedium, meaning restoring balance. The technology involves the use of living green plants for *in situ* risk reduction and/or removal of contaminants from contaminated soil, water, sediments, and air. Specially selected plants or engineered plants are used in the process. Risk reduction can be through a process of removal, degradation of, or containment of a contaminant or a combination of any of these factors. Phytoremediation is an energy-efficient, aesthetically pleasing method of remediating sites with low to moderate levels of contamination and it can be used in conjunction with other more traditional remedial methods [2,130,166].

The process includes mitigating pollutants concentrations in contaminated soils, water or air with plants able to contain, degrade, or eliminate metals, pesticides, solvents, explosives, crude oil, and its derivatives and various other contaminants

from the media that contain them [166,205]. The method may be applied wherever the soil or static water environment has been polluted or is suffering from an ongoing chronic pollution. Examples where phytoremediation has been used successfully include the restoration of abandoned metal-mine, reduction of the impact of sites where polychlorinated biphenyls have been dumped during manufacture, and mitigation of ongoing coal mine discharges.

Phytoremediation causes plants to remove, detoxify, or immobilize environmental contaminants in a growth matrix including soil, water, or sediments through the natural biological, chemical, or physical activities and processes of the plants. Plants are unique organisms with remarkable metabolic and absorption capabilities as well as transport systems that can use up nutrients or contaminants selectively from the growth matrix (soil or water). The process involves growing plants in a contaminated matrix for a required growth period to remove contaminants from the matrix or facilitate immobilization through binding and containment or degradation and detoxification of the pollutants. The plants can be subsequently harvested, processed, and disposed. Plants have evolved a great diversity of genetic adaptations to handle the accumulated pollutants that occur in the environment. It is common practice to grow and in some cases harvest plants on a contaminated site as a passive technique of remediation method. Phytoremediation can be used for river basin management through the hydraulic control of contaminants.

There are several ways in which plants can be used to clean up or remediate contaminated sites. To remove pollutants from the soil, sediments, and or water, plants can break down or degrade organic pollutants or contain and stabilize metal contaminates by acting as filters or traps. The uptake of contaminants in plants occurs primarily through the root system in which the principal mechanism for preventing contaminant toxicity is located. The root system provides an enormous surface area that absorbs and accumulates the water and nutrients essential for growth, as well as other non-essential contaminants. However, it had been observed that using trees, instead of smaller plants, is more effective in treating deeper contamination because tree roots penetrate more deeply into the ground which is quite logical. In addition, deep-lying contaminated groundwater can be treated by pumping the water out of the ground and using plants to treat the contamination. Furthermore, plant roots cause changes at the soil–root interface as they release organic and inorganic compounds. That is, root exudates in the rhizosphere. These root exudates affect the number and activity of the microorganisms, the aggregation, and stability of the soil particles around the root and the availability of the contaminants. Root exudates by themselves can increase that is, mobilize or decrease that is, immobilize directly or indirectly the availability of the contaminants in the root zone, that is, rhizosphere of the plant through changes in soil characteristics, release or organic substances, changes in chemical composition and or increase in plant assisted microbial activity [45,48,63].

One of the main advantages of phytoremediation is that of its relatively low cost compared to other remedial methods such

TABLE 53.1 Types of Phytoremediation Systems

Treatment Method	Mechanism	Media
Rhizofiltration	Uptake of metals in plant roots	Surface water and water pumped through troughs
Phytotransformation	Plant uptake and degradation of organics	Surface water, groundwater
Plant-assisted bioremediation	Enhanced microbial degradation in the rhizosphere	Soils, groundwater within the rhizosphere
Phytoextraction	Uptake and concentration of metals via direct uptake into plant tissue with subsequent removal of the plants	Soils
Phytostabilization	Root exudates cause metals to precipitate and become less bioavailable	Soils, groundwater, mine tailings
Phytovolatilization	Plant evapotranspirates selenium, mercury, and volatile organics	Soils, groundwater
Removal of organics from the air	Leaves take up volatile organics	Air
Vegetative caps	Rainwater is evapotranspirated by plants to prevent leaching contaminants from disposal sites	Soils

TABLE 53.2 Estimates of Phytoremediation Costs versus Costs of Established Technologies

Contaminant	Phytoremediation Costs	Estimated Cost Using Other Technologies	Source
Metals	$80 per cubic yard	$250 per cubic yard	Black [28]
Site contaminated with petroleum hydrocarbons (site size not disclosed)	$70,000	$850,000	Jipson [103]
10 acres lead contaminated land	$500,000	$12 million	Plummer [152]
Radionuclides in surface water	$2 to $6 per thousand gallons treated	None listed	Richman [160]
1 ha to a 15-cm depth (various contaminants)	$2,500 to $15,000	None listed	Cunningham et al. [55]

Source: Adapted from Mitra, N. et al. 2012. *International Journal of Ecosystem,* 2(3), 32–37 and Workshop on Phytoremediation of Organic Contaminants.

as excavation. The cost of phytoremediation has been estimated as $25–$100 per ton of soil and $0.60–$6.00 per 1000 gallons of polluted water with remediation of organics being cheaper than remediation of metals. In many instances, phytoremediation has been found to be half the price of alternative methods. It is pertinent to note that phytoremediation also offers a permanent *in situ* remediation rather than simply translocating the problem. However, like any other process, it has its disadvantages and limitations. For example, the process is dependent on the depth of the roots and the tolerance of the plant to the contaminants. Exposure of animals to plants that act as hyperaccumulators can also be a concern to environmentalists as herbivorous animals may accumulate contaminated particles in their tissues which could in turn affect a whole food web. Furthermore, slow growth and low biomass require a long-term commitment [90,130,169,185] (Tables 53.1 and 53.2).

53.1.1 Principles of Phytoremediation

Phytoremediation is based on the natural ability of certain plants called hyperaccumulations to bioaccumulate, degrade, or render contaminants in soils, water, or air. Phytoremediation has been applied worldwide to mitigate effects of contaminants such as metals, pesticides, solvents, explosives crude oil and its derivatives [84,88,166] plants may break down or degrade organic pollutants, or remove or stabilize metal contaminants through one or of a combination of different phytoremediation

methods to be described later. The basic principle is uptake of contaminants by the roots, biodegradation of some of the contaminants at the root level, accumulation and degradation of the tree level, and volatilization of some of the contaminants by the levels (Figure 53.1).

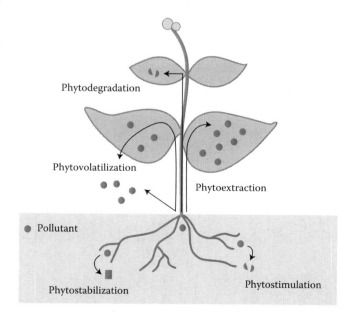

FIGURE 53.1 General principle of phytoremediation.

53.1.2 Bioremediation

Bioremediation is the use of microorganism metabolism to remove pollutants. Microorganisms used to perform the function of bioremediation are known as bioremediators. They can be naturally occurring bacteria, fungi, plants, or genetically modified species.

Bioremediation can occur on its own that is, natural attenuation or intrinsic bioremediation or can be spurred on via addition of fertilizers to increase the bioavailability within the medium that is, biostimulation. Addition of modified microbe strains to the medium to enhance the resident microbe population's ability to break down contaminants had proved very useful. Technologies can be generally classified as *in situ* or *ex situ*. The former method involves treating the contaminated materials at the site while the latter involves the removal of the contaminated materials to be treated elsewhere. Some examples of bioremediation-related technologies include phytoremediation, bioventing, bioleaching land farming, bio reaction, composting bio augmentation rhizofiltration and biostimulation. Bioremediation offers the possibility of destroying and rendering harmless various contaminants using natural biological activity. As such, it uses relatively low cost, low technological techniques; which generally have a high public acceptance and can often be carried out onsite. However, bioremediation has its limitation because the range of contaminants in which it is effective is limited, the time scales involved are relatively long, and the residual contaminant levels achievable may not always be appropriate [174]. Although, the methodologies employed are not technically complex, considerable experience and expertise may be required to design and implement a successful bioremediation program, due to the need to thoroughly assess a site for suitability and to optimize conditions to achieve a satisfactory result. Not all contaminants are easily treated by bioremediation using microorganisms. For example, heavy metals such as cadmium or lead are not easily absorbed or captured by natural organism. In addition, the assimilation of metals such as mercury into the food chain may worsen matters. In these circumstances, phytoremediation may prove very useful because natural plants or transgenic species are able to bioaccumulate these toxins in their aboveground parts which are then harvested for removal [131]. The heavy metals in the harvested biomass may be further concentrated or even recycled for industrial uses [8], Figure 53.2.

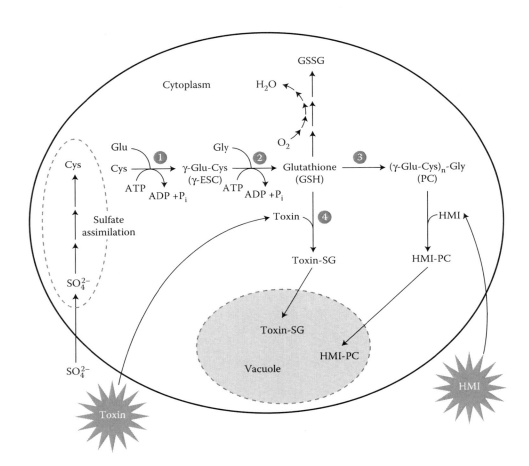

FIGURE 53.2 Mechanism of detoxification of heavy metals, organic pollutants, and oxidative stress in plant cells by glutathione. Cys, cysteine; γ-Glu-Cys, γ-L-glutamyl-L-cysteine; γ-ECS, γ-glutamylcysteine synthetase; GSH, glutathione. (Adapted from Peuke, A.D. and Rennenberg, H. 2005. *EMBO Reports*, 6, 497–501.)

Microorganisms utilized for bioremediation may be indigenous to a contaminated area or they may be isolated from elsewhere and brought to the contaminated site. Contaminant compounds are transformed by living organisms through reactions that take place as a part of their metabolic processes. Usually, biodegradation of a compound occurs because of the actions of multiple organisms. When microorganisms are imported to a containment site to enhance degradation, the process is called biodegradation. For biodegradation to be effective, microorganisms must enzymatically attack the pollutants and convert them to harmless products. Since biodegradation can be effective only where environmental conditions permit microbial growth and activity, its application often involves the manipulation of environmental parameters to allow microbial growth and degradation to proceed at a faster rate. Like other technologies, biodegradation has its limitations, some contaminants such as chlorinated organic or high aromatic hydrocarbons are resistant to microbial attack. Either they are degraded slowly or not at all; therefore, it is not easy to predict the rates of clean-up for a bioremediation exercise.

Recently, hydrocarbon-degrading microorganisms have been found to be ubiquitous in the marine environment [92,212] and biodegradation was shown to be successful in naturally remediating oil contamination associated with several spills that impacted shore has precheminated by permeable marine sediment [36,114,162]. In a study conducted at the Pansacola Beach, in Florida, chemical analysis revealed weathered oil petroleum hydrocarbons C8 to C40 concentrations ranging from 3.1 to 4,500 mg kg^{-1} in beach sands, a total of 24 bacterial strains from 14 genera were isolated from viled beach sands and confirmed as oil-degrading microorganisms. Isolated bacteria include representations of genera with known oil degraders such as *Alcanivorax, Marius bacter, Pseudomonas*, and *Acinetobacter* [107].

Most bioremediation systems are run under anaerobic conditions but running a system under anaerobic conditions may permit microbial organisms to degrade otherwise recalcitrant molecules. For degradation, it is necessary that bacteria and the contaminants should be in contract. This is not easily achieved as neither the microbes for the contaminants are uniformly spread. However, some bacteria are mobile and exhibit chemotactic response and moves towards the contaminant on sensing its presence. Other microbes such as fungi grow in a filamentous form towards the contaminant. It is suffice to note that it is possible to enhance the mobilization of the contaminant utilizing some surfactants such as sodium dodecylsulfate (SDS).

The control and optimization of bioremediation processes is a complex system of many factors which include (1) the existence of a microbial population capable of degrading the pollutants, (2) the availability of contaminants in the microbial population, and (3) the environmental factors which include soil type, temperature, pH, the presence of oxygen or other electron acceptors, and nutrients. Microorganisms can be isolated from almost any environmental conditions. Microbes will adapt and grow at sub zero temperatures as well as in extreme heat, desert conditions, in water, with an excess of oxygen, in aerobic conditions, with the presence of hazardous compounds or on any waste stream.

The main requirements are an energy source and a carbon source. Because of adaptability of microbes and other biological systems, these can be used to degrade or remediate environmental hazards. Effective degradation of pollutants requires that bacteria and the contaminants be in contact. However, this is not easily achieved because neither the microbes nor the contaminants are uniformly spread in the soil. Some bacteria are mobile and exhalant a chemotactic response, sensing the contaminant and moving toward it. Other microbes such as fungi grow in a filamentous form toward the contaminant. However, it is possible to enhance the mobilization of the contaminant utilizing some surfactants such as sodium dodecyl sulfate [198,208,213].

Although the microorganisms are present in contaminated soil, the population is not usually present in adequate quantity to cause effective degradation. Therefore, their growth and activity must be stimulated in what is referred to as biostimulation. Biostimulation usually involves the addition of nutrients and oxygen to help indigenous microorganisms. These nutrients, which include nitrogen, phosphorous and carbon, are the basic building blocks of life and allow microbes to create the necessary enzymes to break down the contaminants. Carbon is the most basic elements of living forms and is needed in greater quantities than other elements. In addition to hydrogen, oxygen, and nitrogen, it constitutes about 95% of the weight of cells. In addition, phosphorus and sulfur contribute the balance of nutrients. The nutritional requirement of carbon to nitrogen ratio is 10:1 and carbon to phosphorus is 3:1. Microbial growth and activity are readily affected by pH, temperature, and moisture. Although microorganisms have also been isolated in extreme conditions, most of them grow optimally over a narrow range, so that it is important to achieve optimal conditions. By implication, it is necessary to assess bioremediation sites for pH, available oxygen, temperature, and nature of substrate structure using soil sample analyses. In a situation whereby the soil is too acidic, the pH can be raised by the addition of lime. Temperature affects biochemical reaction rates and the rates of most reactions double for each 10°C rise in temperature. However, above a certain temperature, the cells die. Plastic covering can be used to enhance solar warming in late spring, summer, and autumn. Furthermore optional moisture level is desirable by irrigation with water. The amount of available oxygen will determine whether the system is aerobic or anaerobic. Hydrocarbons are readily degraded under aerobic conditions, whereas chlorate compounds are degraded only in aerobic ones. To increase the oxygen content of the soil, it is possible to till or sparge air. In some cases, hydrogen peroxide or magnesium peroxide can be introduced to the environment. Soil structure controls the effective delivery of air water and nutrients. To improve soil structure, materials such as gypsum or organic matter can be applied. Low soil permeability can impede movement of water, nutrients and oxygen, hence, soils with cow permeability may not be appropriate for *in situ* cleanup techniques [202].

Aerobic operations indicate the presence of oxygen. Examples of aerobic bacteria recognized for their degradative abilities are *Pseudomonas, Alcaligenes, Sphingomonas, Rhodococcus*, and

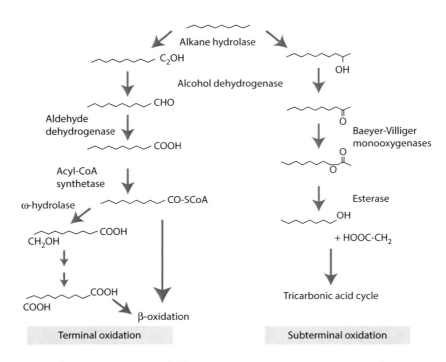

FIGURE 53.3 Aerobic pathways for the degradation of alkanes by terminal and subterminal oxidation. (Adapted from Royo, F. 2009. *Environmental Microbiology*, 11(10), 2477–2490.)

Mycobacterium. These microbes were reported to degrade pesticides and hydrocarbons, alkenes and polyaromatic compounds. Many of these bacteria use the contaminants as the sole source of carbon and energy.

Anaerobic operations indicate the absence of oxygen. Anaerobic bacteria are not as frequently used as aerobic ones. However, there is an increasing interest in anaerobic bacteria used for the bioremediation of polychlorinated biophenyls (PCBs) in river sediments and dechlorination of the solvent trichloroethylene (TCE) and chloroform (Figure 53.3).

Ligninolytic fungi such as the white rot fungus *Phanaerochaete chryosporum* have the ability to degrade an extremely diverse range of persistent or toxic environmental pollutants. Common substrates used includes straw, saw dust, or corn cobs.

Methylotrophs are aerobic bacteria that grow utilizing methane for carbon and energy. The initial enzyme pathway for aerobic degradation; methane mono oxygenase has a broad substrate range and is active against a wide range of compounds, including the chlorinated aliphatics trichloroethylene and 1,2-dichloroethane.

53.1.3 Bioremediation Techniques

Different techniques are employed depending on the degree of saturation and aeration of an area. *In situ* techniques are defined as those that are applied to soil and groundwater at the site with minimal disturbance, while *ex situ* techniques are those that are applied to groundwater at the site, which has been removed from the site via excavation of soil or pumping of water.

53.1.3.1 *In Situ* Bioremediation

These techniques are generally the most desirable options due to lower cost and production of less disturbance because they provide the treatment to the contaminated site, avoiding excavation and transport of contaminants. *In situ* treatment however is limited by the depth of the soil that can be effectively treated. In many soils, effective oxygen diffusion for desirable rates of bioremediation extend to a range of only a few centimeters to about 30 cm into the soil, although depths of 60 cm and greater have been effectively treated in some cases [193,194,196].

The bioremediation process includes the following:

1. Bioventing: This is the most common *in situ* treatment and involves supplying air and nutrients through wells to contaminated soil to stimulate the indigenous bacteria. Bioventing employs low airflow rates and provides only the amount of oxygen necessary for the biodegradation while minimizing volatilization and release of contaminants to the atmosphere. It is quite useful for remediation of simple hydrocarbons and can be used where the contamination is deep under the surface [54].

2. *In situ* biodegradation: This process involves supplying oxygen and nutrients by circulating aqueous solutions through contaminated soils to stimulate naturally occurring bacteria to degrade organic contaminants. It can be used for soil and groundwater. Generally, this technique includes conditions such as the infiltration of water containing nutrients and oxygen or other electron acceptors for groundwater treatment.

3. Biosparging: This is a process of injection of air under pressure below the water table to increase groundwater oxygen concentrations and enhance the rate of biological degradation of contaminants naturally occurring bacteria. Biosparging increases the mixing in the saturated zone and thereby increases the contact between soil and groundwater. The case and low cost of installing small-diameter are injection points allows considerable flexibility in the design and construction of the system.

4. Bioaugmentation: Bioremediation frequently involves the addition of microorganisms indigenous or exogenous to the contaminated sites. Two factors limit the use of added microbial cultures in a land treatment unit (a) nonindigenous cultures rarely compete well enough with an indigenous population to develop and sustain useful population levels and (b) most soils with long-term exposure to biodegradable waste have indigenous microorganisms that effectively degrades if the land treatment unit is well managed [32,53,203].

53.1.3.2 *Ex Situ* Remediation

Ex situ remediation techniques as stated earlier involve the excavation or removal of contaminated soil from the group. Some of these techniques include the following:

1. Land farming: This is a simple technique in which contaminated soil is excavated and spread over a prepared bed and periodically titled until pollutants are degraded. The goal is to stimulate indigenous biodegrative microorganisms and facilitate their aerobic degradation of contaminants. In general, the practice is limited to the treatment of superficial 10–35 cm of soil. Since land farming has the potential to reduce monitoring and maintenance costs as well as cleanup liabilities, it has received much attention as a disposal alternative [35].

2. Composting: This is a technique that involves combining contaminated soil with nonhazardous organic amendment such as manure or agricultural wastes. The presence of these organic materials supports the development of a rich microbial population and elevated temperature characteristic of composting.

3. Biophiles: These are a hybrid of land farming and composting. Essentially engineered cells are constructed as aerated composted piles. Typically, biophiles are used for the treatment of surface contamination with petroleum hydrocarbons. They are a refined version of land farming that tend to control physical losses of the contaminants by teaching and volatilization. Biophiles provide a favorable environment for indigenous aerobic and anaerobic microorganisms.

4. Bioreactions: *Ex situ* treatment of contaminated soil and water can be facilitated by being pumped up from a contaminated plume into slurry reactions or aqueous reactions. Bioremediation in reactors involves the processing of contaminated solid material, which may be soil, sediment, sludge, or water through an engineered containment system. A slurry bioreactor may be defined as a containment vessel and apparatus used to create a three-phase, solid, liquid and gas mixing condition to increase the bioremediation rate of soil-bound and water-soluble pollutants as a water slurry of the contaminated soil and biomass which are usually indigenous microorganisms capable of degrading target contaminants. In general, the rate and extent of biodegradation are greater in a bioreactor system than *in situ* or in solid-phase systems because the contained environment is more manageable and hence more controllable and predictable. Despite the advantages of reactor systems, there are some disadvantages. The contaminated soil requires pretreatment, for example, excavation or alternatively the contaminant can be stripped from the soil via soil washing or physical extraction, for example, vacuum extraction before being placed in a bioreactor [202].

Disadvantages of bioremediation:

- Bioremediation is limited to those compounds that are biodegradable. Not all compounds are susceptible to rapid and complete degradation.
- There are some concerns that the products of biodegradation may be more persistent or toxic than the parent compound.
- Biological processes are often highly specific. Important site factors required for success include the presence of metabolically capable microbial populations, suitable environmental growth conditions, and appropriate levels of nutrients and contaminants.
- It is difficult to extrapolate from bench and pilot-scale studies to full-scale field operations.
- Research is needed to develop and engineer bioremediation technologies that are appropriate for sites with complex mixtures of contaminants that are not evenly dispersed in the environment. Contaminants may be present as solids, liquids, and gases.
- Bioremediation often takes longer than other treatment options, such as excavation and removal of soil or incineration.
- Regulatory uncertainty remains regarding acceptable performance criteria for bioremediation. There is no accepted definition of "clean," evaluating performance of bioremediation is difficult, and there are no acceptable endpoints for bioremediation treatments.

Advantages of bioremediation:

- Bioremediation is a natural process and is therefore perceived by the public as an acceptable waste treatment process for contaminated materials such as soil. Microbes able to degrade the contaminant increase in numbers when the contaminant is present. When the contaminant is degraded, the biodegradative population declines. The residues for the treatment are usually harmless products and include carbon dioxide, water, and cell biomass.

- Theoretically, bioremediation is useful for the complete destruction of wide variety of contaminants. Many compounds that are legally considered to be hazardous can be transformed to harmless products. This eliminates the chance of future liability associated with treatment and disposal of contaminated material.
- Instead of transferring contaminants from one environmental medium to another, for example, from land to water or air, the complete destruction of target pollutants is possible.
- Bioremediation can often be carried out on site, often without causing major disruption of normal activities. This also eliminates the need to transport quantities of waste off site and the potential threats to human health and the environment that can arise during transportation.
- Bioremediation can prove less expensive than other technologies that are used for clean-up of hazardous waste.

53.1.3.3 Mycoremediation

Mycoremediation, a phrase coined by Paul Stamets, a prominent American mycologist, is a form of bioremediation in which fungi is used to degrade or sequester contaminants in the environment. One of the primary roles of fungi in the ecosystem is decomposition, which is facilitated by the mycelium. This is a form of bioremediation in which fungi are used to decontaminate the area utilizing fungal myceli. One of the primary roles of fungi in the ecosystem is decomposition, which is performed by the mycelium, which secretes extracellular enzymes and acids that break down lignum and cellulose, the two main building blocks of plant fiber. The mycelium secretes extracellular enzymes and acids that break down lignin and cellules, the two main building blocks of plant fiber. These are organic compounds composed of long chains of carbon and hydrogen, structurally similar to many organic pollutants.

In a couple of experiments, more than 95% of many of the polycyclic aromatic hydrocarbons (PAHS) were found to be reduced to nontoxic components in the mycelial-inoculated plots, breaking down contaminants into Co_2 and water. Wood-decaying fungi, fungi like oyster mushrooms (*Pleurotus ostreatus*), are particularly effective in breaking down aromatic pollutants, which are toxic components of petroleum as well as chlorinated compound present in certain persistent pesticides [182,188]. The enzymes secreted by many fungi actually stimulate aggressive toxin degradation by other microbes presenting the soil as well as plants. Research findings support the outlook that some fungi can degrade complex toxins such as 4–5 ring PAHs into simpler toxins more bio-available to other organisms. Starting with a supporting role, fungi can trigger a chain of command in the soil ecosystem toward soil health. The key to mycoremediation is determining the right fungus specie to target a specific pollutant. Certain strains have been reported to successfully degrade the nerve gases UX and Sarin [182,188]. The procedure of mycoremediation includes soil tests and rainwater tests from polluted sites for pH, microorganism load and toxin load. Determination of types of fungi and possibly plants to be introduced to the site, that is, inoculation of the site and weekly tests to monitor results of remediation are also part of the procedure.

53.1.3.4 Mycofiltration

Microfiltration is a similar process to mycoremediation. In this scenario, mushroom mycelium mats are used as biological filters. The technique was discovered by Paul Stamets, an American mycologist to control *Escherichia coli* in the water outflow from his property, using a mushroom bed planted in the gulch where the water was leaving the property, Using this technique, the coliform count decreased to nearly undetected levels. It was further observed that the mushroom produced crystalline entities advancing in front of the growing mycelium, which disintegrates when they encountered *E. coli*. As they did so, a chemical signal was sent back to the mycelin that in turn generate what appeared to be a customized macro-crystal which attracted the motile bacteria by the thousands, summarily stunning them. The advancing mycelia then consumed the *E. coli,* effectively eliminating them from the environment. One industrial application of mycofiltration has been to prevent erosion due to water runoff. Its primary application has been on abandoned logging roads. The approach here has been to place bark and wood chips into logging roads and inoculated the wood debris with mycelia of natural fungal species. As the wood chips decompose, the mycelia networks develop and they act as filters to prevent slit-flow. In the process, they also renew top soils spurring the growth of native flora and fauna.

53.1.3.5 Phytoremediation of Poor Air Quality

Initial studies on the detoxifying properties of houseplants to improve air quality were credited to B.C. Wolverton [211]. Experiments by Wolverton and others were carried out on behalf of the National Aviation and Space Administration (NASA) as a means of preserving air quality in space. Approximately 50 plants (mostly houseplants) were observed on their ability to clean the air, ease of cultivation, susceptibility to insect infestation, and transpiration rate (Wolverton et al. [212]). It was also demonstrated that soil microorganisms play a significant role in the detoxification of air along with the plants in that study. Of the plants in this study, English ivy (*Hedera helix*) has been shown to have the greatest capacity of removing formaldehyde, followed by spider plant (*Chlorophytum comosum*), snake plant (*Sansevieria trifasciata*), and aloe (*Aloe barbadenis*). However, spider plant removes the most xylene followed by snake plant and English ivy [210].

In another study, golden pothos (*Epiprenum aureum*) was also shown to remove comparable amounts of formaldehyde to English Ivy [212]. While the latter was the best at removing benzene, the former was third best. Although many of the studies used varying techniques and plant materials and standardized comparison was difficult, the overall summation is that these plants mentioned are some of the most effective at detoxifying while also factoring in ease of propagation. Many other plants were shown to be better at detoxifying than the ones in this study; however' most of these plants are either more expensive to acquire, harder to propagate, or harder to care for. Monkey grass (*Liruope spicata*) is been found to be second overall in the removal of formaldehyde and xylene as well as a strong remover of ammonia from air [210]. Florist's mum (*Chrysanthemum morifolium*) is another potentially readily available good detoxifier of formaldehyde, xylene, and ammonia. Typically, terms, members of the *Ficus* genus and members of the peace lily (Araceae), palm (Aracaceae), and agave (Agavaceae) families have been shown to be the best at detoxifying in general for formaldehyde, xylene, ammonia, and benzene [210,211,212]. Therefore, potting a few of the most effective detoxifying plants may significantly increase the air quality of the average enclosed office and environments like histopathology laboratories and mortuary environments where the combinations of these chemicals are frequently used [105,210].

A summary of the analysis of data of these experiments revealed that English ivy (*Hedera helix*) has by far the best prospect for remediating poor air quality. It is an exotic invasive that can be readily found free in the environment. It is easy to propagate and it is one of the most effective detoxifiers. However, English ivy is also toxic and should not be planted outdoors due to its invasiveness. Snake plant (*Sansevieria trifasciata*) is easy to propagate by division but hard to propagate by cuttings. Spider plant (*Chlorophytum comosum*), aloe (*Aloe barbedensis*), and pothes (*Epiprenum aurem*) are easy to propagate but not as good at toxin removal. Furthermore, many houseplants grow slowly or are hard to propagate and it might be difficult to reproduce a plant that one might find in the store economically; English ivy (*Hedera helix*), spider plant (*Chlorophytum comosum*), and pothos (*Epiprenum aureum*) are notable exceptions. It is, however, pertinent to note that detoxifying plants represent just one method of relieving the problem of indoor air pollution (IAP). A mixture of techniques is most prudent to avoid the effects of VOCs. However, formaldehyde emissions naturally decrease over time and proper air circulation can help remove these gases. Air circulation in combination with carbon filtration also increases the efficacy of plants detoxification [106,211].

53.1.3.6 Plant–Microbe Interaction

Plants normally interact with bacteria by providing resources through sloughed cells, root exudates, and diffusion of oxygen through roots. Exudates and sloughed cells may comprise 7%–27% of a plant's mass [139]. An oxygen diffusion rate of 0.5 mol O_2 m^{-2} soil surface day^{-1} has been observed. Additional nitrogen may be provided by biological fixation [57]. Soybean, clover, and alfalfa in conjunction with *Rhizobium*, for example, adds 50–200 kg of nitrogen per hectare to the rhizosphere each year [51,57].

In an environment polluted by oil spills, for example, plant, animal, bacterial, and fungal abundance and diversity are generally reduced. Further loss of biodiversity and ecosystem function may also occur as a result of disturbance of the complex interactions among the biotic components [189]. Thus, bioremediation

of residual petroleum occurs in an environment with an altered physical character and a reduced biological community. The rate at which the hydrocarbons degrade is a function of the interaction between the hydrocarbons, the altered soil physical properties, the remaining microbial and plant community, and the local climate [21]. The altered soil environment may limit the metabolism of oil by bacteria and fungi changed soil texture may limit biologically available oxygen and moisture. Sufficient nutrients, relative to the high carbon content, may also not be available. Extreme ambient and soil temperature and precipitation will also reduce the rate of oil degradation. Greater reduction in residual petroleum in the presence of vegetation provides evidence for these positive interactions. Therefore, the indirect contribution of plants to oil degradation is potentially significant. The contaminated soil environment may also reduce plant performance; however, reduced soil oxygen and moisture availability in addition to direct oil exposure may reduce plant survival, growth, and reproduction fungal species sensitivity to oil may diminish fungal contribution to hydrocarbon degradation. In addition to direct degradation of oil, fungi may enhance oil degradation indirectly by improving plant performance and the environment for bacteria.

Mycorrhizal fungi augment plant uptake of water and nutrients increasing drought tolerance and plant growth as well [14,190]. Thus, rates of oil degradation are expected to be low because of perturbation of the normal interactions between bacteria, plants, and fungi. Improving restoration via accelerated biodegradation may involve complete understanding of these interactions and perhaps the manipulations the biotic and abiotic components and their interactions of the environment [125,200].

53.1.3.7 Monitoring of Bioremediation

It is of importance that the process of bioremediation be monitored to assess performance. This can be achieved indirectly by measuring the oxidation-reduction potential or redox in soil and groundwater, together with pH, temperature, oxygen content, electron/donor concentrations, and concentrations of breakdown products (e.g., carbon dioxide). It is necessary to sample enough points on and around the contaminated site to be able to determine the contours of equal redox potential. Contouring is usually done using specialized software, for example, using kringing interpolation. If all measurements of redox potential show that the electron acceptors have been used up, it is, in effect, an indicator for total microbial activity. In addition, chemical analysis is also required to determine when the levels of contaminants and their breakdown products have been reduced to blow regulatory limits.

53.1.3.7.1 Phytobial Remediation

Phytobial remediation is a remediation type that combines the use of both microbes and plants in order to eliminate toxic substances from soil and water. Like phytoremediation, plants are grown to help uptake toxic substances and like microbial remediation, microbes are introduced to help degrade toxic substances.

The microbes exist with the plants and combine bioremediation techniques and phytoremediation techniques to attain the best of both traditional forms of remediation. The whole concept of the phytobial process is that it utilizes the rhizodeposition products of the plant root as the energy source for the microorganisms to function [120]. Therefore, the microbes that are utilized must be rhizosphere competent, meaning that they can multiply and thrive in the rhizosphere of the plant [56,61].

53.2 Mechanism and Processes

Vegetation-based remediation shows potential for accumulating, immobilizing, and transforming a low level of persistent contaminants. In national ecosystems, plants acts as filters and metabolize substances generally by nature. This underpins the study of these processes of action. Phytoremediation techniques are classified based on the fate of the contaminants. Plants may break down or degrade organic pollutants, or remove and stabilize metal contaminants. This may be done through one of or a combination of the following methods [46]. Tests performance at the Oak Ridge National Laboratory showed a disappearance of trichloroethylene (TCE) over time. Differences among five different plant species have been observed [218]. Another study confirmed the direct relationship between microbial mineralization of atrazine and the fraction of organic carbon in the soil [142]. Demonstration of phytostimulation has been performed to investigate remediation of chlorinated solvents from groundwater in Fort Worth, Texas, petroleum hydrocarbons from soil and groundwater in Ogden, Utah, petroleum from soil in Portsmith, Virginia, and polyaromatic hydrocarbon (PAHs) from soil in Texas [176]. Degradation by microbes and dense root systems are needed for a successful design. Toxicity and fate of contaminants need to be evaluated prior to implementation of this technology. Vegetation may include trees, grasses, and legumes [135]. It also seems possible that in some cases of cometabolism, energy derived from the oxidation of the nongrowth substrate can be utilized to fix carbon from the growth substrate. The capacity for co-metabolism and fortuitous activity seems to occur most frequently in hydrocarbon utilizing bacteria. In sewage and wastewater treatment processes, biotransformation of pollutants in the absence of bacterial growth also occurs as a result of the activities of endoenzymes present in dead, non-viable bacteria and as a result of exoenzyme excreted by viable bacteria, but the magnitude of such effects has yet to be defined [100,197,198,208,213].

53.2.1 Phytoextraction (Phytoaccumulation)

Phytoextraction is the process where plant roots uptake metal contaminants from the soil and translocate them to their above soil tissues. Since different plants have different abilities to uptake and withstand high levels of pollutants, many different plants may be used. This is of particular importance on sites that have been polluted with more than one type of metal contaminants. Hyperaccumulator plant species, which absorb higher amounts of pollutants than most other species, are used on many sites due to their tolerance of relatively extreme levels of pollution. Once the plants have grown and absorbed the metal pollutant, they are harvested and disposed of safely. This process is repeated several times to reduce contamination to acceptable levels. In some cases, it is possible to recycle the metals through a process known as phytomining. Phytomining is usually reserved for use with precious metals. Two versions of phytoextraction are available and these include

1. Natural hyper accumulation: Where plants naturally take up the contaminants in soil unassisted.
2. Induced or assisted hyperaccumulation: In which a conditioning fluid containing a chelator or another agent is added to soil to increase metal solubility or mobilization so that the plants can absorb them more easily. In many cases, natural hyperaccumulations are metallophyte plants that can tolerate and incorporate high levels of toxic metals.

Applications:

1. Arsenic: Using the sunflower (*Helianthus annuus*) or the Chinese brake fern (*Pteris vittata*) a hyperaccumulator that stores arsenic in its leaves.
2. Cadmium: Using willow (*Salix viminalis*), which possesses specific characteristics like high transport capacity of heavy metals from root to shoot and huge amounts of biomass production has a significant potential as a phytoextractor of cadmium (Cd), zinc (Zn), and copper (Ca). In addition, it can be used for production of bioenergy in the biomass energy power plant [84].
3. Cadmium and zinc: Using alpine pennycress (*Thlaspi caerulescens*) a hyperaccumulator of these metals at levels that would be toxic to many plants. However, the presence of copper seems to impair its growth.
4. Lead: Using Indian mustard (*Brassica juncea*), ragweed (*Ambrosia artemisiifolia*), hemp dogbae (*Apocynum cannabinum*), or poplar trees, which sequester lead in their biomass.
5. Cesium 137 and strontium – 90 were removed from a pond using sunflowers after the Chernobyl accident [2].
6. Mercury, selenium, and organic pollutants such as polychlorinated biphenyls (PCBs) have been removed from soils by transgenic plants containing genes for bacterial enzymes [130].
7. Sodium chloride (desalination): Barley, which are salt tolerant and moderately halophytic, and/or sugar beets are commonly used for salt extraction to reclaim fields that were previously flooded by seawater.
8. Petroleum and hydrocarbons using alfalfa, poplar, juniper, and fescue for soil and groundwater remediation.

The main advantage is environmental friendliness. This is because the traditional methods used for cleaning up heavy meta-contaminated soil disrupt soil structure and reduce soil productivity, whereas phytoextraction can clean up soil without causing any kind of harm to soil quality. Another benefit of phytoextraction is that it is less expensive than any other cleanup process. However, since the process is dependent on plant growth, it takes more time than anthropogenic soil cleanup methods, which is a disadvantage.

53.2.2 Phytostabilization

Phytostabilization is the use of certain plants to immobilize soil and water contaminants. The contaminants are absorbed and accumulated by roots, adsorbed onto the roots, or precipitated in the rhizosphere. This reduces or even prevents the mobility of the contaminants preventing migration into the groundwater or air and reduces the bioavailability of the contaminant, thus preventing spread through the food chain. This technique can also be used to re-establish a plant community or sites that have been denuded due to the high levels of metal contamination. Once a community of tolerant species has been established, the potential for wind erosion and the spread of the pollution is reduced and leaching of the soil contaminants is also reduced. Phytostabilization focuses on long-term stabilization and containment of the pollutant. Unlike phytoextraction, phytostabilization focuses mainly on sequestering pollutants in soil near the roots but not in plants. Tissues pollutants become less bio available and livestock, wildlife, and human exposure is reduced [27].

Applications:

A type of photostabilization process is the use of a vegetative cap to stabilize and contain mine tailing [134]. Another application is the use of hybrid poplar grasses for phytostabilization of heavy metals in the soil.

53.2.3 Phytotransformation

Phytodegradation involves chemical modification of environmental substances as a direct result of plant metabolism, often resulting in their inactivation, degradation (phytodegradation), or immobilization (phytostabilization). *Ex planta* metabolic processes hydrolyze organic compounds into smaller units that can be absorbed by the plant. Some contaminants can be absorbed by the plant and are then broken down by plant enzymes. These smaller pollutant molecules may then be used as metabolites by the plant as it grows, thus becoming incorporated into the plant tissues. Plant enzymes have been identified that break down ammunition wastes, chlorinated solvents such as trichloroethane (TCE), and others that degrade organic herbicides. In the case of organic pollutants such as pesticides, explosives, solvents, industrial chemicals, and other xenobiotic substances, certain plants such as cannas, render these substances non-toxic by their metabolism. In other cases, microorganisms living in association with plant roots may metabolize these substances in soil or water. These complex and recalcitrant compounds cannot be broken down to basic molecules, that is, water and carbon dioxide by plant molecules and, hence, the term phytotransformation

represents a change in chemical structure without complete breakdown of the compound. The term "Green Liver Model" is used to describe phytotransformation as plants behave analogously to the human liver when dealing with these xenobiotic compounds, that is, foreign compounds or pollutants [40]. After uptake of the xenobiotics, plant enzymes increase the polarity of the xenobiotics by adding functional groups such as hydroxyl groups (−OH). This is known as phase I metabolism; similar to the way that the human liver increases the polarity of drugs and foreign compounds (drug metabolism) whereas in the human liver, enzymes such as cytochrome P_{450S} are responsible for the initial reactions. In plants, enzymes such as nitroreductases carry out the same role. In the second stage of phytotransformation, known as phase II metabolism, plant biomolecules such as glucose and amino acids are added to the polarized xenobiotic to further increase the polarity in a process called conjugation. This is again similar to the processes occurring in the human liver where glucuronidation, that is, addition of glucose molecules by uridyl glucuronyl transferases (UGT) class of enzyme of which UGT1A1 is a classical example and glutathione addition reactions occur on reactive centers of the xenobiotic [205,206].

Phase I and II reactions serve to increase the polarity and reduce toxicity of the compounds, although many exceptions to the rule are seen. The increased polarity also allows for easy transport of the xenobiotic along aqueous channels. In the final state of phytotransformation, which is Phase III of the metabolism, a sequestration of the xenobiotics polymerize in a lignin-like manner and develop a complex structure that is sequestered in the plant. This phase is analogous to the Phase III stage of drug metabolism in humans, which is the final phase facilitated by the efflux transporters of which the ABC cassette is a typical example. In phytotransformation, this stage ensures that the xenobiotic is safely stored, and does not affect the functioning of the plant. However, preliminary studies have shown that these plants can be toxic to small animals such as snails, and hence, plants involved in phytotransformation may need to be maintained in a closed enclosure. Hence, the plants reduce toxicity with some exceptions and sequester the xenobiotics in phytotransformation. Trinitrotoluene phytotransformation has been extensively researched and a transformation pathway has been proposed [191].

Breeding programs and genetic engineering are powerful methods for enhancing phytoremediation capabilities or for introducing new capabilities into plants. Genes for phytoremediation may originate from microorganisms or may be transferred from one plant to another variety better adapted to the environmental conditions at the cleanup site. For example, genes encoding a nitroreductase from a bacterium inserted into tobacco showed faster removal of trichlorotoluene (TNT) and enhanced resistance to the toxic effect of the chemical [88]. A mechanism in plants that allows them to grow even when the pollution concentration in the soil is lethal for non-treated plants had been observed [40,185]. Some natural biodegradable compounds such as exogenous polyamines allow the plants to tolerate concentrations of pollutants 500 times higher than untreated plants and to absorb more pollutants.

Applications:

- Use of duckweed to degrade explosives in groundwater.
- Degradation of atrazine by hornwort in aquatic systems.
- Riperian corridor phytodegradation of nitrate in groundwater using hybrid poplar.

53.2.4 Rhizofiltration

This is similar in concept to phytoextraction but is concerned with the remediation of contaminated groundwater rather than remediation of polluted soils. The process entails the use of hydroponically cultivated plant roots to remediate contaminated water of dirt. The pollutants remain absorbed or adsorbed to the roots. The contaminated water is either collected from a waste site and brought to the plants or the plants are planted in the contaminated area, where the roots than take up the water and the contaminants dissolved in it. Many plant species naturally uptake heavy metals and excess nutrients for a variety of reasons which include sequestration, drought resistance, disposal by leaf abscission, interference with other plants, and defense against pathogens and herbivores [33]. Some of these species are better than others and can accumulate extra ordinary amounts of these contaminants. Identification of such plants species has led environmental researchers to realize the potential for using these plants for remediation of contaminated soil and wastewater, Rhizofiltration is very similar to phytoextraction in that it removes contaminants by trapping them into a harvestable plant biomass and they both follow the same basic path to remediation. First, plants are put in contact with the contamination, where they absorb contaminants through their root systems and store them in root biomass and or transport them up into the stem or leaves. The plants continue to absorb contaminants until they are harvested. The plants are then replaced to continue the growth/harvest cycle until satisfactory levels of contaminant are achieved. Both processes are useful in concentrating and precipitating heavy metals than organic contaminants. The major difference between rhizofiltration and phytoextraction is that the former is used for the treatment in aquatic environments; the latter is used in soil remediation.

Rhizofiltration is cost effective for large volumes of water having low concentrations of contaminants that are subjected to stringent standards [135]. It is relatively inexpensive, yet potentially more effective than comparable technologies. For example, the removal of radionuclides from water using sunflower was estimated to cost between $2 and $6 per thousand gallons of water treated, including waste disposal and capital costs [52]. Rhizofiltration has the advantage of being applicable *in situ*, with plants being grown directly in the contaminated water body. This allows for a relatively inexpensive procedure with low capital costs. Operation costs are also low but depend on the contaminant. This treatment method is also aesthetically pleasing and results in a decrease of water infiltration and leaching of contaminants [201]. Furthermore, after harvesting, the crop may be

converted to biofuel briquette, a substitute for fossil fuel However, this treatment has its limit, which is a major disadvantage, any contaminant that is below the rooting depth will not be extracted. In addition, the plants used may not be able to grow in highly contaminated areas. Most importantly, it can take years to reach regulatory levels; this is therefore a long-term maintenance process. There can also be combination of metals and organics, in which case, rhizofiltration treatment will not suffice [192]. Plants grown on polluted water and soils become a potential threat to human and animal health, and therefore, careful attention must be paid to the harvesting process and only non-fodder crop should be chosen for rhizofiltration remediation method [201]

Applications:

- Treatment of surface water, groundwater, industrial and residential effluents, down washes from powerlines, storm waters, acid mine drainage, agricultural runoffs, diluted sludges and radionuclide contaminated solutions [75,77].
- Removal of toxic metals such as Ca^{2+}, Cd^{2+}, Cr^{6+}, Ni^{2+}, Pb^{2+}, and Zn^{2+} from aqueous solutions using various terrestrial plant species [65]. A system to achieve this can consist of a "feeder layer" of soil suspended above a contaminated stream through which plants grow; extending the bulk of their roots into the water. The feeder layer allows the plants to receive fertilizer without contaminating the stream, while simultaneously removing heavy metals from the water [97]
- Low level radioactive contaminants can be removed from liquid streams [70]
- Trees: these are the lowest cost plant type and they can grow on land of marginal quality and have long lifespans with resultant little or no maintenance costs. The most commonly used are willows and poplars, which can grow 6–8 per year and have a high flood tolerance. For deep contamination, hybrid poplars with roots extending 30 ft deep have been used. Their roots penetrate microscopic scale pores in the soil matrix and can cycle 100 of water per day per tree. Those trees act almost like a pump and treat remediation system [192]

53.2.5 Phytovolatilization

This is a process whereby pollutants are removed from soil or water with their release into the air sometimes because of phytotransformation to more volatile and or less polluting substances. There are varying degrees of success with plants as phytovolatilizers with one study showing poplar trees to volatilize up to 90% of the tetrachloro ethylene (TCE) they absorb [218].

Applications:

- Phytoremediation of volatile metals pollutants and organic compounds, chlorinated solvents
- Plants in the mustard family (Brassicaceae) such as Ida Gold had been used for selenium polluted soils

- Primarily used for the removal of mercury, the mercuric ion is transformed into less toxic elemental mercury [93]
- Successfully used in remediation of tritium (3H) a radioactive isotope of hydrogen, it is decayed to stable helium with a half-life of about 12 years [64]. Phytovolatilization is the most controversial of all phytoremediation technologies

Advantages:

- Cost is cheaper compared to conventional methods.
- It is aesthetically pleasing.
- *In situ* application is possible.
- Low impact application.
- Produces less impact products.

Disadvantages:

- Mercury released into the atmosphere is likely to be recycled by precipitation and then deposited back into the ecosystem.
- Need for more research and fieldwork for further knowledge on this remediation strategy.
- It is limited to sites with lower contaminant concentrations.
- It is restricted to sites with contamination as deep as the roots of the plants being used.
- The food chain could be adversely affected by the degradation of chemicals.
- The air could be contaminated by the burning of leaves or limbs of plants contaminating dangerous chemicals.

53.2.6 Phytostimulation

Phytostimulation also referred to as enhanced rhizosphere biodegradation or rhizodegradation or plant assisted bioremediation/degradation is the breakdown of organic contaminants in the soil via enhanced microbial activity in the plant root zone or rhizosphere. Microbial activity is stimulated in the rhizosphere in several ways and these include

- Compounds such as sugars, carbohydrates amino acids, acetates, and enzymes, exuded by the roots enrich indigenous microbe populations.
- Root systems bring oxygen to the rhizosphere, which ensures aerobic transformation.
- Fire root biomass increases available organic carbon.
- Mycorrhizae fungi which grow within the rhizosphere can degrade organic contaminants that cannot be transformed solely by bacteria because of unique enzymatic pathways
- The habitat for increased microbial populations and activity is enhanced by plants [6]

Five enzyme systems in soils and sediments have been investigated by the US EPA Laboratory in Athens, GA. These enzymes include the following:

1. Dehalogenase: Important in dechlorination reactions of chlorinated hydrocarbons

2. Nitroreductase: Required in the first step of nitroaromatic degradation
3. Peroxidase: Important in oxidation reactions
4. Lacase: Breaks aromatic ring structures or organic compounds
5. Nitrilase: Important in oxidation reactions [218]

Applications:

- Removal of organic contaminants such as pesticides, aromatics, and polyaromatic hydrocarbons (PAHs) from soil and sediments
- Removal of chlorinated solvents at pollution sites

Advantages:

- Potentially more cost effective them many other technologies.
- A comparative performance in New Jersey using five different plant species showed that the cost of phytostimulation ranges from $10 to $35 per ton of soil. While other technologies, such as incineration, range from $2000 to $1000 per ton of soil [218].

Disadvantages:

- Applications limited to locations of low level of contamination and in shallow areas
- High levels of contaminations can be toxic to plants [174]

53.2.7 Co-Metabolism

This is a process whereby two compounds are simultaneously degraded. The degradation of the second compound, that is, the secondary substrate is dependent on the presence of the first compound, that is, the primary substrate.

For example, in the process of metabolizing methane, propane or simple sugars, some bacteria such as *Pseudomonas stutzeri OX I* can degrade hazardous chlorinated solvents such as tetrachloro ethylene and trichloroethylene, that they would otherwise be unable to attack. They do this by producing the methane monooxygenase enzyme, which is known to degrade some pollutants, such as chlorinated solvents via co-metabolism. Co-metabolism is thus used as an approach to biological degradation of hazardous solvents. Another example is the scenario where *Mycobacterium vaccae* uses an enzyme to oxidize propane. The same enzyme was discovered to oxidize cyclohexane into cyclohexanol. Thus, cyclohexane is co-metabolized in the presence of propane. This allows for the commensal growth of *Pseudomonas* on cyclohexanol. Pseudomonas can metabolize cyclohexanol but not cyclohexane [22,167].

Applications:

- Most useful in improving the remediation performance at chlorinated solvent sites. For chlorinated ethanes such as trichloro ethane (TCE) and lesser halogenated ethane such as dichloro ethane (DCE) and vinyl chloride (VC).

53.2.8 Hydraulic Control

This is the term given to the use of plants to control the migration of subsurface water through the rapid uptake of large volumes of water by the plants. The plants effectively act as natural hydraulic pumps, which a dense root network has established near the water and can transpire up to 300 gallons of water per day. This mechanism is utilized to decrease the migration of contaminants from surface water into the groundwater, below the water and drinking water supplies. This principle is utilized in the maintenance of the riparian corridors, buffer strips, and vegetative caps.

1. Riparian corridors: A riparian corridor is the transition zone between the land and the watercourse and is a unique plant community consisting of the vegetation growing near a river, stream, lake, lagoon, or other natural body of water. It serves a variety of functions important to people and the environment as a whole by preserving water quality through filtration of sediments from runoff before it enters rivers and stream. In addition, it protects stream banks from erosion, provides a storage area for food waters, provides food and habitat for fish and wildlife, and preserves open space and aesthetic surroundings. The riparian corridor ensures combined utilization of different processes of phytoremediation including phytodegradation, phytovolatilization, and rhizodegradation to control the spread of contaminants and to remediate polluted sites. Riparian strips refer to these uses along the banks of rivers and streams whereas buffer strips refers to their uses along the perimeter of landfills [11].

2. Vegetative cover: Vegetative cover also utilizes the principle of hydraulic control to prevent the spread of pollutants. This entails the use of plants as a cover or cap growing over landfill sites. The standard caps for such sites are usually plastic or clay. Plants used in this manner are not only more aesthetically pleasing, they may also help to control erosion, leaching of contaminants and may be useful in degrading the underlying landfill.

53.2.9 Phytoscreening

As plants are able to translocate and accumulate particular types of contaminants, plants can be used as biosensors of subsurface contamination, thereby allowing investigators to quickly delineable contaminant plumes [40,185]. Chlorinated solvents such as trichloroethylene, have been observed in tree trunks at concentrations related to groundwater concentrations [206]. To ease field implementation of phytoscreening, standard methods have been developed to extract a section of the tree trunk for later laboratory analysis often by using an increment borer [205]. Phytoscreening may lead to more optimized site investigations and reduce contaminated site cleanup costs.

Applications:
Detection of unknown subsurface contamination by chlorinated volatile organic compounds (Cl-VOCs) and petroleum hydrocarbons [185].

53.3 Plants, Bacteria, and Fungi Utilized in Phytoremediation

53.3.1 Fungi

Fungi tend to play more significant roles than bacteria in the biodegradation of hydrocarbons in soils. In aquatic systems, yeasts and filamentous fungi are predominantly associated with sediments and surface films. The enzymatic processes used in PAHs by fungi are similar to those used by mammalian systems. Two principal types of cytochrome P_{450} monooxygenases have been well characterized in yeasts and filamentous fungi. Many fungi oxidize PAHs to dihydrodiols, phenols, conjugates, and other metabolites, but only a few, such as *Phanerochaete chrysosporium*, appear to have the ability to catabolize them completely to carbon dioxide.

Biological treatment technologies for the remediation of soils and groundwater contaminated with organopollutants are widely used for their environmentally friendly impact combined with low cost compared to other treatment alternatives [15,171,199]. An alternative to the biostimulation of indigenous microflora, the so-called practice of bioaugmentation, can favor contaminant degradation when dealing with historically and/or heavily contaminated sites [134]. Sites contaminated by recalcitrant organic compounds have often been shown to be characterized by the concomitant presence of heavy metals. In such a difficult case, the use of filamentous fungi (white-rot fungi, in particular) may give some advantages over bacterial bioaugmentation [73,102]. Fungi display a high ability to immobilize toxic metals by insoluble metal oxalate formation, biosorption, or chelation onto melanin-like polymers. Moreover, due to the low substrate specificity of their degradative enzyme machinery (e.g., lacase, lignin peroxidase, and Manganese peroxidase), fungi are able to perform the breakdown of a wide range of organopollutants in contaminated soils [73,101].

Fungi are the decomposers in the global cycle of life and death. They are usually there to do the work when animal, plant, or non-living object is ready to be broken down into its molecular constituents. Fungi are found in soil, in fresh and seawater, inside the bodies of plants and animals, and traveling through the air as spores. While they are often found functioning together with bacteria and an array of microorganisms, fungi can especially handle breaking down sore of in largest molecules present in nature [71]. Fungi can grow in a multicellular form, with the somatic mycelium extending minute root-like structures through the substrate, or it can be present as unicellular yeast. Some fungi exude extracellular enzymes that allow for digestion of energy sources in their surroundings and further diffusion of these molecules through the substrate towards the fungi [20,131]. One of the largest categories of inputs to the environment in need of decomposition is manufactured hydrocarbon waste. Large reserves of hydrocarbons that were previously stored deep underground are being brought to the surface, altered, and used. A majority of the of the pollution that occurs now involves fossil fuels, whether it is the exhaust and by products of spent fuel or the accumulated polymer plastic made from these same hydrocarbons fossil fuels are composed of PAHs as well as shorter carbon molecules. Many PAHs are naturally occurring [155] in plants and animals and were the raw ingredients that initially decomposed to form fossil fuel reserves. PAHs are building blocks of life, and they are very common on the planet. However, the accumulation and chemical alternations of these PAHs are following a pattern now dominated by the actions of humans [155]. PAHs form when carbon materials are not completely burned for example, sooty exhaust from cars, broiled hamburgers, and burnt toast contain them. Furthermore, large amounts of PAHs are extracted, refined, and transport and thereby causing contamination of the environment frequently all over the world [95].

Fungi possess these decomposing abilities to deal with the array of naturally occurring compounds that serve as potential carbon sources. Hydrocarbon pollutants have similar or analogous molecular structures which enable the fungi to act on them as well. When an area is contaminated, the ability to deal with the contamination and turn it into n energy source is selected for within the fungal population and leads to a population that is better able to metabolize the contaminant [71]. It had been reported that the genes responsible. For PAH degradation are present as many homologous loci within the genome which provides a particularly large pool of mutation and rearrangement possibilities within the gene family [149].

Fungi are especially well suited to PAH degradation relative to other bacterial decomposers for a few reasons. They can degrade high molecular weight PAHs whereas bacteria are at best at degrading smaller molecules [149]). They also function well in nonaqueous environments where hydrophobic PAHs accumulate; a majority of other microbial degradation occurs in aqueous phase. Furthermore, they can function in the very low oxygen conditions that occur in heavily PAH contaminated zones [71]. A review of different studies revealed a list of over 51 fungal species or specie groups capable of degrading PAHs. A wide variety of fungi has evolved effective mechanisms to attack specific PAHs. One reason for this ability lies in the similarity between lignin, a long aromatic family of molecules that is present in wood and PAHs. Lignin is one of the main components of woody tissue in all vascular plants along with cellulose and hemicelluloses. It has been described as the cement in woody tissue that adds strength and flexibility to cellulose, which gives the trees the strength to grow taller towards the light and provides the crunch of vegetables. Fungi produce extracellular enzymes to degrade lignin, which cannot pass through the cell walls of microorganisms. This process of degradation is called mineralization, and the end product is carbon dioxide. Since lignin is comprised of many different aromatic rings in long varied chains, the fungal enzymes for mineralization are non-specific and frequently can also mineralize PAHs [13].

Fungi attack plastic polymers as well; these come in a wide range of structures as lignin and are acted upon by different fungi species for different polymers. The decomposing ability is perhaps more impressive than PAH decomposition. PAHs are naturally occurring although altered and concentrated by human activity, plastic polymers on the other hard are thoroughly transformed by human processes and they are designed to resist degradation. The white rot bacidiomycetes known for lignin degradation specifically *P. ostreatus* could effectively break down polyacrylamide; the super absorbent polymer material in diapers and hygiene products [13].

The list of fungal genera important in oil bioremediation are as follows:

Cellular organisms–Eukaryota–Fungi/Metazoa group–Fungi

1. Ascomycota
   a. Mitosporic Ascomycota
      i. *Helminthosporium*: Was shown to utilize oil components from fuel tanks.
   b. Dothioraceae
      ii. *Aureobasidium*: *A. pullulans* was isolated from fuel tanks.
   c. Saccharomycetales
      i. *Candida*: *Candida* species and strains were shown to degrade petroleum hydrocarbons.
      ii. *Pichia*: Shown to utilize PAHs.
      iii. *Saccharomyces*: Shown to utilize crude oil components.
   d. Mitosporic Hypocreaceae
      i. *Trichoderma*: Shown to utilize phenanthrene.
   e. Mitosporic Hypocreales
      ii. *Fusarium*: Was shown to utilize petroleum components.
   f. Mitosporic Pleosporaceae
      i. *Dendryphiella*: Marine, sea-weed-associated fungi that may have oil-degrading capabilities.
   g. Mitosporic Davidiellaceae
      i. *Cladosporium*: Isolated from oil-degrading consortia.
   h. Halosphaeriaceae
      i. *Corollospora*: Marine, sea-weed-associated fungi that may have oil-degrading capabilities.
      – *Varicosporina*: Marine, sea-weed-associated fungi that may have oil-degrading capabilities.
   i. Lulworthiaceae
      i. *Lulworthia*: Marine, seaweed-associated fungi that may have oil-degrading capabilities.
   j. Mitosporic Trichocomaceae
      i. *Aspergillus*: Shown to utilize pyrene and benzo(*a*) pyrene.
      ii. *Penicillium*: Was isolated from oil-contaminated media.
2. Basidiomycota
   a. Mitosporic Sporidiobolales
      i. *Rhodotorula*: Was tested for degradation of nitrobenzene from fuel tanks.
      ii. *Sporobolomyces*: Isolated from marine samples.
   b. Mitosporic Tremellales
      i. *Trichosporon*: Was tested for degradation of alkyl-substituted aromatic oil components.
   c. Corticiaceae

d. *Phanerochaete*: *P. chrysosporium* has been reported to produce carbon dioxide from benzo(*a*)pyrene, phenanthrene, and fluorene.
e. Pleurotaceae
    i. *Pleurotus*
3. Fungi incertaesedis
    a. Mortierellaceae
        i. *Mortierella*: Abundant in soil; was shown to utilize 2,4-dichlorophenol.
    b. Cunninghamellaceae
        i. *Cunninghamella*: Soil-dwelling, potential oil-biodegraders and opportunistic pathogens.
    c. Mucoraceae
        i. *Mucor*: Soil-dwelling, potential oil-biodegraders.

53.3.2 Bacteria

Interest in the microbial biodegradation of pollutants has intensified in recent years as mankind strives to find sustainable ways to clean up contaminated environments. These bioremediation and bio transformation methods endeavor to harness the astonishing naturally occurring microbial catabolic diversity to degrade, transform or accumulate a huge range of compounds including hydrocarbons, for example, (oil) PCBs PAHs; pharmaceutical substances, radio nuclides, and metals. Major methodological breakthroughs in recent years have enabled detailed genomic, metagenomic proteomic, and bioinformatic and other high through put analyses of environmentally relevant microorganisms providing unprecedented insights into key biodegradative pathways and the ability of organisms to adapt to changing environmental condition. Biological processes play a major role in the removal of contaminants and they take advantage of the astonishing catabolic versatility of microorganisms to degrade/convert such compounds. The burgeoning amount of bacterial genomic data provides unparalleled opportunities for understanding the genetic and molecular bases of the degradation of organic pollutants. Aromatic compounds are among the most recalcitrant of these pollutants and lessons can be learned from the recent genomic studies of *Burkholderia xenovarans CB 400* and *Rhodococcus* sp. strain RHA 1, two of the largest bacterial genomes completely sequenced to data. [147].

Bacterial degradation of aromatic compounds can be divided into three steps:

1. Modification and conversion of the many different compounds into a few central aromatic intermediates (ring-fission substrates); this step is referred as peripheral pathway and involves considerable modification of the ring and/or perhaps elimination of substituent groups.
2. Oxidative ring cleavage by dioxygenases, which are responsible for the oxgenolytic ring cleavage of dihydroxylated aromatic compounds (catechol, protocatechuate, gentisate).
3. Further degradation of the non-cyclic, non-aromatic ring-fission products to intermediates of central metabolic pathways.

It is important to keep in mind that many strains within one species of bacteria usually exist. Usually, only some strains are capable of hydrocarbon degradation and some can cause opportunistic infections in humans and animals. Anaerobic microbial mineralization of recalcitrant organic pollutant is of great environmental significance and involves intriguing novel biochemical reactions. In particular, hydrocarbons and halogenated compounds have long been doubted to be degradable in the absence of oxygen; but the isolation of hitherto unknown anaerobic hydrocarbon degrading and reductively dehalogenating bacteria during the last decades provided ultimate proof for those processes in nature. *Geobacter* species are often the predominant organisms when extracellular electron transfer is an important bioremediation process in subsurface environments. It has been proposed that *Geobacter*-catalyzed bioremediation strategies will provide substantial saving in large-scale *in situ* bioremediation projects for groundwater polluted with uranium and or organic contaminants.

The two elements needed for an efficient utilization of aromatic compounds by bacteria are the enzymes responsible for their degradation and the regulatory elements that control the expression of the catabolic openers to ensure the more efficient output depending on the presence absence of the aromatic compounds or alternative environmental signaling. In general, the regulatory networks that control the operons involves in the catabolism of aromatic compounds are endowed with an extra ordinary degree of plasticity and adaptability. Elucidating such regulating networks pave the wave for a better understanding of the regulatory intricacies that control microbial degradation of aromatic compounds which are key issues for the rationale design of more efficient recombinant biodegraders, bacterial biosensors and biocatalysts. Important physiological events that precede biodegradation by microorganisms include bioavailability or the amount of a substance that is physicochemically accessible to the microorganisms and chemotaxis, or the directed movement of motile organisms towards or away from chemicals in the environment. In addition, mechanisms for the intracellular accumulation of aromatic molecules via various transport mechanisms are also important.

Organic solvents are toxic for microorganisms because they dissolve in the cytoplasmic membranes, a process that alters the membrane's physical structure and renders the cell unable to synthesize ATP. The degree of toxicity varies depending on the chemical and the strain involved, and chemical toxicity correlates with the partition coefficient of the compound in a mixture of oil and water. Microbial tolerance to solvents can be mediated by physical and biochemical barriers. Physical barriers are usually based on increased membrane rigidity through the alternations of the as/trans unsaturated fatty acid ratio, the increase in the saturated unsaturated fatty acid ratio or alterative in the phospholipid head groups. Although these barriers counteract the effect of initial toxicity, long-term resistance is based on the active extrusion of solvents which is mainly mediated by extrusion pumps. Genomic analyses in Gram-negative bacteria have revealed that the resistance—nodulation—cell division (RND) family of efflux pumps is the main group involved in the removal of solvents from the acid. These pumps are made up of three components that span the membranes and extrude solvents from the inner membrane or cytoplasm to the outer medium.

Halomones are a hardy breed of bacteria. They can withstand heat, high salinity, low oxygen, utter darkness, and pressure that would kill most other organisms, these traits enable these microbes to inhabit the deep sandstone formalin that also happen to be useful for hydrocarbon extraction and carbon sequestration using DNA technologies, low diversity microbial community dominated by *Holomonas sulfidaeris* like bacteria that have evolved several strategies to cope with and survive the high pressure, high temperature and nutrient deprived deep sub surface environment were recovered. An analysis of the microbe's metabolism revealed that these bacteria are able to utilize iron and nitrogen from their surroundings and recycle scarce nutrients to meet their metabolic needs. Another member of the same group *Halomonas tiranicae* was so named because it consumes iron superstructure of the Titanic. Perhaps most importantly is the ability of these microbes to metabolized aromatic compounds, a common component of petroleum. By implication, those indigenous microbe would have the adaptive edge in event of hydrocarbon spills and migration. It was thereby suggested that a better understanding of the microbial life of the subterranean world could enhance our ability to explore for and recover oil and gas and to make more environmentally sound choices for subsurface gas storage [62]. Some bacteria utilized in bioremediation all thus listed below:

Cellular organisms—Bacteria

1. Actinobacteria
   a. Micrococcaceae
      i. *Arthrobacter*: *Arthrobacter* spp. was shown to degrade various aromatic hydrocarbons such as phenanthrene and others.
      ii. *Micrococcus*: Isolated from oil-biodegrading consortia in marine environment.
   b. Brevibacteriaceae
      i. *Brevibacterium*: These bacteria were isolated from petroleum-degrading consortia.
   c. Dermabacteraceae
      i. *Brachybacterium*: *B. phenoliresistens* was isolated from an oil-contaminated coastal sand sample.
   d. Dietziaceae
      i. *Dietzia*: Marine hydrocarbon-utilizing bacteria.
   e. Cellulomonadaceae
      i. *Cellulomonas*: Sediment hydrocarbon-utilizing bacteria.
   f. Intrasporangiaceae
      i. *Janibacter*: Implicated in degradation of PAHs in *marine sediments*.
   g. Corynebacteriaceae
      i. *Mycobacterium*: Some species can utilize PAH and other pollutants.
      ii. *Corynebacterium*: Isolated from oil-degrading consortia.
   h. Gordoniaceae
      i. *Gordonia*: Some strains also utilize oil ingredients.
   i. Nocardioidaceae
      i. *Nocardioides*: Most species are *free-living in soil and water*. Some species can utilize PAH and other pollutants.
      ii. *Rhodococcus*: Some species can utilize PAH and other pollutants.
   j. Nocardiaceae
      i. *Nocardia*
      ii. *Smaragdicoccus*
2. Cyanobacteria: Cyanobacteria can play an important role in oil-degrading consortia by not only oxidizing oil components but also by providing microbial community with nitrogen.
3. Bacteroidetes/Chlorobi group
   a. Flavobacteria
      i. *Chryseobacterium*: Were isolated from stable carbazole-degrading consortium with *Achromobacter* and other oil-degrading bacterial communities.
      ii. *Flavobacterium*: Some strains are capable of degrading polycyclic aromatic hydrocarbons and heterocyclics.
      iii. *Yeosuana*: A marine bacterium, *Y. aromativorans* GW1–IT, capable of degrading benzo[*a*]pyrene (Bap).
4. Deinococcus–Thermus
   a. Thermaceae
      i. *Themus*: Aerobic rods found in warm water (40–79C°) such as hot springs, hoot water tanks, and thermally polluted rivers; can degrade crude oil.
5. Thermotogae
   a. Thermotogaceae
      i. *Petrotoga*
6. Firmicutes
   a. Bacillaceae: Endospore-producing; mostly saprophytes from soil, but a few are insect or animal parasites or pathogens.
      i. *Bacillus*: Common in soil; several species (*B. subtilis*, *B. cereus*, and others) were shown to use naphthalene, pyrene, and other aromatics.
      ii. *Geobacillus*: Endospore-forming, thermophilic bacteria capable of utilizing long-chain alkanes.
   b. Staphylococcaceae
      i. *Staphylococcus*: Some species are opportunistic pathogens of humans and animals. Pathways of utilization of phenanthrene and other aromatic compounds by these organisms were studied.
7. Proteobacteria

a. Alphaproteobacteria: Comprised mostly of two major phenotypes: purple non-sulfur bacteria and aerobic bacteriochlorophyll-containing bacterial
  i. Sphingomonadaceae
    A. *Sphingomonas*: Degrade a broad range of substituted aromatic compounds.
    B. *Sphingobium*: Degrade a range of aromatic compounds.
  ii. Rhodobacteriaceae
    A. *Paracoccus*: Hydrocarbon-utilizing bacteria
    B. *Stappia*: Alkaliphilic and halophilic hydrocarbon-utilizing bacteria
    C. *Roseobacter*: Marine hydrocarbon-utilizing bacteria.
  iii. Rhodospirillaceae
    A. *Thalassospira*: A polycyclic aromatic hydrocarbon-degrading marine bacterium.
    B. *Tistrella*: A phenanthrene-degrading marine bacterium.
  iv. Brucellaceae
    A. *Ochrobactrum*: A polycyclic aromatic hydrocarbon-degrading marine bacterium.
    B. Rickettsiales
      – SAR11 cluster
        – Candidatus Pelagibacter
  v. Betaproteobacteria: Comprised of chemoheterotrophs and chemoautotrophs which derive nutrients from decomposition of organic material.
    A. Alcaligenaceae
      – *Achromobacrer*: Were isolated from stable carbazole-degrading consortium with *Chryseobacterium* and other oil-degrading bacterial communities.
      – *Alcaligenes*: Implicated in degradation PAH from oil and other pollutants.
    B. Comamonadaceae
      – *Acidovorax*: Has been found in consortia utilizing heterocyclic aromatics.
      – *Polaromonas*: Has been shown to utilize naphthalene, benzene, toluene.
8. Burkholderiaceae
  a. *Burkholderia*: Found in consortia of microorganisms degrading PAH and other environmental pollutants.
  b. *Ralstonia*: Free-living forms are known to utilize PAHs.
9. Rhodocyclaceae
  a. *Azoarcus*: Gram-negative, facultatively anaerobic bacteria including species which are often associated with grasses and which fix nitrogen as well as species which anaerobically degrade toluene and other mono-aromatic hydrocarbons.
  b. *Thauera*: Gram-negative, rod-shaped bacteria able to anaerobically oxidize and degrade toluene.
10. Delta proteobacteria: Represented by morphologically diverse, anaerobic sulfidogens; some members of this group are considered bacterial predators, having bacteriolytic properties.
  a. Geobacteraceae
    i. *Geobacter*: Anaerobic, metal-reducing bacteria in the family Geobacteraceae. They have the ability to oxidize a variety of organic compounds, including aromatic hydrocarbons.
  b. Desulfobacteraceae
    i. *Desulfobacterium*: Anaerobic, metabolizes C_{12}–C_{20} alkanes.
    ii. *Desulfobacula*: Anaerobic, metabolizes toluene and benzene.
    iii. *Desulfotignum*: A Gram-negative, sulfate-reducing bacterium.
11. Epsilon-proteobacteria: Consists of chemoorganotrophs usually associated with the digestive system of humans and animals.
12. Gamma-proteobacteria: Comprised of facultatively anaerobic and fermentative Gram-negative bacteria.
  a. Piscirickettsiaceae
    i. *Cycloclasticus*: Marine bacteria; play a major role in degrading PAHs from crude oil in marine environment.
  b. Pseudomonadaceae
    ii. *Pseudomonas*: Numerous strains are most studied oil biodegraders; many strains are patented and are included in commercial bioremediation mixtures.
  c. Alteromonadaceae
    i. *Marinobacter*: Implicated in degrading PAHs and other environmental pollutants.
  d. Pseudoalteromonadaceae
    i. *Pseudoalteromonas*: Marine oil-degrading bacteria
  e. Pasteurellaceae
    i. *Pasteurella*: Was shown to degrade fluoranthene
  f. Shewanellaceae
    i. *Shewanella*: Marine organisms frequently isolated from oil-contaminated sites.
  g. Moraxellaceae
    i. *Acinetobacter*: Abilities for bioremediation of oil were documented.
    ii. *Moraxella*: Plasmid-mediated degradation of hydroxylated, methoxylated, and carboxylated benzene derivatives in *Moraxella* spp. were documented.
  h. Halomonadaceae
    i. *Halomonas*: Synonym Deleya; isolated from oil-contaminated soils.
  i. Alcanivoracaceae
    i. *Alcanivorax*: Present in un-polluted seawater in low numbers; principal carbon and energy sources are linear-chain alkanes and their derivatives.
  j. Oceanospirillaceae
    i. *Thalassolituus*: Marine hydrocarbonoclastic alkane-degrading bacteria.

ii. *Oleispira*: Marine hydrocarbonoclastic bacteria.

iii. *Neptunomonas*: Marine hydrocarbonoclastic bacteria.

k. Oleiphilaceae

i. *Oleiphilus*: Marine obligate hydrocarbon-degrading bacteria.

l. Xanthomonadaceae

i. *Rhodanobacter*: *Rhodanobacter* spp. is capable of utilizing benzo[a]pyrene (Bap).

ii. *Stenotrophomonas*: *S. maltophilia* is capable of utilizing various PAH.

iii. *Xanthomonas*: Produce a yellow pigment; some species are pathogenic to plants. Biodegradation of complex PAHs were studied.

iv *Arenimonas*: Was isolated on nutrient agar from a soil sample collected from an oil-contaminated site.

13. Zetaproteobacteria

53.3.3 Plants

Phytoremediation is an emerging technology that uses plant to degrade, extract, contain, or immobilize contaminants such as metals, pesticides, explosives, oil, excess nutrients, and pathogen from soil and water. The technology has been identified as a more cost effective noninvasive and publicly acceptable method of removing environment contaminants than chemical and physical methods. The majority of plants currently used in phytoremediation application including storm water ponds, riparian buffers, rain garders, green roofs, and constructed west lands are herbaceous or nonwoody plants there are current studies to identity woody land scape plant to be incorporated into the system. These plants known as hyperaccumulators are known to have great phytoremediation potentials. To date, more than 400 plant species have been identified as metal hyperaccumulations representing less than 0.2% of all angiosperis ([19,38]. The plant species that have been identified for remediation of soil include either high biomass plants such as willow [112] or those with low biomass but high hyperaccumulating characteristics such as *Thlaspi* and *Arabidopsis* species which have been most extensively studied by the scientific community. Both genera belong to the family of Brassicaceae and Alyssum. *Thlospis* sp. are known to hyper accumulate more than one metal, that is, *T. caerulescens* used for Cd, Ni, Pb, and Zn, *T. goesingense* for Ni and Zn, *T. onchroleucum* for Ni and Zn and *T. rotunfolium* for Ni, Pb, and Zn [217]. Among the genus *Thlaspi*, the hyperaccumulator plant *Thlaspi caerulescens* received much attention and has been extensively studied as potential candidates for Cd and Zn contaminated soils T caerulescens has higher uptake of Cd due to specific rooting strategy and a high uptake rate resulting from the existence in this population of Cd—specific transport channels or carriers in the root membrane [18,69,178].

Rhizo filtration is the removal of pollutants from contaminated wells by accumulation into plant biomass. Several aquatic species have been identified and tested for phytoremediation of heavy metals from the polluted water. These include Sharp dock (*Polygonium amphibium* L), duckweed (*Lemma minor* L), water hyacinth (*Eichhornia crasssipes*), water lettuce (*P. stratiotes*), water drop work (*Oenathe javanica* (BL) DC), calamus (*Lepinonia articulate*), and penny wort (*Hydrocotyle umbellate* L) [217]. The roots of Indian mustard and found to be effective in the removal of Cd, Cr, Cu, Ni, Pb, and Zn and sunflower can remove Pb, U, Cs-137, and Sr-90 from hydroponic solutions [207,214,217]. Duckweed was found to be a good accumulator of Cd, Se, and Cu, a moderate accumulator for Cr, but a poor accumulator of Ni and Pb [215]. Furthermore, aquatic macrophyte (*Eiechhornia crassipes*) was found to be useful in the elimination of Pb from industrial effluents Water hyacinth possesses a well-developed fibrous root system and large biomass and has been successfully used in wastewater treatment systems to improve water quality by reducing the levels of organic and inorganic nutrients. This plant can also reduce the concentrations of heavy metals in acid mine water while exhibiting few signs of toxicity. Water hyacinth accumulates trace elements such as Ag, Pb, Cd, etc. and is efficient for phytoremediation of wastewater polluted with Cd, Cr, Cu, and Se [143,216]. Some wetland plant species including sharp dock, duckweed, water hyacinth, water dropwort, and calamus were shown to have varying properties of phytoremediation. Sharp dock was shown to be a good accumulator of N and P, water hyacinth and duckweed strongly accumulated the two elements with a concentration of 462 and 14200 mg/kg respectively. Water dropwort achieved the highest concentration of Hg whereas the calamus accumulated Pb (512 m/kg) substantially in its roots. The ferns are not left out of the group of bioaccumulators *Pteris* vilta, commonly known as Brake fern has been identified as hyperaccumulator for As contaminated soil and waters. It can accumulate up to 7500 ng As/Kg on a contaminated site without showing toxicity symptoms [123]. Another fan cultivar is available commercially for As phytoremediation and has been successfully used in field trials [169]. In a study of three hydrophytes: *Gladiolus isoetes taiwanenses* Dwvol and *Echinodorus amazonicus*. It was observed that the biomass of all the plants decreased with an increase in Cd concentrations from 5 to 20 mg/L. However, Cd toxic effect was greater in *Isoetes taiwanenses* Dwvol and *Echinodorus amazonicus* than that on *Gladiolus*. In addition, the accumulation of Cd was higher in *Gladiolus* than in the other two plants [113].

In order to cope with heavy metal contaminated soils, various phytoremediation approaches can be applied and these include phytostabilization, phytoimmobilization, and phytoextraction. However the choice will depend on many factors such as plant tolerance to pollutants, soil physics chemical properties, agronomic characteristics of the plant species, climatic conditions such as rainfall and temperature and additional technologies available for the recovery of metals from the harvested plant biomass. The solubility of heavy metals in the polluted soils can be increased using organic and inorganic solvents, thus enhancing the phytoextraction capabilities of many plant species [68]. Other applied enhancement materials include Ethylene diamine tetra acetic acid (EDTA) citric acid, elemental sulfur, or ammonium sulfate. For example, increases greater than 100-fold in Pb concentration in the biomass of crops were reported when EDTA was applied to contaminated soils [27,55]. In a similar fashion, uranium, cadmium, and zinc concentrations were increased by the application of citric acid, elemental sulfur or ammonium sulfate respectively [173]. Apart from the chelating, the plant roots also excrete metal mobilizing substances called phytosiderophores.

Other enhancers of phytoremediation are exudates, which include mugenic and deoxymugenic acids from barley and corn and avenic acids from oats. In addition, plant roots can increase metal bioavailability by exuding protons that acidify the soil and mobilize the metals. The lowering of soil pH decreases the adsorption of heavy metals and increases their concentration in the soil solution. Soil microbes associated with plant roots are also helpful in the phytoextraction of the heavy metals in soils through the degradation of organic pollutants. These include several strains of Bacillus and Pseudomonas, which increase Cd accumulation in *Brassica juncea* seedling [170]. Hyperaccumulators were observed to grow well under high soil moisture content and the biomass of all the tested species was generally greater at higher soil moistures and inhibited at lower soil moistures. It is therefore imperative that for successful phytoremediation of metal polluted soils, a strategy should be developed to combine a rapid screening of plant species possessing the ability to accumulate heavy metals with agronomic practices that enhance shoot biomass production and/or increase metal bioavailability in the rhizosphere. A list of plant hyperaccumulators are shown in Table 53.3.

53.3.4 Genetic Engineering in Phytoremediation

The role of genetic engineering in phytoremediation cannot be over emphasized. A number of bacteria and algae have been genetically engineered to remove a specific heavy metal from contaminated water by over expressing the heavy metal binding protein such as metallothionein along with specific metal transport system. This was first experimented with a mercury (Hg^{++}) transport system [47,116]. Since then, the introduction and over expression of metal binding proteins have been widely exploited to increase the metal binding capacity, tolerance, or accumulation of bacteria and plants. Modification in the biosynthesis of phytochelatins (PCs) in plants has receptly been accomplished to enhance metal accumulation. In addition, various peptides including metal binding amino acids such as histidne cystene residues have been studied for enhanced heavy metal accumulation by bacteria [116,132]. An important pathway by which plants detoxify heavy metals is through sequestration with heavy-metal binding peptide called phytochelations or their precursor; glutathione (GSH). The transgenic Indian mustard (*Brassica juncea*) over expressing gshII (gene encoding glutathione synthase) accumulated more Cd than the wild type and the Cd accumulation and tolerance were correlated with gshII expression level [216]. By implication, transgenic plants expressing glutathione

(GSH) offer great promise for increased capacity to accumulate and tolerate Cd and other heavy metals because PCs are thought to play a role in tolerance to a range of heavy metals. In addition, phytochelatin synthase, an enzyme that catalyzes the biosynthesis of phytochelatins from GSH is also important.

This mechanism protects vascular plants and some algae and fungi and invertebrate from toxic effects of non-essential heavy metals such as As, Cd, or Hg [159]. The heterogenous over expression of phytochelatins gene (PCS) in *Nicotiana glance* R. Graham (shrub tobacco) was observed to cause increased production of phytochelatins (PCs) end higher accumulation of lead (Pb) than in wild type [80]. Furthermore, the expression of the thiosulfate reductase gene from *Salmonella typhimurium* in *Escherichia coli* resulted in the increase in efficiency of removal of heavy metals from solution and accumulation of cadmium up to 150 m in 78% of bacterial cells [21]. Simultaneous expression of genes of *Allium sativum* phytochelatins synthase (ASPCSI) and yeast cadmium factor I (YCF 1) in Arabidopsis to the increased tolerance to Cd and As and higher capacity to accumulate these metals than in corresponding single gene-transgenic strains and wild type. Such combination of modified genes involved in chelation of toxic metals by thiols and vacuolar compartmentalization represents a highly promising new tool for use in phytoremediation efforts [85].

Nicotinamine (NA) a chelator of metals is widely present in higher plants. Expressing a nicotinamine synthase cDNA (TcNAs 1) isolated from the polymetallic hyperaccumulator *Thlaspi caerulescens* in *Arabidopsis thaliana* resulted in 100-fold more NA production and increased nickel tolerance than in the wild type species [151].

53.4 Application of Phytoremediation in Water Reuse

53.4.1 Groundwater

Groundwater is water present below the ground surface that saturates the pore space in the subsurface. At least about half of the world's population depends on groundwater as a source of drinking water. Groundwater is also used by farmers to irrigate crops and by industries to produce every day goods most groundwater is clean but can be polluted or contaminated as a result of human activities or natural conditions. The many diverse activities of man produce innumerable waste materials and by products which when disposed of or stored on land surfaces may percolate into the underlying soil and eventually carried downwards, thereby contaminating the underlying groundwater and jeopardizing its natural quality. Other sources of groundwater pollution include overapplication of fertilizers or pesticides, spills from industrial operations, infiltration from urban runoff, and leakages from landfills. Contaminants found in groundwater include physical, inorganic chemical, organic chemical, bacteriological, and radioactive substances therefore using contaminated groundwater may result in serious health hazards. This underpins the development of various

TABLE 53.3 List of Plant Hyperaccumulators

Contaminant	Accumulation Rates (in mg/kg dry weight)	Binomial Name	English Name	H—Hyperaccumulator or A—Accumulator P—Precipitator T—Tolerant	Notes	References
Al-Aluminum	A-	*Agrostis castellana*	Highland bentgrass	As(A), Mn (A), Pb(A), Zn(A)	Origin Portugal	McCutcheon and Schnoor [127]
Al- Aluminum	1000	*Hordeum vulgare*	Barley	xxx	25 records of plants	Grauace and Horst [83]
Al- Aluminum	xxx	*Hydrangea* spp.	Hydrangea (a.k.a. *Hortensia*)	xxx	xxx	Grauace and Horst [83]
Al- Aluminum	Al concentrations in young leaves, mature leaves, old leaves, and roots were found to be 8.0, 9.2, 14.4, and 10.1 mg gl, respectively [4]	*Melastoma malabathricum* L.	Blue tongue or native lassiandra	P competes with aluminum and reduces uptake [5]	xxx	Grauace and Horst [83]
Al-Aluminum	xxx	*Solidago hispida* (*Solidago canadensis* L.)	Hairy goldenrod	xxx	Origin Canada	McCutcheon and Schnoor [127]
Al-Aluminum	100	*Vicia faba*	Horse bean	xxx	xxx	McCutcheon and Schnoor [127]
Ag-Silver	xxx	*Brassica napus*	Rapeseed plant	Cr, Hg, Pb, Se, Zn	Phytoextraction	Fiegl et al. [72]
Ag-Silver	xxx	*Salix* spp.	Osier spp.	Cr, Hg, Se, petroleum hydrocarbons, organic solvents, MTBE, TCE, and by-products [7]; Cd, Pb, U, Zn (*S. viminalix*) [8]; potassium ferrocyanide (*S. babylonica* L.) [9]	Phytoextrafraction Perchlorate (wetland halophytes)	McCutcheon and Schnoor [127]
Ag-Silver	xxx	*Amanita strobiliformis*	European pine cone Lepidella	Ag(H)	Macrofungi, basidiomycete. Known from Europe, prefers calcareous areas	Borovicka et al. [31]
Ag-Silver	10–1200	*Brassica juncea*	Indian mustard	Ag(H)	Can form alloys of silver–gold–copper	Havar kamp et al. [81]
As-Arsenic	100	*Agrostis capillaris* L.	Common bentgrass, Browntop (=*A. tenuris*)	Al(A), Mn (A), Pb(A), Zn(A)	xxx	McCutcheon and Schnoor [127]
As-Arsenic	H-	*Agrostis castellana*	Highland bentgrass	Al(A), Mn (A), Pb(A), Zn(A)	Origin Portugal	McCutcheon and Schnoor [127]
As-Arsenic	1000	*Agrostis tenerrima* Trin.	Colonial bentgrass	xxx	4 records of plants	Porter and Peterson [153]

(Continued)

TABLE 53.3 (Continued)   List of Plant Hyperaccumulators

Contaminant	Accumulation Rates (in mg/kg dry weight)	Binomial Name	English Name	H—Hyperaccumulator or A—Accumulator P—Precipitator T—Tolerant	Notes	References
As-Arsenic	27,000 (fronds) [12]	Pteris vittata L.	Ladder brake fern or Chinese brake fern	26% of arsenic in the soil removed after 20 weeks' plantation, about 90% as accumulated in fronds [13]	Root extracts reduce arsenate to arsenite [14]	Porter and Peterson [153]; Dwivedi et al. 2010 [67]
As-Arsenic	100–7000	Sarcosphaera coronaria	Pink crown, violet crown-cup, or violet star cup	As(H)	Ectomycorrhizal ascomycete, known from Europe	Borovička 2004 [31]
Be-Beryllium	xxx	xxx	xxx	xxx	No reports found for accumulation	McCutcheon and Schnoor [127]
Cr-Chromium	xxx	Azolla spp.	Mosquito fern, duckweed fern, fairy moss, water fern	xxx	xxx	McCutcheon and Schnoor [127]
Cr-Chromium	H-	Bacopa monnieri	Smooth water hyssop, water hyssop, brahmi, thyme-leafed gratiola,	Cd(H), Cu(H), Hg(A), Pb(A)	Origin India Aquatic emergent species	Gupta et al. [86]
Cr-Chromium	xxx	Brassica juncea L.	Indian mustard	Cd(A), Cr(A), Cu(H), Ni(H), Pb(H), Pb(P), U(A), Zn(H)	Cultivated in agriculture	Bennett et al. [25]
Cr-Chromium	xxx	Brassica napus	Rapeseed plant	Ag, Hg, Pb, Se, Zn	Phytoextraction	
Cr-Chromium	A-	Vallisneria americana	Tape grass	Cd(H), Pb(H)	Native to Europe and North Africa Widely cultivated in the aquarium trade	McCutcheon and Schnoor [127]
Cr-Chromium	1000	Dicoma niccolifera	xxx	xxx	35 records of plants	McCutcheon and Schnoor [127]
Cr-Chromium	Roots naturally absorb pollutants, some organic compounds believed to be carcinogenic [19], in concentrations 10,000 times that in the surrounding water [20]	Eichhornia crassipes	Water hyacinth	Cd(H), Cu(A), Hg(H) [19], Pb(H) [19], Zn(A); also Cs, Sr, U [19,21], and pesticides [22]	Pantropical/subtropical Plants sprayed with 2,4-D may accumulate lethal doses of nitrates [23] The troublesome weed, hence an excellent source of bioenergy [19]	McCutcheon and Schnoor [127]
Cr-Chromium	xxx	Helianthus annuus	Sunflower	xxx	Phytoextraction and rhizofiltration	McCutcheon and Schnoor [127]
Cr	A-	Hydrilla verticillata	Hydrilla	Cd(H), Hg(H), Pb(H)	xxx	McCutcheon and Schnoor [127]
Cr-Chromium	xxx	Medicago sativa	Alfalfa	xxx	xxx	McCutcheon and Schnoor [127]
Cr-Chromium	xxx	Pistia stratiotes	Water lettuce	Cd(T), Hg(H), Cr(H), Cu(T)	xxx	

(Continued)

TABLE 53.3 (*Continued*) List of Plant Hyperaccumulators

Contaminant	Accumulation Rates (in mg/kg dry weight)	Binomial Name	English Name	H—Hyperaccumulator or A—Accumulator P—Precipitator T—Tolerant	Notes	References
Cr-Chromium	xxx	*Salix* spp.	Osier spp.	Ag, Hg, Se, petroleum hydrocarbons, organic solvents, MTBE, TCE, and by-products [7]; Cd, Pb, U, Zn (*S. viminalix*) [8]; potassium ferrocyanide (*S. babylonica* L.) [9]	Phytoextraction Perchlorate (wetland halophytes)	McCutcheon and Schnoor [127]
Cr-Chromium	xxx	*Salvinia molesta*	Karibaweeds or water ferns	Cr(H), Ni(H), Pb(H), Zn(A)	xxx	Strivastav [186]
Cr-Chromium	xxx	*Spirodela polyrhiza*	Giant duckweed	Cd(H), Ni(H), Pb(H), Zn(A)	Native to North America	Strivastav [186]
Cr-Chromium	100	*Jamesbrittenia fodina* (Wild) Hilliard (a.k.a. *Sutera fodina* Wild)	xxx	xxx	xxx	Brook [38]
Cr-Chromium	A-	*Thlaspi caerulescens*	Alpine pennycress, alpine pennygrass	Cd(H), Co(H), Cu(H), Mo, Ni(H), Pb(H), Zn(H)	Phytoextraction *T. caerulescens* may acidify its rhizosphere, which would affect metal uptake by increasing available metals [29]	Lombi et al. [118]
Cu-Copper	9000	*Aeolanthus biformifolius*	xxx	xxx	xxx	Morrison et al. [138]
Cu-Copper	xxx	*Athyrium yokoscense*	(Japanese false spleenwort?)	Cd(A), Pb(H), Zn(H)	Origin Japan	McCtucheon and Schnoor [127]
Cu-Copper	A-	*Azolla filiculoides*	Pacific mosquito fern	Ni(A), Pb(A), Mn(A)	Origin Africa Floating plant	McCtucheon and Schnoor [127]
Cu-Copper	H-	*Bacopa monnieri*	Smooth water hyssop, water hyssop, brahmi, thyme-leafed gratiola	Cd(H), Cr(H), Hg(A), Pb(A)	Origin India Aquatic emergent species	Gupta et al. [86]
Cu-Copper	xxx	*Brassica juncea* L.	Indian mustard	Cd(A), Cr(A), Cu(H), Ni(H), Pb(H), Pb(P), U(A), Zn(H)	Cultivated	Benneth et al. 2000 [25]
Cu-Copper	H-	*Vallisneria americana*	Tape Grass	Cd(H), Cr(A), Pb(H)	Native to Europe and North Africa. Widely cultivated in the aquarium trade.	McCtucheon & Schnoor 2003
Cu-Copper	xxx	*Eichhornia crassipes*	Water hyacinth	Cd(H), Cr(A), Hg(H), Pb(H), Zn(A), also Cs, Sr, U [21], and pesticides [22]	Pantropical/subtropical, the troublesome weed	McCtucheon and Schnoor [127]
Cu-Copper	1000	*Haumania strumrobertii* (Lamiaceae)	Copper flower	xxx	27 records of plants Origin Africa This species' phanerogam has the highest cobalt content Its distribution could be governed by cobalt rather than copper [34]	Baker and Brooks [18]

(*Continued*)

TABLE 53.3 (*Continued*) List of Plant Hyperaccumulators

Contaminant	Accumulation Rates (in mg/kg dry weight)	Binomial Name	English Name	H—Hyperaccumulator or A—Accumulator P—Precipitator T—Tolerant	Notes	References
Cu-Copper	xxx	*Helianthus annuus*	Sunflower	xxx	Phytoextraction with rhizofiltration	Baker and Brooks [18]
Cu-Copper	1000	*Larrea tridentata*	Creosote bush	xxx	67 records of plants Origin United States	Baker and Brooks [18]
Cu-Copper	H-	*Lemna minor*	Duckweed	Pb(H), Cd(H), Zn(A)	Native to North America and widespread worldwide	
Cu-Copper	xxx	*Ocimum centraliafricanum*	Copper plant	Cu(T), Ni(T)	Origin Southern Africa	Howard Williams [96]
Cu-Copper	T-	*Pistia stratiotes*	Water lettuce	Cd(T), Hg(H), Cr(H)	Pantropical Origin South United States Aquatic herb	McCutcheon and Schnoor [127]
Cu-Copper	xxx	*Thlaspi caerulescens*	Alpine pennycress, alpine pennygrass	Cd(H), Cr(A), Co(H), Mo, Ni(H), Pb(H), Zn(H)	Phytoextraction Copper noticeably limits its growth [32]	McCutcheon and Schnoor [127]
Mn-Manganese	A-	*Agrostis castellana*	Highland bentgrass	Al(A), As(A), Pb(A), Zn(A)	Origin Portugal	McCutcheon and Schnoor [127]
Mn-Manganese	xxx	*Azolla filiculoides*	Pacific mosquito fern	Cu(A), Ni(A), Pb(A)	Origin Africa Floating plant	McCutcheon and Schnoor [127]
Mn-Manganese	xxx	*Brassica juncea* L.	Indian mustard	xxx	xxx	Bennetta et al. [25]
Mn-Manganese	xxx	*Helianthus annuus*	Sunflower	xxx	Phytoextraction and rhizofiltration	McCutcheon and Schnoor [127]
Mn-Manganese	1000	*Macadamia neurophylla* (now *Virotia neurophylla* (*Guillaumin*) (P. H. Weston & A. R. Mast)	xxx	xxx	28 records of plants	McCutcheon and Schnoor [127]
Mn-Manganese	200	xxx	xxx	xxx	xxx	McCutcheon and Schnoor [127]
Hg-Mercury	A-	*Bacopa monnieri*	Smooth water hyssop, water hyssop, brahmi, thyme-leafed gratiola	Cd(H), Cr(H), Cu(H), Hg(A), Pb(A)	Origin India Aquatic emergent species	McCutcheon and Schnoor [127]
Hg-Mercury	xxx	*Brassica napus*	Rapeseed plant	Ag, Cr, Pb, Se, Zn	Phytoextraction	McCutcheon and Schnoor [127]
Hg-Mercury	xxx	*Eichhornia crassipes*	Water hyacinth	Cd(H), Cr(A), Cu(A), Pb(H), Zn(A) also Cs, Sr, U [21], and pesticides [22]	Pantropical/subtropical, the troublesome weed	McCutcheon and Schnoor [127]
Hg-Mercury	H-	*Hydrilla verticillata*	Hydrilla	Cd(H), Cr(A), Pb(H)	xxx	McCutcheon and Schnoor [127]
Hg-Mercury	1000	*Pistia stratiotes*	Water lettuce	Cd(T), Cr(H), Cu(T)	35 records of plants	McCutcheon and Schnoor [127]

(Continued)

TABLE 53.3 (*Continued*) List of Plant Hyperaccumulators

Contaminant	Binomial Name	English Name	Accumulation Rates (in mg/kg dry weight)	H—Hyperaccumulator or A—Accumulator P—Precipitator T—Tolerant	Notes	References
Hg-Mercury	*Salix* spp.	Osier spp.	xxx	Ag, Cr, Se, petroleum hydrocarbons, organic solvents, MTBE, TCE, and by-products [7]; Cd, Pb, U, Zn (*S. viminalix*) [8]; potassium ferrocyanide (*S. babylonica* L.) [9]	Phytoextraction Perchlorate (wetland halophytes)	McCutcheon and Schnoor [127]
Mo-molybdenum	*Thlaspi caerulescens* (Brassicaceae)	Alpine pennycress	1500	Cd(H), Cr(A), Co(H), Cu(H), Ni(H), Pb(H), Zn(H)	Phytoextraction	McCutcheon and Schnoor 2003 [127]
Naphthalene	*Festuca arundinacea*	Tall fescue	xxx	xxx	Increases catabolic genes and the mineralization of naphthalene	
Naphthalene	*Trifolium hirtum*	Pink clover, rose clover	xxx	xxx	Decreases catabolic genes and the mineralization of naphthalene	Siciliano et al. [180]
Pb-Lead	*Agrostis castellana*	Highland bentgrass	A-	Al(A), As(H), Mn(A), Zn(A)	Origin Portugal	McCutcheon and Schnoor [127]
Pb-Lead	*Ambrosia artemisiifolia*	Ragweed	xxx	xxx	xxx	Fiegl et al. [72]
Pb-Lead	*Armeria maritima*	Seapink thrift	xxx	xxx	xxx	Fiegl et al. [72]
Pb-Lead	*Athyrium yokoscense*	(Japanese false spleenwort?)	xxx	Cd(A), Cu(H), Zn(H)	Origin Japan	McCutcheon and Schnoor [127]
Pb-Lead	*Azolla filiculoides*	Pacific mosquito fern	A-	Cu(A), Ni(A), Mn(A)	Origin Africa Floating plant	McCutcheon and Schnoor [127]
Pb-Lead	*Bacopa monnieri*	Smooth water hyssop, water hyssop, brahmi, thyme-leafed gratiola	A-	Cd(H), Cr(A), Cu(H), Hg(A)	Origin India Aquatic emergent species	McCutcheon and Schnoor [127]
Pb-Lead	*Brassica juncea*	Indian mustard	H-	Cd(A), Cr(A), Cu(H), Ni(H), Pb(H), Pb(P), U(A), Zn(H)	79 recorded plants Phytoextraction	McCutcheon and Schnoor [127]
Pb-Lead	*Brassica napus*	Rapeseed plant	xxx	Ag, Cr, Hg, Se, Zn	Phytoextraction	McCutcheon and Schnoor [127]
Pb-Lead	*Brassica oleracea*	Ornamental kale, cabbage, broccoli	xxx	xxx	xxx	McCutcheon and Schnoor [127]
Pb-Lead	*Vallisneria americana*	Tape grass	H-	Cd(H), Cr(A), Cu(H)	Native to Europe and North Africa Widely cultivated in the aquarium trade	McCutcheon and Schnoor [127]
Pb-Lead	*Eichhornia crassipes*	Water hyacinth	xxx	Cd(H), Cr(A), Cu(A), Hg(H), Zn(A); also Cs, Sr, U [21], and pesticides [22]	Pantropical/subtropical, the troublesome weed	McCutcheon and Schnoor [127]
Pb-Lead	*Festuca ovina*	Blue sheep fescue	xxx	xxx	xxx	McCutcheon and Schnoor [127]
Pb-Lead	*Helianthus annuus*	Sunflower	xxx	xxx	Phytoextraction and rhizofiltration	McCutcheon and Schnoor [127]
Pb-Lead	*Hydrilla verticillata*	Hydrilla	H-	Cd(H), Cr(A), Hg(H)	xxx	McCutcheon and Schnoor [127]
Pb-Lead	*Lemna minor*	Duckweed	H-	Cd(H), Cu(H), Zn(H)	Native to North America and widespread worldwide	McCutcheon and Schnoor [127]

(*Continued*)

TABLE 53.3 (*Continued*) List of Plant Hyperaccumulators

Contaminant	Accumulation Rates (in mg/kg dry weight)	Binomial Name	English Name	H—Hyperaccumulator or A—Accumulator P—Precipitator T—Tolerant	Notes	References
Pb-Lead	xxx	*Salix viminalis*	Common osier	Cd, U, Zn [8]; Ag, Cr, Hg, Se, petroleum hydrocarbons, organic solvents, MTBE, TCE, and, by-products (S. spp.) [7]; potassium ferrocyanide (S. babylonica L.) [9]	Phytoextraction Perchlorate (wetland halophytes)	Schmidt [173]
Pb-Lead	H-	*Salvinia molesta*	Kariba weeds or water ferns	Cr(H), Ni(H), Pb(H), Zn(A)	Origin India	McCutcheon and Schnoor [127]
Pb-Lead	xxx	*Spirodela polyrhiza*	Giant duckweed	Cd(H), Cr(H), Ni(H), Zn(A)	Native to North America	Srivastav [186]
Pb-Lead	xxx	*Thlaspi caerulescens* (Brassicaceae)	Alpine pennycress, alpine pennygrass	Cd(H), Cr(A), Co(H), Cu(H), Mo(H), Ni(H), Zn(H)	Phytoextraction	Lombi et al. [118]
Pb-Lead	xxx	*Thlaspi rotundifolium*	Round-leaved pennycress	xxx	xxx	Fiegl et al. [72]
Pb-Lead	xxx	*Triticum aestivum*	Common wheat	xxx	xxx	Fiegl et al. [72]
Pb-Lead	A-200	xxx	xxx	xxx	xxx	McCutcheon and Schnoor [127]
Pd-Palladium	xxx	xxx	xxx	xxx	No reports found for accumulation	McCutcheon and Schnoor [127]
Pt-Platinum	xxx	xxx	xxx	xxx	No reports found for accumulation	McCutcheon and Schnoor [127]
Se-Selenium	.012–20	*Amanita muscaria*	Flyagaric	xxx	Cap contains higher concentrations than stalks [40]	McCutcheon and Schnoor [127]
Se-Selenium	xxx	*Brassica juncea*	Indian mustard	xxx	Rhizosphere bacteria enhance accumulation [41]	McCutcheon and Schnoor [127]
Se-Selenium	xxx	*Brassica napus*	Rapeseed plant	Ag, Cr, Hg, Pb, Zn	Phytoextraction	McCutcheon and Schnoor [127]
Se-Selenium	Low rates of Se volatilization from selenate-supplied Muskgrass (10-fold less than from selenite) may be due to a major rate limitation in the reduction of selenate to organic forms of Se in muskgrass	*Chara canescens* Desv. & Lois	Muskgrass	xxx	Muskgrass treated with selenite contains 91% of the total Se in organic forms (selenoethers and diselenides), compared with 47% in Muskgrass treated with selenate [42] 1.9% of the total Se input is accumulated in its tissues; 0.5% is removed via biological volatilization [43]	Li et al. [113]
Se-Selenium	xxx	*Bassia scoparia* (a.k.a. *Kochia scoparia*)	Burningbush, ragweed, summer cypress, fireball, belvedere and Mexican firebrush, Mexican fireweed	U [8], Cr, Pb, Hg, Ag, Zn	Perchlorate (wetland halophytes) Phytoextraction	McCutcheon and Schnoor [127]

(Continued)

TABLE 53.3 (*Continued*) List of Plant Hyperaccumulators

Contaminant	Accumulation Rates (in mg/kg dry weight)	Binomial Name	English Name	H—Hyperaccumulator or A—Accumulator P—Precipitator T—Tolerant	Notes	References
Se-Selenium	xxx	*Salix* spp.	Osier spp.	Ag, Cr, Hg, petroleum hydrocarbons, organic solvents, MTBE, TCE, and byproducts [7] Cd, Pb, U, Zn (*S. viminalis*) [8]; potassium ferrocyanide (*S. babylonica* L.) [9]	Phytoextraction Perchlorate (wetland halophytes)	McCutcheon and Schnoor [127]
Zn-Zinc	A-	*Agrostis castellana*	Highland bentgrass	Al(A), As(H), Mn(A), Pb(A)	Origin Portugal	McCutcheon and Schnoor [127]
Zn-Zinc	xxx	*Athyrium yokoscense*	(Japanese false spleenwort?)	Cd(A), Cu (H), Pb(H)	Origin Japan	McCutcheon and Schnoor [127]
Zn-Zinc	xxx	*Brassicaceae*	Mustards, mustard flowers, crucifers or cabbage family	Hyperaccumulators: Cd, Cs, Ni, Sr	Phytoextraction	McCutcheon and Schnoor [127]
Zn-Zinc	xxx	*Brassica juncea* L.	Indian mustard	Cd(A), Cr (A), Cu(H), Ni(H), Pb (H), Pb(P), U(A).	Larvae of *Pieris brassicae* do not even sample its high-Zn leaves	McCutcheon and Schnoor [127]
Zn-Zinc	xxx	*Brassica napus*	Rapeseed plant	Ag, Cr, Hg, Pb, Se	Phytoextraction	McCutcheon and Schnoor [127]
Zn-Zinc	xxx	*Helianthus annuus*	Sunflower	xxx	Phytoextraction et rhizofiltration	Schmidt [173]
Zn-Zinc	xxx	*Eichhornia crassipes*	Water hyacinth	Cd(H), Cr (A), Cu(A), Hg(H), Pb (H) Also Cs, Sr, U [21], and pesticides [22]	Pantropical/subtropical, the troublesome weed	McCutcheon and Schnoor [127]
Zn-Zinc	xxx	*Salix viminalis*	Common osier	Ag, C, Hg, Se, petroleum hydrocarbons, organic solvents, MTBE, TCE, and by-products [7]; Cd, Pb, U (*S. viminalis*) [8]; potassium ferrocyanide (*S. babylonica* L.) [9]	Phytoextraction Perchlorate (wetland halophytes)	Schmidt [173]
Zn-Zinc	A-	*Salvinia molesta*	Karibaweeds or water ferns	Cr(H), Ni(H), Pb(H), Zn(A)	Origin India	McCutcheonand Schnoor [127]
Zn-Zinc	1400	*Silene vulgaris* (Moench) Garcke (Caryophyllaceae)	Bladder campion	xxx	xxx	Liu et al. [117]
Zn-Zinc	xxx	*Spirodela polyrhiza*	Giant duckweed	Cd(H), Cr (H), Ni(H), Pb (H)	Native to North America	Srivastav [186]
Zn-Zinc	H-10,000	*Thlaspi caerulescens* (Brassicaceae)	Alpine pennycress	Cd(H), Cr (A), Co(H), Cu(H), Mo, Ni(H), Pb (H)	48 records of plants May acidify its own rhizosphere, which would facilitate absorption by solubilization of the metal [29]	McCutcheon and Schnoor [127]
Zn-Zinc	xxx	*Trifolium pratense*	Red clover	Nonmetal accumulator	Its rhizosphere is denser in bacteria than that of *Thlaspi caerulescens*, but *T. caerulescens* has relatively more metal-resistant bacteria [29]	McCutcheon and Schnoor [127]

groundwater bioremediation technologies. Groundwater remediation techniques span biological, chemical, and physical treatment technologies. Most groundwater treatment techniques utilize the combination of technologies. Physical treatment techniques include but not limited to pump and treat, air sparging and dual phase extraction, while chemical treatments include ozone and oxygen gas injection, chemical precipitation, membrane separation, ion exchange, carbon absorption, aqueous chemical oxidation, and surfactant enhanced recovery. The biological treatment techniques, which are the focus of this chapter, include bioaugmentation, bioventing, biosparging, bioslurping, and phytoremediation. In the phytoremediation process, certain plants and trees are planted whose roots absorb contaminants from groundwater over time and are harvested and destroyed. This process can be carried out in areas where the roots can tap the groundwater. Examples of plants used in this process include the Chinese ladder fern (*Pteris vittata*) also known as brake fern, which is a highly effective accumulation of arsenic, an important groundwater contaminant. In addition, genetically altered cottonwood trees are good absorbers of mercury and transgenic Indian mustard plants soak up selenium quite well [109,133,135].

In situ bioremediation of groundwater speeds the natural biodegradation processes that takes place in the water soaked underground region that has below the water table. For sites at which both the soil and groundwater are contaminated, this single technology is effective at treating both. Generally an onsite groundwater bioremediation system consists of an extraction well to remove groundwater from the ground, an above groundwater, and injection wells to return the conditioned groundwater to the subsurface where the microorganisms degrade the contaminants. One limitation of this technology is that differences in underground soil layering and density may cause reinjected conditions groundwater to follow certain preferred flow paths, consequently, the conditioned water may not reach some areas of contamination. Frequently used method of *in situ* groundwater treatment is air sparging which involves pumping air into the groundwater to flush out contaminants Air sparging is used in conjunction with a technology called soil vapor extraction.

The advent of nanotechnology has witnessed the development of many bioremediation products. Example of such products is Nualgi, a nanotechnology micronutrient powder that causes a bloom of Diatom algae in any type of water. Diatoms are naturally present in all water bodies and are at the bottom of the marine food chain where they increase the dissolved oxygen level of water rapidly and consume nitrogen and phosphorus in the water. Nualgi can be mixed into sewage and wastewater where it causes a bloom of diatoms [104]. The process does not require any mechanical mixing which makes it quite simple. Oxygen from diatoms substitutes oxygen from electric aerators in wastewater treatment plants. Nualgi can also be used in natural water bodies and specially build ponds or lagoons fresh and salt waters. Algae is a diverse group of organism sharing only a few characteristics, they do photosynthesize capturing carbon dioxide and producing oxygen they do not make flowers and have simple reproductive structures and they are relatively simple without roots, stems, and leaves. Algae can be microscope that is, microalgae like the unicellular phytoplankton horns, and they can also be very large and referred to as macroalgae such as the giant kelps which can be found in marine waters, freshwater, on trees, as associated with fungi in lichens, boring in stones in high attitude snow, glaciers, geothermal sources, and even deserts. A good example of this green is the seaweed which represents 51% of the total world aquaculture in the marine environment. All algae absorb dissolved nutrients such as nitrogen and phosphorus to grow while trapping Co_2 and released O_2 making them suitable bioremediation candidates of either greenhouse gas emissions or of nutrients from fish farm. However, cultivating algae is relatively expensive [136].

Bioremediation of metal/metalloid—contaminated groundwater makes use of sulfate reducing bacteria (SRB) that reduce sulfate to sulfide while oxidizing a carbon source [124]. The sulfide so generated can remove metals, precipitating then as metal sulfides. The effectiveness of SRB in removing metals from contaminated groundwater depends on the choice on the choice of an appropriate organic carbon source for use by the bacteria the primary consideration when selecting a carbon source is its effect on the extent of microbial activity (biotreatment efficiency) and economic feasibility [80]. SRB oxidize organic matter into bicarbonate anaerobically using sulfate as a terminal electron acceptor according to the reaction:

$$2CH_2O + SO_4^{2-} \rightarrow H_2S + 2HCO_3$$

Where CH_2O represents the organic substrate, the hydrogen sulfide generated may form insoluble complexes with many heavy metals such as Ni, Zn, Cu, [154]. Molasses is a by-product of sugar processing and can be employed as a relatively cheap carbon source. The composition of molasses can be influenced by a number of factors and contains varying amounts of water, sucrose, glucose, muctose, and nonnitrogenous acids [148]. Immobilized microbial cells often outperform their planktonic counterparts in the treatment of polluted water [182]; thus, the immobilization of SRB on solid support matrices could improve their performance during the bioremediation of metal contaminated waters [177,181,187].

53.4.2 Surface Water

Surface waters are waters located above the ground levels and they include streams, rivers, and ponds. Surface water has been shown to have so many beneficial uses and is easily polluted due to easy access. Such uses include support for aquatic life, drinking water supply, agriculture industrial and municipal uses. Pollution of surface water is usually from untreated sewage municipal sources, storm runoff, and industrial effluent discharges. The accumulation of organic pollutants including PCBs, DDT, and heavy metals such as Pb, Cd, Se, and As in the aquatic systems can cause serious problems in the environment and organisms affecting negatively the stability of many aquatic ecosystems and can also cause difficulties for animals and human health. The accumulation of these pollutants is due to intensive anthropogenic activity. These problems of pollution can be partially solved by the application of phytoremediation technologies using algae or aquatic plant to remove pollutants from the environment. A bioremediation technology using *Micrococcus roseus* immobilized in porous spherical beads by an improved polyvinyl alcohol (PVA)–sodium alginate (SA) embedding method was found to be very effective in remediating contaminated surface water. The immobilized cells have an excellent tolerance to pH and temperature changes and were also more resistant to heavy metal stress compared with free cells. The immobilized *M. roseus* possesses an excellent regeneration capacity and could be reused after 180-day's continuous usage. [114]. Polluted surface water can also be remediated in a bioreactor using biofilms on filamentous bamboo in batch and continuous flow modes. The CODcr (chemical oxygen demand, using $K_2Cr_2O_7$ as oxidizer) removal rate of the enhanced systems increased more than 13% relative to a controlled system. In the batch reactor with 4-h cycles, the CODcr was mainly removed by the biofilm on the filamentous bamboo. The biofilm contained diverse organisms including protozoa and metazoa. [43].

53.4.3 Stormwater Runoff

Stormwater runoff from urban areas is a major contributor to groundwater and lake contamination [29,70]. Types of contaminants that are of concern are hydrocarbons, oil and grease, suspended solids, nitrogen, and heavy metals. These are common to areas with roadways; parking lots and other impervious surfaces prevalent in urban areas. The goal of remediation is to remove or change the nature of the pollutants in such a way that they are not harmful to the environment that is receiving the treated waters. Urban contaminants are a leading cause of pollution in storm water runoff. A very viable option to remediate storm water runoff is through phytobial remediation in which the efforts of bioremediation and phytoremediation in water and soil were harnessed through the unique interactions between microorganisms and plant roots. These interactions take place in the plants rhizosphere. Vegetative swalves can be constructed in areas that use this concept to treat waters before they enter other water sources [29].

53.4.4 Heavy Metals in Water Bodies

Heavy metal contamination in aquatic ecosystems due to discharge of industrial effluents may pose a serious threat to human health alkaline precipitation, ion exchange columns, electrochemical removal, filtration, and membrane technologies are the currently available technologies for heavy metal removal. These conventional technologies are not economical and many produce adverse impacts on aquatic ecosystems. Phytoremediation of metals is a cost-effective green technology based on the use of specially selected metal accumulating plants to remove toxic metals from soils and water. Wetland plants are important tools

in the removal of heavy metals from polluted environments [140]. Wetlands plants are preferred over other bio-agents due to their low cost, frequent abundance in the aquatic ecosystem, and ease of handling. The extensive rhizosphere of wetland provides an enriched culture zone for the microbes involved in degradation. The wetland sediment zone provides reducing conditions that are conducive to the metal removal pathway. Constructed wetland proved to be effective for the abatement of heavy metal pollution from acid mine drainage, landfill leachate, thermal power and municipal agricultural, refinery and chloro-alkali effluent. The physicochemical properties of wetlands provide many positive attributes for remediating heavy metals. Some of the wetland plants with high capacity for heavy metal removal include *Typha, Phragmites, Eichhornia, Azolla,* and *Lemna* and other aquatic macrophytes. However, biomass disposal problem and seasonal growth of aquatic macrophytes are some limitations in the transfer of phytoremediation technology from the laboratory to the field. Currently, the disposed biomass of macrophytes is being considered for usage in biofuel production and biomining for rare metals [98,156] Heavy metals that have been identified in the polluted environment include As, Cu, Cd, Pb, Cr, Ni, Hg, and Zn. The presence of any metal may vary from site to site, depending on the source of individual pollutant. Excessive uptake of metals by plants may produce toxicity in human nutrition, and cause acute and chronic diseases. The choice of plants to be used for phytoremediation depends on the target metals while some hyperaccumulators show capacity for more than one metal, others are good at specifics. To date, more than 400 plant species have been identified as metal hyperaccumulators representing less than 0.2% of all angiosperms [38,19]. *Thelaspi* spp., *Arabidopsis* spp., and *Sedium alfredi* spp. belonging to the family of Brassicaceae and *Alyssum* are known for bioaccumulating for Cd, Ni, Pb, and Zn, *T. goesingense* for Ni and Zn, *T. ochroleucum* for Ni and Zn, and *T. rotundi folum* for Ni, Pb, and Zn [217]. The roots of Indian mustard are found to be effective in the removal of Cd, Cr, Cu, Ni, Pb, and Zn and sunflower can remove Pb, U, Cs 137, and Sr-90 from hydroponic solutions [207,214,217]. Duckweed is a good accumulator for Cd, Se, and Cu, a moderate accumulator for Cr, but a poor accumulator of Ni and Pb. Water hyacinth can be used in wastewater treatment systems to improve water quality by reducing the levels of organic and inorganic nutrients. This plant can also reduce the concentrations of heavy metals in acid mine water while exhibiting few signs of toxicity. Water hyacinth accumulating trace elements such as Ag, Pb, Cd, and is efficient for phytoremediation of wastewater polluted with Cd, Cr, Cu, and Se [216]. Sharp duckweed can be used to remediate Hg and calamus for Pb. Brake fern (*Piteris vilta*) is a good hyperaccumulator for As. One fern cultivar is available commercially for As phytoremediation and has been successfully used in field trials [169].

Recently, the use of aquatic plants, especially micro- and macroalgae has received much attention due to their ability to absorb metals, ability to grow both autotrophically and heterotrophically, large surface area/volume ratios, phototaxy phytochelatin expression, and potential for genetic manipulations

[43,108,115,116]. Macroalgae have been used extensively to measure heavy metal pollution and marine environments throughout the world. In recent years, several species of the green algae *Enteromorpha* and *Cladophora* have been utilized to measure heavy metal levels in many parts of the world. [4]. The ability of macroalgae to accumulate metals within their tissues has led to their widespread use as biomonitors of metal availability in marine systems [82,157]. *Chlorophyta* and *Cyanophyta* are hyperabsorbent and hyperaccumulators for arsenic and boron [17] and therefore absorbing and accumulating these elements from their environments into their bodies. The brown algae (Phaeophyta) accumulate metals due to the high levels of sulfated polysaccharides and alginates within their cell walls for which metals show a strong affinity [49,58]. The brown algae *Focus* spp. often dominate the vegetation of heavy metals contaminated habitats. A marine green microalgae, *Platymonas subcordiformis,* was shown to have a very high capacity for strontium uptake; however, very high concentration of strontium cause oxidative damage as evidenced by the increase in lipid peroxidation in the algal cell samples and decrease in growth rate and chlorophyll contents [131]. Duckweed (*Lemnar minor*) is useful in the removal of Cu while the blue green algae *Phormidium* can hyperaccumulate Cd, Zn, Pb, Ni, and Cu [207]. Some algal species may convert mercuric or phenylmercuric ions into metallic mercury, which is then volatilized out of the cell and from the solution [59,60]. The principal mechanism of metallic eathion sequestration involves the formation of complexes between a metal ion and functional groups on the surface or inside the porous structure of the biological material. The carboxyl group of alginate plays a major role in the complexation. Different species of algae and the algae of the same species have different absorption capacity. Further, greater accumulation of Al and B may be related to the high concentrations of pectins in the cell wall. This would provide the plants a strong tolerance to heavy metals and therefore greater phytoextraction [16] (Table 53.4)

Some plant enzymes degrade lignin, which is similar in structure to the PAH. The oyster mushroom, *Pleurotus oestrealus*, can degrade 80%–95% of all PAHs present in soil after 80 days [189]. *Stropharia nogosoannualata* was found to be the most efficient strain of basidiomycete for the removal of a variety of PAHs, doing away with over 85% of them after six weeks. The addition of manganese enhances the performance of the degradation due to the role of manganese peroxidase, one of the extracellular lignolytic enzymes, as an important component of the degradation [189]. A mixture of white rot fungi and bacteria has been suggested to be best for PAH degradation. As the fungi breaks down the largest molecules into low molecular weight PAHs, the bacteria then act on those molecules; a case for phytobial process in remediation. Sometimes, the same fungi that degrade PAHs can also remediate toxic metals. *Fusarium* and *Hypocrea* could degrade one carcinogenic high-weight PAH, pyrene, as well as uptake copper and zinc. There is, however, a caution in the use of fungi associated with PAH degradation, which require them to be handled with care. *Cladophialophora bantiana* in particular

TABLE 53.4 Some Algae and Metals They Accumulate

Species	Metal	Reference
Ascophyllum nodosum	Gold (Au)	Haverkamp et al. [89]
		Kuyucak and Volesky [111]
	Cobalt (Co)	Kuyucak and Volesky [110]
	Nickel (Ni)	Holan and Volesky [94]
	Lead (Pb)	Holan and Volesky [94]
Caulerpa racemosa	Boron (B)	Bursali et al. [41]
Daplmia magna	Arsenic (As)	Irgolic et al. [98]
Fucus vesiculosus	Zinc (Zn)	Forest and Volesky [76]
	Nickel (Ni)	Holan and Volesky [94]
Laminaria japonica	Zinc (Zn)	Forest and Volesky [76]
Micrasterias denticulate	Cadmium (Cd)	Volland et al. [204]
Phormedium bohner	Cromium (Cr)	Dwivedi et al. [66]
Platymonas subcordiformis	Strontium (Sr)	Mei et al. [13]
Sargassum filipendula	Copper (Cu)	Davis et al. [58]
Sargassum fluitans	Copper (Cu)	Davis et al. [58]
	Iron (Fe)	Figueira et al. [74]
	Zinc (Zn)	Forest and Volesky [76]
	Nickel (Ni)	Holan and Volesky [94]
Sargassium natans	Lead (Pb)	Holan and Volesky [94]
Sargassum vulgare	Lead (Pb)	Holan and Volesky [94]
Spirogvra hyaline	Cadmium (Cd), mercury (Hg), lead (Pb), arsenic (As), and cobalt (Co)	Nirmal Kumar and Oommen [145]
Tetraselmis chuil	Arsenic (As)	Irgolic et al. [98]

Source: Adapted from Ben Chekroun, K. and Baghour, M. 2013. *Journal of Material and Environmental Science*, 4(6), 873–880.

causes a severe brain infection, which is fatal without brain surgery and intensive follow-up treatment. It had been observed that many fungal orders' ability to decompose lignin is linked to the ability to infect the human brain and cause neuropathy. Probably the reason is because the chemical make up of the brain is also high in aliphatic and phenolic compounds and lipids, 50% of the brain is composed of aliphatic lipids, which are similarly affected by the nonspecific lignin-degrading enzymes. Furthermore, dopamine, one of the main neurotransmitters throughout the brain, polymerizes into polycyclic aromatic rings and degrades into the same by-products as lignin, making it another target of the fungi enzymes within the brain.

53.4.5 Petroleum Oil Spills and Hydrocarbon Pollutants

One of the largest categories of inputs to the environment in need of decomposition is manufactured hydrocarbon waste. A majority of such pollution that occurs involves fossil fuels. Whether it is the exhaust and by products of spent fuel or the accumulated polymer plastics made from these same hydrocarbons. Fossil fuels are composed of PAHs as well as shorter carbon molecules. Seven of the sixteen PAHs listed as pollutants by the EPA are carcinogenic, teratogenic, and mutagenic. The buildup of waste polymers is also a worldwide problem. Fungi are utilized in the decomposition of petroleum spills because they produce extracellular enzymes.

A very common contaminant of surface water is petroleum oil spills. Petroleum oil is toxic for most life forms and episodic and chronic pollution of the environment by oil causes major ecological perturbations. Marine environments are especially vulnerable because oil spills of coastal regions and the open sea are poorly contained and mitigation is difficult. In addition to pollution through human activities, millions of tons of petroleum enter the marine environment every year from natural seepages. Despite its toxicity, a considerable fraction of petroleum oil entering marine systems is eliminated by the hydrocarbon degrading activities of microbial communities, in particular, by a remarkably recently discovered group of specialists, the so-called hydrocarbonoclastic bacteria (HCB). *Alcanivorax borkumensis,* a paradigm of HCB and probably the most important global oil degrader was the first to be subjected to a functional genomic analysis. HCB also has potential biotechnological applications in the areas of bioplastics and bio catalysts [147].

Until recently, one of the most ignored contaminants of surface waters were pharmaceuticals. Antibiotics in particular are difficult to selectively remove from surface waters by present treatment methods. Bacterial efflux pumps have evolved the ability to discriminately expel antibiotics and other noxious agents via proton and ATP driven pathways. A method whereby light-dependent removal of antibiotics was made possible by engineering the bacterial efflux pump AcrB into a proteovesicle system. A chimeric protein with the requisite proton motive force was created by coupling AcrB to the light-driven proton pump delta-rhodopsin (dR) via a glycoporin, a transmembrane domain. This creates a solar powered protein material capable of selectively capturing antibiotics from bulk solutions using environmental water and direct sunlight. AcrB-dR vesicles removed almost twice as much as the activated carbon treatment, which is the standard method. Altogether, the AcrB-dR system provides an effective means of extracting antibiotics from surface waters as well as potential antibiotic recovery through vesicle solubilization. Another phytoremediation method using bamboos for the treatment of wastewater has been patented by a French company. In this case, wastewater is evenly spread over a bamboo grove and supplies water and nutrient growth for plants. The bamboo grove produces new culms each year and mature culms can be harvested and processed in wood energy industries. It was observed in this technology that 99% of the organic matter and 98% of the nutrients were removed by the soil–bamboo system [15,163,175].

53.4.6 Reduction of Ammonium Hydroxide in Water

Ammonia water or ammonium hydroxide (NH_4OH), also referred to as alkali, is obtained through the absorption of ammonia in demineralized water of very high purity. This reaction is highly exothermic and allows the preparation of a range of ammonia solutions; from 20.5% to 32.5% depending on the required applications. Ammonia water is used in the following processes: yeast manufacturing, nutrient removal, water treatment stations, adjustments of pH in food processing, chemical synthesis, regeneration of ion exchange resins (demineralization), and nitrogen oxide reduction (using GP NOX–A). In the human body, ammonia is normally produced in the intestinal tracts at significantly higher levels than from external exposure. Since ammonia occurs naturally in the environment, humans are regularly exposed to low levels of ammonia through water, food, air, consumer products, and soil. Ammonia is commonly found in surface water and rainwater. Groundwater generally contains low concentrations of ammonia but some deep wells affected by specific geological formations have been shown to have high concentrations of ammonia. The level of ammonia in surface water varies regionally and seasonally and can be affected by localized anthropogenic influences such as runoff from agricultural fields or industrial or sewage treatment discharges. Ammonia may also be added to treated water as part of the disinfection strategy to form chloramines as a secondary disinfectant. In municipal water treatment plants, ammonia can be removed using physicochemical processes such as break point chlorination, ion exchange and membrane filtration and biological treatment (controlled nitrification). At the residential level, some treatment devices utilize osmosis and ion exchange mechanism. There are currently available different words of biological water treatment processes. Most of these operate in a fixed biofilm configuration, which includes a biogrowth support medium for the bacterial activity [10,121,141,161,162]. Other systems operate in a suspended growth mode, where bacteria are hydraulically maintained in suspension within a reactor such as fluidized bed filter [79,81]. A full-scale biological treatment plant consisting of three parallel gravity flow sand filters, each operated with a hydraulic loading rate of 2 gallons per minute (gpm) per square foot (4.9 m/n) was shown to remove 95% of ammonia present in water. The filtered water was chlorinated and had a free chlorine residual of 0.9 mg Cl_2/L and a stable pH (Lytle et al. 2007). A rise in the nitrate—nitrogen concentration (NO_3–N) from below 0.04 mg/L to 1.11 mg/L in the filtered water while no nitrite was detected, an indication of a complete oxidation of ammonia to nitrate through the filters [144].

53.4.7 Wetlands

Wetlands can be defined as areas where dry land meets or is saturated by water. In other words, where the water table is at or near the surface of the land or where the land is covered by shallow water. Wetlands are of two types: (1) natural wetland and (2) constructed or artificial wetlands. Artificial or constructed wetlands are manmade for treatment of wastewater, polluted water, and desalinated water. The natural wetlands are nature's creations and are characterized by water-dependent species of plants and animals, cover 4%–6% of Earth–land surface and are of great economic value. Being one of the most productive environmental service producers on the planet, wetlands deliver products such as fish, timber, wildlife, medicinal plants, drinking water, and energy resources, and perform many vital functions such as storage of water supply, regulation of water table, storm protection and flood mitigation, retention of nutrients, sediments and pollutants, and groundwater recharge or discharge. Wetlands are among the most threatened ecosystems in the world due to their ongoing loss and degradation and this underpins the establishment of the Ramsar Convention on Wetlands of 1971 [158].

53.4.7.1 Natural Wetland

Assessing the phytoremediation potential of wetlands is complex due to variable conditions of hydrology, soil/sediment types, plant species diversity, growing season, and water chemistry. Conclusions about long-term phytoremediation are further complicated by the process of ecological succession in wetlands. Physico-chemical properties of wetlands provide many positive attributes for remediating contaminants the expensive rhizosphere of wetlands herbaceous shrub and tree species provides an enriched culture zone for microbes involved in degradation. Redox conditions in most wetland/soil sediment zones enhance degradation pathways requiring reducing conditions. However, heterogeneity complicates generalizations within and between systems [99]. Contaminant loads in water and sediments including metals, volatile organic compounds VOCs, pesticides and other organohalogens, TNT and other explosives, petroleum hydrocarbons and additives are known to be taken care of by the natural wetland remediating systems. Many of the processes utilized in the course of remediation include rhizofiltration, phytobial remediation, phytoextraction, and actions of plant tissue enzymes [209]. The removal of metals in wetlands occurs through a member of processes which include sedimentation/

coagulation, filtration, plant uptake/removal efficiency, adsorption (binding to sand particles and roots) formation of solid compounds, cation exchange and microbial mediated oxidation [77] vegetation in a wetland provides a substrate (roots stems and leaves) upon which microorganisms can grow as they breakdown organic materials [179]. This community of microorganisms is known as the periphyton. The periphyton and natural chemical processes are responsible for approximately 90% of pollutants and waste breakdown. The plants remove about 7%–10% of pollutants and act as a carbon source for the microbes when they decay. The lists of some plants in natural wetlands that can be selected in the construction of artificial wetlands are thus:

1. Nature trees tolerant of wet soils
   a. Red and silver maple (*Acer rubrum, A. saccarinum*)
   b. River birch (*Betula nigra*)
   c. *Catalpa* spp.
   d. Ash (*Fraxinus* spp.)
   e. Cotton wood (*Populus deltoides*)
   f. Swamp white oak (*Quercus bicolor*)
   g. Sycamores (*Platanus* spp.)
2. Nature shrubs tolerant of wet soils
   a. Red oster dogwood (*Cornus servicea*)
   b. Leather wood (*Dirca plaustris*)
   c. Winterberry (*Ilex verticilata*)
   d. Inkberry (*Ilex glabra*)
   e. Pussy willow (*Salix discolor*)
   f. Shrubby cinquefoil (*Potentilla frushcosa*)
3. Nature herbaceous and flowering plants for sunny moist or boggy conditions
   a. Cattalis (*Typhus* spp.)
   b. Joe–Pye weed (*Eupatorium maculatum*)
   c. Great blue lobella (*Lobella siphilitica*)
   d. Iron weed (*Vernonia noveboracensis*)
   e. Blue flag Iris (*Iris versicolor*)
   f. Boneset (*Eupatorium perfoliatum*)
   g. Cardinal flower (*Lobella cardinalis*)
   h. Golden rods (*Solidago* spp.)
   i. Mersh marigold (*Caltha palustis*)
   j. Swamp milkweed (*Asciepias incarnate*)
   k. *Gentian* spp.
4. Nature herbaceous and flowering plants for shady, moist, or boggy conditions
   a. Bee balm (*Monarda didyma*)
   b. Arrow head (*Sagittaris latifolia*)
   c. False hellebore (*Veratrum viride*)
   d. Turtle head (*Chelone* spp.)
   e. Skunk cabbage (*Symplo carpus foetidue*)
   f. Royal fern (*Osmunda regalis*)
   g. Netted chain fern (*Woodwardia areolat*)
   h. Jack-in-the-pulpit (*Arisaema triphyllum*)
   i. Cinnamon fern (*Osmunda einnamonmea*)
   j. Shield ferns (*Dropteris* spp.)
   k. Laody ferns (*Athyrium* spp.)
5. True bog plants requiring low pH and sun

a. Sundews (*Drosera* spp.)
b. Butter worts (*Pinguicula* spp.)
c. Pitcher plants (*Sarracenia* spp.)

53.4.7.2 Constructed or Artificial Wetlands

Constructed wetlands are decentralized low-energy, low-cost systems to improve water quality. They rely on natural wetland functions, which include plants and microorganisms that uptake and breakdown the wastewater nutrients either aerobically or anaerobically. These systems are responsible for providing multiple benefits like improvement in water quality, water security and reuse, CO_2 reduction, and provides habitat for many plants and animals. It is also act as a source of recreation like rowing and fishing, education, with aesthetic and amenity values [184]. Plants which constitute free floating, emergent and submergent vegetation are the part of the constructed ecosystem to remediate contaminants from municipal, industrial wastewater, metals, acid mine drainage [176].

The microorganism that lures on the exposed surfaces of the aquatic plants and soils remove the dissolved and particulate organic matter and convert them to carbon dioxide and water. The process of active decomposition in the artificial wetlands produces the effluent, which has lower pH and BOD. Some plant species that are commonly used for eliminating contaminants include *Eichhornia cressipes*, *Chara* spp., *Ipomoea aquatic*, *Pistia stratiotes*, *Typha latifolia*, *Brassica juncea*, and *Helianthus annuus*. They are responsible for removal of nutrients such as heavy metals such as cadmium, chromium, copper, iron, nickel, lead, zinc, nitrate, phosphorus, hydrocarbons suspended solids, and COD. Plant selection depends upon plant characteristics such as tolerance to pH and salinity of wastewater, translocation and uptake capabilities, transpiration rate or water use, depth of root zone, native and non-native species and commercial availability. They should be disease, and drought resistant as well as insect and stress tolerant. Growth rate/biomass production and reproduction rate are also important selection criteria that affects the phytoremediation process as well as the designing of the wetland [91,183].

Constructed wetlands are of two basic types (a) subsurface flow and (b) surface flow wetland. The subsurface flow can be further classified as (i) horizontal flow and (ii) vertical flow constructed wetland. In subsurface-flow wetlands, more effluent that is, household wastewater, agricultural, paper mill wastewater, mining runoff, tannery, meat processing waste, storm drains or other water to be cleansed is passed through a gravel generally of limestone or volcanic rock lavastone material or sand medium on which plants are rooted [50]. In subsurface flow system, the effluent may move either horizontally parallel to the surface over vertically, from the planted layer down through the substrate and out. Subsurface horizontal flow wetlands are less hospitable to mosquitoes as there is to water exposed to the surface whose populations can be a problem in surface-flow constructed wetlands. They also have the advantage of requiring less land area for water treatment but are not generally as suitable for wildlife habitat as are surface flow constructed wetlands.

53.4.7.2.1 Surface Flow Wetlands

These wetlands move effluent above the soil in a planted marsh or swamp and thus can be supported by a wider variety of soil types including bay mud and other suly clays. Common phytoremediation plants used in surface-flow wetlands include cattails (*Typha* spp.) sedges, water hyacinth (*Eichhornia crassipes*), and *Pontederia* spp. In subarctic regions, buckbeans (*Menyanthes trifoliata*) and pendant grass (*Arctophila fulva*) have been found to be useful in remediation of heavy metals.

53.4.7.2.2 Tidal-Flow Wetland

These are the latest wetland technology used in the treatment of agricultural and industrial wastewater particularly in heavy load treatment. In this system, organic carbon is primarily oxidized with nitrate, which is produced through a series of flood and drain cycles from one side of the wetland to the other. This process holds a number of benefits over traditional subsurface and surface-flow wetlands, and these benefits include reduced land requirement and increased nitrification capabilities for the treatment of heavy load of contaminants. The combination of physical, chemical and biological processes are at play in wetlands to remove contaminants from wastewater. Theoretically, wastewater treatment within a constructed wetland occurs as it passes through the wetland medium and the plant rhizosphere. A thick film around each root hair is aerobic due to the leakage of oxygen from the rhizomes, roots and rootless [87]. Aerobic and anaerobic microorganisms facilitate decomposition of organic matter. Microbial nitrification and subsequent denitrification releases nitrogen as gas to the atmosphere. Phosphorus is co-precipitated with iron, aluminum, and calcium compound located in the root bed medium [57,87,168]. Suspended solids filter out as they settle in the water column in surface flow wetlands or are physically filtered out by the medium within subsurface flow wetland cells. Harmful bacteria and viruses are reduced by filtration and adsorption by biofilms on the rock media in subsurface flow and vertical flow systems. Nitrogen forms that are of importance in wetland wastewater treatment include organic nitrogen, ammonia, ammonium, nitrate, nitrite, and nitrogen gases. The inorganic forms of nitrogen are essential to plant growth on aquatic system while the wastewater nitrogen removal is important because ammonia is toxic to fish. Further, excessive nitrites in drinking water are thought to cause methemoglobinemia in infants with resultant decrease in blood oxygen transport ability.

Plants used are usually grown on coco peat (Lukmertens) at the time of implantation to water purifying ponds, denutrified soil is they used to prevent unwanted algae and other organisms from taking over. In addition, locally grown bacteria and nonpredatory fish are added to eliminate or reduce pests such as mosquitoes. The bacteria are usually grown locally by submerging straw to support bacteria arriving from the surrounding. Three types of nonpredatory fish are chosen to ensure their co-existence and these include

1. Surface swimming fish, for example, common dace (*Leuciscus leuciscus*), ide (*Leuciscis idus*), and common rudd (*Scardinius erythrophthalmus*).

2. Middle swimmers, for example, common roach (*Bulilus rutilas*)
3. Bottom swimming fish, for example, tench (*Tinca tinca*)

53.4.7.3 Desalination and Post Desalination Treatments

Desalination of agricultural land by phytoextraction has been a long tradition. The most common application of phytoremediation is constructed wetlands is wastewater treatment. There are however recent procedures for desalination and post desalination treatments using the same technology. Seawaters are known to contain salt while some of this salt can be removed through evaporation of water, phytoremediation using constructed wetlands is providing a complimentary process. In addition, salinity in treated wastewater is often increased especially in arid and semi-arid areas and may harm crops irrigated from wetlands, Haplophyte plants are thought to possess high capabilities for accumulating salt examples include *Bassia indica* [179], *Salicornia europaca*, and *Limonium gmelinii* [195] and are therefore very useful in water recovery.

53.5 Discussion

Land and water are precious natural resources on which the sustainability of agriculture and the civilization of humanity rely. Unfortunately, they have been subjected to maximum exploitation and severely degraded or polluted due to anthropogenic activities. The pollution includes point sources such as emission, effluents and solid discharge from industries, vehicle exhaustion and metals from smelting and mining, and nonpoint sources such as soluble salts (natural and artificial) use of pesticides/insecticides, disposal of industrial and municipal wastes in agriculture and excessive use of fertilizers [128,146,172]. Each source of contamination has its own damaging effects to plants and animals and ultimately to human health, but these that add heavy metals to soils and waters are of serious concerns due to their persistence in the environment and carcinogenicity to human beings. They cannot be destroyed biologically but are only transformed from one oxidation state or organic complex to another [78,80]. Therefore, heavy metal pollution poses a great potential threat to the environment and human health. In order to maintain good quality of soils and waters and keep them free from contamination, continuous efforts have been made to develop technologies that are easy to use, sustenance and economically feasible. Physicochemical approaches have been widely used for remedying polluted soil and water, especially at a small scale. However, they experience more difficulties for a large-scale remediation because of high costs and side effects. The use of plant species for cleaning polluted soil and water termed phytoremediation has gained increasing attention since last decade as an emerging cheaper technology. Numerous plant species have been identified and tested for their traits in the uptake and accumulation of different heavy metals. Mechanisms of metal uptake at whole plant and cellular levels have been investigated and progresses have been made in the mechanistic and practical application aspects of phytoremediation [119]. Phytoremediation

is actually a naturally occurring process, discovery of its effectiveness and advances in its application as an innovative treatment technology for contaminated land water and air have been recent. The technology utilizes plants, bacteria, fungi, and algae of various species to remediate contaminated environments. Further, phytoremediation has opened a new vista in environmental microbiology resulting in application of genetic engineering to cultivate transgenic plants, fungi, algae, and bacteria with unusual properties to bioaccumulate and metabolize environmental toxicants, leading to better understanding of plants and bacterial metabolism. Recently, nanotechnology has also been employed to produce such products as "Nualgi," a micronutrient powder that causes a bloom of diatom algae in any type of water where they increase the dissolved oxygen level of water rapidly and consume nitrogen and phosphorus in the water. Nualgi can also be used in natural water bodies and specially built ponds or lagoons fresh and salt water [104].

Every living organism requires water for survival and water covers nearly three-fourths of the surface of the earth. However, almost all of the world's water that is, about 97% is located in the oceans but as might be expected, the high concentration of salt renders the oceans virtually unusable as a source of water for municipal, agricultural, or most industrial needs. However, ocean water is useful in cooling of power plants and as a sink for much of our pollution. While desalination technologies exist, the capital and energy requirements to produce significant quantities of freshwater from such source are generally prohibitive, although some small regions of the world do rely quite heavily on the approach. It is pertinent to note that the sun performs some desalination when it provides the energy needed to evaporate water, leaving the salts behind. Freshwater, lakes and rivers, which are the main source of water for human use, account for just 0.007% of the world's stock of water or about 93,000 km³.

Roughly, 10% of the world's annual runoff is withdrawn for human use each year. While the small figure may suggest ample supplies for the future, that is not always the case. Some areas of the world are inundated with water while others have so little rainfall that human existence is barely possible. Even areas with adequate average precipitation are vulnerable to chaotic variation from year to year. Unless major water storage and conveyance facilities are constructed a region may have plenty of water on the average but not enough to cover needs during dry spells. Furthermore, the geographic distribution of water does not match well the distribution of people on the planet. For example, Asia with 60% of the world's population has only 36% of global runoff, while South America with only 5% of the world's population has 25% of the runoff [126].

Water that has been withdrawn used for some purposes and then returned will somewhat be polluted. Industrial return water may contain different categories of chemical and organic wastes while agricultural returns may contain pesticides fertilizers and salts. Municipal return waters carry human sewage while power plants discharge water with elevated temperature. Different phytoremediation processes have been found applicable for water reuse and these include rhizofiltration, bioaugmentation, bioventing, biosparging, and bioslurping. Contaminants that are removed by phytoremediation include heavy metals, pesticides, organic and inorganic chemicals and some radioactive mater rate/examples of higher plants used in phytoremediation include water hyacinth (*Eichhornia crassipes*), duckweed (*Lemna minor*), and water lettuce (*Pistia stritiotes*) most plants show accumulation, tolerance or hyperaccumulation of only one metal. Water plants show the largest ability to accumulate multiple metals. These include hydrilla (*Hydrilla verticillata*), duckweed (*Lemna minor*), water lettuce (*Pistia stratiotes*), water fern (*Salvinia melesta*), giant duckweed (*Spirodela polyrhiza*), water hyacinth (*Eichhornia crassipes*), water hyssop (*Bacopa monnieri*), water fern (*Azolla filiculoides*), and tape grass (*Vallisneria americana*) [129].

Interest in microbial biodegradation of pollutants has intensified in recent years. Oil and petroleum polluted waters have been remediated using bacteria. Bioremediation of contaminated water sources by sulfate reducing bacteria is proving to be a cost effective process especially if a suitable carbon source and support matrix were available [124]. Furthermore, various types of algae are being used in phytoremediation of heavy metals. These include both the micro and micro algae with varying degrees of metal accumulation. Recently, there has been a growing interest in using algae for biomonitoring eutriphication of organic and inorganic pollutants. By using the chlorophyll formation of the algae for example, it was possible to estimate spectrophotometrically the total nitrogen content in water collected from aquatic systems giving us an idea of eutrophication levels [1,23].

Phytoremediation is an emerging technology that uses plants to degrade, extract, contain, or immobilize contaminants such as metals, pesticides, explosives, oil, excess nutrients, and pathogens from soil and water. Phytoremediation has been identified as a more cost effective noninvasive and publicly acceptable method of removing environmental contaminants than most chemical and physical methods. Two nutrients commonly found in storm water runoff are nitrogen (N) and phosphorus (B). Both of these pollutants are also macronutrients needed for agronomic and horticultural plant growth and are components of all complete fertilizers. Fertilizer application to residential, commercial and municipal lawns and landscape is a major nonpoint source of pollution with potential for reduction via phytoremediation. The majority of plants currently used in phytoremediation including storm water ponds (BMPs) riparian buffers, rain gardens, green roofs, and constructed wetlands are herbaceous non-woody. However, new storm water runoff systems that incorporate woody landscape plants into the systems are being designed for streetscapes and landscapes [165].

However, it must be determined whether phytoremediation can be effective for the site-specific conditions and contamination. In some cases, phytoremediation might not provide adequate protection, from an eco-receptor perspective. For example, contamination that is below ground can be transferred into the leaves and stems of plants that are a food source. Further, in some cases, contaminants are not destroyed in the phytoremediation process; instead, they are transferred from the soil onto the plants and then are transpired into the air, it is also pertinent

to factor potential costs associated with monitoring and maintaining the phytoremediation process. It is also expedient that the development and evaluation of a particular phytoremediation design and long-term performance strategy be performed by a team of specialist in associated field including soil science or agronomy, hydrology, plant biology, environmental engineering regulatory analysis, cost engineering and evaluation, risk assessment and toxicology and landscape architecture.

53.6 Summary and Conclusions

Phytoremediation is currently proving to be an alternative cheaper and environmentally acceptable technology for the bioremediation of contaminated environments including air, land and water. This mode of remediation of polluted environments encompasses other associated techniques including bioremediation and mycoremediation. Overall vascular plants, micro and micro algae, bacteria and fungi which all belong to the plant super family are utilized understanding the fact that biological processes play a major role in the removal of contaminants and the astonishing catabolic versatility of microorganism to degrade and convert such compounds led to new methodological breakthrough in sequencing, genomics, bioteomics, bioinformatics and imaging producing vast amounts of information. In the field of environmental microbiology, genome based global studies opened a new era providing unprecedented *in silico* views as well as clues to the evolution of degradation pathways and to the molecular adaptation strategies to challenging environmental conditions some specific applications of phytoremediation include rhizofiltration for surface water and water pumped through troughs, phytotransformation for surface and groundwater, plant assisted bioremediation for soils, phytostabilization for soils, groundwater and mine tailings, phytovolatilization for soils and groundwater, removal of organics from the air and vegetable caps for soils. Phytoremediation has been employed for water reuse in remediating contaminated surface water, groundwater, urban run- off water, petroleum and hydrocarbon oil spills, reduction of ammonium hydroxide in water removal of pesticides from storm runoff water, removal of heavy metals in water, desalinization and post desalination treatments, natural and constructed wetlands. Some of the advantages of using this mode of treatment include cost-effectiveness, treatment of a wide variety of contaminants, provision of *in situ* treatment, offers permanent solutions and in some instances interim solution, can be integrated into the natural environment and landscaping plans and can be an effective element of a unified treatment—train remediation approach. Some of the disadvantages include dependence on climate and site-specific characteristics, treatment can take longer than other treatment options, it may not be applicable in all contaminated situations [26].

References

1. Abe, K., Takizawa, H., Kumra, S., and Hirano, A. 2004. Characteristics of chlorophyll formation of the aerial

microalgal *Coetastrella striolata* var. *multistriata* and its application for environmental biomonitoring. *Journal of Bioscience and Bioengineering*, 98, 34–39.

2. Adler, T. 1996. Botanical cleanup crews: Using plants to tackle polluted water and soil. *Science News,* July 20.

3. Adriano, D.C. 2001. *Trace Elements in Terrestrial Environments: Biochemistry, Bioavailability and Risks of Metals.* Springer-Verlag, New York.

4. Al-Homaiden, A.A., Al-Ghanayem, A.A., and Areej, H. 2011. Green algae as bioindicators of heavy metal pollution in Wadi Hanifah Stream, Riyadh, Saudi Arabia. *International Journal of Water Resources & Arid Environment*, 1(1), 10–15.

5. American Chemical Society (ACS). 2013. Engineering bacterial efflux pumps for solar-powered bioremediation of surface waters. http://www.pubs.acs.org. Accessed December 15, 2013.

6. Anderson, T.A., Guthrie, E.A., and Walton, B.T. 1993. Bioremediation, in the rhizosphere. *Environmental Science and Technology*, 27(13), 2630–2635.

7. Anderson, C.W.N., Brooks, R.R., Stewart, R.B., and Simcock, R. 1998. Harvesting a crop of gold in plants. *Nature*, 395(6702), 553–554.

8. Anderson, G.C. and Schubert, P. 1981. *Reclamation of Surface Mined Lands for the Production of Forages for Ruminant Animals: An Annotated Bibliography.* West Virginia University, Agricultural and Forestry Experiment Station, Morgantown, WV.

9. Anderson, I.C., Campbell, C.D., and Prosser, J.I. 2003. Diversity of fungi in organic soils under a moorland—Scotspine (*Pinus sylvestris* L.) gradient. *Applied and Environmental Microbiology*, 63, 840–843.

10. Andersson, A., Laurent, P., Kihn, A., Prévost, M., and Servais, P. 2001. Impact of temperature on nitrification in biological activated carbon (BAC) filters used for drinking water treatment. *Water Research*, 35(12), 2923–2934.

11. APHA (American Public Health Association), AWWA (American Water Works Association), and WPFC (Water pollution Control Federation). 1988. *Standard Methods for the Examination of Water and Waste Water*, 19th Edition. American Public Health Association, Washington.

12. Arfi, Y., Buée, M., Marchand, C., Levasseur, A., and Record, E. 2012. Multiple markers pyrosequencing reveals highly diverse and host-specific fungal communities on the mangrove trees *Avicennia marina* and *Rhizophora stylosa*. *FEMS Microbiology Ecology*, 79, 433–444.

13. Auge, R.M., Ebel, R.C., and Deam, X. 1994. Nonhydraulic signaling of soil drying in mycorrhizal maize. *Planta*, 193, 74–82.

14. Ayres, R.U. and Ayres, L.W. (eds.) 2002. *A Handbook of Industrial Ecology*. Edward Elgar Publishing Ltd., Cheltenham, UK.

15. Bagoudou, D., Bois, G. Korboulewsky, N., and Arfi, V. 2009. Initial efficiency of a bamboo grove-based treatment for winery waste water. *Desalination*, 247, 70–78.

16. Baghour, M., Moreno, D.A., Vilora, G., Hernandez, J., Cashtila, N., and Romero, L. 2001. Phyto extraction of Cd and Pb and physiological effect in potato plants (*Solanum tuberosum* var *spunta*). Importance of root temperature. *Journal of Agricultural and Food Chemistry*, 49, 5356–5363.

17. Baker, A.J.M. 1981. Accumulators and excluders—Strategies in the response of plants to heavy metals. *Journal of Plant Nutrition*, 3, 643–654.

18. Baker, A.J.M. and Brooks, R.R. 1989. Terrestrial higher plants which hyperaccumulate metallic elements—A review of their distribution, ecology and phytochemistry. *Biorecovery*, 1(2), 81–126.

19. Baker, K. 2008. *Costs of Reclamation on Southern Appalachian Coal Mines: A Cost-Effectiveness Analysis for Reforestation versus Hayland/Pasture Reclamation.* Virginia Polytechnic Institute and State University. http://scholar.lib.edu/theses/available/etd-06272008-162512.

20. Baldrian, P. 2003. Interactions of heavy metals with white-rot fungi. *Enzyme and Microbial Technology*, 32, 78–91.

21. Bang, S.W., Clark, D.S., and Keahing, J.D. 2000. Engineering hydrogen sulfide production and cadmium removal by expression of the thiosulfate reductase gene (Phs ABG) from salmonella enteric serovar typhimurium in E. *Applied and Environmental Microbiology*, 66, 3939–3944.

22. Beam, H.W. and Perry, J.J. 1974. Microbial degradation and assimilation of n-alkyl-substituted cycloparaffin. *Journal of Bacteriology*, 118(2), 391–399.

23. Bellow, L. 2002. Right as rain: Control water pollution with your own home. *House and Home*, 13(2), 44.

24. Ben Chekroun, K. and Baghour, M. 2013. The role of algae in phytoremediation of heavy metals: A review. *Journal of Materials and Environmental Science*, 4(6), 873–880.

25. Bennett, P.C., Hiebert, F.K., and Choi, W.J. 1996. Microbial colonization and weathering of silicates in petroleum contaminated ground water. *Chemical Geology*, 132, 45–53.

26. Bergkvist, B., Folkeson, L., and Berggren, D. 1989. Fluxes of Cu, Zn, Pb, Cd, Cr, and Ni in temperate forest ecosystems. *Water, Air and Soil Pollution*, 47(3), 217–286.

27. Berti, W.R., Cunningham, S.D., and Cooper, E.M. 1998. Case studies in the field: In-place inactivation and phytorestoration of Pb-contaminated soils. In: Cunningham, S.D. and Vangronsveld, J. (eds.), *Metal-Contaminated Soils: In Situ Inactivation and Phytorestoration*. Springer, Berlin, p. 265.

28. Black, 1995. Absorbing possibilities. Phytoremediation. *Environmental Health Perspectives*, 103, 12.

29. Bonilla-Warford, C. and Zedler, J.B. 2002. Potential for using native plant species in storm water wetlands. *Environmental Management*, 20(3), 385.

30. Books, R.R. and Young, X.H. 1984. Elemental levels and relationships in the endemic serpentine flora of the Great Dyke, Zimbabwe and their significance as controlling factors of the flora. *Taxon*, 33, 392–399.

31. Borovicka, J., Kortba, P., Gryndler, M., Mihaljevic, M., Randa, Z., Rohovec, J., Cajthami, T., Stijve, T., and Dunn,

C.E. 2010. Bioaccumulation of silver in ectomycorrhizal and saprobic macrofungi from pristine and polluted areas. *Science of the Total Environment*, 408(13), 2733–2744. 3.26.

32. Bouchez, T., Patureau, D., Dabert, P., Juretschko, S., Dore, J., Delegenes, P., Moletta, R., and Wagner, M. 2000. Ecological study of bioaugmentation failure. *Environmental Microbiology*, 2, 179–190.

33. Boyd, R.S. 1998. Plants that hyperaccumulate heavy metals. In: Brooks, R.R. (ed.), CAB International, Willingford, UK, pp 181–201.

34. Bradley, F.M. and Ellis, B.W. (eds.) 1992. *Gardening the Indispensable Resource for Every Gardener*, 2nd Edition, Rodale, Emmaus, PA.

35. Bradley, F.M. and Barbara, E. (eds.) 1997. *Rodale's Encyclopedia of Organic Gardening*. Rodale Press, Emmaus, PA.

36. Bragg, J.R., Prince, R.C., Harner, E.J., and Atlas, R.M. 1994. Effectiveness of bioremediation for the Exxon-Valdez oil spill. *Nature*, 368, 413–448.

37. Brooks, R.R. 1998. Geobotany and hyperaccumulators. In: Brooks, R.R. (ed.), *Plants That Hyperaccumulate Heavy Metals: Their Role in Phytoremediation, Microbiology, Archaeology. Mineral Exploration and Phytomining*, CAB International, New York, pp. 55–94.

38. Brooks, R.R. (ed.). 1998. *Plants That Hyperaccumulate Heavy Metals: Their Role in Phytoremediation, Microbiology, Archaeology, Mineral Exploration, and Phytomining*. CAB International, Oxford.

39. Bryant, G. 2006. *Plant Propagation A to Z: Growing Plants for Free*. Firefly Books, Buffalo, NY.

40. Burken, J.G. 2004. Uptake and metabolism of organic compounds: Green-liver model. In: McCutcheon, S.C. and Schnoor, J.L. (eds.), *Phytoremediation: Transformation and Control of Contaminants, A Wiley-Interscience Series of Texts and Monographs*. John Wiley, Hoboken, NJ, p. 59.

41. Bursali, E.A., Cavas, L., Seki, Y., Bozkurt, S.S., and Yurdakoc, M. 2009. The sorption of boron by invasive marine sea weed: *Canlerpa racemosa* var. *cylindrecea*. *Chemical Engineering Journal*, 150, 385–390.

42. Cairney, T. 1993. *Contaminated Land: Problems and Solutions*, 2nd Edition. CRC Press, Boca Raton, FL.

43. Cao, W., Zhang, H., Wang, Y., and Dan, J.Z. 2012. Bioremediation of polluted surface water by using biofilms on filamentatous bamboo. *Ecological Engineering*, 42, 146–149.

44. Carson, R. 2002. *Silent Spring*, 40th Anniversary Edition. Houghton Mifflin, Boston.

45. Carter, C.T. and Ungar, I.A. 2002. Aboveground vegetation, seed bank and soil analysis of a 31 year old forest restoration on coal mine spoil in southeastern Ohio. *The American Midland Naturalist*, 147(1), 44–59.

46. Centre for Public Environment Oversight (CPEO). 2013. Dual Phase Extraction. The Center for Public Environmental Oversight (CPEO). http://www.cpeo.org./techtree/ttde-script/dualphex.htm. Retrieved November 29, 2009.

47. Chen, S. and Wilson, D.B. 1997. Construction and characterization of *Escherichia coli* genetically engineered for bioremediation of Hg (2+) contaminated environments. *Applied and Environmental Microbiology*, 63(6), 2442–2445.

48. Chiboju, D.J, Jadia, M., and Fulekar, H. 2008. Phytoremediation: The application of vermicompost to remove zinc, cadmium, copper, nickel and lead by sunflower plant. *Environmental Engineering and Management Journal*, 7(5), 547–558.

49. Chiarawatchai, N. 2010. Implementation of earthworm-assisted constricted wetlands to treat wastewater and possibility of using alternative plants in constructed wetlands. doku.b.tuharbugde/volltexte/2010/858/pdf/Einzelseiten_ok.pdf.

50. Chowdhury, Z.K., Passantino, L., Summers, S., Work, L., Rossman, L., and Uber, J. 2006. *Assessment of Chloramine and Chlorine Residual Decay in the Distribution System*. American Water Works Association Research Foundation and American Water Works Association, Denver, CO.

51. Colpaert, J.V. 1998. Biological interaction: The Significance of root-microbial symbioses for phytorestoration of metal contaminated soils. In: Vangronsveld, J. and Cunningham, S.D. (eds.), *Metal Contaminated Soils: In situ Inactivation and Phytorestoration*, Springer, Berlin, pp. 75–84.

52. Cooney, C.M. 1996. Sunflowers remove radionuclides from water in ongoing phytoremediation field tests. *Environmental Science and Technology*, 30, 194.

53. CPEO, 2013. Bioslurping the Center for Public Environmental Oversight (CPEO). http://www.epeo.org/techtree/ttdescript/bislurp.htm. Retrieved November 29, 2013.

54. CPEO, 2013. Bioventing The Center for Public Environmental Oversight (CPEO). http:///www.epeo.org./techtree/ttdescript/bioven.htm. Retrieved November 29, 2013.

55. Cunningham, S.D., Anderson, T.A., Schwab, A.P., and Hsu, F.C. 1996. Phytoremediation of soil contaminated with organic pollutants. *Advances in Agronomy*, 56, 55–114.

56. Dannibale, D.L., Rosetto, F., Leonardi, V., Federia, F., and Petruccioli, M. 2006. Role of autochthonous filamentous fungi in bioremediation of soil historically contaminated with aromatic hydrocarbons. *Applied Environmental Microbiology*, 72(1), 28–36.

57. Davies, T.H., Hart, B.T. 1990. Use of aeration to promote nitrification in reed beds treating wastewater. *Advanced Water Pollution Control*, 11, 77–84.

58. Davis, T.A., Volesky, B., and Vierra, R. HSF. 2000. Sargassum seaweed as biosorbent for heavy metals. *Water Research*, 34, 4270–4278.

59. De Filippis, L.F. 1978. The effect of sub lethal concentrations of mercury and Zinc on Chlorella IV characteristics of a general enzyme system for metallic ions. *Zeit Schrift Panzenphysiologie*, 86, 339–352.

60. De Filippis, L.F. and Palaghy, C.K. 1994. In: Rai, L.C., Gaur, J.P. Soeder, C.J. (eds.), *Algae and Water Pollution*. E. Schweizer bartsche verlag buch handling, Stutgart, Germany, p. 31.

61. Devices, W.J., Taridieu, F., and Trego, C.L. 1994. How do chemical signals look in plants that grow in drying soil. *Plant Physiology*, 104, 304–314.

62. Dong, Y., Kumar, C.G., Chia, N., Kin, P-J., Miller, P.A., Price, N.D., Can, Iko, et al. 2013. Halomonas sufideris—dominated microbial community inhabits 91.8 km—deep subsurface Cambrian sand stone reservoir. *Environmental Microbiology*, Doi 10.1111/1462–2920.

63. Duran, N. and Esposito, E. 2000. Potential applications of oxidative enzymes and phenoloxidase-like compounds in wastewater and soil treatment. *Applied Catalysis B*, 28, 83–89.

64. Duschenkor 2003. Trends in phytoremediation of radionuclides. *Plant and Soil*, 249, 167–175.

65. Dushenkov, V., Harry, M., Ilya, R., and Nanda Kumar, P.B.A. 1995. Rhizofiltration: The use of plants to remove heavy metals from aqueous streams. *Environmental Science Technology*, 30, 1239–1245.

66. Dwivedi, S., Srivastava, S., Mishra, S., Kumar, A., Tripathi, R.D., Rao, U.N., Dave, R., Tripathi, P., Charkrabarty, D., and Trivedi, Plc. 2000. Characterization of native microalgal strains for the chromium bioaccumulation potential: Phytoplankton response in polluted habitats. *Journal of Hazardous Materials*, 173, 95–101.

67. Dwivedi, S., Tripathi, R.D., Tripathi, P., Kumar, A., Dave, R., Mishra, S., Singh, R. et al. 2010. Arsenate exposure affects amino acids Mineral nutrient status and antioxidants in rice (*Oryza sativa* L.) genotypes. *Environmental Science and Technology*, 44, 9542–9549.

68. Ebbs, S.D., Lasat, M.M., Brady, D.J., Cornish, J., Gordon, R., and Kochian, I.V. 1997. Phytoextraction of cadmium and zinc from a contaminated soil. *Journal of Environmental Quality*, 26(5), 1424–1430.

69. Elpel, T.J. 2004. *Botany in a Day: The Patterns Method of Plant Identification*. 5th Edition. HOPS Press, Pony, MT.

70. EPA. 1998. *A Citizen's Guide to Phytoremediation, U.S. Environmental Protection Agency, Office of Solid Waste and Emergency Response*. EPA 542-F-98–011, www.epa.gov/. Accessed on October 11, 2013.

71. Fernandez-Luquano of R., Martnez-Su Dendooven L. 2010. Microbial communities to mitigate contamination of PAHs in soil—possibilities and challenges: A review. *Environmental Pollution Science Research*, 10, 11–30.

72. Fiegl, J.L., McDonnell, B.P., Kostel, J.A., Finster, M.E., and Gray, K. 2005. *A Resource Guide. The Phytoremediation of Lead in Urban Residential Soils*. Northwestern University, Evanston, IL.

73. Field, C.B., Malson, P.A., and Money, H.A. 1992. Responses of terrestrial ecosystems to a changing atmosphere: A resource based approach. *Annual Review of Ecology and Systematics*, 23, 201–235.

74. Figueira, M.M., Volesky, B., and Ciminelli, V.S.T. 1997. Assessment of interference in biosorption of heavy metals. *Biotechnology and Bioengineering*, 54, 344–349.

75. Filip, Z. 1978. Decomposition of polyurethan—a garbage landfill leakage water and so microorganisms. *European Journal of Applied Microbiology and Biotechnology*, 5, 225–231.

76. Forest, E. and Volesky, B. 1997. Alginate properties and heavy metals biosorption by seaweed. *Applied Biochemistry and Biotechnology*, 67(3), 215–226.

77. Ganjo, D.G.A. and Khwakaram, A.I. 2010. Hytoremediation of wastewater using some of aquatic macrophytes as biological purifiers for irrigation purposes (removal efficiency and heavy metals FE, Mn, Zn and Cu). *World Academy of Science, Engineering and Technology*, 565–567, 574.

78. Garbisu, C. and Alkorta, I. 2001. Phytoextraction: A cost effective plant-based technology for the removal of metals from the environment. *Bioresource Technology*, 77(3), 229–236.

79. Gauntlett, R. 1981. Removal of ammonia and nitrate in the treatment of potable water. In: Cooper, P. and Atkinson, B. (eds.), *Biological Fluidized Bed Treatment of Water and Wastewater*. Water Research Centre, Stevenage Laboratory, Horwood, London, p. 48.

80. Gisbert, C., Ros, R., de Haro, A., Walker, D.J., Pilar, B.M., Serrano, R., and Avino, J.N. 2003. A plant genetically modified that accumulates Pb is especially promising for phytoremediation. *Biochemical and Biophysical Research*, 303(2), 440–445.

81. Goodall, J.B. 1979. *Biological Removal of Ammonia*. U.S. Environmental Protection Agency, Washington, DC (EPA 570/19-79/020).

82. Gosavi, K., Sammut, J., Gifford, S., and Jankowski, J. 2004. Macroalgal bio-monitors of trace metal contamination in acid sulfate soil aquaculture ponds. *Science of the Total Environment*, 324, 25–39.

83. Grauer, U.E. and Horst, W.J. 1990. Effect of pH and nitrogen source on aluminium tolerance of rye (secale cereale L) and yellow Lupin (Lupinus luteus L). *Plant and Soil*, 127, 13–21.

84. Greger, M. and Landberg, T. 1999. Using of willow in phytoextraction. *International Journal of Phytoremediation*, 1(2), 115–123.

85. Guo, J., Xu, W., and Ma, M. 2012. The assembly of metals chelation by thiols and vacuolar compartmentalization conferred increased tolerance to and accumulation of Cadmium and arsenic in transgenic *Arabidopsis thaliana*. *Journal of Hazardous Materials*, 309(199–200), 309–313.

86. Gupta, M., Singh, S., and Chandra, P. 1994. Uptake and toxicity of metals in *Scirpus lacustris* L and *Bacopa monnieri* L. *Journal of Environmental Science and Health*, 29(10), 2185–2202.

87. Hammer, D.A. 1989. *Constructed Wetlands for Wastewater Treatment Municipal, Industrial and Agricultural*. Lewis Publishers, Boca Raton FL.

88. Hannink, N., Rosser, S.J., French, C.E., Basran, A., Murray, J.A., Nicklin, S., and Bruce, N.C. 2001. Phytodetoxification of TNT by transgenic plants expressing a bacterial nitroreductase. *Nature Biotechnology*, 19(12), 1168–1172.

89. Haverkamp, R.G., Marshall, A.T., and van Agterveld, D. 2007. Pick your carats: Nanoparticles of gold-silver-copper alloy produced in vivo. *Journal of Nanoparticle Research*, 9, 697–700.

90. Hayman, M. and Dupont, R.R. 2001. *Groundwater and Soil Remediation: Process Design and Cost Estimating of Proven Technologies*. ASCE Press, Reston, VA.

91. Hazra, M., Avishek, K., and Pathak, G. 2011. Developing an artificial wetland system for waste water treatment: A designing perspective *IJEP*, 1(1), 8–18.

92. Head, I.M., Jones, D.M., and Rolimg, W.F.M. 2006. Marine microorganisms make a meal of oil. *Nature Reviews Microbiology*, 4, 173–182.

93. Henry, J.R. 2000. In an overview of phytoremediation of lead and mercury. NNEMS Report, Washington, DC, pp. 3–9.

94. Holan, Z.R. and Volesky, B. 1994. Biosorption of lead and nickel by biomass of marine algae. *Biotechnology and Bioengineering*, 43(11), 1001–109.

95. Hong, J.W., Park, J.Y., and Gadd, G.M. 2009. Pyrene degradation and copper and zinc uptake by *Fusarium solani* and *Hypocrea lixii* isolated from petrol station soil. *Journal of Applied Microbiology*, 108, 2030–2040.

96. Howard-Williams, C. 1970. The ecology of *Becium homblei* in Central Africa with special reference to metalliferous soils. *Journal of Ecology*, 58(3), 745–763.

97. Ilya, R., Smith, R.D., and Salt, D.E. 1997. Phytoremediation of metals: Using plants to remove pollutants from the environment. *Current Opinion in Biotechnology*, 8(2), 221–226.

98. Irgolic, K.J., Woolsont, E.A., Stockton, R.A., Newman, R.D., Bottino, N.R., Zingaro, R.A. Kearney, P.C. et al. 1977. Characterization of arsenic compounds formed by Daphniamagna and Tetraselmis Chuii from inorganic arsenate. *Environ Health Perspective*, 19, 61–66.

99. Jahher, A.S., Power, M.A., and Raut, P.D. 2012. Eichhonia crassipes (Mart) solutions in constructed wetlands for the treatment of textile effluent treatment plant. *Journal of Water Research*, 2, 44.

100. Jones, C.G., Lawton, J.H., and Shachak, M. 1994. Organism as ecosystem engineers. *Oikos*, 69, 373–386.

101. Joost, R.E., Olsen, F.J., and Jones, J.H. 1987. Revegetation and minesoil development of coal refuse amended with sewage sludge and limestone. *Journal of Environmental Quality*, 16(1), 65–68.

102. Juhasz, A.L. and Naidu, R. 2000. Bioremediation of high molecular weight polycyclic aromatic hydrocarbons: A review of the microbial degradation of benzo[a]pyrene. *International Biodeterioration & Biodegradation*, 45, 57–88, 44–49.

103. Jupson, E. 1996. ChevRon grows a new remediation technology alfalfa and poplars. *Environmental Business New Press Release*. http://www. Environbiz.com. Accessed on December 10, 2013.

104. Kadambari Consultants. 2013. NUALGI http://www.kadambari.net. Accessed on December 10, 2013.

105. Kent, A., Kobayashi, K., Kaufman, A., Griffis, J., and McConnell, J. 2007. *Using Houseplants to Clean Indoor Air*. Cooperative Extension Service, University of Hawaii at Manoa.

106. Kostiainen, R. 1995. Volatile organic compounds in the indoor air of normal and sick houses. *Atmospheric Environment*, 29(6), 693–702.

107. Kostka, J.E., Prakesh, O., Overhelf, W.A., Stefan, J.G., Frayer, G., Canion, A., Delgardio, J., Jorton, N., Hazen, T.C., and Huettal, M. 2011. Hydrocarbon—Degrading bacteria and the bacterial community response in Gulf of Mexico beach sands impacted by the deep water horizon oil spill. *Applied and Environmental Microbiology*, 77(22), 7962–7974.

108. Kumar, J.I.N., Soni, H., Kumar, R.N., and Bhatt, I. 2008. Macrophytes in phytoremediation of heavy metal contaminated water and sediments in Pariyej community reserve, Guyarat, India. *Turkish Journal of Fisheries and Aquatic Sciences*, 8, 193–200.

109. Kumar, R., Singh, S., and Singh, O.V. 2008. Bioconversion of ligno cellulosic biomass biochemical and molecular perspective. *Journal of Industrial Microbiology and Biotechnology*, 35(5), 377–391.

110. Kuyucak, N. and Volesky, B. 1988. Biosorbents for removal and recovery of metals from industrial solutions. *Biotechnology Letters*, 102, 137–142.

111. Kuyucak, N. and Volesky, B. 1989. Accumulation of cobalt by Marine alga. *Biotechnology and Bioengineering*, 33, 815.

112. Landberg, T. and Greger, M. 1996. Differences in uptake and tolerance to heavy metals in Salix from unpolluted and polluted areas. *Applied Geochemistry*, 11(1–2), 175–180.

113. iH, L., Li, P., Hua, T., Zhang, Y., Xiong, X., and Gong, Z. 2005. Bioremediation of contaminates surface water by immobilized (*Micrococus rosen*). *Environmental Technology*, 26(8), 931–940.

114. Lindstrom, J.E., Prince, R.C., Clark, J.C., Grossman, M.J., Yeager, T.R., Braddock, J.R., and Brown, E.J. 1991. Microbial—populations and hydrocarbon biodegradation potentials in fertilized shoreline sediments affected by the T/V Exxon—Valdez Oil—spill. *Applied and Environmental Microbiology*, 57, 2514–2522.

115. Liu, D.H.F., Liptack, B.G., and Bouis, P.A. 1997. *Environmental Engineers Handbook*, 2nd Edition. Lewis Publishers, Boca Raton, FL.

116. Liu, G.Y. Zhang, Y.X., and Chai, T.Y. 2011. Phytochelatin synthase of *Thlaspi caerulescens* enhanced tolerance and accumulation of heavy metals when expressed in yeast and tobacco. *Plant Cell Reports*, 30(6), 1067–1076.

117. Liu, Y., Yang, S., Tan, S., Lin, Y-M., and Tay, J.H. 2002. Aerobic gramiles: A novel zinc biosorbant. *Letters in Applied Microbiology*, 35, 548–551.

118. Lombi, E., Zhao, F.J., and Dunham, S.J. 2000. Cadmium accumulation in populations of *Thlaspi caerulescens* and *Thlaspi goesingense. New Phytologist*, 145(1), 11–20.

119. Lone, M.I., He, Z., Stoffela, P.J., and Yang, X. 2008. Phytoremediation of heavy metal polluted soils and water. Progresses and perspectives. *Journal of Zhejiang University Science B*, 9(3), 210–220.

120. Lynch, J.M. and Moffat, A.J. 2005. Bioremediation prospects for the future application of innovative applied biological research. *Annals of Applied Biology*, 146(2), 217–221.

121. Lytle, D., Sorg, T., Wang, L., Muhlen, C., Rahrig, M., and French, K. 2007. Biological nitrification in a full scale and pilot scale iron removal drinking water treatment plant. *Journal of Water Supply, Research and Technology–Aqua*, 56(2), 125–136.

122. Ma, L.Q., Komar, K.M., Tu, C., Zhang, W., Cai, Y., and Kennely, E.D. 2001. A fern that hyperaccumulates arsenic. *Nature*, 409(6820): 579.

123. Mai, C., Schormann, W., Majcherczyk, A., and Htterman, A. 2004. Degradation of acrylic copolymers by white—rot fungi. *Applied Microbiology and Biotechnology*, 65, 479–487.

124. Mallick, N. 2002. Biotechnological potential of immobilized algae for wastewater N, P and metal removal: A review. *BioMetals*, 15, 377–390.

125. Mariahal, A.K. and Jaga Deesh, K.S. 2013. Plant-microbe interaction: A potential tool for enhanced bioremediation. In: Arora, N.K. (ed.), *Plant Microbe Symbiosis: Fundamental and Advances*. Springer, New Delhi, pp. 5-395–410.

126. Masters, G.M. 2008. *Introduction to Environmental Engineering and Science*, 2nd Edition, Prentice-Hall, New Jersey, pp. 166–170.

127. McCutcheon, S. and Schnoor, J. 2003. (eds.), *Phytoremediation Transformation and Control of Contaminants*. John Wiley & Sons, Hoboken, NJ.

128. McGrath, S.P., Zhao, F.J., and Lombi, E. 2001. Plant and rhizosphere process involved in phytoremediation of metal-contaminated soils. *Plant Soil*, 232(1/2), 207–214.

129. McIntyre, T.C. 2003. Databases and protocols for plant microorganism selection: Hydrocarbons and metals. In: Schnoor, J.L. and McCutcheon, S.C. (eds.), *Phytoremediation: Transformation and Control of Contaminants* Wiley-Interscience, Hoboken, NJ, pp. 887–904.

130. Meagher, R.B. 2000. Phytoremediation of toxic elemental and organic pollutants. *Current Opinion in Plant Biology*, 3(2), 153–162.

131. Mei, L., Xitao, X., Renhao, X., and Zhili, L. 2006. Effects of strontium-induced stress on marine microalgae platymonas subcordiformis (chlorophyta: volvocales). *Chinese Journal of Oceanology and Limnology*, 24,154.

132. Mejare, M. and Bulow, L. 2001. Metal—Binding proteins and peptides in bioremediation and phytoremediation of heavy metals. *Trends in Biotechnology*, 19(2), 62–61.

133. Mellinger, R.H., Glover, F.W., and Hall, J.G. 1966. *Results of Revegetation of Strip Mine Soil by Soil Conservation Districts in West Virginia*. West Virginia University Agricultural Experiment Station, Morgantown, WV.

134. Mendez, M.O. and Maier, R.M. 2008. Phytostabilization of mine tailings in arid and semiarid environments an emerging remediation technology. *Environmental Health Perspectives*, 116, 278–283.

135. Miller, W.D. 1980. *Waste Disposal Effects on Groundwater: A Comprehensive Survey of the Occurrence and Control of Ground-Water Contamination Resulting from Waste Disposal Particles*. Premier Press, Berkeley, CA.

136. Mitra, N., Rezvan, Z., Seyed, A.M., Gharaie, M., and Hosein, M. 2012. Study of water arsenic and boron pollutants and algae phytoremediation in Three Springs, Iran. *International Journal of Ecosystem*, 2(3), 32–37.

137. Moeller, D.W. 2005. *Environmental Health*, 3rd Edition. Harvard University Press, Cambridge, MA.

138. Morrison, R.S., Brooks, R.R., Reeves, R.D., Malaise, F., Horowiz, P., Aronsoon, M., and Merriam, G.R. 1981. The diverse chemical forms of heavy metal in tissue extracts of some metallophytes from Shaba Province Zaire. *Phytochemistry*, 20, 455–458.

139. Moser, M. and Hase-Wanter, K. 1983. Ecophysiology of mycorrhizal symbioses. In: Lange, O.L., Nobel, P.S., Osmond, C.B. et al. (eds.), *Physiological Plant Ecology M Vol 12 C*. Springer-Verlag, New York, pp. 391–421.

140. Mudgal, V., Madaan, N., and Mudgal, A. 2010. *Heavy Metals in Plants: Phytoremediation: Plants Used to Remediate Heavy Metal Pollution. Agriculture and Biology Journal of North America*, 1(1), 40.

141. Muramoto, S., Udagawa, T., and Okamura, T. 1995. Effective removal of musty odor in the Kanamachi purification plant. *Water Science Technology*, 31(11), 219–222.

142. Nair, D.R. and Schnoor., J.L. 1993. Effect of soil conditions on model parameters and Atrazine mineralization rates. *Water Research*, 28(5), 1199–1205.

143. Ndimele, P.E. and Jimoh, A.A. 2011. *Eichhornia Crassipes* (Water hyacinth) in phytoremediation of heavy metals polluted water at Ologe Lagoon in Lagos, Nigeria. *Research Journal of Environmental Science*, 5, 242–433.

144. Nikolic, V., Milicevic, D., and Milvojevic, A. 2010. *Constructed. Wetland Application in Waste Water Treatment Processes in Serbia*. BALWOIS 2010. Ohrid, Republic of Macedonia, p. 3, May 2010.

145. Nirmal Kumar, J.I. and Oommen, C. 2012. Removal of heavy metals by biosorption using fresh water algae *Spirogyra hyaline. Journal of Environmental Biology*, 33, 27–31.

146. Nriagu, J.O. and Pacyna, J.M. 1988. Quantitative assessment of worldwide contamination of air water and soils by trace metals. *Nature*, 333(6169), 134–139.

147. Ogilvie, L.A. and Hirsch, P.R. (eds.), 2012. *Microbial Ecology Theory—Current Perspectives*. Caster Academic Press, UK.

148. Paturam, J.M. 1989. *By Products of the Cane Sugar Industry an Introduction to Their Industrial Utilization.* Elsevier, Amsterdam.

149. Peng, R., Xiong, A., Xue, Y., Fu, X., Gao, F., Zhao, W., Tian, Y., and Yao, Q. 2008. Microbial biodegradation of polyaromatic hydrocarbons. *FEMS Microbiology Reviews,* 32, 927–955.

150. Peuke, A.D. and Rennenberg, H. 2005. Phytoremediation: Molecular biology, requirements for application, environmental protection, public attention and feasibility. *EMBO Reports,* 6, 497–501.

151. Pianelli, K., Mari, S., Margues, L., Lebrum, M., and Czernic, P. 2005. Nicotianamine over accumulation confers resistance to nickel in *Arabidopsis thaliana. Transgenic Research,* 14, 739–748.

152. Economic Research Service, USDA, 1996. Interest increases in using plants for environmental remediation. In: *Industrial Uses of Agricultural Materials: Situation Outlook Report.*

153. Porter, E.K. and Peterson, P.J. 1975. Arsenic Accumulation by plants in mining waste. *Science of the Total Environment,* 4, 365–371.

154. Poulson, S.R., Colberg, P.J.S., and Drever, J.I. 1997. Toxicity of heavy metals (Ni, Zn) to *Desulfovibrio desulfuricans. Geomicrobiology Journal,* 14(1), 41–49.

155. Prenafeta-Boldu, F.X., Summerbell, R., and Sybren de Hoog, G. 2006. Fungi growing on aromatic hydrocarbons biotechnology's unexpected encounter with biohazard? *FEMS Microbiology Reviews,* 30, 109–130.

156. Rai, P.K. 2008. Heavy metal pollution in aquatic ecosystems and its phytoremediation using wetlands plants: An ecosustainable approach. *International Journal of Phytoremediation,* (2), 131–158.

157. Rainbow, P.S. 1995. Biomonitoring of heavy metal availability in the marine environment. *Marine Pollution Bulletin,* 8(1), 16–19.

158. Ramsar Convention. 1971. http://www. Ramsar.org. Accessed on October 21, 2013.

159. Rea P.A. 2012. Phytochelatin synthase: Of a prottase a peptide polymerase made. *Physiologia Plantarum,* 145(1), 154–164.

160. Richman, M. 1996. Terrestrial plants tested for cleanup of radionucleides explosive residues. *Water Environment and Technology,* 8(5), 17–18.

161. Rittmann, B.E. and Snoeyink, V.L. 1984. Achieving biologically stable drinking water. *Journal of the American Water Works Association,* 76(10), 106–114.

162. Rogalla, F., Ravarini, P., de Larminat, G., and Couttelle, J. 1990. Large scale biological nitrate and ammonia removal. *Water EnvironMental Journal,* 4(4), 319–328.

163. Rosenberg, E., Legman, R., Kushmano. A., Adler. E., Abir. A., and Ron, E.Z. 1996. Oil bioremediation using insoluble nitrogen source. *Journal of Biotechnology,* 15, 51(3), 273–278.

164. Royo, F. 2009. Degradation of alkenes by bacteria. *Environmental Microbiology,* 11(10), 2477–2490.

165. Ruby, M. and Appleton, B. 2010. Using landscape plants for phytoremediation. In: Struck, S. and Lichten, R. (eds.), *Low Impact Development: Redefining Water in the City. Pan America Society of Civil Engineer.* ASCE, Reston, VA, pp. 323–332.

166. Rupassara, S.I., Larson, R.A., Sims, G.K., and Marley, K.A. 2002. Degradation of Atrazine by Hornwort in aquatic systems. *Bioremediation Journal,* 6(3), 217–224.

167. Ryoo, D., Shum, H., Canada, K., Barbieri, P., and Wood, T.K. 2000. Aerobic degradation of tetrachloroethylene by toluene-O-xylene mono oxygenase of *Pseudomonas stutzeri* OX1. *National Biotechnology,* 18, 775–778.

168. Sah, R.N. and Mikkelsen, D.S. 1986. Effects of anaerobic decomposition of organic matter on sorption and transformations of phosphate in drained soils: 1. Effects on phosphate sorption. *Soil Science,* 142, 267–227.

169. Salido, A.L., Hastly, K.L., Lim, J.M., and Butcher, D.J. 2003. Phytoremediation of arsenic and lead in contaminated soils using Chinese brake ferns (*Pteris vittata*) and Indian mustard (*Brassica juncea*). *International Journal of Phytoremediation,* 5(2), 89–103.

170. Salt, D.E., Smith, R.D., and Raskin, I. 1998. Phytoremediation. *Annual Review of Plant Physiology and Plant Molecular Biology,* 49, 643–668.

171. Sasek, V. 2003. Why mycoremediations have not yet come into practice. In: Sasek, V., Glaser, J.A., and Baveye, P. (eds.), *The Utilization of Bioremediation to Reduce Soil Contamination: Problems and Solution.* Kluwer Academic Publishers, New Delhi.

172. Schalscha, E. and Ahumada, I. 1998. Heavy metals in rivers and soils of central Chile. *Water Science and Technology,* 37(8), 251–255.

173. Schmidt, U. 2003. Enhancing phytoremediation: The effect of chemical soil manipulation on mobility, plant accumulation, and leaching of heavy metals. *Journal of Environmental Quality,* 32, 1939–1954.

174. Schnoor, J.L. 1997. Phytoremediation, technology overview report, *Ground-water Remediation Technologies Analysis Centre,* Series E, Vol 1, Oct.

175. Schroder, P., Scheer, C., and Belford, B.J.D. 2001. Metabolism of organic xenobiotics in plants: Conjugating enzymes and metabolic endpoints. *Minerva Biotechnology,* 13, 85–91.

176. Schroder, P., Avino, J.N., Azaizeh, H., Goldhirsh, A.G., Gregorio, S.D., Komives, T., Langergraber, G. et al. 2007. Using phytoremediation Technologies to upgrade waste water treatment in Europe. *Environmental Science and Pollution Research,* 14(7), 495.

177. Schroder, U. 2007. Anodic electron transfer mechanisms in microbial fuel cells and their energy efficiency. *Physical Chemistry Chemical Physics,* 9(21), 2619–2629.

178. Schwartz, C., Echevarria, G., and Morel, J.L. 2003. Phytoextraction of cadmium with *Thlaspi caerulescens. Plant Soil,* 249(1), 27–35.

179. Shelef, O., Gross, A., and Rachmilevitch, S. 2012. The use of *Brassica indica* for salt phytoremediation in constructed wetlands. *Water Research*, 46(13), 3967–3976

180. Siciliano, S.D., Germida, J.J., Bank, K., and Greer, C.W. 2003. Changes in microbial community composition and function during a polyaromatic hydrocarbon phytoremediation field trial. *Applied Environmental Microbiology*, 69, 483–489.

181. Singh, D. and Chauhan, S.K. 2011. Organic acids of crop plants in aluminium detoxification. *Current Science*, 100, 1509–1515.

182. Singh, H. 2006. *Mycoremediation: Fungal Bioremediation.* Wiley Interscience, New York.

183. Singh, A. and Ward, O.P. (eds.) 2004. *Applied Biocremediation and Phytoremediation.* Springer, Berlin.

184. Smith, B.R. 2007. *Constructed Wetlands for Waster Treatment: A Planning & Design Analysis for San Francisco.* Department of City & Regional Planning Department of Landscape Architecture & Environment Planning UC, Berkeley.

185. Sorek, A., Atzmon, N., Dahan, O., Gerstl, Z., Kushisin, L., Laor, Y., Migelgrin, U. et al. 2008. Phytoscreening: The use of trees for discovering subsurface contamination by VOCs. *Environmental Science and Technology*, 42(2), 536–542.

186. Srivastava. M., Ma, L.Q., Singh, N., and Singh, S. 2005. Antioxidant responses of hyper accumulator and sensitive fern species to arsenic. *Journal of Experimental Botany*, 56(415), 1335–1342.

187. Stamets, P. 1999. Earth's natural Internet. *Whole Earth Magazine.*

188. Stamets, P. 2003. Mycofiltration: A novel approach for the bio-transformation of abandoned logging roads. Fungi Perfecti. www.fungi.com. Retrieved November 30, 2013.

189. Steffen, K.T., Hatakka, A., and Hofrichter, M. 2007. Removal and mineralization of polycyclic aromatic hydrocarbons by litter-decomposing basidiomycetous fungi. *Applied Microbiology Biotechnology*, 60, 212–217.

190. Stijvet, Velinga, E.C., and Herrman, A. 1990. Arsenic accumulation in some higher fungi. *Personia*, 14, 161–166.

191. Subramanian, M., Oliver, D.J., and Shanks, J.V. 2006. TNT phytotransformation pathway characteristics in arabidopsis: Role of aromatic hydroxylamines. *Biotechnology Progress*, 22(1), 208–216.

192. Suthersan, S.S. 1997. *Remediation Engineering: Design Concepts.* CRC Press, Boca Raton, FL.

193. Taylor, E.M. and Schuman, G.E. 1988. Fly ash and lime amendment of acidic coal spill to aid revegetation. *Journal of Environmental Quality*, 17(1), 120–124.

194. Thibodean, B. 2006. *The Alternative To Pump And Treat.* http://www.wateronline.com/article.mvc.The-Alternative-To-Pump-And-Treat-0001. Bob Thibodeau, Water Online Magazine.

195. Tikhomirova, N.A., Ushakova, S.A., Kudenko, Y.A., Gribovskaya, I.V., Shklavtsova, E.S., Balnokin, Y.V. et al. 2011. Potential of salt accumulating and salt secreting halophytic plants for recycling sodium chloride in human urine in bioregenerative life support systems. *Advances in Space Research*, 48, 378–82.

196. Topper, K.F. and Sabey. B.R. 1986. Sewage sludge as a coal mine spoil amendment for revegetation in Colorado. *Journal of Environmental Quality*, 15, 44–49.

197. Tsezos, M.M.C. and Cready, R.G.L. 1997. The pot plant testing of the continuous extraction of radio nuclides using immobilized biomass technology. In: Sayler, G. and Fox, R.B. (eds.), *Environmental Biotechnology for Waste Treatment.* Plenim Press, New York, pp. 249–260.

198. UNEP Newsletter and Technical Publications. *Fresh Water Management Series No 2.* http://www. unep.org. Accessed, November 30, 2013.

199. Van de Gast, C., Whiteley, A.S., Starkey, M., Knowles, C.J., and Tompson, I.P. 2003. Bioaugmentation strategies for remediating mixed chemical effluents. *Biotechnology Progress*, 19, 1156–1161.

200. Vassil, A.D., Kapulnik, Y., Raskin, I., and Salt, D.E. 1998. The role of EDTA in lead transport and accumulation by Indian mustard. *Plant Physiology*, 117(2), 447–53.

201. Verma, P., George, K.V., Singh, K.V., Singh, K., Jawakar, A., and Singh, R.N. 2006. Modeling rhizofiltration: Heavy-metal uptake by plant roots. *Environmental Modeling & Assessment*, 11, 387–394.

202. Vidali, M. 2005. Bioremediation: An overview. *Pure and Applied Chemistry*, 73(7), 1167–1172.

203. Vogel, T.M. 1996. Bioaugmentation as a soil bioremediation approach. *Current Opinion in Biotechnology*, 7, 311–316.

204. Volland, S., Schaumloffel, D., Dobritzsch, D., Kraus, S.G.J., and Lutz-Meind, L.U. 2012. Nitrification of phytochelatins in the cadmium-stressed conjugate green alga *Micrasterias denticulate. Chemosphere*, 91(4), 448–454.

205. Vroblesky, D. 2008. *User's Guide to the Collection and Analysis of Tree Cores to Assess the Distribution of Subsurface Volatile Organic Compounds.* www.pubs.usgs. gov/. Accessed on December 20, 2013.

206. Vroblesky, D., Nietch, C., and Morris, J. 1998. Chlorinated ethenes from groundwater in tree trunks. *Environmental Science and Technology*, 33(3), 510–515.

207. Wang, Q., Cui, Y., and Dong, Y. 2002. Phytoremediation of polluted waters potential and prospects of wetland plants. *Acta Biotechnology*, 22(1–2), 199–208.

208. WHO 2003. Ammonia in drinking water. Background document for development of WHO Guidelines for drinking-water quality. World Health Organization, Geneva (WHO/SDE/WSH/03.04/01).

209. Williams, J.B. 2002. Phytoremediation in wetland ecosystems: Progress, problems, and potential. *Critical Reviews in Plant Sciences*, 21(6), 607–635.

210. Wolverton, B.C. and Wolverton, J. 1993. Plants and soil microorganisms: Removal of formaldehyde, xylene, and ammonia from the indoor environment. *Journal of Mississippi Academy of Sciences*, 38(2), 11–15.

211. Wolverton, B.C. and Wolverton, J. 1996. Interior plants: Their influence on airborne microbes inside efficient

buildings. *Journal of the Mississippi Academy of Sciences,* 41 (2), 99–105.

212. Wolverton, B.C., Johnson, A., and Bounds, K. 1989. Interior landscape plants for indoor air pollution abatement: Final report. National Aeronautics and Space Administration.

213. Yakimor, M.M., Timmis, K.N., and Golyshin, P.N. 2007. Obligate oil—Degrading marine bacteria. *Current Opinion in Biotechnology,* 18, 257–266.

214. Zaranyika, M.F. and Ndapwadza, T. 1995. Uptake of Ni, Zn, Fe, Co, Cr, Pb, Cu and Cd by water hyacinth (*Eichhornia crassipes*) in Mukuvisi and Manyame Rivers, Zimbabwe. *Journal of Environmental Science and Health Part A,* 30(1), 157–169.

215. Zayed, A. and Gowthmans, T.N. 1998. Phytoaccumulation of trace elements by wetland plant I, duck weed. *Journal of Environmental Quality,* 27, 715–721.

216. Zhu, Y.L., Pilon–Smits, E.A.H., Jonanin, L., and Terry, N. 1999. Cadmium tolerance and accumulation in Indian mustard, is enhanced by over overexpressing – glutamylcysteine synthetase 1. *Plant Physiology,* 121(4), 1169–1177.

217. Prasad, M.N.V. and Freitas, H.M.D. 2003. Metal hyperaccumulation in plants—Biodiversity prospecting for phytoremediation technology. *Electronic Journal of Biotechnology,* 93(1), 285–321.

218. Miller, R.R. 1996. Phytoremediation. Goundwater remediation technologies analysis centre technology overview Report TO-96-03 National Environmental Technology Application Centre and the University of Pittsburgh, Pennsylvania. http://www.gwrtac.org/pdf/phyto-0.pdf. Accessed on October 25, 2013.

Water Reuse and Artificial Wetlands

54

Treatment Wetlands: Fundamentals

Armando Rivas
Hernández
*Mexican Institute of
Water Technology*

Iván Rivas Acosta
*Mexican Institute of
Water Technology*

Saeid Eslamian
*Isfahan
University of Technology*

PREFACE

Treatment wetlands are used to reduce the existing wastewater pollutant load. Wetlands emulate, in a controlled way, the processes of phytodepuration occurring in natural wetlands. Treatment wetlands are designed, built, and man-operated. They consist of ponds that contain plant species, through which wastewater flows for depuration by means of physical, chemical, and biological processes. Treatment wetlands use has increased over the past years at the international level, mainly due to its low cost of treatment, simplicity of operation, and environmental benefits.

Besides large-scale projects, treatment wetlands are commonly used at smaller scales; for instance, in rural areas at a family level and in medium-sized cities where enough space is available since one of the main constraints is the land availability. Treatment wetlands are also used for sewage sludge treatment in conventional wastewater treatment plants.

54.1 Introduction

Treatment wetlands have taken importance in various regions of the world, such as North America, Europe, and Asia. The beginning of its use dates back to 1960 in Germany on small prototypes, while its formal development began in 1990 [18,37].

Over the years, treatment wetlands have received different names, the most common are constructed wetlands and treatment wetlands [16,18]. The terminology applied to subsurface flow wetlands is particularly unclear. The terminology can be confusing even in English, where terms such as vegetated submerged beds and beds network among others have been used [12]. In this chapter, the term treatment wetlands is used.

Scientific and systematic research extended the application horizons of the wetlands to consider pollution control applications; such research started around 1950 with the work of Seidel. Since 1990, an intensive study and research on treatment wetlands has been developed as evidenced by the high number of manuals and technical reports [12]. During the next following 10 years, there was an increase in the number of wetlands due to the expansion of treatment of different types of wastewater, such as domestic, industrial, and rainwater [15].

In the United States, during 1990, there existed 98 subsurface flow systems and 352 surface flow systems, which were built in the eight following years [18]. The state of Kentucky had 4000 underground flow systems at the family level in 1998; whereas in the United Kingdom, 1000 groundwater flow systems were built in 2009 [18]. In Italy, 150 wetland treatment systems were under operation [24].

During 1991, in the United States, there were more than 200 operating artificial wetlands with different types of effluents, such as municipal wastewater, industrial discharges, and water from agro-food industries [10]. Great Britain, United States, and Australia have been applying treatment wetland technology extensively and this success is mainly because these countries have funded a great amount of scientific research related to wastewater treatment [27].

Despite the progress that has been obtained with the use of this technology, there are not enough case studies and shared experiences related to their successful construction and evaluation. Thus, it is necessary to adapt and develop standard criteria for design, operation, and maintenance. These design guidelines should consider cultural aspects of the local people, as well as economic and climatic conditions that increase the efficiencies of contaminant removal and decrease the treatment costs [18].

Although there have been several years of research and practical experience, there are some fundamental aspects not fully understood about artificial wetlands. An explanation for this is because compared with other technologies such as activated sludge, the treatment wetlands depend on the interaction of many more components [10].

54.2 Classification of Treatment Wetlands

There are diverse and complex criteria to classify wetlands [7,12,17,44]. The Ramsar Convention (formally, the Convention on Wetlands of International Importance) takes a broad approach in determining the wetlands and these are defined as: "areas of marsh, fen, peatland or water, whether natural or artificial, permanent or temporary, with water that is static or flowing, fresh, brackish or salt, including areas of marine water the depth of which at low tide does not exceed six meters" [29]. However, in general, wetlands might be classified in two large types: natural wetlands and treatment wetlands (also known as constructed or artificial wetlands).

On one hand, natural wetlands have two variants: natural with little anthropogenic influence and artificial or induced by human-related activities. The first group includes freshwater and salty marshes, swamps, marshes of cypress trees, including a wide variety of inland, coastal, and marine habitats. The second

group includes artificial or manufactured wetlands that have been developed in areas with municipal and domestic wastewater discharges or agricultural discharges, whose high nutrient content promotes their development.

On the other hand, treatment wetlands could be classified based on the type of flow. In this sense, two types of systems are identified: surface flow and subsurface or underground flow. The latter type is further subdivided into horizontal subsurface flow and vertical subsurface flow.

Wetlands of different types could be combined to conform to hybrid systems [15,44].

In horizontal subsurface systems, the water flow enters the pond through a series of pipes that distribute the flow rate through the filter media (gravel or sand) and water flows slowly beneath the surface toward the collector pipe, located next to the outlet structure. Filter media supports the development of root plants. At the same time, filter media constitutes the surface where biofilms of microorganisms are developed and where most of the debugging occurs. The pollutant removal efficiency in subsurface systems is greater than the systems with inundated flow due to its bigger surface area for the development of the microorganisms and also because the porous medium intervenes effectively in the processes of filtration and sedimentation [3,11].

An artificial wetland of subsurface flow consists of a waterproof channel, which is filled with a solid porous material occupying almost all of its depth. The wastewater flows through the porous medium, and always below the surface of the same. Rocks and gravel are commonly used as a porous medium and emergent vegetation has an essential role for a proper operation.

The filling and the sediments are important because both act as the primary screening barrier, as a supporting structure for plants, and finally as a surface for the growth and development of microbial mass. In this last role, the particle size is important because the smaller the particle sizes, the greater the amount of biofilm will be able to host. However, a seal of the pores or clogging is likely to occur and, thus, corresponding flooding above the subsurface. As a result, it is necessary to optimize the particle size [25].

Figure 54.1 shows a subsurface wetland whose constituents are plants, gravel, pipes, waterproofing, root detail, and biofilms of microorganisms.

In vertical subsurface flow wetlands, the flow rate is distributed on the water surface and it is collected at the bottom of the pond. Under the previous condition exists an efficient oxygen transfer that predominantly supports aerobic condition, which qualifies them as efficient systems to remove organic materials and ammonium. However, because the shortage of anaerobic zones, acceptable rates of denitrification are not obtained. As a result, ammonium is only transformed into nitrate [23,43,45].

The water application occurs intermittently to preserve and encourage aerobic conditions. Also, emergent vegetation is placed in this granular medium. In addition, usually a system of ventilation chimneys is placed to promote aerobic conditions

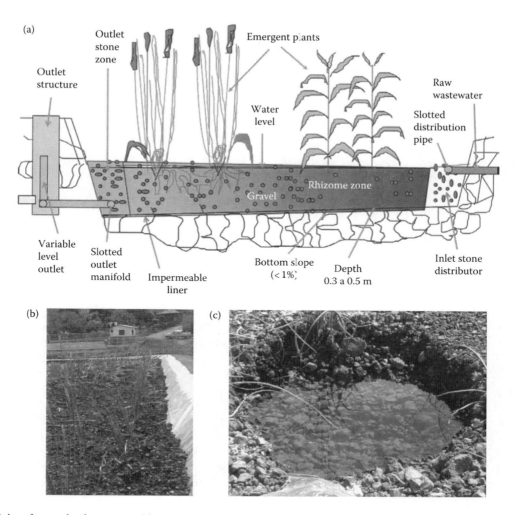

FIGURE 54.1 Subsurface wetland treatment. (a) Cross-section, (b) general appearance, and (c) flooded volume detail.

of the porous medium. Ventilation chimneys consist of screened pipes connected to the outside. Unlike the subsurface horizontal flow wetlands, the substrate consists of several layers, being the finest in the upper part and increasing the gravel diameter toward the bottom [4].

Wetlands for the treatment of sludge in the last decade have taken special importance at the international level, due to its lower costs of treatment, simplicity of operation, and high treatment efficiencies. Therefore, this type of application is widely used for the treatment of sludge from conventional systems, both aerobic and anaerobic.

Sludge treatment wetlands are replacing traditional electromechanical systems that dry out sludge. Also, sludge treatment wetlands are an excellent alternative for the treatment of the sludge generated in small treatment plants, septic tanks, and biodigesters [33].

Uggetti et al. [41] reports high efficiencies of sludge treatment by wetlands, with a reduction of humidity between 16% and 30%, volatile solids decreased between 30% and 49%, and mud mineralization. Uggetti et al. [41] point out that a potential application of sludge treatment wetlands is in agriculture

because their investment, operation, and maintenance costs are reasonable. These types of applications have plenty of potential in small communities. Figure 54.2 shows the general aspect of a wetland of vertical subsurface flow, in the sludge treatment modality. There is accumulation of solids at the surface.

Flooded wetlands or surface systems basically consist of a shallow pond waterproofed with a geomembrane or built in soils with a low hydraulic permeability. They contain floating aquatic plants on their surface and are highly efficient in the removal of nutrients such as phosphorus and ammoniacal nitrogen [11]. In systems of surface flow, water is directly exposed to the atmosphere and preferably circulating through the stems and leaves of plants [12]. Figure 54.3 shows the general aspect of a flooded wetland and Figure 54.4 shows a cross-section.

54.3 Processes Involved in Biochemical Pollutants Removal

As it has been described, treatment wetlands consist of flooded or saturated areas that contain floating, submerged, or emergent

FIGURE 54.2 Sludge treatment wetland.

FIGURE 54.3 Flooded wetland.

plants. For their development, these aquatic plants uptake the nutrients contained in the wastewater. When wastewater flows through the ponds, the pollutants are transformed into gases, biomass (microorganisms and vegetation), and mineral compounds [14]. Treatment wetlands consist of shallow lagoons or channels networks with regional plants, usually less than 1 m high. Depuration processes take place through the interactions between the solid substrate, water, microorganisms, vegetation, and even fauna [13]. Figure 54.5 presents a general appearance of a treatment wetland.

Ponds are usually less than 0.6 m deep and water flows slowly through the porous medium. In its interior there are floating or submerged aquatic plants, such as algae, reeds, vines, lilies, cypress, and bulrushes, among others. Kadlec and Knight [17] suggest installing a broad coverage of plants in order to take advantage of sunlight.

A typical subsurface flow system presents 30 or 70 cm of bedding thickness. In this sense, it is important to have a total depth greater than water depth to ensure a dry zone in the upper part [30]. Wetlands need water to maintain saturated conditions and they normally have a water depth less than 60 cm with emerging plants such as bulrushes and reeds [22].

Oxygen is carried out to the wetland from the atmosphere. Also, oxygen is produced by photosynthesis occurring in the leaves of plants, which is transferred to the root plants and to the water. Then the oxygen becomes dissolved and it is taken up by microorganisms (suspended or attached to all kinds of surfaces such as roots, gravel, and membranes). Major purification processes are sedimentation, filtration, food chains, and assimilation by microorganisms and plants [8].

Wetland vegetation provides surfaces for the formation of bacterial films and facilitates the filtration and adsorption of wastewater components. In addition, plants promote the transfer of oxygen to the water column and control the growth of algae by limiting sunlight penetration [22].

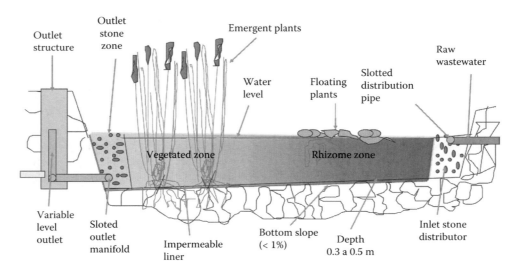

FIGURE 54.4 Flooded wetland cross-section.

FIGURE 54.5 Treatment wetland general appearance.

Treatment wetlands are basically constituted by water, microorganisms, plants, soil or substrate medium, and a hydraulic system.

The treatment wetlands consist of a proper design of a bucket that contains water, soil substrate, and usually emerging plants. These components can be manipulated by constructing a wetland. Other important components that develop naturally in the wetlands are the microorganism communities and aquatic invertebrates [22].

These systems support strong variations in quality and quantity of the water mainly due to their large retention times, which is from 5 to up to 10 days. Such time depends on the quality of water expected in the effluent. It is important to understand the types of pollutants in wastewater in order to determine potential adverse effects due to the presence of toxic substances [17].

Residual water flows slowly through the porous medium (spaces between the gravel, the roots, and accumulated solids). During this process, the flow makes contact with the microorganisms, which are predominantly attached to surfaces. However, also a minor fraction of microorganisms are suspended in the water [17]. Aerobic organisms require dissolved oxygen to decompose into carbonaceous matter. Fernandez et al. [11] point out that also anaerobic bacteria use inorganic oxidized compounds such as NO_3 (nitrates) and SO_4 (sulfates).

The plants release metabolites that microorganisms use as food. Roots, stems, and leaves surfaces offer protection and hosting to microorganisms. This symbiosis results in an efficient and a high-speed process to degrade and remove contaminants [11]. The most used macrophytes in treatment wetland are Typha, Scirpus, and Phragmites. These macrophytes feature several advantages: simple management, resistance to diseases, adaptability to climate, ability to support a wide range of pollutant concentration, and, finally, they are easy to obtain [17].

The body of the plants or biomass takes up substances and this process serves as a method to remove contaminants; for example, heavy metals and another organic components such as phosphorus and potassium. Nitrogen is taken up by plants as a nutrient. Some phosphates enter the tissues of the plants and the insoluble residue is deposited on the wetland floor [5].

Sludge is formed over a complex process. First, partially degraded plant material sediments begin to cover anaerobic sludge on the bottom. Then, this sludge is immobilized by metal sulfites. Within this mossy material (formed by a mixture of degraded plant debris), naturally ion exchange reactions are carried out and absorbed minerals are retained indefinitely [39].

The hydraulic system has as a function the control of flows and water levels. This system is composed of pipelines, pipe collection, and input and output structures. The proper location of the hydraulic structures avoids the generation of preferential flows and dead zones [20].

The objective of the input and output structures is to distribute and control the flow paths in the wetland. Multiple structures are placed at both extremes of the wetland and are essential to ensure a uniform distribution of the input flow. Furthermore, these structures help to avoid "dead zones" where the exchange of water is poor. Under operational conditions, this type of structure improves flow circulation, which results in lower retention times than theoretical retention times [42].

Kadlec and Knight [17] indicate the major contaminants in wastewater that can be reduced or eliminated through treatment wetlands are carbon (organic matter), nitrogen, phosphorus, pathogens (helminthes eggs, bacteria, protozoa, fungi, and viruses), and heavy metals (cadmium, copper, chromium, lead, mercury, selenium, and zinc).

Treatment wetlands must be preceded by a pretreatment, which consists of racks to remove coarse solids, such as garbage, leaves, and floating objects. In addition, this pretreatment removes particles with density greater than the water, for example, sand. It is recommended to install an anaerobic process (a septic tank or a sedimentation tank) in order to reduce the organic load and the concentration of solids [12].

In treatment wetlands, there is an interaction between aquatic plants and microorganisms. The metabolic products of microorganisms degrade organic matter, for instance nitrogen, phosphorus, potassium, and other minerals. These substances are taken up by plants, which in turn provide oxygen to the microorganisms [11].

Oxygen is produced during photosynthesis during the daytime or it may come directly from the atmosphere. Atmospheric oxygen is located near the surface of the bed where it is used by aerobic bacteria. Also, atmospheric oxygen is developed at a slightly aerobic environment in the root zone due to an oxygen transfer. This transfer occurs between existing oxygen in the atmosphere and the oxygen produced during photosynthesis by the plants, which is carried out by means of interconnected aeration in the plants' tissue [11].

Conversely, anaerobic zones contribute with purification processes in the rest of the volume of the wetland [18]. Numerous investigations have shown that the oxygen transport capacity of the macrophytes to the roots is insufficient to ensure an aerobic decomposition; thus, the condition is mostly anaerobic [43].

The oxygen dissolved in the water within the wetland treatment usually is less than 1 mg/L, which favors anaerobic conditions. Because of this, long times of hydraulic retention are required and consequently large land extensions are needed within the treatment processes. Initially, the oxygen enters through convection from the air and it is transported by the aquatic plants through their leaves, stems, and roots. If the oxygen is not fully consumed during the respiration of the roots, the oxygen can enter the water and it could be used by aerobic bacteria to oxidize organic carbon.

The aquatic plants have differences in the ability to transport oxygen. On the one hand, Reddy and DeBusk [31] reference values of oxygen transport in a range between 0.12 and 3.95 mg O_2/g/h depending on the species of aquatic plants. On the other hand, Tyroller et al. [40] report a concentration range between 0.3 and 3.2 O_2/g/day in case of subsurface flow treatment wetland.

Removal mechanisms include several types, such as physical, chemical, biological, and hydraulic. The main ones are of two types: liquid/solid separations and changes of the components in the residual water. The first group includes the processes of sedimentation, filtration, absorption, adsorption, hydrolysis, ion exchange, and leachate. In the second group, removal processes include oxidation/reduction reactions, acid/base changes, precipitation, flocculation, and biochemical reactions under anaerobic/aerobic conditions [11]. The biological aspects include the production of gases (methane, carbon dioxide, and hydrogen sulfide), as well as the fraction that is used by the microorganisms and plants for their development [15].

54.4 Removal of Contaminants

54.4.1 Biochemical Oxygen Demand (BOD)

In addition to sedimentation and filtration, organic matter is reduced also by adsorption, hydrolysis, absorption, and mineralization and production of gases such as methane and carbon dioxide. Organic carbon is used by bacteria as a source of energy and for cell synthesis. Bacteria live scattered across the water in microenvironments, sediments, and roots of the plants [17]. Biodegradation occurs in aerobic and anaerobic conditions or under facultative conditions [14].

54.4.2 Total Suspended Solids (TSS)

Large solids or wastes and floating objects in general are filtered in metallic structures or grids. These types of structures have different openings, from 1 cm up to 3 cm. Sands and other suspended solids are retained on sand removal structures.

The particles are held in a subsurface flow wetland mainly for three reasons: (1) the constrictions of the flow produced by the granular media, (2) low water velocity, and (3) adhesion forces between particles. These three physical processes are known as granular media filtration. In addition, the roots and rhizomes of plants contribute similarly with removal of particles [12]. In this variant of wetlands, the solids are also retained by sedimentation [9]. Cooper et al. [3] indicate that a higher solid reduction occurs in the first meters of the wetland, near the entrance point.

In flooded treatment wetlands, the main mechanism to remove suspended solids is by sedimentation. Sherwood et al. [39] point out that about 90% of the removal usually occurs in 20% of the total area of the system near the entrance.

54.4.3 Nitrogen

Nitrogen in wetlands could be removed by several mechanisms, for instance, volatilization of ammonium, nitrification/denitrification, and by assimilation of plants [39]. In addition, nitrogen in the form of nitrate is used by plants for their growth. The removal of nitrate through biological denitrification requires anaerobic conditions, a proper source of carbon, and an acceptable temperature [5].

The macrophytes assimilate inorganic forms of nitrogen to form structural organic nitrogenous compounds; also, the nitrogen is a macronutrient for plants [11]. Removal of nitrogen by plants is quite small and is effective only when the harvest is well scheduled [43].

The combination of aerobic conditions is required so that the treatment wetlands perform successfully the oxidation of ammonium to nitrate (nitrification). However, also anaerobic conditions are necessary to reduce nitrates to nitrogen gas (denitrification). Therefore, it is advisable to design hybrid systems, in other words, as a combination of wetlands and lagoons of maturation. In the lagoons, the algae produce the required oxygen for nitrification [33].

The organic nitrogen associated with the particles is rapidly eliminated in conjunction with TSS by sedimentation, filtration, and hydrolysis [39]. Large amounts of organic nitrogen are converted into ammonium and released into the water. Ammoniacal nitrogen can be removed by direct volatilization to the atmosphere as ammonia gas, or by ion exchange on the soil particles [18].

54.4.4 Phosphorus

Phosphorus is mainly removed by filtration, sedimentation, assimilation by microorganisms and plants, precipitation/dissolution, adsorption/desorption, and mineralization and soil retention; is used by the macrophytes as orthophosphate [6,21,43]. A systematic maintenance is suggested, in which external vegetation is removed. Such practice avoids natural vegetation entering the wetland [19]. After one or two years of operation, the wetland treatment efficiency to remove phosphorus is reduced. Thus, a complementary physical and chemical treatment such as the use of aluminum (alum) or iron (ferric) salts to precipitate dissolved phosphorous and remove it in a solid (sludge) form is necessary [1,17].

54.4.5 Pathogens

Initially, pathogens enter the wetland treatment because they can be found suspended in the water, and then the populations of microorganisms are reduced in number by several mechanisms. These mechanisms include natural death, predation, sedimentation, and adsorption [17] as well as sedimentation and filtration [39].

54.4.6 Sulfur

The hydrogen sulfide and the sulfur-methylated compounds are volatile and come out of the wetland into the atmosphere [18]. The existing bacteria are involved in the reduction of sulfur to sulfide compounds and during the sulfides oxidation [11]. Wastewater contains sulfur as sulfite or as sulfate, usually in small quantities. Then a weak adsorption in soil occurs and the removal is done mainly by the assimilation of plants, particularly after harvest time [39].

54.4.7 Iron and Manganese

One of the applications of wetlands consists of the removal of iron and manganese, particularly of the wastewater produced by the mining industry. Batty et al. [2] found that a key removal process is iron oxidation to form iron hydroxide sediments, which occurs in the initial stages of the wetlands and ponds.

54.4.8 Heavy Metals

Heavy metals are removed mainly by adsorption, complexation, precipitation, and assimilation by plants. Since organic and inorganic sediments are continuously increasing at a low speed, the adsorption improves [39]. Removal of metals from physical-chemical type processes are cation exchange and formation of chelates with the substrate or with sediment, the union with

humic materials, and precipitation of insoluble salts sulfates or carbonates [11].

54.5 Types of Plants Used

Aquatic plants are classified in three types based on the place where they develop within the water body; they could be emergent, submerged, or floating [11,28].

54.5.1 Emergent or Rooted Species

This type is commonly seen in groundwater flow systems, and the main species are *Arundo donax* (giant cane, giant reed), *Limnobium stoloniferum* (West Indian spongeplant), *Juncus* sp. (reeds), *Hydrocotyle* sp. (Marsh pennywort, Navelwort, dollarweed), *Scirpus* sp. (bulrush), *Alternathera philoxeroides* (alligatorweed), *Salvinia rotundifolia* (salvina), *Typha angustifolia* (lesser bulrush or narrowleaf cattail or lesser reedmace), and *Phragmites communis* (common reed).

54.5.2 Submerged Species

This last group develops exclusively within the water, although they can be used in flooded wetlands. Some examples are *Elodea canadensis* (American or Canadian waterweed or pondweed), *Potamogeton* sp. (stem water grass), and *Elodea canadensis* and *Hydrilla verticillata*.

54.5.3 Floating Species

This kind is used in flooded treatment wetlands and some examples are *Eichhornia crassipes* (water hyacinth), *Lemna minor* (common duckweed or lesser duckweed), *Wolffia* sp. (common watermeal, duckweed), and *Pistia stratiotes* (water lettuce) (Figure 54.6).

One of the most significant functions of the wetland treatment is its capacity to assimilate nutrients, which is directly related to the types of aquatic plants. The assimilation capacity of the plants is based on their speed of growth and harvest time [44]. Plants with a large biomass per unit area have the potential to store a large amount of nutrients. Also, large plants have a great capacity to assimilate a large numbers of traces of elements, some of which are essential for the growth of plants. The demand for these elements may increase when plants are grown in wastewater containing high levels of macronutrients [11].

Water lilies and other aquatic plants can absorb metals such as Cu, Zn, Pb, Cd, Hg, and Ni. Where the plants are not harvested, the vegetal tissue will die and the plants will decompose quickly. During that process, nutrients and metals will be released in the water and the treatment efficiency will be reduced [22]. Also, Lara-Borrero [22] and Rivas et al. [35] pointed out some of the necessary aspects to take into account for the selection of aquatic plants:

- Adaptability to the local climate
- High photosynthetic speed

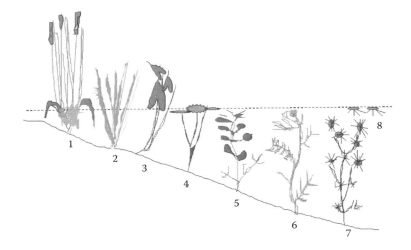

FIGURE 54.6 Classification of aquatic plants: emergent (1–4), submerged (5–7), and floating (8). (Adapted from Odum, E.P. 1972. *Ecología*. Interamericana, México.)

- High transportation capacity or transfer of oxygen
- Wide range of tolerance to contaminant concentration
- High assimilation capacity of pollutants
- Tolerance to adverse weather conditions
- Resistance to insects and diseases
- Ease of control

54.6 Treatment Wetlands Design

The design of wetland treatment requires considering the following:

- Concentration of contaminants in input wastewater
- Required effluent quality
- Type of disposal or reuse according to the existing regulations
- Maximum and minimum water temperature

The most frequent design parameters are flow, BOD, TSS, total nitrogen, total phosphorus, and fecal coliform [9]. Design criteria depend on the concentrations and types of contaminants present. However, it is important to take into account the vegetation types, pretreatment units, and economic aspects (design, construction, operation, and maintenance costs). Wood [46] indicates the following physical and kinetic aspects:

- *Organic surface load.* 75–85 kg/Ha/d in flooded systems and 70–80 in groundwater flow systems
- *Water depth.* 0.30–0.50 m in flooded systems and 0.40–0.70 m in groundwater flow systems
- *Organic hydraulic load.* 760 mm/d in flooded systems and 2–30 in groundwater flow systems
- *Flow rate system.* Modules with less than 10 L/s
- *Retention time.* 6–15 days in flooded systems and 3–7 days in groundwater flow systems
- *Length to width ratio.* Greater than 2:1 and less than 4:1
- *Type of substrate.* Gravel with diameter between 100 and 150 mm

- *Effluent requirements*: BOD, NT, PT, CF, and TSS
- Temperature
- Porosity
- Hydraulic gradient

The equations most commonly used for the design correspond to the equations of Kadlec and the US Environmental Protection Agency. These equations were obtained empirically from the evaluation of hundreds of systems in operation, located primarily in the United States and Europe. These equations consider climatic aspects, variety of vegetal species, different types of filter media, and varied times of operation, among other factors [9,17].

Kadlec and Knight [17] proposed a methodology to estimate the required area, depending on the contaminant to be removed from a model called k–C^* (where k or $k_{A,T}$ represents the flow piston and C^* represents the system background contamination). It is expressed by the following equation:

$$\mathrm{Ln}\left(\frac{C_e - C^*}{C_i - C^*}\right) = -\frac{k_{A,T}}{q} \tag{54.1}$$

where

C_e = Outlet target concentration, effluent (mg/L)
C_i = Inlet concentration, influent (mg/L)
C^* = Background concentration in the bed of aquatic plants (mg/L)
$k_{A,T}$ = first-order dependent areal rate constant at temperature T in °C (m/year)
q = Hydraulic loading rate (m/year)

The k–C^* model could be rearranged to determine the total area as follows:

$$A = \left(\frac{0.0365Q}{k_{A,T}}\right)\mathrm{Ln}\left(\frac{C_e - C^*}{C_i - C^*}\right) \tag{54.2}$$

TABLE 54.1 $k_{A,20}$, θ, and C^* Values

Variable	BOD	TSS	NT	PO$_4$T	CF
Subsurface flow wetland					
$k_{A,20}$ (m/year)	180	1000	27	12	95
θ	1.0	1.0	1.05	1.0	1.0
C^* (mg/L)	$3.5 + 0.053\, C_i$	$5.1 + 0.16\, C_i$	1.50	0.02	10
Surface flow wetland					
$k_{A,20}$ (m/year)	34	1000	22	12	75
θ	1.0	1.0	1.05	1.0	1.0
C^* (mg/L)	$3.5 + 0.053\, C_i$	$5.1 + 0.16\, C_i$	1.50	0.02	300

Note: BOD, biochemical oxygen demand; TSS, total suspended solids; TN, total nitrogen; TPO$_4$, total phosphates; FC, fecal coliform.

where

A = Required wetland area (ha)

Q = Average flow rate (m³/d)

where the variable $k_{A,T}$ is obtained from the following equation:

$$k_{A,T} = k_{A,20}\theta^{(T-20)} \tag{54.3}$$

where

T = wetland water temperature in the coldest month (°C)

$k_{A,20}$ = first-order dependent areal rate constant at 20°C (m/year)

θ = temperature correction factor (1.06)

The values of $k_{A,20}$, θ, and C^* are obtained from Table 54.1.

Subsequently the US Environmental Protection Agency [8] proposed a simplified expression. This equation is commonly used for bed dimensioning of aquatic plants. The removal of BOD$_5$ has as a maximum limit the BOD$_5$ from the bottom of the bed. The value assigned to the coefficients is at the discretion of the designer and the selected value significantly modifies the resulting area. The surface is given by the following equation:

$$A_s = \frac{Q(\operatorname{Ln} C_i - \operatorname{Ln} C_e)}{k_{A,20}d_w\eta} \tag{54.4}$$

where

A_s = wetland surface area (m²)

$k_{A,20} = k_{20}\,(1.06)^{(T-20)}$

$k_{20} = 0.678\,\text{d}^{-1}$

d_w = water depth average, typically 0.20–0.6 m

η = wetland porosity, typically 0.4–0.65 (nondimensional medium porosity)

On the other hand, Cooper et al. [3] proposed the next equation:

$$A_h = \frac{Q(\operatorname{Ln} C_i - \operatorname{Ln} C_e)}{K_{\text{BOD}_5}} \tag{54.5}$$

where

A_h = wetland surface area (m²)

K_{BOD5} = Constant reduction of BOD (Cooper et al. [3] suggest a value of 0.10 in the range of 150–300 mg/L, depending on the temperature), m/day

Experiences at the Mexican Institute of Water Technology (or IMTA by its acronym in Spanish) indicate that the previous design criteria could generate overdesigned systems if specific climatic conditions are uncertain. This increases the overall cost because more land will be necessary and operation expenses will rise. As a result, it is suggested to fully investigate the expected weather conditions [32–35]. Treatment wetlands are technologies that have been widely used for the treatment of different types of wastewater and are used successfully under warm and temperate weather conditions [38].

Other important considerations are as follows [33]:

- *Depth.* Usually the suggested design depth is 0.60 m.
- *Filter media.* This structure consists of three layers with different thickness and size of particles as shown in Table 54.2.
- *Vegetation.* Native plants of the region should be used as much as possible.
- *Channel width.* It is suggested a minimum channel width of 5 m to prevent clogging, reduce the dispersion, and upgrade the pollutants removal. Broader channels present a higher risk of producing preferential flows; the more narrow, the great the clogging risk.
- *Location of inlet and outlet structures.* The installation of several input and output structures generates a better distribution of the water into the ponds, which improves the treatment efficiency.

TABLE 54.2 Suggested Filter Media Specifications

Layer	Thickness (cm)	Diameter of Particles (cm)
Top	10	1–2
Medium	20	2–4
Bottom		4–5

- *Distribution and collection pipes.* Slotted pipes with small holes usually clog easily, so it is recommended to use slots with a minimum diameter of 50.8 mm or slots of 100 mm long and 50.8 mm wide.
- *Slope.* A proper slope ensures an adequate flow velocity. In general, a slope of 0.05% is recommended. Some authors suggest a slope of 1.0%; however, this should be used just in small systems. Regarding the depth near the outlet in large systems, it could be very high.

54.6.1 Pretreatment Units

Grit, sand, and grease removal structures are necessary to prevent clogging in the wetland treatment. The absence of these units in several systems has been identified and this has allowed the entry of solids to treatment ponds causing blockages, preferential flow, stagnant flow, and a decrease in the retention time. Consequently, the efficiency in removing pollutants is reduced [12].

54.7 Operation and Maintenance

The suggested routine operation and maintenance activities are as follows [9,13,17]:

- Measurement and distribution of the flow (monitoring and control)
- Water level control in treatment units
- Cleaning in pretreatment units (grids and sand removal structures)
- Cleaning of pumping sumps (if any)
- Regular harvesting of excess vegetation and replanting of low covered areas
- Removal of vegetation on embankments
- Destruction of mouse nests and anthills
- Cleaning and repair of crossing structures
- Register activities in a record book

Figure 54.7 shows operation and maintenance activities in pretreatment units. Notice the grit removal structure is composed by two alternate channels and Figure 54.8 is an example of a harvesting task.

The resulting dead foliage from plant trimming has many nutrients. Thus, it should be disposed outside the wetland area to avoid water quality deterioration. A systematic pruning (1–2 times per year) allows maintaining plant vigor (development) and therefore a better removal of nutrients [44]. However, trimmed vegetation can be composted and used as fertilizer for crops as long as sewage waters are municipal. The industrial wastewater must be confined first to analyze the existence of toxic materials, such as heavy metals. After harvesting, some plants such as the common reed (*Phragmites communis*) could be used to manufacture crafts [35].

Other activities carried out less frequently are the maintenance of the operation rooms, the protective fences, and the access roads to the treatment system. Furthermore, it is important that operators follow safety and hygiene standards for their own protection, such as the use of gloves, waders, boots, caps, mouth covers, protection lens, and soap. Finally, a first aid kit should be available at any time.

54.8 Applications

Treatment wetlands have been commonly used as municipal wastewater systems. However, there are other uses, for example [12,18,39]:

- Domestic and urban waters.
- Industrial (particularly for industries with residues of the organic type such as sugar mills, hydrocarbons, food processing, chemicals, pharmaceuticals, cosmetics, distilleries, paper, and cattle farms).
- Treatment of acid water, such as acidophiles from mine drainage.

(a)   (b)

FIGURE 54.7 Cleaning the grid (a) and grit removal structures (b).

FIGURE 54.8 Regular harvesting of excess vegetation.

- Removal of heavy metals, for instance, iron and manganese.
- Reduction of nutrients, mainly nitrogen and phosphorus.
- Agricultural wastewater and urban runoff.
- Water quality improvement in rivers and lakes.
- As a final treatment of effluents from other technologies, such as anaerobic systems and electromechanical systems.
- In rural areas where population is scattered, single-family or multi-family wetlands eliminate the construction of expensive drainage and sewerage systems.
- Greywater treatment in urban environments.
- In communities located over hilly terrain, it is difficult to construct single treatment sites due to land availability. In those cases, individual treatment wetlands can be scattered.
- Wildlife protection sites.
- Treatment wetlands could be used in the production of ornamental flowers (Figure 54.9) and forage cultivation.

54.9 Advantages and Disadvantages of Treatment Wetlands

Several advantages and disadvantages of treatment wetlands are pointed out [12,22].

54.9.1 Disadvantages

- Requires large areas of land.
- The investment cost is increased with expensive land.
- Filter media clogs when the operation is not efficient.
- Odor generation in undersized or overloaded systems.
- In cold climates, the speed reduction of contaminants is lower.
- The presence of toxic substances may affect its treatment efficiency.
- The plants of the system are affected during long periods without inflow, for example in schools.

FIGURE 54.9 Ornamental plants development in wetlands of subsurface flow.

- Waterproof deficiencies such as cracks during construction are difficult to correct.
- There is a risk of development of flies and mosquitoes in surface flow wetlands.
- Flow losses occur due to ET, which is higher in dry climates with warm temperatures.
- The salinity in the beds may accumulate after several years of operation.
- Even when treatment wetlands have been applied successfully, there is still a lack of knowledge of the physical, chemical, biological, and hydrological processes involved in the purification of the water.
- Usually the sizing does not include a proper hydraulic design.
- The pond depth is limited by the root penetration.
- Contaminant removal efficiencies are reduced with a systematic pruning of vegetation and with the rehabilitation of areas with accumulated sediments.
- Experienced project engineers are required to determine the equations to be used, the type of plants, the grain size, the most suitable geometry, and finally the location of input and output structures to reduce preferential flows and dead zones.
- The construction process demands careful supervision.

54.9.2 Advantages

- Minimal or no power consumption.
- Low operation and maintenance costs.
- Simple operation and maintenance.
- There is no odor generation when they are well designed and operated.
- There is no noise generation as long as the operation does not involve electro-mechanical equipment.
- Low production of sewage sludge.
- There is low risk of development of flies and mosquitoes in the subsurface wetlands.
- Efficient removal of BOD_5, TSS, pathogens, and nutrients (nitrogen and phosphorus).
- Lower treatment costs than other wastewater treatment systems.
- Flexibility and tolerance to variations in flow rate and organic load.
- Since treatment wetlands constitute new habitats, they protect wildlife areas and contribute to the development and increase of biodiversity.
- Treatment wetlands support higher ranges of variation of hydraulic and organic loading than electromechanical systems (up to 300%).
- Treatment wetlands could be designed to meet different size criteria, starting from one-person facilities, multi-family systems, and even for small cities with land availability.
- Effluents are low in BOD, pathogenic microorganisms, and macronutrients (nitrogen, phosphorus, and potassium).

- The effluent can be reused in multiple applications. For example, in irrigation of green areas and crops, aquaculture, groundwater recharge, and aquatic life protection in rivers and lakes.
- Treatment wetlands can be used in cold, mild, and warm climates.
- Wetlands are aesthetic and environmentally friendly.
- Production of usable products (flowers, forage crops, and raw materials for craft manufacturing.

54.9.3 Removal Efficiencies

It is feasible to obtain concentrations lower than 35 mg/L of suspended solids, although the removal performance depends mainly on the applied organic load and pond depth [12]. Organic matter removal is relatively rapid in the first few meters of the wetland, particularly in temperate climates. However, after the starting point, the removal becomes slower because new residual organic matter from the decomposition of plants and other organisms is generated. Thus, it is not possible to obtain effluent completely free of organic material [39].

Rivas [34] indicates that the global efficiencies reported in the literature are as follows: BOD, COD, and TSS 80%; fecal coliforms 99.9%; ammoniacal nitrogen 65%; phosphorus 80%; color 90%; turbidity 95%; iron 96%; manganese 83%; and pH changes vary from 4% at the entrance and up to 8% at the outlet.

54.10 Summary and Conclusions

In this chapter, treatment wetlands are defined and classified, describing the mechanisms of removing pollutants, the most frequently used design criteria, routine operation, and maintenance activities. Also, their advantages and disadvantages are discussed, as well as different applications. Finally, a case study under operation is evaluated and the obtained economic, social, environmental, and technological benefits are discussed.

At present, there are basically two types of wetlands—natural and those used for the treatment of residual water. In contrast to the natural wetlands, the treatment wetlands are sewage systems that incorporate elements of engineering that are designed and built specifically to meet water quality regulatory standards. Also, treatment wetlands have as their objective the reduction of contaminant content in wastewater. During such processes, flows and water levels are controlled and plant species are harvested. Operation and maintenance activities are carried out with the objective of improving water quality.

Acknowledgment

The second author expresses his sincere gratitude to Mexico's National Council of Science and Technology (CONACyT, by its name in Spanish) for its continuous financial support to foster research activities.

References

1. Arias, C.A., Cabello, A. Brix, H., and Johansen, N.H. 2003. Removal of indicator bacteria from municipal wastewater in an experimental two-stage vertical flow treatment wetland system. *Water Science and Technology*, 48(5), 35–41.

2. Batty, L., Hooley, D., and Younger, P. 2008. Iron and manganese removal in a wetland treatment systems: Rates, processes and implications for management. *Science of the Total Environment*, 394(1), 1–8.

3. Cooper, P.F., Job, G.D., Green, M.B., and Shutes, R.B.E. 1996. *Reed Beds and Constructed Wetlands for Wastewater Treatment*. WRC Publications, Marlow, UK.

4. Delgadillo, O., Camacho, A., Pérez, L., and Andrade, M. 2010. *Depuración de aguas residuales por medio de humedales Artificiales* (*Waste Water Treatment by Artificial Wetlands*). Centro Andino para la Gestión y Uso del Agua (*Andean Center for the Management and Use of Water*) (Centro AGUA) (*Water Center*), Cochabamba, Bolivia.

5. Dinges, R. 1982. *Natural Systems for Water Pollution Control*. Van Rostrand Reinhold Company, New York.

6. Drizo, A., Frost, C.A., Smith, K.A., and Grace, J. 1997. Phosphate and ammonium removal by constructed wetlands with horizontal subsurface flow, using shale as a substrate. *Water Science and Technology*, 35, 95–102.

7. EPA. 1988. *Design Manual. Constructed Wetlands and Aquatic Plant Systems for Municipal Wastewater Treatment*. Cincinnati, OH.

8. EPA 832-R-93-008. 1993. *Subsurface Flow Constructed Wetlands for Wastewater Treatment*. Environmental Protection Agency. United States Office of Water (4204). USA.

9. EPA. 2000. *Manual. Constructed Wetlands Treatment of Municipal Wastewaters*. Cincinnati, OH.

10. Estrada, I.Y. 2010. *Monografía sobre humedales artificiales de flujo subsuperficial (HAFSS) para remoción de metales pesados en aguas residuales*. (*Monograph on Subsurface Flow Constructed Wetlands (SSF) for Removal of Heavy Metals in Wastewater*.) Colombia.

11. Fernández, G.J., De Miguel, B.E., De Miguel, M.J., and Curt, F.D.M. 2004. *Manual de Fitodepuración, filtros de macrófitas en flotación* (*Manual of Fitodepuración, Filters of Macrophytes in Flotation*). Editado por Ayuntamiento de Lorca, Universidad Politécnica de Madrid, Fundación Global Nature, y Obra Social. Edited by Ayuntamiento de Lorca, Polytechnic University of Madrid, Fundación Global Nature, and Social Work.

12. García, J., Motaó, J., and Bayona, J. 2004. *Nuevos Criterios para el Diseño y Operación de Humedales Construidos*. (*New Criteria for the Design and Operation of Constructed Wetlands*.) Ediciones CPET. Barcelona, Spain.

13. García, S.J. and Corzo, H.A. 2008. *Depuración con humedales construidos. Guía práctica de diseño, construcción y explotación de sistemas de humedales de flujo subsuperficial*. (*Treatment with constructed wetlands. Practical Guide for Design, Construction and Operation of Subsurface Flow Constructed Wetland Systems*.) Departamento de Ingeniería Hidráulica, Marítima y Ambiental de la Universidad Politécnica de Catalunya. (Department of Hydraulic Engineering, Maritime and Environmental of the Polytechnic University of Catalunya.)

14. Hammer, D.A. 1989. *Constructed Wetlands for Wastewater Treatment. Municipal, Industrial and Agricultural*. Lewis Publishers, Michigan.

15. Hoffmann, H., Platzer, Ch., Winker, M., and Von Muench, E. 2011. *Revisión técnica de humedales artificiales de flujo subsuperficial para el tratamiento de aguas grises y aguas domésticas*. Alemania. (*Technical Review of Subsurface Flow Constructed Wetlands for Treatment of Gray Water and Domestic Sewage*. Germany.)

16. Kadlec, R.H. 1999. Chemical, physical and biological cycles in treatment wetlands. *Water Science and Technology*, 40(3), 37–44.

17. Kadlec, R.H. and Knight, R.L. 1996. *Treatment Wetlands*. Lewis Publishers, Boca Raton, FL.

18. Kadlec, R.H. and Wallace, S.D. (ed.). 2009. *Treatment Wetlands*. CRC Press, Boca Raton, FL.

19. Kim, S.Y.K and Geary, P.M. 2001. The impact of biomass harvesting on phosphorus uptake by wetland plants. *Water Science and Technology*, 44(11–12), 61–67.

20. Langergraber, G. and Simunek, J. 2005. Modeling variably saturated water flow and multicomponent reactive transport in constructed wetlands. *Vadose Zone Journal*, 4, 924–938.

21. Lantzke, I.R., Heritage, A.D., Pistillo, G., and Mitchell, D.S. 1998. Phosphorus removal rates in bucket size planted wetlands with a vertical hydraulic flow. *Water Research*, 32, 1280–1286.

22. Lara-Borrero, J.A. 1999. *Depuración de aguas residuales municipales con humedales artificiales*. (*Purification of Municipal Sewage with Artificial Wetlands*.) Tesis maestría, Instituto Catalán de Tecnología, (Master thesis, Catalan Institute of technology) UPC, Spain, 114 p.

23. Maina, C.W., Mutua, B.M., and Oduor, S.O. 2011. Evaluating performance of vertical flow constructed wetland under various hydraulic loading rates in effluent polishing. *Journal of Water, Sanitation and Hygiene for Development*, 1(2), 144–151.

24. Masi, F., Bendoricchio, G., Conte, G., Garuti, G., Innocenti, A., Franco, D., Pietrelli, L., Pineschi, G., Pucci, B., and Romagnolli, F. 2000. Constructed wetlands for wastewater treatment in Italy: State-of-the art and obtained results, In: *Proceedings 7th International Conference on Wetland Systems for Water Pollution Control*, Lake Buena Vista, FL. University of Florida, Gainesville and International Water Association.

25. Mena, J., Lourdes Rodriguez, M.L., Núñez, M.J., and Villaseñor, C.J. 2008. *Depuración de aguas residuales con humedales artificiales: Ventajas de los sistemas híbridos*

(Waste Treatment with Artificial Wetlands: Advantages of the Hybrid Systems). Congreso Nacional del Medio Ambiente. Cumbre del Desarrollo Sostenible (National Congress of the Environment. Summit of the Sustainable development). CONAMA.

26. Odum, E.P. 1972. *Ecología.* (Ecology.) Interamericana. México, 639 p.

27. Olmedilla, M. and Rojo, C. 2000. *Función depuradora de los humedales I: una revisión bibliográfica sobre el papel de los micrófitos (Purification Function of Wetlands I: A Bibliographical Review on Macrophytes Role).* Boletín SEHUMED Año IV-Número 14-Junio 2000. Valencia (España). Valencia (Spain).

28. Plaza, D.C. and Vidal, S.G. 2007. *Humedales construidos: una alternativa a considerar para el tratamiento de aguas residuales. (Constructed Wetlands: An Alternative to be Considered for the Treatment of Wastewater.)* Tecnología del agua (Water Technology), Chile.

29. Secretaría de la Convención de Ramsar. 2013. *Manual de la Convención de Ramsar: Guía a la Convención sobre los Humedales (The Ramsar Convention Manual: Guide to the Convention on Wetlands* (Ramsar, Irán, 1971)), 6a. edición. Secretaría de la Convención de Ramsar, Gland (Suiza) (6th. Edition. Secretariat of the Ramsar Convention, Gland (Switzerland).)

30. Reed, S., Crites, R., and Middlebrooks, E. 1995. *Natural Systems for Waste Management and Treatment.* 2nd Edition. McGraw-Hill, New York.

31. Reddy, K.R. and DeBusk, T.A. 1987. State-of-the art utilization of aquatic plants in water pollution control. *Water Science and Technology,* 19(10), 61–79.

32. Rico, M. and Rivas, H.A. 1992. *Sistemas de tratamiento de aguas usando lechos de hidrófitas. (Systems of Water Treatment Using Hydrophyte Beds.)* Informe final. CNA. IMTA. México. Final report. CNA. IMTA. Mexico.

33. Rivas, A., Barceló, Q.I., and Moeller, G.E. 2011. Pollutant removal in a multi-stage municipal wastewater treatment system comprised of constructed wetlands and a maturation pond, in a temperate climate. *Water Science and Technology,* 64(4), 980–987.

34. Rivas, H.A. 1997. Lechos de plantas acuáticas (LPA) para el tratamiento de aguas residuales. *Ingeniería Hidráulica en México,* 3(12), 74–77.

35. Rivas, H.A., Pozo, R.F., and Soto, S.S.I. 2005. *Instalación de humedal para el tratamiento de las aguas residuales de la localidad de Santa Fe de la Laguna, Qluiroga, Mich.*

(Installation of Wetland for Treatment of Wastewater from the Town of Santa Fe de la Laguna, Quiroga, Mich.) Informe final. SEMARNAT, IMTA, Fundación Gonzalo Río Arronte, I.A.P. México. (Final report. SEMARNAT, IMTA, Gonzalo Rio Arronte Foundation, I.A.P., Mexico.)

36. Rivas, H.A. and Ramírez, G.A. 1999. *Diseño y diagnóstico de sistemas de tratamiento de aguas residuales mediante lechos de plantas acuáticas (wetlands). (Design and Diagnosis of Systems of Wastewater Treatment through Beds of Aquatic Plants (Wetlands).)* Informe final. (Final report.) SEMARNAT, CNA, IMTA. México.

37. Rodríguez, P. de A.C. 2003. Humedales construidos. Estado del arte. (II). (Constructed wetlands. State of the art. (II).) *Ingeniería Hidráulica y Ambiental,* 24(3), 42–48.

38. Rousseau, D.P.L. and Hooijmans, T. M. 2009. *Recent Advances in Modelling of Natural Treatment Systems.* Agua 2009, Cali, Colombia.

39. Sherwood, C.R., Crites, R.W., and Middlebrooks, E.J. 1995. *Natural Systems for Waste Management and Treatment.* McGraw-Hill, New York.

40. Tyroller L., Rousseau, D.P.L., Santa, S., and García, J. 2010. Application of the gas tracer method for measuring oxygen transfer rates in subsurface flow constructed wetlands. *Water Research,* 44(14), 4217–4225.

41. Uggetti, E., Ferrer, I., Molist, J., and García, J. 2011. Technical, economic and environmental assessment of sludge treatment wetlands. *Water Research,* 45(2), 573–582.

42. UN-HABITAT. 2008. *Constructed Wetlands Manual.* UN-HABITAT Water for Asian, Kathmandu, Nepal.

43. Vymazal, J. 2008. Constructed wetlands for wastewater treatment: A review. In: M. Sengupta and R. Dalwani (eds.), *Proceedings of Taal2007: The 12th World Lake Conference: 965:980.*

44. Vymazal, J. and Kropfelavá, L. 2008. *Wastewater Treatment in Constructed Wetlands with Horizontal Sub-Surface Flow.* Environmental pollution, Volume 14, pp. 8, Springer, Netherlands.

45. Weedon, C.M. 2010. A decade of compact vertical flow constructed wetlands. *Water Science and Technology,* 62(12), 2790–2800.

46. Wood, A. 1995. Constructed wetlands in water pollution control, Fundamentals to their understanding. *IAWQ. Water Science and Technology,* 32(3), 21–29.

55

Constructed Wetlands: Case Studies

Dongqing Zhang
Nanyang Technological University

Manish Kumar Goyal
Indian Institute of Technology

Vishal Singh
Indian Institute of Technology

Richard M. Gersberg
San Diego State University

PREFACE

This chapter describes the applications of constructed wetlands for the removal of anthropogenic contents from wastewater. Emphasis is given to natural and constructed wetlands for wastewater treatment, especially pharmaceutical waste management. For this purpose, various case studies around the world have been reviewed and two most important case studies are presented to highlight the water treatment process through constructed wetlands. Attention is also paid to the importance of different water treatment processes, important species of plants that work as water purifier, parameters that can influence natural water, and cost–benefit analysis.

55.1 Introduction

55.1.1 Background

Wetlands are zones or areas that lie on a continuum between terrestrial and aquatic environments and demarcation of the boundaries often is not defined. In general terms, wetland is land where water is the dominant factor determining the nature of soil development and the types of plant and animal communities living in the soil and on its surface and the water table is usually at or near the surface or the land is covered by shallow water. It spans a continuum of environments where terrestrial and aquatic systems intergraded. Natural wetlands perform both functions, which are beneficial for both human and wildlife. Cowardin et al. [16] suggested that wetlands must have one or more of the following attributes for classification purposes: (1) at least periodically, the land supports predominantly hydrophytes; (2) the substrate is predominantly undrained hydric soil; and (3) the substrate is nonsoil and is saturated with water or covered by shallow water at some time during the growing season of each year.

The constructed wetlands can be defined as an artificial wastewater treatment system that consists of shallow ponds that have been planted with aquatic plants, and which rely upon natural processes such as chemical, physical, biological, and microbial to treat wastewater. The sallow ponds or channels are generally less than 1 m deep and consist of impervious clay (or synthetic liners), and engineered structures to control the flow direction, liquid detention time, and water level. Inert porous media such as sand, gravel, or rock may be used.

The most important function of wetland is water filtration. As water flows through a wetland, it slows down. In addition, several suspended solid particles become confined and trapped by vegetation and settle out. Other pollutants are transformed to less soluble forms taken up by plants. The plants in wetland also foster the basis surroundings for microorganisms to live there. Through a series of complex processes, these microorganisms also transform and remove pollutants from the water.

Constructed wetlands are generally built on uplands and outside floodplains in order to avoid destruction to natural wetlands, water bodies, and other aquatic resources. Wetlands are frequently built using a number of engineering processes such as excavation, backfilling, grading, diking, and installing water control structures to establish desired hydraulic flow patterns. In case of highly permeable soils, an impervious, compacted clay liner is usually installed and the original soil placed over the liner. Wetland vegetation is then planted or allowed to establish naturally. Constructed wetlands have been used to treat a variety of wastewaters including urban runoff, and municipal, industrial, agricultural, and acid mine drainage. Wetlands have been subjected to wastewater discharges from municipal, industrial, and agricultural sources, and have received agricultural and surface mine runoff, irrigation return flows, urban storm water discharges, leachates, and other sources of water pollution. The actual impacts of such inputs on different wetlands have been quite variable [1].

A constructed wetland system that is specifically engineered for water quality improvement as a primary purpose is termed a constructed wetland treatment system (CWTS). CWTS is a human-induced wetland area that is mostly formulated for water quality improvement through optimizing physical, chemical, and biological processes that is a part of natural wetland systems [68]. Wetlands can provide a cheaper and low-cost alternative system for wastewater treatment, either constructed or natural. Natural characteristics are applied to CWTS with emergent macrophyte stands that duplicate the physical, chemical, and biological processes of natural wetland systems. In the recent decades, CWs have been considered low cost alternatives of domestic and industrial wastewater treatments. The wastewater consists of many organic and inorganic toxicants that are problematic issues because these toxicants have potential to filter in the treatment systems. Thus, various toxicants and heavy metals significantly influence the effectiveness of CWs [50]. However, the toxicant removal processes in CWs is very composite. These processes include sedimentation, flocculation, adsorption, precipitation, coprecipitation, cation and anion exchange,

complexation, oxidation and reduction, microbial activity, and plant uptake based on various biotic and abiotic reactions [58]. However, note that if constructed wetland is used as a treatment system for metal containing domestic wastewater, the treatment design should be done cautiously. Hydrology is also a significant driving force controlling wetland structure and function. It can be altered runoff, and resultant urban runoff changes the magnitude, frequency, and timing of flow to wetlands. These changes have been best observed in riverine wetlands, which are immediately impacted by changes in watershed hydrology [14].

55.1.2 Natural Wetlands versus Constructed Wetlands

It is important to understand the difference between "natural" and "constructed" wetlands to know how this potentially affects the habitat value they provide. In general, constructed wetlands are designed to optimize water quality benefits. Equally, natural wetlands, often restored or enhanced to optimize wildlife habitat, might also provide water quality benefits. A natural wetland is one that "occurs without the aid of humans," while a constructed wetland is one that is "purposely constructed by humans in a non-wetland environment." Natural wetlands (e.g., swamps, bogs, marshes, fens, sloughs, etc.) are being considered as providing many benefits in terms of water quality improvement, flood protection and control measurements, food for wildlife, shoreline erosion protection, habitat for wildlife, and opportunities for recreation and aesthetic appreciation. The functional role of wetlands in improving water quality has been a compelling argument for the preservation of natural wetlands and, recently, the construction of wetlands systems for wastewater treatment.

Constructed wetlands are created from a nonwetland ecosystem or a former terrestrial environment, mainly for the purpose of contaminant or pollutant removal from wastewater. In constructed wetlands, many natural processes such as wetland vegetation, soils, and their associated microbial assemblages are used to assist in treating wastewater in a natural manner. These systems have been designed and operated for a number of purposes including treating wastewater, and using treated wastewater effluent as a water source for the creation and restoration of wetland habitat for wildlife use and environmental enhancement [21].

55.1.3 Types of Constructed Wetlands

Constructed wetland systems can be divided into two general types: (1) the horizontal flow system (HFS) and (2) the vertical flow system (VFS). HFS can be further divided into two general types: (i) free-water surface (FWS) or surface flow (SF) and (ii) sub-surface flow (SSF) systems. In HFS, wastewater is fed at the inlet and flows horizontally through the bed to the outlet. VFS are fed sporadically and drain upright through the bed via a network of drainage pipes [65]. Odum [46] began a study using coastal lagoons for recycling and reuse of municipal wastewaters.

55.1.3.1 Free-Water Surface (FWS)

The FWS systems are used mainly with large wastewater flows for nutrient polishing especially for municipal wastewater treatment (Figure 55.1). The FWS system tends to be rather large with only a few smaller systems in use. The majority of constructed wetland treatment systems are SF or FWS systems. The use of FWS systems started extensively in North America. These types utilize influent waters that flow across a basin or a channel that supports a variety of vegetation, and water is visible at a relatively shallow depth above the surface of the substrate materials.

The common species for FWS constructed wetlands are in Europe: *Phragmitesaustralis* (common reed), North America: *Typha*

FIGURE 55.1 Tropical application of free-water surface flow constructed wetlands in Singapore (Alexandra Constructed Wetlands).

spp. (cattail); Australia and New Zealand: *Phragmitesaustralis, Eleocharissphacelata* (tall spikerush) etc. [62]. FWS constructed wetlands with emergent microphytes function as a land-intensive biological treatment system. Nitrogen is most effectively removed in FWS constructed wetlands by nitrification/denitrification. Because of the atmospheric diffusion and anoxic and anaerobic zones in and near the sediments, FWS constructed wetlands typically have altered zones, especially near the water surface [62].

55.1.3.2 Sub-surface Flow (SSF) System

The SSF technology is mostly used in Northern Europe and vegetated gravel beds are found in Europe, Australia, South Africa, and almost all over the world (Figures 55.2 and 55.3). This is based on the soil. Water flows from one end to the other end through permeable substrates in a vegetated SSF system, which is made of mixture of soil and gravel or crusher rock. It is also called "Root-Zone Method," "Rock-Reed-Filter," or "Emergent Vegetation Bed System." It has a sloped bottom bed to minimize water that flows overland. Wastewater flows through the gravity in a horizontal direction from the root zone of the vegetation about 100–150 mm below the gravel surface. Numerous macro- and microorganisms inhabit the substrates. The treated water can be collected at outlets at the base of the media, typically 0.3–0.6 m below the bed surface [65].

55.1.4 Floating Treatment Wetlands

Floating treatment wetlands (FTWs) are a novel treatment concept that employ rooted, emergent macrophytes growing on a floating mat rather than rooted in the sediments (Figure 55.4) [26]. The roots hang beneath the floating mat and provide a large surface area for biofilm attachment. Thus, they can tolerate the widely fluctuating water depths typical of stormwater

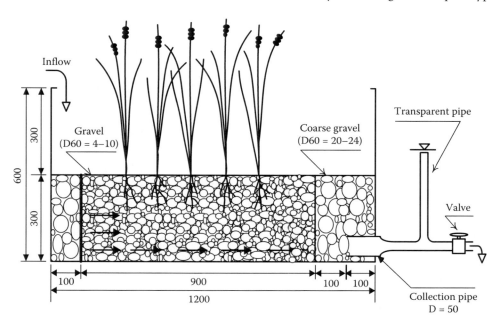

FIGURE 55.2 Typical layout and configuration of horizontal subsurface flow constructed wetlands.

FIGURE 55.3 Horizontal subsurface flow constructed wetlands.

systems, without the risk of the plants drowning. This property then confers FTWs with one main advantage over conventional sediment-rooted wetlands; that is, their ability to cope with variable water depths that are typical of event-driven storm water systems. Additionally, with such a feature, it also enables FTW systems to be designed as detention basins so that large volume storm events may be retained and then released slowly over subsequent days. They represent a means of improving the treatment performance of conventional pond systems by including the beneficial aspects of emergent macrophytes without being constrained by the requirement of shallow water depth.

55.1.5 Advances of Constructed Wetlands

Constructed wetland is a newly emerging cheaper alternative for wastewater treatment using local resources that has been used widely for wastewater and effluent treatment. This system helps in sustainable use of local resources, which is a more

environmentally friendly biological wastewater treatment system. Aesthetically, it is a more landscaped primitive wetland site compared to the conventional wastewater treatment plants. Compared to the other treatment options, constructed wetlands could be constructed at lower costs; however, for this no new technological tools are needed. The system relies on renewable energy sources such as solar and kinetic energy, and wetland plants and microorganisms, which are the active agents in the treatment processes.

In general, to purify the water through wetlands, the plants' physical effects can be brought significantly by the presence of the plants. The plants provide a sufficient surface area for attachment and growth of microbes. The physical components stabilize the surface of the beds and slow down the water flow. In this way, they assist in increasing water transparency through sediment settling and trapping process. Wetlands help in preventing the eutrophication of wetlands plants through the removal and retention of nutrients. The common reed (*Phragmiteskarka*) and cattail typha (*Angustifolia*) are good examples of marsh species that have shown their role in uptaking nutrients. These plants have a large biomass both above in form of the leaves and below in form of the underground stem and roots. The sub-surface plant tissues grow vertically and horizontally, which creates a large surface area for the uptake of nutrients and ions through binding of the soil particles. Apart from this, the macrophytes, which are found in the wetlands, stabilize the surface of the plant beds, provide good conditions for physical filtration, and provide a huge surface area for attached microbial growth. Growth of macrophytes allows for sedimentation and increase in contact time between effluent and plant surface area by the reduction of current velocity, thus, to an increase in the removal of nitrogen [37].

The water hyacinth (*Eichhorniacrassipes*) and duckweed (*Lemna*) are several familiar floating aquatic plants that showed their capability to decrease concentrations of BOD, TSS, total phosphorus, and total nitrogen. However, prolonged presence of *Eichhorniacrassipes* and *Lemna* can lead to deterioration of

FIGURE 55.4 Tropical application of floating treatment wetlands in Kandy, Sri Lanka.

TABLE 55.1   Comparison of Advantages and Disadvantages between Free Water Surface (FWS) and Subsurface Flow (SSF) CWs

	Advantages	Disadvantages
Free water surface (FWS) CWs	Less expensive to construct (per m² basis)	More land requirement
	Simple operation and design	Lower rates of contaminant removal per unit
	Good performance for removal of TSS	Risk of ecological of human exposure
	Offer greater flow control	Odor and insects problems
	More diverse wildlife habitat	
Subsurface flow (SSF) CWs	Higher rates of contaminant removal per unit of land	More expensive for construction
	Less land requirement	Waters containing high suspended solids may cause clogging
	Less expensive for operation	Better P removal performance
	More accessible for maintenance	
	Minimal ecological risk due to absence of an exposure pathway	
	No odor and insects problems because the water level is below the media surface	

TABLE 55.2   Comparison of Advantages and Disadvantages between Vertical and Horizontal Subsurface Flow CWs

	Advantages	Disadvantages
Vertical subsurface flow (HSSF)	Small area requirement	Short flow distances
	Good oxygen supply	Loss of performance consistence in P removal
	Good conditions for nitrification occurrence	Poor conditions for denitrification
	Good NH_4-N removal	Higher technical demand
	Simple hydraulic design	
	High purification performance	
Horizontal subsurface flow (VSSF)	Long flowing distances and good nutrient gradient establishment	Hydraulic design necessary for optimal oxygen supply
	Good removal for BOD_5,COD and TSS	Higher area requirement
	Good NO_3-N removal	
	Nitrification and nitrification possible	
	Formation of humic acid for N and P removal	
	Longer life cycle	

the water quality unless these plants are manually removed on a regular basis. These marshy plants will produce a huger that will delay light penetration to the lower layer of the water column, which could be affected by the survival of the living water organisms [37]. The polluted rivers and other water bodies could also be cleaned using constructed wetland systems. This resultant technology can ultimately be used to regenerate grossly polluted rivers in the country. The constructed wetland treatment plan is broadly applied for various purposes. These water treatment functions include urban and rural runoff management, primary settled and secondary treated sewage treatment through toxicant management from industrial and other wastes, landfill and mining leachate treatment, sludge management, enhancement of in-stream nutrient assimilation, nutrient removal via biomass production, and groundwater recharge. However, the primary purpose of any constructed wetland treatment systems is to treat various wastewater (municipal, industrial, agricultural, and stormwater). The wetland system can also be served as an attractive destination for tourists and local urban dwellers. It can also be served to explore environmental and educational possibilities as a public attraction sanctuary for visitors. The comparison

with different types of wetland systems is shown in Tables 55.1 and 55.2.

55.2 Contaminants Removal in Constructed Wetlands

In wetland systems, complex physical, chemical, and biological processes may occur simultaneously, including volatilization, sorption and sedimentation, phytodegradation, plant uptake, and accumulation, as well as microbial degradation [27]. In wastewater treatment wetlands, the pollutant removal efficiency varies considerably not only from system to system, but also within the same system [53]. Such variability can be traced back to the complex combination of physical, chemical, and biological processes for contaminant removal brought about by the plants, microorganisms, soil matrix, as well as their interactions with each other.

55.2.1 Removal of Total Suspended Solid (TSS)

Wetlands are excellent sediment traps. Stottmeister et al. [53] reported that solids removal mostly occurred in the initial

12%–20% of the cell area of a pilot-scale system studied at Arcata, California. The relevant physical processes include sedimentation, filtration, adsorption onto biofilms, and flocculation/precipitation. Sedimentation and filtration of suspended solids is based on flow retardation that leads to gravity settling of solids. The rate of settling depends on the size, shape, and charge of the particles. Suspended solids, during their settling process can react and bind with various pollutants including organic substances, nutrients, heavy metals, and pathogens, thereby aiding removal of these pollutants from the water column. Accumulation of trapped solids is a major threat to good performance of SSF systems as the solids may clog the soil medium. Therefore, an effective pretreatment is usually necessary for a successful SSF system.

55.2.2 Removal of Organic Matter

Biodegradation of organic compounds involves catabolic activities of both autotrophic and heterotrophic microorganisms. Both aerobic and anaerobic degradation of soluble organic substances are responsible for BOD removal. Aerobic degradation of soluble organic matter is governed by the aerobic heterotrophic bacteria. Ammonifying bacteria also degrade organic compounds containing nitrogen under aerobic conditions [15]. Both bacterial groups consume organics but the faster metabolic rate of the heterotrophs means that they are mainly responsible for the reduction in the BOD of the system. However, many studies have shown that oxygen release from roots of different microphytes is far less than the amount needed for aerobic degradation of the oxygen-consuming substances delivered with sewage [7]. Insufficient supply of oxygen to this group will greatly reduce the performance of aerobic biological oxidation. In most systems designed for the treatment of domestic or municipal sewage, the supply of dissolved organic matter is sufficient and aerobic degradation is generally limited by oxygen availability.

Anaerobic degradation of organics also takes place in a CW in the absence of dissolved oxygen. This is a multistep process carried out by either facultative or obligate anaerobic heterotrophic microorganisms. Anaerobic respiration occurs in the soil zone below the Fe^{3+} reduction zone and the process can be carried out by either facultative or obligate anaerobes. It represents one of the major ways by which high molecular weight carbohydrates are broken down to low molecular weight organic compounds, usually as dissolved organic carbon, which are in turn available to microbes [61]. In the first step, the primary end-products of fermentation are fatty acids, such as acetic, butyric and lactic acids, alcohols, and the gases CO_2 and H_2. Both groups play an important role in decomposition of organic matters [61].

55.2.3 Phosphorus Removal

Although the processes that contribute to nitrogen removal in CWs are complex and include volatilization, plant and microbial uptake, mineralization (ammonification), sorption, desorption, and leaching, nitrification and denitrification have been regarded

as major mechanisms for nitrogen retention [62]. Nitrification is defined as the biological oxidation of ammonium to nitrate with nitrite as an intermediate product in the reaction sequence. Nitrifying bacteria utilize carbon dioxide as a carbon source and oxidize ammonia or nitrite to derive energy. Nitrification is carried out by two types of nitrifying organisms. The first step converts ammonium to nitrite and the second step converts nitrite to nitrate. The nitrification process itself does not remove nitrogen from wastewaters, but nitrification coupled with denitrification appears to be the major removal pathway in many CWs [54,62].

Nitrification occurs in all types of wetlands but has been found to be the limiting step in nitrogen removal. This is because the vast amounts of nitrogen entering the wetlands are in the form of ammonia. Denitrification is most commonly defined as the process in which nitrate is converted into dinitrogen gas via the intermediates nitrite, nitric oxide, and nitrous oxide [29]. From a biochemical viewpoint, denitrification is a bacterial process in which nitrogen oxide (in ionic and gaseous forms) serves as terminal electron acceptors for respiratory electron transport. Electrons are carried from an electron-donating substrate through several carrier systems to a more oxidized N form. The resultant free energy is conserved in ATP, following phosphorylation, and is used by the denitrifying organisms to support respiration. Usually the nitrate concentration of domestic wastewater is low, and thus nitrification is crucial for denitrification to take place. The level of dissolved oxygen is also important when dealing with denitrification, as the process only occurs under anoxic conditions. Various reported studies have shown that nutrients can be taken up directly by roots of microphyte plants. Reddy and DeBusk [48] reported nitrogen standing stock for emergent species in the range of 14–156 g N m^{-2} but the authors indicated that more than 50% of this amount could be stored belowground. Aboveground N standing stock values were reported to be in the range of 22–88 g N m^{-2}.

55.2.4 Climate Change Impact on Snow and Glacier

CWs provide an environment for the interconversion of all forms of phosphorus. Soluble reactive phosphorus is taken up by plants and converted to tissue phosphorus or may be adsorbed to wetland soils and sediments. Organic structural phosphorus may be released as soluble phosphorus if the organic matrix is oxidized. Insoluble precipitates may form under certain circumstances, and may redissolve under altered conditions. Phosphorus transformation in wetlands includes soil accretion, adsorption/desorption, precipitation/dissolution, plant/microbial uptake, fragmentation and leaching, mineralization, and burial [62]. However, it is also well established that the mechanisms that remove P in constructed wetlands include mostly adsorption on substrates, storage in biomass, and the formation and accretion of new sediments and soils.

Adsorption and precipitation of phosphorus is effective in systems where wastewater comes into direct contact with soil substrate. This means that CWs with subsurface flow have a high potential for phosphorus removal via these mechanisms. Plant uptake plays a minor role in phosphorus removal [62]. Most of the phosphorus is taken up by plant roots, and the absorption through leaves and shoots is restricted to submerged species. However, this uptake amount is usually very low. Phosphorus uptake by macrophytes is usually highest during the beginning of the growing season, before maximum growth rate is attained. Biomass increases, however, should not be counted as part of the long-term sustainable phosphorus removal capacity of wetlands. The concentration of phosphorus in the plant tissue varies among species and site, and it varies seasonally as well. Reddy and DeBusk [48] reported the nitrogen standing stock for emergent species in the range of 1.4–37.5 g P m^{-2} with more than 50% of this amount stored belowground. Aboveground P standing stock values were reported in the range of 0.1–11 g P m^{-2}.

55.2.5 Removal of Pharmaceutical and Personal Care Products

Pharmaceuticals and personal care products (PPCPs) in the aquatic environment have attracted increasing concern during the last decade because of their widespread uses and continuous release to the aquatic environment [22,55]. The high polarity and low volatility of most pharmaceuticals results in their easy transportation and discharge into the water compartment [5]. The effect of active pharmaceutical compounds on organisms in the aquatic and terrestrial environment is increasing [23]. These adverse effects may include development of antimicrobial resistance, decrease in plankton diversity, and inhibition of growth of human embryonic cells [49]. In the meantime, the adverse effect caused by chronic aquatic toxicity for pharmaceuticals on

Daphnia, algae, higher aquatic plant, and bacteria has also been demonstrated using low concentrations in chronic test [17].

The environmental exposure scenarios differ substantially between pharmaceuticals used in veterinary and human medicine. Unlike veterinary medicines, which may be introduced to the environment through a variety of direct and indirect exposure, it may therefore be at continuous and low concentrations [18]. Technologies including ozonation, reverse osmosis, and advanced oxidation processes, as well as process optimization in WWTPs (e.g., increasing sludge residence time) [12] do exist to reduce the level of pharmaceutical compounds in wastewater in WWTPs. Most pharmaceutical compounds are refractory to conventional treatment in WWTPs; thus, the pharmaceutical contaminants cannot be effectively and completely removed through conventional WWTPs [31]. Consequently, various kinds of pharmaceutical compounds are released into surface, ground, and coastal waters. The application of CW technology as a cost-effective solution for the elimination of pharmaceuticals has been explored [40]. In CW systems, complex physical, chemical, and biological processes may occur simultaneously, including volatilization, sorption and sedimentation, phytodegradation, plant uptake and accumulation, as well as microbial degradation. A general pharmaceutical removal mechanism in CWs is shown in Figure 55.5. Furthermore, the rate of these processes depends on a variety of design and operational factors such as loading mode (batch or continuous operational mode), the presence of vegetation, soil matrix/substrate, the depth of bed, plant species, organic and hydraulic loading rate, as well as wetland configuration. Table 55.3 shows an overview of the physicochemical properties for the frequently investigated PPCPs in CWs.

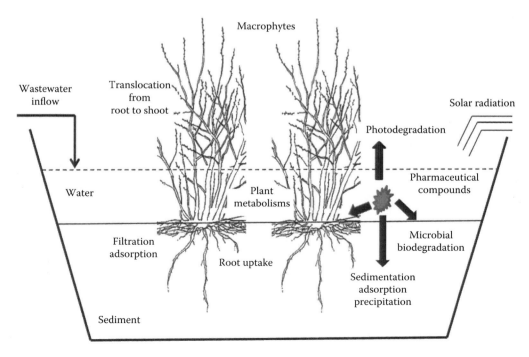

FIGURE 55.5 Pharmaceutical removal mechanisms in constructed wetlands.

TABLE 55.3   Overview of the Physico-Chemical Properties for the Frequently Investigated PPCPs in CWs

Compounds	Use	Chemical Structure	MW^a	S_w^b	pK_a^c	$Log\ k_{ow}^d$	$\log D_{ow}^e$	$\log K_{oc}^f$	H^g
						Physico-Chemical Properties			
Carbamazepine	Antiepileptic		236.28	17.7	13.9	2.45	2.77	3.588	1.08×10^{-10}
Naproxen	Analgesic		230.27	15.9	4.15	3.18	0.18	2.543	3.39×10^{-10}
Ibuprofen	Analgesic		206.29	21.0	4.91	3.97	1.68	2.596	1.50×10^{-7}
Diclofenac	Analgesic		296.16	2.37	4.15	4.57	0.96	2.921	4.73×10^{-12}
Caffeine	Stimulant		194.19	2.16×10^4	10.4	−0.07	−0.55	1	3.58×10^{-11}

(Continued)

TABLE 55.3 (*Continued*) Overview of the Physico-Chemical Properties for the Frequently Investigated PPCPs in CWs

Compounds	Use	Chemical Structure	MW^a	S_w^b	pK_a^c	$Log\,k_{ow}^d$	$\log D_{ow}^e$	$\log K_{oc}^f$	H^g
Salicylic acid	Analgesic		138.12	2240	2.97	2.26	−1.47	1.379	7.34×10^{-9}
Ketoprofen	Analgesic		254.29	51	4.45	3.12	0.54	2.459	2.12×10^{-11}
Clofibric acid	Lipid regulators		214.65	583	3	2.57	−0.42	N.A.	2.19×10^{-8}
Triclosan	Antimicrobial agent		289.55	10	N.A.	4.76	N.A.	N.A.	4.99×10^{-9}
Atenolol	β-Blockers		266.34	1.33×10^4	9.6	0.16	N.A.	N.A.	1.37×10^{-18}
Tonalide	Musk		258.41	1.25	N.A.	5.7	N.A.	3.933	4.22×10^{-5}

(*Continued*)

TABLE 55.3 (*Continued*)   Overview of the Physico-Chemical Properties for the Frequently Investigated PPCPs in CWs

Compounds	Use	Chemical Structure	MW^a	S_w^b	pK_a^c	$\text{Log } k_{ow}^d$	$\log D_{ow}^e$	$\log K_{oc}^f$	H^g
Oxybenzone	Sunscreens		228.25	68.6	N.A.	3.79	N.A.	N.A.	1.5×10^{-8}
Bisphenol A	Phenolic estrogenic compound		228.29	120	N.A.	3.32	N.A.	N.A.	1.0×10^{-11}
Galaxolide	Fragrance		258.4	1.75	N.A.	5.9	N.A.	4.016	1.32×10^{-4}
Methyl-dihydrojasmonate	Fragrance		226.31	280	N.A.	2.98	N.A.	2.153	5.02×10^{-7}

N.A. not available.

a Molecular weight (g mol^{-1}).

b Water solubility (25°C) (mg L^{-1}).

c Ionization constant (pK_a).

d Octanol–water partition coefficient (log k_{ow}).

e Octanol–water distribution coefficient (log D_{ow}).

f Organic carbon partition coefficient (log K_{oc}).

g Henry's law constant (25°C) (Pa m^3 mol^{-1}).

55.2.5.1 Photodegradation

Photodegradation is recognized to be the most vital and prime removal pathway for certain pharmaceutical compounds [2]. Based on the study of pharmaceutical elimination through photodegradation in surface water (e.g., lake, river, open sea, etc.), there have been few studies on the pharmaceutical photodegradation in CWs. Llorens et al. [39] and Matamoros et al. [40,41] investigated that the high removal efficiency of ketoprofen was most probably linked to its fast photodegradation kinetics. In more direct experimental studies of the role of photodegradation, Matamoros et al. [42] evaluated aquatic plants for removing polar microcontaminants in a mesocosm experiment and observed that the reactors planted with *Lemna minor* had a higher mean concentration of diclofenac and triclosan than unplanted control reactors, presumably due to the high plant coverage which blocked the light radiation and consequently reduced the photodegradation of these compounds.

55.2.5.2 Adsorption

Many organic chemicals already have shown their capacity to absorb highly hydrophobic and persistent organic pollutants (POP) in the environment and thus, they accumulate in sediments of CWs [11]. It is notified that the adsorption onto the organic matter present in the granular medium can make hydrophobic compounds more recalcitrant to biodegradation resulting in accumulation in wetland sediments, while hydrophilic contaminants are more easily removed by different processes. In another study, Hijosa-Valsero et al. [28] reported that the pharmaceuticals that are moderately or sometimes highly hydrophilic (log D_{ow} ranging from −2.3 to 3) bind significantly to organic matter.

55.2.5.3 Plant Uptake and Accumulation

The physicochemical characteristic of the organic compound is a typical method for estimating the potential for a plant to assimilate an organic compound [10]. Stottmeister et al. [53] reported that in plant cell membranes, there are no specific transporters for organic compounds. Thus, the uptake of micropollutants is largely dependent on the compound's chemical properties, especially their hydrophobicity. Burken and Schnoor [10] studied that compounds which are highly polar in nature (log K_{ow} < 1) will not be extensively taken up by the plant. However, Matamoros and Salvadó [42] showed that caffeine (log K_{ow} = −0.07) was removed efficiently (83%–99%) by planted CWs (with *L. minor, Salviniamolesta,* and *Elodea canadensis*), while unplanted beds have shown poor removal (0%–30%). Jones and de Voogt [30] studied that phytoremediation process is particularly important for those organic compounds that are recalcitrant or persistent in soils. In this way, in treatment wetlands, the capacities of the higher aquatic plant for phytoremediation could be appreciated.

55.2.5.4 Phytodegradation

Higher aquatic plants can possess metabolic capability through phytodegradation, which uses plants to achieve complete mineralization of the compound to nontoxic end products by plant enzymes or enzyme cofactors. In the processes of metabolic degradation or transformation of organic compounds, plant enzymes act on organic pollutants and mineralize them either completely into inorganic compounds, such as CO_2, H_2O, and Cl_2, or partially into stable intermediates that are stored in the plants [43]. In phytodegradation, plant-associated microbes are employed and rhizodeposition products may stimulate cometabolic degradation of xenobiotics, and the breakdown of organic pollutants occurs either inside the plant tissues or by plant enzymes secreted into the soil [44]. In addition to the important role played by plant uptake and metabolism in the removal of pharmaceutical compounds, the final location of pharmaceuticals and the final form (after metabolic transformation) of the compound could be improved when considering the need for harvesting plants to avoid the reentry of contaminants into the environment after plant senescence and death.

55.2.5.5 Microbial Degradation

In constructed wetlands, physicochemical properties of the toxicants are the main things for nature and the extent of microbial degradation of organic chemicals. The biological degradability of organic compounds is well explained by their chemical structure based on the presence of secondary, tertiary, or quaternary carbon atoms as well as functional groups. Matamoros et al. [41] did a study on the removal efficiencies of 11 PPCPs in two HSSF constructed wetlands and found that while the removal of diclofenac was less than 45%, whereas ibuprofen (80%) and ketoprofen (69%) both have shown significant removal efficiencies. These compounds unexpectedly exhibit enormous variation in rates of biodegradation, though, in the therapeutic class of non-steroidal anti-inflammatory (NSAID), they differ in the aryl functional group. A recent study is suggested to recognize the intermediate transformation products of biodegradation in wastewater. This study indicated that the metabolites of numerous pharmaceuticals are not very different from their parent compounds [33]. Kim et al. [33] also concluded that pharmaceuticals present as microcontaminants in wastewater can be effectively removed by biodegradation alone. Table 55.4 shows the evaluation of the roles of photodegradation, plant uptake, and biodegradation in the removal of selected pharmaceutical compounds in hydroponic microcosms.

TABLE 55.4 Evaluation of the Roles of Photodegradation, Biodegradation, and Plant Uptake (%)

	Photodegradation	Biodegradation	Plant Uptake
Carbamazepine	2	3	22
Naproxen	42	50	5
Diclofenac	77	3	3
Clofibric acid	4	3	53
Caffeine	6	29	65
Ibuprofen	15	34	29
Triclosan	80	19	1
MCPA	5	0	12

55.3 Factors Affecting Contaminants Removal in Constructed Wetlands

Pollutant removal efficiency in constructed treatment wetlands depends on a number of variables including wastewater application rate, organic loading rate, vegetation type, temperature, hydrologic regime, hydraulic detention time (HRT), operational mode, and vegetation type, all which may be highly variable among various systems [4,56]. Most commonly, pollutant removal is often accomplished by manipulating the system's hydraulic conditions and by selecting the appropriate type of dominant vegetation [62]. Hydraulic and hydrologic conditions may strongly influence the biotic community composition, biogeochemical processes, and the fate of pollutants in these treatment wetlands. Therefore, pollutant removal is often accomplished by manipulating the system's hydraulic and hydrologic conditions and by selecting the type of dominant vegetation accordingly [62]. Nevertheless, adverse impact on treatment performance can be expected from low ambient temperatures (especially inhibition of N-removal), peak flows (washout of solids), and clogging of SSF systems.

55.3.1 The Presence of Vegetation

The planted wetlands show better treatment performance as compared to unplanted controls. In plant roots, the plant rhizosphere stimulates microbial densities and activities by giving oxygen through diffusion from roots, which provide a source of carbon compounds for microbes. As per the study done by Gersberg et al. [25], it is observed that the higher aquatic plant within a treatment wetland system shows a higher efficiency of removal for a variety of contaminants as compared to an unplanted system. The removal efficiency vegetation can be explained by: (1) the plant, which can actively translocate oxygen from the shoots to the rhizosphere; (2) the root system, where there exists higher densities of bacteria than away from the plant roots, such that each plant root system is regarded as a mini aerobic/anoxic biological treatment system; and (3) uptake by the plants themselves. In tropical climates, it is found that the plants grow faster and throughout the year. Thus, in this way, the uptake of nutrients can contribute to significantly high removal of nutrients [35].

Kyambadde et al. [35] compared the wastewater treatment efficiencies of CWs in Kampala with respect to an investigation on the effect of plants on contaminant removal. The authors studied that the removal efficiencies in CWs planted with *Cyperus papyrus* of 75.43% NH_4-N, 72.47% TN, and 83.23% TP, and values for CWs planted with *Miscanthidium violaceum* of 64.59% NH_4-N, 69.40% TN, and 48.39% TP. In this, the authors found that both vegetated wetlands gave significant removal efficiencies than those in the unplanted control bed (28%, 25.6%, and 8% for NH_4-N, TN, and total reactive P, respectively). Similarly, Yang et al. [67] conducted a study of the efficiency of contaminant removal by five emergent plant species in Guangzhou, China. Here, authors observed that there was a significant difference in the removal rate of TN and TP, with the average removal efficiencies for TN and NH_4-N in the vegetated wetlands of 75% and 72%, respectively, while those of the unplanted wetlands were markedly lower.

55.3.2 Plant Species

The role of macrophytes on contaminant removal in constructed wetlands is well documented. However, the type of species and how it affects pollutant removal is less known. Different vegetation species may possess different capacities for pollutant uptake and accumulation, as well as more complex effect on the function and structure of bacterial communities especially in the plant rhizosphere.

It is recognized that the type of macrophyte species can influence the treatment efficiency in a treatment wetland [35,67]. Kyambadde et al. [35] did a study on the CWs and found that the plant uptake and storage are the major factors that could be responsible for N and P removal from the medium. He studied that the constructed wetlands planted with *Cyperus papyrus* contributed to 69.5% for N and 88.8% for P of the total N and P removal. However, for a CW planted with *Microbotryum violaceum*, the plants only accounted for 15.8% N and 30.7% P of the total N and P removal.

55.3.3 Hydraulic Retention Time

Hydraulic retention time (HRT), the length of time during which the pollutants are in contact with the substrate and the plants, is well known to be an important factor in determining contaminant removal efficiency [53]. At a low HLR (hydraulic loading rate) (or high HRT), wastewater moves slowly through the system so that the contact time with the substrate and rhizosphere microorganisms is high. However, longer HRTs typically require greater land area and higher capital cost. Konnerup et al. [34] assessed gravel-based HSSF CWs planted with *Canna xgeneralis L. Bailey* in Thailand at four HLRs (55, 110, 220, and 440 mm d[-1]) and found that effluent levels of pollutants were significantly affected by HRT. Bojcevska and Tonderski [4] investigated treatment of sugar factory effluent in a pilot-scale FWS CW system in Kenyaat at HLRs of 75 and 225 mm d[-1] and found that mass removal rates of the tested parameters, that is, TP, NH_4-N, and TSS, were significant affected by mass loading rate ($p < 0.001$), and that there was a linear relationship between the mass removal and mass loading rate.

55.3.4 Operational Mode

A study conducted by Stein et al. [52] suggested that the operation of SSF constructed wetland in batch mode (alternating drain and fill cycles) could improve nitrogen and phosphorus removal efficiency in wastewater wetlands. Wijler and Delwiche [66] first revealed this idea that the alternating periods of submergence and drying of soils might enhance nitrogen (N) loss compared to a continuously flooded condition. The reason behind this is that alternating periods of aerobic and anaerobic soil conditions could facilitate the sequential coupling of nitrification and denitrification. The long-term mechanism of phosphorus (P) retention in wetlands is the adsorption of orthophosphate onto the surfaces of soil minerals, particularly hydrous oxides of iron and aluminum. The sequential nitrification-denitrification and Fe oxyhydroxide formation operations can promote additional oxidized conditions by mass flow of air into pore spaces and it should display improved performance. However, the rates of both in constructed wetlands are affected by oxygen supply.

Recently, Zhang et al. [70] studied the influence of batch versus continuous mode on the removal efficiencies of COD, N, and P in tropical SSF CWs. It this study, it is shown that ammonia removal efficiencies were significantly ($p < 0.05$) enhanced in the planted beds operated in batch mode versus continuous flow (95.2% versus 80.4%, respectively, at the 4-day HRT), and that the presence of plants enhanced this removal significantly ($p < 0.05$) in both types of operational modes. With respect to TP removal, Zhang et al. [70] studied that there is a significant enhancement ($p < 0.05$) of TP removal in batch flow operation (69.6% for planted beds; 39.1% for unplanted beds) as compared to continuous flow operation (46.8% for planted beds; 25.5% for unplanted beds). Any flux of dissolved oxygen produced either by radial oxygen flow away from the plant roots or by drain and fill reaeration may also react with Fe at the metal's surface to convert it to hydrous ferric oxide.

55.3.5 Seasonal Variation

The potential for application of treatment wetland technology in the tropical regions is enormous because most of the developing countries have warm tropical and subtropical climates [20]. It is generally accepted that CWs are more suitable for wastewater treatment in tropical regions than intemperate regions [19]. Tropical regions are characterized by a relatively steady solar energy flux, as well as high humidity and warm temperatures throughout the whole year. Differences inherent to tropical as opposed to temperate environments can have important effects on wetland functions and this in turn will have impact on the use of wetlands for wastewater treatment. As the rates of almost all biological processes are temperature-dependent and increase with increasing temperature, a warm climate is conducive to year-round plant growth and heightened microbiological activity, which in general have a positive effect on treatment efficiency [32,70]. Therefore, the tropical environment should favor the biodegradation of organic matter and nitrification/denitrification, etc. Truu et al. [57] indicated that tropical conditions could enhance the removal of contaminants, as microorganisms living in the CWs usually reach

their optimal activity at warm temperatures (15–25°C). Vymazal [63] reported that the optimum temperature for nitrification in pure cultures ranges from 25°C to 35°C and from 30°C to 40°C in soils. Temperature also has a strong effect on the removal efficiencies of total Kjeldahl nitrogen (TKN) and ammonium. Nitrogen removal rates at a water temperature greater than 15°C are significantly higher than are those observed at lower temperatures [24]. Ammonia volatilization increases 1.3–3.5 times with each 10°C rise in temperature from 0°C to 30°C, and denitrification rates almost double (1.5–2.0) with each 10°C increment [45].

Nutrient removal that can be associated in a wetland is mainly due to biotic, temperature-dependent activity. Consequently, temperature can be considered an important parameter where the pollutant treatment effectiveness of a CW is to be evaluated. In general, because of reduced biotic activity, the effectiveness of treatment in a CW decreases at low temperature. Song et al. [51] found that COD, NH_4-N, and TP removal efficiencies displayed seasonal variations in a full-scale CW system in Shanghai, China. The COD removal was found higher in the spring (65.4%) and summer (66.3%) than in autumn (61.1%) and winter (59.4%), and NH_4-N and TP removal was higher in summer (54.5% for NH_4-N and 35.0% for TP) and autumn (43.3% for NH_4-N and 34.0%) than in the spring (33.7% for NH_4-N and 28.2% for TP) and the winter (32.4% for NH_4-N and 28.9% for TP). Li et al. [36] investigated a CW for treating the eutrophic lake water of Taihu, China, and reported no statistically significant seasonal differences for COD, but did find higher NH_4-N and TN removal effects in autumn and summer as compared to winter (December–February). Here, no statistically significant seasonal variation was detected for NO_3-N ($p > 0.05$).

55.4 Cost, Energy, and Land Requirement

55.4.1 Cost

In Table 55.4, the capital and operational costs for a traditional wastewater treatment plant (WWTP) is compared to CWs by taking the example of developing countries. Here, it is found that the CWs are more affordable as wastewater treatment approaches for small communities, while conventional WWTP and activated sludge processes are found cost-effective in densely populated urban areas. In Table 55.5, one can clearly

TABLE 55.5 Comparison of Cost Requirements between CWs and WWTPs

	Design Capacity	Total Capital Cost	Unit Capital Cost	Treatment Cost	O/M Cost	Energy Cost	Reference
	$m^3 d^{-1}$	US$	US$ m^{-3}	US$ m^{-3}	US$ m^{-3}	US$ m^{-3}	
Conventional WWTP	–	–	246–657	–	0.1151–0.2465	–	Liu et al. [38]
				0.7717	0.6362	0.1036	
CWs	–	–	164–460	–	0.0082–0.039	–	Liu et al. [38]
Dongying, Shandon, China	100,000	8.2 million	82	–	0.012	–	Wang et al. [64]
Bogota Savannah, Colombia	65	14,672	225.72	–	0.0134	–	Arias et al. [3]
Longdao River, Beijing, China	200	32,616	163.08	0.0223	0.014		Chen et al. [13]

observe that the operation and maintenance (O/M) costs for CW systems are much cheaper than those for conventional WWTP and activated sludge processes. Furthermore, CW systems have found tremendously low for O/M cost (US$ 0.0082–0.039 m⁻³ of wastewater), which includes power for pumps, harvesting of vegetation, and insect/pest control when compared to WWTP (US$ 0.1151–0.2465 m⁻³ wastewater). Wang et al. [64] studied the total capital cost of a CW system (Shandong, China), which was US$ 82 m⁻³ d⁻¹, about half of the cost of a conventional activated sludge system. In another study, Chen et al. [13] also found that at a treatment capacity of 200 m⁻³ d⁻¹, the construction cost of the Longdao River CW, which is situated in Beijing, China, was estimated only one-fifth of that for a conventional WWTP.

55.4.2 Land Requirement

CW systems for wastewater treatment generally do need more liberty than that for conventional wastewater treatment systems [6]. It is found that the main barrier for CWs is a high land requirement for expanding the application of CWs, especially in densely populated areas. However, it is also noticed that the CW treatment can be an inexpensive option compared to other options where land is available and/or affordable. According to Vymazal [63], the area required for domestic sewage treatment by HSSF CW is usually about 5 m² PE⁻¹ (PE = population equivalent = 60 BOD₅ d⁻¹), while a VSSF CW requires less land, usually 1–3 m² PE⁻¹. However, in the Asian Cities Program (WAC), which was aimed at developing practical community-based approaches in small and medium sized towns, UN-HABITAT [59] indicated that an area of only1–2 m² PE⁻¹ would be required for a HSSF CW and 0.8–1.5 m² PE⁻¹ would be needed for a VSSF CW.

Li et al. [36], for CWs constructed in Shenzhen, Guangdong Province, China, reported the land requirement was approximately 1.88 m²/m³ and 1.2 m²/PE at 300 L/PE · day. However, with the additional consideration of preliminary treatment, the total land requirement came to 4 m²/m³. There are reports in the literature of treatment wetlands with even lower land requirements. For example, Zhai et al. [69] studied a new type of hybrid CW system consisting of a vertical-baffled flow CW and an HSSF CW to treat municipal wastewater in Southern China, and found the land required for municipal sewage treatment was only 0.70–0.93 m² PE⁻¹ for vegetated wetland beds, which is much smaller than that cited by Vymazal [63] above. Apparently, his hybrid CW system can treat municipal wastewater with similar or even higher removal efficiencies than the conventional CW system but with less land required.

55.4.3 Sustainability

In 1987, the concept of sustainable development was defined at Brundt Land Commission as: "Development that meets the needs of the present without compromising the ability of future generations to meet their own needs" [9]. Energy analysis is an ecological approach, which occurred out of creative combination of thermodynamics and systems ecology and can represent both the environmental and economic values of a given system [8,47]. Therefore, this approach displays a good determination for the comparison of environmental goods, energy quality, and economic valuation of alternative treatment systems. Zhou et al. [71] studied the comparison of the energy and resource consumption between a constructed wetland over Longdao River, Beijing, China in a conventional wastewater treatment system and observed that the ratio of purchased inputs to free inputs for CWs were 3.4, compared with the ratio of 1450 for the conventional treatment system. Similarly, a comparative study performed on the sustainability of original and constructed wetlands in Jiangsu Province, China, found that the energy yield ratio (Yr) of the original wetlands was the highest.

55.5 Applications

55.5.1 Case Study 1: Carolina Bays: A Natural Wastewater Treatment Program

55.5.1.1 Brief about the Study

Carolina bays are strange land features filled with bay trees and other wetland vegetation. The natural forces of wind and artesian water flow caused the formation of lakes, which later filled with vegetation. As per the USEPA [60], over 500,000 of these shallow basins dot the coastal plain from Georgia to Delaware. Many of them occur in the Carolinas, which accounts for their name. Most Carolina bays are swampy or wet areas, and most of the hundreds present in coastal Horry County, South Carolina, are nearly impenetrable jungles of vines and shrubs. The regional water utility, the Grand Strand Water & Sewer Authority (GSWSA), retained CH2M HILL in the late 1970s to evaluate wastewater treatment and disposal options.

To dispose of additional effluent, locations were limited because of serious environmental and recreational concerns. The Waccamaw River flow is recorded critically low and Intracoastal Waterway as well, into which existing facilities discharged, could not incorporate extra loading without unpleasant effects on water quality and consequential impacts on tourism and recreational activities. The USEPA has considered the use of wetlands to be an emerging alternative to conventional treatment processes. As a result, EPA Region IV and the South Carolina Department of Health and Environmental Control awarded an innovative/alternative technologies funding grant for the Carolina bays treatment project, enabling GSWSA to provide expanded collection, treatment, and disposal services at affordable costs.

55.5.1.2 Site Description

To evaluate viable treatment and disposal alternatives, four Carolina bays were selected as treatment sites in the study. A site selection criterion was based on three primary factors: (1) distance from the wastewater source, (2) available treatment area, and (3) environmental sensitivity. The bays chosen for the GSWSA treatment complex are already affected by humans and were the least environmentally sensitive of the bays measured. Carolina Bays 4-A and 4-B are joined along a portion of their margins and encompass about 390 acres of dense, shrubby plant communities with scattered pine trees. This plant association is called "pocosin" after an Indian word describing a bog on a hill. A powerline right-of-way bisects Bay 4-A and also cuts through the southern end of Bay 4-B. The 240-acre Pocosin Bay (Bay 4-C) is also dominated by pocosin vegetation and is filled with up to 15 ft of highly organic peat soils. This bay is being used only as a contingency discharge area and had received the least amount of prior disturbance. Bear Bay (Bay 4-D) covers 170 acres and is dissimilar from the other bays because it is densely forested by pine and hardwood tree species. A large portion of this Carolina bay was cleared for forestry purposes in the mid-1970s but has since been revegetated with a mixture of upland and wetland plant species.

55.5.1.3 Operation and Management

After undergoing conventional primary and secondary treatment processes at the George R. Vereen Wastewater Treatment Plant, the wastewater is slowly released into a Carolina bay for tertiary treatment, rather than directly to recreational surface waters of the area. The plants found in the Carolina bays are naturally adapted to wet conditions, so the addition of a small amount of treated water increases their productivity and, in the process, provides final purification of the wastewater. The treated effluent can be distributed to 700 acres within the four selected Carolina bays through a series of gated aluminum pipes supported on wooden boardwalks. Depending on effluent flow rate and biological conditions in the bays, wastewater flow is alternated among the bays. Here it is observed that the water levels and outflow rates can be partially controlled in Bear Bay. Natural surface outlets in the other three bays were not altered by construction of the project. Ongoing monitoring indicates that significant assimilation is occurring in Bear Bay before the fully treated effluent recharges local groundwater or flows into downstream surface waters.

55.5.1.4 Benefits

The Carolina Bay Natural Land Treatment Program plays an important role in protecting the environment and serves wastewater management needs. The Carolina bays are unique; 98% of the bays in South Carolina have been disturbed by agricultural activities and ditching. The four bays in the treatment program will be maintained in a natural ecological condition. The use of wetlands for treatment can significantly lower the cost of wastewater treatment because the systems rely on plant and animal growth instead of the addition of power or chemicals. The plant communities present in the wetlands naturally adjust to changing water levels and water quality conditions by shifting dominance to those species best adapted to growing under the new conditions. Carolina bays provide a serious refuge for rare plants and animals. Interestingly, black bears still roam the bays' shrub thickets and forested bottom lands just a few miles from the thousands of tourists on South Carolina's beaches. The Carolina Bay Nature Park, to be managed by GSWSA, is currently being planned. The focal point of the park will be an interpretive

visitor center open to the public. This simple structure will be designed and built in harmony with its surroundings on a sand ridge overlooking two Carolina bays.

55.5.2 Case Study 2: Feasibility Studies for Reuse of Constructed Wetlands Treating Simulated Nickel-Containing Groundwater

55.5.2.1 Brief about the Study

Constructed wetlands are considered a low cost treatment option for domestic and industrial wastewater in recent decades. The availability of various toxic heavy metals in wastewater is a serious concern because these heavy metals have potential to accumulate in the treatment systems. Therefore, these heavy metals greatly influence the efficiency of constructed wetlands. Therefore, a feasibility study was proposed for long-term usage of constructed wetlands as treatment systems [50]. Initially, sediment in a constructed wetland was contaminated with simulated nickel-containing groundwater followed by using suitable leaching solution to rejuvenate the heavy metal contaminated sediment. In this, a batch study was performed to recognize the optimum pH for nickel adsorption on sand. The efficiency of different leaching solutions to remove the adsorbed nickel from sand was studied. Following this, a pilot scale study was carried out in a constructed wetland treatment plant (vertical flow) at Anna University, Chennai, India. The amount of nickel solution charged into the control cell (sand) and test cell (sand planted with *Arundodonax*) was 29,000 mg/cell. Here it is seen that the concentration of nickel adsorbed to sand increased from 0.2 mg/kg to 4.34 mg/kg in the control cell. Leaching the wetland with EDTA solution resulted in removal of nickel up to its background concentrations on the sand. Based on the above considerations, it is suggested that the proposed feasibility study can be used to revive the sediment in a constructed wetland for its long-term usage as treatment systems.

55.5.2.2 Site Description

The use of constructed wetlands as a treatment system for domestic wastewater has gained wide acceptance. The metal removal processes in constructed wetlands is very tedious and these processes include a combination of biotic and abiotic reactions such as sedimentation, flocculation, adsorption, precipitation, coprecipitation, cation and anion exchange, complexation, oxidation and reduction, microbial activity, and plant up-take. The pilot scale study was carried out in the constructed wetland treatment plant (vertical flow type) at Anna University, Chennai, India. Sand used in the study was collected from the Chengalpat River, Chennai, India. The wetland plant used in the study was *Arundodonax*, which was collected from the Chengalpat River Basin, Chennai, India. *Arundodonax* is a rhizomatous perennial grass species belonging to the Poaceae family. It reproduces by rhizomes and stem, and also can essentially remain alive throughout the year. It grows in a number of freshwater riparian habitats such as irrigation ditches, streams, lakes, and wetlands. *Arundo* has the ability to survive in a number of different types of soils, ranging from heavy clays to lose sands and gravelly soils. Sandy soil is the most common type of soil in which it is found. The constructed wetland plant has been defined by a control cell and a test cell. The control cell was filled with gravel and sand, whereas the test cell was filled with gravel, sand, and planted with *Arundodonax*. Each cell was fitted with a separate drainage pipe that was directed toward a trench to collect the effluent.

55.5.2.3 Operation and Management

EDTA and hydrochloric acid are purchased in bulk packs from Merck India. The other chemicals that are used in the study were purchased from Merck India of highest purity available. A batch study was performed to identify the optimum pH for nickel adsorption through the sand. The adsorbed metal from the sand was leached using hydrochloric acid (0.1 N) and ethylene diamine tetra acetic acid (EDTA) (0.01 M). Pilot scale study was carried out after the plants were grown for a consistent period. Simulated nickel containing groundwater (hereafter described as nickel solution) was prepared by dissolving hydrated nickel nitrate in groundwater filled in a plastic container of 100 L capacity. The pH of the nickel solution was adjusted to 6 using concentrated hydrochloric acid. EDTA solution was prepared by dissolving it in the groundwater to a concentration of 0.01 M. Here, sand corresponding to each cell was sampled out using a hand steel auger and kept in air-tight polyethylene bags in order to report the nickel adsorption. The auger was washed with water during every single sampling. In each location, sand was sampled out at surface, 30, and 60 cm in depth in order to know about the nickel adsorption at different depths. Root and shoot samples were collected from control cell at the end of the nickel charging phase. The pH of the sand was found to be 7.6 at 1:10 suspensions. The background concentration of nickel in sand and groundwater were found to be 0.2 mg/kg and below detection limits, respectively.

In wetlands, the mobility of trace elements is highly influenced by redox potential and pH of the water sediment system. The leaching efficiencies of hydrochloric acid (0.1 N) and ethylene diamine tetra acetic acid (EDTA) (0.01 M) are found to be 55% and 30.6%, respectively. Since acidic properties of hydrochloric acid affect the sand properties such as porosity and texture, EDTA was used to leach out the nickel from the sand because it is less toxic toward the wetland species and its biodegradability in the system. Analysis of nickel has been done four times during the course of the nickel charging phase. Due to a decline in the pH of the sand, more nickel was adsorbed by the sand. During the leaching phase at the end of every EDTA charging, sand sampling was done as a nickel charging phase to quantify the amount of nickel leached out from the sand.

55.5.2.4 Benefits

As per the given study, it is marked that constructed wetlands are fully able to remove toxic heavy metals and trace metals from the wastewater. It can also be concluded that the proposed feasibility study was efficient in rejuvenating the metal contaminated sediment in constructed wetlands during a short period. Further work was planned to find a breakthrough for nickel and its absorption by the wetland plant in the charging phase during the pilot scale study.

55.6 Summary and Conclusions

Wetland represents the medium where water is the prevailing factor that could be utilized for soil development. Wetlands may improve medium of plant and animal communities living in the soil and on its surface and the water table is usually at or near the surface or the land is covered by shallow water. Based on the previous discussion and findings, it is evaluated that these constructed wetlands can also be called an artificial wastewater treatment system that consists of shallow ponds, which generally depend on natural processes such as chemical, physical, biological, and microbial medium, and are significantly helpful in the treatment of wastewater and other organic chemicals from the water. CWTS is a manufactured wetland area, which is mostly formulated for water quality improvement through optimizing physical, chemical, and biological processes that could be applicable for the treatment of wastewater in a natural basis. As per the above findings, it is found that constructed wetlands are considered low cost alternatives of domestic and industrial wastewater treatments. The VFS- and HFS-based CWTS system can be utilized effectively for the treatment of water in a natural way. The VFS- and HFS-based CWTS have been adopted worldwide. Countries such as China, the United States, India, and other European countries are significantly using CWTS for wastewater treatment.

References

1. Allen, G.H. and Gearheart, R.A. 1988. *Proceedings of a Conference on Wetlands for Wastewater Treatment and Resource Enhancement.* Humbolt State University, Arcata, CA.
2. Andreozzi, R., Marotta, R., and Paxéus, N. 2003. Pharmaceutical in STP effluents and their solar photodegradation in aquatic environment. *Chemosphere*, 50, 1319–1330.
3. Arias, M.E. and Brown, M.T. 2009. Feasibility of using constructed wetlands for municipal wastewater treatment in the Bogotá Savannah, Colombia. *Ecological Engineering*, 35, 1070–1078.
4. Bojcevska, H. and Tonderski, K. 2007. Impact of loads, season, and plant species on the performance of a tropical constructed wetland polishing effluent from sugar factory stabilization ponds. *Ecological Engineering*, 29, 66–76.
5. Breton, R. and Boxall, A. 2003. Pharmaceuticals and personal care products in the environment: Regulatory drivers and research needs. *QSAR and Combinatorial Science*, 22, 399–409.
6. Brissaud, F. 2007. Low technology systems for wastewater perspectives. *Water Sci. Technol.*, 55(7), 1–9.

7. Brix, H. and Schierup, H.-H. 1990. Soil oxygenation in constructed reed beds: The role of macrophyte and soil-atmosphere interface oxygen transport. In: Cooper, P.F. and Findlater, B.C. (eds.), *Constructed Wetlands in Water Pollution Control.* Pergamon Press, Oxford, UK, pp. 53–66.

8. Brown, M.T. and Ugliati, S. 1999. Emergy evaluation of the biosphere and natural capital. *AMBIO*, 28, 486–493.

9. Brundtland Commission. 1987. *Our Common Future.* Oxford University Press, New York, 1987.

10. Burken, J.G. and Schnoor, J.L. 1998. Predictive relationships for uptake of organic contaminants by hybrid poplar trees. *Environmental Science and Technology*, 32, 3379–3385.

11. Campanella, B.F., Bock, C., and Schröder, P. 2002. Phytoremediation to increase the degradation of PCBs and PCDD/Fs Potential and limitations. *Environmental Science and Pollution Research*, 9, 73–85.

12. Carballa, M., Omil, F., Ternes, T., and Lema, J.M. 2007. Fate of pharmaceutical and personal care products (PPCPs) during anaerobic digestion of sewage sludge. *Water Research*, 41, 2139–2150.

13. Chen, Z.M., Chen, B., Zhou, J.B., Li, Z., and Zhou, Y. 2008. A vertical subsurface-flow constructed wetland in Beijing. *Communications in Nonlinear Science and Numerical Simulation*, 13, 1986–1997.

14. Chua Lloyd, H.C., Tan Stephen, B.K., Sim, C.H., and Goyal, M.K. 2012. Treatment of baseflow from an urban catchment by a floating wetland system. *Journal of Ecological Engineering*, 49, 170–180.

15. Cooper, P.F., Job, G.D., Greenand, M.B., and Shutes, R.B.E. 1996. *Reed Beds and Constructed Wetlands for Wastewater Treatment.* WRc, Swindon.

16. Cowardin, L., Carter, V., Golet, F., and LaRoe, E. 1979. *Classification of Wetlands and Deepwater Habitats of the United States.* Office of Biological Services, US Fish and Wildlife Service. Federal Geographic Data Committee, U.S., www.epa.gov/owow/wetlands

17. Crane, M., Watts, C., and Boucard, T. 2006. Chronic aquatic environmental risks from exposure to human pharmaceuticals. *Science of the Total Environment*, 367, 23–41.

18. Daughton, C.G. and Ternes, T.A. 1999. Pharmaceuticals and personal care products in the environment: Agents of subtle change. *Environmental Health Perspectives*, 107, 907–938.

19. Denny, P. 1997. Implementation of constructed wetlands in developing countries. *Water Science and Technology*, 35(5), 27–34.

20. Diemont, S.A.W. 2006. Mosquito larvae density and pollutant removal in tropical wetland treatment systems in Honduras. *Environment International*, 32, 332–341.

21. Ellin, S. 1993. *Constructed Wetlands for Wastewater Treatment and Wildlife Habitat.* United States Environmental Protection Agency. EPA832-R-93-005.

22. Ellis, J.B. 2006. Pharmaceutical and personal care products (PPCPs) in urban receiving waters. *Environmental Pollution*, 144(1), 184–189.

23. Fent, K., Weston, A.A., and Caminada, D. 2006. Ecotoxicology of human pharmaceuticals. *Aquatic Toxicology*, 76, 122–159.

24. García, J., Rousseau, D.P.L., Morató, J., Lesage, E., Matamoros, V., and Bayona, J.M. 2010. Contaminant removal processes in subsurface-flow constructed wetlands: A review. *Critical Reviews in Environmental Science and Technology*, 40, 561–661.

25. Gersberg, R.M., Elkins, B.V., Lyon, S.R., and Goldman, C.R. 1986. Roles of aquatic plants in wastewater treatment by artificial wetland. *Water Research*, 20(3), 363–368.

26. Headley, T.R. and Tanner, C.C. 2011. Constructed wetlands with floating emergent macrophytes: An innovative stormwater treatment technology. *Critical Reviews in Environmental Science and Technology*, 42, 2261–2310.

27. Hijosa-Valsero, M., Matamoros, V., Martín-Villacorta, J., Bécares, E., and Bayona, J.M. 2010. Assessment of full-scale natural systems for the removal of PPCPs from wastewater in small communities. *Water Research*, 44, 1429–1439.

28. Hijosa-Valsero, M., Matamoros, V., Sidrach-Cardona, R., Martín-Villacorta, J., Bécares, E., and Bayona, J.M. 2010. Comprehensive assessment of the design configuration of constructed wetlands for the removal of pharmaceuticals and personal care products from urban wastewaters. *Water Research*, 44, 3669–3678.

29. Jetten, M.S.M., Logemann, S., Muyzer, G.M., Robertson, L.A., DeVries, S., Van Loosdrecht, M.C.M. et al. 1997. Novel principles in the microbialconversion of nitrogen compounds. *Antonie van Leeuwenhoek*, 71, 75–93.

30. Jones, K.C. and de Voogt, P. 1999. Persistent organic pollutants (POPs): State of the science. *Environmental Pollution*, 100, 209–221.

31. Joss, A., Zabczynski, S., Göbel, A., Hoffmann, B., Löffler, D., McArdell, C.S., Ternes, T.A., Thomsen, A., and Siegrist, H. 2006. Biological degradation of pharmaceuticals in municipal wastewater treatment: Proposing a classification scheme. *Water Research*, 40, 1686–1696.

32. Kaseva, M.E. 2004. Performance of a sub-surface flow constructed wetland in polishing pre-treated wastewater—A tropical case study. *Water Research*, 38, 681–687.

33. Kim, S., Weber, A.S., Batt, A., and Aga, D.S. 2008. Removal of pharmaceutical in biological wastewater treatment plants. In: K. Kümmerer (ed.), *Pharmaceuticals in the Environment—Sources, Fate, Effects and Risks.* Springer-Verlag, Berlin, Heidelberg, pp. 349–361.

34. Konnerup, D., Koottatep, T., and Brix, H. 2009. Treatment of domestic wastewater in tropical, subsurface flow constructed wetlands planted with *Canna* and *Heliconia*. *Ecological Engineering*, 35, 248–257.

35. Kyambadde, J., Kansiime, F., Gumaelius, L., and Dalhammar, G. 2004. A comparative study of *Cyperus papyrus* and *Miscanthidiumviolaceum*-based constructed wetlands for wastewater treatment in a tropical climate. *Water Research*, 38, 475–485.

36. Li, L., Li, Y., Biswas, D.K., Nian, Y., and Jiang, G. 2008. Potential of constructed wetlands in treating the eutrophic water: Evidence from Taihu Lake of China. *Bioresource Technology*, 99, 1656–1633.

37. Lim, P.E., Wong, T.F., and Lim, D.V. 2001. Oxygen demand, Nitrogen and Copper removal by free-water-surface and subsurface-flow constructed wetlands under tropical conditions. *Environment International*, 26, 425–431.

38. Liu, D., Ge, Y., Chang, J., Peng, C., Gu, B., Chan, G.Y.S., and Wu, X. 2008. Constructed wetlands in China: Recent developments and future challenges. *Frontiers in Ecology and the Environment*, 7, 261–268.

39. Llorens, E., Matamoros, V., Domingo, V., Bayona, J. M., and García, J. 2009. Water quality improvement in a full-scale tertiary constructed wetland: Effects on conventional and specific organic contaminants. *Science of the Total Environment*, 407, 2517–2524.

40. Matamoros, V., Caselles-Osorio, A., García, J., and Bayona, J.M. 2008a. Behaviour of pharmaceutical products and biodegradation intermediates in horizontal subsurface flow constructed wetland. A microcosm experiment. *Science of the Total Environment*, 394, 171–176.

41. Matamoros, V., García, J., and Bayona, J.M. 2008b. Organic micropollutant removal in a full-scale surface flow constructed wetland fed with secondary effluent. *Water Research*, 42, 653–660.

42. Matamoros, V. and Salvadó, V. 2012. Evaluation of the seasonal performance of a water reclamation pond-constructed wetland system for removing emerging contaminants. *Chemosphere*, 86, 111–117.

43. McCutcheon, S.C. and Schnoor, J.L. 2003. Overview of phytotransformation and control of wastes. In: McCutcheon, S.C. and Schnoor, J.L. (eds.), *Phytoremediation: Transformation and Control of Contaminants*. Wiley, New York, pp. 3–58.

44. Moormann, H., Kuschk, P., and Stottmeister, U. 2002. The effect of rhizodeposition from helophytes on bacterial degradation of phenolic compounds. *Acta Biotechnologica*, 22, 107–112.

45. Ng, W.J. and Gunaratne, G. 2011. Design of tropical constructed wetlands. In: Tanaka, N., Ng, W.J., and Jinadasa, K.B.S.N. (eds.), *Wetlands for Tropical Application*. Wastewater Treatment by Constructed Wetlands, Imperial College Press, London, pp. 69–94.

46. Odum, H.T. 1985. *Self-organization of Estuarine Ecosystems in Marine Ponds Receiving Treated Sewage*. A data report. University of North Carolina. Sea Grant Publications. UNC-SG-85-04.

47. Odum, H.T. 1996. *Environmental Accounting: Energy and Environmental Decision Making*. John Wiley & Sons, New York.

48. Reddy, K.R. and DeBusk, W.F. 1987. Nutrient storage capabilities of aquatic and wetland plants. In: Reddy, K.R. and Smith, W.H. (eds.), *Aquatic Plants for Water Treatment and Resource Recovery*. Magnolia Publishing, Orlando, FL, pp. 337–357.

49. Reinhold, D., Vishwanathan, S., Park, J.J., Oh, D., and Saunders, F.M. 2010. Assessment of plant-driven removal of emerging organic pollutants by duckweed. *Chemosphere*, 80, 687–692.

50. Sivaraman, C., Arulazhagan, P., Walther, D., and Vasudevan, N. 2014. Feasibility studies for reuse of constructed wetlands treating simulated nickel containing groundwater. *Universal Journal of Environmental Research and Technology*, 1(3), 293–300.

51. Song, H.L., Li, X.N., Lu, X.W. and Inamori, Y. 2009. Investigation of microcystin removal from eutrophic surface water by aquatic vegetable bed. *Ecological Engineering*, 35, 1589–1598.

52. Stein, O.R., Hook, P.B., Biederman, J.A., Allen, W.C. and Borden, D.J. 2003. Does batch operation enhance oxidation in subsurface constructed wetlands? *Water Science and Technology*, 48(5), 149–156.

53. Stottmeister, U., Wiebner, A., Kuschk, P., Kappelmeyer, U., Kästner, M., Bederski, O., Müller, R.A. and Moormann, H. 2003. Effects of plants and microorganisms in constructed wetlands for wastewater treatment. *Biotechnology Advances*, 22, 93–117.

54. Tanner, C.C., Kadlec, R.H., Gibbs, M.M., Sukias, J.P.S. and Nguyen, M.L. 2002. Nitrogen processing gradients in subsurface-flow treatment wetlands—Influence of wastewater characteristics. *Ecological Engineering*, 18, 499–520.

55. Ternes, T.A. 1998. Occurrence of drugs in German sewage treatment plants and rivers. *Water Research*, 32, 3245–3260.

56. Trang, N.T.D., Konnerup, D., Schierup, H., Chiem, N.H., Tuan, L.A. and Brix, H. 2010. Kinetics of pollutant removal from domestic wastewater in a tropical horizontal subsurface flow constructed wetland system: Effect of hydraulic loading rate. *Ecological Engineering*, 36, 527–535.

57. Truu, M., Juhanson, J. and Truu, J. 2009. Microbial biomass, activity and community composition in constructed wetlands. *Science of the Total Environment*, 407, 3958–3971.

58. Ujang, Z., Soedjono, E., Salim, M.R. and Shutes, R.B. 2005. Landfill leachate treatment by an experimental subsurface flow constructed wetland in tropical climate countries. *Water Science and Technology*, 52, 243–250.

59. UN-HABITAT. 2008. *Constructed Wetlands Manual*. UN-HABITAT Water for Asian Cities Programme Nepal, Kathmandu, Nepal.

60. USEPA (United States Environmental Protection Agency). 1993. *Carolina Bays: A Natural Wastewater Treatment Program*. EPA Number: 832R93005a.

61. Valiela, I. 1984. *Marine Ecological Processes*. Springer-Verlag, New York.

62. Vymazal, J. 2007. Removal of nutrients in various types of constructed wetlands. *Science of the Total Environment*, 380, 48–65.

63. Vymazal, J. 2011. Constructed wetlands for wastewater treatment: five decades of experience. *Environmental Science and Technology*, 45(1), 61–69.

64. Wang, L., Peng, J., Wang, B. and Cao, R. 2005. Performance of a combined eco-system of ponds and constructed wetlands for wastewater reclamation and reuse. *Water Science and Technology*, 51(12), 315–323

65. Wetland International. 2003. *The Use of Constructed Wetlands for Wastewater Treatment*. Wetlands International, Malaysia Office. www.wetlands.org.

66. Wijler, J. and Delwiche, C.C. 1954. Investigations on the denitrifying process in soil. *Plant Soil*, 5, 55–169.

67. Yang, Q., Chen, Z.H., Zhao, J.G. and Gu, B.H. 2007. Contaminant removal of domestic wastewater by constructed wetlands: Effects of plant species. *Journal of Integrative Plant Biology*, 49 (4), 437–446.

68. Yeh, T.Y., Chou, C.C. and Pan, C.T. 2009. Heavy metal removal within pilot-scale constructed wetlands receiving river water contaminated by confined swine operations. *Desalination*, 249, 368–373.

69. Zhai, J., Xiao, H.W., Kujawa-Roeleveld, K., He, Q. and Kerstens, S.M. 2011. Experimental study of a novel hybrid constructed wetland for water reuse and its application in Southern China. *Water Science and Technology*, 64(11), 2177–2184.

70. Zhang, D.Q., Tan, S.K., Gersberg, R.M., Zhu, J.F., Sadreddini, S. and Li, Y.F. 2012. Nutrient removal in tropical subsurface flow constructed wetlands under batch and continuous flow conditions. *Journal of Environmental Management*, 96, 1–6.

71. Zhou, J.B., Jiang, M.M., Chen, B. and Chen, G.Q. 2009. Energy evaluations for constructed wetland and conventional wastewater treatments. *Communications in Nonlinear Science and Numerical Simulation*, 14, 1781–1789.

Constructed Wetlands: Basic Design

Jatin Kumar Srivastava
*Environmental Research
and Analysis (ERA)*

Atul Bhargava
*Amity University
(Lucknow Campus)*

PREFACE

Wetlands have a natural tendency to remove impurities from water. Constructed wetlands (CWs) are an excellent example of human endeavor to mimic this act of nature capable of treating virtually all classes of wastewaters including municipal, sewage, industrial, urban runoff, agricultural, and waste from smaller communities. Proper design of a constructed wetland helps increase the efficiency of the secondary or sometimes tertiary treatment. The present work is an overview of the totality of the CWs in terms of designing, performance, and mechanisms.

56.1 Introduction

Wetlands are complex, natural, shallow, lentic water regimes dominated by hydrophytes (aquatic plants) providing enormous opportunity for environmental restoration especially the waters. Natural wetlands have been used to provide tertiary treatment to municipal wastewater as a cost-effective alternative to conventional methods [85]. Constructed wetlands (CW) or engineered wetlands are designed to utilize the benefits of natural wetlands and processes for various purposes *ex-situ*. CWs were initially developed 50 years ago in Europe and North America for the treatment of wastewater and have been developing fast over the last few decades and it represents a widely accepted water treatment alternative [62]. Constructed wetlands used for treatment include artificial marshes, ponds, and trenches [96]. Macrophytes (large aquatic plants) are assumed the main biological component of wetlands depending on their biological characteristics such as interspecific competition and ability to tolerate the changing characteristics of the residual water [30]. Aquatic macrophytes, especially the submerged roots, provide substratum to the bacterial community to grow on the submerged roots and stems. In fact, these bacterial communities are instrumental in the removal of colloidal biochemical oxygen demand (BOD) from wastewaters [96]. Constructed wetlands typically are of two types: (1) surface flow (free water surface) and (2) subsurface flow having emergent aquatic plants as the wetland species for the treatment of water. Subsurface flow is further divided into three groups, namely, (a) vertical flow, (b) horizontal flow, and (c) hybrid engineered wetlands [103]. Subsurface flow CWs have merits over surface flow as the latter one causes foul odor and infestation of insects [101]. Voluminous literature is available on the different usages of constructed wetlands explaining the removal of BOD, chemical oxygen demand (COD), suspended solids (SS) efficiency [77,78,91], removal of nutrients especially nitrogenous compounds [6,27,39], and many more. Studies on various combinations of water flow with the variety of effluents treated successfully through the engineered (constructed) wetlands indicate that many opportunities are available to treat wastewater of hazardous nature. The successful treatment of effluents with the maximum removal efficiency of a CW depends largely on the various environmental, hydrodynamic flow, physico-chemical, and biological factors.

56.1.1 Constructed Wetlands versus Conventional Treatment Methods

Engineered or CW systems have certain advantages over the conventional treatment systems because of their convenient establishment at the site of wastewater generation and relatively lower-energy requirements and low-cost systems, especially for final stage treatment of municipal and other urban wastewaters. Moreover, conventional methods of wastewater treatment need the application of mechanized tools to remove impurities and chemicals to bring desired alterations to the wastewater whereas CWs are simple to design and have no technical obligations to follow. Figures 56.1 and 56.2 are the simple diagrammatic representation of a natural treatment wetland and conventional treatment system, respectively. A disadvantage of the CW systems is their relatively slow rate of operation in comparison to conventional wastewater treatment technology [84] and therefore extremely polluted wastewaters (industrial effluents) are preferred to be treated conventionally.

56.1.2 Pollution Removal Efficiency of CWs

Generally, CWs are used as an advanced [96] secondary treatment alternative, which is environmentally sound and occupationally safe. Previous research experiences from all over the world demonstrate the robust nature of the engineered wetlands in the wastewater treatment of various characteristics

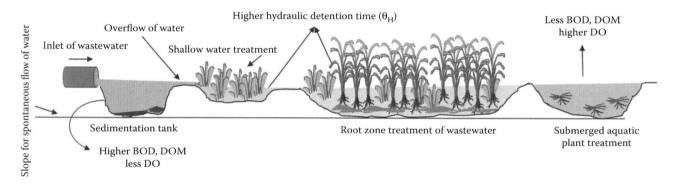

FIGURE 56.1 Simple schematic of natural treatment wetland (free flow).

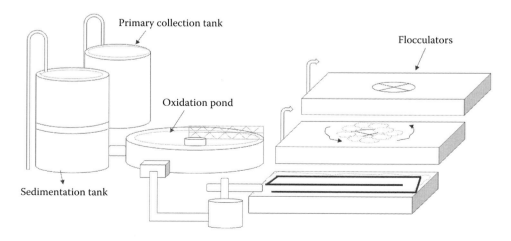

FIGURE 56.2 Simple schematic of conventional wastewater treatment plant.

particularly for municipal waste, sewage, and urban wastewaters [9,24,57,65,76]. Wastewater containing higher organic loading can be treated perfectly with the help of adequately designed wetlands to reduce the values of COD (<87%–89%) in multistage CW [26]; however, reduction up to <60% has been reported in subsurface flow CW by Kaseva [42] and García et al. [24]. There exists differences in COD removal percentage as experienced by researchers all over the world because of the choice of design of CW, environmental factors, performance evaluation strategies, and the species of plants. Reduction of BOD and SS from wastewaters in CW is reportedly very high (≥90%) [24,26,65,91]. CWs have varying efficiency for the removal of nutrient ions, particularly ammonia and total nitrogen. Nutrients are required by the wetland plants for their growth and reproduction. Aquatic plants assimilate nutrient ions in their tissues and provide favorable conditions for microbes to grow [101]; thus, this helps remove excess nutrients from the wastewater. However, reports also indicate the inefficiency of CWs in nutrient removal [9,42,102]. According to Vymazal [102] and other researchers as stated earlier, the removal of nutrients (N and P) is usually low in CWs and never exceeds 50% in cases of municipal wastewater. Furthermore, nutrient removal in CWs depends on the hydraulic loading rate (HLR), flow of water (horizontal or vertical flow) [12], environmental conditions, and availability of oxygen in the voids of media used [35]. Oxygen transfer rate (OTR) plays a crucial role in the activity of microorganisms in the root zone [92], performing various roles in pollution mitigation in a wetland. Comparative study of horizontal sub-surface (HSS) flow constructed wetlands and vertical flow (VF) CWs show higher OTR in VFCWs varying from 50 to 90 gm^{-2}d, whereas OTR is limited in HSSFCW raging between 1 and 8 gm^{-2}d [13]. Nitrification is a microbial process and needs oxygen; therefore, oxidized conditions are necessary in planted CWs [6]. In addition to this, reports of Dunne et al. [18] suggest higher rates (<80%) of phosphorus removal from dairy farmyard wastewater treated in an integrated CW especially in warm seasonal conditions.

56.2 Basic Designing Aspects of CWs

Designing CWs is more logical than a technical endeavor requiring basic skills of civil engineering; however, recent technical advancements have resulted in the development of more efficient systems involving numbers of multidisciplinary inputs without any significant engineering inputs. Several problems encountered during CW operations can be minimized by the application of appropriate selection of the type of wetlands that further depends on the nature of wastewater. Various designs of CWs, contingent with the desired characteristics of the treated wastewater quality, have been worked out all over the world including horizontal surface and subsurface flow, vertical flow, and floating raft systems [84,103]. Subsurface flow systems are more efficient as compared to the other variants of CWs [104]. Whatever the type, all CWs essentially have four major systems: (1) aquatic macrophytes (wetland plants), (2) substrate or media in which plants grow, (3) water column, and (4) microorganisms.

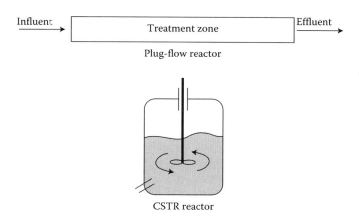

FIGURE 56.3 Simple diagram of plug flow and stirred tank reactors.

In general, CWs for wastewater treatment should be designed as plug-flow reactors with aspect ratios >15 (length to width) [96]. Prior to designing CWs, one must have an understanding of the hydrology and hydraulics (Figure 56.3).

56.2.1 Hydrology and Hydraulics

Hydrology influences the substrate and thereby influences the wetland biota. The residence time of water in wetlands allows the interaction of wetland species and water, which is a determinant of the efficient treatment strategy. Depth and storage are two important hydrologic features of a wetland influenced by landscape; however, in a treatment wetland these are dependent on the inflow to outflow rate and the aspect ratio (length to width) and losses of water such as evapotranspiration (ET). In a constructed wetland, ET is very critical and depends on the diurnal and seasonal cyclic variations [40]. Such loss of water in treatment wetlands disturbs the hydraulics. Wetland hydraulics is referred to as the movement of water through the wetlands. The net effect of any treatment process depends on the water flow characteristics. By definition, CWs are basically the surface flow detention systems and high hydraulic efficiency is required to treat wastewater for the desirable results [69]. Improper hydraulic designing causes issues of water conveyance, water quality, odor, and vector insects [20]. Hydrodynamical properties of the flow such as the velocity, streamlines, and the residence times are important during the designing of CWs [1]. Fundamental to CW's sustainable operation is the proper control of the hydrologic regime of the wetlands and optimal flow hydrodynamics within the wetland [69]. For an effective treatment practice, three principle designs, namely, (1) hydrologic effectiveness, (2) hydrologic efficiency, and (3) facilitation and optimization of water quality treatment processes have been suggested by Wong et al. [107]. In general, hydrologic effectiveness corresponds to the rainfall, relative environmental humidity, and ability of a wetland to detain the wastewater, whereas hydraulic efficiency can be achieved by proper measures of shape and depth (aspect ratio) of the wetland and inflow/outflow structures. Several measures such as vegetation and water depth designs may be introduced for the even distribution of flow throughout the wetland required to facilitate the precise conditions for necessary biochemical processes for wastewater treatment. Furthermore, the affectivity of a treatment wetland largely depends on the basic hydraulic variables such as hydraulic detention time (HDT), hydraulic loading rate (HLR), wetland porosity, etc.

From a practical point of view, wetlands can be thought of as a cross-section of two ideal reactors well known as plug-flow (PF) and continuous stirred tank reactor (CSTR) [40,96] because of the good mixing of water within the substrate of wetland (CSTR) entering and leaving the reactor without any internal mixing (PF). For an affective CW design, these two reactors, despite having different behavior for fluids, facilitate the exact calculations for hydraulic detention time τ_n.

56.2.1.1 Hydraulic detention time (HDT)-τ_n

HDT is the ratio between flow rate and the wetland volume available for water flow.
For free water flow [40]:

$$\tau_n = \frac{V_n}{Q} = \frac{(LWh)_n}{Q} \qquad (56.1)$$

where τ_n = hydraulic detention time, L = wetland length (m), W = wetland width (m), h = water depth (m), $(LWh)_n$ = wetland volume (m³), and Q = flow rate (m³/d).

For subsurface flow [40]:

$$\tau_n = \frac{V_n}{Q} = \frac{\varepsilon(LWh)_n}{Q} \qquad (56.2)$$

where ε = bed medium porosity.

The porosity (ε) of the bed medium of a wetland is the ratio of the empty basin volume to the actual volume occupied by water in a wetland. Reportedly, the value of ε ranges from 0.65 to 0.75 for vegetated wetlands with lower values for dense and mature wetlands [38,71]. Kadlec and Knight [38] reported an average porosity value greater than 0.95, whereas $\varepsilon = 1.0$ can be used as a good approximation. The overall effect of porosity is to reduce the wetland volume available for water flow and storage [20].

56.2.1.2 Hydraulic loading rate (HLR) = q

HLR = qLR or q is defined as the rainfall equivalents irrespective of the type of flow (free surface or subsurface including vertical and horizontal flow). In fact, the hydraulic loading rate is the volume of wastewater that the media filled in wetland system can transmit far enough from the infiltration surface such that it no longer influences the infiltration of additional wastewater. It also implies the addition of the water in a wetland surface without uniform distribution [40]:

$$q = \frac{Q}{A} \qquad (56.3)$$

where Q = water flow rate (m³/d), A = wetland or wetted area (m²).

56.2.1.3 Aspect Ratio

The aspect ratio is defined as the effective ratio of the average length (L) of the major axis and the average width (W) of a wetland. Reports of Wile et al. [106] suggest that treatment efficiency of a constructed wetland (free water surface) is better at higher aspect ratios although reports of García et al. [24] and Bound et al. [4] show no effect of aspect ratio (L:W) on the removal efficiency of the wetland systems. However, the longest aspect ratio with shallow water depth results in a greater efficiency of CWs especially for organic contaminants [24].

56.2.1.4 Substrate Medium

Nutrient ions removal from the wastewater in a CW system largely depends on the substrate media selected. The occurrence and potential importance of diffusion-mediated processes for solute transport in subsurface systems is well recognized. Diffusion is a rate-controlled process; changes in residence time (pore-water velocity) influence the impact of a diffusion-mediated process on solute transport. The mobility of nutrients in soil depends on their solubility in soil pore water and transportation by diffusive processes [68]; however, soil is not supportive in subsurface flow because of the low hydraulic conductivity of soils [5] as the water infiltration influences permeability of nearby soil as the cations present in the soil layers form microaggregates (micelles). The more cations in the soil, the less will be the movement of solutes. In addition, higher porosity increases the water flow through the media while less porous medium increases the saturation of the void space between the media particles resulting in the choking of the media [90] causing the anoxic conditions at the rhizoplane, thereby reducing the nutrient ion removal efficiency of a CW. These voids serve as a flow channels for water in CWs providing the nutrient anions of wastewaters a large surface area (media particles) with which to react. Apart from soil, several other media options are also available that can be used as substrate such as vermiculite, calcite, crushed marbles, sand, and coarse gravel. Sorption of P ions by the sand as substrate is the major P removal mechanism from the wastewaters owing to its large surface area, higher porosity, and Ca content [5]. To achieve maximum reduction of P ions from the wastewater, proportionate mixing of alum-like compounds with the substrate is also commonly recommended.

56.3 Performance of Constructed Wetlands

As mentioned in Section 56.2, the removal efficiency of CWs depends largely on hydraulics, pattern of wastewater flow, environmental conditions, retention time, substrate medium, and the type of vegetation. Table 56.1 presents the plant species commonly used in constructed wetlands. Reed grass (*Glyceria maximus*), yellow flag grass (*Iris pseudacorus*), common reed (*Phragmites australis*), narrow leaved cattail (*Typha angustifolia*), and cattail (*Typha latifolia*) are the commonly used wetland

TABLE 56.1 Common Wetland Plant Species and Their Usage Frequency in Constructed Wetlands

Plant Species	Common Name	Usage Frequency[a] (%)	BOD, COD, SS Removal Efficiency	Nutrient Removal Efficiency
Carex spp.	Sedges	15	++	+
Eichhornia cressipes	Water hyacinth	0.5	+	--
Glyceria maximus	Reed grass	20	+++	+
Iris pseudacorus	Yellow flag	75	+++	++
Juncus effuses	Rushes	28	++	+
Lemna minor	Small duckweed	0.5	+	++
Phragmites australis	Common reed	**90**	+++	+++
Scirpus spp.	Bulrushes	22	++	+++
Typha angustifolia	Narrow leaved cattail	**45**	+++	+++
Typha latifolia	Cattail	**85**	+++	+++

[a] Percentage usage frequency indicate the plant species generally used in the CWs and has been deduced on the basis of available data from 2001–2012 in the research literatures.

species having significant pollution/nutrient removal ability from the wastewater. Detailed discussion on the wetland species is provided in a separate section. Table 56.2 presents the data of the earlier researches that exhibit the type of CW used and their ability to reduce certain specific parameters of wastewater. Although it has been observed that design of a CW and the pattern of water flow influence the removal efficiency, the data of Table 56.2 demonstrate a different conclusion. From Table 56.2, it seems that the removal efficiency of a CW does not entirely depend upon the HLR, which further depends on the particular contaminant of the wastewater. Furthermore, increased hydraulic loads negatively affect CW retention [3]. Table 56.2 also

indicates that simple designed CWs can perform better than the well-organized and engineered wetlands [26,110]. The reduction in COD, BOD, and SS is natural in vegetated wetlands and the designs of CWs having high HRT only enhance the reduction up to a certain extent. Moreover, lower HLR and higher HRT increases the total efficiency of the treatment [88]. It has also been observed that the environmental temperature is decisive as reported by Steer et al. [91]. It is in general agreement that CWs are less efficient in removing nutrients like nitrogen and phosphorus [24,31,42,102]; however, few reports [6,9,108,110]) showed the substantial removal of nitrogen from the wastewater in CWs. Performance of CWs can be a direct function of influent and

TABLE 56.2 Performance of Constructed Wetlands (with Respect to HLR/Time/Nature of Wastewater/Type of CW)

Constructed Wetland	Waste Treated	Treatment Time	HLR	Removal % of Nutrients/ Pollutants	Reference
FWS + SSF	Municipal wastewater	Variable (12 months)	Variable (20–1736 m³/d)	86%–90% reduction in BOD, COD, and nutrients	[28]
FWS ↔ SSF	Recirculating waste-water of aquaculture	2 moths	0.3 m/d	80%–90% reduction in BOD, and COD; 5.4% for P	[55]
VF	Mine wastewater	12 months	4.6 m/d	98%–99% reduction in Pb and Cu concentrations	[76]
SSF	Pretreated Swine Effluent	8 months	6–3.5 cm/d	SS—99%; COD—82%; BOD—92%; TN—24%	[52]
FSW + SSF	Simulated water containing AN	24 months	0.06(FSW); 0.04 (SSF) m/d	AN—63.37 %	[36]
HSFRB	Urban wastewater	9 months	20–45 mm/d	COD—70%–80%; BOD—75%–85%; NH₄-N—40%–50%	[24]
SF	Urban wastewater	5 months	0.064 m³/d	ORP—83.2%; N-NH₄-75.3%	[47]
HF	Tannery wastewater	17 months	3–6 cm/d	COD—41%–73%; BOD—41%–58%	[9]
SSF	Municipal wastewater	6 months	—	78%–79% reduction in pollutant level	[108]
SF-H	Textile wastewater	24 months	13.3 cm/d	72%—Cr(VI); 26%—Cr(III)	[21]
SSF	Urban wastewater	3 months	227 L/m²d	SS—95%; COD—78%; turbidity—95%	[34]

Note: FWS, free water surface; HS, horizontal surface; SF, surface flow; SSF, sub-surface flow; VF, vertical flow; HSFRB, horizontal sub-surface flow reed bed; SF-H, horizontal surface flow; HLR, hydraulic loading rate; ↔, series; ↓, vertical; →, horizontal; +, treatment wetlands used separately. The different units are used to express the HLR.

effluent concentrations. The treatment efficiency of a CW can be calculated as [88]:

$$R = \left(1 - \frac{C_e}{C_i}\right) \times 100 \qquad (56.4)$$

where C_e = effluent concentration in mgl^{-1}; C_i = influent concentration in mgl^{-1}, and R is the removal percentage.

56.3.1 Factors Influencing the Performance of CWs

Naturally vegetated wetlands have inherent ability to reduce the pollutant levels from the water flowing through them; however, the removal efficiency of CWs is largely and essentially influenced by certain environmental factors such as temperature, vegetation density, and seasonal variations and by physical factors such as HRT, HLR (in general), ET, porosity of the substrate, aspect ratio, depth of water bed, and flow of water [1,88,104]. The physical properties of substrate such as porosity influence the conductance of the flow as the void spaces in the media serve as flow channels for water in CWs. In most cases, gravel has been adopted as the best media forming the subsurface. It provides the effective volume for the treatment and poses the least hindrance to the flow velocity [104]. The flow of wastewater in CW systems can be heavily modified by the atmospheric fluxes of precipitation and ET also, as the loss of water through ET slows the flow and concentrates the pollutants, whereas rainfall increased the water flow, reduced the contact time [32,37] and altered the removal efficiency of CWs. In addition, the removal efficiency and volumetric removal rate increase exponentially with the increase of water temperature [36].

Another very important factor that influences the performance of CWs is the flow of wastewater in the system. As stated earlier in the introduction of this chapter, there are two basic types of CWs, namely, (1) vertical flow (VF) and (2) horizontal flow (HF), which further has sub-divisions, namely, (a) surface flow and (b) subsurface flow. In HFs, increasing the aspect ratio of the beds will work like a plug flow and therefore better efficiency [24]. It has been observed by researchers worldwide that subsurface flow (where the water flows in the subsurface layers that are beneath the substrate on which plants grow) to be the best treatment design of a CW [105]. Subsurface flow facilitates (Figures 56.4 and 56.5) the direct contact of wastewater to the roots of plants, the base of treatment wetlands.

Vegetation type is also a factor that sometimes challenges the removal strategy even at the most advanced engineered CWs. The wrong choice of macrophytes may lead to erroneous results and the failure of the treatment system. Table 56.1 presents the most preferred plant species used in CWs, out of which *Phragmites australis, Typha latifolia,* and *Typha angustifolia* are frequently used because of their ability to reduce the level of organic loading, BOD, COD, and nutrient ions such as soluble reactive phosphorus (SRP), for example, 95% of soluble phosphorus can be removed by the application of reed beds [33]. Earlier reports of Dierberg et al. [16] demonstrated substantial (>95%) removal of SRP from the urban storm water runoff by the submerged aquatic macrophytes in a comparatively smaller CW facility.

56.4 Constructed Wetlands and the Urban Wastewater Treatment

The centralized development in most urban parts of the world have increased the migration of people from rural to urban establishments for want of services and better living conditions. Because of this, significantly higher amounts of waste are being generated especially in developing countries. Urban wastewater is a serious challenge before the world community because of the shortage of drinking water. Untreated urban runoff

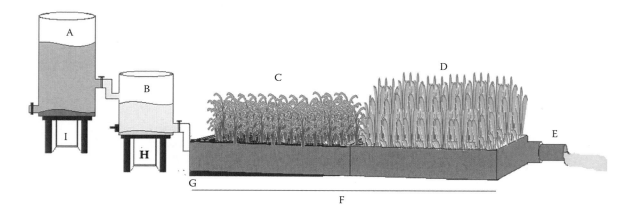

FIGURE 56.4 Horizontal subsurface flow constructed wetlands showing the interlinked primary settling tank and equalization tank. A: primary settling tank, B: secondary settling tank, C: first half CW with different species, D: second half of the CW having another wetland plant species, E: outlet of the treated water, F: horizontal subsurface flow constructed wetland, G: slope given to increase volumetric flow rate, I and H: stands to hold the tanks.

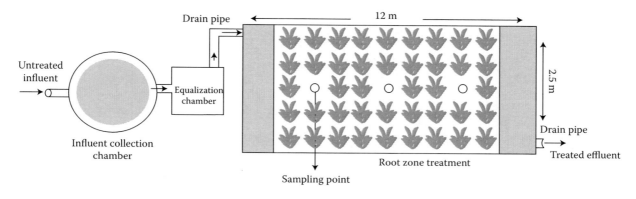

FIGURE 56.5 Simple layout of subsurface flow constructed wetland (pilot scale).

accounted for most of the negative effects observed in rivers, lakes, and other receiving water within urban areas, disrupting the river or any other aquatic ecosystem [95]. The planning of urban water treatment first needs the characterization of pollution load. Urban wastewater typically contains BOD (200 mg/L), COD (450 mg/L), SS (250 mg/L), N-NH$_4^+$ (20–30 mg/L), TKN (50–60 mg/L), and trace elemental pollution. Major parts of urban wastewater are formed by organic substances, pharmaceuticals [73], xenobiotic compounds [10,58], and compounds of benzene [41]. Conventional wastewater treatment plants are not economically feasible whereas the CWs are attracting the attention of the world community to treat urban wastewater cost-effectively having a higher quality of effluent. Generally, HF

constructed wetlands are particularly important for the treatment of urban wastewater generated from a smaller community [24,104]. Figure 56.6 presents a layout of treatment strategy for urban wastewater treatment. Recirculation of the effluent prior to the final discharge is optional, which depends on the level of a particular component of the wastewater. Recirculation is always a preferred choice in cases of highly polluted wastewater and to achieve a higher quality of final discharge from a CW system. Research from all over the world proved the higher efficiency of CWs in treating urban wastewater with the removal percentage of BOD (97%), COD (95%), TSS (85%), N-NH$_4^+$ (39%), and P-PO$_4^{-3}$ (28%–35%). Reports of Chung et al. [11] demonstrate 90% removal of nitrogenous compounds.

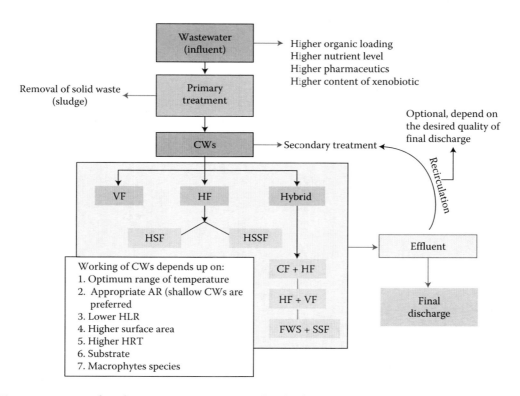

FIGURE 56.6 Treatment strategies for urban wastewater in constructed wetlands systems.

56.4.1 Mechanism of Nutrient Removal from Wastewater in a Constructed Wetland

The ability of wetlands for improving the quality of wastewater has long been recognized in natural wetlands in many parts of the world [62,64] because of their rich floral and microbial diversity. Wetland plants belong chiefly to four group systems, namely, emergent macrophytes (EM) (e.g., *Phragmites australis, Typha latifolia*), floating-leaved macrophytes (FLM) (e.g., *Nuphar luteum*), free-floating macrophytes (FFM) (e.g., *Eichhornia crassipes*), and submerged macrophytes (SM) (e.g., *Myriophyllum spicatum*) [89]. Major mechanisms of nutrient removal in wetlands depend on sedimentation and absorption by the aquatic macrophytes [83]. Macrophytes create conditions for the sedimentation of SS and prevent erosion by reducing the velocity of the water in wetlands. Macrophytes in association with the aquatic microorganisms and periphytons enhance the uptake of nutrients from the water [101]. Periphyton potentially removes metal cations [82] and nutrient anions such as PO_4^{3-} and NO_3^- by direct absorption from the water column [43]. The macrophytes transport approximately 90% of the oxygen available in the rhizosphere. This stimulates both aerobic decomposition of organic matter and promotes the growth of nitrifying bacteria [51,77] (Figure 56.7). The nutrient removal by the four wetland plants systems follows the order as SM > FFM > FLM > EM, indicating the maximum ability of submerged aquatic macrophytes, whereas metal removal follows the order FFM > EM > SAM > FLM [89], although emergent macrophytes because of their biomass accumulation are supposed to accumulate more heavy metals and other nutrient elements. The amount of accumulated nutrients depends on the physiological capacity for further uptake and biomass of aquatic macrophytes, which vary with the species [70]. In most aquatic ecosystems, attention has been focused on the cycling of N and P, most likely to limit primary producers and perhaps heterotrophic microorganisms [74,86]. Aquatic macrophytes act as substratum for the growth of periphyton communities composed of complex assemblages of cyanobacteria, eubacteria, diatoms, and eukaryotic algae [59].

In aquatic systems, interaction of macrophytes with the microbes is common especially at rhizoplane (Figure 56.8). Moreover, the structure of microbial assemblages differs on different plant species depending on the nature and availability of organic carbon and oxygen level at rhizoplane. In water, rooted macrophytes continuously replenish the loss of oxygen because of microbial and chemical consumption by supplying through the plant's interconnected lacunae right from shoot to root where the O_2 is released. This is also known as radial oxygen loss (ROL) [92]. In most wastewaters, single deep branching planctomycete, candidatus *Bocardia anammoxidans* containing anammoxosome having hydroxylamine oxidoreductase a key enzyme responsible for the oxidation of ammonium ions to nitrite/nitrate. Okabe et al. [66] also reported the presence of nitrosomonas like AOB in the roots of common reed, indicating that nitrification occurs under waterlogged conditions as that of CW systems. The ROL at rhizoplane contributes its major role as major electron acceptor to ensure the survival of aerobic life forms; however, in the absence of or in low oxygen levels (as in sediments), several other electron acceptors such as CO_2, CH_4, and NO_3^- support anaerobic life forms. There is a sharp oxic-anoxic interface near the rhizoplane as most of the facultative anaerobes survive at this zone and are critical for water chemistry (Figure 56.7). In an aquatic system, rhizoplane is the site of active nitrification, denitrification, sulfur reduction, iron oxidation, methanogenesis, methanotrophism, and many more bio/physicochemical reactions [89].

56.4.2 Mechanism of Removal of Metal Ions in Constructed Wetlands

The rhizoplane of aquatic plants has different water chemistry than the rest of the water column because of high microbial activity [93]. The aquatic microbial community includes most of the common environmental bacteria and sulfate-reducing

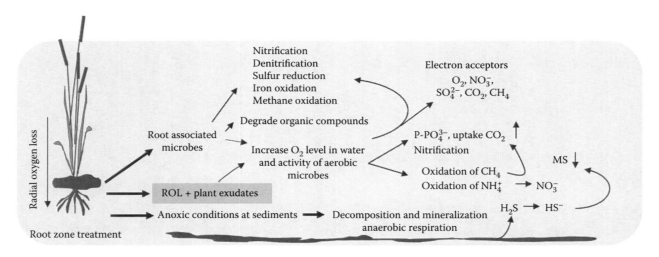

FIGURE 56.7 Mechanism of removal of wastewater components (nutrients and metals) in a CW.

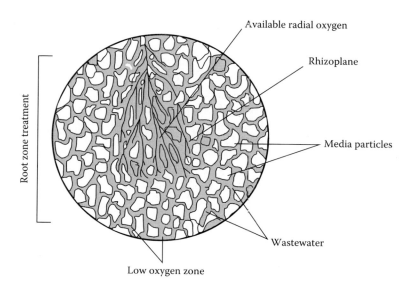

FIGURE 56.8 Root zone treatment and the impact of substrate in a CW system.

bacteria (SRB) are one of them. In natural wetlands, SRB produce sulfide ions. Sulfide anions react with the metal cations in the water column to form metal sulfide precipitate forming the sediments, an important process for the removal of metal ions from water [56,87]. This natural phenomenon has been well exploited in CWs although the water chemistry differs to a large extent in CW systems as most of them are fed with wastewater of varying nature. Biogenic sulfide production is widely used for treating sulfate-containing wastewater such as municipal sewage where the sulfate is reduced to hydrogen sulfide. This is further exploited to remove toxic metal cations from the wastewater by the formation of metal sulfide, a highly stable precipitate [54]. Such reactions can be introduced in the CW systems treating the wastewater containing high metal loading by a combination of the wastewater having more sulfate content such as municipal sewage, urban wastewater, or even molasses. To achieve perfect results in terms of maximum metal removal from the water column, electron donors such as ethanol (molasses), sugar (molasses), lactate, and acetate are usually added, whereas the $SO_4{}^{2-}$ acts as an electron acceptor required by SRB. Furthermore, the metal sulfide precipitation is not only the process; metal reduction is also achieved by the formation of iron plaques on the surface of submerged parts of aquatic plants. Iron plaque is a layer of iron (hydr)-oxide precipitate around the plant parts caused by oxidation of iron by molecular O_2 or by iron oxidizing bacteria (e.g., *Leptospirillum ferroxidans*) [45], which can reduce the metal ions from the water column. Reed plants are particularly well suited for metal removal from water columns [17,75] through the formation of plaque and because of high transpiration rate growing in inundated conditions or in free surface flow CW systems. Plants release excessive metal ions through transpiration, reducing the toxic concentration in the plant tissues of leaves, which is also common to reed plants especially in *Phragmites australis* [8], the most preferred plant species for treatment wetlands.

56.4.3 Removal/Degradation of Pharmaceuticals in CWs

Urban stormwater contains various waste components and it has been observed that pharmaceuticals form the major part [100]. CWs have been observed to substantially reduce the pharmaceuticals from urban wastewater [53]. Pharmaceuticals include any substance or mixture of substances that have the potential for use in the diagnosis, treatment, mitigation, or prevention of a disease, disorder, or abnormal physical state, or its symptoms [19]. Some of these pharmaceuticals ultimately make their way into the environment (water cycle) as the original compound or in a metabolized form by industrial waste, animal husbandry practices, or municipal and domestic wastewater, and they pollute the biosphere [14,25,60,81,97]. Higher concentrations of pharmaceuticals are found in wastewater originating from hospitals and elderly houses [48]. Most of these contaminants are structurally complex organic chemicals that are resistant to biological degradation [12,72]. Although the concentrations have most likely been present in the water for a long time, their levels have only recently been quantified and acknowledged as a potential ecological risk [48]. The persistence of pharmaceutical wastes in aquatic bodies and contamination of drinking water is a risk that can lead to problems with public health [79,80]. With no legally regulated maximum permitted concentrations of pharmaceuticals in the environment, there is a need to remove these micropollutants in a cost-effective manner to reduce their adverse environmental impact. Considering the SF systems alone, the highest removal values were found for caffeine (99.9%), acetaminophen (99.98%), ibuprofen (99.6%), naproxen (99.4%) and triclosan (98%), while the lowest were found for nadolol (80%) and sotalol (20%) [100]. CW systems provide favorable conditions such as nutrients, oxygen, and temperature conditions for the growth of microbes capable of degrading pharmaceuticals.

TABLE 56.3 Examples of Bacterial Removal/Degradation of Pharmaceuticals Present in Urban Wastewater

Pharmaceutical Contaminant	Bacteria	Reference
Thiomersal	*Pseudomonas putida*	[22]
Ibuprofen	*Nitrosomonas europaea*	[94]
Naphthalene	*Bacillus* sp., *Brachybacterium* sp.	[99]
Sulfamethoxazole (SMX)	*Rhodococcus equi*	[49]
Synthetic estrogen 17α-ethinylestradiol (EE2)	*Bacillus subtilis, Pseudomonas aeruginosa, Pseudomonas putida, Rhodococcus equi, Rhodococcus erythropolis, Rhodococcus rhodochrous, Rhodococcus zopfii*	[50]
Acetaminophen	*Delftia tsuruhatensis, Pseudomonas aeruginosa*	[15]

Highly selective and rapid reactions such as advanced oxidation processes (AOPs) or adsorbents such as activated carbon have been utilized for efficient removal of these pharmaceutical micropollutants [29,46,67]. However, the oxidation technologies are characterized by a large ecological footprint due to a high energy use [81] and may also lead to the formation of potentially harmful by-products [48]. Biological techniques can provide an answer to the solution because they are cost-effective as well as more efficient for the removal of micropollutants compared to oxidation technologies [46]. Recently, the use of bacteria has been explored in the removal of pharmaceuticals from contaminated water bodies. Table 56.3 gives an overview of the different types of bacteria used for degradation/removal of pharmaceuticals.

56.5 Hybrid Constructed Wetlands

Hybrid CWs are commonly understood as a combination of two or more types of systems in order to have higher quality of effluent [102]. Most of the time, hybrid systems are required in case of the failure of one kind of CW system in removing a particular component, namely nutrients, toxic organic compounds, and xenobiotic, although having excellent performance in removing other components from the wastewater such as suspended solids. Ávila et al. [2] reported the successful use of hybrid CW systems (VF – HF – FWS) for the treatment of urban wastewater

containing estrogens and dioxins rendering toxicity to the waste. Ninety percent reduction of toxicity was accounted in VF, however; HF and FWS could able to reduce the level of estrogenicity and dioxin-like compounds. HCW systems are rather less efficient in removing the nitrogenous pollution, such as ammonia, which can be stripped by the VFCW systems owing to the increased oxidation. Figure 56.9 present a typical hybrid system made with a combination of VF and HF CW systems. Hybrid systems have been identified and proved to be more efficient for the removal of organic contaminants and in reducing the toxic nature of wastewaters as suggested by the reports of researchers worldwide [2,7,44]. The voluminous literature is too large to be quoted here in the text. Some commonly used hybrid systems are VF – HF; VF – HF – FWS; VF – HF – VF and VF – HF – FWS – P (pond). Hybrid are particularly very useful in the treatment of wastewaters of different natures and characteristics having high pollution loads such as that of hatcheries [61] dairy farmyard [18], heavy metal contaminated wastewaters [21,23], removal of azo-dyes [109], pathogen removal [63], and agricultural wastewater [98].

56.6 Summary and Conclusions

CWs are the result of excellent cost-effective engineering efforts to make use of natural cleaning properties of wetlands. Proper

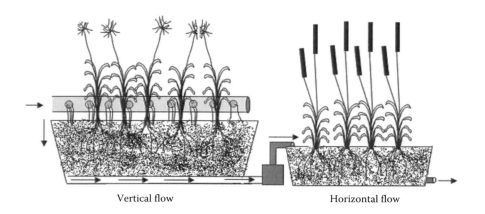

Vertical flow   Horizontal flow

FIGURE 56.9 Hybrid constructed wetlands of VF and HF systems.

designing of CWs along with the selection of appropriate species coupled with the suitable environmental conditions can lead to higher waste reduction efficiency. Advancements are being brought in designing patterns of CWs to make them perfect secondary treatment units. Variety of wastewaters of different natures, namely, municipal, piggery, slaughterhouse, mining fields, and urban wastewater have been successfully treated by different CW systems, whereas subsurface flow CWs have excellent percentage of reduction in almost all the components of wastewater especially urban wastewaters from small communities. In addition, recirculation of the effluent is followed to have higher quality of effluents. Efficiency of CWs depends on certain factors such as HLR, HRT, ambient temperature, evapotranspiration, precipitation, substrate matrix, aspect ratio, and vegetation. Plant species largely influence the treatment process in CWs and the reed beds are supposed to be the best in achieving higher efficiency of CWs.

References

1. Anderssen, B., Mooney, J., and Stricker, J. 1996. Decision support for the design of constructed wetlands. *Applied Mathematical Modelling*, 20, 93–100.
2. Ávila, C., Matamoros, V., Reyes-Contreras, C., Piña, B., Casado, M., Mita, L., Rivetti, C., Barata, C., García, J., and Bayona, J.M. 2014. Attenuation of emerging organic contaminants in a hybrid constructed wetland system under different hydraulic loading rates and their associated toxicological effects in wastewater. *Science of the Total Environment*, 470–471, 1272–1280.
3. Baskerud, B.C. 2003. Clay particle retention in small constructed wetlands. *Water Research*, 37, 3793–3802.
4. Bounds, H.C., Collins, J., Liu, Z., Qin, Z., and Sasek, T.A. 1998. Effects of length-width ratio and stress on a rock plant filter operation. *Small Flow Journal*, 4(1), 4–14.
5. Brix, H., Arias, C.A., and del Bubba, M. 2001. Media selection for sustainable phosphorus removal in subsurface flow constructed wetlands. *Water Science Technology*, 44, 47–54.
6. Brix, H., Arias, C.A., and Johnsen, N. 2003. Experiments in a two stage constructed wetland system: Nitrification capacity and effects of recycling on nitrogen removal. In: J. Vymazal (ed.), *Wetlands-Nutrients, Metals and Mass Cycling*. Backhuys Publishers, Leiden, the Netherlands, pp. 237–258.
7. Brix, H., Koottatep, T., and Laugesen, C.H. 2006. Re-establishment of wastewater treatment in tsunami affected areas of Thailand by the use of constructed wetlands. In: *Proceedings of the of the 10th International Conference on Wetland Systems for Water Pollution Control, MAOTDR 2006*, Lisbon, Portugal, pp. 59–67.
8. Burke, D.J., Weis, J.S., and Weis, P. 2000. Release of metals by the leaves of the salt marsh grasses Spartina alterniflora and Phragmites australis. *Estuarine Coastal and Shelf Science*, 51, 153–159.
9. Calheiros, C.S.C., Rangel, A.O.S.S., and Castro, P.M.L. 2007. Constructed wetland systems vegetated with different plants applied to the treatment of tannery wastewater. *Water Research*, 41(8), 1790–1798.
10. Cheng, S., Vidakovic-Cifrek, Z., Grosse, W., and Karrenbrock, F. 2002. Xenobiotics removal from polluted water by a multifunctional constructed wetland. *Chemosphere*, 48, 415–418.
11. Chung, A.K.C., Wu, Y., Tam, N.F.Y., and Wong, M.H. 2008. Nitrogen and phosphate mass balance in a subsurface flow constructed wetland for treating municipal wastewater. *Ecological Engineering*, 32, 81–89.
12. Cokgor, U.E., Karahan, O., Dogruel, S., and Orlon, D. 2004. Biological treatability of raw and ozonated penicillin formulated effluent. *Journal of Hazardous Material*, 116, 159–166.
13. Cooper, D., Griffin, P., and Cooper, P. 2005. Factors affecting the longevity of sub-surface horizontal flow systems operating as tertiary treatment for sewage effluent. *Water Science Technology*, 51(9), 127–135.
14. Daughton, C.G. and Ternes, T.A. 1999. Pharmaceuticals and personal care products in the environment: Agents of subtle change? *Environmental Health Perspectives Supplements*, 107, 907.
15. De-Gusseme, B., Vanhaecke, L., Verstraete, W., and Boon, N. 2011. Degradation of acetaminophen by *Delftia tsuruhatensis* and *Pseudomonas aeruginosa* in a membrane bioreactor. *Water Research*, 45, 1829–1837.
16. Dierberg, F.E., DeBusk, T.A., Jackson, S.D., Chimney, M.J., and Pietro, K. 2002. Submerged aquatic vegetation-based treatment wetlands for removing phosphorus from agriculture runoff: Response to hydraulic and nutrient loading. *Water Research*, 36, 1409–1422.
17. Drzewiecka, K., Borowiak, K., Mieczek, M., Zawada, I., and Goliński, P. 2010. Cadmium and lead accumulation in two plants of five lakes in Poznan, Poland. *Acta Biologica Cracoviensia*, 52(2), 59–68.
18. Dunne, E.J., Culleton, N., O'Donovan, G., Harrington, R., and Daly, K. 2005, Phosphorus retention and sorption by constructed wetland soils in southeast Ireland. *Water Research*, 39(18), 4355–4362.
19. Enick, O.V. and Moore, M.M. 2007. Assessing the assessments: Pharmaceuticals in the environment. *Environmental Impact Assessment Review*, 27, 707–729.
20. EPA. 1999. Free water surface wetlands for wastewater treatment: A technology assessment. Official Document, Phoenix, AZ.
21. Fibbi, D., Doumett, S., Lepri, L., Checchini, L., Gonnelli, C., Coppini, E., and Del Bubba, M. 2012. Distribution and mass balance of hexavalent and trivalent chromium in a subsurface, horizontal flow (SF-h) constructed wetland operating as post-treatment of textile wastewater for water reuse. *Journal of Hazardous Materials*, 199–200, 209–216.
22. Fortunato, R., Crespo, J.G., and Reis, M.A.M. 2005. Biodegradation of thiomersal containing effluents by

a mercury resistant *Pseudomonas putida* strain. *Water Research*, 15, 3511–3522.

23. Galletti, A., Verlicchi, P., and Ranieri, E. 2010. Removal and accumulation of Cu, Ni and Zn in horizontal subsurface flow constructed wetlands: Contribution of vegetation and filling medium. *Science of the Total Environment*, 408, 5097–510

24. García, J., Aguirre, P., Mujeriego, R., Huang, Y., Ortiz, L., and Bayona, J.M. 2004. Initial contaminant removal performance factors in horizontal flow reed beds used for treating urban wastewater. *Water Research*, 38, 1669–1678.

25. Glassmeyer, S.T., Furlong, E.T., Kolpin, D.W., Cahill, J.D., Zaugg, S.D., Werner, S.L., Meyer, M.T., and Kryak, D.D. 2005. Transport of chemical and microbial compounds from known wastewater discharges: Potential for use as indicators of human fecal contamination. *Environmental Science & Technology*, 39, 5157–5169.

26. Gómez Cerezo, R., Suárez, M.L., and Vidal-Abarca, M.R. 2001. The performance of a multi-stage system of constructed wetlands for urban wastewater treatment in a semi-arid region of SE Spain. *Ecological Engineering*, 16, 501–517.

27. Greenway, M. 2005. The role of constructed wetlands in secondary effluent treatment and water reuse in subtropical Australia. *Ecological Engineering*, 25, 501–509.

28. Greenway, M. and Woolley, A. 1999. Constructed wetlands in Queensland: Performance efficiency and nutrient bioaccumulation. *Ecological Engineering*, 12, 39–55.

29. Gupta, V.K., Gupta, B., Rastogi, A., Agarwal, S., and Nayak, A. 2011. Pesticides removal from waste water by activated carbon prepared from waste rubber tire. *Water Research*, 45, 4047–4055.

30. Hadad, H.R., Maine, M.A., and Bonetto, C.A. 2006. Macrophyte growth in a pilot-scale constructed wetland for industrial wastewater treatment. *Chemosphere*, 63, 1744–1753.

31. Hamersley, M.R., Howes, B.L., White, D.S., Johnke, S., Young, D., Peterson, S.B., and Teal, J.M. 2001. Nitrogen balance and cycling in an ecologically engineered septage treatment system. *Ecological Engineering*, 18, 61–75.

32. Headley, T.R., Davison, L., Huett, D.O., and Müller, R. 2012. Evapotranspiration from subsurface horizontal flow wetlands planted with *Phragmites australis* in sub-tropical Australia. *Water Research*, 46, 345–354.

33. Headly, T.R. Huett, D.O., and Davison, L. 2003. Seasonal variation in phosphorus removal processes within reed beds- mass balance investigations. *Water Science and Technology*, 48(5), 59–66.

34. Herrera-Melián, J.A., González-Bordón, A., Martín-González, M.A., García-Jiménez, P., Carrasco, M., and Araña, J. 2014. Palm tree mulch as substrate for primary treatment wetlands processing high strength urban wastewater. *Journal of Environmental Management*, 139, 22–31.

35. Hu, Y., Zhao, Y., and Rymszewicz, A. 2014. Robust biological nitrogen removal by creating multiple tides in a single bed tidal flow constructed wetland. *Science of the Total Environment*, 470–471, 1197–1204.

36. Jing, S. and Lin, Y. 2004. Seasonal effect on ammonia nitrogen removal by constructed wetlands treating polluted river water in southern Taiwan. *Environmental Pollution*, 127, 291–301.

37. Kadlec, R.H. 1989. Hydrologic factors in wetland water treatment. In: D.A. Hammer (ed.).*Constructed Wetlands for Wastewater Treatment; Municipal, Industrial and Agricultural*. Lewis Publishers, Chelsea, MI, pp. 21–40.

38. Kadlec, R.H. and Knight, R.L. 1996. *Treatment Wetlands*. CRC Press, Boca Raton, FL.

39. Kadlec, R.H., Tanner, C.C., Hally, V.M., and Gibbs, M.M. 2005. Nitrogen spiraling in subsurface-flow constructed wetlands: Implications for treatment response. *Ecological Engineering*, 25, 365–381.

40. Kadlec, R.H. and Wallace, S.D. 2009. *Treatment Wetlands*, 2nd Edition. CRC Press, Boca Raton, FL.

41. Kantawanichkul, S. and Wara-Aswapati, S. 2005. LAS removal by a horizontal flow constructed wetland in tropical climate. In: J. Vymazal (ed.). *Natural and Constructed Wetlands. Nutrients, Metals and Management*. Backhuys Publishers, Leiden, the Netherlands, pp. 261–270.

42. Kaseva, M.E. 2004. Performance of a sub-surface flow constructed wetland in polishing pre-treated wastewater—A tropical case study. *Water Research*, 38, 681–687.

43. Kathiwada, N.R. and Polprasert, C. 1999. Assessment of effective specific surface area for free water surface wetlands. *Water Science Technology*, 40, 83–89.

44. Kato, K., Koba, T., Ietsugu, H., Saigusa, T., Nozoe, T., Kobayashi, S., Kitagawa, K., and Yanagiya, S. 2006. Early performance of hybrid reed bed system to treat milking parlour wastewater in cold climate in Japan. In: *Proceedings of the 10th International Conference on Wetland Systems for Water Pollution Control, MAOTDR 2006*, Lisbon, Portugal, pp. 1111–1118.

45. King, G.M. and Garey, M.A. 1999. Ferric iron reduction by bacteria associated with the roots of freshwater and marine macrophytes. *Applied and Environmental Microbiology*, 65(10), 4393–4398.

46. Kunkel, U. and Radke, M. 2008. Biodegradation of acidic pharmaceuticals in bed sediments: insight from a laboratory experiment. *Environmental Science and Technology*, 42, 7273–7279.

47. Kyambadde, J., Kansiime, F., Gumaelius, L., and Dalhammar, G. 2004. A comparative study of *Cyperus papyrus* and *Miscanthidium violaceum*-based constructed wetlands for wastewater treatment in a tropical climate. *Water Research*, 38, 475–485.

48. Langenhoff, A., Inderfurth, N., Veuskens, T., Schraa, G., Blokland, M., Kujawa-Roeleveld, K., and Rijnaarts, H. 2013. Microbial removal of the pharmaceutical compounds Ibuprofen and diclofenac from wastewater. *Biomedical Research International*, 2013, 325–806.

49. Larcher S. and Yargeau V. 2011. Biodegradation of sulfamethoxazole by individual and mixed bacteria. *Applied Microbiology and Biotechnology*, 91, 211–218.

50. Larcher, S. and Yargeau, V. 2013. Biodegradation of 17α-ethinylestradiol by heterotrophic bacteria. *Environmental Pollution*, 173, 17–22.

51. Lee, B. and Scholz, M. 2007. What is the role of *Phragmites australis* in experimental constructed wetland filters treating urban runoff? *Ecological Engineering*, 29, 87–95.

52. Lee, C., Lee, C., Lee, F., Tseng, S., and Liao, C. 2004. Performance of subsurface flow constructed wetland taking pretreated swine effluent under heavy loads. *Bioresource Technology*, 92, 173–179.

53. Li, Y., Zhu, G., Ng, W.J., and Tan, S.K. 2014. A review on removing pharmaceutical contaminants from wastewater by constructed wetlands: Design performance and mechanism. *Science of the Total Environment*, 468–469, 908–932.

54. Liamlean, W. and Annachhatre, A.P. 2007. Electron donor for biological sulfate reduction. *Biotechnology Advances*, 25(5), 452–463.

55. Lin, Y., Jing, S., and Lee, D. 2003. The potential use of constructed wetlands in a recirculating aquaculture system for shrimp culture. *Environmental Pollution*, 123, 107–113.

56. Machemer, S.D. and Wildeman, T.R. 1992. Adsorption compared with sulphide precipitation as metal removal processes from acid mine drainage in a constructed wetland. *Contaminant Hydrology*, 9, 115–131.

57. Martín, M., Gargallo, S., Hernández-Crespo, C., and Oliver, N. 2013. Phosphorus and nitrogen removal from tertiary treated urban wastewaters by a vertical flow constructed wetland. *Ecological Engineering*, 61, 34–42.

58. Masi, F., Conte, G., Lepri, L., Martellini, T., and Del Bubba, M. 2004. Endocrine Disrupting Chemicals (EDCs) and pathogen removal in a hybrid CW system for a tourist facility wastewater treatment and reuse. In: *Proceedings of the 9th International Conference on Wetland Systems for Water Pollution Control*, ASTEE, France, pp. 461–468.

59. McCormick, P.V. and O'Dell, M.B. 1996. Quantifying periphyton responses to phosphorus in the Florida Everglades: A synoptic-experimental approach. *Journal of North American Benthological Society*, 15, 450–468.

60. Meyer, M.T., Bumgarner, J.E., Varns, J.L., Daughtridge, J.V., Thurman, E.M., and Hostetler, K.A. 2000. Use of radioimmunoassay as a screen for antibiotics in confined animal feeding operations and confirmation by liquid chromatography/mass spectrometry. *Science of the Total Environment*, 248, 181–187.

61. Michael, Jr, J.H. 2003. Nutrients in salmon hatchery wastewater and its removal through the use of a wetland constructed to treat off-line settling pond effluent. *Aquaculture*, 226, 213–225.

62. Mitsch, W.J. and Gosselink, J.G. 1993. *Wetlands*. 2nd Edition. Van Nostrand Reinhold, New York.

63. Morató, J., Codony, F., Sánchez, O., Pérez, L.M., García, J., and Mas, J. 2014. Key design factors affecting microbial community composition and pathogenic organism removal in horizontal subsurface flow constructed wetlands. *Science of the Total Environment*, 481, 81–89.

64. Nahlik, A.M. and Mitsch, W.J. 2006. Tropical treatment wetlands dominated by free-floating macrophytes for water quality improvement in Costa Rica. *Ecological Engineering*, 28, 246–257.

65. Neralla, S., Weaver, R.W., Lesikar, B.J., and Persyn, R.A. 2000. Improvement of domestic wastewater quality by subsurface flow constructed wetlands. *Bioresource Technology*, 75, 19–25.

66. Okabe, S., Nakamura, Y., and Satoh, H. 2012. Community structure and *in situ* activity of nitrifying bacteria in Phragmites root-associated biofilms. *Microbes and Environments*, 27(3), 288–292.

67. Oller, I., Malato, S., and Sánchez-Pérez, J.A. 2011. Combination of advanced oxidation processes and biological treatments for wastewater decontamination: A review. *Science of the Total Environment*, 409, 4141–4166.

68. Peijnenburg, W.J.G.M., Posthuma, L., Eijsakers, H.J.P., and Allen, H.F. 1997. A conceptual framework for implementation of bioavailability of metals for environmental management purposes. *Ecotoxicology and Environmental Safety*, 37(2), 163–172.

69. Persson, J., Somes, N.L.G., and Wong, T.H.F. 1999. Hydraulics efficiency of constructed wetlands and ponds. *Water Science Technology*, 40(3), 291–300.

70. Pieczynska, E. 1990. Lentic aquatic terrestrial ecotones: Their structure functions and importance. In: R.J. Naiman and H. Decamps (Eds.). *The Ecology and Management of Aquatic Terrestrial Ecotones. Man and the Biosphere Series*. The Parthenon Publishing Group, Paris, pp 103–140.

71. Reed, S.C., Crites, R., and Middlebrooks, E.J. 1995. *Natural Systems for Waste Management and Treatment*. 2nd Edition. McGraw-Hill, New York.

72. Ren, N., Chen, Z., Wang, A., Zhang, Z P., and Yue, S. 2008. A novel application of TPAD-MBR system to the pilot treatment of chemical synthesis-based pharmaceutical wastewater. *Water Research*, 42, 3385–3392.

73. Reyes-Contreras, C., Matamorous, V., Ruitz, I., Soto, M., and Bayona, J.M. 2011. Evaluation of PPCPs removal in a combined anaerobic digester-constructed wetland pilot plant treating urban wastewater. *Chemosphere*, 84, 1200–1207.

74. Rosemond, A.D., Pringle, C.M., Ramfrez, A., Paul, M.J., and Meyer, J.L. 2002. Landscape variation in phosphorus concentration and effects on detritus based tropical streams. *Limnology and Oceanography*, 47, 278–289.

75. Samecka-Cymerman, A. and Kempers, A.J. 2001. Concentrations of heavy metals and plant nutrients in water, sediments and aquatic macrophytes of anthropogenic lakes (former open cut brown coal mines) differing in stage of acidification. *Science of the Total Environment*, 281, 87–98.

76. Scholz, M. 2003. Performance predictions of mature experimental constructed wetlands which treat urban water receiving high loads of lead and copper. *Water Research*, 37, 1270–1277.

77. Scholz, M. 2006. *Wetland Systems to Control Urban Runoff.* Elsevier, Amsterdam, the Netherlands.

78. Scholz, M. and Xu, J. 2002. Performance comparison of experimental constructed wetlands with different filter media and macrophytes treating industrial wastewater contaminated with lead and copper. *Bioresource Technology*, 83, 71–79.

79. Schulman, L.J., Sargent, E.V., Naumann, B.D., Faria, E.C., Dolan, D.G., and Wargo, J.P. 2002. A human health risk assessment of pharmaceuticals in the aquatic environment. *Human and Ecological Risk Assessment*, 8, 657–680.

80. Schwab, B.W., Hayes, E.P., Fiori, J.M., Mastrocco, F.J., Roden, N.M., Cragin, D., Meyerhoff, R.D., D'Aco, V.J., and Anderson, P.D. 2005. Human pharmaceuticals in US surface waters: A human health risk assessment. *Regulatory Toxicology and Pharmacology*, 42, 296–312.

81. Schwarzenbach, R.P. Escher, B.I., and Fenner, K. 2006. The challenge of micropollutants in aquatic systems. *Science*, 313, 1072–1077.

82. Scinto, L.J. and Reddy, K.R. 2003. Biotic and abiotic uptake of phosphorous by periphyton in a subtropical freshwater wetland. *Aquatic Botany*, 77, 203–222.

83. Serra, T. Fernando, H.J.S., and Rodríguez, R.V. 2004. Effects of emergent vegetation on lateral diffusion in wetlands. *Water Research*, 38, 139–147.

84. Shutes, R.B.E. 2001. Artificial wetlands and water quality improvement. *Environmental International*, 26, 441–447.

85. Siracusa, G. and La Rasa, A.D. 2006. Design of a constructed wetland for wastewater treatment in a Sicilian town and environmental evaluation using the energy analysis. *Ecological Engineering*, 197, 490–497.

86. Smith, V.H. 1998. Cultural eutrophication of inland, estuarine, and coastal waters. In: M.L. Pace and P.M. Groffman (eds.). *Success, Limitations and Frontiers in Ecosystem Science.* Springer, New York, pp 7–49.

87. Sochacki, A., Surmacz Górska, rska, J., Faure, O., and Guy, B. 2014. Polishing of synthetic electroplating wastewater in microcosm up-flow constructed wetlands: Metals removal mechanisms. *Chemical Engineering Journal*, 242, 43–52.

88. Solano, M.L., Soriano, P., and Ciria, M.P. 2004. Constructed wetlands as a sustainable solution for wastewater treatment in small villages. *Biosystems Engineering*, 87(1), 109–118.

89. Srivastava, J., Gupta, A., and Chandra, H. 2008. Managing water quality with aquatic macrophytes. *Reviews in Environmental Science and Bio/Technology*, 7, 255–266.

90. Srivastava J.K., Kalra, S.J.S., and Shukla, K. 2011. Action of cations on the leachability of the bio-stimulants in a soil column at soil water interface and its impact on surface water quality. *International Journal of Hydrology Science and Technology*, 1(3/4), 252–259.

91. Steer, D., Fraser, L., Boddy, J., and Seibert, B. 2002. Efficiency of small constructed wetlands for subsurface treatment of single family domestic effluent. *Ecological Engineering*, 18, 429–440.

92. Stottmeister, U., Wießner, A., Kuschk, P., Kappelmeyer, U., Kästner, M., Bederski, O., Müller, R.A., and Moormann, H. 2003. Effects of plants and microorganisms in constructed wetlands for wastewater treatment. *Biotechnology Advances*, 22, 93–117.

93. Stout, L.M. and Nüsslein, K. 2010. Biochemical potential of aquatic plant-microbe interactions. *Current Opinion in Biotechnology*, 21, 339–345.

94. Subramanya, N.T. 2003. Biodegradation of bisphenol a and ibuprofen by ammonia oxidizing bacteria. Master's thesis, Texas A & M University.

95. Taebi, A. and Droste, R.L. Pollution loads in urban runoff and sanitary wastewater. *Science of the Total Environment*, 327, 175–184.

96. Tchobanoglous, G. and Schroeder, E.D. 1985. *Water Quality.* Addison-Wesley Publishing Company, USA.

97. Ternes T. 2007. The occurrence of micopollutants in the aquatic environment: A new challenge for water management. *Water Science and Technology*, 55, 327–332.

98. Trang, N.T.D. and Brix, H. 2014. Use of planted biofilters in integrated recirculating aquaculture-hydroponics systems in the Mekong Delta, Vietnam. *Aquaculture Research*, 45(3), 460–469.

99. Velmurugan, A.M. and Arunachalam, C. 2009. Bioremediation of phenol and naphthalene by *Bacillus* species and *Brachybacterium* species isolated from pharma soil sample. *Current World Environment*, 4, 299–306.

100. Verlicchi, P. and Zambello, E. 2014. How efficient are constructed wetlands in removing pharmaceuticals from untreated and treated urban wastewater? A review. *Science of the Total Environment*, 470–471, 1281–1306.

101. Vymazal, J., 2002. The use of sub-surface constructed wetlands for wastewater treatment in the Czech Republic: 10 years experience. *Ecological Engineering*, 18, 633–646.

102. Vymazal, J. 2005. Horizontal sub-surface flow and hybrid constructed wetland systems for wastewater treatment (review). *Ecological Engineering*, 25, 478–490.

103. Vymazal, J. 2006. Constructed wetlands with emergent macrophytes: From experiments to a high quality treatment technology. In: V. Dias and J. Vymazal (eds.). *Proceedings of the 10th International Conference on Wetland Systems for Water Pollution Control*, September 23–29, Lisbon, Portugal, pp. 3–27.

104. Vymazal, J. 2009. The use constructed wetlands with horizontal sub-surface flow for various types of wastewater. *Ecological Engineering*, 35, 1–17.

105. Vymazal, J., Brix, H., Cooper, P.F., Green, M.B., and Haberl, R. (eds.). 1998. *Constructed Wetlands for Wastewater Treatment in Europe.* Backhuys Publishers, Leiden, the Netherlands.

106. Wile, I., Miller, G., and Black. S. 1985. Design and use of artificial wetlands. In: E.R. Kaynor, S. Pelczarski, and J. Benforado (eds.), *Ecological Considerations in Wetlands Treatment of Municipal Waters.* Van Nostrand Reinhold, New York. pp. 26–37.

107. Wong, T.H.F., Breen, P.F., Somes, N.L.G., and Lloyd, S.D. 1998. Managing urban stormwater using constructed wetlands. *Industry Report 98/7, Cooperative Research Centre for Catchment Hydrology*, November.

108. Wu, Y., Chung, A., Tam, N.F.Y., Pi, N., and Wong, M.H. 2008. Constructed mangrove wetland as secondary treatment system for municipal wastewater. *Ecological Engineering*, 34, 137–146.

109. Yadav, A.K., Jena, S., Acharya, B.C., and Mishra, B.K. 2012. Removal of azo dye in innovative constructed wetlands: Influence of iron scrap and sulfate reducing bacterial enrichment. *Ecological Engineering*, 49, 53–58.

110. Yuan, C., Lien, H., Huang, S., Chen, Y., and Fang, T. 2005. Application of subsurface flow constructed wetlands for campus water reuse–A bench-scale system study. *Journal of the Chinese Institute of Environmental Engineering*, 15(4), 243–253.

57

Constructed Wetlands: Pollutant Removal Mechanisms

Christos S. Akratos
University of Patras

PREFACE

Constructed wetlands emulate the processes occurring in natural wetlands and are used to treat various types of wastewaters. Over the years, constructed wetlands have been rather efficient at removing various pollutants and providing final effluents that can be reused for various purposes. This chapter presents basic information on constructed wetlands (types and pollutant removal mechanisms) and the various applications reported for reusing their effluents.

57.1 Introduction

Due to their simplicity and low capital and operational costs, constructed wetlands (CWs) have gained popularity in the treatment of various wastewaters (e.g., municipal, agro-industrial, industrial, urban runoff, agricultural runoff, landfill leachate) [32,64]. CWs use only renewable energy sources (e.g., solar and wind energy) and can operate from tropical to cold climatic zones [32,78]. CW operation is based on several physical, chemical, and biological processes occurring in the complex environment comprising wetland plants (hydrophytes), microorganisms, and porous media.

57.2 Constructed Wetland Types

CWs are usually divided into two main categories based on the system's water flow:

- Free water surface (FWS) CWs
- Subsurface (SSF) CWs

SSF CWs are further categorized into horizontal subsurface flow (HSF) and vertical flow (VF) CWs.

57.2.1 Free Water Surface Constructed Wetlands

FWS CWs (Figure 57.1) are small basins whose bases are sealed (e.g., with geotextile or clay) and contain a layer of soil, or other organic substrate (e.g., compost), up to 60 cm deep, while wastewater flows horizontally across the wetland above the soil layer at depths ranging from 20 to 80 cm [2,18,64,79]. CW vegetation is planted in the soil layer. FWS CWs cannot treat high pollutant loads and are therefore used mainly for municipal wastewater treatment [32]. FWS CWs present high removal efficiencies for organic matter, suspended solids (SS), nitrogen (N), pathogens, and heavy metals, while phosphorus removal efficiencies are significantly lower [64].

FIGURE 57.1 Free water surface flow constructed wetland.

57.2.2 Horizontal Subsurface Flow Constructed Wetlands

HSF CWs (Figure 57.2) are sealed basins containing gravel or other porous media (with sizes ranging from 2.5 to 5 cm) up to 80 cm deep [64]. CW vegetation is planted in the gravel layer and wastewater flows through the pores. As wastewater flows below the substrate's surface, it is not exposed to the atmosphere, thus minimizing any health risks [32]. The bases of HSF CWs usually have a minimum inclination of 1%–3% to ensure wastewater gravitational flow [32,64].

Pollutant removal in HSF CWs occurs when wastewater flows through the porous media and plant roots onto which various microorganisms are attached. The extensive plant root system provides an ideal environment for microorganism growth because it provides significant quantities of dissolved oxygen (DO). DO is used by the microorganisms to degrade pollutants. One crucial detail in HSF CW operation is to ensure the uniform distribution of the wastewater through perforated tubes placed across the CW in order to avoid the occurrence of dead zones [1]. HSF CWs are more efficient bioreactors than FWS as they achieve high removal efficiencies of organic matter, SS, N, P, and heavy metals [1,26,32,64].

57.2.3 Vertical Flow Constructed Wetlands

VF CWs (Figure 57.3) were initially introduced by Seidel [55]. They comprise sealed basins (with geo-membrane or reinforced concrete) that contain several layers of gravel and sand with the

total bed depth varying from 0.45 to 1.20 m [64]. Wastewater floods the VF CW surface at a depth of up to 5 cm through a series of perforated tubes [64]. With this mode of operation, the wastewater spreads throughout the entire CW surface and moves downward by gravity, pushing out the trapped air and sucking fresh air into the bed, thus enhancing aeration. The treated wastewater is collected by a series of perforated pipes located at the bottom of the CWs. Aeration is further enhanced by placing aeration pipes into the substrate layers and by employing a wet-dry operation cycle. This feed method is important in the treatment process because increased oxygen transfer within the bed provides better aerobic conditions for the oxidation of ammonia nitrogen (nitrification) and decomposition of organic matter (OM), compared to HSF CWs [15,17,32,78,79]. However, these conditions do not favor denitrification and the removal of P is limited, mainly due to inadequate contact time between the porous media and the wastewater, as the latter flows downward by gravity [61].

VF CW applications include the treatment of various wastewaters (e.g., landfill leachate, dairy wastewater, food-processing wastewater) [32]. VF CWs are gaining ground against HSF systems as they provide higher oxygen transfer capacity (OTC), and thus they are more effective at organic matter and nitrogen removal [16,64]. Being more effective bioreactors, VF CWs require smaller surface areas (up to 2 m²/pe) than HSF CWs (usually 5–10 m²/pe) [15,64,78].

Besides the common VF CW described previously, other VF CWs have also been developed recently [64]:

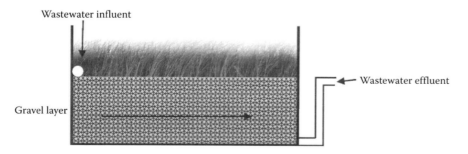

FIGURE 57.2 Horizontal subsurface flow constructed wetland.

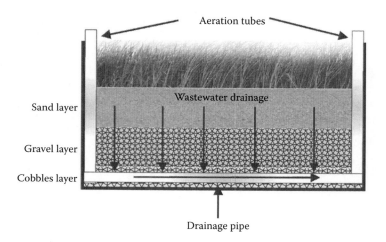

FIGURE 57.3 Vertical flow constructed wetland.

- *The French system:* Its operation is similar to that of the standard VF CW. The only difference is that the upper drainage layer comprises coarser gravel; thus, raw wastewater can be introduced directly into the VF CW without requiring a primary settlement stage [6,7,44].
- *Recirculating VF CWs:* This type of VF CW was developed in order to increase the contact time between the wastewater and the porous media by recalculating a portion of the treated wastewater [70], thus enhancing organic matter removal and denitrification [3,35,60,70].
- *Tidal flow CWs:* This CW has a different feed system from those previously described. Its operation is divided into two stages. During the first stage, the entire VF CW is flooded with wastewater. In the second stage, the VF CW is drained [32,67,69,86]. In this way, the contact time between wastewater and porous media increases [64].
- *Saturated vertical upflow CWs:* Wastewater is introduced at the bottom of the CW and upflows to its surface where it overflows and is collected by a series of pipes [10,11]. The constant saturated conditions in this type of VF CW lead to increased hydraulic residence time and thus to enhanced pollutant removal rates [20].
- *Saturated vertical downflow CWs:* Similar to the VF CW described previously, this type was also designed to provide higher hydraulic residence times and thus improve pollutant removal performance [72]. The only difference between this and the previous type is that the wastewater is introduced at the CW surface and is collected at the bottom.
- *Integrated VF CWs:* This VF CW combines an upflow and a downflow saturated VF CW [13,40,50,82,83].

57.2.4 Hybrid Constructed Wetlands

Various types of CWs have been used recently in hybrid treatment systems in order to improve pollutant removal by exploiting all

their advantages [15,16,64,77,78]. Most commonly, hybrid CWs comprise one VF and one HSF CW [42–45,47,48,56,75,81].

57.3 Pollutant Removal Mechanisms

CWs have been proved to successfully remove various pollutants (i.e., organic matter, nitrogen, phosphorus, heavy metals, and pathogenic microorganisms) through a series of processes (physical, chemical, and biological) that occur in the complex environment formed in the CW [64]. Removal processes are affected by several operating parameters (e.g., HRT, temperature, porous media size and origin, vegetation etc.).

57.3.1 Organic Matter

OM is usually the most abundant pollutant in most wastewaters and occurs as both soluble and particulate fractions. These two fractions are removed by different processes; thus, particulate OM settles in or is filtered by the CW's porous media. Soluble OM is removed by aerobic and anaerobic degradation [24,78]. Usually, OM removal is quicker than the removal of other pollutants in CWs. Aerobic degradation of OM occurs at the surface of FWS CWs and near the roots in all CWs where dissolved oxygen concentrations are high. Anoxic or anaerobic conditions occur in the other areas of CWs. OM degradation occurs through the following reactions [31]:

Aerobically:

$$C_6H_{12}O_6 + 6O_2 \rightarrow 6CO_2 + 6H_2O \qquad (57.1)$$

Anaerobic fermentation:

$$C_6H_{12}O_6 \rightarrow 2CH_3CHOHCOOH \qquad (57.2)$$

Lactic acid:

$$C_6H_{12}O_6 \rightarrow 2CH_3CH_2OH + CO_2 \qquad (57.3)$$

Ethanol:
Methanogenesis:

$$4H_2 + CO_2 \rightarrow CH_4 + 2H_2O \qquad (57.4)$$

$$CH_3COO^- + 4H_2 \rightarrow 2CH_4 + H_2O + OH^- \qquad (57.5)$$

Anaerobically through sulfates reduction:

$$2CH_3CHOCOO^- + SO_4^{2-} + H^+ \rightarrow 2CH_3COO^- \\ + 2CO_2 + 2H_2O + HS^- \qquad (57.6)$$

$$CH_3COO^- + SO_4^{2-} + 2H^+ \rightarrow 2CO_2 + 2H_2O + HS^- \qquad (57.7)$$

Anaerobically through nitrate reduction:

$$C_6H_{12}O_6 + 4NO_3^- \rightarrow 6H_2O + 6CO_2 + 2N_2 + 4e^- \qquad (57.8)$$

Anaerobically through iron reduction:

$$CH_3COO^- + 8Fe^{+3} + 3H_2O \rightarrow 8Fe^{+2} + CO_2 \\ + HCO_3^- + H_2O + 8H^+ \qquad (57.9)$$

Biomass production:

$$COHNS + O + N,P \xrightarrow{\text{bacteria}} \\ CO_2 + NH_3 + C_5H_7NO_2P + \text{products} \qquad (57.10)$$

Biomass

57.3.2 Suspended Solids

Suspended solids (SS) are removed by settlement and filtration. Although SS settlement is achieved in CWs, it does lead to clogging of the porous media. Therefore, to reduce SS load, wastewater is usually subjected to primary settlement before being introduced into the CW.

57.3.3 Nitrogen

Nitrogen (N) is also a pollutant of great interest in wastewaters, as its high concentration in natural water bodies is associated with eutrophication phenomena. N in wastewaters can be found in either organic forms (urea, amino acids, uric acid, purine, pyrimidines) or inorganic forms (ammonia, nitrate, nitrite, and nitrous oxide) [21,24,31,32,52,76].

In CWs, N is mainly removed through the common pathway of ammonification-nitrification-denitrification. A significant portion of organic N is removed with total solids (TS) and the remaining part is transformed into ammonia by the ammonification process. Urea mineralization is described by the following reaction:

$$NH_2CONH_2 + H_2O \rightarrow 2NH_3 + CO_2 \qquad (57.11)$$

Ammonification is affected by a series of parameters (i.e., temperature, pH, C/N ratio, nutrient content, and soil

conditions) [64]. Ammonia is then transformed into nitrate and nitrite (nitrification). Nitrification requires aerobic conditions; however, it can be achieved even in very low DO concentrations (0.3 mg/L) [31]. Similar to ammonification, nitrification is also affected by a series of parameters (i.e., temperature, pH value, water alkalinity, inorganic carbon source, moisture, microbial population, ammonia N concentration, and DO) [15,37,64,76]:

$$NH_4^+ + \frac{3}{2}O_2 \rightarrow NO_2^- + H_2O + 2H^+ \qquad (57.12)$$

$$NO_2^- + \frac{1}{2}O_2 \rightarrow NO_3^- \qquad (57.13)$$

Ammonia is transformed into nitrite by specific microbes (nitrosomonas and nitrobacter), through the following reactions [31]:

$$55NH_4^+ + 76O_2 + 109HCO_3^- \rightarrow C_5H_7NO_2 \\ + 54NO_2^- + 57H_2O + 104H_2CO_3 \qquad (57.14)$$

$$400NO_2^- + NH_4^+ + 4H_2CO_3 + HCO_3^- + 195O_2 \rightarrow \\ C_5H_7NO_2 + 3H_2O + 400NO_3^- \qquad (57.15)$$

In CWs, denitrification takes place simultaneously with nitrification in areas where anoxic conditions prevail, through the following reaction [31]:

$$NO_3^- + 1.08CH_3OH + 0.24H_2CO_3 \rightarrow 0.056C_5H_7NO_2 \\ + 0.47N_2 + 1.68H_2O + HCO_3^- \qquad (57.16)$$

Nitrification is a temperature-dependent process as nitrobacter and nitrosomonas bacteria optimize their function in temperatures above 15°C [34]. Furthermore, in high temperatures, plant vegetation also grows rapidly, providing higher amounts of oxygen for nitrification [34,46,51,74]. In contrast, denitrification requires anoxic conditions and a carbon source, while it is not significantly affected by temperature. Thus, denitrification is limited in VF CWs as they operate under aerobic conditions.

Other N removal mechanisms in CWs include:

- *Plant uptake:* N is an essential element for plant growth; thus, plants tend to absorb N in its anionic forms (ammonia, nitrate, and nitrite). Although plants play an important role in CW operation, N uptake is limited (up to 4%–5% [64]) and depends on a series of parameters (e.g., plant species, plant density).
- *Adsorption:* N can also be removed in CWs by adsorption in the substrate material. The exact contribution of this mechanism to N removal depends on the ion exchange ability of the substrate material [64].
- *Ammonia volatilization:* This process occurs when ammonia is directly removed from the wastewater in gaseous form. In most cases, its contribution to the overall N removal is almost zero as the process occurs in environments with pH values above 9.3 [52,64,76].

- *Anaerobic ammonium oxidation (ANAMMOX):* It was recently discovered that under anaerobic conditions, ammonia can be directly transformed into gaseous N_2 [21] using nitrite and nitrate as electron acceptors [64].
- *Completely autotrophic nitrogen-removal over nitrite (CANON):* This process occurs under aerobic conditions when autotrophic bacteria directly convert ammonia to gaseous N_2 [68].

57.3.4 Phosphorus

Phosphorus (P), similar to N, is a major pollutant as it is also contributes to eutrophication. In wastewaters, P exists in two forms: orthophosphate (OP) and particulate P. In CWs, phosphorus is mainly removed through adsorption, precipitation, plant uptake, and microbial uptake [31]. Plant and microbial uptake only remove ortho-phosphate, while adsorption and sedimentation remove all forms of phosphorus [31]. Phosphorus adsorption is affected by substrate material as substrates containing iron (Fe), aluminum (Al), and calcium (Ca) usually increase P removal [1,4,12,19,31,53,61,62,65,73]. Although adsorption and precipitation are rather effective removal mechanisms during the first years of CW operation, their effectiveness decreases with time due to substrate saturation. Plant and microbial uptake of P is limited in CWs and does not significantly contribute to its removal [64]. It should be noted that P can be released in a CW by biomass decomposition and desorption [64]; therefore, it is essential to remove plant biomass annually.

57.3.5 Heavy Metals

Heavy metals are a major pollutant found mainly in industrial wastewaters and landfill leachates. Heavy metals are rather toxic as they can cause serious health damage in humans and other living organisms. Although heavy metals can be degraded, in CWs they are successfully removed mainly by plant uptake,

adsorption, and precipitation. In CWs, heavy metal adsorption and precipitation follows several pathways [64]:

- Heavy metals bind with organic substances (e.g., humic acids) forming complexes [84].
- Heavy metals precipitate when they form salts with carbonate, bicarbonate, and sulfide ions and hydroxides [24,31,80,84].
- Adsorption by the substrate surface. This process is affected by the substrates cation exchange capacity and wastewater pH [49,54,57,58,78].

The uptake of heavy metals by plant vegetation is also a major removal mechanism in CWs. Almost all plant species (e.g., *Phragmites australis, Typha latifolia, Glyceria fluitans, Juncus effusus, Scirpus* spp.) used in CWs can absorb heavy metal ions [36,41,57,66]. CW plant vegetation first absorbs heavy metals in its root zone and then transfers a small amount to the aerial parts (stem, leaves, etc.) [8,9,22,63,64].

Heavy metals can also be removed by biological processes, as the microorganisms present in a CW can: (1) remove heavy metals by biosorption; (2) use the sulfide reduction mechanism for heavy metal precipitation; (3) cause methylation of heavy metals; and (4) accumulate heavy metals in their cells by forming amorphous minerals [33,71].

57.4 Constructed Wetlands and Urban Water Reuse

CWs have been used to treat various types of wastewaters. Although the majority of publications refer to the extremely low effluent concentration, only a few refer to possible effluent reuse (Table 57.1). CW effluents are mainly used for irrigation purposes [5,26–28,30,59] and stream restoration [14,29], but also to cover household and potable water needs [25,38,85].

Avila et al. [5] treated municipal wastewater from a combined sewer using a hybrid CW system (comprising one VF, one

TABLE 57.1 Constructed Wetlands: Applications for Water Reuse

Reference	Wastewater Type	CW Type	Effluent Applications
Avila et al. [5]	Municipal	Hybrid	• Recharge of aquifers • Irrigation of green areas not accessible to the public
Collins et al. [14]	Stormwater		Stream restoration
Frazer-Williams et al. [23]	Greywater	HSF, VF	
Ghumni et al. [25]	Greywater		Bathing water
Gikas and Tsihrintzis [26]	Household	HSF	Garden irrigation
Gikas and Tsihrintzis [27]	Household	VF	Garden irrigation
Herrera-Melian et al. [28]	Municipal	Hybrid	Irrigation
Jenkis et al. [29]	Stormwater		Stream restoration
Jing et al. [30]	Municipal		Irrigation
Li et al. [39]	Greywater		
Li et al. [38]	Greywater		Household needs
Siracusa and La Rosa [59]	Greywater		Irrigation
Zhang and Tan [85]	Greywater		Household needs

HSF, and one FWS stage), with the aim of using the effluent for groundwater recharge and green area irrigation. This hybrid system proved rather effective as it achieved high removal efficiencies for all pollutants (BOD$_5$: 94%; COD: 85%; SS: 90%; TN: 86%; TP: 35%; *Escherichia coli*: 99%). All pollutant effluent concentrations were below Spanish regulation limits for aquifer recharge, and the irrigation of forests and other green areas not accessible to the public. Herrera-Melian et al. [28] used a hybrid CW system to treat municipal wastewater. This hybrid CW proved rather efficient in wastewater treatment (mean removal rates: 86% for BOD, 80% for COD, 88% for ammonia-N, 96% for SS and turbidity, 24% for phosphate-P, 99.5% for fecal coliforms, and 99.7% for fecal enterococci), while its effluent was reused for irrigation purposes.

Collins et al. [14] examined a variety of wastewater treatment methods to assess their ability in providing effluents that could be used for stream restoration. They set TN removal as the crucial factor and stated that the majority of CWs could provide efficient rates of nitrification and denitrification, in order to produce effluents with low N load. Thus, CW effluents are more suitable for stream restoration than other conventional technologies (e.g., dry ponds and wet ponds). Collins et al. [14] also suggest that CW technology could transform complete natural ecosystems as denitrification could be achieved in the natural wetlands of riparian zones. Nevertheless, attention should be paid to the design of these systems in order to achieve anoxic conditions for denitrification, drainage times are usually increased thus affecting the hydrological regime. Jenkins et al. [29] also examined the use of CWs for stormwater treatment and compared their use with rainwater harvesting. Their results indicate that although rainwater harvesting could reduce a household's potable water demand by up to 36%, CWs provided better hydrological conditions that improved the ecological condition of the urban stream system.

Ghumni et al. [25] report that CWs could be used as a treatment method for greywater before its reuse. CWs appear to achieve high efficiencies in removing organic matter, SS, N, and P, while they were insufficient at pathogen removal, as the effluents were above legislation limits. Nevertheless, when CWs are combined with suitable pretreatment (e.g., settling tank) and post-treatment stages (e.g., photo-oxidation using TiO$_2$ and UV), concentrations of all pollutants (including pathogenic) in the effluent are below European bathing water standards. Li et al. [39] also used CWs in combination with TiO$_2$-based photocatalytic oxidation for greywater treatment. Their results showed that greywater treatment by CWs (HSF and VF) and TiO$_2$-based photocatalytic oxidation could result in a final effluent that meets European bathing water quality standards. Frazer-Williams et al. [23] also used CWs to treat greywater. All CWs tested proved to successfully remove pollutants (removal efficiencies ranges: 87%–93% BOD$_5$, 70%–88% SS, and 2.1%–2.8% log reduction indicator microorganisms). Siracusa and La Rosa [59] designed a CW system to treat greywater and according to their design specification, the final effluent could be used to irrigate orchards, golf courses, decorative landscapes, wildlife habitats, and riparian trees. Furthermore, Li et al. [38] report that apart from their high

performance efficiency, CWs were considered the most environmentally friendly and cost-effective technology for greywater treatment compared with other treatment methods (i.e., rotating biological contactors, sequencing batch reactors, anaerobic sludge blankets, and membrane bioreactors). Zhang and Tan [85] report the use of CWs for greywater treatment in China, where the treated effluents were reused for household needs.

VF and HSF CWs were also used to treat household wastewater in rural areas and their effluents were used to irrigate domestic gardens [26,27]. Both systems were constructed to serve a single household (8 persons for the VF system and 4 persons for the HSF system). Both CWs systems achieved extremely high pollutant removal efficiencies (VF: 96.4% for BOD, 94.4% for COD, 90.8% for TKN, 92.8% for ammonia, 61.6% for OP and 69.8% for TP; HSF: 86.5% for BOD, 84.6% for COD, 83.7% for TKN, 82.2% for ammonia, 63.1% for OP, 63.3% for TP, 79.3% for TSS, and 99.9% for TC), while effluent pollutant concentrations were below the legislation limits for garden water use.

In Taiwan, CW effluent is reused for various applications [30]. CWs are widely used in Taiwan as a cost-effective wastewater treatment method that provides pollutant effluent concentrations below the legislation limits for different reuse purposes [30]:

- *Reuse in rural areas:* Due to the lack of centralized wastewater treatment plants (WWTPs) in Taiwan's rural areas, CWs are extensively used to treat domestic wastewater. Various types of CWs are used (FWS, HSF, and hybrid) resulting in high pollutant removal efficiencies. Pollutant concentrations in the CW effluents are below legislation limits and the waters are usually reused for irrigation purposes.
- *Wastewater treatment and reuse in campuses and hospitals:* Over 47 CWs with total surface areas of up to 500 m² each have been constructed in various schools, university campuses, and hospitals. The treated effluents are reused to irrigate green areas.
- *Wastewater treatment and teuse in ecological recreational areas and green architecture:* To improve life quality in Taiwanese cities, several urban reclamation programs were implemented. A significant number of these programs include the use of CWs in stormwater treatment and these produce effluents used for stream restoration.

57.5 Summary and Conclusions

Although CWs are well established as a wastewater treatment method not only for municipal wastewater but also for a variety of other wastewater types (industrial, agro-industrial, landfill leachate, etc.), data regarding their effluent reuse are rather limited. Nevertheless, CW effluents have been reported to cover from irrigation to household water requirements, thus proving the excellent performance of CWs in wastewater treatment. It should be noted that CWs have been extensively used for greywater treatment and reuse, providing a viable and cost-effective treatment method. Furthermore, another important

CW application is their use for stormwater treatment, as they produce effluents that can recharge streams. The two main advantages of CWs in water reuse are (1) their ability to treat a great variety of wastewater types, and (2) their ability to work together with other treatment methods to produce effluents that meet potable water standards.

References

1. Akratos, C.S. and Tsihrintzis, V.A. 2007. Effect of temperature, HRT, vegetation and porous media on removal efficiency of pilot-scale horizontal subsurface flow constructed wetlands. *Ecological Engineering*, 29, 173–191.

2. Akratos, C.S., Tsihrintzis, V.A., Pechlivanidis, I., Sylaios, G.K., and Jerrentrup, H. 2006. A free-water surface constructed wetland for the treatment of agricultural drainage entering Vassova Lagoon, Kavala, Greece. *Fresenius Environmental Bulletin*, 15, 1553–1562.

3. Arias, C.A., Brix, H., and Marti, E. 2005. Recycling of treated effluents enhances removal of total nitrogen in vertical flow constructed wetlands. *Journal of Environmental Science and Health*, 40, 1431–1443.

4. Arias, C.A., Del Bubba, M., and Brix, H. 2001. Phosphorus removal by sands for use as media in subsurface flow constructed reed beds. *Water Research*, 35, 1159–1168.

5. Ávila, C., José Salas, J., Martín I., Aragón, C., and García, J. 2013. Integrated treatment of combined sewer wastewater and stormwater in a hybrid constructed wetland system in southern Spain and its further reuse. *Ecological Engineering*, 50, 13–20.

6. Boutin, C. and Liénard, A. 2003. Constructed wetlands for wastewater treatment: The French experience. In: *1st International Seminar on the Use of Aquatic Macrophytes for Wastewater Treatment in Constructed Wetlands*, May 5–10, Lisbon, Portugal.

7. Boutin, C., Liénard, A., and Esser, D. 1997. Development of a new generation of reed-bed filters in France: First results. *Water Science and Technology*, 35, 315–322.

8. Bragato, C., Achiavon, M., Polese, R., Ertani, A., Pittarello, M., and Malagoli, M. 2009. Seasonal variations of Cu, Zn, Ni and Cr concentration in *Phragmites australis* (Cav.) Trin. ex Steudel in a constructed wetland of North Italy. *Desalination*, 247, 36–45.

9. Bragato, C., Brix, H., and Malagoli, M. 2006. Accumulation of nutrients and heavy metals in *Phragmites australis* (Cav.) *Trin. ex Steudel* and *Bolboschoenus maritimus* (L.) Palla in a constructed wetland of the Venice lagoon watershed. *Environmental Pollution*, 144, 967–975.

10. Breen, P.F. 1990. A mass balance method for assessing the potential of artificial wetlands for wastewater treatment. *Water Research*, 24, 689–697.

11. Breen, P.F. 1997. The performance of vertical flow experimental wetland under a range of operational formats and environmental conditions. *Water Science and Technology*, 35, 167–174.

12. Brix, H., Arias, C.A., and Del Bubba, M. 2001. Media selection for sustainable phosphorus removal in subsurface flow constructed wetlands. *Water Science and Technology*, 44, 47–54.

13. Chang, J-J., Wu, S-Q., Dai, Y-R., Liang, W., and Wu, Z-B. 2012. Treatment performance of integrated vertical-flow constructed wetland plots for domestic wastewater. *Ecological Engineering*, 44, 152–159.

14. Collins, K.A., Lawrence, T.J., Stander, E.K., Jontos, R.J., Kaushal, S.S., Newcomer T.A., Grimm N.B., and Ekberg M.L.C. 2010. Opportunities and challenges for managing nitrogen in urban stormwater: A review and synthesis. *Ecological Engineering*, 36, 1507–1510.

15. Cooper, P. 1999. A review of the design and performance of vertical-flow and hybrid reed bed treatment systems. *Water Science and Technology*, 40, 1–9.

16. Cooper, P., Griffin, P., Humphries, S., and Pound, A. 1999. Design of a hybrid reed bed system to achieve complete nitrification and denitrification of domestic sewage. *Water Science and Technology*, 40, 283–289.

17. Cooper, P.F., Job, G.D., Green, M.B., and Shutes, R.B.E. 1996. *Reed Beds and Constructed Wetlands for Wastewater Treatment*. Water Research Center Publications, Swindon, UK.

18. Crites, R.W., Middlebrooks, J., and Reed, S.C. 2006. *Natural Wastewater Treatment Systems*. CRC Press, Boca Raton, FL.

19. Drizo, A., Frost, A.C., Grace, J., and Smith, A.K. 1999. Physico-chemical screening of phosphate-removing substrates for use in constructed wetland systems. *Water Research*, 33, 3595–3602.

20. Farahbakshazad, N. and Morrison, G.M. 2003. Phosphorus removal in a vertical up flow constructed wetland system. *Water Science and Technology*, 48(5), 43–50.

21. Faulwetter, J.L., Gagnon, V., Sundberg, C., Chazarenc, F., Burr, M.D., Brisson, J., Camper, A.K., and Stein, O.R. 2009. Microbial processes influencing performance of treatment wetlands: A review. *Ecological Engineering*, 35, 987–1004.

22. Fediuc, E. and Erdei, L. 2002. Physiological and biochemical aspects of cadmium toxicity and protective mechanisms induced in *Phragmites australis* and *Typha latifolia*. *Journal of Plant Physiology*, 159, 265–271.

23. Frazer-Williams, R., Avery, L., Winward, G., Jeffrey, P., Shirley-Smith, C., Liu, S., Memon, F.A., and Jefferson, B. 2008. Constructed wetlands for urban grey water recycling. *International Journal of Environmental Pollution*, 33, 93–109.

24. García, J., Rousseau, D.P.L., Morató, J., Lesage, E., Matamoros, V., and Bayona, J.M. 2010. Contaminant removal processes in subsurface-flow constructed wetlands: A review. *Critical Reviews in Environmental Science and Technology*, 40, 561–661.

25. Ghunmi, L.A., Zeeman, G., Fayyad, M., and van Lier, J.B. 2011. Grey water treatment systems: A review. *Critical Reviews in Environmental Science and Technology*, 41, 657–698.

26. Gikas, G.D. and Tsihrintzis, V.A. 2010. On-site treatment of domestic wastewater using a small-scale horizontal

subsurface flow constructed wetland. *Water Science and Technology*, 62, 603–614.

27. Gikas, G.D. and Tsihrintzis, V.A. 2012. A small-size vertical flow constructed wetland for on-site treatment of household wastewater. *Ecological Engineering*, 44, 337–343.

28. Herrera Melián, J.A., Martín Rodríguez, A.J., Arana, J., González Díaz, O., and González Henríquez, J.J. 2010. Hybrid constructed wetlands for wastewater treatment and reuse in the Canary Islands. *Ecological Engineering*, 36, 891–899.

29. Jenkins, G.A., Greenway, M., and Polson, C. 2012. The impact of water reuse on the hydrology and ecology of a constructed stormwater wetland and its catchment. *Ecological Engineering*, 47, 308–315.

30. Jing, S.R., Lin, Y.F., Shih, K.C., and Lu, H.W. 2008. Applications of constructed wetlands for water pollution control in Taiwan: Review. *Practice Periodical of Hazardous, Toxic, and Radioactive Waste Management*, 12, 249–259.

31. Kadlec, R.H. and Knight, R.L. 1996. *Treatment Wetlands*. CRC Press/Lewis Publishers, Boca Raton, FL.

32. Kadlec, R.H. and Wallace, S.D. 2009. *Treatment Wetlands*, 2nd Edition, CRC Press, Boca Raton, FL.

33. Kosolapov, D.B., Kuschk, P., Vainshtein, M.B., Vatsourina, A.V., Wiebner, A., Kastner, M., and Muller, R.A. 2004. Microbial processes of heavy metal removal from carbon-deficient effluents in constructed wetlands. *Engineering in Life Science*, 4, 403–411.

34. Kuschk, P., Wiebner, A., Kappelmeyer, U., Weibbrodt, E., Kästner, M., and Stottmeister, U. 2003. Annual cycle of nitrogen removal by a pilot-scale subsurface horizontal flow in a constructed wetland under moderate climate. *Water Research*, 37, 4236–4242.

35. Laber, J., Perfler, R., and Haberl, R. 1997. Two strategies for advanced nitrogen elimination in vertical flow constructed wetlands. *Water Science and Technology*, 35, 71–77.

36. Lee, B-H., Scholz, M., and Horn, A. 2005. Constructed wetlands: Treatment of concentrated storm water runoff (Part A). *Environmental Engineering Science*, 23, 191–202.

37. Lee, C.-G., Fletcher, T.D., and Sun, G. 2009. Nitrogen removal in constructed wetlands. *Engineering in Life Sciences*, 9, 11–22.

38. Li, F., Wichmann, K., and Otterpohl, R. 2009. Review of the technological approaches for grey water treatment and reuses. *Science of the Total Environment*, 407, 3439–3449.

39. Li, Z., Gulyas, H., Jahn, M., Gajurel, D.R., and Otterpohl, R. 2004. Greywater treatment by constructed wetlands in combination with TiO_2-based photocatalytic oxidation for suburban and rural areas without sewer system. *Water Science and Technology*, 48, 101–106.

40. Liu, D., Ge, Y., Chang, J., Peng, C., Gu, B., Chan, G.Y.S., and Wu, X. 2008. Constructed wetlands in China: Recent developments and future challenges. *Frontiers in Ecology and the Environment*, 7, 261–268.

41. Marchand, L., Mench, M., Jacob, D.L., and Otte, M.L. 2010. Metal and metalloid removal in constructed wetlands, with emphasis on the importance of plants and standardized measurements: A review. *Environmental Pollution*, 158, 3447–3461.

42. Masi, F., Martinuzzi, N., Bresciani, R., Giovannelli, L., and Conte, G. 2007. Tolerance to hydraulic and organic load fluctuations in constructed wetlands. *Water Science and Technology*, 56, 39–48.

43. Melián, J.A.H., Rodríguez, A.J.M., Araña, J., Díaz, O.G., and Henríquez, J.J.B. 2010. Hybrid constructed wetlands for wastewater treatment and reuse in the Canary Islands. *Ecological Engineering*, 36, 891–899.

44. Molle P., Liénard A., Boutin C., Merlin, G., and Iwema, A. 2005. How to treat raw sewage with constructed wetlands: An overview of the French systems. *Water Science and Technology*, 51, 11–21.

45. Molle, P., Prost-Boucle, S., and Lienard, A. 2008. Potential for total nitrogen removal by combining vertical flow and horizontal flow constructed wetlands: A full-scale experimental study. *Ecological Engineering*, 34, 23–29.

46. Newman, J.M., Clausen, J.C., and Neafsey, J.A. 2000. Seasonal performance of a wetland constructed to process dairy milkhouse wastewater in Connecticut. *Ecological Engineering*, 14, 181–198.

47. O'Hogain, S. 2003. The design, operation and performance of a municipal hybrid reed bed treatment system. *Water Science and Technology*, 48, 119–126.

48. Öövel, M., Tooming, A., Mauring, T., and Mander, Ü. 2008. Schoolhouse wastewater purification in a LWA-filled hybrid constructed wetland in Estonia. *Ecological Engineering*, 29, 17–26.

49. Ouki, S.K. and Kavannagh, M. 1999. Treatment of metals-contaminated wastewaters by use of natural zeolites. *Water Science and Technology*, 39, 115–122.

50. Perfler, R., Laber, J., Langergraber, G., and Haberl, R. 1999. Constructed wetlands for rehabilitation and reuse of surface waters in tropical and subtropical areas—First results from small-scale plots using vertical flow beds. *Water Science and Technology*, 40, 155–162.

51. Reed, S., Middlebrooks, E., and Crites, R. 1995. *Natural Systems for Waste Management and Treatment*. McGraw Hill, New York.

52. Saeed, T. and Sun, G. 2012. A review on nitrogen and organics removal mechanisms in subsurface flow constructed wetlands: Dependency on environmental parameters, operating conditions and supporting media. *Journal of Environmental Management*, 112, 429–448.

53. Sakadevan, K. and Bavor, H.J. 1998. Phosphate adsorption characteristics of soils, slags and zeolite to be used as substrates in constructed wetland systems. *Water Research*, 32, 393–399.

54. Scholz, M. and Xu, J. 2002. Comparison of constructed reed beds with different filter media and macrophytes treating urban stream water contaminated with lead and copper. *Ecological Engineering*, 18, 385–390.

55. Seidel, K. 1966. Reinigung von Gewässern durch höhere Pflanzen. *Deutsche Naturwissenschaft*, 12, 297–298.

56. Serrano, L., de la Varga, D., and Ruiz, I. 2011. Winery wastewater treatment in a hybrid constructed wetland. *Ecological Engineering*, 37, 744–753.

57. Sheoran, A.S. and Sheoran, V. 2006. Heavy metal removal mechanisms of acid mine drainage in wetlands: A critical review. *Minerals Engineering*, 19, 105–116.

58. Sheta, A.S., Falatah, A.M., Al-Sewailem, M.S., Khaled, E.M., and Sallam, A.S.H. 2003. Sorption characteristics of zinc and iron by natural zeolite and bentonite. *Microporous and Mesoporous Materials*, 61, 127–136.

59. Siracusa, G. and La Rosa, A.D. 2004. A surface flow constructed wetland as a measure to maintain the natural habitat of a reserve and a way to reuse wastewaters. *5th International Conference on Environmental Problems in Coastal Regions Incorporating Oil Spill Studies*, Alicante, Spain 10, April 26–28, 2004, 175–182.

60. Sklarz, M.Y., Gross, A., Yakirevich, A. and Soares, M.I.M. 2009. A recirculating vertical flow constructed wetland for the treatment of domestic wastewater. *Desalination*, 246, 617–624.

61. Stefanakis, A.I. and Tsihrintzis, V.A. 2012. Effects of loading, resting period, temperature, porous media, vegetation and aeration on performance of pilot-scale vertical flow constructed wetlands. *Chemical Engineering Journal*, 181–182, 416–430.

62. Stefanakis, A.I. and Tsihrintzis, V.A. 2012. Use of zeolite and bauxite in filter media treating the effluent of vertical flow constructed wetlands. *Microporous and Mesoporous Materials*, 155, 106–116.

63. Stefanakis, A.I. and Tsihrintzis, V.A. 2012. Effect of various design and operation parameters on performance of pilot-scale sludge drying reed beds. *Ecological Engineering*, 38, 65–78.

64. Stefanakis, A.I., Akratos, C.S., and Tsihrintzis, V.A. 2014. *Vertical Flow Constructed Wetlands: Eco-Engineering Systems for Wastewater and Sludge Treatment*. Elsevier, Burlington.

65. Stefanakis, A.I., Akratos, C.S., Gikas, G.D., and Tsihrintzis, V.A. 2009. Effluent quality improvement of two pilot-scale, horizontal subsurface flow constructed wetlands using natural zeolite (clinoptilolite). *Microporous and Mesoporous Materials*, 124, 131–143.

66. Stottmeister, U., Buddhawong, S., Kuschk, P., Wiessner, A., and Mattusch, J. 2006. Constructed wetlands and their performance for treatment of water contaminated with Arsenic and heavy metals. In: Twardowska, I., Allen, H.E., Häggblom, M.M., and Stefaniak, S. (eds.), *Soil and Water Pollution Monitoring, Protection and Remediation*, Springer, Dordrecht, pp. 417–432.

67. Sun, G., Zhao, Q.Y., Allen, S.J., and Cooper, D. 2006. Generating "Tide" in pilot-scale constructed wetlands to enhance agricultural wastewater treatment. *Engineering in Life Sciences*, 6, 560–565.

68. Sun, G. and Austin, D. 2007. A mass balance study on nitrification and deammonification in vertical flow constructed wetlands treating landfill leachate. *Water Science and Technology*, 56, 117–123.

69. Sun, G., Gray, K.R., and Biddlestone, A.J. 1999. Treatment of agricultural wastewater in a pilot-scale tidal flow reed bed system. *Environmental Technology*, 20, 233–237.

70. Sun, G., Gray, K.R., Biddlestone, A.J., Allen, S.J., and Cooper, D.J. 2003. Effect of effluent recirculation on the performance of a reed bed system treating agricultural wastewater. *Process Biochemistry*, 39, 351–357.

71. Vainshtein, M.B., Suzia, N., Kudryashova, E., and Ariskina, E., 2002. New magnet-sensitive structures in bacterial and archaea cells. *Biology of the Cell*, 94, 29–35.

72. Visesmanee, V., Polprasert, C., and Parkpian, P. 2008. Long-term performance of subsurface-flow constructed wetlands treating Cd wastewater. *Journal of Environmental Science and Health Part A*, 43, 765–771.

73. Vohla, C., Kõiv, M., Bavor, H.J., Chazarenc, F., and Mander, Ü. 2011. Filter materials for phosphorus removal from wastewater in treatment wetlands—A review. *Ecological Engineering*, 37, 70–89.

74. Vymazal, J. 2002. The use of sub-surface constructed wetlands for wastewater treatment in the Czech Republic: 10 years experience. *Ecological Engineering*, 18, 633–646.

75. Vymazal, J. 2005. Horizontal sub-surface flow and hybrid constructed wetlands systems for wastewater treatment. *Ecological Engineering*, 25, 478–490.

76. Vymazal, J. 2007. Removal of nutrients in various types of constructed wetlands. *Science of the Total Environment*, 380, 48–65.

77. Vymazal, J. 2011. Constructed wetlands for wastewater treatment: Five decades of experience. *Environmental Science and Technology*, 45, 61–69.

78. Vymazal, J., Brix, H., Cooper, P.F., Green, M.B., and Haberl, R. 1998. *Constructed Wetlands for Wastewater Treatment in Europe*. Backhuys Publishers, Leiden, the Netherlands.

79. Vymazal, J., Greenway, M., Tonderski, K., Brix, H., and Mander, Ü. 2006. Constructed wetlands for wastewater treatment. In: Verhoeven, J.T.A., Beltman, B., Bobbink, R. and Whigham, D.F. (eds.), *Wetlands and Natural Resource Management. Ecological Studies*. Springer-Verlag, Berlin, Germany, Vol. 190, pp. 69–94.

80. Walker, D.J. and Hurl, S. 2002. The reduction of heavy metals in a storm water wetland. *Ecological Engineering*, 18, 407–414.

81. Wen, Y., Xu, C., Liu, G., Chen, Y., and Zhou, Q. 2012. Enhanced nitrogen removal reliability and efficiency in integrated constructed wetland microcosms using zeolite. *Frontiers of Environmental Science & Engineering*, 6, 140–147.

82. Wu, Z-B., Xu, G-L., Zhou, P-J., Zhang, B-Z., Cheng, S-P., Fu, G-P., and He, F. 2004. Removal effects of nitrogen in integrated vertical flow constructed wetland sewage treating system. *Journal of Agro-Environment Science*, 23, 757–760.

83. Xie, X-L., He, F., Xu, D., Dong, J-K., Cheng, S-P., and Wu, Z-B. 2012. Application of large-scale integrated vertical-flow constructed wetland in Beijing Olympic forest park: Design, operation and performance. *Water and Environment Journal*, 26, 100–107.

84. Yeh, T.Y. 2008. Removal of metals in constructed wetlands: A review. *Practice Periodical of Hazardous, Toxic, and Radioactive Waste Management*, 12, 96–101.

85. Zhang, D.Q. and Tan, S.K., 2010. Decentralized wastewater management and its application in an urban area of Beijing, China. *4th International Conference on Bioinformatics and Biomedical Engineering*, June 2010, Chengdu, China, pp. 1–4.

86. Zhao, Q.Y., Sun, G., Lafferty, C., and Allen, S.J. 2004. Optimising the performance of a lab-scale tidal flow reed bed system treating agricultural wastewater. *Water Science and Technology*, 50, 65–72.

XIII

Rainwater Harvesting in Cities

58

Rainwater Use in Urban Areas: Rainwater Quality, Harvesting, Toxicity, Treatment, and Reuse

Elif Burcu Bahadır
Namık Kemal University

Süreyya Meriç
Namık Kemal University

PREFACE

Precipitation is an efficient pathway for removing the gases and particles from the atmosphere. It also plays a significant role in controlling the concentration of these species. Also, rainwater may collect organic air pollutants, such as polycyclic aromatic hydrocarbons (PAHs), phthalate ester (PEs), pesticides, and polychlorinated biphenyls (PCBs), which could be present in air in consequence of different factors such as road traffic, domestic heating, industrial emissions, and agricultural use. The determination of these compounds in rainwater samples consequently may be interesting for a first screening of air quality in urban or industrial areas. Second, rainwater harvesting (RWH) for urban/industrial reuse after proper treatment methods has been a growing issue in many countries in the world as an alternative resource. This chapter is designed to revise quality, toxicity, harvesting, and treatment and reuse guidelines for rainwater by means of urban water reuse. Data showed that a majority of rainwater samples had a neutral or alkaline character. However, major ions and trace metals concentrations are variable. Metals and organic pollutants were detected in almost all samples. Toxicity of rainwater samples varied due to the location and courses of rainwater samples derived; thus, the content of the samples was rich with metals and various organic pollutants with a complex mixture character. Some countries have already developed RWH and reuse guidelines for mostly urban water reuse and greenland watering after a simple or advanced treatment technology depending on rainwater quality by means of chemical and microbiological pollutants.

58.1 Introduction

Among alternative water sources such as greywater, rainwater harvesting (RWH), and urban wastewater reuse, RWH is one of the best methods to solve water scarcity and to establish sustainable water cycles in urban developments [49,58,64,75,122].

However, several studies have reported that RWH may pose a public health risk because of the presence of chemical and microbial pathogens [2,4,64,100]. The microbiological pollutants primarily originate from birds, small mammals, and leaves from overhanging vegetation. Microbiological pollutants show site-specific nature in terms of load and constituents [102].

To determine rainwater quality, it is important to understand the primary sources where pollutants originate [28]. The roof type and enviromental conditions (i.e., the local climate and atmospheric pollution) affect the quality of rainwater. Rainwater quality is related to aquatic toxicity, which is influenced by various inorganics and organic pollutants present in harvested water. Thus, proper treatment methods are needed to provide safe quality before urban reuse.

This chapter is designed to revise quality, toxicity, harvesting and treatment methods, and reuse guidelines for rainwater by means of urban water reuse.

58.2 Rainwater Quality

The pollution level of a site affects the chemical composition of rainwater. Meteorological factors consist of two main processes. The first process is washout of the below cloud atmosphere during precipitation events by raindrops, which scavenge and dissolve particles and gases along their fall. These washout components are mainly of local/regional origin. The second process, called "rain out," is related to the condensation of water vapor on aerosol particles during formation of droplets and gases around the droplets by aqueous-phase reactions. The pollutants of rainwater can be classified in three groups as described in the following sections.

58.2.1 Inorganic Pollutants

58.2.1.1 pH and Minerals

Rainwater chemistry depends on the result of a complex interaction between cloud dynamics and the microphysical processes. Especially, the main sources of constituents strongly affect the composition of rainwater samples: marine, terrigenous, and anthropogenic (industry, traffic, heating, and agriculture) [22]. Rain droplets were seen in higher altitudes where the atmosphere is less loaded with compounds originating from ground-based emissions [15]. The acidity and chemical concentration in rain depend on source strength of the constituents of their physical properties into hydrogeological, chemical transformation during cloud formation, and below scavenging. Volcanic emissions caused low pH of rainwater due to HF and HCl acids [33].

As shown in Table 58.1, the mean pH of rainwater samples is acidic range. The natural acidity of rainwater is often taken to be 5.6, which is pure water in equilibrium with global atmospheric concentration of CO_2 (330 ppm). At a pH level of 6.0 or below, freshwater shrimp cannot survive. At a pH level of 5.5, bottom-dwelling bacterial decomposers begin to die, causing non-decomposed leaf litter and other organic debris to lie on the bottom and deprive plankton of food supply. At a pH level of 4.5 or below, all fish and most frogs and insects die. In addition, acid rain occurs when acidic pollutants, sulfur dioxide (SO_2), and nitrogen oxide (NO_x) precipitate in the form of rain. pH level of the precipitation is below 5.6, the average acidity of "pure" rain. Acid rain affects human life in a variety of ways. Moreover, acid rain damages buildings and historical monuments, and leads to

the release of harmful chemicals, such as aluminum, from rocks and soil into drinking water sources.

An important source of cations was atmospheric sea salt particles, which was directly related to wind speed [33]. Na^+ and Cl^- concentrations are high due to long-range transport and rose by a typhoon through the middle troposphere. Andre et al. reported foundries emission and nuclear power station near a study site as the source of Na^+ and Cl^- [9]. K^+ was originated from terrestrial potassium or anthropogenic aerosols [22,81] and soil, for example, at dry land area of Mirleft of Morocco [67]. High Ca^{2+} concentration in the rainwater samples is due to desert soil, which contains a large fraction of calcite (for instance, Saharan) [6,22,25,61,67]. Dust particles in the atmosphere caused high pH due to it being rich in calcium bicarbonate/carbonate, which is a major buffering agent for acidity generated by sulfuric and nitric acids. Al-Kashman reported that the lowest value of bicarbonate was measured during the coldest month of the rainy season, with a high intensity of precipitation and low dust in the atmosphere [6].

The main sources of NO_3^- in the rainwater samples are industrial activities, traffic emissions, agricultural activities [67,97], and further oxidation of nitrogen oxides to HNO_3^- in the atmosphere [9]. HNO_3^- is originated from especially terrestrial sources or desert dust [22,120]. High concentrations of NH_4^+ are due to high levels of fertilizer use in agricultural activities and human activities [6,22,47,66,107] and local anthropogenic emissions (biomass burning and livestock breeding) [22,47,66].

Table 58.1 reviews pH and minerals in different locations in the world. Some data regarding rainwater quality from Turkey are presented in Table 58.2.

58.2.1.2 Metals

The high values of trace metals (Fe, Al, Zn, Cu, Cd, Pb) in rainwater samples are due to long-range atmospheric transport of anthropogenic activities (industrial activity, coal combustion, and automobile exhaust) and soil dust [5,79]. These trace elements are responsible for the catalytic activity of rainwater [77]. Özsoy and Örnektekin [90] reported that Al concentrations in rainwater samples were high and it caused very high conductivity due to the high content of soluble solids, which is called "red rain." Lee et al. [65] harvested rainwater in Gangneung City, South Korea. They found that Al shows a notable increase to over 200 µg L^{-1} in the spring due to the intense periodic dust storms that can pass over the Gobi Desert in Northern China [65].

Table 58.3 presents a detailed survey of the presence of metals in rainwater collected from different locations as well as Turkey.

58.2.2 Organic Pollutants

Since many organic compounds are present in both the gas and aerosol phase, both the processes of gas and particle scavenging are important paths of atmospheric deposition of these compounds [72,73]. Ugur et al. [114] investigated the activity concentrations of 210Po and 210Pb in rainwater samples. The activity concentrations of 210Po and 210Pb in rainwater vary

TABLE 58.1 pH and Mineral Parameters Measured in Rainwater Samples in Different Locations in the World

Country	pH	EC (µS/cm)	Ca^{+2} (µeq/L)	Mg^{+2}	Na$^+$	K$^+$	NH$_4^+$	HCO$_3^-$	Cl$^-$	NO$_3^-$	SO$_4^{2-}$
Jordan[a]	6.91	95	165.32	93.12	130.56	85.21	75.36	133.65	142.36	67.31	112.36
Italy[b]	5.18	–	70	77	252	17	25	–	322	29	90
Galilee,[c] Israel	–	–	44.7	28	166	3.7	24.3	–	176.3	28	150.3
Tirupati,[d] India	–	–	150.66	55.51	33.08	33.89	20.37	–	33.91	40.84	127.96
Mexico[e]	5.08	–	34.6	3.69	4.52	2.04	95	–	9.29	42.86	76.67
Melle,[f] Belgium	5.19	–	26.87	9.25	36.98	1.96	65.65	–	33.37	31.26	47.32
Spain[g]	6.4	–	57.5	9.8	22.3	4	22.9	–	28.4	20.7	46.1
Tibetan, China[h]	6.59	19.7	65.58	7.43	15.44	14.49	18.13	72.34	19.17	10.37	15.5
Avignon, France[i]	5.2	36.4	117.8	16.7	49.8	13.3	29.2	41	60.2	45.7	77
Kefalonia Island, Greece[j]	8.31	103	189.62	12.34	260.87	61.38	0.56	–	197.18	113.55	83.33
Southern Taiwan[k]	5.32–5.74	–	53.4	32.6	97.1	10.9	50.2	119.6	63.1	15.7	40.5
Newark, NJ[l]	4.6	–	6	3.3	10.9	1.3	24.4	–	10.7	14.4	38.1
Hong Kong, China[m]	–	–	16.2	7	36.9	4.2	22	–	42.4	27.6	70
Shenzhen, China[n]	4.56	–	35.4	3.26	11.2	1.75	33.5	–	20.6	21.9	64.7
Seoul, Korea[o]	4.7	–	34.9	6.9	10.5	3.5	66.4	–	18.2	29.9	70.9
Tokyo, Japan[p]	4.52	–	24.9	11.5	37	2.9	40.4	–	55.2	30.5	50.2
Negev, Israel[q]	7.24	390	1215	188	406	19	47	569	558	103	573
Los Angles, CA[r]	–			–	–	–	22	–	51	54	22
Guangzhou, China[s]	4.49	–	103.6	17	55	32.9	70.6	–	86.8	53.4	163.3
Jinhua, China[t]	4.54	–	56.1	4.4	7.5	5.3	97	–	10.3	37	116.9
Beijing, China[u]	6.18	–	397	160	58.5	40.1	376	–	82.8	174	521
Nanjing, China[v]	5.15	–	295.4	31.7	23	12.1	193.2	–	142.6	39.6	241.8
Gangneung, South Korea[w]	5.3	30	1.6	0.22	1.1	2.1	0.02	–	3	2.2	2.4

[a] Al-Khashman [6].
[b] Le Bolloch and Guerzoni [62].
[c] Herut et al. [43].
[d] Mouli et al. [86].
[e] Baez et al. [12].
[f] Staelens et al. [103].
[g] Avila and Alarcon [10].
[h] Li et al. [69].
[i] Celle-Jeanton et al. [22].
[j] Sazakli et al. [97].
[k] Tsai et al. [109].
[l] Song and Gao [101].
[m] Wai et al. [117].
[n] Huang et al. [47].
[o] Lee et al. [63].
[p] Okuda et al. [89].
[q] Kidron and Starinsky [56].
[r] Kawamura et al. [54].
[s] Cao et al. [21].
[t] Zhang et al. [120].
[u] Hu et al. [46].
[v] Tu et al. [110].
[w] Lee et al. [65].

considerably due to seasonal transport of the radionuclides. The highest activity concentrations in the winter months are due to the wind direction [114].

Volcanic eruptions and natural fires are the major natural sources of PAHs. As anthropogenic sources, many processes can be considered: automobile exhaust emissions and tire degradation, industrial emissions from catalytic cracking, air-blowing of asphalt, coking coal, domestic heating, refuse incineration, power generation, and biomass burning [11,17,18].

Pesticides reach the atmosphere mainly through three pathways: volatilization during and after application, transport of soil particles and dust loaded with pesticides, and losses during pesticide manufacturing processes.

Another group of organic pollutants found in rainwater is phenols and nitrophenols. Leuenberger et al. [68] reported that dinitrophenols induce toxic effects such as inhibiting the growing of plants at 1 and 10 nM concentrations. They can be directly emitted through combustion process in vehicles, especially diesel

TABLE 58.2 pH and Mineral Parameters Measured in Rainwater Samples Collected in Different Areas in Turkey

Location	pH	EC	Ca^{+2}	Mg^{+2}	Na^+	K^+	NH_4^+	Cl^-	NO_3^-	SO_4^{2-}
Mugla[a]	6.9	–	174	–	17	3.5	30	–	23	124
Ankara[b]	6.1	–	210	–	21	19	12	–	62	150
İstanbul[c]	4.81	–	285	–	75.2	57.4	12.8	–	33.4	115.2
Antalya[d]	5.17	–	140	–	450	12.1	50	–	70	113
Rize[e]	6.9	–	–	–	–	–	–	–	71	33
Istanbul[f]	6.96	24.2	32.9	9.17	61.74	3.85	–	–	–	–
Ankara[g]	6.3	–	71.4	9.3	15.6	9.8	86.4	20.4	29.2	48
İzmir[h]	5.64	–	81 ± 86	101 ± 105	117 ± 123	17 ± 23	43 ± 52	117 ± 123	23 ± 32	66 ± 69
Ankara[i]	6.13	–	74	11	16	8.4	64	18	28	56

[a] Demirak et al. [25].
[b] Tuncel and Ungor [112].
[c] Basak and Alagha [14].
[d] Okay et al. [88].
[e] Balcı et al. [13].
[f] Uygur et al. [115].
[g] Topçu et al. [108].
[h] Al-Momani et al. [7].
[i] Tuncer et al. [113].

TABLE 58.3 Metals Measured in Different Locations in the World

Country	Fe (ppb)	Al (ppb)	Zn (ppb)	Pb (ppb)	Cu (ppb)	Cd (ppb)	Ni (ppb)	Cr (ppb)
Jordan[a]	430	324	210	66	73	52	3.5	3.1
Mexico[b]	–	50.7	–	2.48	–	0.41	3.37	0.52
Kefalonia Island, Greece[c]	11	–	10	<2.0	<2.5	0.05	<10	<1.3
Athens, Greece[d]	4.38 ± 2.54	5.87 ± 8.67	33.46 ± 40.81	0.88 ± 1.05	15.41 ± 14.51	0.2 ± 0.14	4.14 ± 3.97	1.29 ± 0.97
N. Jordan[e]	92 ± 104	382 ± 323	6.52 ± 7.84	2.57 ± 2.33	3.08 ± 1.61	0.42 ± 0.63	2.62 ± 2.87	0.77 ± 0.84
Southern Jordan[f]	21.5 ± 32.84	–	32 ± 32.61	51 ± 36.4	40 ± 26.99	42 ± 22.86	1.75 ± 1.36	–
Seoul, South Korea (wooden shingle roof)[g]	154	227	135	10	34	–	–	–
Seoul, South Korea (concrete roof)[g]	160	535	196	14	58	–	–	–
Seoul, South Korea (clay roof)[g]	155	243	313	11	37	–	–	–
Seoul, South Korea (concrete roof)[g]	302	622	428	12	59	–	–	–
Gangneung, South Korea[h]	–	100	60	20	35	ND	–	1
Jeju Island, Korea[i]	30	–	–	–	–	–	–	–
Australia[j]	2.66	–	98	73	2.45	0.01	ND	0.52
Monmouth Country, NJ[k]	3600	1500	66.8	14.6	16.6	–	–	15.2
Mersin, Turkey[l]	743.2 ± 115	484.5 ± 49.5	50.2 ± 6.06	11.36 ± 0.81	3.94 ± 0.27	0.81 ± 0.09	7.23 ± 0.51	5.72 ± 0.43
Ankara, Turkey[m]	750 ± 2370	980 ± 2900	0.03 ± 0.03	19.1 ± 37.6	6.1 ± 9.5	9.5 ± 12.0	3.37 ± 0.60	0.52 ± 0.03
Antalya, Turkey[n]	–	580 ± 758	137 ± 510	10 ± 14	5.9 ± 6.4	4.9 ± 6.4	24 ± 26	9.0 ± 11.6
İstanbul, Turkey[o]	2750	7660	–	1.47	1450	–	0.77	0.58

[a] Al-Khashman [6].
[b] Baez et al. [12].
[c] Sazakli et al. [97].
[d] Kanellopoulou [53].
[e] Al-Momami et al. [7].
[f] Al-Khashman [5].
[g] Lee et al. [64].
[h] Lee et al. [65].
[i] Moon et al. [84].
[j] Morrow et al. [85].
[k] Tuccillo [111].
[l] Özsoy and Örnektekin [90].
[m] Kaya and Tuncel [55].
[n] Al-Momani et al. [7].
[o] Uygur et al. [115].

engines, plastic and chemical industries, or used as reagent production of dyes, drugs, fungicides, and pesticides. Furthermore, phenols and nitrophenols can be directly emitted in atmosphere during photochemical reactions of benzene, toluene, and phenolic compounds with OH and NO_2 radicals [68].

Seyfioğlu et al. collected 27 rain samples; formaldehyde ranged between 10 and 304 µg/L. The significant correlation between HCHO and temperature suggested that measured concentrations were affected by atmospheric photochemical reactions [98].

Organic pollutants measured in rainwater samples collected in different locations in the world are summarized in Table 58.4.

58.2.3 Microbial Pollution

Pathogens are found primarily in the feces of birds and mammals that have access to the rooftop. Most guidelines for rainwater utilization suggest that bacterial pathogens or indicators such as total coliforms and *Escherichia coli* (*E. coli*) are not detectable at counts <1 CFU/100 mL [2,118]. Literature studies have suggested that harvested rainwater used for drinking water should be assessed by monitoring the presence of fecal indicators and bacterial pathogens [2,3,64]. In the samples from the first flush tanks, Lee et al. [64] detected average total coliform counts of 131 CFU/100 mL for the wooden shingle, 197 CFU/100 mL

TABLE 58.4 Organic Pollutants Measured in Rainwater Samples Collected from Different Locations in the World

Organic Pollutants	Location/Time	Organic Pollutants	References
Polycyclic aromatic hydrocarbons (PAH)	Lake Balaton, Tihany, 1995	Naphthalene, fluorene, phenonthrene, anthracene, fluoranthene, pyrene, benz(a)anthracene, chrysene, benzo(b)fluoranthene, benzo (k)fluoranthene, benzo (a)pyrene, dibenz(a,h)antracene, benzo(g,h,i)perylene, indeno(1,2,3-cd) pyrene	Kiss et al. [59]
	Tier, Germany, 1999–2000	Acenaphthylene, acenaphene, fluorene, phenonthrene, anthracene, fluoranthene, pyrene, cyclopenta(c,d)pyrene, benzo(a)antrrhacene, chrysene, benzo(b,k)fluoranthene, benzo(e)pyrene, benzo(a)pyren, indeno(c,d-123) pyrene, dibenzo(a,h)anthracene, benzo(g,h,i)perylene, coronene	Rossi et al. [95]
	Rieti, Italy,1997	Naphtalene, acenaphtlyene, acenaphthlene, fluorene, phenonthrene, anthracene, fluoranthene, pyrene, benzo(a)anthracene, chrysene, benzo(b)fluoranthene, benzo(c)fluoranthene, benzo(a)pyrene, dibenzo(a,h)antracene, benzo(g,h,i) perylene, indeno(1,2,3-cd)pyrene	Guidotti et al. [39]
	BUTAL Bursa, Turkey	Phenanthrene (Phe), anthracene (Ant), fluoranthene (Flt),pyrene (Pyr), benz[a] anthracene (BaA), chrysene (Chr),benzo[b]fluoranthene (BbF), benzo[k] fluoranthene (BkF), enzo[a]pyrene (BaP), indeno[1,2,3-c,d]pyrene (IcdP), dibenz[a,h]anthracene (DahA), and benzo[g,h,i]perylene (BghiP)	Birgül et al. [18]
	Ankara, Turkey	Phenanthrene (Phe), anthracene (Ant), fluoranthene (Flt), pyrene (Pyr), benz[a]anthracene (BaA), chrysene (Chr), benzo[b]fluoranthene (BbF), enzo[a]pyrene (BaP), indeno[1,2.3-c,d]pyrene (IcdP), dibenz[a,h]anthracene (DahA), and benzo[g,h,i]perylene (BghiP)	Gaga et al. [34]
	Albertslund, Denmark	Fluoranthene	Birch et al. [17]
Pesticides	Tier, Germany, 1999–2000	Desisopropyl-atrazin, desthylatrazine, desethyl-terbuthylazine, simazin, atrazine, terbuthylozine, vinclozoline, methyl-paration, dichlofluanid, diethofencaeb, ethyl-paration, triadimefon, penconazole, pyrifenox, procymidone, tebuconazole	Rossi et al. [95]
	Beijing, China, 2006–2007	Dichlorodiphenyltrichloroethanes (DDTs)	Zhang et al. [121]
	Rieti, Italy, 1997	Trifluralin,α-HCH, hexachlorobenzene, propazine, fonotos, diazinon, heptachlor, alachor, terbutryn, pirimiphos-methyl, malathion, metolachor, paration, pendimetalin, heptachloroepoxid, α-endosulfan, ppDDE, ethion, metoxichlor, azinphos-methyl	Guidotti et al. [39]
	Kosetice, Czech Republic, 1997–1999 and 2004–2006	Hexachlorocyclohexanes (HCHs), hexachlorobenzen (HCB), DDT	Dvorska et al. [27]
	Pearl River Delta, China	Dichlorodiphenyltrichloroethane (DDT) anditsmetabolites	Yue et al. [119]
	Braunschweig, Lower Saxony, Germany	Fungicideazoxystrobin, insecticideparathion-ethyl, herbicides	Berenzen et al. [16]
Polychlorinated biphenyls (PCBs)	Kosetice, Czech Republic, 1997–1999 and 2004–2006	PCB52, PCb153	Dvorska et al. [27]
	Rieti, Italy, 1997	2-Monochlorobiphenyl, 2,4′-dichlorobiphenyl, 2,2′5-trichlorobiphenyl, 2,4,4′-trichlorobiphenyl, 2,4,5-trichlorbiphenyl	Guidotti et al. [39]
Phthalates	Rieti, Italy, 1997	Diethylphtalateidiisopropilphtalate, dipropylphtalate, diisobutylphtalate, dibutylphtalate, benzyl-butylphthalate, di-2-ethylhexylphtalate, dioctylphalate	Guidotti et al. [39]

for the concrete tile, 76 CFU/100 mL for the clay tile, and 70 CFU/100 mL for the galvanized steel roofs. The average *E. coli* counts in the first flush tank were 14 CFU/100 mL for the wooden shingle roof, 18 CFU/100 mL for the concrete tile roof, 8 CFU/100 mL for the clay tile roof, and 4 CFU/100 mL for the galvanized steel roof. Of the first flush tank samples, 88% of those from the wooden shingle roof, 77% of those from the concrete tile roof, 92% of those from the clay tile roof, and 92% of those from the galvanized steel roof (none shown) had no measurable levels of enterococci. *Cryptosporidium* spp. was not detected in any of the water samples. The percentages of samples that tested positive for *Pseudomonas* spp. in the first flush tank were 12.5% for the wooden shingle roof, 7.5% for the concrete tile roof, 2% for the clay tile roof, and 0% for the galvanized steel roof. The percentages for *Salmonella* spp. were 5% for the wooden shingle roof and 5% for the concrete tile roof. No *Salmonella* spp. was found in the clay tile or galvanized steel samples. The absence of bacterial pathogens in the water samples from the galvanized steel roof may be due to the high temperatures of these roofs, which reach 75–85°C in the summertime. In addition, galvanized steel concentrates ultraviolet sunlight, which acts as a disinfectant against bacterial pathogens. The highest number of bacterial pathogens was found in the wooden shingle roof samples, mostly likely because of the greater presence and growth of lichens, mosses, and plants on this roofing material [64].

58.2.4 Rainwater Toxicity

Toxic effects of highway runoff (water and sediment) were estimated by numerous researchers on samples collected both at the edge of pavement or immediately downstream of drainage outfalls. Marsalek et al. studied toxicity of combined sewer overflows and stormwater in 15 sites in southern Ontario using a battery of seven bioassays. The highest frequencies of severe and moderate toxicity (19% and 24%, respectively) were found at high runoff sites. Stormwater ponds contributed to toxicity reduction, with respect to both water and sediment downstream ponds [78].

Rouvalis et al. [96] reported that, among 36 rainwater samples collected in Achaia Prefecture, Greece, 52% and 46.7% were found toxic to *Daphnia pulex* organisms in the rural and urban area samples, respectively. Moderate toxicity was correlated to the presence of the insecticide only in the rural areas [99].

Campus parking lot stormwater samples were collected in Clemson University, Georgia. CPLSW was characterized (metals, oil, grease, and general chemistry). Fish (*Primephales promelas*) were more sensitive to the samples than *Ceriodaphnia dubia* with decreased survival in 92% and 15% of the samples, respectively [38,80].

Rosenkrantz et al. [94] used *Hydra hexactinella* to assess the toxicity of stormwater and sediment samples from three retarding basins in Melbourne, Australia. Stormwater from the Avoca St retarding basins resulted in an LC_{50} of 613 mL/L, NOEC and LOEC values of 50 mL/L and 100 mL/L, while the 7-h pulse exposure caused a significant increase in the mean population growth rate compared to the control. Water samples from the

two other retarding basins were found nontoxic to *H. hexactinella* [94].

Hamers et al. collected rainwater samples at 14-day periods in both open and wet only samplers. The esterase inhibiting potency of *Vibrio fisheri* bacteria was found in a sample that collected in an area with intense horticultural activities in June and included high concentrations of dichlorvos, mevinphos, primiphos-methyl, and methiocarb. Maximum individual concentrations of dichlorvos and primiphosmethyl even exceeded the EC_{50} for *Daphnia*, suggesting that pesticides in rainwater pose a risk for aquatic organism [41].

The effects of stormwater runoff sampled during two rain events were determined by exposing *Scenedesmus subspicatus* cells [51]. The runoff from the first rain event had no negative effects to *S. subspicatus*, posing in most cases growth stimulation, whereas the runoff from the second rain event inhibited algae growth.

58.3 Rainwater Harvesting (RWH) and Guidelines

The basic principles of RWH in developing countries correspond in general to the state-of-art that was defined in DIN 1989-1 [26]. Every RWH system consists of waterproof catchment surfaces for collecting the rainwater (e.g., roof or ground surfaces), delivery systems for transporting rainwater from the catchment to appropriate storage tanks (e.g., gutters or surface drains), and the storage tank. Gutters and downpipes are usually made out of plastic or metal, as these materials are most durable [104].

The main design parameters (DPs) of an RWH system are rainfall, catchment area, collection efficiency, tank volume, and water demand [1]. Its operational parameters include rainwater use efficiency (RUE), water saving efficiency (WSE), and cycle number (CN). The sensitivity analysis of a rooftop RWH system DPs to its OPs reveals the ratio of tank volume to catchment area (V/A). The appropriate design value of V/A is varied with D/A. The extra tank volume up to V/A of 0.15–0.2 is also available, if necessary to secure more water [87].

Kahinda and Taigbenu [52] classified RWH according to the type of catchment surface used by: (1) *in situ* RWH (iRWH), where the system uses part of the target area as the catchment area; (2) *ex situ* RWH (xRWH), where the system uses on uncultivated area as its catchment area; and (3) domestic RWH (dRWH), where the system collects water from rooftops, courtyards, and compacted or treated surfaces, and stored it in RWH tanks for domestic uses.

In Sweden, a significant measure of potable water can be saved if rainwater tanks are included as part of a dual water supply solution [116].

In Brazil, potential for using rainwater for saving potable water in residential sectors situated in varied geographic regions ranged from 48% to 100% [37]. A detailed analysis in southeastern Brazil revealed that the potential ranged from 12% to 79% [36].

In Japan, due to the shortage of freshwater, the rainwater utilization engineering was promoted initially in 1963. For example, in Sumida-ku, located in central Tokyo, the technology of roof rainwater collection is promoted and recycled vigorously [74].

Liang and Van Dijk reported economic and financial analysis of RWH in Beijing, China. The results show that the small, medium, and large sizes of RWH systems are all economically feasible. However, the financial feasibility of RWH systems depends on the charge for groundwater and on the size of the RWH systems [70].

Malta Environment and Planning Authority (MEPA) regulates the creation of water collecting surfaces on roofs, calculating the possible size and capacity of tanks according to precise coefficients proportional to the surface of the roof [93].

Zhang et al. investigated rainwater use potential in Sydney, Perth, Darwin, and Melbourne. The economic analysis suggests Sydney was the most appropriate for RWH compared to the other cities [122].

Table 58.5 summarizes RWH applications, while Table 58.6 shows a brief survey of RWH guidelines in different countries.

TABLE 58.5 Rainwater Harvesting (RWH) Systems in Different Locations in the World

References	Location (Average Rain, mm/yr)	Note
Farreny et al. [31]	City located 30 km North Barcelona, Catalonia, Spain (650 mm)	The research is based on a neighborhood of dense social housing (600 inhabitants/ha) with multistory buildings. Four strategies and two scenarios of water prices have been considered. In order to evaluate the cost efficiency of these strategies, the rainwater conveyance, storage, and distribution systems have been designed. The results indicate that RWH strategies in dense urban areas under Mediterranean conditions are economically feasible. However, it is necessary to chose the appropriate scale for rainwater infrastructures in order to make them economically feasible.
Sazakli et al. [97]	The northern part of Kefalonia Island, Greece (198 mm)	The chemical and microbial quality of rainwater samples examined in a 3-year study indicates that Kefalonia Island is unpolluted. In 12 seasonal samplings, 156 rainwater and 144 ground or mixed water samples were collected from ferroconcrete storage tanks (300–1000 m^3 capacity), which are adjacent to cement-paved catchment areas (600–3000 m^2).
Eroksuz and Rahman [29]	In multi-unit buildings in three cities of Australia: Sydney, Newcastle, and Wollongang (1204 mm)	A larger tank size is more appropriate to maximize water savings. The rainwater tank of appropriate size in a multi-unit building can provide significant water in dry years.
Cheng and Liao [23]	Northern Taiwan (2500 mm)	A rainfall station system based on a point concept to one based on spatial concept in order to cope with the problems of rainfall data were converted. A two-step cluster analysis was used to classify the sample areas into several regions. The acquired rainfall level classification represents the homogeneity of rainfall intensity and duration because of the minimum combined difference within a cluster. This rainfall zoning system would contribute to the standardized regional precipitation for the rainwater harvesting application.
Zhang et al. [122]	The Australian cities of Melbourne, Sydney, Perth, and Darwin (800; 1300; 900; 1600)	Small tank sizes were more suited to the city of Melbourne, while medium ones were more likely suitable for Sydney, and large ones for Darwin and Perth.
Sturm et al. [104]	The village of Epyeshona, Northern Namibia	Two scale RWH systems are examined; roof catchments using corrugated iron roofs as rain collection area and ground catchment using treated ground surfaces. The feasibility of the RWH systems was assessed in relation to local socioeconomic conditions. Decentralized techniques of RWH are economically feasible to apply in terms of the roof catchment systems. The ground catchment system, however, needs moderate subsidies to obtain the same benchmark.
Rahman et al. [92]	10 different locations in Greater Sydney, Australia (1241 mm)	A water balance simulation model on daily time scale is developed and water savings, reliability, and financial viability are examined for three different tank sizes—2, 3, and 5 kL. It is found that the average annual water savings from rainwater tanks are strongly correlated with average annual rainfall. A 5-kL tank is preferable to 2- and 3-kL tanks and rainwater tanks should be connected to toilet, laundry, and outdoor irrigation to achieve the best financial outcome for the homeowners.
Jones and Hunt (2010) [50]	Craven Country, the city of Raleigh, the city of Kinston in North Carolina, USA	In Raleigh, a 5300-L cistern collected water from 2400 m2 roof top and the water was used to flush a toilet at a nature education center. In Craven Country a 167-m^2 section of roof top contributed water to an 11,350-L cistern and was used to irrigate gardens around the facility A 19,680 L cistern in Kinston, collected from a 406 m2 section of rooftop, with the collected water used to wash vehicles A simulation tool was developed to model rainwater harvesting system performance based on historic rainfall data and anticipated usage. Calculations were performed on hourly or daily time step depending on the interval of the uploaded rainfall data.

(Continued)

TABLE 58.5 (*Continued*)   Rainwater Harvesting (RWH) Systems in Different Locations in the World

References	Location (Average Rain, mm/yr)	Note
Herrmann and Schmida (1999) [42]	Bochum, Germany (787 mm)	The water balance of a one-family house and multi-story building was calculated.
Liaw and Tsai (2004) [71]	3 cities in Taiwan (range from 1755 to 3350 mm)	Determination of the optimum storage volume of rainwater tanks considering economic aspects.
Tam et al. (2010) [105]	7 cities in Australia (range from 520 to 1597)	Costs of RWH compared to other water supply alternatives.
Ghisi and Mengotti de Oliveira (2007) [37]	Florianopolis, Brazil (1706)	Combination of greywater and RWH systems
Mitchell et al. (2005) [83]	Melbourne, Australia (800)	Role of stormwater as a substitute for potable water
Rahman et al. (2010) [91]	Sydney, Australia (1200)	Determination of the most sustainable RWH scenario for multi-story buildings (LCC).

TABLE 58.6   Guidelines

India

Himachal Pradesh	All commercial and institutional buildings, tourist and industrial complexes, hotels, and so on, existing or coming up and having a plinth area of more than 1000 square meters will have rainwater storage facilities commensurate with the size of roof area. *No objection certificates,* required under different statutes, will be issued to the owners of the buildings, unless they produce satisfactory proof of compliance of the new law. Toilet flush systems will have to be connected with the rainwater storage tank. It has been recommended that the buildings will have rainwater storage facility commensurate with the size of the roof in the open and set back area of the plot at the rate of 0.24 cft./m² of the roof area.
Ahmedabad	In 2002, the Ahmedabad Urban Development Authority (AUDA) made rainwater harvesting mandatory for all buildings covering an area of over 1500 m². According to the rule, for a cover area of over 1500 m², one percolation well is mandatory to ensure groundwater recharge. For every additional 4000 m² cover area, another well needs to be built.
Bangalore	In order to conserve water and ensure groundwater recharge, the Karnataka government in February 2009 announced that buildings constructed in the city would have to compulsorily adopt rainwater harvesting facilities. Residential sites, which exceed an area of 2400 ft² (40 × 60 ft), shall create rain harvesting facilities according to the new law.
Port Blair	In 2007, Port Blair Municipal Council (PBMC) directed all persons related to construction work to provide a proper spout or tank for the collection of rainwater to be utilized for various domestic purposes other than drinking. As per the existing building by-laws 1999, the slab or roof of the building would have to be provided with a proper spout or gutter for collection of rainwater, which would be beneficial for the residents of the municipal area during water crisis. The PBMC had advised all the owners of buildings in the municipal area to comply with the provisions within four months, failing which action would be taken against them by the Council.
Chennai	Rainwater harvesting has been made mandatory in three-storied buildings (irrespective of the size of the rooftop area). All new water and sewer connections are provided only after the installation of rainwater harvesting systems.
Kerala	The Kerala Municipality Building Rules, 1999, was amended by a notification dated January 12, 2004 issued by the Government of Kerala to include rainwater harvesting structures in new construction.
New Delhi	Since June 2001, the Ministry of Urban Affairs and Poverty Alleviation has made rainwater harvesting mandatory in all new buildings with a roof area of more than 100 m² and in all plots with an area of more than 1000 m² that are being developed. The Central Ground Water Authority (CGWA) has made rainwater harvesting mandatory in all institutions and residential colonies in notified areas (south and southwest Delhi and adjoining areas like Faridabad, Gurgaon, and Ghaziabad). This is also applicable to all the buildings in notified areas that have tubewells. The deadline for this was for March 31, 2002.
Indore (Madhya Pradesh)	Rainwater harvesting has been made mandatory in all new buildings with an area of 250 m² or more. A rebate of 6% on property tax has been offered as an incentive for implementing rainwater harvesting systems.
Kanpur (Uttar Pradesh)	Rainwater harvesting has been made mandatory in all new buildings with an area of 1000 m² or more.
Hyderabad (Andhra Pradesh)	Rainwater harvesting has been made mandatory in all new buildings with an area of 300 m² or more. Tentative date for enforcing this deadline was June 2001.
Tamil Nadu	Through an ordinance titled Tamilnadu Muncipal Laws ordinance, 2003, dated July 19, 2003, the government of Tamil Nadu has made rainwater harvesting mandatory for all buildings, both public and private, in the state. The deadline to construct rainwater harvesting structures is August 31, 2003. The ordinance cautions, "Where the rain water harvesting structure is not provided as required, the Commissioner or any person authorised by him in this behalf may, after giving notice to the owner or occupier of the building, cause rain water harvesting structure to be provided in such building and recover the cost of such provision along with the incidental expense thereof in the same manner as property tax." It also warns the citizens on disconnection of water supply connection provided rainwater harvesting structures are not provided.

(Continued)

TABLE 58.6 (*Continued*) Guidelines

Haryana	Haryana Urban Development Authority (HUDA) has made rainwater harvesting mandatory in all new buildings irrespective of roof area. In the notified areas in Gurgaon town and the adjoining industrial areas, all the institutions and residential colonies have been asked to adopt water harvesting by the CGWA. This is also applicable to all the buildings in notified areas having a tubewell. The deadline was for March 31, 2002. The CGWA has also banned drilling of tubewells in notified areas.
Rajasthan	The state government has made rainwater harvesting mandatory for all public establishments and all properties in plots covering more than 500 m² in urban areas.
Mumbai	The state government has made rainwater harvesting mandatory for all buildings that are being constructed on plots that are more than 1000 m² in size. The deadline set for this was October 2002.
Gujarat	The state roads and buildings department has made rainwater harvesting mandatory for all government buildings.
Australia	State governments in the following states in Australia have taken active steps to ensure that the newly constructed houses are designed and built with the latest energy and water efficient designs and products. This initiative is supported by legislation.
Victoria	Since July 2005, new houses and apartments in Victoria must be built to meet the energy efficiency and water management requirements of the 5 Star standard. The 5 Star standard requires: 5 Star energy efficiency rating for the building fabric; water efficient taps and fittings; plus either a rainwater tank for toilet flushing or a solar hot water system.
South Australia	In South Australia, new homes will be required to have a rainwater tank plumbed into the house.
Sydney and New South Wales	In Sydney and New South Wales, the BASIX (Building And Sustainability Index) building regulations call for a 40% reduction in mains water usage. A typical single dwelling design will meet the BASIX target for water conservation if it includes 1. Showerheads, tap fittings, and toilets with at least a 3A rating 2. A rainwater tank or alternative water supply for outdoor water use and toilet flushing and/or laundry
Gold Coast	Construction of 3000-L (800-gallon) rainwater tanks has been made mandatory in the Pimpama Coomera Master Plan area of Gold Coast. This is for all homes and business centers connected to the Class A+ recycled water system (those approved for development after August 29, 2005). The tank should be plumbed to their cold-water washing machine and outdoors faucets.
Queensland	A rebate of up to $1500 for the purchase and installation of home rainwater storages has been offered by the State of Queens in Australia.
Germany	Rain taxes are collected for impervious surface cover on a property that generates runoff directed to the local storm sewer. Therefore, the more rainwater caught and conserved, the less runoff is added to the storm drains. Less runoff allows smaller storm sewers, which, in turn, saves construction and maintenance costs at the site. Thus, people get rain tax reductions by converting their impervious pavement/roof into a porous one
United States	
State of Arizona	The government had announced a one-time tax credit of 25% of the cost of water conservation systems (the maximum limit is $1000) for its residents. The water conservation system is defined as any system, which can harvest residential greywater and/or rainwater. Even the builders are eligible to get the tax credit up to $200 per residence unit constructed with a water conservation system. Any citizen in this state who has purchased a water harvesting system on or after January 1st, 2008, can apply for the Arizona tax credit. There is roughly $250,000 per year allocated for these tax credits.
New Mexico	a. *Santa Fe County*—Residences with 2500 sq ft or more area must install an active rainwater catchment system comprised of cisterns. All commercial developments are required to collect all roof drainage into cisterns to be reused for landscape irrigation. b. *Albuquerque and Bernalillo County*—Residences with 2500 sq ft or more area must install an active rainwater catchment system comprised of cisterns. All commercial developments are required to collect all roof drainage into cisterns to be reused for landscape irrigation.
State of Texas	The 77th Texas Legislature passed in 2001 amended Section 11.32 of the Texas Tax Code to allow taxing units of government the option to exempt from taxation all or a part of the assessed value of the property on which water conservation modifications have been made. *City of Austin*—The residents of the city of Austin can buy rain barrels at subsidized rates and they can avail rebates for the installation of approved cistern systems. Commercial/industrial properties can avail rebates up to $40,000 for the installation of rainwater harvesting and greywater systems. *City of San Antonio*—The citizens are eligible up to 50% rebate for rainwater harvesting projects under San Antonio Water System (SAWS). SAWS will give up to 50% rebate on the cost of new water-saving equipment, including rainwater harvesting systems, to its commercial, industrial, and institutional customers. Rebates are calculated by multiplying acre-feet of water conserved by a set value of $200/acre-foot.

58.4 Rainwater Treatment Systems

RWH applications would generally need proper treatment methods before reuse. Figure 58.1 shows simple rainwater treatment and storage methods.

Mendez et al. examined a pilot-scale and a full-scale roof for rainwater quality and the results demonstrated that rainwater harvested from any of these roofing materials would require treatment if the consumer wanted to meet USEPA primary and secondary drinking water standards or nonpotable water reuse guidelines; at a minimum, first-flush diversion, filtration, and disinfection are recommended. Rainwater harvested from metal roofs tends to have lower concentrations of fecal indicator bacteria as compared to other roofing materials. Although the shingle and green roofs produced water quality comparable in many respects to that from the other roofing materials, their dissolved organic carbon concentrations were very high, which might lead to high concentrations of disinfection byproducts after chlorination [82].

When rainwater needs further purification due to various pollutants and microbial contamination, several treatment approaches are attempted. Table 58.7 illustrates some process modifications proposed in different areas. These alternatives are mostly considered the reuse quality and local abilities to afford treatment costs.

58.5 Rainwater Reuse Systems

The European Union puts priority on water savings, including rainwater harvesting and reuse in buildings [30].

The harvested rainwater is mainly used for nonpotable purposes such as toilet flushing, laundry, car washing, and garden watering [1,116]. Furthermore, the harvested rainwater is also drinkable if specific treatments are taken [8]. The supplementary water can also be used in ecological pools [50,52,57,99]. Figure 58.2 shows individual household RWH systems and reuse options for garden watering and inhouse use.

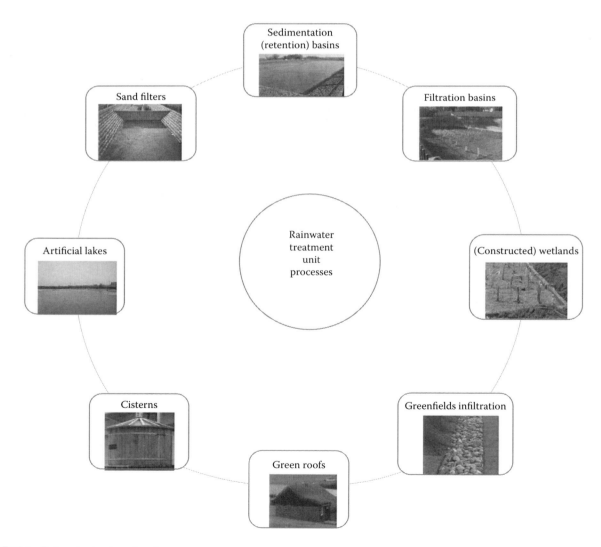

FIGURE 58.1 Rainwater treatment unit processes.

TABLE 58.7 Removal Efficiencies of Different Treatment Systems Applied in RWH Systems

Treatment type	Treatment Yield	References
Solar collector disinfection	20%–30% increase in disinfection efficiency than solar disinfection	Amin and Han [8]
Biofilters	>95% removal for heavy metals	Blecken et al. [19]
Biofilters	70% for nitrogen; 85% for phosphorus; 95% suspended solids	Bratieres et al. [20]
Glass filter strip	56% for nitrogen; 46% for phosphorus; 69% suspended solids	Deletic and Fletcher [24]
ll sorbent	As, Cd, Cr, Cu, Ni, and Zn removal depend on rainwaters heavy metal speciation, surface charge and pH	Genc-Fuhrman et al.[35]
Granular active carbon (GAC)	70% for dissolved organic carbon; 84% for turbidity	Kus et al. [60]
Floating treatment wetlands (FTWs)	34%–42% for turbidity	Tanner and Headley [106]

In China, after chemical processing, rainwater is used to flush the toilet, support indoor hydrant systems, supply cooling tower of air conditioners, irrigate the lawn, pour on outdoor roads, and wash the parking lot [40]. A business center in Guangzhou supplies the cooling water of air conditioners with the collected roof rainwater [123].

In the UK, a study revealed that the average water saving efficiency because of the use of rainwater for toilet flushing was approximately 57% [32].

In Taiwan, 32% of potable water in the residential sector could be replenished by rainwater, mostly for toilet flushing, cleaning, and gardening [23].

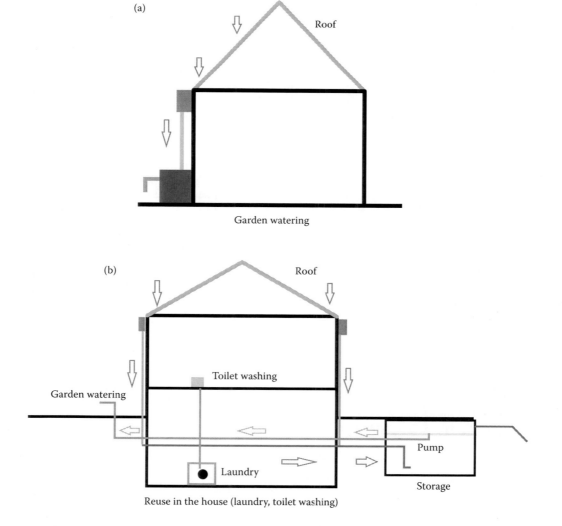

FIGURE 58.2 Rainwater harvesting and reuse in the buildings for garden watering (a) and inhouse use (b).

In the Netherlands, rainwater became more popular economically and an environmentally sustainable water source for splash parks. Man et al. investigated microbial risk assessment using *Legionella pheumophiala* and *Compylobacter jejuni* [76].

Imteaz et al. developed daily a water balance model using daily rainfall data, contributing roof area, rainfall loss factor, available storage volume, tank overflow, and rainwater demand. This water balance model was then used to design an optimum size of domestic rainwater tank to be used in southwest Nigeria. For the tank water, two demand scenarios were assessed: (1) toilet flushing only, and (2) toilet flushing and laundry use [48].

58.6 Summary and Conclusions

Rainwater has been considered an alternative water resource in many countries for various reuse purposes such as green watering, and inhouse use (laundry, wc). Although rainwater quality varies from one location to another, in general, inorganics, organics, and microbial parameters are taken into consideration to choose proper treatment and reuse options. Roof materials affect the quality of rainwater. RHW systems are easily constructed according to norms developed; however, a feasibility study is needed according to present conditions (water resources, current prices) to promote its effective use in an area.

Further to control pollutants in rainwater for reuse, water resources are protected in a way from diffuse pollutions by various pollutants, which would be better to follow them using shorter-term chronic toxicity tests (more sensitive and points of reproductive inhibition, survival, and growth).

References

1. Abdulla, F.A. and Al-Shareef, A.W. 2009. Roof rainwater harvesting systems for household water supply in Jordan. *Desalination*, 243, 195–207.
2. Ahmed, W., Gardner, T., and Toze, S. 2011. Microbiological quality of roof-harvested rainwater and health risks: A review. *Journal of Environmental Quality*, 40, 1–9.
3. Ahmed, W., Goonetilleke, A., and Gardner, T. 2010. Implications of faecal indicator bacteria for the microbiological assessment of roof-harvested rainwater quality in Southeast Queensland, Australia. *Canadian Journal of Microbiology*, 56, 471–479.
4. Ahmed, W., Huygens, F., Goonetilleke, A., and Gardner, T. 2008. Real-time PCR detection of pathogenic microorganisms in roof-harvested rainwater in Southeast Queensland, Australia. *Applied and Environmental Microbiology*, 74(17), 5490–5496.
5. Al-Khashman, O.A. 2005. Study of chemical composition in wet atmospheric precipitation in Eshidiya area, Jordan. *Atmospheric Environment*, 39, 6175–6183.
6. Al-Khashman, O.A. 2009. Chemical characteristics of rainwater collected at a western site of Jordan. *Atmospheric Research*, 91, 53–61.
7. Al-Momani, I.F., Ataman, O.Y., Anwari, M.A., Tuncel, S., Köse, C., and Tuncel, G. 1995. Chemical composition of precipitation near an industrial area at Izmir, Turkey. *Atmospheric Environment*, 29, 1131–1143.
8. Amin, M.T. and Han, M.Y. 2009. Roof-harvested rainwater for potable purposes: Application of solar collector disinfection (SOCO-DIS). *Water Research*, 43, 5225–5235.
9. Andre, F., Jonard, M., and Ponette, Q. 2007. Influence of meteorological factors and polluting environment on rain chemistry and wet deposition in a rural area near Chimay, Belgium. *Atmospheric Environment*, 41, 1426–1439.
10. Avila, A. and Alarcon, M. 1999. Relationship between precipitation chemistry and metrological situations at a rural site in NE Spain. *Atmospheric Environment*, 33, 1663–1667.
11. Baek, S.O., Field, R.A., Goldstone, M.E., Kirk, P.W., Lester, J.N., and Perry, R. 1991. A review of atmospheric polycyclic aromatic hydrocarbons: Sources, fate and behavior. *Water Soil Pollution*, 60, 279–300.
12. Baez, A.P., Belmont, R.D., Garcia, R.M., Torres, M.C., and Padilla, H.G. 2007. Rainwater chemical composition at two sites in Central Mexico. *Atmospheric Research*, 80, 67–85.
13. Balcı, A., Demirak, A., and Tabak, F. 2001. Investigation of the some chemical properties of the rains in Rize (in Turkish). *Journal of Ecology*, 40, 17–19.
14. Basak, B. and Alagha, O. 2004. The chemical composition of rain water over Büyükçekmece Lake Istanbul. *Atmospheric Research*, 71, 275–288.
15. Beiderwieden, E., Wrzesinsky, T., and Klemm, O. 2005. Chemical characterization of fog and rain water collected at the eastern Andes cordillera. *Hydrology and Earth System Sciences*, 2, 863–885.
16. Berenzen, N., Lentzen-Godding, A., Probst, M., Schulz, H., Schulz, R., and Liess, M. 2005. A comparison of predicted and measured levels of runoff-related pesticide concentrations in small lowland streams on a landscape level. *Chemosphere*, 58, 683–691.
17. Birch, H., Mayer, P., Lutzhøft, H.H., and Mikkelsen, P.S. 2012. Partitioning of fluoranthene between free and bound forms in stormwater runoff and other urban discharges using passive dosing. *Water Research*, 46, 6002–6012.
18. Birgül, A., Tasdemir, Y., and Cindoruk, S.S. 2011. Atmospheric wet and dry deposition of polycyclic aromatic hydrocarbons (PAHs) determined using a modified sampler. *Atmospheric Research*, 101, 341–353.
19. Blecken, G.T., Zinger, Y., Deletić, A., Fletcher, T.D., and Viklander, M. 2009. Impact of a submerged zone and a carbon source on heavy metal removal in stormwater biofilters. *Ecological Engineering*, 35, 769–778.
20. Bratieres, K., Fletcher, T.D., Deletić, A., and Zinger, Y. 2008. Nutrient and sediment removal by stormwater biofilters: A large-scale design optimisation study. *Water Research*, 42, 3930–3940.
21. Cao, Y.Z., Wang, S., Zhang, G., Luo, J., and Lu, S. 2009. Chemical characteristics of wet precipitation at an urban

site of Guangzhou, South China. *Atmospheric Research*, 94, 462–469.

22. Celle-Jeanton, H., Travi, Y., Loÿe-Pilot, M., Huneau, F., and Bertrand, G. 2009. Rainwater chemistry at a Mediterranean inland station (Avignon, France): Local contribution versus long-range supply. *Atmospheric Research*, 91, 118–126.

23. Cheng, C.L. and Liao, M.C. 2009. Regional rainfall level zoning for rainwater harvesting systems in northern Taiwan. *Resources, Conservation and Recycling*, 53, 421–428.

24. Deletić, A. and Fletcher, T.D. 2006. Performance of grass filters used for stormwater treatment—A field and modelling study. *Journal of Hydrology*, 317, 261–275.

25. Demirak, A., Balcı, A., Karaoğlu, H., and Tosmur, B. 2006. Chemical characteristics of rainwater at an urban site of South western Turkey. *Environmental Monitoring and Assessment*, 123, 271–283.

26. DIN 1989–1. 2002. *Regenwassernutzungsanlagen Teil 1: Planung, Ausführung, Betrieb und Wartung (Rainwater Harvesting Facilities Part 1: Design, Construction, Operation, and Maintenance)*. DIN Deutsches Institut für Normung e.V., Beuth Verlag GmbH, Berlin (in Germany)

27. Dvorska, A., Lammel, G., and Holoubek, I. 2009. Recent trends of persistent organic pollutants in air in central Europe—Air monitoring in combination with air mass trajectory statistics as a tool to study the effectivity of regional chemical policy. *Atmospheric Environment*, 43, 1280–1287.

28. Egodawatta, P., Thomas, E., and Goonetilleke, A. 2009. Understanding the physical processes of pollutant build-up and wash-off on roof surfaces. *Science of the Total Environment*, 1834–1841.

29. Eroksuz, E. and Rahman, A. 2010. Rainwater tanks in multi-unit buildings: A case study for three Australian cities. *Resources, Conservation and Recycling*, 54, 1449–1452.

30. European Commision. 2009. Study on water performance of buildings. Reference report 070307/2008/520703/ETU/D2.

31. Farreny, R., Gabarrell, X., and Rieradevalla, J. 2011. Cost-efficiency of rainwater harvesting strategies in dense Mediterranean neighbourhoods. *Resources, Conservation and Recycling*, 55, 686–694.

32. Fewkes, A. 1999. The use of rainwater for WC flushing: The field testing of a collection system. *Building and Environment*, 34(6), 765–772.

33. Fujita, S., Takahashi, A., Weng, J., Huang, L., Kim, H., Li, C., Huang, F., and Jeng, F. 2000. Precipitation chemistry in East Asia. *Atmospheric Environment*, 34, 525–537.

34. Gaga, E., Tuncel, G., and Tuncel, S. 2009. Sources and wet deposition fluxes of polycyclic aromatic hydrocarbons (PAHs) in an urban site 1000 meters high in central Anatolia (Turkey). *Environmental Forensics*, 10, 286–298.

35. Genç-Fuhrman, H., Mikkelsen, P.S., and Ledin, A. 2007. Simultaneous removal of As, Cd, Cr, Cu, Ni and Zn from stormwater: Experimental comparison of 11 different sorbents. *Water Research*, 41, 591–602.

36. Ghisi, E., Bressan, D.L., and Martini, M. 2007. Rainwater tank capacity and potential for potable water savings by using rainwater in the residential sector of southeastern Brazil. *Building and Environment*, 42(4), 1654–1666.

37. Ghisi, E. and Mengotti de Oliveira, S. 2007. Potential for potable water savings by combining the use of rainwater and greywater in houses in southern Brazil. *Building and Environment*, 42, 1731–1742.

38. Greenstein, D., Tiefenthaler, L., and Bay, S. 2004. Toxicity of parking lot runoff after application of simulated rainfall. *Bulletin of Environmental Contamination and Toxicology*, 47, 199–206.

39. Guidotti, M., Giovinazzo, R., Cedrone, O., and Vitali, M. 2000. Determination of organic micropollutants in rain water for laboratory screening of air quality in urban environment. *Environment International*, 26, 23–28.

40. Guo, R., Liu, P., and Zhao, X. 2006. Design introduction of water supply and drainage construction drawing of national stadium (I). *Water Supply and Drainage*, 32(4), 83–85.

41. Hamers, T., Smit, M.G.D., Murk, A.J., and Koeman, J.H. 2001. Biological and chemical analysis of the toxic potency of pesticides in rainwater. *Chemosphere*, 45, 609–624.

42. Herrmann, T. and Schmida, U. 1999. Rainwater utilisation in Germany: Efficiency, dimensioning, hydraulic and environmental aspects. *Urban Water*, 1(4), 307–316.

43. Herut, B., Starinsky, A., Katz, A., and Rosenfeld, D. 2000. Relationship between the acidity and the chemical composition of rainwater and climatological conditions a long a transition zone between large deserts and Mediterranean climate. Israel. *Atmospheric Environment*, 34, 1281–1292.

44. http://keeplebanongreen.wordpress.com/tag/water-conservation/.

45. http://www.rainwaterharvesting.org/policy/legislation.htm.

46. Hu, Z.Y., Xu, C.K., Zhou, L.N., Sun, B.H., He, Y.Q., Zhou, J., and Cao, Z.H. 2005. Contribution of Atmospheric Nitrogen Compounds to N Deposition in a Broadleaf Forest of Southern China. *Pedosphere*, 17, 360–365.

47. Huang, X.F., Li, X., He, L.Y., Feng, N., Hu, M., Niu, Y.W., and Zeng, L.W. 2010. 5-Year study of rainwater chemistry in a coastal mega-city in South China. *Atmospheric Research*, 97, 185–193.

48. Imteaz, M.A., Adeboye, O.B., Rayburg, S., and Shanableh, A. 2012. Rainwater harvesting potential for southwest Nigeria using daily water balance model. *Resources, Conservation and Recycling*, 62, 51–55.

49. Imteaz, M.A., Ahsan, A., Naser, J., and Rahman, A. 2011. Reliability analysis of rainwater tanks in Melbourne using daily water balance model. *Resources, Conservation and Recycling*, 56, 80–86.

50. Jones, M.P. and Hunt, W.F. 2010. Performance of rainwater harvesting systems in the southeastern United States. *Resources, Conservation and Recycling*, 54, 623–629.

51. Kaczala, F., Salomon, P.S., Marques, M., Granéli, E., and Hoglanda, W. 2011. Effects from log-yard stormwater runoff on the microalgae *Scenedesmus subspicatus*: Intra-storm magnitude and variability. *Journal of Hazardous Materials*, 185, 732–739.

52. Kahinda, M. and Taigbenu, A.E. 2011. Rainwater harvesting in South Africa: Challenges and opportunities. *Physics and Chemistry of the Earth*, 36, 968–976.

53. Kanellopoulou, E.A. 2001. Determination of heavy metals in wet deposition of Athens. *Global Nest: The International Journal*, 3(1), 45–50.

54. Kawamura, K., Steinberg, S., Ng. L., and Kaplan, I.R. 2001. Wet deposition of low molecular weight mono- and di-carboxylic acids, aldehydes and inorganic species in Los Angeles. *Atmospheric Environment*, 35, 3917–3926.

55. Kaya, G. and Tuncel, G. 1997. Trace element and major ion composition of wet and dry deposition in Ankara, Turkey. *Atmospheric Environment*, 31, 3985–3998.

56. Kidron, G. and Starinsky, A. 2012. Chemical composition of dew and rain in an extreme desert (Negev): Cobbles serve as sink for nutrients. *Journal of Hydrology*, 420–421, 284–291.

57. Kim, H., Han, M., and Lee, J.Y. 2012. The application of an analytical probabilistic model for estimating the rainfall–runoff reductions achieved using a rainwater harvesting system. *Science of the Total Environment*, 424, 213–218.

58. Kim, R.H., Lee, S., Kim, Y.M., Lee, J.H., Kim, S.K., and Kim, J.G. 2005. Pollutants in rainwater runoff in Korea: Their impacts on rainwater utilization. *Environmental Technology*, 26, 411–420.

59. Kiss, G., Gelencsér, A., Krivácsy, Z., and Hlavaya, S. 1997. Occurrence and determination of organic pollutants in aerosol, precipitation, and sediment samples collected at Lake Balaton. *Journal of Chromatography A*, 774, 349–361.

60. Kus, B., Kandasamy, J., Vigneswaran, S., Shon, H., and Moody, G. 2012. Two stage filtration for stormwater treatment: A pilot scale study. *Desalination and Water Treatment*, 45, 361–369.

61. Lara, L.B.L.S., Artaxo, P., Martinellia, L.A., Victoria, R.L., Camargo, P.B., Krusche, A., Ayers, G.P., Ferraz, E.S.B., and Ballester, M.V. 2001. Chemical composition of rainwater and anthropogenic influences in the Piracicaba River Basin, Southeast Brazil. *Atmospheric Environment*, 35, 4937–4945.

62. Le Bolloch, O. and Guerzoni, S. 1995. Acid and alkaline deposition in precipitation on the western coast of Sardinia, central Mediterranean (40°N, 8°E). *Water Air Soil Pollution*, 85, 2155–2160.

63. Lee, B.K., Hong, S.H., and Lee, D.S. 2000. Chemical composition of precipitation and wet deposition of major ions on the Korean peninsula. *Atmospheric Environment*, 34, 563–575.

64. Lee, J.Y., Bak, G., and Han, M. 2012. Quality of roof-harvested rainwater—Comparison of different roofing materials. *Environmental Pollution*, 162, 422–429.

65. Lee, J.Y., Yang, J., Han, M., and Choi, J. 2010. Comparison of the microbiological and chemical characterization of harvested rainwater and reservoir water as alternative water resources. *Science of the Total Environment*, 408, 896–905.

66. Lekouch, I., Mileta, M., Muselli, M., Milimouk-Melnytchouk, I., Šojat, V., Kabbachi, B., and Beysens, D. 2010. Comparative chemical analysis of dew and rain water. *Atmospheric Research*, 95, 224–234.

67. Lekouch, I., Muselli, M., Kabbachi, B., Ouazzani, J., Melnytchouk-Milimouk, I., and Beysens, D. 2011. Dew, fog, and rain as supplementary sources of water in south-western Morocco. *Energy*, 36, 2257–2265.

68. Leuenberger, C., Czuczwa, J., Tremp, J., and Giger, W. 1988. Nitrated phenols in rain: Atmospheric occurrence of phy-totoxic pollutants. *Chemosphere*, 17, 511–515.

69. Li, M., Kang, S., Zhang, Q., and Kaspari, S. 2007. Major ionic composition of precipitation in the Nam Co region, Central Tibetan Plateau. *Atmospheric Research*, 85, 351–360.

70. Liang, X. and Van Dijk, M.P. 2011. Economic and financial analysis on rainwater harvesting for agricultural irrigation in the rural areas of Beijing. *Resources, Conservation and Recycling*, 55, 1100–1108.

71. Liaw, C.H. and Tsai, Y.L. 2004. Optimum storage volume of rooftop rain water harvesting systems for domestic use. *Journal of American Water Resource Association*, 40, 901–12.

72. Ligocki, M.P., Leuenberger, C., and Pankow, J.F. 1985. Trace organic compounds in rain–II. Gas scavenging of neutral organic compounds. *Atmospheric Environment*, 19, 1609–1617.

73. Ligocki, M.P., Leuenberger, C., and Pankow, J.F. 1985. Trace organic compounds in rain–III. Particle scavenging of neutral organic compounds. *Atmospheric Environment*, 19, 1619–1626.

74. Liu, Y. 2005. The raiwanter utilization and subsidy system of Sumida-ku in Tokyo, Beijing. *Water Conservancy*, 6, 44–46.

75. Lye, D.J. 2009. Rooftop runoff as a source of contamination: A review. *Science of the Total Environment*, 407, 5429–5434.

76. Man, D.H., Bouwknegt, M., Van Heijnsbergen, E., Leenen, E.J.T.M., Van Knapena, F., and De Roda Husman, A.M. 2014. Health risk assessment for splash parks that use rainwater as source water. *Water Research*, 54 (2014), 254–261, doi: 10.1016/j.watres.2014.02.010.

77. Manoj, S.V., Mishra, C.D., Sharma, M., Rani, A., Jain, R., Bansal, S.P., and Gupta, K.S. 2000. Iron, manganase and copper concentrations in wet precipitations and kinetics of the oxidation of SO_2 in rain water at urban sites, Jaipur and Kota, in Western India. *Atmospheric Environment*, 34, 4479–4486.

78. Marsalek, J., Rochfort, Q., Mayer, T., Servos, M., Dutka, B., and Brownlee, B. 1999. Toxicity testing for controlling urban wet-weather pollution: Advantages and limitations. *Urban Water*, 1, 91–103.

79. Matsumoto, K., Kawai, S., and Igawa, M. 2005. Dominant factors controlling concentrations of aldehydes in rain, fog, dew water, and in the gas phase. *Atmospheric Environment*, 39, 7321–7329.

80. McQueen, A.D., Johnson, B.M., Rodgers, J.J.H., and English, W.R. 2010. Campus parking lot stormwater runoff: Physicochemical analyses and toxicity tests using Ceriodaphnia dubia and Pimephales promelas. *Chemosphere*, 79, 561–569.

81. Mello, W.Z. 2001. Precipitation chemistry in coast of the metropolitan region of Rio de Janeiro, Brazil. *Environmental Pollution*, 114, 235–242.

82. Mendez, C.B., Klenzendorf, J.B., Afshar, B.R., Simmons, M.T., Barrett, M.E., Kinney, K.A., and Kirisits, M.J. 2011. The effect of roofing material on the quality of harvested rainwater. *Water Research*, 45, 2049–2059.

83. Mitchell, V.G., Taylor, A., Fletcher, T.D., and Deletić, A. 2005. *Storm Water Reuse. Potable Water Substitution for Melbourne*. ISWR Rep. No. 05/12. Monash University, Melbourne, Australia.

84. Moon, S., Lee, J., Lee, B., Park, K., and Jo, Y. 2012. Quality of harvested rainwater in artificial recharge site on Jeju volcanic island, Korea. *Journal of Hydrology*, 414–415, 268–277.

85. Morrow, A.C., Dunstan, R.H., and Coombes, P.J. 2010. Elemental composition at different points of the rainwater harvesting system. *Science of the Total Environment*, 408, 4542–4548.

86. Mouli, P., Mohan, S., and Reddy, S. 2005. Rainwater chemistry at a regional representative urban site: Influence of terrestrial sources on ionic composition. *Atmospheric Environment*, 39, 999–1008.

87. Mun, J.S. and Han, M.Y. 2012. Design and operational parameters of a rooftop rainwater harvesting system: Definition, sensitivity and verification. *Journal of Environmental Management*, 93, 147–153.

88. Okay, C., Akkoyunlu and B.O., Tayan, M. 2002. Composition of wet deposition in Kaynarca. Turkey. *Environmental Pollution*, 118, 401–410.

89. Okuda, T., Iwase, T., Ueda, H., Suda, Y., Tanaka, S., Dokiya, Y., Fushimi, K., and Hosoe, M. 2005. Long-term trend of chemical constituents in precipitation in Tokyo Metroplitan area, Japan, from 1990–2002. *Science of the Total Environment*, 339, 127–141.

90. Özsoy, T. and Örnektekin, S. 2009. Trace elements in urban and suburban rainfall, Mersin, Northeastern Mediterranean. *Atmospheric Research*, 94, 203–219.

91. Rahman, A., Dbais, J., and Imteaz, M. 2010. Sustainability of rainwater harvesting systems in multistorey residential buildings. *American Journal of Engineering and Applied Science*, 3, 889–898.

92. Rahman, A., Keanea, J., and Imteaz, M.A. 2012. Rainwater harvesting in Greater Sydney: Water savings, reliability and economic benefits. *Resources, Conservation and Recycling*, 61, 16–21.

93. Reitano, R. 2011. Water harvesting and water collection systems in Mediterranean area. The case of Malta. *Procedia Engineering*, 21, 81 – 88.

94. Rosenkrantz, R.T., Pollino, C.A., Nugegoda, D., and Baun, A. 2008. Toxicity of water and sediment from stormwater retarding basins to Hydra hexactinella. *Environmental Pollution*, 156, 922–927.

95. Rossi, C., Bierl, R., and Riefstahl, J. 2003. Organic pollutants in precipitation: Monitoring of pesticides and polycyclic aromatic hydrocarbons in the region of Trier (Germany). *Physics and Chemistry of the Earth*, 28, 307–314.

96. Rouvalis, A., Karadima, C., Zioris, I.V., Sakkas, V.A., Albanis, T., and Georgudaki, J.I. 2009. Determination of pesticides and toxic potency of rainwater samples in western Greece. *Ecotoxicology and Environmental Safety*, 72, 828–833.

97. Sazakli, E., Alexopoulos, A., and Leotsinidis, M. 2007. Rainwater harvesting, quality assessment and utilization in Kefalonia Island, Greece. *Water Research*, 41, 2039–2047.

98. Seyfioğlu, R., Odabaşı, M., and Çetin, E. 2005. Formaldehyde deposition in İzmir, Turkey. Paper presented at the *1st International Conference on the Air Pollution and Combustion*, Middle East Technical University; 72, Ankara, Turkey.

99. Sheng, L., Maria, T.S., Ariffin, A.R.M., and Hussein, H. 2011. Integrated sustainable roof design. *Procedia Engineering*, 21, 846–852.

100. Simmons, G., Hope, V., Lewis, G., Whitmore, J., and Gao, W. 2011. Contamination of potable roof-collected rainwater in Auckland, New Zealand. *Water Research*, 35(6), 1518–1524.

101. Song, F. and Gao, Y. 2009. Chemical characteristics of precipitation at metropolitan Newark in the US East Coast. *Atmospheric Environment*, 43, 4903–4913.

102. Spinks, A.T., Coombes, P., Dunstan, R.H., and Kuczera, G. 2003. Water quality treatment processes in domestic rainwater harvesting systems. In: Boyd, M.J., Ball, J.E., Babister, M.K., and Green, J. *Proceedings of the 28th International Hydrology and Water Resources Symposium*, Barton, A.C.T.: Institution of Engineers, Wollongong, Australia, pp. 2.227–2.234.

103. Staelens, J., Schrijver, A.D., Avermaet, P.V., Genouw, G., and Verhoest, N. 2005. A comparison of bulk and wet-only deposition at two adjacent sites in Melle. *Atmospheric Environment*, 39, 7–15.

104. Sturm, M., Zimmermann, M., Schütz, K., Urban, W., and Hartung, H. 2009. Rainwater harvesting as an alternative water resource in rural sites in central northern Namibia. *Physics and Chemistry of the Earth*, 34, 776–785.

105. Tam, V.W.Y., Tam, L., and Zeng, S.X. 2010. Cost effectiveness and tradeoff on the use of rainwater tank: An empirical study in Australian residential decision-making. *Resources, Conservation and Recycling*, 54, 178–186.

106. Tanner, C.C. and Headley, T.R. 2011. Components of floating emergent macrophyte treatment wetlands influencing removal of stormwater pollutants. *Ecological Engineering*, 37, 474–486.

107. Tiwari, S., Chate, D.M., Bisht, D.S., Srivastava, M.K., and Padmanabhamurty, B. 2012. Rainwater chemistry in the North Western Himalayan Region, India. *Atmospheric Research*, 104–105, 128–138.

108. Topçu, S., Incecik, S., and Atimtay, A.T. 2002. Chemical composition of rainwater at EMEP station in Ankara, Turkey. *Atmospheric Research*, 65, 77–92.

109. Tsai, Y., Hsieh, L., Kuo, S., Chen, C., and Wu, P. 2011. Seasonal and rainfall-type variations in inorganic ions and dicarboxylic acids and acidity of wet deposition samples collected from subtropical East Asia. *Atmospheric Environment*, 45, 3535–3547.

110. Tu, J., Wang, H., Zhang, Z., Jin, X., and Li, W. 2005. Trends in chemical composition of precipitation in Nanjing, China, during 1992–2003. *Atmospheric Research*, 73, 283–298.

111. Tuccillo, M.E. 2006. Size fractionation of metals in runoff from residential and highway storm sewers. *Science of the Total Environment*, 355, 288–300.

112. Tuncel, G.S. and Ungor, S. 1996. Rainwater chemistry in Ankara, Turkey. *Atmospheric Environment*, 30, 2721–2727.

113. Tuncer, B., Bayar, B., Yeşilyurt, C., and Tuncel, G. 2001. Ionic composition of precipitation at the Central Anatolia (Turkey). *Atmospheric Environment*, 35, 5989–6002.

114. Ugur, A., Özden, B., and Filizok, I. 2011. Determination of 210Po and 210Pb concentrations in atmospheric deposition in Izmir (Aegean sea-Turkey). *Atmospheric Environment*, 45, 4809–4813.

115. Uygur, N., Karaca, F., and Alagha, O. 2010. Prediction of sources of metal pollution in rainwater in Istanbul, Turkey using factor analysis and long-range transport models. *Atmospheric Research*, 95, 55–64.

116. Villarreal, E.L. and Dixon, A. 2005. Analysis of a rainwater collection system for domestic water supply in Ringdansen, Norrkoping, Sweden. *Building and Environment*, 40, 1174–1784.

117. Wai, K.M., Tanner, P.A., and Tam, C.W.F. 2005. 2-Year study of chemical composition of bulk deposition in a south China coastal city: Comparison with East Asian cities. *Environmental Science & Technology*, 39, 6542–6547.

118. World Health Organization (WHO). 2004. *Guideline for Drinking Water Quality*, 3rd Edition. World Health Organization, Geneva, Switzerland.

119. Yue, Q., Zhang, K., Zhang, B., Li, S., and Zeng, E. 2011. Occurrence, phase distribution and depositional intensity of dichlorodiphenyltrichloroethane (DDT) and its metabolites in air and precipitation of the Pearl River Delta, China. *Chemosphere*, 84, 446–451.

120. Zhang, D.D., Peart, M., Jim, C.Y., He, Y.Q., Li, B.S., and Chen, J.A. 2003. Precipitation chemistry of Lhasa and other remote towns, Tibet. *Atmospheric Environment*, 37, 231–240.

121. Zhang, K., Zhang, B., Li, S., and Zeng, E. 2011. Regional dynamics of persistent organic pollutants (POPs) in the Pearl River Delta, China: Implications and perspectives. *Environmental Pollution*, 159, 2301–2309.

122. Zhang, Y., Chen, D., Chen, L., and Ashbolt, S. 2009. Potential for rainwater use in high-rise buildings in Australian cities. *Journal of Environmental Management*, 91, 222–226.

123. Zhao, Z. and Xu, H. 2012. Study on the supplying system of cooling water of air conditioner based on the urban street rainwater. *Energy Procedia*, 16, 8–13.

59

Water Reuse in Rainwater Harvesting

Saeid Okhravi
Isfahan University of Technology

Saeid Eslamian
Isfahan University of Technology

Jan Adamowski
McGill University

PREFACE

The planning and efficient management of water resources in a catchment has several problems and complexities. The integrated management of water resources is a very important issue for designing managers, investors, and legislators in the areas of water resources issues and water systems management, as it affects the natural and social environment, as well as the economy. Rainwater harvesting is one of the best alternative water resources that can be exploited using conventional approaches because it can be used for public consumption and contributes to sustainable management of freshwater. Rainwater harvesting is an ancient technique that is enjoying a revival in popularity due to the inherent quality of rainwater and the increased interest in reducing consumption of treated water. There are numerous positive benefits to harvesting rainwater. The main benefits are its high water quality due to lack of hardness, low loss of energy, and reduced costs of treatment and purification.

For the sustainable use of water resources, it is critical that rainwater harvesting is used as a water source, as is the case for groundwater and surface water. Rainwater harvesting can contribute to ecosystem protection and human wellbeing, and is especially important when water utilities are unable to obtain sustainable supplies of good quality drinking water. Interest in adapting this practice to urban areas is increasing as it provides the combined benefits of conserving potable water and reducing stormwater runoff. Rainwater harvesting is one of the most important low impact developments (LID) to retain stormwater onsite, where vegetation can filter out or break down pollutants. This chapter covers all the aspects of rainwater reuse in different ways as an integrated stormwater management for solving the unique challenges of individual communities.

59.1 Introduction

Populations in most urban areas are increasing quickly, and supplying sufficient water of good quality is one of the most urgent and significant challenges faced by decision-makers.

There are two solutions to ensure sustainably managing shortages in water supply: finding alternate or additional water sources using conventional approaches, or using the limited available supplies more efficiently. Rainwater harvesting systems are independent systems, which means that users have the responsibility of operation and maintenance. They can also avoid many of the environmental problems that can be caused by conventional large-scale water supply projects that use centralized approaches.

The global population has more than doubled since 1950, and reached 6 billion in 1999. The most recent population forecasts from the United Nations indicate that, under a medium-fertility scenario, the global population is likely to peak at about 8.9 billion in 2050 [31]. Since 1950, the number of people living in urban areas has jumped from 750 million to more than 2.5 billion people. The environmental impacts of urban areas and their water demands are different from those in rural areas. By 2025, the total urban population is expected to double to more than 5 billion, and more than 90% of this increase is expected to occur in developing countries.

Rainwater harvesting is the process of capturing, diverting, and storing rainwater from an impervious surface, such as a roof, or from a pervious surface, such as soil. The collected water can be used for a number of different purposes, including landscape irrigation, drinking and domestic use, aquifer recharge, stormwater retention, and runoff reduction.

The advantages of rainwater harvesting are numerous. Suburban and urban sites typically have a high ratio of impervious areas, including rooftops, parking areas, driveways, and sidewalks, to pervious areas covered with vegetation or bare soil. As a result, a surplus of water runoff is available to harvest and use. Both the local site and the wider community can benefit from implementing rainwater harvesting systems.

The specific advantages of rainwater harvesting include:

- Harvested water is located close to the source and the point of use.
- The systems can be operated and managed by both owners and utilities.
- Rainwater harvesting has few negative environmental impacts compared to other technologies for water resources development.
- Rainwater harvesting decreases flow to stormwater drains and reduces nonpoint source pollution. It also reduces onsite erosion and flooding, as well as the costs of managing runoff.
- Rainwater is relatively clean and the quality is usually acceptable for many purposes, which can therefore reduce consumption of treated water.
- The physical and chemical properties of rainwater are usually superior to groundwater sources that may have been exposed to contamination.

- It promotes self-sufficiency and relieves pressure on other water sources.
- Rainwater harvesting technologies are flexible, and construction, operation, and maintenance are not labor-intensive.
- Rainwater is eco-friendly to landscape plants and gardens. The high nitrogen and low salt content of harvested rainwater can extend the life of landscaping.

Harvesting can occur passively, through storing water directly in the ground (passive water harvesting), or actively through storing water in tanks for later use (active water harvesting) [1].

Active rainwater harvesting systems use equipment to collect, filter, store, and deliver harvested water in residential or small-scale commercial systems, for both irrigation and potable use. The volume of water that can be collected by a rainwater harvesting system depends primarily on two factors: the catchment area and the amount of rainfall. In theory, for every square foot of catchment area, about 0.62 gallons of water can be collected for 1 in. of rainfall. In practice, however, this volume is usually lower because of loss of water in the system; an 80% recovery rate, which translates to about 0.5 gallon per square foot of catchment area, is average. Thus, for example, if a house with a roof area of 2000 ft^2 receives 1 in. of rainfall, the volume of water that the homeowner can expect to realistically collect is about 1000 gallons. Collection systems can range from 55-gallon rain barrels to meet outdoor watering needs to 10,000-gallon or larger tanks to meet domestic and landscape water needs [16]. This chapter focuses on active rainwater harvesting systems for different practical uses.

Passive water harvesting includes systems that use land shaping and other techniques to direct, collect, and infiltrate rainwater directly into the soil where water is put to beneficial uses such as supporting vegetation and managing stormwater.

59.2 Active Rainwater Harvesting System Components

Regardless of its size or complexity, a rainwater harvesting system (Figure 59.1) [22] has six basic components.

1. Catchment surface: The effective catchment area from which rainfall runs off. It may be a roof or an impervious area.
2. Conveyance: Channels water from the catchment surface to the storage tank.
3. Roof washing: Components and filters that remove debris and dust from the captured rainwater before it is stored. This involves first-flush devices.
4. Storage: Cisterns or tanks.
5. Distribution: Delivers the rainwater, either by gravity or a pump.
6. Treatment: For potable systems, filters and other techniques to disinfect the water to make it safe to drink.

Because of the higher risk of contamination, rainwater collected from catchment areas should not be used for potable purposes unless there is a good purification system in place (Figure 59.2) [13].

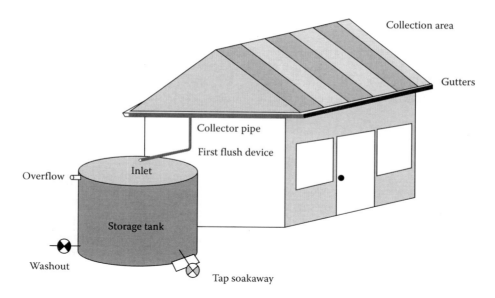

FIGURE 59.1 Typical rainwater harvesting installation. (Adapted from Okhravi, S.S. et al. 2015. *Journal of Flood Engineering*, 5(1).)

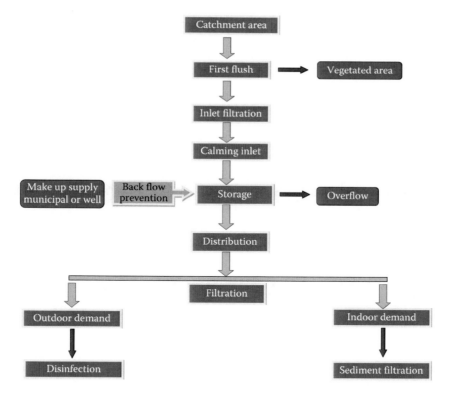

FIGURE 59.2 Active rainwater harvesting system flow chart. (Adapted from Guideline Committee Members. 2009. Georgia Rainwater Harvesting Guidelines, University of Georgia, pp. 31.)

59.3 Water Quality and Treatment Components

Harvested rainwater can be contaminated from both point and nonpoint pollution sources within the catchment. Rainwater quality is evaluated using several physical, chemical, and biological parameters. Pollution from the atmosphere, soil surface, and soil-water interactions are determined. Generally, sediment and toxic materials are the main pollutants. Therefore, it is necessary to adopt environmental and agricultural policies in order to control water quality in catchment systems.

59.3.1 Water Quality Issues

Harvested rainwater quality can be affected by a number of factors, primarily site location and climate, as well as the nature of the catchment system [19]. Water quality is also affected by air pollution and potential contamination of the rooftop water by plants or animals [6].

In terms of physical-chemical parameters, harvested roof water, rainwater, and urban stormwater are generally of a quality that meets the World Health Organization (WHO) guideline values for drinking water (Table 59.1). However, rainwater can have a low pH because of sulfur dioxide, nitrous oxide, and other industrial emissions, and therefore, air quality standards must be reviewed and enforced. In addition, high lead concentrations can sometimes be attributed to the composition of certain roofing materials. Thus, it is recommended that for roof water collection systems, the type of roofing material should be carefully considered.

The aim of this chapter is to outline best-practice techniques for rainwater collection systems in order to obtain good quality water. Information is provided on the following subjects:

1. Types of contaminants in rainwater systems
2. Appropriate rain collection surfaces
3. Appropriate storage vessels
4. Filtration screens
5. First-flush devices
6. Tank maintenance
7. Water quality testing

The majority of current information on rainwater capture focuses on new technologies or water quality [12]. Water from various sites around the world has been tested, with the quality of collected water ranging from very poor to very safe and drinkable.

Several WHO guidelines [7] are relevant to the water quality of rainwater systems. In some cases, the standards are to protect human health, while others are to reduce consumer complaints (e.g., due to poor taste).

59.3.2 Water Treatment

The cleanliness of the roof in a rainwater harvesting system directly affects the quality of the captured water. Treatment of rainwater ensures that the water will be safe to use. This section provides information on prestorage and poststorage treatment.

The goal of prestorage treatment is to clean the rainwater runoff before it goes to the storage tank. It prevents organic matter from accumulating in the storage tank, and it reduces the amount of treatment needed after storage. Diversion and screening can be used for prestorage treatment.

59.3.2.1 Prestorage Treatment

Diversion refers to collecting the first flush of runoff and preventing it from entering the storage container. This first flush of water contains the most debris from the roof surface. While runoff from later in the rainfall event can remove the larger debris, such as leaves, twigs, and blooms that fall on the roof, the first-flush diverter gives the system a chance to relieve itself of the smaller contaminants [26]. A diverter can be installed just before the storage tank or near the downspouts. As it begins to rain, the diverter fills up with the first flush of rain from the roof (Figure 59.3). Once the diverter is full, the cleaner water flows into the tank. The volume of water that is diverted is determined by the length, diameter, and situation of the diverter pipe. Several reports state that 5 to 10 gallons of water per 1000 ft² of roof area should be diverted.

Screening processes aim to remove debris that gathers on the catchment surface, and to assure high-quality water, some filtration is necessary. Debris screens filter out large contaminants and can be installed at several locations along the conveyance network. The gutter debris screen (leaf screen) is the best first defense for keeping debris out of a rainwater harvesting system, and can be placed along the gutter or in the downspout (Figure 59.4). These screens usually consist of 1/4-in. mesh in wire frames fitted along the length of the gutter [8]. Another

TABLE 59.1 World Health Organization Guideline

Parameter	Guideline Value
Fecal coliform of E. coli	Generally, not detectable in a 100 mL sample $< \dfrac{100 \text{ cfu}}{100 \text{ mL}}$ (for outdoor demand)
Aluminum	0.2 mg/L (level likely to result in consumer complaints)
Cadmium	0.003 mg/L
Copper	2 mg/L
Chloride	250 mg/L (level likely to result in consumer complaints)
Fluoride	1.5 mg/L
Iron	0.3 mg/L (level likely to result in consumer complaints)
Lead	0.01 mg/L
Sodium	200 mg/L (level likely to result in consumer complaints)
Sulfate	250 mg/L (level likely to result in consumer complaints)
Turbidity	5 NTU (level likely to result in consumer complaints)
Total dissolved solids	1000 mg/L (level likely to result in consumer complaints)
Zinc	3 mg/L (level likely to result in consumer complaints)

Source: Adapted from Fawell, J.K. et al. 1996. *Guidelines for Drinking Water Quality*, 2nd Edition, Vol. 2. Health Criteria and Other Supporting Information. World Health Organization (WHO), Geneva.

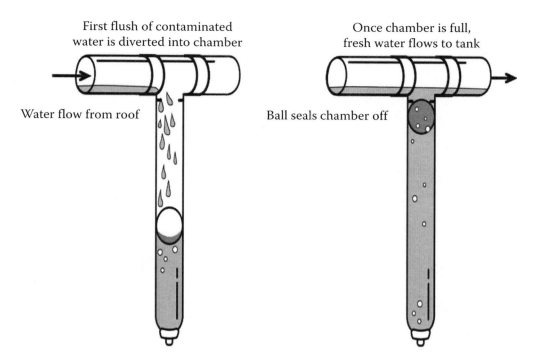

First flush of contaminated water is diverted into chamber

Once chamber is full, fresh water flows to tank

Water flow from roof

Ball seals chamber off

FIGURE 59.3 First-flush diverter. (Adapted from Guideline Committee Members. 2009. Georgia Rainwater Harvesting Guidelines, University of Georgia, pp. 31.)

FIGURE 59.4 Leaf screen. (Adapted from Kinkade-Levario, H. 2007. *Design for Water.* New Society Publishers, Gabriola Island, British Columbia, Canada.)

screen type is the basket screen. The basket screen fits into the inlet of the storage container and catches debris falling into it. The homeowner may need to experiment with various strainer basket screen sizes. It is necessary to clear basket screens after each rain event.

59.3.2.2 After-Storage Treatment

Treatment of the water after storage and before use is critical for both health of the users and maintenance of the system. The level of treatment will depend on the intended use of the water. For nonpotable water systems, treatment generally consists of filtration and disinfection processes in series before distribution to the plumbing system. Some of the common types of treatment are described next.

Filtration is similar to screening but on a smaller scale. The size of the filter is measured in microns, referring to the diameter of the particles that are blocked by the filter.

Turbidity is an indicator of the presence of suspended material in water, such as finely divided organic material, and is measured in nephelometric turbidity units (NTU). High turbidity interferes with disinfection, and therefore it is important to maintain turbidity levels below 10 NTU to improve disinfection effectiveness [3].

The most popular disinfection method in Texas consists of two inline sediment filters: a 5-μm fiber cartridge filter followed by a 3-μm activated charcoal cartridge filter, followed by ultraviolet (UV) light [26]. The disinfection system is located after the pump. Generally, a 5-μm sediment filter is adequate for reducing the concentration of suspended solids for nonpotable indoor uses. A higher level of treatment can be achieved by using a 3-μm sediment filter followed by a 3-μm activated carbon filter.

If larger sized filters are used, small microorganisms, such as bacteria, can pass through, so a disinfection method, such as chlorination, UV light, or ozonation, is required.

Chlorination is a very effective treatment against viruses and bacteria. Unless water pH is over 8.5 (as occurs in some cement

TABLE 59.2   Treatment Techniques

Treatment Method	Location	Result
	Screening	
Leaf screens and strainers	Gutters and downspouts	Prevents leaves and other debris from entering tank
	Settling	
Sedimentation	Within tank	Settles out particulate matter
Activated charcoal	Before tap	Removes chlorine[a]
	Filtering	
Roof washer	Before tank	Eliminates suspended material
Inline/multicartridge	After pump	Sieves sediment
Activated charcoal	After sediment filter	Removes chlorine, improves taste
Slow sand	Separate tank	Traps particulate matter
	Microbiological treatment/Disinfection	
Boiling/distilling	Before use	Kills microorganisms
Chemical treatments (Chlorine or Iodine)	Within tank or at pump (liquid, tablet, or granular)	Kills microorganisms
Ultraviolet light	Before activated charcoal filter	Kills microorganisms
Ozonation	After activated charcoal filter, before tap	Kills microorganisms
Nanofiltration	After activated charcoal filter, before tap	Removes molecules
Reverse osmosis	Before use; polymer membrane (pores 10^{-3}–10^{-6} inch)	Removes ions (contaminants and microorganisms)
	Before use, polymer membrane (pores 10^{-9} in.)	

Source: Adapted from *Texas Manual on Rainwater Harvesting*, third edition. 2005. Texas Water Development Board, Austin, TX.

[a] Should be used if chlorine has been used as a disinfectant.

tanks), chlorine is highly recommended as a treatment to kill bacteria, fungi, and viruses. It is necessary to determine the contact time required for chlorine to kill bacteria. Contact time is estimated based on water pH, temperature, and bacterial load, and is usually 2–5min with a free chlorine residual of 2 mg/L. Contact time increases with pH and decreases with temperature.

UV light is a common disinfection method used in domestic rainwater systems. In order for a UV system to be effective, the water passing through it must be relatively free of particles because any sediment in the water can block pathogens from the light.

In ozonation, ozone gas is introduced to the water to disinfect it. It is usually done at the point where water is used or in the storage tank. Ozone is a colorless gas that disinfects, oxidizes, deodorizes, and decolorizes the water and reduces total organic carbon. As it is a powerful disinfectant and toxic, installation and maintenance of this type of system must be carried out according to the manufacturer's recommended installation guidelines.

A synopsis of treatment techniques is shown in Table 59.2.

59.4 Stormwater Management

Stormwater is any water running off a land surface before it reaches a natural water body. It occurs when the rate of precipitation is greater than the rate at which water can infiltrate, or soak, into the soil, which can occur when the soil is saturated. Runoff remains on the surface and flows into streams, rivers, and eventually large bodies such as lakes or the ocean. Stormwater can transport untreated pollutants, such as sediment, nutrients, and pesticides, into surface water bodies, which has a significant impact on water quality. In addition, stormwater runoff from construction sites can cause soil erosion to occur at a rate 20 times greater than from routine land uses. Learn which sources are regulated and what can be done to control stormwater [14].

Rainwater harvesting offers a small-scale best management practice (BMP) to reduce stormwater runoff and the problems associated with it. By harvesting the rainfall and storing it, the water can be slowly released back into the soil, either through irrigation or direct application [11]. The water then moves into groundwater, providing a steady supply of water to local streams and rivers.

During the past three decades, the practice of stormwater management has evolved. In the mid-1970s, attempts to control runoff flow rates from urban developments were initiated. Early stormwater management plans developed in the 1980s focused on controlling water quantity with the intent of ensuring that runoff from newly developed urban areas did not increase the potential for flooding downstream [32]. Today, with improvements in our understanding of watershed systems and the potential impacts urbanization can have on aquatic ecosystems, rainwater harvesting can be a BMP for stormwater management.

Active rainwater harvesting is a key component of developing a holistic approach for water management in urban environments. A holistic approach to urban water management requires recognizing that all water on a site should be considered part of the site's water balance and should be actively managed to replicate the natural hydrology [33].

59.4.1 Rainwater Harvesting for Stormwater Management

This section focuses on rainwater harvesting for stormwater management. Rainwater harvesting is especially well known as a stormwater management practice in urban areas.

Rainwater harvesting is a BMP for effectively controlling movement of pollutants and preventing degradation of soil and water resources that is compatible with the land use.

Collecting and storing rooftop runoff, and providing a consistent, dedicated, and reliable end use will reduce the volume of runoff and enable the reduction in size of other required stormwater treatment systems on the site [11]. In watersheds where removal of nutrients from stormwater is required, dedicated uses of the collected rainwater or proper treatment/infiltration can reduce these requirements.

BMPs can be divided into two categories: structural and nonstructural. Structural BMPs can be thought of as engineering solutions to stormwater management. They are designed particularly for urban areas, whereas others may be designed for agriculture, forestry, or mining areas. Common examples of structural BMPs usually found within urban areas include stormwater ponds and open channels.

Nonstructural BMPs are designed to reduce the abundance of pollutants in the environment that have the potential to be transported by stormwater runoff. Nonstructural BMPs typically lessen the need for the more costly structural BMPs, and can be achieved through such things as education, management, and development practices. Examples include ordinances and practices associated with land use and comprehensive site planning.

In developed countries in recent years, there has been increased emphasis on green roof design and active rainwater harvesting for in-building nonpotable water uses, such as flushing toilets, and outdoor nonpotable uses, such as landscape irrigation [20]. Assuming a 20% loss due to splash and evaporation, rainfall depth of 1 cm on a single 1000 m² rooftop area can generate about 8 m³ of runoff volume to stormwater drainage systems. The impact can be significant when considering the high density of buildings in urban settings.

Urbanization causes significant changes in the proportion of precipitation that infiltrates into the ground, evaporates back into the atmosphere, and enters drainage features as surface runoff, primarily as a result of the change from land being covered with native vegetation to being covered with paved surfaces (Figure 59.5).

Several studies have documented the impact of rainwater harvesting systems on reduction in runoff volume [11,14]. The larger volume of rainfall retained at the site, the lower potential for phosphorus and other pollutants to be transported by water leaving the site, which impacts water quality in downstream waterways. For active rainwater harvesting, the characteristics of rooftop runoff quality differ from other impervious surfaces such as roads and parking areas, and vary among roofs depending on the roof material. Metals are the most commonly noted contaminants of concern in rooftop runoff [10]. Concentrations of aluminum (Al), manganese (Mn), copper (Cu), lead (Pb), and zinc (Zn) in roof runoff exceeded the national surface water quality standards at least 5% of the time, while Zn and Cu concentrations most often violated standards [4]. Rainwater harvesting can be a technique to reduce the transport of these metals to surface waters.

A rainwater harvesting system considers stormwater runoff to be a resource not a pollutant. In most rainwater harvesting

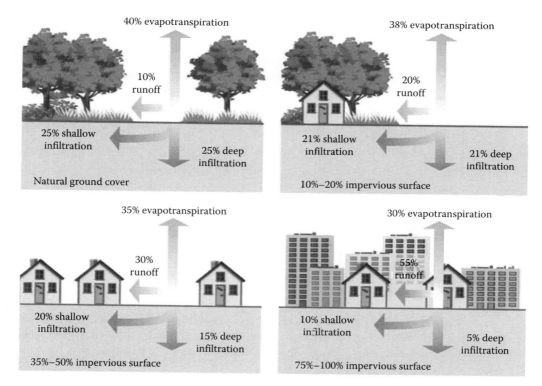

FIGURE 59.5 The impact of urbanization on the hydrologic cycle. (Adapted from Toronto and Region Conservation Authority. 2010. *Low Impact Development Stormwater Management Planning and Design Guide*. Toronto and Region Conservation for Living.)

systems, the volume of potable water saved is equal to the volume of runoff reduced, and both represent a decrease in imports and exports of water from the site. In this sense, rainwater harvesting as an alternative water source and as a BMP are synergistic.

Based on a study by the Pacific Institute, nonpotable water uses, such as landscaping (35%), cooling (15%), laundry (2%), and toilet flushing (12%) (72% of restroom water use is toilet flushing and restroom use represents 16% of overall water use) represent a significant proportion of water use in the commercial, institutional, and industrial sectors [9]. A holistic approach to water management recognizes that these end uses do not require water of drinking water quality (treated water).

59.5 Low Impact Development (LID) Approach

Low impact development (LID) is an innovative stormwater management approach that seeks to mitigate the impacts of increased runoff and stormwater pollution.

LID comprises a set of site design strategies and distributed stormwater management practices that harvest, filter, evapotranspire, and detain runoff close to its source, as well as increase infiltration of rainwater.

Techniques are based on the premise that stormwater management should not be seen as stormwater disposal. Instead of conveying stormwater to large, costly, end-of-pipe facilities located at the bottom of drainage areas, LID addresses stormwater through small, cost-effective landscape features located at the lot level.

LID allows for greater development potential with fewer environmental impacts using smarter designs and advanced technologies that achieve a better balance between conservation, growth, ecosystem protection, and public health [27].

These practices include innovative site design strategies that minimize runoff (i.e., nonstructural LID practices). They also include distributed, small-scale lot level and conveyance practices (i.e., structural LID practices). Examples of LID include:

1. Active rainwater harvesting
2. Green roofs
3. Soakaways and infiltration basins/trenches
4. Bioretention
5. Vegetated filter strips
6. Soil amendments
7. Permeable pavements
8. Swales
9. Perforated pipe systems

All of these approaches attempt to reproduce the predevelopment hydrologic regime through innovative site design and distributed engineering techniques aimed at infiltrating, filtering, evaporating, harvesting, and detaining runoff, as well as preventing pollution.

LID has numerous benefits and advantages over conventional stormwater management approaches. Key elements for LID design can be summarized as follows:

1. Small-scale control using existing natural systems:
   • Viewing conservation of the natural environment as the integrating framework for planning
   • Considering regional and watershed scale contexts, objectives, and targets
   • Identifying stormwater management opportunities and limitations at watershed/sub-watershed and neighborhood scales
   • Recognizing and supporting all environmentally sensitive resources
2. Runoff control:
   • Installing permeable pavements to minimize impervious sites
   • Directing runoff to natural areas and recharging streams, aquifers, and wetlands
   • Incorporating rainwater harvesting systems and green roofs into building designs
   • Preserving existing urban trees
3. Stormwater treatment and conveyance to nearest source area:
   • Using decentralized lot level and conveyance stormwater management practices as part of the treatment approach
   • Flattening slopes, lengthening overland flow paths, and maximizing sheet flow
   • Maintaining natural flow paths by using open drainage [27] (e.g., swales)
4. Customized site design and multifunctional landscapes:
   • Ensuring each site contributes to watershed protection
   • Integrating stormwater management facilities that provide filtration, peak flow attenuation, infiltration and water conservation benefits
   • Designing multifunctional landscaping to reduce runoff and the urban warming effect, and to enhance site aesthetics
5. Maintenance and education:
   • Providing adequate training and funding for municipalities to operate and maintain lot level and conveyance stormwater management practices on public property
   • Teaching property owners, managers and on private property

In short, LID uses environmentally sound technologies, and is an economically sustainable approach for addressing the adverse impacts of urbanization. By managing runoff close to its source though intelligent site design, LID can enhance the local environment, protect public health, and improve community livability—all while saving developers and local governments money.

59.5.1 Considerations of Structural LID Practices

As briefly described above, this chapter focuses on rainwater harvesting systems that reuse stormwater. However, an overview of each LID technology and its implementation is provided, including:

- Description of practice
- Common concerns
- Physical suitability and constraints
- Typical performance

The design templates provide the following:

- Applications
- Typical details
- Design guidance
- BMP sizing
- Design resources
- Design and material specifications
- Construction considerations and sequencing

59.5.1.1 Green Roofs

The use of green roofs represents a truly rare occurrence in modern economies, as it can counter urban climate change, cool buildings, save energy, reduce noise and toxic gas emissions, manage stormwater, conserve biodiversity, and create new markets for existing goods and services.

Green roofs are touted for their benefits to cities, as they improve energy efficiency, reduce urban heat island effects, increase the longevity of roof structures, and create green space for passive recreation or aesthetic and habitat enjoyment. From a water management point of view, they are attractive for impacts on water quality, water balance, and pressure on downstream infrastructure. The green roof acts like a lawn or meadow by storing rainwater in the growing medium and ponding areas. The foliage on the roof also enhances evapotranspiration.

The major components in a green roof are (Figure 59.6)

- The vegetation
- The growing medium
- The drainage system
- The waterproofing

This system is typically sized-based on the available roof area rather than on treatment volume requirements. There are two types of green roofs: intensive and extensive [27]. Intensive green roofs contain greater than 15 cm depth of growing medium, can be planted with deeply rooted plants, and are designed to handle pedestrian traffic. Roof structures supporting intensive green roofs require significantly greater load bearing capacity, thereby increasing their overall cost and the complexity of the design. Extensive green roofs consist of a thin layer of a growing medium used conventionally in building (at a depth of 15 cm or less) with an herbaceous vegetative cover.

59.5.1.2 Soakaways and Infiltration Basins/Trenches

Soakaways are rectangular or circular excavations lined with geotextile fabric and filled with clean granular stone or other void forming material, which receive runoff from a perforated pipe inlet and allow it to infiltrate into the native soil. Soakaways can also be referred to as infiltration galleries, dry wells, or soakaway pits.

An infiltration trench is a gravel or rock-filled trench designed to cause runoff to infiltrate into the ground. Infiltration trenches are essentially long thin soakaways (Figure 59.7).

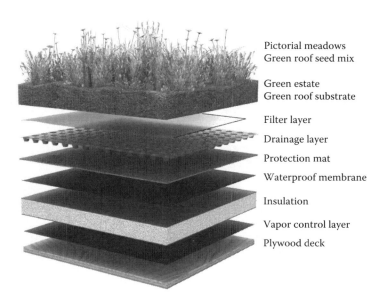

Pictorial meadows
Green roof seed mix

Green estate
Green roof substrate

Filter layer

Drainage layer

Protection mat

Waterproof membrane

Insulation

Vapor control layer

Plywood deck

FIGURE 59.6 Configuration of green roof layers. (Adapted from Toronto and Region Conservation Authority. 2010. *Low Impact Development Stormwater Management Planning and Design Guide.* Toronto and Region Conservation for Living.)

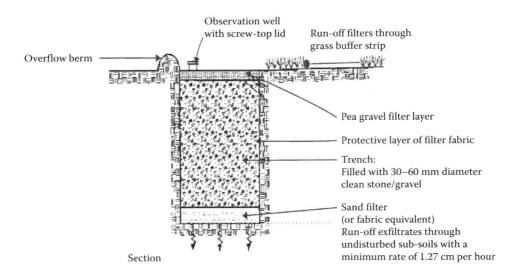

FIGURE 59.7   Schematic of a soakaway. (Adapted from Toronto and Region Conservation Authority. 2010. *Low Impact Development Stormwater Management Planning and Design Guide*. Toronto and Region Conservation for Living.)

There are several factors that must be considered for these systems to be effective, which are described as follows:

- A pretreatment device, such as a swale or filter strip, is recommended upstream of the trench to reduce incoming velocities and coarser sediments.
- The device should be constructed at least 1.5 m above the maximum groundwater level or bedrock layer and only where the groundwater classification allows.
- The trench should be filled with clean stones that can retain the required volume of water to be treated in their void space.
- The stones should be wrapped in a geo-textile, which should be selected based on its durability and should have opening sizes that are adequate to resist clogging.
- The maximum contributing area to infiltration trenches should be less than 5 ha.
- The infiltration device should not be constructed within 5 m of the foundations of buildings or under a road.
- Areas upstream of the trench should be stabilized.
- They must not be constructed near drinking water wells, septic tanks, or drain fields.
- It is advisable to provide inspection tubes at regular intervals along the trench for long infiltration trenches.

59.5.1.3 Bioretention

Bioretention systems treat stormwater by filtering runoff through planted vegetation and percolating (drip-feeding) the runoff through filter media, such as loamy sand. As the water is percolated through the soil, pollutants are captured by fine filtration, absorption, and biological uptake.

Depending on the infiltration rate of the soil and physical constraints, the system may be designed without an underdrain for full infiltration, with an underdrain for partial infiltration, or with an impermeable liner and underdrain for filtration only, which can also be referred to as a biofilter [21].

Bioretention creates a good environment for runoff reduction, filtration, biological uptake, and microbial activity, and provides high pollutant removal. Hence, it can become an attractive landscaping feature with high amenity value and community acceptance.

Finally, bioretention reduces the overall volume of discharge and assists with attenuation of peak discharges through infiltration and temporary storage. It also aids in groundwater recharge and runoff temperature control, and achieves high rates of removal of heavy metals (Cu, Pb, and Zn).

There are six basic components of bioretention, namely (Figure 59.8):

1. Grass buffer strips
2. Sand bed
3. Ponding area
4. Organic/mulch layer
5. Planting soil
6. Plants

59.5.1.4 Vegetated Filter Strips

Vegetated filter strips consist of a permanent, maintained strip of vegetation. They function by slowing the runoff velocity and filtering out suspended sediments and associated pollutants. To function, filter strips require sheet flow across the strip, which can be achieved using level spreaders (Figure 59.9). Frequently, filter strips are designed where runoff is directed from a parking lot into a stone trench, a grass strip, and a longer naturally vegetative strip. They provide good performance at a low cost, as well as providing aesthetics and habitat benefits. Vegetated filter strips can be applied as a first-flush treatment for parking areas where flow is directed toward green space. They can also

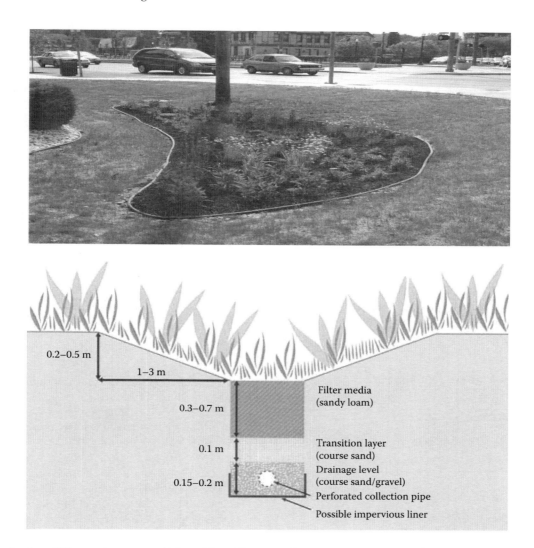

FIGURE 59.8 Section of bioretention system. (Adapted from Toronto and Region Conservation Authority. 2010. *Low Impact Development Stormwater Management Planning and Design Guide.* Toronto and Region Conservation for Living.)

be used to slow the velocity from a level spreader as the flow travels down slope.

Using vegetated filter strips as pretreatment practices to other BMPs is highly recommended. With proper design and maintenance, filter strips can provide relatively high pollutant removal.

According to several studies by ASCE [2], the pollutant removal efficiencies of vegetated filter strips are highly variable (Table 59.3). For this reason, filter strips should be used in conjunction with other water quality BMP (i.e., used as a pretreatment).

59.5.1.5 Soil Amendments

Site preparation typically involves removing and stockpiling the topsoil, compacting of the remaining soil, and then replacing a thin layer of topsoil. This will decrease infiltration, leading to increased runoff, erosion, discharge rate, and downstream volume.

Soil amendments can be used to minimize development impacts on native soils by restoring their infiltration capacity and chemical characteristics. Soil amendments can include not

only compost and mulch but also topsoil, lime, and gypsum (to address nutritional deficiencies and control pH). A thorough soil analysis of the native soil is required to design the ideal mixture.

59.5.1.6 Permeable Pavement

Permeable pavements are alternative paving surfaces that allow stormwater runoff to filter through voids in the pavement surface into an underlying stone reservoir, where it is temporarily stored or infiltrated. A variety of permeable pavement surfaces is available, including pervious concrete, porous asphalt and permeable interlocking concrete pavers. While the specific design may vary, all permeable pavements have a similar structure, consisting of a surface pavement layer, an underlying stone aggregate reservoir layer and a filter layer or fabric installed on the bottom (Figure 59.10). They can be used for low traffic roads, parking lots, driveways, pedestrian plazas, and walkways.

The thickness of the reservoir layer is determined by both a structural and hydrologic design analysis, as it serves to retain

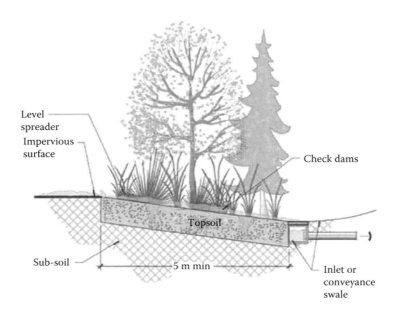

FIGURE 59.9 Vegetated filter strips. (Adapted from Landmark Design Group. http://www.thelandmarkgroup.ca.)

TABLE 59.3 Pollutant Removal Efficiencies of Vegetated Filter Strips

Pollutant	Removal Efficiency (%)
Total suspended solids (TSS)	20–80
Total nitrogen	20–60
Total phosphorus	20–60
Total heavy metals	20–80

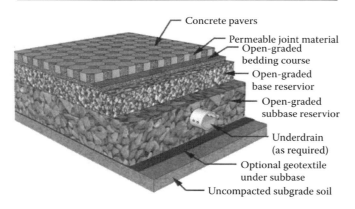

FIGURE 59.10 Typical detail of permeable pavement. (Adapted from U.S. EPA. 2008. Clarification on which stormwater infiltration practices/technologies have the potential to be regulated as "Class V" wells by the Underground Injection Control Program. Water Permits Division and Drinking Water Protection Division. Washington, D.C.)

stormwater and supports traffic loads. Depending on the native soils and physical constraints, the system may be designed with no underdrain for full infiltration, with an underdrain for partial infiltration, or with an impermeable liner and underdrain for a no infiltration or detention and filtration only practice. In fact, in low-infiltration soils, some or all of the filtered runoff is collected in an underdrain and returned to the stormdrain system. If infiltration rates in the native soils permit, permeable pavements can be designed without an underdrain to enable full infiltration of runoff. A combination of these methods can be used to infiltrate a portion of the filtered runoff.

59.5.1.7 Swales

Swales, also known as grassed channels, dry swales, or ditches, are vegetated depressions in the land that accept runoff and cause it to infiltrate. Water enters the swale and is allowed to soak into the base and sides, or flow into another receiving system. They are typically vegetated with grasses and plants that help absorb and infiltrate the stormwater.

Grass swales can also be vegetated open channels that are designed to convey, treat, and attenuate stormwater runoff. Check dams and vegetation in the swale slow the water to allow sedimentation, filtration through the root zone and soil matrix, evapotranspiration, and infiltration into the underlying native soil.

A dry swale is a design variation that incorporates an engineered soil media bed and an optional perforated pipe underdrain system (Figure 59.11). Enhanced grass swales are not capable of providing the same water balance and water quality benefits as dry swales, as they lack their engineered soil media and storage capacity.

Dry swales are similar to bioretention cells in terms of the design of the filter media bed, gravel storage layer, and optional underdrain components. Dry swales may be planted with grasses or be more elaborately landscaped.

59.5.1.8 Perforated Pipe Systems

Conveyance systems designed to artificially infiltrate urban stormwater typically consist of perforated pipes embedded in stone-filled trenches installed within roads.

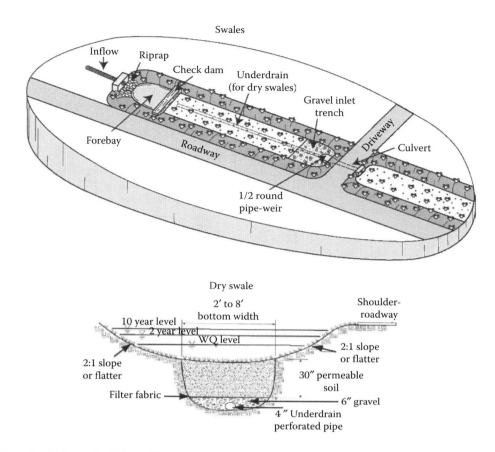

FIGURE 59.11 Schematic of a dry swale. (Adapted from Ontario Ministry of the Environment (OMOE). 2003. *Stormwater Management Planning and Design Manual*. Ontario, Canada.)

Perforated pipe systems can be thought of as long infiltration trenches or linear soakaways that are designed for both conveyance and infiltration of stormwater runoff. They are composed of perforated pipes lined with geotextile fabric that are installed in gently sloping granular stone beds. They allow infiltration of runoff into the gravel bed and underlying native soil while it is being conveyed from source areas or other BMPs to an end-of-pipe facility or receiving water body. Perforated pipe systems are usually combined with other pre-treatment BMPs such as vegetated swales or sediment traps to avoid being clogged with sediment or hydrocarbons. Pollutants are filtered out of the stormwater as it infiltrates into the surrounding soils, and the infiltrated runoff helps to sustain groundwater recharge and maintain baseflows in nearby streams.

Perforated pipe systems can also be referred to as pervious pipe systems, exfiltration systems, clean water collector systems, and percolation drainage systems (Figure 59.12).

The applicability of perforated pipe systems may be limited by several factors, including the permeability of native soils, elevation of the groundwater table, and the slope of the surrounding land. The potential for groundwater contamination must also be considered when selecting suitable sites for perforated pipe systems, as some soluble constituents, such as road de-icing salts,

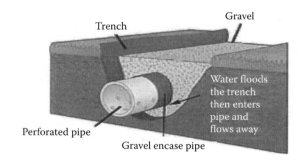

FIGURE 59.12 Schematic of a perforated pipe system. (Adapted from Toronto and Region Conservation Authority. 2010. *Low Impact Development Stormwater Management Planning and Design Guide*. Toronto and Region Conservation for Living.)

are not filtered out of runoff by native soils. Generally, conveyance perforated pipe systems should not be used in areas that are vulnerable to spills of chemicals or hazardous materials.

The gravel beds in which the perforated pipes are installed are typically rectangular excavations with a bottom width of between 600 and 2400 mm. The gravel beds should have gentle slopes of between 0.5% and 1% [27].

59.5.2 Consequences of Implementing the LID Approach

LID has emerged as a highly effective approach to create, retain, or restore natural hydrologic and water quality conditions that may be affected by human alterations, as well as for controlling stormwater pollution and protecting developing watersheds and already urbanized communities throughout the country. There are several specific advantages of using the LID approach in urban communities, as discussed next.

- Several studies have demonstrated that LID is a simple, practical, and universally applicable approach for treating urban runoff [29]. By reproducing predevelopment hydrology, LID effectively reduces runoff and pollutant loads. Researchers have shown the practices to be successful at removing common urban pollutants, including nutrients, metals, and sediment. In addition, the runoff not absorbed by plants could be directed to a vegetated swale next to the sidewalk, where other vegetation will filter pollutants and allow the runoff to infiltrate. The remaining runoff, which is much cleaner and of a lower volume than that generated by conventional development, can be directed to a detention pond or nearby stream. The result is better water quality in local water bodies. Furthermore, since many LID practices cause runoff to infiltrate into groundwater, they help to maintain lower surface water temperatures. LID improves environmental quality, protects public health, and provides a multitude of benefits to the community.

From 2005 to 2008, the US Geological Survey, in a cooperative funding agreement with the Massachusetts Department of Conservation and Recreation, monitored small-scale installations of LID enhancements designed to diminish the effects of storm runoff on the quantity and quality of surface water and groundwater [34]. The monitoring studies examined the effects of replacing an impervious parking-lot surface with a porous surface on groundwater quality; of installing rain gardens and porous pavement in a neighborhood of 3 acres on the quantity and quality of stormwater runoff; and of installing a 3000 ft^2 green roof on the quantity and quality of rainfall-generated roof runoff.

Concentrations of phosphorus, nitrogen, cadmium, chromium, copper, lead, nickel, zinc, and total petroleum hydrocarbons in groundwater were monitored. Enhancing the infiltration of precipitation did not result in discernible increases in concentrations of these potential groundwater contaminants. However, concentrations of dissolved oxygen increased slightly in groundwater profiles following the removal of the impervious asphalt parking-lot surface.

Water quality samples were analyzed for nutrients, metals, total petroleum hydrocarbons, and total coliform and *Escherichia coli* bacteria. A decrease in runoff quantity was observed for storms when precipitation was 0.25 in. or less [34]. Water quality monitoring results were inconclusive; there were no statistically significant differences in concentrations when the pre- and postinstallation period samples were compared.

- Because of its emphasis on natural processes and microscale management practices, LID is often less costly than conventional stormwater controls. The United States Environmental Protection Agency (USEPA) studied different LID sites across the nation and determined that the cost of LID is often less than the cost of conventional development [29]. LID practices can be inexpensive to construct and maintain, and the need to build and maintain conventional treatment practices, such as stormwater ponds, will be reduced and, in some cases, eliminated. In addition, using LID practices to capture runoff onsite can reduce the size of stormwater detention areas, enabling more land to be developed. LID contributes to the creation of a desirable development that often sells faster and at a higher price than equivalent conventional developments.

- LID is not limited to residential areas. Working at a small scale allows the retention volume and water quality control to be tailored to specific site characteristics. Since pollutants vary across land uses and from site to site, the ability to customize stormwater management techniques and the degree of treatment is a significant advantage over conventional management methods. Almost every site and every building can apply some level of site-specific LID and integrated management practices, meaning that any development can become an LID.

- By emulating natural systems and functions, LID offers a simple and effective approach to watershed sensitive development. It makes efficient use of land for stormwater management and therefore interferes less than conventional techniques with other uses of the site. It preserves the landscape and natural features, thereby enhancing the aesthetic worth of a property for homebuyers, property users, and commercial customers. Developers and local governments continue to find that LID saves them money, contributes to public relations and marketing benefits, and improves regulatory expediencies. LID connects people, ecological systems, and economic interests in a desirable way.

Other benefits include habitat enhancement, flood control, improved recreational opportunities, drought impact prevention, and reduction of the urban heat island effect.

59.6 Integrating Reuse of Rainwater Harvesting

Integrated rainwater systems still follow these basic steps, but use modern technology to refine the process with automation and filtration.

Rainwater harvesting is an important element of a well-designed residential site. Rainwater harvesting strategies are most effective when fully integrated with the site's other elements, including vegetation, soils, water flow, solar orientation, and structures (house, driveways, sidewalks, etc.), such that the elements work synergistically to create a productive, self-sustaining site [1].

Rather than rainwater being channeled into a drain, it passes through a mesh filter (to remove leaves and debris) before entering a storage tank. When reused, this water is then automatically pumped back into the building and, after further filtration, is put to use in nonpotable applications, such as toilet flushing, laundry, or commercial washing areas (Figure 59.13).

Functional devices for system control have changed significantly. Traditionally, input and relay devices were used. New instruments, such as web-based information streams, modeling software, and remote access, can now be used to easily control the system.

Float level switches within the tank alert an electronic control device to switch to using the mains supply should the storage tank run empty. The system will always draw on harvested water first. System variations include using a header tank and booster sets, but in essence any integrated rainwater harvesting system follows the same process in its operation.

Rainwater can be used for nearly any purpose that requires water, including landscape use, stormwater control, wildlife and livestock watering, in-home use, and fire protection.

With minimal pretreatment (e.g., gravity filtration or first-flush diversion), the captured rainwater can be used for outdoor nonpotable water uses such as irrigation and washing, or in the building to flush toilets. It is estimated that using rainwater for these applications alone can reduce household municipal water consumption by up to 55% [24]. In some cases, multiple applications of harvested rainwater can be used, allowing significant cost savings.

Due to the different types of water to be reused (reclaimed, rainwater, stormwater, etc.) and the lack of national guidance, use and treatment standards differ between state and local governments. Because of the lack of specific guidance, rainwater and stormwater reuse is often regulated at the same level as reclaimed water, which typically has more clearly defined guidance and standards. Although the general guidance would be similar, it may also differ because of lower levels of initial contamination in harvested rainwater and its potential uses [17].

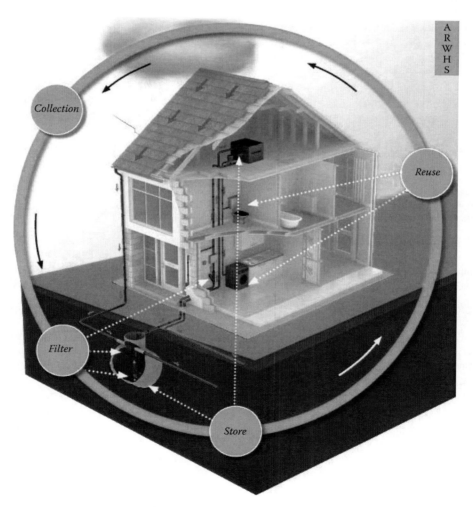

FIGURE 59.13 Integrated active rainwater harvesting systems. (Adapted from College Road North, Aston Clinton, Aylesbury, Buckinghamshire HP22 5EW. Kingspanwater Company website, www.kingspanwater.com.)

59.7 Adaptation Facilities to Climate Change by Rainwater Harvesting

The increasing risk of natural and manufactured catastrophic events is an incentive for decentralized water provision at limited cost.

In Asia, including central, south, east, and southeast Asia, freshwater availability is expected to decrease by 2050, affecting over 1 billion people. The onset of water stress is predicted to come earlier in Africa; by 2020, between 75 and 250 million people are projected to be affected by increased water stress, particularly in the Mediterranean region and the northern and southern parts of the continent.

Currently, 14 out of Africa's 53 countries are classed as water stressed or water scarce. A water-stressed country has a water availability of less than 1700 m³/person-year, and a water-scarce country has less than 1000 m³/person-year. It is estimated that the number of countries in Africa in this situation could double by 2025, to around half of all countries [30].

Rainwater harvesting is an effective water technology for adapting to increased variability in rainfall and water supply. Its decentralized nature allows the owners to benefit from direct management of demand as well as supply. With both modern and indigenous technologies, rainwater harvesting is cost-effective and can release the capital needed in times of disasters of surprising magnitudes. It also reduces greenhouse gas emissions related to water supplies. Rainwater harvesting technology can therefore contribute to both climate change mitigation and adaptation.

Numerous reports and studies describe the contributions of rainwater harvesting to climate change mitigation and adaptation, the most important of which are described next.

59.7.1 Reducing CO_2 Emissions

The Fourth Assessment Report of the Intergovernmental Panel on Climate Change (IPCC) indicated that the expanded use of rainwater harvesting technologies has the potential to reduce emissions by approximately 6 Gt CO_2 equivalent/year in 2030 [15]. Public water systems contribute to climate change through direct emissions of greenhouse gases from water storage reservoirs and water treatment processes and through the material and effective energy uses in the system. Blunt and Holt [25] in Melbourne found that producing potable water using rainwater and other conservation devices generates 0.173 CO_2t/mL (the IEA Energy Sector Carbon Intensity Index [ESCII] tracks how many tonnes of CO_2 are emitted for each unit of energy supplied), while a wastewater treatment plant or wastewater recycling plant generates 0.875 CO_2t/mL. Thus, saving water can reduce greenhouse gas emissions. The greenhouse gas emissions produced by rainwater harvesting as a decentralized water system are far less than those of the current centralized water systems being used.

59.7.2 Providing Adaptation Opportunities for Ecosystems

It is expected that many ecosystem services will be more vulnerable as climate change affects rainfall patterns and increases surface temperatures. Rainwater harvesting will continue to be one way to adapt to these increased changes in water supply and rainfall variability in the future, and, at the same time, enhance ecosystem services.

59.7.3 Helping Communities Adapt to Drought

Rainwater harvesting can help communities adapt to the declining availability of drinking water during droughts. The adoption of rainwater harvesting enables high-quality water to be produced for residential supply, which is particularly important when groundwater is saline, unavailable, or contaminated by pollutants, including agrochemicals. The Central Ground Water Authority has reported that an additional 215 billion m³ of groundwater can be generated by harvesting and recharging only 11% of the surplus runoff. By implementing low-cost rainwater harvesting for drinking and sanitation purposes, communities and individuals can stay healthy during times of crisis events.

59.7.4 Environmental Conservation

Rainwater harvesting can make significant contributions to environmental sustainability. It can mitigate soil erosion, especially in agro-forestry, forests, and agriculture, which is useful for conserving soil, as well as improving water quality for downstream users.

Rainfall volumes are more than adequate to meet the needs of the current population several times over. For example, Kenya would not be categorized as a "water-stressed country" if rainwater harvesting were considered. Overall, Africa's precipitation levels are equivalent to the needs of 9 billion people, or one and half times the current global population.

The chapter arrived at the conclusion that active rainwater harvesting can make a substantial contribution to climate change adaptation and can reduce the vulnerability of water supply in urban areas. There appears to be a significant difference in the response to climate change of active and passive systems for collecting water. Centralized water catchment systems supplying dams were seen as more sensitive, particularly during periods of reduced runoff, and hence potentially more susceptible to failure.

59.8 Summary and Conclusions

Climate change models generally agree that temperature may increase, precipitation patterns may change, and annual precipitation may even decrease. Projected climate change will mean utilities must meet peak seasonal demand in conditions likely to be hotter and drier than in the past. They may be managing stormwater from more intense rainfall events, and facing floods exceeding historic levels. Harvesting rain and stormwater as a way of water reuse produces many potential benefits.

A suite of techniques known today as Green Infrastructure and Low Impact Development collect and manage rainwater/stormwater in ways that capture those benefits. Unlike traditional approaches to stormwater management, these techniques mimic natural predevelopment systems and enhance them to direct water where it can be used and away from where it is a nuisance. A multitude of benefits can accrue including potable water savings, cost savings, reduced flood peaks, water quality management, erosion control, habitat enhancement, reduction of urban heat island effect, and finally helping to reuse and recycle.

Reuse in rainwater harvesting keeps relatively clean water out of the combined sewer system and makes it available for use. This reduces the volume and peak flows of stormwater entering the sewer, thereby reducing flooding and combined sewer overflows and reduces the volume of potable water used for nonpotable applications such as irrigation and toilet flushing.

References

1. Accetturo, A. and Audrey, A. 2012. *Rainwater Harvesting: Guidance toward a Sustainable Water Future.* City of Bellingham, Public Works Department, Washington DC, USA.
2. American Society of Civil Engineers (ASCE). 2000. *National Stormwater Best Management Practices (BMP) Database.* Prepared by the Urban Water Resources Council of ASCE for the U.S. EPA. Office of Science and Technology, Washington, DC.
3. American Water Works Association. 2006. *Residential End Uses of Water.* American Water Works Association Research Foundation, Denver, CO.
4. Chang, M.M., McBroom, W., and Beasley, R.S. 2004. Roofing as a source of nonpoint water pollution. *Journal of Environmental Management,* 73, 307–315.
5. College Road North, Aston Clinton, Aylesbury, Buckinghamshire HP22 5EW. Kingspanwater Company website, www.kingspanwater.com
6. Eslamian, S., Okhravi, S., Fazlolahi, H., and Eslamian, F. 2012. Sustainable management of water resources with techniques of rainwater harvesting in ancient and present. *IWA Specialized Conference on Water and Wastewater Technologies in Ancient Civilizations,* Istanbul, Turkey, March 22–24, 2012.
7. Fawell, J.K., Hickman, J.R., Lurid, U., Mintzand, B., and Pike, E.B. 1996. *Guidelines for Drinking Water Quality,* 2nd Edition, Vol. 2. Health Criteria and Other Supporting Information. World Health Organization (WHO), Geneva.
8. Georgia Amendments to the 2006 International Plumbing, Appendix I, Rainwater Recycling Systems.
9. Gleick, P.H., Haasz, D., Henges-Jeck, C., Srinivasan, V., Wolff, G., Cushing, K.K., and Mann, A. 2003. Waste not, want not: The potential for urban water conservation in California. *Pacific Institute for Studies in Development, Environment, and Security.* Oakland, CA.
10. Göbel, P., Dierkes, C., and Coldewey, W.G. 2007. Storm water runoff concentration matrix for urban areas. *Journal of Contaminant Hydrology,* 91, 26.
11. Gowland, D. and Younos, T. 2008. *Feasibility of Rainwater Harvesting BMP for Stormwater Management SR38–2008.* Virginia Water Resources Research Center, Virginia Tech, Blacksburg, VA.
12. Grady, C. and Younos, T. 2008. *Analysis of Water and Energy Conservation of Rainwater Captures System on a Single Family Home.* Virginia Polytechnic Institute and State University Blacksburg, Virginia.
13. Guideline Committee Members. 2009. Georgia Rainwater Harvesting Guidelines, University of Georgia, pp. 31.
14. Herrmann, T. and Hasse, K. 1997. Ways to get water: Rainwater utilization or long distance water supply. *Water Science Technology.* 36 (8–9).
15. Intergovernmental Panel on Climate Change (IPCC). 2007. *Summary for Policymakers: An Assessment of the Intergovernmental Panel for Climate Change.* Valencia, Spain.
16. Kinkade-Levario, H. 2007. *Design for Water.* New Society Publishers, Gabriola Island, British Columbia, Canada.
17. Kloss, C. and Low Impact Development Center, Inc. 2008. *Managing Wet Weather with Green Infrastructure Municipal Handbook: Rainwater Harvesting Policies.* U.S. Environmental Protection Agency, Washington, DC.
18. Landmark Design Group. http://www.thelandmarkgroup.ca.
19. Langdon, D. and Rawlinson, S. 2007. Sustainability—Managing water consumption. *The Financial Times,* July 10, 2007.
20. Lawson, S., LaBranche-Tucker, A., Otto-Wack, H., Hall, R., Sojka, B., Crawford, E., Crawford, D., and Brand, C. 2009. *Virginia Rainwater Harvesting Manual,* 2nd Edition. The Cabell Brand Center, Salem, VA. www.cabellbrandcenter.org.
21. North Shore City. 2007. *Bioretention Design Guidelines.* Sinclair, Knight and Merz, Auckland, New Zealand.
22. Okhravi, S.S., Eslamian, S.S., Eslamian, F., and Tarkesh Isfahani, S. 2015. Indigenous knowledge as a supportive tool for sustainable development and utilisation of rainwater harvesting systems. *Journal of Flood Engineering,* 5(1–2), 39–50.
23. Ontario Ministry of the Environment (OMOE). 2003. *Stormwater Management Planning and Design Manual.* Ontario, Canada.
24. Reids Heritage Homes, Ontario's Largest Homebuilders. 2007. *Leeds Home Showcased Features.* http://www.ReidsHeritageHomes.com

25. Salas, J.C. 2003. An exploratory study on rainwater harvesting in the Philippines. A Report to the World Bank.

26. *Texas Manual on Rainwater Harvesting,* Third Edition. 2005. Texas Water Development Board, Austin, TX.

27. Toronto and Region Conservation Authority. 2010. *Low Impact Development Stormwater Management Planning and Design Guide.* Toronto and Region Conservation for Living.

28. U.S EPA. 2008. Clarification on which stormwater infiltration practices/technologies have the potential to be regulated as "Class V" wells by the Underground Injection Control Program. Water Permits Division and Drinking Water Protection Division. Washington, DC.

29. U.S. Environmental Protection Agency, Office of Water and Low Impact Development Center. 2000.Low Impact Development (LID); A Literature Review. EPA-841-B-00-005, Washington, DC.

30. UNEP and World Agroforestry Centre. 2006. *Harvesting Rainfall a Key Climate Adaptation Opportunity for Africa.* http://www.UNEP.org.

31. United Nations Environment Programme (UNEP). *Rainwater Harvesting and Utilisation: An Environmentally Sound Approach for Sustainable Urban Water Management: An Introductory Guide for Decision-Makers.* Newsletter and Technical Publications. http://www.UNEP.org.

32. Young, D. 2012. Design and implementation of stormwater infiltration and rainwater harvesting practices: New insights from field evaluations. Toronto and Region Conservation Authority, *Innovative Stormwater Management Conference.* Toronto.

33. Younos, T. and Lawson, S. 2011. Rainwater harvesting: A holistic approach for sustainable water management in built environments. *Low Impact Development Symposium,* Philadelphia, September 26–28, 2011.

34. Zimmerman, M.J., Waldron, M.C., Barbaro, J.R., and Sorenson, J.R. 2010. Effects of low-impact-development (LID) practices on streamflow, runoff quantity, and runoff quality in the Ipswich River Basin, Massachusetts—A summary of field and modeling studies. US Geological Survey Circular, 1361.

60

Rainwater Tanks as a Means of Water Reuse and Conservation in Urban Areas

Ataur Rahman
University of Western Sydney

Saeid Eslamian
Isfahan University of Technology

PREFACE

A rainwater harvesting system (RWHS) is the most popular alternative water supply system in many cities around the world. It has been found that a typical RWHS in an urban area can save 20%–60% of mains water depending on the local rainfall, roof size, number of occupants in the house, and types of water use. To estimate an appropriate size of a rainwater tank for a given application, a rainwater tank model should ideally be adopted, which considers local rainfall, roof area, and water demand data to determine an appropriate tank size. However, for general purposes, standard local council guidelines may be used to select an appropriate tank size. For example, in Sydney, Australia, for a detached house with 4 people, if rainwater is to be used for toilet flushing, washing clothes, and moderate gardening, a 5-kL tank may be adopted. In general, water from an RWHS can safely be used for nonpotable purposes. However, it is not recommended to drink water from an RWHS without treatment, in particular for people who fall in immuno-compromised groups (e.g., children, old people, and cancer patients). A RWHS must be maintained regularly if the water is to be consumed by human beings.

60.1 Introduction

Water is a vital source for the very existence of human beings on Earth. Although the total volume of water on Earth has remained constant since its creation, its variation over space and time, with respect to quantity and quality, presents some unique problems and challenges to all who have an interest in water. The major source of water is the ocean; however, this water cannot be consumed directly due to poor quality. Water in lakes and rivers often needs treatment before it can be consumed by human beings. Only groundwater, rainwater in some areas, and some spring water may be consumed without treatment. In urban areas, due to ever-increasing water demand, urban authorities often struggle to supply adequate water to their residents. In particular, long droughts and climate change related factors have made water supply in urban areas a challenging task. Under this context, water reuse has become popular in many urban areas. There are a number of different modes of water reuse including wastewater reuse, greywater reuse, and stormwater reuse. Rainwater harvesting falls under stormwater reuse. This chapter presents various issues of rainwater harvesting in urban areas including the quantity and quality aspects of the harvested rainwater.

60.2 Rainwater Harvesting Systems in Urban Areas

RWHS can be defined as a facility that collects and stores rainwater to provide water for various potable and nonpotable purposes including irrigation, toilet flushing, and washing of clothing, cars, and hard surface. Rainwater can be collected from different types of surfaces such as house roof, driveway, and car park. However, roof-harvested rainwater is the most common form of alternative water used in urban areas. Harvested water can be stored in a number of ways: (1) metal, concrete, or poly tanks; (2) underground reserve such as a house basement; and (3) gravel pit located below ground. There are several advantages to using roof-harvested rainwater, including (1) reducing the pressure on the mains water supply, (2) reducing stormwater runoff that can

often degrade creek ecosystem health, and (3) providing an alternative water supply during times of water restrictions [3]. A typical RWHS is shown in Figure 60.1.

Many countries have investigated the potential use of roof harvested rainwater including Australia, Canada, Denmark, Germany, India, United Kingdom, South Korea, Japan, New Zealand, Thailand, and the United States. In Australia, about 21% of households source their water from rainwater tanks [1]; however, the dominant source is the public water supply (mains) as can be seen in Figure 60.2. Among Australian states, South Australia has the highest rate of rainwater tank use; 49% of the total households use an RWHS, and the Northern Territory has the lowest rate of rainwater tank use (6%). Use of an RWHS is more common outside capital cities (35%) than in capital cities (12%). In Australian capital cities, the most common reason for installing an RWHS is to save water. About 42% of households having an RWHS state saving water as a reason for installing it (see Figure 60.3). Cost appears to be the most common reason for not having an RWHS (48%). Walton et al. [58] mentioned that there are over 300,000 rainwater tanks in South East Queensland alone, with one in three homes owning a tank.

The Australian Bureau of Statistics (ABS) updated its report on RWHS in 2010, accordingly, about 26% of households used an RWHS as a source of water in 2010, 21% higher than 2007. As per ABS [2], 49% of South Australian households used an RWHS, followed by Queensland (36%) and Victoria (30%). The Northern Territory had the lowest proportion of households (5%) that used a rainwater tank. Between 2007 and 2010,

FIGURE 60.1 A picture of a typical rainwater harvesting system.

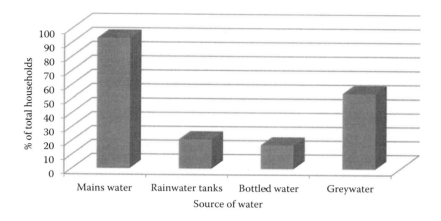

FIGURE 60.2 Water sources in Australian households.

the greatest rise in RWHS use had been for Queensland and Victoria. ABS [2] reported that of households living in a dwelling suitable for installation of an RWHS, the percentage with an RWHS has increased from 24% in 2007 to 32% in 2010. Among all the Australian states, Brisbane had seen the largest increase in households having an RWHS (18% in 2007 to 43% in 2010).

More recently, RWHS has been established as an important alternative source of freshwater in Australian cities. RWHS is regarded as an important component of water sensitive urban design [26], which is a sustainable urban design practice. An RWHS provides multiple benefits, for example, it can save mains water significantly, provide onsite detention effect, and reduce treatable urban runoff volume. The Building Sustainability Index (referred to as BASIX) was introduced by New South Wales Department of Planning in Australia [39]. It is a web-based tool

that assesses the potential performance of new residential dwellings against various sustainability indices. As per BASIX, all new houses in New South Wales must save at least 40% potable water than the average house by adopting various water savings techniques including installation of RWHS.

Use of an RWHS is more common in detached houses in Australia. As reported by ABS [2], about 26% of separate houses are fitted with an RWHS, as opposed to only 6% of semi-detached or townhouses. About 25% of family households had an RWHS compared with only 13% of group households or multi-unit houses.

60.3 Rainwater Harvesting System versus Onsite Detention

In some instances, an RWHS may provide stormwater detention effect, that is, to hold the water for a certain period to reduce flooding in the downstream of an urban catchment [56]. In this case, a much larger tank is needed and there should be an RWHS fitted to most of the houses in an urban catchment. However, it should be noted that this can provide retention effect to only smaller to medium floods (up to an average recurrence interval of about 2 years).

Coombes et al. [11] found that the airspace in a rainwater tank could be used to provide onsite detention effect, which would vary with the type of development and design of the system generally ranging from 32% to 65%. They noted that on the lot scale, a properly designed RWHS could reduce the peak discharge, but an RWHS could only reduce the volume of discharge without affecting the peak flows. Coombes et al. [11] argued that peak discharges at the lot scale had little or no bearing on the floods at a catchment scale, as flooding is a volume-driven process and roof area represents only a smaller fraction of the total area of an urban catchment.

A number of researchers (e.g., [5,25]) reported that the reduction of peak discharge on lot scale with onsite retention such as RWHS would not be significant. Argue and Scott [5] indicated

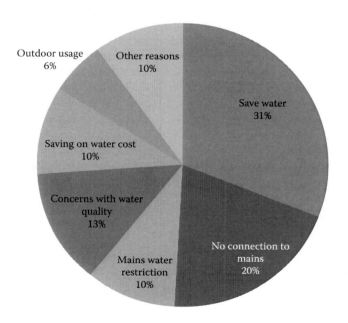

FIGURE 60.3 Percentages of households in Australia using a rainwater harvesting system.

that with a large-scale model (i.e., 14 ha), the onsite detention and onsite retention systems would produce similar hydrographs. They argued that the peak discharge on a lot scale is likely to be larger for onsite retention than onsite detention, but in medium to large catchments, the cumulative effect of volume reduction, under onsite retention, reduces the effect of high peak discharges delivered by individual sites.

60.4 Water Savings from Rainwater Harvesting Systems

Many studies all over the world have demonstrated that a typical RWHS in an urban area can save 20%–60% of mains water depending on the local rainfall, roof size, number of occupants in the house, and types of water use. Some examples are provided next.

Coombes and Kuczera [12] evaluated the performance of 1- to 10-kL rainwater tanks in four Australian capital cities with mains water trickle top-up used to supplement mains water supply for domestic toilet, laundry, hot water, and outdoor use. They found that an individual dwelling with 150 m² roof area and 1- to 5-kL tank size located in Sydney could achieve 10%–58% mains water savings depending on the number of people living in the house. Depending on roof area and number of occupants in the household, the use of RWHS resulted in an annual mains water savings ranging from 18 to 55 kL for a 1-kL rainwater tank and 25 to 144 kL for a 10-kL rainwater tank.

Villarreal and Dixon [57] investigated the water savings potential of an RWHS system from large roof areas in Sweden. They found that 30% of mains water savings can be achieved from a 40-m³ tank if rainwater is used for toilet flushing and washing machine.

Muthukumaran et al. [38] found that use of rainwater inside a purpose-built home in regional Victoria in Australia can save up to 40% of potable water use. Farreny et al. [18] in a study of rainwater harvesting in Spain reported that a sloping smooth roof may harvest up to about 50% more rainwater than flat rough roofs.

Eroksuz and Rahman [17] conducted a research on the use of an RWHS for multi-unit buildings in three cities of New South Wales (Sydney, Newcastle, and Wollongwong), Australia. They found that for a 30-m² roof area per person (for toilet, hot water, laundry, and irrigation use), a 10-kL rainwater tank can provide 21% mains water savings in a multi-unit building. They noted that in order to maximize the water savings, a larger tank would be more appropriate in a multi-unit building, for example, a 30-kL tank provided 40% mains water savings.

Ghisi et al. [22] evaluated the potential for potable water savings by using rainwater for washing vehicles in petrol stations located in Brasilia in Brazil. They found that the average potential for potable water savings by using rainwater was 32%. Imteaz et al. [27–29] demonstrated the water savings from an RWHS that was fitted with a building in Swinburne University of Technology, Melbourne, Australia.

Rahman et al. [43] examined the performance of an RWHS at 10 different locations in the Greater Sydney region, Australia. Three different combinations of water use were considered: (1) toilet and laundry, (2) irrigation, and (3) a combination of toilet, laundry, and irrigation (combined use). Three different tank sizes were considered—2, 3, and 5 kL for a roof area of 200 m² and 4 occupants. For toilet and laundry use, for a 5-kL tank, average annual water savings ranged from 33 (Campbelltown) to 35 kL (Bankstown), with a mean value of 34 kL over all 10 locations. For irrigation use, for a 5-kL tank, average annual water savings ranged from 43 (Sydney) to 57 kL (Penrith), with a mean value of 48 kL. For combined use, for a 5-kL tank, Manly had the highest average annual water savings (69 kL), Richmond had the lowest annual water savings (54 kL), and Parramatta corresponded to the average annual water savings over the 10 locations (59 kL).

In a study in Iran, Mehrabadi et al. [36] assessed the applicability and performance of an RWHS to supply daily nonpotable water. They found that in humid climate, it was possible to supply about 75% of nonpotable water demand by storing rainwater from larger roof areas. Sample et al. [45] assessed the performance of an RWHS in Richmond, Virginia, using storage volume, roof area, irrigated area, and an indoor nonpotable demand, and a storage dewatering goal as independent design variables. Results indicated that land uses that provided larger demands, such as offices, commercial sites, and high-density residential sites, would be better suited than lower-density residential lots.

60.5 Essential Components of a Rainwater Harvesting System

60.5.1 Roof Catchment

As collecting surface, the roof area of a house, factory, or shed is generally used. Rainwater can be collected from most types of roofs, including asbestos roofs, provided they have not been painted with lead-based paints or coated with bitumen-based material [44]. Some types of new tiles and freshly applied acrylic paints may affect the color and taste of rainwater and the first few runoff should be discarded. It is recommended not to use chemically treated timbers and lead flashing in roof catchments. Overflows or discharge pipes from roof-mounted appliances such as evaporative air conditioners or hot water systems should not be allowed to discharge onto the roof catchment area or gutters.

60.5.2 Types of Tanks

Rainwater tanks can vary widely in shape, size, and construction materials. Most of the rainwater tanks are built aboveground as underground tanks often fail due to lateral soil pressure when the tank is empty. A rainwater tank is generally made of polypropylene, concrete, and coated corrugated iron. The size of a rainwater tank depends on water demand, roof catchment area, rainfall in the locality, and the presence of an alternative water supply. If the urban water supply is present, an RWHS is generally used to save potable water and the rainwater tank size is generally smaller; however, in peri-urban and rural areas where urban water supply or groundwater is not available, a much larger rainwater tank is needed to maintain the reliability of the water supply. Each of these factors has a different effect on the size of the rainwater tank to meet a given demand [7,11,50].

For a larger tank size, a polypropylene (PVC) tank is not appropriate as it cannot withstand higher hydrostatic pressure exerted by the water inside the tank when it is nearly full. Concrete is the most preferred material to construct a large tank. In some cases, a series of smaller PVC or metal tanks can be used to meet a large demand, which however is not feasible in urban areas due to space limitation. In terms of visual appeal, a smaller tank is preferred over a very large tank in urban areas. The cost of the tank is often the final deciding factor in terms of the selected tank material and size.

An RWHS consists of an inlet, an overflowing outlet, and an outlet to provide desired water to the users. To facilitate indoor use (such as toilet flushing and washing machine), an RWHS is often fitted with a pump. In urban areas where mains water is available, a top-up system is often included with the RWHS, which allows mains water to enter into the tank when the water level inside the tank falls below a threshold (e.g., 10% of the tank capacity). Other components of an RWHS include a first-flush [16,50] and an optional water treatment system [21,50].

60.5.3 Accessories of a Rainwater Harvesting System

A first-flush device should be used to prevent the first part of the runoff from reaching the tank. This device can reduce the amount of dust, bird droppings, leaves, and debris that have accumulated on the roof from being washed into the tank directly.

A first-flush device diverts the first part of the runoff from the roof (e.g., first 5 min). Griffin et al. [24] showed that the first 30% runoff generated by a storm has a higher contaminant concentration and contains nearly 70% of the total pollutant load. Bucheli et al. [8] found that the runoff from the first 2 mm of rain contains most of the total pollutants on a number of different roof types. The remaining flows for the roofs have a lower contaminant concentration and could be treated by the natural processes within the rainwater tank itself. Bach et al. [6] suggested estimating the first-flush based on statistical analysis of the concentration of pollutants in the first-flush versus the background concentrations rather than a set amount of rainfall as the criterion. Van der Sterren et al. [56] found in a study in Western Sydney, Australia that water quality in the first-flush device was much poorer than the water inside the tank. A similar finding was reported by Kus et al. [32,33].

It is necessary for the rainwater tank to rest on a concrete slab due to the weight of the rainwater tank particularly when it is full. Sometimes, a metal frame is used to support a rainwater tank.

TABLE 60.1 Elements of a Rainwater Harvesting System

Element	Function
Roof and gutter system	To collect water from rainfall and convey to rainwater tank
Down pipe	To allow water entering from the roof and gutter system into the tank
First-flush	To separate first portion of runoff containing high concentration of pollutants
Tank	To hold water
Concrete base or metal frame	To support the weight of rainwater tank
Pump	To make water flowing inside the house for toilet flushing and washing machine, etc.
Leaf eater	To filter out leaves from entering into tank
Tank inlet and outlet	To allow water into the tank from roof and gutter system and to release the water when the tank is full
Screen	To seal all the inlets and outlets so that small animals and insects (e.g., frogs, rats, and mosquitoes) cannot enter into the tank
Tank top-up system	To allow mains water (if existing) to enter into the tank to maintain a minimum water level inside the tank

A leaf eater is a component mounted directly under a roof gutter preventing leaves and debris from flowing through the gutters and into the rainwater tank. A typical leaf eater has a primary and a secondary screen. The primary screen filters out leaves and debris while the secondary screen filters out mosquitoes and vermin. The device is self-cleaning and hence it needs low maintenance.

A tank top-up system enables a dual supply of water. That is, it ensures there is always a supply of water in the tank and tops up when the water level drops below a minimum level. The system utilizes a valve that when activated would introduce mains water to the tank. The tank top-up automatically ceases once the water level reaches the designated minimum level.

Tank screens are generally fitted to all openings to and from the rainwater tank. They provide the last line of defense in keeping pests out of the tank and allowing the water to breathe. Two tank screens are required to be fitted, one for the tank inlet and the other for the overflow outlet. Various elements of an RWHS are listed in Table 60.1.

60.6 Quality of Water from a Rainwater Tank

Quality of roof-harvested rainwater in urban areas depends on a number of factors including surrounding land use, air quality, bird population, roofing and tank materials, rainfall characteristics, and quality and frequency of maintenance of the RWHS.

Generally, water from roof-harvested rainwater is safe to drink. If the tank water is clear, has little smell, and is from a well-maintained system, it is probably safe and unlikely to cause any illness for most users. However, disinfecting the water (e.g., by boiling) before consumption may need to be considered for those who are immuno-compromised such as the very young or very old, cancer patients, people with diabetes, organ transplant recipients, or people who are HIV positive.

Rainwater collected from roof catchments generally contains few pollutants. However, there may be increased pollution by airborne contaminants from very heavy traffic, industrial areas, and construction sites. Collection of rainwater for human consumption (drinking and cooking) in areas affected by very heavy traffic, industry, incinerators, and smelters is not recommended [44]. The microbiological quality of roof-harvested rainwater may not be as good as mains water, but if the collection systems are well maintained, the risk of harmful organisms being present is low [44]. Rainwater does not contain fluoride and hence if the rainwater is the major source of water for drinking and cooking, consumers should obtain necessary fluoride from alternative sources.

Roof-harvested rainwater has not been widely utilized for drinking due to a lack of information on the risk from chemical and microbiological pollutants that could be found in the roof-harvested rainwater. There is also a lack of appropriate guidelines on the use of roof-harvested rainwater for both drinking and nonpotable uses and how the risk from chemical and microbiological pollutants can be managed [3]. Ahmed et al. [3] presented a review of microbiological quality of the roof harvested rainwater and the associated health risk. They noted that several case studies established potential links between gastroenteritis and consumption of untreated roof-harvested rainwater. In the state of Victoria, Australia, 49 rainwater tanks surveyed for the presence of total coliforms, *E. coli*, and enterococci [49] demonstrated that 33% were positive for *E. coli* and 73% positive for enterococci, exceeding the Australian Drinking Water Guidelines of 0 CFU/100 mL [4]. A study in South Korea found that 92% and 72% of roof-harvested rainwater samples were positive for total coliforms and *E. coli*, respectively [35]. Ahmed et al. [3] suggested that health risks assessment models, such as those using quantitative microbial risk assessments, should be used to manage and mitigate health risks associated with drinking and nonpotable uses of roof harvested rainwater. Table 60.2 presents examples causing disease from drinking of untreated roof-harvested rainwater.

The quality of water varies with the location inside the rainwater tank. Previous research has found that the water quality at the surface of the water in the tank is significantly different from that of the bottom of the tank [10]. This difference had been attributed to stratification of water within the tank. Furthermore, a biofilm can form on the air/liquid interface and the solid/liquid interface within a rainwater tank [41,48]. These biofilms were reported to protect natural systems against several pollutants, but were very susceptible to the flow of the liquid [41]. Similar to lakes, chemical and biological gradients can be found in large rainwater tanks [15,48].

Rainwater tank and roof materials largely govern the water quality within the rainwater tank. For example, the galvanized steel roofs and rainwater tanks are likely to add significant concentrations of aluminum and zinc to the water. The lead flashing on the roof can increase the lead concentrations in the tank water [56]. Concrete roof and tanks can cause an increased hardness and conductivity of the water in the rainwater tank.

Van der Sterren et al. [53–55] found in a study in Western Sydney, Australia that topping-up of a rainwater tank with mains water could cause a change in the harvested water quality due to the disturbance of water and mixing with the sediments at the bottom of the tank. They also found that water quality in the tank after heavy storm was poorer compared with the start of the storm. In many instances, the tank water did not meet the standard in terms of biological and chemical parameters and hence they suggested not consuming the rainwater directly from the tank without treatment.

If an RWHS has shown sign of pollution, it can be disinfected using 40 mL of liquid sodium hypochlorite (12.5% available chlorine) or 7 g of granular calcium hypochlorite per 1000 L of water [44]. A chlorine taste and odor may persist for a few days after such disinfection, but this does not make the water unsafe for drinking. It is recommended not to use stabilized chlorine in the RWHS.

60.7 Sizing of a Rainwater Tank

There are different approaches for sizing a rainwater tank; all are based on the water balance principle. Many studies have attempted to develop standard design curves to estimate tank size in a given area for a given application. Some examples are provided here.

Khastagir and Jayasuriya [30] used water demand and roof area to develop a set of dimensionless number curves to obtain the optimum rainwater tank size for a group of suburbs in Melbourne. A paper by Su et al. [51] focused on the development of a relationship between storage and deficit rates for RWHS. Results showed that as the deficit rate increased, so too did the storage size of the tanks. Fewkes [19] studied the performances of rainwater tanks in a house in the UK, which produced a set of dimensionless design curves that enables estimation of the rainwater tank capacity required to achieve a desired performance level given the roof area and demand patterns.

Palla et al. [40] investigated the performances of an RWHS based on 46 sites within the European territory, equally distributed among 5 main climate zones based on the Koppen–Geiger classification. A behavioral model was implemented and

TABLE 60.2   Examples of Reported Cases of Medical Problems Associated with Drinking of Untreated Roof Harvested Rainwater

Country of Occurrence	Types of Disease	Number of People Affected	Reference
Australia	Diarrhea	23	[37]
Australia	Diarrhea, abdominal pain, nausea	27	[20]
New Zealand	Diarrhea	2	[47]
U.S. Virgin Islands	Legionnaires' disease	27	[46]

nondimensional parameters were used to suitably compare the system performance under various environmental (i.e., hydrologic characteristics) and operational (storage capacity) conditions. They found that the antecedent dry weather period was the main hydrologic parameter affecting the system behavior, while rainfall event characteristics (including event rainfall depth, intensity, and duration) had weak correlations.

Campisano et al. [9] examined the performance of domestic RWHS for toilet flushing in 44 Italian sites by using a nondimensional approach characterized by a demand and storage fraction. Regression curves were developed to describe the relationship between the water saving efficiency and the modified storage fraction that allowed the RWHS systems to be sized based on the desired level of water saving performance. The results showed that the system performance was dependent on climatic zones and the intra-annual characteristics of the precipitation regimes.

Ghisi and Schondermark [23] presented an investment feasibility analysis of the RWHS for the residential sector for five towns in Santa Catarina State, southern Brazil. They observed that the ideal tank capacity would be conservative for high rainwater demands and in such cases, an investment feasibility analysis should be carried out in order to obtain a more appropriate tank capacity. It was noted that rainwater use would be economically feasible for most cases, and the higher the rainwater demand, the higher the feasibility.

To simulate the performance of an RWHS, a water balance simulation model can be built based on the following equations, which implements a simple water balance-storage model by assuming that the tank is empty at the beginning. This model accounts for various factors for an RWHS such as tank size, daily rainfall data, losses, daily water demand, mains top-up, and tank spillage. The spillage indicates the overflow from the tank if the tank capacity is exceeded. This approach has been advocated by Su et al. [51] and Rahman et al. [43].

In this modeling, the rainfall is regarded as inflow and the release and possible spillage as outflow. The release is estimated based on the following equations:

$$R_t = D_t \quad \text{if} \quad I_t + S_t \geq D_t \tag{60.1}$$

$$R_t = I_t + S_{t-1} \quad \text{if} \quad I_t + S_{t-1} < D_t \tag{60.2}$$

where D_t is the daily demand (m³) on day t, S_{t-1} is the tank storage at the end of the previous day (m³), R_t is release from rainwater tank (m³), and I_t is inflow (m³). Spill (SP_t) (m³) is calculated from the following equations:

$$SP_t = I_t + S_{t-1} - D_t - S_{MAX} \quad \text{if} \quad I_t + S_{t-1} - D_t > S_{MAX} \tag{60.3}$$

$$SP_t = 0 \quad \text{if} \quad I_t + S_{t-1} - D_t \leq S_{MAX} \tag{60.4}$$

where S_{MAX} is the design storage capacity (m³). The tank storage S_t at the end of day t is calculated using the following equations:

$$S_t = S_{MAX} \quad \text{if} \quad SP_t > 0 \tag{60.5}$$

$$S_t = S_{t-1} + I_t - R_t \quad \text{if} \quad SP_t = 0 \tag{60.6}$$

60.8 Maintenance of a Rainwater Harvesting System

Maintenance of an RWHS is important to receive its intended services, similar to any other plant. However, there have been reports of lack of maintenance of RWHSs. For example, Walton et al. [58] mentioned that few tank owners in South East Queensland, Australia undertake any maintenance of their tanks. This issue was regarded a concern for government and water planners because of the potential implications for public health, strategic water planning, and loss of publicly subsidized assets, if people abandon their tanks due to unsatisfactory performance. The need for the maintenance of RWHS has been highlighted by Cunliffe [13] and en Health Council [16].

Leaves, snails, rodents, spiders, and frogs may enter rainwater tanks where they decompose on the tank floor when dead, which can create bio-film that may contaminate the water inside the tank. Droppings from birds and possums on roof catchment may enter into the tank, where they form *E. coli*, Campylobacter, Cryptosporidium, Giardia, and waterborn worms. Hence, regular cleaning of RWHS is important, in particular when tank water is to be consumed by human beings.

All access points in the RWHS except the inlet and overflow should be sealed. The inlet should incorporate a mesh cover and a strainer to keep out debris and prevent the entry of mosquitoes and other insects. The tank overflow point should also be covered with an insect-proof screen.

Roof catchments feeding an RWHS should be kept clean and clear of leaves and debris. Overhanging tree branches should be removed. Gutters should be regularly inspected and cleaned if leaves and other dirt are deposited. The use of screens/guards should be considered to seal all the openings and all screens should be cleaned regularly. Water ponding in gutters must be prevented as it can provide breeding sites for mosquitoes and could result in eggs being washed into tanks. Tanks should not be allowed to become breeding sites for mosquitoes. If mosquitoes are detected in a tank, the entry points should be located and closed. For most types of tanks, mosquito breeding can be stopped by adding a teaspoon (5 mL) of domestic kerosene [44]. However, kerosene should not be used in AquaplateTM or some plastic tanks. Prevention of mosquito access is the best control option.

It is recommended to examine the tanks for accumulation of sludge at least every 2–3 years. If sludge is covering the bottom of the tank, it should be removed by siphon or by completely emptying and rinsing the tank. Excessive sludge is a sign of inadequate maintenance of the roof and gutters. The maintenance essentials of an RWHS are summarized in Table 60.3.

TABLE 60.3   Maintenance Essentials of a Typical Rainwater Harvesting System

Maintenance	Recommended Frequency
Wash roof and gutter	Once a year
Cut branches of overhanging trees	Once a year
Clean first-flush device, leaf eater, and filter	After a big storm
Remove sludge	Once in 2–3 years
Kill mosquitoes by adding kerosene (not for plastic tanks)	If sign of mosquito breeding is noticed
Inspect screens and openings and seal if needed	Inspect once a year and repair/replace if needed
Disinfect rainwater tank using 40 mL of liquid sodium hypochlorite (12.5% available chlorine) or 7 g of granular calcium hypochlorite per 1000 L of water)	If signs of pollution are noticed

60.9 Economics of Rainwater Harvesting Systems

There have been many studies on the financial benefits of an RWHS. The outcomes of these studies vary with a payback period of 10–60 years depending on various adopted assumptions. Some of these studies are noted next.

A study by Rahman et al. [42] for multi-story buildings in Sydney found that it could be possible to achieve "payback" for the RWHS under some favorable scenarios and conditions. They found that a smaller discount rate is more favorable and the greater the number of users, the higher the benefit-cost ratio for an RWHS. Domenech and Sauri [14] investigated the financial viability of the RWHS in single and multi-family buildings in the metropolitan area of Barcelona, Spain. In single-family households, an expected payback period was found to be between 33 and 43 years depending on the tank size, while in a multifamily building, a payback period was 61 years for a 20-m³ tank. Imteaz et al. [27] found that for commercial tanks connected to large roofs in Melbourne, total construction costs can be recovered within 15–21 years' time depending on the tank size, climatic conditions, and future water price increase rate. Tam et al. [52] investigated the cost effectiveness of RWHS in residential houses around Australia and found that this system can offer notable financial benefit for Brisbane, the Gold Coast, and Sydney due to the relatively higher rainfall in those cities compared to Melbourne. Zhang et al. [59] examined the financial viability of RWHS in high-rise buildings in four capital cities in Australia and found that Sydney has the shortest payback period (about 10 years) followed by Perth, Darwin, and Melbourne. Khastagir and Jayasuriya [31] conducted the financial viability of RWHS in Melbourne, Australia and found that payback periods vary considerably with the tank size and local rainfall.

Kyoungjun and Chulsang [34] showed that rainwater collection would only be feasible in South Korea during six months of the year. They also found that a benefit cost ratio higher than 20% could not be gained due to the cost of water being so inexpensive in South Korea. They suggested that the cost of water supply would need to be increased by a factor of five approximately for the RWHS to become financially viable in South Korea.

Rahman et al. [43] examined the performance of an RWHS in 10 different locations in Greater Sydney, Australia. Three different combinations of water use were considered: (1) toilet and laundry, (2) irrigation, and (3) a combination of toilet, laundry, and irrigation (combined use). Three different tank sizes were considered: 2, 3, and 5 kL. It was found that the benefit cost ratios for an RWHS for all 10 study locations and three tank sizes were smaller than 1.00.

60.10 Summary and Conclusions

A RWHS can save a significant volume of potable water and assist in minimizing pollution of urban waterways. It has been found that a typical RWHS in an urban area can save 20%–60% of mains water depending on the local rainfall, roof size, number of occupants in the house, and types of water use. To estimate an appropriate size of a rainwater tank for a given application, a rainwater tank model should ideally be adopted, which considers local rainfall, roof area, and water demand data to determine an appropriate tank size. However, for general purposes, standard local council guidelines may be used to select an appropriate tank size. For example, in Sydney, Australia, for a detached house with 4 people, if rainwater is to be used for toilet flushing, washing clothes, and moderate gardening, a 5-kL tank may be adopted. In general, water from an RWHS can safely be used for nonpotable purposes. However, it is not recommended to drink water from an RWHS without treatment, in particular for people who fall in immuno-compromised groups (e.g., children, old people, and cancer patients). An RWHS must be maintained regularly if the water is to be consumed by human beings. An RWHS has often been found not to be financially viable in particular where mains water is relatively cheaper such as in Australia. Government authorities should provide subsidy to install an RWHS in such countries as RWHS saves significant volumes of mains water and provides many other environmental benefits.

References

1. ABS. 2007. *Environmental Issues: People's Views and Practices*. Australian Bureau of Statistics (ABS). http://www.abs.gov.au.
2. ABS. 2010. *More Australian Using Rainwater Tanks*. Australian Bureau of Statistics (ABS), Media Release, November 19, 2010 by ABS. http://www.abs.gov.au.
3. Ahmed, W., Gardner, T., and Toze, S. 2011. Microbiological quality of roof-harvested rainwater and health risks: A review. *Journal of Environmental Quality*, 40, 1–9.
4. ADWG. 2004. *Guidelines for Drinking Water Quality in Australia*. National Health and Medical Research Council/Australian Water Resources Council.
5. Argue, J.R. and Scott, P. 2000, On-site stormwater retention (OSR) in residential catchments: A better option? *Proceedings of the 40th Annual Conference of New South*

Wales Floodplain Management Authorities, Parramatta, Australia, May 9-12, 2000, 1–9.

6. Bach, P.M., MacCarthy, T.D., and Deletic, A. 2010. Redefining the stormwater first-flush phenomenon. *Water Research*, 44(8), 2487–2498.

7. Barry, M.E. and Coombes, P.J. 2006. Optimisation of mains trickle top-up supply to rainwater tanks in an urban setting. *Australian Journal of Water Resources*, 10(3), 269–275.

8. Bucheli, T.D., Müller, S.R., Heberle, S., and Schwarzenbach, R.P. 1998. Occurrence and behaviour of pesticides in rainwater, roof runoff, and artificial stormwater infiltration. *Environmental Science and Technology*, 32(22), 3457–3464.

9. Campisano, A., Gnecco, I., Modica, C., and Palla, A. 2013. Designing domestic rainwater harvesting systems under different climatic regimes in Italy. *Water Science and Technology*, 67, 2511–2518.

10. Coombes, P.J. 2002. *Rainwater Tanks Revisited: New Opportunities for Urban Water Cycle Management*. Research Report, The University of Newcastle, Australia.

11. Coombes, P.J., Frost, A., and Kuczera, G. 2001. *Impact of Rainwater Tanks and On-Site Detention Options on Stormwater Management in the Upper Parramatta River Catchment*. Research Report, University of Newcastle, Australia.

12. Coombes, P.J. and Kuczera, G. 2003. A sensitivity analysis of an investment model used to determine the economic benefits of rainwater tanks. *Proceedings 28th International Hydrology and Water Resources Symposium*, Wollongong, Australia, November 10–14, 2003, 243–250.

13. Cuncliffe, D.A. 1998. *Guidance on the Use Rainwater Tanks*. National Environmental Health Forum Monographs, Water Series Number 3. Department of Human Services, Rundle Mall.

14. Domenech, L. and Sauri, D.A. 2010. Comparative appraisal of the use of rainwater harvesting in single and multi-family buildings of the metropolitan area of Barcelona (Spain): Social experience, drinking water savings and economic costs. *Journal of Cleaner Production*, 11, 1–11.

15. Doods, W. 2002. *Freshwater Ecology—Concepts and Environmental Applications*, Elsevier, San Diego, California, USA.

16. en Health Council. 2004. Guidance on use of rainwater tanks. In: Cunliffe, D. (ed.) *Australian Government*, Australian Government Department of Health and Ageing, Canberra, pp. 1–72.

17. Eroksuz, E. and Rahman, A. 2010. Rainwater tanks in multi-unit buildings: A case study for three Australian cities. *Resources, Conservation and Recycling*, 54, 1449–1452.

18. Farreny, R., Morales-Pinzon, T., Guisasola, A., Taya, C., Rieradevall, J., and Gabarrell, X. 2011. Roof selection for rainwater harvesting: Quantity and quality assessments in Spain. *Water Research*, 45, 3245–3254.

19. Fewkes, A. 1999. The use of rainwater for WC flushing: The field testing of a collection system. *Building and Environment*, 34(6), 765–772.

20. Franklin, L.J., Fielding, J.E., Gregory, J., Gullan, L., Lightfood, D., Poznaski, S.Y., and Vally, H. 2009. An outbreak of Salmonella Typhimurium 9 at a school camp linked to contamination of rainwater tanks. *Epidemiology and Infection*, 137, 434–440.

21. Föster, J. 1996. Patterns of roof runoff contamination and their potential implications on practice and regulations of treatment and local infiltration. *Water Science and Technology*, 33(6), 39–48.

22. Ghisi, E., da Fonseca, T., and Rocha, V.L. 2009. Rainwater harvesting in petrol stations in Brasilia: Potential for potable water savings and investment feasibility analysis. *Resources, Conservation and Recycling*, 54, 79–85.

23. Ghisi, E. and Schondermark, P.N. 2013. Investment feasibility analysis of rainwater use in residences. *Water Resources Management*, 27(7), 2555–2576, doi: 10.1007/s11269-013-0303-6.

24. Griffin, D.M.J., Randall, C., and Grizzard, T.J. 1980. Efficient design of stormwater holding basins used for water quality protection. *Water Research*, 14(10), 1540–1554.

25. Herrmann, T. and Schmida, U. 1999. Rainwater utilisation in Germany: Efficiency, dimensioning, hydraulic and environmental aspects. *Urban Water*, 1(4), 307–316.

26. IEAUST. 2006. Urban water harvesting and reuse. In: Wong, T.H.F. (ed.), *Australian Runoff Quality: A Guide to Water Sensitive Urban Design*. The Institution of Engineers Australia (IEAUST), Canberra.

27. Imteaz, M.A., Ahsan, A., Naser, J., and Rahman, A. 2011. Reliability analysis of rainwater tanks in Melbourne using daily water balance model. *Resources, Conservation and Recycling*, 56, 80–86.

28. Imteaz, M.A., Rahman, A., and Ahsan, A. 2012. Reliability analysis of rainwater tanks: A comparison between South-East and Central Melbourne. *Resources, Conservation and Recycling*, 66, 1–7.

29. Imteaz, M.A., Shanableh, A., Rahman, A., and Ahsan, A. 2011. Optimisation of rainwater tank design from large roofs: A case study in Melbourne, Australia. *Resources, Conservation and Recycling*, 55, 1022–1029.

30. Khastagir, A. and Jayasuriya, N. 2008. Optimal sizing of rain water tanks for domestic water conservation. *Journal of Hydrology*, 381, 181–188.

31. Khastagir, A. and Jayasuriya, N. 2011. Investment evaluation of rainwater tanks. *Water Resources Management*, 25, 3769–3784.

32. Kus, B., Kandasamy, J., Vigneswaran, S., and Shon, H.K. 2010. Analysis of first flush to improve the water quality in rainwater tanks. *Water Science and Technology*, 61, 421–428.

33. Kus, B., Kandasamy, J., Vigneswaran, S., Shon, H.K., and Areerachakul, N. 2011. Water quality of membrane filtered rainwater. *Desalination and Water Treatment*, 32(1–3), 208–213.

34. Kyoungjun, K. and Chulsang, Y. 2009. Hydrological modelling and evaluation of rainwater harvesting facilities:

Case study on several rainwater harvesting facilities in Korea. *Journal of Hydrological Engineering*, 14(6), 545–561.

35. Lee, J.Y., Yang, J.S., Han, M., and Choi, J. 2010. Comparison of the microbiological and chemical characterization of harvested rainwater and reservoir water as alternative water sources. *Science of the Total Environment*, 408, 896–905.

36. Mehrabadi, M.H., Saghafian, B., and Fashi, F.H. 2013. Assessment of residential rainwater harvesting efficiency for meeting nonpotable water demands in three climate conditions. *Resources, Conservation and Recycling*, 73, 86–93.

37. Merritt, A., Miles, R., and Bates, J. 1999. An outbreak of campylobacter enteritis on an island resort, north Queensland. *Communicable Diseases Intelligence Journal*, 23, 215–219.

38. Muthukumaran, S., Baskaran, K., and Sexton N. 2011. Quantification of potable water savings by residential water conservation and reuse—A case study. *Resources, Conservation and Recycling*, 55, 945–52.

39. NSWDP. 2005. *The Building Sustainability Index (BASIX)*. New South Wales Department of Planning (NSWDP), http://www.basix.nsw.gov.au/information/index.jsp.

40. Palla, A., Gnecco, I., Lanza, L.G., and Barbera, P.L. 2012. Performance analysis of domestic rainwater harvesting systems under various European climate zones. *Resources, Conservation and Recycling*, 62, 71–80.

41. Percival, S.L., Walker, J.T., and Hunter, P.R. 2000. Microbiological aspects of biofilms and drinking water. In: Vreeland, R.H. (ed.), *Microbiology of Extreme and Unusual Environments*, CRC Press, Boca Raton, FL.

42. Rahman, A., Dbais, J., and Imteaz, M.A. 2010. Sustainability of RWHSs in multistorey residential buildings. *American Journal of Engineering and Applied Science*, 1(3), 889–898.

43. Rahman, A., Keane, J., and Imteaz, M.A. 2012. Rainwater harvesting in greater Sydney: Water savings, reliability and economic benefits. *Resources, Conservation and Recycling*, 61, 16–21.

44. SA Health. 2008. *Water Quality Fact Sheet*. Department Health, Government of South Australia, Australia.

45. Sample, D., Liu, J., and Wang, S. 2013. Evaluating the dual benefits of rainwater harvesting systems using reliability analysis. *Journal of Hydrologic Engineering*, 18(10), 1310–1321.

46. Schlech, W.F., Gorman, G.W., Payne, M.C., and Broome, C.V. 1985. Legionnaires' disease in Caribbean: An outbreak associated with a resort hotel. *Archives of Internal Medicine*, 145, 2076–2079.

47. Simmons, G. and Smith, J. 1997. Roof water potable source of Salmonella infections. *New Zealand Public Health Report*, 4, 1–5.

48. Spinks, A.T., Coombes, P.J., Dunstan, R.H., and Kuczera, G. 2003. Water quality treatment processes in domestic rainwater harvesting systems. *Proceedings 28th International Hydrology and Water Resources Symposium*, Institution of Engineers Australia, Wollongong, Australia, pp. 227–234.

49. Spinks, J., Phillips, S., Robinson, P., and Van Buynder, P. 2006. Bushfires and tank rainwater quality: A cause for concern? *Journal of Water Health*, 4, 21–28.

50. Standards Australia. 2008. HB 230–2008 Rainwater tank design and installation handbook. In: Stevens, D. (ed.), *Australian and New Zealand Standards*, Standards Australia, Sydney, Australia, pp. 1–111.

51. Su, M., Lin, C., Chang, L., Kang, J., and Lin, M. 2009. A probabilistic approach to rainwater harvesting systems design and evaluation. *Resources, Conservation and Recycling*, 53, 393–399.

52. Tam, V.W.Y., Tam, L., and Zeng, S.X. 2009. Cost effectiveness and trade off on the use of rainwater tank: An empirical study in Australian residential decision-making. *Resources, Conservation and Recycling*, 54, 178–86.

53. Van der Sterren, M., Rahman, A., and Dennis, G. 2012. Rainwater harvesting systems in Australia. In: Voudouris, K. (ed.), *Water Quality*, InTech, Croatia.

54. Van der Sterren, M., Rahman, A., and Dennis, G.R. 2012. Implications to stormwater management as a result of lot scale rainwater tank systems: A case study in Western Sydney, Australia. *Water Science and Technology*, 65(8), 1475–1482.

55. Van der Sterren, M., Rahman, A., and Dennis, G. 2013. Quality and quantity monitoring of five rainwater tanks in Western Sydney, Australia. *Journal of Environmental Engineering*, 139, 332–340.

56. Van der Sterren, M., Rahman, A., Shrestha, S., Barker, G., and Ryan, G. 2009. An overview of on-site retention and detention policies for urban stormwater management in the greater Western Sydney region in Australia. *Water International*, 34(3), 362–372.

57. Villarreal, E.L. and Dixon, A. 2005. Analysis of rainwater collection system for domestic water supply in Ringdansen, Norrköping, Sweden. *Building and Environment*, 40(9), 1174–1184.

58. Walton, A., Gardner, J., Sharma, A., Moglia, M., and Tjandraatmadja, G. 2012. *Exploring Interventions to Encourage Rainwater Tank Maintenance*. Urban Water Security Research Alliance Technical Report No. 59, CSIRO, Australia.

59. Zhang, Y., Chen, D., Chen, L., and Ashbolt, S. 2009. Potential for rainwater use in high-rise buildings in Australian cities. *Journal of Environmental Management*, 91, 222–226.

XIV

Water Reuse Specific Applications

61

Groundwater Recharge and Unconventional Water: Design and Management Criteria

Qin Qian
Lamar University

Saeid Eslamian
Isfahan University of Technology

PREFACE

It is known that groundwater is an important source of water supply for urban water use worldwide, especially in acid and semi-arid regions. With unprecedented environmental and global challenges on population growth, the growing pressure on urban water resources is increasing greatly. Recently, using unconventional water as source water to recharge groundwater is expected to become increasingly necessary because more water storage is needed to save water in times of water surplus and water shortage. In this chapter, the artificial ground recharge system, that is, management of aquifer recharge (MAR) techniques, are introduced. The hydrogeological characteristics of the project area, infiltration rate of the soil, the water quality of the source water, and management of soil clogging effect are major design and management criteria. The environmental issues related to water quality of reuse unconventional water and soil aquifer treatment (SAT) for MAR systems are discussed. Case studies are presented to demonstrate the successful projects of the MAR techniques. The main objective is to demonstrate approaches to reuse the unconventional water in urban areas for better water resource management.

61.1 Introduction

Groundwater is the water filled in soil pore spaces and the fractures of rock formations beneath the Earth's surface as shown in Figure 61.1 [30]. The water table is between the saturated zone (the blue area) and unsaturated zone (the greenish area). Water and air are filled in pores in the unsaturated zone, that is, vadose-zone, while, pores in the saturated zone are filled only with water as shown on the close-up diagram of the bottom of the Figure 61.1.

It is known that groundwater is an important source of water supply for urban water use worldwide, especially in acid and semi-arid regions. With unprecedented environmental and global challenges on population growth, the growing pressure on urban water resources is increasing greatly. Groundwater, as a component of the water cycle, is the function of water inputs from precipitation and infiltration from streams, lakes, or other nature water bodies and water outputs through evapotranspiration and runoff. With urbanization, the land is covered with streets, driveways, roofs, and other impermeable surfaces, which produce more runoff, and then in turn decrease the infiltration rate and the groundwater natural recharge inputs.

Recently, using unconventional water as another groundwater recharge resource is expected to become increasingly necessary because more water storage is needed to save water in times of water surplus and water shortage. Such projects have been

817

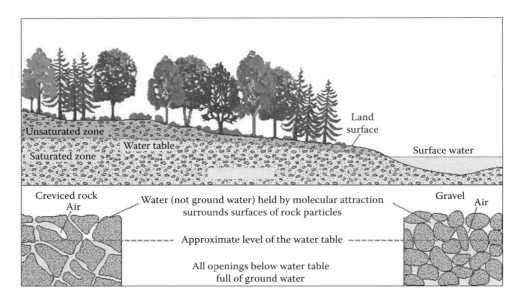

FIGURE 61.1 Groundwater in unsaturated zone and saturated zone. (Adapted from United States Geological Survey (USGS). http://ga.water. usgs.gov/edu/earthgwaquifer.html. Accessed on November 15, 2013.)

practical in areas like most Mediterranean counties [2,17,19], Europe [33] and western United States [9]. Unconventional water may include higher concentrations of organic matter or microorganism content such as municipal wastewater and reclamation water, and it can be water with high saline such as sea and brackish water, industrial and agriculture wastewater [22]. Using unconventional water resources as a viable option to augment drinking water supplies in urban areas has been practiced since 1946 in the United States [34]. Extended studies including spreading-basin experiments, well-injected experiments, hydrological numerical models, and water chemistry analysis have been conducted [34]. In this chapter, artificial groundwater recharge systems and MAR techniques will be introduced, design and management criteria will be discussed, water quality and sustainability issues related to reuse, the unconventional water and SAT for MAR system will be addressed and some case studies will also be presented.

61.2 Artificial Recharge Systems and Management of Aquifer Recharge Techniques

Groundwater recharge with unconventional water plays very important role for groundwater resources management and becomes a great solution to solve issues such as declining groundwater level, saline intrusion, and land subsidence [4,25,29]. The artificial recharge is "the practice of increasing by artificial means that amount of water that enters a ground-water reservoir" [28]. The artificial means include direct recharge of water through spreading basins, pits, and injection or drainage wells, or indirect recharge by infiltration from wells, galleries, and collectors placed near streams. According to the location

of unconventional water recharge, Bouwer [5] defined three engineering systems: surface infiltration system, vadose-zone infiltration system, and wells. The surface infiltration system is designed by ponding or flowing water on the soil surface with basins, furrows, which require permeable surface soils, free clay, or other fine-textured material layers in vadose zones, unconfined and sufficiently transmissive aquifers to achieve high infiltration rates [5]. The vadose-zone infiltration systems are achieved with vertical infiltration means with trenches, shafts, or wells in the vadose zone, which are relatively inexpensive but they clog up easily by accumulation of suspended soils or biomass [5]. Well systems are direct injection water into the aquifer, which is typically used where both surface infiltration systems and vadose zone systems are not suitable [5]. However, the injection water is usually required to be treated in the United States to meet drinking-water quality standards to minimize clogging and protect the groundwater quality. In all three systems, aquifer sediments act as a natural filter removing various contaminants and provide SAT to enhance the aesthetics and public acceptance of potable water from the groundwater where it is recharged by unconventional water [5].

In the early twenty-first century, the water resource management community recognized that the term "artificial" mislead the public regarding water used for recharge as unnatural. The International Association of Hydrogeologists (IAH) has described a new terminology—MAR—to exclude the effects of land clearing, irrigation, and installing water mains to recharge into groundwater by incidental recharge. In urban areas where groundwater is either already overexploited or saline, MAR has the potential to enhance the recharge by indirect reuse of the municipal wastewater, industry wastewater, agricultural wastewater, reclamation water, stormwater/runoff, and sewage effluent.

Dillon [8] defined the major techniques to establish a common glossary. In Figure 61.2, the types of groundwater recharge techniques modified by Voudouris [33] are aquifer storage and recovery (ASR), aquifer storage transfer and recovery (ASTR), bank filtration, dune filtration, infiltration ponds, percolation tanks, rainwater harvesting, SAT, underground dams, sand dams, and recharge releases.

The injected water storage and recovery in the ASR is from the same well; however, the ASTR uses different wells for storage and recovery, which can provide longer residential time for treatment. Bank filtration produces the pressure difference between the surface water body and a well or caisson near or under the water body. After water passes through banks of rivers or lakes to reach the well, higher water quality and recovery can

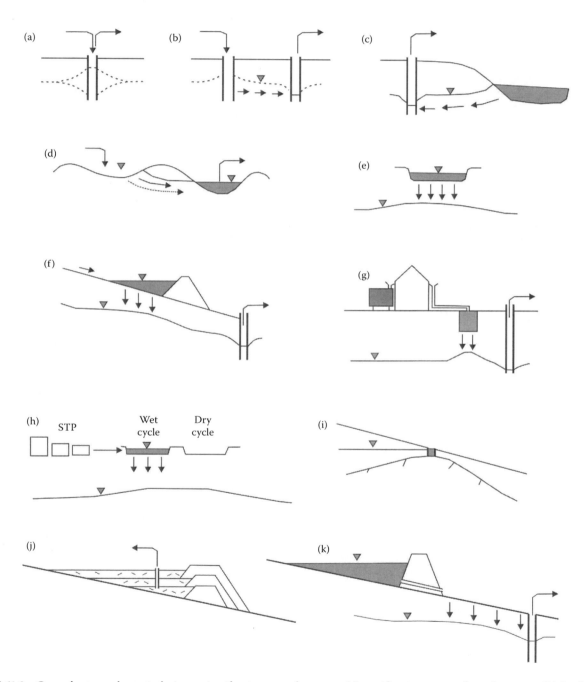

FIGURE 61.2 Groundwater recharge techniques: Aquifer storage and recovery (a), aquifer storage transfer and recovery (b), bank filtration (c), dune filtration (d), infiltration ponds (e), percolation tanks (f), rainwater harvesting (g), soil aquifer treatment (h), underground dams (i), sand dams (j), and recharge releases (k). (Adapted from Dillon, P.J. 2005. *Journal of Hydrology*, 13, 313–316; Modified by Voudouris, K. 2011. *Water*, 3, 964–975.)

be achieved. Ponds are constructed in higher elevations for dune filtration and infiltration pond to increase the infiltration rate. Percolation tank and rainwater harvesting are applied to store the water from runoff before it percolates to the water table. SAT is used as secondary and tertiary treatment of unconventional water through infiltration ponds to remove nutrients and pathogens in passing the unsaturated zone. Sand dams, underground dams, and recharge release are used to save the flood water in arid areas and ephemeral streams. How to apply MAR techniques for water supplies will depend on an understanding of the capabilities and limitations of various techniques to use within the water catchment and aquifer system in relation to needs, existing water infrastructure, space for water harvesting, social and regulatory environment, and skills of personnel [8].

More information about water reuse and water quality guidelines can be found on the IAH Commission on MAR's website: http://www.iah.org/recharge. A database of literature on MAR, an email list for networking, information on forthcoming conferences and workshops, significant recent publications, and downloadable brochures are also included.

61.3 Design and Management

To design and manage a suitable MAR site and techniques, the hydrogeological conditions of the surface soil, unsaturated zone,

and aquifer characteristics play a critical role in transporting and storing management recharge water. The infiltration capacity of the surface soil and unsaturated zone is the first step. The depth of the groundwater level, direction of the groundwater flows, and hydraulic parameters of aquifer need to be investigated. The availability of storage space and moderate permeability of the aquifer can have a good recharge rate to achieve and retain the recharged water in the aquifer for sufficient periods [19]. In practice, older alluvium, buried channels, alluvial fans, dune sands, glacial outwash, fractured, weathered and cavernous rocks, and basaltic rocks are favorable places [19]. Meeting the requirement for the source water quality standard before injection and management clogging in the MAR are also crucial for successful projects.

61.3.1 Infiltration Rates

The infiltration capacity of the soil is an important factor for designing the artificial recharge system. Therefore, the first screening for selecting promising sites is the soil maps and hydrogeological reports [5]. The soil-textural triangle (Figure 61.3) developed by the Soil Survey Staff of the U.S. Department of Agriculture provides evaluations for the textural classification of soil. The hydraulics conductivity values for typical soil [7] are listed in Table 61.1.

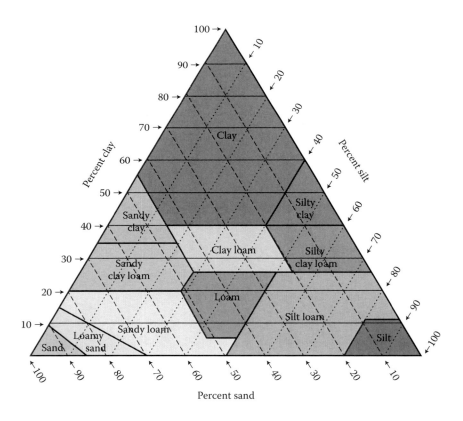

FIGURE 61.3   Soil textural triangle—Based on the triangle, a loamy soil has 40% sand, 20% clay, and 40% silt. A sandy loam has 60% sand, 10% clay, and 30% silt. (Adapted from U.S. Department of Agriculture, http://www.nrcs.usda.gov/wps/portal/nrcs/detail/soils/survey/?cid=nrcs142p2_054167. Accessed on November 15, 2013.)

TABLE 61.1 Typical Values of Saturated Hydraulic Conductivity for Soils

Soil Description	Hydraulic Conductivity K	
	cm/s	ft/s
Clean gravel	$1 \sim 100$	$3 \times 10^{-2} \sim 3$
Sand–gravel mixtures	$10^{-2} \sim 10$	$3 \times 10^{-4} \sim 0.3$
Clean coarse sand	$10^{-2} \sim 1$	$3 \times 10^{-4} \sim 3 \times 10^{-2}$
Find sand	$10^{-3} \sim 10^{-1}$	$3 \times 10^{-5} \sim 3 \times 10^{-3}$
Silty sand	$10^{-3} \sim 10^{-2}$	$3 \times 10^{-5} \sim 3 \times 10^{-4}$
Clayey sand	$10^{-4} \sim 10^{-2}$	$3 \times 10^{-6} \sim 3 \times 10^{-4}$
Silt	$10^{-8} \sim 10^{-3}$	$3 \times 10^{-10} \sim 3 \times 10^{-5}$
Clay	$10^{-10} \sim 10^{-6}$	$3 \times 10^{-12} \sim 3 \times 10^{-8}$

Source: Adapted from Coduto, D.P. 1999. *Geotechnical Engineering: Principles and Practices.* Prentice Hall, Englewood Cliffs, NJ, 759 pp.

Applying Darcy's equation to a soil surface covered by ponded water of negligible depth, the Green-Ampt infiltration model, proposed in 1911 [21], has been developed. If the saturated conductivity K_s in the wetted zone with the homogenous soil, the Green-Ampt equation can be expressed in Equation 61.1.

$$f_p = \frac{K_s(L + S)}{L} \qquad (61.1)$$

where L is the distance from the ground surface to the wetting front and S is the capillary suction at the wetting front. Diffusion-based and empirical equations have been developed and widely used in stormwater management models.

61.3.2 Hydrogeological Studies

To determine the type of MAR and suitable location for MAR, the precise regional hydrogeological setting is necessary to understand before the design stage. The information about the surface soil, unsaturated zone, detailed groundwater level contours, maximum, minimum, and mean depth of the water table, the hydraulic connection of groundwater with surrounding water bodies, and chemical quality of water in different aquifers need to be known precisely.

The hydrologic properties of an unsaturated zone determine the suitability of a particular location for MAR techniques. Typically, areas with high permeability soils, the capacity for horizontal movement of water in the unsaturated zone and the receiving aquifer, no impeding layers, and a thick unsaturated zone are optimal locations [18]. Field experiment and numerical modeling are tools for estimating groundwater recharge rate. Allison et al. [1] used both physical and chemical field experiment methods to estimate recharge in arid and semi-arid areas. Their review indicates that indirect, physical approaches, such as water balance and Darcy flux measurements, are the least successful, while methods using tracers (e.g., Cl, 3H, and 36Cl) have been the most successful in estimating groundwater recharge in dry regions. A three-dimensional intergraded finite-difference numerical code, TOUGH2 [27], is used widely to analyze the

flow of heat, air, water, and nitrate in unsaturated areas under saturated and unsaturated conditions.

To evaluate the effectiveness of a MAR project, the geologic and hydrologic characterization of the aquifer system is essential. A combination of field experiments, laboratory measurements, analytical and numerical simulation methods generally are used to develop the hydrogeological system of MAR projects [26]. Methods including surficial mapping; drilling and core analysis; borehole, surface, and aerial geophysics, sequence stratigraphy; geostatistical analysis of lithology; etc. are used to understand geologic characterizations of the aquifer. The continuous groundwater level and offset distribution could be measured by piezometers or standard electronic monitoring equipment in observation wells in the project area. The hydraulic properties, that is, storage and transmissive properties are determined by laboratory analysis of core samples, borehole geophysics and velocity logs, multi- or single-well aquifer tests [26]. The direction and rate of the saturated flow can be determined with hydraulic properties, distribution of stress, and groundwater head with development of the simulation groundwater transport models.

In addition, for the observation of the possible alteration of the chemical characteristics of the groundwater, chemical analyses should be carried out before and after the recharge application.

61.3.3 Quality of Source Water

Reuse of unconventional water as source water for groundwater recharge is expected to become increasingly important in urban areas. However, the problems such as degradation of the subsurface environment and groundwater due to the transport of pathogenic viruses with recycled water, the clogging effect by suspended solids, and bacteria in the recycled water still remain. Therefore, the quality of source water needs to meet the required standards before injection. A major requirement is silt-free because silt may settle into stagnant water. The chemical and bacteriological analysis of source water and ground water is also needed before the treatment method is determined. Due to treatment costs, the recycled wastewater is usually used for nonpotable purposes such as urban irrigation, power-plant cooling, industrial processing, construction, dust control, fire protection, toilet flushing in commercial buildings, and environmental purposes [5]. When directly injecting water into potable aquifers for potable water use, it usually requires that the drinking water and additional water quality standards are met prior to infiltration. For example, the state of California requires not only reverse osmosis (RO) treatment, but also total organic carbon concentrations of less than 0.5 mg/L as well as total nitrogen concentrations of less than 5 mg N/L before injection [6].

61.3.4 Soil Clogging

One of the most important considerations in design and management of the artificial recharge system is clogging of the infiltration surface, then in turn reduction in infiltration rates.

Clogging is caused by physical, biological, and chemical processes [3]. Physical processes are accumulation of inorganic and organic suspended solids and the downward movement of fine particles in the soil [5]. The typical means to control clogging due to physical processes is presedimentation to settle clay silt and other suspended solids. Periodically removing the mud-cake and dicing or scraping the surface layer as well as installation of a filter on the surface can also prevent this type of clogging. Biological clogging processes include accumulation of algae and bacteria in the water and growth of microorganisms on and in the soil to form biofilms and biomass [5]. Pretreatment of recharge water to remove nutrients and organic carbon can control clogging due to algae growth and other biological clogging. Chemical processes involve precipitation of calcium carbonate, gypsum, phosphate, and other chemicals on and in the soil [5]. An important geochemical process, the calcite dissolution, can cause chemical clogging because the different chemical actions within an aquifer system depend on hydrogeological conditions and water quality [32]. The possible chemical actions in the system need to be investigated to ensure that the hydraulic conductivity can satisfy design requirements.

61.4 Water Quality and Sustainability Issues

Unconventional water in urban areas includes municipal wastewater, sewage effluent, industry wastewater, reclamation water, and seawater. Treated unconventional water is conventionally reused through direct application in irrigation. The existing guidelines and risk estimation provide microbial and chemical limits for such applications. Recently, reuse through a MAR system with a secondary- or tertiary-treatment has been developed. What type of water quality standard might be set in order to provide the proper level of safety associated with the reuse of unconventional water as the groundwater source for MAR systems is still a question. In general, the drinking water quality standard is required for direct aquifer recharge. Advanced tertiary treatment such as reverse osmosis or percolation through the soil is needed for indirect aquifer recharge. Additional water quality standards for emerging disinfection by-products (DBP) using high-dose ultraviolet (UV) radiation after reverse osmosis treatment is also required to achieve N-nitrosodimethylamine (NDMA) concentration of less than 10 ng/L [9].

In any MAR system, an additional water quality improvement can be achieved by allowing the wastewater effluent to percolate the vadose zone and saturated aquifer. The unsaturated zone and aquifer sediments act as a natural filter to remove various contaminants, such as all suspended solids, biodegradable materials, bacteria, viruses, and other microorganisms, to significantly reduce nitrogen, phosphorus, and heavy metals [19]. This process is known as SAT. Many studies demonstrate the performance of SAT systems in removing pathogens, DBP precursors (i.e., TOC), nutrients, and selected

trace organic chemicals [10–16,20,23]. A large number of studies have been conducted to indicate the factors affecting the quality of recovered water and performance of SAT systems. Besides the unconventional water quality, the factors including spreading basin characteristics, subsurface conditions, and the degree of blending with native ground water, operation conditions affect the final recovered water quality [9].

61.5 Case Studies

Many practical applications of MAR technologies are available around the world. In Egypt, the reuse of reclaimed municipal wastewater is being considered as MAR water resources plan. Areas located in western delta fringes, eastern delta fringes, and Nile valley fringes have been identified for MAR application [19]. Many projects used the SAT as a method for tertiary treatment of wastewater. Two projects in New York applied artificial recharge system to treat wastewater from municipal water and sewage water [34]. Van Puffelen [32] identified highly efficient removal of bacteria and viruses in the recharge of heavily polluted river water in the coastal sand dunes of the Netherlands. One of the largest applications is the groundwater replenishment (GWR) system at Orange County, CA, which is an indirect potable reuse project that is initially producing up to 265,000 m³/day of highly treated recycled water with an ultimate capacity of 492,000 m³/day for groundwater recharge and direct injection to protect the groundwater basin from seawater intrusion [24]. Three major components are the advanced water purification facility (AWPF) and pumping station, a 14-mile pipeline connecting the treatment facilities to existing recharge basins, and an expansion of the existing seawater intrusion barrier [24]. The AWPF uses microfiltration, reverse osmosis, and ultraviolet light/advanced oxidation to treat clarified secondary effluent for discharging into the ocean. The effluent from the AWPF is lime stabilized and pumped into injection wells and recharge basins, where blending with groundwater supplies to provide water for 500,000 people [24].

61.6 Summary and Conclusions

MAR techniques play an important role in reuse of the unconventional water in urban areas because MAR systems provide SAT and storage of the effluent and enhance the aesthetics and public acceptance of potable water reuse with very cost-effective solutions. Design and management of MAR systems involves geological, geochemical, hydrological, biological, chemical, physical, and engineering aspects. With inherently heterogeneous soil and underground formation, planning, design, and construction of MAR systems must be piecemeal, first testing for fatal flaws and general feasibility and then proceeding with pilot and small-scale systems until the complete system can be designed and constructed [5]. The water quality standard is essential for different water reuse purposes. The existing regulations and guidance need to be consulted before the MAR design.

References

1. Allison, G.B., Gee, G.W., and Tyler, S.W. 1994. Vadose-Zone techniques for estimating groundwater recharge in arid and semiarid regions. *Soil Science Society of America*, 58 (1), 6–14, doi:10.2136/sssaj1994.03615995005800010002x.

2. Angelakis, A.N., Marecos Do Monte, M.H.F., Bontoux, L., and Asano, T. 1999. The status of wastewater reuse practice in the Mediterranean basin: Need for guidelines. *Water Research*, 33(10), 2201–2217.

3. Beveye, P., Vandevivere, P., Hoyle, B.L., DeLeo, P.C., Sanchez de and Lozada, D. 1998. Environmental impact and mechanisms of the biological clogging of statured soils and aquifer materials. *Critical Reviews in Environmental Science and Technology*, 28(2), 123–191.

4. Bouwer, H. 1996. Issues in artificial recharge. *Water Science and Technology*, 33, 381–390.

5. Bouwer, H. 2002. Artificial recharge of groundwater: Hydrogeology and engineering. *Hydrogeology Journal*, 10, 121–142.

6. California Department of Public Health (CDPH). 2008. Groundwater recharge draft regulation. California CDPH, Division of Drinking Water and Environmental Management, Sacramento, CA. www.cdph.ca.gov/HealthInfo/environhealth/water/Pages/Waterrecycling.aspx. Accessed November 15, 2013.

7. Coduto, D.P. 1999. *Geotechnical Engineering: Principles and Practices*. Prentice Hall, Englewood Cliffs, NJ, 759pp.

8. Dillon, P.J. 2005. Future management of aquifer recharge. *Journal of Hydrology*, 13, 313–316.

9. Drewes, J.E. 2009. Ground water replenishment with recycled water-water quality improvements during managed aquifer recharge. *Ground Water*, 47(4), 502–505.

10. Drewes, J.E. and Fox, P. 1999. Fate of natural organic matter (NOM) during groundwater recharge using reclaimed water. *Water Science and Technology*, 40(9), 241–248.

11. Drewes, J.E. and Fox, P. 2000. Effect of drinking water sources on reclaimed water quality in water reuse systems. *Water Environment Research*, 72(3), 353–362.

12. Drewes, J.E., Heberer, T., Rauch, T., and Reddersen, K. 2003. Fate of pharmaceuticals during groundwater recharge. *Ground Water Monitoring & Remediation*, 23(3), 64–72.

13. Drewes, J.E., Heberer, T., and Reddersen, K. 2002. Fate of pharmaceuticals during indirect potable reuse. *Water Science and Technology*, 46(3), 73–80.

14. Drewes, J.E., Hoppe, C., and Jennings, T. 2006. Fate and transport of *N*-nitrosamines under conditions simulating full-scale groundwater recharge operations. *Water Environment Research*, 78(13), 2466–2473.

15. Drewes, J.E., Quanrud, D., Amy, G., and Westerhoff, P. 2006. Character of organic matter in soil-aquifer treatment systems. *Journal of Environmental Engineering*, 132(11), 1447–1458.

16. Drewes, J.E., Reinhard, M., and Fox, P. 2003. Comparing microfiltration-reverse osmosis and soil-aquifer treatment for indirect potable reuse of water. *Water Research*, 37(15), 3612–3621.

17. Eslamian, S., Amiri M.J., and Abedi-Koupai, J. 2013. Reclamation of unconventional water using nano zero-valent iron particles: An application for groundwater. *International Journal of Water*, 7(1/2), 1–13.

18. Flint, A.L. 2002. The role of unsaturated flow in artificial recharge projects. *U.S. Geological Survey Artificial Recharge Workshop Proceedings*, April 2–4, 2002, Sacramento, CA.

19. Fadlelmawla, A., Abdel-Halem, M., and Vissers, M. 1999. Preliminary plans for artificial recharge with reclaimed wastewater in Egypt. *9th Biennial Symposium on the Artificial Recharge of Groundwater*, Tempe, AZ, June 10–12.

20. Fox, P., Houston, S., Westerhoff, P., Drewes, J.E., Nellor, M., Yanko, E., Baird, R. et al. 2001. *Soil Aquifer Treatment for Sustainable Water Reuse*. American Water Works Association Research Foundation, Denver, CO.

21. Green, W.H. and Ampt, G.A. 1911. Studies on soil physics, 1. The flow of air and water through soils. *Journal of Agricultural Science*, 4, 11–24.

22. Indelicato, S., Tamburino, V., and Zimbone, S.M. 1993. Unconventional water resource use and management. *Etat de l'agriculture en Méditerranée: Ressources en eau: développement et gestion dans les pays méditerranéens*. CIHEAM, Bari, pp. 57–74. (Cah iers Option s Méditerran éen n es; n. 1 (1)).

23. Leenheer, J.A., Rostad, C.E., Barber, L.B., Schroeder, R., Anders, R., and Davisson, M.L. 2001. Nature and chlorine reactivity of organic constituents from reclaimed water in groundwater, Los Angeles County, California. *Environmental Science and Technology*, 35, 3869–3876.

24. Markus, M.R., and Deshmukh, S.S. 2010. An innovative approach to water supply—The Groundwater Replenishment System. *World Environmental and Water Resources Congress 2010*, May 16–20, Rhode Island.

25. Peters, J.H. (ed.). 1998. Artificial recharge of groundwater. In: *Proceedings of the 3rd International Symposium—TISAR 98*, September 21–25, 1998, Balkema, Rotterdam, the Netherland.

26. Phillips, S.P. 2002. The role of saturated flow in artificial recharge projects. *U.S. Geological Survey Artificial Recharge Workshop Proceedings*, April 2–4, 2002, Sacramento, CA.

27. Pruess, K. 1991. *TOUGH2-A General—Purpose Numerical Simulator for Multiphase Fluid and Heat Flow*. Lawrence Berkeley National Laboratory Rep. LBL-29400, University of California, Berkeley, CA, 102p.

28. Todd, D.K. 1959. Annotated bibliography on artificial recharge of ground water through 1954. U.S. Geological Survey Water-Supply Paper 1477, United States Government Printing Office, Washington, DC, 115p.

29. Tsourlos, P., Vargemezis, G., Voudouris, K., Spachos, T., and Stampolidis, A. 2007. Monitoring recycled water injection into a confined aquifer in Sindos (Thessaloniki) using

electrical resistivity tomography (ERT): Installation and preliminary results. In: *Proceedings of the 11th International Congress*, May 2007, Athens, Greece, Vol. 37, Part 2, pp. 580–592.

30. United States Geological Survey (USGS). http://ga.water. usgs.gov/edu/earthgwaquifer.html. Accessed on November 15, 2013.

31. U.S. Department of Agriculture, http://www.nrcs.usda.gov/ wps/portal/nrcs/detail/soils/survey/?cid=nrcs142p2_054167, Accessed on November 15, 2013.

32. Van Puffelen, J. 1982. Artificial groundwater recharge in the Netherlands. *DVWK Bulletin*, 11. (Artificial Groundwater Recharge), 353p.

33. Voudouris, K. 2011. Artificial recharge via boreholes using treated wastewater: Possibilities and prospects. *Water*, 3, 964–975, doi:10.3390/w3040964.

34. Weeks, E.P. 2002. A historical overview of hydrologic studies of artificial recharge in the U. S. Geological Survey. *U.S. Geological Survey Artificial Recharge Workshop Proceedings*, April 2–4, 2002, Sacramento, CA.

62

Water Reuse via Aquifers: Urban Stormwater and Treated Sewage

Declan Page
*Commonwealth Scientific and
Industrial Research Organization
(CSIRO) Land and Water Flagship*

Joanne Vanderzalm
*Commonwealth Scientific and
Industrial Research Organization
(CSIRO) Land and Water Flagship*

PREFACE

Climate change and a growing population, increasingly urbanized, add to the stresses on the world's water resources. To meet the increasing urban water requirements, there is a need to diversify future sources of supply. Desalination of seawater, water recycling, increased use of groundwater, and stormwater and rainwater harvesting are being used internationally to augment urban water supply. However, to date there has been a lag in the uptake of managed aquifer recharge (MAR) for diversifying water sources for urban areas.

This chapter gives a brief overview about the use of MAR primarily for urban areas and reuse.

Recharged water may be sourced from rainwater, stormwater, reclaimed water, mains water, or other aquifers. The number, diversity, and scale of MAR projects is growing internationally, particularly in urban areas, due to water shortages, fewer available dam sites, low costs compared with alternatives where conditions are favorable, and associated benefits of MAR.

Water quality improvements during aquifer storage of recycled waters are increasingly being documented at demonstration sites and operational projects. The growing body of knowledge allows more confident reliance on aquifer treatment.

Recycled water, if stored in an aquifer for a period before recovery as drinking water, provides an additional level of public health protection beyond direct reuse. In urban areas, confined aquifers provide better protection for waters recharged via wells to supplement drinking water supplies. However, unconfined aquifers may generally be used for nonpotable uses to substitute for mains water supplies and, in some cases, provide adequate protection for recovery as drinking water. In Australia, as in the United States over the last 40 years, there is evidence that public acceptance of water recycling via aquifer recharge for drinking water supplies is strong, in marked contrast with water recycling without natural storage and treatment.

62.1 Introduction

This chapter summarizes at an introductory level the relevant information needed to consider water recycling via aquifers, or MAR, as a prospective new water supply for drinking or nonpotable uses in urban areas. It contains information on the sources of urban water for reuse via MAR and outlines the opportunities that MAR may provide. Though the focus of this chapter is primarily for urban areas, the concepts are also equally applicable to rural and regional areas. This chapter does not attempt to fully describe the many technical issues that are covered in the scientific literature accessible from the sources referenced here.

62.2 Definition and Purpose of MAR

MAR, previously known as artificial recharge, is the purposeful recharge of water to aquifers for reuse and/or environmental benefit. Aquifers are permeable geological strata that contain water and are replenished naturally by rain permeating through soil and rock to the aquifer below or by infiltration from streams. However, several human activities may enhance aquifer recharge, which can generally be put into three categories:

1. *Unintentional Recharge*: Such as through land clearing, removing deep-rooted vegetation, deep seepage under irrigation areas and in urban areas by leaks from water infrastructure and sewers, for example, Tula Valley, Mexico [13].
2. *Unmanaged Recharge*: Including stormwater drainage wells and sumps, and septic tank drain fields, generally for disposal of water without thought of recovery or reuse, for example, stormwater drainage in Mount Gambier, Australia [29].
3. *Managed Recharge*: Through mechanisms such as injection wells and infiltration basins for rainwater, urban stormwater, treated sewage, industrial wastewater, mains water, or water from other aquifers that is subsequently recovered for urban water reuse or stored to provide environmental benefit to the aquifer, for example, aquifer storage and recovery (ASR) in Bolivar, Australia [22], bank filtration in Berlin, Germany [10], and dune infiltration in Torreele/St-André, Belgium [28].

This chapter focuses only on the final category of recharge, but acknowledges that frequently there are opportunities to convert from unintentional or unmanaged recharge to managed recharge with the aim of protecting the environment and recovering the stored water for beneficial use. Enhancing natural rates of groundwater recharge via MAR provides an important potential source of water for urban areas where impervious cover has altered the natural recharge regime. This chapter addresses varied application of MAR, but the emphasis is on urban water reuse applications.

MAR can be used to store water from various sources, including urban stormwater, treated sewage, mains water, desalinated seawater, water produced during mining, and rainwater or groundwater from other aquifers. With appropriate pretreatment before recharge and sometimes post-treatment on recovery of the water, it may be used for drinking water supplies [1,10,28], industrial water [19], irrigation [3,13,15], toilet flushing [19], and sustaining ecosystems.

Common reasons for using MAR in water reuse include

- Securing and enhancing urban water supplies
- Improving groundwater quality
- Preventing saltwater intrusion in coastal aquifers
- Providing storage without loss of valuable land surface area
- Reducing evaporation of stored water
- Maintaining environmental flows and groundwater-dependent ecosystems, which improve local amenity, land value, and biodiversity

Consequential benefits of water recycling via aquifers in urban areas may also include

- Improving coastal water quality by reducing nutrient-rich urban discharges
- Mitigating floods and flood damage
- Facilitating urban landscape improvements that increase land value

MAR can play a role in increasing storage capacity to help urban water supplies cope with the runoff variability increasingly being reported due to climate change. It can also assist in harvesting abundant stormwater and treated sewage in urban areas that may be currently underutilized.

62.2.1 Types of MAR

A wide range of methods are in use for recharging water to meet a variety of local conditions, including infiltration techniques to recharge unconfined aquifer and well injection techniques, generally suited to deeper, confined aquifers (Figure 62.1).

There are a large number and growing variety of methods used for MAR internationally. Some international examples of those currently in use are

Aquifer storage and recovery (ASR): Injection of water into a well for storage and recovery from the same well. This is especially useful in brackish aquifers, where storage is the primary goal and water treatment is a smaller consideration. Commonly used to store drinking water in the United States [23], but applied successfully for reuse of treated sewage or urban stormwater in Australia (e.g., Rossdale, Australia [16]).

Aquifer storage, transfer, and recovery (ASTR): Involves injecting water into a well for storage, and recovery from a different well. This is used to achieve additional water treatment in the aquifer by extending residence time in the aquifer beyond that of a single well (e.g., Parafield Gardens, Australia [19]).

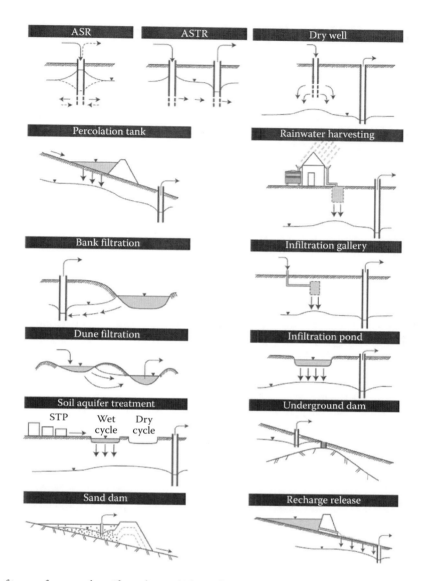

FIGURE 62.1 Schematic of types of managed aquifer recharge. (After Dillon, P. 2005. *Hydrogeology Journal*, 13, 313–316.)

Infiltration ponds: Involve diverting surface water into off-stream basins and channels that allow water to soak through an unsaturated zone to the underlying unconfined aquifer (e.g., Atlantis, South Africa [25]).

Infiltration galleries: Buried trenches (containing polythene cells or slotted pipes) in permeable soils that allow infiltration through the unsaturated zone to an unconfined aquifer (e.g., Floreat, Australia [4]).

Soil-aquifer treatment (SAT): Treated sewage effluent is intermittently infiltrated through infiltration ponds to facilitate nutrient and pathogen removal in passage through the unsaturated zone for recovery by wells after residence in the unconfined aquifer (e.g., Tucson, Arizona [9]).

Percolation tanks, check dams, or recharge weirs: Dams built in ephemeral streams detain water, which infiltrates through the bed to enhance storage in unconfined aquifers and is extracted down-valley (e.g., Gujarat, India [14]).

Rainwater harvesting for aquifer storage: Roof runoff is diverted into a well, sump, or caisson filled with sand or gravel and allowed to percolate to the water table where it is collected by pumping from a well (e.g., Kathmandu Valley, Nepal [27]).

Recharge releases: Dams on ephemeral streams are used to detain flood water and uses may include slow release of water into the streambed downstream to match the capacity for infiltration into underlying aquifers, thereby significantly enhancing recharge (e.g., Little Para River, Australia [7]).

Dry wells: Typically shallow wells where water tables are very deep, allowing infiltration of very high quality water to the unconfined aquifer at depth (e.g., Phoenix, AZ [11])

Bank filtration: Extraction of groundwater from a well or caisson near or under a river or lake to induce infiltration from the surface water body thereby improving and making more consistent the quality of water recovered (e.g., Berlin, Germany [10]).

Dune filtration: Infiltration of water from ponds constructed in dunes and extraction from wells or ponds at lower elevation for water quality improvement and to balance supply and demand (e.g., Torreele/St-André, Belgium [28].

Underground dams: In ephemeral streams where basement highs constrict flows, a trench is constructed across the streambed, keyed to the basement and backfilled with low permeability material to help retain flood flows in saturated alluvium for stock and domestic use (e.g., Kenya [12]).

Sand dams: Built in ephemeral stream beds in arid areas on low permeability lithology, these trap sediment when flow occurs, and following successive floods the sand dam is raised to create an "aquifer," which can be tapped by wells in dry seasons (e.g., Namibia [12] and Kenya [18]).

Selection of suitable sites for MAR and choice of method will depend on the hydrogeology, topography, hydrology, and land uses within a given area. It is common to find similar types of MAR projects clustered in the same geographic area due to shared physical attributes.

MAR is in wide use internationally to enhance urban water supplies, particularly those in semi-arid and arid areas, but also in humid areas, primarily for water quality improvement. The International Association of Hydrogeologists has a Commission on MAR whose website contains information on various applications of MAR (www.iah.org/recharge).

62.2.2 Components of a MAR Project

While MAR operations may appear quite different because of the type of aquifer available for storage and the method used for recharge, MAR systems share seven common elements. These seven components are illustrated for the two main types of aquifers—those that are confined by a low permeability layer, which for MAR requires injecting water via a well (Figure 62.2), and those that are unconfined and allow water to infiltrate through permeable soils, where recharge can be enhanced by basins and galleries (Figure 62.3).

There are many combinations of urban water source, water treatment methods, and recovered water end uses possible and detailing all combinations is beyond the scope of this chapter.

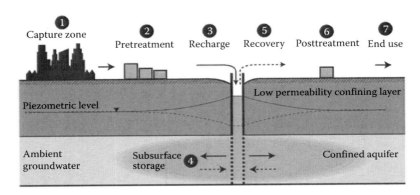

FIGURE 62.2 An example of MAR, ASR in a confined aquifer, showing the seven elements common to each system [17]. Piezometric level is the level of water in a well if a well were constructed. For an unconfined aquifer, this is the water table. During recharge, levels rise and, near recovery wells, levels fall.

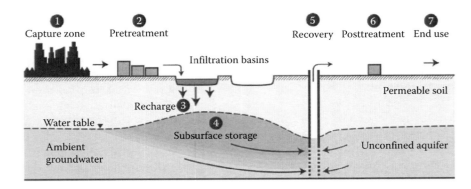

FIGURE 62.3 An example of MAR, SAT in an unconfined aquifer, showing the seven elements common to each system [17]. Piezometric level is the level of water in a well if a well were constructed. For an unconfined aquifer, this is the water table. During recharge, levels rise and, near recovery wells, levels fall.

In general, poorer quality source waters will require a higher level of treatment before recharge in cases where

- The aquifer already contains high quality water
- The water is to be recovered for higher valued uses such as drinking
- The aquifer is fine-grained and there is a need to minimize clogging of the recharge basin, gallery, or well

Passive treatment such as in a wetland may be suitable when urban stormwater is being used to recharge an aquifer with recovery of water for irrigation without any requirement for post-treatment. An example of this can be found in Salisbury, Australia, where the recovered water is used for open space irrigation [19]. However, pretreatment by microfiltration (MF) and granular activated carbon (GAC) filtration was necessary at an ASR site with a very fine-grained aquifer to prevent clogging of the well. This operational constraint was more stringent than those to protect groundwater quality were or for the recovered water to be fit for use in irrigation [20]. No treatment may be necessary where river water of low turbidity is diverted to infiltration basins for enhancing irrigation supplies. Where reclaimed water is used for recharge to recover for drinking water supplies, the source water will require considerable treatment prior to recharge [10,28]. Regulations of source water quality prior to recharge vary considerably between countries.

62.3 Drivers for MAR in Urban Areas

In rural areas, MAR has been used primarily to increase the security of groundwater irrigation supplies and improving the quality of water available for irrigation. However, in urban areas that are facing population growth and climate change, there are often additional drivers for water reuse and MAR. For example, in Salisbury, Australia, MAR (via ASR) was only considered after wetlands were established for flood mitigation, urban amenity, and coastal water quality improvement [19]. The additional costs of injecting and recovering the detained urban stormwater provided a water supply competitive with mains water prices.

For new projects, there may be multiple reasons for introducing MAR at any one site. The combination of benefits as listed above, not just water supply, may determine whether water recycling via MAR should proceed.

With respect to securing and enhancing water supplies, pioneering MAR projects have been established internationally to satisfy immediate nonpotable water supply needs on a commercial basis for agriculture, local government, and industry. However in the future, application of MAR for establishing high-valued drought and emergency supplies could increase as water managers become more confident in the economics and reliability of MAR in relation to alternative water resource management tools [13]. Drought and emergency supplies have high social value but require procedures to support the development of strategic reserves to meet such demands and prevent consumption of available source water and aquifer storage capacity for more immediate needs. A hierarchical approach to allocate these resources could give priority to strategic reserves and substitutional supplies over supplies to meet new water demands. This would help increase the security of urban water supplies, especially in light of climate change and population growth.

62.3.1 Climate Variability

Climate change has been implicated in causing less reliable and more variable rainfall. In the future, many parts of the world will face decreasing mean annual rainfall, shifts in seasonal patterns, more frequent high intensity rainfall, higher temperatures, and higher evaporation rates because of climate change. As a result, to retain the current level of urban water security will require more stable alternative supplies, larger water storages, and a range of demand management measures.

In urban catchments where most runoff is from impervious surfaces, annual runoff is expected to decline by the same proportion as rainfall, although peak storm intensity may actually increase. Hence, the relative efficiency of urban catchments compared to rural catchments will increase as water supplies become more stressed, and MAR could play a role in averting the need to augment future urban stormwater drainage systems.

Climate change will also affect land use, soil cover, erosion, and fire frequency, so water quality from traditional water supply catchments and reservoirs may become increasingly variable. As a result, the water quality advantages of traditional catchments over urban catchments may also change over time.

Bank filtration is one MAR technique that could be adopted by towns currently using river water directly, as a measure to adapt to climate change. This MAR practice is common in Europe where drinking water supplies [10] are drawn from wells in alluvium next to streams rather than the stream itself, as a means of smoothing out water quality variations and prefiltering water.

The need for water storage depends largely on seasonal variations in water sources in addition to interannual variability. For example, cities with prolonged dry periods have a greater need for water storage than wetter cities with more uniform rainfall. However, for very dry cities, the opportunities for urban stormwater harvesting are infrequent and so MAR projects with stormwater in these areas are unlikely. In these locations treated sewage, which has a very stable flow, is likely to be the preferred source of water for MAR. Due to high evaporation rates, cities in this sector are also more likely to prefer subsurface storage of water than dams.

62.3.2 Community Preferences

Community acceptance and use of water recycled via aquifers is a key factor in a successful scheme. One driver for developing MAR with recycled water is its high level of public acceptance, especially for drinking water supplies, with respect to other forms of water recycling. Groundwater recharge with reclaimed water has been practiced since the 1960s in United States for recovery for nonpotable and drinking water supplies. Reclaimed water that has undergone natural treatment is well accepted by the public when recovered water is used for potable purposes [2].

62.4 Availability of Aquifers and Storage Advantages in Urban Areas

An essential prerequisite for MAR is the presence of a suitable aquifer in which to store water. The best aquifers are those that can store and recover large volumes of water because generally there is an economy of scale favoring larger projects. Aquifers that are thick and have uniform hydraulic properties are also preferred to maximize the ability to recover water. Having a very low regional groundwater flow rate through the aquifer also helps to make recovery of recharged water easier. Generally, consolidated aquifers are preferred to unconsolidated ones for ASR due to simpler well construction and ease of maintenance.

There can be both positive and negative aspects of other aquifer attributes in relation to MAR. For example, if aquifers are unconfined (Figure 62.3) and sufficient land is available, infiltration methods may be used and these are generally cheaper than well injection methods. However, stored water is vulnerable to pollution from overlying land uses. Confined aquifer systems are by nature protected from pollution but require wells for access (Figure 62.2). With respect to water quality, the source water quality requirements for turbidity and nutrients to avoid clogging [16] are generally more stringent for recharge wells than for surface infiltration systems such as SAT (Figure 62.3) and depend on the pore sizes in the aquifer, its mineral composition, and the form of construction of the well.

If the ambient groundwater in the storage zone is brackish, the pretreatment requirements for aquifer protection may be less than they would be for a fresh aquifer. However, if groundwater is too saline, the recovery efficiency may be low and render the site nonviable [16,19]. Reactive minerals in aquifers, such as carbonate, can assist in controlling clogging in ASR wells through dissolution, but the same minerals can in some cases also contain metals that are released and impair the quality of recovered water [29,31]. Finally, the oxygen status of the aquifer can also affect water quality [10]. Pathogens [24] and some organic chemicals [15,28] are most effectively removed under aerobic conditions but other organics are only removed under anoxic conditions. The ideal situation is to have variable redox zones in the aquifer so that water is exposed to both conditions to get the best water quality improvement [10].

Where several aquifers are present at one location, interleaved with low permeability layers, this allows choice of one or more with the most favorable characteristics for water storage and MAR. Depending on the degree of aquifer interconnection, it may be possible to store water of different qualities in separate aquifers at the same location.

In other areas, there may be no aquifer, or none with suitable characteristics to allow sufficient storage while ensuring environmental protection. For example,

- Where the aquifer is unconfined and the water table is very shallow. In these locations, water recycling via aquifers could potentially lead to localized urban flooding.
- Where the aquifer is very thin or composed of fine-grained unconsolidated material. These aquifers do not have a high storage volume and will be very susceptible to well clogging [22].
- Where the site is adjacent to a leaky fault or a semiconfining layer containing poor quality water. Here, mixing of the recharged water may lead to recovery of poor quality water unfit for its intended reuse.
- Where the aquifer contains poor quality water and is highly heterogeneous or has a high lateral flow rate. Here, recovery efficiencies are likely to be poor due to water quality constraints.

At these locations, MAR is not feasible. A site that is hydrogeologically complex requires more detailed investigations and more sophisticated risk management [21], which add to the costs, and even though technically feasible, may become economically unviable.

Local hydrogeological knowledge is needed to identify the presence of aquifers and their suitability for MAR. Internationally many countries have been mapping aquifers and the combination of these maps and accompanying hydrogeological reports serve as valuable background information before drilling. Hydrogeological reports generally provide some indication of the level of knowledge of the local aquifers and their degree of uniformity. As aquifer properties vary spatially, it is not generally reliable to extrapolate from one site to predict viability or performance at a nearby site.

Storing water below ground rather than aboveground can have a number of benefits but also some disadvantages. A clear advantage in urban areas is that the land above the storage zone may be used for other uses, particularly if the target aquifer is confined. Even brackish, relatively unutilized aquifers may be used to store freshwater for recovery to meet high-valued uses [16,19]. Although evaporation is eliminated, mixing in a brackish aquifer can result in loss of a similar volume of water to that which would have been lost through evaporation from a surface storage. The rate of recharge and recovery may also restrict the volume of water stored and recovered, which in turn influences the number of recharge systems and recovery wells required for water management.

62.5 Water Sources: Urban Stormwater

Urban stormwater is rainwater that runs off urban surfaces such as roofs, pavements, car parks, roads, gardens, parks, and forests. Much stormwater flows into stormwater drains, creeks, and rivers. Stormwater also forms part of the freshwater that flows to estuaries and the coastal ocean, which is important for maintaining the healthy ecological function of those systems, including the many fish species and associated recreational and commercial fishing activities. While the availability of stormwater is linked to that of rainfall, it is usually an abundant resource in urban areas, but may require treatment and storage before reuse. The availability of stormwater to make useful contributions to urban water supplies is usually not a major constraint and the volume of stormwater runoff from a city is often greater than its entire combined household water use. The primary limitation to stormwater harvesting and reuse in urban areas is the ability to store the water from runoff events for subsequent use. MAR can provide an economic means of storing stormwater in urban areas. Common uses of stormwater recycled via an aquifer include the irrigation of parks and gardens, ovals and golf courses, and other municipal and commercial purposes.

62.6 Water Sources: Treated Sewage

Volumes of treated sewage effluent in urban areas tend to be constant but will require extensive treatment before recycling via aquifers prior to reuse. Infiltration techniques can provide passive treatment during infiltration through the unsaturated zone. SAT, which employs alternating cycles of wetting and drying cycles, can enhance the treatment by providing variable redox conditions and has been shown to be effective for nitrogen removal [9]. Aquifers have advantages with respect to ongoing passive treatment of the water and allowing longer assured residence times before recovery. This form of aquifer water treatment can have an impact on the recovered water quality.

Common uses of treated sewage recycled via an aquifer include the irrigation of parks and gardens, ovals and golf courses, other municipal and commercial purposes, and also drinking water [26].

62.7 Water Quality Considerations

In general, aquifers do not behave like inert systems, but rather as biochemical reactors. It has been found that temperature, redox potential, and nutrient status of water affect the rate of pathogen inactivation, primarily due to their influence on the composition and metabolic activity of microorganisms native to the groundwater. Similarly, these also influence biodegradation of organic chemicals. For example, several endocrine disrupting chemicals are biodegraded in aerobic aquifers in the presence of nutrients but not in anaerobic aquifers [32]. Near a recharge zone, a gradient in temperature, oxygen, and nutrients is likely to occur, and if recharge is intermittent, there will be ongoing changes in microbial composition and activity in space and time.

Rates of inactivation of pathogenic viruses, protozoa, and bacteria can be determined in-situ in aquifers using diffusion cells [6]. Although some aquifers are relatively chemically inert, in general there will also be mineral assemblages that can participate in acid-base, redox, and ion exchange reactions. When groundwater, which has been resident in an aquifer for a long time and reached geochemical equilibrium with the aquifer matrix, receives an influx of surface water with a different composition, it is expected that the imbalance will result in geochemical reactions. This may mobilize metals such as arsenic, iron, and manganese [29] from the aquifer or result in dissolution of calcite and dolomite, leading to chemical weathering of an aquifer and the possibility of structural instability of injection wells in the long term in some aquifers if not managed effectively.

Risk-based assessment and management systems allow for sustainable water quality treatment processes within unsaturated and saturated zones [21]. The American Water Works Association Research Foundation, along with Australian, European, and American partners, has supported much of the research in this area [6], along with the European Community through its RECLAIM WATER projects [8,13,15,19,21,24,26,28].

62.8 Summary and Conclusions

For many years, aquifers have been inadvertently relied on for water recycling. More recently, they have begun to be deliberately used for water recycling via aquifers via MAR. There has been increasing uptake of MAR internationally, not only as a water supply and treatment technology, but also as a sustaining practice in integrated water resources management. Even so, the value of MAR has been dominated in urban areas by the value of subsurface storage capacity rather than for treatment or distribution, as these benefits have not been as well characterized. A range of factors needs to be taken into account in urban water supply, and MAR provides an option that should be considered among all alternatives to determine the best methods for urban water reuse.

References

1. Almulla, A., Hamad, A., and Gadalla, M. 2005. Aquifer storage and recovery (ASR): A strategic cost-effective facility to balance water production and demand for Sharjah. *Desalination*, 174, 193–204.
2. Asano, T., Burton, F., Leverenz, H., Tsuchihashi, R., and Tchobanoglous, G. 2006. *Water Reuse: Issues, Technologies and Applications.* Metcalf & Eddy, McGraw-Hill, New York.
3. Ayuso-Gabella, M.N., Page, D., Masciopinto, C., Aharoni, A., Salgot, M., and Wintgens, T. 2011. Quantifying the effect of managed aquifer recharge on the microbiological human health risks of irrigating crops with recycled water. *Agricultural Water Management*, 99, 93–102.
4. Bekele, E., Toze, S., Patterson, B.M., Fegg, W., Shackleton, M., and Higginson, S. 2013. Evaluating two infiltration gallery designs for managed aquifer recharge using secondary treated wastewater. *Journal of Environmental Management*, 117, 115–120.
5. Dillon, P. 2005. Future management of aquifer recharge. *Hydrogeology Journal*, 13, 313–316.
6. Dillon, P. and Toze, S. (eds.) 2005. *Water Quality Improvements During Aquifer Storage and Recovery.* American Water Works Association Research Foundation Report 91056F.
7. Dillon, P.J. and Liggett, J.A. 1983. An ephemeral stream-aquifer interaction model. *Water Resources Research*, 19(3), 621–626.

8. Ernst, M., Hein, A., Asmin, J., Krauss, M., Fink, G., Hollender, J., Ternes, T., Jørgensen, C., Jekel, M., and McArdell, C.S. 2012. Water quality analysis—Detection, fate, and behavior, of selected trace organic pollutants at managed aquifer recharge sites. In: C. Kazner, T. Wintgens, and P. Dillon (eds.), *Water Reclamation Technologies for Safe Managed Aquifer Recharge.* IWA Publishing, London, UK.

9. Fox, P. 2001. An investigation of soil aquifer treatment for sustainable water reuse. American Water Works Association Research Foundation and American Water Works Association.

10. Greskowiak, J., Prommer, H., Massmann, G., and Nützmann, G. 2006. Modeling seasonal redox dynamics and the corresponding fate of the pharmaceutical residue phenazone during artificial recharge of groundwater. *Environmental Science and Technology,* 40, 6615–6621.

11. Heilweil, V.M., Ortiz, G., and Susong, D.D., 2009, *Assessment of Managed Aquifer Recharge at Sand Hollow Reservoir, Washington County, Utah, Updated to Conditions through 2007.* U.S. Geological Survey Scientific Investigations Report 2009–5050.

12. Ishida, I., Tsuchihara, T., Yoshimoto, S., and Imaizumi, M. 2011. Sustainable use of groundwater with underground dams. *Japan Agricultural Research Quarterly,* 45(1), 51–61.

13. Jiménez, B., Chávez, A., Gibson, R., and Maya, C. 2012. Unplanned aquifer recharge in El Mezquital/Tula Valley, Mexico. In: C. Kazner, T. Wintgens, and P. Dillon (eds.), *Water Reclamation Technologies for Safe Managed Aquifer Recharge.* IWA Publishing, London, UK.

14. Kavuri, M., Boddu, M., and Annamdas, V.G.M. 2011. New methods of artificial recharge of aquifers: A review 2011. *Proceedings of 4th International Perspective on Water Resources & the Environment,* January 4–6, 2011, National University of Singapore (NUS), Singapore.

15. Masciopinto, C., La Mantia, R., Pollice, A., and Giuseppe, L. 2012. Managed aquifer recharge of a karstic aquifer in Nardó, Italy. In: C. Kazner, T. Wintgens, and P. Dillon (eds.), *Water Reclamation Technologies for Safe Managed Aquifer Recharge.* IWA Publishing, London, UK.

16. Miotlinski, K., Dillon, P.J., Pavelic, P., Cook, P.G., Page, D.W., and Levett, K. 2011. Recovery of injected freshwater to differentiate fracture flow in a low-permeability brackish aquifer. *Journal of Hydrology,* 409, 273–282.

17. NRMMC–EPHC–NHMRC. 2009. *Australian Guidelines for Water Recycling, Managing Health and Environmental Risks, Volume 2C—Managed Aquifer Recharge.* Natural Resource Management Ministerial Council, Environment Protection and Heritage Council National Health and Medical Research Council, http://www.environment.gov.au/water/publications/quality/water-recycling-guidelines-mar-24.html A accessed on September 23, 2013.

18. Nzomo Munyao, J., Muinde Munywoki, J., Ikuthu Kitema, M., Ngui Kithuku, D., Mutinda Munguti, J., and Mutiso, S. 2004. *Kituï Sand Dams: Construction and Operation Principles.* Sasol Foundation, Kenya.

19. Page, D., Barry, K., Regel, R., Kremer, S., Pavelic, P., Vanderzalm, J., Dillon, P., Rinck-Pfeifer, S., and Pitman, C. 2012. The aquifer storage, transfer and recovery project in Salisbury, South Australia. In: C. Kazner, T. Wintgens, and P. Dillon (eds.), *Water Reclamation Technologies for Safe Managed Aquifer Recharge.* IWA Publishing, London, UK.

20. Page, D., Miotlinski, K., Dillon P., Taylor, R., Wakelin S., Levett, K., Barry, K., and Pavelic P. 2011. Water quality requirements for sustaining aquifer storage and recovery operations in a low permeability fractured rock aquifer. *Journal of Environmental Management,* 92, 2410–2418.

21. Page, D.W., Ayuso-Gabella, Kopac, I., Bixio, D., Dillon, P., Salgot, M., and Genthe, B. 2012. Risk assessment and risk management in managed aquifer recharge. In: C. Kazner, T. Wintgens, and P. Dillon (eds.), *Water Reclamation Technologies for Safe Managed Aquifer Recharge,* IWA Publishing, London, UK.

22. Pavelic, P., Dillon, P.J., Barry, K.E., Vanderzalm, J.L., Correll, R. L., and Rinck-Pfeiffer, S.M. 2007. Water quality effects on clogging rages during reclaimed water ASR in a carbonate aquifer. *Journal of Hydrology,* 334, 1–16.

23. Pyne, R.D.G. 2005. *Aquifer Storage Recovery: A Guide to Groundwater Recharge through Wells.* 2nd Edition, ASR Systems, Florida, USA.

24. Tandoi, V., Levantesi, C., Toze, S., Böckelmann, U., Divizia, M., Ayuso-Gabella, M.N., Salgot, M. et al. 2012. Water quality analysis—Microbiological hazards. In: C. Kazner, T. Wintgens, and P. Dillon (eds.), *Water Reclamation Technologies for Safe Managed Aquifer Recharge.* IWA Publishing, London, UK.

25. Tredoux G., Cavé L.C., and Bishop, R. 2002. Long-term stormwater and wastewater infiltration into a sandy aquifer, South Africa. *Management of Aquifer Recharge for Sustainability. Proceedings of ISAR-4,* Adelaide, South Australia.

26. Tredoux, G., Genthe, B., Steyn, M., and Germanis, J. 2012. Managed aquifer recharge for potable reuse in Atlantis, South Africa. In: C. Kazner, T. Wintgens, and P. Dillon (eds.), *Water Reclamation Technologies for Safe Managed Aquifer Recharge.* IWA Publishing, London, UK.

27. UN-HABITAT. 2006. Rainwater harvesting and utilisation, Blue Drop Series, Books 1 to 3. United Nations Human Development Programme, Water and Sanitation Infrastructure Branch, Nairobi.

28. van Houtte, E., Cauwenberghs, C., Weemaes, M., and Thoeye, C. 2012. Indirect potable reuse via managed aquifer recharge in the Torreele/St-André project. In: C. Kazner, T. Wintgens, and P. Dillon (eds.), *Water Reclamation Technologies for Safe Managed Aquifer Recharge.* IWA Publishing, London, UK.

29. Vanderzalm, J., Page, D., and Dillon, P. 2011. Application of a risk management framework to a drinking water supply augmented for 100 years by stormwater recharge. *Water Science and Technology,* 63, 719–726.

30. Vanderzalm, J., Sidhu, J., Bekele, E., Ying, G.-G., Pavelic, P., Toze, S., Dillon, P. et al. 2009. *Water Quality Changes During Aquifer Storage and Recovery.* Water Research Foundation, Denver, CO.

31. Vanderzalm, J.L., Dillon, P.J., Barry, K.E., Miotlinksi, K., Kirby, J.K., and Le Gal La Salle, C. 2011. Arsenic mobility and impact on recovered water quality during aquifer storage and recovery using reclaimed water in a carbonate aquifer. *Applied Geochemistry*, 26, 1946–1955.

32. Ying, G.-G., Toze, S., Hanna, J., Yu, X.-Y., Dillon, P.J., and Kookana, R.S. 2008. Decay of endocrine-disrupting chemicals in aerobic and anoxic groundwater. *Water Research*, 42, 1133–1141.

63

Use of Wastewater for Hydroelectric Power Generation

R.K. Saket
*Indian Institute of Technology
(Banaras Hindu University)*

Saeid Eslamian
Isfahan University of Technology

PREFACE

Energy is a fundamental need to daily human life. Whether it is providing lights for our classrooms, refrigeration of our food and medicine, pumps to irrigate our crops, or to run our commercial and industrial enterprises, energy provides the means for economic growth and social and political development. Energy is classified into several types based on the following criteria: (1) primary and secondary energy, (2) commercial and noncommercial energy, and (3) renewable and nonrenewable energy. Renewable energy is obtained from sources that are essentially inexhaustible. It includes wind power, solar power, geothermal energy, tidal power, and hydroelectric power. The most important feature of renewable energy is that it is very clean and does not pollute the environment. This chapter presents use of municipal wastewater for hydroelectric power generation. Generally, most of the urban and industrial wastes generated find their way into land and water bodies, resulting in bad odor, and air and water pollution. Now, this problem is being significantly mitigated through adoption of environmentally friendly waste-to-energy technologies. India has a waste-to-energy potential of 1700 MW. The successful operations of some units, coupled with the recent technological advances, have proven that waste-to-energy projects are commercially viable. Annual and daily flow duration curves for sewage waste water systems (SWWS) have been obtained for design, installation, development, scientific analysis, and reliability evaluation of the hydroelectric power generation system. The hydro potential of wastewater flowing through the sewage system of the Banaras Hindu University campus has been determined to produce annual flow duration and daily flow duration curves by ordering the recorded water flows from maximum to minimum values. Design pressure, the roughness of the pipe's interior surface, method of joining, weight, ease of installation, accessibility to the sewage system, design life, maintenance, weather conditions, availability of material, related cost, and likelihood of structural damage are considered for design of a particular penstock for reliable operation of the power generation system. Micro hydropower plants (MHPP) using municipal wastewater are designed, developed, and practically implemented to provide reliable electric energy to suitable load on the campus of the Banaras Hindu University (BHU), Varanasi, Uttar Pradesh, India during the academic session 2008–2009. Generation reliability evaluation of the developed MHPP using Gaussian distribution approach, safety factor concept, peak load consideration, and Simpson's 1/3rd rule has been described in various published papers [9,11]. This chapter describes basic considerations for development of an MHPP based on municipal wastewater. Performance evaluation, reliability considerations, and voltage stability aspects of the MHPP have not been considered in this chapter.

63.1 Introduction

The international energy outlook in 2004 projected strong growth for worldwide energy demand over the 24-year projection period from 2001 to 2025. The major growth in energy demand is projected in developing countries as 2 billion people lack access to affordable and reliable energy supplies. The rural electrification is a vital program for socioeconomic development of rural areas. The objectives are to trigger economic development and generate employment by providing electricity as an input for productive uses in agriculture and rural industries, and improve the quality of life of the rural people by supplying electricity for lighting of homes, shops, community centers, and public places in all villages [4]. Generally, most of the urban and industrial wastes generated find their way into land and water bodies, resulting in bad odor, and air and water pollution. Now, this problem is being significantly mitigated through adoption of environmentally friendly waste-to-energy technologies. The successful operation of some units, coupled with the recent technological advances, have proven that waste-to-energy projects are commercially viable. The micro hydropower generation system (MHPGS) is one of the popular renewable energy sources in developing countries. MHPGS has obtained increasing interests in the twenty-first century due to their ecological irreproachability and acceptable prices for generating electrical power without producing harmful pollution and greenhouse gases. MHPGS based on urban wastewater (UWW) is considered an environmentally friendly renewable energy source because this can be sized and designed to limit the interference with river flow and canal flow. Most of the MHPGSs operate in isolated mode for rural electrification of remote villages where the population is very small and extension of the distributed grid system is not geographically and financially feasible due to high cost investment in power transmission systems. Small hydroelectric power generation systems (SHPGSs) are relatively small power sources that are appropriate in many cases for individual users or groups of users who are independent of the electricity supply grid. Although this technology is not new, its wide application to small waterfalls and other potential sites like municipal waste water (MWW)-based systems are new [9,11]. SHPGSs are the application of hydroelectric power on a commercial scale serving a small community and are classified by power and size of waterfall. Small hydropower plants (SHPP) can be divided into mini hydro (less than 1000 kW capacity), and micro hydropower systems (MHPS), which have less than 100 kW capacity. Hydroelectric power is the technology of generating electric power from the movement of water through rivers, rivulets, streams, artificially created storage dams, canal drops, and tides. Water is fed via a channel to a turbine where it strikes the turbine blades and causes the shaft to rotate. To generate electricity, the rotating shaft is connected to a generator that converts the motion of the shaft into electrical energy [13]. It has been recognized that SHPP can play a role in improving the energy position in some remote and inaccessible areas.

Generally, in an autonomous MHPS, the small hydropower generators (SHPG) are the main constituents of the system and are designed to operate in parallel with local power grids. The main reasons are to obtain economic benefit of no fuel consumption by micro hydro turbines, enhancement of power capacity to meet the increasing demand, to maintain the continuity of supply in the system, etc. Small/micro hydro is highly fluctuating in nature, will affect the quality of supply considerably, and even may damage the system in the absence of proper control mechanisms. Main parameters to be controlled are the system frequency and voltage, which determine the stability and quality of the supply. In an MHPGS, frequency deviations are mainly due to real power mismatch between generation and demand. Reactive power balance in the hybrid system can be obtained by making use of variable reactive power devices, for example, static volt ampere reactive (VAR) compensator [3]. Comparisons of various penstock materials have been presented considering friction, weight, cost, corrosion, joining, and pressure for reliable operation of the MHPG system. Hybrid power systems are the most attractive option for the electrification of the remote locations. These include high cost because of the system complexity, site-specific design requirements, and the lack of available control system flexibility. Many countries have targets and aspirations for growth in renewable energy. If a new alternative generation technology is introduced that makes a relatively low contribution to the reliability of meeting peak demand, then additional capacity may be needed to provide system margin, and cost is improved on the rest of the system. Quantification of the system costs of additional renewable in 2020 has been presented in Reference 8.

MHPP using MWW neither requires a large dam nor is land flooded. Only wastewater from different parts of the city is collected to generate power, which has minimum environmental impact. After proper chemical treatment, water is provided to farmers for irrigation purpose. MHPGS using MWW of sewage plant can offer a stable, inflation-proof, reliable, economical, and renewable source of electricity. This alternative technology has been appropriately designed, developed, and practically implemented on the campus of the Indian Institute of Technology,

Banaras Hindu University, Varanasi, India. Reuse of MWW can be a stable, inflation-proof, economical, reliable, and renewable energy source of electricity in the power scenario of the twenty-first century [12].

This chapter is organized as follows: Section 63.1 presents an overview on MHPGS based on MWW. Selection criteria and brief descriptions on basic components for designed and developed MHPGS at BHU campus is described in Section 63.2. Selection criteria of penstock pipe materials, water turbines and various generators for different heads are explained in this section. Annual, monthly, and daily flow-duration curves (AFDC, MFDC, and DFDC) have been obtained for design and development of MHPP based on MWW by recording water flow rates in Section 63.3. Generation reliability of constructed MHPP using Gaussian distribution approach, peak load considerations, safety factor concept and Simpson's 1/3rd rule have not been included in this chapter. Reliability aspects have been published in previous publications [15]. Measurement and evaluation of potential power with various head and water flow rates are described in Section 63.4. Results and analytical discussions are presented in Section 63.5. This section concludes with project work as an energy scenario of the twenty-first century. Related bibliography is included in the last section.

63.2 Description of Basic Components of MHPP: An Overview

The principal components of the MWW-based MHPGS are wastewater tank, penstock pipe, water turbine, and induction generator. In hydroelectric power plants, energy from water is utilized to move the turbines, which in turn run the electric generators. The energy of water utilized for power generation may be kinetic or potential. The kinetic energy of water is its energy in motion, while the potential energy is a function of the difference in water head between two points. MWW and self excited induction generator (SEIG)-based MHPGS have been designed, developed, and constructed at Broacha sewage station of the BHU during the session 2008–2009 [6,20,22]. Prototype MHPGS have been designed, installed, and tested using farmland irrigation water flowing in canals in Taiwan. The designed MHP unit consists of a water wheel containing 16 blades, a mechanical shaft, two bearings, a gearbox and a permanent magnet generator (PMG) and inverters. The low rotational speed of the water wheel can be speeded up to required rotational speed of the PMG through the designed gearbox [20]. The following principal components of the developed MWW-based MHPGS have been designed for reliable operation of the system.

63.2.1 Selection of Penstock Pipe

The penstock is the most expensive item in this project, which may cost up to 40% of the total project cost. Several factors should be considered when deciding which material to use for a particular penstock design pressure, that is, the roughness of the pipe's interior surface, method of joining, weight, ease of

installation, accessibility to the site, design life, maintenance, weather conditions, availability, relative cost, and likelihood of structural damage [18]. The most commonly used materials for a penstock are mild steel, high-density polyethylene (HDPE) and unplasticized polyvinyl chloride (uPVC) because of their suitability, availability, and approvability. The uPVC exhibits excellent performance over mild steel and HDPE in terms of least friction losses, weight, corrosion, cost, etc. Comparison of the penstock materials considering friction, weight, corrosion, cost, joining, and pressure have been explained [17]. The uPVC material has been selected and used for design of the penstock for development of this project.

63.2.2 Selection of Turbines

A hydraulic turbine converts the potential energy of water into mechanical energy, which in turn is utilized to run an electric generator to get electric energy. The choice of water turbine depends mainly on the head and MWW flow rate for installation of the MWW and SEIG-based MHPP. The selection also depends on the desired running speed of the generator. To adjust for variations in stream flow, water flow to these turbines is easily controlled by changing nozzle sizes or by using adjustable nozzles. Turbines used in the hydro system can be classified as impulse (Pelton, Turgo, and Cross-flow), reaction (Francis, Propeller, and Kaplan) and water wheels (under-short, breast-shot, and overshot). Groups of the water turbine for various heads available have been proposed for sewage system-based MHPP. The reliability and performance of the MHPGS based on MWW and SEIG depends on the turbines/water wheels efficiency [1]. Kaplan turbine is used for reliable performance, installation, and establishment of this project. For very low heads, bulb turbines are employed these days.

63.2.3 Selection of Generator

Induction and synchronous generators are used in power plants and both are available in three-phase or single-phase systems. Induction generators are generally appropriate for MHPGS [11]. Induction generators offer many advantages over conventional synchronous generators as a source of isolated power supply. Reduced unit cost, ruggedness, brushless (in squirrel cage construction), reduced size, absence of separate DC source and ease of maintenance, and self-protection against severe overloads and short circuits are the main advantages. Capacitors are used for excitation and are popular for smaller systems that generate less than 10–15 kW. Use of an induction generator is increasingly becoming more popular in MHP application because of its simpler excitation system, lower fault level, lower capital cost, and less maintenance requirement. However, one of its major drawbacks is that it cannot generate the reactive power as demanded by the load. Most of the early stage MHPPs are equipped with synchronous generators [5]. SEIG is used at installed MHPGS. In general,

SEIG is applicable for the unregulated prime mover speed. The working principle of SEIG depends on the minimum shaft speed to start generation. Generation voltage of SEIG depends on the shaft speed, residual magnetism, reduced permeability at low magnetization, and value of the capacitor connected to the machine. Reliability of the self-excitation must be very high at generation time. It is achieved either by increasing the speed or by increasing the capacitor value or both. Residual magnetism and permeability of the rotor iron core cannot be changed during operating conditions of the SEIG. However, capacitor value and shaft speed can be changed [7]. The generation of the SEIG depends only on the shaft speed, if maximum values of the capacitors or fixed capacitors are used. At constant parameters of the developed system, such as SEIG torque, gears ratio, water head, and exciting capacitor value, only shaft speed is variable. Reliability condition of the power generation only depends on the shaft speed. Terminal voltage of the SEIG is directly proportional to the shaft speed. If shaft speed of the SEIG is decreased below the particular value of the speed, then generation will stop. This particular value of shaft speed is called the minimum value of speed or threshold speed. This has a particular constant value, where the generation starts and stops. This means that SEIG generates voltage when the shaft speed of the generator is equal or greater than this minimum value of the speed. Similarly, if the shaft speed of the generator is below the minimum value of the threshold speed, then SEIG will not generate any voltage. Many times generation failure was noticed due to lower shaft speeds from the threshold speed of the SEIG. Therefore, evaluation of the success and failure probabilities of the MWW-based MHPGS is required [10].

63.3 Development of Flow Duration Curve

Flow duration curve (FDC) is a plot between flows available during a period versus a fraction of time. If the magnitude on the ordinate is the potential power contained in the stream flow, then the curve is known as power duration curve (PDC). This curve is a very useful tool in the analysis for the development of waterpower. The area under the FDC gives the total quantity of run-off during that period. It is highly useful in the planning and design of water resources projects. This curve plotted on a log-log paper provides a qualitative description of the run-off variability in the stream. In this chapter, annual, monthly, and daily FDCs have been obtained for development and installation of the MHPP by recording water flows from maximum to minimum values. Annual flow duration curve (AFDC) and daily flow duration curve (DLDC) of the developed system have been obtained for performance and reliability evaluation of the MHPGS. The AFDC and DFDC are used to assess the expected availability of water flow, flow variations, and power capability to select the type of turbine and generator [12]. According to FDC, there is a difference in

wastewater flow between summer (March–June) and winter (November–February) cycles and this can affect the power output produced by installed MHPP. These variations have been considered for estimation of the total expected energy generation. Peak load demand has been considered for evaluation of reliability of the MHPGS at different generation capacities, load demand, and atmosphere variations. Output power has been estimated at different flow rates and heads available in the sewage system of BHU. Reuse of MWW of the city can be a stable, inflation-proof, economical, reliable, and renewable energy source in the power scenario of the twenty-first century [16].

According to the above discussion, if SEIG parameter, torque, water head, capacitor value, etc. are constant, then the generation depends only on the shaft speed. SEIG shaft speed depends on the MWW flow rate. It means if the shaft speed is below the threshold speed, then the generation will stop. The threshold speed of the MWW and SEIG-based MHPG system has been calculated from the experimental results. The value of threshold speed of MWW is 3.5 m^3/s for this particular case. Thus, shaft speed of the SEIG due to MWW flow rate lower than the 3.5 m^3/s generation will stop. So, reliability of the generation failure due to lower speed from the threshold speed is examined successfully. For reliability evaluation of the MHPP, ADFDC of each month have been drawn to get an idea of the failure and success generation at particular times. Generation time, repair time, and number of times when generation fails for each month have been discussed successfully for reliability evaluation of the system [19].

In this paragraph, average DFDC and AFDC were obtained by recording the MWW flows from maximum to minimum values. For development of the DFDC of any month of the assessment year, the 15th day of each month is considered for flow rate measurement of 24 duration. DFDC and AFDC of the developed plant were also obtained from the average of three years (from July 2009 to June 2012) for reliability evaluation of the MHPG system as shown in Figures 63.1 and 63.2, respectively. The educational calendar or academic session of BHU starts from July 1 and ends on June 30. Plant data are collected from July 2009 to June 2012. Peak load demand at different working conditions of the generation capacity, load demand, and atmosphere variations have been considered to evaluate the reliability of the MHPG system.

63.3.1 DFDC for July, August, and September

The new academic session of BHU starts in July, which is a rainy season. Therefore, wastewater flow rate mostly reaches the maximum point from July to September as shown in Figure 63.3. During these three months, the average MWW flow rate is above 5 m^3/s as shown in Figure 63.3. The number of failure of generation due to minimum speed and repair time is less. The operating time of the MHPGS is high due to availability of the MWW.

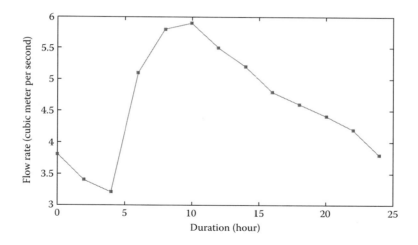

FIGURE 63.1 ADFDC for three years (July 2009–June 2012).

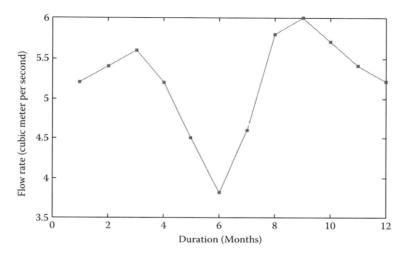

FIGURE 63.2 AAFDC for three years (July 2009–June 2012).

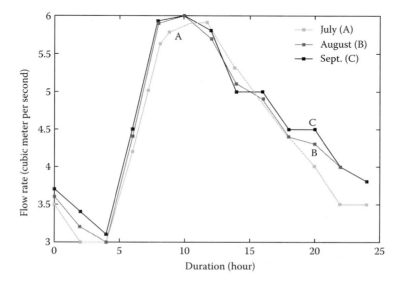

FIGURE 63.3 DFDC for July, August, and September months.

63.3.2 DFDC for October, November, and December

Around October and November, MWW flow rate decreases due to the start of the winter season. Two long holiday periods of Dashahara and Deewali festivals of India in these months decrease MWW flow rate of the sewage plant. The month of December provides a semester break period as well as the coldest atmospheric conditions. Therefore, MWW flow rate decreases and comes down to its deep point. During these three months, the average MWW flow rate decreases from 5 m³/s to below 4 m³/s as shown in Figure 63.4. In October and November, the number of generation failure and repair time are higher than in July, August, and September. The operating time of the MHPGS decreases during these months. Number of failures and repair time of the developed system are high and operating time is low during December.

63.3.3 DFDC for January, February, and March

MWW flow rate increases in January due to starting of the semester from the first week of this month. However, it does not reach up to the maximum point due to winter days. Flow rate increases in February and March but does not reach up to the maximum point due to winter and holidays on Indian Holi festival in these months. The average flow rate decreases from 5 m³/s to below 4 m³/s as shown in Figure 63.5. The number of failure of generation and repair time decreases as compared with the report of December. However, the operating time of the MHPGS increases during these months.

63.3.4 DFDC for April, May, and June

In April, MWW flow rate is standard for electricity generation. Summer vacation of the BHU starts from May to June, so MWW

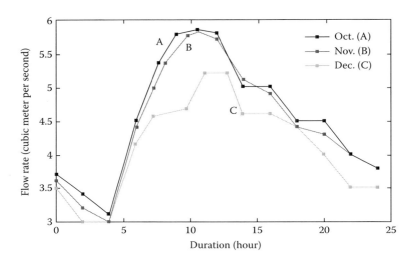

FIGURE 63.4 DFDC for October, November, and December months.

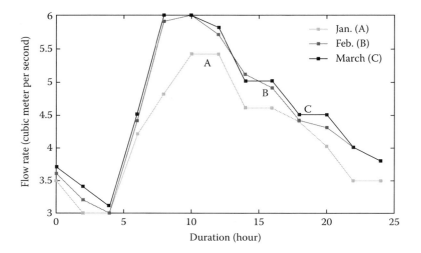

FIGURE 63.5 DFDC for January, February, and March months.

flow rate again decreases during these two months. The average flow rate decreases from approximately 5 to 4 m³/s as shown in Figure 63.6. The number of failure of generation and repair time is approximately low in April. However, the operating time is approximately low in April. Number of generation failure and repair time increases during May and June. However, the operating time of the MHPGS decreases due to unavailability of the MWW during the summer vacation.

Average daily flow-duration curve (ADFDC) from July 2009 to June 2012 and average annual flow-duration curve (AAFDC) of the MWW-based MHPG power plant from July 2009 to June 2012 have been presented as shown in the Figures 63.1 and 63.2, respectively. According to available AFDC of the developed sewage power station, the average value of the flow rate is 4.67 m³/s. Reliability indices for the evaluated MWW flow rates are described in the next section.

63.4 Power Estimation of the Wastewater-Based Hydropower System

63.4.1 Theoretical Power Estimation

The theoretical amount of power in kW (P_{th}) available from a micro hydropower system is directly related to the flow rate in m³/s (Q), water head available in meters (H), and weight density of water in N/m³ (w) as given here:

$$P_{th} = wQH/100 (kW) \qquad (63.1)$$

To calculate the actual power output (P_{act}) from micro hydropower plants, it is required to consider friction losses in the penstock pipes and the efficiency of the turbine and generator. Typically, overall efficiencies for electrical generating systems can vary from 50% to 70% with higher overall efficiencies

occurring in high head systems. Therefore, to determine a realistic power output as shown in Table 63.1, the theoretical power must be multiplied by an overall efficiency factor of 0.5–0.7 depending on the capacity and type of system as given here:

$$P_{act} = QHw\eta/100 (kW) \qquad (63.2)$$

where η = efficiency factor (0.5–0.7). The actual useful or effective output depends on the efficiency of the various parts of the installation.

63.4.2 Experimental Power Estimation Using FDC

63.4.2.1 Development of DFDC for Power Estimation

DFDC has been obtained for hydropower stations based on sewage wastewater from a community of the city recording water flow from maximum to minimum flow. Daily load duration curve and power estimation of the proposed hybrid system are obtained for reliability evaluation of the wastewater-based power plant as shown in Figures 63.4 and 63.5, respectively. From Figure 63.4 it is seen that there is a difference in wastewater flow between summer and winter and this can affect the power output produced by a micro hydropower system. These variations have been considered in the estimation of total energy generation expected from the site with recycle of the water using a solar pumping system [13]. Output power is estimated at different flow rates and heads available in the sewage system of BHU. Recorded daily flow rates of the wastewater at Broacha pumping station of BHU are included in Table 63.2. Seasonal variation in the wastewater flow rates is considered development of the DFDC. DFDCs have been developed for rainy, summer, and winter seasons as shown in Figures 63.7 through 63.9, respectively. These curves are developed for the academic year 2012–2013 for power estimation of the developed system. Brief descriptions on development of DFDC/AFDC are given in Section 63.3.

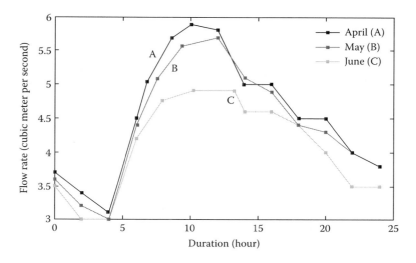

FIGURE 63.6 DFDC for April, May, and June months.

TABLE 63.1 Typical Power Output (W) with Various Head and Water Flow Rates

Head (m)	WasteWater Flow Rate (m³/s)							
	5	10	15	20	40	60	80	100
	Theoretical Power Output (W)							
1	25	49	74	98	196	294	392	490
2	49	98	147	196	392	588	784	980
4	98	196	294	392	784	1176	1568	1960
8	196	392	588	784	1568	2352	3136	3920
10	245	490	735	980	1960	2940	3920	4900
15	368	735	1103	1470	2940	4410	5880	7350
20	490	980	1470	1960	3920	5880	7840	9800
30	735	1470	2205	2940	5880	8820	14,112	17,640
40	980	1960	2940	3920	7840	14,112	18,816	23,520

TABLE 63.2 Daily Wastewater Flow Rate of Broacha Sewage Pump Station

Year 2012–2013 (h)	March 10	April 10	May 10	June 10	July 10	Aug 10	Sept 10	Oct 10	Nov 10	Dec 10
	WasteWater Flow Rates in m³/s									
00.00	11.1	11.5	10	10.5	11.1	10.4	10	9.5	9.5	9
01.00	11.1	11.5	10	10.5	11.1	10.4	10	9.5	9.5	10
02.00	11.1	11.5	12.5	11.8	12.5	12.5	12.3	10.1	10.5	11
03.00	11.1	11.5	12.5	11.8	12.5	12.5	12.3	10.1	10.5	11
04.00	12.5	12.12	12.5	11.8	12.5	12.5	12.3	11	11.5	12
05.00	12.5	12.12	12.5	11.8	12.5	12.5	12.3	11	11.5	12
06.00	12.5	12.12	20	18.2	21.1	19	21.8	11	11.5	12.5
07.00	16	15.4	20	18.2	21.1	19	21.8	13	14	18
08.00	16	15.4	20	18.2	21.1	19	21.8	13	14.1	18.3
09.00	16	15.4	17.4	16	16.7	16.4	17	13	14.1	18.3
10.00	16	15.4	17.4	16	16.7	16.4	17	13	14.1	18.3
11.00	14.3	14.8	17.4	16	16.7	16.4	17	14	14.5	16
12.00	14.3	14.8	17.4	16	16.7	16.4	17	14	14.5	16
13.00	14.3	14.8	14.3	13.8	13.5	14.1	14.5	14.2	14.8	15
14.00	14.3	14.8	14.3	13.8	13.5	14.1	14.5	14.2	14.8	15
15.00	12.5	12.12	14.3	13.8	13.5	14.1	14.5	13.5	13	13.1
16.00	12.5	12.12	14.3	13.8	13.5	14.1	14.5	13.5	13	13.1
17.00	12.5	12.12	16	16.7	15.8	15.4	16	13.5	13	13.1
18.00	12.5	12.12	16	16.7	15.8	15.4	16	13.1	12	12.5
19.00	13.3	13.8	16	16.7	15.8	15.4	16	13.1	12	12.5
20.00	13.3	13.8	14.8	14.3	13.3	13.5	14	13.1	12	12.5
21.00	11.5	11.8	14.8	14.3	13.3	13.5	14	12.5	11.4	11.9
22.00	11.5	11.8	14.8	14.3	13.3	13.5	14	12.5	11.4	11.9
23.00	11.5	10	10.5	11.1	10.4	10	11.8	11	10.5	10.4
24.00	11.5	10	10.5	11.1	10.4	10	11.8	11	10.5	10.4

63.4.2.2 Power Estimation from Developed DFDC

Theoretical power estimation and experimental obtained power at MHPP are given in Tables 63.1 and 63.3, respectively. Table 63.1 includes calculated theoretical power data using Equations 63.1 and 63.2. Power generation capacity data according to availability of the wastewater flow rates during the 10th day of each month are given in Table 63.3. Variations in the generation capacity indicate variations in the developed DFDC. Seasonal variations in the DFDC and power generation are graphically illustrated in Figures 63.7 through 63.12, respectively. These graphs provide power generation capacity of wastewater-based MHPP with seasonal variation in the wastewater flow rates of the sewage power station.

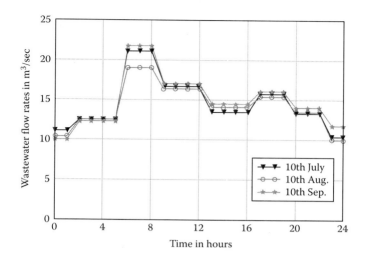

FIGURE 63.7 Flow rates in the rainy season.

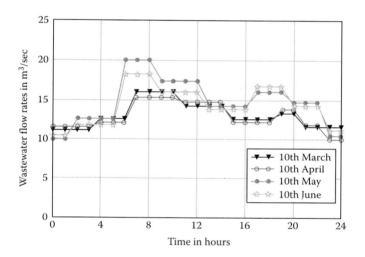

FIGURE 63.8 Flow rates in the summer season.

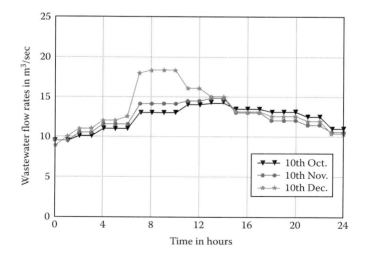

FIGURE 63.9 Flow rates in the winter season.

TABLE 63.3 Output Power Estimation Using DFDC of Developed SPGS

Time (h)	March 10	April 10	May 10	June 10	July 10	Aug 10	Sept 10	Oct 10	Nov 10	Dec 10
	\multicolumn{10}{c}{Estimated power output of developed MHPS (Watts)}									
00.00	82	85	74	77	82	77	74	75	80	70
01.00	82	85	74	77	82	77	74	75	80	70
02.00	82	85	92	87	92	90	84	85	90	80
03.00	82	85	92	87	92	90	84	85	90	80
04.00	92	89	92	87	92	90	84	85	90	80
05.00	92	89	92	87	92	90	160	132	90	155
06.00	92	89	147	134	155	140	160	132	145	155
07.00	118	113	147	134	155	140	160	132	145	155
08.00	118	113	147	134	155	140	160	118	145	155
09.00	118	113	128	118	123	121	125	118	120	120
10.00	118	113	128	118	123	121	125	118	120	120
11.00	105	109	128	118	123	121	125	118	120	120
12.00	105	109	128	118	123	121	125	118	120	120
13.00	105	109	105	102	99	104	107	100	95	104
14.00	105	109	105	102	99	104	107	100	95	104
15.00	92	98	105	102	99	104	107	100	95	104
16.00	92	98	105	102	99	104	107	100	95	104
17.00	92	89	118	123	116	113	118	120	112	115
18.00	98	102	118	123	116	113	118	120	112	115
19.00	98	102	118	123	116	113	118	120	112	115
20.00	98	102	109	105	98	99	103	103	94	99
21.00	85	87	109	105	98	99	103	103	94	99
22.00	85	87	109	105	98	99	103	103	94	99
23.00	85	74	77	82	77	74	87	80	75	83
24.00	85	74	77	82	77	74	87	80	75	83

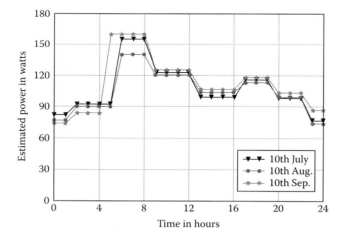

FIGURE 63.10 Estimated power for the rainy season.

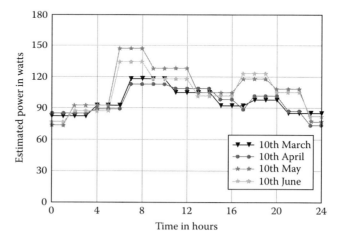

FIGURE 63.11 Estimated power for the summer season.

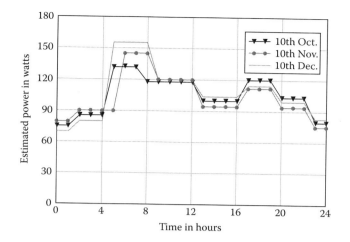

FIGURE 63.12 Estimated power for the winter season.

63.5 Summary and Conclusions

Design aspects of the installed MHPGS have been described successfully in this chapter. Based on the proposed concept and development of experimental MHPP at Banaras Hindu University campus, a number of experimental tests were performed for designing AFDC, DFDC, and SLDC. Generating capacity of installed MHPGS based on MWW at different heads available at Broach sewage station of the BHU has been evaluated successfully. Turbine-generator sets have been recommended for different heads of MWW for reliable operation of the MHPP [9]. MWW flow rate and available head have been measured considering summer/winter and day/night cycles. Typical power output has been estimated and measured at different flow rates and heads of sewage plant for calculation of reliability indices and reliability evaluation of generation capacity. Case studies with several conditions of the configuration of the developed MHPP at BHU campus revealed the good performance of the proposed methodology not only in terms of failure probability evaluation but also in terms of the accuracy of the reliability indices [1,2]. A case study introduced in this chapter has been given to demonstrate the applications of the methodologies. The major portion of these evaluations comes from adequacy analysis based on power flow and optimization related with actions taken to correct the system operation conditions [19]. Future research work in this area is proposed for mini and micro hydropower generation from irrigation canals. Power engineers can also plan for pico and nano hydropower generation from overhead water tanks on buildings to charge small chargeable batteries. Simpson's 3/8th rule is proposed for better approximation and future research work [11].

Acknowledgments

The authors are grateful to Indian Institute of Technology (Banaras Hindu University), Varanasi, Uttar Pradesh (India) for partial financial support for carrying out this project. Corresponding author would like thank to Aanchal, Aakanksha, Siddharth, and Siddhant for providing emotional support during design work of this project. This work is heartily dedicated to sweet memories of the first author's great father parinibutta Ram Das Saket Vardhan, who passed away on Shravan Poornima: August 17, 2008.

References

1. Arya, L.D., Choube, S.C., and Saket, R.K. 2000. Composite system reliability evaluation based on static voltage stability limit. *Journal of the Institution of Engineers (India)*, 80(February), 133–140.

2. Arya, L.D., Choube, S.C., and Saket, R.K. 2001. Generation system adequacy evaluation using probability theory. *Journal of the Institution of Engineers (India)*, 81(March), 170–174.

3. Bansal, R.C., Ahmed, F.Z., and Saket, R.K. 2005, Some issues related to power generation using wind energy conversion system: An overview. *International Journal of Emerging Electric Power System*, 3(2), 01–19.

4. Bhardwaj, A.K., Bansal, R.C., Saket, R.K., and Srivastava, A.K. 2010. Electric power demand forecasting of KAVAL cities. *International Journal of Electrical and Power Engineering*, 4(2), 85–89.

5. Bhatti, T.S., Bansal, R.C., and Kothari D.P. (eds.). 2004. *Small Hydro Power Systems*. Dhanpat Rai and Sons, Delhi, India.

6. Choudhary, R. and Saket, R.K. 2015. A critical review on the self-excitation process and steady state analysis of an SEIG driven by wind turbine. *Renewable and Sustainable Energy Reviews, Science Direct SCI Journal*, 47, 344–353. Web-site: http://www.sciencedirect.com/rser.

7. Fang, Y., Bansal, R.C., Dong, Z.Y., Saket, R.K., and Shakya, J.S. 2011. Wind energy resources: Theory, design and applications. In: A.F. Zobaa and R. C. Bansal (eds.), *Handbook of Renewable Energy Technology, Section I: Wind Energy and Their Applications*, World Scientific Publishing House, Singapore, pp. 3–20.

8. Fraenkel, P., Paish, O., Harvey, A., Brown, A., Edwards, R., and Bokalders, V. 1991. *Micro-Hydro Power: A Guide for Development Workers*. Intermediate Technology Publishers with the Stockholm Environment Institute, London, UK.

9. Saket, R.K. 2008. Design, development and reliability evaluation of micro hydro power generation system based on municipal waste water. In: *Proceedings of the IEEE Electrical Power and Energy International Conference: 2008 (EPEC 2008), IEEE Canada and IEEE Vancouver Section*, The University of British Columbia, Vancouver, BC, Canada, October 6–7, 2008, pp. 1–8.

10. Saket, R.K. 2011. Reliability evaluation of defence support systems. *Innovations in Defence Support Systems*, 2, L.C. Jain, Aidman, E. Abeynayake. C. (eds.), *Socio-Technical Systems Series: Studies in Computational Intelligence*, 338, 239–284, University of South Australia, Australia, Springer Verlag, Berlin, Germany.

11. Saket, R.K. 2013. Design aspects and probabilistic approach for generation reliability evaluation of MWW based micro-hydro power plant. *Renewable and Sustainable Energy Reviews, Elsevier (SCI Science Direct Journal)*, 28, 917–929. DOI: 10.1016j.rser.2013.08.033.

12. Saket, R.K. and Anand Kumar, K.S. 2006. Hybrid micro hydro power generation using municipal waster water and its reliability evaluation. In: *Proceeding of the Greater Mekong Sub region Academic and Research Network (GMSARN) International Conference-6-7 December 2006 on Sustainable Development: Issues and Prospects for the Greater Mekong Sub region*, Asian Institute of Technology, Bangkok, Thailand, pp. 13–20.

13. Saket, R.K., Bansal, R.C., and Anand Kumar, K.S. 2007. Reliability evaluation of hybrid micro hydro power generation using municipal waste water. *GMSARN International Journal on Sustainable Development: Issues and Prospects for the Greater Mekong Sub-Region*, Klong Luang, Pathumthani, Thailand, First Issue of the Biannual Journal, June 2007, 1(1), 13–20.

14. Saket, R.K., Bansal, R.C., and (Col.) Gurmit Singh. 2006. Generation capacity adequacy evaluation based on peak load considerations. *The South Pacific International Journal on Natural Sciences*, 24, 38–44.

15. Saket, R.K., Bansal, R.C., and Col. Gurmit Singh. 2007. Reliability evaluation of power system considering voltage stability and continuation power flow. *Journal of Electrical Systems, Engineering and Scientific Research Groups, France*, (32), 48–60.

16. Saket, R.K., Bansal R.C., and Col. Gurmit Singh. 2009. Power systems component modelling and reliability evaluation of generation capacity. *International Journal of Reliability and Safety*, 03(04), 427–441.

17. Saket, R.K. and Varshney L. 2011. Self excited induction generator and municipal waste water based micro hydro power generation system. *IEEE International Conference on Product Development and Renewable Energy Resources (ICPDRE 2011)*, Chennai (Tamil Nadu), India, pp. 180–185, February 19–20, 2011.

18. Saket, R.K. and Varshney, L. 2012. Self excited induction generator and municipal waste water based micro hydro power generation system. *International Journal of Engineering and Technology*, 04(03), 282–287.

19. Saket, R.K. and Zobaa, A.F. 2009. Power system reliability evaluation using safety factor concept and Simpson's 1/3rd Rule. *The International Federation of Automatic Control (IFAC), 6th IFAC Symposium on Power Plants and Power System Control (PPPSC)*, Vol. 01, Part: 01, Tampare Hall, Finland, July 5–8, 2009, pp. 68–73. DOI: 10.3182/20090705-4–SF–2005. 00014. Elsevier Publishers.

20. Varshney, L. and Saket, R.K. 2012. Power estimation of MWW based generation system using SEIG. *IEEE Students International Conference on Electrical, Electronics and Computer Sciences (IEEE SCEECS 2012)*, Maulana Azad National Institute of Technology, Bhopal (Madhya Pradesh), India, March 01–02, 2012, pp. 130–134.

21. Varshney, L., Saket R.K., and Eslamian S. 2013, Power estimation and reliability evaluation of MWW and SEIG based micro hydro power generation system. *International Journal of Hydrology Science and Technology*, 03(2), 176–191.

22. Varshney, L. and Saket, R.K. 2014. Reliability evaluation of SEIG rotor core magnetization with minimum capacitive excitation for unregulated renewable energy applications in remote areas. *Ain Shams Engineering Journal, Science Direct*, 05(03), 751–757. Web-site: http://www.sciencedirect.com/asej.

64

Application of Nanotechnology in Water Reuse

Saeid Eslamian
Isfahan University of Technology

Seyedeh Matin
Amininezhad
*Islamic Azad University
of Shahreza*

Sayed Mohamad
Amininejad
University of Stuttgart

Jan Adamowski
McGill University

PREFACE

Rapid population growth and droughts have led to water shortages, particularly in arid regions of the world. In response to this scarcity of water, the reuse of treated wastewater has been found to be a good and reliable alternative water source. In addition, the management of water supply and wastewater is becoming increasingly difficult. Hence, the reuse of water and wastewater can improve urban water management.

Untreated and secondary treated wastewater contains a range of pollutants that pose a significant risk to the health of humans and can lead to disease outbreaks. To reduce this risk, several technologies are available to treat polluted effluents.

Nanotechnology is a new generation technology that holds great potential for advancing water and wastewater treatment to improve treatment efficiency as well as to augment water supply through safe use of unconventional water sources.

This chapter reviews the current technologies that are used for reuse of wastewater and the application of nanotechnology in this field.

64.1 Introduction

Humans have practiced wastewater reuse for over 3000 years since the ancient Minoans used wastewater for agricultural irrigation. However, according to Asano and Levine [2], the development of programs for the planned reuse of wastewater began in the 1910s with some of the earliest water reuse systems being developed during the 1920s.

The first system to reuse industrial wastewater was implemented in the 1940s, and in the 1960s Colorado and Florida developed urban wastewater reuse systems. During the last 30 years, research has been focused on the technical barriers and health risks associated with wastewater reuse [2,10]. At the same time, some earlier optimization models have been adapted for water reuse planning [3,10,11–13,15,18]. During the last decade, wastewater reuse has gained much attention in many parts of the world as a means of alleviating the growing pressures on water supplies. From a technical point of view, modern wastewater treatment facilities are able to treat wastewater to the quality required for any purposes [2]. It has been recognized throughout the world that water reuse is an important factor for the optimal planning and efficient use of water resources.

64.2 Requirements for Water Reuse

Wastewater can be reused for several purposes such as agricultural irrigation, industrial processes, groundwater recharge, and even for potable water supply after extended treatment to ensure sustainable and successful wastewater reuse.

The potential public health risk associated with wastewater reuse must be evaluated and minimized, and the treated wastewater must meet the required quality for its specific applications.

In order to meet these requirements, it is necessary to treat the wastewater prior to reuse, and to ensure an appropriate level of disinfection to control pathogens. While a comprehensive overview of wastewater treatment options and public health protection is beyond the scope of this chapter, the following sections provide brief summaries on the basic principles of wastewater treatment and options for minimizing the public health risk and environmental impact of water reuse.

64.2.1 Basic Principles of Wastewater Treatment

In order to reuse wastewater, it is necessary to treat the raw wastewater to meet the specific needs of the planned application and to ensure public safety. In this section, some basic information on wastewater treatment technologies is provided and the terminology is explained.

Wastewater treatment involves three main processes:

- *Physical process*: Impurities are physically removed by screening, sedimentation, filtration, flotation, absorption or adsorption or both, and centrifugation
- *Chemical processes*: Impurities are removed chemically through coagulation, absorption, oxidation-reduction, disinfection, and ion-exchange
- *Biological processes*: Pollutants are removed using biological mechanisms, including aerobic treatment, anaerobic treatment, and photosynthetic processes, such as in oxidation ponds

Conventional wastewater treatment consists of four stages, namely: preliminary, primary, and secondary treatment, and disinfection. In municipal wastewater treatment facilities, a combination of physical, biological, and chemical treatment technologies is used. Preliminary and primary treatments usually consist of physical processes, such as screening for the removal of debris and large solids, and sedimentation. Secondary treatment may utilize biological processes, such as stabilization ponds, trickling filters, oxidation ditches, and activated sludge, which are then followed by sedimentation of the biomass (sludge). Tertiary and advanced treatments are used to remove specific pollutants, such as nitrogen or phosphorus, which cannot be removed by conventional secondary treatments. A summary of the purposes of each treatment process and the technologies that are used is given in Table 64.1.

64.2.2 Minimizing Public Health Risks

The fundamental precondition for wastewater reuse is that it will not cause unacceptable public health risks. Untreated wastewater poses a serious risk of waterborne diseases, such as cholera, typhoid, dysentery, the plague and helminthiasis. In the nineteenth century, large-scale use of untreated wastewater for agriculture triggered epidemics of such waterborne diseases in Europe. With medical and public health advancements, the links between untreated wastewater and diseases have become better understood, and measures to minimize exposure to such pathogens have been introduced.

Raw (untreated) wastewater, however, continues to be used in some regions for direct crop irrigation, despite the clear health risks associated with it. This practice should be discontinued and replaced with irrigation using treated water that meets public health guidelines in order to minimize the exposure of farm workers and consumers to pathogens. For agricultural applications, the World Health Organization (WHO) has published guidelines for the use of wastewater for restricted and unrestricted irrigation. Governments have also developed stringent criteria for agricultural applications. For nonagricultural applications, no global water quality standards exist, and various governments have produced their own standards.

Some of the key pathogens that are found in raw wastewater are summarized in Table 64.2. In addition to pathogens, untreated wastewater may contain chemical substances that are harmful to humans and the environment.

TABLE 64.1 Wastewater Treatment Process

	Preliminary	Primary	Secondary	Tertiary and Advanced
Purpose	Removal of large solids and grit particles	Removal of suspended solids	Biological treatment and removal of common biodegradable organic pollutants	Removal of specific pollutants, such as nitrogen or phosphorus, color, odor, etc.
Sample technologies	Screening, settling	Screening, sedimentation	Percolating/trickling filter, activated sludge Anaerobic treatment Waste stabilization ponds (oxidation ponds)	Sand filtration Membrane bioreactor Reverse osmosis Ozone treatment Chemical coagulation Activated carbon

Source: Adapted from Asano, T. and Levine, A. 1996. *Water Science and Technology*, 33(10–11), 1–14.

TABLE 64.2 Example of Pathogens Associated with Municipal Wastewater

Waterborne bacteria	*Salmonella* sp., *Vibrio cholerae*, Legionellaceae
Protozoa	*Giardia lamblia*, *Cryptosporidium* sp.
Helminths	*Ascaris*, *Toxocara*, *Taenia* (tapeworm), *Ancylostoma* (hookworm)
Viruses	Hepatitis A virus, rotaviruses, enteroviruses

While wastewater reuse has substantial merits, the trade-off between the benefits and potential health risks of applications should be evaluated carefully. These risks can be minimized by the proper treatment, disinfection, and controlled use of reclaimed water. If adequate measures to minimize risk cannot be implemented consistently, wastewater reuse should not be adopted. A further benefit of only using properly treated wastewater for irrigation is that the crops can then be exported to other countries that have strict regulations to minimize the health risks of irrigated crops.

In most cases, disinfection is an essential step prior to wastewater reuse in order to minimize environmental and health risks. The purpose of disinfection is to kill or inactivate pathogenic microorganisms, viruses, and parasites in the treated water. Commonly, disinfection is carried out using strong oxidizers such as chlorine, ozone, and bromine. However, these disinfectants do not inactivate helminth eggs.

64.3 Water Reuse Application

According to the United States (US) Environmental Protection Agency's (EPA) Guidelines for Water Reuse [17], water reuse applications can be classified according to the stringency of the required water quality. From most to least stringent requirements, these categories are briefly introduced here.

64.3.1 Potable Reuse

Potable reuse (i.e., reusing treated wastewater for drinking) has the highest quality requirements of all potential uses of wastewater. Due to the health risk concerns to the public, potable reuse is not widely practiced.

64.3.2 Unrestricted Urban and Recreational Uses, and Agricultural Irrigation of Food Crops

Very high levels of treatment are required in this category. This represents the highest level of water reuse that is currently practiced.

Typical treatment processes include secondary treatment, filtration, and disinfection, with strict restrictions on water quality parameters such as effluent, biochemical oxygen demand (BOD), turbidity, total and/or fecal coliforms, disinfectant residuals, and pH.

64.3.3 Restricted-Access Urban Use, Restricted Recreational Use, and Agricultural Irrigation of Nonfood Crops

This category refers to only providing treated wastewater for reuse by limited populations. In this category, reuse water for irrigation is beneficial because the nutrients in wastewater are good chemical fertilizers. As a result, using reclaimed water to irrigate golf courses and other landscapes is widely practiced. For example, 419 golf courses in Florida were irrigated with 1 mgd (million gallons per day) of reclaimed water [8]. Typical wastewater treatment for reuse in these applications includes secondary treatment and disinfection. Quality requirements are slightly lower than for the previous category.

64.3.4 Industrial Reuse

Less than 20% of the water intake for primary resource and manufacturing industries is consumed and, therefore, there are many opportunities for reuse between industries or in other urban water use sectors. Treated municipal wastewater can be reused for industrial water supply. Furthermore, most industrial water reuse focuses on recycling and process modifications within the boundary of one plant [4]. In water reuse modeling studies, little attention has been given to reusing water between industries. Typical reclamation treatment for municipal wastewater in this category includes secondary treatment and disinfection.

64.4 Nanotechnology and Water Reuse

Recent advances in nanotechnology offer leapfrogging opportunities to develop next-generation water supply systems. Our current water treatment, distribution, and discharge practices, which rely heavily on conveyance and centralized systems, are no longer sustainable. The highly efficient, modular, and multifunctional processes enabled by nanotechnology are envisaged to provide high performance, affordable water and wastewater treatment solutions that are less reliant on large-scale infrastructures [14].

Nanotechnology-enabled water and wastewater treatment have the potential to not only overcome the major challenges faced by existing treatment technologies, but also to provide new treatment capabilities that could allow use of unconventional water sources to become an economic option for expanding the water supply.

Here, we provide an overview of recent advances in nanotechnologies for water and wastewater treatment. The major applications of nanomaterials are critically reviewed based on their functions in unit operation processes. The barriers to their full-scale application and the research needed to overcome these barriers are also discussed.

64.4.1 Nanofiltration

Nanofiltration dates back to the time when reverse osmosis membranes that allowed a reasonable water flux at relatively low pressure were developed. The high pressures traditionally used in reverse osmosis resulted in considerable energy costs, but the quality of the permeate was very good, and often even too good. Thus, membranes that rejected fewer dissolved components, but had higher water permeability were a significant improvement in separation technology. Such "low-pressure reverse osmosis membranes" became known as nanofiltration membranes. By the second half of the 1980s, nanofiltration had become established, and the first applications were reported [5,7]. Since the beginning, the drinking water industry has been the major application area for nanofiltration. The historical reason for this is that nanofiltertation membranes were originally developed for water softening, and to this date, nanofiltration membranes are still sometimes referred to as "softening" membranes [6]. The first nanofiltration plants were developed for water softening, and nanofiltration became an alternative to lime softening. Softening was primarily required for groundwater rather than for surface waters, for which the major problem is usually a high organic matter content. Hardness removal is still one of the major purposes of nanofiltration today. However, the removal of dissolved organic matter soon became an essential part of the process.

64.4.2 Carbon Nanotubes

Since being discovered in 1991 [9], carbon nanotubes have attracted considerable research attention in various scientific communities. Due to their tunable physical, chemical, electrical, and structural properties, carbon nanotubes can inspire innovative technologies to address the problems of water shortage and pollution. Nanotechnologies based on carbon nanotubes have many applications in water treatment, including being used as sorbents, catalysts, filters, or membranes. Cost is often a limiting factor for large-scale applications of carbon nanotubes-based water-treatment materials, as carbon nanotubes are relatively expensive, with current prices being approximately \$75–\$250/g for single-walled carbon nanotubes and approximately \$5–\$25/g for multi-walled carbon nanotubes. However, recent developments have reduced the cost of manufacturing high-quality carbon nanotubes. For example, they can be mass produced using catalytic chemical vapor deposition in a fluidized bed reactor, and a production rate of 595 kg/h can be achieved [1]. Mitsui have projected that the cost of multi-walled carbon nanotubes produced on commercial scales will be approximately \$80/kg, and an eventual price of \$10/lb seems to be achievable. Large-scale manufacturing at low costs may facilitate wider applications of carbon nanotubes. Through rational design and manipulation, they can be incorporated into conventional water treatment materials, opening a new avenue to more efficient water purification and disinfection.

64.4.3 Fullerenes

Fullerenes are molecules composed totally of carbon (C60, C70, etc.). They have a low solubility in water, and therefore, they require surface modification or combination with a stabilization agent to become soluble [16]. Fullerenes are known for their antimicrobial properties that inactivate viruses and bacteria and kill tumor cells. They are not currently used in commercial products as disinfectants because knowledge of their properties and characteristics is limited. One of the main obstacles to using fullerenes in water treatment is that separating and recycling fullerol nanoparticles is difficult. Currently, there is no method that removes these small, light nanoparticles easily and cost-efficiently.

64.5 Summary and Conclusions

Increasing demand for water particularly in developing countries has led to reuse of treated wastewater being seen as a visible alternative water source. To ensure successful wastewater reuse, various water quality requirements must be fulfilled. There are several methods that can be used to achieve these requirements, but nanotechnology as a new generation technology may be hugely influential in the future. Nanotechnology can overcome many of the major challenges faced by existing treatment technologies, and can provide new treatment capabilities that could allow use of unconventional water sources that expand the water supply to become economic. However, certain precautions must be taken to avoid threats to human health or to the environment from nanoparticles.

References

1. Agboola, A.E., Pike, R.W., Hertwig, T.A., and Lou, H.H. 2007. Conceptual design of carbon nanotube processes. *Clean Technologies and Environmental Policy*, 9(4), 289–311.
2. Asano, T. and Levine, A. 1996. Wastewater reclamation, recycling and reuse: Past, present and future. *Water Science and Technology*, 33(10–11), 1–14.
3. Bishop, A. and Hendricks, D. 1971. Water reuse systems analysis. *Journal of the Sanitary Engineering Division*, 97(1), 41–57.
4. Bowman, J.A. 1994. Saving water in Texas industries. *Texas Water Resources*, 20(1), 1–10.
5. Conlon, W.J. and McClellan, S.A. 1989. Membrane softening: Treatment process comes of age. *Journal of the American Water Works Association*, 81(11), 47–51.
6. Duran, F.E. and Dunkelberger, G.W. 1995. A comparison of membrane softening on 3 South Florida groundwaters. *Desalination*, 102(1–3), 27–34.
7. Eriksson, P. 1988. Nanofiltration extends the range of membrane filtration. *Environmental Progress*, 7(1), 58–62.
8. FDEP (Florida Department of Environmental Protection). 2011. *Report by Division of Water Facilities*, Tallahassee, FL.
9. Iijima S. 1991. Helical microtubules of graphitic carbon. *Nature*, 354(6348), 56–58.
10. Jacques G. and Anastasia, P. 1996. Risk analysis of wastewater reclamation and reuse. *Water Science and Technology*, 33(10–11), 297–302.

11. Mulvihill, M.E. and Dracup, J.A. 1974. Optimal timing and sizing of a conjunctive urban water supply and wastewater system with nonlinear programming. *Water Resources Research*, 10(2), 171–175.

12. Ocanas, G. and Mays, L. 1981. A model for water reuse planning. *Water Resources Research*, 17(1), 25–32.

13. Pingry, D.W. and Shaftel, T.L. 1979. Integrated water management with reuse: A programming approach. *Water Resources Research*, 15(1), 8–14.

14. Qu, X.L., Brame, J., Li, Q., and Alvarez, J.J.P. 2013. Nanotechnology for a safe and sustainable water supply: enabling integrated water treatment and reuse. *Accounts of Chemical Research*, 46(3), 834–843.

15. Schwartz, M. and Mays, L. 1983. Models for water reuse and wastewater planning. *Journal of Environmental Engineering*, 109(5), 1128–1147.

16. Spesia, M.B., Milanesio, M.E., and Durantini, E.N. 2007. Synthesis, properties and photodynamic inactivation of Escherichia coli by novel cationic fullerene C(60) derivatives. *European Journal of Medicinal Chemistry*, 43, 853–861.

17. US EPA and US Agency for International Development 1992. *Guidelines for Water Reuse*. EPA/625/R-92/004; US EPA Technology Transfer, Cincinnati, OH.

18. Vieira, J. and Lijklema, L. 1989. Development and application of a model for regional water quality management. *Water Resources*, 23(6), 767–777.

65

Recycled Water in Basin and Farm Scales

Ehsan Goodarzi
Georgia Institute of Technology

Lotfollah Ziaei
Zayandab Consulting Engineers Co.

Saeid Eslamian
Isfahan University of Technology

PREFACE

The available water resources in basins are becoming scarce while demands for water are considerably increasing among various sectors due to economic and population growths. Water deficiency is becoming a main constraint for sustainable regional development and it is the primary motivation in creating water to supply user requirements in particular for agricultural demands within a basin. As agriculture is the largest water user at the global level, the general focus has been on getting higher efficiency by changing irrigation systems or improving irrigation scheduling to reduce the water shortage effects. In addition, significant efforts are being placed to improve water usages efficiency and optimize water consumptions by developing systematic and implementable plans. One way to preserve existing natural water resources is using recycled water. For instance, in the case of farm lands irrigation, water is almost never wholly finished because the amount of water that drains away in the forms of surface runoff and deep percolation can be returned into the system. In other words, the only real loss in the basin scale is evaporation. Based on the classical concepts of water efficiency, all of the wasted water including evaporation, surface runoff, and deep percolation are lost, whereas both surface runoff and deep percolation can be re-entered into the system and added to the surface or groundwater bodies to be used again by downstream users. Considering the return flows as part of available water resources and reuse it, is known as *multiplier effect of water recycling*. To better understand the potential impacts of irrigation interferences at a water basin scale, the multiplier effect of water recycling from an irrigation perspective is studied in the Zayandeh Rud Basin located in the central part of Iran. This study is developed in a general manner to describe how the efficiency of a system can be different from farm to basin scales.

65.1 Introduction

Population increase and socioeconomic rise of developing countries in many parts of the world has escalated the water demands in agricultural, municipal, recreational, and industrial sectors. Increase in those demands in the past decades put severe stresses on the available water resources across the world, particularly in arid and semi-arid regions [6]. Based on the recent studies by the United Nations [23], around one-fifth of world's population have been already affected by water scarcity and they do not have sufficient water resources to meet their needs. In addition, about one-quarter of the world's population is facing economic water deficiency due to lack of necessary hydro-structure to transfer water from sources (e.g., river, lake, and aquifer) and provide their requirements including domestic, industrial, environmental, and agricultural needs. In fact, the nexus between water and energy use seems to be a real issue that needs the attention of decision makers at all levels of governments and international organizations. The water energy nexus and related stresses do not subscribe to jurisdictional and political boundaries recognized nationally or internationally, requires multi-organizational/stakeholders solutions. Hence, optimal management of water resources is imperative to adopt realistic policies to ensure that water is used more efficiently in various sectors [6].

One way to reduce water scarcity is getting more yields per cubic meter of water or increasing water use efficiency. Regarding this issue, identifying effective strategies to gain more productivity by managing nonproductive uses would be very helpful to increase water use efficiency and increase potential savings. A number of important techniques to raise the water use efficiency briefly are improving commercial facility (e.g., toilets and faucets) to use water more efficiently, developing on-farm water management (e.g., using pressurized irrigation systems) in irrigated agriculture as the world's largest water user, pricing water to decrease unnecessary demands, and educating people about water conservation. Based on Seckler et al. [21] most water shortage in the future can be removed if and only if the water efficiency increases without any development in water supply across the world. However, it is important to note that the agricultural sector is the largest freshwater consumer in which 70% of withdrawals from various sources are consumed in this section, while the share of industry and domestic are only 23% and 8%, respectively [5]. Therefore, increasing performance of water usage in the agricultural sector (e.g., increasing irrigation efficiency) is an immediate need to produce more food with less water. However, any increase in irrigation efficiency does not necessarily imply improvement in the whole basin water productivity. In the following sections, we will show how the efficiency in basin can be increased by considering the lost water. The interesting point here is the lost part of irrigation water in the basin scale will remain in the hydrological system and it can be used several times. In other words, excess irrigation water may return to surface or groundwater bodies to enhance the efficiency of the whole system. Based on the classical concept of irrigation efficiency, this amount of excess water is totally wasted and so the estimated efficiency in the basin scale has been underestimated. In other words, the efficiency in the basin scale due to reusing lost amounts of irrigation water in the same part or another part of the system is much higher than the efficiency of the individual use-cycles. Based on the neoclassical irrigation concept, considering the return flows in the system is known as the multiplier effect of water recycling [13].

65.2 Irrigation Efficiency

As clean and safe drinking water become scarcer over the last years, efficiency has received much attention in particular in the field of irrigation as a major water user at the global level. Based on the American Heritage Dictionary, the term *efficiency* is defined as "the ratio of the effective or useful output to the total input in any system." In the case of irrigation, efficiency is known as a way to promote use of water resources and increase limited water supplies in agriculture management and it is a critical parameter in designing irrigation systems and measuring their performance. According to the presented picture for efficiency, irrigation efficiency can be defined as the ratio between water utilized by growing crops and water extracted from a water source and applied by an irrigation system. Therefore, the amount of water that is diverted from a source and not being used to supply growing crops is wasted and results in lower efficiency. Definition of water efficiency may be varied based on the purposes to obtain minimum water requirements. A number of important parameters in the case of water efficiency definitions are presented next.

Water conveyance efficiency (E_c) signifies the effectiveness of water transport systems to deliver water to growing crops. The normal conveyance systems to transfer water from the source of supply to the irrigated field are open channels, pipelines, or natural drainage systems. However, not all of withdrawn water would be delivered to the farm due to transmission losses in the form of seepage and evaporation from channels, overtopping from the channels, deep percolation to soil layers, and leakage in the pipelines. Typically, the losses of unlined and lined channels are much more than losses of pipelines and closed conduits. The conveyance efficiency can be expressed in the following mathematical form:

$$E_c \% = \frac{V_a}{V_t} = \frac{\text{Water delivered to the field (m}^3)}{\text{Water extracted from a source (m}^3)} \times 100 \quad (65.1)$$

Water application efficiency (E_a) represents how well the transported water is applied in the field to meet the crop water needs and it is mainly a function of the applied irrigation method. For example, the average water application efficiency for surface irrigation is about 60%, for sprinkler irrigation 75%, and in the case of drip irrigation is 90%. Water losses reduce water application efficiency in various ways including evaporation surface runoff and droplets, leaks and drift from sprinklers, and deep percolation beneath the root zone. In addition to efficiency, distribution uniformity of water application also is a key factor here in which poor distribution can result in crop stress, while there is high value of water application efficiency. This parameter shows how evenly water is applied in a field and it is a function of a number of variables such as topography, soil permeability, and its hydraulic characteristics.

The water application efficiency can be calculated using the following formula:

$$E_a \% = \frac{V_s}{V_a} = \frac{\text{Irrigation needed by the crop (m}^3)}{\text{Water delivered to the field (m}^3)} \times 100 \quad (65.2)$$

Sometimes, water application efficiency is considered a function of the volume of water stored in the root zone instead of the amount of water needed by the crop, while irrigation can be applied for some other reasons such as frost control, weed germination, or crop cooling [7,8]. In this condition, water storage efficiency concept is more applicable.

Water storage efficiency (E_s) is defined as a fraction of stored water in the crop root zone in the following form:

$$E_s\% = \frac{V_s}{V_r} = \frac{\text{Irrigation needed by the crop (m}^3)}{\text{Root zone storage capacity (m}^3)} \times 100 \quad (65.3)$$

In the case of water storage efficiency, we are trying to maximize water storage in the root zone and minimize the deep percolation. However, there are many difficulties to determine the water storage efficiency due to an inability to obtain the crop root zone and amount of necessary water to recharge the soil root zone depth precisely.

Irrigation efficiency (E_i). In general, water use during irrigation is not only applied for satisfying crop water requirements or crop requirements for evaporation (ET), but also it is applied for other beneficial uses including salt removal or salt leaching, germination of seeds, fertilizer application, crop cooling, etc. In other words, the irrigation efficiency concept is applied to show the effectiveness of irrigation systems when the used water is more than evaporation requirements and cover all water beneficially needs (or conversely, water wasted) to produce the crop. Therefore, this term can be defined as follows:

$$E_i\% = \frac{V_b}{V_a} = \frac{\text{Water used beneficially (m}^3)}{\text{Water delivered to the field (m}^3)} \times 100 \quad (65.4)$$

It is important to note that water losses, which reduce the irrigation efficiency, include surface runoff, weed evaporation, excessive deep percolation, and wind drift [9].

Overall irrigation efficiency (E_o). This parameter shows the efficiency of the whole system in delivering the necessary water to the growing crops, and can be defined as a product of water conveyance and water application efficiencies.

$$E_o\% = [E_c \times E_a] \times 100 \quad (65.5)$$

For example, if during delivering water to the field, 15% of water is lost due to leakage, and the water application efficiency is assumed to be 60%, then the overall efficiency in percentage will be the product of 0.85 and 0.6, which is 51%.

65.2.1 Classical Irrigation Efficiency

Based on the classical concept of irrigation efficiency, the amount of water that should be supplied equals the crop water requirements or the actual evapotranspiration of the crop (ET_a) mines effective precipitation (V_p). Keller et al. [12] defined the net evapotranspiration as follows:

$$ET_{net} = ET_a - V_p \quad (65.6)$$

The effective precipitation in Equation 65.6 is expressed as "that part of the total precipitation on the cropped area, during a specific time period, which is available to meet the potential transpiration requirements in the cropped area" [7,8].

Therefore, the classical irrigation efficiency based on the net evapotranspiration can be calculated as follows:

$$E_i\% = \frac{ET_{net}}{\text{Water delivered to the field}} \times 100 \quad (65.7)$$

EXAMPLE 65.1

Assume in a surface-irrigation system the actual evapotranspiration requirements are 0.8 m, the effective precipitation is 0.35 m, and the amount of water withdrawn is 1.5 m (all values are per unit of area). The irrigation efficiency for the desired field will be calculated as:

$$E_i\% = \frac{ET_a - V_p}{V_a} = \frac{0.8 - 0.35}{1.5} \times 100 = 30\%$$

According to the computations, irrigation efficiency is only 30% and therefore, 70% of supplied water is wasted in this case. However, this is true only from a classical point of view and we will show later that the real irrigation efficiency is more than 30% based on a neoclassical perspective. The neoclassical perspective is important because any change in the irrigation efficiency will affect designing the irrigation system, water-delivery systems, and consequently the total cost of the system in many ways. For instance, usually the water engineers assume a value for irrigation efficiency and then calculate the amount of water that has to be supplied to satisfy the crops requirements. In the case of this example, if we assume the irrigation efficiency is 0.3 m per unit area, and the net evapotranspiration is 0.45 m, the size of the water transport system should be able to handle 0.45/0.3 = 1.5 m. If the efficiency increases from 0.30 to 0.55 m per unit area, the desired water transport system needs to handle 0.81 m; that is, 46% less than the previous one.

In addition to the presented picture of irrigation efficiency, the On-Farm Irrigation Committee of American Society of Civil Engineers defined the irrigation efficiency as the ratio of the beneficially used water to the applied irrigation water (Equation 65.3) [16]. As already explained, the term *beneficial* covers a range of water use including crop cooling, crop ET, soil preparation, leaching for salt control, and any particular cultural operations. However, there are many mathematical formulations to describe efficiency in which differences in those formulas are linked to the ways of considering runoff and deep percolation, estimating efficiency for a season or for an individual irrigation cycle, and calculating efficiency in farm scale or basin scale. It is important to note that reaching 100% irrigation efficiency is not practically possible because there would be instant evaporation losses. Nevertheless, it is roughly possible to get a high efficiency (e.g., about 90%) under the following conditions: (1) the crop is under continuous irrigation, (2) there are no deep percolations, (3) there is no immediate evaporation, and (4) apply all water to crops [22]. Although underwatering crops will result in increasing irrigation efficiency, there are a number of negative side-effects as well as increasing water wasting.

65.2.2 Neoclassical Irrigation Efficiency

All of the presented classical irrigation efficiency concepts are appropriate for design of irrigation systems in the scale of fields or farms. Applying those classical concepts in the context of basin scale will result in poor management of agricultural irrigation systems. The main difference between irrigation efficiency in the farm and basin scales is in considering return flows into the system. In other words, deep percolation and irrigation water runoff are considered as losses in farm scale, while in basin scale, those losses will be added to the available water resources and the next users can use it many times. In general, return flow occurs in two forms: (1) in the form of surface water runoff, and (2) in the form of flow through the groundwater system [14]. However, many irrigation engineers and planners mistakenly ignore the role of return flows and multiplier effect of water recycling as well, and so an inaccurate estimate of real water savings will occur.

Three significant sources of water losses in irrigation process are (1) evaporation from plants and surface runoff, which does not contribute to the ET of the crop, (2) drainage losses that happen in the process of transferring water from the desired sources to the irrigated field or root zone of crops, and (3) spillage losses due to conflicts between water supply and irrigation water demand in the field (e.g., when heavy rain supply is bigger than demands) [13], only drainage and spillage parts are not really lost in the scale of basin and they can be added and continuously cycled to the system, unless they flow to the sinks [21]. However, many planners ignore the return flows in evaluating available water for irrigation. For example, assuming irrigation efficiency based on the classical concept is 40%, which means 60% of the supplied water is wasted through crop evapotranspiration, drainage and spillage losses, or deep percolations. Although 60% of used water for irrigation is wasted, some parts of the lost water can be collected and poured onto downstream fields or be used in the next irrigation round. In other words, the lost water is not necessarily lost in the whole system and recycled water can change the irrigation efficiency. Figure 65.1 shows the amount of excessive water that is added to the groundwater.

Bagley [1] was one of the first people who worked on irrigation efficiency within a basin context and presented the fact that the lost water in a field is not necessarily lost in a basin scale. Bos [2] cleared that a part of wasted water during irrigating a field can be returned to the hydrological system to be used by the next users. Willardson [26] pointed out that increasing irrigation efficiency may have either negative and positive effects in the scale

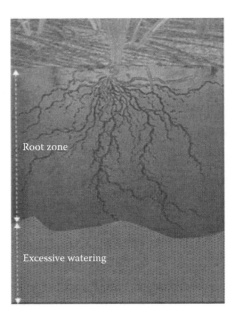

FIGURE 65.1 Schematic view of root zone and excessive watering.

of basin from water quantity and quality perspectives. In another study, Bos and Wolters [3] showed not all of wasted water during diversion from a source to a field is lost, and much of the wasted water is added to the ground or surface water and can be used several times. Palacios-Velez [20] figured out how the lost water can change the irrigation efficiency in the basin [18]. Based on an important study by Jensen [11], he revealed that the standard efficiency concept is misapplied in developing water resources due to ignoring the recycling irrigation losses. Hence, the net irrigation efficiency (E_{ni}) was presented by him as follows:

$$E_{ni} = E_i + I_r(1 - E_i) \qquad (65.8)$$

where E_i is the classical efficiency (Equation 65.6), $1 - E_i$ is the irrigation inefficiency or a fraction of supplied water which does not satisfy crop water requirements (ET_a), and I_r is the fraction of lost water or irrigation inefficiency. Irrigation inefficiency theoretically shows the amount of available water in the hydrological system, which potentially can be reused in the next irrigation cycle.

EXAMPLE 65.2

Assume the evaporation lost is 35% in the previous example, and so there is a potential to return 65% of the lost water to the system. The net irrigation efficiency in this case will be

$$E_{ni} = 0.3 + 0.65(1 - 0.3) = 0.75 \text{ or } 75\%$$

Compare the estimated value of net irrigation efficiency with the classical one shown. There is about 2.5 times increase in the efficacy of system.

65.3 Misunderstanding of FAO's Recommendation

According to the FAO's report provided by Brouwer et al. [4], approximately 40% of the supplied water for irrigation is wasted at the farm level in the forms of surface runoff and deep percolation, while this part of water is not really wasted and can be considered as reusable source. Ignoring this part of water in estimating water balances in the basins is a common mistake between planners and usually results in measuring unreal irrigation efficiency. It is worth pointing out that the percentage of returning water presented by FAO is estimated based on one-time irrigation cycle and it is only applicable in the scale of farm land. In other words, it cannot be used in calculating the efficiency of irrigation in the basin scale as the recycling water is added to the surface and groundwater many times and so the withdrawal water will be more than total the available water. Figure 65.2 shows how the farm and basin scales are different from each other. The following example shows the multiplier effect of water recycling on the irrigation efficiency in basin scale.

Before giving the example, some important definitions related to the inflows and outflows from the basin are briefly presented next.

1. *Committed Water*: This is a part of the lost water from an irrigated area, which is committed to the downstream users for different purposes such as environmental issues or satisfying minimum water flow to keep fish, plants, and wildlife alive.
2. *Uncommitted Water*: As it is clear from its name, this part of water is not committed to any particular purpose and it can be used in the basin for irrigation or urban uses [18].
3. *Closed Basins*: A closed basin is a drainage basin that water cannot flow throughout the basin due to topography of the area and only internal drainage will happen.

In other words, water does not reach the external bodies such as oceans, and evaporation is the only way for the water to escape.

EXAMPLE 65.3

Assume the available water in a closed-drainage basin is 1000 m³ and the total necessary water for cultivated crop plus evaporation losses is 10 m³ per hectare. To estimate the cultivated area, two scenarios can be assumed as follows:

1. Assume the deep percolation equals zero, and so the irrigation efficiency will be 100% as

$$E_i\% = \frac{\text{Beneficial uses (m}^3)}{\text{Water delivered to the field (m}^3)} \times 100$$

$$= \frac{1000}{1000} \times 100 = 100\%$$

and the cultivated area will be

$$\text{Area (ha)} = \frac{\text{Beneficial uses (m}^3)}{\text{Total water requirements per hectare (m}^3/\text{ha})}$$

$$= \frac{1000}{10} = 100 \text{ (ha)}$$

2. Assume the deep percolation is 60%; therefore, the irrigation efficiency is 40%. The cultivated areas based on this irrigation efficiency in different water cycles are presented in Table 65.1. As a case in point, the useable water, cultivated area, and volume of returning flow in the first row of the table is computed as follows.

The total available water in the first cycle is 1000 m³, and by considering 40% irrigation efficiency, the beneficial water would be 40% × 1000 or 400 m³. As the

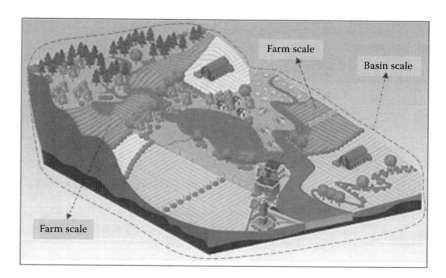

FIGURE 65.2 Schematic view of farm and basin scales. (Adapted from Water Innovation Center. An illustrative example of a potential "Watershed of the Future" for Manitoba. http://www.iisd.org/wic/.)

TABLE 65.1 Cultivated Area and Returning Flows in Example 65.3

Cycle	Total Available Water (m³)	Beneficial Uses (m³)	Cultivated Area (ha)	Return Flow (m³)
1	1000	400	40	600
2	600	240	24	360
3	360	144	14.4	216
4	216	86	8.6	130
5	130	52	5.2	78
6	78	31	3.1	47
7	47	19	1.9	28
8	28	11	1.1	17
9	17	7	0.7	10
10	10	4	0.4	6
11	6	2.4	0.24	3.6
12	3.6	1.4	0.14	2.2
Total	2500	1000	100	1500

necessary water for cultivated crop is 10 m³/ha, the cultivated area is 400/10 or 40 ha, and the amount of return water will be 600 m³.

$$\text{Useable water} = 1000 \times 40\% = 400 \ (\text{m}^3)$$

$$\text{Cultivated area} = \frac{400 \ (\text{m}^3)}{10 \ (\text{m}^3/\text{ha})} = 40 \ (\text{ha})$$

$$\text{Return flow} = 1000 \times 60\% = 600 \ (\text{m}^3)$$

In the second cycle, the return water is considered as available water and so the beneficial water, cultivated area, and return flow are 240 (m³), 24 (m³/ha), and 360 (m³), respectively. As it can be seen from Table 65.1, the total available water at the end of the irrigation cycles is 2500 (m³) in which it is 1500 (m³) more than the real available water in this example, while the total cropped area is the same as the first scenario with 100% irrigation efficiency. Hence, it can be concluded that the total extracted water is 2.5 (2500/1000) times more than the total available water. The ratio 2.5 also is equal to the inverse value of irrigation efficiency or 1/40% = 2.5.

Traditionally, two methods are applied to determine the irrigation efficiency in basins while both of them result in the incorrect efficiency values. The methods concisely are

1. Using the average value of irrigation efficiency based on various field measurements in several parts of a basin
2. Applying the beneficial uses and total demand concepts

Based on the first method, the basin will be divided into a number of parts according to the soil type, cultivation pattern, and used irrigation systems. Then, the irrigation efficiency of basin will be computed as follows:

$$E_{\text{basin}} = \frac{\sum_{i=1}^{n} E_i A_i}{\sum_{i=1}^{n} A_i} \qquad (65.9)$$

where E_{basin} is the basin's efficiency, E_i is the measured efficiency in each region, and A_i is the area of each region. The calculated efficiency based on this method is the average of individual farm's efficiency and not the efficiency in basin scale or the real efficiency. On the other hand, the irrigation efficiency based on the second method is computed as

$$E_{\text{basin}} = \frac{V_n}{V_S + V_G} \times 100 \qquad (65.10)$$

where E_{basin} is the basin's efficiency, V_n is the net water demand, V_G is the water supplied from groundwater resources, and V_S is the water supplied from surface water resources. In this method, the multiplier effect of water recycling in terms of deep percolation and surface runoff are not considered and it is assumed that this part of losses goes away. Hence, the calculated irrigation efficiency using Equation 65.10 also would have unreal value in the basin scale.

Regarding the irrigation efficiency in basins, it is very important to find suitable answers for the following questions:

- What is the real amount of water in basins by considering the multiplier effect of water recycling?
- What is the real irrigation efficiency in the basin scale?
- What are the right amounts, types, and location of losses?

65.4 Multiplier Effect in Zayandeh Rud Basin: Case Study

The Zayandeh Rud River basin is located in the central part of Iran with the Qom desert in the north, the Zagros Mountain Range in the west, and the Iran central desert in the east and south. This basin is the major water source for Isfahan Province, while it receives no precipitation for at least nine months each year. Although this region is recognized as a semi-arid area, Zayandeh Rud River forms one of the most important agricultural regions in central Iran. This river starts in the Zagros Mountains and flows 400 km eastward before ending in the Gavkhouni Marsh, in the southeast of Isfahan City (Figure 65.3).

The basin's area is 26,917 km², altitudes from 1466 to 3874 m, average annual rainfall of 281 mm, annual potential evapotranspiration 1500 mm, total area under irrigation 260,000 ha, on-farm irrigation efficiency range 35%–39%, monthly average temperatures of 3–29°C, and the share of agricultural, industrial, and domestic sections from the total consumed water are 90%, 5%, and 5%, respectively [17]. The rainfall distribution in the basin is not uniform and it is varied between 1700 mm in mountainous western parts and less than 100 mm in eastern flat parts. This basin has always faced severe water shortage. For instance, about 400 years ago, Safavi dynasty tried to transfer water from other neighborhood water sources to the Zayandeh Rud basin.

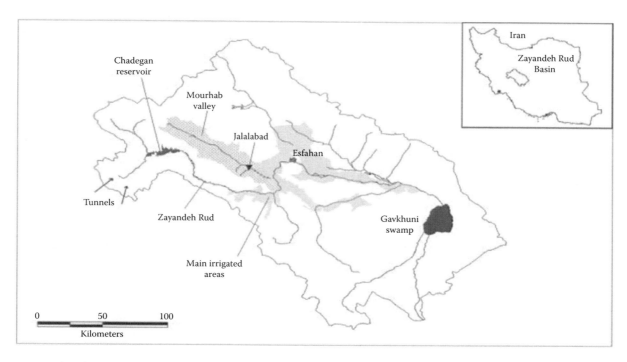

FIGURE 65.3 The schematic view of Zayandeh Rud Basin. (Adapted from Molle, F. and Mamanpoush, A., 2012. *Geoforum*, 43, 285–294.)

However, the first successful pattern of water transition between two basins happened through the *Koohrang-Tunnel 1*, which was built in 1954 with the annual water transferring of 300 MCM. The second and third projects in this case were *Koohrang–Tunnel 2* (1985) and *Cheshemeh langan-Tunnel* (2005) with the annual water transferring of 250 and 120 MCM, respectively. In addition, there are two more ongoing water transition projects for Zayandeh Rud Basin—*Koohrang-Tunnel 3* with the annual water transferring of 250 MCM and *Beheshtabad* with the annual water capacity of 587 MCM. It is important to note that although the desired area has faced severe water shortages over the last decades, some parts of the transferred water to the Zayandeh Rud Basin is conveyed to other neighboring basins that suffer from extreme drought. At the present, the amount of transferring water from Zayandeh Rud to other basins is about 180 MCM, which will increase to 470 MCM by using the *Beheshtabad* project. Based on the presented report by Jamab Consulting Engineering Co. [10], about 90% of the available water in the Zayandeh Rud Basin is spent for agricultural uses, whereas only around 7% of the basin is under cultivation.

65.4.1 Water Consumptions in Zayandeh Rud Basin

The most important water users of Zayandeh Rud River are agricultural, domestic, industrial, and environmental sectors. This basin, with a population of 3.7 million, is a major population area in the central part of Iran. In this case, Zayandeh Rud River provides the most part of the drinking water demand of Isfahan province with the annual average of 420 MCM [24]. Furthermore,

the Zayandeh Rud Basin is one of the most important industrial centers in Iran and it includes many fundamental industries of cement, steel, power stations, etc. The annual water consumption for the industrial section is about 180 MCM. The agricultural segment is the greatest consumer in the basin and almost 90% of available water, about 5035 MCM, has been allocated to this part. However, the recent severe and extensive drought over the last years seriously affected the desired basin with negative impacts on the crop and livestock sectors, which can be result in increasing food prices at the retail level. Regarding the water scarcity issues, agricultural lands within the Zayandeh Rud Basin are reduced from 420,000 to 262,000 ha.

As 90% of the available water in the Zayandeh Rud Basin is consumed by the agricultural sector (in particular in the form of surface irrigation), there are considerable return flows in the basin with the potential to be reused by downstream users. Therefore, the multiplier effects of water recycling in two sub-basins of Zayandeh Rud are presented in the following sections to see how the multiplier effect of recycling water will affect the irrigation efficiency in basin scale. The hydrological characteristics of two desired sub-basins are presented in Table 65.2.

The sub-basin S5 includes *Karvan, Najafabad, Northern Mahyar, Alavijeh-Dahagh, Moorcheh khort,* and *Meimeh-Borkhar* and the sub-basin S6 involves *Koohpayeh* and *Segzy* stations. According to Table 65.2, the total reproducible water in region S5 is 911 MCM, while the annual volume of withdrawals water, which is 2675 MCM, is much more than the reproduced water. This huge difference of 1764 MCM is due to the multiplier effect of water recycling or counting returning water several times in the water cycle. As it can be seen from Table 65.2, the

TABLE 65.2 Hydrological Characteristics of Two Desired Sub-Basins

ID	ARS[a] (MCM)			AWAC[b] (MCM)			SW[c] (MCM)
	Effective rainfall	Surface water	Total	Groundwater	Surface water	Total	
S5	417	494	911	1825	850	2675	1764
S6	149	725	874	770	597	1367	493

[a] Annual reproducible sources.
[b] Annual withdrawals and agricultural consumptions.
[c] Surplus water (multiplier effect).

same trend is happening to the sub-basin S6 in which the difference between these two parts is 493 MCM.

65.4.2 Irrigation Efficiency in Zayandeh Rud Basin

The following section includes the appropriate equations to estimate the irrigation efficiency in the Zayandeh Rud Basin as a closed basin.

$$V_a = V_p - V_e + (V_i - V_o) - \Delta V_g - \Delta V_s \qquad (65.11)$$

where V_a MCM is the real available water diverted from sources, V_p MCM is the effective precipitation, V_e MCM is evaporation losses, V_i MCM is trans-basin diversion (into the basin), V_o MCM is trans-basin diversion (out of the basins), ΔV_g MCM is change in groundwater storage, and ΔV_s MCM is change in surface water storage. The evaporation losses also can be estimated as

$$V_a = V_e + V_n \qquad (65.12)$$

in which V_n MCM is the net water demand.

The total water in the basin (V_c) is the sum of withdrawals of water from ground and surface water sources or

$$V_c = V_G + V_S \qquad (65.13)$$

where V_G MCM is the withdrawals from groundwater, and V_S MCM is withdrawals from surface water. Therefore, the recycled water in the basin can be evaluated using the following equation:

$$V_m = V_c - V_a \qquad (65.14)$$

in which, V_m (MCM) is the reusable water due to the multiplier effect of water recycling.

Based on the presented formulas, the total efficiency of basin is

$$\%E_{t-\text{basin}} = \frac{V_n}{V_a} \times 100 \qquad (65.15)$$

It is important to note that V_n includes all demands in the basin such as agricultural, industrial, domestic, etc. By substituting Equation 65.14 in 65.15, we have

$$\%E_{t-\text{basin}} = \frac{V_n}{V_c - V_m} \times 100 \qquad (65.16)$$

If only the irrigated part is considered in the basin (not including industrial, domestic, and environmental demands), the irrigation efficiency in the basin will be

$$\%E_{i-\text{basin}} = \frac{V_{\text{in}}}{V_a - V_{\text{dw}} - V_{\text{iw}} - V_{\text{ew}}} \times 100 \qquad (65.17)$$

where $E_{i-\text{basin}}$ is the irrigation efficiency in basin scale, V_{in} is irrigation water requirement (not including the other demands), V_{dw} is net drinking water uses (or domestic uses), V_{iw} is industrial water uses, and V_{ew} is the environmental water uses.

It should be noted that land area under cultivation in this basin was 261,257 (ha) with the net irrigation requirement of 5816 m^3/ha in 2006. Hence, the agricultural demands will be 261,257 ha × 5816 m^3/ha = 1519.47 MCM. The amount of water that is needed to irrigate 5000 ha green space area of the city of Isfahan is 8000 m^3/ha. Therefore, the green space demand for the whole area is 40 MCM. In addition, the withdrawals from surface water in the Zayandeh Rud Basin are 1878 MCM. Table 65.3 shows the available water resources, demands, and withdrawals of the Zayandeh Rud Basin in 2006.

Based on the presented information in Table 65.3, the total withdrawals from groundwater (sum of the values in the last column of Table 65.3) equal 3757.717 (MCM). Using all of the presented information, the irrigation efficiency for Zayandeh Rud Basin can be calculated as follows:

$$V_a = 2215 - 57 + (656 - 179) - 38 = 2597 \text{ MCM}$$

The total water demand, which is the sum of values in the second column of Table 65.3, is

$$V_n = 1519.47 + 40 + 84 + 81 + 190 = 1914.47 \text{ (MCM)}$$

Therefore, the evaporation loss is

$$V_e = 2597 - 1914.47 = 682.53 \text{ MCM}$$

The variable V_c will be estimated as

$$V_c = 3757 + 1878 = 5635 \text{ MCM}$$

Therefore, the reusable water due to the multiplier effect of water recycling is

$$V_m = 5635 - 2597 = 3038 \text{ MCM}$$

TABLE 65.3 Available Water Resources, Demands, and Withdrawals of Zayandeh Rud Basin

Available Water Resources	Value (MCM)	Demands	Value (MCM)	Withdrawals	Value (MCM)
Effective rainfall	2215	Agricultural demand	1,519.47	From 22,312 wells	2,980.72
Transbasin diversion (into the basin)	656	Green space demand	40	From 1535 springheads	177.507
Transbasin diversion (out of the basin)	179(−)	Domestic demand	84	From 1508 Qanats	599.49
Evaporation	57(−)	Gavkhooni swamp	81	–	–
Groundwater storage	38(−)	Industrial demands	190	–	–
Total	2597		1,914.47		

Source: Adapted from Jamab Consulting Engineering Co. 1998. *National Comprehensive Water Survey*. Zayandeh-Rud River Basin, Iran.

Finally, the total efficiency of basin will be

$$\%E_{t-basin} = \frac{1914.47}{5635 - 3038} \times 100 = 73.7\%$$

The irrigation efficiency in the basin scale can be evaluated as follows:

$$\%E_{i-basin} = \frac{1519.47 + 40}{2597 - 84 - 190 - 81} \times 100$$
$$= \frac{1559.47}{2597 - 84 - 190 - 81} \times 100 = 69.5\%$$

According to the achieved results, the total basin efficiency and the irrigation efficiency of Zayandeh Rud Basin are 73.7% and 69.5%, respectively. To compare the proposed method with the classical method, both efficiencies are calculated using classical concept as follows:

$$\%E_{t-basin} = \frac{\text{Water used beneficially (m}^3)}{\text{Water delivered to the field (m}^3)} \times 100$$
$$= \frac{1914.47}{5635} \times 100 = 33.96\%$$

and

$$\%E_{i-basin} = \frac{1559.47}{5635 - 84 - 190 - 81} \times 100 = 29.5\%$$

It can simply be concluded that ignoring the multiplier effect of water recycling will result in wrong estimations of available water resources and irrigation efficiency in the basin scale. In other words, the total efficiency of a multiple use-cycle system is considerably greater than the classical efficiency of a single use-cycle because of reusing lost water, which is added to ground or surface water resources. It is also important to note that in an open basin there is less water to be returned rather than in a closed basin and so the multiplier effect of water recycling is not as much as a closed basin. However, reusing the available water still increases the whole basin efficiency in open basins.

The importance of returning flow also is investigated by Kendy and Bredehoeft [14] in the Gallatin Valley, Montana. They studied transient effects of groundwater pumping and surface-water irrigation returns on stream flow, and the impacts of seasonal groundwater pumping on improving irrigation efficiency. Results of their study demonstrated that irrigation efficiency improvements decrease groundwater recharge and discharge to streams

if crop production is not reduced. In addition, they showed that eliminating return flow causes reduction of late-season stream flow, excluding downstream users of late-season water supplies, and shortens wetland hydro-periods at the farm scale. Although improving irrigation efficiency in farm scale might result in more water saving on some farms, it would deprive downstream users from return flow that can be used in different ways.

65.5 Impact of Improving Irrigation Efficiency

Water deficiency is becoming a main constraint for sustainable regional development and effect on the food and energy systems across the world in particular in arid and semi-arid regions. Hence, the general focus is on getting higher efficiency by changing irrigation systems or improving irrigation scheduling, to reduce the water shortage effects. However, increasing the irrigation efficiency does not necessarily result in decreasing the water shortages in closed basins. In the following section, we will briefly explain why getting higher irrigation efficiency is not essentially an appropriate strategy to save more water in the basin scale. In general, the common methods to improve the irrigation efficiency are

1. Improving surface irrigation by decreasing surface runoff and deep percolation below the root zone
2. Replacing surface irrigation with a sprinkler irrigation system

In closed basins, part of surface runoff and deep percolation eventually reach the groundwater and so, the lost water is not necessarily lost in the whole system. In this regard, increasing efficiency of surface irrigation practically would not result in more water savings and also essentially doesn't decrease water deficiency. Although replacing surface irrigation system with the sprinkler reduces deep percolations, the wind drift is a problem here in which the evaporation losses escalate. As discussed previously, deep percolations can be returned to the system and so reduction in percolation cannot be considered as an advantage in the case of increasing water savings. From this point of view, the sprinkler irrigation system only increases the evaporation losses and it is not going to reduce the water shortage in basin scale. However, losses in the sprinkler irrigation system depend on a number of variables such as air and water temperature, humidity, wind condition, applied methods to quantify losses, and considering the aggregated losses over whole fields or only for an individual sprinkler. Kohl et al. [15] reported that losses in a sprinkler system vary in a wide range from 1% to 45% in the literature.

A very important point here is that what was mentioned previously is applicable only for basins that deep percolation can be recycled and reused several times, and it is not applicable in the farm scale. In addition, if there is no possibility to collect deep percolation in a basin, or a major fraction of the salts is leached below the root zone with the deep percolation, increasing irrigation efficiency by using sprinkler or any modern irrigation system can result in more water savings.

65.6 Summary and Conclusions

Population increase and socioeconomic activities have escalated water demands for various purposes and put stress on existing water resources across the world, in particular in arid and semi-arid countries including Iran. Hence, managing the optimum use of water resources is a crucial issue and it is imperative to adopt realistic policies to ensure water is used more efficiently in various sectors. In this chapter, a new look at the water balance problem in basin scale is presented and it is shown that ignoring recycled water misleads in estimating the real water resources and consequently causes confusion in measuring irrigation efficiency. In other words, it is demonstrated that all of the wasted part of irrigation water is not really lost and amounts of water always remain in the hydrological system. Therefore, the recycled part can enhance the efficiency of the whole system, while based on the classical concept, this section has not been considered, and the efficiency mistakenly was underestimated. On this basis, the multiplier effect of water recycling means multiple use-cycle systems where the losses can be added to the available water resources several times and be reused in the same part or another part of the system.

In addition to providing basic concepts of irrigation efficiency, the multiplier effect of water recycling in the Zayandeh Rud Basin from an irrigation point of view is presented as a real case study. The classical and neoclassical methods are applied to measure the irrigation efficiency in the basin scale and the achieved results are compared. It is concluded that using the returned water increases the efficiency in the basin scale up to 40% in two sub-basins of Zayandeh Rud. In other words, the efficiency of a single use-cycle system is much lower than efficiency of a multiple use-cycle system in a closed basin. This also should be noted that improving on-farm irrigation efficiency and changing the irrigation system would result in saving more water if and only if the lost water is evaporated or its quality is too low to be reused.

Symbols

I_r	Irrigation inefficiency
ET_a	Actual evapotranspiration of the crop
ET_{net}	Net evapotranspiration
E_a	Water application efficiency
E_c	Water conveyance efficiency
E_i	Irrigation efficiency
E_{ni}	Net irrigation efficiency
E_o	Overall irrigation efficiency
E_s	Water storage efficiency
$E_{t\text{-basin}}$	Total basin efficiency
V_a	Water delivered to the field
V_r	Root zone storage capacity
V_G	Withdrawals from groundwater
V_S	Withdrawals from surface water
V_b	Water used beneficially
V_c	Sum of withdrawals water from ground and surface water
V_e	Evaporation loss
V_i	Trans-basin diversion (into the basin)
V_m	Reusable water
V_n	Net water demand
V_o	Trans-basin diversion (out of basins)
V_p	Effective precipitation
V_s	Irrigation needed by the crop
ΔV_g	Change in groundwater storage
ΔV_s	Change in surface water storage

References

1. Bagley, J.M. 1965. Effects of competition on efficiency of water use. *Journal of Irrigation and Drainage Division of the American Society of Civil Engineers*, 91(IR1), 69–77.

2. Bos, M.G. 1979. Der Einfluss der Grosse der Bewasserungs einheiten auf die verschienden Bewasserungs wirkungsgrade. *Zeitschrift für Bewasserungs Wirtschaft, Bonn*, 14(1), 139–155.

3. Bos, M.G. and Wolters, W. 1989. Project or overall irrigation efficiency. In irrigation theory and practice. In: J.R. Rydzewski and C.F. Ward (eds.), *Proceedings of the International Conference Held at the University of Southampton*, September 12–15, 1989. Pentech Press, London, UK, pp. 499–506.

4. Brouwer, C., Prins, K., and Heibloem, M. 1989. Irrigation water management, *Training Manual No 5*; Food and Agriculture Organization (FAO), USA.

5. Clay, J. 2004. *World Agriculture and the Environment: A Commodity-by-Commodity Guide to Impacts and Practices*, Island Press, Washington, DC, USA.

6. Goodarzi, E., Ziaei, M., and Shokri, N. 2013. Reservoir operation management by optimization and stochastic simulation. *Journal of Water Supply: Research and Technology—AQUA*, 62(3), 138–154.

7. Howell, T.A. 2002. Irrigation efficiency. Encyclopedia of water science. Marcel Dekker, Inc., New York, USA.

8. ICID Committee on Assembling Irrigation Efficiency Data. 1978. M.G. Bos (Chairman) Standards for the calculation of irrigation efficiencies. *ICID Bulletin* 27(1), 91–101. New Delhi (also published in French, Spanish, Turkish, Arabic, and Persian).

9. Irmac, S., Odhiambo, L.O., Kranz, W.A., and Eisenhauer, D.E. 2011. *Irrigation Efficiency and Uniformity, and Crop Water Use Efficiency*. University of Nebraska, Lincoln, Nebraska.

10. Jamab Consulting Engineering Co. 1998. *National Comprehensive Water Survey*. Zayandeh-Rud River Basin, Iran.

11. Jensen, M.E. 1967. Evaluating irrigation efficiency. *Journal of the Irrigation and Drainage Division American Society of Civil Engineers*, 93(IR1), 83–98.

12. Keller, A., Keller, J., and Seckler, D. 1996. *Integrated Water Resource Systems: Theory and Policy Implications*. International Irrigation Management Institute, Colombo, Sri Lanka.

13. Keller, A.A. and Keller, J. 1995. *Effective Efficiency: A Water Use Efficiency Concept for Allocating Freshwater Resources*. Center for Economic Policy Studies, Winrock International, Arlington, VA.

14. Kendy, E. and Bredehoeft, J.D. 2006. Transient effects of groundwater pumping and surface-water irrigation returns on streamflow. *Water Resources Research*, 42(8), W08415. DOI: 10.1029/2005WR004792.

15. Kohl, K.D., Kohl, R.A., and DeBoer, D.W. 1987. Measurement of low pressure sprinkler evaporation loss. *Transactions of the American Society of Agricultural Engineers*, 30(4), 1071–1074.

16. Kruse, G. 1978. Describing irrigation efficiency and uniformity. *ASCE Journal of the Irrigation and Drainage Division*, 104(1), 35–41.

17. Madani, K. and Mariño, M.A. 2009. System dynamics analysis for managing Iran's Zayandeh-Rud river basin. *Water Resource Management*, 23, 2163–2187.

18. Molden, D. 1997. *Accounting for Water Use and Productivity*. SWIM Paper 1, International Irrigation Management Institute, Colombo, Sri Lanka.

19. Molle, F. and Mamanpoush, A., 2012. Scale, governance and the management of river basins: A case study from Central Iran. *Geoforum*, 43, 285–294.

20. Palacios-Velez, E. 1994. Water use efficiency in irrigation districts. In: H. Garduno and F. Arreguin-Cortes (eds.), *Efficient Water Use*, UNESCO/ROSTLAC, Montevideo, Uruguay.

21. Seckler, D., Molden, D., and Sakthivadivel, R. 2003. The concept of efficiency in water resources management and policy. In: J.W. Kijne, R. Barker, and D.J. Molden (eds.), *Water Productivity in Agriculture: Limits and Opportunities for Improvement*, CABI Publishing and International Water Management Institute, Wallingford, UK/Colombo, Sri Lanka.

22. The Center for Irrigation Technology. 2005. *Irrigation Performance Measurements—Distribution Uniformity and Irrigation Efficiency*.

23. United Nations. 2013. Water Scarcity—International Decade for Action 'Water for Life' 2005–2015. https://www.un.org/waterforlifedecade/scarcity.shtml.

24. Water Engineering Group of IRNCID. 2002. Saline water utilization sustainable agriculture. Iranian National Committee on Irrigation and Drainage (IRNCID) Publishing, Tehran, Iran.

25. Water Innovation Center. An illustrative example of a potential "Watershed of the Future" for Manitoba. http://www.iisd.org/wic/.

26. Willardson, L.S. 1985. Basin-wide impacts of irrigation efficiency. *Journal of Irrigation and Drainage Engineering*, 111(3), 241–246.

Water Reuse in Climatically and Physically Different Regions

Manzoor Qadir
*United Nation University
Institute for Water, Environment
and Health (UNU-INWEH)*

*International Water Management
Institute (IWMI)*

*International Center for
Agricultural Research in the
Dry Areas (ICARDA)*

Toshio Sato
*OYO Corporation
Tottori University*

66

Water Reuse in Arid Zones

PREFACE

Water scarcity is the main driver for water recycling and reuse in countries located in arid zones. Despite multiple benefits from water recycling and reuse for the farmers and communities, there are several constraints to the collection, treatment, and safe and productive use of wastewater in agriculture. Apart from lack of supportive policies, unclear institutional arrangements, critical shortage of skilled human resources, and the public budgets in most countries for water recycling and reuse are inadequate. In addition, limited economic analysis, lack of reuse cost-recovery mechanisms, no or little value for treated wastewater, lack of awareness about the potential of water recycling and reuse, and inefficient irrigation and water management schemes are constraints to water recycling and reuse. Despite these constraints, some countries in arid zones have employed a range of conventional and nonconventional systems and have national standards and regulations in place for water recycling and reuse. Salient features of water recycling and reuse in Tunisia, Jordan, Israel, and Cyprus reveal that although each country has taken a slightly different path, policymakers in these countries consider reuse of reclaimed water an essential aspect of strategic planning and management of water and wastewater. Other countries in arid zones with similar situations can benefit from Tunisia, Jordan, Israel, and Cyprus by transforming wastewater from an environmental burden into an economic asset through aggressive implementation of water recycling and reuse programs.

66.1 Introduction

Given the current demographic trends and future growth projections, as much as 60% of the global population may suffer water scarcity by the year 2025 [9]. In addition to water scarcity, water quality deterioration is expected to intensify in resource-poor countries in dry areas, due to human activity and climate change. It is projected that water scarce countries will have to increasingly rely on alternative water resources such as wastewater to narrow the gap between water demand and supply for agriculture [16,18,19].

There are large gaps between developing and developed countries as well as low- and high-income countries for managing wastewater generated within their boundaries. High-income countries on average treat 70% of the generated wastewater, followed by upper-middle-income countries (38%), lower-middle-income countries (28%), and low-income countries, where only 8% of the wastewater generated is treated [31].

Wastewater treatment and use and/or disposal in the arid and semi-arid regions of developed countries, such as western North America, Australia, and southern Europe, are motivated by stringent effluent discharge regulations and public preferences regarding environmental quality [13]. Treated wastewater is primarily for irrigation, given the increasing competition for water between agriculture and other sectors [31,34]. In developing countries, wastewater treatment is limited, as investments in treatment facilities have not kept pace with persistent increases in population and the consequent increases in wastewater volume in many countries. Thus, much of the wastewater generated is not treated, and much of the untreated wastewater is used for irrigation by small-scale farmers with little ability to optimize the volume or quality of the wastewater they receive [29,32].

Many farmers in water-scarce developing countries irrigate with wastewater because: (1) it is the only water source available for irrigation throughout the year, (2) wastewater irrigation reduces the need for purchasing fertilizer as wastewater is rich in essential nutrients, (3) wastewater irrigation involves less energy cost if the alternative clean water source is deep groundwater, and (4) wastewater enables farmers in peri-urban areas to produce high-value vegetables for sale in local markets.

Irrigation with treated wastewater likely will expand in developed countries, particularly in arid and semi-arid areas, where competition for freshwater supplies will continue to increase [11,24]. Technical solutions and public policies generally are adequate in developed countries to accommodate increases in the treatment and use of wastewater. The same is not true for many developing countries, where treatment facilities already are inadequate, and much of the wastewater used by farmers is not treated.

This chapter provides an overview of the water and wastewater resources in arid zone countries, constraints in water reuse and recycling, and showcases examples from water-scarce countries that have progressed in implementing water recycling and reuse policies and practices. The aim is to provide a number of examples for countries with similar conditions for consideration and implementing water recycling and reuse projects and practices leading to safe and productive use of wastewater in different sectors, particularly in agriculture.

66.2 Water and Wastewater Resources in Arid Zone Countries

Looking at the natural global water cycle that yields an annual renewable water supply of about 7000 m³ per capita, it is evident that there is enough freshwater available every year to fulfill the needs of the present population of this planet. However, in certain regions and countries, the annual renewable supply of water is less than 500 m³ per capita. In addition, the availability of water varies greatly over time in these areas, which results in extreme events. Floods and droughts, for example, occur frequently, sometimes in the same area or neighboring regions. What this contrast illustrates is that freshwater resources and population densities are unevenly distributed across the globe.

Among the arid zones of the world, the Middle East and North Africa (MENA) region is the driest region of the world. The region is home to about 5% of the existing global population, but with only 1% of the world's freshwater resources. The MENA countries depend on seasonal rainfall, have very few rivers—most of them originate in other countries—and often rely on fragile, and sometimes nonrenewable, aquifers. There is already an increasing competition for high-quality water among different water-use sectors. Although agriculture is the dominant user

of freshwater in these countries, it has been yielding its share gradually to nonagricultural uses, that is, household, municipal, and industrial activities. Since the use of freshwater for these activities generates wastewater, the volume of wastewater has been increasing commensurate with rapidly growing population, urbanization, improved living conditions, and economic development.

The phenomenon—less freshwater allocation to agriculture, more freshwater allocation to nonagricultural sectors vis-à-vis increasing volumes of wastewater—is expected to continue and intensify in the countries in the arid zones in the foreseeable future. Most small-scale farmers in urban and peri-urban areas of these countries already depend on wastewater to irrigate a range of crops, often as they have no alternative sources of reliable irrigation water.

In addition to the MENA region, several countries in other regions also encounter a water deficit. The datasets and maps published in recent years show that more and more countries will become water-stressed because of increased water scarcity. In arid zones, water scarcity remains the main driver for water recycling and reuse.

66.3 Constraints to Water Recycling and Reuse

Considering the acute problem of water scarcity and emerging problem of water quality deterioration amid climate change events, the developing countries located in dry areas have an opportunity in terms of planned and beneficial water recycling and reuse. The rate of wastewater treatment is still low in most of these countries while many wastewater treatment plants are plagued by poor operation and maintenance, and if operational, they are operated well beyond their design capacity [2]. These conditions eventually question wastewater treatment processes, quality of treated wastewater, and safety of practices aimed at using wastewater in agriculture [12]. In addition, the regulations prohibiting the agricultural use of untreated or partly treated wastewater exist, but their implementation is not strictly enforced [28].

The major constraints leading to low percentage of water recycling and reuse in countries in dry areas can be: (1) inadequate public budgets in most countries to collect domestic and industrial wastewater separately for treatment per reuse or disposal options; (2) inadequate information on the status of water recycling and reuse or disposal of different forms of wastewater and associated impacts on environmental and human health; (3) incomplete economic analysis of the wastewater treatment and reuse options, usually restricted to financial feasibility analysis rather than a complete financial and cost and benefit evaluation; (4) perceived high cost of developing wastewater collection networks and wastewater treatment plants and low returns without sound assessment and feasibility studies; (5) lack of wastewater treatment and reuse cost-recovery mechanisms including commitment in most countries to support comprehensive wastewater treatment programs; (6) mismatch between water pricing and regional water scarcity with no or little value for treated wastewater resulting in closure of wastewater treatment plants or substantial subsidies to keep the treatment plants in operation; (7) lack of awareness about the potential of water recycling and reuse; (8) general preference for freshwater over wastewater; (9) overall inefficient irrigation and water management schemes undermining the potential of water reuse; (10) lack of supportive policies to promote water recycling and reuse; (11) unclear institutional arrangements and lack of coordination between national agencies and local institutions for wastewater management; and (12) critical shortage of skilled human resources to address the complex issues resulting from wastewater collection, treatment, and reuse systems.

Most developing countries in dry areas are generally characterized by increased population growth and urbanization, improved living conditions, and economic development; all are drivers of increased volumes of wastewater emanating from the domestic and industrial sectors. While these changes are rapid, there is inadequate information on the status of collection and reuse and/or disposal of different forms of wastewater (untreated, partly treated, diluted, and treated) and associated environmental and health impacts [28]. Even in cases where such information is available, there are large differences in the qualitative and quantitative assessment of wastewater because of the different criteria used.

Although the economic impacts of reusing wastewater largely depend on the degree of treatment and the nature of the water reuse options, economic assessments of wastewater treatment and reuse options in the arid region countries are usually restricted to financial feasibility analysis [19,27]. In fact, there are costs and benefits associated with the specific wastewater treatment and reuse systems. There is a need to consider other factors such as centralized and decentralized treatment options, levels of treatment (primary, secondary, and tertiary), intended reuse options, wastewater collection and conveyance infrastructure leading to wastewater treatment plant in locations where such infrastructure does not exist, and transportation options for treated effluent to specific locations for intended reuse. In addition, the opportunity cost of water reuse should be considered under conditions where new uses and moving a given supply of water from one place to another specific location are anticipated. In doing so, the economic analysis therefore needs to consider the implications of wastewater distribution as well as certain restrictions on crop choices based on the quality and quantity of reclaimed water in the anticipated reuse projects.

The perceived high cost of establishing wastewater collection networks and treatment plants capable of satisfactory wastewater treatment is another major constraint leading to uncertainty in terms of adopting comprehensive wastewater treatment and reuse programs. Wastewater treatment facilities and costs vary from location to location and are based on the infrastructure needed, the quality of the wastewater collected and the anticipated quality of the treated wastewater, which may be the result

of primary, secondary, and tertiary treatment. Lee et al. reported that the cost of wastewater treatment may range from 0.46 to 0.74 US\$ m^{-3}, with an average of 0.53 US\$ m^{-3} [21]. The major components of the cost based on these estimates in 2001 include capital (0.10–0.16 US\$ m^{-3}), operation (0.25–0.40 US\$ m^{-3}), maintenance (0.08–0.15 US\$ m^{-3}), and miscellaneous (0.03 US\$ m^{-3}). Estimates from Middle Eastern countries reveal wastewater treatment cost in Saudi Arabia for tertiary treated wastewater at \$ 0.30 m^{-3}, in United Arab Emirates for tertiary treated wastewater at \$0.43 m^{-3}, and in Kuwait for secondary treated wastewater at \$0.18 m^{-3} [15]. These cost calculations are based on projected plant life, interest rate, plant availability, and production capacity around the mid-1990s.

While considering the cost of wastewater and stormwater collection, customer service and billing, stormwater treatment and drainage, property taxes, capital and rehabilitation, corporate services, and planning and engineering aspects of overall wastewater treatment process, the cost could be much higher than reported by Lee et al. [21]. For example, based on the estimates of treating wastewater from Ottawa (Robert O. Pickard Environmental Centre—Wastewater Treatment Plant), the overall wastewater treatment cost may reach CAD\$1.64 per m^3 (1.00 CAD\$ = 0.97 US\$). The cost components consist of wastewater treatment (24.4¢), wastewater and stormwater collection (26.1¢), customer service and billing (7.9¢), stormwater treatment and drainage (6.1¢), property taxes (1.4¢), capital and rehabilitation (75.5¢), corporate services (17.4¢), and planning and engineering (5.0¢) [8]. Based on the analysis of 338 treatment plants in Spain for technical efficiency and cost analysis in wastewater treatment processes, Hernández-Sancho et al. estimated the cost of secondary treatment of wastewater in the range of € 0.40–€ 0.77 m^{-3} (1.00 € = 1.33 US\$ in 2010) [14]. The above estimates reveal that there are differences in the approaches used to calculate the overall cost of wastewater treatment. These range from only considering the cost of reclaimed water recycling (additional treatment, storage, and distribution) to adjusting for other associated costs such as that of the wastewater collection and treatment [20].

The lack of wastewater treatment cost-recovery mechanisms has led to low demand for cost-based reclaimed water when compared with treated or untreated wastewater supplied free of charge to the farmers. The reasons are that both farmers and households have skepticism about the quality of the reclaimed water as they do not have access or means to monitor and verify the quality of water they use. In addition, the availability of untreated wastewater free of charge adds to the complexity of the whole issue and makes it difficult to convince farmers to pay anything for reclaimed water that is not of high quality. For example, despite low prices charged to Tunisian farmers for reclaimed water (0.02 US\$ m^{-3}) compared to approximately fourfold higher conventional water supply costs (0.08 US\$ m^{-3}), the demand for reclaimed water was lower than other water supplies [19]. In most countries, there are no or few farmers or water users associations that can help farmers make use of wastewater in a safe and productive manner.

The mismatch between water pricing and water scarcity is another important constraint affecting the whole process of wastewater treatment in most arid regions in developing countries where water pricing should also consider its scarcity value. This aspect has particular importance in the agricultural sector. According to Kfouri et al., the price of freshwater delivered to irrigators does not reflect even the cost of water supply. At best, some countries, such as Tunisia, charge sufficient rates to cover operation and maintenance costs [5,6]. In general, most arid regions in developing countries do not charge or control groundwater abstractions other than the private cost of pumping and the permitting process.

In general, the lack of commitment on the part of several developing-country governments to advocate and support comprehensive wastewater treatment programs has led to the lack of understanding among farmers and households about the perceived environmental benefits of wastewater treatment and reuse of reclaimed water. Because of the collection and conveyance of wastewater away from urban areas, households do not recognize the benefits of wastewater treatment and reuse amid extreme water scarcity. The governments therefore find it easier to collect fees for connection and wastewater service than for eventual treatment of wastewater. In addition, they do not take into consideration regulatory and monitoring costs. In most arid region developing countries, irrigation and water management schemes are inefficient and do not pay due attention to the potential of reclaimed water as a resource that can be used for irrigation, environment conservation, and other purposes such as groundwater recharge, municipal, recreational, or industrial uses.

66.4 Opportunities for Water Recycling and Reuse

Despite these constraints, some countries in arid zones have employed a range of conventional and nonconventional systems and have national standards and regulations for reuse. Although each country has taken a slightly different path, they also are similar in important ways. Policymakers in these countries consider reuse of reclaimed water an essential aspect of strategic water and wastewater sector planning and management. Salient features of water recycling and reuse in Tunisia, Jordan, Israel, and Cyprus are presented in the following sections.

66.4.1 Tunisia

With actual renewable water resources (ARWR) of 432 m^3 per capita in 2010, water recycling and reuse has been a priority of Tunisia since the early 1980s, when Tunisia launched a nationwide water reuse program to increase the country's usable water resources. Most municipal wastewater receives secondary biological treatment generally through activated sludge and, in some cases, there is tertiary treatment of wastewater in place. In order to promote tourism and protect the environment for tourists, several treatment plants are located along the coast to protect coastal resorts and prevent marine pollution.

Reusing reclaimed water for irrigation in Tunisia is viewed as a method to increase water resources, provide supplemental nutrients, and enhance wastewater treatment in a way that protects coastal areas, water resources, and water receiving bodies sensitive to water quality. Restrictions for water reuse designed to protect public health have received considerable attention in Tunisia and are in line with recommendations of the World Health Organization [35]. The government also supports research studies investigating the use of treated wastewater for groundwater recharge, irrigation of forests and highways, wetlands development, and industrial use.

Tunisian regulations allow the use of secondary treated effluent on all crops except vegetables, whether eaten raw or cooked. Regional agricultural departments supervise the treated wastewater reuse and collect charges from the farmers. In Tunisia, farmers pay for irrigation water on the basis of the volume of water required and the area to be irrigated, and the number of hours corresponding to the contract, at a rate of TND 0.02–0.03 per m^3 (1 TND = US $0.61 in 2013).

Of the annual volume of wastewater generated in Tunisia (0.246 km^3 in 2010), 0.226 km^3 are treated [30,31]. The annual volume of treated wastewater in Tunisia is expected to reach 0.290 km^3 by 2020 [3]. At that point, the expected amount of treated wastewater will be around 18% of the available groundwater resources and could be used where excessive groundwater mining is causing seawater intrusion in coastal aquifers.

In general, there is a strong government support for wastewater reclamation and reuse [4,30]. However, this support has yet to trigger wastewater use at a large scale as groundwater is preferred and being used even though large efforts have been made to provide reclaimed water to the farmers using groundwater. There are a number of issues related to social acceptance, regulations concerning crop choices, and other agronomic considerations that affect these decisions. Farmers in the arid south have expressed their concerns about the long-term impacts of saline wastewater on their crops and soils. In addition, farmers also consider crop restrictions as an impediment as they cannot grow high-value crops such as vegetables with reclaimed water.

Considering these challenges, the Tunisian policy makers have started to pay greater attention through better coordination and demand driven approaches to better plan wastewater reclamation and irrigation with treated effluent in Tunisia. To bridge the gap between the needs of different parties, ensure the achievement of development objects, and preserve the human and natural environment, interdepartmental coordination and follow-up commissions with representatives from the different ministries and their respective departments or agencies, the municipalities and representatives of the water users associations have been set up at national and regional levels [2].

66.4.2 Jordan

With ARWR of 145 m^3 per capita in 2010, Jordan is one of the most water-scarce countries of the world. In terms of establishing wastewater collection networks, treatment of collected wastewater, and use of treated wastewater in agriculture, Jordan is one of the countries where relevant wastewater policy framework and institutional structure exist [23]. The wastewater policy "Jordanian National Wastewater Management Policy," developed more than 15 years ago in 1998, has three major considerations: (1) reclaimed water is to be considered a part of the water budget in the country with no consideration of disposal; (2) water reuse is to be planned on a basin scale; and (3) fees for wastewater treatment may be collected from the water users [25]. The Jordanian wastewater policy is unique and innovative and although the government has not achieved full success in implementing the policy, it represents a different way of thinking about water recycling and reuse [19].

Jordan has implemented an aggressive campaign to rehabilitate and improve wastewater treatment plants. In addition, enforceable standards have been introduced to protect the health of fieldworkers and consumers. In addition, Jordan has extensive research studies on water recycling and reuse. The share of reclaimed water in the total water supply in Jordan is about 13%. There are three categories of water reuse in Jordan: (1) planned direct use within or adjacent to wastewater treatment plants; (2) unplanned reuse of reclaimed water in *wadis*; and (3) indirect reuse after mixing with surface water supplies, which is mainly practiced in the Jordan Valley where reclaimed wastewater provides about half of the irrigation water in the valley.

The planned direct use of reclaimed water is administrated by the Water Authority of Jordan, which has special contracts with the farmers formalizing their rights to use reclaimed water directly at 20 fils per m³ (1000 fils = 1 Jordanian Dinar = US $1.4 in 2014). At the policy level, the Jordanian National Wastewater Management Policy requires that the reclaimed water supply prices cover at least oil and maintenance of the reclaimed water delivery to the farmers [22]. However, reclaimed water supply prices do not include costs incurred on wastewater collection and treatment. Other planned direct reclaimed water users are some private enterprises and experimental pilot projects cosponsored by international donors.

Phase 1 of the As-Samra Wastewater Treatment Plant, the major wastewater treatment facility located 50 km north of Amman, was designed to handle annual wastewater volume of 25 million m³; however, it used to process at least double the amount of wastewater. The cost incurred on Phase 1 was US$ 169 million. Implementing Phase 2 with an estimated cost of US$ 223 million, the treatment plant has been upgraded recently to handle wastewater at a daily volume of 267,000 m³ and an annual volume of 97 million m³. A Private Public Partnership (PPP) model has been used to finance the construction and operation of the treatment plant, with major funding provided by the United States Agency for International Development (USAID). The model is based on a 25-year build-operate-transfer (BOT) approach. Treated wastewater flows into the King Talal Dam, where it is mixed with freshwater from the King Abdullah Canal before being discharged into the Jordan Valley to be used for irrigation.

Treated wastewater constitutes 13% of the ARWR and its contribution is set to rise as conventional wet sanitation and wastewater collection is expected to increase in the coming decades [7]. Jordan has taken the lead in terms of treating part of the domestic wastewater, generally known as greywater, at the household level and making use of it in home gardens to irrigate a range of plant species. It comprises 55%–75% of residential wastewater. The greywater reuse projects have revealed considerable potential of greywater reuse in irrigation at the household level in poor communities. In addition to increasing yields of high-value crops and economical returns, greywater reuse has increased community participation in the national efforts to conserve limited water resources onsite and low-cost greywater treatment and reuse systems [1].

More recently, on February 20, 2014, Jordan launched its first wastewater master plan to help the government determine investment priorities in wastewater services across the country through the year 2035. Referred to as the National Strategic Wastewater Master Plan, it identifies investment needs and priorities for wastewater collection and treatment in every governorate to enable the Ministry of Water and Irrigation to better direct donor and government resources to areas with limited wastewater services or overloaded capacities. The USAID-funded Jordan Institutional Support and Strengthening Programme prepared the plan upon a request from the Ministry of Water and Irrigation.

The master plan will aim at providing wastewater services to all areas with more than 5000 residents as an action plan for investment, development, and donor support through 2035. This will help the water and wastewater sectors as they are under increasing pressure due to massive population growth, high energy costs, and climate change impacts, among other challenges. To move forward with the plan, the Water Authority of Jordan has prepared an accurate and up-to-date map for wastewater information using Geographic Information Systems tools. While this information will allow new projects to save months of time in the planning stage, this will also help in identifying priority areas for wastewater projects.

66.4.3 Israel

Israel is another water-scarce country with ARWR of 240 m³ per capita in 2010. As early as 1953, Israel drafted the standards for water reuse, which have continued to evolve to reflect the latest scientific findings on microbiological and chemical risks [33]. The Water Law of 1959 and policy enacted by the administration up until today define sewage as a "water resource" and an integral part of the water resources of the country and its water budget.

Most of Israel's farmers using reclaimed water are organized into different types of communities and cooperatives. The Ministry of Agriculture provides professional guidance through an efficient extension service. Part of the success of the practice of water reuse in Israel is due to the capacity of the well-organized and informed farmers to adapt quickly to the switch to reclaimed water from other water resources [17]. The reuse of reclaimed water has provided an attractive means of increasing water supply in the country.

On the water reuse front, the institutional set-up of the water sector in Israel has played a pivotal role in facilitating the reuse of reclaimed water in agriculture. The cost of reclaimed water paid by the farmers is about 20% lower than that of the equivalent volume of freshwater. With no provision of private wells, the farmers have no alternate sources of water for irrigation within the zones where treated wastewater is available. This governance structure largely solves problems of reduced reclaimed water demand while ensuring a regular supply of water for irrigation [19].

Israel is a country that has practiced massive water reuse for decades. Most reclaimed water is still dedicated to restricted irrigation, which helps liberate freshwater resources for unrestricted irrigation. However, with long-term use of wastewater, Israel has a substantial risk of salinization of groundwater in the wastewater-irrigated areas. Wastewater produced by the domestic, municipal, and industrial sectors becomes more saline than the water used in these sectors and there are no inexpensive ways to remove salts once they enter sewage. This concern has led to expanding research on developing crops that can withstand higher levels of salts in the growth medium. The Ministry of the Environment has been engaged in a campaign to reduce the addition of salts to sewage since the early 1990s. The country has one of the best-documented and analyzed experiences of water reuse demonstrating much innovation, particularly over the past three decades.

66.4.4 Cyprus

Cyprus is the third largest island in the Mediterranean region and has always been confronted with inadequate volumes of water to meet its agricultural and domestic needs. Its geographical location, semi-arid climate with frequent droughts, economy, and political situation exacerbate its water problems. Cyprus had ARWR of 707 m^3 per capita in 2010. Over 87% of the island's water extractions, 96% of which are coming from groundwater, are for agricultural purposes. The salinization of the groundwater resources by overexploitation of the aquifers has led to the development of alternative water supplies. With desalination as an alternative to meet the water needs, the country has steadily been increasing the use of desalinized water, reaching an annual volume of 49 km^3 [10]. Wastewater reuse is another alternate option with most of the 25 million m^3 of treated wastewater used for agricultural and landscape irrigation [36].

The Sewage Boards at the municipality level are responsible for the collection, treatment, and disposal of wastewater while the Water Development Department undertakes the management and distribution of recycled water. Cyprus has followed effective strategies toward improving the efficient use of water such as improvement in irrigation systems, conservation of groundwater, water pricing, campaigns of water awareness, and reuse of treated wastewater and greywater effluent, among others. This has resulted in a very efficient system of irrigation,

with closed systems, and an overall conveyance efficiency averaging 90%–95%.

In terms of groundwater conservation with treated wastewater for later recovery and reuse, Cyprus has taken several steps. For example, the entire quantity of treated wastewater produced in Paphos, the fourth largest city located in southwest Cyprus, is used for Ezousa aquifer recharge, which is subsequently pumped for irrigation through diversion in an irrigation channel. The Code of Good Agricultural Practice regulates irrigation with reclaimed water. The treated effluent can be applied to all kinds of crops except leafy vegetables, bulbs, and corn eaten raw. The major crops irrigated with treated effluent are citrus trees, olive trees, fodder crops, industrial crops, and cereals. In addition, it is used for landscape and football field irrigation [26].

In the case of Limassol, the largest municipality of Cyprus, there are multiple uses of wastewater generated by the city and subsequently treated. During winter months when the demand for water in agriculture decreases, treated wastewater is pumped to an irrigation dam for irrigation or subject to aquifer recharge. In 2010, about 15% of treated wastewater was used for recharge of the Akrotiri aquifer. The government's long-term plan and water policy aim at fully utilizing reclaimed water and incorporating it into the country's water balance [26]. There are considerations to increase the volume of treated wastewater for aquifer recharge to replenish this depleting aquifer. The implementation of the plan would yield a welfare improvement that would not only increase the economic benefits to all stakeholders in both the short- and long-term, it would also help Cyprus in its efforts to meet the European Union's (EU) Water Framework Directive, WFD (2000/60/EC), requirements by 2015. In compliance with Article 9 of the WFD (2000/60/EC), Cyprus has launched a new water pricing policy to recover the cost of water services. To encourage the use of treated wastewater, it is supplied to the potential users without full cost recovery for different uses at a cost lower than freshwater.

66.5 Summary and Conclusions

Amid water scarcity, the volumes of wastewater generated by the domestic and industrial sectors in countries located in arid zones are increasing due to increased population growth and rapid urbanization, improved living conditions, and economic development. These countries will have to increasingly rely on alternative water resources such as wastewater to narrow the gap between water demand and supply for agriculture or afforestation.

The issues regarding wastewater generation, treatment, and use will intensify in the future, with increasing water scarcity and economic growth. The implementation of research-based technical options for wastewater treatment and reuse in developing countries located in dry areas, supported by flexible policy level interventions and pertinent institutions with skilled human resources, offers great promise for environment and health protection as well as livelihood resilience through agricultural productivity enhancement. This may not be achieved in the next few years. Therefore, interim measures would be needed to address water recycling and reuse to gradually reach a level when most wastewater in these countries would be collected, treated, and used in treated form safely and productively.

References

1. Al-Balawenah, A., Al-Karadsheh, E., and Qadir, M. 2011. Community-based reuse of greywater in home farming. In: *Abstract of the International Conference on Food Security and Climate Change in Dry Areas*, Amman, Jordan, February 1–4, 299–303.

2. Bahri, A. 2008. Case studies in Middle Eastern and North African countries. In: B. Jimenez and T. Asano (eds.), *Water Reuse: An International Survey of Current Practice, Issues and Needs*. IWA Publishing, London, UK.

3. Bahri, A. 2009. *Managing the Other Side of the Water Cycle: Making Wastewater an Asset*. Global Water Partnership Technical Committee (TEC), Background Paper 13, Global Water Partnership.

4. Bahri, A., Basset, C., Oueslati, F., and Brissaud, F. 2001. Reuse of reclaimed wastewater for golf course irrigation in Tunisia. *Water Science and Technology*, 43(10), 117–124.

5. Bazza, M. and Ahmad, M. 2002. A comparative assessment of links between irrigation water pricing and irrigation performance in the Near East. In: *Conference on Irrigation Water Policies: Micro and Macro Considerations*, Agadir, Morocco, June 15–17.

6. Bucknall, J., Kremer, A., Allan, T., Berkoff, J., Abu-Ata, N., Jarosewich-Holder, M., Deichmann, U. et al. 2007. *Making the Most of Scarcity: Accountability for Better Water Management in the Middle East and North Africa*. World Bank, Washington, DC.

7. Carr, G. and Potter, R.B. 2013. Towards effective water reuse: Drivers, challenges and strategies shaping the organisational management of reclaimed water in Jordan. *The Geographical Journal*, 179(1), 61–73.

8. City of Ottawa. 2013. *Wastewater Treatment at Robert O Pickard Environmental Centre—Wastewater Treatment Plant*. http://ottawa.ca/en/residents/water-and-environment/sewers-and-septic-systems/wastewater-treatment. Accessed on July 31, 2013.

9. Cosgrove, W.J. and Rijsberman, F. 2000. *World Water Vision: Making Water Everybody's Business*. World Water Council, World Water Vision, and Earthscan, UK.

10. Eurostat. 2014. *European Commission Eurostat*. http://epp.eurostat.ec.europa.eu/portal/page/portal/statistics/search_database. Accessed on February 20, 2014.

11. Evans, A.E.V., Hanjra, M.A., Jiang, Y., Qadir, M., and Drechsel, P. 2012. Water quality: Assessment of the current situation in Asia. *International Journal of Water Resources Development*, 28(2), 195–216.

12. Grangier, C., Qadir, M., and Singh, M. 2012. Health implications for children in wastewater-irrigated peri-urban Aleppo, Syria. *Water Quality Exposure and Health*, 4(4), 187–195.

13. Hamilton, A.J., Stagnitti, F., Xiong, X., Kreidl, S.L., Benke, K.K., and Maher, P. 2007. Wastewater irrigation the state of play. *Vadose Zone Journal*, 6(4), 823–840.

14. Hernández-Sancho, F., Molinos-Senante, M., and Sala-Garrido, R. 2010. Economic valuation of environmental benefits from wastewater treatment processes: An empirical approach for Spain. *Science of the Total Environment*, 408(4), 953–957.

15. Husain, T. and Ahmed, A.H. 1997. Environmental and economic aspects of wastewater reuse in Saudi Arabia. *Water International*, 22(2), 108–112.

16. Jiménez, B., Drechsel, P., Koné, D., Bahri, A., Raschid-Sally, L., and Qadir, M. 2010. Wastewater, sludge and excreta use in developing countries: An overview. In: P. Drechsel, C.A. Scott, L. Raschid-Sally, M. Redwood and A. Bahri (eds.), *Wastewater Irrigation and Health: Assessing and Mitigating Risks in Low-Income Countries*. Earthscan-International Development Research Centre (IDRC)-International Water Management Institute (IWMI), London, UK, 3–27.

17. Juanicó, M. 2008. Israel as a case study. In: B. Jimenez and T. Asano (eds.), *Water Reuse: An International Survey of Current Practice, Issues and Needs*, IWA Publishing, London, UK, 483–502.

18. Keraita, B., Jimenez, B., and Drechsel, P. 2008. Extent and implications of agricultural reuse of untreated, partly treated and diluted wastewater in developing countries. *CAB Reviews: Perspectives in Agriculture, Veterinary Science, Nutrition and Natural Resources*, 3(058), 1–15.

19. Kfouri, C., Mantovani, P., and Jeuland, M. 2009. Water reuse in the MNA region: Constraints, experiences, and policy recommendations. In: N.V. Jagannathan, A.S. Mohmed, and A. Kremer (eds.), *Water in the Arab World-Management Perspectives and Innovations*, The World Bank, Middle East and North Africa (MNA) Region, World Bank, 447–477.

20. Lazarova, V., Levine, B., Sack, J., Cirelli, G., Jeffrey, P., Muntau, H., Salgot, M., and Brissaud, F. 2001. Role of water reuse for enhancing integrated water management in Europe and Mediterranean countries. *Water Science and Technology*, 43(10), 25–33.

21. Lee, T., Oliver, J.L., Teneere, F., Traners, L., and Valiron, W.F. 2001. Economics and financial aspects of water resources. In: C. Maksimovic and J.A. Tejada-Guibert (eds.), *Frontiers in Urban Water Management: Deadlock or Hope*. IWA Publishing, London, UK, 313–343.

22. McCornick, P.G., Haddadin, N., Rashid, H., and Sabella, R. 2001. *Water reuse in Wadi Zarqa and from Other Amman-Zarqa Sources*, Water Reuse Component, Water Policy Support Project, Ministry of Water and Irrigation, Amman, Jordan.

23. McCornick, P.G., Hijazi, A., and Sheikh, B. 2004. From wastewater reuse to water reclamation: Progression of water reuse standards in Jordan. In: C. Scott, N.I. Faruqui, and L. Rachid-Sally (eds.), *Wastewater Use in Irrigated Agriculture: Confronting the Livelihood and*

Environmental Realities. CABI Publishing—International Water Management Institute (IWMI), Wallingford, UK, 153–162.

24. Murtaza, G., Ghafoor, A., Qadir, M., Owens, G., Aziz, M.A., Zia, M.H., and Saifullah. 2010. Disposal and use of sewage on agricultural lands in Pakistan: A review. *Pedosphere*, 20(1), 23–34.

25. Nazzal, Y.K., Mansour, M., Al-Najjar, M., and McCornick, P.G. 2000. *Wastewater Reuse Law and Standards in the Kingdom of Jordan.* Ministry of Water and Irrigation, Amman, Jordan.

26. Papaiacovou, I. and Papatheodoulou, A. 2013. Integration of water reuse for the sustainable management of water resources in Cyprus. In: V. Lazarova, T. Asano, A. Bahri, and J. Anderson (eds.), *Milestones in Water Reuse: The Best Success Stories.* IWA Publishing, London, 75–82.

27. Qadir, M., Bahri, A., Sato, T., and Al-Karadsheh, E. 2010b. Wastewater production, treatment, and irrigation in Middle East and North Africa. *Irrigation and Drainage Systems*, 24(1–2), 37–51.

28. Qadir, M., Sharma, B.R., Bruggeman, A., Choukr-Allah, R., and Karajeh, F. 2007. Non-conventional water resources and opportunities for water augmentation to achieve food security in water scarce countries. *Agricultural Water Management*, 87(1), 2–22.

29. Qadir, M., Wichelns, D., Raschid-Sally, L., McCornick, P.G., Drechsel, P., Bahri, A., and Minhas, P.S. 2010a. The challenges of wastewater irrigation in developing countries. *Agricultural Water Management*, 97(4), 561–568.

30. Saloua, R. 2012. *National Report: Tunisia. Presented at the First Regional Workshop of the Project Safe Use of Wastewater in Agriculture.* http://www.ais.unwater.org/ais/mod/page/view.php?id=59. Accessed on December 21, 2012; in French.

31. Sato, T., Qadir, M., Yamamoto, S., Endo, T., and Zahoor, A. 2013. Global, regional, and country level need for data on wastewater generation, treatment, and use. *Agricultural Water Management*, 130, 1–13.

32. Simmons, R.W., Qadir, M., and Drechsel, P. 2010. Farm-based measures for reducing human and environmental health risks from chemical constituents in wastewater. In: P. Drechsel, C.A. Scott, L. Raschid-Sally, M. Redwood, and A. Bahri (eds.), *Wastewater Irrigation and Health-Assessing and Mitigating Risks in Low—Income Countries.* Earthscan-International Development Research Centre (IDRC)-IWMI, London, UK, 209–238.

33. Tal, A. 2006. Seeking sustainability: Israel's evolving water management strategy. *Science*, 313(5790), 1081–1084.

34. Toze, S. 2006. Reuse of effluent water—Benefits and risks. *Agricultural Water Management*, 80(1–3), 147–159.

35. WHO (World Health Organization). 2006. *Guidelines for the Safe Use of Wastewater, Excreta and Grey Water: Volume 2. Wastewater Use in Agriculture.* WHO, Geneva, Switzerland.

36. Zimmo, O. and Imseih, N. 2011. Overview of wastewater management practices in Mediterranean countries. *The Handbook of Environmental Chemistry*, 14, 155–181.

67

Water Reuse in Coastal Areas

Johanny A. Perez Sierra
*Institute for Sanitary Engineering,
Water Quality and Solid
Waste Management (ISWA)—
University of Stuttgart*

Saeid Eslamian
Isfahan University of Technology

PREFACE

Water demand has been increasing tremendously in the last few years and water reuse represents a good opportunity for a reliable water supply. Nevertheless, it is important that certain measures be taken in order to protect the health of users and the natural ecosystem. Coastal regions are very vulnerable to water scarcity and water reuse is becoming much more popular.

The treatment scheme in these regions is comparable to the typical state-of-the-art in water treatment—a preliminary, primary, and secondary treatment, joined with an advanced process. This allows the supply of treated water for a myriad variety of users, such as agricultural and garden irrigation, potable and nonpotable water, and recreational uses, among others.

In this chapter, a synopsis of the state of water reuse in the countries of Barbados, Jamaica, and Cyprus is presented. The first two islands belong to the Caribbean region and the latter, to the Mediterranean, European region. It is aimed at providing available information and research in these zones that could serve as examples for similar countries and to overcome water demand by reusing its treated water.

67.1 Introduction

Coastal areas are regions with extreme vulnerability to water scarcity and in many zones, water reuse has been considered the central new water source. Due to the significant risks associated with the reutilization of treated wastewater, more sophisticated technologies are being implemented. In addition, the areas of use are even more diversified.

The islands of Barbados, Jamaica, and Cyprus have been approaching this concept and there are up-to-date cases of its reutilization and application locally. The following is an overview of the potential of water reuse in regions along the seaside and the analysis of some current examples.

67.2 Drivers for Water Reuse in Coastal Areas

Among the parameters that influence the necessity to reutilize reclaimed water can be listed the following as shown in Table 67.1, which differ according to developed and developing countries. The physical aspects arise from the composition of a certain environment, the social ones from the users of the water, and the economic and political aspects from the consequences of the water management strategies.

In coastal areas particularly, there are situations of deficient and low quality flows due to intensive uses upstream; there are also needs of planning for water extraction. Another driver is

TABLE 67.1 Fundamental Wastewater Reuse Drivers

Driver	Developed Countries	Developing Countries
Due to the physical situation		
Lack of water	++ +	++ +
Drought management and to ensure a reliable source of water	++ +	++
Generation of wastewater near agricultural fields demanding water	++ +	++ +
Lack of sanitation inducing unintentional reuse of wastewater		++ +
As a result of water management policies		
For environmental protection to control negative fall-out from the disposal of treated effluents (mainly in coastal and tourist areas or in highly sensitive aquatic ecosystems) water reuse is being practiced	++	
To improve ecological conditions in environments with low water quality by reusing properly treated wastewater	++	
Convenience of reusing wastewater in order to first use water for drinking water supply	++	
Growing recognition among water and wastewater managers of the economic and environmental benefits of using recycled water	+	
For health protection in regions where unintentional reuse is being performed with low water quality	+	
Greater recognition of the environmental and economic costs of water storage facilities such as dams and reservoirs	++	
The growing numbers of successful water recycling projects in the world	+	
As a result of social pressure		
Preference for supporting water reuse programs rather than increasing water prices to transport water from other sources or to cover advanced wastewater treatment costs	++	
Increasing awareness of the environmental impact associated with the overuse of water supplies	+	
To reclaim compounds in reused water (such as nitrogen and phosphorus) at no cost	+	++ +
Community enthusiasm for the concept of water reuse	+	
Economic reasons		
As an option to partially recover treatment costs to meet stringent standards	++	
Economic reasons and water management policies		
To employ reuse as a lower-cost disposal option	++	+
Environmental protection in tourist areas	++	++
Physical, social and economic reasons		
High water demand in local areas, mainly for urban and industrial use	++ +	++ +

Source: Adapted from Jimenez, B. and Asano, T. 2008. In: *Water Reuse: An International Survey of Current Practice Issues and Needs.* IWA Publishing, London, UK. p. 16.

++ + : high; ++ : medium; +: low importance.

the diffuse pollution from agricultural irrigated fields with water that is polluted with salts, nutrients, and pesticides. Besides, groundwater bodies located near the coastline face salty water intrusion. In addition, peak demands in certain periods of the year, especially in summer, provoke stress in water distribution. Finally, there is evacuation of wastewaters from industries in the coastline and sewers to the sea [7].

67.3 Legal Issues and Potential Risks in Water Reuse

According to Jimenez and Asano [5] there is not a general policy to regulate the reutilization of treated wastewater. This is based on the fact that water reuse involves different utilization purposes, which have different quality requirements. This is a relatively new proceeding and many techniques have been created and adapted to certain local conditions, which may make them unsuitable

for other regions. The World Health Organization (WHO) has embraced the responsibility to inform and support decision makers in matters of the safe use of treated wastewater. In 1989, it published a guideline for water reuse in agriculture. Nevertheless, the impact of these guidelines was not very relevant. In 2006, WHO reviewed the previous guideline and joined aspects of American standards, but still its scope is in progress. There is still a need of worldwide-accepted guidelines that will ease the adoption of water reuse habits. Table 67.2 shows the different risks associated with water reuse.

67.4 Scheme of the State-of-the-Art for Wastewater Treatment

Figure 67.1 is a general scheme of the consecutive steps for wastewater treatment, as described in previous chapters, for different reutilization purposes, published by the U.S. Environmental Protection Agency (EPA) [2].

TABLE 67.2 Risks and Hazards Related to Water Reuse

Sector or Element Concerned	Risks	Hazards	Sensitivity (Exposure)
Health	Microbial risks: Cholera, infections, diarrhea, allergies Chemical risk: Intoxication, cancer, poor quality of food products	Pathogens Toxic compounds Emerging pollutants, endocrine disrupters	Exposure (public, users, consumers)
Environment	Eutrophication, groundwater pollution odors Impact of treatment by-products (membrane concentrates, sludge) CO_2 emissions (energy consumption for treatment)	Nitrogen, phosphorus toxic compounds, heavy metals	Depth of aquifer, environmental sensitivity (coastal areas)
Soil and plants	Plant toxicity (salts) Salinization and land degradation (salt water) Accumulation of pollutants in the soil	Salinity, heavy metals	Crop sensitivity, soil fragility
Perception	Visual impact (storage) Odors Social rejection (lack of knowledge, fears) Tensions in case of expropriations	Nuisance	Level of perception, tendency to change
Distribution, equipment	Algae growth Corrosion, biofilms, clogging	Organic matter, nitrogen, phosphorus, suspended solids	Type of irrigation system

Source: Adapted from Plan bleu pour la Méditerranée, Condom, N., Lefebvre, M., and Vandome, L. 2012. Treated wastewater reuse in the Mediterranean: Lessons learned and tools for project development. *Plan Bleu*, Valbonne.

67.5 Potential Uses of Treated Wastewater

The reutilization of reclaimed water will be subject to the source of the water and the treatment scheme. It can be classified as follows [10]:

- *Agriculture irrigation*: The most common type of reutilization nowadays.
- *Landscape irrigation*: This group includes the sprinkling of parks, playgrounds, golf courses, freeway medians, landscaped zones, that is, commerce, offices, industries, and residential zones.
- *Process water*: This refers to the reused water for industrial processes, such as cooling towers and ponds. The uses are diverse and as are the quality requirements, which are subject to each different type of industry.
- *Groundwater recharge*: This is done by spreading basins or by direct injection into aquifers. This category also considers the groundwater recharge by creating a hydraulic impediment to saltwater intrusion on coastal areas.
- *Recreational uses and environmental reconstruction*: It can be used for nonpotable uses on land application, including: augmentation of stream flows, recreational lakes, golf course storage ponds, and wetlands, among others.
- *Nonpotable urban uses*: Due to the costs involved, this application is restricted to cases when the treatment plant is near the point of use. Typical uses include: fire mitigation, air conditioning, toilet flushing, construction water, and flushing of sewers.
- *Potable uses*: This is only possible after the water has been treated under intensive purification systems; and by combining water supply reservoirs or by direct injection in water distribution systems.

67.6 Case Studies

67.6.1 Water Reuse Scheme in Barbados

According to a publication of the United Nations Environment Program (UNEP) [9], there are reports of water reuse in Barbados. In the southeastern coast of this Caribbean island, near to a coral-rich zone is located the "Sam Lord's Castle Hotel." This complex has implemented a wastewater reuse system after the experience of ample water demand for irrigation and a scarcity of freshwater supply. In Figure 67.2 can be identified its location, which belongs to the St. Philip community, one of the driest regions in this island; the location of the different wastewater treatment plants in Barbados also can be seen.

Initially, the hotel was provided with fresh groundwater; however, due to an extensive water demand to irrigate the grass and gardens, saline water intruded into the water extraction points and depleted the resource. Given this situation, it was decided to use the outflows of the aerated sewage treatment plant of the hotel. For that, permission to reuse wastewater as irrigation water was requested to the Ministry of Health and Environment, which accepted the requisition [9]. A packaged wastewater treatment plant (WWTP) was constructed in this hotel by a foreign consulting company and the execution of local workers. The Environmental Engineering Division (EED) of the Ministry of Health and Environment was in charge of the inspection, evaluation, approval, and check out of the performance of this plant. Nevertheless, at the moment of approval, no local regulatory instrument regarding effluent quality was available. Due to this, several terms were specified before permission, which considered the participation of the Town and Country Planning Offices and the Barbados Water Authority (BWA) [9].

In 1997, there were 12 installed wastewater plants in Barbados, with a total treatment capacity of 786,280 gpd. The technologies

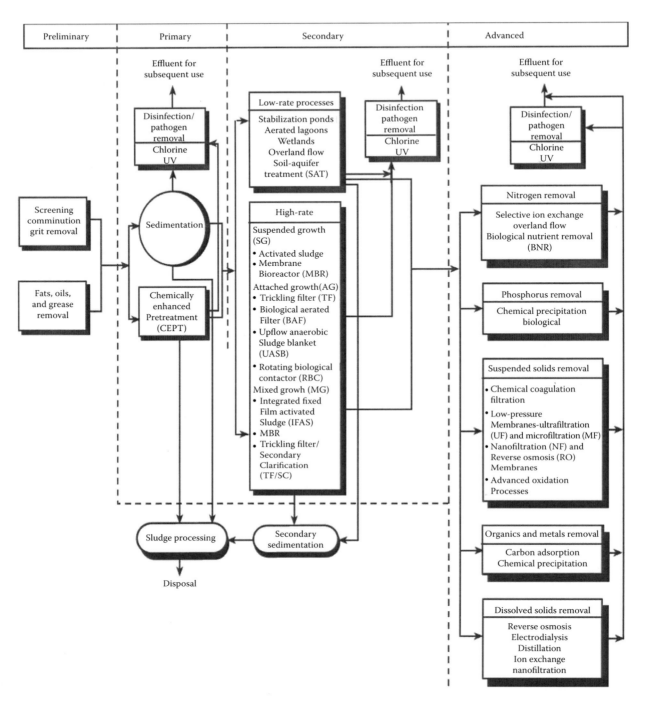

FIGURE 67.1 General outline for wastewater treatment. (Adapted from EPA—U.S. Environmental Protection Agency. 2004. *Guidelines for Water Reuse*. Washington, DC, p. 136.)

implemented included: extended aeration plants, rotating biological contactor, and contact stabilization plants. After the experience of the Sam Lord's Castle Hotel, several touristic facilities were interested in reusing their wastewater effluents for irrigation of their green areas [9]. The UNEP's report [9] indicated that after monitoring the wastewater plant in 1990 in the Sam Lord's Castle Hotel, it was found that the plant had more than 12 years in operation. There appeared to be a lack of maintenance,

no flow meters installed, no laboratories, and especially a lack of know-how among the operators, who basically assessed the quality of the WWTP by the color of the effluent. However, EED carries out some chemical analyses on a monthly basis.

In Figure 67.3 the wastewater treatment scheme and reuse concept in the Sam Lord's Castle Hotel is shown. From results between 1989 and 1990, BOD and TSS removal efficiencies achieved up to 98% and 83%, respectively. Nevertheless, no

FIGURE 67.2 Map of Barbados indicating location of wastewater treatment plants and protection zones. (Adapted from UNEP—United Nations Environment Program. 1997. *Source Book of Alternative Technologies for Freshwater Augmentation in Latin America and the Caribbean*. pp. 289–292.)

information regarding the microbiological disinfection was provided. In any case, the report indicates that a good appearance in the gardens of the hotel can be observed [9].

Finally, in this report the advantages and disadvantages of this technology in Barbados were listed. Among the positive outcomes can be mentioned the avoidance of water pollution by treating the water before discharge, and also by saving costs involved with irrigation. Besides, it reduces the freshwater demand. The main disadvantages are focused on a lack of maintenance and potential risks associated with inefficient disinfection techniques [9].

67.6.2 Water Recycling in Jamaica

Sharif [8] recognized the significant challenges that Jamaica is facing due to severe drought and reduced fresh water sources. This author carried out a study to determine the effects of reclaiming water with floating aquatic species, such as *Eichhornia crassipes, Lemna minor,* and *Pistia stratiotes* for crops production. The quality of the effluent was analyzed after reclamation, where removal of pollutants, TSS, DO, and

nutrients were higher at longer residence times. After using reclaimed water for growing cereal crops, it showed that compared to the plants irrigated with freshwater, the maturity time of crops was reduced, while grain production was increased. It was also found that there was only aggregation of Zn as a heavy metal. The death of bacteria was possible due to the increment in the pH, favored by the growth of algae, when water was in contact with sunlight.

Other reclamation techniques have been aimed at in Jamaica. In November 2010, the International Desalination & Water Reuse Quarterly industry website [4] published that a small wind and solar project designed to operate a reverse osmosis (RO) desalination facility in Jamaica was awarded a US$ 40,000 fund from the Global Environment Facility, in addition to US$ 10,000 offered from the Environmental Foundation of Jamaica. The turbines are expected to generate 2 kW of energy, which will operate the RO desalination system. This project is to be managed by the Caribbean Maritime Institute (CMI).

The legislative context that regulates the water reuse and disposal in Jamaica includes The Natural Resources Conservation Authority Act, 1991, The Beach Control Act, 1956, The Water

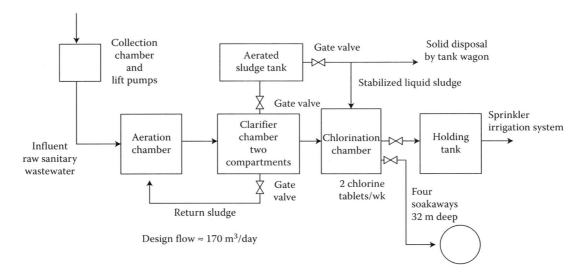

FIGURE 67.3 WWTP scheme and reuse concept at the Sam Lord's Castle Hotel. (Adapted from UNEP—United Nations Environment Program. 1997. *Source Book of Alternative Technologies for Freshwater Augmentation in Latin America and the Caribbean*, pp. 289–292.)

Resources Act, Ministry of Agriculture Act, and National Irrigation Commission Act, among others [6].

67.6.3 Wastewater Reuse in Cyprus

The Mediterranean is characterized as a zone of changeable water availability, which is accentuated in the sunniest periods. This disparity is given due to the irregular precipitation patterns, growing irrigation water demands, elevated temperatures, and pressures of the growing tourist sector. In Cyprus, two sources of water supply are desalination and water reuse [11].

In this country, there are approximately 25 WWTPs running, providing services to the principal cities and municipalities, but there are approximately 175 small-sized WWTPs dealing with wastewater from hotels, military bases, and hospitals. In this country, there are plans of expansion of current WWTPs in order to cope with the regulations of the European Directive 91/271/EC [3].

According to Fatta et al. [3], in Cyprus an amount of 1.5 million m³/y reclaimed water is used for irrigating football fields, parks, hotels, and gardens; while 3.5 million m³/y reclaimed water is used for irrigating crops and the perspectives are to increase this proportion to up to 30 million m³/y reclaimed water for landscape and agricultural irrigation [3].

Thus far, no adverse environmental impact has been observed in Cyprus from the utilization of treated wastewater, since quite strict standards are established in this country [3]. Current water reuse regulations in Cyprus are even more conservative than the WHO guidelines. For water reutilization in urban areas with unlimited general access, the disinfection level is 50 FC/100 mL (80% per time, and a threshold of 100 FC/100 mL). For agricultural irrigation, the standard is set to 200 FC/100 mL up to a maximum of 1000 FC/100 mL, whereas for pastures and

industrial crops, the limit values are 1000 and 3000 FC/100 mL, for each respective case [2].

67.7 Summary and Conclusions

Coastal areas are greatly susceptible to water scarcity; thus, water reuse represents a vital opportunity to supply this precious resource for the different demanding sectors. We have seen that the main drivers to water reutilization are based on social pressures, political initiatives in the form of water management legislations, physical matters which include local hydrological issues, sanitation, and wastewater generation, and finally, as a matter of economical reasons. Since the reclaimed water has different utilization purposes and there have been local procedures that not always can be extrapolated to different conditions, there is no universal water reuse standard. Nevertheless, the WHO has developed and published a series of guidelines in order to support mainly decision makers for the safe usability of treated wastewater and to prevent any risks, particularly of microbiological, chemical, odorous, toxicity, eutrophication, or any other type. The state-of-the-art in water treatment for reuse contemplates a preliminary, primary, and secondary treatment. When the final users have strong water requirements, then advanced treatment procedures are employed, such as UV chlorination, nutrient removal, membrane filtration, and ion exchange, among several other advanced techniques.

We have seen that in Barbados, water reuse has been considered mainly for the touristic sector, especially for garden irrigations. Despite the reuse of treated wastewater, there are no clear policies that regulate its application. The main problems raised thus far refer to a lack of maintenance and potential disinfection risks. In Jamaica, biological water treatment has been successfully assessed, but desalination also has been introduced. In this country, there are specific regulations in terms of water reuse. In Cyprus, there is advanced implementation of water reuse techniques and very strict regulations for its safe application have been established.

References

1. Plan bleu pour la Méditerranée, Condom, N., Lefebvre, M., and Vandome, L. 2012. Treated wastewater reuse in the mediterranean: Lessons learned and tools for project development. *Plan Bleu*, Valbonne.
2. EPA—U.S. Environmental Protection Agency. 2004. *Guidelines for Water Reuse*. EPA—U.S. Environmental Protection Agency, Washington, DC, p. 136.
3. Fatta, D., Arslan Alaton, I., Gokcay, C., Rusan, M.M., Assobhei, O., Mountadar, M., and Papadopoulus, A. 2005. Wastewater reuse: Problems and challenges in cyprus, Turkey, Jordan and Morocco. *European Water*, 11(12), 63–69.
4. International Desalination & Water Reuse Quarterly industry website. 2010. Small Jamaican wind/solar desalination project funded. http://www.desalination.biz/news/news_story.asp?id=5586&channel=0. Accessed on November 28, 2013.
5. Jimenez, B. and Asano, T. (eds.), 2008. Water reclamation and reuse around the world. *Water Reuse: An International Survey of Current Practice Issues and Needs*. IWA Publishing, London, UK, p. 16.
6. Kolbusch, P. 2013. Nutrient management using wastewater and sludge—Jamaica's approach. *Global Conference on Land-Ocean Connection Building Bridges through Partnerships*. National Environment and Planning Agency, Jamaica.
7. Salgot, M. and Tapias, J.C. 2004. Non-conventional water resources in coastal areas: A review on the use of reclaimed water. *Geologica Acta*, 2(2), 121–133.
8. Sharif, M.A.F.M. 2009. Reclamation of wastewater and its use for crop production. PhD thesis. The University of the West Indies at St. Augustine, Trinidad and Tobago.
9. UNEP—United Nations Environment Program. 1997. *Source Book of Alternative Technologies for Freshwater Augmentation in Latin America and the Caribbean*. Unit Sustainable Development and Environment. General Secretariat, Organization of American States (OEA), Washington, DC, pp. 289–292.
10. Urkiaga, A. 2004. Best available technologies for water reuse and recycling. Needed steps to obtain the general implementation of water reuse. *Parque Tecnológico de Bizkaia*. http://technologies.ew.eea.europa.eu/technologies/resourc_mngt/water_use/Anaurkiaga.pdf. Accessed on November 29, 2013.
11. Zimmo, O. and Imseih, N. 2010. Overview of wastewater management practices in mediterranean countries. In: Barcelo, O. and Petrovic, M. (eds.), *Waste Water Treatment and Reuse in the Mediterranean Region, The Handbook of Environmental Chemistry* (2011), Springer-Verlag, Berlin, Heidelberg, vol. 14, pp. 155–181.

68

Water Reuse Sustainability in Cold Climate Regions

Shafi Noor Islam
*University of Brunei
Darussalam (UBD)*

Sandra Reinstädtler
*Brandenburg University of
Technology Cottbus-Senftenberg*

Saeid Eslamian
Isfahan University of Technology

PREFACE

Water is essential for life and total development of human societies in different parts of the world. Geographically, the surface land of the world is divided due to climatic conditions. This is the most critical and potential issue of population distribution, settlement and urbanization, and agricultural development. The civilization, urban development, and agricultural cropping pattern are totally dependent on naval communication and water availability. Soil and fertility development depend on water supply and quality of surface and groundwater. In the cold climate region, the last two decades have witnessed growing water stress, both in terms of water scarcity and quality deterioration. By 2020, water use and reuse is expected to increase by 40%, and 17% more water will be required for food production to meet the needs of the growing population. By 2025, 1.8 billion people will be living in regions with absolute water scarcity; about two out of three people in the world could be living under considerations of water stress.

Therefore, water reuse and proper management strategies are very potential factors for water management in the urban and rural regions. The primary evaluation indicates that for an increased utilization of received wastewater, cleaner arrangements are necessary. The setup of water reuse guidelines is needed. The modern engineering technology setup of water reuse guidelines is needed. The innovation of water reuse approach would be the best practice in the cold climate region toward protecting water resources management sustainability.

68.1 Introduction

Water is a critical resource for life and essential for economic success and improvement of national and regional socioeconomics [16–18,25]. However, its indiscriminate use may lead to the shortage of its precious asset [20]. The concept of water being a never-ending resource with a limitless renewable capacity belongs to the past. Water reuse is a rational practice that contributes to environmental protection [20]. In cold climate regions, there are plenty of water resources compared to other regions of the world, and water has long been considered an inexhaustible public commodity [5]. In general, about 73.3% of the Earth's water is in the oceans and bodies of surface water, the remaining 2.7% is freshwater, 2.1% is tied up in the polar ice caps and in glaciers, leaving only 0.6% to circulate [40]. Greenhouse and soil moisture constitutes 22.4% of the global freshwater, but two-thirds of the groundwater reserve lies below 800 m depth and is beyond human capacity to exploit [40]. The challenging fact is that due to the huge need of water for drinking, irrigation and industrial purposes in the last decades, water scarcity and quality deteriorations almost 70% of the population in cold climate regions are facing water stress issues today [5]. Water supply and sanitation, and management variables to measures of "ecosystems services" and "ecosystems health," which are in turn related to economic

and human health benefits [11,24]. The question for sustainable development is a key concept in the strongest driving force of the water sector [19]. Climatic condition is a potential factor of regional water use, reuse, and management issue. Cold and tropic climate regions are equally facing the problem of sustainable uses and management. The cold climatic regions are the most hostile regions where people are facing huge problems to maintain natural surface and groundwater management. The cold climate region, in general, is the land of water [1] but water is one of the most critical problems in these regions, especially the scarcity of surface freshwater has created serious environmental problems in the cold climate regions. The location of the landward boundary of the coastal zone is a function of three basic geophysical processes: tidal fluctuations, salinity, and risk for cold storm surges. These processes affect the coastal zone of cold climatic regions, which cover an area almost 10% of the world (Figure 68.1). The cold climate coastal zone is the area on both sides of the actual land-sea interface, where the influences of land and water on each side are still determining factors climatically, physiographically, and ecologically [14]. The cold climate water resources play an important role in the socioeconomic development of all coastal regions in the cold climate regions as well as in Sweden and surrounding areas [21]. The coastal region is composed of the land and the sea including estuaries

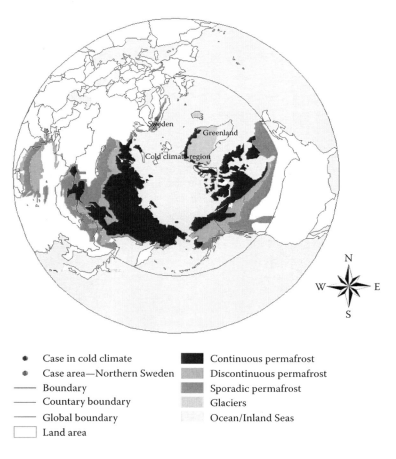

●	Case in cold climate	■	Continuous permafrost
◑	Case area—Northern Sweden	▨	Discontinuous permafrost
—	Boundary	▬	Sporadic permafrost
—	Countary boundary	░	Glaciers
—	Global boundary		Ocean/Inland Seas
▢	Land area		

FIGURE 68.1 Geographical locations and permafrost areas in cold climate regions.

and islands adjacent to the land water interfaces and the coast can multifariously divide into different distinctive features in terms of physiography, seasonality, and use patterns [32,33]. It is at the coast where the heavily sediment-laden river water with very little salinity meets the saline seawater. This mixing creates a unique ecosystem, which has given rise to the species of flora and fauna that exhibit their presence here. Among these, the most important are the northern Nordic region with its unique cold climatic landscape and vegetation and the wildlife, the fisheries dominated by economically important species [21,32]. Cold climatic freshwater availability and supply to the coastal regions and surrounding is playing a potential role to make a balance of cold climatic ecosystems. Therefore, sustainable cold climate water resources management planning is essential for the ecosystem protection and crop production as well as cold climatic community development and their livelihoods' sustainability in the regions. A comprehensive analysis of the various factors leading to cold climate surface, groundwater, and water reuse management sustainability is proposed in this study.

68.2 The Objectives of the Study

The specific objectives of this study are as follows:

- The objective of this is to understand about unique characteristics of the cold climatic region of the world and their potentiality of coastal socioeconomy and regional development and the role of coastal water resource reuse.
- Study seeks the inadequate cold-water reuse and management strategies.
- By introducing the potentiality of GIS, remote sensing (RS) application, and making some practical recommendations for conservation and a better management plan, we can lay the path for future development of ecosystem goods and its services in the cold climate region for livelihood and habitat sustainability.
- Make some applied recommendations for future sustainable water reuse and management policies for cold climate regions and protection and management in the northern hemisphere.

68.3 Geography and Physical Characteristics of Cold Climate Regions

Cold climate regions have been recognized based on their geographical, geological, hydrological, and meteorological characteristic conditions of the regions. In general, it is where the temperature is very cold and human life has to face hardship in the environment. According to Smith, cold climate regions are recognized where the mean temperatures for one month of the year are below +1°C and snow may stay on the ground for a period. The general temperature ranges from as low as −50°C to as high as +40°C, and a range of 90°C. The climate is harsh such as cold snow, wind, and rain resulting in difficulties in

water management and runoff conditions. The potential political land territories of the cold climate regions are Canada, north United States, Greenland, Iceland, north of Norway, Sweden, Finland, and the most northern region of Russia in the North Pole region. On the other hand, southern Argentina, the south part of Chile, southeast corner of Australia, and New Zealand are the territorial part of the cold climate regions (Figure 68.1). The above-mentioned political territorial boundaries are the permanent cold climate regions. In addition, the seasonal cold climate regions are extending gradually from the northern pole to the southern direction in Europe. On the other hand, the permafrost regions are one of the most potential characteristics of cold climate regions and the permafrost dominated regions are also considered cold climate regions (Figure 68.1).

It has been estimated that within the territorial boundaries of the cold climate regions, more than 1 billion people are settled and surviving within the cold climate conditions.

68.3.1 Characteristics and Location of Permafrost in the Cold Climates

Permafrost is a product of cold climates and an important driver of polar desert, arctic tundra, subarctic taiga, and boreal ecosystems and its landscape balance in general: thermal characteristics of the ground directly control or indirectly influence local hydrology, patterns of vegetation as well as landscape, extensive land use, wildlife communities, and ecosystem dynamics [6].

For defining the term of permafrost, it can be stated that permafrost consists of soil, sediment and rock. The temperature conditions are constant for two or more years, staying at or below the freezing point of water, although ice is not always present. Permafrost can be divided in continuous, discontinuous, or sporadic permafrost districts (Figure 68.1). Continuous permafrost is rich of ice and contains more than 50% frozen water in the soil. Some discontinuous or sporadic permafrost grounds are "ice-poor," contain little or no frozen water, and may be covered by a thin or no permanent active ice layer that seasonally thaws during the summer times [30]. In the cold climate regions, the precipitation is generally low. As soil and water remains frozen during most of the year, vegetation is therefore in correlation to the given hydrological circumstances. Often, moisture is limited in the form of deserted areas, tundra, taiga, or boreal landscape parts [30].

Thus far, permafrost soils exist in cold climate regions, which are characterized after the Köppen-Geiger Climate Classification as having a polar climate and the temperature classification of the polar tundra. Permafrost landscapes are further characterized due to extreme temperatures by cryoturbation, which means to have frost churning of soils by freezing and thawing. Cryoturbation creates patterned geomorphic features such as frost heaves, earth hummocks, ice-wedge polygons, cryoturbation steps, non-sorted circles, and non-sorted stripes. In addition, beaded streams are features within cryoturbation, which is having beads along the stream as pools at the "corners" of polygons. Palsas, which are mounds of peat formed by ice lenses and sheet wash rills, are creating the characteristic qualities of cryoturbation within permafrost landscapes. Sheet wash rills are developed while water drains off permafrost ground in small parallel channels. In addition, gelifluction lobes and soil creeps downslope over permafrost depict the appearance of cryoturbation and permafrost [30].

The northern Canada, Alaska, Greenland, Iceland, and northern Scandinavian areas like subarctic peatlands in northern Sweden, partially also subarctic fens in Finland and Norway are bewaring these characteristics. In the northern parts of Russia, permafrost is intensively in place within regions in Siberia and also in North Korea, northern parts of Mongolia, and in the mountainous regions of Nepal and Bhutan as well as the Indian and Chinese parts of the Himalayas (Figure 68.1) [28].

Northern Canada consists of all three possible permafrost conditions of continuous, discontinuous, or sporadic nature, mainly of continuous permafrost. Alaska is divided in a balanced way throughout all of the three sorts of permafrost soils. Greenland mainly consists of continuous permafrost as well as glacial surfaces. Northern parts of European Norway, Finland, and mainly of Sweden have major sporadic permafrost soils. The northern Russian federation along the Arctic Circle (66°33′N) and next to the Arctic Ocean have the highest amounts of permafrost soils in its spatial distribution.

On one hand, a sporadic permafrost condition exists in the southeastern parts (near Finland and Estonia) and on the other hand, next to the continuous permafrost of the northern Arctic Circle is bewaring continuous permafrost. The Russian archipelago Novaya Zemlya is bewaring half of the archipelago with permanent permafrost soils. The uninhabited group of islands of Zemlya Frantsa-Iosifa as well as the island group of Severnaya Zemlya in the Russian high Arctic consists of mainly glaciers and continuous permafrost soils in the tundra and desert vegetative areas. The permafrost areas in North Korea and the northern parts of Mongolia are mixed up by sharing continuous, discontinuous, and sporadic areas of permafrost soil. The mountainous Himalayan regions of Nepal, Bhutan, India, and China inhabit still discontinuous and sporadic permafrost soils (Figure 68.1).

68.3.2 Permafrost and Land Change in Times of Global Warming

Global warming affects nearly all aspects of our physical, social, ecological, economic, and cultural environment and as a consequence will induce challenges for human life, security, infrastructure, ecosystems (services), the worldwide face of our landscapes, and all sorts of spatial planning activities [35]. Climate change can ecologically and in its physical attributes degrade permafrost by affecting its extent and the degree of seasonal thawing. The arctic, subarctic, and boreal regions are already facing drastic changes in ecosystem dynamics and daily lives and livelihood as well as land use manners for the inhabitants are obviously being affected during the last decades [7]. Climate modeling and scenarios are signaling that arctic and subarctic regions are already facing significant changes in

climate [37]. With warmer temperatures especially during the twenty-first century, precipitation and, in particular, winter precipitation is also expected to increase. These changes are likely to have large impacts on the arctic and subarctic environment, ecosystems, and parts of landscapes. Up to that, permafrost is considered a fragile landform characteristic, if it is within a few degrees of thawing. With the example of Denali National Park and its permafrost soils in Alaska, some of the southernmost continuous permafrost in Alaska is represented [30]. These permafrost soils are monitored for research purposes in dealing with a better understanding of the extent of impacts of climate change related temperature rise: in this state of monitoring it was stated that permafrost soils are very susceptible to climate change. Further on, significant landscape change is likely to occur with continued climate warming. So within these monitoring activities in Denali National Park as part of the Central Alaska Network's Inventory and Monitoring Program, it is obvious with its bore hole data that permafrost is one of the "vital signs" of ecosystem and landscape health [30].

The way climate change is continuously degrading permafrost ground can be visualized within the monitoring processes while collecting data about thawing landforms: in general, permafrost soils are displaying different sorts of landform characteristics (and distributions) because of the specific subsurface hydrological circumstances of ice changing to mud slurry and ground subsides. The naturally existing thermokarst terrains are already extending because of hydrological changes and thawing of continuous subsurface ice layers. Increased thermokarst landforms of channels, pits, troughs, potholes, ponds, and leaning trees are occurring.

Landscape-scale interactions and changes in general are typical for climate change impacts [35] as well as plant distribution and productivity are remolded by modifying ground surface (e.g., by increased gelifluction lobes or sheet-wash rills) and the hydrological system into thermokarst terrains [30]. The initialized landscape changes because of thawing permafrost is a continuing and increasing demand. Landscape change means in correlation to climate change and global warming effects to generate unstable development conditions for landscapes in the form of extreme events or in the form of a steady change in micro-macro climatic qualities [34] and so also effecting land use structures and habits.

The important question for water, land use, and landscape management should be in how far climate change is affecting permafrost and the overall system dynamics of water, ecosystem, and landscape balance influencing the sort and intensity of management options in the field of water reuse systems and newly to be invented retention systems for thawing water. Impacts despite manufactured climate change ecosystem responses and permafrost thawing, also non-climatic anthropogenic stressors, changes in land use such as military purposes with extension of fire threats by military training activities, technical installations, and the presence of invasive species are to be included in management plans for adapting and mitigating climate change. Permafrost's reach extends even further than surface hydrology and features. Its distribution is the primary influence governing groundwater flow, recharge, discharge, and hence ecosystem state, landscape evolution as well as infrastructure sustainability. Management for water cycles, water reuse, water quality, and amount for a moderate land use and landscape change has to be developed in an integrative overall procedure.

68.3.3 Data and Methodology

The present study was carried out based on secondary and primary data sources. The primary data collected in 2013 Participatory Rural Appraisal practices were arranged with the local graduate students of Environmental and Applied Science at Halmstad University in Sweden in the coastal region. The information was collected from the northern climatic region of Sweden and the Tromsö city region of Norway. Besides these, a lot of various reports and published articles in journals and conference proceedings have been used for this study [8]. Published materials, reports, and journals were collected from different government and nongovernmental organizations, and university libraries in Germany, Sweden, Denmark, Poland, and France. For secondary data collection, reports on coastal water resource reuse status and management have been used very openly. The research reports from Halmstad Universty, Sweden, and University of Trömso, Norway, and international organizational reports were also used for this study. The collected data were reconstructed, analyzed, and visualized through MS EXCEL, VISIO 32, and ArcGIS 10.1 tools.

68.4 Water Reuse in Cold Climate Regions

Water crisis and scarcity is faced all around the world. In cold climate regions, the surface freshwater crisis is more harmful for life and the natural environment. In general, scarcity occurs when the threat is less than 200 m³ of freshwater available per person per year [20]. Water scarcity is detected in different parts of the world [26] as well as in the cold climate regions such as Sweden, Norway, Finland, Siberia, and the northern parst of Russia, Canada, Greenland, Iceland, southern Argentina, south of Chile, and southeast corner of Australia and New Zealand, the northern part of the United States, and the rest of the Northern Hemisphere regions where cold climate characteristics are seen as in Figure 68.1 [9,26]. According to the demand and necessity of the local and regional level, the specific methods and techniques are developed to reuse the surface water and wastewater reuse in agriculture, irrigation, and industrial purposes. Water reuse or wastewater reuse is a rational practice that contributes to ecosystems and environmental protection in the cold climate regions [20]. Wastewater reuse technique is being implemented to reuse water consumption and the high costs of water treatment. In general, more than an environmental measure, water reuse has an economic impact. Water reuse or wastewater reuse must be a safe, environmentally conscious, responsible, and accountable process, which meets the recycling requirements

and the current legislation and directives of the WHO because irresponsible wastewater reuse may cause various problems in agricultural crops production, industrial products quality, and protection of local and regional ecology of the cold climate regions (Figure 68.1) [20].

The specific sources of surface water in the cold climate regions are included the following sources:

- Rivers
- Streams
- Lakes
- Impoundments (manmade lakes made by demanding, a river or stream)
- Very shallow wells that receive input via precipitation
- Springs affected by precipitation (flow or quantity directly dependent upon precipitation)
- Rain catchments (drainage basins)
- Tundra ponds or muskegs (peat bogs) [38]

Surface water has advantages as a source of potable water. Surface water sources are usually easy to locate. Unlike groundwater, finding surface water does not require a geologist or hydrologist. Normally, surface water is not trained with minerals precipitated from the Earth's strata. Ease of discovery aside, surface water also presents some disadvantages. Surface water sources are easily contaminated with microorganisms that can cause waterborne diseases and are polluted by chemicals that enter from surrounding runoff and upstream discharges. In general, the surface water is highly variable for two main reasons: (1) human interference and (2) natural conditions. In some cases, surface water quickly runs off land surfaces [38].

The maximum contaminated level (MCL) is defined as the maximum permissible level of a contaminant in water at the free-flowing outlet of the ultimate user of a public water system, except in the case of turbidity, where the maximum permissible level is measured at the point of entry to the distribution system. The MCL for a drinking water standard requires a balance between public health benefits and what is technologically and economically feasible. The development of water quality standards or MCLs involves an intensive technological evaluation that includes assessments of

- Occurrences in the environment
- Human exposure in specific and general populations
- Adverse health effects
- Risks to the population
- Methods of decision
- Chemical transformations of the contaminant in drinking water
- Treatment technologies and coasts [36]

In history, it has been treated that the cold climatic regions are the less populated zones in the world. There the household demand for surface and groundwater use was not high in the sixteenth to nineteenth centuries. However, gradually population has increased and economic activities and agricultural activities have been increased dramatically. The global change and climatic condition has risen as a new threat for the socio-economic and industrial development and urbanization. The cold climate regions are less population dense and land use practices. At present, global warming, the growth of population, and urbanization are simultaneously are increasing. Therefore, the demand for freshwater is also increasing all over the world as well as in cold climate regions. According to the demand for freshness and quality, natural water is not sufficient. Therefore, water reuse is necessary for economic, agriculture production, and human wellbeing in cold climate regions and other parts of the world.

68.4.1 Historical Development of Water Reuse and Technology

Indications that wastewater was applied as an alternative water source for agricultural irrigation extended back approximately 5000 years [2]. During the nineteenth century, the introduction of large-scale wastewater carriage systems for discharge into surface waters led to the inadvertent use of sewage and other effluents as components of potable water supplies. Unplanned reuse and development of wastewater treatment plants were the reasons of cholera and typhoid during the 1840s–1850s. This is why proper wastewater reuse and treatment plant is now a question for introducing sophisticated engineering solutions. As global warming, population growth, urbanization, and industrial growth in different regions of the world as well as in the cold climate regions, they are facing the challenge of clean water supply for drinking, agriculture, and industrial uses. Due to the necessity of human well-being in cold climate regions, England progressively introduced water filtration technology during the 1850s and 1860s [2]. In the last quarter of the twentieth century, the promotion of wastewater reuse as a means of supplementing water resources has been recognized in the United States as well as the European Union. The European Communities Commission Directive (91/271/EEC) declared, "Treated wastewater shall be reused whenever appropriate." Disposal routes shall minimize the adverse effects on the environment [13]. Wastewater reuse, when appropriately applied, is considered an example of environmental scientific technology (EST) applications. According to the demand and criteria fulfillment of Agenda 21, technologies are indicated to achieve the following objectives: protect the environment, less polluting, use all resources in a more sustainable manner, recycle more of their wastes and products, and handle residual wastes in a more acceptable manner than the technologies for which they are substitutes. The use of ESTs plays a key role in facilitating freshwater protection and integrated water resource development and management, as recognized in Agenda 21.

According to the historical water cycle management system, the flowing (Figure 68.2) is applicable in the cold climate regions. Traditionally this water cycle system (Figure 68.2) is implemented in the regions, besides the modern water reuse systems are developing according to the demand and necessity to the communities. Water reuses and cycle management systems in cold climate regions are similar to the dry or semi-dry regions of the world. The following cycle scholar model of water is recognized in urban areas and rural regions (Figure 68.2).

68.4.2 Present Demand, Supply, and Management Planning

Most surface water originates directly from precipitation rainfall or snow. Only in the United States' mainland does rainfall average about 4250 billion gallons per day [2]. Of this massive amount, about 60% returns to the atmosphere through evaporation directly from the surface of lakes and rivers and transpiration from plants [9]. This leaves about 1250 billion gallons per day [2,12] to flow across or through the ground to return to the sea [38]. Approximately 326 million cubic miles of water comprise Earth's entire water supply, whereas only 3% of this massive amount of water is fresh, and most of that minute percentage

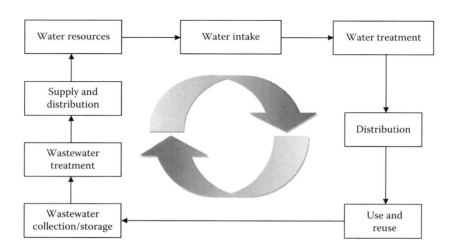

FIGURE 68.2 Close urban and rural water cycle management in cold climate. (Adapted from Van Dijk, M.P. 2007. *Constitution to an International Seminar on Sustainable Urbanization in Tripoli*, Libya, Hotel Bab Africa, June 30–July 1.)

TABLE 68.1 World Water Distribution

Location	Percent of Total (%)
Land area	2.8
Freshwater lake	0.009
Saline water lake and inland seas	0.008
Rivers (average instantaneous volume)	0.0001
Soil moisture	0.005
Groundwater (above depth of 4000 m)	0.61
Icecaps and glaciers	2.14
Atmosphere (water vapor)	0.001
Oceans	97.3
Total all locations (rounded)	100

Source: Adapted from Spellman, F.R. and Drinan, J.E. 2012. *The Drinking Water Handbook*, 2nd Edition, CRC Press, Boca Raton, FL, pp. 1–347.

of freshwater is located in polar ice caps and glaciers. The rest is held in lakes, in flows through soil, and in river and stream systems. Only 0.027% of Earth's freshwater is available for human consumption. For distribution percentages of the Earth's water supply, see Table 68.1.

Although municipal usage of water is only a small fraction of this great volume, the per capita consumption of water in the United States is rather high—as much as 150 gallons per person per day—probably the reason is that public water is relatively inexpensive here. The present population in the cold climate regions is estimated at more than 1 billion, this figure will gradually be increased and the water demand will be increased in cold climate regions. Therefore, proper water use and reuse and sustainable technologies are essential to fulfill the continuously increasing demand of water in cold climate regions. Considering the present scenarios of water demand and necessity in the communities of cold climate regions, the wastewater reuse model has been developed (Figure 68.3) for much public awareness and implementation should start at the microscale level in cold climate regions.

68.4.3 Water Reuse and Pricing

In 2003 Novib, the Dutch branch of Oxfam International, asked what total volume of water is needed to produce a cup of coffee, and similarly for a cup of tea. The water uses during the whole production chain, it was clear from the beginning that the agricultural stages, the stage in which the coffee and tea plants grow and deliver their products, would turn out to be the most water consuming [22]. The average production rate of a cup of coffee is 140 L of water, mostly rainwater for growth of the coffee plant. The real-water content of products is generally negligible compared with the virtual-water content. The global average virtual-water content of wheat, for instance, is 13,000 m³/ton [26], while the real-water content is less than 1 m³/ton [22]. Considering the international trade flows between all major countries of the world and looking at the major agricultural products being treated (285 crop products and 123 livestock products), the actual calculated water use for producing export products amounts to 1250 billion m³/

yr. If the imported countries would like to produce the imported products domestically, they would require a total of 1600 billion m³/yr. This means that the global water savings by trade in agricultural products is estimated at 350 billion m³/yr [22].

In general, there is a need to arrive at a global agreement on water pricing structures that cover the full cost of water use, including investment costs, operational and maintenance costs, a water scarcity rent, and the cost of negative external impacts of water use. Without an international treaty on proper water pricing, it is unlikely that a globally efficient pattern of water use will ever be achieved. The Dublin conference in 1992 [23] was the first time this important issue was raised to the international communities. A global ministerial forum to pursue agreements on this does exist in the regular World Water Forums like Morocco 1997, The Hague 2000, Japan 2003, and Mexico 2006, but these forums have not been used to take up the challenge of making international agreements on the implementation of the principle that water should be considered as a scarce economic good [22].

68.5 Results and Discussion

The potential results of water reuse in contemporary times are raised in cold climate regions, which are very important to solve for the better management of water use and reuse in cold climate regions. Some are discussed in the following subsections.

68.5.1 Water Reuse and Land Use Forms in Cold Climate Regions

Land-use planning is perhaps the most fundamental tool for mainstreaming disasters risk reduction into urban and rural development processes in cold, dry, and semi-dry regions [47]. Indeed, from the twentieth century onward, there was an increasing incidence of problems associated with the management of precipitation across most developed parts even in the cold climate regions of the world. Moreover, in the face of both a changing climate and a more urbanized environment, planning systems were rightly accused of being slow to react to and manage potential problems. Perhaps more seriously it was also suggested that neither the power nor the knowledge existed in order to enable planners to provide the expected level of landscape and land use management and protection [47].

As the land use planning system is concerned with balancing economic, social, and environmental issues, in practice some of the negative effects of runoff should be regulated by controls, either influencing the water and sewerage companies or the built environment toward protection of the traditional land use pattern in the cold climatic countries. In cold climates, the water management approaches are methods of concentrations, conversion, and relocation of waste, such as physiology, chemical, and biological treatment, incinerations, or disposal. The water treatment and quality in development is working as diverse for changing use management approach. The management and agricultural land use and wastewater treatment planning is related to regional economic conditions [27].

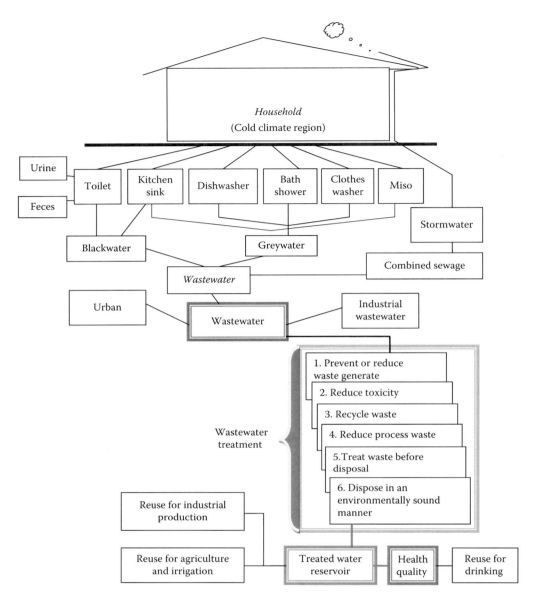

FIGURE 68.3 Model of wastewater treatment plant on a microscale level.

68.5.2 The Importance of Water Reuse for Agriculture and Food Security

Water availability is one of the driving forces of a region for agricultural irrigation, protection of ecosystems, and food production. The availability of natural resources or availability of water resources does not matter. The cold climate regions are recognized as very harsh and sensitive areas in the world [31], where temperature and water resources are shaped in different forms due to variables of seasonality. Therefore, proper water use is a matter of community lifestyle sustainability, agricultural productivity, and industrial production. Quality is dependent on water supply and availability. The landscapes and land use patterns are also dependent on the water availability, supply, and uses in the cold climate regions as well as other regions of the world. In general, the major portions of cold climate regions are

dominated by permafrost soil, which is 1–2 feet layer of ice on soil which is a massive threat for agricultural productivity and management of regional ecosystems and their services.

Permafrost is the frozen ground located 1–2 ft below the surface in cold regions. As permafrost thaws and the soil sinks, structures built on or within the soil are damaged. Although most Alaskans live in permafrost-free areas, an estimated 100,000 Alaskans (about 14% of the population) live in areas sensitive to permafrost degradation [43]. As explained next, the impacts of melting permafrost on transportation, forests, ecosystems, and the economy could have widespread implications for Alaskans.Over the past 50 years, thawing permafrost and increased evaporation have caused a substantial decline in the area of Alaska's closed-basin lakes (lakes without stream inputs and outputs). These surface waters and wetlands provide breeding habitat for millions of waterfowl and shorebirds that winter in the lower 48 states [45]. These wetland ecosystems and the wildlife resources are important to Alaska natives who hunt and fish for food [43,44].

As the climate warms, shrubs are invading the tundra. In some areas, the shrubs are replacing lichens and other tundra vegetation. Lichens are an important winter food source for caribou, and the loss of lichens can lead to declines in the growth and abundance of these animals. Caribou in turn are a critical food source for predators such as bears and wolves, as well as for Alaska natives [45]. Higher temperatures and less summer moisture increase the risks of drought, wildfire, and insect infestation. Alaska's boreal spruce forest declined substantially in recent decades from both fire and insect damage. By mid-century, the average area burned by wildfire each year is likely to double [44].

68.5.3 Legally Binding Declarations and Its Development for Wastewater Reuse

Multifarious purposes in fields of agricultural irrigation, industrial processes, and groundwater recharge are benefited throughout wastewater reuse [15,41]. In the best case, standards, guidelines, criteria, and requirements in general, legal frameworks, and directives are guiding into proper water reuse and protective measurement and management for human and environmental health. Water reuse needs to be monitored for ensuring consistent wastewater reuse application and quality [15,41].

Table 68.2 illustrates different products and water content as a common. But this water content ratio could be changed in different climatic condition and geographical areas. Sustainable and successful wastewater reuse applications are mainly characterized through the requirements of monitoring, evaluating, and

TABLE 68.2 Global Average Virtual Water Content of Selected Products, Per Unit of Product

Product	Virtual Water Content (L)
1 sheet of A4 paper (80 g/m²)	10
1 tomato (70 g)	13
1 potato (100 g)	25
1 slice of bread (30 g)	40
1 orange (100 g)	50
1 apple (100 g)	70
1 glass of beer (250 mL)	75
1 slice of bread (30 g) with cheese (10 g)	90
1 glass of wine (125 mL)	120
1 egg (40 g)	135
1 glass of orange juice (200 mL)	170
1 bag of potato crisps (200 g)	185
1 glass of apple juice (200 mL)	190
1 glass of milk (200 mL)	200
1 hamburger (150 g)	2400
1 pair of shoes (bovine leather)	8000

Source: Adapted from Hoekstra, A.Y. and Chapagain, A.K. 2008. *Globalization of Water—Sharing the Planet's Freshwater Resources.* Blackwell Publishing, Malden, pp. 1–208.

minimizing potential public health risks throughout wastewater reuse and specific water reuse applications for water quality objectives. Enacting the wastewater reuse guidelines in 1918 in California [49], eventually the first standards for the above-mentioned requirements were prepared. The next important historical step for water reclamation and reuse policy was the Water Act, which became law in 1956 [3,4].

In the United States, another milestone event was the passage of the Federal Water Pollution Control Act of 1972 (PL 92–500) "to restore and maintain the chemical, physical and biological integrity of the nation's waters" with the ultimate goal of zero discharge of pollutants into navigable, fishable, and swimmable waters.

To illustrate alternative regulatory practices governing the use of reclaimed wastewater for irrigation, the major microbiological quality guidelines by the WHO [48] and the state of California's Wastewater Reclamation Criteria from 1978 (State of California, 1978) are to be mentioned.

68.6 Water Reuse Sustainability in Cold Climate Regions

The climatic and geosphere condition, the uncontrolled disposal to the environment of municipal, industrial, and agricultural liquid, solid, and gaseous wastes constitute one of the most serious threats to the sustainability of human civilization by contaminating the water, land, and air through global warming. With continuous increasing population and economic growth, industrial development, treatment, and safe disposal of wastewater are essential to preserve public health and reduce intolerable levels of environmental degradation in the cold climate regions [41]. In addition, adequate wastewater management is also required for preventing contamination of water bodies for preserving sources of clean water. Effective wastewater management planning is well established in developed cold regions but is still limited and there are some limitations to water reuses. Now that the requirements for a sustainable wastewater treatment system have been presented, there are several options one can choose from in order to find the most appropriate technology for a particular region. This chapter will discuss a model (Figure 68.3) of microscale level water reuse treatment plant at the community level for cold climate regions.

A plan for a community-scale integrated wastewater treatment facility is proposed for the cold climate regions for water reuse sustainability. Wastewater could be used in the waste treatment plant where the facilities would use biogas produced from the anaerobic digestion of organic waste to generate electricity, heat, light, and carbon dioxide for enhanced plant growth in a controlled environment [10]. In this study, the proposed model (Figure 68.3) could be a popular model that could be used at the community level, as it has been designed based on the raw materials (wastewater) source at the household level.

The importance of wastewater use and reuse in the cold climate regions is essential to demonstrate this potential message in the community level so that all level of inhabitants could be involved in such environmental programs for the better management of their own regional water use and reuse activities. This type of micro-scale level model implementation would be more popular and active at the community level when the policymakers and environmental and climatic scientists make plans to involve the community and stakeholders in any kind of environmental program in cold climate regions [37,45].

Figure 68.3 demonstrates a model flowchart of a wastewater reuse treatment plant at the household level in the community of cold climate regions [39]. The functionality, cost-effectiveness, and easy technology could be used at this model implementation. The effective treatment of wastewater to meet water quality objectives for water reuse applications and to protect public health is a critical element of water reuse systems [39]. The wastewater treatment consists of a combination of physical, chemical, and biological processes and operations to remove solids, organic matter, pathogens, metals, and nutrients from wastewater. In order to increase treatment level, the different degrees of treatment are preliminary, primary, secondary, tertiary, and advanced treatment, which has been set up in the model. The household wastewater, industrial wastewater, and urban wastewater could be stored in micro-scale level and could be treated within this model in six different steps. The first step is the function to reduce or prevent waste to generate; step two is reduce waste toxicity; step three is recycle for waste; step four is reduce the process of waste; step five is treat waste before disposal, and step six is allocated to dispose in an environmentally friendly manner of wastewater treatment. After this, wastewater could be treated as clean water for use in different purposes such as agricultural irrigation, industrial production, and gardening, and other portions of treated wastewater could be used for drinking before monitor health quality of drinking water. This is the micro-scale level model of wastewater treatment plant that could be implemented in the community household level and this type of environmental activity could ensure water reuse sustainability in the community level in cold climate regions.

68.7 Summary and Conclusions

Water is a driving force for community social and economic development all over the world. The world geographical features do not consider the matter of water distribution, but climate is a matter of productivity and development. The cold climate region is an important factor for quality water availability and distribution and supply management and development factors. As it has been analyzed in this study, water resources are the driving forces for the social well-being, economic improvement, industrial production, and agricultural crops production in cold climate regions.

They have the ability to focus tremendous energy and to generate significant creative and economic betterment. In general, the water resources are degrading its quality and quantities due to economic activities, agricultural irrigation, and natural calamities. The hydrological cycle of the cold climate regions are losing the balance capacity for so many reasons such as

permafrost, snow, cold wind, saline water, and natural calamities. The cold climate regions need an adequate interdisciplinary policy framework as a whole, as the regions are very broad and vast areas. Therefore, national and regional water directives or frameworks perhaps would be activated for the whole national territories within the cold climate regions. The methods for the generation of transformation of engineering knowledge and innovation should be disseminated and implemented within the common strategies of international platform to achieve and maintain water resources, agro-farming system development, fishing development, and in general use and reuse of water resources in the cold climate regions.

The following recommendations could be followed in implementation and management stages.

- To build up people's awareness, there will be arranged a yearly local cultural festival on international water day in different parts of the cold climate regions as well as in areas in the cold climate regions such as in Sweden.
- Capacity-building training for the stakeholders and water resource users could be arranged through this local-based training and innovation for wastewater use and reuse tools development process in the cold climate regions where the national government can take proper initiatives for the territorial water resources management initiatives.
- The requirement to ensure the provision of adequate institutional capacity for policy development, delivery, and monitoring; the importance of considering water management infrastructure rather than nature reserves and the need to consider the wise use of permafrost areas within the national territories of cold climate regions.
- Traditionally, water directive, frameworks, and spatial plans for area based and structured on administrative boundaries. This approach fails to address or in many cases even recognize that these boundaries are not functional or commensurate with the environment or boundaries required for the protection of natural water sources for future use and reuse within the national and international directive and framework.

For time series analysis, the satellite and remote sensing imageries could be used to compare the historical changes of agricultural cropping systems, land use pattern, shrinking trends of freshwater sources, and present scenarios measures. As a whole, the developed model of water reuse treatment plant for the microscale level (Figure 68.3) could be implement at the local level in cold climate regions so that the community will be involved in the water reuse treatment process as well as they will be more aware for the future protection of water resources in the cold climate regions.

References

1. ACIA (Arctic Climate Impact Assessment). 2004. *Impacts of a Warming Arctic: Arctic Climate Impact Assessment. Arctic Climate Impact Assessment.* Cambridge University Press, Cambridge, UK, pp. 1–140.

2. Angelakis, A.N. and Spyridakis, S.N. 1996. The status of water resources in Minoan times a preliminary study. In: A.N. Angelakis and A.S. Issar (eds.), *Diachronic Climatic Impacts on Water Resources with Emphasis on Mediterranean Region.* Springer Verlag, Heidelberg, Germany, pp. 161–191.

3. Asano, T. and Levine, A.D. 1998. Wastewater reclamation, recycling, and reuse: An introduction. In: T. Asano (ed.), *Wastewater Reclamation and Reuse.* Technomic Publishing Company, Lancaster, PA, pp. 1–56.

4. Asano, T. (ed.) 1998. *Wastewater Reclamation and Reuse,* Technomic Publishing Company, Lancaster, PA.

5. Bixio, D., Thoeye, C., Koning, J.D., Joksimovic, D., Savic, D., Wintgens, T., and Melin, T. 2006. Water reuse in Europe. *Desalination,* 187, 89–101.

6. Callaghan, T.V. and Björn, L.O. 2004. Effects of changes in climate on landscape and regional processes, and feedbacks to the climate system. November 2004, Royal Swedish Academy of Sciences. *Ambio,* 33(7), 436–447.

7. Callaghan, T.V., Johansson, M. and Brown, R.D. 2011. Multiple effects of changes in arctic snow cover. *Ambio,* 40, 32–45.

8. Cook, H.F. 1998. *The Protection and Conservation of Water Resources—A British Perspective.* John Wiley & Sons, New York, pp. 1–328.

9. Cooley, H. 2009. Water management in a changing climate. In: P.H. Gleick, H. Cooley, M. Cohen, M. Marikawa, J. Morrison, and M. Palaniappan (Authors), *The World's Water 2008-2009. The Biennial Report on Freshwater Resources.* Island Press, Washington, DC, pp. 39–56.

10. Crosby, R.L.J. 1997. A plan for community greenhouse, and waste treatment facility for cold climate regions. In: C. Etnier and B. Guterstam, (eds.), *Ecological Engineering Wastewater Treatment.* Lewis Publishers, New York, pp. 145–152.

11. Costanza, R.L.J., Norton, B.G. and Halskell, B.D. (eds.) 1992. *Ecosystem Health: New Goals for Environmental Management.* Island Press, Washington, DC.

12. Doran, J.W. and Zeiss, M.R. 1997. Soil health and Sustainability: Managing the biotic component of soil quality. *Applied Soil Ecology,* 15(2000), 3–11.

13. ECCD. 1991. Council directive 91/676/EEC of 12 December 1991 concerning the protection of waters against pollution caused by nitrates from agricultural sources. *Official Journal L,* 375(31), 1–8.

14. Fedra, K. and Feoli, E. 1998. GIS technology and spatial analysis in coastal zone management. *EEZ Technology,* 3rd edition, pp. 171–179.

15. GEC, ENEP, 2005. Water and wastewater reuse. An environmentally sound approach for sustainable urban water management. *UN Env. Prog.,* Osaka, Japan, pp. 1–50.

16. Gleick, P. 1998. *The World's Water.* Island Press, Washington, USA.

17. Gleick, P.H., Cooley, H., Cohen, M., Marikawa, M., Morrison, J., and Palaniappan, M. 2009. *The World's Water (2008-2009), The Biennial Report on Freshwater Reuses.* Island Press, Washington, DC, pp. 39–56.

18. Grigg, N.S. 1998. *Water Resources Management Principles, Regulation and Cases.* Mcgraw Hill, Washington, pp. 8–14.

19. Grigg, N.S. 2008. Total water management: Practices for a sustainable future. *Water Alternatives,* Vol. 2, American Water Works Association, Denver, pp. 289–290.

20. Gutterres, M. and Aquim, P.M. 2013. Wastewater reuse focused on industrial applications. In: S.K. Sharma and R. Singh (eds.), *Wastewater Reuse and Management.* Springer Science, pp. 127–140.

21. Hidayati, D. 2000. Coastal management in ASEAN countries. *The Struggle to Achieve Sustainable Coastal Development.* UN University Press, Tokyo, pp. 1–74.

22. Hoekstra, A.Y. and Chapagain, A.K. 2008. *Globalization of Water—Sharing the Planet's Freshwater Resources.* Blackwell Publishing, Malden, pp. 1–208.

23. ICWE (International Conference on Water and Environment). 1992. *The Dublin Statement on Water and Sustainable Development, International Conference on Water and Environment,* January 26–31, 1992, Dublin, Ireland.

24. Islam, S.N. 2007. Salinity instrusion due to fresh water scarcity in the Ganges catchment: A challenge for urban drinking water and mangrove wetland ecosystem in the Sundarbans region, Bangladesh. At the 6th edition of the World Wide Workshop for young environmental Scientists (WWW YES 2007) from April 24–27, 2007 at Domaine de Cherioux, Vitry Sur Seine, Workshop Proceedings, Paris, France, pp. 20–30.

25. Juthi, M.F. Biswas, R.M., and Bahar, M.N. 2009. Assessment of supply water quality in the Chittagong city of Bangladesh. *ARPN Journal of Engineering and Applied Science,* 4(3), 73–80.

26. Kundzewicz, Z.W., Mata, I.J., Arnell, N.W., Döll, P., Katat, P., Jimenez, B., Miller, K.A., Oki, T., Sen, Z., and Shiklomanov, I.A. 2007. Freshwater resources and their management, Climate change 2007: Impacts, adaptation and vulnerability. *Contribution of Working Group II to the Fourth Assessment Report of the Intergovernmental Panel on Climate Change,* Cambridge University Press, Parry, pp. 173–210.

27. Liu, S.X. 2007. Innovative technologies for value-added substance/energy recovery from wastewaters. In: *Food and Agricultural Wastewater Utilization and Treatment.* Blackwell Publishing, London, UK, pp. 247–260.

28. Mackay J.R. 1972. The world of underground ice. *Annals of the Association of American Geographers,* 6, 1–22.

29. Moeller, D.W. 2011. *Environmental Health,* 4th Edition. Harvard University Press, UK, pp. 1–507.

30. National Park Service and U.S. Department of the Interior. 2006. Denali National Park and Preserve: Permafrost Landscapes. Center for Resources, Science, and Learning, http://www.nps.gov/dena/naturescience/upload/Permafrost-Landscapes.pdf. Accessed on February 16, 2014.

31. Nordell, E. 1961. *Water Treatment for Industrial and Other Uses,* 2nd Edition, Reinhold Publishing Corporation, New York, pp. 1–598.

32. Pramanik, M.A.H. 1983. Remote sensing applications to coastal morphological investigations in Bangladesh. Ph.D. Thesis, Jahangimagar University, Savar, Dhaka, Bangladesh.

33. Rahman, A.A. 1988. Bangladesh coastal environment and management. In: M.J. Hasna, H.E. Rashid, and A.A. Rahman, A.A. (eds.), *National Workshop on Coastal Area Resource Development and Management*. Academic Publishers, CARDMA, Dhaka, pp. 1–22.

34. Reinstädtler, S. 2012. Modern 21st century landscapes—Challenges for a climate adaptive landscape change. *Conference Proceeding of Abstracts of the PECSRL—The Permanent European Conference for the Study of the Rural Landscape*, 25. August 20–24, 2012, Leeuwarden & Terschelling, Netherlands (Abstract in conference proceeding).

35. Reinstädtler, S. 2013. Sustaining landscapes—Landscape units for climate adaptive regional planning. *Conference Proceeding of the World Heritage Studies (WHS)*, June 16–19, 2011, IAWHP (International Association for World Heritage Professionals) Cottbus, Germany (article in conference proceeding: http://www.iawhp.com/).

36. Rowe, D.R. and Abden-Magid, I.M. 1995. *Handbook of Wastewater Reclamation and Reuse*, CRC Press, Boca Raton, FL, pp. 1–531.

37. Shur, Y.L. and Jorgenson, M.T. 2007. Patterns of permafrost formation and degradation in relation to climate and ecosystems. *Permafrost and Periglacial Processes*, 18, 7–19.

38. Spellman, F.R. and Drinan, J.E. 2012. *The Drinking Water Handbook*, 2nd Edition, CRC Press, Boca Raton, FL, pp. 1–347.

39. Tao, W.K. and Simpson, J. 1993. The Goddard cumulus ensemble model. Part I: Model description. *Terrestrial, Atmospheric and Oceanic Sciences*, 4, 35–72.

40. Thanh, N.C. and Tam, D.M. 1990. Water systems and the environment. In: N.C. Thanh and A.K. Biswas (eds.), *Environmentally Sound Water Management*, Oxford University Press, Delhi, pp. 1–30.

41. UNEP (United Nations Environmental Programme). 2005. Water and wastewater reuse: An environmentally sound approach for sustainable urban water management. Osaca, Japan. http://www.unep.org.jp/ictc/publications/

42. United Nations. 2006. *The Millennium Development Goals Report 2006*, UN, New York, pp. 1–32.

43. USARC. 2003. *Climate Change, Permafrost, and Impacts on Civil Infrastructure*, U.S. Arctic Research Commission, Arlington, VA, pp. 1–72.

44. USGCRP. 2000. Climate change impacts in the United States: The potential consequences of climate variability and change. *United States Global Change Research Program*, Cambridge University Press, Cambridge, UK.

45. USGCRP. 2009. Global climate change impacts in the United States. In: T.R. Karl, J.M. Melillo, and T.C. Peterson (eds.), *United States Global Change Research Program*. Cambridge University Press, New York.

46. Van Dijk, M.P. 2007. Urban management and institutional change: An integrated approach to achieving ecological cities. *Constitution to an International Seminar on Sustainable Urbanization in Tripoli*, Libya, Hotel Bab Africa, June 30–July 1.

47. White, I. 2010. *Water and the City: Risk, Resilience and Planning for a Sustainable Future*. Routledge, Taylor & Francis Group, New York, pp. 1–202.

48. WHO—World Health Organization, 1989. Health Guidelines for wastewater Use in Agriculture und Aquaculture. World Health Organization Technical Report Series No. 778, Geneva, Switzerland.

49. WHO—World Health Organization, 2001. *Water Quality: Guidelines, Standards and Health: Assessment of Risk and Risk Management for Water-Related Infectious Diseases*. IWA Publishing, London, UK.

50. Wittgren, H.B. and Maehlum, T. 1997. Wastewater treatment wetlands in cold climates. *Journal of Water Science and Technology*, 35(5), 45–53.

Water Reuse Practices in Cold Climates

Kirsten Exall
University of Calgary

Jiri Marsalek
Technical University of Lulea

PREFACE

Although water reuse is practiced more frequently in regions of the world experiencing water scarcity, jurisdictions with lower or no water stress are increasingly turning to water reclamation and reuse as a means of protecting sensitive receiving waters, as well as saving on energy and infrastructure investments. Both stormwater and treated wastewater effluents (reclaimed water) are used in a variety of reuse applications, but cold climate conditions present unique challenges to water quality and reuse. Contaminant inputs and stormwater or wastewater treatment process efficiencies vary during winter months in cold climate regions. Water reuse practices must take into account differences between warm and cold weather operations; for example, there may be shorter irrigation seasons available for agricultural or landscape irrigation applications and greater demand for seasonal storage or alternative disposal options in regions experiencing snow and freezing temperatures in winter. At the same time, cold climates can offer unique opportunities for treatment processes and reuse applications that take advantage of freezing conditions.

69.1 Introduction

Water reuse is practiced all over the globe, but drivers for water reuse often differ in cold climates than in hot and arid regions of the world. In cold climates (defined here as regions regularly experiencing below-freezing temperatures in winter months), water reuse is more likely to be investigated for protection of sensitive receiving waters than as a means of securing additional water supply [2].

Higher concentrations of antibiotics and other contaminants have been detected in receiving waters and sediments in winter, due in part to reduced treatment efficiency, low flow conditions, and cold-water temperatures, which might enhance the persistence of these compounds [22]. Discharges of urban runoff and wastewater effluents may also be more hazardous to fish during winter. Lemly [25] defined Winter Stress Syndrome (WSS) as "a condition of severe lipid depletion in fish brought on by external stressors in combination with normal reductions in feeding and activity during cold weather." Water reuse may be used as a tool in protecting aquatic ecosystems from adverse effects in winter months, by reducing the discharge of effluents to such receiving environments at particularly sensitive times.

Remote communities in cold regions that rely on trucked water supply and long-term storage of effluents can benefit from water reuse as a means to reduce the dependence on the more expensive water supply. There may also be energy savings due to the energy-water nexus. Although additional reclaimed water treatment and distribution may require an energy investment, Stillwell et al. [36] found that adopting water reuse could result in substantial energy savings in potable water collection, treatment, and distribution; the energy investment could be minimized for municipalities already using advanced wastewater treatment processes. Under certain circumstances, economic benefits may also be derived from water reuse, partly from savings on expansion of the water supply and wastewater treatment infrastructure. The incentives for adoption of water reuse may also change over the course of a project. Bischel et al. [6] noted that although such projects may be initiated as a means of increasing disposal capacity and protecting receiving waters, users often come to depend on the reliable water supply over time.

Whatever the initial motivation for adoption of water reclamation and reuse may be, cold climate applications involve additional considerations with respect to seasonal variations in stormwater and wastewater contaminant inputs, treatment efficiencies, and reclaimed water use.

69.2 Contaminant Inputs in Cold Climates

69.2.1 Stormwater Systems

Stormwater quantity and quality vary seasonally in cold climates, where much winter precipitation can fall as snow. The contaminant accumulation and release processes that occur in urban drainage systems during winter have been reviewed in Marsalek et al. [27]; major processes will be summarized here.

Snow along roadsides acts as a sink for contaminants, collecting and storing solids, litter, contaminants from vehicles and winter road maintenance materials, such as road salts and sand, until the snow is removed to a snow disposal site or these materials are released to the stormwater system through snowmelt and rain-on-snow events. Accumulation of contaminants in the snowpack is a dynamic process over the winter, consisting of pollutant influx with deposition and precipitation, and release during intermittent melts or the final snowmelt.

Soluble contaminants are transported within the snowpack by snowpack elution processes, where ice crystals and snowflakes respond to freezing and thawing cycles by metamorphosis, excluding impurities, which are then loosely bound and available to be flushed from the snowpack. Soluble pollutants are collected in meltwater that percolates through the pack, and are eventually flushed from the pack as a highly concentrated pulse of meltwater. High salt levels in the snowpack and meltwater may shift the speciation of metals into the soluble phase. The degree to which soluble pollutants are excluded and washed from the snowpack depends upon the number of freeze–thaw cycles and whether the snowpack receives any additional moisture as rain.

Solids and associated hydrophobic substances, such as polycyclic aromatic hydrocarbons (PAHs), stay in the snowpack until the last 5%–10% of meltwater leaves the snowpack. Such particulate material is filtered or coagulated with other particles

as it moves through the snowpack and remains behind as the soluble component washes through and may only be washed during high flows near the end of the melt. Larger solids remain on the ground as sediment residual after the melt.

The initial stages of melt are generally slow, so the early meltwater runoff can be highly concentrated and toxic, but does not exert a high pollution load. The latter portion of the melt adds both high concentrations and high loads because of wash-off of surfaces and the movement of particulates out of the pack. This process may be affected by rainfall occurring during the melt, which can dilute soluble pollutants, promote the movement of particulates through the pack, and increase flows engaged in wash-off processes. The intensity of the resulting rainfall/melt wash-off and runoff may be augmented because of the low infiltration capacity of the frozen soil substrate [1] and the added volume of water coming from the melting snowpack. Furthermore, the infiltration rates depend on the initial soil moisture at the onset of freezing temperatures and the depth of frost penetration into the soil; small depths (≤ 0.15 m) in soils with low initial moisture exert practically no influence, but larger frost depths in soils with higher initial moisture reduce the infiltration rates to zero [1].

The solids, metals, PAHs, road salts, and sand (grit) that are mobilized from the roadside snowpack or snow disposal sites may accumulate in stormwater management facilities over the winter [26]. The high contaminant loads and presence of road salts alter the mobility and release of solids, metals, and nutrients from such facilities through the winter and spring months (see Reference 27 and references cited therein).

69.2.2 Wastewater Systems

Although less variable than stormwater, seasonal differences have also been observed in the composition of municipal wastewaters, particularly in concentrations of pharmaceuticals and personal care products (PPCPs) [30,35,42]. For example, pain relievers and related medications have been observed at significantly higher concentrations at wastewater treatment plants during winter months, corresponding to periods with higher incidences of the flu and related illnesses [42]. At the same time, influent loads of organic UV filters, such as are used in sunscreens and cosmetics, were found to be lower in Swiss wastewaters during colder months than in warmer ones [3]. Another factor could be reduced flows in sewers, with less infiltration of extraneous waters into sewers resulting from frozen ground during winter months [34].

69.3 Water Reclamation in Cold Climates

69.3.1 Stormwater

Contaminant removals during treatment vary under cold and warm conditions. Some of the challenges to effective contaminant removal experienced in stormwater management in cold climates include frozen filter media, reduced hydraulic efficiency, high chloride loads, temperature-related changes in chemical and biological processes and water density, and dormant biological functions [31]. Roseen et al. [31] examined seasonal performance variations for stormwater management systems in cold climate conditions; filter media frozen to an unspecified depth did not reduce the performance of low impact development (LID) facilities, but conventional stormwater treatment facilities exhibited larger variations in seasonal performance. A much less favorable experience with the effectiveness of a bioretention cell (i.e., an infiltration facility with a vegetated surface cover) in cold climate was reported in Calgary, Canada [21]; infiltration into a partially frozen bioretention cell was significantly impaired.

The road salts that are often used for winter road maintenance in cold climate municipalities impact contaminant removal in stormwater facilities. Both thermal (temperature-induced) and chemical (salt-induced) stratification have been observed in urban stormwater management ponds and in oil and grit separators, which can impact sedimentation efficiency. High salt levels can also shift the speciation of metals into the soluble phase, reducing the removal of such contaminants during sedimentation [27]. The resulting elevated soluble metal concentrations could limit some stormwater reuse applications that depend on higher water quality.

69.3.2 Wastewater

Wastewater treatment, or water reclamation, processes are also very sensitive to temperature changes, although modifications to facility design can mitigate the effects. The kinetics of chemical and biological treatment processes decrease substantially at low temperatures, including biomass growth and activity, nitrification, denitrification, and disinfection processes, while the viscosity of water increases, impeding sedimentation. Temperature correction factors and facility design modifications can mitigate the effects on wastewater treatment processes, but longer reaction times and larger tank volumes are often required [29]. Geography and wind protection play an important role in temperature control in wastewater treatment facilities; Scherfig et al. [33] developed models to predict the temperature changes in biological treatment tanks based on hydraulic residence times, initial tank and influent temperatures, and meteorological parameters. In testing the effects of temperature on various wastewater treatment reactors, Sundaresan et al. [38] observed that chemical oxygen demand (COD) removal and nitrification decreased substantially in fluidized bed, activated sludge, and submerged bed reactors as temperature decreased, but performance quickly rebounded in the latter as temperatures were raised from 5°C to 10°C.

Trace contaminants also display seasonality in removal during wastewater treatment. Sui et al. [37] described seasonal variations in the occurrence and removal of PPCPs in different biological wastewater treatment processes. Most PPCPs had higher concentrations in winter than in summer. Membrane bioreactor (MBR) treatment was found to more effectively remove the more easily biodegradable PPCPs than conventional activated sludge (CAS) or biological nutrient removal (BNR) processes, particularly in winter months. Recalcitrant PPCPs were not removed by any treatment process or in any season [37]. Matamoros et al. [28] evaluated the seasonal performance for removing emerging contaminants of a water reclamation pond-constructed wetland system receiving secondary effluent from a wastewater treatment facility in northeastern Spain. Photodegradable compounds were least removed during winter months, corresponding to the phase with the shortest daylight period. Biodegradable compounds such as ibuprofen and naproxen were more affected by seasonality in the pond than in the constructed wetland; this was assumed to be due to the higher presence of biomass in the wetland, with rooted plants that allowed accumulation of organic matter, transport of oxygen, and development of biofilms [28].

Effluent water quality variations may be mitigated with judicious selection of wastewater treatment processes and adjustments to operations designed to address low temperature challenges. Guidance on the design and operation of wastewater treatment systems for cold climates can be found for waste stabilization ponds and lagoon systems [16,17], and constructed wetlands [20,40,41]. Various process combinations have also been proposed. Hanæus et al. [15] described the improved solids and phosphorus removal efficiency that could be achieved in cold climates by inclusion of chemical precipitation with aluminum salts, iron salts, or lime in conventional wastewater stabilization ponds; such chemically assisted sedimentation ponds (termed *fellingsdams*) have been common since the late 1960s in Scandinavia. Di Trapani et al. [9] described a hybrid activated sludge/biofilm process for upgrading a conventional activated sludge plant for achieving nitrification even at low temperatures and without the need for additional volumes. Bioaugmentation with specialized, cold-adapted microbial communities has been shown to facilitate the rapid start-up and stable long-term performance of biological wastewater treatment processes at low temperatures, although great care must be taken to ensure retention and growth of the introduced microbial cells [14]. Finally, Kim et al. [23] found that biological aerated filtration (BAF) was an effective addition to an existing secondary wastewater treatment plant for the purpose of water reuse in cold regions, achieving nitrification even at 7°C. It was noted that disinfection was still required prior to water reuse.

One treatment process that is unique to cold climate regions is freeze separation, which is also termed freeze crystallization, spray freezing, or Snowfluent. In this technique, wastewater effluents are converted to snow by the use of snow guns. Contaminant removal occurs because as water freezes, impurities are rejected and remain predominantly in the liquid phase while pure ice crystals grow. The freezing process is rarely perfect in practical application, however, and some impurities typically remain associated with the snow and ice crystals [32]. Gao et al. [11] noted that freezing of wastewater was also very effective in inactivating *E. coli* through cell damage. Trickle freeze separation is a variation on spray freezing that has shown promise for treatment of oil sands mine wastewaters [5]. As described above, when the snow begins to melt, soluble contaminants are washed out first; solids and solids-associated contaminants tend to remain behind on the ground under the snowpack. The rate of release of contaminants depends on factors including temperature, rain-on-snow events, snow aging, and structure and materials underneath the snow storage area [32].

69.4 Water Reuse Practices in Cold Climates

Many water reuse applications practiced in warmer climates are applied similarly in cold climate regions. Industrial water reuse practices, for example, whether using reclaimed municipal wastewater or wastewater treated and recycled within the industrial facility, can often be employed in cold regions with similar water quality requirements and water treatment technologies as in warmer climates. Other water reuse applications may require modifications or are uniquely suited to cold climates; this section highlights a number of such applications.

69.4.1 Agricultural and Landscape Irrigation

Reclaimed water demand varies according to regional precipitation patterns and growing seasons. The practice of water reuse for agricultural and landscape irrigation in cold climates differs from that in warmer climates. Cold climates have shorter irrigation seasons, which generally necessitate planning of adequate seasonal storage volume or alternate disposal options for reclaimed water. A recent study of year-round irrigation with reclaimed water from an onsite wastewater treatment operation in Ohio [8] found no significant differences in plant survival or growth in plots subjected to year-round irrigation (including during freezing conditions) and those subjected to irrigation in only nine months of the year. There were, however, indications of ice damage on two evergreen species in the year-round irrigation plots, indicating that careful selection of plant species and irrigation methods is needed. Although the cost of winterizing the irrigation system ($4000) tripled the installation cost of the irrigation system, the authors calculated that in the area of study, which exhibited shallow soils unsuited for septic systems, this still represented substantial savings over the costs to install and operate treatment and storage facilities [8].

In all situations where reclaimed water is used for irrigation, the build-up of heavy metals in soils must be considered. In cold climates where road salts are applied during winter, the salinity of stormwater or reclaimed wastewater used for irrigation must also be carefully monitored along with the resultant soil health [19].

Reclaimed water can also be used in snowmaking on farm fields or landscaped areas. This alternative approach to agricultural and landscape irrigation in cold regions provides winter storage of effluents until the spring melt, when soils and crops may be able to take advantage of the nutrient content of the melt and residual materials [39]. Further discussion of snowmaking as a water reuse application in cold climates is provided next.

69.4.2 Snowmaking

Snowmaking with reclaimed water is practiced in Canada, Australia, and the United States; case studies are described in Maine, Pennsylvania, and California in the 2012 USEPA document, Guidelines for Water Reuse [39]. Important considerations include the potential for human contact, required snow storage, potential volumes and fate of snowmelt in terms of infiltration to underlying ground or the potential for overland flow, and proximity to sensitive surface water bodies, particularly in a drinking water watershed [39].

The water quality of the snowmelt from reclaimed water snowmaking applications may differ from that of the original effluent, due to the transformations that occur during repeated freeze-thaw cycles and the preferential elution of soluble contaminants (discussed previously), as well as the fact that the snowmelt may pick up additional contaminants from the soil during overland flow.

A number of jurisdictions have developed guidelines or regulations relating to reclaimed water use for snowmaking. As the potential exists for human contact and ingestion of snow on ski hills, the British Columbia Reclaimed Water Guideline [4] requires the highest quality of reclaimed water be used for snowmaking for skiing or snowboarding. Clear communication of the use of reclaimed water to snowmaking equipment users, ski hill staff, and skiers is also required in British Columbia, through the marking of equipment and the inclusion of a statement on ski passes that reclaimed water is being used. A moderate or lower exposure potential category of reclaimed water quality may be used for snowmaking in areas where skiing is not intended. Arizona allows snowmaking with Class A reclaimed water, which has undergone secondary treatment, filtration and disinfection meeting turbidity criteria, and containing no detectable fecal coliform bacteria; California and Pennsylvania similarly allow for snowmaking with filtered, disinfected reclaimed water meeting specific turbidity criteria. New Hampshire requires filtration with site-specific nutrient removal depending on snowmelt runoff to surface streams, as well as disinfection [39].

69.4.3 Household Reclaimed Water Use

Isolated communities in cold climates with trucked water supply and long-term storage of effluents in waste stabilization ponds can also benefit from the increased household use of treated greywater to reduce the reliance on costly water supplies. Gunnarsdóttir et al. [13] reviewed the state of wastewater treatment processes in Arctic regions of North America and Europe and found it to be lacking or inadequate in many jurisdictions. The lack of conventional centralized wastewater treatment facilities in the Arctic is in part due to the challenges in designing, constructing, and operating such systems: permafrost conditions, hard rock surfaces, freezing, flooding in the spring, limited quantity of water, high costs of electricity, fuel and transportation, and a settlement pattern with limited accessibility. In many areas of northern Europe, mechanical or chemical treatment processes are preferred over biological treatment as they are less temperature dependent and can occupy a smaller footprint, making them cheaper to construct [13]. In contrast, waste stabilization ponds with seasonal (spring and summer) discharge are the most common form of wastewater treatment in Canada's Northern communities, some with natural wetlands for further effluent polishing [24].

Gunnarsdóttir et al. [13] suggested that costs of construction for new wastewater treatment facilities in northern communities can be minimized by separation of blackwater (human excreta) at the household level and onsite treatment of the remaining greywater, reducing or eliminating the need for expensive secondary sewer collection systems. The treated greywater can then be applied for use in toilet and urinal flushing, as described by Hill [18] in two communities in Canada's North.

The character of greywater depends on the points of collection within a household; low-load greywater excludes kitchen and laundry sources, while high-load greywater includes these sources. The household water savings potential from reusing treated greywater for such nonpotable duties as toilet flushing, garden irrigation, and floor cleaning ranges between 9% and 46% [7]. All greywater may still contain solids, organic matter, nutrients, and various xenobiotics from such products as detergents and dyes [7]; treatment before use within the home is therefore necessary, even in toilet flushing or outdoor irrigation applications. Numerous greywater treatment processes exist and have recently been reviewed [7,12]; performance in regions experiencing extreme cold will likely need additional testing.

69.4.4 Indirect Reuse

Reclaimed water may also be reused indirectly in cold regions as a heat source. Funamizu et al. [10] discussed the urban reuse of the heat energy in wastewater for heating and air conditioning, as well as for snow melting. For snow melting, the snow can be thrown directly into the combined sewer pipe (although this practice may have implications in downstream treatment processes), or the warmed effluent can be diverted to a centralized basin for melting collected snow. The rate of snow melting depends on the flow rate of the effluent, as well as the temperatures, densities, and specific heat capacities of the water and snow [10].

69.5 Summary and Conclusions

Cold climate regions regularly experience freezing temperatures, which not only present unique challenges to wastewater treatment and reclaimed water use, but also offer unique opportunities. Winter conditions in these regions influence both the character and treatability of stormwater and municipal wastewater. Treatment processes can be modified to mitigate temperature effects or even be designed to take advantage of freezing temperatures. Agricultural and landscape irrigation

with reclaimed water can be practiced in cold climates, but the shorter irrigation season often requires seasonal storage or alternate disposal options during winter months. Other applications of water reuse with specific benefits in cold climates include snowmaking, household reclaimed water reuse, and heat recovery.

References

1. Al-Houri, Z.M., Barber, M.E., Yonge, D.R., Ullman, J.L., and Beutel, M.W. 2009. Impacts of frozen soils on the performance of infiltration treatment facilities. *Cold Regions Science and Technology*, 59, 51–57.
2. Angelakis, A.N. and Durham, B. 2008. Water recycling and reuse in EUREAU countries: Trends and challenges. *Desalination*, 218, 3–12.
3. Balmer, M.E., Buser, H.-R., Mueller, M.D., and Pogier, T. 2005. Occurrence of some organic UV filters in wastewater, in surface waters, and in fish from Swiss lakes. *Environmental Science & Technology*, 39, 953–962.
4. BC Ministry of Environment. 2013. *British Columbia Reclaimed Water Guideline.* http://www2.gov.bc.ca/gov/DownloadAsset?assetId=514F9C184A6C432188E204218468F70F&filename=reclaimedwater.pdf. Accessed on November 5, 2014.
5. Beier, N., Sego, D., Donahue, R., and Biggar, K. 2007. Laboratory investigation on freeze separation of saline mine waste water. *Cold Regions Science and Technology*, 48(3), 239–247.
6. Bischel, H.N., Simon, G.L., Frisby, T.M., and Luthy, R.G. 2012. Management experiences and trends for water reuse implementation in Northern California. *Environmental Science & Technology*, 46, 180–188.
7. Boyjoo, Y., Pareek, V.K., and Ang, M. 2013. A review of greywater characteristics and treatment processes. *Water Science and Technology*, 67(7), 1403–1424.
8. Caldwell, H., Mancl, K., and Quigley, M.F. 2007. The effects of year-round irrigation on landscape plant quality and health in Ohio. *Ohio Journal of Science*, 107(4), 76–81.
9. Di Trapani, D., Christensson, M., and Ødegaard, H. 2011. Hybrid activated sludge/biofilm process for the treatment of municipal wastewater in a cold climate region: A case study. *Water Science and Technology*, 63(6), 1121–1129.
10. Funumizu, N., Iida, M., Sakakura, Y., and Takakuwa, T. 2001. Reuse of heat energy in wastewater: Implementation examples in Japan. *Water Science and Technology*, 43(10), 277–285.
11. Gao, W., Smith D.W., and Li, Y. 2006. Natural freezing as a wastewater treatment method: *E. coli* inactivation capacity. *Water Research*, 40(12), 2321–2326.
12. Ghunmi, L.A., Zeeman, G., Fayyad, M., and van Lier, J.B. 2011. Grey water treatment systems: A review. *Critical Reviews in Environmental Science and Technology*, 41(7), 657–698.

13. Gunnarsdóttir, R., Jenssen, P.D., Erland Jensen, P., Villumsen, A., and Kallenborn, R. 2013. A review of wastewater handling in the Arctic with special reference to pharmaceuticals and personal care products (PPCPs) and microbial pollution. *Ecological Engineering*, 50, 76–85.

14. Guo, J., Wang, J., Cui, D., Wang, L., Ma, F., Chang C.-C., and Yang, J. 2010. Application of bioaugmentation in the rapid start-up and stable operation of biological processes for municipal wastewater treatment at low temperatures. *Bioresource Technology*, 101(17), 6622–6629.

15. Hanæus, J., Grönlund, E., and Johansson, E. 2010. Seasonal operation of ponds for chemical precipitation of wastewater. *Journal of Cold Regions Engineering*, 24(4), 98–111.

16. Heaven, S., Lock, A.C., Pak, L.N., and Rspaev, M.K. 2003. Waste stabilisation ponds in extreme continental climates: A comparison of design methods from the USA, Canada, northern Europe and the former Soviet Union. *Water Science and Technology*, 48(2), 25–33.

17. Heinke, G.W., Smith, D.W., and Finch, G.R. 1991. Guidelines for the planning and design of wastewater lagoon systems in cold climates. *Canadian Journal of Civil Engineering*, 18, 556–567.

18. Hill, T.T. 1999. Water reclamation north of 60-wastewater treatment and recycling in NWT and Nunavut. Code 66875, *Proceedings, Canadian Society for Civil Engineering Annual Conference*, 3, Regina, Saskatchewan, Canada, June 2–5, 1999, 379–388.

19. Iskandar, I.K. 1979. The effect of waste water reuse in cold regions on land treatment systems. *Journal of Environmental Quality*, 7(3), 361–368.

20. Jenssen, P.D., Mæhlum, T., Krogstad, T., and Vråle, L. 2005. High performance constructed wetlands for cold climates. *Journal of Environmental Science and Health, Part A*, 40(6–7), 1343–1353.

21. Khan, U., Valeo, C., Chu, A., and van Duin, B. 2012. Bioretention cell efficacy in cold climates: Part 1—Hydrologic performance. *Canadian Journal of Civil Engineering*, 39, 1210–1221.

22. Kim, S.-C. and Carlson, K. 2007. Temporal and spatial trends in the occurrence of human and veterinary antibiotics in aqueous and river sediment matrices. *Environmental Science & Technology*, 41(1), 50–57.

23. Kim, S.W., Park, J.B., and Choi, E. 2007. Possibility of sewage and combined sewer overflow reuse with biological aerated filters. *Water Science and Technology*, 55(1–2), 1–8.

24. Krkosek, W., Ragush, C., Boutilier, L., Sinclair, A., Krumhansl, K., Gagnon, G., Jamieson, R., and Lam, B. 2012. Treatment performance of wastewater stabilization ponds in Canada's Far North. *Cold Regions Engineering* 2012, 612–622.

25. Lemly, A.D. 1996. Winter stress syndrome: An important consideration for hazard assessment of aquatic pollutants. *Ecotoxicology and Environmental Safety*, 34, 223–227.

26. Marsalek, J. 2003. Road salts in urban stormwater: An emerging issue in stormwater management in cold climate. *Water Science and Technology*, 48(9), 61–70.

27. Marsalek, J., Oberts, G., Exall, K., and Viklander, M. 2003. Review of operation of urban drainage systems in cold weather: Water quality considerations. *Water Science and Technology*, 48(9), 11–20.

28. Matamoros, V. and Salvado, V. 2012. Evaluation of the seasonal performance of a water reclamation pond-constructed wetland system for removing emerging contaminants. *Chemosphere*, 86, 111–117.

29. Metcalf and Eddy. 2003. *Wastewater Engineering: Treatment and Reuse*. 4th Edition. McGraw Hill, New York, NY.

30. Nie, Y., Qiang, Z., Zhang, H., and Ben, W. 2012. Fate and seasonal variation of endocrine-disrupting chemicals in a sewage treatment plant with A/A/O process. *Separation and Purification Technology*, 84, 9–15.

31. Roseen, R.M., Ballestero, T.P., Houle, J.J., Avellaneda, P., Briggs, J., Fowler, G., and Wildey, R. 2009. Seasonal performance variations for storm-water management systems in cold climate conditions. *Journal of Environmental Engineering*, 135, 128–137.

32. Sallanko, J. and Haanpää, K-M. 2008. Release of impurities from melting snow made from treated municipal wastewater. *Journal of Cold Regions Engineering*, 22, 54–61.

33. Scherfig, J., Schleisner, L., Brond, S., and Kilde, N. 1996. Dynamic temperature changes in wastewater treatment plants. *Water Environment Research*, 68(2), 143–151.

34. Semadeni-Davies, A. 2004. Urban water management vs. climate change: Impacts on cold region waste water inflows. *Climatic Change*, 64(1–2), 103–126.

35. Smyth, S.A., Lishman, L.A., McBean, E.A., Kleywegt, S., Yang J-J, Svoboda, M.L., Lee H.-B., and Seto, P. 2008. Seasonal occurrence and removal of polycyclic and nitro musks from wastewater treatment plants in Ontario, Canada. *Journal of Environmental Engineering and Science*, 7(4), 299–317.

36. Stillwell, A.S., King, C.W., Webber, M.E., Duncan, I.J., and Hardberger, A. 2011. The energy-water nexus in Texas. *Ecology and Society*, 16(1), 2. http://www.ecologyandsociety.org/vol16/iss1/art2/.

37. Sui, Q., Huang, J., Deng, S., Chen, W., and Yu, G. 2011. Seasonal variation in the occurrence and removal of pharmaceuticals and personal care products in different biological wastewater treatment processes. *Environmental Science & Technology*, 45(8), 3341–3348.

38. Sundaresan, N. and Philip, L. 2008. Performance evaluation of various aerobic biological systems for the treatment of domestic wastewater at low temperatures. *Water Science and Technology*, 58(4), 819–830.

39. U.S. EPA (United States Environmental Protection Agency). 2012. *Guidelines for Water Reuse*. EPA/600/R-12/618, U.S. Environmental Protection Agency, Washington, DC.

40. Werker, A.G., Dougherty, J.M., McHenry, J.L., and Van Loon, W.A. 2002. Treatment variability for wetland wastewater treatment design in cold climates. *Ecological Engineering,* 19, 1–11.

41. Wittgren, H.B. and Mæhlum, T. 1997. Wastewater treatment wetlands in cold climates. *Water Science and Technology,* 35(5), 45–53.

42. Yu, Y., Wu, L., and Chang, A.C. 2013. Seasonal variation of endocrine disrupting compounds, pharmaceuticals and personal care products in wastewater treatment plants. *Science of the Total Environment,* 442, 310–316.

70

Water Reuse in Hilly Urban Areas

Yohannes Yihdego
*Snowy Mountains Engineering
Corporation (SMEC)*

PREFACE

Water reuse systems are an important component of the environmental sustainability program in hilly urban areas. This chapter begins with a review of techniques to reduce overall potable water demand, as reduction of potable water usage is a critical first step in accomplishing the goals of this Handbook. This chapter also provides design guidelines for rainwater harvesting and greywater systems, and points to ideas for utilizing blackwater systems. These recommendations have been informed by research and local, national, and international case studies, and provide strategies for developing water reuse systems in urban areas. However, users of this Handbook should keep in mind that different buildings would have different needs depending on the water sources being used, project budget, aesthetic and space planning concerns, etc. There has been considerable recent activity at many urban area levels in regulatory guidance for rainwater harvesting systems to serve as a guideline for evaluating rainwater harvest systems to be used for toilet and urinal flushing. The guideline outlines the basic requirements for such systems requirements to include rainwater harvesting for nonpotable uses in jurisdiction, and subsequently in plumbing code. These new standards would facilitate implementation of rainwater systems in hilly urban areas to reduce the time and number of steps involved in review of rainwater harvest systems for use in flush fixtures and irrigation. At the core of the regulatory discussion is the responsibility to ensure public health and safety. For such newly emerging systems, it is critical to protect public health and safety while regulatory standards are being drafted. The water reuse systems that currently exist in some urban areas have been approved on a case-by-case basis. In the absence of accepted standards, minimum sanitary requirements have been used at times as the standard to which water must be treated within rainwater harvesting or greywater reuse systems. I would like to acknowledge Prof. Saeid Eslamian for his contribution to this work.

70.1 Introduction

Urban water reuse refers to the use of reclaimed water for nonpotable uses in urban environments. This can include activities such as irrigation, fire protection, window washing, dust control, and toilet flushing. This type of use acknowledges that there are several water uses in urban areas that do not necessarily require water that meets drinking water standards [8]. In order for an area to make use of this type of application, distribution systems must be built to provide the reclaimed water to their place of use. This is a completely separate distribution system from the already common drinking water systems because this is not potable water. Therefore, there is a high extra cost associated with this because most areas do not have this infrastructure in place [14].

Greywater is water discharging from laundry, showers, bathtubs, and kitchen sinks and is about 50%–80% of residential wastewater [3]. Freshwater is becoming a rare resource in the world as a result of a variety of factors, the main ones being pollution and waste management. Studies have shown that in the next 50 years, more than 40% of the world's population will live in countries facing water scarcity especially in parts of Asia and Africa [27]. This has implications for human survival and the socioeconomic development in these countries, which calls for wise management of this resource. With increasing demand for freshwater, there is an urgent need for alternative water resources and optimization of water use efficiency through reuse options. Experiences elsewhere in the world indicate that greywater can be a cost-effective alternative source of water. This is primarily related to its availability and low concentration of pollutants compared to the combined household wastewater [15]. In most cities in developing countries, the increasing urbanization has resulted in development of densely populated peri-settlements (slums). These settlements, inhabited by the poor, have increased the demand for water supply. However, the municipal authorities in most of these cities have limited capacity to provide basic services (adequate safe water, excreta, solid waste and greywater management) to these communities resulting in poor environmental sanitation [27,28].

The populations living in these settlements are therefore vulnerable to health risks given the increasing waste (solid and liquid) generation. In Kampala's (capital of Uganda) peri-urban areas, for example, greywater is indiscriminately disposed of in drainage channels and open spaces resulting in creation of malaria mosquito breeding grounds and smelly stagnant waters, which are a health hazard particularly when children play in them and contaminate shallow groundwater sources. The communities in the peri-urban settlements perceive greywater as being dirty and hence regard its reuse with disaffection. However, appropriate reuse of greywater not only reduces agricultural use of drinking water and water costs, but also increases food security and improves public health [23]. Hence, there is a need to promote its reuse in the peri-urban areas as it can be treated and used close to its origin thereby contributing to the Millennium Development Goals that have health and improving the lives of the people

living in slums as a priority. It is noted that the main barrier for wider and faster dissemination of suitable greywater management systems at the household level in developing countries is the lack of knowledge and experience [17]. Scientific knowledge is sparse regarding greywater characteristics allowing its reuse. In addition, realizing greywater reuse requires operational research, which is pointed out as being important in implementing fundamental public health services effectively in slums [24]. Greywater needs some level of treatment prior to disposal to ensure minimal impacts on the environment. Greywater treatment does not aim at providing water for drinking but may be used for toilet flushing, laundry, lawn irrigation, window and car washing, groundwater recharge, agriculture, or fire extinguishing [18]. This calls for low cost management and treatment systems. Small-scale and onsite treatment of greywater for this purpose includes either one or a combination of septic tanks, soil infiltration, drip irrigation, and constructed wetlands. Septic tanks and soil infiltrations systems, though simple methods for onsite greywater treatment are problematic in the peri-urban areas with high water table due to potential for groundwater contamination. Constructed wetlands, on the other hand, require pretreated greywater, can be expensive and need expertise for design, construction, operation, and monitoring [4,23]. For greywater treatment, the technology of tower gardens was selected as it is a simple, innovative system, which uses greywater for growing vegetables on a small footprint (<1 m^2) and can be easily self-constructed with few local materials. Furthermore, they are easy to operate and maintain and have been found to be appropriate in densely populated settlements, for example, in Kenya, Ethiopia, and South Africa [9]. In other developing nations, a previous study undertaken in peri-urban settlements on tower gardens receiving greywater had a positive uptake, with communities appreciating the benefits of greywater reuse and being able to install these systems on their own after being taught how. However, in this study and all the others, application of tower gardens is still in its infancy necessitating a lot of scientific experimentation to determine their potential to make a real difference in the peri-urban areas [9]. Here, information regarding the greywater characteristics, hydraulic load of a tower garden, greywater treatment potential and impacts on the vegetables and low cost pest control measures is still scarce. In this study, therefore, an investigation into the reuse of greywater for growing vegetables in tower gardens was undertaken in a selected peri-urban settlement in an urban area to provide a scientific basis for application of this option. The specific objectives of the study were (1) to determine the characteristics of the greywater generated by the communities in the settlement, (2) to assess the impacts of the greywater on vegetable growth and soils within the tower gardens, and (3) to ascertain communities' perspectives of this technology and impacts on their livelihood [20].

Reusing rainwater or greywater for toilet flushing or irrigation, as well as using high efficiency and low-flow fixtures and appliances, can have a significant impact on the amount of potable water used in a building, particularly for commercial and institutional buildings where toilet flushing comprises a large percentage of water use.

Rainwater harvesting and greywater systems replace potable water; rainwater-harvesting systems divert water from the urban area combined sewer system, reducing its overall load. This is particularly important in many cities due to the frequency of combined sewer overflows (CSOs) and basement flooding. CSOs are harmful because they pollute our freshwater bodies with raw sewage and pollutants such as motor oil, heavy metal trace elements, fertilizers, and other chemicals, leading to degraded water quality and wildlife habitats. CSOs can be limited by reducing, as much as possible, the amount of water input into the sewer system, particularly during periods of high system strain (such as during large rainfall events). Utilizing cisterns or rainwater harvesting systems reduces the amount of rainfall hitting the ground and entering the sewers in an area. This diminishes the risk of basement, yard, and street flooding, which occurs occasionally during major storm events in the urban area. Lower sewer loads at peak times reduce flood risk and the resultant property damage and costs associated with such events.

There are many ancillary environmental benefits to utilizing water reuse systems: reduced overall demand for potable water conserves freshwater sources, which saves energy used in the treatment and conveyance of municipally supplied water, prevents flooding, protects water quality through diminished CSOs, and improves community water assets for recreation and wildlife habitat. A commitment to sustainability ensures that present needs are met without compromising the ability of future generations to meet their own needs. Conserving our water resources helps to provide for our own needs as well as those of future generations. There are many different measures, which together form a tool kit from which individual measures can be selected as part of a specific design response suiting the characteristics of any development (or redevelopment). Urban water harvesting and reuse is one of those measures. Sustainable approaches to urban water management involve the use of locally generated runoff and wastewater to supplement traditional urban water sources. The incorporation of these water sources in the urban water resource-planning framework reflects the increased scarcity of water sources to meet demands, technological advancements, increased public acceptance, and improved understanding and management of risks including those concerning public health. There is a myriad of methods to utilize runoff and wastewater as a resource [26].

70.2 Applications

Urban water harvesting and reuse schemes can be developed for existing urban areas or new developments and are mainly suitable for nondrinking purposes such as

- Residential uses (including toilet flushing)
- Irrigation of public open spaces (including sporting grounds)
- Industrial uses
- Water features

Harvesting of urban water is possible over a number of scales, from individual domestic allotment level to community scale or industrial precinct development.

Key factors in determining the type and scale of harvesting possible is dependent on

- The proposed water source and quality (i.e., runoff, treated wastewater, etc.)
- The proposed water use (i.e., irrigation)
- The demand pattern and volume (i.e., summer for irrigation)
- The seasonality and volume of water available for harvest (depends on type and source of water)
- The storage options and site constraints (if required)
- Treatment options (if required)
- Objectives for harvesting system (i.e., reduced mains water supply or reduced runoff from site)
- Capital and operational costs including monitoring and maintenance costs

Capture and use of water on site is an environmentally preferable source of alternative water as this method generally does away with the need for piping or pumping. Fewer resources are needed and greenhouse gas emissions are reduced. As urban water harvesting and reuse can be applied at a range of scales and can utilize a range of water sources and storage options, this chapter only provides an overview of the range of options and the factors to be considered.

Most arid regions have highly seasonal rainfall. This seasonal variation in rainfall affects the availability of stormwater and rainwater (or roof runoff). To maintain security of supply when demands are present for these water sources, storage is required. Wastewater has less variation in supply as it is generally dependent on mains water use. Climate in arid region also impacts the demand patterns for water, particularly outdoor uses. This is primarily evident for irrigation with high demand in summer and low demand in winter. Climatic conditions provide challenges, which need to be addressed during the concept design phase. Each available source of water for urban water harvesting and reuse schemes are discussed briefly next [26].

70.2.1 Wastewater

Treated wastewater reuse can provide a relatively constant supply in urban areas because its source is mains water. The production of wastewater is dependent on seasonal and diurnal fluctuations in water use habits. The primary technical disadvantage of wastewater reuse is the level of treatment and associated cost required to achieve the level of water quality necessary for reuse. The principal risk to human health is the inappropriate consumption of wastewater treated for nonpotable uses. In addition, the public perception of treated wastewater reuse and possible health risks needs to be considered [29].

The common sources of greywater production are kitchen, bathroom, and laundry. Greywater disposal infrastructure is poor with the majority of households pouring laundry water in open drains and spaces adjacent to their homes. Laundry greywater was mostly reused to clean houses and then disposed of. Greywater on application onto tower gardens resulted in an increase in the phosphorus content with limited impact on the potassium, organic matter, and nitrogen content of the soils. Tomatoes, collard greens, and plants grown in the tower gardens thrived with the greywater. However, they were attacked by pests necessitating pesticide application. Pesticide use nonetheless has health and environmental impacts. This calls for awareness creation among the communities on the risks of pesticide use and hence importance of protection to safeguard health, crops, animals, and water. In addition, communities should be encouraged to use biological measures for pest control such as growing of onions or garlic on the tower gardens as these limit the use of potentially harmful pesticides. The household occupants where tower gardens were installed appeared to appreciate the technology. However, to create interest and demand to scale up, this greywater reuse option necessitates increased sensitization and social marketing within the community on greywater reuse and associated benefits. There is a need to determine the hydraulic load of a greywater tower to guide the number and size of gardens needed for a particular quantity of generated greywater for optimum performance. Also, further monitoring of this greywater reuse option is needed to ascertain the yield per volume of soil for each tower garden, and impact of greywater application on crop growth and people's livelihoods [5,16].

The interest of the respondents in the gardens is critical to their success. A case in point is where one of the respondents bought a fence to protect his garden from attack by animals. Also differentiating between the chemical content of the greywater and non-greywater is not obvious among the residents in these settlements, which might have also contributed to the decision to apply greywater on the control gardens. The slow uptake of the tower gardens among the residents in the area is most likely a result of the limited sensitization on gray reuse and associated benefits as well as training on installation of the technology. Additionally, no social marketing was undertaken to create interest and demand for the technology. The tower gardens, however, promoted a sense of neighborhood in that the application of the greywater and planted vegetables on maturing were shared with the neighbors without tower gardens.

This study was undertaken to create an understanding of greywater characteristics and to demonstrate a low cost reuse option involving application of small tower gardens for greywater treatment and reuse in a peri-urban settlement in an urban area. To realize this, field surveys, greywater and soil sampling and analysis, and tower garden installation at selected households were undertaken. Often, greywater application to the tower gardens had limited impact on the soil potassium, organic matter, and nitrogen content but increased the phosphorus content [7,21]. The vegetables grown in the greywater towers thrived but were attacked by pests necessitating pest control. The households with the tower gardens appreciated the simplicity of the technology. To create demand for it requires sensitization and social marketing within the community. There is a need to determine the

hydraulic load of a tower garden to guide the number and size for a particular quantity of generated greywater for optimum performance. Also, further monitoring is needed to ascertain the vegetable yield per soil volume and impact on crops [2].

70.2.2 Stormwater

Stormwater can require a similar level of treatment to wastewater and can be a variable source of water that is dependent on rainfall patterns. Stormwater supply may not be available during long dry periods. A backup supply from another water source can be used to maintain continuity of supply. Investigations into the public perception of water reuse show that the past use of water has an effect on how it is viewed [25]. From a health perspective, a study has shown that public perceptions of stormwater reuse are more positive than wastewater reuse [22].

70.2.3 Rainwater

Rainwater captured in rainwater tanks often requires little or no treatment and can be more easily used for a variety of end uses than stormwater and wastewater because of its higher raw water quality. During long dry periods, a rainwater supply may not be available but the provision of a mains water top up or bypass system can ensure continuity of supply. It should be noted that it is possible to blend multiple water sources for recycling [26].

70.3 Water Storage Options

The capacity of any harvesting and reuse scheme is significantly influenced by the size and possible type of storage system.

There are various types of storage systems including

- Rainwater tanks
- Underground storage tanks
- Aboveground storage tanks
- Surface storages (e.g., dams or wetlands)
- Groundwater (e.g., aquifer)

Storages used in urban water reuse schemes can provide a varying level of treatment in addition to other processes included in the treatment train. For example, storage in an aquifer can reduce the number of microorganisms present. Other water storages such as dams or tanks can reduce suspended solids and particulates through settling [26].

70.4 Public Perception

Public perception is a key issue for the design and implementation of water harvesting and reuse projects. In general, as the end use becomes more personal, support for water recycling falls [25]. Investigations have also shown that there is a correlation between the scale of a water harvesting and reuse project and its degree of public acceptance. Water from a person's own home is generally more acceptable than a communal or neighborhood scale water harvesting system. However, acceptance is high again with respect to a large-scale system such as that serving an urban area [22].

70.5 Reducing Freshwater Usage

Before considering the reuse of rainwater and greywater, the most basic strategy to reduce potable water consumption is to use low-flow or high efficiency fixtures and appliances wherever possible. These may include showerheads, restroom or kitchen faucets, toilets, urinals, water fountains, dishwashers, washers, cooling systems, and landscape irrigation systems [26].

70.5.1 Accepted Toilet/Urinal Strategies

The Project Building Certifiers typically installs high efficiency fixtures in its buildings. Such fixtures have been implemented in many different buildings across the urban area and are well accepted by local municipal code and the public. One resource for many of these products is EPA's Water Sense program, which certifies high-efficiency fixtures if they are shown to perform as well or better than their less efficient but code-compliant counterparts, and if they are 20% more water efficient than average products in that category.

High-efficiency toilets: High efficiency toilets have been defined by the EPA as using, on average, 20% less water per flush than the industry-accepted standard of 6.05 L per flush—around 4.84 L per flush maximum. Utilizing high-efficiency toilets can significantly reduce the amount of potable water demand, particularly if more innovative water-saving toilets are used.

Dual flush toilets: Dual flush toilets are commonly used in Europe and Australia and are gaining popularity in the United States as well. A dual flush toilet has two flushing options: a 3.03 L flush for urine only or a 6.05 L full tank flush for solid waste. The two options are typically displayed as two buttons or a variant on a handle. The handle either goes up for one flushing option and down for the other option, or a smaller handle (for the 3.03 L flush) is included in addition to the regularly sized handle. Although many users have reported that using the 3.03 L flush for all flushes causes no significant problems, it is necessary to educate building occupants on how to properly use the system. This should include instructional signage, as well as choosing a toilet with buttons or handles that are straightforward and easy to use [26].

High-efficiency urinals: High-efficiency urinals use on average 1.9 L per flush or less, which is at least half the water used to flush a traditional urinal. Several high-efficiency urinals now on the market use just a pint of water per flush, and have been found to outperform waterless urinals and perform similarly to traditional urinals [26].

70.5.2 Other Toilet/Urinal Strategies

Other fixtures exist that not currently recognized by plumbing and building codes. Use is conditional upon approval of installation.

Waterless urinals: Waterless urinals have been implemented locally over the past several years, as test cases, with mixed results. One cause for failure has been the corrosion of copper piping that drains the urinals. The US Army Corps of Engineers specifically states that drainpipes for waterless urinals should not be made of copper due to corrosion; however, many building codes require copper piping. Another key reason for waterless urinal failures is foreign liquids (such as coffee, juice, and chemicals) being poured down the urinal, which ruins the cartridge in the system. Each time a cartridge is ruined, it must be replaced. Technology for waterless urinals should continue to be explored (particularly for low-use facilities) as the technology evolves to resolve some of these issues. Alternatively, high-efficiency toilets should be implemented to save water when appropriate [26].

Composting toilet: A traditional composting toilet consists of an aerobic processing system that treats human waste with little or no water involved. This is either accomplished through composting or managed aerobic decomposition. Thermophilic bacteria break down the waste, reducing its volume and eliminating potential pathogens. Composted waste is eventually converted to a humus-like material, which may be used as fertilizer, although not suitable for food crops. Moisture must be managed (50% ±10 is desirable) in order to control odor and ensure the effectiveness of the system. Another way to manage odor is to add absorbent material, such as sawdust or peat moss, after each use. This enhances aerobic productivity and absorbs excess liquid [26].

Urine-diverting toilet: Urine diverting toilets use two separate bowls (one for solid waste, one for urine) to harvest urine. Urine is rich in nitrogen and phosphorus, and may be useful as a fertilizer should appropriate refinement technology become available. Such systems depend on occupants using the system properly and instructional signage should be included to ensure proper use. Their use is not allowed in urban areas [26].

A thorough investigation of required approvals and permits should be undertaken as part of the conceptual design of an urban water harvesting and reuse scheme. This would include consultation with relevant

- Local council
- Environment Protection Authority
- Department of Health
- Department for Water
- Natural Resources Management Boards

A proposed urban water harvesting and reuse scheme needs to meet the requirements of a range of legislation including

- Development Act
- Environment Protection Act
- Natural Resources Management Act
- Local Government Act
- Public and Environmental Health Act

A brief description for some of the requirements follows.

70.5.3 Development Act

An urban water harvesting and reuse scheme will generally be part of a larger development. However, whenever an urban water harvesting and reuse scheme is planned, it is advised that the local council be contacted to determine whether

- Development approval is required under the Development Act. The likely issues that a council may want covered in a development application involving an urban water harvesting and reuse scheme include:
  - Compatibility of the proposed scheme with council's objectives, plans, or strategies, including any relevant strategic water management plan or strategy
  - Compatibility of the proposed plan with surrounding land uses (compliance with zoning requirements)
  - Anticipated benefits and impacts associated with scheme construction and operation (including social, environmental, and economic aspects)
  - Consideration of environmental impacts during construction and operation phases
  - How public health and safety risks are addressed
  - Management arrangements (including monitoring and maintenance) for the scheme
  - What (if any) risks and/or financial obligations would be transferred to council if it operates the scheme (e.g., operations, maintenance, monitoring and reporting costs)
  - A management plan for the scheme (including monitoring and maintenance) [26]

70.5.4 Environment Protection Act

Any development, including the construction of an urban water harvesting and reuse scheme, has the potential for environmental impact which can result from vegetation removal, stormwater management, and construction. There is a general environmental duty, required by the Environment Protection Act, to take all reasonable and practical measures to ensure that the activities on a site, including during construction, do not pollute the environment in a way that causes or may cause environmental harm. Aspects of the Environment Protection Act that must be considered when an urban water harvesting and reuse scheme is being considered are discussed next [10].

70.5.5 Water Quality

Water quality is protected under the Environment Protection Act and the associated Environment Protection (Water Quality) Policy. The principal aim of the Water Quality Policy is to achieve the sustainable management of waters by protecting or enhancing water quality while allowing economic and social development. In particular, the policy seeks to

- Ensure that pollution from both diffuse and point sources does not reduce water quality
- Promote best practice environmental management

Through inappropriate management practices, building sites can be major contributors of sediment, suspended solids, concrete wash, building materials, and wastes to the stormwater system. Consequently, all precautions will need to be taken on a site to minimize potential for environmental impact during construction. In addition, the discharge of water into any water body must meet the requirements of the Environment Protection (Water Quality) Policy [1].

70.5.6 Noise

The issue of noise has the potential to cause nuisance during any construction works or ongoing operation (i.e., if pumps are required) of an urban water harvesting and reuse scheme. The noise level at the nearest sensitive receiver should be at least below the Environment Protection (Industrial Noise) Policy allowable noise level when measured and adjusted in accordance with that policy. Reference should be made to the EPA Information Sheets on Construction Noise and Environmental Noise, respectively, to assist in complying with this policy. Air quality may be affected during the construction phase of an urban water harvesting and reuse scheme. Dust generated by machinery and vehicular movement during site works, and any open stockpiling of soil or building materials at a site, must be managed to ensure that dust generation does not become a nuisance offsite [10].

70.5.7 Waste

Any wastes arising from excavation and construction work on a site should be stored, handled, and disposed of in accordance with the requirements of the Environment Protection Act. For example, during construction, all wastes must be contained in a covered waste bin (where possible) or alternatively removed from the site on a daily basis for appropriate offsite disposal [10].

70.6 Legislative Requirements and Approvals

70.6.1 Setting Standards to Protect Health and Safety

Most urban areas set minimum standards that regulate the design and construction of plumbing systems. Approval is also needed for rainwater harvesting and greywater systems for reuse interior to a building.

Most of the plumbing building codes currently do not include standards for water reuse systems, although this may change in the near future given recent activity around urban area legislation. Until these standards are fully established, public health and safety is protected through required review by the urban area Permit Program. A growing number of water reuse systems have been approved to date in many urban areas. State approval is not necessary for rain barrel exterior rainwater harvesting systems for irrigation, as they are not physically tied to the public water system, and therefore are exempt from the state plumbing code. However, cisterns, which store larger quantities of water and have greater maintenance requirements, and may have a connection to city water for supplemental supply, are subject to review under state and city code. Both such systems are widely accepted and approved in many urban areas [10].

At the core of the regulatory discussion is the responsibility to ensure public health and safety. To this end, it is important to have accepted minimum standards for water reuse system operations. Such systems are relatively new in the Midwest region, and water quality and development standards for water reuse systems have yet to be established locally. For such newly emerging systems, it is critical to protect public health and safety while regulatory standards are being drafted. The water reuse systems that currently exist in some urban areas have been approved on a case-by-case basis. Requirements vary, but most include the incorporation of chlorine dosing and/or micron filtration and/or UV filtration into the treatment process. In the absence of accepted standards, Minimum Sanitary Requirements for Bathing Beaches has been used at times as the standard to which water must be treated within rainwater harvesting or greywater reuse systems. In particular, the regulator requires that rainwater or greywater be treated to the minimum sanitary requirements for bathing beaches before it is reused. This means that water reuse systems must remove suspended solids, viruses, and *E. coli* bacteria to a level of less than 235 colony-forming units per 100 mL, and fecal coliform bacteria to a level of less than 500 colony-forming units per 100 mL. The Building Code should contain key provisions when approving a water reuse system. Specifically, CBC requires that only potable water be supplied to plumbing fixtures that provide water for drinking, bathing, or cooking purposes, or for the processing of food, medical, or pharmaceutical products. Potable water also must meet the requirements set forth in the Public Area Sanitary Practice Code. This provision requires that potable water be supplied to all plumbing fixtures, unless expressly permitted, and then have the responsibility to evaluate the proposed system to determine whether it aligns reasonably with safety standards. The institution of regulations and standards related to water reuse systems will eventually resolve the need for individual projects to be considered independently, and will facilitate water harvest system reuse [10].

Water reuse systems are an important component of the environmental sustainability program. This section highlights a review of techniques to reduce overall potable water demand in urban areas, as reduction of potable water usage is a critical first step in accomplishing the goals of this Handbook. This section also provides design guidelines for rainwater harvesting and greywater systems, and points to ideas for utilizing blackwater systems. These recommendations have been informed by research and local, national, and international case studies, and provide strategies for developing water reuse systems. However, users of this Handbook should keep in mind that different buildings would have different needs depending on the water sources being used, project budget, aesthetic and space planning concerns, etc.

Certain activities—whether development or not—require a license granted under the Environment Protection Act. The discharge of stormwater from stormwater infrastructure to underground aquifers in urban areas requires a license. It should be noted that there is no provision in the license for extraction of the water. The Code of Practice for Aquifer Storage and Recovery [12] outlines the requirements of the Environment Protection Authority for the storage of waters in aquifers. It should be noted that the Code of Practice for Aquifer Storage and Recovery is currently under review by the EPA, and a revised draft will cover managed aquifer recharge (MAR).

70.6.2 Guidelines

The guidelines describe methods by which reclaimed water can be used in a sustainable manner without imposing undue risks to public health or the environment [6]. It considers the use of reclaimed water for agricultural, municipal, residential (nonpotable), environmental, and industrial purposes. It provides information on the quality of reclaimed water required for each use, treatment processes, system design, operation, and reliability, site suitability, and monitoring and reporting. These guidelines should be consulted when considering an urban water harvesting and reuse scheme. If groundwater is to be extracted from the aquifer, the proponent must obtain a license from the relevant department. The proponent must also obtain a well construction permit from the Department for Water for any proposed wells (i.e., groundwater bores) that will intersect the water table. The relevant department licenses the discharge of stormwater to underground aquifers wherever an EPA license is not required under the Environment Protection Act. This relevant department also provides the required information and assistance in establishing an urban water harvesting and reuse scheme concerning health issues.

The Environment Protection and Heritage Council and the Natural Resource Management Ministerial Council have developed Australian Guidelines for Water Recycling [13]. The guidelines comprise a risk management framework and specific guidance on managing the health risks and the environmental risks associated with the use of recycled water. Phase one of the guidelines focuses on large-scale treated wastewater to be used for

- Residential garden watering, car washing, toilet flushing, and clothes washing
- Irrigation for urban recreational and open space, and agriculture and horticulture
- Fire protection and firefighting systems
- Industrial uses, including cooling water
- Greywater treated on site (including in high-rise apartments and office blocks) for use for garden watering, car washing, toilet flushing, and clothes washing. The Australian Guidelines for Water Recycling [13] call for a four-step process to prepare the required risk management plan for a recycled water scheme. The guidelines state that a risk management plan should be prepared for every recycled water system. Phase two of guideline development is currently in draft and focuses on three modules:

- Stormwater harvesting and reuse
- Managed aquifer recharge
- Augmentation of drinking water supplies

The guidelines are expected to replace the existing South Australian Reclaimed Water Guidelines (Treated Effluent) [11] and will be the basis for assessment of urban water harvesting and reuse schemes [10].

70.7 Design Process

70.7.1 Overview

There is a range of scales and types of urban water harvesting and reuse schemes that can be designed and installed. The type of scheme can vary from a greywater diversion hose in a household yard for garden irrigation to a community scale dual reticulation system using tertiary treated wastewater. The scope and degree of complexity is dependent on the individual system.

Key drivers for complexity in the systems are

- The number of users
- The quality of the water to be recycled
- The end use

The greater the treatment requirements, the more complex the treatment component and the more involved the monitoring and management systems will need to be.

Water harvesting and reuse schemes can be implemented either in existing urban areas or as part of a new urban development. The project's context will therefore influence the nature of the planning and design process. The key steps in the design process for an urban water harvesting and reuse scheme include

- Assess the site, catchment, and appropriate regulatory requirements
- Identify the objectives and targets
- Identify potential options
- Consult with key stakeholders and relevant authorities
- Evaluate options
- Prepare a detailed design of selected option
- Undertake the approvals process
- Develop an operations, maintenance, and monitoring plan

The design process is likely to be iterative, requiring several rounds of review in earlier stages as new information arises and negotiations progress with stakeholders (including end users) that may alter the objectives and/or available options.

In summary, Water Sensitive Urban Design (WSUD) is an approach to urban planning and design that integrates the management of the total water cycle into the urban development process. It includes

- Integrated management of groundwater, surface runoff (including storm water), drinking water and wastewater to protect water related environmental, recreational and cultural values
- Storage, treatment, and beneficial use of runoff

- Treatment and reuse of wastewater
- Using vegetation for treatment purposes, water-efficient landscaping, and enhancing biodiversity
- Utilizing water saving measures within and outside domestic, commercial, industrial, and institutional premises to minimize requirements for drinking and non-drinking water supplies.

Therefore, WSUD incorporates all water resources, including surface water, groundwater, urban and roof runoff, and wastewater [10].

70.7.2 Assess the Site, Catchment, and Appropriate Regulatory Requirements

WSUD responds to site conditions and land capability and cannot be applied in a standard way. Careful assessment and interpretation of the site conditions is a fundamental part of designing a development that effectively incorporates WSUD. This step identifies and assesses the potential constraints and opportunities of the proposed project site. Potential constraints may include

- Topography
- Land use (including surrounding catchment land use)
- Adjacent land uses (including potential land use conflicts)
- Watercourse characteristics
- Vegetation and other sensitive ecosystems (potential biodiversity impacts)
- Soil characteristics, such as salinity or acid sulfate
- Existing water management infrastructure
- Depth to groundwater, groundwater quality and existing uses of the groundwater in the vicinity
- Statutory or regulatory constraints

This step should identify opportunities for reusing treated stormwater or wastewater, as well as suitable locations for storages. Other aspects of the end users' operations may also be important, such as future development plans or land use changes that may affect longer-term water use patterns. The level of the site and catchment investigation required should match the size and scale of the development and its potential impacts (i.e., larger developments having a greater impact would require greater site investigation). A staged approach to site investigations can be adopted to minimize costs. This involves an initial screening level assessment using readily available information to identify major constraints and opportunities, and then focusing efforts on any identified constraints. An evaluation of the pollutants that may be present within runoff needs to be carried out on a catchment basis, as the quality of runoff for a reuse project is affected by the characteristics of the scheme's catchment. Pollutants will vary according to whether the catchment drains residential, industrial, rural, or a combination of any of these land use types. For example, the risk of chemical pollution in a catchment increases with the extent and nature of industrial uses and paved roads, particularly those with high traffic volumes [10].

The impact of such diffuse pollution sources can be gauged by investigating water quality during wet and dry weather, or by referring to existing water quality data. Similarly, the scheme should investigate the impacts on water quality from any point sources of pollution. The hazard assessment for the scheme may need to consider both diffuse and point sources of pollution. Concentrations of pollutants typically have seasonal or within event patterns, and heavy pollutant loadings can be avoided by being selective in the timing of diversions (e.g., not diverting flow during large floods when treatment systems are often bypassed). Knowledge of the potential pollutant profile helps to define water quality sampling and analysis costs when determining the viability of a project (e.g., if there are any specific industrial activities upstream that contribute particular pollutants such as hydrocarbons).

70.7.3 Identify Objectives and Targets

The design objectives and targets will vary from one location to another and will depend on site characteristics, development form, and the requirements of the receiving ecosystems. It is essential that these objectives are established as part of the conceptual design process and approved by the relevant council prior to commencing the engineering design. Specifying the objectives for an urban water harvesting and reuse scheme is an important step for ensuring that it operates as intended. In developing reuse schemes for a site, broader catchment or regional objectives are important. These could involve specified reductions in

- Mains (potable) water use
- Runoff flow rates and/or volumes
- Runoff pollution loads
- The effective (connected) impervious area of the catchment
- Wastewater disposal volumes

Organizational objectives, government policies, and environmental planning instruments may also provide a strategic context for the project [10].

The most common project objectives will relate to

- Managing public health and safety risks
- Managing environmental risks
- Meeting the requirements of the end user, primarily relating to water quality, quantity, and reliability of supply
- Protecting or enhancing visual amenity or aesthetics

70.7.4 Identify Potential Options

This step identifies various possible layouts for a scheme to meet the project's objectives. Various combinations of measures can be used in a water harvesting and reuse scheme, depending on the nature of the site and the end uses. The design process needs to consider the following components:

- Collection (i.e., swales)
- Storage (i.e., rainwater tanks, wetlands, underground tanks)

- Treatment (i.e., wetlands, wastewater treatment plant)
- Distribution

This step is likely to involve modeling the outcomes from various options and identifying the degree to which each option meets the adopted project objectives. This could be iterative, modeling the influence of a number of key aspects of the project (such as different storage volumes against predicted outcomes), and may include modeling of

- Water balance
- Water pollution and environmental flows
- Water peak flows and flood levels

A risk assessment approach should be utilized during this stage of the process [10].

70.7.5 Identify and Consult with Key Stakeholders

The designer (or applicant) should liaise with civil designers and council officers prior to preceding any further to ensure

- Urban water harvesting and reuse scheme will not result in water damage to existing services or structures
- Access for maintenance to existing services is maintained
- No conflicts arise between the location of services and WSUD devices
- The objectives are consistent with the council's directions for the area

The council will also be able to advise whether

- Development approval is required, and if so, what information should be provided with the development application
- Any other approving authorities should be consulted
- Any specific council requirements need to be taken into consideration

Key stakeholders should also be consulted throughout the planning process (depending on the scale of the scheme), particularly during the setting of project objectives. Their engagement in the scheme from the planning stage will

- Allow for any concerns or misconceptions to be identified and addressed early in the scheme
- Provide opportunities for educating the community and the proponents and build user confidence in the scheme, resulting in greater use of treated water as an alternative to mains water

The key stakeholders will depend on the nature of the scheme [10].

70.7.6 Evaluate Options

The various options identified should be evaluated, taking into account social, economic, and environmental considerations. The evaluation of options should primarily assess how well each option meets the project's objectives. It is likely that during this process trade-offs between objectives may need to be assessed as, for example, it may not be cost-effective to meet all objectives. There is no widely used evaluation technique for urban water harvesting and reuse schemes. This may be partially due to the difficulty in quantifying many of the costs and benefits of such schemes, and where some of the costs and benefits can be attributed to parties not directly involved in the proposed scheme. Possible evaluation techniques include

- Cost-benefit analysis
- Triple bottom line analysis
- Multiple criteria decision analysis

70.7.7 Detailed Design of Selected Option

During the detailed design of the selected scheme, a risk management strategy should be developed. This should identify public health and environmental hazards and an appropriate mix of controls to be implemented during the design and operational phases. As discussed previously, there are several approvals that would generally be required for an urban water harvesting and reuse scheme. Therefore, ensuring that there is adequate time to obtain the approvals is an important part of the process [10].

70.7.8 Maintenance and Monitoring

Appropriate maintenance of urban water harvesting and reuse schemes is important to ensure that the scheme continues to meet its design objectives in the long term and does not present public health or environmental risks. The actual maintenance requirements will depend on the nature of the scheme. Maintenance may include measures relating to each element of a scheme. Protection of treatment, retention and detention systems from contamination is a necessary part of designing an urban water harvesting and reuse system. This includes constructing treatment systems away from flood-prone land, taking care with or avoiding the use of herbicides and pesticides within the surrounding catchment, planting nondeciduous vegetation, and preventing mosquitoes and other pests from breeding in storage ponds. Contingency plans should be developed to cater to the possibility of contaminated water being inadvertently utilized. These plans should focus on

- Determining the duration of recovery pumping required (to extract contaminated water)
- Sampling intervals required
- Managing recovered water

Regular inspections of a scheme are needed to identify any defects or additional maintenance required. The inspections may need to include

- Storages for the presence of cyanobacteria (i.e., algae), particularly during warmer months
- Spillways and creeks downstream of any online storage after a major storm for any erosion
- Water treatment systems

- Distributions systems for faults (e.g., broken pipes)
- Irrigation areas for signs of erosion, under watering, water logging, or surface runoff [10]

70.7.9 Approximate Cost

Due to the variability in the scale and type of urban water harvesting and reuse schemes, it is difficult to provide an indication of the approximate costs of construction and operation of such schemes. However, Kellogg Brown, Root Pty Ltd [19] undertook a review of various scale stormwater-harvesting schemes and was able to develop a relationship between unit production costs against average annual production. For example, it is estimated that a 10 ML stormwater-harvesting scheme would have a water supply cost of $2/kL [10].

70.8 Summary and Conclusions

Wastewater reuse is a proven technology that has been used for more than half a century. It is a drought-proof, renewable supply of water that will help hilly urban communities' water resources from shrinking, and waterways from becoming polluted.

The common sources of greywater in hilly urban areas are bathroom and laundry. Landry greywater has been mostly reused to clean houses and then disposed of.

The most common reasons for establishing a wastewater reuse program is to identify new water sources for increased water demand and to find economical ways to meet increasingly more discharge standards.

The irrigation of public parks, schoolyards, highway medians, and residential landscapes, as well as for fire protection and toilet flushing in commercial and industrial buildings use greywater. The goal of the water reuse in hilly urban areas is to examine the water reuse strategies and identify best practices in water reuse strategies. Water reuse systems have and will continue to become more efficient [10].

References

1. Abu Ghunmi, L., Zeeman, G., van Lier, J., and Fayyad M. 2008. Quantitative and qualitative characteristics of grey water for reuse requirements and treatment alternatives: The case of Jordan. *Water Science Technology*, 58, 1385–1396.
2. Alit, W., Zeev, R., Noam, W., Eilon, A., and Amit, G. 2006. Potential changes in soil properties following irrigation with surfactant-rich greywater. *Ecological Engineering*, 26, 348–354
3. Al-Jayyousi, O.R. and Odeh, R. 2003. Grey water reuse: Towards sustainable management. *Desalination*, 156(1–3), 181–192.
4. American Public Health. 1998. American water works association and water environment federation (APHA/AWWA/WEF). *Standard Methods for the examination of Water and Wastewater*. APHA/AWWA/WEF United Press Inc, Baltimore.

5. Ayers, R.S. and Westcot, D.W. 1994. Water quality for agriculture. Food and Agriculture Organisation (FAO) Irrigation and Drainage Paper 29.

6. Birks, R. and Hills, S. 2007. Characterisation of indicator organisms and pathogens in domestic grey water for recycling. *Environmental Monitoring Assessment*, 129, 61–69.

7. Carden, K., Armitage, N., Winter, K., Sichone, O., Rivett, U., and Kahonde, J. 2007. The use and disposal of grey water in the non-sewered areas of South Africa: Paper l-quantifying the grey water generated and assessing its quality. *Water SA*, 33, 425–432.

8. Christova-Boal, D., Eden, R.E., and McFarlane, S. 1996. An investigation into grey water reuse for urban residential properties. *Desalination*, 106, 391-397.

9. Crosby, C. 2005. Food from used water: Making the previously impossible happen. *The Water Wheel*, (Jan./Feb.), 10–13.

10. Department of Planning and Local Government. 2010. *Water Sensitive Urban Design Technical Manual for the Greater Adelaide Region*. Government of South Australia, Adelaide, Australia.

11. Environment Protection Authority South Australia. 1999. *South Australian Reclaimed Water Guidelines*. Adelaide, South Australia.

12. Environment Protection Authority South Australia. 2004. Code of Practice for Aquifer Storage and Recovery.

13. Environment Protection and Heritage Council. 2006. Australian Guidelines for Water Recycling: Managing Health and Environmental Risks (Phase 1).

14. EPA. 2004. *Guidelines for Water Reuse*. U.S. Environmental Protection Agency. EPA/625/R-04/108.

15. Eriksson, E., Auffarth, K., Henze, M., and Ledin, A. 2002. Characteristics of grey wastewater. *Urban Water*, 4, 85–104.

16. Hernàndez, L.L., Zeeman, G., Temmink, H., and Buisman, C. 2007. Characterisation and biological treatment of grey water. *Water Science Technology*, 56, 193–200.

17. Imhof, B. and Muhlemann, J. 2005. *Grey Water Treatment on Household Level in Developing Countries: A State of The Art Review*. Department of Environmental Sciences at the Swiss Federal Institute of Technology (ETH), Zurich, Switzerland.

18. Jefferson, B., Laine, A.T., Parsons, S., Stephenson, T., and Judd, S. 2000. Technologies for domestic wastewater recycling. *Urban Water*, l, 285–292.

19. Kellogg Brown, Root Pty Ltd. 2004. *Metropolitan Adelaide Stormwater Management Study, Part B—Stormwater Harvesting and Use*. Local Government Association, South Australia.

20. Kulabakoa, N.R., Ssonko, N.K.M., and Kinobe, J. 2011. Grey water characteristics and reuse in tower gardens in peri-urban areas—experiences of Kawaala, Kampala, Uganda. *The Open Environmental Engineering Journal*, 4, 147–154.

21. McGechan, M.B. and Lewis, D.R. 2002. Sorption of phosphorus by soil, part 1: Principles, equations and models. *Bioprocess and Biosystems Engineering*, 82, 1–24.

22. Mitchell, V.G., Hatt, B.E., Deletic, A., Fletcher, T.D., McCarthy, D., and Magyar, M. 2006. *Integrated Stormwater Treatment and Harvesting*. Institute for Sustainable Water Resources, Faculty of Engineering, Monash University. Melbourne.

23. Morel, A. and Diener, S. 2006. *Grey Water Management in Low and Middle-Income Countries, Review of Different Treatment Systems for Households or Neighborhoods*. Swiss Federal Institute of Aquatic Science and Technology.

24. Patel, R.B. and Burke, T.F. 2009. Urbanisation—An emerging humanitarian disaster. *New England Journal of Medicine*, 368, 741–743.

25. Po, M., Kaercher, J.D., and Nancarrow, B. 2003. *Literature Review of Factors Influencing Public Perceptions of Water Reuse*. CSIRO Land and Water, Australia.

26. Public Building Commission of Chicago. 2011. *Water Reuse Handbook*. Board of commissioners, August 2011, p. 75.

27. WHO. 2006. *Guidelines for Safe Use of Wastewater, Excreta and Grey Water Volume 4: Excreta and Grey Water Use in Agriculture*. World Health Organization, Geneva.

28. Yusof, K. and Kwai-Sim, L. 1990. Urbanization and its effects on health in squatter areas. *Journal of Human Ergology*, 19, 171–184.

29. Zìmmo, O.R., Van der Steen, N.P., and Gijzen, H.J. 2003. Comparison of ammonia volatilisation rates in algae and duckweed based waste stabilisation ponds treating domestic wastewater. *Water Research*, 37, 4587–4594.

XVI

Water Reuse Case Studies

71

Feasibility Studies for Water Reuse Systems

Kumari Rina
Central University of Gujarat
Jawaharlal Nehru University

Saeid Eslamian
Isfahan University of Technology

Geetika Tyagi
Space Application Center

Neha Singh
Jawaharlal Nehru University

PREFACE

This chapter describes feasibility of water reuse in an urban environment, the importance of reuse, and the methodology to achieve this. It also emphasizes the barriers and issues in water reuse including guidelines in water reuse. Attention has also been given to the world scenario of water reuse along with life-cycle assessment of water reuse.

71.1 Introduction

Water is essential to human survival. Freshwater scarcity is a growing problem throughout the world. Throughout the world, freshwater resources and population densities are unevenly distributed. As a result, water demands already exceed supplies in regions with more than 40% of the world's population [12]. Along with surface-water, groundwater is an important source of water supply throughout the world. More than 1.5 billion people depend on groundwater sources worldwide, withdrawing approximately 600–700 km³/yr, that is, about 20% of global water withdrawals.

During the past 50 years, groundwater depletion has been reported from isolated pockets to large areas in many countries throughout the world [2]. Globally, freshwater demands are increasing mainly due to population growth, globalization, investment policies, urbanization, and intensification of agriculture and industrialization. Per capita consumption and the demands for the industrial and agricultural sector are increasing. In 2025, nearly one-third of the population of developing countries, some 2.7 billion people, will live in a severe water scarce region. As urban areas continue to grow, pressure on local water supplies will continue to increase. The scarcity of water resources will be one of the most challenging constraints in the coming decades that will militate against economic growth, social justice, and ecological integrity in many developed and developing countries of the world. They will have to reduce the amount of water used in irrigation and transfer it to the domestic, industrial, and environmental purposes. Along with availability of water, industrialization and expanding urbanization has also degraded the water quality in many parts of the world. More water consumption is also directly associated with increased wastewater generation.

Along with water scarcity, pollution of water resources poses a critical challenge in many developing countries. Therefore, ensuring the adequate quantity and quality of water in urban areas is becoming difficult for the authorities to manage water supply, and it is important to look for affordable, implementable and safe solutions to alleviate water problems. Water reclamation, recycling, and reuse address can solve these challenges by creating new sources of high-quality water supplies. There is tremendous potential for reclaimed treated effluent [45]. Although water reclamation and reuse is practiced in many countries around the world, current levels of reuse constitute a small fraction of the total volume of municipal and industrial effluent generated. There are only a few countries that are fully implementing this practice. The wastewater is usually a mixture of domestic and industrial effluents as well as stormwater. Industrial wastewater often contains high concentration levels of metals, metalloids, and volatile or semi-volatile compounds, while domestic wastewater is rich in nutrients and pathogens.

Traditionally, wastewater treatment was mainly confined to pollution abatement, public health protection, and environmental protection by removing biodegradable material, nutrients, and pathogens [39].

71.2 Wastewater Reuse as Environmentally Sound Technologies

When treated appropriately, wastewater reuse is considered an example of environmentally sound technology applications (Agenda 21). It has become an attractive alternative for conserving conventional water resources. It represents a viable, long-term solution to the challenges faced by growing municipal, industrial, and agricultural demands for water.

Throughout the world, water reuse has been developed for two purposes:

1. Potable uses
   a. Direct—Use of reclaimed water for drinking water supply following high levels of treatment
   b. Indirect—After passing through the natural environment
2. Nonpotable uses
   a. Irrigated agriculture
   b. Use for irrigating parks and public places such as golf courses
   c. Use for aquaculture

d. Aquifer recharge (indirect reuse) or uses in industry and urban settlements, etc.

It has numerous benefits:

1. Reduces the demand pressure on the conventional water resources
2. It protects the environment, is less polluting
3. Typically uses less energy than importing water
4. Recycles more of their wastes and products and uses resources in a sustainable manner
5. Avoids construction impact of new supply development
6. Reduces the quantity of wastewater discharged to sensitive surface water body

In developed countries, approximately 73% of the population is provided with wastewater collection and treatment facilities, while in developing countries, the percentage is reduced to only 35% [71].

71.3 What Is a Feasibility Study?

Wastewater reuse has been recognized as a potential strategy in addressing water scarcity [28] and to achieve sustainable water management. Energy efficiency and sustainability are key drivers of water reuse. A feasibility study is defined as an evaluation or analysis of the potential impact of a proposed project or program. Generally, it is carried out to help decision-makers in determining whether to implement a particular project [72]. It is based on extensive research and it contains much data related to financial and operational impact. It also tells about advantages and disadvantages of both the current situation and the proposed plan. In urban areas, only 15%–25% of water diverted or withdrawn is consumed, the rest being returned as wastewater to the urban hydrologic system.

The word "feasible" can be defined in three different ways. The first is "capable of being done, executed or effected"; the second is "capable of being managed, utilized or dealt with successfully"; and the third one is "reasonable, likely" [72]. Most guidelines define "feasibility" with the following meanings:

- The degree to which a given alternative mode, management strategy, designs, or location is economically justified
- The degree to which such an alternative is considered preferable from an environmental or social perspective
- The degree to which eventual construction and operation of such an alternative can be financed and managed

Recently, the concept of "sustainability" has been rapidly implemented in wastewater management [40]. According to the concept of sustainable development, a water-reuse project must comply with environmental, sociocultural, and economic needs [10].

71.4 Wastewater Reuse Terminology

There is some terminology that is generally used in water reuse:

- Wastewater reuse: The use of treated wastewater for a beneficial use.

- Reclaimed water: Wastewater treated or processed to a certain standard for reuse.
- Recycled water: Synonymous with reclaimed water.
- Effluent: The treated water discharged by a wastewater treatment plant.
- Planned reuse: A planned reuse system involves treatment of wastewater to careful standards to be reused for predetermined uses.
- Unplanned reuse: An unplanned reuse system involves inadvertent reuse of wastewater. The most common example of this is downstream river users in rivers where treated wastewater is discharged upstream (although wastewater is highly diluted).

71.5 Categories of Water Reuse

Based on demand and treatment process, water reuse applications can be divided into seven categories.

71.5.1 Agriculture

Irrigation represents the largest user of reclaimed water throughout the world. Using reclaimed water to replace part of the agriculture demand can alleviate local water stress. Furthermore, nutrients contained in the wastewater can also reduce the demand of fertilizers. Effluent from secondary treatment was recommended for irrigating non-food crops, orchards, and vineyards, while effluent from tertiary treatment was recommended for food crop irrigation. To date, California and Florida are the leading states for agricultural water reuse in the United States, reusing 48% and 19% of the total volume of reclaimed water, respectively [23].

Wastewater is being used for irrigation on an estimated 4.5 million ha worldwide [34]. Other estimates suggest that about 200 million farmers irrigate with treated and untreated wastewater, on an estimated 20 million ha [29,57,59].

71.5.2 Landscape Irrigation

It is the second largest user of reclaimed water in industrialized countries and in developing countries as well. It includes the irrigation of parks, playgrounds, golf courses, landscaped areas around commercial, office, and industrial developments, and landscaped areas around residences.

71.5.3 Industrial Activities

It represents the third major use of reclaimed water. Industrial reuse has increased substantially since the early 1990s because of water shortages and increased population pressure [23]. It is primarily used for cooling, boiler makeup water, and other industrial process water [23]. Other industries such as petroleum refineries and chemical manufacturers also require substantial amounts of water. The thermal power sector represents around 50% of the total water withdrawal in the United States in 2005 [74]. These industries, however, do not require water quality as high as for potable supply; they can easily use reclaimed water. Reusing water also helps the industries to reduce cost and improve sustainability. On the other hand, reclaimed water may cause problems such as corrosion, biological growth, and scaling compared with freshwater.

71.5.4 Groundwater Recharge

This is also one of the major applications for water reuse, either via spreading basins or direct injection to groundwater aquifers. Groundwater recharge can alleviate land subsidence and seawater intrusion in coastal areas. It also provides water storage and further treatment for subsequent retrieval and reuse of the reclaimed water. Major planned indirect potable reuse projects have been carried out in places such as Orange County Water District in California and the Occoquan Reservoir in northern Virginia [6] in the United States and in Singapore as "Newater" [54]. The two leading states, Florida and California, use 16% and 15% of the reclaimed water for groundwater recharge, respectively.

71.5.5 Recreational and Environmental Uses

It constitutes one of the largest uses of reclaimed water in industrialized countries and involves nonpotable uses related to land-based water features such as the development of recreational lakes, marsh enhancement, and stream flow augmentation. Manufactured lakes, golf course storage ponds, and water traps can be supplied with reclaimed water. Reclaimed water has been applied to wetlands for a variety of reasons including habitat creation, restoration, and enhancement.

71.5.6 Nonpotable Reuse

In urban areas fire protection, air conditioning, toilet flushing, construction water, and flushing of sanitary sewers are potential users of reclaimed water. Typically, for economic reasons, these uses are incidental and depend on the proximity of the wastewater reclamation plant. In addition, the economic advantages of urban uses can be enhanced by coupling with other ongoing reuse applications such as landscape irrigation.

71.5.7 Potable Reuse

Reclaimed water can also be used either by blending in water supply storage reservoirs or, in the extreme, by direct input of highly treated wastewater into the water distribution system. Public acceptance is another major issue for implementing the direct potable reuse [23]. Planned potable recycling has taken place at Windhoek in Namibia since 1968 [4]. Cloudcroft in New Mexico and Big Springs in Texas in the United States have also recently started reusing water for direct potable use [65].

71.6 Issues, Barriers, and Impediments to Reuse

There are some issues and barriers that presently hinder water reuse; generally, it varies with the region. Some of these points are as follows.

71.6.1 Lack of Available Funding

Although many larger municipalities have constructed water reuse project, lack of funding is probably a major constraint around the world, especially in developing countries [45].

71.6.2 Need of Public Education

An educational campaign is needed to provide information to local public, decision makers, and politicians on costs and benefits of water reuse and about the success stories. Local decision-makers, generally do not consider water reuse as an option when planning for water resource alternatives in areas where the rainfall is abundant.

71.6.3 Public Perception/Public Acceptance

It is also noted throughout the world that nonpotable reuse applications (e.g., landscape irrigation, golf) are generally acceptable, but as reuse moves toward indirect potable reuse, the public generally responded negatively. Respondents in both the United States and UK noted the need for research into the psychological factors that lead to the public's resistance to indirect potable reuse.

71.6.4 Advanced Research

Advanced research is also needed to improve the economics of water reuse by reducing the cost of technologies, to ensure technology transfer, and to promote reuse for specific industrial applications (e.g., cooling in power plants, concrete production, etc.). Most wastewater, especially in arid and semi-arid areas, cannot be used to irrigate golf courses or for industrial applications. It needs to be recycled to serve growing demands of populations. These highly treated reclaimed waters can be used to irrigate edible crops and for indirect potable reuse.

71.6.5 Role of Government

Government has to play a leadership role in assuring adequate water resources for regional areas and to improve water use efficiency systems for which use of reclaimed water is one of the major potential sources. Government has to provide sufficient funding to promote water use efficiency and conservation.

71.6.6 Better Documentation of the Economics of Water Reuse

Although several practitioners have documented the need for a complete accounting of financial and social costs and non-monetizable benefits, such an accounting has yet to be accomplished. A well-written, documented economic treatise on water reuse is needed as a resource.

71.6.7 Chemical and Microbiological Safety

The biggest challenge is to apply treatment technologies that will remove exotic chemicals as well as remove or inactivate microbial pathogens. This must be done on a continuous basis and water utilities must be able to convince the public that the recycled water is chemically and microbiologically safe for the intended applications.

71.7 World Scenario

Throughout the world, studies on wastewater reuse have been carried out. As compared to developed countries, in developing countries wastewater treatment is limited as investment in treatment facilities have not kept pace with persistent increase in population and consequent increase in wastewater volume in many countries. The present text covers the reuse of wastewater study in different regions of the world.

Taking into account the largest water reclamation users (Europe, Israel, California, Japan, and Australia), more than a half of recycled water is used for agricultural purposes. Urban use is the second most used type of recycled water in these regions, immediately followed by groundwater recharge. Reuse of wastewater in different parts of the world is given in Table 71.1.

TABLE 71.1 Reuse of Wastewater in Different Regions of the World

Regions	Water Reuse (Mm³/a)						
	Total	Agricultural	Ground Water Recharge	Industrial	Ecological Use	Urban Use	Domicillary
Europe and Israel	963	674	164	39	48	39	0
California-EUA	434	213	61	22	56	82	0
Japan	206	16	0	16	66	78	29
Australia	166	50	0	66	5	40	5
Total (Mm³/a)	1769	953	224	143	175	239	34
Percentage (%)		54	13	8	10	14	2

TABLE 71.2   Wastewater Generated, Treated, and Used in North America

Country	Wastewater Generated		Wastewater Treated		Treated Wastewater Used	
	Reporting Year	Volume (km³/year)	Reporting Year	Volume (km³/year)	Reporting Year	Volume (km³/year)
Canada	2006	5.395[a]	2006	4.47[a]	–	NA
United States	1995	79.573[b]	1995	56.642[b]	2002	2.345[c]

NA refers to data not available.
[a] Environment Canada [22].
[b] Solley et al. [62].
[c] USEPA [23].

71.7.1 North America

In North America each year wastewater generation is about 85 km³, of which 61 km³ are treated. The annual use of treated wastewater accounts for 2.3 km³, which is only 3.8% of the wastewater treated in the region. Thus, while about 75% of the wastewater generated in North America is treated, only a small portion is used. The amount of wastewater generated and treated in North America is given in Table 71.2.

In the United States, 34.9 bgd (billion gallons per day) (132.1 GL/d) is produced of which only 2.6 bgd (9.8 GL/d) are reused, which represents only 7.4% of total reused waste water [45]. In the United States, California and Florida use notable amounts of reclaimed water in agriculture. An estimated 46% of California's annual reclaimed water use takes place in agriculture. In Florida, the proportion is 44% [16]. Due to increased stress on water resources, wastewater use in Arizona, California, and Texas is motivated.

71.7.2 Latin America

In Latin American countries, only about 20% of generated wastewater undergoes treatment because many Latin American countries do not have well-developed wastewater collection and treatment systems [70]. In 8 of 15 Latin American countries, less than half the population is connected to wastewater collection and treatment systems [70]. In Mexico, an estimated 70,000 and 190,000 ha are irrigated with treated and untreated wastewater, respectively. In Peru, an estimated 1350 and 9346 ha are irrigated with treated and untreated wastewater, respectively. In Argentina and Chile, the areas irrigated with treated wastewater are similar or larger than the areas irrigated with untreated wastewater [33]. In Chile, untreated wastewater was used directly for agricultural purposes until 1992. The 1991 cholera epidemic in Chile made it necessary to revisit the approach of using untreated wastewater for food production in that country [77].

71.7.3 Europe

In Europe, most of the wastewater generated undergoes treatment (71%), due to high public awareness regarding health and environment protection as well as technological advancement for wastewater treatment methods [51,70]. In addition, the legal and regulatory framework for water and wastewater management plays a crucial role in supporting wastewater treatment in the region. In southern Europe, reclaimed wastewater is used predominantly for agricultural irrigation (44% of wastewater reused) whereas in northern Europe, wastewater is used primarily for environmental applications (51% of wastewater reused). Countries that are mainly using wastewater for agriculture include Spain, Italy, Greece, Cyprus, France, and Portugal. From the 964 mm³/yr of treated wastewater that is used in Europe, 347 mm³/yr are accounted for by Spain and 233 mm³/yr by Italy [43]. Cyprus uses 100% of the treated wastewater that is generated (MED WWR WG, 2007) and the main application is irrigation for agriculture [84]. In France, the effluent from the WWTP of Clermont-Ferrand, designed for 400,000 population equivalents, is partially used for crop irrigation by a farmers' union. In Portugal, constructed wetlands have been tested as reclamation technology for irrigation [42] and evaluated as suitable for such purposes, according to limits set by Metcalf and Eddy [44], US EPA [23], Bixio and Wintgens [14], Asano et al. [6], and Marecos do Monte and Albuquerque [41].

According to the country-specific situations, there can be multiple uses of wastewater. Seventy-one percent of volume of wastewater in Spain is used for irrigation, 17% for environmental applications, 7% for recreation, 4% for urban reuse, and 0.3% for industrial purposes [24]. In Portugal, the area irrigated with treated wastewater ranges between 35,000 and 100,000 ha, depending on the storage retention time for wastewater [5]. In Cyprus, 38,200 ha are irrigated with treated wastewater. The primary use of wastewater in Italy also is for irrigation, which covers 28,285 ha [33].

71.7.4 Middle East and North Africa (MENA)

Middle Eastern and North African (MENA) countries are located within arid and semi-arid areas characterized with low availability of freshwater resources [81]. The amount of available freshwater resources is below 500 m³/capita/year in most MENA countries. Therefore, treated wastewater use is essential in the water scarce MENA region.

In the MENA region, the estimated volume of wastewater generation is 22.3 km³ out of which 11.4 km³/year (51% year) is treated. In different MENA regions, efficiency of wastewater treatment is highly variable due to design limitations of treatment plants.

Saudi Arabia collects and treats 672 million m³ of wastewater per day but less than 20% is ultimately reused [1]. Although in 1999 there were 30 major sewage treatment facilities with secondary, tertiary, and advanced levels of treatment and a total design capacity of 1,426,000 m³/day [33], a significant challenge continues to exist in the low overall sewerage rate of 37% [55]. Saudi Arabia intends to increase wastewater use to 65% by 2016 [73].

High-income countries in the region use treated wastewater for agricultural and landscape irrigation. Israel already uses 70% of the wastewater generated in the domestic sector. In Israel, the area irrigated with treated wastewater varies between 28,000 and 65,000 ha [34]. In Kuwait, the use of treated wastewater for landscape irrigation is increasing in urban areas. However, the primary use of treated wastewater is agricultural irrigation (4470 ha), representing 25% of the irrigated area in the country [23]. In the United Arab Emirates, 16,950 ha are irrigated with treated wastewater [33], of which 15,000 ha are in urban forests, public gardens, trees, shrubs, and grasses along roadways [23]. According to FAO [26], an estimated 217,527 ha in Egypt are irrigated with treated wastewater. In Syria, 9000 ha are irrigated with treated wastewater, while 40,000 ha are irrigated with untreated wastewater [33]. In Morocco, about 8000 ha are irrigated with untreated or insufficiently treated wastewater [23]. Israel is characterized by scarcity of water resources, which limits agricultural production possibilities. The amount of water currently recycled is 1900 million cubic meters (mcm) annually, or 260 m³ per capita. This is allocated for urban, industrial, and agricultural uses, leaving 1200 mcm or 63% for current agricultural consumption [32]. The good-quality water resources available for agricultural use tend to decrease as population growth enhances domestic water use, and within the next four decades, treated sewage effluent will become the main source of water for irrigation in Israel.

Of the 1400 mcm (40% of total water supply) that will be used for irrigation in the year 2040 in Israel and the Palestinian autonomous regions, the amount of treated sewage water will be 1000 mcm. Effluent will satisfy 70% of agricultural water demand and will play a dominant role in sustaining agricultural development. Agricultural use of wastewater should be followed by appropriate management techniques to control long-term effects of its constituents. In Israel, the integrated programs for the use of wastewater enable wastewater to account for 20% of the water resources used in agriculture [53]. Wastewater generated, treated, and used in the Middle East and North Africa (MENA) region is given in Table 71.3.

TABLE 71.3   Wastewater Generated, Treated, and Used in Middle East and North Africa (MENA)

Country	Wastewater Generated		Wastewater Treated		Treated Wastewater Used	
	Reporting Year	Volume (km³/year)	Reporting Year	Volume (km³/year)	Reporting Year	Volume (km³/year)
Algeria	2010	0.73	2010	0.15		NA
Bahrain	2010	0.084	2005	0.062	2005	0.016
Egypt	2011	8.5	2011	4.8	2011	0.7
Iran	2010	3.548[a]	2010	0.821[a]	2010	0.328
Iraq	2012	0.58[b]	2012	0.58[b]		NA
Israel	2007	0.5	2007	0.45	2004	0.262
Jordan	2008	0.18	2011	0.115[c]	2012	0.108[c]
Kuwait	2008	0.254	2005	0.25	2002	0.078
Lebanon	2003	0.31	2006	0.004	2005	0.002
Libya	1999	0.546	2009	0.04	2000	0.04
Morocco	2010	0.7	2010	0.124	2008	0.07
Oman	2000	0.09	2006	0.037	2006	0.037
Palestinian Territories	2001	0.071[d]	2001	0.03[d]	1998	0.01
Qatar	2005	0.055[e,*]	2006	0.058	2005	0.043
Saudi Arabia	2000	0.73	2002	0.548	2006	0.166
Syria	2002	1.364	2002	0.55	2003	0.55
Tunisia	2010	0.246	2010	0.226	2001	0.021
Turkey	2010	3.582[f]	2010	2.719[f]	2006	1.0
United Arab Emirates	1995	0.5	2006	0.289	2005	0.248
Yemen	2000	0.074	2009	0.046	2000	0.006

The data are from FAO-AQUASTAT (2012). NA refers to data not available.

Note: * Greater volume of treated wastewater than volume of wastewater generated may be due to respective data from different reporting years.

[a] Domestic wastewater only [64].

[b] Aziz and Aws [9].

[c] Ulimat [68].

[d] PEDCAR [52].

[e] Saloua [58].

[f] TURKSTAT [67].

71.7.5 Sub-Saharan Africa

In sub-Saharan countries, most wastewater goes untreated [82,83]. Untreated wastewater is used for irrigation in the peri-urban zones around Kumasi in Ghana [37], Dakar in Senegal [25], Nairobi in Kenya [17], and Bulawayo in Zimbabwe [48]. In peri-urban Kumasi, about 11,900 ha are irrigated with untreated wastewater [37].

71.7.6 Oceania

In Oceania, about 45% of the 450 wastewater use projects are in agriculture sector [13]. In Australia, annually an estimate of 0.35 km^3 of wastewater is treated. This volume accounts for 19% of the wastewater treated in the country and about 4% of the total water supply [8]. Agriculture is the major sector of water reuse in Australia, where about 20,000 ha are irrigated with treated wastewater [23]. Wastewater use in Australia is more common in inland, rural areas, where rainfall is limited and agricultural demands for irrigation water are notable [3]. According to the Australian Academy of Technological Sciences and Engineering, a total of 166.2 GL/y (43.91 billion gal/y) was reused in 2001–2002, up from 112.9 GL/y (29.83 billion gal/y) during 1996–1999. The proportion of effluent recycled also increased during this period from 7.3% during 1996–1999 to 9.1% in 2001–2002.

In New Zealand, wastewater is used to irrigate golf courses and for industrial applications, but the volumes involved likely are small [3].

71.7.7 Asia

In Asia, only about 32% of the wastewater generated is treated. In Asia, lack of financial resources as well as well-defined policies are the most important constraints [69].

In China, a water reuse project has been constructed in a chemical industrial park in Jiangsu province. This industrial park is one of China's national EIPs and is a pioneer in Jiangsu for having a water recycling plant (WRP). Wastewater reclamation rate in different cities of China is given in Table 71.4.

Japan has adopted a comprehensive strategy for treated wastewater use [27]. Japan's wastewater use strategy is unique because it focuses on meeting urban water needs, rather than providing

water primarily for agricultural uses [23]. In 2009, only 0.2 km of treated wastewater was used in the country. More than half was used for environmental purposes, such as landscape irrigation (27% of treated wastewater), recreation (2%), and river maintenance (29%) [76,81]. Wastewater use in agriculture and industry is not substantial, accounting only for 7% and 1% of the treated wastewater, respectively. In addition, more than 3% of the treated wastewater is used for toilet-flushing [81].

In Vietnam, an estimated 9500 ha are irrigated with untreated wastewater [33]. At least 2% of the agricultural land around most Vietnamese cities is irrigated with wastewater, and much of that land is planted in rice [56]. In Pakistan, there are no clear regulations regarding irrigation of crops with wastewater [75]. It is estimated that 32,500 ha is irrigated with wastewater [21], of which most of the wastewater remains untreated.

In India, per capita yearly surface water availability in the years 1991 and 2001 were 2300 m^3 (6.3 m^3/day) and 1980 m^3 (5.7 m^3/day), respectively, and these are projected to reduce to 1401 and 1191 m^3 by the years 2025 and 2050 [38]. Total water requirement of the country in 2050 is estimated to be 1450 km^3, which is higher than the current availability of 1086 km^3. In order to meet the growing water demand in the country, reuse of treated wastewater will have high potential to tackle the situation of water scarcity. In India, most of the wastewater is used untreated. In 1985, an estimated 73,000 ha were irrigated with direct use of untreated wastewater [63]. Since then, wastewater volumes and wastewater use for irrigation have increased substantially [23].

Due to rapid industrialization and development, there is an increased opportunity for greywater reuse in India [49]. The average greywater generation was 110 lpcd (liter per capita per day), out of which 80 lpcd was contributed from bathing, clothes washing, and wash basin (hand, teeth, face, mouth washing, etc.) and greywater generation from the kitchen was 30 lpcd. Greywater from the kitchen was not considered for treatment and reuse; hence, the average greywater generation was 80 lpcd. Maximum greywater generation was observed to be 100 lpcd and minimum value was 56.25 lpcd. Greywater treatment and reuse systems were constructed in residential schools in Madhya Pradesh, India, and treated greywater (wastewater generated from bathroom, laundry, and kitchen) was used for toilet flushing and irrigating the food crops. Wastewater generated, treated, and used in Asia is given in Table 71.5.

71.8 World Health Organization (WHO) Guidelines for the Safe Use of Wastewater

Both treated and untreated wastewaters are used directly and indirectly (i.e., as fecally contaminated surface water) for irrigation and various other purposes both in developed and developing countries. In places where untreated wastewater or highly contaminated surface water is used for irrigation, health and environmental problems are a matter of concern. The magnitude

TABLE 71.4 Wastewater Reclamation Rate in Different Cities of China

Area	Wastewater Reclamation Rate of Cities in China,[a] 2009	
	Wastewater Reclamation Rate (%)	Wastewater Reuse Rate[b] (%)
Northern cities	16.68	9.37
Central and western cities	3.70	1.89
Southern cities	9.14	5.72
National average	8.59	4.83

[a] Data source: MOHURD (2010).

[b] Wastewater reuse rate = quantity of consumed reclaimed wastewater/quantity of water provided through municipal water supply systems.

TABLE 71.5　Wastewater Generated, Treated, and Used in Asia

Country	Wastewater Generated		Wastewater Treated		Treated Wastewater Used	
	Reporting Year	Volume (km³/year)	Reporting Year	Volume (km³/year)	Reporting Year	Volume (km³/year)
Bangladesh	2000	0.725[a]	–	NA	–	NA
Bhutan	2000	0.004[a]	–	NA	–	NA
Cambodia	2000	1.184[b]	1994	0.0002	–	13.39
China	2009	58.920[c]	2006	17.89	2005	0.45
India	2012	13.999[d]	2012	4.302[d]	2000	0.204[e]
Japan	2009	27.000[f]	2009	14.65[f]	2009	NA
Laos	2000	0.546[b]	–	NA	–	NA
Malaysia	2000	1.403[a]	1995	0.398	–	NA
Maldives	2000	0.004[a]	–	NA	–	NA
Mongolia	2002	0.126[g]	2002	0.083[g]	–	NA
Myanmar	2000	0.017[a]	–	NA	–	NA
Nepal	2006	0.135[h]	2006	0.006[h]	–	NA
Pakistan	2011	6.849[i]	2011	0.548[j]	–	NA
Philippines	2000	7.500[k]	1993	0.01	–	NA
Republic of Korea	2000	6.895[a]	1996	4.18	2008	0.157[l]
Singapore	2000	0.470[a]	–	NA	2008	0.027
Sri Lanka	2000	0.950[a]	–	NA	–	NA[l]
Thailand	2008	5.293	1995	0.035	–	NA
Vietnam	2003	1.1	2009	0.07	2003	0.175

The data are from FAO-AQUASTAT (2012). No data are available on wastewater production, treatment, or use from the following countries: Afghanistan, Brunei, Democratic People's Republic of Korea, Indonesia, Papua New Guinea, and Timor-Leste. NA refers to data not available.

[a] Reference 70.

[b] Values refer to domestic wastewater estimated considering daily per capita water consumption (230 L) and total population in the respected countries in 2000 (Kamal et al. [35]).

[c] MEPPRC [46].

[d] Kaur et al. [36].

[e] Shrivastava and Swarup [60].

[f] References 76 and 81.

[g] Basandorj [11].

[h] Nyachhyon [50].

[i] Murtaza [47].

[j] Calculated based on the estimates that about 8% of the wastewater generated in Pakistan undergoes treatment [47].

[k] Values refer to domestic wastewater only [70].

[l] Jiménez and Asano [33].

of the problem is directly related to chemical constituents in wastewater. Wastewater generally contains higher concentrations of sodium, trace elements, excessive chlorine residuals, and nutrients as well various pathogenic microbes, etc. The types and concentrations of constituents in wastewater depend on the municipal water supply, the influent waste streams (i.e., domestic and industrial contributions), and amount and composition of infiltration in the wastewater collection system. The standards required for the safe use of wastewater and the amount and type of wastewater treatment needed are contentious.

For water reuse in agricultural irrigation and other purposes, WHO [80] has given guidelines to reduce potential risks to public health and promote public acceptance. The US EPA [23] has standards for various nonpotable applications. Many developed (e.g., Australia [19] and United States [23]) and developing countries (e.g., Mexico and Indonesia [15]) have guidelines for water reuse.

The WHO initially published Guidelines for the Safe Use of Wastewater and Excreta in Agriculture and Aquaculture in 1989 (Table 71.6) and later revised it as "Guidelines for the Safe Use of Wastewater, Excreta and Greywater, Volume 2: Wastewater Use in Agriculture" (Table 71.7) [80]. The WHO Guidelines for the safe use of wastewater, excreta, and greywater in agriculture and aquaculture are based on a risk analysis approach, which is recognized internationally as the fundamental methodology underlying the development of food safety standards that both provide adequate health protection and facilitate trade in food.

The rules that govern international trade in food were agreed during the Uruguay Round of Multilateral Trade Negotiations and apply to all members of the World Trade Organization. Guidelines for the international trade of wastewater-irrigated food products should be based on scientifically sound risk assessment and management principles. For application on land, there are still concerns with respect to elevated concentrations

TABLE 71.6 1989 WHO Guidelines for Using Treated Wastewater in Agriculture[a]

Category	Reuse Condition	Exposed Group	Intestinal Nematodes[b] (Arithmetic Mean No. of Eggs per Liter[c])	Fecal Coliforms (Geometric Mean No. per 100 mL[c])	Wastewater Treatment Expected to Achieve
A	Irrigation of crops, likely to be eaten uncooked, sports fields, public parks[d]	Workers, consumers, public	< 1	< 1000	A series of stabilization ponds to achieve the microbiological quality indicated, or equivalent treatment
B	Irrigation of cereal crops, industrial crops, fodder crops, pasture, and trees[e]	Workers	< 1	No standard recommended	Retention in stabilization ponds for 8–10 days or equivalent helminth and fecal coliform removal
C	Localized irrigation of crops in category B if exposure to workers and the public does not occur	None	Not applicable	Not applicable	Pretreatment as required by irrigation technology but not less than primary sedimentation

[a] In specific cases, local epidemiological, sociocultural and environmental factors should be taken into account and the guidelines modified accordingly.

[b] Ascaris and Trichuris species and hookworms.

[c] During the irrigation period.

[d] A more stringent guideline limit (<200 fecal coliforms/100 mL) is appropriate for public lawns, such as hotel lawns, with which the public may come into direct contact.

[e] In the case of fruit trees, irrigation should cease two weeks before fruit is picked, and no fruit should be picked off the ground. Sprinkler irrigation should not be used.

of soluble salts, nutrients, and microbiological quality of the treated effluent.

The comprehensive characterization of wastewater quality provides a means to ensure a basic level of irrigation quality control. It also provides useful baseline information to evaluate impacts from future irrigation. These impacts may relate to changes that occur in community water sources, waste treatment processes, community size, and community or industrial discharge loadings. In addition, the wastewater quality characterization process may also provide an opportunity for community planners and engineering consultants to better evaluate the effectiveness of the treatment process and its ability to eliminate harmful constituents that could normally restrict the potential for irrigation use.

The quality standard for irrigation water is a broader topic spread across an extensive amount of guidelines or regulations that vary in how specific they are for a particular water end use.

71.9 Technology Available for Wastewater Treatment

To meet public demands, specific needs and safety, raw wastewater is treated to make it available for reuse. Technologies to treat recycled water should depend almost entirely on the application, which is called "highest treatment for highest use."

Treatment technology of recycled water must be established based on the expected degree of public contact. For example, if the primary application is irrigation or cooling tower water, sand or dual-media filtration after secondary treatment is sufficient to achieve a state's water quality criteria. However, if the intended application is for indirect potable reuse, sophisticated

technologies such as MF, reverse osmosis, and UV must be employed to ensure chemical and microbiological safety of the reclaimed water.

Wastewater treatment processes can be divided into the following three categories:

1. Physical process: Impurities are removed physically by screening, sedimentation, filtration, flotation, absorption, adsorption or both, and centrifugation.
2. Chemical process: Impurities are removed chemically through coagulation, absorption, oxidation-reduction, disinfection, and ion-exchange.
3. Biological process: Pollutants are removed using biological mechanisms, such as aerobic treatment, anaerobic treatment, and photosynthetic process (oxidation pond).

Conventional wastewater treatment consists of the following stages: preliminary, primary, secondary, and disinfection. Municipal wastewater treatment facilities use combinations of physical, biological, and chemical treatment technologies. Preliminary and primary treatments are usually physical processes, such as screening for the removal of debris and large solids, and sedimentation. A secondary treatment may utilize biological processes, such as stabilization ponds, trickling filter, oxidation ditch, and activated sludge, which is then followed by sedimentation of biomass (sludge). Tertiary and advanced treatment is an additional treatment for higher-level removal of specific pollutants, such as nitrogen or phosphorus, which cannot be removed by conventional secondary treatments. A summary of the purposes and sample technologies of each treatment process is given in Table 71.8. A diagrammatic representation of a low-cost wastewater treatment system is given in Figure 71.1.

TABLE 71.7 WHO Guideline for Irrigation Water Quality

Guideline on Irrigation Water Quality		Westcot /Ayers [78]/Water Quality for Irrigation	WHO [79]/ Wastewater Quality for Irrigation
Guideline			
Irrigation parameter/ type of guideline			
Salinity			
Electrical conductivity (EC)	ds/m	0.7–3	
Sodium adsorption ratio (SAR)			
Sodium (Na)	mg/L	3–9	> 3.0
Total dissolved solids (TDS)	mg/L	450–2000	
Suspended solids	mg/L		
pH		6.5–8	
Pathogenicity			
Intestinal nematodes	eggs/L	–	< 1
	eggs/10 L	–	–
E. coli	CFU/100 mL	–	–
Fecal coliform	CFU/100 mL	–	< 1000
Thermotolerant coliforms	CFU/100 mL	–	–
Total coliform	CFU/100 mL	–	–
Nutrient		–	–
Nitrogen (NO_3-N)	mg/L	–	–
Total nitrogen (TN)	mg/L	–	–
Phosphorus	mg/L	–	–
Heavy metals		–	–
Aluminium (Al)	mg/L	–	–
Arsenic (As)	mg/L	–	–
Beryllium (Be)	mg/L	–	–
Cadmium (Cd)	mg/L	–	–
Cobalt (Co)	mg/L	–	–
Chromium (Cr)	mg/L	–	–
Copper (Cu)	mg/L	–	–
Iron (Fe)	mg/L	–	–
Lithium (Li)	mg/L	–	–
Manganese (Mn)	mg/L	–	–
Molybdenum (Mb)	mg/L	–	–
Nickel (Ni)	mg/L	–	–
Lead (Pb)	mg/L	–	–
Selenium (Se)	mg/L	–	–
Vanadium (V)	mg/L	–	–
Zinc (Zn)	mg/L	–	–

71.10 Methodology to Achieve the Feasibility Study

A successful and thorough feasibility study should address many different aspects such as geological, technical, economical, environmental, sociological, and quality and risks analysis, which help in decision making before the start of any project. Methodology for water reuse feasibility has been described only in a few studies.

To address a complete water reuse feasibility study, a number of aspects have to be considered. However, identifying those different aspects is crucial for a successful water reuse system. The key aspects that should be addressed are

1. Background information
2. Description of the proposed system
3. Life-cycle assessment

71.10.1 Background Information

This is a most important and fundamental step in studying a reuse feasibility study. During data collection, it is necessary to contact the main stakeholders such as water and wastewater agencies, regional environmental agencies, councils and regional governments (land and population projections, funding options), farmers associations, end-users associations, etc. During data collection, some basic records should also to be collected:

- Water supply and demand (local and seasonal)
- Water and wastewater management agencies working in the area
- Regional water and wastewater facilities (in operation and planned)
- Environmental setting: climate, geography and topography, water resources
- Land use/land cover
- Structure and location of potential users
- Water-related socioeconomic facts (water supply restrictions on domestic, industrial, and/or irrigation uses)
- Status of public acceptance of water reuse

Different thematic maps such as locations of the wastewater treatment facility, different land use land cover, and population zones should be prepared with the help of GIS. Within the background information study, characterizing the zone is the most important part. Under this, topography and climate are important.

TABLE 71.8 Wastewater Treatment Process

	Preliminary	Primary	Secondary	Tertiary and Advanced Processes
Purpose	Removal of large solids and grit particles	Removal of suspended Solids	Biological treatment and removal of common biodegradable organic pollutant	Removal of specific pollutant, such as nitrogen or phosphorus, color odor, etc.
Sample technologies	Screening settling	Screening, sedimentation	Percolating/tricking filter, activated sludge anaerobic treatment waste stabilization ponds	Sand filteration, membrane bioreactor, reverse osmosis, ozone treatment, chemical coagulation, activated carbon.

Source: Adapted from Asano, T. and Levine, A. 1998. In T. Asano (ed.), *Wastewater Reclamation and Reuse.* CRC Press, Boca Raton, FL, pp. 1–55.

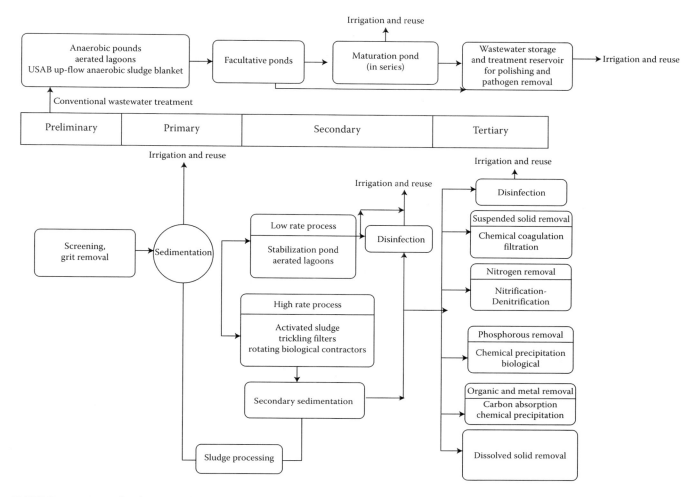

FIGURE 71.1 Generalized municipal wastewater treatment scheme and point of water reuse.

71.10.1.1 Topography and Climate

To evaluate a site for the wastewater treatment system, geography and topography are the most influencing parameters that help in targeting a zone. This should include conditions like location of river basins and wells, urban settlements, agricultural lands, dispersion of the discharged wastewater, etc. They will also influence the evaluation of site possibilities and location of pumping stations, treatment facilities, and distribution lines of the proposed water reuse options. It should also address whether the zone is flat or mountainous, soil type, vegetation, etc.

Along with geography, data on climate of the area is also necessary because it determines the present water resources and future need. Along with annual precipitation, average temperature and evaporation should also be recorded.

71.10.1.2 Water Balance and Water Supply of the Region

Before starting the water reuse treatment system, one should also collect data on water flows, hydrological plans (at regional and national level), water abstraction, seawater intrusion, or pollution (origin, type, and intensity) of surface and groundwater wells of the region. The water supply system will determine the location of the potential reclaimed water users as well as the location of the wastewaters. For the management of a water system in a region, sources and quality characteristics, water consumer trends, average flows, and distribution by sectors (industry, agriculture, and domestic uses) of each type of water supply should be known for that area.

71.10.1.3 Sanitation System

During the establishment of water reuse treatment plant, the sanitation system will also be included as an integrated part of planning. Urban wastewater treatment directive should be followed. Toward the end of the 1990s, around 80% of the European Union population was connected to public sewerage systems and 77% to wastewater treatment plants. Different information should be complied and analyzed under this system such as

• Location and number of water sewage and wastewater treatment plants

- Percentage of population connected to sewerage
- Percentage of the collected wastewater that is treated
- Main characteristics, types of treatment, treatment costs, and cost of the different wastewater treatment plants
- Price paid for sanitation
- Wastewater legislation to be fulfilled

71.10.1.4 Potential Users for the Reclaimed Water

Success of reuse of the wastewater treatment plant depends on identification of potential users of the reclaimed water. For each potential user, volume, frequencies, applications, and required quality have to be compiled. Location and distance between users and treatment plant has to be considered. The location of the greatest water consumers is essential.

Potential demand of reclaimed water are for irrigation (agricultural, landscaping, golf courses), greywater, and cooling water reuse, and other potential reuses distribution of water flows, present and future quantity needs, timing and reliability of needs, water quality needs.

71.10.2 Description of Proposed Water Reuse Options

Before the start of a wastewater treatment plant, description of a proposed system that will depend on the composition of wastewater, requirement of equipments, and impact on the environment must be described.

71.10.2.1 Description of Proposed System

Depending on the composition of the wastewater to be treated and on the required reclaimed water quality, the treatments and systems will be different. Usually, intensive treatments are more expensive, more technological, and require less space compared to extensive ones.

71.10.2.2 Equipment Needs and Costs

The evaluation of investment and operation and maintenance costs is a fundamental phase in the planning of wastewater reuse systems, as it is necessary for an economic comparisons with the other conventional and unconventional water resources. Unfortunately, wastewater treatment costs are not usually well-documented [61].

71.10.2.3 Water Quality and Risk Analysis

Required quality of reclaimed water depends on specific user (agriculture use, industrial uses, and toilet flush). Along with water quality, risk analysis assessment should be an integral part of management and communication. Risk assessment must be based on human health and impact on environment. In order to minimize as much as possible the potential risks associated to water reuse, it is encouraged to carry out a very strict and complete monitoring program together with a guideline of best practices.

71.10.2.4 Advantages and Disadvantages of the Proposed System

Before the start of the feasibility study, advantages or disadvantages of a proposed system must be evaluated. The evaluation must be done on the basic criteria such as

- Resources requirement (land requirement, civil works, installation of pipelines, energy and water requirements, human resources, ease of construction)
- Reliability (quality and changes on the quality of the outlet water, healthy and sanitary issues, safety and risk issues)
- Adaptability (i.e., capacity to treat different flows or inlet loads)
- Capacity to be upgraded (e.g., improvement of the quality changing the membrane cut-off in a membrane bioreactor treatment) or to be enlarged
- Environmental sustainability of the considered alternative
- Economical cost (investment and operation and maintenance costs). Under the O&M costs, labor and energy costs are usually the main items to be analyzed. The need of qualified personnel or a treatment with a very time-intensive demand of operators will highly increase the labor costs associated with a given treatment.

71.10.2.5 Implementation Requirements

Each proposed treatment option in a water reuse project will need some requirements for its implementation such as land requirement, power availability, roads and infrastructures, etc. Furthermore, the treatment system may require a minimum quality in the inlet water, a specific monitoring plan, or restrict the water feed flow.

Reclaimed water distribution is a key aspect to consider in a water reuse project. It is estimated to be 8% of the total construction costs of the project, while maintaining the pipelines and storage tanks is calculated to be 2% of the capital cost.

Accordingly, the different alternatives for reclaimed water distribution and storage have to be firmly evaluated.

When designing the distribution and storage systems, different aspects such as orography, geology, and soil availability need consideration. The precise coordinates and elevation of the proposed systems have to be indicated, specifying if they imply water supply, water storage, pumping, or treatment. In case of dual distribution of tap water and recycled water, their differences and the adopted security measures (risk assessment) need to be pointed out (e.g., distances between their distribution lines, specific materials and elements to be used, etc.).

71.10.3 Life Cycle Assessment (LCA)

Life-cycle assessment (LCA) is a tool to assess the potential impacts on environment and resources used throughout a product's life cycle, that is, from raw material acquisition, via production and use phases, to waste management [30]. International standards, ISO 14040 and ISO 14044, define LCA as follows: The

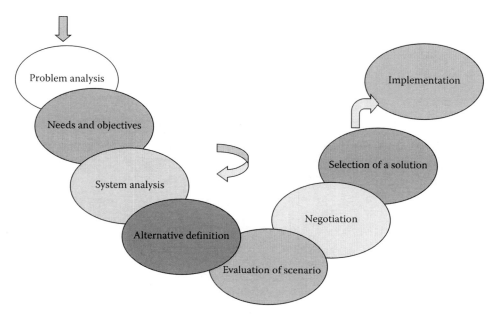

FIGURE 71.2 Life-cycle assessment of water reuse system.

term "product" includes both goods and services [30]. The waste management phase includes disposal as well as recycling. LCA is a comprehensive assessment and it considers all aspects of natural environment and human health. In general, sustainability assessment of any technology is assessed through its effect in social, environmental, and economic dimensions [20]. LCA study involves four phases: goal and scope definition, life-cycle inventory analysis, life cycle impact assessment, and interpretation. The goal and scope definition includes the reasons for carrying out the study, the intended application, and the intended audience [30,31].

LCA covers a diversity of environmental impacts and it may include comparison across impact categories. It can therefore be argued that modeling should ideally be done with the same degree of realism for every impact. On this background, LCA also aims for a comparable way of assessing impacts. Diagrammatic representation of LCA is given in Figure 71.2.

71.10.3.1 Environmental Assessment

Environmental impact assessment is a mandatory tool before the start of any project. It is addressed by the project developer in the pre- and feasibility stages in consultation with other institutions involved in the information on the environmental effects of a project. It tells about design and size or scale of the development. It is intended to identify and assess base related to potential impact and environmentally sensitive areas in the catchment.

During environmental impact assessment studies, the following are the most relevant aspects that must be considered in a water reuse project: environment of the old receiving medium and how it will change with the reuse project, environment of the new receiving medium of the reclaimed water, floral and faunal diversity, natural resource utilization, human health,

hydrology, geomorphology, downstream impact, migration, freshwater requirements, and reserve considerations [72].

When irrigated with treated wastewater for a long time, changes in soil characteristics can be observed. It should be studied time to time. If we are studying irrigation with treated water, changes in the soil characteristics for long periods of time should be studied. Nutrient status of the soil such as nitrogen, phosphorus, sodium, calcium, and boron as well as special attention should be put on risks associated with human health. Issues such as distance between potable and reclaimed water pipe lines, possible consequences in case of a leakage or reclaimed water pipelines breaking, parks and gardens irrigation, and so on, should be studied [72].

71.10.3.2 Economical Assessment

In a feasibility study, economical and financial viability is one of the most important aspects. The economic viability of water reclamation projects is directly linked with the sustainability of projects. The major challenges that are faced during the project implementation are (1) high cost (capital and operational), (2) unavailable or inadequate incentives associate with conservation of water resources and pollution reduction, (3) relatively long-term investment return, and (4) limited level of revenue from recycled water.

The purpose of economic evaluation of recycled water project is to provide the economic rationale and robust evidence base to the decision makers. Development of the economic performance tools could involve quantitative analytical method of measuring, monitoring, and predicting technology performance tools [18]. The cost-effectiveness of reuse projects is directly related to the volume of reclaimed water used: the more water utilized, the more cost-effective it is.

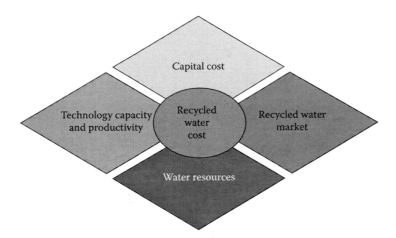

FIGURE 71.3 Drivers of recycled water productivity.

A thorough review of key technical, operational, financial, economic, social, and environmental factors resulted in the three most prominent performance pillars: productivity, efficiency, and reliability. Productivity is a measure that indicates how resources are converted into the products.

In most of cases, there is no single option that is most suitable. There must be two different types of treatment and water reuse applications. To increase the use of reclaimed water, different options could be implemented: a higher price of tap water (high-quality water), a tax for using high-quality waters for nonexigent uses, or funding the use of reclaimed water. Drivers of recycled water are given in Figure 71.3.

Economic efficiency of the water reclamation scheme and technology could be best calculated using the following indicators and their relationships:

- Capital investment cost defined at the end of construction periods
- Annual capital, maintenance and operating costs per volume of produced water
- Annual energy costs per volume of produced water

71.10.3.3 Social Assessment

In order to improve the social acceptance of water reuse, information about the different benefits (environmental, economic, such as impact on tourism and water price) of water reuse and an information session about the different terms (water, wastewater, reclaimed water, water treatment, water quality, etc.) are needed, as information has been proved to have a positive influence on user acceptance [66].

Some of the main issues raised during the consultation program are noise, access problems to different places, damage to structures and properties, the level of treatment of the recycled water, health impact related to the inadvertent consumption of the treated water, possible problems in crop irrigation, price of the recycled water, system connection issues, and when the product will be available.

It was observed that higher level of income and education are positively correlated with a respondent's willingness to use recycled water. Extra information on the advantages of recycled water will have a significant impact on reported degrees of willingness to use recycled water.

71.11 Summary and Conclusions

Throughout the world, communities are facing severe freshwater crisis due to increasing population pressure, intensification of agriculture, industrialization, drought, depletion, and contamination of groundwater. Water reclamation, recycling, and reuse can address these problems. Reclaimed water has enormous potential to solve these problems. Reuse of treated wastewater is a common and rapidly increasing practice, mainly in arid and semi-arid regions around the world. Although water reclamation and reuse is currently being used in many countries around the world, current levels of reuse constitute only a small fraction of the total volume of wastewater generated. As compared to developed countries, developing countries have limited wastewater treatment facilities due to lack of financial resources and well defined polices. In these developing countries, persistent increases in population have not kept pace with investments in treatment facilities. In developed countries, generally technical solutions and public policies are adequate to accommodate increases in the treatment and use of wastewater. The increasing competition between industrial and domestic sectors for limited freshwater resources is also motivating investment in wastewater treatment and use. Water reuse for agricultural irrigation is typically separated into restricted and unrestricted uses. The optimal treatment strategy generally varies with the economic and institutional capacities, wastewater sources and constituents, but should preferably consider the requirements of reuse as well as standards, which are very much essential to maintain public health and the environment. For the implementation of a water reuse project, a feasibility study is a fundamental step. It helps the decision maker to evaluate the potential impact of a proposed project or program. It helps in defining the criteria and methods to assess, calculate, and compare the effect of wastewater reuse programs. Prior identification of the potential users of the reclaimed water and social willingness of the proposal are two main factors that a feasibility study takes into account.

References

1. Al-Musallam, L. 2006. Water and wastewater privatization in Saudi Arabia. *SAWEA 2006 Workshop: Privatization and Outsourcing of Water and Wastewater,* June 7, 2006, Saudi Arabia.

2. Alley, W.M., Healy, R.W., LaBaugh, J.W., and Reilly, T.E. 2002. Flow and Storage in Groundwater Systems. *Science,* 296, 1985. DOI: 10.1126/science.1067123.

3. Anderson, J., Bradley, J. and Radcliffe, J. 2008. Water reuse in Australia and New Zealand. In: B. Jiménez and T. Asano (eds.), *Water Reuse: An International Survey of Current Practice, Issues and Needs.* IWA, London, pp. 105–121.

4. Anderson, J.M. 1996. Current water recycling initiatives in Australia: Scenarios for the 21st century. *Water Science and Technology,* 33, 37–43.

5. Angelakis, A. and Bontoux, L. 2001. Wastewater reclamation and reuse in Eureau countries. *Water Policy,* 3, 47–59.

6. Asano, T., Burton, F.L., Leverenz, H.L., Tsuchihashi, R., and Tchobanoglous, G. 2007. *Water Reuse: Issues, Technologies, and Applications.* McGraw-Hill, New York.

7. Asano, T. and Levine, A. 1998. Wastewater reclamation, recycling and reuse: Introduction. In: T. Asano (ed.), *Wastewater Reclamation and Reuse.* CRC Press, Boca Raton, FL, pp. 1–55.

8. Australian Bureau of Statistics (ABS). 2010. Water Account Australia 2008–2009. http://www.ausstats.abs.gov.au/ausstats/subscriber.nsf/0/D2335EFFE939C9BCCA2577E7001 58B1C/$File/46100 2008-09.

9. Aziz, A.M. and Aws, A. 2012. Wastewater production, treatment and use in Iraq: Country report, Presented at the Second Regional Workshop of the Project "Safe Use of Wastewater in Agriculture," May 16–18, 2012, New Delhi. http://www.ais.unwater.org/ais/mod/page/view.php?id=62. Accessed December 21, 2012.

10. Balkema, A., Preisig, J., Otterpohl, H.A., and Lambert, F.J.D. 2002. Indicators from sustainable assessment of wastewater treatment systems. *Urban Water,* 4, 153–161.

11. Basandorj, D. 2002. Water resource management in Mongolia. In: The National Seminar on Water Quality Improvement, Ulaanbaatar, June 12–13 (Cited by: Dore, G. and Nagpal, T. 2006. Urban transition in Mongolia: Pursuing sustainability in a unique environment. *Environment: Science and Policy for Sustainable Development,* 48, 10–24).

12. Bennett, A.J. 2000. Environmental consequences of increasing production: Some current perspectives. *Agriculture Ecosystem Environment,* 82, 89–95.

13. Bixio, D., De heyder, B., Cikurel, H., Muston, M., Miska, V., Joksimovic, D., Schäfer, A.I., Ravazzini, A., Aharoni, A., Savic, D., and Thoeye, C. 2005. Municipal Wastewater reclamation: Where do we stand? An overview of treatment Technology and management practice. *Water Science and Technology: Water Supply,* 15, 77–85.

14. Bixio, D. and Wintgens, T. (eds.). 2006. *Water Reuse System Management Manual AQUAREC.* Office for Official Publications of the European Communities, Luxembourg.

15. Blumenthal, U., Mara, D.D., Peasey, A., Ruiz-Palacios, G., and Stott, R. 2000. Guidelines for the microbiological quality of treated wastewater used in agriculture: Recommendations for revising WHO guidelines. *WHO Bulletin,* 78(9), 1104–1116.

16. Bryck, J., Prasad, R., Lindley,T., Davis, S., and Carpenter, G. 2008. *National Database of Water Reuse Facilities Summary Report.* Water Reuse Foundation, Alexandria, VA.

17. Cornish, G.A. and Kielen, N.C. 2008. Wastewater irrigation—Hazard or lifeline? Empirical results from Nairobi, Kenya and Kumasi, Ghana. In: C.A. Scott, N.I. Faruqui, and L. Raschid-Sally (eds.), *Wastewater Use in Irrigated Agriculture Confronting. The Livelihood and Environmental Realties.* CABI-IWMI-IDRC, Wallingford, UK, pp. 69–79.

18. Davenport, T.H. 2009. *The Rise of Analytical Performance Management.* Harvard Business Digital, SAS Institute Inc.

19. Dettrick, D. and Gallagher, S. 2002. Environmental Guidelines for the Use of Recycled Water in Tasmania, Department of Primary Industries, Water and Environment, Tasmania, Australia. www.environment.tas.gov.au/file.aspx?id=1886. Accessed on May 3, 2010.

20. Elkington, J. 1998. *Cannibals with Forks—The Triple Bottom Line of 21st Century Business.* New Society Publishers, Canada.

21. Ensink, J.H.J., Mahmood, T., Van der Hoek, W., Raschid-Sally, L., and Amerasinghe, F.P. 2004. A nation-wide assessment of wastewater use in Pakistan: An obscure activity or a vitally important one? *Water Policy,* 6, 1–10.

22. Environment Canada. 2010. *Municipal Water Use Report: Municipal Water Use, 2006 Statistics.* Environment Canada, Gatineau.

23. Environment Protection Agency (EPA). 2004. *Guidelines for Water Reuse.* EPA/625/R-04/108. http://www.epa.gov/ord/.

24. Esteban, R.I., Ortega, E., Batanero, G., and Quintas, L. 2010. Water reuse in Spain: Data overview and costs estimation of suitable treatment trains. *Desalination,* 263, 1–10.

25. Faruqui, N.I., Niang, S. and Redwood, M. 2004. Untreated wastewater use in market gardens: A case study of Dakar, Senegal. In: C.A. Scott, N.I. Faruqui, and L. Raschid-Sally (eds.), *Wastewater Use in Irrigated Agriculture Confronting the Livelihood and Environmental Realties.* CABI-IWMI-IDRC, Wallingford, UK, pp. 113–125.

26. Food and agriculture organization (FAO). 2005. *Irrigation in Africa in Fgures: AQUASTAT Survey—2005.* Water Report 29. FAO, Rome.

27. Funamiu, N., Onitsuka, T. and Hatori, S. 2008. Water reuse in Japan. In: B. Jiménez and T. Asano (eds.), *Water Reuse: An International Survey of Current Practice, Issues and Needs*. IWA, London, pp. 373–386.

28. Hamoda, M.F. 2004. Water strategies and potential of water reuse in the south Mediterranean countries. *Desalination*, 165, 31–41.

29. Hussain, I., Raschid, L., Hanjra, M.A., Marikar, F., and van der Hoek, W. 2001. A framework for analyzing socio-economic, health and environmental impacts of wastewater use in agriculture in developing countries. Working Paper 26, International Water Management Institute, Colombo.

30. ISO. 2006. ISO 14040 International Standard. In: *Environmental Management—Life Cycle Assessment—Principles and Framework*. International Organization for Standardization, Geneva, Switzerland.

31. ISO. 2006. ISO 14044 International Standard. In: *Environmental Management—Life Cycle Assessment—Requirements and Guidelines*. International Organization for Standardization, Geneva, Switzerland.

32. Israel Hydrological Service, Water Commission, 1995. Development, utilization and status of aquifers in Israel, Israel Hydrological Service, Jerusalem (in Hebrew).

33. Jiménez, B. and Asano, T. 2008. *Water Reuse: An International Survey of Current Practice, Issues and Needs*. IWA, London.

34. Jiménez, B. and Asano, T. 2008. Water reclamation and reuse around the world. In: B. Jiménez and T. Asano (eds.), *Water Reuse: An International Survey of Current Practice, Issues and Needs*. IWA, London, pp. 3–26.

35. Kamal, A.S.M., Goyer, K., Koottatep, T., and Amin, A.T.M.N. 2008. Domestic wastewater management in South and Southeast Asia: The potential benefits of a decentralized approach. *Urban Water Journal*, 5, 345–354.

36. Kaur, R., Wani, S.P., Singh, A.K., and Lal, K. 2012. Wastewater production, treatment and Use in India, Presented at the Second Regional Workshop of the Project "Safe Use of Wastewater in Agriculture," May 16–18, 2012, New Delhi. http://www.ais.unwater.org/ais/mod/page/view.php?id=61. Accessed December 21, 2012.

37. Keraita, B., Drechsel, P., Huibers, F., and Raschid-Sally, L. 2002. Wastewater use in informal irrigation in urban and peril-urban areas of Kumasi, Ghana. *Urban Agriculture Magazine*, 8, 11–13.

38. Kumar, R., Singh, R.D. and Sharma, K.D. 2005. Water resources of India. *Current Science*, 89(5), 794–811.

39. Levine, A.D. and Asano, T. 2004. Recovering sustainable water from wastewater. *Environmental Science and Technology Journal*, 38(11), 201A–208A.

40. Lim, S.R., Park, D. and Park, J.M. 2008. Environmental and economic feasibility study of a total wastewater treatment network system. *Journal of Environmental Management*, 88(3), 564–575.

41. Marecos do Monte, H. and Albuquerque, A. 2009. *Wastewater Reuse*. Technical Guide 11 IRAR, Lisbon, Portugal, p. 363 (in Portuguese).

42. Marecos do Monte, H. and Albuquerque, A. 2010. Analysis of constructed wetland performance for irrigation reuse. *Water Science and Technology*, 61(7), 1699–1705.

43. Mediterranean Wastewater Reuse Working Group (MED WWR WG), 2007. Mediterranean Wastewater Reuse Report.

44. Metcalf and Eddy. 2003. *Wastewater Engineering Treatment and Reuse*, 4th Edition. McGraw Hill, New York.

45. Miller, G.W. 2006. Integrated concepts in water reuse: managing global water needs. *Desalination*, 187, 65–75.

46. Ministry of Environment Protection, the People's Republic of China (MEPPRC). 2012. Report on the State of the environment in China 2010. http://english.mep.gov.cn/standards reports/soe/soe2010/

47. Murtaza, G. 2012. Wastewater Production, Treatment and use in Pakistan, Presented at the Second Regional Workshop of the Project "Safe Use of Wastewater in Agriculture", May 16–18, 2012, New Delhi. http://www.ais.unwater.org/ais/mod/page/view.php?id=62. Accessed December 21, 2012.

48. Mutengu, S., Hoko, Z. and Makoni, F.S. 2007. An assessment of the public health Hazard potential of wastewater reuse for crop production. A case of Bulawayo city, Zimbabwe. *Physics and Chemistry of the Earth*, 32, 1195–1203.

49. NEERI. 2007. *Greywater Reuse in Schools—A Guidance Manual*.

50. Nyachhyon, B.L. 2006. Service Enhancement and Development of Sanitary Sewage System In Urban and Semi-Urban Setting in Nepal, Policy Paper 23, Prepared for Economic Policy Network, Ministry of Finance (MOF)/HMGN, ADB Nepal Resident Mission, pp. 86–94 (Cited by: Tuladhar, B., Shrestha, P., and Shrestha, R. 2008. Decentralised wastewater management using constructed wetlands. In: Wicken, J., Verhagen, J., Sijbesma, C., Silva, C.D., and Ryan, P. (eds.), 2008. *Beyond Construction: Use by All. A Collection of Case Studies from Sanitation and Hygiene Promotion Practitioners in South Asia*. IRC, International Water and Sanitation Centre—Water Aid, Delft and London).

51. Organisation for Economic Co-operation and Development (OECD). 2008. OECD Environmental data: Compendium 2006–2008. http://www.oecd.org/.22/55/41878136.pdf.

52. Palestinian Economic Council for Development and Reconstruction (PEDCAR), 2001. Water Sector Strategic Planning. Environmental Authority, Palestine (Cited by: Fatta, D., Salem, Z., Mountadar, M., Assobhei, O., and Loizidou, M. 2004. Urban wastewater treatment and reclamation for agricultural irrigation: the situation in Morocco and Palestine. *Environmentalist*, 24, 227–236).

53. Pedrero, F., Kalavrouziotis, I., Alarcon, J.J., Koukoulakis, P., and Asano, T. 2010. Use of treated municipal wastewater in irrigated agriculture—Review of some practices in Spain and Greece. *Agricultural Water Management*, 97(9), 1233–1241.

54. PUB, 2011. Water for all: NeWater, http://www.pub.gov.sg/water/newater/NEWaterOverview/Pages/default.aspx.

55. Qadir, M., Wichelns, D., Raschid-Sally, L., McCornick, P.G., Drechsel, P., Bahri, A., and Minhas, P.S. 2010. The challenges of wastewater irrigation in developing countries. *Agricultural Water Management*, 97, 561–568.

56. Raschid-Sally, L., Doan, D.T. and Abayawardana, S. 2004. National assessments on Wastewater use in agriculture and an emerging typology: The Vietnam case study. In: C.A. Scott, N.I. Faruqui, and L. Raschid-Sally (eds.), *Wastewater Use in Irrigated Agriculture: Confronting the Livelihood and Environmental Realities*. IWMI/IDRC-CRDI/CABI, Wallingford, UK.

57. Raschid-Sally, L. and Jayakody, P. 2008. *Drivers and Characteristics of Wastewater Agriculture in Developing Countries: Results from a Global Assessment*. Research Report 127. International Water Management Institute, Colombo, pp. 81–90.

58. Saloua, R. 2012. National Report: Tunisia, Presented at the First Regional Workshop of the Project "Safe Use of Wastewater in Agriculture," February 18–19, 2012, Marrakech. http://www.ais.unwater.org/ais/mod/page/view.php?id=59. Accessed December 21, 2012 (in French).

59. Scott, C.A., Faruqui, N.I. and Raschid-Sally, L. 2004. Wastewater use in irrigation agriculture: Management challenges in developing countries. In: C.A. Scott, N.I. Faruqui, and L. Raschid-Sally (eds.), *Wastewater Use in Irrigated Agriculture Confronting the Livelihood and Environmental Realties*. CABI-IWMI-IDRC, Wallingford, UK, pp. 1–10.

60. Shrivastava, P. and Swarup, A. 2001. Management of Wastewater for Environmental Protection of Freshwater Resources: An Approach for Tropical Countries both Developing and Undeveloped, International Conference on Freshwater, 3–7 December, Bonn. (Cited by: Kamal, A.S.M., Goyer, K., Koottatep, T., and Amin, A.T.M.N. 2008. Domestic wastewater management in South and Southeast Asia: The potential benefits of a decentralized approach. *Urban Water Journal*, 5, 345–354).

61. Sipala, S., Mancini, G. and Vagliasindi, F.G.A. 2003. Development of a web based tool for the calculation of costs of different wastewater treatment and reuse scenarios. *Water Supply*, 3(4), 89–96.

62. Solley, W.B., Pirce, R.R., and Perlman, H.A. 1998. Estimated use of water in the United States in 1995. U.S. Geological Survey Circular, 1200. U.S. Geological Survey, Colorado.

63. Strauss, M. and Blumenthal, U.J. 1990. *Use of Human Wastes in Agriculture and Aquaculture: Utilization Practices and Health Perspectives*. International Reference Centre for Waste Disposal (IRCWD) report No 08/90, IRCWD, Dübendorf.

64. Tajrishy, M. 2012. Wastewater production, treatment and use in Iran, Presented at the Second Regional Workshop of the Project "Safe Use of Wastewater in Agriculture", May 16–18, 2012, New Delhi. http://www.ais.unwater.org/ais/mod/page/view.php?id=62. Accessed December 21, 2012.

65. Tchobanoglous, G., Leverenz, H., Nellor, M.H. and Crook, J. 2011. *Direct: Potable Reuse: A Path Forward*. WateReuse Research Foundation, Alexandria, VA.

66. Tsagarakis, K.P. and Georgantzis, N. 2003. The role of information on farmers' willingness to use recycled water for irrigation. *Water Science and Technology Water Supply*, 3(4), 105–113.

67. TURKSTAT, 2012. Environment statistics; Municipal Wastewater Statics. http://www.turkstat.gov.tr/Start.do;jsessionid=hkLSRRjHdxhhKXPhMZLTwYG 0w6SRJ5HLT-2bGZdlpB0L0khfYB9tY!-1381182142. Accessed December 21, 2012.

68. Ulimat, A.A. 2012. Wastewater Production, Treatment and Use in Jordan, Presented at the Second Regional Workshop of the Project "Safe Use of Wastewater in Agriculture," May 16–18, 2012, New Delhi. http://www.ais. unwater.org/ais/mod/page/view.php?id=62. Accessed December 21, 2012.

69. United Nations (UN). 2000. *Wastewater Management Policies and Practices in Asia and the Pacific*. Water Resources Series No 79, UNESCAP, Bangkok.

70. UN. 2012. Population connected to wastewater collecting system. http://unstats.un.org/unsd/environment/wastewater.htm.

71. United States Environmental Protection Agency (USEPA). 2004. *Guidelines for Water Reuse*. EPA/625/R-04/108. USEPA, Washington, DC.

72. Urkiaga, A., Fuentes, L., De las, B.B., Chiru, E., Bodo, B., Hernández, F., and Wintgens, T. 2007. Methodologies for feasibility studies related to wastewater reclamation and reuse projects. *Desalination*, 187, 263–269.

73. USEPA. 2012. *Guidelines for Water Reuse*. USEPA, Cincinnati, OH; National Risk Management Research Laboratory. USAID. (EPA/600/R-12/618), Washington, DC.

74. USGS. 2009. National Water Summary 1983—Hydrologic Events and Issues, U.S. Geological Survey Water-Supply Paper 2250.

75. Van der Hoek, W. 2004. A framework for a global assessment of the extent of wastewater irrigation: The need for a common wastewater typology. In: C.A. Scott, N.I. Faruqui, and L. Raschid-Sally (eds.), *Wastewater Use in Irrigated Agriculture Confronting the Livelihood and Environmental Realties*. CABI-IWMI-IDRC, Wallingford, UK, pp. 11–24.

76. Water Resource Department of Land and Water Bureau (WRDLWB), Ministry of Land, Infrastructure, Transport and Tourism, Japan. 2012. Water Resources in Japan, 2012. http://www.mlit.go.jp/tochimizushigen/mizsei/hakusyo/index5.html.

77. Westcot, D.W. 1997. *Quality Control of Wastewater for Irrigated Crop Production*. Water Reports 10. FAO, Rome. 86.

78. Westcot, D.W. and Ayers, R.S. 1985. Irrigation water quality criteria. In: Pettygrove, G.S. and Asano, T. (eds.), *Irrigation with Reclaimed Municipal Wastewater—A Guidance Manual*. Lewis Publishers, Inc., Chelsea, MI.

79. WHO. 1989. *Guidelines for the Use of Wastewater and Excreta in Agriculture and Aquaculture: Measures for Public Health Protection.* WHO Technical Reporting Service, Rome, Italy. (Executive summary).

80. WHO. 2006. *Guidelines for the Safe Use of Wastewater, Excreta and Greywater, Wastewater Use in Agriculture.* WHO Press, Geneva, Switzerland.

81. World Bank, WB. 2012. The World Bank data; GNI per capita, Atlas method (current US$). http://data.worldbank.org/indicator/NY.GDP.PCAP.CD.

82. World Health Organization (WHO). 2007. *Diarrhea: Why Children Are Still Dying and What Can Be Done.* WHO, Geneva, Switzerland.

83. World Health Organization (WHO). 2008. *The Global Burden of Disease: 2004 Update.* WHO, Geneva, Switzerland.

84. Zachariou, D. 2005. *Implementation of Water Reclamation and Reuse in Cyprus, Drought and Water Deficiency: From Research to Policy Making.* Arid Cluster. http://arid.chemeng.ntua.gr/Project/Default.aspx?t=74.

Wastewater Agriculture in Peri-Urban Areas of Dhaka City

Kamrul Islam
Institute of Water Modeling (IWM)

Subrota Kumar Saha
University of Dhaka

72.1 Introduction

Agriculture represents a key sector in the economies of developing countries and cannot be sustained without sufficient water. In most of these countries, agricultural activities are a major source of livelihood and an essential dimension of local social cohesion and culture. This activity is carried on by small farmers in rural areas, very often with huge constraints. However, it must be remembered that, in the end, the dominant use of water around the world will continue to be water for food security [49].

Food and agriculture production account for 70%–80% of global water use, thereby often limiting the development of other economic sectors as well as threatening the sustainability of aquatic ecosystems [49]. Among them fresh water use is 69% in agriculture, 21% in industry, and 10% in domestic purposes in 2000 [16,49].

While the world population has grown threefold over the last 100 years, water use has expanded six fold over the same period. The world population is predicted to expand from the present 6 billion to nearly 8 billion in 2015, stretching the limits on available freshwater resources. Demands on water resources by municipal and industrial uses are expected to increase dramatically [49].

In many regions of the world, this has placed severe strains on existing resources with resulting environmental impacts. An example is in Perth, Australia, where the major drinking water aquifer is being depleted by a combination of uses such as public drinking water source, horticultural irrigation, and the positioning of pine plantations on a large area of the mound [56].

Saudi Arabia is another example of a country with demonstrated impacts on natural water resources due to increasing demands on groundwater by the agricultural sector [8]. Large decreases in groundwater levels (up to 200 m in some places) have been observed due to overextraction. In a number of countries, for example, Australia, this has been compounded by prolonged periods of drought or seasons of low rainfall. In addition, predicted climate impacts from global warming also point to further stresses on water resources, thus reducing the amount of water available for both irrigation and the environment [50].

In large areas of India and China, groundwater levels are falling by 1–3 m per year, causing intrusion of seawater into aquifers, higher pumping costs, and jeopardizing agricultural production [49].

Wastewater reuse is not a recent invention. There are indicators that wastewater was used for irrigation in ancient Greece and in the Mayan civilization (ca. 3000–1000 BC) [2,3].

During 1950–1960, interests in applying wastewater on land in the western hemisphere increased as wastewater treatment technology advanced and the quality of treated effluents steadfastly improved. Land application became a cost-effective alternative of discharging effluent into surface water bodies [3].

The wastewater used in irrigation can be from different sources. It can be completely untreated municipal or industrial wastewater, mechanically purified wastewater, or particularly or fully purified wastewater treated biologically [12].

Wastewater reuse has recently emerged as a focus of study in the developing countries where its use by urban and peri-urban farming communities is becoming popular [44]. Wastewater may supply organic matter and mineral nutrients that are beneficial to crop production, reduces the cost of fertilizer application and the use of groundwater for agricultural purpose[27]. In arid regions, wastewater is especially valued as an additional resource besides the added benefits from its nutrient contents [22]. However, urban wastewater may also contain hazardous substances including heavy metals and pathogenic microbes. These substances may eventually harm the environment, human health, soil, ground water, and crops. Hence, any decision making related to wastewater reuse should consider both positive and negative aspects [24]. Many countries have developed guidelines and quality criteria for wastewater reuse for agriculture irrigation. Example of these guidelines is the USEPA manual of wastewater reuse [53]. WHO [57] also provides a guideline and quality for the reuse of wastewater for various purposes. In Bangladesh, there are no such guidelines or quality criteria for wastewater reuse in urban and peri-urban areas.

Bangladesh is an agricultural-based country. The agriculture sector contributes about 21.9% of the national GDP [5]. Agriculture provides food security and livelihood support to about 80% of the rural people in Bangladesh. Like other developing countries, agriculture is the highest water use sector in Bangladesh. In 1990, total water withdrawals for agriculture, domestic, and industrial uses were 86%, 2%, and 12%, respectively, of which 73% comes from groundwater sources [15].

More than 3000 industrial units in the city area are discharging the most polluted wastewater into rivers. They are discharging more than 1.3 million cubic meters of wastewater into the rivers everyday compared to the daily discharge of 0.5 million cubic meters of household wastewater (www.thefinancialexpress-bd.com).

Groundwater levels in Dhaka are decreasing at an alarming rate[33]. It is also evident that the rate of water-level drop in the city is about 2.5 m per year in recent years, which was about 1 m per year during the late 1990s [34]. Wastewater agriculture has been a traditional practice in the peri-urban areas of Dhaka city. The domestic sewage and some industrial wastewater of Gulshan, Badda, Rampura, etc. are discharged through the open drain into the Rampura canal and the wastewater of Mirpur Pallobi and Mirpur-11 areas are discharged into the Mirpur Pallobi canal and Kalshi canal, respectively. Wastewater is lifted from these canals for farming of many kinds of vegetables such as cabbage, tomato, celery, cauliflower, arum, brinjal, lady's finger, pumpkin, gourd, radish, etc.

In Bangladesh, reuse of urban wastewater has not started significantly. However, it may have potential benefits for agricultural use in many peri-urban areas of the country. If properly managed, wastewater may be a potential for peri-urban agriculture in Bangladesh.

If wastewater reuse can be managed under an institutional arrangement between the City Corporation and local water users, effluent from this plant can be efficiently used to support agriculture in the peri-urban areas. However, the benefits, risks, and social acceptance should be assessed before

wastewater can be reused in any form in the long run. The present study was conducted to evaluate future wastewater reuse potential and its benefits in agricultural use in the peri-urban areas of Dhaka city.

72.2 Wastewater Reuse, Benefits, and Risk

Wastewater in agriculture has been in practice for decades in countries like Mexico, Vietnam, China, Saudi Arabia, Jordan, India, Pakistan, and Israel [47].

Israel is one of the leading countries in wastewater usage as it expects that 70% of its agricultural water demand in 2040 will be met with treated wastewater. Advantages of wastewater irrigation are the possibility of decreasing the purification level and the derived treatment costs, thanks to the role of soil and crops in acting as a bio-filter. In addition, using the nutrients available in wastewater may diminish fertilization costs [24].

Public health concern is a major constraint because wastewater carries a wide spectrum of pathogenic organisms and poses risk to agricultural workers, crop handlers, and consumers [7,47]. High levels of nitrogen in wastewater may result in nitrate pollution of groundwater and lead to adverse health effects. Accumulation of heavy metals in soils and its uptake by plants is another risk associated with wastewater reuse in agriculture [35].

Guidelines set by the WHO and United Nations Environment Program advise that at least some sort of treatment is necessary before wastewater reuse [7,39].

Urban wastewater reuse is a popular practice in many water shortage countries of the world. Wastewater contains high nutrient values for crop production and may improve food security, employment, and livelihood of many urban poor farmers [48].

Several studies have shown a positive impact of wastewater on crop production [35,46]. Wastewater containing industrial effluent may input high levels of nutrients, heavy metals, and other constitutes that are toxic for plant growth. For example, nitrogen is essential for growth, but in cases of over-application may lead to prolonged vegetative stage making the crop more susceptible to pests and diseases and eventually resulting in lower yields [41]. However, water quality together with climatic conditions, physical and chemical soil properties, and water management practices, to a large extent, determine the maximum crop yield [13].

Reuse of wastewater would have a greater impact on future usable sources of water than any of the technological solutions available for increasing water supply. Water reuse in agriculture is justified on agronomic and economic grounds, but care must be taken to minimize adverse health and environmental impacts [22].

On the other hand, wastewater is seen as a reliable water source where high value crops like vegetables can grow without applying extra chemical fertilizers. Land application of wastewater is often seen as a positive means of low cost wastewater disposal in countries where adequate treatment facilities are absent [40,45].

72.2.1 Agriculture Benefits of Wastewater Reuse

Wastewater provides an opportunity for dry season irrigation water availability and farmers can sell their crops at a price three to five times higher than the Kharif (monsoon) price. The high nutrient load increases crop yields and reduces the need for costly fertilizer inputs.

For a long time, wastewater has been regarded as a problem as it involves hygienic hazards, as well as containing organic matter and eutrophying substances in the form of nitrogen and phosphorus. These substances cause problems in sea, lakes, and streams, but on the other side, they would be valuable for agriculture purposes [14]. Especially the macronutrients nitrogen (N), phosphorus (P), and potassium (K) in urine and feces can be utilized instead of artificial fertilizer [55] produced mostly by fossil resources, which cannot be relied securely in a long-term perspective [42]. According to Jonsson [32], "all nitrogen, phosphorous and potassium from urine and feces can be recycled to agriculture, except for some small losses of nitrogen in the form of ammonia. A lot of energy is conserved since a lot of chemical fertilizers can be replaced by urine and feces." Furthermore, the organic material increases the humus content and thus the water holding capacity of the soil and prevents the degradation of soil fertility [14,20].

Bradford et al. [6] showed that farming practice might alleviate poverty for many urban and peri-urban farmers. Rutkowski et al. [44] presented results from two case studies in the Kirtipur and Bhaktapur municipalities of the Kathmandu Valley and found that water availability generates possibilities to cultivate high-value crops in the peri-urban areas and generates better livelihoods for farmers. Sally et al. [45] showed that year-round availability of water allows multiple crop cultivation cycles, which ensure food security for the poorer households in India, Pakistan, Vietnam, and Ghana, which lift farmers from poverty. Hoek et al. [54] explored that the greater benefit for farmers using wastewater is a reliable water supply, which allows growing high-value vegetables and crops. Wastewater reuse also has an indirect benefit associated with the reduction of pollutants discharged into natural watercourses. Irrigation with untreated wastewater is practiced in many cities in Pakistan because of its high productivity.

Hamoda [22] stated that water reuse resulted in considerable saving of freshwater and led to secure water for irrigation and food sufficiency in water-scarce regions. Water reuse is a cost-effective alternative that would have greater impact on the future than any of the technological solutions available. However, appropriate technologies of wastewater treatment and irrigation systems must be developed and implemented in order to promote wide use of treated wastewater.

Jenssen et al. [30] indicated that the value of wastewater as a source of N, P, and K depends on the nutrient availability, types of crops, and soil fertility. Thus, evaluation of wastewater as a source of nutrient is not possible by simple comparisons of wastewater and fertilizers because the effects of N, P, and K in wastewater are different.

Jayakody et al. [29] presented an assessment of wastewater agriculture in Rajshahi city, which was undertaken as part of the Wastewater Agriculture and Sanitation for Poverty Alleviation in Asia (WASPA) project. It is observed that leafy vegetables are almost exclusively grown in the wastewater irrigation area because of the readily available water from the wastewater canals. Wheat, potato, and jute are grown in the clean water area, but to a much lesser extent in the wastewater areas. Several paddy varieties are grown in both clean and wastewater areas and there is a significant difference between clean water and wastewater yield. The average paddy yields in the wastewater are 3.9 and 4.7 tons/ha in the clean water areas, respectively. Urea is less applied in wastewater areas where as application rates of Muriate of Potash and Triple Super Phosphate are well above the recommended levels in all areas. Fertilizer application rates do not follow any guidelines provided by the DAE.

Fitamo et al. [19] found that secondary treated sewage effluent can efficiently substitute potable water for irrigation and provide economic benefits because of the nitrogen content in STSF water. This corresponds to an annual savings of 179–450 USD per hectare. However, the responses to STSF irrigation and the mineral N-fertilizer economy are dependent on rainfall intensities and irrigation amount applied.

72.2.2 Health Impact of Wastewater Reuse

Feenstra et al. [18] estimated health risks associated with untreated wastewater and showed that wastewater contains far more fecal coliform bacteria and helminth eggs than advised by the WHO. This poses a high health risk to farmers, their families, and crop consumers. It is found that prevalence of diarrheal disease and hookworm infection are very high in the farm community, and these diseases are higher among male farm workers and children than females. For crop consumers, the chance to acquire a hookworm infection seemed slightly increased. The study recommended that appropriate treatment for wastewater and helminthes eggs is needed before use.

Habbari et al. [21] assessed the health risk of wastewater reuse for agricultural purposed in Beni-Mellal, Morocco and said that prevalence of *Ascariasis* infection was approximately five times higher among children in wastewater exposed sites compared with the control site. However, Bradford et al. [6] found that in Hubli-Dharwad, women are the most exposed to the hazards of wastewater (pathogens and organophosphate pesticides). Hired farm laborers are most likely to be women, the poorest social group in the society. Thus, wastewater poses a high risk of geohelminthic infections where adequate treatment of wastewater and public health education are highly recommended.

Cifuentes et al. [9] assessed the risk factors of *Giardia intestinalis* infection in an agricultural population in Mexico. Results showed that long periods of hydraulic retention and partial improvement of wastewater quality did not reduce the health risk. Children 1–14 years of age had the highest prevalence of infection (20%). Individuals with the longest time of exposure to agricultural activities (5 years or more) had the lowest prevalence of infection (6%) compared with those who experienced 1–4 years of exposure time. Individuals from households who purchased vegetables from the market in Mexico City had a higher prevalence of infection with *G. intertinalis* than those who tend to patronize local shops. The study recommended that protective measures for children against protozoan infections might consist of providing primary health care and health education, fostering breastfeeding, safe weaning practices, disinfection of vegetables, safe drinking water containers, and domestic sanitation.

Ensink et al. [13] studied the costs and benefits of untreated wastewater reuse for agriculture. This study focused on health, environment, and socioeconomic aspects of irrigation with untreated wastewater. Results show that although wastewater contains high amounts of fecal coliform bacteria and worm eggs, farmers will take the health risks for direct economic benefits from it. Scott et al. [46] revealed that measured coliform levels in recycling water are not high but show significant evidence of direct health impacts of wastewater irrigation.

Sally et al. [45], Faruqui et al. [17], and Amoah et al. [1] indicated that wastewater reuse causes negative health impacts. The potential health risks include helminthes infections to farmers and crop consumers. Bradford et al. [6] showed that farming practice alleviates poverty for many urban and peri-urban farmers but consumers of their products and the environment are at risk. Untreated wastewater is a major source of pathogenic illness such as anemia and worm infestation. It is also found that urban wastewater contains potentially injurious biomedical wastes including disposable needles and syringes, creating hazardous conditions for the farmers.

Ensink et al. [13] investigated the risks of *Giardia duodenalis* infection due to wastewater irrigation at Faisalabad, Pakistan. The study showed that there is increased risk of *Giardia* infection in wastewater farming households when compared with farming households using regular (non-wastewater) irrigation.

Rutkowski et al. [44] revealed that the quality of wastewater of the Kirtipur and Bhaktapur municipalities of the Kathmandu Valley range from diluted wastewater to raw sewage. Farmers experienced mild to high health impacts from exposure. However, he found that direct contact with wastewater, lack of personal protective measures, and inadequate personal hygiene are the causes of skin ailments in farmers. At the same time, it is said that improved hygiene behaviors in communities engaged in wastewater-fed agriculture and aquaculture benefits public health.

Jamrah et al. [28] revealed that greywater contains significant levels of suspended solids, inorganic constituents, total organic carbon, COD and BOD, total coliforms and *Escherichia coli* bacteria.

72.2.3 Environmental Impact of Wastewater Reuse

Zulu et al. [61] assessed the effect of wastewater reuse in Riceland ecosystem and revealed that water reuse not only helps to meet irrigation water needs, but also aids purification of the agricultural drainage water and preservation of the Riceland ecosystem.

Scott et al. [46] reveled that wastewater recycling tends to concentrate salts in the flows leaving the area. Heavy metal concentrations in bed sediments and in irrigated fields were within Mexican, EU, and US norms. Under the current irrigation practices, the buildup of heavy metals in soils is within EU and US norms.

Hepner [26] indicated that reuse of high quality wastewater minimizes environmental impact and maximizes benefit to crop growth. It is also possible to reduce the extensive and expensive soil and hydrogeologic testing as well as the monitoring requirements if wastewater can be reused for agriculture.

Ensink et al. [13] found that heavy metal buildup in soil was not a significant factor in small cities without major industries. Groundwater contamination is often considered an important negative impact of wastewater reuse. However, when the natural groundwater in an area is saline and unfit for use as drinking water, this is of minor concern.

Hoek et al. [54] revealed that accumulation of heavy metals in soil due to wastewater reuse does not exceed the recommended limit. Thus, there is a need to identify methods to prevent or lower the health risks associated with wastewater reuse for increasing socioeconomic and environmental benefits under the prevailing social and economic conditions.

Davis and Hirji [10] explained that wastewater reuse is a water conservation measure that is increasingly being implemented in water-scarce regions. It also provides additional benefit of protection sensitivity downstream environment. So, wastewater reuse is a viable option that should be considered in water supply projects.

Bradford et al. [6] showed that unregulated and continuous irrigation with wastewater also leads to environment problems such as salinization, phytotoxicity (plant poisoning), and soil structure deterioration (soil clogging), which in India is commonly referred to as sewage sickness. High nutrient content of wastewater greatly increased the incidence of weeds.

Sally et al. [45] analyzed the policy implications of wastewater reuse guidelines in resource-poor situations and found that land application of wastewater is a possible low-cost solution that will improve environmental sanitation and downstream water quality.

Jimenez and Asano [31] said that wastewater adds organic matter and nutrients in soil increasing productivity, which has a great economic and social importance. However, there are also negative impacts such as metal accumulation in soil, soil salinity, and aquifer pollution.

Martijn and Redwood [40] illustrated that nutrient management, crop choice, irrigation methods, health risk regulation, and water rights are the major limitations of adapting safeguards for environmental health regarding wastewater reuse in developing countries.

Toze [51] suggests that several factors that should be considered in water reuse projects include chemical contaminants, soil salinity, and impact on soil structure, which could be controlled through treatment and better farm management practices.

Laturnus et al. [37] studied the degradation and uptake of organic microcontaminants in crops such as barley and carrots grown in agricultural soil amended with anaerobically treated sewage sludge from wastewater treatment plants. This study revealed no immediate risks to soil and crops.

Zaidi [60] reported that collection and treatment of wastewater, disposal of sludge, and use of treated water is a big challenge for environmental security. It is reported that untreated wastewater reuse in agriculture is unacceptable as it may cause soil and groundwater contamination. He also said that wastewater reuse is not economically feasible where an additional distribution system is required and removal of heavy metals from the wastewater will contribute to environmental safety.

Fitamo et al. [19] studied the impact of heavy metals and their potential mobility and bioavailability in soil irrigation with wastewater. The results showed that Zn and V were relatively high, while Hg content was relatively low in the soil. Among heavy metals, Co, Cu, Ni, V, Pb, and Zn showed potential mobility. These heavy metals pose significant health threats through their introduction into the food chain in the wastewater irrigation soils.

Heidarpour et al. [25] assessed the effects of treated effluent on soil chemical properties using two irrigation methods (subsurface and surface irrigation method). Chemical analysis results show that the concentration of chemical constituents in soil layers were influenced by water movement patterns, chemical constituents of irrigation water, and plant uptake mechanisms. The most important concern was the increase of EC in the top soil layer with subsurface irrigation and this might inhibit plant growth. The results also show that wastewater irrigation significantly affected potassium while no significant effect on soil Na, P, and TN (total nitrogen) due to irrigation with wastewater was observed.

Rutkowsi et al. [44] found that the water availability in the dry season in Kirtipur and Bhaktapur municipalities of the Kathmandu Valley increased use to wastewater reuse. However, it has changed the environmental condition downstream of these cities.

72.2.4 Social Acceptability of Wastewater Reuse

Lazarova [38] indicated that sustainability of a wastewater reuse project depends on the public attitude, perception of water quality, and willingness to accept the reuse project. Kretschmer et al. [36] mentioned that social and cultural factors are essential to develop a sustainable water reuse project. Therefore, in the absence of social support, water reuse projects may be unsustainable.

Jamrah et al. [28] conducted a public acceptance survey in Muscat Governorate in the sultanate of Oman and the results show that approximately 76% of the respondents accept the reuse of greywater for gardening, 53% for car washing, and 66% for toilet flushing.

72.2.5 Economic Consideration

Khouri et al. [35] give guidelines for different scenarios, which support an economic evaluation of a planned reuse project in agriculture. Differences between projects where irrigation already exists and wastewater can provide supplemental irrigation and projects without irrigation until now, where the benefits would be, for example, a higher production from existing farms, are being described and mentioned to be considered while evaluating the economic benefit of a reuse project. The difficulty that positive as well as negative effects are not always quantifiable has to be realized and solved (e.g., environmental enhancements from the elimination of wastewater discharges).

The paper of Haruvy [24] demonstrates, in a simple way, how to tackle decision-making questions regarding the economic point of view of the reuse project. Various wastewater reclamation and reuse options (e.g., treatment levels, location of reuse) are compared by computing the net national benefit. A case study in Israel serves as an example. Here, direct benefits as well as costs, indirect benefits, and environmental damage have been considered. As a result, the secondary treatment level and local irrigation (reuse option) was most beneficial, compared to southern conveyance and tertiary or secondary treatment. Still, even this reuse option resulted in much more national benefit than the alternative of river disposal after tertiary treatment.

Baharia [4] points out that there is often a lack of assessment of cost differentials between different alternatives of treatment concepts (treatment for reuse or treatment for sea or river outfall), the necessary transfer systems as well as the cost of freshwater.

The study of Haruvy et al. [24] evaluates the beneficiary and hazardous effects of nutrients by irrigation with secondary water, by using an optimization model. This model determines the monthly optimal treatment level while maximizing the agricultural incomes, in the farmer's point of view. As a result, the added value increases with increasing nitrate concentration, but it is affected by possible hazards to crops. Adaptation of wastewater treatment levels to the regional mix of crops and to crop fertilization demands enhances agricultural incomes. One has to carefully look at the national point of view, as this might lead to different results. There the analysis should also take account of environmental effects, such as nitrate leaching or salinity accumulation.

The presented case study in Xu et al [59] about wastewater reuse on a French island makes a step forward. Here an integrated modeling of technical and economic issues has been realized. The economic submodel estimates the economic efficiency of investments. Furthermore, it makes clear that the evaluation of different scenarios is very important to come to comparable solutions.

72.3 Sample Collection Methods and Techniques

Bangladesh is an agricultural country. Agriculture is more familiar in the peri-urban area of Dhaka city. The study area

TABLE 72.1 GPS Locations of Sampling Points of the Study Area

Sl. No.	Sampling point	GPS Reading	
1	B-1	N 23°45.776′	E 90°26.619′
2	B-2	N 23°45.682	E 90°26.761
3	B-3	N 23°45.786	E 90°26.354
4	B-4	N 23°45.863	E 90°25.881
5	B-5	N 23°45.982	E 90°25.572
6	P-1	N 23°49.784	E 90°21.579
7	P-2	N 23°49.766	E 90°21.554
8	P-3	N 23°49.770	E 90°21.554
9	P-4	N 23°49.747	E 90°21.634
10	P-5	N 23°49.704	E 90°21.622
11	K-1	N 23°49.635	E 90°22.818
12	K-2	N 23°49.630	E 90°22.857
13	K-3	N 23°49.619	E 90°22.927
14	K-4	N 23°49.619	E 90°23.004
15	K-5	N 23°49.617	E 90°23.026

Note: B = Bonosri; P = Pallobi; K = Kalshi.

is mainly the peri-urban area around Dhaka city. The samples are collected from three locations: Banasree (abbreviated as B, between 23°45′40″N and 23°45′58″N latitudes and 90°26′45″E to 90°25′34″E longitudes) on the east side of Dhaka city and other two are in Mirpur, on the north side of Dhaka. Pallabi (abbreviated as P) is located between 23°49′40″N and 23°49′51″N latitudes and 90°21′15″E to 90°21′42″E longitudes. The third place, Kalshi (abbreviated as K) is located between 23°49′30″N and 23°49′40″N latitudes and 90°22′43″E to 90°23′4″E longitudes. The samples are collected from a Jheel (wetland), where the wastewater of Mirpur 11 and Pallabi and Balughat area are discharged (Table 72.1; Figure 72.1).

72.3.1 Sample Collection Techniques

Collection of representative wastewater samples is an important step for evaluating the water quality. Precautions were taken to obtain the representative samples. A total of 15 samples were collected from the exposed sites of three different locations in the

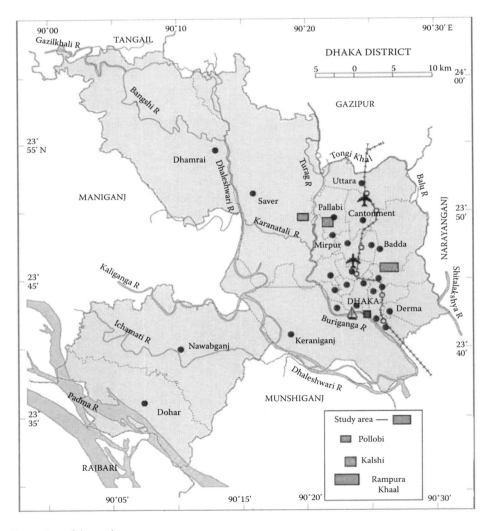

FIGURE 72.1 Sampling points of the study area.

peri-urban areas of Dhaka city. Wastewater samples were collected from the mid-stream of the canal (khaal) and wetland (beels). The samples were collected from 0.5 m below the water surface. Clean and dried 1 L plastic bottles without any contamination were used for sampling. Each bottle was previously washed with detergent and de-ionized water. During the sampling, the bottles were filled with wastewater and then securely sealed with proper leveling and preserved in an ice-filled sample carrier. Finally, the samples were transported to a hydrology laboratory for chemical analysis.

72.4 Result and Discussions

The results of the chemical analysis for all the wastewater samples collected at different places are summarized in Table 72.2.

72.4.1 Physicochemical Properties of Wastewater

The pH value of the study areas range between 7.3 and 8.1, with an average of 7.5, which are slightly alkalic in nature. The pH range within 6.5–7.5 is more suitable for crops [23]. Therefore, the wastewater in the study areas is suitable for crops production in pH context.

EC values of wastewater in all sampling sites of the study areas vary from 773 to 1226 µS/cm. The conductivity range below 750 µS/cm is more suitable; 750–2250 µS/cm is moderately safe for irrigation [58]. Therefore, the overall EC values of the wastewater in the study areas are safe for irrigation purpose.

Total dissolved solids (TDS) contents in wastewater of sampling sites vary from 245 to –481 mg/L and the average value is 416 mg/L. The TDS range below 525 is good for irrigation and 525–1400 is in the permissible level [58]. Therefore, the studied samples are suitable for irrigation with some reservation.

72.4.2 Status of Irrigation Water Chemistry in the Study Area

72.4.2.1 Sodium Hazard

The sodium concentration in wastewater of the study areas varies from 71.89 to 87.01 mg/L and average sodium concentration is 80.66 mg/L. According to irrigation quality standard of WHO and FAO, sodium concentration is within the limit of water quality guidelines for irrigation and shows no restriction for use in crop irrigation.

Potassium concentration in wastewater of the study areas ranges from 12.70 to 17.22 mg/L and average concentration is 14.64 mg/L. Calcium concentration in wastewater of the study areas varies from 42.61 to 50.39 mg/L and average concentration is 47.38 mg/L. Magnesium concentration in wastewater of the study areas ranges from 4.55 to 11 mg/L and average concentration is 8.31 mg/L.

72.4.2.2 Bicarbonate Hazard

Bicarbonate concentration in wastewater of the study areas varies from 167.75 to 381.28 mg/L and average concentration is 325.84 mg/L. Chloride concentration in wastewater of the study areas ranges from 62.12 to 84.31 mg/L and average concentration is 73.36 mg/L. Phosphate concentration in wastewater of the study areas varies from 0.44 to 0.67 mg/L and average concentration is 0.54 mg/L. Sulfate concentration in wastewater of the study areas varies from 0.75 to 6.98 mg/L and average concentration is 2.86 mg/L. In the croplands irrigated with wastewater, nitrate concentrations range from 50 to 130 g m^3, but in the croplands irrigated with groundwater away from the canal, nitrate concentrations are <35 g m^3. Wastewater irrigation was found to control nitrate distribution in groundwater, and ought to be used carefully to protect both soil and groundwater from nitrate pollution.

TABLE 72.2 Wastewater Chemistry of Studied Samples in mg/L

Location	pH	EC (µs)	Temperature (°C)	Na (mg/L)	K (mg/L)	HCO$_3$ (mg/L)	Cl (mg/L)	SO$_4$ (mg/L)	NO$_3$ (mg/L)	PO$_4$ (mg/L)	Fe (mg/L)	Mn (mg/L)	Ca (mg/L)	Mg (mg/L)
B-1	7.9	806	22.9	71.89	12.99	167.75	62.13	2.76	153.75	0.58	1.72	0.23	45.89	9.54
B-2	8.1	1226	25.5	74.37	12.70	366.00	71.00	2.03	178.75	0.59	2.19	0.24	42.61	11.00
B-3	7.5	882	24.5	74.37	13.15	366.00	71.00	3.23	181.25	0.66	1.04	0.19	45.74	10.59
B-4	7.5	855	25.2	76.85	13.73	350.75	71.00	5.43	188.75	0.62	1.16	0.18	45.71	10.73
B-5	7.5	833	25.1	84.28	13.22	381.25	79.88	6.98	180.00	0.62	1.55	0.17	50.10	10.83
P-1	7.4	833	31.9	80.56	13.51	343.13	84.31	3.22	–	0.53	0.32	0.46	47.75	6.99
P-2	7.4	839	31.5	80.56	13.99	327.88	75.44	3.13	–	0.56	0.51	0.38	48.65	6.92
P-3	7.4	838	31.0	80.56	14.32	320.25	71.00	3.26	–	0.51	0.63	0.38	50.39	6.93
P-4	7.5	816	31.0	80.56	13.68	312.63	84.31	3.08	–	0.52	0.74	0.38	48.15	7.00
P-5	7.5	818	31.5	80.56	13.51	327.88	71.00	3.61	–	0.45	0.69	0.38	46.16	6.81
K-1	7.4	809	30.6	87.01	17.22	327.88	71.00	1.01	–	0.45	0.87	0.26	50.21	8.63
K-2	7.7	815	30.3	87.01	17.22	350.75	75.44	2.03	–	0.50	0.72	0.27	48.87	8.61
K-3	7.3	908	30.2	83.79	16.73	381.28	75.44	1.49	–	0.68	0.30	0.37	45.98	8.07
K-4	7.6	785	30.6	83.79	16.89	266.88	71.00	0.97	–	0.46	1.09	0.41	49.28	7.48
K-5	7.5	773	30.8	83.79	16.73	297.38	66.56	0.75	–	0.44	0.56	0.22	45.20	4.55

There is a significant relationship between SAR values and percent sodium for irrigation water and the extent to which sodium is absorbed by the soil. The water required for irrigation is classified based on SAR, percent sodium, and electrical conductivity, which is presented in Table 72.3. This table reveals that wastewater of the study areas is suitable for irrigation. The most important water quality criteria are provided by Richards [43]. The SAR value in wastewater of the study areas ranged from 2.52 to 3.16. According to the quality classification of water irrigation depending upon SAR after Richards [43], wastewaters are excellent to good for irrigation purposes.

72.4.2.3 Doneen Classification

Doneen [11] proposed a classification of water based on the salinity, permeability, and toxicity of irrigation water. Salinity is caused by high solubility of salts, which rapidly accumulate in the soil. The low solubility salts precipitate in the soil as the soil solution increases in the salinity and does not play much role in salinization of soil.

The potential soil salinity (PS) is defined by the concentration of chloride and half of the sulfate ions.

$$PS = Cl^- + 1/2SO_4^{2-}$$

where concentrations of all ions are expressed in meq/L.

The water of study area has been classified for irrigation purposes and the data are presented in Tables 72.4 and 72.5.

72.4.2.4 Wilcox Classification

The PS values of water in the study areas are within 1.77–2.30 ranges, which reveal that all samples are in excellent to good water category for irrigation purposes.

TABLE 72.3 Status of Studied Samples in Terms of Percentage of Sodium, EC, and SAR Value

Location	EC (μs)	Percentage of Sodium	SAR	Quality
B-1	806	52.942	2.522	Permissible
B-2	1226	54.009	2.628	Permissible
B-3	882	53.105	2.576	Permissible
B-4	855	53.866	2.658	Permissible
B-5	833	55.798	2.911	Permissible
P-1	833	57.410	2.932	Permissible
P-2	839	57.537	2.935	Permissible
P-3	838	57.583	2.935	Permissible
P-4	816	57.430	2.932	Permissible
P-5	818	57.537	2.940	Permissible
K-1	809	58.552	3.095	Permissible
K-2	815	58.565	3.096	Permissible
K-3	908	58.035	3.004	Permissible
K-4	785	58.464	3.029	Permissible
K-5	773	60.534	3.163	Doubtful

TABLE 72.4 Water Classes with the Basis on PS Value

Location	PS	Water Quality
B-1	1.77	Excellent to good
B-2	2.02	Excellent to good
B-3	2.02	Excellent to good
B-4	2.02	Excellent to good
B-5	2.28	Excellent to good
P-1	2.40	Excellent to good
P-2	2.15	Excellent to good
P-3	2.02	Excellent to good
P-4	2.40	Excellent to good
P-5	2.02	Excellent to good
K-1	2.02	Excellent to good
K-2	2.15	Excellent to good
K-3	2.15	Excellent to good
K-4	2.02	Excellent to good
K-5	1.90	Excellent to good

Source: After Doneen, L.D. 1962. *Proceedings of the 1961 Biennial Conference on Groundwater Recharge*, 156–163.

TABLE 72.5 Status of Studied Samples in Terms of PI Value

Location	PI	Type
B-1	72.59	Category -3
B-2	99.49	Category -3
B-3	97.58	Category -3
B-4	95.55	Category -3
B-5	99.30	Category -3
P-1	99.30	Category -3
P-2	97.43	Category -3
P-3	96.43	Category -3
P-4	95.36	Category -3
P-5	97.57	Category -3
K-1	95.51	Category -3
K-2	98.30	Category -3
K-3	102.73	Category -3
K-4	89.15	Category -3
K-5	96.53	Category -3

The permeability index (PI) may be expressed as

$$Percent\ PI = \left[\frac{\{Na^+ + 1/2(HCO_{3-})\}}{(Ca^{2+} + Mg^{2+} + Na^+)} \right] \times 100$$

where concentration of all ions are expressed in meq/L.

The Doneen classification diagram reveals that the studied water samples fall in Class III category. It shows that the samples are good for irrigation purposes (Figure 72.2).

However, on the Percent Sodium and Electrical Conductivity (Table 72.3), the samples are in Category I and II, which reveal that they are suitable for irrigation purposes (Figure 72.3).

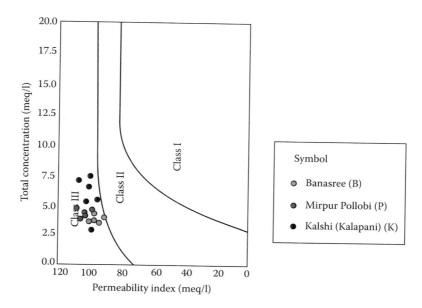

FIGURE 72.2 Permeability index of the study area.

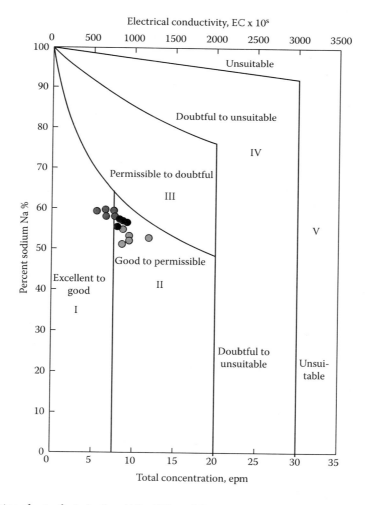

FIGURE 72.3 Quality classification of water for irrigation. (After Wilcox, L.V. 1963. *Factors for Calculation of the Sodium Adsorption Ratio (SAR)*, US Salinity Laboratory Mimeo Report.)

72.4.2.4.1 Index

1. Excellent to good
2. Good to permissible
3. Permissible to doubtful
4. Doubtful to unsuitable
5. Unsuitable

72.4.2.5 USDA Classification

The US Salinity Laboratory Staff [52] has constructed a diagram for classification of irrigation water with SAR as an index of salinity hazard (Figure 72.4). In this diagram, the values of SAR are plotted on an arithmetic scale against electrical conductivity. From the diagram, the samples fall into C_3-S_1, which shows that the wastewater of the study areas is also suitable for irrigation (Table 72.6).

72.5 Summary and Conclusions

Water is one of the most important natural resources. The most important use of water in agriculture is for irrigation, which is a key component to produce enough food. In Dhaka city, with increasing population, the demand of water is alarmingly increasing and thus decreases the natural resources. Wastewater reuse in agriculture would result in considerable savings of ground water and lead to secure water supply required for irrigation. In order to overcome the shortage in food sufficiency in the area in peri-urban areas, farmers consider wastewater a valuable resource because of its high productivity and profitability as well as lack of freshwater. This study shows trends as well as a clear understanding of wastewater use in agriculture. This study shows that the chemistry of wastewater of Dhaka city is suitable for irrigation.

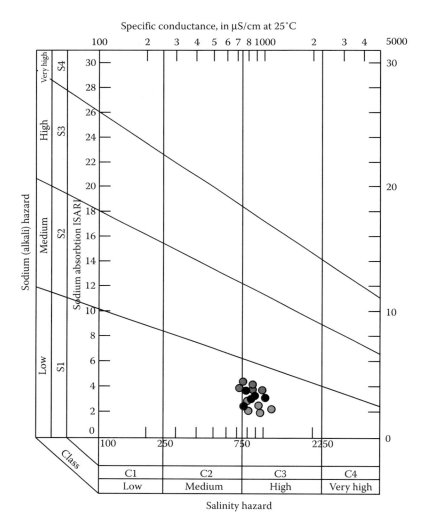

FIGURE 72.4 Suitability of water for irrigation of the study areas. (After US Department of Agriculture (USDA). 1954. *Agriculture Handbook, 60.* USDA, Washington, DC.)

TABLE 72.6 Classification of Water Samples for Irrigation According to USDA

Location	EC (µs)	SAR	Class		Hazards
B-1	806	2.522	C3-S1	S—Moderate–high	A—Low
B-2	1226	2.628	C3-S1	S—Moderate–high	A—Low
B-3	882	2.576	C3-S1	S—Moderate–high	A—Low
B-4	855	2.658	C3-S1	S—Moderate–high	A—Low
B-5	833	2.911	C3-S1	S—Moderate–high	A—Low
P-1	833	2.932	C3-S1	S—Moderate–high	A—Low
P-2	839	2.935	C3-S1	S—Moderate–high	A—Low
P-3	838	2.935	C3-S1	S—Moderate–high	A—Low
P-4	816	2.932	C3-S1	S—Moderate–high	A—Low
P-5	818	2.940	C3-S1	S—Moderate–high	A—Low
K-1	809	3.095	C3-S1	S—Moderate–high	A—Low
K-2	815	3.096	C3-S1	S—Moderate–high	A—Low
K-3	908	3.004	C3-S1	S—Moderate–high	A—Low
K-4	785	3.029	C3-S1	S—Moderate–high	A—Low
K-5	773	3.163	C3-S1	S—Moderate–high	A—Low

Source: Adapted from USDA. 1954. *Agriculture Handbook, 60.* USDA, Washington, DC.

Note: S = Salinity; A = Alkalinity.

The samples are showing SAR values less than four, which indicates that they are excellent to good for irrigation. As percentages of sodium are below 60, this indicates the permissible limit of water quality for irrigation. Based on Wilcox diagram 30% of samples fall in Category I, which is excellent to good, and the other 70% fall in Category II, which is good to permissible class. Based on US salinity diagram, all samples fall in C₃-S₁ category, which shows that the wastewater of the study areas is also suitable for irrigation. The (PS) values are less than three (<3), which reveals that all samples are in excellent to good in water category for irrigation purpose. The Doneen classification diagram based on the permeability index (PI) reveals that all samples in the study areas fall in Class 3. Therefore, it shows that all the samples of wastewater are suitable for irrigation purposes.

It sustains livelihoods of poor peri-urban farming families, contributes to urban food security, helps in solving the urban sanitation problem by preventing pollution of surface water, and makes optimal use of the resources (water and nutrients). The health risks associated with wastewater use in agriculture to farmers and consumers of produce can be reduced by proper irrigation water management and implementation of existing public health measures, even when wastewater treatment is not feasible.

Urban and peri-urban agriculture can enhance food supplies to cities, especially where it has already made its mark, as a cheap and effective source of nutrition, which can be improved at very little marginal cost if officially recognized.

To improve the safety of irrigation of water sources used for agriculture and enhance the direct use of wastewater, it is imperative to separate domestic and industrial discharges in cities, and improve the sewage and septic disposal methods by moving away from ineffective conventional systems.

In addressing health risks, on one hand, authorities have a role to play in planning, financing, and maintaining sanitation and waste disposal infrastructure that is commensurate with their capacities, and which responds to agricultural reuse requirements. On the other hand, as a comprehensive treatment it will remain unlikely in the near future, outsourcing water quality improvements and health risk reduction to the user level and supporting such initiatives through farm tenure security, economic incentives like easy access to credit for safer farming, and social marketing for improving farmer knowledge and responsibility, can lead to reducing public health risks more effectively while maintaining the benefits of urban and peri-urban agriculture. In addition, the following considerations should be kept in mind:

- Farmers should be trained for safe reuse of wastewater from the canals. DPHE should play a major role as the representative organization for public health.
- In the future, a detailed epidemiological study may be conducted at the wastewater reuse site.
- A more detailed study on bioaccumulation to toxic chemicals on soil and crops may be conducted to assess the long-term impact on the human food chain.
- Finally, countries must address the need to develop policies and locally viable practices for safer wastewater use to maintain its benefits for food supply and livelihoods while reducing health and environmental risks.

References

1. Amoah, P., Drechsel, P., and Abaidoo, R.C. 2005. Irrigated urban vegetable production in Ghana: Sources of pathogen contamination and health risk elimination. *Irrigation and Drainage*, 54, 49–61.
2. Angelakis, A.N., Marecos Do Monte, M.H.F., Bontoux, L., and Asano, T. 1999. The status of wastewater reuse practices in the Mediterranean basin: Need for guidelines. *Water Research*, 33(10), 2201–2217.

3. Asano, T. and Levine Audrey, D. 1996. Wastewater reclamation, recycling and reuse: Past, present, and future. *Water Science and Technology*, 33(10–11), 1–14.

4. Bahri, A., 1999. Agricultural reuse of wastewater and global water management. *Water Science and Technology*, 40(4–5), 339–346.

5. Bangladesh Bureau of Statistics. 2008. Preliminary Report of Agri Census-2008, Government of the People's Republic of Bangladesh, Dhaka.

6. Bradford, A., Brook, R., and Hunshal, C.S. 2003. Hubli-Dharwad wastewater irrigation. *Journal of Environment and Urbanization*, 15, 157–170.

7. Blumenthal, U., Mara, D.D., Peasey, A., Ruiz-Palacios, G., and Stott, R. 2000. Guidelines for the microbiological quality of treated wastewater used in agriculture: Recommendations for revising WHO guidelines. *Bulletin of the World Health Organization*, 78(9), 1104–1116.

8. Bushnak, A.A. 2002. Future strategy for water resources management in Saudi Arabia. *A Future Vision for the Saudi Economy Symposium*. October 12–23, 2003, Riyadh.

9. Cifuentes, E., Gomez, M., Blumenthal, U., and Tellez Rajo, M. M. 2000. Risk factors for giardia infection in agriculture village practicing wastewater irrigation in Mexico. *The American Journal of Tropical Medicine and Hygiene*, 62(3), 388–392.

10. Davis, R. and Hirji, R. 2003. Wastewater reuse. *Water Resources and Environment Technical Note F-3*. The World Bank, Washington, DC.

11. Doneen, L.D. 1962. The influence of crop and soil on percolating waters. *Proceedings of the 1961 Biennial Conference on Groundwater Recharge*, 156–163.

12. Donta, A.A. 1997. *Der Boden als Bioreaktor bei der Aufbringungvon Abwasser auf landwirtschaftlich genutzte Flächen*. Veröffentlichungen des Institutes für Siedlungswasserwirtschaft und Abfalltechnik der Universität Hannover, Heft 100.

13. Ensink, J.H.J., van der Hoek, W., and Amerasinghe, F.P. 2006. Giardiaduodenalis infection and wastewater irrigation in Pakistan. *Transactions of the Royal Society of Tropical Medicine and Hygiene*, 100(6), 538–542.

14. Esrey, S., Andersson, I., Hillers, A., and Sawyer, R. 2001. *Closing the Loop Ecological Sanitation for Food Security*. SIDA, Stockholm, Sweden.

15. FAO. 1999. *FAO's Information System on Water and Agriculture*. Food and Agriculture Organization. http://www.fao.org/nr/water/aquastat/countries/bangladesh/index.stm.

16. FAO. 2006. New Gridded Maps of Koeppen's Climate Classification. http://www.fao.org/nr/climpag/globgrids/KC_classification_en.asp.

17. Faruqui, N.I., Niang, S., and Redwood, M. 2004. Untreated wastewater use in market gardens: A case study of Dakar, Senegal. In: C. Scott, N.I. Faruqui, and L. Raschid (eds.), *Wastewater Use in Irrigated Agriculture: Confronting the Livelihood and Environmental Realities*, IWMI-IDRC-CABI, Wallingford, 113–125.

18. Feenstra, S., Hussain, R., and van der Hoek, W. 2000. *Health Risks of Irrigation with Untreated Urban Wastewater in the Southern Punjab, Pakistan*. Institute of Public Health, International Water Management Institute (IWMI), Pakistan Program, Lahore, Pakistan.

19. Fitamo, D., Itana, F., and Olsson, M. 2007. Total contents and sequential extraction of heavy metals in soils irrigated with wastewater, Akaki, Ethiopia. *Journal of Environmental Management*, Springer Science + Business Media, Inc., 39, 178–193.

20. GTZ. 2002. *Deutsche Gesellschaft fur Technische Zusammenarbeit*. Ecosanrecyclingbeats disposal. Eschborn (Germany).

21. Habbari, K., Tifnouti, A., Bitton, G., and Mandil, A. 2000. Gelhelminthicinfections assocaited with raw wastewater reuse for agriculture purposes in Beni-Mellal, Morocco. *Journal of Parasitology International*, 48, 249–254.

22. Hamoda, M.F. 2004. Water strategies and potential of water reuse in the south Mediterranean countries. *Journal of Desalination*, 165, 31–41.

23. Hansen, V.E., Israelsen, O.W., and Stringham, G.E. 1979. *Irrigation Principles and Practices*. John Wiley & Sons, New York.

24. Haruvy, N. 1998. Wastewater reuse: Regional economic considerations. *Journal of Resource, Conservation and Recycling*, 23, 57–66.

25. Heidarpour, M., Fard, B.M., Koupai, J.A., and Malekian R. 2007. The effects of treated wastewater on soil chemical properties using subsurface and surface irrigation methods. *Journal of Agricultural Water Management*, 90, 87–94.

26. Hepner, L.D. 2001. *Reclaimed Wastewater for Agriculture: Signs for the Future*. Department of Agronomy & Environmental Science, Delaware Valley College, Doylestown.

27. Huibers, F.P. and Van Lier, J.B. 2005. Use of wastewater in agriculture: The water chain approach. *Irrigation and Drainage*, 54(Suppl. 1), S-3–S-9.

28. Jamrah, A., Al-Futaisi, A., and Prathapar, S. 2008. Evaluating greywater reusepotential for sustainable water resources management in Oman. *Environmental Monitoring and Assessment*, 137, 315–327.

29. Jayakody, P., Amin, M.M., and Alexandra Clemett, A. 2007. *Wastewater Agriculture in Rajshahi City, Bangladesh*. International Water Management Institute, WASPA Asia Project Report 9.

30. Jenssen, B.H., Boesveld, H., and Rodrigues, M.J. 2005. Some theoretical considerations on evaluating wastewater as a source of N, P and K for crops. *Journal of the International Commission on Irrigation and Drainage*, 54, S35–S47.

31. Jimenez, B. and Asano, T. 2004. Acknowledge all approaches: The global outlook on reuse. *Water*, 21, 32–37.

32. Jönsson, H. 2001. Urine separation-Swedish experiences. *Eco Eng Newsletter*, wstf.go.ke, 1–7.

33. Khan, F.T. 2008. Ground water decrease in Dhaka—Research, the problems and counteractive steps to curb. tanim.blogspot.com/2008/10/groundwater-decrease-in-dhaka-research.html.

34. Khan, S. 2008. *Effective Authority to Save the Rivers a Must.* Reported by Shahiduzzaman Khan, http://www.thefinancialexpressbd.com/search_index.php?page=detail_news&news_id=32448.

35. Khouri, N., Kalbermatten, J.M., and Bartone, C.R. 1994. *The Reuse of Wastewater in Agriculture: A Guide for Planners.* UNDP-World Bank Water and Sanitation Program, The World Bank, Washington, DC.

36. Kretschmer, N., Ribbe, L., and Gaese, H. 2002. Wastewater reuse for agriculture, *Journal of Technology Resource Management & Development*, Scientific Contributions for Sustainable Development, 2, 37–64.

37. Laturnus, F.V., Arnold, K., and Gron, C. 2007. Organic contaminants from sewage sludge applied to agricultural soils. *Environmental Science and Pollution Research*, 14(Special Issue), 53–60.

38. Lazarova, V., Levine, B., Sack, J., Cirelli, G., Jeffrey, P., Muntau, H., Salgot, M., and Brissaud, F. 2000. Role of water reuse for enhancing integrated water management in Europe and Mediterranean countries. *Water Science and Technology*, 43(11), 25–33.

39. Mara, D. and Cairncross, S. 1989. *Guidelines for the Safe Use of Wastewater and Excreta in Agriculture and Aquaculture.* World Health Organization, Geneva, Switzerland.

40. Martijn, E.J. and Redwood, M. 2005. Wastewater irrigation in developing countries-limitation for farmers to adopt appropriate practices. *Journal of the International Commission on Irrigation and Drainage*, John Wiley & Sons, 54, S63–S70.

41. Morishita, T. 1988. Environmental hazards of sewage and industrial effluents onirrigated farmlands in Japan. In: M.B. Pescod and A. Arar (eds.), *Treatment and Use of Sewage Effluent for Irrigation*, Butterworths, Sevenoaks, UK.

42. Palmquist, H. and Jonnsson, H. 2004. Urine, faeces, greywater and biodegradable solid waste as potential fertilizers. *Proceedings of the 2nd International Symposium on Ecological Sanitation*, Lübeck, Germany, p. 828.

43. Richards, L.A. 1954. Diagnosis and improvement of saline and alkali soils. *USDA Agricultural Handbook No. 60.* US Department of Agriculture, Washington, DC.

44. Rutkowski, T., Sally, L.R., and Buecher, S. 2007. Wastewater irrigation in the developing world: Two case studies from the Kathmandu Valley in Nepal. *Journal of Agricultural Water Management*, 88, 83–91.

45. Sally, L.R., Carr, R., and Buechler, S. 2005. Managing wastewater agriculture to improve livelihoods and environment quality in poor countries. *Journal of the International Commission on Irrigation and Drainage*, 54, S11–S22.

46. Scott, C.A., Zarazqa, J.A., and Levine, G. 2000. *Urban-Wastewater Reuse for Crop Production in the Water-short Guanajuato River Basin, Mexico.* IWMI Research Report 41. Colombo, Sri Lanka.

47. Shuval, H.I., Adin, A., Fattal, B., Rawitz, E., and Yekutiel, P. 1986. Wastewater irrigation in developing countries: Health effects and technical solutions. World Bank Technical Paper No. 51. The World Bank, Washington, DC.

48. Siupa, 2000. CGIAR Strategic Initiative on Urban and Peri-Urban Agriculture (SIUPA). http://www.cipotato.org/research/projects/siupa-gl.htm.

49. Tantawi, 2004. Water Potential of Agriculture Forum Universal de les Cultures—Barcelona 2004. 1/13.

50. Toze, S. 2005. *Reuse of Effluent Water—Benefits and Risks CSIRO Land and Water*, CSIRO Centre for Environment and Life Sciences, Wembley, Perth, Australia, www.clw.csiro.au.

51. Toze, S. 2006. Reuse of effluent water-benefits and risks. *Agriculture Water Management*, 80, 147–159.

52. US Department of Agriculture, (USDA). 1954. Diagnosis and improvement of saline and alkali soils. *Agriculture Handbook, 60.* US Department of Agriculture, Washington, DC.

53. USEPA. 1992. *Guidelines for Water Reuse.* United State Environmental Protection Agency, Washington DC.

54. van der Hoek, W., Ul Hassan, M., Ensink, J.H.J., Feenstra, S., Raschid-Sally, L., Munir, S., Aslam, R. et al. 2002. *Urban Wastewater: A Valuable Resource for Agriculture. A Case Study from Haroonabad, Pakistan.* Research Report 63. International Water Management Institute, Colombo, Sri Lanka, p. 21.

55. Vinneras, B., Jonsson, H., Stintzing A.R., and Salomon, E. 2004. *Guidelines on the Use of Urine and Faeces in Crop Production.* EcoSanRes Publication Series. Stockholm Environment Institute, Stockholm, Sweden, 35p.

56. WA State Water Strategy. 2003. *Securing Our Water Future: A State Water Strategy for Western Australia.* Western Australian Government, February 2003. http://www.ourwaterfuture.com.au/community/statewater-strategy.asp.

57. WHO. 1989. Safe use of wastewater and excreta in agriculture and aquaculture measures for public health protection. Prepared by D. Mara, University of Leeds, UK and S. Cairncross, London School of Hygiene and Tropical Medicine, UK.

58. Wilcox, L.V. 1963. Factors for Calculation the Sodium Adsorption Ratio (SAR), US Salinity Laboratory Mimeo Report.

59. Xu, P., Valette, F., Brissaud, F., Fazio, A., and Lazarova, V. 2000. Technical—Economic modeling of integrated water management: Wastewater reuse in a French Island. *3rd International Symposium on Wastewater Reclamation, Recycling and Reuse*, July 3–6, 2000, France.

60. Zaidi, M.K. 2007. *Environmental Aspects of Wastewater Reuse in Book Wastewater Reuse—Risk Assessment, Decision-Making and Environment Security.* Springer, Netherlands, 357–366.

61. Zulu, G., Toyota, M., and Misawa, S. 1996. Characteristics of water reuse and its effects on paddy irrigation system water balance and the Riceland ecosystem. *Journal of Agricultural Water Management*, 31, 269–283.

73

Governance of Nonconventional Water Sources: South Australian Urban Community Views on Ownership, Trust, and Pricing

Zhifang Wu
School of Law, University of South Australia

Ganesh Keremane
School of Law, University of South Australia

PREFACE

Rapid urbanization and growing populations have increased pressure on urban water resources and current climate change predictions have reduced the certainty of traditional supply sources. These have caused a shift in the traditional urban water management regime characterized as being centralized and largely reliant on an engineering approach toward a more sustainable regime that emphasizes the use of new technologies and strategies to increase self-sufficiency by using water sourced from within cities. This means that urban water management now includes services such as water harvesting, water manufacturing, storage, treatment, and distribution, and at times, flood mitigation. However, the integration of these new sources into the urban water supply mix, and the successful implementation of new approaches to urban water management is a multifaceted challenge requiring input beyond the merely technical. Instead, it involves a wide range of factors with varying roles and responsibilities at different levels of administration, generating overlapping concerns among the public agencies managing the resource. Some say that it is a political question and calls for new institutions and organizations to manage this complexity and to consider future generations.

This chapter examines ownership issues within current urban water management regimes in Australia. A case study based on a mail-out survey of the residents of Salisbury City Council was then done, which is renowned worldwide for its water recycling solutions, and for showcasing best practice urban water management. The study found that under current arrangements, there are challenges in achieving sustainable urban water management. The absence of well-defined entitlements and ownership regimes for accessing the produced water, including stormwater, recycled water, and aquifer storage, will require modification of existing legislation and policies in each state, to conform to a nationally consistent framework based on the National Water Initiative principles. It would not be an easy task to accomplish, but a solution may be innovation in governance model as in the case of Salisbury Water.

73.1 Introduction

Rapid urbanization and growing populations have increased pressure on urban water resources and current climate change predictions have reduced the certainty of traditional supply sources. These have caused a shift in the traditional urban water management regime characterized as being centralized and largely reliant on an engineering approach [32] toward a more sustainable regime that emphasizes the use of new technologies and strategies to increase self-sufficiency by using water sourced from within cities [39]. This means that urban water management now includes services such as water harvesting, water manufacturing, storage, treatment and distribution, and, at times, flood mitigation [36]. However, the integration of these new sources into the urban water supply mix and the successful implementation of new approaches to urban water management is a multi-faceted challenge requiring input beyond the merely technical [22]. Instead, it involves a wide range of actors with varying roles and responsibilities at different levels of administration, generating overlapping concerns among the public agencies managing the resource [24,26]. Some say that it is a political question and calls for new institutions and organizations to manage this complexity and to consider future generations [28]. With this background, the goal of this chapter is to examine ownership issues within current urban water management regimes in Australia. In this regard, we focus on one state, South Australia, where the government is supportive of, and engaged in, transitioning South Australia toward becoming a water-sensitive state, and particularly Adelaide into becoming a water-sensitive city [11]. The state already leads the country in a number of key urban water areas, such as treated wastewater reuse, and stormwater harvesting and reuse, and therefore provides opportunities to explore the practical and theoretical resilience of current urban water governance regimes, and to make suggestions for strategies to achieve enhanced governance regimes. The data used in our case study are based on a mail-out survey of the residents of Salisbury City Council, which is renowned worldwide for its water recycling solutions, and for showcasing best practice urban water management.

73.2 Urban Water Management in Australia

The Australian urban water sector has traditionally been heavily reliant on climate-dependent sources of water such as rivers, dams, and reservoirs, which have been dominated by large government monopoly service providers, and characterized by central planning and regulation [21]. However, in the last two decades, the sector has undergone substantial reforms, including corporatization of water utilities, independent price regulation, and industry restructuring. Furthermore, rapid urbanization and increasing population have challenged the limits of traditional water sources that are vulnerable to climate change, as was evident during the 2006 drought [43]. As a result, state governments across Australia are now adopting a holistic approach

to urban water management, particularly urban water supply, which includes incorporating nonconventional sources of water such as desalinated water, stormwater, and wastewater into the supply mix, which had previously only included conventional sources like rivers, dams, and groundwater [11,36]. Additionally, the new holistic approach, termed as Integrated Urban Water Management (IUWM), includes demand management such as water restrictions, water use efficiency, and water conservation, as other additional sources of control of the water supply.

However, too much focus on these measures has been costly to consumers [36]. Previous studies on transitioning to water sensitive cities in Australia [16] have found that institutional capacity is an important issue to be considered while implementing IUWM. This is mainly due to the lack of clarity about roles, responsibilities, and accountabilities within the agencies managing the resources [35]. Therefore, there is a need to reform institutional structures and to adopt a range of policy instruments for IUWM [1] because institutions and rule systems can enable and facilitate the necessary processes for transformation [32]. Our research [19,20,25,45], as well as other research [1,29,35,42], have identified that the property rights and ownership of the captured stormwater, and the reuse of the reclaimed water are not necessarily defined in a precise manner.

Another challenge facing Australia's urban water sector is adapting to climate change. Australian urban water utilities face a significant challenge in designing appropriate demand management and supply augmentation policies in the presence of significant water scarcity and climate variability [43]. Supply augmentation means the creation of additions to water supply infrastructure, such as new dams, desalination plants, or water recycling [43]. While desalination and recycling offer a stable source of water, both involve substantially higher capital and operating costs, ultimately resulting in consumers/households paying higher prices for water and wastewater services [14]. The higher costs have recently been a source of complaint by consumers in South Australia [18]. The Productivity Commission, in its inquiry report [36], noted, "prices for water and wastewater services have increased significantly in recent years and are forecast to rise further in the next few years to finance investment in infrastructure." This contradicts the object of sustainable water management, which is to provide "reliable and affordable water" to users, thereby raising concerns about "affordability" under the current urban water management regime [30]. As a solution, White et al. [43], while discussing the challenges presented by climate change for the urban water industry, suggest that we should be using new economic methods in pricing, cost-benefit analyses, and integrated resource planning, to prepare our cities for climate change. Accordingly, the Council of Australian Governments (COAG) is currently developing a forward work plan to address urban water issues through broad policy reform. Whereas urban water was almost an afterthought for the earlier National Water Initiative, the recent drought and public concern regarding climate change have brought urban water issues centrally onto the national stage [43]. The Commonwealth Government has made a commitment to "support desalination projects, water recycling and major stormwater capturing projects nationwide without adding to climate change and without increasing greenhouse gas emissions" [4].

In the process of including "new sources" into the urban water supply mix, community participation is crucial. Working with a community that does not have alternative water sources as its highest priority requires building participation through a combination of discussions about community outcomes, and more detailed action steps of technology identification, design work, and management [17]. The author further suggests that a lack of community participation results in a wide gap between what is desired from these approaches and what is necessary to get there, and an inability to bridge this gap is the primary reason for the failure of freshwater augmentation projects. Robinson et al. [37] argue that since it is the public who will be served by, and will pay for, these water and wastewater services, the policies on its use and management must include the human dimension. This brings us to the question of users' willingness to pay for the resource in question. This is largely influenced by the tariff structure adopted in a particular scheme/plant. However, the general tendency observed in the case of such freshwater augmentation schemes (reclaimed water in particular) is that users might not be willing to pay more for this resource because it is considered waste. Therefore, the tariff structure should be such that the community being served should perceive it to be appropriate, as well as taking into account the long-term viability of the service provider. Also important is community participation and awareness because, as Tsagarakis and Georgantzís [40] observed in their study, well-informed people, and especially users, are useful in increasing public acceptance and support for water augmentation projects, which may increase their willingness to pay for them.

In addition to the issues discussed above related to producing these new sources of water, there are environmental concerns as well. For example, in Australia there are more than 500 small potable and industrial reverse osmosis plants serving inland communities [10,12], which use brackish groundwater as feed water, as the purpose of these plants is to produce drinking water by removing the salt, which is higher than the acceptable drinking water values in the guidelines [31]. However, this can have a negative impact on the aquifers that contain freshwater [3] and, as well, there are issues related to brine disposal such as government regulations regarding disposal, the lack of information regarding various aspects of concentrate disposal, the cost of disposal, and the environmental impact of different disposal options, which are also very important [2]. All of these relate to the institutional capacity of the public agencies to implement IUWM and to govern urban water. This also means that local government obligations with regard to urban water management have increased over time in Australia [27], and councils are currently responsible for the provision of a wide range of services in their municipalities and make decisions based on a range of economic, financial, ideological, and political factors [5,6,41]. This is exactly the case with the City Council under study—the Salisbury City Council [46].

The Salisbury City Council has pioneered many urban water management initiatives and leads the nation in implementing recycling projects. The Council has created an innovative business model allowing it to separate its core responsibilities as an LGA (local government authority) from its responsibilities to manage the commercial imperatives relating to the produced water.

73.3 Case Study: Implementing IUWM in the City of Salisbury, South Australia

As mentioned earlier, South Australia leads the country in a number of key urban water areas such as treated wastewater reuse, and stormwater harvesting and reuse [11]. Usually the state takes around half of its water supply in an average year from the River Murray. In recent years, this has gone up to nearly 90% of the state's total water supply requirements [13]. However, due to the decline in the flow of water into the river system as a result of the drought years, mismanagement, and overuse of the system, it now appears to be an increasingly unreliable source of water that is in serious decline [11]. The solution is to diversify water supply sources by including new sources, as well as implementing various management measures such as water restrictions, water use efficiency, and water conservation, into the mix [11]. Urban water supply has always been a politically serious endeavor [23] and it is not surprising to see that providing security of water supply for the community, and moving toward more water-sensitive cities, are key electoral issues in South Australia as well as elsewhere in Australia [13]. Furthermore, as the responsibilities of LGAs have increased under the current urban water management regime, many LGAs around Australia have implemented the IUWM approach and the city of Salisbury leads the nation in this process [38]. The work undertaken by the city of Salisbury Council in the northern Adelaide region demonstrates the potential of LGAs to contribute significantly to a major public policy issue [13]. Salisbury's water recycling solutions also set an example and showcase best practice urban stormwater management to audiences across the globe [45].

73.3.1 Description of the Case Study Area

The Salisbury LGA covers an area of 161 km² stretching from the beaches of Gulf St. Vincent to the Adelaide Hills. It is situated 25 km north of the Adelaide CBD, and the terrain is mostly flat with the Little Para River winding its way through the district to the sea. The city has developed 20 strategic stormwater harvesting sites and wetlands. The wetlands are used for storage in conjunction with aquifer storage and recovery (ASR) technology, which allows water to be accessed during the drier periods of the year [7].

The city of Salisbury [8] is now taking stormwater harvesting and recycling to the next level by developing and expanding the distribution network to provide high-quality recycled stormwater including injection in and later recovered from local aquifers by a process known as ASR throughout the Salisbury LGA and beyond. Treated stormwater is distributed via the mains system, and sold to end users, including industrial, commercial, and household facilities. To manage this operation, the City Council decided to form the Salisbury Water Management Board and the Salisbury Water Company, a subsidiary company under the SA *Local Government Act*, 1999. This is a unique structure, and the first of its kind in Australia [34]. The Board is an independent body, and even though it reports to the Council, it can make policy decisions on certain issues independently, such as pricing policies and deciding on applications for water connections. More importantly, significant policy decisions can be made without referring to local government. More details on the governance structure are reported in Wu et al. [45].

According to Stephen Hains [11], the CEO of the city of Salisbury, total investment in the program to date has been nearly $52 million, of which the Council has contributed $17.2 million, the Commonwealth Government $20.3 million, the State Government (principally through its land development agency and connected with its responsibilities as a land owner) $13 million; and others, mainly industrial users, $1.4 million. The Council presently captures around 5 GL of a potential flow through the city of 33 GL, with a plan to increase this by 9 GL over the next five years. Of the total water captured, the Council sells around 1.5 GL, even though there is a strong demand for alternative water supplies in the area. The main customers using this service are industrial users, schools, and other institutions, in addition to new residential subdivisions. However, a major constraint to further sales is the distribution system for the recycled water, which includes direct customer connection, connection to a reticulated network, and water trading. Since the implementation of IUWM encounters a range of institutional, social, economic, and political challenges, we required answers to the following questions:

1. Who do customers think owns the new sources of water (Salisbury Water) and who will/should take responsibility for its quality management?
2. What are the barriers to increased adoption of sustainable urban water management?
3. Does this institutional arrangement/governance model help the implementation of sustainable urban water management?

Accordingly, we decided to conduct a survey of customers who receive Salisbury Water (a term that the Council uses to describe its recycled water) from the Council.

The Council defines Salisbury Water as,

any nonpotable reclaimed, recycled or reused water. This may include recycled stormwater, recycled effluent, native groundwater or any combination of these waters. It may also include any combination of ReWater* with potable

water or rain water. It is NOT suitable for drinking. (http://www.salisbury.sa.gov.au/)

The survey questionnaire included this definition to enable the participants to distinguish it from the mains water that they receive from the SA Water Corporation.

By definition, Salisbury Water can also include any nonpotable reclaimed, recycled, or reused water. The water is not suitable for drinking if it comes from purple (lilac) colored pipes or taps, and can be used for toilet flushing, washing cars, garden irrigation, and filling ornamental ponds, but must not be used for drinking, cooking, or other kitchen purposes, personal washing, evaporative coolers, clothes washing, indoor household cleaning, swimming pools and spas, recreational activities involving contact (e.g., children playing under the sprinklers), and washing of companion animals. Pipes, taps, and water meters that distribute Salisbury Water are purple (or lilac) in color. The city of Salisbury [9] has a ReWater Quality Information Guide that covers appropriate use of Salisbury Water as well as providing guidelines on installation, metering, commissioning of the system, and approved products. All customers of Salisbury Water are required to enter into a Water Supply Agreement, which is an agreement between the owner of the property supplied with Salisbury Water and the city of Salisbury.

The potential customers for the supply of ReWater in Salisbury, which will be supplying recycled stormwater (via the aquifer) to around 16,000 people in the near future, including:

1. 9000 households in Lightsview: The city does not supply water directly to Lightsview residents, but provides supply to a distribution point which is then retailed by the developer to the Lightsview residents. The city has also facilitated a connection to the city of Tea Tree Gully; however, no supply has been taken as yet.
2. 7000 residents in Salisbury, including residents at Mawson Lakes who are currently receiving recycled water from SA Water,[†] which is a mixture of reclaimed wastewater and recycled stormwater.

At this point, the city is directly supplying Salisbury Water to a residential lifestyle village[‡] at the Parafield Gardens Subdivision, which is connected and receiving the water (approximate 154 units) which is supplied through one master meter; while around 216 houses in this subdivision are connected and receiving the water individually [15].

73.3.2 Methodology

We decided to survey the residents in these 370 properties for this study and the City Council agreed to provide us with the residential addresses of their customers. The Total Design Survey Method, comprising a series of precisely laid-out steps, such as

---

* *ReWater* is defined exactly the same as *Salisbury Water* by the City Council (http://www.salisbury.sa.gov.au/Services/Salisbury_Water/Water_Services).

---

† SA Water is a government business enterprise wholly owned by the Government of South Australia.

‡ Residential lifestyle villages refer to retirement homes and villages in South Australia.

personalized questionnaires, incentives, and follow-ups [33], was used to maximize the response rate. Ethics approval was obtained before survey implementation. We eventually received 364 residential addresses of their customers from the City Council. Accordingly, 364 customers received a package which included personalized questionnaires, a cover letter mentioning the objectives of the study, the researcher's contact details, a note about the incentives stating that "Participating in this survey gives you a chance to win 1 of 10 Coles gift vouchers (each valued at \$50)," and a self-addressed reply-paid envelope. The details of the incentives and the sponsorship of the study were printed on the envelopes. The questionnaire included 21 questions, and those wanting to enter into the draw were asked to write down their email address at the end of the questionnaire.

Within four weeks, we received 27 responses. A reminder to follow-up was sent out two weeks later. The questionnaire was attached to the reminder again, in case the recipients did not keep the previous questionnaire. The reminder enabled us to receive an extra 34 responses. Thirty-three letters were returned being noted as either "vacant property" or "property in development." In total, we received 61 responses but identified that one was a double entry and another had many missing answers. Excluding those two entries, the response rate of the study is 59/(364 − 33) = 17.82%. One finding worthy of note is that compared with a community Internet/email survey that we had conducted for a previous study [44–46], the responses to the mail survey had far fewer missing/unanswered questions.

Around 63% of the survey respondents were male, 68% older than 50 years of age, and 57% had household earnings of less than \$1200 per week. About 62% of the respondents were living in the residential lifestyle village, which explains the fact that compared to the Census data, the sample had a greater number of elderly participants.

73.4 Results and Discussions

The survey explored residents' perceptions regarding Salisbury Water in many aspects: ownership issues of the water, that is, who should own Salisbury Water; who should be responsible for the quality of the water; the importance of receiving information about the water supplied; who is trustable in ensuring the water quality; using the water for fit-for-purpose; and pricing of the water. This section reported results of questions and corresponding discussion.

73.4.1 Ownership of Salisbury Water and Responsibility for Quality of the Water

As Salisbury Water may include recycled stormwater, recycled wastewater, native groundwater, or any combination of these, and the ownership of these waters remains unclear, we explored what the customers think about the ownership issues in relation to Salisbury Water. Most respondents considered that the City Council should own Salisbury Water, followed by SA Water and the homeowner (Table 73.1). More males (66.7%) thought

TABLE 73.1 Perceptions of Who Should Own Salisbury Water

Optional Authorities	Frequencies (%)
City Council	61.4
SA Water	10.5
Homeowner	7.0
Federal Govt.	3.5
State Govt.	3.5
Developer	0.0
Not sure	14.0

that City Council should have ownership of Salisbury Water than females (50.0%); while more females (15.0%) thought that it should be owned by homeowners (2.8%). The Crown owns water in law and its management is delegated to state-based institutions who have delegated to City Council of Salisbury in this instance. There is an issue about ownership of produced water that is recycled wastewater, which remains unclear.

Respondents were then asked about their perceptions of who should be responsible for guaranteeing the quality of Salisbury Water. Over one-third of respondents thought it should be City Council, followed by the Health Department and then SA Water (Table 73.2). More males (47.2%) thought that City Council should take this responsibility than females (15.0%), while females (40.0%) favored Health Department responsibility more than males (13.9%). This again illustrates confusion about institutions and suggests that the local supplier will be perceived as responsible whereas under the current arrangements it cannot be.

More than two-thirds of respondents whose household income is less than \$1699 per week thought that Salisbury Water should be owned by City Council, and that City Council should be responsible for the quality of the water. However, more people (42.9%) in the group that has a higher level of household income (\$1700 or more per week) thought that the water should be owned by homeowners rather than by other authorities, and they preferred the Health Department to take responsibility for the quality of the water.

Around 71% of elders (50 years or older) thought that the water should be owned by City Council while only 41% of younger respondents (18–49 years old) thought so. Of those respondents who felt unsure about the issue, most were in the younger age group (33.3%). Respondents who were 50–69 years old had a

TABLE 73.2 Perceptions of Who Should Be Responsible for Guaranteeing the Quality of Salisbury Water

Optional Authorities	Frequencies (%)
City Council	36.8
Health Dept.	22.8
SA Water	15.8
State Govt.	7.0
EPA	7.0
Federal Govt.	3.5
Developer	0.0
Not sure	7.0

TABLE 73.3 Information, Trust, and Customer Satisfaction

	Strongly Disagree (%)	Disagree (%)	Neutral (%)	Agree (%)	Strongly Agree (%)	Average Rating
Receiving information about Salisbury Water is very important to me.	0	3.64	29.09	23.64	43.64	4.07
I have received sufficient information about the Salisbury Water initiative.	3.70	16.67	31.48	22.22	25.93	3.50
I trust my water supplier to ensure that the Salisbury Water to which I have access is healthy and safe.	1.82	5.45	21.82	30.91	40	4.02
I trust in the information about the safety of Salisbury Water given to me by my supplier.	0	5.45	41.82	21.82	30.91	3.78
I am satisfied with the quality of the Salisbury Water that I am using.	1.79	7.14	23.21	35.71	32.14	3.89
I am satisfied with the service of the Salisbury Water supplier.	1.82	9.09	25.45	27.27	36.36	3.87

greater preference for the Health Department being responsible for the quality of water, while those who were either younger or older than 50–69 years had a preference for City Council taking responsibility.

A greater number of the respondents living in the residential lifestyle village (73.5%) considered that City Council should own Salisbury Water than those living in other types of housing (40.9%). However, more respondents living in other types of housing (40.9%) considered that City Council should be responsible for the quality of the water than those living in the residential lifestyle village, who had the same number of respondents supporting the idea that City Council and the Health Department should take this responsibility (both at 32.4%).

73.4.2 Receiving Information about Water Supply and Trust in Water Supplier

Respondents agreed that receiving information about water services is important, but less than half of the respondents considered that they had received sufficient information about the water initiative. Seventy percent of respondents indicated that they trusted their water supplier to ensure that Salisbury Water

was healthy and safe, but only half of them trusted the information about the safety of the water provided by the supplier. More than 60% of respondents were satisfied with the quality of the water, and the service provided by the supplier. Again, this points out that the local authority/supplier is perceived as trustable and responsible in this instance but customers would like to be more informed of the instance (Table 73.3).

73.4.3 Perceptions of Using Salisbury Water

We found that two-thirds of respondents indicated that they had a moderate to high level of knowledge about the water initiatives in the local council area. Older respondents (50 years or older) showed a higher level of knowledge about local water initiatives than younger residents (18–49 years). Respondents who lived in the residential lifestyle village showed higher levels of knowledge than those living in other types of housing (Table 73.4).

About 87% of respondents indicated that they enjoy living in a community that actively contributes to environmental sustainability. About 40% of them claimed that the feeling to do something positive for the environment "motivated me to live in my current place." Approximately 70% of respondents considered

TABLE 73.4 Motivation to Live in Current Place and Perception of Using Salisbury Water

	Strongly Disagree (%)	Disagree (%)	Neutral (%)	Agree (%)	Strongly Agree (%)	Average Rating
I like to live in a community that actively contributes to environmental sustainability.	1.82	3.64	7.27	27.27	60	4.40
The feeling to do something positive for the environment motivated me to live in my current place.	3.64	14.55	40	12.73	29.09	3.49
I need to use lower quality water for some purposes in the future.	1.89	7.55	22.64	24.53	43.40	4.00
Everybody needs to use lower quality water for some purposes in the future.	3.64	3.64	21.82	25.45	45.45	4.05
Using Salisbury Water contributes to a sustainable environment.	0	1.82	5.45	34.55	58.18	4.49
Using Salisbury Water is good for managing water shortages.	0	3.64	12.73	27.27	56.36	4.36
It is fair to ask industries/councils/schools to use Salisbury Water for fit-for-purposes.	0	3.64	14.55	32.73	49.09	4.27
It is fair to ask people to use Salisbury Water for household uses.	7.41	7.41	18.52	24.07	42.59	3.87

that everybody needs to use lower quality water for some purposes in the future. More than 83% of them considered that using Salisbury Water contributes to a sustainable environment and is good for managing water shortages. About 82% of respondents thought it fair to ask industries/councils/schools to use Salisbury Water for fit-for-purposes and less of them (about 67%) thought it fair to ask people to use the water for household uses.

Near to three-quarters of respondents thought that Salisbury Water should be used as a fit-for-purpose supply given that an appropriate quality could be guaranteed, while only 3.5% did not agree with this, and the rest of the respondents (22.8%) were unsure on this issue. A greater number of respondents whose household income was lower (less than $650 per week) and higher ($1700 or more) than those respondents whose household income was at a medium level ($650–1699 per week) agreed that Salisbury Water should be used as a fit-for-purpose supply given that the quality could be guaranteed. Near to four out of five older respondents (50 years or older) agreed with this, while only 58% of the younger respondents (18–49 years old) agreed.

73.4.4 Perceptions of Current Pricing

The pricing of Salisbury Water is set by the City Council of Salisbury in accordance with Section 188 of the Local Government Act 1999, and takes into account the costs of building, operating, and renewing the assets required to capture, treat, and distribute Salisbury Water to customers.

There are currently two charges relating to the usage of Salisbury Water:

1. *Usage Charge*: A rate per kiloliter used. As of July 1, 2012, the usage charge for Salisbury Water is $2.48 per kiloliter. This is $0.97 cheaper than the mains water price.
2. *Residential Supply Charge*: A fixed rate of $12.50 per quarter. This is to cover the cost of cross-connection audits, which are provided at connection and then every 4 years.

Salisbury Water customers are billed quarterly (or half-yearly) depending on their Water Supply Agreement. In addition, there is also a connection fee involved when initially connecting to Salisbury Water. Connection fees are a one-off charge to provide a connection point from the Salisbury Water pipeline to the customers' boundary, and to install a nonpotable meter. Properties supplied with Salisbury Water will always have two water meters consisting of the SA Water main and meter for drinking water and general use, and the Salisbury Water main and meter for toilet flushing, car washing, garden irrigation, and filling of ornamental ponds.

This study explored the perceptions of current customers of Salisbury Water about the fees charged for the water. Of respondents, 40% thought that the current charges for Salisbury Water were "about right," while a greater number of respondents (58.3%) thought that the charge was "too much" or "far too much."

A far greater number of female respondents (81.3%) thought that the fees charged were too high, compared to males (45.2%).

More than half (56.2%) of the respondents thought that a flat fee for every household using Salisbury Water was a bad or a very bad idea, while 41.4% thought it to be a fair or a good idea. Again, a far greater number of females (76.2%) thought that the flat fee was a bad idea compared to males (41.6%).

A greater number of respondents whose household income was more than $1200 per week thought that the current pricing of Salisbury Water was "about right," while a greater number of respondents whose household income was less than $1200 per week thought the charge to be too much. A greater number of respondents whose household income was less than $650 per week and between $1200 and $1699 thought that a flat fee for every household was a good idea; however, more respondents in the other two groups (household income $650–$1199 and $1700 or more per week) thought that it was a bad or a very bad idea. A greater number of respondents in the younger group (49 years or younger) thought that the current pricing of Salisbury Water was "about right" and that a flat fee for every household was fair, while more people in the older group (50 years or older) thought it cost too much and that a flat fee was a bad idea.

Finally, a greater number of respondents living in other types of housing thought that the current pricing was "about right," and that a flat fee for every household was a bad idea; however, more respondents living in the residential lifestyle village thought it cost too much and that a flat fee was fair.

73.5 Summary and Conclusions

The capacity to meet current and future demand is a significant challenge for many urban areas in Australia. Furthermore, uncertainty regarding the future performance of traditional water sources suggests that source diversification should be part of a broader urban water security strategy. However, under current arrangements, there are challenges in achieving sustainable urban water management. The absence of well-defined entitlements for accessing the produced water, including stormwater, recycled water, and aquifer storage, entails the modification of existing legislation and policies to conform to a nationally consistent framework based on the National Water Initiative principles. Given that the production of new water (recycled/desalinated) is capital intensive, and increases the costs borne by households, it is important to ensure that access to water and wastewater services is affordable. It would not be an easy task to accomplish, but given adequate political will and efficient management, it can be achieved. The solution may be innovation in governance model as in the case of Salisbury Water. However, the governance institutions must exist to provide an external regulator.

Acknowledgments

We thank the National Centre for Groundwater Research and Training for funding our study. We would like to show our gratitude to Jennifer McKay for providing valuable suggestions that greatly assisted the research. We are grateful to Mr Colin Pitman and Ms Roseanne Irvine, Salisbury City Projects Department, for their valuable input into this study. We also appreciate the time and contribution of our survey participants.

References

1. Agriculture and Resource Management Council of Australia and New Zealand. 1997. A national framework for improved wastewater reuse and stormwater management in Australia. Occasional Paper Number 3. Task Force on COAG Water Reform, Canberra.

2. Ahmed, M., Shayya, W.H., Hoey, D., and Al-Handaly, J. 2002. Brine disposal from inland desalination plants. *Water International*, 27(2), 194–201.

3. Alley, W.M. 2003. Desalination of ground water: Earth science perspective, U.S. Geological Survey Fact Sheet 075-03. http://pubs.usgs.gov/fs/fs075-03/.

4. Australian Labour Party. 2007. *Labour's National Plan to Tackle the Water Crisis*. Election 2007 Policy Document. Parliament of Australia. Canberra.

5. Bel, G. and Fageda, X. 2009. Factors explaining local privatization: A meta-regression analysis. *Public Choice*, 139(1), 105–119.

6. Biswas, A.K. 2006. Water management for major urban centres. *International Journal of Water Resources Development*, 22(2), 183–197.

7. City of Salisbury. 2012. *Salisbury Annual Report 08/09, House of Assembly, 3rd version, 51st Parliament*. https://www.google.com.au/url?sa=t&rct=j&q=&esrc=s&source=web&cd=2&cad=rja&uact=8&ved=0CCIQFjAB&url=https%3A%2F%2Fwww.parliament.sa.gov.au%2FHouseofAssembly%2FBusinessoftheAssembly%2FRecordsandPapers%2FTabledPapersandPetitions%2FPages%2FTabledPapersandPetitions.aspx%3FTPLoadDoc%3Dtrue%26TPDocType%3D0%26TPP%3D51%26TPS%3D3%26TPItemID%3D593%26TPDocName%3DSalisbury%252B0809.pdf&ei=-OaZVfTUCMTW8gWimruQCw&usg=AFQjCNHAMqrmb95wo09ls6NZLWOT2bGxhg. Retrieved March 13, 2012.

8. City of Salisbury. 2012. *Sustainability That Saves You Money*. http://www.makesgoodbusinesssense.com.au/advantages/sustainability. Retrieved March 12, 2012.

9. City of Salisbury. 2012. *ReWater Quality Information Guide*. http://www.salisbury.sa.gov.au/files/187b7dec-1254-4f48-b18f-9e85010fae04/quality.pdf. Retrieved October 20, 2012.

10. Crisp, G.J. 2012. Desalination and water reuse-sustainably drought proofing Australia. *Desalination and Water Treatment*, 42(1–3), 323–332.

11. Department for Water. 2012. *Water Sensitive Urban Design—Consultation Statement*. http://www.waterforgood.sa.gov.au/homepage/consultation/current-consultation/. Retrieved March 10, 2012.

12. El Saliby, I., Okour, Y., Shon, H.K., Kandasamy, J., and Kim, I.S. 2009. Desalination plants in Australia, review and facts. *Desalination*, 247(1–3), 1–14.

13. Hains, S. 2009. Stormwater: Towards water sensitive cities. *LGMA National Congress, Darwin, May 26, 2009*. https://www.google.com.au/url?sa=t&rct=j&q=&esrc=s&source=web&cd=1&cad=rja&uact=8&ved=0CB0QFjAA&url=http%3A%2F%2Fwww.salisbury.sa.gov.au%2Ffiles%2F1c9524b4-f17b-4650-ad4f-9e850103891b%2Fwatersensitivecities-paper.pdf&ei=-OaZVfTUCMTW8gWimruQCw&usg=AFQjCNGCbDnw9u8Ic4YNZ6H5x1VOgamcXg. Retrieved January 15, 2013.

14. Hughes, N., Hafi, A., and Goesch, T. 2009. Urban water management: Optimal price and investment policy under climate variability. *The Australian Journal of Agricultural and Resource Economics*, 53, 175–192.

15. Irvine, R. 2012. Stormwater Harvesting Project Department in Salisbury LGA, City of Salisbury, personal communication, September 13, 2013.

16. Ison, R.L., Collins, K.B., Bos, J.J., and Iaquinto, B. 2009. Transitioning to water sensitive cities in Australia: A summary of the key findings, issues and actions arising from five national capacity building and leadership workshops. *NUWGP/IWC*. Monash University, Clayton.

17. Jones, K. 2005. Engaging community members in wastewater discussions. *EcoEng Newsletter*, October (11).

18. Kemp, M. 2012. Silver lining to $1.8bn port stanvac desalination plant white elephant. *Advertiser*. http://www.adelaidenow.com.au/news/south-australia/adelaide-desalination-plant-to-be-mothballed/story-e6frea83-1226488293662. Retrieved December 15, 2012.

19. Keremane, G.B. and McKay, J.M. 2008. Water reuse schemes—The role of community social infrastructure. *Water*, 35(1), 35–39.

20. Keremane, G.B., McKay, J.M., and Wu, Z. 2011. No stormwater in my tea cup—An internet survey of residents in three Australian cities. *Water*, 38(2), 118–124.

21. LECG Limited Asia Pacific. 2011. *Competition in the Australian Urban Water Sector*. Waterlines report, National Water Commission, Canberra.

22. Livingston, D., Stenekes, N., Colebatch, H.K., Ashbolt, N.J., and Waite, T.D. 2004. Water recycling and decentralised management: The policy and organisational challenges for innovative approaches. In: T. Daniell (ed.), *Proceedings of the International Conference on Water Sensitive Urban Design: Cities as Catchments*. Adelaide, November 21–25, 2004, pp. 581–592.

23. Lund, J.R. 1988. Metropolitan water market development: Seattle, Washington, 1887–1987. *Journal of Water Resources Planning and Management*, 114(2), 223–240.

24. MacDonald, D.H. and Dyack, B. 2004. *Exploring the Institutional Impediments to Conservation and Water Reuse—National Issues*. CSIRO Land and Water Client Report.

25. McKay, J.M. 2003. Who owns Australia's water?—Elements of an effective regulatory model. *Water Science and Technology*, 48(7), 165–172.

26. McKay, J.M. 2007. Water governance regimes in Australia: Implementing the national water initiative. *Water*, 34(1), 150–156.

27. McKay J.M. 2008. Insubstantial, tenuous and vague laws— The achievement of ecologically sustainable development by water supply business CEO's. *Australian Business Law Review*, 36(6), 432–445.

28. McKay J.M. 2011. Australian water allocation plans and the sustainability objective—conflicts and conflict-resolution measures. *Hydrological Sciences Journal*, 56(4), 615–629.

29. Mitchell, C., Retamal, M., Fane, S., Willets, J., and Davis, C. 2008. Decentralised water systems—Creating conducive institutional arrangements. Paper presented at *Enviro 08*, Melbourne, May 5–7, 2008.

30. National Water Commission. 2007. *Institutional and Regulatory Models for Integrated Urban Water Cycle Management*. NWC, Australian Government. http://www.nwc.gov.au/resources/documents/Institutional-reg-models-integrated-urban-water-cycle-management-PUB-0307.pdf. Retrieved October 11, 2011.

31. NSW Public Works. 2011. *Brackish Groundwater: A Viable Community Water Supply Option?* Waterlines report, National Water Commission, Canberra.

32. Pahl-Wostl, C. 2002. Participative and stakeholder-based policy design, evaluation and modelling processes. *Integrated Assessment*, 3(1), 3–14.

33. Paxson, M.C. 1995. Increasing your survey response rates: Practical instructions from the total-design method. *Cornell Hotel and Restaurant Administration Quarterly*, 36(4), 66–73.

34. Pitman, C. 2012. The General Manager of the Stormwater Harvesting Project in Salisbury LGA, personal communication, March 8, 2012.

35. Productivity Commission. 2008. *Towards Urban Water Reform: A Discussion Paper*. Productivity Commission Research Paper, Melbourne, March.

36. Productivity Commission. 2011. *Australia's Urban Water Sector*. Report No. 55, Final Inquiry Report, Canberra.

37. Robinson, K.G., Robinson, C.H., and Hawkins, S.A. 2005. Assessment of public perception regarding wastewater reuse. *Water Science and Technology: Water Supply*, 5(1), 59–65.

38. Ruiz-Villaverde, A., García-Rubio, M.A., and González-Gómez, F. 2010. Analysis of urban water management in historical perspective: Evidence for the Spanish Case. *International Journal of Water Resources Development*, 26(4), 653–674.

39. Rygaard, M., Binning, P.J., and Albrechtsen, H. 2011. Increasing urban water self-sufficiency: New era, new challenges. *Journal of Environmental Management*, 92(1), 185–194.

40. Tsagarakis, K.P. and Georgantzís, N. 2003. The role of information on farmers willingness to use recycled water for irrigation. *Water Science and Technology: Water Supply*, 3(4), 105–113.

41. Varis, O., Biswas, A.K., Tortajada, C., and Lundqvist, J. 2006. Megacities and water management, international. *Journal of Water Resources Development*, 22(2), 377–394.

42. Ward, J. and Dillon, P. 2011. *Robust Policy Design for Managed Aquifer Recharge*. Waterlines report no. 38, National Water Commission, Canberra.

43. White, S., Noble, K., and Chong, J. 2008. Policy forum: Urban water pricing and supply, reform, risk and reality: Challenges and opportunities for Australian urban water management. *The Australian Economic Review*, 41(4), 428–434.

44. Wu, Z., Hemphill, E., McKay, J.M., and Keremane, G.B. 2013. Investigating psychological factors of behavioural intention of urban residents in South Australia to use treated stormwater for nonpotable purposes. *Journal of Water Reuse and Desalination*, 3(1), pp. 16–25.

45. Wu, Z., McKay, J.M., and Keremane, G.B. 2012. Governance of urban freshwater: Some views of three urban communities in Australia. *Water*, 39(1), 88–92.

46. Wu, Z., McKay, J.M., and Keremane, G.B. 2012. Issues affecting community attitudes and intended nehaviours in stormwater reuse: A case study of Salisbury, South Australia. *Open Access Journal Water 2012*, 4(4), 835–847, doi:10.3390/w4040835.

XVII

Impacts of Climate Change, Mitigation, and Adaptation on Water Reuse

74

Resilience, Not CO$_2$ Mitigation, Is the Imperative

Christopher Monckton
of Brenchley
Science and Public Policy Institute

PREFACE

The issue of how plant and networks for the reclamation of urban water supplies should be re-engineered to proof them against climatic and hydrological cycles altered by anthropogenic global warming has become the wrong question. After up to 18 years 2 months without detectable global warming, opinion among climatologists is migrating toward low-sensitivity scenarios suggested in a growing body of analyses now published in the learned journals. Notwithstanding continuing increases in CO$_2$ emissions, there may well be less than 1 K warming this century. In the short to medium term, global cooling is possible. Any warming below 1.5 K compared with today will scarcely be harmful and, through CO$_2$ fertilization and enhanced drought resilience of trees and plants, may be net-beneficial. Human population growth toward 9–10 billion by mid-century and the growing connectivity and consequent interdependence of complex survival systems, including urban water storage, distribution, and recirculation networks, pose a new challenge: how to build in resilience not only against climate changes in whatever direction they may occur but also against more pressing threats. Threats greater than that of climate change include network overload arising from rapid or ill-planned growth; supply shortage arising from underinvestment and consequent capacity deficit; maintenance failure arising from negligence or municipal bankruptcy; loss of network architecture knowledge arising from unduplicated or corrupted archives or from sudden budget cuts; systemic water loss to leakage arising from insufficiently resolved network monitoring; widespread inadvertent or deliberate pollution or infection of networks arising from lack of engineering for fast emergency isolation or shutdown of affected areas or regions; and partial or total system collapse arising from inadequate robustness against terrorist acts, or from warfare. Set against real problems such as these, the comparatively insignificant impact of slightly warmer worldwide weather on urban water supplies is of little or no account. Managers of water resources should take precautions against budgetary predation by agencies promoting a global-warming mitigation agenda that may well prove as unnecessary as it is costly.

74.1 Introduction

The trend toward living in cities is accelerating everywhere. Furthermore, the global population is now projected to peak at 9–10 billion by 2050, although it is expected to decline sharply thereafter. In large cities, the growing connectivity and consequent close interdependence of complex survival systems, specifically including urban water storage, distribution, disposal, and recirculation networks, poses significant and intriguing challenges for strategic planners, designers, and operators of urban water supply, drainage, and sewerage architectures.

In addition to the usual paper insurances against various risks, physical insurances in the form of redundancy and resilience must be engineered in from the outset in new systems, or retrofitted to legacy systems.

There are many actual or potential threats to the integrity and sustainability of urban water supplies. A nonexhaustive list includes the danger of network overload arising from rapid or ill-planned growth, or from inadequate leak-proofing of distribution systems; theft of water in regions where it is scarce and accordingly worth stealing; supply shortage as a consequence of underinvestment and consequent capacity deficit and bottlenecking; inadequate provision for the growing cost of maintaining aging storage, distribution, and disposal networks; negligence on the part of public authorities whose members are insufficiently tutored in the importance and value of the legacy systems they have inherited from previous generations; the growing propensity of municipalities to fall into bankruptcy and consequently to interrupt management, operational and funding continuity; loss of network architecture knowledge owing to unduplicated, corrupted, or abandoned archives, or to indiscriminate budget cuts; systemic water loss to leakage arising from insufficient resolution or central failure in network monitoring, particularly in aging networks; widespread inadvertent or—in the era of terror—deliberate interruption, pollution, contamination or infection of networks; failure arising from lack of prophylactic engineering for swift and, where appropriate, automated emergency diversion, isolation, or shutdown of affected areas or regions; and the growing risk of partial or total system collapse arising from inadequate robustness of extensive networks that are, by their nature, particularly difficult to defend against terrorist acts and against the collateral effects of warfare.

The risk of long-term damage to the sustainability of sources or of supplies of urban water as a result of anthropogenic global warming must be realistically assessed in the context of real and well-understood risks to water business continuity such as those outlined above.

74.2 Risk Assessment

The prudent chief executive, when first appointed, insists first and foremost on inspecting and verifying the insurances of the business, together with its risk assessments. Every urban water use system should maintain a risk assessment file, with a heading or folder for each identifiable risk, containing a description of the risk, preferably expressed in rigorously quantitative terms; an assessment of the risk, again in quantitative terms, and perhaps using a probability-density function; and a risk management statement identifying all practicable options for containing or neutralizing the risk, and costing each option, together with a timetable and budget for the works necessary for implementation.

At least annually, a summary of the risk assessment, and of the budgetary implications of acting to contain the identified risks and—just as important—of not acting to contain them, should be placed before the board or authority responsible for urban water use systems. Ideally, the water use risk manager should either have a seat on the board or, at the very least, have the unfettered right of access to the chief executive or to the nonexecutive directors.

74.3 Continuity and Sustainability Plans

Risk assessments, which form a necessary part of the board's business continuity and sustainability plans, generally focus chiefly on the short to medium term, while continuity and sustainability plans look to the long term by ensuring that all potential risks are foreseen, planned against, and forestalled in the short term.

Continuity plans ask difficult questions, such as whether the business is sufficiently well justified in social, political, or economic terms to deserve to survive, as well as more straightforward questions such as how to overcome foreseeable risks of shutdown or disaster.

Sustainability plans concern themselves chiefly with ensuring that a sufficient supply of necessary deliverables—whatever consumable raw materials or products are essential to the uninterrupted operation of the urban water use systems—will continue to be delivered in reality, and at a price that the board is willing to afford.

Sustainability plans also consider the impact of the various physical operations of urban water systems on the environment, and of the environment on systems. They will also consider actual or potential scarcities and interruptions in the timely supply of deliverables.

Risk assessments will also take full account of political risk. For the challenge of global warming is that it poses a material risk to supplies if it is a real problem, and a political risk in terms of the taxation and regulation of the business if global warming is not real but the governing power chooses to treat it as though it were real.

No contingency planner can ignore global warming, therefore, even if it is not a real problem.

74.4 Climate-Related Risks and Sustainability Issues

The above sketch of the risks and sustainability issues to which urban water use systems are vulnerable provides a context for the consideration of climate-related risks. The strategic planner and the operational engineer will both be concerned with that uneasy and eternal trade-off: the product of the probability that a foreseeable risk may one day occur and the cost of its occurrence, set against the certain and immediate cost of the insurances and physical prophylactic measures that are necessary to avert or minimize the occurrence of that risk.

To take an extreme example, there is a risk that a meteorite may fall from the sky and do very costly and perhaps terminal damage to water use systems. Though the cost of occurrence is extreme, there are no practicable measures to fend off a meteorite: accordingly, until there are, there is little point in allocating scarce resources in that direction.

Managers of urban water use systems will already be familiar with the usual climate risks: flood, drought, lightning strike, weather-related algal blooms, and suchlike. However, it is now suggested in official quarters that anthropogenic emissions of greenhouse gases are likely to cause systemic change to regional as well as global climate. Yet the true risk the climate question poses to urban water use management may well lie chiefly in the opportunity loss occasioned by the predatory diversion of what may become a damagingly large fraction not only of municipal but also of regional and national budgets toward attempts to mitigate imagined and perhaps imaginary future anthropogenic climate change—attempts that may well prove as unnecessary as they will certainly be ineffectual.

The second-order draft of IPCC ([16]: Figure 11.33) (Figure 74.1: top) followed the general-circulation climate models in projecting global warming of 0.13–0.33 K decade⁻¹ to 2050. The final draft ([16]: Figure 11.25) (Figure 74.1: bottom) rejected the models and

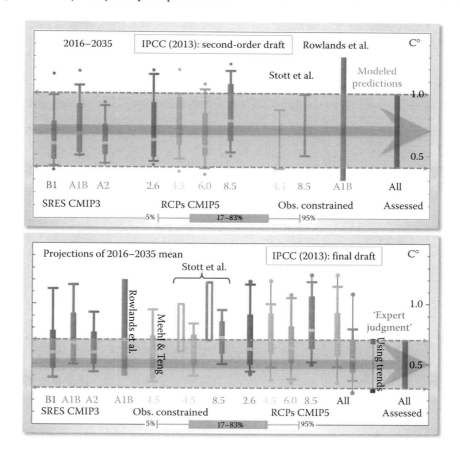

FIGURE 74.1 Upper panel: Models' global warming projections, 2016–2035 versus 1986–2005, against the IPCC's projected interval of 0.4–1.0 K over 30 years, equivalent to 0.13–0.33 K decade⁻¹ (between the gray dotted lines, based on IPCC 2013, 2nd draft, Figure 11.33c). Lower panel: The draft's revised interval 0.3–0.7 K over 30 years or 0.10–2.33 K decade⁻¹. (Based on IPCC. 2013. *Climate Change 2013: The Physical Science Basis. Contribution of Working Group I to the Fifth Assessment Report of the Intergovernmental Panel on Climate Change.* Cambridge University Press, Cambridge, UK, Figure 11.25c.)

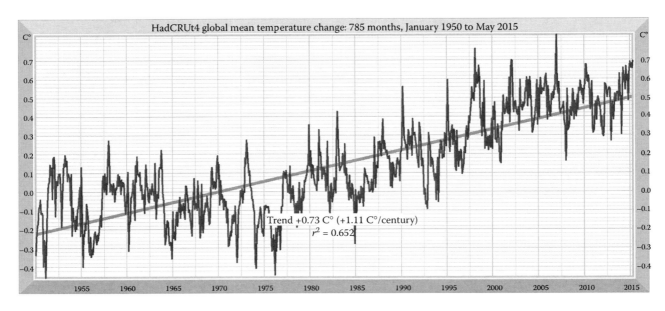

FIGURE 74.2　HadCRUT4 [13] global mean surface temperature anomalies and 0.1 K decade⁻¹ least-squares linear trend, January 1950–May 2015.

sharply reduced the projected interval to 0.10–0.23 K decade⁻¹. Remarkably, IPCC's current best estimate, in the region of 0.5 K decade⁻¹, projects very little more warming over the next 30 years than over the last 30, notwithstanding continuing increases in atmospheric CO_2 concentration.

IPCC ([16], pp. 11–52) says:

> Overall, in the absence of major volcanic eruptions—which would cause significant but temporary cooling—and, assuming no significant future long term changes in solar irradiance, it is *likely* (>66% probability) that the GMST [global mean surface temperature] anomaly for the period 2016–2035, relative to the reference period of 1986–2005, will be in the range 0.3–0.7°C (expert assessment, to one significant figure; *medium confidence*) (Figure 74.2).

Though the IPCC now acknowledges that the models have not proven skillful, even its much-reduced interval of projected temperature change may still be excessive. Observed outturn since 1950 has been little more than 0.1 K decade⁻¹ (Figure based on Reference 23; and see Reference 1).

74.5 The Recent Global Temperature Record

More recently, notwithstanding record increases in atmospheric CO_2 concentration, according to the RSS satellite lower-troposphere dataset [30] there has been no global warming at all for as much as 18 years 6 months (Figure 74.3).

Consider two 60-year periods. Sixty years is the cycle of the influential Pacific Decadal Oscillation. Comparing 60-year periods cancels the largest source of natural short-run temperature variability. From 1894 to 1953, IPCC's methods [11] would have led us to expect 0.1 K anthropogenic warming, yet 0.5 K was observed.

From 1954 to 2013, IPCC's methods would predict 0.9 K, but 0.7 K was observed. Even after removal of the Pacific Decadal Oscillation (PDO) cycle, natural variability unexplained in the models is evident in the record.

74.6 Central England Temperatures and the Sun's Role

The Central England Temperature Record [18,19,27], has been continuously maintained since 1659. It is a reasonable proxy for global temperature change: in 120 years (1894–2013), the Central England record shows warming within 0.01 K of the mean of the three global terrestrial datasets. Central England, and inferentially the world, too, warmed at 0.90 C°/century from 1663 to 1762, before the Industrial Revolution got underway. However, the world has warmed by only 0.73 C° in the past 100 years. If the Central England Temperature Record is accepted as a fair proxy for pre-1850 global temperatures then, contrary to what is sometimes asserted (see, e.g., References 20 and 21, followed by Reference 14), the rate of global warming in the twentieth century may not, after all, be unprecedented in the present millennium.

In Central England, the fastest supra-decadal warming was measured at a rate equivalent to 4.33 C°/century over the four decades from 1694 to 1733 (Figure 74.4). That rate was more than twice the greatest supra-decadal global warming rate in the twentieth century.

Solar activity increased rapidly from the Maunder Minimum of 1645–1715, when the Sun was less active than during the past 11,400 years, to the near Grand Maximum of 1925–1995, during which the Sun was almost more active than at any previous time (Figure 74.5). Solar activity is now declining, and the current solar cycle is not dissimilar to that which led to the Dalton Minimum in the early nineteenth century.

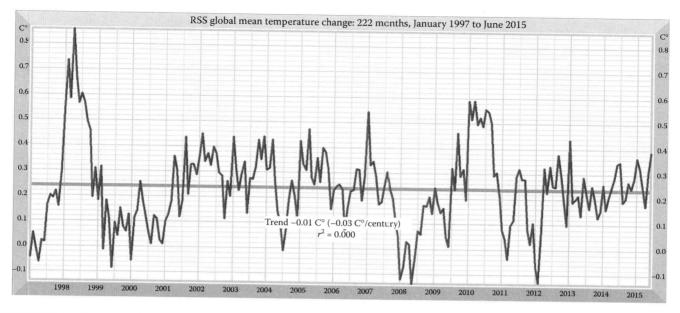

FIGURE 74.3 The RSS satellite dataset shows no global warming for 18 years 6 months.

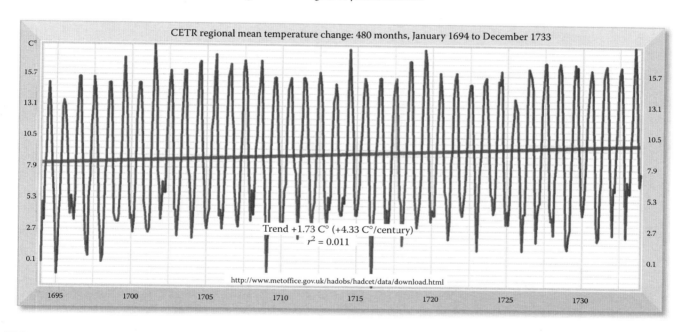

FIGURE 74.4 Central England temperature anomalies and trend, January 1694–December 1733, showing warming at a rate equivalent to 4.33 C°/century.

74.7 Prediction and Observation

In light of the growing divergence between projection and observation, a direct comparison between IPCC's much-reduced near-term global warming projections and observed temperature change since 1990 is presented (Figure 74.6). An over-prediction of 0.17 K has occurred in the nine years since 2005. This unique monthly comparison provides a timely visual verification of the extent to which IPCC near-term projections are proving consistent with observed temperature change.

In Figure 74.7, several official projections are contrasted with observed global warming. Why have the models over-estimated warming? Hot running suggests systemic errors in the current understanding of climate sensitivity. Via intercomparison, these errors may have become integral to all general-circulation models. Possible errors include underestimation of the homeostatic influence of the two bounds of the atmosphere, the ocean (a near-infinite heat-sink) and outer space (an infinite heat-sink); omission of the necessary damping term without which the feedback-amplification has little

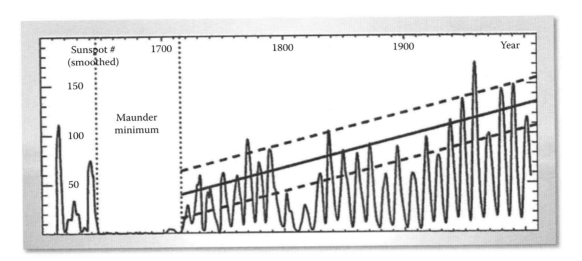

FIGURE 74.5 Growth in solar activity (smoothed sunspot number) from the 70-year maunder minimum (1645–1715), during which solar activity was at its least in the Holocene, to the year 2000. The high activity in the 70 years 1925–1995, peaking in 1960, was a near grand maximum. (Adapted from Hathaway, D.H. and Wilson, R.M. 2004. *Solar Physics*, 224, 5–19.)

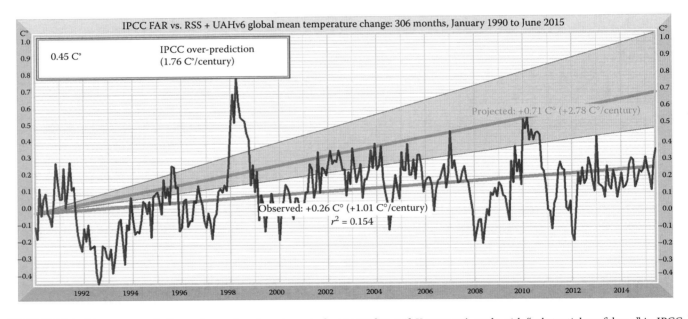

FIGURE 74.6 Near-term projections of warming at a rate equivalent to 2.8 [1.9, 4.2] K century^{-1}, made with "substantial confidence" in IPCC [12], January 1990–June 2015 (shaded region and gray trend line), versus observed anomalies (black spline-curve and trend-line) at 1 K century^{-1} equivalent, taken as the mean of the RSS [30] and UAH [33] satellite monthly mean lower-troposphere temperature anomalies.

meaning in the real climate; overestimation of the forcing from temperature feedbacks, none of which can be measured or otherwise empirically distinguished from the forcings that caused the warming that engendered them; neglect of natural forcings such as the temporary naturally occurring reduction in global cloud cover from 1983 to 2001 [29]; overstatement of the negative forcings from anthropogenic particulate aerosols [24]; and incorrect parameterization of transient influences such as the el Niño/la Niña Southern Oscillations and the great ocean oscillations.

On the empirical evidence of direct temperature measurement to date, the climate problem is not proving to be a problem at all. The question arises: what went wrong in the general-circulation models on whose exaggerated predictions concern about global warming was based? This question was explicitly considered by Monckton of Brenchley et al. [23], who concluded that frank errors in the treatment of temperature feedbacks consequent upon the initial small warming engendered by atmospheric CO_2 enrichment had led to a doubling or tripling of the true rate of anthropogenic warming.

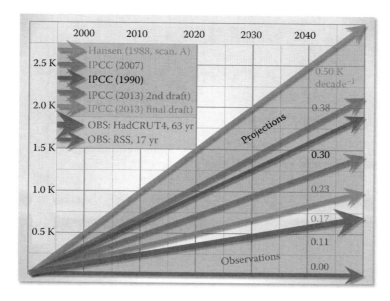

FIGURE 74.7 Five projections of global warming from 1990 to 2050, compared with the linear trends on two observed datasets. James Hansen's 1988 projection [10] is his business-as-usual Scenario A. IPCC projections are mid-range estimates. The trend on the HadCRUT4 monthly global surface temperature anomalies extrapolates to 2050 the warming at 0.11 K decade⁻¹ observed since 1950. The trend on the RSS satellite data has been zero since late 1996—A period of more than 18 years. This trend, too, is extrapolated to 2050.

Now that the IPCC itself has abandoned its original, extreme projections of future global warming, there is no longer any basis for their diversion of expenditures toward attempting to mitigate global warming that may not have been occurring for up to a quarter of a century and is now unlikely to occur at a dangerous rate.

For public officials nervous of questioning what has been presented (improperly) as near-unanimous agreement among climatologists, economic arguments may be more relevant.

To provide a credible basis for the economic appraisal of climate-related risks, a simple model is now presented, allowing economists and other non-specialists in climate physics to input relevant mainstream climatological data, methods, and results into economic models of their own making.

74.8 Climatological Inputs to Mitigation Investment Appraisal

The three conditions precedent to the justification of a climate mitigation strategy are that unabated CO_2-driven global warming may eventually prove damaging; that if the first condition is met, the cost of abating projected warming will be well below the expected benefit in the avoided loss in adapting to its consequences; and that if the second condition is met, the strategy under review will be cost-competitive with other mitigation strategies. It is assumed *ad argumentum* that the first condition is met, though allowance should be made for the possibility that it is not.

A first-order benefit-cost model intended to be adaptable to any CO_2 abatement strategy is described. No attempt is made to provide a complete economic analysis: rather, the focus is on expounding and applying the essential climatological

information without which no mitigation benefit-cost appraisal can be relied on. This approach will give water-use managers competing for scarce funds the necessary ammunition to demonstrate that mitigation of global warming may not be worth the carbon-emitting candle and that, therefore, diversion of public funds away from public infrastructure and toward mitigation would be a futile attempt to solve a problem which (even if it is a problem) cannot be affordably solved.

The model is applied illustratively to two mitigation strategies with the aim of determining quantitatively the extent to which each strategy satisfies the second and third conditions precedent to intervention and is accordingly justifiable in economic terms. The two strategies are the UK's climate mitigation program from 2012 to 2021 and the US Congressional Budget Office's proposed CO_2 capture and storage program.

Uncertainties as to temperature outturn, CO_2 forcing, climate sensitivity, climate impacts, and the influence of mathematical chaos on the climate are addressed by modeling values across a generous interval for twenty-first-century global warming and accordingly for the cost of adaptation to unabated warming.

Temperature-feedback uncertainties are constrained by confining the analysis to short periods of a few decades at most, over which feedbacks will operate only to a limited extent.

For caution, the now-demonstrated absence of scientific consensus as to the extent of humanity's influence on the climate [17] is not taken into account.

74.9 The Model in Outline

Agreement among stakeholders is arguably maximized by adopting a precautionary approach through adopting as normative *ad argumentum* not only the mainstream climatological projections of the models relied upon by the IPCC but also the central economic projections of those economists (such as Reference 32) who argue for an avowedly interventionist stance on grounds of intergenerational equity.

If the model is constructed so that the initial conditions favor intervention, and if intervention even then proves cost-ineffective, it becomes unnecessary either to refine the model by increasing its complexity or to nudge the initial conditions toward a more realistic and hence less interventionist and more politically contentious stance.

74.9.1 Initial Conditions

The climatological initial conditions embody the minimum physical inputs necessary to answer the questions of how much global warming a mitigation strategy may be expected to abate; at what unit cost per Kelvin of warming abated; and what would be the cost of abating all projected global warming over the lifetime of the strategy by measures of equivalent unit abatement cost worldwide.

74.9.2 Climatological Conditions

Climate sensitivity: The global warming ΔT_t to be expected in response to a given proportionate increase in CO_2 concentration over a specified term of years t is for present purposes sufficiently described by the simplified climate-sensitivity relation (Equation 74.1), where ΔT_t, denominated in Kelvin, is the product of three quantities: the reciprocal of the fraction q of total anthropogenic forcing that is driven by CO_2; a time-dependent climate-sensitivity parameter λ_t, which is itself the product of the instantaneous or Planck sensitivity parameter λ_0 and a time-dependent temperature-feedback gain factor G_t; and the CO_2 radiative forcing ΔF_t.

$$\Delta T_t = q^{-1}\lambda_t \Delta F_t = q^{-1}\lambda_0 G_t \Delta F_t \mid \text{Kelvin} \qquad (74.1)$$

Global warming ΔT_t: On business as usual, without mitigation, global warming of 2.8 K from 2000 to 2100 is the mid-range projection in IPCC ([15], table SPM.3). Since the Earth has warmed at a rate well below those projected in all five IPCC *Assessment Reports* and there has been no global warming since 1996 [30], 2.8 K twenty-first-century warming is close to the upper bound.

CO_2 concentration: On business as usual, unmitigated CO_2 concentration over the twenty-first century will attain the annual values in Table 74.A1, derived from the mid-range estimates in IPCC [15].

CO_2 forcing: According to the IPCC, a radiative forcing is an external perturbation in a presumed pre-existing climatic radiative equilibrium, leading to a transient radiative imbalance that will eventually settle toward a new equilibrium at a different global temperature. Experiment and line-by-line radiative transfer analyses have demonstrated that the CO_2 radiative

forcing ΔF_t is sufficiently approximated by the logarithmic relation (Equation 74.2),

$$\Delta F_t = k\ln(C_t/C_0) \mid \text{Watts per square meter} \qquad (74.2)$$

where (C_t/C_0) is a proportionate change in CO_2 concentration, with C_0 the unperturbed value. Myhre et al. [25], followed by IPCC [14], give the coefficient k as 5.35, so that, for example, the CO_2 forcing that arises from doubled concentration is 5.35 ln 2, or 3.708 W m^{-2}. Since Equation 74.2 is logarithmic, the law of diminishing returns applies: each additional CO_2 molecule emitted exercises less forcing than its predecessors.

Planck parameter λ_0: Immediately after a perturbation by an external radiative forcing such as anthropogenically increased CO_2 concentration, the climate sensitivity parameter by which the forcing is multiplied to yield the global temperature response will take its instantaneous or Planck value $\lambda_0 = 0.31$ K W^{-1} m^2 (expressed reciprocally as 3.2 W m^{-2} K^{-1} in IPCC [15], p. 361 fn.).

Sub-centennial sensitivity parameter $\lambda_{1 \le n < 100}$: To allow for the incremental operation of temperature feedbacks, considered by the IPCC to be strongly net-positive, λ_n is projected to increase over time. Taking λ_n as rising linearly from the instantaneous value $\lambda_0 = 0.31$ K W^{-1} m^2 by annual increments 0.013 K W^{-1} m^2 to the centennial value $\lambda_{100} = 0.44$ K W^{-1} m^2 (derived below), Table 74.A2 in the Annex shows indicative annual values of λ_n.

Centennial parameter λ_{100}: The centennial and longer-term values of λ_n allow for longer-term mitigation benefit-cost appraisals. The IPCC projects CO_2 concentration of 713 ppmv in 2100 against 368 ppmv in 2000 (Annex, Table 74.A1), and a mid-range estimate of 2.8 K warming by 2100, of which 0.6 K is pre-committed ([15], table SPM.3), leaving 2.2 K of new warming, of which 70% (Table 74.1), or 1.54 K, is CO_2-driven.

Therefore, the IPCC's implicit centennial climate sensitivity parameter λ_{100} is 1.54 K divided by 5.35 ln(713/368) W m^{-2}, or 0.44 K W^{-1} m^2, representing an increase of 0.13 K W^{-1} m^2 over a century against the Planck value $\lambda_0 = 0.31$ K W^{-1} m^2.

Bicentennial parameter λ_{200}: Examination of the six SRES emissions scenarios for 1900–2100 (Table 74.1) demonstrates the IPCC's implicit bicentennial sensitivity parameter λ_{200} to be 0.50 K W^{-1} m^2 on each scenario.

Equilibrium parameter λ_∞: Dividing the 3.26 K central estimate of climate sensitivity to a CO_2 doubling ([15], p. 798, box 10.2) by the 3.71 W m^{-2} radiative forcing in response to a CO_2 doubling gives the equilibrium sensitivity parameter $\lambda_\infty = 0.88$ K W^{-1} m^2, attained after 1000–3000 years [31].

CO_2 fraction: In Table 74.1, the fraction $q = 0.7$ of total anthropogenic forcing attributable to CO_2 emissions is derivable from each of the six SRES standard emissions scenarios. However, IPCC [16] suggests that $q = 0.83$ is a more realistic value, given that methane concentration is not increasing as rapidly as predicted.

Projected business-as-usual global warming $\Delta T_{1 \le t \le p, \text{bau}}$ over a term of years t is given to reasonable precision by Equation 74.3 as long as $y_t \le 2050$:

$$\Delta t_{t,\text{bau}} = mt \mid K; m = 0.0233 \qquad (74.3)$$

TABLE 74.1 The Bicentennial Sensitivity Parameter and the CO_2 Fraction

Scenario	ΔT_{100}	ΔT_{200}	$\Delta F_{200,all}$	C_{100}	λ_{200}	$\Delta F_{200,CO2}$	q
Units	K	K	W m^{-2}	ppmv	K W^{-1} m^2	W m^{-2}	
A1B	2.8	3.0	6.2	700	0.5	4.5	0.7
A1F1	4.0	4.5	9.1	960	0.5	6.2	0.7
A1T	2.4	2.5	5.1	570	0.5	3.4	0.7
A2	3.4	3.8	8.0	840	0.5	5.5	0.7
B1	1.8	2.0	4.1	520	0.5	2.9	0.7
B2	2.4	2.7	5.6	610	0.5	3.8	0.7
Mean	2.8	3.1	6.3	700	0.5	4.4	0.7

Note: Projected centennial anthropogenic warming ΔT_{100} from 2000 to 2100 ([15], p. 13, table SPM.3), bicentennial warming ΔT_{200}, and total radiative forcings ΔF_{200} from all greenhouse gases for 1900–2100 on all emissions scenarios, and CO_2 concentration C_{100} in 2100 ([15], p. 803, Figure 10.26); and, derived from these, the bicentennial transient-sensitivity parameter $\lambda_{200} = \Delta T_{200}/\Delta F_{200}$; the CO_2 radiative forcing $\Delta F_{200,CO2} = 5.35 \ln(C_{100}/C_{-100})$ from 1900 to 2100, taking C_{-100} as 300 ppmv; and the ratio $q = \Delta F_{200,CO2}/\Delta F_{200}$, all of CO_2 forcing to total greenhouse-gas forcing.

74.9.3 Economic Conditions

Projected global GDP in US \$ is taken as the sum of annual GDP values over the term, where each annual value is the \$63 tn global GDP in 2010 [34] uplifted to account for 3% real annual GDP growth.

Abatement benefit: It has become apparent that the projections of rapid global warming on the interval [5,10] K century^{-1} that formed the basis of the more extreme estimates of the cost of adaptation to unmitigated warming in Stern [32] were excessive. At most, the IPCC's mid-range estimate of 2.8 K century^{-1} will obtain, and warming may be substantially less than this [22].

Stern ([32], Executive Summary, p. ix) says the benefit of CO_2 mitigation, which is the avoided cost of allowing global warming to occur at 2–3 K century^{-1} and paying to adapt fully to its net-adverse consequences, falls on the interval [0.0, 3.0]% of global GDP:

Most formal modeling in the past has used as a starting point a scenario of 2–3 K warming. In this temperature range, the cost of climate change could be equivalent to a permanent loss of around 0%–3% in global world output compared with what could have been achieved in a world without climate change.

The global abatement benefit is assumed to arise at a uniform rate throughout the century, even though the mid-range projection of the near-term global warming rate, equivalent to 1.67 K century^{-1} ([16]: Figure 11.25ab), is well below the mid-range projected warming rate of 2.8 K over the entire century ([15]: table SPM.3).

Uniformity: It will be assumed that climate-damage costs rise *pari passu* with global temperature at 1% of twenty-first-century GDP per 1 K global warming. In practice, up to 1.1 K global warming compared with today's global mean surface temperature (i.e., 2 K compared with the pre-industrial temperature in 1750) may be net-beneficial via CO_2 fertilization, which has already enhanced crop yields as well as net plant primary productivity and drought resistance. However, it is here assumed

that harm will arise at once in response to any increment in CO_2 concentration, however small.

Abatement benchmark: The 1%-of-GDP benchmark cost of abating projected global warming is given by Stern ([32], Executive Summary, p. xiii):

...[the] central estimate is that stabilization of greenhouse gases at levels of 500–550 ppmv CO_2e [mid-range estimate 525 ppmv] will cost, on average, around 1% of annual global GDP by 2050.

Stern's indicative abatement cost falls on his [1,3]%-of-GDP adaptation cost interval, implying that, if global warming does not exceed the IPCC's central estimate this century, the net benefit of mitigation today against that of adaptation later is small.

On business as usual, the IPCC projects a mid-range 514 ppmv CO_2 by 2050 (Annex, Table 74.A2). To this value, 43% of the 236 ppmv increases above the pre-industrial 278 ppmv, or about 101 ppmv, is added to allow for other anthropogenic greenhouse-gas emissions, giving a mid-range business-as-usual estimate 615 ppmv CO_2e.

Accordingly, Stern's benchmark global abatement-cost estimate implies that 1% of global GDP buys an abatement of 90 ppmv CO_2e. Pro rata, the cost of abating the centrally estimated 910 ppmv CO_2e business-as-usual concentration in 2100 by 385 ppmv to keep it stabilized below 525 ppmv CO_2e could reach 4.25% of global GDP by 2100, exceeding Stern's 3%-of-GDP high-end twenty-first-century estimate of the benefit in the avoided cost of adaptation to unmitigated global warming.

Discount rate: Dietz et al. [7], following Stern [32], say,

The discount factor (the relative weight on an investment in consumption at time t, relative to now) in this context is given by Equation 74.4,

$$u'_c e^{-\delta t} \tag{74.4}$$

where u'_c is the marginal utility of consumption per head at time t and δ is a "pure time discount rate"... describing a lower weight on the future, simply *because* it is in the future.... The consumption discount rate r is the rate of fall of the discount factor (Equation 74.5):

$$r = \eta g + \delta \qquad (74.5)$$

where η is the elasticity of marginal utility of consumption and g is per-capita consumption growth....

The most straightforward and defensible interpretation (as argued in the Review) of δ is the probability of existence of the world. In the Review, we took as our base case $\delta = 0.1\%$ per year, which gives roughly a one-in-ten chance of the planet not seeing out this century... Across the infinite horizon of our analysis in Chapter 6, g is on average around 1.3% in a world without climate change, giving an average consumption or social discount rate across the entire period of 1.4% (being lower where the impacts of climate change depress consumption growth.

Under the assumption of a 10% probability of planetary annihilation, extinguishing any consumption growth for all time, the mean value of r will scarcely exceed zero. Accordingly, a zero social discount rate is adopted here. This choice maximally favors the interventionist case, though by depriving future generations of a significant part of their inheritance, perhaps needlessly, it may not favor them.

Immediacy: It is assumed that the benefits of a mitigation strategy will arise at once, although the IPCC's assessment that CO_2 persists in the atmosphere for 40–200 years rules out any significant fall in CO_2 concentration this century.

Casualty: It is assumed that anthropogenic emissions are and will remain the sole cause of the observed increase in CO_2 concentration since 1950.

Exclusions: The model takes no account of the opportunity loss, including the loss to future generations, occasioned by the pre-emptive diversion of otherwise-productive resources to meet the cost of mitigation measures to the extent that they may not be necessary.

No allowance is made for the fact that, in the Western nations where nearly all spending on mitigation occurs, the emissions intensity of energy consumption is significantly lower than in other countries.

Capital and current costs external but essential to a project, such as the provision of spinning-reserve generation for wind turbines on windless days, are excluded. Emissions from project construction and installation, such as those from the burning of lime to make the large quantities of concrete used in the foundations of wind turbines, are ignored.

The model also takes no account of failure by nations to achieve declared emissions abatements. For instance, in April 2013 HM government's climate change advisory committee published a report demonstrating that, although direct CO_2 emissions in the UK had fallen by 20% in the previous two decades, emissions imported in goods from foreign suppliers had resulted in a net 10% increase in the UK's CO_2 emissions over the period. However, it is assumed that expenditure will achieve the stated net reductions in CO_2 emissions.

74.9.4 Case-Specific Initial Conditions

Only two case-specific initial conditions need be established: the fraction of global CO_2 emissions that the strategy, if successful, might abate over its lifetime (typically, the product of the fraction of statewide or national emissions the strategy is expected to abate and the fraction of global emissions represented by statewide or national emissions: Annex, Tables 74.A3 through 74.A5); and the strategy's projected lifetime cost.

74.9.5 Modus Operandi

Once the initial conditions have been established, the algorithm proceeds as follows. The fraction p of global CO_2 emissions expected to be abated determines a new and somewhat abated value $C_{t,\text{aba}}$ of the projected end-of-term business-as-usual CO_2 concentration C_t at y_t, via Equation 74.6:

$$C_{t,\text{aba}} = C_t - p(C_t - C_0)\,\text{ppmv} \qquad (74.6)$$

From the CO_2 concentration growth abated, the CO_2-forcing function (Equation 74.2) is applied to determine how much of the business-as-usual CO_2 radiative forcing that would otherwise have occurred over the term the strategy may abate. In Equation 74.7, ΔF_t is business-as-usual CO_2 forcing at year t, and $\Delta F_{t,\text{aba}}$ is the diminished forcing also at year t:

$$\Delta F_{t,\text{aba}} = k \ln(C_t / C_{t,\text{aba}}) \mid \text{W} \cdot \text{m}^{-2}; \quad k = 5. \qquad (74.7)$$

where appropriate, an approximation of the projected global anthropogenic forcing from all sources that the strategy might abate is obtained by dividing the CO_2 forcing abated by the CO_2 fraction $q = 0.7$. The global warming that the strategy may be expected to abate over the term is the product of the CO_2 forcing abated or, where appropriate, the total anthropogenic forcing abated and the time-dependent sensitivity parameter λ_t.

The unit abatement cost per Kelvin of global warming abated is the ratio of the total project cost over the term and the global warming abated.

The global lifetime abatement cost, the total cost of abating all projected global warming over the term by the adoption worldwide of methods whose unit cost is equivalent to that of the strategy under review, is the product of the unit abatement cost and the projected business-as-usual global warming over the term. It may then be compared with the global abatement benefit, here taken as falling on Stern's interval 1 [0, 3]% of GDP.

The benefit-cost ratio r of a mitigation strategy is the ratio of the global abatement benefit to the global term abatement cost. Where $r \ll 1$, the strategy cannot be justified on climate mitigation grounds alone and, unless other benefits justifying the cost can be found, comparison of its unit abatement cost with those of competing strategies will not be worthwhile. Where $r > 1$, it becomes worthwhile to compare the benefit-cost ratios of competing strategies.

TABLE 74.2 UK Greenhouse Gas Emissions Budget (Mt CO$_2$e), 2012–2021

Period	5-yr Mte CO$_2$e	Mte CO$_2$e yr^{-1}	Years	Mt CO$_2$e
2008–2012	3018	630.6	1	630.6
2013–2017	2782	556.4	5	2782.0
2018–2022	2544	508.8	4	2035.2
2012–2021				5447.8

Note: Budgeted UK anthropogenic greenhouse-gas emissions (Mte CO$_2$e) for the five-year periods 2008–2012, 2013–2017, and 2018–2022, allocated to 2012–2021.

74.10 The United Kingdom's Climate Program: An Illustrative Case Study

The UK emissions abatement target for 2012–2021 is derived from HM government's five-year carbon budgets [5], which limit Britain's CO$_2$ emissions to the values in Table 74.2, which shows budgeted greenhouse-gas emissions as 5448 MtCO$_2$e.

On the assumption that in the absence of mitigation the 1.155% annual exponential rate of decay in emissions observed from 1990 to 2008, largely from replacing coal with gas generation, would have persisted to 2021 (a fair assumption given that by 2015 many large coal-fired plants must be decommissioned to comply with the European Union's Large Combustion Plants Directive), total UK emissions on business as usual from 2012 to 2021 would have been 5726 Mt CO$_2$e (Table 74.3, sum of data in last row).

The implication is that the UK's mitigation program will reduce 2012–2021 emissions by 278 Mt CO$_2$e, of which about five-sixths, or 232 Mt CO$_2$e, is the reduction attributable to CO$_2$ alone.

In 2011, global CO$_2$ emissions were 32.5 Gt CO$_2$, growing at 1 Gt CO$_2$ yr^{-1} [8]. Mean annual emission for 2012–2021, therefore, will be about 38 Gt CO$_2$, to which 10% is added to allow for non-energy emissions, making 42 Gt CO$_2$ yr^{-1}, of which the mean 23.2 Mt annual CO$_2$ abatement arising from the UK government's emissions budget represents 0.055%.

CO$_2$ concentration will rise by 26 ppmv from 390 ppmv in 2011 [26], to 416 ppmv on business as usual in 2021 (Annex, Table 74.A1). The diminished end-of-term CO$_2$ concentration in 2021, assuming that the fractions of global CO$_2$ emissions abated and of CO$_2$ concentration increase abated are identical, is 416 ppmv less 0.055% of the 26 ppmv business-as-usual increase over the term: that is, 415.9863 ppmv.

By Equation 74.7, the CO$_2$ forcing abated is 5.35 times the logarithm of the ratio of the business-as-usual to the diminished end-of-term CO$_2$ concentration: 5.35 ln(416/415.9863) = 0.00177 W m^{-2}. Total anthropogenic forcing abated is the ratio of CO$_2$ forcing abated to the CO$_2$ fraction: thus, 0.00018 W m^{-2}/0.7 is 0.00025 W m^{-2}.

The 0.00006 K (1/17,500 K) global warming the CO$_2$ element in the UK's climate program may abate after 10 years is the product of the CO$_2$ forcing abated by the 10-year climate-sensitivity parameter λ_{10} = 0.323 K W^{-1} m^2 (Annex, Table 74.A2). That concludes the climatological phase, providing input to the benefit-cost analysis.

Peiser [28] provided for the first time a detailed projection of the outturn cost of the UK's greenhouse-gas abatement program under the Climate Change Act (2008) in the decade 2012–2021. Of the $142 bn (£85 billion) cost thus identified, about five-sixths, or $118 bn (£71 bn), is attributable to CO$_2$ emissions reduction alone, for CO$_2$ emissions of 460.7 Mt CO$_2$ in 2011 represented five-sixths of the 553.15 Mt CO$_2$e total emissions for the year ([6], table 2.1).

The UK program's $2066 tn K^{-1} unit abatement cost is the ratio of the $118 billion total cost of the CO$_2$ component to the 0.00006 K CO$_2$-driven warming abated. The global lifetime abatement cost, the product of the unit abatement cost, and the 0.233 K projected business-as-usual decadal global warming, is $482 trillion, or $69,000/head of global population, or 63% of the $766 tn decadal global GDP over the period.

Assuming 1 K twenty-first-century warming, where the benefit in avoided adaptation costs is taken as 1% of GDP the benefit-cost ratio is 0.02, and global mitigation measures of equivalent unit cost would be more than 60 times costlier than adaptation.

Even where twenty-first-century global warming is 2.8 K and the benefit is thus taken 2.8% of GDP, the benefit-cost ratio is only 0.05. However, only a ratio exceeding unity would justify the program in terms of the welfare benefit arising from CO$_2$ mitigation.

74.11 CO$_2$ Capture and Storage in the United States: An Illustrative Case Study

In the United States, several new generating plants purpose-built for *in-situ* CO$_2$ capture and storage (CCS) are planned or under construction. The mean unit cost per MWh of electricity over the lifetime of a power plant, its levelized cost, takes into account the costs of construction, debt service, return on equity investment, taxes, fuel costs, operating expenses, and generating capacity [3].

TABLE 74.3 Actual UK Annual Emissions (Mt CO$_2$e) versus 1.155% Annual Decay Since 1990

1990	774.8	781.6	757.7	735.8	723.6	715.6	735.6	709.1	706.7	675.3	677.9
	774.8	*766.0*	*757.2*	*748.6*	*740.0*	*731.6*	*723.3*	*715.0*	*706.8*	*698.8*	*690.8*
2001	681.2	660.1	666.7	665.6	659.3	655.4	644.7	630.2	576.6	594.1	553.2
	682.9	*675.1*	*667.4*	*659.8*	*652.3*	*644.9*	*637.5*	*630.2*	*623.0*	*616.9*	*608.9*
2012	*602.0*	*595.1*	*588.3*	*581.6*	*574.9*	*568.4*	*561.9*	*555.5*	*549.2*	*542.9*	2021

Note: Actual annual UK CO$_2$ emissions, 1990–2011 (roman typeface), against CO$_2$ emissions declining from 1991 to 2021 at the 1.155% yr^{-1} that prevailed from 1990 to 2008 (italicized), arising chiefly from replacement of coal-fired power stations with gas-fired stations. Budgeted 5448 Mt CO$_2$e total emissions for 2012–2021 (Table 74.2) are 278 Mt CO$_2$e less than the 5726 Mt CO$_2$e (sum of data in last line above) that would have obtained from 2012 to 2021 if the previous 1.155% rate of decay were to continue.

CBO [4] says:

> A large amount of new CCS capacity—installed either at new plants or, through retrofitting, at existing plants—would be needed to reduce costs by enough to achieve DOE's goal.

Accordingly, it is assumed that all of the existing 317 GW coal-fired generating capacity in the United States will be replaced by CCS plants, and, for convenience, that the replacement plants will all be built by 2020, when they will come on stream simultaneously and generate power for 30 years.

It is assumed that, in line with CBO [4], nine-tenths of the CO_2 exhaust from the CCS plants will be successfully captured and stored. Assuming a mean deadweight loss of 30%, coal plants' 33.3% of US CO_2 emissions would increase by 30% (or 10% of US CO_2 emissions) to 43.3%. Of this, nine-tenths, or 39%, will be captured and stored. Deducting back the 10% deadweight loss leaves a net 29% of US CO_2 emissions abated. Since US CO_2 emissions represent 18.7% of global CO_2 emissions, 5.4% of global CO_2 emissions would be abated over the 30-year term.

CBO [4] estimates that at present the cost of building and operating the first commercial CCS power plant would be 74% greater than that of building and operating a conventional plant. The chief reason is the deadweight cost of 30% more coal burned per net MWh delivered.

However, CBO [4] considers that the CCS premium would fall to 35% compared with a conventional plant once 210 GW of CCS capacity had been built and experience had led to economies. It is assumed that the CCS premium will fall in a straight line from 74% to 35% for the first 210 GW of CCS capacity in the United States, and will remain at 35% for the remaining 117 GW capacity.

CBO [4] gives estimates of the total plant construction costs of the first coal-fired CCS plant per kilowatt of net-output capacity and of the levelized electricity cost per megawatt-hour generated (Table 74.4). After 210 GW CCS capacity has been built, "learning by doing" may cut the levelized cost of electricity to $55 MWh^{-1} generated without CCS, and to $74 MWh^{-1} with

TABLE 74.4 Total Pulverized-Coal Plant Construction Costs and Levelized Lifetime Costs

Total Plant Construction Costs 2010$ kW^{-1} net-Output Capacity			Levelized Cost of Generation 2010$ MWh^{-1} Generated		
No CCS	With CCS	Premium (%)	No CCS	With CCS	Premium (%)
$1734	$2790	61	$53.1	$95.4	79
$1919	$3464	81	$57.4	$101.8	77
$1788	$3237	81	$55.9	$97.3	74
$1637	$2895	77	$63.2	$107.7	71
$1888	$3138	66	$65.5	$111.5	70
$1793	$3105	73	$59.0	$102.7	74

Source: CBO. 2012. *Federal Efforts to Reduce the Cost of Capturing and Storing Carbon Dioxide.* Congressional Budget Office, Washington, DC.

Note: Total supercritical pulverized-coal plant construction costs (left: 2010$ kW^{-1} net-output capacity) and levelized lifetime cost of electricity generated (right: 2010$ MWh^{-1} generated), based on estimates from MIT, Global CCS Institute, Carnegie Mellon University, National Energy Technology Laboratory, and Electric Power Research Institute, all recalculated on a uniform basis; and the mean of all five estimates (last row).

CCS, a premium of just 35% in levelized-cost terms, so that the net additional cost of electricity generated by a CCS plant will fall from the initial $43.70 (i.e., $102.70–59.00: Table 74.4) to just $19 MWh^{-1} by the time 210 GW CCS capacity has been built.

Assuming a straight-line reduction in the CCS premium from 74% to 35% over the first 210 GW of CCS capacity installed and a fixed 35% premium thereafter, the mean CCS premium over 30 years if all 317 GW of coal-fired power plants in the United States were replaced with CCS would be $27.25 MWh^{-1}, or $27,250 GWh^{-1}, or $240 million yr^{-1}, or $7.2 bn over the plant's 30-year life, or $2.25 tn for 317 GW.

Once the fraction of global CO_2 emissions abated and the project costs are established, the method of calculation is the same as for the English climate program. Assuming that the benefit in the avoided cost of adaptation to unabated warming is 1% of GDP, that the estimates in CBO [4] are realistic, and that CCS eventually proves practicable, climate mitigation via strategies whose unit cost is similar to that of CCS would cost about twice as much as adaptation, while all other abatement strategies tested using the simple model would cost 1–2 orders of magnitude more than adaptation.

74.12 Summary and Conclusions

The empirical evidence does not suggest that the anthropogenic influence on the climate will be significant enough materially to influence patterns of drought and flood or of water supply or reuse on a global scale. Regionally, there may be some changes, but the climate behaves as a chaotic object, making such changes inherently unpredictable beyond a few days ahead ([14], Section 14.2.2.2). The IPCC, in its special report on extreme weather, confirmed by the *Fifth Assessment Report,* says there is little evidence of a connection between anthropogenic global warming and extreme weather events.

Also, as noted earlier, the IPCC has now accepted that the general-circulation models of future climate on which it had previously relied lack predictive skill, and it has consequently made a sharp cut in its near-term global temperature projections, though it has not yet adjusted its longer-term projections.

Even if the global warming that the IPCC continues to predict for the longer term were to come to pass, and even if it were to cause as much damage as had been predicted and at as much cost as had been predicted, it would be considerably costlier to attempt to mitigate that warming today than to adapt to its consequences the day after tomorrow.

The implications for urban water use management are clear. Encroachment upon water management budgets by municipal, regional, national, and, increasingly, transnational and global authorities in the name of preventing global warming that is not in practice occurring at anything approaching the predicted rate and is unlikely to do so should be firmly resisted, not only on climatological but on environmental and economic grounds.

It remains prudent for urban water-use managers to strive for resilience against climatic changes in whatever directions they may occur. For the time being, though, there is little climatological and no economic case for acting to prevent the modest future warming that is now to be expected in response to our returning to the atmosphere a small fraction of the CO_2 concentration that once prevailed.

Annex of Model Tables

The model tables are a useful reference for non-specialists, for whom the data may not otherwise be accessible. Table 74.A1:

RCP85 concentration-driven model mean annual projected business-as-usual atmospheric CO_2 concentrations from 2001 to 2100 ([16], Figure 12.36). Table 74.A2: Indicative values of the sensitivity parameter for 1–100 years. Table 74.A3: Countries' percentages of global CO_2 emissions, 2008. Table 74.A4: US states' percentages of US and global CO_2 emissions, 2008. Table 74.A5: EU states' percentages of EU and global CO_2 emissions, 2008.

TABLE 74.A1 Projected CO_2 Concentrations (ppmv), 2001–2100

2001	2002	2003	2004	2005	2006	2007	2008	2009	2010
370	372	375	377	379	382	384	387	389	392
2011	**2012**	**2013**	**2014**	**2015**	**2016**	**2017**	**2018**	**2019**	**2020**
394	397	400	402	405	408	411	413	416	419
2021	**2022**	**2023**	**2024**	**2025**	**2026**	**2027**	**2028**	**2029**	**2030**
422	425	428	431	434	438	441	444	447	451
2031	**2032**	**2033**	**2034**	**2035**	**2036**	**2037**	**2038**	**2039**	**2040**
454	458	461	465	468	472	476	479	483	487
2041	**2041**	**2043**	**2044**	**2045**	**2046**	**2047**	**2048**	**2049**	**2050**
491	495	499	503	507	511	515	520	524	528
2051	**2052**	**2053**	**2054**	**2055**	**2056**	**2057**	**2058**	**2059**	**2060**
535	541	548	555	561	568	575	582	589	596
2061	**2062**	**2063**	**2064**	**2065**	**2066**	**2067**	**2068**	**2069**	**2070**
603	610	617	624	632	639	647	654	662	670
2071	**2072**	**2073**	**2074**	**2075**	**2076**	**2077**	**2078**	**2079**	**2080**
677	685	693	701	709	718	726	734	743	751
2081	**2082**	**2083**	**2084**	**2085**	**2086**	**2087**	**2088**	**2089**	**2090**
760	768	777	786	795	804	813	823	832	841
2091	**2092**	**2093**	**2094**	**2095**	**2096**	**2097**	**2098**	**2099**	**2100**
851	860	870	880	890	900	910	920	931	941

Source: Derived from IPCC. 2007. *Climate Change 2007: The Physical Science Basis. Contribution of Working Group I to the Fourth Assessment Report of the Intergovernmental Panel on Climate Change, 2007.* Cambridge University Press, Cambridge, UK. Figure 10.26.

Note: Mid-range projections of mean business-as-usual atmospheric CO_2 concentration (ppmv), 2001–2100.

TABLE 74.A2 Indicative 1–100 Year Climate-Sensitivity Parameters (K W⁻¹ m²)

1 yr	2 yr	3 yr	4 yr	5 yr	6 yr	7 yr	8 yr	9 yr	10 yr
0.311	0.313	0.314	0.315	0.317	0.318	0.319	0.320	0.322	0.323
11 yr	**12 yr**	**13 yr**	**14 yr**	**15 yr**	**16 yr**	**17 yr**	**18 yr**	**19 yr**	**20 yr**
0.324	0.326	0.327	0.328	0.330	0.331	0.332	0.333	0.334	0.336
21 yr	**22 yr**	**23 yr**	**24 yr**	**25 yr**	**26 yr**	**27 yr**	**28 yr**	**29 yr**	**30 yr**
0.337	0.339	0.340	0.341	0.343	0.344	0.345	0.346	0.348	0.349
31 yr	**32 yr**	**33 yr**	**34 yr**	**35 yr**	**36 yr**	**37 yr**	**38 yr**	**39 yr**	**40 yr**
0.350	0.352	0.353	0.354	0.356	0.357	0.358	0.359	0.361	0.362
41 yr	**41 yr**	**43 yr**	**44 yr**	**45 yr**	**46 yr**	**47 yr**	**48 yr**	**49 yr**	**50 yr**
0.363	0.365	0.366	0.367	0.369	0.370	0.371	0.372	0.374	0.375
51 yr	**52 yr**	**53 yr**	**54 yr**	**55 yr**	**56 yr**	**57 yr**	**58 yr**	**59 yr**	**60 yr**
0.376	0.378	0.379	0.380	0.382	0.383	0.384	0.385	0.387	0.388
61 yr	**62 yr**	**63 yr**	**64 yr**	**65 yr**	**66 yr**	**67 yr**	**68 yr**	**69 yr**	**70 yr**
0.389	0.391	0.392	0.393	0.395	0.396	0.397	0.398	0.400	0.401
71 yr	**72 yr**	**73 yr**	**74 yr**	**75 yr**	**76 yr**	**77 yr**	**78 yr**	**79 yr**	**80 yr**
0.402	0.403	0.405	0.406	0.407	0.409	0.480	0.411	0.413	0.414
81 yr	**82 yr**	**83 yr**	**84 yr**	**85 yr**	**86 yr**	**87 yr**	**88 yr**	**89 yr**	**90 yr**
0.415	0.417	0.418	0.419	0.421	0.422	0.423	0.424	0.426	0.427
91 yr	**92 yr**	**93 yr**	**94 yr**	**95 yr**	**96 yr**	**97 yr**	**98 yr**	**99 yr**	**100 yr**
0.428	0.430	0.431	0.432	0.434	0.435	0.436	0.437	0.439	0.440

Note: Indicative 1–100 yr climate sensitivity parameters (K W⁻¹ m²). The sensitivity parameter varies over time to allow for the incremental influence of temperature feedbacks, which, in the IPCC's understanding, are strongly net-positive.

TABLE 74.A3 Percentages of Global CO_2 Emissions by Countries, 2008

China	23.16%	Libya	0.19%	Costa Rica	0.03%
United States	18.68%	Finland	0.19%	Latvia	0.03%
India	5.74%	Portugal	0.19%	Ethiopia	0.02%
Russian Federation	5.63%	Hungary	0.18%	Ivory Coast	0.02%
Japan	3.98%	Bulgaria	0.17%	Panama	0.02%
Germany	2.59%	Norway	0.16%	Tanzania	0.02%
Canada	1.79%	Serbia	0.16%	Netherlands Antilles	0.02%
Iran	1.77%	Trinidad & Tobago	0.16%	Kyrgyzstan	0.02%
United Kingdom	1.72%	Morocco	0.16%	El Salvador	0.02%
South Korea	1.68%	Turkmenistan	0.16%	Cameroon	0.02%
Mexico	1.57%	Azerbaijan	0.16%	Armenia	0.02%
Italy	1.47%	Bangladesh	0.15%	Georgia	0.02%
South Africa	1.44%	Denmark	0.15%	Senegal	0.02%
Saudi Arabia	1.43%	Oman	0.15%	Botswana	0.02%
Indonesia	1.34%	Ireland	0.14%	Ecuador	0.02%
Australia	1.31%	Peru	0.13%	Moldova	0.02%
Brazil	1.30%	Switzerland	0.13%	Cambodia	0.02%
France & Monaco	1.24%	Hong Kong	0.13%	Nicaragua	0.01%
Spain	1.08%	Slovakia	0.12%	Albania	0.01%
Ukraine	1.07%	New Zealand	0.11%	Paraguay	0.01%
Poland	1.04%	Singapore	0.11%	Benin	0.01%
Thailand	0.94%	Cuba	0.10%	Namibia	0.01%
Turkey	0.94%	Bosnia & Herzegovina	0.10%	Mauritius	0.01%
Taiwan	0.85%	Ecuador	0.09%	Uganda	0.01%
Kazakhstan	0.78%	Tunisia	0.08%	Nepal	0.01%
Egypt	0.69%	Yemen	0.08%	New Caledonia	0.01%
Malaysia	0.69%	Angola	0.08%	Tajikistan	0.01%
Argentina	0.63%	Croatia	0.08%	Reunion	0.01%
Netherlands	0.57%	Bahrain	0.07%	Zaire	0.01%
Venezuela	0.56%	Dominican Republic	0.07%	Malta	0.01%
Pakistan	0.54%	Jordan	0.06%	Gabon	0.01%
Utd. Arab Emirates	0.51%	Estonia	0.06%	Suriname	0.01%
Vietnam	0.42%	Slovenia	0.06%	Haiti	0.01%
Uzbekistan	0.41%	Lebanon	0.06%	Mozambique	0.01%
Czech Republic	0.39%	Lithuania	0.05%	Aruba	0.01%
Algeria	0.37%	Sudan	0.05%	Iceland	0.01%
Belgium	0.35%	Bolivia	0.04%	Guadeloupe	0.01%
Iraq	0.34%	Burma	0.04%	Bahamas	0.01%
Greece	0.32%	Jamaica	0.04%	Papua New Guinea	0.01%
Nigeria	0.32%	Guatemala	0.04%	Palestine	0.01%
Romania	0.31%	Macedonia	0.04%	Mauritania	0.01%
Philippines	0.27%	Sri Lanka	0.04%	Montenegro	0.01%
North Korea	0.26%	Mongolia	0.04%	Congo	0.01%
Kuwait	0.25%	Brunei	0.03%	Martinique	0.01%
Chile	0.24%	Luxembourg	0.03%	Madagascar	0.01%
Syria	0.24%	Kenya	0.03%	Zambia	0.01%
Qatar	0.23%	Zimbabwe	0.03%	Burkina Faso	0.01%
Austria	0.22%	Honduras	0.03%	Laos	0.01%
Colombia	0.22%	Ghana	0.03%	Guyana	0.01%
Israel	0.22%	Cyprus	0.03%	All others	0.09%
Belarus	0.21%	Uruguay	0.03%		

Note: Countries ranked by total CO_2 emissions, 2008, based on Boden et al., 2011.

TABLE 74.A4 States' Percentages of US and World CO$_2$ Emissions, 2008

US State	% of US	% World	US State	% of US	% World
Alabama	2.374	0.444	Montana	0.584	0.109
Alaska	0.699	0.131	Nebraska	0.956	0.179
Arizona	1.708	0.319	Nevada	0.657	0.123
Arkansas	1.237	0.231	New Hampshire	0.301	0.023
California	6.592	1.232	New Jersey	2.153	0.402
Colorado	1.684	0.315	New Mexico	1.036	0.194
Connecticut	0.482	0.090	New York	3.046	0.569
Delaware	0.240	0.045	North Carolina	2.343	0.438
District of Columbia	0.058	0.011	North Dakota	0.922	0.172
Florida	4.230	0.790	Ohio	4.333	0.810
Georgia	2.865	0.535	Oklahoma	1.976	0.369
Hawaii	0.354	0.066	Oregon	0.682	0.127
Idaho	0.294	0.055	Pennsylvania	4.519	0.844
Illinois	4.222	0.789	Rhode Island	0.202	0.038
Indiana	3.814	0.713	South Carolina	1.469	0.274
Iowa	1.601	0.299	South Dakota	0.270	0.050
Kansas	1.325	0.248	Tennessee	1.936	0.362
Kentucky	2.717	0.508	Texas	12.447	2.326
Louisiana	3.944	0.737	Utah	1.171	0.219
Maine	0.323	0.060	Vermont	0.108	0.020
Maryland	1.205	0.225	Virginia	1.829	0.342
Massachusetts	1.262	0.042	Washington	1.297	0.242
Michigan	2.935	0.548	West Virginia	1.757	0.328
Minnesota	1.697	0.317	Wisconsin	1.796	0.336
Mississippi	1.121	0.209	Wyoming	0.745	0.139
Missouri	2.482	0.463	Total	100.0	18.68

Source: Based on Boden, T., Marland, G., and Andres, R. 2011. Ranking of the world's countries by 2008 total CO$_2$ emissions from fossil-fuel burning, cement production, and gas flaring. CDIAC, Oak Ridge National Laboratory, doi: 10.3334/CDIAC/00001_V2011, http://cdiac.ornl.gov/trends/emis/top2008.tot; EPA. 2013. *CO$_2$ Consumption from Fossil Fuel Combustion by States, 1990–2013.* http://www.epa.gov/statelocalclimate/documents/pdf/CO2FFC_2011.pdf.

TABLE 74.A5 EU States' Percentages of EU and Global CO$_2$ Emissions, 2011

EU State	% EU	% World	EU State	% E.U.	% World
Austria	1.70	0.22	Italy	11.35	1.47
Belgium	2.70	0.35	Latvia	0.23	0.03
Bulgaria	1.31	0.17	Lithuania	0.39	0.05
Croatia	0.62	0.08	Luxembourg	0.23	0.03
Cyprus	0.23	0.03	Malta	0.08	0.01
Czech Republic	3.01	0.39	Netherlands	4.40	0.57
Denmark	1.16	0.15	Poland	8.03	1.04
Estonia	0.46	0.06	Portugal	1.47	0.19
Finland	1.47	0.19	Romania	2.39	0.31
France	9.58	1.24	Slovakia	0.93	0.12
Germany	20.00	2.59	Slovenia	0.46	0.06
Greece	2.47	0.32	Spain	8.34	1.08
Hungary	1.39	0.18	Sweden	1.24	0.16
Ireland	1.08	0.14	UK	13.28	1.72
			EU total	100	12.95

Source: Based on Boden, T., Marland, G., and Andres, R. 2011. Ranking of the world's countries by 2008 total CO2 emissions from fossil-fuel burning, cement production, and gas flaring. CDIAC, Oak Ridge National Laboratory, doi: 10.3334/CDIAC/00001_V2011, http://cdiac.ornl.gov/trends/emis/top2008.tot.

References

1. Akasofu, S.-I. 2013. On the present halting of global warming. *Climate*, 1, 4–11, doi:10.3390/cli1010004.

2. Boden, T., Marland, G., and Andres, R. 2011. Ranking of the world's countries by 2008 total CO_2 emissions from fossil-fuel burning, cement production, and gas flaring. CDIAC, Oak Ridge National Laboratory, doi: 10.3334/CDIAC/00001_V2011, http://cdiac.ornl.gov/trends/emis/top2008.tot.

3. CBO. 2008. *The Methodology Behind the Levelized Cost Analysis* (supplemental information for nuclear power's role in generating electricity). Congressional Budget Office, Washington, DC.

4. CBO. 2012. *Federal Efforts to Reduce the Cost of Capturing and Storing Carbon Dioxide*. Congressional Budget Office, Washington, DC.

5. DECC. 2013. *UK Greenhouse Gas Emissions: Performance Against Emissions Reduction Targets—2012 Provisional Figures*. https://www.gov.uk/government/uploads/system/uploads/attachment_data/file/211907/Progress_towards_targets_2012_provisional_figures.pdf. Accessed on December 2, 2013.

6. DECC, 2013. *UK Greenhouse Gas Inventory, 1990 to 2011*. Annual Report for Submission Under the Framework Convention on Climate Change, April.

7. Dietz, S., Hope, C., Stern, N., and D. Zenghelis, 2007, Reflections on the stern review (1): A robust case for strong action to reduce the risks of climate change. *World Economics*, 8(1), 121–168.

8. EIA. 2013, *Total Carbon Dioxide Emissions from the Consumption of Energy, 2011*. http://www.eia.gov/cfapps/ipdbproject/IEDIndex3.cfm?tid=90&pid=44&aid=8.

9. EPA. 2013. *CO_2 Consumption from Fossil Fuel Combustion by States, 1990–2013*. http://www.epa.gov/statelocalclimate/documents/pdf/CO2FFC_2011.pdf.

10. Hansen, J., Fung, I., Lacis, A., Rind, D., Lebedeff, S., Ruedy, R., and Russell, G. 1988. Global climate changes as forecast by Goddard Institute for Space Studies Three-Dimensional Model. *Journal of Geophysical Research*, 93(D8), 9341–9364.

11. Hathaway, D.H. and Wilson, R.M. 2004. What the sunspot record tells us about space climate. *Solar Physics*, 224, 5–19.

12. IPCC. 1990. In: J.T. Houghton, G.J. Jenkins, J.J. Ephraums (eds.), *Climate Change—The IPCC Assessment 1990: Report prepared for Intergovernmental Panel on Climate Change by Working Group I*. Cambridge University Press, Cambridge, UK.

13. IPCC. 1995. In: J.T. Houghton, L.G. Meira Filho, B.A. Callander, N. Harris, A. Kattenberg, and K. Maskell (eds.), *Climate Change 1995—The Science of Climate Change: Contribution of WG1 to the Second Assessment Report*. Cambridge University Press, Cambridge, UK.

14. IPCC. 2001. In: J.T. Houghton, Y. Ding, D.J. Griggs, M. Noguer, P.J. van der Linden, X. Dai, K. Maskell, and C.A. Johnson (eds.), *Climate Change 2001: The Scientific Basis. Contribution of Working Group I to the Third Assessment Report of the Intergovernmental Panel on Climate Change*. Cambridge University Press, Cambridge, UK.

15. IPCC. 2007. In: S. Solomon, D. Qin, M. Manning, Z. Chen, M. Marquis, K.B. Avery, M. Tignor, and H.L. Miller (eds.), *Climate Change 2007: The Physical Science Basis. Contribution of Working Group I to the Fourth Assessment Report of the Intergovernmental Panel on Climate Change, 2007*. Cambridge University Press, Cambridge, UK.

16. IPCC. 2013. In: T.F. Stocker, D. Qin, G.-K. Plattner, M. Tignor, S.K. Allen, J. Boschung, A. Nauels, Y. Xia, V. Bex, and P.M. Midgley (eds.), *Climate Change 2013: The Physical Science Basis. Contribution of Working Group I to the Fifth Assessment Report of the Intergovernmental Panel on Climate Change*. Cambridge University Press, Cambridge, UK.

17. Legates, D.R., Soon, W.W.-H., Briggs, W.M., and C.W. Monckton of Brenchley. 2013. Climate consensus and misinformation: A rejoinder to "Agnotology Scientific Consensus, and the Teaching and Learning of Climate Change," *Science Education*, doi:10.1007/s11191-013-9647-9.

18. Manley, G. 1953. The mean temperature of Central England, 1698 to 1952. *Quarterly Journal of the Royal Meteorological Society*, 79, 242–261.

19. Manley, G. 1974. Central England temperatures: Monthly means 1659 to 1973. *Quarterly Journal of the Royal Meteorological Society*, 100, 389–405.

20. Mann, M.E., Bradley, R.S., and Hughes, M.K. 1998. Global-scale temperature patterns and climate forcing over the past six centuries. *Nature*, 392, 779–787.

21. Mann, M.E., Bradley, R.S., and Hughes, M.K. 1999. Northern hemisphere temperatures during the past millennium: Inferences, uncertainties, limitations. *Geophysical Research Letters*, 26, 759–762.

22. Monckton of Brenchley, C.W. 2008. Climate sensitivity reconsidered. *Physics and Society*, 37(3), 6–19.

23. Morice, C.P., Kennedy, J.J., Rayner, N., and Jones, P.D. 2012, Quantifying uncertainties in global and regional temperature change using an ensemble of observational estimates: The HadCRUT4 data set. *Journal of Geophysical Research*, 117, D08101, doi:10.1029/2011JD017187.

24. Murphy, D.M., Solomon, S., Portmann, R.W, Rosenlof, K.H., Forster, P.M., and Wong, T. 2009. An observationally based energy balance for the Earth since 1950. *Journal of Geophysical Research*, 114, D17107, doi: 10.1029/2009/2009JD012105.

25. Myhre, G., Highwood, E.J., Shine, K.P., and Stordal, F. 1998. New estimates of radiative forcing due to well-mixed greenhouse gases. *Geophysical Research Letters*, 25(14), 2715–2718.

26. NOAA. 2014. *Monthly Mean Atmospheric CO_2 Concentration at Mauna Loa*. Hawaii. ftp://ftp.cmdl.noaa.gov/ccg/co2/trends/co2_mm_mlo.txt.

27. Parker, D.E., Legg, T.P., and Folland, C.K. 1992. A new daily Central England temperature series, 1772–1991. *International Journal of Climatology*, 12, 317–342.

28. Peiser, B. 2013. *The UK's Greenhouse-Gas Abatement Program*. Global Warming Policy Foundation, London.

29. Pinker, R.T., Zhang, B., and Dutton, E.G. 2005. Do satellites detect trends in surface solar radiation? *Science*, 308, doi:10.1126/science.1103159.

30. RSS. 2014. Satellite monthly global mean lower-troposphere temperature anomalies. remss.com/data/msu/monthly_time_series/RSS_Monthly_MSU_AMSU_Channel_TLT_Anomalies_Land_and_Ocean_v03_3.txt.

31. Solomon, S., Plattner, G.-K., Knutti, R., and Friedlingstein, P. 2009. Irreversible climate change due to carbon dioxide emissions. *PNAS*, 106(6), 1704–1709, doi:10.1073/pnas.0812721106.

32. Stern, N. 2006. *The Economics of Climate Change: The Stern Review*. Cambridge University Press, Cambridge, UK.

33. UAH. 2014, Satellite MSU monthly global mean lower-troposphere temperature anomalies. vortex.nsstc.uah.edu/data/msu/t2lt/uahncdc.lt.

34. World Bank. 2011. Gross Domestic Product 2009, World Development Indicators. http://siteresources.worldbank.org/DATASTATISTICS/Resources/GDP.pdf.

75

Climate Change Adaptation and Water Reuse

Md Salequzzaman
Khulna University

S.M. Tariqul Islam
Khulna University of Engineering and Technology

Mostafa Shiddiquzzaman
Curtin University of Technology

Saeid Eslamian
Isfahan University of Technology

PREFACE

Climate is changing around the world and it is now an important global environmental, socioeconomic, health, safety, and political issue, especially for those residing in urban areas because of their pure drinking water supply. Water—mainly drinkable freshwater—is one of the scarce resources not only for the survival of human beings, but for all of living beings because pure drinkable water is decreasing rapidly due to anthropogenic pollution and contamination along with human's indiscriminate use that directly or indirectly relates with climate change too. Sustainable climate change adaptation can mitigate unexpected, contaminated, and polluted drinkable pure water supply in urban areas around the world using the key sustainable factor of water reuse, recycle, and resource recovery. This chapter elaborately discusses urban water reuse and how this could be sustainably practiced using various scientific appropriate technologies. At the end, the chapter illustrates water reuse due to climate change in several representative cities around the world.

75.1 Introduction

Water is an essential element for the well-being of humanity and a basic requirement for proper functioning of every living organism and the ecosystems. It is also a critical input for the production and sustainable growth in cultural practices such as agriculture, aquaculture/fisheries, livestock, and forestry. The most important factor is to ensure a sufficient amount of quality water for the system sustainability. This situation becomes worse as climate change hazards are introduced. For example, there is a severe crisis of sufficient amount of quality water (mostly natural and sometimes treated water having no health risks) in the present world. Arnell [7] shows that by 2025, 5 billion people out of 8 billion will experience shortage of quality water. According to UNDP Report, there are some 1.1 billion people in developing countries lacking adequate access to quality water [41] that is further deteriorating by unexpected weather behavior, that is, climatic change like floods and tornadoes such as increasing the frequency of extreme climatic events and the unpredictability of rainfall patterns [25]. This situation again deteriorated by changing the seasonal demands for urban water usages including water for agricultural irrigation, which is reducing the water supply in many cases [10]. The changes of climate variables such as unusual rainfall, more evaporation during summer, salinity intrusion and sea level rise have been influencing the freshwater flow [3]. In addition, large-scale land erosion, loss in biodiversity, flood, storm surges, damage to infrastructures, failure of desired level crop production, fisheries, etc. are some examples of consequence of climate change [3]. Many scientists concluded that shortage of pure drinking water enhances the severe health effects especially for the peoples in developing countries [33].

As global warming is solely responsible for climate change, which reduces the river flows across most of the regions in the world, reducing the sustainable yield of urban and rural water systems [6], such as it causes significant reductions in rainfall across most of southern and eastern Australia [19] where declining river flows and droughts are becoming more recurrent and expanding to other areas. In addition, many peoples who belong to coastal communities in developing countries migrated to urban cities due to climate-induced vulnerabilities like cyclone, floods, hurricane, tornados, and seasonal inundation at habitable areas such as Bangladesh and the Philippines, where at present urban cantres face severe quality water supply for the climatic migrated peoples, known as "climatic refugees" [23]. The scenario of such water scarcity is compounded by climate change induced sea level rise and salinity intrusion in the surface and groundwater. Projected climate scenarios indicate that frequency and severity of natural hazards like cyclone and storm surges would increase [23]. All these disasters are now adding to the vulnerabilities of the concerned local peoples around the world, especially of the coastal communities. This way, climate change leads sea level rise that again aggravates severe sustainable development issues like drainage congestion, water logging and flooding problems especially in the urban and peri-urban areas in many countries around the world [23].

As consequences of these outbreaks, the usable water might have impacted by contamination, pollution, or hazardous situations and rapidly decreasing its usable forms. Thus, water, especially urban water, needs reuse using various engineering tools and appropriate technologies to make a balance "demand and supply of usable forms of waters" [25].

75.2 Cause and Effect: The Climate Change Saga

Presently the atmosphere is at a stage of loading with various greenhouse gases that eventually are collectively causing a net increase in global surface temperature, which is solely by anthropogenic activities [20]. As a reason, climate change is regarded as a common developmental, environmental, socioeconomical, psychological, and political issue for every country and generally regarded to be a major global problem or issue. Although, historically the Earth's climate always had cyclical trends and shows continuous variations through the centuries with constant averages of increased greenhouse gases [22]. For example, temperature record over a period of 100 years shows that the Earth's surface temperature has risen by more than 0.7°C since the 1800s [23] and average global temperature is likely to rise by between 1.8°C and 6.4°C by 2100 AD [24]. Even with good progress on abatement strategies, the temperature rise will most likely be 3.0°C or more over this century. This change and the cause of change have now been a big concern to scientists and social activists all over the world, which are describe briefly in the following points.

Global warming and sea level rise: Warming of the climate system in recent decades is unequivocal, as is now evident from observations of increases in global average air and ocean temperatures, widespread melting of snow and ice, and rising global sea level. Global warming ingress the melting of the polar ice caps leading to a rise in sea levels and flooding in some regions especially areas near the coast—in regions where high temperatures have been the generally norm, like the semi-arid tropics, occurrence of droughts and dry spells will increase [17]. Global warming could lead to the impacts that are abrupt or irreversible, depending on the rate and magnitude of the climate change [4]. Among the victims of climate change, developing countries are now in the most vulnerable position [29]. The effects of this change continue to intensify the potential for social, economic, and overall living standards. Due to the sea level rise and several floods in the coastal belt, vast proportions of people of the world become homeless, many have no choice but to flee from their origin to urban centers; this is known as "climatic migration" [5]. For example, in Bangladesh one-third of the total population is claimed directly by this disaster [35]. This climatic disaster increases almost twofold of water demand over the whole country especially the Jessore-Khulna regions of the urban center.

The impacts: Climate change is expected to increase the frequency and intensity of current hazards and the probability of extreme events and new vulnerabilities with differential spatial and socioeconomic impacts [36]. The study of UNFCCC [42] shows that climate variability can influence peoples' decisions with consequences for their social, economic, political, and personal conditions and effects on their lives and livelihood. Some communities have already been hit by global climate change, especially in recent years in several developing countries including the United States. The negative consequences of climate change in African countries is happening as well, with t frequent floods, droughts, and shift in marginal agricultural systems [13]. The climate change impact on agriculture is believed to be stronger in sub-Saharan Africa [26]. All these are happening by climate change and water crisis/issues become a priority issue, specially reliability and security of urban and rural water systems worldwide [6]. The global climate models suggest that significant shifts in rainfall patterns will occur in the near future [6], where several countries are responding to the impact of the world's most vulnerable communities of El Salvador, Bangladesh, Kenya, Malawi, Bolivia, and the Philippines [29]. Among the victim countries, Bangladesh is situated at the most vulnerable position as the country's proximity to sea level is another natural condition that increases its vulnerability to the effects, and the frequency and severity of these natural disasters are rapidly escalating [29].

Influencing factor of hydrological cycle and water regime: The hydrological cycle around the world is intimately linked with changes in atmospheric temperature and radiation balance. Global warming and climate change observed over the past several decades is consistently associated with changes in a number of components of the hydrological cycle and hydrological systems such as changing precipitation patterns, intensity, and extremes; widespread melting of snow and ice; increasing atmospheric water vapor; increasing evaporation; and changes in soil moisture and runoff [21]. There is significant natural variability of the climatic factor on inter-annual to decadal. Warming ocean waters are likely increasing the power of tropical cyclones (variously known as hurricanes and typhoons), raising the risk of human death, injury, and disease as well as destroying coral reefs and property [42]. The biggest impact may prove to be changes to the availability of freshwater [23]. It is 90% certain that episodes of extreme heat will increase worldwide, leading to increased danger of wildfires, human deaths, and water quality issues [42]. These impacts are intensifying more in developing countries, although they have very different individual circumstances and the specific impacts depend on the climate they experience as well as their geographical, social, cultural, economic, and political situations [42]. Different sectors are affected by these climate change factors at differing degrees, where main sectors includes agriculture, water resources, human health, terrestrial ecosystems and biodiversity, and coastal zones. Further changes in rainfall pattern are likely to lead to severe water shortages and/or flooding that add subsequent impacts in all of these mentioned sectors. This means water is the most critical factor and this needs a balance of demand and supply of water usage.

75.3 Climate Change Diversification Towards Water Regimes

Climate change generates water scarcity and makes the available water unfit for human consumption through contamination, pollution, or other hazards. Water is becoming scarcer globally and several evidences predict that it will become even more scarce in the near future [6,23,33]. The changes of climate variables decrease availability and decline the water quality. Thus, climate change is expected to affect water supply if extreme climatic events and unpredictable rainfall patterns become more prevalent [43]. This situation has increased the demand for pure water/portable water, specifically for urban areas where peoples become more concentrated as one of the effects of climatic refugee. In addition, salinity intrusion, sea level rise, unusual rainfall, tidal flood, and storm surge are threatening safe water

resources for drinking, agricultural, industrial, commercial, and other purposes. Sedimentation of riverbeds and increasing risk of flood are leading to changing ecosystem dynamics, which are influencing the safe water sources, too. In this way, climate change situations limit the access to safe water availability either in terms of drinking or other purposes. Chronic shortage of freshwater is likely to threaten food production, constrain sanitation, hinder the economic development, and damage the ecosystem [3]. Water scarcity and quality deterioration threaten the survival and livelihood of the people itself. Too little water decreases food production, as it limits the water available for farming, which can cause lower crops and ultimately exaggerate vulnerability of poor people.

Usable water availability is principally influenced by three factors: (1) climate (primarily precipitation and evaporation); (2) fossil aquifer availability; and (3) available technology (e.g., desalination or advanced wastewater treatment for reuse—it is a costly process but the costs could significantly decrease in the future with the advancement of technology [34]. Thus, it is perceived that reductions in water availability are expected as a consequence of climate change [8,10]. Water resources are not beyond a climatic disaster. This water resource faces two-dimensional pressures in the climatic change perspective, that is, demand-side pressure and supply-side pressure [10]. The supply-side pressures include climate change (reducing or increasing the amount of water available) and environmental degradation. The demand side pressure is the population growth, urban sprawl, agricultural purposes (irrigation), industrial perspectives (electricity production), and growing environmental demand.

Climate change increases the variability of water supply, leading to floods during some times of the year and droughts in other times and these problems accelerate the vulnerability of people. The population of the world is growing especially in developing countries and thus improvements in water supply have failed (also there has been further deterioration of usable water) to keep pace. Therefore, degradation of water quality is expected to be one of the key impacts of climate change on water resources and water supply [37]. Many scientists stated that projected increases in flooding, drought, decreasing water availability, algal blooms, coastal inundation, and sea level rise have both direct and indirect effects on drinking water quality and limit its availability [31,32]. Observational records and climate projections provide abundant evidence that freshwater resources are vulnerable and have the potential to be strongly impacted by climate change, with wide-ranging consequences for human societies and ecosystems.

However, water availability is one of the most dramatic consequences of climate change for the agricultural sector. However, it is expected to be even more limited in the future as it is related to an increase in air and earth surface temperatures [10]. This phenomenon is important in low-precipitation seasons, and is even greater in dry areas. By the middle of the twenty-first century, annual average river runoff and water availability are projected to increase because of climate change at high latitudes and in some wet tropical areas, and decrease over some dry regions at mid-latitudes and in the dry tropics. Many semi-arid and arid areas (e.g., the Mediterranean Basin, western United States, southern Africa, and northeastern Brazil) are particularly exposed to the impacts of climate change and are projected to suffer a decrease of water resources due to climate change [42]. Climate change affects the function and operation of existing water infrastructure—including hydropower, structural flood defenses, drainage and irrigation systems—as well as water management practices. Adverse effects of climate change on freshwater systems aggravate the impacts of other stresses, such as population growth, changing economic activity, land use change, and urbanization. Globally, water demand will grow in the coming decades, primarily due to population growth and increasing affluence; regionally, large changes in irrigation water demand because of climate change are expected.

75.4 Many Solutions to Climate Change: Adaptation but Confusion

Adapting to climate change will entail adjustments and changes at every level—from community to national and international. Communities must build their resilience, including adopting appropriate technologies while making the most of traditional knowledge, and diversifying their livelihoods to cope with current and future climate stress [3]. Local coping strategies and traditional knowledge needs to be used in synergy with government and local interventions. The choice of adaptation interventions depends on national circumstances [42]. Early strategies for addressing climate change focused almost exclusively on mitigation (i.e., reduction of greenhouse gas concentrations in the atmosphere) [24].

The extent of global warming can be daunting and dispiriting. It is not possible for one person, or even one nation, to slow and reverse climate change. Developing countries are the most vulnerable to climate change impacts because they have fewer resources to adapt: socially, technologically, and financially [23,42]. Adaptation is a process through which societies make themselves better able to cope with an uncertain future. Adapting to climate change involves taking the right measures to reduce the negative effects of climate change (or exploit the positive ones) by making the appropriate adjustments and changes [6]. There are many options and opportunities to adapt. These range from technological options such as increased sea defenses or flood-proof houses on stilts, to behavior change at the individual level, such as reducing water use in times of drought and using insecticide-sprayed mosquito nets. Other strategies include early warning systems for extreme events, better water management, improved risk management, various insurance options, and biodiversity conservation [42].

Good policies, planning, and institutions are essential to ensure that more capital-intensive measures are used in the right circumstances and yield the expected benefits. The adaptation strategy includes both reactive and anticipatory responses to climate change. Reactive responses are those that are implemented

TABLE 75.1 Adaptation Strategy

Reactive adaptation	Anticipatory adaptation
Protection of groundwater resources	Better use of recycled water
Improved management and maintenance of existing water supply systems	Conservation of water catchment areas
Protection of water catchment areas	Improved system of water management
Improved water supply	Water policy reform including pricing and irrigation policies
Groundwater and rainwater harvesting and desalination	Development of flood controls and drought
	Monitoring

Source: UNFCCC. 2007. *Climate Change: Impacts, Vulnerability and Adaptation in Developing Countries.* http://unfccc.int/resource/docs/publications/impacts.pdf.

as a response to an already observed climate impact, whereas anticipatory responses are those that aim to reduce exposure to future risks posed by climate change (Table 75.1).

Adaptation options designed to ensure water supply during average and drought conditions require integrated demand-side as well as supply-side strategies. The former improve water-use efficiency, for example, by recycling water. Supply-side strategies generally involve increases in storage capacity, abstraction from watercourses, and water transfers. Integrated water resources management provides an important framework to achieve adaptation measures across socioeconomic, environmental, and administrative systems. To be effective, integrated approaches must occur at the appropriate scales. Adaptation is more than coping. In well-adapted systems, people are doing well despite changing conditions. They are doing well either because they shift strategies or because the underlying system on which their livelihoods are based are sufficiently resilient and flexible to absorb the impact of changes.

Desalination can greatly aid climate change adaptation, primarily through diversification of water supply and resilience to water quality degradation. Diversification of water supply can provide alternative or supplementary sources of water when current water resources are inadequate in quantity or quality. Desalination technologies also provide resilience to water quality degradation because they can usually produce very pure product water, even from highly contaminated source waters. Increasing resilience to reduced per capita freshwater availability is one of the key challenges of climate change adaptation. Both short-term drought and longer-term climatic trends of decreased precipitation can lead to decreased water availability per capita. These climatic trends are occurring in parallel with population growth, land use change, and groundwater depletion; therefore, rapid decreases in per capita freshwater availability are likely (Table 75.2).

Climate change is a continuous process and it has already destroyed many development efforts in poor countries, forcing susceptible communities to adapt their ways of life to increasingly extreme weather events and changes. Adaptation to climate change must also occur through the prevention and removal of

TABLE 75.2 Some Adaptation Options Based on Water Supply and Demand

Supply-Side	Demand-Side
Prospecting and extraction of groundwater	Improvement of water-use efficiency by recycling water
Increasing storage capacity by building reservoirs and dams	Reduction in water demand for irrigation by changing the cropping calendar, crop mix, irrigation method, and area planted
Desalination of seawater	Reduction in water demand for irrigation by importing agricultural products, i.e., virtual water
Expansion of rainwater storage	Promotion of indigenous practices for sustainable water use
Removal of invasive non-native vegetation from riparian areas	Expanded use of water markets to reallocate water to highly valued uses
Water transfer	Expanded use of economic incentives including metering and pricing to encourage water conservation

Source: IPCC. 2007. *Contribution of Working Group II to the Fourth Assessment Report of the Intergovernmental Panel on Climate Change (IPCC).* Cambridge University Press, Cambridge, UK. www.ipcc.ch/pdf/assessment-report/ar4/wg2/ar4_wg2_full_report.pdf.

maladaptive practices. Maladaptation refers to adaptation measures that do not succeed in reducing vulnerability but increase it instead. Examples of measures that prevent or avoid maladaptation include better management of irrigation systems and removal of laws that can inadvertently increase vulnerability such as destruction of mangroves and relaxation of building regulations on coasts and in floodplains [42].

75.5 Is Water Reuse One of the Best Options?

Water, an important natural resource and an important element of human life, is easily available in the coastal region around the world due to its proximity to the sea. However, the source of drinking water is not good enough. Living standards without quality water cannot be good at all. Good water means water from a safe and pure source. Primary water uses in the urban and peri-urban areas are domestic use (drinking, washing, bathing), agricultural use, and industrial use. The water bodies are also used for subsistence fisheries and capture fisheries by the poor. Farmers use irrigation water from the nearby rivers, canals, lakes, and ponds for their agriculture. Industries consume a large amount of freshwater (groundwater) resources. Future urban expansion and industrial growth are likely to increase the overall water demand. It also is observed that lack of safe potable water during and after an extreme weather event causes suffering to millions of people from waterborne diseases [33]. Such usage of impure water for household affairs causes various health hazards. The problem becomes severe after cyclone, flood, and other natural disasters [2]. Scarcity of drinking water comes as an aftershock of natural disasters to the people.

Climate changes induced natural disasters creating pressure on freshwater and turning freshwater into wastewater. Changes

of climatic factors make diversified impacts on people's livelihood, especially on the water sector. People cannot drink unsafe water without treatment. In the water cycle, wastewater is actually recycled through an extensive and long natural treatment [1]. Before water treatment technology, untreated wastewater is discharged into surface water, transported to the sea, evaporated, and finally comes back as rainfall to reload surface and underground waters. With high-level water treatment technologies, water reuse is now a possibility for decision makers.

Water reuse involves multiple and related technical, financial, economic, and social issues that need to be resolved before embarking on a water reuse process [1]. Direct and indirect water reuse can be applied in the water-stressed area. Indirect water reuse is the preferred approach. In urban areas with limited water supply or stressed water resources, however, direct water reuse can be suitable and sustainable [16]. Provided there is a high standard treatment process in place, direct applications can include street cleaning, watering large gardens, and industrial cooling. To make the water available after climate change impact as an adaptation strategy, water reuse is a great concern.

75.6 Definition of Water Reuse for the Case of Climate Change Adaptation

Water reuse, wastewater reuse, and water recycling all generally mean the same thing: using treated wastewater for a beneficial purpose. The process of treating wastewater prior to reuse is called water reclamation [16]. Water reuse is the use of treated wastewater for beneficial purposes, which increases a community's available water supply and makes it more reliable, especially in times of drought [1]. It may be possible to offset the impacts of environmental flow allocations and climate change by a combination of water savings (reductions in demands) and by introducing new water sources to achieve a new balance between supply and demand [6]. New water sources may include additional river and groundwater sources if not already allocated. New water sources may also include harvesting of urban rainwater and storm water; water recycling (reuse of treated wastewater), or desalination of seawater or saline groundwater sources.

Water reuse is at a number of spatial scales including the household level, where greywater is captured through sink drains onsite and used for garden irrigation and toilet flushing; the community level (decentralized) water reuse, involving the construction of small wastewater treatment plants for potable water reuse and small scale nonpotable water distribution systems; and regional level (centralized) water reuse, involving the upgradation of the appropriate standard for potable water reuse and the installation of large-scale nonpotable water distribution systems [34].

Climate change is one of the barriers to create the disturbance between demand and supply of water [25]. Drought is a major problem in some African countries that are the main victims of drastic climatic change. Most of the adverse climatic activities are led by the fast world countries due to rapid industrialization and modern lifestyles. Anderson [6] shows that climate change impacts will pose major challenges in maintaining adequate water supplies for urban and rural water needs in Australia. Water reuse is a very important issue in the modern world. Water savings is the first condition in the climate adaptation process. Water savings can be possible by reducing the demand of water use. Introducing the new water sources is the best solution for climatic change adaptation.

1. Harvesting of urban rainwater and stormwater and using it through processing
2. Industrial and agricultural wastewater can be reused by filtering process
3. Seawater can be used by desalination process and using it in daily needs (households)
4. Groundwater filled with wastage can be used by filtering process to adapt with drastic climatic changes

75.6.1 Classification of Water Reuse

75.6.1.1 Potable Water Reuse

Potable water reuse is used for drinking water purposes. Wastewater is created from different sectors such as irrigation wastewater (agriculture) and industrial wastewater that can be used for drinking purposes by filtering process. Potable reuse is applicable not only for drinking purposes but also for other applications.*

75.6.1.1.1 Example of Water Reuse (Potable Reuse)

Potable water reuse has been practiced in the USA for decades. For example, Los Angeles (Saline District) has been reusing potable water since 1962. Similar systems are in place in other locations in California and in other states, including Virginia, Texas, Georgia, Arizona, and Colorado. About half the nation's potable reuse systems have come online during the past decade.

75.6.1.2 Nonpotable Water Reuse

Nonpotable water reuse generally contains lower quality than potable water reuse. This system brings dual distribution systems for water reuse. Those are separate systems of pipes for distributing potable and nonpotable water supply processes. Nonpotable reclaimed water can be used for flushing toilets, watering parks or residential lawns, supplying fire hydrants, washing cars and streets, filling decorative fountains, and many other purposes.

75.6.1.2.1 Example of Water Reuse (Nonpotable Reuse)

A dual distribution system is in Grand Canyon Village, Arizona, which has been using reclaimed water for nonpotable uses since 1926. In St. Petersburg, Florida, there are nonpotable reuse systems with large scale water since 1970. Nonpotable water reuse covers 40% of water demand in those areas. City parks, schools, golf courses, residential lawns, fire hydrants, and commercial buildings can use nonpotable water.

75.7 Applicability of Water Reuse for Adapting the Climate Change

Water reuse is one of the most important issues in the recent global aspect of sustainable water management where global warming and huge population growth affect the environment badly. To maintain sustainable use, water savings and water reuse are the most beneficial solutions for meeting the huge demand for water.

1. Sydney Metropolitan Water Plan is targeting to achieve sustainability by 2015 through an adaptation strategy to reuse water. This strategy has been able to save water at 40%. If this process is maintained, it is possible to increase annual water savings of 120–140 Mm³ by existing consumers through a variety of programs including water efficiency plans. The estimated economic savings of water and water reuse is US$100 per person per year. The estimated annual savings is US$ 400 million.
2. By 2025, water availability in nine countries, mainly in eastern and southern Africa, is projected to be less than 1000 mg per person per year. Water reuse is the best solution to make per capita water resource available.
3. In a specific example of South Western Cape, one study shows water supply capacity is decreasing either as precipitation decreases or as potential evaporation increases. This projects a water supply reduction of 0.32% per year by 2020. In this case, there is no alternative of water reuse to adapt in changed adverse climatic situations.
4. In Osaka and Shapporo municipal of Japan, wastewater is reused for thermal energy plants. Reviving of flora and fauna by restoration of river flow is another application of wastewater reuse in Osaka, Japan. The Meguro River, which flows through a residential area in Tokyo, had been abandoned by residents due to the decreasing flow of water and pollution with an unpleasant odor. To solve this issue, the Tokyo Metropolitan Government released water treated with UV radiation into the river. With the drastic improvement in water volume and quality, various living species have returned to the river. After the introduction of highly treated water, many insects and small animal populations have been re-established, and fish such as Japanese trout, striped mullets, and gobies also returned to the river. Biodiversity and environmental amenities have thus been restored effectively with wastewater reuse.
5. Water reuse can be applicable by a water recycling process. Due to a water recycling process, water pollution can be reduced. For example, recycled water may contain higher levels of nutrients, such as nitrogen, than potable water. Application of recycled water for agricultural and landscape irrigation can provide an additional source of nutrients and lessen the need to apply synthetic fertilizers.

---

* Types of Water Reuse, available at http://nas-sites.org/waterreuse/what-is-water-reuse/types-of-water-reuse/.

Water reuse is the primary condition for a sustainable development approach. A recycling approach is one of the most vital treatments to keep pace with the shortage in the supply of water. Wetlands provide many benefits, which include wildlife and wildfowl habitat, water quality improvement, flood diminishment, and fisheries breeding grounds. With the increasing demand of water, sometimes water can be transported or moved from one distance to another. Transportation of water requires a lot of energy and costs. Reusing of water, thus drastically reduce this transportation cost. On the other hand introduction to a water reusing program, pollution should be reduced at a higher level.

75.8 Water Reuse for Adapting Climate Change around the World—Examples

Climate change is increasing water stress for many countries of Africa. An estimated 75–220 million people face severe water shortages by 2020 [42]. Africa will face increasing water scarcity and stress with a subsequent potential increase of water conflicts as almost all of the 50 river basins in Africa are trans-boundary [9,15]. Agricultural production relies mainly on rainfall for irrigation and will be severely compromised in many African countries, particularly for subsistence farmers and in sub-Saharan Africa. Under climate change, much agricultural land will be lost, with shorter growing seasons and lower yields. Egypt is one of the African countries that could be vulnerable to water stress under climate change. The water used in 2000 was estimated at about 70 km³, which is already far in excess of the available resources [18]. A major challenge is to close the rapidly increasing gap between the limited water availability and the escalating demand for water from various economic sectors [40]. The rate of water utilization has already reached its maximum in Egypt, and climate change will exacerbate this vulnerability.

In Asia, global warming could lead to a rise in the snow line and disappearance of many glaciers causing serious impacts on the populations relying on the 7 main rivers in Asia fed by melt water from the Himalayas. Throughout Asia, 1 billion people could face water shortage leading to drought and land degradation by the 2050s [12,14]. Water stress is increasing to over 100 million peoples due to a decrease of freshwater availability in Central, South, East, and Southeast Asia, particularly in large river basins as because of an increasing number and severity of glacial meltrelated floods, slope destabilization followed by decrease in river flows of glaciers disappearance [42]. In Latin America, there is an increase in the number of people experiencing water stress—likely to be 7–77 million by the 2020s. Runoff and water supply in many areas is compromised due to loss and retreat of glaciers. Reduction in water quality in some areas due to an increase in floods and droughts is now a common scenario in the Latin American region [42].

Africa, Asia, and Latin America are currently facing the greatest threats from climate change, both to their people and to local environments. For example in some areas of Africa has a prolonged drought and whereas floods in other areas are exacerbating food insecurity, while in Asia, coastal flooding, sea-level rise and storm surges decreasing livable areas that threatening coastal livelihood and increasing the risk of economic loss and widespread climatic migration. In Latin America, the UNDP report notes [41], melting glaciers in the Andes will decrease long-term water availability in parts of South America, negatively affecting food production and boosting the likelihood of massive flooding and severe weather events. For the same reasons, agricultural productions across the world are decreasing drastically and passing the hardiest time to adapt the climatic changes' consequence. This way climate change is one of the greatest challenges for the present world. The intensity and frequency of climate-induced disasters have increased in recent years. Thus, the adverse effects of climate change undermine the economic development, human security, and people's fundamental rights [41].

In many parts of the world, both in dry and wet climates, integrated water resource management (IWRM) is a key challenge under the present, highly variable climatic condition. Each region requires appropriate IWRM concepts that take both natural resources and socioeconomic conditions into full account. This way vulnerable climate change issues could be linked with the economics and environment directly. The study of Steenbergen and Tuinhof [38] shows the water buffer development during the climate change adaptation. The study shows three techniques (recharge, retention, and reuse) to use the water with the changing pattern of the environment (Figure 75.1)

Recharge → Retention → Reuse

FIGURE 75.1 Recharge, retention, and reuse process of water. (From van Steenbergen, F. and Tuinhof, A. 2009. *Managing the Water Buffer for Development and Climate Change Adaptation. Groundwater Recharge, Retention, Reuse and Rainwater Storage.* MetaMeta Communications, Wageningen, the Netherlands. http://www.bebuffered.com/downloads/3R_managing_the_water_buffer_2010.pdf.)

With the adaptation of recent climatic change, these three techniques are very important tools for better utilization of water resources.

75.8.1 Recharge

First, the wastewater coming from the irrigation and industrial purposes can be used further. Rainwater and floodwater sometimes gather in the low land area. This water can be gathered in some places in rural areas. By adding water to the buffer, recharge contributes to water circulation. Recharge can come from the interception of rain and run-off water (natural recharge), and from increased infiltration of natural processes by manufactured interventions.

Example: Recharge has been done at scale in areas of India, China, Kenya, and Ethiopia—yet is uncommon in other parts of the same countries or other parts of the world. African countries are trying to adapt themselves with climatic change. The roof-top rainwater harvesting systems are one of the vital steps in these countries to reuse the water process.

75.8.2 Retention

After the recharging process, the water is maintained for the next step—retention. Retention is the process where water is kept in some fixed areas as preservation. Retention slows down the lateral flow of groundwater. This helps pond up groundwater and create a large wet buffer in the subsoil. Under such conditions, it is easier to retrieve and circulate water. Retention makes it possible to extend the chain of water uses. Retention process is done by a better drainage process where the groundwater level is increased a lot.

Example: The technique of retention is common in the eastern part of India and Bangladesh. Subsurface dams and sand dams have this same effect of retaining groundwater and creating a huge reservoir by increasing the outflow level. In Bangladesh and the southern part of India, where the shortage of water is severe, creating dams and other small water storage systems, water is retained for sanitation and agricultural irrigation purposes.

75.8.3 Reuse

The most important element is the water reuse process. Reuse is the third element in buffer management. The biggest challenge of 3R is making water revolve as much as possible. Scarcity is resolved not only by managing demand through reduction in use, but also by keeping water in active circulation. It consists of management of evaporation and managing water quality.

Example: In some parts of Bangladesh and the African region, people collect rain water in large pots and drams for the use of irrigation and other domestic purposes. In the saturated zone, reuse is rapid as water that seeps away is quickly picked up and circulated again. In rural areas of Bangladesh and India, water reuse practice is commonly seen in the rural areas mainly where water demand is higher. Water reuse practice is adapted by the agricultural production purposes mainly. It saves transportation

costs and reduces marginal costs of agricultural production. Kenya, China, and Brazil are also adapting the 3R technique of recharge, retention, and reuse.

75.9 Sustainability of Water Reuse for Adapting Climate Change

Water management is a critical component that needs to adapt in the face of both climate and socioeconomic pressures in the coming decades. Changes in water use will be driven by the combined effects of changes in water availability, changes in water demand from land, as well as from other competing sectors including urban, and changes in water management. The impacts of changes in the frequency of floods and droughts or in the quantity, quality, or seasonal timing of water availability could be tempered by appropriate infrastructure investments, and by changes in water and land-use management.

75.9.1 Linkage of Water and Sustainable Development

Water is both a resource and a sector; a key to social development, environmental integrity, and economic growth. As a sector, water requires infrastructure development and operational funds, while as a resource it cuts across sectors and requires integrated approaches to management. Mobilizing water is critical within the post-2015 agenda in order to realize economic and social potential. It is critical for agriculture, ecosystem services, and energy, but water and the services it provides are threatened by climate change, population growth, degrading water quality, and extreme hydrological events (floods and droughts). Water is an issue with utmost priority in Millennium Development Goals (MDGs) and Sustainable Development Goals (SDGs). Different cases regarding sustainability of water reuse throughout the world are mentioned in this chapter briefly.

75.10 Water Reuse Sustainability Case Studies

75.10.1 Sydney

Australia's largest City, Sydney has an integrated water management approach (mainly reuse) to adapt to increasing supply risks and achieve a balance between growing demands and available supplies. For example, the major Sydney water storages (about 44%) of the Hawkesbury-Nepean catchment which has a catchment arera of 22,000 km^2, an average annual rainfall of 890 mm/year and an average annual inflows of about 1300 Mm^3 [38].

75.10.2 Egypt

A major challenge is to close the rapidly increasing gap between the limited water availability and the escalating demand for water from various economic sectors. The rate of water utilization has already reached its maximum for Egypt, and climate change will exacerbate this vulnerability. Agriculture consumes about 85% of the annual total water resource and plays a significant role in the Egyptian national economy, contributing about 20% of GDP [38]. Institutional water bodies in Egypt are working to achieve the following targets by 2017 through the National Improvement Plan.

1. Improving water sanitation coverage for urban and rural areas
2. Wastewater management
3. Optimizing the use of water resources by improving irrigation efficiency and agriculture drainage-water reuse

75.10.3 Yemen

There is also a second dimension to the link between floods and groundwater storage. If there is intensive groundwater development, effective flood storage capacity in the shallow aquifer will increase, as the top layers are no longer saturated. As a result, floods will either not occur or, if they occur, they will do so later in the flood season and less frequently.

The central feature of spate irrigation is the usage of short duration floods that originate from episodic rainfall events in highland catchments. Floods—lasting from a few hours to a few days—are diverted from dry riverbeds and spread gently over the land. The water is used in agriculture, with soil moisture often carefully preserved, as the floods usually arrive ahead of the cultivation season. The floodwater is also used for filling water ponds, for improving rangelands and tree stands, and for recharge. Spate irrigation systems have some of the most spectacular social organizations around [38].

1. Mostly 45% of the farmers use floodwater for irrigation. They recharge this water and use it for irrigation purposes when needed. This could be a better example for water reuse and sustainability [38].
2. Here agriculture is at its most productive. The high water productivity comes from the combined use of floodwater and groundwater, with the spate flows being the main source of recharge. Groundwater in the coastal plains of Yemen is mostly of good quality and hence can be easily reused.

75.10.4 Namibia

By storing surface water in aquifers, a "water bank" can be created through the combination of three elements: recharge, retention, and reuse. Excess water from surface reservoirs is stored in aquifers, thus creating a bridge between years with high rainfall (when there is surplus to be stored) and drought years, making the water supply system resilient to climate variability and, in the long term, climate change [38].

1. In Namibia, extra surface water is gathered within a place and the water is used into the dry season where there is scarcity of water.

2. The sustainability of this water reuse is projected to be retained in the long term. Better water management and extra water buffering are the future challenges for this African country.

75.10.5 Peru

Under limitations, South American nations have developed the necessary capacity to adapt to the local environmental conditions. Such capacity involves their ability to solve some hydraulic problems and foresee climate variations and seasonal rain periods. On the engineering side, their developments included the use of captured rainwater for cropping, filtration, and storage; and the construction of surface and underground irrigation channels, including devices to measure the quantity of water stored. Nasca (southern coast of Peru) has introduced the system of water cropping for underground aqueducts and feeding the phreatic layers [11].

75.10.6 Bangladesh and India

Filtering pond water, harvesting rainwater, and refilling aquifers with rainwater are common water management systems in the coastal region of Bangladesh after cyclone Sidr and Aila [3]. Both ground and rooftop rainwater are harvested. Refilling aquifers with rainwater is an essential source of water in agriculture [30]. A flood-proofed hand pumping system has been introduced in Bahraich, Uttar Pradesh, India. In addition, household water treatment filtering for drinking water is seen. Desalination techniques are also in progress.

75.10.6.1 Small-Scale Riverbank Infiltration in Bangladesh and India

Riverbank infiltration usually makes use of a gallery, a well, or a line of (drilled) wells at a short distance from the river. During groundwater flow from the riverbed to the well, contaminants and pathogens are removed through physical, chemical, and biological processes. River bank infiltration provides potable water without expensive treatment and is a cost-effective solution compared to a surface water treatment or long-distance water conveyance. For small-scale riverbank infiltration, there is the additional advantage that the storage capacity of the sediments around the river bed provide a source of water during the dry period when there is no flow in the river. Quality improvement of the water (compared to the direct use of surface water) is the main advantage of riverbank infiltration schemes [38].

A successful application in a humid area is the riverbank infiltration in Chapai Nawabganj, Bangladesh, where arsenic-free water is pumped from a well near the river. The dug well in Maharashtra shows riverbank infiltration in a dry region from which fluoride-free water is pumped during the whole year to supply a nearby community.

75.10.7 MENA Countries

Drip irrigation is affordable, accessible, and significantly increases water efficiency. Moreover, the costs of environmental degradation related to irrigation in the Middle East and North Africa (MENA) has been estimated to be on the order of US$ 9 billion per year, or 2.1%–7.4% of the range of the MENA countries' GDP [39].

75.10.8 Sub-Saharan Africa

Studies have shown that rainwater collected from corrugated iron roofs is very applicable for drinking if operation, management, and maintenance is carried out. If water is needed for other purposes, like livestock breeding, kitchen usage, and gardening, water quality is not much of an issue. In this case, the family could have a rainwater harvesting tank for drinking and an additional pond for catching surface runoff for other purposes [38]. In sub-Saharan Africa, rainwater harvesting potential = 0.81 × (30 m² × 650 mm)/1000 = 15.600 L per year.

75.10.9 South Africa

In case of dune infiltration, ponds were either excavated or formed through enclosure dikes retaining the recharge water until it has infiltrated through the basin floor. The stormwater runoff was regarded a valuable water source to augment freshwater supplies in the region. In Atlantis, South Africa, dune infiltration is used for drinking water supply and protection of fresh groundwater reserves against intrusion of saline groundwater. Most of the 450 mm mean annual rainfall is received from April to September. Since the soils are mostly sandy, 15%–30% of the rainfall recharges the groundwater [38]. To augment groundwater supplies, artificial recharge through infiltration basins was introduced shortly afterward.

75.10.10 Wadi Wala Dam, Jordan

In the framework of a Jordanian-German technical cooperation project, surface water protection zones have recently been established for both the Wadi Wala and Mujib Dams [27,28]. In the related guideline, zone 2 is defined as a buffer zone of 500 m around zone 1, and zone 1 is identified as a buffer zone of 100 m around a reservoir.

75.10.10.1 Key Factors for Establishing Sustainable Initiatives

1. Planning to meet specific needs and conditions
2. Analyzing economic and financial requirements
3. Selecting options to minimize risk
4. Utilizing institutions and organizations
5. Building capacity: Institutional, financial and human resource capacity building
6. Meeting standards and guidelines
7. Community participation

75.11 Summary and Conclusions

Water reuse and its supply are new fields and definitely options for policy and decisions. Clear policy for pricing, planning,

implementation, and regulation are required for proper water reuse. Alternative water sources must be controlled to remove market-access barriers, and sustainable water and wastewater tariffs should be enforced with a sound pricing mechanism for water reuse. Unified water reuse quality and technology standards, based on applications/classifications for both direct and indirect usage are a prerequisite. The selection of suitable technologies for sustainable water reuse production relies on adequate technical criteria and policy.

For ensuring water is secured for all, safe and sustainable drinking water for everyone, water for agriculture, managing wastewater wisely, minimizing its generation, and pollution are utmost necessary. There is a prescription of protecting groundwater quality through conservation and reuse of wastewater but not by engaging much energy in wastewater plants. Water source must be of sufficient quality for all uses (e.g., drinking, cooking, irrigation, industrialization, etc.). Experts suggest using surface water properly especially through harvesting and conservation of rain and storming water. Adaptation technologies in case of climate change pointed in different literatures are

- *Diversification of water supply*: Exploring alternative sources of water, desalination, post-construction support for community-managed water systems, rainwater collection from ground surfaces, small reservoirs, and micro-catchments, rainwater harvesting from rooftops, water reclamation, and reuse
- *Groundwater recharging*: Rainwater collection from ground surfaces, small reservoirs and micro-catchments, rainwater harvesting from roofs, water reclamation, and reuse
- *Preparation for the upcoming extreme weather events*: Boreholes/tube-wells as a drought intervention for domestic water supply, improving resilience of protected wells to flooding, post-construction support for community-managed water systems and water safety plans
- *Resilience to water quality degradation*: Desalination, household water treatment and safe storage, post-construction support for community-managed water systems, water reclamation and reuse, and water safety plans
- *Storm water control and capturing*: Rainwater collection from ground surfaces, small reservoirs and micro-catchments, rainwater harvesting from roofs
- *Water conservation*: Increasing the use of water-efficient fixtures and appliances, leakage management, detection and repair in piped systems
- *Control water demand*: The demand may be affected by pricing of water reuse technology, but is also driven by reliability and quality of the technology. Overdesign is a common problem. Maintenance and repair downtimes affect water reuse quantity and quality and should be taken into consideration during the design stage of the water reuse project to provide alternative sources of water, in case problems arise during the water reuse production process.

To address those subjects, the issues of capacity, institution, financing, environmental viability, risk management, and community participation should be taken as prior conditions. For sustainable development of the world stated MDGs and SDGs environmental sustainability, especially sustainable management of water in case of natural calamity and water scarcity are taken as extreme concerns. Researchers dealing with water suggest the recharge, retention, and reuse of water not only for drinking and household necessity but also for agricultural and industrial purposes. Considering different socioeconomic and environmental situation, the literature suggests

- In developing countries like Bangladesh and India, rainwater harvesting, pond sand filter, riverbank infiltration, etc. are viable. However, desalination processes and special filtering with low cost may also be viable for middle income families.
- In Latin America and sub-Saharan Africa, rainwater harvesting is suggested enormously. In some countries, construction of dams may be viable.
- However, in South Africa, dune infiltration is common along with rooftop rainwater conservation.
- In developed countries, like America and Japan, they reuse wastewater by processing through radiation. This treated wastewater is reused in agriculture, industry, and atomic energy plants. That is not feasible in the context of developing countries.

After all discussion, it is concluded that sustainable use of water suggests reuse of wastewater and utmost use of surface water with a high importance on rainwater harvesting. In order for life-saving freshwater to be ensured, security of water is a must.

References

1. ADB. 2009. *Water Reuse: Supplementary Water, Asian Development Bank (ADB)*. Philippines. http://www.adb.org/features/water-reuse-supplementary-water-0.
2. Adger, W.N., Huges, T.P., Folke, C., Carpenter, S.R., and Rockstrom, J. 2005. Social-ecological resilience to coastal disasters. *Science AAAS*, 309(5737), 1036–1039.
3. Ahmed, M.F. and Haider, M.Z. 2011. *Climate Change and the Quest of Safe Water Resource—A Study on the South-West Coastal Region of Bangladesh*. Coastal Water Convention (CWC), Khulna University, Khulna, Bangladesh, January 29–30.
4. Ahsan, M.N. 2010. Climate change and socioeconomic vulnerability: Experiences and lessons from South-western Coastal Bangladesh. MSc Thesis, Development Economics Group & Chair Disaster Studies, Wageningen University and Research Centre, the Netherlands. http://edepot.wur.nl/146194.
5. Ahsan, M.N., Ahmed, M.F., Bappy, M.H., Hasan, M.N., and Nahar, N. 2011. Climate change induced vulnerability on living standard—A study on south-western coastal region of Bangladesh. *Journal of Innovation and Development Strategy*, 5(3), 24–28.
6. Anderson, J.M. 2008. Adapting to climate change with water savings and water reuse. *Water Practice and Technology*, 3(2), 30–49.
7. Arnell, N.W. 1999. Climate change and global water resources. *Global Environmental Change*, 9(Supp.1), 31–49.
8. Arnell, N.W. 2004. Climate change impacts on river flows in Britain: The UKCIP02 scenarios. *Journal of the Chartered Institute of Water and Environmental Management*, 18(1), 112–117.
9. Ashton, P.J. 2002. Avoiding conflicts over Africa's water resources. *Ambio*, 31(3), 236–242.
10. Bates, B.C., Kundzewicz, Z.W., Wu, S., and Palutikof, J. 2008. *Climate Change and Water. Technical Paper of the Intergovernmental Panel on Climate Change*. IPCC Secretariat, Geneva.
11. Caran, S.C. and Neely, J.A. 2006. Hydraulic engineering in prehistoric Mexico. *Scientific American*, 8 pp.
12. Christensen, J.H., Hewitson, B., Busuioc, A., Chen, A., Gao, X., Held, I., Jones, R. et al. 2007. Management regional climate projections. In: *Proceedings Fourth Assessment Report of the Intergovernmental Panel on Climate Change 2007*, held in Geneva on September 10–15, 2007. Cambridge University Press. Cambridge, UK.
13. Collier, P., Conway, G., and Venables, T. 2008. Climate change and Africa. *Oxford Review of Economic Policy*, 24(2), 337.
14. Cruz, R.V., Harasawa, H., Lal, M., Wu, S., Anokhin, Y., Punsalmaa, B., Honda, Y., Jafari, M., Li, C., and Huu, N.N. 2007. Chapter 11, *Fourth Assessment Report of the Intergovernmental Panel on Climate Change*. Cambridge University Press, Cambridge, UK, pp. 469–506.
15. de Wit, M. and Stankiewicz, J. 2006. Changes in surface water supply across Africa with predicted climate change. *Science*, 311(5769), 1917–1921, doi: 10.1126/science.1119929.
16. Elliott, M., Armstrong, A., Lobuglio, J., and Bartram, J. 2011. *Technologies for Climate Change—The Water Sector*. UNEP Risoe Centre, Roskilde.
17. Eriksen, S., O'Brien, K., and Rosentrater, L. 2008. *Climate Change in Eastern and Southern Africa: Impacts, Vulnerability and Adaptation*. Department of Sociology and Human Geography. University of Oslo, Norway.
18. Gueye, L., Bzioul, M., and Johnson, O. 2005. *Water and Sustainable Development in the Countries of Northern Africa: Coping with Challenges and Scarcity*. Assessing Sustainable Development in Africa, Africa's Sustainable Development Bulletin, Economic Commission for Africa, Addis Ababa, pp. 24–28.
19. Hennessey, K.J., Page, C., McInnes, K., Jones, R., Barthols, J., Collins, D., and Jones, D. 2004. *Climate Change in New South Wales*. Consultancy Report for the NSW Greenhouse Office by CSIRO and the Australian Bureau of Meteorology, July 2004.

20. Houghton, J.T., Meira Filho, L.G., Callander, B.A., Harris, N., Kattenberg, A., and Maskell, K. 1996. Climate change 1995: The science of climate change. *Contribution of Working Group I to the Second Assessment Report of the Intergovernmental Panel on Climate Change*, Cambridge University Press, Cambridge, England.

21. Huntington, T.G. 2006. Evidence for intensification of the global water cycle: Review and synthesis. *Journal of Hydro-Environment*, 319, 83–95.

22. IPCC. 2001. *Climate Change 2001: Impacts, Adaptation, and Vulnerability.* Cambridge University Press, Cambridge, UK.

23. IPCC. 2007. Climate change 2007: Impacts, adaptation and vulnerability. In: M.L. Parry, O.F. Canziani, J.P. Palutikof, P.J. van der Linden, and C.E. Hanson (eds.), *Contribution of Working Group II to the Fourth Assessment Report of the Intergovernmental Panel on Climate Change (IPCC)*. Cambridge University Press, Cambridge, UK. www.ipcc.ch/pdf/assessment-report/ar4/wg2/ar4_wg2_full_report.pdf.

24. IPCC. 2007. Climate change 2007: Mitigation. In: B. Metz, O.R. Davidson, P.R. Bosch, R. Dave, and L.A. Meyer (eds.), *Contribution of Working Group III to the Fourth Assessment Report of the Intergovernmental Panel on Climate Change (IPCC)*. Cambridge University Press, Cambridge, UK. http://www.ipcc.ch/pdf/assessment-report/ar4/wg3/ar4_wg3_full_report.pdf.

25. Kashyap, A. 2004. Water governance: Learning by developing adaptive capacity to incorporate climate variability and change. *Water Science and Technology*, 19(7), 141–146.

26. Kurukulasuriya, P. and Mendelsohn, R. 2007. A Ricardian analysis of the impact of climate change on African cropland, The World Bank. http://ssrn.com/abstract=1005544.

27. Margane, A., Borgstedt, A., Hamdan, I., Subah, A., and Hajali, Z. 2009. *Delineation of Surface Water Protection Zones for the Wala Dam—Technical Cooperation Project Groundwater Resources Management*, Technical Report No. 12, prepared by BGR & MWI, BGR archive no. 0128313, p. 126, Amman.

28. Margane, A., Subah, A., Hajali, Z., Almomani, T., and Koz, A. 2008. *Delineation of Surface Water Protection Zones for the Mujib Dam—Technical Cooperation Project Groundwater Resources Management*, Technical Report No. 10, prepared by BGR & MWI, BGR archive no. 0128312, p. 132, Amman.

29. Matthews, A. 2009. Bangladesh's climate change emergency. *South Asia Monitor*. Centre for Strategic and International Studies, Washington, DC, p. 136.

30. McNamara, S. 2010. Solving the water crisis in climate-ravaged Bangladesh, United Nations Children's Emergency Fund (UNICEF). www.unicef.org/bangladesh/media_6207.htm.

31. Mirza, M.M.Q. 2002. Global warming and changes in the probability of occurrence of floods in Bangladesh and implications. *Global Environmental Change*, 12(1), 127–138.

32. Mirza, M.M.Q., Warrick, R.A., and Ericksen, N.J. 2003. The implications of climate change on floods of the Ganges, Brahmaputra and Meghna Rivers in Bangladesh. *Climatic Change*, 57(1), 287–318.ss

33. Neelormi, S., Hossain, M.A., and Ahmed, A.U. 2009. Combating water borne diseases under climate change in Bangladesh. In: *Proceedings of Suitable Alternative Solutions, International Science Conference on the Human Dimensions of Global Environmental Change World Conference Centre*, Boon, Germany, April 26–30, 2009.

34. Ray, P.A., Kirshen, P.H., and Watkins Jr., D.W. 2012. Staged climate change adaptation planning for water supply in Amman, Jordan. *Journal of Water Resources and Planning, Manage*, 138(5), 403–411.

35. Saleqazzaman, M., Rahman, M.U., Moniruzzaman, M.M., Kashem, M.A., Salam, M.A., Jahan, S., Tariqul, S.M.T., Islam, and Rokunuzzaman, M. 2009. Climate change induced vulnerabilities and people's precipitation in the south-west region of coastal Bangladesh. In: *Climate Change and the Tasks for Bangladesh*. BAPA & BEN, pp. 149–158.

36. Sharma, B. and Sharma, D. 2008. *Impact of Climate Change on Water Resources and Glacier Melt and Potential Adaptations for Indian Agriculture*. International Water Management Institute, New Delhi.

37. Stakhiv, E.Z. 1998. Policy implications of climate change impacts on water resource management. *Water Policy*, 1(1), 159–175.

38. Steenbergen F. van and Tuinhof, A. 2009. *Managing the Water Buffer for Development and Climate Change Adaptation. Groundwater Recharge, Retention, Reuse and Rainwater Storage*. MetaMeta Communications, Wageningen, the Netherlands. http://www.bebuffered.com/downloads/3R_managing_the_water_buffer_2010.pdf.

39. van Steenbergen, F. and El Haouari, N. 2010. The blind spot in water governance: Conjunctive groundwater use in MENA countries. In: S. Bogdanovich and L. Salome (eds.), *Water Policy and Law in the Mediterranean: An Evolving Nexus*. Faculty of Law Business Academy, Novi Sad, Serbia, pp. 171–189.

40. Stern, N. 2007. *The Economics of Climate Change: The Stern Review*. Cambridge University Press, Cambridge, UK.

41. UNDP. 2004. *A Global Report: Reducing Disaster Risk*. A challenge for development, United Nations Development Program, Bureau for Crisis Prevention and Recovery, New York. www.iamz.ciheam.org/medroplan/archivos/UNDP%20dr _english.pdf.

42. UNFCCC. 2007. *Climate Change: Impacts, Vulnerability and Adaptation in Developing Countries*. http://unfccc.int/resource/docs/publications/impacts.pdf.

43. Ziervogel, G., Shale, M., and Du, M. 2010. Climate change adaptation in a developing country context: The case of urban water supply in Cape Town. *Climate and Development*, 2(2), 94–110.

76

Impact of Climate Change on Drinking Water

Manish Kumar Goyal
Indian Institute of Technology Guwahati

Vishal Singh
Indian Institute of Technology Guwahati

Saeid Eslamian
Isfahan University of Technology

PREFACE

This chapter describes the impact of climate change on surface water, groundwater resources enclosing snow and glacier melting rate. Emphasis is given on vulnerability of climate change on drinking water resources. For this purpose, a study area is chosen and hydrological modeling tool SWAT has been applied. Attention is also paid to the importance of the different scenarios generated for variation of precipitation and temperature parameters to understand the impact of climate change.

76.1 Introduction

Water is a vital component for every human civilization, living organism, and natural habitat. Thus, water is necessary for drinking, cleaning, agriculture, transportation, industry, recreation, and animal husbandry, producing electricity for domestic, industrial, and commercial use. Water has multiple benefits and, hence, the problems created by its excesses, shortages, and quality deterioration, water as a natural resource requires special attention. Several cities especially in the developing world are facing rising pressures on organizations and infrastructure due to population growth and urbanization. Developing countries such as India etc. of particular interest with regard to its future water resources, as it is expected to undergo continued rapid population growth while also being especially sensitive to climate change [4]. Climate change will make drinking water security even more difficult and costly to achieve.

This chapter introduces the background and basic drivers of changes; it then introduces climate change impacts on surface and groundwater resources as well as climate change impacts on snow and glacier melting rate; then discusses vulnerability of drinking water resources to climate change; finally the site information (study area in India) and development of various scenarios using hydrological modeling under altered climate change are discussed.

76.1.1 Background

Water is one of the most precious natural resources on Earth for supporting life. Water is essential for every living organism, natural habitat, and human civilization. The hydrologic cycle circulates enormous quantity of water around the three components of the earth system, that is, atmosphere, lithosphere, and sea. The availability of freshwater resources around the world has little potential especially for human use because 97.5% of all water on Earth is saline water, which cannot be used for drinking purpose and irrigation. Out of the remaining 2.5% freshwater, most of which lies deep and frozen in Antarctica and Greenland in the form of snow and glaciers, only about 0.26% freshwater is available in rivers, lakes, and below the ground in the soils and shallow aquifers, which are readily usable for humanity. Water is used for drinking, cleaning, agriculture, transportation, industry, recreation, animal husbandry, and producing electricity for domestic, industrial, and commercial use. Due to its multiple benefits and the problems created by its excesses, shortages and quality deterioration, water as a resource needs special attention. As per the increasing population of India [33] as well as its all-round development, the utilization of water is also increasing in every sector at a fast pace.

On average, India receives annual precipitation of about 4000 km³. However, there exist considerable spatial and temporal variations in the distribution of rainfall and hence in availability of water in time and space across the country. Precipitation includes rainfall, snowfall, and other processes by which water falls to the land surface such as hail and sleet, etc. The circulation of precipitation and its pattern in India can be easily demonstrated by Indian monsoon. Around 80% of the total precipitation over the country is brought by the southwest (SW) monsoon, which is critical for the availability of freshwater as used for drinking and

irrigation purposes. Changes in climate, the SW monsoon would have a significant impact on agricultural production over the Indian region, climate changes, water resources management, and overall economy of the country. The heavy concentration of rainfall in the monsoon months (June–September) in namely the northeastern states of India, results in scarcity of water in many parts of the country during the non-monsoon periods, namely, Rajasthan and Haryana states of India (Figure 76.1). In view of this, a number of studies have attempted to investigate the trend of climatic variables for the country. These studies have looked at the trends on the country scale, regional scales, and at the individual stations. As per the definition by Troll [74], about 80% of India comes under semi-arid tropics. The seasonal rainfall shows a large variation from 11,000 mm/yr at Cheerapunji located near Shillong on the northeastern part of India to 200 mm/yr near Jaisalmer, in the desert part of Western Rajasthan. An understanding of rainfall variability for the country is necessary to appreciate the impacts of climate change; it is also important for water management [15].

As per the international standard, a country can be categorized as "water stressed" when water availability is less than 1700 m³ per capita per year and "water scarce" if it is less than 1000 m³ per capita per year. The availability of surface water in India is recorded as 2309 and 1902 m³ in the years 1991 and 2001. However, it is also expected that the per capita surface water

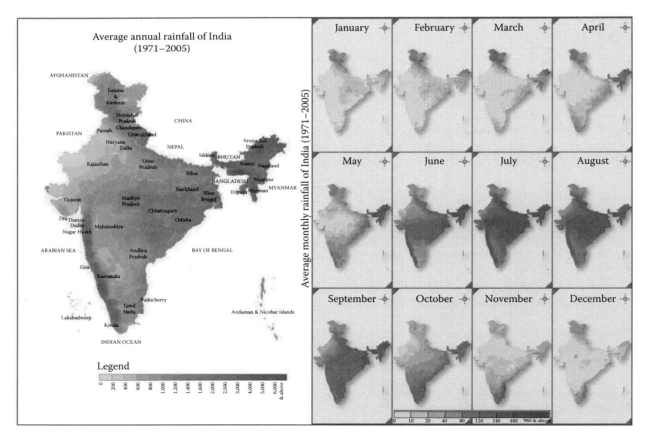

FIGURE 76.1 Average annual rainfall distribution of India.

TABLE 76.1 Water Availability

Facts at a Glance	
Area of the country as % of World Area	2.4%
Population as % of World Population	17.1%
Water as % of World Water	4%
Rank in per capita availability	132
Rank in water quality	122
Average annual rainfall	1160 mm (world average 1110 mm)
Range of distribution	150–11690 mm
Range rainy days	5–150 days, mostly during 15 days in 100 h
Range PET	1500–3500 mm
Per capita water availability (2010)	1588 m³

Source: Water Resources at a Glance 2011, CWC, New Delhi. http://www.cwc.nic.in.

availability is likely to be reduced to 1401 and 1191 m³ by the years 2025 and 2050, respectively. The per capita water availability in the year 2010 was 1588 m³ against 5200 m³ of the year 1951 in the country. India's average annual surface runoff contributed by rainfall and snowmelt is estimated to be about 1869 billion cubic meters (BCM) as per Table 76.1. However, it is estimated that only about 690 BCM or 37% of the surface water resources can actually be mobilized [40]. Out of this total quantity, only 1123 km³ is utilizable (690 km³ from surface water resources and 433 km³ from groundwater resources). The water demand in the year 2000 was 634 km³ and it is likely to be 1093 km³ by the year 2025 [33]. This is because (1) over 90% of the annual flow of the Himalayas rivers occurs over a 4-month period, and (2) potential to capture such resources is complicated by limited suitable storage reservoir sites. Due to fast rise in population and growing economy of the country, there will be continuous increase in demand for water, and it will become scarce in the coming decades (Table 76.2).

TABLE 76.2 India's Water Resources

Sl. No.	Water Resource at a Glance	Quantity (km³)	Percentage
1	Annual precipitation (including snowfall)	4000	100
2	Precipitation during monsoon	3000	75
3	Evaporation+ soil water	2131	53.3
4	Average annual potential flow in rivers	1869	46.7
5	Estimated utilizable water resources	1123	28.1
	Surface water	690	17.3
	Replenishable groundwater	433	10.8
	Current utilization of annual precipitation	634	15.85
	Current utilization of utilizable water	634	56.45
	Storage created of utilizable water	225	20.03
	Storage (under construction) of utilizable water	171	15.22
6	Estimated water need in 2050	1450	129
7	Estimated deficit	327	29
	Interlinking can give us	200	17.8

Source: India-WRIS. 2012. *River Basin Atlas of India.* RRSC-West, NRSC, ISRO, Jodhpur, India.

India's river basins are considered the basic hydrological units for planning and development of water resources. India's water resources wealth has been categorized into six major hydrological zones, namely, water resources regions (Indus drainage, drainage flowing into Arabian Sea, drainage flowing into Bay of Bengal, Brahmaputra drainage, minor rivers draining into other basins/countries such as Myanmar, Bangladesh, and China, and Island drainage) primarily based on drainage of rivers to outlet. Water resource development programs can be well accounted based on the watershed or hydrological unit concepts. The systematic delineation of river basins in India was first attempted by Central Water and Power Corporation (CWPC) under the direction of Dr. A.N. Khosla in 1949 for the entire country. Initially, the whole country was divided into six water resources regions (Watershed Atlas of India, AIS and LUS, 1990). Water resource divisions of India have been classified to highlight the broad drainage pattern of the country [33]. In India, around 12 major river basins having catchment area of 20,000 km² and above are reported. The total catchment area of these rivers is estimated around 25.3 lakh km². The river basins such as Ganga, Meghna and Brahmaputra are recognized as the largest river basins in Indian corresponding to catchment area of about 11.0 lakh km² (more than 43% of the catchment area of all the major rivers in the country) [21]. The other major river basins such as Indus, Mahanadi, Godavari and Krishna correspond to a catchment area of more than 1.0 lakh km². There are 46 medium river basins with catchment area between 2000 and 20,000 km² [33]. The total catchment area of medium river basins is estimated at about 2.5 lakh km². However, all of the major river basins and many medium river basins are inter-state in nature, which covers about 81% of the geographical area of the country (Figure 76.2).

India accounts for largest perennial rivers system in the world; however, most of the inland basins river flow depend on the Indian monsoon. These river channels are the major sources for recharge of groundwater. As per groundwater aspects, India's rechargeable annual groundwater potential has been assessed at around 431 BCM in aggregate terms. The rate of groundwater recharge varies across India as per the variations in rainfall quantity, intensity, and rock type. The physiography of Indian continents varies from rugged mountainous terrains of Himalayas, eastern and western Ghats, and the Deccan plateau to the flat alluvial plains of the river valleys and coastal tracts and the Aeolian desert in the western part. Hydro-geologically it can be divided into four broad provinces [59]. The Indo-Gangetic plains, dune soil of desert region, and coastal alluvium made up of sediments of quaternary to recent age comprise the first group. The second province covers most of western and part of central India and is covered by Late Cretaceous plateau basalt (Deccan trap). The southern and southeastern part of the peninsula, covered by granites and granite gneiss, constitutes the third unit. The Proterozoic, Paleozoic, and Mesozoic sediments distributed all over the country comprise the fourth group. In the sedimentary and alluvial areas, groundwater occurs under semi-confined to confined conditions down to depths of 100 m and more [59].

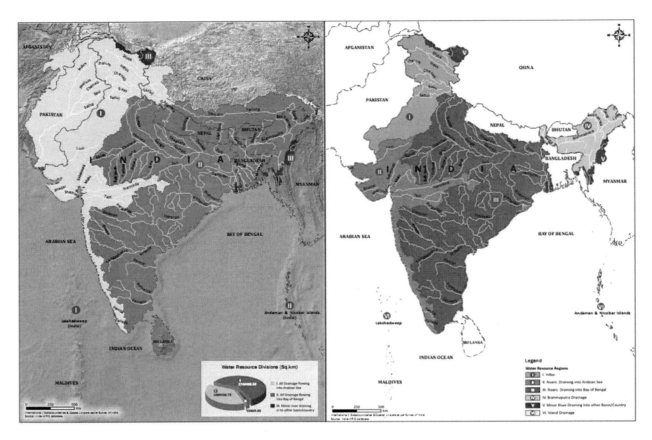

FIGURE 76.2 Major water resource region and river basins of India.

On an all India basis, it is estimated that about 30% of the groundwater potential has been tapped for irrigation and domestic use. The regional situation is very much different and large parts of India have already exploited almost all of their dynamic recharge. As per the recent study done by Kalipada [40], Haryana and Punjab have exploited about 94% of their groundwater resources. Areas with depleting groundwater tables are found in Rajasthan, Gujarat, most of western Uttar Pradesh, and in all of the Deccan states. Occurrence of water availability at about 1000 cubic meters per capita per annum is a common threshold for water indicating scarcity [75]. In the given time, the surface and groundwater resources have transformed in little more than a generation to a situation of water scarcity and limited development options. The surface and groundwater resources were considerably underutilized where significant development probably existed. India faces an increasingly urgent situation: its finite and fragile water resources are stressed and depleting while various sectoral demands are growing rapidly. Increasing population and rapid urbanization also put an additional demand on water resources. Summing up the various sectoral projection reveals a total annual demand for water increasing from 552 BCM in 1997 to 1050 BCM by 2025 [26].

76.1.2 Driver of Changes and Pressures

Many studies, which have been conducted in climate change perspectives, clearly show that climatic change is likely to impact significantly upon freshwater resources availability around the globe. In India, demand for water has already increased manifold over the years due to urbanization, agriculture expansion, increasing population, rapid industrialization, and economic development. At present, changes in cropping patterns and land-use patterns, over-exploitation of water storage, and changes in irrigation and drainage are modifying the hydrological cycle in many climate regions and river basins of India. An assessment of the availability of water resources in the context of future national requirements and expected impacts of climate change and its variability is critical for relevant national and regional long-term development strategies and sustainable development [46]. The demand for water worldwide is water scarcity and it is dependent on many factors such as population growth, urbanization, and industrialization. In India, it is observed that the demand of water is expected to grow by 20%, which could be projected to double from 23.2 trillion liters at present to 47 trillion liters, in which domestic water demand is expected to grow by 40% from 41 to 55 trillion liters, while irrigation will require only 14% more 10 years hence, 592 trillion liters up from 517 trillion liters currently. Table 76.3 has shown that the per capita availability of water has significantly come down and is likely to come down further with the growing population and demand. As per the Ministry of Water Resources, the per capita water availability in 2025 and 2050 is estimated to come down by almost 36% and 60%, respectively, of the 2001 levels [52].

TABLE 76.3 Sectoral Water Demand Situation and Key Drivers (Units in Billion Cubic Meters)

Sector	2002	2030	Water Demand	Key Components
Domestic	34.5	66.4	• Population growth • Increased per capita water consumption	• Absence on regulatory binding on usage and wastage of water
Agriculture	40.86	91.63	• Expansion of the water intensive industries like power, iron and steel, chemical is leading to increase in water demand	• Absence of loss monitoring and subsequent reduction scheme
Industry	606	674	• Domestic food grain demand increasing with increase in population • Demand for water-intensive crops like wheat, rice are increasing substantially poor water management	• Overexploitation of groundwater • Reduction of groundwater levels due to climate change

Source: India's Water Future to 2025–2050: Business as Usual Scenario and Deviations. Ernst & Young Analysis.

76.1.2.1 Increasing Global Annual Temperature

Over the last several years, in many studies there have been observed changes in surface temperature, rainfall, evaporation, and extreme events since the beginning of the twentieth century. The atmospheric concentration of carbon dioxide has increased from about 280 parts per million (ppm) by volume to about 369 ppm and the global temperature of the Earth has increased by about 0.6°C due to deforestation and uneven exploitation of nonrenewable natural resources, namely, coal and petroleum products [34]. The burning of fossil fuels and other uneven human activities are the primary cause for increased concentrations of CO_2 and other greenhouse gases in the atmosphere. Between 1990 and 1999, an estimated 6.3 Gt. C/yr was released due to the combustion of fossil fuels, and another 1.6 Gt. C/yr was released due to the burning of forest vegetation. This was offset by the absorption of 2.3 Gt. C/yr each by growing vegetation and the oceans. This left a balance of 3.3 Gt. C/yr in the atmosphere [16,17]. The global mean sea level has risen by 10–20 cm due to high overland flow and streamflow runoff. There has been a 40% decline in Arctic Sea ice thickness in late summer to early autumn in the past 45–50 years. The frequency of severe floods in large river basins has increased during the twentieth century [63]. Also, synthesis of river-monitoring data reveals that the average annual discharge of freshwater from six of the largest Eurasian rivers to the Arctic Ocean has increased by 7% from 1936 to 1999 [53]. In India, studies by several authors show that there is an increasing trend in surface temperature, no significant trend in rainfall on all-India basis, but decreasing/increasing trends in rainfall at some locations.

76.1.2.2 Hydrological Imbalances

The current climatic uncertainties accelerating the hydrologic cycle, changing precipitation pattern and its intensity, magnitude and timing of runoff. This warmer temperature is increasing melting rate of glaciers consequently [80]. There is clear evidence that Himalayan glaciers have been melting at an unprecedented rate in recent decades. This trend causes major changes in freshwater flow regimes. Further the high intensity

and volume of runoff producing by snow and glacier fed rivers are contributing more water to the sea as well as responsible for rising sea level. On a regional scale, snow cover is important for local water availability, river runoff and groundwater recharge, especially in middle and high latitudes [38]. Moreover, in an environment of increased temperature and evaporation, the lack of available water will decrease soil moisture. Reduction in soil moisture can greatly reduce agricultural yield. In this context, exact knowledge of the snow-covered area is also essential for water resource management and also, information about the water equivalent of snow is important for hydrological modeling and water resource management. Warm air will hold more moisture content and evaporation will be increased as surface moisture. This moisture tends to increase in the atmosphere and consequently rainfall and snowfall events are inclined to be more intense. However, little or no moisture goes into raising the temperature, which could contribute to longer and more severe droughts [73]. In a number of studies, it is projected that increasing temperature and decline in rainfall may reduce net recharge and affect groundwater levels as well [5]. Groundwater has been the mainstay for meeting the domestic needs for drinking water of more than 80% of rural and urban population, besides fulfilling the irrigation needs of around 50% of irrigated agriculture. It has been estimated that 70%–80% of the value of irrigated production in India comes from groundwater irrigation. Climate change has an adverse impact on the Indian groundwater reservoirs. The groundwater level is continuously decreasing due to overdraft and, hence, better management and mitigation strategies for minimizing the threats are necessary [54].

76.1.2.3 Urbanization and Industrialization Impacts

Urbanization is one of the major trends of the twenty-first century in developing countries, there is debate as to whether urbanization will increase or decrease vulnerability to floods and droughts. As per the United Nations report [76], the level of urbanization is estimated to cross the 50% mark in 2005 and by 2025 more than three-fifths of the world's population will live in urban areas [76]. Almost all of this increase in urban population will occur in the developing world and more than half the growth will occur in just two countries, India and China [19]. Over the last 12 years, the urbanization and developmental activities in India have been increased immensely due to increasing population density and their demands especially in urban areas. These high-density urban settlements and corresponding agricultural expansions are creating significant threat to our freshwater reserves as well as increasing hydrological imbalances. Large-scale emigrations from rural areas to urban areas and population growth have been incessant and accelerating these phenomena in most of parts of India, which comes under Ganga, Indus, Narmada, Mahanadi, Godavari, Cauvery, Krishna, and Brahmaputra river basins, where urbanization is increasing at an unprecedented rate especially near river bank areas. Urban agglomeration is causing radical changes in groundwater recharge and modifying the existing mechanisms. A majority of the cities are sited on unconfined or semi-confined aquifers depending on river water and groundwater foremost of their water supply and disposal most of their liquid effluents and solid residues to the rivers and the ground.

There has also been an inevitable rise in waste production. Drainage of surface water has been disrupted as the small natural channels and low-lying areas have been in filled, often with municipal waste. Mishra [48] has studied recently on Ganga basin and found that total water potential of the Ganga basin including surface water potential and groundwater potential is around 525.02 and 170.00 km^3 respectively, in which basin supports approximately 42% of the total population in India. Water tables are declining at approximately an average of 0.20 m per year in many parts of the basin and there is a trend of deteriorating groundwater quality. The greatest hydro-geological impacts of urbanization are found on the flat surfaces of the high terraces and the interfluves. The frequency of extreme hydrological events especially in Central Ganga Plain has been increased due to increase runoff causing more intense local flooding, while droughts during dry weather are deeper and longer. These changes have started showing their impact in Ganga basin on water habitats, exports high concentration of pollution into the rivers, wetlands and reservoirs, destabilizes ecological processes, and handicaps the ecological stability of ecosystems. Srinivasan et al. [69] has studied the Indian growth of urbanization and drinking water vulnerability and found that compact and dense urbanization increase vulnerability to water stress while pre-urban development places less stress on the aquifer. These variations make changes in the water resources of the local environment. These environmental imbalances create serious hazards (heavy rainfall, flash flooding, erosion, warming local climate and reduced biological diversity of flora and fauna) to their local environment [32]. India is increasingly becoming urban. According to the 2001 census, 27.8% of the urban population resides in cities, compared with 25.5% in 1990. The urban population is expected to rise to around 40% by 2020. Ramachandra and Uttam Kumar [57] studied the pattern of growth in Greater Bangalore and its implication on local climate (an increase of ~2–2.5°C during the last decade) and on the natural resources (76% decline in vegetation cover and 79% decline in water bodies).

76.1.2.4 Agricultural Expansion Impacts

The demands for freshwater by industry and especially by agriculture are causing surface water resources to be abstracted and groundwater resources to be depleted in ways which compromise freshwater ecosystem health [67]. The productivity of agricultural, forestry and fisheries systems are dependent on the temporal and spatial distribution of precipitation and evaporation, as well as, especially for the crops, on the availability of freshwater resources for irrigation. Biomass based subsistence agriculture constitutes the main source of rural and urban livelihood. Production systems in marginal areas with respect to water face increased climatic vulnerability and risk under climate change, due to factors such as soil erosion, over-extraction of groundwater and associated salinization, and over-grazing of dryland [23]. While too little water leads to vulnerability of production, too much water can also have deleterious effects on crop productivity, either directly, for example, by affecting soil properties and by damaging plant growth, or indirectly, for example, by harming or delaying necessary farm operations. Heavy precipitation events, excessive soil moisture, and flooding disrupt food production and rural livelihoods worldwide [62]. India is the largest country with a diverse climate. India's wealth mostly depends on the agricultural production. India has largest drainage system around the world. India's agricultural production and dependency mainly corresponds to monsoon rains and a close link exists between climate and water resources. The three main river basins of India, namely, Ganga, Indus, and Brahmaputra contribute about 60% area from the total geographical areas of the country. These river basins are snow fed and originate from Eastern and Western Himalayas. Among the key impacts will be the faster retreat of Himalayan glaciers, frequent floods, and decrease in crop yields. Yield reductions are predicted in wheat and rice due to temperature rise in key growing regions. Until last year, 2009 was the warmest year on record in India since 1901 (+0.913°C above the normal of 24.64°C) now the warmest year is 2010 (+0.93°C) [13]. The quantity of water required for agriculture has increased and will be more severe progressively through the years as more and more areas will bring under irrigation. Since 1947, the irrigated area in India rose from 22.60 to 80.76 mha up to June 1997. Irrigation has played a significant role in India attaining self-sufficiency in food production during the past three decades as per the contribution of surface water and groundwater resources. However, it is likely to become more critical in future in the context of national food security. The requirement of water is projected to decrease to about 68% by the year 2050, due to reasonable utilization, although agriculture will still remain the largest consumer. In order to meet this demand, growth of the existing water resources by expansion of additional sources of water or protection of the existing resources through confiscation additional water in the existing water bodies and their conjunctive use will be needed [46].

During the recent past, developing countries like in India's rural areas, resource development practices have changed in response to population increase and the resultant increased demand on natural resources as well as increasing socioeconomic and political marginalization. This has brought about rapid environmental changes that have reduced the groundwater recharge in the region. About 36% of springs have dried, heads of perennial streams have dried, and water discharge in springs and streams has decreased considerably resulting into severe crisis of water for drinking as well as irrigation during the past 20 years [72]. Due to uneven agricultural expansions, the load on groundwater reserves is continuously increasing. Recent hydrological research regarding northwest India has largely focused on optimizing agricultural yields against the backdrop of increasing water stress. Jalota and Arora [39] used a simple hydro-agricultural model to estimate variability in evaporation and drainage losses under a variety of cropped soils in the state of Punjab. According to the United Nations, globally, nearly 70% of all water consumption is by the agriculture sector, nearly 22% is used by the industries, and 8% is for domestic use. Water demand for irrigation depends on the combined impact of increased temperatures, higher humidity, wind and changes in sunshine hours. Water requirement for thermal energy generation will be increased, depending on future trends in water-use efficiency and the construction of new plants. A major boost to agricultural production and productivity of water in agriculture has been achieved with the green revolution in the 1960s. However, more than 20% of our food production is unsustainable because it depends on overpumping of finite groundwater resources [14].

76.1.2.5 Hydrological Development Impacts

The term sustainable development defines a pattern of resource use that aims to meet the current human needs, while preserving natural resources and ecosystems on which the current and future generations depend. Freshwater sources are of crucial importance to sustain biodiversity, which is increasingly threatened by careless exploitation of natural resources. Many developmental activities are becoming widespread and continuously threatening to our surface as well as groundwater reserves. The hydrological development like construction of dams, reservoirs, hydropower projects, diversion, and overexploitation are affecting the world's freshwater resources. These impacts are responsible for the progressive degradation of inland freshwater systems. Climate change is one of the most important global environmental challenges, with implications for food production, water supply, health, energy, etc. A number of the hydropower development projects are planned to establish by the government of India for meeting the energy demand of the country, though several major and minor projects are already completed. The analysis of climate change impact on the hydrology of high altitude glacierized catchments in the Indian Himalayas is complex due to the high variability in climate, lack of data, large uncertainties in climate change projection and uncertainty about the response of glaciers, which can have significant impact on these developmental activities.

Construction of hydro-electric power plants may also cause ecological impacts on existing river ecosystems and fisheries, induced by changes in flow regime (the hydrograph) and evaporative water losses (in the case of dam-based power-houses). The social disruption and water availability for shipping (water depth) may cause problems on their local environments. Positive effects are flow regulation, flood control, and availability of water for irrigation during dry seasons. Furthermore, hydropower does not require water for cooling (as in the case of thermal power plants) or, as in the case of bio-fuels, for growth. About more than 85% of water reservoirs in India were built for irrigation, flood control, and urban water supply schemes, and many could have small hydropower generation retrofits added without additional environmental impacts. However, In India, the national water policy (NWP) recommended resource planning for a hydrological unit such as a basin or subbasin [51]. This means that all developmental projects in a basin should be formulated within the framework of an overall plan for a basin/subbasin. The NWP further states that there should be an integrated and multidisciplinary approach to the planning, formulation, clearance, and implementation of projects, including catchment and management, environmental and ecological aspects, rehabilitation of affected people and command area development. An approach to river basin planning and management requires the establishment of an appropriate organization at the river basin level for ensuring optimum, all-around, and balanced development of the water resources of a river basin [33]. India has a large number of water resource projects like dams (4169), reservoirs (4089), hydro-electric projects (257), barrages (351) and drinking water supply (168), which are mainly constructed to fulfill the demands of irrigation, drinking, and electricity purposes [33]. However, many water resource development projects have been proposed recently. These development scenarios sometimes create a serious threat to water resources of India. These developmental processes may also disrupt the natural balances of the river flow systems. As per current behavior of climate, these uncertainties would be more serious if effective management plans and precautions are not taken into an account.

76.1.3 Climate Change

Climate change is one of the most important global environmental challenges around the world facing civilization with implications for food production, natural ecosystems, freshwater availability and supply, health, etc. According to the latest scientific assessment, the Earth's climate system has demonstrably affected the freshwater resources on the land at global and regional scales. In India, demand for freshwater has already increased manifold over the years due to urbanization, agriculture expansion, increasing population, rapid industrialization and economic development. At present, changes in cropping pattern and land-use pattern, over-exploitation of water storage and changes in irrigation and drainage are modifying the hydrological cycle in many climate regions and river basins of India. Indus, Ganga and Brahmaputra river basins have been affected abruptly among all the basins because these river basins originate from the Western and Eastern Himalayas. Streamflow generated from these river basins mainly produced by snow and glaciers melting. Apart from this, the other main river basins of India like Narmada, Cauvery, Krishna, Mahanadi, Tapi, and Tungabhadra are mainly dependent on the monsoon rainfall periods. An assessment of the availability of water resources in the context of future national requirements and expected impacts of climate change and its variability is critical for relevant national and regional long-term development strategies and sustainable development. Extreme events from climate alterations and variations pose challenges for water utilities. As per the potential of hydropower in India, various small and large hydropower projects have been proposed by the government of India on the rivers that flow through Indian Himalayan regions. However, the current climatic uncertainties are creating serious devastation to these developmental projects. In India, hydropower projects have been classified in two major types: run of river (ROR) and storage-based schemes. Climate change may influence hydropower generation through changed water availability and distribution. If stream flows are reduced, less power will be generated and vice versa [37]. Further, if the flows are concentrated in fewer months, there will be more chances of spill and less hydropower will be generated. If the precipitation in the higher reaches is reduced, less hydropower can be generated in a cascade system of plants. Hydropower stations with storage will be able to accommodate increased seasonality in inflows, but the ROR schemes will be more vulnerable to climate change. Climate change could also induce a timing mismatch between energy generation and demand. Heavy runoff and precipitation intensity events frequently disturb the hydrological balance result in deterioration of source water quality and increasing the risk of contaminated water supplies [46]. The direct effect of temperature changes on water supplies is also significant in the term of timing of runoff volume. Because of warmer temperatures, freshwater ecosystems have shown significant changes in species composition, organism abundance, productivity, and phenological shifts (including earlier fish migration). Due to warming surface temperature of freshwater lakes and reservoirs, they have exhibited prolonged stratification with decreases in surface layer nutrient concentration and prolonged depletion of oxygen in deeper layers [48]. The most immediate reaction to climate change is expected to be in river and lake water temperatures [29,30].

In Himalayan regions, there would likely be shorter snow accumulation periods especially in lower elevation areas, possibly leading to reduced snow packs, earlier melting, and reduced late summer flows. Warmer temperature during the winter will affect the form of precipitation, with a larger fraction of total precipitation coming as a rain rather than snow. This phenomenon would disturb the real scenarios of snow and glacier hydrology and may cause serious threat to the availability for freshwater in low elevation land areas [38]. These uneven extreme events and climatic uncertainties on snow and glacier aspects have been assessed frequently by various researchers around the world

[61,66]. Effects of climatic variations and changes especially in mountain glaciers and ice caps have been recognized in high volume runoff in downstream flow of the river and channels [41], changing hazard conditions and ocean freshening [9]. The last few years, it has been seen that a variety of changes have emerged in the Indian Himalayan regions, mainly in response to hydrological as well as economical development on the Indian Himalayan basins, and the resultant increased demand for food, fodder, grazing land, water and other natural resources, market forces, and increasing socioeconomic and political marginalization [60,64,70]. As a result, the water resources of the region are diminishing and depleting quickly due to the rapid land use changes and resultant reductions in groundwater recharge [10,83]. By the middle of the twenty-first century, annual average river runoff and water availability are projected to increase because of climate change at high latitudes and in some wet tropical areas, and decrease over some dry regions at mid-latitudes and in the dry tropics. Increased precipitation intensity and variability are projected to increase the risks of flooding and drought in many areas. The frequency of heavy precipitation events (or proportion of total rainfall from heavy falls) will be very likely to increase over most areas during the twenty-first century and consequently affect the source, quantity, and quality of drinking and freshwater on the land. As per the industrialization and urbanization growth of India, water demand will grow in the coming decades, primarily due to population growth and increasing affluence; regionally, large changes in irrigation water demand because of climate change are expected.

76.1.4 Water Use

Water is a limited natural resource but it is widely available resource on the earth in the form of sea water, snow and ice caps, lakes, reservoirs and groundwater beneath the ground. Even though it is a wide presence, the availability and demand of freshwater at many places have high degrees of variations in spatial and temporal domain. In India, it is a challenge to provide water of desired quantity and quality at a desired place. India's water wealth is very diverse and varies from North to West and East to South India. As per the Indian monsoon scenarios, a majority of the annual rain about 70%–90% falls in just 3–4 months (June–September). This leads to too much water and often floods in the wet season especially in North India and too little water and often droughts in the dry season especially in Western India (Rajasthan, Gujarat, and Haryana). At times, enough water may be available but the quality may be so poor that it is of no use without treatment [37]. Each day, a person drinks 3–4 L of water and uses 10–15 L for other essential needs. Globally only about 14% of all water use is for domestic needs (drinking, cooking, washing, etc.) [75]. In India about 90% of official drinking water is from groundwater but many people depend on ponds and tanks. Due to reasons such as growing population, industrializations, etc. per capita water availability is dwindling in most developing countries. At the same time, quality of water is also deteriorating. Hence, as per the current condition, we hear that the world is in the grip of a "water deficit." The present situation shows that it is more a water governance crisis, which has arisen due to mismanagement of water and unscientific policies. The United Nations Development Programme (UNDP) has clearly conveyed the message of water governance crisis through the cover design of the Human Development Report for 2006. Hoekstra and Hung initially introduced the concept of water uses or footprint. The water footprint of a country is defined as the volume of water needed for the production of goods and services consumed by the inhabitants of the country. The volume of water used from domestic water resources to produce goods and services consumed by the inhabitants of the country constitutes the internal water footprint of a nation. The total water footprint of a nation is useful indicator of a nation's call on the global water resources [36].

The utilizable water supply in India is generally used for crop production, industrial purposes, and drinking purposes. At the same time, the water availability in India per capita is less than 1000 m³/yr, which is caused by either the lack of natural water resources or a result of over-exploitation of groundwater resources for irrigation purposes [11]. This other water scarcity reducing strategy can be quantitatively described with the concept of virtual water. This concept defines the virtual water content of a commodity as the volume of water that is actually used to produce the commodity, measured at the place where the commodity is actually produced [5]. From the previous estimates of the water resources in India, the conclusion can be made that precipitation estimates range from 3559 to 4000 km³ (1083–1194 mm), blue water resources estimates range from 1272 to 1650 km³ and green water resources estimates range from 2287 to 2350 km³. Because precipitation is the single input of the calculation of the internal water resources, it determines a large part of the outcome of the internal blue and green water resources. Especially the blue water resources seem to be sensitive to change in precipitation. The green water resources seem to be less sensitive to change in precipitation. The total volume of water used in India for the production of the studied primary crops is 792 billion m³/yr. The total blue water use is 219 billion m³/yr, the total green water use 479 billion m³/yr and the total greywater use billion 95 m³/yr. The blue component refers to the evaporation of groundwater and surface water during the production of a commodity, the green component to the evaporation of rainwater for crop growth [3].

The current climatic uncertainties affect adversely the water availability at a place or region for various uses such as drinking needs, municipal and industrial, irrigation and hydropower. Since the variability of the precipitation and thereby that of the discharge is likely to increase, it will be necessary to be able to regulate the river flows largely. This can be best achieved by creating more water-storage capacity in the country and optimally use the storage space. The existing water infrastructure is inadequate to accommodate the magnitude and temporal patterns of stream flows. However, in India Central Water Commission (Ministry of Water Resources) is the nodal agency which deals with all aspects of water resources of India. India WRIS (water

resource information system of India) portal reports that around 4169 Dams and Reservoirs, 257 Hydro-electric projects (greater than 25 MW) and 169 Drinking-Water supply projects have been established for governing India's water wealth and many projects have been proposed already by the ministry. The total land of India can be largely divided into 20 major river basins [33]. Table 76.4 provides estimates of per capita renewable water resources potential (MCM) of major river basins in India. Rivers are a lifeline of India and have played a significant role in the history of civilizations and economical developmental growths. Table 76.5 shows the annual water requirement in different purposes in India.

76.2 Climate Change Impacts

76.2.1 Climate Change Impacts on Surface Water Resources

Precipitation is a major source of freshwater reserves in India and it accounts 4000 km³ annual rainfall. However, the precipitation pattern varies across the country. The average utilizable surface water accounts around 690 km³. With growing population and industrialization especially in urban areas, the pressure on water resources in increasing and per capita availability of freshwater is continuously decreasing day by day. Apart from this current climatic activities impact badly on our water resources and especially this load is more challenging on the sources of freshwater reserves like rivers, lakes, ponds and snow and glacier reserves. The climate change uncertainties are continuous increasing and similarly it is harnessing our surface water reserves in quality and quantity both aspects [37]. Due to spatial variability in precipitation pattern of India, some places are facing drought and some places are facing flooding problem. Overexploitation of groundwater is leading to reduction of low flows in the rivers and is degrading the river water quality. India has largest river system in the world comprising 22 major river channels along with their several tributaries. Many of them are perennial and some are monsoon dependent. The rivers Ganga, Brahmaputra, and Indus originate from Himalayas and flow throughout the year. The snow and ice melt of the Himalayas and the base flow contributes the stream flow in lean season and managing the environmental flow, though the current climatic changes due to increasing global annual temperature are influencing the natural hydrological system. Average water yield per unit area of the Himalayan Rivers is almost double that of the south peninsular river systems, which indicates the importance of snow and glacier melt contribution from high mountains. Average intensity of mountain glaciations varies from 3.4% for Indus to 3.2% for Ganges, and 1.3% for Brahmaputra [46]. The tributaries of these river systems show maximum intensity of glaciation (2.5%–10.8%) for Indus followed by Ganges (0.4%–10%) and Brahmaputra (0.4%–4%). It demonstrates that the rainfall contributions are greater in the eastern region, while the snow and glacier melt contributions are more important in the western and central Himalayan region.

TABLE 76.4 Basin Wise Water Resource Potential in India

Basin Name	Catchment Area (km²)	Average Water Resource Potential (mcm)	Utilizable Surface Water Resource (mcm)	Life Storage Capacity of Projects (mcm)
Indus Basin	321,289	73,310	46,000	16,568.43
Ganga Basin	861,452	525,020	250,000	60,660.38
Brahmaputra Basin	194,413	537,240	24,000	11,680.56
Barak and others	41,723	48,360	NA	11,680.56
Godavari Basin	312,812	110,540	76,300	6,205.79
Krishna Basin	258,948	78,120	58,000	49,547.52
Cauvery Basin	81,155	21,358	19,000	8,867.02
Subernarekha Basin	29,196	12,370	6,800	2,322.21
Brahmani and Baitarni Basin	51,822	28,480	18,300	5,523.69
Mahanadi Basin	141,589	66,880	50,000	14,207.8
Pennar Basin	55,213	6,900	6,320	4,820.11
Mahi Basin	34,842	11,020	3,100	4,984.03
Sabarmati Basin	21,674	3,810	1,900	1,367.54
Narmada Basin	98,796	45,639	34,500	23,604.6
Tapi Basin	65,145	14,880	14,500	10,255.79
West flowing rivers from Tapi to Tadri	55,940	87,411	11,900	14,732.41
West flowing rivers from Tadri to Kanyakumari	56,177	113,530	24,300	11,553.7
East flowing rivers between Mahanadi and Pennar	86,643	22,520	13,100	3,026.41
West flowing rivers of Kutch and Saurashtra including Luni	321,851	15,100	15,000	5,524.15
Minor rivers draining into Myanmar and Bangladesh	36,202	31,000	NA	312

Source: River Basin Atlas of India 2012. *Ministry of Water Resources.* Government of India.

TABLE 76.5 Annual Water Requirement for Different Uses (in bcm)

	Year					
	2010		2025		2050	
Uses	Low	High	Low	High	Low	High
Irrigation	543	557	561	611	628	807
Domestic	42	43	55	62	90	111
Industry	37	37	67	67	81	81
Power	18	19	31	33	63	70
Inland navigation	7	7	10	10	15	15
Environment-ecology	5	5	10	10	20	20
Evaporation loss	42	42	50	50	76	76
Total	694	710	784	843	973	1180

Source: Adapted from Panwar, S. and Chakrapani, G.J. 2013. *Current Science*, 105(1), 37–46.

In recent decades, the hydrological characteristics of the watersheds in the Himalayan region seem to have undergone substantial changes as a result of extensive land use change (deforestation, agricultural practices, and urbanization), leading to frequent hydrological disasters, enhanced variability in rainfall and runoff, extensive reservoir sedimentation, and pollution of lakes. Climate change and its impact on the hydrological cycle as well as extreme events have posed an additional threat to this mountainous region of the Indian subcontinent. Extreme precipitation events especially in the Himalayas may cause widespread landslides and huge soil erosion. This large amount of sedimentation will affect the storage and seepage capacity of streams and freshwater lakes. The response of hydrological systems, erosion processes and sedimentation in this region could alter significantly due to climate change [35,58]. Many regions of India have suffered unexpected climate change issues such as probable maximum flood in high hilly catchment areas, and flood inundation in lower catchment areas. The flood inundation may affect the availability of freshwater reserves on the ground and beneath the ground. In India, the normal and main sources of drinking water are small hand pumps and tube wells. Many hand pumps and tube wells are submerged in the floodwaters making access to clean drinking water difficult for the initial days especially in the rural areas. People who had moved from their places of original residence experienced great difficulty in accessing drinking water. In some places, people who had taken refuge on raised embankments had to travel by boats to fetch water from hand pumps situated on higher elevations. Several researchers around the world have conducted studies on the effect of climate change on the availability of freshwater resources. McDonald et al. [47] studied the urban growth, climate change, and freshwater availability and found that the freshwater ecosystems in river basins with large populations of urbanites with insufficient water will likely experience flows insufficient to maintain ecological process. Freshwater fish populations will likely be impacted, an issue of special importance in regions such as India's Western Ghats, where there is both rapid urbanization and high levels of fish endemism. As per the study by Vorosmarty et al. [82], the regions of intensive agriculture and dense settlement show high incident threat, as exemplified by much of the United States, virtually all of Europe and large portions of central Asia, the Middle East, the Indian subcontinent and eastern China. Smaller contiguous areas of high incident threat appear in central Mexico, Cuba, North Africa, Nigeria, South Africa, Korea, and Japan. Gosain and Rao [27] projected that the quantity of surface runoff due to climate change would vary across the river basins as well as subbasins in India. However, there is general reduction in the quantity of the available runoff. In India, an increase in precipitation (Pcp) intensity and pattern in the Mahanadi, Brahimani, Ganga, Godavari and Cauvery is projected under climate change scenario. On the other hand, the corresponding total runoff for all these basins does not increase. This may be due to increase in evapotranspiration (ET) rate on account of increased temperature or variation in the distribution of rainfall. In the remaining basins, a decrease in precipitation was noticed. Sabarmati and Luni basins show drastic decrease in precipitation and consequent decrease of total runoff to the tune of two-thirds of the prevailing runoff. This may lead to severe drought conditions in future.

76.2.2 Climate Change Impact on Groundwater Resources

The effects of climate change on subsurface water and groundwater relates to the changes in its recharge and discharge rates plus changes in quantity and quality of water is stored below the ground in the form of water bearing zones called aquifer noticed significantly. Groundwater is a dynamic system. In spite of the national scenario on the availability of groundwater being favorable, there are many areas in the country facing scarcity of water. The development and over-exploitation of groundwater resources in certain parts of the country have raised the concern

and need for judicious and scientific resource management and conservation. Climate change refers to the long-term changes in the components of climate such as temperature, precipitation, evapotranspiration, etc. Climate change affects groundwater recharge rates (i.e., the renewable groundwater resources) and depths of groundwater tables. The groundwater recharges from surface water through seepage and infiltration processes. Thus, the impacts of surface water flow regimes are expected to affect subsurface water as well as groundwater both. Increased precipitation variability may decrease groundwater recharge in humid areas because more frequent. The natural causes of climate change include the Earth's axial and orbital changes, changes in the strength of the Sun, tectonic-plate movements, volcanic eruptions, asteroid collision, and chemical weathering. In semi-arid and arid areas, precipitation variability could increase groundwater recharge. Only high-intensity rainfalls are able to infiltrate fast enough before evaporating, and alluvial aquifers are recharged mainly by inundations due to floods [56]. The growth of urban settlements have largely affected groundwater levels especially in major cities in India. The impact of urbanization on the groundwater regime within a specific urban area depends both on its geographical location and the economic status of the city or even the country [79]. Naik et al. [50] studied significantly on the effect of urbanization on groundwater flow regime in major cities of India.

The groundwater resources supply 80% of domestic needs and more than 45% of total irrigation requirement. The annual Replenishable groundwater resource for the entire country is 433 BCM and net annual groundwater availability is estimated around 399 BCM. However, the annual groundwater draft for irrigation, domestic and industrial was 231 BCM and their Stage of groundwater development for the country as a whole is 58% [33]. The annual replenishable groundwater resource is contributed by two major sources such as rainfall and canal seepage return flow from irrigation, seepage from water bodies and artificial recharge due to water conservation structures. The overall contribution of rainfall to country's annual replenishable groundwater resource is 67% and the share of other sources taken together is 33%. The contribution from other sources such as canal seepage, return flow from irrigation, seepage from water bodies, etc. in annual replenishable resources is more than of 33% in the states of Andhra Pradesh, Delhi, Haryana, Jammu and Kashmir, Jharkhand, Punjab, Tamil Nadu, Uttar Pradesh, Uttaranchal and UT of Pondicherry. Southwest monsoon being the most prevalent contributor of rainfall in the country, about 73% of country's annual replenishable groundwater recharge takes place during the Kharif period of cultivation. Keeping 34 BCM for natural discharge, the net groundwater available for utilization for the entire country is 399 BCM. The annual groundwater draft is 231 BCM, out of which 213 BCM is for irrigation use and 18 BCM for domestic and industrial use. An analysis of groundwater draft figures indicates that in the states of Chhattisgarh, Delhi, Goa, Himachal Pradesh, Jammu and Kashmir, Jharkhand, Kerala, north eastern states of Manipur, Meghalaya, Mizoram, Nagaland and Tripura, Orissa, Sikkim, and Union Territories of Dadra and

Nagar Haveli, Daman and Diu, Lakshadweep and Pondicherry, groundwater draft for domestic and industrial purposes are more than 15% which is comparatively higher than the national average of 8%. In general, the irrigation sector remains the main consumer of groundwater [33].

The per capita water availability of groundwater reserves is continuously declining from 5176 m³ in 1951 to 1820 m³ as on March 1, 2001 [20]. Recent estimates show that 60% of Indians will live in urban areas by 2050, and so high increase in water demand is expected in future. A study conducted on 5723 blocks by the central groundwater board of India (CGWB) in 2004 states that 1615 blocks are semi-critical, critical, or over-exploited. The number of exploited blocks in 1995 was 4%, which increased to 15% by 2004. Groundwater level in Gujarat, Rajasthan, Punjab, Haryana, and Tamil Nadu has shown a critical decline [20]. Groundwater decline has been registered in 289 districts of India. It has been predicted that an average drop in groundwater level by 1 m would increase India's total carbon emissions by over 1% because for the withdrawal of the same amount of water there will be an increase in fuel consumption. Ghats are likely to experience increase in precipitation; however, the increase will show spatiotemporal variability. Due to melting of the Himalayan glaciers, the Indo-Gangetic Plains will experience increased water discharge till 2030s [37,46] but will face gradual reductions thereafter. This increase in precipitation may show higher flooding, devastating major parts of India. At present northern India is losing groundwater at a rate of 549 km³/yr (between April 2002 and June 2008). This rate of high runoff and low recharge will lead to degradation of aquifer in the northern plains of India [45,71]. According to Rangarajan and Athavale [59], the annual replenishable groundwater potential of India, for normal monsoon year based on tritium injection studies, is calculated as 476×10^9 m³/yr. Groundwater is a renewable natural resource and hence it can be replenished by better groundwater management and governance policies. Therefore, spatio-temporal effect of climate change on aquifers should be assessed and based on this risk assessment of each aquifer should be rated and actions and policies should be designed accordingly. Climate change mapping on different resources will give better results and answers about the vulnerability and risks involved over time for a specific area [18]. The new research should also be promoted to obtain better results from the positive effects of climate change, with the aim of reducing the negative effects.

76.2.3 Climate Change Impact on Snow and Glacier

Most of the stable prominent rivers of the world originate from the glaciers. Snow and glacier melt runoff from mountains is the main source of water at the regional scale, with downstream processes, such as hydropower based energy production [81], biodiversity and ecological balance [12], controlled by processes at higher elevations. India has the largest river network in the world and about 70% of the rivers of India are nourishing the Western and Eastern Himalayan glaciers through snow melt

process. Due to anthropogenic activities, namely, deforestation, afforestation, construction of dams, reservoirs, hydroelectric projects (blocking natural flow of rivers), industrialization, burning of fossil fuels (in excess amount) the potential concentration of carbon dioxide (CO_2) and other greenhouse gases (GHGs) is continuously increasing in the atmosphere. The effects of these greenhouse gases result in unbalancing global air mass balance, further enhancing the global average temperature in the lower atmosphere as well as soil surface temperature and reducing carbon sequestration. These morbid changes are showing their negative impacts on the environment mostly due to the absence of less technical efforts and researches against awareness. The Himalayas are an extraordinarily high mountain chain, spanning 2500 km east to west across five countries and encompassing about 15,000 glaciers and 9000 glacial lakes in Bhutan, Nepal, Pakistan, China and India [31]. This mountain range feeds most of the major perennial river systems in the region and is considered the lifeline for approximately 10% of the world's population. The cryosphere of the Hindu Kush-Himalayan region plays a significant role in the regional climatic system, and is a sensitive indicator of global climate change. Climate change is a major driver affecting the cryospheric environment, threatening the freshwater reserves and posing increased risks from climate-induced hazards to the mountain region and its immediate downstream communities. However, there is a marked lack of consistent, detailed, and long-term information for the region, and in particular for glacier and snow cover. The Intergovernmental Panel on Climate Change (IPCC) fourth assessment report shows that the Hindu Kush-Himalayan region comprises a major data gap in terms of any climatic assessment [31]. As per the IPCC, the global annual average temperature has been raised up to 1.6°C in last decades and it will be more severe in the coming days. This warmer temperature consequently increases the melting rate of glaciers. Further the high intensity and volume of runoff producing by snow and glacier fed rivers are contributing more water to the sea as well are responsible for increasing sea level. On a regional scale, snow cover is important for local water availability, river runoff and groundwater recharge, especially in middle and high latitudes [38]. The combined effect of these changes would impact the supply and demand of freshwater of rivers which are fed by snow and glaciers. This mountain range feeds most of the major perennial river systems in the region and is considered the lifeline for approximately 10% of the world's population. Hazard assessment, especially in those river valleys that are known to be potentially at risk for GLOF (glacial lake outburst flash flood) events, is essential in developing the most appropriate responses and mitigation measures. Mitigation measures may help to prevent a GLOF event and/or reduce the severity of its impact [49,65].

Many studies have been carried out on the fluctuation of glaciers in the Indian Himalayan regions and significant changes (mostly retreats) have been recorded in the last three decades. Most of these glaciers have been retreating discontinuously since the post-glacial. Previously, many studies have been carried out on climate monitoring and assessment especially on Brahmaputra basin. Ghosh and Dutta [24] have studied the climatic variability for flood conditions on Brahmaputra basins and highlighted how these parameters are impacting our water balance cycle in snow covered and glaciered areas. The system for real time climate change of uneven climatic impacts on water balance components, that is, downstream movement of river flow, overland flow within the catchment, streamflow generated from rainfall and snow-melt, precipitation pattern, precipitation intensity, temporal and permanent snow cover variation, etc. are areas of concern, which need to be analyzed properly to minimizing their impacts on the environment. The availability of various distributed, stochastic, mathematical and artificial intelligence based models can pro-vide a significant scope in analyzing extreme events simulation and its parameterization under special emphasis of passive and microwave remote sensing. The parameterization and uncer-tainty analysis approach would add more advantage in analyz-ing factors, which are responsible for extreme events and climatic uneven conditions [1]. There are a number of algorithms dealing with how to include different types of uncertainty in calibration procedures such as maximum likelihood- based models (e.g., Reference 68), Bayesian approach [42] and its extension to the generalized likelihood uncertainty estimation (GLUE) approach [8], parameter solution (ParaSol) [78], sequential uncertainty fit-ting (SUFI2) [2] and Markov chain Monte Carlo (MCMC) [42]. These parameterization and uncertainty analysis procedures are fully capable to explore the climatic uncertainties in excess run-off and extreme event analysis.

76.3 Vulnerability of Drinking Water to Climate Change

The surface water availability and shallow groundwater wells depend on the seasonality and interannual variability of stream-flow, and a secured water supply is determined by seasonal low flows. In snow-dominated catchments, higher temperatures lead to reduced streamflow and thus decreased water supply in sum-mer [6]. The factors that determine its water supply potential and level of scarcity can be geo-hydrological; hydrological; techno-logical, managerial, economic and political. At early stages of development, the water needs of a city are generally met through development of local groundwater resources, and diversion of water from lakes, ponds, rivers, and tanks.

Urban drinking water demand is rapidly growing in India due to high growth in urban population and rapid industrial-ization. Drinking water supplies from local water resources including aquifers are falling far short of the high and concen-trated demands in most high dense populated cities like Delhi, Chennai, Kolkata, Mumbai, Bangalore, Hyderabad, and so on. The magnitude of challenge to India's future water resources planning and management would be largely determined not so much by its population growth. The three other factors, namely, (1) the source of this growth whether rural or urban, (2) where this growth is likely to occur whether in water-scarce regions or water rich regions and (3) whether the growth is going to come from increase in urban centers or faster growth of the existing

urban areas. During the period from 1901 to 2001, the average annual compounded growth rate in urban population was 1.4 times higher than that of the total population in the country [28]. The National Water Policy 2002 [28] has prioritized drink-ing water over agriculture and industrial water needs. A sig-nificant share of the water from public reservoirs is normally allocated for domestic water supplies in cities and villages in India during planning. In India, community water supply is the most important requirement and it is about 5% of the total water use. About 7 km³ of surface water and 18 km³ of groundwater are being used for community water supply in urban and rural areas of the country [43]. Different organizations and individuals have given different norms for water supplies in cities and rural areas. The data adopted by the NCIWRD was 220 liters per capita per day (lpcd) for class I cities. For the cities other than class I, the norms are 165 for the 2025 and 220 lpcd for the year 2050. Based on these norms and projection of population, it is estimated that by 2050, water requirements per year for domestic use would be 90 km³ for low demand scenario and 111 km³ for high demand scenario. The estimates indicate that by the year 2025, the water requirement for irrigation would be 561 km³ for low demand scenario and 611 km³ for high demand scenario [43].

76.4 Scenario of the Effect of Climate Change: Application in Drinking Water

India's water resources and management planners are facing significant uncertainties on future demand and availability of water. Vertical processes, evaporation, and precipitation are the important processes that responsible for the water transfer between the hydrosphere and the atmosphere. The water cycle is governed by solar energy mainly through direct vaporiza-tion. Water evaporates from the Earth's surface, vapor is lifted to form clouds and transported in the atmosphere. Water vapor, which has risen from tropical seas to the atmosphere, is carried by winds away from the tropics, where it condenses, releasing latent heat and precipitates over oceans and land. Every year, solar energy lifts about 500,000 km³ of water, evaporating from the Earth's surface, 86% of which (i.e., 430,000 km³) evaporates from the oceanic surface and 14% (i.e., 70,000 km³) from land. About 90% of the volume of water evaporating from oceans pre-cipitates back onto oceans, while 10% is transported to areas over land, where it precipitates [44]. Climate change and its poten-tial hydrological effects are increasingly contributing to this uncertainty. The global annual average temperature is rising day by day mostly by anthropogenic activities. These activities are affecting the radiative forcing into the atmosphere. This will lead to a more vigorous hydrological cycle, with changes in pre-cipitation and evapotranspiration rates. These climate-changing activities have shown their negative effect on global level but this would be more serious and uncertain at region level. These changes will in turn affect water availability and runoff and thus may affect the discharge regime of rivers. The increase of heavy precipitation events, together with higher water temperatures, is likely to exacerbate water quality problems, in particular by accelerating transport of pathogens and other pollutants [44].

Floods and droughts are the main hazardous events in hydrology, which have large impacts on freshwater availability on land and river channels. Besides these quantitative impacts, surface water quality is also affected by climate change. Water pollution through anthropogenic activities is directly linked to urban, industrial or agricultural origin, and climate change could lead to degradation in surface water quality as an indirect consequence of these activities. The climate change determinants especially temperature affect water quality by changing ambient air and soil surface temperature and the increase of extreme hydrological events. The warm climate may increase rate of evaporation and can decrease the water quality level of existing river and lake. Floods and droughts can also modify the water quality by direct effects of dilution or concentration of the dissolved substances. For low river flow rates, the main effect on water quality is as for a temperature increase, a concentration increase of dissolved substances in water but a concentration decrease of dissolved oxygen [55]. For heavy rainfalls and strong hydrologic conditions, runoff and solid material transportation are the main consequences. For countries that are situated in the temperate zone, climate change will decrease the number of rainy days but increase the average volume of each rainfall event [7]. Therefore, drought–rewetting cycles may impact water quality as it enhances decomposition and flushing of organic matter into streams [22].

Climate change may alter the future world's freshwater resources in several aspects, such as freshwater availability, quality, and destructive potential. One of the effects of climate change is that hydrological extremes become more extreme like high intensity rainfall, high rate of evapotranspiration and instant runoff. Many studies have been performed by various researchers around the world for the study of climate change and water resources. Various empirical, mathematical and computer based modeling approaches have been performed to testing and analyzing the effect of climate change on water resources. Regional and global climate scenarios and models are useful tools to produce data inputs for hydrological models in order to understand and predict the potential effects of climate change on water bodies. Considering all these climate change issues in the context of availability of drinking water sources as well as freshwater resources on the land, there is an urgent need to identify the actual problem in appropriate scale in the current trend and initialize new methodology, technology, and tools for implementation as such. The consequential climate change issues which are responsible for the abrupt changes in the freshwater hydrology have been assessed and evaluated by an assortment of meteorological and hydrological parameters, namely, precipitation, temperature, landuse/landcover, soil, slope and water balance components using advance modeling tool, namely, SWAT model (soil and water assessment tool) as per the availability of real time data in the special context of climate change future scenarios.

76.4.1 Study Area and Data Used

To evaluate the climate change impact on the freshwater availability and water balance components of the hydrological cycle, a part of Satluj basin (Rampur to Kasol) has been considered for the current analysis (Figure 76.3). This study area is situated on the Indian Himalayan region and the outlet point of the study area is situated on the snout of Bhakra dam. For this evaluation, various water balance components have been taken into account, namely precipitation, surface runoff, overland flow, evapotranspiration, etc. SWAT model has shown its adaptability in climate change studies for the evaluation of current and future scenarios. In this study, three scenarios have been generated for evaluation of the surface water availability based on the most important hydrological components, that is, precipitation, temperature, evaporation, runoff and water yield (subbasin wise), etc. during the years 1991 to 2008. The current scenario represents the present water resource availability as per the present climatic conditions where as the second and third scenarios have been generated based on the future rainfall intensity (increased 10% and 20%) and temperature conditions (increased as 2.5°C and 4.5°C).

76.4.2 Advance Hydrological Model SWAT

SWAT (soil and water assessment tool) is one of the most recent models developed jointly by the United States Department of Agriculture–Agricultural Research Services (USDA–ARS) and Agricultural Experiment Station in Temple, Texas. It is a physically based, continuous-time, long-term simulation, lumped parameter, deterministic and originated from agricultural models, that can be grouped into eight major divisions: hydrology, weather, sedimentation, soil temperature, crop growth, nutrients, pesticides and agricultural management. SWAT model uses physically based inputs such as weather variables, soil properties, topography, vegetation, and land-management practices as per the catchment's property. The physical processes associated with streamflow, sediment transport, crop growth, nutrient cycling, etc. can be directly modeled into the SWAT. In SWAT, a basin is delineated into subbasins, which are the combinations of the weightage average of the hydrological response units (HRUs). HRUs can be defined by the homogeneous land use and soil types and may either represent different parts of the subbasin with a dominant land use or soil types (also management characteristics). The hydrologic cycle as simulated by SWAT is based on the water balance equation (Equation 76.1):

$$SW_t = SW_a + \sum_{i=1}^{n}(R_{day} - Q_{surf} - E_a - W_{seep} - Q_{gw}) \quad (76.1)$$

where SW_t is the final soil water content (mm H_2O), SW_a the initial soil water content (mm H_2O), t time in days, R_{day} amount of precipitation on day i (mm H_2O), Q_{surf} the amount of surface run-off on day i (mm H_2O), E_a the amount of evapotranspiration on day i (mm H_2O), W_{seep} the amount of percolation and bypass

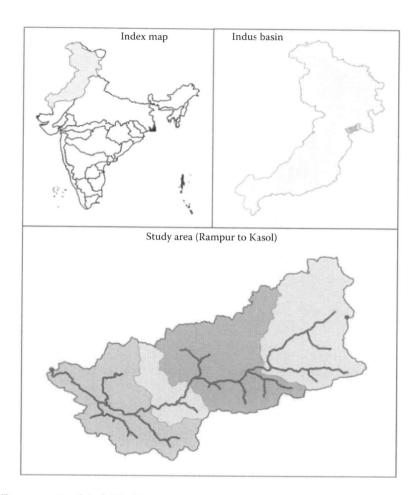

FIGURE 76.3   Study area (Rampur to Kasol, Satluj Basin).

exiting the soil profile bottom on day i (mm H_2O) and Q_{gw} is the amount of return flow on day i (mm H_2O). Surface run-off is computed using a modification of the SCS curve number [77] or the Green and Ampt infiltration method. Surface runoff volume predicted in SWAT using SCS curve number method is given in Equation 76.2:

$$Q_{surf} = \frac{(R_{day} - 0.2S)^2}{(R_{day} + 0.8S)} \qquad (76.2)$$

where Q_{surf} is the accumulated runoff or rainfall excess (mm), R_{day} is the rainfall depth for the day (mm), and S is retention parameter (mm). Runoff will occur when $R_{day} > 0.2S$. The retention parameter varies spatially due to changes in soil, land use, management, and slope and temporally due to changes in soil water content. The retention parameter is defined as (Equation 76.3):

$$S = 25.4\left(\frac{1000}{CN} - 10\right) \qquad (76.3)$$

76.4.3 Selection of Input Parameter

The major inputs, namely, digital elevation model (DEM), land-use/land-cover, soil data, slope map, and hydro-meteorological data like daily rainfall (mm), minimum and maximum daily temperature (°C), wind speed (Km/h), relative humidity (%), and solar radiation have been used for the initial SWAT model set-up. The DEM was used as an initial parameter for generating slope and drainage based on the pour point (generally known as outlet points). The other subbasin parameters such as slope gradient, slope length of the terrain, and stream network characteristics such as channel slope, channel length, and channel width were derived from DEM processing in the SWAT model. For this study, SRTM (shuttle radar topographic mission) DEM, LUC map (downloaded from BHUWAN portal of ISRO, India), daily precipitation, and daily minimum–maximum temperature (IMD 1° gridded data) have been used for set up of the SWAT model.

76.4.4 Analysis and Discussion

Hydrological parameters, namely, precipitation, temperature, and physical parameters, such as landuse/landcover, digital elevation model (DEM) [25], slope, and soil parameters have been used for the generation of the water balance components and scenarios. For the evaluation of the water balance components, SWAT model has been setup based on these parameters. SWAT model generates the outcome, namely, subbasin-wise as

TABLE 76.6 Water Balance Components of the Study Area during Different Scenarios (1991–2008)

Subbasins	Basin Area (km²)	Cumulative Precipitation (mm)	Cumulative Water Yield (mm)	Ratio
		Current Scenarios		
Subbasin 1 (inlet)	878.79	14,350.807	5645.468	0.39
Subbasin 2 (outlet)	672.53	22,158.821	12,422.564	0.56
Subbasin 3	1035.5	14,279.561	5344.541	0.37
Subbasin 4	709.71	14,341.356	4936.507	0.34
		Second Scenarios		
Subbasin 1 (inlet)	878.79	16,508.572	6805.944	0.41
Subbasin 2 (outlet)	672.53	25,488.726	15,261.089	0.60
Subbasin 3	1035.5	16,426.608	6387.237	0.39
Subbasin 4	709.71	16,497.683	6220.931	0.38
		Third Scenarios		
Subbasin 1 (inlet)	878.79	24,755.79	13,219.598	0.53
Subbasin 2 (outlet)	672.53	38,228.459	26,163.223	0.68
Subbasin 3	1035.5	24,632.872	12,704.759	0.52
Subbasin 4	709.71	24,739.467	12,315.438	0.50

well as cumulative on the outlet of the watershed (Table 76.6). The whole study area of Satluj basin (Rampur to Kasol) has been divided into four subbasins or watersheds (Figure 76.3). Three climate-changing scenarios, that is, current, second, and third as per the different climatic conditions, have been generated over the subbasins and at the basin outlet point. The water balance parameters and components have been generated on a daily, monthly, and yearly basis. The SWAT model has been set up initially for the period of 1991–2008. Initially, precipitation and temperature-based scenarios have been generated and it has been analyzed that the incremented precipitation intensity and temperature have shown their influence on the increased rate of evapotranspiration, streamflow runoff, and water yield (Figures 76.4 and 76.5). These climatic uncertainties and changing pattern would create serious threat to the availability of water balance parameters. The high rate of

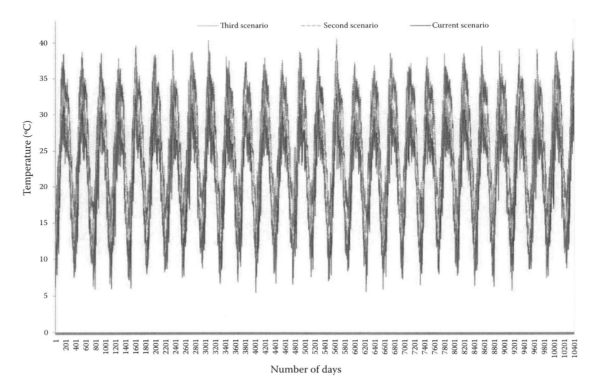

FIGURE 76.4 Daily average temperature at the basin inlet for different scenarios.

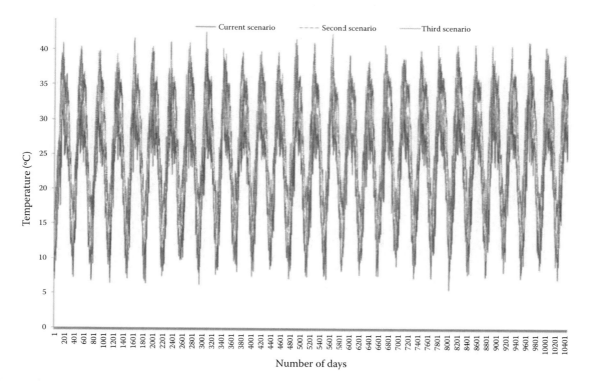

FIGURE 76.5 Daily average temperature at the basin outlet for different scenarios.

evapotranspiration due to increasing temperature will decrease the quantity of freshwater reserves on the land. The changing pattern and intensity of precipitation and its volume will affect the groundwater recharge rate and decrease the quality and quantity of surface water as well as it will increase the sea level that is also a serious threat to the coastal aquifer system and freshwater lakes.

The current climate scenarios for precipitation, evapotranspiration, runoff, and water yield have been generated initially for the period of 1991 to 2008 (Figures 76.6 through 76.14). Based on

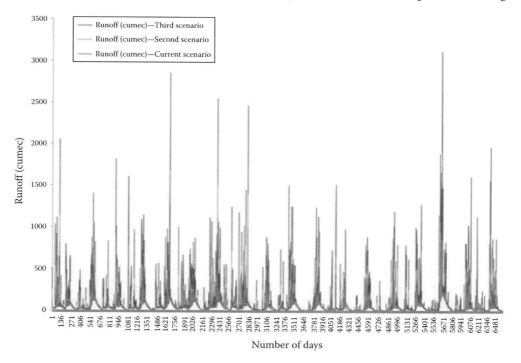

FIGURE 76.6 Model simulated daily runoff at the subbasin outlet for different scenarios.

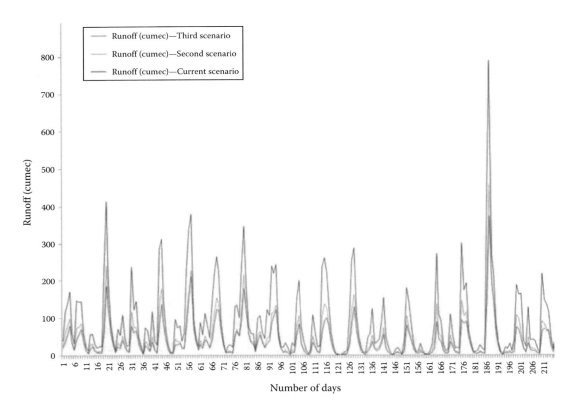

FIGURE 76.7 Model simulated monthly runoff at the subbasin outlet for different scenarios.

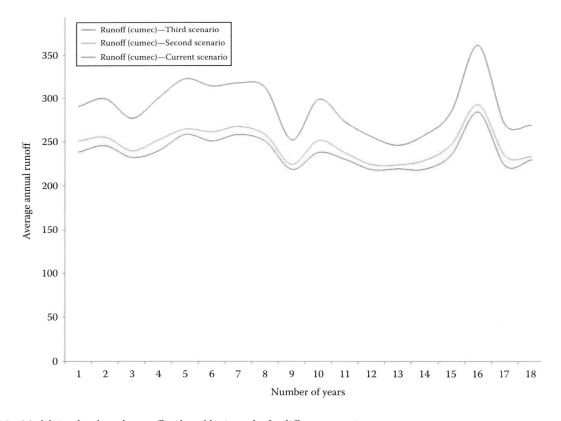

FIGURE 76.8 Model simulated yearly runoff at the subbasin outlet for different scenarios.

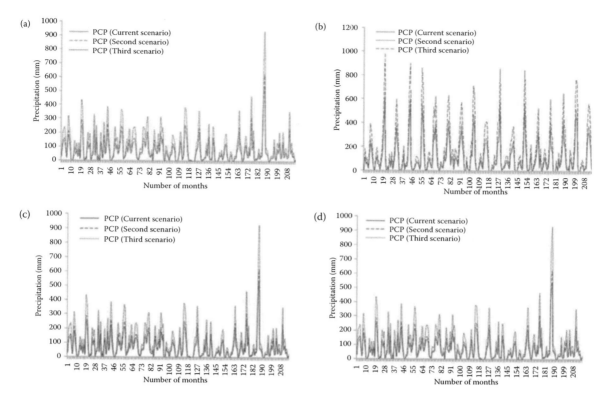

FIGURE 76.9 Subbasin wise model simulated monthly precipitation scenarios: (a) subbasin 1, (b) subbasin 2, (c) subbasin 3, and (d) subbasin 4.

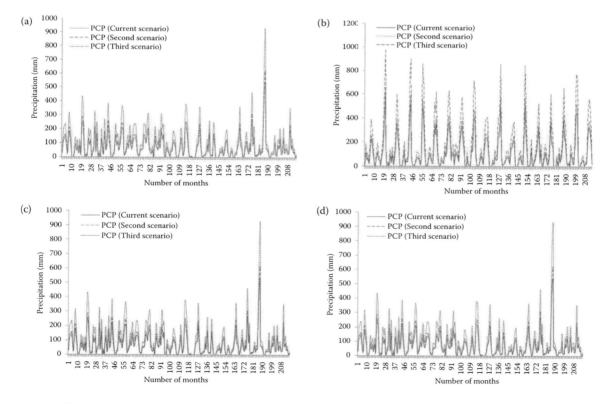

FIGURE 76.10 Subbasin wise model simulated monthly evapotranspiration scenarios: (a) subbasin 1, (b) subbasin 2, (c) subbasin 3, and (d) subbasin 4.

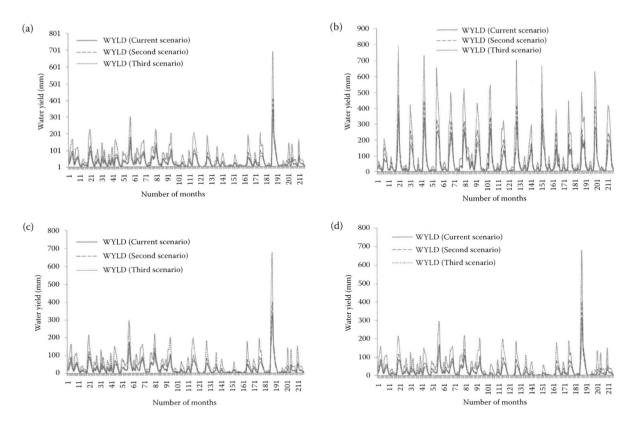

FIGURE 76.11 Subbasin wise model simulated monthly water yield scenarios: (a) subbasin 1, (b) subbasin 2, (c) subbasin 3, and (d) subbasin 4.

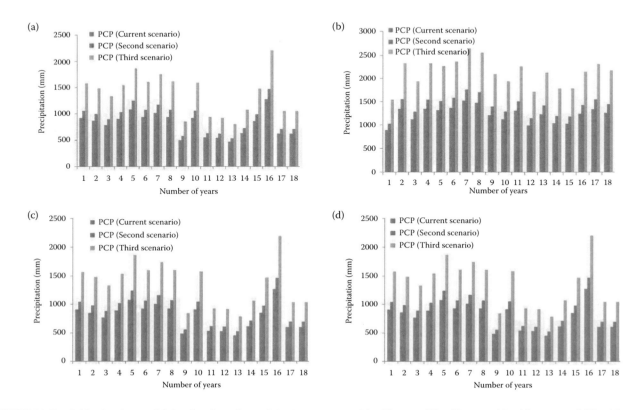

FIGURE 76.12 Subbasin wise model simulated yearly precipitation scenarios: (a) subbasin 1, (b) subbasin 2, (c) subbasin 3, and (d) subbasin 4.

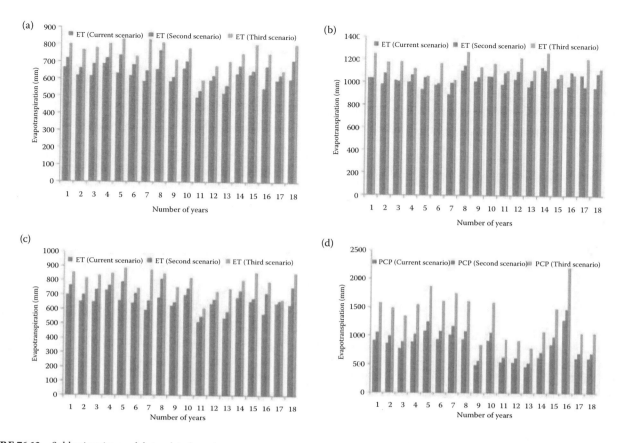

FIGURE 76.13 Subbasin wise model simulated yearly evapotranspiration scenarios: (a) subbasin 1, (b) subbasin 2, (c) subbasin 3, and (d) subbasin 4.

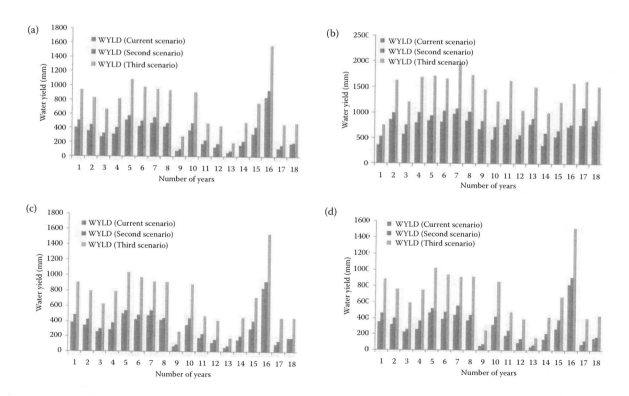

FIGURE 76.14 Subbasin wise model simulated yearly water yield scenarios: (a) subbasin 1, (b) subbasin 2, (c) subbasin 3, and (d) subbasin 4.

the current outcomes of the water balance components, second and third scenarios have been generated in the context of changing climatic conditions. In the second scenario, 10% increment in precipitation and 2.5°C increased temperature have been taken for generating water balance scenarios, similarly for the third scenario 20% increment in the precipitation intensity and 4.5°C increment have been considered as per the different climatic scenarios generated by the IPCC [16]. The IPCCs Special Report on Emissions Scenarios (SRES) was published in 2001 [34], and contains a set of new projections of future greenhouse gas emissions: these projections supersede the IS92 family of projections. The different scenarios generated for the hydrological parameters, namely, precipitation, evapotranspiration, runoff, and water yield clearly show that the increased temperature has shown its effect on precipitation, rate of evapotranspiration, streamflow runoff, and water yield.

76.5 Summary and Conclusions

The aim of this study is to present results of an assessment of the implications of climate change for the global and regional numbers of people living in water-stressed watersheds. The climatic effects have been assessed on the availability and source of freshwater resources using multiple hydrological parameters after generating different scenarios, which are then compared with the current conditions. The outcomes of this study clearly show that the current climatic conditions would be more serious in coming days and it will create a serious threat to the freshwater reserves and sources on the land. Uniquely, this study had been tested if simulated hydrological characteristics would be appropriate for analysis climate change impacts in the watershed level on daily, monthly, and yearly basis. The outcomes of the study allow us to conclude that hydrological factors, namely, precipitation, evapotranspiration, streamflow runoff, and water yield in our hypothesis are significant in estimating climate changing impacts. It had been observed that the increasing annual average temperature impacted the natural streamflow phenomenon in streams and rivers. It is also concluded that the variations in precipitation pattern and intensity are significantly affected by the availability of surface water resources by decreasing the recharge rate of groundwater due to increased water yield. The most important conclusion is that the increasing global annual average temperature would be more critical at the regional or watershed level.

References

1. Abbaspour, K.C. 2011. SWAT-CUP4: SWAT Calibration and Uncertainty Programs—A User Manual. Swiss Federal Institute of Aquatic Science and Technology, Eawag.
2. Abbaspour, K.C., Johnson, A., and Genuchten, V.M.Th. 2004. Estimating uncertain flow and transport parameters using a sequential uncertainty fitting procedure. *Vadose Zone Journal*, 3(4), 1340–1352.
3. Agrawal, G.D. 1999. Diffuse agricultural water pollution in India. *Water Science and Technology*, 39, 33–47.
4. Allan, J.A. 1993. Fortunately there are substitutes for water otherwise our hydro-political futures would be impossible. In: *Priorities for Water Resources Allocation and Management*. ODA, London, pp. 13–26.
5. Allen, D.M., Mackie, D.C., and Wei, M. 2004. Groundwater and climate change: A sensitivity analysis for the Grand Forks aquifer, southern British Columbia, Canada. *Hydrogeology Journal*, 12, 270–290.
6. Barnett, T.P., Malone, R., Pennell, W., Stammer, D., Semtner, B., and Washington, W. 2004. The effects of climate change on water resources in the West: Introduction and overview. *Climatic Change*, 62, 1–11.
7. Bates, B.C., Kundzewicz, Z.W., Wu, S., and Palutikof, J.P. 2008. Climate change and water. Technical paper of the Intergovernmental Panel on Climate change, Geneva. IPCC Secretariat.
8. Beven, K. and Binley, A. 1992. The future of distributed models—Model calibration and uncertainty prediction. *Hydrological Processes*, 6(3), 279–298.
9. Bindoff, N.L., Willebrand, J., Artale, V., Cazenave, A., Gregory, J., Gulev, S., Hanawa, K. et al. 2007. Observations: Oceanic climate change and sea level. In: S. Solomon, D. Qin, M. Manning, Z. Chen, M. Marquis, K.B. Averyt, M. Tignor, and H.L. Miller (eds.), *Climate Change 2007: The Physical Science Basis. Contribution of Working Group I to the Fourth Assessment Report of the Intergovernmental Panel on Climate Change*. Cambridge University Press, Cambridge, United Kingdom.
10. Bisht, B.S. and Tiwari, P.C. 1996. Land use planning for sustainable resource development in Kumaon Lesser Himalaya: A study of Gomti Watershed. *International Journal of Sustainable Development and Ecology*, 3, 23–34.
11. Bobba, A.G., Singh, V.P., and Bengtsson, L. 1997. Sustainable development of water resources in India. *Environmental Management*, 21, 367–393.
12. Brown, L.E., Hannah, D.M., Milner A.M., Soulsby, C., Hodson, A.J., and Brewer, M.J. 2006. Water source dynamics in a glacierized alpine river basin (Taillon-Gabietous, French Pyrenees). *Water Research* 42(8), 1–12. doi: 10.1029/2005wr004268.
13. CDKN Asia. 2012. Agriculture and climate change in India. Climate change and development knowledge network (CDKN). Asia. http://cdkn.org/2012/01/agriculture-and-climate-change-in-india/.
14. CGWB. 2006. *Dynamic Ground Water Resources of India*. Central Ground Water Board, Faridabad.
15. Chaturvedi, M.S. 1985. Water resources of India—An overview. *Sadhana*, 8, 13–38.
16. Climate Change. 2001. The scientific basis, summary for policy makers and technical summary of the working group I report. Intergovernmental Panel on Climate Change, Geneva, Switzerland, 2001.

17. Climate Change. 2001. In: J.T. Houghton, et al. (eds.), *Contribution of Working Group I to the Third Assessment Report of the Intergovernmental Panel on Climate Change.* Cambridge University Press, Cambridge, UK, p. 881.

18. Climate Change Policies in the Asia-Pacific. 2008. Re-uniting climate change and sustainable development. *IGES*, 159–180.

19. Cohen, B. 2004. Urban growth in developing countries: A review of current trends and a caution regarding existing forecasts. *World Development*, 32(1), 23–51.

20. CWC. 2005. Central Water Commission of India. www.cwc.gov.in.

21. Datta, B. and Singh, V.P. 2004. Hydrology. In: V.P. Singh, N. Sharma, C. Shekhar, and P. Ojha (eds.), *The Brahmaputra Basin Water Resources.* Kluwer Academic Publishers, Netherlands, pp. 139–195.

22. Evans, C.D., Monteith, D.T., and Cooper, D.M. 2005. Long-term increases in surface water dissolved organic carbon: Observations, possible causes and environmental impacts. *Environmental Poll*, 137, 55–71.

23. FAO (Food and Agriculture Organization). 2003. World Agriculture Towards 2015/2030. http://www.fao.org/documents/show_cdr.asp?url_file=/docrep/004/y3557e/y3557e00.htm.

24. Ghosh, S. and Dutta, S. 2010. Impact of climate and land use changes on the flood characteristics of the Brahmaputra basin. In: *Proceedings of National Conference on Hydraulics, Water Resources, Coastal and Environmental Engineering (HYDRO-2010)*, December 16–18, 2010, MMU, Mullana, India.

25. GLCF. 2005. *Shuttle Radar Topography Mission (SRTM) Technical Guide.* University of Maryland.

26. Goel, P.S., Datta, P.S., Rama, Sanghal, S.P., Kumar, H., Bahadur, P., Sabherwal, R.K., and Tanwar, B.S. 1975. Tritium tracer studies on ground water recharge in the alluvial deposits of Indo-Gangetic plains of western U.P., Punjab and Haryana. In: R.N. Athavale (ed.), *Proceedings of Indo-German Workshop on Approaches and Methodologies for Development of Ground Water Resources.* NGRI, Hyderabad, pp. 309–322.

27. Gosain, A.K. and Rao, S. 2003. Impacts of climate on water sector. In: P.R. Shukla et al. (eds.), *Climate Change and India: Vulnerability Assessment and Adaptation.* Universities Press (India) Pvt Ltd, Hyderabad, p. 462.

28. Government of India. 2001. Integrated water resource development: A plan for action. Report of the National Commission for integrated water resources development. Ministry of Water Resources, Government of India, New Delhi.

29. Hammond, D. and Pryce, A.R. 2007. Climate change impacts and water temperature. Environment Agency Science Report SC060017/SR, Bristol, UK.

30. Hassan, H., Aramaki, T., Hanaki, K., Matsuo, T., and Wilby, R.L. 1998. Lake stratification and temperature profiles simulated using downscaled GCM output. *Journal of Water Science and Technology*, 38, 217–226.

31. ICIMOD. 2011. *The Status of Glacier in the Hindu-Kush Himalayan Region.* International Centre for Integrated Mountain Development, Kathmandu, Nepal.

32. Ikebuchi, S., Tanaka, K., Ito, Y., Moteki, Q., Souma, K., and Yorozu, K. 2007. Investigation of the effects of Urban Heating on the heavy rainfall event by a cloud resolving model CReSiBUC. *Annuals. of Disease Prevention Research Institute Kyoto University*, No. 50C, 99.

33. India-WRIS. 2012. *River Basin Atlas of India.* RRSC-West, NRSC, ISRO, Jodhpur, India.

34. IPCC. 2001. *Climate Change 2001: The Scientific Basis.* Contribution of Working Group I to the Third Assessment Report of the Intergovernmental Panel on Climate Change. Houghton, J.T., Ding, Y., Griggs, D.J., Noguer, M., van der Linden, P.J., Dai, X., Maskell, K., and Johnson, C.A. (eds). Cambridge University Press, Cambridge, United Kingdom and New York, NY, USA; 881pp.

35. Ives, J. and Messerli, B. 1989. *The Himalayan Dilemma: Reconciling Development and Conservation.* Routledge, London.

36. Jain, S.K. 2011. Population rise and growing water scarcity in India—Revised estimates and required initiatives. *Current Science*, 101, 271–276.

37. Jain, S.K. 2012. Sustainable water management in India considering likely climate and other changes. *Current Science*, 102(2), 177–188.

38. Jain, S.K., Goswami, A., and Saraf, A.K. 2008. Accuracy assessment of MODIS, NOAA and IRS data in snow cover mapping under Himalayan conditions. *International Journal of Remote Sensing*, 29, 5863–5878.

39. Jalota, S.K. and Arora, V.K. 2002. Model-based assessment of water balance components under different cropping systems in north-west India. *Agricultural Water Management*, 57(1), 75–87.

40. Kalipada, C. 2010. *Water Resources of India.* Climate Change Centre Development Alternatives. www.climatechangecentre.net.

41. Kaser, G., Juen, I., Georges, C., Gomez, J., and Tamayo, W. 2003. The impact of glaciers on the runoff and the reconstruction of mass balance history from hydrological data in the tropical Cordillera Blanca, *Peru Journal of Hydrology*, 282, 130–144.

42. Kuczera, G. 1983. Improved parameter inference in catchment models 1. Evaluating parameter uncertainty. *Water Research*, 19(5), 1151–1162.

43. Kumar, R., Singh, R.D., and Sharma, K. D. 2005. Water resources of India. *Current Science*, 89, 794–811.

44. Kundzewicz, Z.W., Mata, L.J., Arnell, N., Doll, P., Kabat, P., Jiménez, B., Miller, K., Oki, T., Sen, Z., and Shiklomanov, I. 2007. Freshwater resources and their management. In: M.L. Parry, O.F. Canziani, J.P. Palutikof, C.E. Hanson, and P.J. van der Linden, (eds.), *Climate Change 2007: Impacts, Adaptation and Vulnerability. Contribution of Working Group II to the Fourth Assessment Report of the Intergovernmental Panel on Climate Change.* Cambridge

University Press, Cambridge, UK, pp. 173–210. http://www.ipcc.ch/pdf/assessment-report/ar4/wg2/ar4-wg2-chapter3.pdf.

45. Mahajan, G., Singh, S., and Chauhan, B.S. 2012. Impact of climate change on weeds in the rice–wheat cropping system. *Current Science*, 102, 1254–1255.

46. Mall, R.K., Gupta, A., Singh, R., Singh, R.S., and Rathore, L.S. 2006. Water resources and climate change: An Indian perspective. *Current Science*, 90(12), 1610–1626.

47. McDonald, R.I., Green, P., Balk, D., Fekete, B., Revenga, C., Todd, M., and Montgomery, M. 2011. Urban growth, climate change, and freshwater availability. *Proceedings of the National Academy of Sciences*, 108(15), 6312–6317.

48. Mishra, A.K. 2011. Impact of Urbanization on the Hydrology of Ganga Basin (India). *Water Resource Management*, 25, 705–719.

49. Mool, P.K., Bajracharya, S.R., and Joshi, S.P. 2001. Inventory of glaciers, glacial lakes, and glacial lake outburst flood monitoring and early warning systems in the Hindu Kush-Himalayan region: Nepal. ICIMOD, Kathmandu, Nepal.

50. Naik, P.K., Tambe, J.A., Dehury, B.N., and Tiwari, A.N. 2008. Impact of urbanization on the groundwater regime in a fast growing city in central India. *Environmental Monitoring and Assessment*, 146, 339–373.

51. National water policy. 2002. *Ministry of Water Resources.* Government of India, New Delhi.

52. PanIIT Conclave. 2010. Water sector in India: Overview and focus areas for the future. kpmg.com/in.

53. Pant, G.B., Rupakumar, K., and Borgaonkar, H.P. 1999. In: Dash, S.K., Bahadur, J. (eds.), *The Himalayan Environment.* New Age International (P) Ltd, New Delhi, pp. 172–184.

54. Panwar, S. and Chakrapani, G.J. 2013. Climate change and its influences on groundwater resources. *Current Science*, 105(1), 37–46.

55. Prathumratana, L., Sthiannopkao, S., and Kim, K.W. 2008. The relationship of climatic and hydrological parameters to surface water quality in the lower Mekong River. *Environment International*, 34, 860–866.

56. Raju, N.J. and Reddy, T.V.K. 2007. Environmental and urbanization affect on groundwater resources in a pilgrim town of Tirupati, Andhra Pradesh, South India. *Journal of Applied Geochemistry*, 9(2), 212–223.

57. Ramachandra T.V. and Kumar, U. 2010. Greater Bangalore: Emerging Urban Heat, Island, *GIS Development*, 14(01), 1–16.

58. Ramakrishnan, P.S. 1999. In: Dash, S.K., Bahadur, J. (eds.), *The Himalayan Environment.* New Age International (P) Ltd, New Delhi, pp. 213–225.

59. Rangarajan, R. and Athavale, R.N. 2000. Annual replenishable ground water potential of India—An estimate based on injected tritium studies. *Journal of Hydrology*, 234, 38–53.

60. Rawat, J.S. 2009. Saving Himalayan Rivers: developing spring sanctuaries in headwater regions. In: Shah, B.L. (ed.), *Natural Resource Conservation in Uttarakhand.* AnkitPrakshan, Haldwani, pp. 41–69.

61. Richard, D. and Gay, M. 2003. Guidelines for scientific studies about glacial hazards. Survey and prevention of extreme glaciological hazards in European mountainous regions. *Glaciorisk Project, Deliverables.* http://glaciorisk.grenoble.cemagref.fr.

62. Rosenzweig, C., Tubiello, F.N., Goldberg, R., Mills, E., and Bloomfield, J. 2002. Increased crop damage in the US from excess precipitation under climate change. *Global Environmental Change*, 12, 197–202.

63. Rupakumar, K., Krishna Kumar, K., and Pant, G.B. 1994. *Geophysical Research Letters*, 21, 677–680.

64. Sharma, E., Bhuchar, S., Xing, M., and Kothyari, B.P. 2007. Land use change and its impact on hydroecological linkages in Himalayan watersheds. *Tropical Ecology*, 48(2), 151–161.

65. Shrestha, A.B., Wake, C.P., Mayewski, P.A., and Dibb, J.E. 1999. Maximum temperature trends in the Himalaya and its vicinity: An analysis based on temperature records from Nepal for the period 1971–94. *Journal of Climate*, 12, 2775–2767.

66. Shrestha, B.B., Nakagawa, H., Kawaike, K., Baba, Y., and Zhang, H. 2010. Glacial Lake Outburst due to Moraine Dam Failure by Seepage and Overtopping with Impact of Climate Change. *Annuals of Disaster PreventionResearch Institute, Kyoto University*, 53B, 569–582.

67. Smakhtin, V. 2008. Basin closure and environmental flow requirements. *Water Resources Development*, 24(2), 227–233.

68. Sorooshian, S. and Gupta, H.V. 1983. Automatic calibration of conceptual rainfall-runoff models: the question of parameter observability and uniqueness. *Water Resource Research*, 19(1), 260–268.

69. Srinivasan, V., Seto, K.C., Emerson, R., and Gorelick, S.M. 2013. The impact of urbanization on water vulnerability: A coupled human–environment system approach for Chennai, India. *Global Environmental Change*, 23(2013), 229–239.

70. Tiwari, P.C. 1995. *Natural Resources and Sustainable Development in Himalaya.* Shree Almora Book Depot, Almora.

71. Tiwari, P.C. 2010. Land use changes and conservation of water resources in Himalayan Headwaters. In: *Proceedings of the 2nd German-Indian Conference on Research for Sustainability: Rnergy & Land Use*, Department of Science & Technology, Government of India, pp. 170–74.

72. Tiwari, P.C. and Joshi, B. 2012. Environmental changes and sustainable development of water resources in the Himalayan Headwaters of India. *Water Resource Manage*, 26, 883–907.

73. Trenberth, K.E. 1999. Short-term climate variations. Recent accomplishments and issues for future progress. *Storms*, 1, 126–141.

74. Troll, C. 1965. Seasonal climates of earth-world maps of climatology. In: Rodenwaldt, E., Jusatz, H. (eds.), *Seasonal Climates of the Earth*. Springer, Berlin, p. 28.

75. UNDP. 2006. *Beyond Scarcity: Power, Poverty and the Global Water Crisis*. Human Development Report 2006 by the United Nations Development Programme, New York.

76. United Nations. 1993. *World Urbanization Prospects: the 1992 Revision*. UN, New York.

77. USDA. 1972. *National Engineering Handbook: Hydrology*. USDA, Soil Conservation Service. U.S. Government Print Office, Washington, DC.

78. Van Griensven, A. and Meixner, T. 2006. Methods to quantify and identify the sources of uncertainty for river basin water quality models. *Water Science and Technology*, 53(1), 51–59.

79. Vazquez-Sune, E., Sanchez-Vila, X., and Carrera, J. 2005. Introductory review of specific factors influencing urban groundwater, an emerging branch of hydrogeology, with reference to Barcelona, Spain. *Hydrogeology Journal*, 13(3), 522–533.

80. Veijalainen, N., Dubrovin, T., and Marttunen, M.V. 2010. Climate change impacts on water resources and lake regulation in the Vuoksi watershed in Finland. *Water Resource Manage*, 24(13), 3437–3459. doi: 10.1007/s11269-010-9614-z.

81. Viviroli, D. and Weingartner, R. 2004. The hydrological significance of mountains: from regional to global scale. *Hydrology and Earth System Sciences*, 8(6), 1016–1029.

82. Vorosmarty, C.J., McIntyre, P.B., Gessner, M.O., Dudgeon, D., Prusevich, A., Green, P., Glidden, S. et al. 2010. Global threats to human water security and river biodiversity. *Nature*, 467, 555.

83. Wasson, R.J., Juyal, N., Jaiswal, M., McCulloch, M., Sarin, M.M., Jain, P., Srivastava, P., and Singhvi, A.K. 2008. The mountain-lowland debate: Deforestation and sediment transport in the upper Ganga catchment. *Journal of Environmental Management*, 88(1), 53–61.

77

The Impact of Climate Change on the Incidence of Infectious Waterborne Disease

Jean O'Dwyer
University of Limerick

Aideen Dowling
University of Limerick

Catherine Adley
University of Limerick

PREFACE

Anthropogenic climate change is currently one of the biggest threats to humankind; with increases in average global temperature predicted to affect natural systems with detrimental consequences to human health. This chapter focuses on climate change and waterborne infectious disease. Meteorological phenomena can play a leading role in waterborne infectious disease epidemiology. The impact of meteorology on waterborne disease is moderated through the effects of rainfall frequency and intensity as well as temperature—both air and water temperatures—both of which are discussed within this section. Also investigated are emerging waterborne infectious diseases, which may prove to be problematic in coming years and present problems to countries unequipped to monitor, report, and moderate diverse waterborne pathogens. This chapter provides a focus on Europe and offers recommendations for adaptability, prevention, and preparation, which can be implemented at a global level.

77.1 Introduction

77.1.1 Climate Change and Human Health

Over the past century, acceleration in economic activity coupled with a staggering increase in fossil fuel consumption has led to an environmental impact of unprecedented proportions; this is termed climate change. There is near unanimous scientific consensus, for example, 97% of publishing scientists agree [9] that rising atmospheric concentrations of greenhouse gases (GHGs) attributed to anthropogenic emissions causing warming (and other climatic changes) at Earth's surface. Climatological research over the past two decades makes clear that Earth's climate will change in response to atmospheric GHG accumulation. The unusually rapid temperature rise (0·5°C) since the mid-1970s is substantially attributable to this anthropogenic increase in GHGs [30]. The Intergovernmental Panel on

Climate Change (IPCC), representing the published results of leading climatological modeling groups around the world, forecasts that global mean temperatures will continue to rise over the twenty-first century if GHG emissions continue unabated. The panel makes their predictions based on representative concentration pathways (RCPs). RCPs are four GHG concentration trajectories adopted by the IPCC for its fifth assessment report (AR5). They describe four possible climate futures, all of which are considered possible depending on how much GHGs are emitted in the years to come. The four RCPs (RCP2.6, RCP4.5, RCP6, and RCP8.5) are named after a possible range of radiative forcing values in the year 2100 (+2.6, +4.5, +6.0, and +8.5 W/m², respectively) [45]. Radiative forcing describes the difference of radiant energy received by the Earth and energy radiated back to space. Under the assumptions of the concentration-driven RCPs and with respect to preindustrial conditions, global temperatures averaged in the period 2081–2100 are projected to likely exceed 1.5°C above preindustrial values for RCP4.5, RCP6.0, and RCP8.5 (high confidence) and are likely to exceed 2°C above preindustrial for RCP6.0 and RCP8.5 (high confidence) [43] as shown in Figure 77.1 [1].

Global warming is not happening uniformly. It is occurring faster during the winter months and the rate of warming is more staggering at higher latitudes. In addition, heat is building up at an unprecedented rate in the ocean; down to 3 km [28] and, as a result, water vapor in the atmosphere is increasing [12]. Sea ice and ice shelves are melting, with September 2012 seeing the lowest sea ice extent ever recorded by satellites at

1.32 million square miles (3.41 million square kilometers) [5]. The result of this heating is a measurable, significant change to the hydrological cycle. There is evidence that the climatic system is experiencing instabilities as extreme weather events, such as prolonged droughts and excessive rainfall (>5 cm/day), which have increased in both intensity and frequency. Some of the projected changes, both globally and specifically in Europe, are detailed in Table 77.1.

The climatic system displays complex interactions of interconnected components, including the atmosphere, hydrosphere, cryosphere, biosphere, and geosphere. Changes in global temperature will have adverse effects on most systems and will undoubtedly have implications in human health. Figure 77.2 summarizes some key pathways in which anthropogenic GHG emission will affect the environment and subsequently human health.

Of particular concern is the effect of a changing climate on the incidence of waterborne infectious disease. Global climate change will interfere with interactions within the hydrologic cycle not only by altering mean meteorological measures, but also by increasing the frequency of extreme events such as excessive precipitation, storm surges, floods, and droughts, many of which have already been seen over the past decade. Structural damage as well as the direct loss of human life because of these extreme weather events is highly publicized in the media. However, a potentially more catastrophic eventuality looms in the form of the increase in transmission, frequency, and dispersal of waterborne infectious diseases.

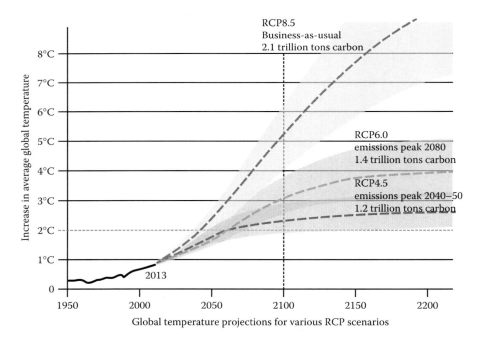

FIGURE 77.1 Global temperature projections for the future based on the representative concentration pathways developed for the intergovernmental panel on climate change report 2005. (Adapted from Architecture 2030, 2012. http://architecture2030.org/files/roadmap_web.pdf. Permission granted December 2, 2013.)

TABLE 77.1   Observed and Global Projections for Climate Change, Globally and in Europe

	Observed and Projected Climate Change Impacts on the Environment and Human Health	
Key Climate Variables	Present Situation	Future Projections
Global temperature	IPCC reports that global average temperature—both land and ocean—has shown an increase of between 0.77°C and 0.80°C in mean temperature as compared to the pre-industrial average.	Further rise in temperature is expected to be between 1.1°C and 6.4°C by 2100.
European temperature	The average temperature for the European Land Area for the decade of 2002–2011 is 1.3°C above preindustrial levels. It was the warmest decade ever recorded with heat waves increasing both in frequency and in length.	Further rise in temperature is expected to be between 2.5°C and 4.0°C by 2100. The largest increases are expected in Northern and Eastern Europe.
Precipitation	Precipitation changes are more spatial and temporal than temperature. Since the 1950s, precipitation has been increasing in Europe throughout the winter but decreasing in parts of southern Europe. In western Europe, intense precipitation events have contributed significantly to the increase.	Most climatic models project a continued increase in precipitation for Northern Europe and decrease in Southern Europe. The number of days with intense precipitation is projected to increase.

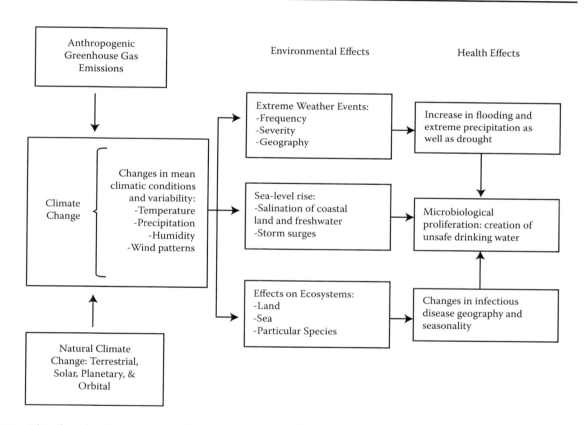

FIGURE 77.2   The effect of anthropogenic emissions of environment and health.

77.2 Waterborne Infectious Diseases

A variety of microorganisms (viruses, bacteria, and protozoans) have the capacity to transmit disease through contact with contaminated water as shown in Table 77.2. The health implications of exposure to pathogenic bacteria, viruses, and protozoa in drinking water are diverse. Worldwide, the most commonly reported manifestation of waterborne disease is in the form of gastrointestinal illness—nausea, vomiting, and diarrhea.

However, in individuals who are more vulnerable to illness, that is, the young, the elderly, and the immunocompromised, the effects can be far more severe. As a result, globally, over 1.5 million people die each year from unsafe water, inadequate sanitation, and insufficient hygiene as a result of diarrheal diseases, schistosomiasis, and others [34].

Transmission of infectious disease is determined by many factors, including extrinsic social, economic, climatic, and ecological conditions [46] and intrinsic human immunity [25]. In

TABLE 77.2   Major Waterborne Disease, Transmission Pathways, and Clinical Features

Organism	Disease	Transmission	Clinical Features
		Protozoa	
Giardia duodenalis	Giardiasis	Fecal oral spread through drinking or recreational water	Diarrhea and abdominal pain, weight loss, and failure to thrive
Cryptosporidium parvum	Cryptosporidiosis	Fecal oral spread through drinking or recreational water	Diarrhea often prolonged
		Bacteria	
Vibrio cholerae	Cholera	Drinking water	Watery diarrhea, may be severe
Salmonella typhi	Typhoid	Drinking Water	Fever, malaise, and abdominal pain with high mortality
Shigella spp.	Shigellosis	Drinking and recreational water	Diarrhea frequently with blood loss
Campylobacter spp.	Camplyobacterosis	Drinking and recreational water	Diarrhea frequently with blood loss
Enterotoxigenic and Enterohemorrhagic *E. coli*		Drinking water	Watery or bloody diarrhea can lead to hemolytic uremic syndrome in children
Legionella	Legionnaires disease	Drinking and recreational water	Watery diarrhea may be severe

relation to urban drinking water, the transmission of waterborne infectious disease can manifest only if the supply has become sufficiently contaminated, or has failed to be treated correctly or the treatment has failed entirely; with an ever-increasing demand on water supplies, these situations are becoming more prevalent. Drinking water may also become contaminated as a result of increases in incidences of heavy rainfall, snow melts, and flooding. These events cause a surge in water flow, causing sewers or wastewater treatment outputs to overflow directly into a surface water body—often this microbiological burden can be too great for water treatment facilities to overcome.

77.3 Meteorology and Waterborne Disease

In terms of climate change, it is important to establish the current impact of water-related illnesses and try to incorporate future predictions to aid policy formulation and improve the adaptive capacity of those nations at greatest risk. To understand the influence of the global meteorological system on water-related disease, we must first distinguish between weather and climate. Weather refers to the short term, usually day to day, meteorological events that take place in local areas. Weather is erratic and variable; changing daily. In contrast, climate usually refers to the "average" conditions for a given location based on its meteorological record (day to day weather over time). While the effects of weather are quite obvious—extreme rainfall events may lead to flooding—changes in climate are more subtle, taking decades to understand and form enough trends to interpret accurately. Accordingly, in relation to our changing climate we must discuss the effects of short-term weather in the form of rainfall and extreme weather events and the more subtle climatic change in temperature. However, while we will treat both meteorological classifications as separate entities, it must not be forgotten that they are interconnected; weather depicting climate and climate influencing weather.

77.4 The Influence on Waterborne Disease

Outbreaks of waterborne disease as a result of the contamination of community water systems have the potential to cause extensive illness among the population; particularly in areas where infrastructure is not up to par. The impact of meteorology on waterborne disease is moderated through the effects of rainfall frequency and intensity as well as temperature; both air and water temperatures. The effects are varied and are pathogen specific, meaning that the risk to human health is entirely a function of local conditions including infrastructure and water treatment capacities. The link between these three parameters must not be understated and is likely to become more evident as climate change scenarios come to fruition. When analyzing the consequences of a changing climate, we must look at both "normal" or "static" weather variables—precipitation, mean air, and water temperature—but, we must also allow some focus on "extreme weather events," which are likely to increase in both frequency and intensity over the coming years. In this chapter, we will discuss the evidence of the impact of climate change on the epidemiology of waterborne disease in relation to three variables:

1. Rainfall
2. Temperature
3. Extreme weather events

77.4.1 Rainfall

One of the most common manifestations of waterborne disease in gastrointestinal illness and many of these enteric diseases that cause this health problem show a seasonal pattern, suggesting that they are sensitive to variations in climatic factors. Levels of precipitation play a pertinent role in influencing the contamination of water and hence increase the vulnerability of drinking water supplies. It is thought that the frequency of heavy precipitation events has increased over many mid-latitude regions since the 1950s, even where there has been a reduction in average precipitation [6]. Of pressing concern is the pollution of water bodies with fecal contamination, particularly from nonpoint sources such as agriculture. Fecal pathogen events in rivers, lakes, and reservoirs have been shown to be associated with rainfall events. Heavy "flash" rainfall or periods of prolonged rainfall can mobilize pathogens within the environment, increasing runoff from agriculture and transporting this microbiologically rich medium into rivers, coastal waters, and groundwater wells [40]. Similarly, during periods of heavy rainfall, a cross-contamination can occur between municipal water and drinking water as water treatment plants may become overwhelmed. This may cause sewage overflow into drinking water supply chains, or may flow directly into local waterways [41]. Lastly, increased levels of rainfall can also lead to flooding. While the structural damages associated with flooding are well reported, the increased exposure to waterborne pathogens is lesser known. This bodes a serious health problem, in both developed and undeveloped countries and can create pandemic outbreaks of infectious disease [15]. Significant research is now available highlighting the links between climate and health. A study based on England and Wales found that 20% of waterborne outbreaks from surface water in the past century were associated with a sustained period of low rainfall, while 10% were associated with heavy rainfall [32]. Similarly, outbreaks of *Cryptosporidium*, *Giardia*, *E. coli*, and other infections have been associated with heavy rainfall events in countries where a public water supply is well regulated [2,10]. Looking to the United States, some of the largest outbreaks of waterborne disease in North America have resulted after notable rainfall events. For example, in May 2000, heavy rainfall in Walkerton, Ontario resulted in approximately 2300 illnesses and 7 deaths after the town's drinking water became contaminated with *E. coli* O157: H7 and *Campylobacter jejuni* [21]. In the United States, from 1948 to 1994, heavy rainfall correlated with more than half of the outbreaks of waterborne diseases [10]. In terms of groundwater aquifers, a recent study in the Republic of Ireland has shown that the strongest predictor of reporting contamination with *E. coli* was rainfall, indicating that contamination was 1.2 times more likely with every 1 mm increase in rainfall [33].

77.4.2 Extreme Weather Events

The impact of extreme weather events (EWEs) on waterborne illness may be widespread and is often a factor in triggering waterborne disease outbreaks [10]. Curriero et al. found that more than half of the reported waterborne disease outbreaks in the United States since the 1950s have followed a period of extreme rainfall. Of these outbreaks, 68% followed severe storms. There is also evidence to support that water-related events like El Nino Southern Oscillation, hurricanes, and typhoons are becoming more frequent, intense, and of greater

duration [29]. Flooding caused by heavy rains or disaster events such as hurricanes or tsunamis can create vast areas of standing water and with it new areas for potential pathogen exposures. It is expected that effects linked to climate change such as elevations in extreme hydrological events may further contribute to water quality impairment by fecal pollution of livestock origin and thus further increase the potential for human exposures and illness. In relation to surface water bodies, an investigation of the effects of wet weather on pathogen and indicator concentrations in an agriculture-intensive watershed in Ontario observed that during storm events, the peak *Campylobacter* concentration arrived earlier than the peak turbidity level. The authors speculated that this was because pathogens are generally in limited supply within a watershed and are therefore more likely to be flushed out of the stream before the turbidity level declines [11]. As mentioned previously, El Niño seasons have been associated with outbreaks of infectious disease; both vector and waterborne, in many areas [18]. El Niño events are predicted to become more common and more severe with an increase in global temperature. As a result of the effects of an El Niño event in 1997, the incidence of infection of patients with diarrheal disease in Lima, Peru, increased significantly due to an increase in temperature [38]. An analysis of the daily hospital data ensued subsequent to the event and was found that the hospital admittance for gastrointestinal symptoms went up by 8% per 1°C increase in temperature [8].

77.4.3 The Impact of Increased Temperature

Transmission of enteric disease can possibly be increased by high temperatures by the direct effect on the growth rate of an organism in the environment [3,4]. The areas affected by drought have increased since the 1970s [23] and this may lead to an increase in exposure to pathogenic microorganisms as a result of pollution concentration in depleted water supplies. In terms of temperature, in general, both salmonella and cholera bacteria, for example, proliferate more rapidly at higher temperatures, salmonella in animal gut and food, and cholera in water [19]. In a technical report produced for the European Centre for Disease Control (ECDC), campylobacteriosis and salmonellosis were cited with the highest frequency in association with air temperature, and campylobacteriosis and non-cholera vibrio infections were reported in association with water temperature.

77.5 Emerging Waterborne Bacterial Pathogens

While climate change is likely to increase the incidence of waterborne infectious disease outbreaks, this is not the only area of concern. Emerging waterborne bacterial pathogens are also an area of immense interest and may burden many countries that have not even been monitoring for these pathogenic organisms to date. Detailed here are emerging pathogens that can be spread through the consumption of drinking water. Worryingly, these organisms do not correlate with the presence of indicator organisms like *E. coli* or coliform bacteria and in most cases there is no satisfactory microbiological indication of their presence in the environment. Further research is needed in order to ascertain the real significance and epidemiology of the diseases caused by these organisms.

77.5.1 *Mycobacterium avium* Complex

The *Mycobacterium avium* complex (MAC) consists of 28 genotypes of two distinct bacterial species: *Mycobacteriumavium* and *Mycobacterium intracellulare*. Like many bacterial pathogens, they are opportunistic and represent the greatest danger to those individuals who are immuocompromised, particularly individuals infected with human immunodeficiency virus (HIV). The organisms have been identified in a diverse range of environmental sources including lakes, streams, rivers springs, soil, piped water supplies, and plants. Importantly, in this context, MAC organisms have been isolated from natural and drinking water distribution systems in the United States [44]. MAC organisms are hardy, surviving under various, normally stressful, conditions. They have been found in water at temperatures of up to 51°C and can grow over a wide pH range. MACs are highly resistant to chlorine and other chemical disinfectants, resulting in standard methods of water treatment not eliminating these organisms from the drinking water distribution system. As MAC organisms grow in biofilms, they represent a real challenge for water quality should they become a common visitor to drinking water systems. The clinical presentation of MAC infections can include a productive cough, fatigue, fever, weight loss, and night sweats.

77.5.2 *Helicobacter pylori*

Helicobacter pylori is commonly associated with gastritis and has been implicated in duodenal and peptic ulcers and stomach cancer. However, most people who become infected by the organism remain asymptomatic [22]. There are no reports that demonstrate *H. pylori* been isolated from an environmental supply, utilizing traditional culture-based method [16]. Instead, molecular methods including fluorescence *in situ* hybridization and polymerase chain reaction (PCR), have been successful in detecting the pathogen. How *H. pylori* is transmitted through human exposure is still not fully understood [27]. The organism has been recovered from saliva, dental plaque, stomach, and fecal samples which could be conclusive of oral–oral or fecal–oral transmission. If this is the case, water may play a significant role in transmission, particularly in situations with improper sanitation and hygiene.

77.5.3 *Aeromonas hydrophyla*

Over the last few years, *A. hydrophila* has been attributed due recognition within the public health sector. It has now been recognized as an opportunistic pathogen, being branded a potential agent of gastroenteritis, septicemia, meningitis, and wound infections [20]. The organism itself is a Gram negative, rod shaped, non-spore forming facultative anaerobic bacilli, native to the family Aeromonadaceae. *A. hydrophila* are abundant in the natural environment and have been isolated from food, drinking water, and aquatic environments [26]. In uncontaminated rivers and lakes, concentrations of *Aeromonas* spp. are usually around 10^2 colony-forming units (CFU)/mL. Groundwater generally contains less than 1 CFU/mL. Drinking water directly leaving the treatment plant has been found to contain between 0 and 10^2 CFU/mL [7]. Drinking water in distribution systems can display higher *Aeromonas* concentrations, due to the growth in biofilms. *Aeromonas* spp. have been found to grow between 5°C and 45°C. *A. hydrophila* is resistant to standard chlorine treatments, possibly persisting inside biofilms [14]. The common routes of infection suggested for *Aeromonas* are the ingestion of contaminated water or food or contact of the organism with a break in the skin. Drinking or natural mineral water can be a possible source of contamination for humans.

77.6 Climate Change and Infectious Disease in Europe

Major causes of diarrheal illness linked to the contamination of drinking water with microorganisms are *Cryptosporidium*, *E. coli*, *Giardia*, *Shigella*, and *Campylobacter*. Cholera is also another pathogenic organism highly associated with drinking water; however, its prevalence in the environment has decreased throughout the late twentieth and early twenty-first century with only 11 confirmed cases in 2006. However, internationally, cholera outbreaks during the warmer months display a seasonal pattern in higher absolute latitudes and climate change might influence the strength, duration, or appearance of such a seasonal pattern [13].

The IPCC predicts that Europe will experience a variance of impacts from global climate change. The variance of impacts is a result of the diverse geographic, demographic, ecological, and socioeconomic conditions in the region. Although the impacts are not uniform across the continent, the annual average temperature and mean precipitation are predicted to experience significant changes overall, with warmer winters predicted for Northern Europe and warmer summers predicted for the South of the continent [17]. Conservative projections foresee global mean air temperatures increasing by 1.8–4.0°C this century, while other models suggest a range of increase of 1.1–6.4°C. The global average temperature is now 0.8°C higher than between 1850 and 1919, while in Europe this average is 1.4°C higher [35]. As to the rate of change, the last decade was the warmest on record, with 2005 being the hottest. The most dramatic increase has been recorded in the Arctic region of Europe, where the temperatures have risen by 3°C over the last 90 years [35]. Northern Europe is also the geographical area with the biggest projected temperature increase according to climate change scenarios [35]. These changes in both temperature and precipitation are likely to influence both water- and foodborne pathogens by altering the effects

TABLE 77.3   Cases of Five Waterborne Diseases Reported in Europe between 1995 and 2010

Year	Campylobacter		Cryptosporidium		VTEC		Giardia		Legionella	
	Number of Cases	Increase/Decrease	Number of Cases	Increase/Decrease	Number of Cases	Increase/Decrease	Number of Cases	Increase/Decrease	Number of Cases	Increase/Decrease
1995	85130.00	0	6814.00	0	3209.00	0	12788.00	0	588.00	0
1996	91285.00	7.23	4070.00	−40.27	3046.00	−5.08	11891.00	−7.01	817.00	38.95
1997	105797.00	15.90	5724.00	40.64	3714.00	21.93	12794.00	7.59	1233.00	50.92
1998	149561.00	41.37	5163.00	−9.80	3597.00	−3.15	11614.00	−9.22	1462.00	18.57
1999	152617.00	2.04	6456.00	25.04	6893.00	91.63	11380.00	−2.01	2263.00	54.79
2000	170065.00	11.43	7833.00	21.33	6847.00	−0.67	10196.00	−10.40	2421.00	6.98
2001	193708.00	13.90	6389.00	−18.43	8675.00	26.70	13833.00	35.67	3763.00	55.43
2002	186780.00	−3.58	4940.00	−22.68	9196.00	6.01	12267.00	−11.32	4791.00	27.32
2003	170218.00	−8.87	8413.00	70.30	9170.00	−0.28	12232.00	−0.29	4503.00	−6.01
2004	182598.00	7.27	6164.00	−26.73	9773.00	6.58	17101.00	39.81	4635.00	2.93
2005	197802.00	8.33	7960.00	29.14	5199.00	−46.80	27240.00	59.29	4094.00	−11.67
2006	175908.00	−11.07	6801.00	−14.56	3262.00	−37.26	16542.00	−39.27	5512.00	34.64
2007	200807.00	14.15	6255.00	−8.03	2908.00	−10.85	14513.00	−12.27	5169.00	−6.22
2008	190579.00	−5.09	7028.00	12.36	3164.00	8.80	18274.00	25.91	5279.00	2.13
2009	198862.00	4.35	8016.00	14.06	3583.00	13.24	16564.00	−9.36	5109.00	−3.22
2010	215058.00	8.14	6605.00	−17.60	3656.00	2.04	16844.00	1.69	5801.00	13.54

of environmental exposure pathways. It is well established that waterborne diseases display a strong correlation with seasonal variation [39], be it through climatic or anthropogenic variation. Although infectious disease outbreaks have been linked to individual weather events, there have been few attempts to detect and attribute temporal trends in infectious diseases to climate change [42]. Many studies have projected future levels of disease spread in response to climate change, but there are currently no means for verifying the accuracy of these models. Unfortunately, waterborne outbreaks have the potential to be rather large and

of mixed etiology, but the actual disease burden in Europe is difficult to approximate and most likely underestimated. Table 77.3 demonstrates the fluctuations in the cases of five water-related infectious diseases for the years of 1995–2010.

Figure 77.3 visually displays the data from Table 77.3, and it can be suggested that there has been an upward trend in reported cases. However, the increases could be attributed to an increase in monitoring regimes, which have become more stringent in recent years. However, it is thought that many cases are still unreported. In 2006, merely 17 waterborne outbreaks were

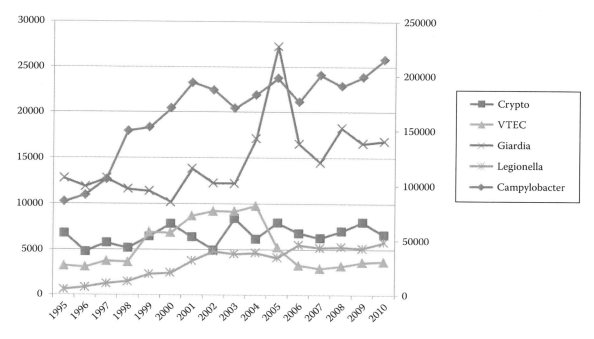

FIGURE 77.3   Graph showing the number of cases of five waterborne diseases as reported to the ECDC from 1995 to 2010.

reported by five countries in Europe, clearly significantly under-reported. They involved a total of 3952 patients, of which 181 were hospitalized, afflicted by a number of causative agents including campylobacter, calicivirus, giardia, and cryptosporidium [40].

Erratic and extreme precipitation events can overwhelm water treatment plants [24] and lead to cryptosporidium outbreaks due to oocysts infiltrating drinking-water reservoirs from springs and lakes and persisting in the water distribution system. In Europe, flooding has rarely been associated with an increased risk of waterborne disease outbreaks, but a few exceptions exist in the UK [37], Finland [31], and Sweden [36]. Projected increases in the intensity and frequency of rainfall in the northern regions could lead to cryptosporidium infiltration in water-treatment and distribution systems [41]. Northern European countries report a more prominent potential increase in climate change risk, as opposed to those from southern European countries, where projected decreases in precipitation could reduce these risks. However, these observations also reflect reporting bias; those countries with better EU cryptosporidium notifications reported a climate change risk, whereas those countries with incomplete (or no) cryptosporidium notifications considered the risk to be low.

77.7 Summary and Conclusions

The purpose of this chapter is to highlight the evidence pertaining to the links between climatic factors and waterborne infectious disease. The effects of temperature, precipitation, and extreme weather events on the incidence of infectious disease has been noted within academic literature and has demonstrably shown to be a key indicator on climate change risks in relation to public health. While intrinsic waterborne infectious disease appears to have been increasing in both Europe and the United States, this is not the only troubling factor. Emerging and re-emerging waterborne infectious diseases are also an area of attention and may prove to be a problematic area of global public health in the years to come. Many studies have described seasonal fluctuations in infectious disease, but few have documented long-term trends associated with a climate-disease interface. A variety of models have been and will continue to be developed with the function of being able to systematically and statistically simulate climatic changes and accordingly predict potential disease outbreaks although very few have successfully controlled for socio-demographic or external environmental indicators. The current gaps in knowledge invite scope for further research in the following areas.

77.7.1 Ongoing Epidemiological Research Highlighting the Links between Climatic Influences and Infectious Disease

A barrier to research in relation to infectious disease as a function of climatology has been the need to develop a relationship between climate and disease patterns across diverse populations and geographical regions. For this barrier to be overcome,

collaborations between research of different nationalities need to be facilitated. Similarly, collaborations between different research fields are paramount to progress. As the subject area requires specialization within many scientific fields, there is a need for climatologists, ecologists, and epidemiologists to combine their knowledge in order to expand the breadth of information being reported and published.

77.7.2 Prioritization of Standardized Global Disease Surveillance

Perhaps the biggest obstacle to coherent reporting of infectious disease outbreaks is the heterogeneity of data collection. It is difficult to ascertain correlation and causation on a global scale as there is an inconsistency in the way data is monitored, collected, and reported. Furthermore, it is near impossible to compare country-specific incidences of infectious disease as the number of incidences may be grossly underestimated if the monitoring and reporting is insufficient. Homogeneity is required across nations in order to establish accurate, representative data that can be used for modeling and epidemiological analysis.

77.7.3 Further Development of Forecasting Models and Improvement of Public Health Infrastructure

Models are an important tool for forecasting the health implications associated with changing climatic conditions. In order for models to be both useful and significant, there must be an integration of multidisciplinary influences including social and environmental considerations. Also needed is training within the public sphere, particularly in relation to emergency response and prevention and control programs. Improved public understanding is paramount to the adaptive capacity of the public to ensure an intelligent response to projected health outcomes of climate change.

References

1. Architecture 2030. 2012. Global temperature projections for various RCP scenarios. http://architecture2030.org/files/roadmap_web.pdf.
2. Atherton, F., Newman, C., and Casemore, D. 1995. An outbreak of waterborne cryptosporidiosis is associated with a public water supply in the UK. *Epidemiological Infection*, 115(1), 123–131.
3. Bentham, G. and Langford, H. 1995. Climate change and the incidence of food poisoning in England and Wales. *International Journal of Biometeorology*, 39(2), 81–86.
4. Bentham, G. and Langford, H. 2001. Environmental temperatures and the incidence of food poisoning in England and Wales. *International Journal of Biometeorology*, 45(1), 22–26.

5. Blunden, J. and Derek, S. 2013. State of the climate in 2012. *Bulletin of the American Meteorological Society*, 94(8), S1–S258.

6. Cann, K.F., Thomas, D.R., Salmon, R.L., Wyn-Jones, A.P., and Kay, D. 2013. Extreme water-related weather events and waterborne disease. *Epidemiology and Infection*, 141(04), 671–686.

7. Chauret, C., Volk, C., Creason, R., Jarosh, J., Robinson, J., and Warnes, C. 2001. Detection of *Aeromonas hydrophila* in a drinking-water distribution system: A field and pilot study. *Canadian Journal of Microbiology*, 47(8), 782–786.

8. Checkley, W., Epstein, L.D., Gilman, R., Figueroa, D., Cama, R., Patz, J., and Black, R. 2000. Effect of El Nino and ambient temperature on hospital admissions for diarrhoeal diseases in Peruvian children. *The Lancet*, 355(9202), 442–450.

9. Cook, J., Nuccitelli, D., Green, S.A., Richardson, M., Winkler, B., Painting, R., and Skuce, A. 2013. Quantifying the consensus on anthropogenic global warming in the scientific literature. *Environmental Research Letters*, 8(2), 24–28.

10. Curriero, F.C., Patz, J., Rose, B., and Lele, S. 2001. The association between extreme precipitation and waterborne disease outbreaks in the United States, 1948–1994. *American Journal of Public Health*, 91(8), 1194–1199.

11. Dorner, S., Anderson, W., Huck, P., Gaulin, T., Candon, H., Slawson, R., and Payment, P. 2007. Pathogen and indicator variability in a heavily impacted watershed. *Journal of Water and Health*, 5(2), 241–257.

12. Easterling, D.R., Meehl, G.A., Parmesan, C., Changnon, S.A., Karl, T.R., and Mearns, L.O. 2000. Climate extremes: Observations, modeling, and impacts. *Science*, 289(5487), 2068–2074.

13. Emch, M., Feldacker, C., Islam, M., and Ali, M. 2008. Seasonality of cholera from 1974 to 2005: A review of global patterns. *International Journal of Health Geographics*, 7(1), 1–13.

14. Fernández, M.C., Giampaolo, B.N., Ibañez, S.B., Guagliardo, M.V., Esnaola, M.M., Conca, L., and Frade, H. 2000. *Aeromonas hydrophila* and its relation with drinking water indicators of microbiological quality in Argentine. *Genetica*, 108(1), 35–40.

15. Fewtrell, L., Kay, D., Watkins, J., Davies, C., and Francis, C. 2011. The microbiology of urban UK floodwaters and a quantitative microbial risk assessment of flooding and gastrointestinal illness. *Journal of Flood Risk Management*, 4(2), 77–87.

16. Giao, M.S., Azevedo, N.F., Wilks, S.A., Vieira, M.J., and Keevil, C.W. 2008. Persistence of *Helicobacter pylori* in heterotrophic drinking-water biofilms. *Applied and Environmental Microbiology*, 74(19), 5898–5904.

17. Giorgi, F., Bi, X., and Pal, J.S. 2004. Mean, inter annual variability and trends in a regional climate change experiment over Europe. I. present-day climate (1961–1990). *Climate Dynamics*, 22(6–7), 733–756.

18. Haines, A. and Patz, J.A. 2004. Health effects of climate change. *Journal of the American Medical Association*, 291(1), 99–103.

19. Hales, S., Kovats, S., and Woodward, A. 2000. What El Niño can tell us about human health and global climate change. *Global Change and Human Health*, 1(1), 66–77.

20. Handfield, M., Simard, P., Couillard, M., and Letarte, R. 1996. *Aeromonas hydrophila* isolated from food and drinking water: Hemagglutination, hemolysis, and cytotoxicity for a human intestinal cell line (HT-29). *Applied and Environmental Microbiology*, 62(9), 3459–3461.

21. Hrudey, S.E., Payment, P., Huck, P.M., Gillham, R.W., Hrudey, E.J. 2003. A fatal waterborne disease epidemic in Walkerton, Ontario: Comparison with other waterborne outbreaks in the developed world. *Journal of Water Science and Technology*, 47(3), 7–14.

22. Hulten, K., Han, S.W., Enroth, H., Klein, P.D., Opekun, A.R., Gilman, R.H., Evans, D.G., Engstrand, L., Graham D.Y., and El-Zaatari, F.A. 1996. *Helicobacter pylori* in the drinking water in Peru. *Gastroenterology*, 110(4), 1031–1035.

23. IPCC. 2007. *Climate Change 2007: Impacts, Adaptation and Vulnerability: Contribution of Working Group II to the Fourth Assessment Report of the Intergovernmental Panel on Climate Change*. Parry, M.L., Canziani, O.F., Palutikof, J.P., van der Linden, P.J., and Hanson, C.E. (eds.). Cambridge University Press, Cambridge, UK.

24. Kistemann, T., Claßen, T., Koch, C., Dangendorf, F., Fischeder, R., Gebel, J., and Exner, M. 2002. Microbial load of drinking water reservoir tributaries during extreme rainfall and runoff. *Applied and Environmental Microbiology*, 68(5), 2188–2197.

25. Koelle, K. and Pascual, M. 2004. Disentangling extrinsic from intrinsic factors in disease dynamics: a nonlinear time series approach with an application tocholera. *The American Naturalist*, 163(6), 901–913.

26. Kussovski, V., Mantareva, V., Angelov, I., Orozova, P., Wöhrle, D., Schnurpfeil, G., and Avramov, L. 2009. Photodynamic inactivation of Aeromonas hydrophilaby-cationic phthalocyanines with different hydrophobicity. *FEMS Microbiology Letters*, 294(2), 133–140.

27. Kusters, J.G., van Vliet, A.H., and Kuipers, E.J. 2006. Pathogenesis of Helicobacter pylori infection. *Clinical Microbiology Reviews*, 19(3), 449–490.

28. Levitus, S., Antonov, J.I., Boyer, T.P., and Stephens, C. 2000. Warming of the world ocean. *Science*, 287(5461), 2225–2229.

29. Lewis, D. and Corvalan, C. 2006. Climate variability and change and their potential health effects in small island states: Information for adaptation planning in the health sector. *Environmental Health Perspectives*, 114(12), 1957–1963.

30. McMichael, A.J., Woodruff, R.E., and Hales, S. 2006. Climate change and human health: Present and future risks. *The Lancet*, 367(9513), 859–869.

31. Miettinen, I., Zacheus, O., Von Bonsdorff, C., and Vartiainen, T. 2001. Waterborne epidemics in Finland in 1998–1999. *Water Science and Technology*, 43(12), 67–71.

32. Nichols, G., Lane, C., Asgari, N., Verlander, N., and Charlett, A. 2009. Rainfall and outbreaks of drinking water related disease and in England and Wales. *Journal of Water and Health*, 7(1), 1–8.

33. O'Dwyer, J., Dowling, A., and Adley, C.C. 2014. Microbiological assessment of private ground water derived potable water supplies in the Mid-West Region of Ireland. *Journal of Water and Health*, 12(2), 310–317.

34. Prüss-Üstün, A., Bos, R., Gore, F., and Bartram, J. 2008. *Safer Water, Better Health: Costs, Benefits and Sustainability of Interventions to Protect and Promote Health.* World Health Organization, Geneva.

35. Räisänen, J., Hansson, U., Ullerstig, A., Döscher, R., Graham, L.P., Jones, C., and Willén, U. 2004. European climate in the late twenty-first century: Regional simulations with two driving global models and two forcing scenarios. *Climate Dynamics*, 22(1), 13–31.

36. Randolph, S. 2001. Tick-borne encephalitis in Europe. *The Lancet*, 358(9294), 1731–1732.

37. Reacher, M., McKenzie, K., Lane, C., Nichols, T., Kedge, I., Iversen, A., and Simpson, J. 2004. Health impacts of flooding in Lewes: A comparison of reported gastrointestinal and other illness and mental health in flooded and non-flooded households. *Communicable Disease and Public Health/PHLS*, 7(1), 39–46.

38. Salazar-Lindo, E., Pinell-Salles, P., Maruy, A., and Chea-Woo, E. 1997. El Niño and diarrhoea and dehydration in Lima, Peru. *The Lancet*, 350(9091), 1597–1598.

39. Schijven, J. and Husman, A. 2005. Effect of climate changes on waterborne disease in The Netherlands. *Water Science and Technology*, 51(5), 79–87.

40. Semenza, J.C. and Menne, B. 2009. Climate change and infectious diseases in Europe. *The Lancet Infectious Diseases*, 9(6), 365–375.

41. Semenza, J.C. and Nichols, G. 2007. Cryptosporidiosis surveillance and water-borne outbreaks in Europe. *European Communicable Disease Bulletin*, 12(5), 13–14.

42. Semenza, J.C., Suk, J.E., Estevez, V., Ebi, K.L., and Lindgren, E. 2012. Mapping climate change vulnerabilities to infectious diseases in Europe. *Environmental Health Perspectives*, 120(3), 385.

43. Stocker, T.F., Qin, D., Plattner, G.K., Tignor, M., Allen, S.K., Boschung, J. and Vasconcellosde Menezes, V. 2013. Climate Change 2013. *The Physical Science Basis. Working Group I Contribution to the Fifth Assessment Report of the Intergovernmental Panel on Climate Change-Abstract for Decision-Makers.* Groupe d'experts intergouvernemental sur l'evolution du climat/Intergovernmental Panel on ClimateChange-IPCC,C/O World Meteorological Organization, Geneva, Switzerland.

44. Von Reyn, C.F., Marlow, J.N., Arbeit, R.D., Barber, T.W., and Falkinham, J.O. 1994. Persistent colonisation of potable water as a source of Mycobacterium avium infection in AIDS. *The Lancet*, 343(8906), 1137–1141.

45. Van Vuuren, D.P., Edmonds, J., Kainuma, M., Riahi, K., Thomson, A., Hibbard, K., and Rose, S.K. 2011. The representative concentration pathways: An overview. *Climatic Change*, 109, 5–31.

46. Weiss, R.A. and McMichael, A.J. 2004. Social and environmental risk factors in the emergence of infectious diseases. *Nature Medicine*, 10, S70-S76.

XVIII

Sustainable Water Reuse and Outlooks

78

Water Reuse and Recreational Waters

Yohannes Yihdego
Snowy Mountains Engineering Corporation (SMEC)

PREFACE

"Recreation" encompasses a wide range of activities, involving different types of water bodies and entailing varying concepts of condition. While the recreational condition of a whitewater stream with a native salmon population will be determined largely by flow levels and condition of fish habitat, for example, the recreational condition of a beach will be assessed more in terms of levels of pathogens and chemical contaminants. Recreation includes all activities relating to sport, pleasure, and relaxation, which for the purpose of these guidelines depend on water resources. There have been many pressures to develop and manage streams and lakes for recreational use in inland areas. The importance of recreation to human well-being highlights the need to establish guidelines that also protect public health. In the absence of proper planning and control, the pressure associated with heavy recreational use can itself rapidly reduce the value of a body of water as a public amenity. Generally, however, the impact of domestic, agricultural, and industrial waste discharges is of greater concern, especially as it affects the microbiological and aesthetic quality of water.

These guidelines have been developed to provide assistance to public health administrators' water authorities, and the general public to assess the suitability of waters for recreational use. They apply to all open waters, that is, fresh and saline inland waters as well as marine and estuarine waters. While the guidelines have not been developed for regulatory purposes, they attempt to provide a benchmark to ensure that recreational waters are safe to use. Several challenges exist in assessing the condition of the nation's recreational waters. Foremost is the lack of a comprehensive national system for collecting data on pathogen levels at beaches, a key concern in assessing the suitability of recreational waters with respect to human health. In addition, data on the types and extent of health effects associated with swimming in contaminated water are limited. The number of occurrences is likely under-reported because individuals may not link common symptoms to exposure to contaminated recreational waters.

I would like to acknowledge Prof. Saeid Eslamian for his contribution to this work.

78.1 Introduction

In considering water reuse, it is important to acknowledge that most tap water has had some contact with treated sewage. Wastewater treatment plants discharge into streams that feed rivers from which other cities pump water for drinking. By the time New Orleans' residents drink the Mississippi, the water has been in and out of more than a dozen cities. More than 200 communities, including Las Vegas, discharge treated wastewater into the Colorado River' which is used as a drinking water source for southern California. In these natural waterways, most contaminates left in the effluent are diluted, biodegraded, and/or photodegraded. In addition, the drinking water treatment process removes anything left over [18].

In a water reuse system, the additional clean up is generally accomplished through technology. Although the additional technologies increase costs of wastewater treatment, reuse can be cost effective if it is designed efficiently, utilized effectively, and marketed correctly. In many areas, the cost of potable water is still less than reclaimed water; however, as the supply of potable water continues to decrease, it will become more cost-effective to use reclaimed water. It is in our best interests to preserve our dwindling water supply by investing in water reuse today [31].

Some advantages to wastewater reuse include:

- Reduces the demands on potable sources of freshwater
- Provision of nutrient-rich wastewaters can increase agricultural production in water-poor areas
- Use of wastewater effluent for irrigation will eliminate the need for nutrient removal at the treatment plant
- The quality of the wastewater as an irrigation water supply is usually superior to that of well water
- Pollution of rivers may be reduced as less wastewater is discharged to the waterways
- Provides better quality water for industrial users

Some disadvantages to wastewater reuse include:

- Reuse may be seasonal in nature (e.g., irrigation, golf course watering)
- Potential for public health problems if treatment system does not work properly
- Not always economically feasible due to the need for a separate distribution system
- May result in groundwater contamination

- May result in over-application of pesticides if users of reuse water for irrigation are not educated about the nutrients already in the water

Perhaps the biggest roadblock to wastewater reuse is the public's aversion to using the water. The phrase "toilet-to-tap" has been used to describe wastewater reuse and in many people's minds, there is an instant "ick" factor when the subject is raised. Of course, "toilet-to-tap" is nowhere near the reality of the situation. Water reuse involves the most state-of-the-art technologies to produce water generally superior to tap water and even bottled water. Aggressive public relations campaigns to educate the community about the actual treatment processes and residual contaminant levels are the best way to combat public aversion [31,40].

There are many different ways in which water can be reused. Most reuse activities fall into one of the following six categories:

- Agricultural reuse
- Direct potable reuse
- Environmental/recreational reuse
- Industrial reuse
- Indirect potable reuse
- Urban reuse

78.1.1 Possible Adverse Effects on Human Health

Greywater contains microorganisms from skin surfaces and dirt; small amounts of urine and feces (e.g., from washing of soiled nappies or bedclothes); and washing and preparation of food. Standing greywater provides an environment in which microorganisms can survive and proliferate [20,47]. As a result, greywater usually contains relatively high numbers of microorganisms, some of which may be capable of causing disease in those who come into contact with the greywater or with plants and crops irrigated with greywater, particularly consumers of greywater-irrigated produce [47].

78.1.2 Possible Adverse Effects on Plant Growth and Yield, and on the Continued Ability of the Soil to Support Plant Growth

Greywater also contains substances that can reduce plant growth or crop yield if present at sufficiently high concentrations. These include salts (e.g., used as bulking agents in detergents), sodium, and boron. Furthermore, extremes of pH, which may be encountered in laundry greywater, can be damaging to plants [20,32]. Some constituents of greywater can change soil properties so that soil becomes progressively less fertile (i.e., less able to support plant growth). Because soil properties change slowly, these tend to be long-term effects, while effects on plant growth and yield are more short term. The major concerns with regard to soil are salinity and sodicity, both of which are related to the concentration of sodium in greywater [42]. Other greywater constituents that may affect soil adversely are detergents, oil and grease, and suspended solids [20,41].

78.1.3 Interdependence of Quality and Quantity of Greywater

The quality of greywater is closely linked to the quantity of greywater generated, both on a household scale and on a community scale. The amount of greywater generated per household varies greatly, being lowest in low-income households and highest in households with in-house taps and an affluent lifestyle [32]. The high housing density in un-sewered settlements generally corresponds to a high overall greywater generation in the settlement, even when the volume of water used per dwelling is relatively low. Combined with the high concentration of pollutants in greywater in informal settlements, the lack of proper waste and wastewater infrastructure and the widespread mingling of different waste streams has led to doubts concerning the possibility of the safe use of greywater in these settlements. The paradox is that greywater offers the greatest potential for improvement in household nutritional status and social functioning in poor rural settlements, and in urban and peri-urban settlements. The challenge, then, lies in identifying conditions and limitations under which greywater could potentially be used beneficially in such settlements.

Many waterways provide an environment for rest and recreational activities both in and around the water. Recreational water activities can have benefits for health and well-being; however, potential health hazards from water bodies will vary depending on the water quality and the type of water contact. All waterways contain microorganisms such as algae, bacteria, and parasites. The type and number of organisms will vary from time to time depending on in-flows and concentrations of contaminants. Agricultural and urban run-off may include chemicals, stormwater, litter, sewage, and animal waste. Water contaminated with fecal material may pose a risk of infection to swimmers and those undertaking contact water sports. Recreational use of water can deliver important benefits to health and well-being. Yet, there may also be adverse health effects associated with recreational use if the water is polluted or unsafe. World Health Organization (WHO) produces international norms on recreational water use and health in the form of guidelines. Environmental and Recreational Reuse refers to a variety of activities that, in general, enhance the overall quality of the environment where they are implemented. This can include activities such as creation or restoration of wetlands [50,52,54,55], creation of water bodies for recreational or aesthetic purposes, and stream augmentation. In particular, reclaimed water's use in wetlands can be particularly important since healthy wetlands can serve many purposes including water purification, water preservation, and wildlife habitats [51,52,56].

Potential health hazards vary depending on the water type and activity. In general, the more contact with the water, the better the water quality must be. The guidelines define physical, chemical, and microbiological limits on water quality for three main categories of recreational activity:

- Primary contact recreation—Where the body can be fully immersed and there is the potential to swallow water, and you are in direct contact with the water. This includes surfing, water skiing, diving, and swimming.
- Secondary contact recreation—Includes activities such as paddling, wading, boating, and fishing in which there is direct contact but the chance of swallowing water is unlikely.
- Passive recreation—Where there is no contact with the water and includes scenic appreciation, walking and picnicking around the water.

78.2 Trends in the Condition of Recreational Waters and Their Effects on Health and the Environment

The nation's rivers, lakes, and coastal waters are used for many different forms of recreation. Some recreational activities take place in or on the water, such as swimming, boating, whitewater rafting, and surfing. Other activities may not involve contact with the water yet may still require water—or be enhanced by proximity to water. Examples include a picnic at the beach, hiking, nature viewing (e.g., bird watching), and hunting (especially waterfowl). People also engage in fishing and shell fishing as recreational activities [17].

In the questions on fresh surface waters and coastal waters, condition is defined as a combination of physical, chemical, and biological attributes of a water body. For recreational waters, condition is more specific, focusing on those physical, chemical, and biological attributes that determine a water body's ability to support recreational activities. The particular attributes necessary to support recreation vary widely, depending on the nature of the activity in question. In a more general sense, however, the components of recreational condition fall into two main categories:

- Attributes that determine whether recreational activities can be enjoyed without unacceptable risk to human health—primarily pathogens and chemical contaminants that can affect the health of humans who are exposed during contact activities such as swimming.
- Attributes associated with ecological systems that support recreation—for example, the status of fish and bird communities, as well as chemical and physical characteristics

that may affect these populations and their habitats. These attributes also contribute to the aesthetic qualities important for recreational activities.

Many stressors affecting the condition of recreational waters fall into the broad category of contaminants. This category includes chemical contaminants, various pathogens (viruses, bacteria, and other parasites or protozoans) that can cause infectious disease, and pollutants such as trash or debris [5,25]. These stressors can come from a variety of point sources and nonpoint sources, and can be discharged or washed directly into recreational waters or carried downstream to lakes or coastal areas. Among the major sources are stormwater and sediment runoff, direct discharge (e.g., from industrial facilities and sewer systems), atmospheric deposition, and recreational activities themselves (e.g., outboard motor exhaust and overboard discharge of sanitary wastes). Some chemicals and pathogens occur naturally, but their abundance may be influenced by other human stressors such as land use and land cover (e.g., paved surfaces and forestry and irrigation practices, which can influence runoff patterns) or by natural stressors such as weather and climate. Land use and land cover can influence recreational conditions in other ways as well [4,23,29,53].

In terms of human health, the stressors that pose the greatest potential risks are chemical and biological contaminants. People can be exposed to these contaminants if they swim in contaminated waters or near stormwater or sewage outfall pipes—especially after a rainfall event. Boating also may pose risks of exposure, although to a lesser extent. For toxic chemical contaminants, the main routes of exposure are through dermal (skin) contact or accidental ingestion. For pathogens, the main route of exposure is by swallowing water, although some infections can be contracted simply by getting polluted water on the skin or in the eyes. In some cases, swimmers can develop illnesses or infections if an open wound is exposed to contaminated water [1,57].

Effects of exposure to chemical and biological contaminants range from minor illnesses to potentially fatal diseases. The most common illness is gastroenteritis, an inflammation of the stomach and the intestines that can cause symptoms such as vomiting, headaches, and diarrhea. Other minor illnesses include ear, eye, nose, and throat infections. While unpleasant, most swimming-related illnesses are indeed minor, with no long-term effects [14]. However, in severely contaminated waters, swimmers can sometimes be exposed to serious and potentially fatal diseases such as meningitis, encephalitis, hepatitis, cholera, and typhoid fever [6,8,16]. Children, the elderly, and people with weakened immune systems are most likely to develop illnesses or infections after coming into contact with contaminated water [2,38].

From an ecological perspective, stressors to recreational waters can affect habitat, species composition, and important ecological processes. For example, changes in land cover (e.g., the removal of shade trees) may cause water temperature to rise above the viable range for certain fish species. Hydro-modifications such as dams may create some recreational opportunities (e.g.,

boating), but they also may impede the migration of fish species such as salmon. Chemical and biological contaminants may harm plants and animals directly, or they may disrupt the balance of the food web. For example, acid deposition may lead to acidification in lakes, while excess nutrients can lead to eutrophic conditions such as low levels of dissolved oxygen, which in turn can harm fish and shellfish populations. Beyond their obvious effects on activities like fishing and nature viewing, stressors such as these also can be detrimental to recreational activities in a more aesthetic sense, as the presence of dead fish or visibly unhealthy plants may diminish one's enjoyment of recreation in or near the water [14].

Ultimately, ecological effects can also impact human health. For example, eutrophic conditions can encourage harmful algal blooms—some of which can produce discomfort or illness when people are exposed through ingestion or skin or eye contact. One well-known type of harmful algal bloom is "red tide," which in humans can cause neurotoxic shellfish poisoning and respiratory irritation [17]. The Report on the Environment presents the best available indicators of information on national conditions and trends in air, water, land, human health, and ecological systems that address what the EPA considers mission critical to protecting our environment and human health. At this time, no National Indicators have been identified to quantify the condition of recreational waters. Individual states monitor certain recreational waters for a set of indicator bacteria and report monitoring results to EPA. However, the methodology and frequency of data collection vary among states, so the data are not necessarily comparable. Challenges and information gaps for developing reliable National Indicators of recreational water conditions are described in more detail in the following section [17].

78.3 Limitations, Gaps, and Challenges

Several challenges exist in assessing the condition of the nation's recreational waters. Foremost is the lack of a comprehensive national system for collecting data on pathogen levels at beaches, a key concern in assessing the suitability of recreational waters with respect to human health. In addition, data on the types and extent of health effects associated with swimming in contaminated water are limited. The number of occurrences is likely under-reported because individuals may not link common symptoms (e.g., gastrointestinal ailments, sore throats) to exposure to contaminated recreational waters.

Another challenge to answering this question is the breadth of the subject. "Recreation" encompasses a wide range of activities, involving different types of water bodies and entailing varying concepts of condition. While the recreational condition of a whitewater stream with a native salmon population will be determined largely by flow levels and condition of fish habitat, for example, the recreational condition of a beach will be assessed more in terms of levels of pathogens and chemical contaminants.

Gaps in assessing the condition of the nation's recreational waters include National Indicators of pathogen levels in recreational waters (rivers, lakes, and coastal beaches), the magnitude of specific stressors—particularly contaminant loadings (biological and chemical)—to recreational waters, harmful algal blooms in recreational waters, and the condition of recreational fish and shellfish populations [17].

78.4 Guidelines for Recreational Use of Water

Greywater treatment and its use for irrigation and other purposes have been reported in relatively high-income, developed countries such as the United States, UK, Australia, Germany, and Sweden [37], and in less developed, low-to-middle income countries such as Costa Rica, Jordan, Malaysia, Mali, Nepal, Palestine, and Sri Lanka [32]. Thus, although individual greywater-use applications are usually small and varied in their design, the concept of greywater use for irrigation to promote water conservation and use of the associated nutrients is not new. Despite this, formal standards or quantitative guidance for the quality of greywater for reuse are not widely available. However, there is an increasing wealth of qualitative guidance obtainable on the Internet, from water authorities, from nongovernmental organizations (NGOs), from manufacturers of greywater treatment systems, and from professionals in the field, regarding precautions to be taken in the application of greywater for irrigation. Examples include guidelines available in Australia [17,40], Sweden [35] and the United States [28]. In South Africa, the National Water Act (NWA) of 1998 is the major piece of legislation addressing the use and disposal of water. The Act makes no specific reference to greywater, but refers to "disposal of waste or water containing waste." Discharge or use of water containing waste requires that the use is listed in a General Authorization (GA) of the Act or alternately requires issue of a license. General Authorizations provided under the NWA were revised in 2004 to allow, among others, limited use of biodegradable industrial wastewater for irrigation. Although greywater is not mentioned among the types of wastewater considered, this is probably the closest that existing legislation comes to providing guidance for quality of greywater intended for irrigation use. Although greywater use for small-scale irrigation is not mentioned, it is considered to be within the spirit of the law. For larger-scale use, either the requirements under the General Authorization (GA) apply as mentioned above, or a license for this use would have to be obtained [21]. Other national legislation and policies that reference wastewater in some manner include the National Building Regulations (NBR) in terms of the National Building Regulations and Building Standards Act, guidelines for the use of wastewater for irrigation [13], and the National Policy on Sanitation. None of these addresses greywater and consequently by implication excludes greywater use from legally recognized water uses. These need to be resolved to clarify the legal position of use of greywater for irrigation. The development of guidance for irrigation use of greywater constitutes a step along the way to introducing greywater use as a formally recognized beneficial use of this waste stream. Guidelines for any form of water use

are intended to address risks associated with that use. With reference to greywater use for small-scale irrigation applications, concerns about risk fall into three categories:

- Possible adverse effects on human health
- Possible adverse effects on plant growth and yield
- Possible adverse effects on the environment, especially on the continued ability of the soil to support plant growth

These guidelines have been developed to assist public health administrators' water authorities, and the general public to assess the suitability of waters for recreational use. They apply to all open waters, that is fresh and saline inland waters as well as marine and estuarine waters. While the guidelines have not been developed for regulatory purposes, they attempt to provide a benchmark to ensure that recreational waters are safe to use.

78.5 Recreation and Health

Recreation includes all activities relating to sport, pleasure, and relaxation which for the purpose of these guidelines depend on water resources. The importance of recreation to human well-being highlights the need to establish guidelines that also protect public health. Apart from establishing guidelines, there are a number of practical measures to be addressed, which can reduce the adverse public health impacts associated with recreation including:

- Education on water safety
- Legislation to control recreational activity
- Water resort management
- Pollution control measures

In the absence of proper planning and control, the pressure associated with heavy recreational use can itself rapidly reduce the value of a body of water as a public amenity. Generally, however, the impact of domestic, agricultural, and industrial waste discharges is of greater concern, especially as it affects the microbiological and aesthetic quality of water [33].

78.6 Primary Contact Recreation

Primary contact recreation is characterized by bodily immersion or submersion where there is a direct contact with the water, and includes activities such as swimming, diving, water skiing, and surfing. There are a number of practical considerations that can be addressed which may help protect the health and well-being of recreationalists [44].

Primary body contact may consist of the following recreational activities depending on the characteristics of the water resource.

- Skiing
- Swimming
- Tubing
- Windsurfing

78.6.1 Health Criteria

People engaged in primary contact recreation may swallow significant amounts of water, absorb toxic chemicals through the skin, or acquire a range of infections. The amount of water that may be accidently swallowed varies considerably but in practice probably does not exceed 100 mL for any individual per day. Depending on the level of control exercised over the open water resource, effort should be made to either warn the public or control access or use when water is found to be heavily polluted. Fecally contaminated water may expose swimmers to a range of infectious gastrointestinal diseases. Norwalk virus has been commonly cited as the most likely disease-causing organism in such waters. The protozoans *Giardia* and *Cryptosporidium* may also be a cause for concern, particularly if farm, animal, and sewage wastes are dumped into streams or lakes [48].

78.6.2 Safety and Aesthetic Criteria

Water should be free of floating or submerged objects, which may cause physical injury. In terms of water quality control, floating debris, oil, grease, scum, foam, and other floating materials originating from waste discharges or of natural origin are of concern. Ideally, water should have a low turbidity and have low color. For many activities, for example, swimming and diving, it is important that the bottom be clearly visible. Formulation of criteria to ensure that water is universally appealing is difficult. Aesthetic preferences are subjective and dependent on cultural conditioning [33].

78.7 Secondary Contact Recreation

Secondary contact recreation includes activities such as paddling activities of children, wading, boating, and fishing in which there is some direct contact with water but where the probability of swallowing water is unlikely. Schedule 2 provides recommended water quality criteria where there is limited body contact [33].

78.7.1 Health Criteria

Since limited skin contact is likely to occur in secondary contact recreation, and the possibility of ingestion of water is less likely, criteria to minimize the risk of internal and external infection are less stringent.

78.7.2 Safety Criteria

Waters should be free of floating and submerged objects that risk the safety of boat users and waders.

78.7.3 Aesthetic Criteria

Secondary body contact may consist of the following recreational activities depending on the characteristics of your water resource.
  Aesthetic use
  Boating

Canoeing
Hunting
Kayaking
Rafting
Similar considerations apply as for primary contact recreation.

78.8 Passive Recreation

Aesthetic enjoyment is the primary consideration of passive recreation. Water provides a focal point for many recreational activities, including scenic appreciation, picnicking, and walking. To maintain the recreational resource, the quality of the water should be consistent with the preservation of flora and fauna, which require the water for their habitat or watering needs. While human wellbeing may be promoted by passive recreation, water is not directly used, so risk to health and safety should not be attributed to water quality problems. Those factors that may degrade the water resource and reduce its recreational amenity are of environmental rather than of public health concern [33].

78.9 Guide to Managing Risks and Uncertainty

Risk, whether expressed qualitatively or quantitatively, indicates the probability of a defined adverse effect occurring in an exposed population. Within the context of the guidance report, the adverse effects and exposed "populations" were as follows:

- Illness in human handlers of greywater and greywater irrigated produce, or human consumers of greywater irrigated produce [49]
- Reduction in plant growth or yield in plants or crops irrigated with greywater
- Environmental degradation, specifically reduction in ability of the soil irrigated with greywater to support plant growth in the long term

Uncertainty refers to the degree of confidence associated with an estimate of risk. In the context of greywater quality, this relates largely to the degree of confidence associated with knowledge of water quality, as once the quality of the greywater is known, suitable steps can be taken to address the risks described above. It should be noted that the baseline of uncertainty associated with greywater use is inherently higher than that associated with, e.g. domestic water use or recreational water use, since greywater is inherently highly variable in quality. Greywater-irrigation implementations are also most likely to occur on a small scale, where frequent monitoring of Greywater quality is likely to be economically and logistically difficult [33].

The conceptual framework of the most recent *WHO Guidelines for the Safe Use of Wastewater, Excreta and Greywater* [47] was applied. These Guidelines recognize that risk exists only where there is both a hazard and exposure to that hazard; thus, risk can be reduced either by improving the quality of the greywater (removing or reducing hazards) or by imposing barriers to exposure (preventing exposure to the hazards) [34]. On this basis, three risk and uncertainty categories were identified among potential users of greywater for irrigation:

- Category 1: Users unable or unwilling to conduct any analyses to characterise greywater quality prior to planning irrigation use and during its implementation.

- Category 2: Users willing and able to conduct limited analyses (minimum analysis) to characterize greywater quality prior to planning irrigation use and during its implementation. Minimum analysis was defined as pH, Electrical conductivity (EC), sodium adsorption ratio (SAR) and fecal coliforms/*E. coli*.

- Category 3: Users willing and able to conduct more extensive analyses to characterize greywater quality prior to planning irrigation use and during its implementation. Full analysis was defined as minimum analysis plus boron, chemical oxygen demand, oil and grease, suspended solids, total inorganic nitrogen and total phosphorus. The higher the magnitude of the hazard (i.e., the poorer the quality of the greywater), the more stringent the required risk management interventions will have to be to protect human health, plants and soil. Risk-management interventions related to exposure were described as barriers that minimize the exposure of human users, plants, or soil to a given hazard, based on the following rationale. As the extent of analysis increases from Category 1 to Category 3–and, by implication, as greywater quality improves as it complies with the quality guidance associated with the analysis—so the magnitude of the hazard decreases, and hence so do the risk-management requirements. Where analysis results are in excess of the quality guidance, either steps can be taken to improve greywater quality, or the more restrictive exposure limits of the preceding category can be accepted. If analysis indicates that the greywater is unsuitable for use, then greywater irrigation should be avoided. As an example of barriers to exposure, the case study on health risks associated with consumption of greywater-irrigated crops demonstrated that personal hygiene and food-preparation methods, such as washing hands and peeling and cooking vegetables, could reduce health risk to acceptable levels [26,27,36]. This makes it possible for consumers to take advantage of improved levels of nutrients in food crops irrigated with mixed domestic greywater [36]. Using this categorization of risk and uncertainty, flowcharts were developed to guide users through decisions associated with each risk-management category, leading to a table of restrictions on use for each category [3,7,15].

The restrictions represent barriers to exposure of human users, plants, or soil to specific hazards. As the knowledge of greywater quality improves (from Category 1 to Category 3), so restrictions apply more to the quality of the greywater and less to the way in which it is used. It is assumed that, where

analysis of greywater is performed, it is used only if quality falls within the quality guidance provided. Thus Category 1 represents almost no knowledge of greywater quality and stringent risk management in terms of exposure barriers, whereas in Category 3 a significant proportion of the risk-management effort is directed at the quality of the greywater and exposure barriers are accordingly less restrictive. The restrictions given in the guidance report for Categories 1 and 3, resemble the precautions given in some existing guidelines for greywater use, such as the "greywater do's and don'ts" published by the State of New South Wales, Australia [19] and the "best management principles" published by the State of Arizona. These were identified as a useful point of departure in developing similar precautions for use elsewhere. More restrictive legislation placing severe limits on the types of greywater use, levels of treatment and permitting of infrastructure, such as those in force in California (California Building Standards Commission, 2010), were considered to be impractical in some nation's context [24].

78.10 Greywater Quality: Guide to Greywater Constituents

The section on greywater quality provides the quality criteria against which measured greywater constituents are compared in risk management Category 2 (minimum analysis) and Category 3 (full analysis) from the preceding section. Both nitrogen and phosphorus are essential plant macronutrients. They are required throughout the lifespan of a plant, but especially at times of growth and development. The guideline ranges recognize that for poor farmers, fertilizer application may be economically infeasible and that greywater used for irrigation is a *de facto* fertilizer. It was shown that mixed domestic greywater from an informal settlement had the potential to increase both growth and yield of crops relative to tap water, although yield appeared to be boosted less than growth [36]. Therefore, the ranges of nitrogen and phosphorus presented in the guidance report consider both the fertilizer potential of greywater and the need to supplement greywater with nutrient-poor irrigation water when nutrient concentrations are likely to exceed the requirements of irrigated plants. While the phosphorus application is high, it was deemed acceptable in view of the fact that most African soils are phosphorus-deficient with a high phosphate fixing capacity [9], the relatively high applications that are required to reach phosphorus levels toxic to plants [39,12] and the generally low risk of groundwater contamination [39]. Where soils are sandy and there are surface water bodies in the vicinity, thereby increasing the risk of leaching to groundwater and surface water, the volume of water should be limited to that which can be absorbed by plants and soil (no leaching). Microbiological constituents, represented by *E. coli*, also deserve further mention. Furthermore, the guidance specifies that where extensive or high-technology treatment would be required to make greywater suitable for irrigation use of any kind, then off-site disposal would be likely to be a safer and cheaper option. The list of constituents was consolidated to include only those considered of greatest relevance to use of greywater for irrigation and congruent with the underlying principles of the guidelines (protection of human health, of plant growth and yield, and of soil fertility).

78.11 Greywater Quality: Mitigation of Greywater Quality

Integrated mitigative practices aim to minimize potential adverse effects of, primarily, physicochemical greywater components (such as EC, SAR, sodium, and boron) as part of plant/crop cultivation. Treatment systems aim to remove, primarily, suspended solids, oil and grease, chemical oxygen demand, and health-related bacteria from greywater. Treatment options were based on pilot projects and all represent forms of biological treatment. These systems may be separate from the irrigation application or integrated with the irrigation application. Suggested mitigative practices address:

- Irrigation method, with subsurface application being the preferred method
- Amelioration of soil, for example, by addition of mulch or gypsum to soil. Kaolinite has also recently been shown to improve soil degraded by long-term application of wastewater [30]
- Leaching of soil to remove excess salt, by application of excess water, preferably, freshwater. This must be balanced against the risk of contaminating surface water or groundwater, depending on local conditions
- Planting of tolerant plant types, particularly plants tolerant of sodium or boron
- Miscellaneous irrigation modifications
- Accepting reduced crop yield

Treatment options included in the guidance report were:

- Mulch tower (separate from irrigation application) [35,45]
- Mulch tower and resorption bed (separate from irrigation application or can be integrated) [35,45]
- Various forms of tower or tube gardens (integrated with irrigation application) [10]

A number of general recommendations arose from discussions of the project team and reference group, based on the reviewed literature and on collective experience. These recommendations apply irrespective of the treatment system used. Greywater-treatment systems should be sufficiently simple and robust to function effectively in rural and peri-urban settlements, preferably without power or piped water and with minimal technical expertise required for maintenance. Treatment systems should use the simplest and most cost-effective technology required to meet the water-quality objectives. Any system that is implemented should be thoroughly tested and proven before implementation to avoid user fatigue caused by system failure and ongoing changes [35].

78.12 Challenges to Sustainable Use of Greywater for Small-Scale Irrigation

Challenges to the uptake of greywater irrigation by users exist at a number of levels. Currently, greywater use falls outside the framework of existing laws governing water and wastewater in many nations. Therefore, the legal status of greywater use needs to be clarified at the national level [35].

78.12.1 Municipalities

A major and very concerning finding from the consultation phases of this project was a lack of awareness and capacity among municipalities with respect to greywater. Response rates to calls for consultation and feedback were typically below 10% after follow-up and most responses received indicated a substantial lack of understanding of greywater and of related issues. There is an urgent need for capacity building at local authority (municipal) level with respect to knowledge of greywater issues, in general, and of the potential of greywater use for irrigation, in particular. A prerequisite for successful implementation is that potential irrigation users of greywater are involved in the planning and execution of projects, and are educated on the risks, benefits, and proper handling of greywater, of irrigated land, and of greywater irrigated crops. This is especially true in low-income settlements where the challenges and barriers to greywater use are greatest. Probably the greatest potential for greywater use lies in including greywater collection, treatment, and use in the planning of new settlements or alterations to existing housing developments. However, it is important to ensure that the greywater technologies introduced have been proven to be sound; that designs are implemented correctly; and that housing management and users are committed to using the facilities properly. Failure in any of these areas has been shown to lead to failure of greywater-use projects [35,46].

78.12.2 Informal Settlements

This is in relation to the extensive challenges facing implementation of greywater management projects in underserviced low-income areas. While informal settlements are, in principle, among those urban and peri-urban settlements that could stand to benefit particularly from greywater use for irrigation, there are many barriers rhat have little to do with greywater. Greywater initiatives in such settlements require concerted upfront input by the local authority to provide a receptive environment. This may appear to be a bleak outlook, particularly in light of capacity problems at the municipal level already mentioned. However, preliminary experiences in a few municipalities (e.g., [36]) have shown it is possible, given political will, focused policies, and cooperation with outside consultants with both social and technical expertise.

78.13 Summary and Conclusions

Recreation is a potential beneficial use of greywater, which can conserve freshwater resources and improve quality of life. The guidance developed in this study represents an important contribution to initiation and management of greywater irrigation implementations in South Africa. Unlike existing South African guidelines for water quality for irrigation, these guidelines are specific to greywater use and to small-scale irrigation. Unlike guidelines for greywater irrigation from other countries, such as Australia [17], United States, and Sweden [11,35], these guidelines are specific to South African conditions, being based on South African studies of greywater [36], South African water quality guidelines, and South African climatic and agricultural data [22]. The latest WHO guidelines [47], although comprehensive and adaptable to local conditions, address primarily minimization of human health risks, and therefore do not extend to all the underlying principles set for this study. By addressing greywater quality and quality mitigation, risk management, and irrigation water quantity, the guidance presented here places greywater irrigation in a broader context than do most existing greywater irrigation guidelines, even though coverage is not as comprehensive for any single topic as, for instance, the WHO guidelines for health risk [47].

The following are necessary to facilitate greywater use:

- Enshrining greywater use as a permitted water use in relevant legislation at the national level
- Capacity building at the provincial and municipal level
- Educating potential users at the settlement or household level, and involvement of potential users in planning greywater use projects at the municipal level
- Integration of greywater-use provisions in new developments at the municipal level
- Municipal and provincial authorities developing an enabling environment in low-income settlements through:
  - Building a sense of community
  - Improving communication between authorities and settlement representatives
  - Providing infrastructure for water, sanitation, wastewater, and solid waste
  - Providing infrastructure and guidance for greywater use in exchange for some form of incentive scheme [35]

References

1. Ahmed, W., Stewart, J., Gardner, T., and Powell D 2008. A real-time polymerase chain reaction assay for quantitative detection of the human-specific enterococci surface protein marker in sewage and environmental waters. *Environmental Microbiology*, 10, 3255–3264.
2. Bartram, J., Corrales, L., Davison, A., Deere, D., Drudy, D., Gordon, B., Howard, G. et al. 2009. *Water Safety Plan Manual: Step-by-Step Risk Management for Drinking-Water Suppliers*. World Health Organization, Geneva.
3. Blignaut, J., Ueckermann, L., and Aronson, J. 2009. Agriculture production's sensitivity to changes in climate in South Africa. *South Africa Journal of Sciences*, 105, 61–68.
4. Boehm, A., Nicolas, J.A., Colford, J.M., Dunbar, L.E., Fleming, L.E., Gold, M.A., Hansel, J.A. et al. 2009. A sea change ahead for recreational water quality criteria. *Journal of Water and Health*, Doi: 10.2166/wh.2009.122.
5. Bofill-Mas, S., Albiñana-Gimenez, N., Clemente-Casares, P., Hundesa, A., Rodriguez-Manzano, J., Allard, A., Calvo, M. et al. 2006. Quantification and stability of human adenoviruses and polyomavirus JCPyV in wastewater matrices. *Applied and Environmental Microbiology*, 72, 7894–7896.
6. Buckalew, D.W., Hartman, L.J., Grimsley, G.A., Martin, A.E., and Register, K.M. 2006. A long term study comparing membrane filtration with Colilert defined substrates in detecting faecal coliforms and Escherichia coli in natural waters. *Journal of Environmental Management*, 80, 191–197.
7. California Building Standards Commission. 2010. 2010 California Plumbing Code, California Code of Regulations, Title 24, Part 5.
8. Chawla, R. and Hunter, P.R. 2005. Classification of bathing water quality based on the parametric calculation of percentiles is unsound. *Water Research*, 39, 4552–4558.
9. Cordell, D., Drangert, J.O., and White, S. 2009. The story of phosphorus: Global food security and food for thought. *Global Environmental Change*, 19, 292–305.
10. Crosby, C. 2005. Food from used water, making the previously impossible happen. *The Water Wheel*, January/February 2005, 10–13.
11. CSBE. 2003. *Gray Water Reuse in Other Countries and its Applicability to Jordan*. Centre for the Study of the Built Environment, Amman, Jordan.
12. Daniels, D., Daniel, T., Carmen, D., Morgan, R., Langston, J., and Vandevender, K. 2011. *Soil Phosphorus Levels: Concerns and Recommendations*. University of Arkansas, Department of Agriculture Cooperative Extension Service http://www.sera17.ext.vt.edu/Documents/Soil_P_Levels_Concerns_and_Recommendations.pdf.
13. Department of National Health and Population Development. 1978. *Guide: Permissible Utilisation and Disposal of Treated Sewage Effluent, Reference No. 11/2/5/3*. Department of National Health and Population Development, Pretoria, South Africa.
14. Dufour, A. and Schaub, S. 2007. The evolution of water quality criteria in the United States, 1922–2003. In: L.J. Wymer (ed.), *Statistical Framework for Recreational Water Quality Criteria and Monitoring. Statistics in Practice*. John Wiley & Sons, Chichester, England, pp. 1–12.
15. Dufour, A.P., Evans, O., Behymer, T.D., and Cantú, R. 2006 Water ingestion during swimming activities in a pool: A pilot study. *Journal of Water Health* 4:425–430.
16. Edge, T.A. and Schaefer, K.A. (eds.). 2006. *Microbial Source Tracking in Aquatic Ecosystems: The State of the Science and an Assessment of Needs*. National Water

Research Institute, Burlington, Ontario. NWRI Scientific Assessment Report Series No. 7 and Linking Water Science to Policy Workshop Series.

17. EPA. 2008. United States Environmental Protection Agency (EPA), Environmental report. http://cfpub.epa.gov/eroe/index.cfm?fuseaction=list.listBySubTopic&ch=47&s=281. Accessed on September 10, 2014.

18. EPA. 2015. Water Recycling and Reuse: The Environmental Benefits. Water Division Region IX- EPA 909-F-98-001. United States Environmental Protection Agency, San Francisco, CA.

19. EPA Victoria. 2008. *The Do's and Don'ts of Reusing Grey Water*. EPA Victoria, Publication No. 884, State of Victoria.

20. Eriksson, E., Auffarth, K., Hen ze, M. and Ledin, A. 2002. Characteristics of greywater. *Urban Water 4*, 85–104.

21. Gravelèt-Blondin, L.R. 2010. Personal consultations with LR Gravelèt-Blondin, Water Lily Consulting, Durban, South Africa. (LR Gravelèt-Blondin previously of Department of Water Affairs and Forestry).

22. Green, G.C. (ed.). 1985. *Estimated Irrigation Requirements of Crops in South Africa, Parts 1 and 2*. Memoirs on the Agricultural Natural Resources of South Africa, No. 2. Soil and Irrigation Research Institute, Department of Agriculture and Water Supply, Pretoria, South Africa.

23. Haugland, R.A., Siefring, S.C., Wymer, L.J., Brenner, K.P., and Dufour, A.P. 2005. Comparison of Enterococcus measurements in freshwater at two recreational beaches by quantitative polymerase chain reaction and membrane filter culture analysis. *Water Research, 39*(4), 559–568.

24. Ishii, S. and Sadowsky, M.J. 2008. Escherichia coli in the environment: Implications for water quality and human health. *Microbes and Environment, 23*, 101–108.

25. Isobe, K.O., Tarao, M., Chiem, N.H., Mihn, L.Y., and Takada, H. 2004. Effect of environmental factors on the relationship between concentrations of coprostanol and fecal indicator bacteria in tropical (Mekong Delta) and temperate (Tokyo) freshwaters. *Applied and Environmental Microbiology, 70*, 814–821.

26. Jackson, S.A.F., Muir, D., and Rodda, N. 2010. Use of domestic greywater for small-scale irrigation of food crops: Health risks. *Proceedings of the 11th Waternet/WARFSA/GWP-SA Symposium*, Victoria Falls, Zimbabwe.

27. Jackson, S.A.F., Rodda, N., and Saluka Zana, S. 2006. Microbiological assessment of food crops irrigated with domestic greywater. *Water SA, 32*, 700–704.

28. Little, V. 2004. *Residential Graywater Reuse Fact Sheet*. Water Conservation Alliance of Southern Arizona. Arizona Department of Water Resources. http://www.watercasa.org/pubs/graywaterguidelines.html.

29. Marsalek, J. and Rochfort, Q. 2004. Urban wet-weather flows: Sources of fecal contamination impacting on recreational waters and threatening drinking-water sources. *Journal of Toxicology and Environmental Health*, Part A 67, 1765–1777.

30. Matiax-Solera, J., Garćia-Irles, L., Morugán, A., Doerr, S.H., Garcia-Orenes, F., Arcenegui, V., and Atanassova, I. 2011. Longevity of soil water repellency in a former wastewater disposal tree stand and potential amelioration. *Geoderma*. DOI: 10.1016/j.geoderma.2011.07.006.

31. McKenzie, C. 2008. Wastewater reuse conserves water and protects waterways. *On Tap*. Winter, 46–51.

32. Morel, A. and Diener, S. 2006. *Greywater Management in Low and Middle-Income Countries, Review of Different Treatment Systems for Households or Neighbourhoods*. Swiss Federal Institute of Aquatic Science and Technology (EAWAG), Dübendorf, Switzerland.

33. National Health and Medical Research Council-Australian Water Resources Council. 1990. Australian Guidelines for recreational use of water. Australian Government Publishing Service, Canberra.

34. National Health and Medical Research Council-Australian Water Resources Council. 2008. Guidelines for managing risks in recreational water. Australian Government Publishing Service, Canberra.

35. Ridderstolpe, P. 2004. *Introduction to Greywater Management*. EcoSanRes Programme, Stockholm Environment Institute, Stockholm, Sweden.

36. Rodda, N., Saluka zana, L., Jackson, S.A.F., and Smith, M.T. 2011. Use of domestic greywater for small-scale irrigation of food crops: Effects on plants and soil. *The Journal of Chemical Physics*, DOI:10.1016/j.pce.2011.08.002.

37. Roesner, L., Qian, Y., Criswell, M., Stromberger, M., and Klein, S. 2006. Long-Term Effects of Landscape Irrigation using Household Greywater: A Literature Review and Synthesis.

38. Roser, D.J., Ashbolt, N.J., Davies, C.M., Glamore, W.C., Hawker, K.M., and Miller, B.M. 2007. Application of TMDL and risk assessment principles for pathogen management at an urban recreational lake. In Watershed Management to Meet Water Quality Standards and TMDLS, 4th Conference Proceedings, March 10–14, 2007. San Antonio, TX. The American Society of Agricultural and Biological Engineers (ASABE), Stephenville, TX, pp. 420–426.

39. Sigua, G.C., Hubbard, R.K., and Coleman, S.W. 2010. Quantifying phosphorus levels in soils, plants, surface water and shallow groundwater associated with bahiagrass-based pastures. *Environmental Science and Pollution Research, 17*(1), 210–219.

40. Standen, R. and McGuckian, R. 2000. Developing irrigation guidelines for wastewater irrigation. *Proceedings of the 63rd Annual Water Engineers and Operators Conference*, September 6–7, Warrambool, Australia.

41. Travis, M.J., Weisbrod, N., and Gross, A. 2008. Accumulation of oil and grease in soils irrigated with grey water and their potential role in soil water repellency. *Science of the Total Environment, 394*, 68–74.

42. Unkovich, M., Stevens, D., Ying, G., and Kelly, J. 2004. *Impacts on Crop Quality from Irrigation with Water Reclaimed from Sewage*. Australian Water Conservation and Reuse Research Program, Australia.

43. Van Averbeke, W., Juma, K.A., and Tshikalange, T.E. 2007. Yield response of African leafy vegetables to nitrogen, phosphorus and potassium: The case of Brassica rapa L. subsp. chinensis and Solanum retroflexum Dun. *Water SA* 33, 355–362.

44. Wade, T.J., Calderon, R.L., Sams, E., Beach, M., Brenner, K.P., Williams, A.H., and Dufour, A.P. 2006. Rapidly measured indicators of recreational water quality are predictive of swimming-associated gastrointestinal illness. *Environmental Health Perspectives* 114(1), 24–28.

45. Whittington-Jones, K. 2007. Construction of Grey Water Treatment Systems for the Scenery Park (Buffalo City Municipality) Pilot Project, Phase 2 Report. Report prepared for EcoSanRes and the Stockholm Environment Institute by Scarab Resource Innovations. South Africa.

46. Whittington-Jones, K.J. and Tandlich, R. 2010. Performance of Pilot Biological Grey Water Treatment Systems in Scenery Park (Buffalo City Municipality, South Africa). Report prepared for the Stockholm Environment Institute and EcoSanRes.

47. WHO. 2006. *Guidelines for the Safe Use of Wastewater, Excreta and Greywater.* World Health Organization, Geneva, Switzerland.

48. Wiedenmann, A. 2007. A plausible model to explain concentration-response relationships in randomized controlled trials assessing infectious disease risks from exposure to recreational waters. In: L.J. Wymer (ed.), *Statistical Framework for Recreational Water Quality Criteria and Monitoring.* Statistics in Practice. John Wiley & Sons, Chichester, England, 153–177.

49. Wiedenmann, A., Krüger, P., Dietz, K., López-Pila, J.M., Szewzyk, R., and Botzenhart, K. 2006. A randomized controlled trial assessing infectious disease risks from bathing in fresh recreational waters in relation to the concentration of Escherichia coli, intestinal enterococci, Clostridium perfringens, and somatic coliphages. *Environmental Health 32 Perspectives* 114, 228–236.

50. Yihdego, Y. and Becht, R. 2013. Simulation of lake–aquifer interaction at Lake Naivasha, Kenya using a three-dimensional flow model with the high conductivity technique and a DEM with bathymetry. *Journal of Hydrology*, 503, 111-122. http://dx.doi.org/10.1016/j.jhydrol.2013.08.034

51. Yihdego, Y. and Webb, J.A. 2011. Modeling of bore hydrograph to determine the impact of climate and land use change in a temperate subhumid region of south-eastern Australia. *Hydrogeology Journal*, 19(4), 877–887.

52. Yihdego, Y. and Webb, J.A. 2012. Modeling of seasonal and long-term trends in lake salinity in South-western Victoria, Australia. *Journal of Environmental management*, 112(2012), 149–159. doi:10.1016/j.jenvman.2012.07.002.

53. Yihdego, Y. and Webb, J.A. 2013. An empirical water budget model as a tool to identify the impact of land-use change in stream flow in southeastern Australia. *Water Resources Management Journal*, 27(14), 4941–4958. http://dx.doi.org/10.1007/s11269-013-0449-2

54. Yihdego Y. and Webb J.A. 2015. Use of a conceptual hydrogeological model and a time variant water budget analysis to determine controls on salinity in Lake Burrumbeet in southeast Australia. *Environmental Earth Sciences Journal.* 73(4), 1587–1600. http://link.springer.com/article/10.1007/s12665-014-3509-x

55. Yihdego, Y, Webb, J.A., and Leahy P., 2014. Modelling of lake level and salinity for Lake Purrumbete in western Victoria, Australia. *Hydrological Sciences Journal.* http://dx.doi.org/10.1080/02626667.2014.975132

56. Yihdego, Y, Webb, J.A., and Leahy P. 2015. Modelling of lake level under climate change conditions: Lake Purrumbete in south eastern Australia. *Environmental Earth Sciences Journal.* 73, 3855–3872. DOI:10.1007/s12665-014-3669-8. http://link.springer.com/article/10.1007/s12665-014-3669-8

57. Yoder. J.S., Hlavsa, M.C., Craun, G.F., Hill, V., Roberts, V., Yu, P.A., Hicks, L.A. et al. 2008. Surveillance for Waterborne Disease Outbreaks Associated with Recreational Use and other Aquatic Facility-associated Health events—United States, 2005–2006, MMWR, Surveillance Summaries 57, No. SS-9.

79

Sustainability and Water Reclamation

Adebayo Oluwole
Eludoyin
Obafemi Awolowo University

PREFACE

One of the important strategies that are aimed at ensuring water sustainability is reclamation of wastewaters for reuse. This chapter provides insights into the concept of water reclamation for the purpose of reuse, in the light of threats of water stress and scarcity, globally. It indicates the usefulness of reclaimed water to offer important cheaper water supply and prevent the environment from receiving untreated and toxic wastes from industry, municipal, and other sources. The chapter identifies the need for good technological development and relevant regulations to ensure that reclaimed water is of desired quality. It also identified the need to sensitize the public that reclaimed waters pass through quality control systems and are certified of required quality before use. The study concluded that water reclamation is an important part of the instruments for water sustainability.

79.1 Introduction

Water sustainability means supplying or being supplied with water for life. It is the continual supply of clean water for human uses and for other living things [31]. The United Nations through the Dublin Water Principles in 1992 recognized water as an economic good that can be sustainably managed with economical, legal, and environmental considerations [32]. Sustainable management of water resources (surface water, groundwater, and wastewater) is aimed at meeting the present human's water requirements without compromising the ability of future generations to meet theirs [7,33]. The Global Water Partnership (GWP) [16] noted that water sustainability is globally important because water is embedded in all aspects of human development—food security, health, and poverty reduction—and in sustaining economic growth in agriculture, industry, and energy generation.

The importance attached to water is stressed by its position as a specific target of the Millennium Goal 7 (ensure environmental sustainability, including the target of halving the number of people without access to water and sanitation). Over 300 million people lack safe water in the sub-Saharan Africa, and over 1.2 billion people worldwide live more than a 15-min walk from a safe water source, many in Africa [21,28]. About 5 million water-borne–related deaths occur, on annual basis, globally, because people have limited access to safe water and adequate sanitation [13,18]. Water-related problems are projected to escalate in many parts of the world [35], probably due to the effect of extreme climate, particularly drought, poor water infrastructure in many developing countries, population increase, and unsustainabe increase in urbanization [27,41].

Many international and local programs have been directed at achieving water sustainability throughout the world, and these include the declaration of the International Drinking Water Supply and Sanitation Decade (IWSSD), Dublin Principle 2, GWP, FairWater, and G8's Evian Water Action Plan, among others. Many of these programs (on water sustainability) have focused on water supply, and have either neglected the other important components of water sustainability or have not significantly incorporated them [2,18,22] (Figure 79.1).

The importance of this chapter is to provide a review on water reclamation as an important process through which water is being recycled for reuse. Discussion of the rationale, importance, and problems of water reclamation makes up the objectives of the chapter.

79.2 What Is Water Reclamation?

Water reclamation is the treatment or processing of wastewater to make it reusable by meeting water quality criteria. Water reclamation is a promising option for water sustenance in urban areas where blackwater (sewage or all in-building wastewater streams, including toilet waste), greywater (all in-building wastewater streams, excluding toilet waste), wash water and rainwater are consistently generated [37]. Reclaimed water can be defined as treated wastewater that has passed through an additional or complementary treatment process if required, following which the quality is appropriate for the use for which it is intended [19]. Reclaimed water has been found useful in many countries of the world, including in Europe and the Americas, as quests for alternative sources of water increase to combat water scarcity problems [24,25,27].

Water reclamation as an option of water sustainability has generated interest over a long time and in different parts of the world. It was probably first practiced in Europe in the Minoan time, approximately 3000–1100 BC for land application in the Ancient Greek and Roman civilizations, as well as in the fourteenth and fifteenth centuries during the Mediterranean civilizations in parts of Europe, especially in Great Britain, Germany, France, and Poland, but the practices have passed through different developmental stages with time, knowledge, technology, and regulations [2,3]. In China, studies [10,11] noted that reclaimed water is significantly used in at least 11 large cities, including Beijing and Shanghai, and that demand for reclaimed water in one of the cities (Beijing Urban Overall Plan, 2004–2020) will increase by 17.6% for urban landscaping, aquatic ecology, street flushing, household nonpotable utilization, and industrial cooling systems. In the landlocked Czech Republic, Janosova [20] advocated for increased used of reclaimed wastewater for crop or vineyard irrigation in the agricultural area in many of the water stressed regions, while Angelakis and Bontoux [1] reviewed wastewater reclamation projects in the EurEau countries (Austria, Belgium, Denmark, Finland, France, Ireland, Italy, Germany, Greece, Luxembourg, the Netherlands, Portugal, Spain, Sweden, the United Kingdom, Iceland, Norway, Switzerland, Bulgaria, Croatia, Cyprus, Estonia, Hungary, Poland, and Romania) and noted spatial differences in the demand for and priority given to water reclamation; it is higher in the southern EurEau countries than in the northern part of the region.

79.3 Rationale for Water Reclamation

An important justification for reclaiming water in most environments is the consideration of the water resource as limited and, perhaps, the most strategically important resource on Earth [23]. Unlike the previous belief that water is a never-ending resource with a limitless renewable capacity, water reclamation (and reuse) are now increasingly being regarded as a necessary tool for water sustainability [5,9,30]. Freshwater makes up less than 1% of the entire water resources in the world, and it is becoming scarce in most countries, including the highly rain-fed countries

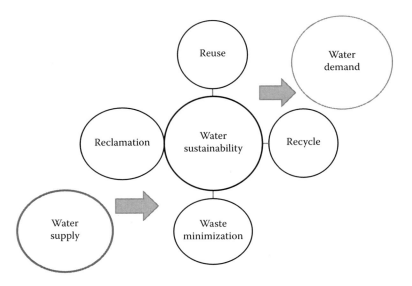

FIGURE 79.1 Targets of sustainable water programs.

[30]. Reliable freshwater supply and the protection of aquatic resources through adequate water management are essential to support all aspects of human life [30]. Water scarcity is associated with the increasing urbanization and rapidly growing urbanization, intensive agricultural practices, effects of climate variability, and extreme events [8,20,36,40]. Water scarcity is experienced in many countries in the world, and it is projected to increase with time if countries do not make efficient plans (Figure 79.2).

79.4 Processes of Urban Water Reclamation

A typical urban water reclamation process often involves collecting and treating urban wastewaters to useable standards (Figure 79.3).

Urban wastewater mainly comes from domestic, municipal, commercial, and institutional sources. Reclaimed wastewaters have been used for toilet flushing (reclaimed household water), urban landscaping and road sweeping (municipal), to replace freshwater as cooling water systems for industrial and municipal heat power plants, as cooling water in industries, and for agricultural irrigation [1,10]. A typical example of community-based wastewater reclamation and reuse is illustrated In Figure 79.4.

Examples of reclamation facilities are found in Sonoma and Monterey Counties in California (for irrigation of edible agricultural crops), Orange County, California (indirect potable wastewater reuse), and in Singapore and Windhoek, Namibia (NEWater facility) [4,15,30,34,39,42]. Wastewater treatment methods range from simple filtration to complex systems, including carbon filtration, softening, reverse osmosis, and strong acid/base ion exchange with microfiltration and ultraviolet-light disinfection, depending on application and the required water quality [3,17]. Wastewater treatment options can be subdivided into preliminary treatment (any physical or chemical process at the wastewater treatment plant that precedes primary treatment, and it functions to protect subsequent treatment units and to minimize operational problems), primary treatment (physical or chemical treatment for the removal of settleable and floatable materials), secondary treatment (processes that use biological and, at times, chemical treatment to accomplish substantial removal of dissolved organics and colloidal materials), and advanced treatment (requiring methods other than those used in conventional treatment). Generally, treatment is a complex process involving several water treatment methods or steps.

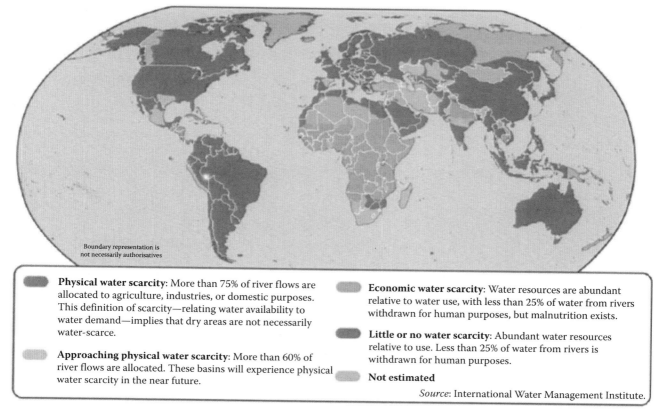

Boundary representation is
not necessarily authorisatives

Physical water scarcity: More than 75% of river flows are allocated to agriculture, industries, or domestic purposes. This definition of scarcity—relating water availability to water demand—implies that dry areas are not necessarily water-scarce.

Approaching physical water scarcity: More than 60% of river flows are allocated. These basins will experience physical water scarcity in the near future.

Economic water scarcity: Water resources are abundant relative to water use, with less than 25% of water from rivers withdrawn for human purposes, but malnutrition exists.

Little or no water scarcity: Abundant water resources relative to use. Less than 25% of water from rivers is withdrawn for human purposes.

Not estimated

Source: International Water Management Institute.

FIGURE 79.2   Projected global water condition by 2025. Countries in Africa, Asia, and Latin America will experience significant water scarcity, despite their natural abundant supply of water. (Adapted from UNEP/GRID-Arendal. 2009. Increased global water stress, Vital Water Graphics 2, UNEP, Norway.)

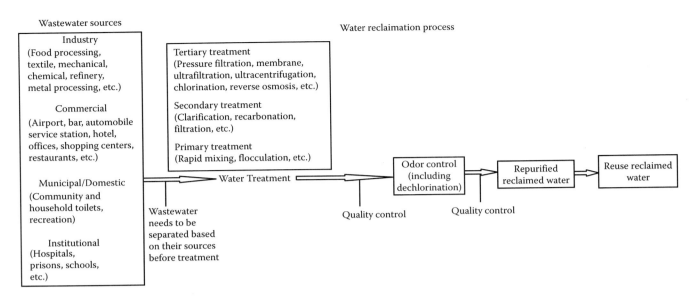

FIGURE 79.3   Typical urban water reclamation process, the quality control is a key step to guarantee the quality of reclaimed water in the wastewater reclamation system. Quality control also includes the monitoring of bioorganics growth and migration. (Adapted from Lu, W. and Leung, A.Y. 2003. *Chemosphere*, 52(9), 1451–1459.)

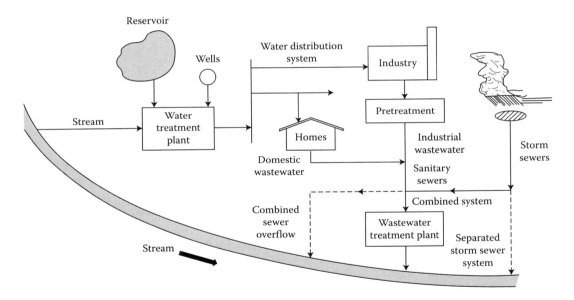

FIGURE 79.4   Typical urban wastewater reclamation and reuse process; arrow indicates direction of water flow. Dotted lines indicate the pathway for the treated effluent from the reclamation cycle.

Sometimes, pretreatment is necessary to improve the efficiency of the final treatment. Information about the different processes is provided in Table 79.1.

79.5 Water Reclamation Issues

79.5.1 Standard Quality

The most significant issue with water reclamation is quality standard of water for various uses. This is probably why reclaimed waters have been used for many nonpotable applications, including toilet flushing, cooling of heat appliances, and irrigation. Concerns on health impacts of using reclaimed wastewater, especially for agricultural irrigation and groundwater recharge, have drawn increasing attention in the last decade [4,6,15,30,34,42], while studies such as Ref. [9] have attracted attention to the need to focus on quality assessment of reclaimed water. Studies in some states of Australia and the United States have shown that although the use of reclaimed water has not been implicated for any significant disease outbreak in any region of the areas, less than 15% of a sampled population in Australia readily agreed to use reclaimed water for

TABLE 79.1 Processes Involved in Water Treatment Phase of Water Reclamation

Treatment Classification	Treatment Process	Application	Advantages and Capabilities	Requirements and Remarks
Preliminary treatment	Equalization	Wastewaters with high variability	Dampers waste variation, reduces chemical requirements, dampens peak flows, and reduces treatment plant sites.	Need large areas, possible septicity, requiring mixing and/or aeration equipment.
	Neutralization	Wastewaters with extreme pH values	Provides the proper conditions for biological, physical, and chemical treatment.	May generate solids.
	Temperature adjustment	Waste streams with extreme temperatures	Provides the proper conditions for biological treatment.	High initial equipment costs.
	Nutrient addition	Nutrient deficient wastes	Optimizes biological treatment.	Possible septicity, requiring mixing and/or aeration equipment.
	Screening	Waste streams containing large solids (wool, rags, etc.)	Prevents pump and pipe clogging; reduces subsequent solids handling.	Maintenance required to prevent screen plugging, ineffective for sticky solids.
	Grit removal	Wastewaters containing significant amounts of large, heavy solids.	Lowers maintenance costs, erosion.	Solids to be disposed of are sometimes offensive.
Primary treatment	Sedimentation	Wastewaters containing settleable suspended solids	Reduces inorganic and organic solid loadings to subsequent biological units. By far the least expensive and most common method of solid–liquid separation. Suitable for treatment of wide variety of wastes. Requires a simpler equipment and operation. Demonstrated reliability as a treatment process.	Possible septicity and odors. Adversely affected by variations in the nature of the waste. Moderately large area requirement.
	Dissolved-air floatation	Wastewaters containing oils, fats, suspended solids, and other floatable matter. Can be used for either clarification or thickening	Removes oils, greases, and suspended solids. Less tank areas than for a sedimentation tank. Higher concentration of solids than for sedimentation. Satisfies immediate oxygen demand. Maintenance of aerobic conditions.	High initial equipment costs. Sophisticated equipment and instrumentation. High power and maintenance cost.
Secondary Treatment	Activated sludge (aeration and secondary sedimentation)	Biologically treatable organic wastes.	Flexible—can adapt to minor pH, organic, and temperature changes. Produces high quality effluent (80%–90% BOD and suspended solids removal). Small area required. Available in package units. The degree of nitrification is controllable. Relatively minor odor problems.	High operating costs (skilled labor, electricity, etc.) Generates solids requiring sludge disposal. Some process alternatives are sensitive to shock loads, and metallic or other poisons. Requires continuous air supply.
	Aerated pond (with secondary sedimentation)	Biologically treatable organic wastes.	Flexible—can adapt to minor pH, organic, and temperature changes. Requires minimum attention. Moderate effluent (80%–90% BOD removal).	Dispersed solids in effluent. Affected by seasonal temperature variations. Operating problems. Moderate power costs. Large area required. No color reduction.
	Aerobic/anaerobic ponds	Biologically treatable organic wastes.	Low construction costs. Non-skilled operation. Moderate quality (80%–95% BOD removal). Removes some nutrients from wastewaters.	Large land area required. Algae in effluent. Possible septicity and odors. Weed growth, mosquito, and insect problems.
	Trickling filter	Biologically treatable organic wastes.	Moderate quality effluent (80%–90% BOD removal). Moderate operating costs (lower than activated sludge and higher than oxidation pond). Good resistance to shock loads.	Clogging of distributors or beds. Snail, mosquito, and insect problems.

(Continued)

TABLE 79.1 (*Continued*)   Processes Involved in Water Treatment Phase of Water Reclamation

Treatment Classification	Treatment Process	Application	Advantages and Capabilities	Requirements and Remarks
	Chemical oxidation	Low flow, high concentration of wastes of known and consistent waste composition, or removal of refractory compounds.	Disinfects effluent. Aids grease removal. Removes taste and odor. Removes organics without producing a residual waste concentrate.	Chemical costs. High initial equipment costs. Skilled operations. Requires handling of hazardous chemicals.
	Chemical mixing flocculation and clarification	Wastewater high in dissolved solids, colloids, metals, or perceptible inorganic and waste containing emulsified oils.	Removes ions, nutrients, colloids, dissolved salts. Recovery of valuable materials. Provides proper conditions for biological treatment.	Sophisticated equipment and instrumentation. Residual salts in effluent. Produces considerable sludge.
	Gravity filtration	Wastewaters with organic or inorganic suspended solids, emulsions, colloids.	Breaks emulsions. Removes suspended solids.	Clogging. Frequent backwashing. High pressure costs.
	Pressure filtration	Wastewater high in suspended solids (i.e., sludge, organic solids).	High solids removal (90%–95%).	High pressure costs. Clogging. High pressure drop (power costs).
	Dissolved air floatation with chemicals	Wastewaters containing oils, fats, colloids and chemically coalesced materials	Produces high degree of treatment Removes oils, greases	High initial equipment costs. High operation cost Sophisticated instrumentation.
	Anaerobic contact	Wastewaters with high BOD and/or high temperature.	Methane recovery. Small area required. Volatile solids destruction.	Heat required. Effluent in reduced chemical form requires further treatment. Requires skilled operation.
Advanced wastewater treatment	Activated carbon adsorption	Wastewater containing trace amounts of organics and color, taste, and odor-producing compounds	Removes nonbiodegradable organics from wastewaters. Removes taste and odor-producing compounds. Reduces color.	High equipment costs. Carbon costs—a. pH adjustment, b. initial carbon, c. makeup carbon. No inorganic removal. Wastes must be solid to prevent clogging. Air pollution potential when regenerating activated carbon. Limited throughput.
	Microstraining filtration	Tertiary treatment	Up to 89% of suspended solids removed. Can produce final effluent of solids less than 10 mg/L.	Very sensitive to solids overloading. Requires automatic controls, absorbent techniques.
	Land treatment	Biologically treatable wastes with low to moderate amounts of toxic substances	Inexpensive. Minimum operator attention, minimum sludge. Crop production. Very high quality effluent and/or discharge.	Large land area required. Possible contamination of potable aquifers. Freezing in winter. Odors in summer under some conditions, usually of minor concern.
	Subsurface disposal (e.g., deep well injection)	Solids free, concentrated wastewaters	Disposal of inorganics and organics. Ultimate disposal of toxic or odorous material.	Subsurface clogging. Groundwater pollution. High maintenance and operation costs. Limited aquifer life. High initial costs.
	Groundwater recharge	Treated wastewaters	Reduces bacteria concentration. Conserves water resources. Prevents saltwater intrusion into potable aquifers.	Possible groundwater contamination. Limited to porous formation.
	Anaerobic digestion	Biodegradable solids	Methane production. Solids stabilization and conditioning. Liquefaction of solids.	Heat required. Process upsets when excess volatile acids.
	Anaerobic digestion	Biodegradable solids. Minimum land required; use of digested sludge as fertilizer or soil conditioner	Methane production. Solids stabilization and conditioning. Liquefaction of solids.	Heat required. Process upsets when excess volatile acids. Odors, skilled labor requirements, and vulnerability to explosion hazard are limitations.
	Aerobic digestion	Biological solids	Relatively little odor. Solids stabilization and conditioning Unsophisticated operation.	Moderate land area required. High energy usage. Reduced dewatering ability.
	Autoclaving	Biological solids	Compact operation. Solids conditioning. Kills microorganisms.	High initial equipment costs. Power costs. Skilled labor. Maintenance cost.
	Aerobic digestion	Biological solids	Relatively little odor. Solids stabilization and conditioning. Unsophisticated operation.	Moderate land area required. High energy usage. Reduced dewatering ability.

potable uses [28,38]. Public opinion is perhaps one of the major drawbacks of water reuse [12], and a proper education will be required to resolve this.

Considering the wide range of the potential of reclaimed waters, it may be difficult to set up the same quality standards for most uses [38]. Water quality standards are available for most water uses to compare before use. For example, Chiaudani and Premazzi [12] documented the quality standards for different water use categories, and in most cases, acceptable standards are documented across countries and international organizations, including the World Health Organization, European Union, etc. Current guidelines and regulations regarding use and reuse of water now acknowledge the use of other water qualities than that of potable water [9,14], and regulations on use of reclaimed water are gradually realizing the importance of balancing quality and quantity with point-of-use availability of water.

79.5.2 Economic Advantage

Reclaimed water is well accepted for its economic advantage, especially when "first-hand" water supply is either scarce or costly. Irrigation, for example, is the highest water-consuming activity in any country, and hence is the first option considered in any reuse planning. For example, water for irrigation practices needs to be significantly available and such demand cannot be met by the available "first-hand" water supply (especially in counties like India, China, and Egypt and most others where at least 90% of available water is used in irrigation), a condition that may make reclaimed water a viable alternative in these countries [10,30].

79.5.3 Environmental Friendliness

Another advantage of water reclamation is its consideration for environmental friendliness in some quarters [29]. Water reclamation prevents discharge of untreated wastes into water bodies, and therefore saves the environment from severe contamination. For example, while nutrients in wastewater can be beneficial to plants when reused in irrigation, they can also be detrimental to the receiving ecosystem when improperly disposed [6].

79.6 Summary and Conclusions

This chapter focused on water reclamation in the context of sustainability, especially in the present threat of water scarcity. Water reclamation and reuse are a water scarcity solution, and are significant because they can increase the available water resource, reduce eutrophication, reduce cost, and sustain human life. The chapter indicated that water reclamation is an essential approach to solving water scarcity problems, water treatments require adequate technology, and water reuse requires education on public opinion, sufficient regulations, and environmental planning.

References

1. Angelakis, A.N. and Bontoux, L. 2001. Wastewater reclamation and reuse in Eureau countries. *Water Policy*, 3(1), 47–59.
2. Angelakis, A.N. and Durham, B. 2008. Water recycling and reuse in EurEau countries: Trends and challenges. *Desalination*, 218(1), 3–12.
3. Angelakis, A.N. and Koutsoyiannis, D. 2003. Urban water engineering and management in Ancient Greece. In: Stewart, B.A. and Howell, T. (Ed.), *The Encyclopedia of Water Science*, Dekker, New York, pp. 999–1007.
4. Barbagallo, S., Cirelli, G.L., and Indelicato, S. 2001. Wastewater reuse in Italy. *Wastewater Reclamation, Recycling and Reuse*, 43(10), 43–50.
5. Beekman, G.B. 1998. Water conservation, recycling and reuse. *International Journal of Water Resources Development*, 14(3), 353–364.
6. Bixio, D., Thoeye, C., Wintgens, T., Ravazzini, A., Miska, V., Muston, M., and Melin, T. 2008. Water reclamation and reuse: Implementation and management issues. *Desalination*, 218(1), 13–23.
7. Burton, I. 1987. Report on reports: Our common future: The world commission on environment and development. *Environment: Science and Policy for Sustainable Development*, 29(5), 25–29.
8. Cairncross, S. 1989. Water supply and sanitation: An agenda for research. *Journal of Tropical Medicine and Hygiene*, 92, 301–314.
9. Casani, S., Rouhany, M., and Knøchel, S. 2005. A discussion paper on challenges and limitations to water reuse and hygiene in the food industry. *Water Research*, 39(6), 1134–1146.
10. Chang, D. and Ma, Z. 2012. Wastewater reclamation and reuse in Beijing: Influence factors and policy implications. *Desalination*, 297, 72–78.
11. Chang, I.S. Chung, C.M., and Han, S.H. 2001. Treatment of oily wastewater by ultrafiltration and ozone. *Desalination*, 133(3), 225–232.
12. Chiaudani, G. and Premazzi, G. 1988. Water quality criteria in environmental management. *Commission of the European Communities. Luxemburg, Environment and Quality of Life Report EUR 11638 EN*, p. 78.
13. De Regt, J. 2005. Water in rural communities. In: *African Water Laws: Plural Legislative Frameworks for Rural Water Management in Africa*. Proceedings of a Workshop Held in Johannesburg, South Africa. January 26–28, 2005.
14. Directive C. 1998. 98/83/EC of November 3, 1998 on the quality of water intended for human consumption. *Official Journal of the European Communities*, 5(98), L330.
15. Durham, B., Rinck-Pfeiffer, S., and Guendert, D. 2003. Integrated water resource management—through reuse and aquifer recharge. *Desalination*, 152(1), 333–338.

16. Global Water Partnership. 2012. Unlocking the door to social development and economic growth: How a more integrated approach to water can help, Policy Brief, Technical Committee.

17. Henze, M. 2006. Wastewater, volumes and composition. In: *Wastewater Treatment: Biological and Chemical Processes*, Henze, M., Harremoes, P., Cour Jansen, J. la, and Arvin, E. (Eds.), Springer-Verlag, New York, 421 pp.

18. Hoko, Z. and Hertle, J. 2006. An evaluation of the sustainability of a rural water rehabilitation project in Zimbabwe. *Physics and Chemistry of the Earth, Parts A/B/C*, 31(15), 699–706.

19. Iglesias, R., Ortega, E., Batanero, G., and Quintas, L. 2010. Water reuse in Spain: Data overview and costs estimation of suitable treatment trains. *Desalination*, 263(1), 1–10.

20. Janosova, B., Miklankova, J., Hlavinek, P., and Wintgens, T. 2006. Drivers for wastewater reuse: Regional analysis in the Czech Republic. *Desalination*, 187(1), 103–114.

21. Jones, J.A. and Van der Walt, I.J. 2004. Challenges for water sustainability in Africa. *GeoJournal*, 61(2), 105–109.

22. Kayaga, S., Mugabi, J., and Kingdom, W. 2013. Evaluating the institutional sustainability of an urban water utility: A conceptual framework and research directions. *Utilities Policy*, 27, 15–27.

23. Lu, W. and Leung, A.Y. 2003. A preliminary study on potential of developing shower/laundry wastewater reclamation and reuse system. *Chemosphere*, 52(9), 1451–1459.

24. Meda, A. and Cornel, P. 2010. Aerated biofilter with seasonally varied operation modes for the production of irrigation water. *Water Science and Technology*, 61(5): 1173–1181.

25. Norton-Brandão, D., Scherrenberg, S. M., and van Lier, J. B. 2013. Reclamation of used urban waters for irrigation purposes—a review of treatment technologies. *Journal of Environmental Management*, 122, 85–98.

26. Pedrero, F., Kalavrouziotis, I., Alarcón, J.J., Koukoulakis, P., and Asano, T. 2010. Use of treated municipal wastewater in irrigated agriculture—Review of some practices in Spain and Greece. *Agricultural Water Management*, 97(9), 1233–1241.

27. Peter, G. and Nkambule, S.E. 2012. Factors affecting sustainability of rural water schemes in Swaziland. *Physics and Chemistry of the Earth, Parts A/B/C*, 50, 196–204.

28. Rodda, J.C. 2001. Water under pressure. *Hydrological Sciences Journal*, 46(6), 841–854.

29. Russell, S. and Lux, C. 2006. Water recycling and the community—Public responses and consultation strategies: A literature review and discussion. *OzAQUAREC WP5 Report*.

30. Salgot, M. 2008. Water reclamation, recycling and reuse: Implementation issues. *Desalination*, 218(1), 190–197.

31. Schnoor, J.L. 2010. Water sustainability in a changing world, The 2010 Clarke Prize Lecture, National Water Research Institute, California.

32. Singh, V.P., Khedun, C.P., and Mishra, A.K. 2014. Water, environment, energy, and population growth: implications for water sustainability under climate change. *Journal of Hydrologic Engineering*, 19(4), 667–673.

33. Sneddon, C., Howarth, R.B., and Norgaard, R.B. 2006. Sustainable development in a post-Brundtland world. *Ecological Economics*, 57(2), 253–268.

34. Tsagarakis, K. and Georgantzs, N. 2003. The role of information on farmers' willingness to use recycled water for irrigation. *Water Supply*, 3(4), 105–113.

35. UNEP/GRID-Arendal. 2009. *Increased Global Water Stress*, Vital Water Graphics 2, UNEP, Norway.

36. United Nations International Strategy for Disaster Reduction Regional Office for Africa, UNISDR. 2012. *Disaster Reduction in Africa*, Special issue on drought 2012.

37. USEPA. 1986. Quality criteria for water, EPA-440/5-86-001, Office of Water Regulations and Standards, Washington, DC.

38. Vigneswaran, S. and Sundaravadivel, M. 2004. Recycle and reuse of domestic wastewater. In: *Encyclopedia of Life Support Systems (EOLSS)*, UNESCO, Eolss Publishers, Oxford, UK, http://www.eolss.net.

39. Wade, M.G. 2006. Integrated concepts in water reuse: Managing global water needs. *Desalination*, 187(1), 65–75.

40. World Health Organization (WHO). 1972. *Health Hazards of Human Environment*. WHO, Geneva.

41. World Health Organization (WHO). 2011. Regional consultation on health of the urban, *Proceedings of the 2010 Regional Consultation of Mumbai, India, Regional Office for South East Asia*, UNFPA, p. 82.

42. Yang, H. and Abbaspour, K.C. 2007. Analysis of wastewater reuse potential in Beijing. *Desalination*, 212(1), 238–250.

80

Sustainable Reuse and Recycling of Treated Urban Wastewater

Atef Hamdy
*CIHEAM/Mediterranean
Agronomic Institute*

Saeid Eslamian
Isfahan University of Technology

PREFACE

Treated and reused treated sewage water is becoming a common source for additional water in some water-scarce regions and many countries have included wastewater reuse in their water planning. Expansion of urban population and increased coverage of domestic water supplies and sewage network will give rise to greater quantities of municipal wastewater, which can become a new water source, particularly, for irrigation. The reuse of such marginal quality waters can be significant in terms of national water budgets, particularly when good quality water is limited.

The increased demand on freshwater resources is a global concern, where in the arid and semi-arid countries it becomes a serious challenge.

Treated sewage effluents are a major component of wastewater. This source of water is considered somewhat attractive, being renewable and accessible in peri-urban areas, a cheap source to be disposed of, and its quantity grows with the expansion of sanitation networks installed for urban or rural communities.

The current overall contribution of treated wastewater to agriculture could hardly exceed 4% of this sector's demand of freshwater. This small percentage, however, should be carefully considered when drafting water master plans, not only for satisfying some irrigation requirements, but also for the fact that reuse of treated effluent is safe, feasible, and an environmentally sound method of wastewater disposal.

Indeed, for most developing countries, those suffering severe shortage in available water resources, reuse of wastewater may have a greater impact on future usable sources of water than any technological solutions available for increasing water such as water harvesting, weather modification, or desalination.

The increasing shift from water management to pollution prevention and/or from wastewater as nuisance to wastewater as a valuable resource is remarkable. The over-riding approach and recommendation is "not to produce an environmental waste but to use technology, to control or to manage the waste" [25]. Therefore, recycling and reuse become the real issue to achieve more efficient and safe irrigation with the wastewater.

Most governments of arid and semi-arid regions have included wastewater reuse in their water resources planning. Policies have been formulated, but few have had the capacity to implement them in their water management practices, in terms of actions to deal with water pollution control and waste disposal. This chapter is focusing on the requirements for successful development and management of wastewater reuse and recycling in agriculture through having updated knowledge and a better acquaintance of the recent techniques and advanced technology in this field.

80.1 Introduction

Demands on water resources for household, commercial, industrial, and agricultural purposes are increasing greatly. The world population will have grown 1.5 times over the second half of the twenty-first century, but the worldwide water usage has been growing at more than three times the population growth. In most countries, human populations are growing while water availability is not. What is available for use, on a per capita basis, therefore, is falling. Out of 100 countries surveyed by the World Resources Institute in 1986, more than half of them were assessed to have low to very low water availability, and quality of water has been the key issue for the low water availability.

The need for increased water requirement for the growing population in the new century is generally assumed, without considering whether available water resources could meet these needs in a sustainable manner. The question about from where the extra water is to come has led to a scrutiny of present water use strategies. A second look at strategies has drawn a picture of making rational use of already available water, which if used sensibly, there could be enough water for all. The new look invariably points at recycle and reuse of wastewater that is being increasingly generated due to rapid growth of population and related developmental activities, including agriculture and industrial productions.

Nowadays, politically and technically the important role of treated wastewater and its reuse could play in reducing the enormous gap between the increasing water demand and limited fragile water supply particularly in many arid and semi-arid developing countries is well recognized. Indeed, for some water-scarce regions including those of the Middle East and Mediterranean, treated wastewater is becoming a common source for additional water and is already included in their master plans. This source of water is considered somewhat attractive being renewable and continuously increasing with time, cheap source, and rich in its content of plant nutrients as well as not affected by climate change.

The reuse of wastewater has been successful for irrigation of a wide array of crops, and increases in crop yields from 10%–30% have been reported [11]. In addition, the reuse of treated wastewater for irrigation and industrial purposes can be used as a strategy to release freshwater for domestic use, and to improve the quality of river waters used for abstraction of drinking water (by reducing disposal of effluent into rivers). Wastewater is used extensively for irrigation in certain countries for example, 67% of total effluent of Israel, 25% in India, and 24% in South Africa is reused for irrigation through direct planning, though unplanned reuse is considerably greater.

The economic, social, and environmental benefits of the reuse of wastewater are quite clear. However, to avoid any environmental degradation and possible health risks, it is needed to adapt an integrated water management, disseminate existing knowledge, and monitor and enforce standards. Equally, it is important to strengthen the capacity of national and local hydrological research institutes to improve their links with those dealing with environmental, economic, and social aspects. Failure in managing and governing the use and recycling of treated wastewater should be counteracted by improving the efficiency of public administration at local levels through the building of responsibilities, combining management and financing functions, improving environmental legislation, and monitoring, dismounting bureaucratic, decentralizing tasks to the lowest levels possible as well as enhancing the skills of the public administration employees.

Nowadays due to the increasing shortage of available water resources, many countries indicate that water supply and sanitation will be one of the main future challenges in a world of growing population and industrialization. The growing awareness of water resource scarcity, the competition for water resources, and the negative impact of contaminated water on human health and environment demand the development of adequate strategies in water management. Next to the development of new management strategies to supply freshwater, the issue of treating and recycling wastewater will play an important role in tackling the existing and occurring problems.

80.1.1 Wastewater History

Wastewater reuse is not a recent invention. There are indicators that wastewater was used for irrigation in ancient Greece and in the Mayan civilization (ca. 3000–1000 BC) [4,6].

During 1950–1960, interests in applying wastewater on land in the western hemisphere as wastewater treatment technology advanced and quality of treated effluents steadfastly improved. Land application became a cost-effective method of discharging effluent into surface water bodies [5].

The potential of wastewater reuse to overcome shortage of freshwater existed in Minoan civilization in ancient Greece, where indications for utilization of wastewater for agricultural irrigation dates back 5000 years. Sewage form practices has been recorded in UK and Germany since the sixteenth and eighteenth centuries, respectively. Irrigation with sewage and their wastewaters has a long history also in India and China.

In the nineteenth century, the discharge of wastewater into surface water bodies led to indirect use of sewage and other wastewaters as unintentional potable water supplies. However, it should be clearly emphasized that such unplanned water reuse coupled with inadequate water and wastewater treatment resulted in catastrophic epidemics of waterborne diseases during the 1840s and 1850s.

For the last three decades, the value of wastewater is becoming increasingly understood in arid and semiarid countries and many countries are now looking forward to ways for improving and expanding wastewater reuse practices [7,10].

80.1.2 The Importance of Wastewater Use in Selected Countries

Wastewater has been recycled in agriculture for centuries as a means of disposal in cities such as Berlin, London, Milan, and Paris [1]. In China, India, and Viet Nam, wastewater has been used to provide nutrients and improve soil quality. In recent years, wastewater has gained importance in water-scarce regions. In Pakistan, 26% of national vegetable production is irrigated with wastewater. Any changes to this practice would reduce the supply of vegetables to cities [15]. In Hanoi, 80% of vegetable production is from urban and peri-urban areas irrigated with wastewater and water from the Red River Delta, which receives drainage effluent from the city [29]. Around Kumasi, Ghana, informal irrigation involving diluted wastewater from rivers and streams occurs on an estimated 11,500 ha, an area larger than the reported extent of formal irrigation in the country [28]. In Mexico, about 260,000 ha are irrigated with wastewater, mostly untreated [34].

In the United States, municipal water reuse accounted for 1.5% of water withdrawals in 2000. California residents reuse 656 million cubic meters of municipal wastewater annually. In Tunisia, reclaimed water accounted for 4.3% of available water resources in 1996 and could reach 11% by 2030. In Israel, wastewater accounted for 15% of water resources in 2000 and could reach 20% by 2010.

80.1.3 What Is Wastewater Reuse?

The term wastewater reuse is often used synonymously with the terms wastewater recycling and wastewater reclamation. Because the general public often does not understand the quality difference between treated and untreated wastewater, many communities have shortened the term to water reuse, which creates a more positive image.

The US Environmental Protection Agency [41] defines wastewater reuse as, "using wastewater or reclaimed water from one application for another application. The deliberate use of reclaimed water or wastewater must be in compliance with applicable rules for a beneficial purpose (landscape irrigation, agricultural irrigation, aesthetic uses, ground water recharge, industrial uses, and fire protection). A common type of recycled water is water that has been reclaimed from municipal wastewater (sewage)."

80.1.4 What Is Water Recycling?

It is reusing treated wastewater for beneficial purposes such as agricultural and landscape irrigation, industrial processes, toilet flushing, and replenishing groundwater basin referred to as

---

BOX 80.1 WATER RECYCLING AND REUSE: DEFINITIONS

Reclaimed water: Treated wastewater suitable for beneficial purposes such as irrigation.

Reuse: The utilization of appropriately treated wastewater (reclaimed water) for some further beneficial purpose.

Recycling: The reuse of treated wastewater.

Potable substitution: The reuse of appropriately treated reclaimed water instead of potable water for nonpotable applications.

Nonpotable reuse: The use of reclaimed water for other than drinking water, for example, irrigation.

Indirect recycling or indirect potable reuse: The use of reclaimed water for potable supplies after a period of storage in surface or in groundwater.

Direct potable reuse: The conversion of wastewater directly into drinking water without any interim storage.

groundwater recharge. Recycled water can satisfy most water demands, as long as it is adequately treated to ensure water quality appropriate for the use.

Although most water recycling projects have been developed to meet notable water demands for nonpotable purposes, a number of projects use recycled water indirectly for potable purposes, such as recharging groundwater aquifers through spreading or injecting recycled water into groundwater supplies to prevent saltwater intrusion in coastal areas.

80.2 Benefits of Reuse and Recycling of Wastewater

In addition to providing dependable, locally controlled water supply, water recycling provides tremendous environmental benefits including the following:

• Decrease diversion of freshwater from sensitive ecosystems

People who reuse water can supplement their demands by using a reliable source of recycled water, which can free considerable amounts of water for the environment and increase flows to vital ecosystems and thereby avoid the diversion of freshwater from sensitive ecosystems that can cause deterioration of water quality and ecosystem health.

• Enhance wetlands and riparian (stream) habitats

Experiences gained demonstrate clearly that streams that have been impaired or dried from water diversion, water flow can be augmented with recycled water to sustain and improve the aquatic and wildlife habitat.

• Prevent and reduce pollution

A clear demonstration for such benefit can be seen with recycled water containing higher levels of nutrients, such as nitrogen. Application of recycled water for agricultural and landscape irrigation can provide an additional source of nutrients and lessen the need to apply synthetic fertilizers [14].

• Saving energy

Recycling water onsite or nearby reduces the energy needed to move water longer distances or pump water from deep within an aquifer. Tailoring water quality to a specific water use also reduces the energy needed to treat water using recycled water that is of lower quality for uses that do not require high quality water, which saves energy and money by reducing treatment requirements.

As water energy demands and environmental needs grow, water recycling will play a greater role in our overall water supply. Future perspectives give the impression that by working together to overcome obstacles, water recycling, along with water conservation and efficiency, can help us to sustainably manage our vital water resources.

In addition to the environmental benefits mentioned above, through the reuse and recycling of wastewater, other benefits could be obtained including:

1. Conservation of freshwater resources for other uses
2. Reduction in the need for wastewater infrastructure for treatment and disposal of sewage
3. Reduction in the need for nitrate and phosphate removal by the treatment system because these elements have added value in agriculture [24,35]
4. Reduction of pollution from disposal of sewage into the environment [21]
5. Combating desertification [3]

However, in spite of the benefits gained by reusing treated wastewater in irrigation, there are associated health risks due to contact with reclaimed wastewater. Consequently, it is of utmost importance to make sure that these risks are minimized and monitored [38].

80.3 Why Wastewater Recycling and Reuse: Motivational Factors

There are several motivational factors for wastewater recycling/reuse, among them the following:

• Reuse of wastewater can be a supplementary source to existing water resources especially in arid and semi-arid climatic regions and thereby having greater opportunities to augment limited primary water sources. This also holds true even in regions where rainfall is adequate. Because of its spatial and temporal variability, water shortages are created [2].
• Costs associated with water supply or wastewater disposal can also make reuse of wastewater an alternative option. Indeed, minimization of infrastructure costs, including total treatment and discharge costs, has been the motivation behind many reuse schemes in Japan.
• Avoidance of environmental problems arising due to discharge of treated/untreated wastewater to the environment is another factor that encourages reuse. While the nutrients in wastewater can assist plant growth when reused for irrigation, their disposal, in extreme cases, is detrimental to ecosystems of the receiving environment. In addition, there may be concerns about the levels of other toxic pollutants in wastewater.
• Scarcity stress and competition climate change are likely to aggravate the scarcity of water that is being driven by other basic forces. On one authoritative view, global warming of 2°C would lead to a situation where "between 100 million and 400 million more people could be at risk of hunger, and 1 to 2 billion more people may no longer have enough water to meet their consumption, hygiene and food needs" [45].

From an environmental perspective, water reuse can reduce demand for freshwater resources, diversify water sources, and enhance reliability of access to resources; it can reduce volume of wastewater discharged into the environment. Decentralized systems can reduce energy required to transport water from the point of production to the point of use and reduce greenhouse gas emissions (due to energy savings) [9,17,21].

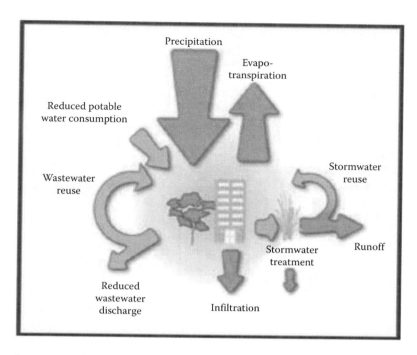

FIGURE 80.1 Sustainable urban water cycle through water reuse. (Adapted from Healthy Waterways. 2011. What Is Water Sensitive Urban Design? http://waterbydesign.com.au/whatiswsud/. Accessed February 2, 2012.)

80.4 Challenges to Reusing and Recycling Water

- *Public acceptance barriers regarding health issues*: The most critical hurdle for water reuse is fears and uncertainties about the health risks, which can severely affect public acceptance. This then impedes the implementing of policies. Unfavorable and uncoordinated regulatory framework can increase such fears and uncertainties [22,43].
- *Technical barriers*: Wastewater treatment for reuse requires such technologies as a membrane and constructed wetlands [25]. Insufficient technical capacity can hinder the installation and maintenance of the wastewater treatment systems. Although high-tech treatment generally produces high-quality reclaimed water, it is not easily a viable option in developing countries [12].
- *Financial barriers*: Initial costs to install wastewater treatment technologies can be expensive, thus making a reuse system unaffordable for households. The unfavorable pricing of freshwater can be a significant hurdle to promote water reuse. If freshwater pricing is too low, it does not drive people to reuse water [43].

80.5 Water Reuse: How It Works

Water reuse can do two things: (1) minimize freshwater demand, and (2) reduce wastewater treatment needs. As a result, water reuse minimizes new water extraction and wastewater effluent, thus enabling the continuous cycle of water in an urban setting.

By minimizing new water inflow and wastewater effluent, water reuse makes the urban water cycle more compact and sustainable. Figure 80.1 shows how water reuse manipulates the directions of unfavorable water flows and creates a cycle of water in an urban setting.

The basic principle of water reuse is to reduce the inefficient mismatch between available water resources and the specific purposes of water use. Although freshwater sources are becoming scarce, it is also true that precious freshwater is inefficiently used for nonpotable purposes, such as irrigation. A 1958 UN Economic and Social Council resolution stated, "No higher quality of water, unless there is surplus of it, should be used for a purpose that can tolerate a lower grade." In order to avoid the inefficient use of precious water sources, an eco-efficient water system is needed that allocates water types to appropriate purposes [42].

80.6 Recognized Wastewater as a Resource

There are several reasons that strongly demonstrate that wastewater should be recognized as an alternative additional water source, among them the following: First, it preserves the high quality, expensive freshwater for the highest value purposes primarily for drinking. The cost of secondary-level treatment for domestic wastewater in MENA, an average of \$US $0.5/m^3$, is cheaper, in most cases much cheaper, than developing new supplies in the region [44]. Second, collecting and treating wastewater protects existing sources of valuable freshwater, the environment in general, and public health. In fact, wastewater treatment and reuse (WWTR) not only protects valuable fresh-water resources, but also it can supplement them, through aquifer recharge. If the true, enormous, benefits of environmental and public health protection were correctly factored into economic analyses, wastewater collection, treatment, and reuse would be one of the highest priorities for scarce public and development funds. Third, if managed properly, treated wastewater can sometimes be a superior source for agriculture than some freshwater sources. It is a constant water source, and nitrogen and phosphorus in the wastewater may result in higher yields than freshwater irrigation, without additional fertilizer application [34].

Another fundamental reason is the prevention of surface water pollution, which could occur when wastewater is not used but is discharged in rivers and lakes. Planned reuse of wastewater for irrigation will greatly help in the elimination of several environmental pollution problems: dissolved oxygen depletion, eutrophication, foaming, fish deaths, etc. [21]. The quantification of the cost of the enormous damage those problems are already causing shows that the treatment and reuse of wastewater is now a must. The actual cost of wastewater treatment is definitely lower than the cost of environmental damage caused by wastewater. In addition, the use of wastewater for irrigation will help in reducing the overpumping and exploitation of groundwater quality being the main source of drinking water supply [13].

One of the economic benefits of wastewater reuse in arid and semi-arid areas is the boosting of agriculture development that would not be possible without a constant and reliable supply of water. This component (agriculture development) is generally neglected in the cost-benefit since the analysis is generally neglected in the operational costs on the one hand, and potential revenues from selling the treated wastewater on the other. Other issues, which are generally overlooked, are the resulting environmental and public health protection even if they are conspicuous, due to the difficulties in assigning economic values to these parameters [24].

In arid and semi-arid regions, improving water productivity and realizing real water savings in poorly performing irrigation systems is an integral part of efforts to bridge the gap between present day and attainable yields [18]. Taking as an example the arid regions of the Mediterranean as those characterized by severe water imbalance uneven rainfall, and at the same time, increased demands for irrigation and domestic water supply as results of expanding urban population and tourist industry have brought water resources availability to be the major constraint for any sustainable development programs. In those arid regions, an appropriate approach for solving water problems could be through much smaller scale planned local water recycling and reuse. Such approaches have become increasingly important for the following two reasons: First, properly treated municipal wastewater often is a significant water resource that can be used for a number of beneficial purposes, such as agricultural and landscape irrigation. Second, discharge of sewage effluent into surface water is becoming increasingly difficult and expensive as treatment requirements become more stringent to protect receiving waters such as rivers, estuaries, and beaches.

80.7 Wastewater as a Potential Water Saving Source

There are several activities not requiring water of potable quality. Reclaimed water may be substituted for conventional resources. Treated wastewater effluents may be reused for irrigation and purposes such as nonagricultural irrigation (parks, green areas, golf courses, etc.), industrial recycling and reuse (cooling water, boiler feed, process water), fire fighting, and groundwater recharge. Therefore, in arid and semi-arid countries, wastewater is considered a potential water saving resource. Taking the case of the Mediterranean region (Table 80.1), the analysis of the presented data indicates that the annual water use in domestic and industrial sectors could reach 83 BMC. Assuming 80% of wastewater will be collected and treated, the annual collected wastewater could reach 66.7 BMC. The existing wastewater reuse is estimated at 0.75 BMC in the Mediterranean countries [16]. The potential treated wastewater for reuse can therefore be estimated at 66 BMC/year in the Mediterranean region. Based on the water demand of year 2025, and assuming that this water could be satisfied, the savings using treated wastewater could reach 70 BMC/year. The cost to achieve this savings is estimated to be 55 billion Euros, which include the need to fill the gap in water supply and sanitation coverage for 25 million people without access to water and to treat the wastewater effluents.

Countries in the Mediterranean region that practice wastewater treatment and reuse include Spain, France, Cyprus, Malta, Tunisia, Israel, Italy, Greece, Portugal, and Egypt. However, only Israel, Cyprus, and Tunisia, and to a certain extent, Jordan, already practice wastewater treatment and reuse as an integral component of their water management and environmental protection strategies.

Agricultural wastewater reuse, however, will not see marked improvement unless restrictions are lifted on pilot wastewater treatment plants with complementary treatment processes. This can only be decided when the stations are functioning with acceptable reliability. This will take a few years of experience [19,20].

80.8 Safe Use of Wastewater

Nonetheless, in all cases, and regardless of the treatment method, technical and organizational measures should be introduced in order to systematically warn those managing the reuse of any breakdowns that may occur in the wastewater treatment plants and to avoid the flow of treated wastewater into the distribution network [33].

Even though the international community recognizes that "safe use of wastewater in agriculture" is an important water resources issue that needs to be addressed, much work still needs to be done to advance the topic in national policies and to implement guidelines and safe practice. The keyword therein is "safe," and one of the essentials is to understand that wastewater is a valuable resource. Addressing the safe use of wastewater

TABLE 80.1 Annual Domestic and Industrial Water Use and Potential Treated Wastewater for Reuse

	Potential Total Irrigation Savings	Potential Total Domestic Savings	Potential Total Industrial/Commercial Savings	Potential Treated Wastewater for Use	Total Potential Water Savings
	M m³/yr	M m³/yr	M m³/yr	M m³/yr	M m³/yr
Syria	1360.0	174.1	3.5	135.9	1673.5
Lebanon	95.0	99.7	1.9	286.3	482.9
Jordan	73.8	71.0	0.4	89.7	234.9
Egypt	4773.0	1079.4	55.5	5108.1	11,016.0
Libya	400.2	161.9	1.2	248.0	811.2
Tunisia	270.6	91..5	1.2	201.1	564.3
Algeria	270.0	368.3	8.4	1138.6	1785.4
Morocco	1016.1	186.5	4.1	553.9	1760.6
Albania	99.4	129.3	0.0	221.4	450.1
Croatia	0.0	121.7	4.8	506.9	633.3
Cyprus	15.6	16.1	0.1	20.1	51.8
France	488.0	1947.9	371.1	26,776.5	29,583.5
Greece	569.4	359.0	2.6	779.2	1710.3
Italy	2537.6	2136.9	1S7.5	16,135.3	21,007.4
Malla	0.7	15.5	0.0	25.3	41.4
Spain	2415.4	1472.1	79.9	7567.3	11,534.7
Turkey	2591.5	1818.8	48.8	6.173.9	10,633.0
Total	16,976	10,250	781	65,968	93,974

Source: Adapted from Cornish, G.A., Mensah, E. and Ghesquiere, P. Water quality and peri-urban irrigation: An assessment of water quality for irrigation and its implications for human health in the peri-urban zone of Kumasi, Ghana. KAR project R7132, H R Wallingford. September, 1999.

in agriculture is an important step. In particular, it needs to be understood that, where water is scarce, a lack of safe practices and guidelines will not prevent the use of wastewater, but will rather result in unsafe practices of use [32].

Water has a precious value and each drop must be accounted for in water-scarce regions such as the Middle East and North Africa. Therefore, although the reuse of municipal wastewater will require more complex management practices and stringent monitoring procedures than when good quality water is used, wastewater has to be reclassified as a renewable water resource rather than waste as it helps increase water availability and, at the same time, reduce environmental degradation [8,30].

80.9 Wastewater and Global Changes

The evidence for climate change is now considered to be unequivocal, and trends in atmospheric carbon dioxide (CO_2), temperature, and sea-level rise are tracking the upper limit of model scenarios elaborated in the Fourth Assessment (AR4) undertaken by the International Panel on Climate Change (IPCC). There remain many scientific questions related to cause and effect that are not yet fully explained, but the probable future costs of climate change are so significant that action now is considered to be a prudent insurance. Current negotiations focus on stabilizing end of century temperatures at no more than 2°C to minimize negative impacts. The criticism that climate science has recently taken does not detract from the reality or the gravity of the clear trends in global climate.

In response to global warming, the hydrological cycle is expected to accelerate as rising temperatures increase the rate of evaporation from land and sea. Thus, rainfall is predicted to rise in the tropics and higher latitudes, but decrease in the already dry semi-arid to arid mid-latitudes and in the interior of large continents. Water-scarce areas of the world will generally become drier and hotter. Both rainfall and temperatures are predicted to become more variable, with a consequent higher incidence of droughts and floods, sometimes in the same place. Anticipation of more drought and extreme rainfall events has impacts on nonexistent or old, inadequate wastewater treatment facilities highlighting the need for infrastructure that can cope with extreme surges of wastewater.

As the global population heads for more than 9 billion people by 2050 (under medium growth projections), the world is rapidly becoming urbanized. Future global food demand is expected to increase by some 70% by 2050, but will approximately double for developing countries. All other things being equal (that is, a world without climate change), the amount of water withdrawn by irrigated agriculture will need to increase by 11% to match the demand for biomass production. The question is how to provide such huge quantities of water, the time most developing arid and semi-arid countries are meeting notable shortage in their available water resources.

Estimates of incremental water requirements to meet future demand for agricultural production under climate change vary from 40% to 100% of the extra water needed without global warming. The amount required as irrigation from ground or surface water depends on the modeling assumptions on the expansion of irrigated area—between 45 and 125 million ha. One consequence of greater future water demand and likely reductions in supply is that the emerging competition between the environment and agriculture for raw water will be much greater, and the matching of supply and demand will be consequently harder to reconcile.

Here come the important roles of reuse and recycling of non-conventional water resources—treated wastewater.

Future perspectives give the evidence that cities will generate increasingly large amounts of effluent that will be recycled for agriculture, subject to water quality and health and safety considerations. The use of wastewater for crop irrigation reduces the use of artificial fertilizers and is thus an important form of nutrient recycling. At an irrigation rate of 1.5 m/year (i.e., 1.5 m^3 of irrigation water per m^2 of field area per year), a typical requirement in a semi-arid climate, treated municipal wastewater can supply 225 kg of nitrogen and 45 kg of phosphorus per hectare per year.

80.10 Wastewater and Climatic Change Relations

The relationship between wastewater and climate change can be seen from three perspectives. First, changing climatic conditions change the volume and quality of water availability in both time and space, thus influencing water usage practices. Second, changes in climate will also require adaptation, in terms of how wastewater is managed. Finally, wastewater treatment results in the emission of greenhouse gases, particularly carbon dioxide (CO_2), methane(CH_4), and nitrous oxide (N_2O).

How wastewater is treated can in turn have an impact on climate change. Wastewater and its treatment generate methane, nitrous oxide, and carbon dioxide. It is worth noting that methane has an impact 21 times greater than the same mass of carbon dioxide. Nitrous oxide is 310 times more potent [1]. Methane emissions from wastewater are expected to increase almost 50% between 1990 and 2020, while estimates of global N_2O emissions from wastewater are incomplete there suggests an increase of 25% between 1990 and 2020 [39]. There is a pressing need to investigate and implement alternatives to current wastewater treatment, which minimize the production of greenhouse gases and power consumption.

In light of rapid global change, communities should plan wastewater management against future scenarios, not current situations. Wastewater management and urban planning lag far behind advancing population growth, urbanization, and climate change. With forward thinking, and innovative planning, effective wastewater management can contribute to the challenges of water scarcity while building ecosystem resilience, thus enabling ecosystem-based adaptation and increased opportunities for solutions to the challenges of current global change scenarios. Population growth and climate change are not uniform in time or space, and so regionally specific planning is essential. Wastewater management must be integrated as part of the solution in existing agreements and actions [18].

80.11 The Challenges of Wastewater Reuse

There have been substantial developments in (waste) water management and treatment technology worldwide during the past decades. In spite of that, in 1997 three billion people on Earth lacked adequate sanitation. In Africa alone, 80 million people are at risk from cholera, and 16 million cases of typhoid infections each year are a result of lack of clean drinking water and adequate sanitation [43]. According to the Kyoto summit in 2003, 2 billion people will not have access to safe drinking water supplies in the year 2015.

The pollution of scarce water resources by untreated or poorly treated wastewater is of great concern worldwide. Contamination of surface water induces algal blooms, fish kill, ecological imbalances, and odor problems, whereas high levels of nitrate make groundwater unfit for potable supplies. Lack of pollution control in general and nitrous emission control in particular might aggravate the availability of scarce water resources, especially in arid and semi-arid areas of the world.

Urban wastewater treatment has received less attention compared to water supply and management. Water scarcity coupled with the bursting seams of our cities and towns have taken a toll on our health and environment. The sewage contamination of our lakes, rivers, and domestic water bodies has reached dangerous levels and is being recognized by leading organizations like the World Bank.

In a developing urban society, the wastewater generation usually averages 30–70 cubic meters per person per year. In a city of 1 million people, the wastewater generated would be sufficient to irrigate approximately 1500–3500 hectare. This urban epidemic needs to be tackled ecologically because of so many pressing issues that are afflicting our waste management process:

- New immigrants to cities have low incomes and cannot afford municipal amenities like waste disposal and sanitary functions.
- In developing countries, approximately 300 million urban residents have no access to sanitation.
- Approximately two-thirds of the population in the developing world has no hygienic means of disposing excreta and an even greater number lack adequate means of disposing of total wastewater.
- It is often an acceptable practice to discharge untreated sewage directly into the bodies of water.

According to the World Bank, "The greatest challenge in the water and sanitation sector over the next two decades will be the implementation of low cost sewage treatment that will at the same time permit selective reuse of treated effluents for agricultural and industrial purposes" [31]. The other two major goals of such treatment are

- The recovery of nutrient and water resources for reuse in agricultural production
- Reducing the overall user-demand for water resources

80.12 Sustainable Water Reuse and Recycling: Major Elements

An important element in sustainable water reuse is the formulation of a framework of realistic, achievable, and enforceable standards for treated wastewater quality and applications. Monitoring and evaluation of water reuse programs and projects are fundamental and thus must overcome in developing countries in arid and semi-arid regions challenges of weak institutions, shortage of trained personnel, lack of monitoring equipment, and the relatively high cost required for monitoring processes.

80.12.1 Monitoring and Evaluation

Monitoring and evaluation of wastewater use programs and projects is a very critical issue, hence, both are the fundamental basis for setting the proper wastewater use and management strategies. Ignoring monitoring evaluation parameters and/or not performing monitoring regularly and correctly could result in serious negative impacts on health, water quality, and environmental and ecological sustainability.

Unfortunately, in many countries that are already using or starting to use treated wastewater as an additional water source, the monitoring and evaluation program aspects are not well developed, and are loose and irregular. This is mainly due to the weak institutions, the shortage of trained personnel capable of carrying the job, lack of monitoring equipment, and the relatively high cost required for monitoring processes.

In the developing countries, two types of monitoring are needed: the first, process control monitoring to provide data to support the operation and optimization of the system in order to achieve successful project performance; the second, compliance monitoring to meet regulatory requirements and not to be performed by the same agency in charge of process control monitoring.

In the developing countries, to avoid failure in wastewater use and to attain the desired success, the monitoring program should be cost-effective, and should provide adequate coverage of the system. Equally so, it must be reliable and timely in order to provide operators and decision-making officials with correct and up-to-date information that allows the application of appropriate remedial measures during critical situations.

80.12.2 Applying Realistic Standards and Regulations

An important element in the sustainable use of wastewater is the formulation of realistic standards and regulations. However, the standards must be achievable and the regulations enforceable.

Unrealistic standards and non-enforceable regulations may do more harm than having no standards and regulations because they create an attitude of indifference toward rules and regulations in general, both among polluters and administrators. In arid and semi-arid countries where wastewater is recognized, additional water source standards, guidelines, and regulations in the majority of developing countries do not consider the reuse aspect as an integrated part of the treatment process; they are only intended to control and protect the quality of water bodies where the reclaimed water is discharged. In reality, in the arid regions of the Near East, North Africa and southern Europe, not all countries have developed guidelines and regulations for reclaimed water use. For those countries, standards and regulations for the reuse should be tailored to match the level of economic and administrative capacity and capability standards should cope with the local prevailing conditions. Some countries have national guidelines for the acceptable use of wastewater for irrigation, many do not. The Guidelines on the Safe Use of Wastewater, Excreta and Greywater in Agriculture and Aquaculture [43] provide a comprehensive framework for risk assessment and management that can be applied at different levels and in a range of socioeconomic circumstances. The main characteristics of the approach proposed by the guidelines are

- The establishment of health-based targets, which allow local authorities to set risk levels that can be handled under the local socio-economic conditions and with the capacities available in a country
- The application of quantitative microbial risk assessment (for pathogenic viruses and bacteria) as a cost-effective way of assessing health risks
- The identification of all risk points along the chain of events from the origin of the wastewater to the consumption of the produce
- The design of a combination of health risk management measures, to be applied along the same chain of events, with the aim of ensuring health protection as a result of incremental risk reduction. Such interventions can include partial wastewater treatment
- Monitoring at all stages to ensure measures are effective, applied correctly, and lead to the desired impact on health.

In many countries, the capacity to apply these guidelines and best practice recommendations are insufficient and need substantial strengthening.

80.12.3 Formulation of National Policies and Strategies

Wastewater management requires the establishment of a clear policy; this policy should be compatible with a number of related sectorial or sub-sectorial policies such as national water management and irrigation policy, national health, sanitation and sewage policy, national agricultural policy, and national environmental protection policy.

Such policy should give guidance on the following issues [4,22]:

- The current and future contribution of treated wastewater to the total national water budget

- Criteria required to achieve maximum benefit of wastewater-reuse for the different water sectorial uses
- Modalities for strengthening the national capacity building in this sector

Ideally, policies of wastewater reuse and strategies for its implementation should be part of water resources planning at the national level. At the local level, individual reuse projects should be part of the overall river basin planning effort.

80.12.4 Verifying Institutional/Administrative Responsibilities and Duties

Safe water treatment, disposal, and reuse are the responsibility of different organizations such as authorities, cooperatives, and communities operating under the jurisdiction of the ministries of agriculture, water resources, and others. The responsibilities of these organizations must be considered and reconciled.

To tackle the range of institutional levels involved and to allocate responsibilities in both treatment and reuse stages, several actions are needed, including the following:

1. A well-defined policy and strategy for the comprehensive management and reuse of treated wastewater is a precondition to success.
2. Many different stakeholders are involved, so roles and responsibilities (who does what) need to be clearly defined, along with mechanisms to ensure the active coordination of the various institutions.
3. Inadequate legislation often hinders the effective reuse of treated wastewater. Integrated legal arrangements can be of great value, along with provisions for active enforcement of all laws and regulations, without exception.
4. A comprehensive plan of action for reusing treated wastewater, with clearly assigned roles, needs to be complemented by periodic reviews and follow-up. Adequate funding is essential.
5. Capacity building is required to analyze staff needs and provide suitable training.
6. More participatory approaches are needed, including raising the awareness of the general public (whose cultural and religious perceptions sometimes regard treated wastewater as impure). Irrigators also need to be involved in the planning and utilization of this resource.
7. More coordination is needed between donors and national institutions involved in wastewater reuse.

To reinforce and help consolidate improved arrangements in countries with many ministries involved, the possible formation of a higher council to create policy and strategies should be considered. This body could oversee implementation and obtain necessary funding.

Where many different laws complicate wastewater reuse, consideration could be given to consolidated legislation that would cover all aspects of water resources planning, management, and utilization.

80.12.5 Public Awareness and Participation

This is the bottleneck governing the wastewater use and its perspective progress. To achieve general acceptance of reuse schemes, it is of fundamental importance to have active public involvement from the planning phase through the full implementation process.

Some observations regarding social acceptance are pertinent. For instance, there may be deep-rooted sociocultural barriers to wastewater reuse. However, to overcome such an obstacle, major efforts are to be carried out by the responsible agencies.

Responsible agencies have an important role to play in providing the concerned public with a clear understanding of the quality of the treated wastewater and how it is to be used; confidence in the local management of the public utilities and in the application of locally accepted technology, give the assurance that the reuse application being considered will involve minimal health risks and minimal detrimental effects on the environment. In this regard, the continuous exchange of information between authorities and public representatives ensures that the adoption of a specific water reuse program will fulfill real user needs and generally recognized community goals for health, safety, ecological concerns program, cost, etc. [2].

80.12.6 Stakeholders and Users Participation

Achieving sustainable reuse and recycling of wastewater implies the full participation of the individuals (stakeholders, users), starting from the planning and designing phase, passing by the operational one and ending with implementation on the ground [19]. Stakeholders from each distinct group have a defined specific role to play in achieving sustainability. However, in spite of the important individual contribution of each group, what are really lacking are the appropriate links and the enabling conditions that facilitate working collectively rather than separately. The cumulative impact of their attitude and actions working as one team will definitely give strong push and straightforward mechanisms toward wastewater reuse sustainability [7].

80.13 Wastewater Management

Managing wastewater is essential for several reasons. First, wastewater is often discharged in places where it cannot be reused, or directly to the sea, thus losing an opportunity for beneficial use. Second, wastewater is often rich in plant nutrients and these and the residual water can both be put into beneficial use through irrigation. Reuse for agriculture, following primary or secondary stage treatment with low-cost ecological technologies, can be a cost-effective and win-win solution in these circumstances [2,28].

Some damage associated with the increasing scarcity of water in the world along with rapid population increase in urban areas gives rise to concern about appropriate water management practices. In the context of trends in urban development, wastewater treatment deserves greater emphasis. Currently, there is a growing awareness of the impact of sewage contamination on rivers and lakes.

Inadequate management and handling of wastewater are behind the following: creating direct or indirect costs caused by illness and mortality; higher costs for producing drinking and industrial water resulting in higher tariffs; and poor water quality, which results in loss of income and loss of valuable biodiversity. The global burden of human disease caused by wastewater pollution of coastal waters has been estimated at 4 million lost person-years annually,. This evidently indicates the important role wastewater management and reuse could play in reducing health hazards and environmental deterioration [23].

The current urban wastewater management system is a linear treatment system that is based on disposal. This traditional system needs to be transformed into a sustainable, closed-loop urban wastewater management system that is based on the conservation of water and nutrient resources. A huge loss of life-supporting resources is the result of failed organic wastewater recovery. For arid and semi-arid regions, there is an urgent need to develop an ecological wastewater management strategy that will result in the reduction of pathogens in surface and groundwater to improve public health.

Successful and sustainable management of wastewater requires a cocktail of innovative approaches that engage the public and private sector at local, national, and transboundary scales. Planning processes should provide an enabling environment for innovation, including at the community level but require government oversight and public management [37].

Irrigation practices with treated municipal wastewater showed, for example, an increase in yield production without the need for additional artificial fertilizers, but carry risks for consumer health—creating costs further down the chain. This again highlights the crosscutting nature of wastewater management that requires collaboration and dialogue between partners, who may not usually talk, for example, farmers, public health officials, municipal and waste managers, planners, and developers. Unmanaged wastewater can be a source of pollution, a hazard for the health of human populations and the environment alike. The Millennium Ecosystem Assessment [39] reported that 60% of global ecosystem services are being degraded or used unsustainably, and highlighted the inextricable links between ecosystem integrity and human health and wellbeing.

Non-appropriate wastewater management, or the lack of, has a direct impact on the biological diversity of aquatic ecosystems, disrupting the fundamental integrity of our life support systems on which a wide range of sectors from urban development to food production and industry depend. It is essential that wastewater management is considered as part of an integrated, ecosystem-based management that operates across sectors and borders. Freshwater and marine countries must adopt a multi-sectorial approach to wastewater management as a matter of urgency, incorporating principles of ecosystems-based management from the water shed, into the sea connecting sectors that will reap immediate benefits from better wastewater management. This implies that wastewater management should reflect the community and ecological needs of each downstream ecosystem and user. Improved ecosystem management, including integrated forestry, livestock, agriculture, wetland, and riparian management, will reduce and mitigate the effects of wastewater entering rivers, lakes, and coastal environments [36,40].

Indeed wastewater management must address not only the urban but also the rural context through sound and integrated ecosystem-based management including, for example, fishers, forests, and agriculture.

On the globe, the experiences gained address clearly the fact that unless wastewater management is given very high priority and dealt with urgently, the financial, environmental, and social costs in terms of human health, mortality, and decreased environmental health are projected to increase dramatically [44].

80.14 Capacity Development— Major Requirements

Capacity development has been defined by the Organization for Economic Cooperation and Development (OECD) as "the process by which individuals, groups, organizations, institutions and societies increase their abilities to: (i) perform core functions, solve problems, define and achieve objectives; and (ii) understand and deal with their development needs in a broad context and in a sustainable manner" [39]. This definition has three important aspects, namely, it

- Indicates that capacity is part of a continuing process
- Ensures that human resources and the way in which they are utilized are central to capacity development
- Recognizes the importance of the overall framework (system) within which individuals and organizations undertake their functions

Capacity for safe wastewater use in agriculture exists at three different, but closely related, levels:

- System level or context in which organizations, groups, and individuals operate
- Organization and group level within the system
- Individual level within organizations and groups

At each of these different levels, there are various dimensions of capacity for the safe wastewater use in agriculture:

- At the system level, dimensions of capacity include the policies, laws, regulations, and standards that provide a framework for safe wastewater use in agriculture, as well as the mechanisms for management, communication, and coordination among the different organizations involved.
- At the organizational level, the mission, structure, operational procedures, and culture of organizations involved in wastewater use in agriculture are important dimensions of capacity, in addition to their human resources, financial resources, information resources, infrastructure, etc.
- At the individual level, knowledge, skills, competencies, experience, and ethics are all part of capacity.

80.15 What Capacities Are Required?

In most developing countries as well as the developed ones, the sustainable use and management of wastewater implies the presence of capacity building supporting programs to develop

- Capacity to position the water sector in the broad social and political context and to deal with externalities
- Capacity to work with economics and financing as these relate to cross-sectorial and cross-boundary competition
- Capacity to create arrangements for, and launch process toward, integrated water resources management (IWRM) as a basis for competition resolution to sectorial water uses conflicts
- Capacity to assess and structure appropriate institutions
- Technical and scientific information system for data sharing
- Flexible "multi-model" education and training packages, maximally geared to address the specific characteristics of different training situations to provide the involved staff with the capability to work in multidisciplinary teams to think strategically and solutions oriented.

Regarding capacity development instruments, emphasis should be notably directed to technical training and education to provide the water sector institutions with human resources of a better capability to utilize, operate, and maintain the infrastructure for a longer time as well as correctly planning, managing, and allocating the available water resources among the competing sectorial water uses. This should be done in parallel with the updating and strengthening of the institutions. Education and training should have a wider scope and be tailored to fulfill the needs of all stakeholders involved in the process. However, needed capacity building in reality goes beyond this and encompasses the wider issues of organizations, which include institutions, rules and regulations, and values of organizations with which the individuals work, on one hand, and the social and economic environment on the other. This emphasizes the importance of having an appropriate framework that links individuals, organizations, and the socioeconomic environment issue and includes activities that need to be engaged to meet the barriers and constraints and assure a sustainable and safe wider use of the wastewater resource.

Well-trained staff cannot function in poorly organized institutions and neither staff nor institutions can achieve their full potential in the absence of an enhancing policy and legal environment. It is doubtless that the strengthening of an organization that has to operate under wrong mandate and terms of reference, or without proper skills and well-trained staff, is quite useless. Evidently, this will only make the organization to continue to do the wrong things more efficiently.

80.16 Experiences Gained and Learned Lessons

In recent years, successful water recycling projects have been implemented in many countries. These are referred to as reuse of agriculture, urban reuse, industrial reuse, and recycling to supplement water resources. This experience has demonstrated the feasibility of water reuse on a large scale and its role in the sustainable management of the world's water. In view of the water recycling projects that have been implemented, a part of the lessons learned and experience gained are as follows:

- The pilot studies have shown that an integrated approach to urban water, sewerage, and stormwater planning can identify opportunities that are not apparent when separate strategies are developed for each service. The pilot studies have shown that both water conservation measures and water reuse are important contributors to environmental water quality improvements, and can also reduce water supply costs. The result is better-integrated, more sustainable solutions and substantial cost savings for local communities. Savings of up to 50% of capital costs have been identified in the pilot studies, but this may be exceptional. It is probably more practical to set a modest target of 15%–20% savings to see if this can be bettered.
- Both project experience and comprehensive health studies have demonstrated the potential to use recycled water to supplement drinking water supplies. Water conservation and beneficial reuse can reduce freshwater diversions from streams and improve downstream water quality. There are many direct and indirect benefits that result from reduced diversions and improved downstream water quality. These benefits should be evaluated and taken into account when assessing the merits of implementing new water reuse projects. Water reuse increases the available supply of water and enables greater human needs to be achieved with less freshwater, thus lessening humanity's impact on the world's water environment. A move from the old "use once and throw away" approach, to a new sustainable "conserve, use wisely, and recycle" water economy will benefit the whole world.
- The selection of technologies should be environmentally sustainable, appropriate to the local conditions, acceptable to the users, and affordable to those who have to pay for them. In developing countries, western technology can be a more expensive and less reliable way to control pollution from human domestic and industrial wastes. Simple solutions that are easily replicated, that allow further upgrading with subsequent development, and that can be operated and maintained by the local community, are often considered the most appropriate and cost effective.
- The use of technology in wastewater management should also be multifaceted and should reflect the needs and capacity of local communities. Incentives should encourage innovative, adaptable approaches to reduce the production of wastewater and potency of its contaminants. The use of green technologies and ecosystem management practices should be used more actively and encouraged, including in rural areas with regard to both water supply and wastewater management.

- The selection of reuse option should be made on a rational basis. Reclaimed water is a valuable but limited water resource; therefore, investment costs should be proportional to the value of the resource. Also, the reuse site must be located as close as possible to the wastewater treatment and storage facilities.
- The lower the financial costs, the more attractive is the technology. However, even a low cost option may not be financially sustainable because this is determined by the true availability of funds provided by the polluter. In the case of domestic sanitation, the people must be willing and able to cover at least the operation and maintenance cost of the total expenses. The ultimate goal should be full cost recovery, although, initially, this may need special financing schemes, such as cross subsidization, revolving funds, and phased investment programs. In this regard, adopting an adequate policy for the pricing of water is of fundamental importance in the sustainability of wastewater reuse systems. The incremental cost basis, which allocates only the marginal costs associated with reuse, seems to be a fair criteria for adoption in developing countries. Marginal cost pricing can reduce successive water use and pollution and can ensure the sustainability of wastewater treatment projects. Setting appropriate tariffs for treated wastewater provides an important incentive mechanism to encourage its reuse; however, it is a very complex issue. Direct benefits of wastewater use are relatively easy to evaluate, whereas the indirect effects are non-monetary issues and, unfortunately, they are not taken into account when performing economic appraisals of projects involving wastewater use for freshwater in water-scarce areas.

80.17 Summary and Conclusions

Out of the above displayed issues, the following main conclusions and recommendations are

- For developing arid and semi-arid countries, water recycling and reuse should be recognized as common practice and an essential method of drought protecting our environment and economies through the development of guidelines and best practice documents on quality criteria, public health, and environmental protection. National or wide water quality and good practice guidelines need to be agreed on to enable water reuse to be implemented for all environmental, social, public health, or economically beneficial applications.
- Pragmatic approaches are needed to protect water quality and achieve sustainable use of wastewater. Many developing countries have adopted legislation and policies to protect water quality and regulate wastewater use. However, the inclusion of unrealistic criteria makes implementation difficult. A more pragmatic approach would combine provisional guidelines with continuing improvements to enhance wastewater quality or the ability to use wastewater in an

environmentally safe manner. Meaningful criteria need to be established in accordance with local, technical, economic, social, and cultural contexts [27]. The implementation of water recycling and reuse requires a global framework that considers institutional, financial, and practical aspects to enable local stakeholders to manage recycled water safely.

- Technical solutions alone cannot provide the increasing population of the arid and semi-arid regions with safe water supply and proper environmental sanitation. Regions need to integrate the technical, institutional, managerial, social, and economic aspects of water resources management. The new approach for sustainable water supply and sanitation depends on local involvement, solutions, and knowledge within an overall framework of water and natural resources planning. Project viability should be based on environmental, social, and economic benefits using whole life, sustainability, and cost-effectiveness tools that provide a fair way of evaluating the benefits. In addition, projects should be able to take advantage of existing or new financial incentives to build skills and confidence in each country.
- Wastewater use cannot be realized unless appropriate capacity development is provided. There is a critical shortage of skilled human resources in developing countries and countries in transition to address the challenges of wastewater management, reuse, and recycling. To be successful and sustainable, wastewater's management must be an integral part of rural and urban development planning across all sectors, and, where feasible, transcending political administrative and jurisdictional borders. This requires a cocktail of innovative approaches that engages the public and private sector at the local level.
- Wise investments in wastewater will generate significant returns, as addressing wastewater as a key step in reducing poverty and sustaining ecosystem services. Instead of being a source of problems, the appropriate management of wastewater will be a positive addition to the environment, which in turn will lead to improved food security, health, and therefore economy.
- Water scientists and engineers must recognize that their task should include learning how to communicate their science in order for their methods to be revealed to other stakeholders in society. Simultaneously, governments, private sector, and civil society movements should seek to incorporate scientific results more systematically in their deliberation and decision-making processes to reach robust solutions.
- Research should connect local knowledge, gender-aware socioeconomic development, culture and policy institutions, and implementing bodies. Research should also focus on the systems beyond the watershed and the conventional concerns of water scientists and managers. In addition, international water research should adopt the constructively engaged IWRAM approach and seek links to education, capacity building, and innovation. In the mean time, interdisciplinary research in a constructively

engaged mode should pay specific attention to strength-ening human capital and implementation capacities and improve the enabling environment within Partner Countries. Bridging the gap between research into the fundamentals and the perception of water users and water policy-makers should be a research focus in its own right.

- Communication and their impact are very challenging and sensitive issues in the water management for sustain-able development purposes. Effective communication of research must be a prime goal and essential component in the research log frame matrix. Indicators of success-ful communication and impact need to be identified with special emphasis on links to major societal constituencies, education, training, and innovation.

- Even though the international community recognizes that "safe use of wastewater in agriculture" is an impor-tant water resources issue that needs to be addressed, much work still needs to be done to advance the topic of international policies and implement guidelines and safe practice. Tested technologies and strategies for the safe use of wastewater in agriculture are available worldwide, but capacities to implement them are still lacking in many countries.

- Finally, a key factor for the success of wastewater treatment and reuse projects is the involvement of the beneficiaries in the decision-making process. Community awareness and involvement in the design, and selection of technol-ogy is crucial to the success of the project. Community involvement ensures proper understanding of needs and local capacities and guarantees ownership of the system. Projects have failed and stations have stopped operating not because of treatment and reuse technology selection, but because of lack of understanding of the capacity of communities to manage and operate the system due to the lack of skills and to cover the required operation and maintenance costs.

References

1. AAEE. 2008. Summary of Pre-conference Workshop at WEFTEC 08, Wastewater Treatment in Tomorrow's Climate Change-Driven World Organized by the American Academy of Environmental Engineers (AAEE) and The Air Quality and Odor Control Committee of WEF, October 18, 2008.

2. Abu-zeid, M. and Hamdy, A. (eds). 2010. Sustainable use and management of treated wastewater. In: *Encyclopaedia on Water Resources Development and Management in Arid and Semi-Arid Regions of the Arab World.* Vol. 9. CHIEAM, Bari, Italy and Arab Water Council, Cairo, Egypt. p. 294.

3. Al-Shreideh, B. 2001. Reuse of treated wastewater in irri-gation and agriculture as a nonconventional water source in Jordan. In: *Proc. Advanced Short Course on Water Saving and Increasing Water Productivity: Challenges and Options.* University of Amman, Jordan, pp. 18.1–18.30.

4. Angelakis, A.N., Marecos de Monte, M.H.F., Bontoux, L., and Asano, T. 1999. The status of wastewater reuse in the Mediterranean basin: Need for guidelines. *Water Research,* 33(10), 2201–2217.

5. Asano, T. 2003. Water from (waste) water—The depend-able water resource. *Water Science and Technology,* 48(8), 23–33.

6. Asano, T. and Levine, A.D. 1998. Wastewater reclama-tion, recycling and reuse: An introduction. In: T. Asano (ed.), *Water Quality Management. Library 10, Wastewater Reclamation and Reuse.* Technomic Publishing Company, Lancaster, PA, pp. 1–56.

7. Bahri, A. 2002. *Wastewater Reuse in the Middle East, North Africa and Mediterranean Countries.* National Research Institute for Agricultural Engineering. Water and Forestry, Tunisia.

8. Barrio, D., Irusta, R., Fatta, D., Papadopoulos, A., and Loizidou, M. 2004. Analysis of best practices and suc-cess stories for sustainable urban wastewater treat-ment and reuse in the agricultural production in the Mediterranean Countries, *Global Symposium on Recycling, Waste Treatment and Clean Technology,* Madrid, Spain, September 26–29, 2004.

9. Bazza, M. 2002. Wastewater recycling and reuse in the near East region: Experiences and issues. In: *Proceedings of IWA–Regional Symposium on Water Recycling in the Mediterranean Region.* Iraklion, Greece, pp. 43–60.

10. CEHA, WHO. 2005. A regional overview of wastewa-ter management and reuse in the eastern Mediterranean region. Regional centre for environmental health activi-ties. CEHA, WHO.

11. Chen, C-L., Kuo, J-F., and Stahl, J.F. 1998, The role of filtra-tion for wastewater reuse. In: T. Asano (ed.), *Wastewater Reclamation and Reuse.* Technomic Publishing Co., Lancaster, PA, pp. 219–262.

12. Choukr-Allah, R. and Hamdy, A. 2005. Wastewater recycling and reuse in some Islamic countries as poten-tial resource for water saving. In: R. Ragab and S. Koo-Oshima (eds), *International Workshop on Environmental Consequences of Irrigation with Poor Quality Waters.* Center for Ecology and Hydrology. Kuala Lumpur, Malaysi, pp. 89–110.

13. Cornish, G.A., Mensah, E., and Ghesquiere, P. Water qual-ity and peri-urban irrigation: An assessment of water qual-ity for irrigation and its implications for human health in the peri-urban zone of Kumasi, Ghana. KAR project R7132, H R Wallingford. September 1999.

14. El-Motaium, R.A. and Abdel Monem, M. 2001. Long term effect of using sewage water for irrigation on nitrate accumulation in soil and plants. *Egyptian Journal of Soil Science,* 41(4), 527–538.

15. Ensink, J.H.J., Mehmood, T., Van der Hoek, W., Raschid-Sally L., and Amerasinghe, F.P. 2004. A nation-wide assess-ment of wastewater use in Pakistan: An obscure activity or a vitally important one? *Water Policy* 6, 197–206.

16. FAO. 1997. Irrigation in the near east region in Figuers: Report No. 9, Rome, Italy.

17. FAO. 2000. Water quality management and pollution control in the Near East: An overview. Proceedings of a Regional Workshop, October 30 to November 2, 2000, Cairo, Egypt (English), FAO, Cairo (Egypt). Regional Office for the Near East, 2002, 195 p.

18. FAO Regional Office for Near East, WHO—Regional Office for the Eastern Mediterranean. 2004. Expert consultation for launching the regional network on wastewater reuse in the near East. In: Proceeding, Cairo, 2003, pp. 160.

19. Fatta, D., Moustakas, K., Loizidou, M., and Assobhei, O. 2005. Waste water treatment technologies and effluent standards applied in the Mediterranean countries for reuse in agriculture. WATAMED, Marrakesh.

20. Fatta, D., Papadopoulos, A., Mentzis, A., Loizidou, M., Sandovar, A., and Barrio, D. 2004. The urgent need for sustainable urban wastewater treatment and reuse in the agricultural production in the Mediterranean countries. The MEDWARE Project, Marrakesh.

21. Frieder, E. 2001. Water reuse an integral part of water resources management: Israel as a case study. *Water Policy,* 3(1), 29–39.

22. Hamdy, A. 1999. Sewage water prospects and challanges for use. In: A. Hamdy and C. Lacerignola (eds), *Mediterranean Water Resources: Major Challenges towards the 21st Century.* CIHEAM/IAM, Bari, Italy March 1999.

23. Hamdy, A. and Lacirignola, C. 2005. Nonconventional water resources as a potential water supply—wastewater and potential use. In: A. Hamdy and C. Laceringola (eds)., *Coping with Water Scarcity in the Mediterranean, What, Why and How,* CHIEAM, MAI-Bari, Italy, pp. 327–351.

24. Hamdy, A. and Ragab, R. 2005. Reuse of treated wastewater in irrigation: Challenges and perspectives. In: R. Ragab and S. Koo-Oshima (eds), *International Workshop of Environmental Consequences of Irrigation with Poor Quality Waters.* Center for Ecology and Hydrology. Kuala Lumpur, Malaysia, pp. 5–28.

25. Hamdy, A. and Sardo, V. 1999. Phytodepuration of urban wastewaters—An appraisal of constructed wetlands efficiency. In: *Proceedings of Special Session on Non-Conventional Water Resources Practices and Management.* Rabat, Morocco.

26. Healthy Waterways. 2011. What is water sensitive urban design? http://waterbydesign.com.au/whatiswsud/. Accessed February 2, 2012.

27. IWMI. 2006. Recycling realities: Managing health risks to make wastewater an asset. Water Policy Briefing 17. Colombo.

28. Keraita, B.N. and Drechsel, P. 2004. Agricultural use of untreated urban wastewater in Ghana. In C.A. Scott, N.I. Faruqui, and L. Raschid-Sally (eds), *Wastewater Use in Irrigated Agriculture.* CABI Publishing, Wallingford, UK.

29. Lai, T.V. 2000. Perspectives of Peri-Urban Vegetable Production in Hanoi. Background paper prepared for the Action Planning Workshop of the CGIAR Strategic Initiative for Urban and Peri-Urban Agriculture (SIUPA), Hanoi, June 6–9. Convened by International Potato Center (CIP), Lima.

30. Lazarova, V. and Bahri, A. 2008. Water reuse practices for agriculture. In: B. Jeminez and T. Asano (eds), *Water Reuse: An International Survey of Current Practice, Issues and Needs.* IWA, London, pp. 199–227.

31. Looker, N. 1998. *Municipal Wastewater Management in Latin America and the Caribbean.* R.J. Burnside International Limited, published for Roundtable on Municipal Water for the Canadian Environment Industry Association.

32. Meda-countries. 2009. Identification and removal of bottlenecks for extended use of wastewater for irrigation or other purposes. RG/2008–10/FTF. Lebanon country report, September 2009, pp. 74.

33. MEDAWARE. 2004. Development of tools and guidelines for the promotion of the sustainable urban wastewater treatment and reuse in the Agricultural production in the Mediterranean countries. Task2: Evaluation of the existing situation related to the operation of urban waste water treatment plants and the disposal with emphasis on the reuse in Agricultural production. MEDAWARE project, pp. 124.

34. Mexico CNA (Comisiòn Nacional del Agua). 2004. Water Statistics. National Water Commission, Mexico City. (In Spanish.)

35. Papadopoulos, I. 1995. Wastewater management for agricultural production and environmental protection in the NER. FAO Regional Office for the Near East, Cairo, Egypt.

36. Scott, C.A., Faruqui, N.I., and L. Raschid-Sally (eds.) 2004. *Wastewater Use in Irrigated Agriculture: Confronting the Livelihood and Environmental Realities.* CAB International, Wallingford, 193 p.

37. Shatnawi, M., Hamdy, A., and Smadi, H. 2010. An overview on urban wastewater: Problems risks and its potential use for irrigation. In: M. Abu-Zeid and A. Hamdy (eds), *Encyclopaedia on Water Resources Development and Management in Arid and Semi Arid Regions of the Arab World.* Vol. 9. pp. 3–53.

38. Toze, S. 2006. Reuse of effluent water—benefits and risks. *Agriculture Water Management,* 80(1–3), 147–159.

39. UNDP. 1998. Capacity Assessment and Development. Technical Advisory Paper No.3. http://magnet.undp.org/Docs/cap/CAPTECH3.htm. Accessed April 13, 2003.

40. UNEP. 2006. Water and wastewater reuse: An environmentally sound approach for sustainable urban water management.

41. US EPA/USAID. 1992 Guidelines for Water Reuse. Washington Technical Report No. 81, September 1992. United States Environmental Protection Agency, Washington, DC.

42. US Geological Survey website "Water Cycle" (December 27, 2011). http://ga.water.usgs.gov/edu/watercycle.html. AccessedFebruary 2, 2012.

43. WHO. 2006. *Guidelines for the Safe Use of Wastewater, Excreta and Gray Water.* Vol. 2. Wastewater use in agriculture. World Health Organization, Geneva, Switzerland.

44. WHO, Regional Office for the Eastern Mediterranean. 2005. A regional overview of wastewater management and reuse in the Eastern Mediterranean region. WHO-EM/CEH/139/E.

45. World Bank. 2009. Development and climate change. Advance Overview of the 2010 World Development Report. World Bank, Washington, DC.

81

Water Reuse Products in Urban Areas

Jay Krishna Thakur
Environment and Information Technology Center—UIZ

Sweety Karmacharya
Eberswalde University for Sustainable Development

Prafull Singh
Amity University

Deepa Gurung
Health and Environmental Management Society (HEMS)

Saeid Eslamian
Isfahan University of Technology

PREFACE

Water scarcity in recent times calls for an alternative to minimize water consumption and economize water use. Correspondingly, wastewater reuse also minimizes the discharge of wastewater into the environment. Recycled wastewater can be conveniently reused for several purposes as long as quality is maintained at the acceptable level for specified purposes. The useful organic carbon and nutrients in wastewater increase its effectiveness of its application in irrigation water. In urban areas, water can be reused in agricultural, domestic, industrial, recreational, and indirect (reuse like aquifer recharge) sectors. Agricultural reuse for crop irrigation, commercial nurseries, and landscape irrigation is practiced in various places. As a domestic application, wastewater can be used within cooling towers to transfer heat in air conditioning systems. In industries, reuse of wastewater for cooling systems make up water, boiler feed water, process wash, and general wash down is in existence. Water use in high-technology manufacturing, such as semiconductor industry is introduced in microchip manufacturing and manufacture of circuit boards. Wastewater can also be successively applied in recreational and landscape impoundments in golf courses and recreational fields as well as in snowmaking. In case there is less preference for direct use of wastewater, it can be indirectly used for aquifer recharge. One of the attractive uses of wastewater can be ascribed to aquaculture, with proper consideration of microbial water quality guidelines, as many suitable species can be cultured in nutrient-rich water. Influences of microbial contamination, salinity, pharmaceutical, and other organic contaminants are some of the issues of concern. However, putting human health risks into focus and effluent quality as per regulatory body requirement, wastewater reuse proves to be an acceptable strategy for sustainable urban water management. Further feasibility study and integrated framework for water management are some of the recommendations.

81.1 Introduction

Water scarcity and demand in arid and semi-arid regions appeal for an alternative water supply. A number of water reuse programs are running around the world in sectors such as agriculture, irrigation, and wetland conservation. Concerns of negative impacts on the environment due to nutrient discharge to surface resources are simultaneously resulting in compulsory reduction in the number of ocean discharges for all seasons of the year [19].

81.2 Wastewater Reuse as Environmentally Sound Technology

When applied appropriately, wastewater reuse is considered an appropriate example of EST applications. ESTs are defined in Chapter 34 of Agenda 21 as technologies that

- Protect the environment
- Are less polluting
- Use all resources in a more sustainable manner
- Recycle more of their wastes and products
- Handle residual wastes in a more acceptable manner than the technologies for which they are substitutes

81.3 Wastewater Reuse Products

Recycled wastewater can serve as a more dependable water source with consistency in quantity and quality, less influenced by droughts and other climatic conditions in comparison to freshwater, with useful substances for some applications. With the application of adequate treatment, wastewater can meet specific needs and purposes, such as toilet flushing, cooling water, and similar other applications. The reuse of wastewater is particularly attractive in arid climates, areas facing demand growth, and those under water stress conditions.

Along with useful organic carbon and nutrients such as nitrogen and phosphorus, the benefit of wastewater use lies in reduced consumption and treatment needs. In various cases, reusing wastewater is less costly than using freshwater because it reduces the infrastructure for advanced wastewater treatment. With the sprawling urbanization and growing urban population, there has been increased water consumption, which requires the additional development of large-scale water resources and associated infrastructure [2].

Reuse is a beneficial use of reclaimed water. Since the user does not have to go to a river to fetch water, this is water byproduct of human sanitation and industrial processes. Reused water can be conveniently used for agricultural, commercial, residential, and industrial processes, which are termed direct reuse. Agricultural irrigation is the largest user product of reclaimed water. Most recycled water will undergo some form of disinfection for the protection of public health.

81.4 Constituents of Water Reuse Products

In general, water reuse products do contain organic and inorganic compounds derived from natural as well as anthropogenic sources. For example, composition of municipal wastewater is approximately 99.9% water and 0.07% of total (dissolved and suspended) solids [17]. Table 81.1 shows the physical, chemical, and biological characteristics of wastewater and their sources [29].

81.5 Guidelines for the Reuse of Water in Urban Areas

The 2004 Guidelines for Water Reuse [18] identified irrigation of golf courses as one of the several typical urban water reuse practices. Combined with data from the Golf Course Superintendents Association of America (GCSAA), AWWA estimated in 2004 that 2900 out of the 18,100 surveyed golf courses were using reclaimed water, a 600% increase from 1994 data. Although most comments were positive, some respondents expressed concern regarding algal problems in ponds, changes in course treatment, and increased turf management.

81.6 Water Reuse Products in Urban Areas

Mostly, nonpotable reuses of water are [42]

- Agriculture reuse, irrigation reuse
- Domestic reuse
- Industrial reuse
- Recreational reuse
- Indirect reuse (e.g., aquifer recharge)

81.6.1 Agriculture/Irrigation Reuse

The main categories of irrigation reuse are agricultural irrigation (crop irrigation, commercial nurseries) and landscape irrigation (parks, playgrounds, golf courses, freeway medians, landscape areas around commercial areas, offices, industrial development, and residential landscape areas). Restricted irrigation reuse is

TABLE 81.1 Physical, Chemical, and Biological Characteristics of Wastewater and Their Sources

Characteristic	Sources
Physical properties:	
Color	Domestic and industrial wastes, natural decay of organic materials
Odor	Decomposing wastewater, industrial wastes
Solids	Domestic water supply, domestic and industrial wastes, soil erosion, inflow/infiltration
Temperature	Domestic and industrial wastes
Chemical constituents:	
Organic:	
Carbohydrates	Domestic, commercial, and industrial wastes
Fats, oils, and grease	Domestic, commercial, and industrial wastes
Pesticides	Agricultural wastes
Phenols	Industrial wastes
Proteins	Domestic, commercial, and industrial wastes
Priority pollutants	Domestic, commercial, and industrial wastes
Surfactants	Domestic, commercial, and industrial wastes
Volatile organic compounds	Domestic, commercial, and industrial wastes
Other	Natural decay of organic materials
Inorganic:	
Alkalinity	Domestic wastes, domestic water supply, groundwater infiltration
Chlorides	Domestic wastes, domestic water supply, groundwater infiltration
Heavy metals	Industrial wastes
Nitrogen	Domestic and agricultural wastes
pH	Domestic, commercial, and industrial wastes
Phosphorus	Domestic, commercial, and industrial wastes, natural runoff
Priority pollutants	Domestic, commercial, and industrial wastes
Sulfur	Domestic water supply; domestic, commercial, and industrial wastes
Gases:	
Hydrogen sulfide	Decomposition of domestic wastes
Methane	Decomposition of domestic wastes
Oxygen	Domestic water supply, surface-water infiltration
Biological constituents:	
Animals	Open watercourses and treatment plants
Plants	Open watercourses and treatment plants
Protists	Domestic wastes, surface-water infiltration, treatment plants
Eubacteria	Domestic wastes, surface-water infiltration, treatment plants
Archaebacteria	Domestic wastes, surface-water infiltration, treatment plants
Viruses	Domestic wastes

Source: Adapted from Metcalf and Eddy. 1991. *Wastewater Engineering. Treatment Disposal Reuse.* McGraw-Hill, New York.

limited to crops that will not be directly consumed by humans (fodder, fiber, and seed crops) and is appropriate for relatively small flows. For this type of reuse, wastewater treatment must effectively remove pathogens and organic matter in order to protect public health and eliminate odors. Sites with steep slopes may not be appropriate for irrigation reuse due to excessive runoff potential. Unrestricted irrigation reuse requires that wastewater be treated to a high quality (turbidity less than 2 NTU) and be disinfected. Golf course and community green spaces prove to be the most common end use of the wastewater reuse products in urban areas.

Worldwide, agriculture receives 67% of total water withdrawal and accounts for 86% of consumption in 2000 [37]. In Africa and Asia, agriculture consumes an estimated 85%–90% of all the freshwater, which is expected to increase 2.5 times by 2025 [46]. Large-scale irrigation projects have accelerated the disappearance of water bodies, such as the Aral Sea, the Iraqi Marshlands, and Lake Chad in West Africa.

For over 4000 years, many countries of Eastern Asia and the Western Pacific practice the application of wastewater containing human excreta to the land, which has maintained soil fertility and remains the only agricultural use option in such areas lacking sewerage facilities. The benefits like freshwater resource conservation, reduced surface water pollution, and reduced requirement for artificial fertilizers, humus build up in the soil preventing soil erosion, and contribution to better nutrition and food security occurs with the wastewater reuse in agriculture [4].

81.6.2 Domestic Reuse

Domestic wastewater from sinks, showers, and washers (grey-water) are treated and reused within cooling-towers to transfer heat in air conditioning systems. In order to check corrosion and biological activities, office cooling system operators usually add anti-corrosion chemicals to cooling water including tolyltriazole, glycols, alcohols, and organic acids. Due to the risk of environmental harm (increased turbidity in water bodies, toxicity to humans, plants, animals and microbes, heavy metal toxicity to aquatic organism) if discharged, heavy metal-based corrosion inhibitors are not widely used now. Studies show that a 1000-ton air conditioner machine evaporates about 30,000 gal/day, discharges about 10,000 gal/day, and makeup of 40,000 gal/day.

While such wastewater should be discharged to sewers, discharges must be approved by the relevant water service provider. Appropriate pretreatment of cooling system wastewater and removing chemical residues should be done prior to discharging to soakage pits. Quality of all releases to soakage pits should also meet national guideline criteria formulated by the government. For example, in Australian cities, storage and handling of toxic chemicals should be in accordance with "AS 4452 The Storage and Handling of Toxic Substances" and this department's note "Toxic and Hazardous Substances."

81.6.3 Industrial Reuse

Several uses of treated wastewater for industrial purposes are available in the sector of power supply, food processing, steel manufacturing, metal finishing, chemistry, and textiles. Industrial reuse of the used water is primarily for cooling system makeup water (replacing the water lost to evaporation in arid climates), boiler-feed water, process water, and general wash down (Figure 81.1). It can also be used for concrete production on construction projects. Table 81.2 shows wastewater reuse potential for industries [40].

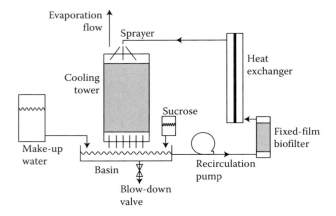

FIGURE 81.1 Schematic diagram of the bench-scale recirculating cooling system with fixed-film biofilter. (Adapted from Meesters, K.P.H. et al. 2003. *Water Research*, 37(3), 525–532.)

TABLE 81.2 Wastewater Reuse Potential for Industries

High Potential	Medium Potential	Low Potential
Pulp and paper	Slaughterhouse	Tanneries and
Cotton textile	Dairy	leather finishing
Pulp and paper	Canning and food processing	Pesticide
Glass and steel	Distillery	Rubber
	Wool textile	Aluminum
	Photographic processing	Explosives
	Chemical	manufacturing
	Fertilizer	Paint manufacturing
	Oil refining	
	Petroleum	
	Electroplating	
	Meat processing	

Source: Adapted from Visvanathan, C. and Asano, T. 2007. *The Potential for Industrial Wastewater Reuse*, Department of Civil and Environmental Engineering, University of California at Davis, California.

81.6.3.1 Cooling System and Water Quality Requirements

Cooling systems are major water consumers in industries, and using reclaimed wastewater for this purpose can contribute significant savings in freshwater consumption. This has been in practice since the early 1970s and is considered a pathway for future freshwater conservation. Various petroleum refineries, petrochemical industries, and chemical plants are fixed with interconnected cooling systems.

Water quality for cooling systems is a concern in the face of problems in circulating cooling systems such as corrosion, scaling, and bio-fouling. The effluent is relatively high in COD, ammonia, phosphorous, total salinity, chlorides, and alkalinity, which are considered important factors while using for cooling makeup water. These constituents are even considerably higher in the arid countries due to the low per capita use of water.

The treatment objective is to remove ammonia and phosphates, excess alkalinity (bicarbonates), and calcium including the reduction of suspended solids (SS) and organic content (COD), which are addressed by high lime precipitation, clarification, ammonia stripping, and pH adjustment. High clarification can be achieved at lime doses that raise the pH to above 10.5 when magnesium hydroxide precipitates, which acts by adsorption–flocculation mechanism and sweeping the flocculants of colloidal phosphates, particulates, and organics. The quality of water to be used in cooling systems must overcome the problems of

- Scaling, corrosion, bio-fouling

Calcium carbonates and phosphates scale formers are specifically higher in reclaimed water while chlorides are general corrosion factors found in wastewater. The problem of bio-fouling in recirculating systems can be addressed by using biocides such as chlorine to reduce the organic substrate concentration to a minimal level.

- Salinity, chlorides and replacement ratio

For each recirculating cooling system, limiting concentration for TDS and chlorides are set. The replacement ratio exists in agricultural reuse schemes, based on different considerations, but is related to salinity and chlorides. The ratio of renovated water makeup consumption is defined as the replacement ratio given by

$$Rr = Ms/Mf$$

where

Rr = replacement ratio
Ms = makeup consumption of renovated wastewater
Mf = makeup consumption of fresh water

Further, Rr depends basically on concentration of chloride in the wastewater and in the freshwater.

$$R = Cs/Cf$$

where

Cs = chloride in wastewater
Cf = chloride in freshwater

Through mass balance consideration, an expression was derived for calculating Rr. This relates the Rr to the ratio of wastewater to freshwater chlorides and to the limiting chloride level in CCW (circulating cooling water), the latter expressed in a normalized manner as CB/CF.

$$Rr = (CB/CF) - 1/(CB/CF) - r$$

81.6.3.2 High Technology Water Reuse

Using water in high-technology manufacturing such as semiconductor industry is a relatively new practice. Two major processes that use water are microchip manufacturing and manufacturing of circuit boards. They rarely utilize reclaimed water. In circuit board manufacturing, water is used primarily for rinsing operations, similar to production of boiler feed water. Both semiconductor and circuit board manufacturing facilities do use reclaimed water for cooling water and site irrigation.

Examples of reuse in high-technology industries include projects by companies such as Intel, which improved the efficiency of the process used to create the ultra-pure water (UPW) required to clean silicon wafers during fabrication. While 2 gallons of water were needed to make 1 gallon of UPW, Intel generates 1 gallon of UPW from between 1.25 and 1.5 gallons. Such reuse approaches enable Intel to harvest as much water from its manufacturing processes as possible. In 2010, Intel internally recycled approximately 2 billion gallons (7.6 MCM) of water, equivalent to 25% of its total water withdrawals for the year [11].

81.6.4 Environmental/Recreational Reuse

Uses such as manufactured wetlands, enhanced natural wetlands, and sustaining stream flows fall under environmental and recreational reuse. An impoundment of reclaimed water where recreation is limited to fishing, boating, and other non-contact

recreational activities constitutes restricted recreational reuse while with unrestricted recreational reuse, reclaimed water is used in an impoundment of water where no limitations are imposed on body-contact recreational activities.

81.6.4.1 Recreational and Landscape Impoundments

Regulation of impoundments that are maintained using reclaimed water is typically in accordance with the potential for contact for that use. For example, in Arizona, the type of recreational impoundments where boating or fishing is attended, requires high quality remediation that includes secondary treatment, filtration, and disinfection falling under class A. Therefore, no detectable fecal coliform organisms are present in four of the last seven daily reclaimed water samples taken, and no single sample contains maximum concentration of fecal coliform organisms exceeding 23/100 mL [41].

81.6.4.2 Golf Course and Recreational Field

Economic development and rapid expansion concentrated in specific destinations have mushroomed the sources of demands for golf courses in urban areas [11]. Scientific study of golf systems reveals that golf courses can serve as places to recycle waste (compost, sludge) and reclaim water in a controlled environment, thus minimizing the negative impacts of both waste and the actual courses [35].

While trying to minimize nonpotable consumption through use of reclaimed water, public health becomes an important matter of concern when such water is used to irrigate residential areas, golf courses, public schoolyards, and parks, and therefore reclaimed water receives treatment and high-level disinfection. Salts and nutrients in reclaimed water are of important concern and the salt sensitivity of the irrigated plants should be considered in case it is applied in irrigated landscape.

Likewise, irrigation of public parks and recreation centers, athletic fields, schoolyards, playing fields, and landscaped areas surrounding public buildings and facilities are the important applications in reuse. Many facilities are required to implement special management practices where reuse is implemented to minimize the potential of cross-connection of water sources. For example, golf courses in San Antonio are required to include a double-check valve on the reclaimed water supply to the property to prevent backflow of reclaimed water into the potable water distribution system.

Some of the practices including the poorest quality reclamation with respect to TDS are produced in the southwest United States, where the greatest golf course reuse occurs:

- Applying extra water to leach excess salts below the turf grass root zone
- Providing adequate drainage
- Modifying turf management practices
- Modifying the root zone mixture
- Blending irrigation waters
- Using amendments

Based on the survey by GCSAA and Environmental Institute for Golf (EIFG) in 2006, an estimated 12% of golf courses in the

United States use reclaimed water, with more courses in the southwest (37%) and southeast (24%) practicing reuse [22].

81.6.4.3 Snowmaking

Making snow is particularly the case where temperature is lower to maintain the snow while natural precipitation will not otherwise support a longer recreation season. Snowmaking using reclaimed water is being done in Maine, Pennsylvania, and California in the United States, Canada, and Australia (e.g., Victoria's Mount Buller Alpine Resort installed in 2008). While states do possess rules and regulations pertaining to snowmaking with reclaimed water, there are no such studies related to human health exposure to snow made with reclaimed water. It becomes necessary to treat the reclaimed water well if snowmelt is to be introduced into the reclaimed water distribution system.

81.6.5 Indirect Reuse: Aquifer Recharge

An attractive option for indirect potable reuse consists of artificial recharge through the wastewater in urban areas. This has been considered attractive for years and has already been implemented in several countries.

Most of the reclaimed water recharge for indirect potable reuse occurs in the United States (e.g., West Basin and Orange County, CA; Mesa and Tucson, AZ). However, the recharge should not degrade the quality of the groundwater nor impose any additional treatment after pumping. In practice, the recharge water reaching the saturated zones of the aquifer should have previously acquired the quality acceptable for drinking purposes [9].

Aquifer storage and recovery is a specific type of practice with the purpose of both augmenting groundwater resources and recovering water in the future for numerous uses. AR and ASR wells are found in areas of the United States that have a high population density and intensive agriculture practice; high dependence and increasing demand of groundwater for drinking and agriculture; and/or limited ground or surface water availability.

81.6 5.1 Recharge for Nonpotable Reuse

The quality of the water extracted from the aquifer should meet the strict standards of microorganisms, organic matters, or heavy metals related to the intended water use for drinking, irrigation, or other purposes.

When recharge is direct, the recycled water should be elevated to meet the standards and limits that are required for the intended uses. In addition, suspended solids and organic matter should be drastically reduced to avoid blockage of the injection wells.

Indirect recharge requires less treated injectant and is easier to implement. Soil aquifer treatment is an appropriate treatment technology to meet required water quality, provided it is properly designed and managed. A detailed investigation of the hydraulic characteristics of the soil layers below the infiltration site is useful for water quality prediction of percolated water. In the cases when highly permeable or heterogeneous onsite soils fail to provide the required treatment, infiltration, percolation through calibrated sand bed filling pits, excavated at the soil surface are used as treatment before infiltration through onsite soil layers [10].

81.7 Reuse Near Urban Settlements

In urban settlements including commercial, residential, and industrial buildings, reclaimed water is reused for various nonpotable purposes such as decorative water features, dust control, fire protection, and toilet flushing. Additionally, irrigation of ornamental landscapes, parks, and golf courses are the various end uses of wastewater in urban areas.

In case the reclaimed water is used for fire protection, additional design issues are to be considered. The volume of storage required to accommodate flow variations can be determined from the daily reclaimed water demand and supply curves.

The US urban reuse is one of the highest volume uses among several major categories with applications in recreational field/golf course irrigation, landscape irrigation, among others such as fire protection and toilet flushing, which are the indicators of the crucial component of water reuse products. Urban reuse is time and again divided into applications that are either accessible or have restricted access, in settings where public access is controlled or restricted by physical or institutional barriers, such as fences or temporal access restriction [11].

81.7.1 Urban and Peri-Urban Agriculture as Wastewater Reuse Product

Urban and peri-urban agriculture is developing wherever land is available close to streams and drains [30] where untreated wastewater is extensively used. Although such practices pose a threat to the health of users and consumers, they are considered to provide important livelihood benefits and perishable food to cities [33]. The major drivers behind such practices include the following:

- Increasing urban water demand and related return flow of used but seldom treated wastewater into the environment and its water bodies, causing pollution of traditional irrigation water sources
- Urban food demand and market incentives favoring food production in city proximity where water sources are usually polluted
- Lack of alternative measures that are cheaper, equally reliable, or safer water sources

Data from a detailed city study in Accra showed that about 200,000 urban dwellers benefit every day from vegetables grown on just 100 ha of land. Among the 53 cities of the developing countries, approximately 0.4 million hectares (Mha) are cultivated with wastewater by a farmer population of 1.1 million with about 4.5 million family dependants. Data of the cities of LDCs (31 cities in this case) show that there are about 1.1 million farmers around these cities who make a living from cultivating 0.4 Mha of land irrigated with wastewater (raw or diluted wastewater, including all those areas that use polluted rivers as the irrigation water source). Some policy recommendations for urban and peri-urban agriculture are

- A very marginal cost is invested to improve an effective source of the nutrition and food supplies of urban and peri-urban areas.
- The WHO [44] guidelines for the safe use of wastewater should be extensively applied as it allows for incremental and adaptive risk reduction, which is more realistic and cost-effective than stressing the need to achieve certain water quality values.
- In addressing health risks, state authorities have a role to play in planning, financing, and maintaining sanitation and waste disposal infrastructure in order to address the health risks.
- Economic incentives like easy access to credit for safer farming, and social marketing for improving farmer knowledge and responsibility, can lead, more effectively, to reduced public health risks while maintaining the benefits of urban and peri-urban agriculture.

Farmers in "poorer" cities tend to face an increasingly polluted water source, which is prevalent in Asia. In African and Asian countries with lower GDP/capita conditions (<USD 2000/year), and where alternative urban employment is not available, high levels of rural-urban migration is prevalent. Second, these migrants are from agricultural backgrounds, which attracts them to use their skills in a similar field. A survey of 12 cities in West Africa showed that in many cities the majority of urban farmers engaged in irrigated agriculture are migrants [30].

Smaller plot sizes in Africa depend not only on access to land and water, but also on security of tenure and on the financial means of farmers. Such a situation is common in the cities falling in the lower range of GDP/capita. Irrigated UPA in many of them has small plot sizes (varies between 0.07 and 1.2 ha, but could sometimes be as small as 0.01 ha), low overall extents of land under urban and peri-urban agriculture (<15,000 ha), and, consequently, lower total extents of wastewater agriculture (Tables 81.3 and 81.4).

Wastewater agriculture is also a significant phenomenon in high- and middle-income countries, where wastewater collection and treatment have yet to take full speed, such as in the cities across Latin America and Asia. However, in cities with higher city poverty index, proportion under wastewater use for urban agriculture appeared to be higher. Higher frequency of use of wastewater in urban irrigation is common in drier and humid areas.

In addressing health risks, state authorities play a role in planning, financing, and maintaining sanitation as well as waste

TABLE 81.3 Extent and Number of Peri-Urban Farmers by Region

Region	No. of Cities with Data	Total Farmers WW	Total WW Area (ha)
		Informal and Formal	Informal and Formal
Subtotal Africa (AF)	9	3550	5100
Subtotal Asia (AS)	19	992,880	214,560
Subtotal Latin America (LA)	8	88,300	142,160
Subtotal Middle East (ME)	3	3320	34,920
Total	39	1,088,050	396,740

TABLE 81.4 Cities with Largest Extents of Wastewater Agriculture

Region	City	Country	Country Population (Millions)	Total WW Area (ha) Informal and Formal	Total Farmers WW Informal and Formal
AS	Ahmedabad	India	2.88	600	No data
AS	Hanoi	Vietnam[a]	3.09	43,778	658,300
AS	Ho Chi Minh	Vietnam[a]	5.55	75,9062[b]	135,000
AS	Kathmandu	Nepal	0.67	5466	19,524
AS	Shijazhuang	China	2.11	11,000	107,000
AS	Zhengzhou	China	2051	1650	25,000
LA	Mexico city/El Mezquital[c]	Mexico	21.3	83,060	73,632
LA	Santafe de Bogota	Columbia	7.03	22,000	3000
LA	Santiago	Chile	5.39	36,500	7300

Source: Adapted from Raschid-Sally, L. and Jayakody, P. 2009. Drivers and characteristics of wastewater agriculture in developing countries: Results from a global assessment, IWMI.

[a] Hanoi and Ho Chi Minh have very large extents of urban and particularly peri-urban agricultural land where irrigation water is often from polluted rivers running through the cities. The farmer numbers are large because of the importance of urban and peri-urban agriculture as a livelihood activity.

[b] Cropped area.

[c] The large volume of wastewater from Mexico City is used to cultivate land in the El Mezquital Valley.

disposal infrastructures, which responds to agricultural reuse requirements. Water quality improvement and health risk reduction to the users and the supportive initiatives through farm tenure security, economic incentives like easy access to credit for safer farming, and social marketing for improving farmer knowledge and responsibility can lead to reducing risks effectively while maintaining the benefits of urban and peri-urban agriculture.

81.7.2 Crops Grown and Irrigation Methods

Across the cities, vegetables and cereals (especially rice) are the two most common crops cultivated with wastewater. Among the

TABLE 81.5 Distribution of Crop Types in Urban Agriculture Grown with Wastewater

	Number of Cities[a]			
Type of Crop	Africa	Asia	Latin America	Middle East
Vegetables	8	16	7	1
Cereals	5	15	5	2
Fodder	1	5	3	0
Other	1	5	3	2

[a] Multiple responses were possible.

type of irrigation method used, furrow, flood, and watering cans appeared to be the most popular types of irrigation (Table 81.5).

81.7.3 Aquaculture

"Water farming" or fish culture and growing aquatic vegetables are various forms of aquaculture. Fishes most frequently grown in aquaculture ponds are carp and tilapia. Local species of the area is grown in India, for example, Indian major carp are mainly grown, such as catla (*Catla catla*), mrigal (*Cirrhina mrigala*), and rohu (*Labeo rohita*) (Figure 81.2).

Kolkata provides the world's largest example of wastewater-fed fisheries [11]. Around 3500 ha of fishponds are fertilized with 550,000 m³/day of untreated wastewater and fish production (Indian major carp, with some tilapia and silver carp) is nearly 20 tonnes/day, equivalent to about 18% of the city's fish demand.

Wastewater-fed (or excreta-fertilized) aquaculture has been practiced for many hundreds of years in China and Indonesia, where several carp species are often grown in the same pond (polyculture), with each species occupying a different ecological niche. A wide range of aquatic vegetables is grown in Asia instead of just watercress, as in most industrialized countries, for example, water spinach (*Ipomoea aquatica*), water chestnut

FIGURE 81.2 Aquaculture practice. (Adapted from Duncan, M. 2003. *Domestic Wastewater Treatment in Developing Countries*. Earthscan, London, UK.)

(*Eleocharis dulcis* and *E. tuberosa*), water bamboo (*Zigania* spp), water calthrop (*Trapa* spp.), and lotus (*Nelumbo nucifera*).

81.7.3.1 Microbial Quality Guidelines

Treated wastewater for aquaculture must be of optimum microbial quality to ensure that the quality of the product does not affect the health of the consumers as well as the pond workers. Human–nematodes infections, which are important criteria for wastewater reuse in agriculture, are not important in aquaculture reuse. Rather, it is the human–trematode infections that are of major concern; among which the most important ones are

- Human schistosomes or blood flukes, principally *Schistosoma mansoni*, *S. haematobium*, and *japonicum*
- Oriental liver fluke, *Clonorichis sinensis*
- Giant intestinal fluke, *Fasciolopis buski*

There are several propositions that sanitary quality of fish grown in wastewater-fed ponds should be assessed by bacterial count per gram of fish muscle [11]. For numbers of both total aerobic bacteria, that is, the "standard plate count" on nutrient agar at 35–37°C and *E. coli*, the quality is interpreted as follows:

- 0–10 per g: Very good
- 11–30 per g: Medium
- 31–50 per g: Poor
- >50 per g: Unacceptable

WHO guidelines for bacterial disease is ≤1000 *E. coli* per 100 ml of aquaculture pond water. The 2004 revised WHO bacterial guideline for *E. coli* numbers in wastewater-fed fish ponds are ≤104/100 mL, rather than ≤1000/100 mL as in the 1989 guidelines—a relaxation of an order of magnitude. In practice,

TABLE 81.6 Distribution of Reclaimed Water in California and Florida

Reuse Category	California (% Use in 2009)	Florida (% Use in 2010)
Irrigation, agricultural	29	11
Urban reuse (landscape irrigation, golf courses)	19	55
Groundwater recharge	5	14
Seawater intrusion barrier	8	–
Industrial reuse	7	13
Natural systems and other uses	23	9
Recreational impoundments	7	–
Geothermal energy	2	

Source: Adapted from Badyal, D. 2009. Municipal Wastewater Recycling Survey.California Water Recycling Funding Program (WRFP). http://www.swrcb.ca.gov/water_issues/programs/grants_loans/water_recycling/munirec.shtml. Retrieved August 23, 2012.

this will make little or no difference to wastewater-fed fish-ponds designed on the basis of a total N loading of 4 kg/ha day (Table 81.6).

81.8 Prevailing Necessity and Opportunity in Acceptance of the Reuse Products

Water reuse must be considered a part of an integrated water resource management plan (IWRMP), which is defined as a sustainable approach to water management, which recognizes its multidimensional character. A cyclic decision-making process is to be established as described by Figure 81.3.

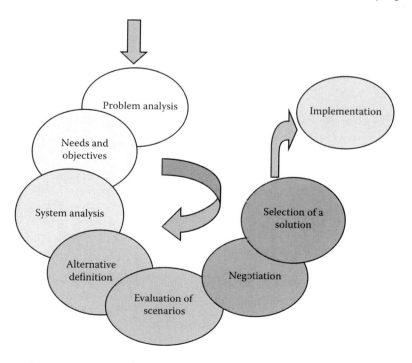

FIGURE 81.3 Cyclic decision-making process to use urban reuse product.

In addition, human health risks must be put to special attention. Issues such as distance between potable and reclaimed water pipelines, possible consequences in case of a leakage or reclaimed water pipelines breaking, parks and gardens irrigation during the night along with other parameters should also be considered.

To date, major emphasis of wastewater reclamation and reuse in the urban areas has been on nonpotable applications, such as agricultural and landscape irrigation, industrial cooling, and in-building applications in large commercial buildings, such as toilet flushing. However, potable reuse raises more public concern due to perceived impression of aesthetics and long-term health concerns. Therefore, water reuse implementation in the future remains to be influenced by diverse debates, such as drought and availability of water; growth versus no growth; urban sprawl, air pollution, and perception of reclaimed water safety and public policy governing sustainable water resources management.

81.9 Effluent Quality

Regulatory bodies such as USEPA establish effluent quality requirements, which are legally binding on authorities responsible for wastewater treatment and are known as effluent quality standards. The regulations that these agencies make are legally binding on the authorities responsible for wastewater treatment [14].

In order to assess the influence of wastewater in a receiving environment, both biotic and abiotic indicators are taken into consideration. In the first group are phytoplankton, macroalgae, angiosperms, benthic invertebrate fauna, and fish fauna. Abiotic indicators are thermal conditions, oxygenation conditions, salinity, acidification status, and nutrient conditions [38].

Treatment to remove fecal bacterial pathogens and human intestinal nematode eggs from the wastewater is essential, but removal to what degree is an important question to answer. They must be removed to the level that does not make the people working in the wastewater-irrigated fields, wastewater products, or those consuming the wastewater-irrigated crops or wastewater-fertilized aquaculture produce prone to illness. The next doubt arises about the minimum requirements for wastewater treatment. The WHO has set up guidelines for the microbiological quality of treated wastewater to be used in agriculture and aquaculture [43], which are as follows [26]:

- For "restricted" irrigation—that is, the irrigation of all crops except salad crops and vegetables eaten uncooked: ≤1 human intestinal nematode eggs/L (reduced to ≤0.1 eggs/L when children under 15 years are exposed), and ≤105 E. coli/100 mL
- For "unrestricted" irrigation—that is, including salad crops and vegetables eaten raw: ≤1 human intestinal nematode eggs/l (also reduced to ≤0.1 eggs/L when children under 15 years are exposed through their fieldworker-parents bringing home "unrestricted" produce directly from the fields), and ≤1000 E. coli/100 mL

The epidemiological and experimental evidence for these guidelines gives risk calculations, based on quantitative micro-

bial risk analysis procedures, to support the E. coli guideline of ≤1000/100 mL for unrestricted irrigation.

- Aquaculture reuse: Zero viable trematode eggs/L of treated wastewater, and ≤1000 E. coli/100 mL of aquaculture pond water

In some regions of the Netherlands, during the water shortage in dry season, reuse of effluent for irrigation is carried out, which is only possible when the effluent quality is sufficient for crop irrigation; chlorine and iron are the limiting substances. Still, the bacterial quality is too bad to meet the standards for cattle and bathing waters [1]. Water boards are also considering an additional treatment of sand filtration after tertiary treatment if effluent can be used for groundwater supply in forest or other nature areas.

81.10 Wastewater Reclamation and Reuse

The foundation of water reuse is built on three principles:

1. Achieving strict water quality standards by adopting appropriate treatment technologies
2. Protecting public health
3. Acquisition of public acceptance

Along with economic consideration, appropriate use and adherence to discharge requirements, public policies, in addition, can be implemented, which promote water conservation and reuse [4]. Among numerous uses, agricultural and irrigation uses are widely used throughout the world with well-established guidelines and agronomic policies.

81.11 Wastewater Treatment Technology

Wastewater treatment processes can be categorized into the following three [40]:

- Physical process: Impurities are removed physically by screening, sedimentation, filtration, flotation, absorption, or adsorption (or both), and centrifugation
- Chemical process: Impurities are removed chemically through coagulation, absorption, oxidation-reduction, disinfection, and ion-exchange
- Biological process: Pollutants are removed using biological mechanisms, such as aerobic treatment, anaerobic treatment, and photosynthetic process (oxidation pond) (Table 81.7).

Billions of dollars are spent for finding water, treating it to high standards, and addressing the ever-increasing needs. Therefore, billions more are spent to collect wastewater, treat it to lower the adverse impacts on environmental and human health, and to address the problems associated with sewage and industrial waste, and then finally dump it into the oceans or other sinks [21].

TABLE 81.7 Wastewater Treatment Process

	Preliminary	Primary	Secondary	Tertiary and Advanced
Purpose	Removal of large solids and grit particles	Removal of suspended solids	Biological treatment and removal of common biodegradable organic pollutants	Removal of specific pollutants, such as nitrogen or phosphorous, color, odor, etc.
Sample technologies	Screening, settling	Screening Sedimentation	Percolating/trickling filter, activated sludge Anaerobic treatment Waste stabilization ponds (oxidation ponds)	Sand filtration Membrane bioreactor Reverse osmosis Ozone treatment Chemical coagulation Activated carbon

Source: Adapted from Asano, T., Smith, R.G. et al. 1984. *Irrigation with Reclaimed Municipal Wastewater: A Guidance Manual.* California State Water Resources Control Board, Sacramento, CA, pp. 2–12.

The intended applications governing the degree of wastewater treatment requires an exact planning and implementation of the water reuse of the reclaimed water. As long as adequate treatment is provided to meet the water quality requirements for the intended use, wastewater or any marginal quality water can be used for any purposes. Wastewater treatment plants consist of a combination of physical, chemical, and biological processes and operations to remove solids, organic matter, and sometimes nutrients from wastewater [36].

81.12 Disadvantage of Wastewater Reuse Product

- Influence of microbial contamination

Agents involved in waterborne diseases from water recycling in urban areas are primarily viral. Though expensive and time consuming, some regulations recommend the use of viral model organisms for the design and monitoring of wastewater disinfection process, in the way the coliform bacteria are used as a model organism for bacterial enteric pathogens. In similar ways, several pathogens such as bacteriophages are used as an indication of inefficient performance of interventions such as chlorination, ozonation, UV doses, membrane bioreactors, and maturation ponds.

- Salinity

In dry climates, when most of the water evaporates, the concentration of chemicals including nitrates, salts, pesticides, disinfecting by-products, fulvic, humic acids, etc. increase in the water percolating down to the ground. The most important concern in such kind of irrigation is the salinity of the effluent and the nitrogen content. Higher TDS of irrigating water ultimately checks the salt or chemical balance in the root zone. Eventually, such water oozes out through natural drains or the sewage effluent discharges in the area where affected groundwater is used as municipal water supply, thereby affecting the public health. If the groundwater were shallow, for example, at a depth of 3 m, it would take the deep percolation about 1.5 years to reach the groundwater, while for a depth of 100 m, it would take about 50 years to reach the groundwater. In addition, urban irrigation can also cause the groundwater level to rise. In an old residential area

of north central Phoenix, Arizona, the groundwater level rose from the depth of about 36 m to the depth of about 15 m in a decade of sewage irrigation. The discharge water from wells was contaminated by local leaking of the underground fuel tanks, which required expensive treatment before it could be discharged to the storm drains. Options include techniques such as blending of wastewater with better quality water and drilling the wells deeper. Higher nitrate is a human and plant health concern while high phosphate may cause deficiency of important micronutrients as mobile copper and zinc in the soil.

- Effects of disinfection in the groundwater

Disinfection byproducts may already be present in the wastewater entering the treatment plants, particularly with the high chloride doses and UV irradiation [25]. Some of them such as trihalomethane and halo acetic acids have been found to be biodegradable in aquifers and recovery wells. Tirhalomethane contains chloroform and bromodichloromethane, which can cause miscarriage in women. Due to their potential of carcinogenicity and toxicity, maximum contaminant levels have been set up, for example, US EPA has set MCL of 80 µg/l for trihalomethane. However, higher nutrient and carbon levels in effluent can be expected to enhance the plant growth as well as bioactivity in the soil, which would later enhance the formation of fulvic and humic acids as the stable products in the soil. Therefore, water to be used below effluent-irrigated is pumped and chlorinated for potable uses and thereby forms a new suite of disinfection byproducts in the water [12].

- Pharmaceuticals and other organic contaminants

While these chemicals may not be directly toxic, they can produce adverse effects on immune and hormone system of both animals and humans, acting as endocrine disruptors [27]. At least 45 chemicals have been identified thus far as endocrine disruptors, including industrial contaminants like dioxins and PCBs, insecticides such as DDT, and herbicides such as atrazines. Not removed by passage through soils, groundwater below the sewage-irrigated water, septic tanks and under agricultural land with the shallow water table (where root activity, decomposing plants, and animals create macropores in the soils), the chemicals move rapidly downward to a greater depth [8].

81.13 Water Reuse and Limitations

The benefits of reusing the wastewater intending to augment water supplies and manage nutrients in treated effluents are the motivators of installation of such reuse programs. Such benefits are reduced nutrient loads to receiving waters due to reuse of the treated wastewater. These drivers center around three categories:

1. Addressing urbanization and water supply scarcity
2. Achieving efficient resource use
3. Environmental and public health protection

Some of issues of water reuse are described next.

81.13.1 Perception on Health

A study conducted in the 53 countries around the world comprising the cities from the developing countries (Table 81.6) reflects the farmers' perception on wastewater irrigation and exposure to health risks. Along with skin infections, gastrointestinal and diarrheal infections are also commonly cited, while a large number of other general illness and respiratory problems were associated with wastewater agriculture (Table 81.8).

81.13.2 Societal Acceptance and Sustainability

To improve the social acceptance of water reuse in urban areas, information about the different benefits (environmental and economic such as impact on tourism and water price) of water reuse and an information session about the different terms (water, wastewater, reclaimed water, water treatment, water quality, etc.) are needed, as information is proved to have a positive influence on user acceptance.

TABLE 81.8 Reported Health Risk Reduction Methods and Perceived Health Problems

Description	Number of Cities Responding Positively out of 53
Type of protection	
Protect feet	20
Protect hands	8
Wash hands	34
No protection	19
Reasons for washing produce	
Reduce contamination	9
Keep produce fresh	23
Clean dirt off produce	23
Farmer attributed perceptions of health problems	
Skin irritation	21
Gastrointestinal/diarrhea	14
Respiratory	6
Other	15

Source: Adapted from Asano, T. 1998. *Wastewater Reclamation and Reuse: Water Quality Management Library*, CRC Press, Boca Raton, FL.

The often-neglected aspect is the public participation while selecting the most appropriate wastewater treatment technology for a particular community, whose preferences and perceptions are important for sustainability concern. The size of community can dictate the type of treatment system selected, its capacity, and thereby, its sustainability.

Increase in population indicates the need for larger plant capacity. Mechanical and lagoon systems are more capable of serving larger populations than land-treatment systems. While mechanical systems are chosen over lagoon systems to service bigger populations, a deciding factor is the land requirement or open space availability.

81.13.3 Nuisance from Odor

Wastewater treatment facilities, regardless of how well they are designed, may generate odor as by-products of wastewater collection and treatment processes. The presence of odor in any wastewater treatment plant is typically an aesthetic problem that usually evokes public intervention and regulatory agency involvements.

81.13.4 Economic Feasibility

Cost effectiveness is directly related to the volume of reclaimed water. Irrigation generally provides the highest potential of water reuse. Although difficult to specify, it could be in the range of a flow corresponding to 10,000–20,000 inhabitant equivalents, or the same water needed to irrigate a golf course or a crop extension of 3,500,000 m² [39].

Financing is another important issue. Subsidies cover planning, technical assistance and research (pilot studies, etc.), and construction costs while they do not cover operation and maintenance costs. In the EU, there are no guidelines yet to quantify non-monetary benefits of projects, and therefore, the grants are provided on a case-by-case basis. Usually, funding water reuse projects are based on political decisions [32].

81.14 Cost of Wastewater Reuse Product

The cost of water reuse involves the extra treatment to reach the reuse quality requirement above and beyond the obligatory baseline treatment in order to protect the community safety, health, and environment; along with the extra conveyance of the effluent to the reuse site. Simultaneously, the benefit of reuse includes the value of freshwater saved and the cost of the alternative safe final disposal of the effluent when reuse is not practiced [23].

Low-cost treatment of wastewater may be an option with stabilization ponds, constructed wetlands, etc. under suitable climatic and geographic conditions. The net cost of treatment may also be reduced through the reuse of biogas for energy and power in the intensive treatment processes, or potentially through the sale of carbon offsets [45].

Although, wastewater can contain heavy metals, organic compounds, and a wide spectrum of pathogens having a negative impact on the environment and human health, recycled water is considered to be the cost-effective means of supplying nutrients for agricultural irrigation as well as avoiding the need for nitrogen removal at wastewater works in sensitive areas. While technical, environmental, and social factors are considered in project planning, monetary factors tend to control the ultimate decision on whether and how a wastewater reuse project is implemented.

In the EU, water reuse is practiced predominantly in arid regions like Greece, Spain, and Italy. Additionally, Israel, Jordan, and Tunisia are among the leading countries in wastewater reuse. In the city of Braunschweig, wastewater has been reused since 1896. By 1996, there were 41 sites in Germany irrigated with domestic wastewater and 33 sites with industrial wastewater reuse including sugar industry (3), starch (7), milk (2), cleaning of vegetables (2), sweets industry (1), distilleries (13), and agricultural cooperatives (5) [15].

81.15 Global Scenario of Water Reuse Products

With respect to the increasing burden on all water resources, both in industrialized and in developing countries, supplementing water resources with reclaimed wastewater cannot be neglected [3]. Worldwide, wastewater reclamation and reuse is estimated to represent a potential extra water resource, approximating 15% of existing water consumption, while on a local basis, this proportion can be significantly higher reaching up to 30% of agricultural irrigation water and 19% of total water supply in Israel in the future. In some Asian cities, including Japan, reclaimed water is used in toilet flushing in high-rise and commercial buildings.

The Mediterranean region includes 60% of the world population with renewable natural resources of less than 1.000 m³ water/inhabitant/year and hence is characterized by the low level and irregularities of water resources throughout the year [24]. Countries of the Southern Mediterranean and Middle East region are facing increasingly serious water shortage problems. Despite its high cost, desalination of brackish and seawater is already under implementation or planned in some countries such as Libya, Egypt, Tunisia, and Algeria. Therefore, several problems appear around the basin including water and soil salinization, desertification, increasing water pollution, and unsustainable land and water use [34].

In Cyprus, about 25 million m³/yr of the collected wastewater is planned to be used for irrigation after tertiary treatment. However, because of the high transportation cost, most of the recycled water, about 55%–60%, will be used for amenity purposes in hotel gardens, parks, golf courses, etc. A net of about 10 million m³ is conservatively estimated to be available for agricultural irrigation, which will allow irrigated agriculture to be expanded by 8%–10%, simultaneously conserving an equivalent amount of water for other sectors [31].

In United Arab Emirates, one of the most water poor countries in the world, groundwater supplies have been conserved by various recycling techniques. The recycled wastewater is used for landscape and horticulture irrigation. Conditions and regulations for the safe use of recycled wastewater for irrigation have been established to check the safety of public health [13].

81.16 Summary and Conclusions

Wastewater reuse practices are considered an essential component of sustainable and integrated urban water resources management. The development of wastewater reclamation and reuse is related to alarming water scarcity, water pollution control measures, protection of the environment, and alternative water resources for the growing population in urban areas. Wastewater reuse is practiced with proper attention to sanitation, public health and environmental protection in the cities of developed countries. With the continuing technological advancement, the reliability of wastewater reuse systems will be widely demonstrated, expanding as an essential element in sustainable urban water management.

Key policy recommendations may be considered in urban wastewater reuse and management:

- Feasibility studies regarding the reuse products must be as complete as possible and they have to detail many different aspects as geological, technical, economic, environmental, social, and water quality, as well as risk issues.
- Wastewater should be considered a potential resource in the overall urban water budget.
- Management must adopt an integrated framework for smooth water supply, stormwater, wastewater, non-point source pollution, and water reuse in urban areas.
- Climate change adaptation and integrated water resources management strategies should be adopted for reclamation and reuse in a sustainable manner.
- Various reuse options should be considered from the outset in the design of treatment plants, as well as in their operation, with corresponding criteria and standards commensurate with the needs of urban dwellers.
- Guidelines and policies that encourage urban and peri-urban communities to determine the most appropriate and cost-effective wastewater treatment solutions, based on local capacities and reuse options, should be adopted.
- Financial stability and sustainability should be ensured by (a) linking waste management with other economic sectors for fast cost-recovery, risk reduction, and sustainable implementation; (b) developing mixed public/private, public/public sector solutions for investment, service delivery, operation, and maintenance; and (c) considering social equity when defining cost-recovery mechanisms.
- EPA 1992: Guidelines for water reuse: Beside the reclaimed water quality guidelines, recommended monitoring and setback distances are given.
- WHO 1989: "Health Guidelines for the Use of Wastewater in Agriculture and Aquaculture," they take into account the treatment process, irrigation system, and the crops to be irrigated. This set of guidelines is controversial but has allowed a real development of wastewater reuse.

- FAO 1985 (Quality criteria) [21] determine the degree of suitability of a given effluent of irrigation [6], for specific values refer to FAO Irrigation and Drainage Paper 47.

References

1. Angelakis, A., Bontoux, L. et al. 2003. Main challenges for water recycling and reuse in EU countries. *Water Supply,* 3(4), 59–68.

2. Aoki, C., Memon, M.A. et al. 2005. *Water and Waste Water Reuse: An Environmentally Sound Approach for Sustainable Urban Water Management Osaka,* Shiga, Japan, UNEP, GEC.

3. Asano, T. 1998. *Wastewater Reclamation and Reuse: Water Quality Management Library,* CRC Press, Boca Raton, FL.

4. Asano, A. and Bahri, A. 2010. Global challenges to wastewater reclamation and reuse. In *Selections from the 2010 World Water Week in Stockholm.* Lundqvist, J. (ed.). pp. 64–72. www.worldwaterweek.org/onthewaterfront.

5. Asano, T., Smith, R.G. et al. 1984. *Irrigation with Reclaimed Municipal Wastewater: A Guidance Manual.* California State Water Resources Control Board, Sacramento, CA, pp. 2–12.

6. Ayers, R.S. and Westcot, D.W. 1985. *Water Quality for Agriculture.* Vol. 29. FAO, Rome.

7. Badyal, D. 2009. Municipal Wastewater Recycling Survey. California Water Recycling Funding Program (WRFP). http://www.swrcb.ca.gov/water_issues/programs/grants_loans/water_recycling/munirec.shtml. Retrieved August 23, 2012.

8. Bouwer, H. 1991. Simple derivation of the retardation equation and application to preferential flow and macro dispersion. *Groundwater,* 29(1), 41–46.

9. Brissaud, F. 2003. Groundwater recharge with recycled municipal wastewater: Criteria for health related guidelines. State of the art report health risks in aquifer recharge using reclaimed water. Water, Sanitation and Health Protection and the Human Environment World Health Organization Geneva and WHO Regional Office for Europe Copenhagen, Denmark 2, 10–15.

10. Brissaud, F., Salgot, M. et al. 1999. Residence time distribution and disinfection of secondary effluents by infiltration percolation. *Water Science and Technology,* 40(4), 215–222.

11. Buras, N., Duek, L. et al. 1987. Microbiological aspects of fish grown in treated wastewater. *Water Research,* 21(1), 1–10.

12. Colborn, T., vom Saal, F.S. et al. 1993. Developmental effects of endocrine-disrupting chemicals in wildlife and humans. *Environmental Health Perspectives,* 101(5), 378.

13. Cooper, I. 2001. *Expanding Sharjah's Green Spaces. Water 21.* Michael Dunn, London, UK, pp. 24–25.

14. Dolan, R.J. 1995. Good science and compliance decisions. *Water Environment Research,* 67(2), 252–252.

15. Donta, A.A. 1997. Der Boden als Bioreaktor bei der Aufbringung von Abwasser auf landwirtschaftlich genutzte Flächen. Veröffentlichungen des Institutes für Siedlungswasserwirtschaft und Abfalltechnik der Universität Hannover (ISAH).

16. Duncan, M. 2003. *Domestic Wastewater Treatment in Developing Countries.* Earthscan, London, UK.

17. Ellis, T.G. 2004. Chemistry of wastewater. http://www.eolss.net/eolsssamplechapters/c06/e6-13-04-05/e6-13-04-05-txt 04.aspx#3._Wastewater_Composition.

18. EPA. 2004. *Guidelines for Water Reuse.* Environmental Protection Agency, Municipal 440 Support Division Office of Wastewater Management Office of Water Washington, DC. Agency for 441 International Development Washington, DC, EPA/625/R-04/108, Cincinnati, OH US EPA/625/R-04, 108, 442.

19. EPA. 2012. Guidelines for Water Reuse, United States Environmental Protection Agency.

20. FAO 1985. Food and Agriculture Organization of the United Nations. Rome, Italy.

21. Gleick, P.H. 2000. A look at twenty-first century water resources development. *Water International,* 25(1), 127–138.

22. Golf Course Superintendents Association of American (GCSAA) and The Environmental Institute for Golf (GCSSAB and EIFG). 2009. Golf Course Environmental Profile, Volume II, Water Use and Conservation Practices on U.S. Golf Courses. GCSAA. Lawrence, KS. http://www.gcsaa.org/Course/Environment/default.aspx. Retrieved on August 23, 2012.

23. Hidalgo, D. and Irusta, R. 2005. The cost of wastewater reclamation and reuse in agricultural production in the Mediterranean countries. *IWA Conference on Water Economics, Statistics and Finance,* Rethymo.

24. Kamizoulis, G., Bahri, A., Brissaud, F., and Angelakis, A.N. 2003. Wastewater recycling and reuse practices in Mediterranean region: Recommended Guidelines. www.med-reunet.com/docs_upload/angelakis_cs.pdf.

25. Lim, R. 2000. Endocrine disrupting compounds in sewage treatment plants (STP) effluent reused in agriculture. *First Symposium on Water Recycling in Australia,* Adelaide.

26. Mara, D. 2013. *Domestic Wastewater Treatment in Developing Countries,* Routledge, Sterling, VA.

27. McGovern, P. and McDonald, H.S. 2003. Endocrine disruptors. *Water Environment & Technology,* 15(1), 35–39.

28. Meesters, K.P.H., Van Groenestijn, J.W., and Gerritse, J. 2003. Biofouling reduction in recirculating cooling systems through biofiltration of process water. *Water Research,* 37(3), 525–532.

29. Metcalf and Eddy. 1991. *Wastewater Engineering. Treatment Disposal Reuse.* McGraw-Hill, New York.

30. Obuobie, E., Keraita, B., Danso, G., Amoah, P., Cofie, O.O., Raschid-Sally, L., and P. Drechsel. 2006. *Irrigated Urban Vegetable Production in Ghana: Characteristics, Benefits and Risks.* IWMI-RUAF-CPWF, Accra, Ghana: IWMI, 150 pp.

31. Papadopoulos, I. 1995. Present and perspective use of wastewater for irrigation in the Mediterranean Basin. *2nd International Symposium on Wastewater Reclamation and Reuse,* A.N. Angelakis, Iraklio, Greece, 2, 735–746.

32. Pirnie, M. 2000. Feasibility study for reuse of wastewater effluent. Final report. Cape May County.

33. Raschid-Sally, L. and Jayakody, P. 2009. Drivers and characteristics of wastewater agriculture in developing countries: Results from a global assessment, IWMI.

34. Salem, Z., Kuhail, Z. et al. 2004. Analysis of best practises and success stories-Medaware project: Report on Task 3, Environmental Quality Authority of Palestine Contribution.

35. Salgot, M., Priestley, G.K. et al. 2012. Golf course irrigation with reclaimed water in the Mediterranean: A risk management matter. *Water*, 4(2), 389–429.

36. Sonune, A. and Ghate, R. 2004. Developments in wastewater treatment methods. *Desalination*, 167, 55–63.

37. UNESCO. 2000. Water use in the world: Present situation/future needs, United Nations Educational, Scientific and Cultural Organization. http://www.unesco.org.

38. Urkiaga, A., De las Fuentes, L. et al. 2006. Methodologies for feasibility studies related to wastewater reclamation and reuse projects. *Desalination*, 187(1), 263–269.

39. Verges, C. 1998. *El agua regenerada como nuevo product en el mercado del abastecimiento y saneamiento de las aguas.* La gestión del agua regenerada. Palamós, Costa Brava, Spain.

40. Visvanathan, C. and Asano, T. 2007. *The Potential for Industrial Wastewater Reuse*, Department of Civil and Environmental Engineering, University of California at Davis, California.

41. Wade, T.J., Sams, E. et al. 2010. Rapidly measured indicators of recreational water quality and swimming-associated illness at marine beaches: A prospective cohort study. *Environmental Health*, 9(66), 1–14.

42. WERF. 2010. Decentralized water resources: Answers to the most frequently asked questions. Water Environment Research Foundation, Alexandria, VA.

43. WHO. 1989. *Health Guidelines for Use of Wastewater in Agriculture and Aquaculture*. World Health Organization, Geneva, Switzerland.

44. WHO. 2006. *Guidelines for the Safe Use of Wastewater, Excreta and Greywater, Volume 1: Policy and Regulatory Aspects*. World Health Organization.

45. Winpenny, J., Heinz, I. et al. 2010. *The Wealth of Waste: The Economics of Wastewater Use in Agriculture*. FAO, Rome. 35.

46. www.webworld.unesco.org. 1999. World water resources and their use. http://webworld.unesco.org/water/ihp/db/shiklomanov/index.shtml. Retrieved November 1, 2013.

82

Conjunctive Use of Water Reuse and Urban Water

Saeid Eslamian
Isfahan University of Technology

Majedeh Sayahi
Isfahan University of Technology

Behnaz Khosravi
Isfahan University of Technology

PREFACE

With the population growth currently and expected in the future, freshwater resources are expected to decrease. Therefore, preserving potable water resources is essential. For nonpotable needs, complex treatment technologies are being instituted to change wastewater into a high-quality product for nonpotable projects. This process is often referred to as wastewater reclamation or water reuse. Water can be a limited resource for many communities. Many water resource professionals believe that reclaiming water after treatment in a modern wastewater treatment plant has an important role in sustainable water resource management. Conjunctive use of water thus far refers to the coordinated and planned use of surface and groundwater. Now it is understood that reuse wastewater is a water resource developed right at the threshold of the urban environment. Surface water, groundwater, and reclaimed wastewater in the watershed are three water resources for water supply. Conjunctive use of water is an integrated management and useful connection between these three water resources. The goal of conjunctive management is to increase the overall water supplies and protect against drought. Problems and costs of disposing wastewater often present new advantages of conjunctive use. The purpose of this chapter is to introduce conjunctive use of groundwater/surface water and reuse water resources for irrigation, industrial and urban water supply in the developing world, and the great potential that planned conjunctive use has as an adaptation strategy to speed up climate change.

82.1 Introduction

Water resources are exigent for agriculture, industry, recreation, and human life. The global renewable water supply is about 7000 m³ per person per year (present population, 2000). Thus, there is sufficient water for at least three times the present world population. In general, lack of water is due to an imbalance between population and precipitation distributions [4]. Urban populations with respect to the sustainable supply of safe and reliable water for residential, commercial, and industrial needs in more developed nations face other challenges [17]. These challenges include different constraints related to water, both from quantity and quality of supply and treatment [17]. Many water resource professionals believe that reclaiming water after it is treated in a modern wastewater treatment plant has an important role for sustainable water resource management [11]. Water supply institutions are searching and considering alternative methods of increasing their available water resources to meet the demands of their customers. One such alternative method is the concept of conjunctive use [12]. The dependence on either surface water or groundwater reservoir alone has inherent limitations. Those limitations include possible restrictions regarding diversions of surface water rights during low flow or drought conditions, evaporative losses of water stored in surface water reservoirs, and limitations due to the nonrenewable nature of certain groundwater supplies [12]. Conjunctive use based on the principle that by using surface water when it is plentiful, recharging aquifers and protecting groundwater supplies in wet years, water will then be used for future objectives in dry years when surface supplies are limited. Specific characteristics of conjunctive use are natural replenishment (e.g., infiltration of precipitation and natural percolation from overlying streams), artificial recharge, return flow from irrigation and sewage, and stream-aquifer interaction [15]. Conjunctive use includes the use of imported water and recycled water for artificial groundwater recharge, water conservation programs, and water banking. As discussed by McClurg [16], most water reuse projects have allowed local districts to substitute recycled water for potable water, for example, to irrigate crops; water city parks, medians, and golf courses; flush toilets in high-rise buildings; and supply industries, increasing the drinking-water supply in a community [16]. The purpose of this overview is to introduce conjunctive use of groundwater/surface water and reuse water resources for irrigation, industrial, and urban water supply in the developing world.

82.2 Methodology

With population growth and higher living standards, demand for good quality municipal, industrial water and even sewage flows have been increased. At the same time, more and more water will be needed for irrigation to meet increasing demands for food for growing populations. Also, more water will be required for environmental concerns. Thus, increased competition for water can be expected. This will need intensive management and international cooperation. Since almost all liquid freshwater on our planet occurs underground, groundwater will be used more and more, and hence, must be protected against depletion and contamination, especially from nonpoint sources like intensive agriculture. Municipal wastewater has the potential to be an important water resource, but its use must be carefully planned and regulated to prevent harmful health effects. As mentioned before, almost all liquid freshwater occurs underground, so its long-term suitability as a source of water is threatened by nonpoint source pollution such as agriculture and aquifer depletion due to groundwater withdrawals in excess of groundwater recharge [4]. The urban water cycle should be managed as a single system in which all urban water flows are recognized as potential resources and where the interconnectedness of water supply, groundwater, stormwater, wastewater, flooding, water quality, wetlands, watercourses, estuaries, and coastal waters is recognized. Water efficiency, reuse, and recycling are integral components of total water cycle management [9,13]. Conjunctive use of water thus far refers to the coordinated and planned use of surface and groundwater. Now it is recognized that reclaimed wastewater is a new water resource, so in the watershed, there

are three water resources for water supply: surface water, groundwater, and reclaimed wastewater. Conjunctive use of water is an integrated management and intelligent use of these three water resources [9]. Precipitation is the input of water to the watershed. The surface water system reactions to rain events are faster than the groundwater system. The water storage capacity of the groundwater system is usually greater than that of the surface water system. In fact, conjunctive use of surface and groundwater, the difference in response time between surface and groundwater systems, and large storage capacity of groundwater systems are utilized. In the holistic water management, these three water resources are used conjunctively. Reclaimed wastewater can be used for artificial groundwater recharge and for surface stream augmentation, and directly for several water uses. The direct use of reclaimed wastewater includes agriculture irrigation, industrial reuse, and urban use. The treatment for wastewater reclamation depends on the water use objectives. One of the effective strategies for development and management of water resources is conjunctive or integrated water use. Conjunctive use programs are designed to increase water supplies. The word conjunctive use has a broad concept now. As shown in Figure 82.1, there are three sources for water supply: surface water, groundwater, and wastewater reclamation/reuse. The new concept of conjunctive use of water is a wise use of total water resources in an integrated manner. Two types of conjunctive use of water will be discussed: conjunctive use of surface water and groundwater and wastewater reclamation/reuse.

82.2.1 Agriculture

Globally, more than 70% of water resources used for agriculture and the major global food production rely on a range of agricultural systems in which water is the critical factor. About 1.5 billion ha, 11%, of the total land surface, is used for agriculture and of this, some 270 million ha are irrigated land. This mere 18% of land under cultivation produces over 40% of the world's staple foods [5].

With the increasing shortage of freshwater for agriculture, the need to use reclaimed water (RW) in agriculture has been increased [4]. The use of recycled water for the irrigation of crops has the benefits in using a resource that would otherwise be discarded and wasted [25]. Prathapar et al. [20] showed that, if needed, an advanced treatment of the excess effluent before aquifer storage and recovery (ASR) is technically possible in Muscat, Oman. However, because of the very high cost of the treatment process, the project will not be beneficial financially. Disposing RW to the sea is not reasonable or prudent either. Therefore, there is a need to avoid RW disposal and maximize the use of it by growing short season crops throughout the year and supplementing RW with groundwater. By doing so, RW will not be excreted to the sea, it need not be injected to the aquifer, stress on groundwater will be reduced, and crop production will be increased [4,20].

Agricultural water management must be integrated with other water management practices. In fact, the act of one user group

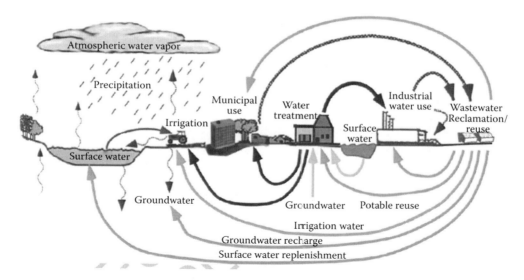

FIGURE 82.1 The cycling of water through the hydrologic cycle. (Adapted from Funamizu, N., and Magara, Y., 2014. *Encyclopedia of Life Support Systems* (EOLSS), UNESCO Publishing, 354–372.)

will affect the water profits of others [4]. Therefore, the decision makers should consider piping RW to areas where groundwater has a good quality to conjunctively use and meet crop water requirements, rather than piping it to areas where groundwater is saline and unsuitable for irrigation. This will prevent disposal of RW to the sea and reduce stress on fresh groundwater zones. On the other hand, piping this RW from sewage treatment plants to non-saline lands where the groundwater is also suitable for irrigation, groundwater quality may complement RW and meet variation in seasonal evaporative demand [1]. By mixing groundwater with the canal water, farmers tend to decrease the risk of soil salinization. So farmers mix groundwater with the canal water in different ratios without sufficient knowledge of the hazards associated with its long-term use [25]. Increasing demands for irrigation water while water resources are limited will lead to reuse and recycling of the available water resources. Today, agriculture drainage, and industrial and domestic wastewaters are reused and recycled for irrigation in many parts of the world [14]. Currently, a system of conjunctive use in many surface irrigation systems has evolved [3]. The potential for conjunctive use in agricultural irrigation varies considerably with "hydrogeological setting," containing such factors as average rainfall and geomorphological position [8].

82.2.2 Landscape

Natural ecosystems are often in association with water or have a strong relationship with it. Lack of water results: soil would parch, forests would wither, and species would die out [5]. The projects done until now indicated that both problems and opportunities exist in using RW for landscape irrigation. Powerful means of water conservation and nutrient recycling are to use recycled wastewater for irrigation in urban landscapes, for reducing the demands of freshwater and lessening

pollution of surface and groundwater. However, using recycled wastewater for irrigation has potential problems associated with it. The challenge of water reuse is to maintain long-term sustainability. Irrigating with wastewater will burn pine needles, symptoms observed in ponderosa pines, because of salts (especially the relatively high Na^+ and high EC) in the treated wastewater. Soil salinity is a function of soil type, management, salinity of water used for irrigation, and the depth of the water table [21]. The reason for concern about possible long-term reductions in soil hydraulic conductivity and infiltration rate in soil with high clay content is the significantly higher soil sodium adsorption ratios (SAR) in RW-irrigated sites compared to surface water irrigated sites provided, although these levels were not high enough to result in short-term soil deterioration. Understanding the responses of urban landscape plants and soils to recycled wastewater irrigation and identifying proper management practices are critical to the long-term success of the water reuse practice. This information is useful to landscape planners and managers to determine what should be monitored and what proactive steps should be taken to minimize any negative effects during planning and managing landscapes receiving recycled wastewater. Using recycled water also decreases the pressures on the environment by reducing the use of environmental waters [25]. The major use of recycled wastewater for landscape is golf courses, mainly because golf courses have intensively managed turf (dense grasses utilize the nutrients in the wastewater) that requires major amounts of water [21]. "The conjunctive of water supply, sewerage and storm water, so that water is used optimally within a catchment resource, state and national policy context. It promotes the coordinated planning, development and management of water, land and related resources (including energy use) that are linked to urban areas and the application of water-sensitive urban design principles in the built urban environment" [7].

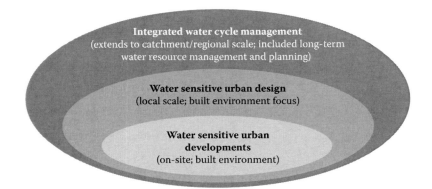

FIGURE 82.2 Managing water resources in an urban development context. (Adapted from Department of the Environment, Water, Heritage and the Arts, 2008. Better urban water management. Australia. http://www.planning.wa.gov.au/.)

This is depicted in Figure 82.2 (from the National Water Initiative) [7].

Turf grasses have a good look, showing salinity damage only on a few sites with poor drainage, heavy soil structure, or shoal water table. However, chronic reduction of conifer trees was often observed under RW irrigation. Ponderosa pines grown on sites irrigated with RW for 5–33 years exhibited 10 times higher needle burn symptoms than those grown on sites irrigated with surface water [21].

82.2.3 Recharge

Iran's groundwater resources have been excessively exploited, often at the expense of density water and land quality, and there is limited room for developing irrigation agriculture [10].

In many parts of the world, exclusively in developing countries, groundwater is thus being massively overabstracted. Consequences of overabstraction are falling water levels and declining well yields, land subsidence, intrusion of saltwater into freshwater supplies, and ecological damages such as drying out wetlands [10]. The purposes of artificial recharge of groundwater include the following:

1. To avoid reduction of groundwater levels due to excessive groundwater withdrawals
2. To protect aquifers against saltwater influx from the ocean
3. To store surface water, including flood or excess water, and reuse water for future use

The benefits of storing water underground are summarized as

1. The price of storing water underground may be less than the cost of equivalent surface reservoirs
2. The aquifer serves as a conditional distribution system and may resolve the need for surface pipe lines or canals
3. Results of storing water in surface reservoirs are evaporation, taste, and odor problems due to algae and other aquatic productivity and pollution, which may be avoided by underground storages
4. In view of appropriate sites for surface reservoirs, they may not be available or environmentally acceptable [9]

5. The stored groundwater serves as a nonevaporating "bank" that can be tapped during subsequent dry periods to sustain consumptive uses or additional stream flows [2]

82.2.4 Case Study of Agriculture

In Australia, the Shepparton Irrigation Region Land and Water Salinity Management Plan has promoted groundwater pumping and reuse for irrigation where groundwater quality and availability allow dilution with canal water (conjunctive water use) to levels that produce minimal production losses from annual and perennial pastures used widely for dairying. In addition, municipal and industrial wastewaters are used on a smaller scale for irrigating pastures (and crops) [23].

In Oman, for conjunctive use of reclaimed water and groundwater projects, three different groundwater qualities (Fresh with EC of 1.0 dS/m, Marginal with EC of 1.5 dS/m, and Saline with EC of 3.0 dS/m) were mixed in four different ratios (25%, 50%, 75%, and 100%) with the canal water. The total 12 combinations of groundwater and canal water were evaluated for their impact on crop production and soil salinity. The results show that mixing fresh groundwater with canal water in any ratio will not make intensive salinity problems except for below average rainfall years [25].

Also in Iran, the impact of irrigation with diluted Caspian Sea water on the growth and yield of barley and on the characteristics of soil was investigated in field plots and in plot experiments during the 2001–2002 growing season. The results indicate that using Caspian Sea water for supplementary irrigation is a viable option and has the potential of reducing pressure on the limited groundwater resources of the region without a significant loss in barley production [22].

82.2.5 Case Study of Landscape

With the growing rate of population in Colorado, increased use of RW is viewed as one approach to maximize the existing water resources and development of Colorado's urban water facilities [21]. While these water quality criteria do help to protect public health and the environment, they do not consider the water chemistry considerations that affect the suitability of treated wastewater for landscape irrigation. The two parameters—salinity and sodicity, as well as other chemical constituents —cause concern among landscape managers. Recycled wastewater chemistry tends to be outstanding in sulfate, bicarbonate, chloride, and sodium. These four ions comprise about 70% of total dissolved salts [21].

82.2.6 Case Study of Industry

The use of a scheme mixing several kinds of unusual waters for reuse in industry, agriculture, and green spaces has been proposed in Iran (2014) so that the seawater, treated water using nanotechnology (NF), and effluent biological systems have been used [18]. The purpose of that research is to find the best mixture for various applications. Determining the best type of mixture by reducing the tested parameters and also analyzing whether up to 5% difference in results of mixing schemes makes sense have been done. For industrial uses, the best mixture is 80% drinking water, 10% raw sewage, and 10% NF effluent, in which SO_4, which is considered an important factor in industry, has the lowest possible percentage. However, the 80% mixture of drinking water, 10% raw sewage, and 10% biological effluent is also convenient for industrial uses and is considered in the scope of the internal and external standards [18].

82.3 Various Aspects of Conjunctive Use

82.3.1 Risk and Health

The World Health Organization (1989) has developed guidelines to avoid use of raw wastewater for irrigation and to ensure that irrigation is reasonably safe from a public health point of view [4]. Water resources and environmental engineering have played an important role in protecting urban public health since the beginning of the Industrial Revolution, which marked the rise of urban populations, growth in wealth, and spread disease due to poor sanitation and inadequate water supply [17]. As the criteria set forth in Regulation 84 for the State of Colorado, in 2000, water treatment professionals typically used parameters that related to human health. These parameters include *E. coli* count, turbidity, total suspended solids (TSS), and nitrogen and phosphorous content to evaluate water quality [21]. In any wastewater reclamation and reuse project, the health protection from the use of reclaimed wastewater is one of the most important aims. It is apparent that raw wastewater has pathogenic microorganisms and toxic materials. The risk of health that is associated with using reclaimed wastewater is decreased by reducing load of pollution and/or limiting contact with the reclaimed wastewater [9]. For using reused waters for agricultural irrigation, there are a number of risk factors. Some risk factors are short term and differ in intensity depending on the potential for human, animal, or environmental contact (e.g., microbial

TABLE 82.1 Threats Arising from Uncontrolled Conjunctive Use of Surface Water and Groundwater Resources in Pakistan

Threat	Main Causes
Groundwater depletion	Unregulated grow of shallow tube wells in areas of fresh groundwater
Soil salinization	Excessive recycling of shallow groundwater leads to salt accumulation in upper layers of soil
	Pumping of poor quality groundwater to compensate for deficiencies in surface water supplies
Deterioration of groundwater quality	Leaching of salt accumulation to groundwater
	Depletion of shallow freshwater overlying saline groundwater
	Lateral intrusion from saline groundwater
Inequity access of water resources	Tail and water users forced to pump excessive amount due to excessive use of surface water resources by head end farmers

Source: Adapted from Murray-Rust, H. 2002. *Conjunctive Water Management for Sustainable Irrigated Agriculture in South*, IWMI, 19–24.

pathogens), while others have longer-term impacts that increase with continued use of reclaimed water (e.g., saline effects on soil) [25]. Table 82.1 lists some of the more immediate threats in Pakistan that are currently being experienced, and where coordinated and planned management of both surface and groundwater resources is needed [19].

82.3.2 Cultural Issues

Survey data shows that the main reason for using recycled wastewater for irrigation is finding an available and reliable water supply [5]. Water quality is also a concern in many areas of the county where pollution sources have degraded water use. Areas with low water quality may require extensive and expensive treatment/remediation [17]. People generally prefer reuse that promotes water conservation, provides environmental protection benefits, protects human health, and cost-effectively treats and distributes a valuable and limited resource [11]. As an example, people in California agree that conjunctive use of water is an economical way for water supply management [15].

82.3.3 Economical Aspects

Storing water underground is a financially better method than dams and reservoirs. The value of a groundwater basin can be evaluated partly in terms of the avoided costs of an equivalent quantity of surface storage capacity. It is especially valuable to have groundwater storage that avoids the costs of surface storage capacity that would be used occasionally for unnecessary uses [2].

Wastewater treatment facilities may realize cost savings due to disposal costs and the sale of the recycled water. Communities can benefit from recycling by deleting or delaying the cost associated with obtaining new sources and facilities for freshwater.

Due to these reasons, currently there are hundreds of successful water reclamation and reuse projects in the United States [21]. In fact, difficulties and costs involved in disposing of wastewater often present new chances for conjunctive use [24]. In golf course irrigation projects, there is generally no economic advantage to using recycled wastewater compared to using ditch water or well water. However, compared to potable water use for landscape irrigation, there is generally an economic advantage for golf courses to use recycled wastewater, although the advantages vary depending on the situation [21].

82.3.4 Educational Issues

We now want to prepare systems that separate nonpotable and potable water supplies. Actually, potable purposes (e.g., direct consumption and food preparation) need small volumes of water, on the order of less than 40 L per person per day (L/capita-day) and much larger volumes of water, ranging from 100 to 400 L/capita-day, is used for nonpotable purposes (e.g., laundry, toilet flushing, bathing, and outdoor water use) [6]. No potable water should be used for other purposes outside of homes and buildings with the use of water to be as efficient as possible [9].

The information that relates to science and technologies, local knowledge and site-specific characteristics, values and interests, local context (e.g., political, social, economic, and environmental landscape), and other information is based on factors that play an important role in shaping perception and the nature of public participation. Furthermore, the uncertainty or incompleteness of information in any of these information categories influences perception about water reuse [11].

NGOs play an important role in bringing competing stakeholders to one platform and reaching a friendly solution. The problem that exists in many developing countries is that although the governments own most of the major sources of surface water, they do not have the ability to manage surface water for irrigation. Several thousands of hectares are water logged every year. Governments also cannot effectively manage groundwater because the farmers think it is their own. Therefore, NGOs play a role in educating the farmers [10].

82.4 Water Rights for Conjunctive Management

Access to a regular safe water supply is a basic human right [5]. A government is not the only factor that can endanger or limit the right to water. Individuals and corporations have the potential to prevent a person's or community's water supply. For example, pollution from factories, farming, or sewage can greatly reduce the quality of water used for drinking. A private individual can refuse access to a river that is needed for washing or a corporation may increase prices for water services without any reason. The duty to protect requires that governments should diligently take all the necessary feasible steps to prevent others from interfering with the right to water. This will usually require a strong regulatory regime that is consistent with other human rights. The Committee on Economic, Social and Cultural Rights has stated that this should include independent monitoring, genuine public participation, and imposition of penalties for noncompliance with standards. Comprehensive regulatory measures will be needed with respect to pollution, disconnection of water supplies, land use, and access to water supplies [5].

Rules governing water use, such as laws defining water rights, are critical to conjunctive water management.

82.5 Summary and Conclusions

There are several factors for enabling conjunctive management by environment.

1. The inflow of water and outflow in an area should be in balance in the long term.
2. Groundwater quality and quantity differ per area and should be recorded.
3. For control of the input and output of water, there are management practices such as irrigation efficiency and land reclamation.

The conjunctive use of surface water and groundwater is ideal in theory, but in practice, several problems occur at the field level during implementation. Ideally, water attributions and applications should be based on the accurate calculations of crop evapotranspiration, precipitation, and salinity build up and should be reviewed yearly. A key to effective conjunctive management is planned investment. Planned investments are required changes or improvements in hardware, software, planning and management capacities, and organizational adjustment. Conjunctive management in an area where unsuitable water quality is more difficult, and to resolve problems needs higher finance. Research in Oman showed that by using RW conjunctively with groundwater, cropping area can be increased (against RW use only) from 694.63 to 2245.31 ha (323% increase) of wheat, 313 to 782 ha (250% increase) of cowpea, and 346 to 754 ha (318% increase) of maize. While the strategy of using conjunctive use achieved acceptable control of soil salinity levels, a concern that it may not be sustainable in the long term arises from the sodicity of the groundwater and waste waters.

Benefits of conjunctive management are

- The reduction in the need for additional surface water reservoirs while adding to the available water supplies
- Offering additional support supplies
- The recovery or decline of decreasing groundwater levels in underlying aquifers
- A reduction in evaporation from surface water reservoirs
- Less impact to the environment
- An increase in available water equipment is realized as compared to the reliance on surface or groundwater sources alone

References

1. AL-Khamisi, S.A., Prathapar, S.A., and Ahmed, M. 2013. Conjunctive use of reclaimed water and groundwater in crop rotations. *Agricultural Water Management,* 116, 228–234.

2. Blomquist, W., Heikkila, T., and Schlager, E. 2001. Institutions and conjunctive water management among three Western States. *Natural Resources Journal,* 41(3), 653–683.

3. Born, N.V.D. 2011. *Conjunctive Water Management: How to Use the Full Potential.* Irrigation and Water Engineering Group Centre for Water and Climate, Wageningen University, the Netherlands.

4. Bouwer, H. 2000. Integrated water management: Emerging issues and challenges. *Agricultural Water Management,* 45, 217–228.

5. Brundtland, G.H. and Vieira de Mello, S. 2003. The right to water. Health and Human Rights Publication Series; No. 3, France.

6. Daigger, G.T. 2011. Sustainable urban water and resource management. *The Bridge,* 41(1), 13–18.

7. Department of the Environment, Water, Heritage and the Arts. 2008. Better urban water management. Australia. http://www.planning.wa.gov.au/.

8. Foster, S., Steenbergen, F.V., Zuleta, J., and Garduño, H. 2010. *Conjunctive Use of Groundwater and Surface Water.* The World Bank, South Asia. Strategic Overview Series No. 2, 3–26.

9. Funamizu, N. and Magara, Y. 2014. Conjunctive use of water, Vol. 1: Environmental and health aspects of water treatment and supply. In: *Encyclopedia of Life Support Systems* (EOLSS), UNESCO Publishing, 354–372.

10. Ghadiri, H., Dordipour, I., Bybordi, M., and Malakouti, M.J. 2006. Potential use of Caspian Sea water for supplementary irrigation in Northern Iran. *Agricultural Water Management,* 79, 209–224.

11. Hartley, T.W. 2005. Public perception and participation in water reuse. *Desalination,* 187, 15–126.

12. Jehn, J.L. 2006. *Conjunctive Use Issues: Optimizing Resources and Minimizing Adverse Effects.* Jehn Water Consultants, Inc., Colorado.

13. Lili, Y., Wentao, J., Xiaoning, Ch., and Weiping, Ch. 2011. An overview of reclaimed water reuse in China. *Journal of Environmental Sciences,* 23(10), 1585–1593.

14. Limaye, S.D. 2014. Conjunctive use of groundwater and surface water in a semiarid hard-rock terrain. In: S. Eslamian (ed.), *Handbook of Engineering Hydrology.* Vol. 1: Fundamentals and Applications. CRC Press, Boca Raton, FL, 41–51.

15. Marino, M.A. 2001. Conjunctive management of surface water and groundwater. *Regional Management of Water Resources,* IAHS Publ. No. 268, 165–173.

16. McClurg, S. 1995. Water recycling. In: *Western Water.* Water Education Foundation, Sacramento, CA, 4–13.

17. Merced County. 2011. Water element. In: 2030 Merced County General Plan. Planning Commission Review Draft. Merced, CA. http://www.co.merced.ca.us/index. aspx?NID=1791.

18. Molaei, H. 2014. Determining the option mixture of industrial wastewater using nano-filter procedure and seawater, municipal water for industrial and agricultural purposes. Master Thesis, College of Agriculture, Isfahan University of Technology, Iran.

19. Murray-Rust, H. 2002. Conjunctive water use and conjunctive water management. *Conjunctive Water Management for Sustainable Irrigated Agriculture in South,* IWMI, 19–24.

20. Prathapar, S.A., Ahmed, M., Abdalla, O., Zekri, S., Al-Jabri, S., Al-Maktoomi, A., and Al-Shuely, M. 2009. Managed Aquifer Recharge of Treated Waste Water in Oman. Report Submitted to Oman Waste Water Company.

21. Qian, Y. 2006. Urban landscape irrigation with recycled wastewater. Completion Report No. 204. Colorado State University, Colorado.

22. Qureshi, A.S. and Masih, I. 2003. Managing soil salinity through conjunctive use of surface water and groundwater: A simulation study. In: ICID Asian Regional Workshop, Sustainable Development of Water Resources and Management and Operation of Participatory Irrigation Organizations, November 10–12, Vol. 1. Taipei, Taiwan, 233–248.

23. Surapaneni, A. and Olsson, K.A. 2002. Sodification under conjunctive water use in the Shepparton irrigation region of northern Victoria: A review. *Australian Journal of Experimental Agriculture,* 42(0), 249–263.

24. The water for food team. 2006. Conjunctive use of groundwater and surface water. *Agricultural and Rural Development Group,* Issue 6. World Bank.

25. Toze, S. 2006. Reuse of effluent water—benefits and risks. *Agricultural Water Management,* 80(1), 147–159.

26. Weinrich, L., Hubler, J.F., and Spatari, S. 2012. Urban water supply: Modeling watersheds and treatment facilities. In: *Handbook of Metropolitan Sustainability,* F. Zeman (ed.), Elsevier Science; Woodhead Publishing, UK, pp. 370–389.

83

Urban Water Reuse Policy

Zareena Begum Irfan
Madras School of Economics

Saeid Eslamian
Isfahan University of Technology

PREFACE

The wastewater needs to be considered as an important component of the water cycles within catchments, if meaningful water management plans are to be implemented within the country. In each landscape, water augmentation has to be considered in conjunction with different wastewater treatment strategies for multiple uses, and should be supported by public policy and social incentives. It can then potentially not only safeguard the downstream users but also provide economic opportunities for alternative uses of wastewater within cities and support the ecosystem services that constitute an integral part of all forms of life. A countrywide approach for wastewater use in agriculture could capture the diversity seen in the Indian context, and could best be done at the state level, by identifying nodal agencies for systematic data collection. Indeed, all states must look at the alternative uses of wastewater for their cities, emphasizing the regional priorities, so that effective wastewater management plans can be developed to face the future with less fresh water. The ongoing dispute between states within India for freshwater as well as for wastewater-turned-freshwater shows the urgency of this matter. This chapter defines urban water reuse policy and categorizes building capacity for urban water reuse.

83.1 Introduction

The relative and absolute growth of the Earth's urban population and areas continues as a major demographic trend. During the 1950s and for the next 30 years, urban populations exploded around the world, and while this rate has slowed down, it is projected that 70% of the world's population will live in urban areas by 2050 [26]. Currently, half the world's population is urban, and with projected population growth being exclusively concentrated in urban areas over the next 30 years, developing regions will have more people living in urban than rural areas by 2030 [26]. Urban growth today is most rapid in developing countries, where cities gain an average of 5 million residents each month [24]. Megacities and metacities—defined by UN Habitat as cities with more than 10 million inhabitants or 20 million inhabitants, respectively—are gaining ground in Asia, Latin America, and Africa, and are spurred by economic development and increased populations [25].

83.1.1 Water Sources in an Urbanized World

In most developing countries, urban growth is inextricably linked with slum expansion and poverty. Sixty-two percent of sub-Saharan Africa's urban population and 43% of south-central Asia's urban population live in slums. In 2000, more than 900 million urban dwellers lived in slums, representing nearly one-third of all urban dwellers worldwide. Though the proportion of the developing world's urban population living in slums declined in the past 10 years, the absolute numbers of slum dwellers has actually grown considerably, and will continue to rise in the near future [26].

City infrastructure has often not kept pace with the massive urban growth, leaving many people, above all those in informal settlements and slums, without adequate access to drinking water and sanitation, which represents one of the major challenges confronting cities today. Having to rely on private vendors for their daily water supply, the urban poor pay up to 50 times more for a liter of water than their richer neighbors [24]. A central component to the adopted international development goals and targets, including most notably the Millennium Development Goals (MDG), is to reduce the share of the population without adequate water and sanitation services [10]. Diarrheal diseases alone are responsible for approximately 1.7 million deaths of children under the age of five every year—a death toll exceeding the combined under-5 mortality burden attributed to malaria and HIV [36]. Investments in drinking water supply and sanitation show a close correspondence with improvement in human health and economic productivity [30].

Invariably, cities consume more ecosystem services than they produce, and create an additional strain on ecosystems through water pollution. As the demand for living space continually increases, concrete and asphalt cover areas that are actually needed for groundwater recharge [22]. While only generating 0.2% of global freshwater supply, urban ecosystems serve 4–5 billion people [30]. City inhabitants benefit from ecosystems that provide services, such as clean water, agricultural products, clean air, and fossil fuels; however, net flows of ecosystem services into cities are increasing even more rapidly than urban populations, and so is the average distance of these flows. By importing goods, urban consumers draw on ecosystem services from other parts of the planet [10].

83.1.2 Water Footprint of Cities

Water is a critical component for human survival, but with the continually increasing demand on this finite resource, we

must find a sustainable balance. By calculating the water footprint, which measures the total volume of water used to produce goods and services that we consume and accounts for the volume of green (rain) and blue (withdrawn) water consumed in the production of agricultural goods from crops and livestock—the major uses of water—as well as the gray (polluted) water generated by agriculture and from household and industrial water use, we can incorporate a more holistic assessment of the demand placed on water resources by humans to calculate water availability [9]. Decision makers and resource managers can use these values to inform discussions on the sustainable and equitable allocation of water [7]. Irrespective of where the goods and services are consumed or produced, the water footprint can evaluate the pressures being placed on ecosystems. Thus far, the water footprint methodology has found little to no application for cities. Such an analysis can, however, help to show how a city populations' high consumption of water and products and services in which water is embedded have an impact on the surrounding rural communities and ecosystems, and at the global level. Going beyond a water footprint, no holistic assessment of a city and its surrounding regions supplying them with goods, services, and water has taken place yet.

83.2 Urban Water Reuse Options Implemented

The uncontrolled growth in urban areas has made planning and expansion of water and sewage systems very difficult and expensive to carry out. In addition, many of those moving to the city have low incomes, making it difficult to pay for any water system upgrades. The problem with the current treatment technologies is that they are not sustainable. The conventional centralized system flushes pathogenic bacteria out of the residential area, using large amounts of water, and often combines the domestic wastewater with rainwater, causing the flow of large volumes of pathogenic wastewater. In fact, the conventional sanitary system simply transforms a concentrated domestic health problem into a diffuse health problem for the entire settlement and/or region. In turn, the wastewater must be treated where the cost of treatment increases as the flow increases. The abuse of water use for diluting human excreta and transporting them away from settled areas is increasingly questioned and being considered unsustainable.

Another reason many treatment systems in developing countries are unsustainable and unsuccessful is that they were simply copied from Western treatment systems without considering the appropriateness of the technology for the culture, land, and climate. Often, local engineers educated in Western development programs supported the choice of the inappropriate systems. Many of the implemented installations were later abandoned due to the high cost of running the system and repairs. On the other hand, conventional systems may even be technologically inadequate to handle the locally produced sewage. For example, in comparison to the United States and Europe, domestic wastewater in arid areas such as the Middle East is up to five times more concentrated in the amount of oxygen demand per volume of sewage.

83.3 Implementation of Water Safety Plans

Framework for Safe Drinking Water and Water Safety Plans. The *Guidelines for Drinking-water Quality* [34] outlines a preventive management framework for safe drinking water that comprises five components (Figure 83.1), three of which combine to form the water safety plan.

Framework for safe drinking water [34]

- Health-based targets (based on an evaluation of health concerns).
- System assessment to determine whether the water supply chain (from source through treatment to the point of consumption) as a whole can deliver water of a quality that meets the health-based targets.
- Operational monitoring of the control measures in the supply chain, which are of particular importance in securing drinking-water safety.
- Management plans (documenting the system assessment and monitoring; describing actions to be taken in normal operation and incident conditions—including upgrade and improvement), documentation, and communication.
- A system of independent surveillance that verifies that the above are operating properly.

A water safety plan, therefore, comprises system assessment and design, operational monitoring, and management plans (including documentation and communication).

Commitment to the Water Safety Plan Approach. While many drinking-water supplies provide adequate and safe drinking-water in the absence of a water safety plan, the formal adoption of a water safety plan and associated commitment to the approach can have a number of benefits. Major benefits of developing and implementing a water safety plan for these supplies include the systematic and detailed assessment and prioritization of hazards and the operational monitoring of barriers or control measures. In addition, it provides for an organized and structured system to minimize the chance of failure through oversight or lapse of management. This process increases the consistency with which safe water is supplied and provides contingency plans to respond to system failures or unforeseeable hazardous events.

For the successful implementation of the water safety plan, management commitment is vital. There are a number of features of water safety plan adoption and implementation that can be attractive to management, including the following:

- Water safety plans represent an approach that demonstrates to the public, health bodies, and regulators that the water supplier is applying best practice to secure water safety
- The benefits that arise from delivering a more consistent water quality and safety through quality assurance systems
- Avoidance of the limitations associated with relying on end-product testing as a means of water safety control
- Potential savings as a result of adopting the water safety plan approach
- Potential for significant improvements in asset management
- Potential for marketing of services, to new and existing customers, of an improved product

Implementation of a pilot water safety plan project, alongside existing water quality management approaches, as a means of demonstrating the feasibility and advantages of the approach may facilitate acceptance of the method.

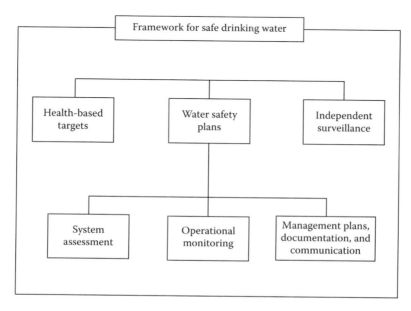

FIGURE 83.1 Framework for safe drinking water.

Development of a Water Safety Plan. A water safety plan essentially consists of three components:

- System assessment
- Operational monitoring
- Management plans, documentation, and communication

Assemble the Water Safety Plan. The preliminary step is to assemble a team to develop the water safety plan. For large supplies, a multi-disciplinary team of key people should be assembled to develop the plan. This should include managers, engineers (operations, maintenance, design, and capital investment), water quality controllers (microbiologists and chemists), and technical staff involved in day-to-day operations. All members of the team should have a good knowledge of the system. Water safety plans for small supplies may be developed generically rather than for individual supplies.

A team leader should be appointed to drive the project and ensure focus. The team leader should have the authority, organizational and interpersonal skills to ensure the project can be implemented. In situations where required skills are unavailable locally, the team leader should explore opportunities for external support. This can include benchmarking or partnering arrangements with other organizations, national or international assistance programs and Internet resources.

It is the team's responsibility to define the scope of the water safety plan. The scope should describe which part of the water supply chain is involved and the general classes of hazards to be addressed. The team should develop each step of the water safety plan. Other desirable features of the water safety plan team include:

- Knowledge of the water supply system and the types of drinking water safety hazards to be anticipated
- Authority to implement any necessary changes to ensure that safe water is produced
- Inclusion of people who are directly involved with the daily operations
- Having sufficient people on the team to allow for a multi-disciplinary approach, but not so many that the team has difficulty in making decisions

Team numbers will vary according to the size of the organization and complexity of the process. The use of subteams is common and might, for example, include water harvesting, water treatment, and distribution operations.

The membership of the team should be periodically reviewed with new or replacement members brought in if required.

Intended Water Use. For general purposes, water safety plans will apply to domestic potable use of drinking water. The expected use of the product should, however, be determined and documented by the water safety plan team. Factors that need to be considered include:

- What consumer education is in place for water use and how is this communicated, including how consumers are notified of potential contamination?
- Who is the water intended for and what is its intended use?

- What special considerations are in place for vulnerable groups such as infants, hospitalized patients, dialysis patients, the elderly, and immunocompromised?

Are there any groups for whom the water is specifically not intended?

- The numbers of people served by different service levels (communal, yard, within-house)
- Socioeconomic status of different communities served

This information is important, as it will be used in the hazard analysis to determine the hazard potential of the water.

EXAMPLE "INTENDED USE" DESCRIPTION

Example 1

Water utility X provides water to the general population. The water supplied is intended for general consumption by ingestion. Dermal exposure to waterborne hazards through washing of bodies and clothes, and inhalation from showering and boiling are also routes for waterborne hazards. Foodstuffs may be prepared with the water. The intended consumers do not include those who are significantly immunocompromised or industries with special water quality needs. These groups are advised to provide additional point-of-use treatment.

Example 2

Utility Y provides water to approximately half the population. The water is intended for general consumption by ingestion. Dermal exposure to waterborne hazards through washing of bodies and clothes, and inhalation from showering and boiling are also routes for waterborne hazards. Foodstuffs may be prepared from the water and market sellers use the water for freshening produce.

About half the population served relies on water supplied from public taps, with a further significant proportion relying on tanker services filled from hydrants. The socio-economic level of the population served by public taps is low and vulnerability to poor health is consequently high.

A significant proportion of the population is HIV positive, which increases vulnerability further. These examples on intended use provides the team with further understanding of the nature of the population served and any particular characteristics that may increase vulnerability to waterborne disease.

83.4 Emerging Dilemmas in Water Recycling

For each control measure, it is important to first define the operational limits (range) which, as part of the overall process train, leads to the supply of water that meets the intended use (including the health targets). However, because it is rarely practical to measure the concentration of hazards directly, some other means of control measure performance needs to be identified and becomes the target of monitoring. Therefore, a relationship between control measure performance, as determined by measurable parameters, and hazard control performance needs to be established. This relationship can be established using theoretical and/or empirical studies. In general, long-term performance data, design specifications, and objective scientific and empirical analysis are likely to be combined.

Not all measurable properties of control measures are suitable for this type of monitoring. Only where the following criteria are satisfied is it possible to define operational limits for control measures:

- Limits for operational acceptability can be defined
- These limits can be monitored, either directly or indirectly (e.g., through surrogates)
- A pre-determined corrective action (response) can be enacted when deviations are detected by monitoring
- The corrective action will protect water safety by bringing the control measure back into specification, by enhancing the barrier or by implementing additional control measures
- The process of detection of the deviation and completion of the corrective action can be completed in a timeframe adequate to maintain water safety

Monitoring Parameters. The parameters selected for operational monitoring should reflect the effectiveness of each control measure, provide a timely indication of performance, and be readily measured and provide opportunity for an appropriate response. Some water quality characteristics can serve as surrogates (or indicators) for characteristics for which testing is more difficult or expensive. Conductivity, for example, is a widely used surrogate for total dissolved solids.

Operational Limits. The water safety plan team should define the operational (or critical) limits for each control measure, based on operational parameters such as chlorine residuals, pH, and turbidity, or observable factors, such as the integrity of vermin-proof screens. The limits need to be directly or indirectly measurable. Current knowledge and expertise, including industry standards and technical data, as well as locally derived historical data, can be used as a guide when determining the limits.

Target or operational limits might be set for the system to run at optimal performance while the term critical limits might be applied when corrective actions are required to prevent or limit the impact of potential hazards on the safety and quality of the water. Limits can be upper limits, lower limits, a range, or an envelope of performance measures. They are usually indicators for which results can be readily interpreted at the time of monitoring and where action can be taken in response to a deviation in time to prevent unsafe water being supplied.

Monitoring. Monitoring relies on establishing the "what," "how," "when," and "who" principles. In most cases, routine monitoring will be based on simple surrogate observations or tests, such as turbidity or structural integrity, rather than complex microbial or chemical tests. The complex tests are generally applied as part of validation and verification activities rather than in monitoring operational or critical limits.

Table 83.1 shows what could be monitored if bacterial contamination of source water is identified as a potential hazard and feral or pest animal control and disinfection are identified as control measures. It can be seen from these examples that the frequency of monitoring will depend on what is being monitored and the likely speed of change.

If monitoring shows that an operational or critical limit has been exceeded, then there is the potential for water to be, or to become, unsafe. The objective is to monitor control measures in a timely manner to prevent the supply of any potentially unsafe water. A monitoring plan should be prepared and a record of all monitoring should to be maintained.

Monitoring plan. The strategies and procedures for monitoring the various aspects of the water supply system should be documented. Monitoring plans should include the following information:

- Parameters to be monitored
- Sampling location and frequency
- Sampling needs and equipment
- Schedules for sampling
- Methods for quality assurance and validation of the sampling results
- Requirements for checking and interpreting the results
- Responsibilities and necessary qualifications of staff
- Requirements for documentation and management of records, including how monitoring results will be recorded and stored
- Requirements for reporting and communication of results

83.5 The Economic Dilemmas in Water Recycling

Water scarcity is one of the key drivers for developing reclaimed water supplies and systems. As part of the overall management of water resources, it is critical to evaluate alternative management strategies for making the most of the existing supplies. Water conservation is an important management consideration for managing the water demand side. On the supply side, the use of

TABLE 83.1 Monitoring Examples

	Animal Control	Disinfection Control
What?	Wild pig densities in catchment must be below 0.5 per km²	Chlorine, pH, temperature and flow must provide for a CT of at least 15 with a turbidity of <5.0 NTU
How?	Scat (animal feces) surveys in spatially stratified transects across the catchment	Measured via telemetry and online probes with alarms
When?	Annually	Telemetry is downloaded automatically and continuously monitored
Who?	Catchment officer	Telemetry engineer

alternative water resources, such as reuse of greywater, rainwater harvesting (where applicable), produced water, and other reuse practices, should also be considered as part of an overall plan.

Integrating water conservation goals and programs into utility water planning is emerging as a priority for communities outside of the traditional water-short regions of the United States. Catalysts for implementing water conservation programs include growing competition for limited supplies, increasing costs and difficulties with developing new supplies, increasing demands that stress existing infrastructure, and growing public support for resource protection and environmental stewardship. As a result of the growing interest in water conservation, one of EPA's most successful partnership programs is WaterSense., which supports water efficiency by developing specifications for water-efficient products and services. The program also provides resources for utilities to help promote their water conservation programs.

In addition to using conservation as a means to utilities to help meet growing water demands, many utilities are also beginning to understand the value of water conservation as a way of saving on costs for both the utility and its customers. Throughout the United States, utilities have experienced quantifiable benefits associated with long-term water conservation programs, including:

- Reduction in operation and maintenance costs resulting from lower use of energy for pumping and less chemical use in treatment and disposal
- Less expensive than developing new sources
- Reduced purchases from wholesalers
- Reduce, defer, or eliminate need for capacity expansions and capital facilities projects

Selecting the appropriate conservation program components includes understanding water use habits of customers, service area demographics, and the water efficiency goals of the utility; some of the most effective practices that encourage conservation include the following:

- Customer education
- Metering
- Rate structures with a volumetric component with rate increases with increased use (tiered rate structure)
- Irrigation efficiency measures
- Time-of-day and day-of-week water limitations
- Seasonal limitations and/or rate structures
- High-efficiency device distribution and rebates.

Since 1991, for example, the Los Angeles Department of Water and Power has installed more than 1 million ultra-low-flush toilets and hundreds of thousands of low-flow showerheads and has provided rebates for high-efficiency washing machines and smart irrigation devices. The city used less water in 2010 than it did in 1990, despite adding more than 700,000 new residents to its service area [17]. While it is clear that potable water resources should be conserved for the reasons above, reclaimed water in some regions of the country is not considered a resource;

rather, it is sometimes viewed as a waste that must be disposed of. With this mindset, customers are sometimes encouraged to use as much reclaimed water as they want, whenever they want. In areas where there are fresh water supply shortfalls or where reclaimed water has become valued as a commodity, however, conservation has also become an important element of reclaimed water management. As a result, reclaimed water is recognized by many states as a resource too valuable to be wasted. The 1995 Substitute Senate Bill 5605 Reclaimed Water Act, passed in the state of Washington, stated that reclaimed water is no longer considered wastewater [28]. The California legislature has declared, "Recycled water is a valuable resource and significant component of California's water supply" [4]. These recent declarations are part of broad statewide objectives to achieve sustainable water resource management. Efficient and effective use can be critical to ensure that the reclaimed water supply is available when there is a demand for it. In addition, storage of reclaimed water can focus on periods of low demand for later use during high-demand periods, thereby stretching available supplies of reclaimed water and maximizing its use. While this practice is sometimes a challenge, it is gaining interest because of recent advances in management practices. Several conservation methods that are used in potable water supply systems are applicable to reclaimed water systems, including volume-based rate structures, limiting irrigation to specific days and hours, incorporation of soil moisture sensors or other controllers that apply reclaimed water when conditions dictate irrigation, and metering. Examples of reclaimed water conservation are prevalent in Florida. Many utilities' reclaimed water availability is limited by seasonal demands that can exceed supply, making conservation and management strategies a necessity. To promote conservation, several utilities have implemented conservation rate structures to encourage efficient use of reclaimed water. In addition, utilities that provide reclaimed water for landscape irrigation, including irrigation for residential lots, medians, parks, and other green space, are promoting efficient use of reclaimed water by limiting the days and hours that users can irrigate. The Loxahatchee River District in Palm Beach County, Florida, has designated irrigation days for residential landscape irrigation reuse customers and can shut off portions of its system on designated non-irrigation days. Port Orange, Florida, retrofitted its entire reuse system with meters so that customers could be charged according to a tiered volumetric rate rather than a flat rate that encouraged excessive use. And the Southwest Florida Water Management District has recognized the importance of conserving reclaimed water to ensure more customers can be served by providing grant funding for reuse programs where efficient use is a criterion for receiving funds.

83.6 Designing Cost-Effective Water Demand Management Programs

Cities are hotspots of consumption—and by this token, they have amazing potential for reducing their water footprint. "Simple" instruments for demand reduction, such as appropriate pricing, techniques like rainwater harvesting or wastewater recycling can have major impacts when implemented and enforced widely on households and industry. Involving marginalized groups into management solutions and implementation is crucial, as the success of Karachi's Orangi Pilot Project clearly demonstrates. Creating a sense of ownership for infrastructure and service provisions also ensures sustainability in their maintenance.

Raising awareness for the sustainable and efficient use of water resources amongst the general public, and especially in agriculture and industry, is essential. Community, industry, and school education programs can raise awareness about the need to conserve water and to bring about long-term changes in water consumption behavior. It must be assured that the targeted community, authority, commercial entity, or any other stakeholder can obtain the necessary information and understand water management practices available for their own needs and local circumstances [14].

Rainwater harvesting has been used extensively to directly recharge groundwater at rates exceeding natural recharge conditions in India. Reports from international organizations focusing on this area indicate that 11 recent projects across Delhi resulted in groundwater level increases from 5 to 10 m in just two years. In fact, the application of rainwater management in India is likely to become one of the most modern in the world [14]. Rainwater harvesting is practiced on a large scale in Chennai, Bangalore, and Delhi where it is included in the state policy and in the building code for new buildings. Collecting rainwater from the abundant rainfalls during the monsoon season is also a feasible option for cities like Karachi and Kolkata.

Treatment and reuse of water from storm water drainage, sewage and other effluents, and industry can greatly supplement local water supplies. Annual reclaimed water volumes total about 2.2 billion m³, based on 2000 and 2001 figures from the World Bank [38]. On a global scale, nonpotable water reuse is currently the dominant method for supplementing supplies for irrigation, industrial cooling, and river flows [14]. For industry, wastewater recycling and reuse can be encouraged through fiscal incentives (subsidies) and pollution taxes [3]. Recovered water or stormwater can also be used to directly recharge groundwater aquifers and thus also create a barrier to saltwater intrusion [14]. Wastewater recycling has been used to recharge Mexico's City overexploited aquifers and halt the City's subsidence since 1992 (environmental norms were implemented to regulate the water quality in 2007) [21]. Recycled wastewater in metropolitan Mexico City is also used to irrigate green areas, fill lakes and canals, and cool industrial processes.

Economic and fiscal incentives and instruments are another possible option to aid the reduction in water demand, effective water basin management by upstream farmers and landowners, or reducing water pollution by industrial users. Water is still perceived as an abundant and free resource and not as an economic good. Due to heavy subsidization, water prices are usually so low that they reflect neither the true economic value nor the costs needed for water provision (infrastructure, treatment, maintenance). In addition, many countries do not meter consumption

and users pay a fixed rate regardless of how much they actually consume. Without adequate price signals, there is no incentive for using water more efficiently or for reducing water consumption. There is evidence from both developed and developing countries that a mix of regulatory and economic/fiscal incentives have lead to 20%–30% reductions in industrial and household water use in the past [3].

Cost recovery for the institutions governing water provision is a major problem in most cities. Incorporating infrastructure maintenance, provision and administrative costs into the water price would lead to better cost recovery, and thus generate needed resources for infrastructure improvements and other water management solutions. Additionally, it would also lead users to economize. An essential component of Buenos Aires' water management plan is the installation of water usage meters across the city. In 2009, only 12.8% of water users had their consumption metered, thereby encouraging the highest per capita water consumption in this report. Through increasing metered connections by 600% (from the current 3000 to 18,000) from 2009–2012, it is estimated that 100,000 m³ of water will be saved daily [5]. Mexico City reported water conservation after installing meters for 90% of its users in 1994 [23].

Integrated River Basin Management (IRBM) is one of the twenty-first century's biggest challenges, but at the same time, it offers a great opportunity for sustainable water resource management in river basins. It refers to a management system where economic, environmental, social administration, and governance is integrated across administrative and regional boundaries. This cross-sectoral approach stems from the recognition that various and competing stakeholders have their own interest in managing water resources. It is increasingly acknowledged that management must become transparent between stakeholders and there has to be communication between administrations and governments regulating water use across the entire river basin. Only in this way, can economic and social benefits derived from water resources be maximized in an equitable manner while preserving and, where necessary, restoring freshwater ecosystems. Mexico administers its water by regional watershed bodies with basin organizations serving as the "technical arm" of broad-based basin councils, incorporating civil society interests (private sector, citizens' groups, etc.) [20]. China's platform for IRBM is the Yangtze Forum, which aims to sustainably manage and develop the Yangtze River. The initiative was launched by authorities of the central government and all relevant provinces, autonomous regions, and municipalities along the Yangtze mainstream and incorporates all stakeholders, as well as domestic and international organizations

Payment for Environmental Services (PES) is a market-based tool that can be integrated into the IRBM approach as a way to create financial incentives for managing natural resources, addressing livelihood issues for the rural poor, and providing sustainable financing for protected areas. The basic principle is that those who "provide" environmental services by conserving natural ecosystems should be compensated by beneficiaries of the service [39]. In the case of freshwater, upstream landowners that protect the watershed (foregoing more lucrative uses of their land, such as agriculture, and incurring costs for implementing conservation measures) are providing a service to downstream users (cities, agriculture, hydropower companies, beverage industry, etc.). Maintaining land in its natural state or implementing conservation techniques and thereby providing environmental services is seldom a more attractive option than its conversion because beneficiaries are not the service provider (the land owner). PES aims to change this by providing incentives for maintaining or restoring land for the desired environmental service [14].

A PES scheme for watershed protection was successfully implemented in the Naivasha Basin in Kenya. It could serve as a model for cities around the world. In Kenya, the Naivasha PES scheme is the first of its kind; the success of the program has generated enormous interest from various government authorities, institutions, and local communities.

83.7 Future Water Supply Augmentation

83.7.1 Cost Consideration

Cities need to reduce water consumption, recycle wastewater, restore adjacent watersheds, and improve engineering solutions to supply water from well-managed ecosystems. The adoption of a multi-sectoral approach to water and wastewater management at the national level is a matter of urgency. This approach should be implemented by incorporating principles of ecosystem-based management extending from the watersheds to the sea, and connecting sectors that will reap immediate benefits from better water and wastewater management. Ecosystem protection, management, and restoration provide a central, effective, sustainable, and economically viable solution to enhancing water supply and quality while mitigating extreme weather events of too much or too little water. Successful and sustainable wastewater management that supports peri-urban agriculture is crucial for reducing water consumption, and requires a mix of innovative approaches that engage the public and private sector at local, national, and trans-boundary scales. Planning processes should provide an enabling multiscale environment for innovation, including at the community level with government oversight and public management.

For cities to better understand their vulnerabilities as well as prepare for the impacts of climate change, they must examine the full suite of potential impacts, both at a regional and local level. Vulnerability and water risk assessments covering the core urban and peri-urban areas, as well as areas that supply water and goods and services that include a complete evaluation of water-related risks such as future water availability, precipitation, drought, runoff patterns, sea level rise, and flooding risks are needed. Local plans should be strengthened by encouraging and, where possible, requiring water and energy utility operators to prepare and update their own site- and system-specific vulnerability assessments that should address utility vulnerability to flooding, drought, and/or sea level rise. More informed

political and financial decisions can be made with access to more diverse information about risks and probabilities. By considering a range of risks, local efforts provide better opportunity for effective long-term adjustment and management.

Local involvement is the key to any vulnerability assessment and adaptation strategy. Proper planning should include not only city personnel, but also representatives from local water and energy utilities, emergency response personnel, natural resource managers, homeowners, businesses, and environmental groups. The businesses, farmers, and food processors (i.e., the supply chain structure) for the city's agricultural and food products, as well as, the city's downstream water users should participate equally in the formulation of adaptation strategies and their implementation.

Innovative financing of water and wastewater infrastructure should incorporate design, construction, operation, maintenance, upgrading, and/or decommissioning. Financing should take the important livelihood opportunities in improving wastewater treatment processes into account, while the private sector can play an important role in operational efficiency under appropriate public guidance, including ecosystem restoration projects.

An inventory of critical infrastructure that is at risk due to flooding, droughts, or sea level rise is also fundamental. So as to inform longer-term planning, construction, funding, and other resiliency goals, the identification of critical facilities at risk (such as roads, hospitals, drinking water supplies and conveyance systems, sewage treatment and conveyance infrastructure) should be prioritized in the short term. Identifying this critical infrastructure should be based on available information and refined as improved data becomes available. The use of green infrastructure and low-impact development in watershed planning offers many benefits and should be encouraged in local planning. Large volumes of stormwater runoff that is discharged through municipal sewer systems can exacerbate storm surges and cause flooding in urban settings. Green infrastructure can capture the runoff, thereby both augmenting water supply and reducing downstream flooding. Low-impact development is a simple and cost-effective green development strategy that can help cities, states, and even individuals meet the water supply challenge. In areas where the groundwater table is too high for infiltration, practices that evaporate or evapotranspire water, like rain gardens or capture-and-use systems (rain barrels and cisterns) can be successfully used. Broad introduction of urban and peri-urban agriculture utilizes otherwise wasted runoff and decreases the reliance on surrounding rural regions for food crops, consequently easing the city's external indirect water footprint impact.

Increasing energy efficiency reduces current and future demand for energy, decreases water consumption related to energy production, and reduces greenhouse gas emissions. Cities should take steps to implement comprehensive and ambitious programs for energy efficiency and saving that promote clean and water-efficient forms of energy such as wind, solar, and geothermal. Solutions for smart water and waste management must be socially and culturally appropriate and acceptable, as well as economically and environmentally viable. Ecosystem protection, management, and restoration are the cheapest, easiest, and most effective ways of improving and securing water supply, filtration, and quality. Education must play a central role in water management and in reducing city's unsustainable demand on water resources.

83.7.2 Sustainability Options

The uncontrolled disposal to the environment of municipal, industrial and agricultural liquid, solid, and gaseous wastes constitutes one of the most serious threats to the sustainability of human civilization by contaminating the water, land, and air and by contributing to global warming. With increasing population and economic growth, treatment and safe disposal of wastewater is essential to preserve public health and reduce intolerable levels of environmental degradation. In addition, adequate wastewater management is also required for preventing contamination of water bodies for the purpose of preserving the sources of clean water. Effective wastewater management is well established in developed countries but is still limited in developing countries. In most developing countries, many people lack access to water and sanitation services. Collection and conveyance of wastewater out of urban neighborhoods is not yet a service provided to all the population, and adequate treatment is provided only to a small portion of the collected wastewater. In slums and peri-urban areas throughout the world, it is common to see raw wastewater flowing in the streets. The inadequate water and sanitation service is the main cause of diseases in developing countries.

In the year 2011, the population of the planet was 7 billion. Population growth forecasts indicate rapid global population growth that will reach 9 billion in 2030. The forecasts also indicate that

- Most of the population growth will occur in developing countries, while the population of developed countries will remain constant at about 1 billion
- A strong migration from rural to urban areas will take place

Considering the expected population growth and the order of priorities in the development of the water and sanitation sector in developing countries—water supply and sewerage first, and only then wastewater treatment—as well as the financial difficulties in these countries, it cannot be assumed that the current low percentage of the coverage of wastewater treatment in these countries will increase in the future, unless a new, innovative strategy is adopted and affordable wastewater treatment options are used.

A key component in any strategy aimed at increasing the coverage of wastewater treatment should be the application of appropriate wastewater treatment technologies that are effective, simple to operate, and low cost (in investment and especially in operation and maintenance). Appropriate technology processes are also more environment-friendly since they consume less energy and thereby have a positive impact on efforts to mitigate the effects of climate change. In addition, with modern design, appropriate technology processes cause less environmental nuisance than conventional processes—for example, they produce lower amounts of excess sludge and their odor problems can be more effectively controlled.

Appropriate technology unit processes include (but are not limited to) the following:

- Preliminary treatment by rotating micro screens
- Vortex grit chambers
- Lagoons treatment (anaerobic, facultative and polishing), including recent developments in improving lagoons performance
- Anaerobic treatment processes of various types, mainly, anaerobic lagoons, upflow anaerobic sludge blanket (UASB) reactors, anaerobic filters and anerobic piston reactor (PAR)
- Physicochemical processes of various types such as chemically enhanced

Primary treatment (CEPT) includes the following:

- Constructed wetlands
- Stabilization reservoirs for wastewater reuse and other purposes
- Overland flow
- Infiltration-percolation
- Septic tanks
- Submarine and large rivers outfalls

Out of these processes, various combinations can be set up. Combinations can also include some other simple processes such as sand filtration and dissolved air floatation (DAF), which are not considered appropriate processes per se but are in fact appropriate processes. One interesting combined process is the generation of effluents suited for reuse in irrigation based on pretreatment by one of the mentioned unit processes followed by a stabilization reservoir.

83.7.3 Building Diverse Water Portfolios

Many countries have the problem of a severe water imbalance. This imbalance in water demand versus supply is due mainly to the relatively uneven distribution of precipitation, high temperatures, increased demands for irrigation, and the impacts of tourism. To alleviate water shortages, serious consideration must be given to wastewater reclamation and reuse. Reclaimed wastewater can be used for a number of options including agricultural irrigation. A wastewater treatment developer must perform an appropriate risk assessment before implementing the reuse of wastewater. Proper consideration of the health risks and quality restrictions must be a part of the assessment. Sourcepoint measures rather than end of pipe solutions are essential. Source-point measures require extensive industrial pretreatment interventions, monitoring and control programs, and incentives for the community to not dispose of any harmful matter into the sewers [32]. For the implementation and promotion of new technology, strategies must include local participation as well as municipal action. Local participation is a positive and important growing trend in government projects. The participation must fit with the local population to meet particular local needs. Local communities can contribute valid indigenous ideas for cost savings in the project. Agreement on key issues between design engineers and the local residents is necessary early on in the project, and if local participation is extensive, capital costs can ultimately be reduced. According to the Inter-American Development Bank, "Citizen participation, properly channeled, generates savings, mobilizes financial and human resources, promotes equity and makes a decisive contribution to the strengthening of society and the democratic system."

There is a strong sense of ownership by members of the community in their projects. This pride in the new development helps to ensure the sustainability of the water supply and sanitation systems. Once the project is implemented, local participation contributes to the community's confidence in the new technology and allows them to take on other challenges such as accessing financial aid for other infrastructure projects. On the governmental level, institutional strengthening is usually needed to assist small to medium-sized cities in dealing with new administrative and financial management responsibilities. One program that has been developed to address the problems associated with decentralization is RIADEL (Local Development Research and Action Network). It is a network for sharing information about local community development in Latin America, including decentralization and the training of social leaders and civil servants.

83.8 Improve Institutional Coordination to Develop Integrated Policies

The rules specifically aimed at water reuse projects, regulations governing utility construction in general also apply. The details of such rules are beyond the scope of this document but can be promulgated by state agencies (including health departments) and local jurisdictions or can be established by federal grant or loan programs.

Once facilities have been constructed, state and local regulations often require monitoring and reporting of performance. To provide production, distribution, and delivery of reclaimed water, as well as payment for it, a range of institutional arrangements can be utilized (Table 83.2).

It is necessary to conduct an institutional inventory to develop a thorough understanding of the institutions with jurisdiction over various aspects of a proposed reuse system. On occasion there is an overlap of agency jurisdiction, which may cause conflict unless steps are taken early in the planning stages to obtain support and delineate roles. The following institutions should be involved or, at a minimum, contacted: federal and state regulatory agencies, administrative and operating organizations, and general units of local (city, town, and county) government.

In developing a viable arrangement, it is critical that both public and private organizations be considered. As access to public funds decrease, the potential for private capital investment increases. It is vital that the agency or entity responsible for financing the project be able to assume bonded or collateralized indebtedness, if such financing is likely, and have accounting and fiscal management structures to facilitate financing. Likewise, the arrangement must designate an agency or entity with contracting power so that agreements can be authorized with other entities in the overall service structure. Additional responsibilities may be assigned to different groups depending on their historical roles and technical and managerial expertise. Close internal coordination between departments and branches of local government, along with a range of legal agreements, will be required to ensure a successful reuse program.

Finally, the relationship between the water purveyor and the water customer must be established, with requirements on both sides to ensure reclaimed water is used safely. Agreements on rates, terms of service, financing for new or retrofitted systems, educational requirements, system reliability or scheduling (for demand management), and other conditions of supply and use reflect the specific circumstances of the individual projects and the customers served. In addition, state laws, agency guidelines, and local ordinances may require customers to meet certain standards of performance, operation, and inspection as a condition of receiving reclaimed water. However, where a system supplies a limited number of users, development of a reclaimed water ordinance may be unnecessary; instead, a negotiated reclaimed water user agreement would suffice. It is worth noting that in some cases, where reclaimed water is still statutorily considered effluent, the agency's permit to discharge wastewater— along with the concomitant responsibilities— may be delegated by the agency to customers whose reuse sites are legally considered to be distributed outfalls of the reclaimed water.

Greywater Policy and Permitting. Key to the viability of small or on-site greywater systems is an effective policy, permitting, and regulatory process to provide adequate treatment of greywater for

TABLE 83.2   Common Institutional Arrangements for Water Reuse

Type of Institutional Arrangement	Production	Wholesale Distribution	Retail Distribution
Separate authorities	Wastewater treatment agency	Wholesale water agency	Retail water entity
Wholesaler/retailer system	Wastewater treatment agency	Wastewater treatment agency	Retail water entity
Joint powers authority (for production and distribution only)	Joint powers authority	Joint powers authority	Retail water entity
Integrated production and distribution	Water/wastewater authority	Water/wastewater authority	Water/wastewater authority

the intended end use. In many states the regulatory system is still designed for large-scale systems; the permitting process for small systems is complex because small systems cross into the purview of various regulatory agencies, which can cause hurdles in the approval process. There are a number of states and local agencies that provide specific regulations or guidance for greywater use, including Arizona, California, Connecticut, Colorado, Georgia, Montana, Nevada, New Mexico, New York, Massachusetts, Oregon, Texas, Utah, Washington, and Wyoming. In addition to the states that have specific policies on greywater use, there are other institutional policies, such as the UPC and the IPC, that are applicable to the implementation of greywater systems.

Land Use and Local Reuse Policy. Most communities in the United States engage in some type of structured planning process whereby the local jurisdiction regulates land use development according to a general plan, sometimes reinforced with zoning regulations and similar restrictions. Developers of approved areas for new development may be required to prepare specific plans that demonstrate sufficient water supply or wastewater treatment capacity. In these contexts, dual-piped systems may be developed at the outset of development. It is important that any reuse project conforms to requirements under the general plan to ensure the project does not face legal challenges on a land use basis. Local planning processes often include public notice and hearings. As the public may have many misconceptions about reclaimed water, it is important for planners to address public concerns or opposition.

The 2004 guidelines identified (use requirements in California) land use and environmental regulation controls used by local government entities to implement and manage reclaimed water systems. Since publication of the 2004 guidelines, many communities and states have implemented more formal water planning processes to meet public health needs for adequate water, wastewater, and reclaimed water services. There are several reasons a utility might create a local policy to require connection to a reclaimed water system, with parallel logic used in many communities to require connection to municipal utilities when reasonably available. The most common reason to require connection is to assure use of the new system, adequate to shift some of the water demand and to pay for the new system or defer new potable main construction. In an integrated water management program, potable water supplies may be limited and require construction of a reclaimed water/dual water system to meet the total demand. Even if reclaimed water is priced lower than the potable supply, the public may not have been adequately informed to understand the benefits of a diversified water system and may resist conversion to reclaimed water.

Mandatory connection to reclaimed water systems is becoming more common. Planning for future use of reclaimed water allows communities to require certain uses to utilize reclaimed water if reasonably available. Because construction cost for retrofit with a dual water system is higher and disruption of other infrastructure is unavoidable, dual water piping can be installed initially with the nonpotable distribution system dedicated to irrigation, cooling towers, or industrial processes. When reclaimed water is available to the development area, a connection to the supply is the only local construction required.

Utilities may also need to secure bonds used for construction with an ordinance requiring connection to a reclaimed water system, thus providing a guarantee of future cash flow to meet bond payments. In addition to state legislative action in California, many utilities have included mandatory connection language. Water Recycling Funding Program Guidelines initially issued in 2004 and amended in July 2008 require loan/grant applicants to include a draft mandatory use ordinance in their application packet [2]. The Marina Coast Water District Ordinance, Title 4, 4.28.030 recycled water service availability, includes the following:

- When recycled water is available to a particular property, as described in Section 1.04.010, the owner must connect to the recycled water system. The owner must bear the cost of completing this connection to the recycled water system.
- New water users who are not required to connect to recycled water because the distance to the nearest recycled water line is greater than the distance provided in Section 1.04.010, shall be required to construct isolated plumbing infrastructure for landscape irrigation or other anticipated nonpotable uses, with a temporary connection to the potable water supply.
- All new private or public irrigation water systems, whether currently anticipating connection to the recycled system or that shall be connected to the potable water system temporarily while awaiting availability of recycled water, shall be constructed of purple polyvinyl chloride (PVC) pipe to the existing district standard specification [11].

Examples of other California utilities with mandatory connection requirements include Dublin San Ramon Services District (DSRSD); Inland Empire Utility Agency; San Luis Obispo Rowland Heights; Cucamonga Valley Water District; and Elsinore Valley Municipal Water District. Florida is another state with mandatory connection requirements; 78 counties, cities, and private utilities responded on their 2011 annual reuse reports that they either require construction of reclaimed water piping in new residential or other developments or require connection to reuse systems when they become available. The Florida communities of Altamonte Springs; Boca Raton; Brevard, Charlotte, Polk, Colombia, Palm Beach, and Seminole Counties; Marco Island; and Tampa are examples. There are no communities in Texas with mandatory connections, but requirements were also found in Yelm, Washington; Cary, North Carolina; and Westminster, Maryland. Along with the mandatory connection requirement, there are also ordinances that promote use of reclaimed water through incentives. The St. Johns River Water Management District, Florida, provides a model water conservation ordinance to cities within the district to promote more water efficient landscape irrigation. The model ordinance includes time-of-day/day-of-week restrictions based on odd-even street address as well as daily irrigation limits of 0.75 in./day (1.9 cm/d). Exemptions may be granted to these limitations. Possible exemptions include using a micro-spray, micro-jet, drip, or bubbler irrigation system; establishing new landscape; or watering in lawn treatment chemicals. The use of water from a reclaimed water system is allowed anytime. The capacity of a reclaimed water system can be strained if customers continue to use reclaimed water beyond the utility capacity to supply it. In Cape Coral, Florida, the city council is considering an ordinance to re-establish an emergency water conservation plan due to a persistent drought since 2007 [1]. The dry-season water demand—and the abuse of reclaimed water—has increased. As much as 42 million gallons (160,000 m³) of reclaimed water are being used each scheduled watering day, and 19 million gallons (72,000 m³) were being used on a day when no watering is allowed. The council is taking a proactive approach to protect the city's water resources, including reclaimed water.

83.9 Key Factors for Establishing Initiatives

83.9.1 Appropriate Treatment Technology

Based on experience from past mistakes in sewage treatment technology, the definition of what is sustainable is clearer. Developers should base the selection of technology upon specific site conditions and financial resources of individual communities.

One approach to sustainability is through decentralization of the wastewater management system. This system consists of several smaller units serving individual houses, clusters of houses, or small communities. Black and greywater can be treated or reused separately from the hygienically more dangerous excreta. Non-centralized systems are more flexible and can adapt easily to the local conditions of the urban area as well as grow with the community as its population increases. This approach leads to treatment and reuse of water, nutrients, and byproducts of the technology (i.e., energy, sludge, and mineralized nutrients) in the direct location of the settlement. Communities must take great care when reusing wastewater, since both chemical substances and biological pathogens threaten public health as well as accumulate in the food chain when used to irrigate crops or in aquaculture. In most cases, industrial pollution poses a greater risk to public health than pathogenic organisms. Therefore, more emphasis is being placed on the need to separate domestic and industrial waste and to treat them individually to make recovery and reuse more sustainable. The system must be able to isolate industrial toxins, pathogens, carbon, and nutrients.

83.9.2 Sustainable Treatment Types

Now that the requirements for a sustainable wastewater treatment system have been presented, there are several options one can choose from in order to find the most appropriate technology for a particular region. This paper will discuss sustainable wastewater treatment systems including:

1. Lagoons/wetlands
2. USAB (anaerobic digesters)
3. SAT technologies

83.9.3 Lagoons and Wetlands

In wetland treatment, natural forces (chemical, physical, and solar) act together to purify the wastewater, thereby achieving wastewater treatment. A series of shallow ponds act as stabilization lagoons, while water hyacinth or duckweed acts to accumulate heavy metals. Multiple forms of bacteria, plankton, and algae act to further purify the water. Wetland treatment technology in developing countries offers a comparative advantage over conventional, mechanized treatment systems because the level of self-sufficiency, ecological balance, and economic viability is greater.

The system allows for total resource recovery [18]. Lagoon systems may be considered a low-cost technology if sufficient, non-arable land is available. However, the requirement of available land is not generally met in big cities. The demand for flat land is high for the expanding urban developments and agricultural purposes. The decision to use wetlands must consider the climate. There are disadvantages to the system that in some locations may make it unsustainable. Some mechanical problems may include clogging with sprinkler and drip irrigation systems, particularly with oxidation pond effluent. Biological growth (slime) in the sprinkler head, emitter orifice, or supply line causes plugging, as do heavy concentrations of algae and suspended solids.

83.9.4 Anaerobic Digestion

Another treatment option available, if there is little access to land, is anaerobic digestion. Anaerobic bacteria degrade organic materials in the absence of oxygen and produce methane and carbon dioxide. The methane can be reused as an alternative energy source (biogas). Other benefits include a reduction of total bio-solids volume of up to 50%–80%, and a final waste sludge that is biologically stable can serve as rich humus for agriculture. So far, anaerobic treatment has been applied in Colombia, Brazil, and India, replacing the more costly activated sludge processes or diminishing the required pond areas. Various cities in Brazil have shown an interest in applying anaerobic treatment as a decentralized treatment system for poor, sub-urban districts. The beauty of the anaerobic treatment technology is that it can be applied on a very small and very large scale. This makes it a sustainable option for a growing community.

83.9.5 Soil Aquifer Treatment

SAT (soil aquifer treatment) is a geopurification system where partially treated sewage effluent artificially recharges the aquifers and is then withdrawn for future use. By recharging through unsaturated soil layers, the effluent achieves additional purification before it is mixed with the natural groundwater. In water scarce areas, treated effluent becomes a considerable resource for improved groundwater sources. The Gaza Coastal Aquifer Management Program includes treated effluents to strengthen the groundwater, in terms of both quantity and quality. With nitrogen reduction in the wastewater treatment plants, the recharged effluent has a potential to reduce the concentration of nitrates in the aquifer. In water scarce areas such as in the Middle East and parts of Southern Africa, wastewater has become a valuable resource that, after appropriate treatment, becomes a commercially realistic alternative for groundwater recharge, agriculture, and urban applications.

SAT systems are inexpensive, efficient for pathogen removal, and are not highly technical to operate. Most of the cost associated with an SAT is for pumping the water from the recovery wells, which is usually $20–50 USD per m³. In terms of reductions, SAT systems typically remove all BOD, TSS, and pathogenic organisms from the waste and tend to treat wastewater to a standard that would generally allow unrestricted irrigation. The biggest advantage of SAT is that it breaks the pipe-to-pipe connection of directly reusing treated wastewater from a treatment plant. This is a positive attribute for those cultures where water reuse is taboo. The pretreatment requirements for SAT vary depending on the purpose of groundwater recharge, sources of reclaimed water, recharge methods, and location. Some may only need primary treatment or treatment in a stabilization pond. However, pretreatment processes should be avoided if they leave high algae concentrations in the recharge water. Algae can severely clog the soil of the Infiltration basin. While the water recovered from the SAT system has much better water quality than the influent, it could still be lower quality than the native groundwater. Therefore, the system should be designed and managed to avoid intrusion into the native groundwater and use only a portion of the aquifer. The distance between infiltration basins and wells or drains should be as large as possible, usually at least 45 to 106 m to allow for adequate soil-aquifer treatment. All the systems described allow for the reuse of treated wastewater in order to have a cyclic, sustainable system. These treated wastewaters provide essential plant nutrients (nitrogen, phosphorus, and potassium) as well as trace nutrients. Phosphorus is an especially important nutrient to recycle, as the phosphorus in chemical fertilizer comes from limited fossil sources.

83.10 Building Capacity for Water and Wastewater Reuse

Successful wastewater reuse projects are designed to reflect specific local conditions, such as water demand, urban growth, climate, socioeconomic characteristics, and cultural preference, as well as institutional and policy frameworks. To do so effectively requires a capacity that is still limited in many developing countries. Capacity building can improve the quality of decision-making and managerial performance in the planning

and implementation of programs, and encompasses the following five elements [24]:

1. *Human resources:* Strengthening people's technical and managerial ability to evaluate limitations of current practice, potential benefits and requirements of wastewater reuse, and fostering their capability to implement new programs
2. *Policy and regulatory framework*: Helping to align or create policy and legal frameworks to facilitate wastewater reuse programs, while ensuring protection of human health and the environment
3. *Institutions*: Supporting national, regional, and local institutions and their enhancement, so that they can determine ways to improve effectiveness in regulating and managing water reuse programs
4. *Financing*: Expanding a range of financing services and opportunities that are available for wastewater reuse initiatives, and improving the capability of utilities and potential EST users to understand and access such services
5. *Participation*: Encouraging civil society to participate in the decision-making process as well as actual implementation of wastewater reuse programs, and to deliver a message to the widest possible audience

These five elements are described in more detail below with examples of real initiatives around the world.

83.10.1 Human Resource Development

Building technical and managerial capacity for operating water and wastewater reuse programs is a critical necessity, due to the variable qualities of source water for wastewater reuse and the complexity of processes. Analytical and problem-solving skills, as well as the ability to maintain and manage technologies, systems and organizations, need to be fostered.

Well-trained personnel, including engineers, scientists and technicians, are necessary for successful water and wastewater recycling projects. In some organizations, resource constraints may force staff with limited training to assume supervisory and management positions, posing a challenge to implementing effective programs. Such problems may be addressed through:

- Carrying out internal human resource development by training courses and on-the-job training
- Developing human capabilities through hiring and retention of qualified personnel

In addition, care should be taken to favor operations that enhance, rather than diminish, employment opportunities, and to utilize reliable mechanisms that can be maintained by a locally trained labor force. Community-level training is also important, as many water reuse and recycling techniques involve actions at a household or shop-floor level. Training materials and methods need to be tailored to meet the needs and qualifications of the target audience.

83.10.2 Policy and Legal Framework Development

Water reuse projects must include regulatory development and implementation to ensure the protection of human health and the environment. Necessary regulations may include permit systems to authorize wastewater discharges, technical specifications on wastewater treatment, reclaimed water quality standards for various applications, and regulations on disposal of waste (sludge, brine, etc.) from treatment. In water scarce areas, water reuse requirements or the installation of a reuse infrastructure may also be introduced. Mechanisms to enforce these regulations are also necessary, including required and voluntary monitoring, inspection programs with adequate staffing, and clear authority to assess and collect fines and penalties. Incentives, such as grants and low-interest loans, flexible permits and priority access to the infrastructure, may also be effective in increasing interest in wastewater reuse.

Example

In Chennai, India, the Chennai Metropolitan Waste and Sewerage Service Board (CMWSSB) began to promote rainwater harvesting and wastewater reuse to alleviate water shortage in industry and households in the mid-1990s. In 1994, the CMWSSB and other regulatory authorities made the installation of suitable rainwater harvesting facilities mandatory for property developers of multistory and special buildings. Recently, this requirement has been extended to individual households for which planning permits are to be submitted. The CMWSSB ensures the installation of such devices by inspection before water/sewer connections are made (CMWSSB).

The evaluation of an institutional support program for water recycling in China by the Japan International Cooperation Agency (JICA) found that policy and legal framework development was crucial for success. With the emergence of a sustainable development policy, various environmental laws, regulations, standards, and their enforcement became more stringent in the 1990s. As a result, industries around the country were compelled to improve their environmental performance, which promoted wastewater treatment and efficient use of water resources, including reuse. At the same time, an increase in water tariffs increased the interest of both industry and citizens in water reuse [8].

83.10.3 Institutional Development and Organizational Management

Studies on managing water supply and sanitation services in developing communities have shown that ensuring the credibility of a responsible agency within its target community, and developing a client-oriented organizational structure are two success factors [19]. While many countries and municipalities may already have institutions for water supply, those for wastewater collection, treatment and disposal may not be organized and managed properly. In some cases, they may not exist at all. A recent analysis revealed that one in ten countries in Asia, Africa, and the Americas do not have a national institution that is identifiably responsible for either urban or rural sanitation [37]. In order to undertake wastewater reclamation projects, it is necessary to examine relevant existing institutions and strengthen them, or to create new ones and assign adequate mandates and responsibilities. It may also be worthwhile establishing collaborative frameworks with other reclamation and reuse programs to achieve a critical mass for service provision.

Example

In the United States, a public utility in the San Francisco Bay area called the Central Contra Costa Sanitary District (CCCSD) began a project in the mid-1990s to modernize its water reclamation facility and to expand its reclaimed water distribution system. In doing so, it established a cooperative agreement with a neighboring utility to expand service provision to a wider area, and the pipeline systems between the two utilities were connected. This collaborative framework was an important factor in achieving 'critical mass' in dealing with stakeholders and a great degree of operational flexibility [6].

83.10.4 Financing

It is costly to build and maintain wastewater treatment plants, and install water distribution lines for reuse. Expanding a range of financial services and opportunities is a key component for promoting water and wastewater recycling. In countries where water and wastewater recycling programs are implemented within a comprehensive water resource development, policy makers may have flexibility in accessing financing. In other cases, technical assistance programs may provide separate funding for water reuse. Locally controlled funds or small-scale financing mechanisms (i.e., microcredit schemes) may also be established to facilitate financing. Along with the introduction of financing mechanisms, a capacity to understand and access such services needs to be fostered among utilities and potential EST users.

Example

Various bi- and multi-lateral cooperation agencies have increasingly provided financing opportunities for water and wastewater reclamation projects. For example, Japan has provided US$1.83 million in total to fund a wastewater reclamation system in Chennai, starting from 1994. The system, which is currently in operation, provides secondary treated sewage, with on-site tertiary treatment, for an oil refinery and fertilizer plant. The service has been expanded to a power plant recently. Chennai Municipal Water and Sewerage Service Board (CMWSSB) hopes to pursue this concept further to make more significant impacts [13].

83.10.5 Raising Public Awareness and Participation

Raising the awareness of the public about water shortages and encouraging their participation in remedial action is crucial in the implementation of wastewater reuse. The issue is of particular importance for water reuse for indirect and direct potable use, including groundwater recharge, as many initiatives have been delayed due to public resistance and legal action. To raise the awareness of stakeholders and ensure that their voices are heard, the decision-making process needs to be participatory, with clearly outlined roles and responsibilities. Proactive public outreach initiatives, such as publications, public announcements, and site visits, are some of the main means to secure wider public acceptance and support.

Civil society organizations usually play an important role in undertaking various activities aimed to raise public awareness. In some countries, local governments and local politicians also take part directly to raise the public awareness of water conservation, better usage to improve public health, and recycling water for secondary uses. Public participation can be scaled-up by bringing the community into the decision-making process. Their participation in the decision-making process also improves public participation in the implementation process. Public participation can be aimed at different objectives including the payment of user charges, conservation, minimizing unaccounted for water rates, recycling and reuse of water, and ownership and operation of the small projects, mainly in slums or peri-urban areas.

Example

The Public Utility Board (PUB) of Singapore opened a public education centre in early 2003 to enhance the public understanding of reclaimed water, called 'NEWater', and other water-related topics. The facility offers multimedia presentations and interactive games, as well as a walk-through of the plant with advanced membrane and ultraviolet technologies. PUB also embarked on an intensive public education program on NEWater in the second half of 2002, with advertisements, posters, leaflets, the broadcast of a documentary, as well as the provision of over 1.5 million bottles of NEWater samples. As a result, an overwhelming majority of Singaporeans have expressed their acceptance of NEWater. An independent poll in October 2002 showed that 82% of citizens indicated that they were prepared to drink NEWater directly, while 16% indicated that they were prepared to drink it indirectly by mixing it with reservoir water [12].

Since 2004, 15 million gallons per day (about 57,000 m³/day) of NEWater has been processed in two factories. While 13 million gallons per day (about 50,000 m³/day) is used for nonpotable usage by industrial and commercial institutions, including wafer fabrication parks, 2 million gallons per day (about 7000 m³/day) are pumped into freshwater reservoirs for indirect potable usage, constituting less than 1% of total water consumption. The plan is to increase the NEWater share in drinking water to 2.5% by 2011 [15].

83.10.6 Opportunities for Innovation in Recycling

A recent development in on-site treatment systems in urban development has been driven largely by the private sector's desire to create more highly sustainable developments through the LEED program. This program area remains small compared to the municipal reuse market. However, it has a growing role for improving water efficiency in new buildings and developments and also for major modifications to existing facilities. A primary driver that compels land developers to consider the implementation of on-site treatment systems is the sustainability accreditation that is promoted and earned through the LEED program. The LEED program was developed by the U.S. Green Building Council (USGBC) in 2000 and represents an internationally-recognized green building certification system. At the time of preparation of this document, the current version of the Rating System Selection Guidance was LEED 2009, originally released in January 2010 and updated in September 2011. The guidance is currently under revision with the new LEED v4 focusing on increasing technical stringency from past versions and developing new requirements for project types such as data centers, warehouses and distribution centers, hotels/motels, existing schools, existing retail, and mid-rise residential buildings. More information is available on the USGBC website [27].

LEED provides building owners/operators with a framework for the selection and implementation of practical, measurable, and sustainable green building design, construction, and operations and maintenance solutions. LEED promotes sustainable building and site development practices through a tiered certification rating system that recognizes projects that implement green strategies for better overall environmental and health performance. The LEED system evaluates new developments, as well as significant modifications to existing buildings, based on a certification point system where applicants may earn up to a maximum of 110 points. LEED promotes a whole-building approach to energy and water sustainability by observance of these seven key areas of the LEED evaluation criteria: (1) sustainable sites, (2) water efficiency, (3) energy and atmosphere, (4) materials and resources, (5) indoor air quality, (6) innovation and design process, and (7) regional-specific priority credits. Developments may qualify for LEED certification designation and points, according to the following qualified certification categories:

- LEED Certified – 40 to 49 points
- LEED Silver – 50 to 59 points
- LEED Gold – 60 to 79 points
- LEED Platinum – 80+ points

On-site treatment systems can comprise a substantial fraction of the certification points with these systems qualifying for up to a maximum number of 11 points through the water efficiency and innovation and design processes in combination with water conservation practices. On-site water treatment systems may qualify for up to 10 points in the water efficiency category through water efficient design, construction, and long-term operation and maintenance features that promote water conservation and efficiency as follows:

- Water Efficient Landscaping – 2 to 4 points
- Innovative Wastewater Technologies – 2 points
- Water Use Reduction – 2 to 4 points

The on-site treatment system must provide water use reductions in conjunction with an associated water conservation program to secure a maximum number of LEED water efficiency points. An on-site treatment system may also help qualify for an Innovation in Design Process maximum credit of one point.

A major sub-category under the Water Efficiency section of the LEED criteria is water use reduction. The water use reduction subcategory determines how much water use can be reduced in and around a LEED-certified development. One item that can receive a score under water reuse is a rainwater (rooftop) harvesting system. The harvested rainwater resource may then be combined with an on-site greywater treatment system, a high-quality wastewater treatment system, or with the use of a municipal reclaimed water system source. The combination of the rainwater harvesting system with either a greywater treatment system, an on-site wastewater treatment system, or a municipal reuse system can together account for a total of up to seven LEED points. While this practice is contrary to the conventional practice of avoiding dilution of biologically degradable material in the sewage that is used by municipal wastewater treatment processes, the on-site treatment system allows multiple objectives of reducing effluent discharges and reducing stormwater runoff while providing water that can be used for nonpotable purposes. The Fay School, located in Southborough, Mass., achieved LEED Gold Certification from the USGBC. The Fay School students now monitor building energy and building water consumption from a digital readout in each new dormitory building. The entire project was developed from the Fay School's interest in sustainable design principles and educates the students on the importance of water efficiency [US-MA-Southborough].

Battery Park City in lower Manhattan, New York City, is a collection of eight high-rise structures with 10 million ft² of floor area that serves 10,000 residents plus 35,000 daily transient workers. Water for toilet flushing, cooling, laundry, and irrigation comes from six on-site treatment systems. On-site systems use MBR technology for biological treatment and UV and ozone for disinfection. Potable water is supplied by New York City and the on-site treatment systems overflow to a combined wastewater/stormwater outfall. All buildings in Battery Park City are LEED certified Gold or Platinum [33].

In an industrial setting, the Frito-Lay manufacturing facility in Casa Grande, Ariz., received a LEED Gold EB (Existing Building) certification with modification to the manufacturing process to incorporate an on-site process water treatment system and addition of 5 MW of on-site photovoltaic power generation [US-AZ-Frito Lay].

Reclaimed water, along with other major alternative water sources, such as harvested rainwater and collected stormwater runoff, offer the opportunity to maximize landscape irrigation and reduce potable water use at many industrial and commercial institutions and at multi-family residential developments. In the south and southwest United States, air conditioning condensate collection and reuse may represent another significant alternate water resource. Onsite treatment systems can be designed to treat municipal wastewater, greywater, harvested rainwater, and stormwater. Regardless of water source selected for use, care must be taken to differentiate pipes on the private side of the municipal utility boxes, appropriately color code on-site pipes, and adopt a cross-connection control program for the different water sources.

83.11 Summary and Conclusions

The responsible management of water resources will become one of the main challenges of the twenty-first century. It will necessitate a concerted effort to both limit our draws on natural water bodies and control the quality of effluents sent back into the environment. Water reuse of all forms should be encouraged as it allows the maximization of water's utility onsite and encourages the treatment of used water prior to discharge. Greywater reuse at the domestic level may well be the simplest form of water reuse and should be investigated as a means to reduce the impact of residential developments on water resources worldwide. However, it excludes toilet wastes and has a low nutrient content, domestic greywater can indeed be highly biologically polluted. High counts of indicator bacteria associated with fecal pollution in greywater studies have been attributed by default to the washing activities that contribute to greywater streams, but their exact source should be investigated. The greywater analyzed in this study contained no heavy metals, but elevated levels of indicator organisms pointed to a strong concentration of pathogenic organisms. Fecal coliform counts in the greywater exceeded [35] water quality standards for unrestricted food crop irrigation by a factor of 10–100. In food tests, however, the contamination of crops by this bacteria-rich irrigation water was not evident. Concentrations of indicator bacteria on food surfaces were low and were not correlated with water quality data. Although the same indicator bacteria present in the greywater were detected on crop surfaces, their numbers were not significantly higher than those found on control crops irrigated with normal tap water. This may indicate some buffering effect of the soil biotic community or signal the insufficiency of chosen indicator bacteria to 54 measure crop contamination by waterborne pathogens. The movement of each pathogenic organism on plant surfaces and into plant tissues will naturally be species-specific and may be difficult to predict with the same methodology employed for water quality testing. More research is needed in the area of pathogen enumeration on crop surfaces so that easy detection methods, sampling regimes, and acceptable contamination limits can be agreed upon. Confirmation of appropriate indicators is also important in order to efficiently investigate the real risk associated with irrigating food crops with fecally contaminated waters of all types.

Assessments of wastewater generation and treatment in many countries have improved within the last 10 years although there are still many sewers ending without treatment plants in rivers as well as with treatment plants with a large enough sewer network to reach treatment capacity. The wastewater generated needs to be treated in order to protect the groundwater and ecosystems, and reduce downstream impacts where many livelihoods are supported. However, treatment levels can also be designed to meet the requirements of end users but this requires adequate discussion at locations where wastewater is to be used. If at the sectoral level, categories of treatment for end use can be agreed upon, and it can be part of the municipal development plan, making effective use of wastewater generated in the cities. Moreover, if annual assessments are made at the city/state level, based on an agreed format, the pollution control boards can perform nationwide projections more effectively, and in a timely manner. With advances made in the IT sector, each nation could well afford to develop an information management system that connects the entire country. However, capacity-building and the infrastructure have to be developed side by side for an overall positive outcome.

Assessments on wastewater irrigated agriculture and livelihood benefits of wastewater are complex. Estimates of potential irrigable land using simple or complex methods have been attempted [16,29].

More methods can be developed by using water-quality parameters, crop types and soil conditions. Modern tools like remote sensing/geographical information system [RS/GIS] and more precise mapping of drainage networks can also provide better overall outcomes that can help assess the nutrient loads leaving the region. The urban planning sector which is currently embarking on GIS-based mapping of municipal areas can make land-use mapping as part of their program of work, to develop baselines, upon which future studies can be modeled. Wastewater irrigation can be a dynamic process in the peri-urban areas, and land-use patterns can change with development and socioeconomic change; therefore, assessments need to involve robust methods to capture this dynamism, spatially and temporally.

Benefits in terms of income generation from wastewater use for marginal farmers were more than evident from many case studies across countries. For many, wastewater agriculture was a primary or secondary income source. Wastewater agriculture is however not without negative externalities, and health impacts on farmers and consumers are of significant concern as reported above. From the developing countries context more studies are required in the areas of wastewater irrigated agriculture, health and food safety, and health economics, specifically at the farm and consumer levels, to capture the diverse settings in which the problems exist. Risk assessment tools like Quantitative Microbial Risk Assessment (QMRA) and Quantitative Chemical Risk Assessment (QCRA) can be used to assess the potential risk, which should then be addressed through multiple barrier approaches with health-based targets for risk reduction [35].

Further, decision makers may find it useful for developing a more holistic national approach for wastewater use in agriculture, with the advantage of feeding national data straight into international databases. Wastewater management and treatment cannot be planned in isolation. They have to be a core part of the strategic plans for water supply and sanitation, irrigation and drainage, energy, and environmental services and other uses [31]. Moreover, it becomes very important to consider these aspects in light of water availability for cities, and to highlight the need for continuous inter-sectoral dialogue and action plans to address the ever-increasing water demands [32]. Integration of water resources development with water services can provide more support for agricultural water management. India being today more urban and peri-urban than rural, it is time safe wastewater use for agriculture was made a priority in its water development agenda.

References

1. Ballaro, C. 2012. Need for water conservation stressed, Water management plan laid out for Cape council, Cape Coral, FL, *Daily Breeze*, April 16, 2012, http://www.cape-coral-daily-breeze.com/page/content.detail/id/529796/Need-for-water-conservation-stre—.html. Retrieved April 2012.

2. Baydal, D. 2009. Municipal Wastewater Recycling Survey. California Water Recycling Funding Program (WRFP). http://www.swrcb.ca.gov/water_issues/programs/grants_loans/water_recycling/munirec.shtml. Retrieved on August 23, 2012.

3. Bhatia, R. and Falkenmark, M. 1992. Water resource policies and the urban poor: Innovative approaches and policy imperatives. Background paper prepared for the ICWE, Dublin, January 26–31, 1992.

4. California State Water Resources Control Board. 2009. Recycled Water Policy. http://www.swrcb.ca.gov/water_issues/programs/water_recycling_policy/. Retrieved July 2012.

5. Garzon, C., Campos, S., Machado, K., Birolo, N., Centeno, C., Jimene, J., and Nuques, C. 2009. Water and Sanitation Program for the Buenos Aires Metropolitan Area and Conurbation. Inter-American Development Bank Loan Proposal. Approval Date: January 1 2009.

6. Hermanowicz, S.W., Sanchez Diaz, E., and Coe, J. 2001. Prospects, problems and pitfalls of urban water reuse: A case study, *Water Science and Technology*, 43(10), 9–16.

7. Hoekstra, A.Y., Chapagain, A.K., Aldaya, M.M., and Mekonnen, M.M. 2011. *The Water Footprint Assessment Manual—Setting the Global Standard*. Water Footprint Network, EarthScan Ltd, London, UK.

8. JICA 2003. Capacity Development and JICA's Activities (in Japanese). JICA, International Development Center (IDJC) and IC Net Ltd.

9. Li, L., Xie G.D., Cao S.Y., Luo Z.H., Humphrey, S., Cheng, S.K., Gai L.Q., et al. 2010. China Ecological Footprint—Biocapacity, cities and development. WWF China.

10. McGranahan, G., Marcotullio, P. Bai, X., Balk, D., Braga, T., Douglas, I., Elmqvist, T. et al. 2005. Chapter 27, Urban systems. In: R. Hassan, R. Scholes, and N. Ash (eds.), *Ecosystems and Human Well-being: Current State and Trends*, Volume 1. Millennium Ecosystem Assessment, Island Press, pp. 795–825.

11. Marina Coast Water District, CA, 2002. Title 4. Recycled Water Chapter 4.28 Recycled Water, Supplement 3-02 to Ordinance 29 Sections 4, 5, and 6, 1995; and Ordinance 27 Section 5, 1994. http://www.mcwd.org/code_4_recycled_water.html. Retrieved April 2012.

12. Ministry of Environment, Singapore. 2003. Speech by Prime Minister Goh Chok Tong at the Official Launch of NEWater, *Environmental News Release* No: 4/2003.

13. Ogura, K. 1999. Reuse of urban waste water for industrial applications in Madras (Chennai), India. *Journal of Water and Wastewater*, 41(2), 48–51, Japanese Version.

14. Pittock, J., Meng, J. Geiger, M., and Chapagain, A.K. 2009. Interbasin water transfers and water scarcity in a changing world—a solution or a pipe dream? WWF Germany.

15. PUB. 2003. Singapore's Experience in Water Reclamation—the NEWater Initiative, International Symposium on Water Resource Management and Green Productivity, Singapore, October 6–9, 2003.

16. Raschid-Sally, L. 2010. The role and place of global surveys for assessing wastewater irrigation. *Irrigation and Drainage Systems*, 24(1–2), 5–21.

17. Rodrigo, D., Lopez Calva, E.J., and Cannan, A. 2010. *Total Water Management*. Environmental Protection Agency. Washington, DC.

18. Rose, G.D. 1999. Community-Based Technologies for Domestic Wastewater Treatment and Reuse: Options for Urban Agriculture. N.C. Division of Pollution Prevention and Environmental Assistance, CFP Report Series: Report 27.

19. Schutte, C.F. 2001. Managing water supply and sanitation services to developing communities: Key success factors. *Water Science and Technology*, 44(6), 155–162.

20. Scott, C. and Banister, J. 2008. The dilemma of water management 'Regionalization' in Mexico under centralized resource allocation. *Water Resources Development*, 24(1), 61–74.

21. Sosa-Rodriguez, F.S. 2010. Impacts of water-management decisions on the survival of a city: From ancient tenochtitlan to modern Mexico City. *Water Resources Development*, 26(4), 677–689.

22. Tortajada, C. 2003. Water management for a megacity: Mexico City metropolitan area. *Ambio*, 32(2), 124–129.

23. Tortajada, C. 2006. Who has access to water? Case study of Mexico City Metropolitan Area. UNDP Human Development Report Office Occasional Paper.

24. UNEP. 2011. Water and Cities Facts and Figures. United Nations Environment Program.

25. UN HABITAT. 2006. *State of The World's Cities 2006/7. The Millennium Development Goals and Urban Sustainability: 30 Years of Shaping the Habitat Agenda*. United Nations Human Settlements Program (UN-HABITAT) & Earthscan Publishing, Sterling, VA.

26. UN HABITAT. 2008. *State of the World's Cities 2010/11. Bridging the Urban Divide*. United Nations Human Settlements Program (UN-HABITAT) and EarthScan Publishing, Sterling, VA.

27. U.S. Green Building Council (USGBC). ND. http://www.usgbc.org/. Accessed July 2012.

28. Van Riper, C., Schlender, G., and Walther, M., 1998. Evolution of water reuse regulations in Washington state. In: *WaterReuse Conference Proceedings*. Denver, CO.

29. Van Rooijen, D.J., Biggs, T.W., Smout, I., and Drechsel, P. 2010. Urban growth, wastewater production and use in irrigated agriculture: A comparative study of Accra, Addis Ababa and Hyderabad. *Irrigation and Drainage Systems*, 24, 53–64.

30. Vörösmarty, C.J., Lévêque, C., Revenga, C., Bos, R., Caudill, C., Chilton, J., Douglas, E.M., et al. 2005. Fresh water. In: R. Hassan, R. Scholes, and N. Ash (eds), *Ecosystems and Human Well-being: Current State and Trends*, Volume 1. Millennium Ecosystem Assessment, Island Press, pp. 165–207.

31. World Bank. 2004. *Water Resources Sector Strategy: Strategic Directions for World Bank Engagement*. Washington, DC.

32. World Bank. 2010. *Sustaining Water of all in a Changing Climate*. World Bank Group Implementation Progress Report of the Water Resources Sector Strategy. Washington, DC.

33. Water Environment Research Foundation. 2009. Case Study: Battery Park Urban Water Reuse. Alexandria, VA.

34. World Health Organization. 2004. *Guidelines for Drinking-water Quality*. Third Edition (incorporating first and second addenda). World Health Organization, Geneva.

35. World Health Organization. 2006. *Guidelines for the Safe Use of Wastewater, Excreta and Greywater*. Volume 2: Wastewater use in agriculture. WHO, Geneva, Switzerland.

36. World Health Organization. 2008. *The Global Burden of Disease: 2004 Update*. WHO, Geneva, Switzerland.

37. World Health Organization and United Nations Children's Fund. 2000 Global Water Supply and Sanitation Assessment 2000 Report, New York.

38. World Water Assessment Programme. 2006. The United Nations World Water Development Report 2: Water—A shared responsibility. UNESCO and Berghahn Books. World Water Assessment Programme.

39. World Wide Fund for Nature. 2006. Payments for Environmental Services—An equitable approach for reducing poverty and conserving nature, Gland, Switzerland.

Urban Water Reuse: Future Policies and Outlooks

Saeid Eslamian
Isfahan University of Technology

Saeid Okhravi
Isfahan University of Technology

Mohammad Naser Reyhani
Isfahan University of Technology

84.1 Introduction

The previous chapters dealt with analysis and interpretation of data by way of focus groups endeavor. In this chapter, a brief summary on the research project about water reuse findings from both the literature review and empirical investigation will be presented.

In this handbook, water recycling is considered as part of integrated resources management. In this process, planners, water utilities, and local and regional governments should establish an outline with future projected water demands. The outline includes forecasts of land use, population, and wastewater. Institutional, legal, and other requirements should be considered in formulating an outline. Clear water supply objectives and alternatives should be specified prior to the economic and financial analysis of each water supply alternative or project. Projects for increasing water conservation, augmenting water supply, and enhancing water quality are often proposed. Refinement of alternatives may be required following economic and financial analyses and prior to final recommendations on the set of water and wastewater management actions.

The role of the environment in supplying materials, services, and waste sinks, which together make the human economy possible, is largely ignored [12]. Current human technology does not have the ability to substitute for major ecological services such as a breathable atmosphere. Increasingly, environmental indicators show that human activity is degrading the capacity of the natural environment to support life on Earth [7,4]. Securing sustainable water for all yields important returns in terms of people's health, food security, and industrial production as well as the health of ecosystems and the services they provide.

Resources are the backbone of every economy and provide raw materials for production of goods and services, and environmental service as basic functions.

Environmental resources can be degraded by pollution and rendered useless, while water reuse plays an important role in water resources, and wastewater and ecosystem management in the world. Water reusing is also sometimes known as purified water, advanced treated water, and recycled or reclaimed water. Water reuse involves using highly treated domestic wastewater for new purposes. By recycling or reusing water, communities can still grow while minimizing or even reducing their impact on the water resources around them. Reusing treated wastewater not only for beneficial aims such as agricultural and landscape irrigation, industrial processes, toilet flushing, replenishing a groundwater basin (groundwater recharge) but also for delaying huge cost of capital investments in the development of new water supplies. Recycled water for landscape irrigation requires less treatment than recycled water for drinking water. No documented cases of human health problems due to contact with recycled water that has been treated to standards, criteria, and regulations have been reported. Reclaimed water systems are continually monitored to ensure the health and welfare of the public and the environment are protected.

Using recycling water is an excellent water supply for

- Irrigating landscaped areas, residences, and urban demands
- Using for industrial and commercial demands for cooling water for power plants and oil refineries and for various processing or washing needs
- Agriculture and pastures for livestock
- Enhancing restoration by creating natural and artificial wetlands and artificial lakes
- Construction activities (e.g., concrete mixing)
- Recharging groundwater
- Environmental and recreational uses as nonpotable urban uses

84.2 Incentives of Reuse

Reused water can satisfy most water demands, as long as it is adequately treated to ensure water quality appropriate for the use. In uses where there is a greater chance of human exposure to the water, more treatment is required. As for any water source that is not properly treated, health problems could appear from drinking or being exposed to recycled water if it contains disease-causing organisms or other contaminants.

The ability to reuse water, regardless of whether the intent is to augment water supplies or manage nutrients in treated effluent, has positive benefits that are also the key motivators for implementing reuse programs [17]. These benefits include improved agricultural production; reduced energy consumption associated with production, treatment, and distribution of water; and significant environmental benefits, such as reduced nutrient loads to receiving waters due to reuse of the treated wastewater. In this fast-changing and highly interconnected world, the problems related to water crisis and conflicts are numerous, complicated, and challenging. Efforts to effectively resolve these problems require a clear vision of the future water availability and demand as well as new ways of thinking, developing, and implementing water planning and management practices. In view of these observation, in 2012 drivers for reuse classified reasons that lead to using reclaimed water into three categories: (1) addressing climate change and water supply scarcity, (2) achieving efficient resource use, and (3) environmental and public health protection [17].

84.2.1 Urbanization, Climate Change, and Water Scarcity

Population growth is an important factor for water-related activities and problems because an increase in population normally causes an increase in water demands in almost all sectors (domestic, industrial, agricultural, energy, and recreation), unless water management practices become more efficient. The estimates by the United Nations (UN) indicate that the world population may increase from 6.7 billion in 2007 to 7.7 billion by 2020 and to 9.2 billion by 2050 [21].

This increase will occur mostly in the developing regions, whose population is projected to increase from 5.4 billion in 2007 to 7.9 billion in 2050. Although these regions are precisely the ones already facing significant water, drought, and sanitation problems, the future situation may likely get much worse.

In view of these observations, some researchers realize that there will not be enough freshwater in the world for humans and ecosystems to survive and that there will be water scarcity, crisis [1,2,5,11,24], and associated conflicts.

As countries develop and populations grow, the potential demand for water is projected to increase by 55% by 2050 [10]. Already by 2025, two-thirds of the world's population could be living in water-stressed countries if current consumption patterns continue [20]. As global water withdrawals continue to rise by approximately 10% every 10 years [18], it is becoming increasingly critical to bring water withdrawals into line with limited renewable levels of ground and surface water. By 2020, water use and reuse is expected to increase by 40%, and 17% more water will be required for food production to meet the needs of the growing population. By 2025, 1.8 billion people will be living in regions with absolute water scarcity; about two out of three people in the world could be living under considerations of water stress. Therefore, water use and reuse is the most potential human demandable issue to survive within a sustainable way [1,6,14].

The hydrological cycle around the world is intimately linked with changes in atmospheric temperature and radiation balance. Global warming and climate change observed over the past several decades is consistently associated with changes in a number of components of the hydrological cycle and hydrological systems such as changing precipitation patterns, intensity, and extremes; widespread melting of snow and ice; increasing atmospheric water vapor; increasing evaporation; and changes in soil moisture and runoff.

Climate change is one of the barriers to create imbalance between demand and supply of water. Climate change poses a serious challenge to social and economic development in all countries and it is expected to increase the frequency and intensity of current hazards and the probability of extreme events and new vulnerabilities with differential spatial and socioeconomic impacts.

Extreme climate events such as aridity, drought, flood, cyclone, and stormy rainfall are expected to remain an impact on human society. In many parts of the world, climate change is a supplementary hazard imposed on long-term water stressed conditions, socioeconomic pressure, and management challenges. These water-stressed conditions are characterized by low, erratic, and poorly distributed rainfall, high evaporation, and excessive runoff and soil losses. The current socioeconomic conditions emanate from historical legacy of apartheid that excluded a majority of people from access to resources including water. They are also expected to generate widespread response to adapt and mitigate the suffering associated with these extremes. Societal and cultural responses to prolonged drought include population dislocation, cultural separation, habitation abandonment, and societal collapse. Clearly, while there is a need to negotiate international commitment to reduce greenhouse gas emissions, it is also important to undertake policies and measures that facilitate adaption to the observed and projected impacts of climate change.

The only way to face increasing water demands by population growth and also climate change concern is water reuse in climate predictions, both for demand projections and for ecological impacts.

Although most water recycling projects have been developed to meet nonpotable water demands, a number of projects use recycled water indirectly for potable purposes. The use of treated greywater (reusable wastewater from residential, commercial, and industrial) at decentralized sites for landscape irrigation and toilet flushing reduces the amount of potable water distributed to these sites, the amount of fertilizer needed, and the volume of wastewater going to septic systems and wastewater treatment plants. Decentralized water reuse systems are being used more in arid regions where long-term drought conditions exist. Successful greywater systems have been operating for many years. They can meet up to 50% of a property's water needs by supplying water for landscaping [9]. Recycling greywater increases infrastructure capacity for new users.

84.2.2 Water–Energy Savings

Energy availability is of fundamental importance for economic and social development. Water and energy are very highly interconnected. A key driver of this increased demand for energy is urbanization; and with more than half of humanity being urbanized, this demand is set to increase [19]. Cities are large consumers of energy, requiring highly concentrated sources of heating, cooling, and transportation systems. Water needed to supply cities is sourced from ever greater distances and frequently needs to be pumped to consumers [3].

Costs associated with water supply or wastewater disposal can also make reuse of wastewater an alternative option. As briefly described, water reuse saves water, energy, and money. Energy efficiency and sustainable development are key parameters that stimulate future people use reusing water. The water-energy nexus identifies that water and energy are mutually dependent—energy production requires large volumes of water and water infrastructure requires large amounts of energy [8].

As the demand for water grows, more water is extracted, treated, and transported sometimes over great distances which can require a lot of energy. If the local source of water is groundwater, the level of groundwater becomes lower and lower as more water is removed. These increases of energy appear by requiring pumping the water to the surface. Recycling water onsite or nearby diminishes the energy needed to move water longer distances or pump water from deep within an aquifer.

Recycling flowback water onsite offers a cost-effective solution that conserves water and energy, saves roads from wear and tear, and keeps production costs down.

Tailoring water quality to a specific water use also reduces the energy needed to treat water. The water quality required to flush a toilet is less stringent than the water quality needed for drinking water and requires less energy to achieve. Using recycled water that is of lower quality for uses that do not require high quality water saves energy and money by reducing treatment requirements.

A large amount of energy is required, first in collecting, extracting, conveying, and distributing water to end users, and second in treating and disposing of the wastewater. Although it requires additional energy to treat wastewater for recycling, the amount of energy required to treat and/or transport other sources of water is generally much greater.

For industry, wastewater recycling and reuse can be encouraged through fiscal incentives (subsidies) and pollution taxes. Indeed minimization of infrastructure costs, including total treatment and discharge costs, has been the motivation behind many reuse schemes.

EPA has developed principles for an energy-water future that incorporate familiar concepts of efficiency, a water-wise energy sector as well as an energy-wise water sector, consideration of wastewater as a resource, and integrated resource planning and recognition of the societal benefits [17].

84.2.3 Environmental Protection

The ecological importance of water recycling is perceived by nations. Around the globe, people are working hard to protect natural resources before their depletion reaches an increased state. People today are "going green" to protect their land and to prove themselves as good citizens. There are many societies concerned about protecting the environment around them. Finally, it is essential to protect water resources and to maintain a balance in the ecosystem.

Environmental concerns over negative impacts from increasing nutrient discharges to waters are resulting in compulsory reduction in the number of water discharges and also side effects of contaminations. From an environmental perspective, water reuse can reduce demand for freshwater resources, diversify water sources, and enhance reliability of access to resources; it can reduce volume of wastewater discharged into the environment. Decentralized systems can reduce energy required to transport water from the point of production to the point of use, and reduce greenhouse gas emissions (due to energy savings). By eliminating effluent discharges for all or even a portion of the year through water reuse, a municipality may be able to avoid or reduce the need for costly nutrient removal treatment processes.

84.3 Benefits of Water Reusing

Plants, wildlife, and fish depend on sufficient water flows to their habitats for living and reproducing. The lack of adequate flow, as a result of diversion for agricultural, urban, and industrial purposes, can cause deterioration of water quality and ecosystem health.

This Handbook covers many aspects of water reuse including application, options, regulations, impacts/benefits, technologies, evaluations, monitoring, management, and financial issues. Collecting all of the information is crucial to conclude. In addition to providing a dependable, locally controlled water supply, water reusing provides tremendous environmental benefits. Some of the most important of them are mentioned next.

- Makes up an additional source of water for various purposes
- Decreases the extraction of water form sources; this may stop being viable as habitats for valuable and endangered wildlife
- Decreases the discharge of effluents; this may be preventing damage and pollution of the ecosystems of sensitive bodies of water
- Creates new wetlands by reusing water
- Enhances and improves the quality of wastewater by constructed wetlands
- Protects the environment by reusing wastewater at power plants
- Makes up an additional source of nutrients and detracts from the need to apply synthetic fertilizers
- Saves energy
- Combats desertification

For further information, by providing an additional source of water, water reusing can lead to decreasing the diversion of water from sensitive ecosystems, decreasing wastewater discharges, reducing and preventing pollution, and creating or enhancing wetlands and riparian habitats.

In some cases, the stimulation for water recycling comes not from a water supply need, but from a need to eliminate or decrease wastewater discharge to lakes, an estuary, or a creek.

However, in spite of the benefits obtained by reusing treated wastewater in irrigation, there are associated health risks due to contact with reclaimed wastewater. Consequently, it is of utmost importance to make sure that these risks are minimized and monitored.

84.4 Sustainable Reuse Policies

Use of treated wastewater has become increasingly important in water resources management for social, environmental, and economic reasons. Despite the many challenges to the sustainable implementation of water resources, evolving conditions across the world will encourage a more systematic consideration and implementation of water reuse. Economic development and population growth continue to increase the demand for high-quality freshwater. However, only a miniscule portion of the global water has a quality safe for human consumption and is easily accessible. Urbanization and the associated increased demand for domestic water and production of wastewater, along with continued reduction in available freshwater, increasing tourism, global warming, and the preferences of a more educated urban population will all play a critical role in affecting plans for

wastewater treatment, increasing demand for locally produced safe and high-value agricultural products, and in encouraging ecological and recreational uses of water. The complexity of sustainable decision making and the challenges of sustainable development in managing water resources are clear.

Wheeler identified five characteristics of successful sustainability policy and practice [22].

84.4.1 A Long-Term Perspective

One of the fundamental concepts of sustainable development is the need to base decisions on a longer-term perspective rather than business as usual. This perspective is necessary because natural systems ebb and flow on a much different time scale than do human-made systems. By focusing on short-term returns on investment, for example, communities may undermine their long-term economic, social, and environmental sustainability. Effective decision making, therefore, requires community leaders to understand and anticipate the long-term implications of current policies and investments.

84.4.2 A Holistic Outlook

Planning for sustainability emphasizes a systems approach to understanding and addressing critical issues. Unlike traditional decision making narrowly focused on one or more sectors or systems within a community, sustainable decision making focuses on relationships among environmental, economic, and social policies and investments. This holistic approach improves understanding of key issues and yields policies and solutions that leverage related responses to achieve integrated community goals. A holistic approach also considers how local policies fit into and inform state, regional, and federal policies.

84.4.3 Acceptance of Limits

Plans and policies promoting sustainability often recognize the concept of limits. For example, many analysts have argued that the planet is physically limited in providing resources and in absorbing pollution and waste. The International Panel on Climate Change suggested such a limit to the amount of greenhouse gases that can be placed in the atmosphere, roughly equivalent to a concentration of 350 parts per million of carbon dioxide. An atmospheric concentration of 392 parts per million has already been reached [22].

84.4.4 A Focus on Place

Discussions around sustainable development often emphasize self-sufficiency and community (or system) resilience. Sustainable decision-making recognizes the uniqueness of individual communities and regions and values their existing and potential contributions to a more sustainable society. Sustainable planning and investment are effective leverage of those local assets to meet local needs. Agriculture and energy are two good examples. In both cases, sustainability policy seeks to leverage existing local assets to achieve greater self-sufficiency. Resilience refers to a community's ability to retain its integrity and to function despite changes and external shocks.

84.4.5 Active Involvement in Problem Solving

Sustainability planning calls for participation of a broad community cross-section, including voices typically under-represented in the policy-making process. Broad active involvement is critical because transitioning to a more sustainable society requires an enormous scale of change. Actively engaging citizenry in problem solving recognizes individuals' power to make a difference and contribute to positive local change. Local involvement also results in local ownership and support of community-based solutions.

84.5 Future and Outlooks of Water Reusing

84.5.1 Surge in Demand for Future

At the global level, withdrawals for irrigation may increase by six times in 2025 when compared to that in 1900. The domination of agricultural sector in water demand is particularly evident in developing regions, such as Africa, Asia, and South America (Table 84.1). The agricultural sector now accounts for over 60% of all water demands, although this figure is also considerably

less than what it used to be during the middle of the last century (over 90% in some regions). This figure will probably further decrease to about 40%, with the exception of Asia where water withdrawals for irrigation are expected to be above 70% in 2025.

Water withdrawals for domestic uses are the highest in Europe and North America, where they may reach more than 40% by 2025 due to a variety of reasons, including increases in temperature and humidity, and decreases in demand for irrigation.

Water demand is expected to increase by 46% by 2035 due to growing population, rising temperatures, longer intervals without rain, and increased evaporation from the soil and water reservoirs [15].

In view of these predictions, scientists have recommended a permanent water source called water reusing. Integrated management of water reusing is a very important issue for designing managers, investors, and legislators in the areas faced with lack in water.

Reusing as described in this meaning is the process to treat wastewater and sewage using advanced technology to produce potable water fit for human consumption. The treated water is first directed to either a groundwater source or a surface water holding area, such as a reservoir, and then treated again at drinking water facilities, before eventually entering the drinking water supply. Most water reusing processes are comprised of a multitiered filtration system, described in Figure 84.1. This concept is represented graphically, which illustrates that water treatment technologies (combined with disinfection) offer a ladder of increasing water quality, and choosing the right level of treatment should be dictated by the end application of the

TABLE 84.1 Ratio of Water Withdrawal and Consumption to Total Water Withdrawal and Consumption by Sectors of Economic Activity (%)

Continent	1950			1995			2025		
	A	I	D	A	I	D	A	I	D
Water withdrawal									
Africa	90.5	7.0	2.6	63.0	8.1	4.4	53.1	18.0	6.0
Asia	93.4	2.4	4.2	80.0	6.9	9.9	72.0	9.5	15.2
Europe	32.2	25.4	41.2	37.4	14.7	44.8	37.2	14.0	45.8
North America	53.5	7.9	36.0	43.5	10.7	41.5	41.4	12.3	41.3
South America	82.4	9.5	7.9	58.6	17.2	15.4	44.2	22.7	23.8
Australia and Oceania	50.0	7.2	39.4	51.0	10.9	23.5	46.8	11.3	26.1
World	78.1	6.3	14.8	66.1	9.1	19.9	60.9	11.6	22.3
Water consumption									
Africa	97.9	1.6	0.5	63.8	1.5	0.8	60.5	3.4	1.3
Asia	98.0	0.7	1.1	91.0	1.5	2.3	88.4	1.8	4.1
Europe	67.7	12.6	15.6	71.4	5.6	15.3	66.8	4.3	22.3
North America	83.5	4.7	3.6	75.1	5.0	7.2	72.4	6.0	7.5
South America	95.0	2.5	1.9	76.4	4.0	3.2	67.4	4.7	8.3
Australia and Oceania	81.3	2.0	9.9	69.1	2.2	3.1	64.1	2.1	6.4
World	94.0	2.2	2.5	84.5	2.4	4.0	81.5	2.7	6.1

Source: Adapted from Shiklomanov, I.A. and Rodda, J.C. 2003. *World Water Resources at the Beginning of the 21st Century.* UNESCO International Hydrology Series, Cambridge University Press.

Note: A: agricultural; I: industrial; D: domestic.

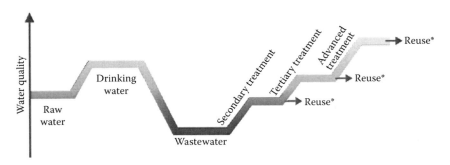

FIGURE 84.1   Achieving any desired level of water quality by kind of treatment technologies. (Adapted from Prugh T. 1999. *Natural Capital and Human Economic Survival*. Lewis Publishers, Boca Raton, FL.)

reclaimed water for achieving economic efficiency and environmental sustainability [17]. In fact, level of treatment is dependent on the end use.

There are numerous case studies that demonstrate the balance of treatment costs along with the intended use of the reclaimed water. Many of these develop reuse in the interest of replacing the use of drinking water for nonpotable applications and meeting future water demands. There are lists of some indirect potable reuse ways presented in Table 84.2. As such, the treatment level required for reclaimed water production depends on the end use. This recognition of "fit for purpose" provides a framework for cost-effective treatment to be applied to a water source sufficient to meet the quality appropriate for the intended use. By selecting appropriate treatment for specific applications, water supply costs can be controlled and the costs for improved wastewater treatment technologies delayed until they are balanced by the benefits. Consideration must also be balanced with the potential for future reuse of higher reclaimed water quality such that these uses are not limited.

TABLE 84.2   Multistep Indirect Potable Reuses Filtration Process

Step	Process	Aim
1 and 2	Initial treatment	Water passes through several screens and sedimentation to remove suspended solids; water at the end of this process is safe for irrigation and other nonconsumption uses
3	Microfiltration	Further filters remaining solids
4	Reverse osmosis	Water pumped through membranes eliminating viruses, bacteria, and protozoa
5	Advanced oxidation	Further disinfection and removal of emerging contaminants using UV or ozone and hydrogen peroxide
6	Fresh water blend	Water blended with surface water reservoirs or added to groundwater
7	Standard water filtration	Water moved from reservoir or groundwater and goes through standard water filtration[a] before being added to potable piping system

[a] Step not required but usually done to address public perception concerns or because recycled water has been blended with less treated water.

84.5.2 Future of Water Reusing

Increasing city populations, the millennium drought, recognition of the limits to water resources, and the gradual awareness of the risks of climate change concentrated the need to look at water reusing as a serious supply option. Shortages of water in the last decades has been determined the most important factor that governments should start investigating the use of water recycling, noting that the world is likely to be out of water. As water energy demands and environmental needs grow, water recycling will play a greater role in overall water supply. State EPAs were formed and started to take an interest in the composition of effluents from wastewater treatment plants [16].

By working together to overcome hindrances, water recycling, along with water conservation and efficiency, can help people to sustainably manage their vital water resources.

Water reusing has proven to be effective and successful in creating a new and dependable water supply without compromising public health. Nonpotable reuse is a widely accepted practice that will continue to grow. Greywater reusing is still an undeveloped technology, and as such its efficiency may improve with time. Advances in water treatment technology in the future may allow heavily contaminated water like greywater to be cleaned to suitable standards for reuse extremely cheaply, but the environmental impacts might still be high and only justifiable on an industrial scale.

However, in many parts of the world, especially advanced countries, the uses of recycled water are expanding in order to accommodate the needs of the environment and growing water supply demands. Advances in wastewater treatment technology and health studies of indirect potable reuse have led many to predict that planned indirect potable reuse will soon become more common. Recycling waste and greywater requires far less energy than treating saltwater using a desalination system.

Reusing purified sewage, also known as reclaimed or recycled water, to boost drinking water supplies has significant potential for helping meet future needs.

Perception of "fit for purpose" has recycled water serving dual piped urban developments, supporting high value horticultural and vegetable crop production, growing pastures for livestock, supporting secondary industries, contributing to recreational water facilities, supporting urban amenities, contributing to the operation of new green city office buildings, being used in food and beverage manufacturing, and replacing environmental water flows earlier removed upstream for consumptive purposes. Treating municipal wastewater and reusing it for drinking water, irrigation, and other applications could significantly increase the nation's water security.

84.6 Summary and Conclusions

There are some significant policy recommendations and considerations for future research concerning expanding water resources and supporting the environment including:

1. Inform people, especially homeowners, about water conservation and water reusing. Shortages of water in the world would have to become much more severe to justify domestic greywater systems in households. It is important, therefore, to inform homeowners how to conserve water and recycle water. This is not only in times of water restrictions but in the ongoing plan to change the effects of climate change.
   In addition, ethical and cultural subjects should be taken into consideration in water reuse and proper management strategies are very much potential factors for water management in the urban and rural regions. The setup of water reuse guidelines, and the ethical, cultural, and engineering aspects should be considered in guideline development processes.

2. Consider setting more aggressive conservation targets. These targets should be followed by maintaining or expanding incentives and creating financing programs that stimulate consumers to retrofit existing homes with the most efficient technologies available.

3. Set up public relations between community and politicians. A more sensitive issue is the extent to which recycling will contribute as a resource for our drinking water systems. There is the technology in some locations, the facilities are in place, but the community and politicians are still not very comfortable. The public water-management sectors in many countries of the world suffer from poor management and/or inadequate investment in the water sector. Private participation in the sector now seems inevitable, and could represent an important means of improving performance in water management and public sanitation, and achieving significant gains in productivity and efficiency.

4. Set up public relations between communities and agencies. While water recycling is a sustainable approach and can be cost-effective in the long term, the treatment of wastewater for reuse and the installation of distribution systems at centralized facilities can be initially expensive compared to such water supply alternatives as imported water, groundwater, or the use of greywater onsite from homes. Institutional barriers, as well as varying agency priorities and public misperception, can make it difficult to implement water recycling projects. Water agencies can also work with the private sector to develop public-private partnerships that can help move the needle on consumer demand.
   Finally, early in the planning process, agencies must reach out to the public to address any concerns and to keep the public informed and involved in the planning process. Engage with a wide range of stakeholders in the region, including the private sector, those from the energy industry, and land use planners to outweigh the options of what our future water portfolio should look like, and to jointly act to implement solutions [25].

5. Establish new technologies. All new single family and duplex residential dwelling units ought to include a separate multiple pipe outlet and drains for all other plumbing fixtures to allow separate discharge of greywater for multiple purposes such as direct irrigation.
   Use data collection systems in billing that can help water managers better assist the residents that are overwatering or have significant leaks.
   Wise investments in wastewater will generate significant returns, as addressing wastewater as a key step in reducing poverty and sustaining ecosystem services. Instead of being a source of problems, the appropriate management of wastewater will be a positive addition to the environment, which in turn will lead to improved food security, health, and therefore economy.

6. Ensure implementation of new green building codes. All greywater systems should be designed and operated according to new green building codes. If future demand estimates at the local and regional level, regard all codes that would require buildings to have more water efficient technologies installed.

7. Support research. Conduct research better documenting the costs and benefits of conservation measures versus the costs of new water reusing projects. More research is also needed for delving into how demand hardening would affect water management efforts in times of drought if more aggressive long-term water conservation methods were put in place. This helps to know some important points that are recognized in this Handbook. There is a remarkable recommendation that water scientists and engineers must recognize that their task should include learning how to communicate their science in order for their methods to be revealed to other stakeholders in society. Simultaneously, governments, private sector, and civil society movements should seek to incorporate scientific results more systematically in their deliberation and decision-making process to reach robust solutions.

In light of rapid global change, communities should plan wastewater management against future scenarios, not current situations. Wastewater management and urban planning lag far behind advancing population growth, urbanization, and climate change. With forward thinking, and innovative planning, effective wastewater management can contribute to the challenges of water scarcity while building ecosystem resilience, thus enabling ecosystem-based adaptation and increased opportunities for solutions to the challenges of current global change scenarios. Protection of water quality from all sources of untreated wastewater, be they domestic, industrial, or agricultural, is a prerequisite for ensuring sustainable development, poverty alleviation, job creation, human and ecosystem health, and people's well-being.

Population growth and climate change are not uniform in time or space, and so regionally specific planning is essential. Wastewater management must be integrated as part of the solution in existing agreements and actions [23]. Given the complex problems we face and the need for efficient and cost-effective strategies, the scientific, engineering, economic, social, and political aspects in an integrated and comprehensive way must be addressed. We are aware of the value and effectiveness of multidisciplinary teams that may include, for example, experts in social sciences, biology, or fisheries.

Opportunities are now available for the development of new technologies to enable more economic treatment of water from lower quality sources. It is expected that breakthroughs in material science will lead to a new generation of materials and systems for water treatment. It is now more important than ever for conservation organizations, governments, civil society, communities, and industries to work together successfully and sustainably meet these needs while protecting nature. The wastewater needs to be considered an important component of the water cycles within catchments, if meaningful water management plans are to be implemented within the country. In each landscape, water augmentation has to be considered in conjunction with different wastewater treatment strategies for multiple uses, and should be supported by public policy and social incentives. It can then potentially not only safeguard the downstream users but also provide economic opportunities for alternative uses of wastewater within cities and support the ecosystem services that constitute an integral part of all forms of life.

References

1. Barlow, M. 2007. *Blue Covenant: The Global Water Crisis and the Coming Battle for the Right to Water*. Black Inc., Melbourne, Australia.
2. De Villiers, M. 1999. *Water Wars: Is the World's Water Running Out?* Weidenfeld & Nicolson Ltd., London, UK.
3. Earle, A. 2013. The role of cities as drivers of international transboundary water management processes, in B.A. Lankford, K. Bakker, M. Zeitoun, and D. Conway (eds.), *Water Security: Principles, Perspectives and Practices*. Earthscan, London.
4. Ekins, P. 1994. The environmental sustainability of economic processes: A framework for analysis. In: J.C.J.M. van den Bergh and J. van der Straaten (eds.), *Toward Sustainable Development*, Island Press, Washington, DC.
5. Gleick, P.H. 1993. *Water in Crisis: A Guide to the World's Fresh Water Resources*. Oxford University Press, Oxford, UK.
6. Gleick, P.H. 1993. Water and conflict: Fresh water and international security. *International Security*, 18: 79–112.
7. Jepson, E.J. 2001. Sustainability and planning: Diverse concepts and close associations. *Journal of Planning Literature*. 15(4), 499–510.
8. National Conference of State Legislatures (NCSL). 2009. Overview of the water-energy nexus in the U.S. http://www.ncsl.org/issues-research/env-res/overviewofthewaterenergynexusintheus.aspx. Retrieved August 2012.
9. National Water Research Institute. 2010. White Paper: Views on the status of water recycling 2030: Recommendations of California's Recycled Water Task Force. Publication Number NWRI-2009-011.
10. OECD. 2012. Environmental Outlook to 2050. Paris.
11. Postel, S.L. 1997. *Last Oasis: Facing Water Scarcity*. W.W. Norton, New York.
12. Prugh T. 1999. *Natural Capital and Human Economic Survival*. Lewis Publishers, Boca Raton, FL.
13. Shiklomanov, I.A. and Rodda, J.C. 2003. *World Water Resources at the Beginning of the 21st Century*. UNESCO International Hydrology Series, Cambridge University Press (extended summary http://webworld.unesco.org/water/ihp/db/shiklomanov/summary/html/summary.html#6.%20Anthropoge).
14. Shiva, V. 2002. *Water Wars: Privatization, Pollution and Profit*. India Research Press, New Delhi, India.
15. Sivakumar, B. 2011. Water crisis: From conflict to cooperation-an overview. *Hydrological Sciences Journal*, 56(4): 531–552.
16. U.S. Environmental Protection Agency (EPA). 2004. Guidelines for Water Reuse. 625/R-04/108. Environmental Protection Agency, Washington, DC.
17. U.S. Environmental Protection Agency (EPA). 2012. Guidelines for Water Reuse. 600/R-12/618. Environmental Protection Agency, Washington, DC.
18. United Nations Environment Programme (UNEP). 2008. *Vital Water Graphics: An Overview of the State of the World's Fresh and Marine Water Resources*, 2nd Edition. Nairobi, Kenya.
19. UN-HABITAT. 2010. *State of the World's Cities 2010–2011 Bridging the Urban Divide*. UN-HABITAT, Nairobi, Kenya.
20. UNESCO. 2009. UN World Water Development Report. Paris.
21. United Nations, Department of Economic and Social Affairs, Population Division, 2007. World population prospects: The 2006 revision highlights. Working Paper No. ESA/P/WP.202, New York.
22. Wheeler, S.M. 2004. *Planning for Sustainability: Creating Livable, Equitable, and Ecological Communities*. Routledge, New York.
23. WHO. 2005. A regional overview of wastewater management and reuse in the Eastern Mediterranean region. Regional Office for the Eastern Mediterranean WHO-EM/CEH/139/E.
24. Wood, C. 2008. *Dry Spring: The Coming Water Crisis of North America*. Vancouver, Canada, Rain Coast Books.
25. http://www.epa.gov/region9/water/recycling. Accessed on December 8, 2014.

Index